Practical Meteorology

An Algebra-based Survey of Atmospheric Science

ROLAND STULL

The University of British Columbia

Vancouver, Canada

Practical Meteorology: An Algebra-based Survey of Atmospheric Science

Dept. of Earth, Ocean & Atmospheric Sciences
University of British Columbia
2020-2207 Main Mall
Vancouver, BC, Canada V6T 1Z4

This work is available at
http://www.eos.ubc.ca/books/Practical_Meteorology/

ISBN-13: 978-0-88865-176-1

Scene is a July 1996 thunderhead over southern Arizona, USA.

•

CONTENTS

Contents Summary

CHAPTER 1 • ATMOSPHERIC BASICS 1

CHAPTER 2 • SOLAR & INFRARED RADIATION 27

CHAPTER 3 • THERMODYNAMICS 53

CHAPTER 4 • WATER VAPOR 87

CHAPTER 5 • ATMOS. STABILITY 119

CHAPTER 6 • CLOUDS 159

CHAPTER 7 • PRECIPITATION PROCESSES 185

CHAPTER 8• SATELLITES & RADAR 219

CHAPTER 9 • WEATHER REPORTS & MAP ANALYSIS 267

CHAPTER 10 • ATMOSPHERIC FORCES & WINDS 289

CHAPTER 11 • GENERAL CIRCULATION 329

CHAPTER 12 • FRONTS & AIRMASSES 389

CHAPTER 13 • EXTRATROPICAL CYCLONES 425

CHAPTER 14 • THUNDERSTORM FUNDAMENTALS 481

CHAPTER 15 • THUNDERSTORM HAZARDS 545

CHAPTER 16 • TROPICAL CYCLONES 603

CHAPTER 17 • REGIONAL WINDS 645

CHAPTER 18 • ATMOSPHERIC BOUNDARY LAYER 687

CHAPTER 19 • POLLUTANT DISPERSION 723

CHAPTER 20 • NUMERICAL WEATHER PREDICTION (NWP) 745

CHAPTER 21 • NATURAL CLIMATE PROCESSES 793

CHAPTER 22 • ATMOSPHERIC OPTICS 833

APPENDIX A • SCIENTIFIC TOOLS 869

PREFACE

I designed this book for students and professionals who want to understand and apply basic meteorological concepts, but who don't need to derive equations.

To make this book accessible to more people, I converted the equations into algebra. With algebraic approximations to the atmosphere, you can see the physical meaning of each term and you can plug in numbers to get usable answers.

No previous knowledge of meteorology is needed — I start from the basics. Your background should include algebra, trig, and classical physics. This book could serve the fields of Atmospheric Science, Meteorology, Environmental Science, Engineering, Air Quality, Climatology, and Geography.

Readers like you asked to see solved examples, to enhance your understanding and speed your ability to apply the concepts to your own situations. To fill this need, I added "Sample Application" boxes for almost every equation in the book.

This book is designed to be both a textbook and a reference. As a textbook, the end of each chapter includes extensive homework exercises in categories inspired by Bloom's taxonomy of learning actions: "Broaden Knowledge & Comprehension"; "Apply"; "Evaluate & Analyze"; and "Synthesize".

Although a hand calculator can be used for some of the homework exercises, other exercises are best solved on a computer spreadsheet such as Excel or using a mathematical program such as MATLAB, R, Mathematica, or Maple. I used Excel for my Sample Applications and most of my graphs.

As a reference, I included in this book many tables, figures and graphs, and have a detailed index. Also, appendices include values of key constants and conversion factors.

The body of the text runs mostly in the inside columns of each page. The outside columns on each page contain the supporting figures, graphs, tables, and sample applications. Other special boxes in these outside columns include supplementary "Info" and "A Scientific Perspective". At the request of some readers, I added "Higher Math" boxes that use calculus, differential equations and other advanced techniques, but you may safely ignore these boxes if you wish.

For instructors, I inserted a bullet next to the most important equations, to help focus the learning. The book contains too much material to cover in one term, so instructors should select the subset of chapters to cover.

I intentionally limited the number of large color photographs and maps in this book, partly to allow it to be inexpensively printed/copied, and partly because most readers can access color images via the internet.

Acknowledgements. I am indebted to the following experts for their suggestions to this book: Phil Austin, Susan Allick Beach, William Beasley, Larry Berg, Allan Bertram, Brian Black, Bob Bornstein, Horst Böttger, Dominique Bourdin, John Cassano, Brian Cheng, Judy Curry, Luca Delle Monache, Xingxiu Deng, Dennis Driscoll, C. Dale Elifrits, David Finley, Jon Foley, Maria Furberg, Charlotte Gabites, Paul Greeley, Josh Hacker, Kit Hayden, Jim Hoke, Ed Hopkins, Dave Houghton, William Hsieh, Phoebe Jackson, Katelyn Janzen, Chris Jeffery, Alison Jolley, Jon Kahl, Scott Krayenhoff, Ian Lumb, Mankin Mak, Jon Martin, Paul Menzel, Stephanie Meyn, Doug McCollor, Mathias Mueller, Laurie Neil, Lorne Nelson, Thomas Nipen, Robert Nissen, Tim Oke, Anders Persson, Richard Peterson, Chris Pielou, Robert Rabin, Curt Rose, Alyson Shave, Scott Shipley, Robert Sica, David Siuta, Zbigniew Sorbjan, John Spagnol, Gert-Jan Steeneveld, David Stensrud, Haizhen Sun, George Taylor, Bruce Thomson, Greg Tripoli, Pao Wang, Greg West, Dave Whiteman, May Wai San Wong, Yongmei Zhou, and Jeff Zong. I apologize for any names that I forgot.

Many students have used earlier drafts, allowing me to fix typos thanks to their careful scrutiny. Any remaining errors are my own.

I thank the faculty and staff at the University of Wisconsin - Madison and the University of British Columbia - Vancouver, who were very supportive while I wrote earlier drafts during my tenures as professor. Storm photographs are reproduced with permission of the copyright holders: Warren Faidley, Gene Moore, and Gene Rhoden. The Space Science and Engineering Center (SSEC) at the University of Wisconsin - Madison granted permission to use their wonderful satellite images. Other images are publicly available from US Government and other sources.

I especially thank my wife Linda for her patience and understanding.

– Roland Stull

INFO • About the Author

Roland Stull holds a Bachelor's degree in Chemical Engineering and a Ph.D. in Atmospheric Science. He is a Certified Flight Instructor (CFI) and a Certified Consulting Meteorologist (CCM) in the USA.

Stull is a professor of Atmospheric Science at the University of British Columbia, Canada. He is a fellow of both the American Meteorological Society (AMS) and the Canadian Meteorological and Oceanographic Society (CMOS). He is author or co-author of over 95 scientific journal papers, and he wrote two single-author books.

1 ATMOSPHERIC BASICS

Contents

Classical Newtonian physics can be used to describe atmospheric behavior. Namely, air motions obey Newton's laws of dynamics. Heat satisfies the laws of thermodynamics. Air mass and moisture are conserved. When applied to a fluid such as air, these physical processes describe **fluid mechanics**. **Meteorology** is the study of the fluid mechanics, physics, and chemistry of Earth's atmosphere.

The **atmosphere** is a complex fluid system — a system that generates the chaotic motions we call **weather**. This complexity is caused by myriad interactions between many physical processes acting at different locations. For example, temperature differences create pressure differences that drive winds. Winds move water vapor about. Water vapor condenses and releases heat, altering the temperature differences. Such feedbacks are nonlinear, and contribute to the complexity.

But the result of this chaos and complexity is a fascinating array of weather phenomena — phenomena that are as inspiring in their beauty and power as they are a challenge to describe. Thunderstorms, cyclones, snow flakes, jet streams, rainbows. Such phenomena touch our lives by affecting how we dress, how we travel, what we can grow, where we live, and sometimes how we feel.

In spite of the complexity, much is known about atmospheric behavior. This book presents some of what we know about the atmosphere, for use by scientists and engineers.

INTRODUCTION

In this book are five major components of meteorology: (1) thermodynamics, (2) physical meteorology, (3) observation and analysis, (4) dynamics, and (5) weather systems (cyclones, fronts, thunderstorms). Also covered are air-pollution dispersion, numerical weather prediction, and natural climate processes.

Starting into the thermodynamics topic now, the state of the air in the atmosphere is defined by its pressure, density, and temperature. Changes of state associated with weather and climate are small perturbations compared to the average (standard)

A SCIENTIFIC PERSPECTIVE • Descartes and the Scientific Method

From René Descartes we get more than the name "Cartesian". In 1637 he published a book *Discours de la Méthode,* in which he defined the principles of the modern scientific method:

- Accept something as true only if you know it to be true.
- Break difficult problems into small parts, and solve each part in order to solve the whole problem.
- Start from the simple, and work towards the complex. Seek relationships between the variables.
- Do not allow personal biases or judgements to interfere, and be thorough.

This method formed the basis of the scientific renaissance, and marked an important break away from blind belief in philosophers such as Aristotle.

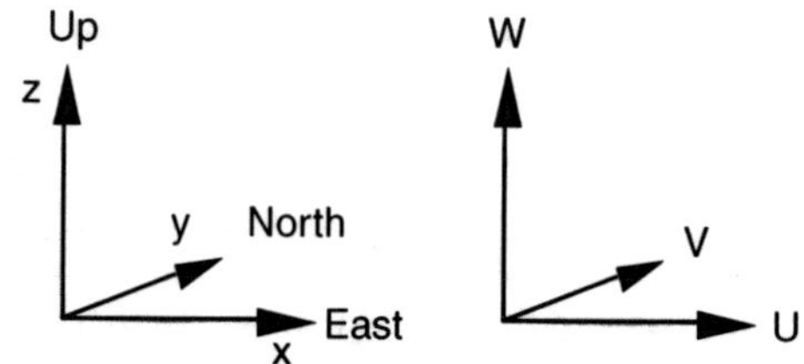

Figure 1.1
Local Cartesian coordinates and velocity components.

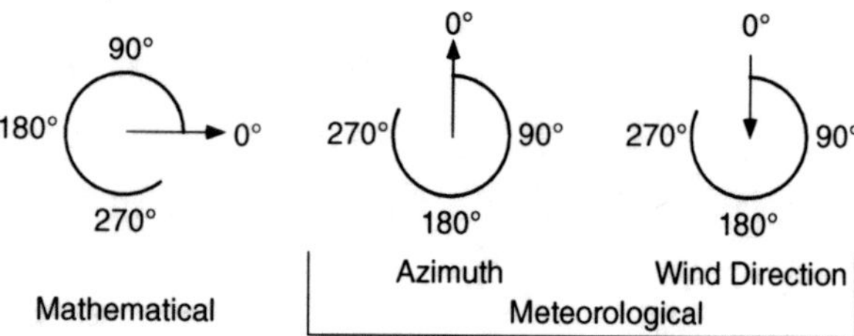

Figure 1.2
Comparison of meteorological and math angle conventions.

INFO • Weather-related Disasters

During 1970 to 2012 there were 8,835 disasters, 1.94 million deaths, and economic losses equivalent to US$ 2.4 trillion due to droughts, temperature extremes, tropical cyclones, floods, and their related health epidemics. Of these totals, storms caused 79% of the disasters, 55% of lives lost, and 86% of economic losses. Individual events included: 300,000 killed in 1970 cyclone Bhola in Bangladesh; 300,000 killed in 1983 drought in Ethiopia; 150,000 killed in drought in Sudan; and 138,866 killed in 1991 cyclone Gorky in Bangladesh. Most of the deaths were in less-developed countries, while most of the economic losses were in the most-developed countries (e.g. US$ 147 billion and $50 billion from hurricanes Katrina and Sandy in the USA). Source: WMO, 2014: "*The Atlas of Mortality and Economic Losses from Weather, Climate and Water Extremes, 1970-2012*".

atmosphere. These changes are caused by well-defined processes.

Equations and concepts in meteorology are similar to those in physics or engineering, although the jargon and conventions might look different when applied within an Earth framework. For a review of basic science, see Appendix A.

METEOROLOGICAL CONVENTIONS

Although the Earth is approximately spherical, you need not always use spherical coordinates. For the weather at a point or in a small region such as a town, state, or province, you can use local right-hand **Cartesian** (rectangular) **coordinates**, as sketched in Fig. 1.1. Usually, this coordinate system is aligned with x pointing east, y pointing north, and z pointing up. Other orientations are sometimes used.

Velocity components U, V, and W correspond to motion in the x, y, and z directions. For example, a positive value of U is a velocity component from west to east, while negative is from east to west. Similarly, V is positive northward, and W is positive upward (Fig. 1.1).

In polar coordinates, horizontal velocities can be expressed as a direction (α), and speed or magnitude (M). Historically, horizontal wind directions are based on the compass, with 0° to the north (the positive y direction), and with degrees increasing in a **clockwise** direction through 360°. Negative angles are not usually used. Unfortunately, this differs from the usual mathematical convention of 0° in the x direction, increasing **counter-clockwise** through 360° (Fig. 1.2).

Historically winds are named by the direction **from** which they come, while in mathematics angles give the direction toward which things move. Thus, a **west** wind is a wind from the west; namely, from 270°. It corresponds to a positive value of U, with air moving in the positive x direction.

Because of these differences, the usual trigonometric equations cannot be used to convert between (U, V) and (α, M). Use the following equations instead, where α is the compass direction from which winds come.

Conversion to Speed and Direction:

$$M = \left(U^2 + V^2\right)^{1/2} \qquad \bullet(1.1)$$

$$\alpha = 90° - \frac{360°}{C} \cdot \arctan\left(\frac{V}{U}\right) + \alpha_o \qquad \bullet(1.2a)$$

where α_o = 180° if $U > 0$, but is zero otherwise. C is the angular rotation in a full circle (C = 360° = 2·π radians).

[NOTE: Bullets • identify key equations that are fundamental, or are needed for understanding later chapters.]

Some computer languages and spreadsheets allow a two-argument arc tangent function (atan2):

$$\alpha = \frac{360°}{C} \cdot \text{atan2}(V, U) + 180° \qquad (1.2b)$$

[CAUTION: in the C and C++ programming languages, you might need to switch the order of U & V.]

Some calculators, spreadsheets or computer functions use angles in degrees, while others use radians. If you don't know which units are used, compute the arccos(–1) as a test. If the answer is 180, then your units are degrees; otherwise, an answer of 3.14159 indicates radians. Use whichever value of C is appropriate for your units.

Conversion to U and V:

$$U = -M \cdot \sin(\alpha) \qquad •(1.3)$$

$$V = -M \cdot \cos(\alpha) \qquad •(1.4)$$

In three dimensions, **cylindrical coordinates** (M, α, W) are sometimes used for velocity instead of Cartesian (U, V, W), where horizontal velocity components are specified by direction and speed, and the vertical component remains W (see Fig. 1.3).

Most meteorological graphs are like graphs in other sciences, with **dependent** variables on the **ordinate** (vertical axis) plotted against an **independent** variable on the **abscissa** (horizontal axis). However, in meteorology the axes are often switched when height (z) is the independent variable. This axis switching makes locations higher in the graph correspond to locations higher in the atmosphere (Fig. 1.4).

Sample Application

Find wind speed and direction, given eastward component 3 m s^{-1}, and northward 4 m s^{-1}.

Find the Answer

(Problem-solving methods are given in Appendix A.)

Given: $U = 3$ m s^{-1}. eastward wind component.
$V = 4$ m s^{-1}. northward wind component.

Find: M = ? m s^{-1}. wind speed
α = ? degrees. wind direction

Sketch:

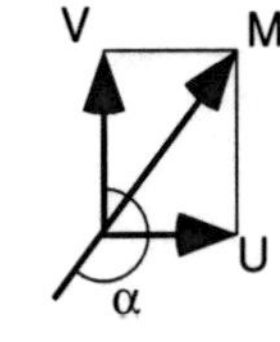

Use eq. (1.1):

$M = [\,U^2 + V^2\,]^{1/2}$
$= [\,(3 \text{ m s}^{-1})^2 + (4 \text{ m s}^{-1})^2\,]^{0.5}$
$= (9 + 16)^{0.5} \cdot [(\text{m s}^{-1})^2]^{0.5}$
$= (25)^{0.5}$ m s^{-1} = **5 m** s^{-1}.

Use eq. (1.2a):

$\alpha = 90° - (360°/C)\cdot\arctan(V/U) + 180°$
$= 90° - (360/360)\cdot\arctan[(4 \text{ m s}^{-1})/(3 \text{ m s}^{-1})] + 180°$
$= 90° - \tan^{-1}(1.333) + 180°$
$= 90° - 53.13° + 180°$ = **216.87°**.

Check: Units OK. Sketch OK. Values physical.
Exposition: Thus, the wind is from the south-southwest (SSW) at 5 m s^{-1}.

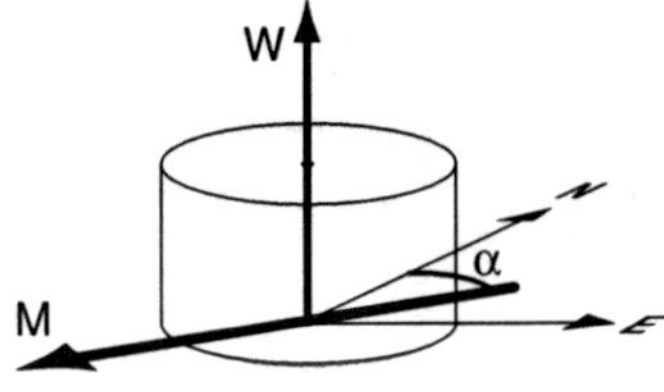

Figure 1.3
Notation used in cylindrical coordinates for velocity.

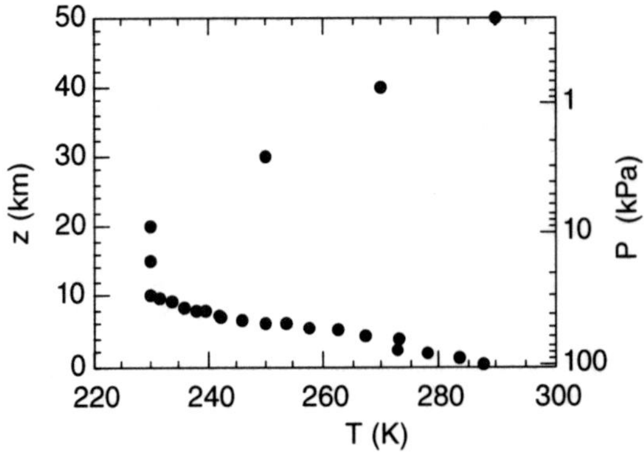

Figure 1.4
Hypothetical temperature T profile in the atmosphere, plotted such that locations higher in the graph correspond to locations higher in the atmosphere. The independent variable can be height z (left axis) or pressure P (right axis).

EARTH FRAMEWORKS REVIEWED

The Earth is slightly flattened into an **oblate spheroid of revolution** (Fig. 1.5). The distance from the center of the Earth to the north (**N**) and south (**S**) poles is roughly 6356.755 km, slightly less than the 6378.140 km distance from the center to

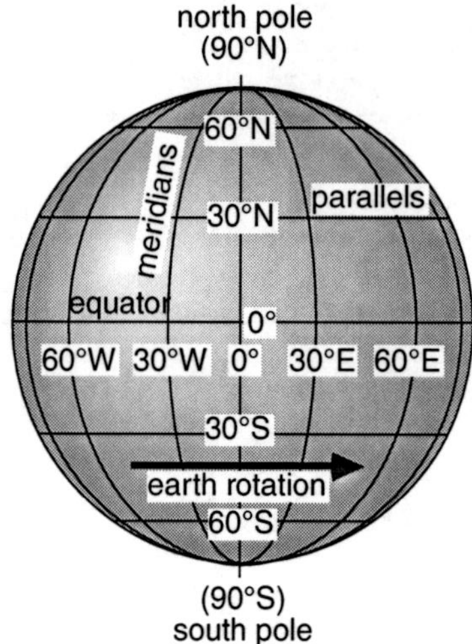

Figure 1.5
Earth cartography.

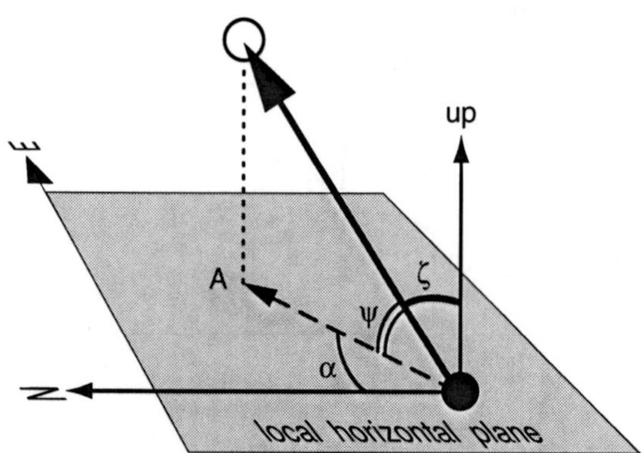

Figure 1.6
Elevation angle Ψ , *zenith angle* ζ , *and azimuth angle* α . *CAUTION: Recall from Fig. 1.2 that azimuth is the compass direction toward the object, while wind direction is the compass direction from which the wind blows.*

> **Sample Application**
> A thunderstorm top is at azimuth 225° and elevation angle 60° from your position. How would you describe its location in words? Also, what is the zenith angle?
>
> **Find the Answer**:
> Given: $\alpha = 225°$
> $\Psi = 60°$.
> Find: Location in words, and find ζ.
> *(continued next page).*

the equator. This 21 km difference in Earth radius causes a north-south cross section (i.e., a **slice**) of the Earth to be slightly elliptical. But for all practical purposes you can approximate the Earth a sphere (except for understanding Coriolis force in the Forces & Winds chapter).

Cartography

Recall that north-south lines are called **meridians**, and are numbered in degrees **longitude**. The **prime meridian** (0° longitude) is defined by international convention to pass through **Greenwich**, Great Britain. We often divide the 360° of longitude around the Earth into halves relative to Greenwich:

- **Western Hemisphere**: 0 – 180°W,
- **Eastern Hemisphere**: 0 – 180°E.

Looking toward the Earth from above the north pole, the Earth rotates counterclockwise about its axis. This means that all objects on the surface of the Earth (except at the poles) move toward the east.

East-west lines are called **parallels**, and are numbered in degrees **latitude**. By convention, the **equator** is defined as 0° latitude; the **north pole** is at 90°N; and the **south pole** is at 90°S. Between the north and south poles are 180° of latitude, although we usually divide the globe into the:

- **Northern Hemisphere:** 0 – 90°N,
- **Southern hemisphere:** 0 – 90°S.

On the surface of the Earth, each **degree of latitude** equals 111 km, or 60 **nautical miles**.

Azimuth, Zenith, & Elevation Angles

As a meteorological observer on the ground (black circle in Fig. 1.6), you can describe the local angle to an object (white circle) by two angles: the **azimuth angle** (α), and either the **zenith angle** (ζ) or **elevation angle** (ψ). The object can be physical (e.g., sun, cloud) or an image (e.g., rainbow, sun dog). By "local angle", we mean angles measured relative to the Cartesian **local horizontal** plane (e.g., a lake surface, or flat level land surface such as a polder), or relative to the local vertical direction at your location. Local **vertical** (**up**) is defined as opposite to the direction that objects fall.

Zenith means "directly overhead". Zenith angle is the angle measured at your position, between a conceptual line drawn to the zenith (up) and a line drawn to the object (dark arrow in Fig. 1.6). The elevation angle is how far above the horizon you see the object. Elevation angle and zenith angle are related by: $\psi = 90° - \zeta$, or if your calculator uses radians, it is $\psi = \pi/2 - \zeta$. Abbreviate both of these forms by $\psi = C/4 - \zeta$, where $C = 360° = 2\pi$ radians.

For the azimuth angle, first project the object vertically onto the ground (*A*, in Fig. 1.6). Draw a conceptual arrow (dashed) from you to *A*; this is the projection of the dark arrow on to the local horizontal plane. Azimuth angle is the compass angle along the local horizontal plane at your location, measured clockwise from the direction to north (N) to the direction to *A*.

Time Zones

In the old days each town defined their own local time. Local noon was when the sun was highest in the sky. In the 1800s when trains and telegraphs allowed fast travel and communication between towns, the railroad companies created standard time zones to allow them to publish and maintain precise schedules. Time zones were eventually adopted worldwide by international convention.

The Earth makes one complete revolution (relative to the sun) in one day. One revolution contains 360° of longitude, and one day takes 24 hours, thus every hour of elapsed time spans 360°/24 = 15° of longitude. For this reason, each time zone was created to span 15° of longitude, and almost every zone is 1 hour different from its neighboring time zones. Everywhere within a time zone, all clocks are set to the same time. Sometimes the time-zone boundaries are modified to follow political or geographic boundaries, to enhance commerce.

Coordinated Universal Time (UTC) is the time zone at the prime meridian. It is also known as **Greenwich Mean Time (GMT)** and **Zulu time (Z)**. The prime meridian is in the middle of the UTC time zone; namely, the zone spreads 7.5° on each side of the prime meridian. UTC is the official time used in meteorology, to help coordinate simultaneous weather observations around the world.

Internationally, time zones are given letter designations A - Z, with Z at Greenwich, as already discussed. East of the UTC zone, the local time zones (A, B, C, ...) are ahead; namely, local time of day is later than at Greenwich. West of the UTC zone, the local time zones (N, O, P, ...) are behind; namely, local time of day is earlier than at Greenwich.

Each zone might have more than one local name, depending on the countries it spans. Most of western Europe is in the Alpha (A) zone, where A = UTC + 1 hr. This zone is also known as Central Europe Time (CET) or Middle European Time (MET). In N. America are 8 time zones P* - W (see Table 1-1).

Near 180° longitude (in the middle of the Pacific Ocean) is the **international date line**. When you fly from east to west across the date line, you lose a day (it becomes tomorrow). From west to east, you gain a day (it becomes yesterday).

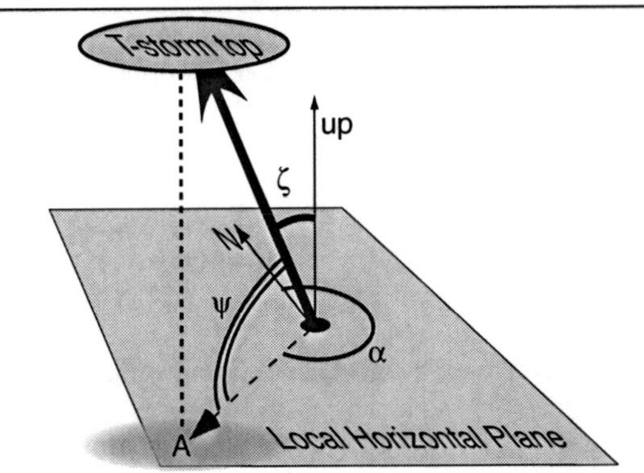

(continuation)

Sketch: (see above)
Because south has azimuth 180°, and west has azimuth 270°, we find that 225° is exactly halfway between south and west. Hence, the object is southwest (SW) of the observer. Also, 60° elevation is fairly high in the sky. So **the thunderstorm top is high in the sky to the southwest of the observer**.

Rearrange equation [$\Psi = 90° - \zeta$] to solve for zenith angle: $\zeta = 90° - \psi = 90° - 60° =$ **30°**.

Check: Units OK. Locations reasonable. Sketch good.
Exposition: This is a bad location for a storm chaser (the observer), because thunderstorms in North America often move from the SW toward the northeast (NE). Hence, the observer should quickly seek shelter underground, or move to a different location out of the storm path.

Table 1-1. Time zones in North America.
ST = standard time in the local time zone.
DT = daylight time in the local time zone.
UTC = coordinated universal time.
For conversion, use:
ST = UTC – α , DT = UTC – β

Zone	Name	α (h)	β (h)
P*	Newfoundland	3.5 (NST)	2.5 (NDT)
Q	Atlantic	4 (AST)	3 (ADT)
R	Eastern	5 (EST)	4 (EDT)
S	Central, and Mexico	6 (CST) 6 (MEX)	5 (CDT) 5
T	Mountain	7 (MST)	6 (MDT)
U	Pacific	8 (PST)	7 (PDT)
V	Alaska	9 (AKST)	8 (AKDT)
W	Hawaii-Aleutian	10 (HST)	9 (HDT)

Sample Application

A weather map is valid at 12 UTC on 5 June. What is the valid local time in Reno, Nevada USA?
Hint: Reno is at roughly 120°W longitude.

Find the Answer:
Given: UTC = 12 , Longitude = 120°W.
Find: Local valid time.

First, determine if standard or daylight time:
Reno is in the N. Hem., and 5 June is after the start date (March) of DT, so it **is** daylight time.
Hint: each 15° longitude = 1 time zone.
Next, use longitude to determine the time zone.
120° / (15° / zone) = 8 zones.
But 8 zones difference corresponds to the Pacific Time Zone. (using the ST column of Table 1-1, for which α also indicates the difference in time zones from UTC)
Use Table 1-1 for Pacific Daylight Time: β = 7 h
PDT = UTC – 7 h = 12 – 7 = **5 am PDT**.

Check: Units OK. 5 am is earlier than noon.
Exposition: In the USA, Canada, and Mexico, 12 UTC maps always correspond to morning of the same day, and 00 UTC maps correspond to late afternoon or evening of the previous day.
Caution: The trick of dividing the longitude by 15° doesn't work for some towns, where the time zone has been modified to follow geo-political boundaries.

INFO • Escape Velocity

Fast-moving air molecules that don't hit other molecules can escape to space by trading their kinetic energy (speed) for potential energy (height). High in the atmosphere where the air is thin, there are few molecules to hit. The lowest escape altitude for Earth is about 550 km above ground, which marks the base of the **exosphere** (region of escaping gases). This equals 6920 km when measured from the Earth's center, and is called the critical radius, r_c.

The escape velocity, v_e, is given by

$$v_e = \left[\frac{2 \cdot G \cdot m_{planet}}{r_c}\right]^{1/2}$$

where $G = 6.67\times10^{-11}$ m^3·s^{-2}·kg^{-1} is the gravitational constant, and m_{planet} is the mass of the planet. The mass of the Earth is 5.975 x 10^{24} kg. Thus, the escape velocity from Earth is roughly v_e = **10,732 m** s^{-1}. Using this velocity in eq. (1.5) gives the temperature needed for average-speed molecules to escape: 9,222 K for H_2, and 18,445 K for the heavier He. Temperatures in the exosphere (upper thermosphere) are not hot enough for average-speed H_2 and He to escape, but some are faster than average and do escape.

Heavier molecules such as O_2 have unreachably high escape temperatures (147,562 K), and have stayed in the Earth's atmosphere, to the benefit of life.

Many countries utilize **Daylight Saving Time** (DT) during their local summer. The purpose is to shift one of the early morning hours of daylight (when people are usually asleep) to the evening (when people are awake and can better utilize the extra daylight). At the start of DT (often in March in North America), you set your clocks one hour ahead. When DT ends in Fall (November), you set your clocks one hour back. The mnemonic "Spring ahead, Fall back" is a useful way to remember.

Times can be written as two or four digits. If two, then these digits are hours (e.g., 10 = 10 am, and 14 = 2 pm). If four, then the first two are hours, and the last two are minutes (e.g., 1000 is 10:00 am, and 1435 is 2:35 pm). In both cases, the hours use a 24-h clock going from 0000 (midnight) to 2359 (11:59 pm).

THERMODYNAMIC STATE

The thermodynamic state of air is measured by its pressure (P), density (ρ), and temperature (T).

Temperature

When a group of molecules (microscopic) move predominantly in the same direction, the motion is called wind (macroscopic). When they move in random directions, the motion is associated with temperature. Higher temperatures T are associated with greater average molecular speeds v :

$$T = a \cdot m_w \cdot v^2 \tag{1.5}$$

where a = 4.0x10^{-5} K·m^{-2} ·s^2 ·mole·g^{-1} is a constant. Molecular weights m_w for the most common gases in the atmosphere are listed in Table 1-2.

[CAUTION: symbol "a" represents different constants for different equations, in this textbook.]

Sample Application

What is the average random velocity of nitrogen molecules at 20°C ?

Find the Answer:
Given: T = 273.15 + 20 = 293.15 K.
Find: v = ? m s^{-1} (=avg mol. velocity)
Sketch:
Get m_w from Table 1-2. Solve eq. (1.5) for v:
$v = [T/a \cdot m_w]^{1/2}$
= [(293.15 K)/(4.0x10^{-5} K·m^{-2}·s^2·mole/g) ·(28.01g/mole)]$^{1/2}$ = **511.5 m** s^{-1}.

Check: Units OK. Sketch OK. Physics OK.
Exposition: Faster than a speeding bullet.

Absolute units such as Kelvin (K) must be used for temperature in all thermodynamic and radiative laws. Kelvin is the recommended temperature unit. For everyday use, and for temperature differences, you can use degrees Celsius (°C).

[CAUTION: degrees Celsius (°C) and degrees Fahrenheit (°F) must always be prefixed with the degree symbol (°) to avoid confusion with the electrical units of coulombs (C) and farads (F), but Kelvins (K) never take the degree symbol.]

At **absolute zero** (T = 0 K = –273.15°C) the molecules are essentially not moving. Temperature conversion formulae are:

$$T_{°F} = [(9/5)\cdot T_{°C}] + 32 \qquad \bullet(1.6a)$$

$$T_{°C} = (5/9)\cdot [T_{°F} - 32] \qquad \bullet(1.6b)$$

$$T_K = T_{°C} + 273.15 \qquad \bullet(1.7a)$$

$$T_{°C} = T_K - 273.15 \qquad \bullet(1.7b)$$

For temperature differences, you can use ΔT(°C) = ΔT(K), because the size of one degree Celsius is the same as the size of one unit of Kelvin. Hence, only in terms involving temperature <u>differences</u> can you arbitrarily switch between °C and K without needing to add or subtract 273.15.

Standard (average) sea-level temperature is

T = 15.0°C = 288 K = 59°F.

Actual temperatures can vary considerably over the course of a day or year. Temperature variation with height is not as simple as the curves for pressure and density, and will be discussed in the Standard Atmosphere section a bit later.

Table 1-2. Characteristics of gases in the air near the ground. Molecular weights are in g mole^{-1}. The volume fraction indicates the relative contribution to air in the Earth's lower atmosphere. EPA is the USA Environmental Protection Agency.

Symbol	Name	Mol. Wt.	Volume Fraction%
Constant Gases (NASA 2015)			
N_2	Nitrogen	28.01	78.08
O_2	Oxygen	32.00	20.95
Ar	Argon	39.95	0.934
Ne	Neon	20.18	0.001 818
He	Helium	4.00	0.000 524
Kr	Krypton	83.80	0.000 114
H_2	Hydrogen	2.02	0.000 055
Xe	Xenon	131.29	0.000 009
Variable Gases			
H_2O	Water vapor	18.02	0 to 4
CO_2	Carbon dioxide	44.01	0.040
CH_4	Methane	16.04	0.00017
N_2O	Nitrous oxide	44.01	0.00003
EPA National Ambient Air Quality Standards (NAAQS. 1990 Clean Air Act Amendments. Rules through 2011)			
CO	Carbon monoxide	28.01	
	(8 h average)		0.0009
	(1 h average)		0.0035
SO_2	Sulfur dioxide	64.06	
	(3 h average)		0.00000005
	(1 h average)		0.0000075
O_3	Ozone (8 h average)	48.00	0.0000075
NO_2	Nitrogen dioxide	46.01	
	(annual average)		0.0000053
	(1 h average)		0.0000100
Mean Condition for Air			
	air	28.96	100.0

Pressure

Pressure P is the force F acting perpendicular (normal) to a surface, per unit surface area A:

$$P = F / A \qquad \bullet(1.8)$$

Static pressure (i.e., pressure in calm winds) is caused by randomly moving molecules that bounce off each other and off surfaces they hit. In a vacuum the pressure is zero.

In the **International System of Units (SI)**, a **Newton** (N) is the unit for force, and m^2 is the unit for area. Thus, pressure has units of Newtons per square meter, or N·m^{-2}. One Pascal (Pa) is defined to equal a pressure of 1 N·m^{-2}. The recommended unit for atmospheric pressure is the **kiloPascal (kPa)**. The average (standard) pressure at sea level is P = 101.325 kPa. Pressure decreases nearly exponentially with height in the atmosphere, below 105 km.

Table 1-3. Standard (average) sea-level **pressure**.

Value	Units
101.325 kPa	kiloPascals (recommended)
1013.25 hPa	hectoPascals
101,325. Pa	Pascals
101,325. N·m^{-2}	Newtons per square meter
101,325 kg$_m$·m^{-1}·s^{-2}	kg-mass per meter per s^2
1.033227 kg$_f$·cm^{-2}	kg-force per square cm
1013.25 mb	millibars
1.01325 bar	bars
14.69595 psi	pounds-force /square inch
2116.22 psf	pounds-force / square foot
1.033227 atm	atmosphere
760 Torr	Torr
Measured as height of fluid in a barometer:	
29.92126 in Hg	inches of mercury
760 mm Hg	millimeters of mercury
33.89854 ft H_2O	feet of water
10.33227 m H_2O	meters of water

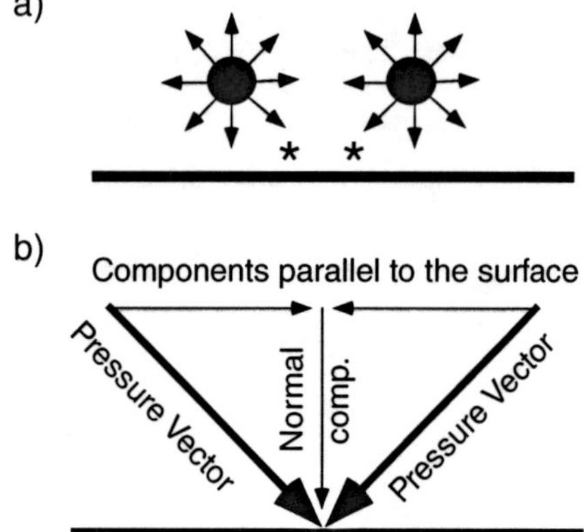

Figure 1.7
*(a) Pressure is isotropic. (b) Dark vectors correspond to those marked with * in (a). Components parallel to the surface cancel, while those normal to the surface contribute to pressure.*

Sample Application

The picture tube of an old TV and the CRT display of an old computer are types of vacuum tube. If there is a perfect vacuum inside the tube, what is the net force pushing against the front surface of a big screen 24 inch (61 cm) display that is at sea level?

Find the Answer

Given: Picture tube sizes are quantified by the diagonal length d of the front display surface. Assume the picture tube is square. The length of the side s of the tube is found from: $d^2 = 2\,s^2$. The frontal surface area is

$A = s^2 = 0.5 \cdot d^2 = 0.5 \cdot (61 \text{ cm})^2 = 1860.5 \text{ cm}^2$
$= (1860.5 \text{ cm}^2)\cdot(1 \text{ m}/100 \text{ cm})^2 = 0.186 \text{ m}^2$.

At sea level, atmospheric pressure pushing against the outside of the tube is 101.325 kPa, while from the inside of the tube there is no force pushing back because of the vacuum. Thus, the pressure difference across the tube face is

$\Delta P = 101.325 \text{ kPa} = 101.325\text{x}10^3 \text{ N m}^{-2}$.

Find: ΔF = ? N,
the net force across the tube.
Sketch:

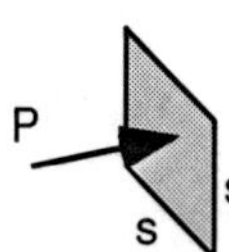

$\Delta F = F_{outside} - F_{inside}$, but $F = P \cdot A$. from eq. (1.8)
$= (P_{outside} - P_{inside})\cdot A$
$= \Delta P \cdot A$
$= (101.325\text{x}10^3 \text{ N m}^{-2})\cdot(0.186 \text{ m}^2)$
$= 1.885\text{x}10^4$ N = <u>**18.85 kN**</u>

Check: Units OK. Physically reasonable.
Exposition: This is quite a large force, and explains why picture tubes are made of such thick heavy glass. For comparison, a person who weighs 68 kg (150 pounds) is pulled by gravity with a force of about 667 N (= 0.667 kN). Thus, the picture tube must be able to support the equivalent of 28 people standing on it!

While kiloPascals will be used in this book, standard sea-level pressure in other units are given in Table 1-3 for reference. Ratios of units can be formed to allow unit conversion (see Appendix A). Although meteorologists are allowed to use **hectoPascals** (as a concession to those meteorologists trained in the previous century, who had grown accustomed to millibars), the prefix "hecto" is non-standard. If you encounter weather maps using millibars or hectoPascals, you can easily convert to kiloPascals by moving the decimal point one place to the left.

In fluids such as the atmosphere, pressure force is **isotropic**; namely, at any point it pushes with the same force in all directions (see Fig. 1.7a). Similarly, any point on a solid surface experiences pressure forces in all directions from the neighboring fluid elements. At such solid surfaces, all forces cancel except the forces normal (perpendicular) to the surface (Fig. 1.7b).

Atmospheric pressure that you measure at any altitude is caused by the weight of all the air molecules above you. As you travel higher in the atmosphere there are fewer molecules still above you; hence, pressure decreases with height. Pressure can also compress the air causing higher density (i.e., more molecules in a given space). Compression is greatest where the pressure is greatest, at the bottom of the atmosphere. As a result of more molecules being squeezed into a small space near the bottom than near the top, ambient pressure decreases faster near the ground than at higher altitudes.

Pressure change is approximately exponential with height, z. For example, if the temperature (T) were uniform with height (which it is not), then:

$$P = P_o \cdot e^{-(a/T)\cdot z} \qquad (1.9a)$$

where a = 0.0342 K m^{-1}, and where average sea-level pressure on Earth is P_o = 101.325 kPa. For more realistic temperatures in the atmosphere, the pressure curve deviates slightly from exponential. This will be discussed in the section on atmospheric structure. *[CAUTION again: symbol "a" represents different constants for different equations, in this textbook.]*

Equation (1.9a) can be rewritten as:

$$P = P_o \cdot e^{-z/H_p} \qquad (1.9b)$$

where H_p = 7.29 km is called the **scale height** for pressure. Mathematically, H_p is the **e-folding distance** for the pressure curve.

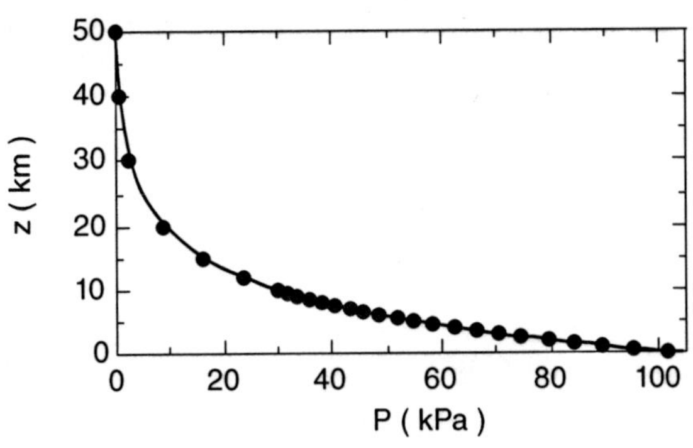

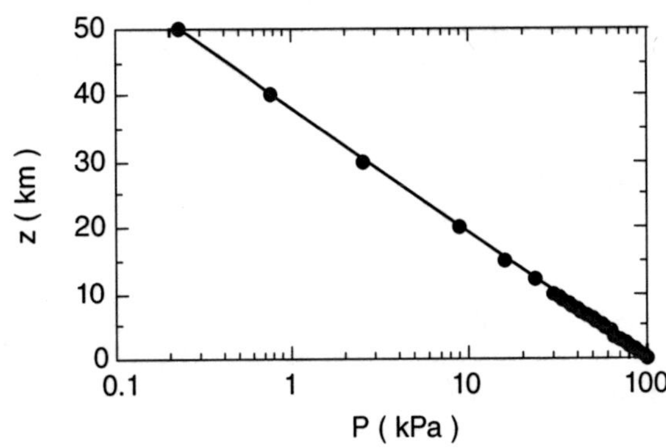

Figure 1.8
Height z vs. pressure P in the atmosphere, plotted on linear (left) and semi-log (right) graphs. See Appendix A for a review of relationships and graphs.

Fig. 1.8 shows the relationship between P and z on linear and semi-log graphs, for $T = 280$ K. [Graph types are reviewed in Appendix A.] In the lowest 3 km of the atmosphere, pressure decreases nearly linearly with height at about (10 kPa)/(1 km).

Because of the monotonic decrease of pressure with height, pressure can be used as a surrogate measure of altitude. (**Monotonic** means that it changes only in one direction, even though the rate of change might vary.) Fig. 1.4 shows such an example, where a reversed logarithmic scale (greater pressure at the bottom of the axis) is commonly used for P. Aircraft also use pressure to estimate their altitude.

In the atmosphere, the pressure at any height z is related to the mass of air above that height. Under the influence of gravity, air mass m has weight $F = m \cdot |g|$, where $|g| = 9.8\ \text{m·s}^{-2}$ is gravitational acceler-

Sample Application

Compare the pressures at 10 km above sea level for average temperatures of 250 and 300 K.

Find the Answer

Given: $z = 10\ \text{km} = 10^4$ m
(a) $T = 250$ K, (b) $T = 300$ K
Find: (a) $P = ?$ kPa, (b) $P = ?$ kPa

(a) Use eq. (1.9a):
P=(101.325kPa)·exp[(–0.0342K m^{-1})·(10^{4}m)/250K]
P = **25.8 kPa**

(b) P=(101.325kPa)·exp[(–0.0342K m^{-1})·(10^{4}m)/300K]
P = **32.4 kPa**

Check: Units OK. Physically reasonable.
Exposition: Pressure decreases slower with height in warmer air because the molecules are further apart.

INFO • e-folding Distance

Some curves never end. In the figure below, curve (a) ends at $x = x_a$. Curve (b) ends at $x = x_b$. But curve (c), the exponentially decreasing curve, asymptotically approaches $y = 0$, never quite reaching it. The area under each of the curves is finite, and in this example are equal to each other.

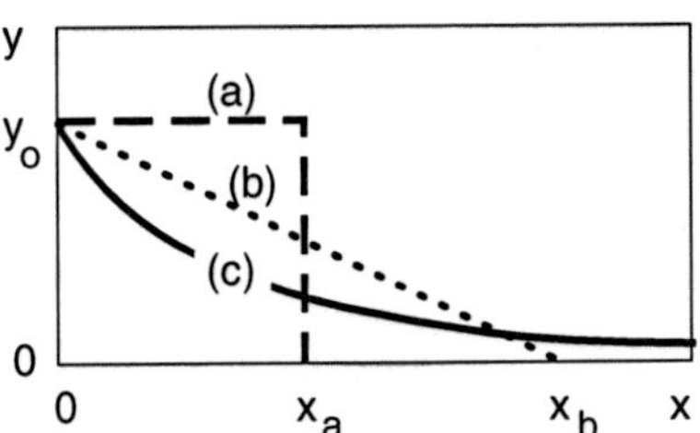

Although the exponential curve never ends, there is another way of quantifying how quickly it decreases with x. That measure is called the e-folding distance (or e-folding time if the independent variable is t instead of x). This is the distance x at which the curve decreases to $1/e$ of the starting value of the dependent variable, where $e = 2.71828$ is the base of natural logarithms. Thus, $1/e = 0.368$.

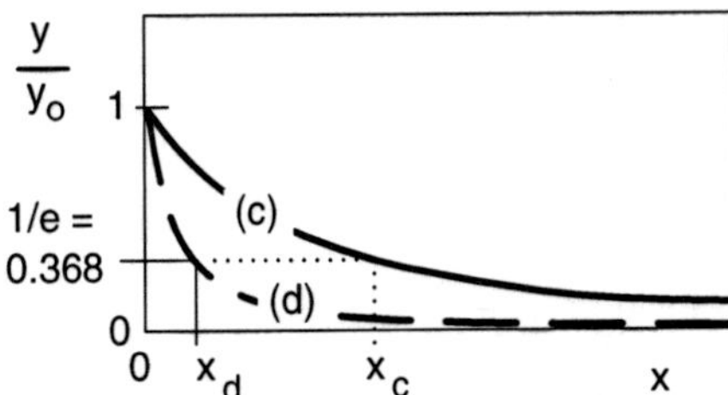

In the example above, both curves (c) and (d) are exponentials, but they drop off at different rates, where x_c and x_d are their respective e-folding distances. Generically, these curves are of the form:

$$y / y_o = e^{-x / x_{efold}} = \exp(-x / x_{efold})$$

Another useful characteristic is that the area A under the exponential curve is $A = y_o \cdot x_{efold}$.

Sample Application

Over each square meter of Earth's surface, how much air mass is between 80 and 30 kPa?

Find the Answer:

Given: P_{bottom} = 80 kPa, P_{top} = 30 kPa, A = 1 m^2
Find: Δm = ? kg

Use eq. (1.11): $\Delta m = [(1\ \text{m}^2)/(9.8\ \text{ms}^{-2})]\cdot(80 - 30\ \text{kPa})\cdot[(1000\ \text{kg}\cdot\text{m}^{-1}\cdot\text{s}^{-2})/(1\ \text{kPa})]$ = **5102 kg**

Check: Units OK. Physics OK. Magnitude OK.
Exposition: About 3 times the mass of a car.

Table 1-4. Standard atmospheric **density** at sea level, for a standard temperature 15°C.

Value	Units
1.2250 kg·m^{-3}.	kilograms per cubic meter (recommended)
0.076474 lb$_m$ ft^{-3}	pounds-mass per cubic foot
1.2250 g liter^{-1}	grams per liter
0.001225 g cm^{-3}	grams per cubic centimeter

Sample Application

At sea level, what is the mass of air within a room of size 5 m x 8 m x 2.5 m ?

Find the Answer

Given: L = 8 m room length, W = 5 m width
H = 2.5 m height of room
ρ = 1.225 kg·m^{-3} at sea level
Find: m = ? kg air mass

H L W

The volume of the room is
$Vol = W\cdot L\cdot H$ = (5m)·(8m)·(2.5m) = 100 m^3.
Rearrange eq. (1.12) and solve for the mass:
$m = \rho\cdot Vol.$ = (1.225 kg·m^{-3})·(100 m^3) = **122.5 kg**.

Check: Units OK. Sketch OK. Physics OK.
Exposition: This is 1.5 to 2 times a person's mass.

Sample Application

What is the air density at a height of 2 km in an atmosphere of uniform temperature of 15°C?

Find the Answer

Given: z =2000 m, ρ_o =1.225 kg m^{-3}, T =15°C =288.15 K
Find: ρ = ? kg m^{-3}

Use eq. (1.13):
ρ=(1.225 kg m^{-3})· exp[(–0.04 K m^{-1})·(2000 m)/288 K]
ρ = **0.928 kg m^{-3}**

Check: Units OK. Physics reasonable.
Exposition: This means that aircraft wings generate 24% less lift, and engines generate 24% less thrust because of the reduced air density.

ation at the Earth's surface. This weight is a force that squeezes air molecules closer together, increasing both the density and the pressure. Knowing that $P = F/A$, the previous two expressions are combined to give

$$P_z = |g|\cdot m_{above\ z} / A \tag{1.10}$$

where A is horizontal cross-section area. Similarly, between two different pressure levels is mass

$$\Delta m = (A/|g|)\cdot(P_{bottom} - P_{top}) \tag{1.11}$$

Density

Density ρ is defined as mass m per unit volume Vol.

$$\rho = m / Vol \qquad \bullet(1.12)$$

Density increases as the number and molecular weight of molecules in a volume increase. Average air density at sea level is given in Table 1-4. The recommended unit for density is kg·m^{-3}.

Because gases such as air are compressible, air density can vary over a wide range. Density decreases roughly exponentially with height in an atmosphere of uniform temperature.

$$\rho = \rho_o\cdot e^{-(a/T)\cdot z} \tag{1.13a}$$

or

$$\rho = \rho_o\cdot e^{-z/H_\rho} \tag{1.13b}$$

where a = 0.040 K m^{-1}, and where average sea-level density is ρ_o = 1.2250 kg·m^{-3}, at a temperature of 15°C = 288 K. The shape of the curve described by eq. (1.13) is similar to that for pressure, (see Fig. 1.9). The scale height for density is H_ρ = 8.55 km.

Although the air is quite thin at high altitudes, it still can affect many observable phenomena: twilight (scattering of sunlight by air molecules) up to

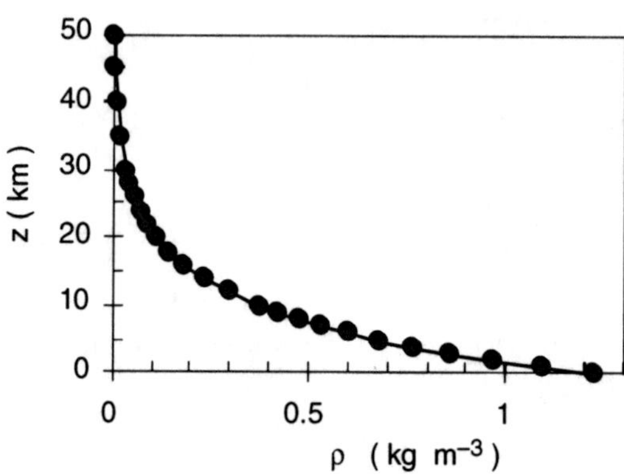

Figure 1.9
Density ρ vs. height z in the atmosphere.

63 km, meteors (incandescence by friction against air molecules) from 110 to 200 km, and aurora (excitation of air by solar wind) from 360 to 500 km.

The **specific volume** (α) is defined as the inverse of density ($\alpha = 1/\rho$). It has units of volume/mass.

ATMOSPHERIC STRUCTURE

Atmospheric structure refers to the state of the air at different heights. The true vertical structure of the atmosphere varies with time and location due to changing weather conditions and solar activity.

Standard Atmosphere

The "1976 U.S. Standard Atmosphere" (Table 1-5) is an idealized, dry, steady-state approximation of atmospheric state as a function of height. It has been adopted as an engineering reference. It approximates the average atmospheric conditions, although it was not computed as a true average.

A **geopotential height**, H, is defined to compensate for the decrease of gravitational acceleration magnitude $|g|$ above the Earth's surface:

$$H = R_o \cdot z \ / (R_o + z) \qquad \bullet(1.14a)$$

$$z = R_o \cdot H \ / (R_o - H) \qquad \bullet(1.14b)$$

where the average radius of the Earth is $R_o = 6356.766$ km. An **air parcel** (a group of air molecules moving together) raised to **geometric height** z would have the same potential energy as if lifted only to height H under constant gravitational acceleration. By using H instead of z, you can use $|g| = 9.8 \text{ m s}^{-2}$ as a constant in your equations, even though in reality it decreases slightly with altitude.

The difference ($z - H$) between geometric and geopotential height increases from 0 to 16 m as height increases from 0 to 10 km above sea level.

Sometimes g and H are combined into a new variable called the **geopotential**, Φ:

$$\Phi = |g| \cdot H \qquad (1.15)$$

Geopotential is defined as the work done against gravity to lift 1 kg of mass from sea level up to height H. It has units of $\text{m}^2 \text{ s}^{-2}$.

HIGHER MATH • Geopotential Height

What is "HIGHER MATH"?

These boxes contain supplementary material that use calculus, differential equations, linear algebra, or other mathematical tools beyond algebra. They are not essential for understanding the rest of the book, and may be skipped. Science and engineering students with calculus backgrounds might be curious about how calculus is used in atmospheric physics.

Geopotential Height

For gravitational acceleration magnitude, let $|g_o| = 9.8 \text{ m s}^{-2}$ be average value at sea level, and $|g|$ be the value at height z. If R_o is Earth radius, then $r = R_o + z$ is distance above the center of the Earth.

Newton's Gravitation Law gives the force $|F|$ between the Earth and an air parcel:

$$|F| = G \cdot m_{Earth} \cdot m_{air\ parcel} \ / \ r^2$$

where $G = 6.67\text{x}10^{-11} \text{ m}^3 \cdot \text{s}^{-2} \cdot \text{kg}^{-1}$ is the gravitational constant. Divide both sides by $m_{air\ parcel}$, and recall that by definition $|g| = |F|/m_{air\ parcel}$. Thus

$$|g| = G \cdot m_{Earth} \ / \ r^2$$

This eq. also applies at sea level ($z = 0$):

$$|g_o| = G \cdot m_{Earth} / \ R_o^2$$

Combining these two eqs. give

$$|g| = |g_o| \cdot [\, R_o \ / \ (R_o + z) \,]^2$$

Geopotential height H is defined as the work per unit mass to lift an object against the pull of gravity, divided by the gravitational acceleration value for sea level:

$$H = \frac{1}{|g_o|} \int_{Z=0}^{z} |g| \cdot dZ$$

Plugging in the definition of $|g|$ from the previous paragraph gives:

$$H = R_o^2 \cdot \int_{Z=0}^{z} (R_o + Z)^{-2} dZ$$

This integrates to

$$H = \left. \frac{-R_o^2}{R_o + Z} \right|_{Z=0}^{z}$$

After plugging in the limits of integration, and putting the two terms over a common denominator, the answer is:

$$H = R_o \cdot z \ / (R_o + z) \qquad (1.14a)$$

Sample Application
Find the geopotential height and the geopotential at 12 km above sea level.

Find the Answer
Given: $z = 12$ km, $R_o = 6356.766$ km
Find: $H = ?$ km, $\Phi = ?$ m^2 s^{-2}

Use eq. (1.14a): $H = (6356.766\text{km})\cdot(12\text{km}) / (6356.766\text{km} + 12\text{km}) =$ **11.98 km**
Use eq. (1.15): $\Phi = (9.8 \text{ m s}^{-2})\cdot(11{,}980 \text{ m}) =$ **$1.17\text{x}10^5$** m^2 s^{-2}
Check: Units OK.
Exposition: $H \le z$ as expected, because you don't need to lift the parcel as high for constant gravity as you would for decreasing gravity, to do the same work.

Sample Application
Find std. atm. temperature & pressure at H=2.5 km.

Find the Answer
Given: $H = 2.5$ km. Find: $T = ?$ K, $P = ?$ kPa
Use eq. (1.16): $T = 288.15 - (6.5\text{K/km})\cdot(2.5\text{km}) =$ **271.9 K**
Use eq. (1.17): $P = (101.325\text{kPa})\cdot(288.15\text{K}/271.9\text{K})^{-5.255877}$
$= (101.325\text{kPa})\cdot 0.737 =$ **74.7 kPa**.
Check: T = –1.1°C. Agrees with Fig. 1.10 & Table 1-5.

Table 1-5. Standard atmosphere.

H (km)	T (°C)	P (kPa)	ρ (kg m^{-3})
-1	21.5	113.920	1.3470
0	15.0	101.325	1.2250
1	8.5	89.874	1.1116
2	2.0	79.495	1.0065
3	-4.5	70.108	0.9091
4	-11.0	61.640	0.8191
5	-17.5	54.019	0.7361
6	-24.0	47.181	0.6597
7	-30.5	41.060	0.5895
8	-37.0	35.599	0.5252
9	-43.5	30.742	0.4664
10	-50.0	26.436	0.4127
11	-56.5	22.632	0.3639
13	-56.5	16.510	0.2655
15	-56.5	12.044	0.1937
17	-56.5	8.787	0.1423
20	-56.5	5.475	0.0880
25	-51.5	2.511	0.0395
30	-46.5	1.172	0.0180
32	-44.5	0.868	0.0132
35	-36.1	0.559	0.0082
40	-22.1	0.278	0.0039
45	-8.1	0.143	0.0019
47	-2.5	0.111	0.0014
50	-2.5	0.076	0.0010
51	-2.5	0.067	0.00086
60	-27.7	0.02031	0.000288
70	-55.7	0.00463	0.000074
71	-58.5	0.00396	0.000064
80	-76.5	0.00089	0.000015
84.9	-86.3	0.00037	0.000007
89.7	-86.3	0.00015	0.000003
100.4	-73.6	0.00002	0.0000005
105	-55.5	0.00001	0.0000002
110	-9.2	0.00001	0.0000001

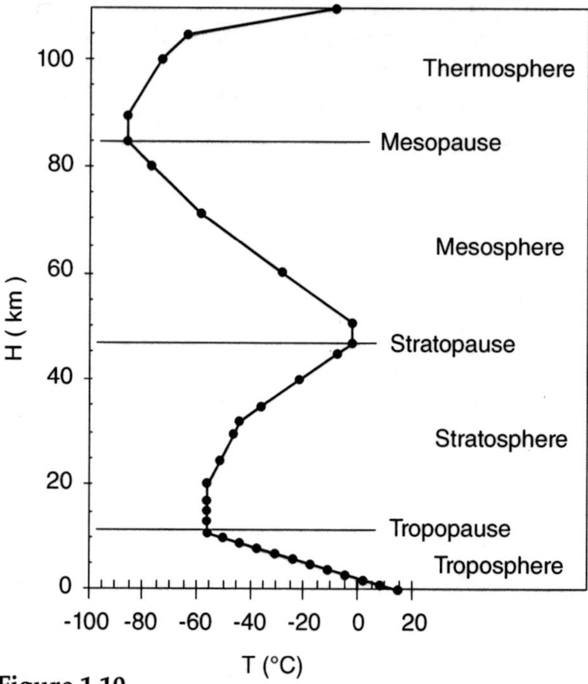

Figure 1.10
Standard temperature T profile vs. geopotential height H.

Table 1-5 gives the standard temperature, pressure, and density as a function of geopotential height H above sea level. Temperature variations are linear between key altitudes indicated in boldface. Standard-atmosphere temperature is plotted in Fig. 1.10.

Below a geopotential altitude of 51 km, eqs. (1.16) and (1.17) can be used to compute standard temperature and pressure. In these equations, be sure to use absolute temperature as defined by $T(\text{K}) = T(°\text{C}) + 273.15$. (1.16)

$T = 288.15 \text{ K} - (6.5 \text{ K km}^{-1})\cdot H$ for $H \le 11$ km

$T = 216.65$ K $11 \le H \le 20$ km

$T = 216.65 \text{ K} + (1 \text{ K km}^{-1})\cdot(H - 20\text{km})$ $20 \le H \le 32$ km

$T = 228.65 \text{ K} + (2.8 \text{ K km}^{-1})\cdot(H - 32\text{km})$ $32 \le H \le 47$ km

$T = 270.65$ K $47 \le H \le 51$ km

For the pressure equations, the absolute temperature T that appears must be the standard atmosphere temperature from the previous set of equations. In fact, those previous equations can be substituted into the equations below to make them a function of H rather than T. (1.17)

$P = (101.325\text{kPa})\cdot(288.15\text{K}/T)^{-5.255877}$ $H \le 11$ km

$P = (22.632\text{kPa})\cdot\exp[-0.1577\cdot(H - 11 \text{ km})]$ $11 \le H \le 20$ km

$P = (5.4749\text{kPa})\cdot(216.65\text{K}/T)^{34.16319}$ $20 \le H \le 32$ km

$P = (0.868\text{kPa})\cdot(228.65\text{K}/T)^{12.2011}$ $32 \le H \le 47$ km

$P = (0.1109\text{kPa})\cdot\exp[-0.1262\cdot(H-47 \text{ km})]$ $47 \le H \le 51$ km

These equations are a bit better than eq. (1.9a) because they do not make the unrealistic assumption of uniform temperature with height.

Knowing temperature and pressure, you can calculate density using the ideal gas law eq. (1.18).

Layers of the Atmosphere

The following layers are defined based on the nominal standard-atmosphere temperature structure (Fig. 1.10).

Thermosphere	$84.9 \le H$ km
Mesosphere	$47 \le H \le 84.9$ km
Stratosphere	$11 \le H \le 47$ km
Troposphere	$0 \le H \le 11$ km

Almost all clouds and weather occur in the **troposphere**.

The top limits of the bottom three spheres are named:

Mesopause	$H = 84.9$ km
Stratopause	$H = 47$ km
Tropopause	$H = 11$ km

On average, the tropopause is lower (order of 8 km) near the Earth's poles, and higher (order of 18 km) near the equator. In mid-latitudes, the tropopause height averages about 11 km, but is slightly lower in winter, and higher in summer.

The three relative maxima of temperature are a result of three altitudes where significant amounts of solar radiation are absorbed and converted into heat. Ultraviolet light is absorbed by ozone near the stratopause, visible light is absorbed at the ground, and most other radiation is absorbed in the thermosphere.

Atmospheric Boundary Layer

The bottom 0.3 to 3 km of the troposphere is called the **atmospheric boundary layer (ABL)**. It is often turbulent, and varies in thickness in space and time (Fig. 1.11). It "feels" the effects of the Earth's surface, which slows the wind due to surface drag, warms the air during daytime and cools it at night, and changes in moisture and pollutant concentration. We spend most of our lives in the ABL. Details are discussed in a later chapter.

Sample Application

Is eq. (1.9a) a good fit to standard atmos. pressure?

Find the Answer

Assumption: Use $T = 270$ K in eq. (1.9a) because it minimizes pressure errors in the bottom 10 km.

Method: Compare on a graph where the solid line is eq. (1.9a) and the data points are from Table 1-5.

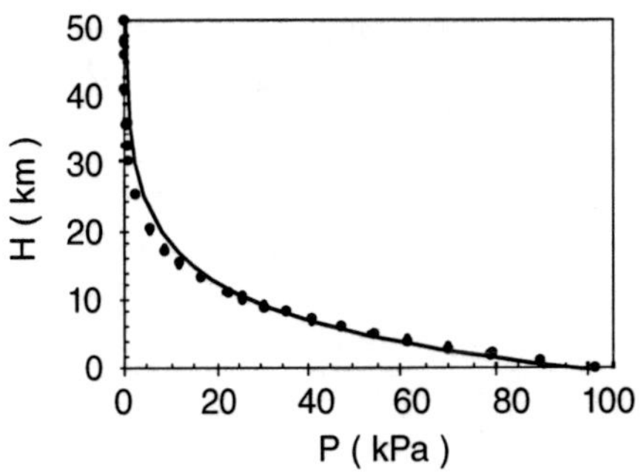

Exposition: Over the lower 10 km, the simple eq. (1.9a) is in error by no more than 1.5 kPa. If more accuracy is needed, then use the hypsometric equation (see eq. 1.26, later in this chapter).

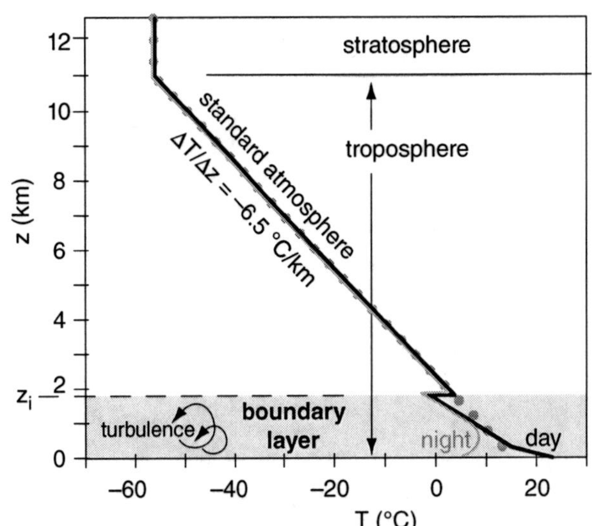

Figure 1.11

Boundary layer (shaded) within the bottom of the troposphere. Standard atmosphere is dotted. Typical temperature profiles during day (black line) and night (grey line) Boundary-layer top (dashed line) is at height z_i.

Sample Application

What is the average (standard) surface temperature for dry air, given standard pressure and density?

Find the Answer:
Given: P = 101.325 kPa, ρ = 1.225 kg·m^{-3}
Find: T = ? K

Solving eq. (1.18) for T gives: $T = P / (\rho \cdot \Re_d)$

$$T = \frac{101.325 \text{ kPa}}{(1.225 \text{ kg}\cdot\text{m}^{-3})\cdot(0.287 \text{ kPa}\cdot\text{K}^{-1}\cdot\text{m}^3\cdot\text{kg}^{-1})}$$

= 288.2 K = <u>15°C</u>

Check: Units OK. Physically reasonable.
Exposition: The answer agrees with the standard surface temperature of 15°C discussed earlier, a cool but pleasant temperature.

Sample Application

What is the absolute humidity of air of temperature 20°C and water vapor pressure of 2 kPa?

Find the Answer:
Given: e = 2 kPa, T = 20°C = 293 K
Find: ρ_v = ? kg$_{\text{water vapor}}$·m^{-3}

Solving eq. (1.19) for ρ_v gives: $\rho_v = e / (\Re_v \cdot T)$
ρ_v = (2 kPa) / (0.4615 kPa·K^{-1}·m^3·kg^{-1} · 293 K)
= <u>**0.0148**</u> kg$_{\text{water vapor}}$·m^{-3}

Check: Units OK. Physically reasonable.
Exposition: Small compared to the total air density.

Sample Application

In an unsaturated tropical environment with temperature of 35°C and water-vapor mixing ratio of 30 g$_{\text{water vapor}}$/kg$_{\text{dry air}}$, what is the virtual temperature?

Find the Answer:
Given: T = 35°C, r = 30 g$_{\text{water vapor}}$/kg$_{\text{dry air}}$
Find: T_v = ? °C

First, convert T and r to proper units
T = 273.15 + 35 = 308.15 K.
r =(30 g$_{\text{water}}$/kg$_{\text{air}}$)·(0.001 kg/g) = 0.03 g$_{\text{water}}$/g$_{\text{air}}$

Next use eq. (1.21):
T_v = (308.15 K)·[1 + (0.61 · 0.03)]
= 313.6 K = <u>**40.6°C**</u>.

Check: Units OK. Physically reasonable.
Exposition: Thus, high humidity reduces the density of the air so much that it acts like dry air that is 5°C warmer, for this case.

EQUATION OF STATE– IDEAL GAS LAW

Because pressure is caused by the movement of molecules, you might expect the pressure P to be greater where there are more molecules (i.e., greater density ρ), and where they are moving faster (i.e., greater temperature T). The relationship between pressure, density, and temperature is called the **Equation of State.**

Different fluids have different equations of state, depending on their molecular properties. The gases in the atmosphere have a simple equation of state known as the **Ideal Gas Law.**

For dry air (namely, air with the usual mix of gases, except no water vapor), the ideal gas law is:

$$P = \rho \cdot \Re_d \cdot T \qquad \bullet(1.18)$$

where $\Re_d$ = 0.287053 kPa·K^{-1}·m^3·kg^{-1}
= 287.053 J·K^{-1}·kg^{-1} .

$\Re_d$ is called the **gas constant** for dry air. Absolute temperatures (K) must be used in the ideal gas law. The total air pressure P is the sum of the partial pressures of nitrogen, oxygen, water vapor, and the other gases.

A similar equation of state can be written for just the water vapor in air:

$$e = \rho_v \cdot \Re_v \cdot T \qquad (1.19)$$

where e is the partial pressure due to water vapor (called the **vapor pressure**), ρ_v is the density of water vapor (called the **absolute humidity**), and the gas constant for pure water vapor is

$\Re_v$ = 0.4615 kPa·K^{-1}·m^3·kg^{-1}
= 461.5 J·K^{-1}·kg^{-1} .

For moist air (normal gases with some water vapor),

$$P = \rho \cdot \Re \cdot T \qquad (1.20)$$

where density ρ is now the total density of the air. A difficulty with this last equation is that the "gas constant" is NOT constant. It changes as the humidity changes because water vapor has different molecular properties than dry air.

To simplify things, a **virtual temperature** T_v can be defined to include the effects of water vapor:

$$T_v = T \cdot [1 + (a \cdot r)] \qquad \bullet(1.21)$$

where r is the water-vapor mixing ratio [r = (mass of water vapor)/(mass of dry air), with units $g_{water\ vapor}/g_{dry\ air}$, see the Water Vapor chapter], a = 0.61 $g_{dry\ air}/g_{water\ vapor}$, and all temperatures are in absolute units (K). In a nutshell, moist air of temperature T behaves as dry air with temperature T_v. T_v is greater than T because water vapor is less dense than dry air, and thus moist air acts like warmer dry air.

If there is also liquid water or ice in the air, then this virtual temperature must be modified to include the **liquid-water loading** (i.e., the weight of the drops falling at their terminal velocity) and **ice loading**:

$$T_v = T \cdot [1 + (a \cdot r) - r_L - r_I] \qquad \bullet(1.22)$$

where r_L is the liquid-water mixing ratio ($g_{liquid\ water}/g_{dry\ air}$), r_I is the ice mixing ratio ($g_{ice}/g_{dry\ air}$), and a = 0.61 ($g_{dry\ air}/g_{water\ vapor}$). Because liquid water and ice are heavy, air with liquid-water and/or ice loading acts like colder dry air.

With these definitions, a more useful form of the ideal gas law can be written for air of any humidity:

$$P = \rho \cdot \Re_d \cdot T_v \qquad \bullet(1.23)$$

where $\Re_d$ is still the gas constant for dry air. In this form of the ideal gas law, the effects of variable humidity are hidden in the virtual temperature factor, which allows the dry "gas constant" to be used (nice, because it really is constant).

HYDROSTATIC EQUILIBRIUM

As discussed before, pressure decreases with height. Any thin horizontal slice from a column of air would thus have greater pressure pushing up against the bottom than pushing down from the top (Fig. 1.12). This is called a **vertical pressure gradient**, where the term **gradient** means change with distance. The net upward force acting on this slice of air, caused by the pressure gradient, is $F = \Delta P \cdot A$, where A is the horizontal cross section area of the column, and $\Delta P = P_{bottom} - P_{top}$.

Also acting on this slice of air is gravity, which provides a downward force (weight) given by

$$F = m \cdot g \qquad \bullet(1.24)$$

where g = – 9.8 m·s^{-2} is the gravitational acceleration. (See Appendix B for variation of g with latitude and altitude.) Negative g implies a negative

Sample Application

In a tropical environment with temperature of 35°C, water-vapor mixing ratio of 30 $g_{water\ vapor}/kg_{dry\ air}$, and 10 $g_{liquid\ water}/kg_{dry\ air}$ of raindrops falling at their terminal velocity through the air, what is the virtual temperature?

Find the Answer:

Given: T = 35°C, r = 30 $g_{water\ vapor}/kg_{dry\ air}$
r_L = 10 $g_{liquid\ water}/kg_{dry\ air}$
Find: T_v = ? °C

First, convert T, r and r_L to proper units
T = 273.15 + 35 = 308.15 K.
r =(30 g_{vapor}/kg_{air})·(0.001 kg/g) = 0.03 g_{vapor}/g_{air}
r_L =(10 g_{liquid}/kg_{air})·(0.001 kg/g) = 0.01 g_{liquid}/g_{air}

Next use eq. (1.22):
T_v = (308.15 K)·[1 + (0.61 · 0.03) – 0.01]
= 310.7 K = <u>37.6°C</u>.

Check: Units OK. Physically reasonable.
Exposition: Compared to the previous Sample Application, the additional weight due to falling rain made the air act like it was about 3°C cooler.

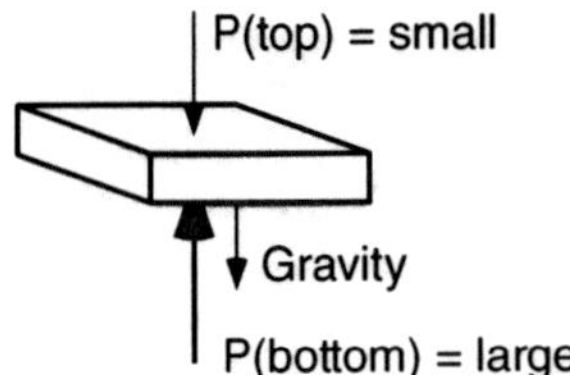

Figure 1.12.
Hydrostatic balance of forces on a thin slice of air.

Sample Application

What is the weight (force) of a person of mass 75 kg at the surface of the Earth?

Find the Answer

Given: m = 75 kg
Find: F = ? N

Sketch:
Use eq. (1.24)
$F = m \cdot g$ = (75 kg)·(– 9.8 m·s^{-2})
= – 735 kg·m·s^{-2} = <u>– 735 N</u>

Check: Units OK. Sketch OK. Physics OK.
Exposition: The negative sign means the person is pulled toward the Earth, not repelled away from it.

Sample Application

Near sea level, a height increase of 100 m corresponds to what pressure decrease?

Find the Answer

Given: $\rho = 1.225\ kg{\cdot}m^{-3}$ at sea level
$\Delta z = 100$ m

Find: $\Delta P = ?$ kPa

Sketch:

P_{top} $\Delta z = 100$ m P_{bottom}

Use eq. (1.25a):

$\Delta P = \rho{\cdot}g{\cdot}\Delta z$
$= (1.225\ kg{\cdot}m^{-3}){\cdot}(-9.8\ m{\cdot}s^{-2}){\cdot}(100\ m)$
$= -1200.5\ kg{\cdot}m^{-1}{\cdot}s^{-2}$
$=$ **-1.20 kPa**

Check: Units OK. Sketch OK. Physics OK.
Exposition: This answer should not be extrapolated to greater heights.

(downward) force. (Remember that the unit of force is $1\ N = 1\ kg{\cdot}m{\cdot}s^{-2}$, see Appendix A). The mass m of air in the slice equals the air density times the slice volume; namely, $m = \rho \cdot (A{\cdot}\Delta z)$, where Δz is the slice thickness.

For situations where pressure gradient force approximately balances gravity force, the air is said to be in a state of **hydrostatic equilibrium**. The corresponding **hydrostatic equation** is:

$$\Delta P = \rho{\cdot}g{\cdot}\Delta z \qquad (1.25a)$$

or

$$\frac{\Delta P}{\Delta z} = -\rho{\cdot}|g| \qquad \bullet(1.25b)$$

The term **hydrostatic** is used because it describes a stationary (static) balance in a fluid (hydro) between pressure pushing up and gravity pulling down. The negative sign indicates that pressure decreases as height increases. This equilibrium is valid for most weather situations, except for vigorous storms with large vertical velocities.

A SCIENTIFIC PERSPECTIVE • Check for Errors

As a scientist or engineer you should always be very careful when you do your calculations and designs. Be precise. Check and double check your calculations and your units. Don't take shortcuts, or make unjustifiable simplifications. Mistakes you make as a scientist or engineer can kill people and cause great financial loss.

Be careful whenever you encounter any equation that gives the change in one variable as a function of change of another. For example, in equations (1.25) P is changing with z. The "change of" operator (Δ) MUST be taken in the same direction for both variables. In this example $\Delta P/\Delta z$ means $[\ P(\text{at } z_2) - P(\text{at } z_1)\ ]\ /\ [\ z_2 - z_1\ ]$. We often abbreviate this as $[\ P_2 - P_1\ ]\ /\ [\ z_2 - z_1\ ]$.

If you change the denominator to be $[\ z_1 - z_2\ ]$, then you must also change the numerator to be in the same direction $[\ P_1 - P_2\ ]$. It doesn't matter which direction you use, so long as both the numerator and denominator (or both Δ variables as in eq. 1.25a) are in the same direction.

To help avoid errors in direction, you should always think of the subscripts by their relative positions in space or time. For example, subscripts 2 and 1 often mean top and bottom, or right and left, or later and earlier, etc. If you are not careful, then when you solve numerical problems using equations, your answer will have the wrong sign, which is sometimes difficult to catch.

HIGHER MATH • Physical Interpretation of Equations

Equations such as (1.25b) are finite-difference approximations to the original equations that are in differential form:

$$\frac{dP}{dz} = -\rho{\cdot}|g| \qquad (1.25c)$$

The calculus form (eq. 1.25c) is useful for derivations, and is the best description of the physics. The algebraic approximation eq. (1.25b) is often used in real life, where one can measure pressure at two different heights [i.e., $\Delta P/\Delta z = (P_2 - P_1)/\ (z_2 - z_1)$].

The left side of eq. (1.25c) describes the infinitesimal change of pressure P that is associated with an infinitesimal local change of height z. It is the vertical gradient of pressure. On a graph of P vs. z, it would be the slope of the line. The derivative symbol "d" has no units or dimensions, so the dimensions of the left side are kPa m^{-1}.

Eq. (1.25b) has a similar physical interpretation. Namely, the left side is the change in pressure associated with a finite change in height. Again, it represents the slope of a line, but in this case, it is a straight line segment of finite length, as an approximation to a smooth curve.

Both eqs. (1.25b & c) state that rate of pressure decrease (because of the negative sign) with height is greater if the density ρ is greater, or if the magnitude of the gravitational acceleration $|g|$ is greater. Namely, if factors ρ or $|g|$ increase, then the whole right hand side (RHS) increases because ρ and $|g|$ are in the numerator. Also, if the RHS increases, then the left hand side (LHS) must increase as well, to preserve the equality of LHS = RHS.

HYPSOMETRIC EQUATION

When the ideal gas law and the hydrostatic equation are combined, the result is an equation called the **hypsometric equation**. This allows you to calculate how pressure varies with height in an atmosphere of arbitrary temperature profile:

$$z_2 - z_1 \approx a \cdot \overline{T_v} \cdot \ln\left(\frac{P_1}{P_2}\right) \qquad \bullet(1.26a)$$

or

$$P_2 = P_1 \cdot \exp\left(\frac{z_1 - z_2}{a \cdot \overline{T_v}}\right) \qquad \bullet(1.26b)$$

where $\overline{T_v}$ is the average virtual temperature between heights z_1 and z_2. The constant $a = \Re_d / |g| = 29.3 \text{ m K}^{-1}$. The height difference of a layer bounded below and above by two pressure levels P_1 (at z_1) and P_2 (at z_2) is called the **thickness** of that layer.

To use this equation across large height differences, it is best to break the total distance into a number of thinner intervals, Δz. In each thin layer, if the virtual temperature varies little, then you can approximate by T_v. By this method you can sum all of the thicknesses of the thin layers to get the total thickness of the whole layer.

For the special case of a dry atmosphere of uniform temperature with height, eq. (1.26b) simplifies to eq. (1.9a). Thus, eq. (1.26b) also describes an exponential decrease of pressure with height.

Sample Application (§)

What is the thickness of the 100 to 90 kPa layer, given [P(kPa), T(K)] = [90, 275] and [100, 285].

Find the Answer

Given: observations at top and bottom of the layer
Find: $\Delta z = z_2 - z_1$
Assume: T varies linearly with z. Dry air: $T = T_v$.

Solve eq. (1.26) on a computer spreadsheet (§) for many thin layers 0.5 kPa thick. Results for the first few thin layers, starting from the bottom, are:

P(kPa)	T_v (K)	$\overline{T_v}$ (K)	Δz(m)
100	285	284.75	41.82
99.5	284.5	284.25	41.96
99.0	284	etc.	etc.

Sum of all Δz = **<u>864.11 m</u>**

Check: Units OK. Physics reasonable.
Exposition: In an aircraft you must climb 864.11 m to experience a pressure decrease from 100 to 90 kPa, for this particular temperature sounding. If you compute the whole thickness at once from $\Delta z = (29.3\text{m K}^{-1})\cdot(280\text{K})\cdot\ln(100/90) = 864.38$ m, this answer is less accurate than by summing over smaller thicknesses.

PROCESS TERMINOLOGY

Processes associated with constant temperature are isothermal. For example, eqs. (1.9a) and (1.13a) apply for an **isothermal** atmosphere. Those occurring with constant pressure are **isobaric**. A line on a weather map connecting points of equal temperature is called an **isotherm**, while one connecting points of equal pressure is an **isobar**. Table 1-6 summarizes many of the process terms.

Sample Application

Name the process for constant density.

Find the Answer:

From Table 1-6: It is an **<u>isopycnal</u>** process.

Exposition: Isopycnics are used in oceanography, where both temperature and salinity affect density.

Table 1-6. Process names. (tendency = change with time)

Name	Constant or equal
adiabat	entropy (no heat exchange)
contour	height
isallobar	pressure tendency
isallohypse	height tendency
isallotherm	temperature tendency
isanabat	vertical wind speed
isanomal	weather anomaly
isentrope	entropy or potential temp.
isobar	pressure
isobath	water depth
isobathytherm	depth of constant temperature
isoceraunic	thunderstorm activity or freq.
isochrone	time
isodop	(Doppler) radial wind speed
isodrosotherm	dew-point temperature
isoecho	radar reflectivity intensity
isogon	wind direction
isogram	(generic, for any quantity)
isohel	sunshine
isohume	humidity
isohyet	precipitation accumulation
isohypse	height (similar to contour)
isoline	(generic, for any quantity)
isoneph	cloudiness
isopleth	(generic, for any quantity)
isopycnic	density
isoshear	wind shear
isostere	specific volume (1/ρ)
isotach	speed
isotherm	temperature

HIGHER MATH • Hypsometric Eq.

To derive eq. (1.26) from the ideal gas law and the hydrostatic equation, one must use calculus. It cannot be done using algebra alone. However, once the equation is derived, the answer is in algebraic form.

The derivation is shown here only to illustrate the need for calculus. Derivations will NOT be given for most of the other equations in this book. Students can take advanced meteorology courses, or read advanced textbooks, to find such derivations.

Derivation of the hypsometric equation:
Given: the hydrostatic eq:

$$\frac{dP}{dz} = -\rho \cdot |g| \tag{1.25c}$$

and the ideal gas law:

$$P = \rho \cdot \Re_d \cdot T_v \tag{1.23}$$

First, rearrange eq. (1.23) to solve for density:

$$\rho = P / (\Re_d \cdot T_v)$$

Then substitute this into (1.25c):

$$\frac{dP}{dz} = -\frac{P \cdot |g|}{\Re_d \cdot T_v}$$

One trick for integrating equations is to separate variables. Move all the pressure factors to one side, and all height factors to the other. Therefore, multiply both sides of the above equation by dz, and divide both sides by P.

$$\frac{dP}{P} = -\frac{|g|}{\Re_d \cdot T_v} dz$$

Compared to the other variables, g and $\Re_d$ are relatively constant, so we will assume that they are constant and separate them from the other variables. However, usually temperature varies with height: $T(z)$. Thus:

$$\frac{dP}{P} = -\frac{|g|}{\Re_d} \cdot \frac{dz}{T_v(z)}$$

Next, integrate the whole eq. from some lower altitude z_1 where the pressure is P_1, to some higher altitude z_2 where the pressure is P_2:

$$\int_{P_1}^{P_2} \frac{dP}{P} = -\frac{|g|}{\Re_d} \cdot \int_{z_1}^{z_2} \frac{dz}{T_v(z)}$$

(continues in next column)

HIGHER MATH • Hypsometric Eq.

(Continuation)

where $|g|/\Re_d$ is pulled out of the integral on the RHS because it is constant.

The left side of that equation integrates to become a natural logarithm (consult tables of integrals).

The right side of that equation is more difficult, because we don't know the functional form of the vertical temperature profile. On any given day, the profile has a complex shape that is not conveniently described by an equation that can be integrated.

Instead, we will invoke the mean-value theorem of calculus to bring T_v out of the integral. The overbar denotes an average (over height, in this context).

That leaves only dz on the right side. After integrating, we get:

$$\ln(P)\Big|_{P_1}^{P_2} = -\frac{|g|}{\Re_d} \cdot \overline{\left(\frac{1}{T_v}\right)} \cdot z\Big|_{z_1}^{z_2}$$

Plugging in the upper and lower limits gives:

$$\ln(P_2) - \ln(P_1) = -\frac{|g|}{\Re_d} \cdot \overline{\left(\frac{1}{T_v}\right)} \cdot (z_2 - z_1)$$

But the difference between two logarithms can be written as the ln of the ratio of their arguments:

$$\ln\left(\frac{P_2}{P_1}\right) = -\frac{|g|}{\Re_d} \cdot \overline{\left(\frac{1}{T_v}\right)} \cdot (z_2 - z_1)$$

Recalling that $\ln(x) = -\ln(1/x)$, then:

$$\ln\left(\frac{P_1}{P_2}\right) = \frac{|g|}{\Re_d} \cdot \overline{\left(\frac{1}{T_v}\right)} \cdot (z_2 - z_1)$$

Rearranging and approximating $\overline{1/T_v} \approx 1/\overline{T_v}$ (which is NOT an identity), then one finally gets the hypsometric eq:

$$(z_2 - z_1) \approx \frac{\Re_d}{|g|} \cdot \overline{T_v} \cdot \ln\left(\frac{P_1}{P_2}\right) \tag{1.26}$$

PRESSURE INSTRUMENTS

Atmospheric-pressure sensors are called **barometers**. Almost all barometers measure the pressure difference between atmospheric pressure on one side of the sensor, and a reference pressure on the other side. This pressure difference causes a net force that pushes against a spring or a weight. For most barometers, the reference pressure is a **vacuum** (zero pressure).

Aneroid barometers use a corrugated metallic can (the **aneroid element**) with a vacuum inside the can. A spring forces the can sides outward against the inward-pushing atmospheric-pressure force. The relative inflation of the can is measured with levers and gears that amplify the minuscule deflection of the can, and display the result as a moving needle on a barometer or a moving pen on a **barograph** (a recording barometer). The scale on an aneroid barometer can be calibrated to read in any pressure units (see Table 1-3).

Mercury (Hg) barometers (developed by Evangelista Torricelli in the 1600s) are made from a U-shaped tube of glass that is closed on one end. The closed end has a vacuum, and the other end is open to atmospheric pressure. Between the vacuum and the air is a column of mercury inside the tube, the weight of which balances atmospheric pressure.

Atmospheric pressure is proportional to the height difference Δz between the top of the mercury column on the vacuum side, and the height on the side of the U-tube open to the atmosphere. Typical Δz scales are **millimeters of mercury (mm Hg)**, **centimeters of mercury (cm Hg)**, or **inches of mercury (in Hg)**. To amplify the height signal, **contra-barometers** (developed by Christiaan Huygens in the 1600s) use mercury on one side of the U-tube and another fluid (e.g., alcohol) on the other.

Because mercury is a poison, modern **Torricelli** (U-tube) barometers use a heavy silicon-based fluid instead. Also, instead of using a vacuum as a reference pressure, they use a fixed amount of gas in the closed end of the tube. All Torricelli barometers require temperature corrections, because of thermal expansion of the fluid.

Electronic barometers have a small can with a vacuum or fixed amount of gas inside. Deflection of the can can be measured by strain gauges, or by changes in capacitance between the top and bottom metal ends of an otherwise non-conductive can. **Digital barometers** are electronic barometers that include analog-to-digital circuitry to send pressure data to digital computers. More info about all weather instruments is in WMO-No. 8 *Guide to Meteorological Instruments and Methods of Observation.*

REVIEW

Pressure, temperature, and density describe the thermodynamic state of the air. These state variables are related to each other by the ideal gas law. Change one, and one or both of the others must change too. Ambient pressure decreases roughly exponentially with height, as given by the hypsometric equation. The vertical pressure gradient is balanced by the pull of gravity, according to the hydrostatic eq.

Density variation is also exponential with height. Temperature, however, exhibits three relative maxima over the depth of the atmosphere, caused by absorption of radiation from the sun. Thermodynamic processes can be classified. The standard atmosphere is an idealized model of atmospheric vertical structure, and is used to define atmospheric layers such as the troposphere and stratosphere. Atmospheric pressure is measured with mercury, aneroid, or electronic barometers.

A SCIENTIFIC PERSPECTIVE • Be Meticulous

Format Guidelines for Your Homework

Good scientists and engineers are not only creative, they are methodical, meticulous, and accurate. To encourage you to develop these good habits, many instructors require your homework to be written in a clear, concise, organized, and consistent format. Such a format is described below, and is illustrated in all the Sample Applications in this book. The format below closely follows steps you typically take in problem solving (Appendix A).

Format:

1. Give the exercise number, & restate the problem.
2. Start the solution section by listing the "Given" known variables (WITH THEIR UNITS).
3. List the unknown variables to find, with units.
4. Draw a sketch if it clarifies the scenario.
5. List the equation(s) you will use.
6. Show all your intermediate steps and calculations (to maximize your partial credit), and be sure to ALWAYS INCLUDE UNITS with the numbers when you plug them into eqs.
7. Put a box around your final answer, or underline it, so the grader can find it on your page amongst all the coffee and pizza stains.
8. Always check the value & units of your answer.
9. Briefly discuss the significance of the answer.

Example:

Problem : What is air density at height 2 km in an isothermal atmosphere of temperature 15°C?

Find the Answer

Given: $z = 2000$ m

$\rho_o = 1.225$ kg m^{-3}

$T = 15°C = 288.15$ K

Find: $\rho = ?$ kg m^{-3}

Use eq. (1.13a): $\rho =$
$(1.225 \text{ kg m}^{-3})\cdot \exp[(-0.040\text{K m}^{-1})\cdot(2000\text{m})/288\text{K}]$

$\rho =$ <u>**0.928**</u> kg m^{-3}

Check: Units OK. Physics reasonable.

Exposition: $(\rho_o - \rho)/\rho_o \approx 0.24$. This means that aircraft wings generate 24% less lift, aircraft engines generate 24% less power, and propellers 24% less thrust because of the reduced air density. This compounding effect causes aircraft performance to decrease rapidly with increasing altitude, until the **ceiling** is reached where the plane can't climb any higher.

Fig. 1.13 shows the Find the Answer of this problem on a computer spreadsheet.

Tips

At the end of each chapter are four types of homework exercises:

- Broaden Knowledge & Comprehension
- Apply
- Evaluate & Analyze
- Synthesize

Each of these types are explained here in Chapter 1, at the start of their respective subsections. I also recommend how you might approach these different types of problems.

One of the first tips is in the "A SCIENTIFIC PERSPECTIVE" box. Here I recommend that you write your exercise solutions in a format very similar to the "Sample Applications" that I have throughout this book. Such meticulousness will help you earn higher grades in most science and engineering courses, and will often give you partial credit (instead of zero credit) for exercises you solved incorrectly.

Finally, most of the exercises have multiple parts to them. Your instructor need assign only one of the parts for you to gain the skills associated with that exercise. Many of the numerical problems are similar to Sample Applications presented earlier in the chapter. Thus, you can try to do the Sample Application first, and if you get the same answer as I did, then you can be more confident in getting the right answer when you re-solve the exercise part assigned by your instructor. Such re-solutions are trivial if you use a computer spreadsheet (Fig. 1.13) or other similar program to solve the numerical exercises.

◇	A	B	C	D	E
1					
2	N21)	What is air density at height			
3		2 km in an atmosphere of			
4		uniform T of 15 °C?			
5					
6		**Given:**	z =	2000	m
7			rho_o=	1.225	kg/m3
8			T =	15	°C
9		**Find:**	rho =	?	kg/m3
10					
11		First convert T to Kelvin.			
12			T =	288.15	K
13		Then use eq (1.13a), where			
14			a =	0.040	K/m
15			rho =	**0.928**	kg/m3
16					
17		**Check:** Units OK. Physics OK.			
18		**Discussion:** This means that			
19		aircraft wings generate			
20		0.928/ 1.225 = 76% of the lift			
21		they would at sea level.			
22					
23	Note to students: the eq used in				
24	cell D15 was: =D7*EXP(-D14*D6/D12)				

Figure 1.13
Example of a spreadsheet used to solve a numerical problem.

HOMEWORK EXERCISES

Broaden Knowledge & Comprehension

These questions allow you to solve problems using current data, such as satellite images, weather maps, and weather observations that you can download through the internet. With current data, exercises can be much more exciting, timely, and relevant. Such questions are more vague than the others, because we can't guarantee that you will find a particular weather phenomenon on any given day.

Many of these questions are worded to encourage you to acquire the weather information for locations near where you live. However, the instructor might suggest a different location if a better example of a weather event is happening elsewhere. Even if the instructor does not suggest alternative locations, you should feel free to search the country, the continent, or the globe for examples of weather that are best suited for the exercise.

Web URL (universal resource locator) addresses are very transient. Web sites come and go. Even a persisting site might change its web address. For this reason, the web-enhanced questions do not usually give the URL web site for any particular exercise. Instead, you are expected to become proficient with internet search engines. Nonetheless, there still might be occasions where the data does not exist anywhere on the web. The instructor should be aware of such eventualities, and be tolerant of students who cannot complete the web exercise.

In many cases, you will want to print the weather map or satellite image to turn in with your homework. Instructors should be tolerant of students who have access to only black and white printers. If you have black and white printouts, use a colored pencil or pen to highlight the particular feature or isopleths of interest, if it is otherwise difficult to discern among all the other black lines on the printout.

You should always list the URL web address and the date you used it from which you acquired the data or images. This is just like citing books or journals from the library. At the end of each web exercise, include a "References" section listing the web addresses used, and any of your own annotations.

A SCIENTIFIC PERSPECTIVE • Give Credit

Part of the ethic of being a good scientist or engineer is to give proper credit to the sources of ideas and data, and to avoid plagiarism. Do this by citing the author and the title of their book, journal paper, or electronic content. Include the international standard book number (isbn), digital object identifier (doi), or other identifying info.

B1. Download a map of sea-level pressure, drawn as isobars, for your area. Become familiar with the units and symbols used on weather maps.

B2. Download from the web a map of near-surface air temperature, drawn is isotherms, for your area. Also, download a surface skin temperature map valid at the same time, and compare the temperatures.

B3. Download from the web a map of wind speeds at a height near the 200 or 300 mb (= 20 or 30 kPa) jet stream level . This wind map should have isotachs drawn on it. If you can find a map that also has wind direction or streamlines in addition to the isotachs, that is even better.

B4. Download from the web a map of humidities (e.g., relative humidities, or any other type of humidity), preferably drawn is isohumes. These are often found at low altitudes, such as for pressures of 850 or 700 mb (85 or 70 kPa).

B5. Search the web for info on the standard atmosphere. This could be in the form of tables, equations, or descriptive text. Compare this with the standard atmosphere in this textbook, to determine if the standard atmosphere has been revised.

B6. Search the web for the air-pollution regulation authority in your country (such as the EPA in the USA), and find the regulated concentrations of the most common air pollutants (CO, SO_2, O_3, NO_2, volatile organic compounds VOCs, and particulates). Compare with the results in Table 1-2, to see if the regulations have been updated in the USA, or if they are different for your country.

B7. Search the web for surface weather station observations for your area. This could either be a surface weather map with plotted station symbols, or a text table. Use the reported temperature and pressure to calculate the density.

B8. Search the web for updated information on the acceleration due to gravity, and how it varies with location on Earth.

B9. Search the web for weather maps showing thickness between two pressure surfaces. One of the most common is the 1000 - 500 mb thickness chart (i.e., the 100 - 50 kPa thickness chart). Comment on how thickness varies with temperature (the most obvious example is the general thickness decrease further away from the equator).

B10. Access from the web an upper-air sounding (e.g., Stuve, Skew-T, Tephigram, etc.) that plots temperature vs. height or pressure for a location near you. We will learn details about these charts later, but for now look at only temperature vs. height. If the sounding goes high enough (up to 100 mb or 10 kPa or so) , can you identify the troposphere, tropopause, and stratosphere.

B11. Often weather maps have isopleths of temperature (isotherm), pressure (isobar), height (contour), humidity (isohume), potential temperature (adiabat or isentrope), or wind speed (isotach). Search the web for weather maps showing other isopleths. (Hint, look for isopleth maps of precipitation, visibility, snow depth, cloudiness, etc.)

Apply

These are essentially "plug & chug" exercises. They are designed to ensure that you are comfortable with the equations, units, and physics by getting hands-on experience using them. None of the problems require calculus.

While most of the numerical problems can be solved using a hand calculator, many students find it easier to compose all of their homework answers on a computer spreadsheet. It is easier to correct mistakes using a spreadsheet, and plotting graphs of the answer is trivial.

Some exercises are flagged with the symbol **(§)**, which means you should use a Spreadsheet or other more advanced tool such as Matlab, Mathematica, or Maple. These exercises have tedious repeated calculations to graph a curve or trend. To do them by hand calculator would be painful. If you don't know how to use a spreadsheet (or other more advanced program), now is a good time to learn.

Most modern spreadsheets also allow you to add objects called text boxes, note boxes or word boxes, to allow you to include word-wrapped paragraphs of text, which are handy for the "Problem" and the "Exposition" parts of the answer.

A spreadsheet example is given in Fig. 1.13. Normally, to make your printout look neater, you might use the page setup or print option to turn off printing of the row numbers, column letters, and grid lines. Also, the borders around the text boxes can be eliminated, and color could be used if you have access to a color printer. Format all graphs to be clear and attractive, with axes labeled and with units, and with tic marks having pleasing increments.

A1. Find the wind direction (degrees) and speed (m s^{-1}), given the (U, V) components:
a. (-5, 0) knots b. (8, -2) m s^{-1}
c. (-1, 15) mi h^{-1} d. (6, 6) m s^{-1}
e. (8, 0) knots f. (5, 20) m s^{-1}
g. (-2, -10) mi h^{-1} h. (3, -3) m s^{-1}

A2. Find the U and V wind components (m s^{-1}), given wind direction and speed:
a. west at 10 knots b. north at 5 m s^{-1}
c. 225° at 8 mi h^{-1} d. 300° at 15 knots
e. east at 7 knots f. south at 10 m s^{-1}
g. 110° at 8 mi h^{-1} h. 20° at 15 knots

A3. Convert the following UTC times to local times in your own time zone:
a. 0000 b. 0330 c. 0610 d. 0920
e. 1245 f. 1515 g. 1800 h. 2150

A4. (i). Suppose that a typical airline window is circular with radius 15 cm, and a typical cargo door is square of side 2 m. If the interior of the aircraft is pressured at 80 kPa, and the ambient outside pressure is given below in kPa, then what are the magnitudes of forces pushing outward on the window and door?

(ii). Your weight in pounds is the force you exert on things you stand on. How many people of your same weight standing on a window or door are needed to equal the forces calculated in part a. Assume the window and door are horizontal, and are near the Earth's surface.
a. 30 b. 25 c. 20 d. 15
e. 10 f. 5 g. 0 h. 40

A5. Find the pressure in kPa at the following heights above sea level, assuming an average T = 250K:
a. -100 m (below sea level) b. 1 km
c. 11 km d. 25 km e. 30,000 ft
f. 5 km g. 2 km h. 15,000 ft

A6. Use the definition of pressure as a force per unit area, and consider a column of air that is above a horizontal area of 1 square meter. What is the mass of air in that column:
a. above the Earth's surface.
b. above a height where the pressure is 50 kPa?
c. between pressure levels of 70 and 50 kPa?
d. above a height where the pressure is 85 kPa?
e. between pressure levels 100 and 20 kPa?
f. above height where the pressure is 30 kPa?
g. between pressure levels 100 and 50 kPa?
h. above a height where the pressure is 10 kPa?

A7. Find the virtual temperature (°C) for air of:

	a.	b.	c.	d.	e.	f.	g.
T (°C)	20	10	30	40	50	0	–10
r (g/kg)	10	5	0	40	60	2	1

A8. Given the planetary data in Table 1-7.

(i). What are the escape velocities from a planet for each of their main atmospheric components? (For simplicity, use the planet radius instead of the "critical" radius at the base of the exosphere.).

(ii). What are the most likely velocities of those molecules at the surface, given the average surface temperatures given in that table? Comparing these answers to part (i), which of the constituents (if any) are most likely to escape? a. Mercury b. Venus c. Mars d. Jupiter e. Saturn f. Uranus g. Neptune h. Pluto

Table 1-7. Planetary data.

Planet	Radius (km)	T_{sfc} (°C) (avg.)	Mass relative to Earth	Main gases in atmos.
Mercury	2440	180	0.055	H_2, He
Venus	6052	480	0.814	CO_2, N_2
Earth	6378	8	1.0	N_2, O_2
Mars	3393	–60	0.107	CO_2, N_2
Jupiter	71400	–150	317.7	H_2, He
Saturn	60330	–185	95.2	H_2, He
Uranus	25560	–214	14.5	H_2, He
Neptune	24764	–225	17.1	H_2, He
Pluto*	1153	–236	0.0022	CH_4, N_2, CO

* *Demoted to a "dwarf planet" in 2006.*

A9. Convert the following temperatures:
- a. 15°C = ?K
- b. 50°F = ?°C
- c. 70°F = ?K
- d. 15°C = ?°F
- e. 303 K = ?°C
- f. 250K = ?°F
- g. 2000°C = ?K
- h. –40°F = ?°C

A10. a. What is the density (kg·m^{-3}) of air, given P = 80 kPa and T = 0 °C ?

b. What is the temperature (°C) of air, given P = 90 kPa and ρ = 1.0 kg·m^{-3} ?

c. What is the pressure (kPa) of air, given T = 90°F and ρ = 1.2 kg·m^{-3} ?

d. Give 2 combinations of pressure and density that have a temperature of 30°C.

e. Give 2 combinations of pressure and density that have a temperature of 0°C.

f. Give 2 combinations of pressure and density that have a temperature of –20°C.

g. How could you determine air density if you did not have a density meter?

h. What is the density (kg·m^{-3}) of air, given P = 50 kPa and T = –30 °C ?

i. What is the temperature (°C) of air, given P = 50 kPa and ρ = 0.5 kg·m^{-3} ?

j. What is the pressure (kPa) of air, given T = –25°C and ρ = 1.2 kg·m^{-3} ?

A11. At a location in the atmosphere where the air density is 1 kg m^{-3}, find the change of pressure (kPa) you would feel if your altitude increases by ___ km.
a. 2 b. 5 c. 7 d. 9 e. 11 f. 13 g. 16
h. –0.1 i. –0.2 j. –0.3 k. –0.4 l. –0.5

A12. At a location in the atmosphere where the average virtual temperature is 5°C, find the height difference (i.e., the thickness in km) between the following two pressure levels (kPa):
a. 100, 90 b. 90, 80 c. 80, 70 d. 70, 60
e. 60, 50 f. 50, 40 g. 40, 30 h. 30, 20
i. 20, 10 j. 100, 80 k. 100, 70 l. 100, 60
m. 100, 50 n. 50, 30

A13. Name the isopleths that would be drawn on a weather map to indicate regions of equal
a. pressure b. temperature
c. cloudiness d. precipitation accumulation
e. humidity f. wind speed
g. dew point h. pressure tendency

A14. What is the geometric height and geopotential, given the geopotential height?
a. 10 m b. 100 m c. 1 km d. 11 km
What is the geopotential height and geopotential, given the geometric height?
e. 500 m f. 2 km g. 5 km h. 20 km

A15. What is the standard atmospheric temperature, pressure, and density at each of the following geopotential heights?
a. 1.5 km b. 12 km c. 50 m d. 8 km
e. 200 m f. 5 km g. 40 km h. 25 km

A16. What are the geometric heights (assuming a standard atmosphere) at the top and bottom of the:
a. troposphere b. stratosphere
c. mesosphere d. thermosphere

A17. Is the inverse of an average of numbers equal to the average of the inverses of those number? (Hint, work out the values for just two numbers: 2 and 4.) This question helps explain where the hypsometric equation given in this chapter is only approximate.

A18(§). Using the standard atmosphere equations, re-create the numbers in Table 1-5 for $0 \le H \le 51$ km.

Evaluate & Analyze

These questions require more thought, and are extensions of material in the chapter. They might require you to combine two or more equations or concepts from different parts of the chapter, or from other chapters. You might need to critically evaluate an approach. Some questions require a numerical answer — others are "short-answer" essays.

They often require you to make assumptions, because insufficient data is given to solve the problem. Whenever you make assumptions, justify them first. A sample solution to such an exercise is shown below.

Sample Application – Evaluate & Analyze (E)

What are the limitations of eq. (1.9a), if any? How can those limitations be eliminated?

Find the Answer

Eq. (1.9a) for P vs. z relies on an average temperature over the whole depth of the atmosphere. Thus, eq. (1.9a) is accurate only when the actual temperature is constant with height.

As we learned later in the chapter, a typical or "standard" atmosphere temperature is NOT constant with height. In the troposphere, for example, temperature decreases with height. On any given day, the real temperature profile is likely to be even more complicated. Thus, eq. (1.9a) is inaccurate.

A better answer could be found from the hypsometric equation (1.26b):

$$P_2 = P_1 \cdot \exp\left(-\frac{z_2 - z_1}{a \cdot \overline{T_v}}\right) \quad \text{with } a = 29.3 \text{ m K}^{-1}.$$

By iterating up from the ground over small increments $\Delta z = z_2 - z_1$, one can use any arbitrary temperature profile. Namely, starting from the ground, set $z_1 = 0$ and $P_1 = 101.325$ kPa. Set $z_2 = 0.1$ km, and use the average virtual temperature value in the hypsometric equation for that 0.1 km thick layer from $z = 0$ to 0.1 km. Solve for P_2. Then repeat the process for the layer between $z = 0.1$ and 0.2 km, using the new T_v for that layer.

Because eq. (1.9a) came from eq. (1.26), we find other limitations.

1) Eq. (1.9a) is for **dry air**, because it uses temperature rather than virtual temperature.

2) The constant "a" in eq. (1.9a) equals = (1/29.3) K m^{-1}. Hence, on a different planet with different gravity and different gas constant, "a" would be different. Thus, eq. (1.9a) is limited to **Earth**.

Nonetheless, eq. (1.9a) is a reasonable first-order approximation to the variation of pressure with altitude, as can be seen by using standard-atmosphere P values from Table 1-5, and plotting them vs. z. The result (which was shown in the Sample Application after Table 1-5) is indeed close to an exponential decrease with altitude.

E1. What are the limitations of the "standard atmosphere"?

E2. For any physical variable that decreases exponentially with distance or time, the e-folding scale is defined as the distance or time where the physical variable is reduced to $1/e$ of its starting value. For the atmosphere the e-folding height for pressure decrease is known as the scale height. Given eq. (1.9a), what is the algebraic and numerical value for atmospheric scale height (km)?

E3(§). Invent some arbitrary data, such as 5 data points of wind speed M vs. pressure P. Although P is the independent variable, use a spreadsheet to plot it on the vertical axis (i.e., switch axes on your graph so that pressure can be used as a surrogate measure of height), change that axis to a logarithmic scale, and then reverse the scale so that the largest value is at the bottom, corresponding to the greatest pressure at the bottom of the atmosphere.

Now add to this existing graph a second curve of different data of M vs. P. Learn how to make both curves appear properly on this graph because you will use this skill repeatedly to solve problems in future chapters.

E4. Does hydrostatic equilibrium (eq. 1.25) always apply to the atmosphere? If not, when and why not?

E5. a. Plug eqs. (1.1) and (1.2a) into (1.3), and use trig to show that $U = U$. b. Similar, but for $V = V$.

E6. What percentage of the atmosphere is above a height (km) of : a. 2 b. 5 c. 11 d. 32 e. 1 f. 18 g. 47 h. 8

E7. What is the mass of air inside an airplane with a cabin size of 5 x 5 x 30 m, if the cabin is pressurized to a cabin altitude of sea level? What mass of outside air is displaced by that cabin, if the aircraft is flying at an altitude of 3 km? The difference in those two masses is the load of air that must be carried by the aircraft. How many people cannot be carried because of this excess air that is carried in the cabin?

E8. Given air of initial temperature 20°C and density of 1.0 kg m^{-3}.

a. What is its initial pressure?
b. If the temperature increases to 30°C in an isobaric process, what is the new density?
c. If the temperature increases to 30°C in an isobaric process, what is the new pressure?
d. For an isothermal process, if the pressure changes to 20 kPa, what is the new density?
e. For an isothermal process, if the pressure

changes to 20 kPa, what is the new T?
f. In a large, sealed, glass bottle that is full of air, if you increase the temperature, what if anything would be conserved (P, T, or ρ)?
g. In a sealed, inflated latex balloon, if you lower it in the atmosphere, what thermodynamic quantities if any, would be conserved?
h. In a mylar (non stretching) balloon, suppose that it is inflated to equal the surrounding atmospheric pressure. If you added more air to it, how would the state change?

E9(§). Starting from sea-level pressure at $z = 0$, use the hypsometric equation to find and plot P vs. z in the troposphere, using the appropriate standard-atmosphere temperature. Step in small increments to higher altitudes (lower pressures) within the troposphere, within each increment. How is your answer affected by the size of the increment? Also solve it using a constant temperature equal to the average surface value. Plot both results on a semi-log graph, and discuss meaning of the difference.

E10. Use the ideal gas law and eq. (1.9) to derive the equation for the change of density with altitude, assuming constant temperature.

E11. What is the standard atmospheric temperature, pressure, and density at each of the following geopotential heights (km)?

a. 75	b. 65	c. 55	d. 45	e. 35
f. 25	g. 15	h. 5	i. –0.5	

E12. The ideal gas law and hypsometric equation are for compressible gases. For liquids (which are incompressible, to first order), density is not a function of pressure. Compare the vertical profile of pressure in a liquid of constant temperature with the profile of a gas of constant temperature.

E13. At standard sea-level pressure and temperature, how does the average molecular speed compare to the speed of sound? Also, does the speed of sound change with altitude? Why?

E14. For a standard atmosphere below $H = 11$ km:
a. Derive an equation for pressure as a function of H.
b. Derive an equation for density as a function of H.

E15. Use the hypsometric equation to derive an equation for the scale height for pressure, H_p.

Synthesize

These are "what if" questions. They are often hypothetical — on the verge of being science fiction. By thinking about "what if" questions you can gain insight about the physics of the atmosphere, because often you cannot apply existing paradigms.

"What if" questions are often asked by scientists, engineers, and policy makers. For example, "What if the amount of carbon dioxide in the atmosphere doubled, then how would world climate change?"

For many of these questions, there is not a single right answer. Different students could devise different answers that could be equally insightful, and if they are supported with reasonable arguments, should be worth full credit. Often one answer will have other implications about the physics, and will trigger a train of related ideas and arguments.

A Sample Application of a synthesis question is presented in the next page. This solution might not be the only correct solution, if it is correct at all.

S1. What if the meteorological angle convention is identical to that shown in Fig. 1.2, except for wind directions which are given by where they blow towards rather than where they blow from. Create a new set of conversion equations (1.1 - 1.4) for this convention, and test them with directions and speeds from all compass quadrants.

S2. Find a translation of Aristotle's *Meteorologica* in your library. Discuss one of his erroneous statements, and how the error might have been avoided if he had following the Scientific Method as later proposed by Descartes.

S3. As discussed in a Sample Application, the glass on the front face of CRT and old TV picture tubes is thick in order to withstand the pressure difference across it. Why is the glass not so thick on the other parts of the picture tube, such as the narrow neck near the back of the TV?

S4. Eqs. (1.9a) and (1.13a) show how pressure and density decrease nearly exponentially with height.
a. How high is the top of the atmosphere?
b. Search the library or the web for the effective altitude for the top of the atmosphere as experienced by space vehicles re-entering the atmosphere.

S5. What is "ideal" about the ideal gas law? Are there equations of state that are not ideal?

S6. What if temperature as defined by eq. (1.5) was not dependent on the molecular weight of the gas. Speculate on how the composition of the Earth's

Sample Application – Synthesize

What if liquid water (raindrops) in the atmosphere caused the virtual temperature to increase [rather than decrease as currently shown by the negative sign in front of r_L in eq. (1.22)]. What would be different about the weather?

Find the Answer

More and larger raindrops would cause warmer virtual temperature. This warmer air would act more buoyant (because warm air rises). This would cause updrafts in rain clouds that might be fast enough to prevent heavy rain from reaching the ground.

But where would all this rain go? Would it accumulate at the top of thunderstorms, at the top of the troposphere? If droplets kept accumulating, they might act sufficiently warm to rise into the stratosphere. Perhaps layers of liquid water would form in the stratosphere, and would block out the sunlight from reaching the surface.

If less rain reached the ground, then there would be more droughts. Eventually all the oceans would evaporate, and life on Earth as we know it would die.

But perhaps there would be life forms (insects, birds, fish, people) in this ocean layer aloft. The reason: if liquid water increases virtual temperature, then perhaps other heavy objects (such as automobiles and people) would do the same.

In fact, this begs the question as to why liquid water would be associated with warmer virtual temperature in the first place. We know that liquid water is heavier than air, and that heavy things should sink. One way that heavy things like rain drops would not sink is if gravity worked backwards.

If gravity worked backwards, then all things would be repelled from Earth into space. This textbook would be pushed into space, as would your instructor. So you would have never been assigned this exercise in the first place.

Life is full of paradoxes. Just be careful to not get a sign wrong in any of your equations — who knows what might happen as a result.

atmosphere might have evolved differently since it was first formed.

S7. When you use a hand pump to inflate a bicycle or car tire, the pump usually gets hot near the outflow hose. Why? Since pressure in the ideal gas law is proportional to the inverse of absolute virtual temperature ($P=\rho \cdot \Re_d/T_v$), why should the tire-pump temperature warmer than ambient?

S8. In the definition of virtual temperature, why do water vapor and liquid water have opposite signs?

S9. How should equation (1.22) for virtual temperature be modified to also include the effects of airplanes and birds flying in the sky?

S10. Meteorologists often convert actual station pressures to the equivalent "sea-level pressure" by taking into account the altitude of the weather station. The hypsometric equation can be applied to this job, assuming that the average virtual temperature is known. What virtual temperature should be used below ground to do this? What are the limitations of the result?

S11. Starting with our Earth and atmosphere as at present, what if gravity were to become zero. What would happen to the atmosphere? Why?

S12. Suppose that gravitational attraction between two objects becomes greater, not smaller, as the distance between the two objects becomes greater.

a. Would the relationship between geometric altitude and geopotential altitude change? If so, what is the new relationship?
b. How would the vertical pressure gradient in the atmosphere be different, if at all?
c. Would the orbit of the Earth around the sun be affected? How?

2 SOLAR & INFRARED RADIATION

Contents

Solar energy powers the atmosphere. This energy warms the air and drives the air motion you feel as winds. The seasonal distribution of this energy depends on the orbital characteristics of the Earth around the sun.

The Earth's rotation about its axis causes a daily cycle of sunrise, increasing solar radiation until solar noon, then decreasing solar radiation, and finally sunset. Some of this solar radiation is absorbed at the Earth's surface, and provides the energy for photosynthesis and life.

Downward **infrared** (IR) radiation from the atmosphere to the Earth is usually slightly less than upward IR radiation from the Earth, causing net cooling at the Earth's surface both day and night. The combination of daytime solar heating and continuous IR cooling yields a **diurnal** (daily) cycle of **net radiation**.

ORBITAL FACTORS

Planetary Orbits

Johannes Kepler, the 17th century astronomer, discovered that planets in the solar system have elliptical orbits around the sun. For most planets in the solar system, the eccentricity (deviation from circular) is relatively small, meaning the orbits are nearly circular. For circular orbits, he also found that the time period Y of each orbit is related to the distance R of the planet from the sun by:

$$Y = a_1 \cdot R^{3/2} \tag{2.1}$$

Parameter $a_1 \approx 0.1996$ d·(Gm)$^{-3/2}$, where d is Earth days, and Gm is gigameters = 10^6 km.

Figs. 2.1a & b show the orbital periods vs. distances for the planets in our solar system. These figures show the duration of a year for each planet, which affect the seasons experienced on the planet.

Orbit of the Earth

The Earth and the moon rotate with a **sidereal** (relative to the stars) period of 27.32 days around their common center of gravity, called the Earth-moon **barycenter**. (Relative to the moving Earth,

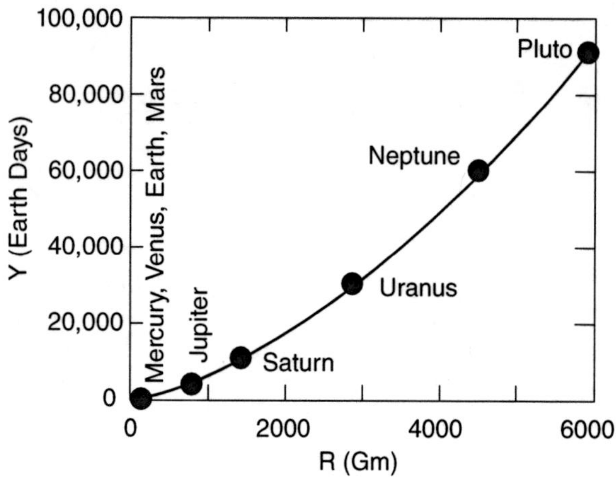

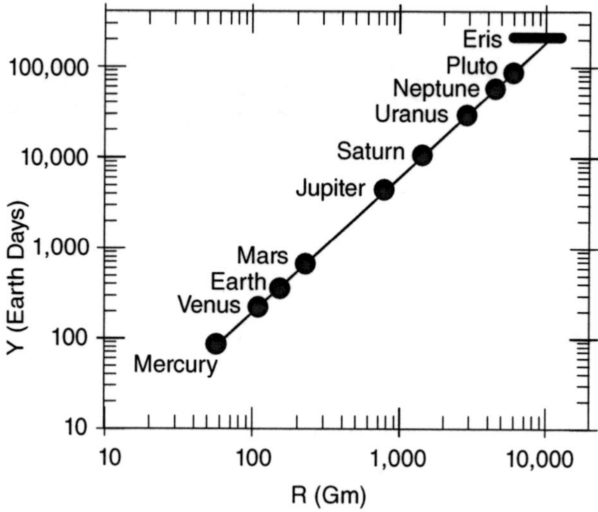

Figure 2.1 a (linear) & b (log-log)
Planetary orbital periods versus distance from sun. Eris (136199) and Pluto (134340) are dwarf planets. Eris, larger than Pluto, has a very elliptical orbit. (a) Linear graph. (b) Log-log graph. (See Appendix A for comparison of various graph formats.)

Sample Application
Verify that eq. (2.1) gives the correct orbital period of one Earth year.

Find the Answer:
Given: R = 149.6 Gm avg. distance sun to Earth.
Find: Y = ? days, the orbital period for Earth

Use eq. (2.1):
$Y = (0.1996 \text{ d}\cdot(\text{Gm})^{-3/2}) \cdot [(149.6 \text{ Gm})^{1.5}]$
= **365.2 days**.

Check: Units OK. Sketch OK. Almost 1 yr.
Exposition: In 365.0 days, the Earth does not quite finish a complete orbit. After four years this shortfall accumulates to nearly a day, which we correct using a leap year with an extra day (see Chapter 1).

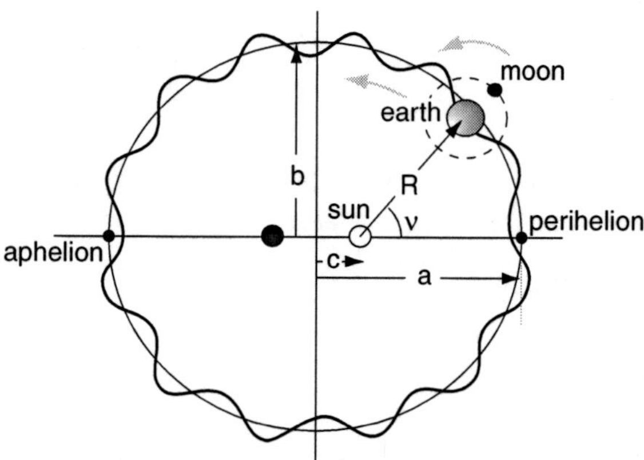

Figure 2.2
Geometry of the Earth's orbit, as viewed from above Earth's North Pole. (Not to scale.) Dark wavy line traces the Earth-center path, while the thin smooth ellipse traces the barycenter path.

the time between new moons is 29.5 days.) Because the mass of the moon (7.35x10^{22} kg) is only 1.23% of the mass of the Earth (Earth mass is 5.9726x10^{24} kg), the barycenter is much closer to the center of the Earth than to the center of the moon. This barycenter is 4671 km from the center of the Earth, which is below the Earth's surface (Earth radius is 6371 km).

To a first approximation, the Earth-moon barycenter orbits around the sun in an **elliptical** orbit (Fig. 2.2, thin-line ellipse) with sidereal period of P = 365.256363 days. Length of the **semi-major axis** (half of the longest axis) of the ellipse is a = 149.598 Gm. This is almost equal to an **astronomical unit** (au), where 1 au = 149,597,870.691 km.

Semi-minor axis (half the shortest axis) length is b = 149.090 Gm. The center of the sun is at one of the foci of the ellipse, and half the distance between the two foci is c = 2.5 Gm, where $a^2 = b^2 + c^2$. The orbit is close to circular, with an **eccentricity** of only about $e \approx c/a$ = 0.0167 (a circle has zero eccentricity).

The closest distance (**perihelion**) along the major axis between the Earth and sun is $a - c$ = 146.96 Gm and occurs at about $d_p \approx$ 4 January. The farthest distance (**aphelion**) is $a + c$ = 151.96 Gm and occurs at about 5 July. The dates for the perihelion and aphelion jump a day or two from year to year because the orbital period is not exactly 365 days. Figs. 2.3 show these dates at the **prime meridian** (Greenwich), but the dates will be slightly different in your own time zone. Also, the dates of the perihelion and aphelion gradually become later by 1 day every 58 years, due to **precession** (shifting of the location of the major and minor axes) of the Earth's orbit around the sun (see the Climate chapter).

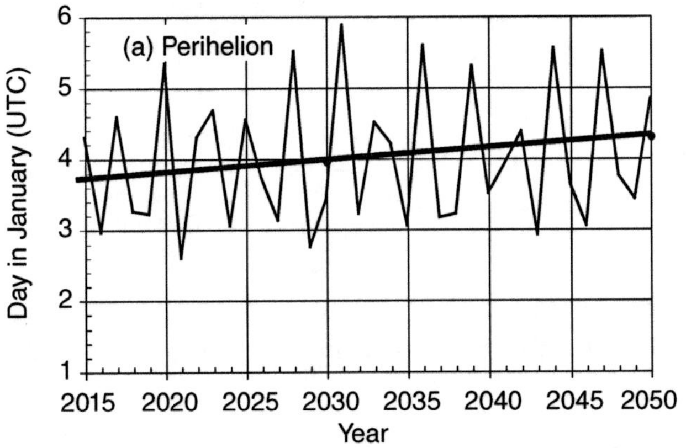

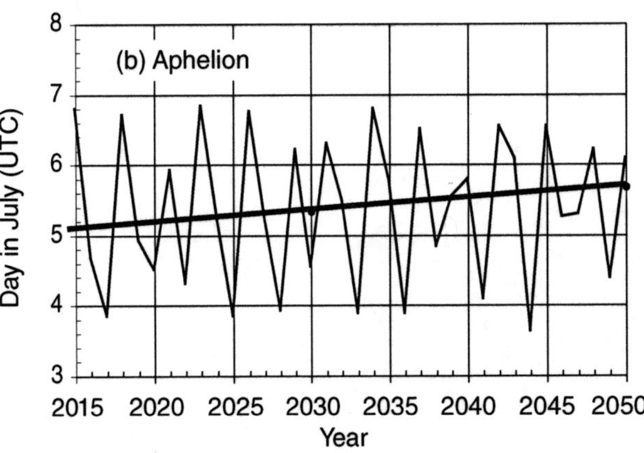

Figure 2.3
Dates (UTC) of the (a) perihelion and (b) aphelion, & their trends (thick line). From the US Naval Observatory. http://aa.usno.navy.mil/data/docs/EarthSeasons.php

Because the Earth is rotating around the Earth-moon barycenter while this barycenter is revolving around the sun, the location of the center of the Earth traces a slightly wiggly path as it orbits the sun. This path is exaggerated in Fig. 2.2 (thick line).

Define a relative **Julian Day**, *d*, as the day of the year. For example, 15 January corresponds to d = 15. For 5 February, d = 36 (which includes 31 days in January plus 5 days in February).

The angle at the sun between the perihelion and the location of the Earth (actually to the Earth-moon barycenter) is called the **true anomaly** ν (see Fig. 2.2). This angle increases during the year as the day d increases from the perihelion day (about d_p = 4; namely, about 4 January). According to **Kepler's second law**, the angle increases more slowly when the Earth is further from the sun, such that a line connecting the Earth and the sun will sweep out equal areas in equal time intervals.

An angle called the **mean anomaly** M is a simple, but good approximation to ν. It is defined by:

$$M = C \cdot \frac{d - d_p}{P} \tag{2.2}$$

where P = 365.256363 days is the (sidereal) orbital period and C is the angle of a full circle (C = 2·π radians = 360°. Use whichever is appropriate for your calculator, spreadsheet, or computer program.)

Because the Earth's orbit is nearly circular, ν ≈ M. A better approximation to the **true anomaly** for the elliptical Earth orbit is

$$\nu \approx M + [2e - (e^3/4)] \cdot \sin(M) + [(5/4) \cdot e^2] \cdot \sin(2M) + [(13/12) \cdot e^3] \cdot \sin(3M) \tag{2.3a}$$

or

$$\nu \approx M + 0.0333988 \cdot \sin(M) + 0.0003486 \cdot \sin(2 \cdot M) + 0.0000050 \cdot \sin(3 \cdot M) \tag{2.3b}$$

for both ν and M in radians, and e = 0.0167.

Sample Application(§)

Use a spreadsheet to find the true anomaly and sun-Earth distance for several days during the year.

Find the Answer

Given: d_p = 4 Jan. P = 365.25 days.
Find: ν = ?° and R = ? Gm.

Sketch: (same as Fig 2.2)
For example, for 15 Feb, d = 46
Use eq. (2.2):
M = (2 · 3.14159)·(46–4)/ 365.256363 = **0.722 radians**
Use eq. (2.3b):
ν =0.722 +0.0333988·sin(0.722) + 0.0003486·sin(1.444) + 0.000005·sin(2.166) = **0.745 radians**
Use eq. (2.4):
R = (149.457Gm)·(1 – 0.0167^2)/[1+0.0167·cos(0.745)]
= (149.457Gm) · 0.99972 / 1.012527 = **147.60 Gm**

Repeating this calculation on a spreadsheet for several days, and comparing M vs. ν and R(M) vs. R(ν) , gives:

Date	d	M (rad)	ν (rad)	R(M) (Gm)	R(ν) (Gm)
4 Jan	4	0	0	146.96	146.96
18 Jan	18	0.241	0.249	147.03	147.04
1 Feb	32	0.482	0.497	147.24	147.25
15 Feb	46	0.722	0.745	147.57	147.60
1 Mar	60	0.963	0.991	148.00	148.06
15 Mar	74	1.204	1.236	148.53	148.60
29 Mar	88	1.445	1.487	149.10	149.18
12 Apr	102	1.686	1.719	149.70	149.78
26 Apr	116	1.927	1.958	150.29	150.36
21 Jun	172	2.890	2.898	151.87	151.88
23 Sep	266	4.507	4.474	149.93	150.01
22 Dec	356	6.055	6.047	147.02	147.03

Check: Units OK. Physics OK.
Exposition: Because M and ν are nearly equal, you can use M instead of ν in eq. (2.4), with good accuracy.

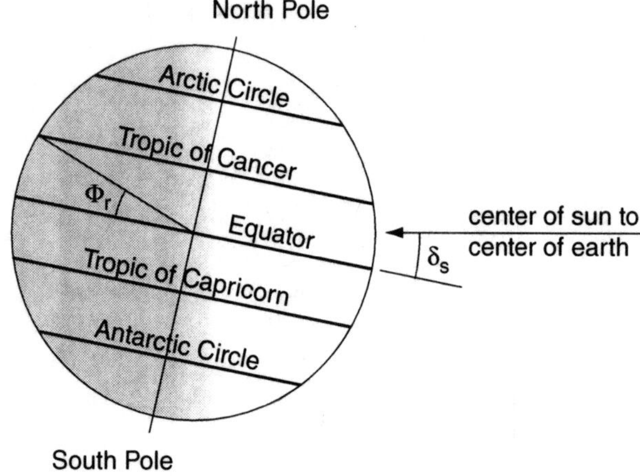

Figure 2.4
Relationship of declination angle δ_S to tilt of the Earth's axis, for a day near northern-hemisphere summer.

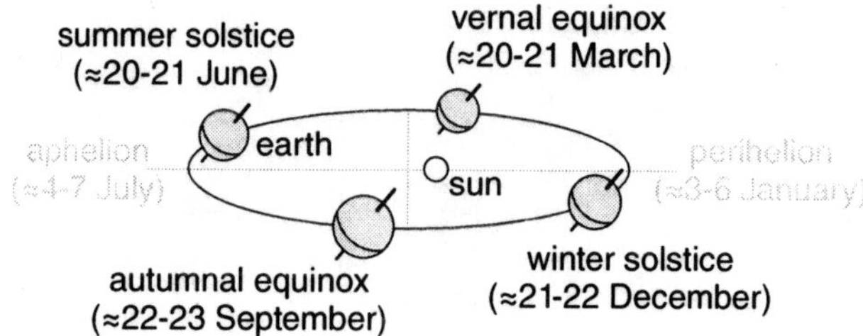

Figure 2.5
Dates (UTC) of northern-hemisphere seasons relative to Earth's orbit.

The distance R between the sun and Earth (actually to the Earth-moon barycenter) as a function of time is

$$R = a \cdot \frac{1 - e^2}{1 + e \cdot \cos(\nu)} \tag{2.4}$$

where e = 0.0167 is eccentricity, and a = 149.457 Gm is the semi-major axis length. If the simple approximation of $\nu \approx M$ is used, then angle errors are less than 2° and distance errors are less than 0.06%.

Table 2-1 . Dates and times (UTC) for northern hemisphere equinoxes and solstices. Format: dd hhmm gives day (dd), hour (hh) and minutes (mm). *From the US Naval Observatory. http://aa.usno.navy.mil/data/docs/EarthSeasons.php*

Year	Spring Equinox (March)	Summer Solstice (June)	Fall Equinox (Sept.)	Winter Solstice (Dec.)
2015	20 2245	21 1638	23 0820	22 0448
2016	20 0430	20 2234	22 1421	21 1044
2017	20 1028	21 0424	22 2002	21 1628
2018	20 1615	21 1007	23 0154	21 2222
2019	20 2158	21 1554	23 0750	22 0419
2020	20 0349	20 2143	22 1330	21 1002
2021	20 0937	21 0332	22 1921	21 1559
2022	20 1533	21 0914	23 0104	21 2148
2023	20 2124	21 1458	23 0650	22 0327
2024	20 0306	20 2051	22 1244	21 0921
2025	20 0901	21 0242	22 1819	21 1503

Seasonal Effects

The tilt of the Earth's axis relative to a line perpendicular to the **ecliptic** (i.e., the orbital plane of the Earth around the sun) is presently Φ_r = 23.44° = 0.409 radians. The direction of tilt of the Earth's axis is not fixed relative to the stars, but wobbles or precesses like a top with a period of about 25,781 years. Although this is important over millennia (see the Climate chapter), for shorter time intervals (up to a century) the precession is negligible, and you can assume a fixed tilt.

The **solar declination angle** δ_S is defined as the angle between the ecliptic and the plane of the Earth's equator (Fig. 2.4). Assuming a fixed orientation (tilt) of the Earth's axis as the Earth orbits the sun, the solar declination angle varies smoothly as the year progresses. The north pole points partially toward the sun in summer, and gradually changes to point partially away during winter (Fig. 2.5).

Although the Earth is slightly closer to the sun in winter (near the perihelion) and receives slightly more solar radiation then, the Northern Hemisphere receives substantially less sunlight in winter because of the tilt of the Earth's axis. Thus, summers are warmer than winters, due to Earth-axis tilt.

The solar declination angle is greatest (+23.44°) on about 20 to 21 June (**summer solstice** in the Northern Hemisphere, when daytime is longest) and is most negative (–23.44°) on about 21 to 22 December (**winter solstice**, when nighttime is longest). The **vernal equinox** (or **spring equinox**, near 20 to 21 March) and **autumnal equinox** (or **fall equinox**, near 22 to 23 September) are the dates when daylight hours equal nighttime hours ("eqi nox" literally translates to "equal night"), and the solar declination angle is zero.

Astronomers define a **tropical year** (= 365.242190 days) as the time from vernal equinox to the next vernal equinox. Because the tropical year is not an integral number of days, the **Gregorian calendar** (the calendar adopted by much of the western world) must be corrected periodically to prevent the seasons (dates of summer solstice, etc.) from shifting into different months.

To make this correction, a **leap day** (29 Feb) is added every 4th year (i.e., **leap years**, are years divisible by 4), except that years divisible by 100 don't have a leap day. However, years divisible by 400 do have a leap day (for example, year 2000). Because of all these factors, the dates of the solstices, equinoxes, perihelion, and aphelion jump around a few days on the Gregorian calendar (Table 2-1 and Figs. 2.3), but remain in their assigned months.

The solar declination angle for any day of the year is given by

$$\delta_S \approx \Phi_r \cdot \cos\left[\frac{C \cdot (d - d_r)}{d_y}\right] \qquad \bullet(2.5)$$

where d is Julian day, and d_r is the Julian day of the summer solstice, and d_y = 365 (or = 366 on a leap year) is the number of days per year. For years when the summer solstice is on 21 June, d_r = 172. This equation is only approximate, because it assumes the Earth's orbit is circular rather than elliptical. As before, $C = 2 \cdot \pi$ radians = 360° (use radians or degrees depending on what your calculator, spreadsheet, or computer program expects).

By definition, Earth-tilt angle (Φ_r = 23.44°) equals the latitude of the **Tropic of Cancer** in the Northern Hemisphere (Fig. 2.4). Latitudes are defined to be positive in the Northern Hemisphere. The **Tropic of Capricorn** in the Southern Hemisphere is the same angle, but with a negative sign. The **Arctic Circle** is at latitude 90° – Φ_r = 66.56°, and the **Antarctic Circle** is at latitude –66.56° (i.e., 66.56°S). During the Northern-Hemisphere summer solstice: the solar declination angle equals Φ_r ; the sun at noon is directly overhead (90° elevation angle) at the Tropic of Cancer; and the sun never sets that day at the Arctic Circle.

Sample Application

Find the solar declination angle on 5 March.

Find the Answer

Assume: Not a leap year.
Given: d = 31 Jan + 28 Feb + 5 Mar = 64.
Find: δ_s = ?°
Sketch:

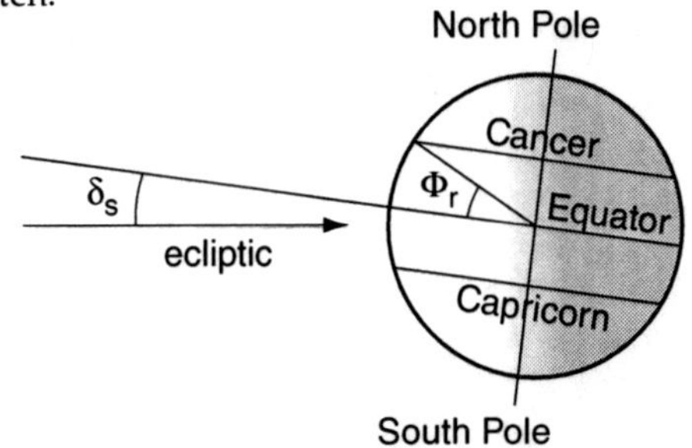

Use eq. (2.5):

δ_s = 23.44° · cos[360°·(64–172)/365]

= 23.44° · cos[–106.521°] = **<u>–6.67°</u>**

Check: Units OK. Sketch OK. Physics OK.
Exposition: On the **vernal equinox** (21 March), the angle should be zero. Before that date, it is winter and the declination angle should be negative. Namely, the ecliptic is below the plane of the equator. In spring and summer, the angle is positive. Because 5 March is near the end of winter, we expect the answer to be a small negative angle.

Sample Application

Find the local elevation angle of the sun at 3 PM local time on 5 March in Vancouver, Canada.

Find the Answer

Assume: Not a leap year. Not daylight savings time. Also, Vancouver is in the Pacific time zone, where $t_{UTC} = t + 8$ h during standard time.

Given: t = 3 PM = 15 h. Thus, t_{UTC} = 23 h.
ϕ = 49.25°N, λ_e = –123.1° (West) for Vancouver.
δ_s = –6.665° from previous Sample Application.

Find: Ψ = ?°

Use eq. (2.6): $\sin(\Psi)$ =
= sin(49.25°)·sin(–6.665°) – cos(49.25°)·cos(–6.665°)·cos[360°·(23h/24h) – 123.1°]
= 0.7576 · (–0.1161) – 0.6528 · 0.9932 · cos(221.9°)
= –0.08793 + 0.4826 = 0.3946
Ψ = arcsin(0.3946) = **23.24°**

Check: Units OK. Physics OK.
Exposition: The sun is above the local horizon, as expected for mid afternoon.

Beware of other situations such as night that give negative elevation angle.

Daily Effects

As the Earth rotates about its axis, the **local elevation angle** Ψ of the sun above the local horizon rises and falls. This angle depends on the latitude ϕ and longitude λ_e of the location:

$$\sin(\Psi) = \sin(\phi)\cdot\sin(\delta_S) - \cos(\phi)\cdot\cos(\delta_S)\cdot\cos\left[\frac{C\cdot t_{UTC}}{t_d} + \lambda_e\right] \qquad \bullet(2.6)$$

Time of day t_{UTC} is in UTC, $C = 2\pi$ radians = 360° as before, and the length of the day is t_d. For t_{UTC} in hours, then t_d = 24 h. Latitudes are positive north of the equator, and longitudes are positive east of the prime meridian. The $\sin(\Psi)$ relationship is used later in this chapter to calculate the daily cycle of solar energy reaching any point on Earth.

[CAUTION: Don't forget to convert angles to radians if required by your spreadsheet or programming language.]

Sample Application(§)

Use a spreadsheet to plot elevation angle vs. time at Vancouver, for 21 Dec, 23 Mar, and 22 Jun. Plot these three curves on the same graph.

Find the Answer

Given: Same as previous Sample Application, except d = 355, 82, 173.

Find: Ψ = ?°. Assume not a leap year.

A portion of the tabulated results are shown below, as well as the full graph.

	Ψ (°)		
t (h)	21 Dec	23 Mar	22 Jun
3	0.0	0.0	0.0
4	0.0	0.0	0.0
5	0.0	0.0	6.6
6	0.0	0.0	15.6
7	0.0	7.8	25.1
8	0.0	17.3	34.9
9	5.7	25.9	44.5
10	11.5	33.2	53.4
11	15.5	38.4	60.5
12	17.2	40.8	64.1
13	16.5	39.8	62.6

(continues in next column)

Sample Application *(continuation)*

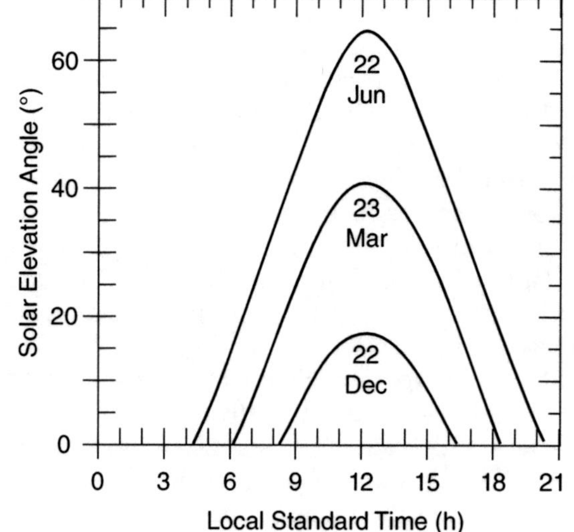

Check: Units OK. Physics OK. Graph OK.
Exposition: Summers are pleasant with long days. The peak elevation does not happen precisely at local noon, because Vancouver is not centered within its time zone.

The local **azimuth angle** α of the sun relative to north is

$$\cos(\alpha) = \frac{\sin(\delta_S) - \sin(\phi)\cdot\cos(\zeta)}{\cos(\phi)\cdot\sin(\zeta)} \tag{2.7}$$

where $\zeta = C/4 - \Psi$ is the **zenith angle**. After noon, the azimuth angle might need to be corrected to be $\alpha = C - \alpha$, so that the sun sets in the west instead of the east. Fig. 2.6 shows an example of the elevation and azimuth angles for Vancouver (latitude = 49.25°N, longitude = 123.1°W) during the solstices and equinoxes.

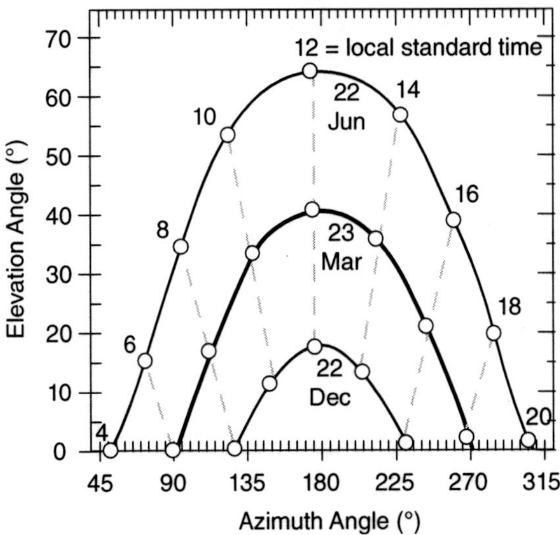

Figure 2.6
Position (solid lines) of the sun for Vancouver, Canada for various seasons. September 21 and March 23 nearly coincide. Isochrones are dashed. All times are Pacific standard time.

Sunrise, Sunset & Twilight

Geometric sunrise and sunset occur when the center of the sun has zero elevation angle. **Apparent sunrise/set** are defined as when the top of the sun crosses the horizon, as viewed by an observer on the surface. The sun has a finite radius corresponding to an angle of 0.267° when viewed from Earth. Also, **refraction** (bending) of light through the atmosphere allows the top of the sun to be seen even when it is really 0.567° below the horizon. Thus, apparent sunrise/set occurs when the center of the sun has an elevation angle of –0.833°.

When the apparent top of the sun is slightly below the horizon, the surface of the Earth is not receiving direct sunlight. However, the surface can still receive indirect light scattered from air molecules higher in the atmosphere that are illuminated by the sun. The interval during which scattered light is present at the surface is called **twilight**. Because twilight gradually fades as the sun moves lower below the horizon, there is no precise definition of the start of sunrise twilight or the end of sunset twilight.

Arbitrary definitions have been adopted by different organizations to define twilight. **Civil twilight** occurs while the sun center is no lower than –6°, and is based on the ability of humans to see objects on the ground. **Military twilight** occurs while the sun is no lower than –12°. **Astronomical twilight** ends when the skylight becomes sufficiently dark to view certain stars, at solar elevation angle –18°.

Table 2-2 summarizes the solar elevation angle Ψ definitions used for sunrise, sunset and twilight. All of these angles are at or below the horizon.

Sample Application(§)
Use a spreadsheet to find the Pacific standard time (PST) for all the events of Table 2-2, for Vancouver, Canada, during 22 Dec, 23 Mar, and 22 Jun.

Find the Answer
Given: Julian dates 355, 82, & 173.
Find: t = ? h (local standard time)
Assume: Pacific time zone: $t_{UTC} = t + 8$ h.

Use eq. (2.8a) and Table 2-2:

	22Dec	23Mar	22Jun
Morning:		PST (h)	
geometric sunrise	8.22	6.20	4.19
apparent sunrise	8.11	6.11	4.09
civil twilight starts	7.49	5.58	3.36
military twilight starts	6.80	4.96	2.32
astron. twilight starts	6.16	4.31	n/a
Evening:			
geometric sunset	16.19	18.21	20.22
apparent sunset	16.30	18.30	20.33
civil twilight ends	16.93	18.83	21.05
military twilight ends	17.61	19.45	22.09
astron. twilight ends	18.26	20.10	n/a

Check: Units OK. Physics OK.
Exposition: During the summer solstice (22 June), the sun never gets below –18°. Hence, it is astronomical twilight all night in Vancouver in mid summer.

During June, Vancouver is on daylight time, so the actual local time would be one hour later.

Table 2-2. Elevation angles for diurnal events.

Event	Ψ (°)	Ψ (radians)
Sunrise & Sunset:		
Geometric	0	0
Apparent	–0.833	–0.01454
Twilight:		
Civil	–6	–0.10472
Military	–12	–0.20944
Astronomical	–18	–0.31416

INFO • Astronomical Values for Time

The constants $a = 2 \cdot e/\omega_E$, $b = [\tan^2(\varepsilon/2)]/\omega_E$, and $c = 2 \cdot \varpi$ in the Equation of Time are based on the following astronomical values: $\omega_E = 2\pi/24\text{h} = 0.0043633$ radians/minute is Earth's rotation rate about its axis, $e = 0.01671$ is the eccentricity of Earth's orbit around the sun, $\varepsilon = 0.40909$ radians = 23.439° is the obliquity (tilt of Earth's axis), $\varpi = 4.9358$ radians = 282.8° is the angle (in the direction of Earth's orbit around the sun) between a line from the sun to the vernal (Spring) equinox and a line drawn from the sun to the moving perihelion (see Fig. 2.5, and Fig. 21.10 in Chapter 21).

Sample Application (§)

Plot time correction vs. day of the year.

Find the Answer:

Given: $d = 0$ to 365

Find: $\Delta t_a = ?$ minutes

Use a spreadsheet. For example, for $d = 45$ (which is 14 Feb), first use eq. (2.2) find the mean anomaly:

$M = 2\pi \cdot (45\text{-}4)/365.25 = 0.705$ rad $= 40.4°$

Then use the **Equation of Time** (2.8b) in

$\Delta t_a = -(7.659 \text{ min}) \cdot \sin(0.705 \text{ rad})$

$\quad + (9.836 \text{ min}) \cdot \sin(2 \cdot 0.705 \text{ rad} + 3.588 \text{ rad})$

$\Delta t_a = -4.96 - 9.46 =$ **<u>–14.4 minutes</u>**

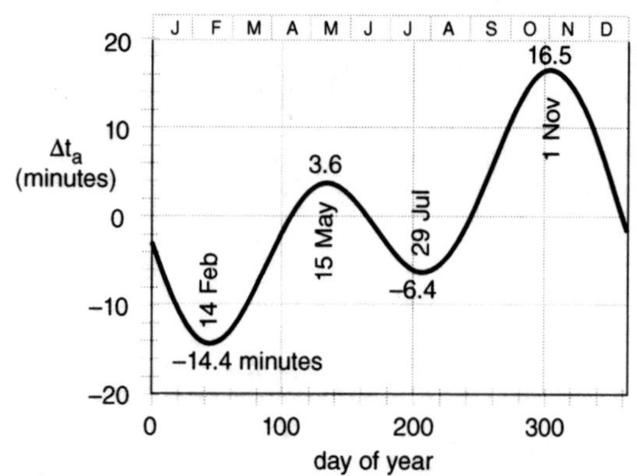

Check: Units OK. Magnitude OK.

Exposition: Because of the negative sign in eq. (2.8c), you need to <u>**ADD**</u> 14.4 minutes to the results of eq. (2.8a) to get the correct sunrise and sunset times.

Approximate (sundial) time-of-day corresponding to these events can be found by rearranging eq. (2.6):

$$t_{UTC} = \frac{t_d}{C} \cdot \left\{ -\lambda_e \pm \arccos\left[\frac{\sin\phi \cdot \sin\delta_s - \sin\Psi}{\cos\phi \cdot \cos\delta_s} \right] \right\} \tag{2.8a}$$

where the appropriate elevation angle is used from Table 2-2. Where the ± sign appears, use + for sunrise and – for sunset. If any of the answers are negative, add 24 h to the result.

To correct the time for the tilted, elliptical orbit of the earth, use the approximate **Equation of Time**:

$$\Delta t_a \approx -a \cdot \sin(M) + b \cdot \sin(2M + c) \tag{2.8b}$$

where $a = 7.659$ minutes, $b = 9.863$ minutes, $c = 3.588$ radians = 205.58°, and where the mean anomaly M from eq. (2.2) varies with day of the year. This time correction is plotted in the Sample Application. The corrected (mechanical-clock) time t_{eUTC} is:

$$t_{eUTC} = t_{UTC} - \Delta t_a \tag{2.8c}$$

which corrects sundial time to mechanical-clock time. Don't forget to convert the answer from UTC to your local time zone.

FLUX

A **flux density,** $\mathbb{F}$, called a **flux** in this book, is the transfer of a quantity per unit area per unit time. The area is taken perpendicular (normal) to the direction of flux movement. Examples with metric (SI) units are mass flux ($\text{kg} \cdot \text{m}^{-2} \text{s}^{-1}$) and heat flux, ($\text{J} \cdot \text{m}^{-2} \cdot \text{s}^{-1}$). Using the definition of a watt ($1 \text{ W} = 1 \text{ J} \cdot \text{s}^{-1}$), the heat flux can also be given in units of ($\text{W} \cdot \text{m}^{-2}$). A flux is a measure of the amount of inflow or outflow such as through the side of a fixed volume, and thus is frequently used in Eulerian frameworks (Fig. 2.7).

Because flow is associated with a direction, so is flux associated with a direction. You must account for fluxes $\mathbb{F}_x$, $\mathbb{F}_y$, and $\mathbb{F}_z$ in the x, y, and z directions, respectively. A flux in the positive x-direction (eastward) is written with a positive value of $\mathbb{F}_x$, while a flux towards the opposite direction (westward) is negative.

The total amount of heat or mass flowing through a plane of area A during time interval Δt is given by:

$$Amount = \mathbb{F} \cdot A \cdot \Delta t \tag{2.9}$$

For heat, $Amount \equiv \Delta Q_H$ by definition.

Fluxes are sometimes written in **kinematic form**, F, by dividing by the air density, ρ_{air} :

$$F = \frac{\mathbb{F}}{\rho_{air}} \qquad (2.10)$$

Kinematic mass flux equals the wind speed, M. Kinematic fluxes can also be in the 3 Cartesian directions: F_x, F_y, and F_z.

[CAUTION: Do not confuse the usage of the symbol M. Here it means "wind speed". Earlier it meant "mean anomaly". Throughout this book, many symbols will be re-used to represent different things, due to the limited number of symbols. Even if a symbol is not defined, you can usually determine its meaning from context.]

Heat fluxes $\mathbb{F}_H$ can be put into kinematic form by dividing by both air density ρ_{air} and the specific heat for air C_p, which yields a quantity having the same units as temperature times wind speed (K·m·s^{-1}).

$$F_H = \frac{\mathbb{F}_H}{\rho_{air} \cdot C_p} \quad \text{for heat only} \qquad (2.11)$$

For dry air (subscript "$_d$") at sea level:

$$\rho_{air} \cdot C_{pd} = 1231 \text{ (W·m}^{-2}) \text{ / (K·m·s}^{-1})$$

$$= 12.31 \text{ mb·K}^{-1}$$

$$= 1.231 \text{ kPa·K}^{-1} .$$

The reason for sometimes expressing fluxes in kinematic form is that the result is given in terms of easily measured quantities. For example, while most people do not have "Watt" meters to measure the normal "dynamic" heat flux, they do have thermometers and anemometers. The resulting temperature times wind speed has units of a kinematic heat flux (K·m·s^{-1}). Similarly, for mass flux it is easier to measure wind speed than kilograms of air per area per time.

Heat fluxes can be caused by a variety of processes. **Radiative fluxes** are radiant energy (electromagnetic waves or photons) per unit area per unit time. This flux can travel through a vacuum. **Advective flux** is caused by wind blowing through an area, and carrying with it warmer or colder temperatures. For example a warm wind blowing toward the east causes a positive heat-flux component F_{Hx} in the x-direction. A cold wind blowing toward the west also gives positive F_{Hx}. **Turbulent fluxes** are caused by eddy motions in the air, while **conductive fluxes** are caused by molecules bouncing into each other.

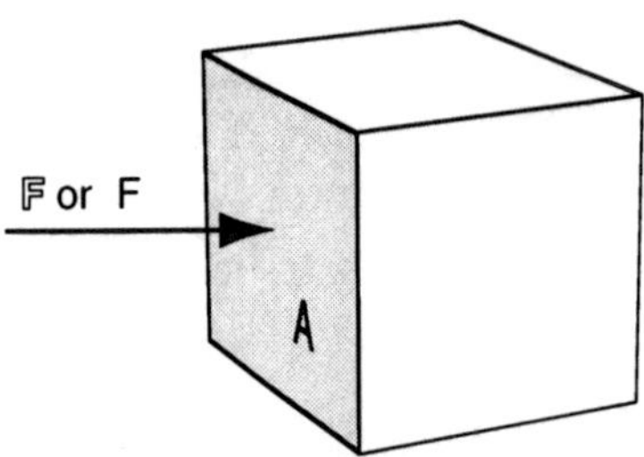

Figure 2.7
Flux F through an area A into one side of a volume.

Sample Application
The mass flux of air is 1 kg·m^{-2}·s^{-1} through a door opening that is 1 m wide by 2.5 m tall. What amount of mass of air passes through the door each minute, and what is the value of kinematic mass flux?

Find the Answer
Given: area A = (1 m) · (2.5 m) = 2.5 m^2
$\mathbb{F}$ = 1 kg·m^{-2}·s^{-1}
Find: (a) Amount = ? kg, and (b) F = ? m·s^{-1}
Sketch: (see Fig. 2.7)

(a) Use eq. (2.9):
Amount = (1 kg·m^{-2}·s^{-1})·(2.5 m^2)·(1 min)·(60 s min^{-1}) = **150 kg**.

(b) Assume: ρ = 1.225 kg·m^{-3} at sea-level
Use eq. (2.10):
F = (1 kg·m^{-2}·s^{-1}) / (1.225 kg·m^{-3})
= **0.82 m·s^{-1}**.

Check: Units OK. Sketch OK. Physics OK.
Exposition: The kinematic flux is equivalent to a very light wind speed of less than 1 m s^{-1} blowing through the doorway, yet it transports quite a large amount of mass each minute.

Sample Application
If the heat flux from the sun that reaches the Earth's surface is 600 W m^{-2}, find the flux in kinematic units.

Find the Answer:
Given: $\mathbb{F}_H$ = 600 W m^{-2}
Find: F_H = ? K·m s^{-1}
Assume: sea level.

Use eq. (2.11)
F_H = (600 W m^{-2}) / [1231 (W·m^{-2}) / (K·m·s^{-1})]
= 0.487 K·m s^{-1}

Check: Units OK. Physics OK.
Exposition: This amount of radiative heat flux is equivalent to an advective flux of a 1 m s^{-1} wind blowing air with temperature excess of about 0.5°C.

Sample Application

Red light has a wavelength of 0.7 µm. Find its frequency, circular frequency, and wavenumber in a vacuum.

Find the Answer

Given: c_o = 299,792,458 m s^{-1}, λ = 0.7 µm
Find: ν = ? Hz, ω = ? s^{-1}, σ = ? m^{-1}.

Use eq. (2.12), solving for ν:
$\nu = c_o/\lambda$ = (3x10^8 m s^{-1}) / (0.7x10^{-6} m cycle^{-1})
= **4.28x10^{14} Hz**.

$\omega = 2\pi\cdot\nu$ = 2·(3.14159)·(4.28x10^{14} Hz)
= **2.69x10^{15}** s^{-1}.

$\sigma = 1/\lambda$ = 1 / (0.7x10^{-6} m cycle^{-1})
= **1.43x10^6** m^{-1}.

Check: Units OK. Physics reasonable.
Exposition: Wavelength, wavenumber, frequency, and circular frequency are all equivalent ways to express the "color" of radiation.

RADIATION PRINCIPLES

Propagation

Radiation can be modeled as electromagnetic waves or as photons. Radiation propagates through a vacuum at a constant speed: c_o = 299,792,458 m·s^{-1}. For practical purposes, you can approximate this **speed of light** as $c_o \approx$ 3x10^8 m·s^{-1}. Light travels slightly slower through air, at roughly c = 299,710,000 m·s^{-1} at standard sea-level pressure and temperature, but the speed varies slightly with thermodynamic state of the air (see the Optics chapter).

Using the wave model of radiation, the **wavelength** λ (m·cycle^{-1}) is related to the **frequency**, ν (Hz = cycles·s^{-1}) by:

$$\lambda\cdot\nu = c_o \tag{2.12}$$

Wavelength units are sometimes abbreviated as (m). Because the wavelengths of light are so short, they are often expressed in units of micrometers (µm).

Wavenumber is the number of waves per meter: σ (cycles·m^{-1}) = $1/\lambda$. An alternative wavenumber definition is radians per meter (= $2\pi/\lambda$). Their units are sometimes abbreviated as (m^{-1}). **Circular frequency** or **angular frequency** is ω (radians s^{-1}) = $2\pi\cdot\nu$. Its units are sometimes abbreviated as (s^{-1}).

Emission

Objects warmer than absolute zero can emit radiation. An object that emits the maximum possible radiation for its temperature is called a **blackbody**. **Planck's law** gives the amount of blackbody monochromatic (single wavelength or color) radiant flux leaving an area, called **emittance** or **radiant exitance**, E_λ*:

$$E_\lambda{}^* = \frac{c_1}{\lambda^5\cdot\left[\exp(c_2/(\lambda\cdot T))-1\right]} \qquad \bullet(2.13)$$

where T is absolute temperature, and the asterisk indicates blackbody. The two constants are:
c_1 = 3.74 x 10^8 W·m^{-2} · µm^4, and
c_2 = 1.44 x 10^4 µm·K.

Eq. (2.13) and constant c_1 already include all directions of exiting radiation from the area. E_λ* has units of W·m^{-2} µm^{-1} ; namely, flux per unit wavelength. For radiation approaching an area rather than leaving it, the radiant flux is called **irradiance**.

Actual objects can emit less than the theoretical blackbody value: $E_\lambda = e_\lambda\cdot E_\lambda{}^*$, where $0 \le e_\lambda \le 1$ is **emissivity**, a measure of emission efficiency.

Sample Application

Find the blackbody monochromatic radiant exitance of green light of wavelength 0.53 µm from an object of temperature 3000 K.

Find the Answer

Given: λ = 0.53 µm, T = 3000 K
Find: E_λ* = ? W·m^{-2}· µm^{-1}.

Use eq. (2.13):

$$E_\lambda{}^* = \frac{c_1}{\lambda^5\cdot\left[\exp(c_2/(\lambda\cdot T))-1\right]}$$

$$E_\lambda{}^* = \frac{(3.74\text{x}10^8\,\text{Wm}^{-2}\mu\text{m}^4)/(0.53\mu\text{m})^5}{\exp\left[(1.44\text{x}10^4\,\text{K}\mu\text{m})/(0.53\mu\text{m}\cdot 3000\text{K})\right]-1}$$

= **1.04 x 10^6** W·m^{-2}· µm^{-1}

Check: Units OK. Physics reasonable.
Exposition: Because 3000 K is cooler than the sun, about 50 times less green light is emitted. The answer is the watts emitted from each square meter of surface per µm wavelength increment.

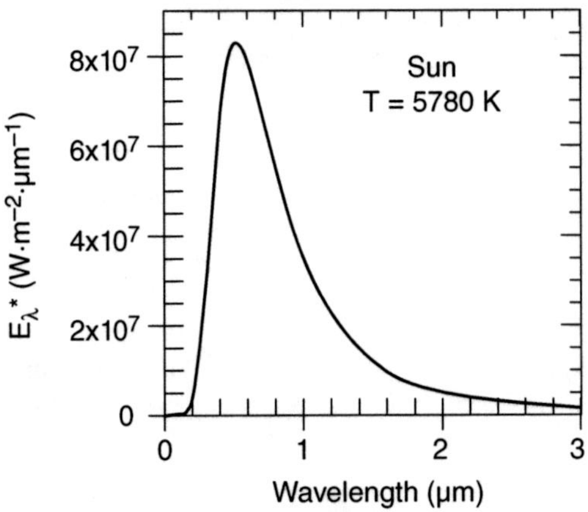

Figure 2.8
Planck radiant exitance, E_λ, from a blackbody approximately the same temperature as the sun.*

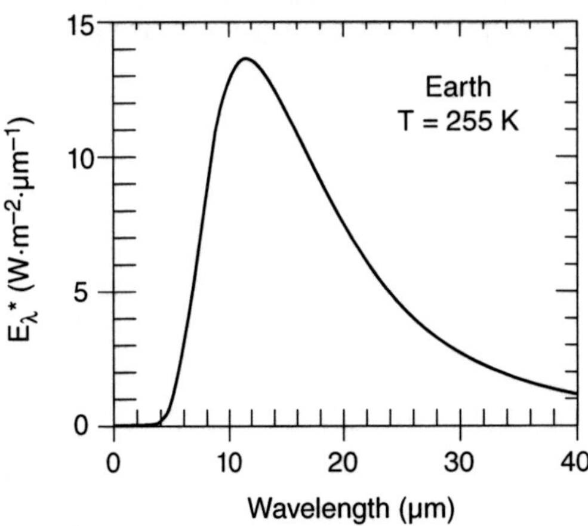

Figure 2.9
Planck radiant exitance, E_λ, from a blackbody approximately the same temperature as the Earth.*

The Planck curve (eq. 2.13) for emission from a blackbody the same temperature as the sun (T = 5780 K) is plotted in Fig. 2.8. Peak emissions from the sun are in the visible range of wavelengths (0.38 – 0.74 µm, see Table 2-3). Radiation from the sun is called **solar radiation** or **short-wave radiation**.

The Planck curve for emission from a blackbody that is approximately the same temperature as the whole Earth-atmosphere system ($T \approx$ 255 K) is plotted in Fig. 2.9. Peak emissions from this idealized average Earth system are in the infrared range 8 to 18 µm. This radiation is called **terrestrial radiation, long-wave radiation**, or infrared (**IR**) radiation.

The wavelength of the peak emission is given by **Wien's law**:

$$\lambda_{max} = \frac{a}{T} \tag{2.14}$$

where a = 2897 µm·K.

The total amount of emission (= area under Planck curve = total emittance) is given by the **Stefan-Boltzmann law**:

$$E^* = \sigma_{SB} \cdot T^4 \qquad \bullet(2.15)$$

where σ_{SB} = 5.67x10⁻⁸ W·m⁻²·K⁻⁴ is the **Stefan-Boltzmann constant**, and E^* has units of W·m⁻². More details about radiation emission are in the Satellites & Radar chapter in the sections on weather satellites.

Table 2-3. Ranges of wavelengths λ of visible colors. Approximate center wavelength is in **boldface**. For more info, see Chapter 22 on Atmospheric Optics.

Color	λ (µm)
Red	0.625 - **0.650** - 0.740
Orange	0.590 - **0.600** - 0.625
Yellow	0.565 - **0.577** - 0.590
Green	0.500 - **0.510** - 0.565
Cyan	0.485 - **0.490** - 0.500
Blue	0.460 - **0.475** - 0.485
Indigo	0.425 - **0.445** - 0.460
Violet	0.380 - **0.400** - 0.425

Sample Application

What is the total radiant exitance (radiative flux) emitted from a blackbody Earth at T = 255 K, and what is the wavelength of peak emission?

Find the Answer

Given: T = 255 K
Find: E^* = ? W·m⁻², λ_{max} = ? µm

Sketch: Earth

Use eq. (2.15):
E^*= (5.67x10⁻⁸ W·m⁻²·K⁻⁴)·(255 K)⁴ = **240 W·m⁻²**.
Use eq. (2.14):
λ_{max} = (2897 µm·K) / (255 K) = **11.36 µm**

Check: Units OK. λ agrees with peak in Fig. 2.9.
Exposition: You could create the same flux by placing a perfectly-efficient 240 W light bulb in front of a parabolic mirror that reflects the light and IR radiation into a beam that is 1.13 m in diameter.

For comparison, the surface area of the Earth is about 5.1x10¹⁴ m², which when multiplied by the flux gives the total emission of 1.22x10¹⁷ W. Thus, the Earth acts like a large-wattage IR light bulb.

HIGHER MATH • Incremental Changes

What happens to the total radiative exitance given by eq. (2.15) if temperature increases by 1°C? Such a question is important for climate change.

Find the Answer using calculus:

Calculus allows a simple, elegant way to find the solution. By taking the derivative of both the left and right sides of eq. (2.15), one finds that:

$$dE^* = 4\cdot\sigma_{SB}\cdot T^3 dT$$

assuming that σ_{SB} is constant. It can be written as

$$\Delta E^* = 4\cdot\sigma_{SB}\cdot T^3\cdot\Delta T$$

for small ΔT.

Thus, a fixed increase in temperature of ΔT = 1°C causes a much larger radiative exitance increase at high temperatures than at cold, because of the T^3 factor on the right side of the equation.

Find the Answer using algebra:

This particular problem could also have been solved using algebra, but with a more tedious and less elegant solution. First let

$$E_2 = \sigma_{SB}\cdot T_2^4 \quad \text{and} \quad E_1 = \sigma_{SB}\cdot T_1^4$$

Next, take the difference between these two eqs:

$$\Delta E = E_2 - E_1 = \sigma\cdot\left[T_2^4 - T_1^4\right]$$

Recall from algebra that $(a^2 - b^2) = (a - b)\cdot(a + b)$

$$\Delta E = \sigma_{SB}\cdot\left(T_2^{\,2} - T_1^{\,2}\right)\cdot\left(T_2^{\,2} + T_1^{\,2}\right)$$

$$\Delta E = \sigma_{SB}\cdot(T_2 - T_1)\cdot(T_2 + T_1)\cdot\left(T_2^{\,2} + T_1^{\,2}\right)$$

Since $(T_2 - T_1) / T_1 << 1$, then $(T_2 - T_1) = \Delta T$, but $T_2 + T_1 \approx 2T$, and $T_2^2 + T_1^2 \approx 2\cdot T^2$. This gives:

$$\Delta E = \sigma_{SB}\cdot\Delta T\cdot 2T\cdot 2T^2$$

or

$$\Delta E \approx \sigma_{SB}\cdot\left[4T^3\Delta T\right]$$

which is identical to the answer from calculus.

We were lucky this time, but it is not always possible to use algebra where calculus is needed.

A SCIENTIFIC PERSPECTIVE • Scientific Laws – The Myth

There are no scientific laws. Some theories or models have succeeded for every case tested so far, yet they may fail for other situations. Newton's "Laws of Motion" were accepted as laws for centuries, until they were found to fail in quantum mechanical, relativistic, and galaxy-size situations. It is better to use the word "relationship" instead of "law". In this textbook, the word "law" is used for sake of tradition.

Einstein said "No amount of experimentation can ever prove me right; a single experiment can prove me wrong."

Because a single experiment can prove a relationship wrong, it behooves us as scientists to test theories and equations not only for reasonable values of variables, but also in the limit of extreme values, such as when the variables approach zero or infinity. These are often the most stringent tests of a relationship.

Example

Query: The following is an approximation to Planck's law.

$$E_\lambda^* = c_1\cdot\lambda^{-5}\cdot\exp[-c_2 / (\lambda\cdot T)] \qquad \text{(a)}$$

Why is it not a perfect substitute?

Find the Answer: If you compare the numerical answers from eqs. (2.13) and (a), you find that they agree very closely over the range of temperatures of the sun and the Earth, and over a wide range of wavelengths. But...

a) What happens as temperature approaches absolute zero? For eq. (2.13), T is in the denominator of the argument of an exponential, which itself is in the denominator of eq. (2.13). If $T = 0$, then $1/T = \infty$. $\text{Exp}(\infty) = \infty$, and $1/\infty = 0$. Thus, $E_\lambda^* = 0$, as it should. Namely, no radiation is emitted at absolute zero (according to classical theory).

For eq. (a), if $T = 0$, then $1/T = \infty$, and $\exp(-\infty) = 0$. So it also agrees in the limit of absolute zero. Thus, both equations give the expected answer.

b) **What happens as temperature approaches infinity?** For eq. (2.13), if $T = \infty$, then $1/T = 0$, and $\exp(0) = 1$. Then $1 - 1 = 0$ in the denominator, and $1/0 = \infty$. Thus, $E_\lambda^* = \infty$, as it should. Namely, infinite radiation is emitted at infinite temperature.

For eq. (a), if $T = \infty$, then $1/T = 0$, and $\exp(-0) = 1$. Thus, $E_\lambda^* = c_1\cdot\lambda^{-5}$, which is not infinity. Hence, this approaches the wrong answer in the ∞ T limit.

Conclusion: Eq. (a) is not a perfect relationship, because it fails for this one case. But otherwise it is a good approximation over a wide range of normal temperatures.

Distribution

Radiation emitted from a spherical source decreases with the square of the distance from the center of the sphere:

$$E_2^* = E_1^* \cdot \left(\frac{R_1}{R_2} \right)^2 \qquad \bullet(2.16)$$

where R is the radius from the center of the sphere, and the subscripts denote two different distances from the center. This is called the **inverse square law**.

The reasoning behind eq. (2.16) is that as radiation from a small sphere spreads out radially, it passes through ever-larger conceptual spheres. If no energy is lost during propagation, then the total energy passing across the surface of each sphere must be conserved. Because the surface areas of the spheres increase with the square of the radius, this implies that the energy flux density must decrease at the same rate; i.e., inversely to the square of the radius.

From eq. (2.16) we expect that the radiative flux reaching the Earth's orbit is greatly reduced from that at the surface of the sun. The solar emissions of Fig. 2.8 must be reduced by a factor of 2.167×10^{-5}, based on the square of the ratio of solar radius to Earth-orbital radius from eq. (2.16). This result is compared to the emission from Earth in Fig. 2.10.

The area under the solar-radiation curve in Fig. 2.10 is the **total** (all wavelengths) **solar irradiance (TSI)**, S_o, reaching the Earth's orbit. We call it an irradiance here, instead of an emittance, because relative to the Earth it is an incoming radiant flux. This quantity was formerly called the **solar constant** but we now know that it varies slightly. The average value of solar irradiance measured at the Earth's orbit by satellites is about

$$S_o = 1366 \ (\pm 7) \ \text{W·m}^{-2} \ . \qquad (2.17)$$

In kinematic units (based on sea-level density), the solar irradiance is roughly S_o = 1.11 K·m s^{-1}.

About ±4 W·m^{-2} of the ±7 W·m^{-2} error bars are due to radiometer calibration errors between the various satellites. Also, during the 11-year sunspot cycle the solar irradiance normally varies by about 1.4 W·m^{-2}, with peak values during sunspot maxima. One example of a longer term variation in solar activity is the **Maunder Minimum** in the late 1600s, during which the solar irradiance was perhaps 2.7 to 3.7 W·m^{-2} less than values during modern solar minima. Such irradiance changes could cause subtle changes (0.3 to 0.4°C) in global climate. See the Climate chapter for more info on solar variability.

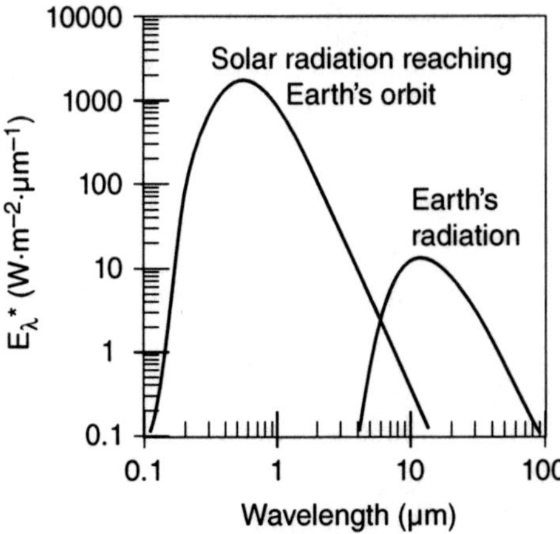

Figure 2.10
Blackbody radiance E reaching top of Earth's atmosphere from the sun and radiance of terrestrial radiation leaving the top of the atmosphere, plotted on a log-log graph.*

Sample Application

Estimate the value of the solar irradiance reaching the orbit of the Earth, given a sun surface temperature (5780 K), sun radius (6.96×10^5 km), and orbital radius (1.495×10^8 km) of the Earth from the sun.

Find the Answer

Given: T_{sun} = 5780 K
$R_{sun} = 6.96 \times 10^5$ km = solar radius
$R_{Earth} = 1.495 \times 10^8$ km = Earth orbit radius
Find: S_o = ? W·m^{-2}
Sketch:

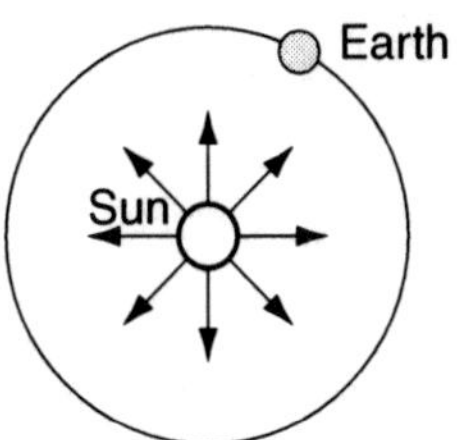

First, use eq. (2.15):
$E_1^* = (5.67 \times 10^{-8} \text{ W·m}^{-2}\text{·K}^{-4})\cdot(5780 \text{ K})^4$
$= 6.328 \times 10^7 \text{ W·m}^{-2}$.
Next, use eq. (2.16), with $R_1 = R_{sun}$ & $R_2 = R_{Earth}$
$S_o = E_2^* = (6.328 \times 10^7 \text{ W·m}^{-2})\cdot$
$(6.96 \times 10^5 \text{ km}/1.495 \times 10^8 \text{ km})^2$ = **1372 W·m^{-2}**.

Check: Units OK, Sketch OK. Physics OK.
Exposition: Answer is nearly equal to that measured by satellites, as given in eq. (2.17). The error is due to a poor estimate of effective sun-surface temperature.

Sample Application(§)

Using the results from an earlier Sample Application that calculated the true anomaly and sun-Earth distance for several days during the year, find the solar radiative forcing for those days.

Find the Answer

Given: R values from previously Sample Application
Find: S = ? W·m^{-2}

Sketch: (same as Fig 2.2)
Use eq. (2.18).

Date	d	R(ν) (Gm)	S (W/m^2)
4 Jan	4	146.96	1418
18 Jan	18	147.04	1416
1 Feb	32	147.25	1412
15 Feb	46	147.60	1405
1 Mar	60	148.06	1397
15 Mar	74	148.60	1386
29 Mar	88	149.18	1376
12 Apr	102	149.78	1365
26 Apr	116	150.36	1354
21 Jun	172	151.88	1327
23 Sep	266	150.01	1361
22 Dec	356	147.03	1416

Check: Units OK. Physics OK.
Exposition: During N. Hemisphere winter, solar radiative forcing is up to 50 W·m^{-2} larger than average.

Sample Application

During the equinox at noon at latitude ϕ =60°, the solar elevation angle is Ψ = 90° – 60° = 30°. If the atmosphere is perfectly transparent, then how much radiative flux is absorbed into a perfectly black asphalt parking lot?

Find the Answer

Given: Ψ = 30° = elevation angle
$E = S_o$ = 1366 W·m^{-2}. solar irradiance
Find: $\mathbb{F}_{rad}$ = ? W·m^{-2}
Sketch:

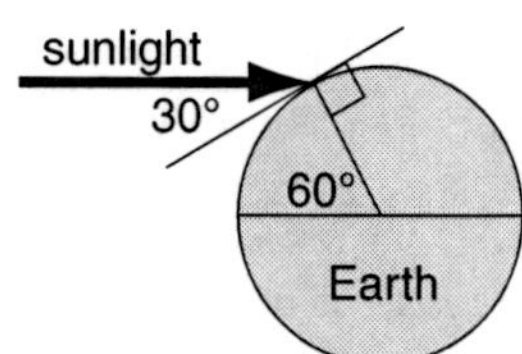

Use eq. (2.19):
$\mathbb{F}_{rad}$ = (1366 W·m^{-2})·sin(30°) = **683 W·m^{-2}**.

Check: Units OK. Sketch OK. Physics OK.
Exposition: Because the solar radiation is striking the parking lot at an angle, the radiative flux into the parking lot is half of the solar irradiance.

According to the inverse-square law, variations of distance between Earth and sun cause changes of the **solar radiative forcing**, S, that reaches the top of the atmosphere:

$$S = S_o \cdot \left(\frac{\overline{R}}{R} \right)^2 \qquad (2.18)$$

where S_o = 1366 W·m^{-2} is the average total solar irradiance measured at an average distance $\overline{R}$ = 149.6 Gm between the sun and Earth, and R is the actual distance between Earth and the sun as given by eq. (2.4). Remember that the solar irradiance and the solar radiative forcing are the fluxes across a surface that is perpendicular to the solar beam, measured above the Earth's atmosphere.

Let irradiance E be any radiative flux crossing a unit area that is perpendicular to the path of the radiation. If this radiation strikes a surface that is not perpendicular to the radiation, then the radiation per unit surface area is reduced according to the **sine law**. The resulting flux, $\mathbb{F}_{rad}$, at this surface is:

$$\mathbb{F}_{rad} = E \cdot \sin(\Psi) \qquad (2.19)$$

where Ψ is the **elevation angle** (the angle of the sun above the surface). In kinematic form, this is

$$F_{rad} = \frac{E}{\rho \cdot C_p} \cdot \sin(\Psi) \qquad •(2.20)$$

where $\rho \cdot C_p$ is given under eq. (2.11).

Average Daily Insolation

The acronym "**insolation**" means "incoming solar radiation" at the top of the atmosphere. The average daily **insolation** $\overline{E}$ takes into account both the solar elevation angle (which varies with season and time of day) and the duration of daylight. For example, there is more total insolation at the poles in summer than at the equator, because the low sun angle near the poles is more than compensated by the long periods of daylight.

$$\overline{E} = \frac{S_o}{\pi} \cdot \left(\frac{a}{R} \right)^2 \cdot \left[h_o{}' \cdot \sin(\phi) \cdot \sin(\delta_s) + \cos(\phi) \cdot \cos(\delta_s) \cdot \sin(h_o) \right] \qquad (2.21)$$

where S_o =1366 W m^{-2} is the solar irradiance, a = 149.457 Gm is Earth's semi-major axis length, R is the actual distance for any day of the year, from eq.

(2.4). In eq. (2.21), h_o' is the sunset and sunrise hour angle in radians.

The hour angle h_o at sunrise and sunset can be found using the following steps:

$$\alpha = -\tan(\phi)\cdot\tan(\delta_S) \tag{2.22a}$$

$$\beta = \min[1, (\max(-1, \alpha)] \tag{2.22b}$$

$$h_o = \arccos(\beta) \tag{2.22c}$$

Eq. (2.22b) truncates the argument of the arccos to be between –1 and 1, in order to account for high latitudes where there are certain days when the sun never sets, and other days when it never rises. [CAUTION. *When finding the arccos, your answer might be in degrees or radians, depending on your calculator, spreadsheet, or computer program. Determine the units by experimenting first with the arccos(0), which will either give 90° or π/2 radians. If necessary, convert the hour angle to units of radians, the result of which is h_o'.*]

Fig. 2.11 shows the average incoming solar radiation vs. latitude and day of the year, found using eq. (2.21). For any one hemisphere, $\overline{E}$ has greater difference between equator and pole during winter than during summer. This causes stronger winds and more active extratropical cyclones in the winter hemisphere than in the summer hemisphere.

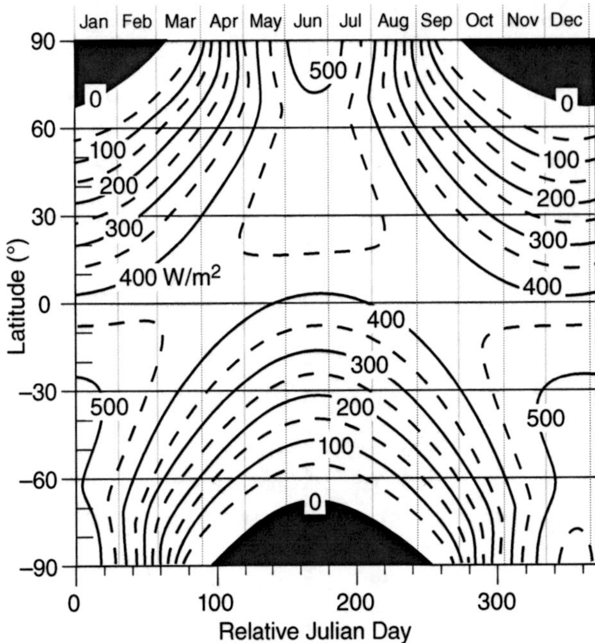

Figure 2.11
Average daily insolation $\overline{E}$ (W m^{-2}) over the globe.

Sample Application
Find the average daily insolation over Vancouver during the summer solstice.

Find the Answer:
Given: $d = d_r = 173$ at the solstice,
$\phi = 49.25°N$, $\lambda_e = 123.1°W$ for Vancouver.
Find: $\overline{E}$ = ? W m^{-2}

Use eq. (2.5): $\delta_S = \Phi_r = 23.45°$
Use eq. (2.2): $M = 167.55°$, and assume $\nu \approx M$.
Use eq. (2.4): $R = 151.892$ Gm.
Use eq. (2.22):
$h_o = \arccos[-\tan(49.25°)\cdot\tan(23.45°)] = 120.23°$
$h_o' = h_o \cdot 2\pi/360° = 2.098$ radians
Use eq. (2.21):

$$\overline{E} = \frac{(1368\text{W}\cdot\text{m}^{-2})}{\pi}\cdot\left(\frac{149\text{Gm}}{151.892\text{Gm}}\right)^2\cdot$$
$$[2.098\cdot\sin(49.25°)\cdot\sin(23.45°) + \cos(49.25°)\cdot\cos(23.45°)\cdot\sin(120.23°)]$$

$\overline{E}$ = (1327 W m^{-2})·[2.098(0.3016)+0.5174]

= **<u>486 W m^{-2}</u>**

Check: Units OK. Physics OK. Agrees with Fig. 2.11.
Exposition: At the equator on this same day, the average daily insolation is less than 400 W m^{-2}.

Absorption, Reflection & Transmission

The emissivity, e_λ, is the fraction of blackbody radiation that is actually emitted (see Table 2-4) at any wavelength λ. The absorptivity, a_λ, is the fraction of radiation striking a surface of an object that is **absorbed** (i.e., stays in the object as a different form of energy). **Kirchhoff's law** states that the **absorptivity** equals the **emissivity** of a substance at each wavelength, λ. Thus,

$$a_\lambda = e_\lambda \tag{2.23}$$

Some substances such as dark glass are semi-**transparent** (i.e., some radiation passes through). A fraction of the incoming (incident) radiation might also be **reflected** (bounced back), and another portion might be **absorbed**. Thus, you can define the efficiencies of reflection, absorption, and transmission as:

$$r_\lambda = \frac{E_{\lambda\ reflected}}{E_{\lambda\ incident}} = \textbf{reflectivity} \tag{2.24}$$

$$a_\lambda = \frac{E_{\lambda\ absorbed}}{E_{\lambda\ incident}} = \textbf{absorptivity} \tag{2.25}$$

Table 2-4. Typical infrared emissivities.

Surface	e	Surface	e
alfalfa	0.95	iron, galvan.	0.13-0.28
aluminum	0.01-0.05	leaf 0.8 µm	0.05-0.53
asphalt	0.95	leaf 1 µm	0.05-0.6
bricks, red	0.92	leaf 2.4 µm	0.7-0.97
cloud, cirrus	0.3	leaf 10 µm	0.97-0.98
cloud, alto	0.9	lumber, oak	0.9
cloud, low	1.0	paper	0.89-0.95
concrete	0.71-0.9	plaster, white	0.91
desert	0.84-0.91	sand, wet	0.98
forest, conif.	0.97	sandstone	0.98
forest, decid.	0.95	shrubs	0.9
glass	0.87-0.94	silver	0.02
grass	0.9-0.95	snow, fresh	0.99
grass lawn	0.97	snow, old	0.82
gravel	0.92	soils	0.9-0.98
human skin	0.95	soil, peat	0.97-0.98
ice	0.96	urban	0.85-0.95

Sample Application

If 500 W m^{-2} of visible light strikes a translucent object that allows 100 W m^{-2} to shine through and 150 W m^{-2} to bounce off, find the transmissivity, reflectivity, absorptivity, and emissivity.

Find the Answer

Given: $E_{\lambda\ incoming} = 500$ W m^{-2},
$E_{\lambda\ transmitted} = 100$ W m^{-2}, $E_{\lambda\ reflected} = 150$ W m^{-2}
Find: $a_\lambda = ?$, $e_\lambda = ?$, $r_\lambda = ?$, and $t_\lambda = ?$

Use eq. (2.26): $t_\lambda = (100 \text{ W m}^{-2}) / (500 \text{ W m}^{-2}) = \underline{0.2}$
Use eq. (2.24): $r_\lambda = (150 \text{ W m}^{-2}) / (500 \text{ W m}^{-2}) = \underline{0.3}$
Use eq. (2.27): $a_\lambda = 1 - 0.2 - 0.3 = \underline{0.5}$
Use eq. (2.23): $e_\lambda = a_\lambda = \underline{0.5}$

Check: Units dimensionless. Physics reasonable.
Exposition: By definition, **translucent** means partly transparent, and partly absorbing.

$$t_\lambda = \frac{E_{\lambda\ transmitted}}{E_{\lambda\ incident}} = \textbf{transmissivity} \qquad (2.26)$$

Values of e_λ, a_λ, r_λ, and t_λ are between 0 and 1.

The sum of the last three fractions must total 1, as 100% of the radiation at any wavelength must be accounted for:

$$1 = a_\lambda + r_\lambda + t_\lambda \qquad (2.27)$$

or

$$E_{\lambda\ incoming} = E_{\lambda\ absorbed} + E_{\lambda\ reflected} + E_{\lambda\ transmitted} \qquad (2.28)$$

For opaque ($t_\lambda = 0$) substances such as the Earth's surface, you find: $a_\lambda = 1 - r_\lambda$.

The reflectivity, absorptivity, and transmissivity usually vary with wavelength. For example, clean snow reflects about 90% of incoming solar radiation, but reflects almost 0% of IR radiation. Thus, snow is "white" in visible light, and "black" in IR. Such behavior is crucial to surface temperature forecasts.

Instead of considering a single wavelength, it is also possible to examine the net effect over a range of wavelengths. The ratio of total reflected to total incoming solar radiation (i.e., averaged over all solar wavelengths) is called the **albedo**, A :

$$A = \frac{E_{reflected}}{E_{incoming}} \qquad \bullet(2.29)$$

The average global albedo for solar radiation reflected from Earth is $A = 30\%$ (see the Climate chapter). The actual global albedo at any instant varies with ice cover, snow cover, cloud cover, soil moisture, topography, and vegetation (Table 2-5). The Moon's albedo is only 7%.

The surface of the Earth (land and sea) is a very strong absorber and emitter of radiation.

Table 2-5. Typical albedos (%) for sunlight.

Surface	A (%)	Surface	A (%)
alfalfa	23-32	forest, decid.	10-25
buildings	9	granite	12-18
clay, wet	16	grass, green	26
clay, dry	23	gypsum	55
cloud, thick	70-95	ice, gray	60
cloud, thin	20-65	lava	10
concrete	15-37	lime	45
corn	18	loam, wet	16
cotton	20-22	loam, dry	23
field, fallow	5-12	meadow, green	10-20
forest, conif.	5-15	potatoes	19

Table 2-5 *(continuation)*. Typical albedos (%).

Surface	A (%)	Surface	A (%)
rice paddy	12	soil, red	17
road, asphalt	5-15	soil, sandy	20-25
road, dirt	18-35	sorghum	20
rye winter	18-23	steppe	20
sand dune	20-45	stones	20-30
savanna	15	sugar cane	15
snow, fresh	75-95	tobacco	19
snow, old	35-70	tundra	15-20
soil, dark wet	6-8	urban, mean	15
soil, light dry	16-18	water, deep	5-20
soil, peat	5-15	wheat	10-23

Within the air, however, the process is a bit more complicated. One approach is to treat the whole atmospheric thickness as a single object. Namely, you can compare the radiation at the top versus bottom of the atmosphere to examine the total emissivity, absorptivity, and reflectivity of the whole atmosphere. Over some wavelengths called **windows** there is little absorption, allowing the radiation to "shine" through. In other wavelength ranges there is partial or total absorption. Thus, the atmosphere acts as a filter. Atmospheric windows and transmissivity are discussed in detail in the Satellites & Radar and Climate chapters.

Beer's Law

Sometimes you must examine radiative **extinction** (reduction of radiative flux) across a short path length Δs within the atmosphere (Fig. 2.12). Let n be the **number density** of radiatively important particles in the air (particles m^{-3}), and b be the **extinction cross section** of each particle (m^2 particle^{-1}), where this latter quantity gives the area of the shadow cast by each particle.

Extinction can be caused by absorption and scattering of radiation. If the change in radiation is due only to absorption, then the absorptivity across this layer is

$$a = \frac{E_{incident} - E_{transmitted}}{E_{incident}} \tag{2.30}$$

Beer's law gives the relationship between incident radiative flux, $E_{incident}$, and transmitted radiative flux, $E_{transmitted}$, as

$$E_{transmitted} = E_{incident} \cdot e^{-n \cdot b \cdot \Delta s} \tag{2.31a}$$

Beer's law can be written using an **extinction coefficient**, k:

$$E_{transmitted} = E_{incident} \cdot e^{-k \cdot \rho \cdot \Delta s} \tag{2.31b}$$

where ρ is the density of air, and k has units of $m^2\, g_{air}^{-1}$. The total extinction across the whole path can be quantified by a dimensionless **optical thickness** (or **optical depth** in the vertical), τ, allowing Beer's law to be rewritten as:

$$E_{transmitted} = E_{incident} \cdot e^{-\tau} \tag{2.31c}$$

To simplify these equations, sometimes a **volume extinction coefficient** γ is defined by

$$\gamma = n{\cdot}b = k{\cdot}\rho \tag{2.32}$$

Visual range (V, one definition of **visibility**) is the distance where the intensity of transmitted light has decreased to 2% of the incident light. It estimates the max distance Δs (km) you can see through air.

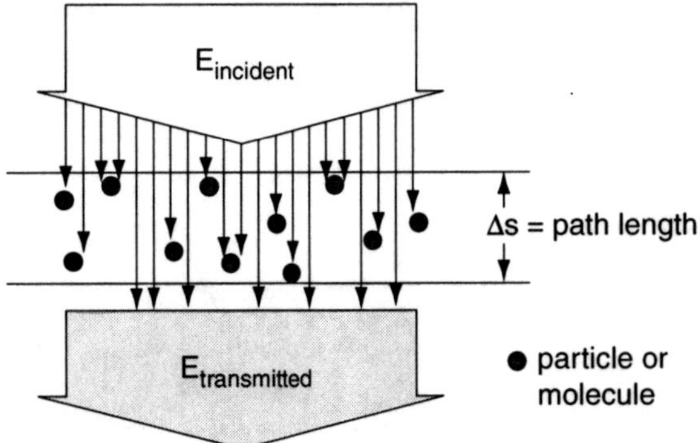

Figure 2.12
Reduction of radiation across an air path due to absorption and scattering by particles such as air-pollution aerosols or cloud droplets, illustrating Beer's law.

Sample Application

Suppose that soot from a burning automobile tire has number density $n = 10^7\ m^{-3}$, and an extinction cross section of $b = 10^{-9}\ m^2$. (a) Find the radiative flux that was not attenuated (i.e., reduction from the solar constant) across the 20 m diameter smoke plume. Assume the incident flux equals the solar constant. (b) Find the optical depth. (c) If this soot fills the air, what is the visual range?

Find the Answer

Given: $n = 10^7\ m^{-3}$, $b = 10^{-9}\ m^2$, $\Delta s = 20$ m
$E_{incident} = S = 1366\ W{\cdot}m^{-2}$ solar constant
Find: E_{tran} =? $W{\cdot}m^{-2}$, τ =? (dimensionless), V =? km
Sketch:

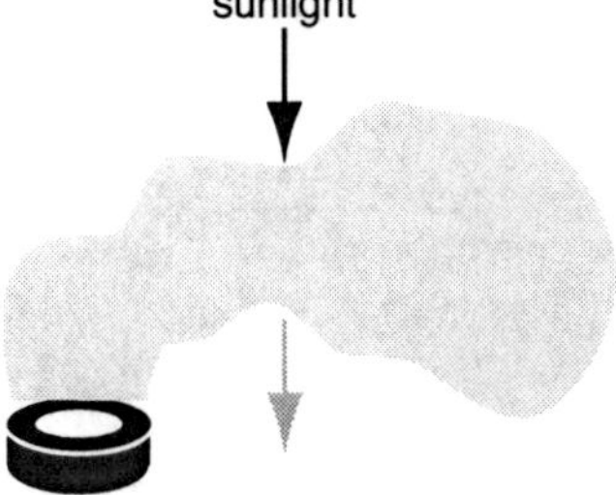

Use eq. (2.32): $\gamma = (10^7\ m^{-3}){\cdot}(10^{-9}\ m^2) = 0.01\ m^{-1}$

(a) Use eq. (2.31a)
E_{tran} = $1366\ W{\cdot}m^{-2} \cdot \exp[-(0.01\ m^{-1}){\cdot}(20\ m)]$
= $1366\ W{\cdot}m^{-2} \cdot \exp[-0.2]$ = $\underline{\mathbf{1118\ W{\cdot}m^{-2}}}$

(b) Rearrange eq. (2.31c):
$\tau = \ln(E_{in}/E_{transmitted}) = \ln(1366/1118) = \underline{\mathbf{0.20}}$

(c) Rearrange eq. (2.31a): $\Delta s = [\ln(E_{incident}/E_{trans})]/\gamma$.
$V = \Delta s = [\ln(1/0.02)]/0.01\ m^{-1} = \underline{\mathbf{0.391\ km}}$

Check: Units OK. Sketch OK. Physics OK.
Exposition: Not much attenuation through this small smoke plume; namely, 1366 – 1118 = 248 $W{\cdot}m^{-2}$ was absorbed by the smoke. However, if this smoke fills the air, then visibility is very poor.

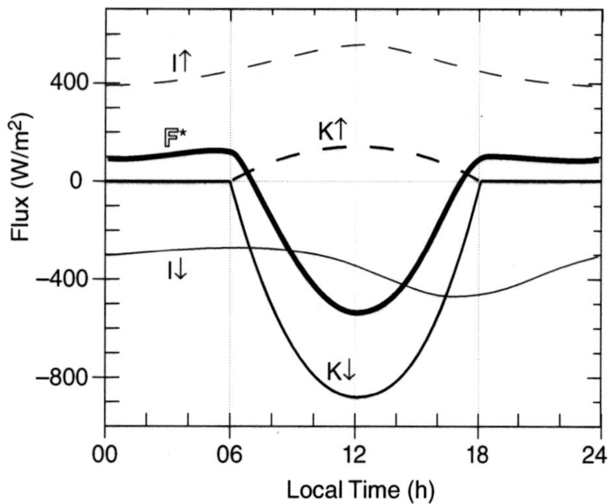

Figure 2.13
Typical diurnal variation of radiative fluxes at the surface. Fluxes are positive upward.

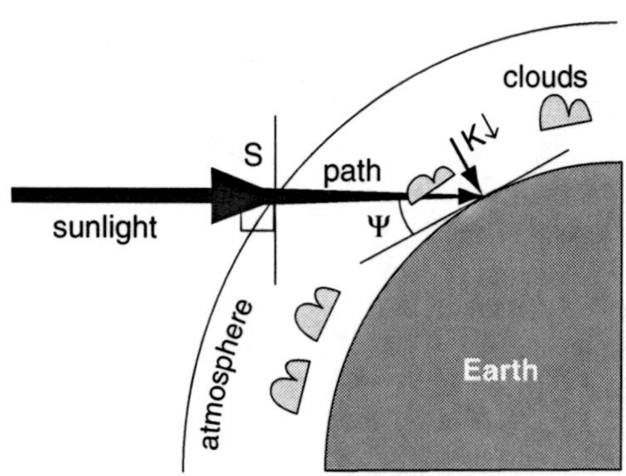

Figure 2.14
Fate of sunlight en route to the Earth's surface.

Sample Application

Downwelling sunlight shines through an atmosphere with 0.8 net sky transmissivity, and hits a ground surface of albedo 0.5, at a time when $\sin(\Psi)$ = 0.3. The surface emits 400 W m^{-2} IR upward into the atmosphere, and absorbs 350 W m^{-2} IR coming down from the atmosphere. Find the net radiative flux.

Find the Answer

Given: $Tr = 0.8$, $S_o = 1366$ W m^{-2}, $A = 0.5$.
$I\uparrow$ = 400 W m^{-2} , $I\downarrow$ = 350 W m^{-2}
Find: $\mathbb{F}^*$ = ? W m^{-2}

Use eq. (2.34):
$K\downarrow$ =–(1366 W m^{-2})·(0.8)· (0.3) = –381 W m^{-2}
Use eq. (2.36): $K\uparrow$= –(0.5)·(–381 W m^{-2})= 164 W m^{-2}
Use eq. (2.33): $\mathbb{F}^*$ =(–381)+(164)+(–350)+(400) W m^{-2}
$\mathbb{F}^*$ = <u>–167</u> W m^{-2}

Check: Units OK. Magnitude and sign OK.
Exposition: Negative sign means net inflow to the surface, such as would cause daytime warming.

SURFACE RADIATION BUDGET

Define $\mathbb{F}^*$ as the **net radiative flux** (positive upward) perpendicular to the Earth's surface. This net flux has contributions (Fig. 2.13) from **downwelling solar** radiation $K\downarrow$, reflected **upwelling solar** $K\uparrow$, **downwelling longwave (IR)** radiation emitted from the atmosphere $I\downarrow$, and **upwelling longwave** emitted from the Earth $I\uparrow$:

$$\mathbb{F}^* = K\downarrow + K\uparrow + I\downarrow + I\uparrow \qquad \bullet(2.33)$$

where $K\downarrow$ and $I\downarrow$ are negative because they are downward.

Solar

Recall that the solar irradiance (i.e., the solar constant) is $S_o \approx 1366 \pm 7$ W·m^{-2} (equivalent to 1.11 K·m s^{-1} in kinematic form after dividing by $\rho \cdot C_p$) at the top of the atmosphere. Some of this radiation is attenuated between the top of the atmosphere and the surface (Fig. 2.14). Also, the sine law (eq. 2.19) must be used to find the component of downwelling solar flux $K\downarrow$ that is perpendicular to the surface. The result for daytime is

$$K\downarrow = -S_o \cdot T_r \cdot \sin(\Psi) \qquad (2.34)$$

where T_r is a **net sky transmissivity**. A negative sign is incorporated into eq. (2.34) because $K\downarrow$ is a downward flux. Eq. (2.6) can be used to find $\sin(\Psi)$. At night, the downwelling solar flux is zero.

Net transmissivity depends on path length through the atmosphere, atmospheric absorption characteristics, and cloudiness. One approximation for the net transmissivity of solar radiation is

$$T_r = (0.6 + 0.2\sin\Psi)(1 - 0.4\sigma_H)(1 - 0.7\sigma_M)(1 - 0.4\sigma_L) \qquad (2.35)$$

where cloud-cover fractions for high, middle, and low clouds are σ_H, σ_M, and σ_L, respectively. These cloud fractions vary between 0 and 1, and the transmissivity also varies between 0 and 1.

Of the sunlight reaching the surface, a portion might be reflected:

$$K\uparrow = -A \cdot K\downarrow \qquad (2.36)$$

where the surface albedo is A.

Longwave (IR)

Upward emission of IR radiation from the Earth's surface can be found from the Stefan-Boltzmann relationship:

$$I\uparrow = e_{IR} \cdot \sigma_{SB} \cdot T^4 \tag{2.37}$$

where e_{IR} is the surface emissivity in the IR portion of the spectrum (e_{IR} = 0.9 to 0.99 for most surfaces), and σ_{SB} is the Stefan-Boltzmann constant (= $5.67\text{x}10^{-8}$ W·m^{-2}·K^{-4}).

However, downward IR radiation from the atmosphere is much more difficult to calculate. As an alternative, sometimes a **net longwave flux** is defined by

$$I^* = I\downarrow + I\uparrow \tag{2.38}$$

One approximation for this flux is

$$I^* = b \cdot (1 - 0.1\sigma_H - 0.3\sigma_M - 0.6\sigma_L) \tag{2.39}$$

where parameter b = 98.5 W·m^{-2} , or b = 0.08 K·m·s^{-1} in kinematic units.

Net Radiation

Combining eqs. (2.33), (2.34), (2.35), (2.36) and (2.39) gives the **net radiation** ($\mathbb{F}^*$, defined positive upward):

$$\mathbb{F}^* = -(1-A) \cdot S \cdot T_r \cdot \sin(\Psi) + I^* \quad \text{daytime} \quad \bullet(2.40a)$$

$$= I^* \quad \text{nighttime} \quad \bullet(2.40b)$$

Sample Application

Find the net radiation at the surface in Vancouver, Canada, at noon (standard time) on 22 Jun. Low clouds are present with 30% coverage.

Find the Answer

Assume: Grass lawns with albedo A = 0.2 .
No other clouds.

Given: σ_L = 0.3

Find: $\mathbb{F}^*$ = ? W·m^{-2}

Use ψ = 64.1° from an earlier Sample Application.
Use eq. (2.35) to find the transmissivity:

T_r = [0.6+0.2·sinψ]·(1–0.4·σ_L)
= [0.6 + 0.2·sin(64.1°)]·[1–(0.4·0.3)]
= [0.80]·(0.88) = 0.686

Use eq. (2.39) to find net IR contribution:

I^* = b·(1 – 0.6·σ_L)
= (98.5 W·m^{-2})·[1–(0.6·0.3)] = 80.77 W·m^{-2}

Use eq. (2.40a):

$\mathbb{F}^*$ = –(1–A) · S · T_r · sin(ψ) + I^*
= –(1–0.2) · (1366 W·m^{-2}) · 0.686 · sin(64.1°)
+ 80.77 W·m^{-2}
=(–674.36 + 80.77) W·m^{-2} = **–593.59 W·m^{-2}**

Check: Units OK. Physics OK.

Exposition: The surface flux is only about 43% of that at the top of the atmosphere, for this case. The negative sign indicates a net inflow of radiation to the surface, such as can cause warming during daytime.

ACTINOMETERS

Sensors designed to measure electromagnetic radiative flux are generically called **actinometers** or **radiometers**. In meteorology, actinometers are usually oriented to measure downwelling or upwelling radiation. Sensors that measure the difference between down- and up-welling radiation are called **net actinometers**.

Special categories of actinometers are designed to measure different wavelength bands:

- **pyranometer** – broadband solar (short-wave) irradiance, viewing a hemisphere of solid angle, with the radiation striking a flat, horizontal plate (Fig. 2.15).
- **net pyranometer** – difference between top and bottom hemispheres for short-wave radiation.

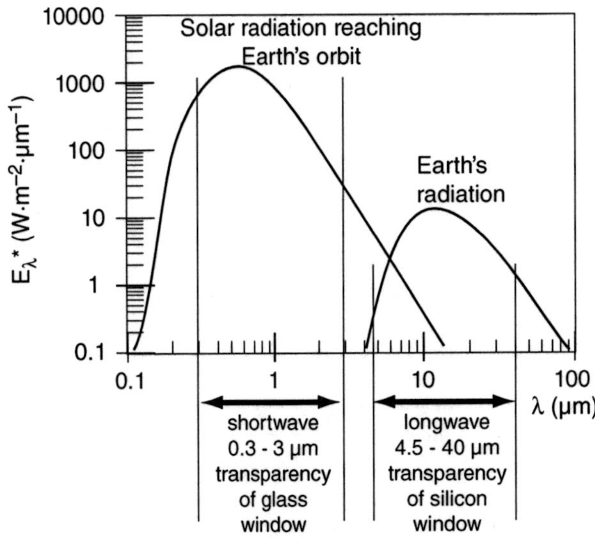

Figure 2.15
The wavelength bands observed by pyranometers (shortwave radiation) and pyrgeometers (longwave radiation) depends on the transparency of the windows used in those instruments.

A SCI. PERSPECTIVE • Seek Solutions

Most differential equations describing meteorological phenomena cannot be solved analytically. They cannot be integrated; they do not appear in a table of integrals; and they are not covered by the handful of mathematical tricks that you learned in math class.

But there is nothing magical about an analytical solution. **Any reasonable solution is better than no solution.** Be creative.

While thinking of creative solutions, also **think of ways to check your answer**. Is it the right order of magnitude, right sign, right units, does it approach a known answer in some limit, must it satisfy some other physical constraint or law or budget?

Example

Find the irradiance that can pass through an atmospheric "window" between wavelengths λ_1 and λ_2.

Find the Answer:

Approach: Integrate Planck's law between the specified wavelengths. This is the area under a portion of the Planck curve.

Check: The area under the whole spectral curve should yield the Stefan-Boltzmann (SB) law. Namely, the answer should be smaller than the SB answer, but should increase and converge to the SB answer as the lower and upper λ limits approach 0 and ∞, respectively.

Methods:

- Pay someone else to get the answer (Don't do this in school!), but be sure to check it yourself.
- Look up the answer in a Table of Integrals.
- Integrate it using the tricks you learned in math class.
- Integrate it using a symbolic equation solver on a computer, such as Mathematica or Maple.
- Find an approximate solution to the full equation. For example, integrate it numerically on a computer. (Trapezoid method, Gaussian integration, finite difference iteration, etc.)
- Find an exact solution for an approximation to the eq., such as a model or idealization of the physics. Most eqs. in this textbook have used this approach.
- Draw the Planck curve on graph paper. Count the squares under the curve between the wavelength bands, and compare to the value of each square, or to the area under the whole curve. (We will use this approach extensively in the Thunderstorm chapter.)
- Draw the curve, and measure area with a planimeter.
- Draw the Planck curve on cardboard or thick paper. Cut out the whole area under the curve. Weigh it. Then cut the portion between wavelengths, & weigh again.
- ...and there are probably many more methods.

- **pyrheliometer** – solar (short wave) direct-beam radiation normal to a flat surface (and shielded from diffuse radiation).
- **diffusometer** – a pyranometer that measures only diffuse solar radiation scattered from air, particles, and clouds in the sky, by using a device that shades the sensor from direct sunlight.
- **pyrgeometer** – infrared (long-wave) radiation from a hemisphere that strikes a flat, horizontal surface (Fig. 2.15).
- **net pyrgeometer** – difference between top and bottom hemispheres for infrared (long-wave) radiation.
- **radiometer** – measure all wavelengths of radiation (short, long, and other bands).
- **net radiometer** – difference between top and bottom hemispheres of radiation at all wavelengths.
- **spectrometers** – measures radiation as a function of wavelength, to determine the spectrum of radiation.

Inside many radiation sensors is a **bolometer**, which works as follows. Radiation strikes an object such as a metal plate, the surface of which has a coating that absorbs radiation mostly in the wavelength band to be measured. By measuring the temperature of the radiatively heated plate relative to a non-irradiated reference, the radiation intensity can be inferred for that wavelength band. The metal plate is usually enclosed in a glass or plastic hemispheric chamber to reduce error caused by heat conduction with the surrounding air.

Inside other radiation sensors are **photometers**. Some photometers use the **photoelectric effect**, where certain materials release electrons when struck by electromagnetic radiation. One type of photometer uses **photovoltaic cells** (also called **solar cells**), where the amount of electrical energy generated can be related to the incident radiation. Another photometric method uses **photoresistor,** which is a high-resistance semiconductor that becomes more conductive when irradiated by light.

Other photometers use **charge-coupled devices** (**CCDs**) similar to the image sensors in digital cameras. These are semiconductor integrated circuits with an array of tiny capacitors that can gain their initial charge by the photoelectric effect, and can then transfer their charge to neighboring capacitors to eventually be "read" by the surrounding circuits.

Simple spectrometers use different filters in front of bolometers or photometers to measure narrow wavelength bands. Higher spectral-resolution spectrometers use **interferometry** (similar to the Michelson inteferometer described in physics books), where the fringes of an interference pattern can be measured and related to the spectral intensities. These are also sometimes called **Fourier-transform spectrometers**, because of the mathematics used to extract the spectral information from the spacing of the fringes.

You can learn more about radiation, including the radiative transfer equation, in the weather-satellite section of the Satellites & Radar chapter. Satellites use radiometers and spectrometers to remotely observe the Earth-atmosphere system. Other satellite-borne radiometers are used to measure the global radiation budget (see the Climate chapter).

REVIEW

The variations of temperature and humidity that you feel near the ground are driven by the diurnal cycle of solar heating during the day and infrared cooling at night. Both diurnal and seasonal heating cycles can be determined from the geometry of the Earth's rotation and orbit around the sun. The same orbital mechanics describes weather-satellite orbits, as is discussed in the Satellites & Radar chapter.

Short-wave radiation is emitted from the sun and propagates through space. It illuminates a hemisphere of Earth. The portion of this radiation that is absorbed is the heat input to the Earth-atmosphere system that drives Earth's weather.

IR radiation from the atmosphere is absorbed at the ground, and IR radiation is also emitted from the ground. The IR and short-wave radiative fluxes do not balance, leaving a net radiation term acting on the surface at any one location. But when averaged over the whole globe, the earth-atmosphere system is approximately in radiative equilibrium.

Instruments to measure radiation are called actinometers or radiometers. Radiometers and spectrometers can be used in remote sensors such as weather satellites.

HOMEWORK EXERCISES

Broaden Knowledge & Comprehension

(Don't forget to cite each web address you use.)

B1. Access a full-disk visible satellite photo image of Earth from the web. What visible clues can you use to determine the current solar declination angle? How does your answer compare with that expected for your latitude and time of year.

B2. Access "web cam" camera images from a city, town, ski area, mountain pass, or highway near you. Use visible shadows on sunny days, along with your knowledge of solar azimuth angles, to determine the direction that the camera is looking.

B3. Access from the web the exact time from military (US Navy) or civilian (National Institute of Standards and Technology) atomic clocks. Synchronize your clocks at home or school, utilizing the proper time zone for your location. What is the time difference between local solar noon (the time when the sun is directly overhead) and the official noon according to your time zone. Use this time difference to determine the number of degrees of longitude that you are away from the center of your time zone.

B4. Access orbital information about one planet (other than Earth) that most interests you (or a planet assigned by the instructor). How elliptical is the orbit of the planet? Also, enjoy imagery of the planet if available.

B5. Access runway visual range reports from surface weather observations (METARs) from the web. Compare two different locations (or times) having different visibilities, and calculate the appropriate volume extinction coefficients and optical thickness. Also search the web to learn how runway visual range (RVR) is measured.

B6. Access both visible and infrared satellite photos from the web, and discuss why they look different. If you can access water-vapor satellite photos, include them in your comparison.

B7. Search the web for information about the sun. Examine satellite-based observations of the sun made at different wavelengths. Discuss the structure of the sun. Do any of the web pages give the current value of the solar irradiance (i.e., the solar constant)? If so, how has it varied recently?

B8. Access from the web daytime visible photos of the whole disk of the Earth, taken from geostationary weather satellites. Discuss how variations in the apparent brightness at different locations (different latitudes; land vs. ocean, etc.) might be related to reflectivity and other factors.

B9. Some weather stations and research stations report hourly observations on the web. Some of these stations include radiative fluxes near the surface. Use this information to create surface net radiation graphs.

B10. Access information from the web about how color relates to wavelength. Also, how does the range of colors that can be perceived by eye compare to the range of colors that can be created on a computer screen?

B11. Search the web for information about albedos and IR emissivities for substances or surfaces that are not already listed in the tables in this chapter.

B12. Find on the web satellite images of either forest-fire smoke plumes or volcanic ash plumes. Compare the intensity of reflected radiation from the Earth's surface as it shines through these plumes with earlier satellite photos when the plumes were not there. Use these data to estimate the extinction coefficient.

B13. Access photos and diagrams from the web that describe how different actinometers are constructed and how they work. Also, list any limitations of these instruments that are described in the web.

Apply

(Students, don't forget to put a box around each answer.)

A1. Given distances R between the sun and planets compute the orbital periods (Y) of:

a. Mercury (R = 58 Gm)
b. Venus (R = 108 Gm)
c. Mars (R = 228 Gm)
d. Jupiter (R = 778 Gm)
e. Saturn (R = 1,427 Gm)
f. Uranus (R = 2,869 Gm)
g. Neptune (R = 4,498 Gm)
h. Pluto (R = 5,900 Gm)
i. Eris: given Y = 557 Earth years, estimate the distance R from the sun assuming a circular orbit. (Note: Eris' orbit is highly eccentric and steeply tilted at 44° relative to the plane of the rest of the solar system, so our assumption of a circular orbit was made here only to simplify the exercise.)

A2. This year, what is the date and time of the:

a. perihelion b. vernal equinox
c. summer solstice d. aphelion
e. autumnal equinox f. winter solstice

A3. What is the relative Julian day for:

a. 10 Jan b. 25 Jan c. 10 Feb d. 25 Feb
e. 10 Mar f. 25 Mar g. 10 Apr h. 25 Apr
i. 10 May j. 25 May k. 10 Jun l. 25 Jun
m. 10 Jul n. 25 Jul o. 10 Aug p. 25 Aug
q. 10 Sep r. 25 Sep s. 10 Oct t. 25 Oct
u. 10 Nov v. 25 Nov w. 10 Dec x. 25 Dec
y. today's date z. date assigned by instructor

A4. For the date assigned from exercise A3, find:

(i) mean anomaly
(ii) true anomaly
(iii) distance between the sun and the Earth
(iv) solar declination angle
(v) average daily insolation

A5(§). Plot the local solar elevation angle vs. local time for 22 December, 23 March, and 22 June for the following city:

a. Seattle, WA, USA
b. Corvallis, OR, USA
c. Boulder, CO, USA
d. Norman, OK, USA
e. Madison, WI, USA
f. Toronto, Canada
g. Montreal, Canada
h. Boston, MA, USA
i. New York City, NY, USA
j. University Park, PA, USA
k. Princeton, NJ, USA
l. Washington, DC, USA
m. Raleigh, NC, USA
n. Tallahassee, FL, USA
o. Reading, England
p. Toulouse, France
q. München, Germany
r. Bergen, Norway
s. Uppsala, Sweden
t. DeBilt, The Netherlands
u. Paris, France
v. Tokyo, Japan
w. Beijing, China
x. Warsaw, Poland
y. Madrid, Spain
z. Melbourne, Australia
aa. Your location today.
bb. A location assigned by your instructor

A6(§). Plot the local solar azimuth angle vs. local time for 22 December, 23 March, and 22 June, for the location from exercise A5.

A7(§). Plot the local solar elevation angle vs. azimuth angle (similar to Fig. 2.6) for the location from exercise A5. Be sure to add tic marks along the resulting curve and label them with the local standard times.

A8(§). Plot local solar elevation angle vs. azimuth angle (such as in Fig. 2.6) for the following location:

a. Arctic Circle b. 75°N c. 85°N d. North Pole
e. Antarctic Circle f. 70°S g. 80°S h. South Pole

for each of the following dates:

(i) 22 Dec (ii) 23 Mar (iii) 22 Jun

A9(§). Plot the duration of evening civil twilight (difference between end of twilight and sunset times) vs. latitude between the south and north poles, for the following date:
a. 22 Dec b. 5 Feb c. 21 Mar d. 5 May
e. 21 Jun f. 5 Aug g. 23 Sep h. 5 Nov

A10. On 15 March for the city listed from exercise A5, at what local standard time is:
a. geometric sunrise
b. apparent sunrise
c. start of civil twilight
d. start of military twilight
e. start of astronomical twilight
f. geometric sunset
g. apparent sunset
h. end of civil twilight
i. end of military twilight
j. end of astronomical twilight

A11. Calculate the Eq. of Time correction for:
a. 1 Jan b. 15 Jan c. 1 Feb d. 15 Feb e. 1 Mar
f. 15 Mar g. 1 Apr h. 15 Apr i. 1 May j. 15 May

A12. Find the mass flux ($kg{\cdot}m^{-2}{\cdot}s^{-1}$) at sea-level, given a kinematic mass flux ($m\ s^{-1}$) of:
a. 2 b. 5 c. 7 d. 10 e. 14 f. 18 g. 21
h. 25 i. 30 j. 33 k. 47 l. 59 m. 62 n. 75

A13. Find the kinematic heat fluxes at sea level, given these regular fluxes ($W{\cdot}m^{-2}$):
a. 1000 b. 900 c. 800 d. 700 e. 600
f. 500 g. 400 h. 300 i. 200 j. 100
k. 43 l. –50 m. –250 n. –325 o. –533

A14. Find the frequency, circular frequency, and wavenumber for light of color:
a. red b. orange c. yellow d. green
e. cyan f. blue f. indigo g. violet

A15(§). Plot Planck curves for the following blackbody temperatures (K):
a. 6000 b. 5000 c. 4000 d. 3000 e. 2500
f. 2000 g. 1500 h. 1000 i. 750 j. 500
k. 300 l. 273 m. 260 n. 250 h. 240

A16. For the temperature of exercise A15, find:
(i) wavelength of peak emissions
(ii) total emittance (i.e., total amount of emissions)

A17. Estimate the value of solar irradiance reaching the orbit of the planet from exercise A1.

A18(§). a. Plot the value of solar irradiance reaching Earth's orbit as a function of relative Julian day.

b. Using the average solar irradiance, plot the radiative flux (reaching the Earth's surface through a perfectly clear atmosphere) vs. latitude. Assume local noon.

A19(§). For the city of exercise A1, plot the average daily insolation vs. Julian day.

A20. What is the value of IR absorptivity of:
a. aluminum b. asphalt c. cirrus cloud
d. conifer forest e. grass lawn f. ice
g. oak h. silver i. old snow
j. urban average k. concrete average
l. desert average m. shrubs n. soils average

A21. Suppose polluted air reflects 30% of the incoming solar radiation. How much ($W\ m^{-2}$) is absorbed, emitted, reflected, and transmitted? Assume an incident radiative flux equal to the solar irradiance, given a transmissivity of:
a. 0 b. 0.05 c. 0.1 d. 0.15 e. 0.2 f. 0.25
g. 0.3 h. 0.35 i. 0.4 j. 0.45 k. 0.5 l. 0.55
m. 0.6 n. 0.65 o. 0.7

A22. What is the value of albedo for the following land use?
a. buildings b. dry clay c. corn d. green grass
e. ice f. potatoes g. rice paddy h. savanna
i. red soil j. sorghum k. sugar cane
l. tobacco

A23. What product of number density times absorption cross section is needed in order for 50% of the incident radiation to be absorbed by airborne volcanic ash over the following path length (km)?
a. 0.2 b. 0.4 c. 0.6 d. 0.8 e. 1.0 f. 1.5 g. 2
h. 2.5 i. 3 j. 3.5 k. 4 l. 4.5 m. 5 n. 7

A24. What fraction of incident radiation is transmitted through a volcanic ash cloud of optical depth:
a. 0.2 b. 0.5 c. 0.7 d. 1.0 e. 1.5 f. 2 g. 3
h. 4 j. 5 k. 6 l. 7 m. 10 n. 15 o. 20

A25. What is the visual range (km) for polluted air that has volume extinction coefficient (m^{-1}) of:
a. 0.00001 b. 0.00002 c. 0.00005 d. 0.0001
e. 0.0002 f. 0.0005 g. 0.001 h. 0.002
i. 0.005 j. 0.01 k. 0.02 l. 0.05

A26. (i) What is the value of solar downward direct radiative flux reaching the surface at the city from exercise A5 at noon on 4 July, given 20% coverage of cumulus (low) clouds.
(ii) If the albedo is 0.5 in your town, what is the reflected solar flux at that same time?

(iii) What is the approximate value of net longwave radiation at that time?

(iv) What is the net radiation at that time, given all the info from parts (i) - (iii)?

A27. For a surface temperature of 20°C, find the emitted upwelling IR radiation (W m^{-2}) over the surface-type from exercise A20.

Evaluate & Analyze

E1. At what time of year does the true anomaly equal:

a. 45° b. 90° c. 135° d. 180° e. 225°
f. 270° g. 315° h. 360°

E2(§) a. Calculate and plot the position (true anomaly and distance) of the Earth around the sun for the first day of each month.

b. Verify Kepler's second law.

c. Compare the elliptical orbit to a circular orbit.

E3. What is the optimum angle for solar collectors at your town?

E4. Design a device to measure the angular diameter of the sun when viewed from Earth. (Hint, one approach is to allow the sun to shine through a pin hole on to a flat surface. Then measure the width of the projected image of the sun on this surface divided by the distance between the surface and the pin hole. What could cause errors in this device?)

E5. For your city, plot the azimuth angle for apparent sunrise vs. relative Julian day. This is the direction you need to point your camera if your want to photograph the sunrise.

E6. a. Compare the length of daylight in Fairbanks, AK, vs Miami, FL, USA.

b. Why do vegetables grow so large in Alaska?

c. Why are few fruits grown in Alaska?

E7. How would Fig. 2.6 be different if daylight (summer) time were used in place of standard time during the appropriate months?

E8(§). Plot a diagram of geometric sunrise times and of sunset times vs. day of the year, for your location.

E9(§). Using apparent sunrise and sunset, calculate and plot the hours of daylight vs. Julian day for your city.

E10. a. On a clear day at your location, observe and record actual sunrise and sunset times, and the duration of twilight.

b. Use that information to determine the day of the year.

c. Based on your personal determination of the length of twilight, and based on your latitude and season, is your personal twilight most like civil, military, or astronomical twilight?

E11. Given a flux of the following units, convert it to a kinematic flux, and discuss the meaning and/or advantages of this form of flux.

a. Moisture flux: $g_{water}{\cdot}m^{-2}{\cdot}s^{-1}$

b. Momentum flux: $(kg_{air}{\cdot}\ m{\cdot}s^{-1}){\cdot}m^{-2}{\cdot}s^{-1}$

c. Pollutant flux: $g_{pollutant}{\cdot}m^{-2}{\cdot}s^{-1}$

E12. a. What solar temperature is needed for the peak intensity of radiation to occur at 0.2 micrometers?

b. Remembering that humans can see light only between 0.38 and 0.74 microns, would the sun look brighter or dimmer at this new temperature?

E13. A perfectly black asphalt road absorbs 100% of the incident solar radiation. Suppose that its resulting temperature is 50°C. How much visible light does it emit?

E14. If the Earth were to cool 5°C from its present radiative equilibrium temperature, by what percentage would the total emitted IR change?

E15(§). Evaluate the quality of the approximation to Planck's Law [see eq. (a) in the " A Scientific Perspective • Scientific Laws — the Myth" box] against the exact Planck equation (2.13) by plotting both curves for a variety of typical sun and Earth temperatures.

E16. Find the solar irradiance that can pass through an atmospheric "window" between $\lambda_1 = 0.3\ \mu m$ and $\lambda_2 = 0.8\ \mu m$. (See the " A Scientific Perspective • Seek Solutions" box in this chapter for ways to do this without using calculus.)

E17. How much variation in Earth orbital distance from the sun is needed to alter the solar irradiance by 10%?

E18. Solar radiation is a diffuse source of energy, meaning that it is spread over the whole Earth rather than being concentrated in a small region. It has been proposed to get around the problem of the inverse square law of radiation by deploying very large mirrors closer to the sun to focus the light as collimated rays toward the Earth. Assuming that all the structural and space-launch issues could be solved, would this be a viable method of increasing energy on Earth?

E19. The "sine law" for radiation striking a surface at an angle is sometimes written as a "cosine law", but using the zenith angle instead of the elevation angle. Use trig to show that the two equations are physically identical.

E20. Explain the meaning of each term in eq. (2.21).

E21. a. Examine the figure showing average daily insolation. In the summer hemisphere during the few months nearest the summer solstice, explain why the incoming solar radiation over the pole is nearly equal to that over the equator.

b. Why are not the surface temperatures near the pole nearly equal to the temperatures near the equator during the same months?

E22. Using Table 2-5 for the typical albedos, speculate on the following:

a. How would the average albedo will change if a pasture is developed into a residential neighborhood.

b. How would the changes in affect the net radiation budget?

E23. Use Beer's law to determine the relationship between visual range (km) and volume extinction coefficient (m^{-1}). (Note that extinction coefficient can be related to concentration of pollutants and relative humidity.)

E24(§). For your city, calculate and plot the noontime downwelling solar radiation every day of the year, assuming no clouds, and considering the change in solar irradiance due to changing distance between the Earth and sun.

E25. Consider cloud-free skies at your town. If 50% coverage of low clouds moves over your town, how does net radiation change at noon? How does it change at midnight?

E26. To determine the values of terms in the surface net-radiation budget, what actinometers would you use, and how would you deploy them (i.e., which directions does each one need to look to get the data you need)?

Synthesize

(Don't forget to state and justify all assumptions.)

S1. What if the eccentricity of the Earth's orbit around the sun changed to 0.2 ? How would the seasons and climate be different than now?

S2. What if the tilt of the Earth's axis relative to the ecliptic changed to 45° ? How would the seasons and climate be different than now?

S3. What if the tilt of the Earth's axis relative to the ecliptic changed to 90° ? How would the seasons and climate be different than now?

S4. What if the tilt of the Earth's axis relative to the ecliptic changed to 0° ? How would the seasons and climate be different than now?

S5. What if the rotation of the Earth about its axis matched its orbital period around the sun, so that one side of the Earth always faced the sun and the other side was always away. How would weather and climate be different, if at all?

S6. What if the Earth diameter decreased to half of its present value? How would sunrise and sunset time, and solar elevation angles change?

S7. Derive eq. (2.6) from basic principles of geometry and trigonometry. This is quite complicated. It can be done using plane geometry, but is easier if you use spherical geometry. Show your work.

S8. What if the perihelion of Earth's orbit happened at the summer solstice, rather than near the winter solstice. How would noontime, clear-sky values of insolation change at the solstices compared to now?

S9. What if radiative heating was caused by the magnitude of the radiative flux, rather than by the radiative flux divergence. How would the weather or the atmospheric state be different, if at all?

S10. Linearize Planck's Law in the vicinity of one temperature. Namely, derive an equation that gives a straight line that is tangent to any point on the Planck curve. (Hint: If you have calculus skills, try using a Taylor's series expansion.) Determine over what range of temperatures your equation gives reasonable answers. Such linearization is sometimes used retrieving temperature soundings from satellite observations.

S11. Suppose that Kirchhoff's law were to change such that $a_\lambda = 1 - e_\lambda$. What are the implications?

S12. What if Wien's law were to be repealed, because it was found instead that the wavelength of peak emissions increases as temperature increases.
a. Write an equation that would describe this. You may name this equation after yourself.
b. What types of radiation from what sources would affect the radiation budget of Earth?

S13. What if the solar "constant" were even less constant than it is now. Suppose the solar constant randomly varies within a range of ±50% of its present value, with each new value lasting for a few years before changing again. How would weather and climate be different, if at all?

S14. What if the distance between sun and Earth was half what it is now. How would weather and climate be different, if at all?

S15. What if the distance between sun and Earth was double what it is now. How would weather and climate be different, if at all?

S16. Suppose the Earth was shaped like a cube, with the axis of rotation perpendicular to the ecliptic, and with the axis passing through the middle of the top and bottom faces of the cube. How would weather and climate be different, if at all?

S17. Suppose the Earth was shaped like a narrow cylinder, with the axis of rotation perpendicular to the ecliptic, and with the axis passing through the middle of the top and bottom faces of the cylinder. How would weather & climate be different, if at all?

S18. Derive eq. (2.21) from the other equations in this chapter. Show your work. Discuss the physical interpretation of the hour angle, and what the effect of truncating it is.

S19. Suppose that the Earth's surface was perfectly reflective everywhere to short-wave radiation, but that the atmosphere absorbed 50% of the sunlight passing through it without reflecting any. What percentage of the insolation would be reflected to space? Also, how would the weather and climate be different, if at all?

S20. Suppose that the atmosphere totally absorbed all short wave radiation that was incident on it, but also emitted an exactly equal amount of short wave radiation as it absorbed. How would weather and climate be different, if at all?

S21. Consider Beer's law. If there are n particles per cubic meter of air, and if a vertical path length in air is Δs, then multiplying the two gives the number of particles over each square meter of ground. If the absorption cross section b is the area of shadow cast by each particle, then multiplying this times the previous product would give the shadowed area divided by the total area of ground. This ratio is just the absorptivity a. Namely, by this reasoning, one would expect that $a = n \cdot b \cdot \Delta s$.

However, Beer's law is an exponential function. Why? What was wrong with the reasoning in the previous paragraph?

S22. What if the atmosphere were completely transparent to IR radiation. How would the surface net radiation budget be different?

S23. Existing radiometers are based on a bolometer, photovoltaic cell, or charge-coupled device. Design a new type of actinometer that is based on a different principle. Hint, think about what is affected in any way by radiation or sunlight, and then use that effect to measure the radiation.

3 THERMODYNAMICS

Contents

Recall from physics that **kinetic energy** relates to the motion of objects, while **potential energy** relates to the attraction between objects. These energies can apply on the **macroscale** — to large-scale objects consisting of many molecules. They can also apply on the **microscale** — to individual molecules, atoms, and subatomic particles. Energy on the microscale is known as **internal energy**, and a portion of internal energy is what we call **heat**.

Energy can change forms between kinetic, potential, and other energy types. It can also change scale. The conversion between microscale and macroscale energies was studied extensively during the industrial revolution to design better engines. This study is called **thermodynamics**.

The field of thermodynamics also applies to the atmosphere. The microscale energy of heat can cause the macroscale motions we call winds. Microscale attractions enable water-vapor molecules to condense into macroscale cloud drops and rain.

In this chapter, we will investigate the interplay between internal energy and macroscale effects in the atmosphere. First, focus on internal energy.

INTERNAL ENERGY

Definitions

In thermodynamics, **internal energy** consists of the sum of microscopic (molecular scale) kinetic and potential energy.

Microscopic kinetic energies include random movement (translation) of molecules, molecular vibration and rotation, electron motion and spin, and nuclear spin. The sum of these kinetic energies is called **sensible energy**, which we humans can sense (i.e., feel) and measure as **temperature**.

Microscopic potential energy is associated with forces that bind masses together. It takes energy to pull two masses apart and break their bonds. This is analogous to increasing the microscopic potential energy of the system. When the two masses snap back together, their microscopic potential energy is released back into other energy forms. Three forms of binding energy are:

- **latent** — bonds between molecules
- **chemical** — bonds between atoms
- **nuclear** — sub-atomic bonds

We will ignore chemical reactions and nuclear explosions here.

Latent energy is associated with **phase change** (solid, liquid, gas). In solids, the molecules are bound closely together in a somewhat rigid lattice. In liquids, molecules can more easily move relative to each other, but are still held close together. In gases, the molecules are further apart and have much weaker bonds.

For example, starting with cold ice (in lower left corner of Fig. 3.1), adding energy causes the temperature to increase (a sensible effect), but only up to the melting point (0°C at standard sea-level pressure). Further addition of energy forces bonds of the solid lattice to stretch, enabling more fluid movement of the molecules. This is melting, a latent effect that occurs with no temperature change. After all the ice has melted, if you add more energy then the liquid warms (a sensible effect), but only up to the boiling point (100°C). Subsequent addition of energy forces further stretching of the molecular bonds to allow freer movement of the molecules; namely, evaporation (a latent effect) with no temperature change. After all the liquid has vaporized, any more energy added increases the water-vapor temperature (a sensible effect).

Fig. 3.1 can also be traversed in the opposite direction by removing internal energy. Starting with hot water vapor, the sequence is cooling of the vapor, condensation, cooling of the liquid, freezing, and finally cooling of the ice.

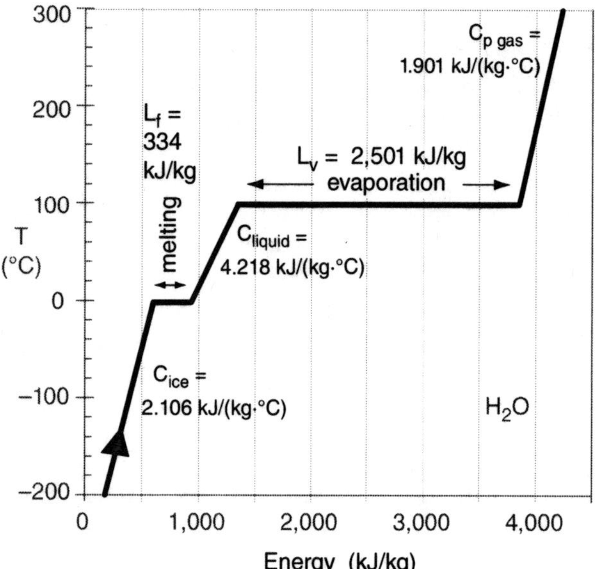

Figure 3.1
Sensible and latent energy for water.

Possession and Transfer of Energy

We cannot ignore the connection between the microscale and the macroscale. A macroscale object such as a cannon ball consists of billions of microscale molecules and atoms, each possessing internal energy. Summing over the mass of all the molecules in the cannon ball gives us the total internal energy (sensible + latent energy) that the cannon ball **possesses.**

Thermal energy transferred to or from the macroscale object can increase or decrease the internal energy it possesses. This is analogous to your bank account, where money transferred (deposited or withdrawn) causes an increase or decrease to the total funds you possess.

Transfer of Heat

Define the **transfer** of **thermal energy** as Δq. It has energy units (J kg^{-1}). In this text, we will refer to Δq as **heat transferred**, although in engineering texts it is just called heat.

Latent Energy Possessed

Define the **latent heat** Q_E as the latent energy (J) possessed by the total mass m of all the molecules in an object.

But usually we are more interested in the change of possessed latent heat ΔQ_E associated with some process that changes the phase of Δm kilograms of material, such as phase change of water:

$$\Delta Q_E = L \cdot \Delta m_{water} \tag{3.1}$$

For example, if we transfer Δq amount of thermal energy into water that is already at 100°C at sea-level pressure, then we can anticipate that the amount of water evaporated will be given by: $\Delta m_{water} = \Delta q \cdot m_{water} / L_v$. [A sample application is on page 86.]

Different materials have different strengths of bonds, so they have different constants of proportionality L (called the **latent heat factor**) between ΔQ_E and Δm. For water (H_2O), those latent heat factors are given in Table 3-1.

Sensible Energy Possessed

The sensible energy (J) possessed by the total mass m of all the molecules in an object (such as air molecules contained in a finite volume) is $m_{air} \cdot C_v \cdot T$, where T is absolute temperature of the air. The constant of proportionality C_v is called the **specific heat at constant volume**. Its value depends on the material. For dry air, $C_{vd\ air} = 717$ J kg^{-1} K^{-1}.

Again, we are interested more in the change of sensible energy possessed. We might suspect that it should be proportional to the change in temperature ΔT, but there is a complication that is best approached using the First Law of Thermodynamics.

FIRST LAW OF THERMODYNAMICS

Definition

Let Δq (J kg^{-1}) be the amount of thermal energy you add to a stationary mass m of air. Some of this energy warms the air (i.e., the internal energy increases). But as air warms, its volume expands by amount ΔV and pushes against the surrounding atmosphere (which to good approximation is pushing back with constant pressure P). Thus, a portion of the thermal energy input does not go into warming the air, but goes into macroscopic movement.

To illustrate, consider a column of air having base of area A and height d. Resting on top of this column is more air, the weight of which causes pressure P at the top of the column. Suppose the volume ($V = A \cdot d$) expansion is all in the vertical, so that $\Delta V = A \cdot \Delta d$. For the column top to rise, it must counteract the downward pressure force from the air above; namely, the top of the column must push up with pressure P as it moves distance Δd.

Recall that **work** W is force times distance ($W = F \cdot \Delta d$). Also, pressure is force per unit area ($P = F/A$) Thus, the work done on the atmosphere by the expanding column is $W = F \cdot \Delta d = P \cdot A \cdot \Delta d = P \cdot \Delta V$.

The **First Law of Thermodynamics** says that energy is conserved, thus the thermal energy input must equal the sum of warming (a microscopic effect) and work done per unit mass (a macroscopic effect):

$$\Delta q = C_v \cdot \Delta T + P \cdot (\Delta V/m) \qquad (3.2a)$$

Apply to the Atmosphere

But $P \cdot (\Delta V/m) = \Delta(P \cdot V/m) - V \cdot \Delta P/m$ (using the product rule of calculus). Also, $P \cdot V/m = P/\rho = \Re \cdot T$ from the ideal gas law, where $\rho = m/V$ is air density and $\Re$ is the gas constant. Thus, $\Delta(P \cdot V/m) = \Delta(\Re \cdot T) = \Re \cdot \Delta T$ because $\Re$ is constant. Using this in eq. (3.2a) gives:

$$\Delta q = C_v \cdot \Delta T + \Re \cdot \Delta T - \Delta P/\rho \qquad (3.2b)$$

By definition for an ideal gas: $C_v + \Re \equiv C_p$, where C_p is the **specific heat of air at constant pressure**. The INFO box on the next page gives:

$$C_p = C_{p\,humid\,air} \approx C_{pd} \cdot (1 + 1.84 \cdot r) \qquad (3.3)$$

where C_{pd} is the dry-air specific heat at constant pressure, and the water-vapor mixing ratio r has units ($g_{water\ vapor} / g_{dry\ air}$). See the Water Vapor chapter for more details on humidity.

Table 3-1. Latent heat factors L for water (H_2O), for the phase-change processes indicated.

Process Name & Direction	L (J kg^{-1})	Process Name & Direction
	vapor	
evaporation ↑	$L_v = 2.5 \times 10^6$	↓ condensation
	liquid	
melting ↑	$L_f = 3.34 \times 10^5$	↓ freezing (fusion)
	solid	
	vapor	
sublimation ↑	$L_d = 2.83 \times 10^6$	↓ deposition
	solid	
↑ requires transfer of thermal energy Δq TO water from the surrounding air.		↓ requires transfer of thermal energy Δq FROM water to the surrounding air.

Sample Application

Suppose 3 kg of water as vapor condenses to liquid. What is the value of latent heat transferred to the air?

Find the Answer

Given: $L_v = 2.5 \times 10^6$ J·kg^{-1}, $m_{vapor} = 3$ kg.
Find: $\Delta Q_E = ?$ J

Apply eq. (3.1):
$\Delta Q_E = (2.5 \times 10^6 \text{ J·kg}^{-1}) \cdot (3 \text{ kg}) = \underline{7{,}500 \text{ kJ}}$

Check: Physics & units are reasonable.
Exposition: Three liters is a small quantity of water (equivalent to 3 mm depth in a bathtub). Yet it represents a large quantity of latent heat.

Sample Application

Find the specific heat at constant pressure for humid air holding 10 g of water vapor per kg of dry air.

Find the Answer

Given: $r = (10\ g_{vapor}) / (1000\ g_{dry\ air}) = 0.01$ g/g
Find: $C_p = ?$ J·kg^{-1}·K^{-1}

Apply eq. (3.3):
$C_p = (1004 \text{ J·kg}^{-1}\text{·K}^{-1}) \cdot [1 + (1.84 \cdot (0.01 \text{ g/g}))]$
$= \underline{1022.5 \text{ J·kg}^{-1}\text{·K}^{-1}}$

Check: Units OK. Magnitude reasonable.
Exposition: Even a modest amount of water vapor can cause a significant increase in specific heat. See Chapter 4 for typical ranges of r in the atmosphere.

INFO • Specific Heat Cp for Air

The specific heat at constant pressure C_p for air is the average of the specific heats for its constituents, weighted by their relative abundance:

$$m_T C_p = m_d \cdot C_{pd} + m_v \cdot C_{pv} \quad (3I.1)$$

where $m_T = m_d + m_v$ is the total mass of air (as a sum of mass of dry air m_d and water vapor m_v), and C_{pd} and C_{pv} are the specific heats for dry air and water vapor, respectively.

Define a mixing ratio r of water vapor as $r = m_v / m_d$. Typically, r is of order 0.01 g/g. Then eq. (3I.1) becomes:

$$C_p = (1 - r) \cdot C_{pd} \cdot [1 + r \cdot C_{pv}/C_{pd}] \quad (3I.2)$$

Given: C_{pd} = 1004 J·kg^{-1}·K^{-1} at 0°C for dry air, and C_{pv} = 1850 J·kg^{-1}·K^{-1} for water vapor, eq. (3I.2) becomes

$$C_p \approx C_{pd} \cdot [1 + 1.84 \cdot r] \quad (3I.3)$$

Even for dry air, the specific heat varies slightly with temperature, as shown in the figure below:

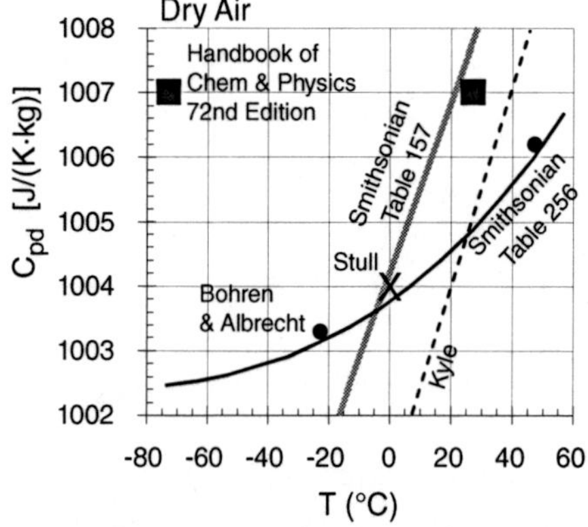

Fig. 3I.1. *Empirical estimates of C_{pd}.*

In this book, we will use C_{pd} = 1004 J·kg^{-1}·K^{-1} , and will approximate it as being constant.

Sample Application

What heat transfer is needed to cause 3 kg of dry air to cool by 10°C?

Find the Answer

Given: C_{pd} = 1004. J·kg^{-1}·K^{-1}, m_{air} = 3 kg , ΔT = –10°C
Find: ΔQ_H = ? J

Apply eq. (3.4b):

$\Delta Q_H = m_{air} \cdot C_{pd} \cdot \Delta T$
= (3 kg)·(1004. J·kg^{-1}·K^{-1})·(–10°C) = **<u>–30.12 kJ</u>**

Check: Physics & units are reasonable.
Exposition: On a hot day, this is the energy your air conditioner must extract from the air in your car.

Appendix B lists some specific heats; e.g.:
$C_{pd\ air}$ = 1004 J·kg^{-1}·K^{-1} for dry air at const. pressure,
$C_{vd\ air}$ = 717 J kg^{-1} K^{-1} for dry air at const. volume,
C_{liq} ≈ 4217.6 J·kg^{-1}·K^{-1} for liquid water at 0°C,
C_{ice} ≈ 2106 J·kg^{-1}·K^{-1} for ice at 0°C,
C_{pv} = 1850 J·kg^{-1}·K^{-1} for pure water vapor at 0°C.

You can combine the first two terms on the right of eq. (3.2b) to give a form of the **First Law of Thermodynamics** that is easier to use for the atmosphere:

$$\underbrace{\Delta q}_{\text{heat transferred}} = \underbrace{C_p \cdot \Delta T}_{\text{enthalpy change}} - \Delta P/\rho \quad (3.2c)$$

Enthalpy vs. Sensible Heat

From our derivation of eq. (3.2c) we saw that the first term on the right includes both the microscopic effect of a temperature change (internal energy or heat possessed) <u>and</u> the macroscopic effect of that same temperature change. Hence, we cannot call that term "heat possessed" — instead we need a new name.

To this end, define **enthalpy** as $h = C_p \cdot T$ with units J kg^{-1}. The corresponding **enthalpy change** is:

$$\Delta h = C_p \cdot \Delta T \quad (3.4a)$$

which is the first term on the right side of eq. (3.4). It is a characteristic <u>possessed</u> by the air.

By tradition, meteorologists often use the word **sensible heat** in place of the word enthalpy. This can be confusing because of the overloading of the word "heat". Here is a table that might help you keep these definitions straight:

Table 3-2. Heat terminology.

Quant. (J/kg)	Character-istic	Terminology	
		Meteorology	Engineering
Δq	transferred	heat transferred	heat
$C_p \cdot \Delta T$	possessed	sensible heat	enthalpy

With this in mind, the heat transferred per unit air mass can be annotated as follows:

$$\underbrace{\Delta q}_{\text{heat transferred}} = \underbrace{C_p \cdot \Delta T}_{\text{sensible heat}} - \Delta P/\rho \quad \bullet(3.2d)$$

This form is useful in meteorology. When rising air parcels experience a decrease in surrounding atmospheric pressure, the last term is non-zero.

The change of **sensible-heat** (ΔQ_H) possessed by air mass m_{air} changing its temperature by ΔT is thus:

$$\Delta Q_H = m_{air} \cdot C_p \cdot \Delta T \quad (3.4b)$$

INFO • Cp vs. Cv

C_v — Specific Heat at Constant Volume

Consider a sealed box of fixed volume V filled with air, as sketched in Fig. 3I.2a below. The number of air molecules (idealized by the little spheres) can't change, so the air density ρ is constant. Suppose that initially, the air temperature T_o is cool, as represented by the short arrows denoting the movement of each molecule in box 3I.2a. When each molecule bounces off a wall of the container, it imparts a small force. The sum of forces from all molecules that bounce off a wall of area A results in an air pressure P_o.

If you add Δq thermal energy to air in the box, the temperature rises to T_2 (represented by longer arrows in Fig. 3I.2b). Also, when each molecule bounces off a wall, it imparts a greater force because it is moving faster. Thus, the pressure P_2 will be larger. This is expected from the ideal gas law under constant density, for which

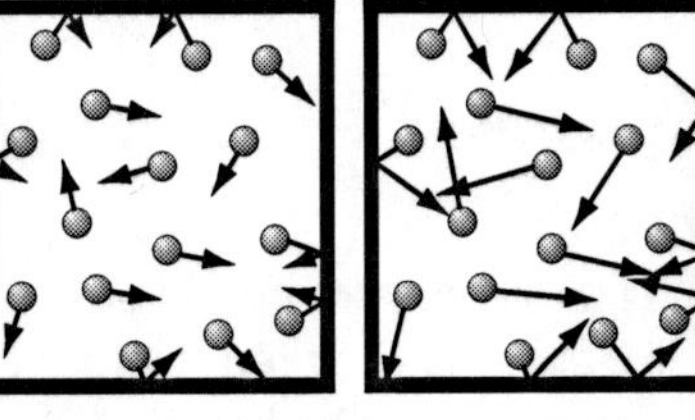

Fig. 3I.2. *Molecules in a fixed volume.*

$$P_o/T_o = P_2/T_2 = \text{constant} = \rho \cdot \Re \tag{3I.4}$$

Different materials warm by different amounts when you add heat. If you know how much thermal energy Δq you added per kilogram of material, and you measure the resulting increase in temperature $T_2 - T_o$, then you can empirically determine the **specific heat at constant volume:**

$$C_v = \Delta q/(T_2 - T_o) \tag{3I.5}$$

which is about $C_v = 717 \text{ J·kg}^{-1}\text{·K}^{-1}$ for dry air.

C_p — Specific Heat at Constant Pressure

For a different scenario, consider a box (Fig. 3I.3c) with a frictionless moveable piston at the top. The weight of the stationary piston is balanced by the pressure of the gas trapped below it. If you add Δq thermal energy to the air, the molecules will move faster (Fig. 3I.3d), and exert greater pressure against the piston and against the other walls of the chamber. But the weight of the piston hasn't changed, so the increased pressure of the gas causes the piston to rise.

But when any molecule bounces off the piston and helps move it, the molecule loses some of its microscopic kinetic energy. (An analogy is when a billiard ball bounces off an empty cardboard box sitting in the middle of the billiard table. The box moves a bit when hit by the ball, and the ball returns more slowly.) The result is that the gas temperature T_1 in Fig. 3I.3e is not as warm as in Figs. 3I.2b or 3I.3d, but is warmer than the initial temperature; namely, $T_o < T_1 < T_2$.

(continues in next column)

(continuation of INFO on Cp vs. Cv)

The molecules spread out within the larger volume in Fig. 3I.3e. Thus, air density ρ decreases, causing fewer molecules near the piston to push against it. The combined effects of decreasing density and temperature cause the air pressure to decrease as the piston rises. Eventually the piston stops rising at the point where the air pressure balances the piston weight, as shown in Fig. 3I.3e.

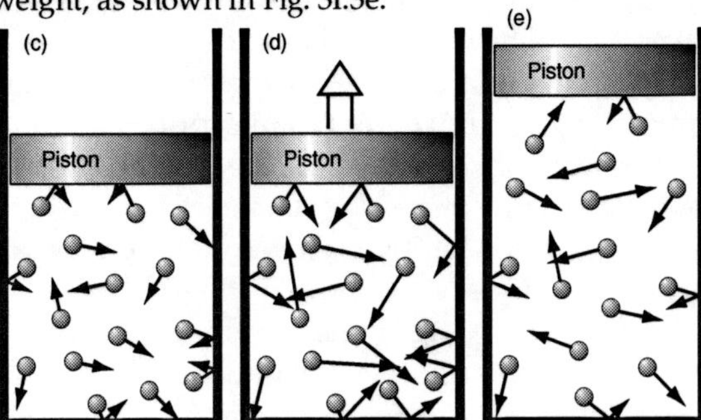

Fig. 3I.3. *Molecules in a constant-pressure chamber.*

Hence, this is an isobaric process (determined by the weight of the piston in this contrived example). The ideal gas law for constant pressure says:

$$\rho_o \cdot T_o = \rho_1 \cdot T_1 = \text{constant} = P/\Re \tag{3I.6}$$

If you know how much thermal energy Δq you added per kilogram of material, and you measure the resulting increase in temperature $T_1 - T_o$, then you can empirically determine the **specific heat at constant pressure:**

$$C_p = \Delta q/(T_1 - T_o) \tag{3I.7}$$

which is about $C_p = 1004 \text{ J·kg}^{-1}\text{·K}^{-1}$ for dry air.

C_p vs. C_v

Thus, in a constant pressure situation, the addition of Δq thermal energy results in less warming [$(T_1 - T_o) < (T_2 - T_o)$] than at constant volume. The reason is that, for constant pressure, some of the random microscopic kinetic energy of the molecules is converted into the macroscale **work** of expanding the air and moving the piston up against the pull of gravity. Such conservation of energy is partly described by the First Law of Thermodynamics.

In the atmosphere, the pressure at any altitude is determined by the weight of all the air molecules above that altitude (namely, the "piston" is all the air molecules above). If you add a small amount of thermal energy to air molecules at that one altitude, then you haven't significantly affected the number of molecules above, hence the pressure is constant. Thus, for most atmospheric situations it is appropriate to use C_p, not C_v, when forecasting temperature changes associated with the transfer of thermal energy into or from an air parcel.

FRAMEWORKS

The First Law of Thermodynamics is a powerful tool that we can apply to different frameworks.

Lagrangian vs. Eulerian

One framework, called **Eulerian**, is fixed relative to a position on the Earth's surface. Thus, if there is a wind, then the air blows through this framework, so we need to be concerned about what heat is carried in by the wind. Weather forecasts for specific points on a map utilize this framework.

Another framework is called **Lagrangian**. It moves with the air — its position is constantly changing. This is handy for investigating what happens to air as it rises or sinks.

Eulerian and Lagrangian frameworks can be used for a variety of budget equations:
- Heat Budget (First Law of Thermo)
- Momentum Budget (Newton's Second Law)
- Moisture Budget (conservation of water),

We will use these frameworks in other chapters too.

Air Parcels

Sometimes a large cluster of air molecules will move together through the atmosphere, as if they were enclosed by a hypothetical balloon about the diameter of two city blocks. We can use a Lagrangian framework that moves with this cluster or "blob", in order to study changes of its temperature, momentum, and moisture.

When these air blobs move through the atmosphere, myriad eddies (swirls of turbulent motion) tend to mix some of the outside air with the air just inside the blob (such as the mixing you see in smoke rising from a campfire). Thus, warmer or colder air could be added to (**entrained** in through the sides of) the blob, and some air from inside could be lost (**detrained**) to the surrounding atmosphere. Also, in the real atmosphere, atmospheric radiation can heat or cool the air blob. These processes complicate the thermodynamic study of real air blobs.

But to gain some insight into the thermodynamics of air, we can imagine a simplified situation where radiative effects are relatively small, and where the turbulent entrainment/detrainment happens only in the outer portions of the air blob, leaving an inner core somewhat protected. This is indeed observed in the real atmosphere. So consider the protected inner core (about the diameter of a city block) as an **air parcel**.

Whenever you see discussions regarding air parcels, you should immediately associated them with Lagrangian frameworks. This is the case for the next section.

HEAT BUDGET OF AN UNSATURATED AIR PARCEL

In this chapter, we consider a special case: unsaturated air parcels, for which no liquid or solid water is involved. The word "**dry**" is used to imply that phase changes of water are not considered. Nonetheless, the air CAN contain water vapor. In the next chapter we include the effects of saturation and possible phase changes in a "**moist**" analysis.

Lagrangian Form of the First Law of Thermo

The pressure of an air parcel usually equals that of its surrounding environment, which decreases exponentially with height. Thus, the last term of eq. (3.2d) will be non-zero for a rising or sinking air parcel as its pressure changes to match the pressure of its environment. But the pressure change with height was given by the hydrostatic equation in Chapter 1: $\Delta P/\rho = -\ |g| \cdot \Delta z$. We can use this to rewrite the First Law of Thermo in the Lagrangian framework of a moving air parcel:

$$\Delta T = -\left(\frac{|g|}{C_p}\right) \cdot \Delta z + \frac{\Delta q}{C_p} \qquad \bullet(3.5)$$

Sample Application

A 5 kg air parcel of initial temperature 5°C rises 1 km and thermally loses 15 kJ of energy due to IR radiation. What is the final temperature of the parcel?

Find the Answer

Given: $\Delta Q = -15{,}000$ J, $m_{air} = 5$ kg, $\Delta z = 1000$ m ,
Find: $T = ?$ K

Convert from energy to energy/mass:
$\Delta q = \Delta Q/m_{air} = (-15000 \text{ J})/(5 \text{ kg}) = -3000 \text{ J kg}^{-1}$.
With lack of humidity info, assume dry air.
$C_p = 1004.\ \text{J·kg}^{-1}\text{·K}^{-1}$ (= units $\text{m}^2\,\text{s}^{-2}\,\text{K}^{-1}$)
Apply eq. (3.5):
$\Delta T = -[(9.8 \text{ m s}^{-2})/\ (1004 \text{ m}^2 \text{ s}^{-2} \text{ K}^{-1})]\cdot(1000\text{m}) + [(-3000 \text{ J kg}^{-1})\ /\ (1004.\ \text{J·kg}^{-1}\text{·K}^{-1})]$
$= (-9.76\ -3.00)\text{ K} = -12.76\ °\text{C}$

Check: Physics & units are reasonable.
Exposition: Because we are working with a temperature difference, recall that 1 K of temperature difference = 1°C of temperature difference. Hence, we could replace the Kelvin units with °C. The net result is that the rising air parcel cools due to both IR radiative cooling and work done on the atmosphere as the parcel rises.

Various processes can cause heat transfer (Δq). The sun could heat the air, or IR radiation could cool the air. Water vapor could condense and release its latent heat back into sensible heat. Exothermic chemical reactions or radioactive decay could occur among air pollutants carried within the parcel. Internal turbulence could dissipate into heat. Molecular conduction in the air is very weak, but turbulence could mix warmer or cooler air into the air parcel. Other processes such as convection and advection (Fig. 3.2) do not change the parcel's temperature, but can move the air parcel along with the heat that it possesses.

Eq. (3.5) represents a **heat budget**. Namely, parcel temperature (which indicates heat possessed) is conserved unless it moves to a different height (where the pressure is different) or if heat is transferred to or from it. Thus, eq. (3.5) and the other First Law of Thermo eqs. (3.2) are also known as **heat conservation** equations.

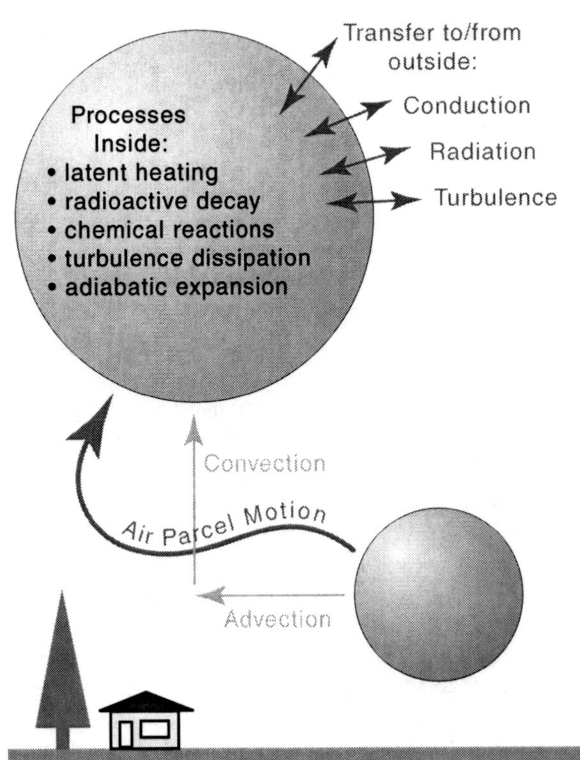

Figure 3.2
Internal and external processes affecting air-parcel temperature.

Lapse-rate Definition

Define the **lapse rate**, Γ, as the amount of temperature <u>decrease</u> with altitude:

$$\Gamma = -\frac{T_2 - T_1}{z_2 - z_1} = -\frac{\Delta T}{\Delta z} \tag{3.6}$$

Note that the lapse rate is the negative of the **vertical temperature gradient** $\Delta T/\Delta z$.

We separately consider the lapse rates inside the air parcel, and in the surrounding environment outside. Inside the air parcel, all the processes illustrated in Fig. 3.2 could apply, causing the parcel's temperature to change with changing altitude. The resulting $\Delta T/\Delta z$ (times –1) defines a **process lapse rate** (Fig. 3.3).

Outside the air parcel, assume the environmental air is relatively stationary. This is the ambient environment through which the air parcel moves. But this environment could have different temperatures at different altitudes, allowing us to define an **environmental lapse rate** (Fig. 3.3). By sampling the ambient air at different heights using weather instruments such as radiosondes (weather balloons) and then plotting T vs. z as a graph, the result is an **environmental sounding** or **vertical temperature profile** of the environment. The environmental sounding changes as the weather evolves, but this is usually slow relative to parcel processes. Thus, the environment is often approximated as being unchanging (i.e., static).

The temperature difference (Fig. 3.3) between the parcel and its environment is crucial for determining parcel buoyancy and storm development. This is our motivation for examining both lapse rates.

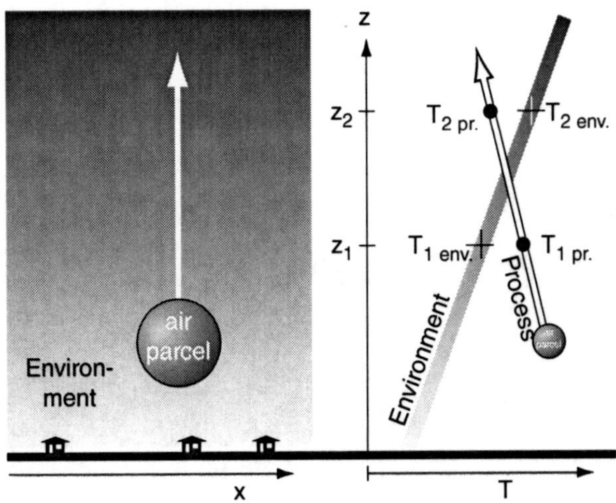

Figure 3.3
Left: Sketch of a physical situation, showing an air parcel moving through an environment. In the environment, darker colors indicate warmer air. Right: Temperature profiles for the environment and the air parcel. The environmental air is not moving. In it, air at height z_1 has temperature $T_{1\ env}$, and air at z_2 has $T_{2\ env}$. The air parcel has an initially warm temperature, but its temperature changes as it rises: becoming $T_{1\ pr}$ at height z_1, and later becoming $T_{2\ pr}$ at height z_2. In the <u>environment</u> of this example, temperature <u>in</u>creases as height increases, implying a <u>negative environmental</u> lapse rate. However, the air <u>parcel's</u> temperature <u>decreases</u> with height, implying a <u>positive parcel</u> lapse rate.

Sample Application

Find the lapse rate in the troposphere for a standard atmosphere.

Find the Answer

Given: Std. Atmos. Table 1-5 in Chapter 1, , where $T = -56.5°C$ at $z = 11$ km, and $T = +15°C$ at $z = 0$ km.

Find: $\Gamma = ?$ °C km^{-1}

Apply eq. (3.6): $\Gamma = -(-56.5 - 15°C) / (11-0 \text{ km})$
$= \underline{+6.5 \text{ °C km}^{-1}}$

Check: Positive Γ, because *T* decreases with *z*.
Exposition: This is the environmental lapse rate of the troposphere. It indicates a static background state.

HIGHER MATH • Adiabatic Lapse Rate in Pressure Coordinates

Start with the First Law of Thermodynamics (eq. 3.2d), but written more precisely using virtual temperature T_v to account for arbitrary concentrations of water vapor in the air. Set $\Delta q = 0$ because adiabatic means no heat transfer:

$$dP = \rho \cdot C_p \cdot dT_v$$

Use the ideal gas law $\rho = P / (\Re_d \cdot T_v)$ to eliminate ρ:

$$dP = \frac{P \cdot C_p \cdot dT_v}{\Re_d \cdot T_v}$$

Group temperature & pressure terms on opposite sides of the eq.:

$$\frac{dP}{P} = \frac{C_p}{\Re_d} \cdot \frac{dT_v}{T_v}$$

Integrate from starting (P_1, T_{v1}) to ending (P_2, T_{v2}):

$$\int_{P_1}^{P_2} \frac{dP}{P} = \frac{C_p}{\Re_d} \cdot \int_{T_{v1}}^{T_{v2}} \frac{dT_v}{T_v}$$

assuming $C_p/\Re_d$ is somewhat constant. The integral is:

$$\ln(P)\Big|_{P_1}^{P_2} = (C_p / \Re_d) \cdot \ln(T_v)\Big|_{T_{v1}}^{T_{v2}}$$

Insert limits of integration. Also: $\ln(a) - \ln(b) = \ln(a/b)$. Thus:

$$\ln\left(\frac{P_2}{P_1}\right) = (C_p / \Re_d) \cdot \ln\left(\frac{T_{v2}}{T_{v1}}\right)$$

Multiply both sides by $\Re_d / C_p$:

$$(\Re_d / C_p) \cdot \ln\left(\frac{P_2}{P_1}\right) = \ln\left(\frac{T_{v2}}{T_{v1}}\right)$$

Use the relationship: $a \cdot \ln(b) = \ln(b^a)$:

$$\ln\left[\left(\frac{P_2}{P_1}\right)^{\Re_d / C_p}\right] = \ln\left(\frac{T_{v2}}{T_{v1}}\right)$$

The anti-log of the equation ($e^{LHS} = e^{RHS}$) yields:

$$\boxed{\left(\frac{P_2}{P_1}\right)^{\Re_d / C_p} = \frac{T_{v2}}{T_{v1}}} \qquad (3.10)$$

Dry Adiabatic Lapse Rate

The word **adiabatic** means zero heat transfer ($\Delta q = 0$). For the protected inner core of air parcels, this means no thermal energy entering or leaving the air parcel from outside (Fig. 3.2). Nonetheless, internal processes are allowed.

For the special case of humid air with no liquid water or ice carried with the parcel (and no water phase changes; hence, a "dry" process), eq. (3.5) gives:

$$\frac{\Delta T}{\Delta z} = -\left(\frac{|g|}{C_p}\right) = -9.8 \text{ K km}^{-1} \qquad \bullet(3.7)$$

Recalling that the lapse rate is the negative of the vertical temperature gradient, we can define a **"dry" adiabatic lapse rate** Γ_d as:

$$\Gamma_d = 9.8 \text{ K km}^{-1} = 9.8 \text{ °C km}^{-1} \qquad (3.8)$$

(Degrees K and °C are interchangeable in this equation for this process lapse rate, because they represent a temperature change with height.)

The HIGHER MATH box at left shows how this dry adiabatic lapse rate can be expressed as a function of pressure *P*:

$$\frac{\Delta T}{T} = \frac{\Re_d}{C_p} \cdot \left(\frac{\Delta P}{P}\right) \qquad (3.9)$$

or

$$\frac{T_2}{T_1} = \left(\frac{P_2}{P_1}\right)^{\Re_d / C_p} \qquad \bullet(3.10)$$

where $\Re_d/C_p = 0.28571$ (dimensionless) for dry air, and where temperatures are in Kelvin.

Sample Application

An air parcel with initial (z, P, T) = (100m, 100 kPa, 20°C) rises adiabatically to (z, P) = (1950 m, 80 kPa). Find its new *T*, & compare eqs. (3.7) & (3.10).

Find the Answer

Given: $P_1 = 100$ kPa, $P_2 = 80$ kPa, $T_1 = 20°C = 293K$
$z_1 = 100$ m, $z_2 = 1950$ m

Find: $T_2 = ?$ °C

First, apply eq. (3.7), which is a function of *z*:
$T_2 = T_1 + (\Delta z)\cdot(-\Gamma_d) = 20°C - (1950-100m)\cdot(0.0098°C/m)$
$= 20°C - 18.1°C = \underline{1.9°C}$.

Compare with eq. (3.10), which is a function of *P*:
$T_2 = (293K) \cdot [(80kPa)/(100kPa)]^{0.28571}$
$T_2 = 293K \cdot 0.9382 = 274.9 \text{ K} = \underline{1.9°C}$

Check: Both equations give the same answer, so either equation would have been sufficient by itself.

Potential-temperature Definition

When an air parcel rises/sinks "dry" adiabatically into regions of lower/higher pressure, its temperature changes due to work done by/on the parcel, even though no thermal energy has been removed/added. Define a new temperature variable called the **potential temperature** θ that is proportional to the sensible heat contained in the parcel, but which is unaffected by work done by/on the parcel.

Namely, potential temperature is constant for an adiabatic process (i.e., $\Delta q = 0$) such as air-parcel ascent. Thus, we can use it as a **conserved variable**. θ can increase/decrease when sensible heat is added/removed. Such **diabatic** (non-adiabatic) heat transfer processes include turbulent mixing, condensation, and radiative heating (i.e., $\Delta q \neq 0$).

Knowing the air temperature T at altitude z, you can calculate the value of potential temperature θ from:

$$\theta(z) = T(z) + \Gamma_d \cdot z \qquad \bullet(3.11)$$

The units (K or °C) of $\theta(z)$ are the same as the units of $T(z)$. There is no standard for z, so some people use height above mean sea level (MSL), while others use height above local ground level (AGL).

If, instead, you know air temperature T at pressure-level P, then you can find the value of θ from:

$$\theta = T \cdot \left(\frac{P_o}{P} \right)^{\Re_d / C_p} \qquad \bullet(3.12)$$

where $\Re_d/C_p$ = 0.28571 (dimensionless) and where temperatures must be in Kelvin. A reference pressure of P_o = 100 kPa is often used, although some people use the local surface pressure instead. In this book we will assume that the surface pressure equals the reference pressure of P_o = 100 kPa and will use z = 0 at that surface, unless stated otherwise.

Both eqs. (3.11) and (3.12) show that $\theta = T$ at $z = 0$ or at $P = P_o$. Thus θ is the actual temperature that an air parcel potentially has if lowered to the reference level adiabatically.

A **virtual potential temperature** θ_v for humid air having water-vapor mixing ratio r but containing no solid or liquid water is defined as:

$$\theta_v = \theta \cdot [1 + (a \cdot r)] \qquad (3.13)$$

where a = 0.61 $g_{air}/g_{water\ vapor}$. If the air contains ice crystals, cloud drops, or rain drops, then virtual potential temperature is given by:

$$\theta_v = \theta \cdot [1 + (a \cdot r) - r_L - r_I] \qquad \bullet(3.14)$$

where r_L is the liquid-water mixing ratio and r_I = ice mixing ratio. Mixing ratio is described in the Wa-

Sample Application

Find θ for air of T = 15°C at z = 750 m?

Find the Answer

Given: T = 15°C, z = 750 m,
Find: θ = ? °C

Apply eq. (3.11), and assume no ice or rain drops.
θ = 15°C + (9.8 °C km^{-1})· (0.75 km) = **22.35 °C**

Check: Physics & units are reasonable.
Exposition: Notice that potential temperatures are warmer than actual air temperatures for $z > 0$.

Sample Application

Find θ for air at P = 70 kPa with T = 10°C?

Find the Answer

Given: P = 70 kPa , T = 10°C = 283 K, P_o = 100 kPa
Find: θ = ? °C

Apply eq. (3.12): θ = (283 K) · [(100 kPa)/(70 kPa)]$^{0.28571}$
θ = 313.6 K = **40.4 °C**

Check: Physics OK. θ is always greater than the actual T, for P smaller than the reference pressure.

Sample Application

What is the virtual potential temperature of air having potential temperature 15°C, mixing ratio 0.008 $g_{water\ vapor}/g_{air}$, and liquid water mixing ratio of: a) 0 ; b) 0.006 $g_{liq.water}/g_{air}$?

Find the Answer

Given: θ =15°C = 288 K, r = 0.008 $g_{water\ vapor}/g_{air}$,
a) r_L = 0 ; b) r_L =0.006 $g_{liq.water}/g_{air}$.
Find: θ_v = ? °C
Abbreviate "water vapor" with "wv" here.

a) Apply eq (3.13): θ_v = (288K)·
[1 + (0.61 g_{air}/g_{wv}) · (0.008 g_{wv}/g_{air})] .
Thus, θ_v = 289.4 K. Or subtract 273 to get Celsius: θ_v = **16.4 °C**.

a) Apply eq (3.14): θ_v = (288K)·
[1 + (0.61 g_{air}/g_{wv}) · (0.008 g_{wv}/g_{air}) – 0.006 $g_{liq.}/g_{air}$] = 287.7 K.
Subtract 273 to get Celsius: θ_v = **14.7°C**.

Check: Physics and units are reasonable.
Exposition: When no liquid water is present, virtual pot. temperatures are always warmer than potential temperatures, because water vapor is lighter than air.

However, liquid water is heavier than air, and has the opposite effect. This is called **liquid-water loading**, and makes the air act as if it were colder.

Sample Application

Given air at P = 70 kPa with T = –1°C, which is either (a) unsaturated, or (b) has $r_s = 5\ g_{water\ vapor}/kg_{air}$ and $r_L = 2\ g_{liq\ water}/kg_{air}$. Find θ_v.

Find the Answer

Given: (a) P = 70 kPa , T = –1°C = 272K,
(b) $r_s = 5\ g_{water\ vapor}/kg_{air} = 0.005\ g_{wv}/g_{air}$
$r_L = 2\ g_{liq\ water}/kg_{air} = 0.002\ g_{liq}/g_{air}$
Find: θ_v = ? °C

(a) Apply eq. (3.12) to find the potential temperature
$\theta = (272\text{K}) \cdot [(100\ \text{kPa})/(70\ \text{kPa})]^{0.28571} = 301\ \text{K}$
= **28°C** .

(b) Apply eq. (3.15):
$\theta_v = (301\ \text{K}) \cdot [\ 1 + (0.61\ g_{air}/g_{wv}) \cdot (0.005\ g_{wv}/g_{air}) - (0.002)\ ] = 301\text{K} \cdot (1.001) = $ **301.3 K**
$\theta_v = 301.3\ \text{K} - 273\ \text{K} = $ **28.3°C**

Check: Physics & units are reasonable.
Exposition: For this example, there was little effect of the vapor and liquid water. However, for situations with greater water vapor or liquid water, the virtual potential temperature can differ by a few degrees, which can be important for estimating thunderstorm intensity.

ter Vapor chapter; it is the ratio of grams of water per gram of air. The θ and θ_v values in the previous three equations must be in units of Kelvin.

An advantage of θ_v is that it can be used to calculate the buoyancy of air parcels that contain water — useful for anticipating storm characteristics. θ_v is constant only when there is no phase changes and no heat transfer; namely, no latent or sensible heat is absorbed or released.

For air that is rising within clouds, with water vapor condensing, it is usually the case that the air is saturated (= 100% relative humidity; see the Water Vapor chapter for details). As a result, the water-vapor mixing ratio r can be replaced with r_s , the saturation mixing ratio.

$$\theta_v = \theta \cdot [1 + (a \cdot r_s) - r_L] \qquad \bullet(3.15)$$

where $a = 0.61\ g_{air}/g_{water\ vapor}$, as before.

However, there are other situations where the air is NOT saturated, but contains liquid water. An example is the non-cloudy air under a cloud base, through which rain is falling at its terminal velocity. For this case, eq (3.14) should be used with an unsaturated value of water-vapor mixing ratio. This situation occurs often, and can be responsible for damaging downbursts of air (see the Thunderstorm chapters).

Why use potential temperature? Because it makes it easier to compare the temperatures of air parcels at two different heights — important for determining if air will buoyantly rise to create thunderstorms. For example, suppose air parcel A has temperature T_A = 20°C at z = 0, while air parcel B has T_B = 15°C at z = 1 km. Parcel A is warmer than parcel B.

Does that mean that parcel A is buoyant (warmer and wants to rise) relative to parcel B? The answer is no, because when parcel B is moved dry adiabatically to the altitude of parcel A, then parcel B is 5°C warmer than parcel A due to adiabatic warming. In fact, you can move parcels A and B to any common altitude, and after considering their adiabatic warming or cooling, parcel B will always be 5°C warmer than parcel A.

The easiest way to summarize this effect is with potential temperature. Using eq. (3.11), we find that θ_A = 20°C and θ_B = 25°C approximately. θ_A and θ_B keep their values (because θ is a conserved variable) no matter to what common altitude you move them, thus θ_B is always 5 °C warmer than θ_A in this illustration.

Another application for potential temperature is to label lines on a thermodynamic diagram, such as described next.

Intro to Thermo Diagrams

Convection is a vertical circulation associated with "warm air rising" and "cold air sinking". Meteorologists forecast the **deep convection** of thunderstorms and their hazards, or the **shallow convection** of thermals that disperse air pollutants.

The phrase "warm air rising" relates to the temperature difference ΔT between an air parcel and its surrounding environment. Air-parcel-temperature variation with altitude can be anticipated using heat- and water-conservation relationships. However, the surrounding environmental temperature profile can have a somewhat arbitrary shape that can be measured by a sounding balloon, but which is not easily described by analytical equations. So it can be difficult to mathematically describe ΔT vs. altitude.

Instead, graphical solutions can be used to estimate buoyancy and convection. We call these graphs **"thermodynamic diagrams"**. In this book, I will abbreviate the name as **"thermo diagram"**.

The diagram is set up so that higher in the diagram corresponds to higher in the atmosphere. In the real atmosphere, pressure decreases approximately logarithmically with increasing altitude, so we often use pressure P along the y-axis as a surrogate for altitude. Along the x-axis is air temperature T. The thin green lines in Fig. 3.4 show the (P, T) basis for a thermo diagram as a semi-log graph.

We can use eq. (3.10) to solve for the "dry" adiabatic temperature change experienced by rising air parcels. These are plotted as the thick orange diagonal lines in Fig. 3.4 for a variety of starting temperatures at P = 100 kPa. These **"dry adiabat"** lines (also known as **isentropes**), are labeled with θ because potential temperature is conserved for adiabatic processes.

INFO • Create Your Own Thermo Diagram

One of the advantages of thermo diagrams is that you do NOT need to calculate adiabatic temperature changes, because they are already calculated and plotted for you for a variety of different starting temperatures. If the starting temperature you need is not already plotted, you can mentally interpolate between the drawn lines as you raise or lower air parcels.

However, it is a useful exercise to see how such a thermo diagram can be created with a tool as simple as a computer spreadsheet.

The green (or dark-grey) items in the spreadsheet below were typed directly as numbers or words. You can follow along on your own spreadsheet. (You don't need to use the same colors — black is OK.)

The orange (or light-grey) numbers were calculated by entering a formula (eq. 3.10) into the bottom leftmost orange cell, and then "filling up" and "filling right" that equation into the other orange cells. But before you fill up and right, be sure to use the dollar sign "$" as shown below. It holds the column ID constant if it appears in front of the ID letter, or holds the row constant if in front of the ID number. Here is the equation for the bottom left orange cell (B12):

= ((B$13+273) * ($A12/A13)^0.28571) - 273

	A	B	C	D
1	Create Your Own Thermo Diagram			
2				
3	P(kPa)	T(degC)	T(degC)	T(degC)
4	10	–152.3	–131.6	–110.9
5	20	–125.9	–100.6	–75.4
6	30	–107.8	–79.5	–51.1
7	40	–93.7	–62.9	–32.1
8	50	–81.9	–49.0	–16.2
9	60	–71.6	–37.1	–2.5
10	70	–62.6	–26.4	9.7
11	80	–54.4	–16.9	20.7
12	90	–46.9	–8.1	30.7
13	100	–40	0	40

Different spreadsheet versions have different ways to create graphs. Select the cells that I outlined with the dark blue rectangle. Click on the Graph button, select the "XY scatter", and then select the option that draws straight line segments without data points.

Under the Chart, Source Data menu, select Series. Then manually switch the columns for the X and Y data for each series — this does an axis switch. On the graph, click on the vertical axis to get the Format Axis dialog box, and select the Scale tab. Check the Logarithmic scale box, and the Values in Reverse Order box. A bit more tidying up will yield a graph with 3 curves similar to Fig. 3.4. Try adding more curves.

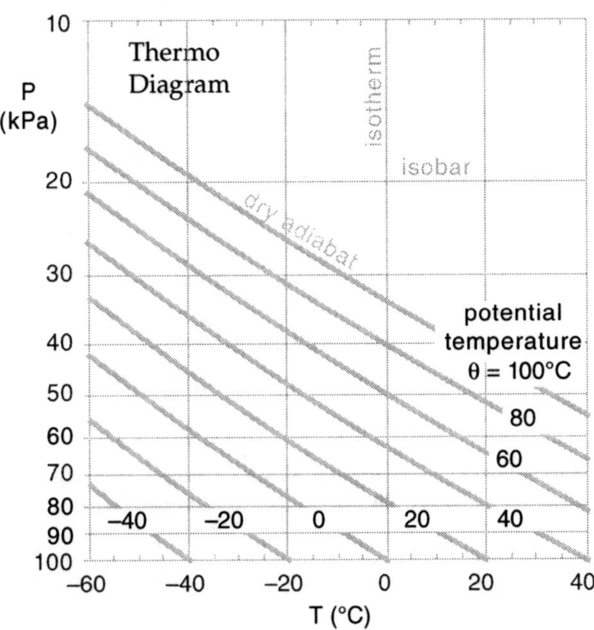

Figure 3.4 (at left)
*Simplified **thermo diagram**, showing isotherms (vertical green lines of constant temperature) and isobars (horizontal green lines of constant pressure). Isobars are plotted logarithmically, but with the scale reversed so that the highest pressure is at the bottom of the graph, corresponding to the bottom of the atmosphere. Dry adiabats are thick orange lines, showing the temperature variation of air parcels rising or sinking adiabatically.*

Sample Application

Given air at P = 70 kPa with T = –1°C. Find θ using the thermo diagram of Fig. 3.4.

Find the Answer

Given: P = 70 kPa , T = –1°C
Find: θ = ? °C

First, use the thermo diagram to find where the 70 kPa isobar and the –1°C isotherm intersect. (Since the –1°C isotherm wasn't drawn on this diagram, we must mentally interpolate between the lines that are drawn.) The adiabat that passes through this intersection point indicates the potential temperature (again, we must interpolate between the adiabats that are drawn). By extending this adiabat down to the reference pressure of 100 kPa, we can read off the temperature 28°C, which corresponds to a potential temperature of θ = <u>28°C</u> .

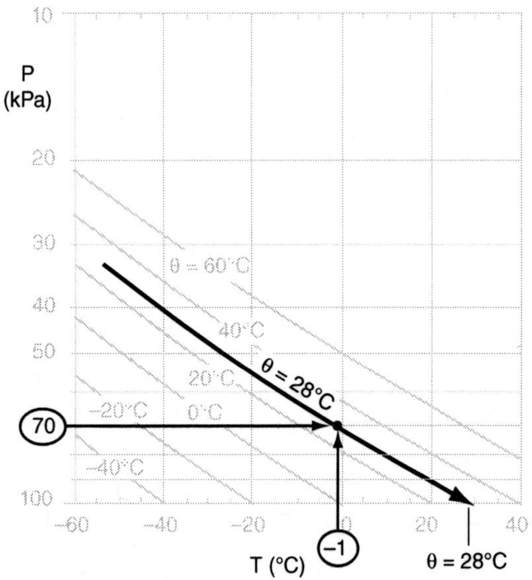

Check: Physics and units are reasonable.
Exposition: This exercise is the same as part (a) of the previous exercise, for which we calculated θ = 301 K = 28°C. Yes, the answers agree.

The advantage of using an existing printed thermo diagram is that we can draw a few lines and quickly find the answer without doing any calculations. So it can make our lives easier, once we learn how to use it.

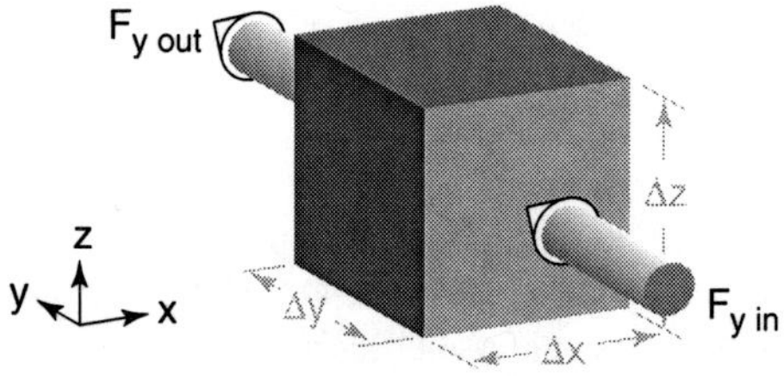

Figure 3.5
If heat flux $F_{y\ in}$ exceeds $F_{y\ out}$, then: (1) heat is deposited in the cube of air, making it hotter, and (2) $\Delta F/\Delta y$ is negative for this case. Similar fluxes can occur across the other faces.

If you know the initial (P, T) of the air parcel, then plot it as a point on the thermo diagram. Move parallel to the orange lines to the final pressure altitude. At that final point, read down vertically to find the parcel's final temperature.

HEAT BUDGET AT A FIXED LOCATION

Eulerian Form of the First Law of Thermo

Picture a cube of air at a fixed location relative to the ground (i.e., an Eulerian framework). By being fixed, the cube experiences only small, slow changes in pressure. As a result, the pressure-change term in the First Law of Thermo (eq. 3.2d) can usually be neglected. What remains is an equation that says thermal energy transferred (Δq) per unit mass causes temperature change: $\Delta T = \Delta q/C_p$.

Dividing this equation by time interval Δt gives a forecast equation for temperature: $\Delta T/\Delta t = (1/C_p)\cdot\Delta q/\Delta t$. A heat flux $\mathbb{F}$ (J m^{-2} s^{-1}, or W m^{-2}) into the volume could increase the temperature, but a heat flux out the other side could decrease the temperature. Thus, with both inflow and outflow of heat, net thermal energy will be transferred into the cube of air if the heat flux <u>de</u>creases with distance s across the cube: $\Delta q/\Delta t = -(1/\rho)\cdot\Delta\mathbb{F}/\Delta s$. The inverse density factor appears because Δq is energy <u>per unit mass</u>.

Heat flux convergence such as this causes warming, while **heat flux divergence** causes cooling. This **flux gradient** (change with flux across a distance) could happen in any of the three Cartesian directions. Thus, the temperature forecast equation becomes:

$$\frac{\Delta T}{\Delta t} = -\frac{1}{\rho\cdot C_p}\left[\frac{\Delta\mathbb{F}_x}{\Delta x} + \frac{\Delta\mathbb{F}_y}{\Delta y} + \frac{\Delta\mathbb{F}_z}{\Delta z}\right] + \frac{\Delta S_o}{C_p\cdot\Delta t} \tag{3.16}$$

where, for example, $\Delta\mathbb{F}_y/\Delta y$ is the change in northward-moving flux $\mathbb{F}_y$ across a north-south distance Δy (Fig. 3.5). Additional heat sources can occur inside the cube at rate $\Delta S_o/\Delta t$ (J kg^{-1} s^{-1}) such as when water vapor already inside the cube condenses into liquid and releases latent heat. The equation above is the Eulerian **heat-budget equation**, also sometimes called a **heat conservation** or **heat balance** equation.

Recall from Chapter 2 that we can define a kinematic flux by $F = \mathbb{F}/(\rho\cdot C_p)$ in units of K m s^{-1} (equivalent to °C m s^{-1}). Thus, eq. (3.16) becomes:

$$\frac{\Delta T}{\Delta t} = -\left[\frac{\Delta F_x}{\Delta x} + \frac{\Delta F_y}{\Delta y} + \frac{\Delta F_z}{\Delta z}\right] + \frac{\Delta S_o}{C_p\cdot\Delta t} \qquad \bullet(3.17)$$

We can also reframe this heat budget in terms of potential temperature, because with no movement of the cube of air itself, then $\Delta T = \Delta\theta$.

$$\frac{\Delta\theta}{\Delta t} = -\left[\frac{\Delta F_x}{\Delta x} + \frac{\Delta F_y}{\Delta y} + \frac{\Delta F_z}{\Delta z}\right] + \frac{\Delta S_o}{C_p \cdot \Delta t} \qquad \bullet(3.18)$$

You may have wondered why, in the previous figure, a $\Delta F_y/\Delta y$ was negative, even though heat was deposited into the cube. The reason is that for gradients, the difference-direction of the denominator must be the same direction as the numerator; e.g.:

$$\frac{\Delta F_y}{\Delta y} = \frac{F_{y\ northside} - F_{y\ southside}}{y_{northside} - y_{southside}} \qquad (3.19)$$

Similar care must be taken for gradients in the x and z directions.

Not only do we need to consider fluxes in each direction in eqs. (3.16 to 3.18), but for any one direction there might be more than one physical process causing fluxes. The other processes that we will discuss next are **conduction** (*cond*), **advection** (*adv*), **radiation** (*rad*), and **turbulence** (*turb*):

$$\frac{\Delta F_x}{\Delta x} = \left.\frac{\Delta F_x}{\Delta x}\right|_{adv} + \left.\frac{\Delta F_x}{\Delta x}\right|_{cond} + \left.\frac{\Delta F_x}{\Delta x}\right|_{turb} + \left.\frac{\Delta F_x}{\Delta x}\right|_{rad} \qquad (3.20)$$

$$\frac{\Delta F_y}{\Delta y} = \left.\frac{\Delta F_y}{\Delta y}\right|_{adv} + \left.\frac{\Delta F_y}{\Delta y}\right|_{cond} + \left.\frac{\Delta F_y}{\Delta y}\right|_{turb} + \left.\frac{\Delta F_y}{\Delta y}\right|_{rad} \qquad (3.21)$$

$$\frac{\Delta F_z}{\Delta z} = \left.\frac{\Delta F_z}{\Delta z}\right|_{adv} + \left.\frac{\Delta F_z}{\Delta z}\right|_{cond} + \left.\frac{\Delta F_z}{\Delta z}\right|_{turb} + \left.\frac{\Delta F_z}{\Delta z}\right|_{rad} \qquad (3.22)$$

In addition to describing these fluxes, we will estimate typical contributions of latent heating as a body source (ΔS_o), allowing us to simplify the full heat budget equation in an Eulerian framework.

Advection of Heat

The AMS *Glossary of Meteorology* (2000) defines advection as transport of an atmospheric property by the mass motion of the air (i.e., by the wind). **Temperature advection** transports heat. Faster winds blowing hotter air causes greater **advective heat flux**:

$$F_{x\ adv} = U \cdot T \qquad (3.23)$$

$$F_{y\ adv} = V \cdot T \qquad (3.24)$$

$$F_{z\ adv} = W \cdot T \qquad (3.25)$$

Sample Application

In the figure below, suppose that the incoming heat flux from the south is 5 W m^{-2}, and the outgoing on the north face of the cube is 7 W m^{-2}. (a) Convert these fluxes to kinematic units. (b) What is the value of the kinematic flux gradient? (c) Calculate the warming rate of air in the cube, assuming the cube has zero humidity and is at a fixed altitude where air density is 1 kg m^{-3}. The cube of air is 10 m on each side.

Find the Answer

Given: $\mathbb{F}_{y\ in}$= 5 W·m^{-2} , $\mathbb{F}_{y\ out}$ = 7 W·m^{-2}, Δy = 10 m
ρ = 1.0 kg m^{-3} ,
Find: a) $F_{x\ right}$ = ? K·m s^{-1}, $F_{x\ left}$ = ? K·m s^{-1}
b) $\Delta F_y/\Delta y$ = ? c) $\Delta T/\Delta t$ = ? K s^{-1}
From Appendix B: C_p = 1004 J·kg^{-1}·K^{-1}
Also, don't forget that 1 W = 1 J s^{-1} .
Diagram:

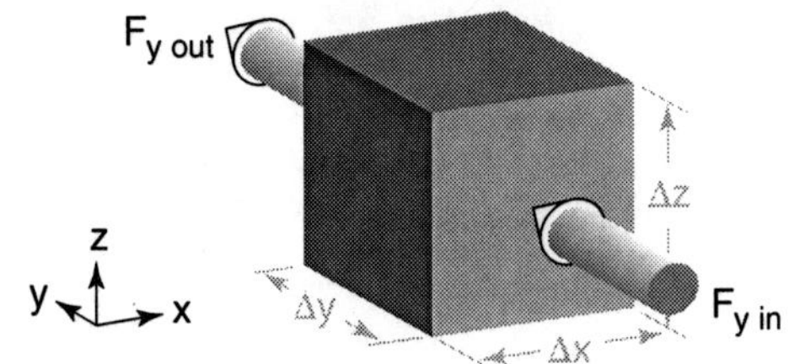

a) Apply eq. (2.11): $F = \mathbb{F} / (\rho \cdot C_p)$

$F_{y\ in}$ = (5 J·s^{-1}·m^{-2}) / [(1 kg m^{-3}) · (1004 J·kg^{-1}·K^{-1}))]
= **4.98x10^{-3} K·m·s^{-1}** .

$F_{y\ out}$ = (7 W·m^{-2}) / [(1 kg m^{-3}) · (1004 J·kg^{-1}·K^{-1}))]
= **6.97x10^{-3} K·m·s^{-1}** .

b) Recall from Chapter 1 that the direction of y is such that y increases toward the north. If we pick the south side as the origin of our coordinate system, then $y_{south\text{-}side}$ = 0 and $y_{north\text{-}side}$ = 10 m. Thus, the kinematic flux gradient (eq. 3.19) is

$$\frac{\Delta F_y}{\Delta y} = \frac{\left[(6.97\times10^{-3}) - (4.98\times10^{-3})\right](\text{K}\cdot\text{m}\cdot\text{s}^{-1})}{[10-0]\ (\text{m})}$$

= **1.99x10^{-4} K·s^{-1}**.

Putting this into eq. (3.21) and then that eq. into eq. (3.17) yields:

$\Delta T/\Delta t$ = **– 1.99x10^{-4} K·s^{-1}**.

Check: Physics & units are reasonable.
Exposition: The cube does not get warmer, it gets colder at a rate of about 0.72°C/hour. The reason is that more heat is leaving than entering, which gave a positive value for the flux gradient.

What happens if either of the two fluxes are negative? That means that heat is flowing from north to south. So the sign is critical in helping us determine the movement and convergence of heat.

Sample Application

The cube of air from Fig. 3.5 has T = 12°C along its south side, but smoothly increases in temperature to 15°C on the north side. This 100 km square cube is advecting toward the north at 25 km/hour. What warming rate at a fixed thermometer can be attributed to temperature advection?

Find the Answer

Given: $V = 25\ \text{km h}^{-1}$, $\Delta T = 15 - 12°\text{C} = 3°\text{C}$,
$\Delta y = 100$ km
Find: $\Delta T/\Delta t$ = ? °C h^{-1} due to advection

Apply eq. (3.27) in eq. (3.21), and apply that in eq. (3.17); namely, $\Delta T/\Delta t = -V \cdot (\Delta T/\Delta y)$
$= -(25\ \text{km h}^{-1}) \cdot [\,3°\text{C} / 100\ \text{km}] = \underline{-0.75\ °\text{C h}^{-1}}$.

Check: Physics and units are reasonable
Exposition: Note that the horizontal temperature gradient is positive (T increases as y increases) and V is positive (south wind), yet this causes negative temperature change (cooling). We call this **cold-air advection**, because colder air is blowing in.

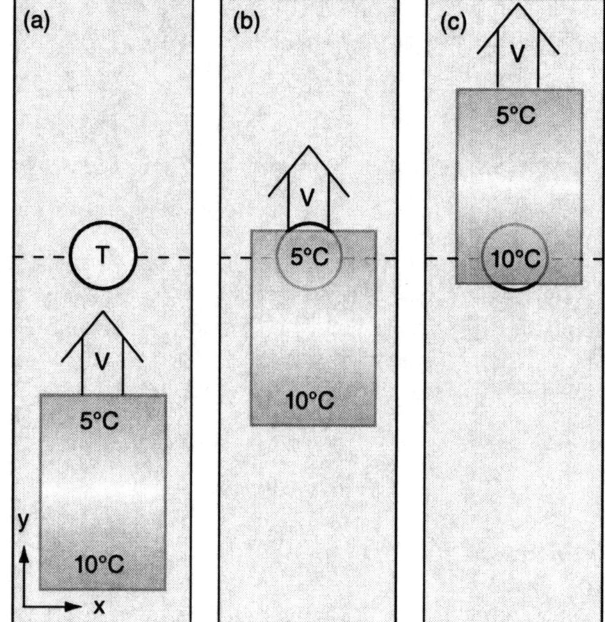

Figure 3.6
Top view of a grass field (green) with a fixed thermometer (T; yellow). Air with temperature gradient is advected north.

Sample Application

Given Fig. 3.6b, except assume that higher in the figure corresponds to higher in the atmosphere (i.e., replace y with z). Suppose that the 5°C air is at a relative altitude that is 500 m higher than that of the 10°C air. If the updraft is 500 m/(10 hours), what is the temperature at the thermometer after 10 hours?

Find the Answer

Given: $\Delta z = 500$ m, $T_{initial} = 5°\text{C}$, $W = 500$ m/(10 h),
$\Delta T/\Delta z = (5{-}10°\text{C})/(500\ \text{m}) = -0.01°\text{C/m}$
Find: T_{final} = ? °C after Δt = 10 h.

Looking at Fig. 3.6c, one might guess that the final air temperature should be 10°C. But Fig. 3.6c does not apply to vertical advection, because there is the added process of adiabatic expansion of the rising air.

The air that is initially 10°C in Fig. 3.6b will adiabatically cool 9.8°C/km of rise. Here, it rises only 0.5 km in the 10 h, so it cools 9.8°C/2 = 4.9°C. Its final temperature is 10°C – 4.9°C = **5.1°C**.

Check: Physics & units reasonable.
Exposition: The equations give the same result. Using eq. (3.28, 3.21 & 3.17): $\Delta T/\Delta t = -W \cdot (\Delta T/\Delta z + \Gamma_d)$.

Since we need to apply this over Δt = 10 h, multiply both sides by Δt: $\Delta T = -W \cdot \Delta t \cdot (\Delta T/\Delta z + \Gamma_d)$.
$\Delta T = -(500\text{m}/10\text{h}) \cdot (10\text{h}) \cdot (-0.01°\text{C/m} + 0.0098°\text{C/m})$
$= -500\text{m} \cdot (-0.0002°\text{C/m}) = +0.1°\text{C}$.
This 0.1°C warming added to the initial temperature of 5°C gives the final temperature = **5.1°C**.

Updrafts also cause heat transport, where buoyant updrafts are called **convection** while non-buoyant updrafts are called **advection**.

To illustrate temperature advection, consider a rectangular air parcel that is colder in the north and warmer in the south (Fig. 3.6). Namely, the temperature gradient $\Delta T/\Delta y$ = negative in this example. A south wind (V = positive) blows the air north toward a thermometer mounted on a stationary weather station. First the cold air reaches the thermometer (Fig. 3.6b). Later, the warm air blows over the thermometer (Fig. 3.6c). So the thermometer experiences warming with time ($\Delta T/\Delta t$ = positive) due to advection. Thus, it is not the advective flux $F_{x\ adv}$, but the gradient of advective flux ($\Delta F_{x\ adv}/\Delta y$) that causes a temperature change.

Although Fig. 3.6 illustrates only horizontal advection in one direction, we need to consider advective effects in all directions, including vertical. For a mean wind with nearly uniform speed:

$$\frac{\Delta F_{x\ adv}}{\Delta x} = \frac{U \cdot (T_{east} - T_{west})}{x_{east} - x_{west}} = U \cdot \frac{\Delta T}{\Delta x} \qquad (3.26)$$

$$\frac{\Delta F_{y\ adv}}{\Delta y} = \frac{V \cdot (T_{north} - T_{south})}{y_{north} - y_{south}} = V \cdot \frac{\Delta T}{\Delta y} \qquad (3.27)$$

$$\frac{\Delta F_{z\ adv}}{\Delta z} = W \cdot \left[\frac{\Delta T}{\Delta z} + \Gamma_d \right] \qquad (3.28)$$

Rising air cools at the dry adiabatic lapse rate of Γ_d = 9.8 °C km^{-1} . Since temperature of a rising air parcel is not conserved, this lapse-rate term must be added to the temperature gradient in the vertical advection equation. This same factor (with no sign changes) works for descending air too.

We can combine eqs. (3.26 - 3.28) with eq. (3.11) to express advection in terms of potential temperature θ:

$$\frac{\Delta F_{x\ adv}}{\Delta x} = U \cdot \frac{\Delta\theta}{\Delta x} \qquad \bullet(3.29)$$

$$\frac{\Delta F_{y\ adv}}{\Delta y} = V \cdot \frac{\Delta\theta}{\Delta y} \qquad \bullet(3.30)$$

$$\frac{\Delta F_{z\ adv}}{\Delta z} = W \cdot \frac{\Delta\theta}{\Delta z} \qquad \bullet(3.31)$$

Sample Application

The potential temperature of the air increases 5°C per 100 km distance east. If an east wind of 20 m s^{-1} is blowing, find the advective flux gradient, and the temperature change associated with this advection.

Find the Answer

Given: $\Delta\theta/\Delta x$ = 5°C/100 km = 5x10^{-5} °C m^{-1}
U = – 20 m s^{-1} (an east wind comes from the east)
Find: $\Delta F/\Delta y$ = ? °C s^{-1}, and $\Delta T/\Delta t$ = ? °C s^{-1}

Apply eq. (3.29): $\Delta F/\Delta x$ = (– 20 m s^{-1})·(5x10^{-5} °C m^{-1})
= **−0.001 °C s^{-1}**
Apply eq. (3.17) neglecting all other terms:
$\Delta T/\Delta t = -\Delta F/\Delta x$ = – (–0.001°C s^{-1}) = **+0.001 °C s^{-1}**

Check: Physics reasonable. Sign appropriate, because we expect warming as the warm air is blown toward us from the east in this example.
Exposition: $\Delta T/\Delta t$ = 3.6°C h^{-1}, a rapid warming rate.

Molecular Conduction & Surface Fluxes

Molecular heat conduction is caused by microscopic-scale vibrations and movement of air molecules transferring some of their microscopic kinetic energy to adjacent molecules. Conduction is what gets heat from the solid soil surface or liquid ocean surface into the air. It also conducts surface heat further underground. Winds are not needed for conduction.

Vertical heat flux due to molecular conduction is:

$$\mathbb{F}_{z\ cond} = -k \cdot \frac{\Delta T}{\Delta z} \qquad (3.32)$$

where k is the molecular conductivity, which depends on the material doing the conducting. The molecular conductivity of air is k = 2.53x10^{-2} W·m^{-1}·K^{-1} at sea-level under standard conditions.

The molecular conductivity for air is small, and vertical temperature gradients are also small in most of the atmosphere, so a good approximation is:

$$\frac{\Delta F_{x\ cond}}{\Delta x} \approx \frac{\Delta F_{y\ cond}}{\Delta y} \approx \frac{\Delta F_{z\ cond}}{\Delta z} \approx 0 \qquad (3.33)$$

But near the ground, large vertical temperature gradients frequently occur in the bottom several mm of the atmosphere (Fig. 3.7). If you have ever walked barefoot on a black asphalt parking lot or road on a hot summer day, you know that the surface temperatures can be burning hot to the touch (hotter than 50°C) even though the air temperatures at the height of your ankles can be 30°C or cooler. This large temperature gradient compensates for the small molecular conductivity of air, to create important vertical heat fluxes at the surface.

Sample Application

Suppose the temperature decreases from 50°C at the Earth's surface to 30°C at 5 mm above ground, as in Fig. 3.7. What is the vertical molecular heat flux?

Find the Answer

Given: ΔT = –20 °C, Δz = 0.005 m
k = 2.53x10^{-2} W·m^{-1}·K^{-1}
Find: $\mathbb{F}_{z\ cond}$ = ? W·m^{-2}

Apply eq. (3.32) :
$\mathbb{F}_{z\ cond}$ = –(2.53x10^{-2} W·m^{-1}·K^{-1}) · [–20°K/(0.005 m)]
= 101.2 W·m^{-2}

Check: Physics and units are reasonable.
Exposition: Although this is a fairly large heat flux into the bottom of the atmosphere, other processes described next (turbulence) can spread this heat over a layer of air roughly 1 km deep.

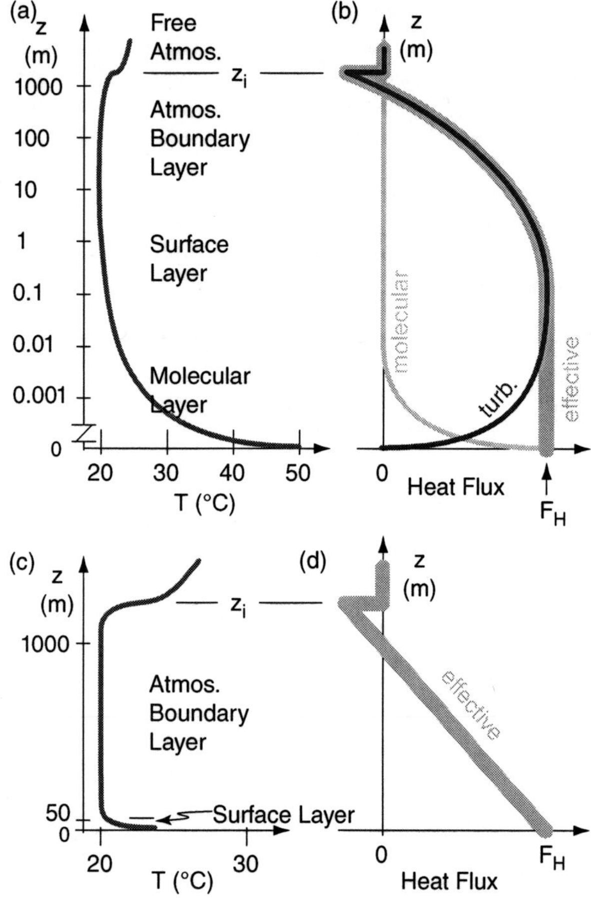

Figure 3.7
Relationship between temperature gradients and heat fluxes. (a & b) logarithmic vertical axis; (c & d) linear axes.

Sample Application

The wind is blowing at 10 m s^{-1} at height 10 m AGL. The 2 m air temperature is 15°C but the surface skin temperature is 30°C. What is the effective surface kinematic heat flux? Assume a surface of medium roughness having $C_H = 0.01$.

Find the Answer

Given: $C_H = 0.01$, $M = 10$ m s^{-1} at $z = 10$ m
T_{sfc} = 30°C , T_{air} = 15°C at z = 2 m
Find: F_H = ? K·m·s^{-1}

Apply eq. (3.35):
$F_H = (1\text{x}10^{-2})\cdot(10 \text{ m s}^{-1})\cdot(30–15°\text{C})$ = **1.5 °C·m·s^{-1}**

Check: Physics and units are reasonable.
Exposition: Recall that the relationship between dynamic and kinematic heat flux is $\mathbb{F}_H = \rho \cdot C_p \cdot F_H$. Thus, the dynamic heat flux is $\mathbb{F}_H \approx$ (1.2 kg m^{-3})·(1004 J kg^{-1} K^{-1})·(**1.5 K·m·s^{-1}**) = 1807. W m^{-2}. This is an exceptionally large surface heat flux — larger than the average solar irradiance of 1366 W·m^{-2}. But such a heat flux could occur where cool air is advecting over a very hot surface.

The bottom layer of the atmosphere that feels the influence of the earth's surface (i.e., the bottom boundary of the atmosphere) is known as the **atmospheric boundary layer** (ABL). This 1 to 2 km thick layer is often turbulent, meaning it has irregular gusts and whorls of motion. Meteorologists have devised an **effective turbulent heat flux** that is the sum of molecular and turbulent heat fluxes (Fig. 3.7), where turbulence is described in the next section. At the surface this effective flux is entirely due to molecular conduction, and above about 5 mm altitude the effective flux is mostly due to turbulence.

Instead of using eq. (3.32) to calculate molecular surface heat fluxes, most meteorologists approximate the effective surface turbulent heat flux, F_H, using what are called **bulk-transfer relationships.**

For windy conditions where most of the turbulence is caused by wind shear (change of wind speed or direction with altitude), you can use:

$$F_H = C_H \cdot M \cdot (\theta_{sfc} - \theta_{air}) \qquad \bullet(3.34)$$

or

$$F_H \cong C_H \cdot M \cdot (T_{sfc} - T_{air}) \qquad \bullet(3.35)$$

where (T_{sfc}, θ_{sfc}) are the temperature and potential temperature at the top few molecules (the skin) of the earth's surface, (T_{air} , θ_{air}) are the corresponding values in the air at 2 m above ground, and the wind speed at altitude 10 m is M. The empirical coefficient C_H is called the **bulk heat-transfer coefficient.** It is dimensionless, and varies from about 2x10^{-3} over smooth lakes or salt flats to about 2x10^{-2} for a rougher surface such as a forest. F_H is a kinematic flux.

For calm sunny conditions, turbulence is created by thermals of warm air rising due to their buoyancy. The resulting convective circulations cause so much stirring of the air that the ABL becomes a well **mixed layer** (ML). For this situation, you can use:

$$F_H = b_H \cdot w_B \cdot (\theta_{sfc} - \theta_{ML}) \qquad \bullet(3.36)$$

or

$$F_H = a_H \cdot w_* \cdot (\theta_{sfc} - \theta_{ML}) \qquad \bullet(3.37)$$

where a_H = 0.0063, is a dimensionless empirical **mixed-layer transport coefficient**, and b_H = 5x10^{-4} is called a **convective transport coefficient.** θ_{ML} is the mid-mixed-layer potential temperature (at height 500 m for a ML that is 1 km thick).

The w_B factor in eq. (3.36) is called the **buoyancy velocity scale** (m s^{-1}):

$$w_B = \left[\frac{|g| \cdot z_i}{T_{v\,ML}} \cdot (\theta_{v\,sfc} - \theta_{v\,ML}) \right]^{1/2} \qquad \bullet(3.38)$$

for a ML of depth z_i , and using gravitational acceleration $|g|$ = 9.8 m s^{-2}. ($\theta_{v\,sfc}$, $\theta_{v\,ML}$) are virtual potential temperatures of the surface skin and in the mid-mixed layer, and T_v is the absolute virtual temperature (Kelvins) in the mid mixed layer. Typical updraft speeds in thermals are of order $0.02 \cdot w_B$. To good approximation, the denominator in eq. (3.38) can be approximated by $\theta_{v\,ML}$ (also in units of K).

Another convective velocity scale w_* is called the **Deardorff velocity**:

$$w_* = \left[\frac{|g| \cdot z_i}{T_v} \cdot F_{Hsfc} \right]^{1/3} \quad \bullet(3.39)$$

for a surface kinematic heat flux of $F_{Hsfc} = F_H$. Often the Deardorff velocity is of order 1 to 2 m·s^{-1}, and the relationship between the two velocity scales is $w_* \approx 0.08 \cdot w_B$.

Later in the chapter, in the section on the Bowen ratio, you will see other formulas you can use to estimate F_H. Bulk transfer relationships can be used for other scalar fluxes at the surface including the moisture flux. For this case, replace temperature or potential-temperature differences with humidity differences between the surface skin and the mixed layer.

Atmospheric Turbulence

Superimposed on the average wind are somewhat-random faster and slower gusts. This **turbulence** is caused by **eddies** in the air that are constantly being created, changing, and dying. They exist as a superposition of many different size swirls (3 mm to 3 km). One eddy might move a cold blob of air out of any fixed Eulerian region, but another eddy might move air that is warmer into that same region. Although we don't try to forecast the heat transported by each individual eddy (an overwhelming task), we instead try to estimate the net heat flux caused by all the eddies. Namely, we resort to a statistical description of the effects of turbulence.

Turbulence in the air is analogous to turbulence in your teacup when you stir it. Namely, turbulence tends to blend all the ingredients into a uniform homogenous mixture. In the atmosphere, the mixing homogenizes individual variables such as potential temperature, humidity, and momentum (wind). The mixing rate depends on the strength of the turbulence, which can vary in space and time. We will focus on mixing of heat (potential temperature) here.

Fair Weather (no thunderstorms)

In a turbulent atmospheric boundary layer (ABL), daytime turbulence caused by convective thermals can transport heat from the sun-warmed Earth's surface and can distribute it more-or-less evenly through the ABL depth. The resulting turbulent heat fluxes decrease linearly with height as shown

Sample Application

What is the value of F_H on a sunny day with no winds? Assume z_i = 3 km, no clouds, dry air, θ_{ML} = 290 K, and θ_{sfc} = 320 K.

Find the Answer

Given: θ_{sfc} = 320 K, θ_{ML} = 290 K, z_i = 3 km,
Find: $F_{z\,eff.sfc.}$ = ? K·m·s^{-1}

If the air is dry, then: $\theta_v = \theta$ (from eq. 3.13).
Apply eqs. (3.38) and (3.36):

$$w_B = \left[\frac{|9.8\text{m}\cdot\text{s}^{-2}|\cdot 3000\text{m}}{290\text{K}} \cdot (320\text{K} - 290\text{K}) \right]^{1/2}$$

$= (3041 \text{ m}^2\text{·s}^{-2})^{1\ 2} = 55.1$ m·s^{-1}

$F_H = (5\text{x}10^{-4})\cdot(55.1 \text{ m·s}^{-1})\cdot(320 \text{ K} - 290 \text{ K})$
= <u>0.83 K·m·s^{-1}</u>

Check: Physics & units are reasonable.
Exposition: Notice how the temperature difference between the surface and the air enters both in the eq. for w_B and again for F_H. Thus, greater differences drive greater surface heat flux.

Sample Application

Given an effective surface kinematic heat flux of 0.67 K·m·s^{-1}, find the Deardorff velocity for a dry, 1 km thick boundary layer of temperature 25°C

Find the Answer

Given: F_H = 0.67 K·m·s^{-1}, z_i = 1 km = 1000 m,
$T_v = T$ (because dry) = 25°C = 298 K.
Find: w_* = ? m s^{-1}

Apply eq. (3.39):
$w_* = [(9.8 \text{ m·s}^{-2})\cdot(1000\text{m})\cdot(0.67 \text{ K·m·s}^{-1})/(298\text{K})]^{1\ 3}$
= <u>2.8</u> m s^{-1}

Check: Physics & units are reasonable.
Exposition: Over land on hot sunny days, warm buoyant thermals often rise with a speed of the same order of magnitude as the Deardorff velocity.

Sample Application

Given the Sample Application at the top of the previous page, what is the value for vertical flux divergence for this calm, sunny ABL?

Find the Answer

Given: $F_H = 0.83$ K·m·s^{-1}, $z_i = 3000$ m
Find: $\Delta F_{z\ turb}/\Delta z = ?$ (K s^{-1})

Apply eq. (3.41):

$$\frac{\Delta F_{z\ turb}}{\Delta z} \approx \frac{-1.2 \cdot F_H}{z_i}$$

$$\frac{\Delta F_{z\ turb}}{\Delta z} \approx \frac{-1.2 \cdot (0.83\ \text{K} \cdot \text{m/s})}{3000\text{m}}$$

$$= \mathbf{-0.000332\ K \cdot s^{-1}}$$

Check: Physics and units are reasonable.
Exposition: Recall from eq. (3.17) that a negative vertical gradient gives a positive warming with time — appropriate for a sunny day. The amount of warming is about 1.2°C/h. You might experience this warming rate over 10 hours on a hot sunny day.

by the thick green line in Fig. 3.7d. This line has a value at the bottom of the ABL as given by the effective surface flux ($F_{z\ bottom} = F_H$), and at the top has a value of ($F_{z\ top} \approx -0.2 \cdot F_H$) on less windy days. Thus, the flux-divergence term for turbulence (during sunny fair weather, within domain $0 < z < z_i$) is:

$$\frac{\Delta F_{z\ turb}}{\Delta z} \approx \frac{F_{z\ top} - F_{z\ bottom}}{z_i} \tag{3.40}$$

$$\frac{\Delta F_{z\ turb}}{\Delta z} \approx \frac{-1.2 \cdot F_H}{z_i} \tag{3.41}$$

for an ABL depth z_i of 0.2 to 3 km.

When no storm clouds are present, the air at $z > z_i$ is often not turbulent during daytime:

$$\frac{\Delta F_{z\ turb}}{\Delta z} \approx 0 \quad \text{above ABL top; for fair weather} \tag{3.42}$$

During clear nights of fair weather, turbulence can be very small over most of the lower 3 km of troposphere, except in the very lowest 100 m where wind shears can still create occasional turbulence.

Stormy Weather

Sometimes horizontal advection can move warm air under colder air. This makes the atmosphere statically unstable, allowing thunderstorms to form. These storms try to undo the instability by overturning the air — allowing the warm air to rise and cold air to sink. But the result is so violently turbulent that much mixing also takes place. The end result can sometimes be an atmosphere with a vertical gradient close to that of the standard atmosphere, as was discussed in Chapter 1. Namely, the atmosphere experiences **moist convective adjustment**, to adjust the initial less-stable lapse rate to one that is more stable.

The **standard atmospheric lapse rate** ($\Gamma_{sa} = -\Delta T/\Delta z$) is 6.5 K km^{-1}. Suppose that the initial lapse rate before the thunderstorm forms is Γ_{ps} ($= -\Delta T/\Delta z$). The amount of heat flux that is required to move the warm air up and cold air down during a storm lifetime of Δt (≈1 h) is:

$$\frac{\Delta F_{z\ turb}}{\Delta z} \approx \frac{z_T}{\Delta t} \cdot \left[\Gamma_{ps} - \Gamma_{sa}\right] \cdot \left(\frac{1}{2} - \frac{z}{z_T}\right) \tag{3.43}$$

where the troposphere depth is z_T (≈11 km). An initially unstable environment gives a positive value for the factor enclosed by square brackets.

Because thunderstorm motions do not penetrate below ground, and assuming no flux above the top of the storm, then the vertical turbulent heat flux must be zero at both the top and bottom of the troposphere, as was sketched in Fig. 3.8. The parabolic

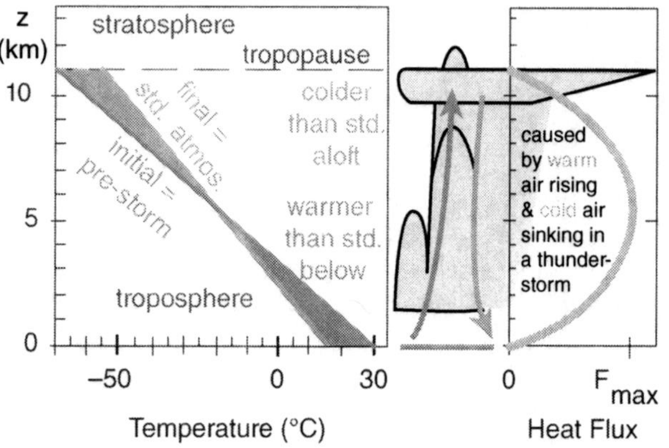

Figure 3.8
If a deep layer of cold air lies above a deep layer of warm air, such as in a pre-thunderstorm environment, then the air is statically unstable. This instability creates a thunderstorm, which not only causes overturning of tropospheric air, but also mixes the air. The final result can differ from storm to storm, but here we assume that the storm dies when the atmosphere has been mixed to the standard (std.)-atmosphere lapse rate of 6.5°C/km.

shape of the heat-flux curve has a maximum value of:

$$F_{\max} = z_T{}^2 \cdot \left[\Gamma_{ps} - \Gamma_{sa} \right] / (8 \cdot \Delta t) \qquad (3.44)$$

The thunderstorm also affects the heat budget via warming at all thunderstorm altitudes where condensation exceeds evaporation. Cooling at the thunderstorm top can be caused by IR radiation from the anvil cloud up into space. These heating and cooling effects should be added to the heat redistribution (heat moved from the bottom to the top of the storm) caused by turbulence.

So far, we focused on vertical flux gradients and the associated heating or cooling. Turbulence can also mix air horizontally, but the net horizontal heat transport is often negligibly small for both fair and stormy weather, because background temperature changes so gradually with distance in the horizontal. Thus, at all locations, a reasonable approximation is:

$$\frac{\Delta F_{x\ turb}}{\Delta x} \approx \frac{\Delta F_{y\ turb}}{\Delta y} \approx 0 \qquad (3.45)$$

Also, at locations with no turbulence there cannot be turbulent heat transport.

Solar and IR Radiation

We will split this topic into short-wave (solar) and long-wave (IR) radiation. Clear air is mostly transparent to solar radiation. Thus, the amount of short-wave radiation entering an air volume nearly equals the amount leaving. No flux gradient means that, to good approximation, you can neglect the direct solar heating of the air. However, sunlight is absorbed at the Earth's surface, which causes surface heat fluxes as already discussed. Sunlight is also absorbed in clouds or smoke, which can cause warming.

IR radiation is more complex, because air strongly absorbs a large portion of IR radiation flowing into a fixed volume, and re-radiates IR radiation outward in all directions. Radiation emission is related to T^4, according to the Stefan-Boltzmann law. In horizontal directions having weak temperature gradients, radiative flux divergence is negligibly small:

$$\frac{\Delta F_{x\ rad}}{\Delta x} \approx \frac{\Delta F_{y\ rad}}{\Delta y} \approx 0 \qquad (3.46)$$

But in the vertical, recall that temperature decreases with increasing altitude. Hence, more radiation would be lost upward from warmer air in the lower troposphere than is returned downward from the colder air aloft, which causes net cooling.

$$\frac{\Delta F_{z\ rad}}{\Delta z} \approx 0.1 \text{ to } 0.2 \text{ (K/h)} \qquad (3.47)$$

Sample Application

Suppose a pre-storm environment has a lapse rate of 9 °C km^{-1}. a) What is the maximum value of vertical heat flux near the middle of the troposphere during a storm lifetime? b) Calculate the vertical flux gradient at 1 km altitude due to the storm.

Find the Answer

Given: Γ_{ps} = 9 K km^{-1},
Find: (a) F_{max} = ? K·m s^{-1} (b) $\Delta F_{z\ turb}/\Delta z$ = ? (K s^{-1})
Assume: Γ_{sa} = 6.5 K km^{-1}, lifetime = Δt = 1 h = 3600 s, z_T = 11 km,

(a) Apply eq. (3.44):
F_{max} =(11,000m)·(11km)·[(9–6.5)(K km^{-1})]/[8·(3600s)]
= **10.5 K m s^{-1}**

(b) Apply eq. (3.43):

$$\frac{\Delta F_{z\ turb}}{\Delta z} \approx \frac{z_T}{\Delta t} \cdot \left[\Gamma_{ps} - \Gamma_{sa} \right] \cdot \left(\frac{1}{2} - \frac{z}{z_T} \right)$$

$$\frac{\Delta F_{z\ turb}}{\Delta z} \approx \frac{11\text{km}}{3600\text{s}} \cdot \left[(9-6.5)\frac{\text{K}}{\text{km}} \right] \cdot \left(\frac{1}{2} - \frac{1\text{km}}{11\text{km}} \right)$$

$\Delta F_{z\ turb}/\Delta z$ = **0.0031 K s^{-1}**

Check: Physics and units are reasonable.
Exposition: The magnitude of the max heat flux due to thunderstorms is much greater than the heat flux due to thermals in fair weather. Thunderstorms move large amounts of heat upward in the troposphere.

Based on Fig. 3.8, we would anticipate that storm turbulence should cool the bottom half of the stormy atmosphere. Indeed, the minus sign in eq. (3.17) combined with the positive sign of answer (b) above gives the expected cooling, not heating.

A SCIENTIFIC PERSPECTIVE • Expert vs. Novice

Expert scientists and engineers often solve problems, organize knowledge, and perceive structure differently than students and other novices.

Problem Solving	Novice	Expert
... is ...	a recall task	a process
... begins with ...	hunt for "the equation"	qualitative analysis
... uses classification based on ...	surface features	deep structure
...tools include ...	"the equation"	graphs, limits, diagrams, conservation laws, units, ...

Organizing Knowledge	Novice	Expert
Memory recall is ...	piecemeal	effortless retrieval of relevant collected facts
Reasoning by ...	jumping to hasty, unfounded conclusions	fast mental scan through a chain of possibilities
Conflicting data, ideas & conclusions are...	not recognized	recognized, pointing to need for more info
Related ideas are...	memorized as separate facts	integrated into a coherent big picture

Structure Perception	Novice	Expert
Cues about the structure are ...	missed	recognized and trigger new lines of thought
Disparate instances are...	separately classified based on surface features	recognized as having the same underlying structure
Tasks are performed...	before thinking about the organization	after data is organized to find structure
Theories that don't agree with data ...	are used without revision	identify ideas ripe for revision

(Paraphrased from Wendy Adams, Carl Wieman, Dan Schwartz, & Kathleen Harper.)

Internal Sources such as Latent Heat

Suppose $\Delta m_{condensing}$ kilograms of water vapor inside the storm condenses into liquid droplets and does not re-evaporate. It would release $L_v \cdot \Delta m_{condensing}$ Joules of latent heat. If this heating is spread vertically through the whole thunderstorm (a gross simplification) of air mass m_{air}, then the heating is:

$$\frac{\Delta So}{C_p \cdot \Delta t} = \frac{L_v}{C_p} \cdot \frac{\Delta m_{condensing}}{m_{air} \cdot \Delta t} \tag{3.48}$$

Because this warming does not require a heat flux across the storm boundaries, we define it as a "source term" that is internal to the thunderstorm. An opposite case of existing suspended cloud droplets that evaporate would yield the same equation, but with opposite sign as indicates net cooling.

In a real thunderstorm, some of the water vapor that initially condensed into cloud droplets can later evaporate. But any precipitation reaching the ground represents condensation that did not re-evaporate. Hence, we can use **rainfall rate** (*RR*) to estimate the internal latent heating rate:

$$\frac{\Delta So}{C_p \cdot \Delta t} = \frac{L_v}{C_p} \cdot \frac{\rho_{liq}}{\rho_{air}} \cdot \frac{RR}{z_{Trop}} \tag{3.49}$$

where the storm is assumed to fill a column of tropospheric air of depth z_{Trop}, liquid-water density is ρ_{liq} = 1000 kg·m^{-3}, latent-heat to specific heat ratio is L_v/C_p = 2500 K·kg$_{air}$·kg$_{liq}^{-1}$, and column-averaged air density is ρ_{air} = 0.689 kg·m^{-3} for z_{Trop} = 11 km.

Combining some of the values in eq. (3.49) gives:

$$\frac{\Delta So}{C_p \cdot \Delta t} = a \cdot RR \tag{3.50}$$

where a = 0.33 K (mm of rain)$^{-1}$, and RR has units [(mm of rain) s^{-1}]. Divide by 3600 for RR in mm h^{-1}.

Simplified Eulerian Net Heat Budget

You can insert the flux-gradient approximations from the previous subsections into eqs. (3.17 or 3.18) for the first law of thermo. Although the result looks complicated, you can simplify it by assuming the following are negligible within a fixed air volume: (1) vertical temperature advection by the mean wind; (2) horizontal turbulent heat transport; (3) molecular conduction; (4) short-wave heating of the air; (5) constant IR cooling.

You then get the following approximate Eulerian net heat-budget equation:

$$\left.\frac{\Delta T}{\Delta t}\right|_{x,y,z} = -\underbrace{\left[U\cdot\frac{\Delta T}{\Delta x}+V\cdot\frac{\Delta T}{\Delta y}\right]}_{advection} \underbrace{-0.1\frac{\text{K}}{\text{h}}}_{radiation}$$

$$\underbrace{-\frac{\Delta F_{z\ turb}(\theta)}{\Delta z}}_{turbulence} + \underbrace{\frac{L_v}{C_p}\cdot\frac{\Delta m_{condensing}}{m_{air}\cdot\Delta t}}_{latent\ heat} \qquad \bullet(3.51)$$

Later in this book you will see similar budget equations for other variables such as water vapor or momentum. In the turbulence term above, the (θ) indicates that this term is for heat flux divergence. Any of the terms on the right-hand side can be zero if the process it represents (advection, radiation, turbulence, condensation) is not active.

The net heat budget is important because you can use it to forecast air temperature at any altitude. Or, if you already know how the air temperature changes with time, you can use the net heat budget to see which processes are most important in causing this change.

The net heat budget applies to a volume of air having a finite mass. For the special case of the Earth's surface (infinitesimally thin; having no mass), you can write a simplified heat budget, as described next.

HEAT BUDGET AT EARTH'S SURFACE

So far, you examined the heat budget for a volume of air, where the volume was fixed (Eulerian) or moving (Lagrangian). Net imbalances of heat flux caused warming or cooling of air in the volume.

But what happens at the Earth's surface, which is infinitesimally thin and thus has zero volume? No heat can be stored in this layer. Hence, the sum of all incoming and outgoing heat fluxes must exactly balance. The net flux at the surface must be zero.

Surface Heat-flux Balance

Recall that fluxes are defined to be positive for heat moving upward, regardless of whether these fluxes are in the soil or the atmosphere.

Relevant fluxes at the surface include:

$\mathbb{F}^*$ = net radiation between sfc. & atmos. (Chapter 2)

$\mathbb{F}_H$ = effective surface turbulent heat flux (the **sensible Heat flux**)

$\mathbb{F}_E$ = effective surface **latent heat flux** caused by Evaporation or condensation (dew formation)

Sample Application

Suppose a thunderstorm rains at rate 4 mm h^{-1}. What is the average heating rate in the troposphere?

Find the Answer

Given: RR = 4 mm·h^{-1}.

Find: $\Delta S_o/(C_p\cdot\Delta t)$ = ? K·h^{-1}

Apply eq. (3.50): $\Delta S_o/(C_p\cdot\Delta t)$ = 0.33 (K mm^{-1})·(4 mm h^{-1}) = **1.32 K·h^{-1}**

Check: Physics and units are reasonable.

Exposition: For fixed Eulerian volumes losing liquid water as precipitation, this heating rate is significant.

Sample Application

For a fixed Eulerian volume, what temperature increase occurs in 2 h if $\Delta m_{cond}/m_{air}$ = 1 g_{water} kg_{air}^{-1}, $F_{H\ sfc}$ = 0.25 K·m·s^{-1} into a 1 km thick boundary layer, U = 0, V = 10 m s^{-1}, and $\Delta T/\Delta y$ = –2°C/100 km. Hint, approximate $L_v/C_p \approx 2.5$ K $(g_{water}\ kg_{air}^{-1})^{-1}$.

Find the Answer

Given: (see above)

Find: ΔT = ? °C over a 2 hour period

For each term in eq. (3.51), multiply by Δt:

$$\text{Lat.Heat\_Source}\cdot\Delta t = \left(2.5\frac{\text{K}\cdot\text{kg}_{air}}{\text{g}_{water}}\right)\cdot\left(1\frac{\text{g}_{water}}{\text{kg}_{air}}\right) = \underline{+2.5°C}$$

$$\text{Turb}\cdot\Delta t = -\frac{-1.2\cdot(0.25\text{K}\cdot\text{m/s})}{1000\text{m}}\cdot(7200\text{s}) = \underline{+2.16°C}$$

$$\text{Adv}\cdot\Delta t = -\left[(10\text{m/s})\cdot\left(\frac{-2°C}{100000\text{m}}\right)\right]\cdot(7200\text{s}) = \underline{+1.44°C}$$

$$\text{Rad}\cdot\Delta t = \left(-0.1\frac{\text{K}}{\text{h}}\right)\cdot(2\text{h}) = \underline{-0.2°C}$$

Combining all the terms gives:

ΔT = (Latent + Turb + Adv + Rad)
= (2.5 + 2.16 + 1.44 – 0.2)°C = **5.9 °C** over 2 hours.

Check: Physics and units are reasonable.

Exposition: For this contrived example, all the terms (except advection in the x direction) were important. Many of these terms can be estimated by looking at weather maps. For example, cloudy conditions might shade the sun during daytime and reduce the surface heat flux. These same clouds can trap IR radiation, causing the net radiative loss to be near zero below cloud base. But if there are no clouds (i.e., no condensation) and no falling precipitation that evaporates on the way down, then the latent-heating term would be zero.

So there is no fixed answer for the Eulerian heat budget — it varies as the weather varies.

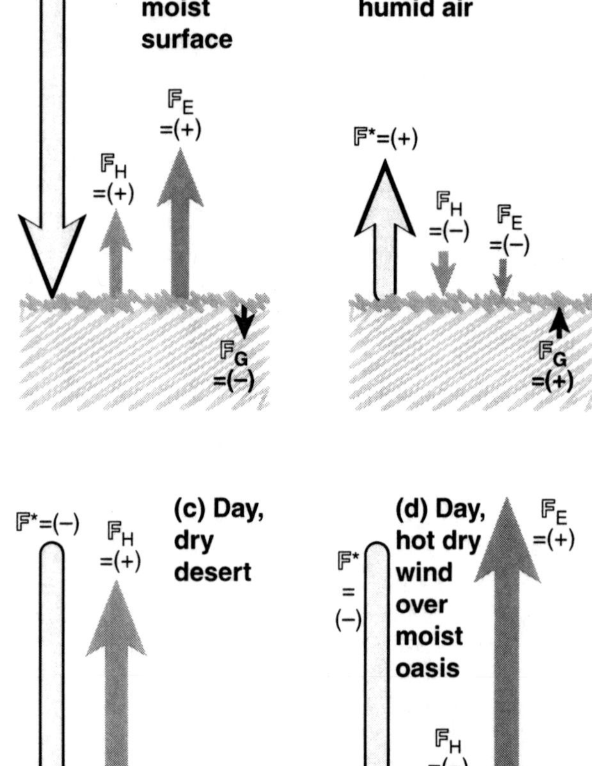

Figure 3.9
Illustration of signs and magnitudes of surface fluxes for various conditions. $\mathbb{F}^*$ *= net radiative flux,* $\mathbb{F}_H$ *= sensible heat flux,* $\mathbb{F}_E$ *= latent heat flux,* $\mathbb{F}_G$ *= conductive heat flux into the ground.*

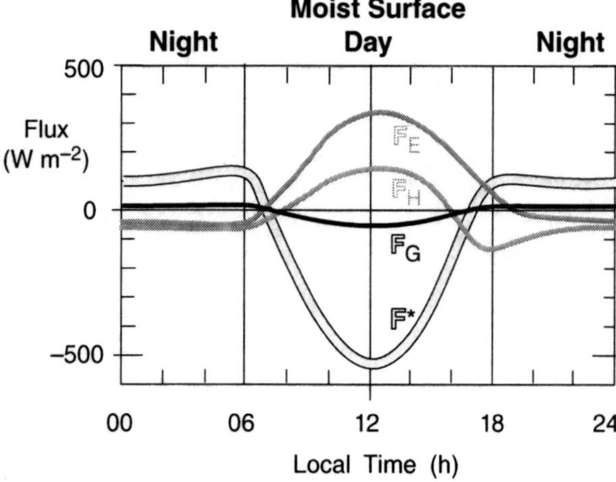

Figure 3.10
Daily variation of terms in the surface heat balance for a moist surface with humid air. Day and night correspond to (a) & (b) of the previous figure.

$\mathbb{F}_G$ = **molecular heat conduction** to/from deeper below the surface (e.g., Ground, oceans).

The surface balance for dynamic heat-fluxes (in units of W m^{-2}) is:

$$0 = \mathbb{F}^* + \mathbb{F}_H + \mathbb{F}_E - \mathbb{F}_G \qquad \bullet(3.52)$$

If you divide by $\rho_{air} \cdot C_p$ to get the balance in kinematic form (in units of K m s^{-1}), the result is:

$$0 = F^* + F_H + F_E - F_G \qquad \bullet(3.53)$$

The first 3 terms on the right are fluxes between the surface and the air above. The last term is between the surface and the Earth below (hence the – sign).

Examples of these fluxes and their signs are sketched in Fig. 3.9 for different surfaces and for day vs. night. For an irrigated lawn or crop, the typical **diurnal cycle** (daily evolution) of surface fluxes is sketched in Fig. 3.10. Essentially, net radiation F^* is an external forcing that drives the other fluxes.

A crude, first-order approximation for dynamic heat flux down into the soil is

$$\mathbb{F}_G \approx X \cdot \mathbb{F}^* \qquad (3.54)$$

with a corresponding kinematic heat flux of:

$$F_G \approx X \cdot F^* \qquad (3.55)$$

with factor X = (0.1, 0.5) for (daytime, nighttime).

There are different options for estimating the other terms in eqs. (3.52 or 3.53). For effective surface sensible heat flux you can use the bulk-transfer relationships already discussed (eqs. 3.34 to 3.37). For latent heat flux at the surface, similar bulk-transfer equations will be given in the Water-Vapor chapter. Another option for estimating latent and sensible heat fluxes at the surface is to utilize the Bowen-ratio, described next.

The Bowen Ratio

Define a **Bowen ratio**, B, as surface sensible-heat flux divided by surface latent-heat flux:

$$B = \frac{\mathbb{F}_H}{\mathbb{F}_E} = \frac{F_H}{F_E} \qquad (3.56)$$

Typical values are: 10 for arid locations, 5 for semi-arid locations, 0.5 over drier savanna, 0.2 over moist farmland, and 0.1 over oceans and lakes.

In the atmospheric **surface layer** (the bottom 10 to 25 m of the troposphere), surface effective sensible heat flux depends on $\Delta\theta/\Delta z$ — the potential-temperature gradient. Namely, $F_H = -K_H \cdot \Delta\theta/\Delta z$, where K_H is an **eddy diffusivity** for heat (see the Atmos. Boundary Layer chapter), z is height above ground, and the negative sign says that the heat flux flows down the local gradient (from hot toward cold air).

An analogous expression for effective surface moisture flux is $F_E = -K_E \cdot \Delta r/\Delta z$, where mixing ratio r is defined in the next chapter as mass of water vapor contained in each kg of dry air. If you approximate the eddy diffusivity for moisture, K_E, as equaling that for heat and if the vertical gradients are measured across the same air layer Δz, then you can write the Bowen ratio as:

$$B = \gamma \cdot \frac{\Delta\theta}{\Delta r} \tag{3.57}$$

for a **psychrometric constant** defined as $\gamma = C_p/L_v = 0.4\ (g_{water\ vapor}/kg_{air}) \cdot K^{-1}$.

Eq. (3.57) is appealing to use in field work because the difficult-to-measure fluxes have been replaced by easy-to-measure mean-temperature and humidity differences. Namely, if you erect a short tower in the surface layer and deploy **thermometers** at two different heights and mount **hygrometers** (for measuring humidity) at the same two heights (Fig. 3.11), then you can compute B. Don't forget to convert the temperature difference to potential-temperature difference: $\Delta\theta = T_2 - T_1 + (0.0098\ K\ m^{-1}) \cdot (z_2 - z_1)$.

With a bit of algebra you can combine eqs. (3.57, 3.56, 3.54, and 3.52) to yield effective surface sensible heat flux in dynamic units (W m^{-2}) as a function of net radiation:

$$\mathbb{F}_H = \frac{-0.9 \cdot \mathbb{F}^*}{\dfrac{\Delta r}{\gamma \cdot \Delta\theta} + 1} \tag{3.58}$$

or kinematic units (K m s^{-1}):

$$F_H = \frac{-0.9 \cdot F^*}{\dfrac{\Delta r}{\gamma \cdot \Delta\theta} + 1} \tag{3.59}$$

A bit more algebra yields the latent heat flux (W m^{-2}) caused by movement of water vapor to or from the surface:

$$\mathbb{F}_E = \frac{-0.9 \cdot \mathbb{F}^*}{\dfrac{\gamma \cdot \Delta\theta}{\Delta r} + 1} \tag{3.60}$$

or in kinematic units (K m s^{-1}):

$$F_E = \frac{-0.9 \cdot F^*}{\dfrac{\gamma \cdot \Delta\theta}{\Delta r} + 1} \tag{3.61}$$

The next chapter shows how to convert latent heat-flux values into waver-vapor fluxes.

If you have found sensible heat flux from eqs. (3.58 or 3.59), then the latent heat flux is easily found from:

$$\mathbb{F}_E = -0.9 \cdot \mathbb{F}^* - \mathbb{F}_H \tag{3.62}$$

or

$$F_E = -0.9 \cdot F^* - F_H \tag{3.63}$$

Sample Application

If the net radiation is –800 W·m^{-2} at the surface over a desert, then find sensible, latent, and ground fluxes.

Find the Answer

Given: $\mathbb{F}^* = -800$ W·m^{-2}
$B = 10$ for arid regions
Find: $\mathbb{F}_H$, $\mathbb{F}_E$ and $\mathbb{F}_G$ = ? W·m^{-2}

Because negative $\mathbb{F}^*$ implies daytime, use $X = 0.1$ in eq. (3.54): $\mathbb{F}_G = 0.1 \cdot \mathbb{F}^* = 0.1 \cdot (-800$ W·m$^{-2}) =$ **−80 W·m^{-2}**

Eqs. (3.52 & 3.56) can be manipulated to give:
$\mathbb{F}_E = (\mathbb{F}_G - \mathbb{F}^*) / (1 + B)$
$\mathbb{F}_H = B \cdot (\mathbb{F}_G - \mathbb{F}^*) / (1 + B)$
Thus,
$\mathbb{F}_E = (-80 + 800$ W·m$^{-2}) / (1 + 10)$
= **65.5 W·m^{-2}**

$\mathbb{F}_H = 10 \cdot (-80 + 800$ W·m$^{-2}) / (1 + 10)$
= **654.5 W·m^{-2}**

Check: Physics and units are reasonable. Also, we should confirm that the result gives a balanced energy budget. So apply eq. (3.52):

$0 = \mathbb{F}^* + \mathbb{F}_H + \mathbb{F}_E - \mathbb{F}_G$???
$0 = -800 + 654.5 + 65.5 + 80$ W·m^{-2} True.

Exposition: Although typical values for the Bowen ratio were given on the previous page, the actual value for any given type of surface depends on so many factors that it is virtually useless when trying to use the Bowen ratio method to predict surface fluxes. However, the field-measurement approach shown in the figure below and in eqs. (3.58 - 3.63) does not require an a-priori Bowen ratio estimate. Hence, this field approach is quite accurate for measuring surface fluxes, except near sunrise and sunset.

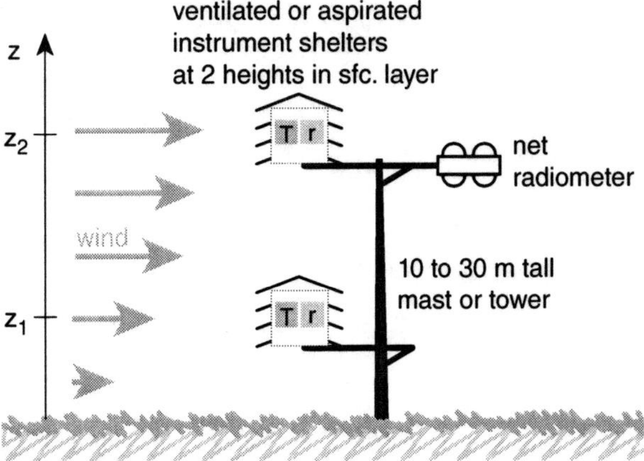

Figure 3.11
Field set-up for getting surface effective sensible and latent heat fluxes using the Bowen-ratio method. (T, r) are (thermometers, hygrometers) that are shielded from sunlight using ventilated instrument shelters. The net radiometer measures $\mathbb{F}^*$.

Sample Application

A Bowen-ratio field site observes the following:

index	z (m)	T (°C)	r (g_{vapor}/kg_{air})
2	15	16	7
1	1	20	12

with, $\mathbb{F}^* = -650$ W·m^{-2}. Find all surface fluxes.

Find the Answer

Given: info above.

Find: surface dynamic fluxes (W·m^{-2}) $\mathbb{F}_E, \mathbb{F}_H, \mathbb{F}_G$ = ?

First step is to find $\Delta\theta$:

$$\Delta\theta = T_2 - T_1 + (0.0098 \text{ K m}^{-1})\cdot(z_2 - z_1)$$
$$= 16 \text{ K} - 20 \text{ K} + (0.0098 \text{ K m}^{-1})\cdot(15\text{m} - 1\text{m})$$
$$= -4 \text{ K} + 0.137 \text{ K} = -3.86 \text{ K}$$

Apply eq. (3.58)

$$\mathbb{F}_H = \frac{-0.9\cdot(-650\text{W}\cdot\text{m}^{-2})}{\dfrac{(-5g_{vap}/kg_{air})}{[0.4(g_{vap}/kg_{air})\cdot\text{K}^{-1}]\cdot(-3.86\text{K})}+1}$$

$\mathbb{F}_H$ = **138 W·m^{-2}**

Next, apply eq. (3.62):

$$\mathbb{F}_E = -0.9\cdot\mathbb{F}^* - \mathbb{F}_H$$
$$= -0.9\cdot(-650 \text{ W}\cdot\text{m}^{-2}) - 138. \text{ W}\cdot\text{m}^{-2}$$
$$= \mathbf{447 \ W\cdot m^{-2}}$$

Finally, apply eq. (3.54): $\mathbb{F}_G = 0.1\cdot\mathbb{F}^*$ = **–65 W m^{-2}**.

Check: Physics & units are reasonable. Also, all the flux terms sum to zero, verifying the balance.

Exposition: The resulting Bowen ratio is B = 138/447 = 0.31, which suggests the site is irrigated farmland.

Sample Application

If wind speed is 30 km h^{-1} and actual air temperature is –25°C, find the wind-chill index.

Find the Answer

Given: M = 30 km h^{-1}, T_{air} = –25°C,

Find: $T_{wind\ chill}$ = ? °C.

Apply eq. (3.64a):

$$T_{wind\ chill} = [0.62\cdot(-25°\text{C}) + 13.1°\text{C}] + [0.51\cdot(-25°\text{C}) - 14.6°\text{C}]\cdot\left(\frac{30\text{km/h}}{4.8\text{km/h}}\right)^{0.16}$$

$$= [-2.4°\text{C}] + [-27.4°\text{C}]\cdot(1.34) = \mathbf{-39.1°C}$$

Check: Physics and units are reasonable. Agrees with a value interpolated from Table 3-1.

Exposition: To keep warm, consider making a fire by burning pages of this book. The book is easier to replace than your fingers, toes, ears, or nose.

APPARENT TEMPERATURE INDICES

Warm-blooded (**homeothermic**) animals including humans generate heat internally via metabolism of the food we eat with the oxygen we breathe. But we also rely on heat transfer with the environment to help maintain an internal core temperature of about 37°C (= 98.6°F). (Our skin is normally cooler — about 33.9°C = 93°F). Heat transfer occurs both via sensible heat fluxes (temperature difference between air and our skin or lungs) and latent heat fluxes (evaporation of moisture from our lungs and of perspiration from our skin).

The temperature we "feel" on our skin depends on the air temperature and wind speed (as they both control the bulk heat transfer between our skin and the environment) and on humidity (is it affects how rapidly perspiration evaporates to cool us).

Define a reference state as being a person walking at speed M_o = 4.8 km h^{-1} through calm, moderately dry air. The actual air temperature is defined to be the temperature we "feel" for this reference state.

The **apparent temperature** is the temperature of a reference state that feels the same as it does for non-reference conditions. For example, faster winds in winter make the temperature feel colder (wind chill) than the actual air temperature, while higher humidities in summer make the air feel warmer (humidex or heat index).

Wind-Chill Temperature

The **wind-chill temperature index** is a measure of how cold the air feels to your exposed face. The official formula, as revised in 2001 by the USA and Canada, for wind chill in °C is:

$$T_{wind\ chill} = (a\cdot T_{air} + T_1) + (b\cdot T_{air} - T_2)\cdot\left(\frac{M}{M_o}\right)^{0.16} \qquad \text{for } M > M_o \quad (3.64a)$$

and

$$T_{wind\ chill} = T_{air} \qquad \text{for } M \le M_o \quad (3.64b)$$

where a = 0.62, b = 0.51, T_1 = 13.1°C, and T_2 = 14.6°C. M is the wind speed measured at the official anemometer height of 10 m. For $M < M_o$, the wind chill equals the actual air temperature. This index applies to non-rainy air.

Fig. 3.12 and Table 3-3 show that faster winds and colder temperatures make us "feel" colder. The data used to create eq. (3.64) was from volunteers in Canada who sat in refrigerated wind tunnels, wearing warm coats with only their face exposed.

At wind chills colder than –27°C, exposed skin can freeze in 10 to 30 minutes. At wind chills colder than –48°C: WARNING, exposed skin freezes in 2 to 5 min. At wind chills colder than –55°C: DANGER, exposed skin freezes in less than 2 minutes. In this danger zone is an increased risk of **frostbite** (fingers, toes, ears and nose numb or white), and **hypothermia** (drop in core body temperature).

Humidex and Heat Index

On hot days you feel warmer than the actual air temperature when the air is more humid, but you feel cooler when the air is drier due to evaporation of your perspiration. In extremely humid cases the air is so uncomfortable that there is the danger of heat stress. Two **apparent temperatures** that indicate this are **humidex** and **heat index**.

The set of equations below approximates Steadman's temperature-humidity index of sultriness (i.e., a **heat index**):

$$T_{heat\ index}(°C) = T_R + [T - T_R]\cdot\left(\frac{RH\cdot e_s}{100\cdot e_R}\right)^p \qquad (3.65a)$$

where e_R = 1.6 kPa is reference vapor pressure, and

$$T_R(°C) = 0.8841\cdot T + (0.19°C) \qquad (3.65b)$$

$$p = (0.0196°C^{-1})\cdot T + 0.9031 \qquad (3.65c)$$

$$e_s(\text{kPa}) = 0.611\cdot\exp\left[5423\left(\frac{1}{273.15} - \frac{1}{(T+273.15)}\right)\right] \qquad (3.65d)$$

The two input variables are T (dry bulb temperature in °C), and RH (the relative humidity, ranging from 0 for dry air to 100 for saturated air). Also, T_R (°C), and p are parameters, and e_s is the saturation vapor pressure, discussed in the Water Vapor chapter. Eqs. (3.65) assume that you are wearing a normal amount of clothing for warm weather, are in the shade or indoors, and a gentle breeze is blowing.

The dividing line between feeling warmer vs. feeling cooler is highlighted with the bold, underlined heat-index temperatures in Table 3-4.

In Canada, a **humidex** is defined as

$$T_{humidex}(°C) = T(°C) + a\cdot(e - b) \qquad (3.66a)$$

where T is air temperature, a = 5.555 (°C kPa^{-1}), b = 1 kPa, and

$$e(\text{kPa}) = 0.611\cdot\exp\left[5418\left(\frac{1}{273.16} - \frac{1}{[T_d(°C)+273.16]}\right)\right] \qquad (3.66b)$$

Table 3-3. The wind-chill-index temperature (°C).

Wind Speed		Actual Air Temperature (°C)					
km·h^{-1}	m·s^{-1}	–40	–30	–20	–10	0	10
60	16.7	–64	–50	–36	–23	–9	5
50	13.9	–63	–49	–35	–22	–8	6
40	11.0	–61	–48	–34	–21	–7	6
30	8.3	–58	–46	–33	–20	–6	7
20	5.6	–56	–43	–31	–18	–5	8
10	2.8	–51	–39	–27	–15	–3	9
0	0	–40	–30	–20	–10	0	10

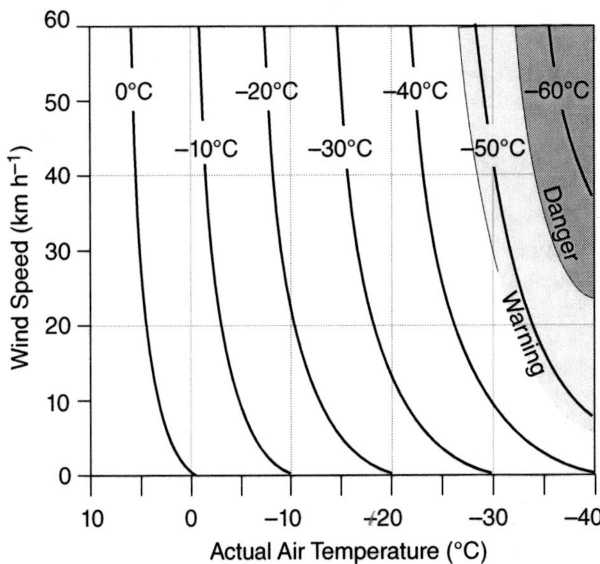

Figure 3.12
For any wind speed M and actual air temperature T read the wind-chill temperature index (°C) from the curves in this graph.

Table 3-4. Heat-index apparent temperature (°C).

Rel. Hum. (%)	Actual Air Temperature (°C)						
	20	25	30	35	40	45	50
100	21	29	41	61			
90	21	29	39	57			
80	21	28	37	52			
70	20	27	35	48			
60	20	26	34	45	62		
50	**<u>19</u>**	25	32	41	55		
40	19	**<u>24</u>**	30	38	49	66	
30	19	24	**<u>29</u>**	36	44	56	
20	18	23	28	**<u>33</u>**	40	48	59
10	18	23	27	32	**<u>37</u>**	**<u>42</u>**	**<u>48</u>**
0	18	22	27	31	36	40	44

Sample Application

Use the equations to find the heat index and humidex for an air temperature of 38°C and a relative humidity of 75% (which corresponds to a dew-point temperature of about 33°C).

Find the Answer

Given: $T = 38°C$, $RH = 75\%$, $T_d = 33°C$

Find: $T_{heat\ index} = ?\ °C$, $T_{humidex} = ?\ °C$

For heat index, use eqs. (3.65):

$$T_R = 0.8841 \cdot (38) + 0.19 = 33.8°C \qquad (3.65b)$$

$$p = 0.0196 \cdot (38) + 0.9031 = 1.65 \qquad (3.65c)$$

$$e_s = 0.611 \cdot \exp[5423 \cdot (\{1/273.15\} - \{1/(38+273.15)\})] = 6.9 \text{ kPa} \qquad (3.65d)$$

$$T_{heat\ index} = 33.8 + [38-33.8] \cdot (0.75 \cdot 6.9/1.6)^{1.65} = \underline{62.9°C} \qquad (3.65a)$$

For humidex, use eqs. (3.66):

$$e = 0.611 \cdot \exp[5418 \cdot (\{1/273.16\} - \{1/(33+273.16)\})] = 5.18 \text{ kPa} \qquad (3.66b)$$

$$T_{humidex} = 38 + 5.555 \cdot (5.18-1) = \underline{61.2°C} \qquad (3.66a)$$

Check: Units are reasonable. Values agree with extrapolation of Tables 3-4 and 3-5.

Exposition: These values are in the danger zone, meaning that people are likely to suffer heat stroke. The humidex and heat index values are nearly equal for this case.

Table 3-5. Humidex apparent temperature (°C)

T_d	Actual Air Temperature T (°C)						
(°C)	20	25	30	35	40	45	50
50							118
45						96	101
40					77	82	87
35				62	67	72	77
30			49	54	59	64	69
25		37	42	47	52	57	62
20	28	33	38	43	48	53	58
15	24	29	34	39	44	49	54
10	21	26	31	36	41	46	51
5	**19**	**24**	**29**	**34**	**39**	**44**	**49**
0	18	23	28	33	38	43	48
–5	17	22	27	32	37	42	47
–10	16	21	26	31	36	41	46

T_d is dew-point temperature, a humidity variable discussed in the Water Vapor chapter.

Humidex is also an indicator of summer discomfort due to heat and humidity (Table 3-3). Values above 40°C are uncomfortable, and values above 45°C are dangerous. **Heat stroke** is likely for humidex ≥ 54°C. This table also shows that for dry air ($T_d \le 5°C$) the air feels cooler than the actual air temperature.

TEMPERATURE SENSORS

Temperature sensors are generically called **thermometers**. Anything that changes with temperature can be used to measure temperature. Many materials expand when warm, so the size of the material can be calibrated into a temperature. Classical **liquid-in-glass thermometers** use either mercury or a dyed alcohol or glycol fluid that can expand from a reservoir or **bulb** up into a narrow tube.

House **thermostats** (temperature controls) often use a **bimetallic strip**, where two different metals are sandwiched together, and their different expansion rates with temperature causes the metal to bend as the temperature changes. Car thermostats use a wax that expands against a valve to redirect engine coolant to the radiator when hot. Some one-time use thermometers use wax that melts onto a piece of paper at a known temperature, changing is color.

Many electronic devices change with temperature, such as resistance of a wire, capacitance of a capacitor, or behavior of various transistors (**thermistors**). These changes can be measured electronically and displayed. **Thermocouples** (such as made by a junction between copper and constantan wires, where constantan is an alloy of roughly 60% copper and 40% nickel) generate a small amount of electricity that increases with temperature. **Liquid crystals** change their orientation with temperature, and can be designed to display temperature.

Sonic thermometers measure the speed of sound through air between closely placed transmitters and receivers of sound. Radio Acoustic Sounder Systems (**RASS**) transmit a loud pulse of sound upward from the ground, and then infer temperature vs. height via the speed that the sound wave propagates upward, as measured by a radio or microwave profiler.

Warmer objects emit more radiation, particularly in the infrared wavelengths. An **infrared thermometer** measures the intensity of these emissions to infer the temperature. Satellite remote sensors also detect emissions from the air upward into space, from which temperature profiles can be calculated (see the Satellites & Radar chapter).

Even thick layers of the atmosphere expand when they become warmer, allowing the **thickness** between two different atmospheric pressure levels to indicate average temperature in the layer.

REVIEW

Three types of heat budgets were covered in this chapter. All depend on the flow rate of energy per unit area (J m^{-2} s^{-1}) into or out of a region. This energy flow is called a flux, where the units above are usually rewritten in their equivalent form (W m^{-2}).

1) One type is a heat balance at the surface of the Earth. The surface has zero thickness — hence no air volume and no mass that can store or release heat. Thus, the input fluxes must exactly balance the output fluxes. Sunlight and IR radiation (see the Radiation chapter) must be balanced by the sum of conduction to/from the ground and effective turbulent fluxes of sensible and latent heat between the surface and the air.

2) Another type is an Eulerian budget for a fixed volume of air. If more heat enters than leaves, then the air temperature must increase (i.e., heat is stored in the volume). Processes that can move heat are advection, turbulence, and radiation. At the Earth's surface, an effective turbulent flux is defined that includes both the turbulent and conductive contributions. Also, heat can be released within the volume if water vapor condenses or radionuclides decay.

3) The third type is a Lagrangian budget that follows a mass of air (called an air parcel) as it rises or sinks through the surrounding environment. This is trickier because the parcel temperature can change even without moving heat into it via fluxes, and even without having water vapor evaporate or condense within it. This adiabatic temperature change is caused by work done on or by the parcel as it responds to the changing pressure as it moves vertically in the atmosphere. For unsaturated (non-cloudy) air, temperature of a rising air parcel decreases at the adiabatic lapse rate of 9.8°C km^{-1}. This process is critical for understanding turbulence, clouds, and storms, as we will cover in later chapters.

The actual air temperature can be measured by various thermometers. Humans feel the combined effects of actual air temperature, wind, and humidity as an apparent air temperature.

HOMEWORK EXERCISES

Broaden Knowledge & Comprehension

B1(§). For an upper-air weather station near you (or for a site specified by your instructor), get recent observation data of T vs. z or T vs. P from the internet, and plot the result on a copy of the thermodynamic diagram from this chapter.

B2. For an upper-air weather station near you (or for a site specified by your instructor), get an already-plotted recent sounding from the internet. Find the background isotherm and isobar lines, and compare their arrangement to the diagram (Fig. 3.4) in this chapter. We will learn more about other thermo-diagram formats in the Atmospheric Stability chapter.

B3. Use the internet to acquire temperatures at your town and also at a town about 100 km downwind of you. Also get the wind speeds in both towns and take an average. Use this average speed to calculate the contribution of advection to the local heating in the air between those two towns.

B4. Use the internet to acquire a weather map or other weather report that shows the observed near-surface air temperature just before sunrise at your location (or at another location specified by your instructor). For the same location, find a map or report of the temperature in mid afternoon. From these two observations, calculate the rate of temperature change over that time period. Also, qualitatively describe which terms in the Eulerian heat budget might be largest. (Hint: if windy, then perhaps advection is important. If clear skies, then heat transfer from the solar-heated ground might be important. Access other weather maps as needed to determine which physical process is most important for the temperature change.)

B5. Use the internet to acquire a local weather map of apparent temperature, such as wind-chill in winter or heat index (or humidex) in summer. If the map covers your location, compare how the air feels to you vs. the apparent temperature on the map.

B6. Use the internet to acquire images of 4 different types of temperature sensors (not 4 models of the same type of sensor).

Apply

A1. Find the change in sensible heat (enthalpy) (J) possessed by 3 kg of air that warms by __°C.
a. 1 b. 2 c. 3 d. 4 e. 5 f. 6 g. 7
h. 8 i. 9 j. 10 k. 11 m. 12

A2. Find the specific heat C_p of humid air having water-vapor mixing ratio ($g_{vapor}/g_{dry\ air}$) of:
a. 0.010 b. 0.012 c. 0.014 d. 0.016 e. 0.018
f. 0.020 h. 0.022 i. 0.024 j. 0.026 k. 0.028
m. 0.030

A3. Find the change in latent heat (J) for condensation of ___ kg of water vapor.
a. 0.2 b. 0.4 c. 0.6 d. 0.8 e. 1.0 f. 1.2 g. 1.4
h. 1.6 i. 1.8 j. 2.0 k. 2.2 m. 2.4

A4. Find the temperature change (°C) of air given the following values of heat transfer and pressure change, assuming air density of 1.2 kg m^{-3}.

	Δq (J kg^{-1})	ΔP (kPa)
a.	500	5
b.	1000	5
c.	1500	5
d.	2000	5
e.	2500	5
f.	3000	5
g.	500	10
h.	1000	10
i.	1500	10
j.	2000	10
k.	2500	10
m.	3000	10

A5. Find the change in temperature (°C) if an air parcel rises the following distances while experiencing the heat transfer values given below.

	Δq (J kg^{-1})	Δz (km)
a.	500	0.5
b.	1000	0.5
c.	1500	0.5
d.	2000	0.5
e.	2500	0.5
f.	3000	0.5
g.	500	1
h.	1000	1
i.	1500	1
j.	2000	1
k.	2500	1
m.	3000	1

A6. Given the following temperature change ΔT (°C) across a height difference of Δz = 4 km, find the lapse rate (°C km^{-1}):
a. 2 b. 5 c. 10 d. 20 e. 30 f. 40 g. 50
h. –2 i. –5 j. –10 k. –20 m. –30

A7. Find the final temperature (°C) of an air parcel with the following initial temperature and height change, for an adiabatic process.

	$T_{initial}$ (°C)	Δz (km)
a.	15	0.5
b.	15	–1.0
c.	15	1.5
d.	15	–2.0
e.	15	2.5
f.	15	–3.0
g.	5	0.5
h.	5	–1.0
i.	5	1.5
j.	5	–2.0
k.	5	2.5
m.	5	–3.0

A8. Using the equations (not using the thermo diagram), find the final temperature (°C) of dry air at a final pressure, if it starts with the initial temperature and pressure as given. (Assume adiabatic.)

	$T_{initial}$ (°C)	$P_{initial}$ (kPa)	P_{final}(kPa)
a.	5	100	80
b.	5	100	50
c.	5	80	50
d.	5	80	100
e.	0	60	80
f.	0	60	50
g.	0	80	40
h.	0	80	100
i.	–15	90	80
j.	–15	90	50
k.	–15	70	50
m.	–15	70	100

A9. Same as previous question, but use the thermo diagram Fig. 3.4.

A10. Given air with temperature and altitude as listed below, use formulas (not thermo diagrams) to calculate the potential temperature. Show all steps in your calculations.

	z (m)	T (°C)
a.	400	30
b.	800	20
c.	1,100	10
d.	1,500	5
e.	2,000	0
f.	6,000	–50
g.	10,000	–90
h.	–30	35
i.	700	3
j.	1,300	–5
k.	400	5
m.	2,000	–20

A11. Same as the previous exercise, but find the virtual potential temperature for humid air. Use a water-vapor mixing ratio of 0.01 $g_{vapor}/g_{dry\ air}$ if the air temperature is above freezing, and use 0.0015 $g_{vapor}/g_{dry\ air}$ if air temperature is below freezing. Assume the air contains no ice or liquid water.

A12. Given air with temperature and pressure as listed below, use formulas (not thermo diagrams) to calculate the potential temperature. Show all steps in your calculations.

	P (kPa)	T (°C)
a.	90	30
b.	80	20
c.	110	10
d.	70	5
e.	85	0
f.	40	–45
g.	20	–90
h.	105	35
i.	75	3
j.	60	–5
k.	65	5
m.	50	–20

A13. Same as previous exercise, but use the thermo diagram Fig. 3.4.

A14. Instead of equations, use the Fig 3.4 to find the actual air temperature (°C) given:

	P(kPa)	θ (°C)
a.	100	30
b.	80	30
c.	60	30
d.	90	10
e.	70	10
f.	50	10
g.	80	–10
h.	50	–10
i.	20	50

A15(§). Use a spreadsheet to calculate and plot a thermo diagram similar to Fig. 3.4 but with: isotherm grid lines every 10°C, and dry adiabats for every 10°C from –50°C to 80°C.

A16. Find the rate of temperature change (°C h^{-1}) in an Eulerian coordinate system with no internal heat source, given the kinematic flux divergence values below. Assume $\Delta x = \Delta y = \Delta z = 1$ km.

	ΔF_x (K·m s^{-1})	ΔF_y (K·m s^{-1})	ΔF_z (K·m s^{-1})
a.	1	2	3
b.	1	2	–3
c.	1	–2	3
d.	1	–2	–3
e.	–1	2	3
f.	–1	2	–3
g.	–1	–2	3
h.	–1	–2	–3

A17. Given the wind and temperature gradient, find the value of the kinematic advective flux gradient (°C h^{-1}).

	V (m s^{-1})	$\Delta T/\Delta y$ (°C 100 km)
a.	5	–2
b.	5	2
c.	10	–5
d.	10	5
e.	–5	–2
f.	–5	2
g.	–10	–5
h.	–10	5

A18. Given the wind and temperature gradient, find the value of the kinematic advective flux gradient (°C h^{-1}).

	W (m s^{-1})	$\Delta T/\Delta z$ (°C km^{-1})
a.	5	–2
b.	5	2
c.	10	–5
d.	10	–10
e.	–5	–2
f.	–5	2
g.	–10	–5
h.	–10	–10

A19. Find the value of the conductive flux $\mathbb{F}_{z\ cond}$ (W m^{-2}) given a change of absolute temperature with height ($T_2 - T_1$ = value below) across a distance ($z_2 - z_1$ = 1 m):

a. –1 b. –2 c. –3 d. –4 e. –5 f. –6 g. –7
h. 1 i. 2 j. 3 k. 4 m. 5 n. 6 o. 7

A20. Find the effective surface turbulent heat flux (°C·m s^{-1}) over a forest for wind speed of 10 m s^{-1}, air temperature of 20°C, and surface temperature (°C) of

a. 21 b. 22 c. 23 d. 24 e. 25 f. 26 g. 27
h. 19 i. 18 j. 17 k. 16 m. 15 n. 14 o. 13

A21. Find the effective kinematic heat flux at the surface on a calm day, for a buoyant velocity scale of 50 m s^{-1}, a mixed-layer potential temperature of 25°C, and with a surface potential temperature (°C) of:

a. 26 b. 28 c. 30 d. 32 e. 34 f. 36 g. 38
h. 40 i. 42 j. 44 k. 46 m. 48 n. 50

A22. Find the effective kinematic heat flux at the surface on a calm day, for a Deardorff velocity of 2

m s^{-1}, a mixed-layer potential temperature of 24°C, and with a surface potential temperature (°C) of:
a. 26 b. 28 c. 30 d. 32 e. 34 f. 36 g. 38
h. 40 i. 42 j. 44 k. 46 m. 48 n. 50

A23. For dry air, find the buoyancy velocity scale, given a mixed-layer potential temperature of 25°C, a mixed-layer depth of 1.5 km, and with a surface potential temperature (°C) of:
a. 27 b. 30 c. 33 d. 36
e. 40 f. 43 g. 46 h. 50

A24. For dry air, find the Deardorff velocity w_* for an effective kinematic heat flux at the surface of 0.2 K·m s^{-1}, air temperature of 30°C, and mixed-layer depth (km) of:
a. 0.4 b. 0.6 c. 0.8 d. 1.0
e. 1.2 f. 1.4 g. 1.6 h. 1.8

A25. Find the value of vertical divergence of kinematic heat flux, if the flux at the top of a 200 m thick air layer is 0.10 K·m s^{-1}, and flux (K·m s^{-1}) at the bottom is:
a. 0.2 b. 0.18 c. 0.16 d. 0.14
e. 0.12 f. 0.10 g. 0.08 h. 0.06

A26. Given values of effective surface heat flux and boundary-layer depth for daytime during fair weather, what is the value of the turbulent-flux vertical gradient?

	F_H (K·m·s^{-1})	z_i (km)
a.	0.25	2.0
b.	0.15	1.5
c.	0.1	1.0
d.	0.03	0.3
e.	0.08	0.3
f.	0.12	0.8
g.	0.15	1.0
h.	0.25	1.5

A27. Given a pre-storm environment where the temperature varies linearly from 25°C at the Earth's surface to –60°C at 11 km (tropopause). What is the value of the vertical gradient of turbulent flux (K s^{-1}) for an altitude (km) of:
a. 0.1 b. 0.5 c. 1 d. 1.5 e. 2 f. 2.5 g. 3
h. 3.5 i. 4 j. 5 k. 6 m. 7 n. 8 o. 11

A28. Find the mid-tropospheric maximum value of heat flux (K·m s^{-1}) for a stormy atmosphere, where the troposphere is 11 km thick, and the air temperature at the top of the troposphere equals the air temperature of a standard atmosphere. But the air temperature (°C) at the ground is:
a. 16 b. 17 c. 18 d. 19 e. 20 f. 21 g. 22
h. 23 i. 24 j. 25 k. 26 m. 27 n. 28 o. 29

A29. Find the latent-heating rate (°C h^{-1}) averaged over the troposphere for a thunderstorm when the rainfall rate (mm h^{-1}) is:
a. 0.5 b. 1 c. 1.5 d. 2 e. 2.5 f. 3 g. 3.5
h. 4 i. 4.5 j. 5 k. 5.5 m. 6 n. 6.5 o. 7

A30. Given below the net radiative flux (W m^{-2}) reaching the surface, find the sum of sensible and latent heat fluxes (W m^{-2}) at the surface. (Hint: determine if it is day or night by the sign of the radiative flux.)
a. –600 b. –550 c. –500 d. –450 e. –400
f. –350 g. –300 h. –250 i. –200 j. –150
k. –100 m. –50 n. 50 o. 100 p. 150

A31. Same as the previous problem, but estimate the values of the sensible and latent heat fluxes (W m^{-2}) assuming a Bowen ratio of:
(1) 0.2 (2) 5.0

A32. Suppose you mounted instruments on a tower to observe temperature T and mixing ratio r at two heights in the surface layer (bottom 25 m of atmosphere) as given below. If a net radiation of –500 W m^{-2} was also measured at that site, then estimate the values of effective surface values of sensible heat flux and latent heat flux.

index	z(m)	T(°C)	r (g_{vap}/kg_{air})
2	10	T_2	10
1	2	20	15

where T_2 (°C) is:
a. 13.5 b. 13 c. 12.5 d. 12 e. 11.5 f. 11
g. 10.5 h. 10 i. 9.5 j. 9 k. 8.5 m. 8

A33. Not only can a stationary person feel wind chill when the wind blows, but a moving person in a calm wind can also feel wind chill, because most important is the speed of the air relative to the speed of the body. If you move at the speed given below through calm air of temperature given below, then you would feel a wind chill of what apparent temperature? Given: M (m s^{-1}), T (°C) .
a. 5, 5 b. 10, 5 c. 15, 5 d. 20, 5 e. 25, 5
f. 30, –10 g. 25, –10 h. 20, –10 i. 15, –10 j. 10, –10

A34(§). Modify eqs. (3.64) to use input and output temperatures in Fahrenheit and wind speeds in miles per hour. Calculate sufficient values to plot a graph similar to Fig 3.12 but in these new units.

A35. Find the heat index apparent temperature (°C) for an actual air temperature of 33°C and a relative humidity (%) of:
a. 5 b. 10 c. 20 d. 30 e. 40 f. 50 g. 60
h. 70 i. 75 j. 80 k. 85 m. 90 n. 90

A36. Find the humidex apparent air temperature (°C) for an actual air temperature of 33°C and a dew-point temperature (°C) of:

a. 32.5 b. 32 c. 31 d. 30 e. 29 f. 28 g. 27
h. 26 i. 25 j. 23 k. 20 m. 15 n. 10 o. 5

Evaluate & Analyze

E1. Assume that 1 kg of liquid water initially at 15°C is in an insulated container. Then you add 1 kg of ice into the container. The ice melts and the liquid water becomes colder. Eventually a final equilibrium is reached. Describe what you end up with at this final equilibrium?

E2. Explain in your own words why the units for specific heat C_p (J·kg^{-1}·K^{-1}) are slightly different than the units for the latent heat factor L (J·kg^{-1}). (Hint: read the INFO box on Internal Energy.)

E3. Explain in your own words why the magnitude of C_p should be larger than the magnitude of C_v. (Hint: read the INFO box on C_p vs. C_v).

E4. Consider the INFO box on C_p vs. C_v, with Fig. 3I.3c representing an initial state at equilibrium. Suppose you add some weight to the piston in Fig (c) causing the piston to become lower to reach a new equilibrium, but no thermal energy is added ($\Delta q = 0$). Describe what would happen to: (a) the molecules on average, (b) the gas temperature in the cylinder, (c) the air density in the cylinder, and (d) the air pressure in the cylinder.

E5. For the First Law of Thermodynamics (eq. 3.4d) which term(s) is are zero for a process that is:

a. adiabatic b. isothermal c. isobaric

E6. Start with eq. (3.4) and use algebra to derive equation (3.5). What did you need to assume to do this derivation? Does the result have any limitations?

E7. For Fig. 3.2, speculate on other processes not listed that might affect the air-parcel temperature.

E8. Using Fig. 3.3, explain in your own words the difference between a process lapse rate and an environmental lapse rate. Can both exist with different values at the same height? Why?

E9. Eq. (3.7) tells us that temperature of an adiabatically rising air parcel will decrease linearly with increasing height. In your own words, explain why you would NOT expect the same process to cause temperature to decrease linearly with decreasing pressure.

E10. If an air parcel rises isothermally (namely, heat is added or subtracted to maintain constant temperature), then what would happen to the potential temperature of the air parcel as it rises?

E11. Chinook winds (also known as foehn winds) consist of air descending down the lee slope of a mountain and then continuing some distance across the neighboring valley or plain. Why are Chinook winds usually warm when they reach the valley? (Hint: consider adiabatic descent of an air parcel.)

E12. In the definition of virtual potential temperature, why do liquid water drops and ice crystals cause the air to act heavier (i.e., colder virtual potential temperature), even though these particles are falling through the air?

E13. First make a photocopy of Fig. 3.4, so that you can keep the original Thermo Diagram clean.
a) On the copy, plot the vertical temperature profile for a standard atmosphere, as defined in Chapter 1. Suppose that this standard profile represents background environmental air.
b) On this same diagram, plat a point representing an air parcel at (P, T) = (100 kPa, 15°C). If you adiabatically lift this parcel to 50 kPa, what is its new temperature?
c) Is the parcel temperature a 50 kPa warmer or colder than the environment at that same pressure?

E14(§). For a standard atmosphere (see Chapt. 1), calculate potential temperature θ at z = 0, 2, 4, 6, 8, 10 km altitudes. Plot θ along the bottom axis and z along the vertical axis.

E15(§). Thermo diagrams often have many different types of lines superimposed. For example, on the background T vs. log-P diagram of Fig. 3.4 is plotted just one type of line: the dry adiabats. Instead of these adiabats, start with the same background of a T vs. log-P diagram, but instead draw lines connecting points of equal height (called **contour** lines). To calculate these lines, use the hypsometric equation from chapter 1 to solve for P vs. (z, T). Do this for the z = 2, 4, 6, 8, 10 km contours, where for any one height, plug in different values of T to find the corresponding values of P that define the contour.

E16. For advection to be a positive contribution (i.e., causing heating) and for wind that is in a positive coordinate direction, explain why the corresponding temperature gradient must be negative.

E17. Suppose that mild air (20°C at 10 m altitude) rests on top of a warm ocean (26°C at the surface), causing **convection** (vertical overturning of the air). If there is no mean horizontal wind, then the effective heat flux at the surface has what value? Assume a mixed layer that is 1200 m thick with average thermodynamic state of r = 0.01 g_{vapor}/g_{air} and θ = 15°C.

E18. Light travels faster in warm air than in cold. Use this info, along with Fig. 3.7, to explain why **inferior mirages** (reflections of the sky) are visible on hot surfaces such as asphalt roads. (Hint: Consider a wave front that is moving mostly horizontally, but also slightly downward at a small angle relative to the road surface, and track the forward movement of each part of this wave front — an optics method known as Huygens' Principle. See details in the atmospheric Optics chapter.)

E19. Under what conditions would eqs. (3.34 - 3.35) be expected to fail? Why?

E20. Use eqs. (3.37) and (3.39) to solve for the heat flux as a function of the temperature difference.

E21. In Fig. 3.8, the heat flux is greatest at the height where there is no change in the vertical temperature profile from before to after a storm. Why should that be the case?

E22. How fast does air temperature change if only if the only thermodynamic process that was active was direct IR cooling?

E23. In a thunderstorm, the amount of water condensation in the troposphere is often much greater than the amount of rain reaching the ground. Why is that, and how might it affect the heat budget averaged over the whole thunderstorm depth?

E24. Eq. (3.51) has what limitations?

E25. Comment on the relative strengths of advective vs. latent heating in an Eulerian system, given V = 5 m s^{-1}, $\Delta T/\Delta y$ = –5°C/1000km, and 1 g/kg of water condenses every 5 minutes.

E26. Create a figures similar to Fig. 3.9, but for:
a) daytime over a white concrete road,
b) nighttime black asphalt road.

E27. It is sometimes said that conductive heat flux into the ground is a response to radiative forcings at the surface. Is that statement compatible with the crude parameterization presented in this book for flux into the ground? Explain.

E28. What is the initial rate of change of average mixed-layer air temperature with horizontal distance downwind if the air is initially 5 °C colder than the water, given that the air blows over the water at speed 15 m s^{-1}? Consider entrainment into the top of the mixed layer, but neglect other heating or cooling processes.

E29. Can the parameterizations (eqs. 3.58 - 3.61) actually give a balanced heat budget? For what types of situations are these parameterizations valid?

E30. (§). Suppose that we used the heat transfer eq. (3.35) as a basis for deriving wind chill. The result might be a different wind-chill relationship:

$$T_{wind\ chill} = T_s + \left(T_{air} - T_s\right)\cdot\left[b + a\cdot\left(\frac{M + M_o}{M_o}\right)^{0.16}\right] + T_c \tag{3.67}$$

where T_s = 34.6°C is an effective skin temperature, and where, $a = 0.5$, $b = 0.62$, T_c = 4.2°C, and M_o = 4.8 km h^{-1}. Plot this equation as a graph similar to Fig. 3.12, and comment on the difference between the formula above and the actual wind-chill formula.

E31. Notice in Fig. 3.12 that the curves bend the most for slow wind speeds. Why might you expect this to be the case?

Synthesize

S1. Describe the change to the ocean if condensation caused cooling and evaporation caused heating of the air. Assume dry air above the ocean.

S2. Suppose that zero latent heat was associated with the phase changes of water. Describe the possible changes to climate and weather, if any?

S3. Describe the change to the atmosphere if rising air parcels became warmer adiabatically while sinking ones became cooler.

S4. Suppose that for each 1 km rise of an air parcel, the parcel mixes with an equal mass of surrounding environmental air. How would the process lapse rate for this rising air parcel be different (if at all) from the lapse rate of an adiabatically rising air parcel (having no mixing).

S5. Macro thermodynamics (the kind we've used in this chapter) considers the statistical state of a large collection of molecules that frequently collide with each other, and how they interact on average with their surroundings. Can this same macro thermodynamics be used in the exosphere, where individual air molecules are very far apart (i.e., have a large mean-free path) and rarely interact? Why? Also, explain if how heat budgets can be used in the exosphere.

S6. Could there be situations where environmental and process lapse rates are equal? If so, give some examples.

S7. Suppose that the virtual potential temperature was not affected by the amount of solid or liquid water in the air. How would weather and climate change, if at all?

S8. The background of the thermo diagram of Fig. 3.4 is an orthogonal grid, where the isotherms are plotted perpendicular to the isobars. Suppose you were to devise a new thermo diagram with the dry adiabats perpendicular to the isobars. On such a diagram, how would the isotherms be drawn? To answer this, draw a sketch of this new diagram, showing the isobars, adiabats, and isotherms. (Do this as a conceptual exercise, not by solving equations to get numbers.)

S9. Describe changes to Earth's surface heat balance if the geological crust was 1 km thick aluminum (an excellent conductor of heat) covering the whole Earth.

S10. Suppose you were on a train moving in a straight line at constant speed. You make measurements of the surrounding environmental air as the train moves down the track.

a) If the environmental air was calm, do you think your measurements are Eulerian, Lagrangian, or neither? Explain.

b) If the environmental air was moving in any arbitrary speed or direction, do you think your measurements are Eulerian, Lagrangian or neither? Explain.

c) Try to create a heat budget equation that works in the framework, given your constant speed of translation of M_o.

S11. Describe how atmospheric structure, climate, and weather would change if the troposphere were completely transparent to all IR radiation, but was mostly opaque to solar radiation.

S12. Describe how errors in surface sensible and latent heat flux estimates would increase as the temperature and humidity differences between the two measurement levels approached zero.

S13. The wind-chill concept shows how it feels colder when it is winder. For situations where the wind chill is much colder than the actual air temperature, to what temperature will an automobile engine cool after it is turned off? Why? (Assume the car is parked outside and is exposed to the wind.)

Sample Application

[This sample applies to eqs. 3.1 and 3.3, but was put here on the last page of the chapter because there was no room for it earlier in the chapter.]

How much dew must condense on the sides of a can of soda for it to warm the soda from 1°C to 16°C?

Hints: Neglect the heat capacity of the metal can. The density of liquid water is 1000 kg·m^{-3}. Assume the density of soda equals that of pure water. Assume the volume of a can is 354 ml (milliliters), where 1 l = 10^{-3} m^3.

Find the Answer

Given: ρ_{water} = 1000 kg·m^{-3}.
C_{liq} = 4200 J·kg^{-1}·K^{-1}
Volume (*Vol*) in Can = 354 ml
L_{cond} = + 2.5x10^6 J·kg^{-1}
ΔT = 15 K

Find: Volume of Condensate

Sketch:

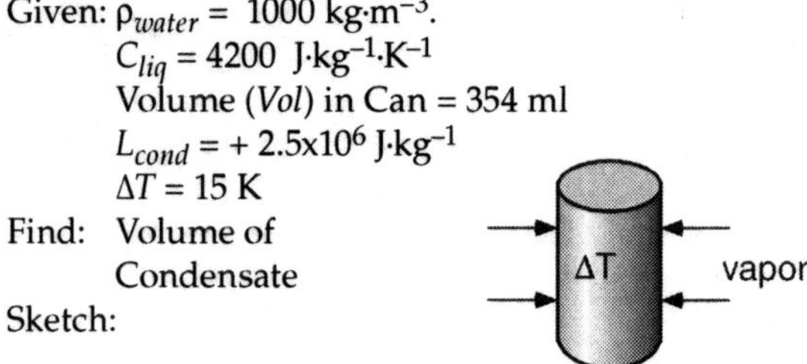

Equate the latent heat release by condensing water vapor (eq. 3.3) with the sensible heat gained by fluid in the can (eq. 3.1)

$$\Delta Q_E = \Delta Q_H$$

$$\rho_{condensate} \cdot (\Delta Vol\ of\ Condensate) \cdot L_{cond} = \rho_{soda} \cdot (Vol\ of\ Can) \cdot C_{liq} \cdot \Delta T$$

Assume the density of condensate and soda are equal, so they cancel. The equation can then be solved for *ΔVolume of Condensate.*

$\Delta Volume\ of\ Condensate = (Vol\ of\ Can) \cdot C_{liq} \cdot \Delta T\ L_{cond}$
= (354 ml)·(4200 J·kg^{-1}·K^{-1})·(15 K) (2.5x10^6 J·kg^{-1})
= **8.92 ml**

Check: Units OK. Sketch OK. Physics OK.

Exposition: Latent heats are so large that an amount of water equivalent to only 2.5% of the can volume needs to condense on the outside to warm the can by 15°C. Thus, to keep your can cool, insulate the outside to prevent dew from condensing.

4 WATER VAPOR

Contents

Water vapor is one of the gases in air. Unlike nitrogen and oxygen which are constant in the bottom 100 km of the atmosphere, water-vapor concentration can vary widely in time and space. Most people are familiar with relative humidity as a measure of water-vapor concentration because it affects our body's moisture and heat regulation. But other humidity variables are much more useful in other contexts.

Storms get much of their energy from water vapor — when water vapor condenses or freezes it releases latent heat. For this reason we carefully track water vapor as it rises in buoyant thermals or is carried by horizontal winds. The amount of moisture available to a storm also regulates the amount of rain or snow precipitating out.

What allows air to hold water as vapor in one case, but forces the vapor to condense in another? This depends on a concept called "saturation".

VAPOR PRESSURE AT SATURATION

Total atmospheric pressure is the sum of the **partial pressures** of all the constituents. For water vapor, the partial pressure is called the **vapor pressure**, e. Vapor pressure has units of kPa.

Gases in the air can mix with any relative proportions. However, for water vapor, there is a critical water-vapor concentration, known as the **saturation humidity**. Above this critical value, water vapor condenses faster than it evaporates, thereby reducing the water-vapor concentration back to the critical value. At this critical value, the air is said to be **saturated**, and the vapor is in equilibrium with liquid water. Let e_s represent the saturation vapor pressure at equilibrium over pure water having a flat surface.

It frequently happens that air holds less than the critical value, and is said to be **unsaturated**. For this case, $e < e_s$.

Although any portion of water vapor can be held in air, it is rare for the vapor pressure to be more than 1% greater than the saturation value. Air having $e > e_s$ is said to be **supersaturated**. Supersaturated air can occur as a transient condition while excess water vapor is condensing onto available surfaces such as on dust particles called cloud condensation

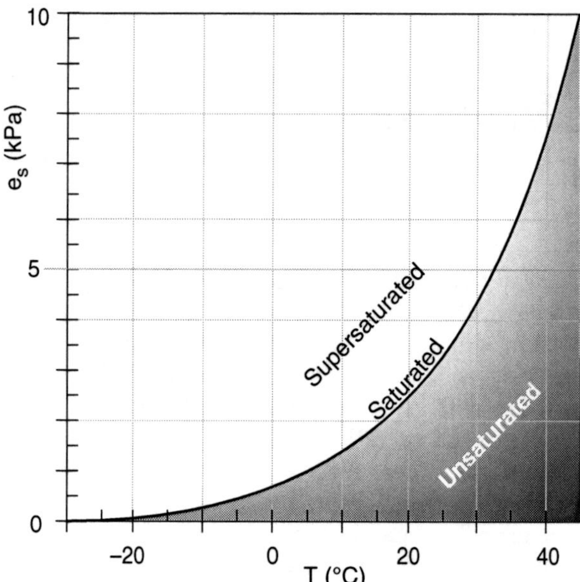

Figure 4.1
Pure-water saturation vapor pressure over a flat water surface.

Sample Application
Air of temperature 30°C has what value of e_s?

Find the Answer
Given: T = 30°C = 303.15 K
Find: e_s = ? kPa for over liq. water. Apply eq. (4.1a).

$$e_s = (0.6113\text{kPa})\cdot\exp\left[(5423\text{K})\cdot\left(\frac{1}{273.15\text{K}} - \frac{1}{303.15\text{K}}\right)\right]$$

$$= (0.6113\text{ kPa})\cdot\exp(1.96473) = \underline{\textbf{4.36 kPa}}$$

Check: Physics reasonable, & agrees with Table 4-1.
Exposition: At saturation for this T, water-vapor partial pressure is only 4.3% of total pressure P≈101.3 kPa.

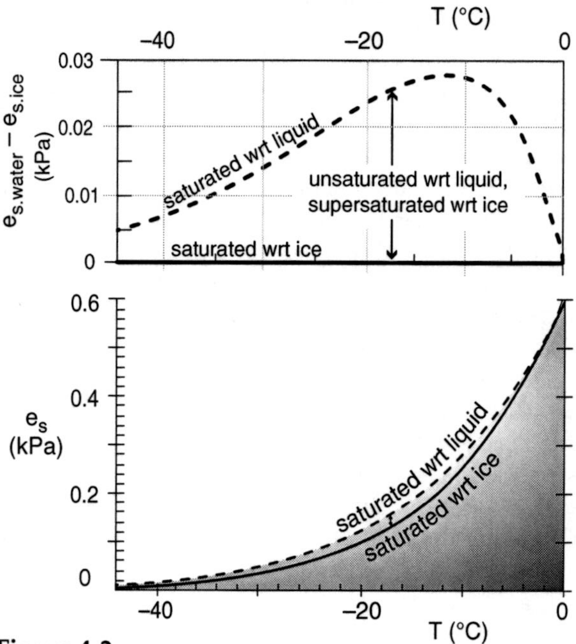

Figure 4.2
Ice & liquid saturation vapor pressures. (wrt = with respect to).

nuclei. However, photographs of air flow over aircraft wings for both subsonic and supersonic flight through humid air indicate that condensation to form cloud droplets occurs almost instantly.

During the equilibrium state (i.e., at saturation) there is a balance between the rate of evaporation from the liquid and the rate of condensation from vapor. Liquid-water temperature controls the rate of evaporation, and **humidity** (water-vapor concentration in air) controls the rate of condensation.

Warmer liquid temperatures cause greater evaporation rates, which allow the humidity in the air to increase until a new balance is attained. The opposite is true for colder temperatures. For a situation where the air and liquid water temperatures are equal, we conclude that colder air has a smaller capacity for holding water vapor than warmer air.

This relationship between saturation vapor pressure and temperature is approximated by the **Clausius-Clapeyron** equation:

$$e_s \approx e_o \cdot \exp\left[\frac{L}{\Re_v}\cdot\left(\frac{1}{T_o} - \frac{1}{T}\right)\right] \qquad \bullet(4.1a)$$

where the water-vapor gas constant is $\Re_v$ = 461 J·K^{-1}·kg^{-1}, T_o = 273.15 K, e_o = 0.6113 kPa, and L is a latent-heat parameter. Temperatures in this equation must have units of Kelvin.

This equation works for saturation over both liquid water and solid water (ice) surface if these surfaces are flat. For liquid water the latent heat of vaporization $L = L_v = 2.5\text{x}10^6$ J·kg^{-1}, giving $L_v/\Re_v$ = 5423 K. For ice the latent heat of deposition $L = L_d = 2.83\text{x}10^6$ J·kg^{-1} and $L_d/\Re_v$ = 6139 K.

The exponentially-shaped curve described by eq. (4.1a) is plotted in Fig. 4.1, with corresponding data values listed Table 4-1. One interpretation of this curve is that as unsaturated humid air is cooled, a temperature is reached at which point the air is saturated. Further cooling forces some water vapor to condense into liquid, creating clouds and rain and releasing latent heat. Hence, the Clausius-Clapeyron equation is important for understanding storms.

In the atmosphere it is possible for liquid water to remain unfrozen at temperatures down to –40°C. Such unfrozen cold water is said to be **supercooled**. The difference between saturation values of water vapor over supercooled liquid water and ice is plotted in Fig. 4.2.

The Clausius-Clapeyron equation also describes the relationship between actual (unsaturated) water-vapor pressure e and dew-point temperature (T_d, to be defined later):

$$e = e_o \cdot \exp\left[\frac{L}{\Re_v} \cdot \left(\frac{1}{T_o} - \frac{1}{T_d}\right)\right] \tag{4.1b}$$

where T_o = 273.15 K and e_o = 0.6113 kPa as before. Use, $L/\Re_v = L_v/\Re_v$ = 5423 K for liquid water, and use $L_d/\Re_v$ = 6139 K for ice.

The latent-heat parameter L varies slightly with temperature. Taking that into account, a different approximation known as **Tetens' formula** has been suggested for saturation vapor pressure e_s as a function of temperature (T, in Kelvins):

$$e_s = e_o \cdot \exp\left[\frac{b \cdot (T - T_1)}{T - T_2}\right] \tag{4.2}$$

where $b = 17.2694$, $e_o = 0.6113$ kPa, $T_1 = 273.15$ K, and $T_2 = 35.86$ K.

Sample Application

How do the vapor pressure values from Tetens' formula differ from those of the Clausius-Clapeyron equation, as a function of temperature?

Find the Answer

Use a spreadsheet to solve eqs. (4.1a) and (4.2) .

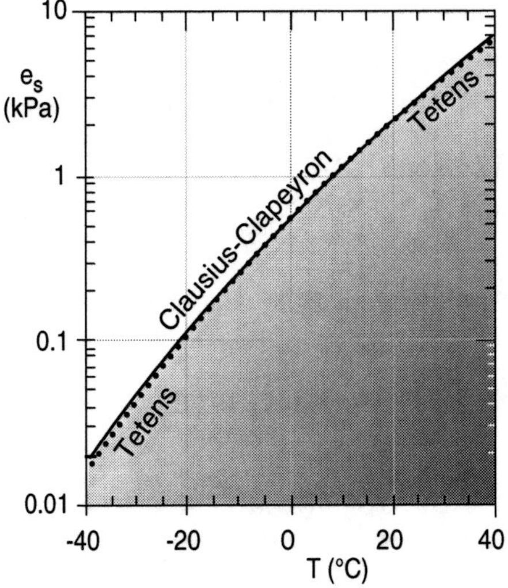

Check: Physics and units are reasonable.

Exposition: The difference between both formulas are very small — often smaller than other uncertainties in measurements of temperature or humidity. So you can be confident using either equation.

HIGHER MATH • Clausius-Clapeyron Eq.

Rudolf Clausius and Benoît Paul Émilie Clapeyron were engineers trying to improve steam engines during the 1800's. They independently made laboratory measurements of water vapor pressure at saturation e_s, and found the following empirical relationship:

$$\frac{de_s}{dT} = \frac{L_v}{T}\left[\frac{1}{\rho_v} - \frac{1}{\rho_L}\right]^{-1}$$

where ρ_L is liquid-water density, and ρ_v is water-vapor density (i.e., **absolute humidity**). The liquid-water density is so much greater than the water-vapor density that the above equation can be approximated by:

$$\frac{de_s}{dT} \cong \frac{L_v}{T}\rho_v$$

The relationship between saturation vapor pressure and the absolute humidity is given by the ideal gas law for water vapor:

$$e_s = \rho_v \cdot \Re_v \cdot T$$

where the water-vapor gas constant is $\Re_v = 4.615\text{x}10^{-4}$ kPa K^{-1} (g m^{-3})$^{-1}$. You can solve the ideal gas law for ρ_v , which you can then use in the previous eq:

$$\frac{de_s}{dT} \cong \frac{L_v \cdot e_s}{\Re_v \cdot T^2}$$

This equation admits a solution if you separate the variables to put all terms involving T on the right, and all those involving e_s on the left:

$$\frac{de_s}{e_s} \cong \frac{L_v}{\Re_v}\frac{dT}{T^2}$$

Next, integrate between some known condition (e_o at T_o) to any arbitrary values of saturation vapor pressure (e_s) and temperature (T):

$$\int_{e_o}^{e_s} \frac{de_s}{e_s} \cong \frac{L_v}{\Re_v}\int_{T_o}^{T}\frac{dT}{T^2}$$

After integrating you get:

$$\ln\left(\frac{e_s}{e_o}\right) \cong -\frac{L_v}{\Re_v}\left[\frac{1}{T} - \frac{1}{T_o}\right]$$

To get eq. (4.1a) take the exponential (i.e., antilog) of both sides and then algebraically rearrange to give:

$$e_s = e_o \cdot \exp\left[\frac{L}{\Re_v} \cdot \left(\frac{1}{T_o} - \frac{1}{T}\right)\right] \tag{4.1}$$

Details are given in the *Atmospheric Thermodynamics* book by C. Bohren and B. Albrecht (1998, Oxford Univ. Press, 402 pp).

INFO • Boiling Point

Liquids boil when the saturation vapor pressure e_s equals the ambient pressure P of the atmosphere:

$$P = e_s$$

Variation of boiling temperature with Altitude

We know e_s as a function of temperature T from the Clausius-Clapeyon equation, so plug that into the right side of the equation above. We also know that ambient atmospheric pressure P decreases exponentially with increasing height z, as was given in Chapter 1, so we can plug that into the left side. This gives

$$P_o \cdot \exp\left[-\frac{z}{H_p}\right] = e_o \cdot \exp\left[\frac{L_v}{\Re_v} \cdot \left(\frac{1}{T_o} - \frac{1}{T}\right)\right]$$

where $P_o = 101.325$ kPa, $H_p \approx 7.29$ km, $e_o = 0.6113$ kPa, $L_v / \Re_v = 5423$ K, and $T_o = 273.15$ K.

Next, divide the left and right sides of the equation by P_o. On the right side, note that $e_o/P_o = \exp[\ln(e_o/P_o)]$. Thus:

$$\exp\left[-\frac{z}{H_p}\right] = \exp\left[\ln\left(\frac{e_o}{P_o}\right) + \frac{L_v}{\Re_v}\left(\frac{1}{T_o}\right) - \frac{L_v}{\Re_v}\frac{1}{T}\right]$$

Create a new constant called a (dimensionless) that is the sum of the first two terms inside the square brackets, because those two terms are constant. Next, take the *ln* of the right and left sides to give:

$$\frac{z}{H_p} = \frac{L_v}{\Re_v}\frac{1}{T} - a$$

Solve this equation for T, which we can re-define as the boiling point $T_{boiling}$:

$$\boxed{T_{boiling} = \frac{L_v / \Re_v}{a + z / H_p}}$$

Knowing that at sea level ($z = 0$) the boiling temperature is 100°C (i.e., $T_{boiling} = 373.15$ K) you can solve for the dimensionless constant, giving : $a = 14.53$.

Exposition:

If you solve this equation for various altitudes, you find that the boiling point decreases by 3.4°C km^{-1}. Thus, $T_{boiling} = 366.35$ K = 93.2°C at 2 km altitude

To soften vegetables to the desired tenderness or to prepare meats to the desired doneness, foods must be cooked at a certain temperature over a certain time duration. Slightly cooler cooking temperatures must be compensated with slightly longer cooking times. Thus, you need to cook boiled foods for longer times at higher altitudes, because boiling happens at a lower temperature.

Table 4-1. Values of humidity variables at saturation (subscript s) over a liquid-water flat surface, for different air temperatures T. The same values also relate the dew-point temperature T_d to the actual humidity. Notation: e = vapor pressure, r = mixing ratio, q = specific humidity, ρ_v = absolute humidity. Note that r and q depend on pressure — this table shows their values for standard sea-level pressure.

		For P = 101.325 kPa		
T	e_s	q_s	r_s	ρ_{vs}
or				
T_d	e	q	r	ρ_v
(°C)	(kPa)	(g kg^{-1})	(g kg^{-1})	(g m^{-3})
–40	0.0203	0.1245	0.1245	0.1886
–35	0.0330	0.2029	0.2029	0.301
–30	0.0528	0.324	0.3241	0.4708
–25	0.0827	0.5079	0.5082	0.7231
–20	0.1274	0.7822	0.7828	1.0914
–15	0.1929	1.1848	1.1862	1.6206
–10	0.2875	1.7666	1.7697	2.3697
–5	0.4222	2.5956	2.6024	3.4151
0	0.6113	3.7611	3.7753	4.8546
5	0.8735	5.3795	5.4086	6.8119
10	1.232	7.6005	7.6587	9.4417
15	1.718	10.62	10.73	12.94
20	2.369	14.67	14.89	17.53
25	3.230	20.07	20.48	23.5
30	4.360	27.21	27.97	31.2
35	5.829	36.58	37.97	41.03
40	7.720	48.8	51.3	53.48
45	10.13	64.66	69.13	69.1
50	13.19	85.18	93.11	88.56
55	17.04	111.7	125.7	112.6
60	21.83	145.9	170.8	142.2

Sample Application

For T = 10°C and **P = 70 kPa**, calculate q_s, r_s, & ρ_{vs}.

Find the Answer

Given: T = 10°C = 283.15 K, P = 70 kPa

Find: q_s = ? g kg^{-1}, r_s = ? g kg^{-1}, ρ_{vs} = ? g m^{-3}

Get e_s = 1.232 kPa from Table 4-1 (independent of P).

Apply eq. (4.8) to get q_s:

q_s = 0.622·(1.232 kPa)/(70 kPa) = 0.0109 g g^{-1} ≈ **11** g kg^{-1}

Apply eq. (4.5) to get r_s:

r_s = [0.622·(1.232 kPa)] / [70 kPa - 1.232 kPa]
= 0.0111 g g^{-1} = **11.1** g kg^{-1}

Apply eq. (4.12) to get ρ_{vs}:

ρ_{vs} = (1232. Pa)/[(461 J·K^{-1}·kg^{-1})·(283.15 K)]
= **0.00944** kg·m^{-3} = **9.44** g·m^{-3} .

Check: Physics and units are reasonable.

Exposition: Table 4-1 could have been used for ρ_{vs} .

MOISTURE VARIABLES

Table 4-2 (continued across several pages) shows most of the moisture variables used in meteorology. Other variables used in the table below include: m = mass, e = vapor pressure, P = total atmospheric pressure, $\Re_d$ = 2.871x10^{-4} kPa·K^{-1}·m^3·g^{-1} is the gas constant for dry air, $\Re_v$ = 4.61x10^{-4} kPa·K^{-1}·m^3·g^{-1} is the gas constant for pure water vapor, P_d is the partial pressure of dry air, ρ_d is the density of dry air (which is a function of pressure, altitude and temperature as given by the ideal gas law), and subscript s denotes saturation.

Table 4-2a. Moisture variables. *(continues on next page)*

Variable name:	Mixing Ratio	Specific Humidity	Absolute Humidity
Symbol:	r	q	ρ_v
Units:	$kg_{water\ vapor}\ kg_{dry\ air}^{-1}$	$kg_{water\ vapor}\ kg_{total\ air}^{-1}$	$kg_{water\ vapor}\ m^{-3}$
Alternative Units:	g kg^{-1} , g g^{-1}, kg kg^{-1}	g kg^{-1} , g g^{-1}, kg kg^{-1}	kg m^{-3}
Defining Equation: (& equation number)	$r = \dfrac{m_{water\ vapor}}{m_{dry\ air}}$ (4.3)	$q = \dfrac{m_{water\ vapor}}{m_{total\ air}}$ (4.6) $q = \dfrac{m_{water\ vapor}}{m_{dry\ air} + m_{water\ vapor}}$	$\rho_v = \dfrac{m_{water\ vapor}}{Volume}$ (4.9)
Relationship to Vapor Pressure:	$r = \dfrac{\varepsilon \cdot e}{P - e}$ (4.4)	$q = \dfrac{\varepsilon \cdot e}{P_d + \varepsilon \cdot e} = \dfrac{\varepsilon \cdot e}{P - e \cdot (1-\varepsilon)}$ (4.7) $q \approx \dfrac{\varepsilon \cdot e}{P}$	$\rho_v = \dfrac{e}{\Re_v \cdot T}$ (4.10) $\rho_v = \dfrac{e \cdot \varepsilon \cdot \rho_d}{P - e} \approx \dfrac{e}{P} \cdot \varepsilon \cdot \rho_d$ (4.11)
If Saturated:	$r_s = \dfrac{\varepsilon \cdot e_s}{P - e_s}$ (4.5)	$q_s = \dfrac{\varepsilon \cdot e_s}{P_d + \varepsilon \cdot e_s}$ (4.8) $q_s = \dfrac{\varepsilon \cdot e_s}{P - e_s \cdot (1-\varepsilon)} \approx \dfrac{\varepsilon \cdot e_s}{P}$	$\rho_{vs} = \dfrac{e_s}{\Re_v \cdot T}$ (4.12) $\rho_{vs} = \dfrac{e_s \cdot \varepsilon \cdot \rho_d}{P - e_s} \approx \dfrac{e_s}{P} \cdot \varepsilon \cdot \rho_d$ (4.13)
Key Constants:	$\varepsilon = \Re_d / \Re_v$ = 0.622 $g_{vapor}\ g_{dry\ air}^{-1}$ = 622 g kg^{-1}	$\varepsilon = \Re_d / \Re_v$ = 0.622 $g_{vapor}\ g_{dry\ air}^{-1}$ = 622 g kg^{-1}	$\Re_v$ = 4.61x10^{-4} kPa·K^{-1}·m^3·g^{-1} ε = 622 g kg^{-1}
Typical Values:	See Table 4-1.	See Table 4-1.	See Table 4-1.
Relevance:	• r is conserved in unsaturated air parcels that move without mixing with their environment. • not affected by heating, cooling, pressure changes. • used in thermo diagrams.	• q is conserved in unsaturated air parcels that move without mixing with their environment. • not affected by heating, cooling, pressure changes.	• easy to measure using absorption of infrared, ultraviolet, or microwave radiation as a function of path length through the air. • is the concentration of water vapor in air.
Notes:	Derivation of eq. (4.4): Given eq. (4.3), divide numerator and denominator by volume. But *m/Volume* is density. Use ideal gas laws for water vapor and for dry air, assuming a common *T*.	Derivation of eq. (4.7) is similar to that for eq. (4.4).	• Eq. (4.10) is the ideal gas law for water vapor. • Eq. (4.11) uses the ideal gas law for dry air to replace temperature *T*. • *T* must be in Kelvin.

Table 4-2b. Moisture variables. *(continuation)*

Variable name:	Relative Humidity	Dewpoint
Symbol:	RH or $RH\%$	T_d
Units:	(dimensionless)	(K)
Alternative Units:	(%)	(°C)
Defining Equation: (& equation number)	$RH = \frac{e}{e_s}$ or $\frac{RH\%}{100\%} = \frac{e}{e_s}$ (4.14a)	"Temperature to which a given air parcel must be cooled at constant pressure and constant water-vapor content in order for saturation to occur."*
Alternative Definitions:	$RH = \frac{q}{q_s} = \frac{\rho_v}{\rho_{vs}} \approx \frac{r}{r_s}$ (4.14b) $\frac{RH\%}{100\%} = \frac{q}{q_s} = \frac{\rho_v}{\rho_{vs}} \approx \frac{r}{r_s}$ (4.14c)	$T_d = \left[\frac{1}{T_o} - \frac{\Re_v}{L}\cdot\ln\left(\frac{e}{e_o}\right)\right]^{-1}$ (4.15a) $T_d = \left[\frac{1}{T_o} - \frac{\Re_v}{L}\ln\left(\frac{r\cdot P}{e_o\cdot(r+\varepsilon)}\right)\right]^{-1}$ (4.15b)
If Saturated:	$RH = 1.0$ or $RH\% = 100\%$ (4.14d)	$T_d = T$
Key Constants:		$e_o = 0.6113$ kPa, $T_o = 273.15$ K $\Re_v/L_v = 1.844\times10^{-4}$ K^{-1}. $\varepsilon = \Re_d/\Re_v = 0.622\ g_{vapor}\ g_{dry\ air}^{-1}$
Typical Values:	$RH = 0.0$ to 1.0 or $RH\% = 0\%$ to 100%	See Table 4-1. $T_d \le T$
Relevance:	• regulates the max possible evaporation into the air. • easy to measure via: (1) capacitance changes across a plastic dielectric; (2) electrical resistance of an emulsion made of carbon powder; or (3) organic-fiber contraction/expansion. • most used by the general public.	• easy to measure via cooling a mirror to the point where condensation (dew) forms on it. The temperature at which this first happens is the dewpoint, as detected by measuring how well a light beam can reflect off the mirror. • a very accurate method for humidity measurement.
Notes:	• it is possible to have relative humidities as high as about 100.5%. This is called **supersaturation** (see the Precipitation Processes chapter).	• T_d also called **dewpoint temperature**. • $(T - T_d)$ = **dewpoint depression** = **temperature dewpoint spread** * Glickman, T. S., 2000: *Glossary of Meteorology*. American Meteorological Society.

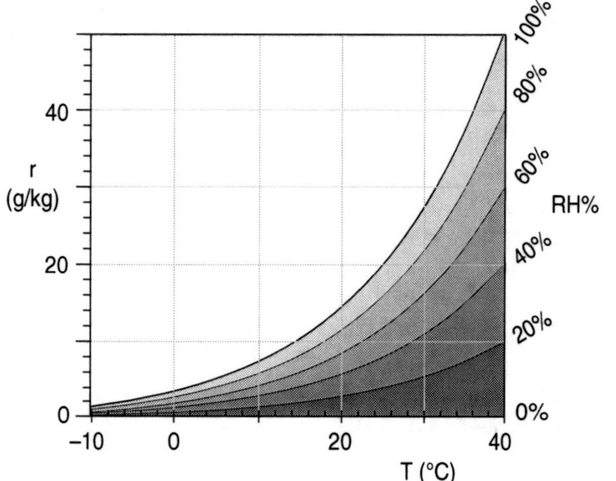

Figure 4.3
Relationship between relative humidity and mixing ratio at P = 101.325 kPa. Grey region is unsaturated. White region is unphysical (except sometimes for 0.5% of supersaturation).

Sample Application
Find T_d and $RH\%$ for r= 10 g kg^{-1}, P= 80 kPa, T=20°C.

Find the Answer
Given: $r = 0.01\ g_{vapor}\ g_{air}^{-1}$, $P = 80$ kPa, $T = 20$°C
Find: T_d = ? °C, and $RH\%$ = ? %

Use eq. (4.15b): T_d = [(1/(273.15K) – (1.844x10^{-4} K^{-1})·
ln{ [(80kPa)·(0.01g$_{vapor}$ g$_{air}^{-1}$)] /
[(0.6113kPa)·((0.01g g^{-1}) +(0.622g g^{-1})] }]$^{-1}$
=[(1/(273.15K)–(1.844x10^{-4}K^{-1})·ln(2.07)]$^{-1}$ = 283.78K.
Thus, T_d = 283.78 – 273.15 = or **10.6°C**

At T_d = 10.6°C, eq. 4.1b gives $e \approx 1.286$ kPa
At T = 20°C, Table 4-1 gives e_s = 2.369 kPa
Use eq. (4.14a): $RH\%$ = 100%·(1.286 kPa/2.369 kPa)
$RH\%$ = **54.3%**

Check: Magnitude and units are reasonable.

Table 4-2c. Moisture variables. *(continuation)*

Variable name:	Lifting Condensation Level (LCL)	Wet-bulb Temperature
Symbol:	z_{LCL}	T_w
Units:	(km)	(K)
Alternative Units:	(m)	(°C)
Defining Equation: (& equation number)	$z_{LCL} = a\cdot(T - T_d)$ (4.16a)	$C_p\cdot(T - T_w) = -L_v\cdot(r - r_w)$ (4.17)
Applications:	$P_{LCL} = P\cdot\left[1 - b\cdot\left(\frac{T - T_d}{T}\right)\right]^{C_p/\Re}$ (4.16b)	$r = r_w - \beta\cdot(T - T_w)$ (4.18a) where $r_w = \frac{\varepsilon}{b\cdot P\cdot\exp\left(\frac{-c\cdot T_w(°C)}{T_w(°C)+\alpha}\right) - 1}$ (4.18b)
If Saturated:	$z_{LCL} = 0$, $P_{LCL} = P$	$T_w = T$, for T = **dry-bulb temperature**
Key Constants:	a = 0.125 km °C^{-1}. Γ_d = 9.8 °C km^{-1} $b = a\cdot\Gamma_d$ = 1.225 (dimensionless) $C_p/\Re$ = 3.5 (dimensionless)	ε = 622 g kg^{-1}, b = 1.631 kPa^{-1}, c = 17.67, α = 243.5°C, β = 0.40224 (g kg^{-1})/°C.
Typical Values:	0 to 5 km	See Figs. 4.4 & 4.5. $T_w \le T$
Relevance:	• for unsaturated air, it is the height (or pressure) to which air must be lifted to become just saturated (i.e., cloudy). • is **cloud-base** altitude for cumulus and other convective clouds.	• easy to measure, by placing a wet wick or sleeve around a thermometer bulb, and then blowing air past the wet bulb (an **aspirated psychrometer**) or moving the wet bulb through the air (a **sling psychrometer**).
Notes:	• does NOT give cloud base for stratiform clouds, because these clouds are caused by advection (nearly horizontal winds). • the LCL is also known as the **saturation level**. For saturated air, it is the height to which air must be lowered to become just unsaturated (eqs 4.16 do NOT apply).	• as water evaporates from the wet wick, the adjacent air temperature drops from T to T_w while the humidity in this air increases from r to r_w, until equilibrium heat balance is reached as described by eq. (4.17). • $(T - T_w)$ is called **wet-bulb depression**. • see **Normand's rule** on later pages.

Sample Application

For an air parcel at P = 90 kPa with T = 25°C and T_d = 8°C, what is the height and pressure of the LCL?

Find the Answer

Given: P = 90 kPa, T = 25°C = 298.15 K, T_d = 8°C .
Find: z_{LCL} = ? km , P_{LCL} = ? kPa

Use eq. (4.16a)

z_{LCL} = (0.125 km °C^{-1})·(25 – 8°C)
= **2.13 km higher than the initial height**

Use eq. (4.16b)

P_{LCL} = (90 kPa)·[1 – 1.225·(25–8°C)/298.15K]$^{3.5}$
= **69.85 kPa**

Check: Physics and units are reasonable.

Exposition: The pressure decreases about 10 kPa for each increase of 1 km of altitude near the surface, the pressure answer is also reasonable. Indeed, an air parcel moving from 90 to 70 kPa rises about 2 km.

Sample Application

You observe a dry-bulb (i.e., normal air) temperature of T = 25°C, and a wet-bulb temperature of 18°C. Use the equations to calculate the mixing ratio for P = 90 kPa. Don't use look-up tables or graphs.

Find the Answer

Given: P = 90 kPa, T = 25°C , T_w = 18°C
Find: r = ? g kg^{-1}

First, solve eq. (4.18b):

$$r_w = \frac{622\text{g/kg}}{(1.631\text{kPa}^{-1})\cdot(90\text{kPa})\cdot\exp\left(\frac{-17.67\cdot 18°C}{18°C + 243.5°C}\right) - 1}$$

r_w = 14.6 g kg^{-1}.

Then solve eq. (4.18a):

r = (14.6g kg^{-1}) – [0.40224 (g kg^{-1})/°C]·(25–18°C)
= **11.78** g kg^{-1}.

Check: Physics and units are reasonable.

Exposition: Knowing r, use other eqs. to find any other humidity variable.

Sample Application

Find the mixing ratio and relative humidity for air temperature of 12°C and wet-bulb temperature of 10°C. Yes, you may use the graphs this time.

Find the Answer

Given: $T = 12°C$, $T_w = 10°C$
Find: $r = ?$ g kg^{-1} , and $RH = ?$ %

Assume $P = 101.3$ kPa, so we can use Figs. 4.4 & 4.5.
The wet-bulb depression is 12 – 10 = 2°C
Use Fig. 4.4. $r = \underline{7}$ g kg^{-1}.
Use Fig. 4.5. $RH = \underline{78\%}$.

Check: Physics and units are reasonable.
Exposition: Much easier than the Sample Applications on the previous page. Notice that in Fig. 4.4, r depends mostly on T_w, because the mixing-ratio lines are mostly horizontal. However, $RH\%$ depends mostly on $T - T_w$, because the lines in Fig. 4.5 are mostly vertical.

More Wet-bulb Temperature Info

The easiest way for you to find humidity from dry and wet-bulb temperature is to look-up the humidity in tables called **psychrometric tables**, which are often published in meteorology books. Figures 4.4 and 4.5 present the look-up information as **psychrometric graphs**, which were computed from the equations in this Chapter.

To create your own psychrometric tables or graphs, first generate a table of mixing ratios in a spreadsheet program, using eqs. (4.18) and (4.19). I assumed a standard sea-level pressure of $P = 101.325$ kPa for the figures here. Then contour the resulting numbers to give Fig. 4.4. Starting with the table of mixing ratios, use eqs. (4.2), (4.5), and (4.14) to create a new table of relative humidities, and contour it to give Fig. 4.5. All of these psychrometric tables and graphs are based on Tetens' formula (see eq. 4.2).

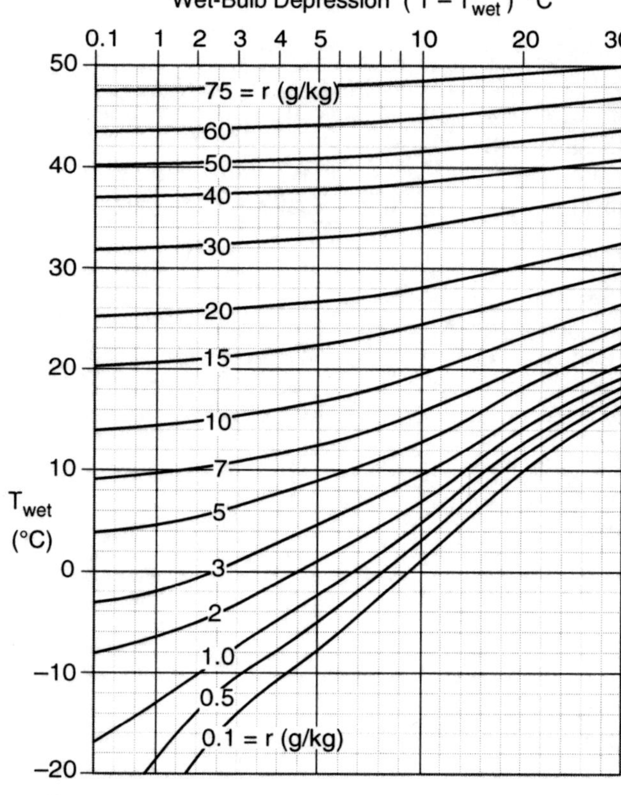

Figure 4.4
Psychrometric graph, to find mixing ratio r from wet and dry-bulb temperatures. Based on P = 101.325 kPa. Caution, the darker vertical lines mark scale changes along the abscissa.

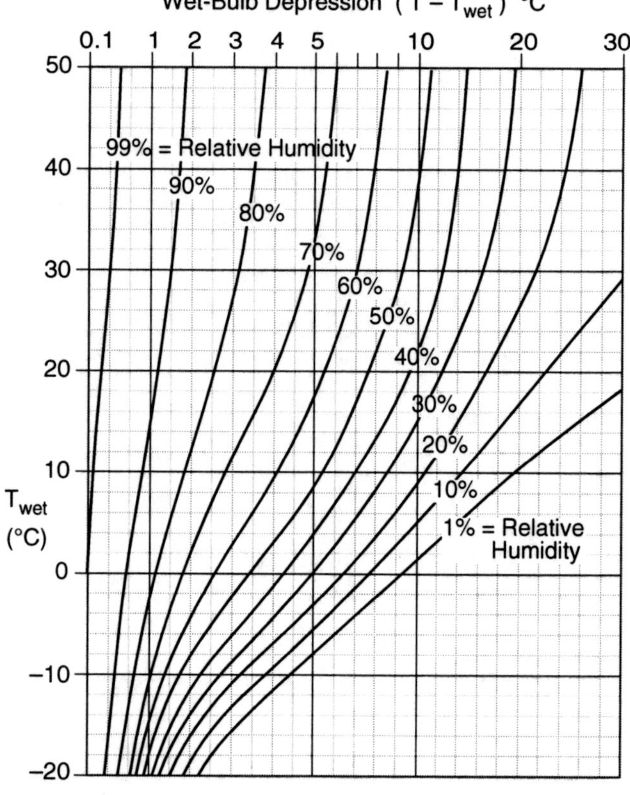

Figure 4.5
Psychrometric graph, to find relative humidity from wet and dry-bulb temperatures. Based on P = 101.325 kPa. Caution, the darker vertical lines mark scale changes along the abscissa.

It is easy to calculate other humidity variables such as mixing ratio or relative humidity from known values of T and T_w. Eqs. (4.18) and Figs 4.4 and 4.5 are examples of this.

However, going the opposite way is more difficult. If you are given other humidity variables it is hard to find the wet-bulb temperature. Namely, to use the equations or figures mentioned above, you would need to iterate to try to converge on the correct answer.

Instead, there are two methods to estimate T_w. One is an empirical approximation (given below), and the other is a graphical method called Normand's Rule (given on the next page).

The empirical approximation for T_w (in °C) at sea level is a function of air temperature T (in °C) and relative humidity $RH\%$ (e.g., using 65.8 to represent 65.8%):

$$T_w \approx T\cdot\text{atan}[0.151977(RH\% + 8.313659)^{1/2}] - 4.686035 + \text{atan}(T + RH\%) - \text{atan}(RH\% - 1.676331) + 0.00391838\cdot(RH\%)^{3/2}\cdot\text{atan}(0.023101\cdot RH\%) \tag{4.19}$$

where the arctangent (atan) function returns values in radians. [CAUTION: If your software returns arctan values in degrees, be sure to convert to radians before you use them in the equation above.]

Sample Application

Given an air temperature of 20°C and a relative humidity of 50%, use the empirical method to estimate the wet-bulb temperature at sea level.

Find the Answer

Given: $T = 20°C$, $RH\% = 50$

Find: $T_w = ?$ °C

Apply eq. (4.19):

$$T_w = 20\cdot\arctan[0.151977\cdot(50 + 8.313659)^{1/2}] + \arctan(20 + 50) - \arctan(50 - 1.676331) + 0.00391838\cdot(50)^{3/2}\cdot\arctan(0.023101\cdot 50) - 4.686035$$

T_w = **<u>13.7 °C</u>**

Check: Units reasonable. Agrees with Fig. 4.5.

Exposition: Although this equation had many terms, it needed to be solved only once. Contrast this with iterative methods, which require repeated solutions of equations in order to converge to an answer.

Sample Application

The air temperature is T = 20°C and the mixing ratio is 7.72 g/kg (which you might have found using a psychrometer with a wet-bulb temperature of 14°C). Use the equations to calculate the relative humidity for P = 100 kPa. Don't use look-up tables or graphs.

Find the Answer

Given: $T = 20°C$, $r = 7.72 \text{ g kg}^{-1}$, $P = 100$ kPa

Find: $RH = ?$ %

First, use Tetens' formula (4.2), with the trick that $\Delta T = T(K) - T_1(K) = [T(°C)+273.15] - 273.15 = T(°C)$, and remembering that for temperature differences: 1°C = 1 K.

$$e_s = 0.611(\text{kPa})\cdot\exp\left[\frac{17.2694\cdot(20\text{K})}{(20 + 273.15)\text{K} - 35.86\text{K}}\right]$$

e_s = 2.34 kPa.

Next, use this in eq. (4.5):

$$r_s = \frac{(622\text{g/kg})\cdot 2.34\text{kPa}}{[101.325 - 2.34]\text{kPa}} = 14.7\text{g/kg}$$

Finally, use eq. (4.14): $RH = 100\%\cdot(r/r_s)$

$RH = 100\% \cdot (7.72/14.7)$ = **<u>52.5%</u>**

Check: Physics and units are reasonable.

Exposition: What a lot of work. If we instead had used psychrometric graph (Fig. 4.5) with T_w = 14°C and $T - T_w$ = 6°C, we would have found almost the same relative humidity much more easily.

Sample Application

For air of $T = 25°C$ and $T_d = 18°C$ at $P = 100$ kPa, find T_w. Use Γ_s = 4.42 °C km^{-1} in Normand's Rule.

Find the Answer

Given: $P = 100$ kPa, $T = 25°C$, $T_d = 18°C$, $\Gamma_s = 4.42$ °C km^{-1}

Find: $T_w = ?°C$

According to Normand's Rule:

First, use eq. (4.16):

$z_{LCL} = (0.125 \text{ km °C}^{-1})\cdot(25 - 18°C) = 0.875$ km.

Next, use eq. (4.20):

$T_{LCL} = 25 - (9.8 \text{ K km}^{-1})\cdot(0.875 \text{ km}) = 16.43$ °C.

Finally, use eq. (4.21):

$T_w = 16.43 + (4.42 \text{ °C km}^{-1})\cdot(0.875 \text{ km}) =$ <u>**20.3°C**</u>.

Check: Physics and units are reasonable.

Exposition: The resulting wet-bulb depression is $(T - T_w) = 4.7°C$.

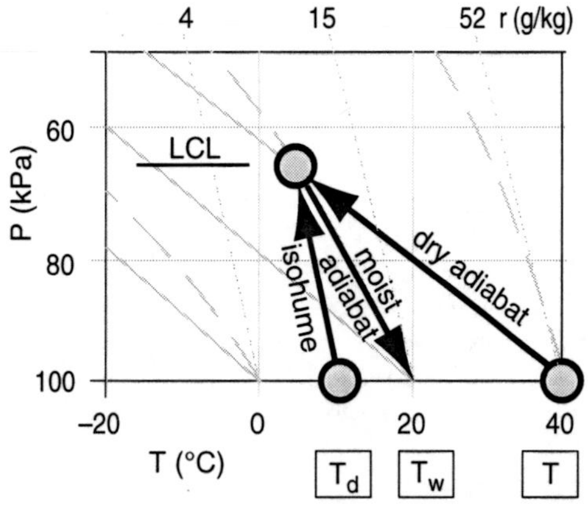

Figure 4.6

Demonstration of Normand's rule on a thermo diagram. It shows how to find wet-bulb temperature T_w, given T and T_d.

INFO • Summary of Humidity Variables

e	= vapor pressure (kPa)
r	= mixing ratio (g kg^{-1})
q	= specific humidity (g kg^{-1})
ρ_v	= absolute humidity (g m^{-3})
RH	= relative humidity (%)
z_{LCL}	= lifting condensation level (km)
T_d	= dewpoint (temperature) (°C)
T_w	= wet-bulb temperature (°C)

Notes:
- subscript s denotes saturation.
- most thermodynamic equations require temperatures to be converted into Kelvin.
- most thermo eqs. require mixing ratio in g g^{-1}.

You can obtain T_w from T_d via **Normand's Rule**:

- First: Find z_{LCL} using eq. (4.16).
- Next: $$T_{LCL} = T - \Gamma_d \cdot z_{LCL} \tag{4.20}$$
- Finally: $$T_w = T_{LCL} + \Gamma_s \cdot z_{LCL} \tag{4.21}$$

where the dry lapse rate is Γ_d = 9.8 K km^{-1} and the moist (saturated) lapse rate is Γ_s (magnitude varies, as explained later in this chapter). If you hypothetically lift an air parcel to its lifting condensation level, then its new temperature is T_{LCL}. The wet-bulb temperature is always constrained between the dry-bulb temperature and the dewpoint: $T_d \le T_w \le T$.

Normand's rule is easy to implement on a thermo diagram. Although isohumes and moist adiabats on thermo diagrams are not introduced until later in this chapter, I demonstrate Normand's rule here for future reference. Follow a dry adiabat up from the given dry-bulb temperature T, and follow an isohume up from the given dew point T_d (Fig. 4.6). At the *LCL* (where these two isopleths cross), follow a moist adiabat back down to the starting pressure to give the wet-bulb temperature T_w.

Converting Between Humidity Variables

Be sure to use g g^{-1} for q and r in the following equations:

$$q = \frac{r}{1+r} \tag{4.22}$$

$$q = \frac{\rho_v}{\rho_d + \rho_v} \tag{4.23}$$

$$q = \left(\frac{\varepsilon \cdot e_o}{P}\right) \cdot \exp\left[\frac{L}{\Re_v} \cdot \left(\frac{1}{T_o} - \frac{1}{T_d}\right)\right] \tag{4.24}$$

$$e = \frac{r}{\varepsilon + r} \cdot P \tag{4.25}$$

$$r = \frac{\rho_v}{\rho_d} \tag{4.26}$$

Also, when humidities are not high:

$$r \approx q \qquad \bullet(4.27a)$$

$$r_s \approx q_s \qquad \bullet(4.27b)$$

TOTAL WATER

Liquid and Solid Water

In clouds, fog, or air containing falling precipitation, one measure of the amount of liquid water in the air is the **liquid water content** (**LWC**). It is defined as

$$\rho_{LWC} = \frac{m_{liq.water}}{Vol} \qquad (4.28)$$

where $m_{liq.water}$ is mass of liquid water suspended or falling through the air, and *Vol* is the air volume. Typical values in cumulus clouds are $0 \le \rho_{LWC} \le 5$ g m^{-3}. It can also be expressed in units of $kg_{liq.water}$ m^{-3}. LWC is the liquid-water analogy to the absolute humidity for water vapor.

Another measure is the **liquid-water mixing ratio**:

$$r_L = \frac{m_{liq.water}}{m_{dry\ air}} \qquad (4.29)$$

where $m_{liq.water}$ is the mass of liquid water that is imbedded as droplets within an air parcel that contains $m_{dry\ air}$ mass of dry air. A similar **ice mixing ratio** can be defined:

$$r_i = \frac{m_{ice}}{m_{dry\ air}} \qquad (4.30)$$

Both mixing ratios have units of $kg_{water}\ kg_{air}^{-1}$, or $g_{water}\ kg_{air}^{-1}$.

Liquid water content is related to liquid-water mixing ratio by

$$r_L = \frac{\rho_{LWC}}{\rho_{air}} \qquad (4.31)$$

where ρ_{air} is air density.

Mixing Ratio of Total Water

The **total-water mixing ratio** r_T is defined as the sum of masses of all phases of water (vapor, liquid, solid) per dry-air mass:

$$r_T = r + r_L + r_i \qquad \bullet(4.32a)$$

where r is **mixing ratio for water-vapor**, r_L is **mixing ratio for liquid-water**, and r_i is **mixing ratio for ice**. Be sure to use common units for all terms in this equation; namely $kg_{water}\ kg_{air}^{-1}$, or $g_{water}\ kg_{air}^{-1}$. Total-water absolute humidity and total-water specific humidity are similarly defined.

Sample Application

Find the liquid water mixing ratio in air at sea level, given a liquid water content of 3 g m^{-3}.

Find the Answer

Given: $\rho_{LWC} = 3\ g_{water}\ m^{-3}$. Sea level.
Find: $r_L = ?\ g_{water}\ kg_{dry\ air}^{-1}$.
Assume standard atmosphere, and use Table 1-5 from Chapter 1 to get: $\rho_{air} = 1.225\ kg_{air}\ m^{-3}$ at sea level.

Use eq. (4.31):
$r_L = (3\ g_{water}\ m^{-3})/(1.225\ kg_{air}\ m^{-3})$
$= 2.45\ g_{water}\ kg_{dry\ air}^{-1}$. = <u>**2.45**</u> g kg^{-1}

Check: Physics, units & magnitude are reasonable.
Exposition: Liquid, solid and water vapor might exist together in a cloud.

Sample Application

At a pressure altitude of 50 kPa inside a thunderstorm, suppose that the air temperature is –5°C, and that each kg of air contains 4 g of water droplets and 2 g of ice crystals suspended in the air. What is the total-water mixing ratio value?

Find the Answer

Given: P = 50 kPa, T = –5°C, r_L = 4 g kg^{-1}, r_i = 2 g kg^{-1}
Find: r_T = ? g kg^{-1}.
Assume cloudy (saturated) air inside the thunderstorm.

We need to solve eq. (4.32c), for which we have everything we need except r_s. To use eq. (4.5) to get r_s, we can read e_s = 0.4222 kPa at T = –5°C from Table 4-1.

Now apply eq. (4.5): $r_s = \varepsilon \cdot e_s/(P-e_s)$ = (622 g·kg^{-1})·(0.4222 kPa)/(50– 0.4222 kPa) = 5.30 g kg^{-1}.

Finally, apply eq. (4.32c):
$r_T = r_s + r_L + r_i$ = 5.3 + 4 + 2 = **11.3** g·kg^{-1}

Check: Physics and units are reasonable.
Exposition: For altitudes where –40°C < T < 0°C, it is often the case in thunderstorms that both liquid water and ice crystals can be present in the air, along with water vapor.

Sample Application

Find the precipitable water in the troposphere if the bottom portion (below P = 50 kPa) has an average total water mixing ratio of 10 g kg^{-1} while the portion of troposphere between 50 and 20 kPa has an average total water mixing ratio of 2 g kg^{-1}.

Find the Answer

Assume: Bottom of troposphere is at 100 kPa.
Given: r_T = 0.010 kg kg^{-1}, P_B = 100 kPa, P_T = 50 kPa,
r_T = 0.002 kg kg^{-1}, P_B = 50 kPa, P_T = 20 kPa
Find: d_W = ? m

Apply eq. (4.33) for the bottom and top segments of the column of tropospheric air, and sum the result.

d_W = [(9.8 m·s^{-2})·(1000 kg$_{water}$·m^{-3})]$^{-1}$ ·
{ (0.010 kg$_{water}$ kg$_{air}$$^{-1}$)·[100 – 50 kPa] +
(0.002 kg$_{water}$ kg$_{air}$$^{-1}$)·[50 – 20 kPa] }
= [1.02x10^{-4} m^2 s^2 kg$_{water}$$^{-1}$] ·
{(0.5 + 0.06) kPa·kg$_{water}$·kg$_{air}$$^{-1}$}
= 5.71x10^{-5} m^2 s^2 kPa kg$_{air}$$^{-1}$.
To convert the units, use Appendix A info:
1 kPa = 1000 kg$_{air}$·m·s^{-2}
Thus, d_w = [(1000 kg$_{air}$·m·s^{-2})/(1 kPa)] ·
(5.71x10^{-5} m^2 s^2 kPa kg$_{air}$$^{-1}$) = **0.057 m**

Check: Physics and units are reasonable.
Exposition: The puddle depth of 5.7 cm is large because the total water mixing ratios were large.

Liquid-water cloud droplets can exist unfrozen in air of temperature less than 0°C. Thus, it is possible for ice and liquid water to co-exist in the same air parcel at the same time, along with water vapor.

Eq. (4.32a) can be simplified if there is no precipitation:

$$r_T = r \qquad \text{if the air is not cloudy} \quad \bullet(4.32b)$$

$$r_T = r_s + r_L + r_i \qquad \text{if the air is cloudy} \quad \bullet(4.32c)$$

By "not cloudy" we mean air that is **unsaturated** (i.e., $r < r_s$). By "cloudy" we mean air that is **saturated** (i.e., $r = r_s$) and has either or both liquid water drops and/or ice crystals suspended in it.

Suppose an air parcel has some total number of water molecules in it. Consider an idealized situation where an air parcel does not mix with its environment. For this case, we anticipate that all the water molecules in the parcel must move with the parcel. It makes no difference if some of these molecules are in the form of vapor, or liquid droplets, or solid ice crystals — all the water molecules must still be accounted for.

Hence, for this idealized parcel with no precipitation falling into or out of it, the amount of total water r_T must be constant. Any changes in r_T must be directly associated with precipitation falling into or out of the air parcel.

Suppose that the air parcel is initially unsaturated, for which case we can solve for the total water using eq. (4.32b). If this air parcel rises and cools and can hold less vapor at saturation (see eq. 4.5), it might reach an altitude where $r_s < r_T$. For this situation, eq. (4.32c) tells us that $r_L + r_i = r_T - r_s$. Namely, we can anticipate that liquid water droplets and/or ice crystals suspended in the air parcel must have formed to maintain the constant total number of water molecules.

Precipitable Water

Consider an air column between (top, bottom) altitudes as given by their respective air pressures (P_T, P_B). Suppose all the water molecules within that column were to fall to the bottom of the column and form a puddle. The depth of this puddle (namely, the **precipitable water**) is

$$d_W = \frac{\overline{r_T}}{|g| \cdot \rho_{liq}} \cdot (P_B - P_T) \qquad \bullet(4.33)$$

where the magnitude of gravitation acceleration is $|g|$ = 9.8 m·s^{-2} , the liquid-water density is ρ_{liq} = 1000 kg·m^{-3}, and a column-average of the total-water mixing ratio is $\overline{r_T}$. For a column where r_T varies with altitude, split it into column segments each having unique r_T average, and sum over all segments.

Precipitable water is sometimes used as a humidity variable. The bottom of the atmosphere is warmer than the mid and upper troposphere, and can hold the most water vapor. In a pre-storm cloudless environment, contributions to the total-column precipitable water thus come mostly from the boundary layer. Hence, precipitable water can serve as one measure of boundary-layer total water that could serve as the fuel for thunderstorms later in the day. See the Thunderstorm chapters for a sample map of precipitable water.

Note that the American Meteorological Society *Glossary of Meteorology* considers only the water vapor in an air column for calculation of precipitable-water depth. However, some satellites can detect total water over a range of altitudes, for which eq. (4.33) would be applicable.

It is possible to have more precipitation reach the ground during a storm than the value of precipitable water. This occurs where moisture advection by the winds can replenish water vapor in a region.

LAGRANGIAN BUDGETS

Moist air parcels have two additional properties that were unimportant for dry air. One is the amount of water in the parcel, which is important for determining cloud formation and precipitation amounts. The second is the latent heat released or absorbed when water changes phase, which is critical for determining the buoyancy of air parcels and the energy of thunderstorms.

Water Budget

Lagrangian Water Conservation

Suppose that the amount of precipitation falling out of the bottom of an air parcel differs from the amount falling into the air parcel from above. This difference gives a net source or sink S^{**}, causing the total water to change inside the air parcel:

$$\frac{\Delta r_T}{\Delta t} = S^{**} \tag{4.34}$$

For situations where $S^{**} = 0$, then total water conservation requires that :

$$(r + r_i + r_L)_{initial} = (r + r_i + r_L)_{final} \qquad \bullet(4.35a)$$

Sample Application

Suppose an air parcel is stationary at sea level. While there, external processes cause it to cool from 30°C to 5°C. If the air is initially unsaturated with humidity of 15 g kg^{-1}, then what is the final disposition of water molecules in the air parcel? There is no precipitation into or out of the air parcel.

Find the Answer

Given: $(T_{initial}, T_{final}) = (30, 5\ °C)$, $r_{initial} = 15$ g kg^{-1}
with $r_{L\ initial} = r_{i\ initial} = 0$ because unsaturated.
Find: $(r, r_L, r_i)_{final} = (?, ?, ?)\ g_{liq}\ kg_{air}^{-1}$

Because the final parcel temperature is warmer than freezing, we can assume **<u>no ice</u>**: $r_{i\,final}$ **<u>= 0</u>**. So this leaves r_{final} and $r_{L\,final}$ to be determined with eq. (4.35b):

$$(15 + 0) = (r_{final} + r_{L\,final})\ \text{g kg}^{-1}$$

Because this problem is set at sea level, we can skip some calculations by using the data in Table 4-1 for the final value of mixing ratio. At final temperature 5°C, Table 4-1 gives a saturation mixing ratio of $r_{s\,final}$ = 5.408 $g_{liq}\ kg_{air}^{-1}$. Since this is less than the initial humidity, we know that the air is saturated with **<u>water-vapor</u>** content: $r_{final} = r_{s\,final}$ = **<u>5.408</u>** $g_{liq}\ kg_{air}^{-1}$.

Finally, solving eq. (4.35b) for $r_{L\,final}$ gives:
$r_{L\,final}$ = 15 g kg^{-1} – 5.408 g kg^{-1} = **<u>9.592</u>** g kg^{-1}

Check: Physics and units are reasonable.
Exposition: Assuming that this final liquid water is suspended in the air as tiny droplets, the result is fog.

Sample Application

Use Fig. 4.7 to answer (**A**) these questions (**Q**).

Find the Answer

Q: What is the saturation mixing ratio for air at P = 30 kPa with T = 20°C?

A: Follow the T = 20°C green isotherm vertically, and the P = 30 kPa green isobar horizontally, to find where they intersect. The saturation mixing ratio (blue diagonal) line that crosses through this intersection is the one labeled: $r_s \approx$ **50** g kg^{-1}.

Q: What is the actual mixing ratio for air at P = 30 kPa with T_d = –20°C?

A: Follow the green –20°C isotherm vertically to where it intersects the horizontal P = 30 kPa isobar. Interpolating between the blue diagonal lines that are adjacent to this intersection gives an actual mixing ratio of $r \approx$ **3** g kg^{-1}.

Q: What is the dew-point temperature for air at P = 60 kPa with r = 0.2 g kg^{-1}?

A: From the intersection of the blue diagonal isohume r = 0.2 g kg^{-1} and the green horizontal isobar for P = 60 kPa, go vertically straight down to find $T_d \approx$ **–40°C**.

If the warm cloud contains no suspended ice crystals, then:

$$(r + r_L)_{initial} = (r + r_L)_{final} \qquad \bullet(4.35b)$$

Namely, an increase in the amount of water in one phase (ice, liquid, vapor) must be compensated by a decrease on other phases in order to satisfy total-water conservation if there are no sources or sinks. For an adiabatic process (i.e., no mixing of air or transfer of precipitation across the boundary of an air parcel), r_T must be conserved. For this reason, isohumes of total water are included on thermo diagrams.

Isohumes on a Thermo Diagram

Thermo diagrams were introduced as Fig. 3.3 in the Thermodynamics chapter. On that diagram, the **state** of the air was represented by two sets of thin solid green lines: isobars (horizontal lines) for pressure and isotherms (vertical lines) for temperature. To that background we will now add another state line: isohumes (thin dotted blue lines) for moisture state of the air (Fig. 4.7).

These isohumes are overloaded with information. As a "state" line, the isohume gives the saturation mixing ratio r_s at any given temperature and

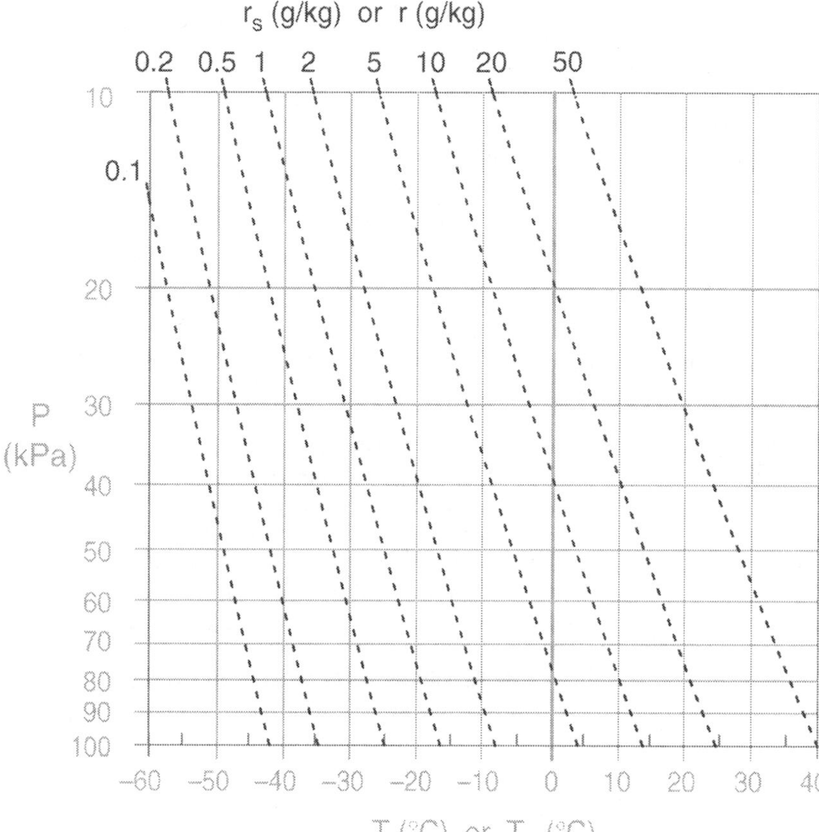

Figure 4.7
Isohumes are dotted blue diagonal lines, isobars are green horizontal lines, and isotherms are green vertical lines on this thermo diagram.

pressure. As a **"process"** line, the isohume shows how total water mixing ratio r_T is conserved for adiabatically rising or sinking air parcels.

To use the thermo-diagram background of P vs. T for isohumes, we need to describe r_s as a function of P and T [abbreviate as $r_s(P, T)$]. Eq. (4.5) gives $r_s(P, e_s)$ and the Clausius-Clapeyron eq. (4.1) gives $e_s(T)$. So combining these two equations gives $r_s(P, T)$.

But to draw any one isohume (i.e., for any one value of r_s), we need to rearrange the result to give $T(P, r_s)$:

$$T = \left[\frac{1}{T_o} - \frac{\Re_v}{L_v} \cdot \ln\left\{ \frac{r_s \cdot P}{e_o \cdot (r_s + \varepsilon)} \right\} \right]^{-1} \qquad (4.36)$$

where T_o = 273.15 K, e_o = 0.6113 kPa, $\Re_v/L$ = 0.0001844 K^{-1}, and ε = 0.622 g g^{-1}. This eq. requires that r_s be in g g^{-1} (not g kg^{-1}). T has units of Kelvin.

Thus, pick any fixed r_s to plot. Then, for a range of P from the bottom to the top of the atmosphere, solve eq. (4.36) for the corresponding T values. Plot these T vs. P values as the isohume line on a thermo diagram. Use a spreadsheet to repeat this calculation for other values of r_s, to plot the other isohumes.

Eqs. (4.36) with (4.15b) are similar. Thus, you can use isohumes of $T(P, r_s)$ to also represent isohumes of $T_d(P, r)$. Namely, you can use isohumes to find the saturation state r_s of the air at any P and T, and you can also use the isohumes to describe the process of how T_d changes when an air parcel of constant r rises or descends to an altitude of different P.

Sample Application

Find the dew-point temperature of air having mixing ratio of 10 g kg^{-1} at an altitude where P = 40 kPa.

Find the Answer

Given: r = 10 g kg^{-1} , P = 40 kPa
Find: T_d = ?°C

Apply eq. (4.15b):

$$T_d = \left[\frac{1}{273.15\text{K}} - 0.000184\text{K}^{-1} \cdot \ln\left\{ \frac{(0.01\text{g/g})\cdot(40\text{kPa})}{(0.6113\text{kPa})\cdot(0.01 + 0.622\text{g/g})} \right\} \right]^{-1}$$

= [(0.003661 K^{-1}) – (0.0001844 K^{-1})·ln{1.035}]$^{-1}$
= [0.003655 K^{-1}]$^{-1}$ = 273.6 K ≈ **0.5°C**

Check: Physics and units are reasonable.
Exposition: You could have saved a lot of time and effort by just looking up the answer in Fig. 4.7. Namely, find the point where the 10 g kg^{-1} mixing-ratio diagonal line intersects the 40 kPa isobar, and then read vertically straight down to find T_d.

We would have found the same numerical answer if we asked for the temperature corresponding to air at 40 kPa that is saturated with r_s = 10 g kg^{-1} with r_L = 0. For this situation, we would have used eq. (4.36).

Heat Conservation for Saturated Air

Moist Adiabatic Lapse Rate

Saturated air is air that is foggy or cloudy, with an amount of water vapor equal to the maximum amount that air can hold given the parcel's temperature. The air parcel contains both vapor and small suspended liquid water droplets or ice crystals. Consider an adiabatic process for which: (1) the parcel does not mix with its surroundings; (2) there is no transfer of heat to/from the parcel from outside; and (3) no liquid (or solid) water falls out of or into the parcel.

An air parcel rising adiabatically has two competing processes that affect its temperature. As for the unsaturated parcel, the saturated parcel expands into regions of lower pressure, doing work on the atmosphere at the expense of thermal energy stored in the parcel. But the colder air parcel can hold less water vapor at saturation than it carried up from the altitude below. So more of the vapor condenses, for which: (1) the latent heating partially offsets the expansion cooling; and (2) the amount of condensed water droplets or ice increases.

The converse occurs for adiabatic descent, where cooling from evaporating liquid-water droplets partially offsets warming due to adiabatic compression.

For air rising across height increment Δz, a saturated parcel has less temperature decrease ΔT than does a dry (unsaturated) parcel. Conversely, for air descending across height increment Δz, a saturated parcel has less temperature increase ΔT than does a dry parcel.

While we previously saw that the dry adiabatic lapse rate was constant ($\Gamma_s = -\Delta T/\Delta z$ = 9.8 °C km^{-1}), we are not so lucky for the saturated lapse rate, which varies with altitude and air temperature.

A saturated air parcel that rises adiabatically has a temperature decrease with increasing height of $-\Delta T/\Delta z = \Gamma_s$, where Γ_s **saturated (or moist) adiabatic lapse rate**. This rate is given by:

$$\Gamma_s = \frac{|g|}{C_p} \cdot \frac{\left(1 + \frac{r_s \cdot L_v}{\Re_d \cdot T}\right)}{\left(1 + \frac{L_v^2 \cdot r_s \cdot \varepsilon}{C_p \cdot \Re_d \cdot T^2}\right)} \qquad (4.37a)$$

Sample Application

What is the value of moist-adiabatic lapse rate at T = 10°C and P = 70 kPa? Do this calculation using both (a) $C_p = C_{p\,dry}$ and (b) the actual moist C_p. (c) Also find $\Delta T/\Delta P$ for a rising saturated air parcel.

Find the Answer

Given: $P = 70$ kPa, $T = 10°C = 283.15$ K,
Find: Γ_s = ? °C km^{-1}

(a) Apply eq. (4.37b). But this eq. needs r_s, which we first can find using eq. (4.5). In turn, that eq. needs e_s, which we can get from Table 4-1: e_s = 1.232 kPa at T = 10°C. Plug this into eq. (4.5): r_s = (0.622 g/g)·(1.232 kPa) / (70kPa – 1.232 kPa) = 0.01114 g/g.
Finally, use this in eq. (4.37b):

$$\Gamma_s = \frac{(9.8\frac{\text{K}}{\text{km}})\cdot\left[1+\frac{(0.01114\text{g/g})\cdot(8711\text{K})}{283.15\text{K}}\right]}{\left[1+\frac{(1.35\times10^7\,\text{K}^2)\cdot(0.01114\text{g/g})}{(283.15\text{K})^2}\right]}$$

Γ_s = (9.8 K km^{-1})·[1.3427] / [2.8758] = **<u>4.58 K km^{-1}</u>**.

(b) Apply eq. (3.3) for saturated air (for which $r = r_s$):
$C_p = C_{p\,dry}$·[1 + 1.84r] = (1004 J·kg^{-1}·K^{-1})·[1+1.84·(0.01114g/g] = 1024.6 J·kg^{-1}·K^{-1}.

Thus: $|g|/C_p$ = 9.565 K km^{-1}
L_v/C_p = 2440. K.
$L_v/\Re_d$ = 8711 K.
When these are applied in eq. (4.37a), the result is:

$$\Gamma_s = \frac{(9.565\frac{\text{K}}{\text{km}})\cdot\left[1+\frac{(0.01114\text{g/g})\cdot(8711\text{K})}{283.15\text{K}}\right]}{\left[1+\frac{(2440\text{K})\cdot(8711\text{K})\cdot(0.01114\text{g/g})\cdot0.622}{(283.15\text{K})^2}\right]}$$

Γ_s = (9.565 K km^{-1})·[1.3427]/[2.837] = **<u>4.53 K km^{-1}</u>**

(c) If we assume $C_p \approx C_{p\,dry}$, then apply eq. (4.38b):

$$\frac{\Delta T}{\Delta P} = \frac{[0.28571\cdot(283.15\text{K})+(2488.4\text{K}\cdot(0.01114\text{g/g}))]}{P\cdot\left[1+[1.35\times10^7\,K^2\cdot(0.01114\text{g/g})/(283.15\text{K})^2]\right]}$$

= [108.62 / (1 + 1.876)] / P = (37.77K) / P

At P = 70 kPa the result is = $\Delta T/\Delta P$ = **<u>0.54 K kPa^{-1}</u>**

Check: Physics and units are reasonable.
Exposition: Don't forget for answers (a) and (b) that lapse rates are the rate of temperature <u>decrease</u> with altitude. Parts (a) and (b) give nearly identical answers, implying that eq. (4.37b) is sufficiently accurate for most applications.

Normally, pressure <u>de</u>creases as altitude increases, thus answer (c) also gives cooling for negative ΔP.

Compare typical saturated adiabatic lapse-rate values of 4 to 7°C km^{-1} to the dry adiabatic lapse rate of 9.8°C km^{-1}.

Instead of a change of temperature with height, this saturated adiabatic lapse rate can be rewritten as a change of temperature ΔT with change of pressure ΔP:

$$\frac{\Delta T}{\Delta P} = \frac{\left[(\Re_d/C_p)\cdot T + (L_v/C_p)\cdot r_s\right]}{P\cdot\left(1+\frac{L_v^2\cdot r_s\cdot\varepsilon}{C_p\cdot\Re_d\cdot T^2}\right)} \tag{4.38a}$$

For the equations above, don't forget that specific heat C_p varies with humidity (see eq. 3.2).

After plugging in the values for the thermodynamic constants and assuming $C_p \approx$ constant, eq. (4.37a) can be simplified as:

$$\Gamma_s = \Gamma_d\cdot\frac{[1+(a\cdot r_s/T)]}{[1+(b\cdot r_s/T^2)]} \qquad \bullet(4.37b)$$

where a = 8711 K, b = 1.35x10^7 K^2, and Γ_d = 9.8 K km^{-1}. Eq. (4.37b) differs from (4.37a) by roughly 1%, so it is often accurate enough for most applications. Use g g^{-1} for mixing ratio, and use Kelvin for temperature. Be aware that r_s is not constant, but is a function of temperature.

Similarly, eq. (4.38a) simplifies to:

$$\frac{\Delta T}{\Delta P} = \frac{[a\cdot T + c\cdot r_s]}{P\cdot[1+(b\cdot r_s/T^2)]} \qquad \bullet(4.38b)$$

with a = 0.28571, b = 1.35x10^7 K^2, and c = 2488.4 K.

Moist Adiabats on a Thermo Diagram

In the Thermodynamics chapter we discussed the process of "dry" adiabatic vertical motion, where "dry" mean <u>un</u>saturated humid air. We had plotted those process lines as the dry adiabats in the thermo diagram of Fig. 3.4.

We can now use eq. (4.38) to calculate and plot the corresponding **moist adiabats** (also called **saturated adiabats**) that apply for saturated (cloudy or foggy) vertical motion. The saturated adiabats are rather complicated to calculate, because the equations above give the slope ($\Delta T/\Delta P$) for the moist adiabat rather than the desired value of T at each P. But don't despair — we can still find the moist-adiabat curves by iterating each curve upward.

Start at P = 100 kPa with some initial value for T. First use these with eq. (4.5) to get r_s. Use these in eq. (4.38) to get $\Delta T/\Delta P$. Then apply that over a small

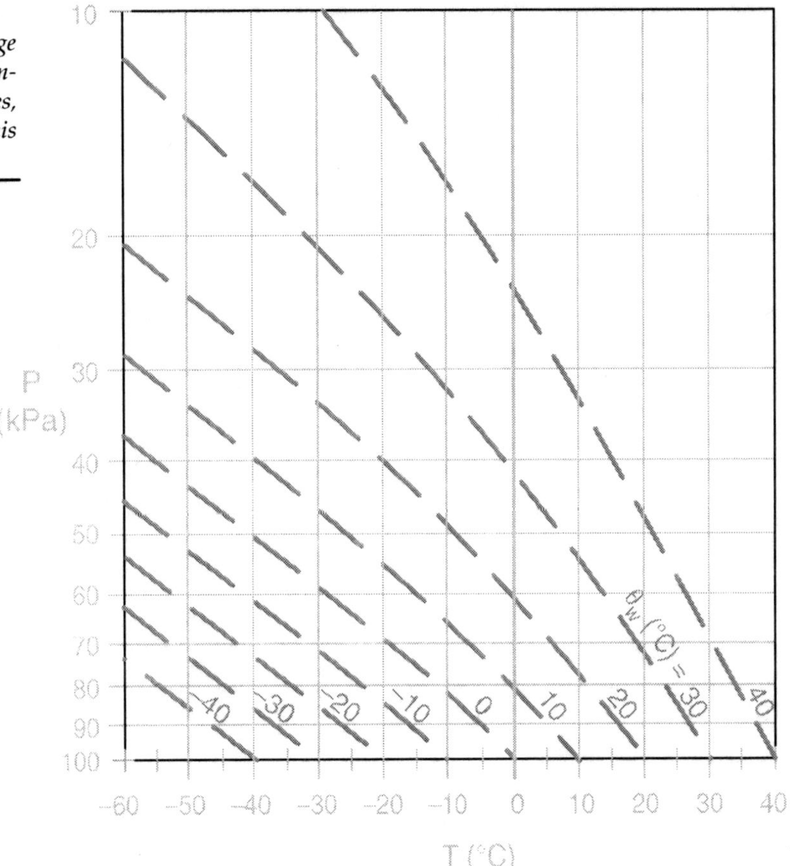

Figure 4.8 (at right)
Moist (saturated) adiabats are thick dashed orange diagonal lines labeled by wet-bulb potential temperature θ_w. Isobars are green horizontal lines, and isotherms are green vertical lines on this thermo diagram.

increment of pressure (such as $\Delta P = P_2 - P_1 = -0.2$ kPa) to solve for the new T_2 at P_2 using:

$$T_2 = T_1 + \frac{\Delta T}{\Delta P} \cdot (P_2 - P_1) \qquad (4.39)$$

Repeat by using the new T_2 at P_2 to find the new r_{s2}, and use all these numbers to solve for a new $(\Delta T/\Delta P)_2$, and use eq. (4.39) again to take the next step. Repeat to iterate your way up the moist adiabat. The result is one of the curves in Fig. 4.8.

To get other curves, start over with a different initial value of T at P = 100 kPa, as shown in Fig. 4.8. Each of those curves is identified by its initial T at the reference pressure of 100 kPa. The next section shows that these labels are called wet-bulb potential temperature θ_w.

INFO • Create Your Own Thermo Diagram — Part 2: Moist Adiabats

As for the dry adiabats, it is useful to see how moist adiabats can be calculated using a computer spreadsheet. The following example is for the moist adiabat that starts at T = 30°C at P = 100 kPa.

	A	B	C	D	E
1	P (kPa)	T (°C)	e_s (kPa)	r_s (g/g)	ΔT/ΔP
2	eq:	(4.39)	(4.1a)	(4.5)	(4.38b)
3	10.0	–70.98	0.0006	0.0000	5.7176
4	10.2	–69.86	0.0007	0.0000	5.6297
5	10.4	–68.75	0.0008	0.0000	5.5440
6	10.6	–67.66	0.0009	0.0001	5.4603
...	...	...	...	...	...
450	99.4	29.82	4.3114	0.0282	0.3063
451	99.6	29.88	4.3270	0.0282	0.3056
452	99.8	29.94	4.3426	0.0283	0.3050
453	100.0	30.00	4.3582	0.0283	0.3043

In row 1, label the variables at the top of the 5 columns as I have done here. In row 2, identify which equation numbers you are using, as documentation for you or others who use your calculations later.

(continues in next column)

INFO • Moist Adiabats (continuation)

In row 3 column A, enter the pressure at the top of the atmospheric column of interest: 10.0 (kPa), shown in red in this example. In row 4 column A type the next pressure 10.2 (kPa), which is the starting pressure plus increment ΔP = 0.2 kPa. Then use automatic series generating methods in your spreadsheet to extend this series down to the point where the pressure is 100.0 (kPa), (on row 453 in my spreadsheet).

Next, in row 453 column B, type in the starting temperature of 30 (°C) for this moist adiabat. In (row, col) = (453, C), use the Clausius-Clapeyron eq. (4.1a) to calculate the saturation vapor pressure for the pressure in column A, and don't forget to convert temperature to Kelvin in your spreadsheet equation. Similarly, use eq. (4.5) to calculate the saturation mixing ratio in cell (453, D). Use eq. (4.38b) to find the moist-adiabat slope in cell (453, E). Again, use Kelvin in your eqs.

For the next row up, but only in cell (452,B), use eq. (4.39) to find the new temperature along the moist adiabat. Be sure to check that the sign is correct (i.e., that temperature is decreasing, not increasing). Then, use the "fill-up" spreadsheet command to fill the eqs. from cells (453, C through E) up one row. Finally, use the fill-up command to fill the eqs. from cells (452, B through E) up to the top pressure level in row 3.

Check: Agrees with Fig. 4.8 for θ_w = 30°C.

Sample Application

What lines do dry & saturated air parcels follow on a thermo diagram if they start at P= 100 kPa, T= 40°C?

Find the Answer

Given: P = 100 kPa, T = 40°C initially.
Plot adiabatic process lines for θ = 40°C & θ_w = 40°C.
Copying the lines from Figs. 3.3 & 4.8.

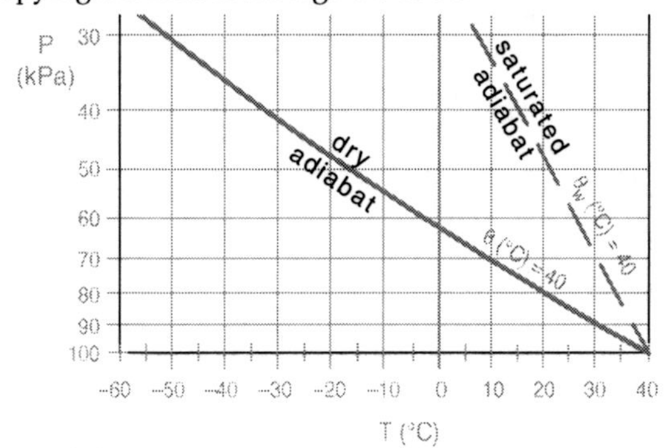

Exposition: Even starting with the same temperature, a rising saturated air parcel becomes warmer than an unsaturated parcel due to latent-heat release.

Sample Application

Air at pressure 80 kPa and T = 0°C is saturated, & holds 2 g kg^{-1} of liquid water. Find θ_e and θ_L.

Find the Answer:

Given: P = 80 kPa, T = 0°C = 273.15K , r_L = 2 g kg^{-1}.
Find: θ_e = ? °C, θ_L = ? °C

First, do preliminary calculations shared by both eqs:
Rearrange eq. (3.12) to give:
$(\theta/T) = (P_o/P)^{0.28571}$ = (100kPa/80kPa)$^{0.28571}$ = 1.066
Thus, θ = 291K ≈ 18°C
At 80 kPa and 0°C, solve eq. (4.5) for r_s = 4.7 g kg^{-1}
Then use eq. (3.2):
$C_p = C_{pd}\cdot(1 + 1.84\cdot r)$ = (1004.67 J·kg^{-1}·K^{-1}) ·[1+1.84·(0.0047 g g^{-1})] = 1013.4 J·kg^{-1}·K^{-1}.
Thus, L_v/C_p = 2467 K/(g$_{water}$ g$_{air}^{-1}$)
and $(L_v/C_p)\cdot(\theta/T)$ ≈ 2630 K/(g$_{water}$ g$_{air}^{-1}$)

Use eq. (4.40):
θ_e = (18°C) + (2630 K/(g$_{water}$ g$_{air}^{-1}$))·(0.0047)
= <u>**30.4 °C**</u>

Use eq. (4.41):
θ_L = (18°C) – (2630 K/(g$_{water}$ g$_{air}^{-1}$))·(0.002)
= <u>**12.7 °C**</u>

Check: Physics, units, & magnitude are reasonable.
Exposition: The answers are easier to find using a thermo diagram (after you've studied the Stability chapter). For θ_e , find the θ value for the dry adiabat that is tangent at the diagram top to the moist adiabat. For θ_L, follow a moist adiabat down to where it crosses the (2 + 4.7 = 6.7 g kg^{-1}) isohume, and from there follow a dry adiabat to P = 100 kPa.

Wet-Bulb, Equivalent, and Liquid-Water Potential Temperatures

Recall from the Thermodynamics chapter that potential temperature θ is conserved during unsaturated adiabatic ascent or descent. However, if an air parcel containing water vapor is lifted above its LCL, then condensation will add latent heat, causing θ to increase. Similarly, if the air contains liquid water such as cloud drops, when it descends some of the drops can evaporate, thereby cooling the air and reducing θ.

However, we can define new variables that are conserved for adiabatic ascent or descent, regardless of any evaporation or condensation that might occur. One is the **equivalent potential temperature** θ_e:

$$\theta_e \approx \theta + \left(\frac{L_v \cdot \theta}{C_p \cdot T}\right)\cdot r \quad \approx \quad \theta + \frac{L_v}{C_{pd}}\cdot r \qquad \bullet(4.40)$$

Another is **liquid water potential temperature**, θ_L:

$$\theta_L \approx \theta - \left(\frac{L_v \cdot \theta}{C_p \cdot T}\right)\cdot r_L \quad \approx \quad \theta - \frac{L_v}{C_{pd}}\cdot r_L \qquad \bullet(4.41)$$

where L_v = 2.5x10^6 J·kg^{-1} is the latent heat of vaporization, C_p is the specific heat at constant pressure for air (C_p is not constant, see the Thermodynamics chapter), T is the absolute temperature of the air, and mixing ratios (r and r_L) have units of (g$_{water}$ g$_{air}^{-1}$). The last approximation in both equations is very rough, with L_v/C_{pd} = 2.5 K·(g$_{water}$/kg$_{air}$)$^{-1}$.

Both variables are conserved regardless of whether the air is saturated or unsaturated. Consider <u>unsaturated</u> air, for which θ is conserved. In eq. (4.40), water-vapor mixing ratio r is also conserved during ascent or descent, so the right side of eq. (4.40) is constant, and θ_e is conserved. Similarly, for unsaturated air, liquid water mixing ratio r_L = 0, hence θ_L is also conserved in eq. (4.41).

For <u>saturated</u> air, θ will increase in a rising air parcel due to latent heating, but r will decrease as some of the vapor condenses into liquid. The two terms in the right side of eq. (4.40) have equal but opposite changes that balance, leaving θ_e conserved. Similarly, the two terms on the right side of eq. (4.41) balance, due to the minus sign in front of the r_L term. Thus, θ_L is conserved.

By subtracting eq. (4.41) from (4.40), we can see how θ_e and θ_L are related:

$$\theta_e \approx \theta_L + \left(\frac{L_v \cdot \theta}{C_p \cdot T}\right)\cdot r_T \qquad (4.42)$$

for a total-water mixing ratio (in g g^{-1}) of $r_T = r + r_L$. Although θ_e and θ_L are both conserved, they are <u>not</u> equal to each other.

We can use θ_e or θ_L to identify and label moist adiabats. Consider an air parcel starting at P = 100 kPa that is saturated but contains no liquid water ($r = r_s = r_T$). For that situation θ_L is equal to its initial temperature T (which also equals its initial potential temperature θ at that pressure). A rising air parcel from this point will conserve θ_L, hence we could label the moist adiabat with this value (Fig. 4.9).

An alternative label starts from same saturated air parcel at P = 100 kPa, but conceptually lifts it to the top of the atmosphere (P = 0). All of the water vapor will have condensed out at that end point, heating the air to a new potential temperature. The potential temperature of the dry adiabat that is tangent to the top of the moist adiabat gives θ_e (Fig. 4.9). [*CAUTION: On some thermo diagrams, equivalent potential temperature is given in units of Kelvin.*]

In other words, θ_L is the potential temperature at the bottom of the moist adiabat (more precisely, at P = 100 kPa), while θ_e is the potential temperature at the top. Either labeling method is fine — you will probably encounter both methods in thermo diagrams that you get from around the world.

Wet-bulb potential temperature (θ_w) can also be used to label moist adiabats. For θ_w, use Normand's rule on a thermo diagram (Fig. 4.10). Knowing temperature T and dew-point T_d at initial pressure P, and plot these points on a thermo diagram. Next, from the T point, follow a dry adiabat up, and from the T_d point, follow an isohume up. Where they cross is the lifting condensation level LCL.

From that LCL point, follow a saturated adiabat back down to the starting altitude, which gives the **wet-bulb temperature** T_w. If you continue to follow the saturated adiabat down to a reference pressure (P = 100 kPa), the resulting temperature is the **wet-bulb potential temperature** θ_w (see Fig. 4.10). Namely, θ_w equals the θ_L label of the moist adiabat that passes through the LCL point. Labeling moist adiabats with values of wet-bulb potential temperature θ_w is analogous to the labeling of dry adiabats with θ, which is why I use θ_w here.

To find θ_e(K) for a moist adiabat if you know its θ_w(K), use:

$$\theta_e = \theta_w \cdot \exp(a_3 \cdot r_s / \theta_w)_o \qquad (4.43a)$$

where a_3 = 2491 K·kg$_{air}$ kg$_{vapor}^{-1}$, and r_s is initial saturation mixing ratio (kg$_{vapor}$ kg$_{air}^{-1}$) at $T = \theta_w$ and P = 100 kPa (as denoted by subscript "o"). You can approximate (4.43a) by

$$\theta_e(\text{K}) \approx a_o + a_1 \cdot \theta_w(°\text{C}) + a_2 \cdot [\theta_w(°\text{C})]^2 \qquad (4.43b)$$

where a_o = 282, a_1 = 1.35, and a_2 = 0.065, for θ_w in the range of 0 to 30°C (see Fig. 4.11). Also the θ_L label for the moist adiabat passing through the LCL equals this θ_w.

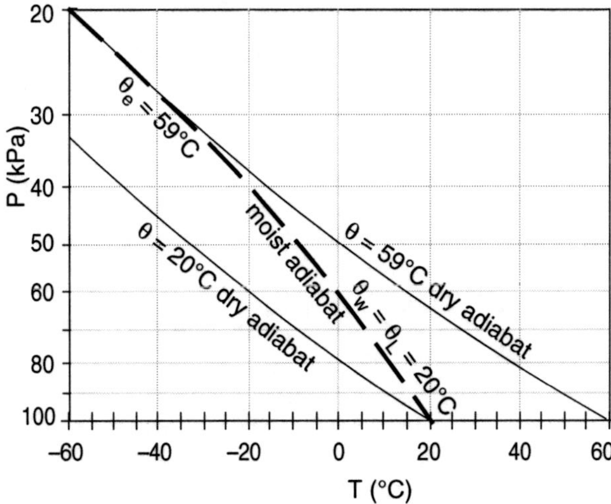

Figure 4.9
Comparison of θ_L, θ_w and θ_e values for the same moist adiabat.

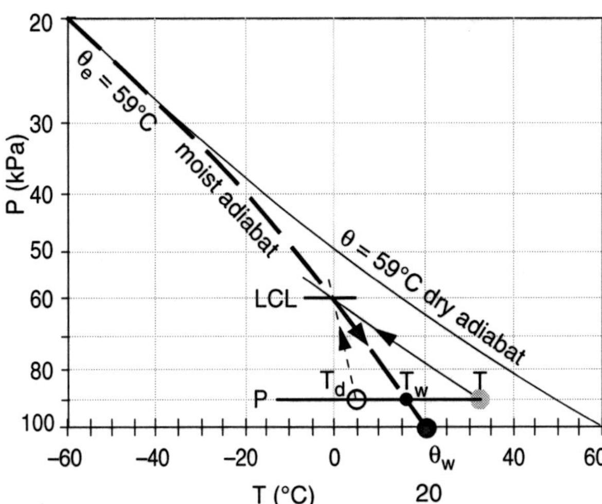

Figure 4.10
Thermo diagram showing how to use Normand's Rule to find wet-bulb potential temperature θ_w, and how it relates to θ_e.

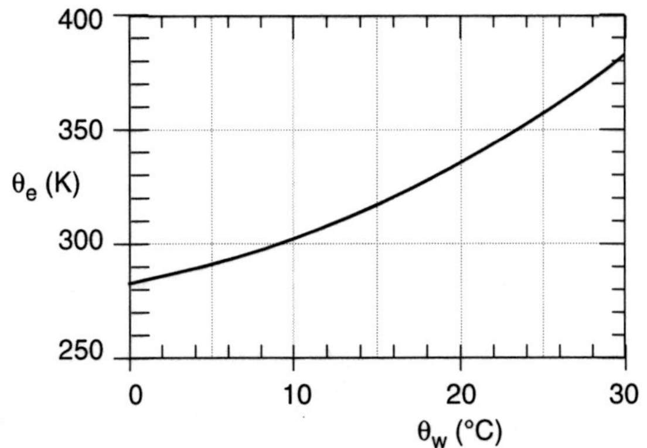

Figure 4.11
Approximate relationship between θ_w and θ_e.

Sample Application

Verify the labels on the moist adiabat that passes through the LCL in Fig. 4.10, given starting conditions $T = 33°C$ and $T_d = 5.4°C$ at $P = 90$ kPa.

Find the Answer

Given: $T = 33°C \approx 306K$, $T_d = 5.4°C \approx 278.4K$
Find: θ_e, θ_w and θ_L labels (°C) for the moist adiabat

First, find the initial θ, using eq. (3.12)

$\theta = T\cdot(P_o/P)^{0.28571} = (306K)\cdot(100kPa/90kPa)^{0.28571}$
$= 315.4K = 42.4°C$

Next, find the mixing ratio using eq. (4.1b) & (4.4):

$e = 0.6114kPa\cdot\exp[5423\cdot(1/273.15 - 1/278.4)] = 0.889kPa$
$r \approx (622g\ kg^{-1})\cdot(0.889kPa)/[90-0.889kPa] \approx 6.2g\ kg^{-1}$

Next, use eq. (3.2) to find $C_p = C_{pd}\cdot(1 + 1.84\cdot r)$

$C_p = (1004.67\ J\cdot kg^{-1}\cdot K^{-1})\cdot[1+1.84\cdot(0.0062\ g\ g^{-1})]$
$C_p = 1016.1\ J\cdot kg^{-1}\cdot K^{-1}$

Solve the more accurate version of eq. (4.40):

$\theta_e = (42.4°C) + \{(2500\ J\ g_{water}^{-1})\cdot(315.4K)/$
$[(1016.1\ J\cdot kg^{-1}\cdot K^{-1})\cdot(306K)]\}\cdot(6.2\ g_{water}\ kg_{air}^{-1})$
$= (42.4°C) + (2.536\ K\cdot kg_{air}\ g_{water}^{-1})\cdot(6.2\ g_{water}\ kg_{air}^{-1})$
$=$ **58.1°C** $= 331.1$ K

The approximate version of eq. (4.40) gives almost the same answer, and is much easier:

$\theta_e = (42.4°C) + (2.5\ K\cdot kg_{air}\ g_{water}^{-1})\cdot(6.2\ g_{water}\ kg_{air}^{-1})$
$=$ **57.9°C**

Eq. (4.43b) is a quadratic eq. that can be solved for θ_w. Doing this, and then plugging in $\theta_e = 331.1$ K gives:

$\theta_w \approx$ **19°C**. , which is the label on the moist adiabat.

Using $\theta_L \approx \theta_w$:

$\theta_L \approx$ **19°C**.

Check: Physics and units are reasonable.
Exposition: These values are within a couple degrees of the labels in Fig. 4.10. Disappointing that they aren't closer, but the θ_e results are very sensitive to the starting point.

Sample Application

For a moist adiabat of $\theta_w = 14°C$, find its θ_e.

Find the Answer

Given: $\theta_w = 14°C = 287$ K
Find: $\theta_e = ?$ K

First get r_s from Fig. 4.7 at $P = 100$ kPa and $T = 14°C$:

$r_s = 10\ g\ kg^{-1} = 0.010\ kg\ kg^{-1}$.

Next, use eq. (4.43a):

$\theta_e = (287K)\cdot\exp[2491(K\cdot kg_{air}\ kg_{vapor}^{-1})\cdot$
$(0.010\ kg_{vapor}\ kg_{air}^{-1}) / (287K)] =$ **313K**

Check: Units are reasonable. Agrees with Fig. 4.11.
Exposition: This $\theta_e = 40°C$. Namely, if a saturated air parcel started with $\theta_w = T = 14°C$, and then if all the water vapor condensed, the latent heat released would warm the parcel to $T = 40°C$.

Sample Application

Suppose a psychrometer at 100 kPa measures dry-bulb and wet-bulb temperatures of 30°C and 15°C. Use a thermo diagram to find the values of T_d and r ?

Find the Answer

Given: $P = 100$ kPa, $T = 30°C$, $T_w = 15°C$.
Find: $r = ?\ g\ kg^{-1}$, $T_d = ?°C$
Hint: We can use the opposite of Normand's rule.

We will learn more about thermo diagrams in the next chapter. So this exercise gives us a preview.

To use Normand's rule in reverse, follow a dry adiabat up from the starting dry-bulb temperature, and follow a saturated adiabat up from the starting wet-bulb temperature. Where they cross, follow the isohume down to the starting altitude to find the dew-point temperature, or follow the isohume up to read the isohume's mixing-ratio value.

Fig. 3.4 shows many dry adiabats, but not the one we want. So I interpolated between the 20 and 40°C dry adiabats, and copied and pasted the result (as the solid dark-orange line for θ = 30°C) onto copy of Fig. 4.7, as shown below. A magenta circle indicates the starting temperature and pressure.

Fig. 4.8 shows many saturated adiabats, but not the one we want. So I interpolated between the 10 and 20°C saturated adiabats, and copied and pasted the result into the figure below as the dark-orange dashed line for $\theta_w = 15°C$. Another magenta circle indicates the starting wet-bulb temperature and pressure.

Those two lines cross (see purple "X" in the figure) almost exactly on the $r =$ **5 g kg^{-1}** isohume (see black circle at top of diagram. Following that same isohume down to $P = 100$ kPa gives $T_d \approx$ **4°C** (circled in black).

Check: Physics and unit are reasonable.
Exposition: The "X" is the location of the lifting condensation level (LCL), which is about 68 kPa in this diagram.

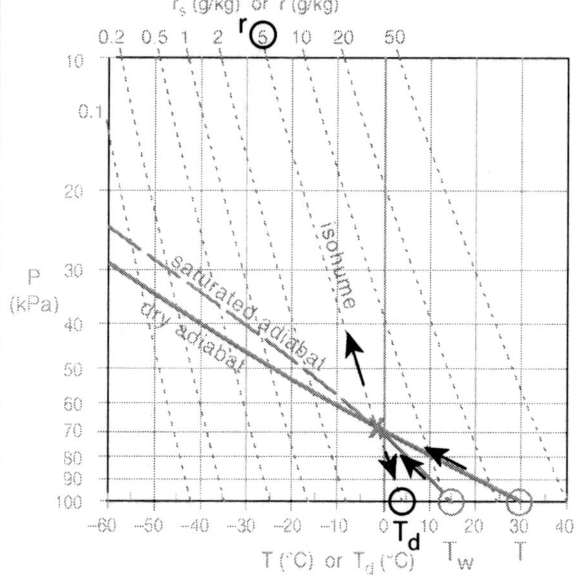

WATER BUDGET AT A FIXED LOCATION

Picture a cube of air at a fixed location relative to the ground (i.e., an **Eulerian** framework). Inflow and/or outflow of water (as vapor, liquid, or solid) can change the total water content r_T of air inside the hypothetical cube. As we did for heat, we can investigate each inflow and outflow process to determine which are significant. The insignificant ones include turbulent transport in the horizontal, molecular conduction (except close to the ground), and mean vertical advection (except in deep convective storms).

What remains is an approximate equation for the total water budget in an Eulerian framework:

•(4.44)

$$\frac{\Delta r_T}{\Delta t} = -\left[U \cdot \frac{\Delta r_T}{\Delta x} + V \cdot \frac{\Delta r_T}{\Delta y}\right] + \left(\frac{\rho_L}{\rho_d}\right)\frac{\Delta Pr}{\Delta z} - \frac{\Delta F_{z\ turb}(r_T)}{\Delta z}$$

storage *horiz. advection* *precipitation* *turbulence*

where precipitation is falling at rate Pr, horizontal advection is caused by wind components (U, V), the turbulent total-water flux (in kinematic units) is $F_{z\ turb}(r_T)$, and (Δx, Δy, Δz, Δt) are the Cartesian dimensions of the cube and the time interval, respectively.

Define **Standard Temperature and Pressure (STP)** as T = 0°C and P = 101.325 kPa. Using air density at STP and liquid water density (ρ_L = 1000 kg_{liq} m^{-3}) from Appendix B, you would find that ρ_L/ρ_d = 836.7 kg_{liq} kg_{air}^{-1}, but this value would increase for hypothetical cubes at higher altitudes where air density is smaller .

In the subsections that follow, we will focus on the individual terms in eq. (4.44).

Horizontal Advection

If more total water (vapor + suspended cloud droplets + suspended ice crystals) is blown out of the cube than blows in, then the total water will decrease during time interval Δt.

Precipitation

Some solid and liquid water particles are large enough that they are not suspended in, and do not move with, the mean winds. Instead, they fall through the air as precipitation. If all the liquid and solid precipitation were collected in a rain gauge and then melted to be all liquid, the resulting water depth is called the **liquid water equivalent**. Thus, 2 mm of liquid precipitation plus 5 cm of snow (which might reduce to 5 mm of water when melted) would give a total liquid water depth of 7 mm.

A SCIENTIFIC PERSPECTIVE• Reasoning

"Reasoning" is the method by which we try to explain how things work. Namely, we try to infer an explanation or a theory that is reasonable. Two types of reasoning are inductive and deductive reasoning.

Characteristics of **inductive reasoning**:
- Going from specific examples to general theories.
- The general theory may be likely, but not certain.
- Often requires a "leap of faith".
- Could apply to situations not yet encountered.
- Can be proven wrong by new contrary evidence.
- Analogous to extrapolation.

Characteristics of **deductive reasoning**:
- Applying a general rule to a specific application.
- Confidence that the result is correct and will work well in new situations if used within the bounds of the general rule.
- Is a sound tool used in engineering.
- Analogous to interpolation.

As a contrived example, what if you observed that as adults, your grandparent was 1.4 m tall, your parent was 1.6 m tall, and you are 1.8 m tall. You could deduce that you are 0.2 m taller than your parent. In fact, you would consider this a "truth" based on the logic of mathematics.

However, you could induce that your unborn child will be 2.0 m tall as an adult. Namely, you saw a pattern from past observations: each offspring was 0.2 m taller than its parent. So extrapolating to your future unborn child, you might expect the child to become 0.2 m taller than you.

Obviously this last prediction carries some uncertainty. You could reduce the uncertainty by collecting more evidence from other families, or by going further back in your family tree. However, many factors over which you have little control could alter the growth of your child — your prediction is uncertain.

This same procedure is done by scientists proposing new theories. Namely, gain confidence that your theory is "universal" in the sense that it works at other places and times in the universe. Such a goal motivated Isaac Newton to design and build telescopes, so he could test his theories of motion by looking at other planets in the universe. The evidence convinced scientists to accept Newton's hypotheses as "laws of motion", yet they were later proven wrong by Einstein.

Let's try some inductive reasoning now. Eq. (4.44) says that change of total water depends on horizontal advection, change of turbulent fluxes in the vertical, and other source terms. Eq. (3.51) said that change of temperature depends on horizontal advection, change of turbulent fluxes in the vertical, and other source terms. By inductive reasoning, we might anticipate that the change of momentum (i.e., wind speed) depends on horizontal advection, change of turbulent fluxes in the vertical, and other source terms. We will check this later, in the Forces & Winds chapter.

Sample Application

Precipitation falls out of the bottom of a cloud at rate 0.5 cm h^{-1}. At 200 m below cloud base, the precipitation rate is 0.4 cm h^{-1}. In this 200 m thick layer of air below cloud base, what is the total-water change rate? Assume ρ_{air} = 1 kg m^{-3}.

Find the Answer

Given: Pr_{bot} = 0.4 cm h^{-1} = 1.11x10^{-6} m s^{-1},
Pr_{top} = 0.5 cm h^{-1} = 1.39x10^{-6} m s^{-1},
ρ_{air} = 1 kg m^{-3}, and
Δz = 200 m

Find: $\Delta r_T/\Delta t$ = ? (g kg^{-1})/s.

Because no other processes are specified, let's neglect them for simplicity. Thus, we can apply eq. (4.44):

$$\Delta r_T/\Delta t = (\rho_L/\rho_d)\cdot(Pr_{top}-Pr_{bot})/(z_{top}-z_{bot}) =$$

$$= \frac{(1.0\times10^6 g_{water}/m^3)}{(1\ kg_{air}/m^3)}\cdot\frac{[(1.39-1.11)\times10^{-6} m/s]}{200m}$$

$$= \underline{0.0014\ (g\ kg^{-1})/s} = 5\ (g\ kg^{-1})/h$$

Check: Physics and units are reasonable.
Exposition: This is typical of **virga**, a weather element where some of the precipitation evaporates before reaching the ground. In virga, the evaporating rain makes the air more humid below cloud base.

If more precipitation falls out of the bottom of the air cube than is falling into the top, then this change of precipitation with altitude would tend to reduce the total water in the cube. Thus, the change of **precipitation rate** (Pr in mm h^{-1} or m s^{-1}) between the top and bottom of the cube of air is important.

Not all the precipitation will reach the ground, because some might evaporate on the way down. The precipitation rate (Pr) at the ground is given a special name: the **rainfall rate** (RR).

Moisture Flux at the Earth's Surface

In the previous chapter we discussed ways to estimate sensible and latent heat fluxes at the Earth's surface. But latent heat flux is tied to the movement of water molecules. Here we show how you can use knowledge of the latent heat flux to estimate the moisture flux at the Earth's surface.

Recall that vertical flux is the movement of something across a horizontal unit area per unit time. Let $\mathbb{F}_E$ be the **latent heat flux** in units of (J·m^{-2}·s^{-1} or W·m^{-2}) and $\mathbb{F}_{water}$ be the **vertical water-vapor flux** in units of (kg_{water}· m^{-2}·s^{-1}). The relationship between these fluxes is:

$$\mathbb{F}_{water} = \mathbb{F}_E / L_v \qquad \bullet(4.45)$$

or

$$\mathbb{F}_{water} = \rho_{air}\cdot(C_p/L_v)\cdot F_E \qquad (4.46)$$

or

$$\mathbb{F}_{water} = \rho_{air}\cdot\gamma\cdot F_E \qquad (4.47)$$

Constants and parameters in these equations are:
F_E = kinematic latent heat flux (K·m·s^{-1});
L_v = 2.5x10^6 J kg_{water}^{-1} = latent heat of vaporization;
γ = C_p/L_v = 0.4 (g_{water} kg_{air}^{-1})·K^{-1}
= **psychrometric constant.**

If you divide these equations by air density (ρ_{air}), then you can get the water flux in kinematic form:

$$F_{water} = \mathbb{F}_{water} / \rho_{air} \qquad (4.48a)$$

For example, eq. (4.45) becomes:

$$F_{water} = \mathbb{F}_E / (\rho_{air}\cdot L_v) = \gamma\cdot F_E \qquad (4.48b)$$

The advantage of **kinematic water flux** (F_{water}) is that its units are similar to (mixing ratio) x (wind speed). Thus, the units are (kg_{water} kg_{air}^{-1})·(m·s^{-1}).

When water from a puddle on the Earth's surface evaporates, the puddle depth decreases. The rate of this depth decrease (mm day^{-1}) is the **evaporation rate**, ($Evap$), which is related to the moisture flux:

$$Evap = \mathbb{F}_{water}/\rho_L = \mathbb{F}_E/(\rho_L\cdot L_v) = a\cdot\mathbb{F}_E \qquad (4.49)$$

$$Evap = (\rho_{air}/\rho_L)\cdot F_{water} = (\rho_{air}/\rho_L)\cdot\gamma\cdot F_E \qquad \bullet(4.50)$$

Sample Application

Suppose the latent heat flux is 250 W·m^{-2}. What are the values of water-vapor flux, kinematic water-vapor flux and evaporation rate? Assume ρ_{air} = 1 kg_{air} m^{-3}.

Find the Answer

Given: ρ_{air} = 1 kg_{air}· m^{-3}, $\mathbb{F}_E$ = 250 W·m^{-2}

Find: $\mathbb{F}_{water}$ = ? kg_{water}·m^{-2}·s^{-1}, $Evap$ = ? mm day^{-1},
F_{water} = ? (kg_{water}· kg_{air}^{-1})·(m^{-2}·s^{-1})

Apply eq. (4.45):

$\mathbb{F}_{water}$ = (250 J·m^{-2}·s^{-1})/(2.5x10^6 J kg^{-1})
= $\underline{0.0001\ kg}_{water}$·$m^{-2}$·$s^{-1}$

Apply eq. (4.48a):

F_{water} = (0.0001 kg_{water}·m^{-2}·s^{-1})/ (1 kg_{air} m^{-3})
= $\underline{0.0001}$ (kg_{water}· kg_{air}^{-1})·(m·s^{-1})

Apply eq. (4.49): $Evap = a\cdot\mathbb{F}_E$
= [0.0346 (m^2 W^{-1})·(mm day^{-1})]·(250 W m^{-2})
= $\underline{8.65\ mm\ day^{-1}}$

Check: Physics and units are reasonable. Namely, the first answer has units of water mass per area per time, the second is like a mixing ratio times velocity, and the third is rate of decreased of water depth in a puddle.
Exposition: If this evaporation rate continues day after day, crops could become desiccated.

For these equations, ρ_L = 1000 kg_{liq} m^{-3}, a = 0.0346 (m^2 W^{-1})·(mm day^{-1}), and a = 4.0x10^{-10} m^3·W^{-1}·s^{-1}.

For windy conditions, another way to estimate the water vapor flux (in kinematic units) is with a **bulk-transfer relationship** similar to eq. (3.35), such as

$$F_{water} = C_H \cdot M \cdot \left(r_{sfc} - r_{air} \right) \qquad (4.51)$$

where the wind speed at 10 m above the surface is M, and the water-vapor mixing ratio in the air at 2 m above the surface is r_{air}. The **bulk-transfer coefficient** for water vapor is roughly the same as the one for heat, C_H, which ranges between 2x10^{-3} (for smooth surfaces) to 2x10^{-2} (for rough surfaces).

In calmer conditions with sunny skies, convective thermals of warm rising air can form, which are effective at transporting moisture in the vertical. The resulting kinematic water-vapor flux is:

$$F_{water} = b_H \cdot w_B \cdot \left(r_{sfc} - r_{ML} \right) \qquad (4.52)$$

for a mid-mixed layer mixing ratio of r_{ML} and a **convective heat-transport coefficient** of b_H = 5x10^{-4} (dimensionless). Eq. (3.38) in the previous chapter gives the expression for buoyancy-velocity scale w_B.

While the two equations above seem physically reasonable, they have a problem in that the mixing ratio at the surface r_{sfc} (literally at the Earth's surface skin) is an abstract concept that is not measurable. For the special case of a river or lake surface or rain-saturated ground, scientists often approximate $r_{sfc} \approx r_s(T_{sfc})$; namely the surface mixing ratio equals the saturation mixing ratio r_s (eq. 4.5) for air at a temperature T_{sfc} equal to that of the surface skin.

Moisture Transport by Turbulence

Recall from the Thermodynamics chapter that turbulence is the quasi-random movement and mixing of air parcels by swirls and eddies — analogous to the effects of an egg beater. By mixing together dry and moist air parcels, turbulence causes a net moisture transport — namely, a **moisture flux**. If the moisture flux out of the top of a layer of air is different than the moisture flux into the bottom, then there is a **moisture flux divergence** $\Delta F_{z\ turb}(r_T)/\Delta z$ that changes the total water content in the layer.

For hot sunny days with light winds, rising thermals create a mixed layer of depth z_i, within which:

$$\frac{\Delta F_{z\ turb}(r_T)}{\Delta z} = \frac{F_{z\ turb\ at\ zi}(r_T) - F_{z\ turb\ at\ surface}(r_T)}{z_i - z_{surface}} \qquad (4.53)$$

where eq. (4.53) says that the local flux divergence equals the average flux divergence across the whole turbulent mixed layer (ML).

Sample Application

A cold front recently passed over your farm field during which it rained hard, but now it is windy with M = 15 m s^{-1}. Air behind the cold front is relatively dry (r_{2m} = 5 g kg^{-1}), but the ground surface is still relatively warm (14°C). Assume your farm is near sea level. What kinematic water-vapor flux do you expect?

Find the Answer

Given: M = 15 m s^{-1}, r_{air} = 5 g kg^{-1} , T_{sfc} = 14°C
Find: F_{water} = ? (kg_{water} kg_{air}^{-1})·(m·s^{-1})

The farm field is probably not perfectly smooth, so let's assume $C_H \approx$ 5x10^{-3} (dimensionless). But assume the recent rain left the soil nearly saturated with water.

To apply eq. (4.51), we need to estimate $r_{sfc} \approx r_s(T_{sfc})$ = r_s (14°C) = 10.0 g kg^{-1} from eqs (4.1) & (4.5). Thus, eq. (4.51) is:

F_{water} = (5x10^{-3})·(15 m s^{-1})·(10.0 – 5 g kg^{-1})
= 0.375 (g_{water}·kg_{air})·(m·s^{-1})
= <u>0.000375</u> (kg_{water} kg_{air}^{-1})·(m·s^{-1})

Check: Physics and units are reasonable.
Exposition: If you apply eq. (4.50) you find this is associated with an evaporation rate of 40 mm day^{-1}.

Sample Application

What is the value of F_{water} on a sunny day with no winds? Assume mixed-layer depth z_i = 3 km, θ_{ML} = 290 K, and θ_{sfc} = 320 K. The air and ground are somewhat dry, so r_{sfc} = 5 g kg^{-1}, and r_{ML} = 3 g kg^{-1}.

Find the Answer

Given: r_{sfc} = 5 g kg^{-1}, r_{ML} = 3 g kg^{-1}, z_i = 3 km,
θ_{sfc} = 320 K, θ_{ML} = 290 K
Find: F_{water} = ? (kg_{water} kg_{air}^{-1})·(m·s^{-1})

Sunny, calm conditions suggests a convective mixed layer, so we can apply eq. (4.52). The temperature info in this sample application is the same as for a sample application in the Surface Fluxes section of Chapter 3, where we found from eq. (3.38) that $w_B \approx$ 55.1 m s^{-1}. Thus, eq. (4.52) is:

F_{water} = (5x10^{-4})·(55.1 m s^{-1})·(5 – 3 g kg^{-1})
= 0.055 (g_{water}·kg_{air})·(m·s^{-1})
= <u>5.5x10^{-5}</u> (kg_{water} kg_{air}^{-1})·(m·s^{-1})

Check: Physics and units are reasonable.
Exposition: If you apply eq. (4.50) you find this is associated with an evaporation rate of 5.83 mm day^{-1} at sea level.

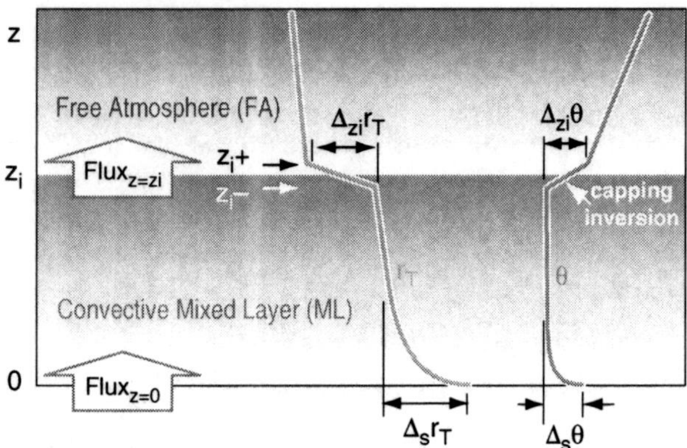

Figure 4.12
Idealized vertical profiles of potential temperature $\Delta\theta$ and total water mixing ratio Δr_T across the turbulent boundary layer (known as a mixed layer; shaded brown in this figure). The differences (jumps: Δ) of these values at the top (z_i) and bottom (s) of the mixed layer can be used to estimate turbulent vertical fluxes in those regions.

Sample Application
Winds are light and solar heating is strong, creating a convective mixed layer (ML) having depth 1.5 km and convective transport of $b_H \cdot w_B$ = 0.03 m s^{-1}. The potential temperature jump near the surface is 7°C while the corresponding jump near the top of the ML is 4°C. The jumps of total water mixing ratio are 6 and –3 g kg^{-1}, respectively. (a) Find the turbulent flux divergence of r_T. (b) If that was the only process active, then at what rate does total water increase in the ML?

Find the Answer
Given: z_i = 1.5 km = 1500 m, $b_H \cdot w_B$ = 0.03 m s^{-1}.
$|\Delta_s\theta|$ = 7°C, $|\Delta_{zi}\theta|$ = 4°C,
$|\Delta_s r_T|$ = 6 g kg^{-1}, $\Delta_{zi} r_T$ = –3 g kg^{-1},
No other processes acting.
Find: (a) $\Delta F_{z\ turb}(r_T)/\Delta z$ = ? (g kg^{-1})/s,
(b) $\Delta r_T/\Delta t$ = ? (g kg^{-1})/s,

(a) Apply eq. (4.56):

$$\frac{\Delta F_{z\ turb}(r_T)}{\Delta z} = -\frac{b_H \cdot w_B}{z_i}\left[0.2 \cdot \left(\Delta_{z_i} r_T\right) \cdot \left|\frac{\Delta_s\theta}{\Delta_{z_i}\theta}\right| + \left|\Delta_s r_T\right|\right]$$

$$= -\frac{(0.03\text{m/s})}{(1500\text{m})}\left[0.2 \cdot \left(-3\frac{\text{g}}{\text{kg}}\right) \cdot \left|\frac{7°\text{C}}{4°\text{C}}\right| + \left(6\frac{\text{g}}{\text{kg}}\right)\right]$$

= **-9.9x10^{-5}** (g kg^{-1})/s

(b) Apply eq. (4.44):
$\Delta r_T/\Delta t$ = – $\Delta F_{z\ turb}(r_T)/\Delta z$ = **9.9x10^{-5}** (g kg^{-1})/s
= 0.36 (g kg^{-1})/hour

Check: Physics and units are reasonable.
Exposition: Entrainment is bringing dry air down into the mixed layer [$F(r_T)$ = –], but that drying rate is offset by moisture flux entering the ML from the Earth's surface, leading to increase of r_T with time.

At the top of the ML, entrainment brings in warmer drier air from just above the ML top (at z_i+) and mixes it with air just below the ML top (at z_i–), see Fig. 4.12. That entrainment is driven by thermals associated with the effective surface heat flux F_H (in kinematic units of K m s^{-1}) and is modulated by the strength of the capping temperature inversion $\Delta_{zi}\theta$, where $\Delta_{zi} = (\ )_{zi+} - (\ )_{zi-}$. Thus,

$$\frac{\Delta F_{z\ turb}(r_T)}{\Delta z} = -\frac{\left[0.2 \cdot (\Delta_{z_i} r_T / \Delta_{z_i}\theta) \cdot F_H\right] + F_{water}}{z_i} \tag{4.54}$$

Also in eq. (4.54) is F_{water}, the effective surface water-vapor flux [units kinematic (kg$_{water}$ kg$_{air}^{-1}$)·(m s^{-1})], which can be rewritten using the psychrometric constant [γ = 0.4 (g$_{water}$ kg$_{air}^{-1}$)·K^{-1}] in terms of the latent heat flux F_E (eq. 4.48), at the surface:

$$\frac{\Delta F_{z\ turb}(r_T)}{\Delta z} = -\frac{0.2 \cdot (\Delta_{z_i} r_T / \Delta_{z_i}\theta) \cdot F_H + \gamma \cdot F_E}{z_i} \tag{4.55}$$

The effective surface sensible and latent heat fluxes in the eq. above can be parameterized using the convective transport coefficient (b_H, see the Thermodynamics chapter) and the buoyancy velocity scale w_B (see eq. 3.38):

$$\frac{\Delta F_{z\ turb}(r_T)}{\Delta z} = -\frac{b_H \cdot w_B}{z_i}\left\{\left[0.2 \cdot \left(\Delta_{z_i} r_T\right) \cdot \left|\frac{\Delta_s\theta}{\Delta_{z_i}\theta}\right|\right] + \left|\Delta_s r_T\right|\right\} \tag{4.56}$$

where $\Delta_s(\ ) = (\ )_{surface} - (\ )_{mid\ mixed\text{-}layer}$

Any one of the four eqs. above can be used to find moisture-flux divergence for $0 < z < z_i$. Above the convective ML (at $z > z_i$) where turbulence is weak or nonexistent:

$$\frac{\Delta F_{z\ turb}(r_T)}{\Delta z} \approx 0 \tag{4.57}$$

During night, turbulence is often weak at most heights (except very close to the surface), allowing you to use eq. (4.57) as a reasonable approximation at z > 0.

Next, consider a different type of mixed layer — one that is mixed mechanically by wind shear on windy days (not by thermals on calm sunny days). Further, suppose that there is such a strong capping temperature inversion at the top of this layer (at height z_i) that there is negligible entrainment (thus near zero moisture flux) into the top of this ML. For this case, the turbulent moisture flux divergence is driven solely by F_{water}, the flux at the Earth's surface: $\Delta F_{z\ turb}(r_T)/\Delta z \approx -F_{water}/z_i$. For example, in winter a humid ML might lose moisture due to condensation of water onto a cold snow-covered landscape. Or a cool dry ML might gain moisture by evaporation from a warm lake.

Thunderstorms are deep convective storms that turbulently mix air over the whole depth of the troposphere. The fuel for such storms is humidity in the prestorm mixed layer. When this ML air is drawn into the storm via the storm updrafts, water vapor condenses and falls out as rain. Meanwhile, drier air from the mid-troposphere is often brought down towards the surface to create a new, but drier, post-storm atmospheric boundary layer (ABL). Some of the falling rain can evaporate into this drier ML, increasing the humidity there.

HUMIDITY INSTRUMENTS

Instruments that measure humidity are called **hygrometers**. Don't confuse the word with "hydrometers", which are used to measure specific gravity of fluids such as battery acid.

Dew-point hygrometers, also known as **chilled-mirror hygrometers**, reflect a beam of light off a tiny metal mirror. When the mirror is cooled to the dew-point temperature, dew forms on the mirror and the light beam scatters instead of reflecting into a detector. Electronics in the instrument cool or heat the mirror to maintain the surface precisely at the dew point temperature, which is provided as an output. These are accurate instruments with relatively slow response. For cold temperatures and low humidities, **frost-point hygrometers** are used instead.

Hair hygrometers use organic fibers such as long hairs, anchored at one end and attached at the other end by amplifying levers to a dial that reads out relative humidity. As the RH increases, the hairs get longer, causing the dial to turn. These are inaccurate, but are inexpensive and are the most common hygrometers for home use. Other hygrometers use other materials that also change their dimensions when they absorb water molecules.

Psychrometers are instruments with two liquid-in-glass thermometers attached to a board or frame. The bulb of one thermometer is surrounded by a sleeve or wick of cloth that is saturated with distilled water, while the other bulb remains dry. After both thermometers are actively ventilated [by whirling the instrument through the air on a hand-held axel (**sling psychrometer**), or by using a spring or electrically driven fan to blow air past the thermometers (**aspirated psychrometer**)], the two thermometers are read to give the **wet and dry-bulb temperatures**. The wet-bulb is cooler than the dry, because of the latent heat absorbed when water evaporates. This thermodynamic information can be used with psychrometric tables or charts (Figs. 4.4 & 4.5) to determine the humidity. These instruments are extremely slow response, but relatively simple. Modern psychrometers replace the liquid-in-glass thermometers with electronic thermometers such as thermistors.

In old **radiosondes** (balloon-borne weather instruments), the electrical resistance across a carbon-coated glass slide was measured. In more humid air, this **carbon-film hygrometer** becomes more resistive. Modern radiosondes often measure the capacitance across a very thin dielectric plastic that is coated on both sides with a porous metallic grid. Both approaches are small and light enough to be carried aloft, but both sensors can be easily contaminated by chemical vapors that change their electrical properties.

Microwave refractometers draw air into a small chamber filled with microwaves. The refraction (bending) of these microwave beams depends on humidity (see the radar section of the Satellites & Radar chapter), and can be measured. These are very fast-response sensors.

Spectral absorption hygrometers, also known as **optical hygrometers**, transmit frequencies of electromagnetic radiation that are strongly absorbed by water vapor. By passing the beam of radiation across a short path of air to a detector, the amount of attenuation can be measured to allow calculation of the absolute humidity. One such instrument, the **Lyman-alpha hygrometer,** uses ultraviolet light of wavelength 0.121567 µm, corresponding to an absorption/emission line of hydrogen. Another, the **krypton hygrometer**, uses emissions at 0.12358 µm, generated by glow tube filled with the noble gas krypton. Other instruments use absorption of infrared light (**infrared hygrometers**). These are all fast-response instruments. See the Satellites & Radar chapter for absorption spectra across the atmosphere.

Some **lidars** (laser radars) have been developed to transmit two neighboring wavelengths of electromagnetic radiation, one of which is affected by water vapor and the other which is not. Such **differential absorption lidars** (**DIAL**) can remotely measure humidity along vertical or slant paths, and can scan the atmosphere to measure the humidity in a volume or in a plane.

Weather radars and other microwave profilers can be used to measure profiles of humidity in the atmosphere, because the speed and/or polarity of microwaves through air depends on humidity.

Some sensors measure path-averaged humidity. One example is the **water-vapor channel on weather satellites**, which measures infrared emissions from water vapor in the air. As discussed in the Satellites & Radar chapter, such emissions come from a layer of air several kilometers thick in the top third of the troposphere. These instruments have the advantage of remotely sampling the atmosphere at locations that are difficult to reach otherwise, such as over the oceans. A disadvantage is that they have difficulty seeing through clouds.

Transmissions from **Global Positioning System (GPS)** satellites are slightly delayed or refracted by humidity along the path of the beam through the atmosphere. However, data from many such crossing beams from the constellation of GPS satellites can be computationally inverted to yield vertical profiles of humidity, similar to the medical X-ray tomography methods used for brain scans.

REVIEW

Many different variables can be used as measures of water-vapor content in air, including: water-vapor partial pressure (known as vapor pressure in meteorology), mixing ratio, absolute humidity, specific humidity, relative humidity, dew-point temperature, lifting condensation level, and wet-bulb temperature. Some of these variables can be used to quantify total water (vapor + liquid + ice).

Formulas exist to convert between these different variables. Some of these humidity variables can be easily measured by instruments called hygrometers, others are useful in conservation equations, while others are commonly known by the general public.

The amount of water actually being held in the air might be less than the maximum amount that could be held at equilibrium, where this equilibrium value is known as the saturation humidity. Cooler air can hold less water vapor at saturation than warmer air — a fact that is critical in understanding why clouds and storms form in rising, cooling air.

By following such a rising or sinking air parcel we can write a Lagrangian moisture budget and calculate adiabatic temperature changes for air that is saturated (foggy or cloudy). A graphical description of this process can be represented as moist adiabats on a thermo diagram. Saturated rising air does not cool as rapidly with altitude as dry air. Thermo diagrams can also include isohume lines that relate humidity state to temperature and pressure.

An alternative frame of reference is Eulerian, which is fixed relative to a location on the Earth's surface. To forecast humidity in such a fixed frame, we can account for the advection of moisture by the winds, the fluxes of moisture due to turbulence, and effects due to precipitation. Clouds and precipitation are discussed in subsequent chapters.

HOMEWORK EXERCISES

Broaden Knowledge & Comprehension

B1. Use the internet to acquire the current humidity at a weather station near you. What type of humidity variable is it?

B2. Use the internet to acquire a current weather map of humidity contours (isohumes) for your region. Print this map and label regions of humid and dry air.

B3. At the time this chapter was written, a web page was available from the National Weather Service in El Paso, Texas, that could convert between different weather variables (search on "El Paso weather calculator"). How do the formulas for humidity on this web page compare to the ones in this chapter?

B4. Use your internet search engine to find additional "weather calculators" that can convert between different units, other than the calculator mentioned in the previous problem. Which one do you like the best? Why?

B5. Use the internet to acquire the company names and model numbers of at least two different instruments for each of 3 different methods for sensing humidity, as was discussed on the previous page.

B6. Two apparent-temperature indices (humidex and heat index) describing heat stress or summer discomfort were presented in the Thermodynamics chapter. Use the internet to acquire journal articles or other information about any two additional indices from the following list:

- **apparent temperature**,
- **discomfort index**,
- **effective temperature**,
- **humisery**,
- **humiture**,
- **index of thermal stress**.
- **livestock weather safety index**,
- **summer simmer index**,
- **temperature-humidity index** (THI),
- **wet-bulb globe temperature**,

B7. Use the internet to acquire a weather map for your region showing isohumes (either at the surface, or at 85 or 70 kPa). For one Eulerian location chosen by your instructor, use the winds and horizontal humidity gradient to calculate the horizontal moisture advection. State if this advection would cause the air to become drier or more humid. Also, what other factors in the Eulerian water balance equation could counteract the advection (by adding or removing moisture to the Eulerian volume)?

B8. A **meteogram** is a graph of a weather variable (such as humidity or temperature) along the vertical axis as a function of time along the horizontal axis. Use the internet to either acquire such a humidity meteogram for a weather station near your location, or create your own meteogram from a sequence of humidity observations reported at different times from a weather station.

B9. Use the internet to search on "**upper air sounding**", where a sounding is a plot of weather variables vs. height or pressure. Some of these web sites allow you to pick the upper-air sounding station of interest (such as one close to you), and to pick the type of sounding plot.

There are several different types of thermo diagram frameworks for plotting soundings, and so far we discussed only one (called an Emagram). We will learn about the other thermo diagrams in the next chapter -- for example, the Stuve diagram looks very similar to an Emagram. Find a web site that allows you to view and plot an emagram or Stuve for a location near you.

Apply

A1. Compare the saturation vapor pressures (with respect to liquid water) calculated with the Clausius-Clapeyron equation and with Tetens' formula, for T (°C):

a. 45 b. 40 c. 35 d. 30 e. 25 f. 20 g. 15
h. 10 i. 5 j. 0 k. –5 l. –10 m. –15 n. –20

A2. Calculate the saturation vapor pressures with respect to both liquid water and flat ice, for T (°C) =

a. –3 b. –6 c. –9 d. –12 e. –15 f. –18
g. –21 h. –24 i. –27 j. –30 k. –35 l. –40

A3. Find the boiling temperature (°C) of pure water at altitudes (km) of:

a. 0.2 b. 0.4 c. 0.6 d. 0.8 e. 1.2 f. 1.4
g. 1.6 h. 1.8 i. 2.2 j. 2.4 k. 2.6 l. 2.8

A4. Calculate the values of e_s (kPa), r (g kg^{-1}), q (g kg^{-1}), ρ_v (g m^{-3}), RH (%), T_d (°C), LCL (km), T_w (°C), r_s (g kg^{-1}), q_s (g kg^{-1}), and ρ_{vs} (g m^{-3}), given the following atmospheric state:

	P (kPa)	T (°C)	e (kPa)
a.	100	45	5
b.	90	35	2
c.	80	25	1
d.	70	15	0.8
e.	80	5	0.5
f.	90	5	0.2
g.	80	15	1
h.	100	25	2.5

A5(§). Some of the columns in Table 4-1 depend on ambient pressure, while others do not. Create a new version of Table 4-1 for an ambient pressure P (kPa) of

a. 95 b. 90 c. 85 d. 80 e. 75 f. 70 g. 65
h. 60 i. 55 j. 50 k. 45 l. 40 m. 35 n. 30

A6. Given the following initial state for air outside your home. If your ventilation/heating system brings this air into your home and heats it to 22°C, what is the relative humidity (%) in your home? Assume that all the air initially in your home is replaced by this heated outside air, and that your heating system does not add or remove water.

T(°C), RH(%)	T(°C), RH(%)
a. 15, 80	g. 5, 90
b. 15, 70	h. 5, 80
c. 15, 60	i. 5, 70
d. 10, 80	j. –5, 90
e. 10, 70	k. –5, 80
f. 10, 60	m. –5, 70

A7. Given the temperatures and relative humidities of the previous exercise, what is the mixing ratio value (g kg^{-1}) for this "outside" air?

A8. For air with temperature and dew-point values given below in °C, find the LCL value (km).

T, T_d	T, T_d
a. 15, 12	g. 5, 4
b. 15, 10	h. 5, 0
c. 15, 8	i. 5, –5
d. 10, 8	j. 20, 15
e. 10, 5	k. 20, 10
f. 10, 2	m. 20, 5

A9. Given the following dry and wet-bulb temperatures, use equations (not graphs or diagrams) to calculate the dew-point temperature, mixing ratio, and relative humidity. Assume P = 95 kPa.

T, T_w	T, T_w
a. 15, 14	g. 5, 4
b. 15, 12	h. 5, 3
c. 15, 10	i. 5, 2
d. 10, 9	j. 20, 18
e. 10, 7	k. 20, 15
f. 10, 5	m. 20, 13

A10. Same as the previous exercise, but you may use the psychrometric graphs (Figs. 4.4 or 4.5).

A11. Given the following temperatures and dew-point temperatures, use the equations of Normand's Rule to calculate the wet-bulb temperature.

T, T_d	T, T_d
a. 15, 12	g. 5, 4
b. 15, 10	h. 5, 0
c. 15, 8	i. 5, –5
d. 10, 8	j. 20, 15
e. 10, 5	k. 20, 10
f. 10, 2	m. 20, 5

A12. Same as the previous exercise, but you may use a thermo diagram to apply Normand's rule.

A13. For air at sea level, find the total-water mixing ratio for a situation where:

a. $T = 3°C$, $r_L = 3$ g kg^{-1}
b. $r = 6$ g kg^{-1}, $r_L = 2$ g kg^{-1}
c. $T = 0°C$, $r_L = 4$ g kg^{-1}
d. $T = 12°C$, $r_L = 3$ g kg^{-1}
e. $r = 8$ g kg^{-1}, $r_L = 2$ g kg^{-1}
f. $r = 4$ g kg^{-1}, $r_L = 1$ g kg^{-1}
g. $T = 7°C$, $r_L = 4$ g kg^{-1}
h. $T = 20°C$, $r_L = 5$ g kg^{-1}

A14. Given air with temperature (°C) and total water mixing ratio (g kg^{-1}) as given below. Find the amount of liquid water suspended in the air: r_L (g kg^{-1}). Assume sea-level.

T, r_T	T, r_T
a. 15, 16	g. 25, 30
b. 15, 14	h. 25, 26
c. 15, 12	i. 25, 24
d. 10, 12	j. 20, 20
e. 10, 11	k. 20, 18
f. 10, 10	m. 20, 16

A15. If the mixing ratio is given below in (g kg^{-1}), then use equations (not figures or graphs) find the dew point (°C), assuming air at $P = 80$ kPa.

a. 28 b. 25 c. 20 d. 15 e. 10 f. 5 g. 3
h. 2 i. 1 j. 0.5 k. 0.3 l. 0.2 m. 0.1 n. 0.05

A16. Same as the previous exercise, but you may use Fig. 4.7.

A17. Given an air parcel starting at 100 kPa with dewpoint (°C) given below, use Fig. 4.7 to find the parcel's final dewpoint (°C) if it rises to a height where $P = 60$ kPa.

a. 40 b. 35 c. 30 d. 25 e. 20 f. 15 g. 10
h. 5 i. 0 j. –5 k. –10 l. –15 m. –20 n. –25

A18. What is the value of relative humidity (%) for air with the following state:

T (°C), T_d (°C), P (kPa)	T (°C), T_d (°C), P (kPa)
a. 15, 10, 90	g. 25, 20, 100
b. 15, 10, 90	h. 25, 15, 100
c. 15, 5, 80	i. 25, 10, 90
d. 10, 0, 80	j. 20, 15, 90
e. 10, 0, 70	k. 20, 10, 80
f. 10, –5, 70	m. 20, 5, 70

A19. Calculate the saturated adiabatic lapse rate (°C km^{-1}) at the temperatures and pressures given below. Use the equations, not the thermo diagram.

T (°C), P (kPa)	T (°C), P (kPa)
a. 15, 90	g. 25, 100
b. 15, 90	h. 25, 100
c. 15, 80	i. 25, 90
d. 10, 80	j. 20, 90
e. 10, 70	k. 20, 80
f. 10, 70	m. 20, 70

A20. Same as the previous exercise, but use equations to find the saturated adiabatic lapse rate as a change of temperature with pressure (°C kPa^{-1}).

A21. For air parcels with initial state as given below, use the thermo diagram (Fig. 4.8) to find the final air-parcel temperature after it is lifted to an altitude where $P = 50$ kPa. Assume the air parcels are saturated at all times.

T (°C), P (kPa)	T (°C), P (kPa)
a. 15, 80	g. 25, 90
b. 15, 70	h. 25, 90
c. 15, 60	i. 25, 80
d. 10, 70	j. 20, 80
e. 10, 60	k. 20, 70
f. 10, 60	m. 20, 60

A22. For air parcels with initial state as given in the previous exercise, use the thermo diagram (Fig. 4.8) to find the final air-parcel temperature after it is lowered to an altitude where $P = 100$ kPa. Assume the air parcels are saturated at all times.

A23. Using Fig. 4.8 and other figures for dry adiabats in the Thermodynamics chapter, determine the values of the liquid-water potential temperature and equivalent potential temperature for the initial air parcel of exercise A21.

A24. Same as the previous exercise, except use equations instead of the thermo diagram.

A25. Given an air parcel that starts at a height where $P = 100$ kPa with $T = 25°C$ and $r = 12$ g kg^{-1} (i.e., it is initially unsaturated). After rising to its final height,

it has an r_L (g kg^{-1}) value listed below. Assuming no precipitation falls out, find the final value for r (g kg^{-1}) for this now-saturated air parcel.
a. 0.5 b. 1 c. 1.5 d. 2 e. 2.5 f. 3 g. 3.5
h. 4 i. 4.5 j. 5 k. 5.5 l. 6 m. 6.5 n. 7

A26. Imagine a horizontally uniform wind given below, which is blowing in to the west side of a fixed cubic domain and blowing out of the east side. The cube is 200 km on each side. The total water mixing ratio for (incoming, outgoing) air is r_T = (12, 8) g kg^{-1}. What is the rate of change of r_T inside the volume due to this advection?
a. 2 b. 4 c. 5 d. 7 e. 10 f. 12 g. 15
h. 18 i. 20 j. 21 k. 23 l. 25 m. 27 n. 30

A27. Imagine a fixed cube of air 200 km on each side. Precipitation is falling at rate 4 mm h^{-1} into the top of this volume, and is falling out of the bottom of the volume at the rate (mm h^{-1}) given below. What is the rate of change of r_T inside the volume due to this precipitation gradient?
a. 6 b. 5.5 c. 5 d. 4.5 e. 4.3 f. 4.2 g. 4.1
h. 4 i. 3.9 j. 3.8 k. 3.7 l. 3.5 m. 3 n. 2.5

A28. Given below the value of latent heat flux (W·m^{-2}) at the surface. Find the kinematic value of latent flux (K·m s^{-1}), the vertical flux of water vapor (kg$_{water}$·m^{-2}·s^{-1}), the vertical flux of water vapor in kinematic form (kg$_{water}$ kg$_{air}$$^{-1}$)·(m s^{-1}), and the evaporation rate (mm d^{-1}).
a. 100 b. 150 c. 200 d. 250 e. 300 f. 350
g. 80 h. 75 i. 70 j. 60 k. 50 l. 40 m. 25

A29. For windy, overcast conditions, estimate the kinematic latent heat flux at the surface, assuming C_H = 5x10^{-3}.

	M (m s^{-1})	r_{sfc} (g kg^{-1})	r_{air} (g kg^{-1})
a.	2	25	10
b.	2	20	10
c.	2	15	10
d.	5	25	10
e.	5	20	10
f.	5	15	10
g.	12	25	10
h.	12	20	10
i.	12	15	10

A30. For sunny, free-convective conditions, estimate the kinematic latent heat flux at the surface.

	w_B (m s^{-1})	r_{sfc} (g kg^{-1})	r_{ML} (g kg^{-1})
a.	5	25	10
b.	5	20	10
c.	5	15	10
d.	12	25	10
e.	12	20	10
f.	12	15	10
g.	25	25	10
h.	25	20	10
i.	25	15	10

A31. Suppose the atmospheric mixed layer (ML) is as sketched in Fig. 4.12, having $|\Delta_s r_T|$ = 8 g kg^{-1}, $|\Delta_s\theta|$ = 5°C, $\Delta_{zi} r_T$ (g kg^{-1}) as given below, and $|\Delta_{zi}\theta|$ = 3°C. The mean winds are calm for this daytime convective boundary layer, for which $b_H w_B$ = 0.02 m s^{-1}. What is the rate of change of total water mixing ratio in the ML due to the turbulence?
a. –10 b. –8 c. –5 d. –3 e. –2 f. –1 g. 0
h. +1 i. 2 j. 4 k. 6 l. 7 m. 9 n. 12

Evaluate & Analyze

E1. When liquid water evaporates into a portion of the atmosphere containing dry air initially, water vapor molecules are added into the volume as quantified by the increased vapor pressure e. Does this mean that the total pressure increases because it has more total molecules? If not, then discuss and justify alternative outcomes.

E2(§). a. Make your own calculations on a spreadsheet to re-plot Fig. 4.2 (both the bottom and top graphs).
b. What factor(s) in the equations causes the vapor pressure curves to differ for the different phases of water?

E3. Suppose that rain drops are warmer than the air they are falling through. Which temperature should be used in the Clausius-Clapeyron equation (T_{air} or T_{water})? Why? (For this exercise, neglect the curvature of the rain drops.)

E4. Notice the column labels in Table 4-1. In that table, why are the same numbers valid for (T, r_s) and for (T_d, r)?

E5(§). a. Plot a curve of boiling temperature vs. height over the depth of the troposphere.
b. The purpose of pressure cookers is to cook boiled foods faster. Plot boiling temperature vs. pressure for pressures between 1 and 2 times $P_{sea\ level}$.

E6(§). Plot r_s vs. T and q_s vs. T on the same graph. Is it reasonable to state that they are nearly equal? What parts of their defining equations allow for this characteristic? For what situations are the differences between the mixing ratio and specific humidity curves significant?

E7(§). On a graph of T vs. T_d, plot curves for different values of RH.

E8. Why is it important to keep thermometers dry (i.e., placing them in a ventilated enclosure such as a **Stevenson screen**) when measuring outside air temperature?

E9(§). For any unsaturated air-parcel, assume you know its initial state (P, T, r). Recall that the LCL is the height (or pressure) where the following two lines cross: the dry adiabat (starting from the known P, T) and the isohume (starting from the known P, r). Given the complexity of the equations for the dry adiabat and isohume, it is surprising that there is such a simple equation (4.16a or b) for the LCL.

Confirm that eq. (4.16) is reasonable by starting with a variety of initial air-parcel states on a thermo diagram, lifting each one to its LCL, and then comparing this LCL with the value calculated from the equation for each initial parcel state. Comment on the quality of eq. (4.16).

E10. Some of the humidity variables [vapor pressure, mixing ratio, specific humidity, absolute humidity, relative humidity, dew-point temperature, LCL, and wet-bulb temperature] have maximum or minimum limits, based on their respective definitions. For each variable, list its limits (if any).

E11. A **swamp cooler** is a common name for an air-conditioning system that lowers the air temperature to the wet-bulb temperature by evaporating liquid water into the air. But this comes with the side-effect of increasing the humidity of the air. Consider the humidex as given in the Thermodynamics chapter, which states that humid air can feel as uncomfortable as hotter dry air. Suppose each cell in Table 3-5 represents a different initial air state. For which subset of cells in that table would a swamp cooler take that initial air state and change it to make the air feel cooler.

E12. We know that $T_w = T$ for saturated air. For the opposite extreme of totally dry air ($r = 0$), find an equation for T_w as a function of (P, T). For a few sample initial conditions, does your equation give the same results you would find using a thermo diagram?

E13(§). Create a table of T_d as a function of (T, T_w). Check that your results are consistent with Figs. 4.4 and 4.5.

E14(§). Do eqs. (4.20) and (4.21) give the same results as using **Normand's Rule** graphically on a thermo diagram (Fig. 4.6)? Confirm for a few different initial air-parcel states.

E15. Derive eqs. (4.23) through (4.27) from eqs. (4.4, 4.7, and 4.10).

E16. Rain falls out of the bottom of a moving cloudy air parcel. (a) If no rain falls into the top of that air parcel, then what does eq. (4.35) tell you? (b) If rain falls out of the bottom at the same rate that it falls into the top, then how does this affect eq. (4.35).

E17(§). Create a thermo diagram using a spreadsheet to calculate isohumes (for r = 1, 3, 7, 10, 30 g kg^{-1}) and dry adiabats (for θ = –30, –10, 10, 30 °C), all plotted on the same graph vs. P on an inverted log scale similar to Figs. 3.3 and 4.7.

E18. Start with Tetens' formula to derive equation (4.36). Do the same, but starting with the Clausius-Clapeyron equation. Compare the results.

E19. It is valuable to test equations at extreme values, to help understand limitations. For example, for the saturated adiabatic lapse rate (eq. 4.37b), what is the form of that equation for T = 0 K, and for T approaching infinity?

E20(§). Create your own thermo diagram similar to Fig. 4.8 using a spreadsheet program, except calculate and plot the following saturated adiabats (θ_w = –30, –10, 10, 30°C).

E21(§). Create a full thermo diagram spanning the domain ($-60 \le T \le 40$°C) and ($100 \le P \le 10$ kPa). This spreadsheet graph should be linear in T and logarithmic in P (with axes reversed so that the highest pressure is at the bottom of the diagram). Plot

- isobars (drawn as thin green solid lines) for P(kPa) = 100, 90, 80, 70, 60, 50, 40, 30, 20
- isotherms (drawn as thin green solid lines) for T(°C) = 40, 20, 0, –20, –40 °C.
- dry (θ, solid thick orange lines) and moist adiabats (θ_w, dashed thick orange lines) for the same starting temperatures as for T.
- isohumes (thin dotted orange lines) for r (g kg^{-1}) = 50, 20, 10, 5, 2, 1, 0.5, 0.2

E22. T and θ are both in Kelvins in eqs. (4.40 and 4.41). Does this mean that these two temperatures cancel each other? If not, then what is the significance of θ/T in those equations?

E23. Consider an air parcel rising adiabatically (i.e., no mixing and no heat transfer with its surroundings). Initially, the parcel is unsaturated and rises dry adiabatically. But after it reaches its LCL, it continues its rise moist adiabatically. Is θ_w or θ_L conserved (i.e., constant) below the LCL? Is it constant above the LCL? Are those two constants the same? Why or why not. Hint, consider the following: In order to conserve total water ($r_T = r + r_L$), r must decrease if r_L increases.

E24. For each of the saturated adiabats in Fig. 4.8, calculate the corresponding value of θ_e (the equivalent potential temperature). Why is it always true that $\theta_e \geq \theta_L$?

E25. a. By inspection of the horizontal advection terms in eq. (4.44), write the corresponding term for vertical advection. b. Which term of that equation could account for evaporation from a lake surface, if the Eulerian cube of air was touching the lake?

E26. Based on the full (un-simplified) Eulerian heat budget equation from the Thermodynamics chapter, create by inspection a full water-balance equation similar to eq. (4.44) but without the simplifications.

E27. Why does the condensation-caused latent heating term in the Eulerian heat balance equation (see the Thermodynamics chapter) have a different form (or purpose) than the precipitation term in the water balance eq. (4.44)?

E28(§). Plot curves kinematic latent flux vs. evaporation rate for different altitudes.

E29. The flux of heat and water due to entrainment at the top of the mixed layer can be written as $F_{H\ zi} = w_e \cdot \Delta_{zi}\theta$ and $F_{water\ zi} = w_e \cdot \Delta_{zi} r_T$, respectively. Also, for free convection (sunny, calm) conditions in the mixed layer, a good approximation is $F_{H\ zi} = 0.02 \cdot F_H$. If the entrainment velocity is w_e, then show how the info above can be used to create eq. (4.54).

E30. Derive eq. (4.56) from (4.55).

E31. Which humidity sensors would be best suited for measuring the rapid fluctuations of humidity in the turbulent boundary layer? Why?

Synthesize

S1. Describe how the formation and evolution of clouds would differ if colder air could hold more water at saturation than warmer air. During a typical daytime summer day, when and at what altitudes would you expect clouds to form?

S2. Describe how isohumes on a thermo diagram would look if saturation mixing ratio depended only on temperature.

S3. Consider the spectral-absorption hygrometers described earlier (e.g., optical hygrometers, Lyman-alpha hygrometer, krypton hygrometer, infrared hygrometers). What principle(s) or law(s) from the Solar and IR Radiation chapter describe the fundamental way that these instruments are able to measure humidity?

S4. Describe the shape (slope and/or curvature) of moist adiabats in a thermo diagram if water-vapor condensation released more latent heat than the cooling associated with adiabatic expansion of the air parcel. Describe any associated changes in climate and weather.

S5. What if the evaporation rate of water from the surface was constant, and did not depend on surface humidity, air humidity, wind speed, or solar heating (i.e., convection). Describe any associated changes in climate and weather.

S6. a. Describe the depth of liquid water in rain gauges at the ground if all the water vapor in the troposphere magically condensed and precipitation out. Assume a standard atmosphere temperature profile, but with a relative humidity of 100% initially (before rainout), and with no liquid or solid water initially suspended in the atmosphere.

b. Oceans currently cover 70.8% of the Earth's surface. If all the water from part (a) flowed into the oceans, describe the magnitude of ocean-depth increase.

c. Do a similar exercise to (a) and (b), but for a saturated standard-atmosphere stratosphere (ignoring the troposphere).

S7. The form of Clausius-Clapeyron equation presented near the beginning of this chapter included both T_o and T as arguments of the exponential function.

a. Use algebra to separate T_o and T into separate exponential functions. Once you have done that, your equation should look like: $e_s = C \cdot \exp[-(L/\Re_v) \cdot (1/T)]$, where C contains the other exponential. Write the expression for C.

b. The **Boltzmann constant** is k_B = 1.3806x10^{-23} J·K^{-1}·molecule^{-1} . This can be used to rewrite the water-vapor gas constant as $\Re_v \approx k_B/m_v$, where m_v represents the mass of an individual water-molecule. Substitute this expression for $\Re_v$ into $e_s = C\cdot\exp[-(L/\Re_v)\cdot(1/T)]$, leaving C as is in this eq.

c. If you didn't make any mistakes in this alternative form for the Clausius-Clapeyron equation, and after you group all terms in the numerator and all terms in the denominator of the argument then your equation should contain a new form for the ratio in the argument of the exponential function. One can interpret the numerator of the argument as the potential energy gained when you pull apart the bond that holds a water molecule to neighboring molecules in a liquid, so as to allow that one molecule to move freely as water vapor. The denominator can be interpreted as the kinetic energy of a molecule as indicated by its temperature.

With that in mind, the denominator is energy available, and numerator is energy needed, for one molecule of water to evaporate. Describe why this ratio is appropriate for understanding saturation vapor pressure as an equilibrium.

S8. Devise an equation to estimate surface water flux that works for a sunny windy day, which reduces to eqs. (4.51 & 4.52) in the limits of zero convection and zero mean wind, respectively.

5 ATMOSPHERIC STABILITY

Contents

A **sounding** is the vertical profile of temperature and other variables in the atmosphere over one geographic location. **Stability** refers to the ability of the atmosphere to be turbulent, which you can determine from soundings of temperature, humidity, and wind. Turbulence and stability vary with time and place because of the corresponding variation of the soundings.

We notice the effects of stability by the wind gustiness, dispersion of smoke, refraction of light and sound, strength of thermal updrafts, size of clouds, and intensity of thunderstorms.

Thermodynamic diagrams have been devised to help us plot soundings and determine stability. As you gain experience with these diagrams, you will find that they become easier to use, and faster than solving the thermodynamic equations. In this chapter, we first discuss the different types of thermodynamic diagrams, and then use them to determine stability and turbulence.

BUILDING A THERMO-DIAGRAM

Components

In the Water Vapor chapter you learned how to compute **isohumes** and **moist adiabats**, and in the Thermodynamics chapter you learned to plot **dry adiabats**. You plotted these isopleths on a background graph having temperature along the horizontal axis and log of pressure along the vertical axis. Figs. 5.1a-d show these diagram components.

When these isopleths are combined on a single graph, the result is called a **thermodynamic diagram** or **thermo diagram** (Fig. 5.1e). At first glance, Fig. 5.1e looks like a confusing nest of lines; however, you can use the pattern-recognition capability of your mind to focus on the components as shown in Fig. 5.1a-d. Your efforts to master thermo diagrams now will save you time in the future.

Several types of thermo diagrams are used in meteorology. They all can show the same information, and are used the same way. The thermo diagram we learned so far is called an **Emagram**. We learned this one first because it was easy to create using a computer spreadsheet.

(a)

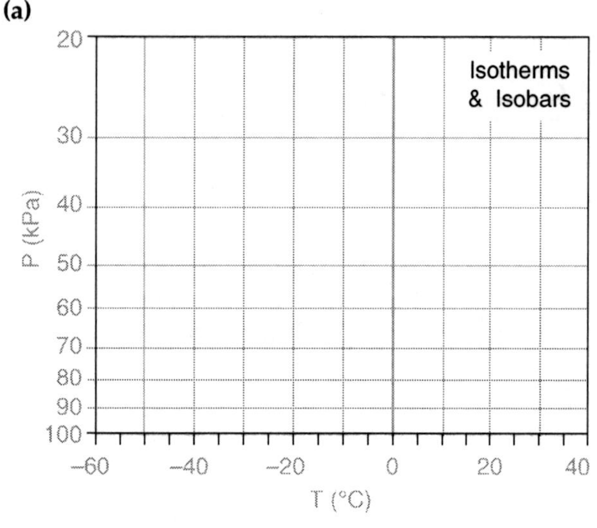

(d)

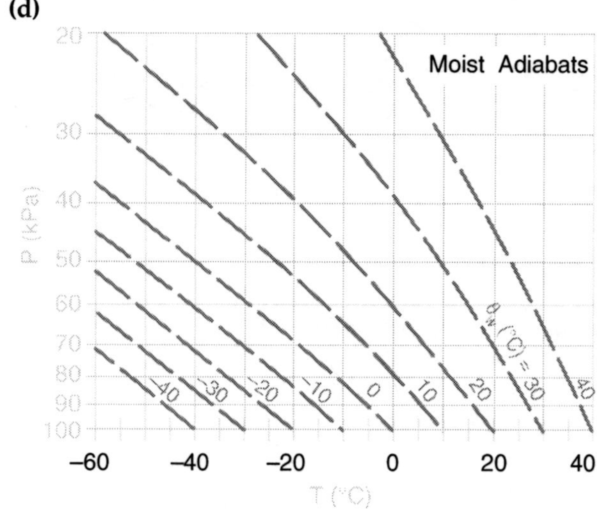

(b)

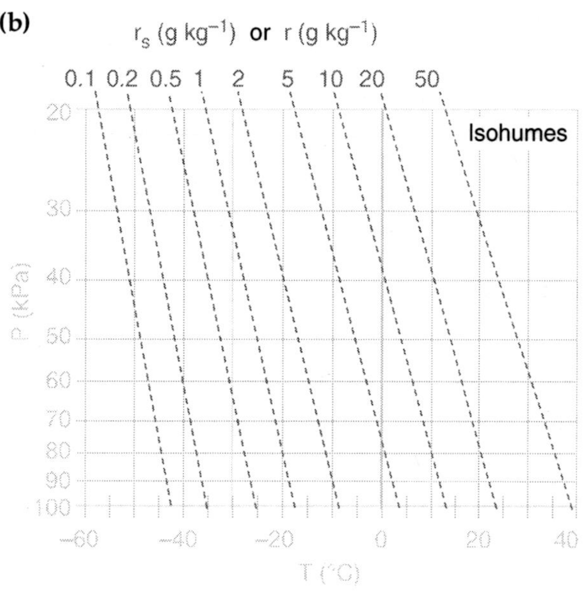

(e)

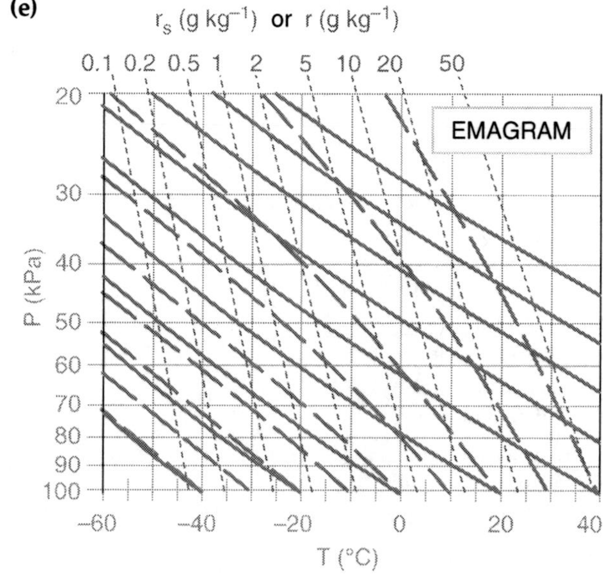

(c)

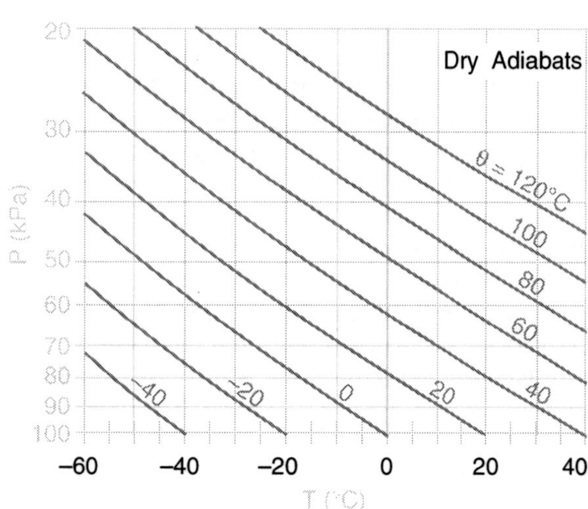

Figure 5.1
Components of an Emagram thermo diagram.
(a) ***Isobars*** *(green thin horizontal lines with logarithmic spacing) and* ***isotherms*** *(green thin vertical lines) are used on all these charts as a common background.*
(b) ***Isohumes*** *(from the Water Vapor chapter) are dotted light-blue lines.*
(c) The dark-orange solid lines are ***dry adiabats****.*
(d) The dark-orange dashed lines are ***moist adiabats****.*
(e) Thermo diagram formed by combining parts (a) through (d).

The variables are: pressure (P), temperature (T), mixing ratio (r), saturated mixing ratio (r_s), potential temperature (θ), and wet-bulb potential temperature (θ_w).

Five or more sets of lines are plotted on every thermo diagram, including the Emagram (Fig. 5.1e). Three sets give the **state** of the air (**isotherms, isobars, isohumes**). Two sets (**dry adiabats** and **moist adiabats**) describe **processes** that change the state. **Height contours** are omitted from introductory thermo diagrams, but are included in full thermo diagrams at the end of this chapter.

Pseudoadiabatic Assumption

In the Water Vapor chapter, we assumed an adiabatic process (no heat transfer or mixing to or from the air parcel) when computing the moist adiabats. However, for any of the thermo diagrams, the moist adiabats can be computed assuming either:

- **adiabatic processes** (i.e., **reversible**, where all liquid water is carried with the air parcels), or
- **pseudoadiabatic processes** (i.e., **irreversible**, where all condensed water is assumed to fall out immediately).

Air parcels in the real atmosphere behave between these two extremes, because small droplets and ice crystals are carried with the air parcel while larger ones precipitate out.

When liquid or solid water falls out, it removes from the system some of the sensible heat associated with the temperature of the droplets, and also changes the heat capacity of the remaining air because of the change in relative amounts of the different constituents. The net result is that an air parcel lifted pseudoadiabatically from 100 kPa to 20 kPa will be about 3°C colder than one lifted adiabatically. This small difference between adiabatic and pseudoadiabatic can usually be neglected compared to other errors in measuring soundings.

Complete Thermo Diagrams

Color thermo diagrams printed on large-format paper were traditionally used by weather services for hand plotting of soundings, but have become obsolete and expensive compared to modern plots by computer. The simplified, small-format diagrams presented so far in this chapter are the opposite extreme – useful for education, but not for plotting real soundings. Also, some weather stations have surface pressure greater than 100 kPa, which is off the scale for the simple diagrams presented so far.

As a useful compromise, full-page thermo diagrams in several formats are included at the end of this chapter. They are optimized for you to print on a color printer.

[*Hint: Keep the original thermo diagrams in this book clean and unmarked, to serve as master copies.*]

INFO • Why so many thermo diagrams?

Meteorological thermo diagrams were originally created as optimized versions of the P vs. α diagrams of classical physics (thermodynamics), where α is the **specific volume** (i.e., α = volume per unit mass = $1/\rho$, where ρ is air density). A desirable attribute of the P vs. α diagram is that when a cyclic process is traced on this diagram, the area enclosed by the resulting curve is proportional to the specific work done by or to the atmosphere. The disadvantage of P vs. α diagram is that the angle between any isotherm and adiabat is relatively small, making it difficult to interpret atmospheric soundings.

Three meteorological thermo diagrams have been devised that satisfy the "area = work" attribute, and are optimized for meteorology to have greater angles between the isotherms and adiabats:

- Emagram,
- Skew-T Log-P Diagram
- Tephigram.

Meteorologists rarely need to utilize the "area = work" attribute, so they also can use any of three additional diagrams:

- Stüve Diagram,
- Pseudoadiabatic (Stüve) Diagram,
- Theta-Height (θ-z) Diagram.

Why are there so many diagrams that show the same things? Historically, different diagrams were devised somewhat independently in different countries. Nations would adopt one as the "official" diagram for their national weather service, and teach only that one to their meteorologists. For example, the tephigram is used in British Commonwealth countries (UK, Canada, Australia, New Zealand). To this day, many meteorologists feel most comfortable with the diagram they learned first.

For many readers, this myriad of diagrams might make an already-difficult subject seem even more daunting. Luckily, all the diagrams show the same thermodynamic state (T, P, r) and process lines (θ, θ_w), but in different orientations. So once you have learned how to read one diagram, it is fairly easy to read the others.

The skill to read diverse thermo diagrams will serve you well when acquiring weather data via the internet, because they can come in any format. The internet is the main reason I cataloged the different thermo diagrams in this book.

Of all these diagrams, the Skew-T and Tephigram have the greatest angle between isotherms and adiabats, and are therefore preferred when studying soundings and stability. These two diagrams look similar, but the Skew-T is growing in popularity because it is easier to create on a computer.

TYPES OF THERMO DIAGRAMS

Emagram

"Emagram" is a contraction for "Energy-per-unit-mass diagram." This semi-log diagram (Fig. 5.3a) has temperature (T) along the abscissa, and pressure (P) decreasing logarithmically upward along the ordinate. The isotherms and isobars are straight lines that are perpendicular to each other, and form the orthogonal basis for this diagram.

Dry adiabats (θ) are nearly-straight, diagonal lines slanted up towards the left, and curve slightly concave upward. Moist adiabats (θ_w) curve concave downward, but become parallel to the dry adiabats at cold temperatures and high altitudes. Isohumes (r) are nearly-straight, nearly-vertical lines (with a slight tilt upward to the left).

Height contours (on the large Emagram at the end of this chapter) are approximate, and are provided as a rough guide to users. These contours are nearly horizontal (tilting downward slightly to the right) and nearly straight (slightly concave upward). They are often not drawn in thermo diagrams, and indeed are not shown in Figs. 5.1 and 5.3.

Stüve & Pseudoadiabatic Diagrams

The Stüve diagram (Fig. 5.3b) has T along the abscissa, and pressure decreasing upward along the ordinate according to $(P/P_o)^{\Re_d/C_{pd}}$, where $\Re_d/C_{pd}$ = 0.28571 (dimensionless) and P_o = 100 kPa. Stüve diagrams look virtually identical to Emagrams, except that the dry adiabats are perfectly straight. These dry adiabats converge to a point at P = 0 kPa and T = 0 K, which is usually well off of the upper left corner of the graph.

Any thermo diagram can be computed using the pseudoadiabatic assumption for the moist adiabats. However, the particular diagram that is known as a "Pseudoadiabatic Diagram" is often a Stüve plot.

Skew-T Log-P Diagram

Fig. 5.3c is a Skew-T. Although T is labeled linearly along the abscissa, the isotherms are parallel, straight, diagonal lines tilting upward to the right. The Skew-T gets its name because the isotherms are not vertical, but skewed. Pressure decreases logarithmically upward along the ordinate, and the isobars are parallel, horizontal, straight lines.

Dry adiabats are diagonal lines slanted up towards the left, with a pronounced curve concave upward. Moist adiabats are more sharply curved concave left near the bottom of the diagram, changing to less curved, concave to the right, as they merge into the dry adiabats at higher altitudes and colder temperatures. Isohumes are almost straight lines, tilting upward to the right.

This diagram is designed so that the isotherms and dry adiabats are nearly (but not perfectly) perpendicular to each other.

Tephigram

The name Tephigram is a contraction of Tee-Phi Diagram. The logarithm of potential temperature θ physically represents the **entropy** (ϕ = Greek letter Phi) change, and is plotted along the ordinate of Fig. 5.2. Temperature T is along the abscissa. Thus, it is a temperature-entropy diagram, or T-ϕ diagram. Isobars are curved green lines in Fig. 5.2.

The range of meteorological interest is shown with the bold rectangle in this figure. When this region is enlarged and tilted horizontally, the result is the tephigram as plotted in Fig. 5.3d.

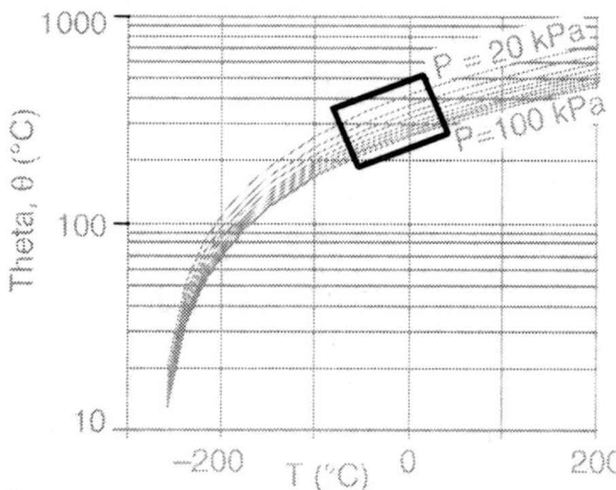

Fig. 5.2
Rectangle shows portion of thermo diagram used for tephigram.

The tephigram looks similar to the Skew-T, with isotherms as straight diagonal lines tilting upward to the right. Dry adiabats are perfectly straight diagonal lines tilting upward left, and are exactly perpendicular to the isotherms. Moist adiabats have strong curvature concave left, and become parallel to the dry adiabats at cold temperatures and high altitudes. Isobars are gently curved (concave down) nearly-horizontal lines. Isohumes are nearly straight diagonal lines tilting upward right.

Theta-Height (θ-z) Diagrams

The orthogonal basis for this diagram are heights (z) plotted along the ordinate and potential temperature (θ) along the abscissa (Fig. 5.3e). Moist adiabats curve to the right of the dry adiabats, but become parallel to the dry adiabats at high altitudes and cold temperatures. Heights are accurate in this diagram, but the isobars are only approximate. Isohumes tilt steeply upward to the right. This diagram is used extensively in boundary-layer meteorology.

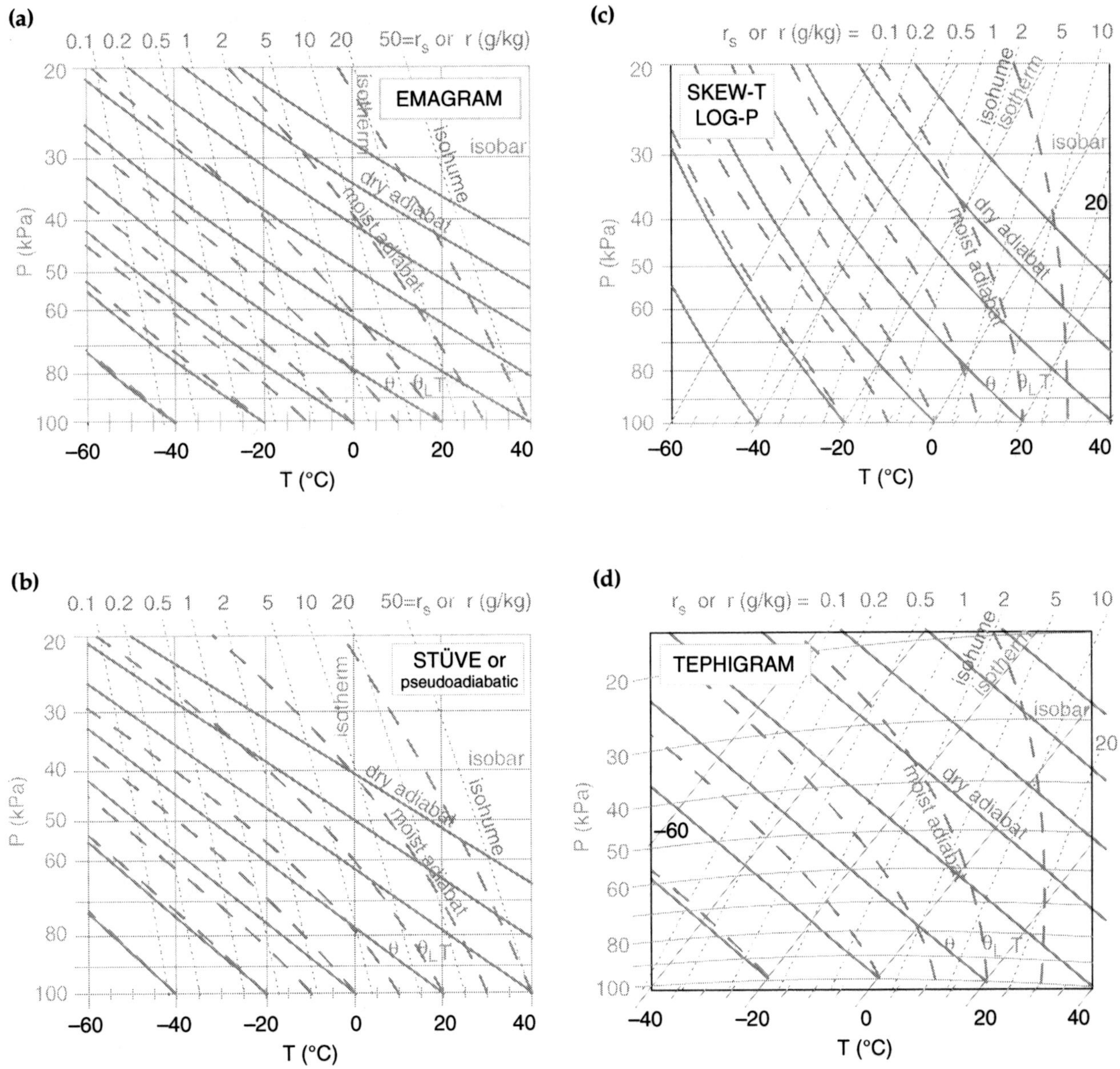

(a)

0.1 0.2 0.5 1 2 5 10 20 50=r_s or r (g/kg)

EMAGRAM

isotherm
isohume
isobar
dry adiabat
moist adiabat
θ
θ_L
T
P (kPa)
T (°C)

(c)

r_s or r (g/kg) = 0.1 0.2 0.5 1 2 5 10

SKEW-T LOG-P

isohume
isotherm
isobar
dry adiabat
moist adiabat
θ
θ_L
T
P (kPa)
T (°C)

(b)

0.1 0.2 0.5 1 2 5 10 20 50=r_s or r (g/kg)

STÜVE or pseudoadiabatic

isotherm
isobar
dry adiabat
moist adiabat
isohume
θ
θ_L
T
P (kPa)
T (°C)

(d)

r_s or r (g/kg) = 0.1 0.2 0.5 1 2 5 10

TEPHIGRAM

isohume
isotherm
isobar
moist adiabat
dry adiabat
θ
θ_L
T
P (kPa)
T (°C)

(e)

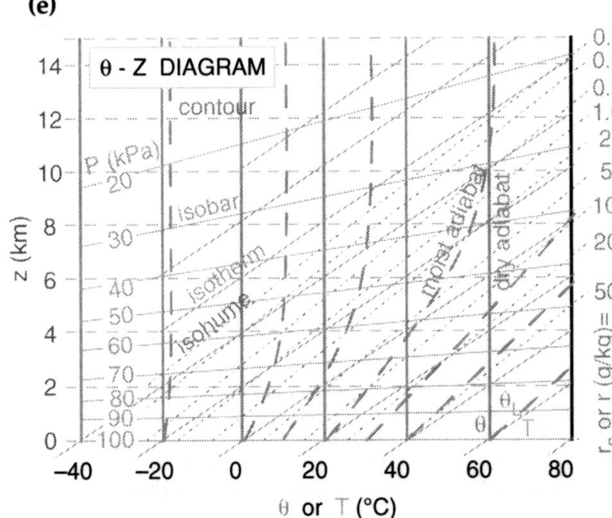

Figure 5.3
Catalog of thermodynamic diagrams. In all diagrams, thick dark-orange lines represent processes, and thin lines (green or blue) represent state. Thick solid dark-orange lines are dry adiabats, and thick dashed dark-orange are moist adiabats. Solid thin green horizontal or nearly-horizontal lines are pressure, and solid thin green vertical or diagonal straight lines are temperature. Isohumes are thin dotted blue lines. In addition, the θ-z diagram has height contours (z) as thin horizontal dashed grey lines.

MORE ON THE SKEW-T

Because of the popularity of the Skew-T, we dissect it in Fig. 5.4, to aid in its interpretation. Tephigram users should find this equally useful, because the only noticeable difference is that the isobars are slightly curved, and increase in slope higher in the tephigram.

The Skew-T diagram is trivial to create, once you have already created an Emagram. The easiest way is to start with a graphic image of an Emagram, and skew it into a parallelogram (Fig. 5.5) using a graphics or drawing program.

(a)

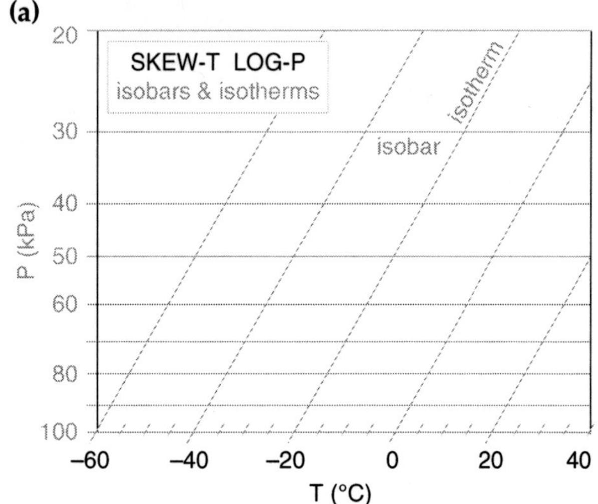

(b)

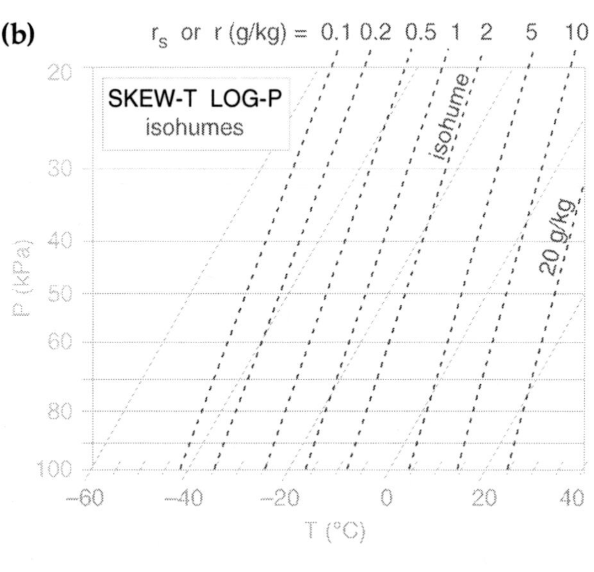

(d)

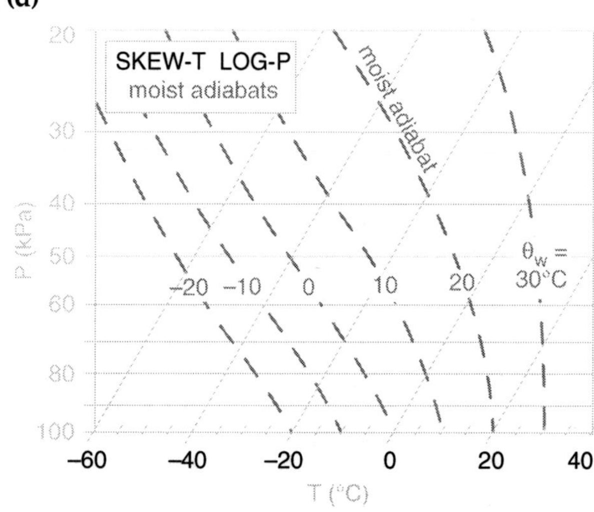

(c)

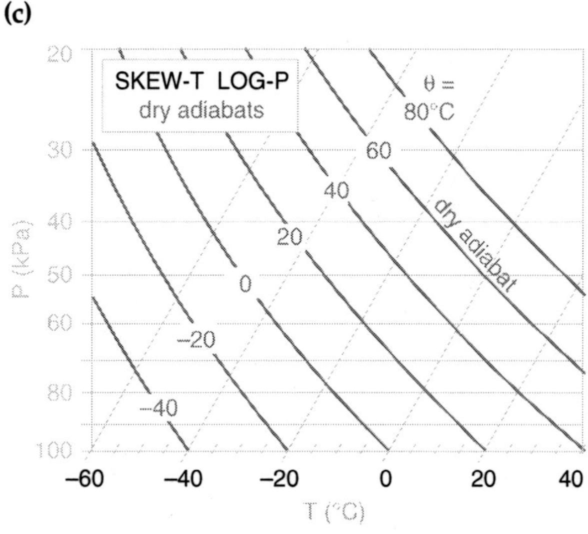

(e)

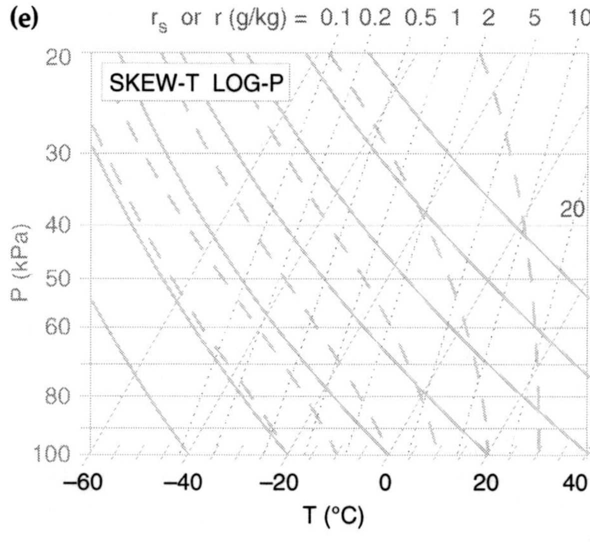

Fig. 5.4
(a)-(d) Components of a Skew-T diagram, drawn to the same scale so they can be superimposed as a learning aid. The state of the air: (a) isobars and isotherms; (b) isohumes as thin dotted lines. Processes: (c) dry adiabats; (d) moist adiabats. (e) Full Skew-T Log-P, shaded lighter to make it easier to write on. The "Skew-T Log-P" diagram is often called a "Skew-T" for short.

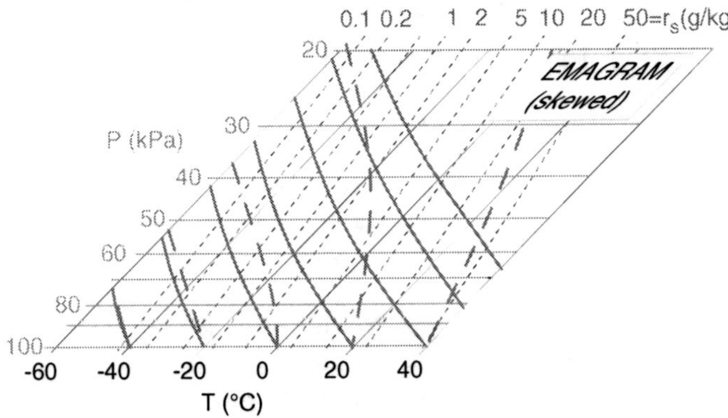

Fig. 5.5
A Skew-T is nothing more than a skewed Emagram.

It is equally simple to draw a Skew-T with a spreadsheet program, especially if you start with the numbers you already calculated to create an Emagram. Let X (°C) be the abscissa of a Cartesian graph, and Y (dimensionless) be the ordinate. Let K be the skewness factor, which is arbitrary. A value of K = 35°C gives a nice skewness (such that the isotherms and adiabats are nearly perpendicular), but you can pick any skewness you want.

The coordinates for any point on the graph are:

$$Y = \ln\left(\frac{P_o}{P}\right) \tag{5.1a}$$

$$X = T + K \cdot Y \tag{5.1b}$$

To draw isotherms, set T to the isotherm value, and find a set of X values corresponding to different heights Y. Plot this one line as an isotherm. For the other isotherms, repeat for other values of T. [Hint: it is convenient to set the spreadsheet program to draw a logarithmic vertical axis with the max and min reversed. If you do this, then use P, not Y, as your vertical coordinate. But you still must use Y in the equation above to find X.]

For the isohumes, dry adiabats, and moist adiabats, use the temperatures you found for the Emagram as the input temperatures in the equation above. For example, the θ = 20°C adiabat has the following temperatures as a function of pressure (recall Thermo Diagrams-Part 1, in the Thermodynamics chapter), from which the corresponding values of X can be found.

P(kPa)= 100 , 80 , 60 , 40 , 20
T(°C) = 20 , 1.9 , –19.8, –47.5 , –88
X = 20 , 9.7 , –1.9 , –15.4 , –31.7

Sample Application

Create a Skew-T diagram with one isotherm (T = 20°C), one dry adiabat (θ = 20°C), one moist adiabat (θ_w = 20°C), and one isohume (r = 5 g/kg). Use as input the temperatures from the Emagram (as calculated in earlier chapters). Use K = 35°C, and P_o = 100 kPa.

Find the Answer

Given: The table of Emagram numbers below.
Use eq. (5.1) in a spreadsheet to find X and Y.

	A		B	C	D	E
1	Create Skew-T from Emagram Numbers					
2						
3			Iso-	Dry	Moist	Iso-
4			therm	Adiabat	Adiabat	hume
5			T=20°C	θ=20°C	θ_w=20C	r=5g/kg
6	Emagram (Given):					
7		P (kPa)	T (°C)	T (°C)	T (°C)	T (°C)
8		100	20	20	20	3.7
9		90	20	11.3	16	2.3
10		80	20	1.9	11.4	0.7
11		70	20	-8.4	6	-1.2
12		60	20	-19.8	-0.3	-3.2
13		50	20	-32.6	-8.8	-5.6
14		40	20	-47.5	-20	-8.5
15		30	20	-65.3	-37	-12
16		20	20	-88	-60	-17
17						
18	Skew-T:					
19		Let K=	35°C		Let P_o=	100kPa
20	P	Y	X	X	X	X
21	100	0.00	20	20	20	3.7
22	90	0.11	23.7.	15.0	19.7	6.0
23	80	0.22	27.8	9.7	19.2	8.5
24	70	0.36	32.5	4.1	18.5	11.3
25	60	0.51	37.9	-1.9	17.6	14.7
26	50	0.69	44.3	-8.3	15.5	18.7
27	40	0.92	52.1	-15.4	12.1	23.6
28	30	1.20	62.1	-23.2	5.1	30.1
29	20	1.61	76.3	-31.7	-3.7	39.3

Using the spreadsheet to plot the result gives:

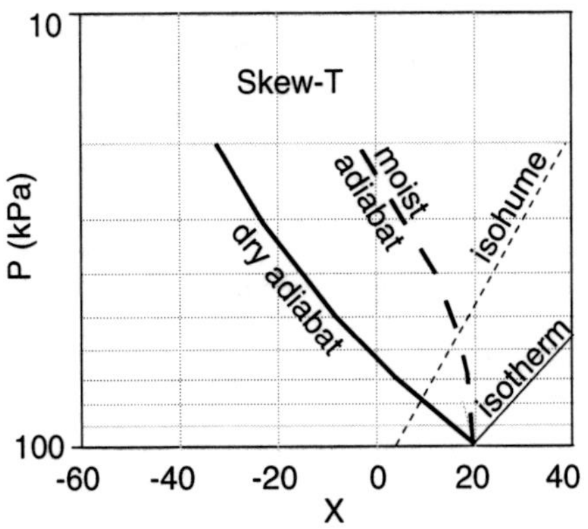

Check: Sketch reasonable. Matches Fig. 5.4e.
Exposition: Experiment with different values of K starting with K = 0°C, and see how the graph changes from an Emagram to a Skew-T.

Sample Application

For each thermo diagram numbered below, identify its name.

1)

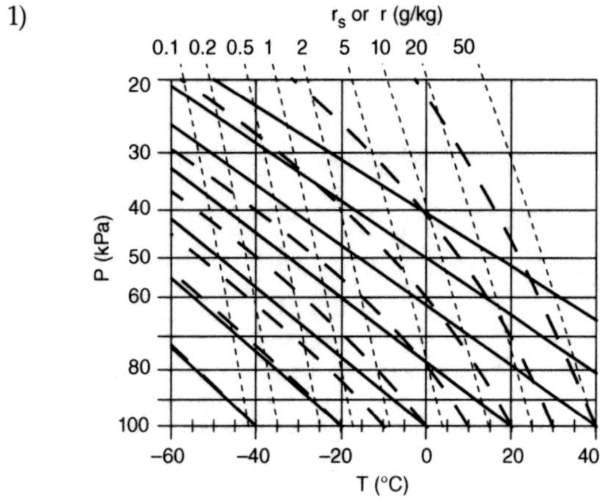

2)

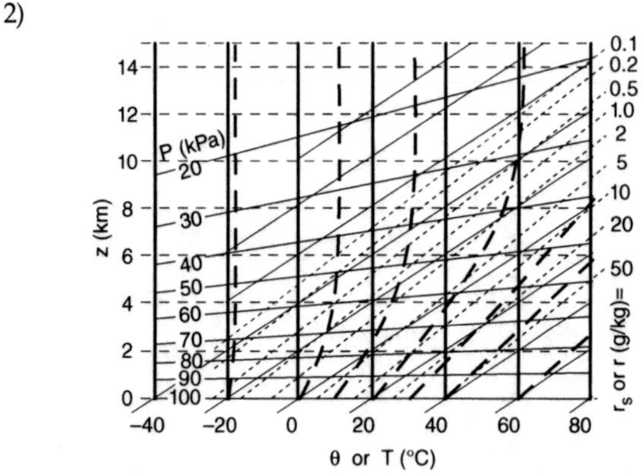

3)

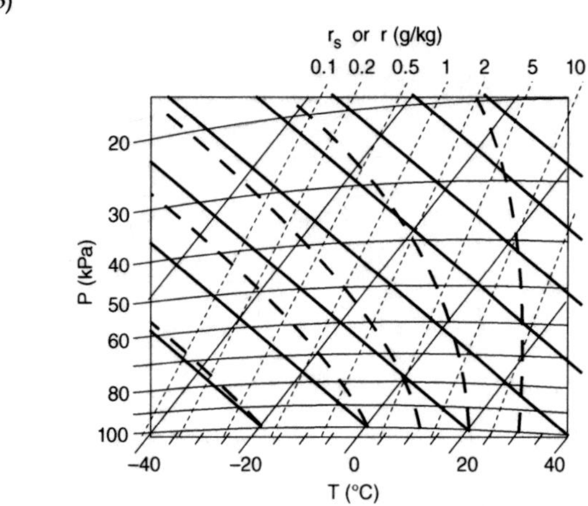

CAUTION: Thermo diagrams printed or published by other agencies do NOT follow any standard format of line style or color. That is the reason these thermo diagrams are shown only with black lines in this sample application — so you don't use the colors as a crutch.]

(continued next page)

GUIDE FOR QUICK IDENTIFICATION OF THERMO DIAGRAMS

If you receive a thermo diagram via the internet without any labels identifying the diagram type, use the following procedure to figure it out:

1) if there are some lines from lower left to upper right, then either Skew-T, Tephigram, or θ-Z .
   a) if the isobars are exactly horizontal and straight, then Skew-T,
   b) else if the strongly curved lines (moist adiabats) asymptotically approach vertical near the top of the diagram, then θ-Z,
   c) else Tephigram,
2) else either Emagram or Stüve .
   a) if dry adiabats are slightly curved, then Emagram,
   b) else Stüve (including Pseudoadiabatic version of Stüve).

When you encounter an unfamiliar thermo diagram, you can take steps to help segregate and identify the various lines. First, identify the isobars, which are horizontal and straight, or nearly so, on virtually all diagrams. Isobars are usually spaced roughly logarithmically with height, and are labeled as either 100, 90, 80, 70 kPa etc., or as 1000, 900, 800, 700 mb or hPa etc. These serve as a surrogate measure of height.

Next, identify the dry and moist adiabats. These are the lines that converge and become parallel at cold temperatures and high altitudes. In all the diagrams, the moist adiabats are the most curved. The adiabats are spaced and labeled roughly linearly, such as 0, 10, 20, 30 °C etc, or often in Kelvin as 273, 283, 293, 303 K, etc. While the dry adiabats are always labeled as potential temperature θ, the moist adiabats can be labeled as either equivalent potential temperature θ_e , liquid-water potential temperature θ_L , or wet-bulb potential temperature θ_w .

An alternative for identifying adiabats is to look at any labeled temperature at the bottom of the graph. Usually at 100 kPa (1000 mb or hPa), you will find that T , θ_w , and θ lines all radiate from the same

Hint: At first glance, a thermo diagram looks like a complicated nest of lines. A way to discriminate one set of lines (e.g., adiabats) from other sets of lines (e.g., isotherms) is to hold the printed diagram so that you can view it from near the edge of the page. For example, if you view an Emagram from the lower right corner, the dry adiabats will stand out as slightly curved lines. Look from different edges to highlight other isopleths.

points. For all diagrams, the T line is rightmost, θ is leftmost, and θ_w is in the middle.

Finally, identify the isotherms and isohumes. These lines cross each other at a small angle, and extend from the bottom to the top of the graph. The isohumes always tilt to the left of the isotherms with increasing height. The isotherms increment linearly (e.g., 0, 10, 20 , 30, °C etc.), while the isohumes usually increment logarithmically (e.g., 1, 2, 5, 10, 20 g kg^{-1}, etc.). Isotherms are either vertical, or tilt upward to the right.

THERMO-DIAGRAM APPLICATIONS

Fig. 5.6 shows a simplified, Skew-T diagram. We use this as the common background for the demonstrations that follow.

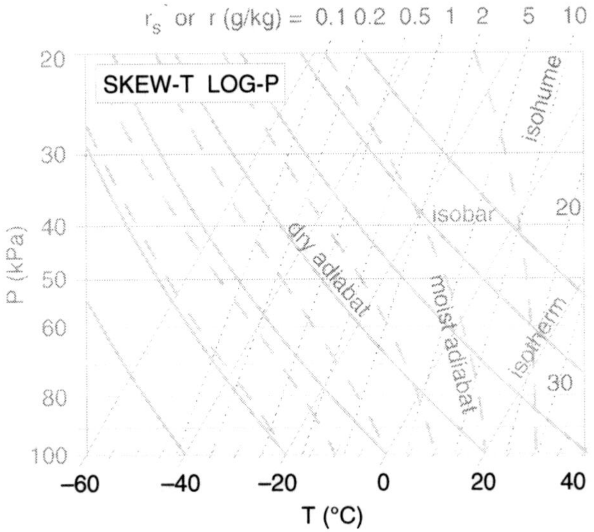

Figure 5.6
Simplified Skew-T diagram. Isopleths are labeled. Colors are lightened to serve as a background for future discussions.

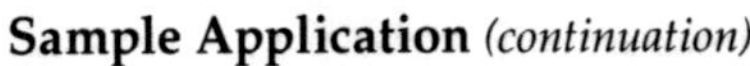

Sample Application *(continuation)*

4)

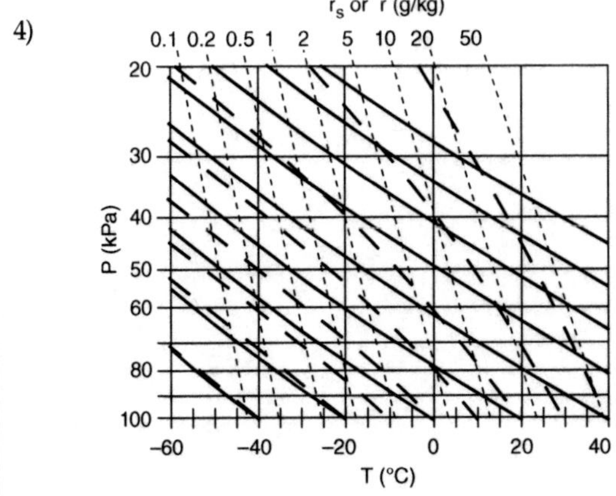

5)

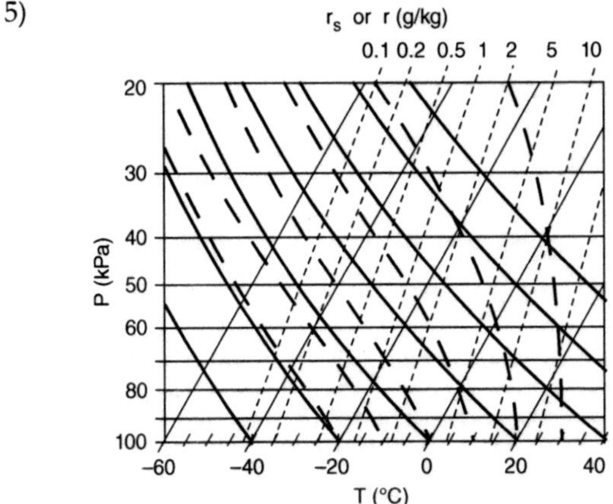

Find the Answer: Use the quick guide.
1) **Stüve**, because satisfies item 2b in the guide.
2) **θ-Z diagram**, because satisfies item 1b.
3) **Tephigram**, because satisfies item 1c.
4) **Emagram**, because satisfies item 2a.
5) **Skew-T**, because satisfies item 1a.

Check: Reasonable, agrees with Fig. 5.3.
Exposition: For practice, pick any one of the diagrams above, and without looking back in Fig. 5.3, try to identify which lines are which.

[Hint: In this book, all thermo diagrams use the same format for the same types of lines: isotherms and isobars are thin solid green, isohumes are thin dotted blue, dry adiabats are thick solid dark-orange, moist adiabats are thick dashed dark-orange, and heights are thin dashed grey (if plotted at all).]

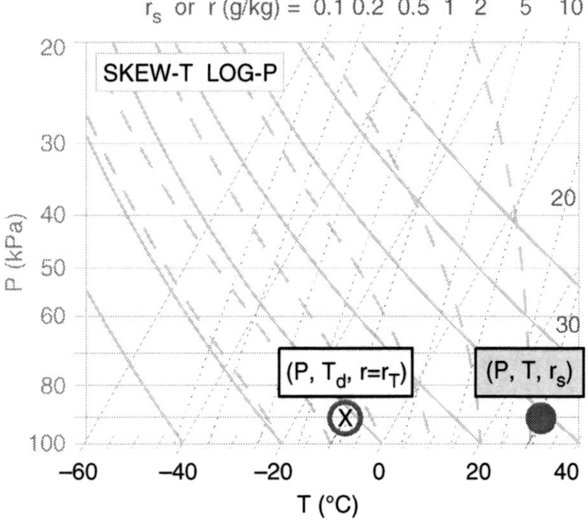

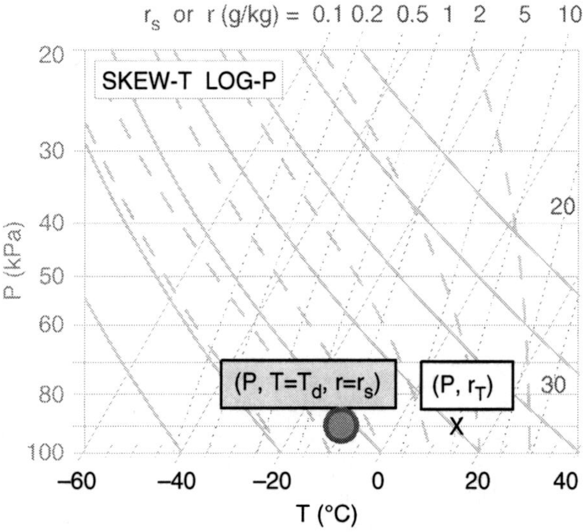

Figure 5.7
In general, three points are needed to indicate one air parcel on a thermo diagram. The red-dot (red-filled circle) indicates air temperature. The blue circle (with no fill) indicates dew-point temperature (as a measure of the amount of water vapor in the air). The X indicates total water content. (a) Example of an unsaturated air parcel plotted on a skew-T diagram. (b) A different air parcel — this one is saturated. (See the Sample Application for details.)

Thermodynamic State

You need three points on a thermodynamic diagram to indicate air-parcel state (pressure, temperature, humidity). For this one air parcel at one height, all three points are always plotted on the same isobar (P).

One point represents T, the actual air temperature (red dot; red filled circle in Fig. 5.7). Its coordinates in the thermo diagram are given by (P, T, r_s). Note that r_s is redundant, because it can be determined from P and T. But we include it for convenience, because we use the r_s value for many calculations.

Another point represents T_d, the dew-point temperature (blue circle outline; no fill). Its coordinates are (P, T_d, r). Again, r is redundant.

The third point gives r_T, the total water mixing ratio in the air parcel (designated by "X" on the thermo diagram). Its coordinates are (P, r_T).

Often, two of these three points coincide. One example is for unsaturated air (Fig. 5.7a) with no rain falling through it. For this case, the temperature and dew-point circles are plotted as described above. However, the X is plotted inside the dew-point circle because the total water in the air parcel equals the water-vapor mixing ratio ($r_T = r$), which is fully specified by the dew point. Namely, the isohume that passes through the dew-point circle indicates the actual amount of water vapor in the air, and there are no other phases of water in that air parcel.

The isohume that passes through the actual air temperature gives the saturation mixing ratio (r_s, the maximum equilibrium amount of moisture that could be held as vapor). A separate symbol is NOT plotted on the thermo diagram for r_s for two reasons. One, it does not indicate the actual state of the air parcel. Two, it is fully specified as the isohume that passes through the temperature point.

For saturated (cloudy or foggy) air parcels (Fig. 5.7b), the two temperature points coincide ($T = T_d$ and $r = r_s$). In this figure, the red-filled circle still represents air temperature, and the blue circle outline still represents the dew-point temperature (for water vapor). However, the total water X is usually to the right of the temperature circles, because clouds contain liquid water and/or ice in addition to water vapor. This is explained later.

[***Tip***: *On these figures, we use large size circles and X's to make them easier to see. However, when you plot them on your own thermo diagram, use smaller circles and X's that you write very precisely on the diagram. Utilize colored pens or pencils if you have them.*]

Processes

Unsaturated (Dry) Lifting

Suppose that an air parcel that is initially unsaturated rises without mixing with the surrounding environment, and with no radiative or other heat transfer to or from it. The rising air parcel cools at the dry adiabatic lapse rate, while carrying all the air and water molecules with it. For this case, mixing ratio (r) and potential temperature (θ) are constant.

Starting from the initial state as plotted by the "old" points on the thermo diagram (Fig. 5.8), determine which dry adiabat is under the center of the red filled circle (temperature), then move the red circle up along this adiabat. Similarly for moisture, determine which isohume is under the center of the blue hollow circle and "X", and follow it up (keeping r constant). Always end the movement of all three points at the same new pressure.

If the points don't lie directly on a drawn isopleth (which is usually the case), then imagine a new isopleth that goes through your point, and which is parallel to (and maintains its relative distance from) the neighboring drawn isopleths. You can do this by eye, or you may use a straight-edge to draw on the diagram a new isopleth that goes through your point of interest. Then, move the air parcel along this imaginary isopleth. Namely, move the temperature point in a direction parallel to the dry adiabats, and move the humidity point parallel to the isohumes.

Descending unsaturated (non-cloudy) air parcels behave similarly, except moving downward along their appropriate isopleths.

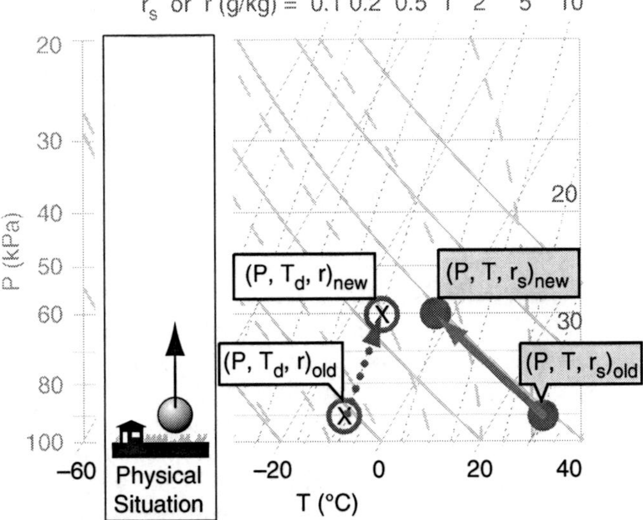

Figure 5.8
Example of the three points (red filled circle for T, open blue circle for T_d, and "X" for r_T) on a skew-T diagram representing the initial (old) state of the air, and how that state changes as the air parcel moves up to a new (lower) pressure level. Note that r_s has changed because the air parcel cooled, but r is unchanged because the air parcel is unsaturated.

Sample Application

(a) What are the values of pressure, temperature, total water mixing ratio, dew point, saturation mixing ratio, actual mixing ratio, and relative humidity of the "old" air parcel in Fig. 5.8a before it starts to rise?

(b) Same question, but at its new altitude.

Find the Answer

Given: Use info in Fig. 5.8 to answer this.
Find: P = ? kPa, T = ? °C, r_T = ? g kg^{-1}, T_d = ? °C, r_s = ? g kg^{-1}, r = ? g kg^{-1}, RH = ? %

(a) Get pressure from the isobar (horizontal green line) that passes through the center of both points. The pressure is displayed along the left side of the thermo diagram. P = **90 kPa**.

Get temperature from the isotherm (green diagonal line) that passes through the center of the red dot. Follow that isotherm diagonally down to the bottom of the graph to read its value. T = **30°C**.

Get total water mixing ratio from the isohume (dotted blue line) that passes through the "X". Read the isohume value at the top of the graph. r_T = **2 g kg^{-1}**.

Get the dew-point temperature from the isotherm (diagonal green line) that passes through the center of the open blue circle. Follow this line diagonally down to the bottom of the thermo diagram to read its label: T_d = **–10°C**.

Get the saturation mixing ratio from the isohume (dotted blue line) that passes through the center of the red circle. It is labeled as: r_s = **30 g kg^{-1}**.

Get the actual mixing ratio from the isohume (dotted blue line) that passes through the center of the open blue circle, and read the label at the top of the graph. r = **2 g kg^{-1}**.

Relative humidity is not displayed on thermo diagrams. But you can find it from $RH\% = (100\%)\cdot(r/r_s)$ = $(100\%)\cdot(2/30)$ = **6.7%** .

(b) Try this on your own, using the same procedure as described above but for the parcel at the new altitude in Fig. 5.8, and then compare with my answers.
P = **60 kPa**, T = **–3°C**, r_T = **2 g kg^{-1}**, T_d = **–15°C**, r_s = **5 g kg^{-1}**, r = **2 g kg^{-1}**, $RH\%$ = **40%**.

Check: Physics and units are reasonable.
Exposition: Although the air was extremely dry initially, as the air parcel was lifted its relative humidity increased. How did this happen, even though the amount of water in the air (r = 2 g kg^{-1}) did not change? The reason was that after the air parcel was lifted to its new altitude, it was much colder, and cold air has a smaller capacity to hold water vapor (r_s decreased).

If you were to continue lifting the parcel, eventually it would cool to the point where $r_s = r$, meaning that the air is saturated. The altitude where this happens indicates cloud base for convective clouds (clouds created by local updrafts of air).

You can get more accurate answers using larger thermo diagrams at the end of this chapter.

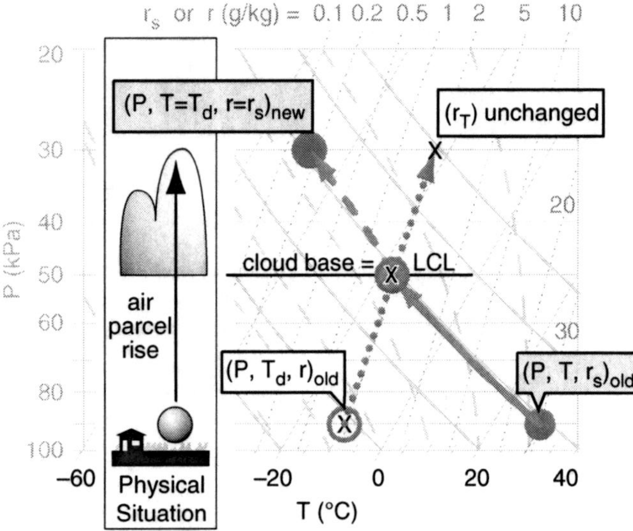

Figure 5.9
Starting with an initial (old) unsaturated air parcel, if you lift it until the temperature and dew-point are equal (namely, where the dry adiabat intersects the isohume), that height is the LCL (lifting condensation level). At this point, the air parcel is just saturated, hence, this marks cloud base for convective clouds because any further lifting causes water vapor to condense onto dust particles — creating liquid cloud droplets. If the air parcel continues to rise up to a new pressure while carrying all its cloud droplets with it, then the three points on the skew-T diagram move as shown in the skew-T diagram. The amount of liquid water (r_L) tied up in cloud droplets at the ending point equals the total-water mixing ratio (r_T) minus the portion that is still held as vapor (r).

Sample Application

Use Fig. 5.9 to estimate P, T, T_d, r, r_s, r_T, r_L or r_{ice} for the parcel: (a) at the LCL, and (b) at P = 30 kPa.

Find the Answer

Given: The dots, circles, and "X's" in Fig. 5.9.
Find: P= ? kPa, T = ?°C, T_d = ?°C,
r = ? g kg^{-1}, r_s = ? g kg^{-1}, r_T = ? g kg^{-1}, r_L = ? g kg^{-1}

(a) At cloud base (i.e., at the LCL), Fig. 5.9 indicates:
P= **50 kPa**, T = **–17°C**, T_d = **–17°C**,
r = **2 g kg^{-1}**, r_s = **2 g kg^{-1}**, r_T = **2 g kg^{-1}**,
$r_L = r_T - r_s$ = **0 g kg^{-1}**

(b) At cloud top (assuming no precipitation):
P= **30 kPa**, T = **–48°C**, T_d = **–48°C**,
r = **0.15 g kg^{-1}**, r_s = **0.15 g kg^{-1}**, r_T = **2 g kg^{-1}**,
$r_{ice} = r_T - r_s$ = 2.0 – 0.15 = **1.85 g kg^{-1}**

Check: Physics and units are reasonable.
Exposition: At temperatures below about –38°C, all hydrometeors are in the form of ice. This is good for aircraft, because ice crystals blow over the wings rather than sticking to them. However, some jet engines have problems when the ice crystals melt and re-freeze on the turbine blades, causing engine problems.

Saturated (Moist) Lifting & Liquid Water

Notice in Fig. 5.8 that the temperature and dew-point curves for the air parcel get closer to each other as the parcel rises. Eventually the dry adiabat and isohume lines cross. The pressure or altitude where they cross (see Fig. 5.9) is defined as the **LCL (lifting condensation level)**, which marks cloud base height for convective clouds (clouds with updrafts from the local region under them, such as cumulus clouds and thunderstorms). This is the altitude where the rising air first becomes saturated — where $T_d = T$, and $r = r_s = r_T$.

Above the LCL the blue circle (T_d) and red dot (T) move together, along a saturated adiabat curve (the orange dashed line that goes through the LCL point). Since saturated air is holding all the vapor that it can, it also means that the actual mixing ratio r equals the saturation value r_s.

If all the condensate (initially as tiny cloud droplets or ice crystals) continues to move with the air parcel and does not fall out as precipitation, then the total-water mixing ratio (r_T) in the parcel is constant. It is easy to estimate the liquid-water mixing ratio (r_L, the relative mass of cloud droplets) as the difference between the total water and the water vapor. In summary, for saturated (cloudy) air: $T = T_d$, $r = r_s$, and $r_L = r_T - r_s$.

Cloudy descending air follows the same lines, but in reverse. Knowing the temperature (red dot with blue circle near the top of Fig. 5.9) and total water content ("X" in that figure) inside the cloud at any one altitude or pressure, then follow the isohume (dotted blue line) down from the "X" and the saturated adiabat (dashed orange line) down from the initial temperature point. As the cloudy air parcel descends it is compressed and warms, allowing more and more of the cloud droplets to evaporate. Eventually, at the LCL, the last droplets have evaporated, so any further descent must follow the dry adiabat (solid orange line) and isohume (dotted blue line) in Fig. 5.9. Namely, the blue circle and the red dot split apart below the LCL.

Precipitation

So far, we assumed that all the liquid water droplets stay inside (move with) the air parcel, meaning r_T = constant. Namely, the "X" in Fig. 5.9 says on one dotted blue line isohume. However, if some of the liquid drops or ice crystals precipitate out of the parcel, then r_T decreases toward r_s, and "X" shifts to a different isohume (Fig. 5.10). The "X" reaches the blue circle ($r_T = r_s$) if all the precipitation falls out.

After precipitation is finished, if there is subsequent ascent or descent, the you must follow the isohume (dotted blue line) that goes through the new location for "X" on the diagram.

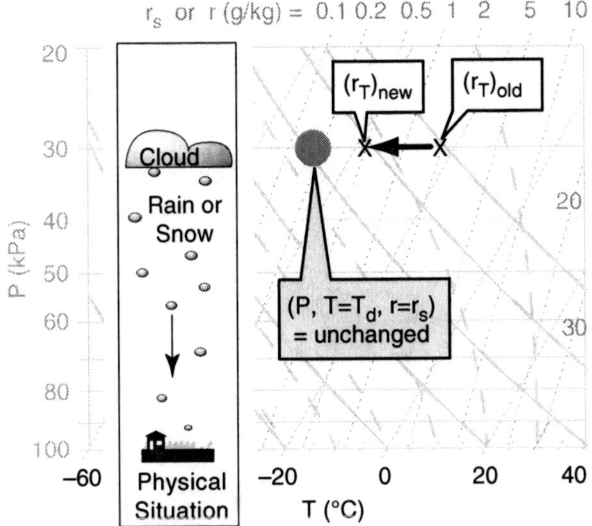

Figure 5.10
Total-water mixing ratio r_T decreases from one isohume (2 g kg^{-1} in this example) to another (0.5 g kg^{-1}) on this skew-T diagram due to precipitation fallout.

Instead of losing water due to precipitation out of an air parcel, sometimes total water is gained when precipitation falls into a parcel from above. Regardless of whether precipitation adds or subtracts water, the process is irreversible, causing a change in the parcel's LCL.

In Fig. 5.10 we illustrate precipitation as happening only while the air parcel is at a fixed pressure or altitude. This is overly simple, as precipitation can also occur while the parcel rises or descends. One way to approximate this on a thermo diagram is to take short adiabatic changes in altitude, and then allow precipitation to occur after each step.

Radiative Heating/Cooling

In the infrared (IR) clouds are blackbodies. They absorb all the IR radiation incident on the outer surface of the cloud, and they emit IR radiation outward in all directions based on the temperature of the cloud droplets on that outer surface.

If the ground and/or atmosphere below cloud base is warmer than the cloud, then there will be more IR radiation absorbed by the cloud base than is re-emitted, and the cloud-base warms. At night, if the ground is colder than cloud base, then the cloud will emit more IR towards the ground than it absorbs, preventing the ground from cooling as much as it would under clear skies, but cooling cloud base.

The atmosphere above the cloud is usually colder than the cloud, and the cloud also "sees" the near-absolute zero temperatures of outer space. Thus, cloud top cools by IR radiation all the time (night and day, see Fig. 5.11). Such cooling creates cold "thermals"

Sample Application

What if ice crystals falling out of the cloud-top air, as illustrated in Fig. 5.10, reduced the total-water mixing ratio from 2 to 0.5 g kg^{-1}. (a) What is its new thermodynamic state at P = 30 kPa? (b) What is its new P_{LCL} if it were to then descend?

Find the Answer

Given: Fig. 5.10, $r_{T\ old}$ = 2 g kg^{-1}, and r_{new} = 0.5 g kg^{-1}
Find: (a) P= ? kPa, T = ?°C, T_d = ?°C, r = ? g kg^{-1}, r_s=? g kg^{-1}, r_T =? g kg^{-1}, r_{ice} =? g kg^{-1}, r_{precip} =? g kg^{-1}
(b) P_{LCL} = ? kPa.

(a) New state at cloud top after precipitation:
P= **30 kPa**, T = **–48°C**, T_d = **–48°C**,
r = **0.15 g kg^{-1}**, r_s = **0.15 g kg^{-1}**, r_T = **0.5 g kg^{-1}**,
$r_{ice} = r_T - r_s$ = 0.5 – 0.15 = **0.35 g kg^{-1}**
r_{precip} = 2 – 0.5 = **1.5 g kg^{-1}**.

(b) New lifting condensation level (LCL):
In Fig. 5.10, from the red dot follow the orange-dashed saturated adiabat downward. (Since no orange dashed lines goes through the red dot, we need to interpolate between neighboring dashed lines left and right of the red dot). Also, from the X for the new r_T = 0.5 g kg^{-1} value, follow the blue-dotted isohume downward. Where those two lines cross is the new LCL. Doing this on a thermo diagram (not shown) gives roughly P_{LCL} = **37 kPa**.

Check: Physics and units are reasonable.
Exposition: Note that the new LCL is at a higher altitude (lower pressure) than the LCL in Fig. 5.9, because precipitation fallout left the air drier.

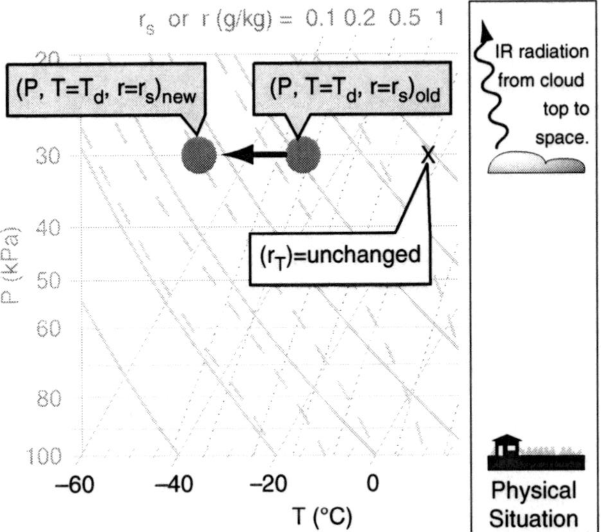

Figure 5.11
Infrared (IR) radiation to space from cloud top can cool the cloud-top air. Such cooling lowers the saturation mixing ratio and forces more water vapor to condense, thus increasing the amount of condensate.

Sample Application

Suppose IR radiation caused the cloud-top thermodynamic state to change as shown in Fig. 5.11, where $T_{new} \approx -71°C$ and no precipitation falls out. What is the air parcel's new thermodynamic state, and what will be its new P_{LCL}?

Find the Answer

Given: data in Fig. 5.11

Find: (a) P= ? kPa, T = ?°C, T_d = ?°C, r = ? g kg^{-1}, r_s=? g kg^{-1}, r_T =? g kg^{-1}, r_{ice} =? g kg^{-1}, r_{precip} =? g kg^{-1}

(b) P_{LCL} = ? kPa.

(a) Approximate new state, base on visual interpretation of Fig. 5.11:

P= **30 kPa**, T = **–71°C**, T_d = **–71°C**,

$r \approx$ **0 g kg^{-1}**, $r_s \approx$ **0 g kg^{-1}**, r_T = **2.0 g kg^{-1}**,

$r_{ice} = r_T - r_s = 2 - 0 =$ **2 g kg^{-1}**

(b) New lifting condensation level (LCL):

In Fig. 5.11, from the red dot follow the orange-dashed saturated adiabat downward. Also, from the X for r_T = 2 g kg^{-1} value, follow the blue-dotted isohume downward. Where those two lines cross is the new LCL. Doing this on a thermo diagram (such as Fig. 5.11) gives roughly $P_{LCL} \approx$ **80 kPa**.

Check: Physics and units are reasonable.

Exposition: The lifting condensation level is at lower altitude (higher pressure) than it was originally.

NOTE: For this and for previous Sample Applications, you might find a slightly different answer if you used a different thermo diagram. This is normal. These slight differences in the new air-parcel state are usually neglected, because there is even greater uncertainty in measured old values of air-parcel state.

in the top 2 cm of cloud-top air that sink lower into the cloud and spread the cooling effect over more of the cloud. This same effect occurs in fog, which tends to make fog denser and more difficult to dissipate.

During daytime light from the sun can reach the clouds, where some is absorbed by the cloud droplets and some is scattered back out (making clouds appear white). The portion absorbed in the tops and sides of clouds is spread over a thicker regions of cloud that extends about 2 m into the cloud from the outer surface. This means that heating due to sunlight is spread over a thicker layer of cloud, as opposed to IR cooling which happens only at the immediate cloud surface.

Recall that the word "adiabatic" means no heat is transferred between the air parcel and the surrounding environment. Radiation is not adiabatic, and hence is called **diabatic**. In Fig. 5.11, the diabatic IR cooling causes the red dot and blue circle to move together ($T = T_d$) to colder temperatures, while the total-water mixing ratio "X" is unchanged because radiation does not create or destroy water molecules. Nonetheless, as temperature drops, r_s (= r) decreases, causing liquid r_L or solid r_{ice} water to increase (according to $r_{L\,new}$ or $r_{ice\,new} = r_T - r_{s\,new}$.) in the absence of precipitation.

Solar heating of the cloud would cause diabatic warming on the thermo diagram.

Radiation is important in cloud evolution and cloud dynamics, particularly for the anvils of mesoscale convective systems (thunderstorms), hurricanes, and frontal stratus clouds (see the Clouds chapter).

INFO • Homeostasis and LeChatelier's Principle

Many systems in nature are in equilibrium as a result of a balance between opposing forcings. For such a system, if one of the forcings is changed, the system responds in a way to reach a new equilibrium that partially undoes the changed forcing.

In chemistry, this is known as **LeChatelier's Principle**. For example, if two reagents are in equilibrium in a beaker, then if you add more of one of the reagents, a reaction occurs that reduces the increase of concentration of the added reagent.

For global climate, the principle is known as **homeostasis**. For example, increased solar radiation will cause the earth-system temperature to increase, thereby causing increased cooling due to outgoing IR radiation (due to the Stefan-Boltzmann law) which partially compensates the original heating.

Many external influences (particularly solar heating) create instabilities in the atmosphere. The atmosphere responds by generating motions on a wide range of scales. These motions transport heat that partially undoes the effect of the external influence. We experience these responses as the global circulation, weather systems and storms.

Sample Application

Suppose an air parcel has an initial state of (P, T, T_d) = (95 kPa, 26°C, 3°C). It rises to an altitude where P = 35 kPa, which represents cloud top. While there, precipitation removes water amount Δr_T = 4 g kg^{-1}, and IR cooling lowers the temperature by ΔT = –18°C. If the air parcel then returns back to its original pressure, what is its final thermodynamic state? Use the full-size skew-T from the end of this chapter for this exercise. Explain all steps.

Find the Answer

Given: $(P, T, T_d)_{initial}$ = (95 kPa, 26°C, 3°C). P_{top} = 35 kPa. Δr = –4 g kg^{-1}, and ΔT = –18°C.
Find: $(P, T, T_d)_{final}$ = (95 kPa, ?°C, ?°C)

[1] The red dot and blue circle show the initial temperature and dew-point of the air parcel, respectively. Interpolating between the 4 and 6 g kg^{-1} isohumes, it appears that the initial dew point corresponds to roughly $r \approx$ 5 g kg^{-1}, which is plotted as the line with large blue dots with upward arrow.
[2] From [1] follow the dry adiabat for T (solid orange line), and follow the isohume for T_d (dotted blue line). They cross at an LCL of about $P_{LCL} \approx$ 68 kPa (given by the horizontal black line through that LCL point, and interpolating between the horizontal green lines to find the pressure). This is cloud base for ascending air.
[3] As the air parcel continues to rise to cloud top at P = 35 kPa, the temperature and dew-point follow the saturated adiabat (dashed orange line). Thus, at cloud top, $T = T_d \approx$ –38°C (which is the isotherm (diagonal green line) that goes through temperature point [3] at the tip of the upward arrowhead. The total-water mixing ratio is conserved and still follows the 5 g kg^{-1} isohume.
[4] Precipitation removes 4 g kg^{-1} from the parcel, leaving r_T = 5 –4 = 1 g kg^{-1} as indicated by the dotted blue isohume. But the parcel air temperature is unchanged from point [3].
[5] Infrared (IR) radiative cooling decreases the temperature by 18°C, causing a new cloud-top temperature of $T = T_d \approx$ –38 – 18 = – 56°C. The corresponding r_s value is so small that it is below 0.1 g kg^{-1}. But the total water content is unchanged from point [4].
[6] As this air is saturated (because $r_s < r_T$), the descent follows a saturated adiabat (dashed orange line) down to where the saturated adiabat crosses the isohume (dotted blue line). This intersection marks the LCL for descending air, which is at about P_{LCL} = 60 kPa (interpolate between the horizontal green line to the black line through this point).
[7] Below this cloud base, T follows the dry adiabat (solid orange line), ending with $T \approx$ 14°C at P = 95 kPa. The dew-point follows the 1 g kg^{-1} isohume (dotted blue line), giving a final $T_d \approx$ –17°C. **$(P, T, T_d)_{final}$ = (95 kPa, 14°C, –17°C)**

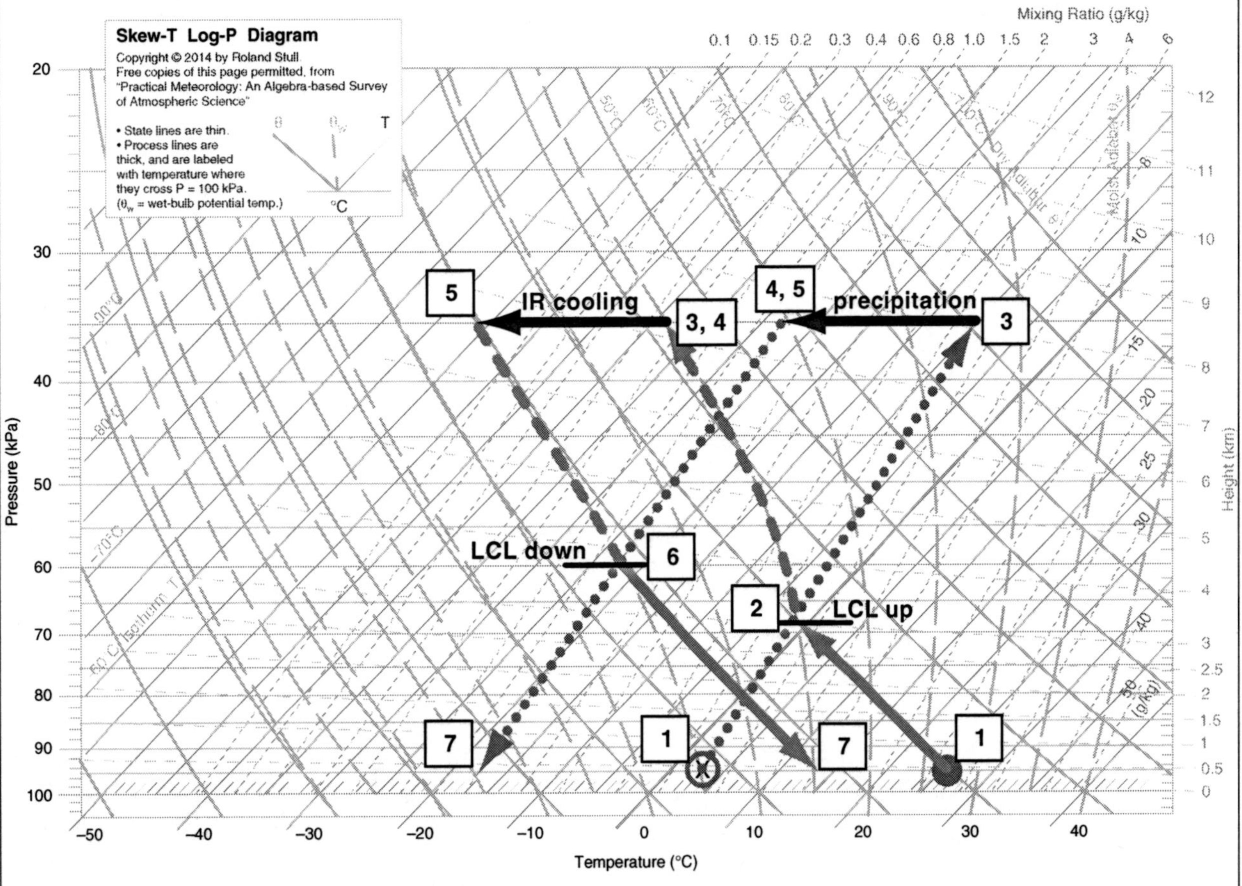

Sample Application

Suppose you made a sounding by releasing a radiosonde balloon, and it reported the following data. Plot it on a skew-T log-P diagram.

P (kPa)	*T* (°C)	T_d (°C)
20	–36	–80
30	–36	–60
40	–23	–23
50	–12	–12
70	0	–20
80	8	–10
90	10	10
100	10	10

Find the Answer

Here I used red for temperature (*T*) and blue for dewpoint (T_d). Always connect the data points with straight lines.

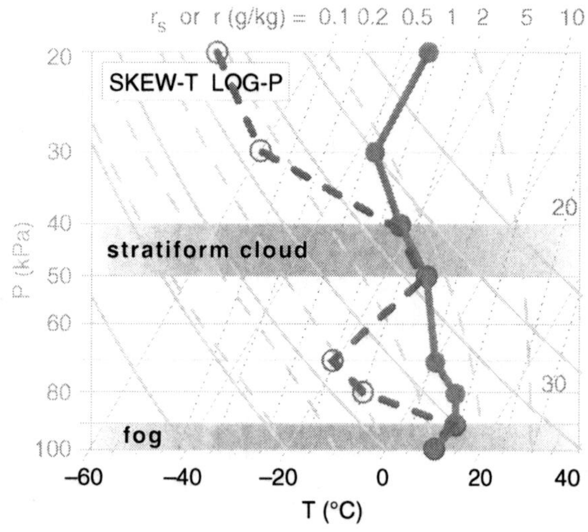

Check: Sketches OK. Also, dew point is never greater than temperature, as required.

Exposition: There are two levels where dew-point temperature equals the actual air temperature, and these two levels correspond to saturated (cloudy) air. We would call the bottom layer fog, and the elevated one would be a mid-level stratiform (layered) cloud. In the Clouds chapter you will learn that this type of cloud is called **altostratus**. (Note, from the sounding it is impossible to diagnose the detailed structure of that cloud, so it could be **altocumulus**.)

There are also two layers of dry air, where there is a large spread between *T* and T_d. These layers are roughly 80 to 60 kPa, and 35 to 20 kPa.

AN AIR PARCEL & ITS ENVIRONMENT

Consider the column of air that is fixed (Eulerian framework) over any small area on the globe. Let this air column represent a stationary **environment** through which other things (aircraft, raindrops, air parcels) can move.

For an air parcel that moves through this environment, we can follow the parcel (Lagrangian framework) to determine how its thermodynamic state might change as it moves to different altitudes. As a first approximation, we often assume that there is no mixing between the air parcel and the surrounding environment.

Soundings

As stated at the start of this chapter, the vertical profile of environmental conditions is called an **upper-air sounding**, or just a **sounding**. The word "sounding" is used many ways. It is the:

- activity of collecting the environment data (as in "to make a sounding").
- data that was so collected (as in "to analyze the sounding data").
- resulting plot of these data on a thermo diagram (as in "the sounding shows a deep unstable layer").

A **sonde** is an expendable weather instrument with built-in radio transmitter that can be attached to a platform that moves up or down through the air column to measure its environmental state. Helium-filled latex weather balloons carry **radiosondes** to measure thermodynamic state (*P, T,* T_d) or carry **rawinsondes** measure (*z, P, T,* T_d*, U, V*) by also utilizing GPS or other navigation signals. **Rocketsondes** are lofted by small sounding rockets, and **dropsondes** are dropped by parachute from aircraft. **Unmanned aerial vehicles (UAVs)** and conventional aircraft can carry weather sensors. Radar and satellites are **remote sensors** that can also measure portions of soundings.

The sounding represents a snapshot of the state of the air in the environment, such as in the Sample Application on this page. The plot of any dependent variable vs. height or pressure is called a **vertical profile**. The negative of the vertical temperature gradient ($-\Delta T_e/\Delta z$) is defined as the **environmental lapse rate**, where subscript $_e$ means environment.

Normally, temperature and dew point are plotted at each height. Other chemicals such as smoke particles or ozone can also be plotted as a sounding. It is difficult to measure density, so instead its value at any height is calculated from the ideal gas law. For cloudy air, the T_d points coincide with the *T* points. Most sondes have some imprecision, so we infer clouds in any layer where T_d is near *T*.

Straight lines are drawn connecting the temperature points (see INFO box on **mandatory** and **significant levels**), and separate straight lines are drawn for the humidity points for unsaturated air. Liquid or ice mixing ratios are not usually measured by radiosondes, but can be obtained by **research aircraft** flying slant ascent or descent soundings.

Buoyant Force

For an object such as an air parcel that is totally immersed in a fluid (air), buoyant force per unit mass (F/m) depends on the density difference between the object (ρ_o) and the surrounding fluid (ρ_f):

$$\frac{F}{m} = \frac{\rho_o - \rho_f}{\rho_o} \cdot g = g' \tag{5.2}$$

where gravitational acceleration (negative downward) is g = –9.8 m·s^{-2}. If the immersed object is **positively buoyant** (F = positive) then the buoyant force pulls upward. **Negatively buoyant** objects (F = negative) are pulled downward, while **neutrally buoyant** objects have zero buoyant force.

Recall that weight is the force caused by gravity acting on a mass ($F = m{\cdot}g$). Since buoyancy changes the net force (makes the weight less), the net effect on the object is that of a **reduced gravity** (g') as defined in eq. (5.2). Thus $F = m{\cdot}g'$.

Identify the air parcel (subscript p) as the object, and the environment (subscript e) as the surrounding fluid. Because both air parcels and the surrounding environment are made of air, use the ideal gas law ($P = \rho\, \Re\, T_v$) to describe the densities in eq. (5.2).

$$\frac{F}{m} = \frac{T_{ve} - T_{vp}}{T_{ve}} \cdot g = g' \tag{5.3a}$$

where the pressure inside the parcel is assumed to always equal that of the surrounding environment at the same height, allowing the pressures from the ideal gas law to cancel each other in eq. (5.3a).

Virtual temperature T_v (in Kelvins) is used to account for the effects of both temperature and water vapor on the buoyancy. For relatively dry air, $T_v \approx T$, giving:

$$\frac{F}{m} \approx \frac{T_p - T_e}{T_e} \cdot |g| = g' \tag{5.3b}$$

Namely, parcels warmer than their environment tend to rise. Colder parcels tend to sink. As air parcels move vertically, their temperatures change as given by adiabatic (dry or saturated) lapse rates, so you need to repeatedly re-evaluate the buoyancy of the parcel relative to its new local environment. Thermo diagrams are extremely handy for this, as sketched in Fig. 5.12.

INFO • Mandatory & Significant Levels

Rising rawinsondes record the weather at $\Delta z \approx 5$ m increments, yielding data at about 5000 heights. To reduce the amount of data transmitted to weather centers, straight-line segments are fit to the sounding, and only the end points of the line segments are reported. These points, called **significant levels**, are at the kinks in the sounding (as plotted in the previous Sample Application). Additional **mandatory levels** at the surface and 100, 92.5, 85, 70, 50, 40, 30, 25, 20, 15, 10, 7, 5, 3, 2, & 1 kPa are also reported, to make it easier to analyze **upper-air charts** at these levels.

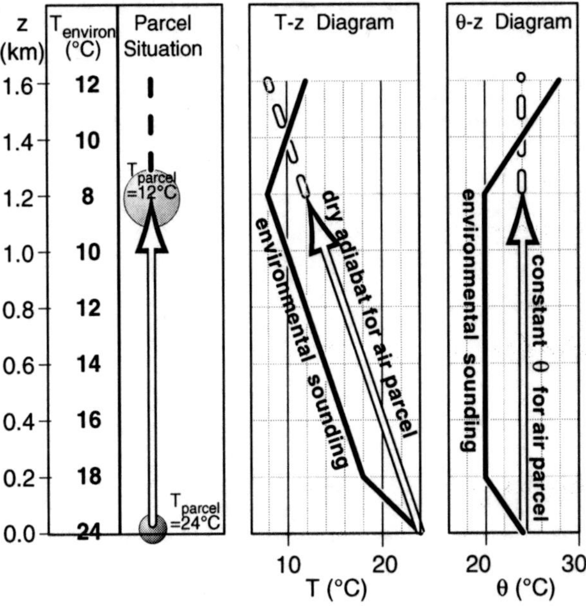

Figure 5.12
Buoyancy at any one height (z) depends on the temperature (T) or potential-temperature (θ) difference between the environment and the air parcel. A thermo diagram aids this, by allowing us to see the background environment and the air-parcel temperature at every height. In this example for unsaturated air lifted from the surface, the parcel is positively buoyant up to z = 1.4 km. Above that altitude, the parcel is colder than the environment and is negatively buoyant.

Sample Application

When the air parcel in Fig. 5.12 reaches 1 km altitude, what is its buoyant force/mass? The air is dry.

Find the Answer

Given: T_e = 10°C +273 = 283 K, T_p = 14°C +273 = 287 K
Find: F/m = ? m·s^{-2}

Apply eq. (5.3a) for dry air:
F/m = (287 K – 283 K)·(9.8 m·s^{-2}) / 283 K
= <u>0.14 m·s^{-2}</u> = 0.14 N/kg (see units in Appendix A)

Check: Physics and units are reasonable.
Exposition: Positive buoyancy force favors rising air.

Sample Application

a) Use a thermo diagram to estimate the potential-temperature values for each point in the environmental sounding, and plot your answers a new, linear graph of θ vs. *P*. Assume dry air. Environmental sounding: [*P*(kPa), *T*(°C)] = [100, 20], [95, 10], [85, 0], [75, 0], [70, –4]

b) An air parcel at [*P*(kPa), *T*(°C)] = [100, 23] moves to a new height of 80 kPa. Plot its path on your θ vs. *P* graph, and find its new buoyant force/mass at 80 kPa.

Find the Answer

Given: The *T* vs. *P* data above. Dry, thus $T_v = T$.
Find: (a) Estimate θ from thermo diag. & plot vs. *P*.
(b) Plot parcel rise on the same graph.
Calculate buoyant F/m at P = 80 kPa.

Method

a) From the large thermo diagrams at the chapter end, the environment potential temperature sounding is:
[*P*(kPa), θ(°C)]
[100,20]
[95,15]
[85,15]
[75,25]
[70,26]

See blue line in the figure:

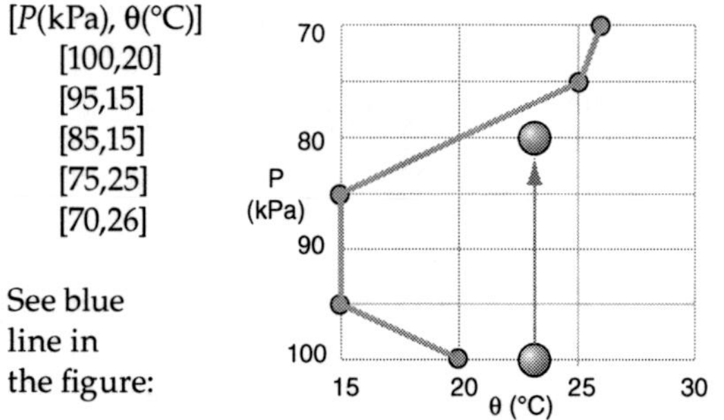

b) The parcel initial state is [*P*(kPa), θ(°C)] = [100,23]. See red line in figure. At *P* = 80 kPa, the figure shows Δθ = 23 – 20°C = 3°C = 3 K. Because dry air: $\Delta\theta_v = \Delta\theta$.
$T_v = T \approx 10°C = 283$ K as approximate average *T*.
Apply eq. (5.3d):
$F/m = (3K/283K)\cdot(9.8 \text{ m s}^{-2}) = 0.10 \text{ m s}^{-2} =$ **<u>0.10 N kg⁻¹</u>**

Check: Physics and units are reasonable.
Exposition: The upward buoyancy force causes the air parcel to continue to rise, until it hits the blue line.

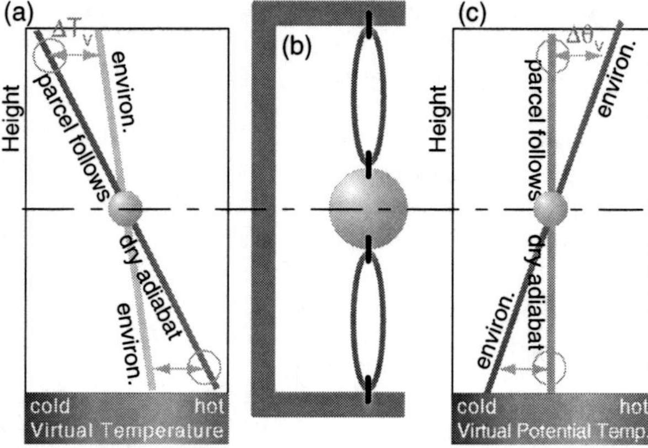

Figure 5.13
(a) In a statically stable environment, a buoyantly neutral air parcel (blue) becomes colder than the environment when lifted (top grey circle) and warmer if lowered (bottom grey circle). (b) The opposing forces act on the parcel similar to a weight suspended between two elastic bands. (c) Same concept using θ_v.

Sometimes it is more convenient to use virtual potential temperature θ_v (eqs. 3.13, 3.14) to calculate buoyant force, because adiabatic parcel motion has constant $\theta_{v\,p}$. Buoyant force per mass is then:

$$\frac{F}{m} = \frac{\theta_{ve} - \theta_{vp}}{T_{ve}} \cdot g = g' \qquad (5.3c)$$

where gravitational acceleration (g = –9.8 m s^{-2}) is a negative number in eq. (5.3c) because it acts downward. Subscript *e* indicates "environment".

Many meteorologists prefer to use the magnitude of gravitational acceleration as a positive number:

$$\frac{F}{m} = \frac{\theta_{vp} - \theta_{ve}}{T_{ve}} \cdot |g| = g' \qquad (5.3d)$$

which more obviously shows that upward buoyancy force occurs for a parcel warmer then the environment at its same altitude or pressure.

Picture the physical situation in the left third of Fig. 5.12 on the previous page. Suppose you grab a small piece of the environmental air near the ground and lift it. As the parcel rises, it cools adiabatically. It is moving through a somewhat static environment, and thus is encountering different surrounding (environmental) temperatures at different altitudes (represented by the colored background). Namely, both the parcel temperature and the environmental temperature are different at different altitudes.

We need to know these different temperatures at each altitude in order to determine the temperature difference and the resulting parcel buoyancy from eqs. (5.3). To make our lives easier, we use thermo diagrams on which we can plot the background environment, and which show dry adiabats as relate to parcel rise. Two examples are the emagram and θ-Z diagrams in Fig. 5.12. These both represent the same parcel and same environment.

Brunt-Väisälä Frequency

Suppose the ambient environment lapse rate Γ is less the dry adiabatic lapse rate (Γ_d = 9.8°C km^{-1}). Namely, temperature does not decrease as fast with increasing height as Γ_d (see Fig. 5.13a).

Suppose you start with an air parcel having the same initial temperature as the environment (the blue sphere in Fig. 5.13a) at any initial altitude. If you push the air parcel down, it warms at the adiabatic lapse rate, which makes it warmer than the surrounding environment at its new height. This positive buoyant force would tend to push the parcel upward. The parcel's inertia would cause it to overshoot upward past its initial height, where adiabatic cooling would create a negative buoyant force that reverses the parcel's upward motion.

In summary, a parcel that is displaced vertically from its initial altitude will oscillate up and down. The environment that enables such an oscillation is said to be statically stable. The oscillation is similar to that experienced by a vertically displaced weight suspended between two elastic bands (Fig. 5.13b).

The air parcel oscillates vertically with frequency (radians s^{-1})

$$N_{BV} = \sqrt{\frac{|g|}{T_v} \cdot \left(\frac{\Delta T_v}{\Delta z} + \Gamma_d \right)} \qquad \bullet(5.4a)$$

where N_{BV} is the Brunt-Väisälä frequency, and $|g|$ = 9.8 m s^{-2} is the magnitude of gravitational acceleration. Virtual temperature T_v must be used to account for water vapor, which has lower density than dry air. If the air is relatively dry, then $T_v \approx T$. Use absolute units (K) for T_v in the denominator.

The same frequency can be expressed in terms of virtual potential temperature θ_v (see Fig. 5.13c):

$$N_{BV} = \sqrt{\frac{|g|}{T_v} \cdot \frac{\Delta \theta_v}{\Delta z}} \qquad \bullet(5.4b)$$

The oscillation period is

$$P_{BV} = \frac{2\pi}{N_{BV}} \qquad (5.5)$$

If the static stability weakens (as the environmental $\Delta T/\Delta z$ approaches $-\Gamma_d$), then the frequency decreases and the period increases toward infinity. The equations above are idealized, and neglect the damping of the oscillation due to air drag (friction).

Sample Application

For a dry standard-atmosphere defined in Chapter 1, find the Brunt-Väisälä frequency & period at z = 4 km.

Find the Answer

Given: $\Delta T/\Delta z = -6.5$ K/km from eq. (1.16).
$T_v = T$ because dry air.
Use Table 1-5 at z = 4 km to get: $T = -11°C = 262$ K
Find: N_{BV} = ? rad s^{-1}, P_{BV} = ? s

Apply eq. (5.4a)

$$N_{BV} = \sqrt{\frac{(9.8\text{ms}^{-2})}{262\text{K}} \cdot \left(-6.5 + 9.8\frac{\text{K}}{\text{km}}\right) \cdot \left(\frac{1\text{km}}{1000\text{m}}\right)}$$

$$= [\,1.234\text{x}10^{-4}\ \text{s}^{-2}]^{1/2} = \underline{0.0111\ \text{rad s}^{-1}}$$

Apply eq. (5.5):

$$P_{BV} = \frac{2\pi \text{ radians}}{0.0111 \text{ radians/s}} = \underline{\mathbf{565.5\ s}} = 9.4 \text{ min}$$

Check: Physics and units are reasonable.

Exposition: The Higher Math box at right shows that N_{BV} must have units of (radians s^{-1}) because it is the argument of a sine function. But meteorologists often write the units as (s^{-1}), where the radians are implied.

HIGHER MATH • Brunt-Väisälä Frequency

Create the Governing Equations

Consider a scenario as sketched in Fig. 5.13a, where the environment lapse rate is Γ and the dry adiabatic lapse rate is Γ_d. Define the origin of a coordinate system to be where the two lines cross in that figure, and let z be height above that origin. Consider dry air for simplicity ($T_v = T$). Based on geometry, Fig. 5.13a shows that the temperature difference between the two lines at height z is $\Delta T = (\Gamma - \Gamma_d)\cdot z$.

Eq. (5.3b) gives the corresponding vertical force F per unit mass m of the air parcel:

$$F/m = |g|\cdot\Delta T/T_e = |g|\cdot[(\Gamma - \Gamma_d)\cdot z]/T_e$$

But $F/m = a$ according to Newton's 2nd law, where acceleration $a \equiv d^2z/dt^2$. Combining these eqs. gives:

$$\frac{d^2 z}{dt^2} = \frac{|g|}{T_e}\cdot(\Gamma - \Gamma_d)\cdot z$$

Because T_e is in Kelvin, it varies by only a small percentage as the air parcel oscillates, so approximate it with the average environmental temperature $\overline{T_e}$ over the vertical span of oscillation.

All factors on the right side of the eq. above are constant except for z, so to simplify the notation, use a new variable $N_{BV}{}^2 \equiv -|g|\cdot(\Gamma - \Gamma_d)/\overline{T_e}$ to represent those constant factors. [This definition agrees with eq. (5.4a) when you recall that $\Gamma = -\Delta T/\Delta z$.]

This leaves us with a second-order differential equation (a hyperbolic differential eq.) known as the wave equation:

$$\frac{d^2 z}{dt^2} = -N_{BV}^2 \cdot z \qquad (5a)$$

Solve the Differential Equation

For the wave equation, try a wave solution:

$$z = A\cdot\sin(f\cdot t) \qquad (5b)$$

with unknown amplitude A and unknown frequency of oscillation f (radians s^{-1}). Insert (5b) in (5a) to get:

$$-f^2 \cdot A \cdot \sin(f \cdot t) = -N_{BV}^2 \cdot A \cdot \sin(f \cdot t)$$

After you cancel the $A\cdot\sin(f\cdot t)$ from each side, you get:

$$f = N_{BV}$$

Using this in eq. (5b) gives an air parcel height that oscillates in time:

$$z = A \cdot \sin(N_{BV} \cdot t) \qquad (5c)$$

Exposition

The amplitude A is the initial distance that you displace the air parcel from the origin. If you plug eq. (5.5) into (5c) you get: $z = A\cdot\sin(2\pi\cdot t/P_{BV})$, which gives one full oscillation during a time period of $t = P_{BV}$.

FLOW STABILITY

Stability is a characteristic of how a system reacts to small disturbances. If the disturbance is damped, the system is said to be **stable**. If the disturbance causes an amplifying response (irregular motions or regular oscillations), the system is **unstable**.

For **fluid-flow stability** we will focus on turbulent responses spanning the smallest eddies to deep thunderstorms. The stability characteristics are:

- **Unstable** air becomes, or is, **turbulent** (irregular, gusty, stormy).
- **Stable** air becomes, or is, **laminar** (non-turbulent, smooth, non-stormy).
- **Neutral** air has no tendency to change (disturbances neither amplify or dampen).

Flow stability is controlled by ALL the processes (buoyancy, inertia, wind shear, rotation, etc.) acting on the flow. However, to simplify our understanding of flow, we sometimes focus on just a subset of processes. If you ignore all processes except buoyancy, then you are studying **static stability**. If you include buoyancy and wind-shear processes, then you are studying **dynamic stability**.

Static Stability

One way to estimate static stability is by taking a small piece of the environment and hypothetically displacing it as an air parcel a small distance vertically from its starting point, assuming the surrounding environment is horizontally large and is quasi-stationary. Another is to lift whole environmental layers. We will look at both methods.

Parcel Method for Static Stability

Will a displaced air parcel experience buoyancy forces that push it in the same direction it was displaced (i.e., an amplifying or unstable response), or will buoyancy forces tend to push the parcel back to its starting (equilibrium) height (a stable response)?

The answer to this question is tricky, because the region of unstable response can span large vertical regions. For example, a parcel that is warmer than its environment will keep rising over large distances so long as it remains warmer than its surrounding environment at the same altitude as the parcel.

This type of buoyancy motion, called **convection**, stirs the air and generates turbulence over the whole span of its rise. Namely, a thin region that is locally stable might become turbulent if an air parcel moves through it from some distant source. Such distant effects are known as **nonlocal turbulence**, and must be considered when determining stability.

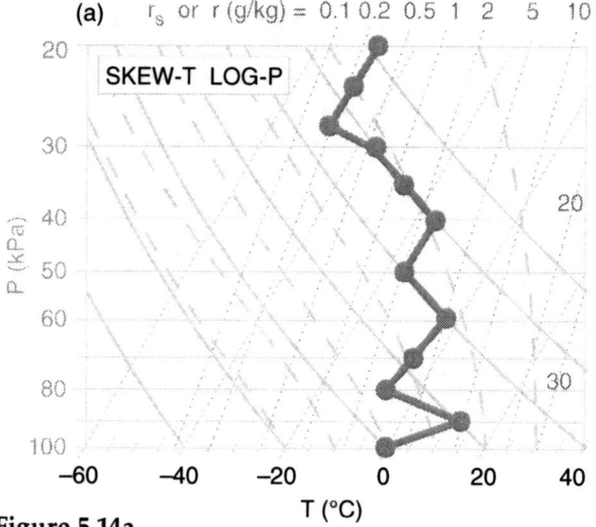

Figure 5.14a
Static stability determination. Step 1: Plot the sounding. (This sounding is contrived to illustrate all stabilities.)

Step 1: Graphical methods are usually best for determining static stability, in order to account for nonlocal effects. Plot the environmental sounding on a thermo diagram (Fig. 5.14a). Any type of thermo diagram will work — they all will give the same stability determination.

Step 2: At every apex (kink in the sounding), hypothetically displace an air parcel from that kink a small distance upward. If the parcel is unsaturated, follow a dry adiabat upward to determine the parcel's new temperature. If saturated, follow a moist adiabat. If that displaced parcel experiences an upward buoyant force (i.e., is warmer than its surrounding environment at its displaced height), then it is locally unstable, so write the letter "U" just above that apex (Fig 5.14b). If the upward displaced

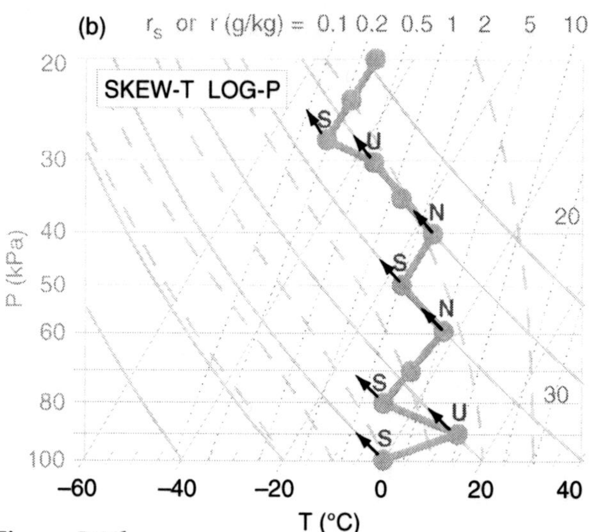

Figure 5.14b
Static stability determination. Step 2: Upward displacements. (Assume this contrived sounding is unsaturated.)

parcel experiences no force (i.e., is nearly the same temperature as the environment at its displaced height), then write "N" just above that apex because it is locally neutral. If the upward displaced parcel experiences a downward force (i.e., is cooler than the environment at its displaced height), then write "S" just above the apex because it is locally stable.

Step 3: Do a similar exercise at every apex, but hypothetically displacing the air parcel downward (Fig. 5.14c). If the parcel is cooler than its environment at its displaced height and experiences a downward force, then it is locally unstable, so write "U" just below that apex. If the parcel is the same temperature as the environment at its displaced height, write "N" just below the apex. If the parcel is warmer than the environment at its displaced height, then write "S" just below the apex.

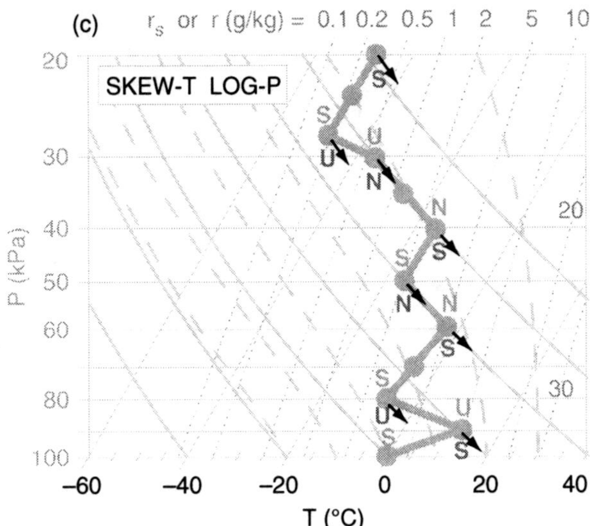

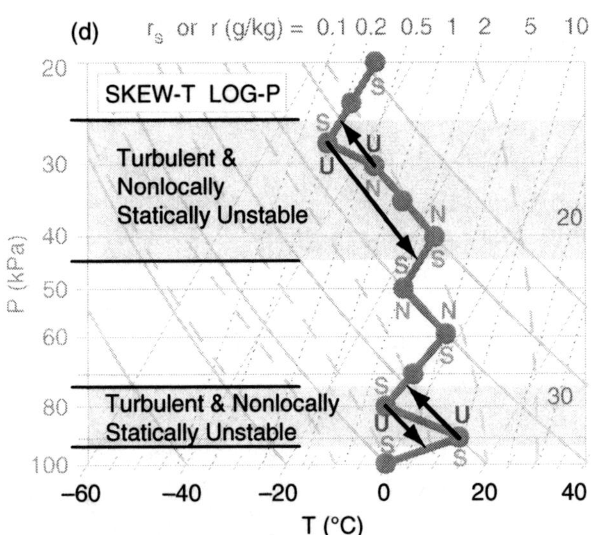

Figure 5.14c & d
Static stability determination. (c) Step 3: Downward displacement. (d) Step 4: Identify unstable regions (shaded yellow).

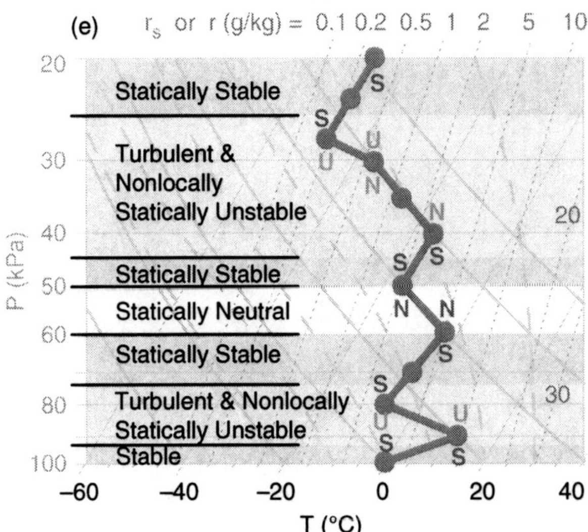

Figure 5.14e
Static stability determination. Step 5: Identify statically stable regions (shaded light blue) and neutral regions (no shading).

Step 4: From every apex with U above, draw an arrow that follows the adiabat upward until it hits the sounding (Fig. 5.14d). From every apex with U below, draw an arrow that follows the adiabat downward until it hits the sounding. All domains spanned by these "U" arrows are turbulent and are nonlocally statically unstable. Any vertically overlapping "U" regions should be interpreted as one contiguous nonlocally unstable and turbulent region.

Step 5: Any regions outside of the unstable regions from the previous step are stable or neutral, as indicated by the "S" or "N" letters next to those sounding line segments (Fig. 5.14e).

Within statically **neutral** subdomains, the vertical gradients of temperature T or potential temperature θ are:

$$\frac{\Delta T}{\Delta z} \approx -\begin{cases}\Gamma_d & \text{if unsaturated} \\ \Gamma_s & \text{if saturated}\end{cases} \qquad \bullet(5.6a)$$

or

$$\frac{\Delta \theta}{\Delta z} \approx \begin{cases}0 & \text{if unsaturated} \\ \Gamma_d - \Gamma_s & \text{if saturated}\end{cases} \qquad \bullet(5.6b)$$

where $\Gamma_d = 9.8$ K km^{-1} is the dry adiabatic lapse rate, and Γ_s is the saturated (or moist) adiabatic lapse rate (which varies, but is always a positive number).

In statically **stable** subdomains, the temperature does not decrease with height as fast as the adiabatic rate (including isothermal layers, and **temperature inversion** layers where temperature increases with height). Thus:

$$\frac{\Delta T}{\Delta z} > -\begin{cases}\Gamma_d & \text{if unsaturated} \\ \Gamma_s & \text{if saturated}\end{cases} \qquad \bullet(5.7a)$$

or

$$\frac{\Delta \theta}{\Delta z} > \begin{cases}0 & \text{if unsaturated} \\ \Gamma_d - \Gamma_s & \text{if saturated}\end{cases} \qquad \bullet(5.7b)$$

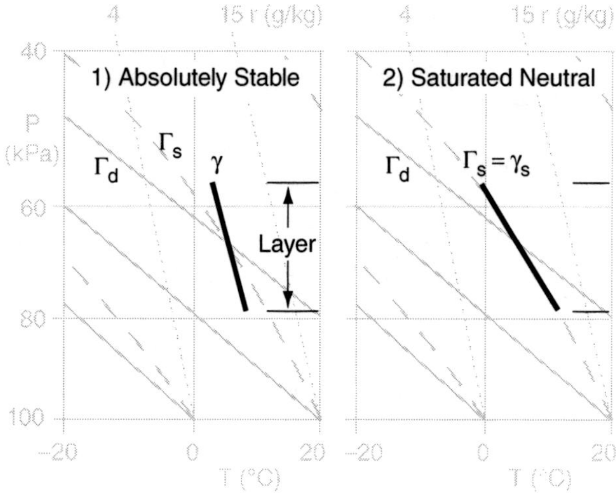

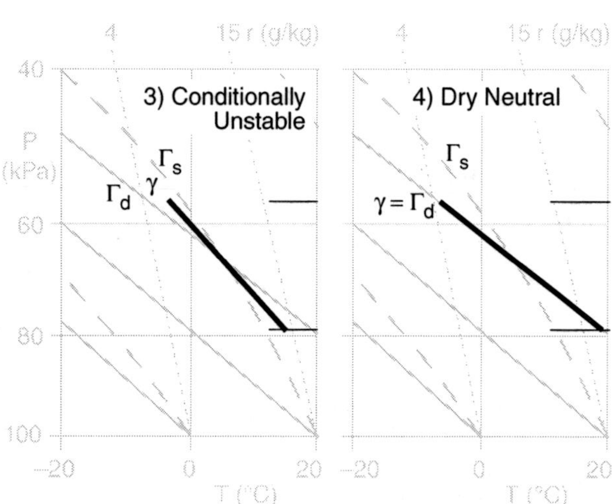

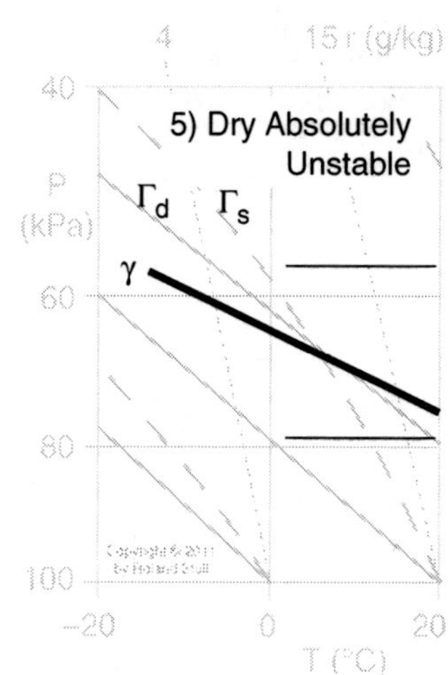

Figure 5.15

Layer stabilities, shown on a portion of an Emagram. The thick solid line is environmental air temperature within one layer of air (i.e., a line segment from a complete sounding), and the thick dotted line is environmental dew-point temperature. Γ_d and Γ_s are the dry and saturated adiabatic lapse rates.

Layer Method for Static Stability

Sometimes a whole layer of environmental air is lifted or lowered by an outside process. Synoptic-scale warm and cold fronts (see the Fronts & Airmasses chapter) and coherent clusters of thunderstorms (**mesoscale convective complexes**, see the Thunderstorm chapters) are examples of layer-lifting processes. In these cases we cannot use the parcel method to determine static stability, because it assumes a quasi-stationary environment.

Lapse Rate Names

The lapse rate γ of an environmental layer is defined as the temperature decrease with height:

$$\gamma = -\Delta T / \Delta z \tag{5.8}$$

Lapse rates are named as shown in Table 5-1, where isothermal and inversion are both subadiabatic.

Table 5-1. Names of environmental lapse rates, where Γ_d is the dry adiabatic lapse rate.

Name	Layer Lapse Rate	Fig.
adiabatic	$\gamma = \Gamma_d = 9.8°C\ km^{-1}$	5.15(4)
superadiabatic	$\gamma > \Gamma_d$	5.15(5)
subadiabatic	$\gamma < \Gamma_d$	5.15(1)
• **isothermal**	T = constant with z	
• **inversion**	T = increases with z	

Layer Stability Classes: Five classes of layer stability are listed in Table 5-2. For a layer of air that is already saturated (namely, it is a layer of clouds, with $T = T_d$), then the environmental lapse rate γ, which is still defined by the equation above, is indicated as γ_s to remind us that the layer is saturated. As before, the word "dry" just means unsaturated here, so there could be moisture in the air.

Table 5-2. Layer static stability, where $\Gamma_d = 9.8°C\ km^{-1}$ is the dry adiabatic lapse rate, and Γ_s is the saturated adiabatic lapse rate. Γ_s is always $\le \Gamma_d$.

Name	Layer Lapse Rate
1) absolutely stable	$\gamma < \Gamma_s$
2) saturated neutral	$\gamma_s = \Gamma_s$
3) conditionally unstable	$\Gamma_s < \gamma < \Gamma_d$
4) dry neutral	$\gamma = \Gamma_d$
5) dry absolutely unstable	$\gamma > \Gamma_d$

Fig. 5.15 illustrates these stabilities. The layer of air between the thin horizontal lines has a linear lapse rate shown by the thick line. The slope of this thick line relative to the slopes of the dry Γ_d and moist Γ_s adiabats determines the layer stability class.

Conditional Stability: One of the types of layer stability listed in Table 5-2 is conditional stability. A layer of conditionally unstable air has lapse rate between the dry and moist adiabatic lapse rates. This layer is stable if the air is unsaturated, but is unstable if the air is saturated (cloudy). Hence, the "condition" refers to whether the air is cloudy or not.

Conditionally unstable air, if over a deep enough layer, is a favorable environment for thunderstorms to grow. A special set of **thunderstorm stability indices** has been developed to determine if thunderstorms will form, and how intense they might be. These are discussed in the Thunderstorm chapters of this book. The presence of thunderstorms indicate violently unstable air that extends throughout the depth of the troposphere.

When a whole layer is lifted, the temperature at each level within that layer will change according to whether those levels are saturated or not. Fig. 5.16 shows a sample sounding, and how temperatures change following dry adiabats until reaching their local LCL, above which they follow the moist adiabat. This particular example shows how a layer that starts as conditionally unstable can become absolutely unstable if the bottom of the layer reaches saturation sooner than the top.

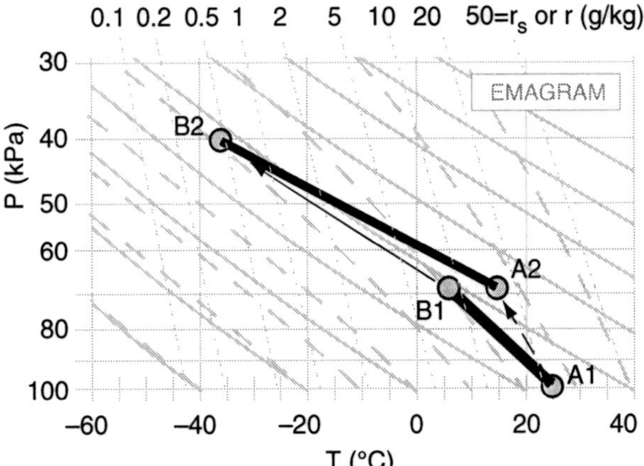

Figure 5.16
Example of lifting a layer to determine layer stability. Conditionally unstable layer (thick black line between points A1 and B1 on this Emagram) is lifted 30 kPa. Suppose that the bottom of the layer is already saturated, but the top is not. Upon lifting, the bottom of the layer will follow a moist adiabat from A1 to A2, while the top follows a dry adiabat from B1 to B2. The layer at its new height (black line A2 to B2) is now absolutely unstable.

Dynamic Stability

Dynamic stability considers both buoyancy and wind shear to determine whether the flow will become turbulent. **Wind shear** is the change of wind speed or direction with height, and can be squared to indicate the kinetic energy available to cause turbulence.

The ratio of buoyant energy to shear-kinetic energy is called the **bulk Richardson number**, *Ri*, which is dimensionless:

$$Ri = \frac{|g|\cdot(\Delta T_v + \Gamma_d \cdot \Delta z)\cdot \Delta z}{T_v \cdot\left[(\Delta U)^2 + (\Delta V)^2\right]} \qquad \bullet(5.9a)$$

and

$$Ri = \frac{|g|\cdot \Delta\theta_v \cdot \Delta z}{T_v \cdot\left[(\Delta U)^2 + (\Delta V)^2\right]} \qquad \bullet(5.9b)$$

where [$\Delta\theta_v$, ΔU, ΔV, ΔT_v] are the change of [virtual potential temperature, east-west wind component, north-south wind component, virtual temperature] across a layer of thickness Δz. As before, Γ_d = 9.8 ·K km^{-1} is the dry adiabatic lapse rate, and T_v must be in absolute units (K) in the denominator of eq. (5.9).

Sample Application

Given these data, is the atmosphere turbulent?

z (km)	T_v (°C)	U (m s^{-1})	V (m s^{-1})
5.5	–23	50	0
5.0	–20	40	8

Find the Answer

Given: ΔT = –3°C, ΔV = –8, ΔU = 10 m s^{-1}, Δz = 0.5 km, where all vertical differences are (top – bottom)

Find: Turbulent (yes/no) ?

First, examine static stability.

$\Delta\theta_v = \Delta T_v + \Gamma_d \cdot \Delta z$
$= (-3°C) + (9.8\ °C\ km^{-1})\cdot(0.5\ km) = 1.9°C$

Because the top of the layer has warmer θ_v than the bottom, the air is locally statically stable. No nonlocal info to alter this conclusion. Thus, static stability does not make the air turbulent. But need to consider wind.

Apply eq. (5.9b). Use average T_v = –21.5+273 = 251.5 K

$$Ri = \frac{(9.8 m\cdot s^{-2})\cdot(1.9K)\cdot(500m)}{(251.5K)\cdot\left[(10)^2 + (-8)^2 (m/s)^2\right]} = \underline{0.226}$$

This *Ri* below the critical value of 0.25, therefore the flow is turbulent because of dynamic instability.

Check: Physics are reasonable. *Ri* is dimensionless.
Exposition: Turbulence tends to mix the air, which reduces wind shears and temperature differences, and thus undoes the dynamic instability that caused it, as expected by LeChatelier's Principle (see INFO box).

Sample Application

For the following sounding, determine the regions of turbulence.

z (km)	T (°C)	U (m s^{-1})	V (m s^{-1})
2	0	15	0
1.5	0	12	0
1.2	2	6	4
0.8	2	5	4
0.1	8	5	2
0	12	0	0

Find the Answer:

Given: The sounding above.

Find: a) Static stability (nonlocal parcel apex method), b) dynamic stability, & (c) identify turbulence.

Assume dry air, so $T = T_v$.

Method: Use spreadsheet to compute θ and Ri. Note that Ri applies to the layers between sounding levels.

z (km)	θ (°C)	z_{layer} (km)	Ri
2	19.6		
1.5	14.7	1.5 to 2.0	9.77
1.2	13.8	1.2 to 1.5	0.19
0.8	9.8	0.8 to 1.2	55.9
0.1	9.0	0.1 to 0.8	5.25
0	12	0 to 0.1	–0.358

Next, plot these results:

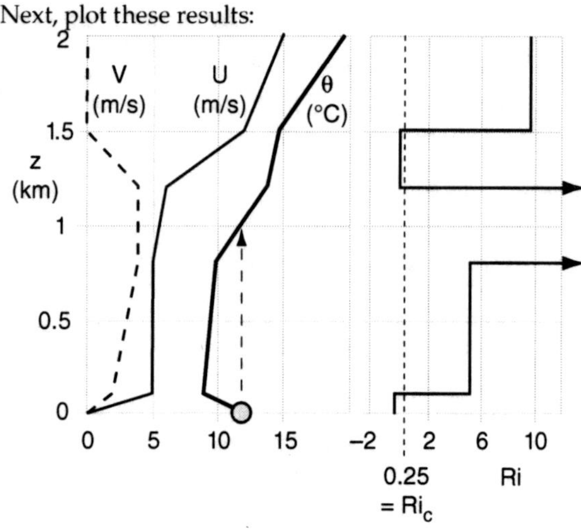

a) Static Stability: Unstable & turbulent for z = 0 to 1 km, as shown by nonlocal air-parcel rise in the θ sounding.

b) Dynamic Stability: Unstable & turbulent for z = 0 to 0.1 km, & for z = 1.2 to 1.5 km, where $Ri < 0.25$.

c) Turbulence exists where the air is statically OR dynamically unstable, or both. Therefore: **Turbulence at 0 - 1 km, and 1.2 to 1.5 km.**

Check: Physics, sketch & units are reasonable.

Exposition: At z = 0 to 1 km is the mixed layer (a type of atmos. boundary layer). At z = 1.2 to 1.5 is **clear-air turbulence** (**CAT**) and perhaps K-H waves.

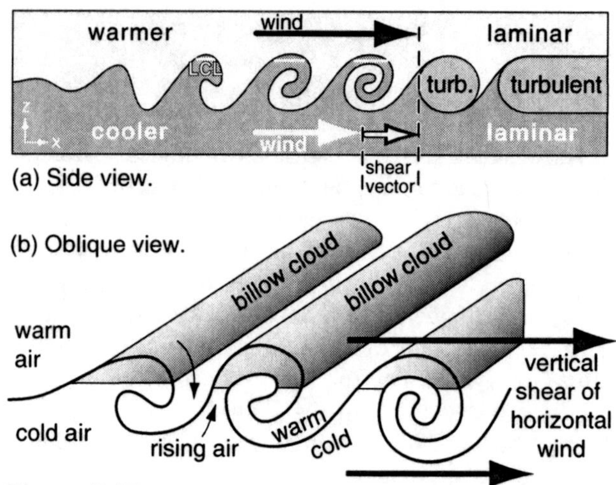

Figure 5.17
(a) Side view of Kelvin-Helmholtz waves growing in a dynamically unstable region: wind shear across a statically stable layer. (b) View showing parallel cloud bands called billow clouds

For the special case of statically stable air:

$$Ri = \frac{N_{BV}{}^2 \cdot (\Delta z)^2}{(\Delta U)^2 + (\Delta V)^2} \tag{5.9c}$$

N_{BV} is undefined for air that is not statically stable.

For thin layers of atmosphere, the system is **dynamically unstable** if

$$Ri < Ri_c \qquad \bullet(5.10)$$

Ri_c = 0.25 is the **critical Richardson number**.

When statically stable laminar flow is just becoming dynamically unstable due to an increase in wind shear, the dynamically unstable layer behaves like breaking ocean waves (Fig. 5.17) in slow motion. These waves are called **Kelvin-Helmholtz waves**, or **K-H waves** for short. As these waves break, the dynamically unstable layer becomes turbulent, and is known as **clear-air turbulence** (**CAT**) if outside thunderstorms and above the boundary layer.

If the bottom layer of cool air is humid enough, then clouds can form in any wave crest that is higher than its lifting condensation level (LCL). Since these clouds form narrow parallel bands perpendicular to the shear vector, the clouds look like billows of an accordion, and are called **billow clouds** (Fig. 5.17).

Existence of Turbulence

Statically stable flows can be dynamically unstable and can become turbulent if the wind shear is strong enough.

Nonlocally statically unstable flow always becomes turbulent regardless of the wind shear. Thus, either dynamic instability or nonlocal static instability are sufficient to enable turbulence.

However, laminar flow exists only if the layer of air is stable both dynamically and statically.

FINDING TROPOPAUSE HEIGHT AND MIXED-LAYER DEPTH

Tropopause

Recall from Chapter 1 (Fig 1.10) that the standard-atmosphere temperature decreases with height within the troposphere, but is isothermal with height in the bottom part of the stratosphere. Isothermal layers are very strongly statically stable. The strong and deep statically stable layer in the stratosphere discourages vertical motion through it, which is why most of our weather (including thunderstorms and hurricanes) is trapped below it. The base of the stratospheric isothermal layer is the tropopause.

Real soundings are more complex than the simple standard atmosphere, and can have many different stable layers (and isothermal layers) at different heights (Fig. 5.18). Nonetheless, if the radiosonde balloon rises high enough before bursting, then it often can enter the stratosphere. To locate the tropopause, we just need to look for the bottom of the very thick isothermal layer that is near a pressure altitude of 40 to 20 kPa.

Strong static stability can be found by regions where potential-temperature lines (**isentropes**) are packed close together. The strongly stable stratosphere has tighter packing of isentropes than the less-stable troposphere, such as sketched in the idealized vertical cross section of Fig. 5.19. Thus, you can locate the tropopause at the bottom of the region of tight isentrope packing, as sketched in Fig. 5.19. [*CAUTION*: *The tropopause usually does NOT follow any single isentrope.*]

To illustrate the relationship between stability and isentrope spacing, suppose a rawinsonde (i.e., a helium-filled weather balloon) is released from near the ground and rises along the orange line in Fig. 5.19. While it rises, it measures temperature and pressure of the surrounding environment, allowing one to calculate the vertical profile of potential temperature θ.

It starts in cold air near the ground, where the large spacing between isentropes shows that the static stability is weak near the ground. At about 4 km above the ground it passes through a frontal inversion, where static stability is stronger as shown by the close spacing of the two isentropes (θ = 260 & 270 K). Higher, in the mid troposphere, the larger spacing indicates weaker stability. Starting at about 10 km altitude the sonde enters a region of close spacing of isentropes, indicating that the sonde has entered the stratosphere. The bottom of this strongly stable region is the tropopause.

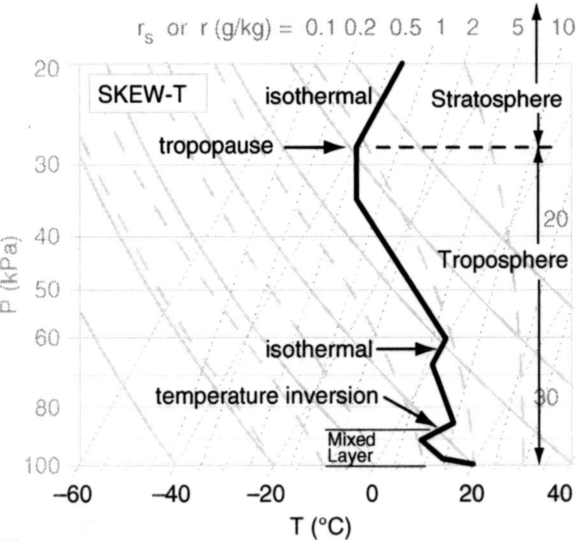

Figure 5.18
Sample atmospheric sounding having two isothermal layers and one temperature inversion.

> **Sample Application**
> What is the tropopause height (or pressure) in the sounding of Fig. 5.18, and how is it identified?
>
> **Find the Answer**
> Given: the plotted sounding.
> Find: P_{Trop} = ? kPa
>
> The base of the deep isothermal layer near the top of this sounding is at P_{Trop} ≈ **28 kPa**. The other isothermal and inversion layers are too low, and too thin.
>
> **Check**: Sketch OK.
> **Exposition**: The tropopause is lower near the poles and higher near the equator. It is lower in winter, higher in summer (see Table 6-1 in the Cloud chapter), but varies considerably from day to day.

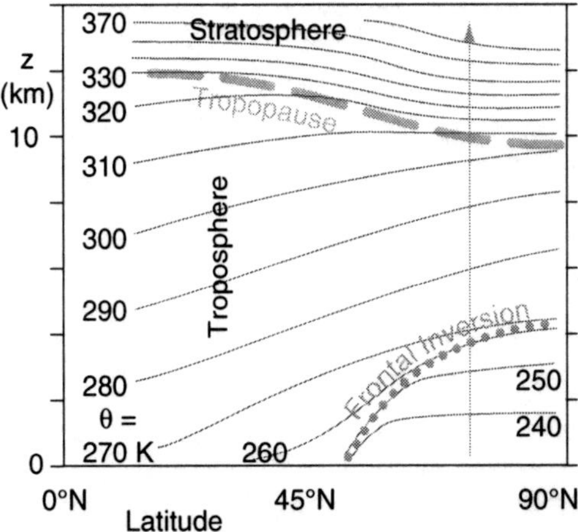

Figure 5.19
Idealized vertical cross section. Isentropes (connecting points of equal θ) are dark blue. Frontal inversion is dotted brown. Tropopause is dashed green. Red arrow indicates rawinsonde.

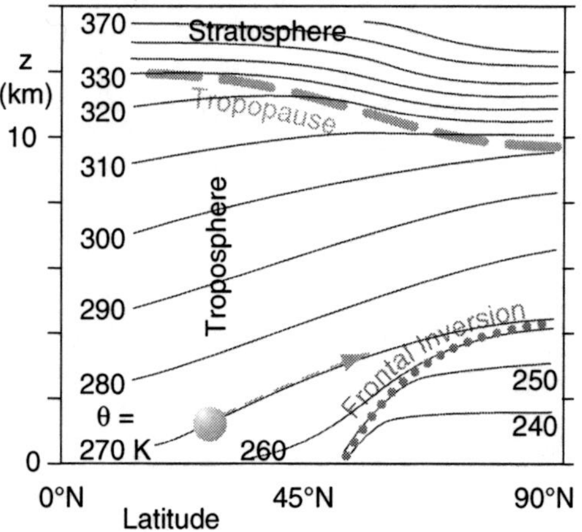

Figure 5.20
Same as Fig. 5.19, but now the cyan sphere represents an air parcel of potential temperature θ = 270 K, which rides along the environmental 270 K isentrope as the parcel is advected toward the North Pole (90°N). Thick green dashed line indicates the tropopause. Thin blue dashed line shows the air-parcel path.

As air parcels advect (move) horizontally, they also tend to move vertically to stay on a surface of constant potential temperature (i.e., an **isentropic surface**), assuming no heat transfer to/from the parcel. The reason is buoyancy. For example, the parcel at 30°N illustrated in Fig. 5.20 has a potential temperature of θ = 270 K, so it would try to stay on the 270 K isentrope (dark blue line in that figure).

If it were to stray slightly [above, below] that surface, it would be surrounded by air that has [warmer, cooler] potential temperature, causing a buoyant force that would tend to [lower, raise] the parcel back to the altitude of the 270 K isentrope. **Isentropic analysis** is a method of plotting weather variables on constant θ surfaces as a way to estimate dynamics under adiabatic conditions.

Sample Application

Plot the following sounding on a Skew-T diagram. If the ground heats an air parcel to the following state [P=100 kPa, T = 25°C], then find the pressure of the mixed-layer top. Sounding: [P (kPa), T (°C)] = [100,22] , [99,19] , [85,7] , [80,5] , [75,5] , [60,–5] , [50,–10] , [30,–40] , [20,–40] .

Find the Answer

Given: the data above.
Find: P_i (pressure in kPa at mixed-layer top z_i)
See Skew-T below. Red dot and arrow represent the heated near-surface air parcel. Black line is the plotted environmental sounding.

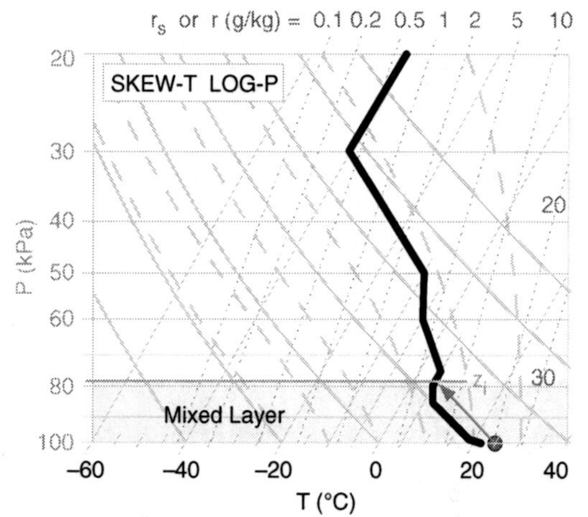

From the diagram, $P_i \approx$ **78 kPa**.

Check: Sketch agrees with larger thermo diagrams.
Exposition: The yellow-shaded region is nonlocally unstable. It is the turbulent mixed layer.

Mixed-Layer

When the sun heats the ground to be warmer than the adjacent air, or when cold air advects over warmer ground, some of the heat is conducted into the thin layer of air touching the ground (see the Thermodynamics chapter). This warm near-surface air wants to rise as bubbles or vertical plumes called **thermals**. You can sometimes see birds soaring in the updrafts of these thermals.

The thermals create vertical **convective circulations** that stir the bottom portion of the troposphere into a **well-mixed layer**. This **convective mixed layer** is evidence of nonlocal static instability in the boundary layer — the bottom 300 to 3000 m of the troposphere.

The **mixed layer** (ML) is usually capped by a statically stable layer or **temperature inversion** (where temperature increases with height), which prevents the thermals from rising through it and traps pollutants below it. Thus, the ML depth z_i is important for air-quality studies.

To find z_i from a sounding, use the nonlocal methods discussed earlier in this chapter. (1) Draw the measured sounding on a thermo diagram. (2) Estimate the temperature of the near-surface air parcel and plot it on the same diagram. (3) Lift this parcel adiabatically until it hits the environmental sounding, which defines z_i (or P_i if using pressure as the vertical coordinate). θ vs. z thermo diagrams (see end of this chapter) are handy for ML studies.

A less accurate, but quick approach, is to look for the temperature inversion (or similar layer that is statically stable) that is near the ground, but not touching it. A **temperature inversion** is where T increases with height. The top of the mixed layer is usually somewhere within this temperature inversion. In Fig. 5.18, the average height of the temperature inversion is about z_i = 87 kPa. See the Boundary Layer and Air Pollution chapters for details.

> **INFO • A holistic approach to stability**
>
> The nonlocal approach is not new. Thunderstorm forecasters usually use the whole sounding, and consider air parcels rising from the surface to the top of the storm to estimate the potential severity of storms. They don't focus on only the local lapse rate in the mid troposphere to estimate thunderstorm intensity.
>
> "Lapse rate" (subadiabatic, adiabatic, superadiabatic) is **not** synonymous with "static stability" (stable, neutral, unstable). If the purpose of a stability definition is to determine whether flow becomes turbulent, then you must consider nonlocal effects within the whole sounding.

REVIEW

Thermo diagrams are useful for estimating thermodynamic state (pressure, temperature, and moisture) of the air without having to solve equations. There are many different types of thermo diagrams: Emagram, Skew-T, Stüve (or Pseudoadiabatic Diagram), Tephigram, and θ-Z diagram. They all serve the same purpose, and all have isotherms, isobars, isohumes, dry and moist adiabats, and height contours (in some cases).

You can also track state changes caused by physical processes acting on the air. In this chapter you read about processes such as radiative heating and cooling, evaporation and condensation, and precipitation falling into or out of an air parcel. You also learned how to estimate cloud-base altitude from the lifting condensation level, and cloud-top altitude from nonlocal stability and buoyancy. You saw how potential temperature is a convenient way to describe the state of vertically moving air parcels.

Thermo diagrams are often used to plot upper-air soundings made with rawinsonde balloons or calculated from computerized forecasts. Soundings plotted on thermo diagrams allow you to find the depth of the mixed layer and the top of the troposphere (i.e., the tropopause). They also allow you to calculate important parameters such as the Brunt-Väisälä frequency and the Richardson number.

Air flow can change from laminar to turbulent if the flow is unstable. Instability can be due to the vertical variation of temperature alone (as quantified by static stability), or be due to the combined effects of wind and temperature variations with height (as quantified by dynamic stability). By plotting soundings and analyzing stability you can anticipate turbulence such as causes rapid dispersion of air pollutants and transport of heat, moisture, and momentum.

HOMEWORK EXERCISES

Broaden Knowledge & Comprehension

B1. Use the internet to acquire a map of rawinsonde sounding locations for your part of the world, and determine which four sounding locations are closest to your school.

B2. Use the internet to acquire **WMO (World Meteorological Organization)** and **ICAO (International Civil Aviation Organization)** identification codes for the four sounding locations closest to your school or business. For each of those four locations, what is the latitude, longitude, topographic elevation, and other info about the site.

B3. Use the internet to acquire different thermo diagrams (e.g., Emagrams, skew-Ts, Stüve diagrams, tephigrams, etc) for the most recent rawinsonde sounding made at the launch site closest to your school or business. Which web site (URL) has the diagram that is clearest or easiest to read?

B4. Use the internet to acquire today's surface weather data of (P, T, and T_d) for the a weather station that is near you. After plotting those on a skew-T diagram, use graphical methods to estimate the lifting condensation level. How does this height compare to any cumulus cloud bases that might be reported from the same station?

B5. Use the internet to acquire rawinsonde data in text form (i.e., a table of numbers) for an upper-air station near you, or a different station that your instructor assigns. Then plot this sounding by hand on a large thermo diagram (skew-T, unless your instructor indicates a different diagram). Determine the LCL for an air parcel rising from the surface.

B6. Use the internet to acquire a sequence of 4 successive soundings from a single rawinsonde station assigned by your instructor. Describe the evolution of the air at that location (e.g., warming, cooling, getting more humid, getting windier, etc.).

B7. Use the internet to acquire an already-plotted sounding near your location <u>prior</u> to a rain storm. From the plotted sounding, determine the temperature and mixing ratio near the ground, and then move that hypothetical air parcel upward to a pressure of 40 kPa. Use the thermo diagram to estimate the final liquid-water mixing ratio of the parcel. Can you relate the actual precipitation depth that was observed from this storm to your hypothetical liquid-water mixing ratio? Would you get a more-realistic answer if you stopped the parcel at a different pressure?

B8. In the Thunderstorm chapters you will learn additional ways to use soundings on thermo diagrams to help estimate thunderstorm strength, tornado likelihood, and other storm characteristics. As a preview of this application, find on the internet an already-plotted sounding that also shows wind information or Convective Available Potential Energy (CAPE) or other stability indices. See if that web site has a link to a legend that defines those terms. If so, use that legend to interpret the stability indices on the sounding, and explain what they mean.

B9. Use the internet to acquire a sounding for a nearby location (or other site that your instructor assigns) that you can output in both plotted graph and text form. Use the plotted sounding to determine the static stability at all heights, using the apex parcel method. Use the text data to calculate Brunt-Väisälä frequency and bulk Richardson number in the stable layers. Combine your dynamic and static stability info to suggest which regions in the sounding are turbulent. [Hints" (1) For any stable layer with uniform lapse rate within it, use only the temperatures and winds at the top and bottom of the layer to calculate these variables. (2) Ignore any moisture indicated by the sounding, and assume $T_v \approx T$].

B10. Use the internet to acquire a plotted sounding from early in the morning. Suppose that as the day evolves, the near-surface dew-point temperature doesn't change, but the near-surface air temperature becomes warmer. Considering nonlocal static stability effects, how warm would the near-surface air parcel need to become later that day in order for it to be positively buoyant (dry below the LCL, saturated above the LCL) up to a height where P = 60 kPa? Draw that air-parcel path on the sounding diagram, and discuss where cloud base and cloud top would be.

B11. Use the internet to acquire a plotted sounding near you. Estimate the pressures at the boundary-layer top and at the tropopause. Draw these locations on your thermo diagram, and discuss how you found them based on your sounding.

B12. Use the internet to acquire images of Kelvin-Helmholtz waves and billow clouds. Print two examples of each, along with their URLs.

Apply

A1. Identify the thermo diagram type.

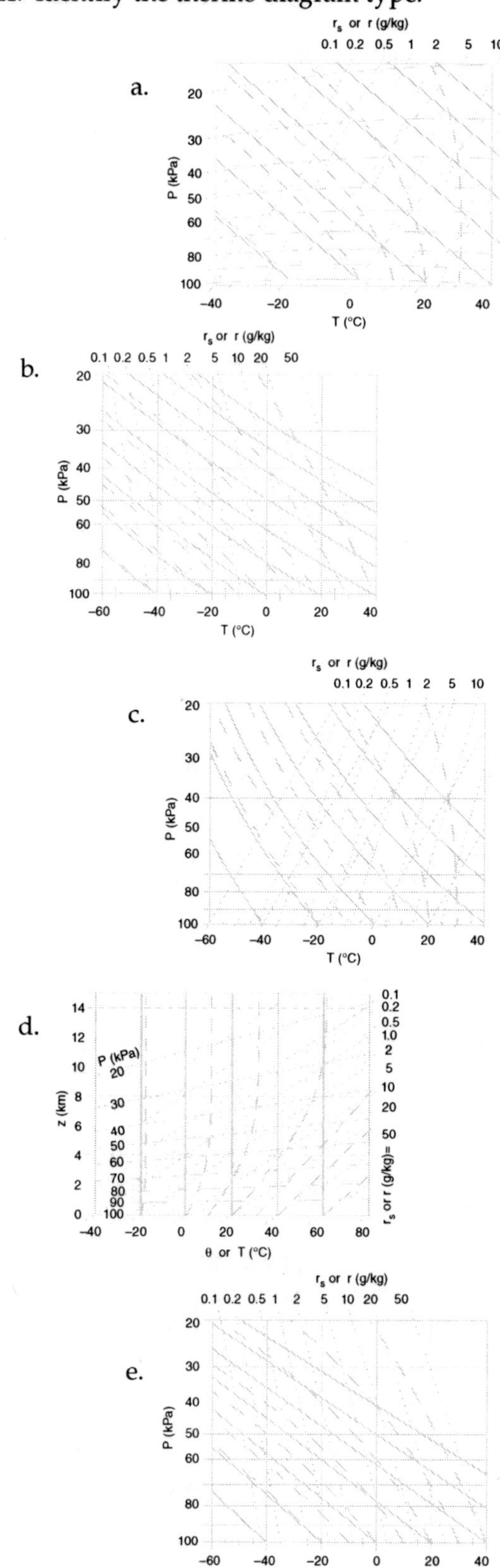

A2. On copies of ALL 5 thermo diagrams from A1, find and label one of the:
a. dry adiabats b. isohumes c. isobars
d. isentropes e. saturated adiabats
f. isotherms g. moist adiabats

A3(§). Use a spreadsheet to create the Skew-T, as given in the Sample Application in the Skew-T section of this Chapter, except do it for isopleths listed below. Describe the results of your experiments with different values of the coefficient K, which determines the amount of skewness.

	T, θ, & θ_w (°C)	r (g kg^{-1})
a.	30	50
b.	25	20
c.	15	10
d.	10	2
e.	5	1
f.	0	0.5
g.	–5	0.2
h.	–10	0.1
i.	–15	15
j.	–20	25

A4. Make a copy of the page holding the thermo diagrams for exercise A1. Then, on all these diagrams, plot the one thermodynamic state given below. (Note: these are single air parcels, not components of a sounding.)

	P(kPa)	T(°C)	T_d(°C)
a.	50	0	–15
b.	80	–5	–10
c.	35	–25	–30
d.	75	25	16
e.	60	15	–10
f.	90	20	10
g.	65	–25	–40
h.	85	–10	–30
i.	70	10	5
j.	80	15	10
k.	55	–10	–15

A5. Using thermodynamic state of the air parcel given in the previous exercise, plot it on the end-of-chapter large thermo diagram specified by your instructor (if none specified, use the skew-T diagram). Hint: Use a copy of a blank thermo diagram so you can keep the master copy clean.

A6. Plot the one thermodynamic state from exercise A4 on one large thermo diagram specified by your instructor (if none specified, use the skew-T). For that plotted point, use the thermo diagram (not equations) to find the values of the following items:
i) mixing ratio
ii) potential temperature
iii) wet-bulb temperature
iv) wet-bulb potential temperature
v) saturation mixing ratio
vi) LCL
vii) relative humidity
viii) equivalent potential temperature

A7. Same as A6, but use only an Emagram or Stüve.

A8. Starting with one assigned air-parcel with state from question A4:
i) lower it to a pressure of 100 kPa, and find its new temperature and dew point. Use only a tephigram or Skew-T.
ii) lower it to a pressure of 100 kPa, and find its new temperature and dew point. Use only a Stüve or Emagram.
iii) raise it to a pressure of 20 kPa, and find its new temperature and dew point. Use only a tephigram or Skew-T.
iv) raise it to a pressure of 20 kPa, and find its new temperature and dew point. Use only a Stüve or Emagram.

A9. Which thermo diagrams from Fig. 5.3 have
a. straight isobars?
b. straight dry adiabats?
c. straight moist adiabats?
d. have the greatest angle between isotherms and dry adiabats?
e. moist adiabats that asymptotically approach dry adiabats near the top of the diagram?
f. moist adiabats that asymptotically approach dry adiabats near the left of the diagram?

A10. Use the data below as an air-parcel initial state, and plot it on the specified large end-of-chapter thermo diagram. Assume this air parcel is then lifted to a final height where P = 25 kPa. Find the final values of the following variables T_d, r_s, T, r_L, and r_T.

	P(kPa)	**T(°C)**	**T_d(°C)**	**Thermo Diagram**
a.	95	30	15	Stüve
b.	95	30	25	Tephigram
c.	95	20	10	Skew-T
d.	95	15	–5	Skew-T
e.	100	25	10	Emagram
f.	100	25	5	Tephigram
g.	100	30	20	Emagram
h.	100	15	5	Stüve
i.	90	20	5	Tephigram
j.	90	15	5	Emagram
k.	90	5	–5	Skew-T
l.	90	10	0	Stüve

A11. Using the same thermo diagram as specified in exercise A10, find the pressure at the LCL for an air parcel that started with the conditions as specified in that exercise.

A12. Starting with the FINAL state of the air parcel specified from exercise A10, allow all of the liquid water to rain out (except if the liquid water from the final state of A10 was greater than 1 g kg^{-1}, then allow only 1 g kg^{-1} of liquid water to rain out). Then lower the air parcel back to its starting pressure, and determine its new temperature and new dewpoint. Use the specified thermo diagram from A10.

A13. Starting with the FINAL state of the air parcel specified from exercise A10, allow radiative cooling to change the air parcel temperature by $\Delta T = -8°C$. Then lower the air parcel back to its starting pressure, and determine its new temperature and dew point. Use the specified thermo diagram from A10.

<u>Sounding Data</u>. The data here gives environmental conditions (i.e, the ambient sounding). Use this sounding for the questions that follow. Given $V = 0$.

P(kPa)	T(°C)	T_d(°C)	U (m s^{-1})
20	–25	–55	45
30	–25	–50	50
40	–20	–20	44
45	–15	–15	40
50	–10	–24	30
70	12	–20	30
80	19	3	20
90	21	15	7
99	29	17	7
100	33	20	2

Definitions: **Significant levels** are altitudes where the sounding has a bend or kink. In between significant levels you must draw straight lines.
Layers are the regions of air that are in between two neighboring levels.

A14. Using the end-of-chapter large thermo diagram specified below, plot the sounding data from the previous column on that diagram. For temperatures, use solid filled circles (red if available), and then connect those temperature-sounding points from bottom to top of the atmosphere with straight line segments. For dewpoints, use an open circles (blue perimeter if possible), and connect those points with dashed straight line segments.
a. Emagram b. Tephigram c. Skew-T
d. θ-Z diagram e. Stüve

A15. a. For each pressure and temperature pair in the sounding data, calculate the corresponding potential temperature. Then plot a graph of potential temperature vs. pressure, using linear scales for both graph axes.
b. Use the result from (a) and the humidity information from the sounding data to find virtual potential temperature, and then plot it vs. pressure using linear scales for both axes.

A16. Given the sounding data. Suppose that you create an air parcel at the pressure-level (kPa) indicated below, where that parcel has the same initial thermodynamic state as the sounding at that pressure. Then move that parcel up through the environment to the next higher significant level (i.e., next lower pressure). What is the value of the buoyant force/mass acting on the parcel at its new level?
a. 100 b. 99 c. 90 d. 80 e. 70 f. 50 g. 45
h. 40 i. 30

A17. Given the sounding data. Find N_{BV} and P_{BV} for an air parcel that starts in the middle of the layer indicated below, and for which its initial displacement and subsequent oscillation is contained within that one layer. Use the layer with a bottom pressure-level (kPa) of:
a. 90 b. 80 c. 70 d. 50 e. 45 f. 40 g. 30

A18. Using the sounding data such as plotted for exercise A14, using the apex parcel method to find static stability for the one environmental layer, where the bottom of that one layer is at P (kPa) =
a. 100 b. 98 c. 87 d. 80 e. 70 f. 60 g. 40
For this exercise you may assume the air is dry. Also, the stability in any one layer could depend on nonlocal effects from other layers.

A19. Same as A18, but find the layer stability using the sounding data.

A20. Using the sounding data, calculate Richardson number, which you can then use to find a layer's dynamic stability, for the layer that has a layer base at level P (kPa) =
a. 100 b. 98 c. 87 d. 80 e. 70 f. 60 g. 40

A21. Indicate which portions of the sounding above are likely to be turbulent, and explain why.

A22. Using the sounding data:
a. Does the sounding data indicate the presence of a mixed layer near the ground? What is its corresponding depth, z_i ?
b. Determine the height of the tropopause.
c. Locate any layer (stratiform) clouds that might be present in the environment.

A23. With the sounding data:

a. Compare the LCL pressure heights found using a thermo diagram vs. that found using the equations from the Water Vapor chapter, for an air parcel starting with the initial state as given by the sounding data at 100 kPa.

b. On the same thermo diagram from part (a), at what pressure altitude does the air parcel from (a) stop rising (i.e., what is the pressure at cloud top)? At cloud top in the cloud, find the mixing ratios for water vapor and liquid water.

c. Suppose that all the liquid water that you found in part (b) were to precipitate out, after which the air parcel descends back to a pressure altitude of 100 kPa. Compare the initial and final relative humidities of the air parcel (i.e., before rising from 100 kPa, and after returning to 100 kPa).

A24. At P = 100 kPa, z = 0 km, an air parcel has initial values of T = 30°C and T_d = 0°C. Use each of the large thermo diagrams at the end of this chapter to find the LCL. How do these LCL values compare to the theoretical value from the equation in the Water Vapor chapter?

Evaluate & Analyze

E1. If an unsaturated air parcel rises, why does its humidity follow an isohume (constant r) instead of an isodrosotherm (constant T_d)?

E2. Suppose a rising dry air parcel cools at a rate different than Γ_d. What process(es) could cause this difference? Why?

E3. From the hypsometric equation one expects that colder air has thinner thickness. So why do the height contour lines on the large Emagram tilt down to the right, rather than down to the left?

E4. The large thermo diagrams at the end of the chapter have mixing-ratio labels near the top. Why don't these label values increase linearly?

E5. Dry adiabats are perfectly straight lines on the Stüve diagram, but have slight curvature on the Emagram (see the large diagrams at the end of the chapter). If the dry adiabatic lapse rate is constant, why are the lines curved in the Emagram?

E6. On a thermo diagram, why are three points needed to represent any single air parcel?

E7. Could you determine the complete thermodynamic state of an air parcel given only its LCL and its initial (P, T)? Explain.

E8. The "air parcel" method of static stability assumes that the ambient environment doesn't change as an air parcel moves through it. Under what conditions is this assumption good? When would the assumption fail?

E9. Given an air parcel of virtual temperature T_v, embedded in a uniform environment of the same T_v. Why does it have zero buoyant force, even though gravity is trying to pull the air parcel down?

E10. Can stationary metal vehicles float in air analogous to how metal ships can float on the ocean? If so, how should the metal aircraft be designed?

E11. Recall equations (5.2 & 5.3a) for buoyant force, based on difference between the air parcel (object) and its environment (surrounding fluid). Why does the equation using density difference have a different sign than the equation using virtual temperature?

E12. Compare the slope of the saturated adiabats on the Emagram and the θ-Z diagram. Why do they tilt in opposite directions on these two diagrams?

E13. To compare the many different thermo diagrams at the end of this chapter, first make a copy of each large blank diagram. Then, on each one, plot the same air parcel (P, T, T_d) = (90 kPa, 20°C, 10°C). If the parcel is lifted to a pressure of 50 kPa, find the final thermodynamic state of the parcel, and compare the answers from the different diagrams.

E14. The sounding in Fig. 5.14e has lines connecting the data points. For each line segment, label its lapse rate (subadiabatic, adiabatic, superadiabatic). Is there a perfect relationship between lapse rate and the static stability indicated in that figure? If not, why not?

E15. Discuss the nature of the Brunt-Väisälä frequency in air that is statically unstable.

E16. The equation for the Brunt-Väisälä oscillation period contains 2π. Why?

E17. a. Check the units of the factors in the Richardson number equations (5.9) to confirm that the Richardson number is indeed dimensionless.

b. What is the value of Richardson number for statically unstable air? What does this value imply about turbulence?

c. What is the value of Richardson number for an air layer with no wind shear? What does this value imply about turbulence?

E18. Although the large thermo diagrams at the end of this chapter include pressures greater than 100 kPa, most of the smaller diagrams earlier in the chapter have a max pressure of 100 kPa at the bottom of the diagram. For these smaller diagrams, what methods can you use to plot pressures that are more than 100 kPa?

E19. If mixing ratio is conserved in an unsaturated rising air parcel, why isn't saturated mixing ratio conserved?

E20(§). (This exercise is lengthy.)
For pressures in the range of 100 to 10 kPa, use a computer spreadsheet to create a:

a. Stüve diagram. [Hint: use $(P_o/P)^{\Re_d/C_{pd}}$ as vertical coordinate.]

b. Skew-T diagram.

c. Tephigram. [Hint: plot it in tilted form, as in Fig. 5.2, but with moist adiabats and isohumes added to the other lines in that figure. This is a very difficult exercise.]

d. θ-Z diagram. [Hint, plot all the isopleths except the isobars.]

E21(§). Use a computer spreadsheet to create an Emagram for a wider range of pressures (120 kPa to 5 kPa) than I had plotted in my small thermo diagrams. Plot only 1 dry adiabat (θ = 0°C) and 1 isohume (r = 10 g kg^{-1}) for this exercise.

E22. How would the air parcel line in Fig. 5.12 be different if the rising parcel entrains a small amount of environmental air as it rises? Sketch this new path on a similar diagram.

E23. For a layer of isothermal air, what is the sign of the Richardson number?

E24. How are the gradient, bulk, and flux Richardson numbers related? The chapter on the Atmospheric Boundary Layer defines the flux Richardson number, and also describes K-theory, a tool you might find useful in answering this question.

E25. Create a new conceptual algorithm <u>different</u> from the "Guide" in this chapter to help you identify different thermo diagram types. Test it to ensure that it doesn't falsely identify some diagrams.

E26(§). Use a spreadsheet to reproduce Fig. 5.2, using the relationship for potential temperature as a function of T and P as given in the Heat chapter.

E27. All the dry adiabats in a Stüve diagram converge to a point above the top left of the diagram. Explain why or why not the dry adiabats in an Emagram converge to a point.

E28. Suppose that turbulent mixing in a dynamically unstable, but statically stable, environment causes both $\Delta T/\Delta z$ and $\Delta U/\Delta z$ to be reduced by the same fraction, b. Show how the Richardson number increases as b decreases. Use this to explain why turbulence acts to reduce the dynamic instabilities that caused it, analogous to LeChatelier's principle. (Hint: Assume V = 0 for simplicity.)

E29. Fig. 5.10 indicates state changes due to fallout of precipitation.

a. Can the opposite happen? Namely, can precipitation fall INTO an air parcel, with all of the water staying in the air? If so, give an example and show how that process would be plotted on a thermo diagram.

b. What happens if rain from above falls through an air parcel with zero or partial evaporation? Namely, the rainfall does not change during its passage through the air parcel. Indicate this on a thermo diagram.

Synthesize

S1. What if you multiply the numerator and denominator of equation (5.9c) by 0.5·m, where m is air-parcel mass. Describe how the numerator and denominator could be interpreted as potential energy and kinetic energy, respectively.

S2. Use the internet to uncover a brief history of Archimedes. Where did he live? What discoveries did he make, and which were relevant to meteorology? What is he most famous for? What was his role in the wars that were waged at that time?

S3. What if water-vapor condensation caused cooling instead of warming. Describe any possible changes in climate and weather. Draw a rough thermodynamic diagram by eye for this physical situation.

S4. Describe any possible changes in climate and weather if buoyancy force was a function of only air-parcel temperature and not on temperature difference?

S5. Fig. 5.17 showed Kelvin-Helmholtz (K-H) waves that form and break in the atmosphere. These are a type of interfacial waves that form on interfaces between dense (cold) air and less-dense (warm) air. Another type of wave called "internal wave" can exist in statically stable air. Use the internet to learn more about internal waves — waves that can move vertically as well as horizontally. Write a brief summary of internal waves.

S6. Describe how the Richardson number would be different if vertical velocity were included in the denominator. Speculate on why it is not included?

S7. Describe any changes in climate and weather that might occur if only Earth's air warmed due to global warming without any associated change in surface temperature. Justify your hypothesis in terms of static stability.

S8. Suppose a nuclear war happened on Earth, and that many of the explosions caused small-diameter dust particles to be blown into the stratosphere where they would settle out very slowly. This situation is called nuclear winter. Using the principles you learned in the radiation chapter, describe how stratospheric and lower-tropospheric temperatures might change under this thick layer of dust. Describe any changes in climate and weather associated with the resulting changes in atmospheric static stability.

S9. An isentropic chart shows the altitude (either z or P) of a constant θ surface. Describe how you would use upper-air soundings from different weather stations to get the data needed to draw an isentropic chart.

S10. Devise a new type of thermo diagram that has log(r) along the abscissa, and log(P) along the ordinate with scale reversed (max and bottom and min at top). Plot the isotherms and dry adiabats in this diagram. Name the diagram after yourself, and explain the virtues and utility of your diagram.

S11. What if T_d could be greater than T in a sounding. Explain how that might be possible, and describe how convective clouds might be different, if at all.

S12. Suppose the tropopause was touching the ground everywhere on Earth. How would the weather or climate be different, if at all?

LARGE-SIZE THERMO DIAGRAMS

You have our permission to freely reproduce the following seven thermo diagrams for your own personal use, or for education (but not for commercial resale), so long as you retain the author, title, and publisher citations in the copies. The diagrams are printed in faded colors so your own plotted soundings will be easier to see. We recommend that you do NOT write on the following seven pages, so that they remain as clean master copies.

These diagrams share common line formats: thin lines representing state of the air; thick lines representing adiabatic processes that change the state.

- Isotherms (temperature: T) and isobars (pressure: P) are thin solid green lines,
- Isohumes (mixing ratio: r) are thin dotted blue,
- Dry adiabats (potential temperature: θ) are thick solid orange lines,
- Moist adiabats (wet-bulb potential temperature: θ_w) are thick dashed orange lines,
- Contours (height: z) are very thin dashed grey lines (if plotted at all).

The height contours are only approximate. Potential temperatures are based on P_o = 100 kPa, where we set z = 0 for simplicity at this reference pressure.

The line labels also follow a common format for most of the diagrams in this book:

- Isobar labels are along the left axis. Units: kPa.
- Isohume labels are along top axis (and right axis in one case). Units: g kg^{-1}.
- Height labels are along the right axis. Units: km.
- Isotherms and both dry and moist adiabats are labeled at the bottom. Units: °C.

For this last item, the three isopleths sprout from each temperature label like branches from a bush. The branches are always in the same order, although their angles differ from graph to graph. Dry adiabats are the left branch, moist adiabats are the center branch, and isotherms are the right branch.

Two versions of the Skew-T and θ-Z Diagram are included. The standard version spans the range P = 105 to 20 kPa. The other version [with "(ABL)" added to the title] is designed for **atmospheric boundary-layer** and air-pollution work, and gives more detail in the bottom third of the troposphere.

On the next pages are these thermo diagrams:

- Emagram
- Stüve Diagram
- Skew-T Log-P Diagram
- Tephigram
- Theta-Z Diagram
- Theta-Z Diagram (ABL)
- Skew-T Log-P Diagram (ABL).

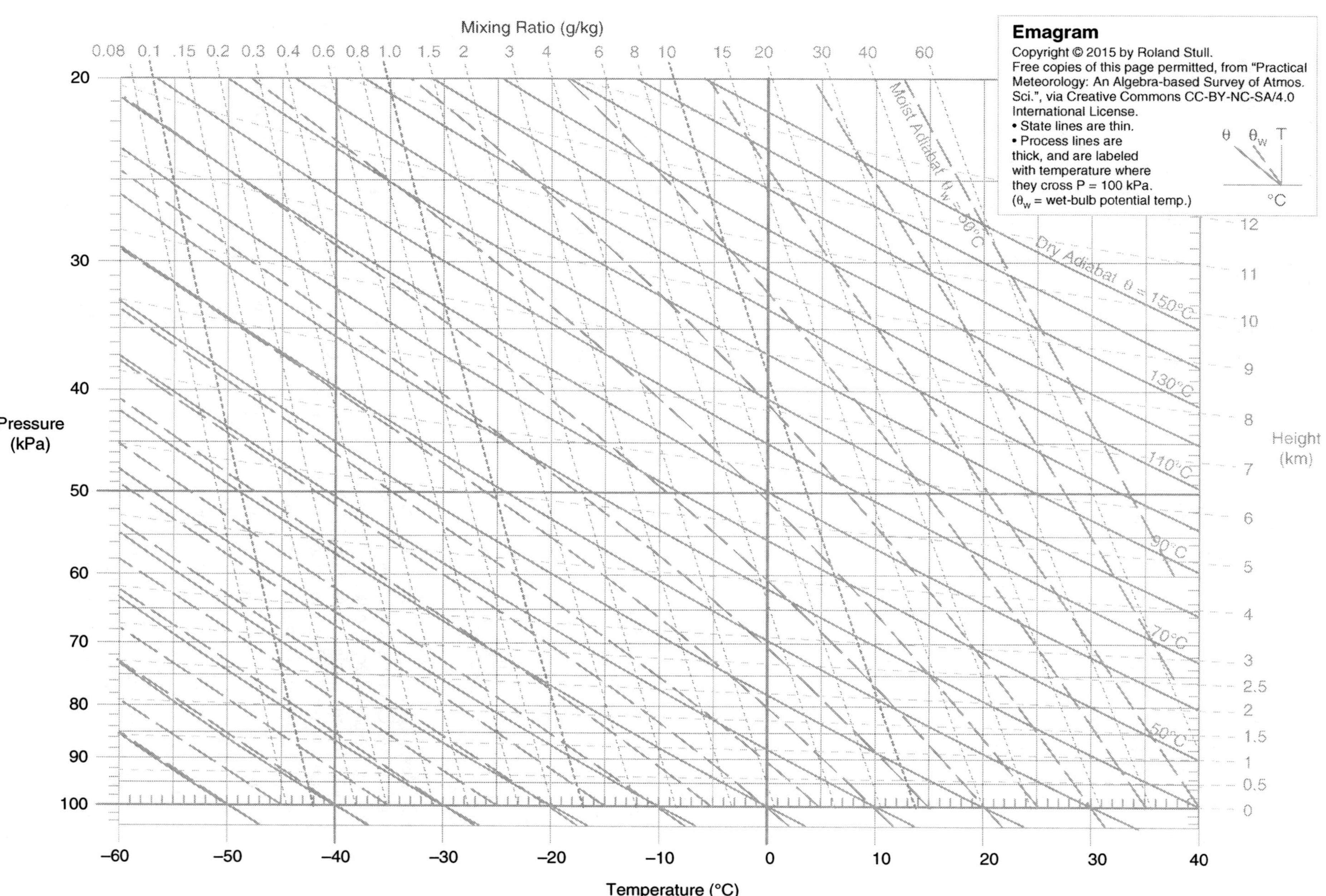

Emagram
Copyright © 2015 by Roland Stull.
Free copies of this page permitted, from "Practical Meteorology: An Algebra-based Survey of Atmos. Sci.", via Creative Commons CC-BY-NC-SA/4.0 International License.
• State lines are thin.
• Process lines are thick, and are labeled with temperature where they cross P = 100 kPa.
(θw = wet-bulb potential temp.)
θ θw T
°C
Mixing Ratio (g/kg)
0.08 0.1 .15 0.2 0.3 0.4 0.6 0.8 1.0 1.5 2 3 4 6 8 10 15 20 30 40 60
Pressure (kPa)
20 30 40 50 60 70 80 90 100
Height (km)
12 11 10 9 8 7 6 5 4 3 2.5 2 1.5 1 0.5 0
Moist Adiabat θw = 50°C
Dry Adiabat θ = 150°C
130°C
110°C
90°C
70°C
50°C
–60 –50 –40 –30 –20 –10 0 10 20 30 40
Temperature (°C)

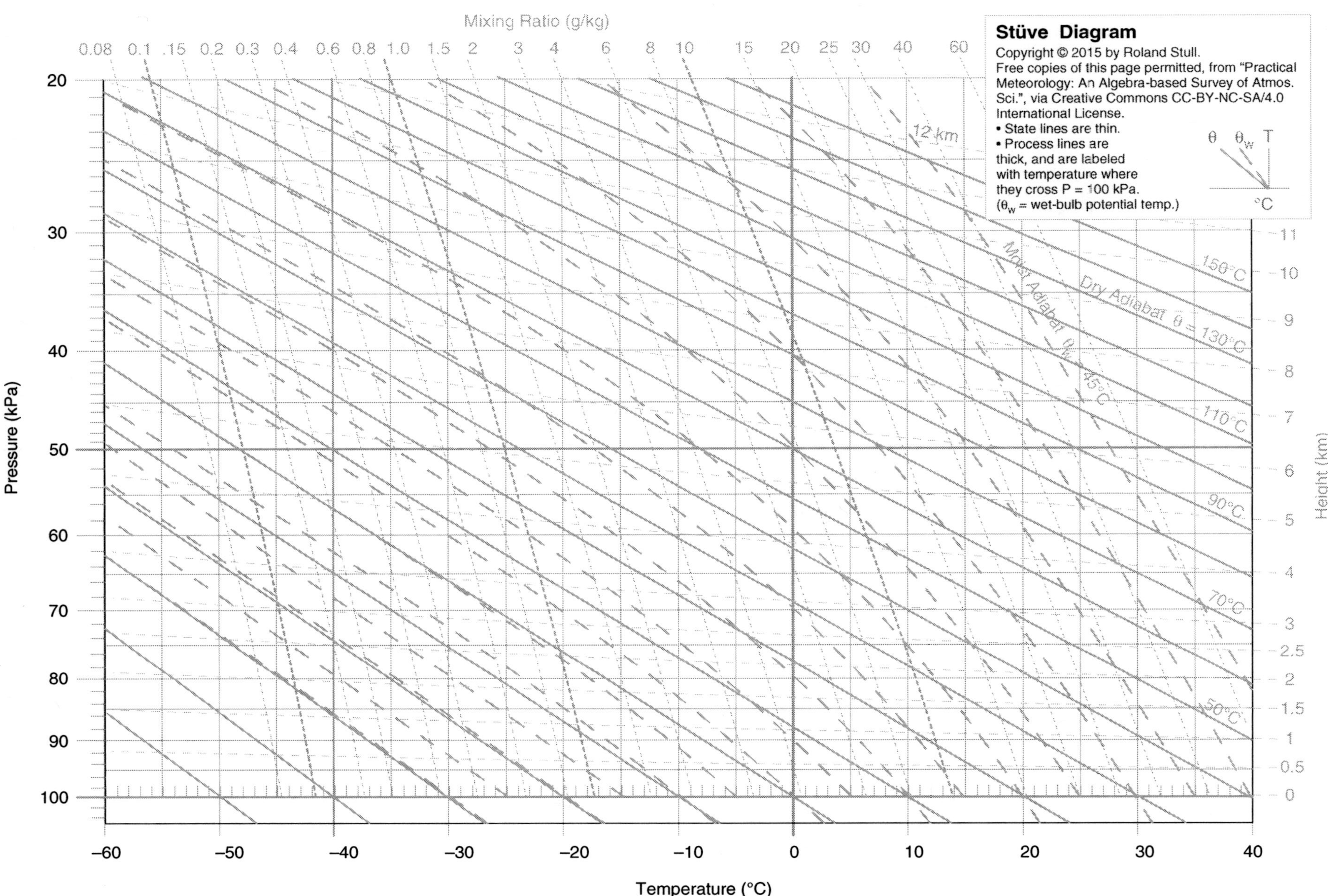
Stüve Diagram
Copyright © 2015 by Roland Stull.
Free copies of this page permitted, from "Practical Meteorology: An Algebra-based Survey of Atmos. Sci.", via Creative Commons CC-BY-NC-SA/4.0 International License.
• State lines are thin.
• Process lines are thick, and are labeled with temperature where they cross P = 100 kPa.
(θ_w = wet-bulb potential temp.)
θ θ_w T
°C
Mixing Ratio (g/kg)
0.08 0.1 .15 0.2 0.3 0.4 0.6 0.8 1.0 1.5 2 3 4 6 8 10 15 20 25 30 40 60
Pressure (kPa)
20 30 40 50 60 70 80 90 100
Temperature (°C)
−60 −50 −40 −30 −20 −10 0 10 20 30 40
Height (km)
11 10 9 8 7 6 5 4 3 2.5 2 1.5 1 0.5 0
12 km
150°C
Dry Adiabat θ = 130°C
110°C
90°C
70°C
50°C
Moist Adiabat θ_w = 45°C

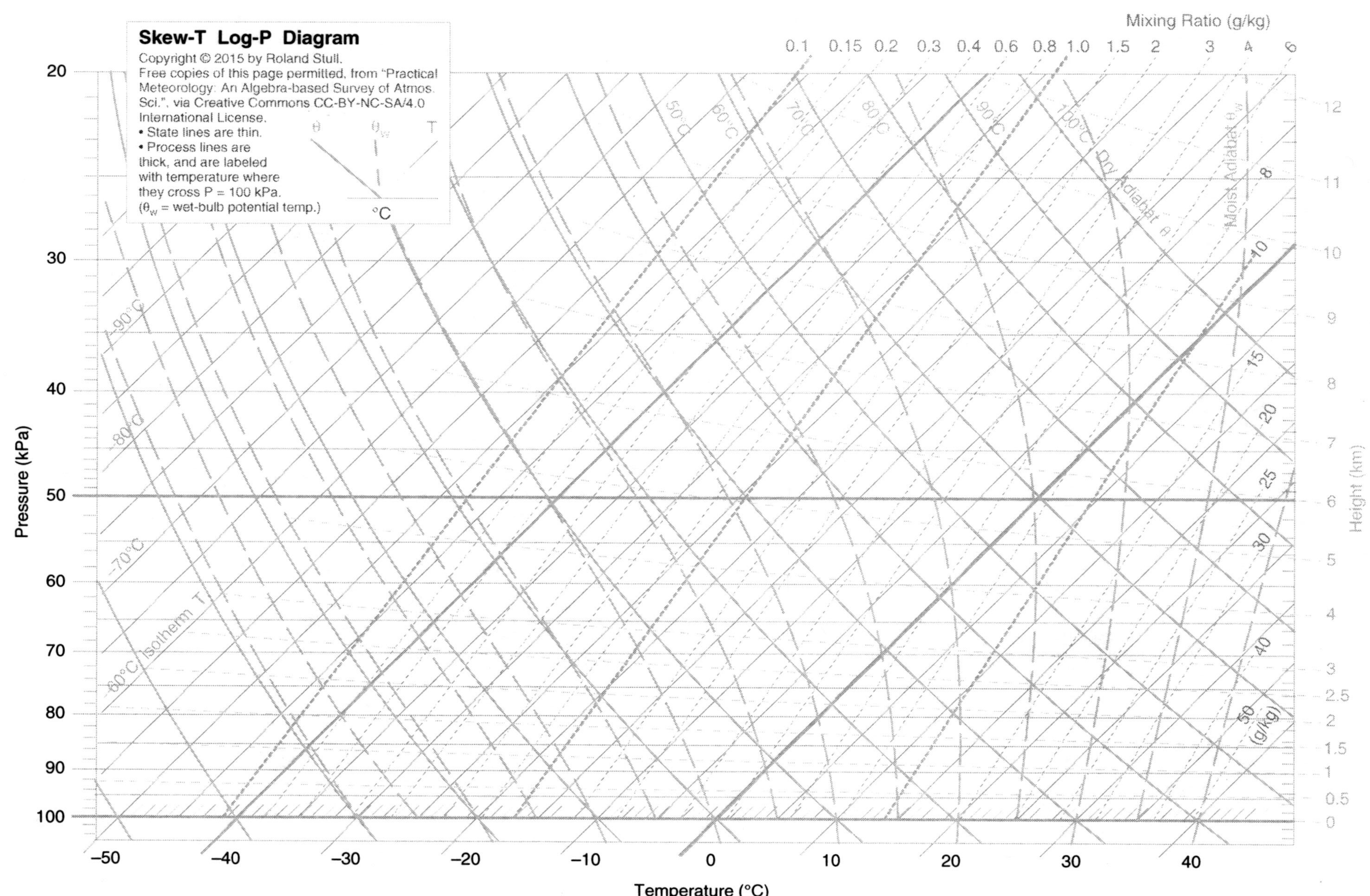

Skew-T Log-P Diagram
Copyright © 2015 by Roland Stull.
Free copies of this page permitted, from "Practical Meteorology: An Algebra-based Survey of Atmos. Sci.", via Creative Commons CC-BY-NC-SA/4.0 International License.
• State lines are thin.
• Process lines are thick, and are labeled with temperature where they cross P = 100 kPa.
(θw = wet-bulb potential temp.)
θ
θw
T
°C
Mixing Ratio (g/kg)
0.1 0.15 0.2 0.3 0.4 0.6 0.8 1.0 1.5 2 3 4 6
8 10 15 20 25 30 40 50 (g/kg)
Dry Adiabat θ
Moist Adiabat θw
50°C 60°C 70°C 80°C 90°C 100°C
–90°C –80°C –70°C –60°C
Isotherm T
Pressure (kPa)
20 30 40 50 60 70 80 90 100
Height (km)
12 11 10 9 8 7 6 5 4 3 2.5 2 1.5 1 0.5 0
Temperature (°C)
–50 –40 –30 –20 –10 0 10 20 30 40

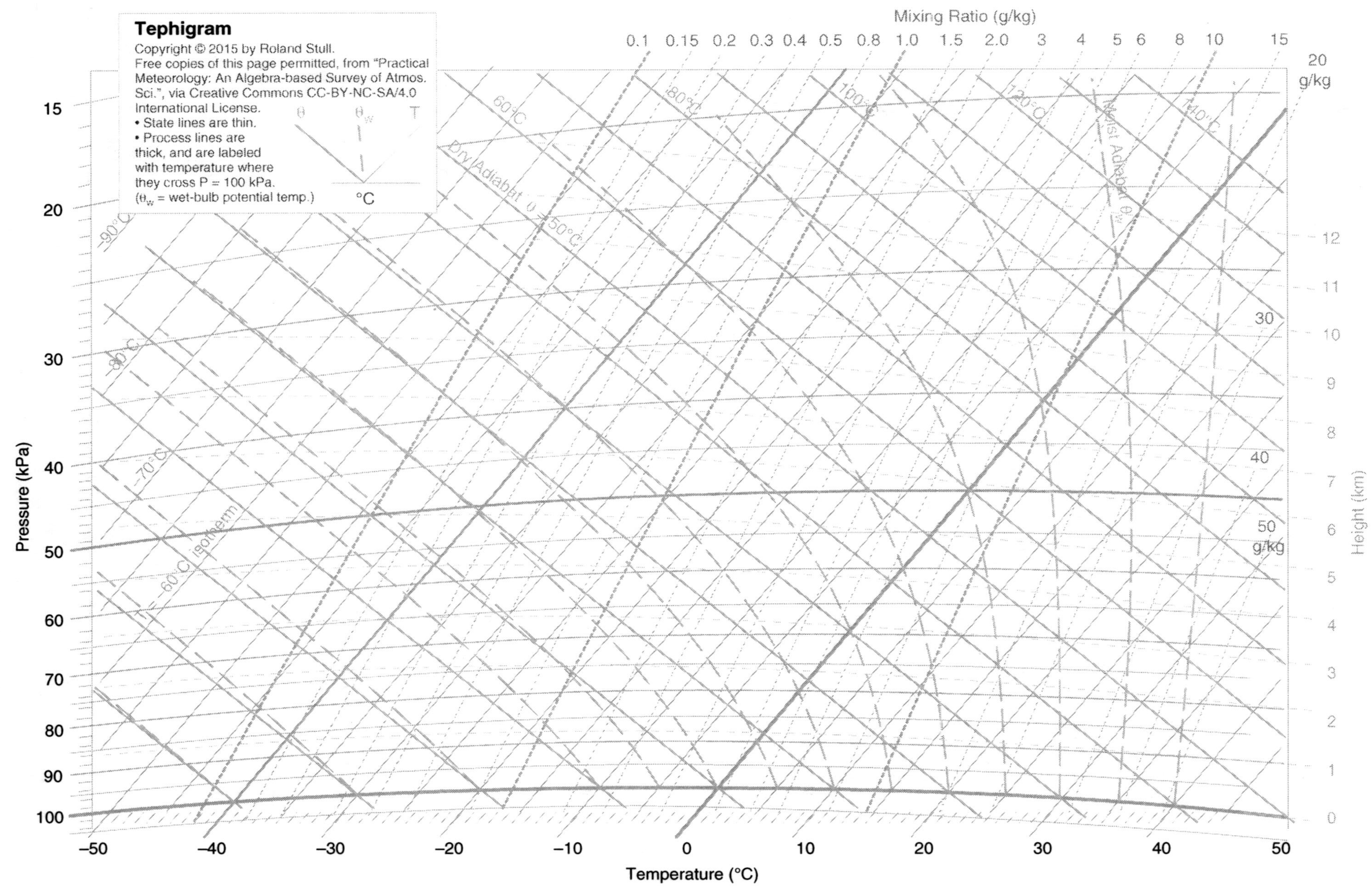
Tephigram
Copyright © 2015 by Roland Stull.
Free copies of this page permitted, from "Practical Meteorology: An Algebra-based Survey of Atmos. Sci.", via Creative Commons CC-BY-NC-SA/4.0 International License.
• State lines are thin.
• Process lines are thick, and are labeled with temperature where they cross P = 100 kPa.
(θ_w = wet-bulb potential temp.)
θ
θ_w
T
°C
Mixing Ratio (g/kg)
0.1
0.15
0.2
0.3
0.4
0.5
0.8
1.0
1.5
2.0
3
4
5
6
8
10
15
20 g/kg
30
40
50 g/kg
Dry Adiabat θ = 50°C
Moist Adiabat θ_w
60°C
80°C
100°C
120°C
140°C
–90°C
–80°C
–70°C
–60°C isotherm
Pressure (kPa)
15
20
30
40
50
60
70
80
90
100
Temperature (°C)
–50
–40
–30
–20
–10
0
10
20
30
40
50
Height (km)
12
11
10
9
8
7
6
5
4
3
2
1
0

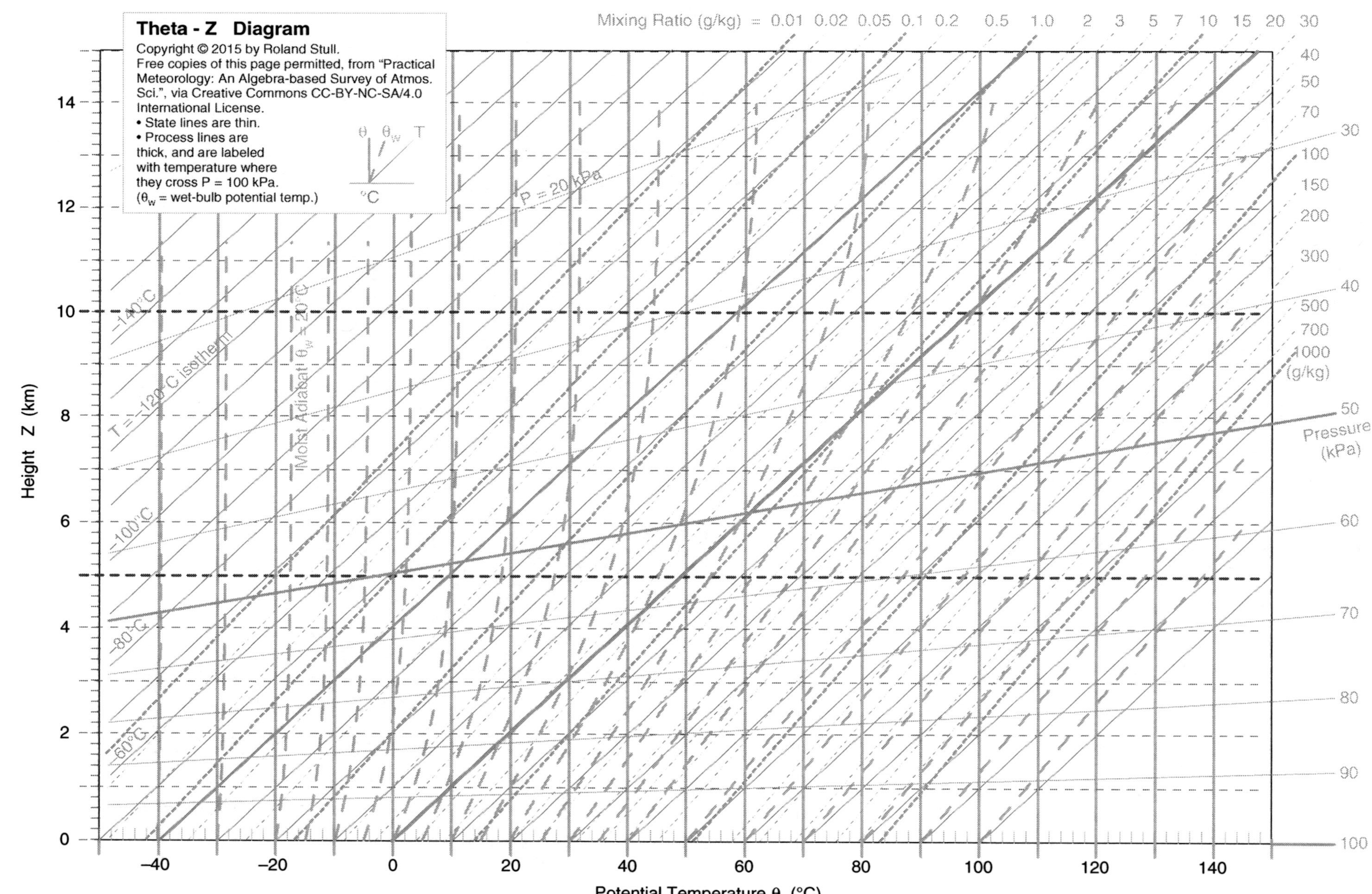
Theta - Z Diagram
Copyright © 2015 by Roland Stull.
Free copies of this page permitted, from "Practical Meteorology: An Algebra-based Survey of Atmos. Sci.", via Creative Commons CC-BY-NC-SA/4.0 International License.
• State lines are thin.
• Process lines are thick, and are labeled with temperature where they cross P = 100 kPa.
(θ_w = wet-bulb potential temp.)
θ θ_w T
°C
Mixing Ratio (g/kg) = 0.01 0.02 0.05 0.1 0.2 0.5 1.0 2 3 5 7 10 15 20 30 40 50 70 100 150 200 300 500 700 1000 (g/kg)
Pressure (kPa)
30
40
50
60
70
80
90
100
P = 20 kPa
Moist Adiabat θ_w = 20°C
T = –120°C isotherm
–140°C
–100°C
–80°C
–60°C
Height Z (km)
0
2
4
6
8
10
12
14
Potential Temperature θ (°C)
–40
–20
0
20
40
60
80
100
120
140

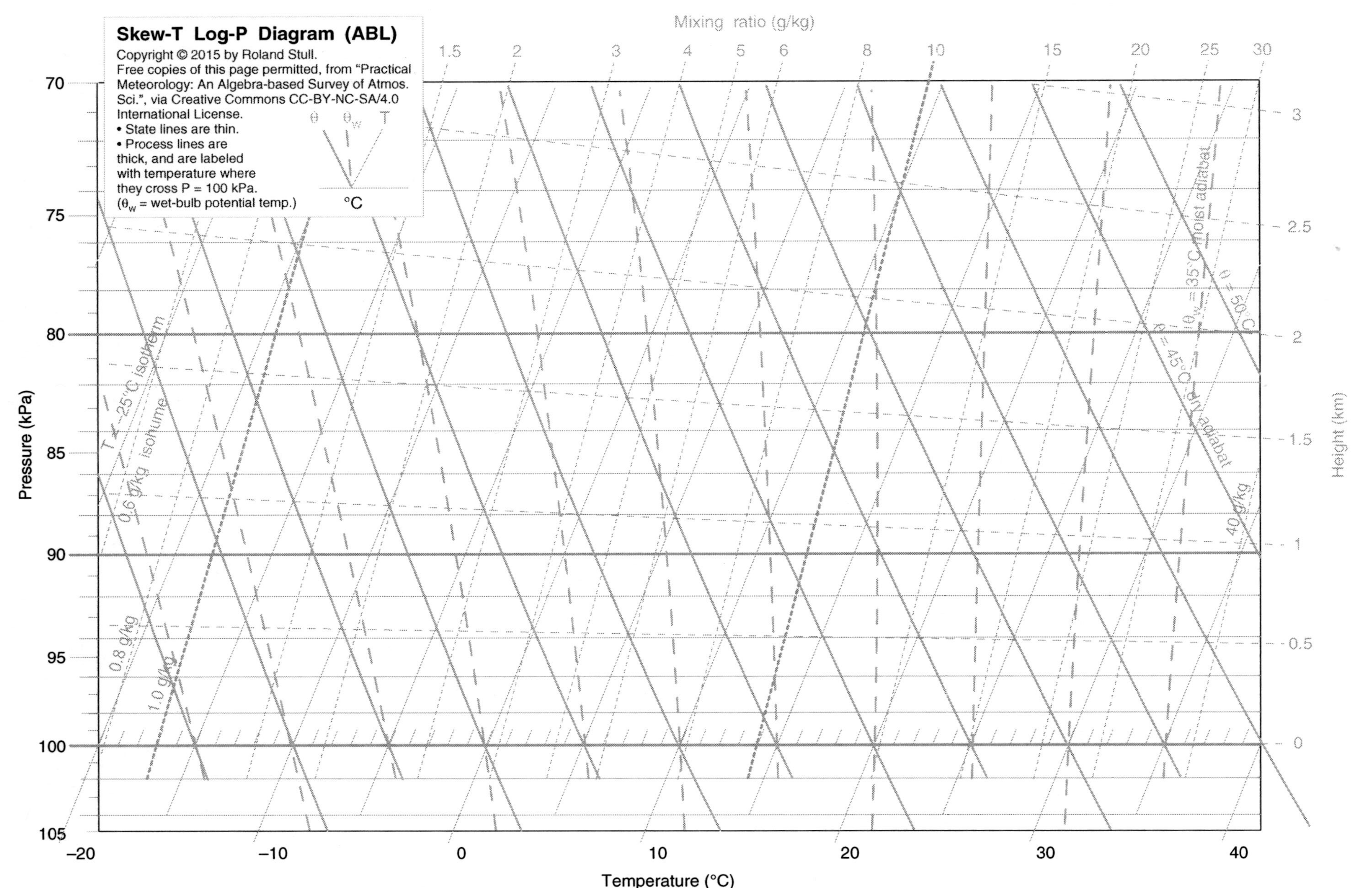

Skew-T Log-P Diagram (ABL)
Copyright © 2015 by Roland Stull.
Free copies of this page permitted, from "Practical Meteorology: An Algebra-based Survey of Atmos. Sci.", via Creative Commons CC-BY-NC-SA/4.0 International License.
• State lines are thin.
• Process lines are thick, and are labeled with temperature where they cross P = 100 kPa.
(θ_W = wet-bulb potential temp.)
θ
θ_W
T
°C
Mixing ratio (g/kg)
1.5
2
3
4
5
6
8
10
15
20
25
30
Pressure (kPa)
70
75
80
85
90
95
100
105
Temperature (°C)
−20
−10
0
10
20
30
40
Height (km)
3
2.5
2
1.5
1
0.5
0
T = −25°C isotherm
0.6 g/kg isohume
0.8 g/kg
1.0 g/kg
θ_W = 35°C moist adiabat
θ = 50°C
θ = 45°C dry adiabat
40 g/kg

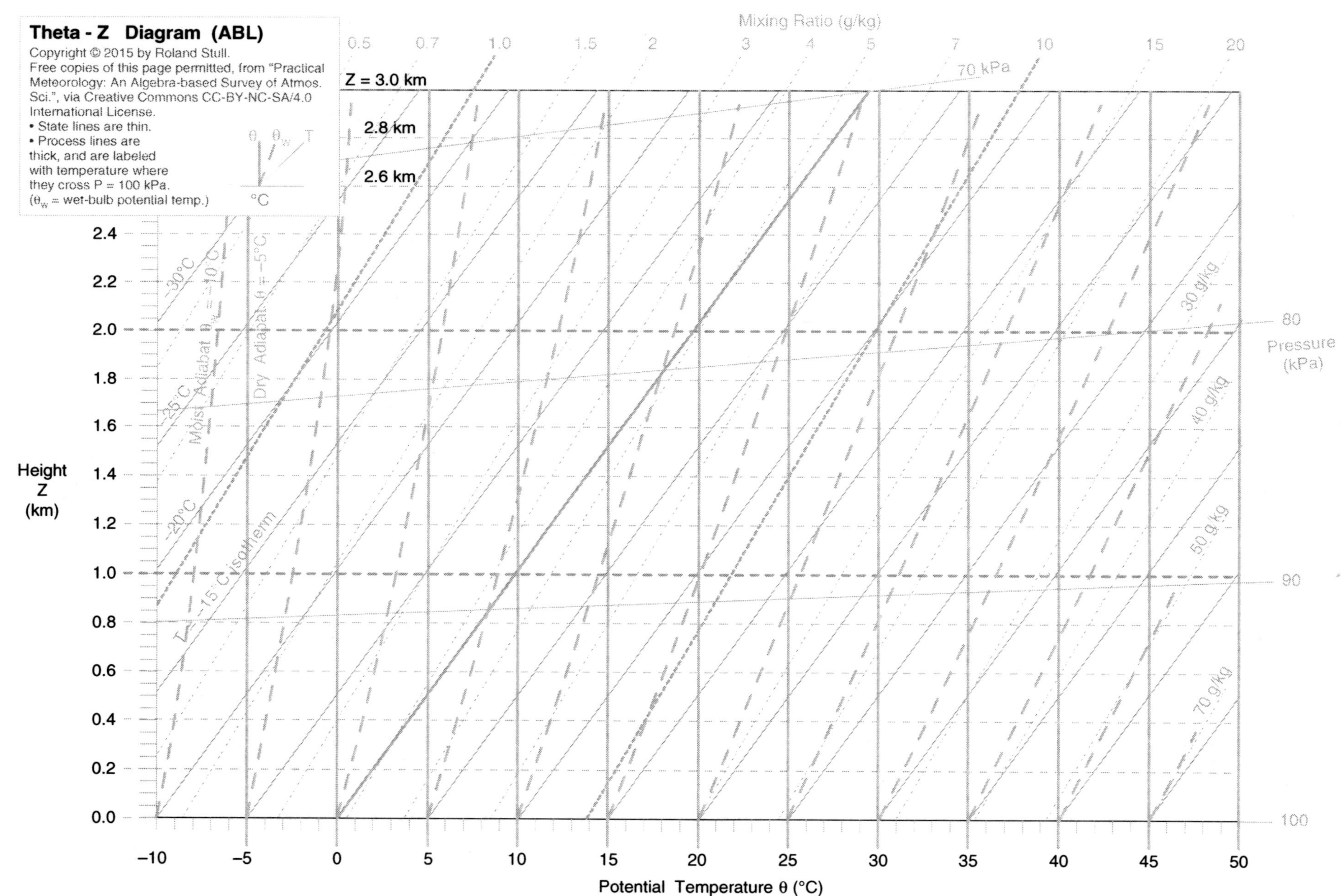

Theta - Z Diagram (ABL)
Copyright © 2015 by Roland Stull.
Free copies of this page permitted, from "Practical Meteorology: An Algebra-based Survey of Atmos. Sci.", via Creative Commons CC-BY-NC-SA/4.0 International License.
• State lines are thin.
• Process lines are thick, and are labeled with temperature where they cross P = 100 kPa.
(θ_w = wet-bulb potential temp.)
θ θ_w T
°C
Mixing Ratio (g/kg)
0.5
0.7
1.0
1.5
2
3
4
5
7
10
15
20
70 kPa
Z = 3.0 km
2.8 km
2.6 km
-30°C
-25°C
-20°C
Moist Adiabat θ_w = -10°C
Dry Adiabat θ = -5°C
T = -15°C isotherm
30 g/kg
40 g/kg
50 g/kg
70 g/kg
80
90
100
Pressure (kPa)
Height Z (km)
2.4
2.2
2.0
1.8
1.6
1.4
1.2
1.0
0.8
0.6
0.4
0.2
0.0
–10
–5
0
5
10
15
20
25
30
35
40
45
50
Potential Temperature θ (°C)

6 CLOUDS

Contents

Clouds have immense beauty and variety. They show weather patterns on a global scale, as viewed by satellites. Yet they are made of tiny droplets that fall gently through the air. Clouds can have richly complex fractal shapes, and a wide distribution of sizes.

Clouds form when air becomes saturated. Saturation can occur by adding water, by cooling, or by mixing; hence, Lagrangian water and heat budgets are useful. The buoyancy of the cloudy air and the static stability of the environment determine the vertical extent of the cloud.

Fogs are clouds that touch the ground. Their location in the atmospheric boundary layer means that turbulent transport of heat and moisture from the underlying surface affects their formation, growth, and dissipation.

PROCESSES CAUSING SATURATION

Clouds are saturated portions of the atmosphere where small water droplets or ice crystals have fall velocities so slow that they appear visibly suspended in the air. Thus, to understand clouds we need to understand how air can become saturated.

Cooling and Moisturizing

Unsaturated air parcels can reach saturation by three processes: cooling, adding moisture, or mixing. The first two processes are shown in Fig. 6.1, where saturation is reached by either **cooling** until the temperature equals the dew point temperature, or **adding moisture** until the dew point temperature is raised to the actual ambient temperature.

The temperature change necessary to saturate an air parcel by cooling it is:

$$\Delta T = T_d - T \tag{6.1}$$

Whether this condition is met can be determined by finding the actual temperature change based on the first law of thermodynamics (see the Heat Budgets chapter).

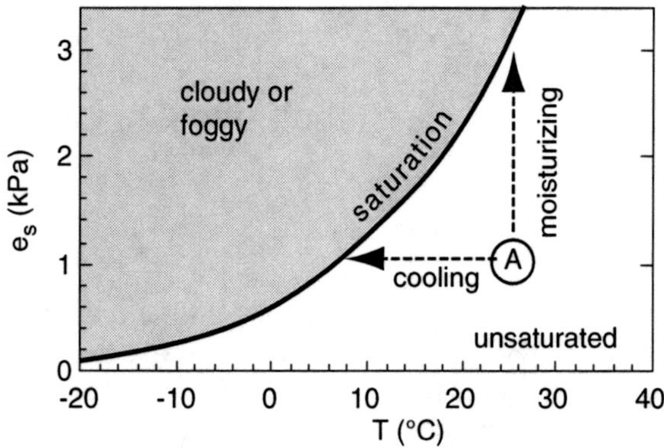

Figure 6.1
Unsaturated air parcel "A" can become saturated by the addition of moisture, or by cooling. The curved line is the saturation vapor pressure from the Water Vapor chapter.

The moisture addition necessary to reach saturation is

$$\Delta r = r_s - r \tag{6.2}$$

Whether this condition is met can be determined by using the moisture budget to find the actual humidity change (see the Water Vapor chapter).

In the real atmosphere, sometimes both cooling and moisturizing happen simultaneously. Schematically, this would correspond to an arrow from parcel *A* diagonally to the saturation line of Fig. 6.1.

Clouds usually form by adiabatic cooling of rising air. Air can be rising due to its own buoyancy (making **cumuliform** clouds), or can be forced up over hills or frontal boundaries (making **stratiform** clouds). Once formed, infrared radiation from cloud top can cause additional cooling to help maintain the cloud.

Sample Application

Air at sea level has a temperature of 20°C and a mixing ratio of 5 g kg^{-1}. How much cooling OR moisturizing is necessary to reach saturation?

Find the Answer

Given: $T = 20°C$, $r = 5$ g kg^{-1}

Find: $\Delta T = ?$ °C, $\Delta r = ?$ g kg^{-1}.

Use Table 4-1 because it applies for sea level. Otherwise, solve equations or use a thermo diagram.

At $T = 20°C$, the table gives $r_s = 14.91$ g kg^{-1}.

At $r = 5$ g kg^{-1}, the table gives $T_d = 4$ °C.

Use eq. (6.1): $\Delta T = 4 - 20 = \underline{-16°C}$ needed.

Use eq. (6.2): $\Delta r = 14.91 - 5 = \underline{9.9 \text{ g kg}^{-1}}$ needed.

Check: Units OK. Physics OK.

Exposition: This air parcel is fairly dry. Much cooling or moisturizing is needed.

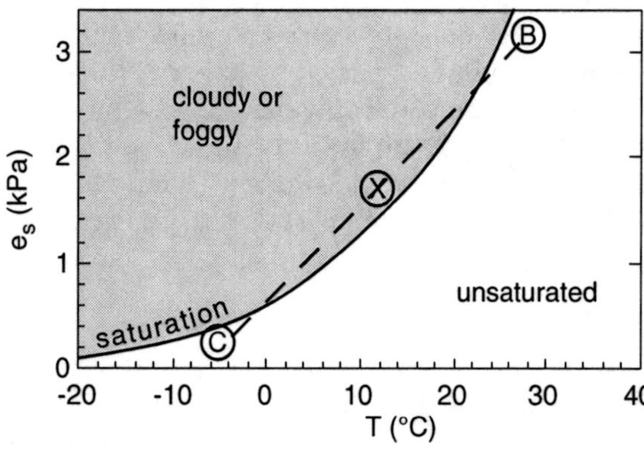

Figure 6.2
Mixing of two unsaturated air parcels B and C, which occurs along a straight line (dashed), can cause a saturated mixture X. The curved line is the saturation vapor pressure from the Water Vapor chapter.

Mixing

Mixing of two unsaturated parcels can result in a saturated mixture, as shown in Fig. 6.2. Jet contrails and your breath on a cold winter day are examples of clouds that form by the mixing process.

Mixing essentially occurs along a straight line in this graph connecting the thermodynamic states of the two original air parcels. However, the saturation line (given by the Clausius-Clapeyron equation) is curved, so a mixture can be saturated even if the original parcels are not.

Let m_B and m_C be the original masses of air in parcels *B* and *C*, respectively. The mass of the mixture (parcel X) is :

$$m_X = m_B + m_C \tag{6.3}$$

The temperature and vapor pressure of the mixture are the weighted averages of the corresponding values in the original parcels:

$$T_X = \frac{m_B \cdot T_B + m_C \cdot T_C}{m_X} \tag{6.4}$$

$$e_X = \frac{m_B \cdot e_B + m_C \cdot e_C}{m_X} \tag{6.5}$$

Specific humidity or mixing ratio can be used in place of vapor pressure in eq. (6.5).

Instead of using the actual masses of the air parcels in eqs. (6.3) to (6.5), you can use the relative portions that mix. For example, if the mixture consists of 3 parts B and 2 parts C, then you can use $m_B = 3$ and $m_C = 2$ in the equations above.

Sample Application

Suppose that the state of parcel B is T = 30°C with e = 3.4 kPa, while parcel C is T = –4°C with e = 0.2 kPa. Both parcels are at P = 100 kPa. If each parcel contains 1 kg of air, then what is the state of the mixture? Will the mixture be saturated?

Find the Answer

Given: B has T = 30°C, e = 3.4 kPa, P = 100 kPa
C has T = –4°C, e = 0.2 kPa, P = 100 kPa
Find: T = ? °C and e = ? kPa for mixture (at X).

Use eq. (6.3): m_X = 1 + 1 = 2 kg
Use eq. (6.4): T_X =
[(1kg)·(30°C) + (1kg)·(–4°C)]/(2kg) = **13°C**.
Use eq. (6.5): e_X =
[(1kg)·(3.4 kPa) + (1kg)·(0.2 kPa)]/(2kg) = **1.8 kPa**
P hasn't changed. P = **100 kPa**.

Check: Units OK. Physics OK.
Exposition: At T = 13°C, Table 4-1 gives e_s = 1.5 kPa. Thus, the mixture **is saturated** because its vapor pressure exceeds the saturation vapor pressure. This mixture would be cloudy/foggy.

CLOUD IDENTIFICATION

You can easily find beautiful photos of all the clouds mentioned below by pointing your web-browser search engine at "cloud classification", "cloud identification", "cloud types", or "International Cloud Atlas". You can also use web search engines to find images of any named cloud. To help keep the cost of this book reasonable, I do not include any cloud photos.

Cumuliform

Clouds that form in updrafts are called **cumuliform** clouds. The small and medium-size ones look like cotton balls, turrets on castles, or cauliflower. The largest cumuliform clouds — thunderstorms — have tops that look like an anvil or like the mushroom cloud of an atom bomb. Cumuliform clouds of all sizes have an aspect ratio of about one; namely, the ratio of cloud diameter to distance of cloud top above ground is roughly one.

Thicker clouds look darker when viewed from underneath, but when viewed from the side, the cloud sides and top are often bright white during daytime. The individual clouds are often surrounded by clearer air, where there is compensating **subsidence** (downdrafts).

Cumulus clouds frequently have cloud bases within 1 or 2 km of the ground (in the boundary layer). But their cloud tops can be anywhere within the troposphere (or lower stratosphere for the strongest thunderstorms). Cumuliform clouds are associated with **vertical** motions.

Cumuliform clouds are named by their thickness, not by the height of their base (Fig. 6.3). Starting from the largest (with highest tops), the clouds are **cumulonimbus (thunderstorms)**, **cumu-**

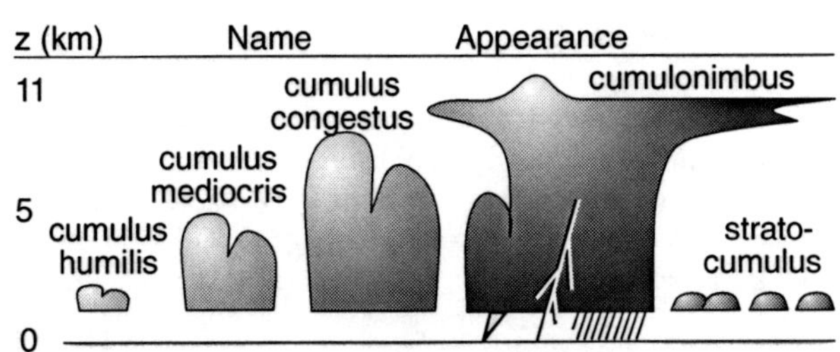

Figure 6.3
Cloud identification: **cumuliform.** *These are lumpy clouds caused by convection (updrafts) from the surface.*

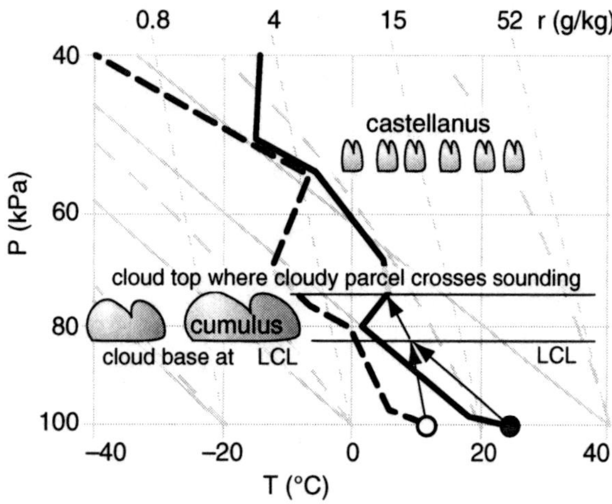

Figure 6.4
Characteristic sounding for cumulus humilis and altocumulus castellanus clouds. Solid thick black line is temperature; dashed black line is dew point; LCL = lifting condensation level. The cumulus humilis clouds form in thermals of warm air rising from the Earth's surface. The altocumulus castellanus clouds are a special stratiform cloud (discussed later in this chapter) that are not associated with thermals rising from the Earth's surface.

Sample Application
Use the sounding plotted here. Where is cloud base and top for an air parcel rising from the surface?

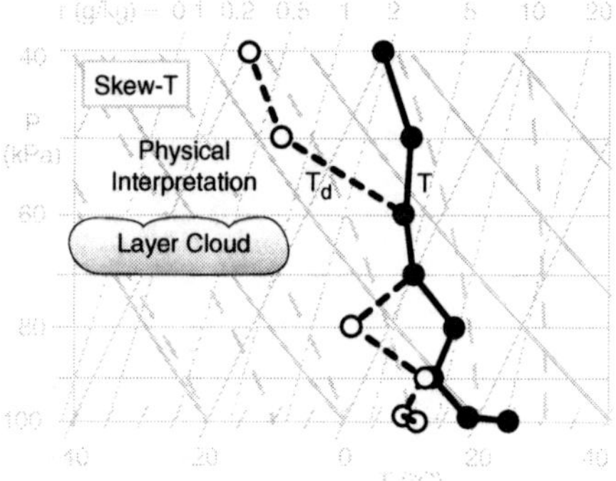

Find the Answer
Given: sounding above.
Find: P_{base} = ? kPa, P_{top} = ? kPa

First, plot the sounding on a large thermo diagram from the Atmospheric Stability chapter. Then conceptually lift a parcel dry adiabatically from the surface until it crosses the isohume from the surface. That LCL is at P_{base} = 80 kPa.

However, the parcel never gets there. It hits the environment below the LCL, at roughly P = 83 kPa. Neglecting any inertial overshoot of the parcel, it would have zero buoyancy and stop rising 3 kPa below the LCL. Thus, there is **no cloud** from rising surface air.

Check: Units OK. Physics OK.
Exposition: This was a trick question. Also, the presence of mid-level stratiform clouds is irrelevant.

lus congestus (**towering cumulus**), **cumulus mediocris** and **cumulus humilis** (**fair-weather cumulus**).

Cumuliform clouds develop in air that is **statically unstable**. This instability creates convective updrafts and downdrafts that tend to undo the instability. Cumuliform clouds can form in the tops of warm updrafts (**thermals**) if sufficient moisture is present. Hence, cumulus clouds are **convective clouds**. Internal buoyancy-forces (associated with latent heat release) enhance and support the convection and updrafts, thus we consider these clouds to be dynamically **active**.

If the air is continually destabilized by some external forcing, then the convection persists. Some favored places for destabilization and small to medium cumulus clouds are:

- behind cold fronts,
- on mostly clear days when sunshine warms the ground more than the overlying air,
- over urban and industrial centers that are warmer than the surrounding rural areas,
- when cold air blows over a warmer ocean or lake.

Cold fronts trigger deep cumuliform clouds (thunderstorms) along the front, because the advancing cold air strongly pushes up the warmer air ahead of it, destabilizing the atmosphere and triggering the updrafts (see Thunderstorm chapters).

Also mountains can trigger all sizes of cumulus clouds including thunderstorms. One mechanism is **orographic lift,** when horizontal winds hit the mountains and are forced up. Another mechanism is **anabatic circulation**, where mountain slopes heated by the sun tend to organize the updrafts along the mountain tops (Regional Winds chapter).

Once triggered, cumulus clouds can continue to grow and evolve somewhat independently of the initial trigger. For example, orographically-triggered thunderstorms can persist as they are blown away from the mountain.

Knowing the environmental sounding and a thermodynamic state of an air parcel starting near the ground, you can use a thermo diagram (Fig. 6.4) to find **cloud base** (at the lifting condensation level, LCL) and **cloud top** (at the equilibrium level, EL). See the Atmospheric Stability chapter for an explanation on how to use thermo diagrams.

Stratiform

Stratiform clouds are horizontal cloud layers. They span wide regions (Fig. 6.5), and have an appearance similar to a blanket or sheet. Stratiform clouds are often associated with warm fronts. Names of stratiform clouds, starting from the lowest, darkest layers, are **nimbostratus, stratus, altostratus, altocumulus, cirrostratus, cirrocumulus,** and **cirrus.**

Layered clouds are often grouped (high, middle, low) by their relative altitude or level within the troposphere. However, troposphere thickness and **tropopause height** vary considerably with latitude (high near the equator, and low near the poles, see Table 6-1). It also varies with season (high during summer, low during winter).

Thus, low, middle and high clouds can have a range of altitudes. Table 6-1 lists cloud levels and their altitudes as defined by **World Meteorological Organization (WMO)**. These heights are only approximate, as you can see from the overlapping values in the table.

High, layered clouds have the prefix "cirro" or "cirrus". The cirrus and cirrostratus are often wispy or have diffuse boundaries, and indicate that the cloud particles are made of ice crystals. In the right conditions, these ice-crystal clouds can cause beautiful **halos** around the sun or moon. See the last chapter for a discussion of Atmospheric Optics.

Mid-level, layered clouds have the prefix "alto". These and the lower clouds usually contain liquid water droplets, although some ice crystals can also be present. In the right conditions (relatively small uniformly sized drops, and a thin cloud) you can see an optical effect called **corona**. Corona appears as a large disk of white light centered on the sun or moon (still visible through the thin cloud). Colored fringes surround the perimeter of the white disk (see the Atmospheric Optics chapter).

Altocumulus and cirrocumulus are layers of lumpy clouds. The edges of these small cloud lumps are often sharply defined, suggesting that they are predominantly composed of liquid water droplets. The lumpiness is not caused by thermals rising from the Earth's surface, but instead are caused by smaller-diameter turbulent eddies generated locally within the clouds.

Low altitude stratiform clouds include **stratus** and **nimbostratus**. In North America, nimbostratus clouds often have low bases and are considered to be a low cloud. However, as we will see in the international cloud classification section, nimbostratus clouds are traditionally listed as mid-level clouds. In this book, we will treat **nimbostratus** as stratiform rain clouds with low cloud base.

The prefix "**nimbo**" or suffix "**nimbus**" originally designated a precipitating cloud, but such a meaning is no longer prescribed in the international cloud atlas. **Nimbo**stratus usually have light to moderate rain or snow over large horizontal areas, while cumulo**nimbus** (thunderstorm) clouds have heavy rain (or snow in winter) and sometimes hail

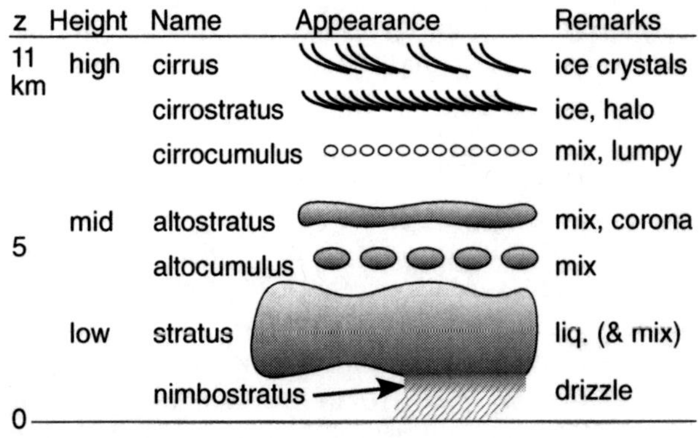

Figure 6.5
Cloud identification: **stratiform.** *Layered clouds caused by nearly horizontal advection of moisture by winds. "Mix" indicates a mixture of liquid and solid water particles. Heights "z" are only approximate — see Table 6–1 for actual ranges.*

Table 6-1. Heights (z) and pressures (P) of clouds and the tropopause. Tropopause values are average typical values, while cloud heights (z) are as defined by the WMO. Pressures are estimated from the heights. sfc. = Earth's surface.

Region:	Polar	Mid-latitude	Tropical
Tropopause:			
z (km)	8	11	18
P (kPa)	35	22	8
High Clouds:			
z (km)	3 - 8	5 - 13	6 - 18
P (kPa)	70 - 35	54 - 16	47 - 8
Middle Clouds:			
z (km)	2 - 4	2 - 7	2 - 8
P (kPa)	80 - 61	80 - 41	80 - 35
Low Clouds:			
z (km)	sfc. - 2	sfc. - 2	sfc. - 2
P (kPa)	P_{sfc} - 80	P_{sfc} - 80	P_{sfc} - 80

INFO • Cloud Deck

Layer clouds, especially those with cloud bases at low altitude, are sometimes called "**cloud decks**".

in small areas or along narrow paths on the ground called **swaths** (e.g., hail swaths or snow swaths).

Smooth-looking, non-turbulent stratiform clouds include stratus (low-altitude, thick, you cannot see the sun through it), altostratus (mid-altitude, modest thickness, you can see the sun faintly through this cloud) and cirrostratus (high altitude, diffuse, allows bright sunlight to cause shadows). None of these clouds (even the lowest ones) are associated with thermals rising from the Earth's surface.

For most **stratiform clouds**, the external forcing is horizontal **advection** (movement by the mean wind), where humid air is blown up the gently sloping surface of a warm front. The source of these air parcels are often 1000 km or more from the clouds. In these warm-frontal regions the air is often **statically stable**, which suppresses vertical motions. Because these clouds are not driven by their own positive buoyancy, we consider them to be **passive** clouds.

This process is illustrated in Fig. 5.20. Let the circle in that figure represent a humid air parcel with potential temperature 270 K. If a south wind were tending to blow the parcel toward the North Pole, the parcel would follow the 270 K isentrope like a train on tracks, and would ride up over the colder surface air in the polar portion of the domain. The gentle rise of air along the isentropic surface creates sufficient cooling to cause the condensation.

Stratiform clouds can be inferred from soundings, in the layers above the boundary layer where environmental temperature and dew point are equal (i.e., where the sounding lines touch). Due to inaccuracies in some of the sounding instruments, sometimes the T and T_d lines become close and parallel over a layer without actually touching. You can infer that these are also stratiform cloud layers.

However, you can<u>not</u> estimate cloud base altitude by the LCL of near-surface air from under the

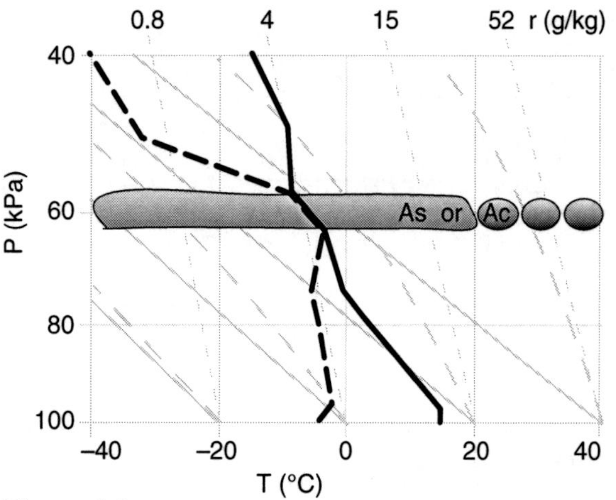

Figure 6.6
Characteristic sounding for stratiform clouds. Shown are altostratus (As) and altocumulus (Ac); mid-level stratiform clouds.

cloud, because stratiform clouds don't form from air that rises from near the Earth's surface. For altocumulus and cirrocumulus clouds, don't let the suffix "cumulus" in fool you — these clouds are still primarily advective, layer clouds.

Stratocumulus

Stratocumulus clouds are low-altitude layers of lumpy clouds, often covering 5/8 or more of the sky (Fig. 6.3). They don't fit very well into either the stratiform or cumuliform categories. They are often turbulently coupled with the underlying surface. Thus, their cloud bases can be estimated using the LCL of near-surface air.

Air circulations in "**stratocu**" can be driven by:
1) wind-shear-generated turbulence (known as **forced convection**),
2) IR radiative cooling from cloud top that creates blobs of cold air that sink **(free convection)**, and
3) advection of cool air over a warmer surface.

Others

There are many beautiful and unusual clouds that do not fit well into the cumuliform and stratiform categories. A few are discussed here: castellanus, lenticular, cap, rotor, banner, contrails, fumulus, billow clouds, pyrocumulus, pileus, and fractus. You can find pictures of these using your web browser.

Other clouds associated with thunderstorms are described in the Thunderstorm chapters. These include funnel, wall, mammatus, arc, shelf, flanking line, beaver tail, and anvil.

Clouds in unstable air aloft

Two types of clouds can be found in layers of statically or dynamically unstable air not associated with the ground: castellanus and billow clouds.

Castellanus clouds look like a layer of small-diameter castle turrets (Fig. 6.4). When a layer of relatively warm air advects under a layer of cooler air, the interface between the two layers aloft can become **statically unstable**. [The advection of air from different sources at different altitudes is called **differential advection**.]

If these castellanus form just above the top of the boundary layer, they are called **cumulus castellanus**. When slightly higher, in the middle of the troposphere, they are called **altocumulus castellanus**. Altocumulus castellanus are sometimes precursors to thunderstorms (because they indicate an unstable mid-troposphere), and are a useful clue for storm chasers.

Billow clouds (discussed in the Atmos. Stability chapter) are a layer of many parallel, horizontal lines of cloud that form in the crests of **Kelvin-**

Helmholtz (**K-H**) waves (Fig. 5.17). They indicate a layer of turbulence aloft caused by wind shear and dynamic instability. When similar turbulent layers form with insufficient moisture to be visible as billow clouds, the result is a layer of **clear-air turbulence** (**CAT**). Pilots try to avoid both CAT and K-H waves. Sometimes instead of a layer of billows, there will be only a narrow band of breaking wave clouds known as **K-H wave clouds**.

Clouds associated with mountains

In mountainous regions with sufficient humidity, you can observe lenticular, cap, rotor, and banner clouds. The Regional Winds chapter covers others.

Lenticular clouds have smooth, distinctive lens or almond shapes when viewed from the side, and they are centered on the mountain top or on the crest of the lee wave (Fig. 6.7). They are also known as **mountain-wave clouds** or **lee-wave clouds**, and are passive clouds that form in hilly regions.

If a lenticular cloud forms directly over a mountain, it is sometimes called a **cap cloud**. A cap cloud can form when air that is statically stable is blown toward a mountain or other terrain slope. As the air is forced to rise by the terrain, the air cools adiabatically and can reach saturation if the air is sufficiently humid. As the air descends down the lee side of the mountain, the air warms adiabatically and the cloud droplets evaporate.

However, the static stability allows the air to continue to oscillate up and down as it blows further downwind. Lenticular clouds can form in the crest of these vertical oscillations (called **mountain waves**) to the lee of (downwind of) the mountain.

They are a most unusual cloud, because the cloud remains relatively stationary while the air blows through it. Hence, they are known as **standing lenticular**. The uniformity of droplet sizes in lenticular clouds create beautiful optical phenomena called **iridescence** when the sun appears close to the cloud edge. See the Regional Winds chapter for mountain wave details, and the Atmospheric Optics chapter for more on optical phenomena.

Rotor clouds are violently turbulent balls or bands of ragged cloud that rapidly rotate along a horizontal axis (Fig. 6.7). They form relatively close to the ground under the crests of mountain waves (e.g., under standing lenticular clouds), but much closer to the ground. Pilots flying near the ground downwind of mountains during windy conditions should watch out for, and avoid, these hazardous clouds, because they indicate severe turbulence.

The **banner cloud** is a very turbulent streamer attached to the mountain top that extends like a banner or flag downwind (Fig. 6.7). It forms on the lee side at the very top of high, sharply pointed,

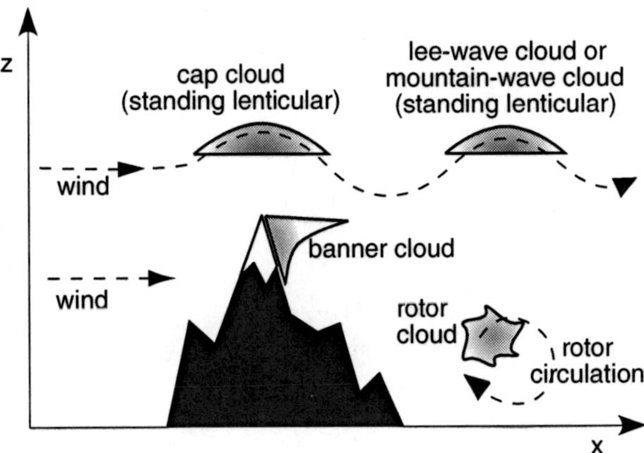

Figure 6.7
Some clouds caused by mountains, during strong winds.

mountain peaks during strong winds. As the wind separates to flow around the mountain, low pressure forms to the lee of the mountain peak, and counter-rotating vortices form on each side of the mountain. These work together to draw air upward along the lee slope, causing cooling and condensation. These strong turbulent winds can also pick up previously fallen ice particles from snow fields on the mountain surface, creating a **snow banner** that looks similar to the banner cloud.

Clouds due to surface-induced turbulence or surface heat

The most obvious clouds formed due to surface-induced turbulence are the **cumuliform** (convective) clouds already discussed. However, there are three others that we haven't covered yet: pyrocumulus, pileus and fractus.

A **pyrocumulus** is a cumulus cloud that forms in the smoke of a fire, such as a forest fire or other wild fire. One of the combustion products of plant material is water vapor. As this water vapor is carried upward by the heat of the fire, the air rises and cools. If it reaches its LCL, the water vapor can condense onto the many smoke particles created by the fire. Some pyrocumulus can become thunderstorms.

Pyrocumulus clouds can also be created by geothermal heat and moisture sources, including volcanoes and geysers. Lightning can often be found in volcanic ash plumes.

The **pileus cloud** looks like a thin hat just above, or scarf around, the top of the rising cumulus clouds (Fig. 6.8). It forms when a layer of stable, humid air in the middle of the troposphere is forced upward by cumuliform cloud towers (cumulus congestus) rising up from underneath. Hence, there is an indirect influence from surface heating (via the cumuliform cloud). Pileus are very short lived, because the cloud towers quickly rise through the pileus and engulf them.

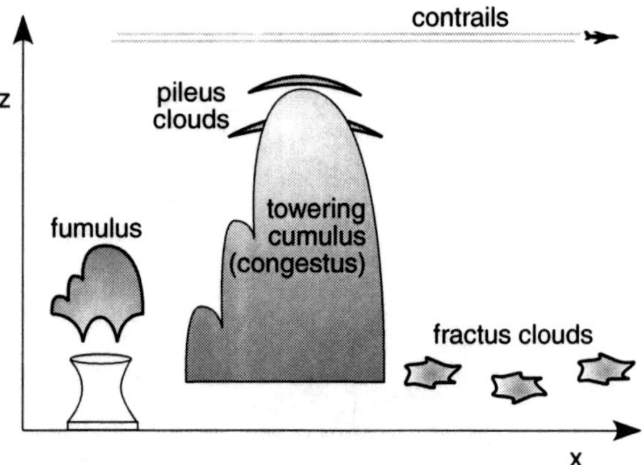

Figure 6.8
Other clouds.

INFO • Strato- & Mesospheric Clouds

Although almost all our clouds occur in the troposphere, sometimes higher-altitude thin stratiform (cirrus-like) clouds can be seen. They occur poleward of 50° latitude during situations when the upper atmosphere temperatures are exceptionally cold.

These clouds are so diffuse and faint that you cannot see them by eye during daytime. These clouds are visible at night for many minutes near the end of evening twilight or near the beginning of morning twilight. The reason these clouds are visible is because they are high enough to still be illuminated by the sun, even when lower clouds are in the Earth's shadow.

Highest are **noctilucent clouds**, in the mesosphere at heights of about 85 km. They are made of tiny H_2O ice crystals (size ≈ 10 nm), and are nucleated by meteorite dust. Noctilucent clouds are **polar mesospheric clouds (PMC)**, and found near the polar mesopause during summer.

Lower (20 to 30 km altitude) are **polar stratospheric clouds (PSC)**. At air temperatures below –78°C, nitric acid trihydrate (**NAT-clouds**) can condense into particles. Also forming at those temperatures can be particles made of a supersaturated mixture of water, sulphuric acid, and nitric acid (causing **STS clouds**).

For temperatures colder than –86°C in the stratosphere, pure H_2O ice crystals can form, creating PSCs called **nacreous clouds**. They are also known as **mother-of-pearl clouds** because they exhibit beautiful iridescent fringes when illuminated by the sun. These stratospheric clouds form over mountains, due to mountain waves that propagate from the troposphere into the stratosphere, and amplify there in the lower-density air. They can also form over extremely intense tropospheric high-pressure regions.

Fractus clouds are ragged, shredded, often low-altitude clouds that form and dissipate quickly (Fig. 6.8). They can form during windy conditions in the turbulent, boundary-layer air near rain showers, under the normal nimbostratus or cumulonimbus cloud base. These clouds do not need mountains to form, but are often found along the sides of mountains during rainy weather. The falling rain from the cloud above adds moisture to the air, and the updraft portions of turbulent eddies provide the lifting to reach condensation.

Sometimes fractus clouds form in non-rainy conditions, when there is both strong winds and strong solar heating. In this case, the rising thermals lift air to its LCL, while the intense turbulence in the wind shear shreds and tears apart the resulting cumulus clouds to make **cumulus fractus**.

Anthropogenic Clouds

The next two clouds are **anthropogenic** (man-made). These are contrails and fumulus.

Fumulus is a contraction for "fume cumulus". They form in the tops of thermal plumes rising from cooling towers or smokestacks (Fig. 6.8). Modern air-quality regulations often require that industries scrub the pollutants out of their stack gasses by first passing the gas through a scrubber (i.e., a water shower). Although the resulting effluent is much less polluted, it usually contains more water vapor, and thus can cause beautiful white clouds of water droplets within the rising exhaust plume. Similarly, cooling towers do their job by evaporating water to help cool an industrial process.

Contrail is a contraction for "condensation trail", and is the straight, long, narrow, horizontal pair of clouds left behind a high-altitude aircraft (Fig. 6.8). Aircraft fuel is a hydrocarbon, so its combustion in a jet engine produces carbon dioxide and water vapor. Contrails form when water vapor in the exhaust of high-altitude aircraft mixes with the cold environmental air at that altitude (see mixing subsection earlier in this chapter). If this cold air is already nearly saturated, then the additional moisture from the jet engine is sufficient to form a cloud. On drier days aloft, the same jet aircraft would produce no visible contrails.

Regardless of the number of engines on the aircraft, the exhaust tends to be quickly entrained into the horizontal **wing-tip vortices** that trail behind the left and right wing tips of the aircraft. Hence, jet contrails often appear initially as a pair of closely-spaced, horizontal parallel lines of cloud. Further behind the aircraft, environmental wind shears often bend and distort the contrails. Turbulence breaks apart the contrail, causes the two clouds to merge into one contrail, and eventually mixes enough dry ambient air to cause the contrail to evaporate and disappear.

Contrails might have a small effect on the global-climate heat budget by reflecting some of the sunlight. Contrails are a boon to meteorologists because they are a clue that environmental moisture is increasing aloft, which might be the first indication of an approaching warm front. They are a bane to military pilots who would rather not have the enemy see a big line in the sky pointing to their aircraft.

CLOUD ORGANIZATION

Clouds frequently become organized into patterns during stormy weather. This organization is discussed in the chapters on Fronts, Midlatitude Cyclones, Thunderstorms, and Tropical Cyclones. Also, sometimes large-scale processes can organize clouds even during periods of fair weather, as discussed here.

Cloud streets or **cloud lines** are rows of fair-weather cumulus-humilis clouds (Fig. 6.9) that are roughly parallel with the mean wind direction. They form over warm land in the boundary layer on sunny days, and over water day or night when cold air advects over warmer water. Light to moderate winds in the convective boundary layer cause **horizontal roll vortices**, which are very weak counter-rotating circulations with axes nearly parallel with the mean wind direction. These weak circulations sweep rising cloud-topped thermals into rows with horizontal spacing of roughly twice the boundary-layer depth (order of 1 km).

Mesoscale cellular convection (MCC) can also form in the boundary layer, but with a much larger horizontal scale (order of 5 to 50 times the boundary layer depth; namely, 10 to 100 km in diameter). These are so large that the organization is not apparent to observers on the ground, but is readily visible in satellite images. **Open cells** consist of a honeycomb or rings of cloud-topped updrafts around large clear regions (Fig. 6.9). **Closed cells** are the opposite — a honeycomb of clear-air rings around cloud clusters. Cloud streets changing into MCC often form when cold continental air flows over a warmer ocean. [*WARNING: Abbreviation* ***MCC*** *is also used for* ***Mesoscale Convective Complex****, which is a cluster of deep thunderstorms.*]

A SCIENTIFIC PERSPECTIVE • Cargo Cult Science

"In the South Seas there is a cargo cult of people. During the war they saw airplanes land with lots of good materials, and they want the same thing to happen now. So they've arranged to make things like runways, to put fires along the sides of the runways, to make a wooden hut for a man to sit in, with two wooden pieces on his head like headphones and bars of bamboo sticking out like antennas (he's the controller) and they wait for the airplanes to land. They're doing everything right. The form is perfect. It looks exactly the way it looked before. But it doesn't work. No airplanes land. So I call these things cargo cult science, because they follow the apparent precepts and forms of scientific investigation, but they're missing something essential, because the planes don't land."

"Now it behooves me, of course, to tell you what they're missing. ... It's a kind of scientific integrity, a principle of scientific thought that corresponds to a kind of utter honesty... For example, if you're doing an experiment, you should report everything that you think might make it invalid – not only what you think is right about it: other causes that could possibly explain your results; and things you thought of that you've eliminated by some other experiment, and how they worked – to make sure the other fellow can tell they have been eliminated."

"Details that could throw doubt on your interpretation must be given, if you know them. You must do the best you can – if you know anything at all wrong, or possibly wrong – to explain it. If you make a theory, ... then you must also put down all the facts that disagree with it, as well as those that agree with it. There is also a more subtle problem. When you have put a lot of ideas together to make an elaborate theory, you want to make sure when explaining what it fits, that those things it fits are not just the things that gave you the idea for the theory; but that the finished theory makes something else come out right, in addition."

"In summary, the idea is to try to give all the information to help others to judge the value of your contribution; not just the information that leads to judgement in one particular direction or another."

– R. P. Feynman, 1985: *"Surely You're Joking, Mr. Feynman!". Adventures of a Curious Character.* Bantam.

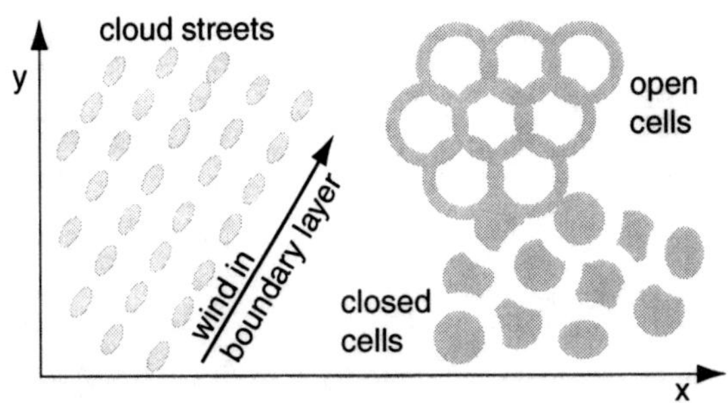

Figure 6.9
Organized boundary-layer clouds, such as are seen in a satellite image. Grey = clouds.

CLOUD CLASSIFICATION

Clouds are classified by their shapes (i.e., their **morphology**). The classification method introduced in 1803 by Luke Howard is still used today, as approved by the World Meteorology Organization (WMO).

WMO publishes an *International Cloud Atlas* with photos to help you identify clouds. As mentioned, you can easily find similar photos for free by pointing your web search engine at "cloud classification", "cloud identification", "cloud types", or "International Cloud Atlas".

The categories and subcategories of clouds are broken into:

- **genera** - main characteristics of clouds
- **species** - peculiarities in shape & structure
- **varieties** - arrangement and transparency
- **supplementary features** and **accessory clouds** - attached to other (mother) clouds
- **mother clouds** - clouds with attachments
- **meteors** - precipitation (ice, water, or mixed)

Most important are the genera.

Genera

The ten cloud **genera** are listed in Table 6-2, along with their official abbreviations and symbols as drawn on weather maps. These genera are mutually exclusive; namely, one cloud cannot have two genera. However, often the sky can have different genera of clouds adjacent to, or stacked above, each other.

Table 6-2. Cloud genera.

Genus	Abbreviation	WMO Symbol	USA Symbol
cirrus	Ci		
cirrostratus	Cs		
cirrocumulus	Cc		
altostratus	As		
altocumulus	Ac		
nimbostratus	Ns		
stratus	St		
stratocumulus	Sc		
cumulus	Cu		
cumulonimbus	Cb		

Species

Subdividing the genera are cloud **species**. Table 6-3 lists the official WMO species. Species can account for:

- forms (clouds in banks, veils, sheets, layers, etc.)
- dimensions (horizontal or vertical extent)
- internal structure (ice crystals, water droplets, etc.)
- likely formation process (orographic lift, etc.)

Species are also mutually exclusive. An example of a cloud genus with specie is "Altocumulus castellanus (Ac cas)", which is a layer of mid-level lumpy clouds that look like castle turrets.

Table 6-3. WMO cloud species. (Ab. = abbreviation)

Specie	Ab	Description
calvus	cal	the top of a deep Cu or Cb that is starting to become fuzzy, but no cirrus (ice) anvil yet
capillatus	cap	Cb having well defined streaky cirrus (ice) anvil
castellanus	cas	small turrets looking like a crenellated castle. Turret height > diameter. Can apply to Ci, Cc, Ac, Sc.
congestus	con	very deep Cu filling most of troposphere, but still having crisp cauliflower top (i.e., no ice anvil). Often called **Towering Cu** (TCU).
fibratus	fib	nearly straight filaments of Ci or Cs.
floccus	flo	small tufts (lumps) of clouds, often with virga (evaporating precip.) falling from each tuft. Applies to Ci, Cc, Ac, Sc.
fractus	fra	shredded, ragged, irregular, torn by winds. Can apply to Cu and St.
humilis	hum	Cu of small vertical extent. Small flat lumps.
lenticularis	len	having lens or almond cross section, often called mountain wave clouds. Applies to Cc, Ac, Sc.
mediocris	med	medium size Cu
nebulosus	neb	Cs or St with veil or layer showing no distinct details
spissatus	spi	thick Ci that looks grey.
stratiformis	str	spreading out into sheets or horizontal layers. Applies to Ac, Sc, Cc.
uncinus	unc	hook or comma shaped Ci

Varieties

Genera and species are further subdivided into **varieties** (Table 6-4), based on:

• transparency (sun or moon visible through cloud)
• arrangement of visible elements

These varieties are NOT mutually exclusive (except for translucidus and opacus), so you can append as many varieties to a cloud identification that apply. For example, "cumulonimbus capillatus translucidus undulatus" (Cb cap tr un).

Table 6-4. WMO cloud varieties. Ab. = Abbreviation.

Variety	Ab.	Description
duplicatus	du	superimposed cloud patches at slightly different levels. Applies to Ci, Cs, Ac, As, Sc.
intortus	in	Ci with tangled, woven, or irregularly curved filaments
lacunosus	la	honeycomb, chessboard, or regular arrangement of clouds and holes. Applies to Cc, Ac, and Sc.
opacus	op	too thick for the sun or moon to shine through. Applies to Ac, As, Sc, St.
perlucidus	pe	a layer of clouds with small holes between elements. Applies to Ac, Sc.
radiatus	ra	very long parallel bands of clouds that, due to perspective, appear to converge at a point on the horizon. Applies to Ci, Ac, As, Sc, and Cu.
translucidus	tr	layer or patch of clouds through which sun or moon is somewhat visible. Applies to Ac, As, Sc, St.
undulatus	un	cloud layers or patches showing waves or undulations. Applies to Cc, Cs, Ac, As, Sc, St.
vertebratus	ve	Ci streaks arranged like a skeleton with vertebrae and ribs, or fish bones.

Supplementary Features

Table 6-5. WMO cloud supplementary features, and the **mother** clouds to which they are attached. (Ab. = abbreviation.)

Feature	Ab.	Description
arcus	arc	a dense horizontal roll cloud, close to the ground, along and above the leading edge of Cb gust-front outflows. Usually attached to the Cb, but can separate from it as the gust front spreads. Often has an arch shape when viewed from underneath, and a curved or arc shape in satellite photos.
incus	inc	the upper portion of a Cb, spread out into an (ice) anvil with smooth, fibrous or striated appearance.
mamma	mam	hanging protuberances of pouch-like appearance. Mammatus clouds are often found on the underside of Cb anvil clouds.
praecipitatio	pra	precipitation falling from a cloud and reaching the ground. Mother clouds can be As, Ns, Sc, St, Cu, Cb.
tuba	tub	a cloud column or funnel cloud protruding from a cloud base, and indicating an intense vortex. Usually attached to Cb, but sometimes to Cu.
virga	vir	visible precipitation trails that evaporate before reaching the ground. They can hang from Ac, Cu, Ns or Cb clouds.

Accessory Clouds

Smaller features attached to other clouds.

Table 6-6. WMO **accessory clouds**, and the **mother** clouds to which they are attached. (Ab.=abbreviation.)

Name	Ab.	Description
pannus	pan	ragged shreds, sometimes in a continuous layer, below a mother cloud and sometimes attached to it. Mother clouds can be As, Ns, Cu, Cb.
pileus	pil	a smooth thin cap, hood, or scarf above, or attached to, the top of rapidly rising cumulus towers. Very transient, because the mother cloud quickly rises through and engulfs it. Mother clouds are Cu con or Cb.
velum	vel	a thin cloud veil or sheet of large horizontal extent above or attached to Cu or Cb, which often penetrate it.

Table 6-7. Sky cover. Oktas= eighths of sky covered.

Sky Cover (oktas)	Symbol	Name	Abbr.	Sky Cover (tenths)
0	○	Sky **Clear**	SKC	0
1	⦶	**Few*** Clouds	FEW*	1
2	◔			2 to 3
3		**Scattered**	SCT	4
4	◑			5
5		**Broken**	BKN	6
6	◕			7 to 8
7				9
8	●	**Overcast**	OVC	10
un-known	⊗	Sky **Obscured**	**	un-known

* *"Few" is used for (0 oktas) < coverage ≤ (2 oktas).*

** *See text body for a list of abbreviations of various obscuring phenomena.*

Sample Application

Use a spreadsheet to find and plot the fraction of clouds ranging from X = 50 to 4950 m width, given ΔX = 100 m, S_X =0.5, and L_X = 1000 m.

Find the Answer

Given: ΔX = 100 m, S_X =0.5, L_X = 1000 m.
Find: $f(X)$= ?

Solve eq. (6.6) on a spreadsheet. The result is:

X (m)	f(X)
50	2.558×10^{-8}
150	0.0004
250	0.0068
350	0.0252
450	0.0495
550	0.0710
etc.	etc.

Sum of all f = 0.999

Check: Units OK. Physics almost OK, but the sum of all f should equal 1.0, representing 100% of the clouds. The reason for the error is that we should have considered clouds even larger than 4950 m, because of the tail on the right of the distribution. Also smaller ΔX would help.

Exposition: Although the dominant cloud width is about 800 m for this example, the long tail on the right of the distribution shows that there are also a small number of large-diameter clouds.

SKY COVER (CLOUD AMOUNT)

The fraction of the sky (celestial dome) covered by cloud is called **sky cover, cloud cover**, or **cloud amount**. It is measured in eights (**oktas**) according to the World Meteorological Organization. Table 6-7 gives the definitions for different cloud amounts, the associated symbol for weather maps, and the abbreviation for aviation weather reports (**METAR**).

Sometimes the sky is **obscured**, meaning that there might be clouds but the observer on the ground cannot see them. Obscurations (and their abbreviations) include: **mist** [BR; horizontal visibilities ≥ 1 km (i.e., ≥ 5/8 of a statute mile)], **fog** [FG; visibilities < 1 km (i.e., < 5/8 statute mile)], **smoke** (FU), **volcanic ash** (VA), **sand** (SA), **haze** (HZ), **spray** (PY), widespread **dust** (DU).

For aviation, the altitude of cloud base for the lowest cloud with coverage ≥ 5 oktas (i.e., lowest broken or overcast clouds) is considered the **ceiling**. For obscurations, the **vertical visibility** (VV) distance is reported as a ceiling instead.

CLOUD SIZES

Cumuliform clouds typically have diameters roughly equal to their depths, as mentioned previously. For example, a fair weather cumulus cloud typically averages about 1 km in size, while a thunderstorm might be 10 km.

Not all clouds are created equal. At any given time the sky contains a spectrum of cloud sizes that has a **lognormal distribution** (Fig. 6.10, eq. 6.6)

$$f(X) = \frac{\Delta X}{\sqrt{2\pi}\cdot X \cdot S_X} \cdot \exp\left[-0.5 \cdot \left(\frac{\ln(X/L_X)}{S_X}\right)^2\right] \quad (6.6)$$

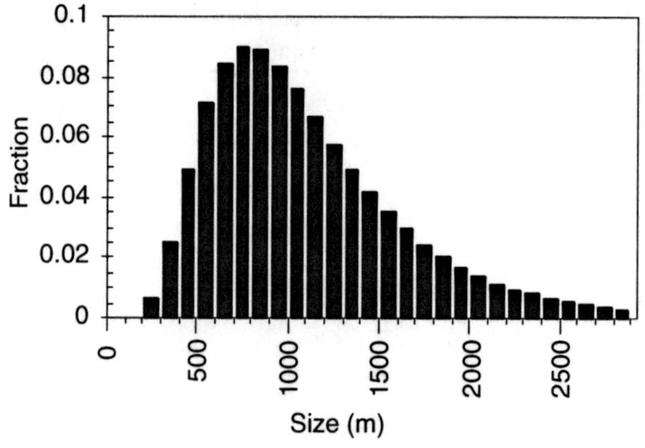

Figure 6.10
Lognormal distribution of cloud sizes.

where X is the cloud diameter or depth, ΔX is a small range of cloud sizes, $f(X)$ is the fraction of clouds of sizes between $X–0.5\Delta X$ and $X+0.5\Delta X$, L_X is a location parameter, and S_X is a dimensionless spread parameter. These parameters vary widely in time and location.

According to this distribution, there are many clouds of nearly the same size, but there also are a few clouds of much larger size. This causes a skewed distribution with a long tail to the right (see Fig. 6.10).

FRACTAL CLOUD SHAPES

Fractals are patterns made of the superposition of similar shapes having a range of sizes. An example is a dendrite snow flake. It has arms protruding from the center. Each of those arms has smaller arms attached, and each of those has even smaller arms. Other aspects of meteorology exhibit fractal geometry, including lightning, turbulence, and clouds.

Fractal Dimension

Euclidian geometry includes only integer dimensions; for example 1-D, 2-D or 3-D. Fractal geometry allows a continuum of dimensions; for example $D = 1.35$.

Fractal dimension is a measure of space-filling ability. Common examples are drawings made by children, and newspaper used for packing cardboard boxes.

A straight line (Fig. 6.11a) has fractal dimension $D = 1$; namely, it is one-dimensional both in the Euclidian and fractal geometry. When toddlers draw with crayons, they fill areas by drawing tremendously wiggly lines (Fig. 6.11b). Such a line might have fractal dimension $D \approx 1.7$, and gives the impression of almost filling an area. Older children succeed in filling the area, resulting in fractal dimension $D = 2$.

A different example is a sheet of newspaper. While it is flat and smooth, it fills only two-dimensional area (neglecting its thickness), hence $D = 2$. However, it takes up more space if you crinkle it, resulting in a fractal dimension of $D \approx 2.2$. By fully wadding it into a tight ball of $D \approx 2.7$, it begins to behave more like a three-dimensional object, which is handy for filling empty space in cardboard boxes.

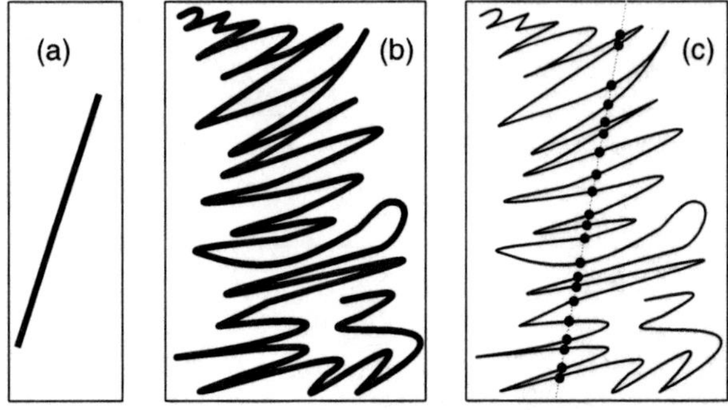

Figure 6.11
(a) A straight line has fractal dimension D = 1. (b) A wiggly line has fractal dimension 1 < D < 2. (c) The zero-set of the wiggly line is a set of points with dimension 0 < D < 1.

INFO • Cloud and Sky Photography using High Dynamic Range (HDR)

Most people take photos with their cell phone or tablet. You cannot easily add special lenses and filters to these devices. Instead, to take better cloud photos, use special software on your device to capture **high dynamic range (HDR)** images.

When you take an HDR photo, your camera takes three photos in rapid succession of the same scene. One photo uses a good average exposure for the whole scene (1.N). Another is underexposed (1.U), but captures the sky and clouds better. The third is overexposed (1.O), to capture darker ground and shadows. The software automatically combines the best parts of the three photos into one photo (2). Finally, after you enhance the saturation, contrast, and sharpness, the image (3) is much better than the normal photo (1.N).

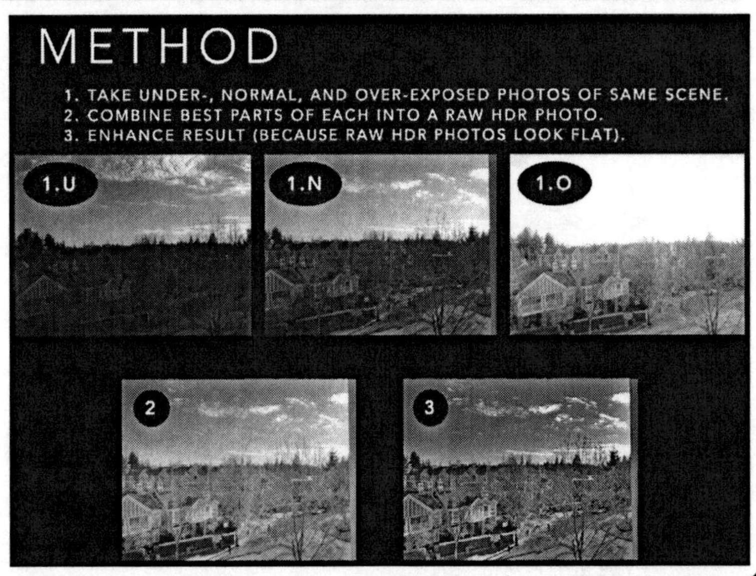

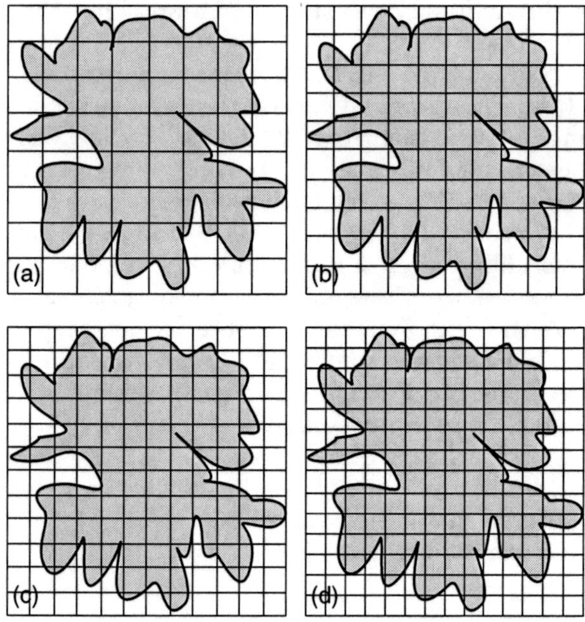

Figure 6.12
Cloud shadow overlaid with small tiles. The number of tiles M per side of domain is: (a) 8, (b) 10, (c) 12, (d) 14.

Sample Application

Use Fig. 6.12 to measure the fractal dimension of the cloud shadow.

Find the Answer

For each figure, the number of boxes is:

	M	N
(a)	8	44
(b)	10	53
(c)	12	75
(d)	14	88

(You might get a slightly different count.)
Plot the result on a log-log graph, & fit a straight line.

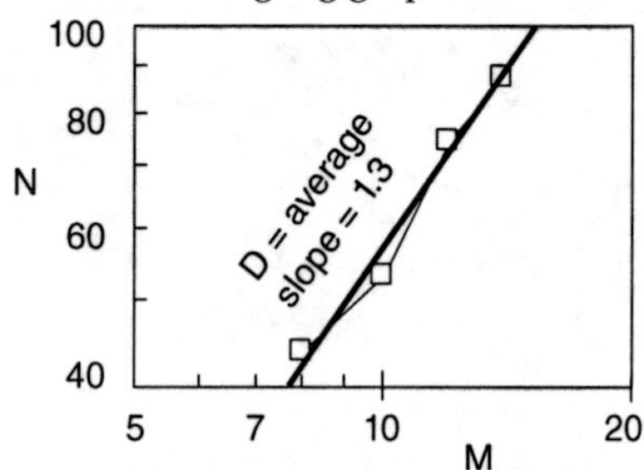

Use the end points of the best-fit line in eq. (6.7):

$$D = \frac{\log(100/40)}{\log(15/7.5)} = \underline{\mathbf{1.32}}$$

Check: Units OK. Physics OK.

Exposition: About right for a cloud. The original domain size relative to cloud size makes no difference. Although you will get different M and N values, the slope D will be almost the same.

A **zero set** is a lower-dimensional slice through a higher-dimensional shape. Zero sets have fractal dimension one less than that of the original shape (i.e., $D_{zero\ set} = D - 1$). Sometimes it is easier to measure the fractal dimension of a zero set, from which we can calculate the dimension of the original.

For example, start with the wad of paper having fractal dimension $D = 2.7$. It would be difficult to measure the dimension for this wad of paper. Instead, carefully slice that wad into two halves, dip the sliced edge into ink, and create a print of that inked edge on a flat piece of paper. The wiggly line that was printed (Fig. 6.11b) has fractal dimension of $D = 2.7 - 1 = 1.7$, and is a zero set of the original shape. It is easier to measure.

To continue the process, slice through the middle of the print (shown as the grey line in Fig. 6.11c). The wiggly printed line crosses the straight slice at a number of points (Fig. 6.11c); those points have a fractal dimension of $D = 1.7 - 1 = 0.7$, and represent the zero set of the print. Thus, while any one point has a Euclidian dimension of zero, the set of points appears to partially fill a 1-D line.

Measuring Fractal Dimension

Consider an irregular-shaped area, such as the shadow of a cloud. The perimeter of the shadow is a wiggly line, so we should be able to measure its fractal dimension. Based on observations of the perimeter of real-cloud shadows, $D = 1.35$. If we assume that this cloud shadow is a zero set of the surface of the cloud, then the cloud surface dimension is $D = D_{zero\ set} + 1 = 2.35$.

A **box-counting** method can be used to measure fractal dimension. Put the cloud-shadow picture within a square domain. Then tile the domain with smaller square boxes, with M tiles per side of domain (Fig. 6.12). Count the number N of tiles through which the perimeter passes. Repeat the process with smaller tiles.

Get many samples of N vs. M, and plot these as points on a log-log graph. The fractal dimension D is the slope of the best fit straight line through the points. Define subscripts 1 and 2 as the two end points of the best fit line. Thus:

$$D = \frac{\log(N_1 / N_2)}{\log(M_1 / M_2)} \tag{6.7}$$

This technique works best when the tiles are small. If the plotted line is not straight but curves on the log-log graph, try to find the slope of the portion of the line near large M.

FOG

Types

Fog is a cloud that touches the ground. The main types of fog are:
- upslope
- radiation
- advection
- precipitation or frontal
- steam

They differ in how the air becomes saturated.

Upslope fog is formed by adiabatic cooling in rising air that is forced up sloping terrain by the wind. Namely, it is formed the same way as clouds. As already discussed in the Water Vapor chapter, air parcels must rise or be lifted to their lifting condensation level (LCL) to form a cloud or upslope fog.

Radiation fog and advection fog are formed by cooling of the air via conduction from the cold ground. **Radiation fog** forms during clear, nearly-calm nights when the ground cools by IR radiation to space. **Advection fog** forms when initially-unsaturated air advects over a colder surface.

Precipitation fog or **frontal fog** is formed by adding moisture, via the evaporation from warm rain drops falling down through the initially-unsaturated cooler air below cloud base.

Steam fog occurs when cold air moves over warm humid surfaces such as unfrozen lakes during early winter. The lake warms the air near it by conduction, and adds water by evaporation. However, this thin layer of moist warm air near the surface is unsaturated. As turbulence causes it to mix with the colder air higher above the surface, the mixture becomes saturated, which we see as steam fog.

Idealized Fog Models

By simplifying the physics, we can create mathematical fog models that reveal some of the fundamental behaviors of different types of fog.

Advection Fog

For formation and growth of advection fog, suppose a fogless mixed layer of thickness z_i advects with speed M over a cold surface such as snow covered ground or a cold lake. If the surface potential temperature is θ_{sfc}, then the air potential temperature θ cools with downwind distance according to

$$\theta = \theta_{sfc} + (\theta_o - \theta_{sfc}) \cdot \exp\left(-\frac{C_H}{z_i} \cdot x\right) \qquad (6.8)$$

Sample Application

Fog formation: A layer of air adjacent to the surface (where P = 100 kPa) is initially at temperature 20°C and relative humidity 68%. (a) To what temperature must this layer be cooled to form **radiation** or **advection** fog? (b) To what altitude must this layer be lifted to form **upslope** fog? (c) How much water must be evaporated into each kilogram of dry air from falling rain drops to form **frontal** fog? (d) How much evaporation (mm of lake water depth) from the lake is necessary to form **steam** fog throughout a 100 m thick layer? [Hint: Use eqs. from the Water Vapor chapter.]

Find the Answer

Given: P = 100 kPa, T = 20°C, RH = 68%, Δz = 100 m
Find: a) T_d=?°C, b) z_{LCL}=?m, c) r_s=?g kg^{-1} d) d=?mm

Using Table 4-1: e_s = 2.371 kPa.
Using eq. (4.14): $e = (RH/100\%) \cdot e_s$
= (68%/100%)·(2.371 kPa) = 1.612 kPa

(a) Knowing e and using Table 4-1: T_d = **14°C**.

(b) Using eq. (4.16):
$z_{LCL} = a \cdot [T - T_d]$ = (0.125 m K^{-1})·
[(20+273)K – (14+273)K] = **0.75 km**

(c) Using eq. (4.4), the initial state is:
$r = \varepsilon \cdot e/(P-e)$ = 0.622 · (1.612 kPa) / (100 kPa – 1.612 kPa) = 0.0102 g g^{-1} = 10.2 g kg^{-1}
The final mixing ratio at saturation (eq. 4.5) is:
$r_s = \varepsilon \cdot e_s/(P-e_s)$ = 0.622·(2.371 kPa) / (100 kPa – 2.371 kPa) = 0.0151 g g^{-1} = 15.1 g kg^{-1}.
The amount of additional water needed is
$\Delta r = r_s - r$ = 15.1 – 10.2 = **4.9** g_{water} kg_{air}^{-1}

(d) Using eqs. (4.11) & (4.13) to find absolute humidity
$\rho_v = \varepsilon \cdot e \cdot \rho_d / P$
= (0.622)·(1.612 kPa)·(1.275 kg·m^{-3})/(100 kPa)
= 0.01278 kg·m^{-3}
$\rho_{vs} = \varepsilon \cdot e_s \cdot \rho_d / P$
= (0.622)·(2.371 kPa)·(1.275 kg·m^{-3})/(100 kPa)
= 0.01880 kg·m^{-3}
The difference must be added to the air to reach saturation: $\Delta\rho = \rho_{vs} - \rho_v$ =
$\Delta\rho$ = (0.01880–0.01278) kg·m^{-3} = 0.00602 kg·m^{-3}
But over A =1 m^2 of surface area, air volume is
$Vol_{air} = A \cdot \Delta z$ = (1 m^2)·(100 m) = 100 m^3.
The mass of water needed in this volume is
$m = \Delta\rho \cdot Vol_{air}$ = 0.602 kg of water.
But liquid water density ρ_{liq} = 1000 kg·m^{-3}: Thus,
$Vol_{liq} = m / \rho_{liq}$ = 0.000602 m^3
The depth of liquid water under the 1 m^2 area is
$d = Vol_{liq} / A$ = 0.000602 m = **0.602 mm**

Check: Units OK. Physics OK.
Exposition: For many real fogs, cooling of the air and addition of water via evaporation from the surface happen simultaneously. Thus, a fog might form in this example at temperatures warmer than 14°C.

HIGHER MATH • Advection Fog

Derivation of eq. (6.8):

Start with the Eulerian heat balance, neglecting all contributions except for turbulent flux divergence:

$$\frac{\partial\theta}{\partial t} = -\frac{\partial F_{z\ turb}(\theta)}{\partial z}$$

where θ is potential temperature, and F is heat flux. For a mixed layer of fog, F is linear with z, thus:

$$\frac{\partial\theta}{\partial t} = -\frac{F_{z\ turb\ zi}(\theta) - F_{z\ turb\ sfc}(\theta)}{z_i - 0}$$

If entrainment at the top of the fog layer is small, then $F_{z\ turb\ zi}(\theta) = 0$, leaving:

$$\frac{\partial\theta}{\partial t} = \frac{F_{z\ turb\ sfc}(\theta)}{z_i} = \frac{F_H}{z_i}$$

Estimate the flux using bulk transfer eq. (3.21). Thus:

$$\frac{\partial\theta}{\partial t} = \frac{C_H \cdot M \cdot (\theta_{sfc} - \theta)}{z_i}$$

If the wind speed is roughly constant with height, then let the Eulerian volume move with speed M:

$$\frac{\partial\theta}{\partial t} = \frac{\partial\theta}{\partial x}\frac{\partial x}{\partial t} = \frac{\partial\theta}{\partial x} \cdot M$$

Plugging this into the LHS of the previous eq. gives:

$$\frac{\partial\theta}{\partial x} = \frac{C_H \cdot (\theta_{sfc} - \theta)}{z_i}$$

To help integrate this, define a substitute variable $s = \theta - \theta_{sfc}$, for which $\partial s = \partial\theta$. Thus:

$$\frac{\partial s}{\partial x} = -\frac{C_H \cdot s}{z_i}$$

Separate the variables: $\frac{ds}{s} = -\frac{C_H}{z_i}dx$

Which can be integrated (using the prime to denote a dummy variable of integration):

$$\int_{s'=s_o}^{s} \frac{ds'}{s'} = -\frac{C_H}{z_i}\int_{x'=0}^{x} dx'$$

Yielding:

$$\ln(s) - \ln(s_o) = -\frac{C_H}{z_i} \cdot (x - 0)$$

or

$$\ln\left(\frac{s}{s_o}\right) = -\frac{C_H}{z_i} \cdot x$$

Taking the antilog of each side (i.e., exp):

$$\frac{s}{s_o} = \exp\left(-\frac{C_H}{z_i} \cdot x\right)$$

Upon rearranging, and substituting for s:

$$\theta - \theta_{sfc} = (\theta - \theta_{sfc})_o \cdot \exp\left(-\frac{C_H}{z_i} \cdot x\right)$$

But θ_{sfc} is assumed constant, thus:

$$\boxed{\theta = \theta_{sfc} + (\theta_o - \theta_{sfc}) \cdot \exp\left(-\frac{C_H}{z_i} \cdot x\right)} \qquad (6.8)$$

where θ_o is the initial air potential temperature, C_H is the heat transfer coefficient (see the Heat Budgets chapter), and x is travel distance over the cold surface. This assumes an idealized situation where there is sufficient turbulence caused by a brisk wind speed to keep the boundary layer well mixed.

Advection fog forms when the temperature drops to the dew-point temperature T_d. At the surface (more precisely, at z = 10 m), $\theta \approx T$. Thus, setting $\theta \approx T = T_d$ at saturation and solving the equation above for x gives the distance over the lake at which fog first forms:

$$x = \frac{z_i}{C_H} \cdot \ln\left(\frac{T_o - T_{sfc}}{T_d - T_{sfc}}\right) \qquad (6.9)$$

Surprisingly, neither the temperature evolution nor the distance to fog formation depends on wind speed.

For example, advection fog can exist along the California coast where warm humid air from the

Sample Application

Air of initial temperature 5°C and depth 200 m flows over a frozen lake of surface temperature –3°C. If the initial dew point of the air is –1°C, how far from shore will **advection** fog first form?

Find the Answer

Given: T_o = 5°C, T_d = –1°C,
T_{sfc} = –3 °C, z_i =200 m

Find: x = ? km.

Assume smooth ice: C_H = 0.002 .

Use eq. (6.9): x = (200 m/0.002)·
ln[(5–(–3))/(–1–(–3))] = **138.6 km**

Sketch, where eq. (6.8) was solved for T vs. x:

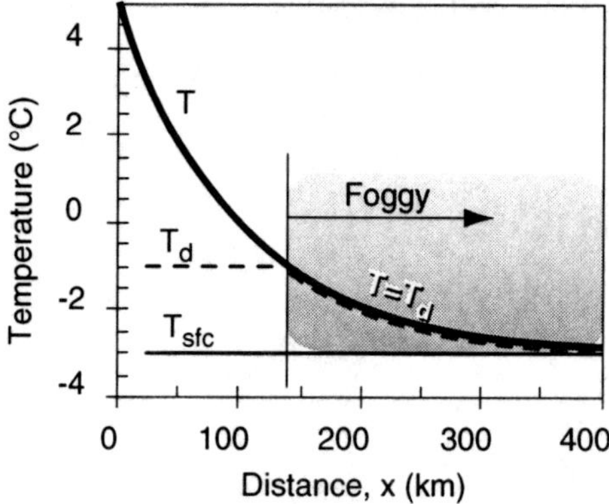

Check: Units OK. Physics OK.

Exposition: If the lake is smaller than 138.6 km in diameter, then no fog forms. Also, if the dew-point temperature needed for fog is colder than the surface temperature, then no fog forms.

west blows over the cooler "Alaska current" coming from further north in the Pacific Ocean.

Advection fog, once formed, experiences radiative cooling from fog top. Such cooling makes the fog more dense and longer lasting as it can evolve into a well-mixed radiation fog, described in the next subsection.

Dissipation of advection fog is usually controlled by the synoptic and mesoscale weather patterns. If the surface becomes warmer (e.g., all the snow melts, or there is significant solar heating), or if the wind changes direction, then the conditions that originally created the advection fog might disappear. At that point, dissipation depends on the same factors that dissipate radiation fog. Alternately, frontal passage or change of wind direction might blow out the advection fog, and replace the formerly-foggy air with cold dry air that might not be further cooled by the underlying surface.

Radiation Fog

For formation and growth of radiation fog, assume a stable boundary layer forms and grows, as given in the Atmospheric Boundary Layer chapter. For simplicity, assume that the ground is flat, so there is no drainage of cold air downhill (a poor assumption). If the surface temperature T_s drops to the dew-point temperature T_d then fog can form (Fig. 6.13). The fog depth is the height where the nocturnal temperature profile crosses the initial dew-point temperature T_{do}.

The time t_o between when nocturnal cooling starts and the onset of fog is

$$t_o = \frac{a^2 \cdot M^{3/2} \cdot (T_{RL} - T_d)^2}{(-F_H)^2} \tag{6.10}$$

where a = 0.15 $m^{1/4} \cdot s^{1/4}$, T_{RL} is the residual-layer temperature (extrapolated adiabatically to the surface), M is wind speed in the residual layer, and F_H is the average surface kinematic heat flux. Faster winds and drier air delay the onset of fog. For most cases, fog never happens because night ends first.

Once fog forms, evolution of its depth is approximately

$$z = a \cdot M^{3/4} \cdot t^{1/2} \cdot \ln\left[(t / t_o)^{1/2} \right] \tag{6.11}$$

where t_o is the onset time from the previous equation. This equation is valid for $t > t_o$.

Liquid water content increases as a saturated air parcel cools. Also, visibility decreases as liquid water increases. Thus, the densest (lowest visibility)

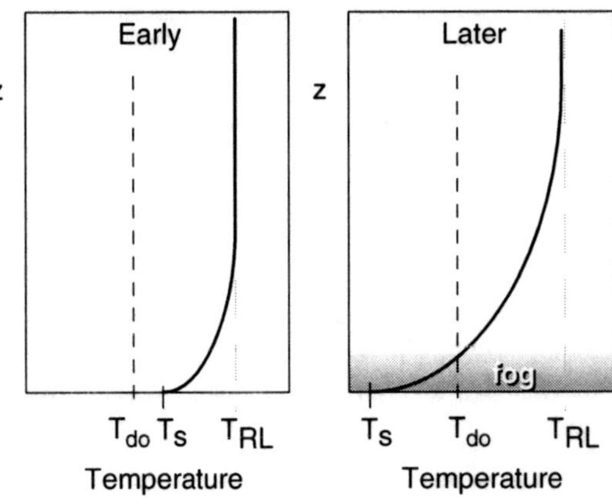

Figure 6.13
Stable-boundary-layer evolution at night leading to radiation fog onset, where T_s is near-surface air temperature, T_{do} is original dew-point temperature, and T_{RL} is the original temperature. Simplified, because dew formation on the ground can decrease T_d before fog forms, and cold-air downslope drainage flow can remove or deposit cold air to alter fog thickness.

Sample Application

Given a residual layer temperature of 20°C, dew point of 10°C, and wind speed 1 m s^{-1}. If the surface kinematic heat flux is constant during the night at –0.02 K·m s^{-1}, then what is the onset time and height evolution of **radiation** fog?

Find the Answer

Given: T_{RL} = 20°C, T_d = 10°C, M = 1 m s^{-1}
F_H = –0.02 K·m s^{-1}
Find: t_o = ? h, and z vs. t.

Use eq. (6.10):

$$t_o = \frac{(0.15\text{m}^{1/4} \cdot \text{s}^{1/4})^2 \cdot (1\text{m/s})^{3/2} \cdot (20-10°\text{C})^2}{(0.02°\text{C} \cdot \text{m/s})^2}$$

= **1.563 h**

The height evolution for this wind speed, as well as for other wind speeds, is calculated using eq. (6.11):

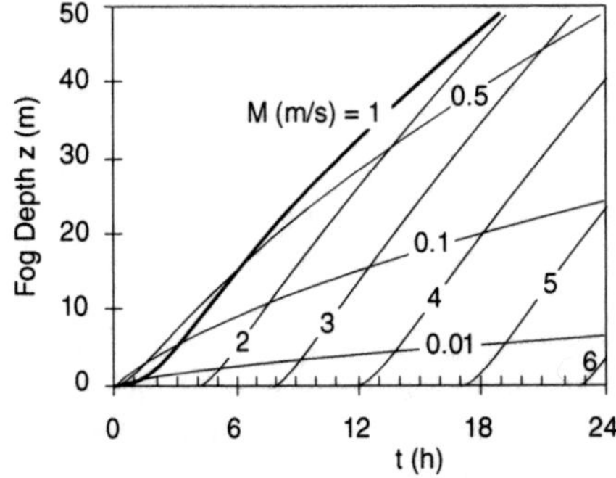

Check: Units OK. Physics OK.
Exposition: Windier nights cause later onset of fog, but stimulates rapid growth of the fog depth.

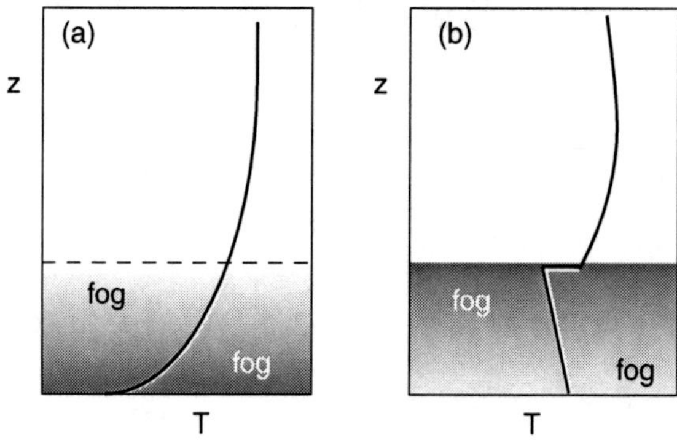

Figure 6.14
(a) Stratified fog that is more dense and colder at ground. (b) Well-mixed fog that is more dense and colder at the top due to IR radiative cooling.

> **Sample Application**
> Initially, total water is constant with height at 10 g kg^{-1}, and wet-bulb potential temperature increases with height at rate 2°C/100m in a stratified fog. Later, a well-mixed fog forms with depth 100 m. Plot the total water and θ_W profiles before and after mixing.
>
> **Find the Answer**
>
> 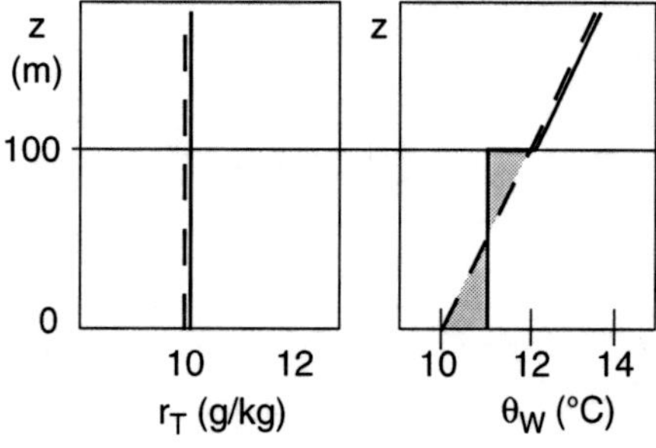
>
>
> **Check**: Units OK. Physics OK.
> **Exposition**: Dashed lines show initial profiles, solid are final profiles. For θ_W, the two shaded areas must be equal for heat conservation. This results in a temperature jump of 1°C across the top of the fog layer for this example. Recall from the Water Vapor and Atmospheric Stability chapters that actual temperature within the saturated (fog) layer decreases with height at the moist adiabatic lapse rate (lines of constant θ_W).

part of the fog will generally be in the coldest air, which is initially at the ground.

Initially, fog density decreases smoothly with height because temperature increases smoothly with height (Fig. 6.14a). As the fog layer becomes optically thicker and more dense , it reaches a point where the surface is so obscured that it can no longer cool by direct IR radiation to space.

Instead, the height of maximum radiative cooling moves upward into the fog away from the surface. Cooling of air within the nocturnal fog causes air to sink as cold thermals. Convective circulations then turbulently mix the fog.

Very quickly, the fog changes into a **well-mixed fog** with wet-bulb-potential temperature θ_W and total-water content r_T that are uniform with height. During this rapid transition, total heat and total water averaged over the whole fog layer are conserved. As the night continues, θ_W decreases and r_L increases with time due to continued radiative cooling.

In this fog the actual temperature decreases with height at the moist adiabatic lapse rate (Fig. 6.14b), and liquid water content increases with height. Continued IR cooling at fog top can strengthen and maintain this fog. This fog is densest near the top of the fog layer.

Dissipation of Well-Mixed (Radiation and Advection) Fogs

During daytime, solar heating and IR cooling are both active. Fogs can become less dense, can thin, can lift, and can totally dissipate due to warming by the sun.

Stratified fogs (Fig. 6.14a) are optically thin enough that sunlight can reach the surface and warm it. The fog albedo can be in the range of A = 0.3 to 0.5 for thin fogs. This allows the warm ground to rapidly warm the fog layer, causing evaporation of the liquid drops and dissipation of the fog.

For optically-thick well-mixed fogs (Fig. 6.14b), albedoes can be A =0.6 to 0.9. What little sunlight is not reflected off the fog is absorbed in the fog itself. However, IR radiative cooling continues, and can compensate the solar heating. One way to estimate whether fog will totally dissipate at time t in the future is to calculate the sum Q_{Ak} of accumulated cooling and heating (see the Atmospheric Boundary Layer chapter) during the time period $(t - t_o)$ since the fog first formed:

$$Q_{Ak} = F_{H.night} \cdot (t - t_o) + (1 - A) \cdot \frac{F_{H.\max} \cdot D}{\pi} \cdot \left[1 - \cos\left(\frac{\pi \cdot (t - t_{SR})}{D} \right) \right] \tag{6.12}$$

where $F_{H.night}$ is average nighttime surface kinematic heat flux (negative at night), and $F_{H.max}$ is the amplitude of the sine wave that approximates surface kinematic sensible heat flux due to solar heating (i.e., the positive value of insolation at local noon). D is daylight duration hours, t is hours after sunset, t_o is hours after sunset when fog first forms, and t_{SR} is hours between sunset and the next sunrise.

The first term on the right should be included only when $t > t_o$, which includes not only nighttime when the fog originally formed, but the following daytime also. This approach assumes that the rate $F_{H.night}$ of IR cooling during the night is a good approximation to the continued IR cooling during daytime.

The second term should be included only when $t > t_{SR}$, where t_{SR} is sunrise time. (See the Atmospheric Boundary Layer Chapter for definitions of other variables). When Q_{Ak} becomes positive, fog dissipates (neglecting other factors such as advection, and assuming no fog initially).

While IR cooling happens from the very top of the fog, solar heating occurs over a greater depth in the top of the fog layer. Such heating underneath cooling statically destabilizes the fog, creating convection currents of cold thermals that sink from the fog top. These upside-down cold thermals continue to mix the fog even though there might be net heating when averaged over the whole fog. Thus, any radiatively heated or cooled air is redistributed and mixed vertically throughout the fog by convection.

Such heating can cause the bottom part of the fog to evaporate, which appears to an observer as **lifting** of the fog (Fig. 6.15a). Sometimes the warming during the day is insufficient to evaporate all the fog. When night again occurs and radiative cooling is not balanced by solar heating, the bottom of the elevated fog lowers back down to the ground (Fig. 6.15b).

In closing this section on fogs, please be aware that the equations above for formation, growth, and dissipation of fogs were based on very idealized situations, such as flat ground and horizontally uniform heating and cooling. In the real atmosphere, even gentle slopes can cause katabatic drainage of cold air into the valleys and depressions (see the Regional Winds chapter), which are then the favored locations for fog formation. The equations above were meant only to illustrate some of the physical processes, and should not to be used for operational fog forecasting.

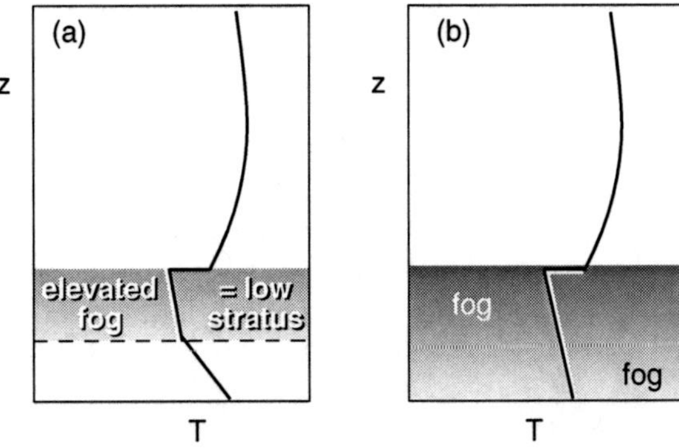

Figure 6.15
(a) Heating during the day can modify the fog of Fig. 6.14b, causing the base to ***lift*** *and the remaining elevated fog to be less dense (b) If solar heating is insufficient to totally dissipate the fog, then nocturnal radiative cooling can re-strengthen it.*

Sample Application(§)

When will fog dissipate if it has an albedo of A = (a) 0.4; (b) 0.6; (c) 0.8 ? Assume daylight duration of D = 12 h. Assume fog forms at t_o = 3 h after sunset. Given: $F_{H\ night} = -0.02$ K·m s^{-1}, and $F_{H\ max\ day} = 0.2$ K·m s^{-1}. Also plot the cumulative heating.

Find the Answer

Given: (see above) Let t = time after sunset.
Use eq. (6.12), & solve on a spreadsheet:

(a) t = 16.33 h = 4.33 h after sunrise = **10:20 AM**
(b) t = 17.9 h = 5.9 h after sunrise = **11:54 AM**
(c) **Fog never dissipates**, because the albedo is so large that too much sunlight is reflected off of the fog, leaving insufficient solar heating to warm the ground and dissipate the fog.

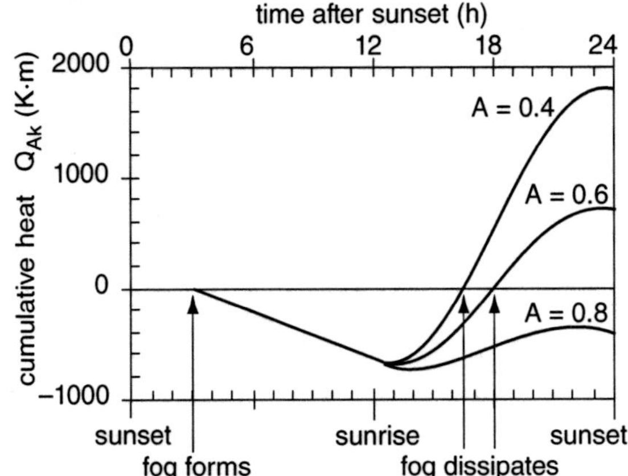

Check: Units OK. Physics OK.
Exposition: Albedo makes a big difference for fog dissipation. There is evidence that albedo depends on the type and number concentration of fog nuclei.

REVIEW

Condensation of water vapor occurs by cooling, adding moisture, or mixing. Cooling and moisturizing are governed by the Eulerian heat and water budgets, respectively. Turbulent mixing of two nearly-saturated air parcels can yield a saturated mixture. We see the resulting droplet-filled air as clouds or fog.

If clouds are buoyant (i.e., the cloud and subcloud air is statically unstable), thermals of warm air actively rise to form cumuliform clouds. These have a lognormal distribution of sizes, and have fractal shapes. Active clouds are coupled to the underlying surface, allowing thermo diagrams to be used to estimate cloud base and top altitudes.

If the clouds are not buoyant (i.e., the cloud layer is statically stable), the clouds remain on the ground as fog, or they are forced into existence as stratiform clouds by advection along an isotropic surface over colder air, such as at a front.

Clouds can be classified by their altitude, shape, and appearance. Special symbols are used on weather maps to indicate cloud types and coverage.

While almost all clouds are created in rising air and the associated adiabatic cooling, fogs can form other ways. IR radiative cooling creates radiation fogs, while cooling associated with advection of humid air over a cold surface causes advection fogs. Precipitation and steam fog form by adding moisture to the air that touches a liquid water surface (e.g., raindrops, or a lake), followed by mixing with the surrounding cooler air to reach saturation.

Although fog forms in a layer of cold air resting on the ground, that cold-air can be either statically stable or unstable depending on whether continued cooling is being imposed at the bottom or top of the layer, respectively. Heat and moisture budget equations can be used to forecast fog onset, development, and dissipation, but only for some idealized situations.

HOMEWORK EXERCISES

Broaden Knowledge & Comprehension

B1. Search the web or consult engineering documents that indicate the temperature and water vapor content of exhaust from aircraft jet engines. What ambient temperature and humidity of the atmosphere near the aircraft would be needed such that the mixture of the exhaust with the air would just become saturated and show a jet contrail in the sky. (You might need to utilize a spreadsheet to experiment with different mixing proportions to find the one that would cause saturation first.)

B2. Search the web for cloud-classification images corresponding to the clouds in Figs. 6.3 and 6.5. Copy the best example of each cloud type to your own page to create a cloud chart. Be sure to also record and cite the original web site for each photo.

B3. Search the web for thermo diagrams showing soundings for cases where different cloud types were present (e.g., active, passive, fog). Estimate cloud base and cloud top altitudes from the sounding, and compare with observations. Cloud base observations are available from the METAR surface reports made at airports. Cloud top observations are reported by aircraft pilots in the form of PIREPS.

B4. Search the web to identify some of the instruments that have been devised to determine cloud-base altitude?

B5. Search the web for cloud images corresponding to the "Other Clouds" discussed near the start of this chapter. Copy the best example of each cloud type to your own page. Be sure to also record and cite the original web site for each cloud photo.

B6. Search the web for satellite and ground-based images of cloud streets, open cells, and closed cells. Create a image table of these different types of cloud organization.

B7. Search the web for a surface weather map that plots surface observations for locations near you. For 3 of the stations having clouds, interpret (describe in words) the sky cover and cloud genera symbols (using Tables 6-7 and 6-2) that are plotted.

B8. Table 6-2 shows only a subset of the symbols used to represent clouds on weather maps. Search the web for map legends or tables of symbols that give a more complete list of cloud symbols.

B9. a. Search the web for fractal images that you can print. Use box counting methods on these images to find the fractal dimension, and compare with the fractal dimension that was specified by the creator of that fractal image.

b. Make a list of the URL web addresses of the ten fractal images that you like the best (try to find ones from different research groups or other people).

B10. Search the web for satellite images that show fog. (Hints: Try near San Francisco, or try major river valleys in morning, or try over snowy ground in spring, or try over unfrozen lakes in Fall). For the fog that you found, determine the type of fog (radiation, advection, steam, etc.) in that image, and justify your decision.

B11. a. Search the web for very high resolution visible satellite images of cumulus clouds or cloud clusters over an ocean. Display or print these images, and then trace the edges of the clouds. Find the fractal dimension of the cloud edge, similar to the procedure that was done with Fig. 6.12.

B12. Search the web for a discussion about which satellite channels can be used to discriminate between clouds and fog. Summarize your findings.

B13. Search the web for methods that operational forecasters use to predict onset and dissipation times of fog. Summarize your findings. (Hint: Try web sites of regional offices of the weather service.)

Apply

A1. Given the following temperature and vapor pressure for an air parcel. (i) How much moisture Δe (kPa) must be added to bring the original air parcel to saturation? (ii) How much cooling ΔT (°C) must be done to bring the original parcel to saturation? Original [T (°C) , e (kPa)] =

a. 20, 0.2 b. 20, 0.4 c. 20, 0.6 d. 20, 0.8 e. 20, 1.0
f. 20, 1.5 g. 20, 2.0 h. 20, 2.2 i. 10, 0.2 j. 10, 0.4
k. 10, 0.6 l. 10, 0.8 m. 10, 1.0 n. 30, 1.0 o. 30, 3.0

A2. On a winter day, suppose your breath has T = 30°C with T_d = 28°C, and it mixes with the ambient air of temperature T = –10°C and T_d = –11°C. Will you see your breath? Assume you are at sea level, and that your breath and the environment mix in proportions of (breath : environment):

a. 1:9 b. 2:8 c. 3:7 d. 4:6 e. 5:5
f. 6:4 g. 7:3 h. 8:2 i. 9:1

A3(§). A jet aircraft is flying at an altitude where the ambient conditions are P = 30 kPa, T = –40°C, and T_d = –42°C. Will a visible contrail form? (Hint, assume that all possible mixing proportions occur.)

Assume the jet exhaust has the following conditions: [T (°C) , T_d (°C)] =

a. 200, 180 b. 200, 160 c. 200, 140 d. 200, 120
e. 200, 100 f. 200, 80 g. 200, 60 h. 200, 40
i. 400, 375 j. 400, 350 k. 400, 325 l. 400, 300
m. 400, 275 n. 400, 250 o. 400, 200 p. 400, 150

A4. Given the following descriptions of ordinary clouds. (i) First classify as cumuliform or stratiform. (ii) Then name the cloud. (iii) Next, draw both the WMO and USA symbols for the cloud. (iv) Indicate if the cloud is made mostly of liquid water or ice (or both). (v) Indicate the likely altitude of its cloud base and top. (vi) Finally, sketch the cloud similar to those in Figs. 6.3 or 6.5.

a. Deep vertical towers of cloud shaped like a gigantic mushroom, with an anvil shaped cloud on top. Flat, dark-grey cloud base, with heavy precipitation showers surrounded by non-precipitating regions. Can have lightning, thunder, hail, strong gusty winds, and tornadoes. Bright white cloud surrounded by blue sky when viewed from the side, but the cloud diameter is so large that when from directly underneath it might cover the whole sky.

b. Sheet of light-grey cloud covering most of sky, with sun or moon faintly showing through it, casting diffuse shadows on ground behind buildings and people.

c. Isolated clouds that look like large white cotton balls or cauliflower, but with flat bases. Lots of blue sky in between, and no precipitation. Diameter of individual clouds roughly equal to the height of their tops above ground.

d. Thin streaks that look like horse tails, with lots of blue sky showing through, allowing bright sun to shine through it with crisp shadows cast on the ground behind trees and people.

e. Thick layer of grey cloud with well poorly-defined cloud base relatively close to the ground, and widespread drizzle or light rain or snow. No direct sunlight shining through, and no shadows cast on the ground. Gloomy.

f. Isolated clouds that look like small, white cotton balls or popcorn (but with flat bases), with lots of blue sky in between. Size of individual clouds roughly equal to their height above ground.

g. Thin uniform veil covering most of sky showing some blue sky through it, with possibly a halo around a bright sun, allowing crisp shadows cast on the ground behind trees and people.

h. Layer of grey cloud close to the ground, but lumpy with some darker grey clouds dispersed among thinner light grey or small clear patches in a patchwork or chessboard pattern. Not usually precipitating.

i. Thin veil of clouds broken into very small lumps, with blue sky showing through, allowing bright sun to shine through it with crisp shadows cast on the ground behind trees and people.

j. Sheet of cloud covering large areas, but broken into flat lumps, with sun or moon faintly showing through the cloudy parts but with small patches of blue sky in between the lumps, casting diffuse shadows behind buildings and people.

k. Deep towers of cloud with bright white sides and top during daytime, but grey when viewed from the bottom. Clouds shaped like stacks of ice-cream balls or turrets of cotton balls with tops extending high in the sky, but with flat bases relatively close to the ground. Usually no precipitation.

l. Thick layer of grey cloud with well defined cloud base relatively close to the ground. No direct sunlight shining through, and no shadows cast on the ground. No precipitation.

A5. Use a thermo diagram to plot the following environmental sounding:

P (kPa)	T (°C)
20	–15
25	–25
35	–25
40	–15
45	–20
55	–15
70	0
80	9
85	6
95	15

Determine cloud activity (active, passive, fog, none), cloud-base height, and cloud-top height, for the conditions of near-surface (P = 100 kPa) air parcels given below: [T (°C) , T_d (°C)] =

a. 20, 14	b. 20, 7	c. 20, 17	d. 20, 19
e. 15, 15	f. 30, 0	g. 30, 10	h. 30, 24
i. 25, 24	j. 25, 20	k. 25, 16	l. 25, 12
m. 25, 10	n. 25, 6	o. 15, 10	p. 22, 22

A6. The buoyancy of a cloudy air parcel depends on its virtual temperature compared to that of its environment. Given a saturated air parcel of temperature and liquid-water mixing ratio as listed below. What is its virtual temperature, and how would it compare to the virtual temperature with no liquid water?

[T(°C), r_L(g kg^{-1})] = a. 20, 1 b. 20, 2 c. 20, 5
d. 20, 10 e. 10, 1 f. 10, 3 g. 10, 7 h. 10, 12
i. 0, 2 j. 0, 4 k. 0, 8 l. 0, 15 m. 5, 1
n. 5, 2 o. 5, 4 p. 5, 6 q. 5, 8 r. 5, 10

A7. On a large thermo diagram from the end of the Atmospheric Stability chapter, plot a hypothetical sounding of temperature and dew-point that would be possible for the following clouds.

a. cirrus b. cirrostratus c. cirrocumulus
d. altostratus e. altocumulus f. stratus
g. nimbostratus h. fog

A8. Name these special clouds.

a. Parallel bands of clouds perpendicular to the shear vector, at the level of altocumulus.
b. Parallel bands of cumulus clouds parallel to the wind vector in the boundary layer.
c. Two, long, closely-spaced parallel cloud lines at high altitude.
d. Clouds that look like flags or pennants attached to and downwind of mountain peaks.
e. Look like altocumulus, but with more vertical development that causes them to look like castles.
f. Clouds that look like breaking waves.
g. Look like cumulus or cumulus mediocris, but relatively tall and small diameter causing them to look like castles.
h. Clouds that look lens shaped when viewed from the side, and which remain relatively stationary in spite of a non-zero wind.
i. A low ragged cloud that is relatively stationary and rotating about a horizontal axis.
j. Ragged low scattered clouds that are blowing rapidly downwind.
k. Low clouds that look like cotton balls, but forming above industrial cooling towers or smoke stacks.
l. A curved thin cloud at the top of a mountain.
m. A curved thin cloud at the top of a rapidly rising cumulus congestus.

A9. Discuss the difference between cloud genera, species, varieties, supplementary features, accessory clouds, mother clouds, and meteors.

A10. Given the cloud genera abbreviations below, write out the full name of the genus, and give the WMO and USA weather-map symbols.

a. Cb b. Cc c. Ci d. Cs e. Cu
f. Sc g. Ac h. St i. As j. Ns

A11. For the day and time specified by your instructor, fully classify the clouds that you observe in the sky. Include all of the WMO-allowed classification categories: genera, specie, variety, feature, accessory cloud, and mother cloud. Justify your classification, and include photos or sketches if possible.
[HINT: When you take **cloud photos**, it is often useful to include some ground, buildings, mountains, or trees in the photo, to help others judge the scale of the cloud when they look at your picture. Try to avoid pictures looking straight up. To enhance the cloud image, set the exposure based on the cloud brightness, not on the overall scene brightness. Also, to make the cloud stand out against a blue-sky background, use a polarizing filter and aim the camera at right angles to the sunbeams. Telephoto lenses are extremely helpful to photograph distant clouds, and wide-angle lenses help with widespread nearby clouds. Many cloud photographers use zoom lenses that span a range from wide angle to telephoto. Also, always set the camera to focus on infinity, and never use a flash.]

A12. Draw the cloud coverage symbol for weather maps, and write the METAR abbreviation, for sky cover of the following amount (oktas).
a. 0 b. 1 c. 2 d. 3 e. 4
f. 5 g. 6 h. 7 i. 8 j. obscured

A13.(§). Use a spreadsheet to find and plot the fraction of clouds vs. size. Use ΔX = 100 m, with X in the range 0 to 5000 m.
(i) For fixed S_x = 0.5, plot curves for L_X (m) =
a. 200 b. 300 c. 400 d. 500 e. 700 f. 1000 g. 2000
(ii) For fixed L_X = 1000 m, plot curves for S_X =
h. 0.1 i. 0.2 j. 0.3 k. 0.4 l. 0.5 m. 0.6 n. 0.8

A14. In Fig. 6.12, divide each tile into 4 equal quadrants, and count N vs. M for these new smaller tiles to add data points to the solved-example figure for measuring fractal dimension. Do these finer-resolution tiles converge to a different answer? Do this box-counting for Fig. 6.12 part:
a. (a) b. (b) c. (c) d. (d)

A15. Use the box counting method to determine the fractal dimensions in Fig. 6.11b. Use M =
a. 4 b. 5 c. 6 d. 7 e. 8 f. 9
g. 10 h. 12 i. 14 j. 16 k. 18 l. 20

A16. For air starting at sea level:
(i) How high (km) must it be lifted to form upslope fog?
(ii) How much water ($g_{liq}\ kg_{air}^{-1}$) must be added to cause precipitation fog?
(iii) How much cooling (°C) is necessary to cause radiation fog?
Given the following initial state of the air [T (°C) , RH (%)] =
a. 10, 20 b. 10, 40 c. 10, 60 d. 10, 80 e. 10, 90
f. 20, 20 g. 20, 40 h. 20, 60 i. 20, 80 j. 20, 90
k. 0, 20 l. 0, 40 m. 0, 60 n. 0, 80 o. 0, 90

A17(§). In spring, humid tropical air of initial temperature and dew point as given below flows over colder land of surface temperature 2 °C. At what downwind distance will advection fog form? Also plot the air temperature vs. distance. Assume z_i = 200 m, and C_H = 0.005 .
The initial state of the air is [T (°C), T_d (°C)] =
a. 20, 15 b. 20, 10 c. 20, 5 d. 20, 0 e. 20, –5
f. 20, –10 g. 20, –15 h. 10, 8 i. 10, 5 j. 10, 2
k. 10, 0 l. 10, –2 m. 10, –5 n. 10, –8 o. 10, –10

A18(§). Given F_H = –0.02 K·m·s^{-1} . (i) When will radiation fog form? (ii) Also, plot fog depth vs. time.
Given residual-layer initial conditions of [T (°C) , T_d (°C) , M (m s^{-1})] =
a. 15, 13, 1 b. 15, 13, 2 c. 15, 13, 3 d. 15, 13, 4
e. 15, 10, 1 f. 15, 10, 2 g. 15, 10, 3 h. 15, 10, 4
i. 15, 8, 1 j. 15, 8, 2 k. 15, 8, 3 l. 15, 8, 4
m. 15, 5, 1 n. 15, 5, 2 o. 15, 5, 3 p. 15, 5, 4

A19(§). (i) When will a well-mixed fog dissipate? Assume: albedo is 0.5, fog forms 6 h after sunset, and daylight duration is 12 h. (ii) Also, plot the cumulative heat vs. time.
Given the following values of surface kinematic heat flux: [$F_{H.night}$ (K·m s^{-1}) , $F_{H.max\ day}$ (K·m s^{-1})] =
a. –0.02, 0.15 b. –0.02, 0.13 c. –0.02, 0.11
d. –0.02, 0.09 e. –0.02, 0.07 f. –0.02, 0.17
g. –0.015, 0.15 h. –0.015, 0.13 i. –0.015, 0.11
j. –0.015, 0.09 k. –0.015, 0.07 l. –0.015, 0.17

Evaluate & Analyze

E1. What processes in the atmosphere might simultaneously cool and moisturize unsaturated air, causing it to become saturated?

E2. Cumulus humilis clouds often have flat bases at approximately a common altitude over any location at any one time. Cumulus fractus clouds do not have flat bases, and the cloud-base altitudes vary widely over any location at one time. What causes this difference? Explain.

E3. Can you use darkness of the cloud base to estimate the thickness of a cloud? If so, what are some of the errors that might affect such an approach? If not, why not?

E4. Fig. 6.4 shows that the LCL computed from surface air conditions is a good estimate of cloud-base altitude for cumulus humilis clouds, but not for cloud-base altitude of altocumulus castellanus. Why?

E5. What methods could you use to estimate the altitudes of stratiform clouds by eye? What are the pitfalls in those methods? This is a common problem for weather observers.

E6. Should stratocumulus clouds be categorized as cumuliform or stratiform? Why?

E7. List all the clouds that are associated with turbulence (namely, would cause a bumpy ride if an airplane flew through the cloud).

E8. What do pileus clouds and lenticular clouds have and common, and what are some differences?

E9. Cloud streets, bands of lenticular clouds, and Kelvin-Helmholtz billows all consist of parallel rows of clouds. Describe the differences between these clouds, and methods that you can use to discriminate between them.

E10. The discrete cloud morphology and altitude classes of the official cloud classification are just points along a continuum of cloud shapes and altitudes. If you observe a cloud that looks halfway between two official cloud shapes, or which has an altitude between the typical altitudes for high, middle, or low, clouds, then what other info can you use as a weather observer to help classify the cloud?

E11 (full term project). Build your own cloud chart with your own cloud photos taken with a digital camera. Keep an eye on the sky so you can try to capture as many different cloud types as possible. Use only the best one example of each cloud type corresponding to the clouds in Figs. 6.3 and 6.5. Some clouds might not occur during your school term project, so it is OK to have missing photos for some of the cloud types.

E12. List all of the factors that might make it difficult to see and identify all the clouds that exist above your outdoor viewing location. Is there anything that you can do to improve the success of your cloud identification?

E13(§). a. On a spreadsheet, enter the cloud-size parameters (S_X, L_X, ΔX) from the Sample Application into separate cells. Create a graph of the lognormal cloud distribution, referring to these parameter cells.

b. Next, change the values of each of these parameters to see how the shape of the curve changes. Can you explain why L_X is called the location parameter, and S_X is called the spread parameter? Is this consistent with an analytical interpretation of the factors in eq. (6.6)? Explain.

E14. The box-counting method can also be used for the number of points on a straight line, such as sketched in Fig. 6.11c. In this case, a "box" is really a fixed-length line segment, such as increments on a ruler or meter stick.

a. Using the straight line and dots plotted in Fig. 6.11c, use a centimeter rule to count the number of non-overlapping successive 1 cm segments that contain dots. Repeat for half cm increments. Repeat with ever smaller increments. Then plot the results as in the fractal-dimension Sample Application, and calculate the average fractal dimension. Use the zero-set characteristics to find the fractal dimension of the original wiggly line.

b. In Fig. 6.11c, draw a different, nearly vertical, straight line, mark the dots where this line crosses the underlying wiggly line, and then repeat the dot-counting procedure as described in part (a). Repeat this for a number of different straight lines, and then average the resulting fractal dimensions to get a more statistically-robust estimate.

E15. a. Crumple a sheet of paper into a ball, and carefully slice it in half. This is easier said than done. (A head of cabbage or lettuce could be used instead.) Place ink on the cut end using an ink pad, or by dipping the paper wad or cabbage into a pan with a thin layer of red juice or diluted food coloring in the bottom. Then make a print of the result onto a flat piece of paper, to create a pattern such as shown in Fig. 6.11b. Use the box counting method to find the fractal dimension of the wiggly line that was printed. Using the zero-set characteristic, estimate the fractal dimension of the crumpled paper or vegetable.

b. Repeat the experiment, using crumpled paper wads that are more tightly packed, or more loosely packed. Compare their fractal dimensions.

(P.S. Don't throw the crumpled wads at the instructor. Instructors tend to get annoyed easily.)

E16. Can precipitation fog form when cold raindrops fall through warmer air? Explain.

E17. A fog often forms near large waterfalls. What type of fog is it usually, and how does it form?

E18. In this chapter we listed some locations and seasons where advection fog is likely. Describe 3 other situations (locations and seasons) when advection fog would be likely. (If possible, use locations close to home.)

E19. Derive eq. (6.10), based on the exponential temperature profile for a stable boundary layer (see the Atmospheric Boundary Layer chapter). State and justify all assumptions. [This requires calculus.] Be critical of any simplifications that might not be appropriate for real fogs.

E20. Derive eq. (6.11) from (6.10), and justify all assumptions. Be critical of any simplifications that might not be appropriate for real fogs.

E21. Derive eq. (6.12) using info from the Atmospheric Boundary Layer chapter, and justify all assumptions. Be critical of any simplifications that might not be appropriate for real fogs.

E22. Use the data in the Sample Application for fog dissipation, but find the critical albedo at which fog will just barely dissipate.

E23. During possible frost events, some orchard and vineyard owners try to protect their fruit from freezing by spraying a mist of water droplets or burning smudge pots to make smoke in and above their plants. Why does this method work, and what are its limitations?

Synthesize

S1. What if the saturation curve in Fig. 6.1 was concave down instead of concave up, but that saturation vapor pressure still increased with increasing temperature. Describe how the cooling, moisturizing, and mixing to reach saturation would be different.

S2. Suppose that descending (not ascending) air cools adiabatically. How would cloud shapes be different, if at all? Justify.

S3. Cloud classification is based on morphology; namely, how the cloud looks. A different way to classify clouds is by the processes that make the clouds. Devise a new scheme to classify clouds based on cloud processes; name this scheme after yourself; and make a table showing which traditional cloud names fall into each category of your new scheme, and justify.

S4. Clouds in the atmospheres of other planets in our solar system have various compositions:

- Venus: sulfuric acid
- Mars: water
- Jupiter: ammonia, sulfur, water
- Saturn: ammonia, ammonia hydrosulfide, water
- Uranus & Neptune: methane.

Would the clouds on these other planets have shapes different than the clouds on Earth? If so, then develop a cloud classification for them, and explain why. If not, why not?

S5. Utilize the information in (a) and (b) below to explain why cloud sizes might have a lognormal distribution.

a. The **central-limit theorem** of statistics states that if you repeat an experiment of adding *N* random numbers (using different random numbers each time), then there will be more values of the sum in the middle of the range than at the extremes. That is, there is a greater probability of getting a middle value for the sum than of getting a small or large value. This probability distribution has the shape of a **Gaussian curve** (i.e., a **bell curve** or a **"normal" distribution**; see the Air Pollution chapter for examples).

For anyone who has rolled **dice**, this is well known. Namely, if you roll one die (consider it to be a **random number generator** with a uniform distribution) you will have an equal chance of getting any of the numbers on the die (1, 2, 3, 4, 5, or 6).

However, if you roll two dice and sum the numbers from each die, you have a much greater chance of getting a sum of 7 than of any other sum (which is exactly half way between the smallest possible sum of 2, and the largest possible sum of 12). You have slightly less chance of rolling a sum of 6 or 8. Even less chance of rolling a sum of 5 or 9, etc.

The reason is that 7 has the most ways (6 ways) of being created from the two dice (1+6, 2+5, 3+4, 4+3, 5+2, 6+1). The sums of 6 and 8 have only 5 ways of being generated. Namely, 1+5, 2+4, 3+3, 4+2, and 5+1 all sum to 6, while 2+6, 3+5, 4+4, 5+3, 6+2 all sum to 8. The other sums are even less likely.

b. The logarithm of a product of numbers equals the sum of the logarithms of those numbers.

S6. Build an instrument to measure relative sizes of cumulus clouds as follows. On a small piece of clear plastic, draw a fine-mesh square grid like graph paper. Or take existing fine-mesh graph paper and make a transparency of it using a copy machine. Cut the result to a size to fit on the end of a short tube, such as a toilet-paper tube.

Hold the open end of the tube to your eye, and look through the tube toward cumulus clouds. [CAUTION: Do NOT look toward the sun.] Do this over relatively flat, level ground, perhaps from the roof of a building or from a window just at tree-top level. Pick clouds of medium range, such that the whole cloud is visible through the tube.

For each cloud, record the relative diameter (i.e., the number of grid lines spanned horizontally by the cloud), and the relative height of each cloud base above the horizon (also in terms of number of grid lines). Then, for each cloud, divide the diameter by the cloud-base height to give a normalized diameter. This corrects for perspective, assuming that cumulus cloud bases are all at the same height.

Do this for a relatively large number of clouds that you can see during a relatively short time interval (such as half an hour), and then count the clouds in each bin of normalized cloud diameter.

a. Plot the result, and compare it with the lognormal distribution of Fig. 6.10.

b. Find the L_X and S_X parameters of the lognormal distribution that best fit your data. (Hint: Use trial and error on a spreadsheet that has both your measured size distribution and the theoretical distribution on the same graph. Otherwise, if you know statistics, you can use a method such a Maximum Likelihood to find the best-fit parameters.)

S7. Suppose that you extend Euclidian space to 4 dimensions, to include time as well as the 3 space dimensions. Speculate and describe the physical nature of something that has fractal dimension of 3.4 .

S8. For advection fog, eq. (6.8) is based on a well-mixed fog layer. However, it is more likely that advection fog would initially have a temperature profile similar to that for a stable boundary layer (Atmospheric Boundary Layer chapter). Derive a substitute equation in place of eq. (6.8) that includes this better representation of the temperature profile. Assume for simplicity that the wind speed is constant with height, and state any other assumptions you must make to get an answer. Remember that any reasonable answer is better than no answer, so be creative.

S9. Suppose that fog was transparent to infrared (longwave) radiation. Describe how radiation fog formation, growth, and dissipation would be different (if at all) from real radiation fogs?

S10. Suppose that clouds were transparent to solar radiation, and couldn't shade the ground. Describe and explain the possible differences in cloud morphology, coverage, duration, and their effects on weather and climate, if any.

S11. Find a current weather map (showing only fronts and winds) for your continent. Use your knowledge to circle regions on the map where you expect to find different types of fog and different types of clouds, and label the fog and cloud types.

7 PRECIPITATION PROCESSES

Contents

Hydrometeors are liquid and ice particles that form in the atmosphere. Hydrometer sizes range from small cloud droplets and ice crystals to large hailstones. **Precipitation** occurs when hydrometeors are large and heavy enough to fall to the Earth's surface. **Virga** occurs when hydrometers fall from a cloud, but evaporate before reaching the Earth's surface.

Precipitation particles are much larger than cloud particles, as illustrated in Fig. 7.1. One "typical" raindrop holds as much water as a million "typical" cloud droplets. How do such large precipitation particles form?

The **microphysics** of cloud- and precipitation-particle formation is affected by super-saturation, nucleation, diffusion, and collision.

Supersaturation indicates the amount of excess water vapor available to form rain and snow.

Nucleation is the formation of new liquid or solid hydrometeors as water vapor attaches to tiny dust particles carried in the air. These particles are called cloud condensation nuclei or ice nuclei.

Diffusion is the somewhat random migration of water-vapor molecules through the air toward existing hydrometeors. Conduction of heat away from the droplet is also necessary to compensate the latent heating of condensation or deposition.

Collision between two hydrometeors allows them to combine into larger particles. These processes affect liquid water and ice differently.

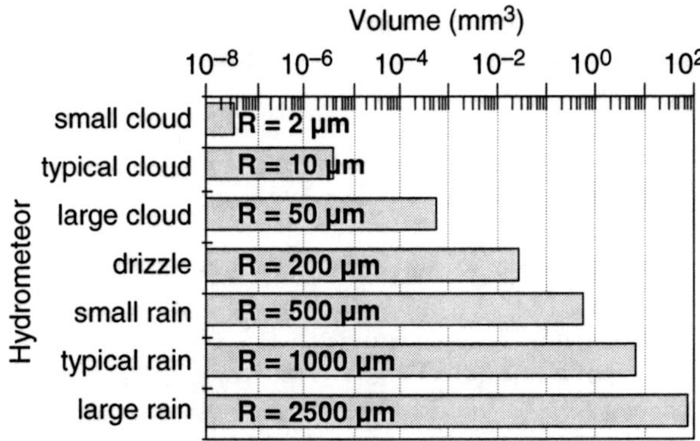

Figure 7.1
Drop volumes and radii, R.

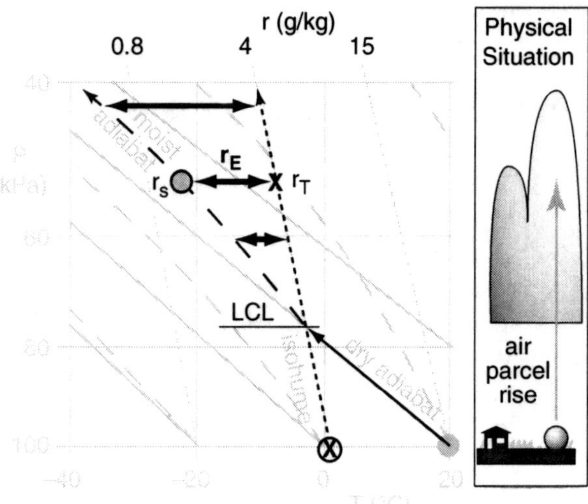

Figure 7.2
Thermo diagram (emagram) showing how excess-water mixing ratio r_E (thick line with arrows at each end) increases with increasing height above cloud base (LCL) for an air parcel rising adiabatically. r_T = total water mixing ratio; r_s = saturation mixing ratio for water vapor. r_E is the amount of water that can form hydrometeors.

Sample Application
An air parcel starts at 100 kPa with temperature 20°C and mixing ratio 4 g kg^{-1}. Use a thermo diagram to find the adiabatic value of excess-water mixing ratio when the parcel reaches a pressure of 60 kPa, and find the max available supersaturation.

Find the Answer
Given: Initially $r = r_T = 4$ g kg^{-1}, $T = 20°C$,
$P = 100$ kPa
Finally: $P = 60$ kPa.
Find: $r_E = ?$ g kg^{-1}, $S_A = ?$ (dimensionless)

Use a full size thermo diagram from the Atmospheric Stability chapter. Starting with the initial conditions, and following the air parcel up dry-adiabatically from the surface, we find the LCL is at $P = 75$ kPa and $T = -3°C$, which marks cloud-base. Above the LCL, follow the moist adiabat, which gives $T \approx -15°C$ at $r_s = 2$ g kg^{-1} at $P = 60$ kPa. Assuming no entrainment and mixing, $r_T = 4$ g kg^{-1}.

Use eq. (7.5): $r_E = (4 - 2)$ g kg^{-1} = <u>**2 g**</u> kg^{-1}.
Then use eq. (7.7): $S_A = r_E / r_s$ = <u>**1**</u> = <u>**100%**</u>.

Check: Units OK. Physics OK. Agrees with Figs. 7.2 & 7.3)
Exposition: A supersaturation of 100% corresponds to a relative humidity of $RH = 200\%$ from eq. (7.2). Thus, there is twice as much water in the air as can be held at equilibrium. This excess water vapor drives condensation. For real clouds, we expect r_E in the range of 0.5 to 1.5 g kg^{-1}, based on the curves in Fig. 7.3.

SUPERSATURATION AND WATER AVAILABILITY

Supersaturation

When there is more water vapor in the air than it can hold at equilibrium, the air is said to be **supersaturated**. Supersaturated air has relative humidity RH greater than 100%. Define this excess relative humidity as a **supersaturation** fraction S:

$$S = RH - 1 \qquad \bullet(7.1)$$

or as a supersaturation percentage $S\%$:

$$S\% = 100\% \cdot S = RH\% - 100\% \qquad \bullet(7.2)$$

Using the definition of relative humidity from the Water Vapor chapter, rewrite supersaturation in terms of total vapor pressure e and saturation vapor pressure e_s:

$$S = \frac{e}{e_s} - 1 \qquad \bullet(7.3)$$

for $e > e_s$. It can also be approximated using mixing ratios:

$$S \approx \frac{r}{r_s} - 1 \qquad \bullet(7.4)$$

for $r > r_s$.

Water Availability

Suppose that an initially-unsaturated air parcel has total water mixing ratio of r_T. If the air parcel rises adiabatically (i.e., no mixing with the environment, and no precipitation in or out), then it cools and its saturation mixing ratio decreases. You can use a thermo diagram (from the Atmospheric Stability chapter) to find these changing mixing ratios, as sketched here in Fig. 7.2.

Above the parcel's lifting condensation level (LCL) the saturation mixing ratio r_s is less than r_T. This supersaturated air has an **excess water mixing ratio** r_E of:

$$r_E = r_T - r_s \qquad \bullet(7.5)$$

The excess water is available to condense onto liquid drops and to deposit onto ice crystals:

$$r_L + r_i = r_E \qquad (7.6)$$

where r_L is liquid-water mixing ratio, and r_i is ice mixing ratio. At temperatures T in the range $-40 < T < 0°C$, supercooled liquid water and ice crystals can co-exist in the air.

Available supersaturation fraction is defined as:

$$S_A = \frac{r_E}{r_s} = \frac{r_T}{r_s} - 1 \qquad (7.7)$$

As hydrometeors grow and remove water vapor, r becomes less than r_T, and S becomes less than S_A. Droplets and ice crystals stop growing by condensation and deposition when they have consumed all the available supersaturation (i.e., when $r \rightarrow r_s$ and $S \rightarrow 0$).

The adiabatic estimate of r_E increases with height above the LCL, as sketched in Figs. 7.2 and 7.3. However, in many real clouds, **diabatic** (i.e., non-adiabatic) processes such as entrainment and mixing of clear drier air into the top and sides of the cloud cause the total-water mixing ratio to not be conserved with height. As a result, typical amounts of excess-water mixing ratio are less than adiabatic (dashed lines in Fig. 7.3).

Beware that the adiabatic r_E curve varies from situation to situation, depending on the thermodynamic initial conditions of the air parcel. So Fig. 7.3 is just one example (based on the Fig. 7.2 scenario).

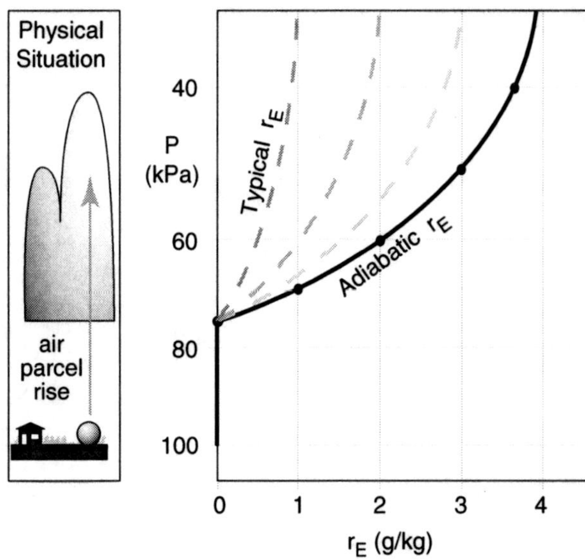

Figure 7.3
Example of the increase of excess water mixing ratio r_E vs. height above cloud base for the scenario in the previous figure. The adiabatic value (solid line) represents the max likely value in non-precipitating clouds, while the dashed curves show ranges of actual max values. P is pressure.

Number and Size of Hydrometeors

Suppose we partition the available excess water equally between all hydrometeors (for example, for all liquid water droplets). In this way, we can estimate the average radius R for each droplet due only to condensation (i.e., before collisions between droplets allow some to merge and grow into larger drops):

$$R = \left[\frac{3}{4\pi} \cdot \frac{\rho_{air}}{\rho_{water}} \cdot \frac{r_E}{n} \right]^{1/3} \qquad \bullet(7.8)$$

where excess-water mixing ratio r_E is in kg_{water} kg_{air}^{-1}, ρ is density, and n is the number density of hydrometeors (the count of hydrometeors per cubic meter of air). Typical values are R = 2 to 50 µm, which is small compared to the 1000 µm separation between droplets, and is too small to be precipitation.

This is an important consideration. Namely, even if we ignore the slowness of the diffusion process (described later), the hydrometeors stop growing by condensation or deposition before they become precipitation. The reason is that there are too many hydrometeors, all competing for water molecules, thus limiting each to grow only a little.

The number density n of hydrometeors is initially controlled by the number density of nuclei upon which they can form, as described next.

Sample Application

Within a cloud, suppose air density is 1 kg m^{-3} and the excess water mixing ratio is 4 g kg^{-1}. Find the final drop radius for hydrometeor counts of (a) 10^8 m^{-3}, and (b) 10^9 m^{-3}.

Find the Answer

Given: $r_E = 0.004\ kg_{water}\ kg_{air}^{-1}$, $\rho_{air} = 1$ kg m^{-3}
n = (a) 10^8 m^{-3}, and (b) 10^9 m^{-3}.
Find: R = ? µm
Assume: ρ_{water} = 1000 kg m^{-3}, as listed in Appendix A.

Use eq. (7.8). Part (a),:

$$R = \left[\frac{3 \cdot (1 kg_{air}/m^3) \cdot (0.004 kg_{water}/kg_{air})}{4\pi \cdot (10^3 kg_{water}/m^3) \cdot (10^8 m^{-3})} \right]^{1/3}$$

$= 2.12\text{x}10^{-5}$ m = **21.2 µm**.

(b) Similarly, R = **9.8 µm**.

Check: Units OK. Physics OK.
Exposition: Both of these numbers are well within the range of "typical" cloud droplets. Thus, the final drop size is NOT large enough to become precipitation.

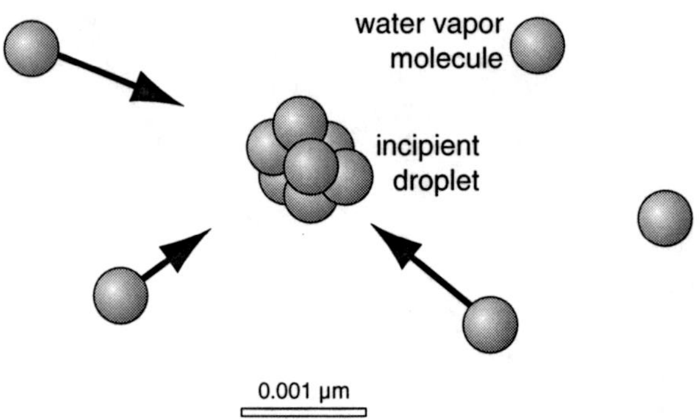

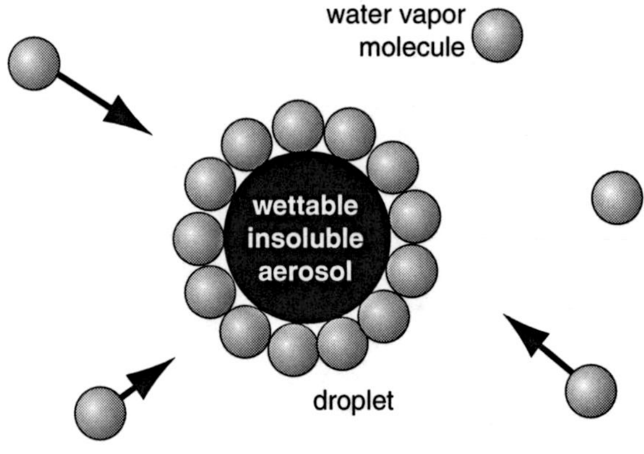

Figure 7.4
Sketch of differences between (a) homogeneous nucleation of cloud droplets and (b) heterogeneous nucleation around a wettable insoluble aerosol particle.

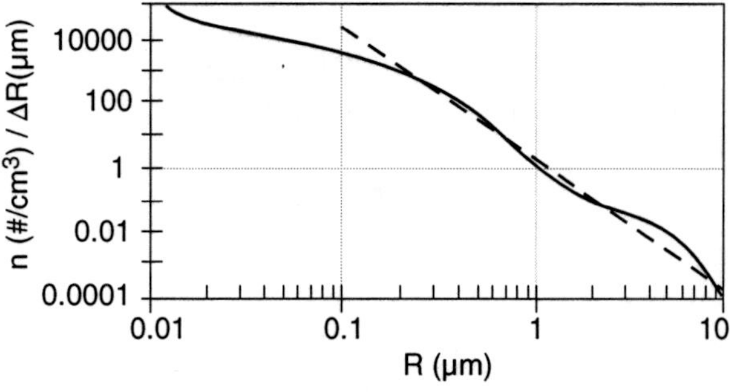

Figure 7.5
CCN particle count n vs. radius R, for c = 2,000,000 $\mu m^3 \cdot m^{-3}$.

NUCLEATION OF LIQUID DROPLETS

Nucleation (the creation of new droplets) in clean air is called **homogeneous nucleation** (Fig. 7.4a). We will show that homogeneous nucleation is virtually impossible in the real atmosphere and can be neglected. Nucleation of cloud droplets by water vapor condensing on tiny dust particles in the air is called **heterogeneous nucleation** (Fig. 7.4b). Even with heterogeneous nucleation, there is a barrier to droplet formation that must first be overcome.

Cloud Condensation Nuclei (CCN)

An **aerosol** is any tiny solid or liquid particle suspended in the air. The subset of aerosol particles that can nucleate cloud droplets are called **cloud condensation nuclei** (**CCN**). To nucleate a droplet, a solid aerosol either must be soluble in water (such as various salt particles), or be sufficiently large in diameter (radius > 0.1 µm) and have a wettable surface (i.e., be **hydrophilic**).

Boundary-layer air over oceans has smaller concentrations of aerosols than continental air. Over oceans, of the 150 to 1000 total aerosol particles cm^{-3} of air, only about 90 to 200 particles cm^{-3} are CCN at normal values of relative humidity ($RH \approx 101\%$) inside clouds. Over continents, of the 2,000 to 70,000 total aerosol particles cm^{-3} of air, only about 200 to 700 particles cm^{-3} are CCN. At higher relative humidities, larger percentages of aerosols act as CCN. Exceptionally clean air over the Arctic can have only 30 CCN particles cm^{-3}, while over industrial cities the CCN count can approach 10^6 particles cm^{-3}.

CCN particles can form when pollutant gases (of molecular size: 10^{-4} to 10^{-3} µm) in the air cluster to form **ultrafine aerosols** (size 10^{-3} to 10^{-2} µm) or are oxidized in the presence of sunlight. Over the oceans, sulfate and sulfuric acid CCN can form this way from gases such as dimethyl sulfide and methane sulfonic acid, produced by **phytoplankton** (microscopic drifting plant life in the ocean).

Further condensation of more pollutant gases and **coagulation** (sticking together) cause the aerosols to quickly grow to 0.01 to 1 µm size, called **fine aerosols**. Beyond this size range they grow more slowly. As a result, aerosols in this size range tend to accumulate — a process called **accumulation mode**. At larger 1 to 10 µm sizes, **coarse mode** CCN can form by other processes, such as strong winds that pick up fine dirt from the ground.

Small nuclei are much more abundant than larger ones (thin wiggly line in Fig. 7.5). Instead of a smooth decrease in number of particles as their size increases, the aerosol curve often has two or three peaks corresponding to the ultrafine, accumulation, and coarse modes.

Over continental regions, the **number density** (n = count of particles per volume of air) of particles with radius between $R - 0.5 \cdot \Delta R$ and $R + 0.5 \cdot \Delta R$ can be approximated by:

$$n(R) = c \cdot R^{-4} \cdot \Delta R \tag{7.9}$$

for particles larger than 0.2 µm, and for small ΔR. Constant c depends on the total concentration of particles. This distribution, called the **Junge distribution**, is the dashed straight line in Fig. 7.5.

Curvature and Solute Effects

Both droplet curvature and chemical composition affect the evaporation rate, which affect the fate of the droplet.

Curvature Effect

The evaporation rate from the curved surface of a droplet is greater than that from a flat water surface, due to surface tension. But droplet growth requires condensation to exceed evaporation. Thus, to be able to grow, smaller droplets need greater RH in the air than larger drops. The resulting equilibrium RH in the air as a function of droplet radius R is described by **Kelvin's equation**

$$RH\% = 100\% \cdot \exp\left[\frac{2 \cdot \sigma}{\rho_m \cdot k_B \cdot T \cdot R}\right] \tag{7.10a}$$

where σ = 0.076 N m^{-1} is **surface tension** of pure water at 0°C, ρ_m = 3.3x10^{28} molecules m^{-3} is the number density of water molecules in liquid at 0°C, k_B = 1.38x10^{-23} J·K^{-1}·molecule^{-1} is Boltzmann's constant, and T is absolute temperature (in Kelvin) of the droplet. Kelvin's equation can be abbreviated as

$$RH\% = 100\% \cdot \exp\left[\frac{c_1}{T \cdot R}\right] \tag{7.10b}$$

where c_1 = 0.3338 µm·K = $2\sigma/(\rho_m \cdot k_B)$.

Fig. 7.6 illustrates the curvature effect described by Kelvin's equation, where conditions above the line allow droplets to grow. For example, droplets of radius smaller than 0.005 µm need an environment having RH > 128% to grow. But larger droplets of radius 0.1 µm need only RH > 101% to grow.

For homogeneous nucleation in clean air (no aerosols), incipient droplets form when several water-vapor molecules merge (Fig. 7.4a). The resulting droplet has extremely small radius (≈0.001 µm), causing it to evaporate quickly, given the typical humidities in clouds of ≈101% (grey shaded band in Fig. 7.6). How-

Sample Application

If c = 5x10^7 µm^3·m^{-3} for the Junge distribution, then how many CCN are expected within radii ranges of (a) 0.45 - 0.55 µm and (b) 0.95 - 1.05 µm?

Find the Answer

Given: R = 0.5 µm & 1.0 µm, ΔR= 0.1 µm for both ranges, and c = 5x10^7 µm^3·m^{-3}

Find: n = ? # m^{-3}

Use eq. (7.9):

(a) n = (5x10^{7}µm^3·m^{-3})·(0.5 µm)$^{-4}$ ·(0.1µm) = **8 x 10^7** m^{-3}.

(b) n = (5x10^7 µm^3·m^{-3})·(1 µm)$^{-4}$ ·(0.1 µm) = **5 x 10^6** m^{-3}.

Check: Units OK. Physics OK.

Exposition: Doubling the particle radius reduces the number density by more than tenfold. If each CCN nucleates a cloud droplet, then each m^3 of cloudy air contain tens of millions of cloud droplets. Concentrations in real clouds can be 10s to 1000s of times greater.

Sample Application

What humidity in a cloud at 0°C is needed to allow a droplet of radius 0.03 µm to grow?

Find the Answer:

Given: R = 0.03 µm. T = 273 K

Find: RH = ? %

Use eq. (7.10b)

$RH\%$ = 100% · exp[(0.334 µm·K) / (273K · 0.03 µm)] = **104.16%**

Check: Units OK. Magnitude OK.

Exposition: A stable droplet this small is unlikely in a real cloud, where typically RH ≈ 101% or less.

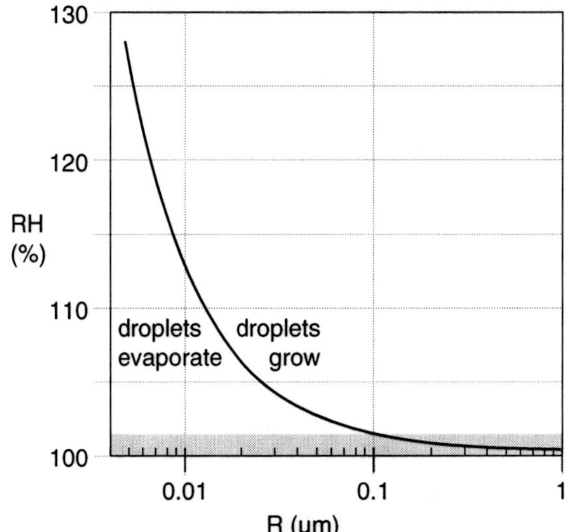

Figure 7.6

Curvature effect, showing the relative humidity (RH) needed in air for a droplet of radius R to grow. The curve is given by Kelvin's equation, for pure water at 0°C. Shaded region highlights the range of RH typically found in clouds.

Table 7-1. Properties of some solutes. M_s is molecular weight, *i* is approximate ion count.

Solute	Chemistry	M_s	i
salt	NaCl	58.44	2
ammonium sulfate	$(NH_4)_2SO_4$	132.13	3
hydrogen peroxide	H_2O_2	34.01	2
sulfuric acid	H_2SO_4	98.07	3
nitric acid	HNO_3	63.01	2

Sample Application

Find the equilibrium relative humidity over a droplet of radius 0.2 µm, temperature 20°C, containing 10^{-16} g of ammonium sulfate.

Find the Answer

Given: $R = 0.2$ µm, $T = 293$ K, $m_s = 10^{-16}$ g
Find: $RH\% = 100\% \cdot (e_s^*/e_s) = ?\ \%$

From Table 7-1 for ammonium sulfate:
$M_s = 132.13$, and $i = 3$.
Use eq. (7.11):

$$\frac{e_s^*}{e_s} \approx \frac{\exp\left(\dfrac{0.3338\text{K}\cdot\mu\text{m}}{(293\text{K})\cdot(0.2\mu\text{m})}\right)}{1+\dfrac{(4.3\times10^{12}\mu\text{m}^3\cdot\text{g}^{-1})\cdot 3\cdot(10^{-16}\text{g})}{(132.13)\cdot(0.2\mu\text{m})^3}}$$

$= 1.00571 / (1+0.00122) = 1.00448$
$RH\% = 100\% \cdot (e_s^*/e_s) = $ **<u>100.448%</u>**

Check: Units OK. Physics OK.
Exposition: Fig. 7.7b gives a value of about 100.49% for a temperature of 0°C. Thus, warmer temperatures require less supersaturation of water vapor in the air to prevent the droplet from vaporizing.

Sample Application

In fog at 10°C, the vapor pressure is 1.4 kPa. Find the supersaturation fraction and percentage.

Find the Answer

Given: $T = 10$°C, $e_s^* = 1.4$ kPa
Find: $S = ?$, and $S\% = ?$

First, use Table 4-1 in the Water Vapor chapter to find the saturation vapor pressure at 10°C: $e_s = 1.233$ kPa.
Next, use eq. (7.12):
$S = (1.4 \text{ kPa} / 1.233 \text{ kPa}) - 1 = 1.135 - 1 = $ **<u>0.135</u>**
$S\% = 100\% \cdot S = $ **<u>13.5%</u>**

Check: e_s^* is indeed $> e_s$, thus supersaturated.
Exposition: Unrealistically large, given typical supersaturations in clouds and fog of order S%=1%.

ever, for heterogeneous nucleation the small number of water molecules can coat the outside of the aerosol particle (Fig. 7.4b), with a resulting radius that is relatively large. Droplets formed by heterogeneous nucleation grow and remove water-vapor molecules from the air, thereby lowering the *RH* and eliminating the chance for homogeneous nucleation.

Solute Effect

Solutions (i.e., water containing dissolved chemicals) evaporate water molecules at a slower rate than does pure water. Solutions occur when condensation occurs on impurities such as certain cloud condensation nuclei (CCN) that dissolve in the nascent water droplet. This can partially compensate the curvature effect.

Recall that the saturation vapor pressure e_s over a flat surface of pure water was given in the Water Vapor chapter by the Clausius-Clapeyron equation. The two opposing effects of curvature and solute can be combined into one equation (**Köhler equation**) for the ratio of actual saturation vapor pressure e_s^* in equilibrium over a solution with a curved surface, to vapor pressure over a flat surface of pure water e_s:

$$RH = \frac{e_s^*}{e_s} \approx \frac{\exp\left(\dfrac{c_1}{T\cdot R}\right)}{1+\dfrac{c_2\cdot i\cdot m_s}{M_s\cdot R^3}} \qquad \bullet(7.11)$$

where *RH* is the relative humidity fraction, *T* is absolute temperature, *R* is drop radius, *i* is number of ions per molecule in solution (called the **van't Hoff factor**), m_s is mass of solute in the droplet, and M_s is molecular weight of solute. The two parameters are: $c_1 = 0.3338$ K·µm, and $c_2 = 4.3\text{x}10^{12}$ µm^3·g^{-1}. Table 7-1 gives properties for common atmospheric solutes.

In eq. (7.11) the relative humidity *RH* can be greater than 1, corresponding to a relative-humidity percentage ($RH\% = 100\% \cdot e_s^*/e_s$) that is greater than 100%. Similar to eq. (7.3), **supersaturation** relative to the hydrometeor can be defined as a fraction:

$$S = (e_s^*/e_s) - 1 \qquad \bullet(7.12)$$

or as a percentage, $S\% = 100\% \cdot S$:

$$S\% = 100\% \cdot [(e_s^*/e_s) - 1] \qquad \bullet(7.13)$$

Thus, the left hand side of eq. (7.11) can be easily rewritten as supersaturation.

The numerator of eq. (7.11) describes the **curvature effect**, and together with the left hand side is the **Kelvin equation**. The denominator describes the **solute effect** of impurities in the water. Eq. (7.11) was solved in a spreadsheet to produce the **Köhler**

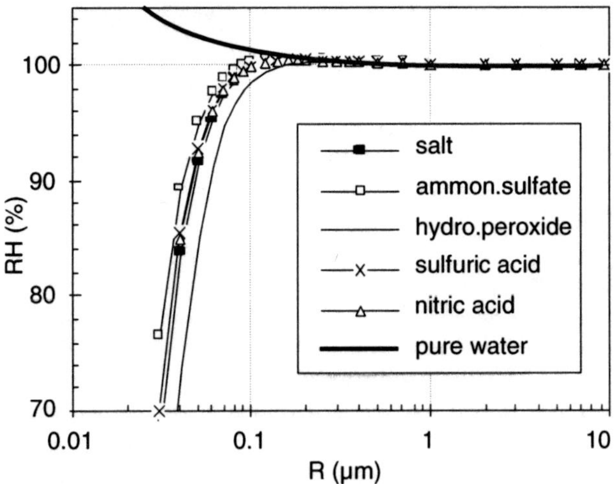

Figure 7.7a
Köhler curves, showing equilibrium relative humidities over droplets of different radius with various solutes. T = 0°C. Solute mass = 10^{-16} g.

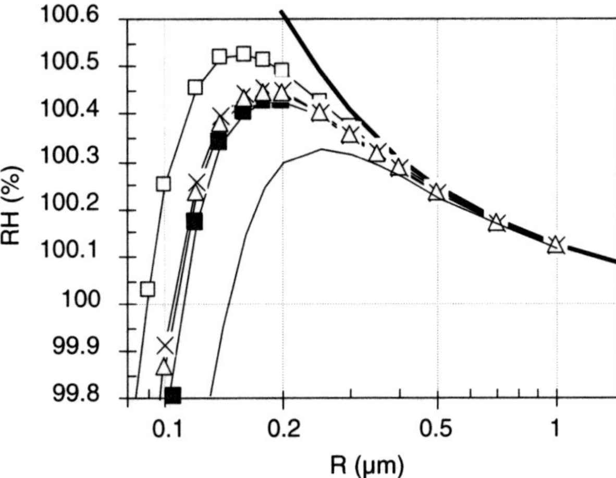

Figure 7.7b
Blow-up of Fig. 7.7a.

curves in Fig. 7.7. Using these Köhler curves, we can study nucleation.

First, the curve in Fig. 7.7 for pure water increases exponentially as drop radius becomes smaller. This curve was already shown in Fig. 7.6, computed from Kelvin's equation. It was used to explain why homogeneous nucleation is unlikely, because when the first several water-vapor molecules come together to form a condensate, their droplet radius is so small that the droplet instantly explodes.

Second, solutions of some chemicals can form small droplets even at humidities of less than 100%. Such **hygroscopic** (water attracting) pollutants in the air will grow into droplets by taking water vapor out of the air.

Third, if humidities become even slightly greater than the peaks of the Köhler curves, or if droplet radius becomes large enough (see INFO box), then

INFO • Droplet Growth

To help interpret Fig. 7.7b, consider a droplet containing just one chemical such as salt, in a cloud having a known humidity, such as RH = 100.3%. This is redrawn below.

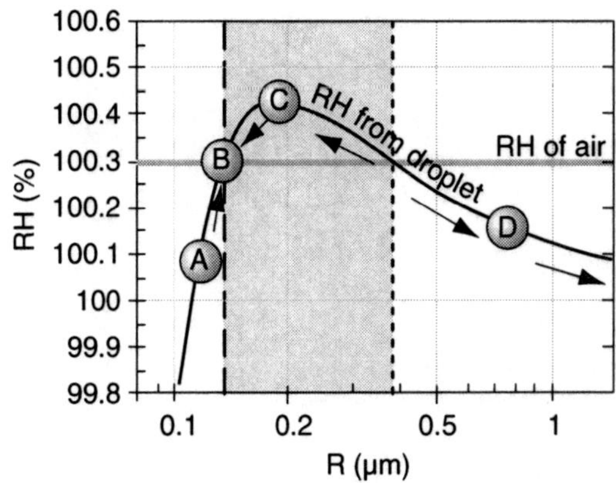

Figure 7.a. *Interpretation of a Köhler curve.*

Think of the Köhler curve as the RH associated with the droplet, which is trying to drive <u>evaporation from</u> the droplet. The RH in air (horizontal grey line) is trying to drive <u>condensation to</u> the droplet.

If $RH_{droplet} > RH_{air}$, then evaporation > condensation, and the droplet radius becomes smaller as water molecules leave the droplet. This is the situation for any droplet in the light grey band; namely, droplet C suffers net evaporation, causing its radius to decrease toward B.

If $RH_{droplet} < RH_{air}$, then evaporation < condensation, and droplet radius increases due to net condensation. Droplet D has this state, and as its radius increases it moves to the right in the graph, causing $RH_{droplet}$ to decrease further and driving even faster droplet growth. Activated droplets such as these continue growing (shown by right-pointing arrows) until they consume the excess humidity (driving the RH_{air} down toward 100%). Such droplet growth is a first stage in formation of precipitation in warm clouds.

Droplet A also has $RH_{droplet} < RH_{air}$, and would tend to increase in radius due to net condensation. But in this part of the Köhler curve, increasing radius causes increasing $RH_{droplet}$, and causes the net condensation to diminish until the droplet radius reaches that at B.

Thus, the vertical dotted line in Fig. 7.a is an unstable equilibrium. Namely, a droplet on the Köhler curve at the dotted line would either grow or shrink if perturbed slightly from its equilibrium point.

However, the vertical dashed line is a stable equilibrium point. Droplets approach this radius and then stop growing. Namely, they stay as small haze particles, and do not grow into larger cloud or precipitation particles.

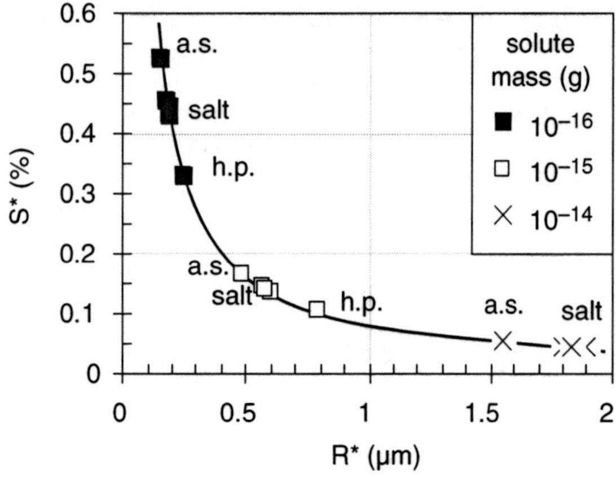

Figure 7.8
Critical values of supersaturation S vs. critical drop radius R*, for various solute masses and chemicals. a.s. = ammonium sulfide, salt = sodium chloride, h.p. = hydrogen peroxide. Nitric acid and sulfuric acid are near the salt data points.*

Sample Application

Find the critical radius and supersaturation value for 10^{-15} g of ammonium sulfate at 0°C.

Find the Answer

Given: $m_s = 10^{-15}$ g, $T = 273$ K
Find: $R^* = ?$ µm, $S\%^* = ?$ %.

Use eq. (7.14) & Table 7-1. $R^* =$

$$\sqrt{\frac{(3.8681\times10^{13} K^{-1}\cdot g^{-1}\mu m^2)\cdot 3\cdot(10^{-15} g)\cdot(273K)}{132.13}}$$

R^* = **0.49 µm**

Use eq. (7.15) & Table 7-1. $S^* =$

$$S^* = \sqrt{\frac{(1.278\times10^{-15}\text{K}^3\cdot\text{g})\cdot 132.13}{3\cdot(10^{-15}\text{g})\cdot(273\text{K})^3}} = 0.00166$$

$S\%^*$ = **0.166%**

Check: Units OK. Physics OK.
Exposition: Agrees with the point (white square) plotted in Fig. 7.8 for "a.s.". Thus, the larger mass nucleus needs less supersaturation to become activated.

For a parcel of rising air with increasing supersaturation, the larger nuclei will become activated first, followed by the smaller nuclei if the parcel keeps cooling and if the excess vapor is not removed by the larger nuclei first.

droplets can grow unimpeded. CCN reaching this state are said to be **activated**. Growth of droplets from activated nuclei continues until enough vapor (i.e., r_E) is consumed to reduce the supersaturation back toward 100%. Pure droplets cannot form or co-exist in an environment with neighboring solution droplets, because of the low supersaturation remaining in the air after much of the vapor has condensed out onto the solution droplets. That is another reason why homogeneous nucleation can be neglected for practical purposes.

Fourth, although not shown in these curves, the equations allow droplets to form at lesser supersaturations if the mass of dissolved solute is greater. Hence, larger CCN can grow into droplets earlier and can grow faster than smaller CCN.

Critical Radius

The location of the peak of the Köhler curves marks the barrier between the larger, activated droplets that can continue to grow, from the smaller **haze** droplets that reach an equilibrium at small size. The drop radius R^* at this peak is called the **critical radius**, and the corresponding **critical supersaturation** fraction is $S^* = e_s^*/e_s - 1$. They are given by:

$$R^* = \sqrt{\frac{c_3 \cdot i \cdot m_s \cdot T}{M_s}} \qquad \bullet(7.14)$$

and

$$S^* = \sqrt{\frac{c_4 \cdot M_s}{i \cdot m_s \cdot T^3}} \qquad \bullet(7.15)$$

where $c_3 = 3.8681\text{x}10^{13}$ µm^2·K^{-1}·g^{-1}, and $c_4 = 1.278\text{x}10^{-15}$ K^3·g. For critical supersaturation as a percentage, use $S\%^* = 100\%\cdot S^*$.

Critical conditions are plotted in Fig. 7.8 for various masses of different chemicals. Obviously S^* is inversely related to R^*, so as solute mass increases, smaller supersaturations are necessary to reach the critical point, and at that point the droplets will be larger. Also, notice that different chemicals will grow to different sizes, which is one factor causing a range of drop sizes in the cloud.

Haze

For conditions left of the peak on any Köhler curve (i.e., $R < R^*$), CCN rapidly grow into small droplets that stop growing at an equilibrium size determined by the humidity, temperature, and solute (see previous INFO box). These small droplets are called **haze** droplets. Thus, tiny droplets can exist even at relative humidities below 100%.

Sample Application

For 10^{-16} g of ammonium sulfate at 0°C, how does haze droplet radius change as RH increases from 70 to 80%? Also, how many molecules are in each aerosol?

Find the Answer

Assume: Same conditions as in Fig. 7.7a.
Given: RH = 70%, 80%.
Find: R = ? µm, n = ? molecules

Solve eq. (7.11), or use Fig. 7.7a. I will use the Fig.
$R \approx$ **0.027 µm** at 70%; $R \approx$ **0.032 µm** at 80%.

The number of molecules is $n = \rho_m \cdot [(4/3) \cdot \pi \cdot R^3]$
n = **2.72x10^6** molecules; n = **4.53x10^6** molecules

Check: Units OK. Physics OK.
Exposition: Haze particles indeed become larger as relative humidity increases, thereby reducing visibility. Scattering of light by this size of particles is called **Mie scattering** (see the Atmos. Optics chapter).

Haze droplets are aerosols. When a tiny, dry CCN grows to its equilibrium size by the condensation of water molecules, this process is called **aerosol swelling**. Aerosol swelling is responsible for reducing **visibility** in polluted air as humidities increase above about 75%.

Even haze particles contain many water molecules. Liquid water contains about $\rho_m = 3.3\text{x}10^{28}$ molecules m^{-3}. Thus, the smallest haze particles of radius 0.02 µm contain roughly $n = \rho_m \cdot [(4/3) \cdot \pi \cdot R^3]$ = 1.1 million molecules.

The word **smog** is a contraction of "smoke" and "fog", which is a reasonable lay description of haze. Many urban smogs are a stew of ingredients including ozone, volatile hydrocarbons such as evaporated gasoline, and various oxides of nitrogen. These react in the atmosphere, particularly in the presence of sunlight, to create sulfates, nitrates, and hydrogen peroxide CCNs. Aerosol swelling and reduced visibilities are quite likely in such urban smogs, particularly when the air is humid.

Activated Nuclei

For conditions to the right of the peak on any Köhler curve (i.e., $R > R^*$), CCN are activated and can continue growing. There is no equilibrium that would stop their growth, assuming sufficient water vapor is present. These droplets can become larger than haze droplets, and are called **cloud droplets**.

Because atmospheric particles consist of a variety of chemicals with a range of masses, we anticipate from the Köhler curves that different CCN will become activated at different amounts of

HIGHER MATH • Critical Radius

Derivation of the critical radius, eq. (7.14).

The critical radius is at e_s^*/e_s = maximum. But at the maximum, the slope is zero: $d(e_s^*/e_s)/dR = 0$. By finding this derivative of eq. (7.11) with respect to R, and setting it to zero, we can solve for R at the maximum. This is R^* by definition.

The right side of eq. (7.11) is of the form a/b. A rule of calculus is:

$$d(a/b)/dR = [b \cdot (da/dR) - a \cdot (db/dR)] \,/\, b^2.$$

Also, a is of the form $a = \exp(f)$, for which another rule is:

$$da/dR = a \cdot (df/dR).$$

Combining these 2 rules and setting the whole thing to zero gives:

$$0 = \frac{a}{b}\left[\frac{df}{dR} - \frac{1}{b}\frac{db}{dR}\right]$$

But (a/b) is just the original right side of eq. (7.11), which we know is close to 1.0 at the max, not close to 0. Thus, the eq. above equals 0 only if:

$$\frac{df}{dR} = \frac{1}{b}\frac{db}{dR}$$

Plugging in for f and b and differentiating yields:

$$\frac{-c_1}{T \cdot R^2} = \frac{1}{\left[1 + \dfrac{c_2 \cdot i \cdot m_s}{M_s \cdot R^3}\right]} \frac{(-3) \cdot c_2 \cdot i \cdot m_s}{M_s \cdot R^4}$$

Multiply both sides by $(-R^4 \cdot T/c_1)$:

$$R^2 = (3 \cdot T / c_1) \cdot \frac{\left[\dfrac{c_2 \cdot i \cdot m_s}{M_s}\right]}{\left[1 + \dfrac{c_2 \cdot i \cdot m_s}{M_s \cdot R^3}\right]}$$

Multiply the numerator and denominator of the right side by $M_s \,/\, (c_2 \cdot i \cdot m_s)$, which gives:

$$R^2 = (3 \cdot T / c_1) \cdot \frac{1}{\left[\dfrac{M_s}{c_2 \cdot i \cdot m_s} + \dfrac{1}{R^3}\right]}$$

By plugging in typical values, we can show that the $1/R^3$ term is small enough to be negligible compared to the other term in square brackets. This leaves:

$$R^2 \cong \left(\frac{3 \cdot c_2}{c_1}\right) \cdot \left(\frac{T \cdot i \cdot m_s}{M_s}\right)$$

Define $c_3 = 3 \cdot c_2 / c_1$, set $R = R^*$, and take the square root of both sides to get the final answer:

$$\boxed{R^* \cong \sqrt{\frac{c_3 \cdot i \cdot m_s \cdot T}{M_s}}} \qquad (7.14)$$

Sample Application

How many nuclei would be activated in continental air of supersaturation percentage 0.5%? Also, how much air surrounds each droplet, and what is the distance between drops?

Find the Answer

Given: $S = 0.005$
Find: n_{CCN} = ? particles m^{-3}, Vol = ? mm^3, x = ? mm.

(a) Use eq. (7.16):
$n_{CCN} = (6\text{x}10^8 \text{ m}^{-3})\cdot(0.5)^{0.5}$ = **$\underline{4.24\text{x}10^8}$** m^{-3}
(b) Also:
$Vol = 1/n_{CCN}$ = **$\underline{2.36 \text{ mm}^3 \text{ droplet}^{-1}}$**
(c) Using eq. (7.17): $x = Vol^{1/3}$ = **$\underline{1.33 \text{ mm}}$**.

Check: Units OK. Physics OK.
Exposition: If all of these nuclei become cloud droplets, then there are over 40 million droplets within each cubic meter of cloud. But there is a relatively large distance between each drop.

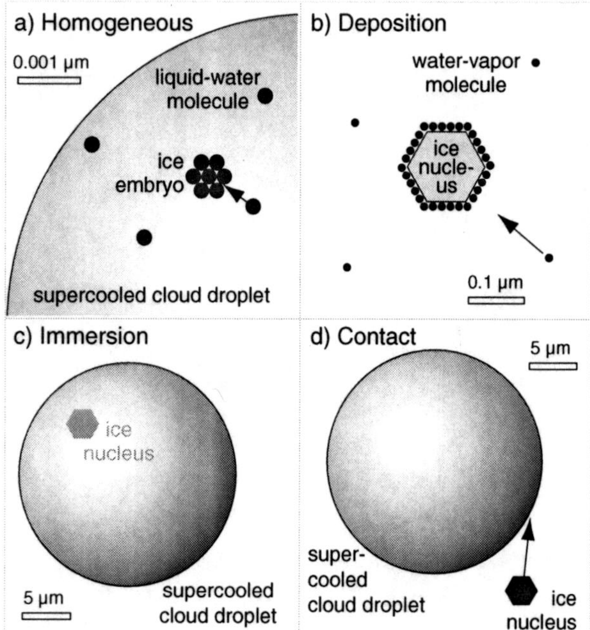

Figure 7.9
Sketch of (a) homogeneous, and (b-d) heterogeneous ice nucleation processes.

supersaturation. The number density n_{CCN} (# of CCN per m^3) activated as a function of supersaturation <u>fraction</u> S is roughly:

$$n_{CCN} = c \cdot (100 \cdot S)^k \tag{7.16}$$

but varies widely. In maritime air, $c \approx 1\text{x}10^8$ m^{-3} and $k \approx 0.7$. In continental air $c \approx 6\text{x}10^8$ m^{-3} and $k \approx 0.5$. The number of activated nuclei is in the range of 10^8 to 10^9 m^{-3}, which is usually just a small fraction of the total number of particles in the air.

The distance x between cloud droplets is on the order of 1 mm, and is given by

$$x = n_{CCN}^{-1/3} \tag{7.17}$$

NUCLEATION OF ICE CRYSTALS

Ice cannot survive at temperatures above 0°C at normal atmospheric pressures. Below that temperature, ice crystals can exist in equilibrium with air that is supersaturated with respect to ice. The saturation curve for ice was plotted in Fig. 4.2, and is close to, but slightly below, the curve for supercooled liquid water.

There is a thermodynamic barrier to ice formation, analogous to the barrier for droplet growth. This barrier can be overcome with either very cold temperatures (colder than –40°C), high supersaturation in the air, or by the presence of ice nuclei.

Processes

Homogeneous freezing nucleation is the name for the spontaneous freezing that occurs within supercooled liquid-water droplets as temperature decreases to near –40°C. No impurities are needed for this. Instead, ice embryos form by chance when clusters of water molecules happen to come together with the correct orientations (Fig. 7.9a). At –40°C, the cluster needs only about 250 molecules. At slightly warmer temperatures the critical embryo size is much larger, and thus less likely to occur. Because larger supercooled droplets contain greater numbers of molecules, they are more likely to form an ice embryo and freeze.

Heterogeneous freezing nucleation can occur a variety of ways (Fig. 7.9b-d), but all require an impurity (Table 7-2), which is generically called an **ice nucleus**. These processes are described next.

Deposition nucleation (Fig. 7.9b) occurs when water vapor deposits directly on a **deposition nucleus**. While the solute effect was important for liquid droplet nucleation, it does not apply to ice nucleation because salts are excluded from the ice-crystal lattice as water freezes. Thus, the size and crystal structure of the ice nucleus is more important. Deposition is unlikely on particles of size 0.1 µm or less. Colder temperatures and greater supersaturation increases deposition nucleation.

Immersion freezing (Fig. 7.9c) occurs for those liquid drops already containing an undissolved ice nucleus called a **freezing nucleus**. Each nucleus has a critical temperature below 0°C at which it triggers freezing. Thus, as external processes cause such a contaminated drop to cool, it can eventually reach the critical temperature and freeze.

Larger drops contain more freezing nuclei, and have a greater chance of containing a nucleus that triggers freezing at warmer temperatures (although still below 0°C). To freeze half of the drops of radius R, the temperature must drop to T, given statistically by

$$T \approx T_1 + T_2 \cdot \ln(R/R_o) \tag{7.18}$$

where R_o = 5 µm, T_1 = 235 K, and T_2 = 3 K.

Condensation freezing is a cross between deposition nucleation and immersion freezing. In this scenario, which occurs below 0°C, nuclei are more attractive as condensation nuclei than as deposition nuclei. Thus, supercooled liquid water starts to condense around the nucleus. However, this liquid water immediately freezes due to the immersion-nucleation properties of the nucleus.

Contact freezing (Fig. 7.9d) occurs when an uncontaminated supercooled liquid drop happens to hit an external **contact nucleus**. If the droplet is cooler than the critical temperature of the ice nucleus, then it will freeze almost instantly. This also happens when (supercooled) **freezing rain** hits and instantly freezes on trees and power lines. Ice crystals in the air are also good contact nuclei for supercooled water.

Ice Nuclei

Only substances with similar molecular structure as ice can serve as ice nuclei. Such substances are said to be **epitaxial** with ice.

Natural ice nuclei include fine particles of clay such as kaolinite stirred up from the soil by the wind. Certain bacteria and amino acids such as l-leucine and l-tryptophan from plants also can nucleate ice. Combustion products from forest fires contain many ice nuclei. Also, ice crystals from one cloud can fall or blow into a different cloud, triggering continued ice formation.

Other substances have been manufactured specifically to seed clouds, to intentionally change the amount or type (rain vs. hail) of precipitation. Silver

Table 7-2. Listed are the warmest ice nucleation threshold temperatures (°C) for substances that act as ice nuclei. Processes are: 1 contact freezing, 2 condensation freezing, 3 deposition, and 4 immersion freezing. (x = not a process for this substance.)

Substance	Process			
	1	2	3	4
silver iodide	–3	–4	–8	–13
cupric sulfide	–6	x	–13	–16
lead iodide	–6	–7	–15	x
cadmium iodide	–12	x	–21	x
metaldehyde	–3	–2	–10	x
1,5-dihydroxynaphlene	–6	–6	–12	x
phloroglucinol	x	–5	–9	x
kaolinite	–5	–10	–19	–32

Sample Application

What cloud temperature is needed to immersion-freeze half of the droplets of 100 µm radius?

Find the Answer

Given: R = 100 µm. Find: T = ? K

Use eq. (7.18)

T = 235K + (3K)·ln(100µm/5µm) = 244K = <u>–29°C</u>

Check: Units OK. Physics OK.

Exposition: Smaller droplets can remain unfrozen at much colder temperatures than larger drops, which are thus available to participate in the WBF precipitation growth process described later.

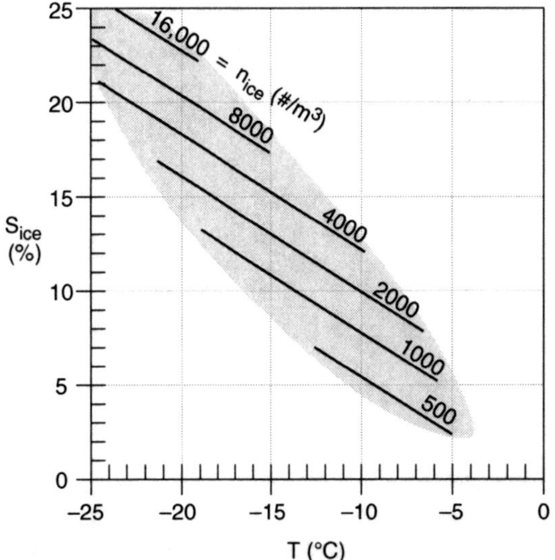

Figure 7.10
Number density n_{ice} of active ice nuclei per m³ of air as a function of ice supersaturation S and temperature T. Shaded region encloses typically observed combinations of S and T.

Sample Application

Given a supersaturation gradient of 1% per 2 µm near a droplet in a cloud at 4 km altitude where saturated mixing ratio is 1.5 g kg^{-1} and the diffusivity is 2x10^{-5} m^2·s^{-1}. Find (a) the mixing-ratio gradient, (b) the kinematic moisture flux, and (c) the dynamic moisture flux.

Find the Answer:

Given: $\Delta S\%/\Delta x$ = 1% / 2 µm, r_s = 0.0015 kg$_{water}$/kg$_{air}$
D = 2x10^{-5} m^2·s^{-1}.
Find: (a) $\Delta r/\Delta x$ = ? (kg$_{water}$/kg$_{air}$)/m , (b) F = ? (kg$_{water}$/kg$_{air}$)·(m s^{-1}) , (c) $\mathbb{F}$ = ? kg$_{water}$·m^{-2}·s^{-1}

First, convert units in the saturation gradient:
$\Delta S/\Delta x$ = [1%/2 µm]·(10^{6}µm/1m)/100% = 5000 m^{-1} .
Next, use eq. (7.21):
$\Delta r/\Delta x$ = (0.0015kg/kg)·(5000m^{-1})
= **7.5** (kg$_{water}$/kg$_{air}$)·m^{-1}.

Then use eq. (7.20):
F = –(2x10^{-5} m^2·s^{-1})·(7.5 (kg/kg)·m^{-1})
= **–1.5x10^{-4}** (kg$_{water}$/kg$_{air}$)·(m s^{-1}).

Finally, use the definition of kinematic flux from the Solar & IR Radiation chapter: $F = \mathbb{F} / \rho_{air}$. As a first guess for density, assume a standard atmosphere at z = 2 km, and use Table 1-5 in Chapter 1 to estimate ρ_{air} ≈ 0.82 kg$_{air}$ m^{-3} . Thus: $\mathbb{F}$ = (ρ_{air})·(F)
= (0.82kg$_{air}$ m^{-3})·[–1.5x10^{-4}(kg$_{water}$/kg$_{air}$)·(m s^{-1})]
= **–1.23x10^{-4}** kg$_{water}$·m^{-2}·s^{-1} .

Check: Units OK. Physics OK.
Exposition: Diffusive fluxes are very small, which is why it can take a couple hours for droplets to grow to their maximum drop radius. The negative sign means water vapor flows from high to low humidity.

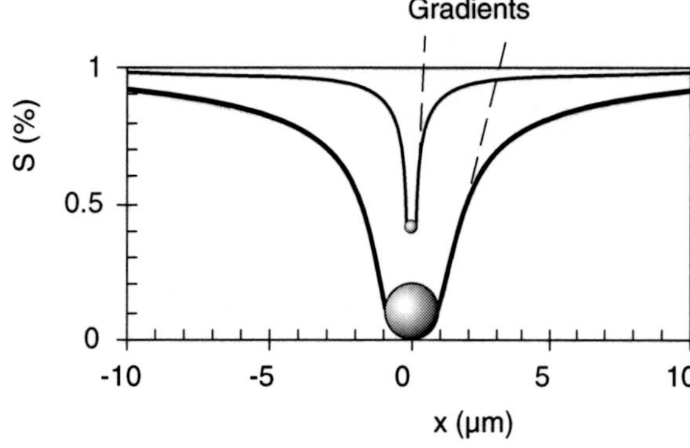

Figure 7.11
Humidity gradients in supersaturated environment near growing droplets. S% is supersaturation percentage, x is distance. The nucleus for both droplets is salt of mass 10^{-16} g. Background supersaturation is 1%, the small droplet has radius 0.2 µm, and the large drop 1.0 µm. Equilibrium supersaturation adjacent to the droplets is taken from Fig. 7.7b. Nearest neighbor droplets are roughly 1000 µm distant.

iodide has been a popular chemical for cold-cloud seeding. Other cold-cloud-seeding chemicals are lead iodide, cupric sulfide, phloroglucinol, and metaldehyde. Table 7-2 lists some ice nuclei and their critical temperatures. Because contact nucleation occurs at the warmest temperatures for most substances, it is the most likely process causing ice nucleation.

Air usually contains a mixture of ice nuclei having a variety of ice nucleation processes that become active at different temperatures and supersaturations. The number density n_{ice} of active ice nuclei per cubic meter of air is shown in Fig. 7.10.

LIQUID DROPLET GROWTH BY DIFFUSION

In a supersaturated environment, condensation onto a growing droplet removes water vapor from the adjacent air (Fig. 7.11). This lowers the humidity near the droplet, creating a humidity gradient down which water vapor can diffuse.

Diffusion is the process where individual water-vapor molecules meander through air via Brownian motion (i.e., random walk). The net direction of diffusion is always down the humidity gradient toward drier air.

The diffusive moisture flux $\mathbb{F}$ in kg$_{water}$·m^{-2}·s^{-1} is

$$\mathbb{F} = -D \cdot \frac{\Delta \rho_v}{\Delta x} \qquad \bullet(7.19)$$

where x is distance, D is diffusivity, and ρ_v is absolute humidity (water-vapor density) in kg$_{water}$ m^{-3}. To rewrite this in kinematic form, divide by dry-air density ρ_{air} to give:

$$F = -D \cdot \frac{\Delta r}{\Delta x} \qquad \bullet(7.20)$$

where r is water-vapor mixing ratio, and kinematic flux F has units of mixing ratio times velocity [(kg$_{water}$ kg$_{air}$$^{-1}$)·(m s^{-1})]. Larger gradients cause larger fluxes, which cause droplets to grow faster.

The mixing ratio gradient is $\Delta r/\Delta x$. This is related to the supersaturation gradient $\Delta S/\Delta x$ by:

$$\frac{\Delta r}{\Delta x} = \frac{\Delta[(1+S)\cdot r_s]}{\Delta x} \approx r_s \cdot \frac{\Delta S}{\Delta x} \qquad (7.21)$$

where r_s is the saturation mixing ratio over a flat water surface. For this expression, the supersaturation fraction is defined by eq. (7.4), and supersaturation percentage by eq. (7.2).

Larger drops create a more gentle gradient than smaller drops. The humidity profile is given by:

$$S \approx S_\infty + \frac{R}{|x|}\cdot(S_R - S_\infty) \tag{7.22}$$

where S is supersaturation fraction at distance x from the center of the drop, S_∞ is background supersaturation at a large distance from the droplet, S_R is equilibrium supersaturation adjacent to the drop, and R is droplet radius. Eq. (7.22) was solved on a spreadsheet to create Fig. 7.11.

The **diffusivity** D is approximately

$$D = c\cdot\frac{P_o}{P}\cdot\left(\frac{T}{T_o}\right)^{1.94} \tag{7.23}$$

where c = 2.11x10^{-5} m^2·s^{-1} is an empirical constant, P_o = 101.3 kPa, and T_o = 273.15 K. This molecular diffusivity for moisture is similar to the thermal conductivity for heat, discussed in the Heat Budgets chapter.

During droplet growth, not only must water vapor diffuse through the air toward the droplet, but heat must conduct away from the drop. This is the latent heat that was released during condensation. Without conduction of heat away from the drop, it would become warm enough to prevent further condensation, and would stop growing.

Droplet radius R increases with the square-root of time t, as governed by the combined effects of water diffusivity and heat conductivity:

$$R \approx c_4\cdot\left(D\cdot S_\infty\cdot t\right)^{1/2} \tag{•(7.24)}$$

where S_∞ is the background supersaturation fraction far from the drop. Also, dimensionless constant c_4 is:

$$c_4 = (2\cdot r_\infty\cdot \rho_{air}/\rho_{liq.water})^{1/2} \tag{7.25}$$

where background mixing ratio is r_∞. Small droplets grow by diffusion faster than larger droplets, because of the greater humidity gradients near the smaller drops (Fig. 7.11 and eq. 7.22). Thus, the small droplets will tend to catch up to the larger droplets.

The result is a drop size distribution that tends to become **monodisperse**, where most of the drops have approximately the same radius. Also, eq. (7.24) suggests that time periods of many days would be necessary to grow rain-size drops by diffusion alone. But real raindrops form in much less time (tens of minutes), and are known to have a wide range of sizes. Hence, diffusion can<u>not</u> be the only physical process contributing to rain formation.

Sample Application

Find the water vapor diffusivity at P = 100 kPa and T = –10°C.

Find the Answer

Given: P = 100 kPa and T = 263 K.
Find: D = ? m^2·s^{-1}.

Use eq. (7.23):

$$D = (2.11\times10^{-5}\,\text{m}^2\text{s}^{-1})\left(\frac{101.3\text{kPa}}{100\text{kPa}}\right)\left(\frac{263\text{K}}{273.15\text{K}}\right)^{1.94}$$

= **<u>1.99x10</u>**$^{-5}$ m^2·s^{-1}.

Check: Units OK. Physics OK.
Exposition: Such a small diffusivity means that a large gradient is needed to drive the vapor flux.

Sample Application

Find and plot drop radius vs. time for diffusive growth, for the same conditions as the previous Sample Application. Assume 1% supersaturation.

Find the Answer

Given: P = 100 kPa, T = 263 K, D = 2x10^{-5} m^2·s^{-1}.
Find: R(µm) vs. t

First get ρ_{air} from the ideal gas law:

$$\rho_{air} = \frac{P}{\Re_d\cdot T} = \frac{100\text{kPa}}{(0.287\text{kPa·K}^{-1}\text{·m}^3\text{·kg}^{-1})\cdot(263\text{K})}$$

= 1.325 kg·m^{-3}.

$r_s \approx$ 1.8 g kg^{-1} = 0.0018 kg kg^{-1} from thermo diagram in Ch. 5. But supersaturation S = 1% = 0.01 = [r_∞ / r_s] – 1.
Thus r_∞ = [1 + S]·r_s
= 1.01·(0.0018 kg kg^{-1}) = 0.00182 kg kg^{-1}.

Using this in eq. (7.25): c_4 =
[2·(0.00182 kg$_w$/kg$_a$)·(1.325 kg$_a$·m^{-3})/(1000 kg$_w$·m^{-3})]$^{1/2}$
c_4 = 0.0022 (dimensionless)

Finally solve eq. (7.24) on a spreadsheet:
$R \approx$ 0.0022 · [(2x10^{-5} m^2 s^{-1}) · (0.01) ·t]$^{1/2}$

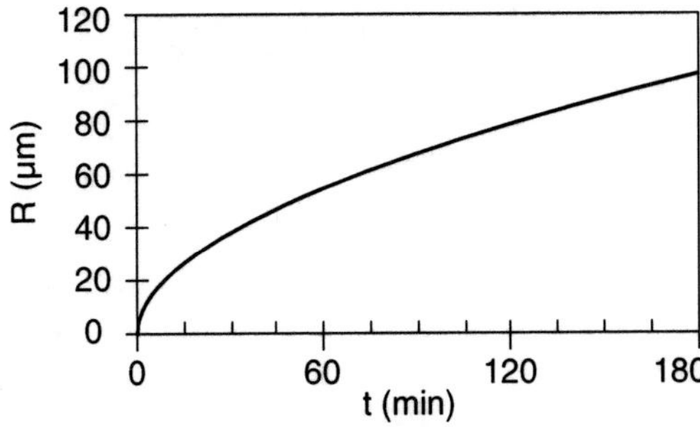

Check: Units OK. Physics OK.
Exposition: The droplet radius increases with the square root of time — fast initially and slower later. After 3 hours, it has a size on the borderline between cloud and rain drops. Thus, it is virtually impossible to grow full-size rain drops solely by diffusion.

INFO • Cubic Ice

Fourteen **phases** of ice have been identified, and are labeled using Roman numerals I–XIV (see Fig. 7.b). More phases might be discovered in the future. Each phase is a preferred arrangement of molecules having uniform chemical composition and physical state.

At normal atmospheric temperatures and pressures, ice I is most prevalent. However, it comes in two variants: **hexagonal ice (Ih)**, and **cubic ice (Ic)**. Ice Ih is the form that is thermodynamically stable in the troposphere. Both forms of ice I have a tetrahedral arrangement of water molecules.

Hexagonal ice Ih forms crystals that are hexagonal plates, hexagonal columns with flat ends, hexagonal columns with pyramidal ends, and **dendrites** (snowflakes with 6 arms). Samples of these crystal shapes have been collected in the atmosphere, and are frequently observed. This is the normal ice that we see.

Cubic ice Ic is believed to be able to form as cubes, square columns capped by pyramids, and octahedrons (equal to two pyramids with their bases merged). Natural crystals of ice Ic have been detected in the lower stratosphere, but never been successfully captured in the lower troposphere partly because it is metastable with respect to ice Ih, and at warmer temperatures ice Ic rapidly converts to Ih. Cubic ice has been created in the lab under atmospheric conditions, and its existence in the atmosphere has been inferred from certain halos observed around the sun (because ice crystals act like prisms; see the Optics chapter).

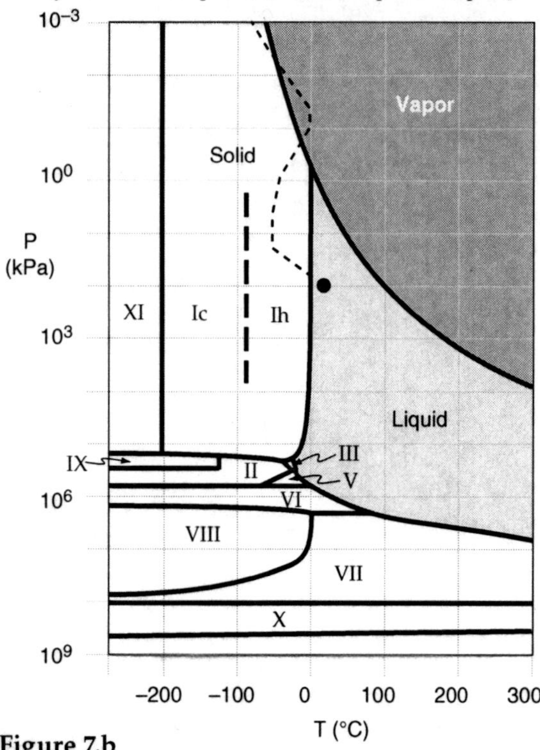

Figure 7.b
Phase diagram for water, as a function of temperature (T) and pressure (P). Standard atmosphere is thin dotted line, with circle at Earth-surface standard conditions. (Not all phases are plotted.) See exercise U26 and Table 7-7.

ICE GROWTH BY DIFFUSION

Ice Crystal Habits

In the troposphere, the normal ice crystal shape that forms from direct deposition of water vapor is hexagonal (see INFO Box at left). The particular hexagonal shape that grows depends on temperature and supersaturation (Fig. 7.12). These shapes are called **habits**. Supersaturation is sometimes given as **water-vapor density excess** $\rho_{ve} = \rho_v - \rho_{vs}$, where ρ_v is absolute humidity and ρ_{vs} is the saturation value of absolute humidity.

As ice crystals fall and move by wind and turbulence, they pass through regions of different temperature and vapor-density excess in the cloud. This allows individual crystals to grow into complex combinations of habits (Fig. 7.13). For example, a crystal that starts growth as a column might later be capped on each end by large plates. Because each crystal travels through a slightly different path through the cloud, each snowflake has a unique shape.

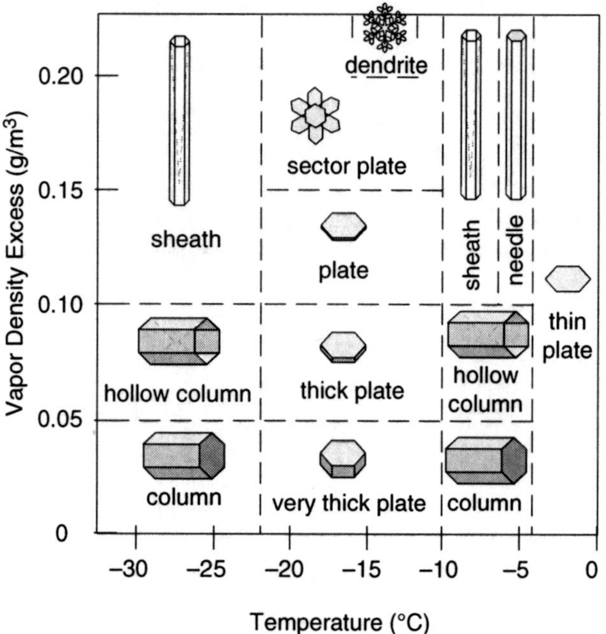

Figure 7.12
Ice crystal habits (idealized). Sheaths are hollow needles, and needles are long, narrow, solid columns. Outside the range of this figure, solid-column rosettes are found at temperatures below −40°C.

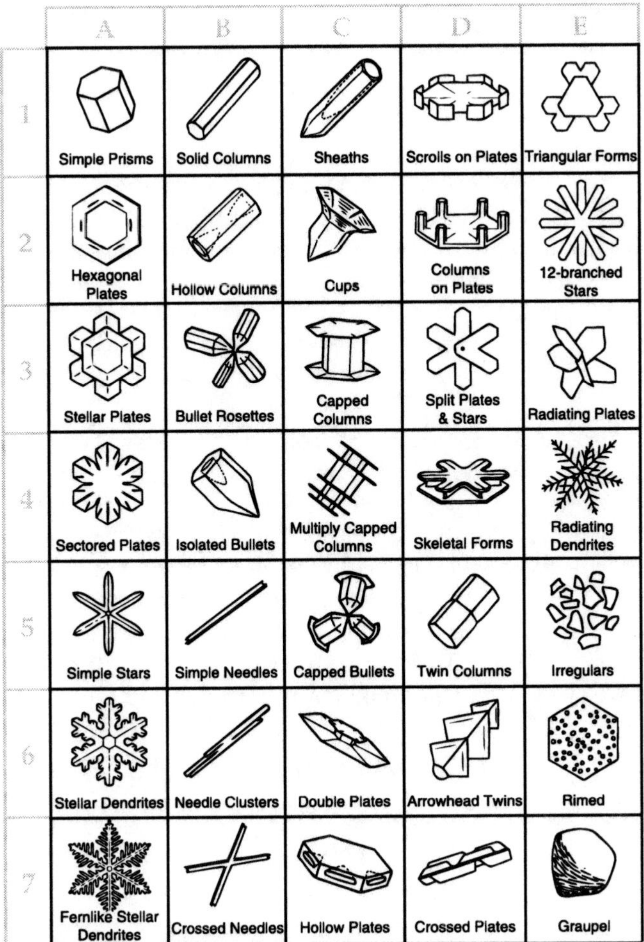

Figure 7.13
Some of the observed ice crystal shapes. [Courtesy of Kenneth Libbrecht, www.SnowCrystals.com.]

Sample Application
For each snow crystal class in Table 7-A (see INFO Box at right), identify (by row and column label) all crystal shapes from Fig. 7.13 that are in that class.

Find the Answer

<u>Class</u> <u>Shapes</u>
1a Columns: 1A, 1B, 2B, 5D
1b Needles: 1C, 5B, 6B, 7B
1c Plates: 1E, 2A, 3A, 3D, 4A, 4D, 6C, 7C, 7D
1d Stellars: 2E, 3A, 3D, 4A, 4D, 4E, 5A, 6A,
1e Irregular: 1D, 1E, 2C, 2D, 3C, 3E, 4C, 5C, 5E, 6D
1f Graupel: 6E, 7E
1g Hail: (none)
1h Ice Pellets: (none)

Exposition: Identifying ice-crystal class is somewhat subjective. Different experts might give a slightly different classification than the solution above.

INFO • Snow Grain Classification

Snow avalanches are often associated with weak snow layers buried under stronger layers of snow. Field observations of snow crystal shape and size are important for detecting the different snow layers. The International Commission on Snow and Ice (ICSI) developed in 1990 a standard symbology (Table 7-A & Table 7-B) to use when logging snow data.

Table 7-A. Morphological (shape-based) classification of precipitation particles. T = temperature, ρ_{ve} = excess water vapor density.
[Colbeck et al, 1990: *The International Classification for Seasonal Snow on the Ground*, ICSI. 37pp. Available from http://www.crrel.usace.army.mil/techpub/CRREL_Reports/reports/Seasonal_Snow.pdf]

	Name	Symbol	Shape & Formation
1a	Columns	▭	Short, prismatic crystal, solid or hollow. See Fig. 7.12 for T & ρ_{ve} conditions.
1b	Needles	⟷	Needle-like, approximately cylindrical. See Fig. 7.12 for T & ρ_{ve} conditions.
1c	Plates	⬡	Plate-like, mostly hexagonal. See Fig. 7.12 for T & ρ_{ve} conditions.
1d	Stellars, Dendrites	✳	Six-fold start-like, planar or spatial. See Fig. 7.12 for T & ρ_{ve} conditions.
1e	Irregular Crystals	[symbol]	Clusters of very small crystals. Polycrystals growing at varying environmental conditions.
1f	Graupel	[symbol]	Heavily rimed particles. Caused by accretion of supercooled water.
1g	Hail	▲	Laminar internal structure, translucent or milky, glazed surface. Growth by accretion of supercooled water.
1h	Ice Pellets	⚠	Transparent, mostly small spheroids. Frozen rain.

Table 7-B. Snow-grain classification (ICSI).

Term	Size (mm)
Very fine	< 0.2
Fine	0.2 to 0.5
Medium	0.5 to 1.0
Coarse	1.0 to 2.0
Very coarse	2.0 to 5.0
Extreme	> 5.0

Sample Application

Given a mixed-phase cloud (i.e., having both ice crystals and supercooled liquid water droplets) at –14°C that is saturated with respect to water (and thus supersaturated with respect to ice; see Fig. 4.2 in the Water Vapor chapter). If the water vapor diffusivity is 1.5×10^{-5} m^2 s^{-1}, then what is the relative mass of ice crystals after 1 hour of growth, for (a) a 3-D crystal and (b) a 2-D crystal that is 15 µm thick?

Find the Answer

Given: $D = 1.5 \times 10^{-5}$ m^2 s^{-1}, t = 1 h = 3600 s, T = 259K
d = 15 µm = 1.5×10^{-5} m.
Find: m = ? kg
Assume: initial mass m is negligible

First, use Fig. 4.2 in the Water Vapor chapter to estimate the supersaturation. The insert in that fig. shows that $(e_{water} - e_{ice}) \approx 0.0275$ kPa at T = –14°C, and $e_{ice} \approx 0.175$ kPa. The supersaturation is:

$S = (e_{water} - e_{ice}) \,/\, e_{ice}$ = (0.0275 kPa)/(0.175 kPa)
S = 0.157 (dimensionless)

Use the ideal gas law for water vapor (eq. 4.10) to estimate vapor density from vapor pressure (which for this case equals $e_{water} \approx 0.20$ kPa, from previous paragraph):

$\rho_v = e_{water}/(\Re_v \cdot T) =$
$= (0.2 \text{ kPa}) \,/\, [(0.461 \text{ kPa·K}^{-1}\text{·m}^3\text{·kg}^{-1})\cdot(259 \text{ K})]$
$= 1.68 \times 10^{-3}$ kg m^{-3}

(a) Use eq. (7.26). The factor in parenthesis is
$(\rho_v{}^3/\rho_i)^{1/2} = [(1.68\times10^{-3} \text{ kg m}^{-3})^3/(916.8 \text{ kg m}^{-3})]^{1/2}$
$= 2.27\times10^{-6}$ kg m^{-3}
and the term in square brackets of eq. (7.26) is
$[\,] = [(1.5\times10^{-5} \text{ m}^2 \text{ s}^{-1})\cdot(0.157)\cdot(3600\text{s})]^{3/2}$
$[\,] = (8.478\times10^{-3} \text{ m}^2)^{3/2} = 7.8\times10^{-4} \text{ m}^3$

Thus, solving eq. (7.26):
$m = 11.85\cdot(2.27\times10^{-6} \text{ kg m}^{-3})\cdot(7.8\times10^{-4} \text{ m}^3)$
$= 2.1\times10^{-8}$ kg = **2.1×10^{-5} g**

(b) Use eq. (7.27):

$$m = \frac{5.09}{1.5\times10^{-5}\text{m}} \cdot \left(\frac{(0.00168\text{kg/m}^3)^2}{916.8\text{kg/m}^3} \right) \cdot [8.478\times10^{-3}\text{m}^2]^2$$

$= 7.51\times10^{-8}$ kg = **7.51×10^{-5} g**

Check: Units OK. Physics OK. Agrees with Fig. 7.14.
Exposition: Typically observed ice-crystal mass is about 3×10^{-5} g. Typical **snowflakes** that fall to Earth are often aggregates of hundreds of ice crystals stuck together, with a total mass of about 3 mg snowflake^{-1}.

Growth Rates

Because of the diversity of shapes, it is better to measure crystal size by its mass m rather than by some not-so-representative radius. Rate of growth by diffusion depends on crystal habit.

Columns and very thick plates have **aspect ratio** (height-to-width ratio) of roughly 1. If the aspect ratio remains constant during growth, then the growth equation is:

$$m \approx c_3 \cdot (\rho_v{}^3 \,/\, \rho_i)^{1/2} \cdot [D \cdot S \cdot t]^{3/2} \qquad (7.26)$$

where c_3 = 11.85 (dimensionless), ρ_v is the density of water vapor (=absolute humidity, see eq. 4.10), ρ_i is ice density (= 916.8 kg m^{-3} at 0°C), D is diffusivity, S is supersaturation fraction, and t is time. If the crystal were spherical with radius R, then its mass would be $m = \rho_{liq.water} \cdot (4\cdot\pi/3)\cdot R^3$. Taking the cube root of both sides of eq. (7.26) gives an equation similar to eq. (7.24). Thus growth rate of a 3-D crystal is very similar to growth of a liquid droplet.

For 2-D growth, such as dendrites or plates of constant thickness d, the growth equation changes to

$$m \approx \frac{c_2}{d} \cdot \left(\frac{\rho_v{}^2}{\rho_i} \right) \cdot [D \cdot S \cdot t]^2 \qquad (7.27)$$

where c_2 = 5.09 (dimensionless). For 1-D growth of needles and sheaths of constant diameter, the growth equation is

$$m \propto \exp\left[(D \cdot S \cdot t)^{1/2} \right] \qquad (7.28)$$

These three growth rates are sketched in Fig. 7.14.

Evidently 2-D crystals increase mass faster than 3-D ones, and 1-D crystals increase mass faster still. Those crystals that gain the mass fastest are the ones that will precipitate first.

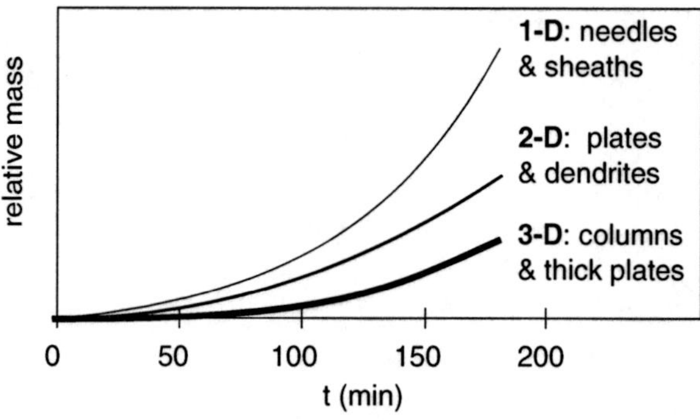

Figure 7.14
Relative growth rates of crystals of different habits.

The Wegener-Bergeron-Findeisen (WBF) Process

Recall from the Water Vapor chapter that ice has a lower saturation vapor pressure than liquid water at the same temperature. Fig. 7.15 shows an enlargement of the saturation vapor-pressure curves for liquid water and ice.

Suppose that initially (time 1, on the time line in Fig. 7.15) there are only supercooled liquid water droplets in a cloudy air parcel. These droplets exist in a supersaturated environment and therefore grow as the excess water vapor diffuses toward the droplets. As the air parcel rises and cools within the cloud, some ice nuclei might become activated at time 2, causing ice crystals to form and grow. The excess water vapor now deposits on both the liquid and solid hydrometeors.

Both the ice crystals and liquid droplets continue to grow, because both are in a supersaturated environment (time 3). However, the ice crystal grows a bit faster because it is further from its ice saturation line (i.e., more supersaturated) than the liquid droplet is from liquid saturation line.

As both hydrometeors grow, water vapor is removed from the air, reducing the supersaturation. Eventually, near point 4 on the time line, so much vapor has been consumed that the relative humidity has dropped below 100% with respect to liquid water. Hence, the liquid droplet begins to evaporate into the unsaturated air. However, at point 4 the ice crystal continues to grow because the air is still supersaturated with respect to ice.

The net result is that the ice crystals grow at the expense of the evaporating liquid droplets, until the liquid droplets disappear (point 5). This is called the **Wegener-Bergeron-Findeisen (WBF)** process.

The difference between ice and liquid saturation vapor pressures is greatest in the range –8°C to –16°C, as shown in Fig. 7.16 (from the insert in Fig. 4.2). This is the temperature range where we expect the maximum effect from the WBF growth process, also known as the **cold-cloud process** because temperatures below freezing are needed.

If a large number of ice nuclei exist in the air, then a large number of ice crystals will form that are each too small to precipitate. For a very small number of ice nuclei, those few ice crystals will rapidly grow and precipitate out, leaving behind many small liquid cloud droplets in the cloud. Both of these scenarios lead to relatively little precipitation.

Only with a medium concentration (1 to 10) ice nuclei per liter (compared to about a million liquid droplets in the same volume) will the ice nuclei be able to scavenge most of the condensed water before precipitating out. This scenario causes the maximum precipitation for the WBF processes. But a

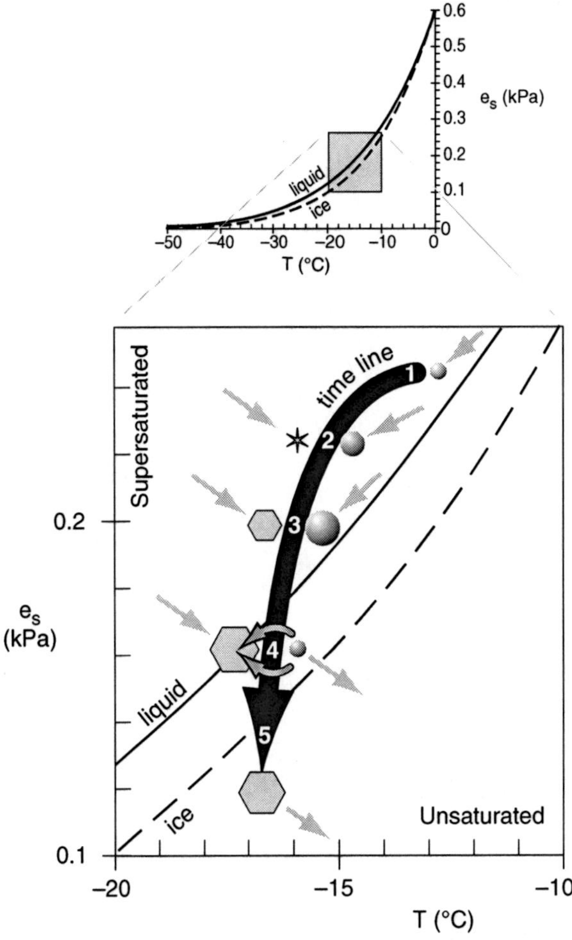

Figure 7.15
Enlargement of part of the saturation curve from Fig. 4.2, illustrating the WBF (cold-cloud) ice-growth process in a rising, cooling air parcel. Shown are saturation vapor pressure (e_s) vs. temperature (T) over liquid water and ice. Spheres represent cloud droplets, and hexagons represent ice crystals. Small grey arrows indicate movement of water vapor toward or away from the hydrometeors. Thick black arrow indicates time evolution.

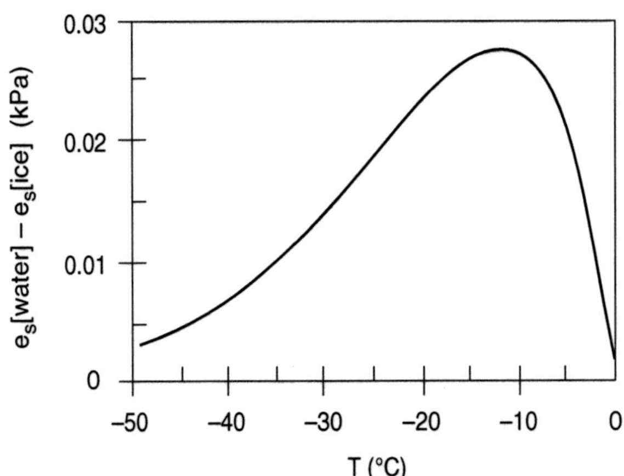

Figure 7.16
Saturation vapor pressure difference over water vs. over ice (copied from Fig. 4.2).

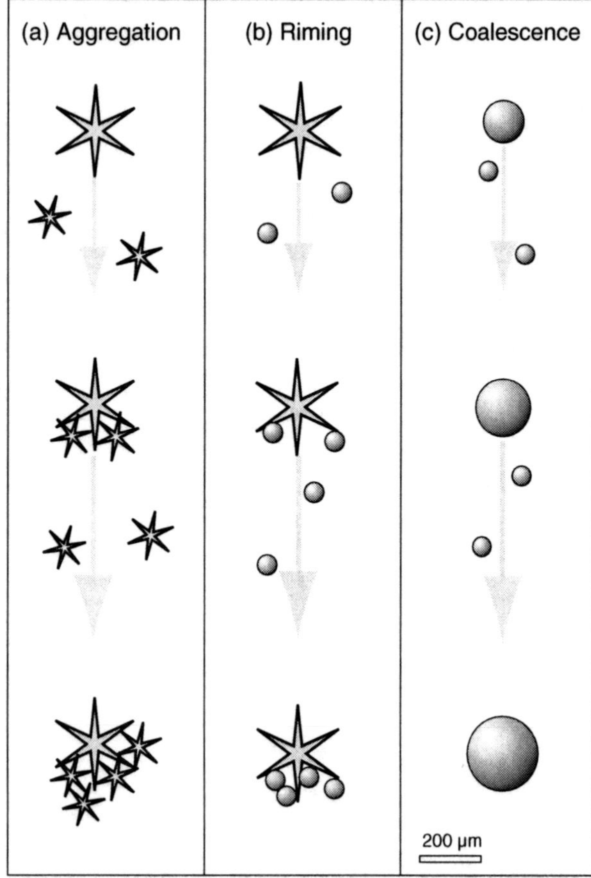

Figure 7.17
Collision and collection of hydrometeors in (a) very cold clouds with no liquid water; (b) cold clouds with both supercooled liquid-water droplets that freeze quickly when they hit ice crystals; and (c) warm clouds. Grey arrows show the fall velocity of the larger hydrometeor relative to the smaller ones.

Sample Application
What updraft wind is needed to keep a typical cloud droplet (R = 10 µm) from falling?

Find the Answer
Given: R = 10 µm
Find: w_T = ? m s^{-1}, and use: $w_{up} = -w_T$

Use eq. (7.29):
w_T = (–1.19x10^8 m^{-1} s^{-1})·(10x10^{-6} m)^2 =
= –0.012 m s^{-1} = **–1.2 cm** s^{-1}
w_{up} = **1.2 cm** s^{-1}

Check: Units OK. Physics OK.
Exposition: The required updraft velocity is positive 1.2 cm s^{-1}, which is a very gentle movement of air. Recalling that most clouds form by adiabatic cooling within updrafts of air, these updrafts also keep cloud droplets aloft.

restriction on this precipitation formation process is that it happens only in cold clouds (clouds colder than 0°C).

As was discussed in the nucleation sections, there are indeed fewer ice nuclei than CCN, hence the cold-cloud process can be an important first step in getting hydrometeors large enough to begin to fall out of the cloud as precipitation.

The cold-cloud process can occur even in summer, but higher in the troposphere where the air is colder. As these ice particles fall into warmer air at lower altitude, they melt into raindrops to create summer convective rain showers (see Fig. 7.21 later in this chapter).

COLLISION AND COLLECTION

Larger hydrometeors (ones with a greater mass/drag ratio) fall the fastest. As a result, different hydrometeors move at different speeds, allowing some to **collide** (hit each other). Not all collisions result in the merging of two hydrometeors. Those particles that do merge form a particle that is even heavier, falls faster, and collides with even more particles (Fig. 7.17). Hence, this positive feedback can cause hydrometeors to rapidly grow large enough to precipitate.

Terminal Velocity of Hydrometeors

Everything including cloud and rain drops are pulled by gravity. The equilibrium velocity resulting when gravity balances frictional drag is called the **terminal velocity**.

Cloud Droplets

For particles of radius R < 40 µm, which includes most cloud droplets and aerosols, **Stokes Drag Law** gives the terminal velocity w_T as

$$w_T \approx -k_1 \cdot R^2 \tag{7.29}$$

where k_1 = 1.19x10^8 m^{-1}·s^{-1} . The negative sign on indicates the droplets are falling.

When drops fall at their terminal velocity, the gravitational pull on the drops is transmitted by frictional drag to the air. In other words, the weight of the air includes the weight of the drops within it. Hence, droplet-laden air is heavier than cloud-free air, and behaves as if it were colder (see virtual temperature, eq. 1.22). Falling rain also tends to drag air with it.

Typical terminal velocities of these smallest droplets and aerosols are mm s^{-1} to cm s^{-1} relative to the air. However, the rising air in clouds often has updraft velocities (cm s^{-1} to m s^{-1}) that are greater than the terminal velocity of the particles. The net result is that cloud droplets and aerosols are carried upward inside the cloud.

Rain Drops

Rain drops are sufficiently large and fall fast enough that Stokes drag law is not appropriate. If raindrops were perfect spheres, then

$$w_T \approx -k_2 \cdot \left(\frac{\rho_o}{\rho_{air}} \cdot R \right)^{1/2} \qquad (7.30)$$

where $k_2 = 220\ m^{1/2} \cdot s^{-1}$, $\rho_o = 1.225\ kg \cdot m^{-3}$ is air density at sea level, and ρ_{air} is air density at the drop altitude. Again, the negative sign in the equation means a downward velocity.

However, the larger raindrops become flattened as they fall due to the drag (see polarimetric radar section of the Satellites & Radar chapter). They do not have a tear-drop shape. This flattening increases air drag even further, and reduces their terminal velocity from that of a sphere. Fig. 7.18 shows raindrop terminal velocities. For the smallest drops, the curve has a slope of 2, corresponding to Stokes law. For intermediate sizes R = 500 to 1000 µm, the slope is 0.5, which corresponds to eq. (7.30). At the larger sizes about R = 2.5 mm, the terminal velocity curve has near zero slope as the droplet becomes so deformed that it begins to look like a parachute. Drops larger than about 2.5 mm radius tend to break up. The largest raindrops rarely exceed 4 mm radius.

Let R be the **equivalent radius** of a sphere having the same volume as the deformed drop. An empirical curve for terminal velocity (relative to air) over range $20 \le R \le 2500$ µm is:

$$w_T = -c \cdot \left[w_o - w_1 \cdot \exp\left(\frac{R_o - R}{R_1} \right) \right] \qquad \bullet(7.31)$$

where $w_o = 12\ m\ s^{-1}$, $w_1 = 1\ m\ s^{-1}$, $R_o = 2500$ µm, and $R_1 = 1000$ µm. This curve gives a maximum terminal velocity of 11 m s^{-1} for the largest drops. The density correction factor is $c = (\rho_{70kPa}/\rho_{air})^{1/2} \approx (70\ kPa/P)^{1/2}$, where P is ambient pressure. Rain falls faster where the air is thinner (less dense).

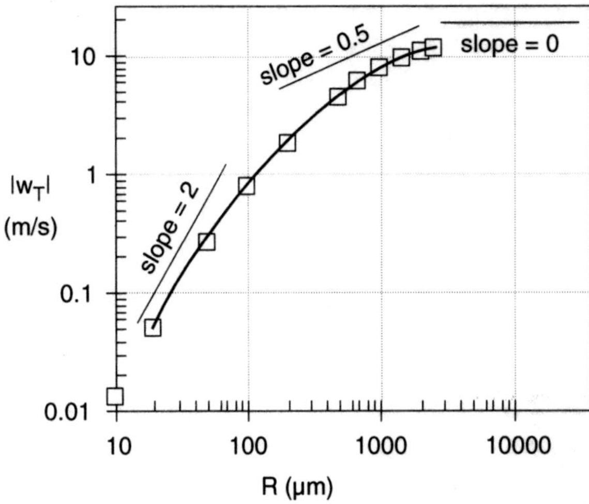

Figure 7.18
Raindrop terminal speed $|w_T|$ at P = 70 kPa and T = 0°C. R is the equivalent radius of a spherical drop. Curve is eq. (7.31).

Sample Application

Find the terminal velocity of a droplet of equivalent radius 1500 µm, at P = 70 kPa.

Find the Answer

Given: R = 1500 µm, c = 1 at P = 70 kPa.
Find: w_T = ? m s^{-1}.

Use eq. (7.31):
$w_T = -1 \cdot [(12 m\ s^{-1}) - (1 m\ s^{-1}) \cdot \exp\{(2500µm-1500µm)/(1000µm)\}]$
$= \underline{-9.3\ m\ s^{-1}}$

Check: Units OK. Physics OK. Agrees with Fig. 7.18. Negative sign means falling downward.
Exposition: This $w_T \approx 34$ km h^{-1}. Updrafts in thunderstorms are fast, and keep these large drops aloft.

Sample Application

Plot the terminal velocity of hailstones vs. equivalent diameter (0.01 to 0.1 m) at 5 km altitude.

Find the Answer

Given: $z = 5$ km
Find: w_T (m s^{-1}) vs. D (m), where $D = 2{\cdot}R$.

Assume: std. atmosphere. Thus $\rho_{air} = 0.7361$ kg m^{-3} at $z = 5$ km, from Chapter 1.
Assume: $C_D = 0.55$, and $\rho_i = 900$ kg m^{-3} , calm air.
Use a spreadsheet to solve eq. (7.32).

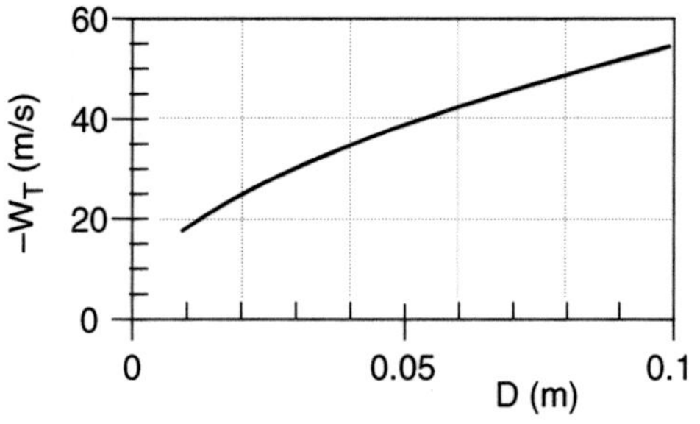

Check: Magnitudes agree with plot in the Thunderstorm chapters. Sign is negative (implying downward fall).

Exposition: Massive hailstones falling at these fast velocities can kill people, strip foliage and small branches off trees, destroy crops, and kill animals.

Hailstones

The terminal velocity w_T of a hailstone relative to the air is approximated by:

$$w_T = -\left[\frac{8}{3}\frac{|g|}{C_D}\frac{\rho_i}{\rho_{air}}R\right]^{1/2} \qquad \bullet(7.32)$$

where $|g|$ = 9.8 m s^{-2} is gravitational acceleration magnitude, $C_D \approx 0.55$ (dimensionless) is a **drag coefficient** of the hailstone through air, $\rho_i \approx 900$ kg m^{-3} is the density of the hailstone, ρ_{air} is air density, and R is hailstone radius. The negative sign means the hailstone falls downward.

The drag coefficient varies between 0.4 and 0.8, because hailstones have different shapes, surface roughnesses, and tumblings. The hailstone ice density can be less than the density of pure ice, because of varying amounts of imbedded air bubbles. Air density decreases with increasing altitude (see Chapter 1); hailstones fall faster in thinner air. For non-spherical hailstones, R is taken as the equivalent radius of a sphere that has the same volume as the actual hailstone.

Large hailstones form only in thunderstorms with strong updrafts. Thus, the hailstone terminal velocity relative to the ground is the sum of the air updraft speed (a positive number) and the hailstone terminal velocity relative to the air (a negative number).

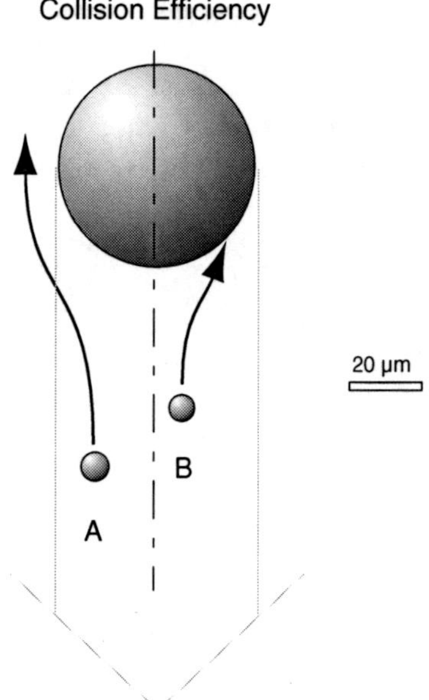

Figure 7.19
Smaller droplets A & B are both in the path of the falling larger drop. Relative to the large drop, small droplet A is pushed aside by the air flowing around the large drop, and does not collide. But droplet B, closer to the centerline of the larger drop, does collide.

Collection & Aggregation Processes

Warm-cloud Process

The merging of two liquid droplets (Fig. 7.17c) is called **coalescence**. This is the only process for making precipitation-size hydrometeors that can happen in warm clouds (clouds warmer than 0°C), and is thus called the **warm-cloud process**.

When droplets of different size approach each other, they do not always merge. One reason is that the smaller droplet partly follows the air as it flows around the larger droplet, and thus may not collide with the larger drop (Fig. 7.19). This is quantified by a **collision efficiency** (E), which is small ($0.02 < E < 0.1$) when the smaller droplet is very small ($2 < R < 5$ µm). But if both droplets are relatively large (such as when the smaller droplet has $R > 10$ µm, and the larger droplet has $R > 30$ µm), then efficiencies can be $0.5 \le E \le 1$.

Even if two droplets collide, they might not merge because a thin layer of air can be trapped between the two droplets (Fig. 7.20b). For this situation, the two droplets bounce off of each other and do not coalesce. The **coalescence efficiency** (E') is very small ($0.1 \le E' \le 0.3$) when both drops are large ($300 \le R \le 500$ µm). The efficiency is greater ($E' > 0.8$) when both droplets are small ($R < 150$ µm, as in Fig. 7.20a).

The product of both efficiencies is the **collection efficiency**: $E_c = E \cdot E'$. The maximum efficiency possible is 1.0, but usually efficiencies are smaller.

Cold-cloud Processes

When ice particles collide and stick to other ice particles (Fig. 7.17a), the process is called **aggregation**. Such aggregation is aided if the colliding particles are dendrites, for which the snowflake arms can interlock. Also, if the ice particles are warmer than –5°C, then the ice surface becomes sticky, allowing multiple ice crystals to aggregate into soft little irregular clumps of snow.

The growth of ice particles by collection and instant freezing of supercooled liquid droplets (Fig. 7.17b) in mixed-phase clouds is called **accretion** or **riming**. Hydrometeors that become so heavily rimed as to completely cover and mask the original habit are called **graupel**. Graupel has the consistency of a sugar cube (i.e., many separate solid grains stuck together), but often in the shape of a cone or a sphere. For an aggregate to be called graupel, it's diameter must be no larger than 5 mm.

If the collected water does not freeze instantly upon contacting the ice particle, but instead flows around it before freezing, then **hail** can form. Graupel contains a lot of air trapped between the frozen droplets on the graupel, and thus is often softer and less dense than hail. See the Thunderstorm chapters for more information about hail.

Precipitation Formation

Warm Clouds

How do terminal velocity and collection efficiency relate to the formation of large, precipitation-size particles in warm clouds? Recall that: (1) the atmosphere has an excessively large number of CCN; (2) this causes the available condensate in a rising cooling air parcel to be partitioned into a large number of small droplets (droplets too small to rain out); (3) droplets tend to become monodisperse (nearly the same size) due to diffusion; (4) droplets of the same size have the same terminal velocity, and thus would be unlikely to collide with each other as they are kept aloft in the updraft; and (5) with no collisions we would not expect larger precipitation drops to form in warm clouds.

Yet warm-cloud rain happens quite nicely in the real atmosphere, especially in the tropics. Why?

Five factors can help make warm-cloud rain:

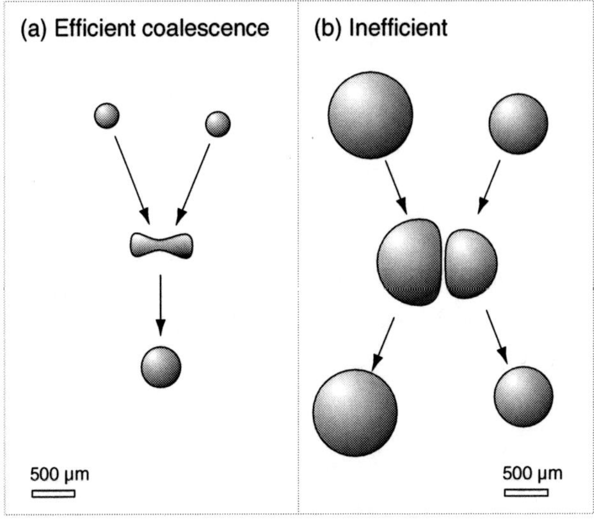

Figure 7.20
Colliding small droplets in (a) coalesce into a single drop. Colliding larger droplets in (b) do not coalesce, because of a film of air trapped between them that cannot completely escape before the two droplets bounce off of each other.

INFO • Meteors and Meteorology

Ancient Greeks defined "meteors" as anything in the sky. They were particularly concerned about missiles the gods might toss down, such as bits of rock, ice, or lightning bolts.

Only much later did scientists discriminate between missiles from space (bits of rock called **meteoroids**) and missiles from the atmosphere (bits of just about anything else from the sky). But by then "**meteorology**" was firmly entrenched as the name for **atmospheric science**.

According to the *Glossary of Meteorology*, meteorologists study the following meteors:

- **hydrometeors** – wet: clouds, rain, snow, fog, dew, frost, etc.
- **lithometeors** – dry: dust, sand, smoke, haze
- **igneous meteors** – lightning, corona
- **electrometeors** – lightning (again), thunder
- **luminous meteors** – rainbows, halos, etc.

Except for "hydrometeors", these terms are seldom used any more.

• First, by random chance a small number of collisions do occur, which starts to broaden the spectrum of drop sizes. This broadens the range of terminal velocities to allow more collisions, which accelerates via positive feedback (with help from breakup of larger drops).
• Second, not all CCN are the same size — some are called **giant CCN** (particles > 3 μm radius, with a wettable surface) and can create a small number of larger cloud droplets that fall relative to (and collide with) the other cloud droplets.
• Third, turbulence can entrain outside clear air into the top and sides of a cloud, causing some droplets to partly evaporate, thereby broadening the spectrum of droplet sizes, again allowing collisions.
• Fourth, IR radiation from individual drops near cloud top and sides can cool the drops slightly below the ambient air temperature, allowing greater condensation growth of those drops relative to interior drops.
• Fifth, electrical charge build-up in cumuliform clouds (see the Thunderstorm chapters) can draw droplets together of different charge, and can cause sparks between nearby droplets to allow them to coalesce more efficiently.

Cold Clouds

In cold clouds (T < 0°C), the smaller number of ice nuclei in the atmosphere allows the available condensate to deposit onto a small number of larger ice particles. Even in mixed-phase clouds, the WBF process can remove water molecules from the large number of droplets and deposit them onto a small number of ice crystals. Thus, the ice crystals are larger, and can fall as precipitation. Also, the crystals often have a wide range of sizes and shapes so they can collide and aggregate easily, which also creates large-enough particles to fall as precipitation.

Cumuliform clouds including thunderstorms can be deep enough to have their bases in warm air and their tops in cold air (Fig. 7.21). Thus, ice particles can grow to large size (order of 1-5 mm) via the WBF process, aggregation, and riming in the cold part of the cloud, and then melt into large raindrops as they pass through warmer air closer to the ground. Most rain from mid-latitude thunderstorms forms this way. See the Thunderstorm chapters for more information about heavy rain.

Also, on those rare occasions when thunderstorms can be triggered in late Fall or early Spring, boundary-layer temperatures can be cold enough to allow snow from thunderstorms to reach the ground as large snow clusters (snow balls) without totally melting, accompanied by lightning and thun-

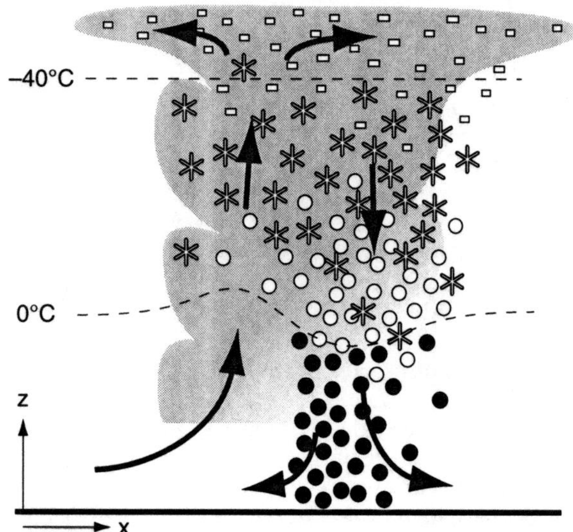

Figure 7.21
Thunderstorms can have cold-cloud tops and warm-cloud bases, allowing efficient growth of ice precipitation that can melt into large raindrops on the way down.

der from the storm. This is sometimes nicknamed **thundersnow**.

Regardless of whether clouds are warm or cold, a simple rule of precipitation is that thicker clouds can cause heavier precipitation rates with larger size drops. The main reason is that hydrometeors take a longer time to fall the greater distance through thicker clouds, giving them more time to grow.

PRECIPITATION CHARACTERISTICS

Rain-Drop Size Distribution

Small raindrops outnumber large ones. Classically, the rain-drop spectrum has been fit by an exponential function, known as the **Marshall-Palmer distribution**:

$$N = \frac{N_o}{\Lambda} \exp(-\Lambda \cdot R) \qquad \bullet(7.33)$$

where N is the number of drops of radius greater than R within each cubic meter of air, and $N_o \approx 1.6\text{x}10^7$ m^{-4}. Parameter $\Lambda(\text{m}^{-1}) = 8200 \cdot (RR)^{-0.21}$, where RR is rainfall rate in mm h^{-1}. Fig. 7.22 shows the Marshall-Palmer distribution. While there are on the order of 1000 drops of drizzle size and larger in each cubic meter of air, there are only tens of drops of typical raindrop size and larger.

Fig. 7.23 shows rain-drop distributions for a variety of rain rates. Larger drop diameters are associated with heavier rainfall rates. Drop diameters in the range of 4 to 8 mm diameters have been observed for the heaviest rain events. Because large liquid drops tend to break up into smaller ones as they fall, it is possible that some of the largest-diameter drops reported in Fig. 7.23 consisted of still-melting graupel or aggregated/rimed snow.

Beware that these curves vary significantly from storm to storm. Many thunderstorms have much greater numbers of large 1000 to 2500 µm radius drops. Real clouds are very heterogeneous in their generation of rain, causing a patchiness in the rainfall rates. This patchiness causes the raindrop size distributions to continually change with time and place; namely, they are not the simplistic, steady distributions given above. This patchiness is also observed by weather radar (see the Satellites & Radar chapter), and in the rainfall observed by rain gauges on the ground. You probably also experienced it when driving or walking through rain.

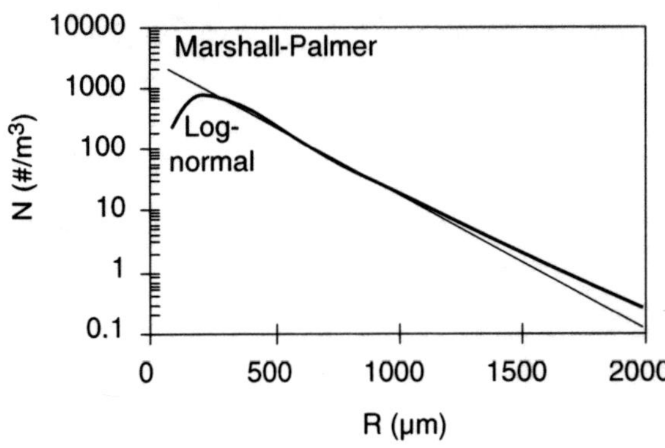

Figure 7.22
Rain drop size distributions. Rainfall rate was RR = 10 mm h^{-1}.

> **Sample Application**
> Find the number of raindrops larger than 1000 µm, using the parameters given below.
>
> **Find the Answer**
> Given: RR = 10 mm h^{-1}, N_o = 1.6x10^7 m^{-4} , R = 1000 µm
> Find: N = ? m^{-3} .
>
> First, find Λ = 8200·(10 mm h^{-1})$^{-0.21}$ = 5056 m^{-1} .
> Next, use eq. (7.33) for Marshall-Palmer:
> N = [(1.6x10^7 m^{-4})/(5056 m^{-1})]·exp[–(5056 m^{-1})·(10^{-3} m)]
> = **20.2 m^{-3}** .
>
> **Check**: Units OK. Physics OK. Agrees with Fig. 7.22.
> **Exposition**: Don't forget that N is like 1 – cumulative distribution. It doesn't give the count of drops of size equal to R, but counts all drops of size R and greater.

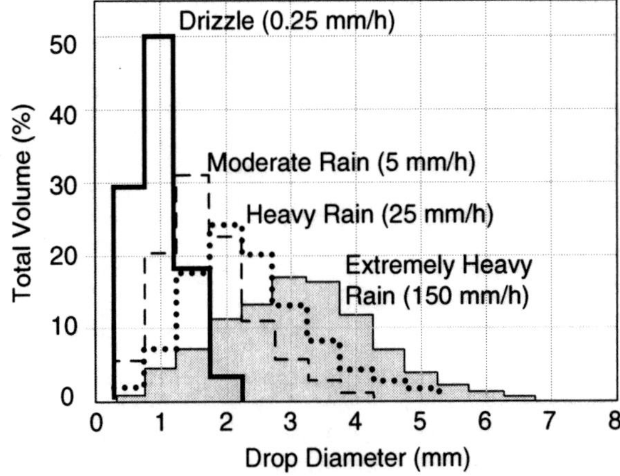

Figure 7.23
Typical distribution of rain-drop sizes during different rain rates. Area under each curve equals 100% of the total volume of rain accumulated in the rain gauge for that rain rate.

Table 7-3. Rain intensity criteria (from USA *Fed. Meteor. Handbook No. 1*, Sep 2005), and their corresponding weather-map symbols and Meteorological Aviation Report (METAR) codes for continuous, non-freezing rain.

Rain Intensity	Rainfall rate		Symbol on Map / METAR code
	inches h^{-1}	≈mm h^{-1}	
heavy	> 0.30	> 7.6	• • • • +RA
moderate	0.11 – 0.30	2.6 – 7.6	• • • RA
light	0* – 0.10	0* to 2.5	• • –RA
trace	< 0.005	<0.1	

Sample Application

Is the record rainfall envelope valid for these other actual observations: (a) 206 mm/20 min at Curtea-de-Arges, Romania on 7 July 1889; (b) 304.8 mm/42 min at Holt, MO, USA on 22 July 1947?

Find the Answer

Given: (a) d_{obs} = 206 mm, P_{rain} = 20 minutes = 0.333 h
(b) d_{obs} = 304.8 mm , P_{rain} = 42 min = 0.7 h

Find: if d_{obs} is approx. equal or less than $d_{max\ rain}$

Use eq. (7.34): (a) $d_{max\ rain}$ = <u>209.6 mm</u>
(b) $d_{max\ rain}$ = <u>303.7 mm</u>

Exposition: Yes, envelope is approx. valid for both.

Table 7-4. Drizzle and snow intensity criteria (from USA *Fed. Meteor. Handbook No. 1,* Sep 2005), and their weather-map symbols and Meteorological Aviation Report (METAR) codes for continuous precipitation.

Precip. Intensity	Visibility (x_v)		Symbol on Map / METAR Code	
	miles	≈km	Drizzle	Snow
heavy	$x_v \le 0.25$	$x_v \le 0.4$	, , , , +DZ	* * * * +SN
moderate	$0.25 < x_v \le 0.5$	$0.4 < x_v \le 0.8$	, , , DZ	* * * SN
light	$x_v > 0.5$	$x_v > 0.8$	, , –DZ	* * –SN

Rainfall Rates

Sometimes **rainfall rate** (i.e., **precipitation intensity**) is classified as **light** (or **slight**), **moderate**, or **heavy**. Different countries set different thresholds for these rainfall categories. Table 7-3 shows rainfall intensity criteria used in the USA to determine weather symbols in station plots on surface weather maps. A **trace** amount of precipitation is a very small amount of rain that might wet the ground, but is too small to be detected in a rain gauge (i.e., < 0.1 mm). See the Thunderstorm chapters for more information about very heavy rain and downpours from thunderstorms.

Drizzle is precipitation of very small drops (diameter < 0.5 mm) that are closely spaced and uniform. Although precipitation rates (mm h^{-1}) from drizzle are usually very small, drizzle can reduce visibility (Table 7-4).

World-record rainfall total accumulated depth $d_{max.rain}$ in a rain gauge over any storm period P_{rain} is approximately contained under the following envelope:

$$d_{max.rain} = a \cdot P_{rain}^{1/2} \tag{7.34}$$

where $a = 363\ \text{mm}_{rain} \cdot \text{h}^{-1/2}$. For example, on 25 May 1920 Fussen, Germany received 126 mm in 8 minutes. On 18 July 1942 Smethport, PA, USA received 780 mm in 6 hours. On 15 March 1952, Cilaos, La Re Union Island received 1,830 mm in 1 day. Cherrapunji, India received 2,493 mm in 48 h on 15-16 Jun 1995, and 9,300 mm for the month of July 1861, and received 26,470 mm in the year ending 31 July 1861.

Snowfall Rates & Snow Accumulation

Snowfall rates in the US are classified using the same visibility criterion as drizzle (Table 7-4). Low visibility due to heavy snowfall or blowing loose snow, when accompanied by strong winds, is classified as a **blizzard** if it persists for 3 to 4 hours (Table 7-5). If you are outside in a blizzard, you could easily get disoriented and not be able to return to a shelter, because of **white-out** conditions, where the snow makes the ground and sky look uniformly white so you cannot discern any features or landmarks.

When it is snowing, the precipitation rate is usually measured as **liquid-water equivalent** in units of mm h^{-1}. Namely, it is the precipitation rate after all precipitation is melted. Heated rain gauges accomplish this.

You can also estimate the snowfall rate by periodically measuring the depth of snow on the ground using a meter stick or other metric. However, melting of the snow on warm ground, and compression of snow by the weight of snow above cause large errors in these estimates. Thus, liquid-water equivalent is used instead as a more accurate measure.

New fallen snow has a density roughly 10% of that of liquid water. Thus, a rough first guess is that new-fallen snow depth is about 10 times the liquid-water equivalent depth.

Actual snow densities vary widely, as listed in Table 7-6. The density of freshly falling dry snow is very small because of the air between branches of each ice crystal, and because of air trapped between ice crystals as they accumulate on the ground. After snow has fallen, **metamorphosis** takes place where the tips of the crystals evaporate and redeposit near the crystal centers. Such snow gradually changes into **snow grains** (similar to sugar grains), and becomes more compact and dense. Snow can be further modified by partial melting and refreezing (on a diurnal cycle, and also on an annual cycle for glacier snow). The weight of additional snow on top can further compact deeper older snow.

Piste is the name for a ski run where the snow has been compacted by grooming machines. Density and strength can be increased by mechanically chopping and compacting the snow, by adding liquid water (that later freezes), and by adding chemicals such as nitrate fertilizers or urea.

Table 7-5. Blizzard criteria.

Weather Condition	Criteria (all must be met)	
	USA	Canada
visibility	≤ 0.25 mile	≤ 1 km
wind speed	≥ 35 mi h^{-1}	≥ 40 km h^{-1}
duration	≥ 3 h	≥ 4 h

Table 7-6. Snow density.

Density (kg m^{-3})	Characteristics
50 - 100	Fresh falling snow.
100 - 200	New top snow. Uncompacted. Called "**powder**" by skiers.
200 - 300	Settled snow on ground. Self-compacted after several days.
300 - 500	Compacted snow by grooming machines. Some target densities (kg m^{-3}) for groomed ski slopes are: 450 for cross-country (**nordic**) tracks, 530-550 for snowboard and downhill (**alpine**) runs, and 585 - 620 for slalom. Also forms naturally in deep layers of snow, such as during glacier formation.
500 - 550	Called "**névé**". Snow that has been partially melted, refrozen, & compacted.
550 - 830	Called "**firn**". Naturally compacted and aged over 1 year. A form of ice still containing air channels, observed during glacier formation.
830 - 917	Ice with bubbles, typical in the top 1000 m of old glaciers.
917	Solid ice (no bubbles). Typical of glacier ice below 1000 m depth.

Precipitation Distribution

Combining rain-gauge data over land with satellite observations over oceans gives the annual precipitation distribution shown in Fig. 7.24. The heaviest rain is in the tropics, where the warm sea surface causes copious amounts of evaporation (Fig. 7.25), where the warm air can hold a large amount of precipitable water, and where the general circulation contains updrafts. Rain is suppressed at 30° north and south due to extensive regions of downdraft in the Hadley-cell circulation. This circulation is discussed in the General Circulation chapter.

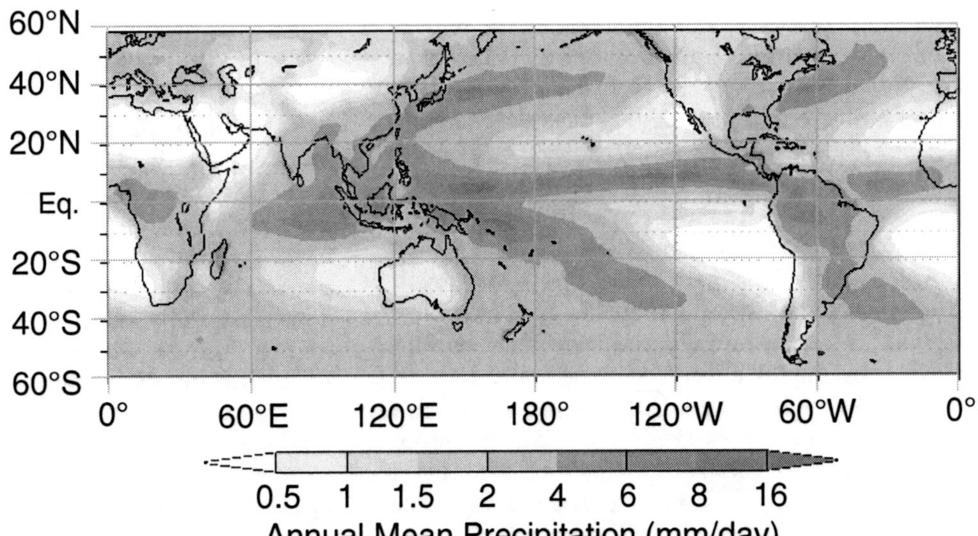

Figure 7.24
Annual mean precipitation during years 1979 to 2000.
[Adapted from image courtesy of US Climate Prediction Center (NOAA/ National Weather Service, National Centers for Environmental Prediction). http://www.cpc.ncep.noaa.gov/products/precip/CWlink/wayne/annual.precip.html]

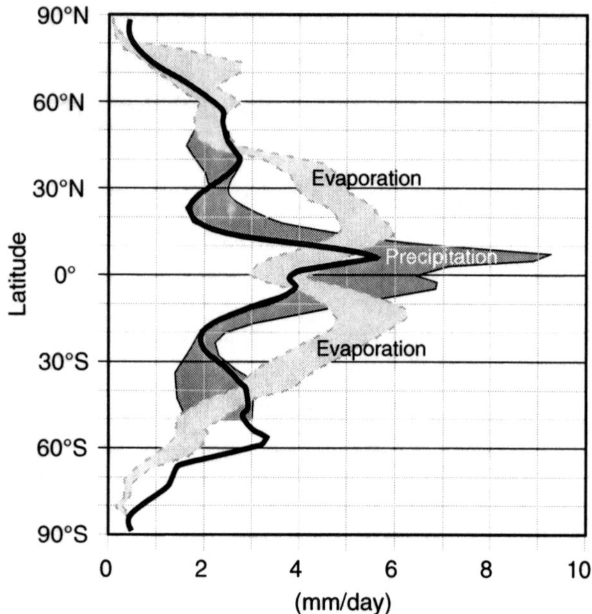

Figure 7.25
Zonally averaged rates of evaporation (light grey with dotted outline) and precipitation (dark grey with solid black outline). Estimated from a combination of satellite observations, numerical simulations, and surface observations. Spread of the grey areas indicates uncertainty (disagreement among the different methods). Data courtesy of NASA GSFC and the Global Precipitation Climatology Project (plotted as the thick black line).

PRECIPITATION MEASUREMENT

The simplest precipitation instrument is a **rain gauge**, which is a cylindrical bucket into which the rain falls. By using a measuring stick to manually read the water depth in the bucket at successive times such as every hour, you can determine rainfall rate. For greater sensitivity, a funnel can be placed over the bucket to collect rain faster, but the depth of water in the bucket must be reduced by the ratio of the horizontal cross-section areas of the bucket to the funnel opening. To get the **liquid-water equivalent** of the snowfall, some rain gauges are heated to melt snow, and others are painted black to passively melt snow by absorbing sunlight. Some gauges are surrounded by a segmented **wind shield** to reduce errors due to blowing precipitation.

Automated rain gauges exist. **Weighing rain gauges** weigh the rain-filled bucket over successive intervals, inferring rain accumulation by weight increase, knowing the density of liquid water.

Tipping-bucket rain gauges direct the captured rain into a tiny bucket on one side of a lever. When the bucket gets full, it tips the lever like a seesaw (teeter-totter), emptying that bucket while simultaneously moving under the funnel an empty bucket from the other end of the lever. Each tip can be counted digitally, and the frequency of tips during an hour gives the rainfall rate.

An **evaporative rain gauge** has two metal plates, one above the other, each oriented horizontally (one facing up, and the other facing down). Each plate is heated electrically to maintain the same specified temperature warmer than ambient air. Precipitation falling on the hot top plate evaporates quickly, thereby removing heat from that plate. By measuring the amount of extra electricity needed to keep the top plate at the same temperature as the bottom, and knowing the latent heat of vaporization, the rainfall rate can be determined.

Attenuation rain gauges have a light beam that shines horizontally across an open air path exposed to precipitation. The attenuation of the light beam is related to precipitation intensity, but errors can be due to air pollution, fog, and different absorption cross-sections of liquid vs. solid precipitation.

A **disdrometer** measures size distribution of rain drops via the momentum imparted to a horizontal plate by each falling drop. Another method is a particle imager that sends light from an array of light-emitting diodes to an array of tiny photodetectors. Each hydrometeor casts a shadow that can be detected, where the size and the shape of the shadows are used together to estimate precipitation rate, hydrometeor type and size. A **Knollenberg probe** uses this imaging method, and can be mounted on aircraft flying through clouds and precipitation.

A **liquid-water content (LWC) probe** consists of an electrically heated wire. When mounted on an aircraft flying through a cloud, the rain and cloud droplets evaporate upon hitting the hot wire. By measuring the electrical power needed to maintain a constant wire temperature against the evaporative cooling, the LWC can be inferred.

Snow amount on the ground can be measured by placing a liquid antifreeze-filled thin-skin metal **snow pillow** on the ground before the winter snow season. As snow accumulates during the season, the weight of the snow squeezes the pillow and displaces some of the fluid. Pressure sensors measure the weight of the displaced fluid to infer snow weight.

Downward-pointing **ultrasonic snow-depth sensors** mounted on a tall pole measure the travel time for an emitted pulse of sound to reach the ground and echo back to the sensor on the pole. This gives the distance between the top of the snow and the sensor, which can be subtracted from the sensor height above bare ground to give the snow depth. Similar sensors use travel time for IR or visible light pulses.

Remote sensors (see the Satellites & Radar chapter) can also be used to measure rain rate or accumulation. Ground-based **weather radar** actively emits microwaves, and can estimate rainfall rate from the echo intensity and polarization characteristics of the microwave signal that is scattered back to the radar from the precipitation particles. **Passive microwave sensors** on some weather satellites can measure the **brightness temperature** of the minute amounts of microwaves emitted from the Earth's surface and atmosphere. With this info one can infer the atmospheric total water content in a column of the atmosphere (used to estimate **tropical rainfall** over the oceans), and can infer snow depth on the ground (over high-latitude regions).

REVIEW

Cloud droplets that form on cloud condensation nuclei (CCN) overcome a formation barrier caused by the surface tension of the curved surface. However, because there are so many CCN between which the available water is partitioned, the result is a large number of very small droplets. These drops grow slowly by diffusion, and develop a monodisperse droplet-size distribution. Such a distribution reduces droplet collisions, and does not favor droplet growth into precipitation hydrometeors. Hence, we get pretty clouds, but no rain.

Warm-cloud ($T > 0°C$) rain can happen in the tropics, particularly over oceans where there are fewer CCN allow formation of a smaller number of larger drops. Several other processes can cause the droplet sizes to have more diversity, resulting in different terminal velocities for different drops. This encourages collision and coalescence to merge smaller droplets into ones that are large enough to precipitate out.

In clouds colder than 0°C, ice nuclei trigger ice crystals to grow. Ice crystals can exist in the air along with supercooled liquid drops. Because of the difference between liquid and ice saturation humidities, the ice can grow at the expense of evaporating liquid droplets. If the ratio of water to ice hydrometeors is about a million to one, then most of the water will be transferred to ice crystals, which are then heavy enough to fall as precipitation.

As larger ice particles fall and hit smaller supercooled liquid droplets, the droplets can freeze as rime onto the ice crystals, causing the hydrometeors to grow even faster. This process can create graupel and hail. Also, ice crystals can aggregate (collide and stick together) to make larger clumps of snow. Most rain at midlatitudes results from melted snow that form from this "cold cloud" process.

HOMEWORK EXERCISES

Broaden Knowledge & Comprehension

B1. Search the web for any journal articles, conference papers, or other technical reports that have pictures of droplet or ice-crystal growth or fall processes (such as photos taken in vertical wind tunnels).

B2. Can you find any satellite photos on the web showing haze? If so, which satellites and which channels on those satellites show haze the best? Do you think that satellites could be used to monitor air pollution in urban areas?

B3. Search the web for microphotographs of ice crystals and snow flakes with different habits. What habits in those photos were not given in the idealized Fig. 7.12?

B4. Find on the web a clear microphotograph of a dendrite snow flake. Print it out, and determine the fractal dimension of the snow flake. (Hint, see the Clouds chapter for a discussion of fractals.)

B5. Search various government air-pollution web sites (such as the U.S. Environmental Protection Agency: http://www.epa.gov/) for sizes of aerosol pollutants. Based on typical concentrations (or on concentrations specified in the air-quality standards) of these pollutants, determine the number density, and compare with Fig. 7.5.

B6. Search the web for information about surface tension, and how it relates to **Gibbs free energy**.

B7. Search the web for information about snow and snowflakes.

B8. Search the web for climate statistics of actual annual precipitation last year worldwide (or for your country or region), and compare with Fig. 7.24.

B9. Search the web for photos and diagrams of precipitation measurement instruments, and discuss their operation principles.

Apply

A1. Using Fig. 7.1, how many of the following droplets are needed to fill a large rain drop.

a. small cloud droplet
b. typical cloud droplet
c. large cloud droplet
d. drizzle droplet

e. small rain droplet
f. typical rain droplet

A2. Find the supersaturation fraction and supersaturation percentage, given relative humidities (%) of: a. 100.1 b. 100.2 c. 100.4 d. 100.5 e. 100.7 f. 101 g. 101.2 h. 101.8 i. 102 j. 102.5 k. 103.3 l. 104.0 m. 105 n. 107 o. 110

A3. For air at T = –12°C, find the supersaturation fraction, given a vapor pressure (kPa) of:
a. 0.25 b. 0.26 c. 0.28 d. 0.3 e. 0.31 f. 0.38 g. 0.5 h. 0.7 i. 0.9 j. 1.1 k. 1.2 l. 1.4 m. 1.6 n. 1.8 o. 1.9
Hint. Get saturation vapor pressure from the Water Vapor chapter.

A4. For air at P = 80 kPa and T = –6°C, find the supersaturation percentage, given a mixing ratio (g kg^{-1}) of: a. 6 b. 5.8 c. 5.6 d. 5.4 e. 5.2 f. 5.1 g. 5 h. 4.9 i. 4.7 j. 4.5 k. 4.3 l. 4.1 m. 3.8 n. 3.6
Hint. Get saturation mixing ratio from a thermo diagram (at the end of the Atmospheric Stability chapter).

A5. For the previous problem, assume the given mixing ratios represent total water mixing ratio. Find the excess water mixing ratio (g kg^{-1}).

A6. For an air parcel with excess water mixing ratio of 10 g kg^{-1} at a geopotential height of 5 km above mean sea level, find the average radius (µm) of the hydrometeor assuming growth by condensation only, given a hydrometeor number density (# m^{-3}) of: a. 1x10^8 b. 2x10^8 c. 3x10^8 d. 4x10^8 e. 5x10^8 f. 6x10^8 g. 7x10^8 h. 8x10^8 i. 9x10^8 j. 1x10^9 k. 2x10^9 l. 3x10^9 m. 4x10^9 n. 5x10^9 o. 8x10^9

A7. If c = 5x10^6 µm^3·m^{-3}, use the Junge distribution to estimate the number density of CCN (# m^{-3}) within a ΔR = 0.2 µm range centered at R (µm) of:
a. 0.2 b. 0.3 c. 0.4 d. 0.5 e. 0.6 f. 0.8 g. 1.0 h. 2 i. 3 j. 4 k. 5 l. 6 m. 8 n. 10

A8. For pure water at temperature –20°C, use Kelvin's equation to find the equilibrium RH% in air over a spherical droplet of radius (µm):
a. 0.005 b. 0.006 c. 0.008 d. 0.01 e. 0.02 f. 0.03 g. 0.04 h. 0.05 i. 0.06 j. 0.08 k. 0.09 l. 0.1 m. 0.2 n. 0.3 o. 0.5

A9 (§). Produce Köhler curves such as in Fig. 7.7b, but only for salt of the following masses (g) at 0°C:
a. 10^{-18} b. 10^{-17} c. 10^{-16} d. 10^{-15} e. 10^{-14} f. 10^{-13} g. 10^{-12} h. 10^{-11} i. 5x10^{-18} j. 5x10^{-17} k. 5x10^{-16} l. 5x10^{-15} m. 5x10^{-14} n. 5x10^{-13}

A10(§). Produce Köhler curves for a solute mass of 10^{-16} g of salt for the following temperatures (°C):
a. –35 b. –30 c. –25 d. –20 e. –15 f. –10 g. –5 h. 2 i. 5 j. 10 k. 15 l. 20 m. 25 n. 30

A11. Find the critical radii (µm) and supersaturations at a temperature of –10°C, for
a. 5x10^{-17} g of hydrogen peroxide
b. 5x10^{-17} g of sulfuric acid
c. 5x10^{-17} g of nitric acid
d. 5x10^{-17} g of ammonium sulfate
e. 5x10^{-16} g of hydrogen peroxide
f. 5x10^{-16} g of sulfuric acid
g. 5x10^{-16} g of nitric acid
h. 5x10^{-16} g of ammonium sulfate
j. 5x10^{-15} g of hydrogen peroxide
k. 5x10^{-15} g of sulfuric acid
l. 5x10^{-15} g of nitric acid
m. 5x10^{-15} g of ammonium sulfate
n. 5x10^{-14} g of hydrogen peroxide
o. 5x10^{-14} g of sulfuric acid
p. 5x10^{-14} g of nitric acid
q. 5x10^{-14} g of ammonium sulfate

A12. For the nuclei of the previous exercise, find the equilibrium haze droplet radius (µm) for the following relative humidities (%):
(i) 70 (ii) 72 (iii) 74 (iv) 76 (v) 78 (vi) 80 (vii) 82 (viii) 84 (ix) 86 (x) 88 (xi) 90 (xii) 92 (xiii) 94 (xiv) 96 (xv) 98

A13. How many CCN will be activated in maritime air at supersaturations (%) of
a. 0.2 b. 0.3 c. 0.4 d. 0.5 e. 0.6 f. 0.8 g. 1.0 h. 2 i. 3 j. 4 k. 5 l. 6 m. 8 n. 10

A14. Find the average separation distances (µm) between cloud droplets for the previous problem.

A15. What temperature is needed to immersion-freeze half the droplets of radius (µm)
a. 10 b. 20 c. 30 d. 40 e. 50 f. 60 g. 70 h. 80 i. 90 j. 125 k. 150 l. 200 m. 200 n. 300

A16. Estimate the number density of active ice nuclei for the following combinations of temperature and supersaturation [T(°C), S_{ice}(%)].
a. –5, 3 b. –5, 5 c. –10, 5 d. –10, 10 e. –15, 10 f. –10, 12 g. –15, 12 h. –20, 13 i. –15, 18 j. –20, 20 k. –25, 20 l. –20, 23 m. –25, 23 n. –23, 25

A17. For a supersaturation gradient of 1% per 2 µm, find the kinematic moisture flux due to diffusion. Given $T = -20°C$, and $P = 80$ kPa. Use D ($m^2 \cdot s^{-1}$) of:
a. $1x10^{-6}$ b. $2x10^{-6}$ c. $3x10^{-6}$ d. $4x10^{-6}$
e. $5x10^{-6}$ f. $6x10^{-6}$ g. $7x10^{-6}$ h. $8x10^{-6}$
i. $9x10^{-6}$ j. $1x10^{-5}$ k. $3x10^{-5}$ l. $4x10^{-5}$
m. $5x10^{-5}$ n. $6x10^{-5}$ o. $7x10^{-5}$ p. $8x10^{-5}$

A18 (§). Compute and plot supersaturation (%) vs. distance (µm) away from drops of the following radii , given a background supersaturation of 0.5%
a. 0.15 µm containing 10^{-16} g of salt
b. 0.15 µm containing 10^{-16} g of ammon.sulfate
c. 0.15 µm containing 10^{-16} g of sulfuric acid
d. 0.15 µm containing 10^{-16} g of nitric acid
e. 0.15 µm containing 10^{-16} g of hydrogen perox.
f. 0.3 µm containing 10^{-16} g of salt
g. 0.3 µm containing 10^{-16} g of ammon.sulfate
h. 0.3 µm containing 10^{-16} g of sulfuric acid
i. 0.3 µm containing 10^{-16} g of nitric acid
j. 0.3 µm containing 10^{-16} g of hydrogen perox.
k. 0.5 µm containing 10^{-16} g of salt
l. 0.5 µm containing 10^{-16} g of ammon.sulfate
m. 0.5 µm containing 10^{-16} g of sulfuric acid
n. 0.5 µm containing 10^{-16} g of nitric acid
o. 0.5 µm containing 10^{-16} g of hydrogen perox.
p. 2 µm containing 10^{-16} g of salt
q. 1 µm containing 10^{-16} g of ammon.sulfate
r. 1 µm containing 10^{-16} g of sulfuric acid
s. 1 µm containing 10^{-16} g of nitric acid
t. 1 µm containing 10^{-16} g of hydrogen perox.

A19. Find the diffusivity ($m^2 \cdot s^{-1}$) for water vapor, given [P(kPa) , T(°C)] of:
a. 80, 0 b. 80, –5 c. 80, –10 d. 80, –20
e. 70, 0 f. 70, –5 g. 70, –10 h. 70, –20
i. 60, 0 j. 60, –5 k. 60, –10 l. 60, –20
m. 50, 0 n. 50, –5 o. 50, –10 p. 50, –20

A20 (§). For the previous exercise, plot droplet radius (µm) vs. time (minutes) for diffusive growth.

A21. What phase (I - XIV) of ice is expected at the following locations in a standard atmosphere?
a. Earth's surface b. mid-troposphere
c. tropopause d. mid stratosphere
e. stratopause f. mid-mesosphere
g. mesopause

A22. What phase (I - XIV) of ice is expected for the following conditions of [P(kPa) , T(°C)]?
a. 1, –250 b. 1, –150 c. 1, –50 d. 1, 50
e. 10^3, –250 f. 10^3, –150 g. 10^3, –50 h. 10^3, 50
i. $5x10^5$, –250 j. $5x10^5$, –150 k. $5x10^5$, –30
l. $5x10^5$, 50 m. 10^7, –250 n. 10^7, –50 o. 10^7, 50

A23. What crystal habit could be expected for the following combinations of [ρ_{ve} ($g\ m^{-3}$) , T(°C)]:
a. 0.22, –25 b. 0.22, –20 c. 0.22, –13
d. 0.22, –8 e. 0.22, –5 f. 0.22, –2
g. 0.12, –25 h. 0.12, –20 i. 0.12, –13
j. 0.12, –8 k. 0.12, –5 l. 0.12, –2
m. 0.08, –25 n. 0.08, –20 o. 0.08, –13
p. 0.08, –8 q. 0.08, –5 r. 0.08, –2

A24. Suppose the following ice crystals were to increase mass at the same rate. Find the rate of increase with time of the requested dimension.
a. effective radius of column growing in 3-D
b. diameter of plate growing in 2-D
c. length of needle growing in 1-D

A25 (§). Given $D = 2x10^{-5}\ m^2 \cdot s^{-1}$ and $\rho_v = 0.003$ $kg \cdot m^{-3}$, plot ice-crystal mass (g) vs. time (minutes) for 3-D growth such as a hexagonal column. Use the following supersaturation fraction:
a. 0.001 b. 0.002 c. 0.003 d. 0.004 e. 0.005
f. 0.006 g. 0.007 h. 0.008 i. 0.009 j. 0.01
k. 0.012 l. 0.014 m. 0.016 n. 0.018 o. 0.020

A26 (§). Same as the previous problem, but for 2-D growth of a thin flat plate of thickness 10 µm.

A27 (§). Use the Clausius-Clapeyon equation from the Water Vapor chapter to calculate the saturation vapor pressure over liquid water and ice for $-50 \le T \le 0°C$, and use that data to calculate and plot the difference. Namely, reproduce Fig. 7.16 with your own calculations.

A28. Find the terminal velocity of cloud droplets of radius (µm): a. 0.2 b. 0.4 c. 0.6 d. 0.8
e. 1.0 f. 2 g. 3 h. 4 i. 5 j. 7
k. 10 l. 15 m. 20 n. 30 o. 40 p. 50

A29. Find the terminal velocity of rain drops of radius (µm): a. 100 b. 150 c. 200 d. 300
e. 400 f. 500 g. 600 h. 700 i. 800
j. 900 k. 1000 l. 1200 m. 1500 n. 2000

A30. Calculate the terminal velocity of hailstones of radius (cm): a. 0.25 b. 0.5 c. 0.75 d. 1 e. 1.25
f. 1.5 g. 1.75 h. 2 i. 2.5 j. 3 k. 3.5
l. 4 m. 4.5 n. 5 o. 5.5 p. 6

A31. What type of "meteor" is:
a. a rainbow b. lightning c. corona
d. dust e. a cloud f. a halo g. sand
h. rain i. smoke j. snow k. fog
l. haze m. dew n. frost

A32. For a Marshall-Palmer rain-drop size distribution, if the rainfall rate is
(i) 10 mm h^{-1}, or (ii) 20 mm h^{-1},
how many droplets are expected of radius (μm) greater than:
a. 100 b. 200 c. 300 d. 400 e. 500
f. 700 g. 1000 h. 1200 i. 1500 j. 2000

A33. What is the rain intensity classification and the weather map symbol for rainfall rates (mm h^{-1}) of:
a. 0.02 b. 0.05 c. 0.1 d. 0.2 e. 0.5
f. 1.0 g. 2 h. 3 i. 4 j. 5
k. 6 l. 7 m. 8 n. 9 o. 10

A34. For precipitation in the form of
(i) drizzle, or (ii) snow,
what is the precipitation intensity classification and weather map symbol for visibility (km) of:
a. 0.1 b. 0.2 c. 0.3 d. 0.4 e. 0.5 f. 0.6 g. 0.7
h. 0.8 i. 0.9 j. 1.0 k. 1.2 l. 1.5 m. 2 n. 5

A35. For a liquid-water equivalent precipitation value of 5 cm, find the snow depth if the snow density (kg m^{-3}) is: a. 50 b. 75 c. 100 d. 150 e. 200
f. 250 g. 300 h. 350 i. 400 j. 450 k. 500
l. 550 m. 600 n. 650 o. 700 p. 800 q. 900

A36. Find the mean annual precipitation for the following locations, given their longitudes, latitudes:
a. 120°W, 50°N b. 120°W, 25°N c. 120°W, 10°N
d. 120°W, 10°S e. 120°W, 30°S f. 60°W, 0°N
g. 60°W, 20°N h. 60°W, 40°N i. 0°W, 50°N
j. 0°W, 25°N k. 0°W, 5°N l. 0°W, 20°S
m. 120°E, 25°S n. 120°E, 0°N o. 120°E, 30°N

A37. What are the values of zonally averaged evaporation and precipitation rates at latitude:
a. 70°N b. 60°N c. 50°C d. 40°N e. 30°N
f. 20°N g. 10°N h. 0° i. 10°S j. 20°S
k. 30°S m. 40°S n. 50°S o. 60°S p. 70°S

Evaluate & Analyze

E1. If saturation is the maximum amount of water vapor that can be held by air at equilibrium, how is supersaturation possible?

E2. Fig. 7.2 shows how excess-water mixing ratio can increase as a cloudy air parcel rises. Can excess-water mixing ratio increase with time in a cloudy or foggy air parcel that doesn't rise? Explain.

E3. What can cause the supersaturation S in a cloud to be less than the available supersaturation S_A?

E4. Fig. 7.3 applies to cumulus clouds surrounded by clear air. Would the curves be different for a uniform stratus layer? Why?

E5. An air parcel contains CCN that allow 10^9 m^{-3} hydrometeors to form. If the air parcel starts at P = 100 kPa with T = 20°C and T_d = 14°C, find the average radius of cloud droplets due to condensation only, after the air parcel rises to P (kPa) of:
a. 85 b. 82 c. 80 d. 78 e. 75 f. 72 g. 70
h. 67 i. 63 j. 60 k. 58 l. 54 m. 50 n. 45
Hint. Use a thermo diagram to estimate r_E.

E6. Derive eq. (7.8), stating and justifying all assumptions. (Hint: consider the volume of a spherical drop.)

E7. Rewrite eq. (7.8) for hydrometeors that form as cubes (instead of spheres as was used in eq. 7.8). Instead of solving for radius R, solve for the width s of a side of a cube.

E8. In Fig. 7.5, consider the solid curve. The number density (# cm^{-3}) of CCN between any two radii R_1 and R_2 is equal to the average value of $n/\Delta R$ within that size interval times ΔR (=$R_2 - R_1$). This works best when R_1 and R_2 are relatively close to each other. For larger differences between R_1 and R_2, just sum over a number of smaller intervals. Find the number density of CCN for droplets of the following ranges of radii (μm):
a. $0.02 < R < 0.03$ b. $0.03 < R < 0.04$
c. $0.04 < R < 0.05$ d. $0.05 < R < 0.06$
e. $0.06 < R < 0.08$ f. $0.08 < R < 0.1$
g. $0.2 < R < 0.3$ h. $0.3 < R < 0.4$
i. $0.4 < R < 0.5$ j. $0.5 < R < 0.6$
k. $0.6 < R < 0.8$ l. $0.8 < R < 1$
m. $0.1 < R < 0.2$ n. $1 < R < 2$
o. $0.1 < R < 1$ p. $0.1 < R < 10$
q. $0.02 < R < 0.1$ r. $0.02 < R < 10$

E9. a. Find a relationship between the number density of CCN particles n, and the corresponding mass concentration c (μg·m^{-3}), using the dashed line in Fig. 7.5 and assuming that the molecular weight is M_s.

b. Use Table 7-1 to determine the molecular weight of sulfuric acid (H_2SO_4) and nitric acid (HNO_3), which are two contributors to acid rain. Assuming tiny droplets of these acids are the particles of interest for Fig. 7.5, find the corresponding values or equations for mass concentration of these air pollutants.

E10. a. From the solid curve in Fig. 7.5, find the number density of CCN at R = 1 μm, and use this result to determine the value and units of parameter c in the CCN number density eq. (7.9)

b Plot the curve resulting from this calibrated eq. (7.9) on a linear graph, and on a **log-log graph**.

c. Repeat (a) and (b), but for $R = 0.1$ µm.

d. Repeat (a) and (b), but for $R = 0.01$ µm.

e. Why does eq. (7.9) appear as a straight line in Fig. 7.5? What is the slope of the straight line in Fig. 7.5? [Hint: for log-log graphs, the **slope** is the number of decades along the vertical axis (ordinate) spanned by the line, divided by the number of decades along the horizontal axis (abscissa) spanned by the line. **Decade** means a factor of ten; namely, the interval between major tic marks on a logarithmic axis.]

f. For number densities less than 1 cm^{-3}, what does it mean to have less than one particle (but greater than zero particles)? What would be a better way to quantify such a number density?

g. As discussed in the Atmospheric Optics chapter, air molecules range in diameter between roughly 0.0001 and 0.001 mm. If the dashed line in Fig. 7.5 were extended to that small of size, would the number density indicated by the Junge distribution agree with the actual number density of air molecules? If they are different, discuss why.

E11. Show how the Köhler equation reduces to the Kelvin equation for CCNs that don't dissolve.

E12 (§). Surface tension σ for a cloud droplet of pure water squeezes the droplet, causing the pressure inside the droplet to increase. The resulting pressure difference ΔP between inside and outside the droplet can be found from the **Young-Laplace equation** ($\Delta P = \sigma \cdot dA/dV$, where A is surface area and V is volume of the droplet). For a spherical droplet of radius R this yields:

$$\Delta P = 2 \cdot \sigma \,/\, R$$

Plot this pressure difference (kPa) vs. droplet radius over the range $0.005 \le R \le 1$ µm.

E13. a. Using the equation from the previous exercise, combine it with Kelvin's equation to show how the equilibrium relative humidity depends on the pressure excess inside the droplet.

b. If greater pressure inside the drop drives a greater evaporation rate, describe why RH greater than 100% is needed to reach an equilibrium where condensation from the air balances evaporation from the droplet.

E14. Using the full Köhler equation, discuss how supersaturation varies with:

a. temperature.

b. molecular weight of the nucleus chemical.

c. mass of solute in the incipient droplet.

E15. Consider a Köhler curve such as plotted in the INFO box on Droplet Growth. But let the RH = 100.5% for air. Starting with an aerosol with characteristics of point A on that figure, (a) discuss the evolution of drop size, and (b) explain why haze particles are not possible for that situation.

E16. Considering Fig. 7.7, which type of CCN chemical would allow easier formation of cloud droplets: (a) a CCN with larger critical radius but lower peak supersaturation in the Köhler curve; or (b) a CCN with smaller critical radius but higher supersaturation? Explain.

E17. What is so special about the critical radius, that droplets larger than this radius continue to grow, while smaller droplets remain at a constant radius?

E18. The Kelvin curve (i.e., the Köhler curve for pure water) has no critical radius. Hence, there is no barrier droplets must get across before they can grow from haze to cloud droplets. Yet cloud droplets are easier to create with heterogeneous nucleation on solute CCN than with homogeneous nucleation in clean air. Explain this apparent paradox.

E19. In air parcels that are rising toward their LCL, aerosol swelling increases and visibility decreases as the parcels get closer to their LCL. If haze particles are at their equilibrium radius by definition, why could they be growing in the rising air parcel? Explain.

E20. How high above the LCL must air be lifted to cause sufficient supersaturation to activate CCN for liquid-droplet nucleation?

E21. Suppose a droplet contained all the substances listed in Table 7-2. What is the warmest temperature (°C) that the droplet will freeze due to:

a. contact freezing b. condensation freezing

c. deposition freezing d. immersion freezing

E22. Discuss the differences in nucleation between liquid water droplets and ice crystals, assuming the air temperature is below freezing.

E23. Discuss the differences in abundance of cloud vs. ice nuclei, and how this difference varies with atmospheric conditions.

E24. Can some chemicals serve as both water and ice nuclei? For these chemicals, describe how cloud particles would form and grow in a rising air parcel.

E25. If droplets grow by diffusion to a final average radius given by eq. (7.8), why do we even care about the diffusion rate?

E26. In the INFO box on Cubic Ice, a phase diagram was presented with many different phases of ice. The different phases have different natural crystal shapes, as summarized in the table below. In the right column of this table, draw a sketch of each of the shapes listed. (Hint, look in a geometry book or on the internet for sketches of geometric shapes.)

Table 7-7. Crystal shapes for ice phases.

Shape	Phases	Sketch
hexagonal	Ih	
cubic	Ic, VII, X	
tetragonal	III, VI, VIII, IX, XII	
rhombic	II, IV	
ortho-rhombic	XI	
monoclinic	V	

E27. The droplet growth-rate equation (7.24) considers only the situation of constant background supersaturation. However, as the droplet grows, water vapor would be lost from the air causing supersaturation to decrease with time. Modify eq. (7.24) to include such an effect, assuming that the temperature of the air containing the droplets remains constant. Does the droplet still grow with the square root of time?

E28. If ice particles grew as spheres, which growth rate equation would best describe it? Why?

E29. Verify that the units on the right sides of eqs. (7.26) and (7.27) match the units of the left side.

E30. Manipulate the mass growth rate eq. (7.26) to show that the effective radius of the ice particle grows as the square root of time. Show your work.

E31. Considering the different mass growth rates of different ice-crystal shapes, which shape would grow fastest in length of its longest axis?

E32. If the atmosphere were to contain absolutely no CCN, discuss how clouds and rain would form, if at all.

E33. In Fig. 7.15 the thick black line follows the state of a rising air parcel that is cooling with time. Why does that curve show e_s decreasing with time?

E34. If both supercooled liquid droplets and ice crystals were present in sinking cloudy air that is warming adiabatically, describe the evolutions of both types of hydrometeors relative to each other.

E35. On one graph similar to Fig. 7.18, plot 3 curves for terminal velocity. One for cloud droplets (Stokes Law), another for rain drops, and a third for hail. Compare and discuss.

E36. Can ice crystals still accrete smaller liquid water droplets at temperatures greater than freezing? Discuss.

E37. Large ice particles can accumulate smaller supercooled liquid water drops via the aggregation process known as accretion or riming. Can large supercooled liquid water drops accumulate smaller ice crystals? Discuss.

E38. Is it possible for snow to reach the ground when the atmospheric-boundary-layer temperature is warmer than freezing? Discuss.

E39. Compare CCN size spectra with raindrop size spectra, and discuss.

E40. For warm-cloud precipitation, suppose that all 5 formation factors are working simultaneously. Discuss how the precipitation drop size distribution will evolve with time, and how it can create precipitation.

E41. Is it possible to have blizzard conditions even with zero precipitation rate? Discuss.

E42. Precipitation falling out of a column of atmosphere implies that there was net latent heating in that column. Use the annual mean precipitation of Fig. 7.24 to discuss regions of the world having the greatest latent heating of the atmosphere.

E43. According to Fig. 7.25, some latitudes have an imbalance between evaporation and precipitation. How can that be maintained?

E44. Suppose that a disdrometer gives you information on the size of each hydrometeor that falls, and how many of each size hydrometer falls per hour. Derive an equation to relate this information to total rainfall rate.

Synthesize

S1. What if the saturation vapor pressure over supercooled water and ice were equal.

a. Discuss the formation of clouds and precipitation.

b. Contrast with those processes in the real atmosphere.

c. Discuss how the weather and climate might change, if at all.

S2. In Fig. 7.15, suppose that the saturation vapor pressure over ice were greater, not less, than that over water. How would the WBF process change, if at all? How would precipitation and clouds change, if at all?

S3. What if you were hired to seed warm ($T > 0°C$) clouds (i.e., to add nuclei), in order to create or enhance precipitation. Which would work better: (a) seeding with 10^5 salt particles cm^{-3}, each with identical of radius 0.1 µm; or (b) seeding with 1 salt particle cm^{-3}, each with identical radius of 0.5 µm; or (c) seeding with a range of salt particles sizes? Discuss, and justify.

S4. What if you were hired to seed cold ($T < 0°C$) clouds (i.e., to add nuclei), in order to create or enhance precipitation. Would seeding with water or ice nuclei lead to the most precipitation forming most rapidly? Explain and justify.

S5. Is it possible to seed clouds (i.e., add nuclei) in such a way as to reduce or prevent precipitation? Discuss the physics behind such weather modification.

S6. What if all particles in the atmosphere were **hydrophobic** (i.e., repelled water). How would the weather and climate be different, if at all?

S7. What if the concentration of cloud nuclei that could become activated were only one-millionth of what currently exists in the atmosphere. How would the weather and climate change, if at all?

S8. Eq. (7.29) indicates that smaller droplets and aerosol particles fall slower. Does Stoke's law apply to particles as small as air molecules? What other factors do air molecules experience that would affect their motion, in addition to gravity?

S9. What if Stoke's law indicated that smaller particles fall faster than larger particles. Discuss the nature of clouds for this situation, and how Earth's weather and climate might be different.

S10. What if rain droplet size distributions were such that there were more large drops than small drops. Discuss how this could possibly happen, and describe the resulting weather and climate.

S11. Suppose that large rain drops did not break up as they fell. That is, suppose they experienced no drag, and there was no upper limit to rain drop size. How might plant and animal life on Earth have evolved differently? Why?

S12. What if cloud and rain drops of all sizes fell at exactly the same terminal velocity. Discuss how the weather and climate might be different.

S13. What if condensation and deposition absorbed latent heat (i.e., caused cooling) instead of releasing latent heat. How would clouds, precipitation, weather and climate be different, if at all.

S14. Weather modification is as much a social issue as a scientific/technical issue. Consider a situation of **cloud seeding** (adding nuclei) to enhance precipitation over arid farm land in county X. If you wanted to make the most amount of money, would you prefer to be the:

a. meteorologist organizing the operation,
b. farmer employing the meteorologist,
c. company insuring the farmer's crop,
d. company insuring the meteorologist, or
e. lawyer in county Y downwind of county X, suing the meteorologist, farmer, and insurance companies?

Justify your preference.

S15. Suppose that you discovered how to control the weather via a new form of cloud seeding (adding nuclei). Should you ...

a. keep your results secret and never publish or utilize them, thereby remaining impoverished and unknown?

b. publish your results in a scientific journal, thereby achieving great distinction?

c. patent your technique and license it to various companies, thereby achieving great fortune?

d. form your own company to create tailored weather, and market weather to the highest bidders, thereby becoming a respected business leader?

e. modify the weather in a way that you feel is best for the people on this planet, thereby achieving great power?

f. allow a government agency to hold hearings to decide who gets what weather, thereby achieving great fairness and inefficiency?

g. give your discovery to the military in your favorite country, thereby expressing great patriotism? (Note: the military will probably take it anyway, regardless of whether you give it willingly.)

Discuss and justify your position. (Hint: See the "A SCIENTIFIC PERSPECTIVE" box at the end of this chapter before you answer this question.)

A SCIENTIFIC PERSPECTIVE • Consequences

The scenario of exercise S15 is not as far-fetched as it might appear. Before World War II, American physicists received relatively little research funding. During the war, the U.S. Army offered a tremendous amount of grant money and facilities to physicists and engineers willing to help develop the atomic bomb as part of the Manhattan Project.

While the work they did was scientifically stimulating and patriotic, many of these physicists had second thoughts after the bomb was used to kill thousands of people at the end of the war. These concerned scientists formed the "Federation of Atomic Scientists", which was later renamed the "Federation of American Scientists" (FAS).

The FAS worked to discourage the use of nuclear weapons, and later addressed other environmental and climate-change issues. While their activities were certainly worthy, one has to wonder why they did not consider the consequences before building the bomb.

As scientists and engineers, it is wise for us to think about the moral and ethical consequences before starting each research project.

8 SATELLITES & RADAR

Contents

To understand and predict the weather, we first must measure it. **In-situ** or **direct** weather instruments must physically touch, or be exposed to, the air being measured. Examples include **thermometers** (temperature), **barometers** (pressure), **hygrometers** (humidity), **anemometers** (wind speed), **wind vanes** (wind direction), **pyranometers** (solar radiation), and **rain gauges** (precipitation).

Remote sensors infer the weather conditions by detecting the characteristics of waves propagating from distant regions. The waves can be electromagnetic (light, infrared, microwaves, etc.) or sound. **Active** remote instrument systems such as **radar** (RAdio Detection And Ranging) transmit their own waves toward the object and then receive the signal bouncing back to the sensors. **Passive** ones, such as some satellite sensors, receive waves naturally emanating from the object.

Clouds, precipitation, and air molecules can totally or partially **absorb** electromagnetic radiation (Fig. 8.1a), **scatter** it into many directions (Fig. 8.1b), or **reflect** it (Fig. 8.1c). Objects also **emit** radiation (Fig. 8.1d) according to Planck's law. Interactions of radiation with the Earth, air, and clouds create the **signals** that satellites and radar use.

This chapter covers the basics of **weather satellites** and **radar**. Other remote-sensor systems, not covered here, include **lidar** (LIght Detection And Ranging), and **sodar** (SOund Detection And Ranging).

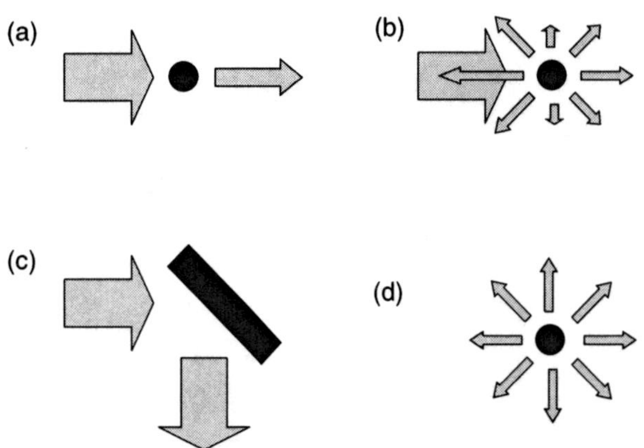

Figure 8.1
(a) Partial absorption, (b) scattering, (c) reflection, and (d) emission of electromagnetic radiation (arrows) by objects (black).

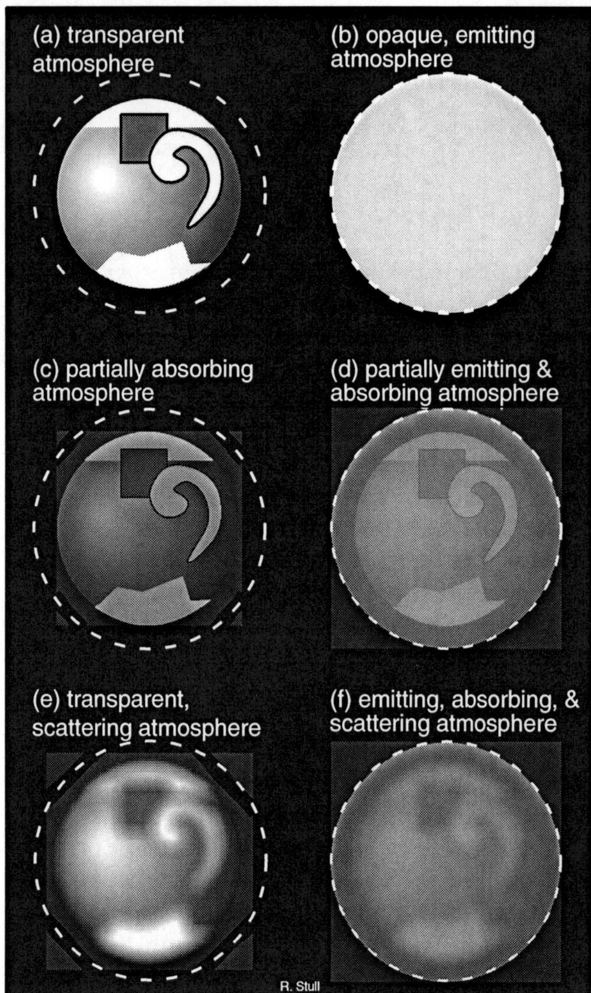

Figure 8.2
Illustration of visibility of the Earth and atmosphere as viewed by satellite. Figure (d) is overly simplistic, because variations in atmospheric constituents in the mid and upper atmosphere will cause atmospheric emissions (glowing) to be uneven.

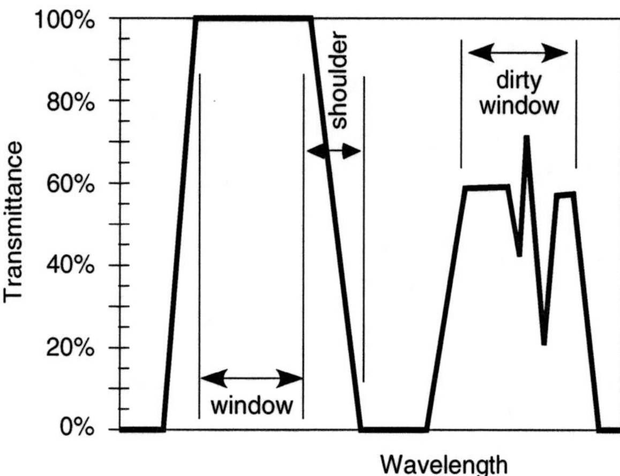

Figure 8.3
Windows in the electromagnetic spectrum.

RADIATIVE TRANSFER for Satellites

Signals

Weather satellites have sensors called **radiometers** that passively measure upwelling electromagnetic radiation from the Earth and atmosphere. **Infrared** (**IR, long-wave**) and **microwave** radiation are emitted by the Earth, atmosphere, clouds, and the sun (see the Radiation chapter). **Visible** light (**short-wave** or **solar radiation**) is emitted by the sun and reflected and absorbed by the Earth system. Additional portions of the **electromagnetic spectrum** are useful for remote sensing.

What the satellite can "see" in any one wavelength depends on the transparency of the air at that wavelength. A perfectly transparent atmosphere allows the upwelling radiation from the Earth's surface or highest cloud top to reach the satellite. Thus, wavelengths for which the air is transparent (Fig. 8.2a) are good for observing clouds and land use.

If air molecules strongly absorb upwelling radiation at another wavelength, then none of the signal at that wavelength from the Earth and clouds will reach the satellite (i.e., an opaque atmosphere). But according to Kirchhoff's law (see the Radiation chapter), absorptivity equals the emissivity at that wavelength. This atmosphere will emit its own spectrum of radiation according to Planck's law, causing the atmosphere to glow like an infrared light bulb (Fig. 8.2b). Wavelengths with this characteristic are good for observing the top of the atmosphere, but are bad for remote sensing of the Earth and clouds.

For other wavelengths, the atmosphere partially absorbs the upwelling radiation, causing the Earth and clouds to look dimmer (Fig. 8.2c). But this usually never happens alone, because Kirchhoff's law says that the atmosphere will also partially emit in the same wavelengths. The result is a dim view of the Earth, partially masked by a dimly glowing atmosphere (Fig. 8.2d).

For wavelengths scattered by air molecules, the signal from the Earth and clouds becomes blurred (Fig. 8.2e). For some wavelengths, this blurring is so extreme that no useful signal reaches the satellite, other than noise from all the scattered light rays. Finally, there are other wavelengths where all of the processes happen: atmospheric scattering, absorption, and emission (Fig. 8.2f).

Transmittance and Windows

Of the electromagnetic energy that is upwelling through any height, the percentage of it that comes out the top of the atmosphere is called **transmittance**. Transmittance varies with wavelength.

Portions of the spectrum where transmittances are large are called **windows** (Fig. 8.3), by analogy to visible light passing through clear glass windows. At wavelengths near the window, there can be **shoulder** regions where transmittance rapidly changes. Portions of the spectrum having partial transmittance are sometimes called **dirty windows**. By designing satellite-borne radiometers that are sensitive to different window and non-window wavelengths, you can measure different characteristics of the Earth and atmosphere.

Figures 8.4 (next 2 pages) show the transmittance at different wavelengths. Different gases in the atmosphere have different molecular vibration and rotation modes, causing them to absorb at discrete wavelengths called **absorption lines**. In Fig. 8.4, the windows are regions with transmittance of about 80% or higher.

These transmittance curves are not physical laws and are not constant, but can change slightly with atmospheric conditions. The **absorption bands** (i.e., non-window regions) shift wavelength very slightly with temperature and pressure.

The amount of absorption and transmission depend strongly on the concentration of absorbing gas along the path length of the radiation (see Beer's Law, in the Radiation chapter). Some gas concentrations vary with season (carbon dioxide CO_2), some vary hourly depending on the weather (water vapor H_2O, ozone O_3), while others are relatively constant. Additional gases indicated in these figures are molecular oxygen (O_2), methane (CH_4), carbon monoxide (CO), and nitrous oxide (N_2O).

Water vapor is a major absorber, so more humid conditions and deeper moist layers cause greater absorption. Recall from Chapter 1 that most of our storms and most of the atmosphere's humidity are trapped within the troposphere. Thus, transmittance is weakest in the tropics (high humidity and deep troposphere) and strongest near the poles (low absolute humidity and shallow troposphere). At mid latitudes, transmittance is greatest in winter (low humidity, shallow troposphere) and weakest in summer (higher humidity and deeper troposphere). Transmittance can easily vary by plus or minus 20% between these different locations and seasons in some portions of the spectrum, especially for wavelengths greater than 5 µm.

Another factor that reduces transmittance is **scattering** by air molecules and **aerosols** (e.g., air-pollution particles). Scattering increases (causing transmittance to decrease) as wavelength gets shorter (dashed curve in Fig. 8.4a). For cleaner air, the dashed curve is higher and transmittance is greater, but the opposite occurs for heavily polluted, aerosol-laden air. The visible light portion of the spectrum is a **dirty window** region. Fig. 8.4 shows that the atmosphere is clearer in some of the IR windows than in the visible light portion of the spectrum that you see every day with your eyes.

So far, we examined atmospheric transmittance (the left column of images in Fig. 8.2). Next, we look at atmospheric emissions (right column of Fig. 8.2).

Sample Application

Wavelength 1.85 µm has: (a) what transmittance, and (b) corresponds to which sketch in Fig. 8.2 of Earth visibility as viewed from space?

Find the Answer

Given: λ = 1.85 µm

Find: (a) transmittance = ? % , (b) Earth visibility = ?

(a) From Fig. 8.4b, transmittance = 10% approx., mostly due to strong absorption by water vapor.

(b) It would look like Fig. 8.2b in the infrared.

Check: (no easy check for this)

Exposition: Most water vapor is in the troposphere. Satellites would see it glow like an IR light bulb.

Planck's Law & Brightness Temperature

In the Radiation chapter, Planck's law allowed computation of total energy flux radiating from a blackbody object per unit wavelength ($W \cdot m^{-2} \cdot \mu m^{-1}$) as a function of temperature. Emissions from a flat surface are in all directions, illuminating a hemisphere around the object.

But a satellite cannot measure the total radiation coming out of the Earth or atmosphere — it can measure only the portion of radiation that happens to be coming toward the satellite within the solid angle intercepted by the radiometer. Assuming the radiation is uniform in all directions (i.e., **isotropic**), then the portion of radiative flux per unit wavelength λ per unit **steradian** (sr) of solid angle is Planck's Law equation divided by π:

$$B_\lambda(T) = \frac{c_{1B} \cdot \lambda^{-5}}{\exp\left(\dfrac{c_2}{\lambda \cdot T}\right) - 1} \qquad \bullet(8.1)$$

where B is the **blackbody radiance** in units of $W \cdot m^{-2} \cdot \mu m^{-1} \cdot sr^{-1}$, and

$$c_{1B} = 1.191\,042\,82 \times 10^8 \ W \cdot m^{-2} \cdot \mu m^4 \cdot sr^{-1}$$

$$c_2 = 1.438\,775\,2 \times 10^4 \ \mu m \cdot K .$$

Thus, $c_{1B} = c_1/\pi$, where c_1 was from eq. (2.13). Don't forget that T must be in units of Kelvin. A steradian is the solid angle with vertex at the center of a sphere

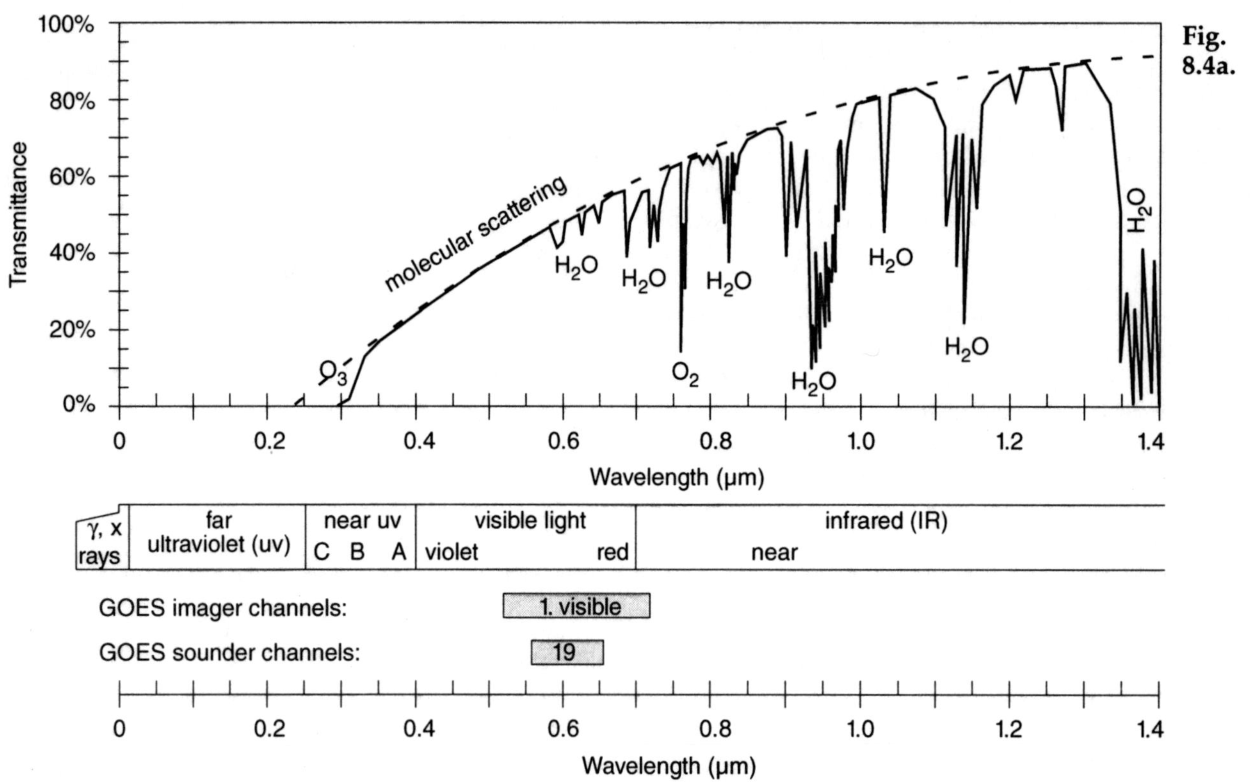

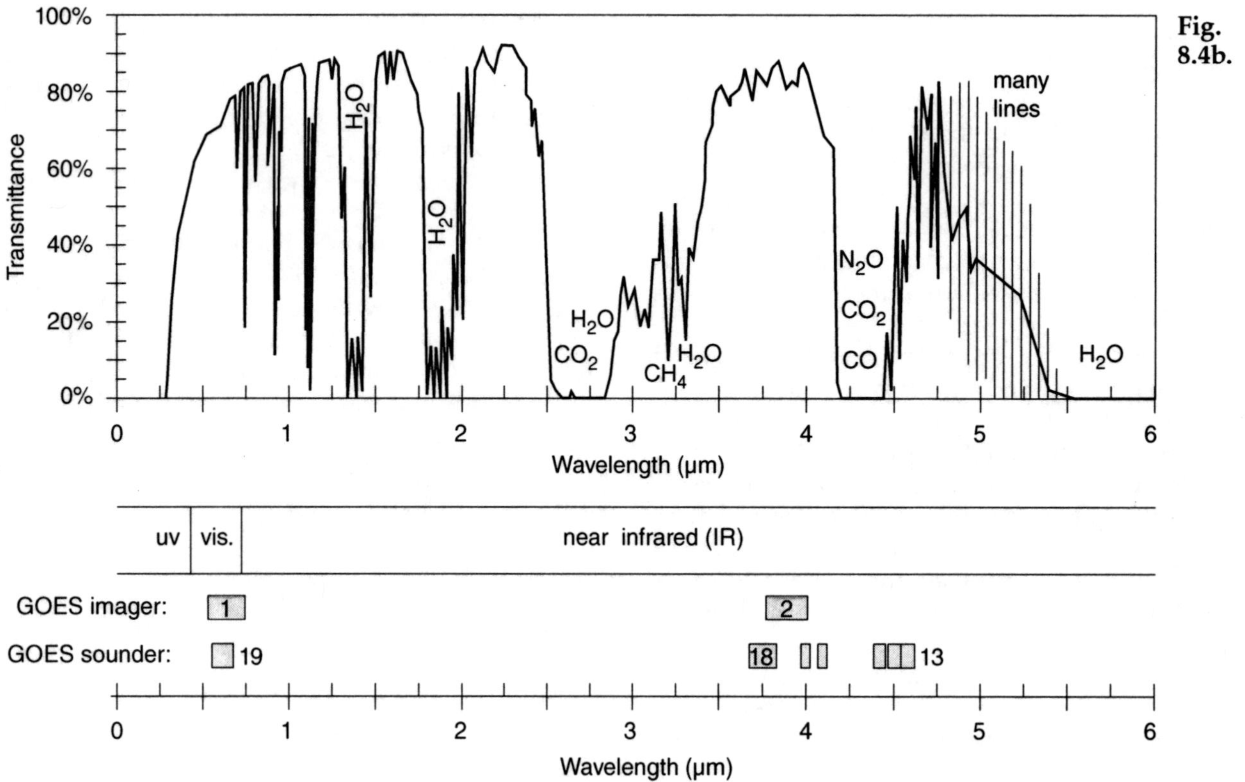

Figure 8.4 (this page and next)
Atmospheric transmittance of electromagnetic radiation. Wavelength bands: (a) 0 to 1.4 μm; (b) 0 to 6 μm; (c) 5 to 30 μm; and (d) 30 μm to 100 cm (logarithmic scale). For regions of strong absorption (i.e., low transmittance) the dominant absorbing chemical is given. The name of the spectral bands and satellite radiometer channels are indicated.

Fig. 8.4c.

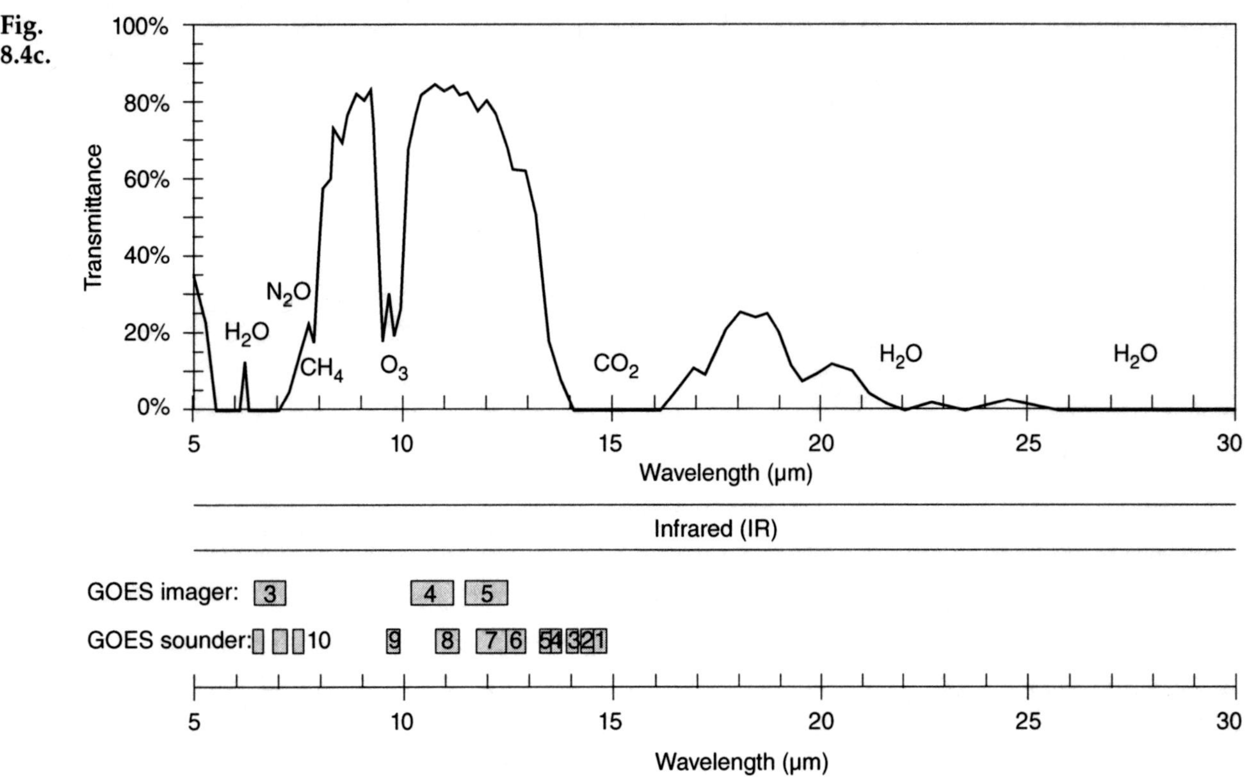

Fig. 8.4d.

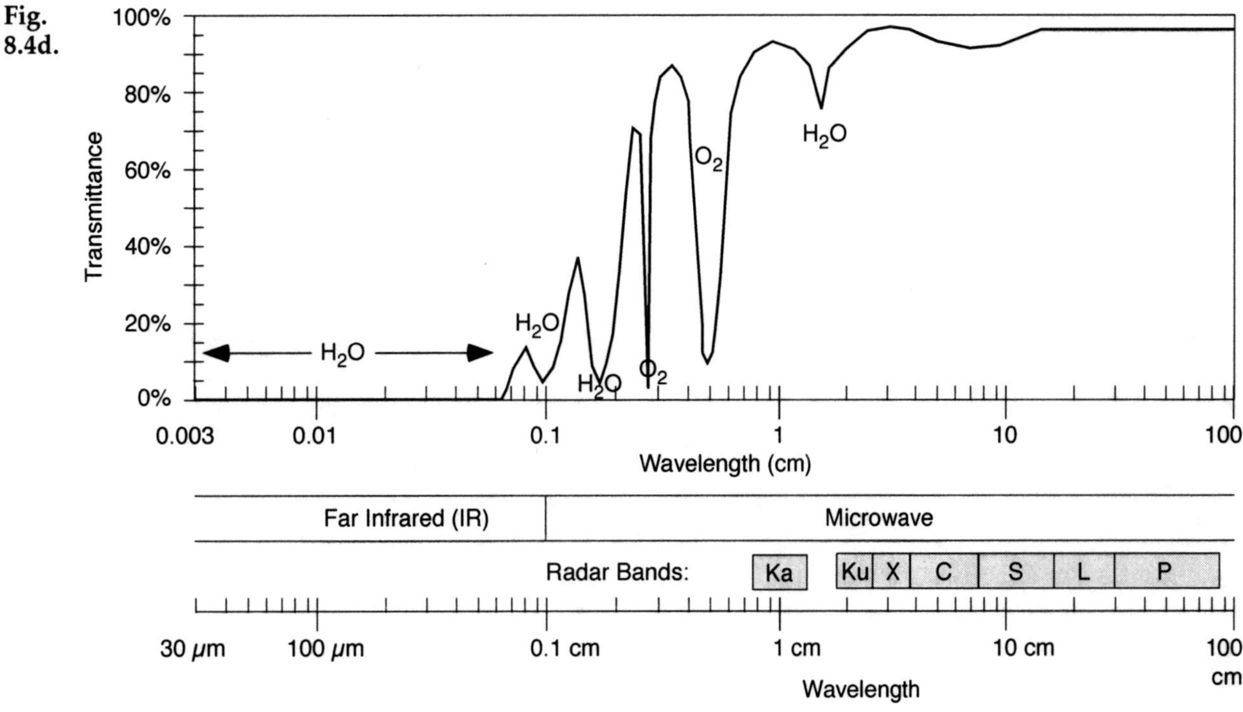

Figure 8.4 (continuation)
See Chapter 2 for wavelengths emitted by sun and Earth. Upwelling radiation can be either terrestrial radiation emitted from the atmosphere and the Earth's surface, or solar radiation reflected upward from the Earth's surface and from within the atmosphere.

Sample Application

What is the blackbody radiance at wavelength 10.9 µm from a cloud top of temperature –20°C ?

Find the Answer

Given: T = –20°C = 253 K, λ = 10.9 µm
Find: $B_\lambda(T)$ = ? W·m^{-2}·µm^{-1}·sr^{-1}

Use eq. (8.1):

$$B_\lambda(T) = \frac{(1.191\text{x}10^8 \text{W}\cdot\text{m}^{-2}\mu\text{m}^4\text{sr}^{-1})\cdot(10.9\mu\text{m})^{-5}}{\left[\exp\left(\frac{1.44\text{x}10^4\mu\text{m}\cdot\text{K}}{(10.9\mu\text{m})\cdot(253\text{K})}\right)\right]-1}$$

= **4.22** W·m^{-2}·µm^{-1}·sr^{-1}

Check: Units OK. Physics OK.
Exposition: 10.9 µm is an IR wavelength in an atmospheric window region of the spectrum (Fig. 8.4). Thus, these emissions from the cloud would be absorbed only little by the intervening atmosphere, and could be observed by satellite.

Sample Application

A satellite measures a radiance of 1.1 W·m^{-2}·µm^{-1} ·sr^{-1} at wavelength 6.7 µm. What is the brightness temperature?

Find the Answer

Given: $B_\lambda(T)$= 1.1 W·m^{-2}·µm^{-1}·sr^{-1}, λ = 6.7 µm
Find: T_B = ? K

Use eq. (8.2) :

$$T_B = \frac{(1.44\text{x}10^4\mu\text{m}\cdot\text{K}) / (6.7\mu\text{m})}{\ln\left(1+\frac{1.191\text{x}10^8\text{W}\cdot\text{m}^{-2}\mu\text{m}^4\text{sr}^{-1}\cdot(6.7\mu\text{m})^{-5}}{1.1\text{W}\cdot\text{m}^{-2}\mu\text{m}^{-1}\text{sr}^{-1}}\right)}$$

= **239 K**

Check: Units OK. Physics OK.
Exposition: 6.7 µm is an IR wavelength in a water-vapor absorption (=emission) part of the spectrum (Fig. 8.4). The atmosphere is partly opaque in this region, so the radiation received at the satellite was emitted from the air. This brightness temperature is about –34°C. Such temperatures are typically found in the upper troposphere. Thus, we can infer from the standard atmosphere that this satellite channel is "seeing" the upper troposphere.

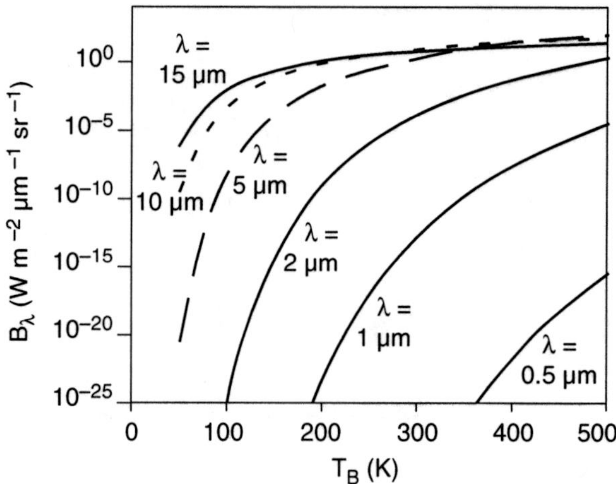

Figure 8.5
Planck's law emissions vs. brightness temperature.

(of radius r) that encompasses an area of r^2 on the surface of the sphere; 4π sr cover the whole surface.

Fig. 8.5 shows Planck's law plotted differently; namely, blackbody radiance vs. temperature. Blackbody radiance increases monotonically with increasing temperature. Hotter objects emit greater radiation (assuming a blackbody emissivity of 1.0).

You can also use Planck's law in reverse. Plug a measured radiance into eq. (8.1), and solve for temperature. This temperature is called the **brightness temperature** (T_B), which is the temperature of a hypothetical blackbody that produces the same radiance as the measured radiance:

$$T_B = \frac{c_2 / \lambda}{\ln\left(1+\frac{c_{1B}\cdot\lambda^{-5}}{B_\lambda}\right)} \qquad \bullet(8.2)$$

Radiative Transfer Equation

Recall from Fig. 8.2 that surface emissions might be partially or totally absorbed by the atmosphere before reaching the satellite. The atmosphere emits its own radiation, some of which might also be lost by absorption before reaching the satellite.

These effects are summarized by the **radiative transfer equation**:

$$L_\lambda = B_\lambda(T_{skin})\cdot\hat{\tau}_{\lambda,sfc} + \sum_{j=1}^{n} e_\lambda(z_j)\cdot B_\lambda(T_j)\cdot\hat{\tau}_{\lambda,j} \qquad (8.3)$$

where L_λ is the radiance at wavelength λ that exits the top of the atmosphere and can be observed by satellite. T_{skin} is the temperature of the top few molecules of the Earth's surface, NOT the standard meteorological "surface" temperature measured 2 m above ground. $\hat{\tau}$ is transmittance; e is emissivity. This equation is called **Schwarzschild's eq.**

The first term on the right hand side (RHS) gives the blackbody emissions from the Earth's surface, reduced by the overall transmittance $\hat{\tau}_{\lambda,sfc}$ between the surface and the top of the atmosphere (from Fig. 8.4). Namely, satellites can see the Earth's surface at wavelengths for which the air is not totally opaque.

The second term on the RHS is a sum over all atmospheric layers (j = 1 to n), representing the different heights z_j in the atmosphere. The net emissions from any one layer j are equal to the emissivity $e_\lambda(z_j)$ of the air at that height for that wavelength, times the blackbody emissions. However, the resulting radiance is reduced by the transmittance $\hat{\tau}_{\lambda,j}$ between that height and the top of the atmosphere.

Transmittance $\hat{\tau}$ = 0 if all of the radiation is absorbed before reaching the top of the atmosphere. Transmittance = 1 – **absorptance**; namely, $\hat{\tau}$ =1 – a.

To help understand the second term, consider a hypothetical or "toy" profile of transmittance (Fig. 8.6). For the air below 5 km altitude, $\hat{\tau}$ = 0, thus $a = (1 - \hat{\tau}) = 1$. But if a = 1, then e = 1 from Kirchhoff's Law (see the Radiation chapter). Hence, the layers of air below 5 km are efficient emitters of radiation, but none of this radiation reaches the top of the atmosphere because it is all absorbed by other air molecules along the way, as indicated by zero transmissivity. So a satellite cannot "see" the air at this range of heights.

Above 10 km altitude, although the atmosphere is transparent in this toy profile, it has zero emissivity. So no radiation is produced from the air at these altitudes, and again the satellite cannot see this air.

But for heights between 5 and 10 km, the atmosphere is partially emitting (has nonzero emissivity), and the resulting upwelling radiation is not totally absorbed (as indicated by nonzero transmissivity). Thus, radiation from this layer can reach the satellite. The satellite can see through the atmosphere down to this layer, but can't see air below this layer.

Thus, with the right wavelength, satellites can measure brightness temperature at an elevated layer of cloudless air. At other wavelengths with different transmittance profiles, satellites can measure temperature in different layers, at the top of the atmosphere, at cloud top, or at the Earth's surface.

In general, heights where transmittance changes are heights for which the air can be remotely observed. After a bit of math, the **radiative transfer equation** can be rewritten in terms of transmittance change $\Delta\hat{\tau} = [\hat{\tau}_{\text{top of layer j}} - \hat{\tau}_{\text{bottom of layer j}}]$:

$$L_\lambda = B_\lambda(T_{skin})\cdot\hat{\tau}_{\lambda,sfc} + \sum_{j=1}^{n} B_\lambda(T_j)\cdot\Delta\hat{\tau}_{\lambda,j} \qquad (8.4)$$

If the full atmospheric depth is black (i.e., no surface emissions to space), then $\sum_{j=1}^{n}\Delta\hat{\tau}_{\lambda,j} = 1$.

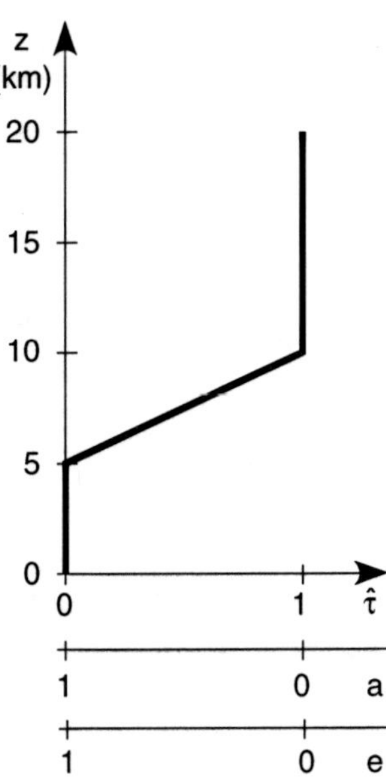

Figure 8.6
Hypothetical variation of atmospheric transmittance ($\hat{\tau}$), absorptance (a), and emissivity (e) with height (z).

Sample Application

Given the IR transmittance profile of Fig. 8.6, for λ = 6.7 µm. Suppose the vertical temperature profile in the atmosphere is:

z (km)	T (°C)
15 to 20	–40
10 to 15	–30
5 to 10	–20
0 to 5	0

and the Earth's surface (skin) temperature is 15°C. Find the upwelling IR radiation at the top of the atmosphere.

Find the Answer

Given: the data above.
Find: L_λ = ? W·m^{-2}·µm^{-1}·sr^{-1} .

You can save some work by thinking about the problem first. From Fig. 8.6, the transmittance $\hat{\tau}_{\lambda,sfc}$ is 0 at the surface (z = 0). Thus, none of the surface emissions will reach the top of the atmosphere, so T_{skin} is irrelevant, and the first term on the RHS of eq. (8.4) is zero.

Also, as discussed, the satellite can't see the layers between 0 to 5 km and 10 to 20 km for this toy profile, so the temperatures in these layers are also irrelevant. Thus, all terms in the sum in eq. (8.4) are zero except the one term for the one layer at 5 to 10 km.

In this layer, the absolute temperature is T = 273 - 20°C = 253 K. Across this layer, the change of transmittance is $\Delta\hat{\tau}$ = [1 – 0]. Thus, the only nonzero part of eq. (8.4) is:

$L_\lambda = 0 + B_\lambda(253\text{ K}) \cdot [1 - 0]$

where you can solve for B_λ using eq. (8.1).

$L_\lambda = B_\lambda = \underline{\mathbf{1.82}}$ W·m^{-2}·µm^{-1}·sr^{-1} .

Check: Units OK. Physics OK.
Exposition: A lot of work was saved by thinking first.

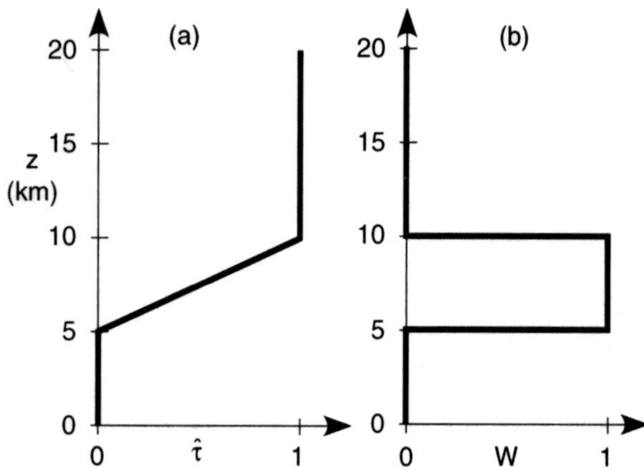

Figure 8.7
Hypothetical transmittance $\hat{\tau}$ and weighting function $W(z)$.

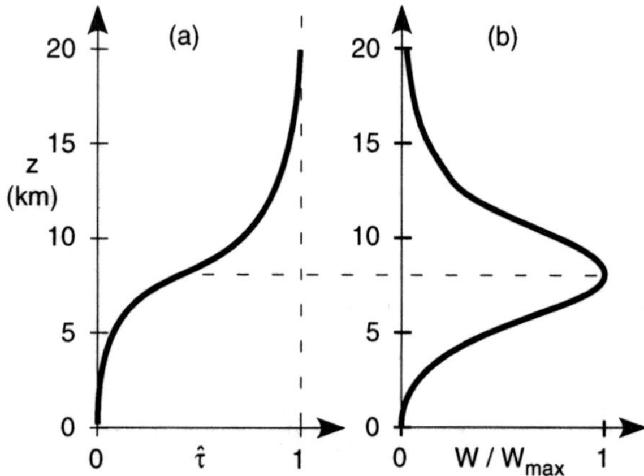

Figure 8.8
Transmittance $\hat{\tau}$ and weighting function $W(z)$ at $\lambda = 6.7\ \mu m$. This is the GOES-8 imager **water-vapor channel** *(IR channel 3). The weight is normalized by its maximum value.*

Weighting Functions

In eq. (8.4) the factor $\Delta\hat{\tau}$ within the sum acts as a weight that determines the relative contribution of each layer to the total radiance out the top of the atmosphere at that wavelength. Use the symbol $W_{\lambda,j}$ for these weights (i.e., $W_{\lambda,j} = \Delta\hat{\tau}_{\lambda,j}$). With this trivial notation change, the **radiative transfer equation** is:

$$L_\lambda = B_\lambda(T_{skin})\cdot\hat{\tau}_{\lambda,sfc} + \sum_{j=1}^{n} B_\lambda(T_j)\cdot W_{\lambda,j} \qquad \bullet(8.5)$$

where $\Sigma_j W_j = 1$ for any one wavelength.

For the toy profile of Fig. 8.7a, the corresponding vertical profile of weights, called the **weighting function**, is shown in Fig. 8.7b. The weights are proportional to the slope of the transmittance line. (Any vertical line segment in Fig. 8.7a has zero slope, remembering that the independent height variable is along the ordinate in this meteorological graph).

In reality, the weighting function for any wavelength is a smooth curve. Fig. 8.8 shows the curve at $\lambda = 6.7\ \mu m$, for which water vapor is the emitter.

In essence, the weighting function tells you the dominant height-range seen by a satellite channel. For Fig. 8.8, all moist layers in the height range of roughly 5 to 12 km (mid to upper troposphere) are blurred together to give one average moisture value. Hence, this **water-vapor channel** can see tops of deep thunderstorms, but not the boundary layer.

By utilizing many wavelengths with weighting functions that peak at different heights, the satellite can focus on different overlapping height ranges. Heights of peak W vary where $\hat{\tau}$ in Fig. 8.4 varies with λ. Figs. 8.9 show the weighting functions for the sounder channels (Fig. 8.4) on the GOES-8 satellite. Newer satellites use interferometer methods, giving a nearly infinite suite of weighting functions that peak at a wide range of different altitudes.

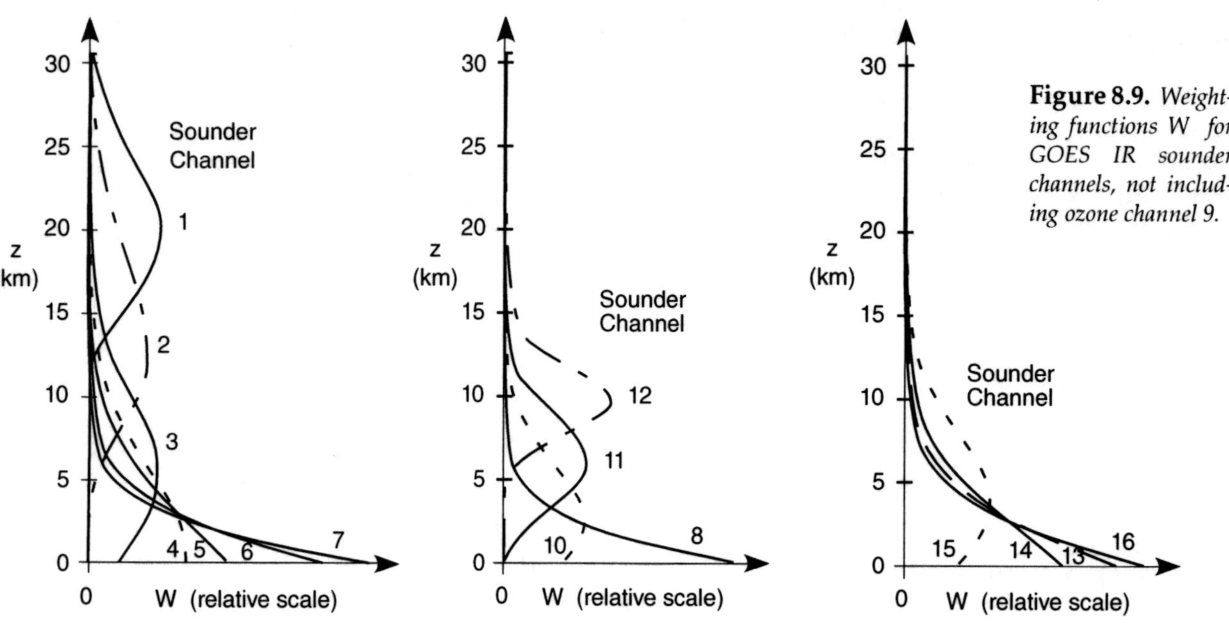

Figure 8.9. *Weighting functions W for GOES IR sounder channels, not including ozone channel 9.*

WEATHER SATELLITES

Orbits

Artificial satellites such as weather satellites orbiting the Earth obey the same orbital mechanics as planets orbiting around the sun. For satellites in near-circular orbits, the pull by the Earth's gravity f_G balances centrifugal force f_C :

$$f_G = \frac{G \cdot M \cdot m}{R^2} \qquad \bullet(8.6)$$

$$f_C = \left(\frac{2\pi}{t_{orbit}}\right)^2 \cdot m \cdot R \qquad \bullet(8.7)$$

where R is the distance between the center of the Earth and the satellite, m is the mass of the satellite, M is the mass of the Earth (5.9742x10^{24} kg), and G is the gravitational constant (6.6742x10^{-11} N·m^2·kg^{-2}). See Appendix B for lists of constants.

Solve for the orbital time period t_{orbit} by setting $f_G = f_C$:

$$t_{orbit} = \frac{2\pi \cdot R^{3/2}}{\sqrt{G \cdot M}} \qquad \bullet(8.8)$$

Orbital period does not depend on satellite mass, but increases as satellite altitude increases.

Weather satellite orbits are classified as either **polar-orbiting** or **geostationary** (Fig. 8.10). Polar-orbiters are **low-Earth-orbit** (**LEO**) satellites.

Geostationary Satellites

Geostationary satellites are in high Earth orbit over the equator, so that the orbital period matches the Earth's rotation. Relative to the fixed stars, the Earth rotates 360° in 23.934 469 6 h, which is the duration of a **sidereal day.** With this orbital period, geostationary satellites appear parked over a fixed point on the equator. From this vantage point, the satellite can take a series of photographs of the same location, allowing the photos to be combined into a repeating movie called a **satellite loop**.

Disadvantages of geostationary satellites include: distance from Earth is so great that large magnification is needed to resolve smaller clouds; many satellites must be parked at different longitudes for imagery to cover the globe; imaging is interrupted during nights near the equinoxes because the solar panels are in darkness — eclipsed by the Earth; and polar regions are difficult to see.

Satellites usually have planned lifetimes of about 3 to 5 years, so older satellites must be continually replaced with newer ones. Lifetimes are limited partly because of the limited propellant storage needed to make orbital corrections. Satellites are also hurt by tiny meteoroids that frequently hit the satellite

Sample Application

At what (a) distance above the Earth's center, & (b) altitude above the Earth's surface, must a geostationary satellite be parked to have an orbital period of exactly one sidereal day? Use Appendix B for Earth constants.

Find the Answer

Given: t_{orbit} = 23.934 469 6 h = 86,164 s = <u>sidereal</u> day.
M = 5.973 6 x 10^{24} kg
G = 6.674 28 x 10^{-11} m^3·s^{-2}·kg^{-1}
Find: (a) R = ? km, (b) z = ? km

(a) Rearrange eq. (8.8):
$R = (t_{orbit} / 2\pi)^{2/3} \cdot (G \cdot M)^{1/3}$ = **<u>42,167.5 km</u>**
(b) From this subtract Earth radius at equator (R_o = 6,378 km) to get height above the surface:
$z = R - R_o$ = **<u>35,790 km</u>**

Check: Units OK. Physics OK.
Exposition: Compares well with real satellites. As of 2 July 2012, the GOES-15 satellite was at R = 42,168.07 km. It was slightly too high, orbiting slightly too slowly, causing it to gradually get behind of the Earth's rotation. Namely, it drifts 0.018°/day toward the west relative to the Earth.

Such drift is normal for satellites, which is why they carry propellant to make orbital adjustments, as commanded by tracking stations on the ground.

For a <u>calendar</u> day (24 h from sun overhead to sun overhead), the Earth must rotate 360.9863°, because the position of the sun relative to the fixed stars changes as the Earth moves around it.

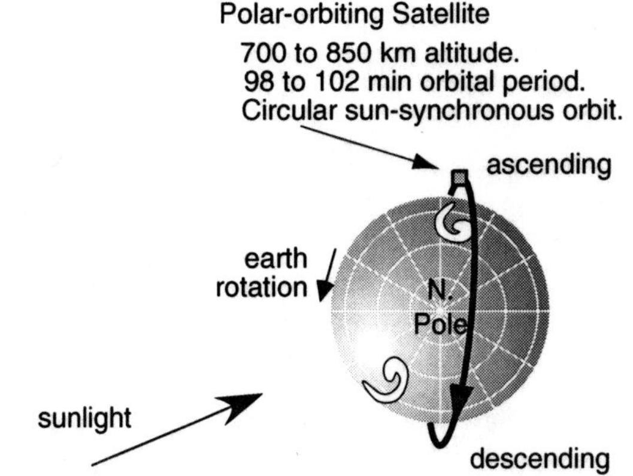

Figure 8.10
Sketch (to scale) of geostationary and polar-orbiting weather satellite orbits.

Figure 8.11
Example of Earth disk image from Meteosat-8 (MSG-1), which was parked at 0° longitude. It began routine operations on 29 Jan 2004. Copyright © 2004 by EUMETSAT. Used with permission. http://oiswww.eumetsat.org/IDDS-cgi/listImages As of July 2012, Meteosat-10 (MSG-3) was being moved to 0° longitude to take over primary duties of full-disk imaging in early 2013.

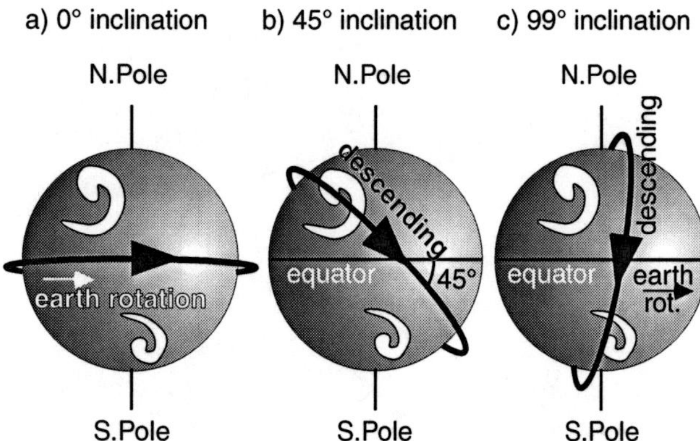

Figure 8.12
Examples of inclinations of satellite orbits. Sun-synchronous satellites use orbit (c).

at high speed, and by major solar storms. For this reason, most meteorological satellite agencies try to keep an in-orbit spare satellite nearby.

The USA has a series of **Geostationary Operational Environmental Satellite** (**GOES**). They usually park one satellite at 75°W to view the N. American east coast and western Atlantic, and another at 135°W to view the west coast and eastern Pacific — named **GOES-East** and **GOES-West**.

The European Organization for the Exploitation of Meteorological Satellites (EUMETSAT) operate **Meteosat** satellites (Fig. 8.11). They try to keep one parked near 0° longitude, to view Europe, Africa, the Mediterranean Sea, and the eastern Atlantic Ocean.

The Japan Meteorological Agency (JMA) operates Multi-functional Transport Satellites (**MTSAT**), with one parked at 145°E to give a good view of Japan and approaching Pacific typhoons. The China Meteorological Administration has a series of **FengYun** (FY-2, "Wind & cloud") geostationary satellites parked at 86.5°E and 105°E. Russia's Geostationary Operational Meteorological Satellite (**GOMS**) program has an Elektro-L satellite parked over the Indian Ocean at 76°E. The India Space Research Organization (ISRO) operates **INSAT** satellites in the 60° to 95°E range of longitudes.

Thus, there are usually sufficient geostationary satellites around the equator to view all parts of the Earth except the poles.

Polar Orbiting Satellites

If geostationary positioning is not required, then weather satellites could be placed at any altitude with any orbital inclination. However, there is a special altitude and inclination that allows satellites to view the Earth at roughly the same local time every day. Advantages are consistent illumination by the sun, lower altitude to better resolve the smaller clouds, and good views of high latitudes.

To understand this special orbit, consider the following. When the orbital plane of the satellite is along the Earth's equator, AND the direction of satellite orbit is the same as the direction of Earth's rotation, then the orbit is defined to have 0° **inclination** (Fig. 8.12a). Greater inclination angles (Fig. 8.12b) indicate greater tilt of the orbit relative to the equator. For inclinations greater than 90°, the satellite is orbiting opposite to the Earth's rotation (Fig. 8.12c). For an inclined orbit, the **ascending node** is the side of the orbit where the satellite crosses the equator northbound (behind the Earth in Fig. 8.12b & c). The **descending node** is where it crosses the equator southbound (in front of the Earth in Fig. 8.12b & c).

Polar orbiting weather satellites are designed so that the locations of the ascending and descending nodes are **sun synchronous**. Namely, the satellite always observes the same local solar times on every orbit. For example, Fig. 8.13a shows a satellite orbit with descending node at about 10:20 AM local time. Namely, the local time at city A directly under the satellite when it crosses the equator is 10:20 AM.

For this sun-synchronous example, 100 minutes later, the satellite has made a full orbit and is again over the equator. However, the Earth has rotated 25.3° during this time, so it is now local noon at city A. However, city B is now under the satellite (Fig. 8.13b), where its local time is 10:20 AM. 100 minutes later, during the next orbit, city C is under the satellite, again at 10:20 AM local time (Fig. 8.13c).

For the satellite orbits in Fig. 8.13, on the back side of the Earth, the satellite always crosses the equator at 10:20 PM local time during its ascension node.

Sun-synchronous polar-orbiting satellites are nicknamed by the time of day when they cross the equator during daylight. It does not matter whether this daylight crossing is during the ascent or descent part of the orbit. For the example of Fig. 8.13, this is the **morning** or **AM satellite**.

Many countries have polar orbiting weather satellites. The USA's **Polar Orbiting Environmental Satellites (POES)** are designated **NOAA**-X, where X is the satellite ID number. NOAA-19, launched in Feb 2009, is the last POES. Each satellite has a design life of about 4 years in space. A **Suomi National Polar-orbiting Partnership (NPP)** satellite was launched in Oct 2011 as a transition to future **Joint Polar Satellite System (JPSS)** satellites.

For the polar orbit to remain sun synchronous during the whole year, the satellite orbit must precess 360°/year as the Earth orbits the sun; namely, 0.9863° every day. This is illustrated in Fig. 8.14. Aerospace engineers, astronomers and physicists devised an ingenious way to do this without using their limited supply of onboard propellant. They take advantage of the pull of the solar gravity and the resulting slight tidal bulge of the "solid" Earth toward the sun. As the Earth rotates, this bulge (which has a time lag before disappearing) moves eastward and exerts a small gravitational pull on the satellite in the direction of the Earth's rotation.

This applies a torque to the orbit to cause it to gradually rotate relative to the fixed stars, so the orbit remains synchronous relative to the sun. The combination of low Earth orbit altitude AND inclination greater than 90° gives just the right amount of precession to maintain the sun synchronous orbit.

The result is that polar-orbiting weather satellites are usually placed in low Earth orbit at 700 to 850 km altitude, with short orbital period of 98 to 102 minutes, and inclination of 98.5° to 99.0°. Polar orbiting satellites do not go directly over the poles, but intentionally miss the poles by 9°. This is still close enough to get good images of the poles.

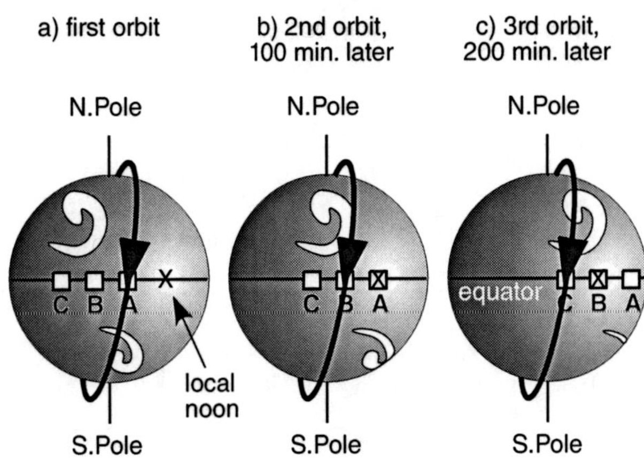

Figure 8.13
Rotation of the Earth under a sun-synchronous satellite orbit. X marks local noon.

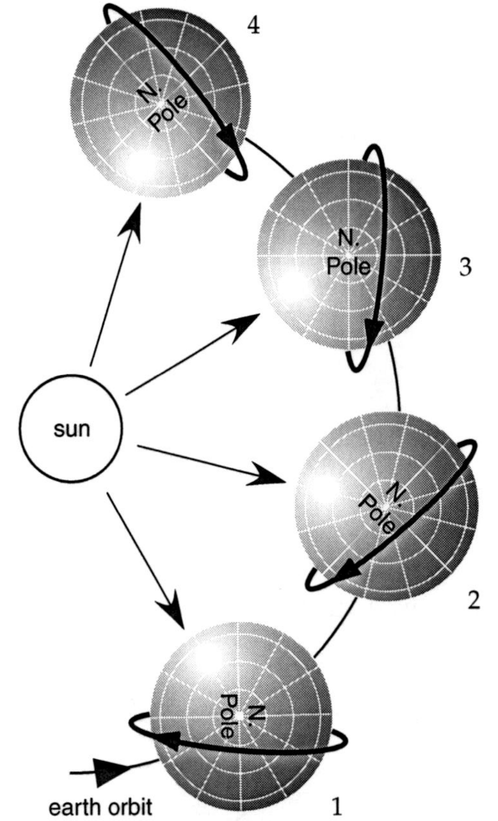

Figure 8.14
Precession of polar satellite orbit (thick lines) as the Earth orbits around the sun (not to scale). The page number at the top of this textbook page can represent a "fixed star".

Table 8-1. Imager channels on USA GOES weather satellites. All channels except #3 are in atmospheric windows. Channel wavelengths vary slightly from satellite to satellite. •Indicates important channels.

Channel #	Name	Center Wave-length (µm)	Wave-length Range (µm)
1 •	visible	0.65	0.52 - 0.71
2	short-wave IR window	3.9	3.73 - 4.07
3 •	water vapor	6.7	5.8 - 7.3
4 •	infrared (long-wave IR window	10.7	10.2 - 11.2
5	IR dirty window		
	(for GOES 10-11)	12	11.5 - 12.5
	(for GOES 12-15)	13.3	13.0 - 13.7

Table 8-2. Imager channels on European MSG-3 (Meteosat-10) weather satellite.

Channel #	Name	Center Wave-length (µm)	Wave-length Range (µm)
1	VIS 0.6 (visible green)	0.635	0.56 - 0.71
2	VIS 0.8 (visible red)	0.81	0.74 - 0.88
3	NIR 1.6 (near IR)	1.64	1.50 - 1.78
4	IR 3.9	3.90	3.48 - 4.36
5	WV 6.2 (water vapor: high trop.)	6.25	5.35 - 7.15
6	WV 7.3 (water vapor: mid-trop.)	7.35	6.85 - 7.85
7	IR 8.7	8.70	8.30 - 9.10
8	IR 9.7 (ozone)	9.66	9.38 - 9.94
9	IR 10.8	10.80	9.80 - 11.8
10	IR 12.0	12.00	11.0 - 13.0
11	IR 13.4 (high-troposphere)	13.40	12.4 - 14.4
12	HRV (high-resolution visible)	broad-band	0.4 - 1.1

INFO • Some Other Satellite Systems

Scatterometer sensors on satellites can detect capillary waves on the ocean, allowing near-surface wind speeds to be estimated. Passive and active Special Sensor Microwave Imagers (**SSM/I**) can retrieve precipitation and precipitable water over the ocean. Combining a series of observations while a satellite moves allows small onboard antennas to act larger, such as via **synthetic aperture radar** (SAR).

Imager

Modern weather satellites have many capabilities, one of which is to digitally photograph (make images of) the clouds, atmosphere, and Earth's surface. Meteorologists use these photos to help identify and locate weather patterns such as fronts, thunderstorms and hurricanes. Pattern-recognition programs can also use sequences of photos to track cloud motions, thereby inferring the winds at cloud-top level. The satellite instrument system that acquires the digital data to construct these photos is called an **imager**.

As of year 2012, USA geostationary (GOES) weather satellites have 5 imager channels (wavelength bands) for viewing the Earth system (Fig. 8.4). Most of the spectral bands were chosen specifically to look through different transmittance windows to "see" different atmospheric and cloud features. These channels are summarized in Table 8-1.

Imager channels for the European Meteosat-10 are listed in Table 8-2. This satellite has 12 channels. Included are more visible channels to better discern colors, including vegetation greenness (important for weather and climate modeling). Future USA satellites will also have more channels. The discussion below is for the most-used GOES imager channels.

Visible

Visible satellite images (GOES channel 1) show what you could see with your eyes if you were up in space. All cloud tops look white during daytime, because of the reflected sunlight. In cloud-free regions the Earth's surface is visible. At night, special low-light visible-channel imagers on some satellites can see city lights, and see the clouds by moonlight, Without this feature, visible images are useless at night.

Infrared (IR)

Infrared satellite images (GOES channel 4) use long wavelengths in a transmittance window, and can clearly see through the atmosphere to the surface or the highest cloud top. There is very little solar energy at this wavelength to be reflected from the Earth system to the satellite; hence, the satellite sees mostly emissions from the Earth or cloud. The advantage of this channel is it is useful both day and night, because the Earth never cools to absolute zero at night, and thus emits IR radiation day and night.

Images made in this channel are normally grey shaded such that colder temperatures look whiter, and warmer looks darker. But in the troposphere, the standard atmosphere gets colder as height increases (Chapter 1). Thus, white colored clouds in this image indicate high clouds (cirrus, thunderstorm anvils, etc.), and darker grey clouds are low clouds (stratus, fog tops, etc.) Medium grey shading implies middle clouds (altostratus, etc.).

Fig. 8.15 demonstrates the principles behind this IR shading. At any one spot in the field of view, (a) a radiance L is measured by the satellite radiometer — for example: 7.6 W·m^{-2}·µm^{-1}·sr^{-1}, as shown by the dashed line. The picture element (**pixel**) in the image that corresponds to this location is shaded darker (b) for greater L values, mimicking photographic film that becomes darker when exposed to more light. Not knowing the emissivity of the emitting object viewed at this spot, you can (c) assume a black body, and then use the Planck curve (d) for this IR channel to infer (e) brightness temperature T_B (= 283 K in this example). But for any normal temperature profile in the troposphere, such as the standard atmosphere (f), warmer temperatures are usually (g) closer to the ground (z = 0.9 km in this example).

The net result for this IR window channel is that the darker shading (from b, redrawn in h) corresponds to lower clouds (i). Similarly, following the dotted curve, lesser values of observed radiance correspond to colder temperatures and higher clouds, and are shown as whiter pixels.

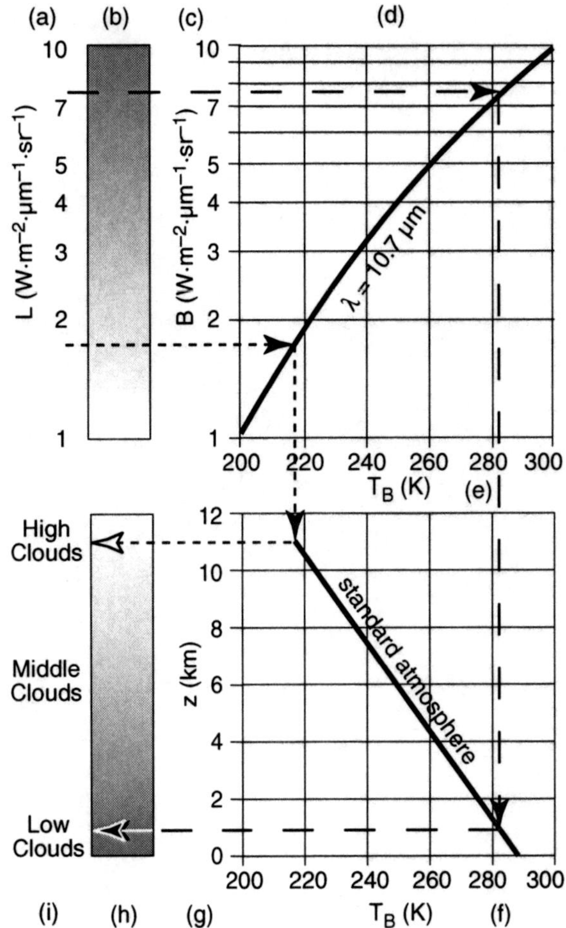

Figure 8.15
Imaging principles for the IR window channel.

Water-vapor (WV)

Water-vapor images are obtained by picking a wavelength (channel 3) that is NOT in a window. In this part of the spectrum, water (as vapor, liquid, or ice) in the atmosphere can absorb radiation. If little water is present in the mid to upper troposphere (Fig. 8.8b), then most of the IR radiation from the Earth can reach the satellite. The warm brightness temperature associated with emissions from the Earth's surface is displayed as dark grey in a water-vapor satellite image — indicating drier air aloft.

For higher concentrations of water in air, most of the surface emissions do not reach the satellite because they are absorbed by the water in the mid to upper troposphere. Kirchhoff's law tells us that this atmospheric layer is also an effective emitter. The colder brightness temperatures associated with strong emissions from this cold layer of air are displayed as light grey — indicating moist air aloft.

Water-vapor images are useful because: (1) they provide data day and night; (2) animations of image sequences show the movement of the air, regardless of whether clouds are present or not; and (3) they give average conditions over a thick layer in the upper troposphere. Because of item (2), pattern recognition programs can estimate average winds in the upper troposphere by tracking movement of blobs of humid air, with or without clouds being present.

Other Channels

Channels 2 and 5 are used less by forecasters, but they do have some specialized uses. Channel 2 sees both reflected solar IR and emitted terrestrial IR, and can help detect fog and low stratus clouds. It is sometimes called the **fog channel**. It can also help to discriminate between water-droplet and ice-crystal clouds, and to see hot spots such as forest fires.

Channel 5 is near the IR longwave window of channel 4, but slightly shifted into a shoulder region (or dirty window) where there are some emissions from low-altitude water vapor. Computerized images of the difference between channels 4 & 5 can help identify regions of greater humidity in the boundary-layer, which is useful for forecasting storms.

Image Examples & Interpretation

Figures 8.16a-c show visible (Vis), infrared (IR) and water vapor (WV) images of the same scene. You can more successfully interpret cloud type when you use and compare all three of these image channels. The letters below refer to labels added to the images. Extra labels on the images are used for a Sample Application and for homework exercises.

Figure 8.16a
Visible satellite image.

To aid image interpretation, letter labels a-z are added to identical locations in all 3 satellite images.

Figure 8.16b
Infrared (IR) satellite image.

[Images a-c courtesy of Space Science & Engineering Center, Univ. of Wisconsin-Madison. Valid time for images: 00 UTC on 16 June 2004.]

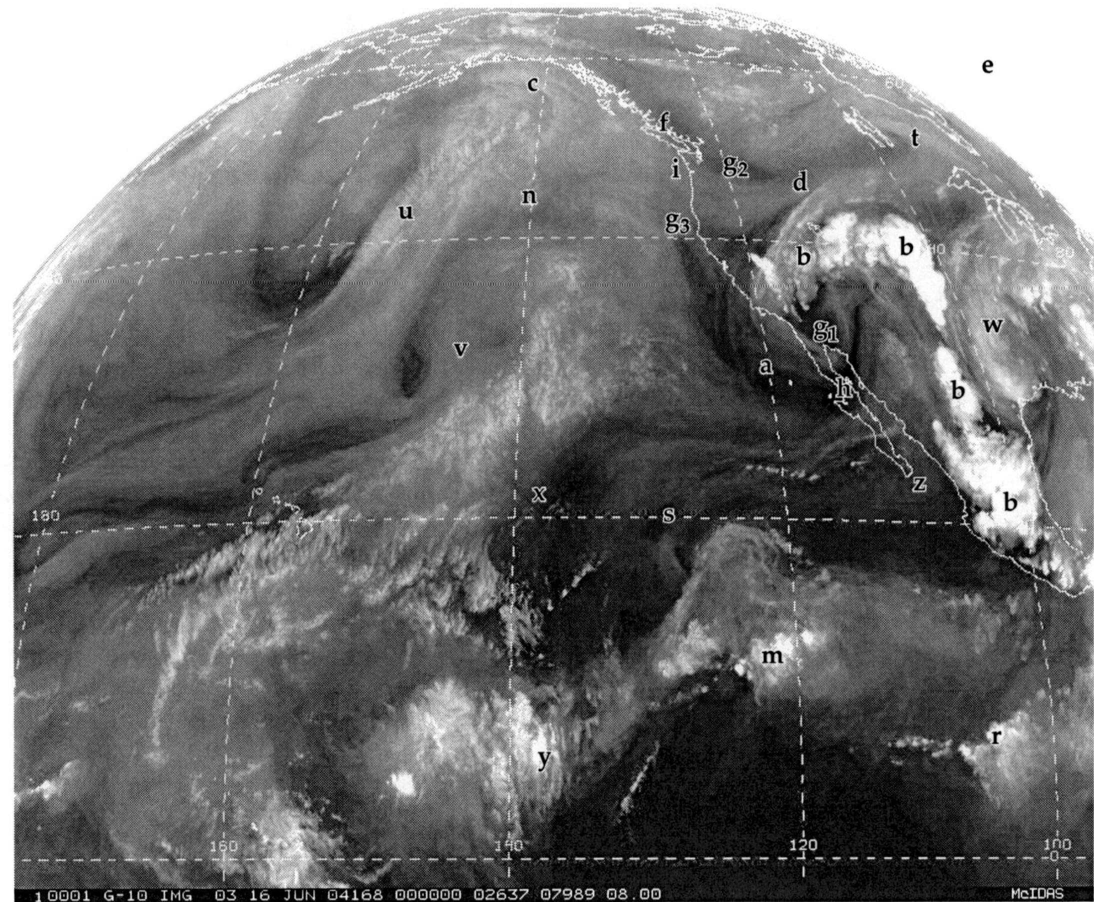

Figure 8.16c
Water-vapor satellite image.

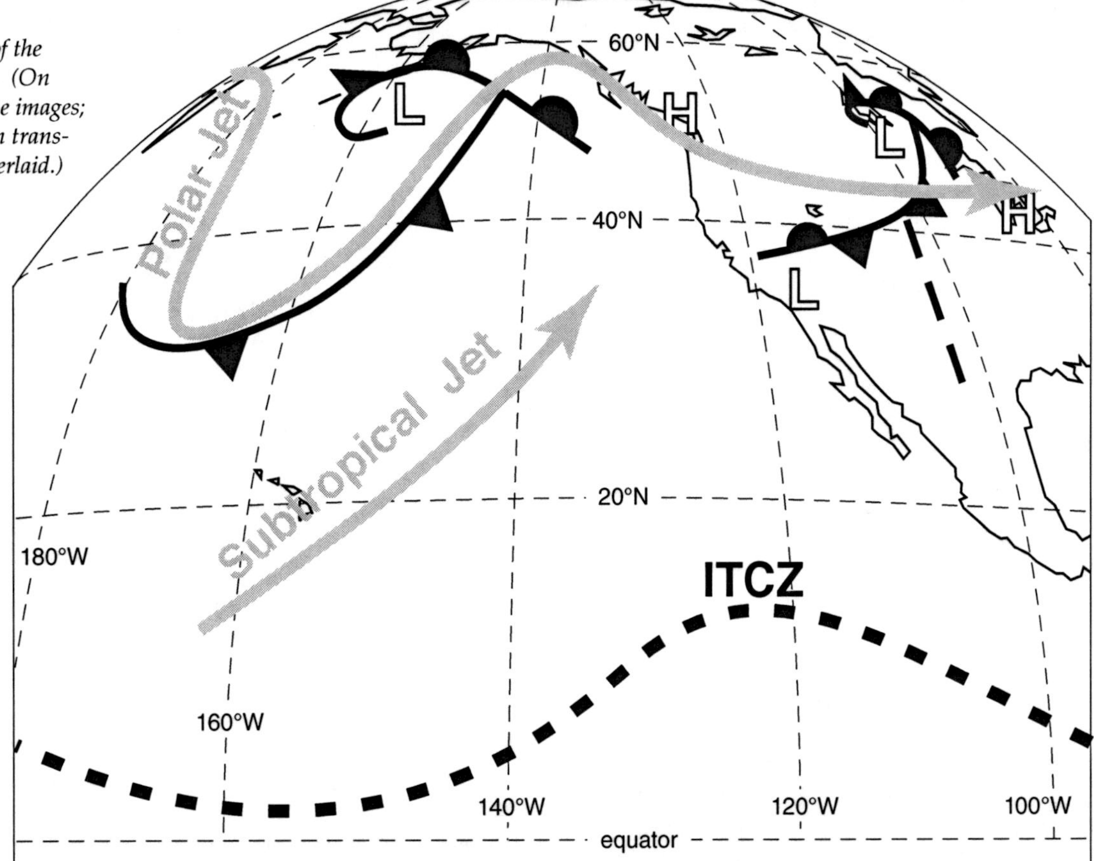

Figure 8.16d
Interpretation of the satellite images. (On same scale as the images; can be copied on transparency and overlaid.) Symbols and acronyms will be explained in later chapters.

a. Fog or low stratus:
Vis: White, because it is a cloud.
IR: Medium to dark grey, because low, warm tops.
WV: Invisible, because not in upper troposphere. Instead, WV shows amount of moisture aloft.

b. Thunderstorms:
Vis: White, because it is a cloud.
IR: Bright white, because high, cold anvil top.
WV: Bright white, because copious amounts of water vapor, rain, and ice crystals fill the mid and upper troposphere. Often the IR and WV images are enhanced by adding color to the coldest temperatures and most-humid air, respectively, to help identify the strongest storms.

c. Cirrus, cirrostratus, or cirrocumulus:
Vis: White, because cloud, although can be light grey if cloud is thin enough to see ground through it.
IR: White, because high, cold cloud.
WV: Medium to light grey, because not a thick layer of moisture that is emitting radiation.

d. Mid-level cloud tops:
Could be either a layer of altostratus/altocumulus, or the tops of cumulus mediocris clouds.
Vis: White, because it is a cloud.
IR: Light grey, because mid-altitude, medium-temperature.
WV: Medium grey. Some moisture in cloud, but not a thick enough layer in mid to upper troposphere to be brighter white.

e. Space:
Vis: Black (unless looking toward sun).
IR: White, because space is cold.
WV: White, because space is cold.

f. Snow-capped Mountains (not clouds):
Vis: White, because snow is white.
IR: Light grey, because snow is cold, but not as cold as high clouds or outer space.
WV: Maybe light grey, but almost invisible, because mountains are below the mid to upper troposphere. Instead, WV channel shows moisture aloft.

g. Land or Water Surfaces (not clouds):
g_1 is in very hot desert southwest in summer, g_2 is in arid plateau, and g_3 is Pacific Ocean.
Vis: Medium to dark grey. Color or greyshade is that of the surface as viewed by eye.
IR: g_1 is black, because very hot ground.
g_2 is dark grey, because medium hot.
g_3 is light grey, because cool ocean.
WV: Light grey or invisible, because below mid to upper troposphere. Instead, sees moisture aloft.

h. Tropopause Fold or Dry Air Aloft.
Vis: Anything.
IR: Anything.
WV: Dark grey or black, because very dry air in the upper troposphere. Occurs during tropopause folds, because dry stratospheric air is mixed down.

i. High Humidities Aloft.
Vis: Anything.
IR: Anything.
WV: Light grey. Often see meandering streams of light grey, which can indicate a jet stream. (Might be hard to see in this copy of a satellite image.)

"Image Interpretation" means the use of satellite images to determine weather features such as fronts, cyclones, thunderstorms and the global circulation. This is a very important part of manual weather forecasting. Whole books are devoted to the subject, and weather forecasters receive extensive training in it. In this book, overviews of image interpretation of cyclones, fronts, and thunderstorms are covered later, in the chapters on those topics.

Sample Application
Determine cloud type at locations "m" and "n" in satellite images 8.16a-c.

Find the Answer
Given: visible, IR, and water vapor images
Find: cloud type

m: vis: White, therefore cloud, fog, or snow.
IR: White, thus high cloud top (cirrus or thunderstorm, but not fog or snow).
wv: White, thus copious moisture within thick cloud layer. Thus, not cirrus.
Conclusion: **thunderstorm**.

n: vis: White or light grey, thus cloud, fog, or snow. (Snow cover is unlikely on unfrozen Pacific).
IR: Medium grey, roughly same color as ocean. Therefore warm, low cloud top.
wv: Medium grey (slightly darker than surrounding regions), therefore slightly drier air aloft. But gives no clues regarding low clouds.
Conclusion: **low clouds or fog**

Check: Difficult to check or confirm now. But after you learn synoptics you can check if the cloud feature makes sense for the weather pattern that it is in.
Exposition: This is like detective work or like a medical diagnosis. Look at all the clues, and rule out the clouds that are not possible. Be careful and systematic. Use other info such as the shape of the cloud or its position relative to other clouds or relative to mountains or oceans. Interpreting satellite photos is somewhat of an art, so your skill will improve with practice.

However, for future reference, Fig. 8.16d shows my interpretation of the previous satellite photos. This particular interpretation shows only some of the larger-scale features. See the Fronts, Cyclones, and General Circ. chapters for symbol definitions.

Sounder

The sounder radiometer measures radiances at different wavelength channels (Table 8-3) that have different weighting functions (Fig. 8.9), in order to estimate vertical temperature profiles (i.e., temperature soundings). These weighting functions peak at different altitudes, allowing us to estimate a **sounding** (temperatures at different altitudes). We will examine the basics of this complex **retrieval** process.

There is a limit to our ability to retrieve sounding data, as summarized in two corollaries. **Corollary 1** is given at right. To demonstrate it, we will start with a simple weighting function and then gradually add more realism in the subsequent illustrations.

Consider the previous idealized transmittance profile (Fig. 8.7), but now divide the portion between $z = 5$ and 10 km into 5 equal layers. As shown in Fig. 8.17a, the change in transmittance across each small layer is $\Delta\hat{\tau} = 0.2$ (dimensionless); hence, the weight (Fig. 8.17b) for each layer is also $W = 0.2$. Assume that this is a crude approximation to sounder channel 3, with central wavelength of $\lambda = 14.0$ µm.

Suppose the "actual" temperature of each layer, from the top down, is $T = -20, -6, -14, -10$, and 0°C, as illustrated by the data points and thin line in Fig. 8.17c. (We are ignoring the portions of the sounding below 5 km and above 10 km, because this weighting function cannot "see" anything at those altitudes, as previously discussed.)

Using Planck's Law (eq. 8.1), find the blackbody radiance from each layer from the top down: $B =$ 3.88, 4.82, 4.27, 4.54, and 5.25 $W \cdot m^{-2} \cdot \mu m^{-1} \cdot sr^{-1}$. Weight each by $W = 0.2$ and then sum according to radiative transfer eq. (8.5) to compute the weighted average. This gives the radiance observed at the satellite: $L =$ 4.55 $W \cdot m^{-2} \cdot \mu m^{-1} \cdot sr^{-1}$. The surface (skin) term in eq. (8.5) was neglected because the transmittance at the surface is zero, so no surface information reaches the satellite for this idealized transmittance profile.

This satellite-observed radiance is communicated to ground stations, where automatic computer programs retrieve the temperature using eq. (8.2). When we do that, we find $T_B = 263.18$ K, or $\underline{T = -9.82°C}$. This is plotted as the thick line in Fig. 8.17c.

Detailed temperature-sounding structure is not retrieved by satellite, because the retrieval can give only one piece of temperature data per weighting function. Vertically broad weighting functions tend to cause significant smoothing of the retrieved temperature sounding.

Table 8-3. Sounder channels on GOES weather satellites. (Also see Fig. 8.4.)

Channel #	Center Wave-length (µm)	Channel #	Center Wave-length (µm)
1	14.7	11	7.0
2	14.4	12	6.5
3	14.0	13	4.6
4	13.7	14	4.5
5	13.4	15	4.4
6	12.7	16	4.1
7	12.0	17	4.0
8	11.0	18	3.8
9	9.7	19	0.6
10	7.5		

Retrieval Corollary 1: The sounder can retrieve (at most) one piece of temperature data per channel. The temperature it gives for that channel is the average brightness temperature weighted over the depth of the weighting function.

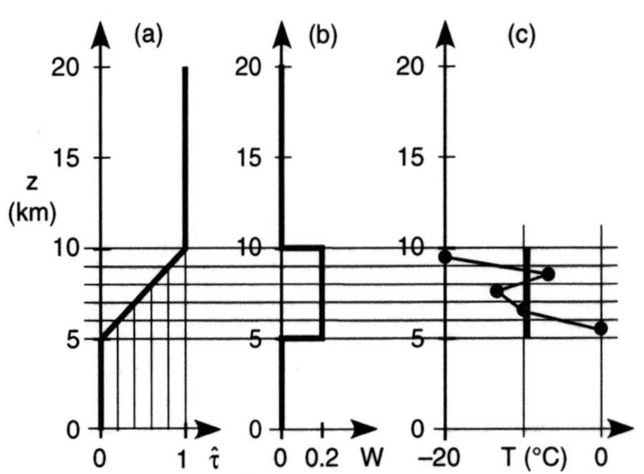

Figure 8.17
Retrieval of temperature from one channel (idealized). (a) transmittance; (b) weighting function; (c) original (thin line with data points) and satellite retrieved (thick line) temperatures.

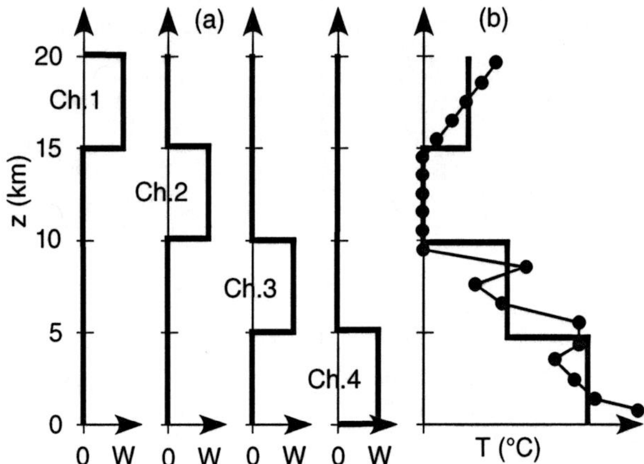

Figure 8.18
(a) Weighting functions for four idealized channels (Ch. 1 – 4). (b) Actual (thin line) and retrieved (thick line) temperature sounding.

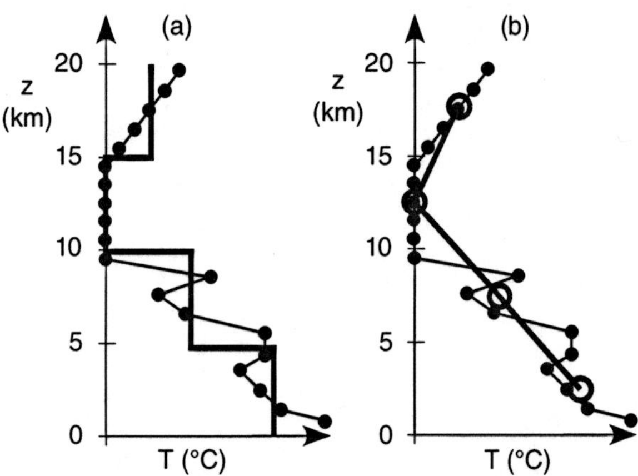

Figure 8.19
Retrieved soundings (a, thick line), are usually plotted as (b) single data points (open circles) for each channel, connected by straight lines.

Retrieval Corollary 2: If weighting functions from different channels have significant overlap in altitude and similar shapes, then they do not provide independent observations of the atmosphere. For this situation, if there are also measurement errors in the radiances or the weighting functions, then the sounding can retrieve **fewer** than one piece of temperature data per sounder channel. (See the "Higher Math" box later in this section for a demonstration.)

Illustration of Retrieval Corollary 1 (Non-overlapping Weighting Functions)

Consider a slightly more realistic illustration of a perfect (idealized) case where the weighting functions do not overlap vertically between different channels (Fig. 8.18a). The relationship between actual temperatures (thin line) and the resulting temperature retrievals are sketched. Namely, the weighting functions are independent of each other, allowing the four channels to retrieve four independent temperatures, as plotted by the thick line (Fig. 8.18b). The thick line is the retrieved sounding.

Instead of plotting the retrieved sounding as a sequence of vertical line segments as shown in Fig. 8.19a, it is often plotted as data points. For our four independent channels, we would get four data points (large, open circles), and the resulting sounding line is drawn by connecting the circles (Fig. 8.19b). The retrieved sounding (thick line in Fig. 8.19b) does a good job of capturing the gross-features of the temperature profile, but misses the fine details such as sharp temperature inversions.

Illustration of Retrieval Corollary 2 (Overlapping Weighting Functions)

With non-overlapping weighting functions the sounding-retrieval process was easy. For more realistic overlapping weighting functions, it becomes very difficult, as summarized in **Retrieval Corollary 2**, given in the left column.

We can first study this as a **forward problem**, where we pretend we already know the temperature profile and want to find the radiances that the satellite would see. This approach is called an **Observing System Simulation Experiment (OSSE)**, used by instrument designers to help anticipate the radiances arriving at the satellite, so that they can fix problems before the satellite is launched. We anticipate that the radiance received in one channel depends on the temperatures at many heights. Easy!

Later, we will approach this more realistically; i.e., as an **inverse problem** where we have satellite-measured radiances and want to determine atmospheric temperatures. The inverse problem for overlapping weighting functions requires us to solve a set of coupled nonlinear equations. Nasty!

To illustrate the forward problem, suppose that idealized weighting functions of Fig. 8.20 and Table 8-4 approximate the actual GOES weighting functions for sounder channels 1 – 4. For any one channel, the sum of the weights equals one, as you can check from the data in the figure. Each weighting function peaks at a different height. For simplicity, look at only the atmospheric contribution to the radiances and ignore the surface (skin) term.

For this forward example, suppose the temperatures for each layer (from the top down) are $T = -40$, -60, -30, and $+20$°C, as plotted in Fig. 8.21. Namely, we are using a coarse-resolution T profile, because we already know from Corollary 1 that retrieval methods cannot resolve anything finer anyway.

For each channel, we can write the radiative transfer equation (8.5). To simplify these equations, use $\lambda = 1, 2, 3, 4$ to index the wavelengths of sounder channels 1, 2, 3, 4. Also, use $j = 1, 2, 3, 4$ to index the four layers of our simplified atmosphere, from the top down. The radiative transfer equation for our simple 4-layer atmosphere, without the skin term, is:

$$L_\lambda = \sum_{j=1}^{4} B_\lambda(T_j) \cdot W_{\lambda,j} \tag{8.9}$$

After expanding the sum, this equation can be written for each separate channel as:

$$L_1 = \tag{8.10a}$$
$$B_1(T_1) \cdot W_{1,1} + B_1(T_2) \cdot W_{1,2} + B_1(T_3) \cdot W_{1,3} + B_1(T_4) \cdot W_{1,4}$$

$$L_2 = \tag{8.10b}$$
$$B_2(T_1) \cdot W_{2,1} + B_2(T_2) \cdot W_{2,2} + B_2(T_3) \cdot W_{2,3} + B_2(T_4) \cdot W_{2,4}$$

$$L_3 = \tag{8.10c}$$
$$B_3(T_1) \cdot W_{3,1} + B_3(T_2) \cdot W_{3,2} + B_3(T_3) \cdot W_{3,3} + B_3(T_4) \cdot W_{3,4}$$

$$L_4 = \tag{8.10d}$$
$$B_4(T_1) \cdot W_{4,1} + B_4(T_2) \cdot W_{4,2} + B_4(T_3) \cdot W_{4,3} + B_4(T_4) \cdot W_{4,4}$$

j: layer 1 layer 2 layer 3 layer 4

Because of the wide vertical spread of the weights, the radiance in each channel depends on the temperature at many levels, NOT just the one level at the peak weight value. But the radiative transfer equations are easy to solve; namely, given T and W, it is straight forward to calculate the radiances L, because we need only solve one equation at a time. I did this on a spreadsheet — the resulting radiances for each channel are in Table 8-5.

Now consider the more realistic **inverse** problem. To find the temperature T for each layer, knowing the radiance L from each sounder channel, you must solve the whole set of coupled equations (8.10a-d). These eqs. are nonlinear in temperature, due to the Planck function B. The number of equations equals the number of sounder channels. The number of terms in each equation depends on how finely discretized are the sounder profiles from Fig. 8.9,

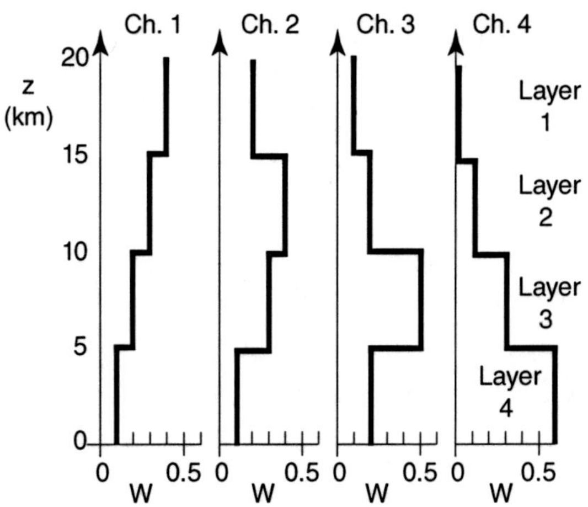

Figure 8.20
Idealized weighting functions for sounder channels Ch. 1 – 4 .

Table 8-4. Idealized sounder weights $W_{\lambda,j}$.

Vector	Channel (λ)	Layer in Atmosphere (j)			
		1 (top)	2	3	4 (bottom)
A	1	0.4	0.3	0.2	0.1
B	2	0.2	0.4	0.3	0.1
C	3	0.1	0.2	0.5	0.2
D	4	0	0.1	0.3	0.6

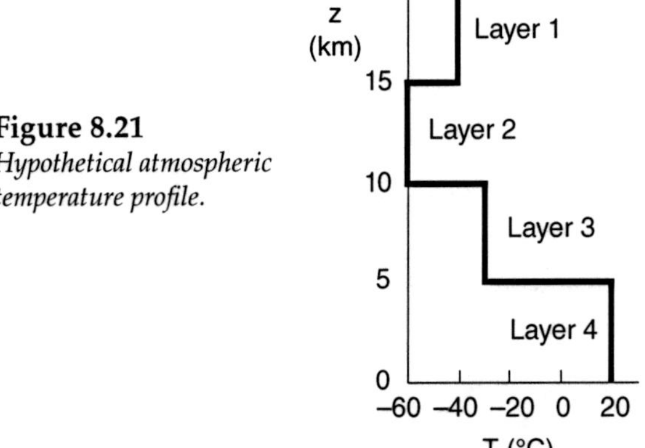

Figure 8.21
Hypothetical atmospheric temperature profile.

Table 8-5. Solution of the forward radiative transfer equation for the 4-layer illustrative atmosphere. The actual wavelengths λ for each channel were copied from Table 8-3.

Channel	λ (µm)	L (W·m^{-2} ·µm^{-1} ·sr^{-1})
1	14.7	2.85
2	14.4	2.87
3	14.0	3.64
4	13.7	5.40

Sample Application

Given a 4-channel sounder with weighting functions in Fig. 8.20 & Table 8-4. The satellite observed radiances are: $L_1 = 2.85$, $L_2 = 2.87$, $L_3 = 3.64$, & $L_4 = 5.4$ W·m^{-2} ·µm^{-1}·sr^{-1} . Retrieve the temperature sounding.

Find the Answer:

Given: *L(obs)* above; weights *W* from Table 8-4.
Find: Temperatures T_1 to T_4 (K), where $()_1$ is top layer.

I did this manually by trial and error on a spreadsheet — a bit tedious, but it worked.

• First, guess a starting sounding of *T(guess)* = –20°C everywhere (= 253K). See Table 8-6.
• Use eqs. (8.10) to compute radiances *L(guess)* for each channel.
• For Ch.1: $error^2_1 = (L_{1\ guess} - L_{1\ obs})^2$; etc. for Ch.2-4.
• Compute *Sum of error*^{2} = $error^2_1 + error^2_2$ + etc.
This initial total error was very large (= 3.7255).
• Experiment with different values of T_4 (temperature of layer 4) to find the "best" value that gives the least *Sum of error*^{2} so far. Then, keeping this "best" T_4 value, experiment with T_3 , finding its best value. Proceed similarly for layers 2 & 1. This completes iteration 1, as shown in Table 8-6.
• Keeping the best T_1 through T_3, experiment with T_4 again — you will find a different "best" value. Then do layers 3, 2, & 1 in succession. This ends iteration 2.
• Keep iterating for layers 4, 3, 2, 1. Boring. But each time, the *Sum of error*^{2} becomes less and less.

I eventually got tired and gave up. The answer is *T(final guess)* in Table 8-6.

Check: The actual *T* is given in Fig. 8.21, & listed here:

Height:	1 (top)	2	3	4 (bot.)
T (actual)	–40	–60	–30	+20
T (final guess)	<u>–45</u>	<u>–50.1</u>	<u>–33.6</u>	<u>+20.1</u>

Exposition: Not a perfect answer, which we know only because this exercise was contrived from an earlier illustration where we knew the actual temperature.

Why was it not perfect? It was very difficult to get the temperatures for layers 1 and 2 to converge to a stable solution. Quite a wide range of temperature values for these layers gave virtually the same error, so it was difficult to find the best temperatures. This is partly related to the close similarity in weighting functions for Channels 1 and 2, and the fact that they had large spread over height with no strong peak in any one layer. The radiances from these two channels are not independent of each other, resulting in a solution that is almost singular (not well behaved in a mathematical sense; not allowing a solution).

These difficulties are typical. See the "Info Box" on the next page.

Table 8-6. Approximate solution to the Sample Applic. inverse problem. *T* (°C) is air temperature. *L* (W·m^{-2}·µm^{-1}·sr^{-1}) is radiance. Error = *L(obs) – L(guess)*.

Height:	1 (top)	2	3	4 (bot.)
T (initial guess)	–20	–20	–20	–20
T (1st iteration)	–32.9	–42.9	–43.3	+2.5
T (final guess)	<u>–45</u>	<u>–50.1</u>	<u>–33.6</u>	<u>+20.1</u>

Channel:	1	2	3	4
L (obs)	2.85	2.87	3.64	5.40
L (from T initial guess)	3.70	3.78	3.88	3.95
error^2	0.7189	0.8334	0.0591	2.114
sum of error^2	3.7255			
L (from T final guess)	2.84	2.93	3.60	5.39
error^2	0.00009	0.00367	0.00155	0.00019
sum of error^2	0.0055			

which is related to the number of retrieval altitudes. From Retrieval Corollary 1 there is little value to retrieve more altitudes than the number of sounder channels. For example, the current GOES satellite has 19 channels; hence, we need to solve a coupled set of 19 equations, each with 19 nonlinear terms.

Solving this large set of coupled nonlinear equations is tricky; many different methods are used by government forecast centers and satellite institutes. Here is a simple approach that you can solve on a spreadsheet, which gives an approximate solution:

Start with an initial guess for the temperature of each layer. The better the first guess, the quicker you will converge toward the best answer. Use those temperatures to solve the much easier forward problem; namely, calculate the radiances L_λ from each eq. (8.10) separately. Calculate the squared error ($L_{\lambda\ calc} - L_{\lambda\ obs}$)^2 between the <u>calculated</u> radiances and the <u>observed</u> radiances from satellite for each channel, and then sum the errors to get an overall measure of the quality of the guessed temperature sounding.

Next, try to reduce the overall error by modifying the temperature guesses. For example, vary the guessed temperature for only one atmospheric layer, until you find the temperature that gives the least total error. Then do the same for the next height, and continue doing this for all heights. Then repeat the whole process, from first height to last height, always seeking the minimum error. Keep repeating these steps (i.e., keep **iterating**), until the total error is either zero, or small enough (considering errors in the measured radiances).

Some of the difficulties in sounding retrievals are listed in the Info Box below. In spite of these difficulties, satellite-retrieved soundings are steadily improving, and make an important positive contribution to weather-forecast quality.

INFO • Satellite Retrieval Difficulties

- Radiance is an average from a deep layer, often overlapping with other layers.
- Radiance observations (in different channels) are not independent of each other.
- It is difficult to separate the effects of temperature and water-vapor variations in radiance signal.
- There is not a unique relationship between the spectrum of outgoing radiance and atmospheric temperature and humidity profiles.
- Temperatures are nonlinearly buried within the Planck function (but linear approximations are often used).
- Radiance observations have errors, caused by instrument errors, sampling errors, interference by clouds, and errors in the estimation of the weighting functions.
- Because of all these factors, there are an infinite number of temperature profiles that all satisfy the observed radiances within their error bars. Thus, statistical estimates must be used.
- To help pick the best profile, a good first guess and boundary conditions are critical. (**Retrieval Corollary 3: The retrieved profile looks more like the first guess than like reality.**)
- Satellite-retrieved soundings are most useful in regions (such as over the oceans) lacking other in-situ observations, but in such regions it is difficult to provide a good first guess. Often numerical weather forecasts are used to estimate the first guess, but such forecasts usually deviate quite significantly from reality over ocean data-voids.

HIGHER MATH • Info Projection

By representing weighting functions as vectors, inner (dot •) products show how much information from one vector is contained in (projected on) the other vectors. This helps demonstrate satellite Retrieval Corollary 2: The amount of retrieved info can be <u>less</u> than the number of weighting functions.

To illustrate, consider a 2-layer atmosphere and let each height represent an orthogonal axis in 2-D space. The discrete weight values at each height within any one weighting function give corresponding coordinates for that vector in the 2-D space.

For example, row 1 of the Fig. at right shows weighting functions for two channels, *A* and *B*, which are represented as vectors in the last Fig. of that row. For this special case of non-overlapping weighting functions, *A* and *B* give completely independent

(continues in the next column)

Higher Math • Info Projection *(continuation)*

pieces of info, as indicated by the orthogonal vectors. The total information value = 2.

Row 2 of the Fig illustrates two identical weighting functions. Their vector representations perfectly coincide in the 2-D space. While the one weighting function gives us information, the second function tells us nothing new. Thus, total info value = 1.

Row 3 of the Fig. shows two different, but broadly overlapping, weighting functions. When plotted in vector space, we see that much of *B* projects onto *A*. Namely, much of *B* tells us nothing new. The only new contribution from vector *B* is the component that is orthogonal to *A*. For this example, only 30% of *B* gives new info (B_{new}); therefore, the total info value is 1.3 (= 1.0 from *A* + 0.3 from *B*).

In a sum of squares sense, the fraction of vector *B* = (W_{B1}, W_{B2}, W_{B3}, ...) that is NOT explained by (i.e., not projected onto) vector *A* = (W_{A1}, W_{A2}, W_{A3}, ...) is:

$$f = 1 - \frac{(A \bullet B)^2}{|A|^2 \; |B|^2} = 1 - \frac{(\Sigma_j \; W_{A,j} \; W_{B,j})^2}{(\Sigma_j \; W_{A,j}^2)(\Sigma_j \; W_{B,j}^2)} \tag{8.11}$$

For higher-order vector spaces, we can use (8.11) to successively find the fraction of vector *C* that provides new info; namely, the portion of *C* that is orthogonal both to *A* and to the new-info part of *B*.

The vector representing the new-info part of *B* (the portion of *B* not projecting on *A*) is:

$$B_{new} = B_{\perp A} = B - \frac{(A \bullet B)}{|A|^2} A \tag{8.12}$$

Using this info projection method on the weights of Fig. 8.20 (with vectors *A* to *D* identified in Table 8-4) give: *A*: 1 , B_{new}: 0.19 , $C_{additional\ contrib.\ to\ Bnew} = 0.267$, $C_{new} = 0.258$, $D_{contrib.\ to\ Bnew} = 0.115$, $D_{contrib.\ to\ Cnew} = 0.303$, $D_{new} = 0.419$.

The total info value sums to <u>2.55</u>. Thus, the 4 weighting functions give fewer than 4 independent pieces of info about the temperature profile.

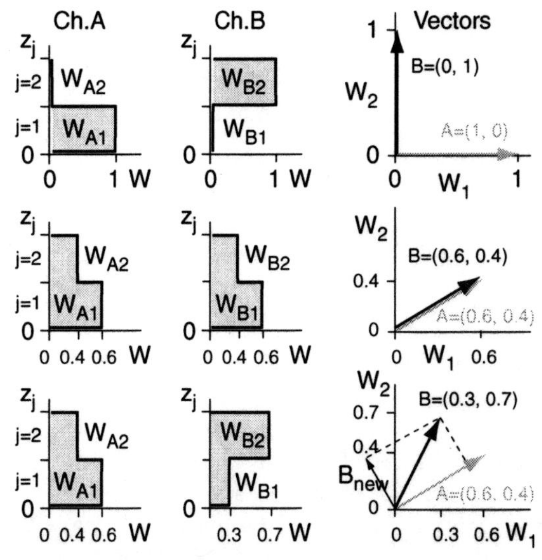

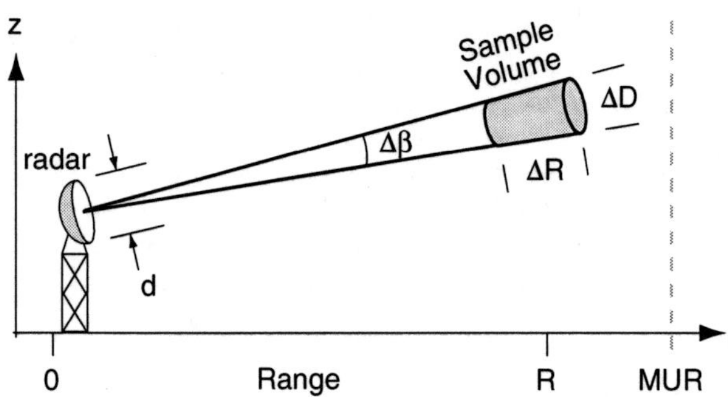

Figure 8.22
Radar beam. MUR is maximum unambiguous range.

WEATHER RADARS

Fundamentals

Weather radars are active sensors that emit pulses of very intense (250 - 1000 kW) microwaves generated by magnetron or klystron vacuum tubes. These transmitted pulses, each of Δt = 0.5 to 10 μs duration, are reflected off a parabolic **antenna dish** that can rotate and tilt, to point the train of pulses toward any azimuth and elevation angle in the atmosphere. **Pulse repetition frequencies** (PRF) are of order 50 to 2000 pulses per second.

The microwaves travel away from the radar at the speed of light through the air ($c \approx 3\text{x}10^8$ m s^{-1}), focused by the antenna dish along a narrow beam (Fig. 8.22). The angular thickness of this beam, called the **beamwidth** $\Delta\beta$, depends on the wavelength λ of the microwaves and the diameter (d) of the parabolic antenna:

$$\Delta\beta = a \cdot \lambda / d \tag{8.13}$$

where a = 71.6°. Larger-diameter antennae can focus the beam into narrower beamwidths. For many weather radars, $\Delta\beta < 1°$.

The volume sampled by any one pulse is shaped like a slightly tapered cylinder (i.e., the frustum of a cone; Fig. 8.22). Typical pulse lengths ($c \cdot \Delta t$) are 300 to 500 m. However, the sampled length ΔR is half the pulse length, because of the round trip the energy must travel in and out of the sample volume.

$$\Delta R = c \cdot \Delta t / 2 \tag{8.14}$$

The diameter (ΔD) is:

$$\Delta D = R \cdot \Delta\beta \tag{8.15}$$

for beamwidth $\Delta\beta$ in radians (= degrees · π/180°). $\Delta D \approx$ 0.1 km at close range (R), and increases to 10 km at far range. Thus, the resolution of the radar decreases as range increases.

A very small amount (10^{-5} to 10^{-15} W) of the transmitted energy is **scattered** back towards the radar when the microwave pulses hit objects such as **hydrometeors** (rain, snow, hail), insects, birds, aircraft, buildings, mountains, and trees. The radar dish collects these weak **returns** (**echoes**) and focuses them onto a detector. The resulting signal is amplified and digitally processed, recorded, and displayed graphically.

The **range** (radial distance) R from the radar to any target is easily calculated by measuring the round-trip time t between transmission of the pulse and reception of the scattered signal:

$$R = c \cdot t / 2 \tag{•8.16}$$

Sample Application
Given a 10 cm wavelength radar with a 9 m diameter antenna dish, find the beamwidth.

Find the Answer:
Given: λ = 10 cm = 0.1 m , d = 9 m.
Find: $\Delta\beta$ = ?°

Use eq. (8.13): $\Delta\beta$ = (71.6°)·(0.1m)/(9m) = **0.8°**

Check: Units OK. Physics OK. Reasonable value.
Exposition: This large dish and wavelength are used for the USA WSR-88D operational weather radars.

Sample Application
Find the round-trip travel time to a target at R=20 km.

Find the Answer:
Given: R = 20 km = $2\text{x}10^4$ m, $c = 3 \times 10^8$ m s^{-1}.
Find: t = ? μs

Rearrange eq. (8.16): $t = 2\,R / c$
$t = 2 \cdot (2\text{x}10^4\text{m}) / (3\text{x}10^8\text{m s}^{-1}) = 1.33\text{x}10^{-4}$ s = **133 μs**

Check: Units OK. Physics OK.
Exposition: The typical duration of an eye blink is 100 to 200 ms. Thus, the radar pulse could make 1000 round trips to 20 km in the blink of an eye.

Also, the azimuth and elevation angles to the target are known from the direction the radar dish was pointing when it sent and received the signals. Thus, there is sufficient information to position each target in 3-D space within the volume scanned.

Weather radar cannot "see" each individual rain or cloud drop or ice crystal. Instead, it sees the average energy returned from all the hydrometeors within a finite size pulse subvolume. This is analogous to how your eyes see a cloud; namely, you can see a white cloud even though you cannot see each individual cloud droplet.

Weather radars look at three characteristics of the returned signal to help detect storms and other conditions: **reflectivity**, **Doppler shift**, and **polarization**. These will be explained in detail, after first covering a few more radar fundamentals.

Maximum Range

The maximum range that the radar can "see" is limited by both the **attenuation** (absorption of the microwave energy by intervening hydrometeors) and pulse-repetition frequency. In heavy rain, so much of the radar energy is absorbed and scattered that little can propagate all the way through (recall Beer's Law from the Radiation chapter). The resulting **radar shadows** behind strong targets are "blind spots" that the radar can't see.

Even with little attenuation, the radar can "listen" for the return echoes from one transmitted pulse only up until the time the next pulse is transmitted. For those radars where the microwaves are generated by klystron tubes (for which every pulse has exactly the same frequency, amplitude, and phase), any echoes received after this time are erroneously assumed to have come from the second pulse. Thus, any target greater than this **maximum unambiguous range** (**MUR**, or R_{max}) would be erroneously displayed a distance R_{max} closer to the radar than it actually is (Fig. 8.35). MUR is given by:

$$R_{max} = c \,/\, [\, 2 \cdot PRF \,] \qquad \bullet(8.17)$$

Magnetron tubes produce a more random signal that varies from pulse to pulse, which can be used to discriminate between subsequent pulses and their return signals, thereby avoiding the MUR problem.

Scan and Display Strategies

Modern weather radars are programmed to automatically sweep 360° in azimuth α, with each successive scan made at different elevation angles ψ (called **scan angles**). For any one elevation angle, the radar samples along the surface of a cone-shaped region of air (Fig. 8.23). When all these scans are merged into one data set, the result is called a **volume scan**. The radar repeats these volume scans roughly every 5 to 10 minutes to sample the air around the radar.

Data from volume scans can be digitally sliced and displayed on computer in many forms. Animations of these displays over time are called **radar loops**. Typical 2-D displays are:

Sample Application

A 5 cm wavelength radar with 5 m diameter antenna dish transmits 1000 pulses per second, each pulse lasting 1 μs. (a) What are the sample-volume dimensions for a pulse received 100 μs after transmission? (b) Is the range to this sample volume unambiguous?

Find the Answer:
Given: $\lambda = 5$ cm, $d = 5$ m, $PRF = 1000\ s^{-1}$,
$\Delta t = 1\ \mu s$, $t = 100\ \mu s$.
Find: (a) ΔR = ? m, and ΔD = ? m. (b) R_{max} = ? km.

(a) Use eq. (8.14): $\Delta R = (3\times10^8 m/s)\cdot(10^{-6}s)/2$ = **<u>150 m</u>**.
To use eq. (8.15) for ΔD, we first need R and $\Delta\beta$.
From eq. (8.13): $\Delta\beta = (71.6°)\cdot(0.05m)/(5m) = 0.72°$
From eq. (8.16): $R = (3\times10^8 m/s)\cdot(10^{-4}s)/2 = 15$ km.
Use eq. (8.15): $\Delta D = (15000m)\cdot(0.72°)\cdot\pi/(180°)$= **<u>188 m</u>**.

(b) Use eq. (8.17): $R_{max}=(3\times10^8 m/s)/[2\cdot 1000\ s^{-1}]=150$km
But R = 15 km from (a).
<u>Yes, range IS unambiguous</u> because $R < R_{max}$.

Check: Units OK. Physics OK.
Exposition: If the range to the rain shower had been R = 160 km, then the target would have appeared on the radar display at an erroneous range of $R - R_{max}$ = 160 – 150 km = 10 km from the radar, and would be superimposed on any echoes actually at 10 km range.

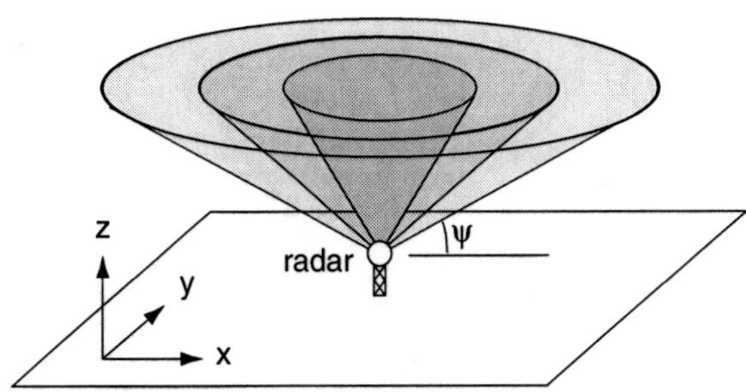

Figure 8.23
Scan surfaces at different elevation angles ψ.

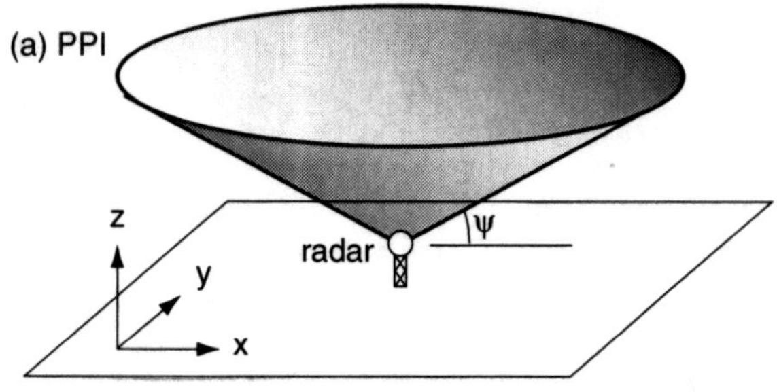

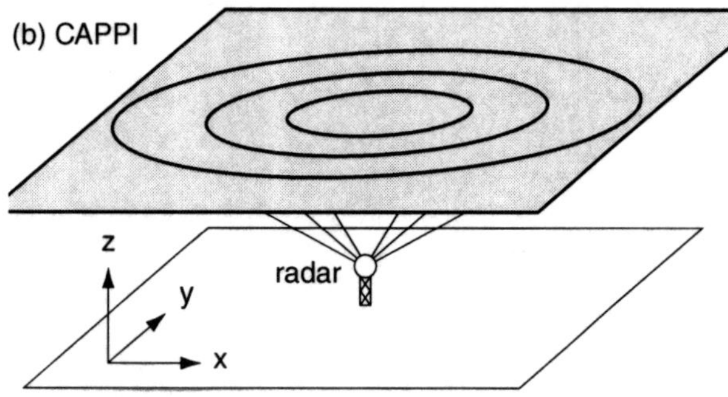

Figure 8.24
(a) Plan-Position Indicator (PPI) display. (b) Constant Altitude Plan Position Indicator (CAPPI) display.

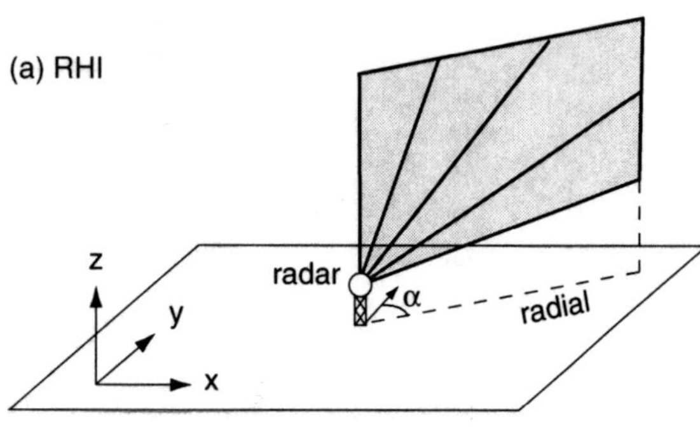

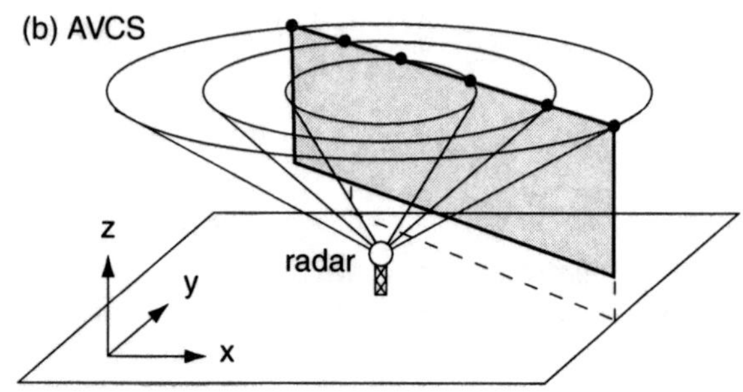

Figure 8.25
(a) Range Height Indicator (RHI) display. (b) Arbitrary Vertical Cross Section (AVCS) display.

- **PPI** (**plan position indicator**), which shows the radar echoes around 360° azimuth, but at only one elevation angle. Namely, this data is from a cone that spans many altitudes (Fig. 8.24a). These displays are often superimposed on background maps showing towns, roads and shorelines.

- **CAPPI** (**constant-altitude plan position indicator**), which gives a horizontal slice at any altitude (Fig. 8.24b). These displays are also superimposed on background maps (Fig. 8.26). At long ranges, the CAPPI is often allowed to follow the PPI cone from the lowest elevation-angle scan.

- **RHI** (**range-height indicator**, see Fig. 8.25a), which is a vertical slice along a fixed azimuth (α) line (called a **radial**) from the radar. It is made by physically keeping the antenna dish pointed at one azimuth while stepping the elevation angle up and down.

- **AVCS** (**arbitrary vertical cross section**) which gives a vertical slice in any horizontal direction through the atmosphere (Fig. 8.25b).

Radar Bands

In the microwave portion of the electromagnetic spectrum are wavelength (λ) bands that have been assigned for use by radar (Fig. 8.4d). Each band is given a letter designation:

- **L band**: λ = 15 to 30 cm. Used in air traffic control and to study **clear-air turbulence** (**CAT**).
- **S band**: λ = 7.5 to 15 cm. Detects precipitation particles, insects, and birds. Long range (roughly 500 km) capabilities, but requires a large (9 m diameter) antenna dish.
- **C band**: λ = 3.75 to 7.5 cm. Detects precipitation particles and insects. Shorter range (roughly 250 km because the microwave pulses are more quickly attenuated by the precipitation), but requires less power and a smaller antenna dish.
- **X band**: λ = 2.5 to 3.75 cm. Detects tiny cloud droplets, ice crystals and precipitation. Large attenuation; therefore very short range. The radar sensitivity decreases inversely with the 4th power of range.
- **K band** is split into two parts:
  Ku band: λ = 1.67 to 2.5 cm
  Ka band: λ = 0.75 to 1.11 cm.
  Detects even smaller particles, but has even shorter range. (The gap at 1.11 to 1.67 cm is due to very strong absorption by a water-vapor line, causing the atmosphere to be opaque at these wavelengths; see Fig. 8.4d).

The US National Weather Service's nationwide network of **NEXRAD** (Next Generation Radar) Weather Surveillance Radars (**WSR-88D**) uses the S band (10 cm) to detect storms and estimate precipitation. The Canadian Meteorological Service uses C band (5 cm) radars. Some North American TV stations also have their own C-band weather radars. Europe is moving to a C-band standard for weather radars, although some S band radars are also used in Spain. Weather-avoidance radars on board commercial aircraft are X band, to help alert the pilots to stormy weather ahead. Police radars include X and K bands, while microwave ovens use S band (12.2 cm). All bands are used for weather research.

Beam Propagation

A factor leading to erroneous signals is **ground clutter**. These are undesired returns from fixed objects (tall towers, trees, buildings, mountains). Usually, ground clutter is found closest to the radar where the beam is still low enough to hit objects, although in mountainous regions ground clutter can occur at any range where the beam hits a mountain. Ground-clutter returns are often strong because these targets are large. But these targets do not move, so they can be identified as clutter and filtered out. However, swaying trees and moving traffic on a highway can confuse some ground-clutter filters.

In a vacuum, the radar pulses would propagate at the speed of light in a straight line. However, in the atmosphere, the denser and colder the air, the slower the speed. A measure of this speed reduction is the **index of refraction**, n:

$$n = c_o / c \qquad \bullet(8.18)$$

where c_o = 299,792,458 m s^{-1} is the speed of electromagnetic radiation in a vacuum, and c is the speed in air. The slowdown is very small, giving index of refraction values on the order of 1.000325.

To focus on this small change, a new variable called the **refractivity,** N, is defined as

$$N = (n - 1) \times 10^6 \qquad \bullet(8.19)$$

For example, if n = 1.000325, then N = 325. For radio waves, including microwaves:

$$N = a_1 \cdot \frac{P}{T} - a_2 \cdot \frac{e}{T} + a_3 \cdot \frac{e}{T^2} \qquad \bullet(8.20)$$

where P is atmospheric pressure at the beam location, T is air temperature (in Kelvin), e is water vapor pressure, a_1 = 776.89 K kPa^{-1}, a_2 = 63.938 K kPa^{-1}, and a_3 = 3.75463x10^6 K^2 kPa^{-1}. In the first term on the RHS, P/T is proportional to air density, according to the ideal gas law. The other two terms account for

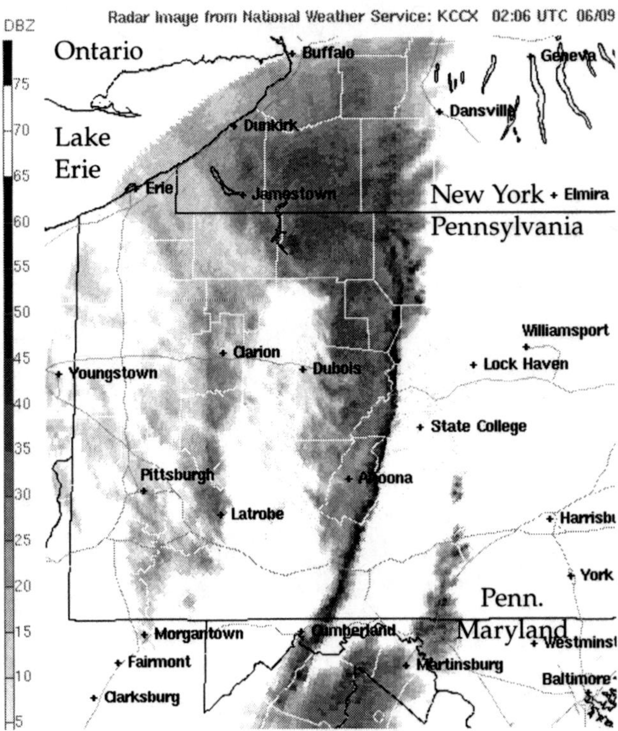

Figure 8.26
CAPPI display of radar reflectivity (related to rainfall intensity) of a violent squall line of thunderstorms (dark curved line through center of image) sweeping across Pennsylvania in 2003. [Courtesy of the US National Weather Service.]

Sample Application
Find the refractivity and speed of microwaves through air of pressure 90 kPa, temperature 10°C, and relative humidity 80%.

Find the Answer:
Given: P = 90 kPa, T = 10°C = 283 K, RH = 80%
Find: N (dimensionless) and c (m·s^{-1})

First, convert RH into vapor pressure. Knowing T, get the saturation vapor pressure from Table 4-1 in the Water Vapor chapter: e_s = 1.233 kPa. Then use eq. (4.14): $e = RH \cdot e_s$ = 0.80·(1.233 kPa). e = 0.986 kPa

Next, use P, T, and e in eq. (8.20):
N= (776.89 K/kPa)·(90kPa)/(283K)
– (63.938 K/kPa)·(0.986kPa)/(283K)
+ (3.75463x10^6 K^2/kPa)·(0.986kPa)/(283K)^2
N = 247.067 – 0.223 + 46.224 = **293.068**

Solve eq. (8.18) for c, and plug in n from (8.19):
$c = c_o$ / [1 + Nx10^{-6}] = (299,792,458 m/s) / 1.000293
c = **299,704,624.2 m s^{-1}**

Check: Units OK. Physics OK.
Exposition: The military carefully monitors and predicts vertical profiles of refractivity to determine whether the signal from their air-defense radars would get ducted or trapped, which would prevent them from detecting enemy aircraft sneaking in above the duct.

(a)

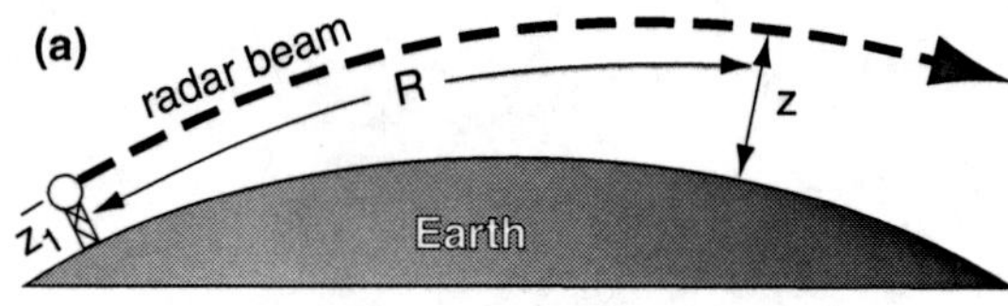

(b)

(c)

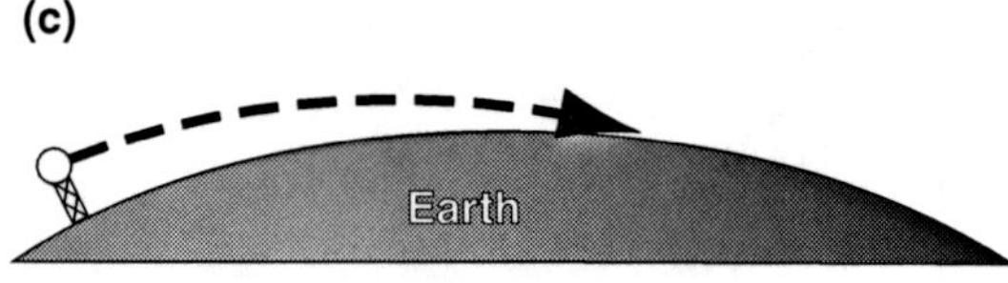

Figure 8.27
(a) Standard refraction of radar beam centerline in Earth's atmosphere. (b) Ducting or trapping over a smooth sea-surface. (c) Trapping over a rough surface.

Sample Application(§)
For a radar on a 10 m tower, plot the centerline height of the beam vs. range for standard refraction, for a 1° elevation angle.

Find the Answer:
Given: $\Delta N/\Delta z = -39\ \text{km}^{-1}$, $z_1 = 10\ \text{m} = 0.01\ \text{km}$,
$R_o = 6371$ km
Find: z (km) vs. R (km)

$\Delta N/\Delta z = -39\ \text{km}^{-1}$. Thus $\Delta n/\Delta z = -39\text{x}10^{-6}\ \text{km}^{-1}$.
Use eq. (8.21):
$R_c = (6371\ \text{km}) / [1+(6356.766\ \text{km})(-39\text{x}10^{-6}\ \text{km}^{-1})]$
$= (6371\ \text{km}) \cdot 1.33 = 8473\ \text{km}$
Use eq. (8.22): $z = 0.01\ \text{km} - 8473\ \text{km}$
$+ [\ R^2 + (8473\ \text{km})^2 + 2 \cdot R \cdot (8473\ \text{km}) \cdot \sin(1°)\ ]^{1/2}$
For example, z = **2.35 km** at R = 100 km. Using a spreadsheet for many different R values gives:

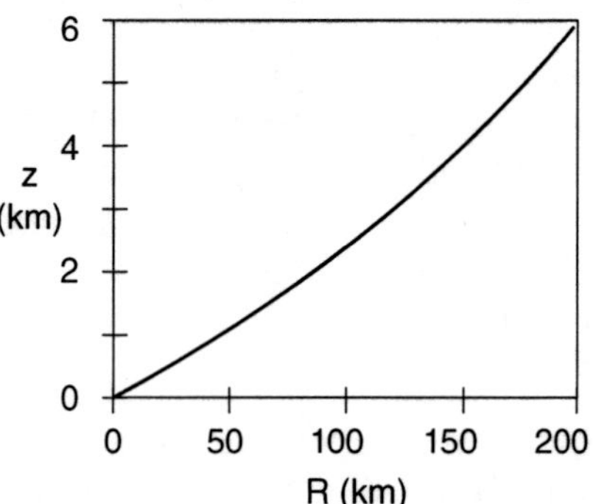

Check: Units OK. Physics OK. Graph reasonable.
Exposition: Why does the line curve up with increasing range? Although the radar beam bends downward, the Earth's surface curves downward faster.

the polarization of microwaves by water vapor (polarization is discussed later). The parameters in this formula have been updated to include current levels of CO_2 (order of 375 ppm) in the atmosphere.

Although small, the change in propagation speed through air is significant because it causes the microwave beam to **refract** (bend) toward the denser, colder air. Recall from chapter 1 that density decreases nearly exponentially with height, which results in a vertical gradient of refractive index $\Delta n/\Delta z$. This causes microwave beams to bend downward. But the Earth's surface also curves downward relative to the starting location. When estimating the beam height z above the Earth's surface at slant range R, you must include both effects.

One way to estimate this geometrically is to assume that the beam travels in a straight line, but that the Earth's surface has an effective radius of curvature R_c of

$$R_c = \frac{R_o}{1 + R_o \cdot (\Delta n / \Delta z)} = k_e \cdot R_o \qquad (8.21)$$

where $R_o = 6371$ km is the average Earth radius. The height z of the center of the radar beam can then be found from the **beam propagation equation**:

$$z = z_1 - R_c + \sqrt{R^2 + R_c^{\ 2} + 2R \cdot R_c \cdot \sin\psi} \qquad (8.22)$$

knowing the height z_1 of the radar above the Earth's surface, and the radar elevation angle ψ.

In the bottom part of the atmosphere, the vertical gradient of refractive index is roughly $\Delta n/\Delta z = -39\text{x}10^{-6}\ \text{km}^{-1}$ (or $\Delta N/\Delta z = -39\ \text{km}^{-1}$) in a standard atmosphere. The resulting **standard refraction** of the radar beam gives an effective Earth radius of approximately $R_c \approx (4/3) \cdot R_o$ (Fig. 8.27a). Thus, $k_e = 4/3$ in eq. (8.21) for standard conditions.

Superimposed on this first-order effect are second-order effects due to temperature and humidity variations that cause changes in refractivity gradient (Table 8-7). For example, when a cool humid boundary layer is capped by a hot dry temperature inversion, then conditions are right for a type of **anomalous propagation** called **superrefraction.** This is where the radar beam has a smaller radius of curvature than the Earth's surface, and thus will bend down toward the Earth.

With great enough superrefraction, the beam hits the ground and is absorbed and/or causes a ground-

Table 8-7. Radar beam propagation conditions.

$\Delta N/\Delta z$ (km^{-1})	Refraction
> 0	subrefraction
0 to –79	normal range
–39	standard refraction
–79 to –157	superrefraction
< –157	trapping or ducting

clutter return at large range if the Earth's surface is rough there (Fig. 8.27c). However, for smooth regions of the Earth's surface such as a waveless ocean, the beam can "bounce" repeatedly, which is called **ducting** or **trapping** (Fig. 8.27b).

When cooler, moister air overlies warmer drier air, then **subrefraction** occurs, where the radius of curvature is larger than for standard refraction. This effect is limited by the fact that atmospheric lapse rate is rarely statically unstable over large depths of the mid and upper troposphere.

Reflectivity

The Radar Equation

Recall that the radar transmits an intense pulse of microwave energy with power P_T (≈ 750 kW for WSR-88D radars). Particles in the atmosphere scatter a miniscule amount of this energy back to the radar, resulting in a small received power P_R. The ratio of received to transmitted power is explained by the **radar equation**:

$$\left[\frac{P_R}{P_T}\right] = [b]\cdot\left[\frac{|K|}{L_a}\right]^2\cdot\left[\frac{R_1}{R}\right]^2\cdot\left[\frac{Z}{Z_1}\right] \qquad \bullet(8.23)$$

$$\begin{bmatrix}\text{relative}\\ \text{received}\\ \text{power}\end{bmatrix} = \begin{bmatrix}\text{equip-}\\ \text{ment}\\ \text{effects}\end{bmatrix}\cdot\begin{bmatrix}\text{atmos-}\\ \text{pheric}\\ \text{effects}\end{bmatrix}\cdot\begin{bmatrix}\text{range}\\ \text{inverse}\\ \text{square}\end{bmatrix}\cdot\begin{bmatrix}\text{target}\\ \text{reflect-}\\ \text{ivity}\end{bmatrix}$$

where each factor in brackets is dimensionless.

The most important parts of this equation are the **refractive-index magnitude** $|K|$, the range R between the radar and target, and the **reflectivity factor** Z. These are summarized below. The Info Box has more details on the radar equation.

$|K|$ (dimensionless): Liquid drops are more efficient at scattering microwaves than ice crystals. $|K|^2 \approx 0.93$ for liquid drops, and $|K|^2 \approx 0.208$ for ice (assuming spherical ice particles of the same mass as the actual snow crystal).

R (km): Returned energy from more distant targets is weaker than from closer targets, as given by the inverse-square law (see Radiation chapter).

Z (mm^6 m^{-3}): Larger numbers (N) of larger-diameter (D) drops in a given volume (Vol) of air will scatter more microwave energy. This **reflectivity factor** is given by:

$$Z = \frac{\sum^{N} D^6}{Vol.} \qquad \bullet(8.24)$$

where the sum is over all N drops, and where each drop could have a different D.

INFO • Deriving the Radar Equation

The radar transmits a burst of power P_T (W). As this pulse travels range R to a target drop, the energy flux (W m^{-2}) diminishes as it spreads to cover spherical surface area $4\pi R^2$, according to the inverse square law (see the Radiation chapter). But the parabolic antenna dish focuses the energy, resulting in antenna gain G relative to the normal spherical spread of energy. While propagating to the drop, some of the energy is lost along the way, as described by attenuation factor L_a. The net energy flux E_i (W m^{-2}) reaching the drop is $E_i = P_T \cdot G \,/\, (4\pi R^2 \cdot L_a)$ (R1)

If that incident "power per unit area" hits a drop that has **backscatter cross-sectional** area σ, then power P_{BS} (W) gets scattered in all directions:

$$P_{BS} = E_i \cdot \sigma \qquad (R2)$$

This miniscule burst of return power spreads to cover a spherical surface area, and is again attenuated by L_a while en route distance R back to the radar. The returning energy flux E_R (W m^{-2}) at the radar is

$$E_R = P_{BS} \,/\, (4\pi R^2 \cdot L_a) \qquad (R3)$$

Eq. (R3) is similar to (R1), except that the drop does not focus the energy, therefore there is no gain G.

But when this flux is captured by the antenna with area A_{ant} and focused with gain G onto a detector, the resulting received power per drop P_R' (W) is

$$P_R' = E_R \cdot A_{ant} \cdot G \qquad (R4)$$

The effective antenna area depends on many complex antenna factors, but the net result is:

$$A_{ant} = \lambda^2 \,/\, [8\pi\cdot\ln(2)] \qquad (R5)$$

The radar sample volume is much larger than a single drop, thus the total returned power P_R is the sum over N drops:

$$P_R = \sum^{N} P_R' \qquad (R6)$$

These N drops are contained in the sample volume Vol of air (Fig. 8.22). Assuming a cylindrical shape of length ΔR (eq. 8.14) and diameter ΔD (eq. 8.15), gives

$$Vol = \pi R^2 \cdot (\Delta\beta)^2 \cdot c \cdot \Delta t \,/\, 8 \qquad (R7)$$

where $\Delta\beta$ is beam width (radians), c is microwave speed through air, and Δt is the radar pulse duration.

For hydrometeor particle diameters D less than a third of the radar wavelength λ, the backscatter cross-section area σ is well approximated by a relationship known as **Rayleigh scattering**:

$$\sigma = \pi^5 \cdot |K|^2 \cdot D^6 \,/\, \lambda^4 \qquad (R8)$$

where $|K|$ is the **refractive index magnitude** of the drop or ice particle.

Combining these 8 equations, and using eq. (8.24) to define the reflectivity factor Z, gives the traditional form of the **radar equation**: (R9)

$$P_R = \left[\frac{c\cdot\pi^3}{1024\cdot\ln(2)}\right]\cdot\left[\frac{P_T\cdot G^2\cdot(\Delta\beta)^2\cdot\Delta t}{\lambda^2}\right]\cdot\left[\left(\frac{|K|}{L_a\cdot R}\right)^2\cdot Z\right]$$

P_R = [*constant*] · [*radar characteristics*] · [*atmos. char.*]

When the right side is multiplied by Z_1/Z_1 and rearranged into dimensionless groups, the result is the version of the radar equation shown as eq. (8.23).

Sample Application

For a 5 cm radar, the energy scattered from a 1 mm diameter drop is how much greater than that from a 0.5 mm drop?

Find the Answer:

Given: $\lambda = 5$ cm, $D_1 = 0.5$ mm, $D_2 = 1$ mm

Find: $\sigma_{scat2} / \sigma_{scat1}$ (dimensionless)

For Rayleigh scattering [see Info Box eq. (R8)]:

$\sigma_{scat2} / \sigma_{scat1} = (D_2/D_1)^6 = 2^6 = \underline{\mathbf{64}}$.

Check: Units OK. Physics OK.

Exposition: Amazingly large difference. Double the drop size and get 64 times the scattered energy.

Sample Application

WSR-88D radar detects a 40 dBZ rain shower 20 km from the radar, with no other rain detected. Transmitted power was 750 kW, what is the received power?

Find the Answer:

Given: Echo = 40 dBZ, $R = 20$ km, $P_T = 750{,}000$ W,

$|K|^2 = 0.93$ for rain (i.e., liquid), $b = 14{,}255$,

$R_1 = 2.17\text{x}10^{-10}$ km for WSR-88D.

Find: $P_R = ?$ W

Assume: $L_a = 1.0$ (no attenuation), because there are no other rain showers between the radar and target.

Use eq. (8.27) and rearrange to solve for Z/Z_1 :

$Z/Z_1 = 10^{(\text{dBZ}/10)} = 10^{(40/10)} = 10^4$.

Use the radar equation (8.23): $P_R =$

$(7.5\text{x}10^5\text{W})\cdot(14{,}255)\cdot[(0.93)\cdot(2.17\text{x}10^{-10}\text{km})/(20\text{km})]^2\}\cdot[10^4]$

$P_R = \underline{\mathbf{1.09 \times 10^{-8}}}$ W .

Check: Units OK. Physics OK.

Exposition: With so little power coming back to the radar, very sensitive detectors are needed. WSR-88D radars can detect signals as weak as $P_R \approx 10^{-15}$ W.

Normal weather radars use the same parabolic antenna dish to receive signals as to transmit. But such an intense pulse is transmitted that the radar antenna electrically "rings" after the transmit impact for 1 μs, where this "ringing" sends strong power as a false signal back toward the receiver. To compensate, the radar must filter out any received energy during the first 1 μs before it is ready to detect the weak returning echoes from true meteorological targets. For this reason, the radar cannot detect meteorological echoes within the first 300 m or so of the radar.

Even when looking at the same sample volume of rain-filled air, the returned power varies considerably from pulse to pulse. To reduce this noise, the WSR-88D equipment averages about 25 sequential pulses together to calculate each smoothed reflectivity value that is shown on the radar display for each sample volume.

The other factors in the radar equation are:

$Z_1 = 1\ \text{mm}^6\ \text{m}^{-3}$ is the **reflectivity unit factor**.

$R_1 = \text{sqrt}(Z_1 \cdot c \cdot \Delta t/\lambda^2)$ (8.25)

is a **range factor** (km), where $c \approx 3\text{x}10^8$ m s^{-1} is microwave speed through air. For WSR-88D radar, wavelength $\lambda = 10$ cm and pulse duration is $\Delta t \approx 1.57$ μs, giving $R_1 \approx 2.17\text{x}10^{-10}$ km.

L_a is a dimensionless **atmospheric attenuation factor** ($L_a \ge 1$) accounting for one-way losses by absorption and scattering as the microwave pulse travels between the radar and the target drop. $L_a = 1$ means no attenuation, and increasing values of L_a imply increasing attenuation. For example, the signal returning from a distant rain shower is diminished (appears weaker than it actually is) if it travels through a nearby shower en route to the radar. If you don't know L_a, then assume $L_a \approx 1$.

$b = \pi^3 \cdot G^2 \cdot (\Delta\beta)^2/[1024 \cdot \ln(2)]$ (8.26)

is a dimensionless **equipment factor**. For WSR-88D, the antenna gain is $G \approx 45.5$ dB $= 35{,}481$ (dimensionless), and the beam width is $\Delta\beta \approx 0.95° = 0.0161$ radians, giving $b \approx 14{,}255$.

Of most interest to meteorologists is the reflectivity factor Z, because a larger Z is usually associated with heavier precipitation. Z varies over a wide range of magnitudes, so **decibels** (dB) of Z (namely, **dBZ**) are often used to quantify radar reflectivity:

$$\text{dBZ} = 10 \cdot \log(Z/Z_1) \qquad \bullet(8.27)$$

where this is a common logarithm (base 10). Although dBZ is dimensionless, the suffix "dBZ" is added after the number (e.g., 35 dBZ).

dBZ can be calculated from radar measurements of P_R/P_T by using the log form of the radar eq.: •(8.28)

$$dBZ = 10\left[\log\left(\frac{P_R}{P_T}\right) + 2\log\left(\frac{R}{R_1}\right) - 2\log\left|\frac{K}{L_a}\right| - \log(b)\right]$$

Although the R/R_1 term compensates for most of the decrease echo strength with distance, atmospheric attenuation L_a still increases with range. Thus dBZ still decreases slightly with increasing range.

Reflectivity or **echo intensity** is the term used by meteorologists for dBZ, and this is what is displayed in **radar reflectivity images**. Typical values are –28 dBZ for haze, –12 dBZ for insects in clear air, 25 to 30 dBZ in dry snow or light rain, 40 to 50 dBZ in heavy rain, and up to 75 dBZ for giant hail.

Rainfall Rate Estimated by Radar Reflectivity

A tenuous, but useful, relationship exists between rainfall rate and radar echo intensity. Both increase with the number and size of drops in the air. However, this relationship is not exact. Complicating factors on the rainfall rate RR include the drop terminal velocity, partial evaporation of the drops while falling, and downburst speed of the air containing the drops. Complicating factors for the echo intensity include the bright-band effect (see next page), and unknown backscatter cross sections for complex ice-crystal shapes.

Nonetheless, an empirically tuned approximation can be found:

$$RR = a_1 \cdot 10^{(a_2 \cdot dBZ)} \qquad \bullet(8.29)$$

where a_1 = 0.017 mm h^{-1} and a_2 = 0.0714 dBZ^{-1} (both a_2 and dBZ are dimensionless) are the values for the USA WSR-88D radars. This same equation can be written as:

$$Z = a_3 \cdot RR^{a_4} \qquad (8.30)$$

where a_3 = 300, and a_4 = 1.4, for RR in mm h^{-1} and Z in mm^6 m^{-3}. Equations such as (8.29) and (8.30) are called **Z-R relationships**.

To simplify storm information presented to the public, radar reflectivities are sometimes binned into **categories** or **levels** with names or numbers representing **rainfall intensities.** For example, aircraft pilots and controllers use the terms in Fig. 8.28. TV weathercasters often display the intensity categories in different colors. Different agencies in different countries use different thresholds for precipitation categories. For example, in Canada moderate precipitation is defined as 2.5 to 7.5 mm h^{-1}.

Z-R relationships can never be perfect, as is demonstrated here. Within an air parcel of mass 1 kg, suppose that m_L = 5 g of water has condensed and falls out as rain. If that 5 g is distributed equally among N = 1000 drops, then the diameter D of each drop is 2.12 mm, from:

$$D = \left[\frac{6 \cdot m_L}{\pi \cdot N \cdot \rho_{liq}} \right]^{1/3} \qquad (8.31)$$

where ρ_{liq} = 10^3 kg m^{-3} is the density of liquid water. However, if the same mass of water is distributed among N = 10,000 droplets, then D = 0.98 mm. When these two sets of N and D values are used in eq. (8.24), the first scenario gives 10 times the reflectivity Z as the second, even though they have identical total mass of liquid water m_L and identical rainfall amount. Nonetheless, Z-R relationships are useful at giving a first approximation to rainfall rate.

Sample Application

If radar reflectivity is 35, what is the ATC precipitation intensity term? Also, estimate the rainfall rate.

Find the Answer

Given: reflectivity = 35 dBZ
Find: ATC term and RR = ? mm h^{-1}.

Use Fig. 8.28: ATC weather radar term = **moderate**
Use eq. (8.29):
RR = (0.017mm h^{-1})·10$^{[0.0714(1/dBZ)\cdot 35dBZ]}$ = **5 mm h^{-1}**.

Check: Units OK. Physics OK.
Exposition: Heavy and extreme echoes are often associated with thunderstorms, which pilots avoid.

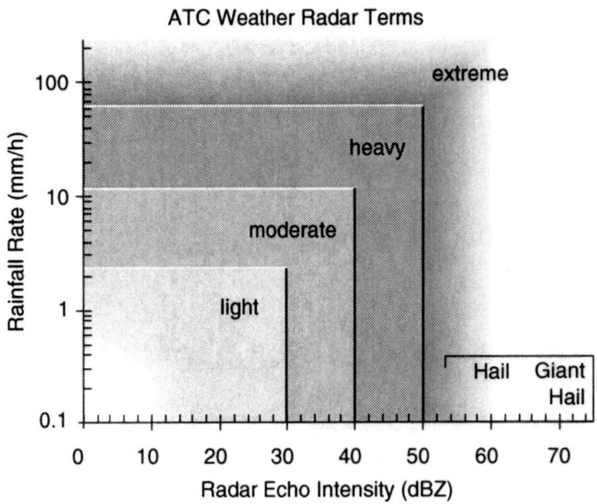

Figure 8.28
Precipitation-intensity terms as a function of weather-radar echo reflectivity, as used by the USA Air Traffic Control (ATC). Rainfall rate is approximate, based on eq. (8.29).

Sample Application

Air parcels A and B both have liquid-water mixing ratio 3 g kg^{-1}, but A has 1000 active liquid-water nuclei and B has 5,000. Compare the dBZ from both parcels.

Find the Answer:

Given: r_L = 3 g kg^{-1}, ρ_L = 10^3 kg m^{-3}
N_A = 1000 , N_B = 5000 .
Find: dBZ$_A$ and dBZ$_B$ (dimensionless)
Assume air density is ρ = 1 kg/m^3. Assume *Vol.*=1 m^3
Assume equal size droplets form on each nucleus, so that $\Sigma D^6 = N \cdot D^6$.

Use eq. (8.31):
D_A=[6·(0.003kg)/(π·1000·(10^3 kg m^{-3})]$^{1/3}$
= 0.00179 m = 1.79 mm

(continues on next page)

Sample Application
(continuation)

D_B=[6·(0.003kg)/(π·5000·(10³ kg m⁻³)]^{1/3}$ = 0.00105 m = 1.05 mm

$$D_B=[6\cdot(0.003\text{kg})/(\pi\cdot 5000\cdot(10^3\text{ kg m}^{-3})]^{1/3}$$
$$= 0.00105 \text{ m} = 1.05 \text{ mm}$$

Use eq. (8.24), but using $N\cdot D^6$ in the numerator:

$$Z_A = 1000\cdot(1.79\text{ mm})^6 /(1\text{ m}^3) = 3.29\text{x}10^4\text{ mm}^6\text{ m}^{-3}$$
$$Z_B = 5000\cdot(1.05\text{ mm})^6 /(1\text{ m}^3) = 6.70\text{x}10^3\text{ mm}^6\text{ m}^{-3}$$

Use eq. (8.27):

$$dBZ_A = 10\cdot\log[\ (3.29\text{x}10^4\text{ mm}^6\text{ m}^{-3})/(1\text{ mm}^6\text{ m}^{-3})\ ]$$
$$= \underline{\mathbf{45.2\ dBZ}}$$
$$dBZ_B = 10\cdot\log[\ (6.70\text{x}10^3\text{ mm}^6\text{ m}^{-3})/(1\text{ mm}^6\text{ m}^{-3})\ ]$$
$$= \underline{\mathbf{38.3\ dBZ}}$$

Check: Units OK. Physics OK.

Exposition: Thus, the same amount of liquid water (3 g kg⁻¹) in each parcel would lead to substantially different radar reflectivities. This is the difficulty of *Z-R* relationships. But not all is lost -- the polarization radar section shows how more accurate rainfall estimates can be made using radar.

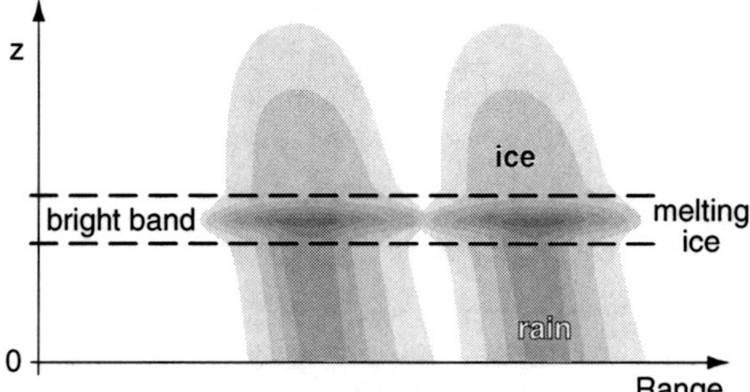

Figure 8.29
RHI sketch of the bright band in two precipitation cells.

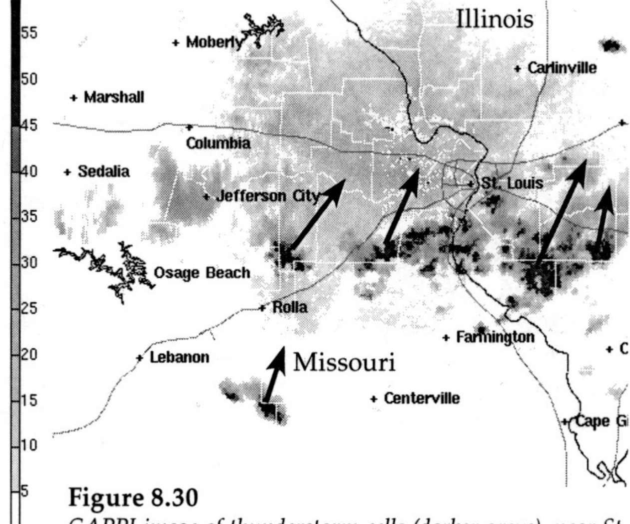

Figure 8.30
CAPPI image of thunderstorm cells (darker greys), near St. Louis, MO, on 22 July 2004. Vectors indicate cell movement. (Courtesy of the US NWS.)

Bright Band

Ice crystals scatter back to the radar only about 20% of the energy scattered by the same amount of liquid water (as previously indicated with the $|K|^2$ factor in the radar equation). However, when ice crystals fall into a region of warmer air, they start to melt, causing the solid crystal to be coated with a thin layer of water. This increases the reflectivity of the liquid-water-coated ice crystals by a factor of 5, causing a very strong radar return known as the **bright band.** Also, the wet outer coating on the ice crystals cause them to stick together if they collide, resulting in a smaller number of larger-diameter snowflake clusters that contribute to even stronger returns.

As the ice crystals continue to fall and completely melt into liquid drops, their diameter decreases and their fall speed increases, causing their drop density and resulting reflectivity to decrease. The net result is that the bright band is a layer of strong reflectivity at the melting level of falling precipitation. The reflectivity in the bright band is often 15 to 30 times greater than the reflectivity in the ice layer above it, and 4 to 9 times greater than the rain layer below it. Thus, Z-R relationships fail in bright-band regions.

In an RHI display of reflectivity (Fig. 8.29) the bright band appears as a layer of stronger returns. In a PPI display the bright band is donut shaped — a hollow circle of stronger returns <u>around the radar</u>, with weaker returns both inside and outside the circle.

Hail

Hail can have exceptionally high radar reflectivities, of order 60 to 75 dBZ, compared to typical maxima of 50 dBZ for heavy rain. Because hailstone size is too large for microwave Rayleigh scattering to apply, the normal Z-R relationships fail. Some radar algorithms diagnose hail for reflectivities > 40 dBZ at altitudes where the air is colder than freezing, with greater chance of hail for reflectivities ≥ 50 dBZ at altitudes where temperature ≤ –20°C.

Other Uses for Reflectivity Data

Radar reflectivity images are used by meteorologists to **identify storms** (thunderstorms, hurricanes, mid-latitude cyclones), track storm movement, find echo-top height and thunderstorm features, indicate likelihood of hail, and estimate rain rate and flooding potential. For example, Fig. 8.30 shows isolated thunderstorm cells, while Fig. 8.26 shows an organized squall line of thunderstorms. The radar image typically presented by TV weather briefers is the reflectivity image.

Clear-air reflectivities from bugs can help identify cold fronts, dry lines, thunderstorm outflow (gust fronts), sea-breeze fronts, and other boundaries (discussed later in this book). Weather radar can also track bird migration. Clear-air returns are often very weak, so radar **clear-air scan** strategies use a slower azimuth sweep rate and longer pulse duration to allow more of the microwave energy to hit the target and to be scattered back to the receiver.

Doppler Radar

Large **hydrometeors** include rain drops, ice crystals, and hailstones. Hydrometeor velocity is the vector sum of their fall velocity through the air plus the air velocity itself. Doppler radars measure the component of hydrometeor velocity that is away from or toward the radar (i.e., **radial velocity**, Fig. 8.31). Hence, Doppler radars can detect wind components associated with gust fronts and tornadoes.

Radial Velocities

Radars transmit a microwave signal of known frequency ν. But after scattered from moving hydrometeors, the microwaves that return to the radar have a different frequency ν + Δν. Hydrometeors moving toward the radar cause higher returned frequencies while those moving away cause lower. The frequency change Δν is called the **Doppler shift**.

Doppler-shift magnitude can be found from the Doppler equation:

$$|\Delta\nu| = \frac{2 \cdot |M_r|}{\lambda} \qquad \bullet(8.32)$$

where λ is the wavelength (e.g., S-band radars such as WSR-88D use λ = 10 cm), and M_r is the average radial velocity of the hydrometeors relative to the radar. Knowing the speed of light $c = 3\text{x}10^8$ m s^{-1}, you can find the frequency as a function of wavelength:

$$\nu = c / \lambda \quad . \qquad \bullet(8.33)$$

However, the <u>**frequency shift**</u> is so miniscule that it is very difficult to measure (see Sample Application).

> **Sample Application**
> An S-band radar would measure what Doppler-shift magnitude for a tornadic speed of 90 m s^{-1}?
>
> **Find the Answer**
> Given: M_r = 90 m/s, assume λ = 10 cm for S-band.
> Find: Δν = ? s^{-1}
> Use eq. (8.32): |Δν| = [2·(90 m s^{-1})] / (0.1 m) = <u>**1800**</u> s^{-1}
>
> **Check**: Units & physics OK.
> **Exposition**: The Doppler-shift magnitude is less than one part per million, relative to the original transmitted frequency (eq. 8.33) of ν = c/λ = (3x10^8 m s^{-1}) / (0.1 m) = 3x10^9 s^{-1}.

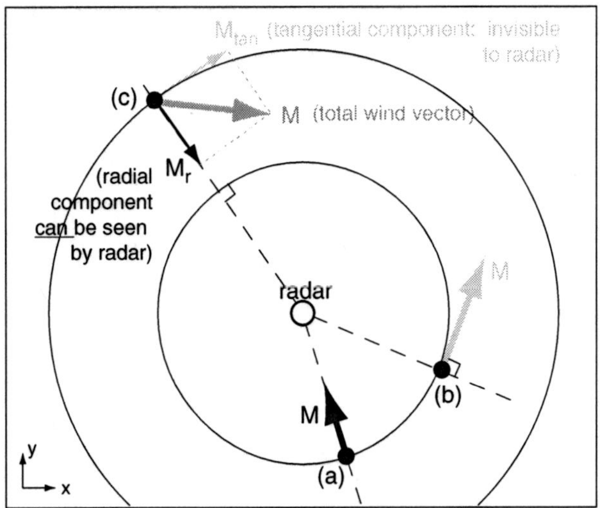

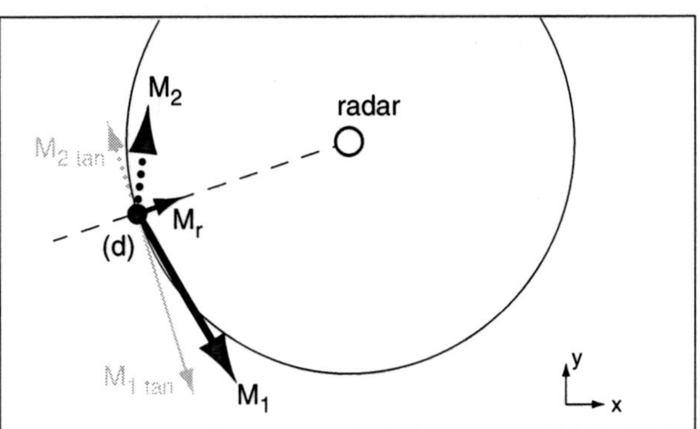

Figure 8.31
Illustration of which velocities can be seen by Doppler radar. For any location where you know the wind vector, mentally draw a circle (centered on the radar) that passes through the location, and draw a radial line (shown dashed in these figs.) from the radar to the location. The radar can see only the component of the wind vector along the radial line, and cannot see the component along the circle. Examples: (a) At point (a) the total wind vector M is along the radial line, hence the radar sees the whole wind. (b) At point (b) the total wind vector M is tangent to the circle, hence the radar sees zero wind there. (c) At point (c) the total wind vector M has a radial component M_r that the radar sees, and a tangential component M_{tan} that is invisible. Hence, only a portion of the total wind is seen there.

Conversely, if you don't know the actual wind but have Doppler radar observations, beware that an infinite number of true wind vectors can create the radial component seen by the radar. For example, at point (d) two completely different wind vectors M_1 and M_2 have the same radial component M_r. Hence, the radar would give you only M_r but you would have no way of knowing the total wind vector. The solution to this dilemma is to have two Doppler radars (i.e., **dual Doppler**) *at different locations that can both scan the same wind location from different directions.*

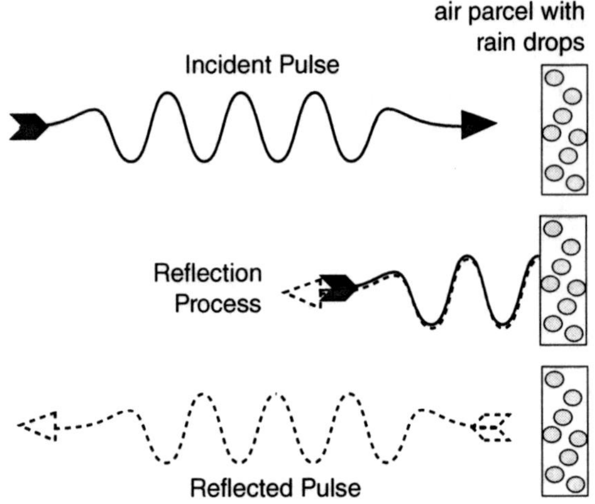

Figure 8.32
A pulse of microwaves scatters off a stationary, droplet-laden air parcel that coincides with the radar sample volume. The same shading (solid for incident pulse, dashed for reflected pulse) is used in the next two figures.

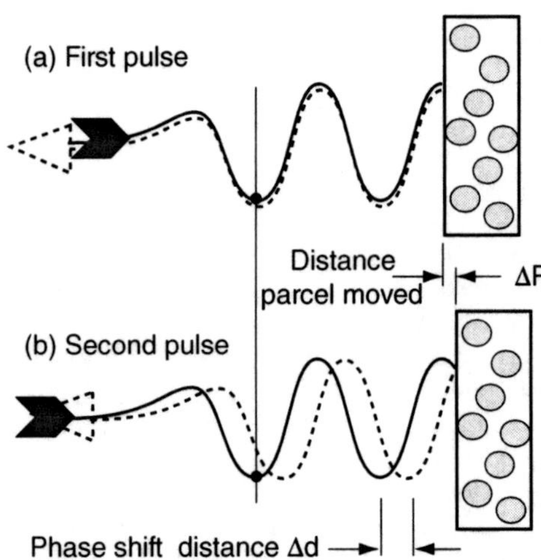

Figure 8.33
Phase shift distance is twice the distance moved by the air parcel.

Table 8-8. Display convention for Doppler velocities.		
Sign of M_r	**Radial Direction Relative to Radar**	**Display Color**
Positive	AWAY	Red & Orange
Zero	(none)	White, Gray or Black
Negative	TOWARD	Blue & Green

Instead, Doppler radars detect air motion by measuring the **phase shift** of the microwaves. Consider a pulse of microwaves that scatter from a droplet-laden air parcel (Fig. 8.32). When that parcel is at any one location and is illuminated by a radar pulse, there is a measurable phase difference (difference in locations of the wave troughs) between the incident wave and the scattered wave. Fig. 8.33a illustrates a phase difference of zero; namely, the troughs coincide.

If the parcel has moved to a slightly different distance from the radar when the next pulse hits, then there will be a different phase shift compared to that of the first pulse (Fig. 8.33b). As illustrated, the phase shift Δd is just twice the radial distance ΔR that the air parcel moved. You can demonstrate this to yourself by drawing a perfect wave on a thin sheet of tracing paper, folding the paper horizontally a certain distance ΔR from the crest, and then comparing the distance Δd between incident and folded troughs or crests. Knowing the time between successive pulses (= 1/PRF), the radial velocity M_r is:

$$M_r = \Delta R \cdot \text{PRF} = \text{PRF} \cdot \Delta d / 2 \qquad (8.34)$$

Table 8-8 gives the display convention for PPI or CAPPI displays of Doppler radial velocities in North America. Slower speeds are sometimes displayed with less color saturation (lighter, paler, closer to white). The color convention is modeled after the astronomical "red shift" for stars moving away from Earth. In directions from the radar that are perpendicular to the mean wind, there is no radial component of wind (Fig. 8.31); hence, this appears as a white or light grey line (Fig. 8.38a) of zero radial velocity (called the **zero isodop**). In some countries, the zero isodop is displayed with finite width as a black "no-data" line, because the ground-clutter filter eliminates the slowest radial velocities.

Maximum Unambiguous Velocity

The phase-shift method imposes a **maximum unambiguous velocity** $M_{r\,max}$ the radar can measure, given by:

$$M_{r\,max} = \lambda \cdot \text{PRF} / 4 \qquad \bullet(8.35)$$

where λ is wavelength. Fig. 8.34 shows the reason for this limitation. From the initial pulse to the second pulse, if the air parcel moved a quarter wavelength (λ/4) AWAY from the radar (Fig. 8.34b), then the scattered wave is a half wavelength out of phase from the incident wave. However, if the air parcel moved a quarter wavelength TOWARD the radar (Fig. 8.34c), then the phase shift is also half a wavelength. Namely, the AWAY and TOWARD velocities give exactly the same phase-shift signal at this critical speed, and the radar phase detector cannot distinguish between them.

Velocities slightly faster than $M_{r\,max}$ AWAY from the radar erroneously appear to the phase detector as fast velocities TOWARD the radar. Similarly, velocities faster than $M_{r\,max}$ TOWARD the radar are erroneously folded back as fast AWAY velocities. The false, displayed velocities $M_{r\,false}$ are:

For $M_{r\,max} < M_r < 2M_{r\,max}$:

$$M_{r\,false} = M_r - 2\,M_{r\,max} \qquad (8.36a)$$

For $(-2M_{r\,max}) < M_r < (-M_{r\,max})$:

$$M_{r\,false} = M_r + 2\,M_{r\,max} \qquad (8.36b)$$

Speeds greater than $2|M_{r\,max}|$ fold twice.

For greater max velocities $M_{r\,max}$, you must operate the Doppler radar at **greater** PRF (see eq. 8.35). However, to observe storms at greater range R_{max}, you need **lower** PRF (see eq. 8.17). This trade-off is called the **Doppler dilemma** (Fig. 8.35). Operational Doppler radars use a compromise PRF.

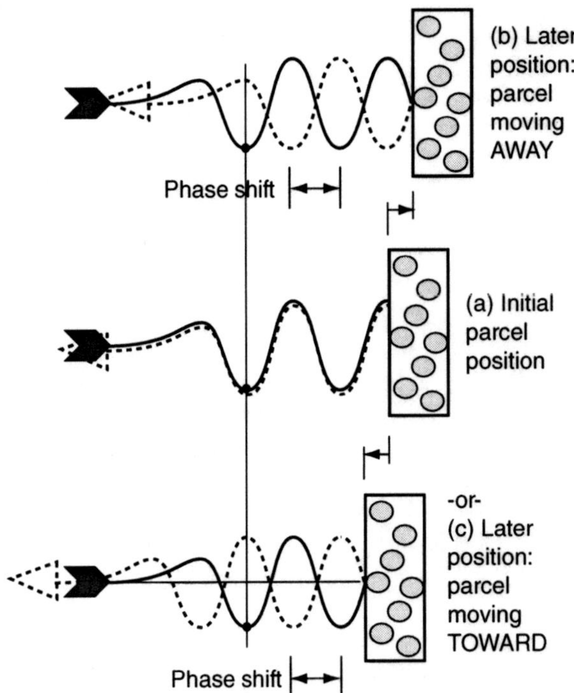

Figure 8.34
Air movement of distance equal to quarter a wavelength toward and away from the radar can yield identical phase shifts, making velocity interpretation ambiguous (i.e., erroneous).

Sample Application

A C-band radar emitting 2000 microwave pulses per second illuminates a target moving 30 m s^{-1} away from the radar. What is the displayed radial velocity?

Find the Answer:
Given: PRF = 2000 s^{-1}, M_r = +30 m s^{-1}.
Assume λ = 5 cm because C-band.
Find: $M_{r\,max}$ = ? m s^{-1} and $M_{r\,displayed}$ = ? m s^{-1}

Use eq. (8.35):
$M_{r\,max}$ = (0.05 m)·(2000 s^{-1})/4 = **25 m/s**

But $M_r > M_{r\,max}$, therefore velocity folding.
Use eq. (8.36a):
$M_{r\,false}$ = (30 m s^{-1}) – 2·(25 m s^{-1}) = **–20 m s^{-1}**,
where the negative sign means toward the radar.

Check: Units OK. Physics OK.
Exposition: This is a very large velocity in the wrong direction. Algorithms that automatically detect **tornado vortex signatures (TVS)**, look for regions of fast AWAY velocities adjacent to fast TOWARD velocities. However, this Doppler velocity-folding error produces false regions of adjacent fast opposite velocities, causing false-alarms for tornado warnings.

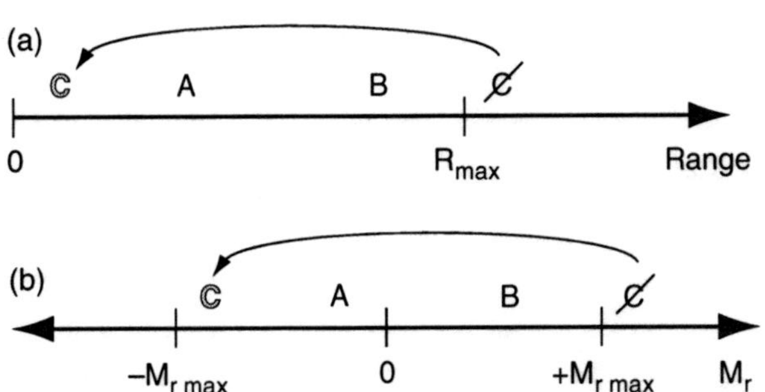

Figure 8.35
(a) Reflectivity targets at ranges A and B from the radar appear in their proper locations. Target C, beyond the max unambiguous range (R_{max}), is not displayed at its actual location, but appears as a false echo **C** *on the radar display closer to the radar.*
(b) Air parcels with negative or positive radial velocities A and B are displayed accurately. Parcel C with positive velocity greater than $M_{r\,max}$ is falsely displayed **C** *with large negative velocity.*

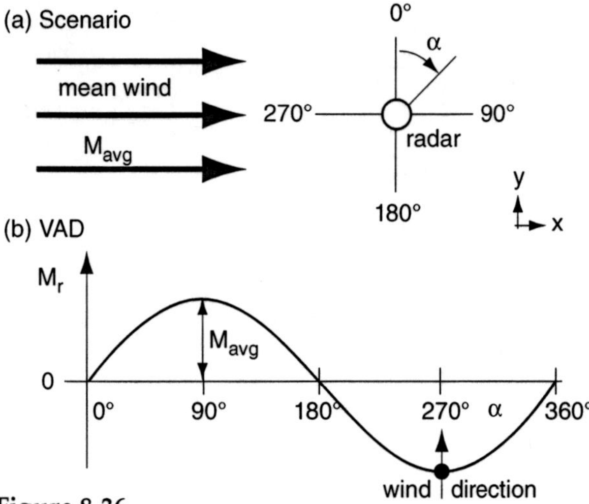

Figure 8.36
(a) Scenario of mean winds relative to radar. (b) Velocity-azimuth display (VAD) of radial wind component.

Sample Application
From a 1 km thick layer of air near the ground, the average radial velocity is -2 m s^{-1} at range 100 km from the radar. What is the vertical motion at the layer top?

Find the Answer:
Given: $R = 100$ km , $\Delta z = 1$ km , $M_{r\,avg} = -2$ m s^{-1}.
Find: $w = ?$ m s^{-1} at $z = 1$ km.
Assume: $w_{bot} = 0$ at the ground (at $z = 0$).

Use eq. (8.37): $\Delta w = -2\cdot(-2 \text{ m s}^{-1}) \cdot (1 \text{ km}) / (100 \text{ km})$
$\Delta w = +0.04$ m s^{-1}.
But $\Delta w = w_{top} - w_{bot}$. Thus: $w_{top} =$ **+0.04 m s^{-1}**

Check: Units OK. Physics OK.
Exposition: This weak upward motion covers a broad 100 km radius circle, and would enhance storms.

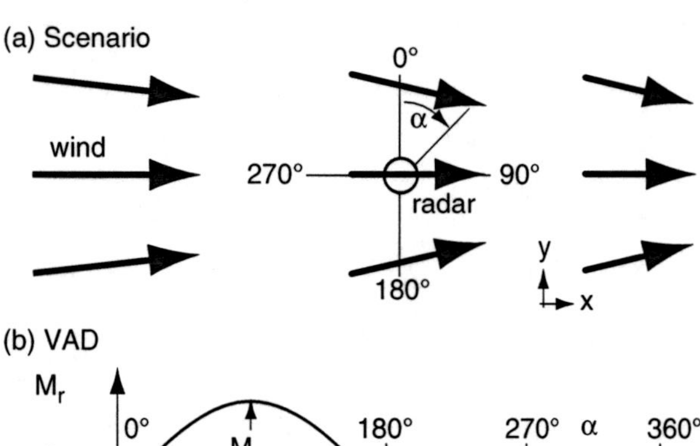

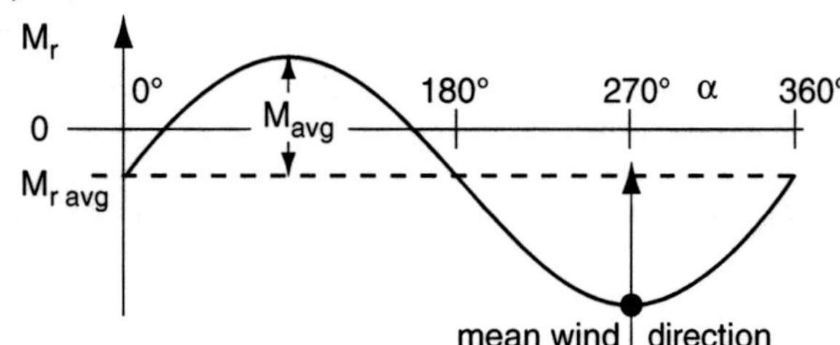

Figure 8.37
(a) Scenario of mean winds AND horizontal convergence.
(b) Velocity-azimuth display (VAD) of radial wind component.

Velocity Azimuth Display (VAD)

A plot of radial velocity M_r (at a fixed elevation angle and range) vs. azimuth angle α is called a **velocity-azimuth display (VAD)**, which you can use to measure the mean horizontal wind speed and direction within that range circle. Fig. 8.36a illustrates a steady mean wind blowing from west to east through the Doppler radar sweep area. The resulting radial velocity component measured by the radar is a sine wave on the VAD (Fig. 8.36b).

Mean horizontal wind speed M_{avg} is the amplitude of the sine wave relative to the average radial velocity $M_{r\ avg}$ (=0 for this case), and the meteorological wind direction is the azimuth at which the VAD curve is most negative (Fig. 8.36b). Repeating this for different heights gives the winds at different altitudes, which you can plot as a wind-profile sounding or as a hodograph (explained in the Thunderstorm chapters). The measurements at different heights are made using different elevation angles and ranges (eq. 8.22).

You can also estimate mean <u>vertical</u> velocity using horizontal divergence determined from the VAD. Picture the scenario of Fig. 8.37a, where there is convergence (i.e., negative divergence) superimposed on the mean wind. Namely, west winds exist everywhere, but the departing winds east of the radar are slower than the approaching winds from the west. Also, the winds have a convergent north-south component. When plotted on a VAD, the result is a sine wave that is displaced in the vertical (Fig. 8.37b). Averaging the slower AWAY radial winds with the faster TOWARD radial winds gives a negative average radial velocity $M_{r\ avg}$, as shown by the dashed line in Fig. 8.37b. $M_{r\ avg}$ is a measure of the divergence/convergence, while M_{avg} still measures mean wind speed.

Due to mass continuity (discussed in the Atmospheric Forces & Winds chapter), an accumulation of air horizontally into a volume requires upward motion of air out of the volume to conserve mass (i.e., mass flow in = mass flow out). Thus:

$$\Delta w = \frac{-2\cdot M_{r\ avg}}{R}\cdot \Delta z \qquad (8.37)$$

where Δw is change in vertical velocity across a layer of thickness Δz, and R is range from the radar at which the radial velocity M_r is measured.

Upward vertical velocities (associated with horizontal **convergence** and negative $M_{r\ avg}$) enhance cloud and storm development, while downward vertical velocities (**subsidence**, associated with horizontal **divergence** and positive $M_{r\ avg}$) suppresses clouds and provides fair weather. Knowing that $w = 0$ at the ground, you can solve eq. (8.37) for

w at height Δz above ground. Then, starting with that vertical velocity, you can repeat the calculation using average radial velocities at a higher elevation angle to find w at this higher altitude. Repeating this gives the vertical velocity profile.

A single Doppler radar cannot measure the full horizontal wind field, because it can "see" only the radial component of velocity. However, if the scan regions of two nearby Doppler radars overlap, then in theory the horizontal wind field can be calculated from this **dual-Doppler** information, and the full 3-D winds can be inferred using mass continuity.

Identification of Storm Characteristics

While the VAD approach averages the winds within the whole scan circle, you can also use the smaller scale patterns of winds to identify storm characteristics. For example, you can use Doppler information to help detect and give advanced warning for **tornadoes** and **mesocyclones** (see the Thunderstorm chapters). Doppler radar cannot see the whole rotation inside the **thunderstorm** — just the portions of the vortex with components moving toward or away from the radar. But you can use this limited information to define a **tornado vortex signature** (**TVS**, Fig. 8.38b) with the brightest red and blue pixels next to each other, and with the zero isodop line passing through the tornado center and following a radial line from the radar.

Hurricanes also exhibit these rotational characteristics, as shown in Fig. 8.39. Winds are rotating counterclockwise around the eye of this Northern Hemisphere hurricane. The zero isodop passes through the radar location in the center of this image, and also passes through the center of the eye of the hurricane. See the Tropical Cyclone chapter for details.

Cold-air **downbursts** from thunderstorms hit the ground and **diverge** (spread out) as damaging horizontal **straight-line winds** (Fig. 8.38c). The leading edge of the outflow is the **gust front**. Doppler radar sees this as neighboring regions of opposite moving radial winds, similar to the TVS but with the zero isodop perpendicular to the radial from the radar, and with the blue region always closer to the radar than the red region.

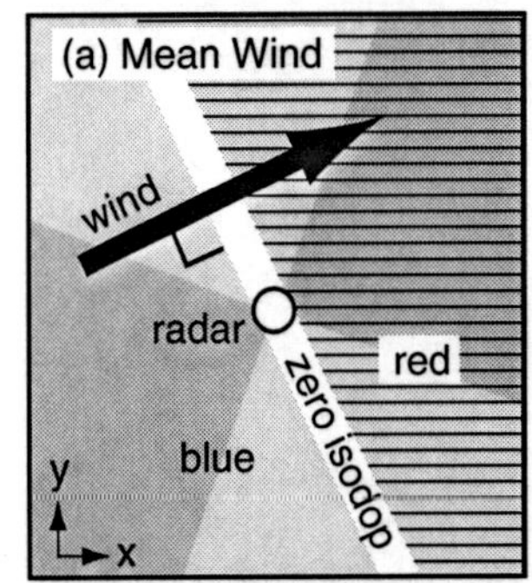

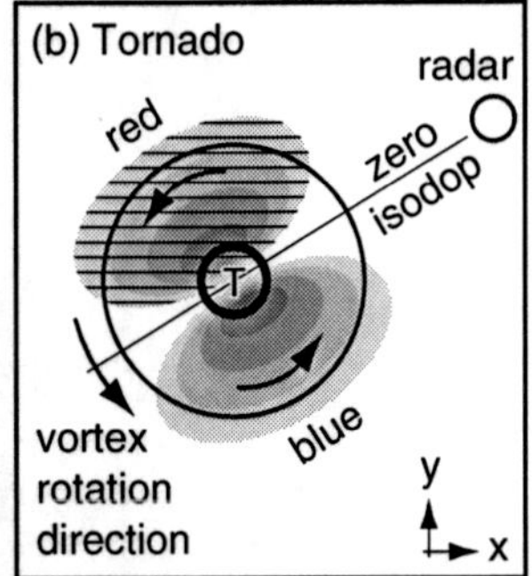

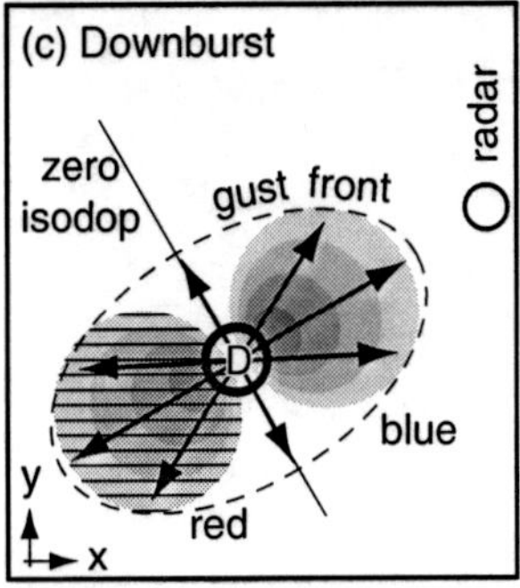

Figure 8.38
PPI or CAPPI radar image interpretation. The white circle indicates the radar location. Hatched shaded regions represent red, and indicate wind components AWAY from the radar. Other shaded regions represent blue, and indicate TOWARD the radar. Darker shadings represent more vivid colors, and mean faster radial winds. Actual wind directions are shown by the black arrows. (a) Uniform mean wind in the domain. (b) Tornado (T) vortex, surrounded by weaker mesocyclone vortex (large circle). (c) Downburst (D) and gust front (dashed line) at the leading edge of outflow air.

Sample Application

A downburst and outflow are centered directly over a Doppler radar. Sketch the Doppler display.

Find the Answer:

Given: Air diverging in all directions from radar
Sketch: Doppler radar display appearance

At every azimuth from the radar, air is moving AWAY. Thus red would be all around the radar, out to the gust front. The fastest outflow is likely to be closest to the radar.

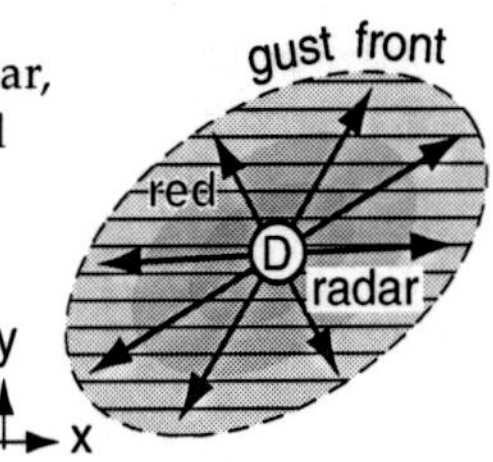

Check: Sketch consistent with display definitions.

Exposition: Meteorologists at the radar might worry about their own safety, and about radome damage. The **radome** is the spherical enclosure that surrounds the radar antenna, to protect the radar from the weather. Also, a wet radome causes extra attenuation, making it difficult to see distant precipitation echoes.

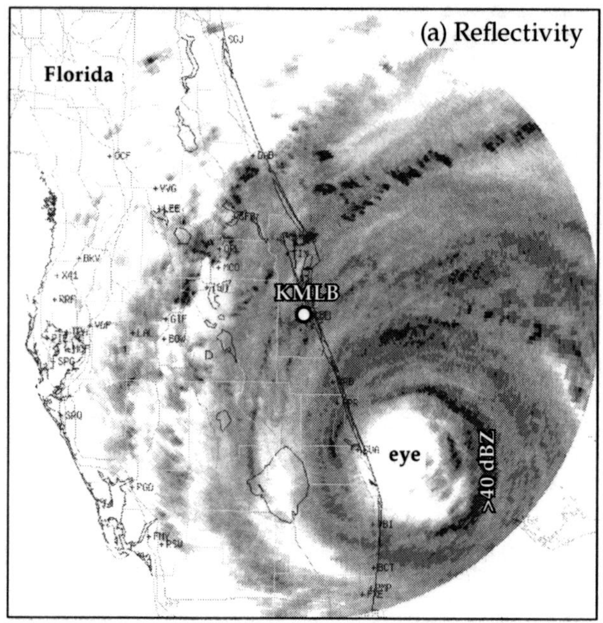

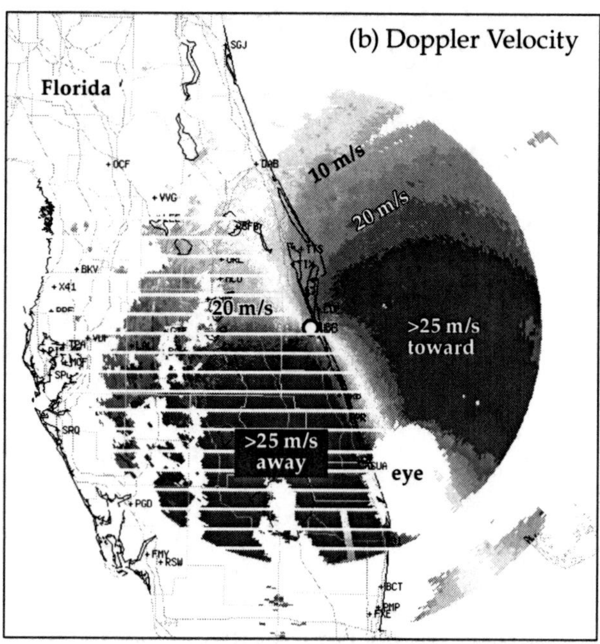

Figure 8.39
Radar images of Hurricane Francis at 0300 UTC on 5 September 2004 as the eye is approaching the coast of Florida, USA. Hurricane intensity at this time was category 2 on the Saffir-Simpson scale, although it had been category 4 before reaching the coast. The Melbourne, FL, radar KMLB is in the center of the radar image. (a) PPI image where darker shading represents greater reflectivity (dBZ). (b) Doppler radial velocities (in the original image velocities toward the radar were colored blue, and away velocities were red). In this black and white image, the AWAY winds are shown with white hatching superimposed. [Adapted from US National Weather Service data as displayed by the Research Applications Program division of the US National Center for Atmospheric Research.]

Sample Application
Explain the "S" curve in the zero isodop of Fig. 8.39b.

Find the Answer:
Given: Doppler velocities in Fig. 8.39b.
Find: Explanation for "S" curve of zero isodop.

For a PPI display, the radar is observing the air along the surface of an inverted scan cone (Fig. 8.24a). Hence, targets at closer range to the radar correspond to targets at lower altitude. Curvature in the zero isodop indicates vertical shear of the horizontal winds (called **wind shear** for short).

Close to the radar the zero isodop is aligned northwest to southeast. The wind direction at low altitude is perpendicular to this direction; namely, from the **northeast at low altitude**.

Further from the radar (but still on the radar side of the eye), the zero isodop is south-southeast of the radar. The winds at this range are perpendicular to a straight line drawn from the radar to the zero isodop; namely, from the **east-northeast at mid altitudes**.

Exposition: As you will see in the Winds & Hurricane chapters, boundary-layer winds feel the effect of drag against the ground, and have a direction that spirals in toward the eye. Above the boundary layer, in the middle of the troposphere, winds circle the eye.

Spectrum Width

Another product from Doppler radars is the **spectrum width**. This is the variance of the Doppler velocities within the sample volume, and is a measure of the intensity of atmospheric **turbulence** inside storms. Also, the Doppler signal can be used with reflectivity to help eliminate ground clutter. For high elevation angles and longer pulse lengths, the spectrum width is also large in bright bands, because the different fall velocities of rain and snow project onto different radial velocities. Strong **wind shear** within sample volumes can also increase the spectrum width.

Difficulties

Besides the maximum unambiguous velocity problem, other difficulties with Doppler-estimated winds are: (1) ground clutter including vehicles moving along highways; (2) lack of information in parts of the atmosphere where there are no scatterers (i.e., no rain drops or bugs) in the air; and (3) scatterers that move at a different speed than the air, such as falling raindrops or migrating birds.

Polarimetric Radar

Because Z-R relationships are inaccurate and do not give consistent results, other methods are used to improve estimates of precipitation type and rate. One method is to transmit microwave pulses with different **polarizations**, and then compare their received echo intensities. Radars with this capability are called **polarimetric**, **dual-polarization**, or **polarization-diversity** radars.

Recall that electromagnetic radiation consists of perpendicular waves in the electric and magnetic fields. **Polarization** is defined by the direction of the electric field. Radars can be designed to switch between two or three polarizations: **horizontal**, **vertical**, and **circular** (see Fig. 8.40). Air-traffic control radars use circular polarization, because it partially filters out the rain, allowing a clearer view of aircraft. Most weather radars use horizontal and vertical linear polarizations.

Horizontally polarized pulses get information about the horizontal dimension of the precipitation particles (because the energy scattered depends mostly on the particle horizontal size). Similarly, vertically polarized pulses get information about the vertical dimension. By alternating between horizontal and vertical polarization with each successive transmitted pulse, algorithms in the receiver can use echo differences to estimate the average shape of precipitation particles in the sample volume.

Only the smallest cloud droplets are spherical (Fig. 8.41a, equal width and height). These would give returns of equal magnitude for both horizontal and vertical polarizations. However, larger rain drops become more **oblate** (Fig. 8.41b, flattened on the bottom and top) due to air drag when they fall. The largest drops are shaped like a hamburger bun (Fig. 8.41c). This gives greater reflectivity factor Z_H from the horizontally polarized pulses than from the vertically polarized ones Z_V. The amount of oblateness increases with the mass of water in the drop. There is some evidence that just before the largest drops break up, they have a parachute or jelly-fish-like shape (Fig. 8.41d).

Some hydrometeor shapes change the polarization of the scattered pulse. For example, clockwise circular polarized transmitted pulses are scattered with counterclockwise circular polarization by any rain drops that are nearly spherical. Tumbling, partially melted, irregular-shaped, ice crystals can respond to horizontally polarized pulses by returning a portion of the energy in vertical polarization. Different shapes of ice crystals have different reflectivity factors in the vertical and horizontal.

Four variables are often analyzed and displayed on special radar images to help diagnose hydrometeor structure and size. These are:

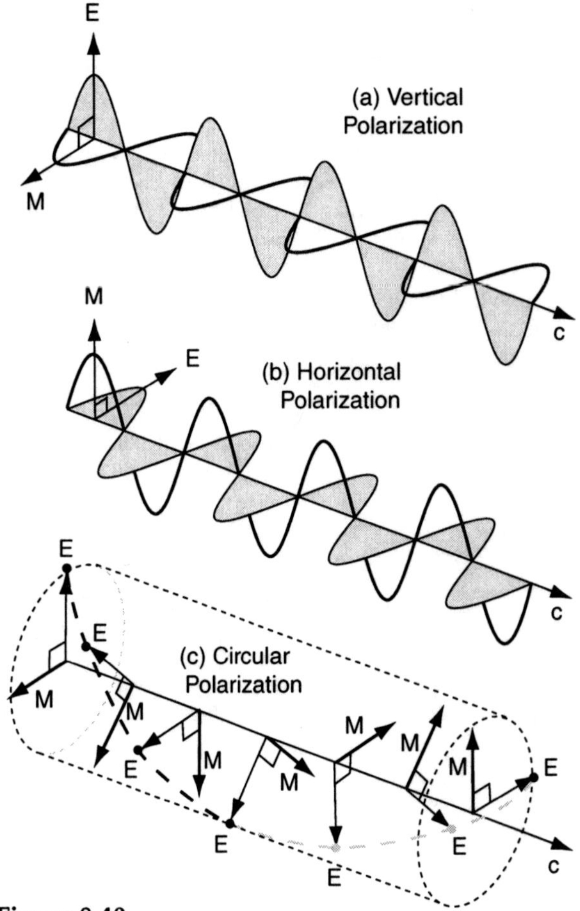

Figure 8.40
Polarization of microwaves, where E = electric field, M = magnetic field, and c = wave propagation at speed of light. (a) Vertical; (b) Horizontal; and (c) Circular polarization.

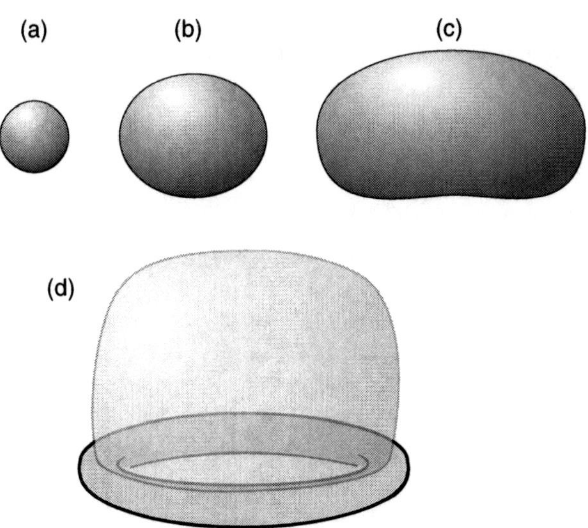

Figure 8.41
Enlargement of raindrop shapes vs. diameter D (approximate). (a) D < 1 mm; (b) 1 mm < D < 3 mm ; (c) 3 mm < D; (d) large (D ≈ 4 mm) jellyfish-like shape just before drop breaks up..

Sample Application

Polarization radar measures reflectivity factors of Z_{HH} = 1000 dBZ , Z_{VV} = 398 dBZ , and Z_{VH} = 0.3 dBZ. (a) Find the differential reflectivity and the linear depolarization ratio. (b) Discuss characteristics of the hydrometeors. (c) What values of specific differential phase and co-polar correlation coefficient would you expect? Why?

Find the Answer:

Given: Z_{HH} = 1000 dBZ, Z_{VV} = 398 dBZ , and Z_{VH} = 0.3 dBZ

Find: Z_{DR} = ? dB, LDR = ? dB

(a) Use eq. (8.38): Z_{DR} = 10·log(1000 dBZ/398 dBZ) = **+4.0** dB

Use eq. (8.39): LDR = 10·log(0.3 dBZ/1000 dBZ) = **–35.2** dB

(b) **Discussion**: The positive value of Z_{DR} indicates a hydrometeor with greater horizontal than vertical dimension, such as **large, oblate rain drops**. The very small value of *LDR* supports the inference of rain, rather than snow or hail.

(c) I would expect K_{DP} = **2 to 4 °/km**, because of large oblate rain drops.

I would also expect ρ_{HV} = **0.95**, because **rain intensity is light to moderate**, based on eq. (8.27):

dBZ = 10·log(Z_{HH}/Z_1)
= 10·log(1000dBZ/1dBZ)
= **30 dBZ**, which is then used to find the intensity classification from Fig. 8.28.

Check: Units OK. Results reasonable for weather.

Exposition: See item (b) above.

In eq. (8.27) I used Z_{HH} because the horizontal drop dimension is the one that is best correlated with the mass of the drop, as illustrated in Fig. 8.41.

- **Differential Reflectivity**: Z_{DR} (unit: dB)

$$Z_{DR} = 10\cdot\log(Z_{HH}/Z_{VV}) \qquad \bullet(8.38)$$

where the first and second subscripts of reflectivity Z give the polarization (H = horizontal; V = vertical) of received and transmitted pulses, respectively.
Typical values: –2 to +6 dB, where negative and positive values indicate vertical or horizontal orientation of the hydrometeor's major axis.
Useful with Z for improving rain-rate estimates, and to identify radar attenuation regions.

- **Linear Depolarization Ratio**. LDR (unit: dB)

$$LDR = 10\cdot\log(Z_{VH}/Z_{HH}) \qquad \bullet(8.39)$$

Typical values: > –10 dB for ground clutter
–15 dB for melting snowflakes.
–20 to –26 dB for melting hail
< –30 dB for rain
Useful for estimating micrometeor type or habit. Can also help remove ground clutter, and identify the bright band.

- **Specific Differential Phase**: K_{DP}
(unit: degrees of phase shift per 1 km path)
Relative amount of phase shift (shift of the crests of the electromagnetic waves) in the returned echoes from the horizontally vs. vertically transmitted pulses.
Typical values: –1 to +6 °/km, where positive values indicate oblate (horizontally dominant) hydrometeors. Tumbling hail appears symmetric on average, resulting in K_{DP} = 0.
Improves accuracy of rain rate estimate, and helps to isolate the portion of echo associated with rain in a rain/hail mixture.

- **Co-polar Correlation Coefficient**: ρ_{HV}
(unit: dimensionless, in range –1 to +1)
Measures the amount of similarity in variation of the time series of received signals from the horizontal vs. vertical polarizations.
Typical values of $|\rho_{HV}|$:
0.5 means very large hail
0.7 to 0.9 means drizzle or light rain
0.90 to 0.95 means hail, or bright-band mix.
≥ 0.95 means rain, snow, ice pellets, graupel
0.95 to 1.0 means rain
Useful for estimating the heterogeneity of micrometeors (mixtures of different-shaped ice crystals and rain drops) in the sampling volume, and radar beam attenuation.

One suggestion for an improved estimate of rainfall rate RR with S-band polarimetric radar is:

$$RR = a_o \cdot K_{DP}^{a_1} \cdot Z_{DR}^{a_2} \qquad \bullet(8.40)$$

where a_o = 90.8 mm h^{-1} (see Sample Application "check" for discussion of units), a_1 = 0.89, and $a_2 = -1.69$. Such improved rainfall estimates are crucial for **hydrometeorologists** (experts who predict drought severity, river flow and flood potential).

Based on all the polarimetric information, fuzzy-logic algorithms are used at radar sites to classify the hydrometeors and to instantly display the result on computer screens for meteorologists. Hydrometeor Classification Algorithms (HCA) for WSR-88D radars use different algorithms for summer (warm mode) and winter (cold mode) precipitation.

- Warm Mode: big drops, hail, heavy rain, moderate rain, light rain, no echoes, birds/insects, anomalous propagation.
- Cold Mode: convective rain, stratiform rain, wet snow, dry snow, no echoes, birds/insects, anomalous propagation.

Combinations of polarimetric variables can also help identify graupel and estimate hail size, estimate attenuation of the radar signal, identify storms that might become electrically active (lots of lightning), identify regions that could cause hazardous ice accumulation on aircraft, detect tornado debris clouds, and correct for the bright band.

Sample Application

Polarization radar measures a differential reflectivity of 5 dB and specific differential phase of 4°/km. Estimate the rainfall rate.

Find the Answer:
Given: Z_{DR} = 5 dB , K_{DP} = 4°/km.
Find: RR = ? mm h^{-1}

Use eq. (8.40):
RR = (90.8 mm h^{-1}) · (4°/km)$^{0.89}$ · (5 dB)$^{-1.69}$
= (90.8 mm h^{-1})· 3.43 · 0.066 = **20 mm h^{-1}**

Check: Units don't match. The reason is that the units of a_o are not really [mm/h], but are [(mm h^{-1}) · (°/km)$^{-0.89}$ · (dB)$^{+1.69}$]. These weird units resulted from an empirical fit of the equation to data, rather than from first principles of physics.
Exposition: This corresponds to heavy rain (Fig. 8.28), so hydrometeorologists would try to forecast the number of hours that this rain would continue to estimate the **storm-total accumulation** (= rainfall rate times hours of rain). If excessive, they might issue a flood warning.

Phased-Array Radars & Wind Profilers

Phased-array radars steer the radar beam not by rotating the antenna dish, but by using multiple transmitter heads that can transmit at slightly different times. Picture a building (the rectangle in Fig. 8.42a) as viewed from above. On one side of the building are an array of microwave transmitters (black dots in the Fig). When any transmitter fires, it emits a wave front shown by the thin semicircle. If all transmitters fire at the same time, then the superposition of all the waves yield the most energy in an effective wave front (thick gray line) that propagates perpendicularly away from the transmitter building, as shown by the gray arrow. For the example in Fig. 8.42a, the effective wave front moves east.

But these transmitters can also be made to fire sequentially (i.e., slightly out of phase with each other) instead of simultaneously. Suppose transmitter 1 fires first, and then a couple nanoseconds later transmitter 2 fires, and so on. A short time after transmitter 5 fires, its wave front (thin semicircle) has had time to propagate only a short distance, however the wave fronts from the other transmitters have been propagating for a longer time, and are further away from the building (Fig. 8.42b). The superposition of

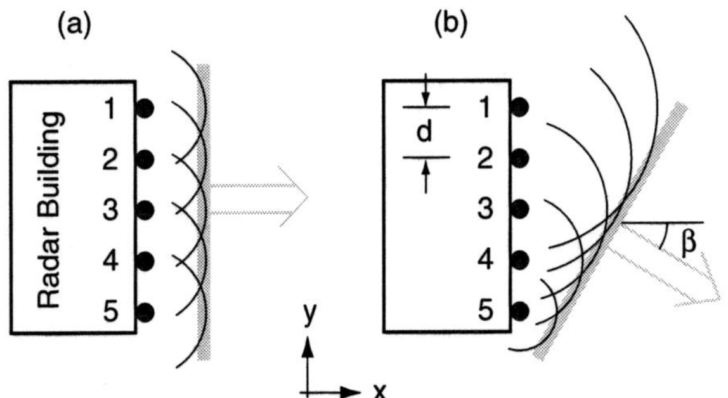

Figure 8.42
Phased array radar concept.

Sample Application

What time delay is needed in phased array radar to steer the beam 30° away from the line normal (perpendicular) to the radar face, if the transmitter elements are 1 m apart?

Find the Answer:
Given: $\beta = 30°$, $d = 1$ m , $c = 3\text{x}10^8$ m s^{-1}
Find: Δt = ? s

Use eq. (8.41): $\Delta t = [(1\text{m})/(3\text{x}10^8 \text{ m s}^{-1})]\cdot\sin(30°)$
$\Delta t = 1.67\text{x}10^{-9}$ s = **1.67 ns**.

Check: Units OK. Physics OK. Agrees with Fig. 8.42.
Exposition: The larger the angle β, the smaller the effective size of the transmitter face, and the less focused is the beam. This is a disadvantage of phased-array radars used for meteorology: the sample volume changes not only with range but also with direction β. Therefore, the algorithms used to interpret meteorological characteristics (reflectivity, Doppler velocity, spectrum width, polarimetric data) must be modified to account for these variations.

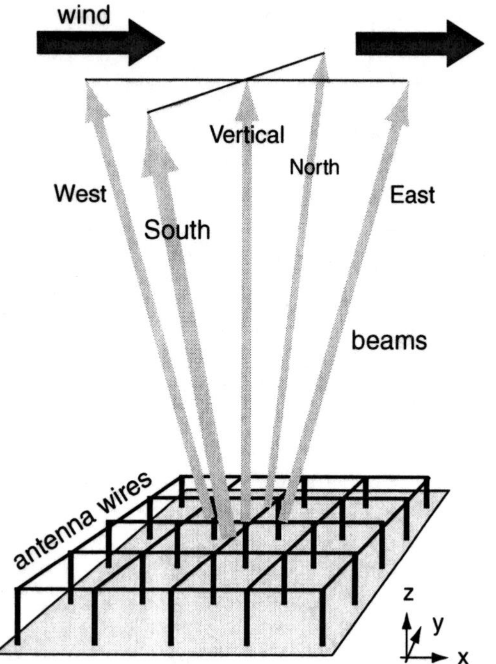

Figure 8.43
Wind profiler concept.

the individual wave fronts now yields an effective wave front (thick gray line) that is propagating toward the southeast.

Let Δt be the time delay between firing of subsequent transmitters. For transmitter elements spaced a distance d from each other, the time delay needed to give an effective propagation angle β (see Fig. 8.42b) is:

$$\Delta t = (d/c)\cdot\sin(\beta) \tag{8.41}$$

where $c \approx 3\text{x}10^8$ m s^{-1} is the approximate speed of microwaves through air.

By changing Δt, the beam (grey arrow) can be steered over a wide range of angles. Negative Δt steers the beam toward the northeast, rather than southeast in this example. Also, by having a 2-D array of transmitting elements covering the whole side of the building, the beam can also be steered upward by firing the bottom elements first.

Two advantages of this approach are: (1) fewer moving parts, therefore less likelihood of mechanical failure; and (2) the beam angle can be changed instantly from one extreme to the other. However, to see a full 360° azimuth around the radar site, phased-array transmitting elements must be designed into all 4 walls of the building. While the military have used phased array radars for decades to detect aircraft and missiles, these radars are just beginning to be used in civilian meteorology.

Another phased-array application to meteorology is the **wind profiler**. This observes horizontal wind speed and direction at many heights by measuring the Doppler shifts in the returned signals from a set of radio beams tilted slightly off from vertical (Fig. 8.43).

Some wind profilers look like a forest of TV antennas sticking out of the ground, others like window blinds laying horizontally, others like a large trampoline, and others like a grid of perpendicular clothes lines on poles above the ground (Fig. 8.43). Each line is a radio transmitter antenna, and the effective beam can be steered away from vertical by phased (sequential) firing of these antenna wires.

Wind profilers work somewhat like radars — sending out pulses of radio waves and determining the height to the clear-air targets by the time delay between transmitted and received pulses. The targets are turbulent eddies of size equal to half the radio wavelength. These eddies cause refractivity fluctuations that scatter some of the radio-wave energy back to receivers on the ground.

Wind profilers with 16 km altitude range use Very High Frequencies (**VHF**) of about ν = 50 MHz (λ = c / ν = 6 m). Other **tropospheric wind profilers** use an Ultra High Frequency (**UHF**) near 400 MHz (λ = 0.75 m = 75 cm). Smaller **boundary-layer wind profilers** for sampling the lower troposphere use an ultra high frequency near 1 GHz (λ = 30 cm).

Normally, wind profilers are designed to transmit in only 3, 4 or 5 nearly-vertical, fixed beam directions (Fig. 8.43). The fixed zenith angle of the slightly-tilted beams is on the order of ζ = 14° to 24°.

The radial velocities M_{east} and M_{west} measured in the east- and west-tilted beams, respectively, are affected by the U component of horizontal wind and by vertical velocity W. Positive W (upward motion; thick gray vector in Fig. 8.44) projects into positive (away from wind profiler) components of radial velocity for both beams (thin gray vectors). However, positive U (thick black vector) wind contributes positively to M_{east}, but negatively (toward the profiler) to M_{west} (thin black vectors).

Assume the same average U and W are measured by both beams, and design the profiler so that all tilted beams have the same zenith angle. Thus, the two radial velocities measured can be subtracted and added to each other, to yield the desired two wind components:

$$U = \frac{M_{east} - M_{west}}{2 \cdot \sin\zeta} \qquad \bullet(8.42a)$$

$$W = \frac{M_{east} + M_{west}}{2 \cdot \cos\zeta} \qquad \bullet(8.42b)$$

Use similar equations with radial velocities from the north- and south-tilted beams to estimate V and W. Fig. 8.45 shows sample wind-profiler output.

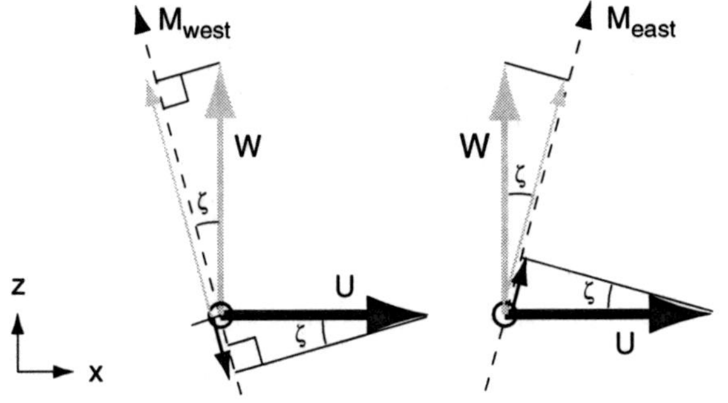

Figure 8.44
Projections of horizontal wind U and vertical velocity W to the radial wind directions M_{west} and M_{east} along wind-profiler west and east tilted beams.

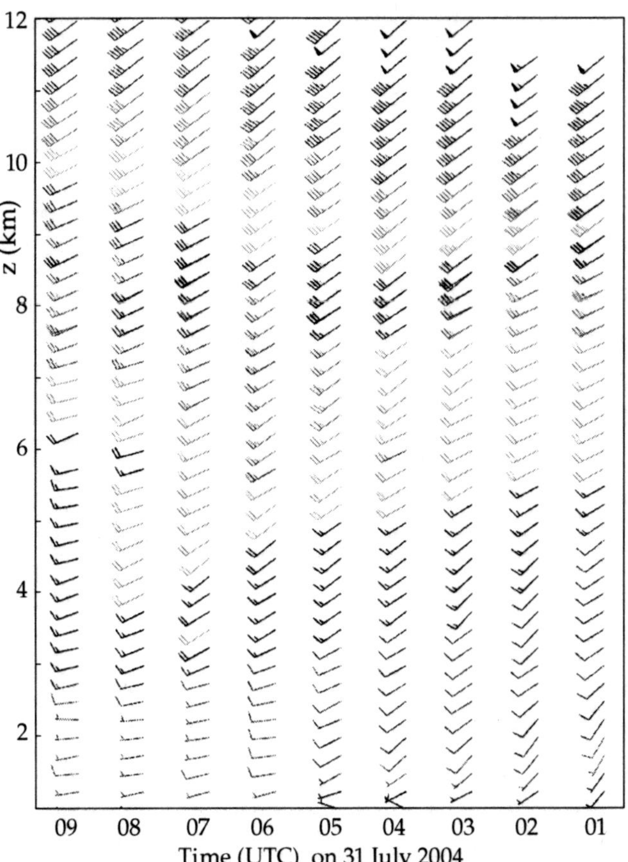

Figure 8.45
Wind profiler observations for Wolcott, IN. Arrow shows horizontal wind direction. Each short wind barb = 5 m/s, long = 10 m/s, & pennant = 50 m/s. (E.g., wind at z = 10 km, t = 02 UTC is 45 m/s from southwest.) Courtesy of US National Oceanic and Atmospheric Admin. (NOAA) Profiler Network (NPN).

Sample Application

A wind profiler with 20° beam tilt uses the Doppler shift to measure radial velocities of 3.326 m s^{-1} and –3.514 m s^{-1} in the east- and west-tilted beams, respectively, at height 4 km above ground. Find the vertical and horizontal wind components at that height.

Find the Answer:
Given: M_{east}= 3.326 m s^{-1}, M_{west}= –3.514 m s^{-1}, ζ = 20°
Find: U = ? m s^{-1} and W = ? m s^{-1}.

Use eq. (8.42a):
U = [(3.326 m s^{-1}) – (–3.514 m s^{-1})]/[2·sin(20°)]=**10 m s^{-1}**
Use eq. (8.42b):
W=[(3.326 m s^{-1})+(–3.514 m s^{-1})]/[2·cos(20°)]=**–0.1 m s^{-1}**

Check: Units OK. Physics OK.
Exposition: The positive U component indicates a west wind (i.e., wind FROM the west) at z = 4 km. Negative W indicates subsidence (downward moving air). These are typical of fair-weather conditions under a strong high-pressure system. Typical wind profilers make many measurements each minute, and then average their results over 5 minutes to 1 hour.

REVIEW

Passive sensors on weather satellites observe many different wavelengths of electromagnetic radiation upwelling from the Earth-atmosphere-cloud system. Imager sensors take high-quality digital photographs of clouds and air motion, for which the most important wavelength channels are visible, IR, and water-vapor. Sounder radiometers remotely probe different depths in the atmosphere.

For reflected visible sunlight and for wavelengths in IR windows, the satellite sees down to the highest cloud top, or to the ground if no clouds block the view. In IR opaque regions, the satellite cannot see through the atmosphere, and instead measures radiation emitted from the air.

For this latter situation, atmospheric emissions at any one wavelength come from a broad range of altitudes, as defined by weighting functions. By observing many different wavelengths with weighting functions that peak at different altitudes, the data can be inverted to retrieve vertical profiles of temperature. Retrievals are difficult, but provide useful remotely-sensed soundings over the oceans having insufficient in-situ observations.

Two favorite satellite orbits are geostationary (high altitude) and sun-synchronous polar orbiting (low altitude). Weather features and storms can be interpreted from satellite images and movie loops. In IR images, high (cold) clouds are often displayed with the lighter grays and white, while lower clouds are darker gray.

Weather radars are active sensors that transmit strong pulses of microwaves, and then measure the echoes that bounce back from precipitation. Stronger echoes (i.e., large scattering) show regions of more and larger rain and snow particles, which are used to estimate rainfall rate and storm intensity. Movement of these echoes show the storm track.

When the wind carries the hydrometeors in a radial direction relative to the radar, the echo frequency is shifted slightly due to the Doppler effect. Doppler radars analyze this radial velocity by measuring the phase shift of the echoes. Certain wind patterns yield characteristic signatures in the Doppler wind field, which are used to detect incipient tornadoes and other damaging winds.

Polarimetric weather radars transmit and receive microwaves with different polarizations (usually horizontal and vertical). Hydrometeors with different shapes return polarized signals differently, allowing better estimation of rainfall rates and precipitation type. Phased-array radars need no moving parts to can scan the weather. Wind profilers measure horizontal wind speed and directions.

HOMEWORK EXERCISES

Broaden Knowledge & Comprehension

B1. Search the web for graphs of transmittance for wavelengths greater than 100 cm (like Figs. 8.4), and find the names of these regions of the electromagnetic spectrum. Comment on the possibility for remote sensing at these longer wavelengths.

B2. Search the web for graphs of weighting functions for one of the newer weather satellites (for the imager or for the sounder), as specified by your instructor. Discuss their advantages and disadvantages compared to the weighting functions given in this book.

B3. Search the web for current orbital characteristics for the active weather satellite(s) specified by your instructor. Also, look for photos or artist drawings of these satellites.

B4. Search the web for data to create tables that list the names and locations (longitude for GOES; equator crossing times for POES) currently active weather satellites around the world. Discuss any gaps in worldwide weather-satellite coverage.

B5. Search the web for images from all the different channels (for the imager or for the sounder) for one of the satellites, as specified by your instructor. Discuss the value and utility of each image.

B6. Search the web for satellite loops that cover your location. Compare these images with the weather you see out the window, and make a short-term forecast based on the satellite loop.

B7. Search the web to find the most recent set of visible, IR, and water-vapor images that covers your area. Print these images; label cloud areas as (a), (b), etc.; and interpret the clouds in those images. For the cloud area directly over you, compare you satellite interpretation with the view out your window.

B8. Search the web for thermo diagrams showing temperature soundings as retrieved from satellite sounder radiance data. Compare one of these soundings with the nearest rawinsonde (in-situ) sounding (by searching on "upper air" soundings).

B9. Search the web for tutorials on satellite-image interpretation; learn how to better interpret one type of cloud system (e.g., lows, fronts, thunderstorms); and write a summary tutorial for your classmates.

B10. Search the web for galleries of classical images showing the best examples of different types of clouds systems (e.g., hurricanes, squall lines).

B11. Search the web for tutorials on radar-image interpretation, and learn how to better interpret one type of echo feature (e.g., tornadoes, thunderstorms). Write a summary tutorial for your classmates.

B12. Search the internet for weather radar imagery at your location or for a radar location assigned by your instructor).

a. List the range of products provided by this site. Does it include reflectivity, Doppler velocity, polarimetric data, derived products?

b. Are the radar images fresh or old?

c. Does the site show sequences of radar reflectivity images (i.e., radar loops)? If so, compare the movement of the whole line or cluster of storms with the movement of individual storm cells.

B13. Sometimes precipitation rate or accumulated precipitation amount estimates are provided at some weather radar web sites. For such a site (or a site assigned by your instructor), compare actual rain-gauge observations of rainfall with the radar-estimated amounts. Explain why they might differ.

B14. Sometimes weather radar can detect bats, birds, bugs, and non-precipitating clouds. Find an example of such a display, and provide its web address.

B15. What percentage of your country is covered by weather radars? Find and print a map showing this coverage, if possible.

B16. Although most weather radar is inside protective radomes, sometimes you can find photos on the internet that show the view inside the radome, or which shows the radar before the radome was installed. Print a photo of such a radar dish and its associated equipment. List specifications for that radar, including its transmitted power, horizontal scan rate, elevation angles, etc.

B17. Search the web to compare radar bands used for weather radar, police radar & microwave ovens.

B18. Search the web for examples of radar ducting or trapping.

B19. Search the web for other radar relationships to estimate rainfall rate, including other Z-R relationships and ones that use polarimetric data.

B20. Discuss the bright-band phenomenon by searching the web for info and photos.

B21. Search the web for info & photos on rain droplet & ice crystal shape, as affects polarimetric radar.

B22. Find real-time imagery of polarimetric radar products such as Z_{DR}, LDR, K_{DP}, and ρ_{HV}, and discuss their value in weather interpretation.

B23. Search the web for photos of phased-array radars and wind profilers. Compare and discuss.

B24. Search the web for locations in the world having "wind profiler" sites. Examine and discuss real-time data from whichever site is closest to you.

B25. Compare high-resolution satellite images from Earth-observing satellites (Aqua, Terra, and newer) with present-generation weather satellite imagery.

Apply

A1. Using Fig. 8.4, identify whether the following wavelengths (μm) are in a window, dirty window, shoulder, or opaque part of the transmittance spectrum, and identify which sketch in Fig. 8.2 shows how the Earth would look at that wavelength. [Hint: transmittance of ≥ 80% indicates a window.]

a. 0.5   b. 0.7   c. 0.95   d. 1.25   e. 1.33
f. 1.37   g. 1.6   h. 2.3   i. 2.4   j. 5.0

A2. Find the blackbody radiance for the following sets of [λ (μm), T (°C)]:

a. 14.7, –60   b. 14.4, –60   c. 14.0, –30
d. 13.7, 0   e. 13.4, 5   f. 12.7, 15
g. 12.0, 25   h. 11.0, –5   i. 9.7, –15

A3. For the wavelengths in the previous problem, identify the closest GOES sounder channel #, and the altitude of the peak in the weighting function.

A4. Find the brightness temperature for the following wavelengths (μm), given a radiance of 10^{-15} W·m^{-2}·μm^{-1}·sr^{-1} :

a. 0.6   b. 3.8   c. 4.0   d. 4.1   e. 4.4
f. 4.5   g. 4.6   h. 6.5   i. 7.0   j. 7.5

A5(§). Given the following temperature sounding:

z (km)	T (°C)	z (km)	T (°C)
15 to 20	–50	5 to 10	–5
10 to 15	–25	0 to 5	+5
		Earth skin	+20

Find the radiance at the top of the atmosphere for the following transmittance profiles & wavelengths:
λ (μm): (a) 7.0 , (b) 12.7 , (c) 14.4 , (d) 14.7

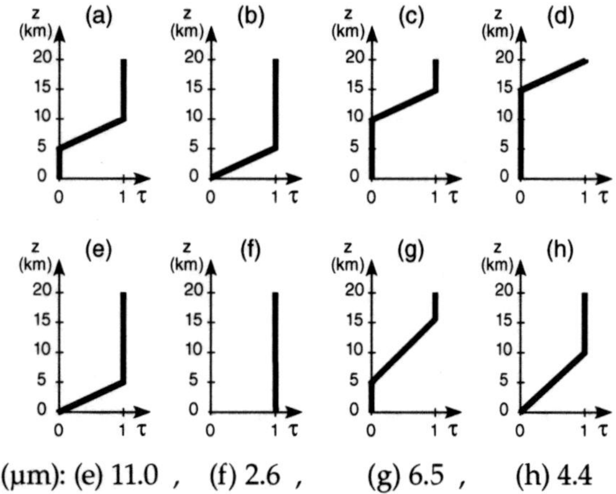

λ (μm): (e) 11.0 , (f) 2.6 , (g) 6.5 , (h) 4.4

A6. For the transmittances of the previous exercise, plot the weighting functions.

A7. For the following altitudes (km) above the Earth's surface, find the satellite orbital periods:
a. 40,000 b. 600 c. 800 d. 2,000 e. 5,000
f. 10,000 g. 15,000 h. 20,000 i. 30,000 j. 50,000

A8. What shade of grey would the following clouds appear in visible, IR, and water-vapor satellite images? a. cirrus b. cirrocumulus c. cirrostratus
d. altocumulus e. altostratus
f. stratus g. nimbostratus h. fog
i. cumulus humilis j. cumulus congestus
k. cumulonimbus l. stratocumulus

A9(§). Do an OSSE to calculate the radiances observed at the satellite, given the weighting functions of Fig. 8.20, but with the following atmospheric temperatures (°C): Layer 1 = –30, Layer 2 = –70, Layer 3 = –20, and Layer 4:
a. 40 b. 35 c. 30 d. 25 e. 20 f. 15 g. 10 h.5

A10. Find the beamwidth angle for a radar pulse for the following sets of [wavelength (cm) , antenna dish diameter (m)]:
a. [20, 8] b. [20, 10] c. [10, 10] d. [10, 5] e. [10, 3]
f. [5, 7] g. [5, 5] h. [5, 2] i. [5, 3] j. [3, 1]

A11. What is the name of the radar band associated with the wavelengths of the previous exercise?

A12. Find the range to a radar target, given the round-trip (return) travel times (μs) of:
a. 2 b. 5 c. 10 d. 25 e. 50
f. 75 g. 100 h. 150 i. 200 j. 300

A13. Find the radar max unambiguous range for pulse repetition frequencies (s^{-1}) of:
a. 50 b. 100 c. 200 d. 400 e. 600
f. 800 g. 1000 h. 1200 i. 1400 j. 1600

A14. Find the Doppler max unambiguous velocity for a radar with pulse repetition frequency (s^{-1}) as given in the previous exercise, for radar with wavelength of: (i) 10 cm (ii) 5 cm

A15. Determine the size of the radar sample volume at a range of 30 km for a 10 cm radar with 5 m diameter antenna dish and pulse duration (μs) of:
a. 0.1 b. 0.2 c. 0.5 d. 1.0 e. 1.5 f. 2 g. 3 h. 5

A16(§). Calculate and plot the microwave refractivity vs. height using P and T of a standard atmosphere, but with vapor pressures (kPa) of:
a. 0 b. 0.05 c. 0.1 d. 0.2 e. 0.5
f. 1.0 g. 2 h. 5 i. 10 j. 20
[Note: Ignore supersaturation issues.]

A17(§). For the previous exercise, plot the vertical gradient $\Delta n/\Delta z$ of refractive index vs. height within the troposphere; determine the average vertical gradient in the troposphere; find the radius of curvature of the radar beam; and find the k_e beam curvature factor.

A18(§). For a radar with 0.5° elevation angle mounted on a 10 m tower, calculate and plot the height of the radar-beam centerline vs. range from 0 to 500 km, for the following beam curvature factors k_e, and name the type of propagation:
a. 0.5 b. 0.8 c. 1.0 d. 1.25 e. 1.33
f. 1.5 g. 1.75 h. 2 i. 2.5 j. 3

A19. Use the simple Z-R relationship to estimate the rainfall rate and the descriptive intensity category used by pilots and air traffic controllers, given the following observed radar echo dBZ values:
a. 10 b. 35 c. 20 d. 45
e. 58 f. 48 g. 52 h. 25

A20. Find reflectivity dBZ for rain 10 km from a WSR-88D, if received power (in 10^{-14} W) is: a. 1 b.2 c. 4 d. 6 e. 8 f. 10 g. 15 h. 20 i. 40 j. 60 k. 80

A21. Estimate the total rainfall accumulated in 1 hour, given the radar reflectivity values below:

Time (min)	dBZ	Time (min)	dBZ
0 - 10	15	29 - 30	50
10 - 25	30	30 - 55	18
25 - 29	43	55 - 60	10

A22. For an S-band radar (wavelength = 10 cm), what is the magnitude of the Doppler frequency shift, given radial velocities (m s^{-1}) of:
a. –110 b. –85 c. –60 d. –20
e. 90 f. 65 g. 40 h. 30

A23. Given a max unambiguous velocity of 25 m s^{-1}, what velocities would be displayed on a Doppler radar for rain-laden air moving with the following real radial velocities (m s^{-1})?

a. 26 b. 28 c. 30 d. 35 e. 20 f. 25 g. 55
h. –26 i. –28 j. –30 k. –35 l. –20

A24. Given a max unambiguous range of 200 km, at what range in the radar display does a target appear if its actual range is:

a. 205 b. 210 c. 250 d. 300 e. 350
f. 400 g. 230 h. 240 i. 390 j. 410

A25. Given the following sets of [average radial velocity (m s^{-1}) , range (km)], find the change of vertical velocity across a change of height of 1 km.

a. [–3, 100] b. [–2, 200] c. [–1, 50] d. [–4, 50]
e. [3, 100] f. [2, 200] g. [1, 50] h. [4, 50]

A26. Given polarimetric radar observations of Z_{VV} = 500 and Z_{VH} = 1.0, find the differential reflectivity and linear depolarization ratio for Z_{HH} values of:

a. 100 b. 200 c. 300 d. 400 e. 500
f. 600 g. 700 h. 800 i. 900 j. 1000

A27. Find the rainfall rate for the following sets of [K_{DP} (°/km) , Z_{DR} (dB)] as determined from polarimetric radar:

a. [1, 2] b. [1, 3] c. [1, 4] d. [1, 5]
e. [2, 2] f. [2, 6] g. [3, 2] h. [3, 4]
i. [3, 6] j. [5, 2] k. [5, 4] l. [5, 6]

A28. A wind profiler has transmitters spaced 0.5 m apart. Find the beam zenith angle for Δt (ns) of:

a. 0.2 b. 0.5 c. 1.0 d. 2 e. 3 f. 5

A29. For a wind profiler with 17° beam tilt, find U and W given these sets of [M_{east}(m s^{-1}), M_{west}(m s^{-1})]:

a. [4, –4] b. [4, –3] c. [4, –5] d. [–4, 4]
e. [–4, 3] f. [–4, 5] g. [1, 1] h. [–1, –1]

Evaluate & Analyze

E1(§). Create blackbody radiance curves similar to Fig. 8.5, but for the following satellite channels. Does a monotonic relationship exists between brightness temperature and blackbody radiance?

a. GOES imager channels 2 - 5.
b. Meteosat-8 imager channels 3 – 6
c. Meteosat-8 imager channels 7 - 11
d. GOES sounder channels 1 - 4
e. GOES sounder channels 5 - 8
f. GOES sounder channels 9 - 12
g. GOES sounder channels 13 - 16
h. GOES sounder channels 17 - 18
i. Meteosat-8 imager channels 1, 2, and 12

E2. For what situations would the brightness temperature NOT equal the actual temperature?

E3. a. Which GOES sounder channels can see the Earth's surface?

b. For the channels from part (a), how would the brightness temperature observed by satellite be affected, if at all, by a scene that contains scattered clouds of diameter smaller than can be resolved. For example, if the channel can see pixels of size 1 km square, what would happen if some of that 1 km square contained cumulus clouds of diameter 300 m, and the remaining pixel area was clear?

E4. Consider the radiative transfer equation. For an opaque atmosphere, $\hat{\tau}_{\lambda\,sfc} = 0$. What happens to the radiation emitted from the Earth's surface?

E5. Knowing the relationship between transmittance profile and weighting function, such as sketched in Figs. 8.7 or 8.8, sketch the associated transmittance profile for the following GOES sounder channels. [Hint, use the weights as sketched in Fig. 8.9 .]

a. 1 b. 2 c. 3 d. 4 e. 5
f. 7 g. 10 h. 11 i. 12 j. 15

E6(§). Given the following temperature sounding:

z (km)	T (°C)	z (km)	T (°C)
15 to 20	–50	5 to 10	–5
10 to 15	–25	0 to 5	+5
		Earth skin	+20

Find the radiance at the top of the atmosphere for the following transmittance profiles & wavelengths:

λ (μm): (a) 7.0 , (b) 12.7 , (c) 14.4 , (d) 4.4

[CAUTION: Transmittance ≠ 0 from the surface.]

E7. For the transmittances of the previous exercise, plot the weighting functions.

E8. a. How low can a LEO satellite orbit without the atmosphere causing significant drag? [Hint, consider the exosphere discussion in the escape-velocity Info box in Chapter 1.]

b. What is the orbital period for this altitude?

c. If the satellite is too low, what would likely happen to it?

E9. a. What is the difference in orbital altitudes for geostationary satellites with orbital periods of 1 calendar day (24 h) and 1 sidereal day?
b. Define a sidereal day, explain why it is different from a calendar day, and discuss why the sidereal day is the one needed for geostationary orbital calculations.

E10. a. Discuss the meanings and differences between geostationary and sun-synchronous orbits.
b. Can satellite loops be made with images from polar orbiting satellites? If so, what would be the characteristics of such a loop?
c. What is the inclination of a satellite that orbits in the Earth's equatorial plane, but in the opposite direction to the Earth's rotation?
d. What are the advantages and disadvantages of sun-synchronous vs. geostationary satellites in observing the weather?

E11. What shade of gray would the following clouds appear in visible, IR, and water-vapor satellite images, and what pattern or shape would they have in the images?
a. jet contrail
b. two cloud layers: cirrus & altostratus
c. two cloud layers: cirrus & stratus
d. three cloud layers: cirrus, altostratus, stratus
e. two cloud layers: fog and altostratus
f. altocumulus standing lenticular
g. altocumulus castellanus
h. billow clouds
i. fumulus
j. volcanic ash clouds

E12. a. Using Fig. 8.8, state in words the altitude range that the water-vapor channel sees.
b. Sometimes water-vapor satellite loops show regions becoming whiter with time, even though there is no advection of water vapor visible in the loop. What might cause this?
c. Why is very dense fog with 100% relative humidity invisible in water-vapor satellite image?

E13(§). Re-do the temperature sounding retrieval exercise of Table 8-6 and its Sample Application. Iterate through each layer 3 times, and then discuss the answer and its errors. Start with an initial temperature (°C) guess at all heights of:
a. –60 b. –50 c. –40 d. –30 e. 0
f. Instead of an isothermal initial guess, use a different initial guess for each layer based on the standard atmosphere near the center of the layer.

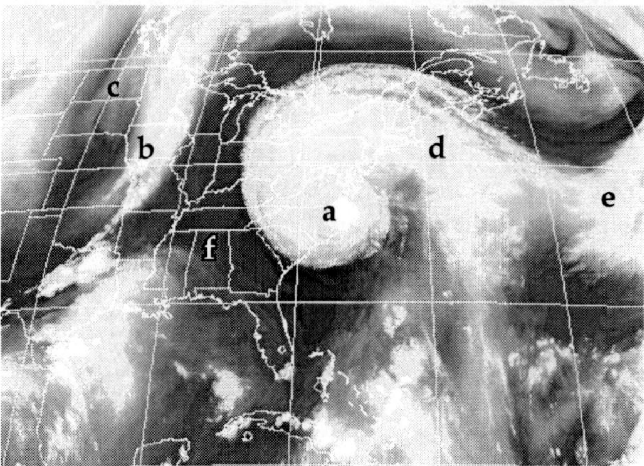

Top: visible. Middle: IR. Bottom: water vapor.
(Images courtesy of Space Science & Engineering Center.)

E14. For locations (a) to (f), interpret the following satellite images, taken by GOES-12 at 2115 UTC on 18 Sept 2003, and justify your interpretation.

E15(§). Re-do the temperature sounding retrieval exercise of Table 8-6 and its Sample Application. Iterate through each layer 3 times, and save your answer as a baseline result. Start with the same initial temperature (–45 °C) guess at all heights.

Then repeat the process with the following errors added to the radiances before you start iterating, and discuss the difference between these new results and the results from the previous paragraph. What is the relationship between radiance error measured by satellite and temperature sounding error resulting from the data inversion?

a. Channel 1, add 5% error to the radiance.
b. Channel 1, subtract 5% error from radiance.
c. Channel 2, add 5% error to the radiance.
d. Channel 2, subtract 5% error from radiance.
e. Channel 3, add 5% error to the radiance.
f. Channel 3, subtract 5% error from radiance.
g. Channel 4, add 5% error to the radiance.
h. Channel 4, subtract 5% error from radiance.

E16(§). For Doppler radar, plot a curve of R_{max} vs. $M_{r\ max}$ for a variety of PRFs between 100 and 2000 s^{-1}, identify the problems with the interpretation of velocities and ranges above and below your curve, and discuss the Doppler dilemma. Use the following Doppler radar wavelengths (cm) for your calculations, and identify the name of their radar band:

a. 20 b. 10 c. 5 d. 3 e. 2 f. 1

E17. Discuss the advantages and disadvantages of PPI, CAPPI, RHI, and AVCS displays for observing:

a. thunderstorms b. hurricanes c. gust fronts
d. low-pressure centers e. fronts

E18(§). For P = 90 kPa, calculate a table of values of $\Delta N/\Delta z$ for different values of temperature gradient $\Delta T/\Delta z$ along the column headers, and different values of $\Delta e/\Delta z$ along the row headers. Then indicate beam propagation conditions (superrefraction, etc.) in each part of the table. Suggest when and where the worst anomalous propagation would occur.

E19. Which is more important in creating strong radar echoes: a larger number of small drops, or a small number of larger drops? Why?

E20. Derive eq. (8.31) using the geometry of a spherical drop.

E21. For radar returns, explain why is the log(range) includes a factor of 2 in eq. (8.28).

E22. By what amount does the radar reflectivity factor Z change when dBZ increases by a factor of 2?

E23. Hong Kong researchers have found that $Z = a_3 \cdot RR^{a_4}$, where $a_3 = 220 \pm 12$ and $a_4 = 1.33 \pm 0.03$, for RR in mm h^{-1} and Z in mm^6 m^{-3}. Find the corresponding equation for RR as a function of dBZ.

E24. Why (and under what conditions) might a weather radar NOT detect cells of heavy rain, assuming that the radar is in good working condition?

E25. How might the bright band look in the vertical velocities measured with a wind profiler? Why?

E26. Given an RHI radar display showing a bright band (see Fig. a below). If the radar switches to a PPI scan having elevation angle ψ shown in (a), sketch the appearance of this bright band in Fig. b. Assume the max range shown in (a) corresponds to the range circle drawn in (b).

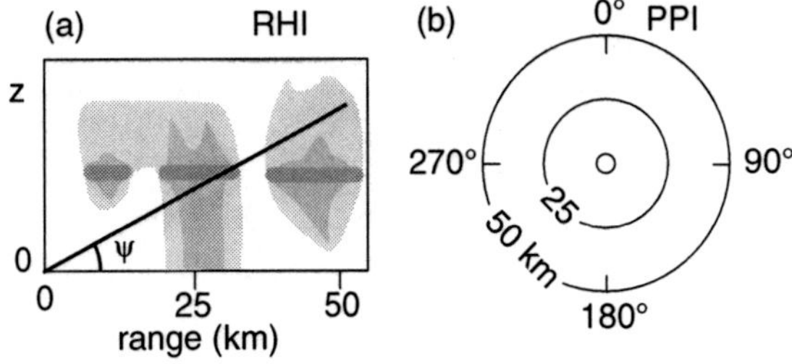

Fig. RHI radar display (a), showing echoes from precipitation regions (hatched). Bright-band echoes are the thick grey horizontal lines. Draw your answer in the blank PPI display of Fig. b.

E27. What design changes would you suggest to completely avoid the limitation of the max unambiguous velocity?

E28. Interpret the VAD display below to diagnose mean wind speed, direction, & convergence, if any.

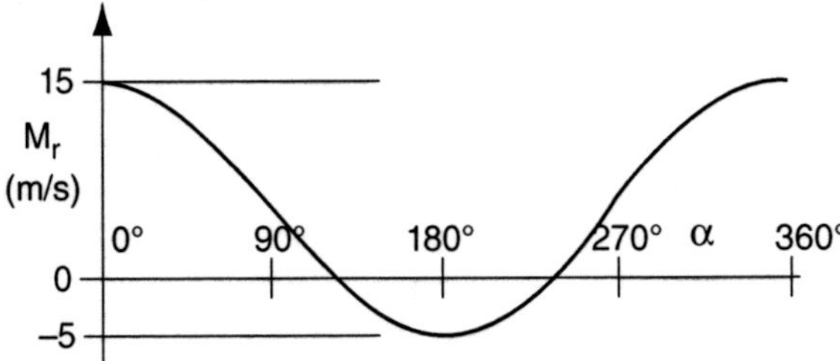

E29(§). Create a table of rainfall rate values, where each column is a different K_{DP} value (within typical range), and each row is a different Z_{DR} value. Draw the isohyets by hand in that table (i.e., draw lines connecting points of equal rainfall rate).

E30. Discuss the advantages and disadvantages of using larger zenith angle for the beams in a wind profiler.

E31. a. Given the vectors drawn in Fig. 8.44, first derive a equations for M_{east} and M_{west} as the sum of the wind vectors projected into the east and west beams, respectively. Then solve those coupled equations to derive eqs. (8.42).
b. Write eqs. similar to eq. (8.42), but for the V and W winds based on M_{north} and M_{south}.

E32. Interpret these weather conditions as observed on a Doppler radar display of radial velocity.

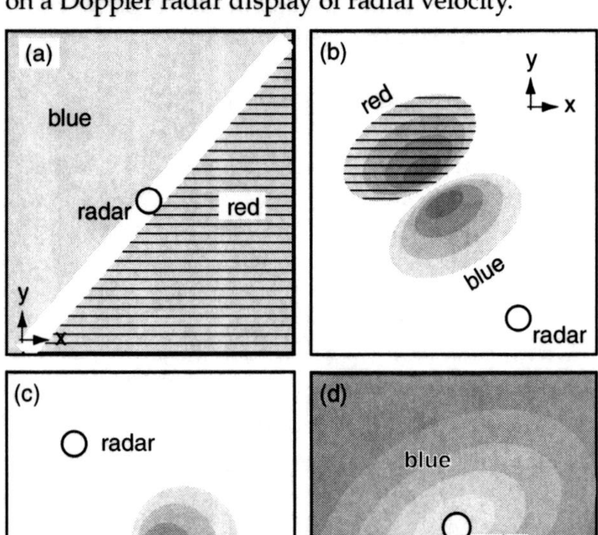

Synthesize

S1. What if substantial cloud coverage could occur at any depth in the stratosphere as well as in the troposphere. Discuss how clouds at different altitudes would look in visible, IR, and water-vapor satellite images. Are there any new difficulties?

S2. What if molecular scattering by air was significant at all wavelengths. How would that affect satellite images and sounding retrievals, if at all?

S3. Suppose that Planck's law and brightness temperature were not a function of wavelength. How would IR satellite image interpretation be affected?

S4. If satellite weighting functions have about the same vertical spreads as plotted in Fig. 8.9, discuss the value of adding more and more channels.

S5. What if gravity on Earth were twice as strong as now. What would be the altitude and orbital period of geostationary satellites?

S6. What if the Earth were larger diameter, but had the same average density as the present Earth. How large would the diameter have to be so that an orbiting geostationary weather satellite would have an orbit that is zero km above the surface? (Neglect atmospheric drag on the satellite.)

S7. Determine orbital characteristics for geostationary satellites over every planet in the solar system.

S8. What if you could put active radar and lidar on weather satellites. Discuss the advantages and difficulties. (Note, this is actually being done.)

S9. Devise a method to eliminate both the max-unambiguous range and max-unambiguous velocity of Doppler radar.

S10. What if pressure and density were uniform with height in the atmosphere. How would radar beam propagation be affected?

S11. Compile (using web searches) the costs of developing, launching, operating, and analyzing the data from a single weather satellite. Compare with analogous costs for a single rawinsonde site.

S12. What if radar reflectivity was proportional only to the number of hydrometeors in a cloud, and not their size. How would radar-echo displays and rainfall-rate calculations be affected?

S13. What if Doppler radars could measure only tangential velocity rather than radial velocity. Discuss how mean wind, tornadoes, and downburst/gust-fronts would look to this radar.

S14. **Radio Acoustic Sounding Systems (RASS)** are wind profilers that also emit loud pulses of sound waves that propagate vertically. How can that be used to also measure the temperature sounding?

9 WEATHER REPORTS & MAP ANALYSIS

Contents

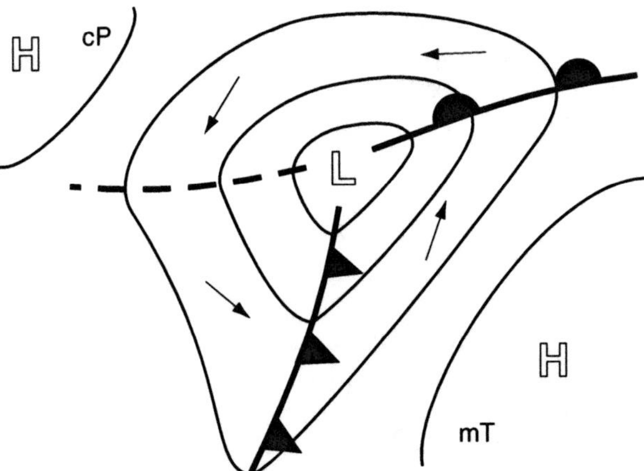

Figure 9.1
Idealized surface weather map showing high (H) and low (L) pressure centers, isobars (thin lines), and fronts (heavy solid lines) in the N. Hemisphere. Vectors indicate near-surface wind. Dashed line is a trough of low pressure. cP indicates a continental polar airmass; mT indicates a maritime tropical airmass.

Surface weather charts summarize weather conditions that can affect your life. Where is it raining, snowing, windy, hot or humid? More than just plots of raw weather reports, you can analyze maps to highlight key features including airmasses, centers of low- and high-pressure, and fronts (Fig. 9.1). In this chapter you will learn how to interpret some surface weather reports, and how to analyze surface weather maps.

SEA-LEVEL PRESSURE REDUCTION

Near the bottom of the troposphere, pressure gradients are large in the vertical (order of 10 kPa km^{-1}) but small in the horizontal (order of 0.001 kPa km^{-1}). As a result, pressure differences between neighboring surface weather stations are dominated by their relative station elevations z_{stn} (m) above sea level.

However, horizontal pressure variations are important for weather forecasting, because they drive horizontal winds. To remove the dominating influence of station elevation via the vertical pressure gradient, the reported station pressure P_{stn} is extrapolated to a constant altitude such as mean sea level (MSL). Weather maps of **mean-sea-level pressure** (P_{MSL}) are frequently used to locate high- and low-pressure centers at the bottom of the atmosphere.

The extrapolation procedure is called **sea-level pressure reduction**, and is made using the hypsometric equation:

$$P_{MSL} = P_{stn} \cdot \exp\left(\frac{z_{stn}}{a \cdot \overline{T_v^*}}\right) \qquad (9.1)$$

where $a = \Re_d / |g| = 29.3$ m K^{-1}, and the average air virtual temperature $\overline{T_v}$ is in Kelvin.

A difficulty is that $\overline{T_v}$ is undefined below ground. Instead, a fictitious average virtual temperature is invented:

$$\overline{T_v^*} = 0.5 \cdot [T_v(t_o) + T_v(t_o - 12\text{ h}) + \gamma_{sa} \cdot z_{stn}] \qquad (9.2)$$

where $\gamma_{sa} = 0.0065$ K m^{-1} is the standard-atmosphere lapse rate for the troposphere, and t_o is the time of the observations at the weather station. Eq. (9.2) attempts to average out the diurnal cycle, and it also extrapolates from the station to halfway toward sea level to try to get a reasonable temperature.

Sample Application

Phoenix Arizona (elevation 346 m MSL) reports dry air with T = 36°C now and 20°C half-a-day ago. P_{stn} = 96.4 kPa now. Find P_{MSL} (kPa) at Phoenix now.

Find the Answer

Given: T(now) = 36°C, T(12 h ago) = 20°C,
z_{stn} = 346 m, P(now) = 96.4 kPa. Dry air.
Find: P_{MSL} = ? kPa

$T_v \approx T$, because air is dry. Use eq. (9.2) : $\overline{T_v^*}$ =
= 0.5·[(36°C) + (20°C) + (0.0065 K m^{-1})·(346 m)]
= 29.16·°C (+ 273.15) = 302.3 K
Use eq. (9.1):
P_{MSL} = (96.4 kPa)·exp[(346 m)/((29.3 m K^{-1})·(302.3 K))]
= (96.4 kPa)·(1.03984) = **100.24 kPa**

Check: Units OK. Physics OK. Magnitude OK.
Discus.: P_{MSL} can be significantly different from P_{stn}

Sample Application

Interpret the following METAR code:
METAR KSJT 160151Z AUTO 10010KT 10SM TS FEW060 BKN075 28/18 A2980 RMK AO2 LTG DSNT ALQDS TSB25 SLP068 T02780178

Hint: see the METAR section later in this chapter.

Find the Answer:

Weather conditions at KSJT (San Angelo, Texas, USA) observed at 0151 UTC on 16th of the current month by an automated station: Winds are from the 100° at 10 knots. Visibility is 10 statute miles or more. Weather is a thunderstorm. Clouds: few clouds at 6000 feet AGL, broken clouds at 7500 feet AGL. Temperature is 28°C and dewpoint is 18°C. Pressure (altimeter) is 29.80 inches Hg. REMARKS: Automated weather station type 2. Distant (> 10 statute miles) lightning in all quadrants. Thunderstorm began at 25 minutes past the hour. Sea-level pressure is 100.68 kPa. Temperature more precisely is 27.8°C, and dewpoint is 17.8°C.

Exposition: As you can see, codes are very concise ways of reporting the weather. Namely, the 3 lines of METAR code give the same info as the 12 lines of plain-language interpretation.

You can use online web sites to search for station IDs. More details on how to code or decode METARs are in the Federal Meteor. Handbook No. 1 (2005) and various online guides. The month and year of the observation are not included in the METAR, because the current month and year are implied.

I am a pilot and flight instructor, and when I access METARs online, I usually select the option to have the computer give me the plain-language interpretation. Many pilots find this the easiest way to use METARs. After all, it is the weather described by the code that is important, not the code itself. However, meteorologists and aviation-weather briefers who use METARs every day on the job generally memorize the codes.

METEOROLOGICAL REPORTS & OBSERVATIONS

One branch of the United Nations is the **World Meteorological Organization (WMO)**. Weather-observation standards are set by the WMO. Also, the WMO works with most nations of the world to coordinate and synchronize weather observations. Such observations are made simultaneously at specified **Coordinated Universal Times (UTC)** to allow meteorologists to create a **synoptic** (snapshot) picture of the weather (see Chapter 1).

Most manual upper-air and surface synoptic observations are made at 00 and 12 UTC. Fewer contries make additional synoptic observatios at 06 and 18 UTC

Weather Codes

One of the great successes of the WMO is the international sharing of real-time weather data via the **Global Telecommunication System (GTS)**. To enable this sharing, meteorologists in the world have agreed to speak the same weather language. This is accomplished by using **Universal Observation Codes** and abbreviations. Definitions of some of these codes are in:

World Meteorological Organization: 1995 (revised 2014): *Manual on Codes. International Codes Vol. 1.1 Part A - Alphanumeric Codes.* WMO No. 306 (http://www.wmo.int/pages/prog/www/WMOCodes.html)

Federal Meteorological Handbook No. 1 (Sept 2005): *Surface Weather Observations and Reports.* FCM-H1-2005 (http: www.ofcm.gov/fmh-1/fmh1.htm).

Sharing of real-time data across large distances became practical with the invention of the electric telegraph in the 1830s. Later developments included the teletype, phone modems, and the internet. Because weather codes in the early days were sent and received manually, they usually consisted of human-readable abbreviations and contractions.

Modern table-driven code formats (**TDCF**) are increasingly used to share data. One is **CREX** (Character form for the Representation and EXchange of data). Computer binary codes include **BUFR** (Binary Universal Form for the Representation of meteorological data) and **GRIB** (Gridded Binary).

However, there still are important sets of **alphanumeric codes** (letters & numbers) that are human writable and readable. Different alphanumeric codes exist for different types of weather observations and forecasts, as listed in Table 9-1. We will highlight one code here — the **METAR**.

Table 9-1. List of alphanumeric weather codes.

Name	Purpose
SYNOP	Report of surface observation from a fixed land station
SHIP	Report of surface observation from a sea station
SYNOP MOBIL	Report of surface observation from a mobile land station
METAR	Aviation routine weather report (with or without trend forecast)
SPECI	Aviation selected special weather report (with or without trend forecast)
BUOY	Report of a buoy observation
RADOB	Report of ground radar weather observation
RADREP	Radiological data report (monitored on a routine basis and/or in case of accident)
PILOT	Upper-wind report from a fixed land station
PILOT SHIP	Upper-wind report from a sea station
PILOT MOBIL	Upper-wind report from a mobile land station
TEMP	Upper-level pressure, temperature, humidity and wind report from a fixed land station
TEMP SHIP	Upper-level pressure, temperature, humidity and wind report from a sea station
TEMP DROP	Upper-level pressure, temperature, humidity and wind report from a dropsonde released by carrier balloons or aircraft
TEMP MOBIL	Upper-level pressure, temperature, humidity and wind report from a mobile land station
ROCOB	Upper-level temperature, wind and air density report from a land rocketsonde station
ROCOB SHIP	Upper-level temperature, wind and air density report from a rocketsonde station on a ship
CODAR	Upper-air report from an aircraft (other than weather reconnaissance aircraft)
AMDAR	Aircraft report (Aircraft Meteorological DAta Relay)
ICEAN	Ice analysis
IAC	Analysis in full form
IAC FLEET	Analysis in abbreviated form
GRID	Processed data in the form of grid-point values
GRAF	Processed data in the form of grid-point values (abbreviated code form)

Table 9-1 (continued). Alphanumeric codes.

Name	Purpose
WINTEM	Forecast upper wind and temperature for aviation
TAF	Aerodrome forecast
ARFOR	Area forecast for aviation
ROFOR	Route forecast for aviation
RADOF	Radiological trajectory dose forecast (defined time of arrival and location)
MAFOR	Forecast for shipping
TRACK-OB	Report of marine surface observation along a ship's track
BATHY	Report of bathythermal observation
TESAC	Temperature, salinity and current report from a sea station
WAVEOB	Report of spectral wave information from a sea station or from a remote platform (aircraft or satellite)
HYDRA	Report of hydrological observation from a hydrological station
HYFOR	Hydrological forecast
CLIMAT	Report of monthly values from a land station
CLIMAT SHIP	Report of monthly means and totals from an ocean weather station
NACLI, CLINP, SPCLI, CLISA, INCLI	Report of monthly means for an oceanic area
CLIMAT TEMP	Report of monthly aerological means from a land station
CLIMAT TEMP SHIP	Report of monthly aerological means from an ocean weather station
SFAZI	Synoptic report of bearings of sources of atmospherics (e.g., from lightning)
SFLOC	Synoptic report of the geographical location of sources of atmospherics
SFAZU	Detailed report of the distribution of sources of atmospherics by bearings for any period up to and including 24 hours
SAREP	Report of synoptic interpretation of cloud data obtained by a meteorological satellite
SATEM	Report of satellite remote upper-air soundings of pressure, temperature and humidity
SARAD	Report of satellite clear radiance observations
SATOB	Report of satellite observations of wind, surface temperature, cloud, humidity and radiation

METAR and SPECI

METAR stands for Meteorological Aviation Report. It contains a routine (hourly) observations of surface weather made at a manual or automatic weather station at an airport. It is formatted as a text message using codes (abbreviations, and a specified ordering of the data blocks separated by spaces) that concisely describe the weather.

Here is a brief summary on how to read METARs. Grey items below can be omitted if not needed.

Format

[**METAR** or SPECI] [corrected] [weather station ICAO code] [day, time] [report type] [wind direction, speed, gusts, units] [direction variability] [prevailing visibility, units] [minimum visibility, direction] [runway number, visual range] [current weather] [lowest altitude cloud coverage, altitude code] [higher-altitude cloud layers if present] [temperature/dewpoint] [units, sea-level pressure code] [supplementary] **RMK** [remarks].

Example (with remarks removed):

```
METAR KTTN 051853Z 04011G20KT 1 1/4SM
R24/6200FT VCTS SN FZFG BKN003 OVC010
M02/M03 A3006 RMK...
```

Interpretation of the Example Above

Routine weather report for Trenton-Mercer Airport (NJ, USA) made on the 5th day of the current month at 1853 UTC. Wind is from 040° true at 11 gusting to 20 knots. Visibility is 1.25 statute miles. Runway visual range for runway 24 is 6200 feet. Nearby thunderstorms with snow and freezing fog. Clouds are broken at 300 feet agl, and overcast at 1000 ft agl. Temperature minus 2°C. Dewpoint minus 3°C. Altimeter setting is 30.06 in. Hg. Remarks...

SPECI

If the weather changes significantly from the last routine METAR report, then a special weather observation is taken, and is reported in an extra, unscheduled SPECI report. The SPECI has all the same data blocks as the METAR plus a plain language explanation of the special conditions.

The criteria that trigger SPECI issuance are:
Wind direction: changes >45° for speeds ≥ 10 kt.
Visibility: changes across threshold: 3 miles, 2 miles, 1 mile, 0.5 mile or instrument approach minim.
Runway visual range: changes across 2400 ft.
Tornado, Waterspout: starts, ends, or is observed.
Thunderstorm: starts or ends.
Hail: starts or ends.
Freezing precipitation: starts, changes, ends.
Ceiling: changes across threshold: 3000, 1500, 1000, 500, 200 (or lowest approach minimum) feet.
Clouds: when layer first appears below 1000 feet.
Volcanic eruption: starts.

Details of METAR / SPECI Data Blocks

Corrected: COR if this is a corrected METAR.
Weather Station **ICAO** Code is a 4-letter ID specified by the Internat. Civil Aviation Organization.
Day, Time: 2-digit day within current month, 4-digit time, 1-letter time zone (Z = UTC. Chapter 1).
Type: AUTO=automatic; (blank)=routine; NIL= missing.
Wind: 3-digit direction (degrees relative to true north, rounded to nearest 10 degrees). VRB=variable. 2- to 3-digit speed. (000000=calm). G prefixes gust max speed. Units (KT=knots, KMH=kilometers per hour, MPS=meters per second).
Direction Variability only if > 60°. Example: 010**V**090, means variable direction between 010° and 090°.
Prevailing Visibility: 4 digits in whole meters if units left blank. If vis < 800 m, then round down to nearest 50 m. If 800 ≤ vis < 5000 m, then round down to nearest 100 m. If 5000 ≤ vis < 9999 m, then round down to nearest 1000 m. Else "9999" means vis ≥ 10 km. In USA: number & fraction, with SM=statute miles. NDV = no directional variations.
Minimum Visibility: 4 digits in whole meters if units are blank & 1-digit (a point from an 8-point compass)
Runway Visual Range (RVR): **R**, 2-digit runway identifier, (if parallel runways, then: L=left, C=center, R=right), **/** , 4-digit RVR. Units: blank=meters, FT=feet. If variable RVR, then append optional: 4 digits, **V**, 4 digits to span the range of values. Finally, append optional tendency code: U=up (increasing visibility), N=no change, D-down (decreasing visibility).
Weather: see Tables in this chapter for codes. 0 to 3 groups of weather phenomena can be reported.
Clouds: 3-letter coverage abbreviation (see Table 9-10), 3-digit cloud-base height in hundreds of feet agl. TCU=towering cumulus congestus, CB = cumulonimbus. If no clouds, then whole cloud block replaced by CLR=clear or by SKC=sky clear. NSC= no significant clouds below 5000 ft (1500 m) with no thunderstorm and good visibility. NCD if no clouds detected by an automated system.
Higher Cloud Layers if any: 2nd lowest clouds reported only if ≥ SCT. 3rd lowest only if ≥ BRN.
Note: if visibility > 10 SM and no clouds below 5,000 ft (1500 m) agl and no precipitation and no storms, then the visibility, RVR, weather, & cloud blocks are omitted, and replaced with CAVOK, which means ceiling & visibility are OK (i.e., no problems for visual flight). (Not used in USA.)
Temperature/Dew-point: rounded to whole °C. Prefix M=minus.

Sea-level Pressure: 4 digits. Unit code prefix: A = altimeter setting in inches mercury, for which last 2 digits are hundredths. Q = whole hectoPascals hPa). Example: Q1016 = 1016 hPa = 101.6 kPa.

Supplementary: Can include: RE recent weather; WS wind shear; W sea state; runway state (SNOCLO=airport closed due to snow); trend, significant forecast weather (NOSIG=no change in significant weather, NSW=nil significant weather)

Remarks: **RMK**. For details, see the manuals cited three pages earlier.

Although you can read a METAR if you've memorized the codes, it is easier to use on-line computer programs to translate the report into plain language. Consult other resources and manuals to learn the fine details of creating or decoding METARs.

Weather-Observation Locations

Several large governmental centers around the world have computers that automatically collect, test data quality, organize, and store the vast weather data set of coded and binary weather reports. For example, Figs. 9.2 to 9.12 show locations of weather observations that were collected by the computers at the European Centre for Medium-Range Weather Forecasts (**ECMWF**) in Reading, England, for a six-hour period centered at 00 UTC on 30 Mar 2015.

The volume of weather data is immense. There are many millions of locations (manual stations, automatic sites, and satellite obs) worldwide that report weather observations near 00 UTC. At ECMWF, many hundreds of gigabytes (GB) of weather-observation data are processed and archived every day. The locations for some of the different types of weather-observation data are described next.

Surface observations (Fig. 9.2) include manual ones from land (SYNOP) and ship (SHIP) at key synoptic hours. Many countries also make hourly observations at airports, reported as METARs.

Surface automatic weather-observation systems make more frequent or nearly continuous reports. Examples of automatic surface weather stations are **AWOS** (Automated Weather Observing System), and **ASOS** (Automated Surface Observing System) in the USA. Those automatic reports that are near the synoptic hours are also included in Fig. 9.2.

Both moored and drifting buoys (BUOY; Fig. 9.3) also measure near-surface weather and ocean-surface conditions, and relay this data via satellite.

Small weather balloons (Fig. 9.4) can be launched manually or automatically from the surface to make **upper-air soundings**. As an expendable **radiosonde** package is carried aloft by the helium-filled

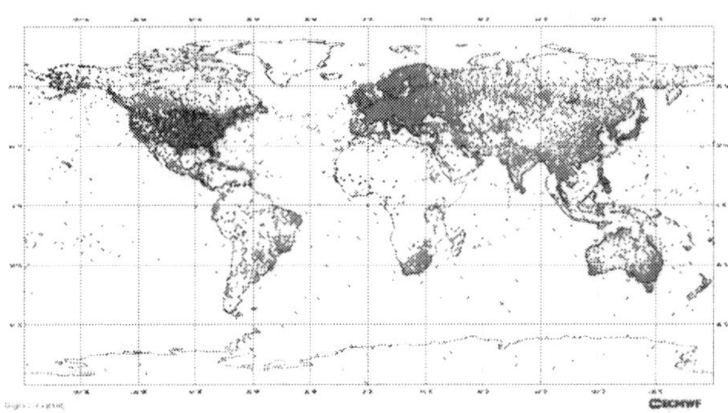

Figure 9.2
Surface data locations for observations of temperature, humidity, winds, clouds, precipitation, pressure, and visibility collected by synoptic weather stations on land and ship. Valid: 00 UTC on 30 Mar 2015. Number of observations: 36024 METAR (land) + 23742 SYNOP (land) + 376079 SHIP = 63526 surface obs. (From ECMWF.)

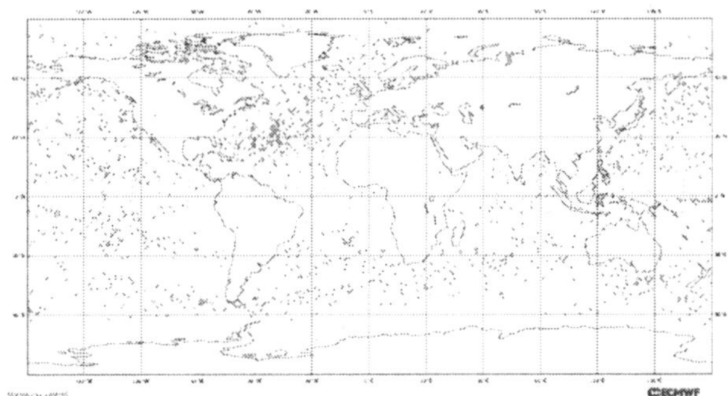

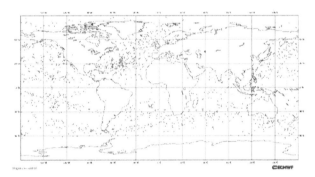

Figure 9.3
Surface data locations for temperature and winds collected by drifting and moored BUOYs. Valid: 00 UTC on 30 Mar 2015. Number of observations: 9114 drifters + 716 moored = 8830 buoys. (From ECMWF.)

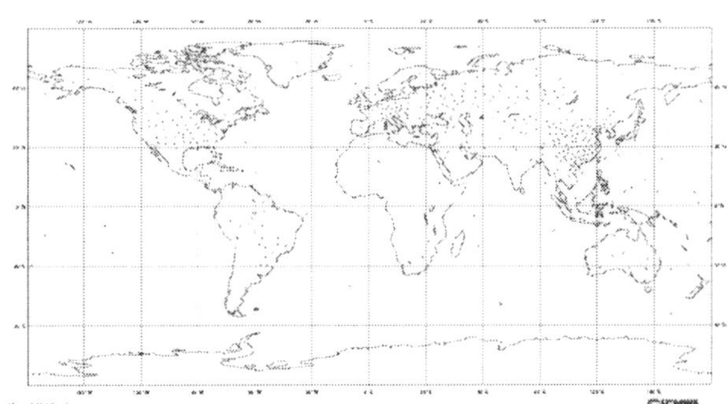

Figure 9.4
Upper-air sounding locations for temperature, pressure, and humidity collected by rawinsonde balloons launched from land and ship, and by dropsondes released from aircraft. Valid: 00 UTC on 30 Mar 2015. Number of observations: 596 land (TEMP) + 1 ship (TEMP SHIP) + 0 dropsondes (TEMP DROP) = 597 soundings. Extra dropsondes are often dropped over oceans at hurricanes, typhoons, and strong winter storms. (From ECMWF.)

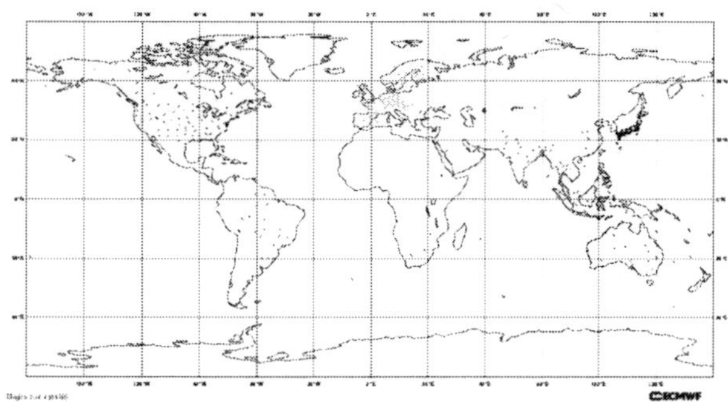

Figure 9.5
Upper-air data locations for winds collected by: PILOT balloons, ground-based wind profilers, and Doppler radars. Valid: 00 UTC on 30 Mar 2015. Number of observations: 324 pilot/rawinsonde balloons + 3158 microwave wind profilers = 3482 wind soundings. (From ECMWF.)

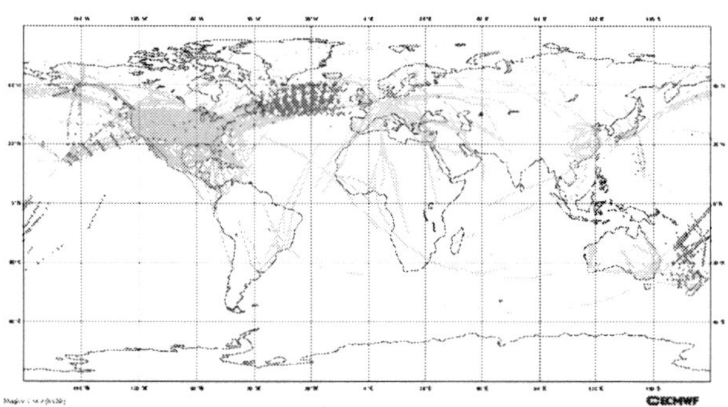

Figure 9.6
Upper-air data locations for temperature and winds collected by commercial aircraft: AIREP manual reports (black), and AMDAR & ACARS (grey) automated reports. Valid: 00 UTC on 30 Mar 2015. Number of observations: 2254 AIREP + 17661 AMDAR + 156136 ACARS = 176051 aircraft observations, most at their cruising altitude of 10 to 15 km above sea level. (From ECMWF.)

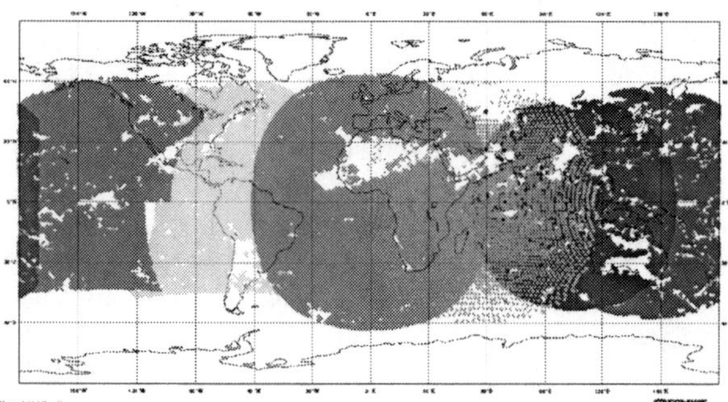

Figure 9.7
Upper-air data locations for winds collected by geostationary satellites (SATOB) from the USA (GOES), Europe (METEOSAT), and others around the world. Based atmospheric motion vectors (AMV) of IR cloud patterns. Similar satellite observations are made using water vapor and visible channels. Valid: 00 UTC on 30 Mar 2015. Number of observations: 442475. (From ECMWF.)

latex balloon, and later as it descends by parachute, it measures temperature, humidity, and pressure. These radiosonde observations are called **RAOBs**.

Some radiosondes include additional instruments to gather navigation information, such as from **GPS** (Global Positioning Satellites). These systems are called **rawinsondes**, because the winds can be inferred by the change in horizontal position of the sonde. When a version of the rawinsonde payload is dropped by parachute from an aircraft, it is called a **dropsonde.**

Simpler weather balloons called **PIBALs** (Pilot Balloons) carry no instruments, but are tracked from the ground to estimate winds (Fig. 9.5). Most balloon soundings are made at 00 and 12 UTC.

Remote sensors on the ground include weather radar such as the **NEXRAD** (Weather Surveillance Radar WSR-88D). Ground-based microwave **wind profilers** (Fig. 9.5) automatically measure a vertical profile of wind speed and direction. **RASS** (Radio Acoustic Sounding Systems) equipment uses both sound waves and microwaves to measure virtual temperature and wind soundings.

Commercial aircraft (Fig. 9.6) provide manual weather observations called Aircraft Reports (**AIREPS**) at specified longitudes as they fly between airports. Many commercial aircraft have automatic meteorological reporting equipment such as **ACARS** (Aircraft Communication and Reporting System), **AMDAR** (Aircraft Meteorological Data Relay), & **ASDAR** (Aircraft to Satellite Data Relay).

Geostationary satellites are used to estimate tropospheric winds (Fig. 9.7) by tracking movement of clouds and water-vapor patterns. Surface winds over the ocean can be estimated from polar orbiting satellites using **scatterometer** systems (Fig. 9.8) that measure the scattering of microwaves off the sea surface. Rougher sea surface implies stronger winds.

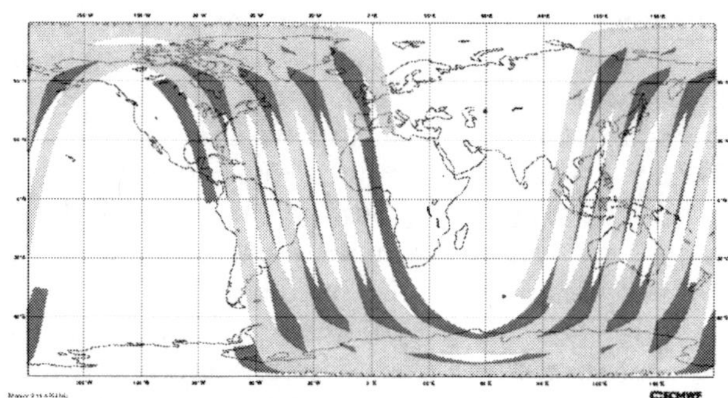

Figure 9.8
Surface-wind estimate locations from microwave scatterometer measurements of sea-surface waves by the polar-orbiting satellites. Valid: 00 UTC on 30 Mar 2015. Number of observations: 526159. (From ECMWF.)

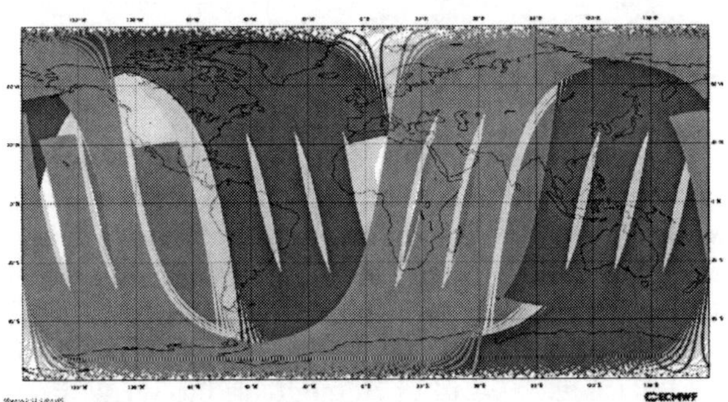

Figure 9.9
Temperature-sounding (SATEM) locations from radiation measurements by polar-orbiting satellites using the AMSU (Advanced Microwave Sounding Unit). Satellites: several NOAA satellites, Aqua, and MetOP. Valid: 00 UTC on 30 Mar 2015. Number of observations: 612703. (From ECMWF.)

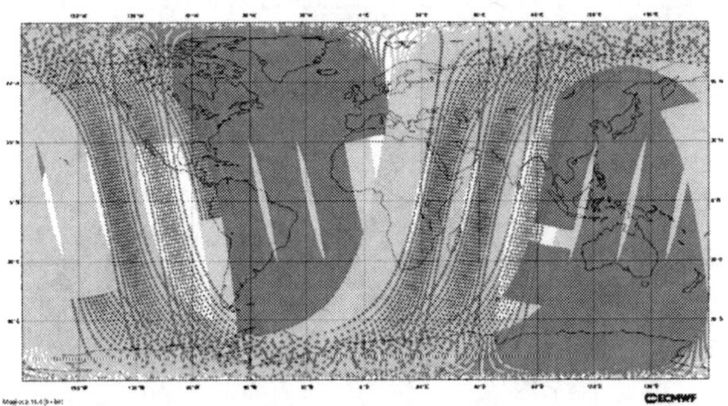

Figure 9.10
Temperature-sounding (SATEM) locations from high-spectral-resolution infrared radiation measurements by polar-orbiting satellites, using HIRS (High-resolution Infrared Radiation Sounder). Valid: 00 UTC on 30 Mar 2015. Number of observations: 5394127. (From ECMWF.)

Satellites radiometrically estimate air-temperature to provide remotely-sensed upper-air automatic data (Fig. 9.9). One system is the **AMSU** (Advanced Microwave Sounding Unit), currently flying on NOAA 15, 16, 17, 18, Aqua, and the European MetOp satellites.

Higher spectral-resolution soundings (Fig. 9.10) are made with the **HIRS** (High-resolution Infrared Radiation Sounding) system on a polar-orbiting satellites.

Estimates of air density can also be made as signals from Global Positioning System (**GPS**) satellites are bent as they pass through the atmosphere to other satellites (Fig. 9.11). Other techniques (not shown) use ground-based sensors to measure the refraction and delay of GPS signals.

Polar orbiting satellites can also be used to estimate atmospheric motion vectors (AMV) from the movement of IR cloud patterns. These can give upper-air wind data over the Earth's poles (Fig. 9.12) — regions not visible from geostationary satellites.

Many more satellite products are used, beyond the ones shown here. Radiance measurements from geostationary satellites are used to estimate temperature and humidity conditions for numerical forecast models via variational data assimilation in three or four dimensions (**3DVar** or **4DVar**). Tropospheric precipitable water can be estimated by satellite from the amount of microwave or IR radiation emitted from the troposphere.

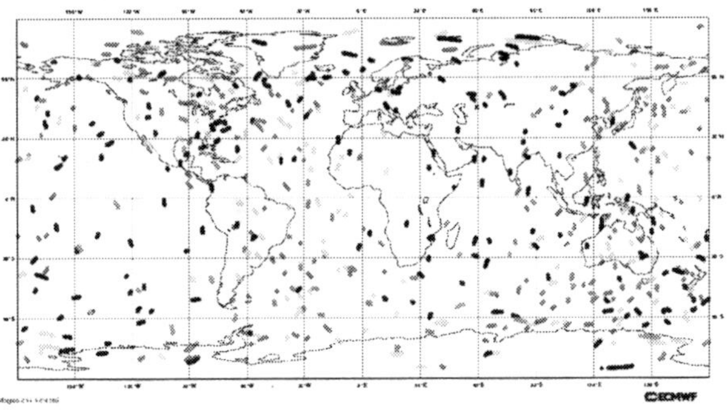

Figure 9.11
Air density estimates are made using Global Positioning System (GPS) Radio Occultation (GPS-RO). Valid: 00 UTC on 30 Mar 2015. Number of observations: 81236. (From ECMWF.)

These synoptically reported data give a snapshot of the weather, which can be analyzed on **synoptic weather maps**. The methods used to analyze the weather data to create such maps are discussed next.

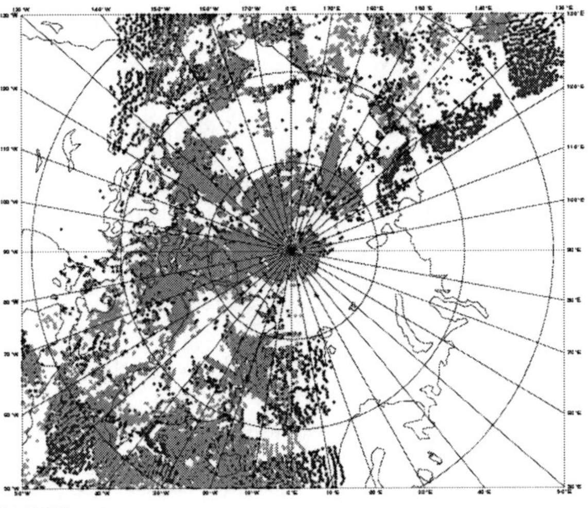

Figure 9.12
Atmospheric motion vector (AMV) locations from IR observations by polar satellites, over the N. Pole. Valid: 00 UTC on 30 Mar 2015. Number of observations: 33171. (From ECMWF.)

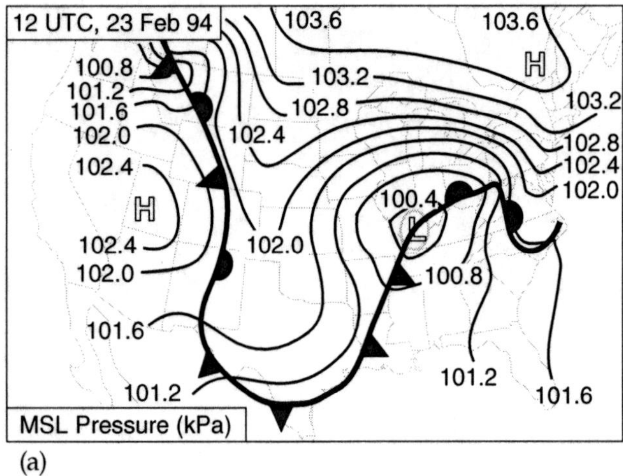

(a)

(b)

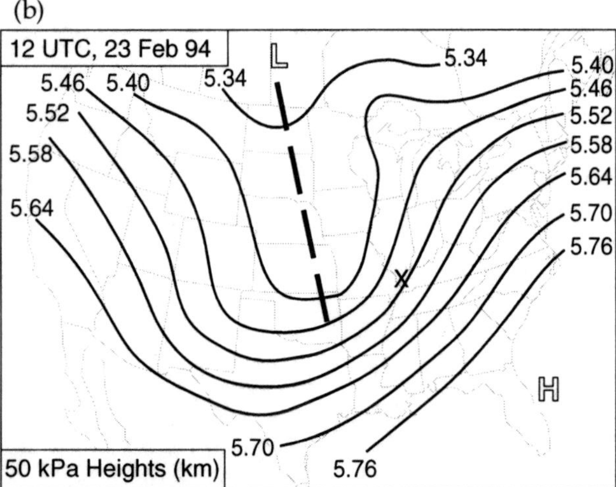

Figure 9.13
Examples of synoptic weather maps, which give a snapshot of the weather at an instant in time. Fig. 9.13a shows an example of a synoptic weather map for pressure at the bottom of the troposphere, based on surface weather observations. It shows pressure reduced to mean sea-level (MSL), fronts, and high (H) and low (L) pressure centers. Fig. 9.13b shows an upper-air synoptic map for geopotential heights of the 50 kPa isobaric surface (a surrogate for pressure) near the middle of the troposphere valid at the same time, using data from weather balloons, aircraft, satellite, and ground-based remote sensors. It also indicates high and low centers, and the trough axis (dashed line). By studying both maps, you can get a feeling for the three-dimensional characteristics of the weather.

Three steps are needed to create such maps. First, the weather data must be observed and communicated to central locations. Second, the data is tested for quality, where erroneous or suspect values are removed. These steps were already discussed.

Third, the data is analyzed, which means it is integrated into a coherent picture of the weather. This last step often involves interpolation to a grid (if it is to be analyzed by computers), or drawing of isopleths and identification of weather features (lows, fronts, etc.) if used by humans.

[Based on original analyses by Jon Martin.]

SYNOPTIC WEATHER MAPS

Weather observations that were taken **synoptically** (i.e., simultaneously) at many weather stations worldwide can be drawn on a weather map. For any one station, the weather observations include many different variables. So a shorthand notation called a **station-plot model** was devised to use symbols or glyphs for each weather element, and to write those data around a small circle representing the station location.

But the raw numbers and glyphs plotted on a map at hundreds of stations can be overwhelming. So computers or people can **analyze** the map to create a coherent picture that integrates together all the weather elements, such as in Fig. 9.13.

The resulting **synoptic-weather map** shows scales of weather (see table in the Forces & Winds chapter) that are called **synoptic-scale** weather. The field of study of these weather features (fronts, highs, lows, etc.) is called **synoptics**, and the people who study and forecast these features are **synopticians.**

Station Plot Model

On weather maps, the location of each weather station is circled, and that station's weather data is plotted in and around the circle. The standardized arrangement of these data is called a **station plot model** (Fig. 9.14). Before the days of computerized geographic information systems (GIS), meteorologists had to rely on abbreviated codes to pack as much data around each plotted weather station as possible. These codes are still used today.

Unfortunately, different weather organizations/countries use different station plot models and different codes. Here is how you interpret the surface station plot models for USA and Canada:

• **T** is a two-digit **temperature** in whole degrees (°C in most of the world; °F if plotted by the USA).

Example: **18** means 18°F if plotted by a USA organization, but means 18°C if plotted by a Canadian organization.

[CAUTION: On weather maps produced in the USA, temperatures at Canadian weather stations are often converted to °F before being plotted. The opposite happens for USA stations plotted on Canadian weather maps — they are first converted to °C. You should always think about the temperature value to see if it is reasonable for the units you assume.]

• **T_d** is a two-digit **dew-point temperature** in whole degrees (°C in most of the world; °F if plotted by the USA).

Example: **14** means 14°F if plotted by a USA organization, but means 14°C if plotted by a Canadian organization. [Same CAUTION as for **T**.]

• **P** is the 3 least-significant digits of a 4- or 5-digit metric **mean sea-level pressure** value. To the left of the 3 digits, prefix either "9" or "10", depending on which one gives a value closest to standard sea-level pressure. For kPa, insert a decimal point two places from the right. For hPa, insert a decimal point one place from the right.

Example: **041** means 100.41 kPa, or 1004.1 hPa. New example: **986** means 99.86 kPa, or 998.6 hPa.

[CAUTION: Some organizations report **P** in inches of mercury (in. Hg.) instead of hPa. P_{MSL} (in. Hg.) is an **altimeter setting**, used by aircraft pilots.]

• **past wx** is a glyph (Table 9-2) for **past weather** in the past hour (or past 6 hours for Canada). It is blank unless different from present weather.

• **wx** is a glyph for **present weather** (at time of the weather observation). Tables 9-3 show the commonly used weather glyphs. Examples:

means snow shower; means thunderstorm with moderate rain.

• **vis** is the code for **visibility** (how far you can see objects).

In the USA, visibility is in statute miles.
(a) if visibility ≤ 3 1/8 miles, then *vis* can include a fraction.
(b) if 3 1/8 < visibility < 10 miles, then *vis* does not include a fraction.
(c) if 10 ≤ visibility, then *vis* is left blank.

Example: **2 1/4** means visibility is 2 1/4 statute miles. New example: **8** means visibility is 8 miles.

In Canada, visibility is in kilometers, but is coded into a two-digit *vis* code integer as follows:
(a) if $vis \le 55$, then visibility (km) = $0.1 \cdot vis$
(b) if $56 \le vis \le 80$, then visibility (km) ≈ $vis - 50$
(c) if $81 \le vis$, then visibility (km) ≈ $5 \cdot (vis - 74)$

Examples: **24** means 2.4 km. **66** means 16 km. **82** means 40 km. See Fig. 9.15 for a graph.

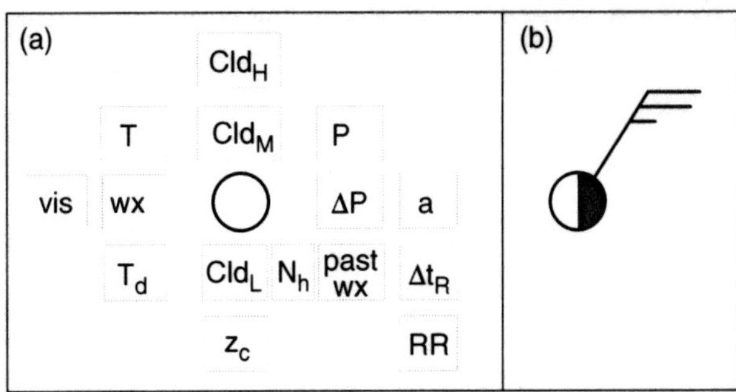

Figure 9.14
Station plot model. (a) Fixed fields (see tables). (b) Total cloud coverage represented by partially filled circle, and wind direction and speed. The circle represents the station location. Normally, (a) and (b) are plotted superimposed.

Table 9-2. Past or recent weather glyphs and codes (past wx).

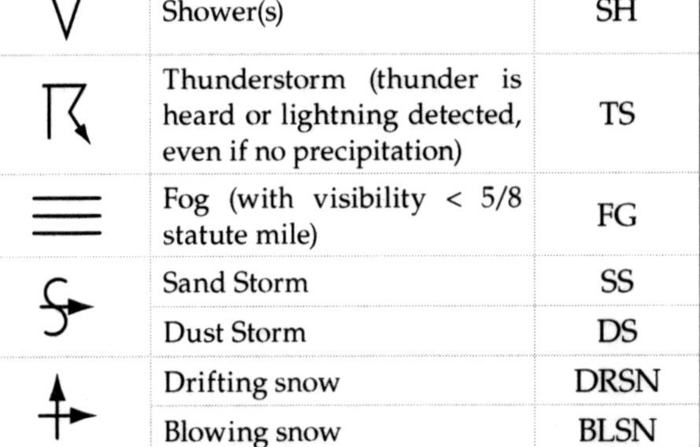

Glyph	Meaning	METAR
	Drizzle	DZ
	Rain	RA
	Snow	SN
	Shower(s)	SH
	Thunderstorm (thunder is heard or lightning detected, even if no precipitation)	TS
	Fog (with visibility < 5/8 statute mile)	FG
	Sand Storm	SS
	Dust Storm	DS
	Drifting snow	DRSN
	Blowing snow	BLSN

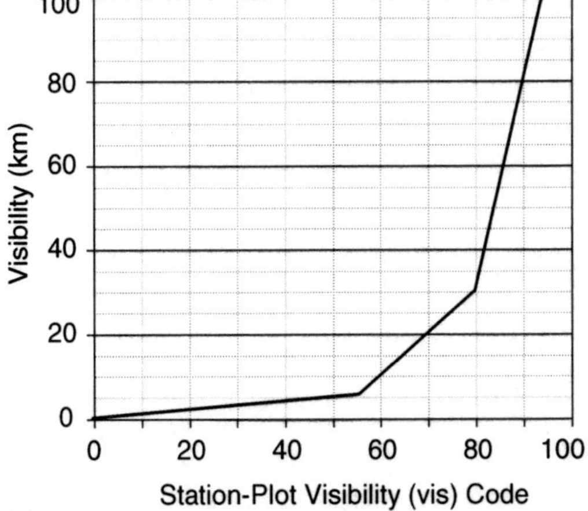

Figure 9.15
Visibility (vis) code for Canada.

Table 9-3a. Basic weather (wx) symbols and codes.

Glyph	Meaning	METAR
Precipitation:		
	Drizzle ‡	DZ
	Rain ‡	RA
	Snow ‡	SN
	Hail (large, diameter ≥ 5 mm)	GR
	Graupel (snow pellets, small hail, size < 5 mm)	GS
	Ice Pellets (frozen rain, called sleet in USA)	PL
	Ice Crystals ("diamond dust")	IC
	Snow Grain	SG
	Ice Needles	
	Unknown Precipitation (as from automated station)	UP
Obscuration:		
	Fog (with visibility < 5/8 statute mile) ‡	FG
	Mist (diffuse fog, with visibility ≥ 5/8 statute mile)	BR
	Haze	HZ
	Smoke	FU
	Volcanic Ash	VA
	Sand in air	SA
	Dust in air	DU
	Spray	PY
Storms & Misc.:		
	Squall	SQ
	Thunderstorm (thunder is heard or lightning detected, even if no precipitation)‡	TS
	Lightning	
	Funnel Cloud	FC
	Tornado or Waterspout	+FC
	Dust Devil (well developed)	PO
	Sand Storm	SS
	Dust Storm	DS

‡ Can be used as a "Past Weather" glyph.

Table 9-3b. Weather-code modifiers.

Glyph	Meaning	METAR prefix
Intensity, Proximity, or Recency: (Grey box is placeholder for a precipitation glyph from Table 9-3a. For example, ❜❜ means light drizzle.)		
	Light	–
	Moderate	(blank)
	Heavy	+
; ;	Intermittent and light ; moderate ; heavy	(no code for intermittent)
( )	In the vicinity. In sight, but not at the weather stn.	VC
	Virga (precip. in sight, but not reaching the ground).	VIRGA
]	In past hour, but not now	
		Increased during past hour, and occurring now	
		Decreased during the past hour, and occurring now	
Descriptor:		
	Shower (slight)	–SH
	Shower (moderate)	SH
	Shower (heavy)	+SH
	Thunderstorm	TS
	Thunderstorm (heavy)	+TS
	Freezing. (if light, use left placeholder only) ***	FZ
	Blowing (slight)**	–BL
	Blowing (moderate)**	BL
	Blowing (strong, severe)**	+BL
	Drifting (low) (For DU, SA, SN raised < 2 m agl)	DR
	Shallow*	MI
– –	Partial*	PR
	Patchy*	BC

(* Prefixes for fog FG only.) (** Prefixes only for DU, SA, SN or PY.)
(*** Prefixes only for FG, DZ, or RN.)

Table 9-4. High Clouds (Cld_H).

Glyph	Meaning
	Cirrus (scattered filaments, "mares tails", not increasing).
	Cirrus (dense patches or twisted sheaves of filament bundles).
	Cirrus (dense remains of a thunderstorm anvil).
	Cirrus (hook shaped, thickening or spreading to cover more sky).
	Cirrus and cirrostratus increasing coverage or thickness, but covering less than half the sky.
	Cirrus and cirrostratus covering most of sky, and increasing coverage or thickness.
	Cirrostratus veil covering entire sky.
	Cirrostratus, not covering entire sky.
	Cirrocumulus (with or without smaller amounts of cirrus and/or cirrostratus).

Table 9-5. Mid Clouds (Cld_M).

Glyph	Meaning
	Altostratus (thin, semitransparent).
	Altostratus (thick), or nimbostratus.
	Altocumulus (thin).
	Altocumulus (thin, patchy, changing, and/or multi-level).
	Altocumulus (thin but multiple bands or spreading or thickening).
	Altocumulus (formed by spreading of cumulus).
	Multiple layers of middle clouds (could include altocumulus, altostratus, and/or nimbostratus).
	Altocumulus castellanus (has turrets or tuffs).
	Altocumulus of chaotic sky (could include multi-levels and dense cirrus).

• **Cld_H** is a glyph for high clouds (see Table 9-4).

• **Cld_M** is a glyph for mid-level clouds (Table 9-5).

• **Cld_L** is a glyph for low clouds (see Table 9-6).

• **N_h** is fraction of sky covered by low and mid clouds only. Units: oktas (eighths).

This can differ from the total sky coverage (see Table 9-10), which is indicated by the shading inside the station circle.

• **z_c** is a single-digit code for cloud-base height of lowest layer of clouds (Table 9-7).

• **ΔP** is 2 digits giving pressure change in the past 3 hours, prefixed with + or –. Units are hundredths of kPa. Example: –28 is a pressure decrease of 0.28 kPa or 2.8 hPa.

Table 9-6. Low Clouds (Cld_L).

Glyph	Meaning
	Cumulus (Cu) humilis. Fair-weather cumulus. Little vertical development.
	Cumulus mediocris. Moderate to considerable vertical development.
	Cumulus congestus. Towering cumulus. No anvil top.
	Stratocumulus formed by the spreading out of cumulus.
	Stratocumulus. (Not from spreading cu)
	Stratus.
	Scud. Fractostratus or fractocumulus, often caused by rain falling from above.
	Cumulus and stratocumulus at different levels (not cause by spreading of Cu.
	Cumulonimbus. Thunderstorm. Has anvil top that is glaciated (contains ice crystals, and looks fibrous).

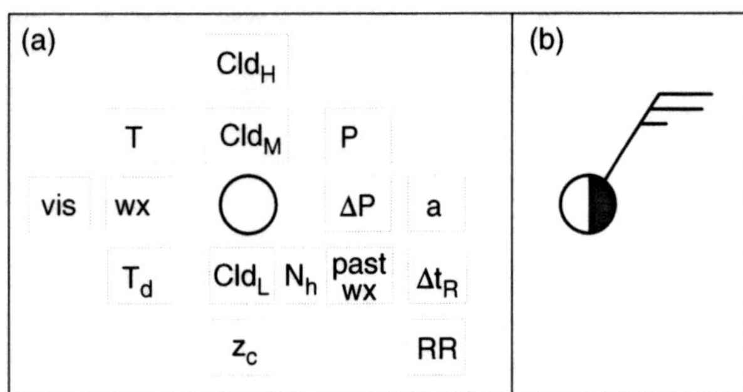

Figure 9.14 (again)
Station plot model.

Table 9-7. Codes for cloud-base height (z_c).		
Code	**meters agl**	**feet agl**
0	0 to 49	0 to 149
1	50 to 99	150 to 299
2	100 to 199	300 to 599
3	200 - 299	600 to 999
4	300 - 599	1,000 to 1,999
5	600 to 999	2,000 to 3,499
6	1,000 to 1,499	3,500 to 4,999
7	1,500 to 1,999	5,000 to 6,499
8	2,000 to 2,499	6,600 to 7,999
9	≥ 2,500	≥ 8,000

Table 9-8. Symbols for pressure change (barometric tendency) during the past 3 hours. (a)	
Glyph	**Meaning**
	Rising, then falling
	Rising, then steady or rising more slowly
	Rising steadily or unsteadily
	Falling or steady, later rising; or Rising slowly, later rising more quickly
	Steady
	Falling, then rising, but ending same or lower
	Falling, then steady or falling more slowly
	Falling steadily or unsteadily
	Steady or rising, then falling; or Falling, then falling more quickly

• <u>**a**</u> is a glyph representing the pressure change (barometric tendency) during the past 3 hours (Table 9-8). It mimics the trace on a barograph.

• <u>**Δt_R**</u> is a single-digit code that gives the number of hours ago that precipitation began or ended.

0 means no precipitation
1 means 0 to 1 hour ago
2 means 1 to 2 hours ago
3 means 2 to 3 hours ago
4 means 3 to 4 hours ago
5 means 4 to 5 hours ago
6 means 5 to 6 hours ago
7 means 7 to 12 hours ago
8 means more than 8 hours ago

• <u>**RR**</u> is the accumulated precipitation in past 6 hours. In the USA, the units are hundredths of inches. For example: 45 means 0.45 inches.

• <u>**wind**</u> is plotted as a direction shaft with barbs to denote speed (Fig. 9.14b). Table 9-9, reproduced from the Forces & Winds chapter, explains how to interpret it.

• <u>**total clouds**</u> is indicated by the portion of the station plot circle that is blackened (Fig. 9.14b). Table 9-10, reproduced from the Cloud chapter, explains its interpretation.

Table 9-9. Interpretation of wind barbs.		
Symbol	**Wind Speed**	**Description**
◎	calm	two concentric circles
	1 - 2 speed units	shaft with no barbs
	5 speed units	a half barb (half line)
	10 speed units	each full barb (full line)
	50 speed units	each pennant (triangle)

• The total speed is the sum of all barbs and pennants. For example, indicates a wind from the west at speed 75 units. Arrow tip is at the observation location.
• CAUTION: Different organizations use different speed units, such as knots, m s^{-1}, miles h^{-1}, km h^{-1}, etc. Look for a legend to explain the units. When in doubt, assume knots — the WMO standard. To good approximation, 1 m s^{-1} ≈ 2 knots.

In the next subsection you will learn how to analyze a weather map. You can do a **hand analysis** (manual analysis) by focusing on just one meteorological variable. For example, if you want to analyze temperatures, then you should focus on just the temperature data from the station plot for each weather station, and ignore the other plotted data. This is illustrated in Fig. 9.16, where I have highlighted the temperatures to make them easier to see.

[*CAUTION: Do not forget that the plotted temperature represents the temperature at the station location (namely, at the plotted station circle), not displaced from the station circle as defined by the station plot model.*]

Table 9-10. Sky cover. Oktas=eighths of sky covered.

Sky Cover (oktas)	Symbol	Name	Abbr.	Sky Cover (tenths)
0	○	Sky **Clear**	SKC	0
1	⦶	**Few*** Clouds	FEW*	1
2	◔			2 to 3
3	◔	**Scattered**	SCT	4
4	◑			5
5	◕	**Broken**	BKN	6
6	◕			7 to 8
7	●			9
8	●	**Overcast**	OVC	10
unknown	⊗	Sky **Obscured**		unknown

* "Few" is used for (0 oktas) < coverage ≤ (2 oktas).

Sample Application

Decode the (a) station plot and the (b) METAR below, and compare the information they contain.

(a) Station Plot Example

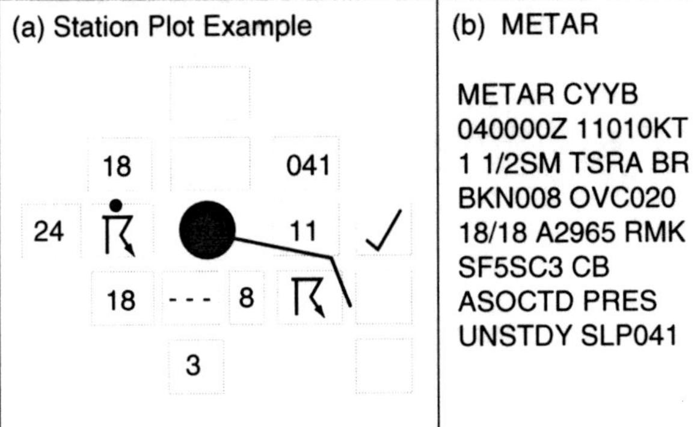

(b) METAR

METAR CYYB
040000Z 11010KT
1 1/2SM TSRA BR
BKN008 OVC020
18/18 A2965 RMK
SF5SC3 CB
ASOCTD PRES
UNSTDY SLP041

(c) Translation of METAR (Info from Stn Plot underlined)

Meteorological Aviation Report for North Bay (CYYB) ON, Canada on 4th day of the month at 0000 UTC.
Wind from 110° true at 10 knots.
Visibility 1.5 statuate miles (= 2.4 km).
Present weather is thunderstorm with moderate rain and mist.
Cloud coverage: broken clouds with base at 800 feet above ground, overcast with base at 2000 feet.
Temperature is 18°C, and dew point is 18°C.
Altimeter is 29.65 inches of mercury.

Remarks:
Stratus fractus clouds with 5/8 coverage, and Stratocumulus clouds with 3/8 coverage, both associated with cumulonimbus.
Sea-level pressure is unsteady 100.41 kPa.
[Additional info in station plot, but not in METAR: past weather was thunderstorm; pressure first decreased, then increased with net increase of 0.11 kPa in 3 hr.]

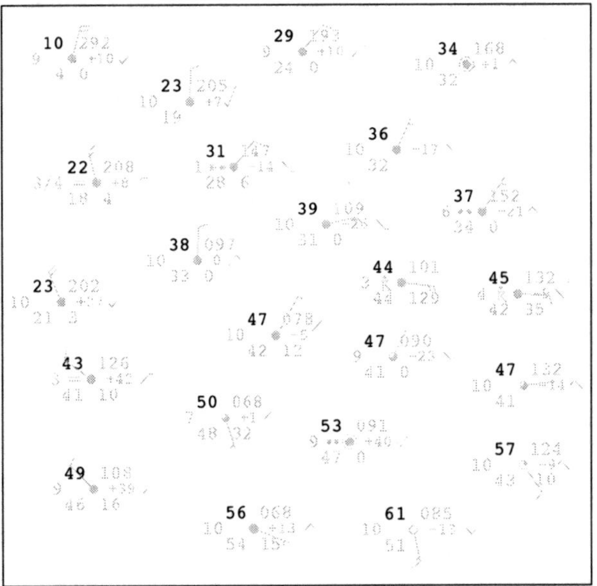

Figure 9.16
A surface weather map with temperatures highlighted. Units: T and T_d (°F), visibility (miles), speed (knots), pressure and 3-hour tendency (see text), 6-hour precipitation (hundredths of inches). Extracted from a "Daily Weather Map" courtesy of the US National Oceanic and Atmospheric Administration (NOAA), National Weather Service (NWS), National Centers for Environmental Prediction (NCEP), Hydrometeorological Prediction Center (HPC). The date/time of this map is omitted to discourage cheating during map-analysis exercises, and the station locations are shifted slightly to reduce overlap.

(a)

north						
4	3	2	2	3	4	5
16	15	7	5	6	6	7
20	20	20	20	12	10	9
23	25	27	30	28	20	13
25	27	30	33	31	28	26
27	28	29	29	29	29	28

south; west — east

Map Analysis, Plotting & Isoplething

You might find the amount of surface-observation data such as plotted in Fig. 9.16 to be overwhelming. To make the plotted data more comprehensible, you can simplify the weather map by drawing **isopleths** (lines of equal value, see Table 1-6).

For example, if you analyze temperatures, you draw **isotherms** on the weather map. Similarly, if you analyze pressures you draw the **isobars**, or for humidity you draw **isohumes**.

Also, you can identify features such as fronts and centers of low and high pressure. Heuristic models of these features allow you to anticipate their evolution (see chapters on Fronts & Airmasses and Extratropical Cyclones).

Most weather maps are analyzed by computer. Using temperature as an example, the synoptic temperature observations are interpolated by the computer from the irregular weather-station locations to a regular grid (Fig. 9.17a). Such a grid of numbers is called a **field** of data, and this particular example is a **temperature field**. A discrete temperature field such as stored in a computer array approximates the continuously-varying temperature field of the real atmosphere. The gridded field is called an **analysis**.

Regardless of whether you manually do a **hand analysis** on irregularly-spaced data (as in Fig. 9.16), or you let the computer do an **objective analysis** on a regularly-spaced grid of numbers (as in Fig. 9.17a), the next steps are the same for both methods.

Continuing with the temperature example of Fig. 9.17, draw isotherms connecting points of equal temperature (Fig. 9.17b). The following rules apply to any line connecting points of equal value (i.e., **isopleths**), not just to isotherms:

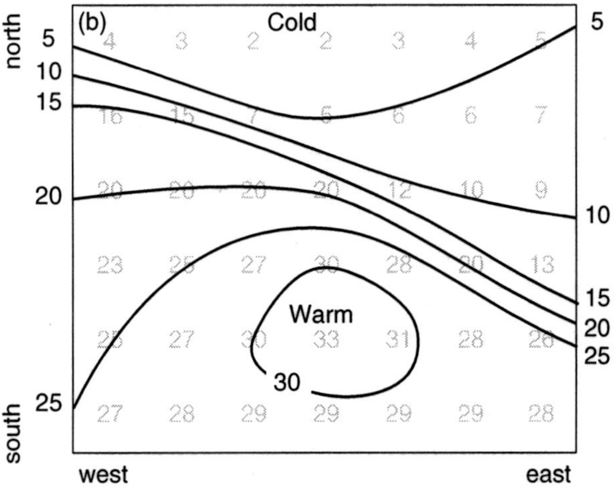

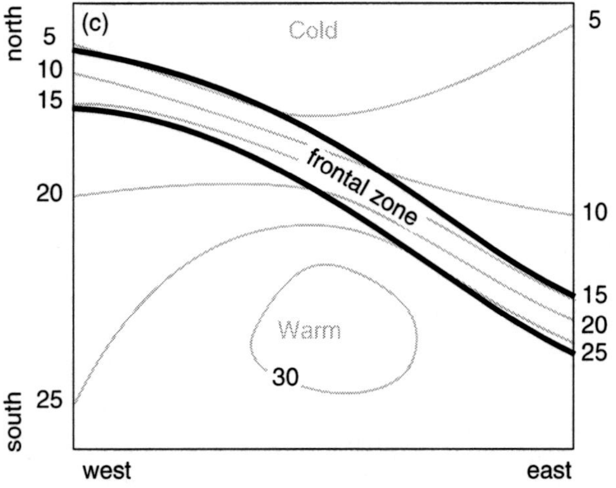

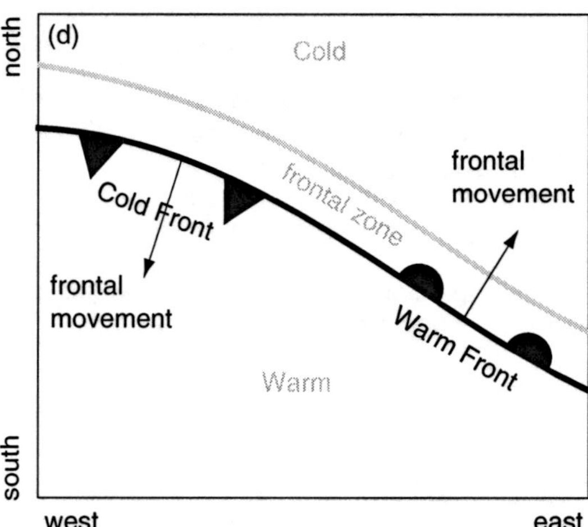

Figure 9.17
Weather map analysis: a) temperature field, with temperature in (°C) plotted on a Cartesian map; b) isotherm analysis; c) frontal zone analysis; d) frontal symbols added. [Note: The temperature field in this figure is for a different location and day than the data that was plotted in Fig. 9.16.]

• draw isopleths at regular intervals (such as every 2°C or 5°C for isotherms)
• interpolate where necessary between locations (e.g., the 5°C isotherm must be equidistant between gridded observations of 4°C & 6°C)
• isopleths never cross other isopleths of the same variable (e.g., isotherms can't cross other isotherms, but isotherms can cross isobars)
• isopleths never end in the middle of the map
• label each isopleth, either at the edges of the map (the only places where isopleths can end), or along closed isopleths
• isopleths have no kinks, except sometimes at fronts or jets

Finally, label any relative maxima and minima, such as the warm and cold centers in Fig. 9.17b.

You can identify **frontal zones** as regions of tight isotherm packing (Fig. 9.17c), namely, where the isotherms are closer together. Note that no isotherm needs to remain within a frontal zone.

Finally, always draw a heavy line representing the front on the warm side of the frontal zone (Fig. 9.17d), regardless of whether it is a cold, warm, or stationary front.

Frontal symbols are drawn on the side of the frontal line toward which the front moves. Draw semicircles to identify warm fronts (for cases where cold air retreats). Draw triangles to identify a cold front (where cold air advances). Draw alternating triangles and semicircles on opposite sides of the front to denote a stationary front, and on the same side for an occluded front. Fig. 9.18 summarizes frontal symbols, many of which will be discussed in more detail in the Airmasses and Fronts chapter.

In Fig. 9.17d, there would not have been enough information to determine if the cold air was advancing, retreating, or stationary, if I hadn't added arrows showing frontal movement. When you analyze fronts on real weather maps, determine their movement from successive weather maps at different times, by the wind direction across the front, or by their position relative to low-pressure centers.

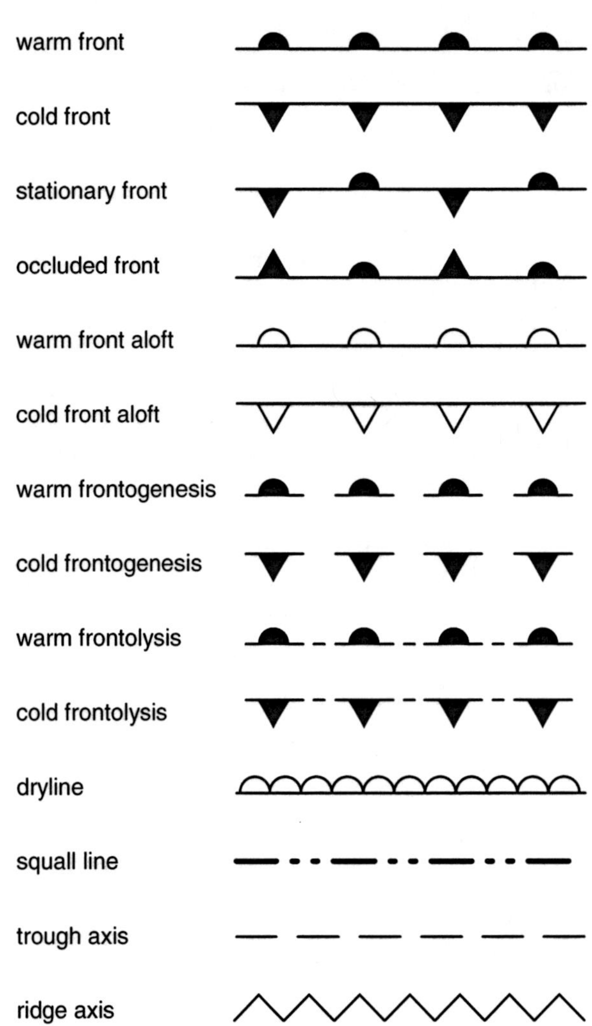

Figure 9.18
Glyphs for fronts, other airmass boundaries, and axes. The suffix "genesis" implies a forming or intensifying front, while "lysis" implies a weakening or dying front. A stationary front is a frontal boundary that does not move very much. Occluded fronts and drylines will be explained in the Fronts chapter.

REVIEW

Hundreds of thousands of weather observations are simultaneously made around the world at standard observation times. Some of these weather observations are communicated as alphanumeric codes such as the METAR that can be read and decoded by humans. A station-plot model is often used to plot weather data on weather maps. Map analysis is routinely performed by computer, but you can also draw isopleths and identify fronts, highs, lows, and airmasses by hand.

HOMEWORK EXERCISES

Broaden Knowledge & Comprehension

B1. Access a surface weather map that shows station plot information for Denver, Colorado USA. Decode the plotted pressure value, and tell how you can identify whether that pressure is the actual station pressure, or is the pressure reduced to sea level.

B2. Do a web search to identify 2 or more suggestions on how to reduce station pressure to sea level. Pick two methods that are different from the method described in this chapter.

B3. Search the web for maps that show where METAR weather data are available. Print such a map that covers your location, and identify which 3 stations are closest to you.

B13. Search the web for sites that give the ICAO station ID for different locations. This is the ID used to indicate the name of the weather station in a METAR.

B4. Access the current METAR for your town, or for a nearby town assigned by your instructor. Try to decode it manually, and write out its message in words. Compare your result with a computer decoded METAR if available.

B5. Search the web for maps that show where weather observations are made today (or recently), such as were shown in Figs. 9.2-13.16. Hint: If you can't find a site associated with your own country's weather service, try searching on "ECMWF data coverage" or "Met Office data coverage" or "FNMOC data coverage".

B6. For each of the different sensor types discussed in the section on Weather Observation Locations, use the web to get photos of each type of instrument: rawinsonde, dropsonde, AWOS, etc.

B7. Search the web for a history of ocean weather ships, and summarize your findings.

B8. Access from the web a current plotted surface weather map that has the weather symbols plotted around each weather station. Find the station closest to your location (or use a station assigned by the instructor), and decode the weather data into words.

B9 Access simple weather maps from the web that print values of pressure or temperature at the weather stations, but which do not have the isopleths drawn. Print these, and then draw your own isobars or isotherms. If you can do both isobars and isotherms for a given time over the same region, then identify the frontal zone, and determine if the front is warm, cold, or occluded. Plot these features on your analyzed maps. Identify highs and lows and airmasses.

B10. Use the web to access surface weather maps showing plotted station symbols, along with the frontal analysis. Compare surface temperature, wind, and pressure along a line of weather stations that crosses through the frontal zone. How do the observations compare with your ideas about frontal characteristics?

Apply

A1. Find the pressure "reduced to sea level" using the following station observations of pressure, height, and virtual temperature. Assume no temperature change over the past 12 hours.

	P (kPa)	z (m)	T_v(°C)
a.	102	–30	40
b.	100	20	35
c.	98	150	30
d.	96	380	30
e.	94	610	20
f.	92	830	18
g.	90	980	15
h.	88	1200	12
i.	86	1350	5
j	84	1620	5
k.	82	1860	2

A2. Decode the following METAR. Hint: It is not necessary to decode the station location; just write its ICAO abbreviation followed by the decoded METAR. Do not decode the remarks (RMK).

a. KDFW 022319Z 20003KT 10SM TS FEW037
SCT050CB BKN065 OVC130 27/20 A2998
RMK AO2 FRQ LTGICCG TS OHD MOV E-NE

b. KGRK 022317Z 17013KT 4SM TSRA BKN025
BKN040CB BKN250 22/21 A3000
RMK OCNL LTGCCCG SE TS OHD-3SE MOV E

c. KSAT 022253Z 17010KT 10SM SCT034 BKN130
BKN250 28/23 A2998 RMK AO2 RAE42
SLP133 FEW CB DSNT NW-N P0001 T02830233

d. KLRD 022222Z 11015KT M1/4SM TSRA FG
OVC001 24/23 A2998 RMK AO2 P0125 PRESRR

e. KELD 022253Z AUTO 14003KT 4SM RA BR OVC024 23/21 A3006 RMK AO2 TSB2153E12 SLP177 T02280211 $

f. KFSM 022311Z 00000KT 10SM TSRA SCT030 22/21 A3004 RMK AO2 P0000

g. KLIT 022253Z 08009KT 7SM TS FEW026 BKN034CB OVC060 27/22 A3003 RMK AO2 RAB28E45 SLP169 OCNL LTGICCC OHD TS OHD MOV N P0000 T02720217

h. KMCB 022315Z AUTO 34010KT 1/4SM +TSRA FG BKN005 OVC035 24/23 A3009 RMK AO2 LTG DSNT ALQDS P0091 $

i. KEET 022309Z AUTO 05003KT 3SM -RA BR SCT024 BKN095 23/23 A3005 RMK AO2 LTG DSNT N AND E AND SW

j. KCKC 022314Z AUTO 20003KT 2SM DZ OVC003 13/11 A3009 RMK AO2

k. CYQT 022300Z 20006KT 20SM BKN026 OVC061 16/11 A3006 RMK SC5AC2 SLP185

l. CYYU 022300Z 23013KT 15SM FEW035 BKN100 BKN200 BKN220 23/08 A3013 RMK CU2AS2CC1CI1 WND ESTD SLP207

m. CYXZ 022300Z 00000KT 15SM -RA OVC035 14/11 A3021 RMK SC8 SLP239

n. KETB 022325Z AUTO 10007KT 009V149 10SM -RA CLR 19/11 A3021 RMK AO2

o. CYWA 022327Z AUTO 33004KT 9SM RA FEW027 FEW047 BKN069 19/12 A3016

A3. Translate into words a weather glyph assigned from Table 9-11.

A4 For a weather glyph from Table 9-11, write the corresponding METAR abbreviation, if there is one.

A5. Using the station plot model, plot the weather observation data around a station circle drawn on your page for one METAR from exercise A2, as assigned by your instructor.

A6. Using the USA weather map in Fig. 9.19, decode the weather data for the weather station labeled (a) - (w), as assigned by your instructor.

A7. Photocopy the USA weather map in Fig. 9.19 and analyze it by drawing isopleths for:

Table 9-11. Weather-Glyph Exercises.

a	b	c	d	e	f
g	h	i	j	k	l
m	n	o	p	q	r
s	t	u	v	w	x
y	z	z0	z1	z2	z3
aa	ab	ac	ad	ae	af
ag	ah	ai	aj	ak	al
am	an	ao	ap	aq	ar
as	at	au	av	aw	ax
ay	az	ba	bb	bc	bd

a. temperature (isotherms) every 5°F
b. pressure (isobars) every 0.4 kPa
c. dew point (isodrosotherms) every 5°F
d. wind speed (isotachs) every 5 knots
e. pressure change (isallobar) every 0.1 kPa

A8. Using the Canadian weather map of Fig. 9.20, decode the weather data for the station labeled (a) - (z), as assigned by your instructor.

A9. Photocopy the Canadian weather map of Fig. 9.20, and analyze it by drawing isopleths for the following quantities.
a. temperature (isotherms) every 2°C
b. dew point (isodrosotherms) every 2°C
c. pressure change (isallobar) every 0.1 kPa
d. total cloud coverage (isonephs) every okta
e. visibility every 5 km

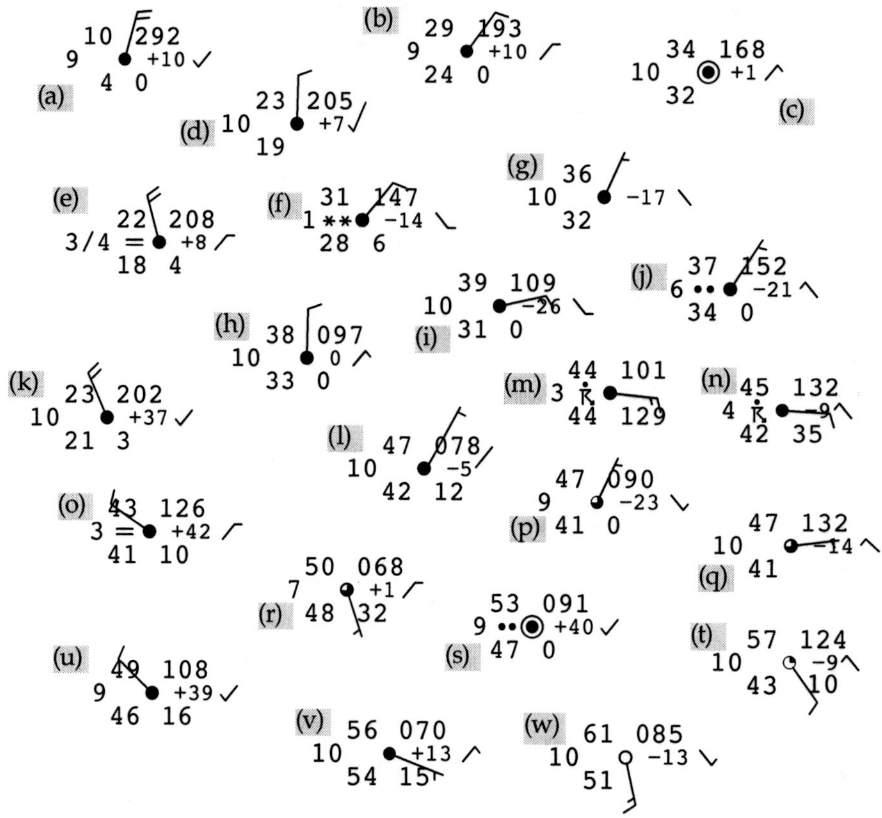

Figure 9.19
USA surface weather map. Units: T and T_d (°F), visibility (miles), speed (knots), pressure and 3-hour tendency (see text), 6-hour precipitation (hundredths of inches). Extracted from a "Daily Weather Map" courtesy of the US National Oceanic and Atmospheric Administration (NOAA), National Weather Service (NWS), National Centers for Environmental Prediction (NCEP), Hydrometeorological Prediction Center (HPC). The date/time of this map is omitted to discourage cheating during map-analysis exercises, and the station locations are shifted slightly to reduce overlap.

8	9	11	12	13	14	15
7	9	11	13	14	15	16
8	9	14	18	20	20	18
9	9	16	19	22	24	24
10	11	18	20	22	25	25
12	14	19	20	22	24	25
14	18	19	21	22	23	24
18	19	20	21	22	23	23

Figure 9.21-i
Temperature (°C). (These figures are show out of order because there was space on this page for it.)

9.5	9.4	9.4	9.5	9.6	9.7	9.8
9.4	9.3	9.2	9.3	9.4	9.6	9.7
9.5	9.2	8.9	9.2	9.3	9.4	9.6
9.5	9.4	9.1	9.4	9.5	9.5	9.6
9.6	9.4	9.2	9.5	9.6	9.7	9.7
9.6	9.4	9.4	9.6	9.7	9.8	9.9
9.6	9.4	9.6	9.7	9.8	9.9	0.0
9.6	9.6	9.7	9.8	9.9	0.0	0.1

Figure 9.21-ii
Pressure (kPa). [The first 1 or 2 digits of the pressure are omitted. Thus, 9.5 on the chart means 99.5 kPa, while 0.1 means 100.1 kPa.]

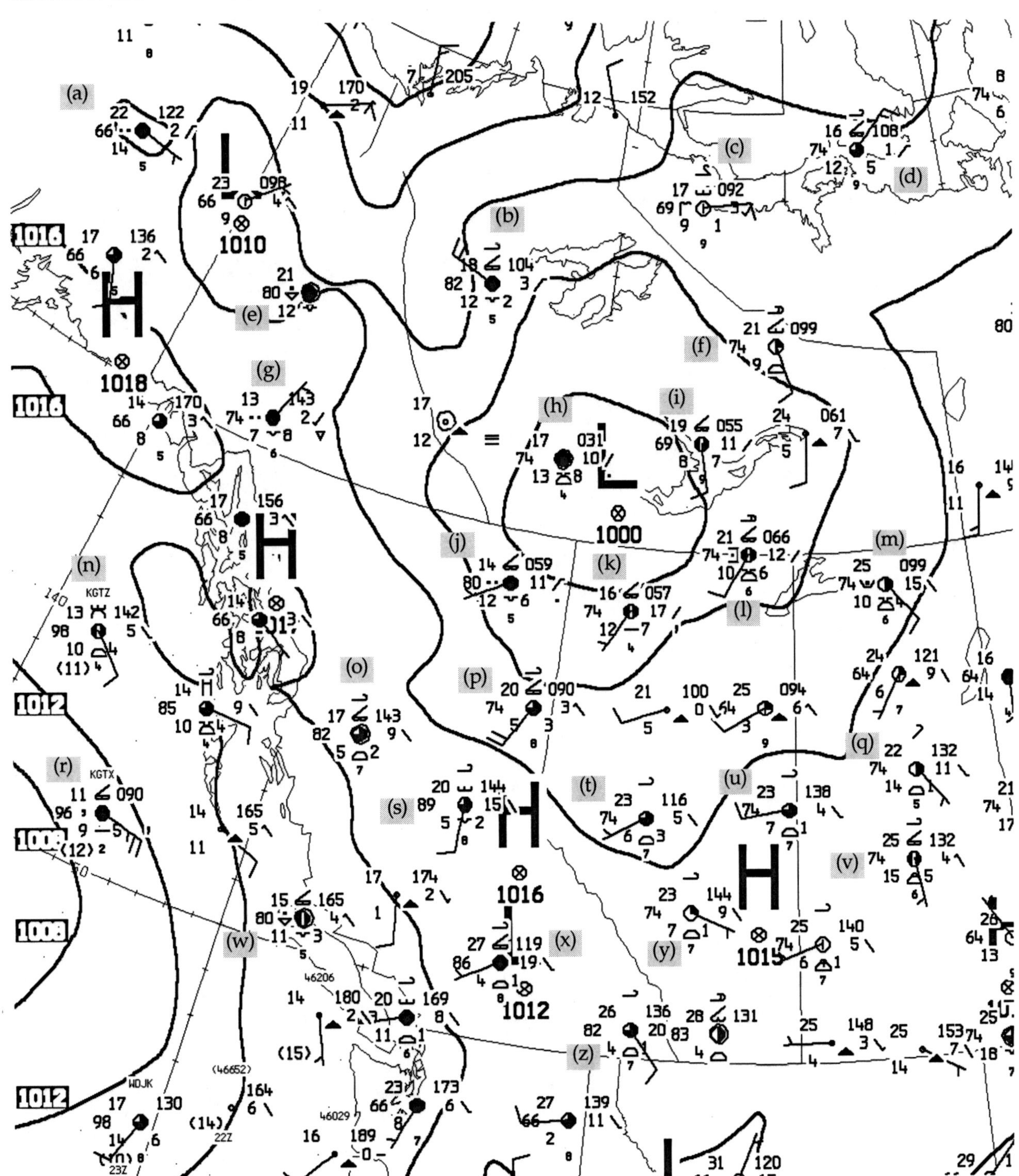

Figure 9.20
Canadian weather map courtesy of Environment Canada. http://www.weatheroffice.gc.ca/analysis/index_e.html

Figure 9.21
(see previous page)

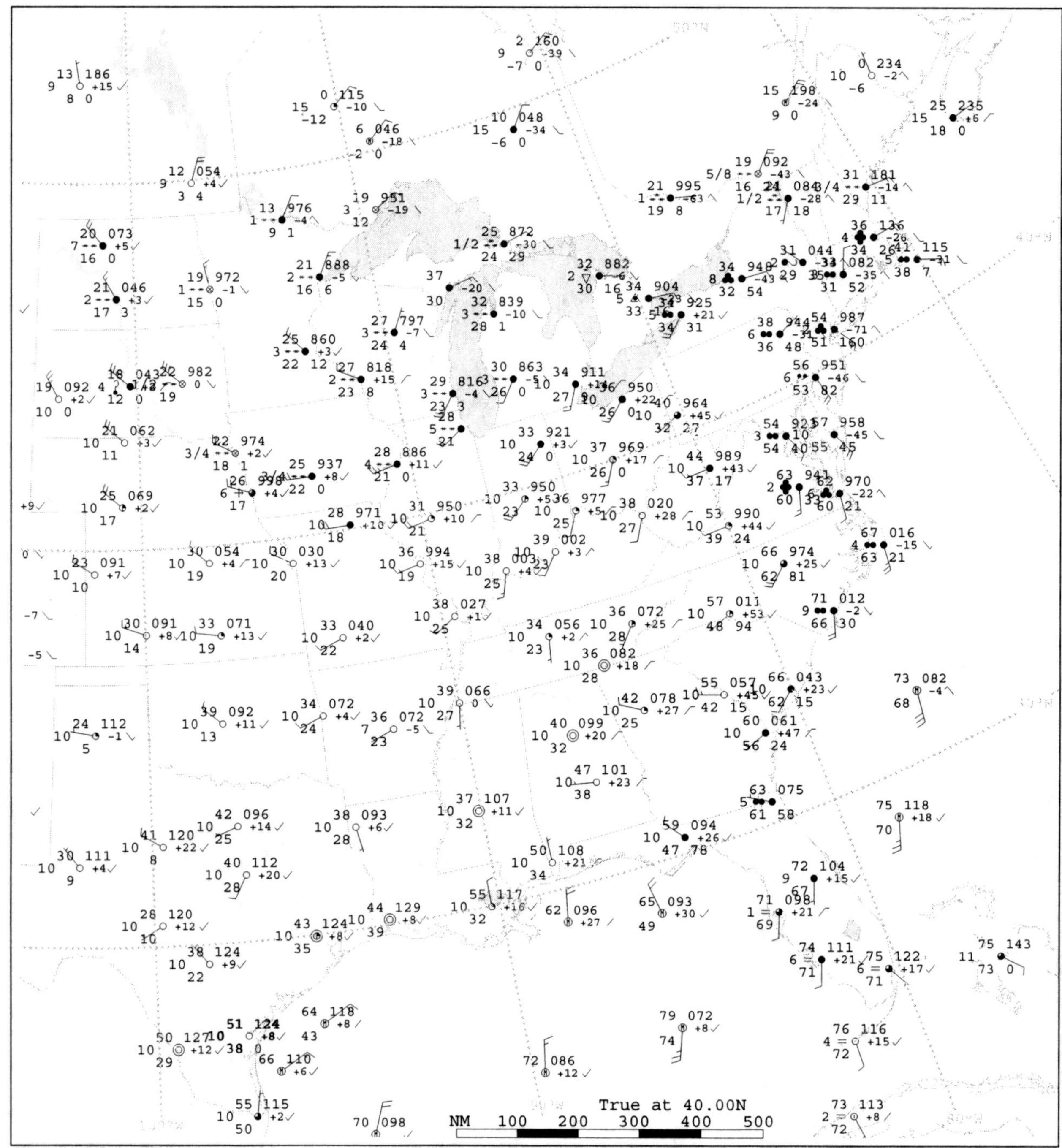

Figure 9.22

A surface weather map of central and eastern N. America. Units: T and T_d (°F), visibility (miles), speed (knots), pressure and 3-hour tendency (see text), 6-hour precipitation (hundredths of inches). Extracted from a "Daily Weather Map" courtesy of the US National Oceanic and Atmospheric Administration (NOAA), National Weather Service (NWS), National Centers for Environmental Prediction (NCEP), Hydrometeorological Prediction Center (HPC). The date/time of this map is omitted to discourage cheating during map-analysis exercises.

A10. Both of the weather maps of Fig. 9.21 correspond to the same weather. Do the following work on a photocopy of these charts:

a. Draw isotherms and identify warm and cold centers. Label isotherms every 2°C.
b. Draw isobars every 0.2 kPa and identify high and low pressure centers.
c. Add likely wind vectors to the pressure chart.
d. Identify the frontal zone(s) and draw the frontal boundary on the temperature chart.
e. Use both charts to determine the type of front (cold, warm), and draw the appropriate frontal symbols on the front.
f. Indicate likely regions for clouds and suggest cloud types in those regions.
g. Indicate likely regions for precipitation.
h. For which hemisphere are these maps?

Evaluate & Analyze

E1. Are there any locations in the world where you could get a reasonable surface weather map without first reducing the pressure to mean sea level? Explain.

E2. What aspects of mean sea level reduction are physically unsound or weak? Explain.

E3. One of the isoplething instructions was that an isopleth cannot end in the middle of the map. Explain why such an ending isopleth would imply a physically impossible weather situation.

E4. Make a photocopy of the surface weather map (Fig. 9.22) on the next page, and analyze your copy by drawing isobars (solid lines), isotherms (dashed lines), high- and low-pressure centers, airmasses, and fronts.

Synthesize

S1. Suppose that you wanted to plot a map of thickness of the 100 to 50 kPa layer (see the General Circulation chapter for a review of thickness maps). However, in some parts of the world, the terrain elevation is so high that the surface pressure is lower than 100 kPa. Namely, part of the 100 to 50 kPa layer would be below ground.

Extend the methods on sea-level pressure reduction to create an equation or method for estimating thickness of the 100 to 50 kPa layer over high ground, based on available surface and atmospheric sounding data.

S2 What if there were no satellite data? How would our ability to analyze the weather change?

S3. What if only satellite data existed? How would our ability to analyze the weather change?

S4. a. Suppose that all the weather observations over land were accurate, and all the ones over oceans had large errors. At mid-latitudes where weather moves from west to east, discuss how forecast skill would vary from coast to coast across a continent such as N. America.

b. How would forecast skill be different if observations over oceans were accurate, and over land were inaccurate?

S5. Pilots flying visually (VFR) need a certain minimum visibility and cloud ceiling height. The **ceiling** is the altitude of the lowest cloud layer that has a coverage of broken or overcast. If there is an obscuration such as smoke or haze, the ceiling is the vertical visibility from the ground looking up.

Use the web to access pilot regulations for your country to learn the ceiling and visibility needed to land VFR at an airport with a control tower. Then translate those values into the codes for a station plot model, and write those values in the appropriate box relative to a station circle.

S6. In Fig. 9.4, notice that west of N. America is a large data-sparse region over the N.E. Pacific Ocean. This region, shown in Fig. 9.23 below, is called the **Pacific data void**. Although there is buoy and ship data near the surface, and aircraft data near the tropopause, there is a lack of mid-tropospheric data in that region.

Although Figs. 9.2 - 9.12 show lots of satellite data over that region, satellites do not have the vertical coverage and do not measure all the meteorological variables needed to use as a starting point for accurate weather forecasts.

Suppose you had an unlimited budget. What instruments and instrument platforms (e.g., weather ships, etc.) would you deploy to get dense spatial coverage of temperature, humidity, and winds in the Pacific data void? If you had a limited budget, how would your proposal be different?

[Historical note: Anchored weather ships such as one called **Station Papa** at 50°N 145°W were formerly stationed in the N.E. Pacific, but all these ships were removed due to budget cutbacks. When they were removed, weather-prediction skill over large parts of N. America measurably decreased, because mid-latitude weather moves from west to east. Namely, air from over the data void regions moves over North America.]

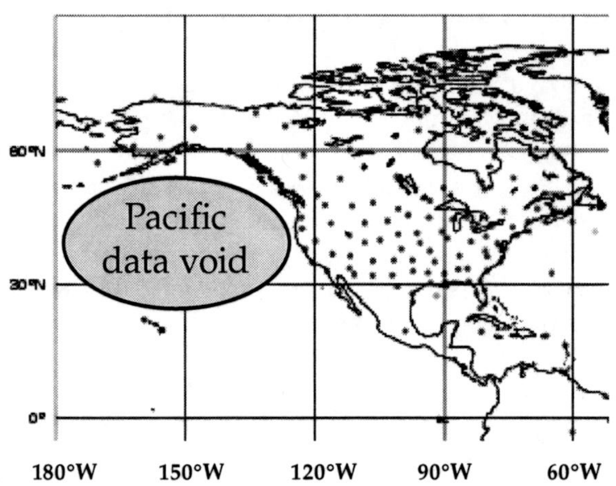

Figure 9.23
The "Pacific data void" & upper-air sounding locations (dots).

A SCIENTIFIC PERSPECTIVE • Creativity in Engineering

"Just as the poet starts with a blank sheet of paper and the artist with a blank canvas, so the engineer today begins with a blank computer screen. Until the outlines of a design are set down, however tentatively, there can be no appeal to science or to critical analysis to judge or test the design. Scientific, rhetorical or aesthetic principles may be called on to inspire, refine and finish a design, but creative things do not come of applying the principles alone. Without the sketch of a thing or a diagram of a process, scientific facts and laws are of little use to engineers. Science may be the theatre, but engineering is the action on the stage."

"Designing a bridge might also be likened to writing a sonnet. Each has a beginning and an end, which must be connected with a sound structure. Common bridges and so-so sonnets can be made by copying or mimicking existing ones, with some small modifications of details here and there, but such are not the creations that earn the forms their reputation or cause our spirits to soar. Masterpieces come from a new treatment of an old form, from a fresh shaping of a familiar genre. The form of the modern suspension bridge — consisting of a deck suspended from cables slung over towers and restrained by anchorages — existed for half a century before John Roebling proposed his Brooklyn Bridge, but the fresh proportions of his Gothic-arched masonry towers, his steel cables and diagonal stays, and his pedestrian walkway centered above dual roadways produced a structure that remains a singular achievement of bridge engineering. Shakespeare's sonnets, while all containing 14 lines of iambic pentameter, are as different from one another and from their contemporaries as one suspension bridge is from another."

– Henry Petroski, 2005: Technology and the humanities. *American Scientist*, **93**, p 305.

10 ATMOSPHERIC FORCES & WINDS

Contents

Winds power our wind turbines, push our sailboats, cool our houses, and dry our laundry. But winds can also be destructive — in hurricanes, thunderstorms, or mountain downslope windstorms. We design our bridges and skyscrapers to withstand wind gusts. Airplane flights are planned to compensate for headwinds and crosswinds.

Winds are driven by forces acting on air. But these forces can be altered by heat and moisture carried by the air, resulting in a complex interplay we call weather. Newton's laws of motion describe how forces cause winds — a topic called **dynamics**.

Many forces such as pressure-gradient, advection, and frictional drag can act in all directions. Inertia creates an apparent centrifugal force, caused when centripetal force (an imbalance of other forces) makes wind change direction. Local gravity acts mostly in the vertical. But a local horizontal component of gravity due to Earth's non-spherical shape, combined with the contribution to centrifugal force due to Earth's rotation, results in a net force called Coriolis force.

These different forces are present in different amounts at different places and times, causing large variability in the winds. For example, Fig. 10.1 shows changing wind speed and direction around a low-pressure center. In this chapter we explore forces, winds, and the dynamics that link them.

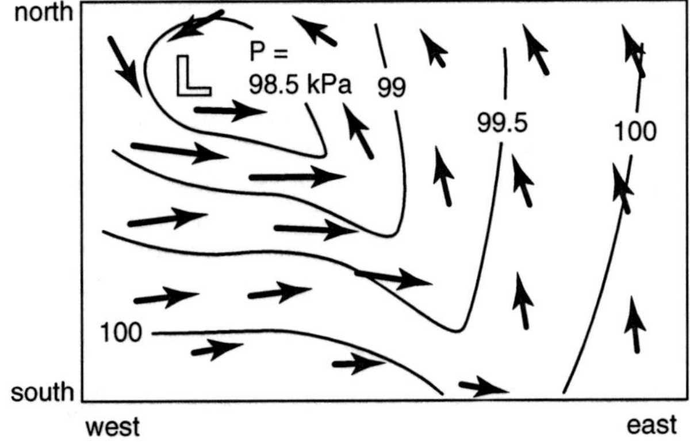

Figure 10.1
Winds (arrows) around a low-pressure center (L) in the N. Hemisphere. Lines are isobars of sea-level pressure (P).

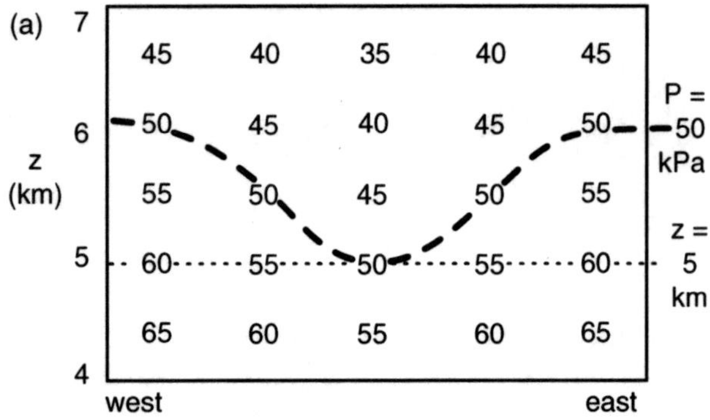

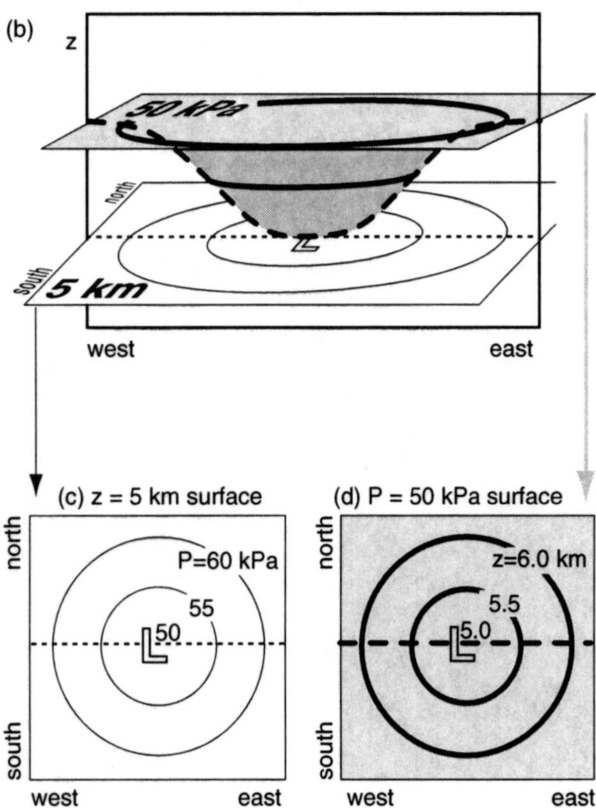

Figure 10.2
Sketch of the similarity of (c) pressures drawn on a constant height surface to (d) heights drawn on a constant pressure surface. At any one height such as z = 5 km (shown by the thin dotted line in the vertical cross section of Fig. a), pressures at one location on the map might differ from pressure at other locations. In this example, a pressure of 50 kPa is located midway between the east and west limits of the domain at 5 km altitude. Pressure always decreases with increasing height z, as sketched in the vertical cross section of (a). Thus, at the other locations on the cross section, the 50 kPa pressure will be found at higher altitudes, as sketched by the thick dashed line. This thick dashed line and the corresponding thin dotted straight line are copied into the 3-D view of the same scenario is sketched in Fig. b. "L" indicates the cyclone center, having low pressure and low heights.

WINDS AND WEATHER MAPS

Heights on Constant-Pressure Surfaces

Pressure-gradient force is the most important force because it is the only one that can drive winds in the horizontal. Other horizontal forces can alter an existing wind, but cannot create a wind from calm air. All the forces, including pressure-gradient force, are explained in the next sections. However, to understand the pressure gradient, we must first understand pressure and its atmospheric variation.

We can create weather maps showing values of the pressures measured at different horizontal locations all at the same altitude, such as at mean-sea-level (MSL). Such a map is called a **constant-height map**. However, one of the peculiarities of meteorology is that we can also create maps on other surfaces, such as on a surface connecting points of equal pressure. This is called an **isobaric map**. Both types of maps are used extensively in meteorology, so you should learn how they are related.

In Cartesian coordinates (x, y, z), z is height above some reference level, such as the ground or sea level. Sometimes we use geopotential height H in place of z, giving a coordinate set of (x, y, H) (see Chapter 1).

Can we use pressure as an alternative vertical coordinate instead of z? The answer is yes, because pressure changes monotonically with altitude. The word **monotonic** means that the value of the dependent variable changes in only one direction (never decreases, or never increases) as the value of the independent variable increases. Because P never increases with increasing z, it is indeed monotonic, allowing us to define **pressure coordinates** (x, y, P).

An isobaric surface is a conceptual curved surface that connects points of equal pressure, such as the shaded surface in Fig. 10.2b. The surface is higher above sea level in high-pressure regions, and lower in low-pressure regions. Hence the height contour lines for an isobaric surface are good surrogates for pressures lines (isobars) on a constant height map. Contours on an isobaric map are analogous to elevation contours on a topographic map; namely, the map itself is flat, but the contours indicate the height of the actual surface above sea level.

High pressures on a constant height map correspond to high heights of an isobaric map. Similarly, regions on a constant-height map that have tight **packing** (close spacing) of isobars correspond to regions on isobaric maps that have tight packing of height contours, both of which are regions of strong pressure gradients that can drive strong winds. This one-to-one correspondence of both types of maps (Figs. 10.2c & d) makes it easier for you to use them interchangeably.

Isobaric surfaces can intersect the ground, but two different isobaric surfaces can never intersect because it is impossible to have two different pressures at the same point. Due to the smooth monotonic decrease of pressure with height, isobaric surfaces cannot have folds or creases.

We will use isobaric charts for most of the upper-air weather maps in this book when describing upper-air features (mostly for historical reasons; see INFO box). Fig. 10.3 is a sample weather map showing height contours of the 50 kPa isobaric surface.

Plotting Winds

Symbols on weather maps are like musical notes in a score — they are a shorthand notation that concisely expresses information. For winds, the symbol is an arrow with feathers (or barbs and pennants). The tip of the arrow is plotted over the observation (weather-station) location, and the arrow shaft is aligned so that the arrow points toward where the wind is going. The number and size of the feathers indicates the wind speed (Table 10-1, copied from Table 9-9). Fig. 10.3 illustrates wind barbs.

Table 10-1. Interpretation of wind barbs.

Symbol	Wind Speed	Description
◎	calm	two concentric circles
	1 - 2 speed units	shaft with no barbs
	5 speed units	a half barb (half line)
	10 speed units	each full barb (full line)
	50 speed units	each pennant (triangle)

• The total speed is the sum of all barbs and pennants. For example, indicates a wind from the west at speed 75 units. Arrow tip is at the observation location.
• CAUTION: Different organizations use different speed units, such as knots, m s^{-1}, miles h^{-1}, km h^{-1}, etc. Look for a legend to explain the units. When in doubt, assume knots — the WMO standard. For unit conversion, a good approximation is 1 m s^{-1} ≈ 2 knots.

Sample Application
Draw wind barb symbol for winds from the:
(a) northwest at 115 knots; (b) northeast at 30 knots.

Find the Answer
(a) 115 knots = 2 pennants + 1 full barb + 1 half barb.
(b) 30 knots = 3 full barbs

(a) (b)

Check: Consistent with Table 10-1.
Exposition: Feathers (barbs & pennants) should be on the side of the shaft that would be towards low pressure if the wind were geostrophic.

INFO • Why use isobaric maps?

There are five reasons for using isobaric charts.

1) During the last century, the **radiosonde** (a weather sensor hanging from a free helium balloon that rises into the upper troposphere and lower stratosphere) could not measure its geometric height, so instead it reported temperature and humidity as a function of pressure. For this reason, **upper-air charts** (i.e., maps showing weather above the ground) traditionally have been drawn on isobaric maps.

2) Aircraft altimeters are really pressure gauges. Aircraft assigned by air-traffic control to a specific "altitude" above 18,000 feet MSL will actually fly along an isobaric surface. Many weather observations and forecasts are motivated by aviation needs.

3) Air pressure is created by the weight of air molecules. Thus, every point on an isobaric map has the same mass of air molecules above it.

4) An advantage of using equations of motion in pressure coordinates is that you do not need to consider density, which is not routinely observed.

5) Numerical weather prediction models sometimes use pressure as the vertical coordinate.

Items (1) and (5) are less important these days, because modern radiosondes use **GPS (Global Positioning System)** to determine their (x, y, z) position. So they report all meteorological variables (including pressure) as a function of z. Also, some of the modern weather forecast models do not use pressure as the vertical coordinate. Perhaps future weather analyses and numerical predictions will be shown on constant-height maps.

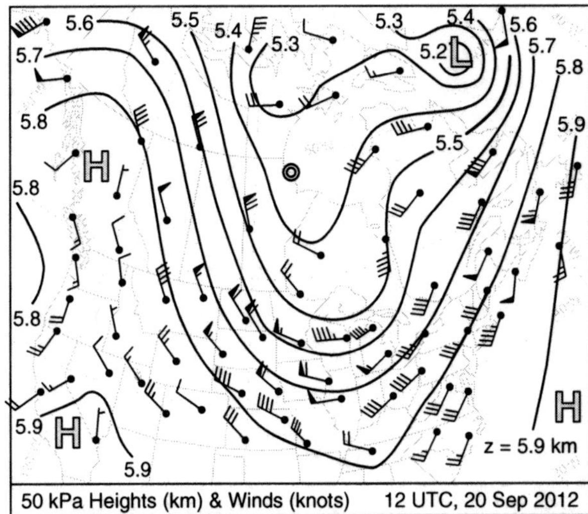

Figure 10.3
Winds (1 knot ≈ 0.5 m s^{-1}) and heights (km) on the 50 kPa isobaric surface. The relative maxima and minima are labeled as H (high heights) and L (low heights). Table 10-1 explains winds.

INFO • Newton's Laws of Motion

Isaac Newton's published his laws in Latin, the language of **natural philosophy** (science) at the time (1687). Here is the translation from Newton's *Philosophiæ Naturalis Principia Mathematica* ("Mathematical Principles of Natural Philosophy"):

"**Law I**. Every body perseveres in its state of being at rest or of moving uniformly straight forward, except inasmuch as it is compelled by impressed forces to change its state.

"**Law II**. Change in motion is proportional to the motive force impressed and takes place following the straight line along which that force is impressed.

"**Law III**. To any action, there is always a contrary, equal reaction; in other words, the actions of two bodies each upon the other are always equal and opposite in direction.

"**Corollary 1**. A body under the joint action of forces traverses the diagonal of a parallelogram in the same time as it describes the sides under their separate actions."

Sample Application

If a 1200 kg car accelerates from 0 to 100 km h^{-1} in 7 s, heading north, then: (a) What is its average acceleration? (b) What vector force caused this acceleration?

Find the Answer

Given: $\vec{V}_{initial} = 0$, $\vec{V}_{final} = 100 \text{ km h}^{-1} = 27.8 \text{ m s}^{-1}$
$t_{initial} = 0$, $t_{final} = 7$ s. Direction is north.
$m = 1200$ kg.

Find: (a) $\vec{a} = ?$ m·s^{-2} , (b) $\vec{F} = ?$ N

(a) Apply eq. (10.2): $\vec{a} = \dfrac{\Delta \vec{V}}{\Delta t}$

$= (27.8 - 0 \text{ m s}^{-1}) / (7 - 0 \text{ s}) =$ **<u>3.97</u>** m·s^{-2} <u>to the north</u>

(b) Apply eq. (10.1): $\vec{F} = (1200 \text{ kg}) \cdot (3.97 \text{ m·s}^{-2})$
= <u>**4766 N to the north**</u>
where 1 N = 1 kg·m·s^{-2} (see Appendix A).

Check: Physics and units are reasonable.
Exposition: My small car can accelerate from 0 to 100 km in 20 seconds, if I am lucky. Greater acceleration consumes more fuel, so to save fuel and money, you should accelerate more slowly when you drive.

NEWTON'S 2ND LAW

Lagrangian

For a Lagrangian framework (where the coordinate system follows the moving object), **Newton's Second Law of Motion** is

$$\vec{F} = m \cdot \vec{a} \qquad \bullet(10.1)$$

where $\vec{F}$ is a force vector, m is mass of the object, and $\vec{a}$ is the acceleration vector of the object. Namely, the object accelerates in the direction of the applied force.

Acceleration is the velocity $\vec{V}$ change during a short time interval Δt:

$$\vec{a} = \frac{\Delta \vec{V}}{\Delta t} \qquad (10.2)$$

Plugging eq. (10.2) into (10.1) gives:

$$\vec{F} = m \cdot \frac{\Delta \vec{V}}{\Delta t} \qquad (10.3a)$$

Recall that **momentum** is defined as $m \cdot \vec{V}$. Thus, if the object's mass is constant, you can rewrite Newton's 2nd Law as **Lagrangian momentum budget**:

$$\vec{F} = \frac{\Delta(m \cdot \vec{V})}{\Delta t} \qquad (10.3b)$$

Namely, this equation allows you to forecast the rate of change of the object's momentum.

If the object is a collection of air molecules moving together as an **air parcel**, then eq. (10.3a) allows you to forecast the movement of the air (i.e., the **wind**). Often many forces act simultaneously on an air parcel, so we should rewrite eq. (10.3a) in terms of the net force:

$$\frac{\Delta \vec{V}}{\Delta t} = \frac{\vec{F}_{net}}{m} \qquad (10.4)$$

where $\vec{F}_{net}$ is the vector sum of all applied forces, as given by Newton's Corollary 1 (see the INFO box).

For situations where $\vec{F}_{net}/m = 0$, eq. (10.4) tells us that the flow will maintain constant velocity due to **inertia**. Namely, $\Delta \vec{V} / \Delta t = 0$ implies that $\vec{V}$ = constant (not that $\vec{V} = 0$).

In Chapter 1 we defined the (U, V, W) wind components in the (x, y, z) coordinate directions (positive toward the East, North, and up). Thus, we can split eq. (10.4) into separate **scalar** (i.e., non-vector) equations for each wind component:

$$\frac{\Delta U}{\Delta t} = \frac{F_{x\,net}}{m} \qquad \bullet(10.5a)$$

$$\frac{\Delta V}{\Delta t} = \frac{F_{y\,net}}{m} \qquad \bullet(10.5b)$$

$$\frac{\Delta W}{\Delta t} = \frac{F_{z\,net}}{m} \qquad \bullet(10.5c)$$

where $F_{x\,net}$ is the sum of the x-component of all the applied forces, and similar for $F_{y\,net}$ and $F_{z\,net}$.

From the definition of Δ = final – initial, you can expand $\Delta U/\Delta t$ to be $[U(t+\Delta t) - U(t)]/\Delta t$. With similar expansions for $\Delta V/\Delta t$ and $\Delta W/\Delta t$, eq. (10.5) becomes

$$U(t+\Delta t) = U(t) + \frac{F_{x\,net}}{m} \cdot \Delta t \qquad \bullet(10.6a)$$

$$V(t+\Delta t) = V(t) + \frac{F_{y\,net}}{m} \cdot \Delta t \qquad \bullet(10.6b)$$

$$W(t+\Delta t) = W(t) + \frac{F_{z\,net}}{m} \cdot \Delta t \qquad \bullet(10.6c)$$

These are forecast equations for the wind, and are known as the **equations of motion**. The Numerical Weather Prediction (NWP) chapter shows how the equations of motion are combined with budget equations for heat, moisture, and mass to forecast the weather.

Eulerian

While Newton's 2nd Law defines the fundamental dynamics, we cannot use it very easily because it requires a coordinate system that moves with the air. Instead, we want to apply it to a fixed location (i.e., an **Eulerian** framework), such as over your house. The only change needed is to include a new term called **advection** along with the other forces, when computing the net force F_{net} in each direction. All these forces are explained in the next section.

But knowing the forces, we need additional information to use eqs. (10.6) — we need the initial winds [U(t), V(t), W(t)] to use for the first terms on the right side of eqs. (10.6). Hence, to make numerical weather forecasts, we must first observe the current weather and create an **analysis** of it. This corresponds to an **initial-value problem** in mathematics.

Average horizontal winds are often 100 times stronger than vertical winds, except in thunderstorms and near mountains. We will focus on horizontal forces and winds first.

Sample Application

Initially still air is acted on by force $F_{y\,net}/m = 5\text{x}10^{-4}$ m·s^{-2}. Find the final wind speed after 30 minutes.

Find the Answer

Given: $V(0) = 0$, $F_{y\,net}/m = 5\text{x}10^{-4}$ m·s^{-2}, $\Delta t = 1800$ s
Find: $V(\Delta t) = ?$ m s^{-1}. Assume: $U = W = 0$.

Apply eq. (10.6b): $V(t+\Delta t) = V(t) + \Delta t \cdot (F_{y\,net}/m)$
$= 0 + (1800\text{s})\cdot(5\text{x}10^{-4}\ \text{m·s}^{-2}) =$ **0.9 m s^{-1}**.

Check: Physics and units are reasonable.
Exposition: This wind toward the north (i.e., from 180°) is slow. But continued forcing over more time could make it faster.

A SCIENTIFIC PERSPECTIVE • Creativity

As a child at Woolsthorpe, his mother's farm in England, Isaac Newton built clocks, sundials, and model windmills. He was an average student, but his schoolmaster thought Isaac had potential, and recommended that he attend university.

Isaac started Cambridge University in 1661. He was 18 years old, and needed to work at odd jobs to pay for his schooling. Just before the plague hit in 1665, he graduated with a B.A. But the plague was spreading quickly, and within 3 months had killed 10% of London residents. So Cambridge University was closed for 18 months, and all the students were sent home.

While isolated at his mother's farm, he continued his scientific studies independently. This included much of the foundation work on the laws of motion, including the co-invention of calculus and the explanation of gravitational force. To test his laws of motion, he built his own telescope to study the motion of planets. But while trying to improve his telescope, he made significant advances in optics, and invented the reflecting telescope. He was 23 - 24 years old.

It is often the young women and men who are most creative — in the sciences as well as the arts. Enhancing this creativity is the fact that these young people have not yet been overly swayed (perhaps misguided) in their thinking by the works of others. Thus, they are free to experiment and make their own mistakes and discoveries.

You have an opportunity to be creative. Be wary of building on the works of others, because subconsciously you will be steered in their same direction of thought. Instead, I encourage you to be brave, and explore novel, radical ideas.

This recommendation may seem paradoxical. You are reading my book describing the meteorological advances of others, yet I discourage you from reading about such advances. You must decide on the best balance for you.

HORIZONTAL FORCES

Five forces contribute to net horizontal accelerations that control horizontal winds: **pressure-gradient force** (PG), **advection** (AD), **centrifugal force** (CN), **Coriolis force** (CF), and **turbulent drag** (TD):

$$\frac{F_{x\,net}}{m} = \frac{F_{x\,AD}}{m} + \frac{F_{x\,PG}}{m} + \frac{F_{x\,CN}}{m} + \frac{F_{x\,CF}}{m} + \frac{F_{x\,TD}}{m} \tag{10.7a}$$

$$\frac{F_{y\,net}}{m} = \frac{F_{y\,AD}}{m} + \frac{F_{y\,PG}}{m} + \frac{F_{y\,CN}}{m} + \frac{F_{y\,CF}}{m} + \frac{F_{y\,TD}}{m} \tag{10.7b}$$

Centrifugal force is an apparent force that allows us to include inertial effects for winds that move in a curved line. Coriolis force, explained in detail later, includes the gravitational and compound centrifugal forces on a non-spherical Earth. In the equations above, force per unit mass has units of N kg^{-1}. These units are equivalent to units of acceleration (m·s^{-2}, see Appendix A), which we will use here.

Advection of Horizontal Momentum

Advection is not a true force. Yet it can cause a change of wind speed at a fixed location in Eulerian coordinates, so we will treat it like a force here. The wind moving past a point can carry **specific momentum** (i.e., momentum per unit mass). Recall that momentum is defined as mass times velocity, hence specific momentum equals the velocity (i.e., the wind) by definition. Thus, the wind can move (**advect**) different winds to your fixed location.

This is illustrated in Fig. 10.4a. Consider a mass of air (grey box) with slow U wind (5 m s^{-1}) in the north and faster U wind (10 m s^{-1}) in the south. Thus, U decreases toward the north, giving $\Delta U/\Delta y$ = *negative*. This whole air mass is advected toward the north over a fixed weather station "O" by a south wind (V = *positive*). At the later time sketched in Fig. 10.4b, a west wind of 5 m s^{-1} is measured at "O". Even later, at the time of Fig. 10.4c, the west wind has increased to 10 m s^{-1} at the weather station. The rate of increase of U at "O" is larger for faster advection (V), and is larger if $\Delta U/\Delta y$ is more negative.

Thus, $\Delta U/\Delta t = -V \cdot \Delta U/\Delta y$ for this example. The advection term on the RHS causes an acceleration of U wind on the LHS, and thus acts like a force per unit mass: $\Delta U/\Delta t = F_{x\,AD}/m = -V \cdot \Delta U/\Delta y$.

You must always include advection when momentum-budget equations are written in Eulerian frameworks. This is similar to the advection terms in the moisture- and heat-budget Eulerian equations that were in earlier chapters.

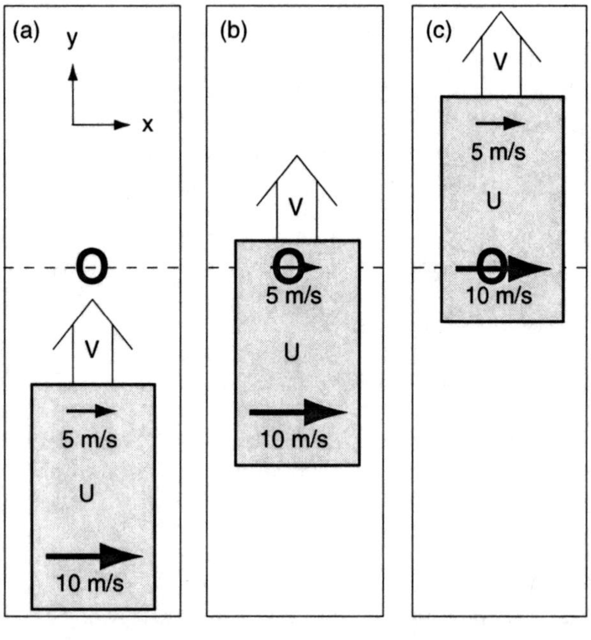

Figure 10.4
Illustration of V advection of U wind. "O" is a fixed weather station. Grey box is an air mass containing a gradient of U wind. Initial state (a) and later states (b and c).

For advection, the horizontal force components are

$$\frac{F_{x\,AD}}{m} = -U\cdot\frac{\Delta U}{\Delta x} - V\cdot\frac{\Delta U}{\Delta y} - W\cdot\frac{\Delta U}{\Delta z} \qquad \bullet(10.8a)$$

$$\frac{F_{y\,AD}}{m} = -U\cdot\frac{\Delta V}{\Delta x} - V\cdot\frac{\Delta V}{\Delta y} - W\cdot\frac{\Delta V}{\Delta z} \qquad \bullet(10.8b)$$

Recall that a **gradient** is defined as change across a distance, such as $\Delta V/\Delta y$. With no gradient, the wind cannot cause accelerations.

Vertical advection of horizontal wind ($-W\cdot\Delta U/\Delta z$ in eq. 10.8a, and $-W\cdot\Delta V/\Delta z$ in eq. 10.8b) is often very small outside of thunderstorms.

Horizontal Pressure-Gradient Force

In regions where the pressure changes with distance (i.e., a **pressure gradient**), there is a force from high to low pressure. On weather maps, this force is at right angles to the height contours or isobars, directly from high heights or high pressures to low. Greater gradients (shown by a tighter packing of isobars; i.e., smaller spacing Δd between isobars on weather maps) cause greater pressure-gradient force (Fig. 10.5). Pressure-gradient force is independent of wind speed, and thus can act on winds of any speed (including calm) and direction.

For pressure-gradient force, the horizontal components are:

$$\frac{F_{x\,PG}}{m} = -\frac{1}{\rho}\cdot\frac{\Delta P}{\Delta x} \qquad \bullet(10.9a)$$

$$\frac{F_{y\,PG}}{m} = -\frac{1}{\rho}\cdot\frac{\Delta P}{\Delta y} \qquad \bullet(10.9b)$$

where ΔP is the pressure change across a distance of either Δx or Δy, and ρ is the density of air.

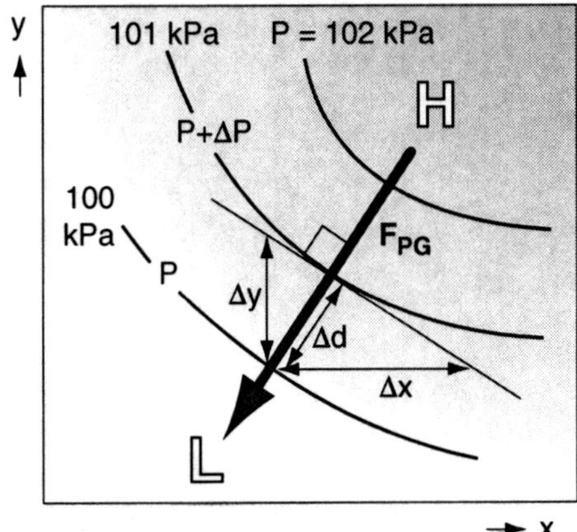

Figure 10.5
The dark arrow shows the direction of pressure-gradient force F_{PG} from high (H) to low (L) pressure. This force is perpendicular to the isobars (solid curved lines).

Sample Application

Minneapolis (MN, USA) is about 400 km north of Des Moines (IA, USA). In Minneapolis the wind components (U, V) are (6, 4) m s^{-1} , while in Des Moines they are (2, 10) m s^{-1}. What is the value of the advective force per mass?

Find the Answer

Given: (U, V) = (6, 4) m s^{-1} in Minneapolis,
(U, V) = (2, 10) m s^{-1} in Des Moines
Δy = 400 km, Δx = is not relevant
Find: $F_{x\,AD}/m$ =? m·s^{-2} , $F_{y\,AD}/m$ =? m·s^{-2}

Use the definition of a gradient:
$\Delta U/\Delta y = (6-2\text{ m s}^{-1})/400{,}000\text{ m} = 1.0\text{x}10^{-5}\text{ s}^{-1}$
$\Delta U/\Delta x$ = not relevant, $\Delta U/\Delta z$ = not relevant,
$\Delta V/\Delta y = (4-10\text{ m s}^{-1})/400{,}000\text{ m} = -1.5\text{x}10^{-5}\text{ s}^{-1}$
$\Delta V/\Delta x$ = not relevant, $\Delta V/\Delta z$ = not relevant
Average $U = (6+2\text{ m s}^{-1})/2 = 4\text{ m s}^{-1}$
Average $V = (4+10\text{ m s}^{-1})/2 = 7\text{ m s}^{-1}$

Use eq. (10.8a):
$F_{x\,AD}/m = -(7\text{m s}^{-1})\cdot(1.0\text{x}10^{-5}\text{ s}^{-1})$
= **$-7\text{x}10^{-5}$** m·s^{-2}
Use eq. (10.8b):
$F_{y\,AD}/m = -(7\text{m s}^{-1})\cdot(-1.5\text{x}10^{-5}\text{ s}^{-1})$
= **$1.05\text{x}10^{-4}$** m·s^{-2}

Check: Physics and units are reasonable.
Exposition: The slower U winds from Des Moines are being blown by positive V winds toward Minneapolis, causing the U wind speed to decrease at Minneapolis. But the V winds are increasing there because of the faster winds in Des Moines moving northward.

Sample Application

Minneapolis (MN, USA) is about 400 km north of Des Moines (IA, USA). In (Minneapolis, Des Moines) the pressure is (101, 100) kPa. Find the pressure-gradient force per unit mass? Let ρ = 1.1 kg·m^{-3}.

Find the Answer

Given: P =101 kPa @ x = 400 km (north of Des Moines).
P =100 kPa @ x = 0 km at Des Moines. ρ = 1.1 kg·m^{-3}.
Find: $F_{y\,PG}/m$ = ? m·s^{-2}

Apply eq. (10.9b):

$$\frac{F_{y\,PG}}{m} = -\frac{1}{(1.1\text{kg}\cdot\text{m}^{-3})}\cdot\frac{(101{,}000-100{,}000)\text{Pa}}{(400{,}000-0)\text{m}}$$

= **$-2.27\text{x}10^{-3}$** m·s^{-2}.
Hint, from Appendix A: 1 Pa = 1 kg·m^{-1}·s^{-2}.

Check: Physics and units are reasonable.
Exposition: The force is from high pressure in the north to low pressure in the south. This direction is indicated by the negative sign of the answer; namely, the force points in the negative y direction.

Sample Application

If the height of the 50 kPa pressure surface decreases by 10 m northward across a distance of 500 km, what is the pressure-gradient force?

Find the Answer

Given: $\Delta z = -10$ m, $\Delta y = 500$ km, $|g| = 9.8$ m·s^{-2}.
Find: $F_{PG}/m = ?$ m·s^{-2}

Use eqs. (10.11a & b):
$F_{x\ PG}/m = 0$ m·s^{-2}, because $\Delta z/\Delta x = 0$. Thus, $F_{PG}/m = F_{y\ PG}/m$.

$$\frac{F_{y\,PG}}{m} = -|g| \cdot \frac{\Delta z}{\Delta y} = -\left(9.8\frac{\text{m}}{\text{s}^2}\right)\cdot\left(\frac{-10\text{m}}{500{,}000\text{m}}\right)$$

F_{PG}/m = **0.000196 m·s^{-2}**

Check: Physics, units & sign are reasonable.
Exposition: For our example here, height decreases toward the north, thus a hypothetical ball would roll downhill toward the north. A northward force is in the positive y direction, which explains the positive sign of the answer.

Table 10-2. To apply centrifugal force to separate Cartesian coordinates, a (+/–) sign factor s is required.

Hemisphere	For winds encircling a Low Pressure Center	For winds encircling a High Pressure Center
Southern	–1	+1
Northern	+1	–1

Sample Application

500 km east of a high-pressure center is a north wind of 5 m s^{-1}. Assume N. Hemisphere. What is the centrifugal force?

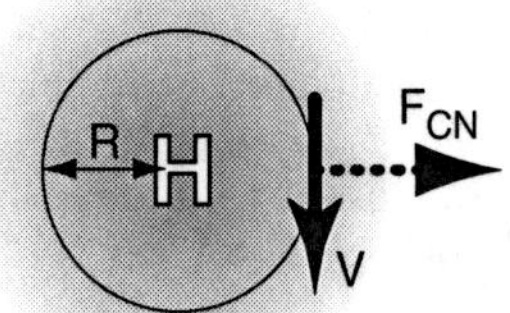

Find the Answer

Given: $R = 5\text{x}10^5$ m,
$U = 0$, $V = -5$ m s^{-1}
Find: $F_{x\ CN}/m = ?$ m·s^{-2}.

Apply eq. (10.13a). In Table 10-2 find $s = -1$.

$$\frac{F_{xCN}}{m} = -1 \cdot \frac{(-5\text{m/s})\cdot(5\text{m/s})}{5\times10^5} = \underline{5\text{x}10^{-5}\ \text{m·s}^{-2}}.$$

Check: Physics and units OK. Agrees with sketch.
Exposition: To maintain a turn around the high-pressure center, other forces (the sum of which is the centripetal force) are required to pull toward the center.

If pressure increases toward one direction, then the force is in the opposite direction (from high to low P); hence, the negative sign in these terms.

Pressure-gradient-force magnitude is

$$\left|\frac{F_{PG}}{m}\right| = \left|\frac{1}{\rho}\cdot\frac{\Delta P}{\Delta d}\right| \tag{10.10}$$

where Δd is the distance between isobars.

Eqs. (10.9) can be rewritten using the hydrostatic eq. (1.25) to give the pressure gradient components as a function of spacing between height contours on an isobaric surface:

$$\frac{F_{x\,PG}}{m} = -|g|\cdot\frac{\Delta z}{\Delta x} \tag{10.11a}$$

$$\frac{F_{y\,PG}}{m} = -|g|\cdot\frac{\Delta z}{\Delta y} \tag{10.11b}$$

for a gravitational acceleration magnitude of $|g| = 9.8$ m·s^{-2}. Δz is the height change in the Δx or Δy directions; hence, it is the slope of the isobaric surface. Extending this analogy of slope, if you conceptually place a ball on the isobaric surface, it will roll downhill (which is the pressure-gradient force direction). The magnitude of pressure-gradient force is

$$\left|\frac{F_{PG}}{m}\right| = \left|g\cdot\frac{\Delta z}{\Delta d}\right| \tag{10.12}$$

where Δd is distance between height contours.

The one force that makes winds blow in the horizontal is pressure-gradient force. All the other forces are a function of wind speed, hence they can only change the speed or direction of a wind that already exists. The only force that can start winds blowing from zero (calm) is pressure-gradient force.

Centrifugal Force

Inertia makes an air parcel try to move in a straight line. To get its path to turn requires a force in a different direction. This force, which pulls toward the inside of the turn, is called **centripetal** force. Centripetal force is the result of a net imbalance of (i.e., the nonzero vector sum of) other forces.

For mathematical convenience, we can define an apparent force, called **centrifugal** force, that is opposite to centripetal force. Namely, it points outward from the center of rotation. Centrifugal-force components are:

$$\frac{F_{xCN}}{m} = +s\cdot\frac{V\cdot M}{R} \qquad \bullet(10.13a)$$

$$\frac{F_{yCN}}{m} = -s\cdot\frac{U\cdot M}{R} \qquad \bullet(10.13b)$$

where $M = (U^2 + V^2)^{1/2}$ is wind speed (always positive), R is radius of curvature, and s is a sign factor from Table 10-2 as determined by the hemisphere (North or South) and synoptic pressure center (Low or High).

Centrifugal force magnitude is proportional to wind speed squared:

$$\left|\frac{F_{CN}}{m}\right| = \frac{M^2}{R} \qquad (10.14)$$

Coriolis Force

An object such as an air parcel that moves relative to the Earth experiences a **compound centrifugal force** based on the combined tangential velocities of the Earth's surface and the object. When combined with the non-vertical component of gravity, the result is called Coriolis force (see the INFO box on the next page). This force points 90° to the right of the wind direction in the Northern Hemisphere (Fig. 10.6), and 90° to the left in the S. Hemisphere.

The Earth rotates one full revolution (2π radians) during a sidereal day (i.e., relative to the fixed stars, $P_{sidereal}$ is a bit less than 24 h, see Appendix B), giving an angular rotation rate of

$$\Omega = 2\cdot\pi / P_{sidereal} \qquad \bullet(10.15)$$

$$= 0.729\ 211\ 6 \times 10^{-4} \text{ radians s}^{-1}$$

The units for Ω are often abbreviated as s^{-1}. Using this rotation rate, define a **Coriolis parameter** as:

$$f_c = 2\cdot\Omega\cdot\sin(\phi) \qquad \bullet(10.16)$$

where ϕ is latitude, and $2{\cdot}\Omega = 1.458423\text{x}10^{-4}\ s^{-1}$. Thus, the Coriolis parameter depends only on latitude. Its magnitude is roughly $1\text{x}10^{-4}\ s^{-1}$ at mid-latitudes.

The Coriolis force in the Northern Hemisphere is:

$$\frac{F_{x\,CF}}{m} = f_c\cdot V \qquad \bullet(10.17a)$$

$$\frac{F_{y\,CF}}{m} = -f_c\cdot U \qquad \bullet(10.17b)$$

In the Southern Hemisphere the signs on the right side of eqs. (10.17) are opposite. Coriolis force is zero under calm conditions, and thus cannot create a wind. However, it can change the direction of an existing wind. Coriolis force cannot do work, because it acts perpendicular to the object's motion.

The magnitude of Coriolis force is:

$$|\ F_{CF}/m\ | \approx 2\cdot\Omega\cdot|\sin(\phi)\cdot M| \qquad (10.18a)$$

or

$$|\ F_{CF}/m\ | \approx |\ f_c\cdot M\ | \qquad (10.18b)$$

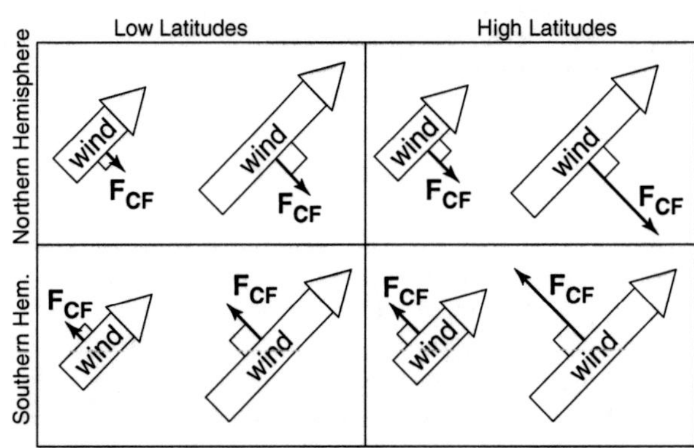

Figure 10.6
Coriolis force (F_{CF}) vs. latitude, wind-speed, and hemisphere.

Sample Application (§)
a) Plot Coriolis parameter vs. latitude.
b) Find F_{CF}/m at Paris, given a north wind of 15 m s^{-1}.

Find the Answer:
a) Given: ϕ = 48.874°N at Paris.
Find f_c (s^{-1}) vs. ϕ(°) using eq. (10.16). For example:
$f_c = (1.458\text{x}10^{-4}\ s^{-1})\cdot\sin(48.874°) = 1.1\text{x}10^{-4}\ s^{-1}$.

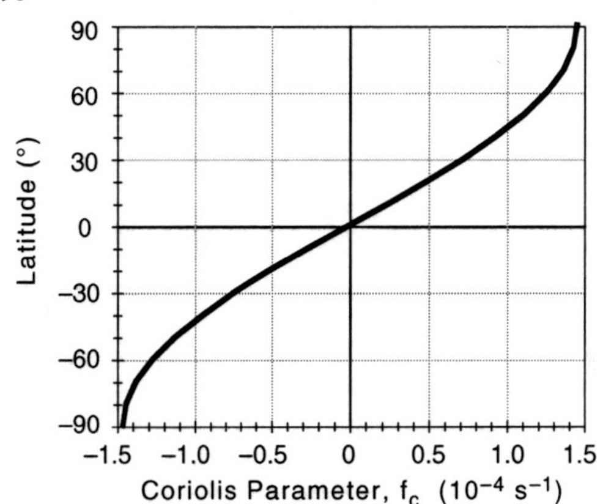

b) Given: $V = -15$ m s^{-1}. Find: F_{CF}/m= ? m s^{-2}
Assume $U = 0$ because no info, thus $F_{y\,CF}/m$= 0.
Apply eq. (10.17a):
$F_{x\,CF}/m = (1.1\text{x}10^{-4}\ s^{-1})\cdot(-15\text{ m s}^{-1})$ = **$-1.65\text{x}10^{-3}$ m s^{-2}**
Exposition: This Coriolis force points to the west.

INFO • Coriolis Force in 3-D

Eqs. (10.17) give only the dominant components of Coriolis force. There are other smaller-magnitude Coriolis terms (labeled *small* below) that are usually neglected. The full Coriolis force in 3-dimensions is:

$$\frac{F_{x\,CF}}{m} = f_c\cdot V - 2\Omega\cdot\cos(\phi)\cdot W \qquad (10.17c)$$

[small because often W<<V]

$$\frac{F_{y\,CF}}{m} = -f_c\cdot U \qquad (10.17d)$$

$$\frac{F_{z\,CF}}{m} = 2\Omega\cdot\cos(\phi)\cdot U \qquad (10.17e)$$

[small relative to other vertical forces]

INFO • On Coriolis Force

Gaspar Gustave Coriolis explained a compound centrifugal force on a rotating non-spherical planet such as Earth (Anders Persson: 1998, 2006, 2014).

Basics

On the rotating Earth an imbalance can occur between gravitational force and centrifugal force.

For an object of mass m moving at tangential speed M_{tan} along a curved path having radius of curvature R, **centrifugal force** was shown earlier in this chapter to be $F_{CN}/m = (M_{tan})^2/R$. In Fig 10.a the object is represented by the black dot, and the center of rotation is indicated by the X.

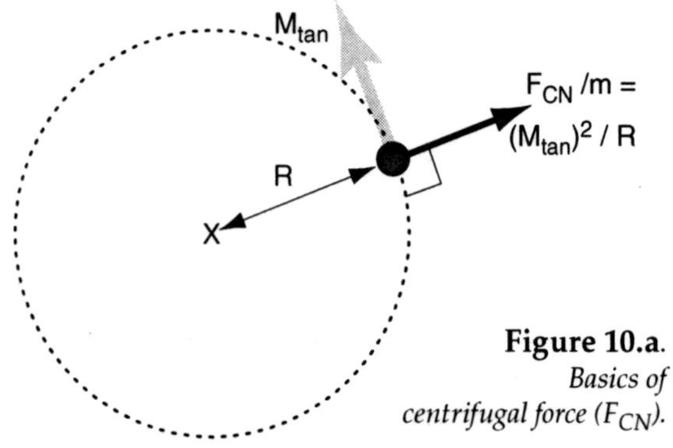

Figure 10.a. *Basics of centrifugal force (F_{CN}).*

The Earth was mostly molten early in its formation. Although gravity tends to make the Earth spherical, centrifugal force associated with Earth's rotation caused the Earth to bulge slightly at the equator. Thus, Earth's shape is an ellipsoid (Fig. 10.b).

The combination of gravity F_G and centrifugal force F_{CN} causes a net force that we feel as **effective gravity** F_{EG}. Objects fall in the direction of effective gravity, and it is how we define the local vertical (V) direction. Perpendicular to vertical is the local "horizontal" (H) direction, along the ellipsoidal surface. An object initially at rest on this surface feels no net horizontal force. [Note: Except at the poles and equator, F_G does not point exactly to Earth's center, due to gravitational pull of the equatorial bulge.]

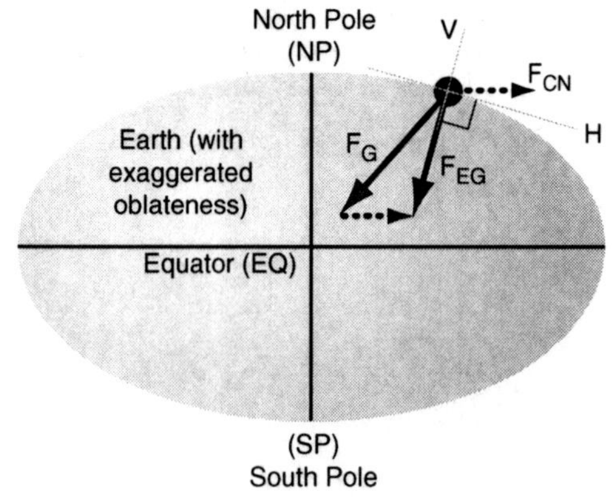

Figure 10.b. *Earth cross section (exaggerated).*

(continues in next column)

INFO • On Coriolis Force (continuation)

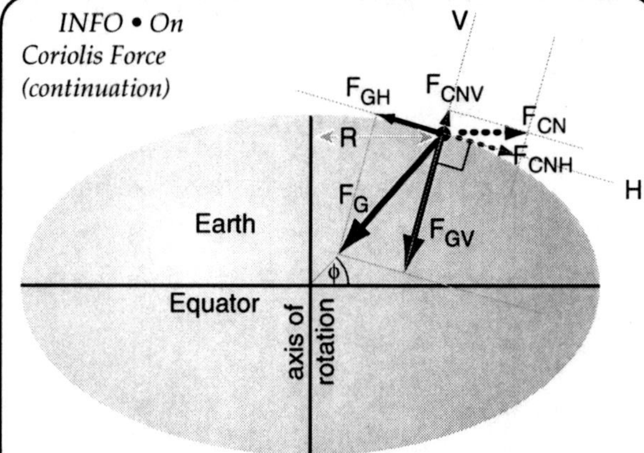

Figure 10.c. *Horizontal & vertical force components.*

Split the vectors of true gravity into local vertical F_{GV} and horizontal F_{GH} components. Do the same for the centrifugal force (F_{CNV}, F_{CNH}) of Earth's rotation (Fig. 10.c). Total centrifugal force F_{CN} is parallel to the equator (EQ). Thus, for an object at latitude ϕ, you can use trig to show $F_{CNH} \approx F_{CN} \cdot \sin(\phi)$.

Objects at Rest with respect to Earth's Surface

Looking down towards the north pole (NP), the Earth turns counterclockwise with angular velocity Ω = 360°/(sidereal day) (Fig. 10.d). Over a time interval Δt, the amount of rotation is $\Omega \cdot \Delta t$. Any object (black dot) at rest on the Earth's surface moves with the Earth at tangential speed $M_{tan} = \Omega \cdot R$ (grey arrow), where $R = R_o \cdot cos(\phi)$ is the distance from the axis of rotation. R_o = 6371 km is average Earth radius.

But because the object is at rest, its horizontal component of centrifugal force F_{CNH} associated with movement following the curved latitude (called a parallel) is the same as that for the Earth, as plotted in Fig. 10.c above. But this horizontal force is balanced by the horizontal component of gravity F_{GH}, so the object feels no net horizontal force.

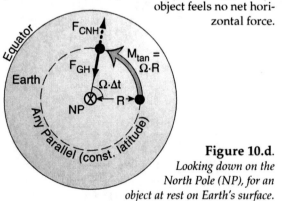

Figure 10.d. *Looking down on the North Pole (NP), for an object at rest on Earth's surface.*

Objects Moving East or West relative to Earth

Suppose an object moves with velocity M due east relative to the Earth. This velocity (thin white arrow in Fig. 10.e) is relative to Earth's velocity, giving the object a faster total velocity (grey arrow), causing greater centrifugal force and greater F_{CNH}. But F_{GH} is constant.

(continues in next column)

INFO • On Coriolis Force (continuation)

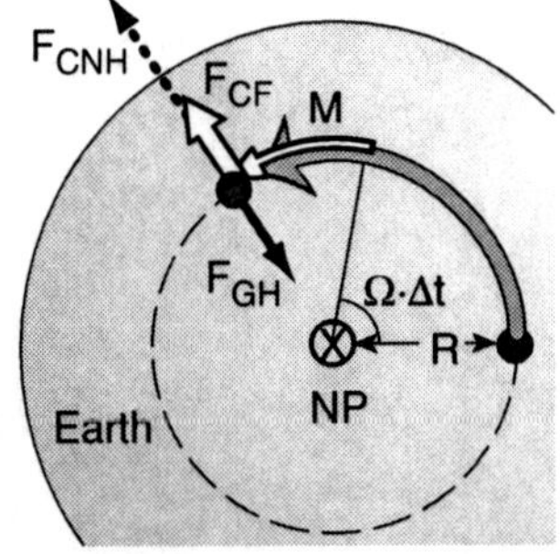

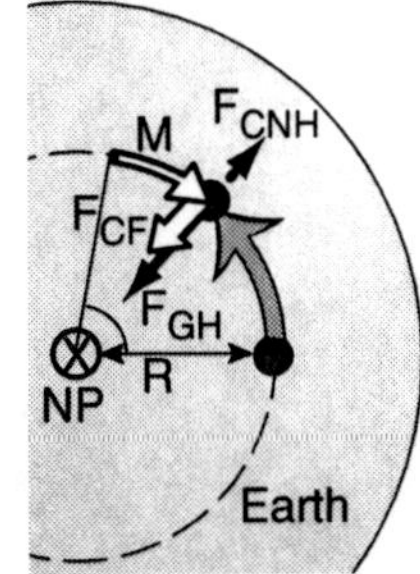

Figure 10.e. *Eastward moving object.*

Figure 10.f. *Westward moving object.*

Horizontal force F_{CNH} does NOT balance F_{GH}. The thick white arrow (Fig. 10.e) shows that the force difference F_{CF} is to the right relative to the object's motion M. F_{CF} is called **Coriolis force.**

The opposite imbalance of F_{CNH} and F_{GH} occurs for a westward-moving object (thin white arrow), because the object has slower net tangential velocity (grey arrow in Fig. 10.f). This imbalance, Coriolis force F_{CF}, is also to the right of the relative motion M.

Northward-moving Objects

When an object moves northward at relative speed M (thin white arrow in Fig. 10.g) while the Earth is rotating, the path traveled by the object (thick grey line) has a small radius of curvature about point X that is displaced from the North Pole. The smaller radius R causes larger centrifugal force F_{CNH} pointing outward from X.

Component $F_{CNH\text{-}ns}$ of centrifugal force balances the unchanged horizontal gravitational force F_{GH}. But there remains an unbalanced east-west component of centrifugal force $F_{CNH\text{-}ew}$ which is defined as Coriolis force F_{CF}. Again, it is to the right of the relative motion vector M of the object.

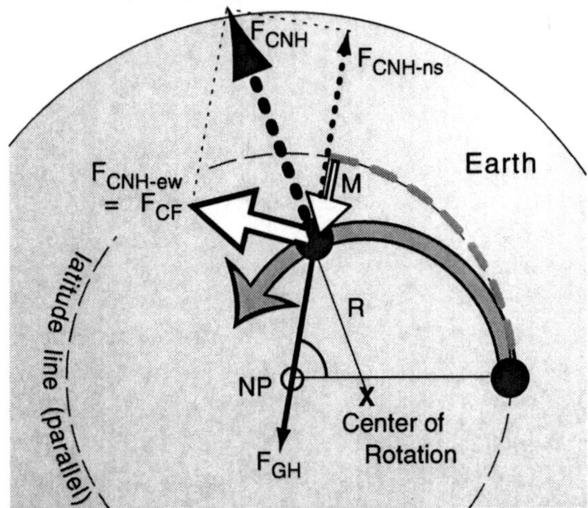

Figure 10.g. *Northward moving object.*

(continues in next column)

INFO • Coriolis Force (continuation)

Objects moving south have a Coriolis force to the right due to the larger radius of curvature. Regardless of the direction of motion in the Northern Hemisphere, Coriolis force acts 90° to the right of the object's motion relative to the Earth. When viewing the Southern Hemisphere from below the south pole, the Earth rotates clockwise, causing a Coriolis force that is 90° to the left of the relative motion vector.

Coriolis-force Magnitude Derivation

From Figs. 10.c & d, see that an object at rest (subscript R) has

$$F_{GH} = F_{CNH} \equiv F_{CNHR} \qquad \text{(C1)}$$

and

$$M_{tan\ rest} = \Omega \cdot R \qquad \text{(C2)}$$

From Fig. 10.e, Coriolis force for an eastward-moving object is defined as

$$F_{CF} \equiv F_{CNH} - F_{GH}$$

Apply eq. (C1) to get

$$F_{CF} = F_{CNH} - F_{CNHR}$$

or

$$F_{CF} = \sin(\phi) \cdot [F_{CN} - F_{CNR}] \qquad \text{(from Fig. 10.c)}$$

Divide by mass m, and plug in the definition for centrifugal force as velocity squared divided by radius:

$$F_{CF} / m = \sin(\phi) \cdot [\, (M_{tan})^2/R - (M_{tan\ rest})^2/R \,]$$

Use $M_{tan} = M_{tan\ rest} + M$, along with eq. (C2):

$$F_{CF} / m = \sin(\phi) \cdot [\, (\Omega{\cdot}R{+}M)^2/R - (\Omega{\cdot}R)^2/R \,]$$

$$F_{CF} / m = \sin(\phi) \cdot [(2{\cdot}\Omega{\cdot}M) + (M^2/R)]$$

The first term is usually much larger than the last, allowing the following approximation for Coriolis force per mass:

$$F_{CF} / m \approx 2{\cdot}\Omega{\cdot}\sin(\phi) \cdot M \qquad (10.18)$$

Define a Coriolis parameter as $f_c \equiv 2{\cdot}\Omega{\cdot}\sin(\phi)$. Thus,

$$F_{CF} / m \approx f_c \cdot M$$

HIGHER MATH • Apparent Forces

In vector form, centrifugal force/mass for an object at rest on Earth is $-\boldsymbol{\Omega} \times (\boldsymbol{\Omega} \times \mathbf{r})$, and Coriolis force/mass is $-2\boldsymbol{\Omega} \times \mathbf{V}$, where vector $\boldsymbol{\Omega}$ points along the Earth's axis toward the north pole, $\mathbf{r}$ points from the Earth's center to the object, $\mathbf{V}$ is the object's velocity relative to Earth, and $\times$ is the vector cross product.

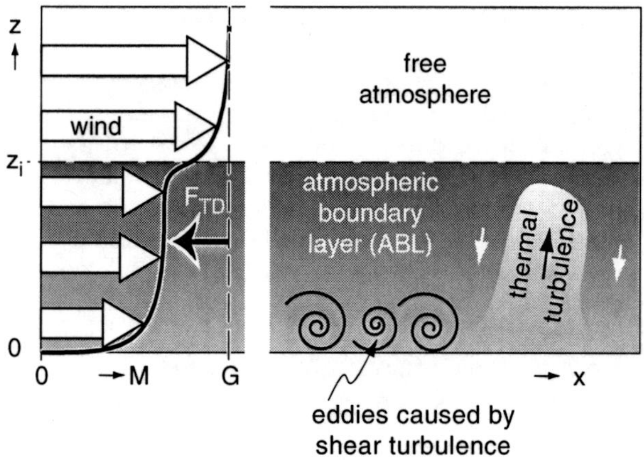

Figure 10.7
Wind speed M (curved black line with white highlights) is slower than geostrophic G (vertical dashed line) because of turbulent drag force F_{TD} in the atmospheric boundary layer.

Turbulent-Drag Force

Surface elements such as pebbles, blades of grass, crops, trees, and buildings partially block the wind, and disturb the air that flows around them. The combined effect of these elements over an area of ground is to cause resistance to air flow, thereby slowing the wind. This resistance is called **drag.**

At the bottom of the troposphere is a layer of air roughly 0.3 to 3 km thick called the **atmospheric boundary layer** (**ABL**). The ABL is named because it is at the bottom boundary of the atmosphere. Turbulence in the ABL mixes the very-slow near-surface air with the faster air in the ABL, reducing the wind speed M throughout the entire ABL (Fig. 10.7).

The net result is a drag force that is normally only felt by air in the ABL. For ABL depth z_i the drag is:

$$\frac{F_{x\,TD}}{m} = -w_T \cdot \frac{U}{z_i} \qquad \bullet(10.19a)$$

$$\frac{F_{y\,TD}}{m} = -w_T \cdot \frac{V}{z_i} \qquad \bullet(10.19b)$$

where w_T is called a turbulent **transport velocity.**

The total magnitude of turbulent drag force is

$$\left|\frac{F_{TD}}{m}\right| = w_T \cdot \frac{M}{z_i} \qquad (10.20)$$

and is always opposite to the wind direction.

For statically **unstable** ABLs with light winds, where a warm underlying surface causes thermals of warm buoyant air to rise (Fig. 10.7), this convective turbulence transports drag information upward at rate:

$$w_T = b_D \cdot w_B \qquad (10.22)$$

where dimensionless factor b_D = 1.83x10^{-3}. The **buoyancy velocity scale,** w_B, is of order 10 to 50 m s^{-1}, as is explained in the Heat Budget chapter.

For statically **neutral** conditions where strong winds M and **wind shears** (changes of wind direction and/or speed with height) create eddies and mechanical turbulence near the ground (Fig. 10.7), the transport velocity is

$$w_T = C_D \cdot M \qquad (10.21)$$

where the **drag coefficient** C_D is small (2x10^{-3} dimensionless) over smooth surfaces and is larger (2x10^{-2}) over rougher surfaces such as forests.

In fair weather, turbulent-drag force is felt only in the ABL. However, thunderstorm turbulence can mix slow near-surface air throughout the troposphere. Fast winds over mountains can create mountain-wave drag felt in the whole atmosphere (see the Regional Winds chapter).

Sample Application

What is the drag force per unit mass opposing a U = 15 m s^{-1} wind (with V = 0) for a: (a) statically neutral ABL over a rough forest; & (b) statically unstable ABL having convection with w_B = 50 m s^{-1}, given z_i = 1.5 km.

Find the Answer

Given: $U = M$ = 15 m s^{-1}, z_i = 1500 m,
C_D = 2x10^{-2}, w_B = 50 m s^{-1}.
Find: $F_{x\,TD}/m$ = ? m·s^{-2}.

(a) Plugging eq. (10.21) into eq. (10.19a) gives:

$$\frac{F_{x\,TD}}{m} = -C_D \cdot M \cdot \frac{U}{z_i} = -(0.02) \cdot \frac{(15\text{m/s})^2}{1500\text{m}}$$

$$= \underline{\mathbf{-3x10^{-3}}} \text{ m·s}^{-2}.$$

(b) Plugging eq. (10.22) into eq. (10.19a) gives:

$$\frac{F_{x\,TD}}{m} = -b_D \cdot w_B \cdot \frac{U}{z_i}$$

$$= -(0.00183) \cdot (50\text{m/s}) \cdot \frac{(15\text{m/s})}{1500\text{m}}$$

$$= \underline{\mathbf{-9.15x10^{-4}}} \text{ m·s}^{-2}.$$

Check: Physics and units are reasonable.

Exposition: Because the wind is positive (blowing toward the east) it requires that the drag be negative (pushing toward the west). Shear (mechanical) turbulence and convective (thermal/buoyant) turbulence can both cause drag by diluting the faster winds higher in the ABL with slower near-surface winds.

Summary of Forces

Table 10-3. Summary of forces.

Item	Name of Force	Direction	Magnitude ($N\ kg^{-1}$)	Horiz. (H) or Vert. (V)	Remarks ("item" is in column 1; H & V in col. 5)
1	**gravity**	down	$\left\lvert\frac{F_G}{m}\right\rvert = \lvert g\rvert = 9.8\ m{\cdot}s^{-2}$	V	**hydrostatic equilibrium** when items 1 & 2V balance
2	**pressure gradient**	from high to low pressure	$\left\lvert\frac{F_{PG}}{m}\right\rvert = \left\lvert g\cdot\frac{\Delta z}{\Delta d}\right\rvert$	V & H	the only force that can drive horizontal winds
3	**Coriolis** (compound)	90° to right (left) of wind in Northern (Southern) Hemisphere	$\left\lvert\frac{F_{CF}}{m}\right\rvert = 2{\cdot}\Omega{\cdot}\lvert\sin(\phi){\cdot}M\rvert$	H*	**geostrophic wind** when 2H and 3 balance (explained later in horiz. wind section)
4	**turbulent drag**	opposite to wind	$\left\lvert\frac{F_{TD}}{m}\right\rvert = w_T\cdot\frac{M}{z_i}$	H*	**atm. boundary-layer wind** when 2H, 3 and 4 balance (explained in horiz. wind section)
5	**centrifugal** (apparent)	away from center of curvature	$\left\lvert\frac{F_{CN}}{m}\right\rvert = \frac{M^2}{R}$	H*	centripetal = opposite of centrifugal. **Gradient wind** when 2H, 3 and 5 balance
6	**advection** (apparent)	(any)	$\left\lvert\frac{F_{AD}}{m}\right\rvert = \left\lvert -M{\cdot}\frac{\Delta U}{\Delta d} - ...\right\rvert$	V & H	neither creates nor destroys momentum; just moves it

**Horizontal is the direction we will focus on. However, Coriolis force has a small vertical component for zonal winds. Turbulent drag can exist in the vertical for rising or sinking air, but has completely different form than the boundary-layer drag given above. Centrifugal force can exist in the vertical for vortices with horizontal axes. Note: units $N\ kg^{-1} = m{\cdot}s^{-2}$.*

EQUATIONS OF HORIZONTAL MOTION

Combining the forces from eqs. (10.7, 10.8, 10.9, 10.17, and 10.19) into Newton's Second Law of Motion (eq. 10.5) gives simplified equations of horizontal motion:

•(10.23a)

$$\frac{\Delta U}{\Delta t} = -U\frac{\Delta U}{\Delta x} - V\frac{\Delta U}{\Delta y} - W\frac{\Delta U}{\Delta z} - \frac{1}{\rho}\cdot\frac{\Delta P}{\Delta x} + f_c\cdot V - w_T\cdot\frac{U}{z_i}$$

•(10.23b)

$$\underbrace{\frac{\Delta V}{\Delta t}}_{tendency} = \underbrace{-U\frac{\Delta V}{\Delta x} - V\frac{\Delta V}{\Delta y} - W\frac{\Delta V}{\Delta z}}_{advection} \underbrace{- \frac{1}{\rho}\cdot\frac{\Delta P}{\Delta y}}_{pressure\ gradient} \underbrace{- f_c\cdot U}_{Coriolis} \underbrace{- w_T\cdot\frac{V}{z_i}}_{turbulent\ drag}$$

These are the forecast equations for wind.

For special conditions where steady winds around a circle are anticipated, centrifugal force can be included.

The terms on the right side of eqs. (10.23) can all be of order $1\text{x}10^{-4}$ to $10\text{x}10^{-4}\ m{\cdot}s^{-2}$ (which is equivalent to units of $N\ kg^{-1}$, see Appendix A for review). However, some of the terms can be neglected under special conditions where the flow is less complicated. For example, near-zero Coriolis force occurs near the equator. Near-zero turbulent drag exists above the ABL. Near-zero pressure gradient is at low- and high-pressure centers.

Other situations are more complicated, for which additional terms should be added to the equations of horizontal motion. Within a few mm of the ground, **molecular friction** is large. Above mountains during windy conditions, **mountain-wave drag** is large. Above the ABL, cumulus clouds and thunderstorms can create strong **convective mixing**.

For a few idealized situations where many terms in the equations of motion are small, it is possible to solve those equations for the horizontal wind speeds. These theoretical winds are presented in the next section. Later in this chapter, equations to forecast vertical motion (W) will be presented.

Table 10-4. Names of idealized steady-state horizontal winds, and the forces that govern them.

$$0 = -U\frac{\Delta U}{\Delta x} \quad -\frac{1}{\rho}\cdot\frac{\Delta P}{\Delta x} \quad +f_c\cdot V \quad -w_T\cdot\frac{U}{z_i} \quad +s\frac{V\cdot M}{R}$$

Forces:	*pressure gradient*	*Coriolis*	*turbulent drag*	*centri-fugal*
Wind Name				
Geostrophic	•	•		
Gradient	•	•		•
Atm.Bound. Layer	•	•	•	
ABL Gradient	•	•	•	•
Cyclostrophic	•			•
Inertial		•		•
Antitriptic	•		•	

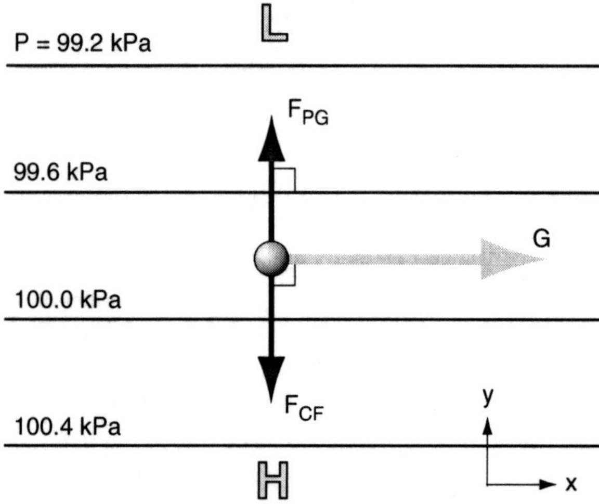

Figure 10.8
Idealized weather map for the Northern Hemisphere, showing geostrophic wind (G, grey arrow) caused by a balance between two forces (black arrows): pressure-gradient force (F_{PG}) and Coriolis force (F_{CF}). P is pressure, with isobars plotted as thin black lines. L and H are low and high-pressure regions. The small sphere represents an air parcel.

Sample Application
Find geostrophic wind components at a location where ρ = 1.2 kg m^{-3} and f_c = 1.1x10^{-4} s^{-1}. Pressure decreases by 2 kPa for each 800 km of distance north.

Find the Answer
Given: f_c =1.1x10^{-4} s^{-1}, ΔP= –2 kPa, ρ=1.2 kg m^{-3}, Δy=800 km.
Find: (U_g, V_g) = ? m s^{-1}

But $\Delta P/\Delta x = 0$ implies $V_g = 0$. For U_g, use eq. (10.26a):

$$U_g = \frac{-1}{(1.2\text{kg/m}^3)\cdot(1.1\times10^{-4}\text{s}^{-1})}\cdot\frac{(-2\text{kPa})}{(800\text{km})} = \underline{\mathbf{18.9\ m\ s^{-1}}}$$

Check: Physics & units OK. Agrees with Fig. 10.10.
Exposition: As the pressure gradient accelerates air northward, Coriolis force turns it toward the east.

HORIZONTAL WINDS

When air accelerates to create wind, forces that are a function of wind speed also change. As the winds continue to accelerate under the combined action of all the changing forces, **feedbacks** often occur to eventually reach a final wind where the forces balance. With a zero net force, there is zero acceleration.

Such a final, equilibrium, state is called **steady state**:

$$\frac{\Delta U}{\Delta t} = 0, \quad \frac{\Delta V}{\Delta t} = 0 \qquad \text{•(10.24)}$$

Caution: Steady state means no further change to the non-zero winds. Do not assume the winds are zero.

Under certain idealized conditions, some of the forces in the equations of motion are small enough to be neglected. For these situations, theoretical steady-state winds can be found based on only the remaining larger-magnitude forces. These theoretical winds are given special names, as listed in Table 10-4. These winds are examined next in more detail. As we discuss each theoretical wind, we will learn where we can expect these in the real atmosphere.

Geostrophic Wind

For special conditions where the only forces are Coriolis and pressure-gradient (Fig. 10.8), the resulting steady-state wind is called the geostrophic wind, with components (U_g, V_g). For this special case, the only terms remaining in, eqs. (10.23) are:

$$0 = -\frac{1}{\rho}\cdot\frac{\Delta P}{\Delta x} + f_c\cdot V \qquad (10.25a)$$

$$0 = -\frac{1}{\rho}\cdot\frac{\Delta P}{\Delta y} - f_c\cdot U \qquad (10.25b)$$

Define $U \equiv U_g$ and $V \equiv V_g$ in the equations above, and then solve for these wind components:

$$U_g = -\frac{1}{\rho\cdot f_c}\cdot\frac{\Delta P}{\Delta y} \qquad \text{•(10.26a)}$$

$$V_g = +\frac{1}{\rho\cdot f_c}\cdot\frac{\Delta P}{\Delta x} \qquad \text{•(10.26b)}$$

where f_c = (1.4584x10^{-4} s^{-1})·sin(*latitude*) is the Coriolis parameter, ρ is air density, and $\Delta P/\Delta x$ and $\Delta P/\Delta y$ are the horizontal pressure gradients.

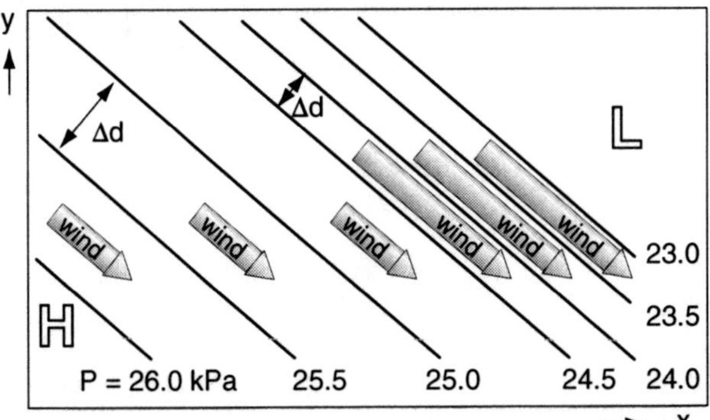

Figure 10.9
Isobars (black lines) that are more closely spaced (i.e., tightly packed) cause stronger geostrophic winds (arrows), for N. Hemisphere.

Real winds are nearly geostrophic at locations where isobars or height contours are relatively straight, for altitudes above the atmospheric boundary layer. Geostrophic winds are fast where isobars are packed closer together. The geostrophic wind direction is parallel to the height contours or isobars. In the (N., S.) hemisphere the wind direction is such that low pressure is to the wind's (left, right), see Fig. 10.9.

The magnitude G of the geostrophic wind is:

$$G = \sqrt{U_g^{\ 2} + V_g^{\ 2}} \qquad (10.27)$$

If Δd is the distance between two isobars (in the direction of greatest pressure change; namely, perpendicular to the isobars), then the magnitude (Fig. 10.10) of the geostrophic wind is:

$$G = \left| \frac{1}{\rho \cdot f_c} \cdot \frac{\Delta P}{\Delta d} \right| \qquad \bullet(10.28)$$

Above sea level, weather maps are often on isobaric surfaces (constant pressure charts), from which the geostrophic wind (Fig. 10.10) can be found from the height gradient (change of height of the isobaric surface with horizontal distance):

$$U_g = -\frac{|g|}{f_c} \cdot \frac{\Delta z}{\Delta y} \qquad \bullet(10.29a)$$

$$V_g = +\frac{|g|}{f_c} \cdot \frac{\Delta z}{\Delta x} \qquad \bullet(10.29b)$$

where the Coriolis parameter is f_c, and gravitational acceleration is $|g|$ = 9.8 m·s^{-2}. The corresponding magnitude of geostrophic wind on an isobaric chart is:

$$G = \left| \frac{g}{f_c} \cdot \frac{\Delta z}{\Delta d} \right| \qquad \bullet(10.29c)$$

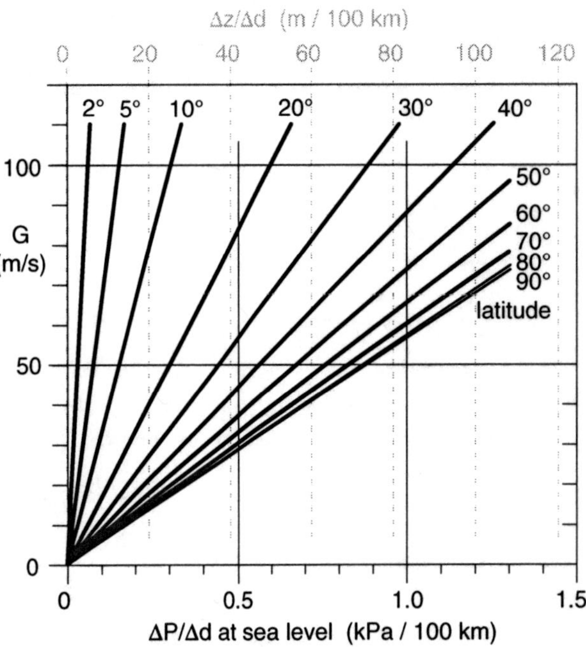

Figure 10.10
Variation of geostrophic wind speed (G) with horizontal pressure gradient (ΔP/Δd) at sea level. Top scale is height gradient of any isobaric surface.

Sample Application
Find the geostrophic wind for a height increase of 50 m per 200 km of distance toward the east. Assume, f_c = 0.9x10^{-4} s^{-1}.

Find the Answer
Given: Δx = 200 km, Δz = 50 m, f_c = 0.9x10^{-4} s^{-1}.
Find: G = ? m s^{-1}

No north-south height gradient, thus $U_g = 0$.
Apply eq. (10.29b) and set $G = V_g$:

$$V_g = +\frac{|g|}{f_c} \cdot \frac{\Delta z}{\Delta x} = \left(\frac{9.8\text{m s}^{-2}}{0.00009\text{s}^{-1}}\right) \cdot \left(\frac{50\text{m}}{200{,}000\text{m}}\right) = \underline{\mathbf{27.2\ m\ s^{-1}}}$$

Check: Physics & units OK. Agrees with Fig. 10.10.
Exposition: If height increases towards the east, then you can imagine that a ball placed on such a surface would roll downhill toward the west, but would turn to its right (toward the north) due to Coriolis force.

INFO • Approach to Geostrophy

How does an air parcel, starting from rest, approach the final steady-state geostrophic wind speed G sketched in Fig. 10.8?

Start with the equations of horizontal motion (10.23), and ignore all terms except the tendency, pressure-gradient force, and Coriolis force. Use the definition of

continues on next page

INFO • Appr. to Geostrophy *(continuation)*

geostrophic wind (eqs. 10.26) to write the resulting simplified equations as:

$$\Delta U / \Delta t = -f_c \cdot (V_g - V)$$
$$\Delta V / \Delta t = f_c \cdot (U_g - U)$$

Next, rewrite these as forecast equations:

$$U_{new} = U_{old} - \Delta t \cdot f_c \cdot (V_g - V_{old})$$
$$V_{new} = V_{old} + \Delta t \cdot f_c \cdot (U_g - U_{new})$$

Start with initial conditions (U_{old}, V_{old}) = (0, 0), and then iteratively solve the equations on a spreadsheet to forecast the wind.

For example, suppose ΔP = 1 kPa, $f_c = 10^{-4}$ s^{-1}, Δx = 500 km, ρ = 1 kg m^{-3}, where we would anticipate the wind should approach (U_g, V_g) = (0, 20) m s^{-1}. The actual evolution of winds (U, V) and air parcel position (X, Y) are shown in Figs. below.

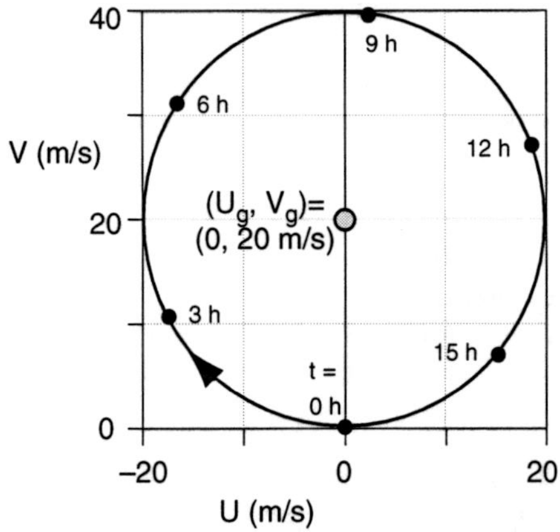

Surprisingly, the winds never reach geostrophic equilibrium, but instead rotate around the geostrophic wind. This is called an **inertial oscillation**, with period of $2 \cdot \pi / f_c$. For our case, the period is 17.45 h. Twice this period is called a **pendulum day**.

The net result in the figure below is that the wind indeed moves at the geostrophic speed of 20 m s^{-1} to the north (≈ 1250 km in 17.45 h), but along the way it staggers west and east with an additional **ageostrophic** (non-geostrophic) part.

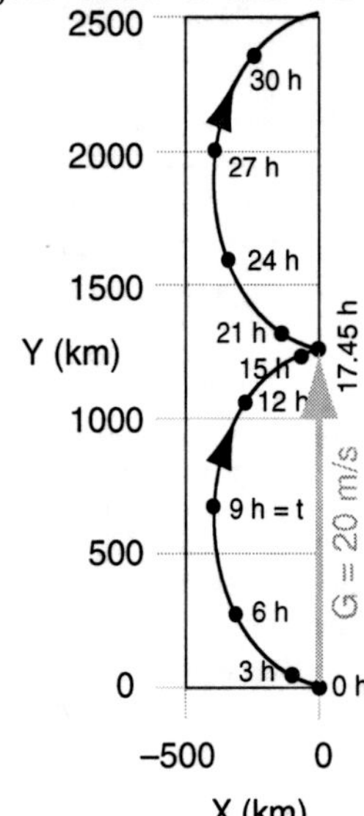

Inertial oscillations are sometimes observed at night in the atmospheric boundary layer, but rarely higher in the atmosphere. Why not? (1) The ageostrophic component of wind (wind from the East in this example) moves air mass, and changes the pressure gradient. (2) Friction damps the oscillation toward a steady wind.

If the **geopotential** $\Phi = |g| \cdot z$ is substituted in eqs. (10.29), the resulting geostrophic winds are:

$$U_g = -\frac{1}{f_c} \cdot \frac{\Delta \Phi}{\Delta y} \tag{10.30a}$$

$$V_g = \frac{1}{f_c} \cdot \frac{\Delta \Phi}{\Delta x} \tag{10.30b}$$

Gradient Wind

If there is no turbulent drag, then winds tend to blow parallel to isobar lines or height-contour lines even if those lines are curved. However, if the lines curve around a low-pressure center (in either hemisphere), then the wind speeds are **subgeostrophic** (i.e., slower than the theoretical geostrophic wind speed). For lines curving around high-pressure centers, wind speeds are **supergeostrophic** (faster than theoretical geostrophic winds). These theoretical winds following curved isobars or height contours are known as **gradient winds**.

Gradient winds differ from geostrophic winds because Coriolis force F_{CF} and pressure-gradient force F_{PG} do not balance, resulting in a non-zero net force F_{net}. This net force is called centripetal force, and is what causes the wind to continually change direction as it goes around a circle (Figs. 10.11 & 10.12). By describing this change in direction as causing an apparent force (centrifugal), we can find the equations that define a steady-state gradient wind:

$$0 = -\frac{1}{\rho} \cdot \frac{\Delta P}{\Delta x} + f_c \cdot V + s \cdot \frac{V \cdot M}{R} \tag{10.31a}$$

$$0 = \underbrace{-\frac{1}{\rho} \cdot \frac{\Delta P}{\Delta y}}_{\text{pressure gradient}} \underbrace{- f_c \cdot U}_{\text{Coriolis}} \underbrace{- s \cdot \frac{U \cdot M}{R}}_{\text{centrifugal}} \tag{10.31b}$$

Because the gradient wind is for flow around a circle, we can frame the governing equations in radial coordinates, such as for flow around a low:

$$\frac{1}{\rho} \cdot \frac{\Delta P}{\Delta R} = f_c \cdot M_{tan} + \frac{M_{tan}^2}{R} \tag{10.32}$$

where R is radial distance from the center of the circle, f_c is the Coriolis parameter, ρ is air density, $\Delta P / \Delta R$ is the radial pressure gradient, and M_{tan} is the magnitude of the tangential velocity; namely, the gradient wind.

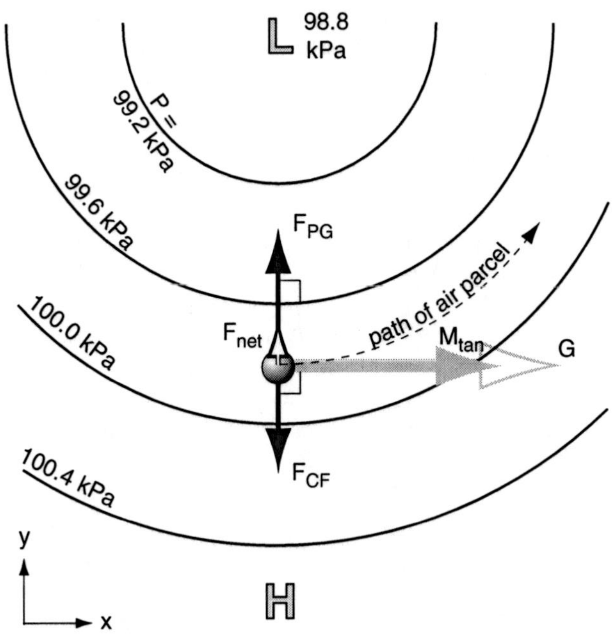

Figure 10.11
Forces (dark arrows) that cause the gradient wind (solid grey arrow, M_{tan}) to be slower than geostrophic (hollow grey arrow) when circling around a low-pressure center (called a cyclone in the N. Hem.). The short white arrow with black outline shows centripetal force (the imbalance between the other two forces). Centripetal force pulls the air parcel (grey sphere) inward to force the wind direction to change as needed for the wind to turn along a circular path.

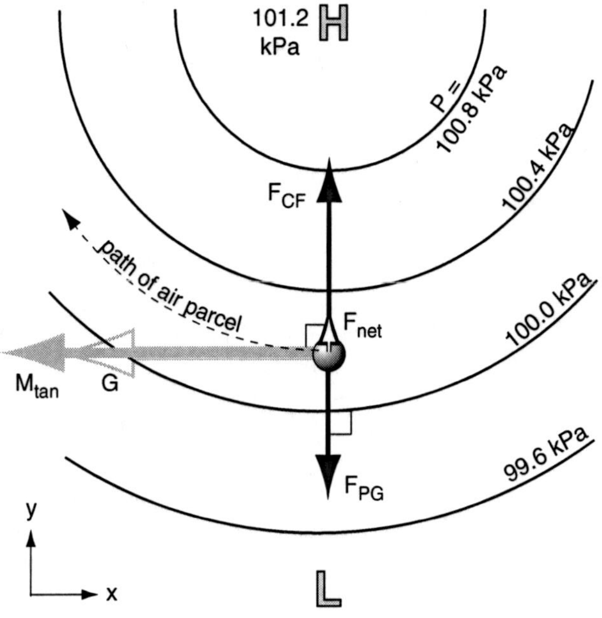

Figure 10.12
Forces (dark arrows) that cause the gradient wind (solid grey arrow, M_{tan}) to be faster than geostrophic (hollow grey arrow) for an air parcel (grey sphere) circling around a high-pressure center (called an anticyclone in the N. Hemisphere).

By re-arranging eq. (10.32) and plugging in the definition for geostrophic wind speed G, you can get an implicit solution for the gradient wind M_{tan}:

$$M_{tan} = G \pm \frac{M_{tan}^{\ 2}}{f_c \cdot R} \tag{10.33}$$

In this equation, use the + sign for flow around high-pressure centers, and the – sign for flow around lows (Fig. 10.13).

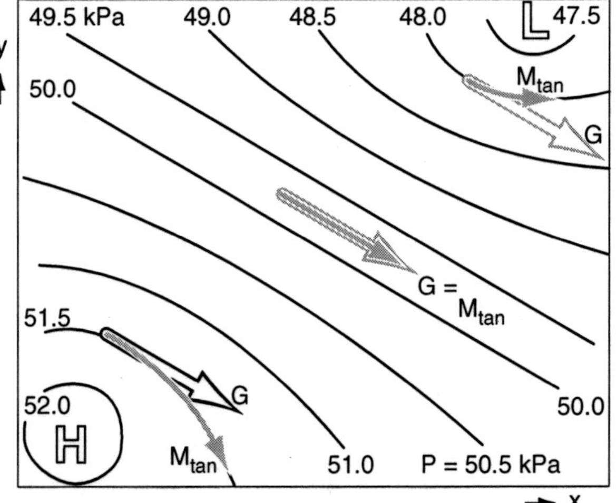

Figure 10.13
Comparison of gradient winds M_{tan} vs. geostrophic wind G for flows around low (L) and high (H) pressures. N. Hemisphere.

Sample Application
What radius of curvature causes the gradient wind to equal the geostrophic wind?

Find the Answer
Given: $M_{tan} = G$ Find: $R = ?$ km

Use eq. (10.33), with $M_{tan} = G$: $G = G \pm G^2/(f_c \cdot R)$
This is a valid equality $G = G$ only when the last term in eq. (10.33) approaches zero; i.e., in the limit of $\underline{R = \infty}$.

Check: Eq. (10.33) still balances in this limit. **Exposition**: Infinite radius of curvature is a straight line, which (in the absence of any other forces such as turbulent drag) is the condition for geostrophic wind.

Sample Application

If the geostrophic wind around a low is 10 m s^{-1}, then what is the gradient wind speed, given $f_c = 10^{-4}$ s^{-1} and a radius of curvature of 500 km? Also, what is the curvature Rossby number?

Find the Answer

Given: $G = 10$ m s^{-1}, $R = 500$ km, $f_c = 10^{-4}$ s^{-1}
Find: M_{tan} = ? m s^{-1}, Ro_c = ? (dimensionless)

Use eq. (10.34a)

$$M_{tan} = 0.5\cdot(10^{-4}\text{s}^{-1})\cdot(500000\text{m})\cdot\left[-1+\sqrt{1+\frac{4\cdot(10m/s)}{(10^{-4}\text{s}^{-1})\cdot(500000\text{m})}}\right] = \underline{\mathbf{8.54\ m\ s^{-1}}}$$

Use eq. (10.35):

$$Ro_c = \frac{(10\text{m/s})}{(10^{-4}\text{s}^{-1})\cdot(5\times10^{5}\text{m})} = \underline{\mathbf{0.2}}$$

Check: Physics & units are reasonable.
Exposition: The small Rossby number indicates that the flow is in geostrophic balance. The gradient wind is indeed slower than geostrophic around this low.

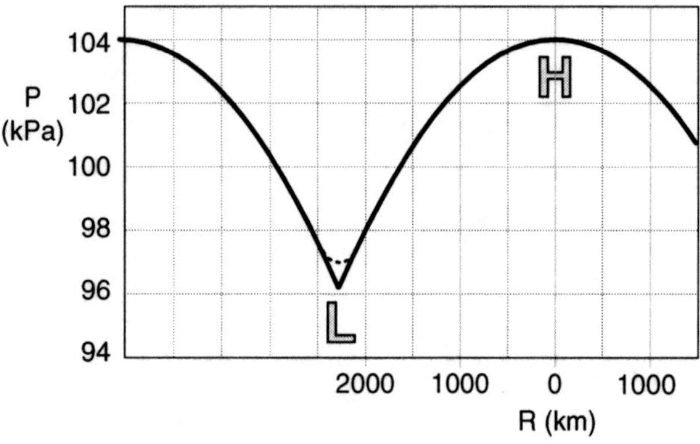

Figure 10.14
Illustration of how mean sea-level pressure P can vary with distance R from a high-pressure (H) center. The anticyclone (i.e., the high) has zero horizontal pressure gradient and calm winds in its center, with weak pressure gradient (ΔP/ΔR) and gentle winds in a broad region around it. The cyclone (i.e., the low) can have steep pressure gradients and associated strong winds close to the low center (L), with a pressure cusp right at the low center. In reality (dotted line), turbulent mixing near the low center smooths the cusp, allowing a small region of light winds at the low center surrounded by stronger winds. Although this graph was constructed using eq. (10.37b), it approximates the pressure variation along the cross section shown in the next figure.

Eq. (10.33) is a quadratic equation that has two solutions. One solution is for the gradient wind M_{tan} around a **cyclone** (i.e., a low):

$$M_{tan} = 0.5\cdot f_c\cdot R\cdot\left[-1+\sqrt{1+\frac{4\cdot G}{f_c\cdot R}}\right] \qquad \bullet(10.34a)$$

The other solution is for flow around an **anticyclone** (i.e., a high):

$$M_{tan} = 0.5\cdot f_c\cdot R\cdot\left[1-\sqrt{1-\frac{4\cdot G}{f_c\cdot R}}\right] \qquad \bullet(10.34b)$$

To simplify the notation in the equations above, let

$$Ro_c = \frac{G}{f_c\cdot R} \qquad (10.35)$$

where we can identify (Ro_c) as a "curvature" **Rossby number** because its length scale is the radius of curvature (R). When Ro_c is small, the winds are roughly geostrophic; namely, pressure gradient force nearly balances Coriolis force. [CAUTION: In later chapters you will learn about a Rossby radius of deformation, which is distinct from both Ro_c and R.]

For winds blowing around a low, the gradient wind is:

$$M_{tan} = \frac{G}{2\cdot Ro_c}\cdot\left[-1+\left(1+4\cdot Ro_c\right)^{1/2}\right] \qquad (10.36a)$$

and for winds around a high) the gradient wind is:

$$M_{tan} = \frac{G}{2\cdot Ro_c}\cdot\left[1-\left(1-4\cdot Ro_c\right)^{1/2}\right] \qquad (10.36b)$$

where G is the geostrophic wind.

While the differences between solutions (10.36a & b) appear subtle at first glance, these differences have a significant impact on the range of winds that are physically possible. Any value of Ro_c can yield physically reasonable winds around a low-pressure center (eq. 10.36a). But to maintain a positive argument inside the square root of eq. (10.36b), only values of $Ro_c \le 1/4$ are allowed for a high.

Thus, strong radial pressure gradients with small radii of curvature, and strong tangential winds can exist near low center. But only weak pressure gradients with large radii of curvature and light winds are possible near high-pressure centers (Figs. 10.14 and 10.15). To find the maximum allowable horizontal variations of height z or pressure P near anticyclones, use $Ro_c = 1/4$ in eq. (10.35) with G from (10.29c) or (10.28):

$$z = z_c - \left(f_c^{\,2}\cdot R^2\right)/\left(8\cdot|g|\right) \qquad \bullet(10.37a)$$

or

$$P = P_c - \left(\rho\cdot f_c^{\,2}\cdot R^2\right)/8 \qquad \bullet(10.37b)$$

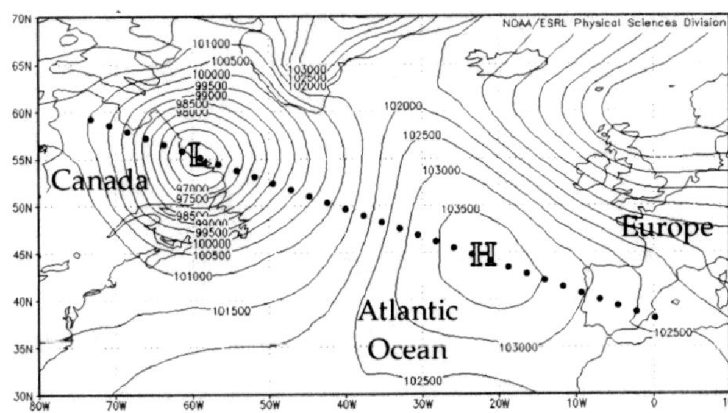

Figure 10.15
Illustration of strong pressure gradients (closely-spaced isobars) around the low-pressure center (L) over eastern Canada, and weak pressure gradients (isobars spaced further apart) around the high (H) over the NE Atlantic Ocean. NCEP reanalysis of daily-average mean sea-level pressure (Pa) for 5 Feb 2013. Pressures in the low & high centers were 96.11 & 104.05 kPa. Pressure variation along the dotted line is similar to that plotted in the previous figure. [Courtesy of the NOAA/NCEP Earth Systems Research Laboratory. http://www.esrl.noaa.gov/psd/data/gridded/data.ncep.reanalysis.html]

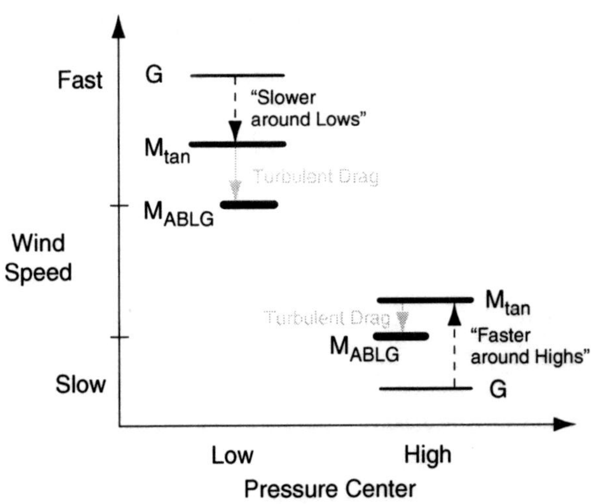

Figure 10.16
Relative magnitudes of different wind speeds around low- and high-pressure centers. G = geostrophic wind, M_{tan} = gradient wind speed, M_{ABLG} = atmospheric-boundary-layer gradient wind speed. G is smaller in highs than in lows, because it is not physically possible to have strong pressure gradients to drive strong steady-state winds at high centers.

where the center pressure in the high (anticyclone) is P_c, or for an isobaric surface the center height is z_c, the Coriolis parameter is f_c, $|g|$ is gravitational acceleration magnitude, ρ is air density, and the radius from the center of the high is R (see Fig. 10.14).

Figs. 10.14 and 10.15 show that pressure gradients, and thus the geostrophic wind, can be large near low centers. However, pressure gradients, and thus the geostrophic wind, must be small near high centers. This difference in geostrophic wind speed G between lows and highs is sketched in Fig. 10.16. The slowdown of gradient wind M_{tan} (relative to geostrophic) around lows, and the speedup of gradient wind (relative to geostrophic) around highs is also plotted in Fig. 10.16. The net result is that gradient winds, and even atmospheric boundary-layer gradient winds M_{ABLG} (described later in this chapter), are usually stronger (in an absolute sense) around lows than highs. For this reason, low-pressure centers are often windy.

Atmospheric-Boundary-Layer Wind

If you add turbulent drag to winds that would have been geostrophic, the result is a **subgeostrophic** (slower-than-geostrophic) wind that crosses the isobars at angle (α) (Fig. 10.17). This condition is found in the atmospheric boundary layer (ABL) where the isobars are straight. The force balance at steady state is:

$$0 = -\frac{1}{\rho}\cdot\frac{\Delta P}{\Delta x} + f_c\cdot V - w_T\cdot\frac{U}{z_i} \tag{10.38a}$$

$$0 = -\frac{1}{\rho}\cdot\frac{\Delta P}{\Delta y} - f_c\cdot U - w_T\cdot\frac{V}{z_i} \tag{10.38b}$$

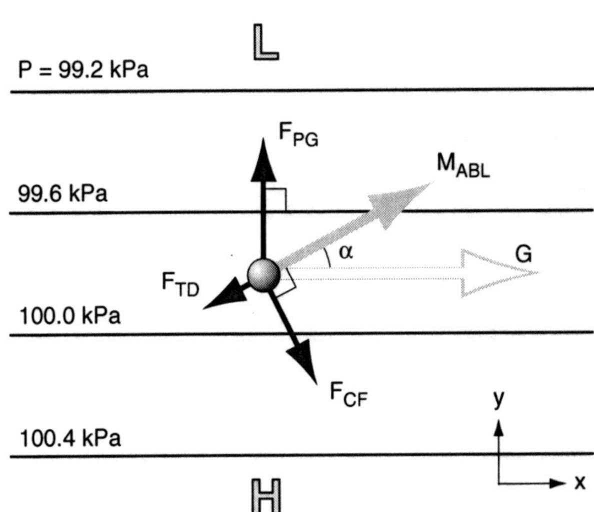

Figure 10.17
Balance of forces (black arrows) creating an atmospheric-boundary-layer wind (M_{ABL}, solid grey arrow) that is slower than geostrophic (G, hollow grey arrow). The grey sphere represents an air parcel. Thin black lines are isobars. L and H are low and high-pressure centers.

Sample Application

For statically neutral conditions, find the winds in the boundary layer given: z_i = 1.5 km, U_g = 15 m s^{-1}, V_g = 0, f_c = 10^{-4} s^{-1}, and C_D = 0.003. What is the cross-isobar wind angle?

Find the Answer

Given: z_i = 1.5 km, U_g = 15 m s^{-1}, V_g = 0, f_c = 10^{-4} s^{-1}, C_D = 0.003.

Find: V_{ABL} =? m s^{-1}, U_{ABL} =? m s^{-1}, M_{ABL} =? m s^{-1}, α = ? °

First: $G = (U_g^2 + V_g^2)^{1/2}$ = 15 m s^{-1}. Now apply eq.(10.41)

$$a = \frac{0.003}{(10^{-4}\text{s}^{-1})\cdot(1500\text{m})} = 0.02 \text{ s/m}$$

Check: $a \cdot G$ = (0.02 s m^{-1})·(15 m s^{-1}) =0.3 (is < 1. Good.)

U_{ABL}=[1–0.35·(0.02s m^{-1})·(15m s^{-1})]·(15m s^{-1})≈**13.4m s^{-1}**

V_{ABL}=[1–0.5·(0.02s m^{-1})·(15m s^{-1})]·(0.02s m^{-1})·(15m s^{-1})·(15m s^{-1}) ≈ **3.8 m s^{-1}**

$$M_{ABL} = \sqrt{U_{ABL}^2 + V_{ABL}^2} = \sqrt{13.4^2 + 3.8^2} = \underline{\mathbf{13.9 \text{ m s}^{-1}}}$$

Isobars are parallel to the geostrophic wind. Thus, the cross-isobar angle is:

$\alpha = \tan^{-1}(V_{ABL}/U_{ABL}) = \tan^{-1}(3.8/13.4)$ = **15.8°**.

Check: Physics & units are reasonable.

Exposition: Drag both slows the wind (13.4 m s^{-1}) in the boundary layer below its geostrophic value (15 m s^{-1}) and turns it at a small angle (15.8°) towards low pressure. Given N. Hem. (because of the positive Coriolis parameter), the ABL wind direction is 254.2°.

Sample Application

For statically unstable conditions, find winds in the ABL given V_g = 0, U_g = 5 m s^{-1}, w_B = 50 m s^{-1}, z_i = 1.5 km, b_D = 1.83x10^{-3}, and f_c = 10^{-4} s^{-1}. What is the cross-isobar wind angle?

Find the Answer

Given: (use convective boundary layer values above)

Find: M_{ABL} =? m s^{-1}, V_{ABL} =? m s^{-1}, U_{ABL} =? m s^{-1}, α =?°

Apply eqs. (10.42):

$$c_1 = \frac{(1.83\times10^{-3})\cdot(50\text{m/s})}{(10^{-4}\text{s}^{-1})\cdot(1500\text{m})} = 0.61 \text{ (dimensionless)}$$

$c_2 = 1/[1+(0.61)^2]$ = 0.729 (dimensionless)

U_{ABL} = 0.729·[(5m s^{-1}) – 0] = **3.6 m s^{-1}**

V_{ABL} = 0.729·[0 + (0.61)·(5m s^{-1})] = **2.2 m s^{-1}**

Use eq. (10.40):

$M_{ABL} = [U^2_{ABL} + V^2_{ABL}]^{1/2}$ = **4.2 m s^{-1}**

$\alpha = \tan^{-1}(V_{ABL}/U_{ABL}) = \tan^{-1}(2.2/3.6)$ = **31.4°**

Check: Physics & units are reasonable.

Exposition: Again, drag slows the wind and causes it to cross the isobars toward low pressure. The ABL wind direction is 238.6°.

Namely, the only forces acting for this special case are pressure gradient, Coriolis, and turbulent drag (Fig. 10.17).

Replace U with U_{ABL} and V with V_{ABL} to indicate these winds are in the ABL. Eqs. (10.38) can be rearranged to solve for the ABL winds, but this solution is **implicit** (depends on itself):

$$U_{ABL} = U_g - \frac{w_T \cdot V_{ABL}}{f_c \cdot z_i} \qquad (10.39a)$$

$$V_{ABL} = V_g + \frac{w_T \cdot U_{ABL}}{f_c \cdot z_i} \qquad (10.39b)$$

where (U_g, V_g) are geostrophic wind components, f_c is Coriolis parameter, z_i is ABL depth, and w_T is the turbulent transport velocity.

You can **iterate** to solve eqs. (10.39). Namely, first you guess a value for V_{ABL} to use in the right side of the first eq. Solve eq. (10.39a) for U_{ABL} and use it in the right side of eq. (10.39b), which you can solve for V_{ABL}. Plug this back into the right side of eq. (10.39a) and repeat this procedure until the solution **converges** (stops changing very much). The magnitude of the boundary-layer wind is:

$$M_{ABL} = [U^2_{ABL} + V^2_{ABL}]^{1/2} \qquad (10.40)$$

For a statically **neutral** ABL under windy conditions, then $w_T = C_D \cdot M_{ABL}$, where C_D is the drag coefficient (eq. 10.21). For most altitudes in the neutral ABL, an approximate but explicit solution is:

$$U_{ABL} \approx (1 - 0.35 \cdot a \cdot U_g) \cdot U_g - (1 - 0.5 \cdot a \cdot V_g) \cdot a \cdot V_g \cdot G \qquad \bullet(10.41a)$$

$$V_{ABL} \approx (1 - 0.5 \cdot a \cdot U_g) \cdot a \cdot G \cdot U_g + (1 - 0.35 \cdot a \cdot V_g) \cdot V_g \qquad \bullet(10.41b)$$

where the parameter is $a = C_D/(f_c \cdot z_i)$, G is the geostrophic wind speed and a solution is possible only if $a \cdot G < 1$. If this condition is not met, or if no reasonable solution can be found using eqs. (10.41), then use the iterative approach described in the next section, but with the centrifugal terms set to zero. Eqs. (10.41) do not apply to the **surface layer** (bottom 5 to 10% of the neutral boundary layer).

If the ABL is statically **unstable** (e.g., sunny with slow winds), use $w_T = b_D \cdot w_B$ (see eq. 10.22). Above the surface layer there is an exact solution that is explicit:

$$U_{ABL} = c_2 \cdot [U_g - c_1 \cdot V_g] \qquad \bullet(10.42a)$$

$$V_{ABL} = c_2 \cdot [V_g + c_1 \cdot U_g] \qquad \bullet(10.42b)$$

where $c_1 = \frac{b_D \cdot w_B}{f_c \cdot z_i}$, and $c_2 = \frac{1}{[1 + c_1^2]}$.

The factors in c_1 are given in the "Forces" section.

In summary, both wind-shear turbulence and convective turbulence cause drag. Drag makes the ABL wind slower than geostrophic (subgeostrophic), and causes the wind to cross isobars at angle α such that it has a component point to low pressure.

ABL Gradient (ABLG) Wind

For curved isobars in the atmospheric boundary layer (ABL), there is an imbalance of the following forces: Coriolis, pressure-gradient, and drag. This imbalance is a centripetal force that makes ABL air spiral outward from highs and inward toward lows (Fig. 10.18). An example was shown in Fig. 10.1.

If we devise a centrifugal force equal in magnitude but opposite in direction to the centripetal force, then the equations of motion can be written for spiraling flow that is steady over any point on the Earth's surface (i.e., NOT following the parcel):

$$0 = -\frac{1}{\rho}\cdot\frac{\Delta P}{\Delta x} + f_c\cdot V - w_T\cdot\frac{U}{z_i} + s\cdot\frac{V\cdot M}{R} \qquad (10.43a)$$

$$0 = -\frac{1}{\rho}\cdot\frac{\Delta P}{\Delta y} - f_c\cdot U - w_T\cdot\frac{V}{z_i} - s\cdot\frac{U\cdot M}{R} \qquad (10.43b)$$

pressure gradient — *Coriolis* — *turbulent drag* — *centrifugal*

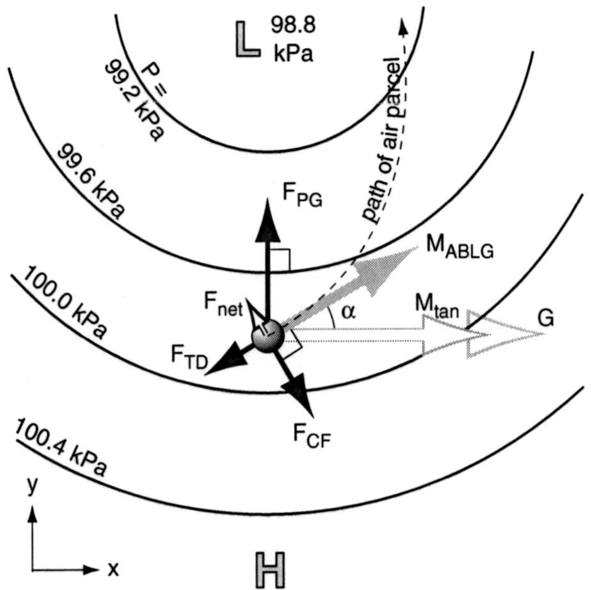

Figure 10.18
Imbalance of forces (black arrows) yield a net centripetal force (F_{net}) that causes the atmospheric-boundary-layer gradient wind (M_{ABLG}, solid grey arrow) to be slower than both the gradient wind (M_{tan}) and geostrophic wind (G). The resulting air-parcel path crosses the isobars at a small angle α toward low pressure.

We can anticipate that the ABLG winds should be slower than the corresponding gradient winds, and should cross isobars toward lower pressure at some small angle α (see Fig. 10.19).

Lows are often overcast and windy, implying that the atmospheric boundary layer is statically neutral. For this situation, the transport velocity is given by:

$$w_T = C_D\cdot M = C_D\cdot\sqrt{U^2+V^2} \qquad (10.21 \text{ again})$$

Because this parameterization is nonlinear, it increases the nonlinearity (and the difficulty to solve), eqs. (10.43).

Highs often have mostly clear skies with light winds, implying that the atmospheric boundary layer is statically unstable during sunny days, and statically stable at night. For daytime, the transport velocity is given by:

$$w_T = b_D\cdot w_B \qquad (10.22 \text{ again})$$

This parameterization for w_B is simple, and does not depend on wind speed. For statically stable conditions during fair-weather nighttime, steady state is unlikely, meaning that eqs. (10.43) do not apply.

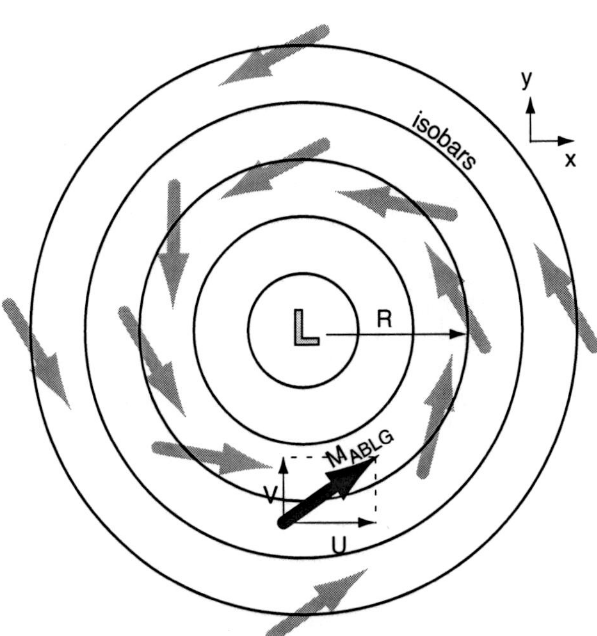

Figure 10.19
Tangential ABLG wind component (U) and radial ABLG wind component (V) for the one vector highlighted as the thick black arrow. N. Hemisphere.

Sample Application

If G = 10 m s^{-1} at R = 400 km from the center of a N. Hem. cyclone, C_D = 0.02, z_i = 1 km, and f_c = 10^{-4} s^{-1}, then find the ABLG wind speed and components.

Find the Answer

Given: (see the data above)
Find: M_{BLG} = ? m s^{-1}, U_{BLG} = ? m s^{-1}, V_{BLG} = ? m s^{-1},

Use a spreadsheet to iterate (as discussed in the INFO box) eqs. (10.44) & (1.1) with a time step of Δt = 1200 s. Use $U = V = 0$ as a first guess.

G (m/s)=	10		z_i (km)=	1	
R (km)=	400		f_c (s^{-1})=	0.0001	
C_D =	0.02		Δt(s)=	1200	

Iteration Counter	U_{ABLG} (m s^{-1})	V_{ABLG} (m s^{-1})	M_{ABLG} (m s^{-1})	ΔU_{ABLG} (m s^{-1})	ΔV_{ABLG} (m s^{-1})
0	0.00	0.00	0.00	0.000	1.200
1	0.00	1.20	1.20	0.148	1.165
2	0.15	2.37	2.37	0.292	1.047
3	0.44	3.41	3.44	0.408	0.861
4	0.85	4.27	4.36	0.480	0.640
5	1.33	4.91	5.09	0.502	0.420
...					
29	4.16	4.33	6.01	-0.003	0.001
30	4.16	4.33	6.01	-0.002	0.001

The evolution of the iterative solution is plotted at right as it approaches the final answer of U_{ABLG} = **4.16 m s^{-1}**, V_{ABLG} = **4.33 m s^{-1}**, M_{ABLG}= **6.01 m s^{-1}**, where (U_{ABLG} , V_{ABLG}) are (tangential, radial) parts.

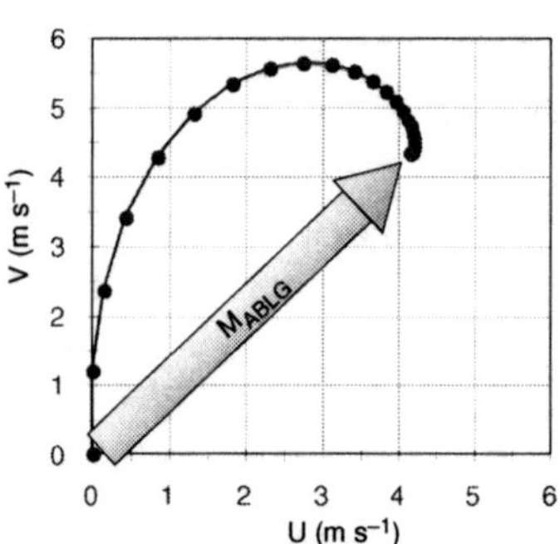

Check: Physics & units are reasonable. You should do the following "what if" experiments on the spreadsheet to check the validity. I ran experiments using a modified spreadsheet that relaxed the results using a weighted average of new and previous winds.

(a) As R approaches infinity and C_D approaches zero, then M_{ABLG} should approach the geostrophic wind G. I got $U_{ABLG} = G$ = 10 m s^{-1}, V_{ABLG} = 0.

(b) For finite R and 0 drag, then M_{ABLG} should equal the gradient wind M_{tan}. I got U_{ABLG} = 8.28 m s^{-1}, V_{ABLG} = 0.

(c) For finite drag but infinite R, then M_{ABLG} should equal the atmospheric boundary layer wind M_{ABL}. I got U_{ABLG} = 3.91 m s^{-1}, V_{ABLG} = 4.87 m s^{-1}. Because this ABL solution is based on the full equations, it gives a better answer than eqs. (10.41).

Exposition: If you take slightly larger time steps, the solution converges faster. But if Δt is too large, the iteration method fails (i.e., blows up).

Nonlinear coupled equations (10.43) are difficult to solve analytically. However, we can rewrite the equations in a way that allows us to iterate numerically toward the answer (see the INFO box below for instructions). The trick is to not assume steady state. Namely, put the tendency terms ($\Delta U/\Delta t$, $\Delta V/\Delta t$) back in the left hand sides (LHS) of eqs. (10.43). But recall that $\Delta U/\Delta t = [U(t+\Delta t) - U]/\Delta t$, and similar for V.

For this iterative approach, first re-frame eqs. (10.43) in cylindrical coordinates, where (U, V) are the (tangential, radial) components, respectively (see Fig. 10.19). Also, use G, the geostrophic wind definition of eq. (10.28), to quantify the pressure gradient.

For a cyclone in the Northern Hemisphere (for which s = +1 from Table 10-2), the atmospheric boundary layer gradient wind eqs. (10.43) become:

$$M = (U^2 + V^2)^{1/2} \qquad \text{(1.1 again)}$$

(10.44a)

$$U(t+\Delta t) = U + \Delta t \cdot \left[f_c \cdot V - \frac{C_D \cdot M \cdot U}{z_i} + s \frac{V \cdot M}{R} \right]$$

(10.44b)

$$V(t+\Delta t) = V + \Delta t \cdot \left[f_c \cdot (G - U) - \frac{C_D \cdot M \cdot V}{z_i} - s \frac{U \cdot M}{R} \right]$$

where (U, V) represent (tangential, radial) parts for the wind vector south of the low center. These coupled equations are valid both night and day.

INFO • Find the Answer by Iteration

Equations (10.44) are difficult to solve analytically, but you can iterate as an alternative way to solve for the ABLG wind components. Here is the procedure.

(1) Make an initial guess for (U, V), such as (0, 0).
(2) Use these (U, V) values in the right sides of eqs. (10.44) and (1.1), and solve for the new values of [$U(t+\Delta t)$, $V(t+\Delta t)$].
(3) In preparation for the next iteration, let $U = U(t+\Delta t)$, and $V = V(t+\Delta t)$.
(4) Repeat steps 2 and 3 using the new (U, V) on the right hand sides .
(5) Keep iterating. Eventually, the [$U(t+\Delta t)$, $V(t+\Delta t)$] values stop changing (i.e., reach steady state), giving $U_{ABLG} = U(t+\Delta t)$, and $V_{ABLG} = V(t+\Delta t)$.

Because of the repeated, tedious calculations, I recommend you use a spreadsheet (see the Sample Application) or write your own computer program.

As you can see from the Sample Application, the solution spirals toward the final answer as a **damped inertial oscillation**.

For daytime fair weather conditions in anticyclones, you could derive alternatives to eqs. (10.44) that use convective parameterizations for atmospheric boundary layer drag.

Because eqs. (10.44) include the tendency terms, you can also use them for non-steady-state (time varying) flow. One such case is nighttime during fair weather (anticyclonic) conditions. Near sunset, when vigorous convective turbulence dies, the drag coefficient suddenly decreases, allowing the wind to accelerate toward its geostrophic equilibrium value. However, Coriolis force causes the winds to turn away from that steady-state value, and forces the winds into an **inertial oscillation**. See a previous INFO box titled *Approach to Geostrophy* for an example of undamped inertial oscillations.

During a portion of this oscillation the winds can become faster than geostrophic (**supergeostrophic**), leading to a low-altitude phenomenon called the **nocturnal jet**. See the Atmospheric Boundary Layer chapter for details.

Cyclostrophic Wind

Winds in tornadoes are about 100 m s^{-1}, and in waterspouts are about 50 m s^{-1}. As a tornado first forms and tangential winds increase, centrifugal force increases much more rapidly than Coriolis force. Centrifugal force quickly becomes the dominant force that balances pressure-gradient force (Fig. 10.20). Thus, a steady-state rotating wind is reached at much slower speeds than the gradient wind speed.

If the tangential velocity around the vortex is steady, then the steady-state force balance is:

$$0 = -\frac{1}{\rho}\cdot\frac{\Delta P}{\Delta x} + s\cdot\frac{V\cdot M}{R} \tag{10.45a}$$

$$0 = -\underbrace{\frac{1}{\rho}\cdot\frac{\Delta P}{\Delta y}}_{\text{pressure gradient}} - s\cdot\underbrace{\frac{U\cdot M}{R}}_{\text{centrifugal}} \tag{10.45b}$$

You can use cylindrical coordinates to simplify solution for the cyclostrophic (tangential) winds M_{cs} around the vortex. The result is:

$$M_{cs} = \sqrt{\frac{R}{\rho}\cdot\frac{\Delta P}{\Delta R}} \tag{10.46}$$

where the velocity M_{cs} is at distance R from the vortex center, and the radial pressure gradient in the vortex is $\Delta P/\Delta R$.

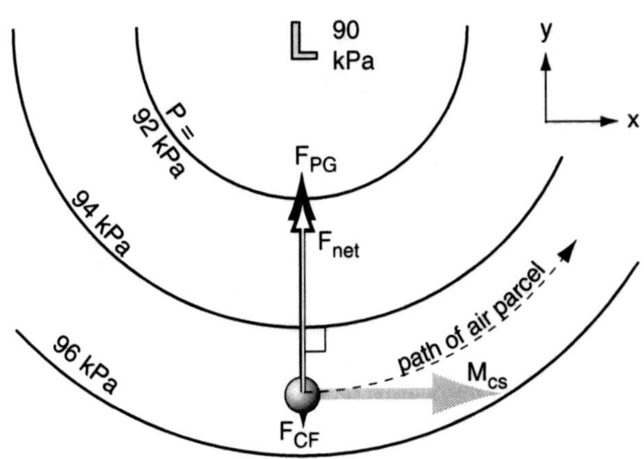

Figure 10.20
Around tornadoes, pressure gradient force F_{PG} is so strong that it greatly exceeds all other forces such as Coriolis force F_{CF}. The net force (F_{net}) pulls the air around the tight circle at the cyclostrophic wind speed (M_{cs}).

Sample Application
A 10 m radius waterspout has a tangential velocity of 45 m s^{-1}. What is the radial pressure gradient?

Find the Answer
Given: M_{cs} = 45 m s^{-1}, R = 10 m.
Find: $\Delta P/\Delta R$ = ? kPa m^{-1}.

Assume cyclostrophic wind, and ρ = 1 kg m^{-3}.
Rearrange eq. (10.46):

$$\frac{\Delta P}{\Delta R} = \frac{\rho}{R}\cdot M_{cs}^2 = \frac{(1\text{kg/m}^3)\cdot(45\text{m/s})^2}{10\text{m}}$$

$\Delta P/\Delta R$ = 202.5 kg·m^{-1}·s^{-2} / m = **0.2 kPa m^{-1}**.

Check: Physics & units are reasonable.
Exposition: This is 2 kPa across the 10 m waterspout radius, which is 1000 times greater than typical synoptic-scale pressure gradients on weather maps.

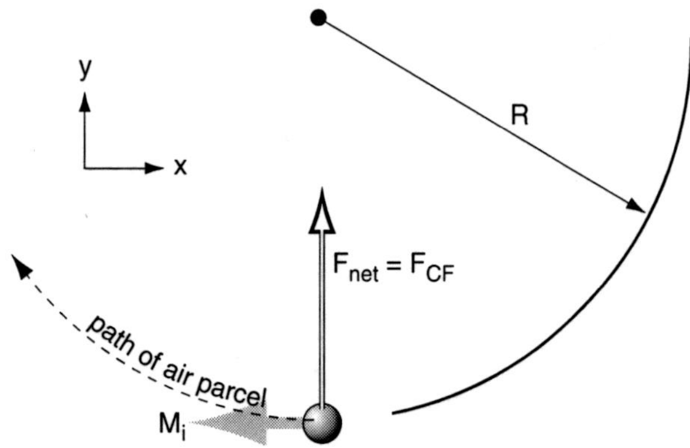

Figure 10.21
Coriolis force (F_{CF}, thick black arrow, behind the white arrow) on an air parcel (grey ball), creating an anticyclonic inertial wind M_i, grey arrow). R is radius of curvature. White arrow is net force F_{net}.

Sample Application
For an inertial ocean current of 5 m s^{-1}, find the radius of curvature and time period to complete one circuit. Assume a latitude where $f_c = 10^{-4}$ s^{-1}.

Find the Answer
Given: $M_i = 5$ m s^{-1}, $f_c = 10^{-4}$ s^{-1}.
Find: R = ? km, *Period* = ? h

Use eq. (10.48): $R = -(5 \text{ m s}^{-1}) / (10^{-4} \text{ s}^{-1})$ = **-50 km**
Use *Period* = $2\pi/f_c$ = 62832 s = **17.45 h**

Check: Units & magnitudes are reasonable.
Exposition: The tracks of drifting buoys in the ocean are often **cycloidal**, which is the superposition of a circular inertial oscillation and a mean current that gradually translates (moves) the whole circle.

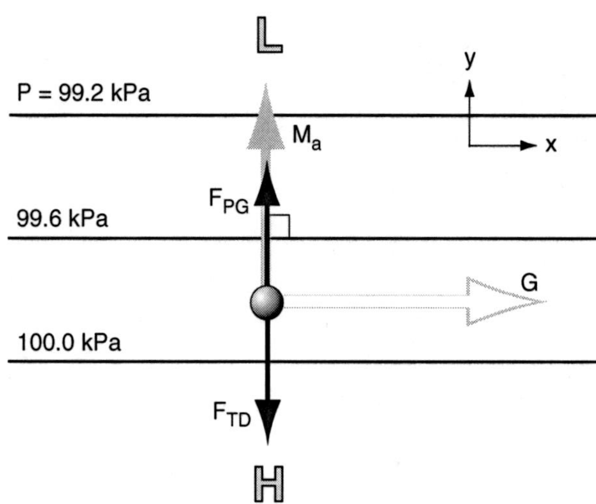

Figure 10.22
Balance of forces (F, black arrows) that create the antitriptic wind M_a (grey arrow). G is the theoretical geostrophic wind. F_{TD} is turbulent drag, and F_{PG} is pressure-gradient force.

Recall from the Gradient Wind section that anticyclones cannot have strong pressure gradients, hence winds around highs are too slow to be cyclostrophic. Around cyclones (lows), cyclostrophic winds can turn either counterclockwise or clockwise in either hemisphere, because Coriolis force is not a factor.

Inertial Wind

Steady-state **inertial motion** results from a balance of Coriolis and centrifugal forces in the absence of any pressure gradient:

$$0 = f_c \cdot M_i + \frac{M_i^2}{R} \tag{10.47}$$

where M_i is inertial wind speed, f_c is the Coriolis parameter, and R is the radius of curvature. Since both of these forces depend on wind speed, the inertial wind cannot start itself from zero. It can occur only after some other force first causes the wind to blow, and then that other force disappears.

The inertial wind coasts around a circular path of radius R,

$$R = -\frac{M_i}{f_c} \tag{10.48}$$

where the negative sign implies anticyclonic rotation (Fig. 10.21). The time period needed for this **inertial oscillation** to complete one circuit is *Period* = $2\pi/f_c$, which is half of a **pendulum day** (see *Approach to Geostrophy* INFO Box earlier in this chapter).

Although rarely observed in the atmosphere, inertial oscillations are frequently observed in the ocean. This can occur where wind stress on the ocean surface creates an ocean current, and then after the wind dies the current coasts in an inertial oscillation.

Antitriptic Wind

A steady-state antitriptic wind M_a could result from a balance of pressure-gradient force and turbulent drag:

$$0 = -\frac{1}{\rho} \cdot \frac{\Delta P}{\Delta d} - w_T \cdot \frac{M_a}{z_i} \tag{10.49}$$

where ΔP is the pressure change across a distance Δd perpendicular to the isobars, w_T is the turbulent transport velocity, and z_i is the atmospheric boundary-layer depth.

This theoretical wind blows perpendicular to the isobars (Fig. 10.22), directly from high to low pressure:

$$M_a = \frac{z_i \cdot f_c \cdot G}{w_T} \tag{10.50}$$

For free-convective boundary layers, $w_T = b_D \cdot w_B$ is not a function of wind speed, so M_a is proportional to G. However, for windy forced-convection boundary layers, $w_T = C_D \cdot M_a$, so solving for M_a shows it to be proportional to the square root of G.

This wind would be found in the atmospheric boundary layer, and would occur as an along-valley component of "long gap" winds (see the Regional Winds chapter). It is also sometimes thought to be relevant for thunderstorm cold-air outflow and for steady sea breezes. However, in most other situations, Coriolis force should not be neglected; thus, the atmospheric boundary-layer wind and BL Gradient winds are much better representations of nature than the antitriptic wind.

Summary of Horizontal Winds

Table 10-5 summarizes the idealized horizontal winds that were discussed earlier in this chapter.

On real weather maps such as Fig. 10.23, isobars or height contours have complex shapes. In some regions the height contours are straight (suggesting that actual winds should nearly equal geostrophic or boundary-layer winds), while in other regions the height contours are curved (suggesting gradient or boundary-layer gradient winds). Also, as air parcels move between straight and curved regions, they are sometimes not quite in equilibrium. Nonetheless, when studying weather maps you can quickly estimate the winds using the summary table.

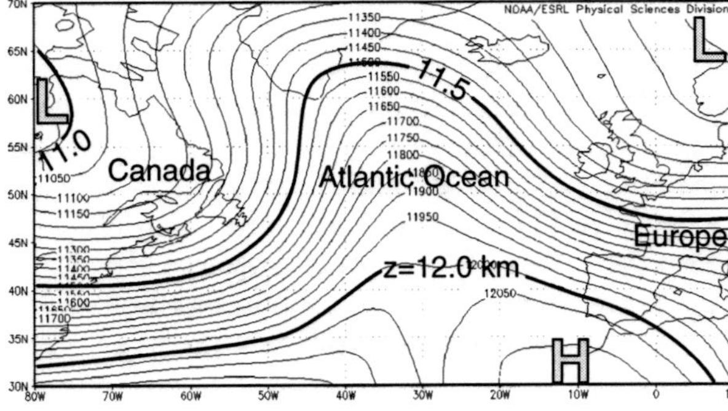

Figure 10.23
One-day average geopotential heights z (thick lines in km, thin lines in m) on the 20 kPa isobaric surface for 5 Feb 2013. Close spacing (tight packing) of the height contours indicate faster winds. This upper-level chart is for the same day and location (Atlantic Ocean) as the mean-sea-level pressure chart in Fig. 10.15. [Courtesy of NOAA/NCEP Earth System Research Lab. http://www.esrl.noaa.gov/psd/data/gridded/data.ncep.reanalysis.html]

Sample Application

In a 1 km thick convective boundary layer at a location where $f_c = 10^{-4}$ s^{-1}, the geostrophic wind is 5 m s^{-1}. The turbulent transport velocity is 0.02 m s^{-1}. Find the antitriptic wind speed.

Find the Answer

Given: $G = 5$ m s^{-1}, $z_i = 1000$ m, $f_c = 10^{-4}$ s^{-1}, $w_T = 0.02$ m s^{-1}

Find: $M_a = ?$ m s^{-1}

Use eq. (10.50):

$M_a = (1000\text{m})\cdot(10^{-4}\text{ s}^{-1})\cdot(5\text{m s}^{-1}) / (0.02\text{ m s}^{-1})$

$= \underline{25 \text{ m s}^{-1}}$

Check: Magnitude is too large. Units reasonable.

Exposition: Eq. (10.50) can give winds of $M_a > G$ for many convective conditions, for which case Coriolis force would be expected to be large enough that it should not be neglected. Thus, antitriptic winds are unphysical. However, for forced-convective boundary layers where drag is proportional to wind speed squared, reasonable solutions are possible.

INFO • The Rossby Number

The **Rossby number** (Ro) is a dimensionless ratio defined by

$$Ro = \frac{M}{f_c \cdot L} \quad \text{or} \quad Ro = \frac{M}{f_c \cdot R}$$

where M is wind speed, f_c is the Coriolis parameter, L is a characteristic length scale, and R is radius of curvature.

In the equations of motion, suppose that advection terms such as $U \cdot \Delta U / \Delta x$ are order of magnitude M^2/L, and Coriolis terms are of order $f_c \cdot M$. Then the Rossby number is like the ratio of advection to Coriolis terms: $(M^2/L) / (f_c \cdot M) = M/(f_c \cdot L) = Ro$. Or, we could consider the Rossby number as the ratio of centrifugal (order of M^2/R) to Coriolis terms, yielding $M/(f_c \cdot R) = Ro$.

Use the Rossby number as follows. If $Ro < 1$, then Coriolis force is a dominant force, and the flow tends to become geostrophic (or gradient, for curved flow). If $Ro > 1$, then the flow tends not to be geostrophic.

For example, a midlatitude cyclone (low-pressure system) has approximately $M = 10$ m s^{-1}, $f_c = 10^{-4}$ s^{-1}, and $R = 1000$ km, which gives $Ro = 0.1$. Hence, midlatitude cyclones tend to adjust toward geostrophic balance, because $Ro < 1$. In contrast, a tornado has roughly $M = 50$ m s^{-1}, $f_c = 10^{-4}$ s^{-1}, and $R = 50$ m, which gives $Ro = 10{,}000$, which is so much greater than one that geostrophic balance is not relevant.

Table 10-5. Summary of horizontal winds**.

Item	Name of Wind	Forces	Direction	Magnitude	Where Observed
1	**geostrophic**	pressure-gradient, Coriolis	parallel to straight isobars with Low pressure to the wind's left*	faster where isobars are closer together. $G=\left\lvert\frac{g}{f_c}\cdot\frac{\Delta z}{\Delta d}\right\rvert$	aloft in regions where isobars are nearly straight
2	**gradient**	pressure-gradient, Coriolis, centrifugal	similar to geostrophic wind, but following curved isobars. Clockwise* around Highs, counterclockwise* around Lows.	slower than geostrophic around Lows, faster than geostrophic around Highs	aloft in regions where isobars are curved
3	**atmospheric boundary layer**	pressure-gradient, Coriolis, drag	similar to geostrophic wind, but crosses isobars at small angle toward Low pressure	slower than geostrophic (i.e., **subgeostrophic**)	near the ground in regions where isobars are nearly straight
4	**atmospheric boundary-layer gradient**	pressure-gradient, Coriolis, drag, centrifugal	similar to gradient wind, but crosses isobars at small angle toward Low pressure	slower than gradient wind speed	near the ground in regions where isobars are curved
5	**cyclostrophic**	pressure-gradient, centrifugal	either clockwise or counterclockwise around strong vortices of small diameter	stronger for lower pressure in the vortex center	tornadoes, waterspouts (& sometimes in the eye-wall of hurricanes)
6	**inertial**	Coriolis, centrifugal	anticyclonic circular rotation	coasts at constant speed equal to its initial speed	ocean-surface currents

* *For Northern Hemisphere. Direction is opposite in Southern Hemisphere.* ** *Antitriptic winds are unphysical; not listed here.*

HORIZONTAL MOTION

Equations of Motion — Again

The geostrophic wind can be used as a surrogate for the pressure-gradient force, based on the definitions in eqs. (10.26). Thus, the **equations of horizontal motion** (10.23) become:

$$\frac{\Delta U}{\Delta t}=-U\frac{\Delta U}{\Delta x}-V\frac{\Delta U}{\Delta y}-W\frac{\Delta U}{\Delta z}+f_c\cdot\left(V-V_g\right)-w_T\cdot\frac{U}{z_i} \quad •(10.51a)$$

$$\underbrace{\frac{\Delta V}{\Delta t}}_{tendency}=\underbrace{-U\frac{\Delta V}{\Delta x}-V\frac{\Delta V}{\Delta y}-W\frac{\Delta V}{\Delta z}}_{advection}\underbrace{-f_c\cdot\left(U}_{Coriolis}\underbrace{-U_g\right)}_{pressure\ gradient}\underbrace{-w_T\cdot\frac{V}{z_i}}_{turbulent\ drag} \quad •(10.51b)$$

For winds turning around a circle, you can add a term for centrifugal force, which is an artifice to account for the continual changing of wind direction caused by an imbalance of the other forces (where the imbalance is the centripetal force).

The difference between the actual and geostrophic winds is the **ageostrophic wind** (U_{ag}, V_{ag}). The term in eqs. (10.51) containing these differences indicates the **geostrophic departure**.

$$U_{ag}=U-U_g \quad •(10.52a)$$

$$V_{ag}=V-V_g \quad •(10.52b)$$

Scales of Horizontal Motion

A wide range of horizontal scales of motion (Table 10-6) are superimposed in the atmosphere: from large global-scale circulations through extra-tropical cyclones, thunderstorms, and down to swirls of turbulence.

The troposphere is roughly 10 km thick, and this constrains the vertical scale of most weather phenomena. Thus, phenomena of large horizontal scale will have a constrained vertical scale, causing them to be similar to a pancake. However, phenomena with smaller horizontal scale can have aspect ratios (width/height) of about one; namely, their characteristics are **isotropic**.

Larger-scale meteorological phenomena tend to exist for longer durations than smaller-scale ones. Fig. 10.24 shows that time scales τ and horizontal length scales λ of many meteorological phenomena nearly follow a straight line on a log-log plot. This implies that

$$\tau/\tau_o = (\lambda/\lambda_o)^b \qquad (10.53)$$

where $\tau_o \approx 10^{-3}$ h, $\lambda_o \approx 10^{-3}$ km, and $b \approx 7/8$.

In the next several chapters, we cover weather phenomena from largest to smallest horiz. scales:

- Chapter 11 General Circulation (planetary)
- Chapter 12 Fronts & Airmasses (synoptic)
- Chapter 13 Extratropical Cyclones (synoptic)
- Chapter 14 Thunderstorm Fundam. (meso β)
- Chapter 15 Thunderstorm Hazards (meso γ)
- Chapter 16 Tropical Cyclones (meso α & β)
- Chapter 17 Regional Winds (meso β & γ)
- Chapter 18 Atm. Boundary Layers (microscale)

Although hurricanes are larger than thunderstorms, we cover thunderstorms first because they are the building blocks of hurricanes. Similarly, midlatitude cyclones often contain fronts, so fronts are covered before extratropical cyclones.

Table 10-6. Horizontal scales of motion in the troposphere.

Horizontal Size	Scale Designation	Name	
40,000 km	macro α	planetary scale	
4,000 km	macro β	synoptic scale*	
700 km			
300 km	meso α	mesoscale**	
30 km	meso β		
3 km	meso γ		
300 m	micro α	microscale***	boundary-layer turbulence
30 m	micro β		surface-layer turbulence
3 m	micro γ		inertial subrange turbulence
300 mm	micro δ		
30 mm			fine-scale turbulence
3 mm	viscous	dissipation subrange	
0.3 μm		mean-free path between molec.	
0.003 μm	molecular	molecule sizes	
0			

Note: Disagreement among different organizations.
*Synoptic: AMS: 400 - 4000 km; WMO: 1000 - 2500 km.
**Mesoscale: AMS: 3 - 400 km; WMO: 3 - 50 km.
***Microscale: AMS: 0 - 2 km; WMO: 3 cm - 3 km.
where AMS = American Meteorological Society, and WMO = World Meteorological Organization.

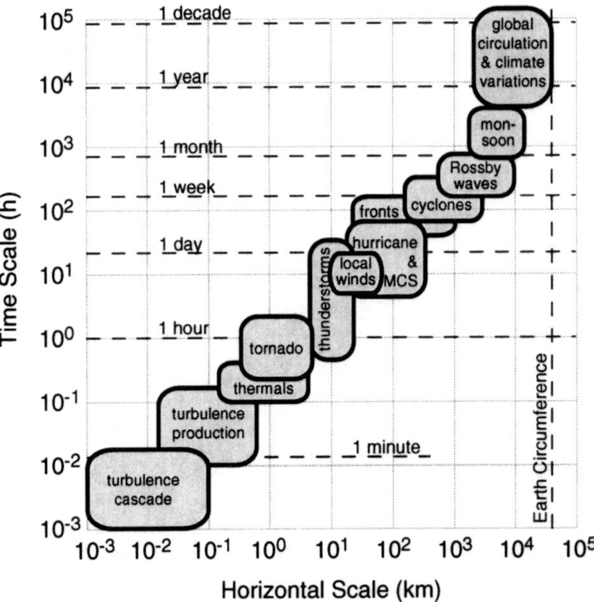

Figure 10.24
Typical time and spatial scales of meteorological phenomena.

VERTICAL FORCES AND MOTION

Forces acting in the vertical can cause or change vertical velocities, according to Newton's Second Law. In an Eulerian framework, the **vertical component of the equations of motion** is:

$$\underbrace{\frac{\Delta W}{\Delta t}}_{tendency} = \underbrace{-U\frac{\Delta W}{\Delta x} - V\frac{\Delta W}{\Delta y} - W\frac{\Delta W}{\Delta z}}_{advection} \underbrace{- \frac{1}{\rho}\frac{\Delta P}{\Delta z}}_{pressure\ gradient} \underbrace{- |g|}_{gravity} \underbrace{- \frac{F_{z\,TD}}{m}}_{turb.\ drag} \qquad (10.54)$$

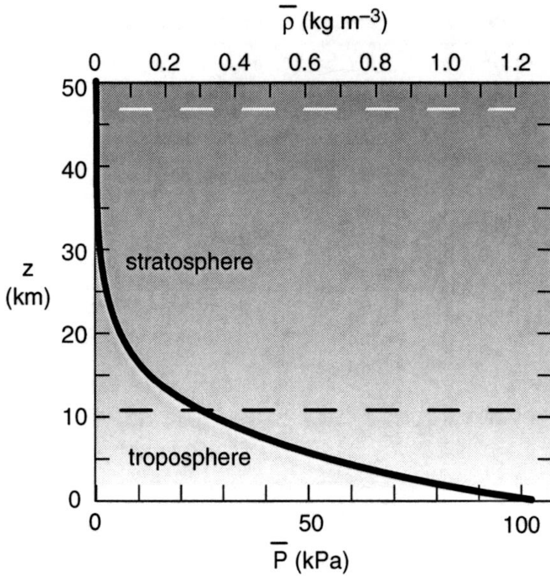

Figure 10.25
Background state, showing change of mean atmospheric pressure $\overline{P}$ and mean density $\overline{\rho}$ with height z, based on a standard atmosphere from Chapter 1.

Sample Application

Suppose your neighborhood has a background environmental temperature of 20°C, but at your particular location the temperature is 26°C with a 4 m s^{-1} west wind and no vertical velocity. Just 3 km west is an 5 m s^{-1} updraft. Find the vertical acceleration.

Find the Answer

Given: T_e = 273+20 = 293 K, $\Delta\theta$ = 6°C, U= 4 m s^{-1}
$\Delta W/\Delta x$= (5 m s^{-1} – 0) / (–3,000m – 0)

Find: $\Delta W/\Delta t$ = ? m·s^{-2}

Assume: Because W = 0, there is zero drag. Because the air is dry: $T_v = T$. Given no V info, assume zero.

Apply eq. (10.59): $\frac{\Delta W}{\Delta t} = -U\frac{\Delta W}{\Delta x} + \frac{\theta_p - \theta_e}{\overline{T}_e}\cdot|g|$

$\Delta W/\Delta t$ = –(4 m s^{-1})·(–5m·s^{-1}/3,000m)+(6/293)·(9.8m·s^{-2})
= 0.0067 + 0.20 = **0.21 m·s^{-2}**

Check: Physics & units are reasonable.
Exposition: Buoyancy dominated over advection for this example. Although drag was zero initially because of zero initial vertical velocity, we must include the drag term once the updraft forms.

where the vertical acceleration given in the left side of the equation is determined by the sum of all forces/mass acting in the vertical, as given on the right. For Cartesian directions (x, y, z) the velocity components are (U, V, W). Also in this equation are air density (ρ), pressure (P), vertical turbulent-drag force ($F_{z\ TD}$), mass (m), and time (t). Magnitude of gravitational acceleration is $|g|$ = 9.8 m·s^{-2}. Coriolis force is negligible in the vertical (see the INFO box on *Coriolis Force in 3-D*, earlier in this chapter), and is not included in the equation above.

Recall from Chapter 1 that our atmosphere has an extremely large pressure gradient in the vertical, which is almost completely balanced by gravity (Fig. 10.25). Also, there is a large density gradient in the vertical. We can define these large terms as a **mean background state** or a **reference state** of the atmosphere. Use the overbar over variables to indicate their average background state. Define this background state such that it is exactly in **hydrostatic balance** (see Chapter 1):

$$\frac{\Delta \overline{P}}{\Delta z} = -\overline{\rho}\cdot|g| \qquad (10.55)$$

However, small deviations in density and pressure from the background state can drive important **non-hydrostatic** vertical motions, such as in thermals and thunderstorms. To discern these effects, we must first remove the background state from the full vertical equation of motion. From eq. (10.54), the gravity and pressure-gradient terms are:

$$\frac{1}{\rho}\left[-\frac{\Delta P}{\Delta z} - \rho|g|\right] \qquad (10.56)$$

But total density ρ can be divided into background ($\overline{\rho}$) and deviation (ρ') components: $\rho = \overline{\rho} + \rho'$. Do the same for pressure: $P = \overline{P} + P'$. Thus, eq. (10.56) can be expanded as:

$$\frac{1}{(\overline{\rho}+\rho')}\left[-\frac{\Delta \overline{P}}{\Delta z} - \frac{\Delta P'}{\Delta z} - \overline{\rho}|g| - \rho'|g|\right] \qquad (10.57)$$

The first and third terms in square brackets in eq. (10.57) cancel out, due to hydrostatic balance (eq. 10.55) of the background state.

In the atmosphere, density perturbations (ρ') are usually much smaller than mean density. Thus density perturbations can be neglected everywhere except in the gravity term, where $\rho'|g|/(\overline{\rho}+\rho') \approx (\rho'/\overline{\rho})\cdot|g|$. This is called the **Boussinesq approximation**.

Recall from the chapters 1 and 5 that you can use virtual temperature (T_v) with the ideal gas law in place of air density (but changing the sign because low virtual temperatures imply high densities):

$$-\frac{\rho'}{\bar{\rho}}\cdot|g| = \frac{\theta_v'}{\bar{T}_v}\cdot|g| = \frac{\theta_{v\,p}-\theta_{v\,e}}{\bar{T}_{v\,e}}\cdot|g| = g' \quad (10.58)$$

where subscripts p & e indicate the air parcel and the environment surrounding the parcel, and where g' is called the **reduced gravity**. The virtual potential temperature θ_v can be in either Celsius or Kelvin, but units of Kelvin must be used for T_v and T_{ve}.

Combining eqs. (10.54) & (10.58) yields:

$$\frac{\Delta W}{\Delta t} = -U\frac{\Delta W}{\Delta x} - V\frac{\Delta W}{\Delta y} - W\frac{\Delta W}{\Delta z} \quad \bullet(10.59)$$

$$-\frac{1}{\bar{\rho}}\frac{\Delta P'}{\Delta z} + \frac{\theta_{vp}-\theta_{ve}}{\bar{T}_{ve}}\cdot|g| - \frac{F_{z\,TD}}{m}$$

Terms from this equation will be used in the Regional Winds chapter and in the Thunderstorm chapters to explain strong vertical velocities.

When an air parcel rises or sinks it experiences resistance (turbulent drag, $F_{z\,TD}$) per unit mass m as it tries to move through the surrounding air. This is a completely different effect than air drag against the Earth's surface, and is not described by the same drag equations. The nature of $F_{z\,TD}$ is considered in the chapter on Air Pollution Dispersion, as it affects the rise of smoke-stack plumes. $F_{z\,TD} = 0$ if the air parcel and environment move at the same speed.

CONSERVATION OF AIR MASS

Due to random jostling, air molecules tend to distribute themselves uniformly within any volume. Namely, the air tends to maintain its **continuity**. Any additional air molecules entering the volume that are not balanced by air molecules leaving (Fig. 10.26) will cause the air density (ρ, mass of air molecules in the volume) to increase, as described below by the **continuity equation**.

Continuity Equation

For a fixed Eulerian volume, the **mass budget equation** (i.e., the **continuity equation**) is:

$$\frac{\Delta\rho}{\Delta t} = \left\{-U\frac{\Delta\rho}{\Delta x} - V\frac{\Delta\rho}{\Delta y} - W\frac{\Delta\rho}{\Delta z}\right\} - \rho\left[\frac{\Delta U}{\Delta x} + \frac{\Delta V}{\Delta y} + \frac{\Delta W}{\Delta z}\right] \quad (10.60)$$

The terms in curly braces { } describe advection. With a bit of calculus one can rewrite this equation as:

INFO • Eötvös Effect

When you move along a path at constant distance R above Earth's center, gravitational acceleration appears to change slightly due to your motion. The measured gravity $|g_{obs}| = |g| - a_r$, where:

$$a_r = 2\cdot\Omega\cdot\cos(\phi)\cdot U + (U^2 + V^2)/R$$

The first term is the vertical component of Coriolis force (eq. 10.17e in the INFO box on p.297), and the last term is centrifugal force as you follow the curvature of the Earth. Thus, you feel lighter traveling east and heavier traveling west. This is the **Eötvös effect**.

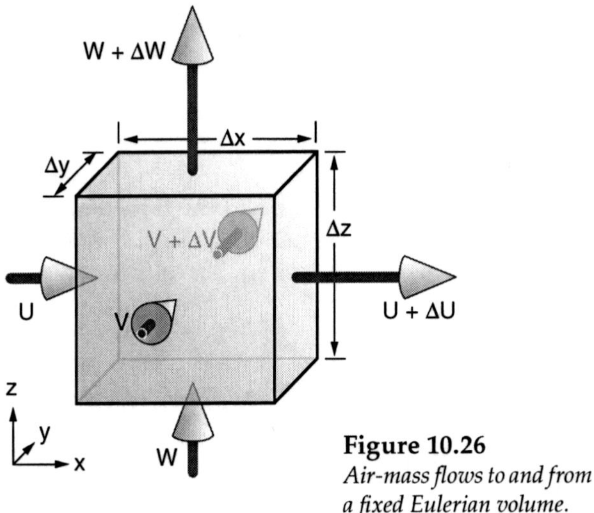

Figure 10.26
Air-mass flows to and from a fixed Eulerian volume.

Sample Application

Hurricane-force winds of 60 m s^{-1} blow into an north-facing entrance of a 20 m long pedestrian tunnel. The door at the other end of the tunnel is closed. The initial air density in the tunnel is 1.2 kg m^{-3}. Find the rate of air density increase in the tunnel.

Find the Answer

Given: $V_{N.\,entrance} = -60$ m s^{-1}, $V_{S.\,entrance} = 0$ m s^{-1}, ρ = 1.2 kg m^{-3}, $\Delta y = 20$ m,

Find: $\Delta\rho/\Delta t$ = ? kg·m^3·s^{-1} initially.

Use eq. (10.60), with $U = W = 0$ because the other walls, roof, and floor prevent winds in those directions:

$$\frac{\Delta\rho}{\Delta t} = -\rho\frac{V_{N.entr.} - V_{S.entr.}}{\Delta y_{tunnel}} = -\left(1.2\frac{kg}{m^3}\right)\frac{(-60-0)m/s}{(20)m}$$

$\Delta\rho/\Delta t$ = <u>+3.6</u> kg·m^{-3}·s^{-1} .

Check: Physics & units are reasonable.

Exposition: As air density increases, so will air pressure. This pressure might be sufficient to blow open the other door at the south end of the pedestrian tunnel, allowing the density to decrease as air escapes.

$$\frac{\Delta\rho}{\Delta t} = -\left[\frac{\Delta(\rho U)}{\Delta x} + \frac{\Delta(\rho V)}{\Delta y} + \frac{\Delta(\rho W)}{\Delta z}\right] \quad (10.61)$$

where U, V, and W are the wind components in the x, y, and z directions, respectively, and t is time.

When you calculate wind gradients, be sure to take the wind and space differences in the same direction. For example: $\Delta U/\Delta x = (U_2 - U_1)/(x_2 - x_1)$.

Incompressible Idealization

Mean air density changes markedly with altitude, as was sketched in Fig. 10.25. However, at any one altitude the density changes only slightly due to local changes in humidity and temperature. For non-tornadic, non-thunderstorm conditions where Fig. 10.25 is valid, we can make a reasonable simplifying idealization that density is constant ($\Delta\rho \approx 0$) at any one altitude. Namely, air behaves as if it is **incompressible**.

If we make this idealization, then the advection terms of eq. (10.60) are zero, and the time-tendency term is zero. The net result is **volume conservation**, where volume outflow equals volume inflow:

$$\frac{\Delta U}{\Delta x} + \frac{\Delta V}{\Delta y} + \frac{\Delta W}{\Delta z} = 0 \quad •(10.62)$$

Fig. 10.26 illustrates such **incompressible continuity**. Can you detect an error in this figure? It shows more air leaving the volume in each coordinate direction than is entering — impossible for incompressible flow. A correct figure would have changed arrow lengths, to indicate net <u>in</u>flow in one or two directions, balanced by net <u>out</u>flow in the other direction(s).

As will be explained in the last section of this chapter, **divergence** is where more air leaves a volume than enters (corresponding to positive terms in eq. 10.62). **Convergence** is where more air enters than leaves (corresponding to negative terms in eq. 10.62). Thus, volume (mass) conservation of incompressible flow requires one or two terms in eq. (10.62) to be negative (i.e., convergence), and the remaining term(s) to be positive (i.e., divergence) so that their sum equals zero.

Horizontal divergence (D) is defined as

$$D = \frac{\Delta U}{\Delta x} + \frac{\Delta V}{\Delta y} \quad (10.63)$$

Negative values of D correspond to convergence. Plugging this definition into eq. (10.62) shows that vertical velocities increase with height where there is horizontal convergence:

$$\frac{\Delta W}{\Delta z} = -D \quad (10.64)$$

Boundary-Layer Pumping

Consider an extratropical cyclone, where the boundary-layer gradient wind spirals in toward the low-pressure center. Those spiraling winds consist of a tangential component following the isobars as they encircle the low center, and a radial component having inflow velocity V_{in} (Fig. 10.27).

But volume inflow ($2\pi R \cdot \Delta z \cdot V_{in}$) through the sides of the cylindrical volume of radius R and height Δz must be balanced by net volume outflow ($\pi R^2 \cdot \Delta W/\Delta z$) through the top and bottom. Equating these incompressible flows gives:

$$\frac{2 \cdot V_{in}}{R} = \frac{\Delta W}{\Delta z} \quad •(10.65a)$$

Thus, for horizontal inflow everywhere (positive V_{in}), one finds that ΔW must also be positive.

If a cylinder of air is at the ground where $W = 0$ at the cylinder bottom, then W at the cylinder top is:

$$W = (2 \cdot V_{in} \cdot \Delta z) / R \quad (10.65b)$$

Namely, extratropical cyclones have rising air, which causes clouds and rain due to adiabatic cooling. This forcing of a broad updraft regions by horizontal-wind drag around a cyclone is known as **boundary-layer pumping** or **Ekman pumping**.

For atmospheric boundary-layer gradient (ABLG) winds around anticyclones (highs), the opposite occurs: horizontal outflow and a broad region of descending air (**subsidence**). The subsidence causes adiabatic warming, which evaporates any clouds and creates fair weather.

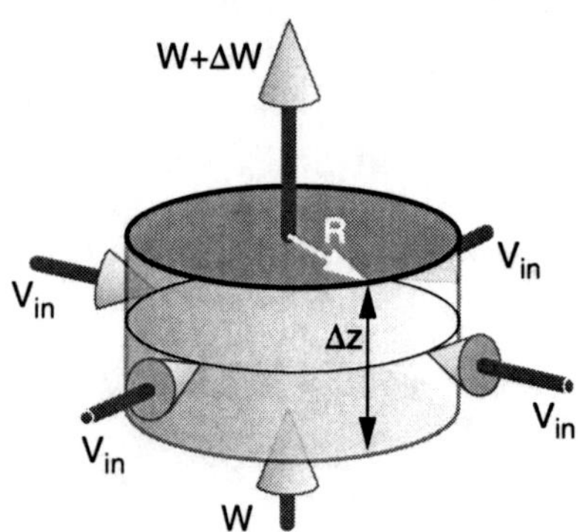

Figure 10.27
Volume conservation for an idealized cylindrical extratropical cyclone.

Recall from the ABLG wind section that an analytical solution could not be found for V_{ABLG} (which is the needed V_{in} for eq. 10.65). Instead, we can approximate $V_{in} \approx V_{ABL}$ for which an analytical solution exists. But V_{ABL} is always larger than V_{ABLG} for flow around cyclones, so we must be aware that our analytical answer will always give winds that are slightly faster than occur around lows in nature.

To solve for V_{ABL}, we need to make an assumption about the static stability of the atmospheric boundary layer. Because cyclones generally have overcast skies and strong winds, we can safely assume neutral stability. In this case, eq. (10.41b) gives the cross-isobaric inflow velocity.

Use V_{ABL} for V_{in} in eq. (10.65b) and solve for W (which we will call W_{ABL} — the vertical velocity at the atmospheric boundary-layer top, as sketched in Fig. 10.28):

$$W_{ABL} = \frac{2 \cdot b \cdot C_D}{f_c} \cdot \frac{G^2}{R} \qquad \bullet(10.66)$$

with geostrophic wind G, radius of curvature R, Coriolis parameter f_c, and drag coefficient C_D for statically neutral boundary conditions. For flow over land, $C_D \approx 0.005$.

Eq. (10.41b) can be used to find $b = \{ 1 - 0.5 \cdot [C_D \cdot G / (f_c \cdot z_i)] \}$ for an atmospheric boundary layer of thickness z_i. If you don't know the actual **atmospheric boundary-layer depth**, then a crude approximation for cyclones (not valid for anticyclones) is :

$$z_i \approx \frac{G}{N_{BV}} \qquad (10.67)$$

In this approximation, you must use a Brunt-Väisälä frequency N_{BV} that is valid for the statically stable air in the troposphere above the top of the statically neutral atmospheric boundary layer. For this special approximation: $b = \{ 1 - 0.5 \cdot [C_D \cdot N_{BV} / f_c] \}$. A required condition for a physically realistic solution is $[C_D \cdot N_{BV} / f_c] < 1$.

You can interpret eq. (10.66) as follows. Stronger pressure gradients (which cause larger geostrophic wind G), larger drag coefficients, and smaller radii of curvature cause greater atmospheric boundary-layer pumping W_{ABL}.

Although the equations above allow a complete approximate solution, we can rewrite them in terms of a **geostrophic relative vorticity**:

$$\zeta_g = \frac{2 \cdot G}{R} \qquad (10.68)$$

which indicates air rotation. Vorticity is introduced later in this chapter, and is covered in greater detail in the General Circulation chapter.

Sample Application

Atmospheric boundary-layer gradient winds have an inflow component of 3 m s^{-1} at radius 500 km from the center of a midlatitude cyclone. What is the updraft speed?

Find the Answer

Given: R = 500 km, V_{in} = 3 m s^{-1}. Let $\Delta z = z_i$ = 1 km
Find: W = ? m s^{-1}

Apply eq. (10.65b):

$$W = 2 \cdot V_{in} \cdot \frac{\Delta z}{R} = 2 \cdot (3 \text{m/s}) \cdot \frac{1\text{km}}{500\text{km}}$$

$$= \underline{0.012 \text{ m s}^{-1}}$$

Check: Physics & units are reasonable.
Exposition: This updraft speed of 1.2 cm/s, while slow, can cause significant lifting over many hours. Nonetheless, this updraft is slow enough to enable us to approximate the atmosphere as being hydrostatic.

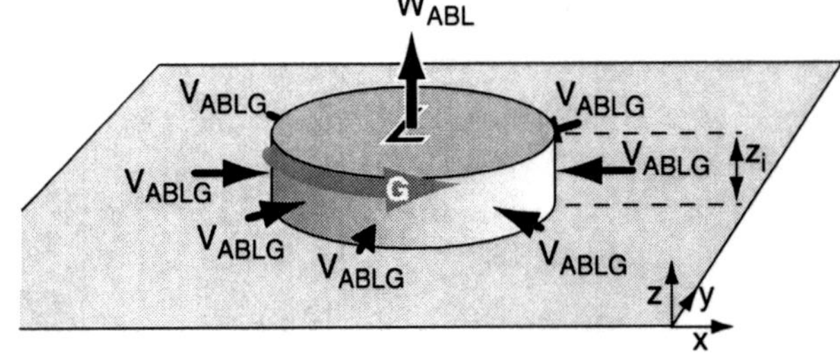

Figure 10.28
Application of atmospheric boundary-layer gradient winds V_{ABLG} to estimate the vertical velocity W_{ABL} due to atmospheric boundary-layer pumping around a low-pressure center (L). G = geostrophic wind.

Sample Application

At 500 km from the center of a midlatitude cyclone at latitude where $f_c = 0.0001$ s^{-1}, the pressure gradient can drive a 15 m s^{-1} geostrophic wind. Assume a standard atmosphere static stability above the top of the atmospheric boundary layer (ABL), and a drag coefficient of 0.004 at the bottom. Find the Ekman pumping updraft speed out of the atmospheric boundary-layer top. Also, what are the geostrophic relative vorticity, the depth of the ABL, and the internal Rossby deformation radius?

Find the Answer

Given: $f_c = 0.0001$ s^{-1}, $R = 5\text{x}10^5$ m, $G = 15$ m s^{-1}, $N_{BV} = 0.0113$ s^{-1} (from a previous Sample Application using the standard atmos.), $C_D = 0.004$,

Find: $W_{ABL} = ?$ m s^{-1}, $\zeta_g = ?$ s^{-1}, $z_i = ?$ m, $\lambda_R = ?$ km

For depth of the troposphere, assume $z_T = 11$ km.

Apply eq. (10.67):

$z_i \approx G/N_{BV} = (15 \text{ m s}^{-1})/(0.0113 \text{ s}^{-1}) = \underline{1327 \text{ m}}$

Apply eq. (10.68):

$$\zeta_g = \frac{2\cdot(15\text{m/s})}{5\times10^5\text{m}} = \underline{\textbf{6x10}}^{-5} \text{ s}^{-1}$$

Apply eq. (10.70):

$$\lambda_R \approx \frac{(15\text{m/s})}{(0.0001\text{s}^{-1})}\cdot\frac{11\text{km}}{1.327\text{km}} = \underline{\textbf{1243 km}}$$

We need to check to ensure that $[C_D \cdot N_{BV}/f_c] < 1$.

$[0.004\cdot(0.0113\text{s}^{-1})/(0.0001\text{s}^{-1})] = 0.452 < 1$.

Thus, we can expect our approximate solution should work for this case.

Apply eq. (10.71): $W_{ABL} =$

$$0.004\cdot\frac{(1.327\text{km})}{(11\text{km})}\cdot(1.243\times10^6\text{m})\cdot(6\times10^{-5}\text{s}^{-1})$$
$$\cdot\left[1-0.5\cdot(0.004)\cdot\frac{1243\text{km}}{11\text{km}}\right]$$

$= (0.036 \text{ m s}^{-1}) \cdot [0.774] = \underline{\textbf{0.028 m s}^{-1}}$

Check: Physics and units are reasonable.

Exposition: The updraft speed 2.8 cm s^{-1} is slow, but over many hours can cause significant lifting. As the rising air cools adiabatically, clouds form and latent heat is released due to condensation. Hence, clouds and bad weather are often associated with midlatitude cyclones.

Eq. (10.66) can be modified to use geostrophic vorticity. The resulting Ekman pumping at the atmospheric boundary layer top in a midlatitude cyclone is:

$$W_{ABL} = C_D \cdot \frac{G}{f_c} \cdot \zeta_g \cdot \left[1 - 0.5 \frac{C_D \cdot N_{BV}}{f_c}\right] \tag{10.69}$$

The first four factors on the right side imply that larger drag coefficients (i.e., rougher terrain with more trees or buildings) and stronger pressure gradients (as indicated by larger geostrophic wind) driving winds around smaller radii of curvature (i.e., larger geostrophic vorticity) at lower latitudes (i.e., smaller f_c) create stronger updrafts. Also, stronger static stabilities (i.e., larger Brunt-Väisälä frequency NBV) in the troposphere above atmospheric boundary-layer top reduce updraft speed by opposing vertical motion.

One can write an **internal Rossby deformation radius** based on the eq. (10.67) approximation for depth z_i of the atmospheric boundary layer:

$$\lambda_R \approx \frac{G}{f_c} \cdot \frac{z_T}{z_i} \tag{10.70}$$

where tropospheric depth is z_T. Internal and external Rossby deformation radii are described further in the General Circulation and Fronts & Airmasses chapters, respectively.

The Rossby deformation radius can be used to write yet another expression for Ekman pumping vertical velocity out of the top of the atmospheric boundary layer:

$$W_{ABL} = C_D \cdot \frac{z_i}{z_T} \cdot \lambda_R \cdot \zeta_g \cdot \left[1 - 0.5 \cdot C_D \cdot \frac{\lambda_R}{z_T}\right] \tag{10.71}$$

KINEMATICS

Kinematics is the study of patterns of motion, without regard to the forces that cause them. We will focus on horizontal divergence, vorticity, and deformation. All have units of s^{-1}.

We have already encountered horizontal **divergence**, D, the spreading of air:

$$D = \frac{\Delta U}{\Delta x} + \frac{\Delta V}{\Delta y} \tag{10.72}$$

Figure 10.29a shows an example of pure divergence. Its sign is positive for divergence, and negative for **convergence** (when the wind arrows point <u>toward</u> a common point).

Vorticity describes the rotation of air (Fig. 10.29b). The relative vorticity, ζ_r, about a locally vertical axis is given by:

$$\zeta_r = \frac{\Delta V}{\Delta x} - \frac{\Delta U}{\Delta y} \qquad (10.73)$$

The sign is positive for counterclockwise rotation (i.e., cyclonic rotation in the N. Hemisphere), and negative for clockwise rotation. Vorticity is discussed in greater detail in the General Circulation chapter. Neither divergence nor vorticity vary with rotation of the axes — they are **rotationally invariant**.

Two types of **deformation** are stretching deformation and shearing deformation (Figs. 10.29c & d). **Stretching deformation**, F_1, is given by:

$$F_1 = \frac{\Delta U}{\Delta x} - \frac{\Delta V}{\Delta y} \qquad (10.74)$$

The axis along which air is being stretched (Fig. 10.29c) is called the **axis of dilation** (x axis in this example), while the axis along which air is compressed is called the **axis of contraction** (y axis in this example).

Shearing deformation, F_2, is given by:

$$F_2 = \frac{\Delta V}{\Delta x} + \frac{\Delta U}{\Delta y} \qquad (10.75)$$

As you can see in Fig. 10.29d, shearing deformation is just a rotated version of stretching deformation. The **total deformation**, F, is:

$$F = \left[F_1^{\,2} + F_2^{\,2} \right]^{1/2} \qquad (10.76)$$

Deformation often occurs along fronts. Most real flows exhibit combinations of divergence, vorticity, and deformation.

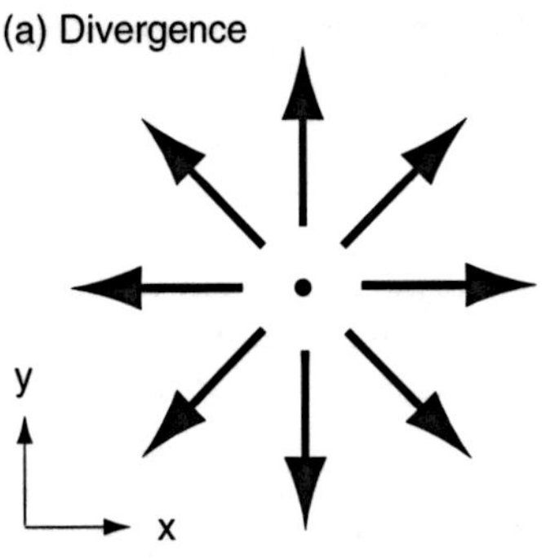

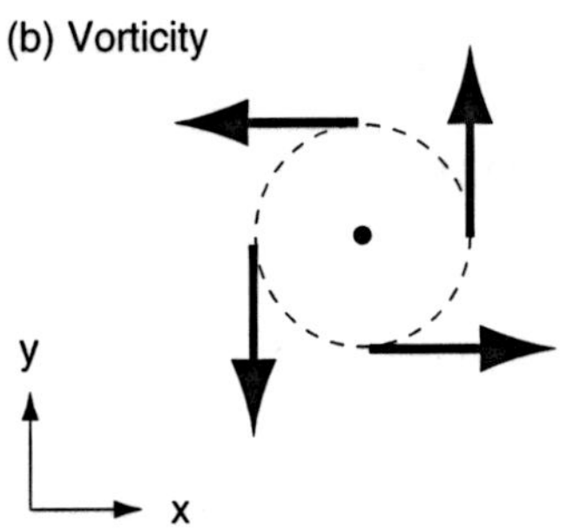

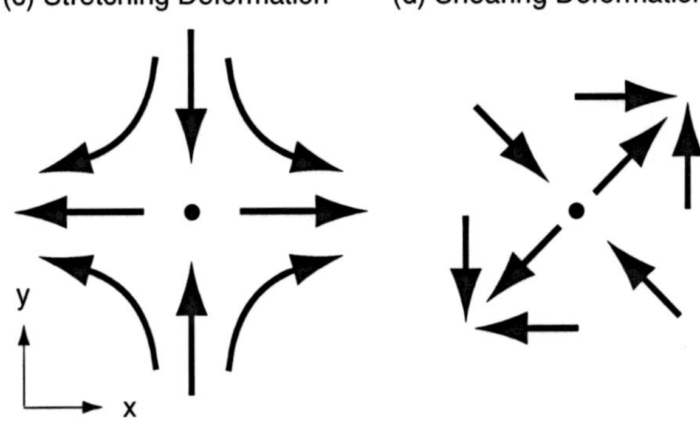

Figure 10.29
Kinematic flow-field definitions. Black arrows represent wind velocity.

MEASURING WINDS

For weather stations at the Earth's surface, wind direction can be measured with a **wind vane** mounted on a vertical axel. **Fixed vanes** and other shapes can be used to measure wind speed, by using strain gauges to measure the minute deformations of the object when the wind hits it.

The generic name for a wind-speed measuring device is an **anemometer**. A **cup anemometer** has conic- or hemispheric-shaped cups mounted on spokes that rotate about a vertical axel. A **propellor**

anemometer has a propellor mounted on a horizontal axel that is attached to a wind vane so it always points into the wind. For these anemometers, the rotation speed of the axel can be calibrated as a wind speed.

Other ways to measure wind speed include a **hot-wire** or **hot-film anemometer**, where a fine metal wire is heated electrically, and the power needed to maintain the hot temperature against the cooling effect of the wind is a measure of wind speed. A **pitot tube** that points into the wind measures the dynamic pressure as the moving air stagnates in a dead-end tube. By comparing this dynamic pressure with the static pressure measured by a different sensor, the pressure difference can be related to wind speed.

Sonic anemometers send pulses of sound back and forth across a short open path between two opposing transmitters and receivers of sound. The speed of sound depends on both temperature and wind speed, so this sensor can measure both. Tracers such as smoke, humidity fluctuations, or clouds can be **tracked** photogramatically from the ground or from remote sensors such as laser radars (**lidars**) or satellites, and the wind speed then estimated from the change of position of the tracer between successive images.

Measurements of wind vs. height can be made with **rawinsonde balloons** (using a GPS receiver in the sonde payload to track horizontal drift of the balloons with time), **dropsondes** (like rawinsondes, only descending by parachute after being dropped from aircraft), **pilot balloons** (carrying no payload, but being tracked instead from the ground using radar or theodolites), **wind profilers**, **Doppler weather radar** (see the Satellites & Radar chapter), and via anemometers mounted on aircraft.

REVIEW

According to Newton's second law, winds are driven by forces. The pressure-gradient creates a force, even in initially calm (windless) conditions. This force points from high to low pressure on a constant altitude chart (such as at sea-level), or points from high to low heights on an isobaric chart (such as the 50 kPa chart). Pressure-gradient force is the main force that drives the winds.

Other forces exist only when there is already a wind. One example is turbulent drag against the ground, which pushes opposite to the atmospheric boundary-layer wind direction. Another example is Coriolis force, which is related to centrifugal force of winds relative to a rotating Earth.

If all the forces vector-sum to zero, then there is no net force and winds blow at constant speed. Theoretical winds based on only a small number of forces are given special names. The geostrophic wind occurs when pressure-gradient and Coriolis forces balance, causing a wind that blows parallel to straight isobars. For curved isobars around lows and highs, the imbalance between these two forces turns the wind in a circle, with the result called the gradient wind. Similar winds can exist in the atmospheric boundary layer, where turbulent drag of the air against the Earth's surface slows the wind and causes it to turn slightly to cross the isobars toward low pressure.

Waterspouts and tornadoes can have such strong winds that pressure-gradient force is balanced by centrifugal force, with the resulting wind speed known as the cyclostrophic wind. In oceans, currents can inertially flow in a circle.

The two most important force balances at mid-latitudes are hydrostatic balance in the vertical, and geostrophic balance in the horizontal.

Conservation of air mass gives the continuity equation, for which an incompressible approximation can be used in most places except in thunderstorms. Mechanisms that cause motion in one direction (horizontal or vertical) will also indirectly cause motions in the other direction as the air tries to maintain continuity, resulting in a circulation.

Kinematics is the word that describes the behavior and effect of winds (such as given by the continuity equation) without regard to the forces that cause them. The word **dynamics** describes how forces cause winds (as given by Newton's 2nd law).

HOMEWORK EXERCISES

Some of these questions are inspired by exercises in Stull, 2000: *Meteorology for Sci. & Engr. 2nd Ed.*, Brooks/Cole, 528 pp.

Broaden Knowledge & Comprehension

For all the exercises in this section, collect information off the internet. Don't forget to cite the web sites you use.

B1. a. Find a weather map showing today's sea-level pressure isobars near your location. Calculate pressure-gradient force (N) based on your latitude and the isobar spacing (km/kPa).

b. Repeat this for a few days, and plot the pressure gradient vs. time.

B2. Get 50 kPa height contour maps (i.e., 500 hPa heights) over any portion of the Northern Hemisphere. In 2 locations at different latitudes having straight isobars, compute the geostrophic wind speed. In 2 locations of curved isobars, compute the gradient wind speed. How do these theoretical winds compare with wind observations near the same locations?

B3. Similar exercise B2, but for 2 locations in the Southern Hemisphere.

B4. a. Using your results from exercise B2 or B3, plot the geostrophic wind speed vs. latitude and pressure gradient on a copy of Fig. 10.10. Discuss the agreement or disagreement of your results vs. the lines plotted in that figure.

b. Using your results from exercise B4 or B5, show that gradient winds are indeed faster than geostrophic around high-pressure centers, and slower around low-pressure centers.

B5. Discuss surprising insights regarding Isacc Newton's discoveries on forces and motion.

B6. Get a map of sea-level pressure, including isobar lines, for a location or date where there are strong low and high-pressure centers adjacent to each other. On a printed copy of this map, use a straight edge to draw a line connecting the low and high centers, and extend the line further beyond each center. Arbitrarily define the high center as location $x = 0$. Then, along your straight line, add distance tic marks appropriate for the map scale you are using. For isobars crossing your line, create a table that lists each pressure P and its distance x from the high. Then plot P vs. x and discuss how it compares with Fig. 10.14. Discuss the shape of your curve in the low- and high-pressure regions.

B7. Which animations best illustrate Coriolis Force?

B8. a. Get a map of sea-level pressure isobars that also shows observed wind directions. Discuss why the observed winds have a direction that crosses the isobars, and calculate a typical crossing angle..

b. For regions where those isobars curve around cyclones or anticyclones, confirm that winds spiral into lows and out of highs.

c. For air spiraling in toward a cyclone, estimate the average inflow radial velocity component, and calculate W_{BL} based on incompressible continuity.

B9. For a typhoon or hurricane, get a current or past weather map showing height-contours for any one isobaric level corresponding to an altitude about 1/3 the altitude of the storm (ie., a map for any pressure level between 85 to 60 kPa). At the eye-wall location, use the height-gradient to calculate the cyclostrophic wind speed. Compare this with the observed hurricane winds at that same approximate location, and discuss any differences.

B10. Get a 500 hPa (= 50 kPa) geopotential height contour map that is near or over the equator. Compute the theoretical geostrophic wind speed based on the height gradients at 2 locations on that map where there are also observed upper-air wind speeds. Explain why these theoretical wind speeds disagree with observed winds.

Apply

A1. Plot the wind symbol for winds with the following directions and speeds:

a. N at 5 kt b. NE at 35 kt c. E at 65 kt
d. SE at 12 kt e. S at 48 kt f. SW at 105 kt
g. W at 27 kt h. NW at 50 kt i. N at 125 kt

A2. How fast does an 80 kg person accelerate when pulled with the force given below in Newtons?

a. 1 b. 2 c. 5 d. 10 e. 20 f. 50
g. 100 h. 200 i. 500 j. 1000 k. 2000

A3. Suppose the following force per mass is applied on an object. Find its speed 2 minutes after starting from rest.

a. 5 N kg^{-1} b. 10 m·s^{-2}
c. 15 N kg^{-1} d. 20 m·s^{-2}
e. 25 N kg^{-1} f. 30 m·s^{-2}
g. 35 N kg^{-1} h. 40 m·s^{-2}
i. 45 N kg^{-1} j. 50 m·s^{-2}

A4. Find the advective "force" per unit mass given the following wind components (m s^{-1}) and horizontal distances (km):

a. U=10, ΔU=5, Δx=3
b. U=6, ΔU=–10, Δx=5
c. U=–8, ΔV=20, Δx=10
d. U=–4, ΔV=10, Δx=–2
e. V=3, ΔU=10, Δy=10
f. V=–5, ΔU=10, Δy=4
g. V=7, ΔV=–2, Δy=–50
h. V=–9, ΔV=–10, Δy=–6

A5. Town A is 500 km west of town B. The pressure at town A is given below, and the pressure at town B is 100.1 kPa. Calculate the pressure-gradient force/mass in between these two towns.

a. 98.6 b. 98.8 c. 99.0 d. 99.2 e. 99.4
f. 99.6 g. 99.8 h. 100.0 i. 100.2 j. 100.4
k. 100.6 l. 100.8 m. 101.0 n. 101.2 o. 101.4

A6. Suppose that $U = 8$ m s^{-1} and $V = -3$ m s^{-1}, and latitude = 45° Calculate centrifugal-force components around a:

a. 500 km radius low in the N. hemisphere
b. 900 km radius high in the N. hemisphere
c. 400 km radius low in the S. hemisphere
d. 500 km radius high in the S. hemisphere

A7. What is the value of f_c (Coriolis parameter) at:

a. Shanghai
b. Istanbul
c. Karachi
d. Mumbai
e. Moscow
f. Beijing
g. São Paulo
h. Tianjin
i. Guangzhou
j. Delhi
k. Seoul
l. Shenzhen
m. Jakarta
n. Tokyo
o. Mexico City
p. Kinshasa
q. Bangalore
r. New York City
s. Tehran
t. (a city specified by your instructor)

A8. What is the magnitude and direction of Coriolis force/mass in Los Angeles, USA, given:

	U (m s^{-1})	V (m s^{-1})
a.	5	0
b.	5	5
c.	5	–5
d.	0	5
e.	0	–5
f.	–5	0
g.	–5	–5
h.	–5	5

A9. Same wind components as exercise A8, but find the magnitude and direction of turbulent drag force/mass in a statically neutral atmospheric boundary layer over an extensive forested region.

A10. Same wind components as exercise A8, but find the magnitude and direction of turbulent drag force/mass in a statically unstable atmospheric boundary layer with a 50 m/s buoyant velocity scale.

A11. Draw a northwest wind of 5 m s^{-1} in the S. Hemisphere on a graph, and show the directions of forces acting on it. Assume it is in the boundary layer.

a. pressure gradient b. Coriolis
c. centrifugal d. drag

A12. Given the pressure gradient magnitude (kPa/1000 km) below, find geostrophic wind speed for a location having $f_c = 1.1\text{x}10^{-4}$ s^{-1} and $\rho = 0.8$ kg m^{-3}.

a. 1 b. 2 c. 3 d. 4 e. 5
f. 6 g. 7 h. 8 i. 9 j. 10
k. 11 m. 12 n. 13 o. 14 p. 15

A13. Suppose the height gradient on an isobaric surface is given below in units of (m km^{-1}). Calculate the geostrophic wind at 55°N latitude.

a. 0.1 b. 0.2 c. 0.3 d. 0.4 e. 0.5
f. 0.6 g. 0.7 h. 0.8 i. 0.9 j. 1.0
k. 1.1 m. 1.2 n. 1.3 o. 1.4 p. 1.5

A14. At the radius (km) given below from a low-pressure center, find the gradient wind speed given a geostrophic wind of 8 m s^{-1} and given $f_c = 1.1\text{x}10^{-4}$ s^{-1}.

a. 500 b. 600 c. 700 d. 800 e. 900
f. 1000 g. 1200 h. 1500 i. 2000 j. 2500

A15. Suppose the geostrophic winds are $U_g = -3$ m s^{-1} with $V_g = 8$ m s^{-1} for a statically-neutral boundary layer of depth $z_i = 1500$ m, where $f_c = 1.1\text{x}10^{-4}$ s^{-1}. For drag coefficients given below, what is the atmos. boundary-layer wind speed, and at what angle does this wind cross the geostrophic wind vector?

a. 0.002 b. 0.004 c. 0.006 d. 0.008 e. 0.010
f. 0.012 g. 0.014 h. 0.016 i. 0.018 j. 0.019

A16. For a statically unstable atmos. boundary layer with other characteristics similar to those in exercise A15, what is the atmos. boundary-layer wind speed, at what angle does this wind cross the geostrophic wind vector, given w_B (m s^{-1}) below?

a. 75 b. 100 c. 50 d. 200 e. 150
f. 225 g. 125 h. 250 i. 175 j. 275

A17(§). Review the Sample Application in the "Atmospheric Boundary Layer Gradient Wind" section. Re-do that calculation for M_{ABLG} with a different parameter as given below:

a. $z_i = 1$ km b. $C_D = 0.003$ c. $G = 8$ m s^{-1}
d. $f_c = 1.2\text{x}10^{-4}$ s^{-1} e. $R = 2000$ km
f. $G = 15$ m s^{-1} g. $z_i = 1.5$ km h. $C_D = 0.005$
i. $R = 1500$ km j. $f_c = 1.5\text{x}10^{-4}$ s^{-1}

Hint: Assume all other parameters are unchanged.

A18. Find the cyclostrophic wind at radius (m) given below, for a radial pressure gradient = 0.5 kPa m^{-1}:

a. 10 b. 12 c. 14 d. 16 e. 18
f. 20 g. 22 h. 24 i. 26 j. 28 k 30

A19. For an inertial wind, find the radius of curvature (km) and the time period (h) needed to complete one circuit, given $f_c = 10^{-4}$ s^{-1} and an initial wind speed (m s^{-1}) of:
a. 1 b. 2 c. 3 d. 4 e. 6 f. 7 g. 8 h. 9
i. 10 j. 11 k. 12 m. 13 n. 14 o. 15

A20. Find the antitriptic wind for the conditions of exercise A15.

A21. Below is given an average inward radial wind component (m s^{-1}) in the atm. boundary layer at radius 300 km from the center of a cyclone. What is the average updraft speed out of the atm. boundary-layer top, for a boundary layer that is 1.2 km thick?
a. 2 b. 1.5 c. 1.2 d. 1.0
e. –0.5 f. –1 g. –2.5 h. 3 i. 0.8 j. 0.2

A22. Above an atmospheric boundary layer, assume the tropospheric temperature profile is $\Delta T/\Delta z = 0$. For a midlatitude cyclone, estimate the atm. boundary-layer thickness given a near-surface geostrophic wind speed (m s^{-1}) of:
a. 5 b. 10 c. 15 d. 20 e. 25 f. 30
g. 35 h. 40 i. 3 j. 8 k. 2 l. 1

A23(§). For atm. boundary-layer pumping, plot a graph of updraft velocity vs. geostrophic wind speed assuming an atm. boundary layer of depth 0.8 km, a drag coefficient 0.005 . Do this only for wind speeds within the valid range for the atmos. boundary-layer pumping eq. Given a standard atmospheric lapse rate at 30° latitude with radius of curvature (km) of:
a. 750 b. 1500 c. 2500 d. 3500 e. 4500
f. 900 g. 1200 h. 2000 i. 3750 j. 5000

A24. At 55°N, suppose the troposphere is 10 km thick, and has a 10 m s^{-1} geostrophic wind speed. Find the internal Rossby deformation radius for an atmospheric boundary layer of thickness (km):
a. 0.2 b. 0.4 c. 0.6 d. 0.8 e. 1.0
f. 1.2 g. 1.5 h. 1.75 i. 2.0 j. 2.5

A25. Given $\Delta U/\Delta x = \Delta V/\Delta x = (5\ m\ s^{-1}) / (500\ km)$, find the divergence, vorticity, and total deformation for ($\Delta U/\Delta y$, $\Delta V/\Delta y$) in units of (m s^{-1})/(500 km) as given below:
a. (–5, –5) b. (–5, 0) c. (0, –5) d. (0, 0) e. (0, 5)
f. (5, 0) g. (5, 5) h. (–5, 5) i. (5, –5)

Evaluate & Analyze

E1. Discuss the relationship between eqs. (1.24) and (10.1).

E2. Suppose that the initial winds are unknown. Can a forecast still be made using eqs. (10.6)? Explain your reasoning.

E3. Considering eq. (10.7), suppose there are no forces acting. Based on eq. (10.5), what can you anticipate about the wind speed.

E4. We know that winds can advect temperature and humidity, but how does it work when winds advect winds? Hint, consider eqs. (10.8).

E5. For an Eulerian system, advection describes the influence of air that is blown into a fixed volume. If that is true, then explain why the advection terms in eq. (10.8) is a function of the wind gradient (e.g., $\Delta U/\Delta x$) instead of just the upwind value?

E6. Isobar packing refers to how close the isobars are, when plotted on a weather map such as Fig. 10.5. Explain why such packing is proportional to the pressure gradient.

E7. Pressure gradient has a direction. It points toward low pressure for the Northern Hemisphere. For the Southern Hemisphere, does it point toward high pressure? Why?

E8. To help you interpret Fig. 10.5, consider each horizontal component of the pressure gradient. For an arbitrary direction of isobars, use eqs. (10.9) to demonstrate that the vector sum of the components of pressure-gradient do indeed point away from high pressure, and that the net direction is perpendicular to the direction of the isobars.

E9. For centrifugal force, combine eqs. (10.13) to show that the net force points outward, perpendicular to the direction of the curved flow. Also show that the magnitude of that net vector is a function of tangential velocity squared.

E10. Why does $f_c = 0$ at the equator for an air parcel that is stationary with respect to the Earth's surface, even though that air parcel has a large tangential velocity associated with the rotation of the Earth?

E11. Verify that the net Coriolis force is perpendicular to the wind direction (and to its right in the N. Hemisphere), given the individual components described by eqs. (10.17).

E12. For the subset of eqs. (10.1 - 10.17) defined by your instructor, rewrite them for flow in the Southern Hemisphere.

E13. Verify that the net drag force opposes the wind by utilizing the drag components of eqs. (10.19). Also, confirm that drag-force magnitude for statically neutral conditions is a function of wind-speed squared.

E14. How does the magnitude of the turbulent-transport velocity vary with static stability, such as between statically unstable (convective) and statically neutral (windy) situations?

E15. Show how the geostrophic wind components can be combined to relate geostrophic wind speed to pressure-gradient magnitude, and to relate geostrophic wind direction to pressure-gradient direction.

E16. How would eqs. (10.26) for geostrophic wind be different in the Southern Hemisphere?

E17. Using eqs. (10.26) as a starting point, show your derivation for eqs. (10.29).

E18. Why are actual winds finite near the equator even though the geostrophic wind is infinite there? (Hint, consider Fig. 10.10).

E19. Plug eq. (10.33) back into eqs. (10.31) to confirm that the solution is valid.

E20. Plug eqs. (10.34) back into eq. (10.33) to confirm that the solution is valid.

E21. Given the pressure variation shown in Fig. 10.14. Create a mean-sea-level pressure weather map with isobars around high- and low-pressure centers such that the isobar packing matches the pressure gradient in that figure.

E22. Fig. 10.14 suggests that any pressure gradient is theoretically possible adjacent to a low-pressure center, from which we can further infer that any wind speed is theoretically possible. For the real atmosphere, what might limit the pressure gradient and the wind speed around a low-pressure center?

E23. Given the geopotential heights in Fig. 10.3, calculate the theoretical values for gradient and/or geostrophic wind at a few locations. How do the actual winds compare with these theoretical values?

E24. Eq. 10.39 is an "implicit" solution. Why do we say it is "implicit"?

E25. Determine the accuracy of explicit eqs. (10.41) by comparing their approximate solutions for ABL wind against the more exact iterative solutions to the implicit form in eq. (10.39).

E26. No explicit solution exists for the neutral atmospheric boundary layer winds, but one exists for the statically unstable ABL? Why is that?

E27. Plug eqs. (10.42) into eqs (10.38) or (10.39) to confirm that the solution is valid.

E28(§). a. Create your own spreadsheet that gives the same answer for ABLG winds as in the Sample Application in the ABLG-wind section.

b. Do "what if" experiments with your spreadsheet to show that the full equation can give the gradient wind, geostrophic wind, and boundary-layer wind for conditions that are valid for those situations.

c. Compare the results from (b) against the respective analytical solutions (which you must compute yourself).

E29. Photocopy Fig. 10.13, and enhance the copy by drawing additional vectors for the atmospheric boundary-layer wind and the ABLG wind. Make these vectors be the appropriate length and direction relative to the geostrophic and gradient winds that are already plotted.

E30. Plug the cyclostrophic-wind equation into eq. (10.45) to confirm that the solution is valid for its special case.

E31. Find an equation for cyclostrophic wind based on heights on an isobaric surface. [Hint: Consider eqs. (10.26) and (10.29).]

E32. What aspects of the *Approach to Geostrophy* INFO Box are relevant to the inertial wind? Discuss.

E33. a. Do your own derivation for eq. (10.66) based on geometry and mass continuity (total inflow = total outflow).

b. Drag normally slows winds. Then why does the updraft velocity increase in eq. (10.66) as drag coefficient increases?

c. Factor b varies negatively with increasing drag coefficient in eq. (10.66). Based on this, would you change your argument for part (b) above?

E34. Look at each term within eq. (10.69) to justify the physical interpretations presented after that equation.

E35. Consider eq. (10.70). For the internal Rossby deformation radius, discuss its physical interpretation in light of eq. (10.71).

E36. What type of wind would be possible if the only forces were turbulent-drag and Coriolis. Discuss.

E37. Derive equations for Ekman pumping around anticyclones. Physically interpret your resulting equations.

E38. Rewrite the total deformation as a function of divergence and vorticity. Discuss.

Synthesize

S1. For zonal (east-west) winds, there is also a vertical component of Coriolis force. Using your own diagrams similar to those in the INFO box on Coriolis Force, show why it can form. Estimate its magnitude, and compare the magnitude of this force to other typical forces in the vertical. Show why a vertical component of Coriolis force does not exist for meridional (north-south) winds.

S2. On Planet Cockeyed, turbulent drag acts at right angles to the wind direction. Would there be anything different about winds near lows and highs on Cockeyed compared to Earth?

S3. The time duration of many weather phenomena are related to their spatial scales, as shown by eq. (10.53) and Fig. 10.24. Why do most weather phenomena lie near the same diagonal line on a log-log plot? Why are there not additional phenomena that fill out the relatively empty upper and lower triangles in the figure? Can the distribution of time and space scales in Fig. 10.24 be used to some advantage?

S4. What if atmospheric boundary-layer drag were constant (i.e., not a function of wind speed). Describe the resulting climate and weather.

S5. Suppose Coriolis force didn't exist. Describe the resulting climate and weather.

S6. Incompressibility seems like an extreme simplification, yet it works fairly well? Why? Consider what happens in the atmosphere in response to small changes in density.

S7. The real Earth has locations where Coriolis force is zero. Where are those locations, and what does the wind do there?

S8. Suppose that wind speed $M = c{\cdot}F/m$, where c = a constant, m = mass, and F is force. Describe the resulting climate and weather.

S9. What if Earth's axis of rotation was pointing directly to the sun. Describe the resulting climate and weather.

S10. What if there was no limit to the strength of pressure gradients in highs. Describe the resulting climate, winds and weather.

S11. What if both the ground and the tropopause were rigid surfaces against which winds experience turbulent drag. Describe the resulting climate and weather.

S12. If the Earth rotated half as fast as it currently does, describe the resulting climate and weather.

S13. If the Earth had no rotation about its axis, describe the resulting climate and weather.

S14. Consider the Coriolis-force INFO box. Create an equation for Coriolis-force magnitude for winds that move:
a. westward b. southward

S15. What if a cyclostrophic-like wind also felt drag near the ground? This describes conditions at the bottom of tornadoes. Write the equations of motion for this situation, and solve them for the tangential and radial wind components. Check that your results are reasonable compared with the pure cyclostrophic winds. How would the resulting winds affect the total circulation in a tornado? As discoverer of these winds, name them after yourself.

S16. What if $F = c{\cdot}a$, where c = a constant not equal to mass, a = acceleration, and F is force. Describe the resulting dynamics of objects such as air parcels.

S17. What if pressure-gradient force acted parallel to isobars. Would there be anything different about our climate, winds, and weather maps?

S18. For a free-slip Earth surface (no drag), describe the resulting climate and weather.

S19. Anders Persson discussed issues related to Coriolis force and how we understand it (see *Weather*, 2000.) Based on your interpretation of his paper, can Coriolis force alter kinetic energy and momentum of air parcels, even though it is only an apparent force? Hint, consider whether Newton's laws would be violated if your view these motions and forces from a fixed (non-rotating) framework.

S20. If the Earth was a flat disk spinning about the same axis as our real Earth, describe the resulting climate and weather.

S21. Wind shear often creates turbulence, and turbulence mixes air, thereby reducing wind shear. Considering the shear at the ABL top in Fig. 10.7, why can it exist without mixing itself out?

S22. Suppose there was not centrifugal or centripetal force for winds blowing around lows or highs. Describe the resulting climate, winds and weather.

S23. Suppose advection of the wind by the wind were impossible. Describe the resulting climate and weather.

11 GENERAL CIRCULATION

Contents

A spatial imbalance between radiative inputs and outputs exists for the earth-ocean-atmosphere system. The earth loses energy at all latitudes due to outgoing infrared (IR) radiation. Near the tropics, more solar radiation enters than IR leaves, hence there is a net input of radiative energy. Near Earth's poles, incoming solar radiation is too weak to totally offset the IR cooling. The net result is **differential heating**, creating warm equatorial air and cold polar air (Fig. 11.1a).

This imbalance drives the global-scale **general circulation** of winds. Such a circulation is a fluid-dynamical analogy to **Le Chatelier's Principle** of chemistry. Namely, an imbalanced system reacts in a way to partially counteract the imbalance. The continued destabilization by radiation causes a general circulation of winds that is unceasing.

You might first guess that buoyancy causes air in the warm regions to rise, while cold air near the poles sink (Fig. 11.1b). Instead, the real general circulation has three bands of circulations in the Northern Hemisphere (Fig. 11.1c), and three in the Southern. In this chapter, we will identify characteristics of the general circulation, explain why they exist, and learn how they work.

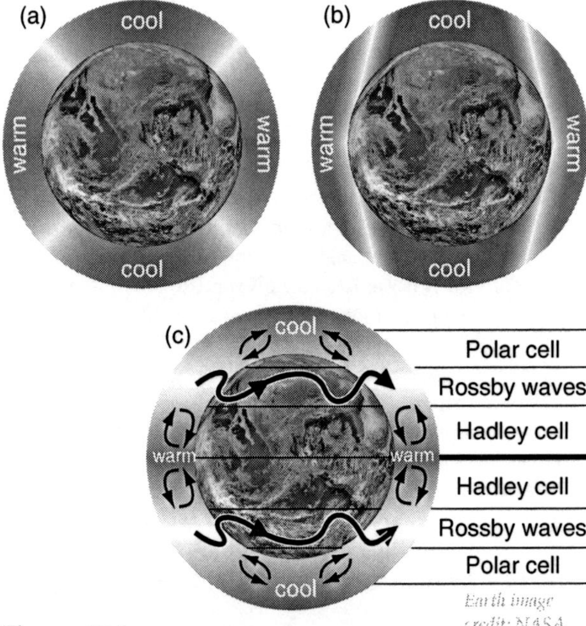

Figure 11.1
Radiative imbalances create (a) warm tropics and cold poles, inducing (b) buoyant circulations. Add Earth's rotation, and (c) three circulation bands form in each hemisphere.

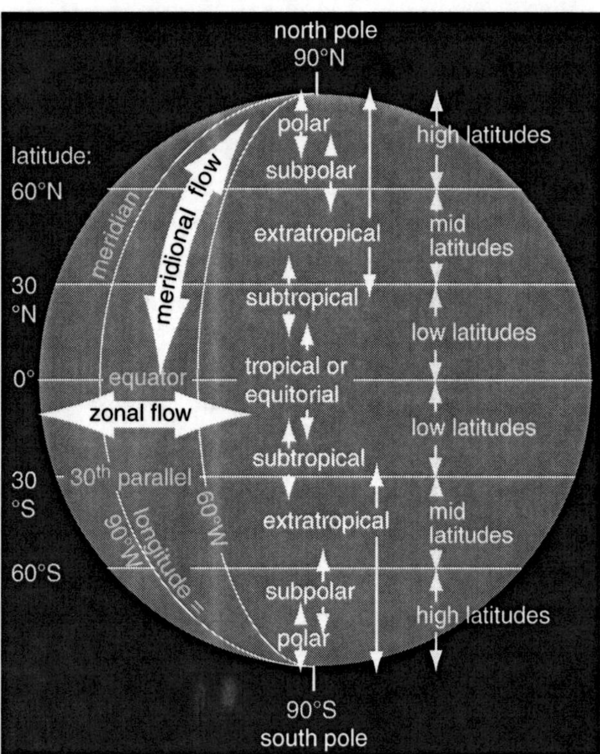

Figure 11.2
Global key terms.

> **A SCIENTIFIC PERSPECTIVE • Idealizations of Nature**
>
> Natural atmospheric phenomena often involve the superposition of many different physical processes and scales of motion. Large-scale average conditions are said to be caused by **zeroth-order** processes. Dominant variations about the mean are controlled by **first-order** processes. Finer details are caused by **higher-order** processes.
>
> Sometimes insight is possible by stripping away the higher-order processes and focusing on one or two lower orders. Equations that describe such simplified physics are known as **toy models**. With the right simplifications, some toy models admit analytical solutions. Compare this to the unsimplified physics, which might be too complicated to solve analytically (although numerical solutions are possible).
>
> A rotating spherical Earth with no oceans and with uniform temperature is one example of a zeroth-order toy model. A first-order toy model might add the north-south temperature variation, while neglecting east-west and continent-ocean variations. With even more sophistication, seasonal or monthly variations might be explained. We will take the approach in this chapter to start with zeroth-order models to focus on basic climate concepts, and then gradually add more realism.
>
> John Harte (1988) wrote a book demonstrating the utility of such toy models. It is *"Consider a Spherical Cow"*, by University Science Books. 283 pages.

KEY TERMS

Lines of constant latitude are called **parallels**, and winds parallel to the parallels are identified as **zonal flows** (Fig. 11.2). Lines of constant longitude are called **meridians**, and winds parallel to the meridians are known as **meridional flows**.

Between latitudes of 30° and 60° are the **mid-latitudes**. **High latitudes** are 60° to 90°, and **low latitudes** are 0° to 30°. Each 1° of latitude = 111 km.

Tropics, **subtropics**, **subpolar**, and **polar** regions are as shown in Fig. 11.2. Regions not in the tropics are called **extratropical**; namely, poleward of about 30°N and about 30°S.

For example, **tropical cyclones** such as **hurricanes** are in the tropics. Low-pressure centers (**lows**, as indicated by 𝕃 on weather maps) outside of the tropics are called **extratropical cyclones**.

In many climate studies, data from the months of June, July, and August (**JJA**) are used to represent conditions in N. Hemisphere summer (and S. Hemisphere winter). Similarly, December, January, February (**DJF**) data are used to represent N. Hemisphere winter (and S. Hemisphere summer).

A SIMPLIFIED DESCRIPTION OF THE GLOBAL CIRCULATION

Consider a hypothetical rotating planet with no contrast between continents and oceans. The **climatological average** (average over 30 years; see the Climate chapter) winds in such a simplified planet would have characteristics as sketched in Figs. 11.3. Actual winds on any day could differ from this climatological average due to transient weather systems that perturb the average flow. Also, monthly-average conditions tend to shift toward the summer hemisphere (e.g., the circulation bands shift northward during April through September).

Near the Surface

Near-surface average winds are sketched in Fig. 11.3a. At low latitudes are broad bands of persistent easterly winds ($U \approx -7$ m s^{-1}) called **trade winds**, named because the **easterlies** allowed sailing ships to conduct transoceanic trade in the old days.

These trade winds also blow toward the equator from both hemispheres, and the equatorial belt of convergence is called the **intertropical convergence zone (ITCZ)**. On average, the air at the ITCZ is hot and humid, with low pressure, strong upward air motion, heavy convective (thunderstorm) precipitation, and light to calm winds except in thunderstorms. This **equatorial trough** (low-pressure belt) was called the **doldrums** by sailors whose sailing ships were becalmed there for many days.

At 30° latitude are belts of high surface pressure called **subtropical highs** (Fig. 11.3a). In these belts are hot, dry, cloud-free air descending from higher in the troposphere. Surface winds in these belts are also calm on average. In the old days, becalmed sailing ships would often run short of drinking water, causing horses on board to die and be thrown overboard. Hence, sailors called these miserable places the **horse latitudes**. On land, many of the world's deserts are near these latitudes.

In mid-latitudes are transient centers of low pressure (**mid-latitude cyclones, L**) and high pressure (**anticyclones, H**). Winds around lows **converge** (come together) and circulate **cyclonically** — counterclockwise in the N. Hemisphere, and clockwise in the S. Hemisphere. Winds around highs **diverge** (spread out) and rotate **anticyclonically** — clockwise in the N. Hemisphere, and counterclockwise in the S. Hemisphere. The cyclones are regions of bad weather (clouds, rain, high humidity, strong winds) and fronts. The anticyclones are regions of good weather (clear skies or fair-weather clouds, no precipitation, dry air, and light winds).

The high- and low-pressure centers move on average from west to east, driven by large-scale winds from the west. Although these **westerlies** dominate the general circulation at mid-latitudes, the surface winds are quite variable in time and space due to the sum of the westerlies plus the transient circulations around the highs and lows.

Near 60° latitude are belts of low surface pressure called **subpolar lows**. Along these belts are light to calm winds, upward air motion, clouds, cool temperatures, and precipitation (as snow in winter).

Near each pole is a climatological region of high pressure called a **polar high**. In these regions are often clear skies, cold dry descending air, light winds, and little snowfall. Between each polar high (at 90°) and the subpolar low (at 60°) is a belt of weak easterly winds, called the **polar easterlies**.

Upper-troposphere

The stratosphere is strongly statically stable, and acts like a lid to the troposphere. Thus, vertical circulations associated with our weather are mostly trapped within the troposphere. These vertical circulations couple the average near-surface winds with the average upper-tropospheric (near the tropopause) winds described here (Fig. 11.3b).

In the tropics is a belt of very strong equatorial high pressure along the tops of the ITCZ thunder-

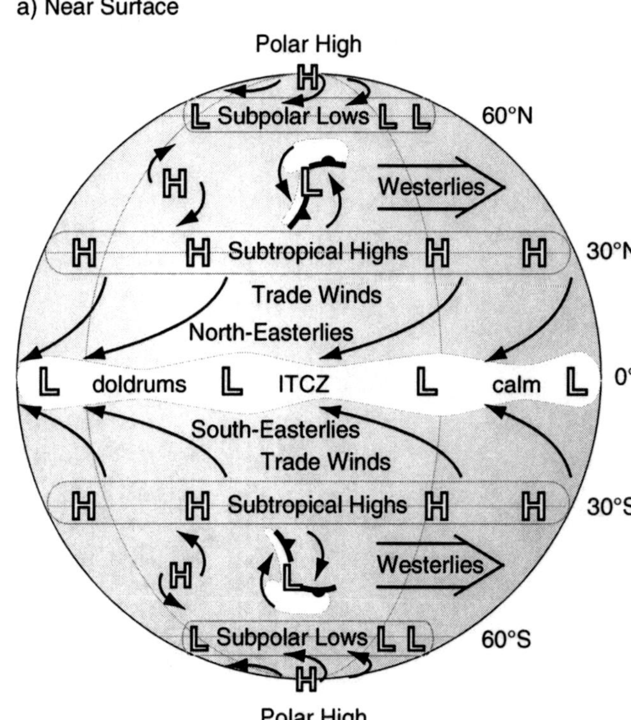

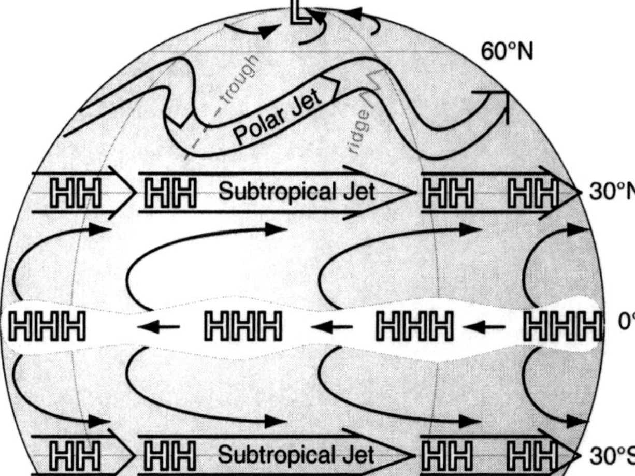

Figure 11.3
Simplified global circulation in the troposphere: (a) near the surface, and (b) near the tropopause. H and L indicate high and low pressures, and HHH means very strong high pressure. White indicates precipitating clouds.

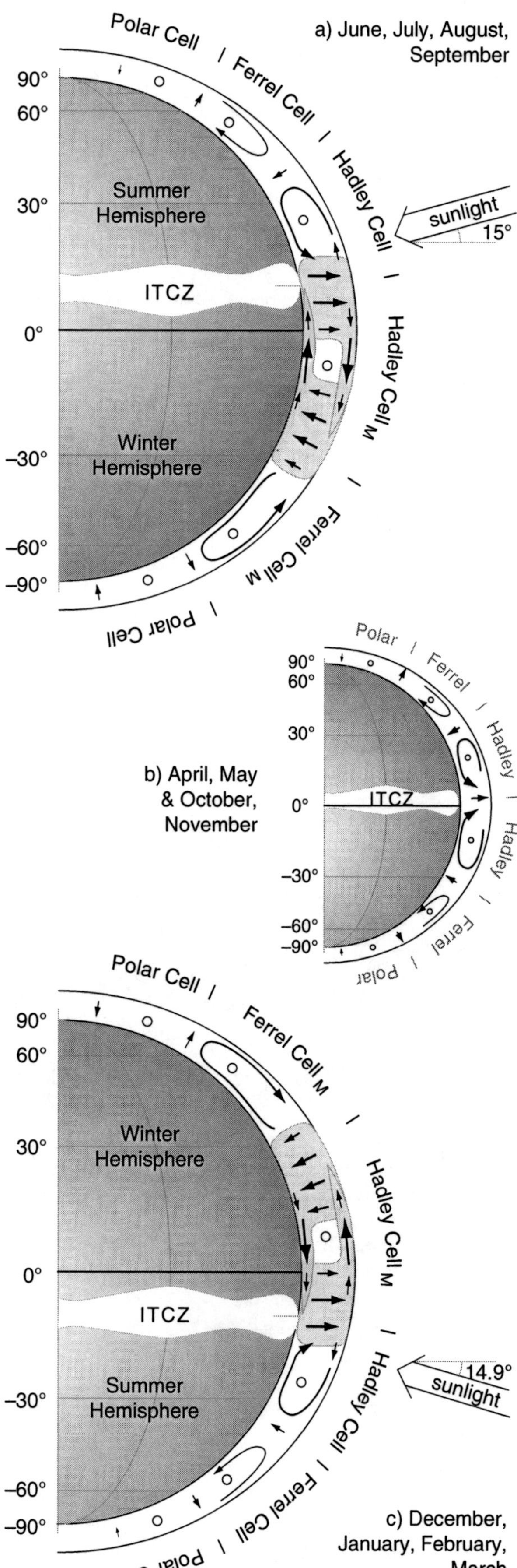

storms. Air in this belt blows from the east, due to easterly inertia from the trade winds being carried upward in the thunderstorm convection. Diverging from this belt are winds that blow toward the north in the N. Hemisphere, and toward the south in the S. Hemisphere. As these winds move away from the equator, they turn to have an increasingly westerly component as they approach 30° latitude.

Near 30° latitude in each hemisphere is a persistent belt of strong westerly winds at the tropopause called the **subtropical jet**. This jet meanders north and south a bit. Pressure here is very high, but not as high as over the equator.

In mid-latitudes at the tropopause is another belt of strong westerly winds called the **polar jet**. The centerline of the polar jet meanders north and south, resulting in a wave-like shape called a **Rossby wave**, as sketched in Fig. 11.1c. The equatorward portions of the wave are known as low-pressure **troughs**, and poleward portions are known as high-pressure **ridges**. These ridges and troughs are very transient, and generally shift from west to east relative to the ground.

Near 60° at the tropopause is a belt of low to medium pressure. At each pole is a low-pressure center near the tropopause, with winds at high latitudes generally blowing from the west causing a cyclonic circulation around the **polar low**. Thus, contrary to near-surface conditions, the near-tropopause average winds blow from the west at all latitudes (except near the equator).

Vertical Circulations

Vertical circulations of warm rising air in the tropics and descending air in the subtropics are called **Hadley cells** or **Hadley circulations** (Fig. 11.4). At the bottom of the Hadley cell are the trade winds. At the top, near the tropopause, are divergent winds. The updraft portion of the Hadley circulation is often filled with thunderstorms and heavy precipitation at the ITCZ. This vigorous convection in the troposphere causes a high tropopause (15 - 18 km altitude) and a belt of heavy rain in the tropics.

The summer- and winter-hemisphere Hadley cells are strongly asymmetric (Fig. 11.4). The major Hadley circulation (denoted with subscript "M") crosses the equator, with rising air in the summer hemisphere and descending air in the winter hemisphere. The updraft is often between 0° and

Figure 11.4 *(at left)*
Vertical cross section of Earth's global circulation in the troposphere. (a) N. Hemisphere summer. (b) Transition months. (c) S. Hemisphere summer. The major (subscript $_M$) Hadley cell is shaded light grey. Minor circulations have no subscript.

15° latitudes in the summer hemisphere, and has average core vertical velocities of 6 mm s^{-1}. The broader downdraft is often found between 10° and 30° latitudes in the winter hemisphere, with average velocity of about –4 mm s^{-1} in downdraft centers. Connecting the up- and downdrafts are meridional wind components of 3 m s^{-1} at the cell top and bottom.

The major Hadley cell changes direction and shifts position between summer and winter. During June-July-August-September, the average solar declination angle is 15°N, and the updraft is in the Northern Hemisphere (Fig. 11.4a). Out of these four months, the most well-defined circulation occurs in August and September. At this time, the ITCZ is centered at about 10°N.

During December-January-February-March, the average solar declination angle is 14.9°S, and the major updraft is in the Southern Hemisphere (Fig. 11.4c). Out of these four months, the strongest circulation is during February and March, and the ITCZ is centered at 10°S. The major Hadley cell transports significant heat away from the tropics, and also from the summer to the winter hemisphere.

During the transition months (April-May and October-November) between summer and winter, the Hadley circulation has nearly symmetric Hadley cells in both hemispheres (Fig. 11.4b). During this transition, the intensities of the Hadley circulations are weak.

When averaged over the whole year, the strong but reversing major Hadley circulation partially cancels itself, resulting in an annual average circulation that is somewhat weak and looks like Fig. 11.4b. This weak annual average is deceiving, and does not reflect the true movement of heat, moisture, and momentum by the winds. Hence, climate experts prefer to look at months JJA and DJF separately to give **seasonal averages**.

In the winter hemisphere is a **Ferrel cell**, with a vertical circulation of descending air in the subtropics and rising air at high latitudes; namely, a circulation opposite to that of the major Hadley cell. In the winter hemisphere is a modest **polar cell**, with air circulating in the same sense as the Hadley cell.

In the summer hemisphere, all the circulations are weaker. There is a minor Hadley cell and a minor Ferrel cell (Fig. 11.4). Summer-hemisphere circulations are weaker because the temperature contrast between the tropics and poles are weaker.

Monsoonal Circulations

Monsoon circulations are continental-scale circulations driven by continent-ocean temperature contrasts, as sketched in Figs. 11.5. In summer, high-pressure centers (anticyclones) are over the relative-

a) June, July, August

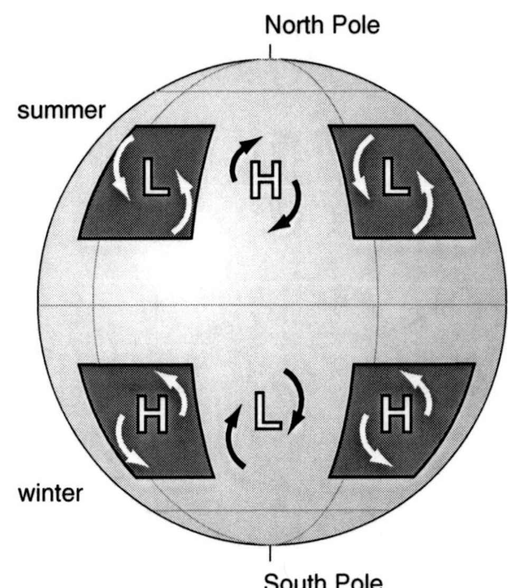

b) December, January, February

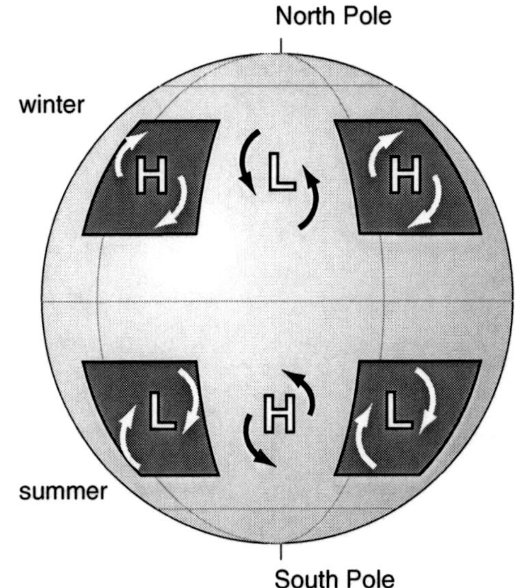

Figure 11.5
Idealized seasonal-average monsoon circulations near the surface. Continents are shaded dark grey; oceans are light grey. H and L are surface high- and low-pressure centers.

ly warm oceans, and low-pressure centers (cyclones) are over the hotter continents. In winter, low-pressure centers are over the cool oceans, and high-pressure centers are over the colder continents.

These monsoon circulations represent average conditions over a season. The actual weather on any given day can be variable, and can deviate from these seasonal averages.

Our Earth has a complex arrangement of continents and oceans. As a result, seasonally-varying monsoonal circulations are superimposed on the seasonally-varying planetary-scale circulation to yield a complex and varying global-circulation pattern.

At this point, you have a descriptive understanding of the global circulation. But what drives it?

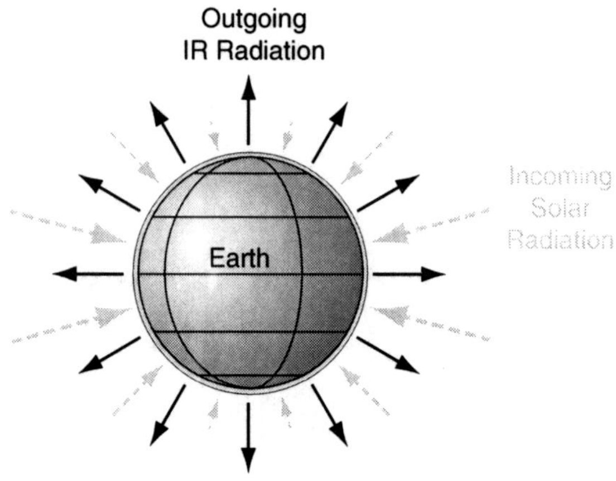

Figure 11.6
Annual average incoming solar radiation (grey dashed arrows) and of outgoing infrared (IR) radiation (solid black arrows), where arrow size indicates relative magnitude. [Because the Earth rotates and exposes all locations to the sun at one time or another, the incoming solar radiation is sketched as approaching all locations on the Earth's surface.]

RADIATIVE DIFFERENTIAL HEATING

The general circulation is driven by differential heating. Incoming solar radiation (**insolation**) nearly balances the outgoing infrared (IR) radiation when averaged over the whole globe. However, at different latitudes are significant imbalances (Fig. 11.6), which cause the differential heating.

Recall from the Solar & Infrared Radiation chapter that the flux of solar radiation incident on the top of the atmosphere depends more or less on the cosine of the latitude, as shown in Fig. 11.7. The component of the incident ray of sunlight that is perpendicular to the Earth's surface is small in polar regions, but larger toward the equator (grey dashed arrows in Figs. 11.6 and 11.7). The incoming energy adds heat to the Earth-atmosphere-ocean system.

Heat is lost due to infrared (IR) radiation emitted from the Earth-ocean-atmosphere system to space. Since all locations near the surface in the Earth-ocean-atmosphere system are relatively warm compared to absolute zero, the Stefan-Boltzmann law from the Solar & Infrared Radiation chapter tells us that the emission rates are also more or less uniform around the Earth. This is sketched by the solid black arrows in Fig. 11.6.

Thus, at low latitudes, more solar radiation is absorbed than leaves as IR, causing net warming. At high latitudes, the opposite is true: IR radiative losses exceed solar heating, causing net cooling. This differential heating drives the global circulation.

The general circulation can't instantly eliminate all the global north-south temperature differences. What remains is a meridional temperature gradient — the focus of the next subsection.

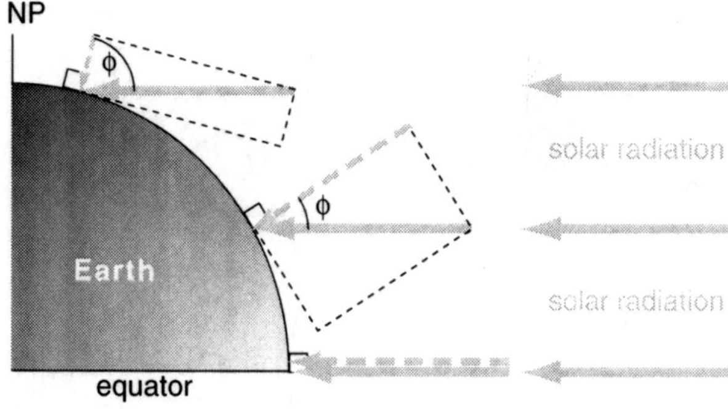

Figure 11.7
Of the solar radiation approaching the Earth (thick solid grey arrows), the component (dashed grey arrow) that is perpendicular to the top of the atmosphere is proportional to the cosine of the latitude ϕ (during the equinox).

North-South Temperature Gradient

To create a first-order toy model, neglect monthly variations, monsoonal variations and mountains. Instead, focus on surface temperatures averaged around separate latitude belts and over one year. Those latitude belts near the equator are warmer, and those near the poles are colder (Fig. 11.8a). One equation that roughly approximates the variation of zonally-averaged surface temperature T with latitude ϕ is:

$$T \approx a + b \cdot \left[\cos^3 \phi \cdot \left(1 + \frac{3}{2} \cdot \sin^2 \phi \right) \right] \tag{11.1}$$

where $a \approx -12°C$ and $b \approx 40°C = 40$ K.

The equation above applies only to the surface. At higher altitudes the north-south temperature difference is smaller, and even becomes negative in the stratosphere (i.e., warm over the poles and cold over the tropics). To account for this altitude variation, b cannot be a constant. Instead, use:

$$b \approx b_1 \cdot \left(1 - \frac{z}{z_T} \right) \tag{11.2}$$

where average tropospheric depth is $z_T \approx 11$ km, parameter $b_1 = 40°C = 40$ K, and z is height above the surface.

A similarly crude but useful generalization of parameter a can be made so that it too changes with altitude z above sea level:

$$a \approx a_1 - \gamma \cdot z \tag{11.3}$$

where $a_1 = -12°C$ and $\gamma = 3.14$ °C km^{-1}.

In Fig. 11.8a, notice that the temperature curve looks like a flattened cosine wave. The flattened curve indicates somewhat uniformly warm temperatures between ±30° latitude, caused the strong mixing and heat transport by the Hadley circulations.

With uniform tropical temperatures, the remaining change to colder temperature is pushed to the mid-latitude belts. The slope of the Fig. 11.8a curve is plotted in Fig. 11.8b. This slope is the north-south **(meridional) temperature gradient:**

$$\frac{\Delta T}{\Delta y} \approx -b \cdot c \cdot \cos^2 \phi \cdot \sin^3 \phi \tag{11.4}$$

where y is distance in the north-south direction, $c = 1.18 \times 10^{-3}$ km^{-1} is a constant valid at all heights, and b is given by eq. (11.2) which causes the gradient to change sign at higher altitudes (e.g., at z = 15 km).

Because the meridional temperature gradient results from the interplay of differential radiative heating and advection by the global circulation, let us now look at radiative forcings in more detail.

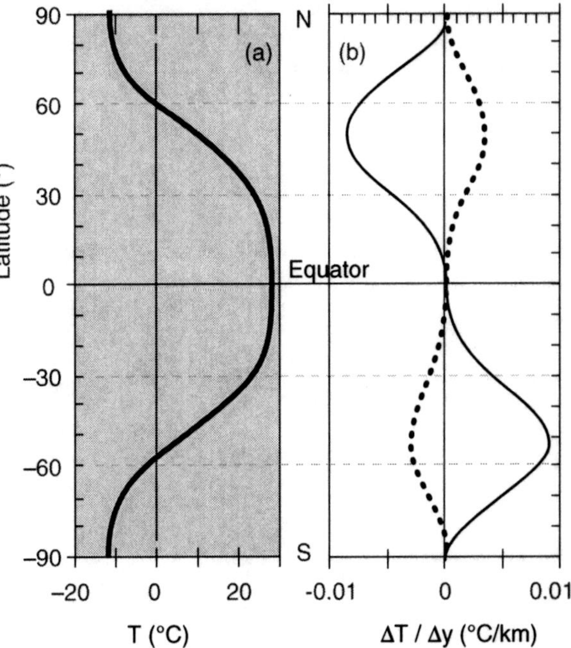

Figure 11.8
Idealized variation of annual-average temperature (a) and meridional temperature gradient (b) with latitude, averaged around latitude belts (i.e., zonal averages). Solid curve represents Earth's surface (= sea-level because the toy model neglects terrain) and dotted curve is for 15 km altitude above sea level.

Sample Application

What is the value of annual zonal average temperature and meridional temperature gradient at 50°S latitude for the surface and for z = 15 km.

Find the Answer

Given: $\phi = -50°$, (a) $z = 0$ (surface), and (b) z = 15 km
Find: T_{sfc} = ? °C, $\Delta T/\Delta y$ = ? °C km^{-1}

a) z = 0. Apply eq. (11.2): $b = (40°C)\cdot(1–0) = 40°C$.
Apply eq. (11.3): $a = (–12°C)–0°C = –12°C$.
Apply eq. (11.1) :
$T_o \approx -12°C + (40°C) \cdot \left[\cos^3(-50°) \cdot \left(1 + \frac{3}{2} \cdot \sin^2(-50°) \right) \right]$ = **7.97 °C**

Apply eq. (11.4):
$\frac{\Delta T}{\Delta y} \approx -(40°C) \cdot (1.18 \times 10^{-3}) \cdot \cos^2(-50°) \cdot \sin^3(-50°)$ = **0.0087°C km^{-1}**

b) z = 15 km.
Apply eq. (11.2): $b = (40°C)\cdot[1–(15/11)] = –14.55°C$.
Apply (11.3): $a = (–12°C)–(3.14°C/km)\cdot(15km) = –59.1°C$.
Apply eq. (11.1) :
$T_{15km} \approx -59.1°C - 14.6°C \cdot \left[\cos^3(-50°) \cdot \left(1 + \frac{3}{2} \cdot \sin^2(-50°) \right) \right]$ = **–66.4°C**

Apply eq. (11.4) at z = 15 km:
$\frac{\Delta T}{\Delta y} \approx +(14.55°C) \cdot (1.18 \times 10^{-3}) \cdot \cos^2(-50°) \cdot \sin^3(-50°)$ = **–0.0032 °C km^{-1}**

Check: Phys. & units reasonable. Agrees with Fig. 11.8 for Southern Hemisphere.
Exposition: Cold at z = 15 km even though warm at z = 0. Gradient signs would be opposite in N. Hem.

HIGHER MATH • Derivation of the North-South Temperature Gradient

The goal is to find $\partial T/\partial y$ for the toy model.

a) First, expand the derivative: $$\frac{\partial T}{\partial y} = \frac{\partial T}{\partial \phi} \cdot \frac{\partial \phi}{\partial y} \qquad \text{(a)}$$

We will look at factors $\partial T/\partial\phi$ and $\partial\phi/\partial y$ separately:

b) Factor $\partial\phi/\partial y$ describes the meridional gradient of latitude. Consider a circumference of the Earth that passes through both poles. The total latitude change around this circle is $\Delta\phi = 2\pi$ radians. The total circumference of this circle is $\Delta y = 2\pi R$ for average Earth radius of $R = 6371$ km. Hence:

$$\frac{\partial \phi}{\partial y} = \frac{\Delta \phi}{\Delta y} = \frac{2\pi}{2\pi \cdot R} = \frac{1}{R} \qquad \text{(b)}$$

c) For factor $\partial T/\partial\phi$, start with eq. (11.1) for the toy model:

$$T \approx a + b \cdot \left[\cos^3 \phi \cdot \left(1 + \frac{3}{2} \cdot \sin^2 \phi\right)\right] \qquad \text{(11.1)}$$

and take its derivative vs. latitude: $\frac{\partial T}{\partial \phi} = b \cdot \left(\frac{3}{2}\right) \cdot$

$$\left[(2\sin\phi \cdot \cos\phi)\cos^3\phi - 3\left(\frac{2}{3} + \sin^2\phi\right)\cos^2\phi \cdot \sin\phi\right]$$

Next, take the common term ($\sin\phi \cdot \cos^2\phi$) out of [ ]:

$$\frac{\partial T}{\partial \phi} = b\left(\frac{3}{2}\right)\sin\phi \cdot \cos^2\phi \cdot \left[2\cos^2\phi - 2 - 3\sin^2\phi\right]$$

Use the trig. identity: $\cos^2\phi = 1 - \sin^2\phi$.
Hence: $2\cos^2\phi = 2 - 2\sin^2\phi$. Substituting this in into the previous full-line equation gives:

$$\frac{\partial T}{\partial \phi} = b\left(\frac{3}{2}\right)\sin\phi \cdot \cos^2\phi \cdot \left[-5\sin^2\phi\right] \qquad \text{(c)}$$

d) Plug equations (b) and (c) back into eq. (a):

$$\frac{\partial T}{\partial y} = -b \cdot \left(\frac{15}{2} \cdot \frac{1}{R}\right) \cdot \sin^3\phi \cdot \cos^2\phi \qquad \text{(d)}$$

Define:
$$c = \left(\frac{15}{2} \cdot \frac{1}{R}\right) = \left(\frac{15}{2} \cdot \frac{1}{6371\text{km}}\right) = 1.177 \times 10^{-3} \text{ km}^{-1}$$

Thus, the final answer is:

$$\boxed{\frac{\Delta T}{\Delta y} \approx \frac{\partial T}{\partial y} = -b \cdot c \cdot \sin^3\phi \cdot \cos^2\phi} \qquad \text{(11.4)}$$

Check: Fig. 11.8 shows the curves that were calculated from eqs. (11.1) and (11.4). The fact that the sign and shape of the curve for $\Delta T/\Delta y$ is consistent with the curve for $T(y)$ suggests the answer is reasonable.

Alert: Eqs. (11.4) and (11.1) are based on a highly idealized "toy model" of the real atmosphere. They were designed only to illustrate first-order effects.

Global Radiation Budgets

Incoming Solar Radiation

Because of the tilt of the Earth's axis and the change of seasons, the actual flux of incoming solar radiation is not as simple as was sketched in Fig. 11.6. But this complication was already discussed in the Solar & Infrared Radiation chapter, where we saw an equation to calculate the incoming solar radiation (**insolation**) as a function of latitude and day. The resulting insolation figure is reproduced below (Fig. 11.9a).

If you take the spreadsheet data from the Solar & Infrared Radiation chapter that was used to make this figure, and average rows of data (i.e., average over all months for any one latitude), you can find the annual average insolation E_{insol} for each latitude (Fig. 11.9b). Insolation in polar regions is amazingly large.

The curve in Fig. 11.9b is simple, and in the spirit of a toy model can be nicely approximated by:

$$E_{insol} = E_o + E_1 \cdot \cos(2\,\phi) \qquad \text{(11.5)}$$

where the empirical parameters are $E_o = 298$ W m^{-2}, $E_1 = 123$ W m^{-2}, and ϕ is latitude. This curve and the data points it approximates are plotted in Fig. 11.10.

But not all the radiation incident on the top of the atmosphere is absorbed by the Earth-ocean-atmosphere system. Some is reflected back into space from snow and ice on the surface, from the oceans, and from light-colored land. Some is reflected from

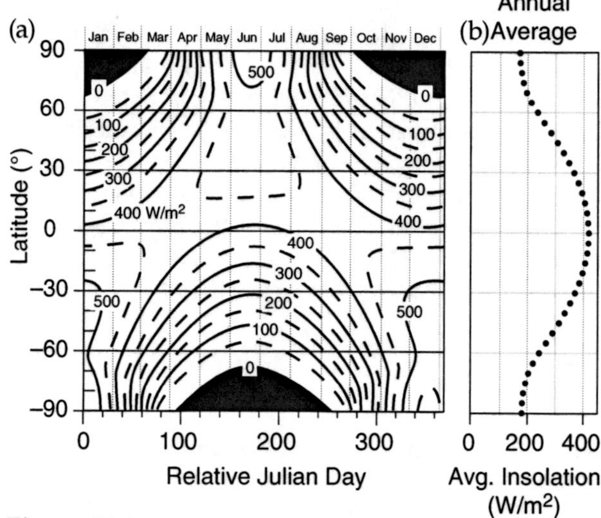

Figure 11.9
(a) Solar radiation (W m^{-2}) incident on the top of the atmosphere for different latitudes and months (copied from the Solar & Infrared Radiation chapter). (b) Meridional variation of insolation, found by averaging the data from the left figure over all months for each separate latitude (i.e., averages for each row of data).

cloud top. Some is scattered off of air molecules. The amount of insolation that is NOT absorbed is surprisingly constant with latitude at about $E_2 \approx 110$ W m^{-2}. Thus, the amount that IS absorbed is:

$$E_{in} = E_{insol} - E_2 \qquad (11.6)$$

where E_{in} is the incoming flux (W m^{-2}) of solar radiation absorbed into the Earth-ocean-atmosphere system (Fig. 11.10).

Outgoing Terrestrial Radiation

As you learned in the Satellites & Radar chapter, infrared radiation emission and absorption in the atmosphere are very complex. At some wavelengths the atmosphere is mostly transparent, while at others it is mostly opaque. Thus, some of the IR emissions to space are from the Earth's surface, some from cloud top, and some from air at middle altitudes in the atmosphere.

In the spirit of a toy model, suppose that the net IR emissions are characteristic of the absolute temperature T_m near the middle of the troposphere (at about $z_m = 5.5$ km). Approximate the outbound flux of radiation E_{out} (averaged over a year and averaged around latitude belts) by the Stefan-Boltzmann law (see the Solar & Infrared Radiation chapter):

$$E_{out} \approx \varepsilon \cdot \sigma_{SB} \cdot T_m^4 \qquad (11.7)$$

where the effective emissivity is $\varepsilon \approx 0.9$ (see the Climate chapter), and the Stefan-Boltzmann constant is $\sigma_{SB} = 5.67\text{x}10^{-8}$ W·m^{-2}·K^{-4}. When you use $z = z_m = 5.5$ km in eqs. (11.1 - 11.3) to get T_m vs. latitude for use in eq. (11.7), the result is E_{out} vs. ϕ, as plotted in Fig. 11.10.

Net Radiation

For an air column over any square meter of the Earth's surface, the radiative input minus output gives the net radiative flux:

$$E_{net} = E_{in} - E_{out} \qquad (11.8)$$

which is plotted in Fig. 11.10 for our toy model.

Sample Application

Estimate the annual average solar energy absorbed at the latitude of the Eiffel Tower in Paris, France.

Find the Answer

Given: $\phi = 48.8590°$ (at the Eiffel tower)
Find: $E_{in} = ?$ W m^{-2}

Use eq. (11.5):
E_{insol} = (298 W m^{-2}) + (123 W m^{-2})·cos(2 · 48.8590°)
= (298 W m^{-2}) – (16.5 W m^{-2}) = 281.5 W m^{-2}
Use eq. (11.6):
E_{in} = 281.5 W m^{-2} – 110.0 W m^{-2} = **171.5 W m^{-2}**

Check: Units OK. Agrees with Fig. 11.10.
Exposition: The actual annual average E_{in} at the Eiffel Tower would probably differ from this zonal avg.

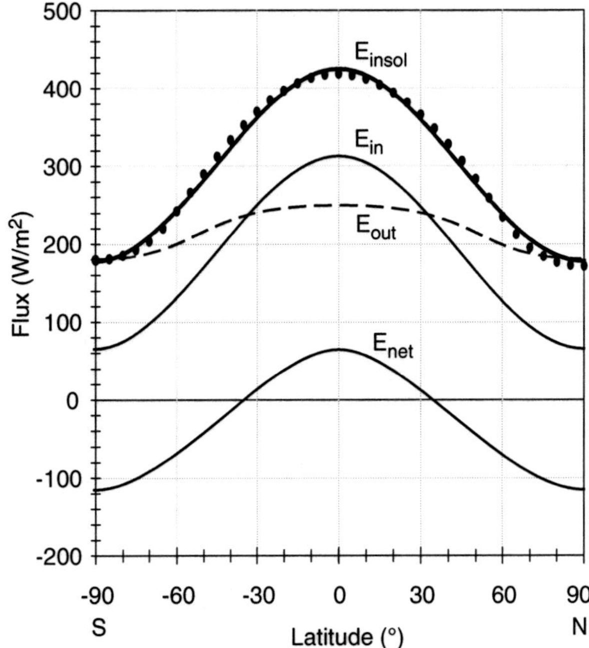

Figure 11.10
Data points are insolation vs. latitude from Fig. 11.9b. Eq. 11.5 approximates this insolation E_{insol} (thick black line). E_{in} is the solar radiation that is absorbed (thin solid line, from eq. 11.6). E_{out} is outgoing terrestrial (IR) radiation (dashed; from eq. 11.7). Net flux $E_{net} = E_{in} - E_{out}$. Positive E_{net} causes heating; negative causes cooling.

Sample Application. What is E_{net} at the Eiffel Tower latitude?
Find the Answer. Given: $\phi = 48.8590°$, $z = 5.5$ km, $E_{in} = 171.5$ W m^{-2} from previous Sample Application
Find: $E_{net} = ?$ W m^{-2}
Apply eq. (11.2): b = (40°C)·[1 – (5.5 km/11 km)] = 20°C
Apply eq.(11.3): a=(-12°C)–(3.14°C km^{-1})·(5.5 km) = -29.27°C
Apply eq. (11.1): T_m = –18.73 °C = 254.5 K
Apply eq.(11.7): E_{out}=(0.9)·(5.67x10^{-8}W·m^{-2}·K^{-4})·(254.5 K)^4 = 213.8 W m^{-2}
Apply eq. (11.8): E_{net} = (171.5 W m^{-2}) – (213.8 W m^{-2}) = **–42.3 W m^{-2}**
Check: Units OK. Agrees with Fig. 11.10
Exposition: The net radiative heat loss at Paris latitude must be compensated by winds blowing heat in.

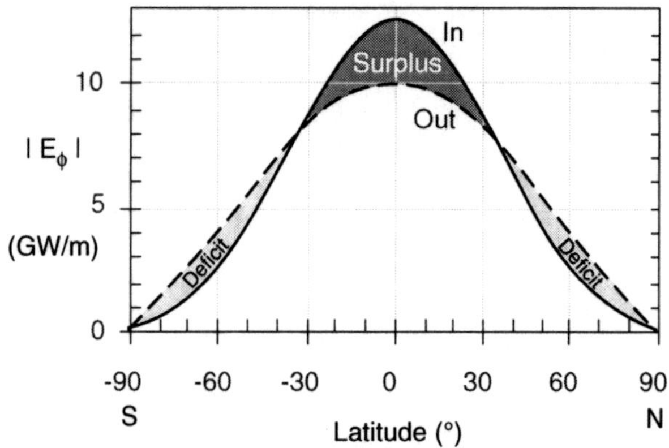

Figure 11.11
Zonally-integrated radiative forcings for absorbed incoming solar radiation (solid line) and emitted net outgoing terrestrial (IR) radiation (dashed line). The surplus balances the deficit.

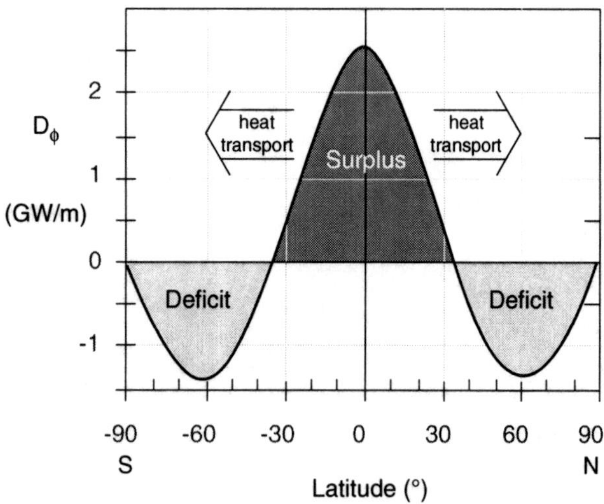

Figure 11.12
Net (incoming minus outgoing) zonally-integrated radiative forcings on the Earth. Surplus balances deficits. This differential heating imposed on the Earth must be compensated by heat transport by the global circulation; otherwise, the tropics would keep getting hotter and the polar regions colder.

> **Sample Application**
> From the previous Sample Application, find the zonally-integrated differential heating at $\phi = 48.859°$.
>
> **Find the Answer**
> Given: $E_{net} = -42.3$ W m^{-2} , $\phi = 48.859°$ at Eiffel Tower
> Find: D_ϕ = ? GW m^{-1}
>
> Combine eqs. (11.8-11.10): $D_\phi = 2\pi \cdot R_{Earth} \cdot \cos(\phi) \cdot E_{net}$
> $= 2(3.14159)\cdot(6.357\text{x}10^6\text{m})\cdot\cos(48.859°)\cdot(-42.3 \text{ W m}^{-2})$
> $= -1.11\text{x}10^9$ W m^{-1} = **<u>-1.11 GW m^{-1}</u>**
>
> **Check**: Units OK. Agrees with Fig. 11.12.
> **Exposition**: Net radiative heat loss at this latitude is compensated by warm Gulf stream and warm winds.

Radiative Forcing by Latitude Belt

Do you notice anything unreasonable about E_{net} in Fig. 11.10? It appears that the negative area under the curve is much greater than the positive area, which would cause the Earth to get colder and colder — an effect that is not observed.

Don't despair. Eq. (11.8) is correct, but we must remember that the circumference $[2\pi \cdot R_{Earth} \cdot \cos(\phi)]$ of a parallel (a constant latitude circle) is smaller near the poles than near the equator. ϕ is latitude and the average Earth radius R_{Earth} is 6371 km.

The solution is to multiply eqs. (11.6 - 11.8) by the circumference of a parallel. The result

$$E_\phi = 2\pi \cdot R_{Earth} \cdot \cos(\phi) \cdot E \tag{11.9}$$

can be applied for $E = E_{in}$, or $E = E_{out}$, or $E = E_{net}$. This converts from E in units of W m^{-2} to E_ϕ in units of W m^{-1}, where the distance is north-south distance. You can interpret E_ϕ as the power being transferred to/from a one-meter-wide sidewalk that encircles the Earth along a parallel.

Figs. (11.11) and (11.12) show the resulting incoming, outgoing, and net radiative forcings vs. latitude. At most latitudes there is nonzero net radiation. We can define E_{net} as **differential heating** D_ϕ :

$$D_\phi = E_{\phi\, net} = E_{\phi\, in} - E_{\phi\, out} \tag{11.10}$$

Our despair is now quelled, because the surplus and deficit areas in Figs. 11.11 and 11.12 are almost exactly equal in magnitude to each other. Hence, we anticipate that Earth's climate should be relatively steady (neglecting global warming for now).

General Circulation Heat Transport

Nonetheless, the imbalance of net radiation between equator and poles in Fig. 11.12 drives atmospheric and oceanic circulations. These circulations act to undo the imbalance by removing the excess heat from the equator and depositing it near the poles (as per Le Chatelier's Principle). First, we can use the radiative differential heating to find how much global-circulation heat transport is needed. Then, we can examine the actual heat transport by atmospheric and oceanic circulations.

The Amount of Transport Required

By definition, the meridional transport at the poles is zero. If we sum D_ϕ from Fig. 11.12 over all latitude belts from the North Pole to any other latitude ϕ , we can find the total transport Tr required for the global circulation to compensate all the radiative imbalances north of that latitude:

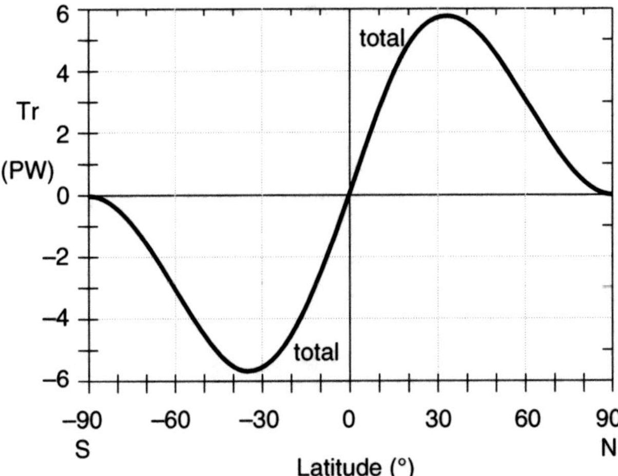

Figure 11.13
Required heat transport Tr by the global circulation to compensate radiative differential heating, based on a simple "toy model". Agrees very well with achieved transport in Fig. 11.14.

$$Tr(\phi) = \sum_{\phi_o=90°}^{\phi} (-D_\phi) \cdot \Delta y \tag{11.11}$$

where the width of any latitude belt is Δy.

[Meridional distance Δy is related to latitude change $\Delta\phi$ by: Δy(km) = (111 km/°) · $\Delta\phi$ (°) .]

The resulting "needed transport" is shown in Fig. 11.13, based on the simple "toy model" temperature and radiation curves of the past few sections. The magnitude of this curve peaks at about 5.6 PW (1 petaWatt equals 10^{15} W) at latitudes of about 35° North and South (positive *Tr* means northward transport).

Transport Achieved

Satellite observations of radiation to and from the Earth, estimates of heat fluxes to/from the ocean based on satellite observations of sea-surface temperature, and in-situ measurements of the atmosphere provide some of the transport data needed. Numerical forecast models are then used to tie the observations together and fill in the missing pieces. The resulting estimate of heat transport achieved by the atmosphere and ocean is plotted in Fig. 11.14.

Ocean currents dominate the total heat transport only at latitudes 0 to 17°, and remain important up to latitudes of ± 40°. In the atmosphere, the Hadley circulation is a dominant contributor in the tropics and subtropics, while the Rossby waves dominate atmospheric transport at mid-latitudes.

Knowing that global circulations undo the heating imbalance raises another question. How does the differential heating in the atmosphere drive the winds in those circulations? That is the subject of the next three sections.

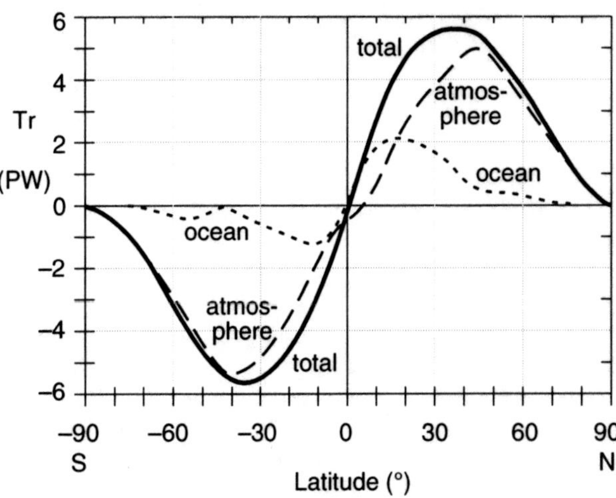

Figure 11.14
*Meridional heat transports: Satellite-observed total (solid line) & ocean estimates (dotted). Atmospheric (dashed) is found as a residual. 1 PW = 1 petaWatt = 10^{15} W. [Data from K. E. Trenberth and J. M. Caron, 2001: "J. Climate", **14**, 3433-3443.]*

Sample Application

What total heat transports by the atmosphere and ocean circulations are needed at 50°N latitude to compensate for all the net radiative cooling between that latitude and the North Pole? The differential heating as a function of latitude is given in the following table (based on the toy model):

Lat (°)	D_ϕ (GW m^{-1})	Lat (°)	D_ϕ (GW m^{-1})
90	0	65	–1.380
85	–0.396	60	–1.403
80	–0.755	55	–1.331
75	–1.049	50	–1.164
70	–1.261	45	–0.905

Find the Answer

Given: ϕ = 50°N. D_ϕ data in table above.
Find: Tr = ? PW

Use eq. (11.11). Use sidewalks (latitude belts) each of width $\Delta\phi$ = 5°. Thus, Δy (m) = (111,000 m/°) · (5°) = 555,000 m is the sidewalk width. If one sidewalk spans 85 - 90°, and the next spans 80 - 85° etc, then the values in the table above give D_ϕ along the edges of the sidewalk, not along the middle. A better approximation is to average the D_ϕ values from each edge to get a value representative of the whole sidewalk. Using a bit of algebra, this works out to:

Tr = – (555000 m)· [(0.5)·0.0 – 0.396 – 0.755 – 1.049 – 1.261 – 1.38 – 1.403 – 1.331 – (0.5)·1.164] (GW m^{-1})

Tr = (555000 m)·[8.157 GW m^{-1}] = **4.527 PW**

Check: Units OK (10^6 GW = 1 PW). Agrees with Fig. 11.14. **Exposition**: This northward heat transport warms all latitudes north of 50°N, not just one sidewalk. The warming per sidewalk is $\Delta Tr/\Delta\phi$.

A SCIENTIFIC PERSPECTIVE • Residuals

If something you cannot measure contributes to things you can measure, then you can estimate the unknown as the **residual** (i.e., difference) from all the knowns. This is a valid scientific approach. It was used in Fig. 11.14 to estimate the atmospheric portion of global heat transport.

CAUTION: When using this approach, your residual not only includes the desired signal, but it also includes the sums of all the errors from the items you measured. These errors can easily accumulate to cause a "noise" that is larger than the signal you are trying to estimate. For this reason, **error estimation** and **error propagation** (see Appendix A) should always be done when using the method of residuals.

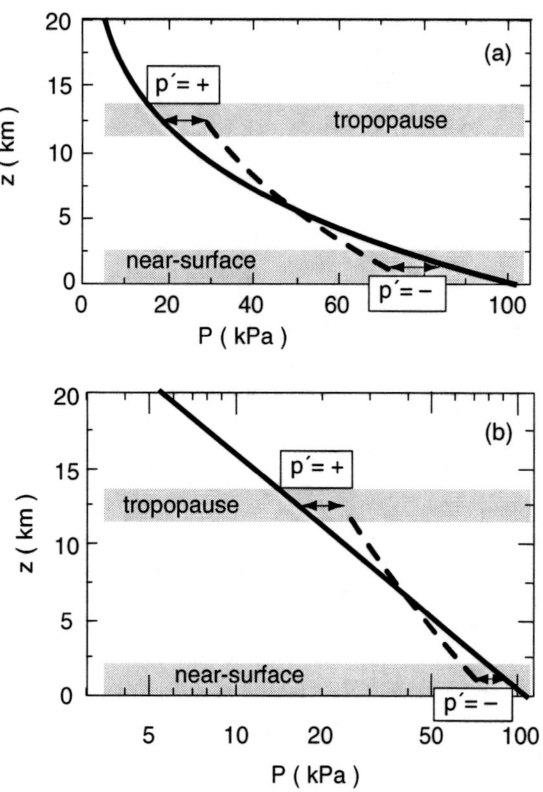

Figure 11.15
Background hydrostatic pressure (solid line), and non-hydrostatic column of air (dashed line) with pressure perturbation p′ that deviates from the hydrostatic pressure $P_{hydrostatic}$ at most heights z. In this example, even though p′ is positive (+) near the tropopause, the total pressure in the column ($P_{column} = P_{hydrostatic} + p'$) at the tropopause is still less than the surface pressure. The same curve is plotted as (a) linear and as (b) semilog.

PRESSURE PROFILES

The following fundamental concepts can help you understand how the global circulation works:

- non-hydrostatic pressure couplets due to horizontal winds and vertical buoyancy,
- hydrostatic thermal circulations,
- geostrophic adjustment, and
- the thermal wind.

The first two concepts are discussed in this section. The last two are discussed in subsequent sections.

Non-hydrostatic Pressure Couplets

Consider a background reference environment with no vertical acceleration (i.e., **hydrostatic**). Namely, the pressure-decrease with height causes an upward pressure-gradient force that exactly balances the downward pull of gravity, causing zero net vertical force on the air (see Fig. 1.12 and eq. 1.25).

Next, suppose that immersed in this environment is a column of air that might experience a different pressure decrease (Fig. 11.15); i.e., **non-hydrostatic pressures**. At any height, let $p' = P_{column} - P_{hydrostatic}$ be the deviation of the actual pressure in the column from the theoretical hydrostatic pressure in the environment. Often a positive p' in one part of the atmospheric column is associated with negative p' elsewhere. Taken together, the positive and negative p's form a **pressure couplet**.

Non-hydrostatic p' profiles are often associated with non-hydrostatic vertical motions through Newton's second law. These non-hydrostatic motions can be driven by horizontal convergence and divergence, or by buoyancy (a vertical force). These two effects create opposite pressure couplets, even though both can be associated with upward motion, as explained next.

Horizontal Convergence/Divergence

If external forcings cause air near the ground to converge horizontally, then air molecules accumulate. As density ρ increases according to eq. (10.60), the ideal gas law tells us that p' will also become positive (Fig. 11.16a).

Positive p' does two things: it (1) decelerates the air that was converging horizontally, and (2) accelerates air vertically in the column. Thus, the pressure perturbation causes **mass continuity** (horizontal inflow near the ground balances vertical outflow).

Similarly, an externally imposed horizontal divergence at the top of the troposphere would lower the air density and cause negative p', which would also accelerate air in the column upward. Hence, we expect upward motion (W = positive) to be driven by a p' couplet, as shown in Fig. 11.16a.

Buoyant Forcings

For a different scenario, suppose air in a column is positively buoyant, such as in a thunderstorm where water-vapor condensation releases lots of latent heat. This vertical buoyant force creates upward motion (i.e., warm air rises, as in Fig. 11.16b).

As air in the thunderstorm column moves away from the ground, it removes air molecules and lowers the density and the pressure; hence, p' is negative near the ground. This suction under the updraft causes air near the ground to horizontally converge, thereby conserving mass.

Conversely, at the top of the troposphere where the thunderstorm updraft encounters the even warmer environmental air in the stratosphere, the upward motion rapidly decelerates, causing air molecules to accumulate, making p' positive. This pressure perturbation drives air to diverge horizontally near the tropopause, causing the outflow in the anvil-shaped tops of thunderstorms.

The resulting pressure-perturbation p' couplet in Fig. 11.16b is opposite that in Fig. 11.16a, yet both are associated with upward vertical motion. The reason for this pressure-couplet difference is the difference in driving mechanism: imposed horizontal convergence/divergence vs. imposed vertical buoyancy.

Similar arguments can be made for downward motions. For either upward or downward motions, the pressure couplets that form depend on the type of forcing. We will use this process to help explain the pressure patterns at the top and bottom of the troposphere.

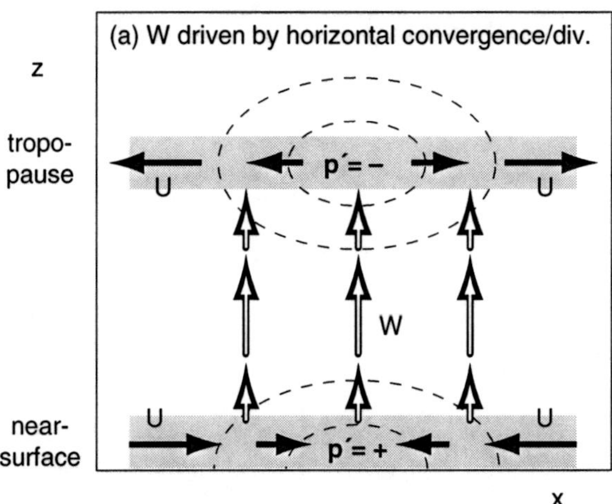

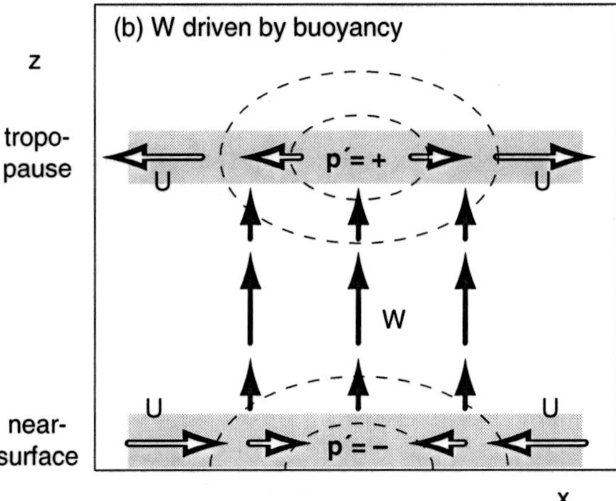

Figure 11.16
(a) Vertical motions driven by horizontal convergence or divergence. (b) Vertical motions driven by buoyancy. In both figures, black arrows indicate the cause (the driving force), and white arrows are the effect (the response). p' is the pressure perturbation (deviation from hydrostatic), and thin dashed lines are isobars of p'. U and W are horizontal and vertical velocities. In both figures, the responding flow (white arrows) is driven down the pressure-perturbation gradient, away from positive p' and toward negative p'.

Hydrostatic Thermal Circulations

Cold columns of air tend to have high surface pressures, while warm columns have low surface pressures. Figs. 11.17 illustrate how this happens.

Consider initial conditions (Fig. 11.17i) of two equal columns of air at the same temperature and with the same number of air molecules in each column. Since pressure is related to the mass of air above, this means that both columns A and B have the same initial pressure (100 kPa) at the surface.

Next, suppose that some process heats one column relative to the other (Fig. 11.17ii). Perhaps condensation in a thunderstorm cloud causes latent heating of column B, or infrared radiation cools column A. After both columns have finished expanding or contracting due to the temperature change, they will reach new hydrostatic equilibria for their respective temperatures.

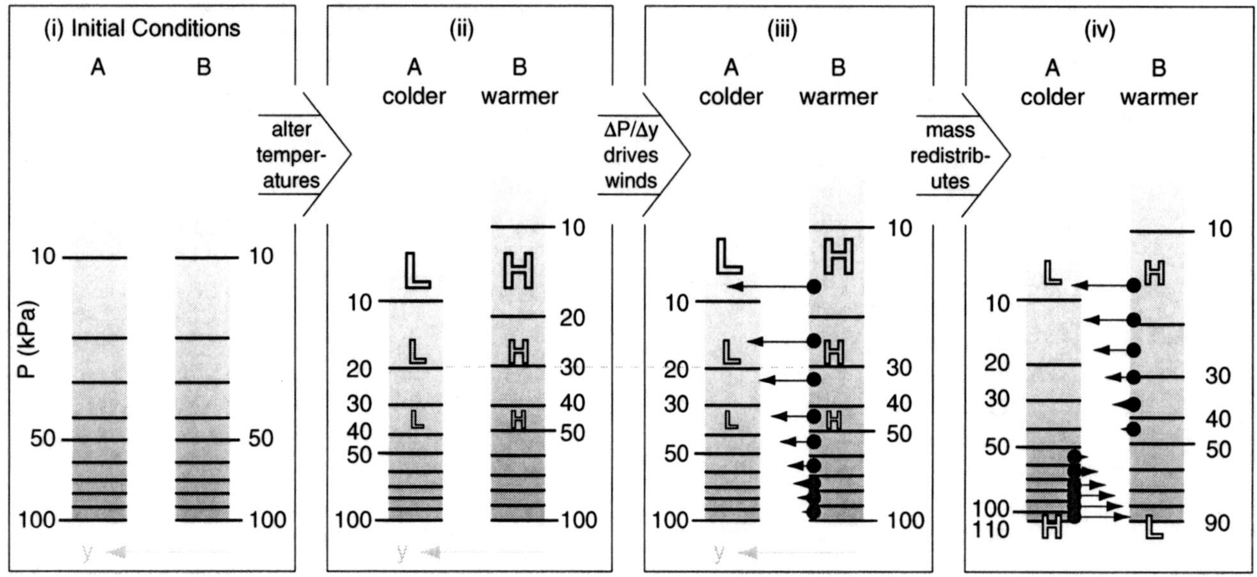

Figure 11.17

Formation of a thermal circulation. The response of two columns of air that are heated differently is that the warmer air column develops a low pressure perturbation at the surface and a high pressure perturbation aloft. Response of the cold column is the opposite. Notation: H = high pressure perturbation, L = low; black dots represents air parcels; thin arrows are winds.

The hypsometric equation (see Chapter 1) says that pressure decreases more rapidly with height in cold air than in warm air. Thus, although both columns have the same surface pressure because they contain the same number of molecules, the higher you go above the surface, the greater is the pressure difference between warm and cold air. In Fig. 11.17ii, the printed size of the "H" and "L" indicate the relative magnitudes of the high- and low-pressure perturbations p'.

The horizontal pressure gradient $\Delta P/\Delta y$ aloft between the warm and cold air columns drives horizontal winds from high toward low pressure (Fig. 11.17iii). Since winds are the movement of air molecules, this means that molecules leave the regions of high pressure-perturbation and accumulate in the regions of low. Namely, they leave the warm column, and move into the cold column.

Since there are now more molecules (i.e., more mass) in the cold column, it means that the surface pressure must be greater (H) in the cold column (Fig. 11.17iv). Similarly, mass lost from the warm column results in lower (L) surface pressure. This is called a **thermal low.**

A result (Fig. 11.17iv) is that, near the surface, high pressure in the cold air drives winds toward the low pressure in warm air. Aloft, high pressure-perturbation in the warm air drives winds towards low pressure-perturbation in the cold air. The resulting **thermal circulation** causes each column to gain as many air molecules as they lose; hence, they are in mass equilibrium.

This equilibrium circulation also transports heat. Air from the warm air column mixes into the cold column, and vice versa. This intermixing reduces the temperature contrast between the two columns, causing the corresponding equilibrium circulations to weaken. Continued destabilization (more latent heating or radiative cooling) would be needed to maintain the circulation.

The circulations and mass exchange described above can be realized at the equator. At other latitudes, the exchange of mass is often slower (near the surface) or incomplete (aloft) because Coriolis force turns the winds to some angle away from the pressure-gradient direction. This added complication, due to geostrophic wind and geostrophic adjustment, is described next.

GEOSTROPHIC WIND & GEOSTROPHIC ADJUSTMENT

Ageostrophic Winds at the Equator

Air at the equator can move directly from high (H) to low (L) pressure (Fig. 11.18 - center part) under the influence of pressure-gradient force. Zero Coriolis force at the equator implies infinite geostrophic winds. But actual winds have finite speed, and are thus **ageostrophic** (not geostrophic).

Because such flows can happen very easily and quickly, equatorial air tends to quickly flow out of highs into lows, causing the pressure centers to neutralize each other. Indeed, weather maps at the equator show very little pressure variations zonally. One exception is at continent-ocean boundaries, where continental-scale differential heating can continually regenerate pressure gradients to compensate the pressure-equalizing action of the wind. Thus, very small pressure gradients can cause continental-scale (5000 km) monsoon circulations near the equator. Tropical forecasters focus on winds, not pressure.

If the large-scale pressure is uniform in the horizontal near the equator (away from monsoon circulations), then the horizontal pressure gradients disappear. With no horizontal pressure-gradient force, no large-scale winds can be driven there. However, winds can exist at the equator due to inertia — if the winds were first created geostrophically at nonzero latitude and then coast across the equator.

But at most other places on Earth, Coriolis force deflects the air and causes the wind to approach geostrophic or gradient values (see the Forces & Winds chapter). Geostrophic winds do not cross isobars, so they cannot transfer mass from highs to lows. Thus, significant pressure patterns (e.g., strong high and low centers, Fig. 11.18) can be maintained for long periods at mid-latitudes in the global circulation.

Definitions

A **temperature field** is a map showing how temperatures are spatially distributed. A **wind field** is a map showing how winds are distributed. A **mass field** represents how air mass is spatially distributed. But we don't routinely measure air mass.

In the first Chapter, we saw how pressure at any one altitude depends on (is a measure of) all the air mass above that altitude. So we can use the **pressure field** at a fixed altitude as a surrogate for the mass field. Similarly, the **height field** on a map of constant pressure is another surrogate.

But the temperature and mass fields are coupled via the hypsometric equation (see Chapter 1). Namely, if the temperature changes in the horizon-

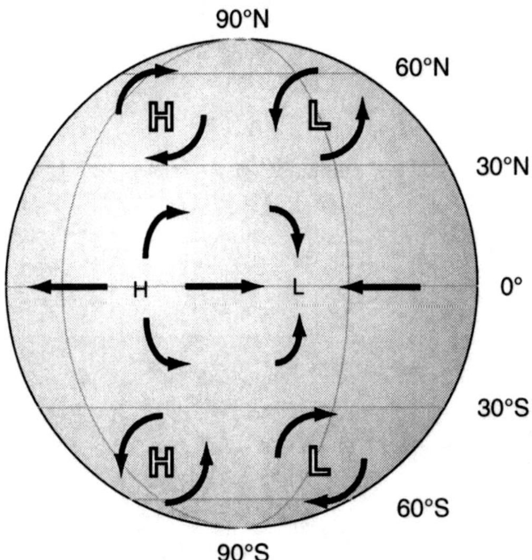

Figure 11.18
At the equator, winds flow directly from high (H) to low (L) pressure centers. At other latitudes, Coriolis force causes the winds to circulate around highs and lows. Smaller size font for H and L at the equator indicate weaker pressure gradients.

A SCIENTIFIC PERSPECTIVE • The Scientific Method Revisited

"Like other exploratory processes, [the scientific method] can be resolved into a dialogue between fact and fancy, the actual and the possible; between what could be true and what is in fact the case. The purpose of scientific enquiry is not to compile an inventory of factual information, nor to build up a totalitarian world picture of Natural Laws in which every event that is not compulsory is forbidden. We should think of it rather as a logically articulated structure of justifiable beliefs about a Possible World — a story which we invent and criticize and modify as we go along, so that it ends by being, as nearly as we can make it, a story about real life."

- by Nobel Laureate Sir Peter Medawar (1982) *Pluto's Republic*. Oxford Univ. Press.

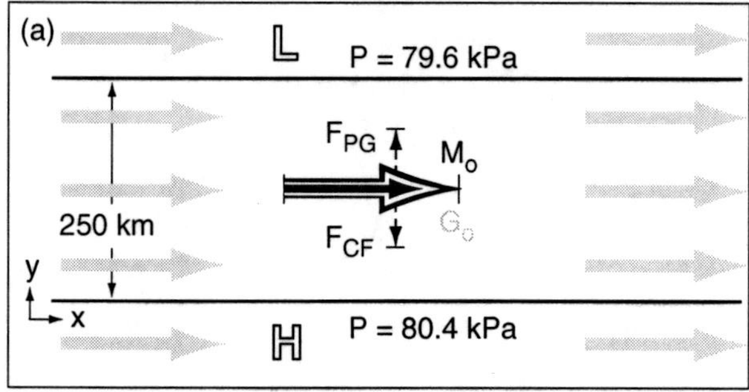

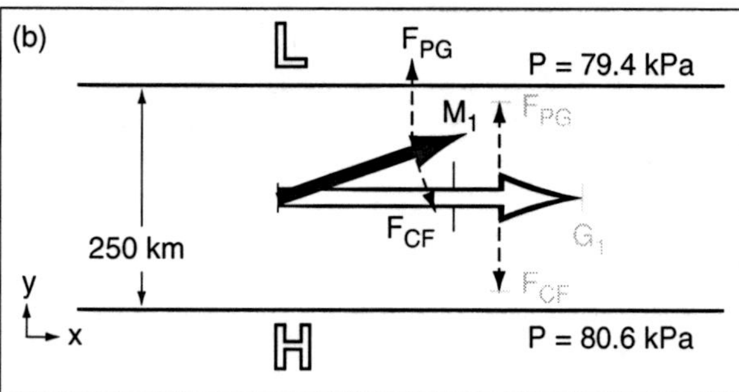

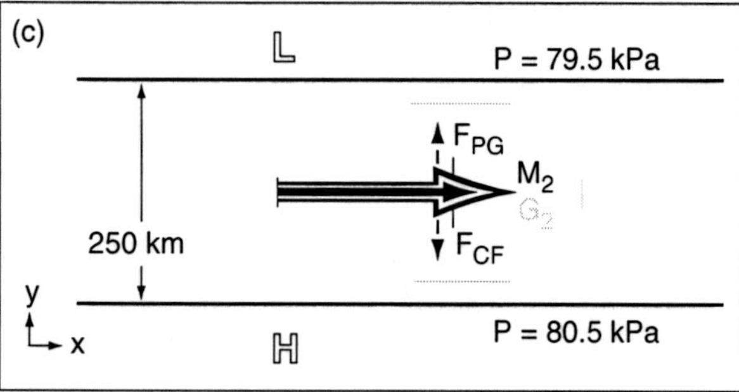

Figure 11.19
Example of geostrophic adjustment in the N. Hemisphere (not at equator). (a) Initial conditions, with the actual wind M (thick black arrow) in equilibrium with (equal to) the theoretical geostrophic value G (white arrow with black outline). (b) Transition. (c) End result at a new equilibrium. Dashed lines indicate forces F. Each frame focuses on the region of disturbance.

Sample Application
Find the internal Rossby radius of deformation in a standard atmosphere at 45°N.

Find the Answer
Given: $\phi = 45°$. Standard atmosphere from Chapter 1:
$T(z = Z_T = 11 \text{ km}) = -56.5°C$, $T(z=0) = 15°C$.
Find: $\lambda_R = ?$ km

First, find $f_c = (1.458\text{x}10^{-4} \text{ s}^{-1})\cdot\sin(45°) = 1.031\text{x}10^{-4} \text{ s}^{-1}$
Next, find the average temperature and temperature difference across the depth of the troposphere:
$T_{avg} = 0.5\cdot(-56.5 + 15.0)°C = -20.8°C = 252$ K
$\Delta T = (-56.5 - 15.0)$ °C $= -71.5°C$ across $\Delta z = 11$ km
(continued on next page)

tal (defined as a **baroclinic** atmosphere), then pressures at any fixed altitude much also change in the horizontal. Later in this chapter, we will also see how the wind and temperature fields are coupled (via the thermal wind relationship).

Geostrophic Adjustment - Part 1

The tendency of non-equatorial winds to approach geostrophic values (or gradient values for curved isobars) is a very strong process in the Earth's atmosphere. If the actual winds are not in geostrophic balance with the pressure patterns, then both the winds and the pressure patterns tend to change to bring the winds back to geostrophic (another example of Le Chatelier's Principle). **Geostrophic adjustment** is the name for this process.

Picture a wind field (grey arrows in Fig. 11.19a) initially in geostrophic equilibrium ($M_o = G_o$) at altitude 2 km above sea level (thus, no drag at ground). We will focus on just one of those arrows (the black arrow in the center), but all the wind vectors will march together, performing the same maneuvers.

Next, suppose an external process increases the horizontal pressure gradient to the value shown in Fig. 11.19b, with the associated faster geostrophic wind speed G_1. With pressure-gradient force F_{PG} greater than Coriolis force F_{CF}, the force imbalance turns the wind M_1 slightly toward low pressure and accelerates the air (Fig. 11.19b).

The component of wind M_1 from high to low pressure horizontally moves air molecules, weakening the pressure field and thereby reducing the theoretical geostrophic wind (G_2 in Fig. 11.19c). Namely, the **wind field changed the mass (i.e., pressure) field**. Simultaneously the actual wind accelerates to M_2. Thus, the **mass field also changed the wind field**. After both fields have adjusted, the result is $M_2 > M_o$ and $G_2 < G_1$, with $M_2 = G_2$. These changes are called **geostrophic adjustments**.

Defined a "disturbance" as the region that was initially forced out of equilibrium. As you move further away from the disturbance, the amount of geostrophic adjustment diminished. The e-folding distance for this reduction is

$$\lambda_R = \frac{N_{BV}\cdot Z_T}{f_c} \qquad \bullet(11.12)$$

where λ_R is known as the **internal Rossby deformation radius.** The Coriolis parameter is f_c, tropospheric depth is Z_T, and the Brunt-Väisälä frequency is N_{BV} (see eq. on page 372). λ_R is roughly 1300 km.

Use eq. (11.12) as follows. If size or wavelength λ of the initial disturbance is large (> λ_R), then the wind field experiences the greatest adjustment. For smaller scales, the pressure and temperature fields experience the greatest adjustment.

Sample Application *(continuation)*
Find the Brunt-Väisälä frequency (see the Atmos. Stability chapter). $N_{BV} = \sqrt{\frac{(9.8\text{m/s})}{252\text{K}}\left[\frac{-71.5\text{K}}{11000\text{m}} + 0.0098\frac{\text{K}}{\text{m}}\right]} = 0.0113\text{s}^{-1}$
where the temperature differences in square brackets can be expressed in either °C or Kelvin.
Finally, use eq. (11.12): λ_R = (0.0113 s^{-1})·(11 km) / (1.031x10^{-4} s^{-1}) = **1206 km**

Check: Physics, units & magnitude are reasonable.
Exposition: When a cold-front over the Pacific approaches the steep mountains of western Canada, the front feels the influence of the mountains 1200 to 1300 km before reaching the coast, and begins to slow down.

THERMAL WIND EFFECT

Recall that horizontal temperature gradients cause vertically varying horizontal pressure gradients (Fig. 11.17), and that horizontal pressure gradients drive geostrophic winds. We can combine those concepts to see how horizontal temperature gradients drive vertically varying geostrophic winds. This is called the **thermal wind effect**.

This effect can be pictured via the slopes of isobaric surfaces (Fig. 11.20). The hypsometric equation from Chapter 1 describes how there is greater thickness between any two isobaric (constant pressure) surfaces in warm air than in cold air. This causes the tilt of the isobaric surfaces to change with altitude. But as described by eq. (10.29), tilting isobaric surface imply a pressure-gradient force that can drive the geostrophic wind (U_g, V_g).

The relationship between the horizontal temperature gradient and the changing geostrophic wind with altitude is known as the **thermal wind effect**. After a bit of manipulation (see the Higher Math box on the next page), one finds that:

$$\frac{\Delta U_g}{\Delta z} \approx \frac{-|g|}{T_v \cdot f_c} \cdot \frac{\Delta T_v}{\Delta y} \quad \bullet(11.13a)$$

$$\frac{\Delta V_g}{\Delta z} \approx \frac{|g|}{T_v \cdot f_c} \cdot \frac{\Delta T_v}{\Delta x} \quad \bullet(11.13b)$$

where the Coriolis parameter is f_c, virtual absolute temperature is T_v, and gravitational acceleration magnitude is $|g|$ = 9.8 m·s^{-2}. Thus, the meridional temperature gradient causes the zonal geostrophic winds to change with altitude, and the zonal temperature gradient causes the meridional geostrophic winds to change with altitude.

Above the atmospheric boundary layer, actual winds are nearly equal to geostrophic or gradient winds. Thus, eqs. (11.13) provide a first-order estimate of the variation of actual winds with altitude.

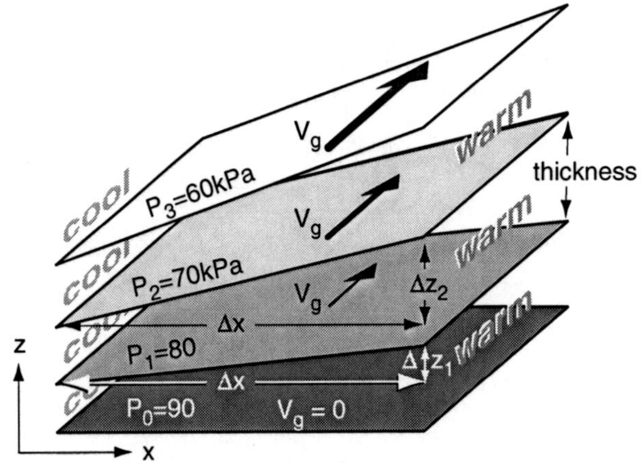

Figure 11.20
Isobaric surfaces are shaded brown. A zonal temperature gradient causes the isobaric surfaces to tilt more and more with increasing altitude. Greater tilt causes stronger geostrophic winds, as plotted with the black vectors for the N. Hemisphere. Geostrophic winds are reversed in S. Hemisphere.

Sample Application
While driving north a distance of 500 km, your car thermometer shows the outside air temperature decreasing from 20°C to 10°C. How does the theoretical geostrophic wind change with altitude?

Find the Answer
Given: Δy = 500 km, ΔT = 10°C – 20°C = –10°C,
average T = 0.5·(10+20°C) = 15°C = 288 K.
Find: $\Delta U_g/\Delta z$ = ? (m s^{-1})/km
Given lack of other data, for simplicity assume:
f_c = 10^{-4} s^{-1}, and $T_v = T$ (i.e., air is dry).

Apply eq. (11.13a):
$$\frac{\Delta U_g}{\Delta z} \approx \frac{-(9.8\text{m/s}^2)}{(288\text{K})\cdot(10^{-4}\text{s}^{-1})} \cdot \frac{(-10°\text{C})}{(500\text{km})} = 6.81 \text{ (m s}^{-1}\text{)/km}$$

Check: Physics & units reasonable. Agrees with Fig.
Exposition: Even if the wind were zero at the ground, the answer indicates that a west wind will increase in speed by 6.81 m s^{-1} for each 1 km of altitude gained.

Given no info about temperature change in the x direction, assume uniform temperature, which implies $\Delta V_g/\Delta z$ = 0 from eq. (11.13b). Namely, the south wind is constant with altitude.

HIGHER MATH • Thermal Wind Effect

Problem: Derive Thermal Wind eq. (11.13a).

Find the Answer: Start with the definitions of geostrophic wind (10.26a) and hydrostatic balance (1.25b):

$$U_g = -\frac{1}{\rho \cdot f_c}\frac{\partial P}{\partial y} \quad \text{and} \quad \rho \cdot |g| = -\frac{\partial P}{\partial z}$$

Replace the density in both eqs using the ideal gas law (1.20). Thus:

$$\frac{U_g \cdot f_c}{T_v} = -\frac{\Re_d}{P}\frac{\partial P}{\partial y} \quad \text{and} \quad \frac{|g|}{T_v} = -\frac{\Re_d}{P}\frac{\partial P}{\partial z}$$

Use $(1/P)\cdot\partial P = \partial \ln(P)$ from calculus to rewrite both:

$$\frac{U_g \cdot f_c}{T_v} = -\Re_d \frac{\partial \ln(P)}{\partial y} \quad \text{and} \quad \frac{|g|}{T_v} = -\Re_d \frac{\partial \ln(P)}{\partial z}$$

Differentiate the left eq. with respect to z:

$$\frac{\partial}{\partial z}\left(\frac{U_g \cdot f_c}{T_v}\right) = -\Re_d \frac{\partial \ln(P)}{\partial y\, \partial z}$$

and the right eq. with respect to y:

$$\frac{\partial}{\partial y}\left(\frac{|g|}{T_v}\right) = -\Re_d \frac{\partial \ln(P)}{\partial y\, \partial z}$$

But the right side of both eqs are identical, thus we can equate the left sides to each other:

$$\frac{\partial}{\partial z}\left(\frac{U_g \cdot f_c}{T_v}\right) = \frac{\partial}{\partial y}\left(\frac{|g|}{T_v}\right)$$

Next, do the indicated differentiations, and rearrange to get the exact relationship for thermal wind:

$$\frac{\partial U_g}{\partial z} = -\frac{|g|}{T_v \cdot f_c}\frac{\partial T_v}{\partial y} + \frac{U_g}{T_v}\left(\frac{\partial T_v}{\partial z}\right)$$

The last term depends on the geostrophic wind speed and the lapse rate, and has magnitude of 0 to 30% of the first term on the right. If we neglect the last term, we get the approximate thermal wind relationship:

$$\frac{\partial U_g}{\partial z} \approx -\frac{|g|}{T_v \cdot f_c}\frac{\partial T_v}{\partial y} \qquad (11.13a)$$

Exposition: A **barotropic** atmosphere is when the geostrophic wind does not vary with height. Using the exact equation above, we see that this is possible only when the two terms on the right balance.

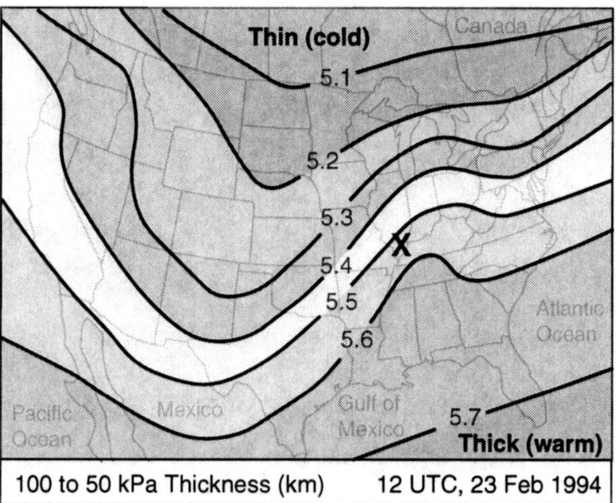

Figure 11.21
Thickness chart based on a US National Weather Service 24-hour forecast. The surface cyclone center is at the "X".

Definition of Thickness

In Fig. 11.20, focus on two isobaric surfaces, such as P = 90 kPa (dark brown in that figure), and P = 80 kPa (medium brown). These two surfaces are at different altitudes z, and the altitude difference is called the "thickness". For our example, we focused on the "90 to 80 kPa thickness". For any two isobaric surfaces P_1 and P_2 having altitudes z_{P1} and z_{P2}, the thickness is defined as

$$TH = z_{P2} - z_{P1} \qquad \bullet(11.14)$$

The hypsometric equation from Chapter 1 tells us that the thickness is proportional to the average absolute virtual-temperature within that layer. Colder air has thinner thickness. Thus, horizontal changes in temperature must cause horizontal changes in thickness.

Weather maps of "100 to 50 kPa thickness" such as Fig. 11.21 are often created by forecast centers. Larger thickness on this map indicates warmer air within the lowest 5 km of the atmosphere.

Thermal-wind Components

Suppose that we use thickness TH as a surrogate for absolute virtual temperature. Then we can combine eqs. (11.14) and (10.29) to yield:

$$U_{TH} = U_{G2} - U_{G1} = -\frac{|g|}{f_c}\frac{\Delta TH}{\Delta y} \qquad \bullet(11.15a)$$

$$V_{TH} = V_{G2} - V_{G1} = +\frac{|g|}{f_c}\frac{\Delta TH}{\Delta x} \qquad \bullet(11.15b)$$

where U_{TH} and V_{TH} are **components of the thermal wind,** $|g|$ = magnitude of gravitational-acceleration, f_c = Coriolis parameter, (U_{G1}, V_{G1}) are geostrophic-wind components on the P1 isobaric surface, and (U_{G2}, V_{G2}) are geostrophic-wind components on the P_2 isobaric surface.

The horizontal vector defined by (U_{TH}, V_{TH}) is the difference between the geostrophic wind vector on the P_2 surface and the geostrophic wind vector on the P_1 surface, as Fig. 11.22 demonstrates. The corresponding **magnitude of the thermal wind** M_{TH} is:

$$M_{TH} = \sqrt{U_{TH}^{\ 2} + V_{TH}^{\ 2}} \tag{11.16}$$

To illustrate this, consider two isobaric surfaces P_2 = 50 kPa (shaded blue in Figure 11.22) and P_1 = 100 kPa (shaded red). The P_1 surface has higher height to the east (toward the back of this sketch). If you conceptually roll a ball bearing down this red surface to find the direction of the pressure gradient, and then recall that the geostrophic wind in the N. Hemisphere is 90° to the right of that direction, then you would anticipate a geostrophic wind G_1 direction as shown by the red arrow. Namely, it is parallel to a constant height contour (dotted red line) pointing in a direction such that low heights are to the vector's left.

Suppose cold air in the north (left side of this sketch) causes a small thickness of only 4 km between the red and blue surfaces. Warm air to the south causes a larger thickness of 5 km between the two isobaric surfaces. Adding those thicknesses to the heights of the P_1 surface (red) give the heights of the P_2 surface (blue). The blue P_2 surface tilts more steeply than the red P_1 surface, hence the geostrophic wind G_2 is faster (blue arrow) and is parallel to a constant height contour (blue dotted line).

Projecting the G_1 and G_2 vectors to the ground (green in Fig. 11.22), the vector difference is shown in yellow, and is labeled as the thermal wind M_{TH}. It is parallel to the contours of thickness (i.e., perpendicular to the temperature gradient between cold and warm air) pointing in a direction with cold air (thin thicknesses) to its left (Fig. 11.23).

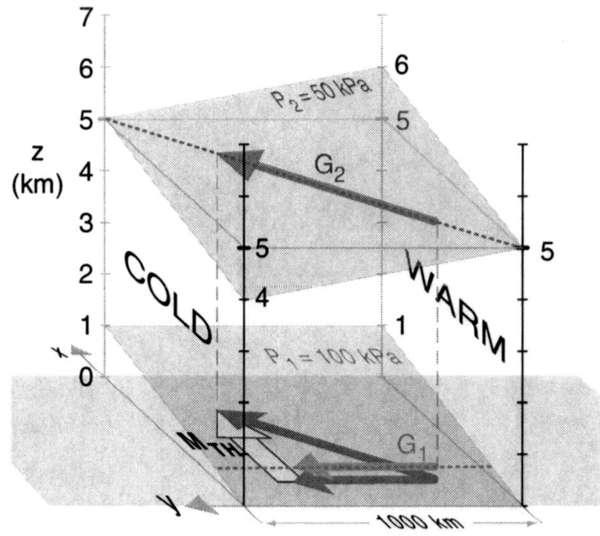

Figure 11.22
Relationship between the thermal wind M_{TH} and the geostrophic winds G on isobaric surfaces P. View is from the west northwest, in the Northern Hemisphere.

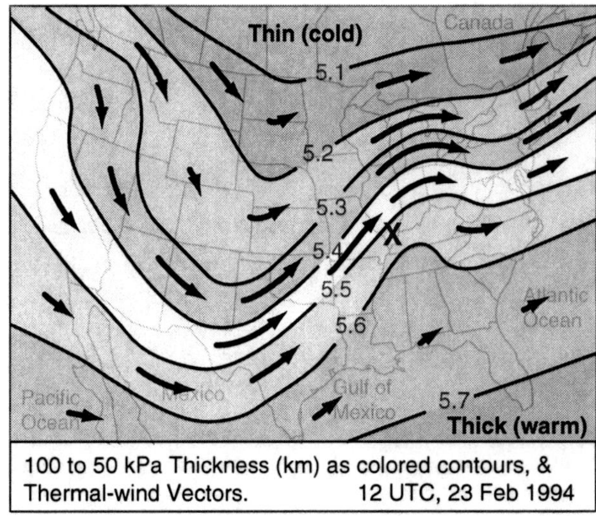

Figure 11.23
Arrows indicate thermal-wind vectors, where longer arrows indicate thermal winds that are stronger. Colored contours are 100 to 50 kPa thicknesses.

Sample Application. For Fig. 11.22, what are the thermal-wind components. Assume $f_c = 1.1\text{x}10^{-4}\ \text{s}^{-1}$.

Find the Answer. Given: TH_2 = 4 km, TH_1 = 5 km, Δy = 1000 km from the figure, $f_c = 1.1\text{x}10^{-4}\ \text{s}^{-1}$.
Find: U_{TH} = ? m s^{-1}, V_{TH} = ? m s^{-1}

Apply eq. (11.15a): $U_{TH} = \dfrac{-|g|}{f_c}\dfrac{\Delta TH}{\Delta y} = \dfrac{-(9.8\text{ms}^{-2})\cdot(4-5)\text{km}}{(1.1\times10^{-4}\text{s}^{-1})\cdot(5000\text{km})}$ = **17.8** m s^{-1}

Check: Physics & units are reasonable. Positive sign for U_{TH} indicates wind toward positive x direction, as in Fig.
Exposition: With no east-to-west temperature gradient, and hence no east-to-west thickness gradient, we would expect zero north-south thermal wind;, hence, V_{TH} = 0.

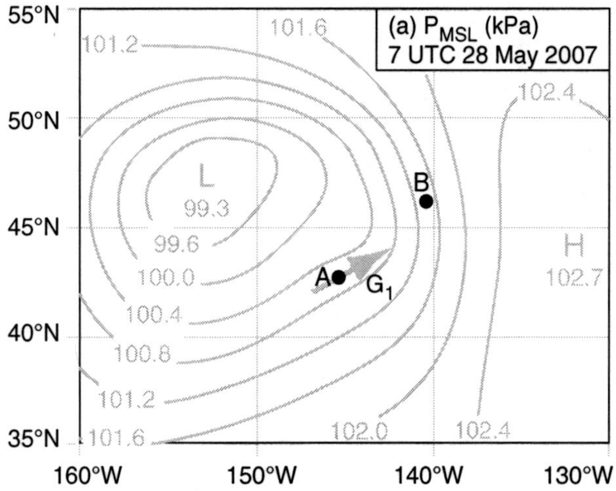

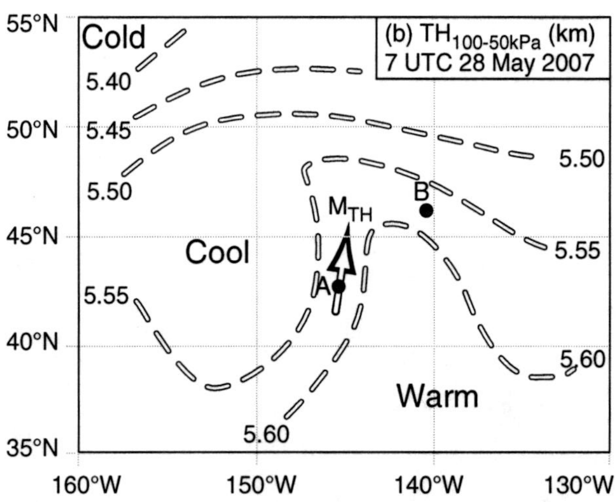

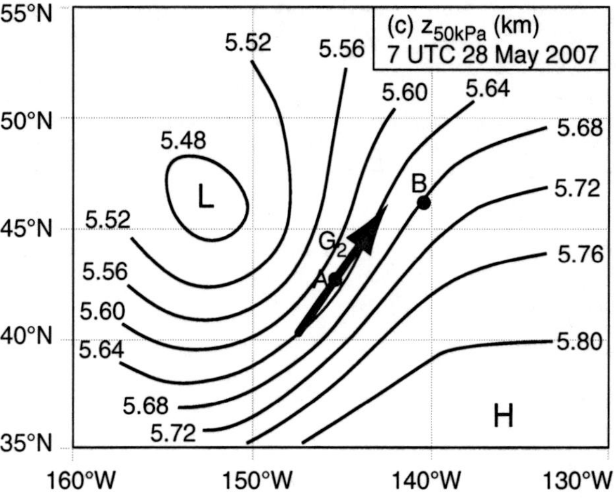

Figure 11.24
Weather maps for a thermal-wind case-study. (a) Mean sea-level pressure (kPa), as a surrogate for height of the 100 kPa surface. (b) Thickness (km) of the layer between 100 kPa to 50 kPa isobaric surfaces. (c) Geopotential heights (km) of the 50 kPa isobaric surface.

Thermal-wind magnitude is stronger where the thickness gradient is greater. Thus, regions on a weather map (Fig. 11.23) where thickness contours are closer together (i.e., have **tighter packing**) indicates faster thermal winds. The relationship between thermal winds and thickness contours is analogous to the relationship between geostrophic winds and height contours. But never forget that no physically realistic wind can equal the thermal wind, because the thermal wind represents the difference or shear between two geostrophic winds.

Nonetheless, you will find the thermal-wind concept useful because it helps you anticipate how geostrophic wind will change with altitude. Actual winds tend towards being geostrophic above the atmospheric boundary layer, hence the thermal-wind concept allows you to anticipate real wind shears.

Case Study

Figs. 11.24 show how geostrophic winds and thermal winds can be found on weather maps, and how to interpret the results. These maps may be copied onto transparencies and overlain.

Fig. 11.24a is a weather map of pressure at sea level in the N. Hemisphere, at a location over the northeast Pacific Ocean. As usual, L and H indicate low- and high-pressure centers. At point A, we can qualitatively draw an arrow (grey) showing the theoretical geostrophic (G_1) wind direction; namely, it is parallel to the isobars with low pressure to its left.

Recall that pressures on a constant height surface (such as at height $z = 0$ at sea level) are closely related to geopotential heights on a constant pressure surface. So we can be confident that a map of 100 kPa heights would look very similar to Fig. 11.24a.

Fig. 11.24b shows the 100 - 50 kPa thickness map, valid at the same time and place. The thickness between the 100 and the 50 kPa isobaric surfaces is about 5.6 km in the warm air, and only 5.4 km in the cold air. The white arrow qualitatively shows the thermal wind M_{TH}, as being parallel to the thickness lines with cold temperatures to its left.

Fig. 11.24c is a weather map of geopotential heights of the 50 kPa isobaric surface. L and H indicate low and high heights. The black arrow at A shows the geostrophic wind (G_2), drawn parallel to the height contours with low heights to its left.

If we wished, we could have calculated quantitative values for G_1, G_2, and M_{TH}, utilizing the scale that 5° of latitude equals 555 km. [*ALERT: This scale does not apply to longitude, because the meridians get closer together as they approach the poles. However, once you have determined the scale (map mm : real km) based on latitude, you can use it to good approximation in any direction on the map.*]

Back to the thermal wind: if you add the geostrophic vector from Fig. 11.24a with the thermal wind vector from Fig. 11.24b, the result should equal the geostrophic wind vector in Fig. 11.24c. This is shown in the Sample Application.

Although we will study much more about weather maps and fronts in the next few chapters, I will interpret these maps for you now.

Point A on the maps is near a cold front. From the thickness chart, we see cold air west and northwest of point A. Also, knowing that winds rotate counterclockwise around lows in the N. Hemisphere (see Fig. 11.24a), I can anticipate that the cool air will advance toward point A. Hence, this is a region of **cold-air advection**. Associated with this cold-air advection is **backing of the wind** (i.e., turning counterclockwise with increasing height), which we saw was fully explained by the thermal wind.

Point B is near a warm front. I inferred this from the weather maps because warmer air is south of point B (see the thickness chart) and that the counterclockwise winds around lows are causing this warm air to advance toward point B. **Warm air advection** is associated with **veering of the wind** (i.e., turning clockwise with increasing height), again as given by the thermal wind relationship. I will leave it to you to draw the vectors at point B to prove this to yourself.

Thermal Wind & Geostrophic Adjustment - Part 2

As geostrophic winds adjust to pressure gradients, they move mass to alter the pressure gradients. Eventually, an equilibrium is approached (Fig. 11.25) based on the combined effects of geostrophic adjustment and the thermal wind. This figure is much more realistic than Fig. 11.17(iv) because Coriolis force prevents the winds from flowing directly from high to low pressure.

With these concepts of:
- differential heating,
- nonhydrostatic pressure couplets due to horizontal winds and vertical buoyancy,
- hydrostatic thermal circulations,
- geostrophic adjustment, and
- the thermal wind,

we can now go back and explain why the global circulation works the way it does.

Sample Application

For Fig. 11.24, qualitatively verify that when vector M_{TH} is added to vector G_1, the result is vector G_2.

Find the Answer

Given: the arrows from Fig. 11.24 for point A. These are copied and pasted here.

Find: The vector sum of $G_1 + M_{TH}$ = ?

Recall that to do a vector sum, move the tail of the second vector (M_{TH}) to be at the arrow head of the first vector (G_1). The vector sum is then the vector drawn from the tail of the first vector to the tip (head) of the second vector.

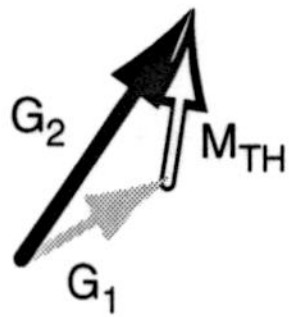

Check: Sketch is reasonable.

Exposition: The vector sum indeed equals vector G_2, as predicted by the thermal wind.

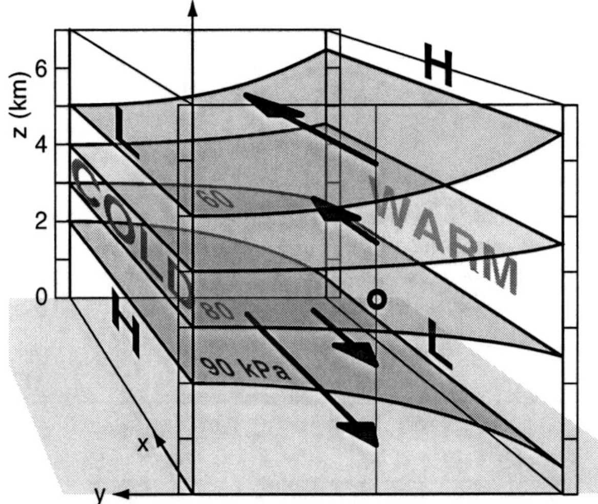

Figure 11.25

Typical equilibrium state of the pressure, temperature and wind fields after it has adjusted geostrophically. Isobaric surfaces are shaded with color, and recall that high heights of isobaric surfaces correspond to regions of high pressure on constant altitude surfaces. Black arrows give the geostrophic wind vectors.

Consider the red-shaded isobaric surface representing P = 60 kPa. That surface has high (H) heights to the south (to the right in this figure), and low (L) heights to the north. In the Northern Hemisphere, the geostrophic wind would be parallel to a constant height contour in a direction with lower pressure to its left. A similar interpretation can be made for the purple-shaded isobaric surface at P = 90 kPa. At a middle altitude in this sketch there is no net pressure gradient (i.e., zero slope of an isobaric surface), hence no geostrophic wind.

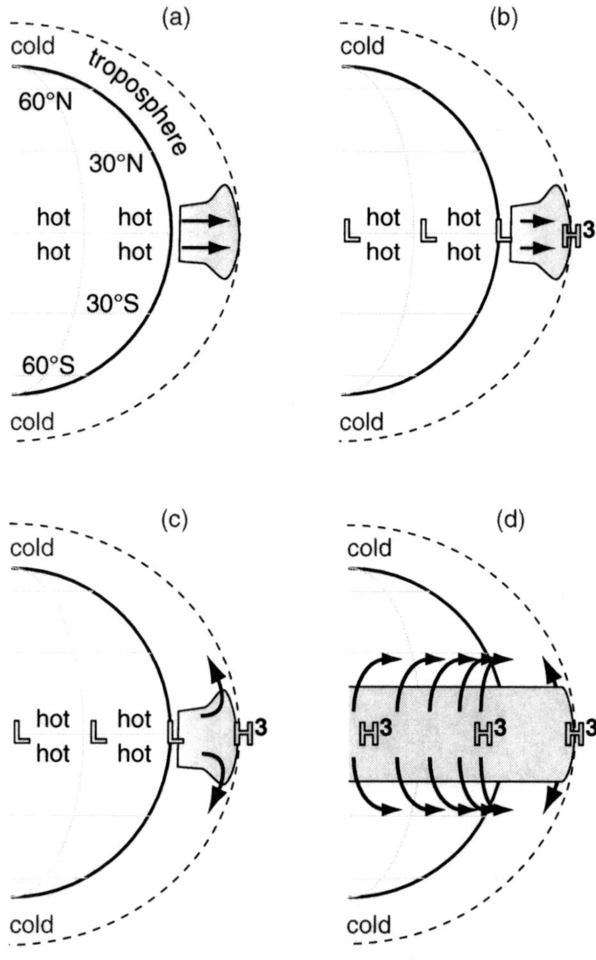

Figure 11.26
Application of physical concepts to explain the general circulation (see text). White-filled H and L indicate surface high and low pressure regions; grey-filled H and L are pressure regions near the tropopause. (continued on next pages)

A SCIENTIFIC PERSPECTIVE • Model Sensitivity

CAUTION: Whenever you find that a model has high **sensitivity** (i.e., the output result varies by large magnitude for small changes in the input parameter), you should be especially wary of the results. Small errors in the parameter could cause large errors in the result. Also, if the real atmosphere does not share the same sensitivity, then this is a clue that the model is poorly designed, and perhaps a better model should be considered.

Models are used frequently in meteorology — for example: numerical weather prediction models (Chapter 20) or climate-change models (Chapter 21). Most researchers who utilize models will perform careful **sensitivity studies** (i.e., compute the output results for a wide range of parameter values) to help them gauge the potential weaknesses of the model.

EXPLAINING THE GENERAL CIRCULATION

Low Latitudes

Differential heating of the Earth's surface warms the tropics and cools the poles (Fig. 11.26a). The warm air near the equator can hold large amounts of water vapor evaporated from the oceans. Buoyancy force causes the hot humid air to rise over the equator. As the air rises, it cools and water vapor condenses, causing a belt of thick thunderstorm clouds around the equator (Fig. 11.26a) with heavy tropical precipitation.

The buoyantly forced vertical motion removes air molecules from the lower troposphere in the tropics, and deposits the air near the top of the troposphere. The result is a pressure couplet (Fig. 11.26b) of very high perturbation pressure p' (indicated with HHH or H^3 on the figures) near the tropopause, and low perturbation pressure (L in the figures) at the surface.

The belt of tropical high pressure near the tropopause forces air to diverge horizontally, forcing some air into the Northern Hemisphere and some into the Southern (Fig. 11.26c). With no Coriolis force at the equator, these winds are driven directly away from the high-pressure belt.

But as these high-altitude winds move away from the equator, they are increasingly affected by Coriolis force (Fig. 11.26d). This causes winds moving into the Northern Hemisphere to turn to their right, and those moving into the Southern Hemisphere to turn to their left.

But as the winds move further and further away from the equator, they are turned more and more to the east, creating the subtropical jet (Fig. 11.26e) at about 30° latitude. Coriolis force prevents these upper-level winds from getting further away from the equator than about 30° latitude (north and south), so the air accumulates and the pressure increases in those belts. When simulations of the general circulation impose a larger Coriolis force (as if the Earth spun faster), the convergence bands occur closer to the equator. For weaker Coriolis force, the convergence is closer to the poles. But for our Earth, the air converges at 30° latitude.

This pressure perturbation p' is labeled as HH or H^2 in Fig. 11.26e, to show that it is a positive pressure perturbation, but not as strong as the H^3 perturbation over the equator. Namely, the horizontal pressure gradient between H^3 and H^2 drives the upper-level winds to diverge away from the equator.

The excess air accumulated at 30° latitudes cannot go up into the stratosphere in the face of very strong static stability there. The air cannot go further poleward because of Coriolis force. And the air cannot move equatorward in the face of the strong upper-level winds leaving the equator. The only remaining path is downward at 30° latitude (Fig. 11.26f) as a nonhydrostatic flow. As air accumulates near the ground, it causes a high-pressure perturbation there — the belt of subtropical highs labeled with H.

The descending air at 30° latitude warms dry-adiabatically, and does not contain much moisture because it was squeezed out earlier in the thunderstorm updrafts. These are the latitudes of the subtropical deserts (Fig. 11.26h), and the source of hot airmasses near the surface.

Finally, the horizontal pressure gradient between the surface subtropical highs near 30° latitude and the equatorial lows near 0° latitude drives the surface winds toward the equator. Coriolis force turns these winds toward the west in both hemispheres (Fig. 11.26g), resulting in the easterly trade winds (winds from the east) that converge at the ITCZ.

The total vertical circulation in the tropics and subtropics we recognize as the Hadley Cell (labeled h.c. in Fig. 11.26f). This vertical circulation (a **thermally-direct circulation**) is so vigorous in its vertical mixing and heat transport that it creates a deeper troposphere in the tropics than elsewhere (Fig. 11.4). Also, the vigorous circulation spreads and horizontally mixes the radiatively warmed air somewhat uniformly between ±30° latitude, as sketched by the flattened temperature curve in Fig. 11.8a.

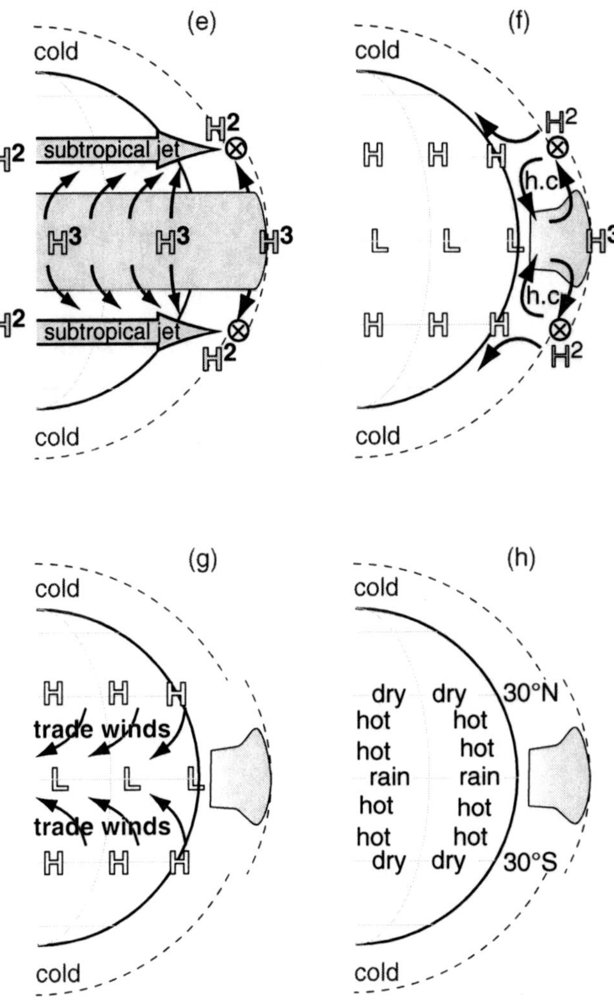

Figure 11.26 (continuation)
Explanation of low-latitude portion of the global circulation. The dashed line shows the tropopause. The "x" with a circle around it (representing the tail feathers of an arrow) indicates the axis of a jet stream that goes into the page.

INFO • The Trade Inversion

Descending air in the subtropical arm of the Hadley circulation is hot and dry. Air near the tropical sea surface is relatively cool and humid. Between these layers is a strong temperature inversion called the **trade inversion** or **passat inversion**. This statically stable layer (between the dashed lines in the Figure) creates a lid to the tropical convection below it. The inversion base is lowest (order of 500 m) in the subtropics, and is highest (order of 2,500 m) near the ITCZ. Fair-weather cumulus clouds (**trade cumuli**) between the lifting condensation level (LCL) and the trade inversion are shallowest in the subtropics and deeper closer to the ITCZ.

By capping the humid air below it, the trade inversion allows a latent-heat fuel supply to build up, which can be released in hurricanes and ITCZ thunderstorms.

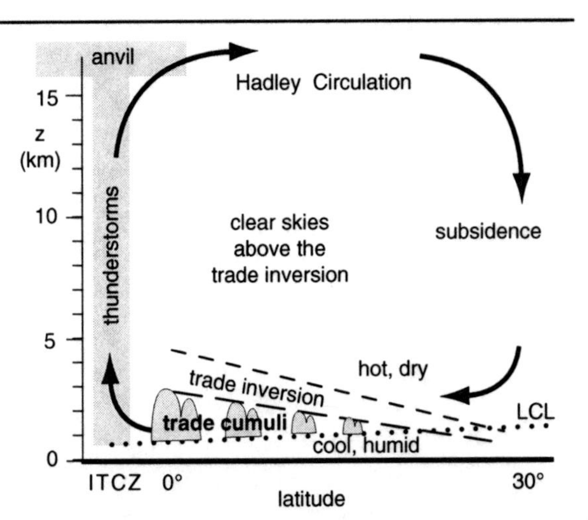

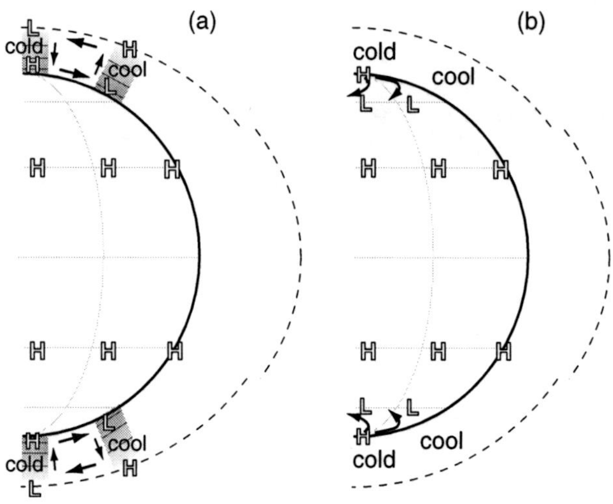

Figure 11.27
Explanation of high-latitude portion of general circulation.

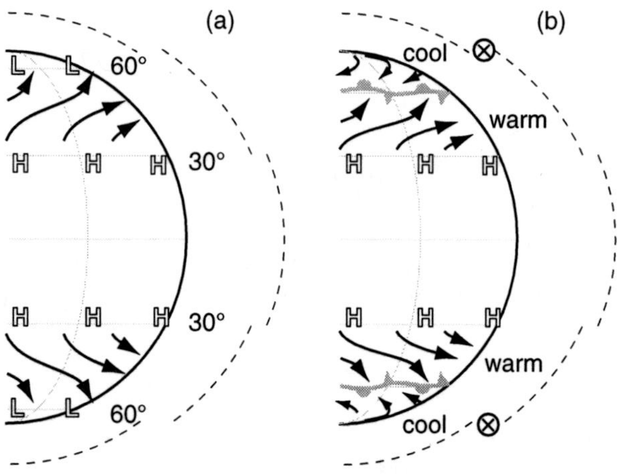

Figure 11.28
Explanation of mid-latitude flow near the Earth's surface. H and L indicate belts of high and low pressure, black arrows are average surface winds, and the polar front is shaded grey. The circle with "x" in it represents the tail feathers of the jet-stream wind vector blowing into the page.

High Latitudes

As sketched in Fig. 11.8a, air temperature is very cold at the poles, and is cool at 60° latitude. The temperature difference between 60° and 90° latitudes creates opposite north-south pressure gradients and winds at the top and bottom of the troposphere, due to the thermal circulation effect (Fig. 11.17). A vertical cross section of this thermal circulation (Fig. 11.27a) shows a weak polar cell. Air generally rises near 60°N and descends near the pole.

At the poles are surface high-pressure centers, and at 60° latitudes are belts of subpolar lows at the surface. This horizontal pressure gradient drives equatorward winds, which are turned toward the west in both hemispheres due to Coriolis force (Fig. 11.27b). Namely, the winds become geostrophic, and are known as **polar easterlies**.

At the top of the shallow (6 to 8 km thick) troposphere are poleward winds that are turned toward the east by Coriolis force. These result in an upper-level westerly flow that circulates around the upper-level polar low (Fig. 11.3b).

Mid-latitudes

Recall that the Hadley cell is unable to mix heat beyond about ±30° latitude. This leaves a very strong meridional temperature gradient in mid-latitudes (Fig. 11.8) throughout the depth of the troposphere. Namely, the temperature change between the equator (0°) and the poles (90°) has been compressed to a latitude band of about 30 to 60° in each hemisphere.

Between the subtropical high-pressure belt near 30° latitude and the subpolar low-pressure belt near 60° latitude is a weak meridional pressure gradient near the Earth's surface. This climatological-average pressure gradient drives weak boundary-layer winds from the west in both hemispheres (Fig. 11.28a), while the drag of the air against the surface causes the wind to turn slightly toward low pressure.

Near the subpolar belt of low pressure is a region of surface-air convergence, with easterly winds from the poles meeting westerly winds from mid-latitudes. The boundary between the warm subtropical air and the cool polar air at the Earth's surface is called the **polar front** (Fig. 11.28b) — a region of even stronger horizontal temperature gradient.

Recall from the hypsometric equation in Chapter 1 that the height difference (i.e., the thickness) between two isobaric surfaces increases with increasing temperature. As a result of the meridional temperature gradient, isobaric surfaces near the top of the troposphere in mid-latitudes are much more steeply sloped than near the ground (Figs. 11.29 & 11.32). This is related to the thermal-wind effect.

In the Northern Hemisphere this effect is greatest in winter (Fig. 11.32), because there is the greatest temperature contrast between pole and equator. In the Southern Hemisphere, the cold Antarctic continent maintains a strong meridional temperature contrast all year.

Larger pressure gradients at higher altitudes drive stronger winds. The core of fastest westerly winds near the tropopause (where the pressure-gradient is strongest) is called the **polar jet stream**, and is also discussed in more detail later in this chapter. Thus, the climatological average winds throughout the troposphere at mid-latitudes are from the west (Fig. 11.30a) in both hemispheres.

Although the climatological average polar-jet-stream winds are straight from west to east (as in Fig. 11.30a), the actual flow on any single day is unstable. Two factors cause this instability: the variation of Coriolis parameter with latitude (an effect that leads to **barotropic instability**), and the increase in static stability toward the poles (an effect that leads to **baroclinic instability**). Both of these instabilities are discussed in more detail later.

These instabilities cause the jet stream to meander meridionally (north-south) as it continues to blow from the west (Fig. 11.30b). The meanders that form in this flow are called **Rossby waves**. Regions near the tropopause where the jet stream meanders equatorward are called **troughs**, because the lower pressure on the north side of the jet stream is brought equatorward there. Poleward meanders of the jet stream are called **ridges**, where higher pressure extends poleward. The locations of Rossby-wave troughs and ridges usually propagate toward the east with time, as will be explained in detail later in this chapter.

Recall that there is a subtropical jet at roughly 30° latitude associated with the Hadley Cell. Thus, in each hemisphere are a somewhat-steady subtropical jet and an unsteady polar jet (Fig. 11.30b). Both of these jets are strongest in the winter hemisphere, where there is the greatest temperature gradient between cold poles and hot equator.

Troughs and ridges in the jet stream are crucial in creating and killing cyclones and anticyclones near the Earth's surface. Namely, they cause the extremely large weather variability that is normal for mid-latitudes. The field of **synoptic meteorology** comprises the study and forecasting of these variable systems, as discussed in the Airmasses, Fronts, and Extratropical Cyclone chapters.

Figs. 11.31 and 11.32 show actual global pressure patterns at the bottom and top of the troposphere. The next section explains why the patterns over the oceans and continents differ.

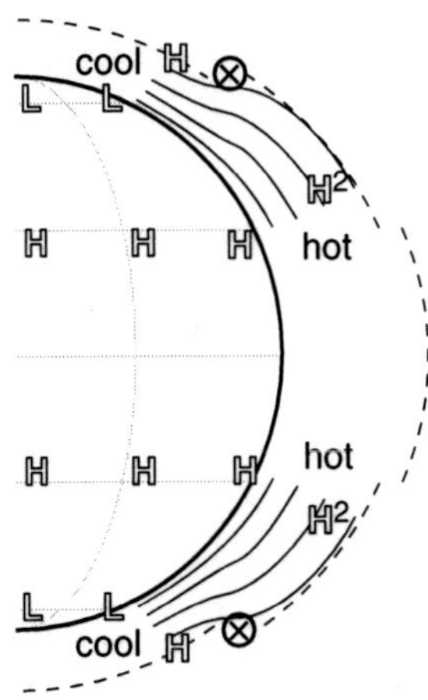

Figure 11.29
Isobaric surfaces (thin solid black lines) are spaced further apart in hot air than in cool air. Regions of steeper slope of the isobars have stronger pressure gradient, and drive faster winds. The jet-stream axis is shown with ⊗ .

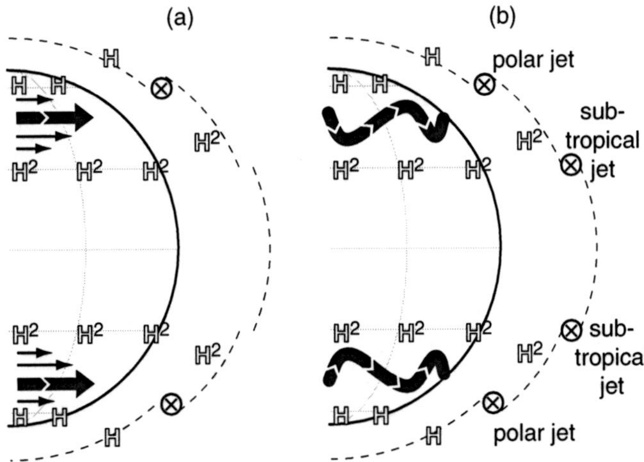

Figure 11.30
Mid-latitude flow near the top of the troposphere. The thick black arrow represents the core or axis of the jet stream: (a) average, (b) snapshot. (See caption in previous two figures for legend.)

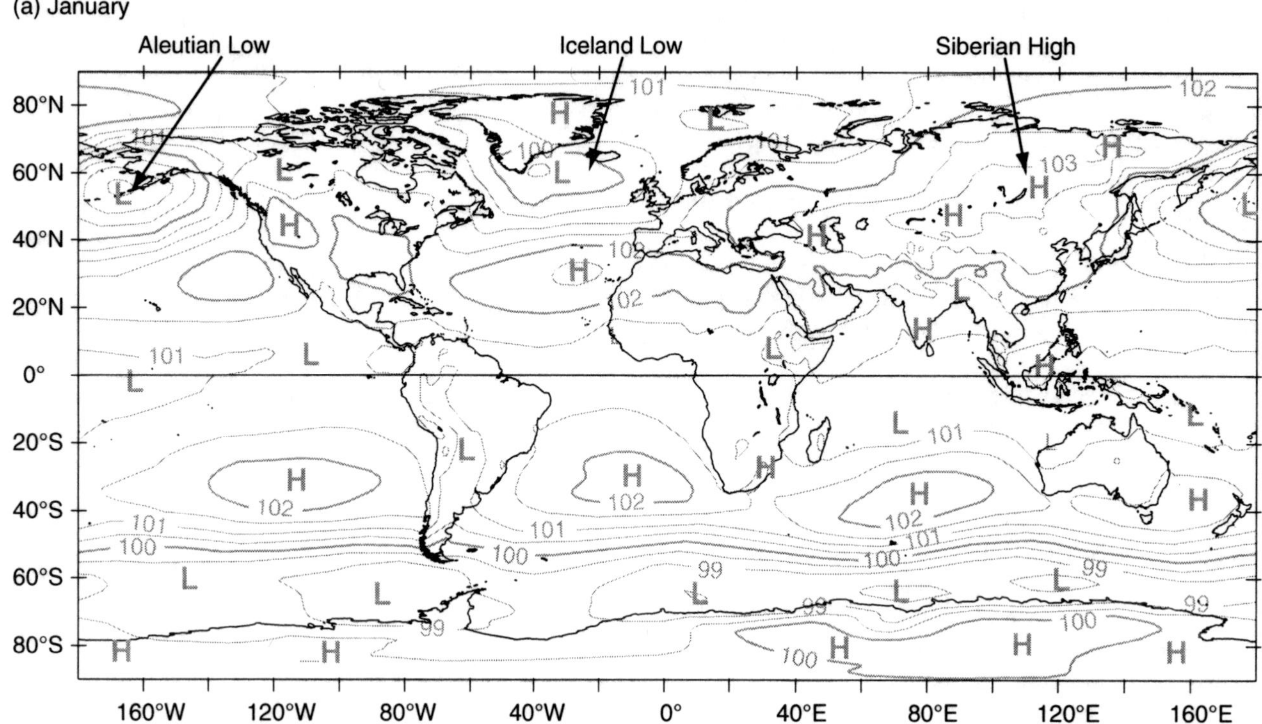

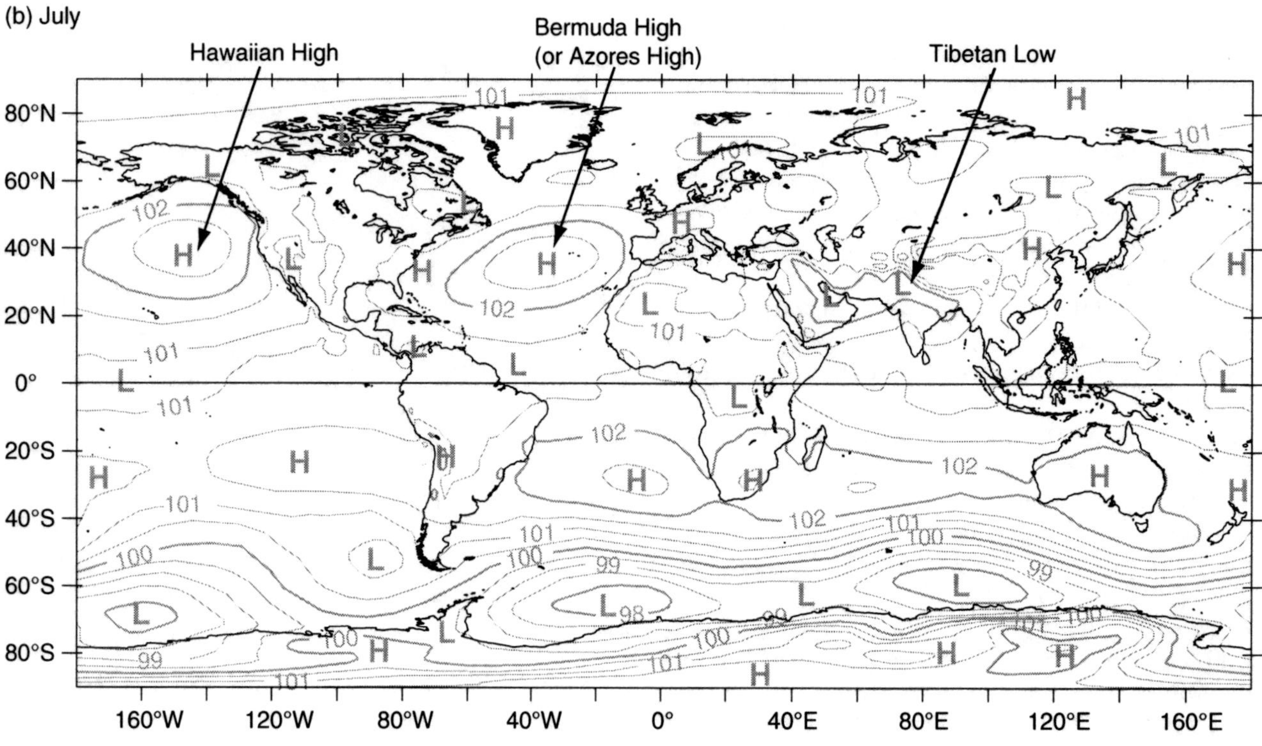

Figure 11.31
Mean sea-level pressure (kPa) for (a) January 2001, and (b) July 2001. [European Centre for Medium Range Weather forecasts (EC-MWF) ERA-40 data used in this analysis have been provided by ECMWF, and have been obtained from the ECMWF data server.]

(a) January

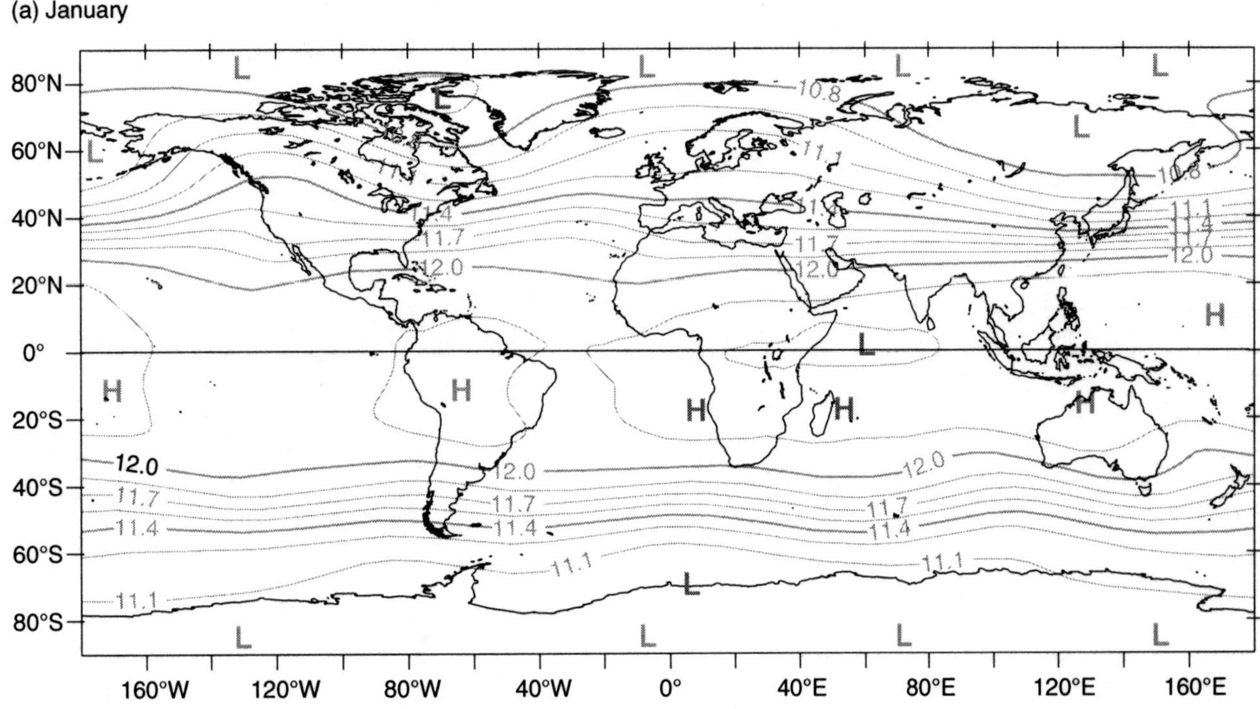

(b) July

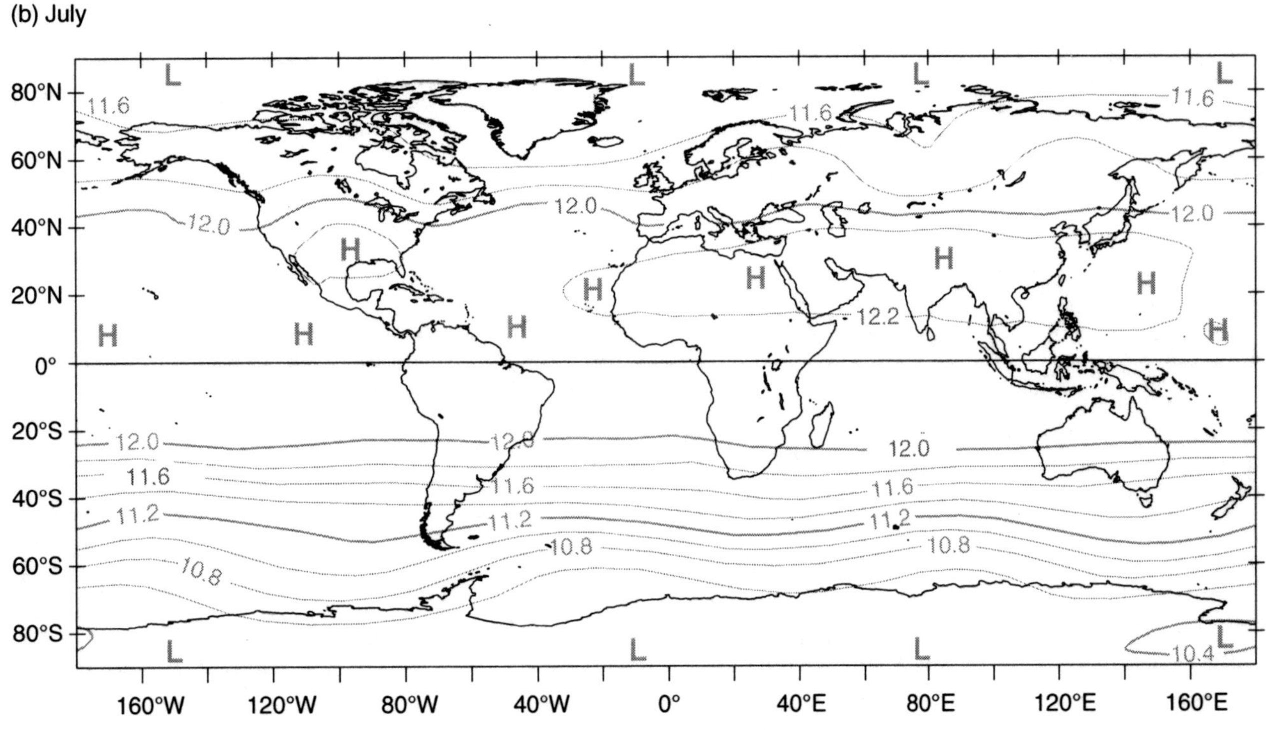

Figure 11.32
Geopotential height (km) of the 20 kPa isobaric surface (near the *tropopause) for (a) January 2001, and (b) July 2001. [ECMWF ERA-40 data used in this analysis have been provided by ECMWF, and have been obtained from the ECMWF data server.]*

Monsoon

Recall from the Heat Budgets chapter that the temperature change of an object depends on the mass of material being heated or cooled, and on the material's specific heat. If you put the same amount of heat into objects of similar material but differing mass, the smaller masses will warm the most.

Rocks and soil on continents are opaque to sunlight, and are good insulators of heat. Thus, sunlight directly striking the land surface warms only a very thin top layer (mm) of molecules, causing this thin layer to get quite warm. Similarly, longwave (infrared) radiative heat loss at night causes the very thin top layer to become very cold. Namely, there is a large **diurnal** (daily) temperature contrast. Also, because there are more daylight hours in summer and more nighttime hours in winter, continental land surfaces tend to become hot in summer and cold in winter.

Water in the oceans is partially transparent and sometimes turbulent, allowing sunlight to be absorbed and spread over a thick layer (meters to tens of meters) of molecules. Also, water has a large specific heat (see the Heat Budgets chapter), hence a large input of heat causes only a small temperature change. Thus, ocean surfaces have very small diurnal temperature changes, and have only a medium amount of seasonal temperature changes.

The net result is that during summer, continents warm faster than the oceans. During winter, continents cool faster than the oceans.

Consider a cold region next to a warm region. Over the cold surface, the near-surface air cools and develops a high-pressure center with anticyclonically rotating winds, as explained by the thermal circulation sketched in Fig. 11.17. Over warm surfaces, the thermal circulation causes near-surface air to warm and develop a low-pressure center with cyclonic winds. As already mentioned, this is called a **thermal low**.

Combining the effects from the previous two paragraphs with the strong tendency of the winds to become geostrophic (or gradient) yields the near-surface monsoonal flows shown in Fig. 11.5. Opposite pressure patterns and circulations would occur near the top of the troposphere. The regions near surface lows tend to have rising air and abundant clouds and rain. Regions near surface highs tend to have dry fair weather with few clouds.

Monsoon circulations occur over every large continent and ocean (Fig. 11.31). Some are given names. Over the Atlantic in summer, winds on the south and west sides of the monsoonal **Bermuda High** (also called the **Azores High**) steer Atlantic hurricanes northward as they near North America. Over the North Pacific in summer is the **Hawaiian High** or **Pacific High** which provides cool northerly breezes and months of fair weather to the west coast of North America.

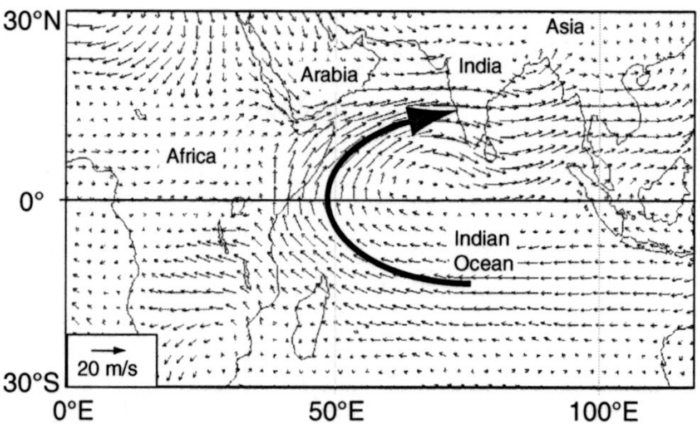

Figure 11.33
Monsoon winds near India on 85 kPa isobaric surface (at z ≈ 1.5 km), averaged over July 2001. Thin arrow length shows wind speed (see legend). Thick arrow shows cross-equatorial flow.

The summer low over northern India is called the **Tibetan Low**. It helps drive strong cross-equatorial flow (Fig. 11.33) that brings the much needed monsoon rains over India. Ghana in West Africa also receives a cross-equatorial monsoon flow.

Winter continental highs such as the **Siberian High** over Asia are formation locations for cold airmasses. Over the North Atlantic Ocean in winter is the **Icelandic Low**, with an average circulation on its south side that steers mid-latitude cyclones toward Great Britain and northern Europe. The south side of the winter **Aleutian Low** over the North Pacific brings strong onshore flow toward the west coast of North America, causing many days of clouds and rain.

The actual global circulation is a superposition of the zonally averaged flows and the monsoonal flows (Fig. 11.31). Also, a snapshot or satellite image of the Earth on any given day would likely deviate from the one-month averages presented here. Other important aspects of the global circulation were not discussed, such as conversion between available potential energy and kinetic energy. Also, monsoons and the whole global circulation are modulated by El Niño / La Niña events and other oscillations, discussed in the Climate chapter.

In the previous sections, we have described characteristics of the global circulation in simple terms, looked at what drives these motions, and explained dynamically why they exist. Some of the phenomena we encountered deserve more complete analysis, including the jet stream, Rossby waves with their troughs and ridges, and some aspects of the ocean currents. The next sections give details about how these phenomena work.

JET STREAMS

In the winter hemisphere there are often two strong jet streams of fast west-to-east moving air near the tropopause: the **polar jet stream** and the **subtropical jet stream** (Figs. 11.34 & 11.35).

The subtropical jet is centered near 30° latitude in the winter hemisphere. This jet: (1) is very steady; (2) meanders north and south a bit; (3) is about 10° latitude wide (width ≈1,000 km); and (4) has seasonal-average speeds of about 45 m s^{-1} over the Atlantic Ocean, 55 to 65 m s^{-1} over Africa and the Indian Ocean, and 60 to 80 m s^{-1} over the western Pacific Ocean. The **core** of fast winds near its center is at 12 km altitude (Fig. 11.35). It is driven by outflow from the top of the Hadley cell, and is affected by both Coriolis force and angular-momentum conservation.

The polar jet is centered near 50 to 60° latitude in the winter hemisphere. The polar jet: (1) is extremely variable; (2) meanders extensively north and south; (3) is about 5° latitude wide; and (4) has widely varying speeds (25 to 100 m s^{-1}). The core altitude is about 9 km. This jet forms over the polar front — driven by thermal-wind effects due to the strong horizontal temperature contrast across the front.

When meteorological data are averaged over a month, the subtropical jet shows up clearly in the data (e.g., Fig. 11.36) because it is so steady. However, the polar jet disappears because it meanders and shifts so extensively that it is washed out by the long-term average. Nonetheless, these transient meanders of the polar jet (troughs and ridges in the Rossby waves) are extremely important for mid-latitude cyclone formation and evolution (see the Extratropical Cyclone chapter).

In the summer hemisphere both jets are much weaker (Figs. 11.35b and 11.36), because of the much weaker temperature contrast between the equator and the warm pole. Core wind speeds in the subtropical jet are 0 to 10 m s^{-1} in N. Hemisphere summer, and 5 to 45 m s^{-1} in S. Hemisphere summer. This core shifts poleward to be centered near 40° to 45° latitude. The polar jet is also very weak (0 to 20 m s^{-1}) or non-existent in summer, and is displaced poleward to be centered near 60° to 75° latitude.

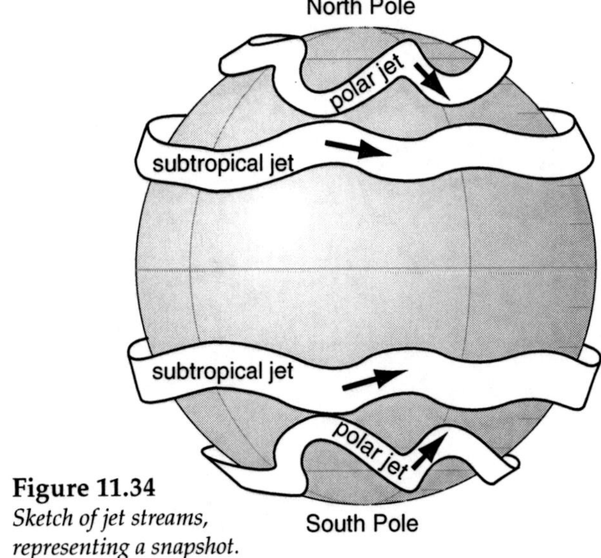

Figure 11.34
Sketch of jet streams, representing a snapshot.

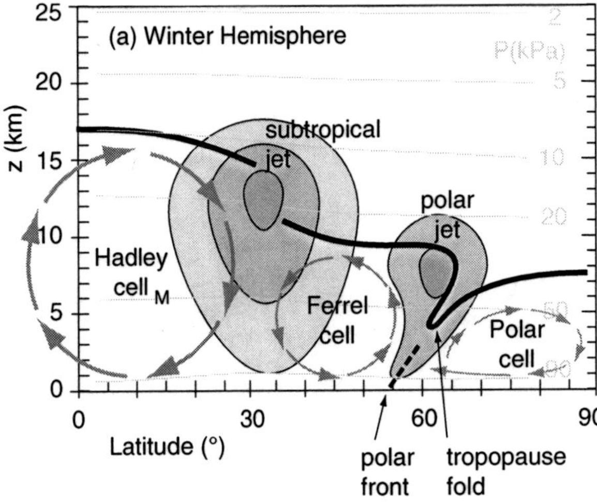

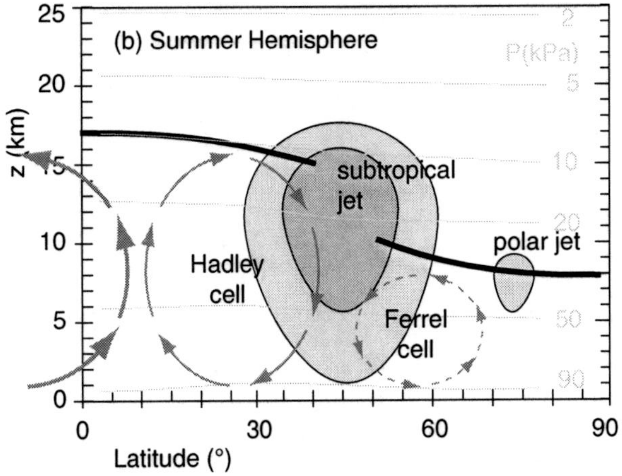

Figure 11.35
Simplified vertical cross-section. Thick solid line is the tropopause. Darker shading indicates faster winds from the west (perpendicular to the page). This is a snapshot, not a climatological average; hence, the polar jet and polar front can be seen.

INFO • Jet Stream Aspect Ratio

Jet streams in the real atmosphere look very much like the thin ribbons of fast-moving air, as sketched in Fig. 11.34. Jet vertical thickness (order of 5 to 10 km) is much smaller than their horizontal width (order of 1000 to 2000 km). Namely, their aspect ratio (width/thickness) is large. Figures such as 11.35 are intentionally distorted to show vertical variations better.

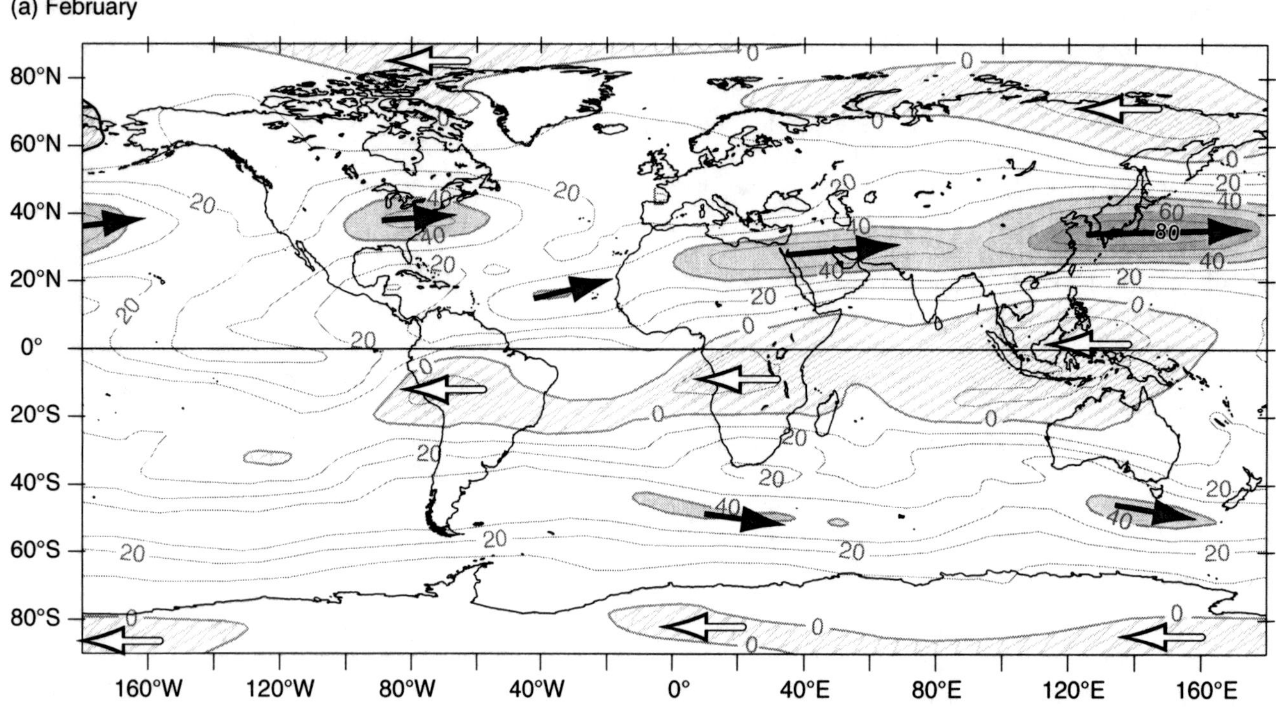

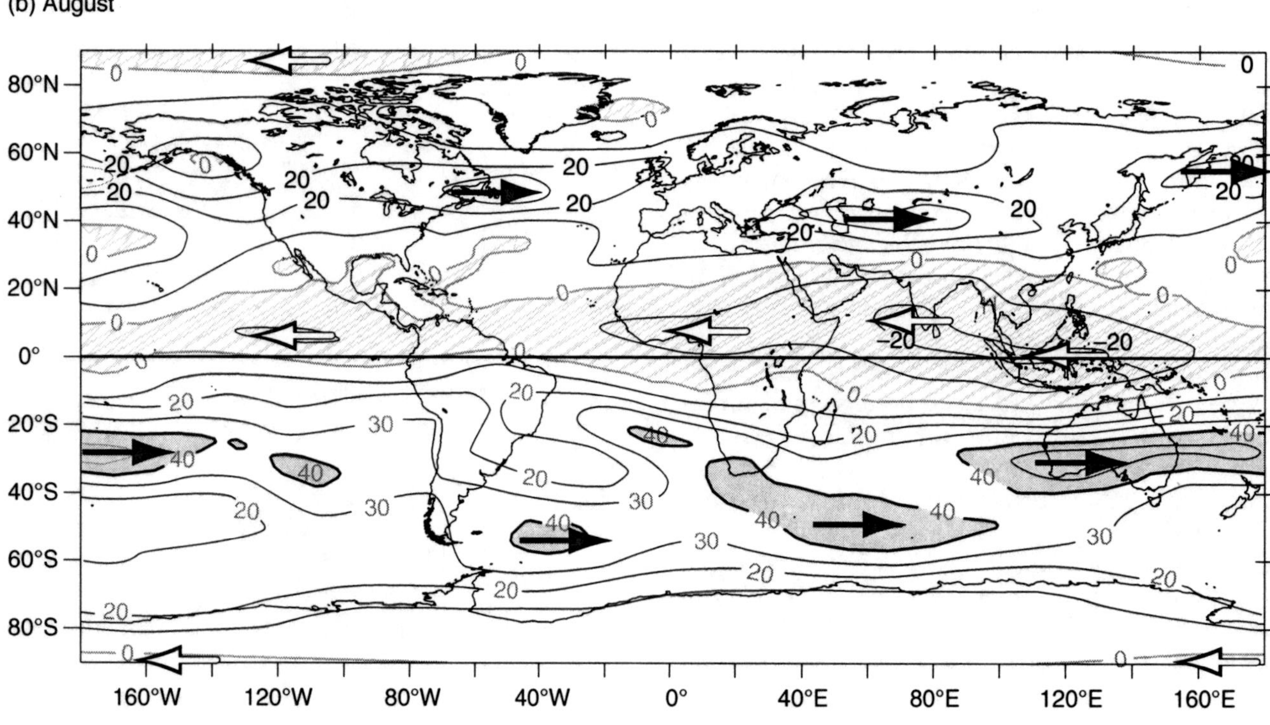

Figure 11.36

Zonal (U) component of winds (m s^{-1}) at 20 kPa isobaric surface (near the tropopause) for (a) February, and (b) August 2001. Contour interval is 10 m s^{-1}. All negative (east) winds are lightly hatched, and are indicated with white arrows. Positive (west) winds are indicated with black arrows, and are shaded at 40, 60, and 80 m s^{-1}. February and August are shown because the jets are stronger than during January and July, respectively. In 2001, polar easterlies were observed at 20 kPa, contrary to the longer-term climate average of polar westerlies as sketched in Fig. 11.3b. [ECMWF ERA-40 data used in this analysis have been provided by ECMWF, and have been obtained from the ECMWF data server.]

Baroclinicity & the Polar Jet

First, consider how temperature varies with height and latitude (Fig. 11.37a). At any altitude in the troposphere you will find a horizontal temperature gradient between colder poles and warmer equator. According to the hypsometric relationship, the thickness between two isobaric surfaces is smaller in the colder (polar) air and greater in the warmer (equatorial) air (Fig. 11.37b). Hence, isobaric surfaces tilt in the horizontal, which drives a geostrophic wind (Fig. 11.37c).

The greatest tilt is near 30° latitude at the tropopause, and the associated pressure gradient drives the fastest winds (**jet-stream core**) there, as expected due to the thermal wind. Because the isobars cross the isotherms (and isobars also cross the isopycnics — lines of equal density), the atmosphere is said to be **baroclinic**. It is this **baroclinicity** associated with the meridional temperature gradient that creates the west winds of the jet stream.

Notice that the troposphere is deeper near the equator than near the poles. Thus, the typical lapse rate in the troposphere, applied over the greater depth, causes colder temperatures at the tropopause over the equator than over the poles (Fig. 11.37a). In the stratosphere above the equatorial tropopause, the air is initially isothermal, but at higher altitudes the air gets warmer. Meanwhile, further north in the stratosphere, such as at latitude 60°, the air is isothermal over a very large depth. Thus, the meridional temperature gradient reverses in the bottom of the stratosphere, with warmer air over 60° latitude and colder equatorial air.

The associated north-south thickness changes between isobaric surfaces causes the meridional pressure gradient to decrease, which you can see in Fig. 11.37b as reductions in slopes of the isobars. The reduced pressure gradient in the lower stratosphere causes wind speeds to decrease with increasing altitude (Fig. 11.37c), leaving the **jet max** at the tropopause. Near the jet core is a region where the tropopause has a break or a fold, as is covered in the Fronts and Extratropical Cyclone chapters.

You can apply the concepts described above to the toy model of eq. (11.1) and Fig. 11.8. Using that model with eqs. (11.2, 11.4 & 11.13) gives

$$U_{jet} \approx \frac{|g| \cdot c \cdot b_1}{2\Omega \cdot T_v} \cdot z \cdot \left(1 - \frac{z}{2 \cdot z_T}\right) \cdot \cos^2(\phi) \cdot \sin^2(\phi) \tag{11.17}$$

where ϕ is latitude, $|g| = 9.8$ m s^{-2}, U_g has been relabeled as U_{jet}, T_v is average absolute virtual temperature, $b_1 \approx 40$ K, $c = 1.18\text{x}10^{-3}$ km^{-1}, $z_T \approx 11$ km is average depth of the troposphere, and assuming $U_{jet} = 0$ at z = 0. The factor $2{\cdot}\Omega = 1.458\text{x}10^{-4}$ s^{-1} comes from the Coriolis-parameter definition $f_c = 2{\cdot}\Omega{\cdot}\sin\phi$.

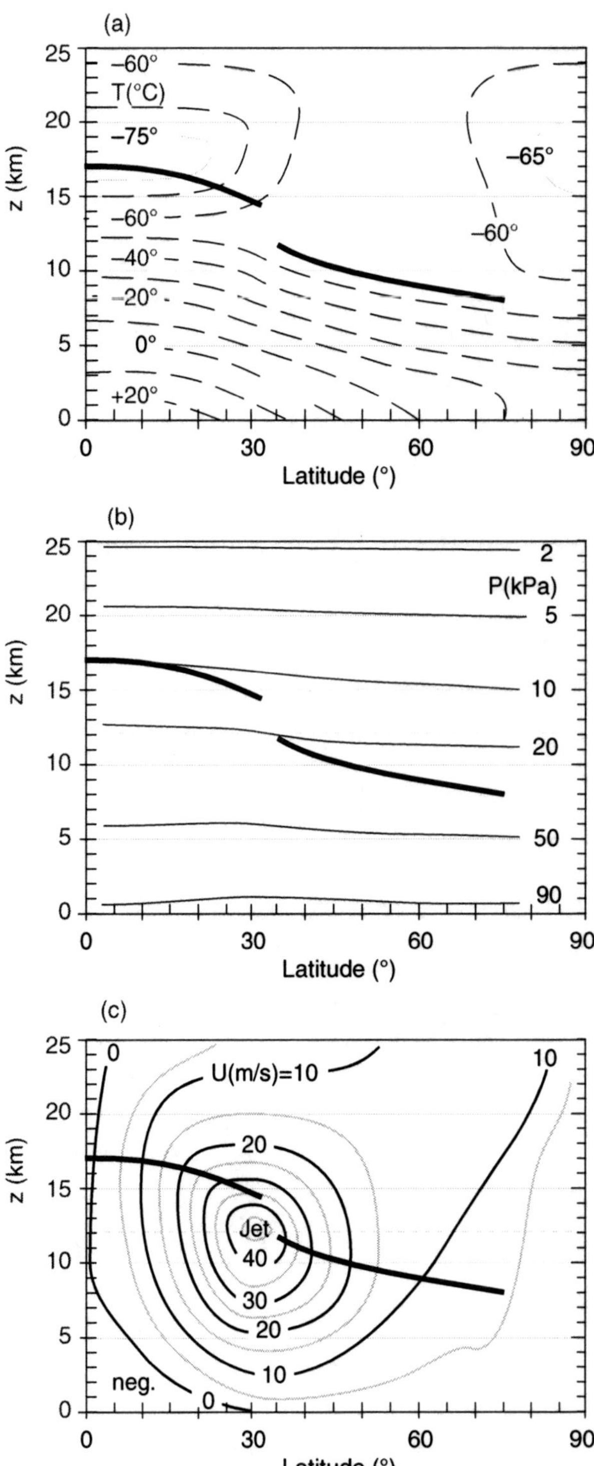

Figure 11.37
Vertical cross sections through the atmosphere on January 2003. (a) isotherms T (°C), (b) isobars P (kPa), (c) isotachs of the zonal wind U (m s^{-1}). The tropopause is indicated by a heavy solid line. For this Northern Hemisphere case, the jet-stream winds in (c) are from the west (i.e., toward the reader). You can superimpose copies of these three figures to see the relationship between the **temperature field**, **mass field** *(i.e., pressures), and* **wind field**.

Sample Application

Use the toy-model temperature eqs. (11.1 & 11.4) to estimate the speed of the geostrophic wind for latitude 60° at the top of the troposphere. Assume $T_v = -5°C$.

Find the Answer

Given: $\phi = 60°$, $T_v = 268$ K, $z = z_T$.
Find: U_{jet} = ? m s^{-1}

From Fig. 11.39 estimate troposphere top $z_T \approx 9$ km. Apply eq. (11.17), which already incorporates toy model eq. (11.4).

$$U_{jet} \approx \frac{(9.8\text{m}\cdot\text{s}^{-2})\cdot(1.18\times10^{-3}\text{km}^{-1})\cdot(40\text{K})}{(1.458\times10^{-4}\text{s}^{-1})\cdot(268\text{K})}\cdot$$

$$(9\text{km})\cdot(1-(1/2))\cdot\cos^2(60°)\cdot\sin^2(60°)$$

$= \underline{10.}$ m s^{-1}

Check: Physics & units are reasonable.

Exposition: Agrees with Fig. 11.37c. During winter the poles are colder relative to the equator, driving faster jet-stream winds than summer. Peak winds in the jet core can reach 100 m s^{-1}, although 3-month-average speeds are typically 40 m s^{-1}.

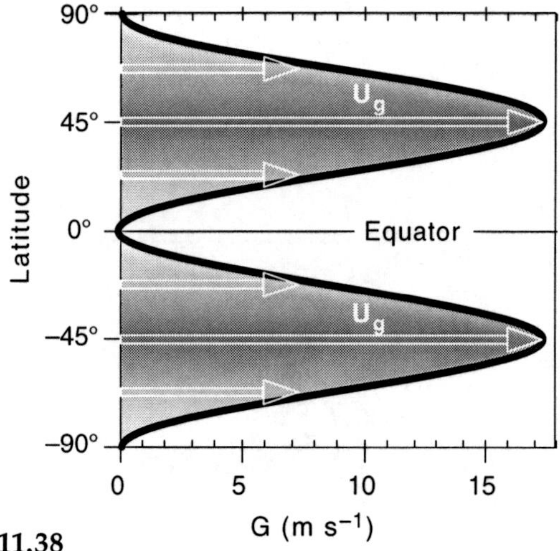

Figure 11.38
Idealized profile of 11 km altitude jet-stream winds vs. latitude.

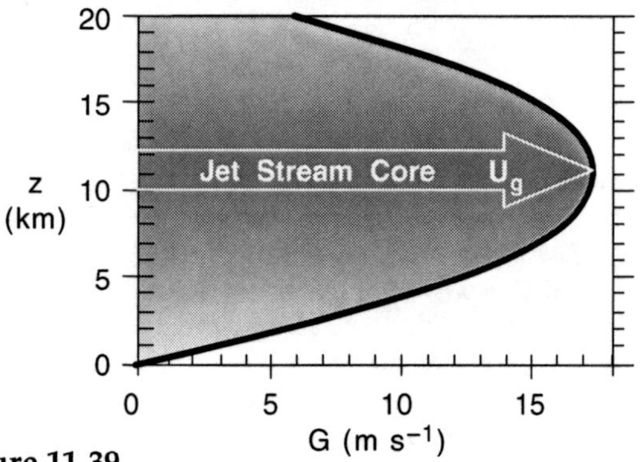

Figure 11.39
Idealized profile of 45°N latitude jet-stream winds vs. altitude.

The thermal-wind relationship tells us that the meridional variation of temperature between the cold poles and warm equator will drive westerly jet-stream winds in both hemispheres. Namely, the opposite Coriolis-parameter signs in the two hemispheres are canceled by the opposite meridional temperature-gradient signs, yielding positive values for U_g in the two hemispheres.

Although Fig. 11.8 indicates the largest meridional temperature gradients are at roughly 50° latitude (north & south) in our toy model, the meridional variation of the Coriolis parameter conspires to make the fastest jet-stream winds at 45° latitude (north & south) according to eq. (11.17). Solving eq. (11.27) for a tropopause height of z_T = 11 km with an average virtual temperature of –20°C in the troposphere gives peak jet-stream wind speeds (Figs. 11.38 & 11.39) of about 17.25 m s^{-1}.

While our simple toy model is insightful because it mimics the main features of the global circulation and allows an analytical solution that you can solve on a spreadsheet, it is <u>too</u> simple. Compared to Fig. 11.37, the toy jet-stream speeds are too slow, are located too far poleward, and don't give fast-enough winds near the ground. Also, actual locations in both hemispheres shift a bit to the (north, south) in northern hemisphere (summer, winter).

Angular Momentum & Subtropical Jet

Angular momentum can influence the <u>subtropical</u> jet, and is defined as mass times velocity times radius of curvature. Suppose that initially there is air moving at some zonal velocity U_s relative to the Earth's surface at some initial (source) latitude ϕ_s. Because the Earth is rotating, the Earth's surface at the source latitude is moving toward the east at velocity U_{Es}. Thus, the total eastward speed of the air parcel relative to the Earth's axis is $(U_s + U_{Es})$.

As sketched in Fig. 11.40, move the air to some other (destination) latitude ϕ_d, assuming that no other forces are applied. Conservation of angular momentum requires:

$$m\cdot(U_s + U_{Es})\cdot R_s = m\cdot(U_d + U_{Ed})\cdot R_d \qquad \bullet(11.18)$$

where U_d represents the new zonal air velocity relative to the Earth's surface at the destination latitude, U_{Ed} is the tangential velocity of the Earth's surface at the destination, and m is air mass.

For latitude ϕ at either the source or destination, the radius is R_s or R_d = $R_E\cdot\cos(\phi)$, where average Earth radius is R_E = 6371 km. Similarly, tangential velocities at either the source or destination are U_s or $U_d = \Omega\cdot R_\phi = \Omega\cdot R_E\cdot\cos(\phi)$, for an Earth angular velocity of $\Omega = 0.729\text{x}10^{-4}$ s^{-1}.

Solving these equations for the destination air velocity U_d relative to Earth's surface gives:

$$U_d = [\Omega \cdot R_E \cdot \cos(\phi_s) + U_s] \cdot \frac{\cos(\phi_s)}{\cos(\phi_d)} - \Omega \cdot R_E \cdot \cos(\phi_d) \tag{11.19}$$

As we already discussed, winds at the top of the Hadley cell diverge away from the equator, but cannot move beyond 30° latitude because Coriolis force turns the wind. When we use eq. (11.19) to predict the zonal wind speed for typical trade-wind air that starts at the equator with U_s = –7 m s^{-1} and ends at 30° latitude, we find unrealistically large wind speeds (125 m s^{-1}) for the subtropical jet (Fig. 11.41). Actual typical wind speeds in the subtropical jet are of order 40 to 80 m s^{-1} in the winter hemisphere, and slower in the summer hemisphere.

The discrepancy is because in the real atmosphere there is no conservation of angular momentum due to forces acting on the air. Coriolis force turns the wind, causing air to accumulate and create a pressure-gradient force to oppose poleward motion in the Hadley cell. Drag due to turbulence slows the wind a small amount. Also, the jet streams meander north and south, which helps to transport slow angular momentum southward and fast angular momentum northward. Namely, these meanders or synoptic-scale eddies cause mixing of zonal momentum.

Next, the concept of vorticity is discussed, which will be useful for explaining how troughs and ridges develop in the jet stream. Then, we will introduce a way to quantify circulation, to help understand the global circulation.

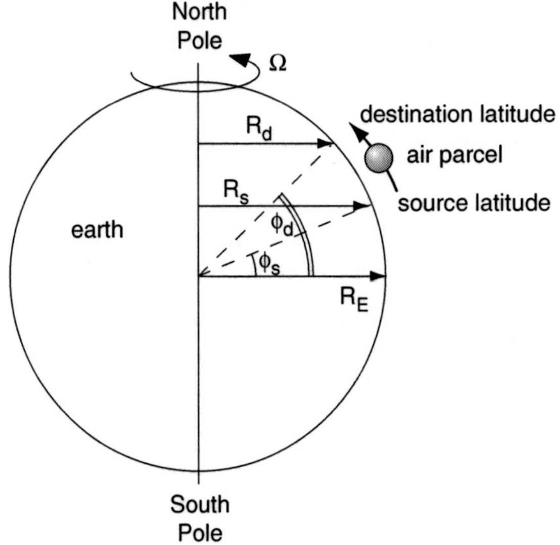

Figure 11.40
Geometry for angular-momentum calculations affecting an air parcel that moves toward the north.

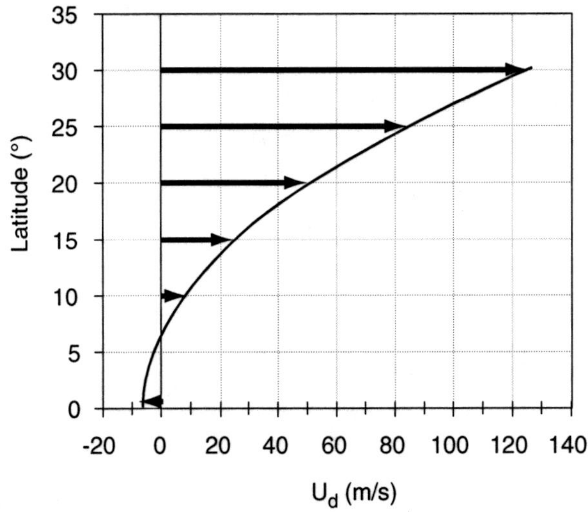

Figure 11.41
Computed zonal wind speed U_d at various destination latitudes for typical trade-wind air that starts at the equator with U_s = –7 m s^{-1}, for the unrealistic case of conservation of angular momentum. Actual zonal winds are not this fast at 30° latitude.

Sample Application

For air starting at the equatorial tropopause, what would be its zonal velocity at 20°N if angular momentum were conserved?

Find the Answer

Given: ϕ_d = 20°N, ϕ_s = 0°, $\Omega \cdot R_E$ = 463 m s^{-1}
Find: U_d = ? m s^{-1}
Assume: no turbulence; U_s = –7 m s^{-1} easterlies.

Use eq. (11.19): U_d =

$$[(463\text{m/s})\cos(0°) + U_s] \frac{\cos(0°)}{\cos(20°)} - (463\text{m/s})\cos(20°)$$

U_d = = **50.2** m s^{-1}

Check: Physics & units are reasonable. Agrees with Fig. 11.41
Exposition: Actual winds are usually slower, because of turbulent drag and other forces..

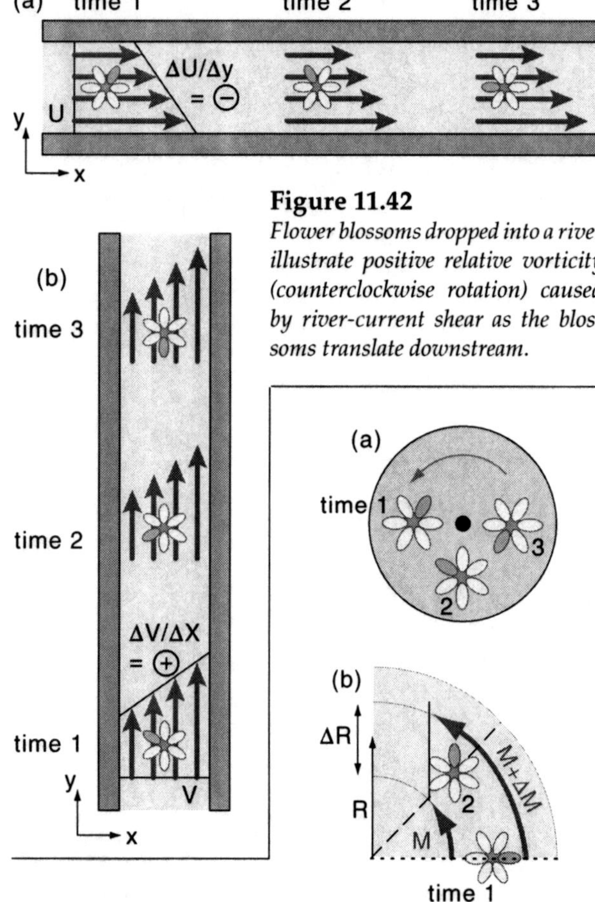

Figure 11.42
Flower blossoms dropped into a river illustrate positive relative vorticity (counterclockwise rotation) caused by river-current shear as the blossoms translate downstream.

Figure 11.43
(a) Vorticity associated with solid-body rotation, illustrated with a flower blossom glued to a solid turntable at time 1. (b) Vorticity caused by radial shear of tangential velocity M.

TYPES OF VORTICITY

Relative-vorticity Definition

Counterclockwise rotation about a local vertical axis defines positive vorticity. One type of vorticity ζ_r, called **relative vorticity**, is measured relative to both the location of the object and to the surface of the Earth (also see the Forces & Winds chapter).

Picture a flower blossom dropped onto a straight river, where the river current has shear (Fig. 11.42). As the floating blossom drifts (translates) downstream, it also spins due to the velocity shear of the current. Both Figs. 11.42a & b show counterclockwise rotation, giving positive relative vorticity as defined by:

$$\zeta_r = \frac{\Delta V}{\Delta x} - \frac{\Delta U}{\Delta y} \qquad \bullet(11.20)$$

for (U, V) positive in the local (x, y) directions.

Next, consider a flower blossom glued to a solid turntable, as sketched in Fig. 11.43a. As the table turns, so does the orientation of the flower petals relative to the center of the flower. This is shown in Fig. 11.43b. Solid-body rotation requires vectors that start on the dotted line and end on the dashed line. But in a river (or atmosphere), currents can have additional radial shear of the tangential velocity. Thus, relative vorticity can also be defined as:

$$\zeta_r = \frac{\Delta M}{\Delta R} + \frac{M}{R} \qquad \bullet(11.21)$$

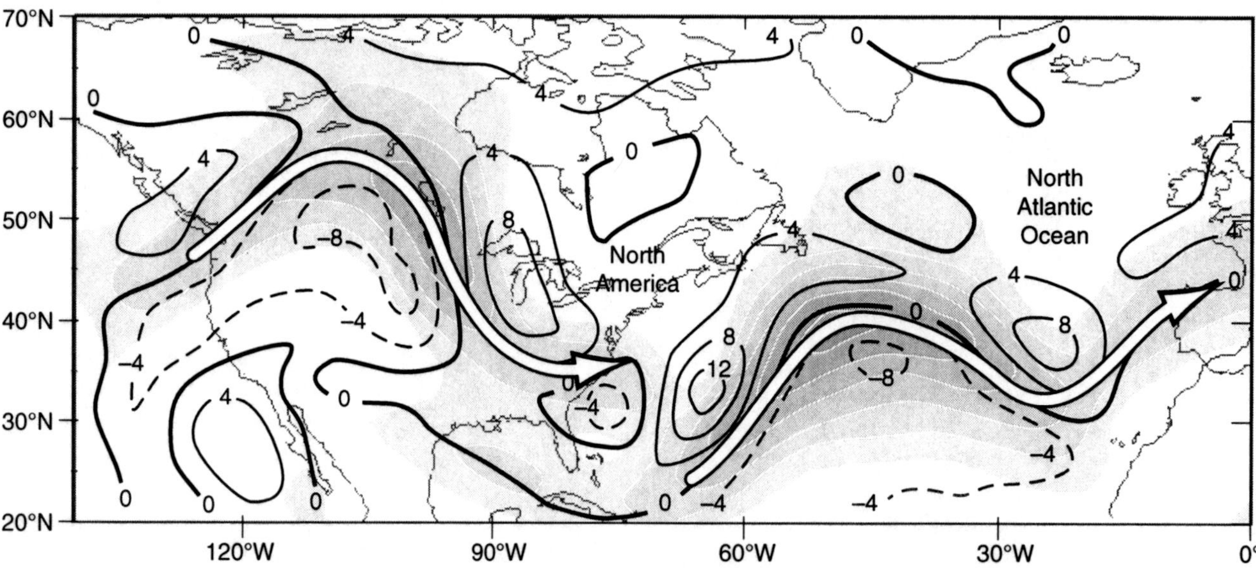

Figure 11.44
Weather map example of relative vorticity (contour lines, with units of 10^{-5} s^{-1}) near the tropopause (at the 20 kPa isobaric surface) at 12 UTC on 5 January 2001. White arrows show jet stream axis of fastest winds over North America and the Atlantic Ocean. Shading gives wind speeds every 10 m s^{-1} from 30 m s^{-1} (lightest grey) to over 80 m s^{-1} (darkest grey). (Based on NCEP/NCAR 40-year reanalysis data, utilizing the plotting tool by Christopher Godfrey, the Univ. of Oklahoma School of Meteorology.)

INFO • Solid Body Relative Vorticity

To get eq. (11.22) you can begin with eq. (11.21):

$$\zeta_r = \frac{\Delta M}{\Delta R} + \frac{M}{R}$$

For solid-body rotation, $\Delta M / \Delta R$ exactly equals M/R, as is illustrated in the figure at right. Hence:

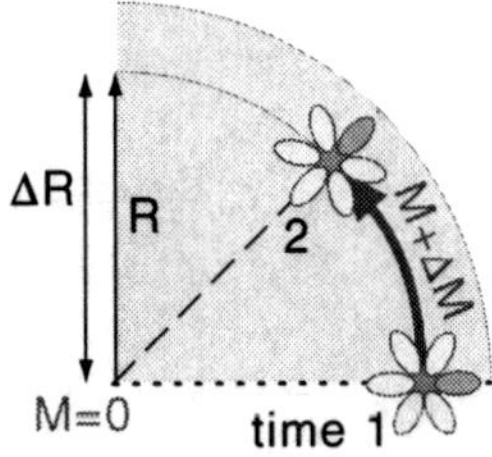

$$\zeta_r = \frac{(M-0)}{R} + \frac{M}{R} = \boxed{\frac{2M}{R}}$$

which is eq. (11.22).

For pure solid-body rotation (where the tangential current- or wind-vectors do indeed start at the same dotted line and end at the same dashed line in Fig. 11.43b), eq. (11.21) can be rewritten as

$$\zeta_r = \frac{2M}{R} \qquad \bullet(11.22)$$

Relative vorticity has units of s^{-1}. A quick way to determine the sign of the vorticity is to curl the fingers of your right hand in the direction of rotation. If your thumb points up, then vorticity is positive. This is the **right-hand rule**. Fig. 11.44 shows relative vorticity of various signs, where vorticity was created by both shear and curvature.

Absolute-vorticity Definition

Define an absolute vorticity ζ_a as the sum of the vorticity relative to the Earth plus the rotation of the Earth relative to the so-called fixed stars:

$$\zeta_a = \zeta_r + f_c \qquad \bullet(11.23)$$

where $f_c = 2\Omega \cdot \sin(\phi)$ is the Coriolis parameter, $2\Omega = 1.458\text{x}10^{-4}\ s^{-1}$ is the angular velocity of the Earth relative to the fixed stars, and latitude is ϕ. Absolute vorticity has units of s^{-1}.

Potential-vorticity Definition

Define a potential vorticity ζ_p as absolute vorticity per unit depth Δz of the rotating air column:

$$\zeta_p = \frac{\zeta_r + f_c}{\Delta z} = \text{constant} \qquad \bullet(11.24)$$

This is the most important measure of vorticity because it is constant (i.e., is conserved) for flows having no latent or radiative heating, and no turbulent drag. Potential vorticity units are $m^{-1}\ s^{-1}$.

Sample Application

At 50°N is a west wind of 100 m s^{-1}. At 46°N is a west wind of 50 m s^{-1}. Find the (a) relative, & (b) absolute vorticity.

Find the Answer

Given: U_2 =100 m s^{-1}, U_1 =50 m s^{-1}, $\Delta\phi$ =4°lat. V =0
Find: ζ_r = ? s^{-1} , ζ_a = ? s^{-1}

(a) Use eq. (11.20): Appendix A: 1°latitude = 111 km
ζ_r = – (100 – 50 m s^{-1}) / (4.4x10^5 m) = **–1.14x10**$^{-4}$ s^{-1}

(b) Average ϕ = 48°N. Thus, f_c = (1.458x10^{-4} s^{-1})·sin(48°)
f_c = 1.08x10^{-4} s^{-1}. then, use eq. (11.23):
ζ_a = (–1.14x10^{-4}) + (1.08x10^{-4}) = **–6x10**$^{-6}$ s^{-1} .

Check: Physics and units are reasonable.
Exposition: Shear vorticity from a strong jet stream, but its vorticity is opposite to the Earth's rotation.

Sample Application

Given a N. Hemisphere high-pressure center with tangential winds of 5 m s^{-1} at radius 500 km. What is the value of relative vorticity?

Find the Answer

Given: $|M|$ = 5 m s^{-1} (anti-cyclonic), R = 500,000 m.
Find: ζ_r = ? s^{-1}

In the N. Hem., anti-cyclonic winds turn clockwise, so using the right-hand rule means your thumb points down, so vorticity will be negative. Apply eq. (11.22):

ζ_r = – 2 · (5 m s^{-1}) / (5x10^5 m) = **–2x10**$^{-5}$ s^{-1}

Check: Physics and units are reasonable.
Exposition: Anticyclones often have smaller magnitudes of relative vorticity than cyclones (Fig. 11.44).

Sample Application

A hurricane at latitude 20°N has tangential winds of 50 m s^{-1} at radius 50 km from the center averaged over a 15 km depth. Find potential vorticity, assuming solid-body rotation for simplicity.

Find the Answer

Assume the shear is in the cyclonic direction.
Given: Δz = 15,000 m, ϕ = 20°N, M = 50 m s^{-1}, ΔR = 50,000 m.
Find: ζ_p = ? $m^{-1} \cdot s^{-1}$

Apply eqs. (11.22) with (11.24):

$$\zeta_p = \frac{\frac{2 \cdot (50\text{m/s})}{5 \times 10^4 \text{m}} + (1.458 \times 10^{-4} \text{s}^{-1}) \cdot \sin(20°)}{15000\text{m}}$$

= (2x10^{-3}+5x10^{-5})/15000 m = **1.37x10**$^{-7}$ $m^{-1} \cdot s^{-1}$

Check: Physics & units are reasonable.
Exposition: Positive sign due to cyclonic rotation.

Sample Application

If $\zeta_a = 1.5\text{x}10^{-4}\ \text{s}^{-1}$, $\rho = 0.7$ kg m^{-3} and $\Delta\theta/\Delta z = 4$ K km^{-1}, find the isentropic potential vorticity in PVU.

Find the Answer

Given: $\zeta_a = 1.5\text{x}10^{-4}\ \text{s}^{-1}$, $\rho = 0.7$ kg m^{-3}, $\Delta\theta/\Delta z = 4$ K km^{-1}

Find: ζ_{IPV} = ? PVU

Apply eq. (11.26):

$$\zeta_{IPV} = \frac{(1.5\times10^{-4}\text{s}^{-1})}{(0.7\text{kg}\cdot\text{m}^{-3})}\cdot\left(4\text{K}\cdot\text{km}^{-1}\right)$$

$= 8.57\text{x}10^{-7}$ K·m^2·s^{-1}·kg^{-1} = 0.857 PVU

Check: Physics & units are reasonable.

Exposition: As expected for tropospheric air, the answer has magnitude less than 1.5 PVU.

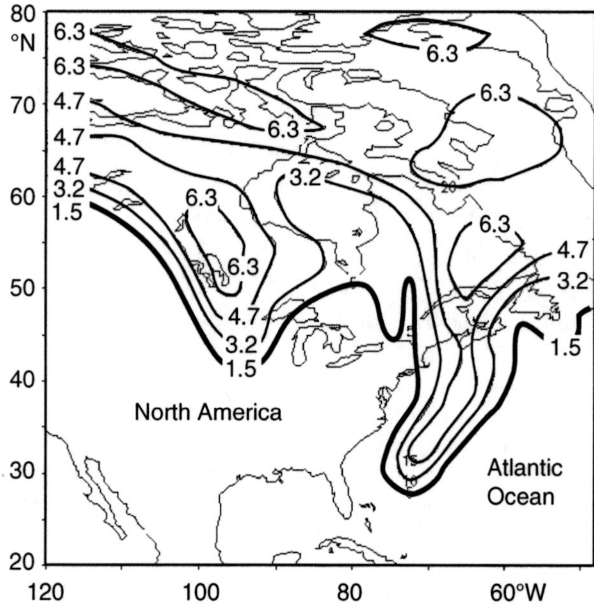

Figure 11.45

Example of isentropic potential vorticity on the 315 K isentropic surface, at 00 UTC on 5 January 2001. Units are PVU. Values greater than 1.5 are in stratospheric air. Because the tropopause is at lower altitude near the poles than at the equator, the 315 K potential temperature surface crosses the tropopause; so it is within the troposphere in the southern part of the figure and in the stratosphere in the northern part. Tropopause folds are evident by the high PVU values just west of the Great Lakes, and just east of the North American coastline. (Based on NCEP/NCAR 40-year reanalysis data, with initial plots produced using the plotting tool by Christopher Godfrey, the University of Oklahoma School of Meteorology.)

Rewrite the potential vorticity using eqs. (11.24 & 11.21):

$$\underset{shear}{\frac{\Delta M}{\Delta R}} + \underset{curvature}{\frac{M}{R}} + \underset{planetary}{f_c} = \underset{stretching}{\zeta_p \cdot \Delta z} \qquad \bullet(11.25)$$

The initial values of absolute vorticity and rotating-air depth determine the value for the constant ζ_p. In order to preserve the equality in the equation above while preserving the constant value of ζ_p, any increase of depth Δz of the rotating layer of air must be associated with greater relative vorticity (air spins faster) or larger f_c (rotating air moves poleward).

Isentropic Potential Vorticity Definition

By definition, an isentropic surface connects locations of equal entropy. But non-changing entropy corresponds to non-changing potential temperature θ in the atmosphere (i.e., adiabatic conditions). If you calculate the absolute vorticity ($\zeta_r + f_c$) on such a surface, then it can be used to define an isentropic potential vorticity (IPV):

$$\zeta_{IPV} = \frac{\zeta_r + f_c}{\rho}\cdot\left(\frac{\Delta\theta}{\Delta z}\right) = \frac{\zeta_a}{\rho}\cdot\left(\frac{\Delta\theta}{\Delta z}\right) \qquad (11.26)$$

or

$$\zeta_{IPV} = -|g|\cdot(\zeta_r + f_c)\cdot\frac{\Delta\theta}{\Delta p} \qquad (11.27)$$

where air pressure is p, air density is ρ, gravitational acceleration magnitude is $|g|$, and $\Delta\theta/\Delta z$ indicates the static stability.

Define a **potential vorticity unit (PVU)** such that 1 PVU = 10^{-6} K·m^2·s^{-1}·kg^{-1}. The stratosphere has lower density and greater static stability than the troposphere, hence stratospheric air has IPV values that are typically 100 times larger than for tropospheric air. Typically, $\zeta_{IPV} > 1.5$ PVU for stratospheric air (a good atmospheric cross-section example is shown in the Extratropical Cyclone chapter).

For idealized situations where the air moves adiabatically without friction while following a constant θ surface (i.e., isentropic motion), then ζ_{IPV} is conserved, which means you can use it to track air motion. Also, stratospheric air does not instantly lose its large IPV upon being mixed downward into the troposphere.

Thus, IPV is useful for finding tropopause folds and the accompanying intrusions of stratospheric air into the troposphere (Fig. 11.45), which can bring down toward the ground the higher ozone concentrations and **radionuclides** (radioactive atoms from former atomic-bomb tests) from the stratosphere.

Because isentropic potential vorticity is conserved, if static stability ($\Delta\theta/\Delta z$) weakens, then eq. (11.26) says absolute vorticity must increase to maintain constant IPV. For example, Fig. 11.46 shows isentropes for flow from west to east across the

Rocky Mountains. Where isentropes are spread far apart, static stability is low. Because air tends to follow isentropes (for frictionless adiabatic flow), a column of air between two isentropes over the crest of the Rockies will remain between the same two isentropes as the air continues eastward.

Thus, the column of air shown in Fig. 11.46 becomes stretched on the lee side of the Rockies and its static stability decreases (same $\Delta\theta$, but spread over a larger Δz). Thus, absolute vorticity in the stretched region must increase. Such increased cyclonic vorticity encourages formation of low-pressure systems (extratropical cyclones) to the lee of the Rockies — a process called **lee cyclogenesis**.

HORIZONTAL CIRCULATION

Consider a closed shape of finite area (Fig. 11.47a). Pick any starting point on the perimeter, and hypothetically travel counterclockwise around the perimeter until you return to the starting point. As you travel each increment of distance Δl, observe the local winds along that increment, and get the average tangential component of wind velocity M_t.

The horizontal **circulation** C is defined as the product of this tangential velocity times distance increment, summed over all the increments around the whole perimeter:

$$C = \sum_{i=1}^{n} (M_t \cdot \Delta l)_i \qquad (11.28)$$

where i is the index of each increment, and n is the number of increments needed to complete one circuit around the perimeter. Take care that the sign of M_t is such that it is positive if the tangential wind is in the same direction as you are traveling, and negative if opposite. The units of circulation are $m^2 \cdot s^{-1}$.

If we approximate the perimeter by Cartesian line segments (Fig. 11.47b), then eq. (11.28) becomes:

$$C = \sum_{i=1}^{n} (U \cdot \Delta x + V \cdot \Delta y)_i \qquad (11.29)$$

The sign of Δx is (+) if you travel in the positive x-direction (toward the East), and (–) if opposite. Similar rules apply for Δy (+ toward North).

To better understand circulation, consider idealized cases (Figs. 11.48a & b). For winds of tangential velocity M_t rotating counterclockwise around a circle of radius R, the circulation is $C = 2\pi R \cdot M_t$. For clockwise circular rotation, the circulation is $C = -2\pi R \cdot M_t$, namely, the circulation value is negative. From these two equations, we see that a fast speed around a small circle (such as a tornado) can give

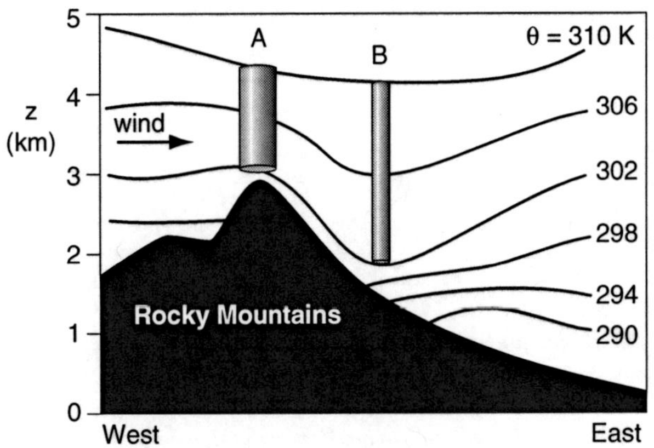

Figure 11.46
Wind blowing from west to east (A) over the Rocky Mountains creates mountain waves and downslope winds, which cause greater separation (B) between the 302 and 310 K isentropes to the lee of the mountains. This greater separation implies reduced static stability and vertical stretching.

Sample Application

Given Fig. 11.46. (a) Estimate $\Delta\theta/\Delta z$ at A and B. (b) if the initial absolute vorticity at A is 10^{-4} s^{-1}, find the absolute vorticity at B.

Find the Answer

Given: $\zeta_a = 10^{-4}$ s^{-1} at A, $\theta_{top} = 310$ K, $\theta_{bottom} = 302$ K.
Find: (a) $\Delta\theta/\Delta z$ = ? K km^{-1} at A and B.
(b) ζ_a = ? s^{-1} at B.

$\Delta\theta$ = 310 K – 302 K = 8 K at A & B. Estimate the altitudes at the top and bottom of the cylinders in Fig. 11.46.
A: z_{top} = 4.4 km, z_{bottom} = 3.1 km. Thus Δz_A = 1.3 km.
B: z_{top} = 4.2 km, z_{bottom} = 1.9 km. Thus Δz_B = 2.3 km.

(a) $\Delta\theta/\Delta z$ = 8K / 1.3 km = <u>6.15 K km^{-1}</u> at A.
$\Delta\theta/\Delta z$ = 8K / 2.3 km = <u>3.48 K km^{-1}</u> at B.

(b) If initial (A) and final (B) IPV are equal, then rearranging eq. (11.26) and substituting $(\zeta_r + f_c) = \zeta_a$ gives:
$\zeta_{aB} = \zeta_{aA} \cdot (\Delta z_B/\Delta z_A)$
$= (10^{-4}$ s$^{-1}) \cdot [(2.3$ km$)/(1.3$ km$)]$ = <u>1.77x 10^{-4}</u> s^{-1}

Check: Sign, magnitude & units are reasonable.
Exposition: Static stability $\Delta\theta/\Delta z$ is much weaker at B than A. Thus absolute vorticity at B is much greater than at A. If the air flow directly from west to east and if the initial relative vorticity were zero, then the final relative vorticity is $\zeta_r = 0.77 \times 10^{-4}$ s^{-1}. Namely, to the lee of the mountains, cyclonic rotation forms in the air where none existed upwind. Namely, this implies cyclogenesis (birth of cyclones) to the lee (downwind) of the Rocky Mountains.

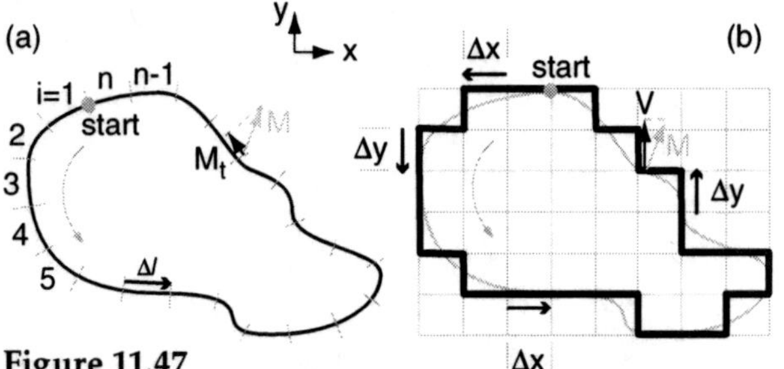

Figure 11.47
Method for finding the circulation. M is the wind vector. M_t and V are projections onto the perimeter. (a) Stepping in increments of Δl around an arbitrary shape. (b) Stepping around a Cartesian (gridded) approximation to the shape in (a).

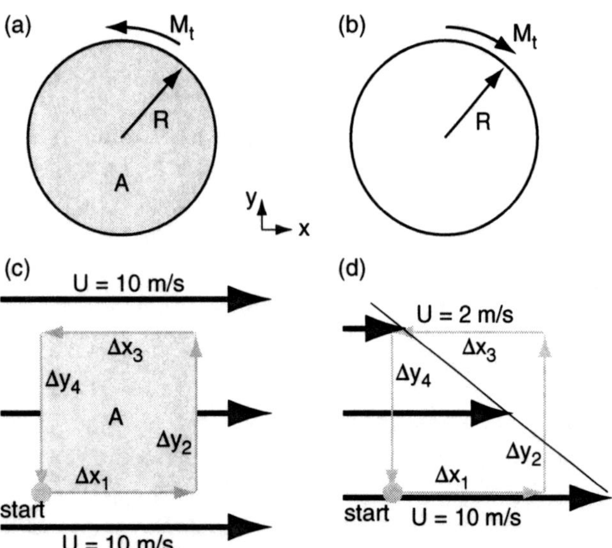

Figure 11.48
Circulation examples. (a) Counterclockwise rotation around a circle. (b) Clockwise rotation around a circle. (c) Uniform wind. (d) Uniform shear. Area A enclosed by circulation is shaded.

Sample Application
Find the horizontal circulation for Fig. 11.48d. Assume $\Delta x = \Delta y = 1$ km. Relate to shear and rel. vorticity.

Find the Answer
Given: U_{bottom}=10 m s^{-1}, U_{top}=2 m s^{-1}, V=0,
Δx=Δy=1 km. Find: C = ? m^2 s^{-1}, ζ_r = ? s^{-1}

Use eq. (11.29) from start point:
$C = (U\cdot\Delta x)_1 + (V\cdot\Delta y)_2 + (U\cdot\Delta x)_3 + (V\cdot\Delta x)_4$
(10m s^{-1})·(1km) + 0 + (2m s^{-1})·(–1km) + 0
C = 8 (m s^{-1})·km = **<u>8000 m</u>^{2} s^{-1}**
Use eq. (11.30): with area $A = \Delta x\cdot\Delta y$ = 1 km^2
$U_{shear} = \Delta U/\Delta y = (U_{top}-U_{bottom})/\Delta y$
= [(2 – 10 m s^{-1})/(1 km)] = –8 (m s^{-1})/km
C = [0 – U_{shear}]·A = –[–8 (m s^{-1})/km]· (1 km^2)
= 8 (m s^{-1})·km = **<u>8000 m</u>^{2} s^{-1}**.
Use eq. (11.31): $\zeta_r = C/A$ = 8 (m s^{-1})/km = **<u>0.008 s</u>$^{-1}$** .

Check: Physics, magnitude & units are reasonable.
Exposition: Strong shear. Large circ. Large vorticity.

the same circulation magnitude as a slower speed around a larger circle (e.g., a mid-latitude cyclone).

Consider two more cases (Figs. 11.48c & d). For a circuit in a constant wind field of any speed, the circulation is C = 0. For a circuit within a region of uniform shear such as $\Delta U/\Delta y$, the circulation is $C = -(\Delta U/\Delta y)\cdot(\Delta y\cdot\Delta x)$. Comparing these last two cases, we see that the wind speed is irrelevant for the circulation, but the wind shear is very important.

In the last equation above, $(\Delta y\cdot\Delta x) = A$ is the area enclosed by the circulation of Fig. 11.48d. In general, for uniform U and V <u>shear</u> across a region, the horizontal circulation is:

$$C = \left(\frac{\Delta V}{\Delta x} - \frac{\Delta U}{\Delta y}\right)\cdot A \tag{11.30}$$

But the term in parentheses is the relative vorticity ζ_r. This gives an important relationship between horizontal circulation and vorticity:

$$C = \zeta_r \cdot A \tag{11.31}$$

Vorticity is defined at any one point in a fluid, while circulation is defined around a finite-size area. Thus, eq. (11.31) is valid only in the limit as A becomes small, or for the special case of a fluid having uniform vorticity within the whole circulation area.

The horizontal circulation C defined by eq. (11.30 & 11.31) is also known as the **relative circulation** C_r. An **absolute circulation** C_a can be defined as

$$C_a = (\zeta_r + f_c)\cdot A \tag{11.32}$$

where f_c is the Coriolis parameter. The absolute circulation is the circulation that would be seen from a fixed point in space looking down on the atmosphere rotating with the Earth.

For the special case of a frictionless **barotropic atmosphere** (where isopycnics are parallel to isobars), **Kelvin's circulation theorem** states that C_a is constant with time.

For a more realistic **baroclinic atmosphere** containing horizontal temperature gradients, the **Bjerknes circulation theorem:**

$$\frac{\Delta C_r}{\Delta t} = -\sum_{i=1}^{n}\left(\frac{\Delta P}{\rho}\right)_i - f_c \cdot \frac{\Delta A}{\Delta t} \tag{11.33}$$

says relative circulation varies with the torque applied to the fluid (via the component of pressure forces in the direction of travel, summed around the perimeter of the circulation area) minus the Earth's rotation effects in a changing circulation area. The units of $\Delta P/\rho$ are J kg^{-1}, which are equivalent to the m^2·s^{-2} units of $\Delta C_r/\Delta t$. The pressure term in eq. (11.33) is called the **solenoid term**.

EXTRATROPICAL RIDGES & TROUGHS

The atmosphere is generally warm near the equator and cool near the poles. This meridional temperature gradient drives a west-to-east wind having increasing speed with increasing altitude within the troposphere, as described by the thermal-wind effect. The resulting fast wind near the tropopause is called the jet stream. To first order, we would expect this jet stream to encircle the globe (Fig. 11.49a) along the zone between the warm and cool airmasses, at roughly 50 to 60° latitude in winter.

However, this flow is unstable, allowing small disturbances to grow into large north-south meanders (Fig. 11.49b) of the jet stream. These meanders are called **Rossby waves** or **planetary waves**. Typical wavelengths are 3 - 4 Mm. Given the circumference of a parallel at those latitudes, one typically finds 3 to 13 waves around the globe, with a normal **zonal wavenumber** of 7 to 8 waves.

The equatorward region of any meander is called a **trough** (pronounced like "troff") and is associated with low pressure or low geopotential height. The poleward portion of a meander is called a **ridge**, and has high pressure or height. The turning of winds around troughs and ridges are analogous to the turning around closed lows and highs, respectively. The trough center or **trough axis** is labeled with a dashed line, while the **ridge axis** is labeled with a zig-zag symbol (Fig. 11.49b).

Like many waves or oscillations in nature, Rossby waves result from the interplay between inertia (trying to make the jet stream continue in the direction it was deflected) and a restoring force (acting opposite to the deflection). For Rossby waves, the restoring force can be explained by the conservation of potential vorticity, which depends on both the Coriolis parameter and the layer thickness (related to layer static stability). **Baroclinic instability** considers both restoring factors, while **barotropic instability** is a simpler approximation that considers only the Coriolis effect.

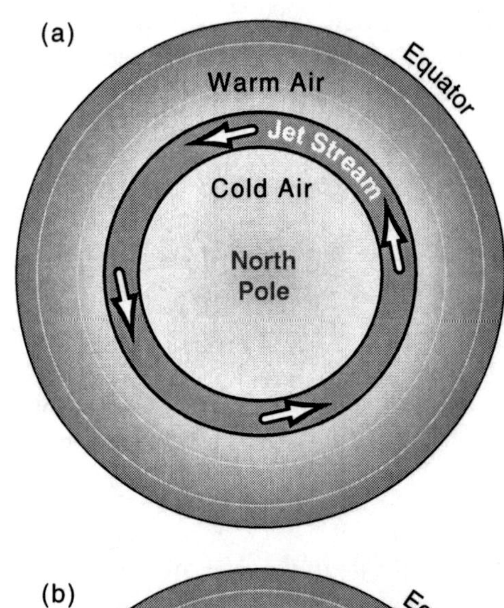

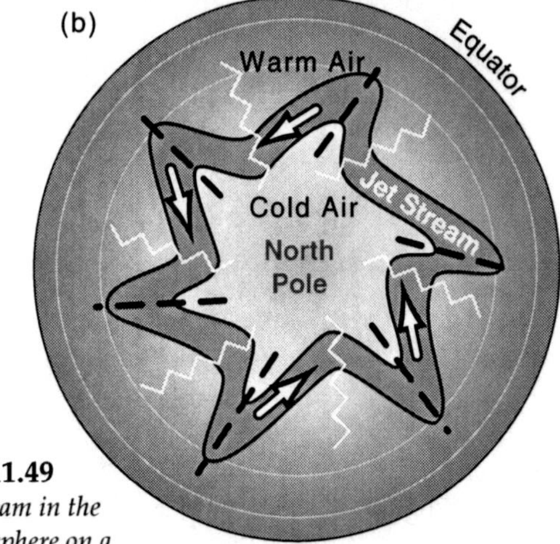

Figure 11.49
(a) Jet stream in the N. Hemisphere on a world with no instabilities. (b) Jet stream with barotropic or baroclinic instabilities, creating a meandering jet. Troughs are marked with a black dashed line, and ridges with a white zig-zag line.

Barotropic Instability

Consider tropospheric air of constant depth Δz ($\approx$ 11 km). For this situation, the conservation of potential vorticity can be written as

$$\left[\frac{M}{R}+f_c\right]_{initial} = \left[\frac{M}{R}+f_c\right]_{later} \qquad (11.34)$$

where jet-stream wind speed M divided by radius of curvature R gives the relative vorticity, and f_c is the Coriolis parameter (which is a function of latitude).

(Location **a** in Fig. 11.50) Consider a jet stream at

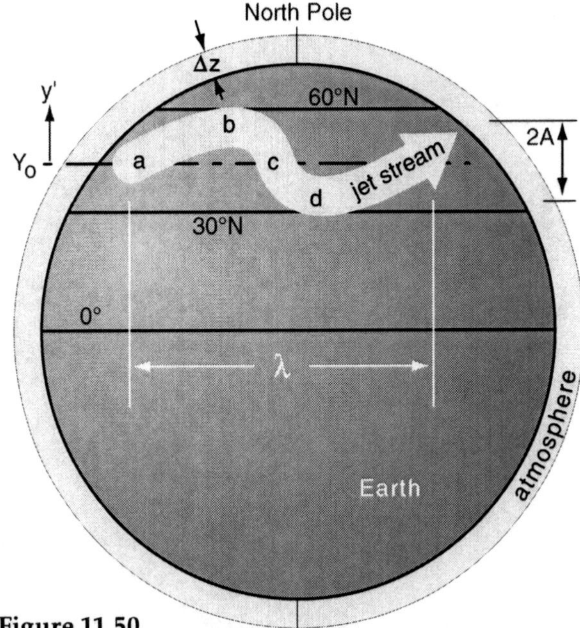

Figure 11.50
Illustration of barotropic instability.

HIGHER MATH • Planetary-wave Vorticity

Suppose the jet-stream axis oscillates north or south some distance y relative an arbitrary reference latitude as the air flows toward the east (x):

$$y = A \cdot \sin(2\pi \cdot x / \lambda) \qquad \text{(a)}$$

where the wavelength is λ and the meridional amplitude of the wave is A. The jet speed along this wave is M, and the associated wind components (U, V) depend on the local slope s of the wave at location x. From geometry: $U^2 + V^2 = M^2$ and $s = V/U$, thus:

$$U = M \cdot (1+s^2)^{-1/2} \quad \& \quad V = M \cdot s \cdot (1+s^2)^{-1/2} \qquad \text{(b)}$$

To find slope s from eq. (a), take the derivative of y:

$$s = \partial y / \partial x = (2\pi A / \lambda) \cdot \cos(2\pi \cdot x / \lambda) \qquad \text{(c)}$$

Next, change eq. (11.20 from finite difference to partial derivatives:

$$\zeta_r = \partial V / \partial x - \partial U / \partial y \qquad \text{(d)}$$

But you can expand the last term as follows:

$$\zeta_r = \partial V / \partial x - (\partial U / \partial x) \cdot (\partial x / \partial y)$$

where the last factor is just one over the slope:

$$\zeta_r = \partial V / \partial x - (\partial U / \partial x) \cdot (1 / s) \qquad \text{(e)}$$

Combine eqs. (e, c, & b) to get the desired relative vorticity:

$$\zeta_r = \frac{-2 \cdot M \cdot A \cdot \left(\frac{2\pi}{\lambda}\right)^2 \cdot \sin\left(\frac{2\pi x}{\lambda}\right)}{\left[1 + \left(\frac{2\pi A}{\lambda}\right)^2 \cdot \cos^2\left(\frac{2\pi x}{\lambda}\right)\right]^{3/2}} \qquad \text{(f)}$$

Fig. 11.a illustrates this for a wave with A = 1500 km, λ = 6000 km, & M = 40 m s^{-1}.

Exposition: Some calculus books give equations for the sine-wave radius of curvature R. Using that in $|\zeta_r| = 2M/R$ would give a similar equation for vorticity. The largest vorticities are concentrated near the wave crest and trough, allowing meteorologists to use positive vorticity to help find troughs.

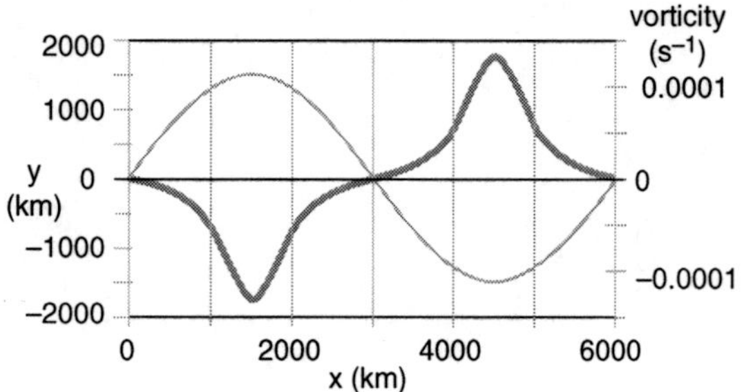

Figure 11.a.
Jet-stream path (blue) and associated vorticity (thick red).

initial latitude Y_o moving in a straight line from the southwest. At that latitude it has a certain value of the Coriolis parameter, but no relative vorticity (M/R = 0, because $R = \infty$ for a straight line). But f_c increases as the air moves poleward, thus the M/R term on the right side of eq. (11.34) must become smaller than its initial value (i.e., it becomes negative) so that the sum on the right side still equals the initial value on the left side.

(Location **b**) We interpret negative curvature as anticyclonic curvature (clockwise turning in the N. Hemisphere). This points the jet stream equatorward.

(Location **c**) As the air approaches its starting latitude, its Coriolis parameter decreases toward its starting value. This allows the flow to become a straight line again at location c. But now the wind is blowing from the northwest, not the southwest.

(Location **d**) But as the air overshoots equatorward, f_c gets smaller, requiring a positive M/R (cyclonic curvature) to maintain constant potential vorticity. This turns the jet stream back toward its starting latitude, where the cycle repeats. The flow is said to be **barotropically unstable**, because even pure, non-meandering zonal flow, if perturbed just a little bit from its starting latitude, will respond by meandering north and south.

This north-south (meridional) oscillation of the west-to-east jet stream creates the wavy flow pattern we call a **Rossby wave** or a **planetary wave**. Because the restoring force was related to the change of Coriolis parameter with latitude, it is useful to define a beta parameter as

$$\beta = \frac{\Delta f_c}{\Delta y} = \frac{2 \cdot \Omega}{R_{earth}} \cdot \cos\phi \qquad \bullet\text{(11.35)}$$

where the average radius of the Earth is R_{Earth} =6371 km. For $2 \cdot \Omega / R_{Earth}$ = 2.29x10^{-11} m^{-1}·s^{-1}, one finds that β is roughly (1.5 to 2)x10^{-11} m^{-1}·s^{-1}.

The wave path in Fig. 11.50 can be approximated with a simple cosine function:

$$y' \approx A \cdot \cos\left[2\pi \cdot \left(\frac{x' - c \cdot t}{\lambda}\right)\right] \qquad \text{(11.36)}$$

where the displacement distance of the Rossby wave from Y_o (its starting latitude) is y'. Let x' be the eastward distance from the start of the wave. The position of the wave crests move at phase speed c with respect to the Earth. The wavelength is λ and its amplitude is A (see Fig. 11.50). The primes indicate the deviations from a mean background state.

Barotropic Rossby waves of the jet stream have wavelengths of about $\lambda \approx$ 6000 km and amplitudes of about $A \approx 1665$ km, although a wide range of both is possible. Typically 4 to 5 of these waves can fit around the earth at mid-latitudes (where the circumference of a latitude circle is $2\pi \cdot R_{Earth} \cdot \cos\phi$, where ϕ is latitude.

The waves move through the air at **intrinsic phase speed** c_o:

$$c_o = -\beta \cdot \left(\frac{\lambda}{2\pi} \right)^2 \qquad \bullet(11.37)$$

The negative sign means that the wave crest propagate toward the west through the air.

However, the air in which the wave is imbedded is itself moving toward the east at wind speed U_o. Thus, relative to the ground, the **phase speed** c is:

$$c = U_o + c_o \qquad \bullet(11.38)$$

Given typical values for U_o, the total phase speed c relative to the ground is positive. Such movement toward the east is indeed observed on weather maps.

Barotropic Rossby waves can have a range of different wavelengths. But eq. (11.37) says that different wavelength waves move at different intrinsic phase speeds. Thus, the different waves tend to move apart from each other, which is why eq. (11.37) is known as a **dispersion relationship**.

Short waves have slower intrinsic phase speed toward the west, causing the background wind to blow them rapidly toward the east. **Long waves**, with their faster intrinsic phase speed westbound end up moving more slowly toward the east relative to the ground, as illustrated in Fig. 11.51.

(a) Short Waves

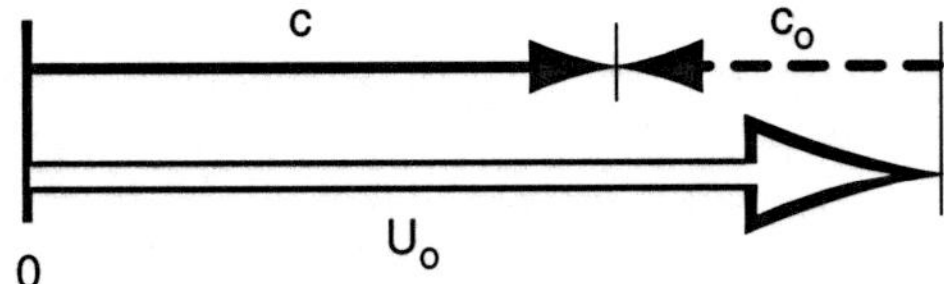

(b) Long Waves

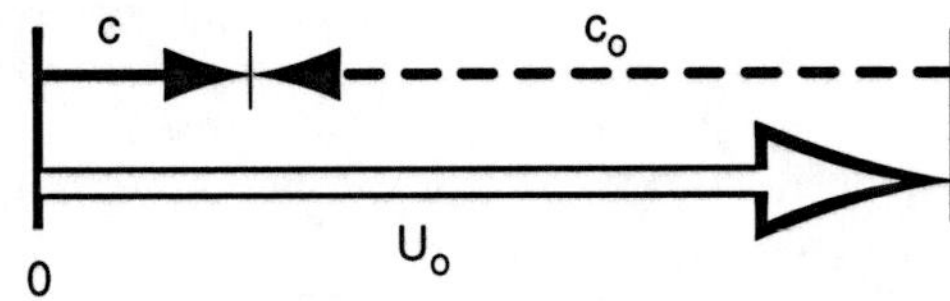

HIGHER MATH • The Beta Plane

Here is how you can get β using the definition of the Coriolis parameter f_c (eq. 10.16):

$$f_c = 2\,\Omega \sin\phi$$

where ϕ is latitude.

Since y is the distance along the perimeter of a circle of radius R_{Earth}, recall from geometry that

$$y = R_{Earth} \cdot \phi$$

for ϕ in radians.

Rearrange this to solve for ϕ, and then plug into the first equation to give:

$$f_c = 2\,\Omega \sin(y/R_{Earth})$$

By definition of β, take the derivative to find

$$\beta = \frac{\partial f_c}{\partial y} = \frac{2 \cdot \Omega}{R_{earth}} \cdot \cos\left(\frac{y}{R_{earth}} \right)$$

Finally, use the second equation above to give:

$$\boxed{\beta = \frac{2 \cdot \Omega}{R_{earth}} \cdot \cos\phi} \qquad \bullet(11.35)$$

For a small range of latitudes, β is nearly constant. Some theoretical derivations assume constant beta, which has the same effect as assuming that the earth is shaped like a cone. The name for this lamp-shade shaped surface is the **beta plane**.

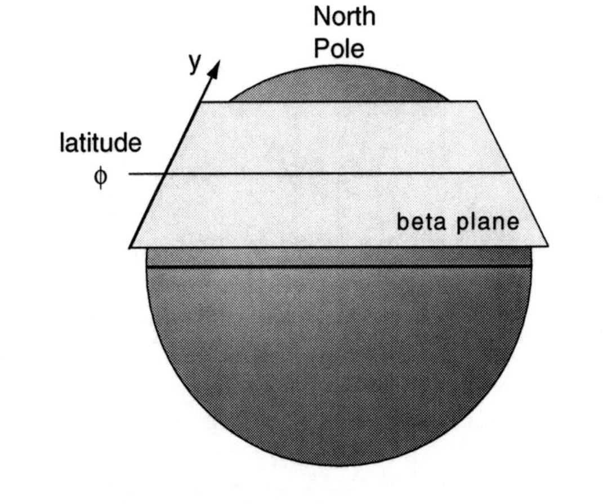

Figure 11.51 *(at left)*
Sum of large background west wind U_o with smaller intrinsic Rossby-wave phase speed c_o from the east gives propagation speed c of Rossby waves relative to the ground.

Sample Application (§)

The jet stream meanders north and south with barotropic wavelength of 8000 km and amplitude of 1200 km relative to reference latitude 50°N. Winds in this jet are 40 m s^{-1}. Calculate the following for a barotropic wave: beta parameter, phase speeds, and wave translation distance in 12 h. Also, plot the initial and final wave streamlines $y'(x')$ for x' of 0 to 11,000 km.

Find the Answer

Given: U_o = 40 m s^{-1}, ϕ =50°, A =1200 km, λ =8000 km, for t = 0 to 12 h

Find: β = ? m^{-1}·s^{-1}, c_o = ? m s^{-1}, c = ? m s^{-1}, $D = c\cdot\Delta t$? km translation distance, $y'(x')$ = ? km.

Apply eq. (11.35):

$$\beta = 2.29\times10^{-11}\cdot\cos(50°) = \mathbf{1.47x10^{-11}}\ \text{m}^{-1}\cdot\text{s}^{-1}.$$

Next, apply eq. (11.37):

$$c_o = -(1.47\times10^{-11}\text{m}^{-1}\text{s}^{-1})\left(\frac{8\times10^6\,\text{m}}{2\pi}\right)^2 = \underline{-23.9}\ \text{m s}^{-1}$$

Then apply eq. (11.38):

$$c = (50-23.9)\ \text{m s}^{-1} = \underline{16.1}\ \text{m s}^{-1}$$

Wave crest translation distance in 12 h is

$$D = c\cdot\Delta t = (16.1\ \text{m s}^{-1})\cdot(12\text{h}\cdot3600\text{s/h}) = \underline{697.1\ \text{km}}$$

Finally solve & plot eq. (11.36) for t = 0 to 12 h:

$$y' \approx (1200\text{km})\cdot\cos\left[2\pi\cdot\left(\frac{x'-(16.1\text{m/s})\cdot t}{8\times10^6\text{m}}\right)\right]$$

with conversions between m & km, and for s & h.

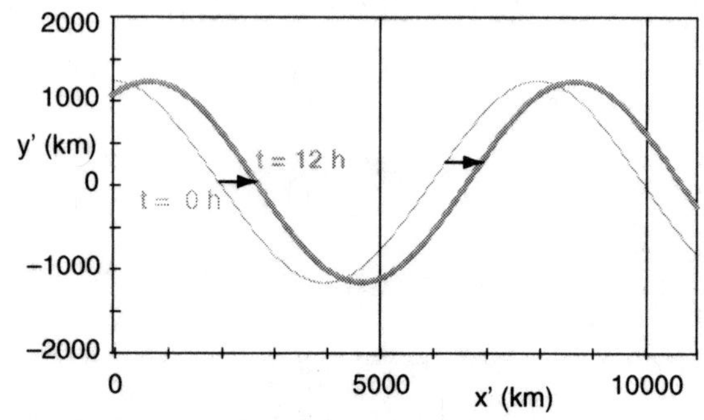

Check: Physics & units are reasonable.

Exposition: The thin blue streamlines plotted above show the path of the 40 m s^{-1} jet stream, but this path gradually shifts toward the east (thick red streamlines).

Although this Rossby-wave phase speed is much slower than a jet airliner, the wave does not need to land and refuel. Thus, during 24 hours, this long wave could travel about 1,400 km — roughly half the distance between San Francisco, CA & Chicago, IL, USA. Even longer waves can be stationary, and some extremely long waves can **retrograde** (move in a direction opposite to the background jet-stream flow; namely, move toward the west).

Sample Application (§)

Similar to the previous Sample, but for a short wave with 500 km amplitude and 2000 km wavelength.

Find the Answer

Given: Same, but λ = 2000 km, A = 500 km.

Apply similar equations (not shown here) as before, yielding: β = **1.47x10**$^{-11}$ m^{-1}·s^{-1}, c_o = $\underline{-1.5}$ m s^{-1}, c = 38.5 m s^{-1}, and D = 1664 km.

The short-wave streamline plot:

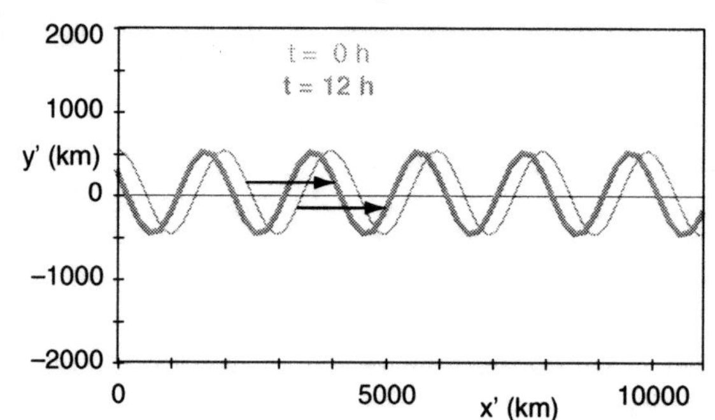

Check: Physics & units reasonable.

Exposition: Short waves move faster than long ones.

Sample Application (§)

Similar to the previous two Sample Applications, but superimpose the long and short waves.

Find the Answer

The combined long- and short-wave streamline plot:

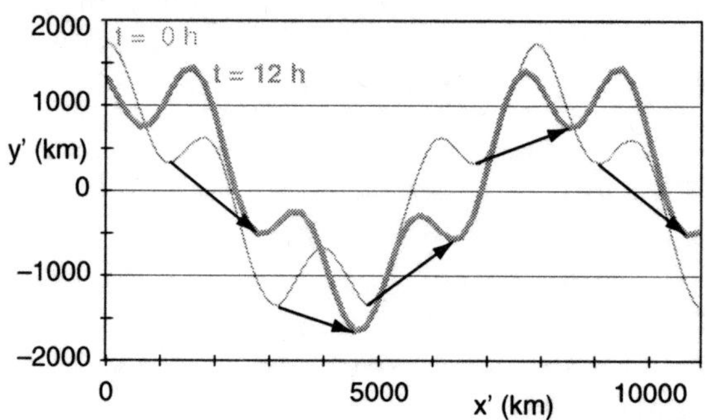

Check: Physics & units reasonable.

Exposition: The short waves move rapidly along the long-wave streamline similar to trains on a track, except that this long-wave track gradually shifts east.

The short waves travel very fast, so they are in and out of any city very quickly. Thus, they cause rapid changes in the weather. For this reason, weather forecasters pay particular attention to short-wave troughs to avoid surprises. Although short waves are sometimes difficult to spot visually in a plot of geopotential height, you can see them more easily on plots of 50 kPa vorticity. Namely, each short-wave trough has a noticeable positive vorticity that can be highlighted or colorized on a forecast map.

Baroclinic Instability

Fig. 11.52 illustrates baroclinic instability using a toy model with thicker atmosphere near the equator and thinner atmosphere near the poles. This mimics the effect of static stability, with cold strongly stable air near the poles that restricts vertical movement of air, vs. warm weakly stable air near the equator that is less limiting in the vertical (see Fig. 5.20 again).

Qualitative View

As before, use potential-vorticity conservation:

$$\left[\frac{f_c + (M/R)}{\Delta z}\right]_{initial} = \left[\frac{f_c + (M/R)}{\Delta z}\right]_{later} \quad (11.39)$$

where M is wind speed, the Coriolis parameter is f_c, and the radius of curvature is R. You must include the atmospheric thickness Δz because it varies south to north.

For baroclinic waves, follow the jet stream as you had done before for barotropic waves, from location **a** to location **d**. All the same processes happen as before, but with an important difference. As the air moves toward location **b**, not only does f_c increase, but Δz decreases. But Δz is in the denominator, hence both f_c and Δz tend to increase the potential vorticity. Thus, the curvature M/R must be even more negative to compensate those combined effect. This means the jet stream turns more sharply.

Similarly, at location **d**, f_c is smaller and Δz in the denominator is larger, both acting to force a sharper cyclonic turn. The net result is that the combined restoring forces are stronger for baroclinic situations, causing tighter turns that create a shorter overall wavelength λ than for barotropic waves.

Quantitative View

The resulting north-south displacement y' for the baroclinic wave is:

$$y' \approx A \cdot \cos\left(\pi \cdot \frac{z}{Z_T}\right) \cdot \cos\left[2\pi \cdot \left(\frac{x' - c \cdot t}{\lambda}\right)\right] \quad (11.40)$$

where the tropospheric depth is Z_T ($\approx$ 11 km), the meridional amplitude is A, and where c is phase speed, λ is wavelength, x is distance East, and t is time. Notice that there is an additional cosine factor. This causes the meridional wave amplitude to be zero the middle of the troposphere, and to have opposite signs at the top and bottom.

[ALERT: this is an oversimplification. Waves in the real atmosphere aren't always 180° out of phase between top and bottom of the troposphere. Nonetheless, this simple approach gives some insight into the workings of baroclinic waves.]

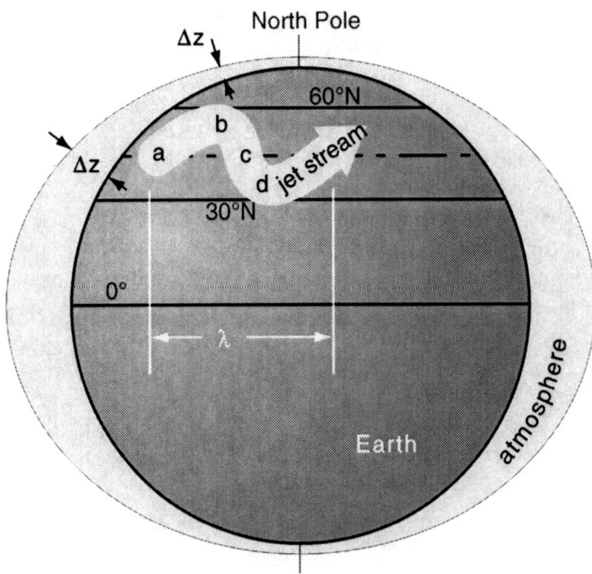

Figure 11.52
Illustration of baroclinic instability, where the effect of static stability is mimicked with at atmosphere that is thicker near the equator and thinner near the poles. The thickness change is exaggerated in this figure..

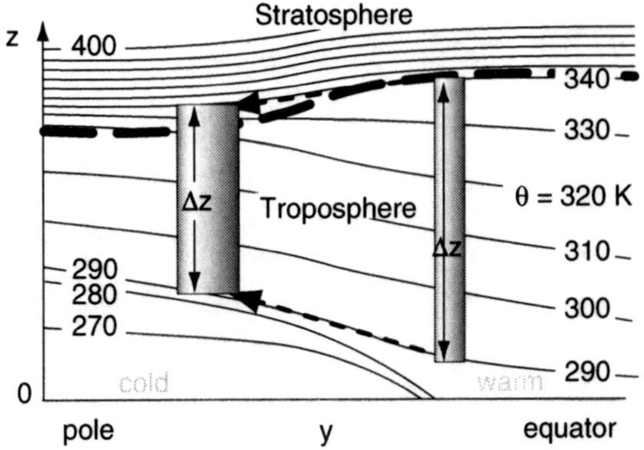

Figure 5.20 (again)
Illustration of how the north-south variation of static stability can be interpreted as a change of effective depth of the atmosphere. Vertical cross section through the atmosphere, showing isentropes (lines of constant potential temperature θ). The depth Δz of column of air on the right will shrink as the column moves poleward, because air tends to follow isentropes during adiabatic processes. Thus, the same Δθ between top and bottom of the air columns spans a shorter vertical distance for the poleward column, meaning that static stability and Brunt-Väisälä frequency are greater there.

Sample Application(§)

The jet stream meanders north and south with amplitude of 900 km relative to reference latitude 50°N. Winds in this jet are 40 m s^{-1}. Temperature decreases 50°C across 12 km thick troposphere, with average temperature –20°C. Calculate the following for a baroclinic wave: Brunt-Väisälä frequency, internal Rossby radius, dominant wavelength, phase speeds, and wave translation distance in 3 h. Also, plot the initial and final wave streamlines $y'(x')$ at both z = 0 and z = 12 km for x' ranging 0 to 5,000 km. β = 1.47x10^{-11} m^{-1}·s^{-1}.

Find the Answer

Given: U_o = 40 m s^{-1}, ϕ =50°, A =900 km, T_{avg} = 253 K, $\Delta T/\Delta z$ = –50K/(12 km), β = 1.47x10^{-11} m^{-1}·s^{-1}, for z=1 & 12 km, t = 0 & 3 h

Find: N_{BV} = ? s^{-1}, λ_R = ? km, λ = ? km, c_o = ? m s^{-1}, c = ? m s^{-1}, $y'(x')$ = ? km.

Get the N_{BV} eq. from the right column of this page.

$$N_{BV} = \left[\frac{(9.8\text{m s}^{-2})}{253\text{K}}\left(\frac{-50\text{K}}{1.2\times10^4\text{m}} + 0.0098\frac{\text{K}}{\text{m}}\right)\right]^{1/2}$$

N_{BV} = **0.0148 s^{-1}**

Apply eq. (10.16):

$$f_c = (1.458\times10^{-4}\text{s}^{-1})\cdot\sin(50°) = 1.117\text{x}10^{-4}\ \text{s}^{-1}$$

Apply eq. (11.12):

$$\lambda_R = \frac{(0.0148\text{s}^{-1})\cdot(12\text{km})}{1.117\times10^{-4}\text{s}^{-1}} = \mathbf{1590.\ km}$$

Apply eq. (11.43):

$\lambda = \lambda_d = 2.38\ \lambda_R$ = **3784. km**

Apply eq. (11.41):

$$c_o = \frac{-(1.47\times10^{-11}\text{m}^{-1}\text{s}^{-1})}{\pi^2\cdot\left[\frac{4}{(3.784\times10^6\text{m})^2} + \frac{1}{(1.59\times10^6\text{m})^2}\right]}$$

= **–2.21** m s^{-1}

Apply eq. (11.42):

c = (40 – 2.21)m s^{-1} = **37.8** m s^{-1}

Use eq. (11.40) for z = 0, 12 km, and t = 0, 3 h:

$$y' \approx (900\text{km})\cos\left(\frac{\pi\cdot z}{12\text{km}}\right)\cos\left[2\pi\left(\frac{x'-(37.8\text{m/s})\cdot t}{3784\text{km}}\right)\right]$$

The results are plotted at right:

Check: Physics & units are reasonable.

Exposition: The Rossby wave at z = 12 km is indeed 180° out of phase from that at the surface. Namely, an upper-level ridge is above a surface trough. In the mid-troposphere (z ≈ 6 km) the baroclinic wave has zero amplitude (not shown). *(continues in right column)*

The intrinsic phase speed c_o for the baroclinic wave is:

$$c_o = \frac{-\beta}{\pi^2\cdot\left[\frac{4}{\lambda^2}+\frac{1}{\lambda_R^2}\right]} \qquad \bullet(11.41)$$

where eq. (11.35) gives β, wavelength is λ, and eq. (11.12) gives the internal Rossby deformation radius λ_R.

The influence of static stability is accounted for in the **Brunt-Väisälä frequency** N_{BV}, which is a factor in the equation for Rossby deformation radius λ_R (eq. 11.12). Recall from the Atmospheric Stability chapter that $N_{BV} = [\ (|g|/T_v)\cdot(\Gamma_d + \Delta T/\Delta z)\ ]^{1/2}$, where gravitation acceleration is $|g|$ = 9.8 m s^{-2}, absolute virtual temperature is T_v (where $T_v = T$ for dry air), dry adiabatic lapse rate is Γ_d = 9.8 °C km^{-1} = 0.0098 K m^{-1}, and $\Delta T/\Delta z$ is the change of air temperature with height.

As before, the phase speed relative to the ground is:

$$c = U_o + c_o \qquad \bullet(11.42)$$

Although many different wavelengths λ are possible, the dominant wavelength (i.e., the wave for which amplitude grows the fastest) is roughly:

$$\lambda_d \approx 2.38\cdot\lambda_R \qquad (11.43)$$

where λ_R is the internal Rossby radius of deformation (eq. 11.12). Typical values are λ_d ≈ 3 to 4 Mm.

Sample Application *(continuation)*

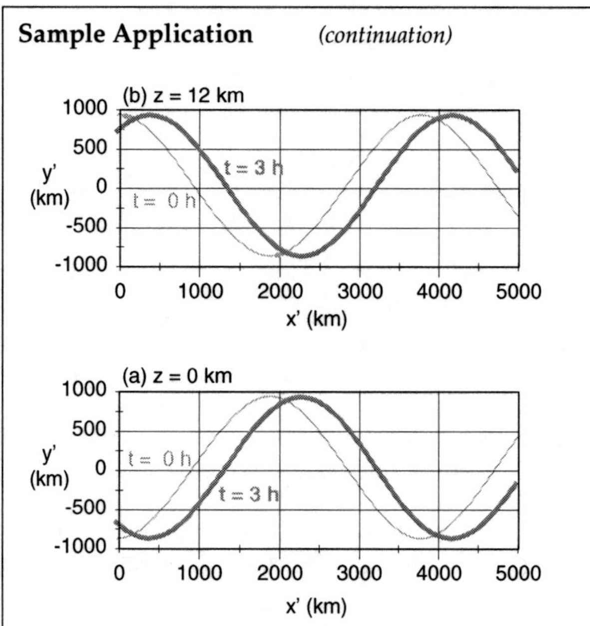

Fig. Baroclinic waves at two heights and two times, for the sample application at left.

INFO • Baroclinic Wave Characteristics

Baroclinic waves perturb many variables relative to their average background states. Represent the perturbation by $e' = e - e_{background}$ for any variable e. Variables affected include:

y' = meridional streamline displacement north,
(u', v', w') = wind components,
θ' = potential temperature,
p' = pressure, and
η' = vertical displacement.

Independent variables are time t and east-west displacement x' relative to some arbitrary location.

Let:

$$a = \pi \cdot z / Z_T \tag{11.44}$$

$$b = 2\pi \cdot (x' - c \cdot t) / \lambda \tag{11.45}$$

Thus:

$$\begin{aligned} y' &= \hat{Y} \cdot \cos(a) \cdot \cos(b) \\ \eta' &= \hat{\eta} \cdot \sin(a) \cdot \cos(b) \\ \theta' &= -\hat{\theta} \cdot \sin(a) \cdot \cos(b) \\ p' &= \hat{P} \cdot \cos(a) \cdot \cos(b) \\ u' &= \hat{U} \cdot \cos(a) \cdot \cos(b) \\ v' &= -\hat{V} \cdot \cos(a) \cdot \sin(b) \\ w' &= -\hat{W} \cdot \sin(a) \cdot \sin(b) \end{aligned} \tag{11.46}$$

Each of the equations above represents a wave having amplitude is indicated by the factor with the caret (^) over it, where the symbols with carets are always positive. For Northern Hemispheric baroclinic waves, the amplitudes are:

$$\begin{aligned} \hat{Y} &= A \\ \hat{\eta} &= \frac{A \cdot \pi \cdot f_c \cdot (-c_o)}{Z_T} \cdot \frac{1}{N_{BV}^2} \\ \hat{\theta} &= \frac{A \cdot \pi \cdot f_c \cdot (-c_o)}{Z_T} \cdot \frac{\theta_o}{g} \\ \hat{P} &= A \cdot \rho_o \cdot f_c \cdot (-c_o) \\ \hat{U} &= \left[\frac{A \cdot 2\pi \cdot (-c_o)}{\lambda}\right]^2 \cdot \frac{1}{A \cdot f_c} \\ \hat{V} &= \frac{A \cdot 2\pi \cdot (-c_o)}{\lambda} \\ \hat{W} &= \frac{A \cdot 2\pi \cdot (-c_o)}{\lambda} \cdot \frac{\pi \cdot (-c_o) \cdot f_c}{Z_T \cdot N_{BV}^2} \end{aligned} \tag{11.47}$$

where ρ_o is average density of air at height z, and intrinsic phase speed c_o is a negative number. A (= north-south displacement) depends on the initial disturbance. *(continues in next column)*

INFO • Baroclinic Wave *(continuation)*

Any atmospheric variable can be reconstructed as the sum of its background and perturbation values; for example: $U = U_{background} + u'$. Background states are defined as follows.

$U_{background} = U_g$ is the geostrophic (jet-stream) wind.
$V_{background} = W_{background} = 0$.
$P_{background}$ decreases with increasing height according to the hydrostatic equation (Chapter 1).
$\theta_{background}$ increases linearly as altitude increases as was assumed to create a constant value of N_{BV}.
$Y_{background}$ corresponds to the latitude of zero-perturbation flow, which serves as the reference latitude for calculation of f_c and β.

Background state for η is the altitude z in eq. (11.44).

All variables listed at left interact together to describe the wave. The result is sketched in Fig. 11.53. Although the equations at left look complicated, they are based on a simplified description of the atmosphere. They neglect clouds, turbulence, latent heating, meridional wave propagation, and nonlinear effects. Nonetheless, the insight gained from this simple model helps to explain the behavior of many of the synoptic weather patterns that are covered in the Extratropical Cyclones chapter.

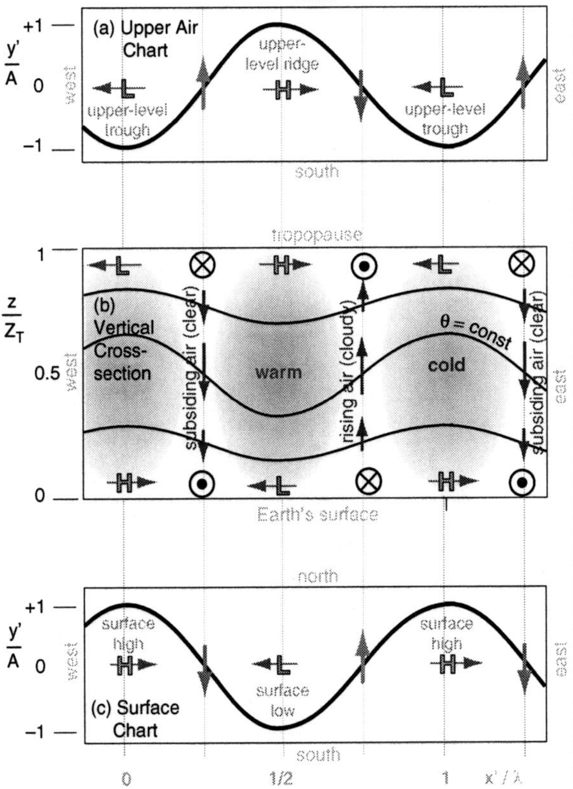

Figure 11.53
Characteristics of a N. Hemisphere baroclinic wave, based on quasi-geostrophic theory. H & L are high and low pressure. Circle-dot and circle-X are flow out-of and into the page, respectively. (after Cushman-Roisin, 1994: "Intro. to Geophysical Fluid Dynamics". Prentice Hall.)

Sample Application

Given $A = 900$ km, $\lambda = 3784$ km, $c_o = -2.21$ m s^{-1}, $Z_T = 12$ km, $f_c = 1.117\text{x}10^{-4}$ s^{-1}, $\rho_o = 1$ kg m^{-3}, $N_{BV} = 0.0148$ s^{-1}, and $|g|/\theta_o = 0.038$ m s^{-2} K^{-1}. Find the amplitudes for all baroclinic wave variables.

Find the Answer

Given: $A = 900$ km, $\lambda = 3784$ km, $c_o = -2.21$ m s^{-1},
$Z_T = 12$ km, $f_c = 1.117\text{x}10^{-4}$ s^{-1}, $\rho_o = 1$ kg m^{-3},
$N_{BV} = 0.0148$ s^{-1}, and $|g|/\theta_o = 0.038$ m s^{-2} K^{-1}.

Find: $\hat{Y}, \hat{\eta}, \hat{\theta}, \hat{P}, \hat{U}, \hat{V}, \hat{W}$

Apply eqs. (11.47). $\hat{Y} = A = 900$ km.

$\hat{\eta} = (900\text{km})\cdot\pi\cdot(1.117\text{x}10^{-4}\ \text{s}^{-1})\cdot(2.21\ \text{m s}^{-1})\ /\ [\ (12\ \text{km})\cdot(0.0148\ \text{s}^{-1})^2\ ] = 265.5$ m

$\hat{\theta} = 1.53$ K

$\hat{P} = 0.222$ kPa

$\hat{U} = 0.109$ m s^{-1}

$\hat{V} = 3.30$ m s^{-1}

$\hat{W} = 0.000974$ m s^{-1} $= 3.5$ m h^{-1}

Check: Physics & units are reasonable.

Exposition: Compared to the 40 m s^{-1} jet-stream background winds, the U perturbation is small. In fact, many of these amplitudes are small. They would be larger for larger A and for shorter wavelengths and a shallower troposphere.

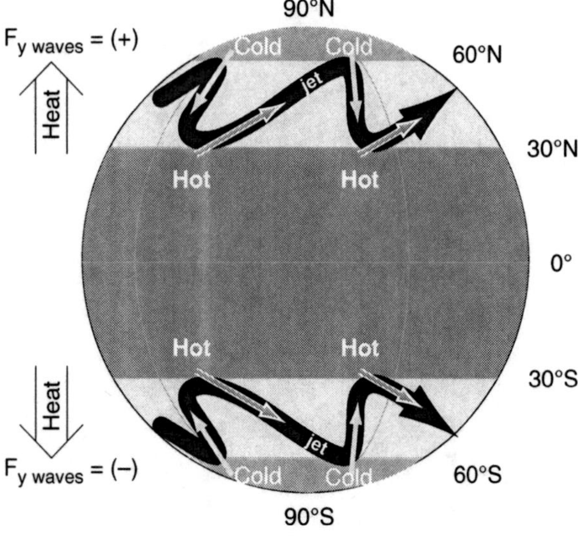

Figure 11.54
Sketch of how Rossby waves (thick black lines) transport heat poleward in midlatitudes. F_y is the meridional heat transport.

Meridional Transport by Rossby Waves

Heat Transport Meridionally

As the jet stream meanders north and south while encircling the earth with its west-to-east flow, its temperature changes in response to the regions it flows over (Fig. 11.54). For example, ridges in the planetary wave are where the jet stream is closest to the poles. This air is flowing over colder ground, and is also getting little or no direct solar heating because of the low sun angles (particularly in winter). As a result, the air becomes colder when near the poles than the average jet-stream temperature.

The opposite temperature change occurs in the troughs, where the jet stream is closest to the equator. The air near troughs is flowing over a warmer Earth surface where significant amounts of heat are moved into the jet stream via convective clouds and solar radiation. Define T' as the amount that the temperature deviates from the average value, so that positive T' means warmer (in the troughs), and negative T' means colder (in the ridges) than average.

Define v' as deviation in meridional velocity relative to the mean ($V = 0$). Positive v' occurs where meandering air has a component toward the north, and negative v' means a component toward the south.

Northward movement ($v' = +$) of warm air ($T' = +$) in the N. Hemisphere contributes to a positive (northward) heat flux $v'T'$. This is a kinematic flux, because units are K·m/s. Similarly, southward movement ($v' = -$) of cold air ($T' = -$) also contributes to a positive heat flux $v'T'$ (because negative times negative = positive). Adding all the contributions from N different parts of the meandering jet stream gives an equation for the mean meridional heat flux F_y caused by Rossby waves:

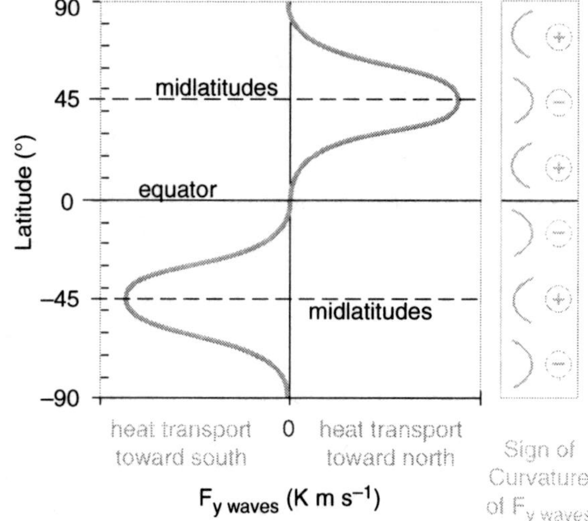

Figure 11.55
Heat flux (magenta line) toward the poles due to Rossby waves Green icons show whether the curvature of the heat-flux line is positive or negative.

$$F_{y\ waves} = (1/N)\cdot\Sigma(v'\cdot T')$$

This flux is (positive, negative) in the (N., S.) Hemispheres. Thus, for both hemispheres, Rossby waves move heat from the tropics toward the poles.

The largest magnitudes of v' and T' occur at midlatitudes, where the jet stream has the largest meridional wind speeds co-located in the region of greatest meridional temperature gradient (Fig. 11.8). Fig. 11.55 shows a sketch of $F_{y\ waves}$ vs. latitude. When this line is concave to the right, it indicates positive curvature (*Curv*). Concave to the left is negative curvature. Curvature of $F_{y\ waves}$ will be used later.

Momentum Transport Meridionally

Even if the atmosphere were calm with respect to the Earth's surface, the fact that the Earth is rotating implies that the air is also moving toward the east. Latitude circles near the equator have much larger circumference than near the poles, hence air near the equator must be moving faster toward the east.

As Rossby waves move some of the tropical air poleward (positive v' in N. Hem.; negative v' in S. Hem.), conservation of U angular momentum requires the speed-up (positive U-wind perturbation: u') as the radius from Earth's axis decreases. Similarly, polar air moving equatorward must move slower (negative u' relative to the Earth's surface).

The north-south transport of U momentum is called the average kinematic momentum flux $\overline{u'v'}$:

$$\overline{u'v'} = (1/N)\cdot\Sigma(u'\cdot v')$$

For example, consider a N. Hemisphere Rossby-wave. Poleward motion (positive v' ; see brown arrows in Fig. 11.56) transports faster U winds (i.e., positive u'), causing positive $u'v'$. Similarly, equatorward motion (negative v' ; see yellow arrows) transports slower U winds (i.e., negative u'), so again the product is positive $u'v'$. Averaging over all such segments of the jet stream gives positive $\overline{u'v'}$ in the N. Hemisphere, which you can interpret as transport of zonal momentum toward the N. Pole by Rossby waves. In the Southern Hemisphere, $\overline{u'v'}$ is negative, which implies transport of U momentum toward the S. Pole.

To compensate for the larger area of fast U winds (brown in Fig. 11.56) relative to the smaller area of slower U winds (yellow in Fig. 11.56), one can multiply the angular momentum by $a = \cos(\phi_s)/\cos(\phi_d)$:

$$a\cdot u' \approx \Omega\cdot R_{earth}\cdot\left[\frac{\cos^2\phi_s}{\cos\phi_d} - \cos\phi_d\right]\cdot\frac{\cos\phi_s}{\cos\phi_d} \qquad (11.48)$$

where subscript s represents source location, d is destination, and $\Omega\cdot R_{Earth} = 463.4$ m s^{-1} as before.

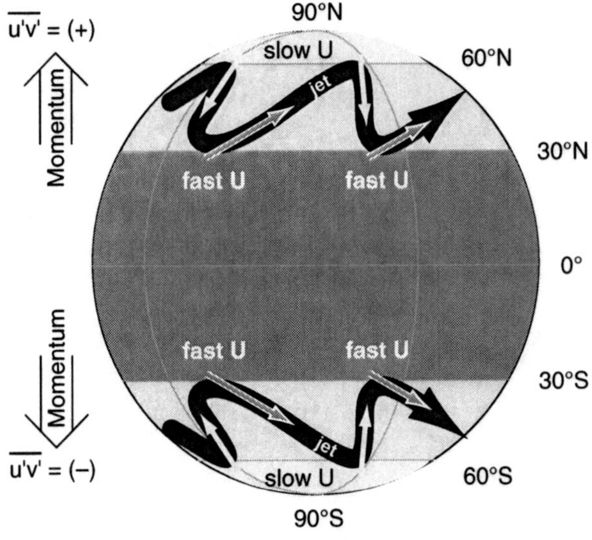

Figure 11.56
Sketch of how Rossby waves (thick black line) transport U momentum toward the poles.

Sample Application

Find the weighted zonal velocity perturbation for air arriving at 50°N from 70°N.

Find the Answer

Given: ϕ_s = 70°N, ϕ_d = 50°N
Find: $a\cdot u'$ = ? m s^{-1}

Apply eq. (11.48):

$$a\cdot u' = (463.4\text{m/s})\cdot\left[\frac{\cos^2(70°)}{\cos(50°)} - \cos(50°)\right]\frac{\cos(70°)}{\cos(50°)}$$

= **– 113.6** m s^{-1}

Check: Physics reasonable, but magnitude too large.
Exposition: The negative sign means air from 70°N is moving slower from the west than any point at 50°N on the Earth's surface is moving. Thus, relative to the Earth, the wind is from the east.

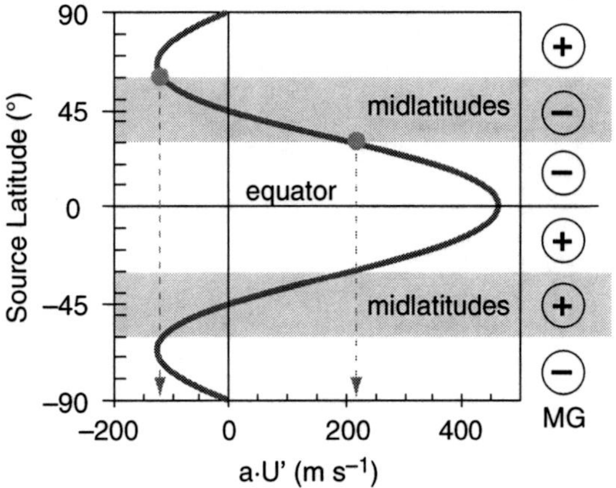

Figure 11.57
Weighted zonal velocity for air that arrives at midlatitude destination 45° from different source latitudes, where the weight is proportional to the amount of air in the source region.

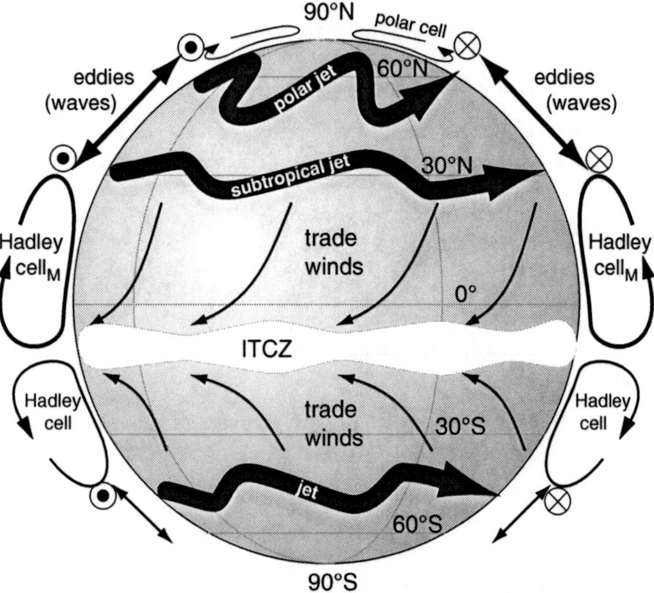

Figure 11.58
Sketch of 3-band global circulation for February (N. Hemisphere winter): 1) a Hadley cell of vertical circulation at low latitudes, 2) Rossby waves with horizontally meandering jet stream at midlatitudes, and 3) a weak polar cell of vertical circulation at high latitudes. For the jet axes, the dot-circle is wind out of the page, and x-circle is wind into the page.

INFO • Torques on the Earth

During high-wind episodes in one of the circulation bands, temporary changes in wind-drag torques are large enough to make measurable changes in the Earth's rotation rate — causing the length of a day to increase or decrease 1 - 3 μs over periods of months. In addition, external influences (lunar and solar tides, solar wind, geomagnetic effects, space dust) cause the Earth to spin ever more slowly, causing the length of a day to increase 1.4 ms/century at present.

For example, consider the weighted zonal velocity $a{\cdot}U'$ that reaches destination latitude 45° from other source latitudes, as sketched in Fig. 11.57. Air starting from 30° has much larger magnitude $|a{\cdot}U'|$ than does air starting from 60°. (*ALERT: Angular-momentum conservation gives unrealistically large velocities, but is qualitatively informative.*)

The change of weighted Rossby-wave momentum flux with latitude is

$$MG = \Delta \overline{u'v'} \,/\, \Delta y \qquad (11.49)$$

where *MG* is the north-south gradient of zonal momentum. From Fig. 11.57 we infer that *MG* is (negative, positive) in (N., S.) Hemisphere midlatitudes. This meridional gradient implies that excess zonal momentum from the tropics is being deposited at midlatitudes by the Rossby waves.

Rossby waves that transport *U* momentum poleward have a recognizable rounded-sawtooth shape, as sketched in Figs. 11.56 & 11.58. Specifically, the equatorward-moving portions of the jet stream are aligned more north-south (i.e., are more meridional), and sometimes even tilt backwards (toward the west as it moves toward the equator). The poleward-moving portions of the jet are more zonal (west to east).

THREE-BAND GLOBAL CIRCULATION

In the chapter introduction, it was stated that Coriolis force causes the thermally-driven planetary circulation to break down into 3 latitude-bands (Fig. 11.58) in each hemisphere. These bands are: (1) a strong, **direct**, asymmetric, vertical-circulation Hadley cell in low latitudes (0° - 30°); (2) a band of mostly horizontal Rossby waves at mid-latitudes (30° - 60°); and a weak direct vertical circulation cell at high latitudes (60° to 90°). Fig. 11.58 includes more (but not all) of the details and asymmetries explained in this chapter.

The circulation bands work together to globally transport atmospheric heat (Fig. 11.14), helping undo the differential heating that was caused by solar and IR radiation. The Earth-atmosphere-ocean system is in near equilibrium thermally, with only extremely small trends over time related to global warming.

The circulation bands also work together in the meridional transport of zonal momentum. The trade winds, blowing opposite to the Earth's rotation, exert a **torque** (force times radius) that tends to slow the Earth's spin due to frictional drag against the land and ocean surface. However, in mid-latitudes, the westerlies dragging against the Earth's surface and against mountains apply an opposite torque, tending to accelerate the Earth's spin. On the long term, the opposite torques nearly cancel each other. Thus, the whole Earth-atmosphere-ocean system maintains a near-equilibrium spin rate.

A Metric for Vertical Circulation

Getting back to atmospheric circulations, one can define the strength *CC* of a vertical circulation cell as:

$$CC = \left[\frac{f_c^2}{N_{BV}^2} \cdot \frac{\Delta V}{\Delta z} \right] - \frac{\Delta w}{\Delta y} \qquad (11.50)$$

Direct circulation cells are ones with a vertical circulation in the direction you would expect if there were no Coriolis force. The units for circulation are s^{-1}.

Using the major Hadley cell as an example of a direct circulation (Fig. 11.59), note that *w* decreases as *y* increases; hence, $\Delta w/\Delta y$ is negative. Similarly, $\Delta V/\Delta z$ is positive. Thus, eq. (11.50) gives a positive *CC* value for direct circulations, and a negative value for **indirect circulations** (having an opposite rotation direction).

Effective Vertical Circulation

When the forecast equations for momentum and heat are applied to eq. (11.50), the result allows you to anticipate the value of *CC* for a variety of situations— even situations where vertical cells are not dominant:

$$\underset{circulation}{CC} \propto \underset{radiation}{-\frac{\Delta E_{net}}{\Delta y}} + \underset{wave\text{-}heat}{Curv(F_{y\ wave})} + \underset{wave\text{-}momentum}{\frac{\Delta MG}{\Delta z}} \qquad (11.51)$$

In this equation are factors and terms that were discussed earlier in this chapter. For example, you can use Fig. 11.10 to estimate E_{net}, the differential heating due to radiation, and how it varies with *y*. The sign of the curvature (*Curv*) of the Rossby-wave heat flux $F_{y\ wave}$ was shown in Fig. 11.55. If we assume that the meridional gradient of zonal momentum $MG \approx 0$ near the ground, then $\Delta MG/\Delta z$ has the same sign as *MG*, as was sketched in Fig. 11.57.

With this information, you can estimate the sign of *CC* in different latitude bands. Namely, you can anticipate direct and indirect circulations. For the Northern Hemisphere, the results are:

<u>*circulation*</u> ∝ <u>*radiation*</u> + <u>*wave-heat*</u> + <u>*wave-momentum*</u> = <u>*total*</u>
CC_{polar} ∝ positive + positive + positive = positive
CC_{midlat} ∝ positive + negative +negative = negative
$CC_{tropics}$ ∝ positive + positive + positive = positive

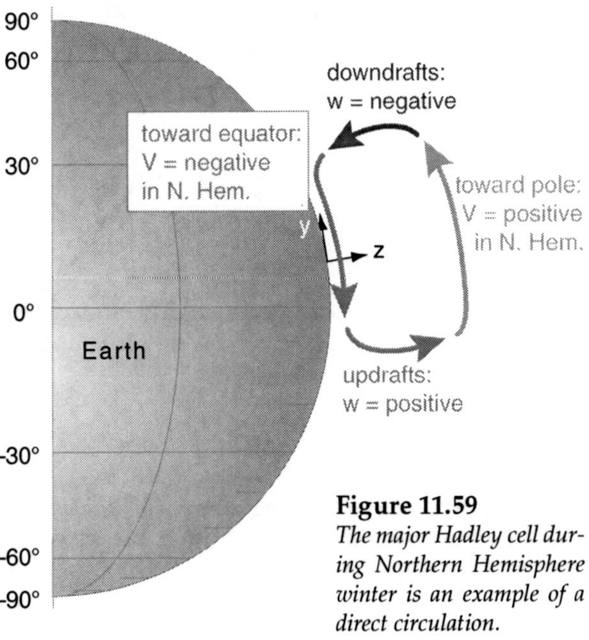

Figure 11.59
The major Hadley cell during Northern Hemisphere winter is an example of a direct circulation.

Sample Application

Suppose the Hadley cell updraft and downdraft velocities are 6 and –4 mm s^{-1}, respectively, and the meridional wind speeds are 3 m s^{-1} at the top and bottom of the cell. The major Hadley cell is about 17 km high by 3900 km wide, and is centered at about 10° latitude. Temperature in the tropical atmosphere decreases from about 25°C near the surface to –77°C at 17 km altitude. Find the vertical cell circulation.

Find the Answer

Given: Δw = –10 mm s^{-1} = –0.01 m s^{-1} , Δy = 3.9x10^3 km
ΔV = 6 m s^{-1}, Δz = 17 km, ϕ = 10°, ΔT = –102°C.
Find: *CC* = ? s^{-1}.

First, use eq. (10.16):
$f_c = (1.458\text{x}10^{-4}\ s^{-1})\cdot\sin(10°) = 2.53\text{x}10^{-5}\ s^{-1}$
Next, for the Brunt-Väisälä frequency, we first need:
T_{avg} = 0.5*(25 – 77)°C = –26°C = 247 K
In the tropics $\Delta T/\Delta z$ = –6 °C km^{-1} = –0.006 K m^{-1}
Then use eq. (5.4), and assume $T_v = T$:

$$N_{BV} = \sqrt{\frac{9.8\text{m/s}^2}{247\text{K}}(-0.006 + 0.0098)\frac{\text{K}}{\text{m}}} = 0.0123\ s^{-1}$$

Finally, use eq. (11.50):

$$CC = \left[\left(\frac{2.53\times10^{-5}\text{s}^{-1}}{0.0123\text{s}^{-1}} \right)^2 \cdot \frac{6\text{m/s}}{17000\text{m}} \right] - \frac{-0.01\text{m/s}}{3.9\times10^6\text{m}}$$

CC = 1.493x10^{-9} + 2.564x10^{-9} s^{-1} = **<u>4.06x10^{-9}</u>** s^{-1}

Check: Physics & units are reasonable.
Exposition: Both terms contribute positively to the circulation of the major Hadley cell during Northern Hemisphere winter.

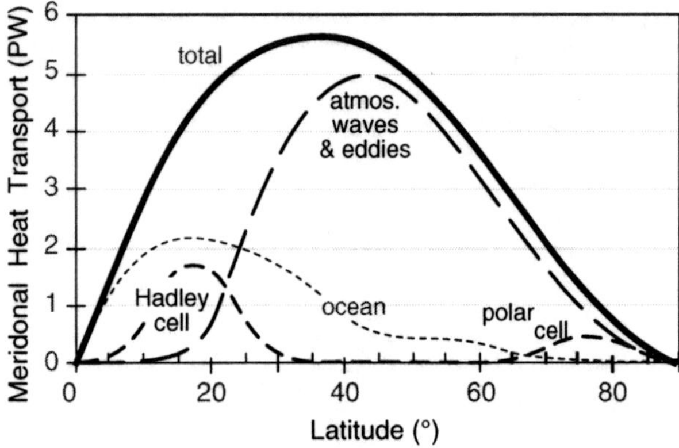

Figure 11.60
Contributions of direct circulations (medium dashed line, for Hadley cell and polar cell) and indirect circulations (long dashed line, for Rossby waves and eddies) to total meridional heat transport Tr in the N. Hemisphere

Rossby waves so efficiently transport heat and momentum at midlatitudes that the effective vertical circulation is negative. This indirect circulation is called the **Ferrel cell**.

Fig. 11.60 redraws the N. Hemisphere portion of Fig. 11.14, qualitatively highlighting the relative contributions of the direct (Hadley and polar cells) and indirect (Ferrel cell/Rossby waves) atmospheric circulations to the total meridional heat transport. At mid-latitudes, the main circulation feature is the Rossby-wave meanders of the jet stream near the tropopause, and the associated low- and high-pressure centers near the surface. High- and low-latitudes are dominated by direct vertical circulations.

Ocean currents also contribute to global heat redistribution. Although ocean-circulation details are not within the scope of this book, we will introduce one ocean topic here — the Ekman spiral. This describes how wind drag can drive some ocean currents, including hurricane storm surges.

EKMAN SPIRAL OF OCEAN CURRENTS

Frictional drag between the atmosphere and ocean enables winds to drive ocean-surface currents. Coriolis force causes the surface current to be 45° to the right of the wind direction in the Northern Hemisphere. Drag between that surface-current and deeper water drives deeper currents that are slower, and which also are to the right of the current above. The result is an array of ocean-current vectors that trace a spiral (Fig. 11.61) called the **Ekman spiral**.

The equilibrium horizontal ocean-current components (U, V, for a rotated coordinate system having U aligned with the wind direction) as a function of depth (z) are:

$$U = \left[\frac{u_{*water}^{\;2}}{(K \cdot f_c)^{1/2}}\right] \cdot \left[e^{z/D} \cdot \cos\left(\frac{z}{D} - \frac{\pi}{4}\right)\right] \tag{11.52a}$$

$$V = \left[\frac{u_{*water}^{\;2}}{(K \cdot f_c)^{1/2}}\right] \cdot \left[e^{z/D} \cdot \sin\left(\frac{z}{D} - \frac{\pi}{4}\right)\right] \tag{11.52b}$$

where z is negative below the ocean surface, f_c = Coriolis parameter, and u_{*water} is a **friction velocity** (m s^{-1}) for water. It can be found from

$$u_{*water}^{\;2} = \frac{\rho_{air}}{\rho_{water}} \cdot u_{*air}^{\;2} \tag{11.53}$$

where the density ratio $\rho_{air}/\rho_{water} \approx 0.001195$ for sea water. The friction velocity (m s^{-1}) for air can be approximated using **Charnock's relationship**:

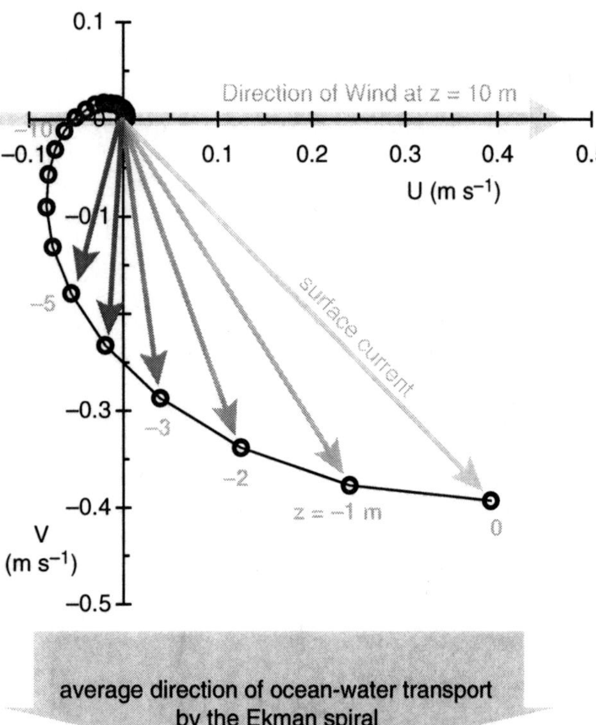

Figure 11.61
Example of an Ekman spiral in the ocean. Blue arrows show ocean currents at different depths (numbers in green). U current direction is aligned with the near-surface wind direction. Red arrow shows the wind direction, but speed is much greater. This example is based on a 10 m s^{-1} wind at 45° latitude.

$$u_{*air}{}^2 \approx 0.00044 \cdot M^{2.55} \tag{11.54}$$

where M is near-surface wind speed (at z = 10 m) in units of m s^{-1}.

The **Ekman-layer depth** scale is

$$D = \sqrt{\frac{2 \cdot K}{f_c}} \tag{11.55}$$

where K is the ocean eddy viscosity (a measure of ability of ocean turbulence to mix momentum). One approximation is $K \approx 0.4\ |z|\ u_{*water}$. Although K varies with depth, for simplicity in this illustration I used constant K corresponding to its value at z = –0.2 m, which gave $K \approx 0.001$ m^2 s^{-1}.

The average water-mass transport by Ekman ocean processes is 90° to the (right, left) of the near-surface wind in the (N., S.) Hemisphere (Fig. 11.61). This movement of water affects sea-level under hurricanes (see the Tropical Cyclone chapter).

REVIEW

The combination of solar radiation input and IR output from Earth causes polar cooling and tropical heating. In response, a global circulation develops due to buoyancy, pressure, and geostrophic effects, which transports heat from the tropics toward the polar regions, and which counteracts the radiative differential heating.

Near the equator, warm air rises and creates a band of thunderstorms at the ITCZ. The updrafts are a part of the Hadley-cell direct vertical circulation, which moves heat away from the equator. This cell cannot extend beyond about 30° to 35° latitude because Coriolis force turns the upper-troposphere winds toward the east, creating a subtropical jet near 30° latitude at the tropopause. Near the surface are the trade-wind return flows from the east.

A strong meridional temperature gradient remains at mid-latitudes, which drives westerly winds via the thermal-wind effect, and creates a polar jet at the tropopause. But instabilities of the jet stream cause meridional meanders called Rossby waves.

These waves are very effective at moving heat poleward, leaving only a very weak indirect-circulating Ferrel cell in midlatitudes. Near the poles is a weak direct-circulation cell. Ocean currents can be driven by the overlying winds in the atmosphere. All these ceaseless global circulations in both the atmosphere and the ocean can move enough momentum and heat to keep our planet in near equilibrium.

Sample Application

For an east wind of 14 m/s at 30°N, graph the Ekman spiral.

Find the Answer

Given: M = 14 m·s^{-1}, ϕ = 30°N
Find: $[U, V]$ (m s^{-1}) vs. depth z (m)

Using a relationship from the Forces and Winds chapter, find the Coriolis parameter: $f_c = 7.29\text{x}10^{-5}$ s^{-1} .

Next, use Charnock's relationship. eq. (11.54):
$u_{*air}{}^2 = 0.00044\ (14^{\ 2.55}) = 0.368$ m^2 s^{-2} .

Then use eq. (11.53):
$u_{*water}{}^2 = 0.001195 \cdot (0.368\ \text{m}^2\ \text{s}^{-2}) = 0.00044\ \text{m}^2\ \text{s}^{-2}$

Estimate K at depth 0.2 m in the ocean from
$K \approx 0.4 \cdot (0.2\ \text{m}) \cdot [0.00044\ \text{m}^2\ \text{s}^{-2}]^{1/2} = 0.00168\ \text{m}^2\ \text{s}^{-1}$

Use eq. (11.55):
$D = [2 \cdot (0.00168\ \text{m}^2\ \text{s}^{-1})\ /\ (7.29\text{x}10^{-5}\ \text{s}^{-1})\ ]^{1/2} = 6.785$ m

Use a spreadsheet to solve for (U, V) for a range of z, using eqs. (11.52). I used z = 0, –1, –2, –3 m etc.

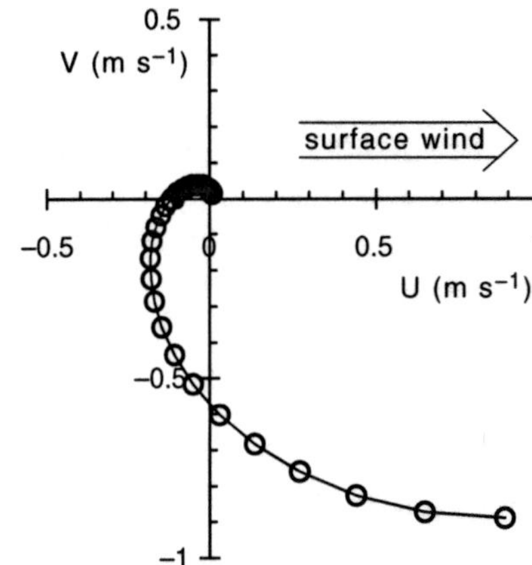

Finally, rotate the graph 180° because the wind is from the East.

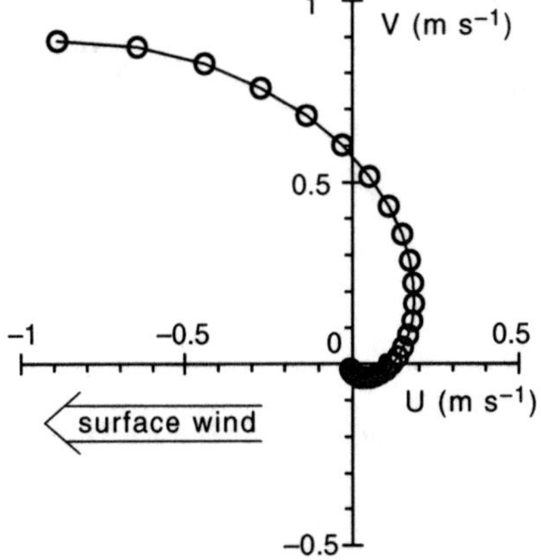

Check: Physics & units are reasonable.
Exposition: Net transport of ocean water is toward the north for this East wind case.

HOMEWORK EXERCISES

Broaden Knowledge & Comprehension

B1. From hemispheric weather maps of winds near the tropopause (which you can access via the internet), identify locations of major global-circulation features including the jet stream, monsoon circulations, tropical cyclones and the ITCZ.

B2. Same as the previous exercise, except using water-vapor or infrared image loops from geostationary satellites to locate the features.

B3. From the web, find a rawinsonde sounding at a location in the trade-wind region, and confirm the wind reversal between low and high altitudes.

B4. Use a visible, whole-disk image from a geostationary satellite to view and quantify the cloud-cover fraction as a function of latitude. Speculate on how insolation at the Earth's surface is affected.

B5. Download a series of rawinsonde soundings for different latitudes between the equator and a pole. Find the tropopause from each sounding, and then plot the variation of tropopause height vs. latitude.

B6. Download a map of sea-surface temperature (SST), and discuss how SST varies with latitude.

B7. Most satellite images in the infrared show greys or colors that are related to brightness temperature (see the legend in whole-disk IR images that you acquired from the internet). Use these temperatures as a function of latitude to estimate the corresponding meridional variation of IR-radiation out. Hint, consider the Stefan-Boltzmann law.

B8. Download satellite-derived images that show the climatological average incoming and outgoing radiation at the top of the atmosphere. How does it relate to the idealized descriptions in this chapter?

B9. Download satellite-derived or buoy & ship-derived ocean currents for the global oceans, and discuss how they transport heat meridionally, and why the oceanic transport of heat is relatively small at mid to high latitudes in the N. Hemisphere.

B10. Use a satellite image to locate a strong portion of the ITCZ over a rawinsonde site, and then download the rawinsonde data. Plot (compute if needed) the variation of pressure with altitude, and discuss how it does or doesn't deviate from hydrostatic.

B11. Search the web for sites where you can plot "reanalysis data", such as the NCEP/NCAR reanalysis or any of the ECMWF reanalyses. Pick a month during late summer from some past year in this database, and plot the surface pressure map. Explain how this "real" result relates to a combination of the "idealized" planetary and monsoonal circulations.

B12. Same as B11, but for monthly average vertical cross sections that can be looped as movies. Display fields such as zonal wind, meridional wind, and vertical velocity, and see how they vary over a year.

B13. Capture a current map showing 85 kPa temperatures, and assume that those temperatures are surrogates for the actual average virtual temperature between 100 and 70 kPa. Compute the thermal wind magnitude and direction for a location assigned by your teacher, and see if this theoretical relationship successfully explains the wind shear between 100 and 70 kPa. Justify your reasoning.

B14 Capture a current map showing the thickness between 100 and 50 kPa, and estimate the thermal wind direction and magnitude across that layer.

B15. Use rawinsonde soundings from stations that cross the jet stream. Create your own contour plots of the jet-stream cross section for (a) heights of key isobaric surfaces; (b) potential temperature; and (c) wind magnitude. Compare your plots with idealized sketches presented in this chapter.

B16. What are the vertical and horizontal dimensions of the jet stream, based on weather maps you acquire from the internet.

B17. Acquire a 50 kPa vorticity chart, and determine if the plotted vorticity is isentropic, absolute, relative, or potential. Where are positive-vorticity maxima relative to fronts and foul weather?

B18. Calculate the values for the four types of vorticity at a location identified by your instructor, based on data for winds and temperatures. Namely, acquire the raw data used for vorticity calculations; do not use vorticity maps captured from the web.

B19. For the 20 kPa geopotential heights, use the wavy pattern of height contours and their relative packing to identify ridges and troughs in the jet stream. Between two troughs, or between two ridges, estimate the wavelength of the Rossby wave. Use that measured length as if it were the dominant wavelength to estimate the phase speed for baroclinic and barotropic waves.

B20. Confirm that the theoretical relationship between horizontal winds, temperatures, vertical velocities, and heights for baroclinic waves is consistent with the corresponding weather maps you acquire from the internet. Explain any discrepancies.

B21. Confirm the three-band nature of the global circulation using IR satellite image movie loops. In the tropics, compare the motion of low (warm) and high (cold) clouds, and relate this motion to the trade winds and Hadley circulation. In mid-latitudes, find the regions of meandering jet stream with its corresponding high and low-pressure centers. In polar regions, relate cloud motions to the polar cell.

B22. Are the ocean-surface current directions consistent with near-surface wind directions as observed in maps or animations acquired from the internet, given the dynamics describe for the Ekman spiral?

Apply

A1(§). For the "toy" model, make a graph of zonally-averaged temperature (°C) vs. latitude for the altitude (km) above ground level (AGL) given here:
a. 0.5 b. 1 c. 1.5 d. 2 e. 2.5 f. 3 g. 3.5
h. 4 i. 4.5 j. 5 k. 5.5 l. 6 m. 6.6 n. 7
o. 8 p. 9 q. 10 r. 11 s. 12 t. 13 u. 14

A2(§). For the "toy" model, make a graph of zonally-averaged $\Delta T/\Delta y$ (°C km^{-1}) vs. latitude for the altitude (km AGL) given here:
a. 0.5 b. 1 c. 1.5 d. 2 e. 2.5 f. 3 g. 3.5
h. 4 i. 4.5 j. 5 k. 5.5 l. 6 m. 6.6 n. 7
o. 8 p. 9 q. 10 r. 11 s. 12 t. 13 u. 14

A3. Estimate the annual average insolation (W m^{-2}) at the following latitude:
a. 90° b. 85° c. 80° d. 75° e. 70° f. 65° g. 60°
h. 55° i. 50° j. 45° k. 40° l. 35° m. 30° n. 25°
o. 20° p. 15° q. 10° r. 5° s. equator

A4. Estimate the annual average amount of incoming solar radiation (W m^{-2}) that is absorbed in the Earth-ocean-atmosphere system at latitude:
a. 90° b. 85° c. 80° d. 75° e. 70° f. 65° g. 60°
h. 55° i. 50° j. 45° k. 40° l. 35° m. 30° n. 25°
o. 20° p. 15° q. 10° r. 5° s. equator

A5. Using the idealized temperature near the middle of the troposphere (at z = 5.5 km), estimate the outgoing infrared radiation (W m^{-2}) from the atmosphere at the following latitude:
a. 90° b. 85° c. 80° d. 75° e. 70° f. 65° g. 60°
h. 55° i. 50° j. 45° k. 40° l. 35° m. 30° n. 25°
o. 20° p. 15° q. 10° r. 5° s. equator

A6. Using the results from the previous two exercises, find the net radiation magnitude (W m^{-2}) that is input to the atmosphere at latitude:
a. 90° b. 85° c. 80° d. 75° e. 70° f. 65° g. 60°
h. 55° i. 50° j. 45° k. 40° l. 35° m. 30° n. 25°
o. 20° p. 15° q. 10° r. 5° s. equator

A7. Using the results from the previous exercise, find the latitude-compensated net radiation magnitude (W m^{-2}; i.e., the differential heating) at latitude:
a. 90° b. 85° c. 80° d. 75° e. 70° f. 65° g. 60°
h. 55° i. 50° j. 45° k. 40° l. 35° m. 30° n. 25°
o. 20° p. 15° q. 10° r. 5° s. equator

A8. Assuming a standard atmosphere, find the internal Rossby deformation radius (km) at latitude:
a. 90° b. 85° c. 80° d. 75° e. 70° f. 65° g. 60°
h. 55° i. 50° j. 45° k. 40° l. 35° m. 30° n. 25°
o. 20° p. 15° q. 10° r. 5° s. equator

A9. Given the following virtual temperatures at your location (20°C) and at another location, find the change of geostrophic wind with height [(m s^{-1})/km]. Relative to your location, the other locations are:

	Δx(km)	Δy(km)	T_v(°C)		Δx(km)	Δy(km)	T_v(°C)
a.	0	100	15	k.	100	0	15
b.	0	100	16	l.	100	0	16
c.	0	100	17	m.	100	0	17
d.	0	100	18	n.	100	0	18
e.	0	100	19	o.	100	0	19
f.	0	100	21	p.	100	0	21
g.	0	100	22	q.	100	0	22
h.	0	100	23	r.	100	0	23
i.	0	100	24	s.	100	0	24
j.	0	100	25	t.	100	0	25

A10. Find the thermal wind (m s^{-1}) components, given a 100 to 50 kPa thickness change of 0.1 km across the following distances:
Δx(km) = a. 200 b. 250 c. 300 d. 350
e. 400 f. 450 g. 550 h. 600 i. 650
Δy(km) = j. 200 k. 250 l. 300 m. 350
n. 400 o. 450 p. 550 q. 600 r. 650

A11. Find the magnitude of the thermal wind (m s^{-1}) for the following thickness gradients:

	ΔTH(km) / Δx(km)	&	ΔTH(km) / Δy(km)
a.	–0.2 / 600	and	– 0.1 / 400
b.	–0.2 / 400	and	– 0.1 / 400
c.	–0.2 / 600	and	+ 0.1 / 400
d.	–0.2 / 400	and	+ 0.1 / 400
e.	–0.2 / 600	and	– 0.1 / 400
f.	–0.2 / 400	and	– 0.1 / 400
g.	–0.2 / 600	and	+ 0.1 / 400
h.	–0.2 / 400	and	+ 0.1 / 400

A12. For the toy model temperature distribution, find the wind speed (m s^{-1}) of the jet stream at the following heights (km) for latitude 30°:
a. 0.5 b. 1 c. 1.5 d. 2 e. 2.5 f. 3 g. 3.5
h. 4 i. 4.5 j. 5 k. 5.5 l. 6 m. 6.6 n. 7
o. 8 p. 9 q. 10 r. 11 s. 12 t. 13 u. 14

A13. If an air parcel from the starting latitude 5° has zero initial velocity relative to the Earth, then find its U component of velocity (m s^{-1}) relative to the Earth when it reaches the following latitude, assuming conservation of angular momentum.
a. 0° b. 2° c. 4° d. 6° e. 8° f. 10° g. 12°
h. 14° i. 16° j. 18° k. 20° l. 22° m. 24° n. 26°

A14. Find the relative vorticity (s^{-1}) for the change of (U , V) wind speed (m s^{-1}), across distances of Δx = 300 km and Δy = 600 km respectively given below.
a. 50, 50 b. 50, 20 c. 50, 0 d. 50, –20 e. 50, –50
f. 20, 50 g. 20, 20 h. 20, 0 i. 20, –20 j. 20, –50
k. 0, 50 l. 0, 20 m. 0, 0 n. 0, –20 o. 0, –50
p. –20, 50 q. –20, 20 r. –20, 0 s. –20, –20 t. –20, –50
u. –50, 50 v. –50, 20 x. –50, 0 y. –50, –20 z. –50, –50

A15. Given below a radial shear ($\Delta M/\Delta R$) in [(m s^{-1})/km] and tangential wind speed M (m s^{-1}) around radius R (km), find relative vorticity (s^{-1}):
a. 0.1, 30, 300 b. 0.1, 20, 300 c. 0.1, 10, 300
d. 0.1, 0, 300 e. 0, 30, 300 f. 0, 20, 300
g. 0, 10, 300 h. –0.1, 30, 300 i. –0.1, 20, 300
j. –0.1, 10, 300 k. –0.1, 0, 300

A16. If the air rotates as a solid body of radius 500 km, find the relative vorticity (s^{-1}) for tangential speeds (m s^{-1}) of:
a. 10 b. 20 c. 30 d. 40 e. 50 f. 60 g. 70
h. 80 i. 90 j. 100 k. 120 l. 140 m. 150

A17. If the relative vorticity is $5\text{x}10^{-5}$ s^{-1}, find the absolute vorticity at the following latitude:
a. 90° b. 85° c. 80° d. 75° e. 70° f. 65° g. 60°
h. 55° i. 50° j. 45° k. 40° l. 35° m. 30° n. 25°
o. 20° p. 15° q. 10° r. 5° s. equator

A18. If absolute vorticity is $5\text{x}10^{-5}$ s^{-1}, find the potential vorticity (m^{-1}·s^{-1}) for a layer of thickness (km) of: a. 0.5 b. 1 c. 1.5 d. 2 e. 2.5 f. 3 g. 3.5
h. 4 i. 4.5 j. 5 k. 5.5 l. 6 m. 6.6 n. 7
o. 8 p. 9 q. 10 r. 11 s. 12 t. 13 u. 14

A19. The potential vorticity is $1\text{x}10^{-8}$ m^{-1}·s^{-1} for a 10 km thick layer of air at latitude 48°N. What is the change of relative vorticity (s^{-1}) if the thickness (km) of the rotating air changes to:
a. 9.5 b. 9 c. 8.5 d. 8 e. 7.5 f. 7 g. 6.5
h. 10.5 j. 11 k. 11.5 l. 12 m. 12.5 n. 13

A20. If the absolute vorticity is $3\text{x}10^{-5}$ s^{-1} at 12 km altitude, find the isentropic potential vorticity (PVU) for a potential temperature change of ___ °C across a height increase of 1 km.
a. 1 b. 2 c. 3 d. 4 e. 5 f. 6 g. 6.5
h. 7 i. 8 j. 9 k. 10 l. 11 m. 12 n. 13

A21. Find the horizontal circulation associated with average relative vorticity $5\text{x}10^{-5}$ s^{-1} over area (km^2):
a. 500 b. 1000 c. 2000 d. 5000 e. 10,000
f. 20,000 g. 50,000 h. 100,000 i. 200,000

A22. For the latitude given below, what is the value of the beta parameter (m^{-1} s^{-1}):
a. 90° b. 85° c. 80° d. 75° e. 70° f. 65° g. 60°
h. 55° i. 50° j. 45° k. 40° l. 35° m. 30° n. 25°
o. 20° p. 15° q. 10° r. 5° s. equator

A23. Suppose the average wind speed is 60 m s^{-1} from the west at the tropopause. For a barotropic Rossby wave at 50° latitude, find both the intrinsic phase speed (m s^{-1}) and the phase speed (m s^{-1}) relative to the ground for wavelength (km) of:
a. 1000 b. 1500 c. 2000 d. 2500 e. 3000
f. 3500 g. 4000 h. 4500 i. 5000 j. 5500
k. 6000 l. 6500 m. 7000 n. 7500 o. 8000

A24. Plot the barotropic wave (y' vs x') from the previous exercise, assuming amplitude 2000 km.

A25. Same as exercise A23, but for a baroclinic Rossby wave in an atmosphere where air temperature decreases with height at 4°C km^{-1}.

A26(§). Plot the baroclinic wave (y' vs x') from the previous exercise, assuming amplitude 2000 km and a height (km):
(i) 2 (ii) 4 (iii) 6 (iv) 8 (v) 10

A27. What is the fastest growing wavelength (km) for a baroclinic wave in a standard atmosphere at latitude:
a. 90° b. 85° c. 80° d. 75° e. 70° f. 65° g. 60°
h. 55° i. 50° j. 45° k. 40° l. 35° m. 30° n. 25°
o. 20° p. 15° q. 10° r. 5° s. equator

A28. For the baroclinic Rossby wave of exercise A25 with amplitude 2000 km, find the wave amplitudes of the:
(i). vertical-displacement perturbation
(ii). potential-temperature perturbation
(iii). pressure perturbation
(iv). U-wind perturbation
(v). V-wind perturbation
(vi). W-wind perturbation

A29(§). For a vertical slice through the atmosphere, plot baroclinic Rossby-wave perturbation amount for conditions assigned in exercise A28.

A30. Find the latitude-weighted $a{\cdot}u'$ momentum value (m s^{-1}) for air that reaches destination latitude 50° from source latitude: a. 80° b. 75° c. 70° d. 65° e. 60° f. 55° g. 45° h. 40° i. 35°

A31. Suppose the ____ cell upward and downward speeds are ___ and ___ mm s^{-1}, respectively, and the north-south wind speeds are 3 m s^{-1} at the top and bottom of the cell. The cell is about __ km high by ___ km wide, and is centered at about ___ latitude. Temperature in the atmosphere decreases from about 15°C near the surface to –57°C at 11 km altitude. Find the vertical circulation.

cell	W_{up} (mm s^{-1})	W_{down} (mm s^{-1})	Δz (km)	Δy (km)	ϕ (°)
a. Hadley	6	–4	17	3900	10
b. Hadley	4	–4	15	3500	10
c. Hadley	3	–3	15	3500	5
d. Ferrel	3	–3	12	3000	45
e. Ferrel	2	–2	11	3000	45
f. Ferrel	2	–2	10	3000	50
g. polar	1	–1	9	2500	75
h. polar	1	–1	8	2500	75
i. polar	0.5	–0.5	7	2500	80

A32. Find the friction velocity at the water surface if the friction velocity (m s^{-1}) in the air (at sea level for a standard atmosphere) is:
a. 0.05 b. 0.1 c. 0.15 d. 0.2 e. 0.25
f. 0.3 g. 0.35 h. 0.4 i. 0.45 j. 0.5
k. 0.55 l. 0.6 m. 0.65 n. 0.7 o. 0.75

A33. Find the Ekman-spiral depth scale at latitude 50°N for eddy viscosity (m^2 s^{-1}) of:
a. 0.0002 b. 0.0004 c. 0.0006 d. 0.0008 e. 0.001
f. 0.0012 g. 0.0014 h. 0.0016 i. 0.0018 j. 0.002
k. 0.0025 l. 0.003 m. 0.0035 n. 0.004 o. 0.005

A34(§). Create a graph of Ekman-spiral wind components (U, V) components for depths from the surface down to where the velocities are near zero, for near-surface wind speed of 8 m s^{-1} at 40°N latitude.

Evaluate & Analyze

E1. During months when the major Hadley cell exists, trade winds cross the equator. If there are no forces at the equator, explain why this is possible.

E2. In regions of surface high pressure, descending air in the troposphere is associated with dry (non-rainy) weather. These high-pressure belts are where deserts form. In addition to the belts at ±30° latitude, semi-permanent surface highs also exist at the poles. Are polar regions deserts? Explain.

E3. The subtropical jet stream for Earth is located at about 30° latitude. Due to Coriolis force, this is the poleward limit of outflow air from the top of the ITCZ. If the Earth were to spin faster, numerical experiments suggest that the poleward limit (and thus the jet location) would be closer to the equator. Based on the spins of the other planets (get this info from the web or a textbook) compared to Earth, at what latitudes would you expect the subtropical jets to be on Jupiter? Do your predictions agree with photos of Jupiter?

E4. Horizontal divergence of air near the surface tends to reduce or eliminate horizontal temperature gradients. Horizontal convergence does the opposite. Fronts (as you will learn in the next chapter) are regions of strong local temperature gradients. Based on the general circulation of Earth, at what latitudes would you expect fronts to frequently exist, and at what other latitudes would you expect them to rarely exist? Explain.

E5. In the global circulation, what main features can cause mixing of air between the Northern and Southern Hemispheres? Based on typical velocities and cross sectional areas of these flows, over what length of time would be needed for the portion $1/e$ of all the air in the N. Hemisphere to be replaced by air that arrived from the S. Hemisphere?

E6. In Fig. 11.4, the average declination of the sun was listed as 14.9° to 15° for the 4-month periods listed in those figures. Confirm that those are the correct averages, based on the equations from the Solar & Infrared Radiation chapter for solar declination angle vs. day of the year.

E7. Thunderstorms are small-diameter (15 km) columns of cloudy air from near the ground to the tropopause. They are steered by the environmental winds at an altitude of roughly 1/4 to 1/3 the troposphere depth. With that information, in what direction would you expect thunderstorms to move as a function of latitude (do this for every 10° latitude)?

E8. The average meridional wind at each pole is zero. Why? Also, does your answer apply to instantaneous winds such as on a weather map? Why?

E9. Can you detect monsoonal (monthly or seasonal average) pressure centers on a normal (instantaneous) weather map analysis or forecast? Explain.

E10. Figs. 11.3a & 11.5a showed idealized surface wind & pressure patterns. Combine these and draw a sketch of the resulting idealized global circulation including both planetary and monsoon effects.

E11. Eqs. (11.1-11.3) represent an idealized ("toy model") meridional variation of zonally averaged temperature. Critically analyze this model and discuss. Is it reasonable at the ends (boundaries) of the curve; are the units correct; is it physically justifiable; does it satisfy any budget constraints (e.g., conservation of heat, if appropriate), etc. What aspects of it are too simplified, and what aspects are OK?

E12. (a) Eq. (11.4) has the 3rd power of the sine times the 2nd power of the cosine. If you could arbitrarily change these powers, what values would lead to reasonable temperature gradients ($\Delta T/\Delta y$) at the surface and which would not (Hint: use a spreadsheet and experiment with different powers)?

(b) Of the various powers that could be reasonable, which powers would you recommend as fitting the available data the best? (Hint: consider not only the temperature gradient, but the associated meridional temperature profile and the associated jet stream.) Also, speculate on why I chose the powers that I did for this toy model.

E13. Concerning differential heating, Fig. 11.9 shows the annual average insolation vs. latitude. Instead, compute the average insolation over the two-month period of June and July, and plot vs. latitude. Use the resulting graph to explain why the jet stream and weather patterns are very weak in the summer hemisphere, and strong in the winter hemisphere.

E14. At mid- and high-latitudes, Fig. 11.9 shows that each hemisphere has one full cycle of insolation annually (i.e., there is one maximum and one minimum each year).

But look at Fig. 11.9 near the equator.

a. Based on the data in this graph (or even better, based on the eqs. from the Solar & Infrared Radiation chapter), plot insolation vs. relative Julian day for the equator.

b. How many insolation cycles are there each year at the equator?

c. At the equator, speculate on when would be the hottest and coldest "seasons".

d. Within what range of latitudes near the equator is this behavior observed?

E15. Just before idealized eq. (11.6), I mentioned my surprise that E_2 was approximately constant with latitude. I had estimated E_2 by subtracting my toy-model values for E_{insol} from the actual observed values of E_{in}. Speculate about what physical processes could cause E_2 to be constant with latitude all the way from the equator to the poles.

E16. How sensitive is the toy model for E_{out} (i.e., eq. 11.7) to the choice of average emission altitude z_m? Recall that z_m, when used as the altitude z in eqs. (11.1-11.3), affects T_m. Hint: for your sensitivity analysis, use a spreadsheet to experiment with different z_m and see how the resulting plots of E_{out} vs. latitude change. (See the "A SCIENTIFIC PERSPECTIVE" box about model sensitivity.)

E17(§). Solve the equations to reproduce the curves in figure: a. 11.10 b. 11.11 c. 11.12 d. 11.13

E18. We recognize the global circulation as a response of the atmosphere to the instability caused by differential heating, as suggested by **LeChatelier's Principle**. But the circulation does not totally undo the instability; namely, the tropics remain slightly warmer than the poles. Comment on why this remaining, unremoved instability is required to exist, for the global circulation to work.

E19. In Fig. 11.12, what would happen if the surplus area exceeded the deficit area? How would the global circulation change, and what would be the end result for Fig. 11.12?

E20. Check to see if the data in Fig. 11.12 does give zero net radiation when averaged from pole to pole.

E21. The observation data that was used in Fig. 11.14 was based on satellite-measured radiation and differential heating to get the total needed heat transport, and on estimates of heat transport by the oceans. The published "observations" for net atmospheric heat transport were, in fact, estimated as the difference (i.e., residual) between the total and the ocean curves. What could be some errors in this atmosphere curve? (Hint: see the A SCIENTIFIC PERSPECTIVE box about Residuals.)

E22. Use the total heat-transport curve from Fig. 11.60. At what latitude is the max transport? For that latitude, convert the total meridional heat-flux value to horsepower.

E23. For Fig. 11.15, explain why it is p' vs. z that drive vertical winds, and not P_{column} vs. z.

E24. a. Redraw Figs. 11.16 for downdraft situations.

b. Figs. 11.16 both show updraft situations, but they have opposite pressure couplets. As you already found from part (a) both pressure couplets can be associated with downdrafts. What external information (in addition to the pressure-couplet sign) do you always need to decide whether a pressure couplet causes an updraft or a downdraft? Why?

E25. a. For the thermal circulation of Fig. 11.17(iv), what needs to happen for this circulation to be maintained? Namely, what prevents it from dying out?

b. For what real-atmosphere situations can thermal circulations be maintained for several days?

E26. a. Study Fig. 11.18 closely, and explain why the wind vectors to/from the low- and high-pressure centers at the equator differ from the winds near pressure centers at mid-latitudes.

b. Redraw Fig. 11.5a, but with continents and oceans at the equator. Discuss what monsoonal pressures and winds might occur during winter and summer, and why.

E27. a. Redraw Fig. 11.19, but for the case of geostrophic wind decreasing from its initial equilibrium value. Discuss the resulting evolution of wind and pressure fields during this geostrophic adjustment.

b. Redraw Fig. 11.19, but for flow around a low-pressure center (i.e., look at gradient winds instead of geostrophic winds). Discuss how the wind and pressure fields adjust when the geostrophic wind is increased above its initial equilibrium value.

E28. How would the vertical potential temperature gradient need to vary with latitude for the "internal Rossby radius of deformation" to be invariant? Assume constant troposphere depth.

E29. In the Regional Winds chapter, gap winds and coastally-trapped jets are explained. Discuss how these flows relate to geostrophic adjustment.

E30. At the top of **hurricanes** (see the Tropical Cyclones chapter), so much air is being continuously pumped to the top of the troposphere that a high-pressure center is formed over the hurricane core there. This high is so intense and localized that it violates the conditions for gradient winds; namely, the pressure gradient around this high is too steep (see the Forces & Winds chapter).

Discuss the winds and pressure at the top of a hurricane, using what you know about geostrophic adjustment. Namely, what happens to the winds and air mass if the wind field is not in geostrophic or gradient balance with the pressure field?

E31. In the thermal-wind relationship (eqs. 11.13), which factors on the right side are constant or vary by only a small amount compared to their magnitude, and which factors vary more (and are thus more important in the equations)?

E32. In Fig. 11.20, how would it change if the bottom isobaric surface were tilted; namely, if there were already a horizontal pressure gradient at the bottom?

E33. Draw a sketch similar to Fig. 11.20 for the thermal-wind relationship for the Southern Hemisphere.

E34. In maps such as Fig. 11.21, explain why thickness is related to average temperature.

E35. Redraw Fig. 11.22 for the case cold air in the west and warm air in the east. Assume no change to the bottom isobaric surface.

E36. Copy Fig. 11.24. a. On your copy, draw the G_1 and G_2 vectors, and the M_{TH} vector at point B. Confirm that the thermal wind relationship is qualitatively satisfied via vector addition. Discuss why point B is an example of veering or backing.

b. Same as (a) but calculate the actual magnitude of each vector at point B based on the spacing between isobars, thickness contours, or height contours. Again, confirm that the thermal wind relationship is satisfied. (1° latitude = 111 km)

E37. Using a spreadsheet, start with an air parcel at rest at the tropopause over the equator. Assume a realistic pressure gradient between the equator and 30° latitude. Use dynamics to solve for acceleration of the parcel over a short time step, and then iterate over many time steps to find parcel speed and position. How does the path of this parcel compare to the idealized paths drawn in Fig. 11.26d? Discuss.

E38. In the thunderstorms at the ITCZ, copious amounts of water vapor condense and release latent heat. Discuss how this condensation affects the average lapse rate in the tropics, the distribution of heat, and the strength of the equatorial high-pressure belt at the tropopause.

E39. Summarize in a list or an outline all the general-circulation factors that make the mid-latitude weather different from tropical weather.

E40. Explain the surface pressure patterns in Figs. 11.31 in terms of a combination of idealized monsoon and planetary circulations.

E41. Figs. 11.31 show mid-summer and mid-winter conditions in each hemisphere. Speculate on what the circulation would look like in April or October.

E42. Compare Figs. 11.32 with the idealized planetary and monsoon circulations, and discuss similarities and differences.

E43. Based on Figs. 11.32, which hemisphere would you expect to have strong subtropical jets in both summer and winter, and which would not. What factors might be responsible for this difference?

E44. For the Indian monsoon sketched in Fig. 11.33, where are the updraft and downdraft portions of the major Hadley cell for that month? Also, what is the relationship between the trade winds at that time, and the Indian monsoon winds?

E45. What are the dominant characteristics you see in Fig. 11.34, regarding jet streams in the Earth's atmosphere? Where don't jet streams go?

E46. In Figs. 11.35, indicate if the jet-stream winds would be coming out of the page or into the page, for the: a) N. Hemisphere, (b) S. Hemisphere.

E47. Although Figs. 11.36 are for different months than Figs. 11.32, they are close enough in months to still both describe summer and winter flows.

a. Do the near-tropopause winds in Figs. 11.36 agree with the pressure gradients (or height gradients) in Figs. 11.32?

b. Why are there easterly winds at the tropopause over/near the equator, even though there is negligible pressure gradient there?

E48. Describe the mechanism that drives the polar jet, and explain how it differs from the mechanism that drives the subtropical jet.

E49. In Fig. 11.37b, we see a very strong pressure gradient in the vertical (indicated by the different isobars), but only small pressure gradients in the horizontal (indicated by the slope of any one isobar). Yet the strongest average winds are horizontal, not vertical. Why?

E50. Why does the jet stream wind speed decrease with increasing height above the tropopause?

E51. a. Knowing the temperature field given by the toy model earlier in this chapter, show the steps needed to create eq. (11.17) by utilizing eqs. (11.2, 11.4 and 11.13). b. For what situations might this jet-wind-speed equation not be valid? c. Explain what each term in eq. (11.17) represents physically.

E52. Why does an air parcel at rest (i.e., calm winds) near the equator possess large angular momentum? What about for air parcels that move from the east at typical trade wind speeds?

E53. At the equator, air at the bottom of the troposphere has a smaller radius of curvature about the Earth's axis than at the top of the troposphere. How significant is this difference? Can we neglect it?

E54. Suppose that air at 30° latitude has no east-west velocity relative to the Earth's surface. If that air moves equatorward while preserving its angular momentum, which direction would it move relative to the Earth's surface? Why? Does it agree with real winds in the general circulation? Elaborate.

E55. Picture a circular hot tub of 2 m diameter with a drain in the middle. Water is initially 1.2 m deep, and you made rotate one revolution each 10 s. Next, you pull the plug, allowing the water depth to stretch to 2.4 m as it flows down the drain. Calculate the new angular velocity of the water, neglecting frictional drag. Show your steps.

E56. In eq. (11.20), why is there a negative sign on the last term? Hint: How does the rotation direction implied by the last term without a negative sign compare to the rotation direction of the first term?

E57. In the Thunderstorm chapters, you will learn that the winds in a portion of the tornado can be irrotational. This is surprising, because the winds are traveling so quickly around a very tight vortex. Explain what wind field is needed to gave **irrotational winds** (i.e., no relative vorticity) in air that is rotating around the tornado. Hint: Into the wall of a tornado, imagine dropping a neutrally-buoyant small paddle wheel the size of a flower. As this flower is translated around the perimeter of the tornado funnel, what must the local wind shear be at the flower to cause it to not spin relative to the ground? Redraw Fig. 11.43 to show what you propose.

E58. Eq. (11.25) gives names for the different terms that can contribute toward vorticity. For simplicity, assume Δz is constant (i.e., assume no stretching). On a copy of Fig. 11.44, write these names at appropriate locations to identify the dominant factors affecting the vorticity max and min centers.

E59. If you were standing at the equator, you would be rotating with the Earth about its axis. However, you would have zero vorticity about your vertical axis. Explain how that is possible.

E60. Eq. (11.26) looks like it has the <u>absolute</u> vorticity in the numerator, yet that is an equation for a form of potential vorticity. What other aspects of that equation make it like a potential vorticity?

E61. Compare the expression of horizontal circulation C with that for vertical circulation CC.

E62. Relate Kelvin's circulation theorem to the conservation of potential vorticity. Hint: Consider a constant Volume = $A \cdot \Delta z$.

E63. The jet stream sketched in Fig. 11.49 separates cold polar air near the pole from warmer air near the equator. What prevents the cold air from extending further away from the poles toward the equator?

E64. If the Coriolis force didn't vary with latitude, could there be Rossby waves? Discuss.

E65. Are baroclinic or barotropic Rossby waves faster relative to Earth's surface at midlatitudes? Why?

E66. Compare how many Rossby waves would exist around the Earth under barotropic vs. baroclinic conditions. Assume an isothermal troposphere at 50°N.

E67. Once a Rossby wave is triggered, what mechanisms do you think could cause it to diminish (i.e., to reduce the waviness, and leave straight zonal flow).

E68. In Fig. 11.50 at point (4) in the jet stream, why doesn't the air just continue turning clockwise around toward points (2) and (3), instead of starting to turn the other way?

E69. Pretend you are a newspaper reporter writing for a general audience. Write a short article describing how baroclinic Rossby waves work, and why they differ from barotropic waves.

E70. What conditions are needed so that Rossby waves have zero phase speed relative to the ground? Can such conditions occur in the real atmosphere?

E71. Will Rossby waves move faster or slower with respect to the Earth's surface if the tropospheric static stability increases? Why?

E72. For a baroclinic wave that is meandering north and south, consider the northern-most point as the wave crest. Plot the variation of this crest longitude vs. altitude (i.e., x vs. z). Hint: consider eq. (11.40).

E73. Use tropopause-level Rossby-wave trough-axes and ridge-axes as landmarks. Relative to those landmarks, where east or west is: (a) vertical velocity the greatest; (b) potential-temperature deviation the greatest; and (c) vertical displacement the greatest?

E74. In Figs. 11.51 and 11.53 in the jet stream, there is just as much air going northward as there is air going southward across any latitude line, as required by mass conservation. If there is no net mass transport, how can there be heat or momentum transport?

E75. For the Southern Hemisphere: (a) would a direct circulation cell have positive or negative CC? (b) for each term of eq. (11.51), what are their signs?

E76. Compare definitions of circulation from this chapter with the previous chapter, and speculate on the relevance of the static stability and Earth's rotation in one or both of those definitions.

E77. Consider a cyclonic air circulation over an ocean in your hemisphere. Knowing the relationship between ocean currents and surface winds, would you anticipate that the near-surface wind-driven ocean currents are diverging away from the center of the cyclone, or converging toward the center? Explain, and use drawings. Note: Due to mass conservations, horizontally diverging ocean surface waters cause upwelling of nutrient-rich water toward the surface, which can support ocean plants and animals, while downwelling does the opposite.

Synthesize

S1. Describe the equilibrium general circulation for a non-rotating Earth.

S2. Circulations are said to **spin-down** as they lose energy. Describe general-circulation spin-down if Earth suddenly stopped spinning on its axis.

S3. Describe the equilibrium general circulation for an Earth that spins three times faster than now.

S4. Describe the **spin-up** (increasing energy) as the general circulation evolves on an initially non-rotating Earth that suddenly started spinning.

S5. Describe the equilibrium general circulation on an Earth with no differential radiative heating.

S6. Describe the equilibrium general circulation on an Earth with cold equator and hot poles.

S7. Suppose that the sun caused radiative cooling of Earth, while IR radiation from space caused warming of Earth. How would the weather and climate be different, if at all?

S8. Describe the equilibrium general circulation for an Earth with polar ice caps that extend to 30° latitude.

S9. About 250 million years ago, all of the continents had moved together to become one big continent called **Pangaea**, before further plate tectonic movement caused the continents to drift apart. Pangaea spanned roughly 120° of longitude (1/3 of Earth's circumference) and extended roughly from pole to pole. Also, at that time, the Earth was spinning faster, with the solar day being only about 23 of our present-day hours long. Assuming no other changes to insolation, etc, how would the global circulation have differed compared to the current circulation?

S10. If the Earth was dry and no clouds could form, how would the global circulation differ, if at all? Would the tropopause height be different? Why?

S11. Describe the equilibrium general circulation for an Earth with tropopause that is 5 km high.

S12. Describe the equilibrium general circulation for an Earth where potential vorticity isn't conserved.

S13. Describe the equilibrium general circulation for an Earth having a zonal wind speed halfway between the phase speeds of short and long barotropic Rossby waves.

S14. Describe the equilibrium general circulation for an Earth having long barotropic Rossby waves that had slower intrinsic phase speed than short waves.

S15. Describe the nature of baroclinic Rossby waves for an Earth with statically unstable troposphere.

S16. Describe the equilibrium general circulation for an Earth where Rossby waves had no north-south net transport of heat, momentum, or moisture.

S17. Describe the equilibrium general circulation for an Earth where no heat was transported meridionally by ocean currents.

S18. Describe the equilibrium ocean currents for an Earth with no drag between atmosphere and ocean.

S19. Suppose there was an isolated small continent that was hot relative to the surrounding cooler ocean. Sketch a vertical cross section in the atmosphere across that continent, and use thickness concepts to draw the isobaric surfaces. Next, draw a plan-view map of heights of one of the mid-troposphere isobaric surfaces, and use thermal-wind effects to sketch wind vectors on this same map. Discuss how this approach does or doesn't explain some aspects of monsoon circulations.

S20. If the Rossby wave of Fig. 11.50 was displaced so that it is centered on the equator (i.e., point (1) starts at the equator), would it still oscillate as shown in that figure, or would the trough of the wave (which is now in the S. Hem.) behave differently? Discuss.

S21. If the Earth were shaped like a cylinder with its axis of rotation aligned with the axis of the real Earth, could Rossby waves exist? How would the global circulation be different, if at all?

S22. In the subtropics, low altitude winds are from the east, but high altitude winds are from the west. In mid-latitudes, winds at all altitudes are from the west. Why are the winds in these latitude bands different?

S23. What if the Earth did not rotate? How would the Ekman spiral in the ocean be different, if at all?

12 FRONTS & AIRMASSES

Contents

A high-pressure center, or **high** (H), often contains an **airmass** of well-defined characteristics, such as cold temperatures and low humidity. When different airmasses finally move and interact, their mutual border is called a **front**, named by analogy to the battle fronts of World War I.

Fronts are usually associated with low-pressure centers, or **lows** (L). Two fronts per low are most common, although zero to four are also observed. In the Northern Hemisphere, these fronts often rotate counterclockwise around the low center like the spokes of a wheel (Fig. 12.1), while the low moves and evolves. Fronts are often the foci of clouds, low pressure, and precipitation.

In this chapter you will learn the characteristics of anticyclones (highs). You will see how anticyclones are favored locations for airmass formation. Covered next are fronts in the bottom, middle, and top of the troposphere. Factors that cause fronts to form and strengthen are presented. This chapter ends with a special type of front called a dry line.

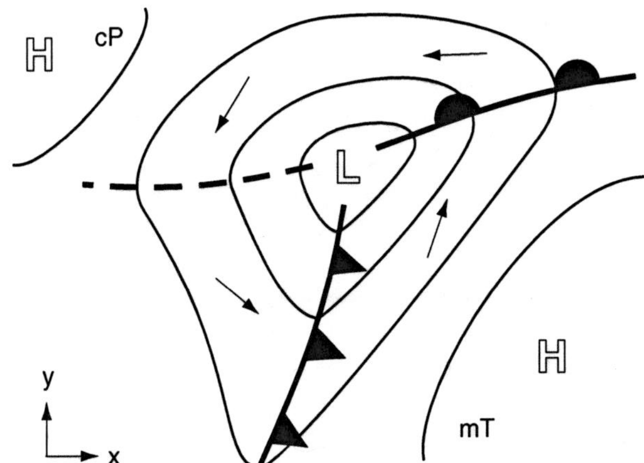

Figure 12.1
Idealized surface weather map (from the Weather Reports & Map Analysis chapter) for the N. Hemisphere showing high (H) and low (L) pressure centers, isobars (thin lines), a warm front (heavy solid line with semicircles on one side), a cold front (heavy solid line with triangles on one side), and a trough of low pressure (dashed line). Vectors indicate near-surface wind. cP indicates a continental polar airmass; mT indicates a maritime tropical airmass.

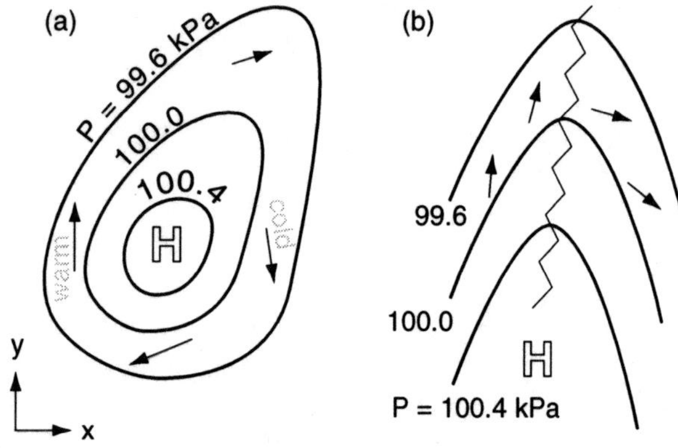

Figure 12.2
Examples of isobars plotted on a sea-level pressure map. (a) High-pressure center. (b) High-pressure ridge in N. Hemisphere mid-latitudes. Vectors show surface wind directions.

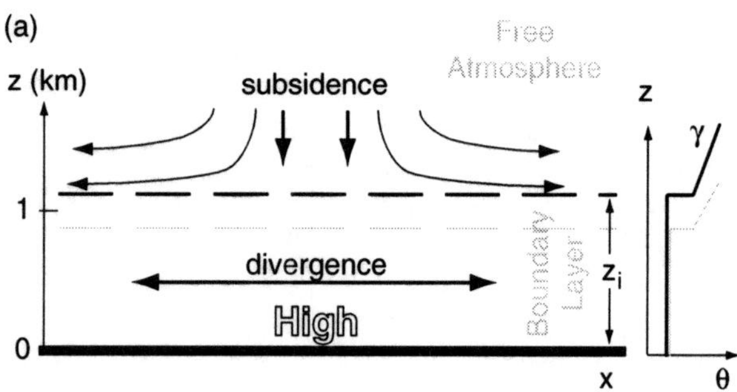

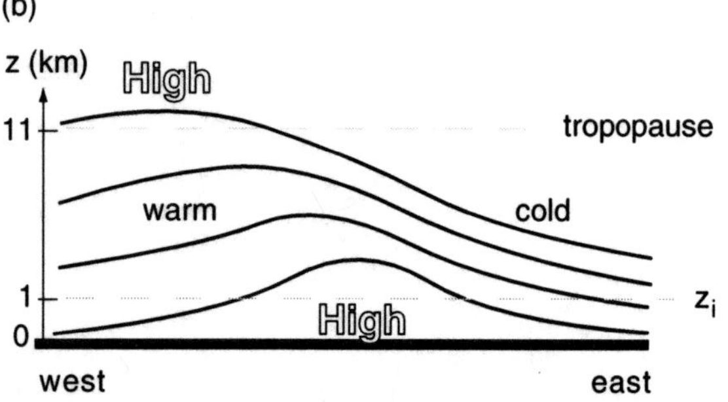

Figure 12.3
(a) Left: vertical circulation above a surface high-pressure center in the bottom half of the troposphere. Black dashed line marks the initial capping inversion at the top of the boundary layer. Grey dashed line shows the top later, assuming no turbulent entrainment into the boundary layer. Right: idealized profile of potential temperature, θ, initially (black line) and later (grey). The boundary-layer depth z_i is on the order of 1 km, and the potential-temperature gradient above the boundary layer is represented by γ.
(b) Tilt of high-pressure ridge westward with height, toward the warmer air. Thin lines are height contours of isobaric surfaces. Ridge amplitude is exaggerated in this illustration.

ANTICYCLONES OR HIGHS

Characteristics & Formation

High-pressure centers, or **highs**, are identified on constant altitude (e.g., sea-level) weather maps as regions of relative maxima in **pressure**. The location of high-pressure center is labeled with "H" (Fig. 12.2a). High centers can also be found on upper-air isobaric charts as relative maxima in **geopotential height** (see the Atmos. Forces & Winds chapter, Fig. 10.2).

When the pressure field has a relative maximum in only one direction, such as east-west, but has a horizontal pressure-gradient in the other direction, this is called a high-pressure **ridge** (Fig. 12.2b). The **ridge axis** is labeled with a zigzag line.

The column of air above the high center contains more air molecules than neighboring columns. This causes more weight due to gravity (see Chapter 1), which is expressed in a fluid as more pressure.

Above a high center is often downward motion (**subsidence**) in the mid-troposphere, and horizontal spreading of air (**divergence**) near the surface (Fig. 12.3a). Subsidence impedes cloud development, leading to generally clear skies and fair weather. Winds are also generally calm or light in highs, because gradient-wind dynamics of highs require weak pressure gradients near the high center (see the Atmos. Forces & Winds chapter).

The diverging air near the surface spirals outward due to the weak pressure-gradient force. Coriolis force causes it to rotate clockwise (anticyclonically) around the high-pressure center in the Northern Hemisphere (Fig. 12.2a), and opposite in the Southern Hemisphere. For this reason, high-pressure centers are called **anticyclones**.

Downward advection of dry air from the upper troposphere creates dry conditions just above the boundary layer. Subsidence also advects warmer potential temperatures from higher in the troposphere. This strengthens the temperature inversion that caps the boundary layer, and acts to trap pollutants and reduce visibility near the ground.

Subsiding air cannot push through the capping inversion, and therefore does not inject free-atmosphere air directly into the boundary layer. Instead, the whole boundary layer becomes thinner as the top is pushed down by subsidence (Fig. 12.3a). This can be partly counteracted by entrainment of free atmosphere air if the boundary layer is turbulent, such as for a convective mixed layer during daytime over land. However, the entrainment rate is controlled by turbulence in the boundary layer (see the Atmos. Boundary Layer chapter), not by subsidence.

Five mechanisms support the formation of highs at the Earth's surface:

• Global Circulation: Planetary-scale, semi-permanent highs predominate at 30° and 90° latitudes, where the global circulation has downward motion (see the General Circulation chapter). The **subtropical highs** centered near 30° North and South latitudes are 1000-km-wide belts that encircle the Earth. **Polar highs** cover the Arctic and Antarctic. These highs are driven by the global circulation that is responding to differential heating of the Earth. Although these highs exist year round, their locations shift slightly with season.

• Monsoons: Quasi-stationary, continental-scale highs form over cool oceans in summer and cold continents in winter (see the General Circulation chapter). They are seasonal (i.e., last for several months), and form due to the temperature contrast between land and ocean.

• Transient Rossby waves: Surface highs form at mid-latitudes, east of high-pressure ridges in the jet stream, and are an important part of mid-latitude weather variability (see the General Circulation and Extratropical Cyclone chapters). They often exist for several days.

• Thunderstorms: Downdrafts from thunderstorms (see the Thunderstorm chapters) create **meso-highs** roughly 10 to 20 km in diameter at the surface. These might exist for minutes to hours.

• Topography/Surface-Characteristics: Meso-highs can also form in mountains due to blocking or channeling of the wind, mountain waves, and thermal effects (anabatic or katabatic winds) in the mountains. Sea-breezes or lake breezes can also create meso-highs in parts of their circulation. (See the Regional Winds chapter.)

The actual pressure pattern at any location and time is a superposition of all these phenomena.

Vertical Structure

The location difference between surface and upper-tropospheric highs (Fig. 12.3b) can be explained using gradient-wind and thickness concepts.

Because of barotropic and baroclinic instability, the jet stream meanders north and south, creating troughs of low pressure and ridges of high pressure, as discussed in the General Circulation chapter. **Gradient winds** blow faster around ridges and slower around troughs, assuming identical pressure gradients. The region east of a ridge and west of a trough has fast-moving air entering from the west, but slower air leaving to the east. Thus, horizontal convergence of air at the top of the troposphere adds more air molecules to the whole tropospheric column at that location, causing a surface high to form east of the upper-level ridge.

West of surface highs, the anticyclonic circulation advects warm air from the equator toward the poles (Figs. 12.2a & 12.3b). This heating west of the surface high causes the **thickness** between isobaric surfaces to increase, as explained by the hypsometric equation. Isobaric surfaces near the top of the troposphere are thus lifted to the west of the surface high. These high heights correspond to high pressure aloft; namely, the upper-level ridge is west of the surface high.

The net result is that high-pressure regions tilt westward with increasing height (Fig. 12.3b). In the Extratropical Cyclone chapter you will see that deepening low-pressure regions also tilt westward with increasing height, at mid-latitudes. Thus, the mid-latitude tropospheric pressure pattern has a consistent phase shift toward the west as altitude increases.

AIRMASSES

An **airmass** is a widespread (of order 1000 km wide) body of air in the bottom third of the troposphere that has somewhat-uniform characteristics. These characteristics can include one or more of: temperature, humidity, visibility, odor, pollen concentration, dust concentration, pollutant concentration, radioactivity, cloud condensation nuclei (CCN) activity, cloudiness, static stability, and turbulence.

Airmasses are usually classified by their temperature and humidity, as associated with their source regions. These are usually abbreviated with a two-letter code. The first letter, in lowercase, describes the humidity source. The second letter, in upper-

Sample Application

A "**cA**" airmass has what characteristics?

Find the Answer

Given: cA airmass.
Find: characteristics

Use Table 12-1: cA = continental Arctic
Characteristics: **Dry and very cold**.

Check: Agrees with Fig. 12.4.
Exposition: Forms over land in the arctic, under the polar high. In Great Britain, the same airmass is labeled as Ac.

Table 12-1. Airmass abbreviations. **Boldface** indicates the most common ones.

Abbr.	Name	Description
c	**continental**	Dry. Formed over land.
m	**maritime**	Humid. Formed over ocean.
A	Arctic	Very cold. Formed in the polar high.
E	Equatorial	Hot. Formed near equator.
M	Monsoon	Similar to tropical.
P	**Polar**	Cold. Formed in subpolar area.
S	Superior	A warm dry airmass having its origin aloft.
T	**Tropical**	Warm. Formed in the subtropical high belt.
k	colder than the underlying surface	
w	warmer than the underlying surface	
Special (regional) abbreviations.		
AA	Antarctic	Exceptionally cold and dry.
r	returning	As in "rPm" returning Polar maritime [Great Britain]

Note: Layered airmasses are written like a fraction, with the airmass aloft written above a horizontal line and the surface airmass written below. For example, just east of a dryline you might have:

$$\frac{cT}{mT}$$

case, describes the temperature source. Table 12-1 shows airmass codes. [*CAUTION: In Great Britain, the two letters are reversed.*]

Examples are **maritime Tropical** (**mT**) airmasses, such as can form over the Gulf of Mexico, and **continental Polar** (**cP**) air, such as can form in winter over Canada.

After the weather pattern changes and the airmass is blown away from its genesis region, it flows over surfaces with different relative temperatures. Some organizations append a third letter to the end of the airmass code, indicating whether the moving airmass is (**w**) warmer or (**k**) colder than the underlying surface. This coding helps indicate the likely static stability of the air and the associated weather. For example, "mPk" is humid cold air moving over warmer ground, which would likely be statically unstable and have convective clouds and showers.

Creation

An airmass can form when air remains stagnant over a surface for sufficient duration to take on characteristics similar to that surface. Also, an airmass can form in moving air if the surface over which it moves has uniform characteristics over a large area. Surface high-pressure centers favor the formation of airmasses because the calm or light winds allow long residence times. Thus, many of the **airmass genesis** (formation) regions (Fig. 12.4) correspond to the planetary- and continental-scale high-pressure regions described in the previous section.

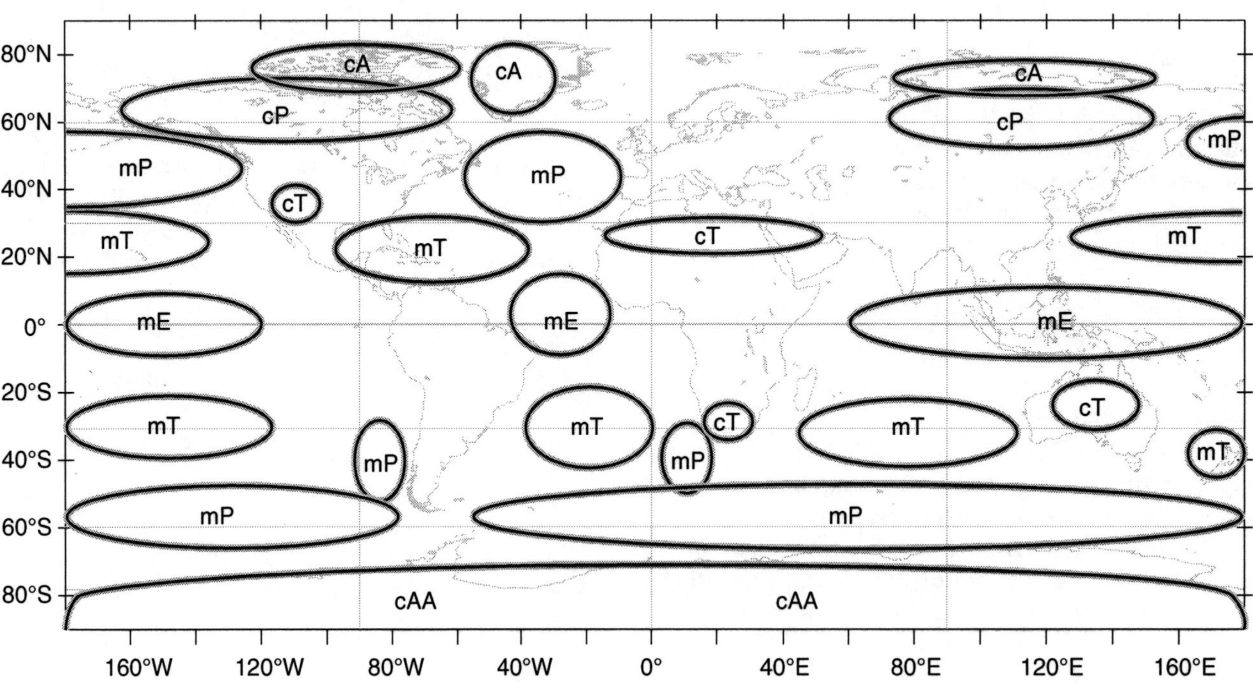

Figure 12.4
Airmass formation regions (symbols are defined in Table 12-1). Some meteorologists change all "mE" to "mT", and "cAA" to "cA".

Airmasses form as boundary layers. During their residence over a surface, the air is modified by processes including radiation, conduction, divergence, and turbulent transport between the ground and the air.

Warm Airmass Genesis

When cool air moves over a warmer surface, the warm surface modifies the bottom of the air to create an evolving, **convective mixed layer (ML).** Turbulence — driven by the potential temperature difference $\Delta\theta_s$ between the warm surface θ_{sfc} and the cooler airmass θ_{ML} — causes the ML depth z_i to initially increase (Fig. 12.5). This is the depth of the new airmass. A heat flux from the warm surface into the air causes θ_{ML} to warm toward θ_{sfc}. θ_{ML} is the temperature of the new airmass as it warms.

Synoptic-scale divergence β and subsidence w_s, which is expected in high-pressure airmass-genesis regions, oppose the ML growth. Changes within the new airmass are rapid at first. But as airmass temperature gradually approaches surface temperature, the turbulence diminishes and so does the rate of ML depth increase. Eventually, the ML depth begins to decrease (Fig. 12.5) because the reduced turbulence (trying to increase the ML thickness) cannot counteract the relentless subsidence.

A "toy model" describing the atmospheric boundary-layer processes that create a warm airmass is given in the INFO Box. The nearby Sample Application box uses this toy model to find the evolution of the warm airmass depth (i.e., the ML depth z_i) and its potential temperature θ_{ML} evolution. This is the solution that was plotted in Fig. 12.5.

The e-folding time (see Chapter 1) for the θ_{ML} to approach θ_{sfc} is surprisingly constant — about 1 to 2 days. As a result, creation of this warm **tropical** airmass is nearly complete after about a week. That is how long it takes until the airmass temperature nearly equals the surface temperature (Fig. 12.5). The time τ to reach the peak ML thickness is typically about 1 to 4 days (see another INFO box).

Cold Airmass Genesis

When air moves over a colder surface such as arctic ice, the bottom of the air first cools by conduction, radiation, and turbulent transfer with the ground. Turbulence intensity then decreases within the increasingly statically-stable boundary layer, reducing the turbulent heat transport to the cold surface.

However, direct radiative cooling of the air, both upward to space and downward to the cold ice surface, chills the air at rate 2°C day^{-1} (averaged over a 1 km thick boundary layer). As the air cools below the dew point, water-droplet clouds form. Continued radiative cooling from cloud top allows ice crystals

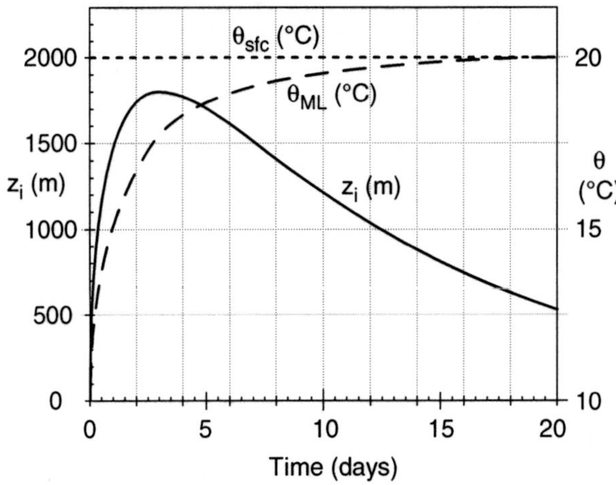

Figure 12.5
Genesis of a warm airmass after cold air comes to rest over a warmer surface. This is an example based on toy-model equations in the INFO Box. Airmass potential temperature is θ_{ML} and depth is z_i. Imposed conditions for this case-study example are: large-scale divergence $\beta = 10^{-6}$ s^{-1}, potential temperature gradient in vertical $\gamma_o = 3.3$ K km^{-1}, initial near-surface air potential temperature $\theta_{ML} = \theta_o = 10$°C, and surface temperature $\theta_{sfc} = 20$°C.

A SCIENTIFIC PERSPECTIVE • Math Clarity

In math classes, you might have learned how to combine many small equations into a single large equation that you can solve. For meteorology, although we could make such large single-equation combinations, we usually cannot solve them.

So there is no point in combining all the equations. Instead, it is easier to see the physics involved by keeping separate equations for each physical process. An example is the toy model given in the INFO Box on the next page for warm airmass genesis. Even though the many equations are coupled, it is clearer to keep them separate.

INFO • Warm Airmass Genesis

Modeling warm airmass creation (genesis) is an exercise in atmospheric boundary-layer (ABL) evolution. Since we do not cover ABLs in detail until a later chapter, the details are relegated to this INFO Box. You can safely skip them now, and come back later after you have studied ABLs.

Define a relative potential temperature θ based on a reference height (z) at the surface (z = 0). Namely, $\theta \approx T + \Gamma_d \cdot z$, where Γ_d = 9.8 °C km^{-1} is the dry adiabatic lapse rate.

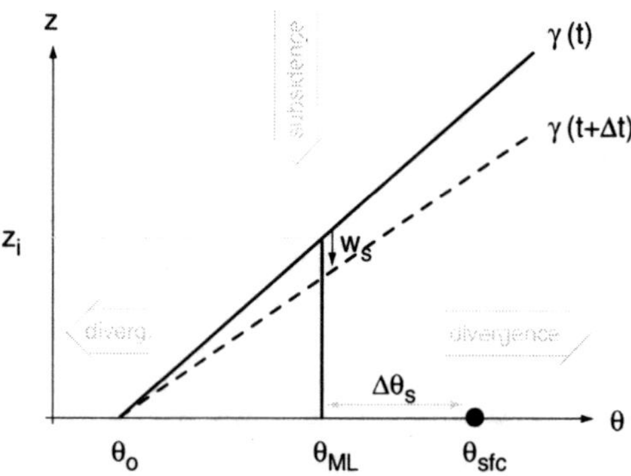

Suppose that a cool, statically stable layer of air initially (subscript *o*) has a near-surface temperature of θ_o and a linear potential-temperature gradient $\gamma_o = \gamma = \Delta\theta/\Delta z$ before it comes to rest over a warm surface of temperature θ_{sfc} (see Fig. above). We want to predict the time evolution of the depth (z_i) and potential temperature (θ_{ML}) of this new airmass. Because this airmass is an ABL, we can use boundary-layer equations to predict z_i (the height of the convective mixed-layer ,ML), and θ_{ML} (the ML potential temperature).

ML depth increases by amount Δz_{ie} during a time interval Δt due to thermodynamic encroachment (i.e., warming under the capping sounding). This is an entrainment process that adds air to the ML through the ML top. Large scale **divergence** $\beta = \Delta U / \Delta x + \Delta V / \Delta y$ removes air horizontally from the ML and causes a subsidence velocity of magnitude w_s at the ML top:

$$w_s = \beta \cdot z_i^* \qquad (7)$$

Thus, a change in ML depth results from the competition of these two terms:

$$z_i = z_i^* + \Delta z_{ie} - w_s \cdot \Delta t \qquad (12)$$

where the asterisk * indicates a value from the previous time step.

The amount of heat ΔQ (as an incremental accumulated kinematic heat flux) transferred from the warm surface to the cooler air during time interval Δt under light-wind conditions depends on the temperature difference at the surface $\Delta\theta_s$ and the intensity of turbulence, as quantified by a buoyancy velocity scale w_b:

$$\Delta Q = b \cdot w_b \cdot \Delta\theta_s \cdot \Delta t \qquad (10)$$

where $b = 5\text{x}10^{-4}$ (dimensionless) is a convective heat transport coefficient (see the Heat chapter).

(continues in next column)

INFO *(continuation)*

The buoyancy velocity scale is:

$$w_b = \left[\frac{|g|}{\theta_{ML}^*} \cdot z_i^* \cdot \Delta\theta_s \right]^{1/2} \qquad (9)$$

where $|g|$ = 9.8 m s^{-2} is gravitational acceleration.

That heat goes to warming θ_{ML}, which by geometry adds a trapezoidal area under the γ curve :

$$\Delta z_{ie} = \left[\frac{2 \cdot \Delta Q}{\gamma} + \left(z_i^*\right)^2 \right]^{1/2} - z_i^* \qquad (11)$$

Knowing the entrainment rate, we get $\Delta\theta_{ML}$ geometrically from where it intercepts the γ curve:

$$\theta_{ML} = \theta_{ML}^* + \gamma \cdot \Delta z_{ie} \qquad (13)$$

With this new ML temperature, we can update the surface temperature difference

$$\Delta\theta_s = \theta_{sfc} - \theta_{ML}^* \qquad (8)$$

Knowing the large-scale divergence, we can also update the potential temperature profile in the air above the ML:

$$\gamma = \gamma_o \cdot \exp(\beta \cdot t) \qquad (6)$$

All that remains is to update the time variable:

$$t = t^* + \Delta t \qquad (5)$$

You might have noticed that some of the equations above are initially singular, when the ML has zero depth. So, for the first small time step ($\Delta t \approx$ 6 minutes = 360 s), you should use the following special equations in the following order:

$$t = \Delta t \qquad (1)$$

$$\Delta\theta_s = \theta_{sfc} - \theta_o \qquad (2)$$

$$z_i = \Delta\theta_s \cdot \left[\frac{2 \cdot b \cdot \Delta t}{\gamma_o} \cdot \left(\frac{|g|}{\theta_o} \right)^{1/2} \right]^{2/3} \qquad (3)$$

$$\theta_{ML} = \theta_o + \gamma_o \cdot z_i \qquad (4)$$

Then, for all the subsequent time steps, use the set of equations (5 to 13 in the order as numbered) to find the resulting ML evolution. Repeat eqs. (5 to 13) for each subsequent step. As the solution begins to change more gradually, you may use larger time steps Δt. The result is a toy model that describes warm airmass formation as an evolving convective boundary layer.

The Sample Application on the next page shows how this can be done with a computer spreadsheet. First, you need to specify the imposed constants γ_o, θ_o, β and θ_{sfc}. Next, initialize the values of: $\gamma = \gamma_o$, $z_i = 0$, and $\theta_{ML} = \theta_o$ at $t = 0$. Then, solve the equations for the first step. Finally, continue iterating for subsequent time steps to simulate warm airmass genesis.

These simulations show that greater divergence causes a shallower ML that can warm faster. Greater static stability in the ambient environment reduces the peak ML depth.

Sample Application (§)

Air of initial ML potential temperature 10°C comes to rest over a 20°C sea surface. Divergence is 10^{-6} s^{-1}, and the initial $\Delta\theta/\Delta z = \gamma = 3.3$ K km^{-1}. Find and plot the warm airmass evolution of potential temperature and depth.

Find the Answer

Given: $\theta_o = 10°C = 283K$, $\theta_{sfc} = 20°C$, $\gamma_o = 3.3$ K km^{-1}, $\beta = 10^{-6}$ s^{-1}.
Find: $\theta_{ML}(t) = ?$ °C, $z_i(t) = ?$ m

For the first time step of Δt = 6 min (=360 s), use eqs. (1 to 4 from the INFO Box on the previous page). For subsequent steps, repeatedly use eqs. (5 to 13 from that same INFO Box).

t (s)	t (d)	γ (K m^{-1})	w_s (m s^{-1})	$\Delta\theta_s$ (°C)	w_b (m s^{-1})	ΔQ (K·m)	Δz_{ie} (m)	z_i (m)	θ_{ML} (°C)
0	0.000							0	10
360	0.004			10				74	10.2
720	0.008	0.00330	0.00007	9.75	5.0	8.8	29.8	104	10.3
1080	0.013	0.00330	0.00010	9.66	5.9	10.3	26.4	131	10.4
1440	0.017	0.00330	0.00013	9.57	6.6	11.3	24.0	155	10.5
. . .									
259200	3.00	0.00428	0.00179	2.41	12.0	104.3	13.6	1789	17.7
270000	3.13	0.00432	0.00179	2.35	11.9	150.8	19.4	1789	17.7
280800	3.25	0.00437	0.00179	2.26	11.7	142.8	18.2	1788	17.8
. . .									
1684800	19.50	0.01779	0.00057	0.14	1.6	4.9	0.5	542	19.9
1728000	20.00	0.01858	0.00054	0.13	1.5	4.4	0.4	519	19.9

Sample results from the computer spreadsheet are shown above. The final answer is plotted in Fig. 12.5.

Check: Units OK. Physics OK. Fig. 12.5 reasonable.
Exposition: I used small time steps of Δt = 6 minutes initially, and then as Δz_{ie} became smaller, I gradually increased past Δt = 3 h to Δt = 12 h. The figure shows rapid initial modification of the cold, statically stable airmass toward a warm, unstable airmass. Maximum z_i is reached in about τ = 3.06 days (see INFO box), in agreement with Fig. 12.5.

INFO • Time of Max Airmass Thickness

The time τ to reach the peak ML thickness for warm airmass genesis is roughly

$$\tau \approx c \cdot \left(\frac{\theta_o}{|g| \cdot \beta \cdot \gamma_o \cdot e^{\beta \cdot \tau}} \right)^{1/3} \quad \text{(c)}$$

where c = 140 (dimensionless), and the other variables are defined in the text. τ is typically about 1 to 4 days. Any further lingering of the airmass over the same surface temperature results in a loss of airmass thickness due to divergence.

Equation (c) is an implicit equation; namely, you need to know τ in order to solve for τ. Although this equation is difficult to solve analytically, you can iterate to quickly converge to a solution in about 5 steps in a computer spreadsheet. Namely, start with $\tau = 0$ as the first guess and plug into the right side of eq. (c). Then solve for τ on the left side. For the next iteration, take this new τ and plug it in on the right, and solve for an updated τ on the left. Repeat until the value of τ converges to a solution; namely, when $\Delta\tau/\tau < \varepsilon$ for ε = 0.01 or smaller.

Sample Application (§)

For the conditions of the previous Sample Application, find the time τ that estimates when the new warm airmass has maximum thickness.

Find the Answer

Given: $\theta_o = 10°C = 283K$, $\theta_{sfc} = 20°C$,
$\gamma_o = 3.3$ K km^{-1}, $\beta = 10^{-6}$ s^{-1}.
Find: $\tau = ?$ days

Use eq. (c) from the INFO Box at left. Start with $\tau = 0$. The first iteration is:

$$\tau = 140 \cdot \left[\frac{283\text{K}}{(9.8\text{m/s}^2)\cdot(10^{-6}\text{s}^{-1})\cdot(0.0033\text{K/m})\cdot e^0} \right]^{1/3}$$

= 288498 s = 3.339 days
Subsequent iterations give: τ =
3.033 -> 3.060 -> 3.057 -> 3.058 -> **3.058 days**

Check: Units OK. Physics OK. Agrees with Fig. 12.5.
Exposition: Convergence was quick. From Fig. 12.5, the actual time of this peak thickness was between 3 and 3.125 days, so eq. (c) does a reasonable job.

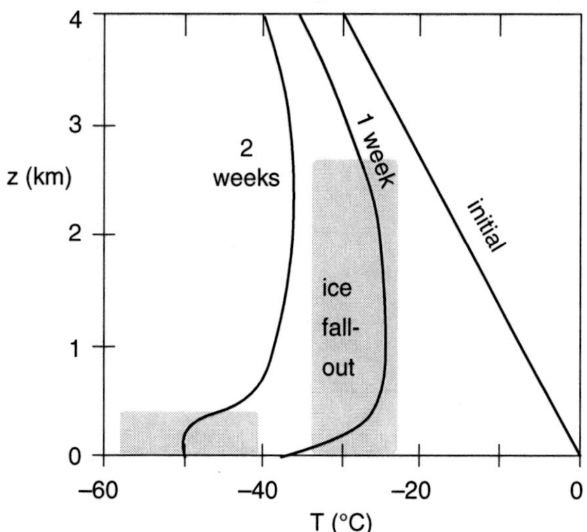

Figure 12.6
Genesis of a continental-polar air mass over arctic ice. The cloud/fog regions are shaded.

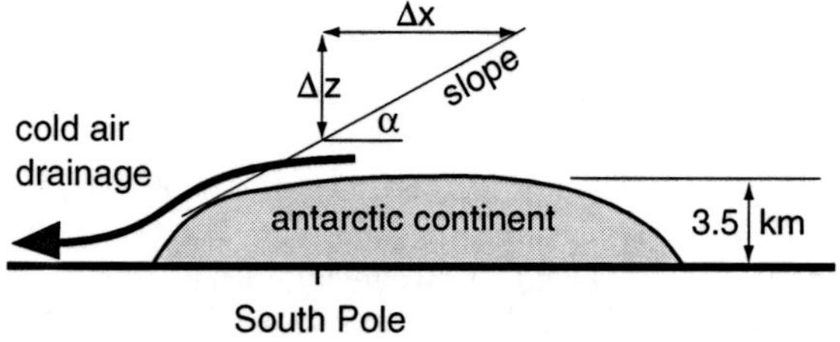

Figure 12.7
Cold katabatic winds draining from Antarctica.

> **Sample Application**
> Find the slope force per unit area acting on a katabatic wind of temperature –20°C with ambient air temperature 0°C. Assume a slope of $\Delta z/\Delta x = 0.1$.
>
> **Find the Answer**
> Given: $\Delta\theta = 20$ K, $T_e = 273$ K, $\Delta z/\Delta x = 0.1$
> Find: $F_{x\,S}/m = ?$ m·s^{-2}
>
> Use eq. (12.1):
>
> $$\frac{F_{x\,S}}{m} = \frac{(9.8\text{m}\cdot\text{s}^{-2})\cdot(20\text{K})}{273\text{K}}\cdot(0.1)$$
> $= \underline{\mathbf{0.072\ m{\cdot}s^{-2}}}$
>
> **Check**: Units OK. Physics OK.
> **Exposition**: This is two orders of magnitude greater than the typical synoptic forces (see the Dynamics chapter). Hence, drainage winds can be strong.

to grow at the expense of evaporating liquid droplets, changing the cloud into an ice cloud.

Radiative cooling from cloud top creates cloudy "thermals" of cold air that sink, causing some turbulence that distributes the cooling over a deeper layer. Turbulent entrainment of air from above cloud top down into the cloud allows the cloud top to rise, and deepens the incipient airmass (Fig. 12.6).

The ice crystals within this cloud are so few and far between that the weather is described as cloudless ice-crystal precipitation. This can create some spectacular **halos** and other optical phenomena in sunlight (see the Atmospheric Optics chapter), including sparkling ice crystals known as **diamond dust**. Nevertheless, infrared radiative cooling in this cloudy air is much greater than in clear air, allowing the cooling rate to increase to 3°C day^{-1} over a layer as deep as 4 km.

During the two-week formation of this **continental-polar** or **continental-arctic** airmass, most of the ice crystals precipitate out leaving a thinner cloud of 1 km depth. Also, subsidence within the high pressure also reduces the thickness of the cloudy airmass and causes some warming to partially counteract the radiative cooling.

Above the final fog layer is a nearly isothermal layer of air 3 to 4 km thick that has cooled about 30°C. Final air-mass temperatures are often in the range of –30 to –50 °C, with even colder temperatures near the surface.

While the Arctic surface consists of relatively flat sea-ice (except for Greenland), the Antarctic has mountains, high ice-fields, and significant surface topography (Fig. 12.7). As cold air forms by radiation, it can drain downslope as a **katabatic wind** (see the Regional Winds chapter). Steady winds of 10 m s^{-1} are common in the Antarctic interior, with speeds of 50 m s^{-1} along some of the steeper slopes.

As will be shown in the Regional Winds chapter, the buoyancy force per unit mass on a surface of slope $\Delta z/\Delta x$ translates into a quasi-horizontal slope-force per mass of:

$$\frac{F_{x\,S}}{m} = \frac{|g|\cdot\Delta\theta}{T_e}\cdot\frac{\Delta z}{\Delta x} \qquad \bullet(12.1)$$

$$\frac{F_{y\,S}}{m} = \frac{|g|\cdot\Delta\theta}{T_e}\cdot\frac{\Delta z}{\Delta y} \qquad \bullet(12.2)$$

where $|g| = 9.8$ m·s^{-2} is gravitational acceleration, and $\Delta\theta$ is the potential-temperature difference between the draining cold air and the ambient air above. The ambient-air absolute temperature is T_e.

The sign of these forces should be such as to accelerate the wind downslope. The equations above work when the magnitude of slope $\Delta z/\Delta x$ is small, because then $\Delta z/\Delta x = \sin(\alpha)$, where α is the slope angle of the topography.

The katabatic wind speed in the Antarctic also depends on turbulent drag force against the ice surface, ambient pressure-gradient force associated with synoptic weather systems, Coriolis force, and turbulent drag caused by mixing of the draining air with the stationary air above it. At an average drainage velocity of 5 m s^{-1}, air would need over 2 days to move from the interior to the periphery of the continent, which is a time scale on the same order as the inverse of the Coriolis parameter. Hence, Coriolis force cannot be neglected.

Katabatic drainage removes cold air from the genesis regions and causes turbulent mixing of the cold air with warmer air aloft. The resulting cold-air mixture is rapidly distributed toward the outside edges of the antarctic continent.

One aspect of the global circulation is a wind that blows around the poles. This is called the **polar vortex**. Katabatic removal of air from over the antarctic reduces the troposphere depth, enhancing the persistence and strength of the antarctic polar vortex due to potential vorticity conservation.

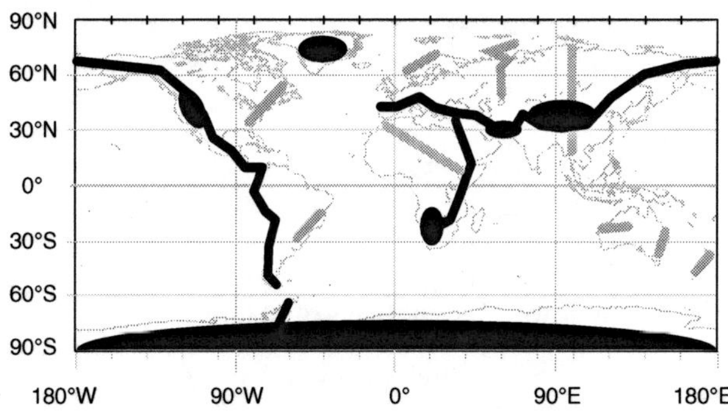

Figure 12.8
Locations of major mountain ranges (black lines), lesser ranges (grey lines), and large plateaus of land or ice (black ovals).

Movement

Airmasses do not remain stationary over their birth place forever. After a week or two, a transient change in the weather pattern can push the airmass toward new locations.

When airmasses move, two things can happen:
(1) As the air moves over surfaces with different characteristics, the airmass begins to change. This is called **airmass modification**, and is described in the next subsection.
(2) An airmass can encounter another airmass. The boundary between these two airmasses is called a **front**, and is a location of strong gradients of temperature, humidity, and other airmass characteristics. Fronts are described in detail later.

Tall mountain ranges can strongly **block** or **channel** the movement of airmasses, because airmasses occupy the bottom of the troposphere. Fig. 12.8 shows a simplified geography of major mountain ranges.

For example, in the middle of North America, the lack of any major east-west mountain range allows the easy movement of cold polar air from Canada toward warm humid air from the Gulf of Mexico. This sets the stage for strong storms (see the Extratropical Cyclone chapter and the chapters on Thunderstorms).

The long north-south barrier of mountains (Rockies, Sierra Nevada, Cascades, Coast Range) along the west coast of North America impedes the easy entry of Pacific airmasses toward the center of that continent. Those mountains also help protect the west coast from the temperature extremes experienced by the rest of the continent.

In Europe, the mountain orientation is the opposite. The Alps and the Pyrenees are east-west mountain ranges that inhibit movement of Mediterranean airmasses from reaching northward. The lack of major north-south ranges in west and central Europe allows the easy movement of maritime airmasses from the Atlantic to sweep eastward, bringing cool wet conditions.

One of the greatest ranges is the Himalaya Mountains, running east-west between India and China. Maritime tropical airmasses moving in from the Indian Ocean reach these mountains, causing heavy rains over India during the monsoon. The same mountains block the maritime air from reaching further northward, leaving a very dry Tibetan Plateau and Gobi Desert in its **rain shadow**.

The discussion above focused on blocking and channeling by the mountains. In some situations air can move over mountain tops (Fig. 12.9). When this happens, the airmass is strongly modified, as described next.

Modification

As an airmass moves from its origin, it is modified by the new landscapes under it. For example, a polar airmass will warm and gain moisture as it moves equatorward over warmer vegetated ground. Thus, it gradually loses its original identity.

Via Surface Fluxes

Heat and moisture transfer at the surface can be described with **bulk-transfer relationships** such as eq. (3.34). If we assume for simplicity that wind-induced turbulence creates a well-mixed airmass of quasi-constant thickness z_i, then the change of

Sample Application

An mP airmass initially has T = 5°C & RH = 100%. Use a thermo diagram to find T & RH at: 1 Olympic Mtns (elevation ≈ 1000 m), 2 Puget Sound (0 m), 3 Cascade Mtns (1500 m), 4 the Great Basin (500 m), 5 Rocky Mtns (2000 m), 6 the western Great Plains (1000 m).

Find the Answer

Given: Elevations from west to east (m) =
0, 1000, 0, 1500, 500, 2000, 1000
Initially RH=100%. Thus $T_d = T$ = 5°C
Find: T (°C) & RH (%) at surface locations 1 to 6.
Assume: All condensation precipitates out. No additional heat or moisture transfer from the surface. Start with air near sea level, where P_{SL} = 100 kPa.

Use Fig. 12.9 and an emagram from the Stability chapter. On the thermo diagram, the air parcel follows the following route: 0 - 1 - 2 - 1 - 3 - 4 - 3 - 5 - 6.

Initially (point 0), $T_d = T$ = 5°C. Because this air is already saturated, it would follow a saturated adiabat from point 0 to point 1 at z = 1 km, where the still-saturated air has $T_d = T$ = –1°C.

If all condensates precipitate out, then air would descend dry adiabatically (with T_d following an isohume) from point 1 to 2, giving T = 9°C and T_d = 1°C.

When this unsaturated air rises, it first does so dry adiabatically until it reaches its LCL at point 1. This now-cloudy air continues to rise toward point 3 moist adiabatically, where it has $T_d = T$ = –4°C. Etc.

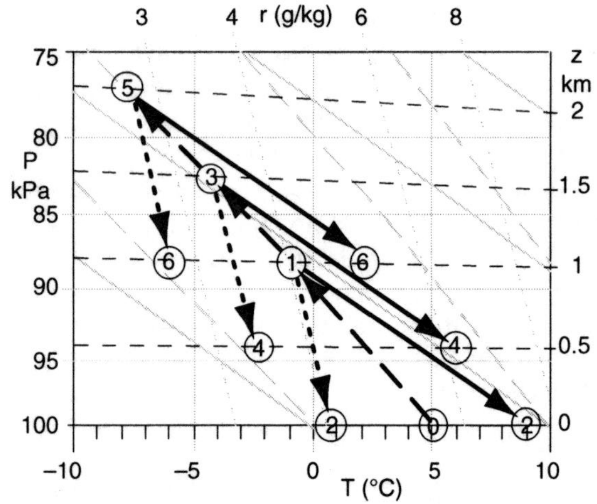

Results, where Index = circled numbers in Fig above.

Index	z (km)	T (°C)	T_d (°C)	RH (%)
0	0	5	5	100
1	1	–1	–1	100
2	0	9	1	55
3	1.5	–4	–4	100
4	0.5	6	–2	54
5	2	–7	–7	100
6	1	2	–6	53

Check: Units OK. Physics OK. Figure OK.
Exposition: The airmass has lost its maritime identity by the time it reaches the Great Plains, and little moisture remains. Humidity for rain in the plains comes from the southeast (not from the Pacific).

airmass potential temperature θ_{ML} with travel-distance Δx is:

$$\frac{\Delta\theta_{ML}}{\Delta x} \approx \frac{C_H \cdot (\theta_{sfc} - \theta_{ML})}{z_i} \tag{12.3}$$

where $C_H \approx 0.01$ is the bulk-transfer coefficient for heat (see the chapters on Heat and on the Atmospheric Boundary Layer).

If the surface temperature is horizontally homogeneous, then eq. (12.3) can be solved for the airmass temperature at any distance x from its origin:

$$\theta_{ML} = \theta_{sfc} - (\theta_{sfc} - \theta_{ML\,o}) \cdot \exp\left(-\frac{C_H \cdot x}{z_i} \right) \tag{12.4}$$

where $\theta_{ML\,o}$ is the initial airmass potential temperature at location x = 0.

Sample Application

A polar airmass with initial θ = –20°C and depth = 500 m moves southward over a surface of 0°C. Find the initial rate of temperature change with distance.

Find the Answer

Given: θ_{ML} = –20°C, θ_{sfc} = 0°C, z_i = 500 m
Find: $\Delta\theta_{ML}/\Delta x$ = ? °C km^{-1}. Assume no mountains.

Use eq. (12.3):

$$\frac{\Delta\theta_{ML}}{\Delta x} \approx \frac{(0.01)\cdot[0°C - (-20°C)]}{(500m)} = \mathbf{0.4\ °C\ km^{-1}}$$

Check: Units OK. Physics OK.
Exposition: Neither this answer nor eq. (12.4) depend on wind speed. While faster speeds give faster position change, they also cause greater heat transfer to/from the surface (eq. 3.34). These 2 effects cancel.

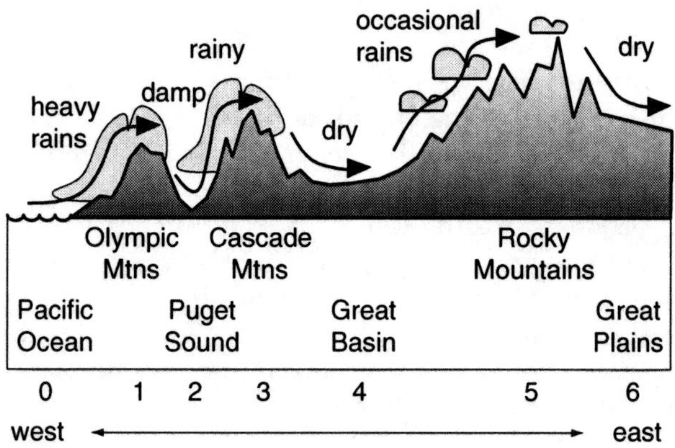

Figure 12.9
Modification of a Pacific airmass by flow over mountains in the northwestern USA. (Numbers are used in the Sample Application.)

Via Flow Over Mountains

If an airmass is forced to rise over mountain ranges, the resulting condensation, precipitation, and latent heating will dry and warm the air. For example, an airmass over the Pacific Ocean near the northwestern USA is often classified as maritime polar (mP), because it is relatively cool and humid. As the prevailing westerly winds move this airmass over the Olympic Mountains (a coastal mountain range), the Cascade Mountains, and the Rocky Mountains, there is substantial precipitation and latent heating (Fig. 12.9).

SURFACE FRONTS

Surface fronts mark the boundaries between airmasses at the Earth's surface. They usually have the following attributes:

- strong horizontal temperature gradient
- strong horizontal moisture gradient
- strong horizontal wind gradient
- strong vertical shear of the horizontal wind
- relative minimum of pressure
- high vorticity
- **confluence** (air converging horizontally)
- clouds and precipitation
- high static stability
- kinks in isopleths on weather maps

In spite of this long list of attributes, fronts are usually labeled by the surface temperature change associated with frontal passage.

Some weather features exhibit only a subset of attributes, and are not labeled as fronts. For example, a **trough** (pronounced like "trof") is a line of low pressure, high vorticity, clouds and possible precipitation, wind shift, and confluence. However, it often does not possess the strong horizontal temperature and moisture gradients characteristic of fronts.

Another example of an airmass boundary that is often not a complete front is the **dryline**. It is discussed later in this chapter.

Recall from the Weather Reports and Map Analysis chapter that fronts are always drawn on the warm side of the frontal zone. The frontal symbols (Fig. 12.10) are drawn on the side of the frontal line toward which the front is moving. For a stationary front, the symbols on both sides of the frontal line indicate what type of front it would be if it were to start moving in the direction the symbols point.

Fronts are three dimensional. To help picture their structure, we next look at horizontal and vertical cross sections through fronts.

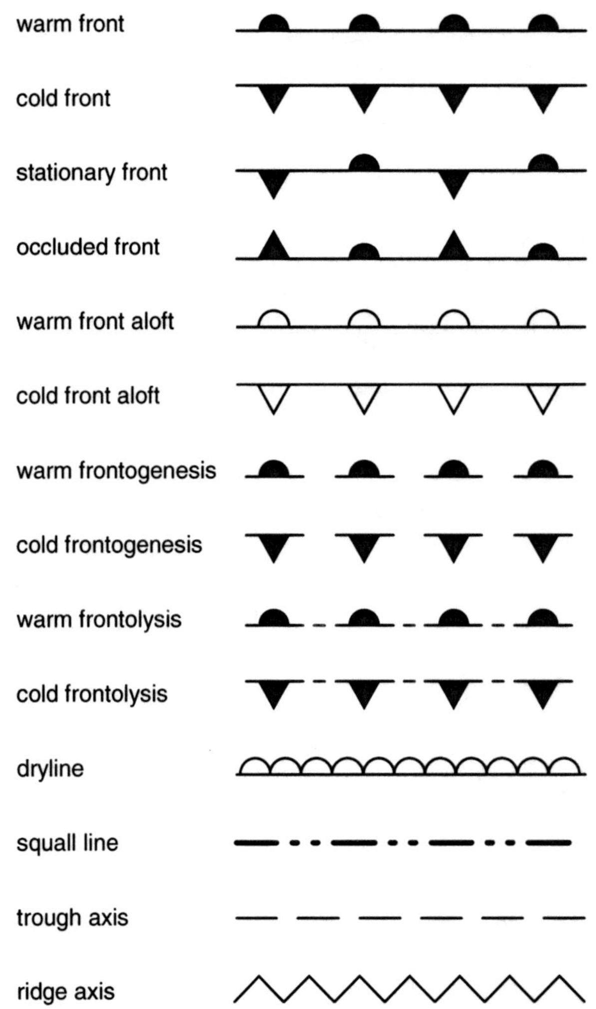

Figure 12.10
Glyphs for fronts, other airmass boundaries, and axes (copied from the Weather Reports and Map Analysis chapter). The suffix "genesis" implies a forming or intensifying front, while "lysis" implies a weakening or dying front. A stationary front is a frontal boundary that doesn't move very much. Occluded fronts and drylines will be explained later.

> **INFO • Bergen School of Meteorology**
>
> During World War I, Vilhelm Bjerknes, a Norwegian physicist with expertise in radio science and fluid mechanics, was asked in 1918 to form a Geophysical Institute in Bergen, Norway. Cut-off from weather data due to the war, he arranged for a dense network of 60 surface weather stations to be installed. Some of his students were C.-G. Rossby, H. Solberg, T. Bergeron, V. W. Ekman, H. U. Sverdrup, and his son Jacob Bjerknes.
>
> Jacob Bjerknes used the weather station data to identify and classify cold, warm, and occluded fronts. He published his results in 1919, at age 22. The term "front" supposedly came by analogy to the battle fronts during the war. He and Solberg also later explained the life cycle of cyclones.

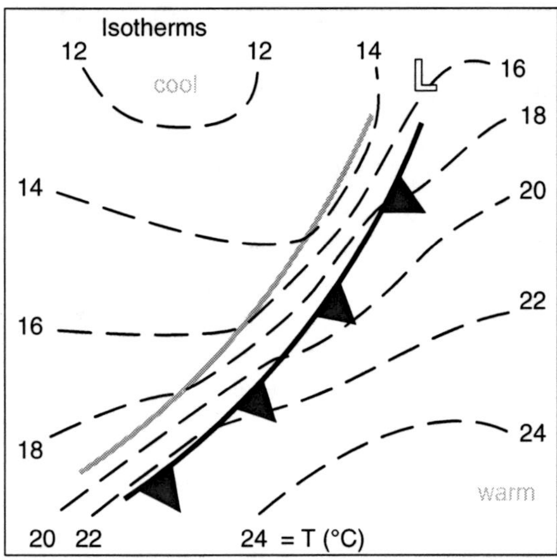

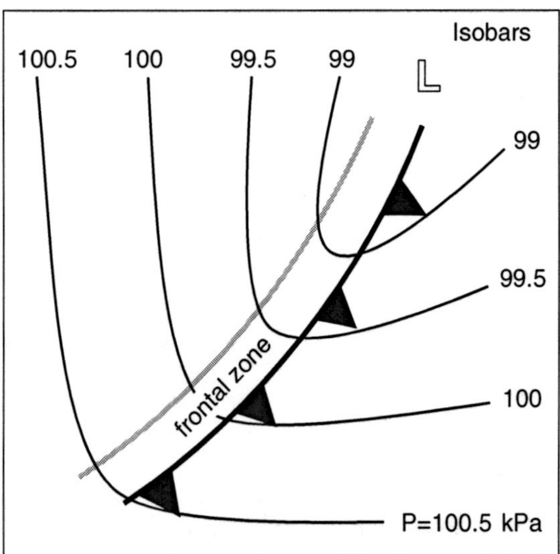

Horizontal Structure

Cold Fronts (Fig. 12.11)

In central N. America, winds ahead of cold fronts typically have a southerly component, and can form strong low-level jets at night and possibly during day. Warm, humid, hazy air advects from the south.

Sometimes a **squall line** of thunderstorms will form in advance of the front, in the warm air. These squall lines can be triggered by wind shear and by the kinematics (advection) near fronts. They can also consist of thunderstorms that were initially formed on the cold front, but progressed faster than the front.

Along the front are narrow bands of towering cumuliform clouds with possible thunderstorms and scattered showers. Along the front the winds are stronger and gusty, and pressure reaches a relative minimum. Thunderstorm anvils often spread hundreds of kilometers ahead of the surface front.

Winds shift to a northerly direction behind the front, advecting colder air from the north. This air is often clean with excellent visibilities and clear blue skies during daytime. If sufficient moisture is present, scattered cumulus or broken stratocumulus clouds can form within the cold airmass.

As this airmass consists of cold air advecting over warmer ground, it is statically unstable, convective, and very turbulent. However, at the top of the airmass is a very strong stable layer along the frontal inversion that acts like a lid to the convec-

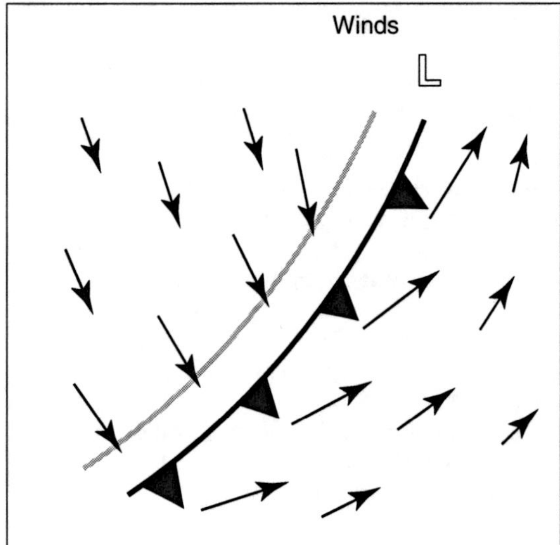

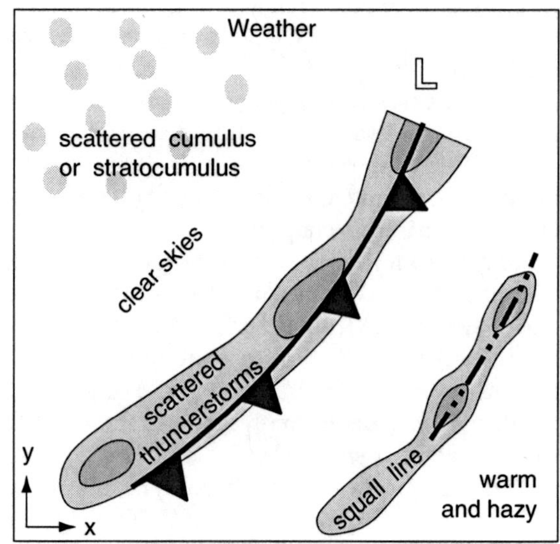

Figure 12.11 (both columns) *Cold front horizontal structure (maps can be overlain).*

tion. Sometimes over ocean surfaces the warm moist ocean leads to considerable post-frontal deep convection.

The idealized picture presented in Fig. 12.11 can differ considerably in the mountains.

Warm Fronts (Fig. 12.12)

In central N. America, southeasterly winds ahead of the front bring in cool, humid air from the Atlantic Ocean, or bring in mild, humid air from the Gulf of Mexico.

An extensive deck of stratiform clouds (called a **cloud shield**) can occur hundreds of kilometers ahead of the surface front. In the cirrostratus clouds at the leading edge of this cloud shield, you can sometimes see halos, sundogs, and other optical phenomena. The cloud shield often wraps around the poleward side of the low center.

Along the frontal zone can be extensive areas of low clouds and fog, creating hazardous travel conditions. Nimbostratus clouds cause large areas of drizzle and light continuous rain. Moderate rain can form in multiple **rain bands** parallel to the front. The pressure reaches a relative minimum at the front.

Winds shift to a more southerly direction behind the warm front, advecting in warm, humid, hazy air. Although heating of air by the surface might not be strong, any clouds and convection that do form can often rise to relatively high altitudes because of weak static stabilities throughout the warm airmass.

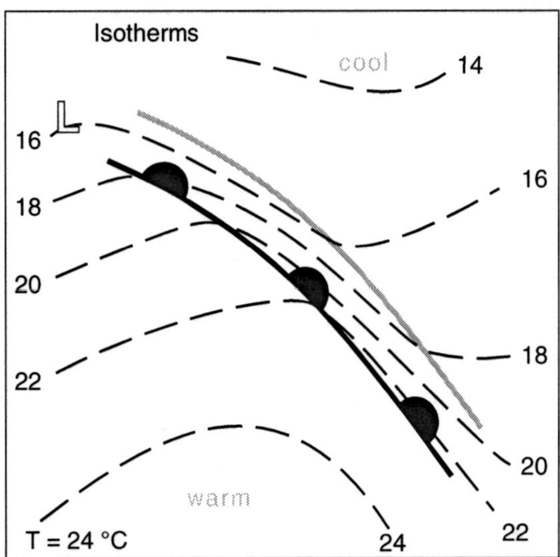

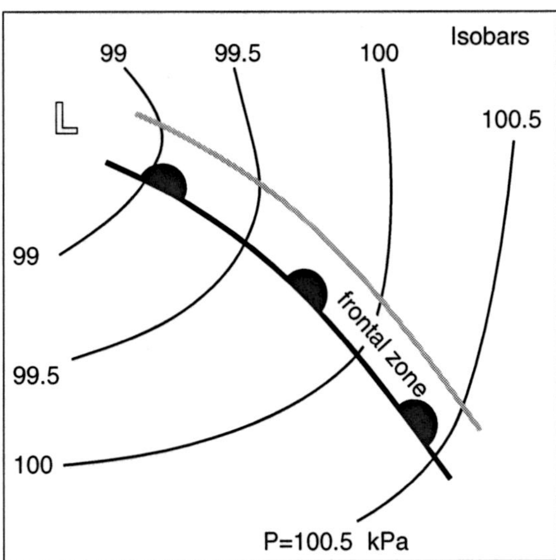

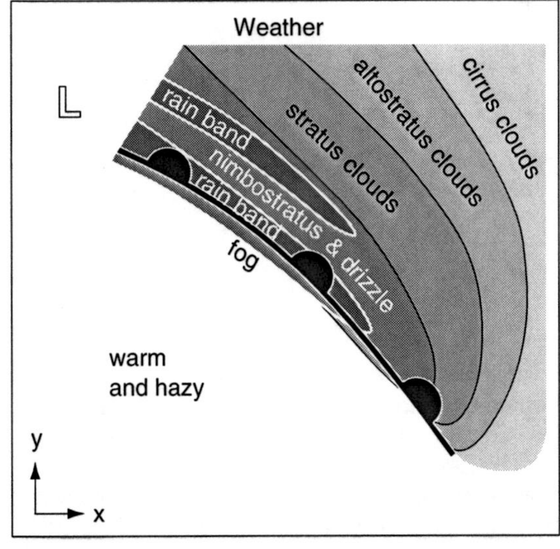

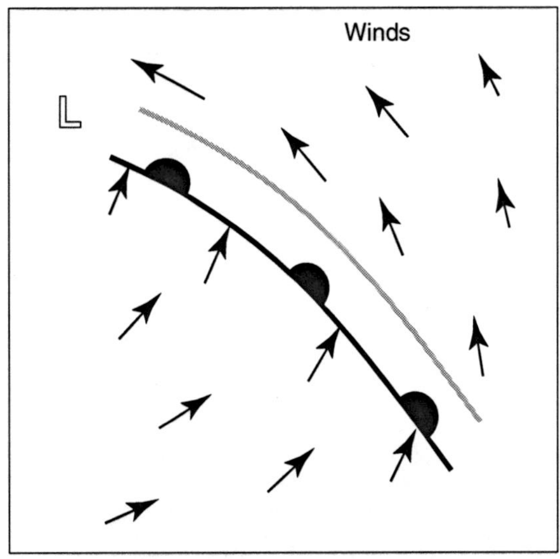

Figure 12.12 (both columns) *Warm front structure (maps can be overlain).*

Sample Application

Given the plotted surface weather data below, analyze it for temperature (50°F and every 5°F above and below) and pressure (101.2 kPa and every 0.4 kPa above and below). Identify high- and low-pressure centers and fronts. Discuss how the winds, clouds and weather compare to the descriptions in Figs. 12.11 & 12.

Find the Answer

First, using methods shown in the Weather Reports & Map Analysis chapter, draw the isotherms (°F):

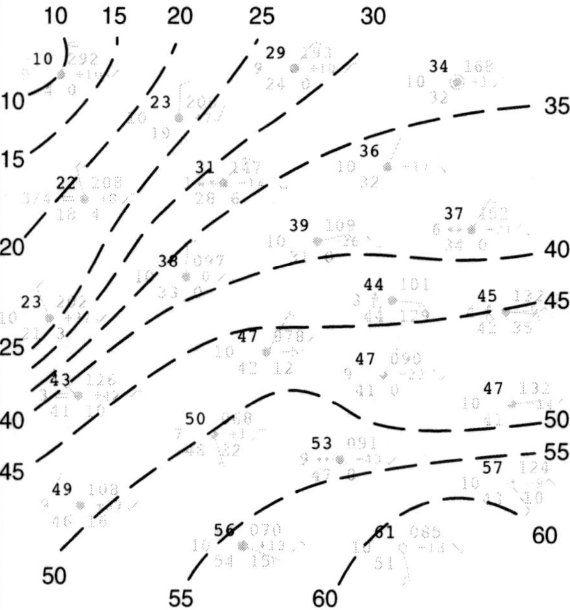

Next, analyze it for pressure. Recall that the plotted pressure is abbreviated. We need to prefix 9 or 10 to the left of the pressure code, and insert a decimal point two places from the right (to get kPa). Choose between 9 and 10 based on which one results in a pressure closest to standard sea-level pressure 101.3 kPa.

(continues in next column)

Sample Application *(continued)*

For the pressure data in this Sample Application, every prefix is 10. For example, the plotted pressure code 097 means 100.97 kPa. Similarly, 208 means 102.08 kPa. By analyzing pressures, we get the following map of isobars (kPa):

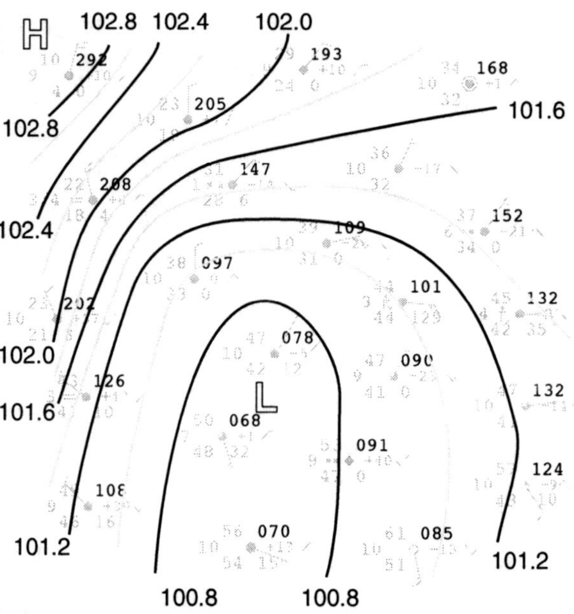

Next, overlay the isobars and isotherms, and find the frontal zones (drawn with the thick black and grey lines) and fronts (thick black line).

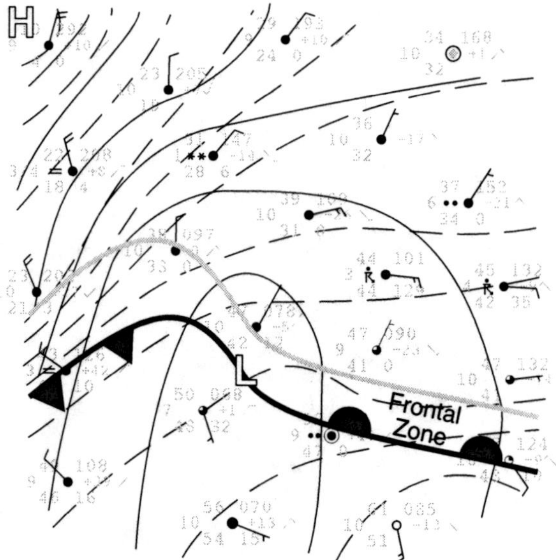

Check: Looks reasonable.

Exposition: Winds are generally circulating counterclockwise around the low center (except the very light winds, which can be sporadic). Overcast skies cover most of the region, with some thunderstorms and rain north of the warm front, and snow further northwest of the low where the air is colder.

Vertical Structure

Suppose that radiosonde observations (**RAOBs**) are used to probe the lower troposphere, providing temperature profiles such as those in Fig. 12.13a. To locate fronts by their vertical cross section, first convert the temperatures into potential temperatures θ (Fig. 12.13b). Then, draw lines of equal potential temperature (**isentropes**). Fig. 12.13c shows isentropes drawn at 5°C intervals. Often isentropes are labeled in Kelvin.

In the absence of diabatic processes such as latent heating, radiative heating, or turbulent mixing, air parcels follow isentropes when they move adiabatically. For example, consider the θ = 35°C parcel that is circled in Fig. 12.13b above weather station B. Suppose this parcel starts to move westward toward C.

If the parcel were to be either below or above the 35°C isentrope at its new location above point C, buoyant forces would tend to move it vertically to the 35°C isentrope. Such forces happen continuously while the parcel moves, constantly adjusting the altitude of the parcel so it rides on the isentrope.

The net movement is westward and upward along the 35°C isentrope. Air parcels that are forced to rise along isentropic surfaces can form clouds and precipitation, given sufficient moisture. Similarly, air blowing eastward would move downward along the sloping isentrope.

In three dimensions, you can picture **isentropic surfaces** separating warmer θ aloft from colder θ below. Analysis of the flow along these surfaces provides a clue to the weather associated with the front. Air parcels moving adiabatically must follow the "topography" of the isentropic surface. This is illustrated in the Extratropical Cyclones chapter.

At the Earth's surface, the boundary between cold and warm air is the **surface frontal zone**. This is the region where isentropes are packed relatively close together (Figs. 12.13b & c). The top of the cold air is called the **frontal inversion** (Fig. 12.13c). The frontal inversion is also evident at weather stations C and D in Fig. 12.13a, where the temperature increases with height. Frontal inversions of warm and cold fronts are gentle and of similar temperature change.

Within about 200 m of the surface, there are appreciable differences in frontal slope. The cold front has a steeper nose (slope ≈ 1 : 100) than the warm front (slope ≈ 1 : 300), although wide ranges of slopes have been observed.

Fronts are defined by their temperature structure, although many other quantities change across the front. Advancing cold air at the surface defines

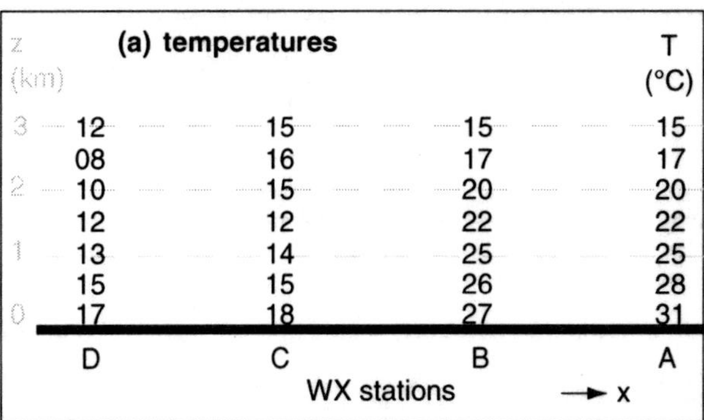

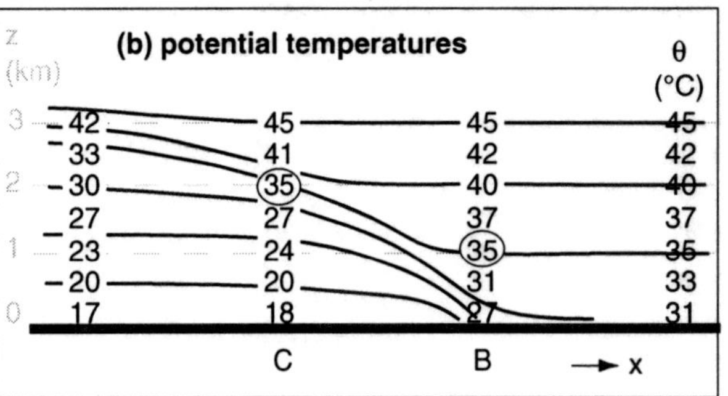

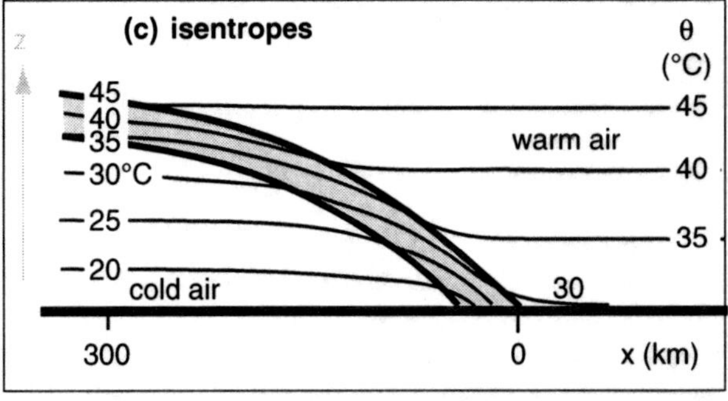

Figure 12.13
Analysis of soundings to locate fronts in a vertical cross section. Frontal zone / frontal inversion is shaded in bottom figure, and is located where the isentropes are tightly packed (close to each other).

Sample Application

What weather would you expect with a warm katafront? (See next page.)

Find the Answer & Exposition

Cumuliform clouds and showery precipitation would probably be similar to those in Fig. 12.16a, except that the bad weather would move in the direction of the warm air at the surface, which is the direction the surface front is moving.

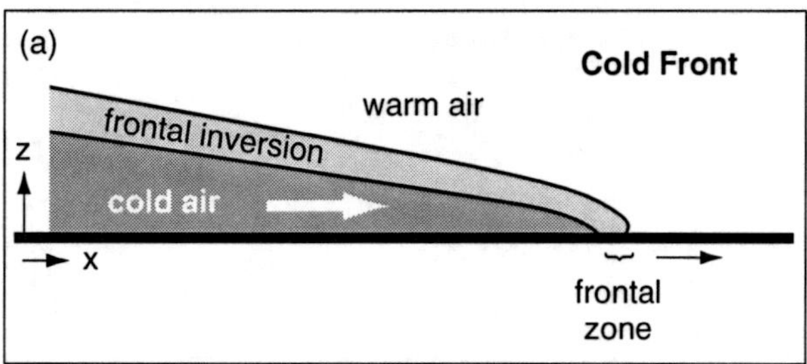

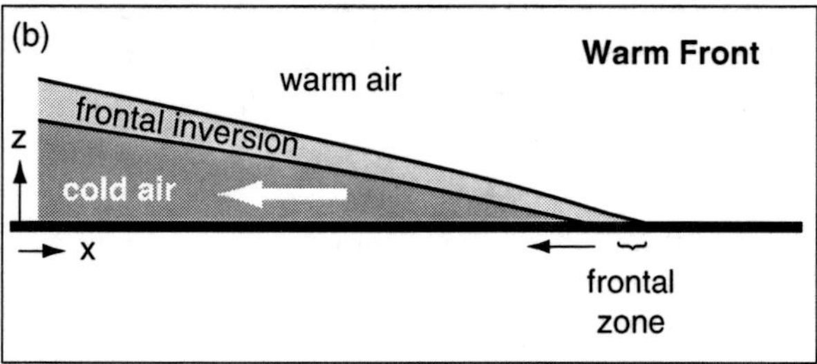

Figure 12.14
Vertical structure of fronts, based on cold air movement.

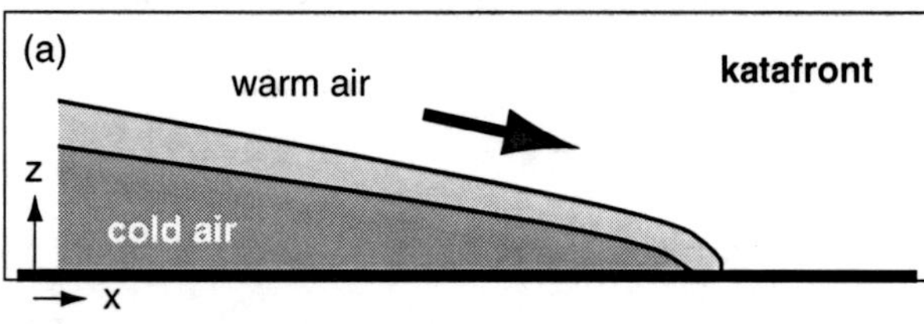

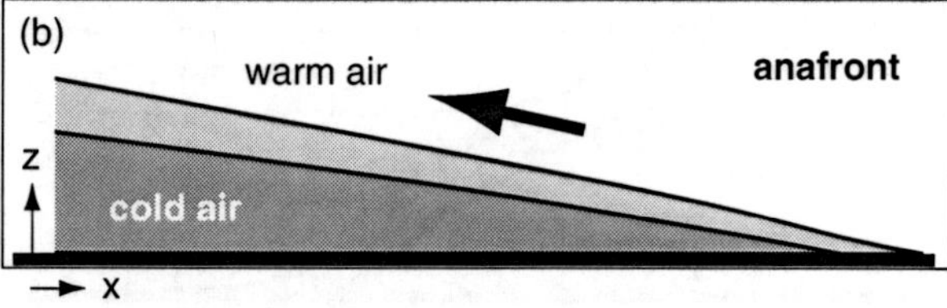

Figure 12.15
Vertical structure of fronts, based on overlying-air movement.

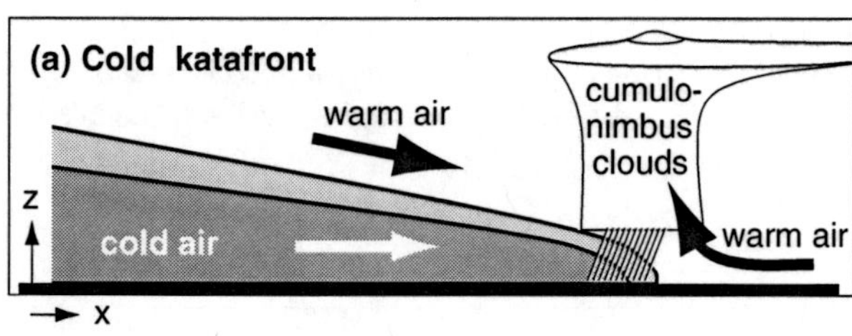

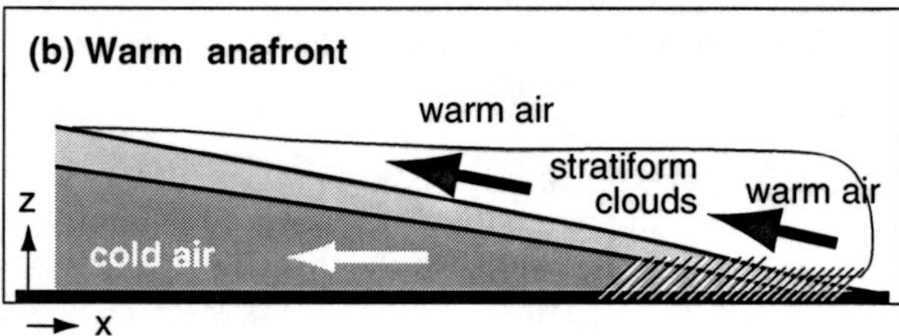

Figure 12.16
Typical fronts in central N. America.

the **cold front**, where the front moves toward the warm airmass (Fig. 12.14a). Retreating cold air defines the **warm front**, where the front moves toward the cold airmass (Fig. 12.14b).

Above the frontal inversion, if the warm air flows down the frontal surface, it is called a **katafront**, while warm air flowing up the frontal surface is an **anafront** (Fig. 12.15). It is possible to have cold katafronts, cold anafronts, warm katafronts, and warm anafronts.

Frequently in central N. America, the cold fronts are katafronts, as sketched in Fig. 12.16a. For this situation, warm air is converging on both sides of the frontal zone, forcing the narrow band of cumuliform clouds that is typical along the front. It is also common that warm fronts are anafronts, which leads to a wide region of stratiform clouds caused by the warm air advecting up the isentropic surfaces (Fig. 12.16b).

A **stationary front** is like an anafront where the cold air neither advances nor retreats.

GEOSTROPHIC ADJUSTMENT – PART 3

Winds in the Cold Air

Why does cold dense air from the poles not spread out over more of the Earth, like a puddle of water? Coriolis force is the culprit, as shown next.

Picture two air masses initially adjacent (Fig. 12.17a). The cold airmass has initial depth H and uniform virtual potential temperature θ_{v1}. The warm airmass has uniform virtual potential temperature $\theta_{v1} + \Delta\theta_v$. The average absolute virtual temperature is $\overline{T_v}$. In the absence of rotation of the coordinate system, you would expect the cold air to spread out completely under the warm air due to buoyancy, reaching a final state that is horizontally homogeneous.

However, on a rotating Earth, the cold air experiences Coriolis force (to the right in the Northern Hemisphere) as it begins to move southward. Instead of flowing across the whole surface, the cold air spills only distance a before the winds have turned 90°, at which point further spreading stops (Fig. 12.17b).

At this quasi-equilibrium, pressure-gradient force associated with the sloping cold-air interface balances Coriolis force, and there is a steady geostrophic wind U_g from east to west. The process of approaching this equilibrium is called **geostrophic adjustment**, as was discussed in the previous chapter. Real atmospheres never quite reach this equilibrium.

At equilibrium, the final spillage distance a of the front from its starting location equals the **external Rossby-radius of deformation**, λ_R:

$$a = \lambda_R = \frac{\sqrt{|g| \cdot H \cdot \Delta\theta_v / \overline{T_v}}}{f_c} \qquad \bullet(12.5)$$

where f_c is the Coriolis parameter and $|g| = 9.8 \text{ m s}^{-2}$ is gravitational acceleration magnitude.

The geostrophic wind U_g in the cold air at the surface is greatest at the front (neglecting friction), and exponentially decreases behind the front:

$$U_g = -\sqrt{|g| \cdot H \cdot (\Delta\theta_v / \overline{T_v})} \cdot \exp\left(-\frac{y+a}{a}\right) \qquad \bullet(12.6)$$

for $-a \le y \le \infty$. The depth of the cold air h is:

$$h = H \cdot \left[1 - \exp\left(-\frac{y+a}{a}\right)\right] \qquad \bullet(12.7)$$

which smoothly increases to depth H well behind the front (at large y).

Figs. 12.17 are highly idealized, having airmasses of distinctly different temperatures with a sharp interface in between. For a fluid with a smooth continuous temperature gradient, geostrophic adjustment occurs in a similar fashion, with a final equilibrium state as sketched in Fig. 12.18. The top of this diagram represents the top of the troposphere, and the top wind vector represents the jet stream.

This state has high surface pressure under the cold air, and low surface pressure under the warm air (see the General Circulation chapter). On the cold side, isobaric surfaces are more-closely spaced in height than on the warm side, due to the hypsometric relationship. This results in a pressure reversal aloft, with low pressure (or low heights) above the cold air and high pressure (or high heights) above the warm air.

Horizontal pressure gradients at low and high altitudes create opposite geostrophic winds, as indicated in Fig. 12.18. Due to Coriolis force, the air represented in Fig. 12.18 is in equilibrium; namely, the cold air does not spread any further.

This behavior of the cold airmass is extremely significant. It means that the planetary-scale flow, which is in approximate geostrophic balance, is unable to complete the job of redistributing the cold air from the poles and warm air from the tropics. Yet some other process must be acting to complete the job of redistributing heat to satisfy the global energy budget (in the General Circulation chapter).

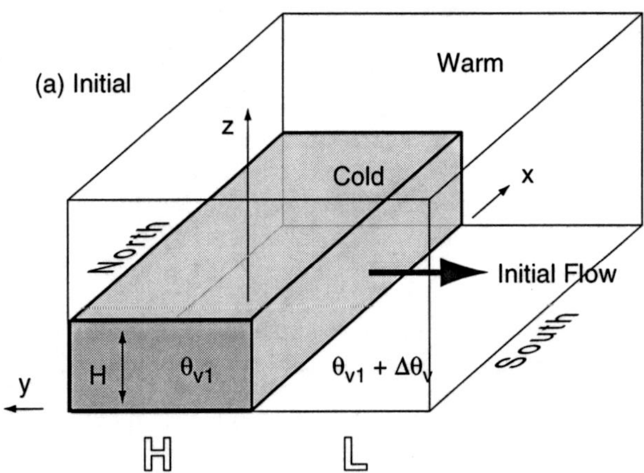

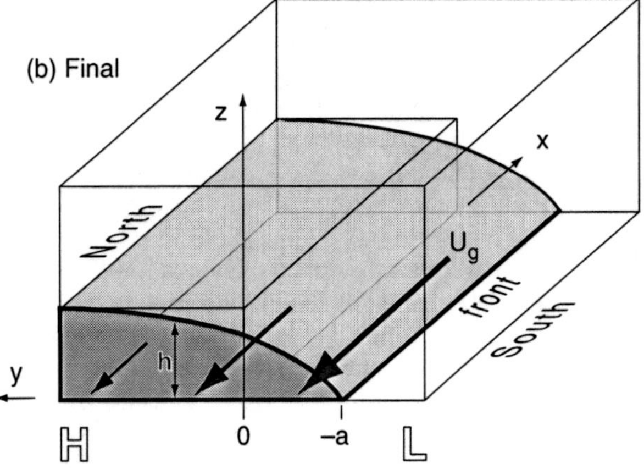

Figure 12.17
Geostrophic adjustment of a cold front. (a) Initial state. (b) Final state is in dynamic equilibrium, which is never quite attained in the real atmosphere.

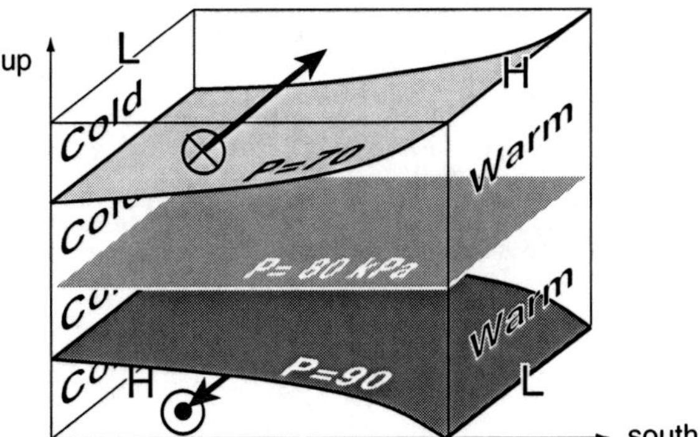

Figure 12.18
Final dynamic state after geostrophic adjustment within an environment containing continuous temperature gradients. Arrows represent geostrophic wind. Shaded areas are isobaric surfaces. H and L indicate high and low pressures relative to surrounding pressures at the same altitude. Dot-circle represents the tip of an arrow pointing toward the reader, x-circle represents the tail feathers of an arrow pointing into the page.

HIGHER MATH • Geostrophic Adjustment

We can verify that the near-surface (frictionless) geostrophic wind is consistent with the sloping depth of cold air. The geostrophic wind is related to the horizontal pressure gradient at the surface by:

$$U_g = -\frac{1}{\rho \cdot f_c} \cdot \frac{\Delta P}{\Delta y} \qquad (10.26a)$$

Assume that the pressure at the top of the cold airmass in Figs. 12.17 equals the pressure at the same altitude h in the warm airmass. Thus, surface pressures will be different due to only the difference in weight of air below that height.

Going from the top of the sloping cold airmass to the bottom, the vertical increase of pressure is given by the hydrostatic eq: $\Delta P_{cold} = -\rho_{cold} \cdot |g| \cdot h$

A similar equation can be written for the warm air below h. Thus, at the surface, the difference in pressures under the cold and warm air masses is:

$$\Delta P = -|g| \cdot h \cdot (\rho_{cold} - \rho_{warm})$$

multiplying the RHS by $\overline{\rho} / \overline{\rho}$, where $\overline{\rho}$ is an average density, yields:

$$\Delta P = -|g| \cdot h \cdot \overline{\rho} \cdot [(\rho_{cold} - \rho_{warm}) / \overline{\rho}]$$

As was shown in the Buoyancy section of the Stability chapter, use the ideal gas law to convert from density to virtual temperature, remembering to change the sign because the warmer air is less dense. Also, $\Delta T_v = \Delta\theta_v$. Thus: $\Delta P = -|g| \cdot h \cdot \overline{\rho} \cdot [(\theta_{v\ warm} - \theta_{v\ cold}) / \overline{T_v}]$ where this is the pressure change in the negative y direction.

Plugging this into eq. (10.26a) gives:

$$U_g = -\frac{|g| \cdot (\Delta\theta_v / T_v)}{f_c} \cdot \frac{\Delta h}{\Delta y}$$

which we can write in differential form:

$$\boxed{U_g = -\frac{|g| \cdot (\Delta\theta_v / T_v)}{f_c} \cdot \frac{\partial h}{\partial y}} \qquad (b)$$

The equilibrium value of h was given by eq. (12.7):

$$h = H \cdot \left[1 - \exp\left(-\frac{y+a}{a} \right) \right] \qquad (12.7)$$

Thus, the derivative is:

$$\frac{\partial h}{\partial y} = \frac{H}{a} \cdot \exp\left(-\frac{y+a}{a} \right)$$

Plugging this into eq. (b) gives:

$$U_g = -\frac{|g| \cdot (\Delta\theta_v / T_v)}{f_c} \cdot \frac{H}{a} \cdot \exp\left(-\frac{y+a}{a} \right) \qquad (c)$$

But from eq. (12.5) we see that:

$$f_c \cdot a = f_c \cdot \lambda_R = \sqrt{|g| \cdot H \cdot (\Delta\theta_v / T_v)} \qquad (12.5)$$

Eq. (c) then becomes:

$$\boxed{U_g = -\sqrt{|g| \cdot H \cdot (\Delta\theta_v / T_v)} \cdot \exp\left(-\frac{y+a}{a} \right)} \qquad (12.6)$$

Thus, the wind is consistent (i.e., geostrophically balanced) with the sloping height.

That other process is the action of cyclones and Rossby waves. Many small-scale, short-lived cyclones are not in geostrophic balance, and they can act to move the cold air further south, and the warm air further north. These cyclones feed off the potential energy remaining in the large-scale flow, namely, the energy associated with horizontal temperature gradients. Such gradients have potential energy that can be released when the colder air slides under the warmer air (see Fig. 11.1).

Sample Application

A cold airmass of depth 1 km and virtual potential temperature 0°C is imbedded in warm air of virtual potential temperature 20°C. Find the Rossby deformation radius, the maximum geostrophic wind speed, and the equilibrium depth of the cold airmass at $y = 0$. Assume $f_c = 10^{-4}\ s^{-1}$.

Find the Answer

Given: $\overline{T_v} = 280$ K, $\Delta\theta_v = 20$ K, $H = 1$ km, $f_c = 10^{-4}\ s^{-1}$.

Find: $a = ?$ km, $U_g = ?$ m s^{-1} at $y = -a$, and $h = ?$ km at $y = 0$.

Use eq. (12.5):

$$a = \frac{\sqrt{(9.8\text{m}\cdot\text{s}^{-2})\cdot(1000\text{m})\cdot(20\text{K})/(280\text{K})}}{10^{-4}\text{s}^{-1}}$$

= **265 km**

Use eq. (12.6):

$$U_g = -\sqrt{(9.8\text{m}\cdot\text{s}^{-2})(1000\text{m})(20\text{K})/(280\text{K})} \cdot \exp(0)$$

= **–26.5 m s^{-1}**

Use eq. (12.7):

$$h = (1\text{km}) \cdot [1 - \exp(-1)] = \underline{\mathbf{0.63\ km}}$$

Check: Units OK. Physics OK.

Exposition: Frontal-zone widths on the order of a = 200 km are small compared to lengths (1000s km).

Winds in the Warm Over-riding Air

Across the frontal zone is a stronger-than-background horizontal temperature gradient. In many fronts, the horizontal temperature gradient is strongest near the surface, and weakens with increasing altitude.

The thermal-wind relationship tells us that the geostrophic wind will increase with height in strong horizontal temperature gradients. If the frontal zone extends vertically over a large portion of the troposphere, then the wind speed will continue to increase with height, reaching a maximum near the tropopause.

Thus, jet streams are associated with frontal zones. The jet blows parallel to the frontal zone, with greatest wind speeds on the warm side of the frontal zone. If the cold air is advancing as a cold front, then this jet is known as a **pre-frontal jet**.

This is illustrated in Fig. 12.19. Plotted are isentropes, isobars, isotachs, and the frontal zone. The cross-frontal direction is north-south in this figure, causing a pre-frontal jet from the West (blowing into the page, in this diagram).

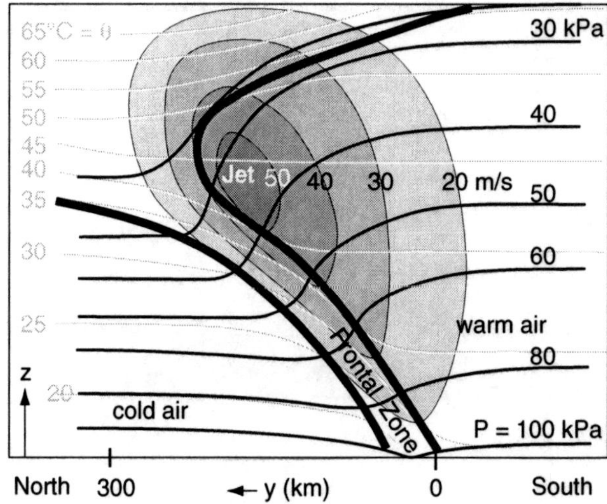

Figure 12.19
Vertical section across a cold front in the N. Hemisphere. Thick lines outline the frontal zone in the troposphere, and show the tropopause in the top of the graph. Medium black lines are isobars. Thin grey lines are isentropes. Shaded areas indicate isotachs, for a jet that blows from the West (into the page).

Frontal Vorticity

Combining the cold-air-side winds from Fig. 12.17b and warm-air-side winds from Fig. 12.19 into a single diagram yields Fig. 12.20. In this sketch, the warm air aloft and south of the front has geostrophic winds from the west, while the cold air near the ground has geostrophic winds from the east.

Thus, across the front, $\Delta U_g/\Delta y$ is negative, which means that the relative vorticity (eq. 11.20) of the geostrophic wind is positive (ζ_r = +) at the front (grey curved arrows in Fig. 12.20). In fact, cyclonic vorticity is found along fronts of any orientation. Also, stronger density contrasts across fronts cause greater positive vorticity.

Also, frontogenesis (strengthening of a front) is often associated with horizontal convergence of air from opposite sides of the front (see next section). Horizontal convergence implies vertical divergence (i.e., vertical stretching of air and updrafts) along the front, as required by mass continuity. But stretching increases vorticity (see the chapters on Atmos. Stability, General Circulation, and Extratropical Cyclones). Thus, frontogenesis is associated with updrafts (and associated clouds and bad weather) and with increasing relative vorticity.

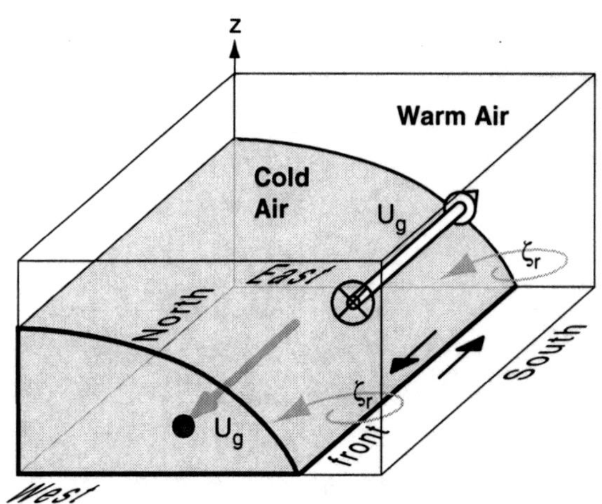

Figure 12.20
Sketch of cold front, combining winds in the cold air (thick straight grey arrow) from Fig. 12.17b with the winds in the warm air (thick white arrow) from Fig. 12.19. These winds cause shear across the front, shown with the black arrows. This shear is associated with positive (cyclonic) relative vorticity of the geostrophic wind (thin curved grey arrows).

INFO • The Polar Front

Because of Coriolis force, cold arctic air cannot spread far from the poles, causing a quasi-permanent frontal boundary in the winter hemisphere. This is called the **polar front**.

It has a wavy irregular shape where some segments advance as cold fronts, other segments retreat as warm fronts, some are stationary, and others are weak and cause gaps in the front.

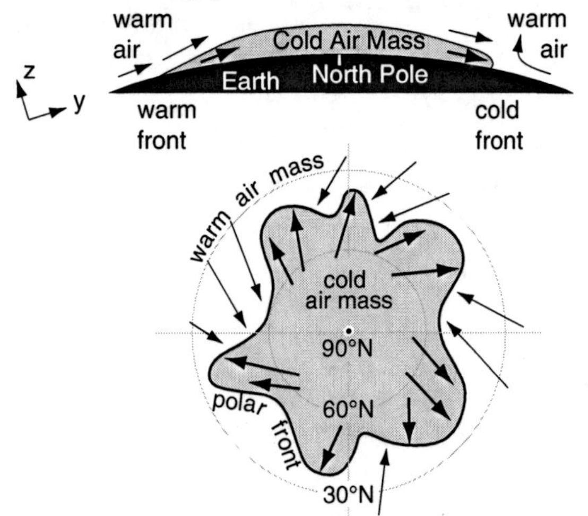

Figure a. *Top: Vertical slice through the North Pole. Bottom: View from space looking down on N. Pole.*

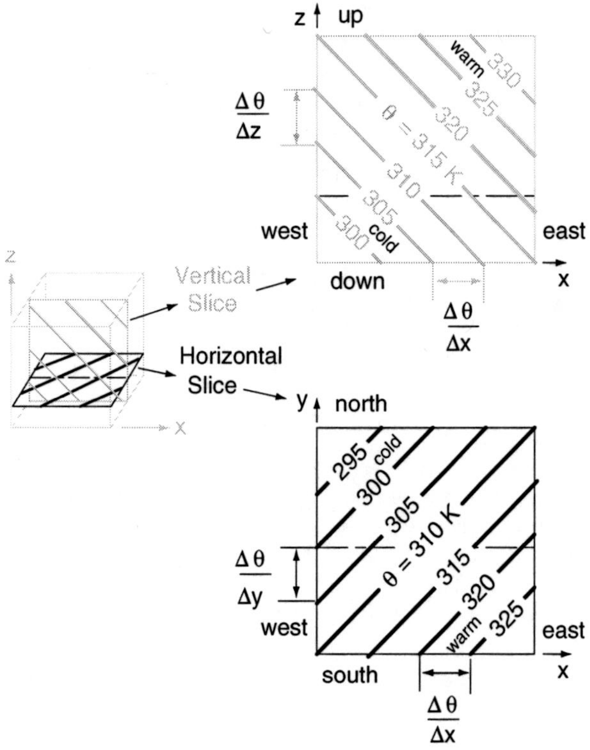

Figure 12.21
Vertical and horizontal slices through a volume of atmosphere, showing initial conditions prior to frontogenesis. The thin horizontal dashed lines show where the planes intersect.

FRONTOGENESIS

Fronts are recognized by the change in temperature across the frontal zone — greatest at the surface. Hence, the horizontal temperature gradient (temperature change per distance across the front) is one measure of frontal strength. Usually potential temperature is used instead of temperature to simplify the problem when vertical motions can occur.

Physical processes that tend to increase the potential-temperature gradient are called **frontogenetic** — literally they cause the birth or strengthening of the front. Three classes of such processes are kinematic, thermodynamic, and dynamic.

Kinematics

Kinematics refers to motion or advection, with no regard for driving forces. This class of processes cannot create potential-temperature gradients, but it can strengthen or weaken existing gradients.

From earlier chapters, we saw that radiative heating causes north-south temperature gradients between the equator and poles. Also, the general circulation causes the jet stream to meander, which creates transient east-west temperature gradients along troughs and ridges. The standard-atmosphere also has a vertical gradient of potential temperature in the troposphere (θ increases with z). Thus, it is fair to assume that temperature gradients often exist, which could be strengthened during kinematic frontogenesis.

To illustrate kinematic frontogenesis, consider an initial potential-temperature field with uniform gradients in the x, y, and z directions, as sketched in Fig. 12.21. The gradients have the following signs (for this particular example):

$$\frac{\Delta\theta}{\Delta x} = + \qquad \frac{\Delta\theta}{\Delta y} = - \qquad \frac{\Delta\theta}{\Delta z} = + \tag{12.8}$$

Namely, potential temperature increases toward the east, decreases toward the north, and increases upward. There are no fronts in this picture initially.

We will examine the subset of advections that tends to create a cold front aligned north-south. Define the strength of the front as the potential-temperature gradient across the front:

$$\text{Frontal Strength} = FS = \frac{\Delta\theta}{\Delta x} \qquad \bullet(12.9)$$

The change of frontal strength with time due to advection is given by the kinematic frontogenesis equation:

•(12.10)

$$\frac{\Delta(FS)}{\Delta t} = -\left(\frac{\Delta\theta}{\Delta x}\right)\cdot\left(\frac{\Delta U}{\Delta x}\right) - \left(\frac{\Delta\theta}{\Delta y}\right)\cdot\left(\frac{\Delta V}{\Delta x}\right) - \left(\frac{\Delta\theta}{\Delta z}\right)\cdot\left(\frac{\Delta W}{\Delta x}\right)$$

Strengthening Confluence Shear Tilting

Confluence

Suppose there is a strong west wind U approaching from the west, but a weaker west wind departing at the east (Fig. 12.22 top). Namely, the air from the west almost catches up to air in the east.

For this situation, $\Delta U/\Delta x = -$, and $\Delta\theta/\Delta x = +$ in eq. (12.10). Hence, the product of these two terms, when multiplied by the negative sign attached to the confluence term, tends to strengthen the front [$\Delta(FS)/\Delta t = +$]. In the shaded region of Fig. 12.22, the isentropes are packed closer together; namely, it has become a frontal zone.

Shear

Suppose the wind from the south is stronger on the east side of the domain than the west (Fig. 12.23 top). This is one type of wind shear. As the isentropes on the east advect northward faster than those on the west, the potential-temperature gradient is strengthened in-between, creating a frontal zone.

While the shear is positive ($\Delta V/\Delta x = +$), the northward temperature gradient is negative ($\Delta\theta/\Delta y = -$). Thus, the product is positive when the preceding negative sign from the shear term is included. Frontal strengthening occurs for this case [$\Delta(FS)/\Delta t = +$].

Tilting

If updrafts are stronger on the cold side of the domain than the warm side, then the vertical potential-temperature gradient will be tilted into the horizontal. The result is a strengthened frontal zone (Fig. 12.24).

The horizontal gradient of updraft velocity is negative in this example ($\Delta W/\Delta x = -$), while the vertical potential-temperature gradient is positive ($\Delta\theta/\Delta z = +$). The product, when multiplied by the negative sign attached to the tilting term, yields a positive contribution to the strengthening of the front for this case [$\Delta(FS)/\Delta t = +$].

While this example was contrived to illustrate frontal strengthening, for most real fronts, the tilting term causes weakening. Such **frontolysis** is weakest near the surface because vertical motions are smaller there (the wind cannot blow through the ground).

Tilting is important and sometimes dominant for upper-level fronts, as described later.

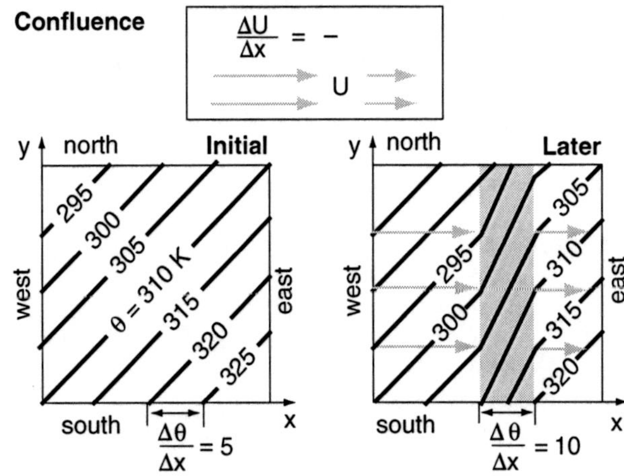

Figure 12.22
Confluence strengthens the frontal zone (shaded) in this case. Arrow tails indicate starting locations for the isotherms.

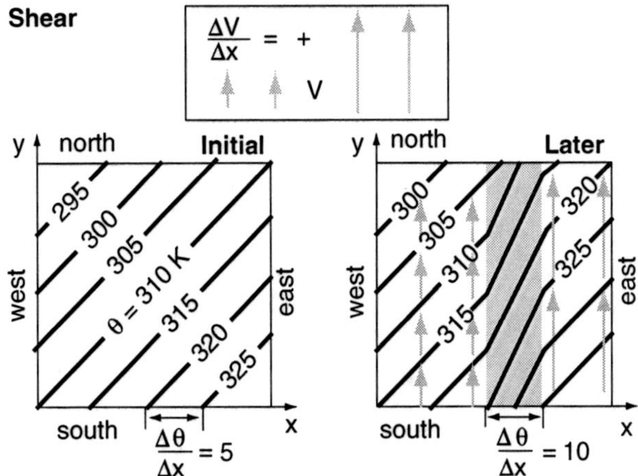

Figure 12.23
Shear strengthens the frontal zone (shaded) in this case. Arrow tails indicate starting locations for the isotherms.

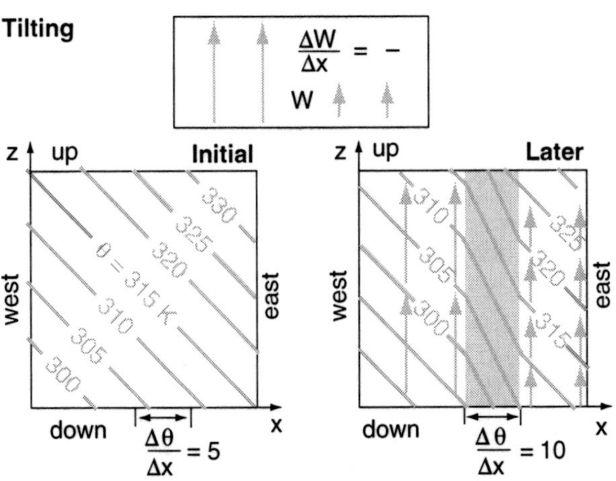

Figure 12.24
Tilting of the vertical temperature gradient into the horizontal strengthens the frontal zone (shaded) in this illustration.

Sample Application

Given an initial environment with $\Delta\theta/\Delta x$ = 0.01°C km^{-1}, $\Delta\theta/\Delta y$ = –0.01°C km^{-1}, and $\Delta\theta/\Delta z$ = 3.3°C km^{-1}. Also, suppose that $\Delta U/\Delta x$ = – 0.05 (m/s) km^{-1}, $\Delta V/\Delta x$ = 0.05 (m/s) km^{-1}, and $\Delta W/\Delta x$ = 0.02 (cm/s) km^{-1}. Find the kinematic frontogenesis rate.

Find the Answer

Given: (see above)

Find: $\Delta(FS)/\Delta t$ = ? °C·km^{-1}·day^{-1}

Use eq. (12.10):

$$\frac{\Delta(FS)}{\Delta t} = -\left(0.01\frac{°C}{km}\right)\cdot\left(-0.05\frac{m/s}{km}\right) - \left(-0.01\frac{°C}{km}\right)\cdot \left(0.05\frac{m/s}{km}\right) - \left(3.3\frac{°C}{km}\right)\cdot\left(0.0002\frac{m/s}{km}\right)$$

= +0.0005 + 0.0005 – 0.00066 °C·m·s^{-1}·km^{-2}

= **+0.029 °C·km^{-1}·day^{-1}**

Check. Units OK. Physics OK.

Exposition: Frontal strength $\Delta\theta/\Delta x$ nearly tripled in one day, increasing from 0.01 to 0.029 °C km^{-1}.

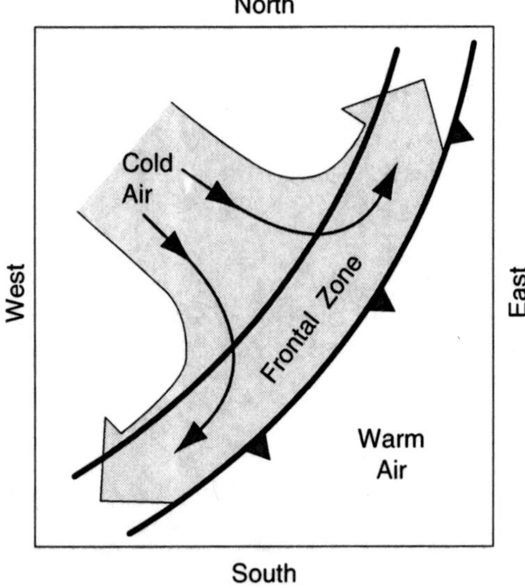

Figure 12.25

An illustration of the near-surface horizontal air-flow pattern at a frontal zone.

Deformation

The previous figures presented idealized kinematic scenarios. Often in real fronts the flow field is a more complex combination of scenarios. For example, Fig. 12.25 shows a **deformation** (change of shape) flow field in the cold air, with **confluence** (→ ← coming together horizontally) of air perpendicular to the front, and **diffluence** (← → horizontal spreading of air) parallel to the front.

In such a flow field both convergence and shear affect the temperature gradient. For example, consider the two identical deformation fields in Figs. 12.26a & b, where the only difference is the angle of the isentropes in a frontal zone relative to the **axis of dilation** (the line toward which confluence points, and along which diffluence spreads).

For initial angles less than 45° (Fig. 12.26a & a'), the isentropes are pushed closer together (frontogenesis) and tilted toward a shallower angle. For initial angles greater than 45°, the isentropes are spread farther apart (frontolysis) and tilted toward a shallower angle. Using this info to analyze Fig. 12.25 where the isentropes are more-or-less parallel to the frontal zone (i.e., initial angle << 45°), we would expect that flow field to cause frontogenesis.

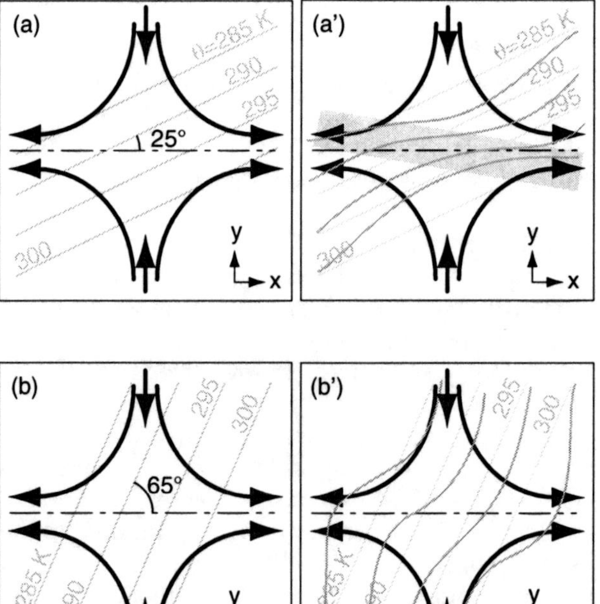

Figure 12.26

Black arrows are wind, and grey arrows are isentropes. Dashed line is the axis of dilation. (a) and (a') are initial and final states for shallow initial angle, showing frontogenesis. The shaded area in (a') highlights the strengthened frontal zone. (b) and (b') are initial and final states for steep initial angle, showing frontolysis. (after J. Martin, 2006: "Mid-Latitude Atmospheric Dynamics: A First Course. Wiley.)

Thermodynamics

The previous kinematic examples showed **adiabatic** advection (potential temperature was conserved while being blown with the wind). However, **diabatic** (non-adiabatic) thermodynamic processes can heat or cool the air at different rates on either side of the domain. These processes include radiative heating/cooling, conduction from the surface, turbulent mixing across the front, and latent heat release/absorption associated with phase changes of water in clouds.

Define the diabatic warming rate (DW) as:

$$\text{Diabatic Warming Rate} = DW = \frac{\Delta\theta}{\Delta t} \qquad \bullet(12.11)$$

If diabatic heating is greater on the warm side of the front than the cold side, then the front will be strengthened:

$$\frac{\Delta(FS)}{\Delta t} = \frac{\Delta(DW)}{\Delta x} \qquad \bullet(12.12)$$

In most real fronts, **turbulent mixing** between the warm and cold sides weakens the front (i.e., causes **frontolysis**). **Conduction** from the surface also contributes to frontolysis. For example, behind a cold front, the cold air blows over a usually-warmer surface, which heats the cold air (i.e., airmass modification) and reduces the temperature contrast across the front. Similarly, behind warm fronts, the warm air is usually advecting over cooler surfaces.

Over both warm and cold fronts, the warm air is often forced to rise. This rising air can cause **condensation** and cloud formation, which strengthen fronts by warming the already-warm air.

Radiative cooling from the tops of stratus clouds reduces the temperature on the warm side of the front, contributing to frontolysis of warm fronts. Radiative cooling from the tops of post-cold frontal stratocumulus clouds can strengthen the front by cooling the already-cold air.

Sample Application

A thunderstorm on the warm side of a 200 km wide front rains at 2 mm h^{-1}. Find the frontogenesis rate.

Find the Answer

Given: $RR = 0$ at $x = 0$,
and $RR = 2$ mm h^{-1} at $x = 200$ km
Find: $\Delta(FS)/\Delta t$ = ? °C·km^{-1}·day^{-1}

From the Heat chapter:
$\Delta\theta/\Delta t = (0.33°\text{C mm}^{-1})\cdot RR = 0.66°\text{C h}^{-1}$.

Combine eqs. (12.11) & (12.12): $\Delta(FS)/\Delta t = \Delta(\Delta\theta/\Delta t)/\Delta x$

$\Delta(FS)/\Delta t = [(0.66°\text{C h}^{-1})\cdot(24 \text{ h day}^{-1}) - 0] / [200 \text{ km} - 0]$
$= (15.84°\text{C day}^{-1})/(200\text{km}) =$ **0.079 °C·km^{-1}·day^{-1}**

Check: Units OK. Physics OK. Magnitude good.
Exposition: This positive value indicates thermodynamic frontogenesis.

Dynamics

Kinematics and thermodynamics are insufficient to explain observed frontogenesis. While kinematic frontogenesis gives doubling or tripling of frontal strength in a day (see previous Sample Applications), observations show that frontal strength can increase by a factor of 15 during a day. Dynamics can cause this rapid strengthening.

Because fronts are long and narrow, we expect along-front flow to tend toward geostrophy, while across front flows could be ageostrophic. We can anticipate this by using the **Rossby number** (see

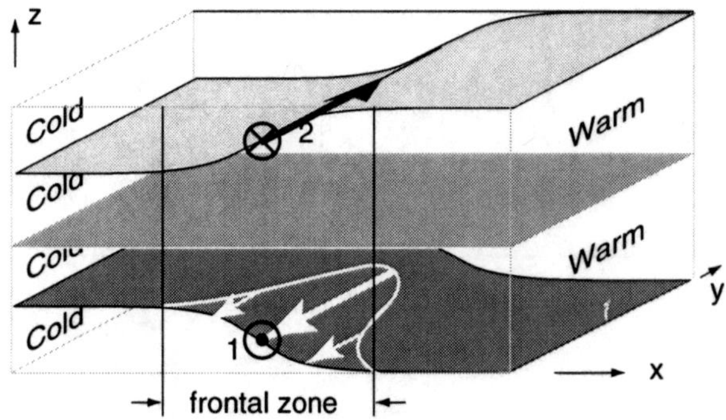

Figure 12.27
Isobaric surfaces near a hypothetical front, before being altered by dynamics. Vectors show equilibrium geostrophic winds, initially (state "o"). Dot-circle (1) represents the tip of an arrow pointing toward the reader; x-circle (2) represents the tail feathers of an arrow pointing into the page.

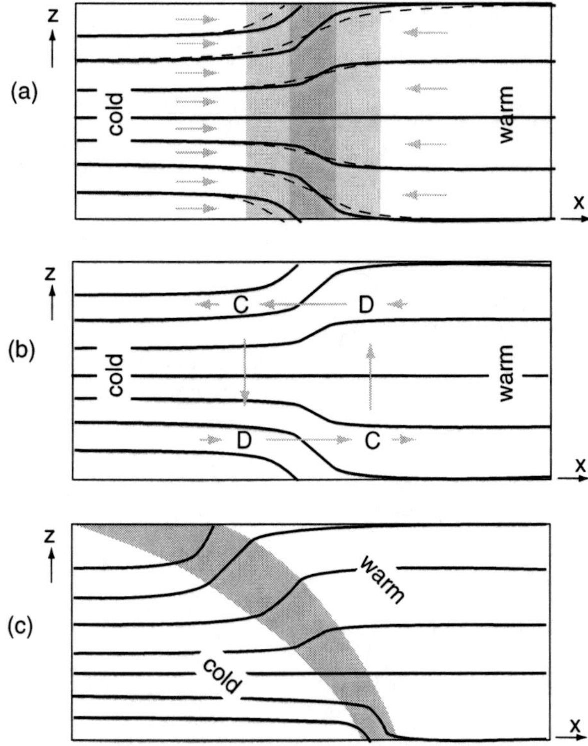

Figure 12.28
Vertical cross section showing dynamic strengthening of a front. The frontal zone is shaded; lines are isobars; and arrows are ageostrophic winds. D and C are regions of horizontal divergence and convergence. (a) Initial state (thin dashed lines) modified (thick lines) by confluence (arrows). (b) Ageostrophic circulation, called a Sawyer-Eliassen circulation. (c) Final equilibrium state.

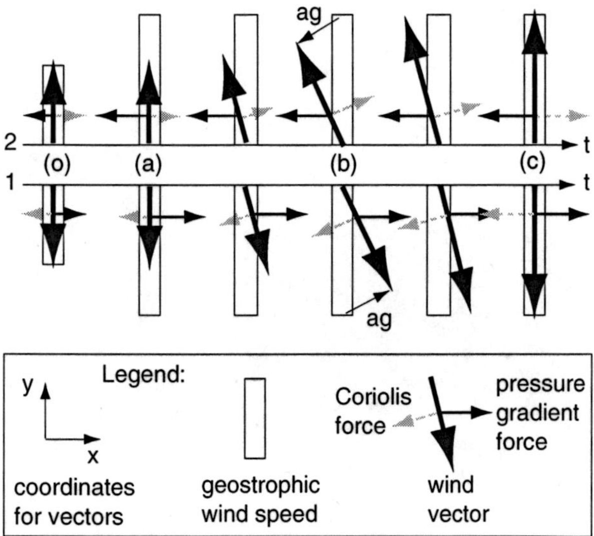

Figure 12.29
Time lines labeled 1 & 2 are for the vectors 1 & 2 in Fig. 12.27. Initial state (o) is given by Fig. 12.27. Later states (a)-(c) correspond to Figs. 12.28(a)-(c). The ageostrophic wind (ag) at time (b) is also indicated. States (o) and (c) are balanced.

INFO Box in the Forces & Winds chapter). Fronts can be of order 1000 km long, but of order 100 km wide. Thus, the Rossby number for along-front flow is of order $Ro =$ 0.1 . But the across-front Rossby number is of order $Ro = 1$. Recall that flows tend toward geostrophy when $Ro < 1$. Thus, ageostrophic dynamics are anticipated across the front, as are illustrated next.

Picture an initial state in geostrophic equilibrium with winds parallel to the front, as sketched in Fig. 12.27. This figure shows a special situation where pressure gradients and geostrophic winds exist only midway between the left and right sides. Zero gradients and winds are at the left and right sides. A frontal zone is in the center of this diagram.

Suppose some external forcing such as kinematic confluence due to a passing Rossby wave causes the front to strengthen a small amount, as sketched in Fig. 12.28a. Not only does the potential-temperature gradient tighten, but the pressure gradient also increases due to the hypsometric relationship.

The increased pressure gradient implies a different, increased geostrophic wind. However, initially, the actual winds are slower due to inertia, with magnitude equal to the original geostrophic speed.

While the actual winds adjust toward the new geostrophic value, they temporarily turn away from the geostrophic direction (Fig. 12.29) due to the imbalance between pressure-gradient and Coriolis forces. During this transient state (b), there is a component of wind in the x-direction (Fig. 12.28b). This is called **ageostrophic flow**, because there is no geostrophic wind in the x-direction.

Because mass is conserved, horizontal convergence and divergence of the U-component of wind cause vertical circulations. These are thermally direct circulations, with cold air sinking and warm air rising and moving over the colder air. The result is a temporary cross-frontal, or **transverse circulation** called a **Sawyer-Eliassen circulation**. The updraft portion of the circulation can drive convection, and cause precipitation.

The winds finally reach their new equilibrium value equal to the geostrophic wind. In this final state, there are no ageostrophic winds, and no cross-frontal circulation. However, during the preceding transient stage, the ageostrophic cross-frontal circulation caused extra dynamic confluence near the surface, which adds to the original kinematic confluence to strengthen the surface front. The transverse circulation also tilts the front (Fig. 12.28c).

In summary, a large and relatively steady geostrophic wind blows parallel to the front (Fig. 12.27). A weak, transient, cross-frontal circulation can be superimposed (Fig. 12.28b). These two factors are also important for upper-tropospheric fronts, as described later.

OCCLUDED FRONTS AND MID-TROPOSPHERIC FRONTS

When three or more airmasses come together, such as in an **occluded front**, it is possible for one or more fronts to ride over the top of a colder airmass. This creates lower- or mid-tropospheric fronts that do not touch the surface, and which would not be signaled by temperature changes and wind shifts at the surface. However, such **fronts aloft** can trigger clouds and precipitation observed at the surface.

Occluded fronts occur when cold fronts catch up to warm fronts. What happens depends on the temperature and static-stability difference between the cold advancing air behind the cold front and the cold retreating air ahead of the warm front.

Fig. 12.30 shows a **cold front occlusion**, where very cold air that is very statically stable catches up to, and under-rides, cooler air that is less statically stable. The warm air that was initially between these two cold airmasses is forced aloft. Most occlusions in interior N. America are of this type, due to the very cold air that advances from Canada in winter.

Observers at the surface would notice stratiform clouds in advance of the front, which would normally signal an approaching warm front. However, instead of a surface warm front, a surface occluded front passes, and the surface temperature decreases like a cold front. The trailing edge of cool air aloft marks the warm front aloft.

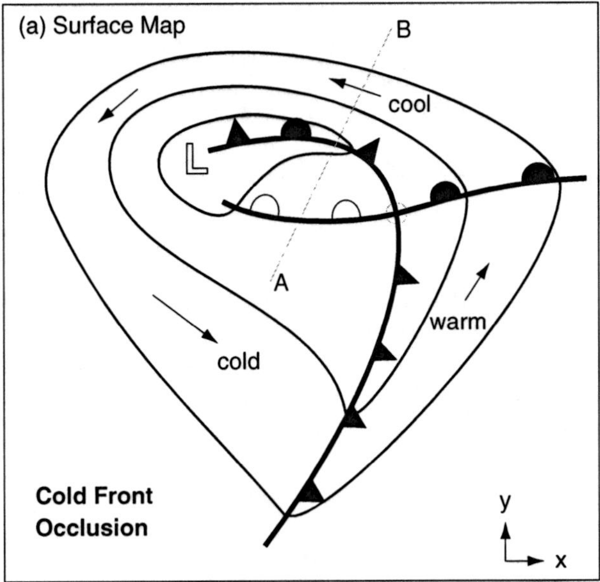

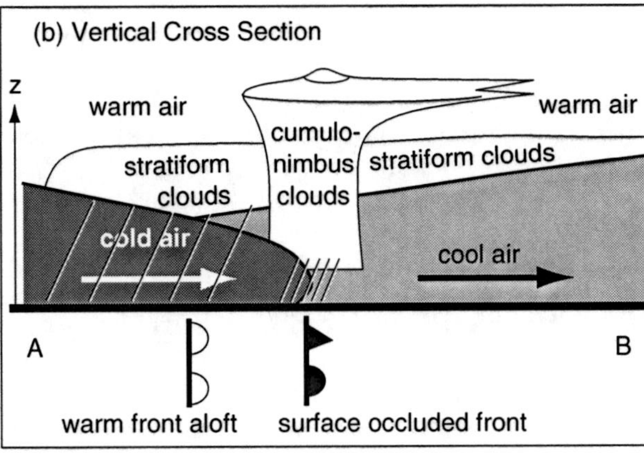

Figure 12.30
Cold front occlusion. (a) Surface map showing position of surface cold front (dark triangles), surface warm front (dark semicircles), surface occluded front (dark triangles and semicircles), and warm front aloft (white semicircles). (b) Vertical cross section along slice A-B from top diagram. Diagonal lines = rain.

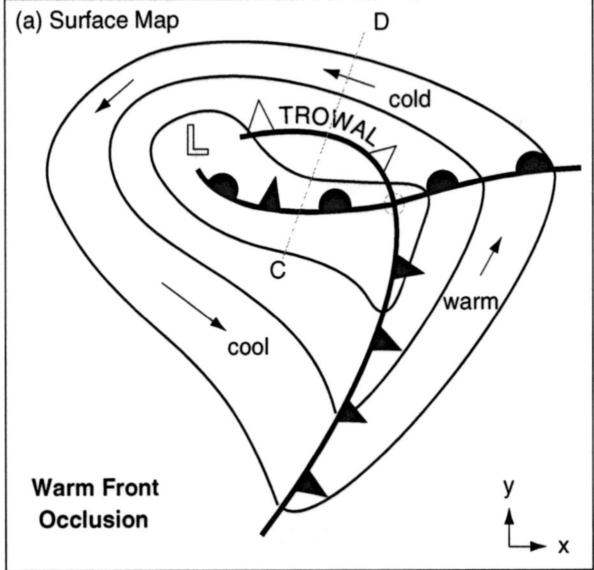

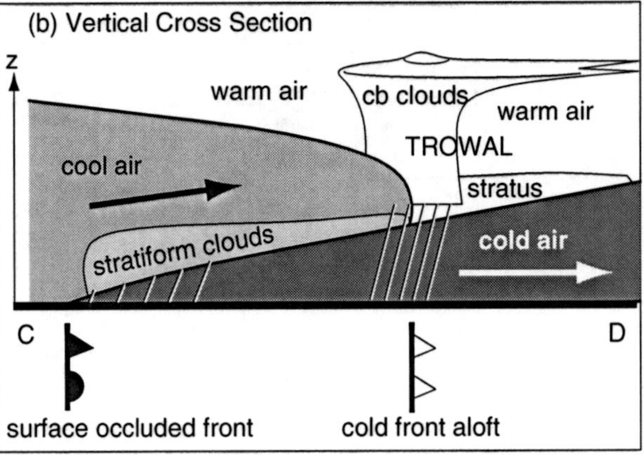

Figure 12.31
Warm front occlusion. (a) Surface map. Symbols are similar to Fig. 12.30, except that white triangles denote a cold front aloft. (b) Vertical cross section along slice C-D from top diagram. TROWAL = trough of warm air aloft. The open grey circle shows the triple-point location (meeting of 3 surface fronts).

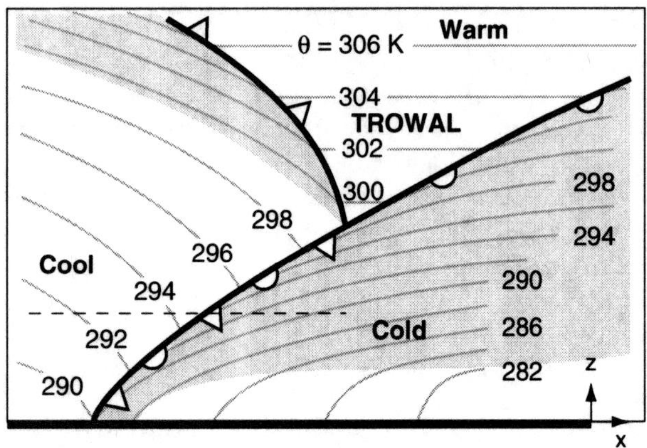

Figure 12.32
Idealized vertical cross-section through a warm front occlusion, showing the bottom 5 km of the troposphere. Dark grey lines are isentropes. Light grey shading indicates strong static stability (regions where the isentropes are packed closer together in the vertical). TROWAL = trough of warm air aloft.

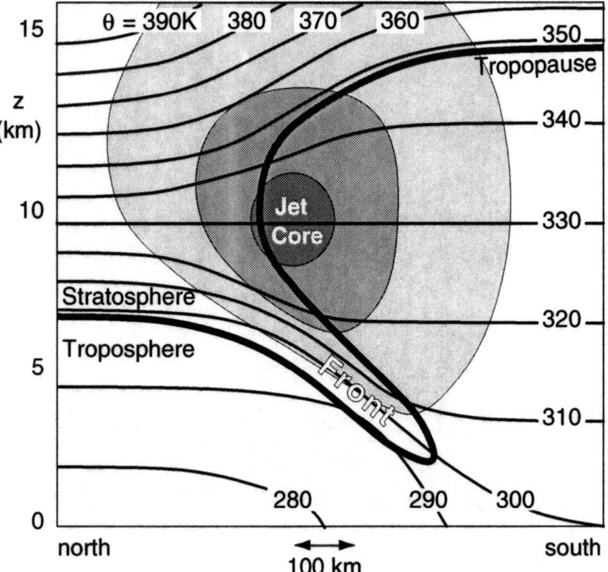

Figure 12.33
Idealized vertical cross-section through the jet stream in N. Hemisphere. Shading indicates west-to-east speeds (into the page): light ≥ 50 m s^{-1}, medium ≥ 75 m s^{-1}, dark ≥ 100 m s^{-1}. Thin lines are isentropes (K). Thick line is the tropopause. "Front" indicates the upper-tropospheric front, where stratospheric air is penetrating into the troposphere.

The location in Figs. 12.30 and 12.31 where the cold, warm, and occluded fronts intersect at the surface is called the **triple point**.

Fig. 12.31 shows a **warm front occlusion**, where cool air that is less statically stable catches up to, and over-rides, colder air that is more statically stable, forcing aloft the warm air that was in-between. Most occlusions in Europe and the Pacific Northwest USA are this type, due to mild cool air that advances from over the cool ocean during winter.

Observers at the surface notice stratiform clouds in advance of the front. But before the surface front arrives, there can be showers or thunderstorms associated with the cool front aloft. Later, a surface occluded front passes with widespread drizzle, and the surface temperature warms.

Fig. 12.32 shows how the static stability relates to the type of occlusion. In this sketch of a warm front occlusion, look at the static stability across the occluded front, at the altitude of the dashed line. To the east of the front in the cold airmass, the air is strongly stable (as shown by the tight packing of the isentropes in the vertical). To the west in the cool airmass, the air is less statically stable (i.e., greater vertical spacing between isentropes). If this had been a cold frontal occlusion instead, the greatest static stability would have been west of the front.

The wedge of warm air (Fig. 12.32) pushed up between the cool and cold airmasses is called a **TROWAL**, an acronym that means "**trough of warm air aloft**." This TROWAL, labeled in Fig. 12.31a, touches the ground at the triple point, but tilts toward higher altitudes further north. Under the TROWAL can be significant precipitation and severe weather at the surface (Fig. 12.31b) — hence it is important for weather forecasting.

The previous sketches are "text-book" examples of prototypical situations. In real life, more complex maps and cross sections can occur. Sometimes more than three airmasses can be drawn together in a low, causing multiple cold or warm fronts, and multiple fronts aloft.

UPPER-TROPOSPHERIC FRONTS

Upper-tropospheric fronts are also called **upper-level fronts**, and are sometimes associated with **folds in the tropopause**. A cross-section through an idealized upper-level front is sketched in Fig. 12.33.

In the lower troposphere, the south-to-north temperature gradient creates the jet stream due to the thermal wind relationship (see the General

Circulation chapter). A reversal of the meridional temperature gradient above 10 km in this idealized sketch causes wind velocities to decrease above that altitude. Within the stratosphere, the isentropes are spaced closer together, indicating greater static stability.

South of the core of the jet stream, the tropopause is relatively high. To the north, it is lower. Between these two extremes, the tropopause can wrap around the jet core, and fold back under the jet, as sketched in Fig. 12.33. Within this fold the isentropes are tightly packed, indicating an **upper-tropospheric front**. Sometimes the upper-level front connects with a surface front (Fig. 12.19).

Stratospheric air has unique characteristics that allow it to be traced. Relative to the troposphere, stratospheric air has high ozone content, high radioactivity (due to former nuclear bomb testing), high static stability, low water-vapor mixing ratio, and high **isentropic potential vorticity**.

The tropopause fold brings air of stratospheric origin down into the troposphere. This causes an injection of ozone and radioactivity into the troposphere. The dry air in the tropopause fold behind the cold front is often clearly visible in water-vapor satellite images of the upper troposphere.

The heavy line in Fig. 12.33 corresponds roughly to the 1.5 PVU (potential vorticity units) isopleth, and is a good indicator of the tropopause. Above this line, PVU values are greater than 1.5, and increase rapidly with height to values of roughly 10 PVU at the top of Fig. 12.33. Below this line, the PVU gradient is weak, with typical values of about 0.5 PVU at mid-latitudes. PVU is negative in the Southern Hemisphere, but of similar magnitude.

Sawyer-Eliassen dynamics help create tropopause folds. Picture a tropopause that changes depth north to south, as in Figs. 11.35 or 11.37 of the General Circulation chapter, but does not yet have a fold. If the jet core advects colder air into the region, then the thickness between the 10 and 20 kPa isobaric surfaces (Fig. 11.37b) will decrease as described by the hypsometric equation. This will cause greater slope (i.e., greater height gradient) of the 10 kPa surface near the jet core. This temporarily upsets geostrophic balance, causing an ageostrophic circulation above the jet from south to north.

This transient flow continues to develop into a cross jet-stream direct circulation that forms around the jet as sketched in Fig. 12.34 (similar to Fig. 12.28b). Below the fold is an indirect circulation that also forms due to a complex geostrophic imbalance. Both circulations distort the tropopause to produce the fold, as sketched in Fig. 12.34, and strengthen the upper-level front due to the kinematic tilting term.

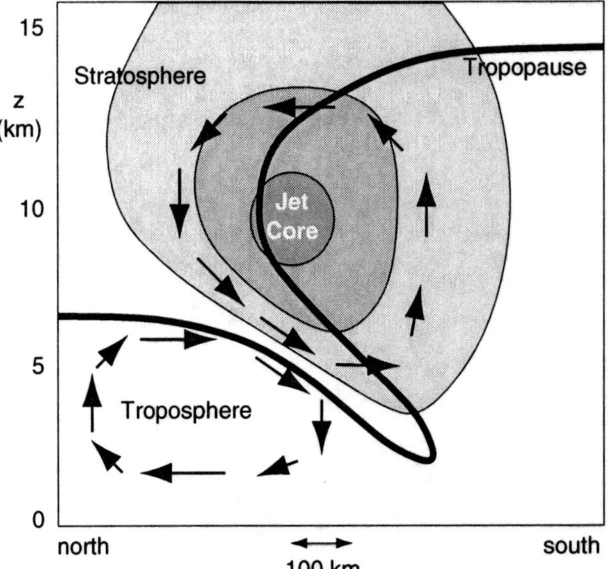

Figure 12.34
Transverse (cross-jet-stream) circulations (arrows) that dynamically form an upper-tropospheric front (i.e., a tropopause fold).

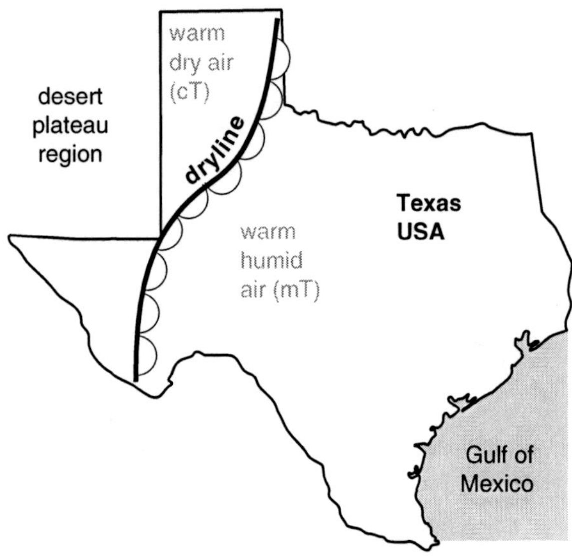

Figure 12.35
Typical location of dryline in the southwestern USA.

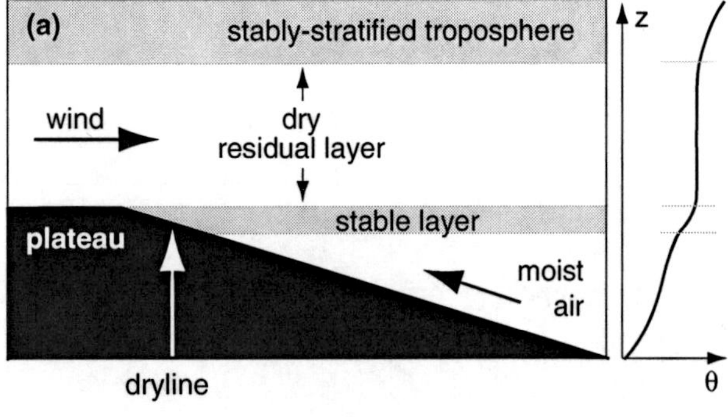

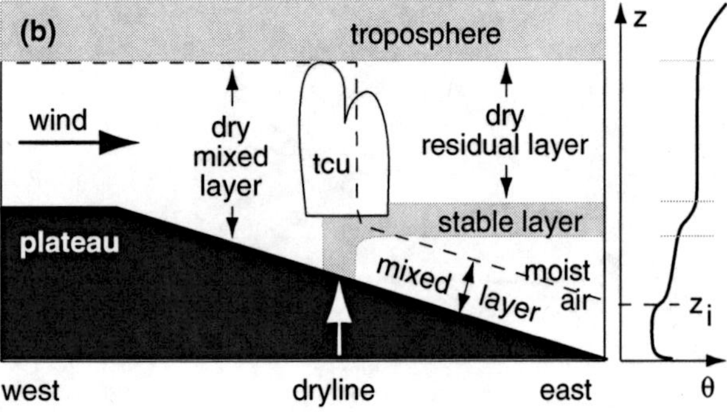

Figure 12.36
Idealized vertical cross section through a dryline (indicated by white arrow). Winds are indicated with black arrows. (a) Early morning. (b) Mid afternoon, with a convective mixed layer of thickness z_i. tcu = towering cumulus clouds. At right of each figure is the potential temperature θ sounding for the east side of the domain.

DRYLINES

In western Texas, USA, there often exists a boundary between warm humid air to the east, and warm dry air to the west (Fig. 12.35). This boundary is called the **dryline**. Because the air is hot on both sides of the boundary, it cannot be labeled as a warm or cold front. The map symbol for a dryline is like a warm front, except with open semicircles adjacent to each other, pointing toward the moist air.

Moist air on the east side comes from the Gulf of Mexico, while the dry air comes from the semi-arid plateaus of Mexico and the Southwestern USA. Drylines are observed during roughly 40 to 50% of all days in spring and summer, in that part of the world. Similar drylines are observed in other parts of the world.

During midday through afternoon, convective clouds are often triggered along the drylines. Some of these cloud bands grow into organized thunderstorm squall lines that can propagate east from the dryline.

Drylines need not be associated with a wind shift, convergence, vorticity, nor with low pressure. Hence, they do not satisfy the definition of a front. However, sometimes drylines combine with troughs to dynamically contribute to cyclone development.

Drylines tend to move eastward during the morning, and return westward during evening. This diurnal cycle is associated with the daily development of a convective mixed layer, interacting with the sloping terrain, via the following mechanism.

Over the high plateaus to the west, deep hot turbulent mixed layers can form in the late afternoon. By night and early morning, after turbulence decays, this thick dry layer of hot air over the plateau becomes a non-turbulent residual layer with nearly adiabatic lapse rate (Fig. 12.36a). Westerly winds advect this dry air eastward, where it overrides cooler, humid air at lower altitudes. Between that elevated residual layer and the moister air below is a strong stable layer such as a temperature inversion.

After sunrise, the warm ground causes a convective mixed-layer to form and grow. The mixed-layer depth is shown by the dashed line in Fig. 12.36b. Because the ground is higher in the west, the top of the mixed layer is able to reach the dry residual-layer air in the west earlier than in the east. When this dry air is entrained downward into the air below, it causes the surface humidity to drop.

Meanwhile, further east, the top of the mixed layer has yet to reach the dry air aloft, so the humidity is still high there. The dryline separates this moist surface air in the east from the dry air to the west.

As the day continues, the mixed layer in the moist air deepens, allowing the region of dry-air entrainment to spread eastward. This causes the dryline to move eastward as the day progresses. At night, convective turbulence ceases, a stable boundary layer develops, and prevailing low-altitude easterlies advect moist air back toward the west (Fig. 12.36a).

If we assume that initially the inversion is level with respect to sea level, then using geometry, we can see that the progression speed of the dryline eastward $\Delta x/\Delta t$ is directly related to the rate of growth of the mixed layer $\Delta z_i/\Delta t$:

$$\frac{\Delta x}{\Delta t} = \frac{\Delta z_i \,/\, \Delta t}{s} \tag{12.13}$$

where $s = \Delta z/\Delta x$ is the terrain slope.

Suppose the early morning stable boundary layer has a linear profile of potential temperature $\gamma = \Delta\theta/\Delta z$, then the mixed-layer growth proceeds with the square-root of accumulated daytime heating Q_{Ak} (see the Atmospheric Boundary Layer chapter):

$$z_i = \left(\frac{2 \cdot Q_{Ak}}{\gamma}\right)^{1/2} \tag{12.14}$$

As a result, the distance Δx that a dryline moves eastward as a function of time is:

$$\Delta x = \frac{1}{s} \cdot \left(\frac{2 \cdot Q_{Ak}}{\gamma}\right)^{1/2} \tag{12.15}$$

Time is hidden in the cumulative heat term Q_{Ak}.

Sample Application (§)

Suppose the surface heat flux during the day is constant with time at 0.2 K·m s^{-1} in the desert southwest USA. The early morning sounding has a potential temperature gradient of γ = 10 K km^{-1}. The terrain slope is 1:400. Plot the dryline position with time.

Find the Answer

Given: γ = 10 K km^{-1}, $s = 1/400$, F_H = 0.2 K·m s^{-1},
Find: $\Delta x(t)$ = ? km.

For constant surface flux, cumulative heating is:

$$Q_{Ak} = F_H \cdot \Delta t$$

where Δt is time since sunrise.

Thus, eq. (12.15) can be rewritten as:

$$\Delta x = \frac{1}{s} \cdot \left(\frac{2 \cdot F_H \cdot \Delta t}{\gamma}\right)^{1/2}$$

$$= \frac{400}{1} \cdot \left[\frac{2 \cdot (0.2\text{K} \cdot \text{m/s}) \cdot \Delta t}{(0.01\text{K/m})}\right]^{1/2}$$

Plotting this with a spreadsheet program:

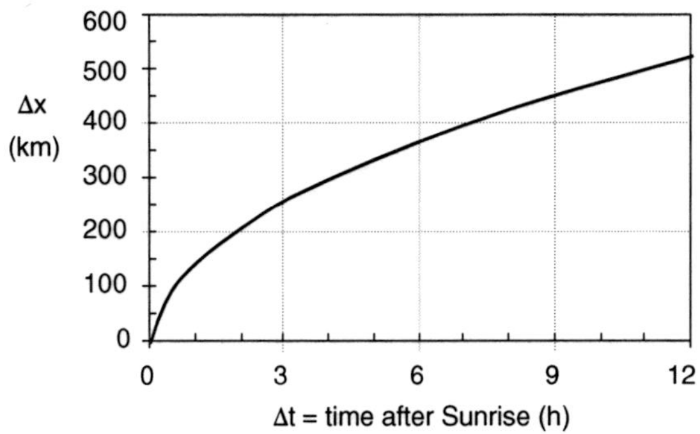

Check: Units OK. Physics OK. Graph reasonable.
Exposition: The dryline moves to the east with the square-root of time. Thus, the further east it gets, the slower it goes.

REVIEW

Surface fronts mark the boundaries between airmasses. Changes in wind, temperature, humidity, visibility, and other meteorological variables are frequently found at fronts. Along frontal zones are often low clouds, low pressure, and precipitation. Fronts rotate counterclockwise around the lows in the Northern Hemisphere as the lows move and evolve.

Fronts strengthen or weaken due to advection by the wind (kinematics), external diabatic heating (thermodynamics), and ageostrophic cross-frontal circulations (dynamics). One measure of frontal strength is the temperature change across it. The spread of cold air behind a front is constrained by Coriolis force, as winds adjust toward geostrophic.

Airmasses can form when air resides over a surface sufficiently long. This usually happens in high-pressure centers of light wind. Airmasses over cold surfaces take longer to form than those over warmer surfaces because turbulence is weaker in the statically stable air over the cold surface. As airmasses move from their source regions, they are modified by the surfaces and terrain over which they flow.

Upper-tropospheric fronts can form as folds in the tropopause. Drylines form over sloping terrain as a boundary between dry air to the west and moist air to the east; however, drylines do not usually possess all the characteristics of fronts.

HOMEWORK EXERCISES

Broaden Knowledge & Comprehension

B1. Monitor the weather maps on the web every day for a week or more during N. Hemisphere winter, and make a sketch of each low center with the fronts extending (similar to Fig. 12.1). Discuss the variety of arrangements of warm, cold, and occluded fronts that you have observed during that week.

B2. Same as B1, but for the S. Hemisphere during S. Hemisphere winter.

B3. On a surface weather map, identify high-pressure centers, and identify ridges. Also, on a 50 kPa (500 hPa) chart, do the same. Look at the ratio of ridges to high centers for each chart, and identify which chart (surface or 50 kPa) has the largest relative number of high centers, and which has the largest relative number of ridges.

B4. From weather maps showing vertical velocity w, find typical values of that vertical velocity near the center of highs. (If $\omega = \Delta P/\Delta t$ is used instead of w, it is acceptable to leave the units in mb s^{-1}). Compare this vertical velocity to the radial velocity of air diverging away from the high center in the boundary layer.

B5. Use a sequence of surface weather maps every 6 or 12 h for a week, where each map spanning a large portion of the globe. Find one or more locations that exhibits the following mechanism for formation of a surface high pressure:

a. global circulation b. monsoon
c. Rossby wave d. thunderstorm
e. topographic

B6. Do a web search to find surface weather maps that also include airmass abbreviations on them. Print one of these maps, and suggest how the labeled airmass moved to its present location from its genesis region (where you will need to make a reasonable guess as to where its genesis region was).

B7. Use weather maps on the web to find a location where air has moved over a region and then becomes stationary. Monitor the development of a new airmass (or equivalently, modification of the old airmass) with time at this location. Look at temperature and humidity.

B8. For a weather situation of relatively zonal flow from west to east over western N. America, access upper air soundings for weather stations in a line more-or-less along the wind direction. Show how the sounding evolves as the air flows over each major mountain range. How does this relate to airmass modification?

B9. Access upper air soundings for a line of RAOB stations across a front. Use this data to draw vertical cross-sections of:

a. pressure or height b. temperature
c. humidity d. potential temperature

B10. Same as previous exercise, but identify the frontal inversions (or frontal stable layers) aloft in the soundings.

B11. Access a sequence of weather maps that cover a large spatial area, with temporal coverage every 6 or 12 h for a week. Find an example of a front that you can follow from beginning (frontogenesis) through maturity to the end (frontolysis). Discuss how the front moves as it evolves, and what the time scale for its evolution is.

B12. Print from the web a weather map that shows both the analyzed fronts and the station plot data. On this map, draw your own analysis of the data to show where the frontal zones are. Discuss how the weather characteristics (wind, pressure, temperature, weather) across these real fronts compare with the idealized sketches of Figs. 12.11 and 12.12.

B13. Access a sequence of surface weather maps from the web that show the movement of fronts. Do you see any fronts that are labeled as stationary fronts, but which are moving? Do you see any fronts labeled as warm or cold fronts, but which are not moving? Are there any fronts that move backwards compared to the symbology labeling the front (i.e., are the fronts moving opposite to the direction that the triangles or semicircles point)? [Hint: often fronts are designated by how they move relative to the low center. If a cold front, for example, is advancing cyclonically around a low, but the low is moving toward the west (i.e., backwards in mid-latitudes), then relative to people on the ground, the cold front is retreating.]

B14. Access weather map data that shows a strong cold front, and compare the winds on both sides of that front to the winds that you would expect using geostrophic-adjustment arguments. Discuss.

B15. For a wintertime situation, access a N. (or S.) Hemisphere surface weather map from the web, or access a series of weather maps from different agencies around the world in order to get information for the whole Hemisphere. Draw the location of the polar front around the globe.

B16. Access weather maps from the web that show a strengthening cold or warm front. Use the other weather data on this map to suggest if the strengthening is due mostly to kinematic, thermodynamic, or dynamic effects.

B17. Access a sequence of weather maps that shows the formation and evolution of an occluded front. Determine if it is a warm or cold occlusion (you might need to analyze the weather map data by hand to help you determine this).

B18. Use the web to access a 3-D sketch of a TROWAL. Use this and other web sites to determine other characteristics of TROWALs.

B19. Download upper-air soundings from a station under or near the jet stream. Use the data to see if there is a tropopause fold. [Hint: assuming that you don't have measurements of radioactivity or isentropic potential vorticity, use mixing ratio or potential temperature as a tracer of stratospheric air.]

B20. Search the web for surface weather maps that indicate a dryline in the S.W. USA. If one exists, then search the web for upper-air soundings just east and just west of the dryline. Plot the resulting soundings of potential temperature and humidity, and discuss how it relates to the idealized sketch of a dryline in this textbook.

Apply

A1. Identify typical characteristics of the following airmass:

a. cAA b. cP c. cT d. cM e. cE
f. mA g. mP h. mT i. mM j. mE
k. cEw l. cAk m. cPw n. mPk o. mTw

A2. List all the locations in the world where the following airmasses typically form.

a. cAA b. cA c. cP d. mP e. mT
f. cT g. mE

A3.(§) Produce a graph of warm airmass depth z_i and airmass potential temperature θ_{ML} as a function of time, similar to Fig. 12.5. Use all the same conditions as in that figure (γ = 3.3 K km^{-1}, $\beta = 10^{-6}$ s^{-1}, initial $\Delta\theta s$ = 10°C; see the Sample Application on the subsequent page) for genesis of a warm airmass, except with the following changes:

a. initial $\Delta\theta s$ = 5°C b. initial $\Delta\theta s$ = 15°C
c. initial $\Delta\theta s$ = 20°C d. initial $\Delta\theta s$ = 15°C

e. $\beta = 0\ s^{-1}$ f. $\beta = 10^{-7}\ s^{-1}$
g. $\beta = 5 \times 10^{-7}\ s^{-1}$ h. $\beta = 5 \times 10^{-6}\ s^{-1}$
i. $\beta = 10^{-5}\ s^{-1}$ j. $\gamma = 5\ K\ km^{-1}$
k. $\gamma = 4\ K\ km^{-1}$ l. $\gamma = 3\ K\ km^{-1}$
m. $\gamma = 2\ K\ km^{-1}$ n. $\gamma = 1\ K\ km^{-1}$

A4. Same as the previous exercise, but find the time scale τ that estimates when the peak airmass depth occurs for warm airmass genesis.

A5. For the katabatic winds of Antarctica, find the downslope buoyancy force per unit mass for this location:

Location	$\Delta z/\Delta x$	$\Delta\theta$ (K)	T_e (K)
a. Interior	0.001	40	233
b. Interior	0.002	35	238
c. Interior	0.003	30	243
d. Intermediate	0.004	27	245
e. Intermediate	0.005	25	248
f. Intermediate	0.006	24	249
g. Intermediate	0.007	23	250
h. Coast	0.008	22	251
i. Coast	0.009	21	252
j. Coast	0.010	20	253
k. Coast	0.011	19	254

A6.(§) An Arctic airmass of initial temperature –30°C is modified as it moves at speed 10 m s^{-1} over smooth warmer surface of temperature 0°C. Assume constant airmass thickness. Find and plot the airmass (mixed-layer) temperature vs. downwind distance for an airmass thickness (m) of:
a. 100 b. 200 c. 300 d. 400 e. 500
f. 600 g. 700 h. 800 i. 900 j. 1000
k. 1200 l. 1400 m. 1500

A7. Find the external Rossby radius of deformation at 60° latitude for a cold airmass of thickness 500 m and $\Delta\theta$ (°C) of:
a. 2 b. 4 c. 6. d. 8 e. 10 f. 12 g. 14
h. 16 i. 18 j. 20 k. 22 l. 24 m. 26 n. 28
Assume a background potential temperature of 300K.

A8.(§) Find and plot the airmass depth and geostrophic wind as a function of distance from the front for the cases of the previous exercise. Assume a background potential temperature of 300 K.

A9. Suppose that $\Delta\theta/\Delta x$ = 0.02°C km^{-1}, $\Delta\theta/\Delta y$ = 0.01°C km^{-1}, and $\Delta\theta/\Delta z$ = 5°C km^{-1}. Also suppose that $\Delta U/\Delta x$ = – 0.03 (m/s) km^{-1}, $\Delta V/\Delta x$ = 0.05 (m/s) km^{-1}, and $\Delta W/\Delta x$ = 0.02 (cm/s) km^{-1}. Find the kinematic frontogenesis contributions from:
a. confluence b. shear c. tilting
d. and find the strengthening rate.

A10. A thunderstorm on the warm side of a 300 km wide front rains at the following rate (mm h^{-1}). Find the thermodynamic contribution to frontogenesis.
a. 0.2 b. 0.4 c. 0.6 d. 0.8 e. 1.0 f. 1.2 g. 1.4
h. 1.6 i. 1.8 j. 2.0 k. 2.5 l. 3.0 m. 3.5 n. 4.0

A11. Plot dryline movement with time, given the following conditions. Surface heat flux is constant with time at kinematic rate 0.2 K·m s^{-1}. The vertical gradient of potential temperature in the initial sounding is γ. Terrain slope is $s = \Delta z/\Delta x$.

	γ (K km^{-1})	s		γ (K km^{-1})	s
a.	8	1/500	b.	8	1/400
c.	8	1/300	d.	8	1/200
e.	12	1/500	f.	12	1/400
g.	12	1/300	h.	12	1/200
i.	15	1/500	j.	15	1/400

Evaluate & Analyze

E1. Would you expect there to be a physically- or dynamically-based upper limit on the number of warm fronts and cold fronts that can extend from a low-pressure center? Why?

E2. Sketch a low with two fronts similar to Fig. 12.1, but for the Southern Hemisphere.

E3. What are the similarities and differences between a ridge and a high-pressure center? Also, on a weather map, what characteristics would you look for to determine if a weather feature is a ridge or a trough?

E4. In high-pressure centers, is the boundary-layer air compressed to greater density as subsidence pushes down on the top of the mixed layer? Explain.

E5. In Fig. 12.3b, use the hypsometric relationship to explain why the ridge shifts or tilts westward with increasing altitude. Hint: consider where warm and cold air is relative to the surface high.

E6. How would Fig. 12.3b be different, if at all, in Southern Hemisphere mid-latitudes. Hint: Consider the direction that air rotates around a surface high, and use that information to describe which side of the high (east or west) advects in warm air and which advects in colder air.

E7. Consider the global-circulation, monsoon, and topographic mechanisms that can form high-pressure regions. Use information from the chapters on General Circulation, Extratropical Cyclones, and Regional Winds to identify 3 or more regions in the world where 2 or more of those mechanisms are superimposed to help create high-pressure regions. Describe the mechanisms at each of those 3 locations, and suggest how high-pressure formation might vary during the course of a year.

E8. Out of all the different attributes of airmasses, why do you think that the airmass abbreviations listed in Table 12-1 focus on only the temperature (i.e., AA, A, P, T, M, E) and humidity (c, m)? In other words, what would make the relative temperature and humidity of airmasses be so important to weather forecasters? Hint: consider what you learned about heat and humidity in earlier chapters.

E9. Fig. 12.4 shows the genesis regions for many different airmasses. At first glance, this map looks cluttered. But look more closely for patterns and describe how you can anticipate where in the world you might expect the genesis regions to be for a particular airmass type.

E10. In this chapter, I described two very different mechanisms by which warm and cold airmasses are created. Why would you expect them to form differently? Hint: consider the concepts, not the detailed equations.

E11. Fig. 12.5 shows how initially-cold air is transformed into a warm airmass after it becomes parked over a warm surface such as a tropical ocean. It is not surprising that the airmass temperature θ_{ML} asymptotically approaches the temperature of the underlying surface. Also not surprising: depth z_i of the changed air initially increases with time.

But more surprising is the decrease in depth of the changed air at longer times. Conceptually, why does this happen? Why would you expect it to happen for most warm airmass genesis?

E12. Using the concept of airmass conservation, formulate a relationship between subsidence velocity at the top of the boundary layer (i.e., the speed that the capping inversion is pushed down in the absence of entrainment) to the radial velocity of air within the boundary layer. (Hint: use cylindrical coordinates, and look at the geometry.)

E13. For the toy model in the INFO Box on Warm Airmass Genesis, ΔQ is the amount of heat (in kinematic units) put into the mixed layer (ML) from the surface during one time step. By repeating over many time steps, we gradually put more and more heat into the ML. This heat is uniformly mixed in the vertical throughout the depth z_i of the ML, causing the airmass to warm.

Yet at any time, such as 10 days into the forecast, the amount of additional heat contained in the ML (which equals its depth times its temperature increase since starting, see Fig. 12.5) is less than the accumulated heat put into the ML from the surface (which is the sum of all the ΔQ from time zero up to 10 days).

That discrepancy is not an error. It describes something physical that is happening. What is the physical process that explains this discrepancy, and how does it work?

E14. In the Sample Application on warm airmass genesis, I iterate with very small time step at first, but later take larger and larger time steps. Why would I need to take small time steps initially, and why can I increase the time step later in the simulation? Also, if I wanted to take small time steps for the whole duration, would that be good or bad? Why?

E15. Suppose that cold air drains katabatically from the center of Antarctica to the edges. Sketch the streamlines (lines that are parallel to the flow direction) that you would anticipate for this air, considering buoyancy, Coriolis force, and turbulent drag.

E16.(§) Assume the katabatic winds of Antarctica result from a balance between the downslope buoyancy force, Coriolis force (assume 70°S latitude), and turbulent drag force at the surface (assume neutral boundary layer with C_D = 0.002). The surface is smooth and the depth of the katabatic layer is 100 m. Neglect entrainment drag at the top of the katabatic flow. The slope, potential temperature difference, and ambient temperature vary according to the table from exercise N5. Find the katabatic wind speed at the locations in the table from exercise N5.

E17. Where else in the world (Fig. 12.8) would you anticipate orographic airmass modification processes similar to those shown in Fig. 12.9? Explain.

E18. For airmass modification (i.e., while an airmass is blowing over a different surface), why do we describe the heat flux from the surface using a bulk transfer relationship ($C_H \cdot M \cdot \Delta\theta$), while for warm airmass genesis of stationary air we used a buoyancy-velocity approach ($b \cdot w_b \cdot \Delta\theta$)?

E19. The Eulerian heat budget from the Heat chapter shows how air temperature change ΔT is related to the heat input over time interval Δt. If this air is also moving at speed M, then during that same time interval, it travels a distance $\Delta x = M \cdot \Delta t$. Explain how you can use this information to get eq. (12.3), and why the wind speed M doesn't appear in that eq.

E20. Look at Fig. 12.9 and the associated Sample Application. In the Sample Application, the air passes twice through the thermodynamic state given by point 1 on that diagram: once when going from point 0 to points 2, and the second time when going from points 2 to point 3.

Similarly, air passes twice through the thermodynamic state at point 3 on the thermo diagram, and twice through point 5.

But air can achieve the same thermodynamic state twice only if certain physical (thermodynamic) conditions are met. What are those conditions, and how might they NOT be met in the real case of air traversing over these mountain ranges?

E21. Background: Recall that a frontal zone separates warmer and cooler airmasses. The warm airmass side of this zone is where the front is drawn on a weather map. This is true for both cold and warm fronts.

Issue: AFTER passage of the cold front is when significant temperature decreases are observed. BEFORE passage of a warm front is when significant warming is observed.

Question: Why does this difference exist (i.e., AFTER vs. BEFORE) for the passage of these two fronts?

E22. Overlay Fig. 12.11 with Fig. 12.12 by aligning the low-pressure centers. Do not rotate the images, but let the warm and cold fronts extend in different directions from the low center. Combine the information from these two figures to create a new, larger figure showing both fronts at the same time, extending from the same low. On separate copies of this merged figure, draw:

a. isotherms b. isobars c. winds d. weather

E23. Suppose you saw from your barometer at home that the pressure was falling. So you suspect that a front is approaching. What other clues can you use (by standing outside and looking at the clouds and weather; NOT by looking at the TV, computer, or other electronics or weather instruments) to determine if the approaching front is a warm or cold front. Discuss, along with possible pitfalls in this method.

E24. Draw new figures (a) - (d) similar to (a) - (d) in Figs. 12.11 and 12.12, but for an occluded front.

E25. Use the columns of temperature data in Fig. 12.13a, and plot each column as a separate sounding on a thermo diagram (See the Atmos. Stability chapter for blank thermo diagrams that you can photocopy for this exercise. Use a skew-T diagram, unless your instructor tells you otherwise). Describe how the frontal zone shows up in the soundings.

E26. Draw isentropes that you might expect in a vertical cross-section through a warm front.

E27. Other than drylines, is it possible to have fronts with no temperature change across them? How would such fronts be classified? How would they behave?

E28. What clouds and weather would you expect with a cold anafront?

E29. Why should the Rossby radius of deformation depend on the depth of the cold airmass in a geostrophic adjustment process?

E30. For geostrophic adjustment, the initial outflow of cold air turns, due to Coriolis force, until it is parallel to the front. Why doesn't it continue turning and point back into the cold air?

E31. For geostrophic adjustment, what is the nature of the final winds, if the starting point is a shallow cylinder of cold air 2 km thick and 500 km radius?

E32. Starting with Fig. 12.18, suppose that ABOVE the bottom contoured surface the temperature is horizontally uniform. Redraw that diagram but with the top two contoured surfaces sloped appropriately for the new temperature state.

E33. By inspection, write a kinematic frontogenesis equation (similar to eq. 12.10), but for an east-west aligned front.

E34. Figs. 12.22 - 12.24 presume that temperature gradients already exist, which can be strengthened by kinematic frontogenesis. Is that presumption valid for Earth's atmosphere? Justify.

E35. For the fronts analyzed in the Sample Application in the section on Surface Fronts – Horizontal Structure, estimate and compare magnitudes of any kinematic, thermodynamic, and dynamic processes that might exist across those fronts, based on the plotted weather data (which includes info on winds, precipitation, etc.). Discuss the relative importance of the various mechanisms for frontogenesis.

E36. Speculate on which is more important for dynamically generating ageostrophic cross-frontal circulations: the initial magnitude of the geostrophic wind or the change of geostrophic wind.

E37. What happens in an occluded front where the two cold air masses (the one advancing behind the cold front, and the one retreating ahead of the warm front) have equal virtual temperature?

E38. Draw isentropes that you expect in a vertical cross-section through a cold front occlusion (where a cold front catches up to a warm front), and discuss the change of static stability across this front.

E39. What types of weather would be expected with an upper-tropospheric front that does not have an associated surface front? [Hint: track movement of air parcels as they ride isentropic surfaces.]

E40. Draw a sketch similar to Fig. 12.34 showing the transverse circulations, but for a vertical cross-section in the Southern Hemisphere.

E41. Would you expect drylines to be possible in parts of the world other than the S.W. USA? If so, where? Justify your arguments.

Synthesize

S1. Fig. 12.2 shows a ridge that is typical of mid-latitudes in the N. Hemisphere. Sketch a ridge for:

a. Southern Hemisphere mid-latitudes
b. N. Hemisphere tropics
c. S. Hemisphere tropics

Hint: Consider the global pressure patterns described in the General Circulation chapter.

S2. What if there was no inversion at the top of the boundary layer? Redraw Fig. 12.3a for this situation.

S3. What if the Earth had no oceans? What mechanisms could create high-pressure centers and/or high-pressure belts?

S4. For your location, rank the importance of the different mechanisms that could create highs, and justify your ranking.

S5. Airmasses are abbreviated mostly by the relative temperature and humidity associated with their formation locations. Table 12-1 also describes two other attributes: returning airmasses, and airmasses that are warmer or colder than the underlying surface. What one additional attribute would you wish airmasses could be identified with, to help you to predict the weather at your location? Explain.

S6. Fig. 12.4 shows many possible genesis regions for airmasses. But some of these regions would not likely exist during certain seasons, because of the absence of high-pressure centers. For the hemisphere (northern or southern) where you live, identify how the various genesis regions appear or disappear with the seasons. Also, indicate the names for any of the high-pressure centers. For those that exist year round, indicate how they shift location with the seasons. (Hint: review the global circulation info in the General Circulation chapter.)

S7. What if airmasses remained stationary over the genesis regions forever? Could there be fronts and weather? Explain.

S8. Would it be possible for an airmass to become so thick that it fills the whole troposphere? If so, explain the conditions needed for this to occur.

S9. What if boundary-layer processes were so slow that airmasses took 5 times longer to form compared to present airmass formation of about 3 to 5 days? How would the weather and global circulation be different, if at all?

S10. Cold-air drainage from Antarctica is so strong and persistent that it affects the global circulation. Discuss how this affect is captured (or not) in the global maps in the General Circulation chapter.

S11. What if Antarctica was flat, similar to the Arctic? How would airmasses and Earth's climate be different, if at all?

S12. Consider Fig. 12.8. How would airmass formation and weather be different, if at all, given:

a. Suppose the Rocky Mountains disappeared, and a new dominant mountain range (named after you), appeared east-west across the middle of N. America.

b. Suppose that the Alps and Pyrenees disappeared and were replaced by a new dominant north-south mountain range (named after you) going through Europe from Copenhagen to Rome.

c. Suppose no major mountain ranges existed.

S13. Suppose a cross-section of the terrain looked like Fig. 12.9, except that the Great Basin region were below sea level (and not flooded with water). Describe how airmass modification by the terrain would be different, if at all.

S14. What if precipitation did not occur as air flows over mountain ranges? Thus, mountains could not modify air masses by this process. How would weather and climate be different from now, if at all?

S15. A really bad assumption was made in eq. (12.3); namely, z_i = constant as the airmass is modified. A better assumption would be to allow z_i to change with time (i.e., with distance as it blows over the new surface). The bad assumption was made because it allowed us to describe the physics with a simple eq. (12.3), which could be solved analytically to get eq. (12.4). This is typical of many physical problems, where it is impossible to find an exact analytical solution to the full equations describing true physics (as best we know it).

So here is a philosophical question. Is it better to approximate the physics to allow an exact analytical solution of the simplified problem? Or is it better to try to get an approximate (iterated or graphical solution) to the exact, more-complicated physics? Weigh the pros and cons of each approach, and discuss.

S16. For a (a) cold front, or a (b) warm front, create station plot data for a weather station 100 km ahead of the front, and for another weather station 100 km behind the front. Hint: use the sketches in Figs. 12.11 and 12.12 to help decide what to plot.

S17. Suppose the width of frontal zones were infinitesimally small, but their lengths remained unchanged. How would weather be different, if at all?

S18. What if turbulence were always so intense that frontal zones were usually 1000 km in width? How would weather be different, if at all?

S19. Does the geostrophic adjustment process affect the propagation distance of cold fronts in the real atmosphere, for cold fronts that are imbedded in the cyclonic circulation around a low-pressure center? If this indeed happens, how could you detect it?

S20. The boxes at the top of Figs. 12.22 - 12.24 show the associated sign of key terms for kinematic frontogenesis. If those terms have opposite signs, draw sketches similar to those figures, but showing frontolysis associated with:

a. confluence b. shear c. tilting.

S21. Suppose that condensation of water vapor caused air to cool. How would the thermodynamic mechanism for frontogenesis work for that situation, if at all? How would weather be near fronts, if at all?

S22. Suppose that ageostrophic motions were to experience extremely large drag, and thus would tend to dissipate quickly. How would frontogenesis in Earth's atmosphere be different, if at all?

S23. After cold frontal passage, cold air is moving over ground that is still warm (due to its thermal inertia). Describe how static stability varies with height within the cold airmass 200 km behind the cold front. Draw thermo diagram sketches to illustrate your arguments.

S24. Suppose a major volcanic eruption injected a thick layer of ash and sulfate aerosols at altitude 20 km. Would the altitude of the tropopause adjust to become equal to this height? Discuss.

S25. Suppose the slope of the ground in a dryline situation was not as sketched in Fig. 12.36. Instead, suppose the ground was bowl-shaped, with high plateaus on both the east and west ends. Would drylines exist? If so, what would be their characteristics, and how would they evolve?

13 EXTRATROPICAL CYCLONES

Contents

A synoptic-scale weather system with **low pressure** near the surface is called a **"cyclone"** (Fig. 13.1). Horizontal winds turn **cyclonically** around it (clockwise/counterclockwise in the Southern/ Northern Hemisphere). Near the surface these turning winds also spiral towards the low center. Ascending air in the cyclone can create clouds and precipitation.

Tropical cyclones such as hurricanes are covered separately in a later chapter. **Extratropical cyclones** (cyclones outside of the tropics) are covered here, and include transient **mid-latitude cyclones** and **polar cyclones**. Other names for extratropical cyclones are **lows** or **low-pressure centers** (see Table 13-1). Low-altitude convergence draws together airmasses to form fronts, along which the bad weather is often concentrated. These lows have a short life cycle (a few days to a week) as they are blown from west to east and poleward by the polar jet stream.

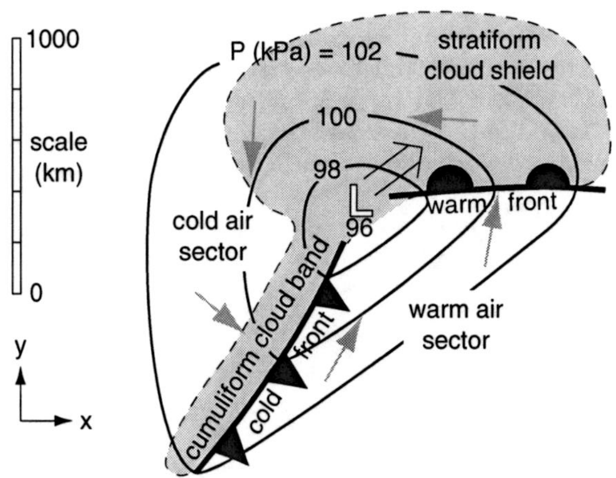

Figure 13.1
Components of a typical extratropical cyclone in the N. Hemisphere. Light grey shading shows clouds. Grey arrows are near-surface winds. Thin black lines are isobars (kPa). Thick black lines are fronts. The double-shaft arrow shows movement of the low center **L**.

Table 13-1. Cyclone names. "Core" is storm center. *T* is relative temperature.

Common Name in N. Amer.	Formal Name	Other Common Names	T of the Core	Map Symbol
low	extra-tropical cyclone	mid-latitude cyclone	cold	L
		low-pressure center		
		storm system*		
		cyclone *(in N. America)*		
hurricane	tropical cyclone	typhoon *(in W. Pacific)*	warm	
		cyclone *(in Australia)*		

(Often used by TV meteorologists.)*

CYCLONE CHARACTERISTICS

Cyclogenesis & Cyclolysis

Cyclones are born and intensify (**cyclogenesis**) and later weaken and die (**cyclolysis**). During cyclogenesis the (1) vorticity (horizontal winds turning around the low center) and (2) updrafts (vertical winds) increase while the (3) surface pressure decreases.

The intertwined processes that control these three characteristics will be the focus of three major sections in this chapter. In a nutshell, updrafts over a synoptic-scale region remove air from near the surface, causing the air pressure to decrease. The pressure gradient between this low-pressure center and the surroundings drives horizontal winds, which are forced to turn because of Coriolis force. Frictional drag near the ground causes these winds to spiral in towards the low center, adding more air molecules horizontally to compensate for those being removed vertically. If the updraft weakens, the inward spiral of air molecules fills the low to make it less low (cyclolysis).

Cyclogenesis is enhanced at locations where one or more of the following conditions occur:

(1) east of mountain ranges, where terrain slopes downhill under the jet stream.
(2) east of deep troughs (and west of strong ridges) in the polar jet stream, where horizontal divergence of winds drives mid-tropospheric updrafts.
(3) at frontal zones or other baroclinic regions where horizontal temperature gradients are large.
(4) at locations that don't suppress vertical motions, such as where static stability is weak.
(5) where cold air moves over warm, wet surfaces such as the Gulf Stream, such that strong evaporation adds water vapor to the air and strong surface heating destabilizes the atmosphere.
(6) at locations further from the equator, where Coriolis force is greater.

If cyclogenesis is rapid enough (central pressures dropping 2.4 kPa or more over a 24-hour period), the process is called **explosive cyclogenesis** (also nicknamed a **cyclone bomb**). This can occur when multiple conditions listed above are occurring at the same location (such as when a front stalls over the Gulf Stream, with a strong amplitude Rossby-wave trough to the west). During winter, such cyclone bombs can cause intense cyclones just off the east coast of the USA with storm-force winds, high waves, and blizzards or freezing rain.

INFO • Southern Hemisphere Lows

Some aspects of mid-latitude cyclones in the Southern Hemisphere are similar to those of N. Hemisphere cyclones. They have low pressure at the surface, rotate cyclonically, form east of upper-level troughs, propagate from west to east and poleward, and have similar stages of their evolution. They often have fronts and bad weather.

Different are the following: warm tropical air is to the north and cold polar air to the south, and the cyclonic rotation is clockwise due to the opposite Coriolis force. The figure below shows an idealized extra-tropical cyclone in the S. Hemisphere.

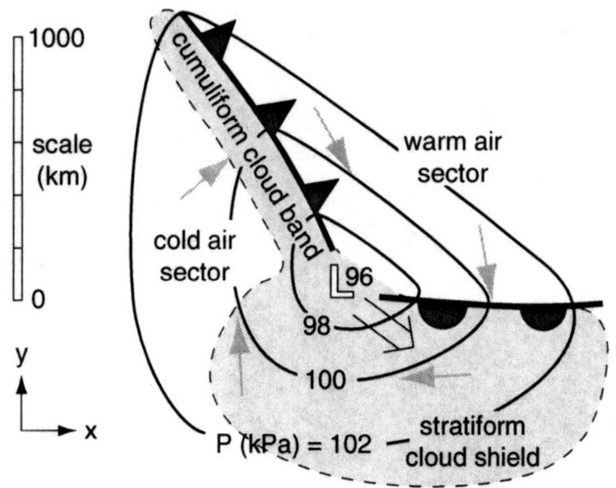

Figure a.
Sketch of mid-latitude cyclone in the Southern Hemisphere.

Cyclone Evolution

Although cyclones have their own synoptic-scale winds circulating around the low-pressure center, this whole system is blown toward the east by even larger-scale winds in the general circulation such as the jet stream. As a study aid, we will first move with the cyclone center as it evolves through its 1-day to 2-week life cycle of cyclogenesis and cyclolysis. Later, we will see where these low centers form and move due to the general circulation.

One condition that favors cyclogenesis is a **baroclinic** zone — a long, narrow region of large temperature change across a short horizontal distance near the surface. Frontal zones such as stationary fronts (Fig. 13.2a) are regions of strong baroclinicity.

Above (near the tropopause) and parallel to this baroclinic zone is often a strong jet stream (Fig. 13.2b), driven by the thermal-wind effect (see the chapters on General Circulation, and Fronts & Airmasses). If conditions are right (as discussed later in this chapter), the jet stream can remove air molecules from a column of air above the front, at location "D" in Fig. 13.2b. This lowers the surface pressure under location "D", causing **cyclogenesis** at the surface. Namely, under location "D" is where you would expect a surface low-pressure center to form.

The resulting pressure gradient around the surface low starts to generate lower-tropospheric winds that circulate around the low (Fig. 13.3a, again near the Earth's surface). This is the **spin-up** stage — so named because vorticity is increasing as the cyclone intensifies. The winds begin to advect the warm air poleward on the east side of the low and cold air equatorward on the west side, causing a kink in the former stationary front near the low center. The kinked front is wave shaped, and is called a **frontal wave**. Parts of the old front advance as a warm front, and other parts advance as a cold front. Also, these winds begin to force some of the warmer air up over the colder air, thereby generating more clouds.

If jet-stream conditions continue to be favorable, then the low continues to intensify and mature (Fig. 13.3b). As this cyclogenesis continues, the central pressure drops (namely, the cyclone **deepens**), and winds and clouds increase as a **vortex** around the **low center**. Precipitation begins if sufficient moisture is present in the regions where air is rising.

The advancing cold front often moves faster than the warm front. Three reasons for this are: (1) The Sawyer-Eliassen circulation tends to push near-surface cold air toward warmer air at both fronts. (2) Circulation around the vortex tends to deform the frontal boundaries and shrink the warm-air region to a smaller wedge shape east and equatorward of the low center. This wedge of warm air is called the **warm-air sector** (Fig. 13.1). (3) Evaporating precipi-

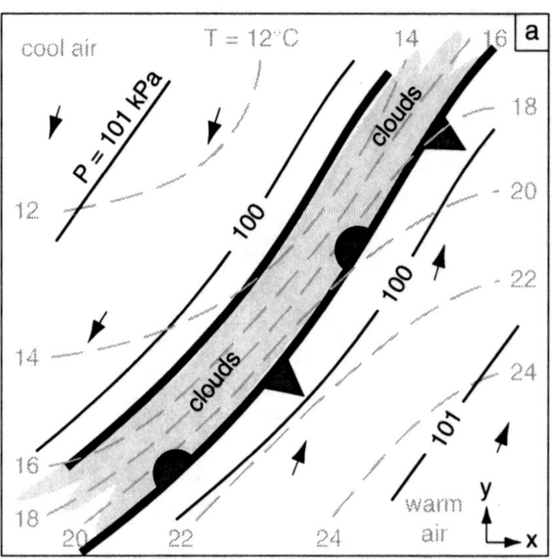

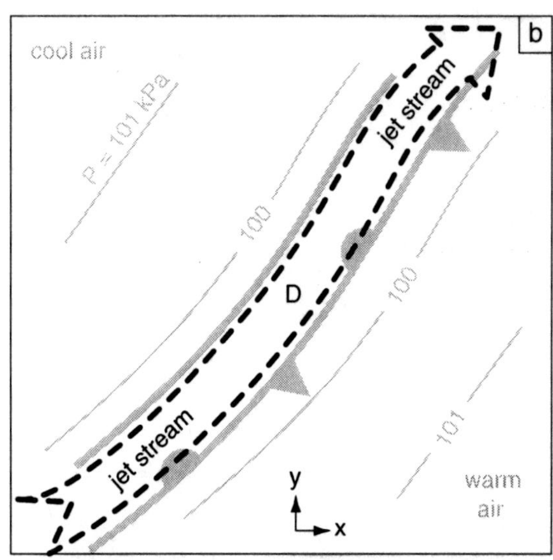

Figure 13.2
Initial conditions favoring cyclogenesis in N. Hemisphere. (a) Surface weather map. Solid thin black lines are isobars. Dashed grey lines are isotherms. The thick black lines mark the leading and trailing edges of the frontal zone. Grey shading indicates clouds. Fig. 13.3 shows subsequent evolution. (b) Upper-air map over the same frontal zone, where the dashed black arrow indicates the jet stream near the tropopause (z ≈ 11 km). The grey lines are a copy of the surface isobars and frontal zone from (a) to help you picture the 3-D nature of this system.

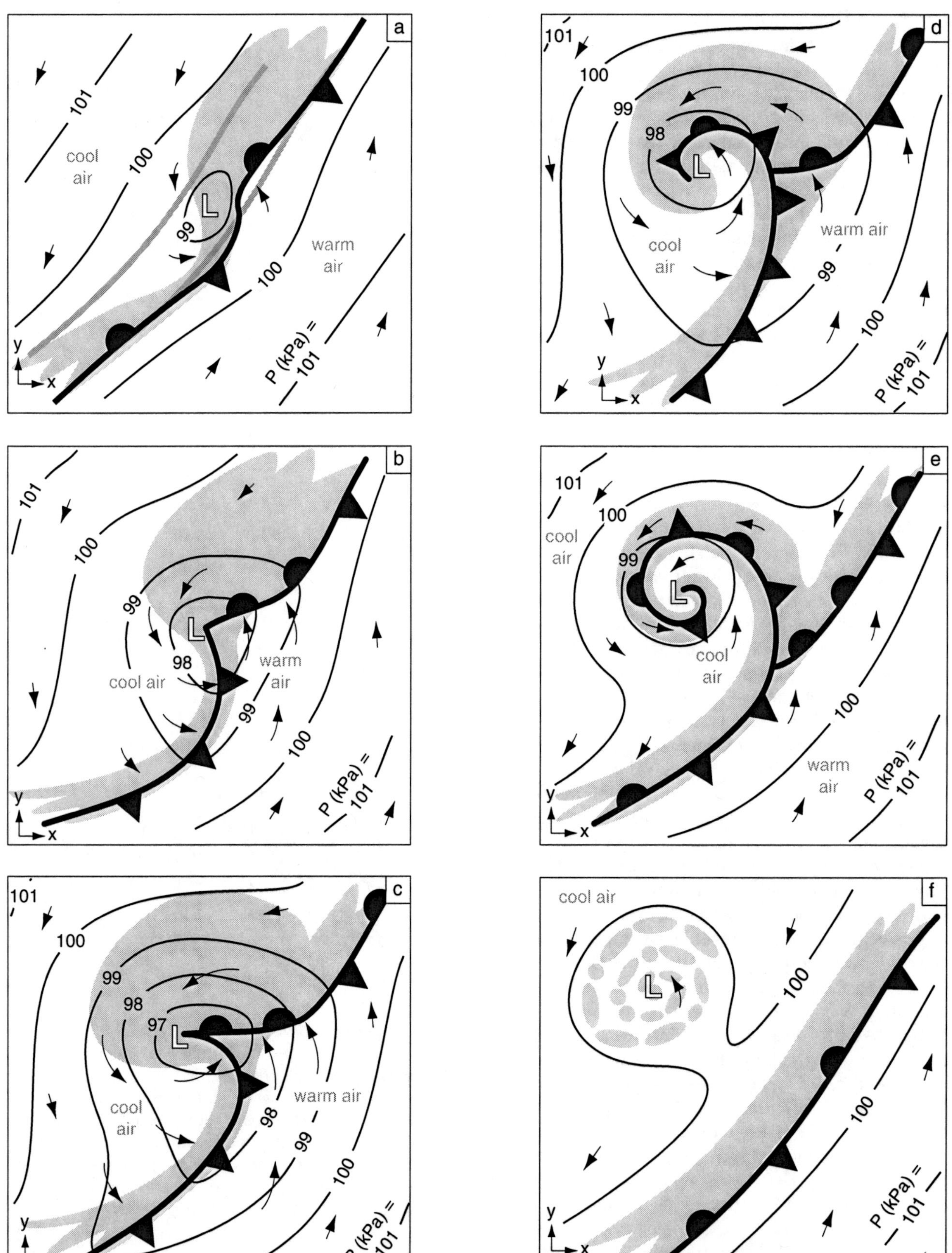

Figure 13.3
Extratropical cyclone evolution in the N. Hemisphere, including cyclogenesis (a - c), and cyclolysis (d - f). These idealized surface weather maps move with the low center. Grey shading indicates clouds, solid black lines are isobars (kPa), thin arrows are near-surface winds, L *is at the low center, and medium grey lines in (a) bound the original frontal zone. Fig. 13.2a shows the initial conditions.*

tation cools both fronts (enhancing the cold front but diminishing the warm front). These combined effects amplify the frontal wave.

At the peak of cyclone intensity (lowest central pressure and strongest surrounding winds) the cold front often catches up to the warm front near the low center (Fig. 13.3c). As more of the cold front overtakes the warm front, an occluded front forms near the low center (Fig. 13.3d). The cool air is often drier, and is visible in satellite images as a **dry tongue** of relatively cloud-free air that begins to wrap around the low. This marks the beginning of the **cyclolysis** stage. During this stage, the low is said to **occlude** as the occluded front wraps around the low center.

As the cyclone occludes further, the low center becomes surrounded by cool air (Fig. 13.3e). Clouds during this stage spiral around the center of the low — a signature that is easily seen in satellite images. But the jet stream, still driven by the thermal wind effect, moves east of the low center to remain over the strongest baroclinic zone (over the warm and cold fronts, which are becoming more stationary).

Without support from the jet stream to continue removing air molecules from the low center, the low begins to **fill** with air due to convergence of air in the boundary layer. The central pressure starts to rise and the winds slow as the vorticity **spins down**.

As cyclolysis continues, the low center often continues to slowly move further poleward away from the baroclinic zone (Fig. 13.3f). The central pressure continues to rise and winds weaken. The tightly wound spiral of clouds begins to dissipate into scattered clouds, and precipitation diminishes.

But meanwhile, along the stationary front to the east, a new cyclone might form if the jet stream is favorable (not shown in the figures).

In this way, cyclones are born, evolve, and die. While they exist, they are driven by the baroclinicity in the air (through the action of the jet stream). But their circulation helps to reduce the baroclinicity by moving cold air equatorward, warm air poleward, and mixing the two airmasses together. As described by **Le Chatelier's Principle**, the cyclone forms as a response to the baroclinic instability, and its existence partially undoes this instability. Namely, cyclones help the global circulation to redistribute heat between equator and poles.

Figures 13.3 are in a moving frame of reference following the low center. In those figures, it is not obvious that the warm front is advancing. To get a better idea about how the low moves while it evolves over a 3 to 5 day period, Figs. 13.4 show a superposition of all the cyclone locations relative to a fixed frame of reference. In these idealized figures, you can more easily see the progression of the low center, the advancement of the warm fronts and the advancement of cold air behind the cold fronts.

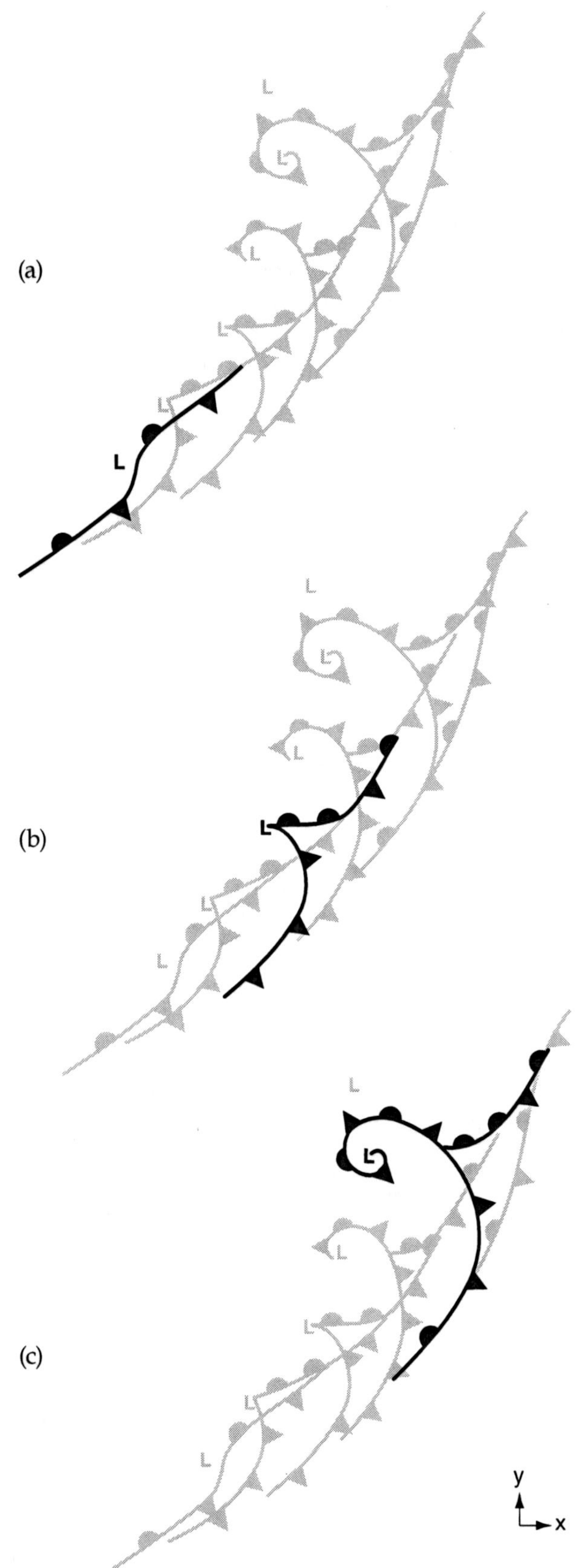

Figure 13.4
Illustration of movement of a low while it evolves. Fronts and low centers are from Figs. 13.3. Every second cyclone location is highlighted in a through c.

Cyclone Tracks

Extratropical cyclones are steered by the global circulation, including the prevailing westerlies at mid-latitudes and the meandering Rossby-wave pattern in the jet stream. Typical **storm tracks** (cyclone paths) of low centers are shown in Fig. 13.5. Multi-year climate variations (see the Climate chapter) in the global circulation, such as associated with the El Niño / La Niña cycle or the North Atlantic Oscillation (NAO), can alter the cyclone tracks. Mid-latitude cyclones are generally stronger, translate faster,

Figure 13.5 (below)
Climatology of extratropical cyclone tracks (lines with arrows) for (a) January and (b) July. Other symbols represent genesis and decay regions, as explained in the text. Circled symbols indicate stationary cyclones.

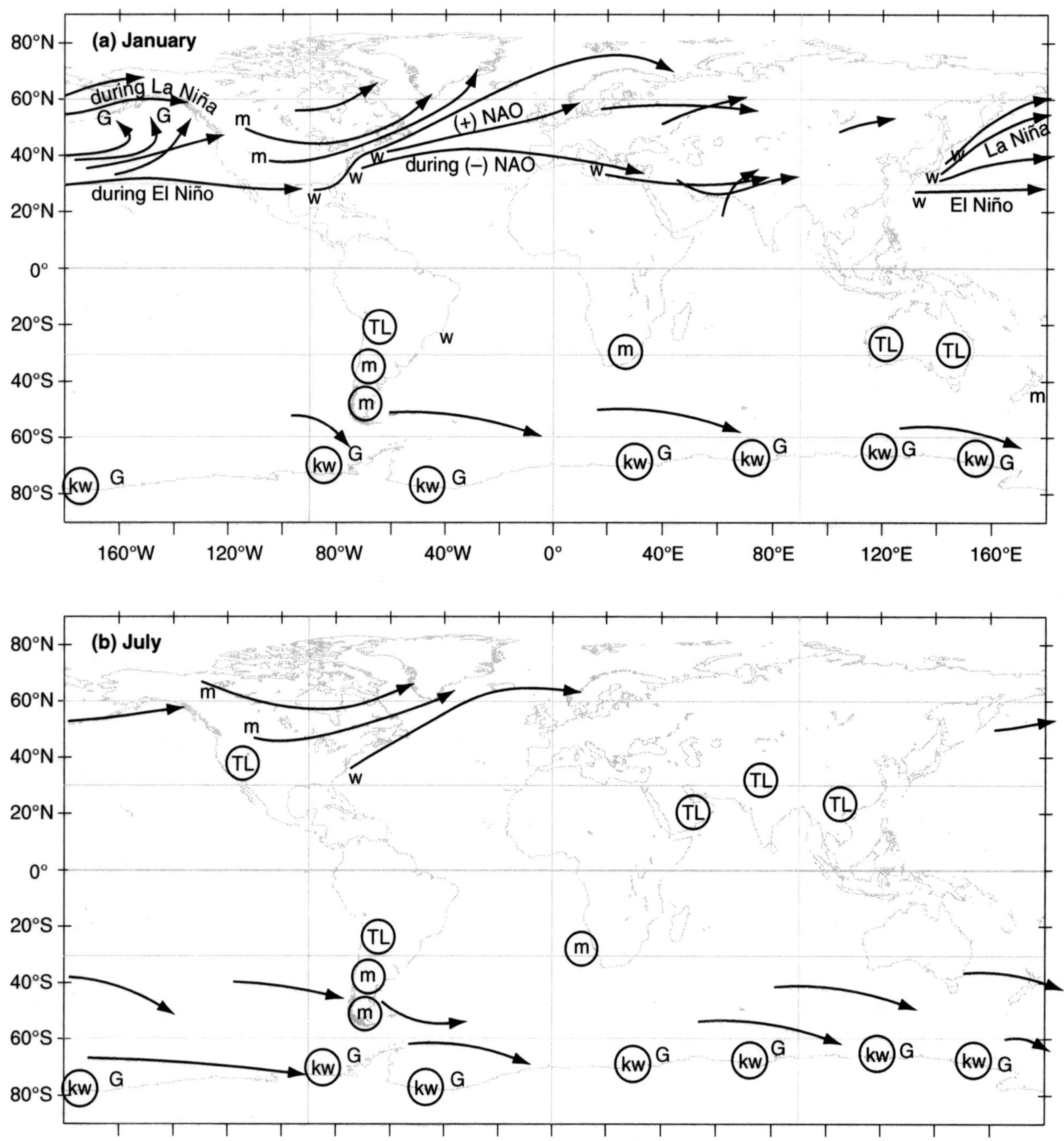

and are further equatorward during winter than in summer.

One favored cyclogenesis region is just east of large mountain ranges (shown by the "m" symbol in Fig. 13.5; see **Lee Cyclogenesis** later in this chapter). Other cyclogenesis regions are over warm ocean **boundary currents** along the western edge of oceans (shown by the symbol "w" in the figure), such as the **Gulf Stream** current off the east coast of N. America, and the **Kuroshio Current** off the east coast of Japan. During winter over such currents are strong sensible and latent heat fluxes from the warm ocean into the air, which adds energy to developing cyclones. Also, the strong wintertime contrast between the cold continent and the warm ocean current causes an intense baroclinic zone that drives a strong jet stream above it due to thermal-wind effects.

Cyclones are often strengthened in regions under the jet stream just east of troughs. In such regions, the jet stream steers the low center toward the east and poleward. Hence, cyclone tracks are often toward the northeast in the N. Hemisphere, and toward the southeast in the S. Hemisphere.

Cyclones in the Northern Hemisphere typically evolve during a 2 to 7 day period, with most lasting 3 - 5 days. They travel at typical speeds of 12 to 15 m s^{-1} (43 to 54 km h^{-1}), which means they can move about 5000 km during their life. Namely, they can travel the distance of the continental USA from coast to coast or border to border during their lifetime. Since the Pacific is a larger ocean, cyclones that form off of Japan often die in the Gulf of Alaska just west of British Columbia (BC), Canada — a cyclolysis region known as a **cyclone graveyard** (G).

Quasi-stationary lows are indicated with circles in Fig. 13.5. Some of these form over hot continents in summer as a monsoon circulation. These are called **thermal lows** (TL), as was explained in the General Circulation chapter in the section on Hydrostatic Thermal Circulations. Others form as quasi-stationary lee troughs just east of mountain (m) ranges.

In the Southern Hemisphere (Fig. 13.5), cyclones are more uniformly distributed in longitude and throughout the year, compared to the N. Hemisphere. One reason is the smaller area of continents in Southern-Hemisphere mid-latitudes and subpolar regions. Many propagating cyclones form just north of 50°S latitude, and die just south. The region with greatest cyclone activity (cyclogenesis, tracks, cyclolysis) is a band centered near 60°S.

These Southern Hemisphere cyclones last an average of 3 to 5 days, and translate with average speeds faster than 10 m s^{-1} (= 36 km h^{-1}) toward the east-south-east. A band with average translation speeds faster than 15 m s^{-1} (= 54 km h^{-1}; or > 10°

INFO • North American Geography

To help you interpret the weather maps, the map and tables give state and province names.

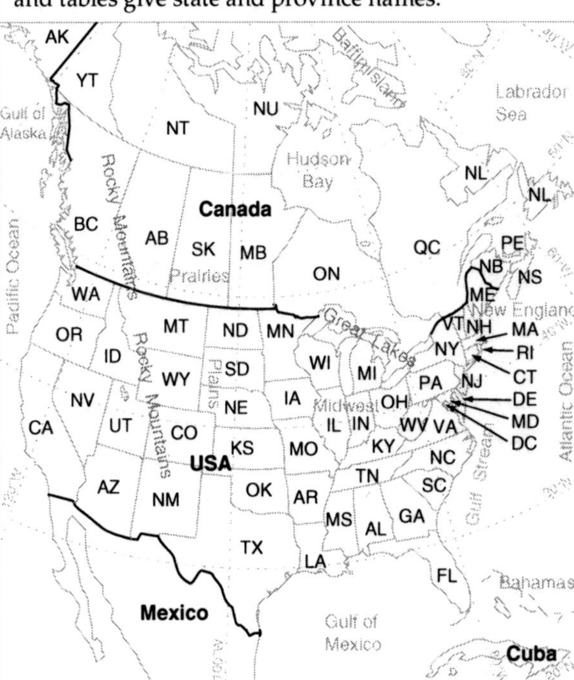

Figure b.

Canadian Postal Abbreviations for Provinces:

AB	Alberta	NT	Northwest Territories
BC	British Columbia	NU	Nunavut
MB	Manitoba	ON	Ontario
NB	New Brunswick	PE	Prince Edward Isl.
NL	Newfoundland & Labrador	QC	Quebec
NS	Nova Scotia	SK	Saskatchewan
		YT	Yukon

USA Postal Abbreviations for States:

AK	Alaska	MD	Maryland	OK	Oklahoma
AL	Alabama	ME	Maine	OR	Oregon
AR	Arkansas	MI	Michigan	PA	Pennsylvania
AZ	Arizona	MN	Minnesota	RI	Rhode Isl.
CA	California	MO	Missouri	SC	South Carolina
CO	Colorado	MS	Mississippi	SD	South Dakota
CT	Connecticut	MT	Montana	TN	Tennessee
DE	Delaware	NC	North Carolina	TX	Texas
FL	Florida	ND	North Dakota	UT	Utah
GA	Georgia	NE	Nebraska	VA	Virginia
HI	Hawaii	NH	New Hampshire	VT	Vermont
IA	Iowa	NJ	New Jersey	WA	Washington
ID	Idaho	NM	New Mexico	WI	Wisconsin
IL	Illinois	NV	Nevada	WV	West Virginia
IN	Indiana	NY	New York	WY	Wyoming
KS	Kansas	OH	Ohio	DC	Wash. DC
KY	Kentucky				
LA	Louisiana				
MA	Massachusetts				

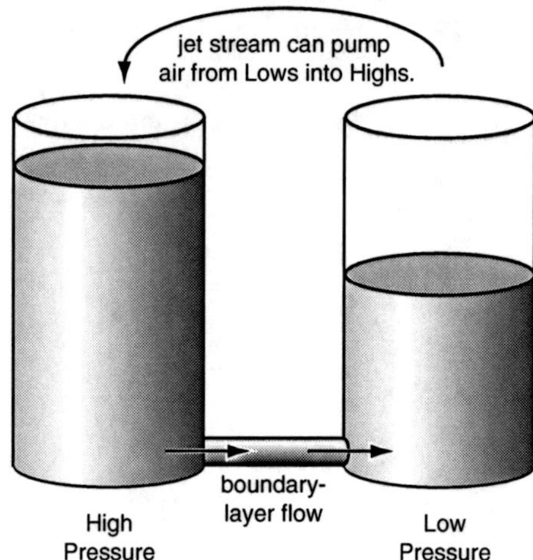

Figure 13.6
Two tanks filled with water to different heights are an analogy to neighboring high and low pressure systems in the atmosphere.

longitude day^{-1}) extends from south of southwestern Africa eastward to south of western Australia. The average track length is 2100 km. The normal **cyclone graveyard** (G, cyclolysis region) in the S. Hemisphere is in the **circumpolar trough** (between 65°S and the Antarctic coastline).

Seven stationary centers of enhanced cyclone activity occur around the coast of Antarctica, during both winter and summer. Some of these are believed to be a result of fast **katabatic** (cold downslope) winds flowing off the steep Antarctic terrain (see the Fronts & Airmasses chapter). When these very cold winds reach the relatively warm unfrozen ocean, strong heat fluxes from the ocean into the air contribute energy into developing cyclones. Also the downslope winds can be channeled by the terrain to cause cyclonic rotation. But some of the seven stationary centers might not be real — some might be caused by improper reduction of surface pressure to sea-level pressure. These seven centers are labeled with "kw", indicating a combination of katabatic winds and relatively warm sea surface.

Stacking & Tilting

Lows at the bottom of the troposphere always tend to kill themselves. The culprit is the boundary layer, where turbulent drag causes air to cross isobars at a small angle from high toward low pressure. By definition, a low has lower central pressure than the surroundings, because fewer air molecules are in the column above the low. Thus, boundary-layer flow will always move air molecules toward surface lows (Fig. 13.6). As a low fills with air, its pressure rises and it stops being a low. Such **filling** is quick enough to eliminate a low in less than a day, unless a compensating process can remove air more quickly.

Such a compensating process often occurs if the axis of low pressure **tilts** westward with increasing height (Fig. 13.7). Recall from the **gradient-wind** discussion in the Atmospheric Forces and Winds chapter that the jet stream is slower around troughs than ridges. This change of wind speed causes divergence aloft; namely, air is leaving faster than it is arriving. Thus, with the upper-level trough shifted west of the surface low (L), the divergence region (D) is directly above the surface low, supporting cyclogenesis. Details are explained later in this Chapter. But for now, you should recognize that a westward tilt of the low-pressure location with increasing height often accompanies cyclogenesis.

Conversely, when the trough aloft is **stacked** vertically above the surface low, then the jet stream

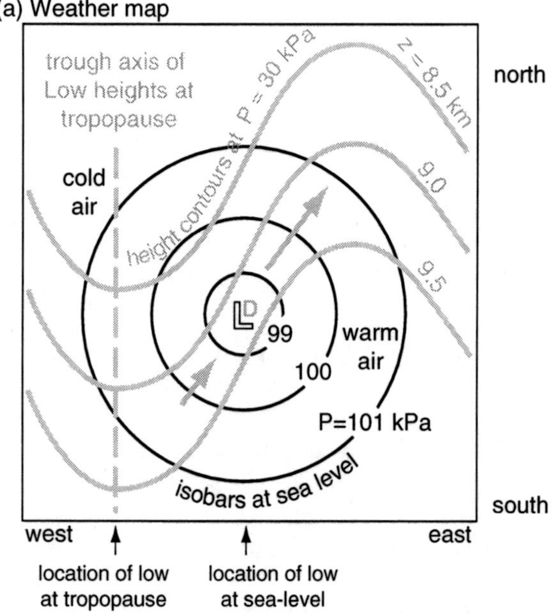

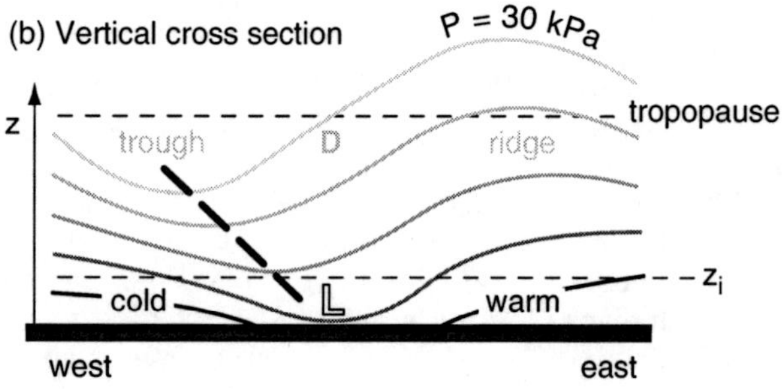

Figure 13.7
(a) Two N. Hemisphere weather maps superimposed: (thin black lines) sea-level pressure, and (grey lines) 30 kPa heights. Jet-stream winds (thick grey arrows) follow the height contours (b) East-west vertical cross section through middle of (a). Heavy dashed line is trough axis. L indicates low center at surface, and D indicates divergence aloft. (Pressure and height variations are exaggerated.)

is not pumping air out of the low, and the low fills due to the unrelenting boundary-layer flow. Thus, vertical stacking is associated with cyclolysis.

Other Characteristics

Low centers often move parallel to the direction of the isobars in the warm sector (Fig. 13.1). So even without data on upper-air steering-level winds, you can use a surface weather map to anticipate cyclone movement.

Movement of air around a cyclone is three-dimensional, and is difficult to show on two-dimensional weather maps. Fig. 13.8 shows the main streams of air in one type of cyclone, corresponding to the snapshot of Fig. 13.3b. Sometimes air in the **warm-air conveyor belt** is moving so fast that it is called a low-altitude **pre-frontal jet**. When this humid stream of air is forced to rise over the cooler air at the warm front (or over a mountain) it can dump heavy precipitation and cause flooding.

Behind the cold front, cold air often descends from the mid- or upper-troposphere, and sometimes comes all the way from the lower stratosphere. This dry air **deforms** (changes shape) into a **diffluent** (horizontally spreading) flow near the cold front.

To show the widespread impact of a Spring midlatitude cyclone, a case-study is introduced next.

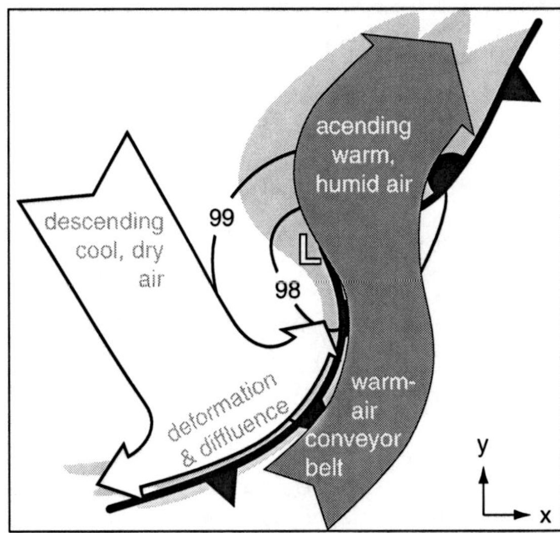

Figure 13.8
Ascending and descending air in a cyclone. Thin black lines with numbers are isobars (kPa). Thick black lines are fronts.

MIDLATITUDE CYCLONE EVOLUTION — A CASE STUDY

Summary of 3 to 4 April 2014 Cyclone

An upper-level trough (Fig. 13.10a) near the USA Rocky Mountains at 00 UTC on 3 April 2014 propagates eastward, reaching the midwest and Mississippi Valley a day and a half later, at 12 UTC on 4 April 2014. A surface low-pressure center forms east of the trough axis (Fig. 13.10b), and strengthens as the low moves first eastward, then north-eastward.

Extending south of this low is a dry line that evolves into a cold front (Figs. 13.10b & 13.11), which sweeps into the Mississippi Valley. Ahead of the front is a squall line of severe thunderstorms (Figs. 13.11 & 13.12). Local time there is Central Daylight Time (CDT), which is 6 hours earlier than UTC.

During the 24 hours starting at 6 AM CDT (12 UTC) on 3 April 2014 there were a total of 392 storm reports recorded by the US Storm Prediction Center (Fig. 13.9). This included 17 tornado reports, 186 hail reports (of which 23 reported large hailstones greater than 5 cm diameter), 189 wind reports (of which 2 had speeds greater than 33 m/s). Next, we focus on the 12 UTC 4 April 2014 weather (Figs. 13.13 - 13.19).

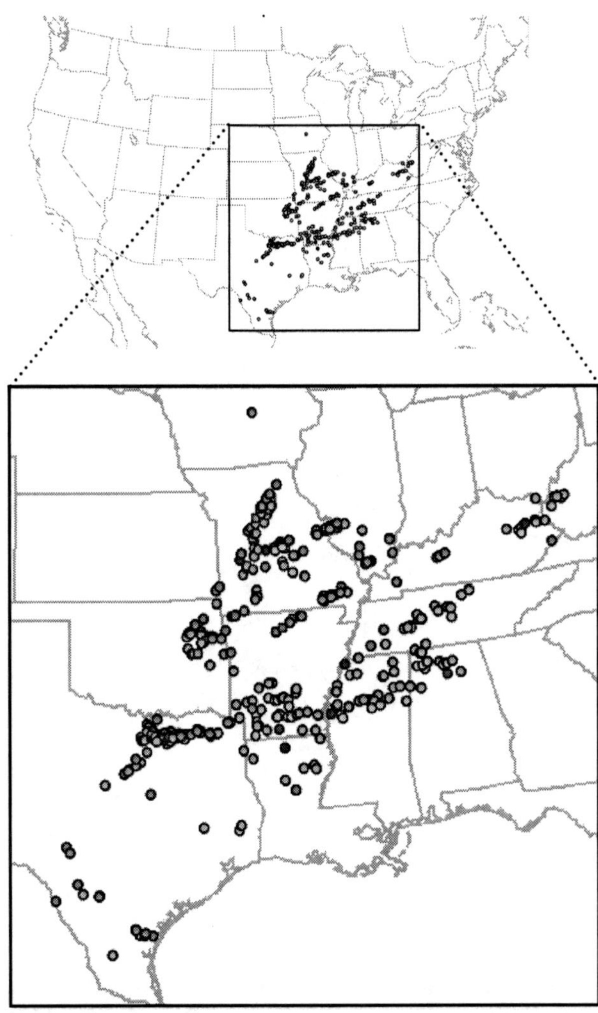

Figure 13.9
Storm reports (dots) during 06 CDT 3 Apr to 06 CDT 4 Apr 2014 [where 06 Central Daylight Time (CDT) = 12 UTC]. Legend: Red = tornado, green = hail, blue = wind reports.

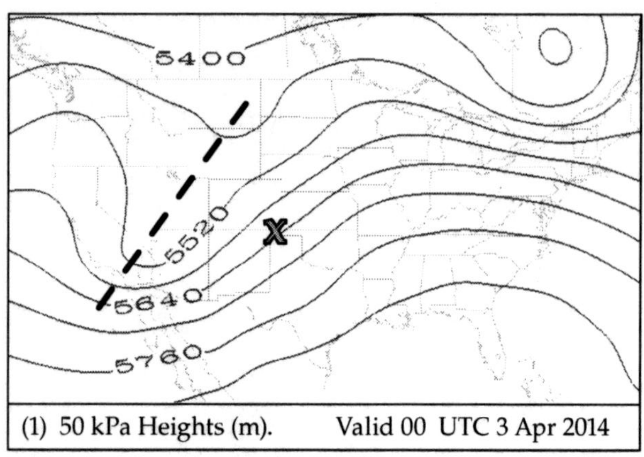

(1) 50 kPa Heights (m). Valid 00 UTC 3 Apr 2014

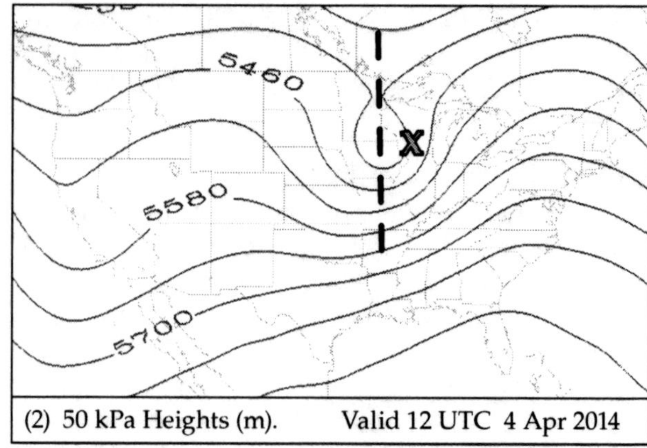

(2) 50 kPa Heights (m). Valid 12 UTC 4 Apr 2014

Figure 13.10a

Evolution of geopotential height contours (m) of the 50 kPa isobaric surface (known as "50 kPa heights") during a day and a half for the case-study cyclone. Height contour interval is 60 m. "X" marks the surface-low location at the valid times of the two different maps. The thick dashed line shows the axis of the low-pressure trough (i.e., the "trough axis").

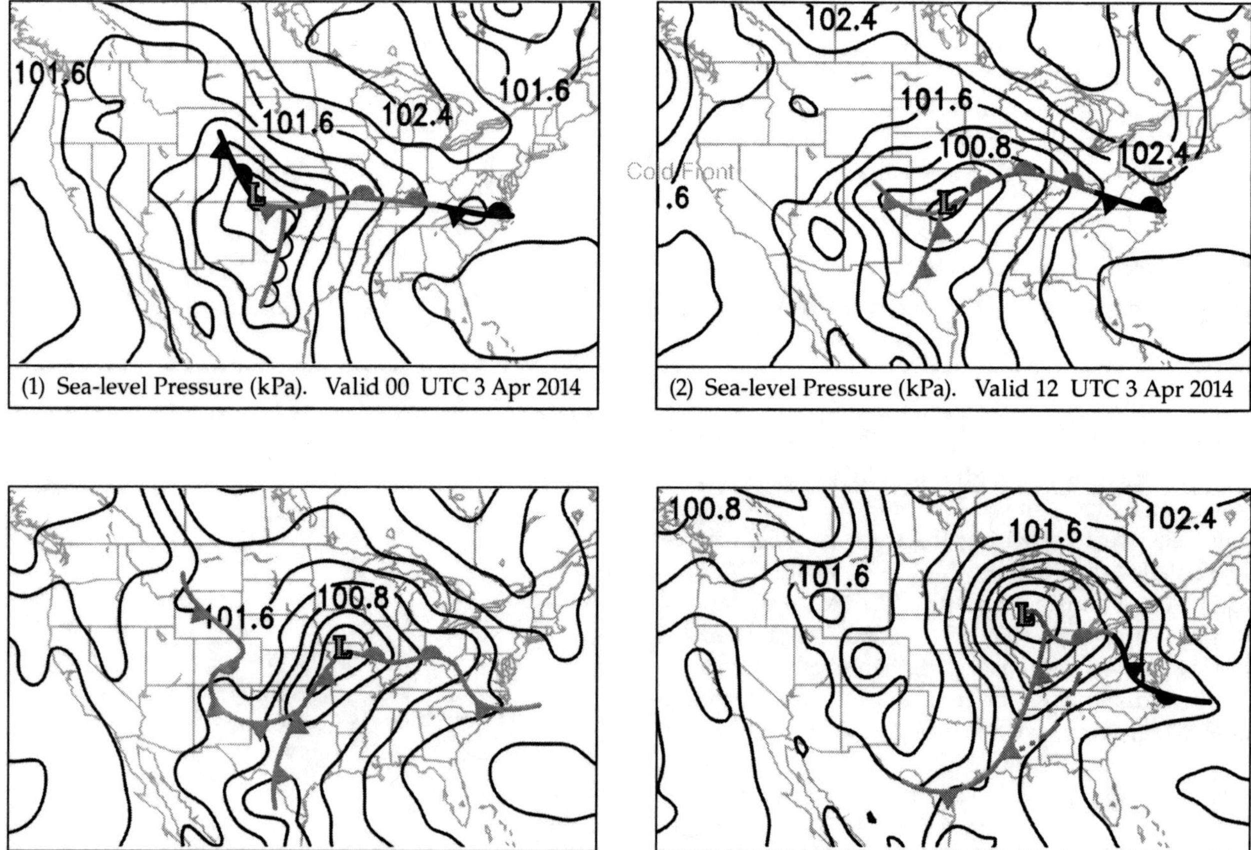

(1) Sea-level Pressure (kPa). Valid 00 UTC 3 Apr 2014

(2) Sea-level Pressure (kPa). Valid 12 UTC 3 Apr 2014

(3) Sea-level Pressure (kPa). Valid 00 UTC 4 Apr 2014

(4) Sea-level Pressure (kPa). Valid 12 UTC 4 Apr 2014

Figure 13.10b

Evolution of mean-sea-level (MSL) pressure (kPa) and surface fronts every 12 hours during a day and a half, from 00 UTC 3 April to 12 UTC 4 April 2014. Isobar contour interval is 0.4 kPa. "L" marks the location of the surface low-pressure center of the case-study cyclone. The central pressure of the low every 12 hours in this sequence was: 99.4, 99.9, 100.0, and 99.7 kPa. Image (1) also shows a dry line in Texas, and image (4) shows a squall line (of thunderstorms) in the southeast USA ahead of the cold front.

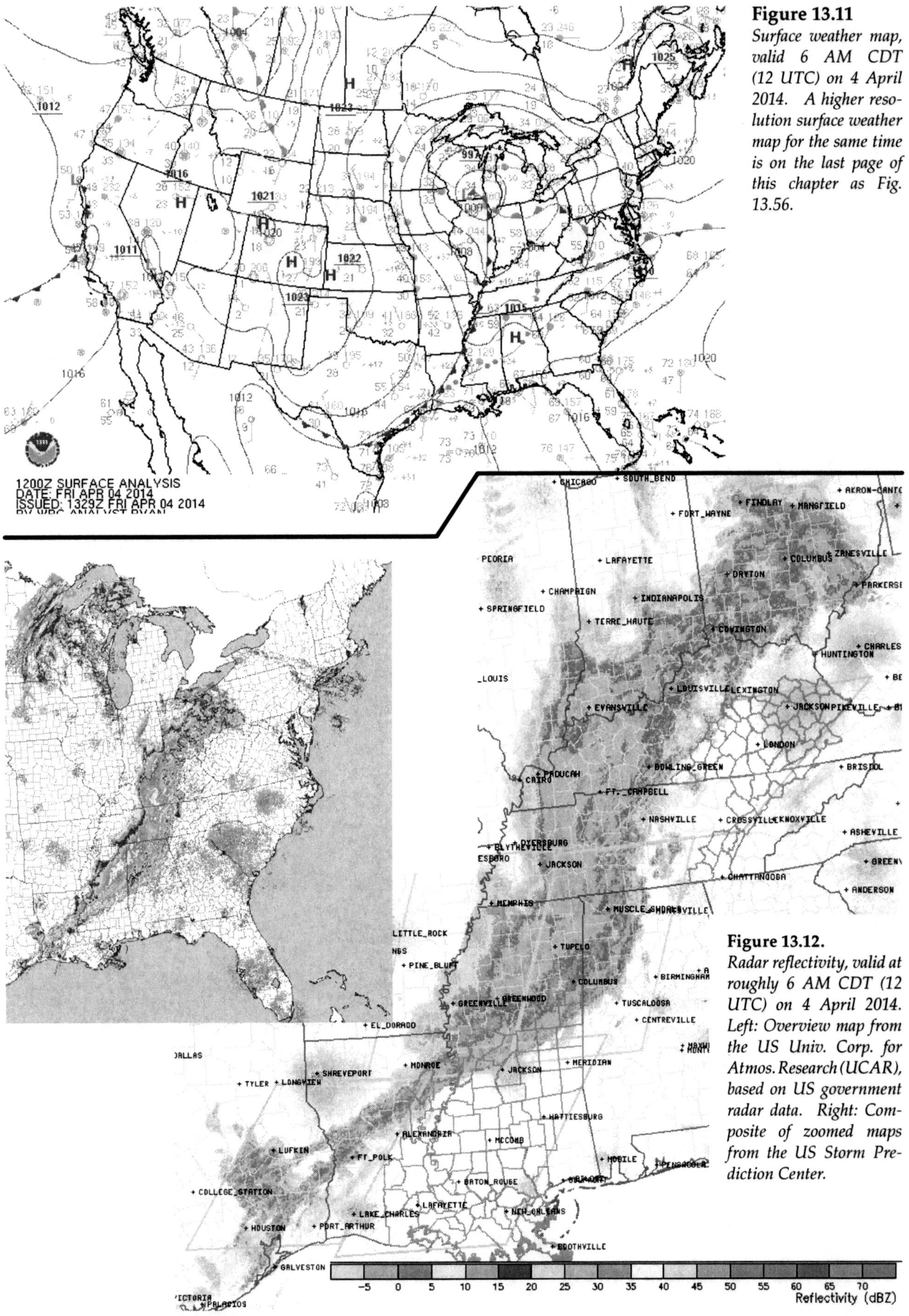

Figure 13.11
Surface weather map, valid 6 AM CDT (12 UTC) on 4 April 2014. A higher resolution surface weather map for the same time is on the last page of this chapter as Fig. 13.56.

Figure 13.12.
Radar reflectivity, valid at roughly 6 AM CDT (12 UTC) on 4 April 2014. Left: Overview map from the US Univ. Corp. for Atmos. Research (UCAR), based on US government radar data. Right: Composite of zoomed maps from the US Storm Prediction Center.

INFO • Isosurfaces & Their Utility

Lows and other synoptic features have five-dimensions (3-D spatial structure + 1-D time evolution + 1-D multiple variables). To accurately analyze and forecast the weather, you should try to form in your mind a multi-dimensional picture of the weather. Although some computer-graphics packages can display 5-dimensional data, most of the time you are stuck with flat 2-D weather maps or graphs.

By viewing multiple 2-D slices of the atmosphere as drawn on weather maps (Fig. c), you can picture the 5-D structure. Examples of such 2-D maps include:
- uniform height maps
- isobaric (uniform pressure) maps
- isentropic (uniform potential temperature) maps
- thickness maps
- vertical cross-section maps
- time-height maps
- time-variable maps (meteograms)

Computer animations of maps can show time evolutions. In Chapter 1 is a table of other iso-surfaces.

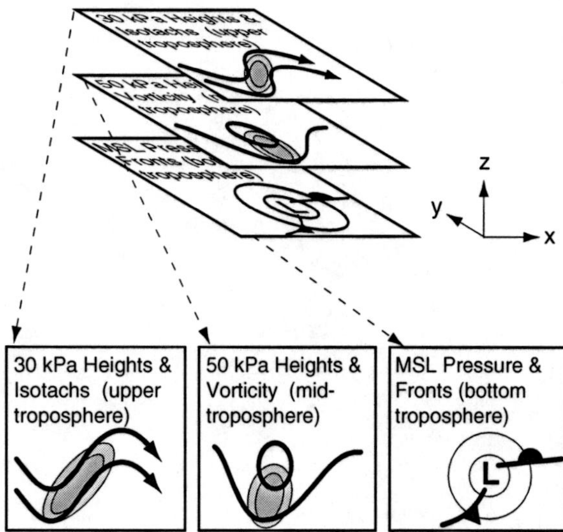

Figure c
A set of weather maps for different altitudes helps you gain a 3-D perspective of the weather. MSL = mean sea level.

Height

Mean-sea-level (**MSL**) maps represent a uniform height of $z = 0$ relative to the ocean surface. For most land areas that are above sea level, these maps are created by extrapolating atmospheric conditions below ground. (A few land-surface locations are below sea level, such as Death Valley and the Salton Sea USA, or the Dead Sea in Israel and Jordan).

Meteorologists commonly plot air pressure (reduced to sea level) and fronts on this uniform-height surface. These are called "MSL pressure" maps.

(continues in next column)

INFO • Isosurfaces *(continuation)*

Pressure

Recall that pressure decreases monotonically with increasing altitude. Thus, lower pressures correspond to higher heights.

For any one pressure, such as 70 kPa (which is about 3 km above sea level on average), that pressure is closer to the ground (i.e., less than 3 km) in some locations and is further from the ground in other locations, as was discussed in the Forces and Winds chapter. If you conceptually draw a surface that passes through all the points that have pressure 70 kPa, then that **isobaric surface** looks like rolling terrain with peaks, valleys (troughs), and ridges.

Like a topographic map, you could draw contour lines connecting points of the same height. This is called a "70 kPa height" chart. Low heights on an isobaric surface correspond to low pressure on a uniform-height surface.

Back to the analogy of hilly terrain, suppose you went hiking with a thermometer and measured the air temperature at eye level at many locations within a hilly region. You could write those temperatures on a map and then draw isotherms connecting points of the same temperature. But you would realize that these temperatures on your map correspond to the hilly terrain that had ridges and valleys.

You can do the same with isobaric charts; namely, you can plot the temperatures that are found at different locations on the undulating isobaric surface. If you did this for the 70 kPa isobaric surface, you would have a "70 kPa isotherm" chart. You can plot any variable on any isobaric surface, such as "90 kPa isohumes", "50 kPa vorticity", "30 kPa isotachs", etc.

You can even plot multiple weather variables on any single isobaric map, such as "30 kPa heights and isotachs" or "50 kPa heights and vorticity" (Fig. c). The first chart tells you information about jet-stream speed and direction, and the second chart can be used to estimate cyclogenesis processes.

Thickness

Now picture two different isobaric surfaces over the same region, such as sketched a the top of Fig. c. An example is 100 kPa heights and 50 kPa heights. At each location on the map, you could measure the height difference between these two pressure surfaces, which tells you the thickness of air in that layer. After drawing isopleths connecting points of equal thickness, the resulting contour map is known as a "100 to 50 kPa thickness" map.

You learned in the General Circulation chapter that the thermal-wind vectors are parallel to thickness contours, and that these vectors indicate shear in the geostrophic wind. That chapter also showed that the 100-50 kPa thickness is proportional to average temperature in the bottom half of the troposphere.

(continues on next page)

INFO • Isosurfaces (continued)

Potential Temperature

On average, potential temperature θ increases toward the equator and with increasing height. Day-to-day variability is superimposed on that average. Over any region for any valid time you can create a surface called an isentropic surface that follows any desired potential temperature, such as the θ = 310 K isentropic surface shaded in blue in Fig. d. below.

As was done for isobaric surfaces, you can also plot contours of the height of that surface above mean sea level (e.g., "310 K heights"). Or you can plot other weather variables on that surface, such as "310 K isohumes".

If you were to look straight down from above the top diagram in Fig. d, you would see a view such as

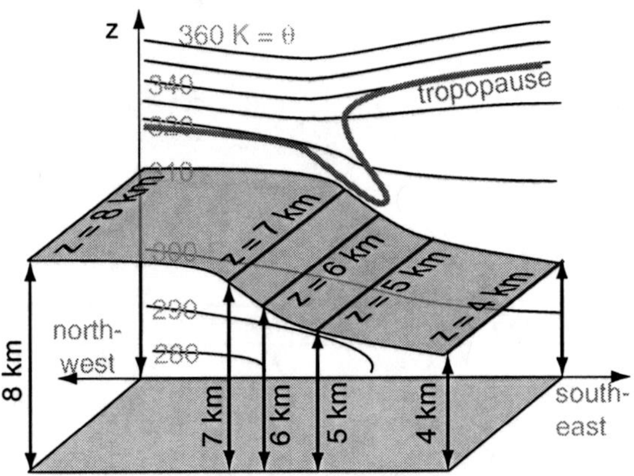

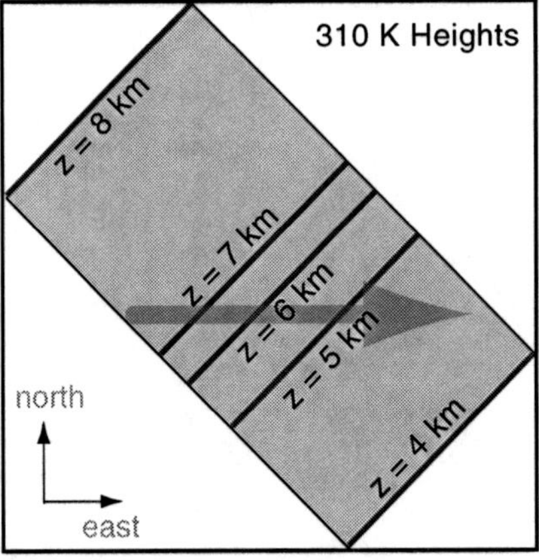

Figure d
Lines of uniform potential temperature (thin black lines) are sketched in the background of the 3-D diagram at top. Each line has a corresponding isentropic surface that goes through it, such as sketched for θ = 310 K in both figures (shaded in blue). The heights (z) of this surface above MSL are plotted as contour lines in the isentropic chart at bottom. The thick grey arrow represents a hypothetical wind vector.

(continues in next column)

INFO • Isosurfaces (continued)

shown in the bottom of Fig. d. This bottom diagram is the isentropic chart that would be presented as a weather map.

For adiabatic processes, unsaturated air parcels that are blown by the wind tend to follow the isentropic surface that corresponds to the parcel's potential temperature. The reason is that if the parcel were to stray off of that surface, then buoyant forces would move it back to that surface.

For example, if an air parcel has θ = 310 K and is located near the tail of the arrow in the bottom of Fig. d, then as the parcel moves with the wind it will the 310 K isentropic surface. In this illustration, the parcel descends (and its temperature would warm adiabatically while its θ is constant). Because of adiabatic warming and cooling, winds descending along an isentrope will warm and be cloud free, while winds rising along isentropes will cool and become cloudy.

Turbulence, condensation and radiation are not adiabatic (i.e., are **diabatic**), and cause the parcel's θ to change. This would cause the parcel to shift to a different isentropic surface (one that matches the parcel's new θ).

Potential Vorticity (PVU)

Recall the definition of isentropic potential vorticity from the General Circulation chapter. That chapter also defined potential vorticity units (PVU) for this variable. PVUs are very large in the stratosphere, and small in the troposphere, with the tropopause often at about 1.5 PVU.

Thus, contour plots of the height of the 1.5 PVU surface approximate the altitude of the tropopause at different locations. The tropopause could be relatively low (at $z \approx 6$ km MSL, at $P \approx 35$ kPa) near the poles, and relatively high near the equator ($z \approx 15$ - 18 km MSL, and $P \approx 10$ kPa). This contour plot can indicate features such as tropopause folds where stratospheric air can be injected into the troposphere.

Surface

The "**surface weather map**" shows weather at the elevation of the Earth's surface. Namely, it follows the terrain up and down, and is not necessarily at mean sea level.

Rules

It is impossible to have two different pressures, or two different potential temperatures, at the same point in space at any instant. For this reason, isobaric surfaces cannot cross other isobaric surfaces (e.g., the 70 kPa and 60 kPa isobaric surfaces cannot intersect). Similar rules apply for isentropic surfaces. But isobaric surfaces can cross isentropes, and they both can intersect the ground surface.

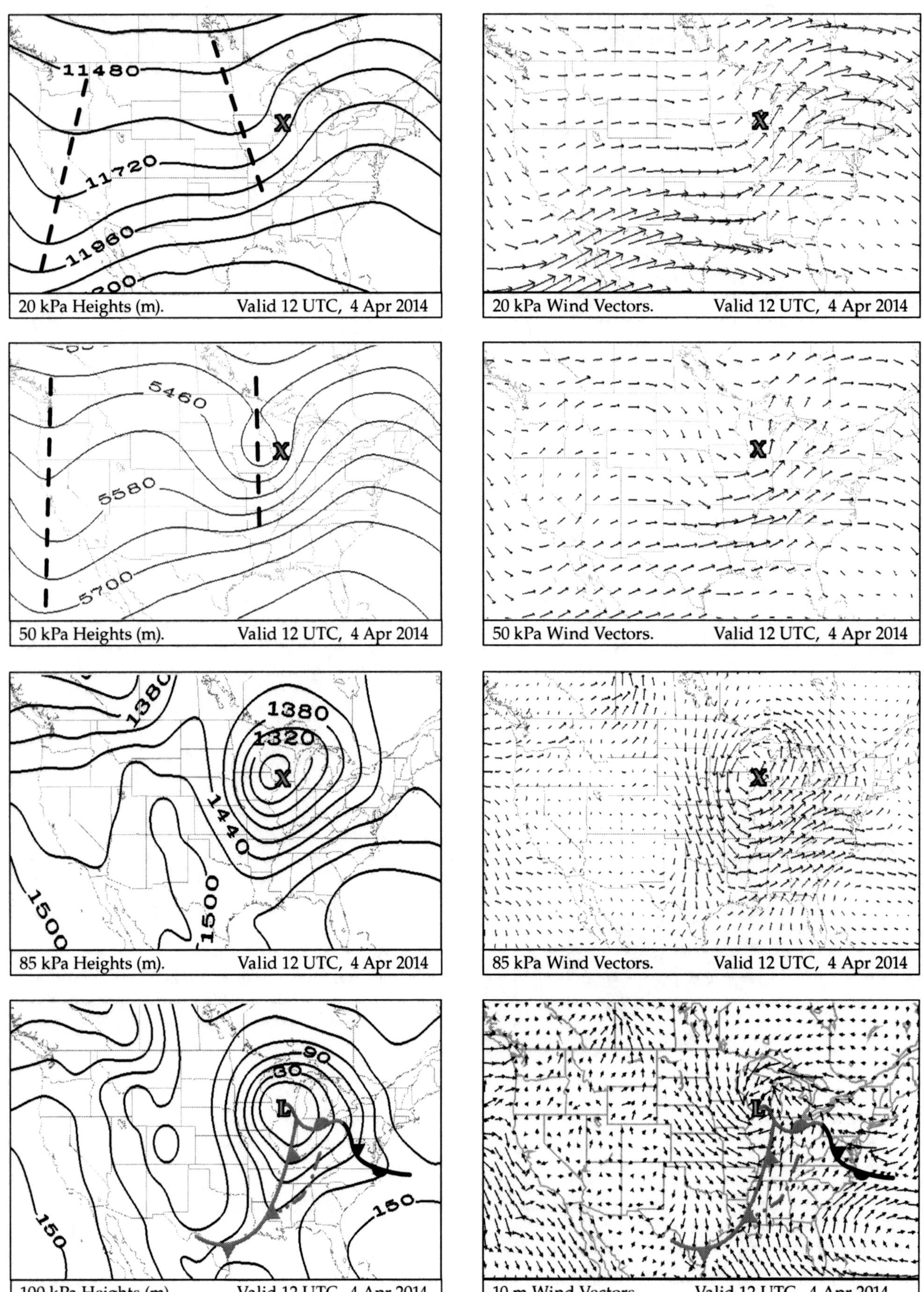

Figure 13.13. *Geopotential heights (left column) and wind vectors (right). Maps higher on the page are for higher in the atmos.*

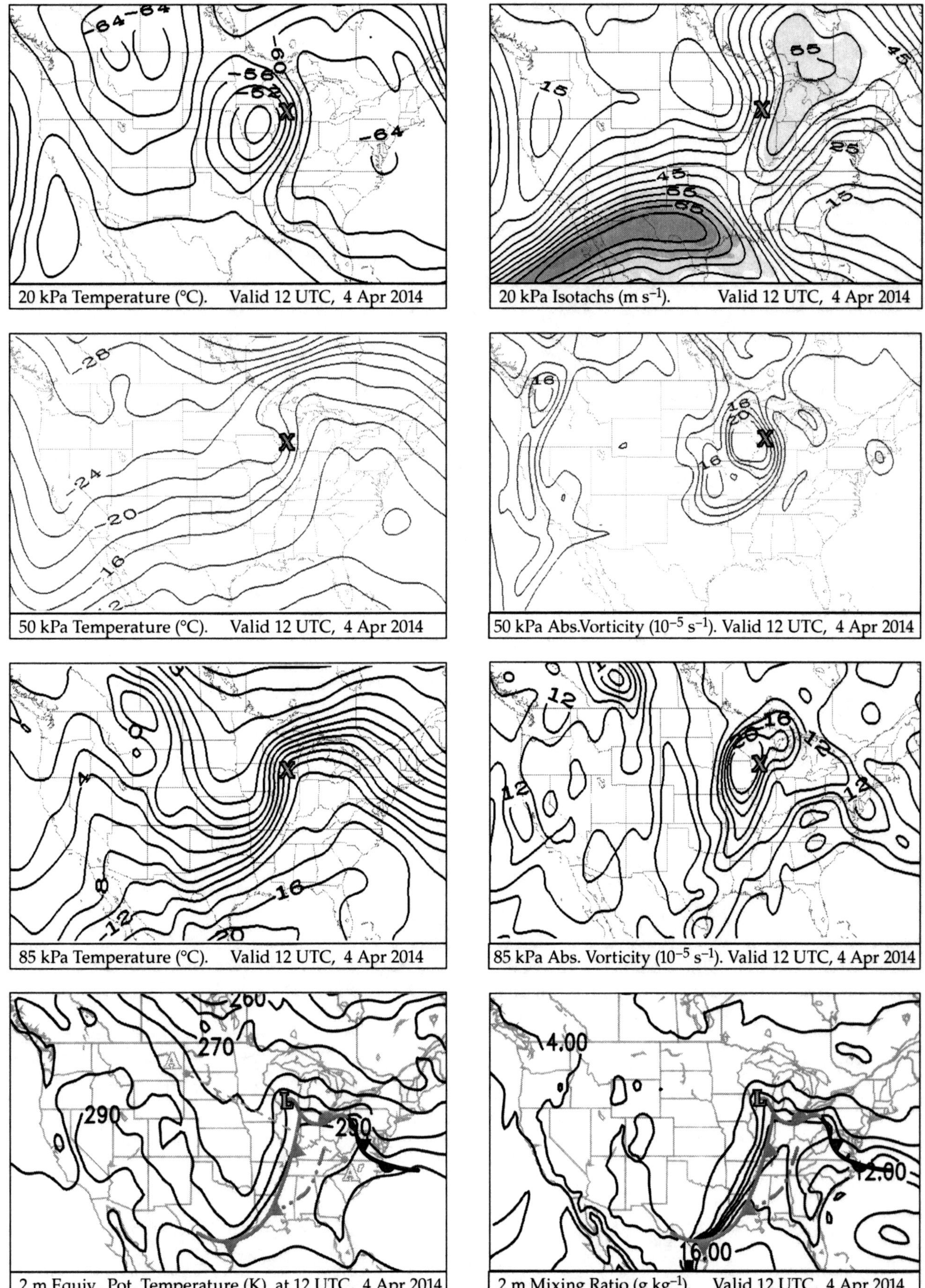

Figure 13.14. *Temperature (left column) and other variables (right). "L" and "X" indicate location of the surface low center.*

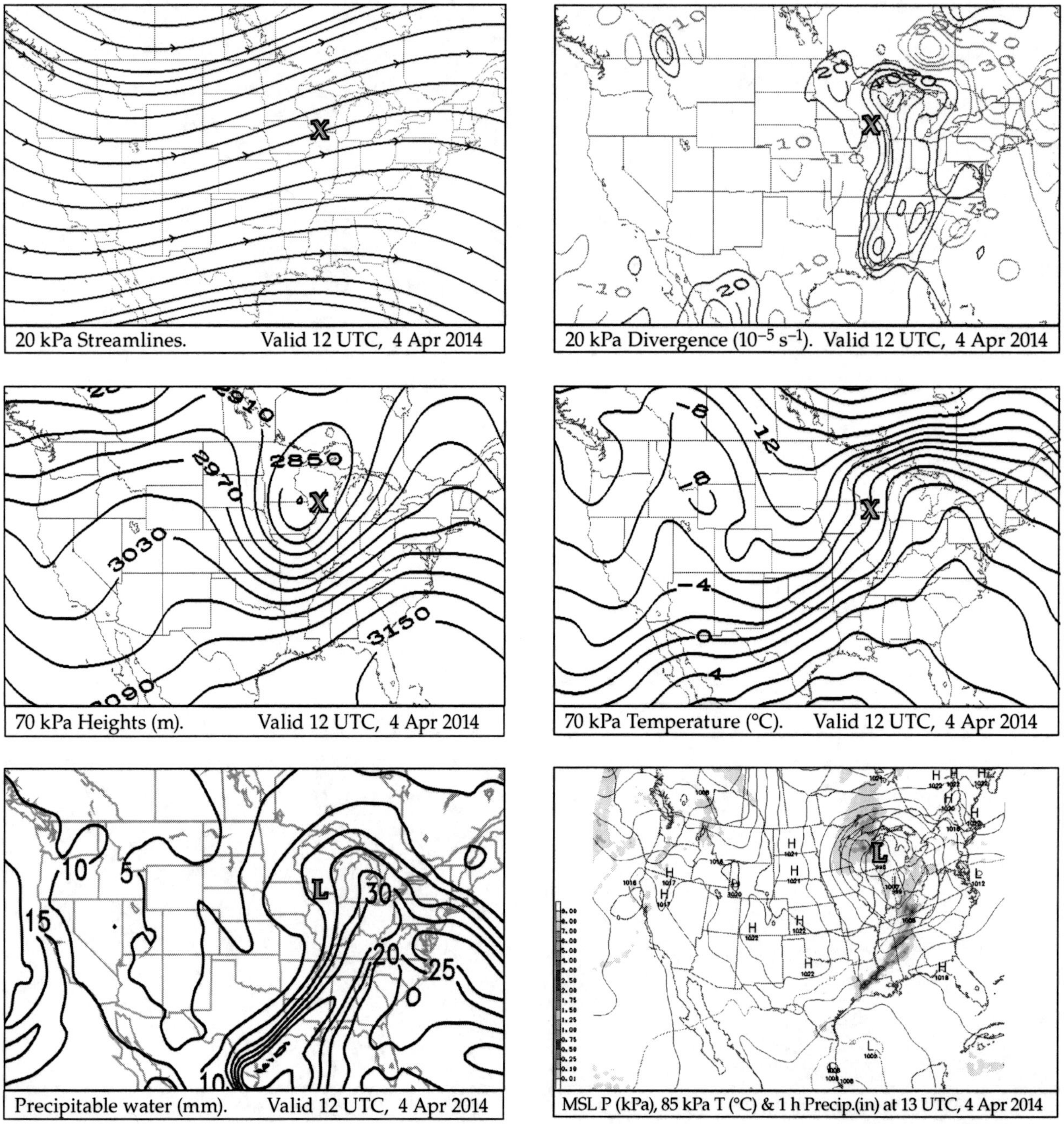

20 kPa Streamlines. Valid 12 UTC, 4 Apr 2014

20 kPa Divergence (10^{-5} s^{-1}). Valid 12 UTC, 4 Apr 2014

70 kPa Heights (m). Valid 12 UTC, 4 Apr 2014

70 kPa Temperature (°C). Valid 12 UTC, 4 Apr 2014

Precipitable water (mm). Valid 12 UTC, 4 Apr 2014

MSL P (kPa), 85 kPa T (°C) & 1 h Precip.(in) at 13 UTC, 4 Apr 2014

Figure 13.15. *Maps higher on the page are for higher in the atmosphere. "L" and "X" indicate location of* surface *low-pressure center.*

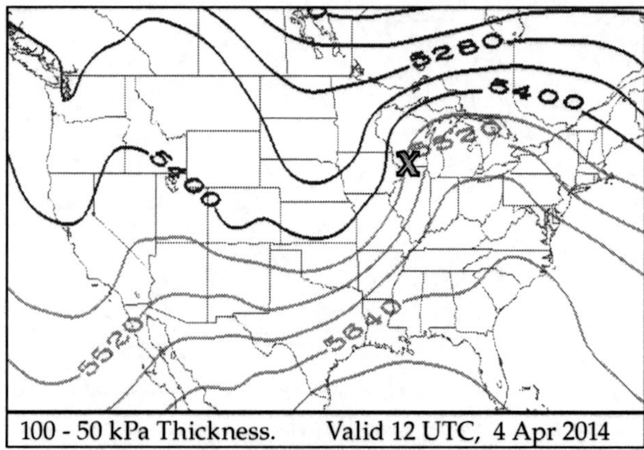

100 - 50 kPa Thickness. Valid 12 UTC, 4 Apr 2014

Figure 13.16.
Thickness (m) between the 100 kPa and 50 kPa isobaric surfaces.

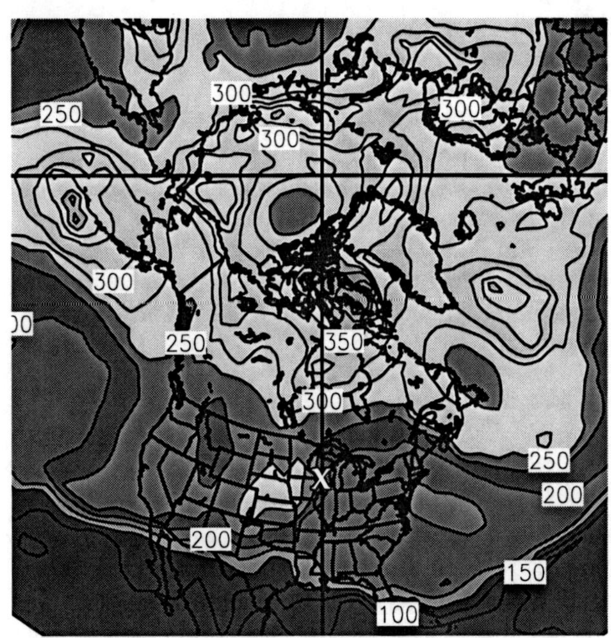

Fig. 13.17
Tropopause pressure (hPa), valid 4 Apr 2014 from NCEP operational analysis. "X" indicates surface-low location. Thanks to NOAA Earth System Research Lab. [100 hPa = 10 kPa.]

Fig. 13.18 (below)
Hemispheric plot of 20 kPa height contours (unit interpretation: 1220 = 12.2 km), valid 4 Apr 2014 from NCEP operational analysis. The North Pole is at the center of the map. Thanks to NOAA Earth System Research Lab.

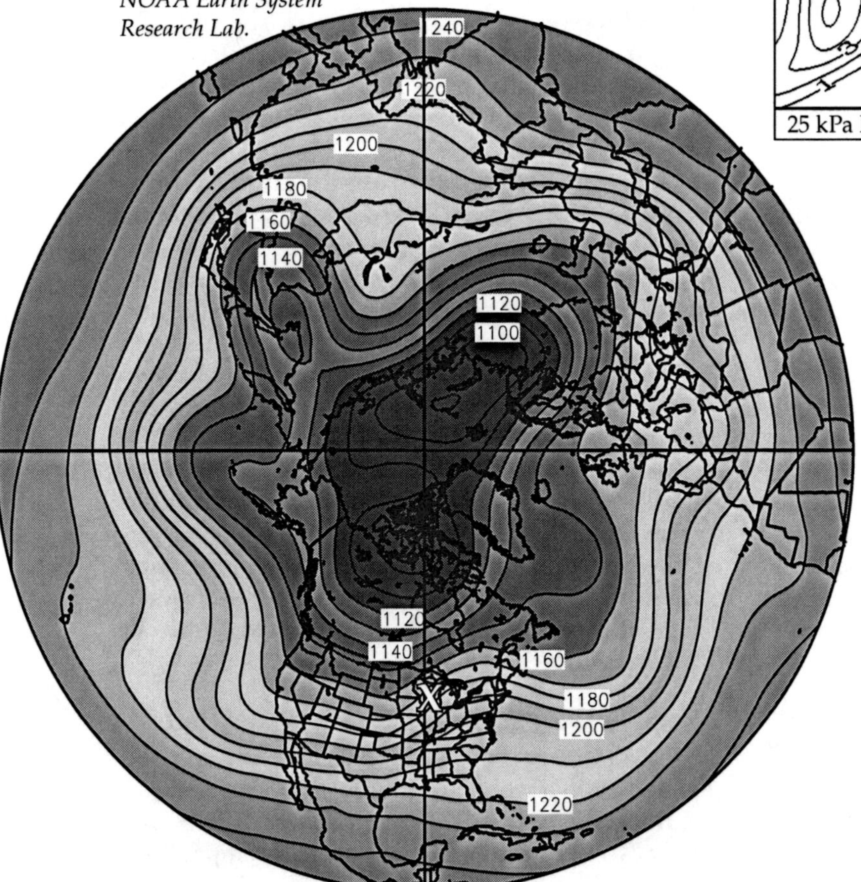

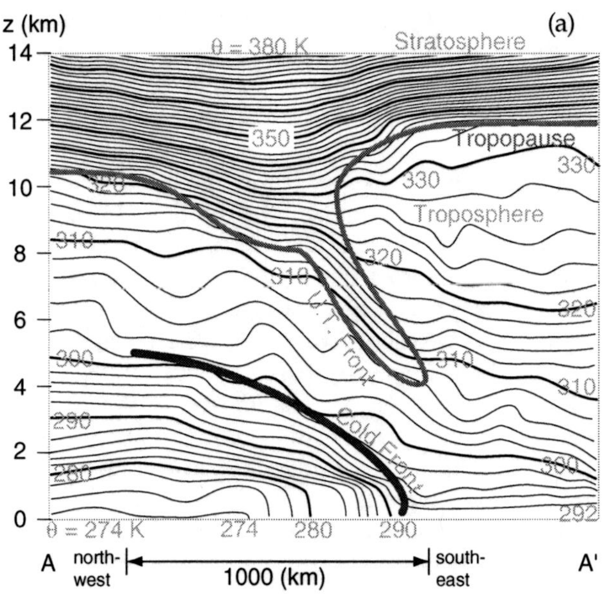

(b)

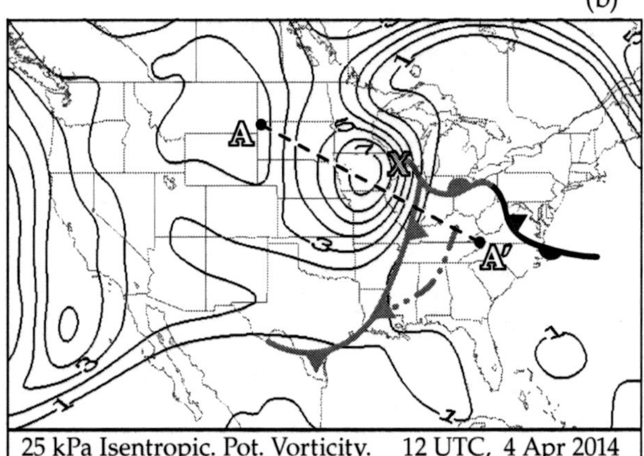

Fig. 13.19 (above)
(a) Vertical cross section through the atmosphere, along the dashed diagonal line shown in (b). Black lines are isentropes (lines of constant potential temperature θ). Red line indicates the tropopause, separating the troposphere from the stratosphere. An Upper Tropospheric (U.T.) Front corresponds to a tropopause fold, where stratospheric air is closer to the ground. Below the U.T. Front is a surface-based Cold Front, where the cold air to the left of the dark-blue line is advancing from left to right.

(b) Isopleths are of isentropic potential vorticity in potential vorticity units (PVU). These are plotted on the 25 kPa isobaric surface. Values greater than 1.5 PVU usually are associated with stratospheric air. The diagonal straight dotted line shows the cross section location plotted in (a). The "bulls eye" just west of the surface low "X" indicates a tropopause fold, where stratospheric air descends closer to the ground. Surface fronts are also drawn on this map.

Weather-map Discussion for this Case

As recommended for most weather discussions, start with the big picture, and progress toward the details. Also, start from the top down.

Hemispheric Map - Top of Troposphere

Starting with the planetary scale, Fig. 13.18 shows **Hemispheric 20 kPa Geopotential Height Contours**. It shows five long Rossby-wave troughs around the globe. The broad trough over N. America also has two short-wave troughs superimposed — one along the west coast and the other in the middle of N. America. The jet stream flows from west to east along the height contours plotted in this diagram, with faster winds where the contours are packed closer together.

Next, zoom to the synoptic scale over N. America. This is discussed in the next several subsections using Figs. 13.13 - 13.15.

20 kPa Charts — Top of Troposphere (z ≈ 11.5 km MSL)

Focus on the top row of charts in Figs. 13.13 - 13.15. The thick dashed lines on the **20 kPa Height** contour map show the two short-wave trough axes. The trough over the central USA is the one associated with the case-study cyclone. This trough is west of the location of the surface low-pressure center ("X"). The **20 kPa Wind Vector** map shows generally westerly winds aloft, switching to southwesterly over most of the eastern third of the USA. Wind speeds in the **20 kPa Isotach** chart show two jet streaks (shaded in yellow) — one with max winds greater than 70 m s^{-1} in Texas and northern Mexico, and a weaker jet streak over the Great Lakes.

The **20 kPa Temperature** chart shows a "bulls-eye" of relatively warm air (–50°C) aloft just west of the "X". This is associated with an intrusion of stratospheric air down into the troposphere (Fig. 13.19). The **20 kPa Divergence** map shows strong horizontal divergence (plotted with the blue contour lines) along and just east of the surface cold front and low center.

50 kPa Charts — Middle of Troposphere (z ≈5.5 km MSL)

Focus on the second row of charts in Figs. 13.13 - 13.14. The **50 kPa Height** chart shows a trough axis closer to the surface-low center ("X"). This low-pressure region has almost become a "closed low", where the height contours form closed ovals. **50 kPa Wind Vectors** show the predominantly westerly winds turning in such a way as to bring colder air equatorward on the west side of the low, and bringing warmer air poleward on the east side of the low (shown in the **50 kPa Temperature** chart). The **50 kPa Absolute Vorticity** chart shows a bulls-eye of positive vorticity just west of the surface low.

70 kPa Charts — (z ≈ 3 km MSL)

The second row of charts in Fig. 13.15 shows a closed low on the **70 kPa Height** chart, just west of the surface low. At this altitude the warm air advection poleward and cold-air advection equatorward are even more obvious east and west of the low, respectively, as shown on the **70 kPa Temperature** chart.

85 kPa Charts — (z ≈ 1.4 km MSL)

Focus on the third row of charts in Figs. 13.13 - 13.14. The **85 kPa Height** chart shows a deep closed low immediately to the west of the surface low. Associated with this system is a complete counterclockwise circulation of winds around the low, as shown in the **85 kPa Wind Vector** chart.

The strong temperature advection east and west of the low center are creating denser packing of isotherms along the frontal zones, as shown in the **85 kPa Temperature** chart. The cyclonically rotating flow causes a large magnitude of vorticity in the **85 kPa Absolute Vorticity** chart.

100 kPa and other Near-Surface Charts

Focus on the last row of charts in Figs. 13.13-13.14. The approximate surface-frontal locations have also been drawn on most of these charts. The **100 kPa Height** chart shows the surface low that is deep relative to the higher pressures surrounding it. The **10 m Wind Vectors** chart shows sharp wind shifts across the frontal zones.

Isentropes of **2 m Equivalent Potential Temperature** clearly demark the cold and warm frontal zones with tightly packed (closely spaced) isentropes. Recall that fronts on weather maps are drawn on the warm sides of the frontal zones. Southeast of the low center is a humid "warm sector" with strong moisture gradients across the warm and cold fronts as is apparent by the tight isohume packing in the **2 m Water Vapor Mixing Ratio** chart.

Next, focus on row 3 of Fig. 13.15. The high humidities also cause large values of **Precipitable water** (moisture summed over the whole depth of the atmosphere), particularly along the frontal zones. So it is no surprise to see the rain showers in the **MSL Pressure, 85 kPa Temperature and 1-h Precipitation** chart.

100 to 50 kPa Thickness

Fig. 13.16 shows the vertical distance between the 100 and 50 kPa isobaric surfaces. Namely, it shows the thickness of the 100 to 50 kPa layer of air. This thickness is proportional to the average temperature in the bottom half of the troposphere, as described by the hypsometric eq. The warm-air sector (red isopleths) southeast of the surface low, and the cold air north and west (blue isopleths) are apparent. Recall that the thermal wind vector (i.e., the vertical shear of the geostrophic wind) is parallel to the thickness lines, with a direction such that cold air (thin thicknesses) is on the left side of the vector.

Tropopause and Vertical Cross Section

Fig. 13.17 shows the pressure altitude of the tropopause. A higher tropopause would have lower pressure. Above the surface low ("X") the tropopause is at about the 20 kPa (= 200 hPa) level. Further to the south over Florida, the tropopause is at even higher altitude (where $P \approx 10$ kPa = 100 hPa). North and west of the "X" the tropopause is at lower altitude, where $P = 30$ kPa (=300 hPa) or greater. Globally, the tropopause is higher over the subtropics and lower over the sub-polar regions.

Fig. 13.19 shows a vertical slice through the atmosphere. Lines of uniform potential temperature (isentropes), rather than absolute temperature, are plotted so as to exclude the adiabatic temperature change associated with the pressure decrease with height. Tight packing of isentropes indicates strong static stability, such as in the stratosphere, **upper-tropospheric (U.T.) fronts** (also called **tropopause folds**), and surface fronts.

Special thanks to Greg West and David Siuta for creating many of the case-study maps in Figs. 13.10 - 13.19 and elsewhere in this chapter.

In the next sections, we see how dynamics can be used to explain cyclone formation and evolution.

LEE CYCLOGENESIS

Recall from the General Circulation chapter that the west-to-east jet stream can meander poleward and equatorward as **Rossby waves,** due to barotropic and baroclinic instability. Such waves in the upper-air (jet-stream) flow can create mid-latitude cyclones at the surface, as shown in Figs. 13.6 & 13.7.

One trigger mechanism for this instability is flow over high mountain ranges. The Rossby wave triggered by such a mountain often has a trough just downwind of (ie., to the "lee" of) mountain ranges. East of this trough is a favored location for cyclogenesis; hence, it is known as **lee cyclogenesis**. Because the mountain location is fixed, the resulting Rossby-wave trough and ridge locations are stationary with respect to the mountain-range location.

INFO • Multi-field Charts

Most of the weather maps presented in the previous case study contained plots of only one field, such as the wind field or height field. Because many fields are related to each other or work together, meteorologists often plot multiple fields on the same chart.

To help you discriminate between the different fields, they are usually plotted differently. One might use solid lines and the other dashed. Or one might be contoured and the other shaded (see Fig. e, Fig. 13.7a, or Fig. 13.15-right map). Look for a legend or caption that describes which lines go with which fields, and gives units.

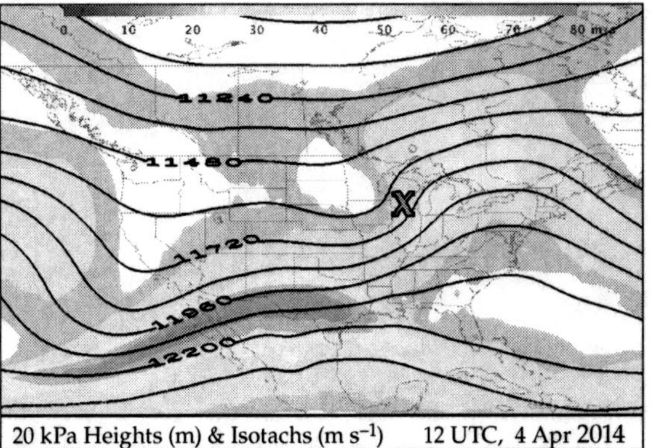

Fig. e.
Geopotential heights (lines) and wind speed (color fill).

INFO • Synoptics

Synoptic meteorology is the study and analysis of weather maps, often with the aim to forecast the weather on horizontal scales of 400 to 4000 km. **Synoptic weather maps** describe an instantaneous state of the atmosphere over a wide area, as created from weather observations made nearly simultaneously.

Typical weather phenomena at these **synoptic scales** include cyclones (Lows), anticyclones (Highs), and airmasses. Fronts are also included in synoptics because of their length, even though frontal zones are so narrow that they can also be classified as mesoscale. See Table 10-6 and Fig. 10.24 in the Atmospheric Forces & Winds chapter for a list of different atmospheric scales.

The material in this chapter and in the previous one fall solidly in the field of **synoptics**. People who specialize in synoptic meteorology are called **synopticians**.

The word "synoptics" is from the Greek "synoptikos", and literally means "seeing everything together". It is the big picture.

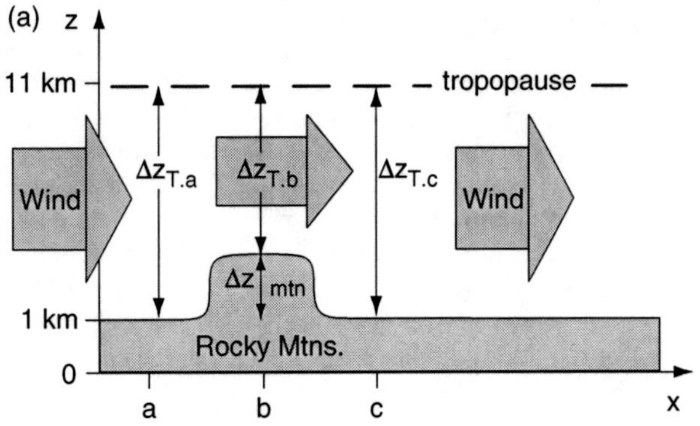

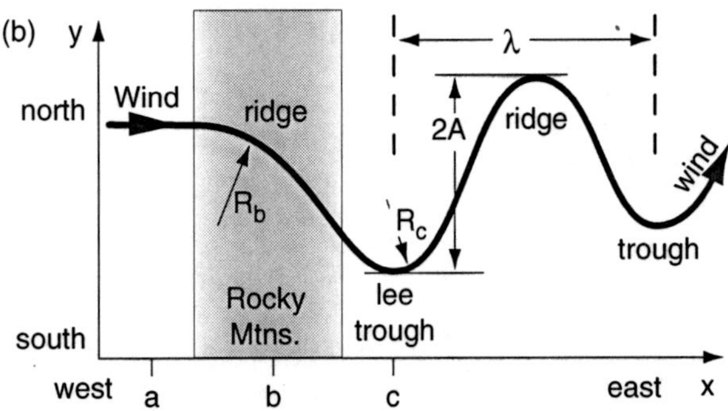

Figure 13.20
Cyclogenesis to the lee of the mountains. (a) Vertical cross section. (b) Map of jet-stream flow. "Ridge" and "trough" refer to the wind-flow pattern, not the topography.

Sample Application

What amplitude & wavelength of terrain-triggered Rossby wave would you expect for a mountain range at 48°N that is 1.2 km high? The upstream depth of the troposphere is 11 km, with upstream wind is 19 m s^{-1}.

Find the Answer

Given: ϕ = 48°N, Δz_{mtn} = 1.2 km, Δz_T = 11 km, M = 19 m s^{-1}.

Find: A = ? km , λ = ? km

Use eq. (13.4):

A = [(1.2 km) / (11 km)] · (6371 km) = **695 km**

Next, use eq. (13.2) to find β at 48°N:

$\beta = (2.294\text{x}10^{-11}\ \text{m}^{-1}\cdot\text{s}^{-1}) \cdot \cos(48°) = 1.53\text{x}10^{-11}\ \text{m}^{-1}\cdot\text{s}^{-1}$

Finally, use eq. (13.1):

$$\lambda \approx 2\cdot\pi\cdot\left[\frac{19\text{m}\cdot\text{s}^{-1}}{1.53\times10^{-11}\text{m}^{-1}\cdot\text{s}^{-1}}\right]^{1/2} = \underline{\mathbf{6990\ km}}$$

Check: Physics and units are reasonable.

Exposition: Is this wave truly a planetary wave? Yes, because its wavelength (6,990 km) would fit 3.8 times around the Earth at 48°N (circumference = $2\cdot\pi\cdot R_{Earth}\cdot\cos(48°)$ = 26,785 km). Also, the north-south meander of the wave spans $2A$ = 12.5° of latitude.

Stationary Rossby Waves

Consider a wind that causes air in the troposphere to blow over the Rocky mountains (Fig. 13.20). Convective clouds (e.g., thunderstorms) and turbulence can cause the Rossby wave amplitude to decrease further east, so the first wave after the mountain (at location c in Fig. 13.20) is the one you should focus on.

These Rossby waves have a dominant wavelength (λ) of roughly

$$\lambda \approx 2\cdot\pi\cdot\left[\frac{M}{\beta}\right]^{1/2} \qquad (13.1)$$

where the mean wind speed is M. As you have seen in an earlier chapter, β is the northward gradient of the Coriolis parameter (f_c):

$$\beta = \frac{\Delta f_c}{\Delta y} = \frac{2\cdot\Omega}{R_{earth}}\cdot\cos\phi \qquad \bullet(13.2)$$

Factor $2{\cdot}\Omega = 1.458\text{x}10^{-4}\ \text{s}^{-1}$ is twice the angular rotation rate of the Earth. At North-American latitudes, β is roughly 1.5 to $2\text{x}10^{-11}\ \text{m}^{-1}\ \text{s}^{-1}$.

Knowing the mountain-range height (Δz_{mtn}) relative to the surrounding plains, and knowing the initial depth of the troposphere (Δz_T), the Rossby-wave amplitude A is:

$$A \approx \frac{f_c}{\beta}\cdot\frac{\Delta z_{mtn}}{\Delta z_T} \qquad (13.3)$$

Because β is related to f_c, we can analytically find their ratio as $f_c/\beta = R_{Earth}\cdot\tan(\phi)$, where the average radius (R_{Earth}) of the Earth is 6371 km. Over North America the tangent of the latitude ϕ is $\tan(\phi) \approx 1$. Thus:

$$A \approx \frac{\Delta z_{mtn}}{\Delta z_T}\cdot R_{earth} \qquad (13.4)$$

where $2A$ is the Δy distance between the wave trough and crest.

In summary, the equations above show that north-south Rossby-wave amplitude depends on the height of the mountains, but does not depend on wind speed. Conversely, wind speed is important in determining Rossby wavelength, while mountain height is irrelevant.

Potential-vorticity Conservation

Use conservation of potential vorticity as a tool to understand such mountain lee-side Rossby-wave triggering (Fig. 13.20). Create a "toy model" by assuming wind speed is constant in the Rossby wave, and that there is no wind shear affecting vorticity. For this situation, the conservation of potential vorticity ζ_p is given by eq. (11.25) as:

$$\zeta_p = \frac{(M/R) + f_c}{\Delta z} = \text{constant} \qquad \bullet(13.5)$$

For this toy model, consider the initial winds to be blowing straight toward the Rocky Mountains from the west. These initial winds have no curvature at location "a", thus $R = \infty$ and eq. (13.5) becomes:

$$\zeta_p = \frac{f_{c.a}}{\Delta z_{T.a}} \qquad (13.6)$$

where $\Delta z_{T.a}$ is the average depth of troposphere at point "a". Because potential vorticity is conserved, we can use this fixed value of ζ_p to see how the Rossby wave is generated.

Let Δz_{mtn} be the relative mountain height above the surrounding land (Fig. 13.20a). As the air blows over the mountain range, the troposphere becomes thinner as it is squeezed between mountain top and the tropopause at location "b": $\Delta z_{T.b} = \Delta z_{T.a} - \Delta z_{mtn}$. But the latitude of the air hasn't changed much yet, so $f_{c.b} \approx f_{c.a}$. Because Δz has changed, we can solve eq. (13.5) for the radius of curvature needed to maintain $\zeta_{p.b} = \zeta_{p.a}$.

$$R_b = \frac{-M}{f_{c.a} \cdot (\Delta z_{mtn} / \Delta z_{T.a})} \qquad (13.7)$$

Namely, in eq. (13.5), when Δz became smaller while f_c was constant, M/R had to also become smaller to keep the ratio constant. But since M/R was initially zero, the new M/R had to become negative. Negative R means anticyclonic curvature.

As sketched in Fig. 13.20, such curvature turns the wind toward the equator. But equatorward-moving air experiences smaller Coriolis parameter, requiring that R_b become larger (less curved) to conserve ζ_p. Near the east side of the Rocky Mountains the terrain elevation decreases at point "c", allowing the air thickness Δz to increase back to is original value.

But now the air is closer to the equator where Coriolis parameter is smaller, so the radius of curvature R_c at location "c" becomes positive in order to keep potential vorticity constant. This positive vorticity gives that cyclonic curvature that defines the lee trough of the Rossby wave. As was sketched in Fig. 13.7, surface cyclogenesis could be supported just east of the lee trough.

Sample Application

Picture a scenario as plotted in Fig. 13.20, with 25 m s^{-1} wind at location "a", mountain height of 1.2 km, troposphere thickness of 11 km, and latitude 45°N. What is the value of the initial potential vorticity, and what is the radius of curvature at point "b"?

Find the Answer

Given: $M = 25$ m s^{-1}, $\Delta z_{mtn} = 1.2$ km, $R_{initial} = \infty$,
$\Delta z_T = 11$ km, $\phi = 45$°N.
Find: $\zeta_{p.a} = ?$ m^{-1}·s^{-1}, $R_b = ?$ km

Assumption: Neglect wind shear in the vorticity calculation.

Eq. (10.16) can be applied to get the Coriolis parameter
$f_c = (1.458\times10^{-4}\ \text{s}^{-1})\cdot\sin(45°) = 1.031\times10^{-4}\ \text{s}^{-1}$

Use eq. (13.6):

$$\zeta_p = \frac{1.031\times10^{-4}\text{s}^{-1}}{11\text{km}} = \underline{\mathbf{9.37x10^{-9}}}\ \text{m}^{-1}\ \text{s}^{-1}$$

Next, apply eq. (13.7) to get the radius of curvature:

$$R_b = \frac{-(25\text{m/s})}{(1.031\times10^{-4}\text{s}^{-1})\cdot(1.2\text{km}/11\text{km})} = \underline{\mathbf{-2223.\ km}}$$

Check: Physics and units are reasonable.
Exposition: The negative sign for the radius of curvature means that the turn is anticyclonic (clockwise in the N. Hemisphere). Typically, the cyclonic trough curvature is the same order of magnitude as the anticyclonic ridge curvature. East of the first trough and west of the next ridge is where cyclogenesis is supported.

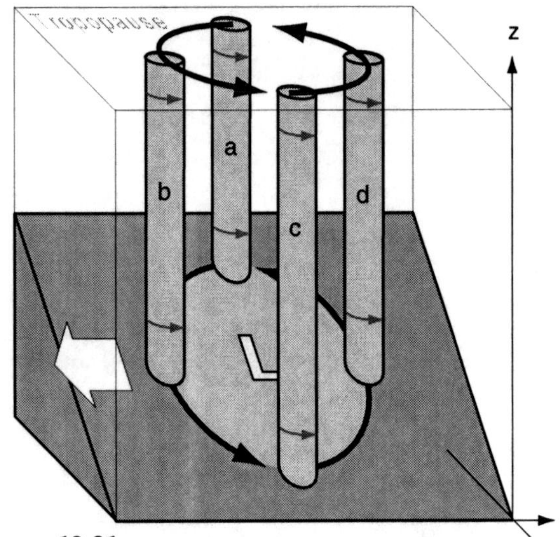

Figure 13.21
A low-pressure (L) center above terrain that slopes downward to the east, for the Northern Hemisphere.

Lee-side Translation Equatorward

Suppose an extratropical cyclone (low center) is positioned over the east side of a mountain range in the Northern Hemisphere, as sketched in Fig. 13.21. In this diagram, the green circle and the air above it are the cyclone. Air within this cyclone has positive (cyclonic) vorticity, as represented by the rotating blue air columns in the figure. The locations of these air columns are also moving counterclockwise around the common low center (L) — driven by the synoptic-scale circulation around the low.

As column "a" moves to position "b" and then "c", its vertical extent Δz stretches. This assumes that the top of the air columns is at the tropopause, while the bottom follows the sloping terrain. Due to conservation of potential vorticity ζ_p, this stretching must be accompanied by an increase in relative vorticity $\Delta\zeta_r$:

$$\Delta\zeta_r = 2 \cdot R \cdot \alpha \cdot \zeta_p \qquad (13.8)$$

R is cyclone radius and $\alpha = \Delta z/\Delta x$ is terrain slope.

Conversely, as column "c" moves to position "d" and then "a", its vertical extent shrinks, forcing its relative vorticity to decrease to maintain constant potential vorticity. Hence, the center of action of the low center shifts (translates) equatorward (white arrow in Fig. 13.21) along the lee side of the mountains, following the region of increasing ζ_r.

A similar conclusion can be reached by considering conservation of isentropic potential vorticity (IPV). Air in the bottom of column "a" descends and warms adiabatically en route to position "c", while there is no descent warming at the column top. Hence, the static stability of the column decreases at its equatorward side. This drives an increase in relative vorticity on the equatorward flank of the cyclone to conserve IPV. Again, the cyclone moves equatorward toward the region of greater relative vorticity.

Sample Application

The cyclone of Fig. 13.21 has $\zeta_p = 1\text{x}10^{-8}\ \text{m}^{-1}\cdot\text{s}^{-1}$ and R = 600 km. The mountain slope is 1:500. Find the relative-vorticity change on the equatorward side.

Find the Answer

Given: $\zeta_p = 1\text{x}10^{-8}\ \text{m}^{-1}\cdot\text{s}^{-1}$, R = 600 km, α = 0.002
Find: $\Delta\zeta_r = ?\ \text{s}^{-1}$.
Assume constant latitude in Northern Hemisphere.

Use eq. (13.8):
$\Delta\zeta_r = 2 \cdot (600{,}000\ \text{m}) \cdot (0.002) \cdot (1\text{x}10^{-8}\ \text{m}^{-1}\cdot\text{s}^{-1})$
$= \underline{2.4\text{x}10^{-5}}\ \text{s}^{-1}$.

Check: Physics and units are reasonable.
Exposition: A similar decrease is likely on the poleward side. The combined effect causes the cyclone to translate equatorward to where vorticity is greatest.

SPIN-UP OF CYCLONIC ROTATION

Cyclogenesis is associated with upward motion, decreasing surface pressure, and increasing vorticity (i.e., spin-up). You can gain insight into cyclogenesis by studying all three characteristics, even though they are intimately related (Fig. 13.22). Let us start with vorticity.

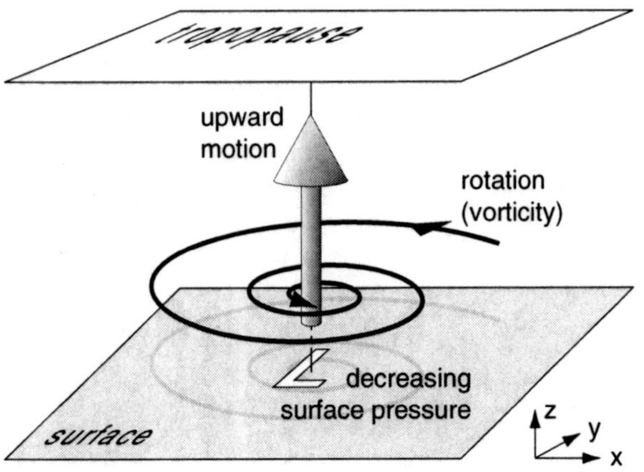

Figure 13.22
Attributes of cyclogenesis. Updrafts remove air molecules from near the ground, which lowers surface pressure. The pressure gradient drives winds, which rotate due to Coriolis force.

The equation that forecasts change of vorticity with time is called the **vorticity tendency equation**. We can investigate the processes that cause cyclogenesis (**spin up**; positive-vorticity increase) and cyclolysis (**spin down**; positive-vorticity decrease) by examining terms in the vorticity tendency equation. Mountains are not needed for these processes.

Vorticity Tendency Equation

The change of relative vorticity ζ_r over time (i.e., the **spin-up** or **vorticity tendency**) can be predicted using the following equation:

•(13.9)

$$\underbrace{\frac{\Delta\zeta_r}{\Delta t}}_{tendency} = \underbrace{-U\frac{\Delta\zeta_r}{\Delta x} - V\frac{\Delta\zeta_r}{\Delta y}}_{horiz.\ advection} \underbrace{- W\frac{\Delta\zeta_r}{\Delta z}}_{vert.\ advect.} \underbrace{+ f_c\frac{\Delta W}{\Delta z}}_{stretching} \underbrace{- V\frac{\Delta f_c}{\Delta y}}_{beta\ effect}$$

$$\underbrace{+\frac{\Delta U}{\Delta z}\cdot\frac{\Delta W}{\Delta y}}_{tilting\ (A)} \underbrace{- \frac{\Delta V}{\Delta z}\cdot\frac{\Delta W}{\Delta x}}_{(B)} \underbrace{- \frac{\Delta W}{\Delta x}\cdot\frac{\Delta W}{\Delta y}}_{(C)} \underbrace{+ \zeta_r\frac{\Delta W}{\Delta z}}_{stretching\ of\ rel.\ vort.} \underbrace{- C_D\frac{M}{z_i}\zeta_r}_{turbulent\ drag}$$

Positive vorticity tendency indicates cyclogenesis.

Vorticity Advection: If the wind blows air of greater vorticity into your region of interest, then this is called **positive vorticity advection** (**PVA**). **Negative vorticity advection** (**NVA**) is when lower-vorticity air is blown into a region. These advections can be caused by vertical winds and horizontal winds (Fig. 13.23a).

Stretching: Consider a short column of air that is spinning as a vortex tube. Horizontal convergence of air toward this tube will cause the tube to become taller and more slender (smaller diameter). The taller or **stretched** vortex tube supports cyclogenesis (Fig. 13.23b). Conversely, horizontal **divergence** shortens the vortex tube and supports cyclolysis or anticyclogenesis. In the first and second lines of eq. (13.9) are the stretching terms for Earth's rotation and relative vorticity, respectively. Stretching means that the top of the vortex tube moves upward away from (or moves faster than) the movement of the bottom of the vortex tube; hence $\Delta W/\Delta z$ is positive for stretching.

Beta Effect: Recall from eq. (11.35) in the General Circulation chapter that we can define $\beta = \Delta f_c/\Delta y$. Beta is positive in the N. Hemisphere because the Coriolis parameter increases toward the north pole (see eq. 13.2). If wind moves air southward (i.e., V = negative) to where f_c is smaller, then relative vorticity ζ_r becomes larger (as indicated by the negative sign in front of the beta term) to conserve potential vorticity (Fig. 13.23c).

Tilting Terms: (A & B in eq. 13.9) If the horizontal winds change with altitude, then this shear causes vorticity along a horizontal axis. (C in eq. 13.9) Neighboring up- and down-drafts give horizontal shear of the vertical wind, causing vorticity along a horizontal axis. (A-C) If a resulting horizontal vortex tube experiences stronger vertical velocity on one end relative to the other (Fig. 13.23d), then the tube will tilt to become more vertical. Because spinning about a vertical axis is how we define vorticity, we have increased vorticity via the tilting of initially horizontal vorticity.

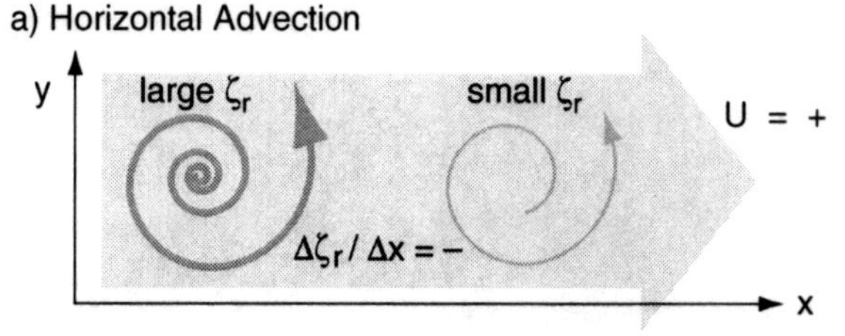

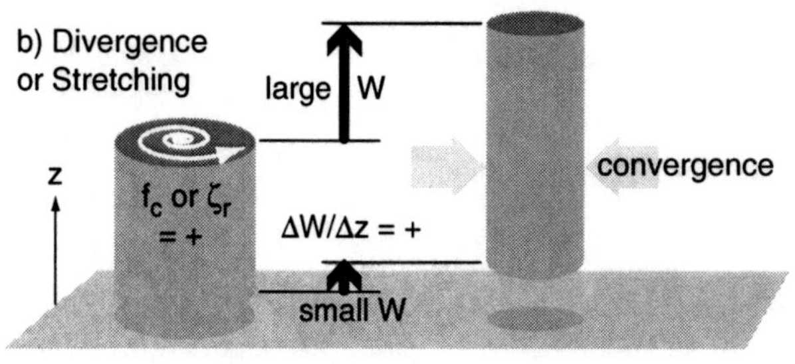

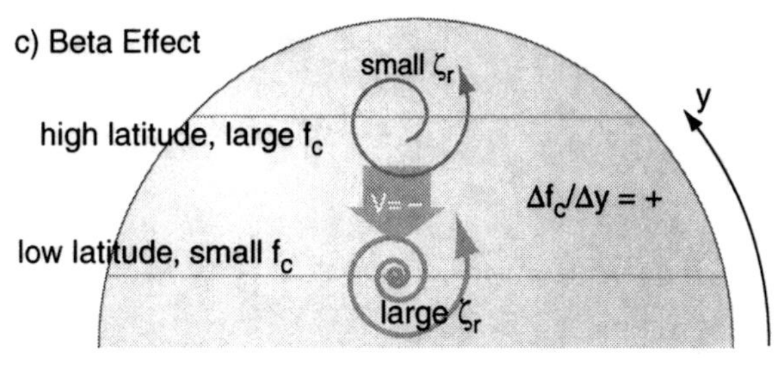

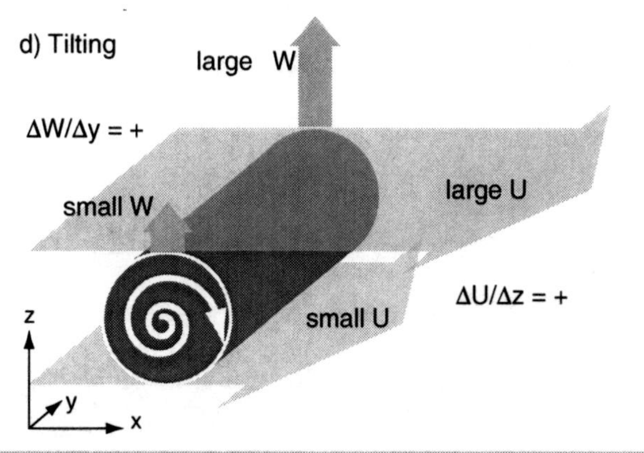

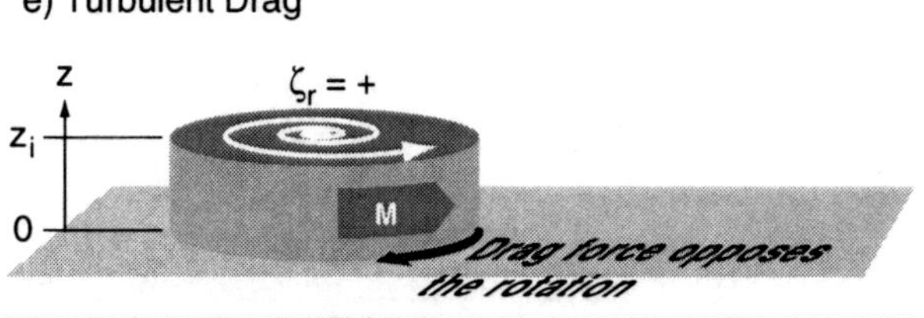

Figure 13.23
Illustration of processes that affect vertical vorticity (see eq. 13.9). An additional drag process is shown in the next column.

f) Turbulent Drag (part 2)

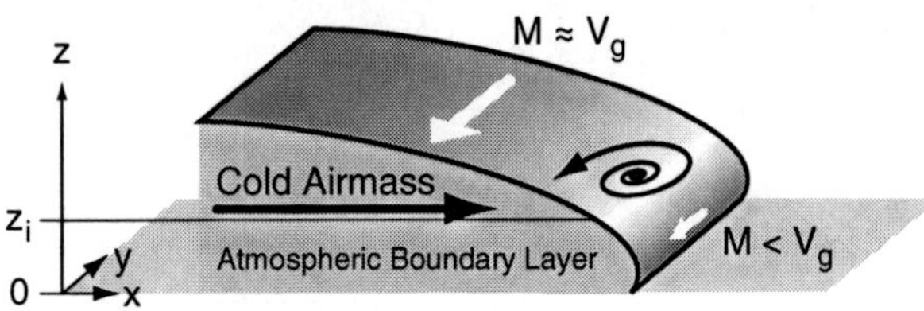

Figure 13.23 *(part 2)*
Illustration of a turbulent-drag process that causes spin-up.

Turbulence in the atmospheric boundary layer (ABL) communicates frictional forces from the ground to the whole ABL. This **turbulent drag** acts to slow the wind and decrease rotation rates (Fig. 13.23e). Such **spin down** can cause cyclolysis.

However, for cold fronts drag can increase vorticity. As the cold air advances (black arrow in Fig. 13.23f), Coriolis force will turn the winds and create a geostrophic wind V_g (large white arrow). Closer to the leading edge of the front where the cold air is shallower, the winds M subgeostrophic because of the greater drag. The result is a change of wind speed M with distance x that causes positive vorticity.

All of the terms in the vorticity-tendency equation must be summed to determine net spin down or spin up.

You can identify the action of some of these terms by looking at weather maps.

Fig. 13.24 shows the wind vectors and absolute vorticity on the 50 kPa isobaric surface (roughly in the middle of the troposphere) for the case-study cyclone. **Positive vorticity advection** (PVA) occurs where wind vectors are crossing the vorticity contours from high toward low vorticity, such as highlighted by the dark box in Fig. 13.24. Namely, higher vorticity air is blowing into regions that contained lower vorticity. This region favors cyclone spin up.

Negative vorticity advection (NVA) is where the wind crosses the vorticity contours from low to high vorticity values (dark oval in Fig. 13.24). By using the absolute vorticity instead of relative vorticity, Fig. 13.24 combines the advection and beta terms.

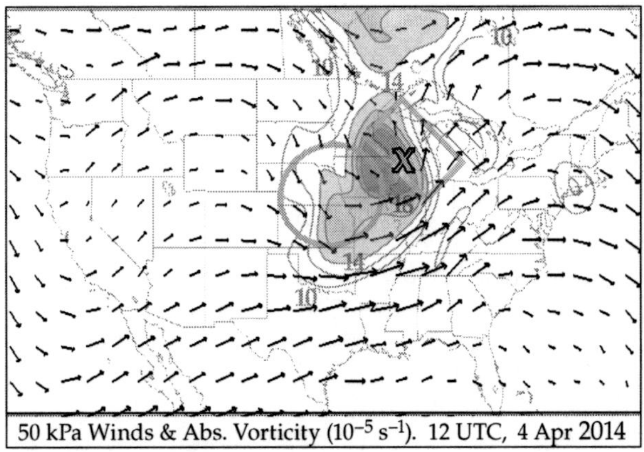

Figure 13.24
Absolute vorticity (shaded) and winds (vectors) at 50 kPa, highlighting regions of positive (green rectangle) and negative (red oval) vorticity advection for the case-study storm.

Fig. 13.25 shows vertical velocity in the middle of the atmosphere. Since vertical velocity is near zero at the ground, regions of positive vertical velocity at 50 kPa must correspond to stretching in the bottom half of the atmosphere. Thus, the updraft regions in the figure favor cyclone spin-up (i.e., cyclogenesis).

In the bottom half of the troposphere, regions of stretching must correspond to regions of convergence of air, due to mass continuity. Fig. 13.26 shows the **divergence** field at 85 kPa. Negative divergence corresponds to convergence. The regions of low-altitude convergence favor cyclone spin-up.

Low-altitude spin-down due to turbulent drag occurs wherever there is rotation. Thus, the rotation of 10 m wind vectors around the surface low in Fig. 13.13 indicate vorticity that is spinning down. The tilting term will also be discussed in the Thunderstorm chapters, regarding tornado formation.

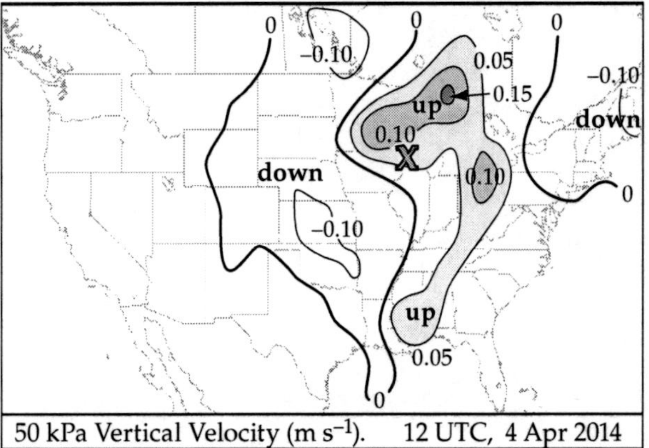

Figure 13.25
Vertical velocity (m s^{-1}) near the middle of the troposphere (50 kPa) for the case-study storm. At this altitude, a value of w = 0.10 m s^{-1} corresponds to ω = –0.68 Pa s^{-1}.

Quasi-Geostrophic Approximation

Above the boundary layer (and away from fronts, jets, and thunderstorms) the terms in the second line of the vorticity equation are smaller than those in the first line, and can be neglected. Also, for synoptic scale, extratropical weather systems, the winds are almost geostrophic (**quasi-geostrophic**).

These weather phenomena are simpler to analyze than thunderstorms and hurricanes, and can be well approximated by a set of equations (quasi-geostrophic vorticity and omega equations) that are less complicated than the full set of **primitive equations** of motion (Newton's second law, the first law of thermodynamics, continuity, and ideal gas law).

As a result of the simplifications above, the vorticity forecast equation simplifies to the following **quasi-geostrophic vorticity equation**:

$$\underbrace{\frac{\Delta\zeta_g}{\Delta t}}_{spin-up} = \underbrace{-U_g\frac{\Delta\zeta_g}{\Delta x} - V_g\frac{\Delta\zeta_g}{\Delta y}}_{horizontal\ advection} \underbrace{-V_g\frac{\Delta f_c}{\Delta y}}_{beta} \underbrace{+f_c\frac{\Delta W}{\Delta z}}_{stretching} \qquad \bullet(13.10)$$

where the relative geostrophic vorticity ζ_g is defined similar to the relative vorticity of eq. (11.20), except using geostrophic winds U_g and V_g:

$$\zeta_g = \frac{\Delta V_g}{\Delta x} - \frac{\Delta U_g}{\Delta y} \qquad \bullet(13.11)$$

For solid body rotation, eq. (11.22) becomes:

$$\zeta_g = \frac{2\cdot G}{R} \qquad \bullet(13.12)$$

where G is the geostrophic wind speed and R is the radius of curvature.

The prefix "quasi-" is used for the following reasons. If the winds were perfectly geostrophic or gradient, then they would be parallel to the isobars. Such winds never cross the isobars, and could not cause convergence into the low. With no convergence there would be no vertical velocity.

However, we know from observations that vertical motions do exist and are important for causing clouds and precipitation in cyclones. Thus, the last term in the quasi-geostrophic vorticity equation includes W, a wind that is not geostrophic. When such an **ageostrophic** vertical velocity is included in an equation that otherwise is totally geostrophic, the equation is said to be **quasi-geostrophic**, meaning partially geostrophic. The quasi-geostrophic approximation will also be used later in this chapter to estimate vertical velocity in cyclones.

Within a quasi-geostrophic system, the vorticity and temperature fields are closely coupled, due to

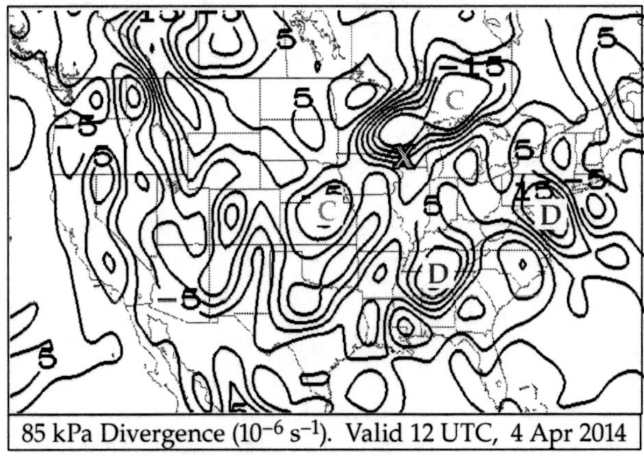

85 kPa Divergence (10^{-6} s^{-1}). Valid 12 UTC, 4 Apr 2014

Figure 13.26
Horizontal divergence ($D = \Delta U/\Delta x + \Delta V/\Delta y$) for the case-study storm. C = horizontal convergence (= – D).

Sample Application
Suppose an initial flow field has no geostrophic relative vorticity, but there is a straight north to south geostrophic wind blowing at 10 m s^{-1} at latitude 45°. Also, the top of a 1 km thick column of air rises at 0.01 m s^{-1}, while its base rises at 0.008 m s^{-1}. Find the rate of geostrophic-vorticity spin-up.

Find the Answer
Given: $V = -10$ m s^{-1}, $\phi = 45°$, $W_{top} = 0.01$ m s^{-1}, $W_{bottom} = 0.008$ m s^{-1}, $\Delta z = 1$ km.
Find: $\Delta\zeta_g/\Delta_t = ?$ s^{-2}

First, get the Coriolis parameter using eq. (10.16):
$f_c = (1.458\text{x}10^{-4}\ \text{s}^{-1})\cdot\sin(45°) = 0.000103\ \text{s}^{-1}$
Next, use eq. (13.2):

$$\beta = \frac{\Delta f_c}{\Delta y} = \frac{1.458\times10^{-4}\text{s}^{-1}}{6.357\times10^{6}\text{m}}\cdot\cos 45°$$
$$= 1.62\text{x}10^{-11}\ \text{m}^{-1}\cdot\text{s}^{-1}$$

Use the definition of a gradient (see Appendix A):

$$\frac{\Delta W}{\Delta z} = \frac{W_{top} - W_{bottom}}{z_{top} - z_{bottom}} = \frac{(0.01 - 0.008)\text{m/s}}{(1000 - 0)\text{m}}$$
$$= 2\text{x}10^{-6}\ \text{s}^{-1}$$

Finally, use eq. (13.10). We have no information about advection, so assume it is zero. The remaining terms give:

$$\underbrace{\frac{\Delta\zeta_g}{\Delta t}}_{spin-up} = \underbrace{-(-10\text{m/s})\cdot(1.62\times10^{-11}\text{m}^{-1}\cdot\text{s}^{-1})}_{beta}$$
$$\underbrace{+(0.000103\text{s}^{-1})\cdot(2\times10^{-6}\ \text{s}^{-1})}_{stretching}$$
$$=(1.62\text{x}10^{-10} + 2.06\text{x}10^{-10})\ \text{s}^{-2} = \underline{3.68\text{x}10^{-10}}\ \text{s}^{-2}$$

Check: Units OK. Physics OK.
Exposition: Even without any initial geostrophic vorticity, the rotation of the Earth can spin-up the flow if the wind blows appropriately.

HIGHER MATH • The Laplacian

A Laplacian operator ∇^2 can be defined as

$$\nabla^2 A = \frac{\partial^2 A}{\partial x^2} + \frac{\partial^2 A}{\partial y^2} + \frac{\partial^2 A}{\partial z^2}$$

where A represents any variable. Sometimes we are concerned only with the horizontal (H) portion:

$$\nabla_H^2 (A) = \frac{\partial^2 A}{\partial x^2} + \frac{\partial^2 A}{\partial y^2}$$

What does it mean? If $\partial A/\partial x$ represents the slope of a line when A is plotted vs. x on a graph, then $\partial^2 A/\partial x^2 = \partial[\,\partial A/\partial x\,]/\partial x$ is the change of slope; namely, the curvature.

How is it used? Recall from the Atm. Forces & Winds chapter that the geostrophic wind is defined as

$$U_g = -\frac{1}{f_c}\frac{\partial \Phi}{\partial y} \qquad V_g = \frac{1}{f_c}\frac{\partial \Phi}{\partial x}$$

where Φ is the geopotential ($\Phi = |g|\cdot z$). Plugging these into eq. (13.11) gives the geostrophic vorticity:

$$\zeta_g = \frac{\partial}{\partial x}\left(\frac{1}{f_c}\frac{\partial \Phi}{\partial x}\right) + \frac{\partial}{\partial y}\left(\frac{1}{f_c}\frac{\partial \Phi}{\partial y}\right)$$

or

$$\boxed{\zeta_g = \frac{1}{f_c}\nabla_H^2(\Phi)} \qquad (13.11b)$$

This illustrates the value of the Laplacian — as a way to more concisely describe the physics.

For example, a low-pressure center corresponds to a low-height center on an isobaric sfc. That isobaric surface is concave up, which corresponds to positive curvature. Namely, the Laplacian of $|g|\cdot z$ is positive, hence, ζ_g is positive. Thus, a low has positive vorticity.

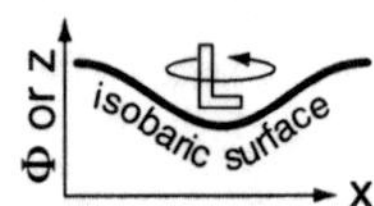

Figure f.

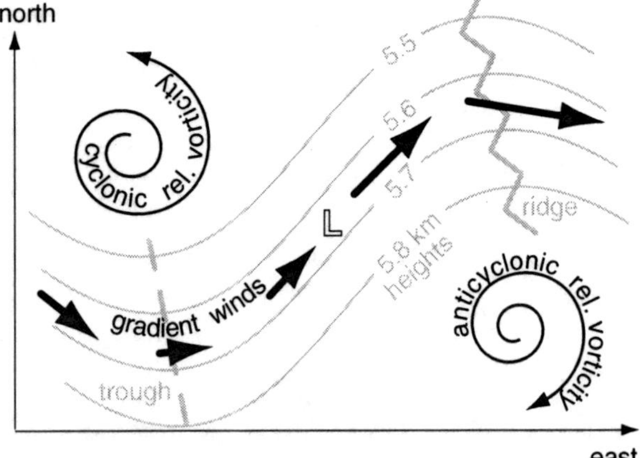

Figure 13.27
An idealized 50 kPa chart with equally-spaced height contours, as introduced by J. Bjerknes in 1937. The location of the surface low L is indicated.

the dual constraints of geostrophic and hydrostatic balance. This implies close coupling between the wind and mass fields, as was discussed in the General Circulation and Fronts chapters in the sections on geostrophic adjustment. While such close coupling is not observed for every weather system, it is a reasonable approximation for synoptic-scale, extratropical systems.

Application to Idealized Weather Patterns

An idealized weather pattern ("toy model") is shown in Fig. 13.27. Every feature in the figure is on the 50 kPa isobaric surface (i.e., in the mid troposphere), except the **L** which indicates the location of the surface low center. All three components of the geostrophic vorticity equation can be studied.

Geostrophic and gradient winds are parallel to the height contours. The trough axis is a region of cyclonic (counterclockwise) curvature of the wind, which yields a large positive value of geostrophic vorticity. At the ridge is negative (clockwise) relative vorticity. Thus, the **advection** term is positive over the **L** center and contributes to spin-up of the cyclone because the wind is blowing higher positive vorticity into the area of the surface low.

For any fixed pressure gradient, the gradient winds are slower than geostrophic when curving cyclonically ("slow around lows"), and faster than geostrophic for anticyclonic curvature, as sketched with the thick-line wind arrows in Fig. 13.27. Examine the 50 kPa flow immediately above the surface low. Air is departing faster than entering. This imbalance (divergence) draws air up from below. Hence, W increases from near zero at the ground to some positive updraft speed at 50 kPa. This **stretching** helps to spin-up the cyclone.

The **beta** term, however, contributes to spin-down because air from lower latitudes (with smaller Coriolis parameter) is blowing toward the location of the surface cyclone. This effect is small when the wave amplitude is small. The sum of all three terms in the quasigeostrophic vorticity equation is often positive, providing a net spin-up and intensification of the cyclone.

In real cyclones, contours are often more closely spaced in troughs, causing relative maxima in jet stream winds called jet streaks. Vertical motions associated with horizontal divergence in jet streaks are discussed later in this chapter. These motions violate the assumption that air mass is conserved along an "isobaric channel". Rossby also pointed out in 1940 that the gradient wind balance is not valid for varying motions. Thus, the "toy" model of Fig. 13.27 has weaknesses that limit its applicability.

ASCENT

You can also use updraft speed to estimate cyclone strength and the associated clouds. For the case-study cyclone, Fig. 13.25 shows upward motion near the middle of the troposphere (at 50 kPa).

Recall from classical physics that the definition of vertical motion is $W = \Delta z/\Delta t$, for altitude z and time t. Because each altitude has an associated pressure, define a new type of vertical velocity in terms of pressure. This is called **omega** (ω):

$$\omega = \frac{\Delta P}{\Delta t} \qquad \bullet(13.13)$$

Omega has units of Pa s^{-1}.

You can use the hypsometric equation to relate W and ω:

$$\omega = -\rho \cdot |g| \cdot W \qquad \bullet(13.14)$$

for gravitational acceleration magnitude $|g|$ = 9.8 m·s^{-2} and air density ρ. The negative sign in eq. (13.14) implies that updrafts (positive W) are associated with negative ω. As an example, if your weather map shows ω = –0.68 Pa s^{-1} on the 50 kPa surface, then the equation above can be rearranged to give W = 0.1 m s^{-1}, where an mid-tropospheric density of $\rho \approx$ 0.69 kg·m^{-3} was used.

Use either W or ω to represent vertical motion. Numerical weather forecasts usually output the vertical velocity as ω. For example, Fig. 13.28 shows upward motion (ω) near the middle of the troposphere (at 50 kPa).

The following three methods will be employed to study ascent: the continuity equation, the omega equation, and Q-vectors. Near the tropopause, horizontal divergence of jet-stream winds can force mid-tropospheric ascent in order to conserve air mass as given by the continuity equation. The almost-geostrophic (quasi-geostrophic) nature of lower-tropospheric winds allows you to estimate ascent at 50 kPa using thermal-wind and vorticity principles in the omega equation. Q-vectors consider ageostrophic motions that help maintain quasi-geostrophic flow. These methods are just different ways of looking at the same processes, and they often give similar results.

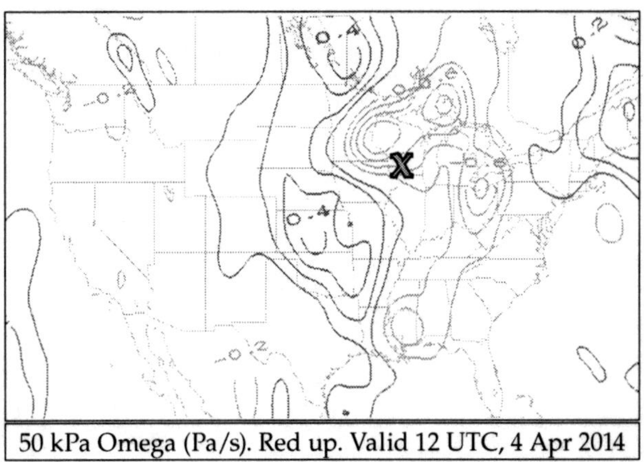

50 kPa Omega (Pa/s). Red up. Valid 12 UTC, 4 Apr 2014

Figure 13.28
Vertical velocity (omega) in pressure coordinates, for the case-study cyclone. Negative omega (colored red on this map) corresponds to updrafts.

Sample Application
At an elevation of 5 km MSL, suppose (a) a thunderstorm has an updraft velocity of 40 m s^{-1}, and (b) the subsidence velocity in the middle of an anticyclone is –0.01 m s^{-1}. Find the corresponding omega values.

Find the Answer
Given: (a) W = 40 m s^{-1}. (b) W = – 0.01 m s^{-1}. z = 5 km.
Find: ω = ? kPa s^{-1} for (a) and (b).

To estimate air density, use the standard atmosphere table from Chapter 1: ρ = 0.7361 kg m^{-3} at z = 5 km.

Next, use eq. (13.14) to solve for the omega values:
(a) ω = –(0.7361 kg m^{-3})·(9.8 m s^{-2})·(40 m s^{-1})
= –288.55 (kg·m^{-1}·s^{-2})/s = –288.55 Pa s^{-1}
= **–0.29 kPa/s**

(b) ω = –(0.7361 kg m^{-3})·(9.8 m s^{-2})·(–0.01 m s^{-1})
= 0.0721 (kg·m^{-1}·s^{-2})/s = 0.0721 Pa s^{-1}
= **7.21x10^{-5} kPa s^{-1}**

Check: Units and sign are reasonable.
Exposition: CAUTION. Remember that the sign of omega is opposite that of vertical velocity, because as height increases in the atmosphere, the pressure decreases. As a quick rule of thumb, near the surface where air density is greater, omega (in kPa s^{-1}) has magnitude of roughly a hundredth of W (in m s^{-1}), with opposite sign.

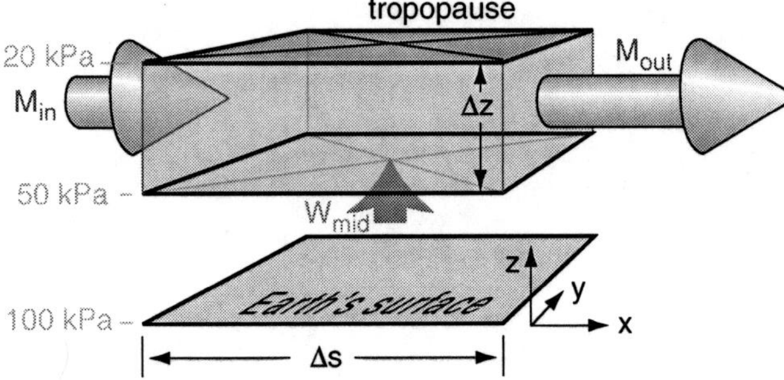

Figure 13.29
For air in the top half of the troposphere (shaded light blue), if air leaves faster (M_{out}) than enters (M_{in}) horizontally, then continuity requires that this upper-level horizontal divergence be balanced by ascent W_{mid} in the mid troposphere.

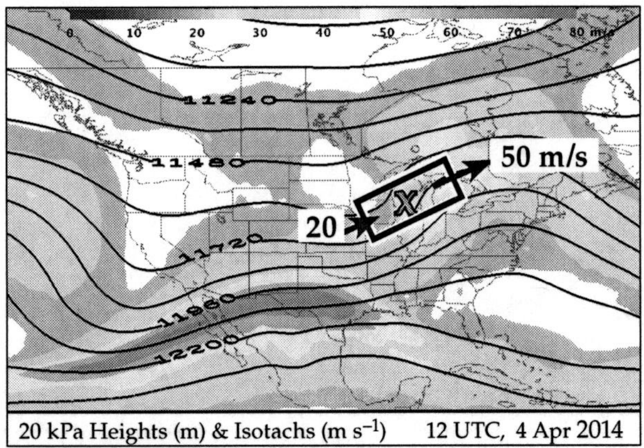

Figure 13.30
Over the surface cyclone (X) is a region (box) with faster jet-stream outflow than inflow (arrows). Isotachs are shaded.

Sample Application
Jet-stream inflow winds are 50 m s^{-1}, while outflow winds are 75 m s^{-1} a distance of 1000 km further downstream. What updraft is induced below this 5 km thick divergence region? Assume air density is 0.5 kg m^{-3}.

Find the Answer
Given: M_{in} = 50 m s^{-1}, M_{out} = 75 m s^{-1}, Δs = 1000 km, Δz = 5 km.
Find: W_{mid} = ? m s^{-1}

Use eq. (13.17):
$W_{mid} = [M_{out} - M_{in}]\cdot(\Delta z/\Delta s)$
= [75 - 50 m s^{-1}]·[(5km)/(1000km)] = 0.125 m s^{-1}

Check: Units OK. Physics unreasonable, because the incompressible continuity equation assumes constant density — a bad assumption over a 5 km thick layer.
Exposition: Although this seems like a small number, over an hour this updraft velocity would lift air about 450 m. Given enough hours, the rising air might reach its lifting condensation level, thereby creating a cloud or enabling a thunderstorm.

Continuity Effects

Horizontal divergence ($D = \Delta U/\Delta x + \Delta V/\Delta y$) is where more air leaves a volume than enters, horizontally. This can occur at locations where jet-stream wind speed (M_{out}) exiting a volume is greater than entrance speeds (M_{in}).

Conservation of air mass requires that the number of air molecules in a volume, such as the light blue region sketched in Fig. 13.29, must remain nearly constant (neglecting compressibility). Namely, volume inflow must balance volume outflow of air.

Net vertical inflow can compensate for net horizontal outflow. In the troposphere, most of this inflow happens a mid-levels ($P \approx 50$ kPa) as an upward vertical velocity (W_{mid}). Not much vertical inflow happens across the tropopause because vertical motion in the stratosphere is suppressed by the strong static stability. In the idealized illustration of Fig. 13.29, the inflows [(M_{in} times the area across which the inflow occurs) plus (W_{mid} times its area of inflow)] equals the outflow (M_{out} times the outflow area).

The continuity equation describes volume conservation for this situation as

$$W_{mid} = D\cdot\Delta z \qquad (13.15)$$

or

$$W_{mid} = \left[\frac{\Delta U}{\Delta x} + \frac{\Delta V}{\Delta y}\right]\cdot\Delta z \qquad (13.16)$$

or

$$W_{mid} = \frac{\Delta M}{\Delta s}\cdot\Delta z \qquad (13.17)$$

where the distance between outflow and inflow locations is Δs, wind speed is *M*, the thickness of the upper air layer is Δz, and the ascent speed at 50 kPa (mid tropospheric) is W_{mid}.

Fig. 13.30 shows this scenario for the case-study storm. Geostrophic winds are often nearly parallel to the height contours (solid black curvy lines in Fig. 13.30). Thus, for the region outlined with the black/white box drawn parallel to the contour lines, the main inflow and outflow are at the ends of the box (arrows). The isotachs (shaded) tell us that the inflow (≈ 20 m s^{-1}) is slower than outflow (≈50 m s^{-1}).

We will focus on two processes that cause horizontal divergence of the jet stream:
- Rossby waves, a planetary-scale feature for which the jet stream is approximately geostrophic; and
- jet streaks, where jet-stream accelerations cause non-geostrophic (ageostrophic) motions.

Rossby Waves

From the Forces and Winds chapter, recall that the gradient wind is faster around ridges than troughs, for any fixed latitude and horizontal pressure gradient. Since Rossby waves consist of a train of ridges and troughs in the jet stream, you can anticipate that along the jet-stream path the winds are increasing and decreasing in speed.

One such location is east of troughs, as sketched in Fig. 13.31. Consider a hypothetical box of air at the jet stream level between the trough and ridge. Horizontal wind speed entering the box is slow around the trough, while exiting winds are fast around the ridge. To maintain mass continuity, this horizontal divergence induces ascent into the bottom of the hypothetical box. This ascent is removing air molecules below the hypothetical box, creating a region of low surface pressure. Hence, **surface lows** (extratropical cyclones) **form east of jet-stream troughs**.

We can create a toy model of this effect. Suppose the jet stream path looks like a sine wave of wavelength λ and amplitude $\Delta y/2$. Assume that the streamwise length of the hypothetical box equals the diagonal distance between the trough and ridge

$$\Delta s = d = \left[(\lambda/2)^2 + \Delta y^2\right]^{1/2} \qquad (13.18)$$

Knowing the decrease/increase relative to the geostrophic wind speed G of the actual gradient wind M around troughs/ridges (from the Forces and Winds chapter), you can estimate the jet-stream wind-speed increase as:

$$\Delta M = 0.5 \cdot f_c \cdot R \cdot \left[2 - \sqrt{1 - \frac{4 \cdot G}{f_c \cdot R}} - \sqrt{1 + \frac{4 \cdot G}{f_c \cdot R}}\right] \qquad (13.19)$$

For a simple sine wave, the radius-of-curvature R of the jet stream around the troughs and ridges is:

$$R = \frac{1}{2\pi^2} \cdot \frac{\lambda^2}{\Delta y} \qquad (13.20)$$

Combining these equations with eq. (13.17) gives a toy-model estimate of the vertical motion:

$$W_{mid} = \frac{\dfrac{f_c \cdot \Delta z \cdot \lambda^2}{4\pi^2 \cdot \Delta y}\left[2 - \sqrt{1 - \dfrac{8\pi^2 G \cdot \Delta y}{f_c \cdot \lambda^2}} - \sqrt{1 + \dfrac{8\pi^2 G \cdot \Delta y}{f_c \cdot \lambda^2}}\right]}{[(\lambda/2)^2 + \Delta y^2]^{1/2}} \qquad (13.21)$$

For our case-study cyclone, Fig. 13.30 shows a short-wave trough with jet-stream speed increasing from 20 to 50 m/s across a distance of about 1150 km. This upper-level divergence supported cyclogenesis of the surface low over Wisconsin ("X" on the map).

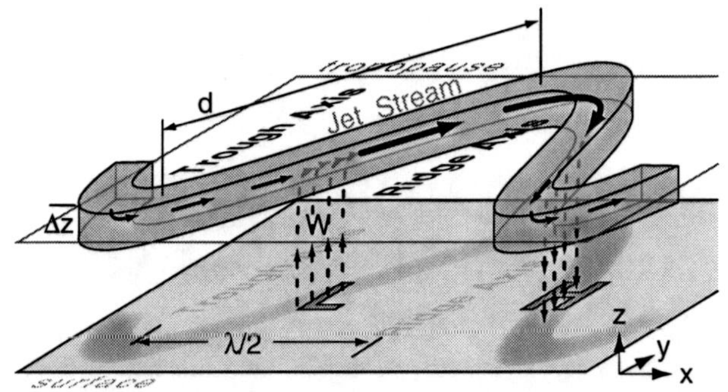

Figure 13.31
Sketch showing how the slower jet-stream winds at the trough (thin lines with arrows) are enhanced by vertical velocity (W, dotted lines) to achieve the mass balance needed to support the faster winds (thick lines with arrows) at the ridge. (N. Hem.)

Sample Application

Suppose a jet stream meanders in a sine wave pattern that has a 150 km north-south amplitude, 3000 km wavelength, 3 km depth, and 35 m s^{-1} mean geostrophic velocity. The latitude is such that $f_c = 0.0001$ s^{-1}. Estimate the ascent speed under the jet.

Find the Answer

Given: $\Delta y = 2 \cdot (150\text{ km}) = 300$ km, $\lambda = 3000$ km, $\Delta z = 3$ km, $G = 35$ m s^{-1}, $f_c = 0.0001$ s^{-1}

Find: $W_{mid} = ?$ m s^{-1}

Apply eq. (13.20):

$$R = \frac{1}{2\pi^2} \cdot \frac{(3000\text{km})^2}{(300\text{km})} = \underline{\mathbf{1520\ km}}$$

Simplify eq. (13.19) by using the curvature Rossby number from the Forces and Winds chapter:

$$\frac{G}{f_c \cdot R} = Ro_c = \frac{35\text{m/s}}{(0.0001\text{s}^{-1}) \cdot (1.52 \times 10^6\text{m})} = 0.23$$

Next, use eq. (13.19), but with Ro_c:

$$\Delta M = \frac{35\text{m/s}}{2(0.23)} \cdot \left[2 - \sqrt{1 - 4(0.23)} - \sqrt{1 + 4(0.23)}\right]$$

$$= 76.1(\text{m/s}) \cdot [2 - 0.283 - 1.386] = \underline{\mathbf{25.2}}\ \text{m s}^{-1}$$

Apply eq. (13.18):

$$d = [\ (1500\text{km})^2 + (300\text{km})^2\ ]^{1/2} = \underline{\mathbf{1530\ km}}$$

Finally, use eq. (13.17):

$$W_{mid} = \left(\frac{25.2\text{m/s}}{1530\text{km}}\right) \cdot 3\text{km} = \underline{\mathbf{0.049}}\ \text{m s}^{-1}$$

Check: Physics, units, & magnitudes are reasonable.

Exposition: This ascent speed is 5 cm s^{-1}, which seems slow. But when applied under the large area of the jet stream trough-to-ridge region, a large amount of air mass is moved by this updraft.

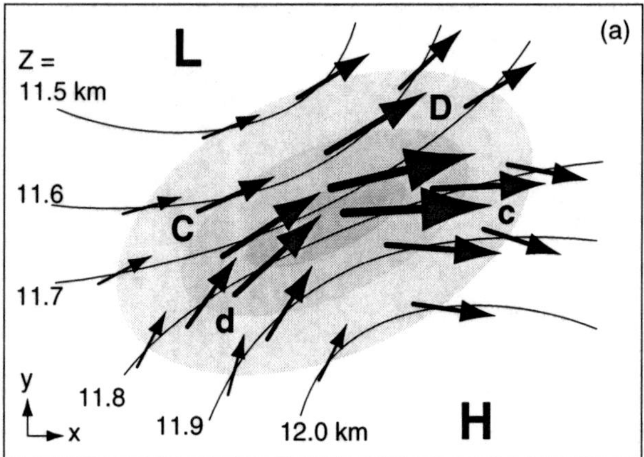

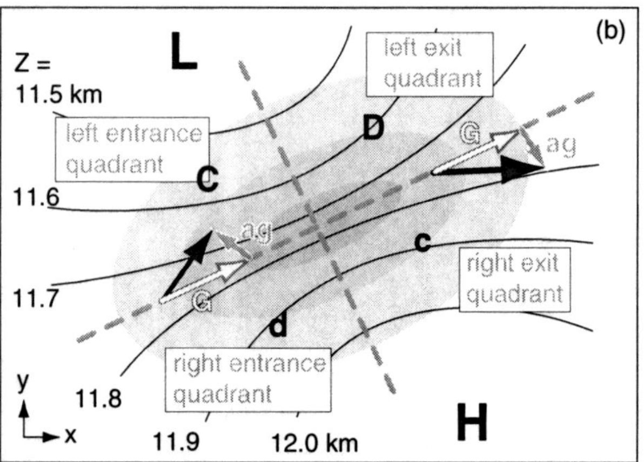

Figure 13.32
Horizontal divergence (D = strong, d = weak) and convergence (C = strong, c = weak) near a jet streak. Back arrows represent winds, green shading indicates isotachs (with the fastest winds having the darkest green), thin curved black lines are height contours of the 20 kPa isobaric surface, L & H indicate low & high height centers. Geostrophic (G) winds are parallel to the isobars, while (ag) indicates the ageostrophic wind component. Tan dashed lines parallel and perpendicular to the jet axis divide the jet streak into quadrants.

Sample Application

A west wind of 60 m s^{-1} in the center of a jet streak decreases to 40 m s^{-1} in the jet exit region 500 km to the east. Find the exit ageostrophic wind component.

Find the Answer

Given: ΔU = 40–60 m s^{-1} = –20 m s^{-1}, Δx = 500 km.
Find: V_{ag} = ? m s^{-1}

Use eq. (13.25), and assume $f_c = 10^{-4}$ s^{-1}.
The average wind is U = (60 + 40 m s^{-1})/2 = 50 m s^{-1}.
V_{ag} = [(50 m s^{-1}) / (10^{-4} s^{-1})] · [(–20 m s^{-1}) / (5x10^5 m)]
= **–20** m s^{-1}

Check: Physics and units are reasonable. Sign OK.
Exposition: Negative sign means V_{ag} is north wind.

Jet Streaks

The jet stream does not maintain constant speed in the **jet core** (center region with maximum speeds). Instead, it accelerates and decelerates as it blows around the world in response to changes in horizontal pressure gradient and direction. The fast-wind regions in the jet core are called **jet streaks**. The response of the wind to these speed changes is not instantaneous, because the air has inertia.

Suppose the wind in a weak-pressure-gradient region had reached its equilibrium wind speed as given by the geostrophic wind. As this air coasts into a region of stronger pressure gradient (i.e., tighter packing of the isobars or height contours), it finds itself slower than the new, faster geostrophic wind speed. Namely, it is **ageostrophic** (not geostrophic) for a short time while it accelerates toward the faster geostrophic wind speed.

When the air parcel is too slow, its Coriolis force (which is proportional to its wind speed) is smaller than the new larger pressure gradient force. This temporary imbalance turns the air at a small angle toward lower pressure (or lower heights). This is what happens as air flows into a jet streak.

The opposite happens as air exits a jet streak and flows into a region of weaker pressure gradient. The wind is temporarily too fast because of its inertia, so the Coriolis force (larger than pressure-gradient force) turns the wind at a small angle toward higher pressure.

For northern hemisphere jet streams, the wind vectors point slightly left of geostrophic while accelerating, and slightly right while decelerating. Because the air in different parts of the jet streak have different wind speeds and pressure gradients, they deviate from geostrophic by different amounts (Fig. 13.32a). As a result, some of the wind vectors converge in speed and/or direction to make horizontal convergence regions. At other locations the winds cause divergence. The jet-stream divergence regions drive cyclogenesis near the Earth's surface.

For an idealized west-to-east, steady-state jet stream with no curvature, the U-wind forecast equation (10.51a) from the Atmospheric Forces and Wind chapter reduces to:

$$0 = -U\frac{\Delta U}{\Delta x} + f_c\left(V - V_g\right) \qquad (13.22)$$

Let (U_{ag}, V_{ag}) be the **ageostrophic wind** components

$$V_{ag} = V - V_g \qquad •(13.23)$$

$$U_{ag} = U - U_g \qquad •(13.24)$$

Plugging these into eq. (13.22) gives for a jet stream from the west:

$$V_{ag} = \frac{U}{f_c} \cdot \frac{\Delta U}{\Delta x} \qquad \bullet(13.25)$$

Similarly, the ageostrophic wind for south-to-north jet stream axis is

$$U_{ag} = -\frac{V}{f_c} \cdot \frac{\Delta V}{\Delta y} \qquad \bullet(13.26)$$

For example, consider the winds approaching the jet streak (i.e., in the **entrance region**) in Fig. 13.32a. The air moves into a region where U is positive and increases with x, hence V_{ag} is positive according to eq. (13.25). Also, V is positive and increases with y, hence, U_{ag} is negative. The resulting ageostrophic entrance vector is shown in blue in Fig. 13.32b. Similar analyses can be made for the jet-streak **exit regions**, yielding the corresponding ageostrophic wind component.

When considering a jet streak, imagine it divided into the four quadrants sketched in Fig. 13.32 (also Fig. 13.35). The combination of speed and direction changes cause strong divergence in the **left exit quadrant**, and weaker divergence in the **right entrance quadrant**. These are the regions where cyclogenesis is favored under the jet. Cyclolysis is favored under the convergence regions of the right exit and left entrance regions.

This ageostrophic behavior can also be seen in the maps for a different case-study (Fig. 13.33). This figure overlays wind vectors, isotachs, and geopotential height contours near the top of the troposphere (at 20 kPa). The broad area of shading shows the jet stream. Embedded within it are two relative speed maxima (one over Texas, and the other over New England) that we identify as jet streaks. Recall that if winds are geostrophic (or gradient), then they should flow parallel to the height contours.

In Fig. 13.33 the square highlights the exit region of the Texas jet streak, showing wind vectors that cross the height contours toward the right. Namely, inertia has caused these winds to be faster than geostrophic (**supergeostrophic**), therefore Coriolis force is stronger than pressure-gradient force, causing the winds to be to the right of geostrophic. The oval highlights the entrance region of the second jet, where winds cross the height contours to the left. Inertia results in slower-than-geostrophic winds (**subgeostrophic**), causing the Coriolis force to be too weak to counteract pressure-gradient force.

Consider a vertical slice through the atmosphere, perpendicular to the geostrophic wind at the jet exit region (Fig. 13.32). The resulting combination of ageostrophic winds (M_{ag}) induce mid-tropospheric ascent (W_{mid}) and descent that favors cyclogenesis and cyclolysis, respectively (Fig. 13.34). The weak, vertical, cross-jet flow (orange in Fig. 13.34) is a **secondary circulation**.

INFO • Ageostrophic right-hand rule

If the geostrophic winds are accelerating, use your right hand to curl your fingers from vertical toward the direction of acceleration (the acceleration vector). Your thumb points in the direction of the ageostrophic wind.

This right-hand rule also works for deceleration, for which case the direction of acceleration is opposite to the wind direction.

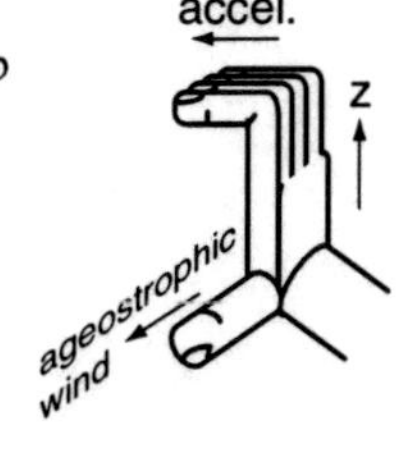

Figure g.

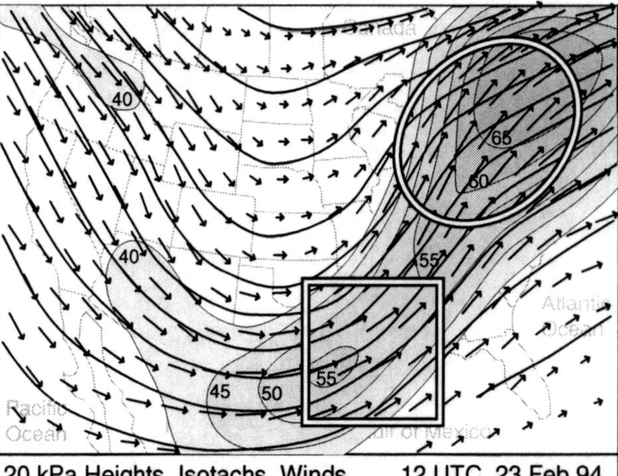

Figure 13.33 (Not the 2014 case study.)
Superposition of the 20 kPa charts for geopotential heights (medium-thickness black curved lines), isotachs in m s^{-1} (shading), and winds (vectors). The scale for winds and the values for the height contours are identical to those in Figs. 13.17. Regions of relatively darker shading indicate the jet streaks. White/black square outlines the exit region from a small jet streak over Texas, and white/black oval outlines the entrance region for a larger jet streak over the northeastern USA.

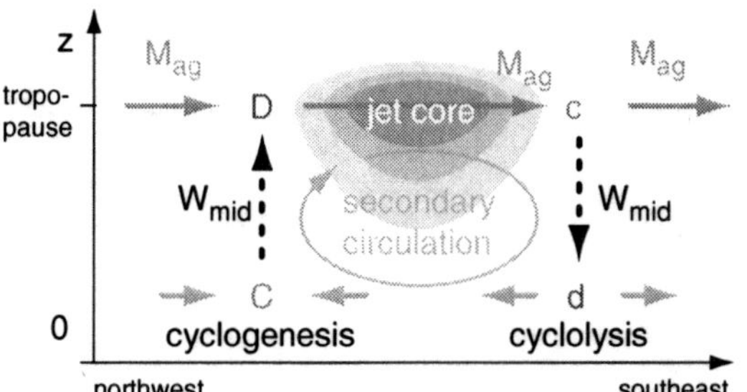

Figure 13.34
Vertical slice through the atmosphere at the jet-streak exit region, perpendicular to the average jet direction. Viewed from the west southwest, the green shading indicates isotachs of the jet core into the page. Divergence (D) in the left exit region creates ascent (W, dotted lines) to conserve air mass, which in turn removes air from near the surface. This causes the surface pressure to drop, favoring cyclogenesis. The opposite happens under the right exit region, where cyclolysis is favored.

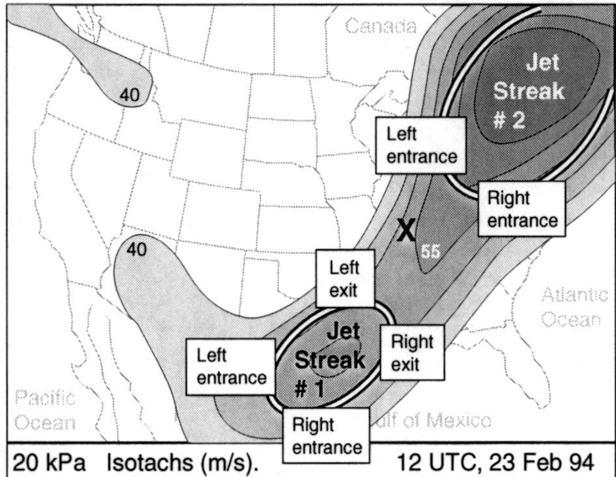

Figure 13.35 (Not the 2014 case study.)
Entrance and exit regions of jet streaks (highlighted with ovals). X marks the surface low center.

Sample Application

A 40 m s^{-1} in a jet core reduces to 20 m s^{-1} at 1,200 km downstream, at the exit region. The jet cross section is 4 km thick and 800 km wide. Find the mid-tropospheric ascent. $f_c = 10^{-4}$ s^{-1}.

Find the Answer

Given: $\Delta U = 20 - 40 = -20$ m s^{-1}, $U = 40$ m s^{-1},
$f_c = 10^{-4}$ s^{-1}, $\Delta z = 4$km, $\Delta y = 800$ km, $\Delta x = 1.2\text{x}10^6$ m.
Find: $W_{mid} = ?$ m s^{-1}

Use eq. (13.27):

$$W_{mid} = \left|\frac{(40\text{m/s})\cdot(-20\text{m/s})}{(10^{-4}\text{s}^{-1})\cdot(1.2\times10^6\text{m})}\right| \cdot \frac{(4\text{km})}{(800\text{km})} = \underline{0.033}\text{ m s}^{-1}$$

Check: Physics and units are reasonable.
Exposition: This small ascent speed, when active over a day or so, can be important for cyclogenesis.

INFO • Sutcliffe Development Theorem

To help forecast cyclogenesis, Sutcliffe devised

$$D_{top} - D_{bottom} = -\frac{1}{f_c}\left[U_{TH}\frac{\Delta\zeta_{gc}}{\Delta x} + V_{TH}\frac{\Delta\zeta_{gc}}{\Delta y}\right]$$

where **divergence** is $D = \Delta U/\Delta x + \Delta V/\Delta y$, column geostrophic vorticity is $\zeta_{gc} = \zeta_{g\ top} + \zeta_{g\ bottom} + f_c$, and ($U_{TH}$, V_{TH}) are the thermal-wind components.

This says that if the vorticity in an air column is positively advected by the thermal wind, then this must be associated with greater air divergence at the column top than bottom. When combined with eq. (13.15), this conclusion for upward motion is nearly identical to that from the Trenberth omega eq. (13.29).

The secondary circulation in the jet <u>exit</u> region is opposite to the Hadley cell rotation direction in that hemisphere; hence, it is called an **indirect** circulation. In the jet <u>entrance</u> region is a **direct** secondary circulation.

Looking again at the 23 Feb 1994 weather maps, Fig. 13.35 shows the entrance and exit regions of the two dominant jet streaks in this image (for now, ignore the smaller jet streak over the Pacific Northwest). Thus, you can expect divergence aloft at the left exit and right entrance regions. These are locations that would favor cyclogenesis near the ground. Indeed, a new cyclone formed over the Carolinas (under the right entrance region of jet streak # 2). Convergence aloft, favoring cyclolysis (cyclone death), is at the left entrance and right exit regions.

You can estimate mid-tropospheric ascent (W_{mid}) under the right entrance and left exit regions as follows. Define $\Delta s = \Delta y$ as the north-south half-width of a predominantly west-to-east jet streak. As you move distance Δy to the side of the jet, suppose that V_{ag} gradually reduces to 0. Combining eqs. (13.25) and (13.16) with V_{ag} in place of M, the mid-tropospheric ascent driven by the jet streak is

$$W_{mid} = \left|\frac{U\cdot\Delta U}{f_c\cdot\Delta x}\right|\cdot\frac{\Delta z}{\Delta y} \tag{13.27}$$

Omega Equation

The **omega equation** is the name of a diagnostic equation used to find vertical motion in pressure units (omega; ω). We will use a form of this equation developed by K. Trenberth, based on quasi-geostrophic dynamics and thermodynamics.

The full omega equation is one of the nastier-looking equations in meteorology (see the HIGHER MATH box). To simplify it, focus on one part of the full equation, apply it to the bottom half of the troposphere (the layer between 100 to 50 kPa isobaric surfaces), and convert the result from ω to W.

The resulting approximate omega equation is:

$$W_{mid} \cong \frac{-2\cdot\Delta z}{f_c}\left[U_{TH}\frac{\overline{\Delta\zeta_g}}{\Delta x} + V_{TH}\frac{\overline{\Delta\zeta_g}}{\Delta y} + V_{TH}\frac{\beta}{2}\right] \tag{•(13.28)}$$

where W_{mid} is the vertical velocity in the mid-troposphere (at P = 50 kPa), Δz is the 100 to 50 kPa thickness, U_{TH} and V_{TH} are the thermal-wind components for the 100 to 50 kPa layer, f_c is Coriolis parameter, β is the change of Coriolis parameter with y (see eq. 13.2), ζ_g is the geostrophic vorticity, and the overbar represents an average over the whole depth of the layer. An equivalent form is:

$$W_{mid} \cong \frac{-2 \cdot \Delta z}{f_c} \left[M_{TH} \frac{\Delta\left(\zeta_g + (f_c/2)\right)}{\Delta s} \right] \qquad \bullet(13.29)$$

where s is distance along the thermal wind direction, and M_{TH} is the thermal-wind speed.

Regardless of the form, the terms in square brackets represent the advection of vorticity by the thermal wind, where vorticity consists of the geostrophic relative vorticity plus a part of the vorticity due to the Earth's rotation. The geostrophic vorticity at the 85 kPa or the 70 kPa isobaric surface is often used to approximate the average geostrophic vorticity over the whole 100 to 50 kPa layer.

A physical interpretation of the omega equation is that greater upward velocity occurs where there is greater advection of cyclonic (positive) geostrophic vorticity by the thermal wind. Greater upward velocity favors clouds and heavier precipitation. Also, by moving air upward from the surface, it reduces the pressure under it, causing the surface low to move toward that location and deepen.

Weather maps can be used to determine the location and magnitude of the maximum upward motion. The idealized map of Fig. 13.36a shows the height (z) contours of the 50 kPa isobaric surface, along with the trough axis. Also shown is the location of the surface low and fronts.

At the surface, the greatest vorticity is often near the low center. At 50 kPa, it is often near the trough axis. At 70 kPa, the vorticity maximum (**vort max**) is usually between those two locations. In Fig. 13.36a, the darker shading corresponds to regions of greater cyclonic vorticity at 70 kPa.

Fig. 13.36b shows the thickness (Δz) of the layer of air between the 100 and 50 kPa isobaric surfaces. Thickness lines are often nearly parallel to surface fronts, with the tightest packing on the cold side of the fronts. Recall that thermal wind is parallel to the thickness lines, with cold air to the left, and with the greatest velocity where the thickness lines are most tightly packed. Thermal wind direction is represented by the arrows in Fig. 13.36b, with longer arrows denoting stronger speed.

Advection is greatest where the area between crossing isopleths is smallest (the INFO Box on the next page explains why). This rule also works for advection by the thermal wind. The dotted lines represent the isopleths that drive the thermal wind. In Fig. 13.36 the thin black lines around the shaded areas are isopleths of vorticity. The solenoid at the smallest area between these crossing isopleths indicates the greatest vorticity advection by the thermal wind, and is outlined by a rectangular box. For this

Sample Application

The 100 to 50 kPa thickness is 5 km and $f_c = 10^{-4}$ s^{-1}. A west to east thermal wind of 20 m s^{-1} blows through a region where avg. cyclonic vorticity decreases by 10^{-4} s^{-1} toward the east across a distance of 500 km. Use the omega eq. to find mid-tropospheric upward speed.

Find the Answer

Given: $U_{TH} = 20$ m s^{-1}, $V_{TH} = 0$, $\Delta z = 5$ km, $\Delta\zeta = -10^{-4}$ s^{-1}, $\Delta x = 500$ km, $f_c = 10^{-4}$ s^{-1}.

Find: $W_{mid} = ?$ m s^{-1}

Use eq. (13.28):

$$W_{mid} \cong \frac{-2 \cdot (5000\text{m})}{(10^{-4}\text{s}^{-1})} \left[(20\text{m/s}) \frac{(-10^{-4}\text{s}^{-1})}{(5 \times 10^5 \text{m})} + 0 + 0 \right] = \underline{0.4} \text{ m s}^{-1}$$

Check: Units OK. Physics OK.

Exposition: At this speed, an air parcel would take 7.6 h to travel from the ground to the tropopause.

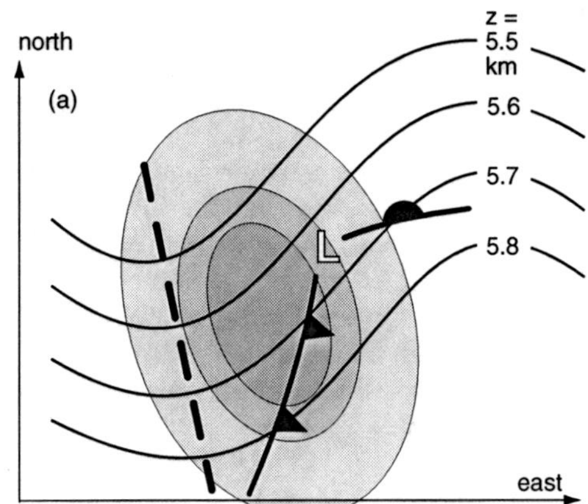

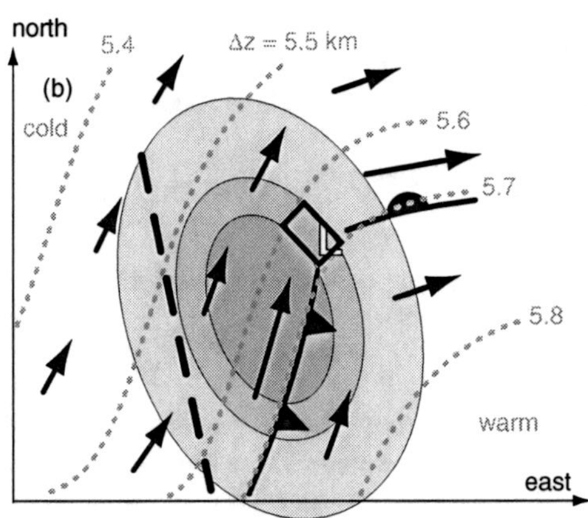

Figure 13.36

(a) Weather at three different pressure heights: (1) 50 kPa heights (solid lines) and trough axis (thick dashed line); (2) surface low pressure center (L) and fronts; (3) 70 kPa vorticity (shaded).
(b) Trough axis, surface low and fronts, and vorticity shading are identical to Fig. (a). Added are: 100 to 50 kPa thickness (dotted lines), thermal wind vectors (arrows), and region of maximum positive vorticity advection by the thermal wind (rectangular box). It is within this box that the omega equation gives the greatest updraft speed, which support cyclogenesis.

HIGHER MATH • The Omega Eq.

Full Omega Equation

The omega equation describes vertical motion in pressure coordinates. One form of the quasi-geostrophic omega equation is:

$$\left\{\nabla_p^2 + \frac{f_o^{\ 2}}{\sigma}\frac{\partial^2}{\partial p^2}\right\}\omega = \frac{-f_o}{\sigma}\cdot\frac{\partial}{\partial p}\left[-\vec{V}_g \bullet \vec{\nabla}_p\left(\zeta_g + f_c\right)\right]$$

$$-\frac{\Re}{\sigma \cdot p}\cdot\nabla_p^2\left[-\vec{V}_g \bullet \vec{\nabla}_p T\right]$$

where f_o is a reference Coriolis parameter f_c at the center of a beta plane, σ is a measure of static stability, V_g is a vector geostrophic wind, $\Re$ is the ideal gas law constant, p is pressure, T is temperature, ζ_g is geostrophic vorticity, and • means vector dot product.

$\vec{\nabla}_p(\,) = \partial(\,)/\partial x|_p + \partial(\,)/\partial y|_p$ is the **del operator**,

which gives quasi-horizontal derivatives along an isobaric surface. Another operator is the **Laplacian**:

$$\nabla^2{}_p(\,) = \partial^2(\,)/\partial x^2\Big|_p + \partial^2(\,)/\partial y^2\Big|_p \ .$$

Although the omega equation looks particularly complicated and is often shown to frighten unsuspecting people, it turns out to be virtually useless. The result of this equation is a small difference between very large terms on the RHS that often nearly cancel each other, and which can have large error.

Trenberth Omega Equation

Trenberth developed a more useful form that avoids the small difference between large terms:

$$\left\{\nabla_p^2 + \frac{f_o^{\ 2}}{\sigma}\frac{\partial^2}{\partial p^2}\right\}\omega = \frac{2f_o}{\sigma}\cdot\left[\frac{\partial \vec{V}_g}{\partial p} \bullet \vec{\nabla}_p\left(\zeta_g + (f_c/2)\right)\right]$$

For the omega subsection of this chapter, we focus on the vertical (pressure) derivative on the LHS, and ignore the Laplacian. This leaves:

$$\frac{f_o^{\ 2}}{\sigma}\frac{\partial^2 \omega}{\partial p^2} = \frac{2f_o}{\sigma}\cdot\left[\frac{\partial \vec{V}_g}{\partial p} \bullet \vec{\nabla}_p\left(\zeta_g + (f_c/2)\right)\right]$$

Upon integrating over pressure from p = 100 to 50 kPa:

$$\frac{\partial \omega}{\partial p} = \frac{-2}{f_o}\cdot\left[\overline{\vec{V}_{TH} \bullet \vec{\nabla}_p\left(\zeta_g + (f_c/2)\right)}\right]$$

where the definition of thermal wind V_{TH} is used, along with the mean value theorem for the last term.

The hydrostatic eq. is used to convert the LHS: $\partial\omega/\partial p = \partial W/\partial z$. The whole eq. is then integrated over height, with $W = W_{mid}$ at $z = \Delta z$ (= 100 - 50 kPa thickness) and $W = 0$ at $z = 0$.

This gives W_{mid} =

$$\frac{-2\cdot\Delta z}{f_c}\left[U_{TH}\frac{\overline{\Delta\left(\zeta_g + (f_c/2)\right)}}{\Delta x} + V_{TH}\frac{\overline{\Delta\left(\zeta_g + (f_c/2)\right)}}{\Delta y}\right]$$

But f_c varies with y, not x. The result is eq. (13.28).

INFO • Max Advection on Wx Maps

One trick to locating the region of maximum advection is to find the region of smallest area between crossing isopleths on a weather (wx) map, where one set of isopleths must define a wind.

For example, consider temperature advection by the geostrophic wind. Temperature advection will occur only if the winds blow across the isotherms at some nonzero angle. Stronger temperature gradient with stronger wind component perpendicular to that gradient gives stronger temperature advection.

But stronger geostrophic winds are found where the isobars are closer together. Stronger temperature gradients are found where the isotherms are closer together. In order for the winds to cross the isotherms, the isobars must cross the isotherms. Thus, the greatest temperature advection is where the tightest isobar packing crosses the tightest isotherm packing. At such locations, the area bounded between neighboring isotherms and isobars is smallest.

This is illustrated in the surface weather map below, where the smallest area is shaded to mark the maximum temperature advection. There is a jet of strong geostrophic winds (tight isobar spacing) running from northwest to southeast. There is also a front with strong temperature gradient (tight isotherm spacing) from northeast to southwest. However, the place where the jet and temperature gradient together are strongest is the shaded area.

Each of the odd-shaped tiles (**solenoids**) between crossing isobars and isotherms represents the same amount of temperature advection. But larger tiles imply that temperature advection is spread over larger areas. Thus, greatest temperature flux (temperature advection per unit area) is at the smallest tiles.

This approach works for other variables too. If isopleths of vorticity and height contours are plotted on an upper-air chart, then the smallest area between crossing isopleths indicates the region of maximum **vorticity advection** by the geostrophic wind. For vorticity advection by the **thermal wind**, plot isoplets of vorticity vs. **thickness contours**.

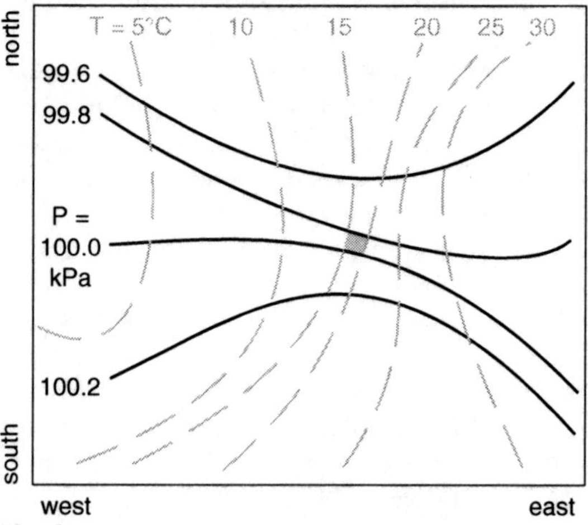

Fig. h. *Solid lines are isobars. Grey dashed lines are isotherms. Greatest temperature advection is at shaded tile.*

Figure 13.37 (At right. Not the 2014 case study.) *Superposition of the vorticity chart (grey lines and shading) at 85 kPa with the chart for thickness (thick black lines) between the 100 and 50 kPa isobaric surfaces, for a 1994 event. The thermal wind (arrows) blows parallel to the thickness lines with cold air to its left. The white box highlights a region of positive vorticity advection (PVA) by the thermal wind, where updrafts, cyclogenesis, and bad weather would be expected.*

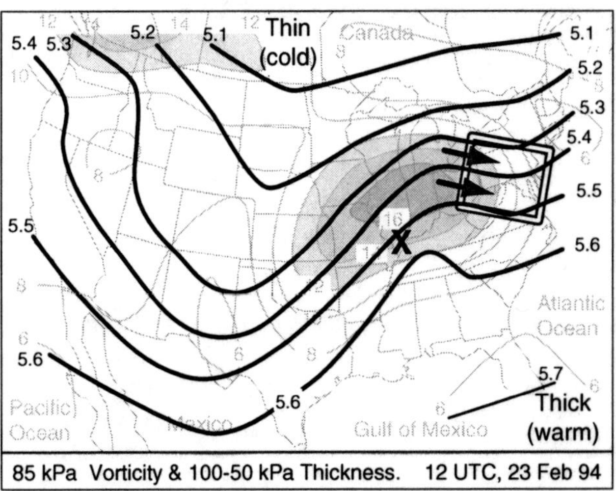

particular example, the greatest updraft would be expected within this box.

Be careful when you identify the smallest area. In Fig. 13.36b, another area equally as small exists further south-south-west from the low center. However, the cyclonic vorticity is being advected away from this region rather than toward it. Hence, this is a region of negative vorticity advection by the thermal wind, which would imply downward vertical velocity and cyclolysis or anticyclogenesis.

To apply these concepts to the 1994 event, Fig. 13.37 superimposes the 85 kPa vorticity chart with the 100 - 50 kPa thickness chart. The white box highlights a region of small solenoids, with the thermal wind blowing from high towards low vorticity. Hence, the white box outlines an area of **positive vorticity advection (PVA)** by the thermal wind, so anticipate substantial updrafts in that region. Such updrafts would create bad weather (clouds and precipitation), and would encourage cyclogenesis in the region outlined by the white box.

Near the surface low center (marked by the X in Fig. 13.37) is weak negative vorticity advection. This implies downdrafts, which contribute to cyclolysis. This agrees with the actual cyclone evolution, which began weakening at this time, while a new cyclone formed near the Carolinas and moved northward along the USA East Coast.

The Trenberth **omega equation** is heavily used in weather forecasting to help diagnose synoptic-scale regions of updraft and the associated cyclogenesis, cloudiness and precipitation. However, in the derivation of the omega equation (which we did not cover in this book), we neglected components that describe the role of ageostrophic motions in helping to maintain geostrophic balance. The INFO box on the Geostrophic Paradox describes the difficulties of maintaining geostrophic balance in some situations — motivation for Hoskin's Q-vector approach described next.

INFO • The Geostrophic Paradox

Consider the entrance region a jet streak. Suppose that the thickness contours are initially zonal, with cold air to the north and warm to the south (Fig. i(a)). As entrance winds (black arrows in Fig. i(a) converge, warm and cold air are advected closer to each other. This causes the thickness contours to move closer together (Fig. i(b), in turn suggesting tighter packing of the height contours and faster geostrophic winds at location "X" via the thermal wind equation. But the geostrophic wind in Fig. i(a) is advecting slower wind speeds to location "X".

Paradox: advection of the geostrophic wind by the geostrophic wind seems to undo geostrophic balance at "X".

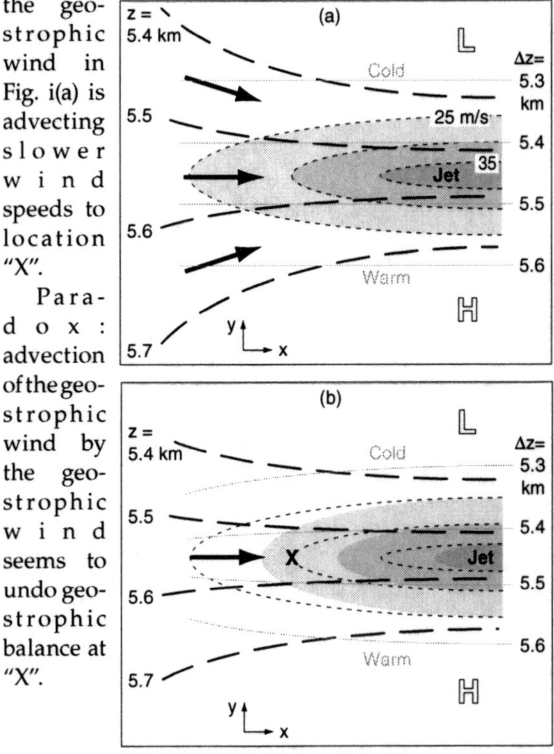

Fig. i. *Entrance region of jet streak on a 50 kPa isobaric surface. z is height (black dashed lines), Δz is thickness (thin grey lines), shaded areas are wind speeds, with initial isotachs as dotted black lines. L & H are low and high heights. (a) Initially. (b) Later.*

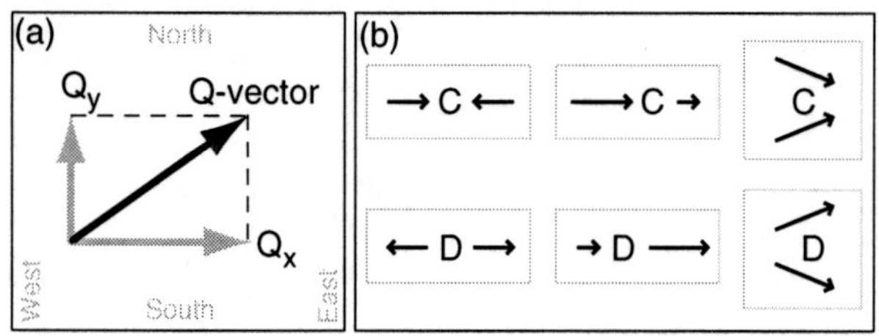

Figure 13.38
(a) Components of a Q-vector. (b) How to recognize patterns of vector convergence (C) and divergence (D) on weather maps.

Sample Application

Given the weather map at right showing the temperature and geostrophic wind fields over the NE USA. Find the Q-vector at the "X" in S.E. Pennsylvania. Side of each grid square is 100 km, and corresponds to $G = 5$ m s^{-1} for the wind vectors.

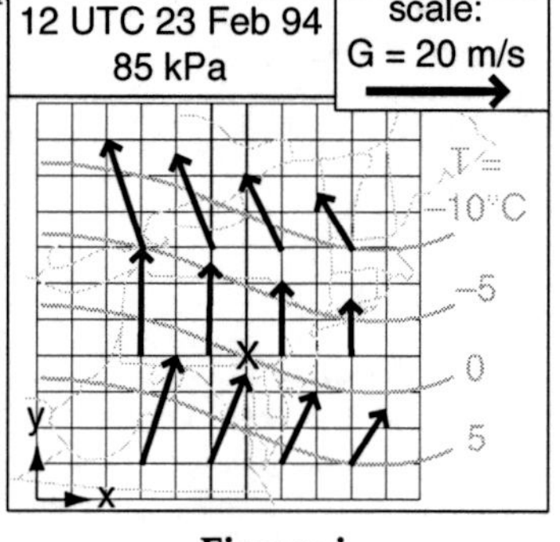

Figure j

Find the Answer

Given: P = 85 kPa, G (m s^{-1}) & T (°C) fields on map.
Find: Q_x & Q_y = ? m^2·s^{-1}·kg^{-1}

First, estimate U_g, V_g, and T gradients from the map.
$\Delta T/\Delta x = -5°\text{C}/600\text{km}$, $\Delta T/\Delta y = -5°\text{C}/200\text{km}$,
$\Delta U_g/\Delta x = 0$, $\Delta V_g/\Delta x = (-2.5\text{m s}^{-1})/200\text{km}$
$\Delta U_g/\Delta y = (-5\text{m s}^{-1})/300\text{km}$, $\Delta V_g/\Delta y = 0$,
$\Re/P = 0.287/85 = 0.003376$ m^3·kg^{-1}·K^{-1}

Use eq. (13.30): $Q_x = -(0.003376 \text{ m}^3\cdot\text{kg}^{-1}\cdot\text{K}^{-1})\cdot$
$[(0)\cdot(-8.3) + (-12.5)\cdot(-25)]\cdot 10^{-12}$ K·m^{-1}·s^{-1}
Q_x = **$-1.06\text{x}10^{-12}$** m^2·s^{-1}·kg^{-1}

Use eq. (13.31): $Q_y = -(0.003376 \text{ m}^3\cdot\text{kg}^{-1}\cdot\text{K}^{-1})\cdot$
$[(-16.7)\cdot(-8.3) + (0)\cdot(-25)]\cdot 10^{-12}$ K·m^{-1}·s^{-1}
Q_y = **$-0.47\text{x}10^{-12}$** m^2·s^{-1}·kg^{-1}

Use eq. (13.32) to find Q-vector magnitude:
$|Q| = [(-1.06)^2 + (-0.47)^2]^{1/2} \cdot 10^{-12}$ K·m^{-1}·s^{-1}
$|Q|$ = **$1.16\text{x}10^{-12}$** m^2·s^{-1}·kg^{-1}

Check: Physics, units are good. Similar to Fig. 13.40.
Exposition: The corresponding Q-vector is shown at right; namely, it is pointing from the NNE because both Q_x and Q_y are negative. There was obviously a lot of computations needed to get this one Q-vector. Luckily, computers can quickly compute Q-vectors for many points in a grid, as shown in Fig. 13.40. Normally, you don't need to worry about the units of the Q-vector. Instead, just focus on Q-vector convergence zones such as computers can plot (Fig. 13.41), because these zones are where the bad weather is.

Q-Vectors

Q-vectors allow an alternative method for diagnosing vertical velocity that does not neglect as many terms.

Defining Q-vectors

Define a horizontal Q-vector (units m^2·s^{-1}·kg^{-1}) with x and y components as follows:

$$Q_x = -\frac{\Re}{P}\left[\left(\frac{\Delta U_g}{\Delta x}\cdot\frac{\Delta T}{\Delta x}\right)+\left(\frac{\Delta V_g}{\Delta x}\cdot\frac{\Delta T}{\Delta y}\right)\right] \qquad (13.30)$$

$$Q_y = -\frac{\Re}{P}\left[\left(\frac{\Delta U_g}{\Delta y}\cdot\frac{\Delta T}{\Delta x}\right)+\left(\frac{\Delta V_g}{\Delta y}\cdot\frac{\Delta T}{\Delta y}\right)\right] \qquad (13.31)$$

where $\Re = 0.287$ kPa·K^{-1}·m^3·kg^{-1} is the gas constant, P is pressure, (U_g, V_g) are the horizontal components of geostrophic wind, T is temperature, and (x, y) are eastward and northward horizontal distances. On a weather map, the Q_x and Q_y components at any location are used to draw the Q-vector at that location, as sketched in Fig. 13.38a. Q-vector magnitude is

$$|Q| = \left(Q_x^{\,2} + Q_y^{\,2}\right)^{1/2} \qquad (13.32)$$

Estimating Q-vectors

Eqs. (13.30 - 13.32) seem non-intuitive in their existing Cartesian form. Instead, there is an easy way to estimate Q-vector direction and magnitude using weather maps. First, look at direction.

Suppose you fly along an isotherm (Fig. 13.39) in the direction of the thermal wind (in the direction that keeps cold air to your left). Draw an arrow describing the geostrophic wind vector that you observe at the start of your flight, and draw a second arrow showing the geostrophic wind vector at the end of your flight. Next, draw the vector difference, which points from the head of the initial vector to the head of the final vector. The Q-vector direction points 90° to the right (clockwise) from the geostrophic difference vector.

The magnitude is

$$|Q| = \frac{\Re}{P}\left|\frac{\Delta T}{\Delta n}\cdot\frac{\Delta \vec{V}_g}{\Delta s}\right| \qquad (13.33)$$

where Δn is perpendicular distance between neighboring isotherms, and where the temperature difference between those isotherms is ΔT. Stronger baroclinic zones (namely, more tightly packed isotherms) have larger temperature gradient $\Delta T/\Delta n$. Also, Δs

is distance of your flight along one isotherm, and ΔV_g is the magnitude of the geostrophic difference vector from the previous paragraph. Thus, greater changes of geostrophic wind in stronger baroclinic zones have larger Q-vectors. Furthermore, Q-vector magnitude increases with the decreasing pressure P found at increasing altitude.

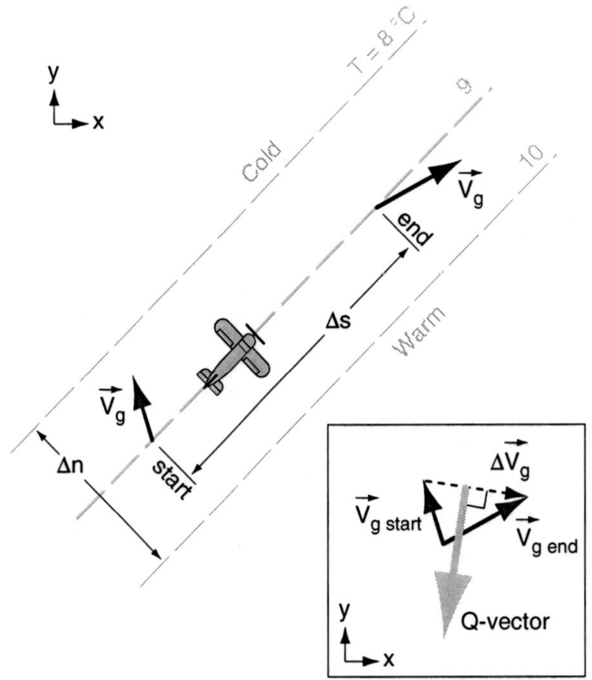

Figure 13.39
Illustration of natural coordinates for Q-vectors. Dashed grey lines are isotherms. Aircraft flies along the isotherms with cold air to its left. Black arrows are geostrophic wind vectors. Grey arrow indicates Q-vector direction (but not magnitude).

Using Q-vectors / Forecasting Tips

Different locations usually have different Q-vectors, as sketched in Fig. 13.40 for a 1994 event. Interpret Q-vectors on a synoptic weather map as follows:

- Updrafts occur where Q-vectors converge. (Fig. 13.41 gives an example for the 1994 event).
- Subsidence (downward motion) occurs where Q-vectors diverge.
- Frontogenesis occurs where Q-vectors cross isentropes (lines of constant potential temperature) from cold toward warm.
- Updrafts in the TROWAL region ahead of a warm occluded front occur during cyclolysis where the along-isentrope component of Q-vectors converge.

Using the tricks for visually recognizing patterns of vectors on weather maps (Fig. 13.38b), you can identify by eye regions of convergence and divergence in Fig. 13.40. Or you can let the computer analyze the Q-vectors directly to plot Q-vector convergence and divergence (Fig. 13.41). Although Figs. 13.40 and 13.41 are analysis maps of current weather, you can instead look at Q-vector forecast maps as produced automatically by numerical weather prediction models (see the NWP chapter) to help you forecast regions of updraft, clouds, and precipitation.

Remember that Q-vector convergence indicates regions of likely synoptic-scale upward motion and associated clouds and precipitation. Looking at Fig. 13.41, see a moderate convergence region running from the western Gulf of Mexico up through eastern Louisiana and southern Mississippi. It continues as a weak convergence region across Alabama and Georgia, and then becomes a strong convergence region over West Virginia, Virginia and Maryland. A moderate convergence region extend northwest toward Wisconsin.

This interpretation agrees with the general locations of radar echoes of precipitation for this 1994 event. Note that the frontal locations need not correspond to the precipitation regions. This demonstrates the utility of Q-vectors — even when the updrafts and precipitation are not exactly along a front, you can use Q-vectors to anticipate the bad-weather regions.

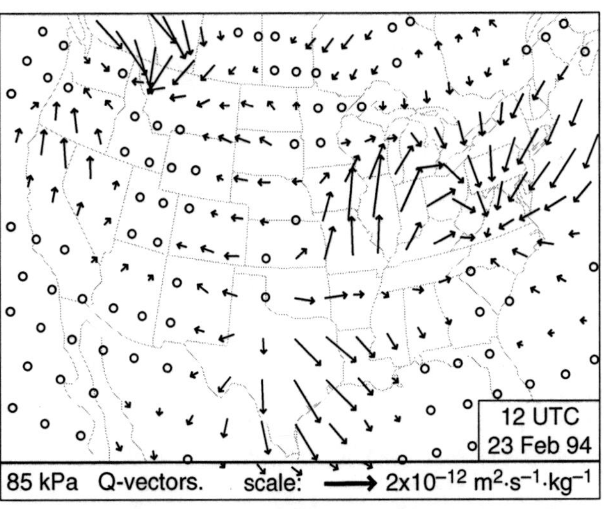

Figure 13.40 (Not the 2014 case study.)
*Weather map of Q-vectors. (**o** means small magnitude.)*

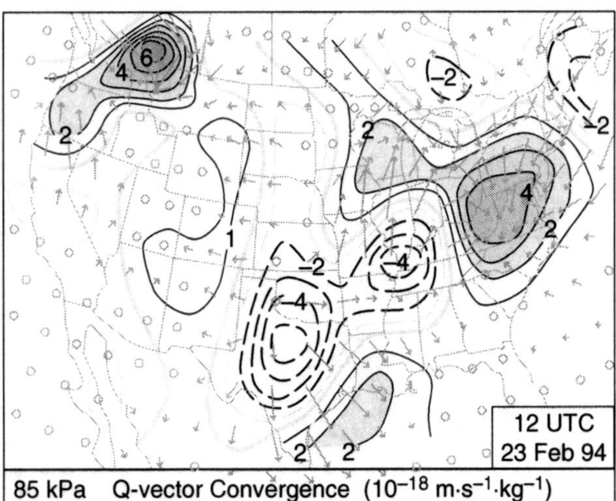

Figure 13.41 (Not the 2014 case study.)
Convergence of Q-vectors (shaded). Divergence (dashed lines).

> **Sample Application**
> Discuss the nature of circulations and anticipated frontal and cyclone evolution, given the Q-vector divergence region of southern Illinois and convergence in Maryland & W. Virginia, using Fig. 13.41.
>
> **Find the Answer**
> Given: Q-vector convergence fields.
> Discuss: circulations, frontal & cyclone evolution
>
> **Exposition**: For this 1994 case there is a low center over southern Illinois, right at the location of maximum divergence of Q-vectors in Fig. 13.41. This suggests that: (1) The cyclone is entering the cyclolysis phase of its evolution (synoptic-scale subsidence that opposes any remaining convective updrafts from earlier in the cyclones evolution) as it is steered northeastward toward the Great Lakes by the jet stream. (2) The cyclone will likely shift toward the more favorable updraft region over Maryland. This shift indeed happened.
>
> The absence of Q-vectors crossing the fronts in western Tennessee and Kentucky suggest no frontogenesis there.
>
> Between Maryland and Illinois, we would anticipate a mid-tropospheric ageostrophic wind from the east-northeast. This would connect the updraft region over western Maryland with the downdraft region over southern Illinois. This circulation would move air from the warm-sector of the cyclone to over the low center, helping to feed warm humid air into the cloud shield over and north of the low.

Also, along the Texas Gulf coast, the Q-vectors in Fig. 13.40 are crossing the cold front from cold toward warm air. Using the third bullet on the previous page, you can anticipate frontogenesis in this region.

Resolving the Geostrophic Paradox

What about the ageostrophic circulations that were missing from the Trenberth omega equation? Fig. 13.41 suggests updrafts at the Q-vector convergence region over the western Gulf of Mexico, and subsidence at the divergence region of central Texas. Due to mass continuity, expect an ageostrophic circulation of mid-tropospheric winds from the southeast toward the northwest over the Texas Gulf coast, which connects the up- and down-draft portions of the circulation. This ageostrophic wind moves warm pre-frontal air up over the cold front in a **direct circulation** (i.e., a circulation where warm air rises and cold air sinks).

But if you had used the 85 kPa height chart to anticipate geostrophic winds over central Texas, you would have expected light winds at 85 kPa from the northwest. These opposing geostrophic and ageostrophic winds agree nicely with the warm-air convergence (creating thunderstorms) for the cold katafront sketch in Fig. 12.16a.

Similarly, over West Virginia and Maryland, Fig. 13.41 shows convergence of Q-vectors at low altitudes, suggesting rising air in that region. This updraft adds air mass to the top of the air column, increasing air pressure in the jet streak right entrance region, and tightening the pressure gradient across the jet entrance. This drives faster geostrophic winds that counteract the advection of slower geostrophic winds in the entrance region. Namely, the ageostrophic winds as diagnosed using Q-vectors help prevent the Geostrophic Paradox (INFO Box).

> **HIGHER MATH • Q-vector Omega Eq.**
>
> By considering the added influence of ageostrophic winds, the Q-vector omega equation is:
>
> $$\left\{\nabla_p^2 + \frac{f_o^{\,2}}{\sigma}\frac{\partial^2}{\partial p^2}\right\}\omega = \frac{-2}{\sigma}\cdot\left[\frac{\partial Q_1}{\partial x}+\frac{\partial Q_2}{\partial y}\right] + \frac{f_o\cdot\beta}{\sigma}\frac{\partial V_g}{\partial p} - \frac{R/C_p}{\sigma\cdot P}\nabla^2(\Delta Q_H)$$
>
> The left side looks identical to the original omega equation (see a previous HIGHER MATH box for an explanation of most symbols). The first term on the right is the convergence of the Q vectors. The second term is small enough to be negligible for synoptic-scale systems. The last term contributes to updrafts if there is a local maximum of sensible heating ΔQ_H.

TENDENCY OF SEA-LEVEL PRESSURE

Because cyclones are associated with low surface pressure, processes that lower the sea-level pressure (i.e., **deepen** the low) favor cyclogenesis. On isobaric surfaces such as 100 kPa, a deepening cyclone corresponds to **falling** geopotential heights.

Conversely, processes that cause rising sea-level pressure (i.e., **filling** the low) cause cyclolysis, or even anticyclogenesis. On an isobaric surface, this corresponds to **rising** geopotential heights.

For an isobaric surface, the change of geopotential heights with time is called the **height tendency** (Fig. 13.42). The corresponding time variation of pressure on a constant height surface (e.g., sea-level) is known as **pressure tendency**. Given the close relationship between geopotential heights and pressures (recall Fig. 10.2 in the Atmospheric Forces and Winds chapter), falling heights correspond to falling pressures, both of which favor cyclogenesis.

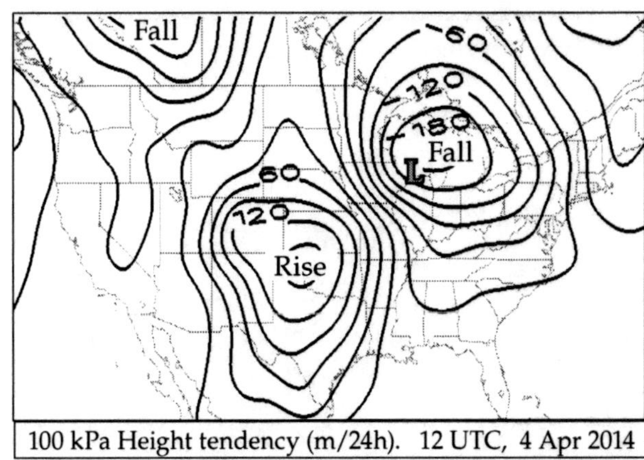

Figure 13.42
Change of geopotential height with time near the surface, for the case-study storm. Negative regions indicate where heights (and surface pressures) are decreasing (falling); namely, regions of cyclogenesis. Height rises favor anticyclogenesis.

Mass Budget

Because sea-level pressure depends on the weight of all the air molecules above it, a falling surface pressure must correspond to a removal of air molecules from the air column above the surface. An accounting of the total number of molecules in an air column is called a **mass budget**.

Imagine a column of air over 1 m^2 of the Earth's surface, as sketched in Fig. 13.43a. Suppose there is a weightless leaf (grey rectangle in that figure) that can move up and down with velocity W_{mid} in response to movement of air molecules in the column. Pick two arbitrary heights above and below the leaf, and consider the air densities at these heights. Because air is compressible, the density (ρ_2) below the leaf is greater than the density (ρ_1) above the leaf. But we will focus mostly on how the densities at these fixed altitudes change for the following scenarios.

Scenario of Fig. 13.43a: Air is pumped into the bottom half of the column, while an equal amount of air molecules are pumped out of the top. Since mass_out = mass_in, the total mass in the column is constant. Therefore the surface pressure (P_{sfc}) is constant, and the two densities do not change. But the leaf is pushed upward (W_{mid} = up) following the flow of air molecules upward in the column.

What could cause analogous inflows and outflows in the real atmosphere: horizontal convergence of wind just above the surface, and divergence aloft in the jet stream.

Scenario of Fig 13.43b: Horizontal divergence of air aloft removes air molecules from the top of the air column, with no flow in or out of the bottom. As

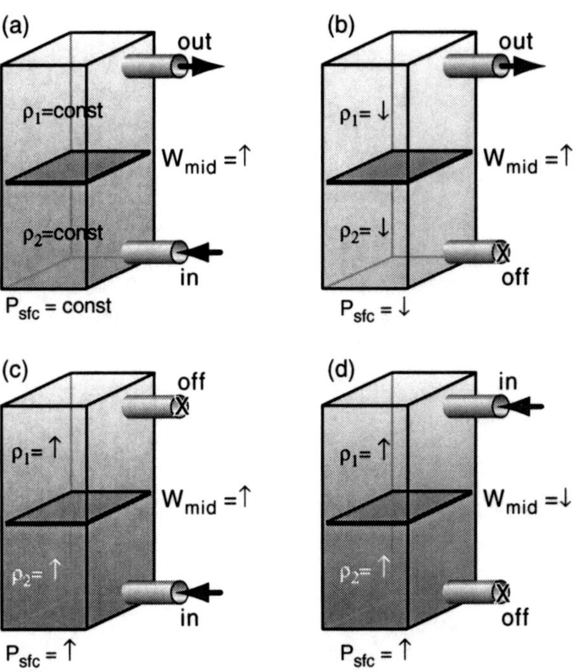

Figure 13.43
(a) Column of air (blue shading) with a leaf (grey sheet) in the middle. (b) Leaf location changes after some air is withdrawn from the top. Assume that the weightless leaf moves up and down with the air.

Sample Application

If the surface pressure is 100 kPa, how much air mass is in the whole air column above a 1 meter squared surface area?

Find the Answer

Given: $A = 1\ m^2$, $P_s = 100$ kPa
Find: $m = ?$ kg

Rearrange eq. (13.34) to solve for m:
$m = P_s \cdot A/|g| = [(100\ kPa)\cdot(1\ m^2)] / (9.8\ m\ s^{-2})$

But from Appendix A: $1\ Pa = 1\ kg \cdot m^{-1} \cdot s^{-2}$, thus:
$m = [(10^5\ kg \cdot m^{-1} \cdot s^{-2})\cdot(1\ m^2)] / (9.8\ m\ s^{-2})$
$= \underline{10.2 \times 10^3\ kg} = 10.2$ Mg

Check: Physics and units are reasonable.
Exposition: This calculation assumed that gravitational acceleration is approximately constant over the depth of the atmosphere.

Eq. 13.34 can be used for the pressure at any height in the atmosphere, but only if m represents the mass of air above that height. For example, if the tropopause is at pressure 25 kPa, then the mass of air above the tropopause is one quarter of the previous answer; namely, 2.55 Mg over each square meter.

Subtracting this value from the previous answer shows that of the total 10.2 Mg of mass in the atmosphere above a square meter, most of the air (7.65 Mg) is within the troposphere.

air molecules are evacuated from the column, the leaf moves upward, densities decrease, and surface pressure decreases because the fewer air molecules in the column cause less weight.

Scenario of Fig. 13.43c: Low-level convergence causes inflow, with no flow in or out of the top. As more molecules are pumped into the column, densities and surface pressure increase, and the leaf is pushed upward.

Scenario of Fig 13.43d: Horizontal convergence aloft causes inflow at the top of the column, with no flow in or out of the bottom. As more molecules enter the column, densities and surface pressure increase, but the leaf is pushed downward.

Pressure at sea level (P_s) is related to total column air mass (m) by:

$$P_s = \frac{|g|}{A} \cdot m \qquad \bullet(13.34)$$

where the column bottom surface as area (A) and gravitational acceleration is $|g| = 9.8\ m \cdot s^{-2}$. If the column mass changes with time, then so must the surface pressure:

$$\frac{\Delta P_s}{\Delta t} = \frac{|g|}{A} \cdot \frac{\Delta m}{\Delta t} \qquad \bullet(13.35)$$

Suppose that a hypothetical column of height z contains constant-density air. You can relate density to mass by

$$m = \rho \cdot Volume = \rho \cdot A \cdot z \qquad (13.36)$$

Thus, $\Delta m/\Delta t$ causes $\Delta z/\Delta t$, where $\Delta z/\Delta t$ is vertical velocity $W_{surrogate}$. Plugging the previous two equations into eq. (13.34) gives a way to estimate the tendency of surface pressure as a function of the motion of our hypothetical leaf $W_{surrogate}$:

$$\frac{\Delta P_s}{\Delta t} = \pm |g| \cdot \rho(z) \cdot W_{surrogate}(z) \qquad (13.37)$$

assuming we know the density $\rho(z)$ of the inflow or outflow air occurring at height z. Use the negative sign in eq. (13.37) if it is driven at the top of the troposphere, and positive sign if driven at the bottom.

Consider the following four processes that can cause inflow or outflow to/from the column:

- Advection
- Boundary-layer pumping
- Upper-level divergence
- Diabatic heating

We can estimate a $W_{surrogate}$ for the last three processes, and then add all the processes to estimate the net pressure tendency.

Advection moves a low-pressure region from one location to another. If you know the wind direction and speed M_c that is blowing the low-pressure air column toward you, and if you know ($\Delta P_s/\Delta s$) how surface pressure changes with distance *s* between your location and the advecting column, then you can estimate the pressure tendency due to advection by $\Delta P_s/\Delta t|_{advection} = -M_c \cdot (\Delta P_s/\Delta s)$.

Recall from the Atmospheric Forces and Winds chapter that frictional drag in the boundary layer causes the horizontal wind to spiral inward toward a low-pressure center. Mass continuity requires that this horizontal inflow be balanced by vertical outflow W_{BL} out of the top of the boundary layer. This **boundary-layer pumping** vertical velocity can be used with air density ρ_{BL} at the boundary-layer top in the pressure tendency equation:

$$\Delta P_s/\Delta t|_{B.L.Pumping} = -|g| \cdot \rho_{BL} \cdot W_{BL} .$$

Earlier in this chapter we saw how jet-stream curvature and jet-streak processes can cause **upper-level divergence** (U.L.Diverg.), which can remove mass from an air column and cause cyclogenesis. The W_{mid} from that section, along with the average mid-level air density ρ_{mid}, can be used to give:

$$\Delta P_s/\Delta t|_{U.L.Diverg.} = -|g| \cdot \rho_{mid} \cdot W_{mid} .$$

Non-adiabatic heating (**diabatic heating**) can be caused by solar or IR radiation, by condensation of water vapor, and other factors. The effect of latent heating on the surface pressure is described next.

Latent Heating

Water vapor might condense into drops or deposit into ice crystals at some heights within an air column — releasing latent heat and warming the column. If some of these precipitation particles evaporate before reaching the ground, then they absorb latent heat and cool the column.

The precipitation that does reach the ground is related to the net amount of condensational heating during time interval Δt by

$$\frac{\Delta T_v}{\Delta t} = \frac{a}{\Delta z} \cdot \frac{L_v}{C_p} \cdot \frac{\rho_{liq}}{\rho_{air}} \cdot RR \qquad (13.38)$$

where *RR* is rainfall rate (mm h^{-1}), T_v is average air-column virtual temperature, $a = 10^{-6}$ km mm^{-1}, Δz is depth of the air column (km), the ratio of latent heat of vaporization to specific heat of air is L_v/C_p = 2500 K·kg$_{air}$ /kg$_{liq}$, and where (ρ_{air} , ρ_{liq}) are air and liquid-water densities, respectively, with ρ_{liq} = 1000 kg m^{-3}).

But the hypsometric equation relates pressure changes to temperature changes (Fig. 13.44). Thus, eq. (13.38) and the hypsometric equation can be merged to give:

Sample Application

Divergence of air at the top of the troposphere removes air molecules from the top of a tropospheric column, causing a 0.1 m s^{-1} updraft at height 8 km above ground level (AGL). No other processes add or remove air mass. What is the corresponding surface pressure tendency?

Find the Answer

Given: z = 8 km, $W_{surrogate}$(8 km) = 0.1 m s^{-1}
Find: $\Delta P_s/\Delta t$ = ? kPa s^{-1}

For air density at z = 8 km, assume a standard atmosphere. Use ρ = 0.5252 kg m^{-3}.

Use eq. (13.37) with negative sign because the forcing is at the top of the troposphere:

$\Delta P_s/\Delta t$ = – (9.8 m s^{-2})·(0.5252 kg m^{-3})·(0.01 m s^{-1})
= – 0.0515 kg·m^{-1}·s^{-3} = –0.0515 Pa s^{-1}
= **<u>– 5.15x10^{-5} kPa s^{-1}</u>**

Check: Physics and units are reasonable.
Exposition: The corresponding hourly pressure tendency is $\Delta P_s/\Delta t$ = –0.185 kPa h^{-1}. If this rapid deepening of the cyclone were to continue for 24 h, we would classify this explosive cyclogenesis as a **cyclone bomb**.

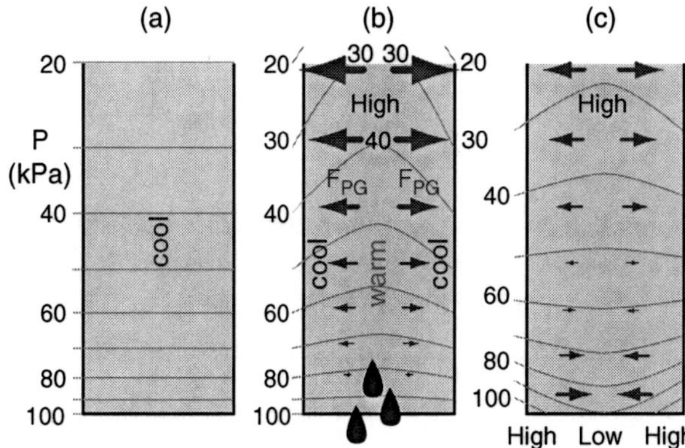

Figure 13.44
Illustration of diabatic heating. (a) Initial condition of cool air in the column. Green lines are isobars. (b) A deep cloud causes warming in the center of the air column due to latent heating, with precipitation falling out of the bottom of the column (represented by black rain drops). But the outside of the air column is not heated (contains no clouds) and remains cool. The thickness between isobaric surfaces is thicker in warm air than cool. This causes pressure gradient forces (F_{PG}, purple arrows) from warm to the cold air. These forces drive winds that remove air molecules from high pressure into lower pressure. (c) The resulting reduction of air mass in the warm core reduces sea-level pressure (Low; i.e., cyclogenesis), and adds mass into the cool-air regions (creating sea-level highs).

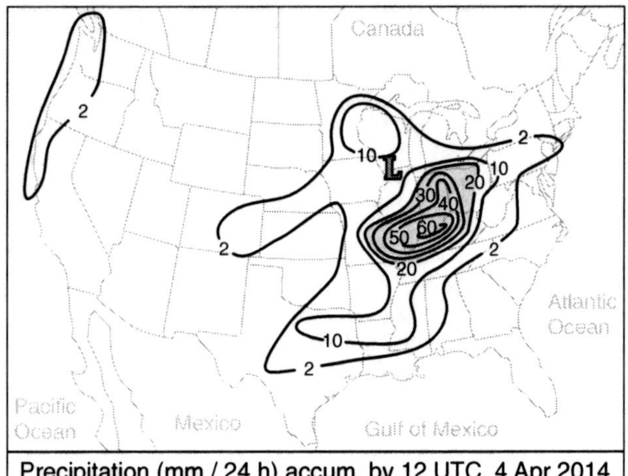

Figure 13.45
Precipitation (liquid equivalent) measured with rain gauges.

$$\frac{\Delta P_s}{\Delta t} = -\frac{|g|}{\overline{T_v}} \cdot \frac{L_v}{C_p} \cdot \rho_{liq} \cdot RR \qquad \bullet(13.39)$$

for an air column with average virtual temperature (Kelvin) of $\overline{T_v}$ (≈ 300 K), and where gravitational acceleration magnitude is $|g|$ = 9.8 m·s^{-2}. Although latent heating and cooling might occur at different heights within the air column, eq. (13.39) describes the net column-average effect.

For a typical value of $\overline{T_v}$, eq. (13.39) reduces to:

$$\frac{\Delta P_s}{\Delta t} \approx -b \cdot RR \qquad \bullet(13.40)$$

with factor $b = \frac{|g|}{\overline{T_v}} \cdot \frac{L_v}{C_p} \cdot \rho_{liq} \approx 0.082 \text{ kPa mm}_{\text{rain}}^{-1}$.

You can estimate rainfall rate with weather radar, or you can measure it with rain gauges. Fig. 13.45 shows measured precipitation liquid-equivalent depth (after melting any snow) for the case-study storm.

Sample Application

For the maximum contoured precipitation rate for the case-study storm (in Fig. 13.45), find the diabatic heating contribution to sea-level pressure tendency.

Find the Answer

Given: RR = 60 mm/24 h
Find: $\Delta P_s/\Delta t$ = ? kPa h^{-1}

First, convert RR from 24 h to 1 hr:
RR = 2.5 mm h^{-1}

Use eq. (13.40):
$\Delta P_s/\Delta t = -(0.082 \text{ kPa/mm}_{\text{rain}})\cdot(2.5 \text{ mm h}^{-1})$
$= \underline{\mathbf{-0.205\ kPa}}$ h^{-1}

Check: Physics, magnitude & units are reasonable.
Exposition: This deepening rate corresponds to 4.9 kPa day^{-1} — large enough to be classified as a cyclone "bomb".

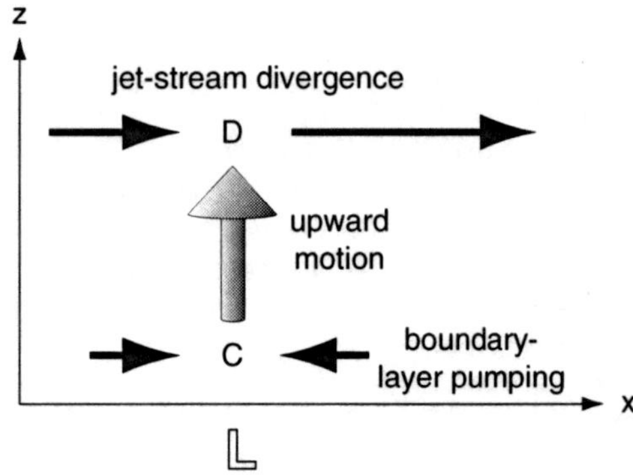

Figure 13.46
Sketch of coupling between convergence (C) in the boundary layer and divergence (D) in the upper atmosphere. Arrows represent winds. L is location of low-pressure center at surface.

Net Pressure Tendency

The previous heuristic models for horizontal advection (horiz. adv.) and boundary-layer pumping (B.L.Pumping) and upper-level divergence (U.L.Diverg.) and latent heating can be combined within the framework of Fig. 13.46 to give an equation for **sea-level net pressure tendency**:

•(13.41)

$$\frac{\Delta P_s}{\Delta t} = -M_c \frac{\Delta P_s}{\Delta s} + |g| \cdot \rho_{BL} \cdot W_{BL} - |g| \cdot \rho_{mid} \cdot W_{mid} - b \cdot RR$$

tendency = horiz. adv. & B.L.Pumping & U.L.Diverg. & heating

where (ρ_{BL}, ρ_{mid}) and (W_{BL}, W_{mid}) are the average air densities and vertical velocities at boundary-layer top and mid-troposphere, respectively. The air-column horizontal translation speed is M_c (defined as positive for the average movement direction along path s). RR is rainfall rate at the surface, $|g|$ = 9.8 m·s^{-2} is gravitational-acceleration magnitude, and b ≈ 0.082 kPa/mm$_{\text{rain}}$. Cyclogenesis occurs when $\Delta P_s/\Delta t$ is negative.

We can create a toy model of cyclone evolution using the equation above. Initially (time A in Fig. 13.47a) there is no extratropical cyclone. But if there is a Rossby wave in the jet stream, then we can anticipate horizontal divergence aloft (at the altitude of the tropopause) at a horizontal location east of the upper-level trough. As this upper-level divergence removes air molecules from the air column underneath it, sea-level pressure begins to decrease (time B in Fig. 13.47).

But once a low-center forms near the surface, the combination of pressure-gradient force, Coriolis force, and frictional drag start to create a boundary-layer gradient wind that spirals into towards the low. The resulting inflow and boundary-layer pumping tends to fill (weaken) the low. So the only way the cyclone can continue to intensify is if the upper-level divergence in the jet stream is greater than the convergence in the boundary layer (times B to C in Fig. 13.47). At this time, cyclogenesis continues and sea-level pressure continues to drop.

Rising air in the cyclone (Fig. 13.46) cools adiabatically, eventually creating clouds and precipitation. The condensation creates diabatic heating, which enhances buoyancy to create faster updrafts and continued cyclogenesis (time C in Fig. 13.47).

Recall from Fig. 13.3 that circulation of air masses around the cyclone center reduces the horizontal temperature gradients there, leaving the strongest temperature gradients eastward and equatorward of the low center. Because these temperature gradients are what drives the jet-stream via the thermal-wind, it means that the Rossby wave shifts eastward relative to the low center (times C to D in Fig. 13.47).

Latent heating in the cloudy updrafts continues to support cyclogenesis. Meanwhile, the upper-level divergence of the jet stream decreases. Eventually the sum of upper-level divergence and diabatic processes is insufficient to counterbalance the continued boundary-layer pumping. At this point (time D in Fig. 13.47), cyclogenesis ceases, and the sea-level pressure stops dropping. The cyclone has reached its mature stage and is still strong (low surface pressures, strong winds, heavy precipitation), but will not be getting stronger.

The strong circulation around the low continues to cause strong boundary-layer pumping, which ceaselessly tends to weaken the cyclone. Without the jet-stream support aloft, the air column begins to fill with molecules — cyclolysis has begun.

As the cyclone occludes (Figs. 13.3e & f), the Rossby wave shifts so far eastward that the jet-stream is causing upper-level convergence over the weakening low (time E in Fig. 13.47), thereby helping to fill it even faster with air molecules.

Eventually, the central pressure rises to equal the surrounding pressures (i.e., no horizontal pressure gradient). Winds decrease, condensation and precipitation end, and the circulation spins-down. By time F in Fig. 13.47, the cyclone has disappeared, and anticyclogenesis (creation of surface high-pressure) has begun.

During the three-day life cycle, the cyclone acted to move cold air equatorward and warm air poleward to reduce the baroclinic instability that had created it (as per Le Chatelier's principle).

Sample Application

A cyclone experiences the following processes:
- Rainfall of 2 mm h^{-1}.
- Advection due to a 15 m/s west wind across a horizontal gradient of $\Delta P/\Delta x$ of 0.5 kPa/300 km.
- Upper-level divergence causing W_{mid} = 0.04 m s^{-1}.
- Boundary-layer pumping W_{BL} = 0.02 m s^{-1}.

Given: $\rho_{mid} \approx$ 0.5 kg·m^{-3}, $\rho_{BL} \approx$ 1.112 kg·m^{-3}

What is the sea-level pressure tendency?.

Find the Answer

Given: $\Delta P/\Delta s$ = 0.5 kPa/300 km, RR = 2 mm h^{-1}, W_{mid} = 0.02 m s^{-1}, W_{BL} = 0.02 m s^{-1},

Find: $\Delta P_s/\Delta t$ = ? kPa h^{-1}

Apply eq. (13.41): $\Delta P_s/\Delta t$ =

$$-\left(15\frac{\text{m}}{\text{s}}\right)\left(\frac{0.5\text{kPa}}{3\times10^5\text{m}}\right)+\left(9.8\frac{\text{m}}{\text{s}^2}\right)\left(1.112\frac{\text{kg}}{\text{m}^3}\right)\left(0.02\frac{\text{m}}{\text{s}}\right)$$
$$-\left(9.8\frac{\text{m}}{\text{s}^2}\right)\left(0.5\frac{\text{kg}}{\text{m}^3}\right)\left(0.04\frac{\text{m}}{\text{s}}\right)-\left(0.084\frac{\text{kPa}}{\text{mm}}\right)\left(2\frac{\text{mm}}{\text{h}}\right)$$

$\Delta P_s/\Delta t$ = (–2.5 + 21.8 – 19.6 – 4.7)x10^{-5} kPa s^{-1}

tendency = horiz. adv. & B.L.Pumping & U.L.Diverg. & heating

= – 5 x10^{-5} kPa s^{-1} = **<u>– 0.18 kPa</u>** h^{-1}

Check: Physics, magnitude & units are reasonable.

Exposition: The negative sign implies cyclogenesis. In the B.L.pumping and U.L.Diverg. terms I divided by 1000 to convert Pa into kPa. For the rainfall term I converted from kPa h^{-1} to kPa s^{-1}.

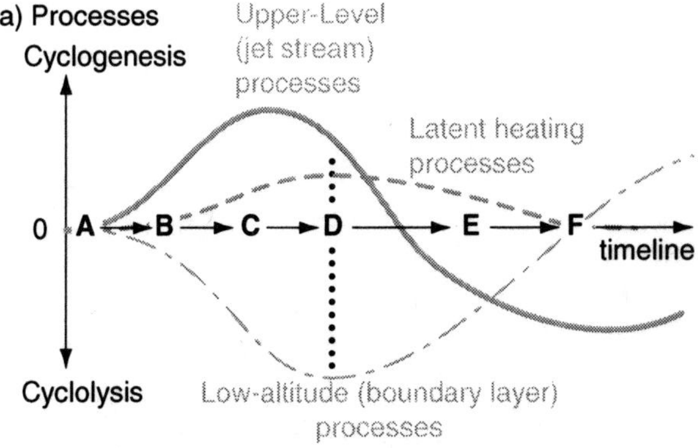

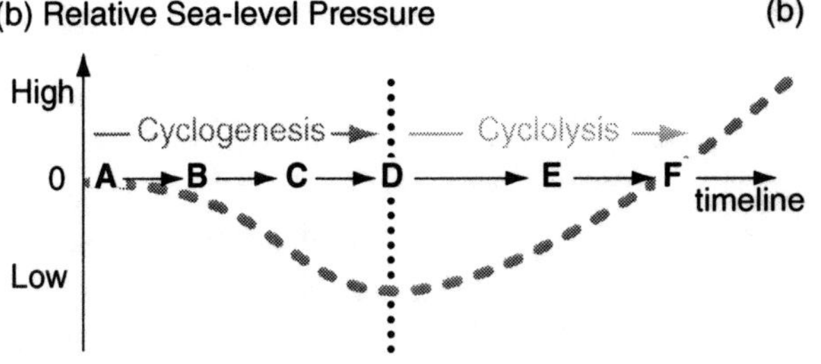

Figure 13.47
Processes that cause extratropical cyclone evolution.

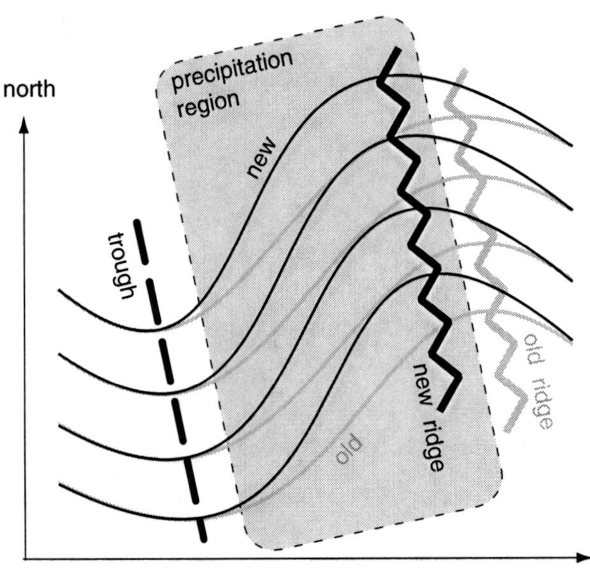

Figure 13.48
50 kPa chart showing cloud and precipitation region in the upper troposphere, causing latent heating and a westward shift of the ridge axis.

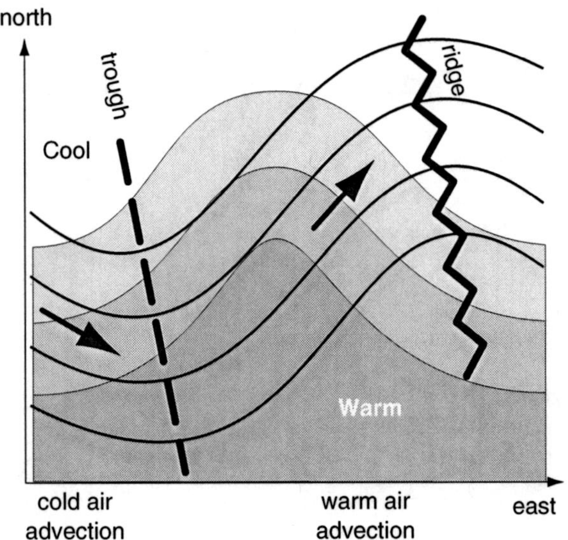

Figure 13.49
50 kPa chart showing a temperature field (shaded) that is 1/4 wavelength west of the wave in the height contours (solid lines).

CYCLONE SELF DEVELOPMENT

Up to this point, cyclogenesis has been treated as a response to various external imposed forcings. However, some positive feedbacks allow the cyclone to enhance its own intensification. This is often called **self development**.

Condensation

As discussed in the quasi-geostrophic vorticity subsection, divergence of the upper-level winds east of the Rossby-wave trough (Fig. 13.48) causes a broad region of upward motion there. Rising air forms clouds and possibly precipitation if sufficient moisture is present. Such a cloud region is sometimes called an **upper-level disturbance** by broadcast meteorologists, because the bad weather is not yet associated with a strong surface low.

Latent heating of the air due to condensation enhances buoyancy and increases upward motion. The resulting **stretching** enhances spin-up of the vorticity, and the upward motion withdraws some of the air away from the surface, leaving lower pressure. Namely, a surface low forms.

Diabatic heating also increases the average temperature of the air column, which pushes the 50 kPa pressure surface upward (i.e., increasing its height), according to the hypsometric relationship. This builds or strengthens a ridge in the upper-level Rossby wave west of the initial ridge axis.

The result is a shortening of the wavelength between trough and ridge in the 50 kPa flow, causing tighter turning of the winds and greater vorticity (Fig. 13.48). Vorticity advection also increases.

As the surface low strengthens due to these factors (i.e., divergence aloft, vorticity advection, precipitation, etc.), more precipitation and latent heating can occur. This positive feedback shifts the upper-level ridge further west, which enhances the vorticity and the vorticity advection. The net result is rapid strengthening of the surface cyclone.

Temperature Advection

Cyclone intensification can also occur when warm air exists slightly west from the Rossby-wave ridge axis, as sketched in Fig. 13.49. For this situation, warm air advects into the region just west of the upper-level ridge, causing ridge heights to increase. Also cold air advects under the upper-level trough, causing heights to fall there.

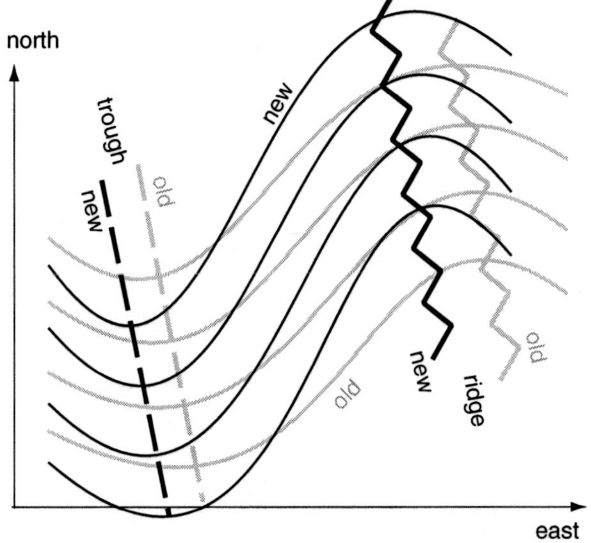

Figure 13.50
50 kPa chart showing the westward shift and intensification of north-south wave amplitude caused by differential temperature advection.

The net result is intensification of the Rossby-wave amplitude (Fig. 13.50) by deepening the trough and strengthening the ridge. Stronger wave amplitude can cause stronger surface lows due to enhanced upper-level divergence.

Propagation of Cyclones

For a train of cyclones and anticyclones along the mid-latitude baroclinic zone (Fig. 13.51a), a Q-vector analysis (Fig. 13.51b) suggests convergence of Q-vectors east of the low, and divergence of Q-vectors west of the low (Fig. 13.51c). But convergence regions imply updrafts with the associated clouds and surface-pressure decrease — conditions associated with cyclogenesis. Thus, the cyclone (𝕃) in Fig. 13.51c would tend to move toward the updraft region indicated by the Q-vector convergence.

The net result is that cyclones tend to propagate in the direction of the thermal wind, i.e., parallel to the thickness contours.

Creation of Baroclinic Zones

Cyclones and anticyclones tend to create or strengthen baroclinic zones such as fronts. This works as follows.

Consider a train of lows and highs in a region with uniform temperature gradient, as shown in Fig. 13.52a. The rotation around the lows and highs tend to distort the center isotherms into a wave, by moving cold air equatorward on the west side of the lows and moving warm air poleward on the east side.

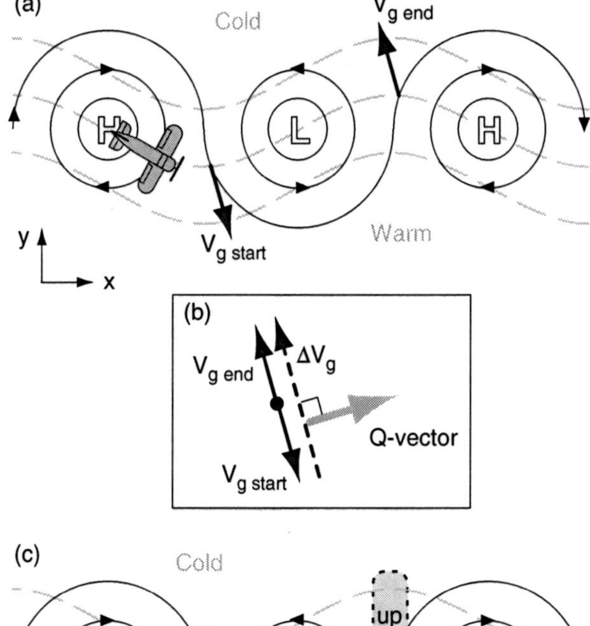

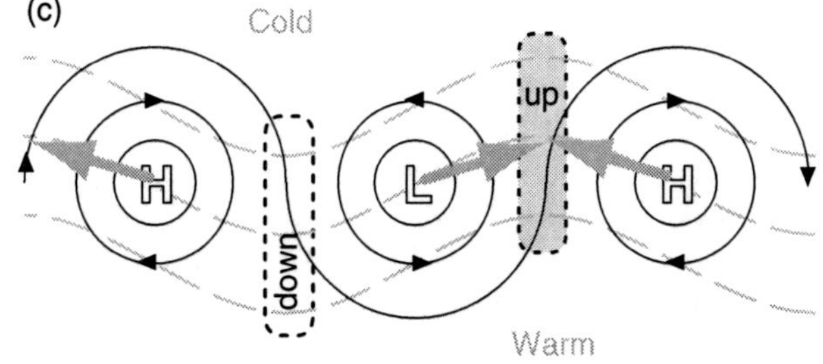

Figure 13.51
Using Q-vectors to estimate cyclone propagation. Grey dashed lines are isotherms (or thickness contours). Thin black lines are isobars (or height contours). (a) Airplane is flying in the direction of the thermal wind. Black arrows are geostrophic wind vectors encountered by the airplane on either side of the cyclone (𝕃). (b) Vector difference (black dashed line) between starting and ending geostrophic-wind vectors (drawn displaced to the right a bit so you can see it) for the portion of aircraft flight across the low. Q-vector (grey thick arrow) is 90° to the right of the dashed vector. (c) Q-vector from (b) is copied back to the cyclone. Similar analyses can be done to find the Q-vectors for the anticyclones (ℍ). Convergence of Q-vectors (grey shaded box with dotted outline) indicates region of updraft. Divergence of Q-vectors (white box with dotted outline) indicates region of downdraft. (for N. Hemisphere.)

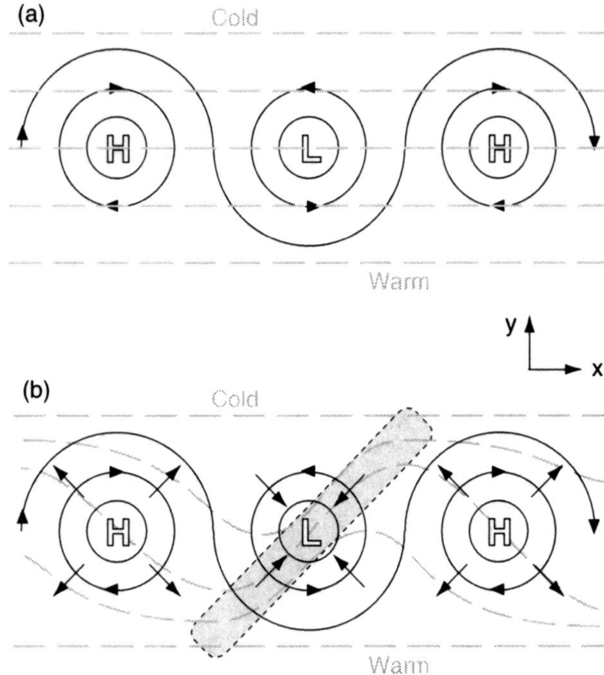

Figure 13.52
Rotational and divergent wind components (thin black arrows) and isotherms (dashed grey lines) in the lower troposphere. (a) Initial train of highs (H) and lows (L) in a uniform temperature gradient in the N. Hemisphere. (b) Later evolution of the isotherms into frontal zones (shaded rectangle is a baroclinic zone).

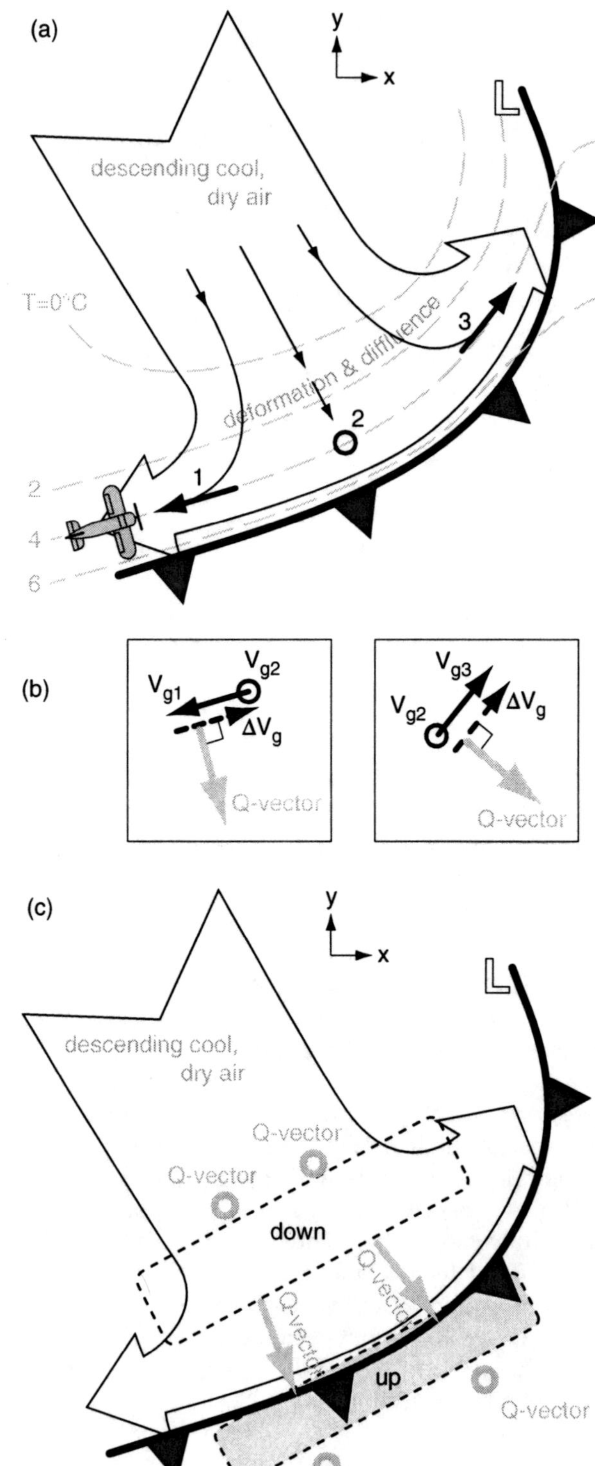

Figure 13.53
Using Q-vectors to locate regions of upward and downward motion due to diffluence of air behind a cold front. (a) Thin black lines are wind direction. Dashed grey lines are isotherms, along which an imaginary airplane flies. Thick black arrows show geostrophic wind vectors, with "o" representing zero wind. (b) Estimation of Q-vectors between points 1 and 2, and also between points 2 and 3. (c) The Q-vectors from (b) plotted in the baroclinic zone, with near-zero Q-vectors (grey "o") elsewhere. Q-vector convergence in grey shaded region suggests updrafts.

In addition, convergence into lows pulls the isotherms closer together, while divergence around highs tends to push isotherms further apart. The combination of rotation and convergence/divergence tends to pack the isotherms into frontal zones near lows, and spread isotherms into somewhat homogeneous airmasses at highs (Fig. 13.52b).

Much of the first part of this chapter showed how cyclones can develop over existing baroclinic regions. Here we find that cyclones can help create those baroclinic zones — resulting in a positive feedback where cyclones modify their environment to support further cyclogenesis. Thus, cyclogenesis and frontogenesis often occur simultaneously.

Propagation of Cold Fronts

Recall from Fig. 13.8 that the circulation around a cyclone can include a deformation and diffluence region of cold air behind the cold front. If the diffluent winds in this baroclinic zone are roughly geostrophic, then you can use Q-vectors to analyze the ageostrophic behavior near the front.

Fig. 13.53a is zoomed into the diffluence region, and shows the isotherms and geostrophic wind vectors. By flying an imaginary airplane along the isotherms and noting the change in geostrophic wind vector, estimate the Q-vectors at the front as drawn in Fig. 13.53b. Further from the front, the Q-vectors are near zero.

Thus, Q-vector convergence is along the leading edge of the cold front (Fig. 13.53c), where warm air is indeed rising over the front. Behind the cold front is Q-vector divergence, associated with downward air motion of cool dry air from higher in the troposphere (Fig. 13.8).

REVIEW

Extratropical cyclones (**lows**) are horizontally large (thousands of km in diameter) relatively thin (11 km thick) storms in mid latitudes. They undergo an evolution of intensification (cyclogenesis) and weakening (cyclolysis) within a roughly three-day period while they translate eastward and poleward. The storm core is generally cold, which implies that the upper-level Rossby wave trough is generally west of the cyclone.

Cyclones rotate (counterclockwise/clockwise) in the (Northern/Southern) Hemisphere around a center of low sea-level pressure. Bad weather (clouds, precipitation, strong winds) are often concentrated in narrow frontal regions that extend outward from the low-pressure center. Cyclone strength can be inferred from its vorticity (rotation), ascent (updrafts), and sea-level pressure.

Weather maps are used to study **synoptic-scale storm systems** such as mid-latitude cyclones and fronts. Many case-study maps were presented in this chapter for the 4 April 2014 cyclone, and a few maps were for a similar cyclone during 23 February 1994.

A SCIENTIFIC PERSPECTIVE • Uncertainty and Truth in Science

The Supreme Court of the USA ruled that: "There are no certainties in science. Scientists do not assert that they know what is immutably 'true' — they are committed to searching for new, temporary theories to explain, as best they can, phenomena." They ruled that science is "a process for proposing and refining theoretical explanations about the world that are subject to further testing and refinement."

[*CAUTION: The definition, role, and activities of science cannot be defined or constrained by a legal court or religious inquisition. Instead, science is a philosophy that has gradually developed and has been refined by scientists. The "A SCIENTIFIC PERSPECTIVE" boxes in this book can help you to refine your own philosophy of science.*]

INFO • Landfalling Pacific Cyclones

Mid-latitude cyclones that form over the warm ocean waters east of Japan often intensify while approaching the dateline (180° longitude). But by the time they reach the eastern North Pacific ocean they are often entering the cyclolysis phase of their evolution. Thus, most landfalling cyclones that reach the Pacific Northwest coast of N. America are already occluded and are spinning down.

The cyclone labeled L_1 in Fig. 13.54 has just started to occlude. Satellite images of these systems show a characteristic tilted-"T" (⟋) shaped cloud structure (grey shaded in Fig. 13.54), with the cold front, occluded front, and a short stub of a warm front.

As the cyclone translates further eastward, its translation speed often slows and the low center turns northward toward the cold waters in the Gulf of Alaska — a cyclone graveyard where lows go to die. In the late occluded phase, satellite images show a characteristic "cinnamon roll" cloud structure, such as sketched with grey shading for cyclone L_2.

For cyclone L_2, when the cold front progresses over the complex mountainous terrain of the Pacific Northwest (British Columbia, Washington, Oregon), the front becomes much more disorganized and difficult to recognize in satellite and surface weather observations (as indicated with the dashed line over British Columbia).

The remaining portion of L_2's cold front still over the Pacific often continues to progress toward the southeast as a "headless" front (seemingly detached from its parent cyclone L_2).

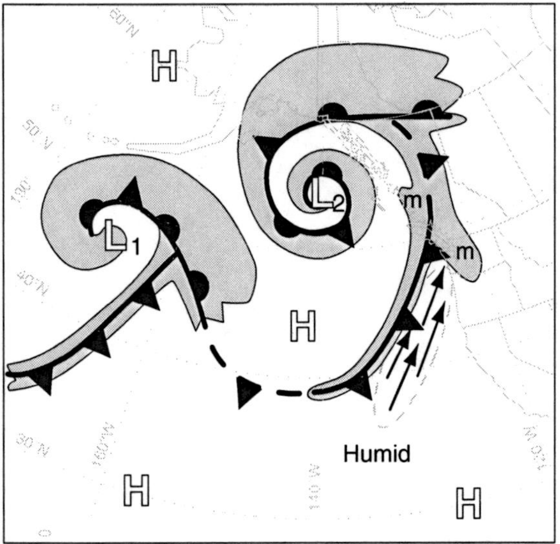

Figure 13.54
Sketch of occluding mid-latitude cyclones approaching the Pacific Northwest coast of N. America.

(continues in next column)

INFO • Pacific Cyclones *(continued)*

Sometimes there is a strong **"pre-frontal jet"** of fast low-altitude winds just ahead of the cold front, as shown by the black arrows in Fig. 13.54. If the source region of this jet is in the humid sub-tropical air, then copious amounts of moisture can be advected toward the coast by this **atmospheric river**. If the source of this jet is near Hawaii, then the conveyor belt of moist air streaming toward N. America is nicknamed the "**Pineapple Express**".

When this humid air hits the coast, the air is forced to rise over the mountains. As the rising air cools adiabatically, clouds and orographic precipitation form over the mountains (indicated by "m" in Fig. 13.54). Sometimes the cold front stalls (stops advancing) while the pre-frontal jet continues to pump moisture toward the mountains. This atmospheric-river situation causes extremely heavy precipitation and flooding.

Fig. 13.55 shows an expanded view of the Vancouver, Seattle, Victoria region (corresponding to the lower-right "m" from Fig. 13.54). Low-altitude winds split around the Olympic Mountains, only to converge (thick dashed line) in a region of heavy rain or snow called the **Olympic Mountain Convergence Zone** (also known as the **Puget Sound Convergence Zone** in the USA). The windward slopes of mountain ranges often receive heavy orographic precipitation (dotted ovals), while in between is often a **rain shadow** of clearer skies and less precipitation.

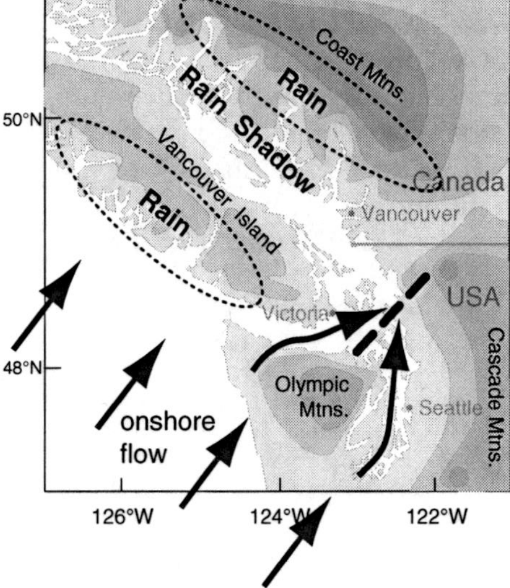

Figure 13.55
Zoomed view of the Pacific Northwest. Pacific Ocean, sounds, and straits are white, while higher terrain is shaded darker. Arrows represent low-altitude wind.

HOMEWORK EXERCISES

Broaden Knowledge & Comprehension

For all the exercises in this section, collect information from the internet. Don't forget to cite the web site URLs that you use.

B1. Collect weather maps for a new case-study cyclone assigned by your instructor. Use these maps to explain the processes involved in the evolution of this cyclone.

B2. What are some web sites that provide data on damage, deaths, and travel disruptions due to storms?

B3. Same as exercise B1, but use the maps to create a sketch of the 3-D nature of the storm, and how the storm dynamics affect its evolution.

B4. Find weather maps of common cyclogenesis locations and storm tracks.

B5. Get weather maps that display how a cyclone is affected as it moves over large mountain ranges, such as the North American Coast Range, Rocky Mountains, Cascade Mountains, or other significant range in the world (as assigned by your instructor).

B6. Draw on a map the path of a cyclone center, and encircle regions on that map experiencing bad weather (heavy rains, blizzards, windstorms, etc.).

B7. Get maps showing cyclone bombs, and discuss their life cycle.

B8. Get upper-air weather maps (such as at 30 or 20 kPa) showing wind isotachs or geopotential height contours. Use these maps to identify Rossby wave lengths and amplitudes.

B9. Using the maps from exercise B8, measure the radius-of-curvature of troughs, and combine with wind-speed information to calculate vorticity.

B10. Get geopotential height and temperature maps for the 70 or 85 kPa isobaric surfaces. Use the method of crossing isopleths to identify which point on the map has the largest temperature advection.

B11. Get geopotential height and vorticity maps on the 50 kPa isobaric surface. Identify points on the map that have the largest positive vorticity advection and largest negative vorticity advection. Identify a location that supports cyclogenesis.

B12. Get upper-air maps of vorticity on the 70 kPa isobaric surface, and a map of 100 to 50 kPa thickness. For a location near a cyclone, estimate the thermal wind. Next, identify a point on the map having largest advection of vorticity by that thermal wind. Recall that ascent and cyclogenesis is supported at those max advection locations, as described by the omega equation.

B13. Get an upper-air map that shows contours of the height or pressure of a surface of constant potential temperature (i.e., an isentropic surface). Depending on the potential temperature you pick, indicate which part of the isentropic surface might be in the stratosphere (if any). Given typical wind directions (from other maps or information), indicate if the air is like ascending or descending along the isentropic surface.

B14. Get a map of isotachs for the 20 or 30 kPa isobaric surface. Label the quadrants of relative to the jet-streak axis. Suggest locations where cyclogenesis is favored. At one of those regions, calculate vertical velocity in the mid-troposphere that would be caused by divergence in the jet-streak winds.

B15. Get a map of geopotential height on an isobaric surface of 20 or 30 kPa. Comment on how the packing of heights changes as the jet stream flows in troughs and ridges.

B16. Capture a radar reflectivity image (perhaps a composite from many radars) that show the distribution of dBZ around an extratropical cyclone. Comment on the amount of sea-level pressure tendency due to latent-heating in different parts of the cyclone.

B17. The case-study in this textbook showed weather maps of typical variables and fields often employed by meteorologists. Get and comment on the value of weather maps of other types of fields or variables, as you can acquire from national weather services or weather research centers.

B18. Find a web site that produces maps of Q-vectors or Q-vector divergence. Print this map and a normal surface weather map with fronts, and discuss how you would anticipate the cyclone to evolve based on a Q-vector analysis.

Apply

A1. For latitude 50°N, find the approximate wavelength (km) of upper-atmosphere (Rossby) waves triggered by mountains, given an average wind speed (m s^{-1}) of:
a. 20 b. 25 c. 30 d. 35 e. 40 f. 45 g. 50
h. 55 i. 60 j. 65 k. 70 l. 75 m. 80 n. 85

A2. Find the rate of increase of the β parameter (i.e., the rate of change of Coriolis parameter with distance north) in units of $m^{-1}s^{-1}$ at the following latitude (°N):
a. 40 b. 45 c. 50 d. 55 e. 60 f. 65 g. 70
h. 80 i. 35 j. 30 k. 25 l. 20 m. 15 n. 10

A3. Given a tropospheric depth of 12 km at latitude 45°N, what is the meridional (north-south) amplitude (km) of upper-atmosphere (Rossby) waves triggered by mountains, given an average mountain-range height (km) of:
a. 0.4 b. 0.6 c. 0.8 d. 1.0 e. 1.2 f. 1.4 g. 1.6
h. 1.8 i. 2.0 j. 2.2 k. 2.4 l. 2.6 m. 2.8 n. 3.0

A4. How good is the approximation of eq. (13.4) to eq. (13.3) at the following latitude (°N)?
a. 40 b. 45 c. 50 d. 55 e. 60 f. 65 g. 70
h. 80 i. 35 j. 30 k. 25 l. 20 m. 15 n. 10

A5. For a troposphere of depth 12 km at latitude 43°N, find the potential vorticity in units of $m^{-1}s^{-1}$, given the following:
[wind speed (m s^{-1}) , radius of curvature (km)]
a. 50 , 500
b. 50 , 1000
c. 50 , 1500
d. 50 , 2000
e. 50 , –500
f. 50 , – 1000
g. 50 , –1500
h. 50 , -2000
i. 75 , 500
j. 75 , 1000
k. 75 , 1500
l. 75 , 2000
m. 75 , –1000
n. 75 , – 2000

A6. For air at 55°N with initially no curvature, find the potential vorticity in units of $m^{-1}s^{-1}$ for a troposphere of depth (km):
a. 7.0 b. 7.5 c. 8.0 d. 8.5 e. 9.0 f. 9.5 g. 10.0
h. 10.5 i. 11 j. 11.5 k. 12 l. 12.5 m. 13 n. 13.5

A7. When air at latitude 60°N flows over a mountain range of height 2 km within a troposphere of depth 12 km, find the radius of curvature (km) at location "C" in Fig. 13.20 given an average wind speed (m s^{-1}) of: a. 20 b. 25 c. 30 d. 35 e. 40 f. 45 g. 50
h. 55 i. 60 j. 65 k. 70 l. 75 m. 80 n. 85

A8. Regarding equatorward propagation of cyclones on the eastern slope of mountains, if a cyclone of radius 1000 km and potential vorticity of $3x10^{-8}$ $m^{-1}s^{-1}$ is over a slope as given below ($\Delta z/\Delta x$), find the change in relative vorticity (s^{-1}) between the north and south sides of the cyclone.
a. 1/500 b. 1/750 c. 1/1000 d. 1/1250
e. 1/1500 f. 1/1750 g. 1/2000 h. 1/2250
i. 1/2500 j. 1/2750 k. 1/3000 l. 1/3250

A9. Recall from the Atmospheric Forces and Winds chapter that incompressible mass continuity implies that $\Delta W/\Delta z = -D$, where D is horizontal divergence. Find the Coriolis contribution to the stretching term (s^{-2}) in the relative-vorticity tendency equation, given the average 85 kPa divergence in Fig. 13.26 for the following USA state:
a. AZ b. WI c. KS d. KY e. VA f. CO g. KA
h. AB i. MD j. ID k. central TX l. NY m. IN

A10. Find the spin-up rate (s^{-2}) of quasi-geostrophic vorticity, assuming that the following is the only non-zero characteristic:
a. Geostrophic wind of 30 m s^{-1} from north within region where geostrophic vorticity increases toward the north by $6x10^{-5}$ s^{-1} over 500 km distance.
b. Geostrophic wind of 50 m s^{-1} from west within region where geostrophic vorticity increases toward the east by $8x10^{-5}$ s^{-1} over 1000 km distance.
c. A location at 40°N with geostrophic wind from the north of 25 m s^{-1}.
d. A location at 50°N with geostrophic wind from the south of 45 m s^{-1}.
e. A location at 35°N with vertical velocity increasing 0.5 m s^{-1} with each 1 km increase in height.
f. A location at 55°N with vertical velocity decreasing 0.2 m s^{-1} with each 2 km increase in height.

A11. Find the value of geostrophic vorticity (s^{-1}), given the following changes of (U_g, V_g) in m s^{-1} with 500 km distance toward the (north, east):
a. (0, 5) b. (0, 10) c. (0, –8) d. (0, –20)
e. (7, 0) f. (15, 0) g. (–12, 0) h. (-25, 0)
i. (5, 10) j. (20, 10) k. (–10, 15) l. (–15, –12)

A12. Find the value of geostrophic vorticity (s^{-1}), given a geostrophic wind speed of 35 m s^{-1} with the following radius of curvature (km). Assume the air rotates similar to a solid-body.
a. 450 b. –580 c. 690 d. –750 e. 825 f. –988
g. 1300 h. –1400 i. 2430 j. –2643 k. 2810
l. –2900 m. 3014 n. –3333

A13. What is the value of omega (Pa s^{-1}) following a vertically-moving air parcel, if during 1 minute its pressure change (kPa) is:
a. –2 b. –4 c. –6 d. –8 e. –10 f. –12 g. –14
h. –16 i. –18 j. –20 k. 0.00005 l. 0.0004
m. 0.003 n. 0.02

A14. At an altitude where the ambient pressure is 85 kPa, convert the following vertical velocities (m s^{-1}) into omega (Pa s^{-1}):
a. 2 b. 5 c. 10 d. 20 e. 30 f. 40 g. 50
h. –0.2 i. –0.5 j. –1.0 k. –3 l. –5 m. – 0.03

A15. Using Fig. 13.15, find the most extreme value horizontal divergence (10^{-5} s^{-1}) at 20 kPa over the following USA state:
a. MI b. WI c. IL d. IN e. TN f. GA
g. MS h. AB i. KY j. PA k. NY l. SC

A16. Find the vertical velocity (m s^{-1}) at altitude 9 km in an 11 km thick troposphere, if the divergence (10^{-5} s^{-1}) given below occurs within a 2 km thick layer within the top of the troposphere.
a. 0.2 b. 1 c. 1.5 d. 2 e. 3 f. 4 g. 5 h. 6
i. –0.3 j.–0.7 k. –1.8 l. –2.2 m. –3.5 n. –5

A17. Jet-stream inflow is 30 m s^{-1} in a 4 km thick layer near the top of the troposphere. Jet-stream outflow (m s^{-1}) given below occurs 800 km downwind within the same layer. Find the vertical velocity (m s^{-1}) at the bottom of that layer
a. 35 b. 40 c. 45 d. 50 e. 55 f. 60 g. 65
h. 70 i. 30 j. 25 k. 20 l. 15 m. 10 n. 5

A18. Find the diagonal distance (km) from trough to crest in a jet stream for a wave of 750 km amplitude with wavelength (km) of:
a. 1000 b. 1300 c. 1600 d. 2000 e. 2200
f. 2500 g. 2700 h. 3000 i. 3100 j. 3300
k. 3800 l. 4100 m. 4200 n. 4500

A19. Given the data from the previous exercise, find the radius (km) of curvature near the crests of a sinusoidal wave in the jet stream.

A20. Find the gradient-wind speed difference (m s^{-1}) between the jet-stream speed moving through the anticyclonic crest of a Rossby wave in the N. Hemisphere and the jet-stream speed moving through the trough. Use data from the previous 2 exercises, and assume a geostrophic wind speed of 75 m s^{-1} for a wave centered on latitude 40°N.

A21. Given the data from the previous 3 exercises. Assuming that the gradient-wind speed difference calculated in the previous exercise is valid over a layer between altitudes 8 km and 12 km, where the tropopause is at 12 km, find the vertical velocity (m s^{-1}) at 8 km altitude.

A22. Suppose that a west wind enters a region at the first speed (m s^{-1}) given below, and leaves 500 km downwind at the second speed (m s^{-1}). Find the north-south component of ageostrophic wind (m s^{-1}) in this region. Location is 55°N.
a. (40, 50) b. (30, 60) c. (80, 40) d. (70, 50)
e. (40, 80) f. (60, 30) g. (50, 40) h. (70, 30)
i. (30, 80) j. (40, 70) k. (30, 70) l. (60, 20)

A23. Use the ageostrophic right-hand rule to find the ageostrophic wind direction for the data of the previous problem.

A24. Using the data from A22, find the updraft speed (m s^{-1}) into a 4 km thick layer at the top of the troposphere, assuming the half-width of the jet streak is 200 km.

A25. Suppose that the thickness of the 100 - 50 kPa layer is 5.5 km and the Coriolis parameter is 10^{-4} s^{-1}. A 20 m s^{-1} thermal wind from the west blows across a domain of x dimension given below in km. Across that domain in the x-direction is a decrease of cyclonic relative vorticity of 3x10^{-4} s^{-1}. What is the value of mid-tropospheric ascent velocity (m s^{-1}), based on the omega equation?
a. 200 b. 300 c. 400 d. 500 e. 600
f. 700 g. 800 h. 900 i. 1000 j. 1100
k. 1200 l. 1300 m. 1400 n. 1400

A26. On the 70 kPa isobaric surface, $\Delta U_g/\Delta x$ = (4 m s^{-1})/(500 km) and $\Delta T/\Delta x$ = ___°C/(500 km), where the temperature change is given below. All other gradients are zero. Find the Q-vector components Q_x, Q_y, and the magnitude and direction of Q.
a. 1 b. 1.5 c. 2 d. 2.5 e. 3 f. 3.5 g. 4
h. 4.5 i. 5 j. 5.5 k. 6 l. 6.5 m. 7 n. 7.5

A27. Find Q-vector magnitude on the 85 kPa isobaric surface if the magnitude of the horizontal temperature gradient is 5°C/200 km, and the magnitude of the geostrophic-wind difference-vector component (m s^{-1}) along an isotherm is __ /200 km, where __ is:
a. 1 b. 1.5 c. 2 d. 2.5 e. 3 f. 3.5 g. 4
h. 4.5 i. 5 j. 5.5 k. 6 l. 6.5 m. 7 n. 7.5

A28. Find the mass of air over 1 m^2 of the Earth's surface if the surface pressure (kPa) is:
a. 103 b. 102 c. 101 d. 99 e. 98 f. 97 g. 96
h. 95 i. 93 j. 90 k. 85 l. 80 m. 75 n. 708

A29. Assume the Earth's surface is at sea level. Find the vertical velocity (m s^{-1}) at height 3 km above ground if the change of surface pressure (kPa) during 1 hour is:

a. –0.5 b. –0.4 c. –0.3 d. –0.2 e. –0.1 f. 0.1
g. 0.2 h. 0.4 i. 0.6 j. 0.8 k. 1.0 l. 1.2 m. 1.4

A30. Given the rainfall (mm) accumulated over a day. If the condensation that caused this precipitation occurred within a cloud layer of thickness 6 km, then find the virtual-temperature warming rate (°C day^{-1}) of that layer due to latent heat release.

a. 1 b. 50 c. 2 d. 45 e. 4 f. 40 g. 5
h. 35 i. 7 j. 30 k. 10 l. 25 m. 15 n. 20

A31. For the data in the previous exercise, find the rate of decrease of surface pressure (kPa) per hour, assuming an average air temperature of 5°C.

Evaluate & Analyze

E1. Compare the similarities and differences between cyclone structure in the Northern and Southern Hemisphere?

E2. In the Cyclogenesis & Cyclolysis section is a list of conditions that favor rapid cyclogenesis. For any 3 of those bullets, explain why they are valid based on the dynamical processes that were described in the last half of the chapter.

E3. For a cyclone bomb, are the winds associated with that cyclone in geostrophic equilibrium? Hints: consider the rate of air-parcel acceleration, based on Newton's 2nd Law. Namely, can winds accelerate fast enough to keep up with the rapidly increasing pressure gradient?

E4. Make a photocopy of Fig. 13.3. For each one of the figure panels on this copy, infer the centerline position of the jet stream and draw it on those diagrams.

E5. Create a 6-panel figure similar to Fig. 13.3, but for cyclone evolution in the Southern Hemisphere.

E6. Justify the comment that cyclone evolution obeys Le Chatelier's Principle.

E7. Refer back to the figure in the General Circulation chapter that sketches the position of mountain ranges in the world. Use that information to hypothesize favored locations for lee cyclogenesis in the world, and test your hypothesis against the data in Fig. 13.5.

E8. Why are there no extratropical cyclone tracks from east to west in Fig. 13.5?

E9. Contrast the climatology of cyclone formation and tracks in the Northern vs. Southern Hemisphere (using the info in Fig. 13.5), and explain why there is a difference in behaviors based on the dynamical principles in the last half of the chapter.

E10. Justify why the tank illustration in Fig. 13.6 is a good analogy to atmospheric flow between cyclones and anticyclones.

E11. Fig. 13.7 shows the axis of low pressure tilting westward with increasing height. Explain why this tilt is expected. (Hints: On which side of the cyclone do you expect the warm air and the cold air? The hypsometric equation in Chapter 1 tells us how fast pressure decreases with height in air of different temperatures.)

E12. Redraw Fig. 13.7 a & b for the Southern Hemisphere, by extending your knowledge of how cyclones work in the Northern Hemisphere.

E13. Regarding stacking and tilting of low pressure with altitude, make a photocopy of Fig. 13.3, and on this copy for Figs. b and e draw the likely position of the trough axis near the top of the troposphere. Justify your hypothesis.

E14. Fig. 13.8 shows a warm-air conveyor bring air from the tropics. Assuming this air has high humidity, explain how this conveyor helps to strengthen the cyclone. Using dynamical principles from the last half of the chapter to support your explanation.

E15. Fig. 13.10a has coarse temporal resolution when it shows the evolution of the case-study cyclone. Based on your knowledge of cyclone evolution, draw two weather maps similar to Fig. 13.10a, but for 12 UTC 3 Apr 2014 and 00 UTC 4 Apr 2014.

E16. Speculate on why the hail reports in Fig. 13.9 are mostly in different regions than the wind-damage reports.

E17. From the set of case-study maps in Figs. 13.9 through 13.19, if you had to pick 3 maps to give you the best 5-D mental picture of the cyclones, which 3 would you pick? Justify your answer.

E18. Of the following isosurfaces (height, pressure, thickness, and potential temperature), which one seems the most peculiar (unusual, illogical) to you?

What questions would you want to ask to help you learn more about that one isosurface?

E19. Starting with a photocopy of the 4 height charts in the left column of Figs. 13.13, use a different color pen/pencil for each isobaric surface, and trace all the height contours onto the same chart. Analyze the tilt with height of the axis of low pressure, and explain why such tilt does or does not agree with the state of cyclone evolution at that time.

E20. Compare and contrast the 85 kPa temperature map of Fig. 13.14 with the thickness map of Fig. 13.16 for the case-study storm. Why or why not would you expect them to be similar?

E21. Compare the wind vectors of Fig. 13.13 with the heights in Fig. 13.13. Use your understanding of wind dynamics to explain the relationship between the two maps.

E22. In the vertical cross section of Fig. 13.19a, why are the isentropes packed more closely together in the stratosphere than in the troposphere? Also, why is the tropopause higher on the right side of that figure? [Hints: Consider the standard atmosphere temperature profile from Chapter 1. Consider the General Circulation chapter.]

E23. Describe the relationship between the surface values of isentropes in Fig. 13.19a and the temperature values along cross-section A - A' in Fig. 13.14 (bottom left).

E24. In Fig. 13.20, how would lee cyclogenesis be affected if the tropopause perfectly following the terrain elevation? Explain.

E25. In Fig. 13.20, speculate on why the particular set of Rossby waves discussed in that section are known as "stationary" waves.

E26. Can "stationary" Rossby waves such as in Fig. 13.20 occur near the equator? If so, what are their characteristics?

E27. Given zonal flow of the whole troposphere (12 km depth) hitting a semi-infinite plateau at 32°N latitude. Use math to explain the flow behavior over the plateau, assuming a 2 km plateau height above sea level. Assume the tropopause height doesn't change. Would the triggering of cyclones, and the wavelength of planetary waves be different? Why? Can you relate your answer to weather over the Tibetan Plateau?

E28. Considering the terrain height changes, would anticyclones be triggered upwind of mountain ranges, analogous to lee-side cyclogenesis? Justify your answer, and discuss how you can confirm if this happens in the real atmosphere.

E29. Suppose that all of North America is nearly at sea level, except for a 2 km deep valley that is 500 km wide, running north-south across the center of North America (assume the valley is not filled with water). Explain what Rossby waves would be triggered, and the associated weather downwind.

E30. Summer tropopause height is higher, and jet-stream winds are slower, than in winter. Explain the seasonal differences you would expect in terrain-triggered Rossby waves, if any.

E31. If extratropical cyclones tend to propagate equatorward on the lee side of mountain ranges, is there any geographic feature that would cause these storms to propagate poleward? Justify.

E32. Fig. 13.22 highlights 3 important attributes of cyclones that are discussed in greater detail in the last half of the chapter. Speculate on why we study these attributes separately, even though the caption to that figure discusses how all 3 attributes are related.

E33. For cyclogenesis we focused on three attributes: vertical velocity, vorticity, and pressure-tendency at sea level. What attributes would you want to focus on to anticipate cyclolysis? Explain.

E34. Three vorticity tilting terms are given in eq. (13.9), but just the first tilting term was illustrated in Fig. 13.23d. Draw figures for the other two terms.

E35. Except for the last term in the full vorticity tendency equation, all the other terms can evaluate to be positive or negative (i.e., gain or loss of relative vorticity). What is it about the mathematics of the turbulent drag term in that equation that always make it a loss of relative vorticity?

E36. Based on what you learned from Fig. 13.24, what tips would you teach to others to help them easily find regions of PVA and NVA.

E37. Make a diagram that shows vertical advection of vorticity, similar to the drawing in Fig. 13.23a.

E38. If drag were the only non-zero term on the right side of eq. (13.9), then how would vorticity change with time if initially the flow had some amount of positive vorticity?

E39. Devise a tilting term for vertical vorticity that is tilted into horizontal vorticity.

E40. For what situations might the quasi-geostrophic approximation be useful, and for what situations would it be inappropriate?

E41. The quasi-geostrophic vorticity equation includes a term related to vorticity advection by the geostrophic wind. The omega equation has a term related to vorticity advection by the thermal wind. Contrast these terms, and how they provide information about cyclogenesis.

E42. Employ Figs. 13.9 - 13.19 to estimate as many vorticity-tendency terms as reasonably possible for that case study event, on the 50 kPa isobaric surface at the location of the "X".

E43. Create a "toy model" similar to Fig. 13.27, but focus on the vorticity effects east of the ridge axis. Use this to help explain anticyclogenesis.

E44. If nothing else changes except latitude, explain the relationship between Rossby-wave radius-of-curvature and latitude.

E45. Why does jet-stream curvature contribute to surface cyclogenesis east of the jet trough axis rather than west of the trough axis?

E46. In the Forces and Winds chapter you saw that the horizontal pressure gradient and wind speeds are weak. Does this physical constraint influence the possible strength of the jet-stream curvature effect for cyclogenesis?

E47. At the "X" in Fig. 13.30, use the information plotted on that map to estimate how the jet-stream curvature and jet-streak processes influence changes to sea-level pressure and ascent speed in the middle of the troposphere.

E48. Make a photo copy of Fig. 13.18. Using the jet-stream curvature and jet-streak information that you can estimate from the height contours, draw on your copy the locations in the Northern Hemisphere where cyclogenesis is favored.

E49. For the jet-streak illustration of Fig. 13.32b, explain why two of the quadrants have weaker convergence or divergence than the other two quadrants.

E50. Consider a steady, straight jet stream from west to east. Instead of a jet streak of higher wind speed imbedded in the jet stream, suppose the jet streak has lower wind speed imbedded in the jet stream. For the right and left entrance and exit regions to this "slow" jet streak, describe which ones would support cyclogenesis at the surface.

E51. Does the "Ageostrophic right-hand rule" work in the Southern Hemisphere too? Justify your answer based on dynamical principles.

E52. Eq. (13.25) was for a west wind, and eq. (13.26) was for a south wind. What method would you use to estimate ageostrophic wind if the wind was from the southwest?

E53. Fig. 13.34 suggests low-altitude convergence of air toward cyclones (lows), rising motion, and high-altitude divergence. Is this sketch supported by the case-study data from Figs. 13.15, 13.25, and 13.26?

E54. The green arrows showing near-surface winds are plotted in Fig. 13.34 as a component of the secondary circulation. What drives these near-surface winds? (Hint: Recall that winds are driven by forces, according to Newton.)

E55. a. Use Figs. 12.12 from the Fronts & Airmasses chapter to find the location where horizontal temperature advection is greatest near a warm front. b. Similar questions but using Fig. 12.11 for cold fronts. Hint: Use the technique shown in the INFO box for max advection in Chapter 13.

E56. Discuss how Fig. 13.37 relates to the omega equation, and how the figure and equation can be used to locate regions that favor cyclogenesis.

E57. Create a new form for the omega equation, based on:

a. the change of geostrophic wind with height;

b. the temperature change in the horizontal.

Hint: The horizontal-temperature gradient is related to the geostrophic-wind vertical gradient by the thermal-wind relationship.

E58. What steps and assumptions must you make to change eq. (13.28 into eq. (13.29)?

E59. Would the omega equation give any vertical motion for a situation having zero temperature gradient in the horizontal (i.e., zero baroclinicity)? Why?

E60. What role does inertia play in the "geostrophic paradox"?

E61. Suppose that the 85 kPa geostrophic wind vectors are parallel to the height contours in Fig. 13.13 for the case-study storm. Use that information along with the 85 kPa isotherms in Fig. 13.14 to estimate the direction of the Q-vectors at the center of the following USA state:

a. IL b. IA c. MO d. KS e. AR

E62. Use the Q-vector approach to forecast where cyclogenesis might occur in the Pacific Northwest USA (in the upper left quadrant of Figs. 13.40 and 13.41).

E63. In Fig. 13.43 we showed have the vertical speeds of the "leaf" in the air column could be a surrogate for changes in mass and surface pressure. Discuss the pros and cons of that approach.

E64. If rainfall rate (RR) affects surface pressure, and weather-radar echo intensity (dBZ) can be used to estimate rainfall rate, then devise an equation for surface-pressure change as a function of dBZ.

E65. Considering surface-pressure tendency, what cyclogenesis information can be gained from Doppler velocities measured by weather radar? State the limitations of such an approach.

E66. Are there situations for which cyclolysis (quantified by sea-level pressure tendency) might be caused by latent heating. Justify your arguments.

E67. In Fig. 13.47, describe the dynamics that makes the time of maximum convergence in the boundary layer occur <u>after</u> the time of maximum divergence in the jet stream, during cyclone evolution.

E68. Suppose that the temperature wave in Fig. 13.49 was a quarter of a wavelength east of the height wave. Would the flow differ from that sketched in Fig. 13.50? Speculate on how likely it is that the temperature wave is shifted this way.

E69. Do you think that anticyclones could self develop? Explain what processes could make this happen.

E70. Explain why the Q-vector analysis of Fig. 13.51 indicates the propagation of cyclones is toward the east. Also explain why this relates to self-propagation rather than relating to cyclogenesis driven by the jet-stream flow.

E71. Use Fig. 13.52 to explain why fronts are associated with cyclones and not anticyclones. The same figure can be used to explain why airmasses are associated with anticyclones. Discuss.

E72. If global baroclinicity is absent (e.g., no air-temperature gradient between the equator and the poles), could there be cyclogenesis? Why?

E73. Explain how the up- and down-couplet of air motion in Fig. 13.53c (as diagnosed using Q-vectors), works in a way to strengthen and propagate the cold front.

E74. Use a Q-vector analysis to speculate on the dynamics needed to cause <u>warm</u> fronts to strengthen and propagate.

E75. Synthesize a coherent description of the dynamics of the case-study cyclone, based on information from the weather maps that help you estimate ascent, spin-up, and sea-level pressure tendency.

E76. Compare and contrast the Pacific cyclones of Fig. 13.54 with the case-study cyclone of Figs. 13.13 - 13.15.

Synthesize

S1. Consider Fig. 13.6. What if frictional drag is zero at the bottom of the atmosphere (in the boundary layer). Describe differences in the resulting climate and weather.

S2. What if the case-study cyclone of Figs. 13.9 - 13.19 occurred in February rather than April, what broad aspects of the storm data would change, if at all?

S3. What if there were no mountain ranges oriented south-north in North America. Describe differences in the resulting climate and weather.

S4. What if all south-north mountain ranges disappeared in North America, and were replaced by one west-east mountain range. Describe differences in the resulting climate and weather.

S5. Suppose western North America was cold, and eastern North America was warm. Describe the orientation of baroclinicity, the jet stream, and differences in resulting climate and weather relative to our actual climate and weather.

S6. Describe changes to Rossby waves and cyclogenesis for an Earth that rotates twice as fast as the real Earth.

S7. How would you numerically solve (iterate) the quasi-geostrophic omega and vorticity equations to step forward in time to forecast those variables.

Next, describe how you could use those forecasts to estimate the corresponding temperature and wind. Finally, describe the pros and cons of using these quasi-geostrophic equations instead of using the forecast equations for momentum, heat, water, continuity, and the ideal gas law (as is done in modern numerical weather prediction).

S8. Accelerations and direction changes of the jet stream create regions of horizontal convergence and divergence that support cyclogenesis and anticyclogenesis. What if this happened on a planet where the tropospheric and stratospheric static stability were nearly the same. Describe the resulting differences of climate and weather on that planet compared to Earth.

S9. Describe changes in climate and weather that might be expected of stratospheric static stability extended all the way to the Earth's surface.

S10. Describe possible changes to climate and weather if there were no mid-latitude cyclones.

S11. Suppose the sun turned off, but radioactive decay of minerals in the solid earth caused sufficient heat to keep the Earth-system temperature the same as now. Describe resulting changes to the jet stream, Rossby waves, and weather.

S12. Re-read this chapter and extract all the forecasting tips to create your own concise synoptic-weather forecast guide.

Fig. 13.56.
Surface weather map for the case-study cyclone, from www.hpc.ncep.gov/dailywxmap/ .

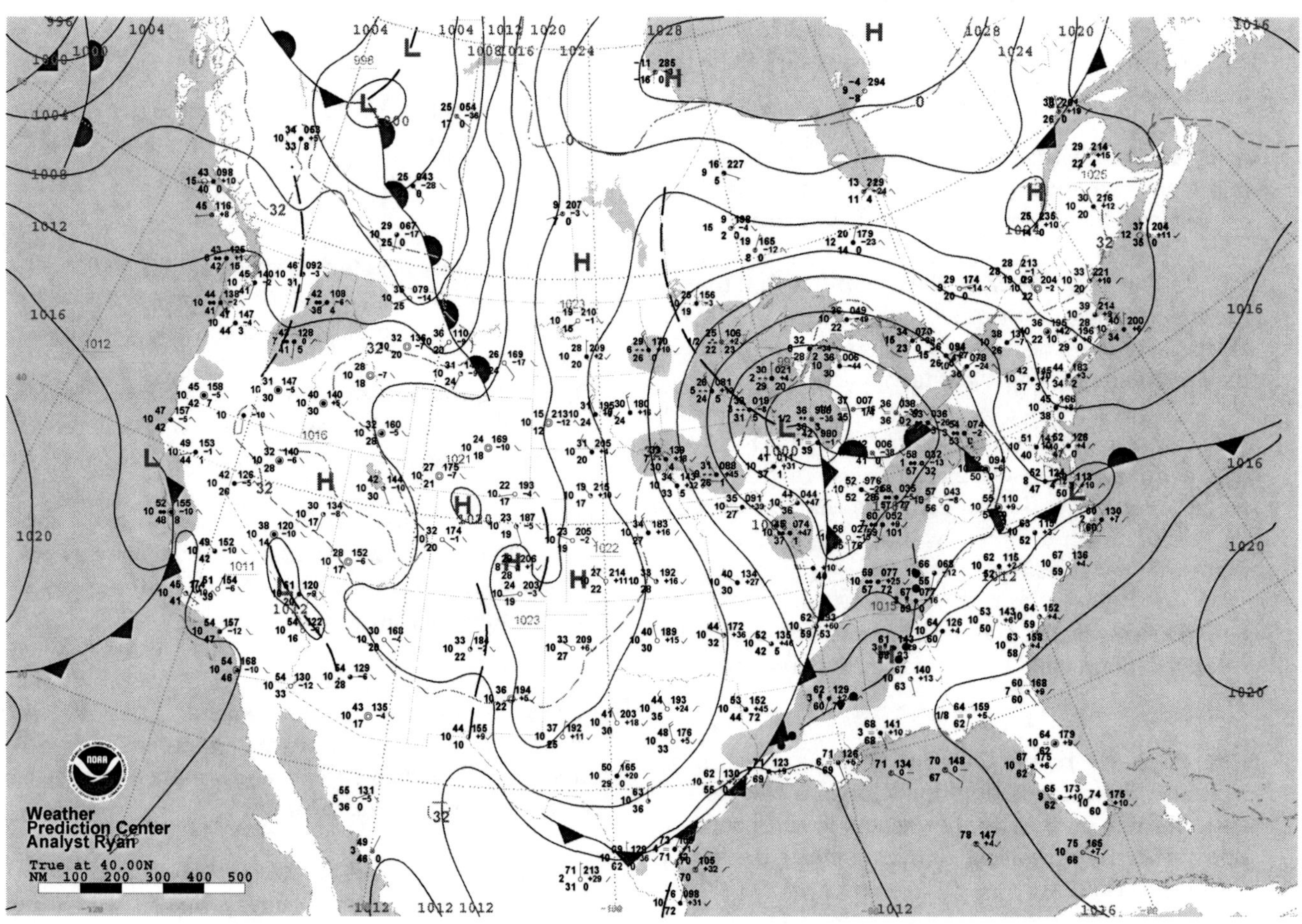

14 THUNDERSTORM FUNDAMENTALS

Contents

Thunderstorm characteristics, formation, and forecasting are covered in this chapter. The next chapter covers thunderstorm hazards including hail, gust fronts, lightning, and tornadoes.

THUNDERSTORM CHARACTERISTICS

Thunderstorms are **convective clouds** with large vertical extent, often with tops near the tropopause and bases near the top of the boundary layer. Their official name is **cumulonimbus** (see the Clouds Chapter), for which the abbreviation is **Cb**. On weather maps the symbol ⌈⟨ represents thunderstorms, with a dot •, asterisk *, or triangle ▵ drawn just above the top of the symbol to indicate rain, snow, or hail, respectively. For severe thunderstorms, the symbol is ⌈⟩.

© Gene Rhoden / weatherpix.com

Figure 14.1
Air-mass thunderstorm.

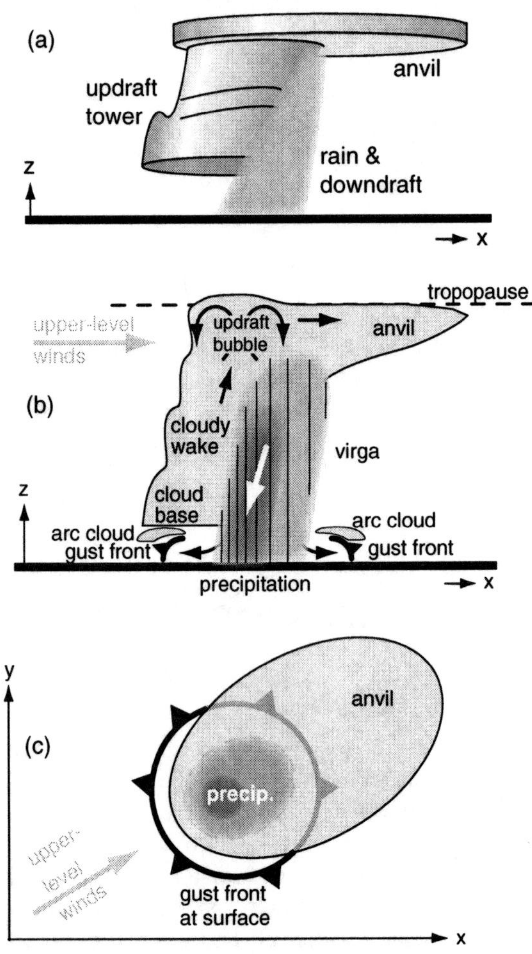

Figure 14.2
(a) Sketch of a basic (airmass) thunderstorm in its mature stage. (b) Vertical slice through the storm. Light shading indicates clouds, medium and dark shadings are moderate and heavy precipitation, and arrows show air motion. (c) Horizontal composite, showing the anvil at storm top (as viewed from above by satellite), the precipitation in the low-to-middle levels (as viewed by radar), and the gust front of spreading winds at the surface.

© Gene Rhoden / weatherpix.com

Figure 14.3
Photo of supercell thunderstorm.

Appearance

A mature thunderstorm cloud looks like a mushroom or anvil with a relatively large-diameter flat top. The simplest thunderstorm (see Figs. 14.1 & 14.2) has a nearly vertical stem of diameter roughly equal to its depth (of order 10 to 15 km). The large top is called the **anvil**, **anvil cloud**, or **thunderhead**, and has the official name **incus** (Latin for anvil). The anvil extends furthest in a direction as blown by the upper-tropospheric winds.

If the thunderstorm top is just starting to spread out into an anvil and does not yet have a fibrous or streaky appearance, then you identify the cloud as **cumulonimbus calvus** (see the Clouds Chapter). For a storm with a larger anvil that looks strongly **glaciated** (i.e., has a fibrous appearance associated with ice-crystal clouds), then you would call the cloud a **cumulonimbus capillatus**.

Within the stem of a mature thunderstorm is the cloudy **main updraft tower** topped by an **updraft bubble** (Fig. 14.2b). When this rising air hits the tropopause, it spreads to make the anvil. Also in the stem is a downdraft with precipitation. When the downdraft air hits the ground it spreads out, the leading edge of which is called the **gust front**. When viewed from the ground under the storm, the main updraft often has a darker cloud base, while the rainy region often looks not as dark and does not have a well-defined cloud base.

Not all cumulonimbus clouds have lightning and thunder. Such storms are technically not thunderstorms. However, in this book we will use the word **thunderstorm** to mean any cumulonimbus cloud, regardless of whether it has lightning.

More complex thunderstorms can have one or more updraft and downdraft regions. The most severe, long-lasting, less-frequent thunderstorms are **supercell thunderstorms** (Figs. 14.3 & 14.4).

Clouds Associated with Thunderstorms

Sometimes you can see other clouds attached to thunderstorms, such as a funnel, wall, mammatus, arc, shelf, flanking line, scud, pileus, dome, and beaver tail (Fig. 14.4). Not all thunderstorms have all these associated clouds.

Arc clouds (official name **arcus**, Fig. 14.2b) or **shelf clouds** form near the ground in boundary-layer air that is forced upward by undercutting cold air flowing out from the thunderstorm. These cloud bands mark the leading edge of **gust-front** outflow from the **rear-flank downdraft** (Fig. 14.4), usually associated with the **flanking line**. Often the undersides of arc clouds are dark and turbulent-looking, while their tops are smooth. Not all gust fronts have arc clouds, particularly if the displaced air is dry. See the Thunderstorm Hazards chapter.

The **beaver tail** (Fig. 14.4) is a smooth, flat, narrow, low-altitude cloud that extends along the boundary between the inflow of warm moist air to the thunderstorm and the cold air from the rain-induced **forward flank downdraft** (**FFD**).

A **dome** of **overshooting** clouds sometimes forms above the anvil top, above the region of strongest updraft. This is caused by the inertia of the upward moving air in the main updraft, which overshoots above its neutrally buoyant **equilibrium level** (**EL**). Storms that have **overshooting tops** are often more violent and turbulent.

The **flanking line** is a band of cumuliform clouds that increase from the medium-size **cumulus mediocris** (**Cu med**) furthest from the storm to the taller **cumulus congestus** (**Cu con**) close to the **main updraft**. Cumulus congestus are also informally called **towering cumulus** (**TCu**). The flanking line forms along and above the gust front, which marks the leading edge of colder outflow air from the **rear-flank downdraft** (**RFD**).

If humid layers of environmental air exist above rapidly rising cumulus towers, then **pileus** clouds can form as the environmental air is pushed up and out of the way of the rising cumulus clouds. Pileus are often very short lived because the rising cloud tower that caused it often keeps rising through the pileus and obliterates it.

The most violent thunderstorms are called **supercell** storms (Figs. 14.3 & 14.4), and usually have a quasi-steady rotating updraft (called a **mesocyclone**). The main thunderstorm updraft in supercells sometimes has curved, helical **cloud striations** (grooves or ridges) on its outside similar to threads of a screw (Fig. 14.4a). Supercells can produce intense **tornadoes** (violently rotating columns of air), which can appear out of the bottom of an isolated cylindrical lowering of the cloud base called a **wall cloud**. The portion of the tornado made visible by cloud droplets is called the **funnel cloud**, which is the name given to tornadoes not touching the ground. Tornadoes are covered in the next chapter. Most thunderstorms are not supercell storms, and most supercell storms do not have tornadoes.

Attached to the base of the wall cloud is sometimes a short, horizontal cloud called a **tail cloud**, which points towards the **forward flank precipitation** area. Ragged cloud fragments called **scud** (**cumulus fractus**) often form near the tip of the tail cloud and are drawn quickly into the tail and the wall cloud by strong winds.

The wall cloud and tornado are usually near the boundary between cold downdraft air and the low-level warm moist inflow air, under a somewhat **rain-free cloud base**. In mesocyclone storms, you can see the wall cloud rotate by eye, while rotation

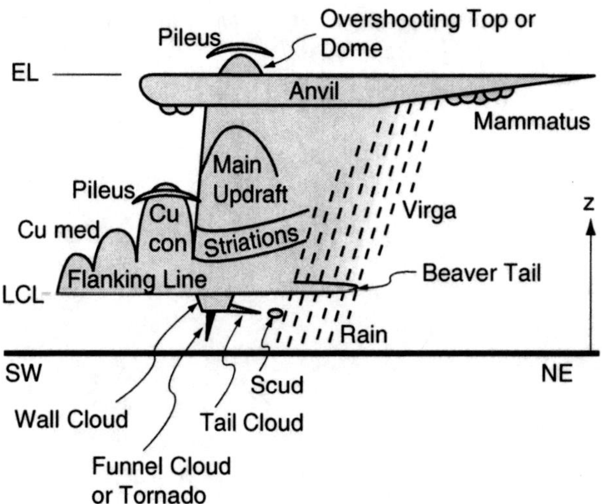

Figure 14.4a
Sketch of a classic supercell thunderstorm (Cb) as might be viewed looking toward the northwest in central North America. The storm would move from left to right in this view (i.e., toward the northeast). Many storms have only a subset of the features cataloged here. Cu med = cumulus mediocris; Cu con = cumulus congestus; LCL = lifting condensation level; EL = equilibrium level (often near the tropopause, 8 to 15 km above ground); NE = northeast; SW = southwest.

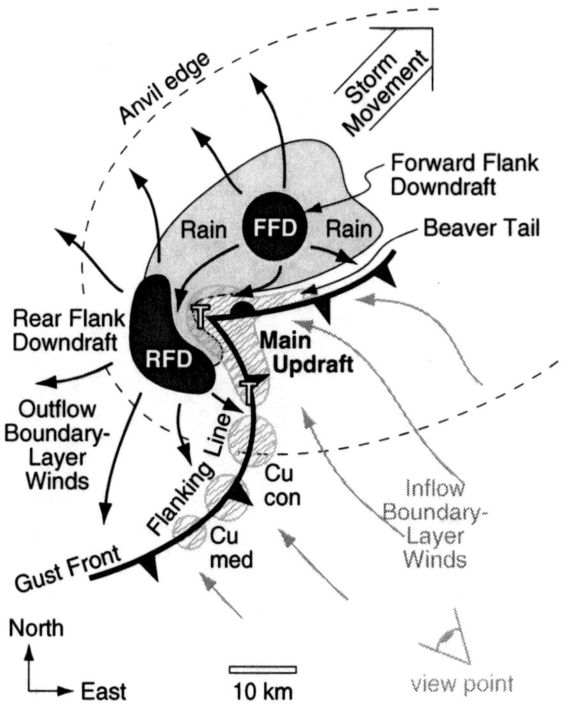

Figure 14.4b
Plan view (top down) sketch of a classic supercell thunderstorm. T indicates possible tornado positions; RFD = Rear Flank Downdraft; FFD = Forward Flank Downdraft. Regions of updraft are hatched with grey; downdrafts are solid black; rain is solid light grey. Low surface pressure is found under the updrafts (especially near the "T" locations), and high pressure under the downdrafts. Vectors are near-surface winds: cold are black, warm are grey. From the view point you see Fig. 14.4a.

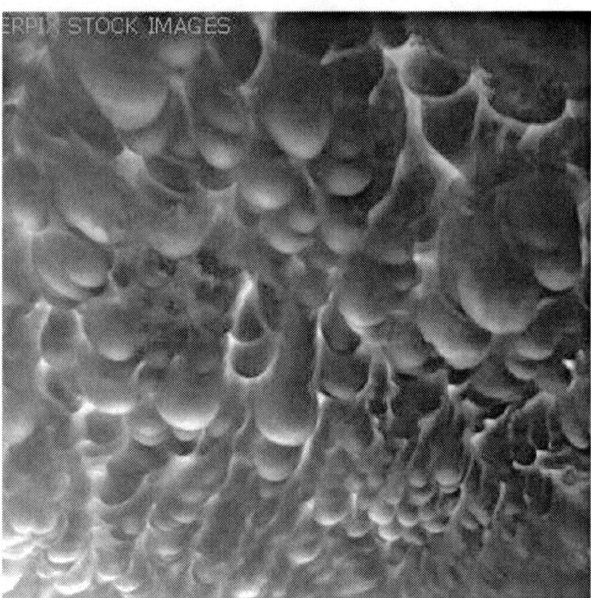

© Gene Rhoden / weatherpix.com

Figure 14.5
Mammatus clouds.

Sample Application
Identify the thunderstorm features, components, and associated clouds that are in Fig. 14.3.

Find the Answer
Given: Photo 14.3, reproduced here.
Find: Cloud features.

Method: Compare photo 14.3 to diagram 14.4a.

Base photo: © Gene Rhoden / weatherpix.com

Exposition: This cloud is a supercell thunderstorm, because it looks more like the cloud sketched in Fig. 14.4a than the simple single-cell air-mass thunderstorm in Fig. 14.2. The main updraft is tilted from lower left to upper right, a sign of wind shear in the environment. If a tornado exists under the wall cloud, it is too small to see in this figure. Given the orientation of the cloud features (similar to those in Fig 14.4a), the photographer was probably looking toward the northwest, and the thunderstorm is moving toward the northeast (to the right in this figure).

of the larger-diameter supercell is slower and is apparent only in time-lapse movies. Non-rotating wall clouds also occur with non-rotating thunderstorms (i.e., non-supercell storms).

The **rain** portion of a thunderstorm is often downwind of the main updraft, in the forward portion of the storm. Cloud edges and boundaries are often not distinguishable in the rain region. Although thunderstorms are rare in cold weather, when they occur they can cause very heavy snowfall instead of rain. Further downwind of the heavy rain area is a region of **virga**, where the precipitation (rain or snow) evaporates before reaching the ground.

Mammatus clouds look like sacks or protuberances hanging from the bottom of layered clouds, and are sometimes found on the underside of thunderstorm anvils (Fig. 14.5). No correlation exists between the presence of mammatus clouds and the intensity of thunderstorms (i.e., hail, tornadoes, etc.). Not all thunderstorms have mammatus clouds.

One of many theories for mammatus clouds is that the thunderstorm updraft pumps air into the anvil with slightly cooler virtual temperature than environmental air. This anvil air is statically unstable, so the cold air begins sinking as upside-down thermals. Because these sinking air parcels are laden with cloud droplets or ice crystals, the air warms at only the moist adiabatic rate. Thus, evaporation of some of the water helps keep the mammatus clouds cooler than their environment, and the liquid water loading of the remaining droplets contributes to cooler virtual temperatures, so they keep sinking.

In an environment with weak wind shear, the **anvil cloud** in weak thunderstorms can be symmetrical, looking like a mushroom cap (Fig. 14.1). However, for larger, longer lasting thunderstorms in the presence of strong winds aloft, the ice crystals brought up by the main updraft will spread out asymmetrically (Fig. 14.2) in an anvil that can be 100 km wide (crosswind) and 300 km long (downwind).

If you want to chase and photograph isolated storms in N. America, the best place to be is well to the side of the storm's path. As most storms move northeast in N. America, this corresponds to your being southeast of the thunderstorm. This location is relatively safe with good views of the storm and the tornado (see "A SCI. PERSPECTIVE • Be Safe").

For non-chasers overtaken by a storm, the sequence of events they see is: cirrus anvil clouds, mammatus clouds (if any), gust front (& arc cloud, if any), rain, hail (if any), tornado (if any), and rapid clearing with sometimes a rainbow during daytime.

Cells & Evolution

The fundamental structural unit of a thunderstorm is a **thunderstorm cell**, which is of order 10 km in diameter and 10 km thickness (i.e., its **aspect ratio** = depth/width = 1). Its evolution has three stages: (1) towering-cumulus; (2) mature; and (3) dissipation, as sketched in Fig. 14.6. This life cycle of an individual cell lasts about 30 to 60 min.

The towering cumulus stage consists of only updraft in a rapidly growing cumulus congestus tower. It has no anvil, no downdrafts, and no precipitation. It is drawing warm humid boundary-layer air into the base of the storm, to serve as fuel. As this air rises the water vapor condenses and releases latent heat, which adds buoyancy to power the storm.

In the mature stage (Fig. 14.7), the thunderstorm has an anvil, both updrafts and downdrafts, and heavy precipitation of large-size drops. This often is the most violent time during the storm, with possible lightning, strong turbulence and winds. The precipitation brings cooler air downward as a downburst, which hits the ground and spreads out. In the absence of wind shear, two factors conspire to limit the lifetime of the storm (Fig. 14.6): (1) the storm runs out of fuel as it consumes nearby boundary-layer air; and (2) the colder outflow air chokes off access to adjacent warm boundary-layer air.

In the dissipating stage, the storm consists of downdrafts, precipitation, and a large **glaciated** (containing ice crystals) anvil. The storm rains itself out, and the precipitation rate diminishes. The violent aspects of the storm are also diminishing. With no updraft, it is not bringing in more fuel for the storm, and thus the storm dies.

The last frame in Fig. 14.6 shows the anvil-debris stage, which is not one of the official stages listed in severe-weather references. In this stage all that remains of the former thunderstorm is the anvil and associated virga, which contains many ice crystals that fall and evaporate slowly. No precipitation reaches the ground. The anvil remnant is not very turbulent, and drifts downwind as a thick cirrostratus or altostratus cloud as it gradually disappears.

Meanwhile, the cold outflow air spreading on the ground from the former storm might encounter fresh boundary-layer air, which it lifts along its leading edge (gust front) to possibly trigger a new thunderstorm (labeled "Cu con 2" in the last frame of Fig. 14.6). This process is called **storm propagation**, where the first storm (the "mother") triggers a daughter storm. This daughter storm can go through the full thunderstorm life cycle if conditions are right, and trigger a granddaughter storm, and so forth.

Some thunderstorms contain many cells that are in different stages of their evolution. Other storms have special characteristics that allow the cell to be maintained for much longer than an hour. The types of thunderstorms are described next.

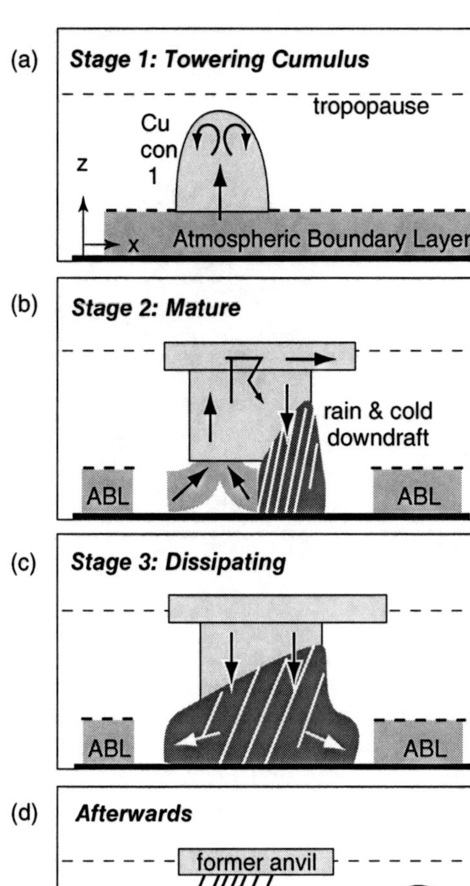

Figure 14.6
Phases of thunderstorm cell evolution. Light grey shading represents clouds, medium grey is pre-storm boundary layer (ABL) of warm humid air that serves as fuel for the storm. Dark shading is colder outflow air. Diagonal white lines represent precipitation. Arrows show air motion. Cu con = cumulus congestus.

A SCIENTIFIC PERSPECTIVE • Be Safe

In any scientific or engineering work, determine the possible hazards to you and others before you undertake any action. If you are unable to increase the safety, then consider alternative actions that are safer. If no action is relatively safe, then cancel that project.

For example, chasing thunderstorms can be very hazardous. Weather conditions (heavy rainfall, poor visibility, strong winds, hail, lightning, and tornadoes) can be hazardous. Also, you could have poor road conditions and a limited ability to maneuver to get out of harm's way in many places.

Your safety and the safety of other drivers and pedestrians must always be your first concern when storm chasing. Some storm-chase guidelines are provided in a series of "A SCIENTIFIC PERSPECTIVE" boxes later in this and the next chapters.

(continues in later pages)

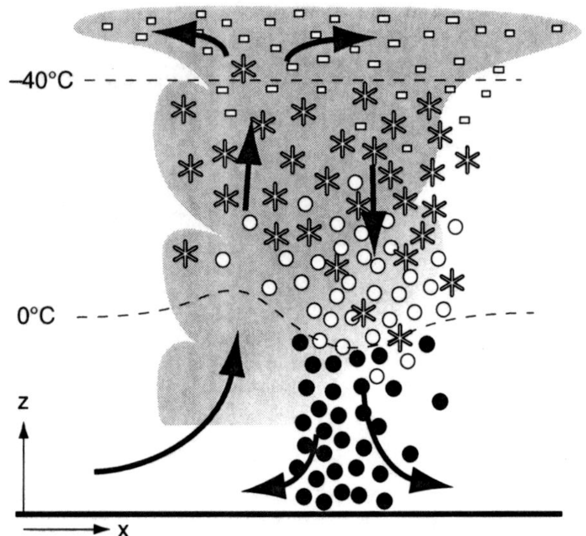

Figure 14.7
Airmass thunderstorm in its mature stage during summer. Grey shading represents cloud droplets (except above the –40°C isotherm, where all hydrometeors are frozen), black dots represent rain, white dots are graupel, dendrite shapes are snow, and small white rectangles are small ice crystals. Arrows show winds, and dashed lines give air temperature.

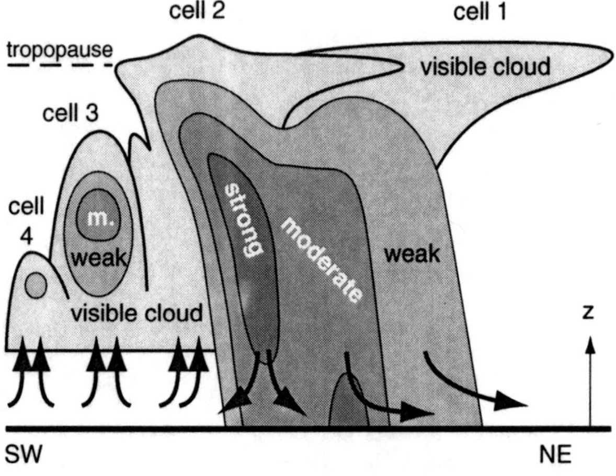

Figure 14.8
Multicell thunderstorm. Cell 1 is the oldest cell, and is in the dissipating stage. Cell 2 is mature. Cell 3 is in the cumulus stage. Cell 4 is the newest and smallest, in an early cumulus stage. Shaded regions indicate where we would see clouds and rain by eye. Medium, dark, and very dark shading indicate weak, moderate (m.), and strong radar echoes inside the clouds, which generally correspond to light, moderate, and heavy precipitation. Arrows show low-altitude winds.

Thunderstorm Types & Organization

Types of thunderstorm organization include:

- basic storms
  - single cell (airmass)
  - multicell
  - orographic
- mesoscale convective systems (MCS)
  - squall line
  - bow echo
  - mesoscale convective complex (MCC)
  - mesoscale convective vortex (MCV)
- supercells
  - classic (CL)
  - low precipitation (LP)
  - high precipitation (HP)

Basic Storms

Single-cell Thunderstorms. An **airmass thunderstorm** (Figs. 14.1, 14.2 & 14.7) is often a single-cell thunderstorm with evolution very similar to Fig. 14.6. It can form in the middle of a warm humid airmass, usually in late afternoon. Satellite images of airmass thunderstorms show that several often exist at the same time, all with somewhat symmetric anvils, and scattered within the airmass analogous to mushrooms in a meadow. These storms can be triggered by thermals of rising warm air (heated at the Earth's surface) that happen to be buoyant enough to break through the capping inversion at the boundary-layer top.

The lifetime of airmass thunderstorms are about the same as that of an individual cell (30 to 60 minutes), but they can spawn daughter storms as sketched in Fig. 14.6. Each can produce a short-duration (15 minutes) rain shower of moderate to heavy intensity that covers an area of roughly 5 - 10 km radius. Airmass thunderstorms that produce short-duration severe weather (heavy precipitation, strong winds, lightning, etc.) during the mature stage are called **pulse storms**.

Multicell Thunderstorms. Most thunderstorms in North America are **multicell storms** (Fig. 14.8). Within a single cloud mass that by eye looks like a thunderstorm, a weather radar can see many (2 to 5) separate cells (Fig. 14.9) by their precipitation cores and their winds. In a multicell storm, the different cells are often in different stages of their life cycle. Given weak to moderate wind shear in the environment, the downburst and gust front from one of the mature cells (cell 2 in Fig. 14.8) can trigger new adjacent cells (namely, cells 3 and 4).

In central North America where warm humid boundary-layer winds often blow from the southeast in the pre-storm environment, new cells often

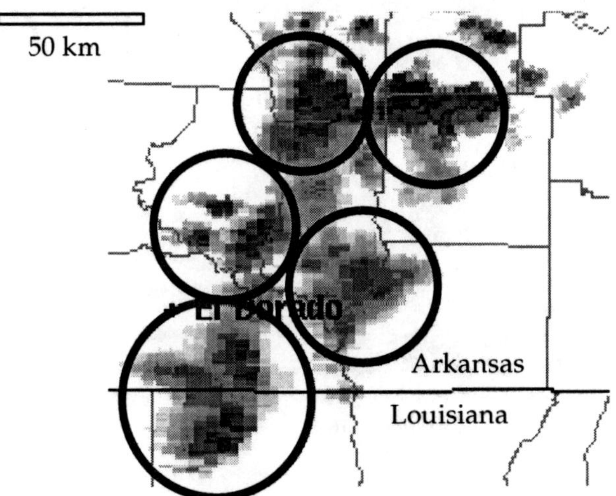

Figure 14.9.
Radar reflectivity PPI image of a multicell thunderstorm in south-central USA, 28 June 2001. Darkest grey indicates > 50 dBZ; namely, the heaviest precipitation. Cells are circled. Courtesy of the US National Weather Service, NOAA.

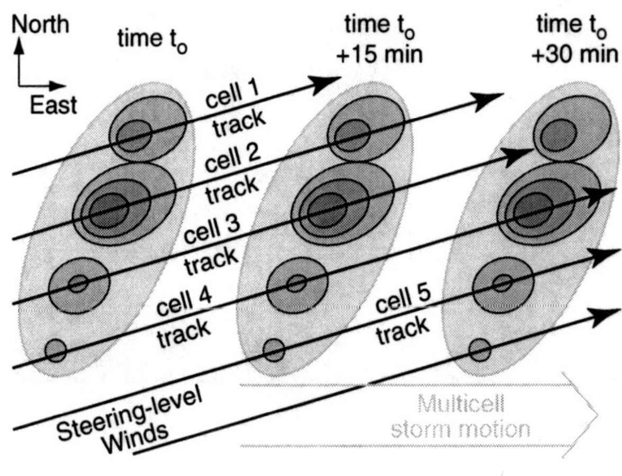

Figure 14.10
Diagram of multicell thunderstorm motion. Grey shading indicates visible clouds and radar echoes corresponding to Fig. 14.8. Black arrows show tracks of individual cells. Double-arrow in grey shows motion of the multicell cloud mass. Each light-grey large oval shows position of the cloud mass every 15 min.

form on the south flank of the storm (closest to the fuel supply), while the northeastward-moving mature cells decay as their fuel supply is diminished (Fig. 14.10). The resulting multicell thunderstorm appears to move to the right of the **steering-level winds** (normal or average winds in the bottom 6 km of the troposphere), even though individual cells within the storm move in the direction of the normal steering-level winds (Fig. 14.10).

Orographic Thunderstorms. Single or multicell storms triggered by mountains or hills are called **orographic thunderstorms**. With the proper environmental wind shear (Fig. 14.11), low-altitude upslope winds can continuously supply humid boundary-layer air. If the upper-level winds are from the opposite direction (e.g., from the west in Fig. 14.11), then these storms can remain stationary over a mountain watershed for many hours, funneling heavy rains into canyons, and causing devastating **flash floods**. One example is the Big Thompson Canyon flood in the Colorado Rockies, which killed 139 people in 1976 (see the Exposition in the Thunderstorm Hazards chapter).

Sometimes if winds are right, thunderstorms are triggered over the mountains, but then are blown away to other locations. For example, orographic thunderstorms triggered over Colorado and Wyoming in late afternoon can move eastward during the evening, to hit the USA Midwest (e.g., Wisconsin) between midnight and sunrise.

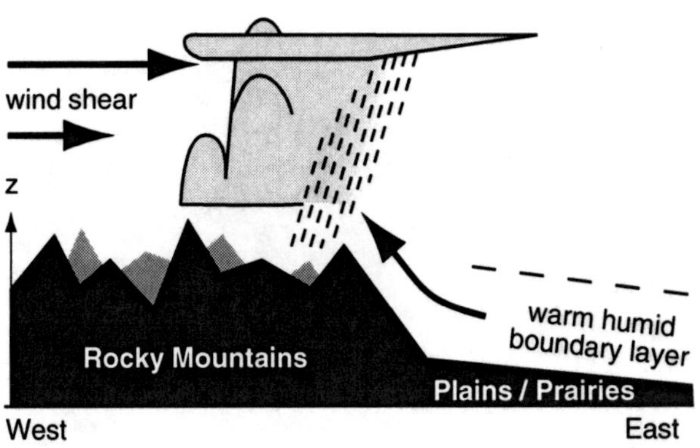

Figure 14.11
Orographic thunderstorm, stationary over the mountains.

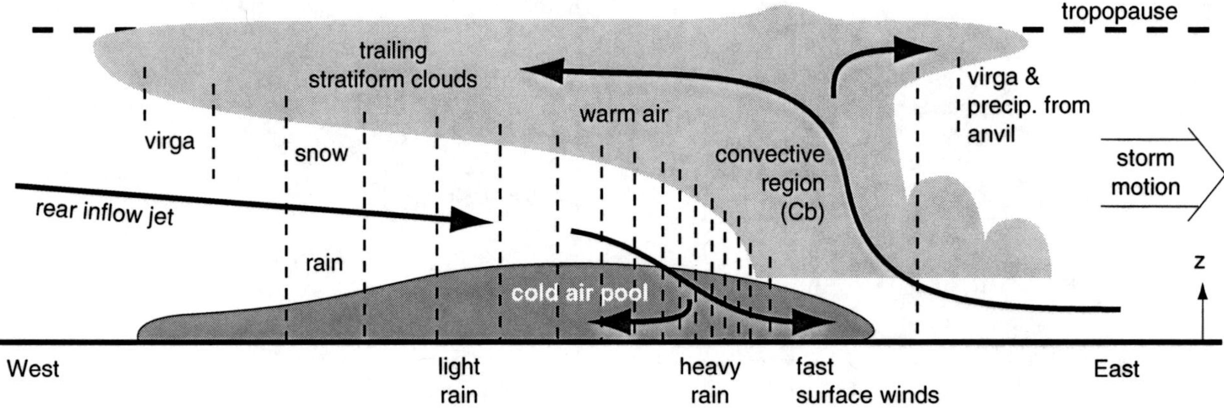

Figure 14.12
Vertical cross section of a mesoscale convective system (MCS).

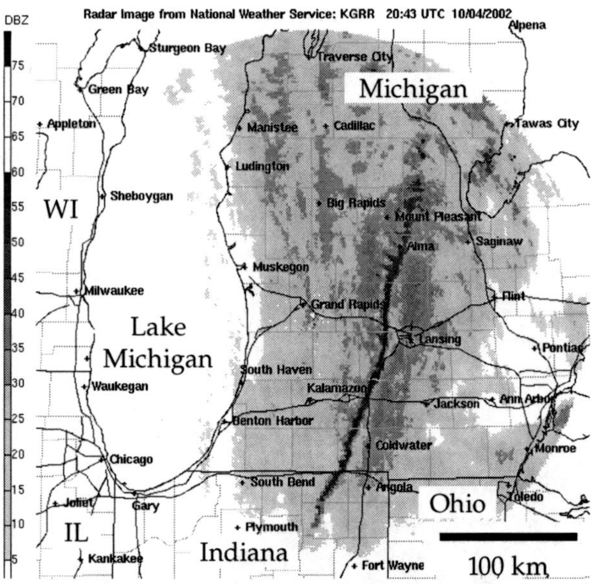

Figure 14.13.
Radar reflectivity PPI image of a mesoscale convective system (MCS) in southern Michigan, USA, on 4 Oct 2002. Dark shading indicates a line of heavy thunderstorm rain associated with radar reflectivities > 50 dBZ. Ahead and behind this squall line are broad regions of lighter, stratiform precipitation with imbedded convective cells. Courtesy of the US National Weather Service, NOAA.

Mesoscale Convective Systems

Another type of multicell-storm organization is a **mesoscale convective system (MCS)**. This often has a narrow line of thunderstorms with heavy precipitation, followed by additional scattered thunderstorm cells (Figs. 14.12 & 14.13). The anvils from all these thunderstorms often merge into a single, large stratiform **cloud shield** that trails behind. This trailing **stratiform** region has widespread light to moderate precipitation that contributes about 1/3 of the total MCS rainfall.

These systems can be triggered by the terrain (many form just east of the Rocky Mountains), by synoptic systems (such as cold fronts), or by the gust fronts from a smaller number of thunderstorms. To survive, they need large convective instability and large wind shear over a deep layer of the environment. MCSs are responsible for 30 to 70% of precipitation over central N. America during the warm season. Some MCSs reach their peak intensity about midnight.

Ice crystals falling from the trailing stratiform clouds melt or partially sublime, and the resulting rain then continues to partially evaporate on the way down. This produces a broad region of chilled air (due to latent heats of melting, sublimation, and evaporation) that descends to the ground and accumulates as a **cold-air pool** at the surface. The extra weight of the cold air produces a mesoscale high pressure (**meso-high**) at the surface in this region (Fig. 14.14).

Meanwhile, warm humid boundary-layer air ahead of the MCS is drawn into the storm by the combined effects of buoyancy and wind shear. This air rises within the imbedded thunderstorm cells, and releases latent heat as the water vapor condenses. The resulting heating produces a layer of relatively warm air at mid and upper levels in the MCS (Fig. 14.14). The top of the anvil is cooled by IR radiation to space.

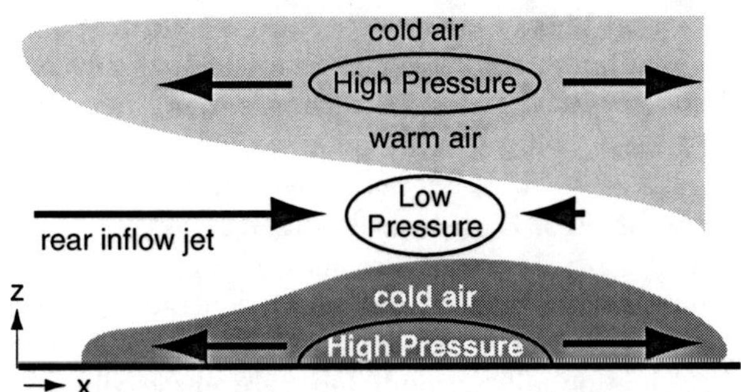

Figure 14.14
Zoomed view of the left half of Fig. 14.12, showing the rear portion of an MCS. Variation of pressure (relative to a standard atmosphere) with altitude caused by layers of warm and cold air in the stratiform portion of an MCS.

The resulting sandwich of relatively warmer and cooler layers causes a mesoscale low-pressure region (a **meso-low**) at mid altitudes (Fig. 14.14) and high pressure aloft (see Sample Application). The horizontal pressure-gradient force at mid altitudes causes air to converge horizontally. The thunderstorm updraft partially blocks mid-altitude inflow from the front of the storm, so the majority of inflow is from the rear. This is called a **rear inflow jet** (**RIJ**).

As the RIJ approaches the core of the MCS, it passes through the precipitation falling from above, and is cooled. This cooling allows the RIJ to descend as it blows toward the front of the storm, causing damaging **straight-line winds** at the ground. Large damaging straight-line wind events are called **derechos** (see a later INFO Box).

Sample Application

Plot MCS thermodynamics:

(a) Assume a standard atmosphere in the environment outside the MCS, and calculate $P_{std.atm.}$ vs. z.

(b) Inside the MCS, assume the temperature T_{MCS} differs from the $T_{std.atm.}$ by:
–12 K for $9 < z \le 11$ km due to rad. cooling at anvil top;
+8 K for $4 \le z \le 9$ km due to latent heating in cloud;
–12 K for $0 \le z \le 2$ km in the cold pool of air;
0 K elsewhere. Use P_{sfc} = 101.8 kPa in the MCS.
Plot $\Delta T = T_{MCS} - T_{std.atm.}$ vs. z.

(c) Calculate P_{MCS} vs. z in the MCS.

(d)Plot pressure difference $\Delta P = P_{MCS} - P_{std.atm.}$ vs. z

Find the Answer:
Given: T and P data listed above within MCS, and std. atm. in the environment outside the MCS.
Find & plot: ΔT (K) vs. z (km); and ΔP (kPa) vs. z (km)

Method:
(a) In a spreadsheet, use standard-atmosphere eq. (1.16) to find $T_{std.atm.}$ vs. z, assuming $H = z$.

(Continues in next column.)

Sample Application. *(Continuation)*

For this standard atmosphere outside the MCS, use hypsometric eq. (1.26b) to find P vs. z by iterating upward in steps of Δz = 250 m from $P_1 = P_{sfc}$ = 101.325 kPa at $z_1 = 0$. Assume $T_{v\,avg} = T_{avg} = 0.5 \cdot (T_1 + T_2)$ in eq. (1.26b) for simplicity (since no humidity information was given in the problem). Thus (with a = 29.3 m·K^{-1}):

$$P_2 = P_1 \cdot \exp[\,(z_1 - z_2) \,/\, (a \cdot 0.5 \cdot (T_1 + T_2)\,]\qquad(1.26b)$$

(b) Plot ΔT vs. z, from data given in the problem.

(c) Find $T_{MCS}(z) = T_{std.atm.}(z) + \Delta T(z)$. Do the similar hypsometric P calculations as (a), but inside the MCS.

(d) Find $\Delta P = P_{MCS} - P_{std.atm.}$ from (c) and (a), and plot.

Spreadsheet segments:

	(a)		(b)	(c)		(d)
	Std. Atmos.			MCS		ΔP
z (m)	T (K)	P (kPa)	ΔT (K)	T (K)	P (kPa)	(kPa)
0	288.2	101.3	–12	276.2	101.8	0.5
250	286.5	98.3	–12	274.5	98.7	0.4
500	284.9	95.4	–12	272.9	95.7	0.2
750	283.3	92.6	–12	271.3	92.7	0.1
. . .	. . .	. . .	. . .	. . .	. . .	. . .
2000	275.2	79.5	–12	263.2	79.0	–0.5
2250	273.5	77.1	0	273.5	76.6	–0.5
. . .	. . .	. . .	. . .	. . .	. . .	. . .
3750	263.8	63.7	0	263.8	63.3	–0.4
4000	262.2	61.7	8	270.2	61.3	–0.4
. . .	. . .	. . .	. . .	. . .	. . .	. . .
9000	229.7	30.8	8	237.7	31.3	0.5
9250	228.0	29.6	–12	214.0	30.1	0.5
. . .	. . .	. . .	. . .	. . .	. . .	. . .
11000	216.7	22.7	–12	204.7	22.7	0.0
11250	215.0	21.8	0	215.0	21.8	0.0

(Note: The actual spreadsheet carried more significant digits.)

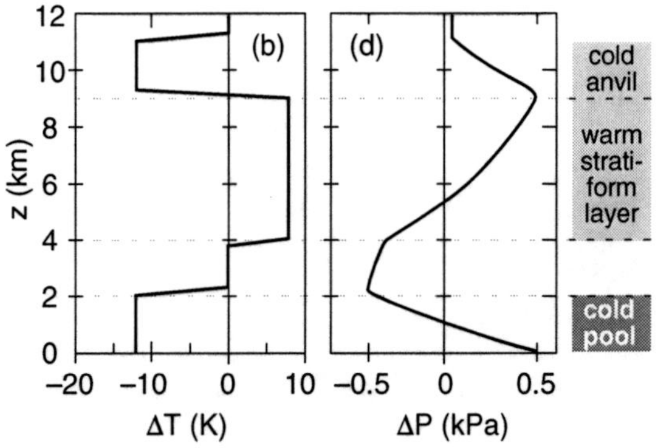

Check: Units OK. Standard atmosphere P agrees with Table 1-5. Graphs are reasonable.

Exposition: High pressure at the surface causes the cold-pool air to spread out, as sketched in Figs. 14.12 & 14.14. The lowest pressure is above the cold pool and below the warm stratiform clouds. This low pressure (relative to the environment) sucks in air as a rear inflow jet. The deep layer of warm stratiform clouds eventually causes high pressure aloft, causing the cloudy air to spread out in the anvil. The cold anvil top then brings the pressure back to ambient, so that above the anvil there are no pressure differences.

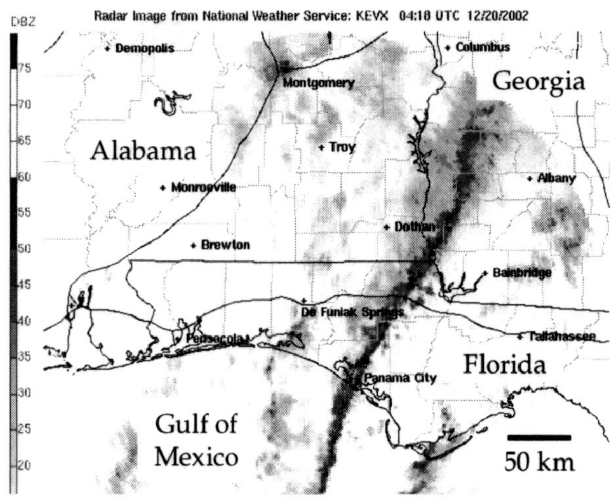

Figure 14.15
Radar reflectivity PPI image of squall line in Florida, Georgia and Alabama, USA, on 20 December 2002. Darkest shading corresponds to thunderstorms with reflectivity > 50 dBZ. Courtesy of the US National Weather Service, NOAA.

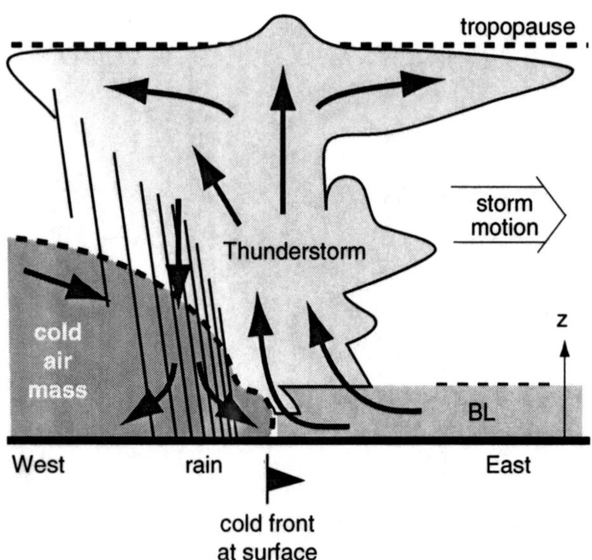

Figure 14.16
Vertical cross section through a squall-line thunderstorm, triggered by a cold front. Boundary-layer (BL) pre-storm air is warm and humid; cold air mass is colder and drier. Dashed lines indicate base of stable layers such as temperature inversions.

Squall Line. One type of MCS is a **squall line**, where a linear triggering mechanism such as a cold front, dry line or gust front causes many neighboring thunderstorms to merge into a long narrow line (Figs. 8.26 & 14.15). Squall lines can be many hundreds (even a thousand) kilometers long, but are usually only 15 to 400 km wide. They can persist for several hours to several days if continuously triggered by the advancing front. Heavy thunderstorm precipitation falls immediately behind the cold front at the surface, and sometimes has weaker stratiform precipitation behind it. Squall line storms can also have prodigious lightning, hail, damaging winds and tornadoes (although the tornadoes tend to be weaker than supercell-thunderstorm tornadoes).

Fig. 14.16 illustrates flow characteristics in a vertical cross section perpendicular to the squall-line axis. Environmental wind shear along the advancing cold front allows the thunderstorms to persist in the mature stage for the lifetime of the squall line. Cold, dry air descends from the west in the cold air mass from behind the squall line. This cold descent is enhanced by cool downburst air in the precipitation region. Both effects help advance the leading edge of the cold front at the ground. Warm, humid boundary-layer air blowing from the east ahead of the front is lifted by the undercutting denser cold advancing air, thereby continuously feeding the main updraft that moves with the advancing front.

A well defined arc or shelf cloud often exists along the leading edge of the cold front, associated with violent straight-line winds. If the gust front from the downburst advances eastward faster than the cold front, then a second line of thunderstorms may be triggered along the gust front in advance of the cold front or the squall line might separate from the front.

Bow Echo. For MCSs with strong buoyancy in the updraft air and moderate to strong wind shear in the bottom 2 to 3 km of the environment, the cold pool becomes deeper and colder, and rear-inflow-jet (RIJ) wind speeds can become stronger. The RIJ pushes forward the center portion of the squall line, causing it to horizontally bend into a shape called a **bow echo** (Fig. 14.17). Bow-echo lines are 20 to 200 km long (i.e., smaller than most other MCSs) and are often more severe than normal MCSs.

Vorticity caused by wind shear along the advancing cold pool causes counter-rotating vortices at each end of the bow-echo line. These are called **line-end vortices** or **bookend vortices** (Fig 14.17), which rotate like an eggbeater to accelerate the RIJ and focus it into a narrower stream. For bow echoes lasting longer than about 3 h, Coriolis force enhances the northern cyclonic vortex in the

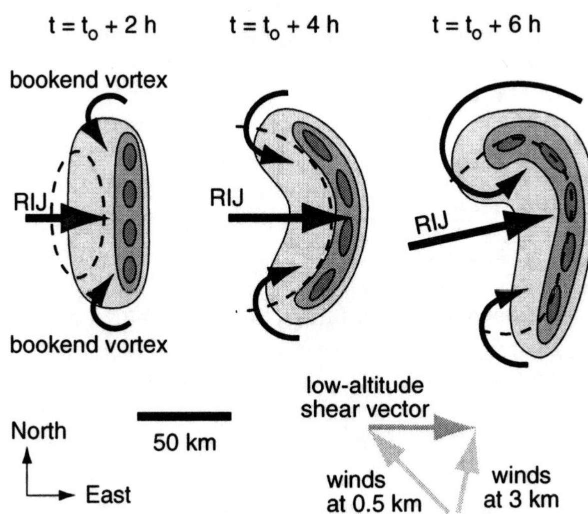

Figure 14.17
Evolution of a bow-echo line of thunderstorms. Shading indicates radar reflectivity, dashed line outlines cold air pool at surface, RIJ is rear inflow jet, t is time, and t_o is formation time.

Northern Hemisphere and diminishes the southern anticyclonic vortex, causing the bow echo to become asymmetric later in its evolution.

Strong straight-line winds near the ground from a bow echo can damage buildings and trees similar to an EF2 tornado (see the "Tornado" section in the next chapter for a description of tornado damage scales). If a sequence of bow echoes causes wind damage along a 400 km path or greater and lasts about 3 hours or more, with wind speeds ≥ 26 m s^{-1} with gusts ≥ 33 m s^{-1}, then the damaging wind event is called a **derecho** (see INFO Box 3 pages later).

Mesoscale Convective Complex (MCC). If the size of the MCS cloud shield of merged anvils is large (diameter ≥ 350 km), has an overall elliptical or circular shape, and lasts 6 to 12 h, then it is called a **mesoscale convective complex** (**MCC**). An additional requirement to be classified as an MCC is that the top of the cloud shield, as seen in IR satellite images, must have a cold brightness-temperature ≤ -33°C. Within this cold cloud shield must be a smaller region (diameter ≥ 250 km) of higher, colder cloud top with brightness temperature ≤ –53°C.

MCCs can produce heavy rain over large areas, with the most intense rain usually happening first. MCCs are most often observed at night over a statically stable boundary layer, which means that the fuel for these storms often comes from the warm humid air stored in the residual boundary layer from the day before. They can be triggered by: weak warm frontal zones, weak mid-tropospheric short waves, and are often associated with low-level jets of wind.

Sample Application

Estimate the speed of the rear inflow jet (RIJ) wind that is drawn into the MCS mid-tropospheric low-pressure region, for the previous Sample Application. The MCS has horizontal radius of 175 km, and is located in Oklahoma (latitude 35°N) .

Find the Answer

Given: ΔP = –0.5 kPa at height z = 2 km, from previous Sample Application.
R = 175 km = radius of meso-low in mid-troposphere, latitude ϕ = 35°N

Find: M_{RIJ} = ? m s^{-1}

Assume the RIJ is geostrophic ($M_{RIJ} = U_g$). To use eq. (10.26a) for the geostrophic wind, we need:
$\Delta P/\Delta y$, ρ and f_c .

Consider the region just south of the low center.
$\Delta P/\Delta y$ = (–0.5 kPa)/(175 km) = –0.00286 kPa·km^{-1}

Assuming std. atmosphere. Use Table 1-5 at 2 km
ρ = 1.0 kg·m^{-3} .

Use eq. 10.16 to find the Coriolis parameter:
f_c = (1.458x10^{-4} s^{-1})·sin(35°) = 0.836x10^{-4} s^{-1}.

Combine these data in eq. (10.26a):
U_g= –[(1.0 kg·m^{-3})·(0.836x10^{-4} s^{-1})]$^{-1}$·(–0.00286 Pa·m^{-1})
= 34.2 m s^{-1}

Thus,
$M_{RIJ} = U_g$ = **34.2** m s^{-1}

Check: Units OK. Magnitude reasonable. Sign OK.
Exposition: This implies a RIJ from the west, in the region just south of the mid-tropospheric meso-low center. If this RIJ descends down to the surface due to the influence of precipitation cooling, it could cause damaging straight-line wind gusts that exceed the threshold to be classified as a derecho.

One caution: the geostrophic wind assumes steady state. In a rapidly intensifying MCS, steady-state might not be the best assumption. If the wind were accelerating, then it could have an ageostrophic component pointing towards the low center.

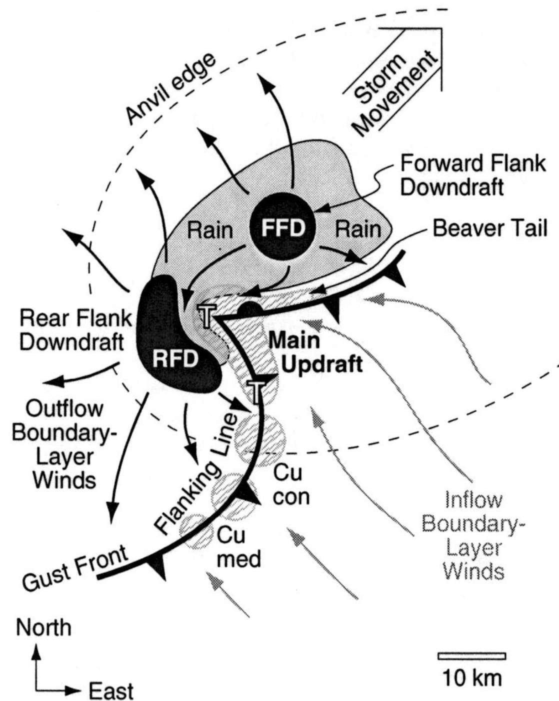

Figure 14.18
Plan view (top down) sketch of a classic supercell in the N. Hemisphere (copied from Fig. 14.4b). T indicates possible tornado positions; RFD = Rear Flank Downdraft; FFD = Forward Flank Downdraft. Regions of updraft are hatched with grey; downdrafts are solid black; rain is solid light grey. Low surface pressure is found under the updrafts (especially near the "T" locations), and high pressure under the downdrafts. Vectors are surface winds: cold are black, warm are grey.

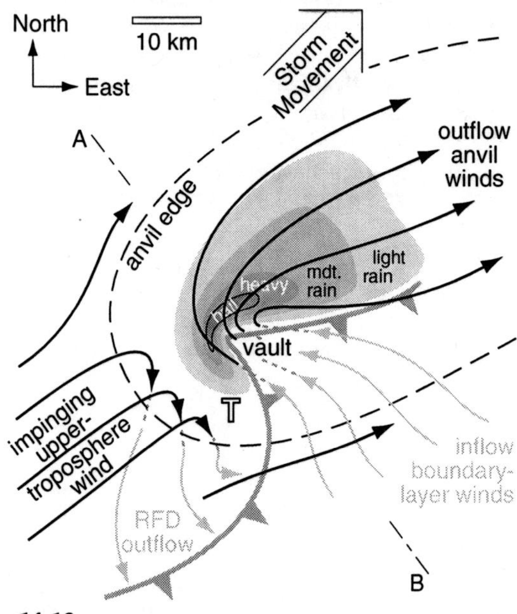

Figure 14.19
Plan view of classic (CL) supercell in the N. Hemisphere. Low altitude winds are shown with light-grey arrows, high-altitude winds with black arrows, and ascending/descending air with short-dashed lines. T indicates tornado location.

Mesoscale Convective Vortex (MCV). Between midnight and sunrise, as the static stability of the boundary layer increases, the thunderstorms within MCSs often weaken and die. Thunderstorms are more difficult to trigger then, because cold nocturnal boundary-layer air resists rising. Also, above the nocturnal layer is a thin residual layer that has only a small capacity to hold the fuel supply of warm humid air from the previous day's mixed layer. As the MCS rains itself out, all that is left by the next morning are mid-level stratiform clouds.

However, the mid-tropospheric mesoscale low-pressure region (Fig. 14.14) in the MCS has existed for about half a day by this time, and has developed cyclonic rotation around it due to Coriolis force. This mesoscale rotation is often visible (by satellite) in the mid-level (former MCS) stratiform clouds, and is called a **mesoscale convective vortex** (**MCV**). The stable boundary layer beneath the MCV reduces surface drag, allowing the rotation to persist for many more hours around the weak low.

Although these weakly rotating systems are not violent and often don't have a circulation at the surface, they can be tracked by satellite during the morning and early afternoon as they drift eastward. They are significant because they can enhance new thunderstorm formation via weak convergence and upward motion. Thus, by evening the dying MCV can change into a new MCS.

Supercell Thunderstorms

A violent thunderstorm having a rotating updraft and persisting for hours is called a **supercell storm** (Figs. 14.3 & 14.4). Its downdraft and surface features (Fig 14.18, a duplicate of Fig. 14.4b) are organized in a way to trigger new thunderstorms nearby along a **flanking line**, and then to entrain these incipient updrafts (cumulus congestus clouds) into the main supercell updraft to give it renewed strength. The main, persistent, rotating updraft, or **mesocyclone**, is 3 to 10 km in diameter, and can extend throughout the whole depth of the thunderstorm.

Supercell storms are responsible for the most violent tornadoes (EF3 - EF5, as defined in the next chapter) and hail, and can have damaging straight-line winds and intense rain showers causing flash floods. To support supercells, the pre-storm environment must have a deep layer of convectively unstable air, and strong wind shear in the bottom 6 km of the troposphere.

Although supercells can have a wide range of characteristics, they are classified into three categories: **low-precipitation** (**LP**) storms, medium precipitation or **classic** (**CL**) supercells, and **high-precipitation** (**HP**) storms. Storms that fall between these categories are called **hybrid storms**.

Classic Supercells. Figures 14.3 and 14.4 show characteristics of a classic supercell. One way to understand how supercells work is to follow air parcels and hydrometeors (precipitation particles) as they flow through the storm.

Warm humid boundary-layer air flowing in from the southeast (Fig. 14.19) rises in the main supercell updraft, and tiny cloud droplets form once the air is above its LCL. So much water vapor condenses that large amounts of latent heat are released, causing large buoyancy and strong updrafts (often ≥ 50 m s^{-1}). Cloud droplets are carried upward so quickly in the updraft that there is not yet time for larger precipitation particles to form. Hence, the updraft, while visible by eye as a solid cloud, appears only as a **weak echo region (WER)** on radar PPI scans because there are no large hydrometeors. This region is also known as an **echo-free vault** (Fig. 14.19).

As the air reaches the top of the troposphere, it encounters stronger winds, which blows these air parcels and their cloud particles downwind (to the northeast in Fig. 14.19). Since the whole storm generally moves with the mean wind averaged over the bottom 6 km of troposphere (to the northeast in Fig. 14.19), this means that the air parcels in the top part of the storm are flowing to the storm's forward flank. This differs from MCSs, where the updraft air tilts back toward the storm's trailing (rear) flank.

While rising, the air-parcel temperature drops to below freezing, and eventually below –40°C near the top of the thunderstorm (similar to Fig. 14.7), at which point all hydrometeors are frozen. Various microphysical processes (cold-cloud Bergeron process, collision, aggregation, etc.) cause larger-size hydrometeors to form and precipitate out. The larger, heavier hydrometeors (hail, graupel) fall out soonest, adjacent to the updraft (to the north in Fig. 14.19). Medium size hydrometeors are blown by the upper-level winds farther downwind (northeast), and fall out next. Lighter hydrometeors fall more slowly, and are blown farther downwind and fall in a larger area.

In summer, all of these ice particles (except for some of the hail) melts while falling, and reaches the ground as rain. The greatest precipitation rate is close to the updraft, with precipitation rates diminishing farther away. The very smallest ice particles have such a slow fall velocity that they have a long residence time in the top of the troposphere, and are blown downwind to become the anvil cloud.

Meanwhile, the falling precipitation drags air with it and causes it to cool by evaporation (as will be explained in more detail in the next chapter). This causes a downdraft of air called the **forward flank downdraft (FFD**, see Fig. 14.18). When this cold air hits the ground, it spreads out. The leading edge of

Sample Application

If a Doppler radar were located near the words "10 km" in the bottom right corner of Fig. 14.18, then sketch the resulting color patterns on the Doppler PPI wind display for near-surface winds.

Find the Answer:

Given: Near-surface winds from Fig. 14.18 are copied below, with radar located southeast of the storm center.

Find: Sketch the Doppler wind display.

Assume sufficient number-density of insects to allow Doppler radar returns even in rain-free areas.

Hatched grey corresponds to red colors, for winds away from the radar. Plain grey corresponds to blues and greens, for winds toward the radar. White is the zero isodop line. Darker shading near the "T" is to suggest a tornado vortex signature.

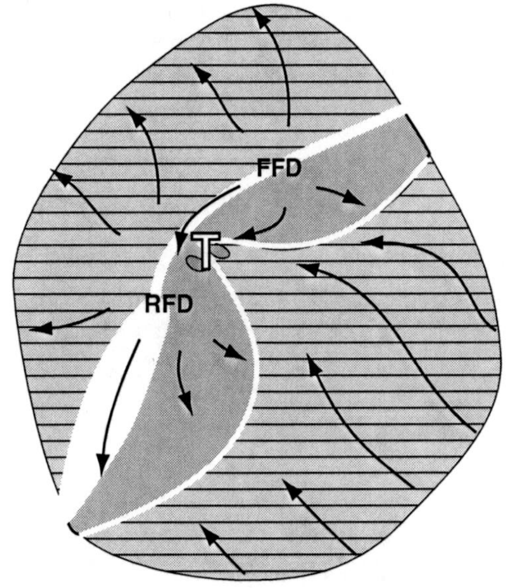

Check: Compares well with info in Chapter 8.

Exposition: Fig. 14.19 already shows the radar reflectivity display. By interpreting the reflectivity and Doppler info together, you can create in your mind a coherent picture of the storm.

The gust fronts show up clearly in the Doppler display. The tornado-vortex signature in the Doppler display and the hook echo in the reflectivity display both suggest the presence of a tornado.

Finally, real thunderstorms are imbedded in an environment that is not usually calm, so there would also be red and blue colors outside of the regions sketched here.

INFO • Derecho

A **derecho** is a hazardous event of very strong straight-line (non-tornadic) horizontal winds (≥ 26 m s^{-1}) often causing widespread (≥ 400 km length) damage at the surface. It is associated with clusters of downbursts of air from a single moving mesoscale convective system (MCS), which causes localized destruction that sweeps through the event area during several hours to over a day. Within the event area, the wind intensity and damage can be variable.

The word "derecho" is based on a Spanish word for "straight-ahead" or "direct". Derechos occur in the prairies east of the Rocky Mountains (see Fig. below), and were first reported over a century ago. Although derechos can occur during any month, they are most frequent in April through August, with peak frequencies during May and July. Peak wind speeds over 65 m s^{-1} have been rarely observed.

In the USA, an average of 21 derechos occur each year, killing 9 and injuring 145 people annually. Most of the deaths are people in cars and boats, while most of the injuries are people in mobile homes and cars. Derechos can blow down trees, destroy mobile homes, and damage other structures and cars.

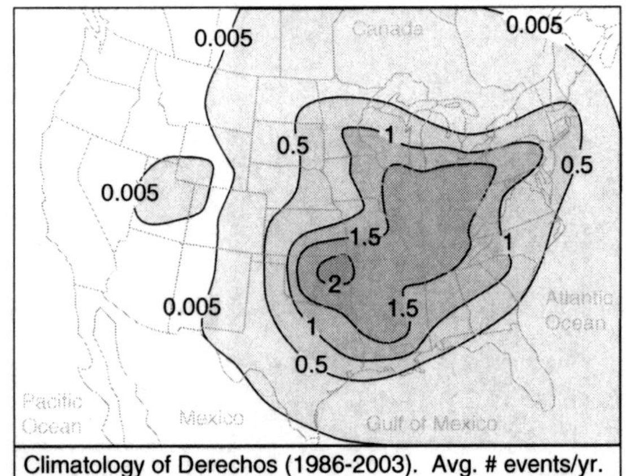

Figure 14.a.
Derecho climatology.

Based on data from Ashley & Mote, 2005: Derecho hazards in the United States. Bull. Amer. Meteor. Soc., 86, 1577-1592.

the spreading cold air is called the (forward-flank) **gust front**, and is often indicated with cold-frontal symbols. This gust front helps to deflect the boundary-layer air upward into the main updraft, and can create a **beaver-tail cloud**, as described earlier.

Due to rotation of the whole updraft (the origin of rotation will be discussed later), the precipitation region is often swept counterclockwise around the updraft for N. Hemisphere supercells. This creates a characteristic curve of the precipitation region that is seen on weather radar as a **hook echo** around the echo-free vault (Fig. 14.19).

In the S. Hemisphere (and for a small portion of N. Hemisphere supercells — see Fig. 14.63), the updrafts rotate clockwise. These supercells can still produce hook echoes, but with the hook projecting in a direction different than that sketched in Fig. 14.19).

The supercell storm is so tightly organized that it acts as an obstacle to fast winds in the upper troposphere. When these ambient winds from the west or southwest hit the supercell, some are deflected downward, creating the **rear-flank downdraft (RFD)**. As the dry upper-tropospheric air moves down adjacent to the cloudy updraft, it entrains some cloud droplets, which quickly evaporate and cool the air. This negative buoyancy enhances the RFD such that when the air hits the ground it spreads out, the leading edge of which is a (rear-flank) **gust front**.

When warm, humid boundary-layer air flowing into the storm hits the rear-flank gust front, it is forced upward and triggers a line of cumulus-cloud growth called the **flanking line** of clouds. The inflow and circulation around the supercell draws these new cumulus congestus clouds into the main updraft, as previously described. At the cusp of the two gust fronts, there is low surface pressure under the updraft. This **meso-low** (see Table 10-6) and the associated gust fronts look like mesoscale versions of synoptic-scale cyclones, and are the surface manifestation of the **mesocyclone**.

Strong **tornadoes** are most likely to form in one of two locations in a supercell. One is under the largest cumulus congestus clouds just as they are being entrained into the main updraft, due to the rotation between the RFD and the updraft. These form under the rain-free cloud-base portion of the supercell, and are often easy to view and chase. The other location is under the main updraft at the cusp where the two gust fronts meet, due to rotation from the parent mesocyclone. Tornadoes are discussed in more detail in the next chapter.

Low Precipitation Supercells (Fig. 14.20). In drier regions of North America just east of the Rocky Mountains, LCLs are much higher above ground. However, mountains or dry lines can cause sufficient lift to trigger high-base supercell storms. Also, if strong environmental wind shear causes a strongly tilted updraft, then incipient precipitation particles are blown downwind away from the rich supply of smaller supercooled cloud droplets in the updraft, and thus can't grow large by collection/accretion. Most of the hydrometeors that form in these storms evaporate before reaching the ground; hence the name **low-precipitation (LP) supercells**.

With fewer surrounding low clouds to block the view, LP storms often look spectacular. You can often see the updraft, and can see evidence of its rotation due to cloud striations, curved inflow bands and possibly a rotating wall cloud. If present, the beaver tail is easily visible. Although rainfall is only light to moderate, LP storms can produce great amounts of large-diameter hail. Those LP supercells having exceptionally strong updrafts produce anvils that look like a mushroom cloud from a nuclear bomb explosion.

High Precipitation Supercells (Fig. 14.21). In very warm humid locations such as southeastern North America, some supercell storms can produce widespread heavy precipitation. These are known as **high-precipitation (HP) supercells**. Precipitation falls from most of the cloud-base areas, often filling both the FFD and the RFD. So much rain falls upstream in the rear-flank region of the storm that the main mesocyclone is sometimes found downstream (forward) of the main precipitation region.

Some storms have curved inflow bands and cloud striations on the updraft that you can see from a distance. Also, the smooth-looking beaver-tail clouds are most common for HP storms, near the forward-flank gust front. Much more of the surrounding sky is filled with low and mid-level clouds, which together with low visibilities makes it difficult to get good views of this storm.

As this type of storm matures, the heavy-rain region of the hook echo can completely wrap around the updraft vault region. Storm chasers call this region the **bear's cage**, because of the high danger associated with tornadoes and lightning, while being surrounded by walls of torrential rain and hail with poor visibility and hazardous driving/chasing conditions.

Now that we are finished covering the descriptive appearance and types of thunderstorms, the next five sections examine the meteorological conditions needed to form a thunderstorm.

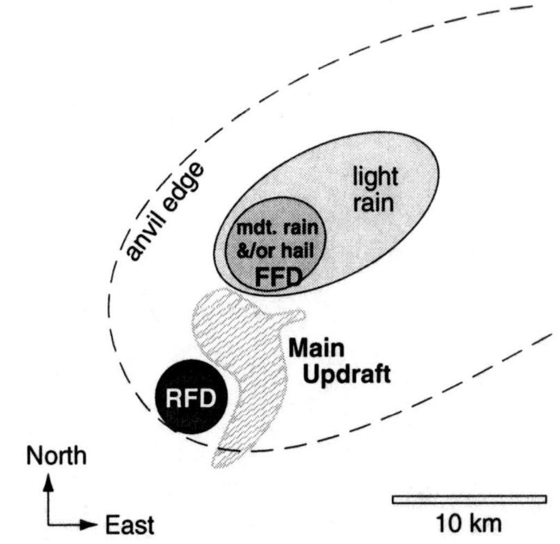

Figure 14.20
Low-precipitation (LP) supercell thunderstorm in the N. Hemisphere. FFD = forward flank downdraft; RFD = rear flank downdraft. Often little or no precipitation reaches the ground. The hatched grey region indicates updrafts.

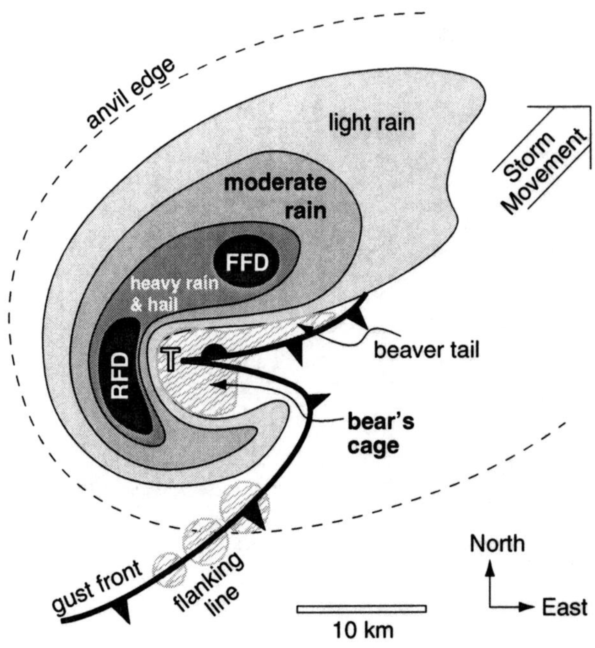

Figure 14.21
High-precipitation (HP) supercell thunderstorm in the N. Hemisphere. FFD = forward flank downdraft; RFD = rear flank downdraft. Hatched grey regions indicate updrafts.

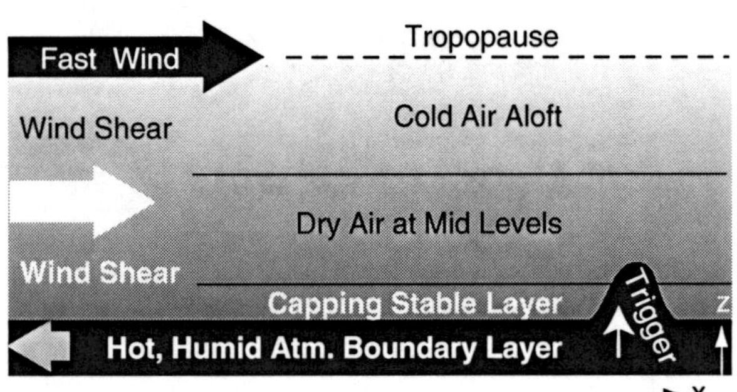

Figure 14.22
Atmospheric conditions favorable for formation of strong thunderstorms.

THUNDERSTORM FORMATION

Favorable Conditions

Four environmental (pre-storm) conditions are needed to form the strong **moist convection** that is a severe thunderstorm:

1) **high humidity** in the boundary layer,
2) nonlocal conditional **instability**,
3) strong **wind shear**,
4) a **trigger mechanism** to cause **lifting** of atmospheric boundary-layer air.

Fig. 14.22 illustrates these conditions. Meteorologists look for these conditions in the pre-storm environment to help forecast thunderstorms. We will examine each of these conditions in separate sections. But first, we define key convective altitudes that we will use in the subsequent sections to better understand thunderstorm behavior.

Key Altitudes

The existence and strength of thunderstorms depends partially on layering and stability in the pre-storm environment. Thus, you must first obtain an **atmospheric sounding** from a **rawinsonde balloon** launch, **numerical-model** forecast, aircraft, dropsonde, satellite, or other source. Morning or early afternoon are good sounding times, well before the thunderstorm forms (i.e., **pre-storm**). The environmental sounding data usually includes temperature (T), dew-point temperature (T_d), and wind speed (M) and direction (α) at various heights (z) or pressures (P).

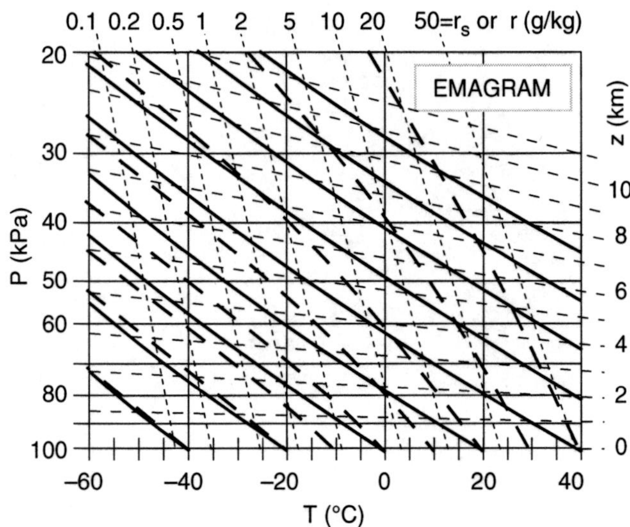

Figure 14.23
Thermo diagram, with approximate height contours (z) added as nearly horizontal thin dashed lines (labeled at right). Sounding data plotted on this diagram will enable you to anticipate storm formation and strength.

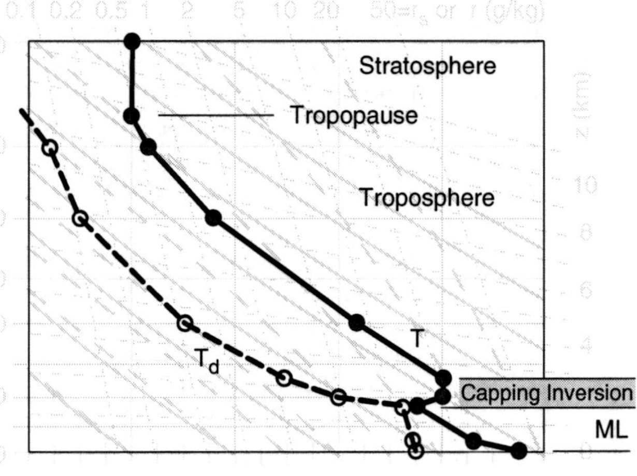

Figure 14.24
Environmental sounding of temperature (T) and dew-point temperature (T_d) plotted on a thermo diagram. The mixed layer (ML), capping inversion (Cap), tropopause, troposphere, and stratosphere are identified. This example shows an early-afternoon pre-storm environment.

Plot the environmental sounding on a thermo diagram such as the Emagram of Fig. 14.23, being sure to connect the data points with straight lines. Fig. 14.24 shows a typical pre-storm (early afternoon) sounding, where T is plotted as black-filled circles connected with a thick black solid line, and T_d is plotted as white-filled circles connected with a thick black dashed line. The mixed layer (ML; i.e., the daytime turbulently well-mixed boundary layer), capping inversion, and tropopause are identified in Fig. 14.24 using the methods in the Atmospheric Stability chapter.

Daytime solar heating warms the Earth's surface. The hot ground heats the air and evaporates soil moisture into the air. The warm humid air parcels rise as thermals in an unstable atmospheric boundary layer. If we assume each rising air parcel does not mix with the environment, then its temperature decreases dry adiabatically with height initially, and its mixing ratio is constant.

Fig. 14.25 shows how this process looks on a thermo diagram. The parcel temperature starts near the ground from the solid black circle and initially follows a **dry adiabat** (thin black solid line with arrow). The dew point starts near the ground from the open circle and follows the **isohume** (thin black dotted line with arrow).

Typically, the rising parcel hits the statically stable layer at the top of the mixed layer, and stops rising without making a thunderstorm. This capping temperature-inversion height z_i represents the average **mixed-layer (ML) top**, and is found on a thermo diagram where the dry adiabat of the rising parcel first crosses the environmental sounding (Fig. 14.25). In pressure coordinates, use symbol P_i to represent the ML top.

If the air parcel were to rise a short distance above z_i, it would find itself cooler than the surrounding environment, and its negative buoyancy would cause it sink back down into the mixed layer. On many such days, no thunderstorms ever form, because of this strong lid on top of the mixed layer.

But suppose an external process (called a trigger) pushes the boundary-layer air up through the capping inversion (i.e., above z_i) in spite of the negative buoyancy. The rising air parcel continues cooling until it becomes saturated. On a thermo diagram, this saturation point is the **lifting condensation level (LCL)**, where the dry adiabat first crosses the isohume for the rising parcel.

As the trigger mechanism continues to push the reluctant air parcel up past its LCL, water vapor in the parcel condenses as clouds, converting latent heat into sensible heat. The rising cloudy air parcel thus doesn't cool as fast with height, and follows a **moist adiabat** on a thermo diagram (Fig. 14.25).

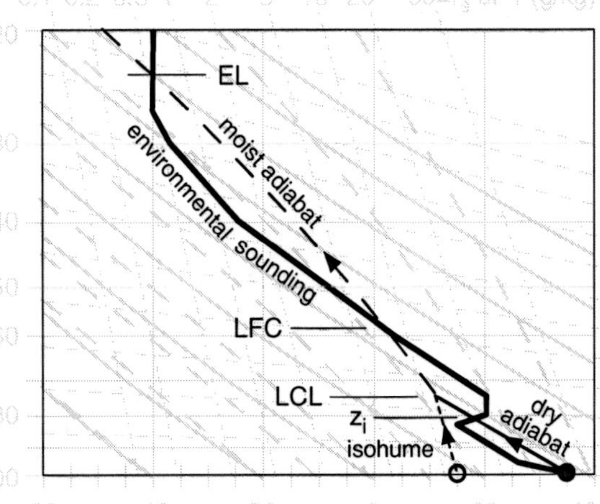

Figure 14.25
Afternoon pre-storm environmental air from near the surface (indicated with the circles), is hypothetically lifted as an air parcel to the top of the thermo diagram. Identified are the average mixed-layer height (z_i), the lifting condensation level (LCL), the level of free convection (LFC), and the equilibrium level (EL).

INFO • Cap vs. Capping Inversion

The strongly stable layer at the top of the daytime boundary layer (i.e., at the top of the mixed layer) is called the "**capping inversion,**" as sketched in the figure below (a modification of Fig. 14.25).

The region near the top of the mixed layer where the rising air parcel is colder than the environment is called the "**cap.**" It is this region between z_i and the LFC that opposes or inhibits the rise of the air parcel, and which must be overcome by the external forcing mechanism. See the section later in this chapter on "Triggering vs. Convective Inhibition" for more details.

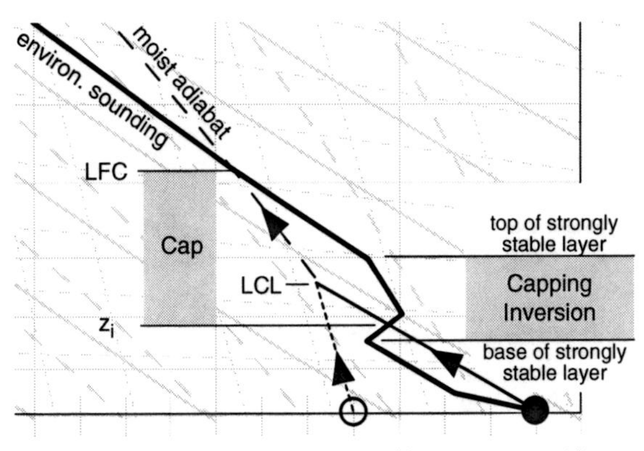

Figure 14.b
Emagram with idealized pre-storm environmental sounding (thick solid line). Circles and lines with arrows show the rise of an air parcel from near the surface.

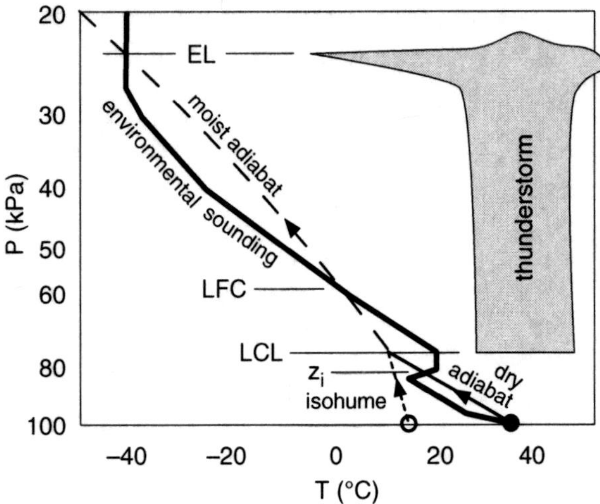

Figure 14.26
From Fig. 14.25, we infer that a thunderstorm could form with cloud top and base as sketched, if successfully triggered. Thunderstorm air parcels follow the thin diagonal solid and dashed lines, while the surrounding air is shown with the thick line.

Sample Application

Plot this sounding on a full-size **skew-T** diagram (Atm. Stab. chapter), and estimate the pressure altitudes of the ML top, LCL, LFC, and EL for an air parcel rising from near the surface.

P (kPa)	T (°C)	T_d (°C)
100	30	20.
96	25	.
80	10	.
70	15	.
50	−10	.
30	−35	.
20	−35	.

Find the Answer

Given: Data in the table above.
Find: P_i = ? kPa (= ML top), P_{LCL} = ? kPa, P_{LFC} = ? kPa, P_{EL} = ? kPa.

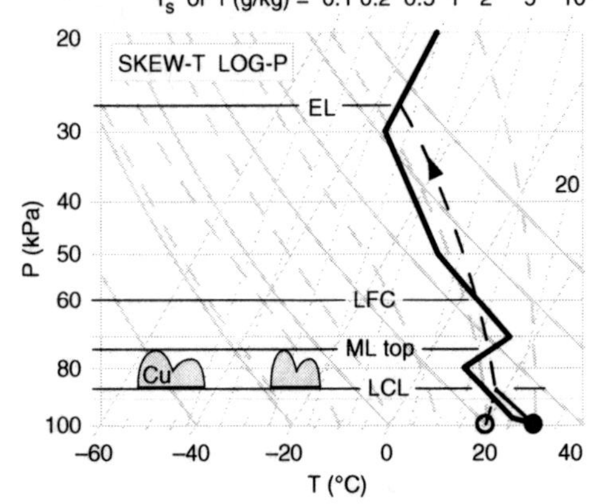

After plotting the air-parcel rise, we find:
P_i = **74 kPa**, P_{LCL} = **87 kPa**,
P_{LFC} = **60 kPa**, P_{EL} = **24 kPa**.

Exposition: The LCL for this case is below the ML top. Thus, the ML contains scattered fair-weather cumulus clouds (cumulus humilis, Cu). If there is no external trigger, the capping inversion prevents these clouds from growing into thunderstorms.

Even so, this cloudy parcel is still colder than the ambient environment at its own height, and still resists rising due to its negative buoyancy.

For an atmospheric environment that favors thunderstorm formation (Fig. 14.25), the cloudy air parcel can become warmer than the surrounding environment if pushed high enough by the trigger process. The name of this height is the **level of free convection** (**LFC**). On a thermo diagram, this height is where the moist adiabat of the rising air parcel crosses back to the warm side of the environmental sounding.

Above the LFC, the air parcel is positively buoyant, causing it to rise and accelerate. The positive buoyancy gives the thunderstorm its energy, and if large enough can cause violent updrafts.

Because the cloudy air parcel rises following a moist adiabat (Fig. 14.25), it eventually reaches an altitude where it is colder than the surrounding air. This altitude where upward buoyancy force becomes zero is called the **equilibrium level** (**EL**) or the **limit of convection** (**LOC**). The EL is frequently near (or just above) the tropopause (Fig. 14.24), because the strong static stability of the stratosphere impedes further air-parcel rise.

Thus, thunderstorm cloud-base is at the LCL, and thunderstorm anvil top is at the EL (Fig. 14.26). In very strong thunderstorms, the rising air parcels are so fast that the inertia of the air in this updraft causes an **overshooting dome** above the EL before sinking back to the EL. The most severe thunderstorms have tops that can penetrate up to 5 km above the tropopause, due to a combination of the EL being above the tropopause (as in Fig. 14.26) and inertial overshoot above the EL.

For many thunderstorms, the environmental air between the EL and the LFC is **conditionally unstable**. Namely, the environmental air is unstable if it is cloudy, but stable if not (see the Atmospheric Stability chapter). On a thermo diagram, this is revealed by an environmental lapse rate between the dry- and moist-adiabatic lapse rates.

However, it is not the environmental air between the LFC and EL that becomes saturated and forms the thunderstorm. Instead, it is air rising above the LFC from lower altitudes (from the atmospheric boundary layer) that forms the thunderstorm. Hence, this process is a **nonlocal conditional instability** (**NCI**).

Frequently the LCL is below z_i. This allows fair-weather cumulus clouds (cumulus humilis) to form in the top of the atmospheric boundary layer (see the Sample Application at left). But a trigger mechanism is still needed to force this cloud-topped mixed-layer air up to the LFC to initiate a thunderstorm.

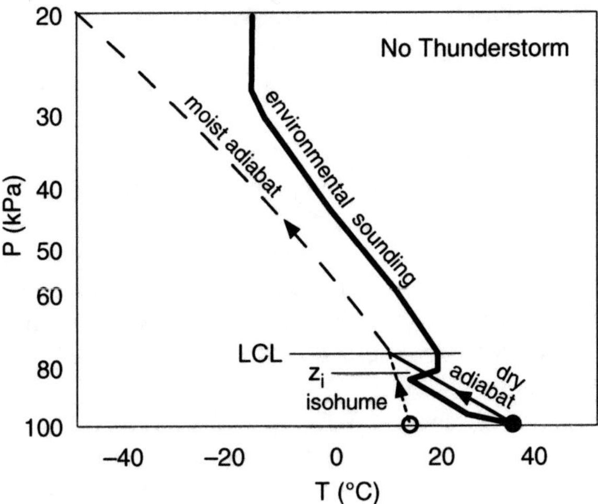

Figure 14.27
An environmental sounding that does not favor thunderstorms, because the environmental air aloft is too warm.

In many situations, no LFC (and also no EL) exists for an air parcel that is made to rise from the surface. Namely, the saturated air parcel never becomes warmer than the environmental sounding (such as Fig. 14.27). Such soundings are NOT conducive to thunderstorms.

HIGH HUMIDITY IN THE ABL

One of the four conditions needed to form **convective storms** such as thunderstorms is high humidity in the atmospheric boundary layer (ABL). Thunderstorms draw in pre-storm ABL air, which rises and cools in the thunderstorm updraft. As water vapor condenses, it releases latent heat, which is the main energy source for the storm (see the Sample Application). Thus, the ABL is the fuel tank for the storm. In general, stronger thunderstorms form in **moister warmer ABL air** (assuming all other factors are constant, such as the environmental sounding, wind shear, trigger, etc.).

The **dew-point temperature** T_d in the ABL is a good measure of the low-altitude humidity. High dew points also imply high air temperature, because $T \geq T_d$ always (see the Water Vapor chapter). Higher temperatures indicate more sensible heat, and higher humidity indicates more latent heat. Thus, high dew points in the ABL indicate a large fuel supply in the ABL environment that can be tapped by thunderstorms. Thunderstorms in regions with $T_d \geq 16°C$ can have heavy precipitation, and those in regions with $T_d \geq 21°C$ can have greater severity.

Sample Application
How much **energy does an air-mass thunderstorm** release? Assume it draws in atmospheric boundary layer (ABL, or BL) air of T_d = 21°C & depth Δz_{BL} = 1 km (corresponding roughly to ΔP_{BL} = 10 kPa), and that all water vapor condenses. Approximate the cloud by a cylinder of radius R = 5 km and depth Δz = 10 km with base at P = 90 kPa & top at P = 20 kPa.

Find the Answer
Given: $R = 5$ km, $\Delta z = 10$ km, $T_d = 21°C$,
$\Delta z_{BL} = 1$ km, $\Delta P_{BL} = 10$ kPa,
$\Delta P_{storm} = 90 - 20 = 70$ kPa
Find: $\Delta Q_E = ?$ (J)

Use eq. (3.3) with $L_v = 2.5\text{x}10^6$ J·kg^{-1}, and eq. (4.3)
$$\Delta Q_E = L_v \cdot \Delta m_{water} = L_v \cdot r \cdot \Delta m_{air} \quad \text{(a)}$$
where $r \approx 0.016$ kg$_{water}$·kg$_{air}^{-1}$ from thermo diagram.

But eq. (1.8) $\Delta P = \Delta F/A$ and eq. (1.24) $\Delta F = \Delta m \cdot g$ give:
$$\Delta m_{air} = \Delta P \cdot A/g \quad \text{(b)}$$
where A = surface area = πR^2.

For a cylinder of air within the ABL, but of the same radius as the thunderstorm, use eq. (b), then (a):
Δm_{air} = (10 kPa)·π·(5000m)^2/(9.8 m s^{-2})= 8x10^8 kg$_{air}$
ΔQ_E=(2.5x10^6 J·kg^{-1})(0.016 kg$_{water}$·kg$_{air}^{-1}$)·
(8x10^8 kg$_{air}$) = 3.2x10^{15} J

Given the depth of the thunderstorm (filled with air from the ABL) compared to the depth of the ABL, we find that $\Delta P_{storm} = 7 \cdot \Delta P_{BL}$ (see Figure).

Thus, $\Delta Q_{E\ storm} = 7\ \Delta Q_{E\ BL}$

$\Delta Q_{E\ storm}$ = **2.24x10^{16}** J

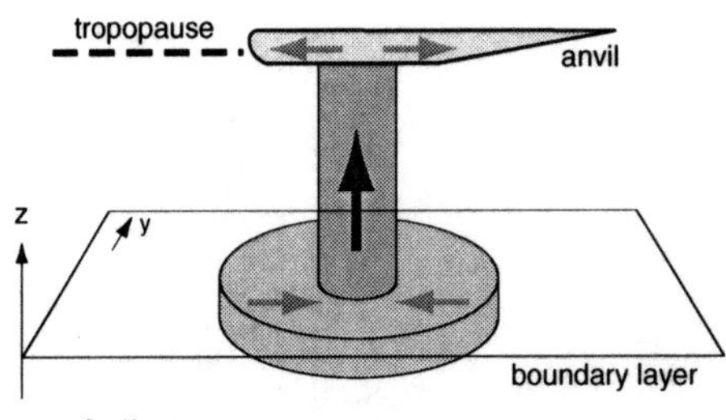

Check: Physics OK. Units OK. Values approximate.
Exposition: A **one-megaton nuclear bomb** releases about 4x10^{15} J of heat. This hypothetical thunderstorm has the power of 5.6 one-megaton bombs.

Actual heat released in a small thunderstorm is about 1% of the answer calculated above. Reasons include: T_d < 21°C; storm entrains non-ABL air; and not all of the available water vapor condenses. However, supercells continually draw in fresh ABL air, and can release more heat than the answer above. Energy released differs from energy available (CAPE). For another energy estimate, see the "Heavy Rain" section of the next chapter.

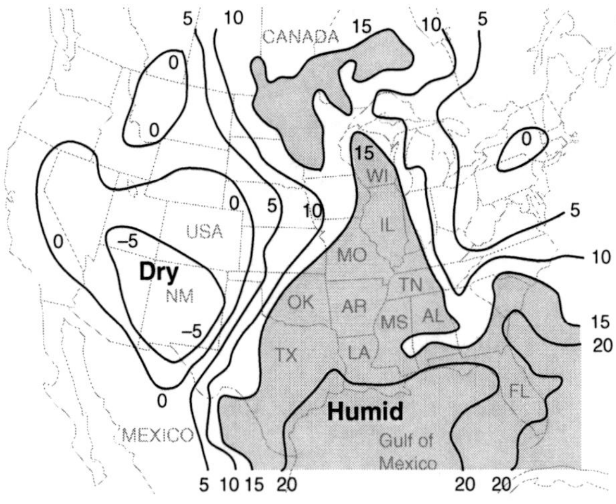

Figure 14.28
Surface weather map valid at 22 UTC on 24 May 2006 showing isodrosotherms (lines of equal dew-point, T_d, in °C) over N. America. Moistest air is shaded, and highlights the greatest fuel supply for thunderstorms. A dry line (sharp decrease in humidity) exists in west Texas (TX) and Oklahoma (OK).

Note, this is a different case study than was used in the previous chapter. The Extratropical Cyclone chapter used a Winter storm case, while here we use an early Spring severe-storm case.

On weather maps, lines of equal dew-point temperature are called **isodrosotherms** (recall Table 1-6). Fig. 14.28 shows an example, where the isodrosotherms are useful for identifying regions having warm humid boundary layers.

Most of the weather maps in this chapter and the next chapter are from a severe weather case on 24 May 2006. These case-study maps are meant to give one example, and do not show average or climatological conditions. The actual severe weather that occurred for this case is described at the end of this chapter, just before the Review.

An alternative moisture variable is **mixing ratio**, r. Large mixing-ratio values are possible only if the air is warm (because warm air can hold more water vapor at saturation), and indicate greater energy available for thunderstorms. For example, Fig. 14.29 shows **isohumes** of mixing ratio.

Wet-bulb temperature T_w, wet-bulb potential temperature θ_w, equivalent potential temperature θ_e, or **liquid-water potential temperature θ_L** also indicate moisture. Recall from Normand's rule in the Water Vapor chapter that the wet-bulb potential temperature corresponds to the moist adiabat that passes through the LCL on a thermo diagram. Also in Water Vapor chapter is a graph relating θ_w to θ_e.

For afternoon thunderstorms in the USA, pre-storm boundary layers most frequently have wet-bulb potential temperatures in the θ_w = 20 to 28°C range (or θ_e in the 334 to 372 K range, see example in Fig. 14.30). For supercell thunderstorms, the boundary-layer average is about θ_w = 24°C (or θ_e = 351 K), with some particularly severe storms (strong tornadoes or large hail) having $\theta_w \geq$ 27°C (or $\theta_e \geq$ 366 K).

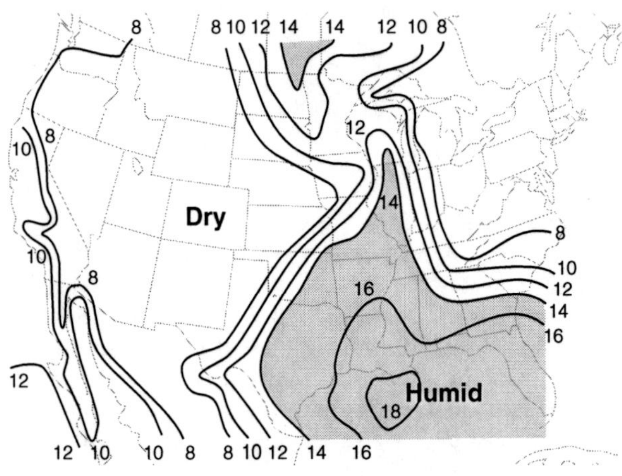

Figure 14.29
Similar to the previous weather map, but for isohumes of mixing ratio (g·kg^{-1}).

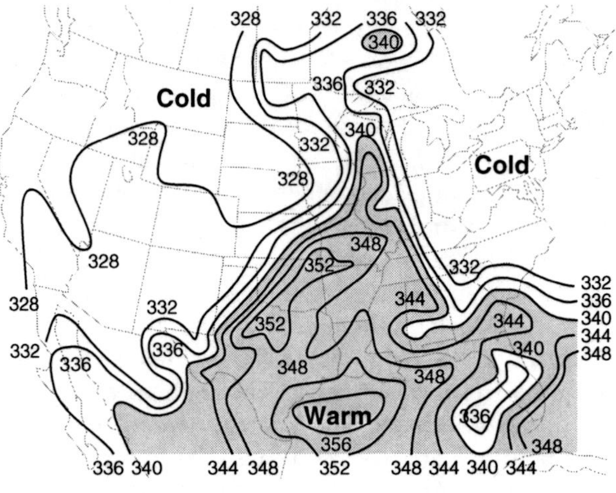

Figure 14.30
Similar to previous figure, but for equivalent potential temperature θ_e in Kelvin. Larger θ_e indicates warmer, moister air.

Precipitable water gives the total water content in a column of air from the ground to the top of the atmosphere (see example in Fig. 14.31). It does not account for additional water advected into the storm by the inflow winds, and thus is not a good measure of the total amount of water vapor that can condense and release energy.

Mean-layer lifting condensation level (**ML-LCL**) is the average of the LCL altitudes for all air parcels starting at heights within the bottom 1 km of atmosphere (i.e., ≈ boundary layer). Lower ML-LCL values indicate greater ABL moisture, and favor stronger storms (Figs. 14.32 & 14.33).

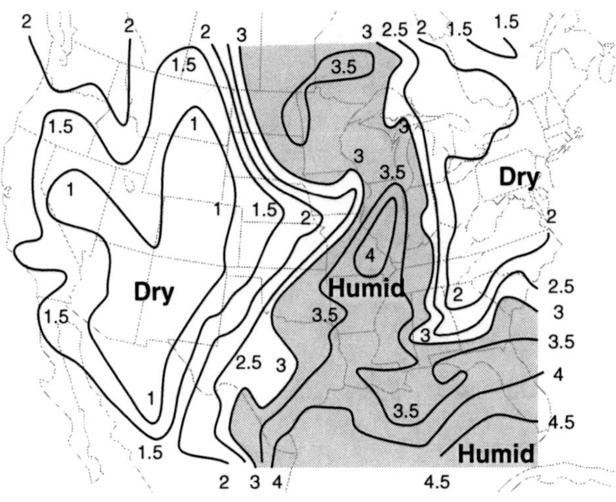

Figure 14.31
Similar to previous figure, but for precipitable water in cm.

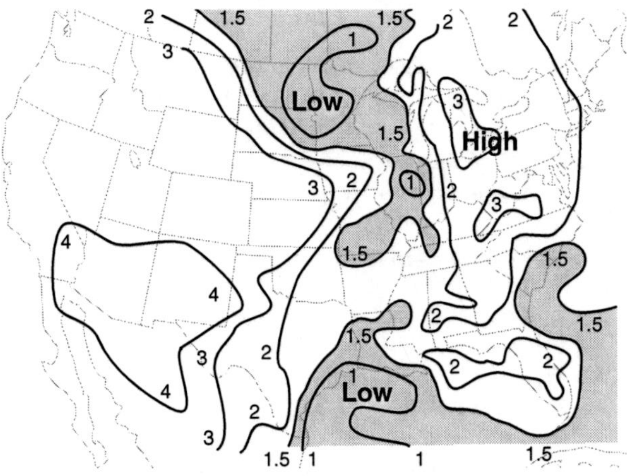

Figure 14.32
Similar to previous figure, but for mean-layer lifting condensation level (ML-LCL) heights in km above ground level. Lower ML-LCL heights (suggesting more intense storms) are shaded.

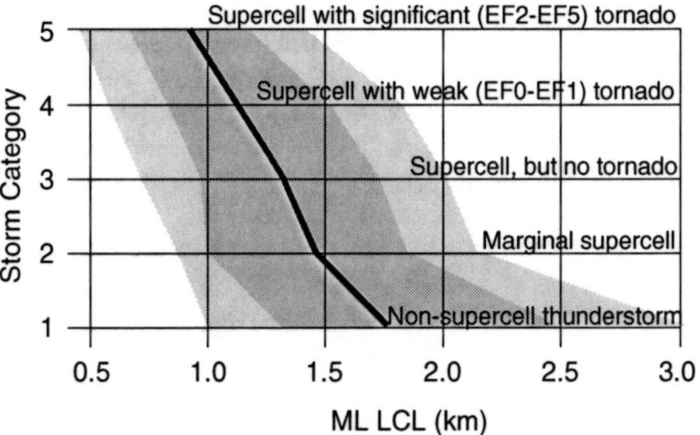

Figure 14.33
Statistical relationship between storm category (as labeled along the horizontal lines) and the mean-layer lifting condensation level (ML-LCL). A marginal supercell is one with weak (< 20 m s^{-1}) or short-duration (< 30 min) cyclonic shear. EF is the Enhanced Fujita scale tornado intensity. Thick line is median (50 percentile) of about 500 observed thunderstorms in the central USA. Dark grey spans 25 through 75 percentiles (i.e., the interquartile range), and light grey spans 10 through 90 percentiles. Lower ML-LCL values generally correspond to greater storm intensities. Caution: There is significant overlap of ML-LCLs for all storm categories, implying that ML-LCL cannot sharply discriminate between storm severities.

See the INFO box on the next page to learn about the median, interquartile range, and percentiles.

INFO • Median, Quartiles, Percentiles

Median, quartiles, and percentiles are statistical ways to summarize the location and spread of experimental data. They are a robust form of **data reduction**, where hundreds or thousands of data are represented by several summary statistics.

First, **sort** your data from the smallest to largest values. This is easy to do on a computer. Each data point now has a **rank** associated with it, such as 1st (smallest value), 2nd, 3rd, ... nth (largest value). Let $x_{(r)}$ = the value of the rth ranked data point.

The middle-ranked data point [i.e., at $r = (1/2)\cdot(n+1)$] is called the **median**, and the data value x of this middle data point is the median value ($q_{0.5}$). Namely,

$$q_{0.5} = x_{(1/2)\cdot(n+1)} \qquad \text{for } n = \text{odd}$$

If n is an even number, there is no data point exactly in the middle, so use the average of the 2 closest points:

$$q_{0.5} = 0.5\cdot [x_{(n/2)} + x_{(n/2)+1}] \qquad \text{for } n = \text{even}$$

The median is a measure of the **location** or center of the data

The data point with a rank closest to $r = (1/4)\cdot n$ is the **lower quartile** point:

$$q_{0.25} = x_{(1/4)\cdot(n+1)}$$

The data point with a rank closest to $r = (3/4)\cdot n$ is the **upper quartile** point:

$$q_{0.75} = x_{(3/4)\cdot(n+1)}$$

These last 2 equations work well if n is large (≥100, see below). The **interquartile range (IQR)** is defined as IQR = $q_{0.75} - q_{0.25}$, and is a measure of the **spread** of the data. (See the Sample Application nearby.)

Generically, the variable q_p represents any **quantile**, namely the value of the ranked data point having a value that exceeds portion p of all data points. We already looked at p = 1/4, 1/2, and 3/4. We could also divide large data sets into hundredths, giving **percentiles**. The lower quartile is the same as the 25th percentile, the median is the 50th percentile, and the upper quartile is the 75th percentile.

These **non-parametric statistics** are **robust** (usually give a reasonable answer regardless of the actual distribution of data) and **resistant** (are not overly influenced by **outlier** data points). For comparison, the mean and standard deviation are NOT robust nor resistant. Thus, for experimental data, you should use the median and IQR.

To find quartiles for a small data set, split the ranked data in half, and look at the lower and upper halves separately.

Lower half of data: If n = odd, consider those data points ranked less than or equal to the median point. For n = even, consider points with values less than the median value. For this subset of data, find its median, using the same tricks as in the previous paragraph. The resulting data point is the **lower quartile**.

Upper half of data: For n = odd, consider those data points ranked greater than or equal to the original median point. For n = even, use the points with values greater than the median value. The median point in this data subset gives the **upper quartile**.

Sample Application (§)

Suppose the z_{LCL} (km) values for 9 supercells (with EF0-EF1 tornadoes) are:

1.5, 0.8, 1.4, 1.8, 8.2, 1.0, 0.7, 0.5, 1.2

Find the median and interquartile range. Compare with the mean and standard deviation.

Find the Answer:

Given: data set listed above.

Find: $q_{0.5}$ = ? km, IQR = ? km,
Mean$_{zLCL}$ = ? km, σ_{zLCL} = ? km

First sort the data in ascending order:

Values (z_{LCL}):	0.5,	0.7,	0.8,	1.0,	1.2,	1.4,	1.5,	1.8,	8.2
Rank (r):	1	2	3	4	5	6	7	8	9
Middle:					^				

Thus, the median point is the 5th ranked point in the data set, and corresponding value of that data point is <u>median = $q_{0.5}$ = $z_{LCL(r=5)}$ = **1.2 km**</u>.

Because this is a small data set, use the special method at the bottom of the INFO box to find the quartiles.

Lower half:

Values:	0.5,	0.7,	0.8,	1.0,	1.2
Subrank:	1	2	3	4	5
Middle:			^		

Thus, the lower quartile value is $q_{0.25}$ = 0.8 km

Upper half:

Values:	1.2,	1.4,	1.5,	1.8,	8.2
Subrank:	1	2	3	4	5
Middle:			^		

Thus, the upper quartile value is $q_{0.75}$ = 1.5 km

The IQR = $q_{0.75} - q_{0.25}$ = (1.5km – 0.8km) = <u>**0.7 km**</u>

Using a spreadsheet to find the mean and standard deviation:

Mean$_{zLCL}$ = <u>**1.9 km**</u> , σ_{zLCL} = <u>**2.4 km**</u>

Check: Values reasonable. Units OK.

Exposition: The original data set has one "wild" z_{LCL} value: 8.2 km. This is the **outlier**, because it lies so far from most of the other data points.

As a result, the mean value (1.9 km) is not representative of any of the data points; namely, the center of the majority of data points is not at 1.9 km. Thus, the mean is not robust. Also, if you were to remove that one outlier point, and recalculate the mean, you would get a significantly different value (1.11 km). Hence, the mean is not resistant. Similar problems occur with the standard deviation.

However, the median value (1.2 km) is nicely centered on the majority of points. Also, if you were to remove the one outlier point, the median value would change only slightly to a value of 1.1 km. Hence, it is robust and resistant. Similarly, the IQR is robust and resistant.

INSTABILITY, CAPE & UPDRAFTS

The second requirement for convective-storm formation is instability in the pre-storm sounding. **Nonlocal conditional instability** (**NCI**) occurs when warm humid atmospheric boundary layer (ABL) air is capped by a temperature inversion, above which is relatively cold air. The cold air aloft provides an environment that gives more buoyancy to the warm updraft air from below, allowing stronger thunderstorms. The "nonlocal" aspect arises because air from below the cap becomes unstable above the cap. The "condition" is that the ABL air must first be lifted past the cap (i.e., past its LCL and LFC) for the instability to be realized.

The capping inversion traps the warm humid air near the ground, allowing sensible and latent heat energy to build up during the day as the sun heats the ground and causes evaporation. Without this cap, smaller cumulus clouds can withdraw the warm humid air from the boundary layer, leaving insufficient fuel for thunderstorms. Thus, the cap is important — it prevents the fuel from leaking out.

Convective Available Potential Energy

Thunderstorms get their energy from the buoyancy associated with latent-heat release when water vapor condenses. The **Convective Available Potential Energy** (CAPE) is a way to estimate this energy using a thermo diagram. **CAPE** is proportional to the shaded area in Fig 14.34; namely, the area between LFC and EL altitudes that is between the environmental sounding and the moist adiabat of the rising air parcel.

To explain this, use the definition of buoyancy force per unit mass $F/m = |g|\cdot(T_{vp} - T_{ve})/T_{ve}$ that was covered in the Atmospheric Stability chapter. T_{vp} is the virtual temperature of the air parcel rising in the thunderstorm, T_{ve} is virtual temperature in the surrounding environment at the same altitude as the thunderstorm parcel, and $|g|$ = 9.8 m s^{-2} is gravitational acceleration magnitude.

Recall from basic physics that **work** equals force times distance. Let $(\Delta E/m)$ be the incremental work per unit mass associated with a thunderstorm air parcel that rises a small increment of distance Δz. Thus, is $\Delta E/m = (F/m)\cdot \Delta z$, or:

$$\frac{\Delta E}{m} \cong \Delta z \cdot |g| \frac{(T_{vp} - T_{ve})}{T_{ve}} \tag{14.1}$$

Recall from Chapter 1 that virtual temperature includes the effects of both water vapor and liquid water. Water vapor is less dense than air ($T_v > T$), thus increasing the buoyant energy. Liquid- and

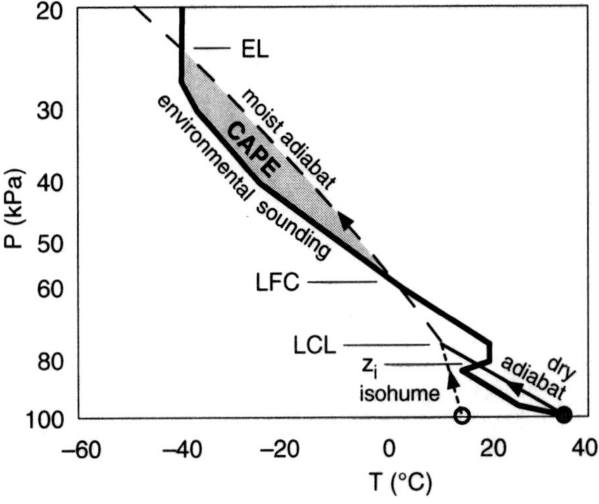

Figure 14.34
Surface-based Convective Available Potential Energy (CAPE) is the grey shaded area for an afternoon pre-storm environment.

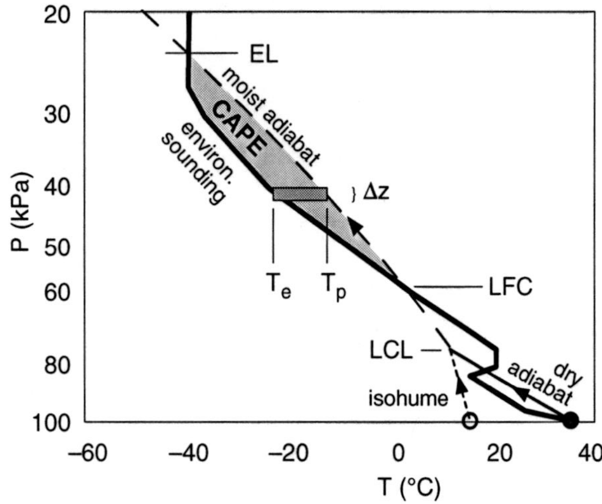

Figure 14.35
Dark shaded rectangle of incremental height Δz and width $T_p - T_e$ shows the portion of total CAPE area associated with just one thin layer of air.

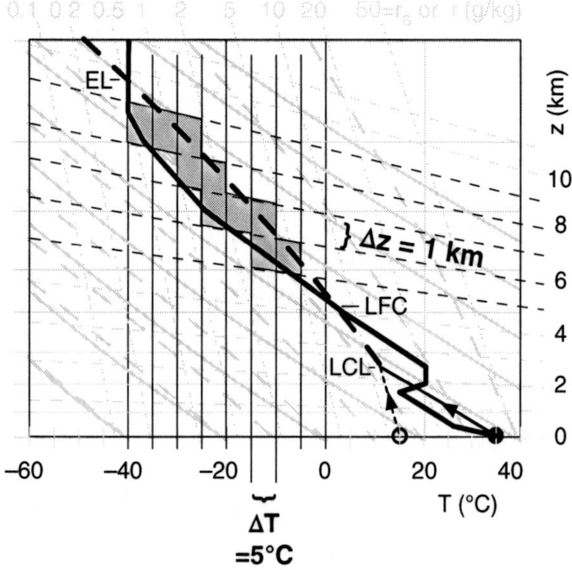

Figure 14.36
Approximating the CAPE area by non-overlapping tiles, each of area Δz by ΔT. In this example, Δz = 1 km, and ΔT = 5°C.

solid-water hydrometeors (cloud droplets, rain, and snow) falling at their terminal velocity are heavier than air ($T_v < T$), thus decreasing the upward buoyant energy. Both are important for thunderstorms, but are often difficult to determine.

Instead, you can use the approximation $T_v \approx T$, which gives:

$$\frac{\Delta E}{m} \cong |g| \frac{\Delta z \cdot (T_p - T_e)}{T_e} = |g| \cdot (Increm.Area) / T_e \qquad (14.2)$$

where T_p is air-parcel temperature, T_e is environmental temperature, and the incremental area (*Increm. Area*) is shown Fig. 14.35 as the dark-grey rectangle.

Adding all the incremental rectangles between the LFC and the EL gives a total area (light shading in Fig. 14.35) that is proportional to CAPE.

$$CAPE \cong |g| \cdot (Total.Area) / T_e \qquad (14.3)$$

or

$$CAPE \cong |g| \sum_{LFC}^{EL} (T_p - T_e) \cdot \Delta z / T_e \qquad \bullet(14.4)$$

The units of CAPE are J·kg^{-1}. These units are equivalent to m^2·s^{-2} ; namely, velocity squared. The temperature in the denominator of eqs. (14.3 & 14.4) must be in Kelvin. Both numerator temperatures must have the same units: either Kelvin or °C.

The shape of the CAPE area is usually not simple, so calculating the area is not trivial. At severe-weather forecast centers, computers calculate CAPE automatically based on the pre-storm sounding. By hand, you can use a simple graphical method by first plotting the sounding and the surface-parcel rise on a thermo diagram, and then using whatever height and temperature increments are plotted on the background diagram to define "bricks" or "tiles", each of known size.

Namely, instead of using long narrow rectangles, each of different width, as sketched in Fig. 14.35, you can cover the complex shaded area with smaller, non-overlapping tiles, each of equal but arbitrary size ΔT by Δz (such as ΔT = 5°C and Δz = 1 km, see Fig. 14.36). Count the number of tiles, and multiply the result by the area of each tile ($\Delta T \cdot \Delta z$ = 5 °C·km in this example). When tiling the CAPE area, try to compensate for small areas that are missed by the tiles by allowing some of the other tiles to extend slightly beyond the boundaries of the desired shaded area by roughly an equal area (e.g., see the Sample Application). Smaller-size tiles (such as ΔT = 1°C and Δz = 0.2 km) give a more accurate answer (and are recommended), but are more tedious to count.

Sample Application
For the sounding in Fig 14.36, estimate the CAPE.

Find the Answer
Given: Fig 14.36.
Find: CAPE = ? $J{\cdot}kg^{-1}$

Each tile has size ΔT = 5°C by Δz = 1000 m. The area of each tile is 5,000 °C·m, and there are 10 tiles. The total area is: *Area* = 10 x 5000 °C·m = 50,000 K·m , [where I took advantage of ΔT(°C) = ΔT(K)].

By eye, the average T_e in the CAPE region is about –25°C = 248 K. Use eq. (14.3):

CAPE = [9.8 $m{\cdot}s^{-2}$ / 248K]·(50,000 K·m) =
= 1976 $m^2{\cdot}s^{-2}$ = **1976 $J{\cdot}kg^{-1}$**.

Check: Physics and units OK. Figure OK.
Exposition: This is a moderate value of CAPE that could support supercell storms with tornadoes.

On many thermo diagrams such as the emagram of T vs. ln(P) used here, the height contours and isotherms are not perpendicular; hence, the rectangles look like trapezoids (see the Sample Application that employed Fig. 14.36). Regardless of the actual area within each trapezoid in the plotted graph, each trapezoid represents a contribution of ΔT by Δz toward the total CAPE area.

To replace heights with pressures in the equation for CAPE, use the hypsometric equation from Chapter 1, which yields:

$$CAPE = \Re_d \cdot \sum_{LFC}^{EL} (T_p - T_e) \cdot \ln\left(\frac{P_{bottom}}{P_{top}}\right) \qquad \bullet(14.5)$$

where $\Re_d$ = 287.053 $J{\cdot}K^{-1}{\cdot}kg^{-1}$ is the gas constant for dry air, P_{bottom} and P_{top} are the bottom and top pressures of the incremental rectangle, and the sum is still over all the rectangles needed to tile the shaded CAPE area on the sounding (Fig. 14.37). Again, the temperature difference in eq. (14.5) in °C is equal to the same value in Kelvin.

All of the CAPE figures up until now have followed a rising air parcel that was assumed to have started from near the surface. This is called **Surface-Based CAPE** (SBCAPE), which is often a good method for the mid-afternoon pre-storm soundings that have been shown so far.

CAPE values vary greatly with location and time. By finding the CAPE for many locations in a region (by using rawinsonde observations, or by using forecast soundings from numerical weather prediction models), you can write the CAPE values on

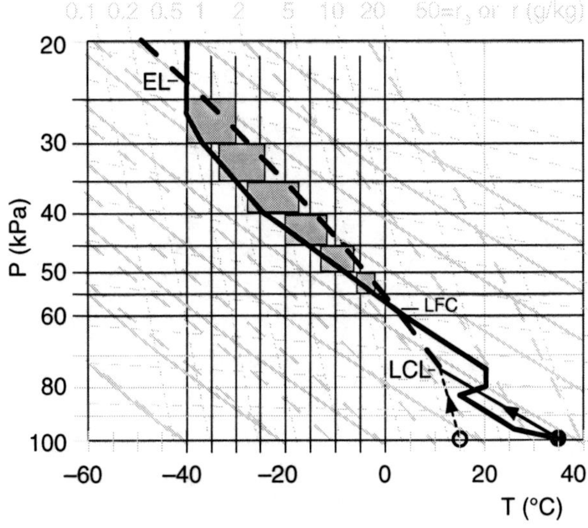

Figure 14.37
Approximating the CAPE area by non-overlapping tiles, each of area ΔP by ΔT. In this example, ΔP = 5 kPa, and ΔT varies.

Sample Application
For the sounding in Fig. 14.37, estimate the CAPE using pressure rather than height increments.

Find the Answer:
Given: Fig. 14.37.
Find: CAPE = ? $J{\cdot}kg^{-1}$

Use eq. (14.5):
CAPE = [287 J/(K·kg)] · {
[–30°C – (–40°C)] · ln(30kPa/25kPa)] +
[–24°C – (–33°C)] · ln(35kPa/30kPa)] +
[–17°C – (–27°C)] · ln(40kPa/35kPa)] +
[–12°C – (–20°C)] · ln(45kPa/40kPa)] +
[–7°C – (–13°C)] · ln(50kPa/45kPa)] +
[–2°C – (–6°C)] · ln(55kPa/50kPa)] }

CAPE = [287 $J{\cdot}(K{\cdot}kg)^{-1}$] ·
{ 10K · 0.182 + 9.2K · 0.154 + 10K · 0.134 +
8K · 0.118 + 6K · 0.105 + 4K · 0.095 }

CAPE = [287 $J{\cdot}(K{\cdot}kg)^{-1}$] · 6.53 K = **1874 $J{\cdot}kg^{-1}$** .

Check: Physics and units OK.
Exposition: Theoretically, we should get exactly the same answer as we found in the previous Sample Application. Considering the coarseness of the boxes that I used in both Sample Applications, I am happy that the answers are as close as they are. I found the height-tiling method easier than the pressure-tiling method.

a weather map and then draw isopleths connecting lines of equal CAPE, such as shown in Fig. 14.38.

Other parcel-origin assumptions work better in other situations. If the pre-storm sounding is from earlier in the morning, then the forecast max surface temperature for that afternoon (along with the dew point forecast for that time) is a better choice for the rising air-parcel initial conditions (Fig. 14.39). This is also a type of SBCAPE.

Another way is to average the conditions in the bottom 1 km (roughly 10 kPa) of the environmental sounding to better estimate ABL conditions. Thus, the initial conditions for the rising air parcel represents the mean layer (ML) conditions, or the mixed-layer (ML) conditions (Fig. 14.40). CAPE calculated this way is called **Mean Layer CAPE** (MLCAPE). Fig. 14.41 shows a case-study example of MLCAPE.

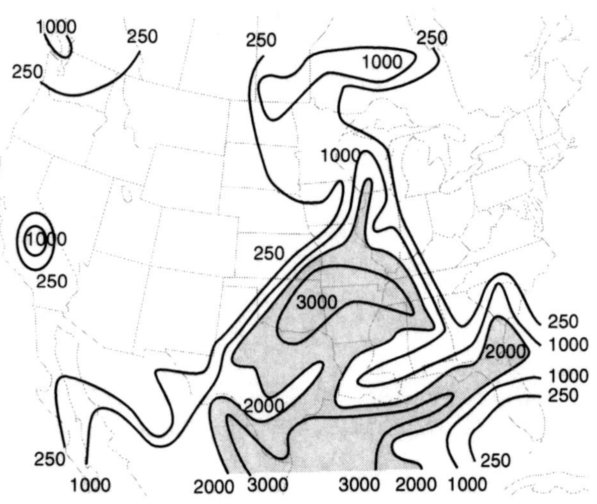

Figure 14.38
Weather map of surface-based CAPE (SBCAPE) in J·kg^{-1} over N. America. Valid 22 UTC on 24 May 2006. Shaded region highlights larger SBCAPE values, where more intense thunderstorms can be supported.

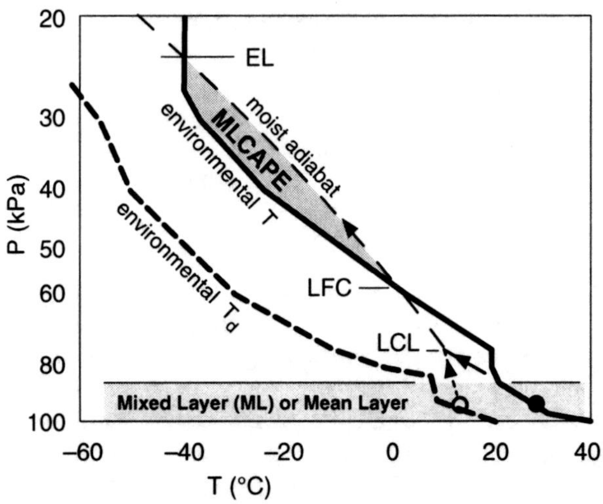

Figure 14.40
Another method for estimating CAPE is to use air-parcel initial conditions equal to the average conditions (filled and open circles) in the boundary layer (shaded). This is called the mean layer CAPE (MLCAPE).

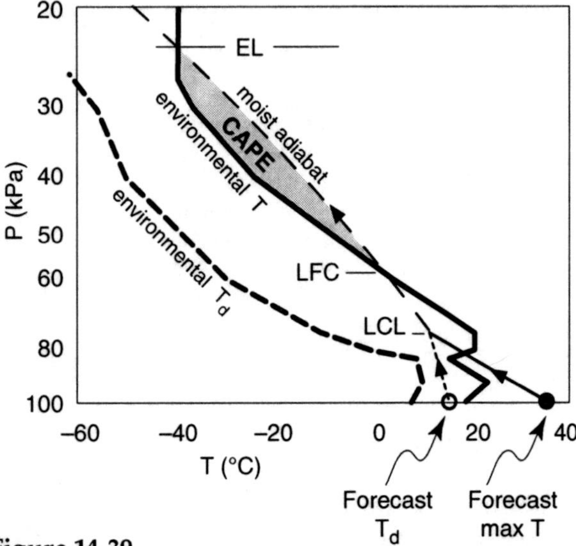

Figure 14.39
Shown is one method for estimating CAPE from an early morning sounding. Instead of using actual surface environmental conditions for the air parcel, you can use the forecast maximum near-surface air temperature (max T) for later that day, and assume no change in environmental conditions above the ML.

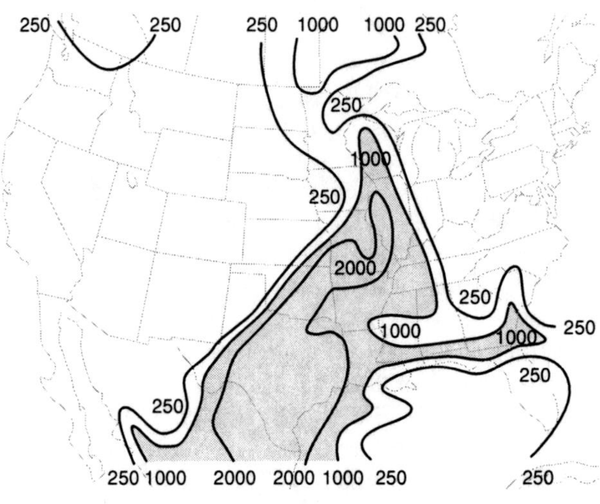

Figure 14.41
Weather map similar to Fig. 14.38, but for mean-layer CAPE (MLCAPE) in J·kg^{-1}.

Early studies of thunderstorm occurrence vs. MLCAPE lead to forecast guidelines such as shown in Table 14-1. However, MLCAPE is not a sharp discriminator of thunderstorm intensity, as shown by the large overlap of storm categories in Fig. 14.42.

Yet another way is to calculate many different CAPEs for air parcels that start from every height in the bottom 30 kPa of the pre-storm environmental sounding, and then select the one that gives the greatest CAPE values. This is called the **Most Unstable CAPE** (MUCAPE). This method works even if thunderstorm updrafts are triggered by an elevated source, and also works for pre-storm soundings from any time of day. Although it is too tedious to compute these multiple CAPEs by hand, it is easily automated on a computer (Figs. 14.43 & 14.44). MUCAPE is always ≥ SBCAPE.

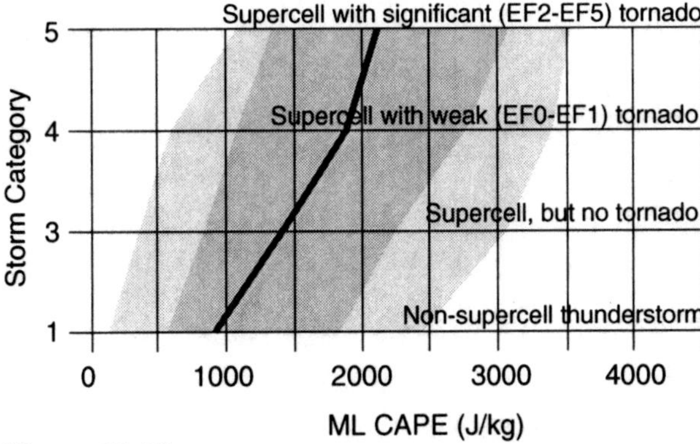

Figure 14.42
Statistics of thunderstorm intensity vs. MLCAPE, based on several hundred paired soundings and thunderstorms in central N. America. Dark line is median (50 percentile); dark shading spans the interquartile range (25 to 75 percentiles); light shading spans the 10 to 90 percentile range.

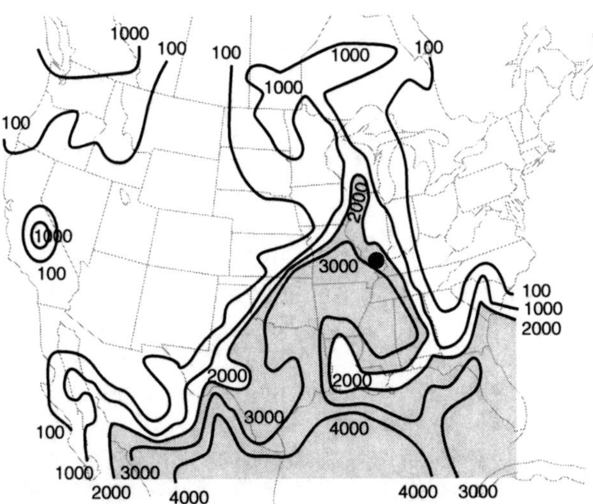

Figure 14.43
Weather map similar to Fig. 14.38 over the USA, but for most-unstable CAPE (MUCAPE) in J·kg^{-1}.

Table 14-1. Thunderstorm (CB) intensity guide.

MLCAPE (J·kg^{-1})	Stability Description	Thunderstorm Activity
0 - 300	mostly stable	little or none
300 - 1000	marginally unstable	weak CB
1000 - 2500	moderately unstable	moderate CB likely; severe CB possible
2500 - 3500	very unstable	severe CB likely, possible tornado
3500 & greater	extremely unstable	severe CB, tornadoes likely

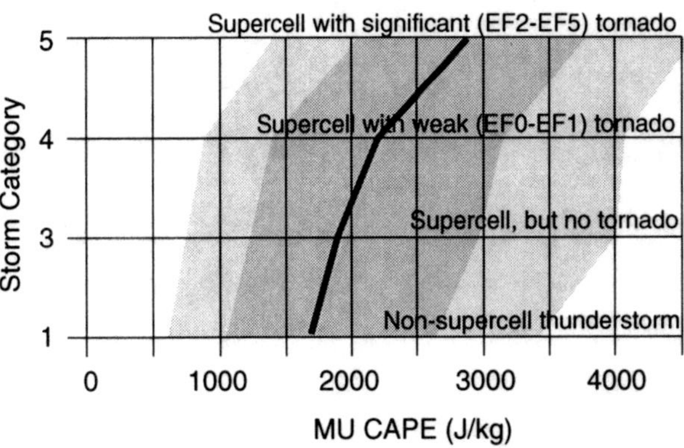

Figure 14.44
Statistics of thunderstorm intensity vs. MUCAPE, similar to Fig. 14.42.

Sample Application

Assume a thunderstorm forms at the dot in Fig. 14.43, where MUCAPE ≈ 2700 J·kg^{-1}. Estimate the possible intensity of this storm. Discuss the uncertainty.

Find the Answer

Given: MUCAPE ≈ 2,700 J·kg^{-1} at the dot (•) in Fig. 14.43, located in southern Illinois.
Find: Possible storm intensity. Discuss uncertainty.

Use Fig. 14.44 to estimate intensity. For MUCAPE of 2,700 J·kg^{-1}, the median line is about half way between "supercell with **weak tornado** (EF0-EF1)", and "supercell with **significant tornado** (EF2-EF5)." So we might predict a supercell with an EF1 - EF2 tornado.

Exposition: Although there is some uncertainly in picking the MUCAPE value from Fig. 14.43, there is even greater uncertainty in Fig 14.44. Namely, within the interquartile range (dark shading) based on hundreds of past storms used to make this figure, there could easily be a supercell with no tornado, or a supercell with a significant tornado. There is even a chance of a non-supercell thunderstorm.

Thus, MUCAPE is not capable of sharply distinguishing thunderstorm intensity.

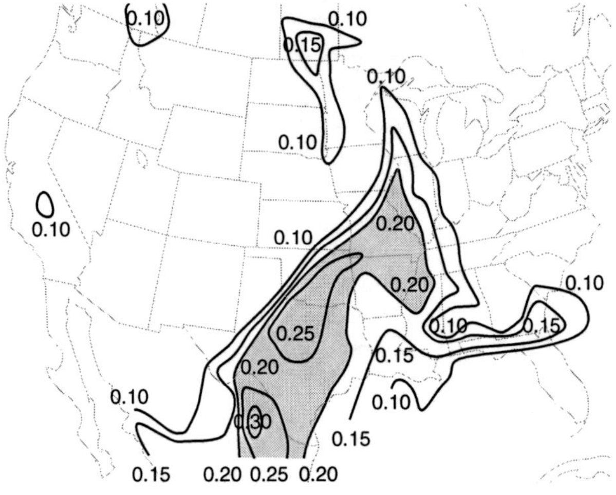

Figure 14.45
Weather map similar to Fig. 14.38 over the USA, but for normalized CAPE (nCAPE) in m·s^{-2}.

Sample Application
For the sounding in Fig. 14.36 (CAPE = 1976 J·kg^{-1}) find thunderstorm intensity, normalized CAPE, max updraft velocity, and likely updraft velocity.

Find the Answer
Given: Fig. 14.36, with CAPE = 1976 J·kg^{-1}
Find: Intensity = ? , $nCAPE$ = ? m·s^{-2},
w_{max} & $w_{max\ likely}$ = ? m s^{-1}

a) From Table 14-1, **moderate thunderstorms**.

b) By eye using Fig. 14.36, the bottom and top of the CAPE region are $z_{LFC} \approx 4.2$ km, and $z_{EL} = 9.6$ km.
Use eq. (14.6):
$nCAPE$=(1976 J·kg^{-1})/[9600–4200m]=**0.37**m·s^{-2}.

c) Use eq. (14.7):
$w_{max} = [2 \cdot (1976\ m^2\ s^{-2})\ ]^{1/2} = w_{max} =$ **63** m s^{-1} .

d) Use eq (14.8):
$w_{max\ likely} = (63\ m\ s^{-1})/2 =$ **31** m s^{-1} .

Check: Physics and units OK. Figure OK.
Exposition: Because the LFC is at such a high altitude, the CAPE area is somewhat short and fat, as indicated by the small value of nCAPE. Thus, strong low-altitude updrafts and tornadoes are possible. This thunderstorm has violent updrafts, which is why aircraft would avoid flying through it.

The shape of the CAPE area gives some information about the storm. To aid shape interpretation, a **normalized CAPE (nCAPE)** is defined as

$$nCAPE = \frac{CAPE}{z_{EL} - z_{LFC}} \qquad \bullet(14.6)$$

where z_{EL} is height of the equilibrium level, z_{LFC} is height of the level of free convection, and the units of nCAPE are m·s^{-2} or J·(kg·m)$^{-1}$. Tall, thin CAPE areas (i.e., nCAPE ≤ 0.1 m·s^{-2}) often suggest heavy precipitation, but unlikely tornadoes. Short, wide CAPE area (i.e., nCAPE ≥ 0.3 m·s^{-2}) in the mid to lower part of the sounding can result in thunderstorms with strong, low-altitude updrafts, which cause vertical stretching of the air, intensification of rotation, and greater chance of tornadoes. See Fig. 14.45.

A word of caution. CAPE gives only an estimate of the strength of a thunderstorm if one indeed forms. It is a necessary condition, but not a sufficient condition. To form a thunderstorm, there must also be a process that triggers it. Often no thunderstorms form, even in locations having large CAPE. If thunderstorms are triggered, then larger CAPE indicates greater instability, stronger updrafts, and a chance for more violent thunderstorms. It is a useful, but not perfect, forecast tool, as evident by the lack of sharpness in the statistics (Figs. 14.42, & 14.44).

Updraft Velocity

You can also use CAPE to estimate the updraft speed in thunderstorms. Recall from basic physics that kinetic energy per unit mass is $KE/m = 0.5{\cdot}w^2$, where w is updraft speed. Suppose that all the convective available potential energy could be converted into kinetic energy; namely, CAPE = KE/m. Combining the two equations above gives

$$w_{max} = \sqrt{2{\cdot}CAPE} \qquad (14.7)$$

which gives unrealistically large speed because it neglects frictional drag. Studies of actual thunderstorm updrafts find that the most likely max updraft speed is

$$w_{max\ likely} \approx w_{max} / 2 \qquad \bullet(14.8)$$

Air in an updraft has inertia and can overshoot above the EL. Such **penetrative convection** can be seen by eye as mound or turret of cloud that temporarily overshoots above the top of the thunderstorm anvil. You can also see it in some satellite images. As stated before, such turrets or domes give a good clue to storm spotters that the thunderstorm is probably violent, and has strong updrafts. Such strong updrafts are felt as severe or extreme **turbulence** by aircraft.

WIND SHEAR IN THE ENVIRONMENT

Wind shear, the change of horizontal wind speed and/or direction with height, is the third requirement for thunderstorm formation. The wind shear across a layer of air is the vector difference between the winds at the top of the layer and winds at the bottom (Fig. 14.46), divided by layer thickness Δz. Shear has units of (s^{-1}).

Using geometry, shear can be expressed via its components as:

$$\frac{\Delta U}{\Delta z} = \frac{U_2 - U_1}{z_2 - z_1} \tag{14.9}$$

$$\frac{\Delta V}{\Delta z} = \frac{V_2 - V_1}{z_2 - z_1} \tag{14.10}$$

or

$$\text{Shear Magnitude} = \frac{\left[(\Delta U)^2 + (\Delta V)^2\right]^{1/2}}{\Delta z} \tag{14.11a}$$

Shear Direction:

$$\alpha_{shear} = 90° - \frac{360°}{C} \arctan\left(\frac{\Delta V}{\Delta U}\right) + \alpha_o \tag{14.11b}$$

where subscript 2 is the layer top, and subscript 1 is layer bottom. Eq. (1.2) was used to find the direction α of the shear vector, but with ΔU in place of U, and ΔV in place of V. $\alpha_o = 180°$ if $\Delta U > 0$, but is zero otherwise. C is the angular rotation in a circle (360° or 2π radians, depending on the output from "arctan").

For thunderstorm forecasting, meteorologists look at shear across many layers at different heights in the atmosphere. To make this easier, you can use layers of equal thickness, such as Δz = 1 km. Namely, look at the shear across the bottom layer from z = 0 to z = 1 km, and the shear across the next layer from z = 1 to z = 2 km, and so forth.

When studying shear across layers of equal thickness, meteorologists often use the **vector wind difference** (ΔU, ΔV) as a surrogate measure of **vector wind shear**. We will use this surrogate here. This surrogate, and its corresponding **wind-difference magnitude**

$$\Delta M = \left(\Delta U^2 + \Delta V^2\right)^{1/2} \tag{14.12}$$

has units of wind speed (m s^{-1}).

Thunderstorms can become intense and long-lasting given favorable wind shear in the lower atmosphere. Under such conditions, humid boundary-layer air (thunderstorm fuel) can be fed into the moving storm (Fig. 14.47). Another way to picture this process is that wind shear causes the storm to move away from locations of depleted boundary-layer fuel into locations with warm, humid bound-

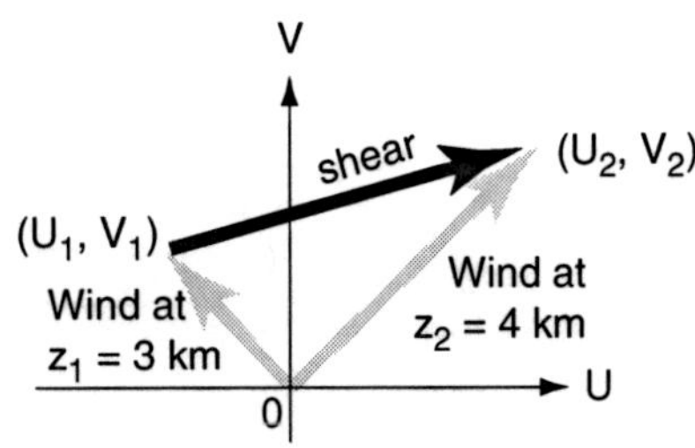

Figure 14.46
Wind difference (black arrow) between two altitudes is the vector difference between wind (long grey arrow) at the top altitude minus wind at the bottom (short grey arrow). We will use vector wind difference as a surrogate for vector wind shear.

Sample Application

Given a horizontal wind of (U, V) = (1, 2) m s^{-1} at 2 km altitude, and a wind of (5, –3) m s^{-1} at 3 km, what are the wind-difference & shear magnitudes & direction?

Find the Answer:

Given: (U, V) = (1, 2) m s^{-1} at z = 2 km
(U, V) = (5, –3) m s^{-1} at z = 3 km
Find: ΔM = ? (m s^{-1}), α = ? (°)
Shear Magnitude = ? (s^{-1}),

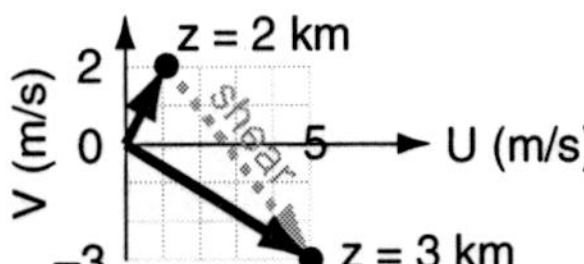

Use eq. (14.12) to find the wind difference magnitude:
ΔU = (5 – 1) = 4 m s^{-1}. ΔV = (–3 – 2) = –5 m s^{-1}.
ΔM = [(4 m $s^{-1})^2$ + (–5 m $s^{-1})^2$ $]^{1/2}$ =
[16+25 m^2 $s^{-2}]^{1/2}$ = <u>6.4</u> m s^{-1}

Use eq. (14.11b) to find shear direction:
α = 90°–arctan(–5/4)+180° = 90°–(–51.3°)+180° = <u>**321.3°**</u>
which is the direction that the wind shear is coming <u>from</u>.

Layer thickness is Δz = 3 km – 2 km = 1 km = 1000 m.
Use eq. (14.11a) to find shear magnitude, rewritten as:
Shear Mag.= $\Delta M/\Delta z$ = (6.4 m s^{-1})/(1000 m) = <u>**0.0064 s**</u>$^{-1}$

Check: Units OK. Magnitude & direction agree with figure.

Exposition: There is both **speed shear** and **directional shear** in this example. It always helps to draw a figure, to check your answer.

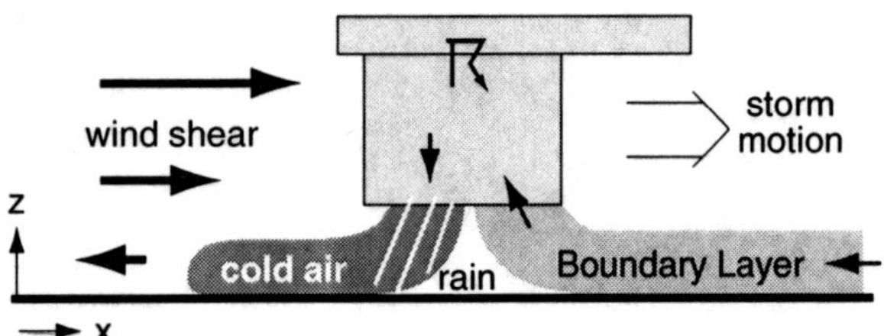

Figure 14.47
Wind shear allows long-lasting strong thunderstorms, such as supercell storms and bow-echo thunderstorm lines.

ary layers. The storm behaves similar to an upright vacuum cleaner, sucking up fresh boundary-layer air as it moves, and leaving behind an exhaust of colder air that is more stable.

In an environment with wind but no wind shear, the thunderstorms would last only about 15 minutes to 1 h, because the thunderstorm and boundary-layer air would move together. The storm dies after it depletes the fuel in its accompanying boundary layer (Fig. 14.6).

Shear can also control the direction that a thunderstorm moves. As will be shown later, a clockwise turning wind shear vector can create a dynamic vertical pressure gradient that favors thunderstorms that move to the right of the mean steering-level wind direction.

Mesocyclones can also be created by wind shear. A mesocyclone is where the whole thunderstorm rotates. A mesocyclone can be a precursor to strong tornadoes, and is discussed in detail in the next chapter. Strong shear also enhances MCSs such as bow-echo thunderstorms. Thus, there are many ways that **upper-air wind** (winds above the surface) and wind shear affect thunderstorms.

Hodograph Basics

One way to plot upper-air wind vs. height is to draw wind symbols along the side of a thermo diagram (Fig. 14.48). Recall from the Atmospheric Forces and Winds chapter that wind direction is from the tail end (with the feathers) toward the other end of the shaft, and more barbs (feathers or pennants) indicate greater speed. The tip of the arrow is at the altitude of that wind.

Another way to plot winds is to collapse all the wind vectors from different heights into a single polar graph, which is called a **hodograph**. For example, Fig. 14.49 shows winds at 1 km altitude blowing from the southeast at 15 m s^{-1}, winds at 3 km altitude blowing from the south at 25 m s^{-1}, and winds at 5 km altitude blowing from the southwest at 35 m s^{-1}. Both **directional shear** and **speed shear** exist.

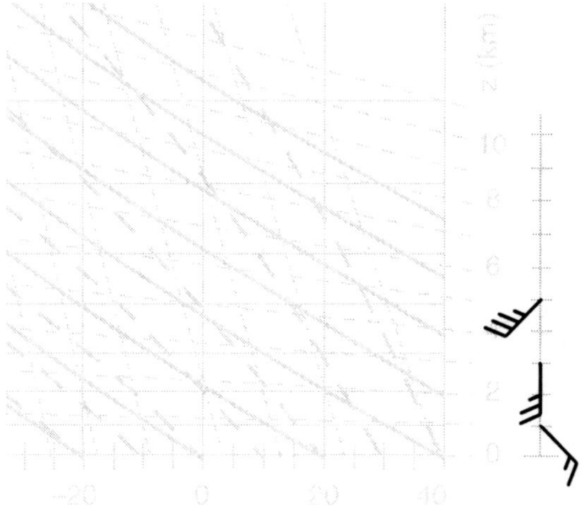

Figure 14.48
Winds at different heights, indicated with wind barbs adjacent to a thermo diagram. The same speeds are indicated by vectors in the next figure. Wind-barb speed units are m s^{-1}. (Look back at Table 10-1 for wind-barb definitions.)

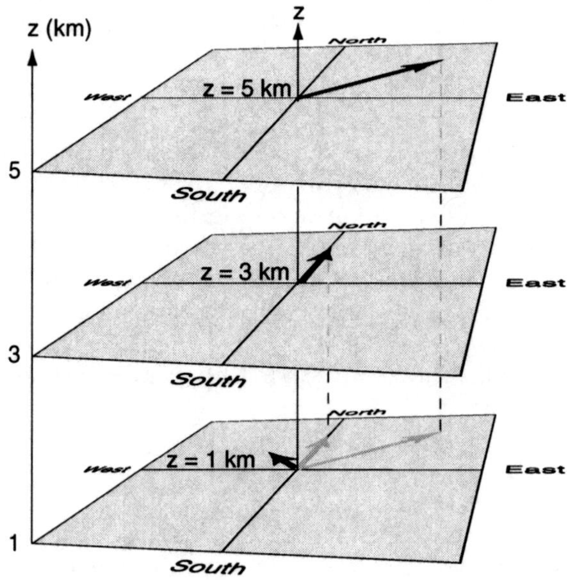

Figure 14.49
Projection of wind vectors (black arrows) from different altitudes onto the bottom plane (grey arrows). These wind vectors correspond to the wind vectors in the previous figure.

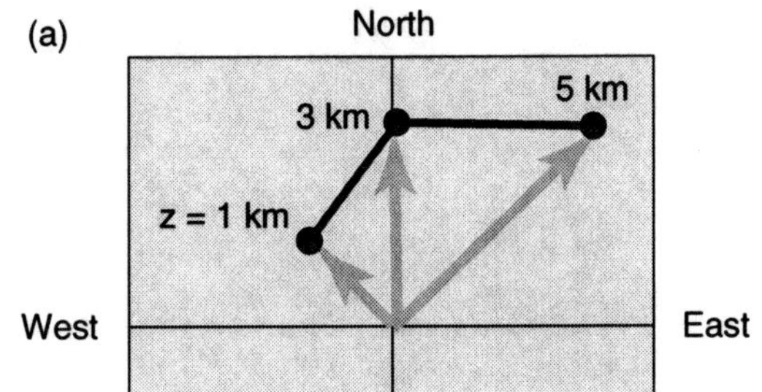

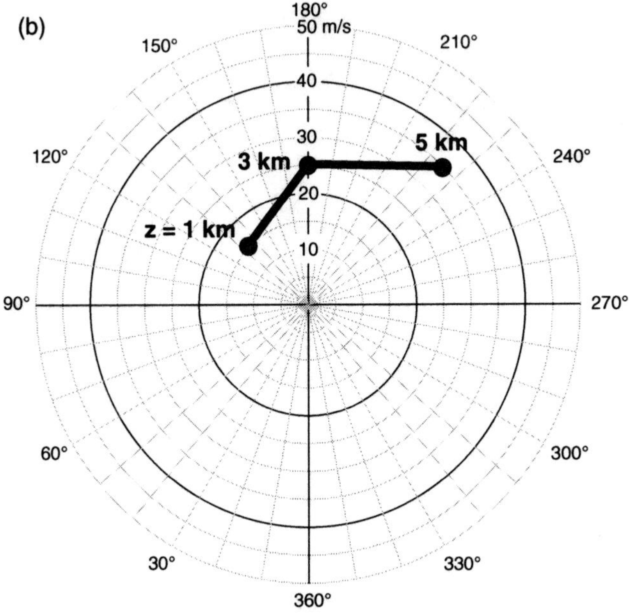

Figure 14.50
(a) Using the projected vectors from the previous figure, draw dots at the end of each vector, and connect with straight lines. When that line and dots are plotted on polar graph paper (b), the result is a hodograph.

If you project those wind vectors onto a single plane, you will get a graph that looks like the bottom of Fig. 14.49. Next, draw a dot at the end of each projected vector (Fig. 14.50a), and connect the dots with straight lines, going sequentially from the lowest to the highest altitude. Label the altitudes above ground level (AGL) next to each dot.

Plot the resulting dots and connecting lines on polar graph paper, but omit the vectors (because they are implied by the positions of the dots). The result is called a **hodograph** (Fig. 14.50b). Fig. 14.51 is a large blank hodograph that you can copy and use for plotting your own wind profiles. Coordinates in this graph are given by radial lines for wind direction, and circles are wind speed.

Notice that the compass angles labeled on the blank hodograph appear to be backwards. This is not a mistake. The reason is that winds are specified

Sample Application

Use the table of winds below to plot a hodograph.

z (km)	dir. (°)	M (m s^{-1})	z (km)	dir. (°)	M (m s^{-1})
0	0	0	4	190	15
1	130	5	5	220	20
2	150	10	6	240	30
3	170	13			

Find the Answer

Plot these points on a copy of Fig. 14.51, and connect with straight lines. Label each point with its altitude.

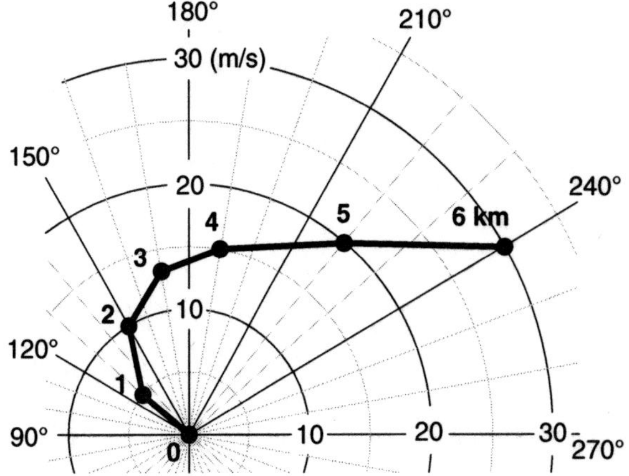

Check: The wind at 6 km, for example, is from the west southwest (240°), implying that the tip of the wind vector would be pointing toward the east northeast. This agrees with the location of the 6 km dot.

Exposition: This profile shows both speed and directional shear. Also, these winds **veer** (turn clockwise) with increasing altitude. From the thermal wind Exposition in Chapter 11, recall that veering winds are associated with warm-air advection, which is useful for bringing warmer air into a thunderstorm.

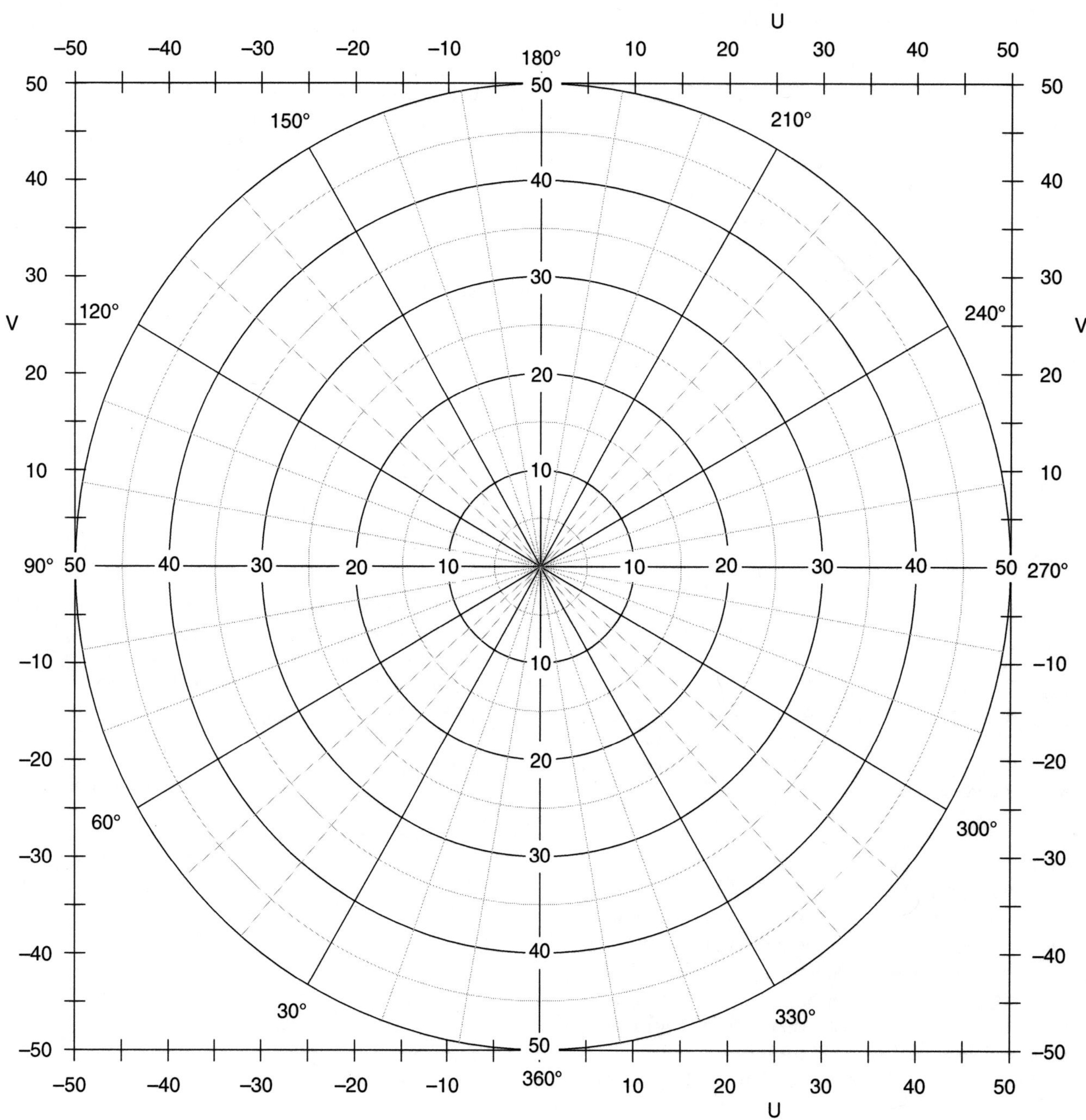

Figure 14.51
Blank hodograph for you to copy and use. Compass angles are direction winds are from. Speed-circle labels can be changed for different units or larger values, if needed.

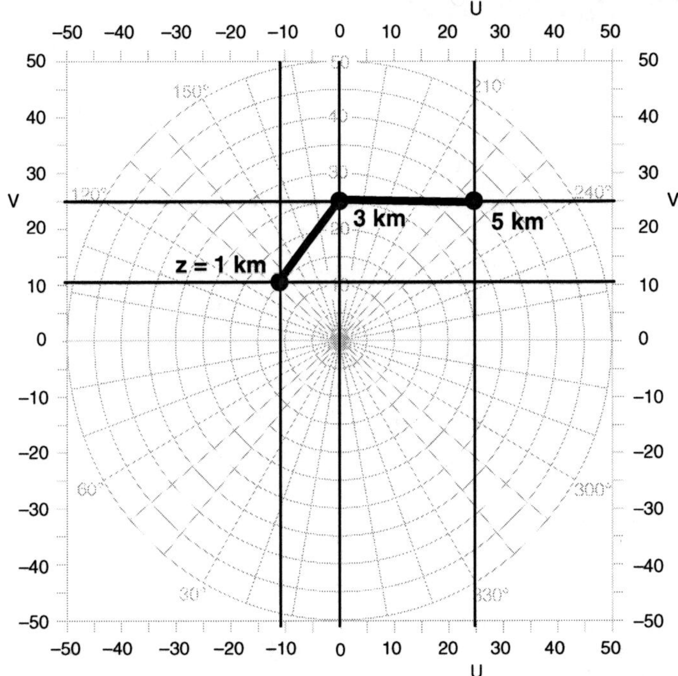

Figure 14.52
Example of determination of U and V Cartesian wind components from a hodograph. Legend: speeds are in (m s^{-1}).

by where they come <u>from</u>. For example, the wind at 1 km altitude is from the southeast; namely, from a compass direction of 135° (Fig. 14.50). The south wind at 3 km is from 180°, and the southwest wind at 5 km is from 235°. Using the "from" direction and the wind speed to specify a point on the hodograph, the result is a hodograph that implies wind vectors pointing from the origin to the correct directions.

Often you need to use the U and V wind components to determine thunderstorm characteristics. You could calculate these using eqs. (1.3) and (1.4) from Chapter 1. Alternately, you can pick them off from the coordinates of the hodograph. As demonstrated in Fig. 14.52, for any hodograph point, use a straight-edge to draw a horizontal line from that point to the ordinate to determine V, and draw a vertical line from the point to the abscissa to get U.

Sample Application

For the hodograph of Fig. 14.52, find the (U, V) Cartesian coordinate for each of the 3 data points.

Find the Answer

Given: Hodograph Fig. 14.52 with data points at z = 1, 3, and 5 km.
Find: (U, V) = ? (m s^{-1}) for each point.

Hint: The hodograph has the U velocities labeled at both top and bottom to make it easier to draw perfectly vertical lines thru any data point. For the same reason, V velocities are labeled at both left and right.

Method: Draw horizontal and vertical lines through each data point, and then pick off the U and V values by eye. Results:

For z = 1 km: <u>U = –11</u> m s^{-1}, <u>V = 11</u> m s^{-1}
For z = 3 km: <u>U = 0</u> m s^{-1}, <u>V = 25</u> m s^{-1}
For z = 5 km: <u>U = 25</u> m s^{-1}, <u>V = 25</u> m s^{-1}

Check: Signs and magnitudes are consistent with directions and speeds.
Exposition: Be careful to use the proper units. The legend in the Fig. 14.52 caption tells us the units for wind speed.

CAUTION: Different agencies use different units (m s^{-1}, knots, miles·h^{-1}, km·h^{-1}) and sometimes forget to state the units in the hodograph legend or caption.

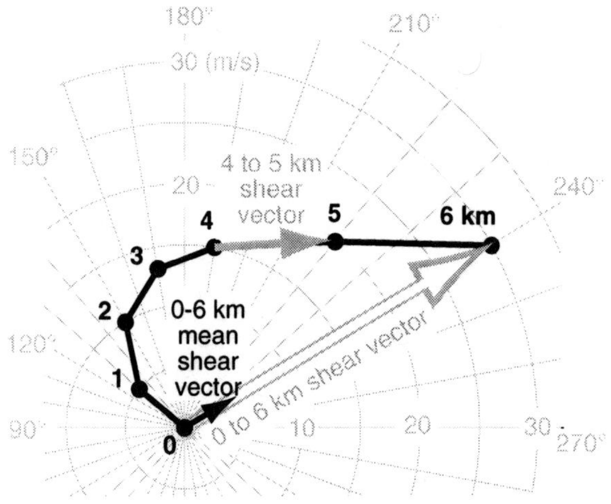

Figure 14.53
Example of mean shear vector between 0 to 6 km, shown by the small black arrow. There are six 1 km layers in this example. Vector wind differences are used as surrogates for shear.

Sample Application
Using data from the previous Sample Application, mathematically get the mean shear vector coordinates $(\Delta U, \Delta V)_{ms}$, magnitude ($\Delta M_{ms}$), and direction ($\alpha_{ms}$).

Find the Answer
Given: Wind vectors at 6 and 0 km altitude:
$M = 30$ m s^{-1} and $\alpha = 240°$ at $z = 6$ km,
$M = 0$ m s^{-1} and $\alpha = 0°$ at $z = 0$ km.
Let: subscript "*ms*" = "mean shear"
Find: $(\Delta U, \Delta V)_{ms}$ = ? (m s^{-1}),
ΔM_{ms} = ? (m s^{-1}), α_{ms} = ? (°)

First, find (U, V) components from eqs. (1.3) & (1.4):
$U_{6km} = -M \cdot \sin(\alpha) = -(30\text{m s}^{-1}) \cdot \sin(240°) = 26$ m s^{-1}
$V_{6km} = -M \cdot \cos(\alpha) = -(30\text{m s}^{-1}) \cdot \cos(240°) = 15$ m s^{-1}
and
$U_{0km} = -M \cdot \sin(\alpha) = -(0\text{m s}^{-1}) \cdot \sin(0°) = 0$ m s^{-1}
$V_{0km} = -M \cdot \cos(\alpha) = -(0\text{m s}^{-1}) \cdot \cos(0°) = 0$ m s^{-1}

Next, use eqs. (14.13 & 14.14):
$\Delta U_{ms} = (26 - 0 \text{ m s}^{-1})/6 = \underline{4.3}$ m s^{-1}
$\Delta V_{ms} = (15 - 0 \text{ m s}^{-1})/6 = \underline{2.5}$ m s^{-1}

Then, use $(\Delta U, \Delta V)$ in eq. (14.15):
$\Delta M_{ms} = [(\Delta U_{ms})^2 + (\Delta V_{ms})^2]^{1/2}$
$= [(4.3 \text{ m s}^{-1})^2 + (2.5 \text{ m s}^{-1})^2]^{1/2}$
$= [24.74 \text{ (m s}^{-1})^2]^{1/2} = \underline{5}$ m s^{-1}

Finally, use $(\Delta U, \Delta V)$ in eq. (14.16):
$\alpha_{ms} = 90° - \arctan(\Delta V_{ms}/\Delta U_{ms}) + 180°$
$= 90° - \arctan(2.5/4.3) + 180°$
$= 90° - 30° + 180° = \underline{\mathbf{240°}}$

Check: Units OK. Agrees with Fig. 14.53 black arrow.
Exposition: The only reason the mean shear direction equaled the wind direction at 6 km was because the surface wind was zero. Normally they differ.

Using Hodographs

Wind and wind-shear between the surface and 6 km altitude affects the dynamics, evolution, and motion of many N. American thunderstorms. You can use a hodograph to determine key wind-related quantities, including the:
- local shear across a single layer of air
- mean wind-shear vector
- total shear magnitude
- mean wind vector (normal storm motion)
- right & left moving supercell motions
- storm-relative winds.

Shear Across a Single Layer

Local shear between winds at adjacent wind-reporting altitudes is easy to find using a hodograph. Comparing Figs. 14.46 with 14.50, you can see that the hodograph line segment between any two adjacent altitudes is equal to the local shear across that one layer. For example, the solid grey arrow in Fig. 14.53 (one of the line segments of the hodograph) shows the local shear across the 4 to 5 km layer.

Mean Wind Shear Vector

The **mean wind-shear vector** across multiple layers of air of equal thickness is the vector sum of the local shear vectors (i.e., all the hodograph line segments for those layers), divided by the number of layers spanned. Namely, it is a vector drawn from the bottom-altitude wind point to the top-altitude wind point (white arrow with grey border in Fig. 14.53), then divided by the number of layers to give the mean (solid black arrow). For Fig. 14.53, the mean 0 - 6 km shear vector is 5 m s^{-1} from the west southwest. This graphical method is easy to use.

To find the same mean wind-shear vector mathematically, use the U and V components for the wind at $z = 0$, and also for the wind at $z = 6$ km. The mean shear (subscript $_{ms}$) vector coordinates $(\Delta U, \Delta V)_{ms}$ for the 0 to 6 km layer is given by:

$$\Delta U_{ms} = (U_{6km} - U_{0km})/6 \qquad \bullet(14.13)$$

$$\Delta V_{ms} = (V_{6km} - V_{0km})/6 \qquad \bullet(14.14)$$

You can use these coordinates to plot the mean shear vector on a hodograph. Find the magnitude of this mean shear vector using eq. (14.15), and the direction using eq. (14.16). Namely:

$$\Delta M_{ms} = \left(\Delta U_{ms}^{\ 2} + \Delta V_{ms}^{\ 2}\right)^{1/2} \qquad (14.15)$$

$$\alpha_{ms} = 90° - \frac{360°}{C} \cdot \arctan(\Delta V_{ms} / \Delta U_{ms}) + \alpha_o \qquad (14.16)$$

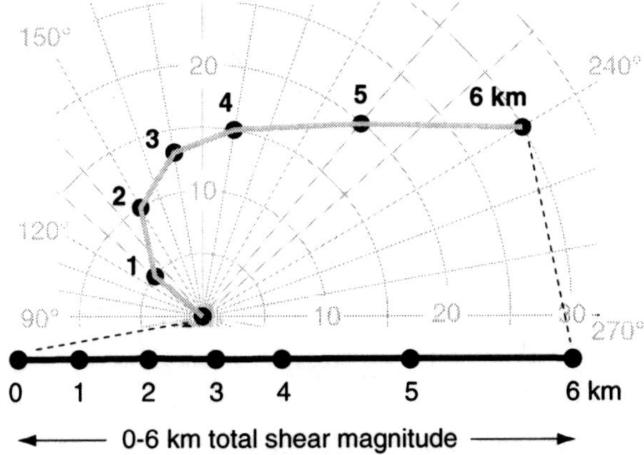

Figure 14.54
Example of total shear magnitude (black line), found as the sum of the shear magnitudes from the individual layers (grey line segments) between 0 to 6 km altitude. Vector wind differences are used as surrogates for shear.

where $C = 360°$ or 2π radians (depending on the units returned by your calculator or spreadsheet), and $\alpha_o = 180°$ if $\Delta U_{ms} > 0$, but is zero otherwise. As before, these wind-difference values are surrogates for the true shear.

Total Shear Magnitude

The **total shear magnitude** (TSM) across many layers, such as between 0 to 6 km, is the algebraic sum of the lengths of the individual line segments in the hodograph. For a simple graphic method that gives a good estimate of TSM, conceptually lay flat the grey line from Fig. 14.54 without stretching or shrinking any of the line segments. The length of the resulting straight line (black line in Fig. 14.54) indicates the total shear magnitude. For this example, the total shear magnitude is roughly 45 m s^{-1}, comparing the length of the black line to the wind-speed circles in the graph. Again, wind difference is used as a surrogate for shear.

To find the total shear magnitude mathematically, sum over all the individual layer shear magnitudes (assuming layers of equal thickness):

$$TSM = \sum_{i=1}^{n} \left[(U_i - U_{i-1})^2 + (V_i - V_{i-1})^2 \right]^{1/2} \quad \bullet(14.17)$$

where n is the total number of layers, and subscripts i and i–1 indicate top and bottom of the ith layer. For the Fig. 14.54 example, n = 6, and each layer is 1 km thick.

Although two different environments might have the same total wind shear (TSM), the distribution of that shear with height determines the types of storms possible. For example, Fig. 14.55 shows two hodographs with exactly the same total shear

Sample Application(§)

Mathematically calculate the total shear magnitude for the hodograph of Fig. 14.53 (i.e., using data table from the Sample Application 4 pages ago).

Find the Answer

Given: The data table from 4 pages ago, copied below.
Find: TSM = ? m s^{-1}

Use a spreadsheet to do these tedious calculations: First, use eqs. (1.3) and (1.4) to find U and V components of the winds. Next, find the wind differences $\Delta U = U_i - U_{i-1}$, and $\Delta V = V_i - V_{i-1}$. Finally, use these in eq. (14.17).

All speed and surrogate-shear units below are (m s^{-1}) except for direction α (degrees) and height z (km).

z	α	M	U	V	ΔU	ΔV	Shear
6	240	30	26.0	15.0	13.1	-0.3	13.1
5	220	20	12.9	15.3	10.3	0.5	10.3
4	190	15	2.6	14.8	4.9	2.0	5.2
3	170	13	-2.3	12.8	2.7	4.1	5.0
2	150	10	-5.0	8.7	-1.2	5.4	5.6
1	130	5	-3.8	3.2	-3.8	3.2	5.0
0	0	0	0.0	0.0	Sum=TSM=		44.2

Check: Units OK.
Exposition: The graphical estimate of 45 m s^{-1} by eye from Figure 14.54 was very close to the calculated value here.

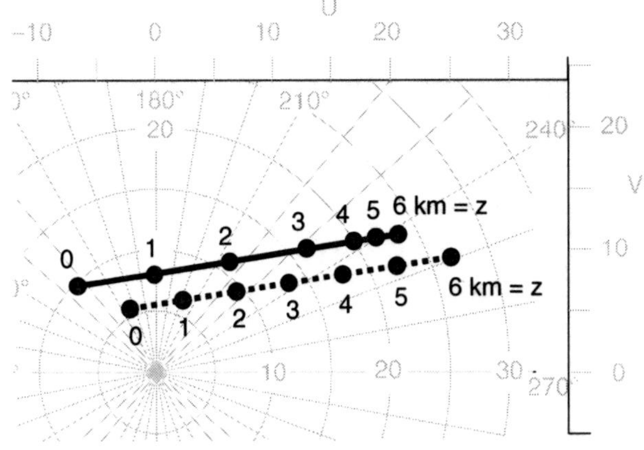

Figure 14.55
Two lines on this hodograph have exactly the same total length which means they have the same total shear magnitude. But that shear distributed differently with depth z for each line.

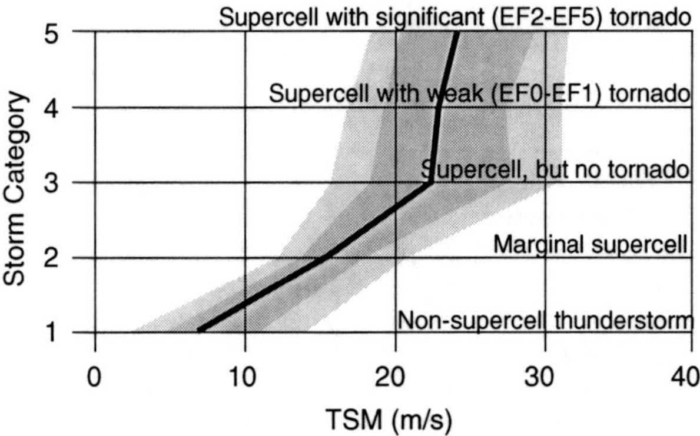

Figure 14.56
Observation frequency of thunderstorms of various intensities vs. environmental total shear magnitude (TSM) in the 0 to 6 km layer of atmosphere. Black line is the median (50 percentile) of several hundred observations in central N. America; dark grey shading spans 25 to 75 percentiles (the interquartile range); and light grey spans 10 to 90 percentiles.

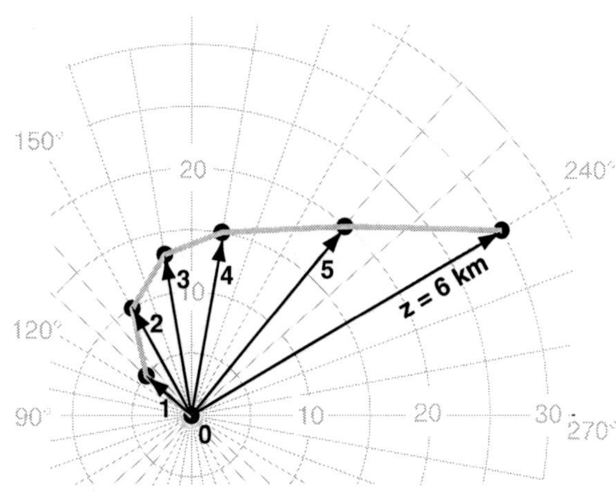

Figure 14.57
Wind vectors (black) associated with points on a hodograph.

Sample Application
How many fence posts are needed to support a straight fence with 6 sections?

Find the Answer: 7 (see figure)

Exposition: Only 6 posts are needed for a closed loop.

magnitude, but the solid line has most of that shear in the bottom 3 km (as seen by the spacing between height (z) points along the hodograph), favoring **squall lines** and **bow echoes**. The dashed line has evenly distributed shear through the bottom 6 km, favoring **supercells**. Fig. 14.56 verifies that greater TSM values support supercells.

Mean Environmental Wind (Normal Storm Motion)

Airmass thunderstorms often translate in the direction and speed of the **mean environmental wind** vector in the bottom 6 km of the atmosphere. These are sometimes called **steering-level winds** or **normal winds**. To find the 0 to 6 km **mean wind** (i.e., NOT the shear) first picture the wind vectors associated with each point on the hodograph (Fig. 14.57). Next, vector sum all these winds by moving them tail to head, as shown in Fig. 14.58).

Finally, divide the vector sum by the number of wind points to find the **mean wind vector**, as indicated with the "X". [CAUTION: As you can see in Fig. 14.58, the 6 layers of air are bounded by 7 wind-vector points, not forgetting the zero wind at the ground. Thus, the vector sum in Fig. 14.58 must be divided by 7, not by 6.] For this example, the mean wind (i.e., the location of the "X") is from about 203° at 11 m s^{-1}. An arrow from the origin to the "X" gives the forecast **normal motion for thunderstorms**.

An easier way to approximate this mean wind vector without doing a vector addition is to estimate (by eye) the center of mass of the area enclosed by

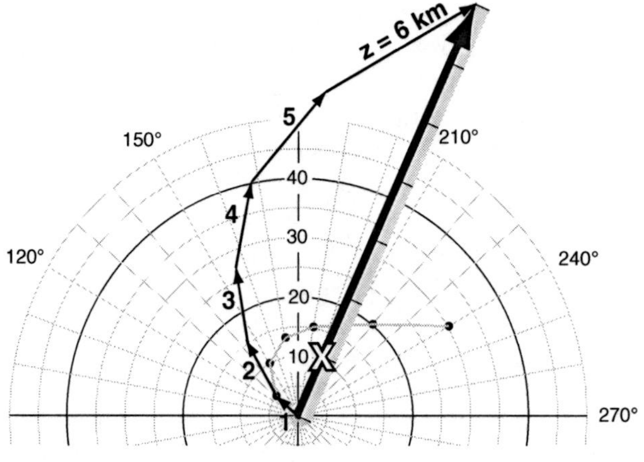

Figure 14.58
Individual wind vectors (thin black arrows), added as a vector sum (thick black arrow), divided by the number of wind vectors (7 in this example, not forgetting the zero wind at the ground), gives the mean wind (X) in the 0 to 6 km layer of air.

the original hodograph (shaded black in Fig. 14.59; NOT the area enclosed by the re-positioned vectors of Fig. 14.58). Namely, if you were to: (1) trace that area onto a piece of cardboard, (2) cut it out, & (3) balance it on your finger tip, then the balance point (shown by the "X" in Fig. 14.59) marks the center of mass. "X" indicates the normal motion of thunderstorms, and is easy to estimate graphically using this center-of-mass estimate.

To mathematically calculate the **mean wind** (indicated with overbars) in the bottom 6 km of the atmosphere, first convert the vectors from speed and direction to their Cartesian components. Then sum the U velocities at all equally-spaced heights, and separately sum all the V velocities, and divide those sums by the number of velocity heights to get the components $(\overline{U}, \overline{V})$ of the mean wind.

$$\overline{U} = \frac{1}{N}\sum_{j=0}^{N} U_j \qquad \bullet(14.18)$$

$$\overline{V} = \frac{1}{N}\sum_{j=0}^{N} V_j \qquad \bullet(14.19)$$

where N is the number of wind levels (not the number of shear layers), and j is the altitude index. For winds every 1 km from $z = 0$ to $z = 6$ km, then $N = 7$. [CAUTION: Don't confuse the mean wind with the mean shear.] Then, use eqs. (1.1) and (1.2) [similar to eqs. (14.15 & 14.16), but for mean wind instead of mean shear] to convert from Cartesian coordinates to mean speed and direction $(\overline{M}, \overline{\alpha})$.

In summary the normal motion of thunderstorms is $(\overline{U}, \overline{V})$, or equivalently $(\overline{M}, \overline{\alpha})$. This corresponds to the center-of-mass "X" estimate of Fig. 14.59.

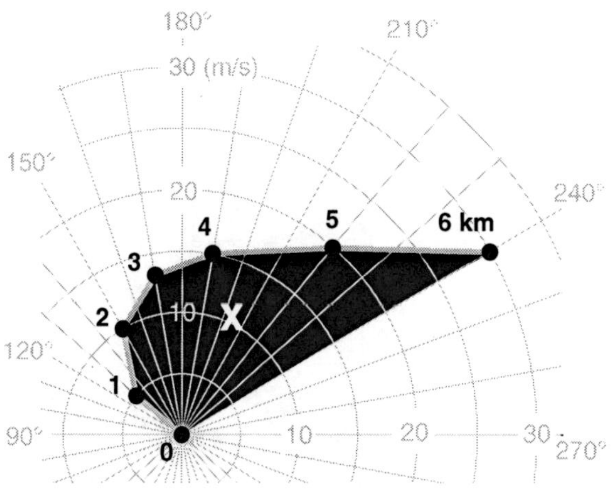

Figure 14.59
Mean wind ("X") in the 0 to 6 km layer of air, estimated as a center of mass of the area (black) enclosed by the hodograph. "X" indicates the normal motion of thunderstorms.

Sample Application

Mathematically calculate the mean wind components, magnitude, and direction for the hodograph of Fig. 14.57 (i.e., using data table from the Sample Application about 5 pages ago).

Find the Answer:
Given: The data table from 5 pages ago, copied below.
Find: $(\overline{U}, \overline{V}) = ?$ (m s^{-1}), $(\overline{M}, \overline{\alpha}) = ?$ (m s^{-1}, °)

Use a spreadsheet to do these tedious calculations: First, use eqs. (1.3) and (1.4) to find U and V components of the winds. Next, average the U's, and then the V's, using eqs. (14.18 & 14.19). Finally, use eqs. (1.1) and (1.2) to covert to speed and direction.

All velocities below are (m s^{-1}), direction is in degrees, and height z in km.

z	direction	M	U	V
6	240	30	26.0	15.0
5	220	20	12.9	15.3
4	190	15	2.6	14.8
3	170	13	-2.3	12.8
2	150	10	-5.0	8.7
1	130	5	-3.8	3.2
0	0	0	0.0	0.0
Mean wind components (ms^{-1})=			**4.3**	**10.0**
Mean wind magnitude (m s^{-1}) =			**10.9**	
Mean wind direction (°) =			**203.5**	

Check: Units OK. Magnitudes and signs reasonable.
Exposition: The graphical estimate of mean wind speed 11 m s^{-1} and direction 203° by eye from Figs. 14.58 and 14.59 were very close to the calculated values here. Namely, expect normal thunderstorms to move from the south southwest at about 10.9 m s^{-1} in this environment. Thus, the center-of-mass estimate by eye saves a lot of time and gives a reasonable answer.

A SCIENTIFIC PERSPECTIVE • Be Safe
(part 2)

Charles Doswell's web page recommends storm-chase guidelines, which I paraphrase here:

The #1 Threat: Being on the Highways
1. Avoid chasing alone.
2. Be very alert to standing water on the roads.
3. Avoid chasing in cities if at all possible.
4. Don't speed.
5. Pull fully off the road when you park.
6. Use your turn signals.
7. Slow down in poor visibility (rain; blowing dust).
8. Plan where to get fuel; don't let your tank get low.
9. Avoid unpaved roads (very slippery when wet).
10. Make your vehicle visible to other vehicles.

(continues in the next chapter).

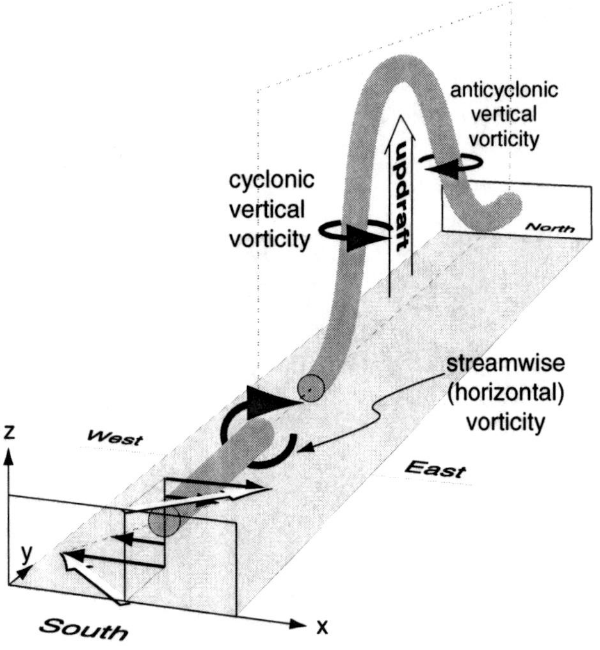

Figure 14.60
Counter-rotating mesocyclones formed when a convective updraft tilts the horizontal vorticity associated with environmental wind shear. Near-surface environmental winds from the southeast, and winds at 6 km altitude from the southwest, are white arrows. Dark cylinders represent vorticity axis.

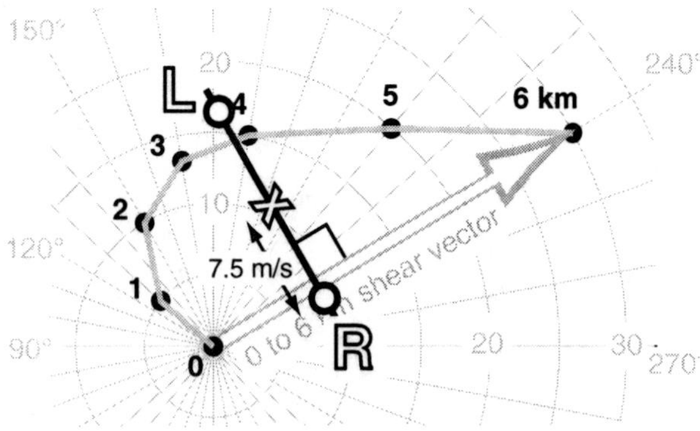

Figure 14.61
Bunker's internal dynamics technique for finding the movement of right-moving (R) and left-moving (L) supercell thunderstorms.

Supercell Storm Motion

Supercell thunderstorms have rotating updrafts, and to get that rotation they need sufficient environmental wind shear. The environmental total shear magnitude typically must be TSM ≥ 25 m s^{-1} and be distributed relatively evenly across the lower troposphere (from the surface to 6 km) to support supercells. Total shear magnitudes less than 15 m s^{-1} often are too small for supercells. In between those two shears, thunderstorms may or may not evolve into supercells.

One reason that strong shear promotes supercells is that horizontal vorticity associated with vertical shear of the horizontal wind in the environment can be tilted into vertical vorticity by the strong convective updraft of a thunderstorm (Fig. 14.60). This causes counter-rotating vortices (mesocyclones) on the left and right sides of the updraft. These two mesocyclones are deep (fill the troposphere) and have diameters roughly equal to the tropospheric depth. One is cyclonic; the other is anticyclonic.

Sometimes supercell thunderstorms split into two separate storms: **right-moving** and **left-moving supercells**. Namely, the cyclonic and anticyclonic mesocyclones support their own supercells, and move right and left of the "normal" motion as would have been expected from the mean steering-level wind.

Bunker's **Internal Dynamics (ID)** method to forecast the movement of these right and left-moving supercells is:

(1) Find the 0 to 6 km mean wind (see "X" in Fig. 14.61), using methods already described. This would give normal motion for thunderstorms. But supercells are not normal.

(2) Find the 0.25 to 5.75 km layer shear vector (approximated by the 0 to 6 km layer shear, shown with the white arrow in Fig. 14.61).

(3) Through the mean wind point (X), draw a line (black line in Fig. 14.61) perpendicular to the 0 to 6 km shear vector.

(4) On the line, mark 2 points (white in black circles): one 7.5 m s^{-1} to the right and the other 7.5 m s^{-1} to the left of the "X".

(5) The point that is right (clockwise) from the mean wind "X" gives the average motion forecast for the right (R) moving storm.

(6) The point that is left (counter-clockwise) from the mean wind "X" gives the average forecast motion of the left (L) moving storm.

Not all storms follow this simple rule, but most supercell motions are within a 5 m s^{-1} radius of error around the R and L points indicated here. For the example in Fig. 14.61, motion of the right-moving supercell is 8 m s^{-1} from 246°, and the left mover is 17 m s^{-1} from 182°.

Only for straight hodographs (Fig. 14.62a) are the right and left-moving supercells expected to exist simultaneously with roughly equal strength (Fig. 14.63a). Hodographs that curve clockwise (Fig. 14.62b) with increasing height favor right-moving supercell thunderstorms (Fig. 14.63b). For this situation, the left-moving storm often dissipates, leaving only a right-moving supercell.

Hodographs that curve counterclockwise (Fig. 14.62c) favor left-moving supercells (Fig. 14.63c), with the right-moving storm often dissipating. In the plains/prairies of North America, most hodographs curve clockwise (as in Figs. 14.61 and 14.62b), resulting in right-moving supercell thunderstorms that are about ten times more abundant than left-moving storms. The opposite is true in the S. Hemisphere for midlatitude thunderstorms.

In summary, supercell thunderstorms often do **not** move in the same direction as the mean environmental (steering-level) wind, and do not move at the mean wind speed. The methods that were discussed here are useful to forecast movement of storms that haven't formed yet. However, once storms form, you can more accurately estimate their motion by tracking them on radar or satellite. Regardless of the amounts of shear, CAPE, and moisture in the pre-storm environment, thunderstorms won't form unless there is a trigger mechanism. Triggers are discussed in the next main section (after the Bulk Richardson Number subsection).

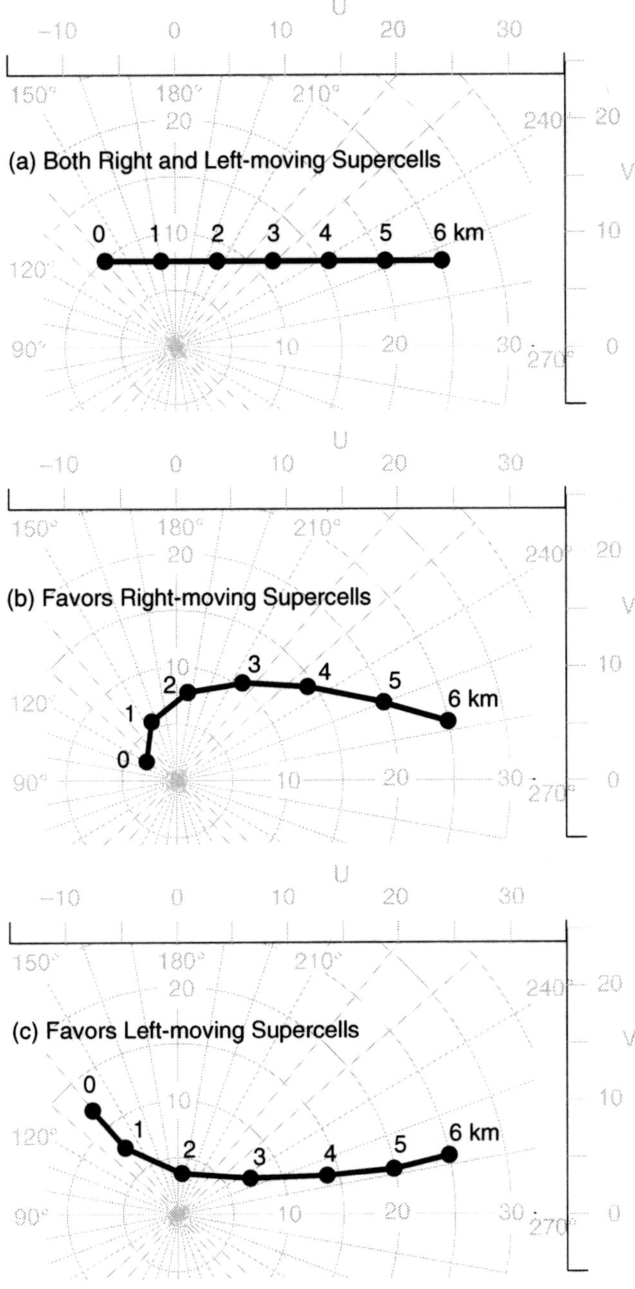

Figure 14.62
Shapes of hodograph that favor right and left-moving supercell thunderstorms: (a) straight hodograph; (b) hodograph showing clockwise curvature with increasing height; (c) hodograph showing counter-clockwise curvature with increasing height.

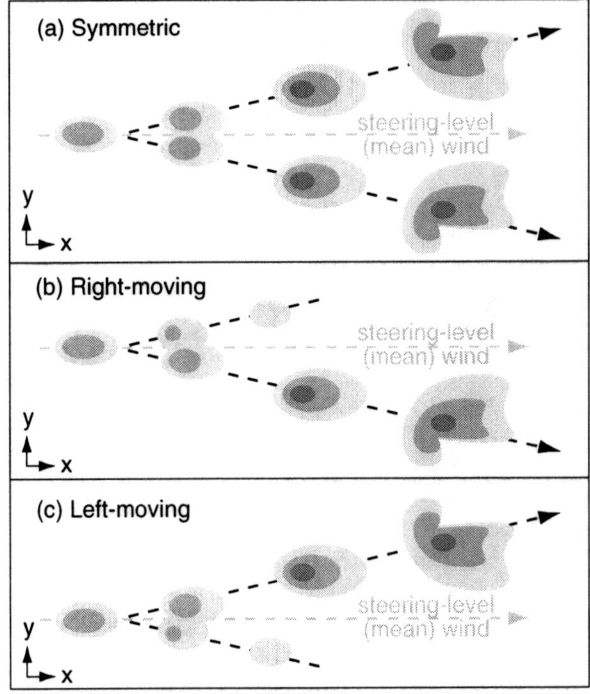

Figure 14.63
Sketch of sequence of radar reflectivity echoes over time (left to right) for: (a) symmetric supercell split; (b) dominant right-moving supercells; and (c) dominant left-moving supercells. Note the different orientation of the hook echo in the left-moving supercell. The steering-level wind is the "normal" wind.

Sample Application

Given the wind profile in the table below:

z (km)	dir. (°)	speed (ms^{-1})	z (km)	dir. (°)	speed ($m\ s^{-1}$)
0	150	15	4	270	15
1	160	8	5	270	22
2	210	5	6	270	30
3	260	8			

(a) Plot the hodograph;
(b) Graphically find the mean wind-shear vector for the 0 to 6 km layer;
(c) Graphically find the total shear magnitude.
(d) Does this hodograph favor bow echoes or supercells? Why?
(e) Graphically find the 0 to 6 km mean wind.
(f) Graphically find the motions for the right and left-moving supercell thunderstorms.
(g) Which is favored: right or left-moving supercell thunderstorms? Why?

Find the Answer

(a) Plot the hodograph (see curved black line below):

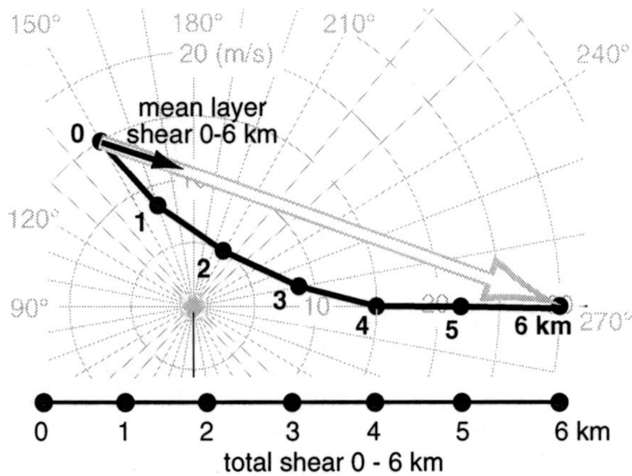

(b) The 0 to 6 km shear vector is shown with the white arrow. Dividing by 6 gives a mean layer shear (black arrow) of about **7 m s^{-1} from the west northwest**.

(c) The total shear magnitude, shown at the bottom of the hodograph above, is about **42 m s^{-1}**.

(d) **Supercells** are favored, because shear is evenly distributed throughout the bottom 6 km of atmosphere. Also, the total shear magnitude is greater than 25 m s^{-1}, as required for supercells.

(e) The answer is a vector, drawn from the origin to the "X" (see hodograph in next column). Namely, the mean wind is from about **250° at 10 m s^{-1}**. Vector method roughly agrees with center-of-mass method, done by eye.

(Continues in next column.)

Sample Application *(continuation)*

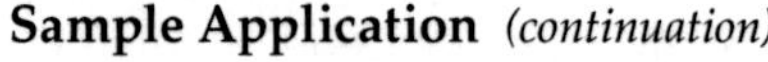
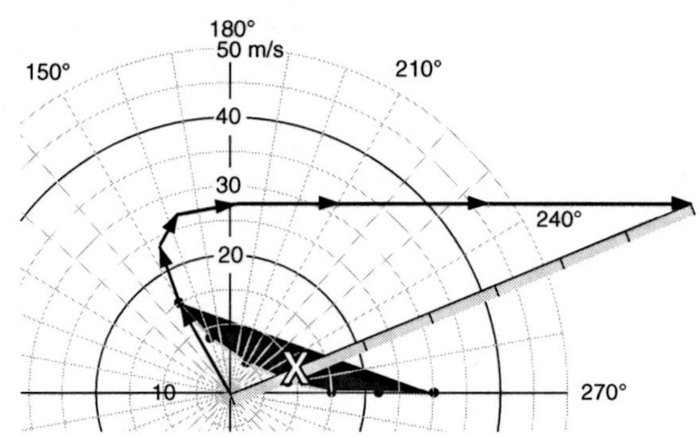

Thus, for normal thunderstorms (and for the first supercell, before it splits into right and left-moving supercells) forecast them to move from the west southwest at about 10 m s^{-1}. For both this example and Fig. 14.59, the center-of-mass method gives a better estimate if we look at the hodograph area between z = 0.5 and 5.5 km.

(f) Bunker's Internal Dynamics method is shown graphically on the hodograph below. The left supercell (L) is forecast to move at **16** m s^{-1} **from 229°**, and the right supercell (R) at **8** m s^{-1} **from 298°**.

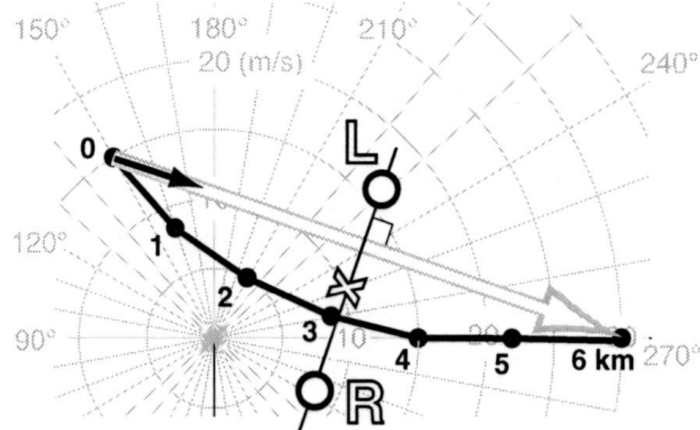

(g) **Left-moving supercells** are favored, because the hodograph curves counterclockwise with increasing height. This storm moves faster than the mean wind (i.e., faster than the "normal" thunderstorm steering-level wind).

Exposition: This hodograph shows **veering winds** (winds turning clockwise with increasing altitude), implying warm-air advection according to the thermal-wind relationship. Even though the wind vectors veer clockwise, the hodograph curvature is counterclockwise (implying dominant left-moving supercell storms), so be careful not to confuse these two characteristics.

Winds alone are not sufficient to forecast whether thunderstorms will occur. Other key ingredients are instability and abundant moisture, as previously discussed. Also needed is sufficient lifting by a trigger mechanism to overcome the capping inversion.

Bulk Richardson Number

Thunderstorm type depends both on the amounts of instability and wind shear in the pre-storm environmental sounding. The **bulk Richardson number (BRN)** is the ratio of nonlocal instability (i.e., the CAPE) in the mid to upper part of the troposphere to shear (ΔM) in the lower half of the troposphere:

$$BRN = \frac{CAPE}{0.5 \cdot (\Delta M)^2} \qquad \bullet(14.20)$$

Table 14-2 summarizes the utility of the BRN. The BRN is dimensionless, because the units of CAPE ($J \cdot kg^{-1}$) are the same as ($m^2 \cdot s^{-2}$) (see Appendix A).

In the denominator, the shear magnitude (ΔM) is given by:

$$\Delta M = \left[(\Delta U)^2 + (\Delta V)^2 \right]^{1/2} \qquad \bullet(14.21)$$

where

$$\Delta U = \overline{U} - U_{SL} \qquad (14.22a)$$

$$\Delta V = \overline{V} - V_{SL} \qquad (14.22b)$$

Eqs. (14.22) are the shear components between the mean wind ($\overline{U}, \overline{V}$) in the 0 to 6 km layer, and the surface-layer winds (U_{SL}, V_{SL}) estimated as an average over the bottom 0.5 km of the atmosphere. Fig. 14.64 shows how to estimate this shear magnitude from a hodograph.

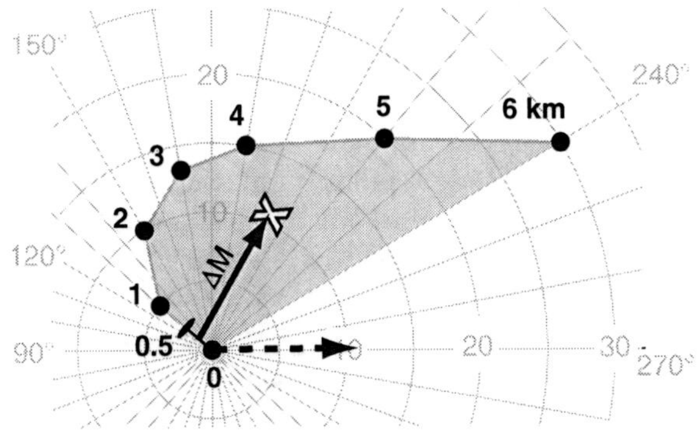

Figure 14.64
Solid black arrow shows shear ΔM between the average winds in the lowest 0.5 km of air, and the mean wind (X) in the lowest 6 km. Dashed arrow is just the solid arrow moved (without stretching) to the origin, so that the speed circles indicate the shear magnitude. $\Delta M = 11$ m s^{-1} in this example. ΔM is important in the bulk Richardson number (BRN), and BRN shear.

Table 14-2. Thunderstorm type determination using the bulk Richardson number (BRN)

BRN	Thunderstorm Type
< 10	(unlikely to have severe thunderstorms)
10 - 45	Supercells
45 - 50	Supercells and/or multicells
> 50	Multicells

Sample Application

For the previous Sample Application, assume the associated CAPE = 3000 J·kg^{-1}. (a) Find the BRN shear. (b) Find the BRN. (c) Are tornadic supercells likely?

Find the Answer

Given: CAPE = 3000 J·kg^{-1}
hodograph = previous Sample Application
Find: BRN shear = ? m^2·s^{-2}
BRN = ? (dimensionless)
Yes/no: thunderstorms? supercells? tornadoes?

(a) On the hodograph from the previous Sample Application, put the tail of a vector (solid black arrow) a quarter of the way from the $z = 0$ point to the $z = 1$ point on the hodograph). Put the arrowhead on the "X" mean wind. Then measure the vector's length (see dashed arrow), which is about $\Delta M = 18$ m s^{-1}.

Then use eq. (14.23):
BRN Shear = 0.5·(18 m s^{-1})^2 = <u>162</u> m^2 s^{-2}

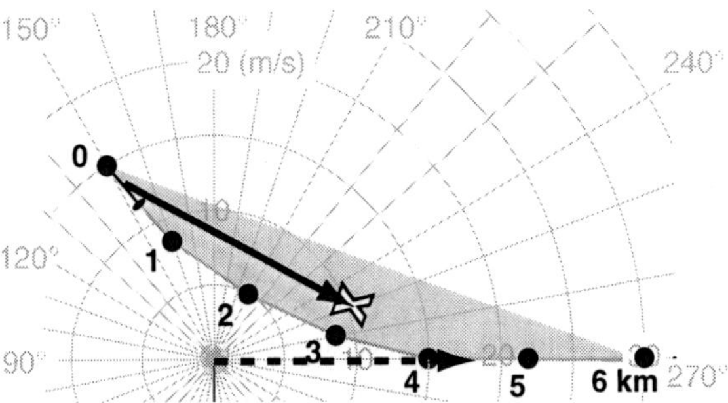

(b) Use eq. (14.20):
BRN = (3000 m^2 s^{-2}) / (162 m^2 s^{-2}) = **18.5**

(c) Assume CAPE ≈ MLCAPE. Then from Table 14-1:
<u>Severe thunderstorms likely, possible tornado.</u>
Next, use Table 14-2: Thunderstorm type = <u>supercell</u>.
Finally, use Table 14-3: <u>Tornadoes are likely</u>.

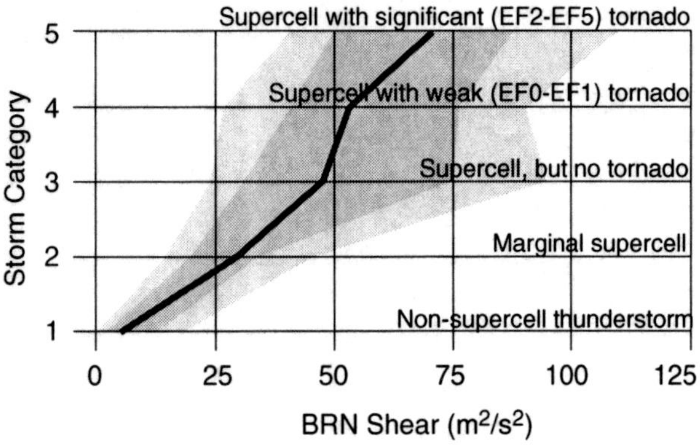

Figure 14.65
Statistics of bulk Richardson number shear (BRN shear) as a discriminator for thunderstorm and tornado intensity, based on several hundred storms in central N. America. Light grey spans 10 to 90 percentiles of the observations; dark grey spans the interquartile range of 25 to 75 percentiles; and the black line is the median (50 percentile).

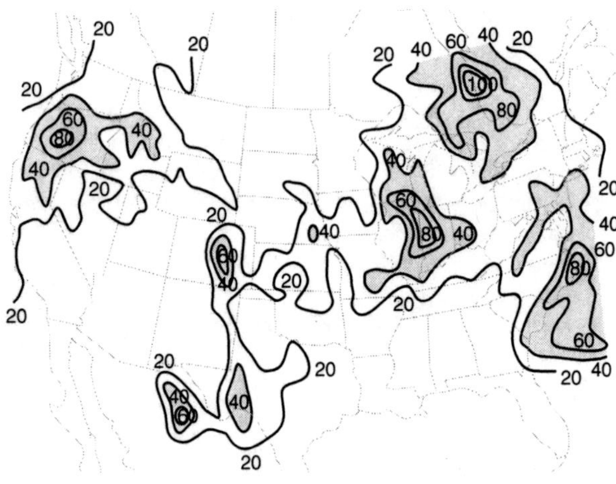

Figure 14.66
Weather map similar to Fig. 14.38, but for bulk Richardson number shear (BRN shear) in $m^2 \cdot s^{-2}$. Unlike the smooth variations in CAPE across N. America, the BRN shear values have a more local structure and are highly variable.

Table 14-3. If supercell thunderstorms form, then the bulk-Richardson-number shear (BRN Shear) suggests whether they might have tornadoes.

BRN Shear ($m^2 \cdot s^{-2}$)	Tornadoes
25 - 50	possible, but less likely
50 - 100	more likely

The denominator of the BRN, known as the **bulk-Richardson-number shear (BRN Shear)**, is

$$\text{BRN Shear} = 0.5 \cdot (\Delta M)^2 \qquad \bullet(14.23)$$

which has units of ($m^2 \cdot s^{-2}$). BRN shear can help indicate which supercells might be tornadic (see Table 14-3). Although BRN shear is statistically sharper in its ability to discriminate between thunderstorms and tornadoes of different severity, there is still significant overlap in thunderstorm/tornado categories as was found by verification against several hundred thunderstorms in central N. America (Fig. 14.65).

A case study weather map of BRN shear is analyzed in Fig. 14.66. As before, you should NOT use BRN shear alone as an indicator of tornado likelihood, because additional conditions must also be satisfied in order to allow tornadic thunderstorms to form. Some of these other conditions are instability, and a trigger mechanism.

TRIGGERING VS. CONVECTIVE INHIBITION

The fourth environmental requirement for thunderstorm formation is a **trigger** mechanism to cause the initial **lifting** of the air parcels. Although the capping inversion is needed to allow the fuel supply to build up (a good thing for thunderstorms), this cap also inhibits thunderstorm formation (a bad thing). So the duty of the external trigger mechanism is to lift the reluctant air parcel from z_i to the LFC. Once triggered, thunderstorms develop their own circulations that continue to tap the boundary-layer air.

The amount of external forcing required to trigger a thunderstorm depends on the strength of the cap that opposes such triggering. One measure of cap strength is convective inhibition energy.

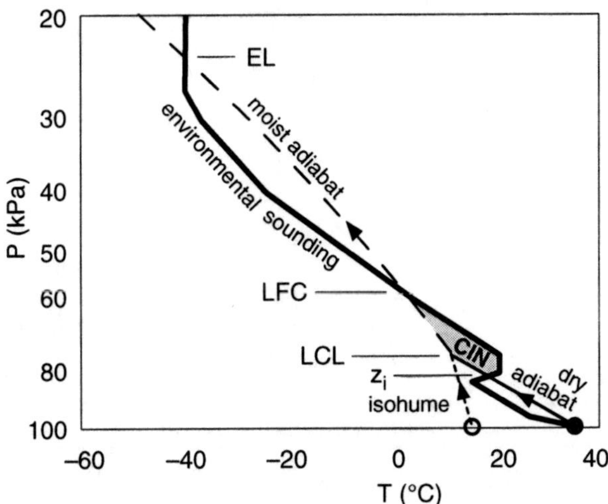

Figure 14.67 (at left)
Convective Inhibition (CIN) is proportional to the area (shaded) where the rising air parcel is colder than the environment for an early afternoon pre-storm sounding.

Convective Inhibition (CIN)

Between the mixed-layer top z_i and the level of free convection LFC, an air parcel lifted from the surface is colder than the environment. Therefore, it is negatively buoyant, and does not want to rise. If forced by a trigger mechanism to rise above z_i, then the trigger process must do work against the buoyant forces within this cap region.

The total amount of work needed is proportional to the area shaded in Fig. 14.67. This work per unit mass is called the **Convective Inhibition** (**CIN**). The equation for CIN is identical to the equation for CAPE, except for the limits of the sum. Another difference is that CIN values are negative. CIN has units of J·kg^{-1}.

For the afternoon sounding of Fig. 14.67, CIN corresponds to the area between z_i and LFC. For a morning sounding, a forecast of the afternoon's high temperature provides the starting point for the rising air parcel, where z_i is the corresponding forecast for mixed-layer top (Fig. 14.68). Thus, for both Figs. 14.67 and 14.68, you can use:

$$CIN = \sum_{z_i}^{LFC} \frac{|g|}{T_{ve}} (T_{vp} - T_{ve}) \cdot \Delta z \qquad \bullet(14.24)$$

where T_v is virtual temperature, subscripts p and e indicate the air parcel and the environment, $|g|$ = 9.8 m·s^{-2} is the magnitude of gravitational acceleration, and Δz is a height increment. Otherwise, CIN is usually found by summing between the surface (z = 0) and the LFC (Fig. 14.69).

$$CIN = \sum_{z=0}^{LFC} \frac{|g|}{T_{ve}} (T_{vp} - T_{ve}) \cdot \Delta z \qquad (14.25)$$

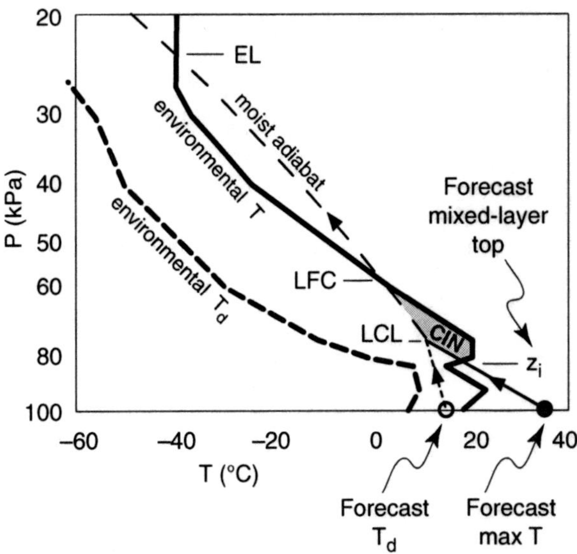

Figure 14.68
CIN based on max surface temperature forecasts and the early morning pre-storm sounding.

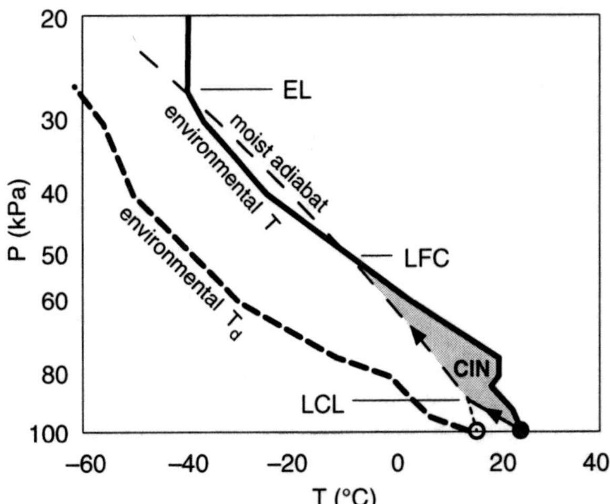

Figure 14.69
CIN between the surface and the LFC, for an early morning pre-storm sounding.

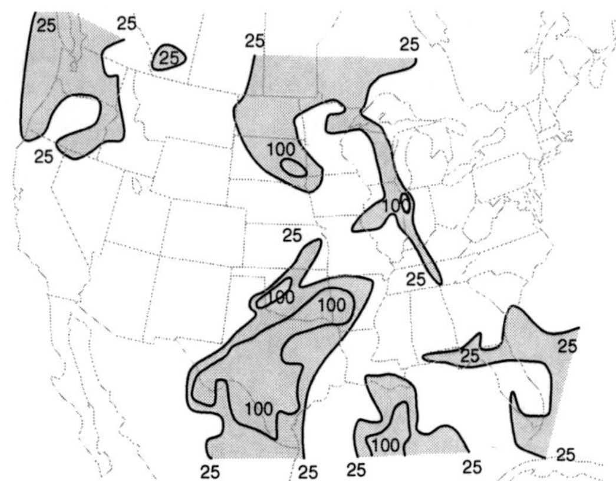

Figure 14.70
Weather map similar to Fig. 14.38 over USA, but for magnitude of mean-layer convective inhibition energy (ML CIN) in J·kg^{-1}.

Table 14-4. Convective Inhibition (CIN)

CIN Value (J·kg^{-1})	Interpretation
> 0	No cap. Allows weak convection.
0 to –20	Weak cap. Triggering easy. Air-mass thunderstorms possible.
–20 to –60	Moderate cap. Best conditions for CB. Enables fuel build-up in boundary layer. Allows most trigger mechanisms to initiate storm.
–60 to –100	Strong cap. Difficult to break. Need exceptionally strong trigger.
< –100	Intense cap. CB triggering unlikely.

Sample Application

Estimate the convective inhibition (CIN) value for the sounding of Fig. 14.25. Are thunderstorms likely?

Find the Answer

Given: Sounding of Fig. 14.25, as zoomed below.
Find: CIN = ? J·kg^{-1}.

Cover the CIN area with small tiles of size Δz = 500 m and ΔT = –2°C (where the negative sign indicates the rising parcel is colder than the environmental sounding). Each tile has area = $\Delta T \cdot \Delta z$ = –1,000 K·m. I count 12 tiles, so the total area is –12,000. Also, by eye the average environmental temperature in the CIN region is about T_e = 12°C = 285 K.

Use eq. (14.26):
CIN = [(9.8 m s^{-2})/(285 K)]·(–12,000 K·m) = **–413 J·kg^{-1}**.
From Table 14-4, **thunderstorms are unlikely**.

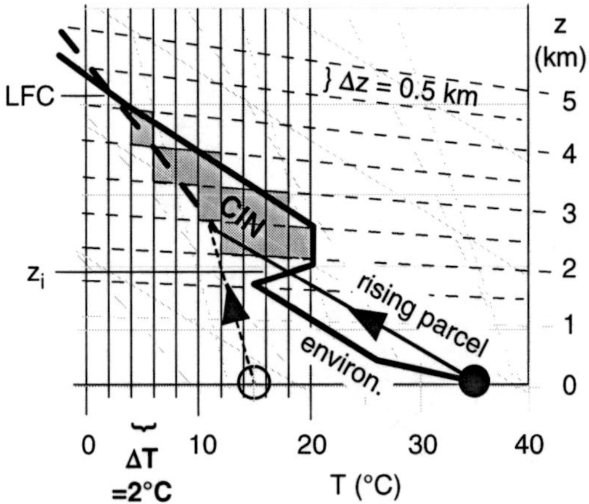

Check: By eye, the CIN area of Fig. 14.25 is about 1/5 the size of the CAPE area. Indeed, the answer above has about 1/5 the magnitude of CAPE (1976 J·kg^{-1}), as found in previous solved exercises. Units OK too.

Exposition: The negative CIN implies that work must be done ON the air to force it to rise through this region, as opposed to the positive CAPE values which imply that work is done BY the rising air.

As before, if the virtual temperature is not known or is difficult to estimate, then often storm forecasters will use T in place of T_v in eq. (14.24) to get a rough estimate of CIN:

$$CIN = \sum_{z_i}^{LFC} \frac{|g|}{T_e}(T_p - T_e)\cdot\Delta z \qquad \bullet(14.26)$$

although this can cause errors as large as 35 J·kg^{-1} in the CIN value. Similar approximations can be made for eq. (14.25). Also, a **mean-layer CIN (ML CIN)** can be used, where the starting air parcel is based on average conditions in the bottom 1 km of air.

CIN is what prevents, delays, or inhibits formation of thunderstorms. Larger magnitudes of CIN (as in Fig. 14.69) are less likely to be overcome by trigger mechanisms, and are more effective at preventing thunderstorm formation. CIN magnitudes (Table 14-4) smaller than about 60 J·kg^{-1} are usually small enough to allow deep convection to form if triggered, but large enough to trap heat and humidity in the boundary layer prior to triggering, to serve as the thunderstorm fuel. CIN is the **cap** that must be broken to enable thunderstorm growth.

Storms that form in the presence of large CIN are less likely to spawn tornadoes. However, large CIN can sometimes be circumvented if an elevated trigger mechanism forces air-parcel ascent starting well above the surface.

The area on a thermo diagram representing CIN is easily integrated by computer, and is often reported with the plotted sounding. **[CAUTION: often the magnitude of CIN is reported (i.e., as a positive value).]** CIN magnitudes from many sounding stations can be plotted on a weather map and analyzed, as in Fig. 14.70. To estimate CIN by hand, you can use the same tiling method as was done for CAPE (see the Sample Application at left).

Thunderstorms are more likely for smaller values of the difference $\Delta z_{cap} = z_{LFC} - z_{LCL}$, which is the nonlocally stable region at the bottom of the storm. Namely, a thinner cap on top of the ABL might allow thunderstorms to be more easily triggered.

Triggers

Any external process that forces the boundary-layer air parcels to rise through the statically stable cap can be a trigger. Some triggers are:

- Boundaries between airmasses:
    - cold, warm, or occluded fronts,
    - gust fronts from other thunderstorms;
    - sea-breeze fronts,
    - dry lines,
- Other triggers:
    - mountains,
    - small regions of high surface heating.
    - vertical oscillations called buoyancy waves,

If one airmass is denser than an adjacent airmass, then any convergence of air toward the dividing line between the airmasses (called an **airmass boundary**) will force the less-dense air to rise over the denser air. A cold front is a good example (Fig. 14.71), where cold, dense air advances under the less-dense warm air, causing the warm air to rise past z_i.

Surface-based **cold** and **warm fronts** are synoptically-forced examples of airmass boundaries. Upper-level synoptic fronts (with no signature on a surface weather map) can also cause lifting and trigger thunderstorms. Fig. 14.72 shows the frontal analysis for the 24 May 2006 case study that has been presented in many of the preceding sections. These frontal zones could trigger thunderstorms.

Sea-breeze (see the Regional Winds chapter) or **lake-breeze fronts** create a similar density discontinuity on the mesoscale at the boundary between the cool marine air and the warmer continental air during daytime, and can also trigger convection. If a cold **downburst** from a thunderstorm hits the ground and spreads out, the leading edge is a **gust front** of cooler, denser air that can trigger other thunderstorms (via the propagation mechanism already discussed). The **dry line** separating warm dry air from warm humid air is also an airmass boundary (Figs. 14.71 & 14.72), because dry air is more dense than humid air at the same temperature (see definition of virtual temperature in Chapter 1).

These airmass boundaries are so crucial for triggering thunderstorms that meteorologist devote much effort to identifying their presence. Lines of thunderstorms can form along them. Synoptic-scale fronts and dry lines are often evident from weather-map analyses and satellite images. Sea-breeze fronts and gust fronts can be seen by radar as convergence

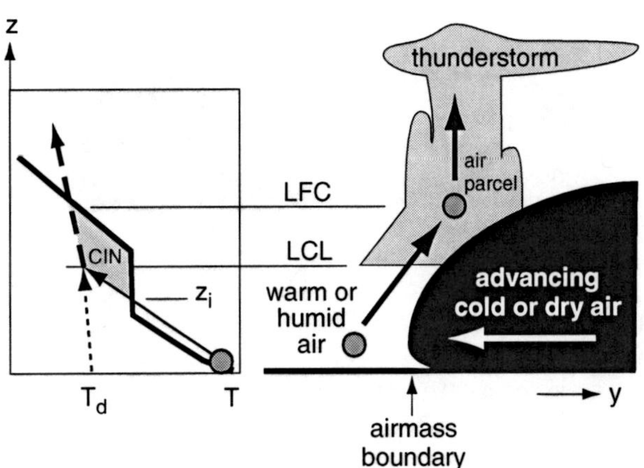

Figure 14.71
Warmer or moister (less-dense) air forced to rise over advancing colder or drier (more-dense) air can trigger thunderstorms if the air parcels forced to rise reach their LFC. The portion of thunderstorm clouds between the LCL and the LFC can have a smooth or laminar appearance. Above the LFC the clouds look much more convective, with rising turrets that look like cauliflower.

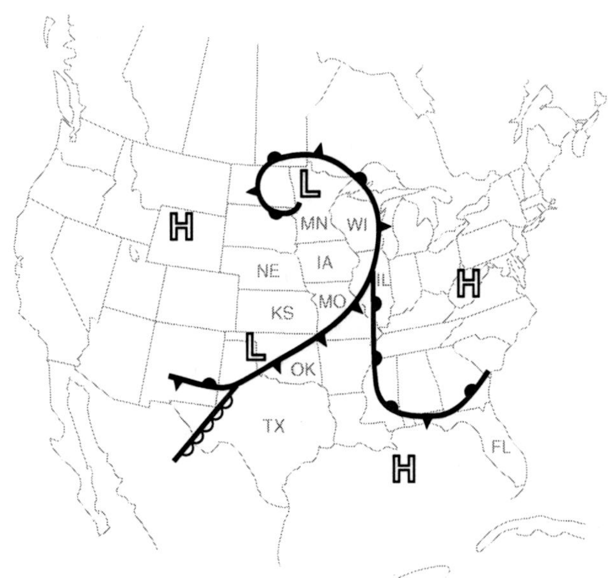

Figure 14.72
Synoptic frontal analysis over N. America at 22 UTC on 24 May 2006. An occluded front extends from a low (L) in Minnesota (MN) into Illinois (IL), where the warm and cold fronts meet. The cold front extends from IL through Oklahoma (OK) and Texas (TX), and merges with a dry line in SW Texas.

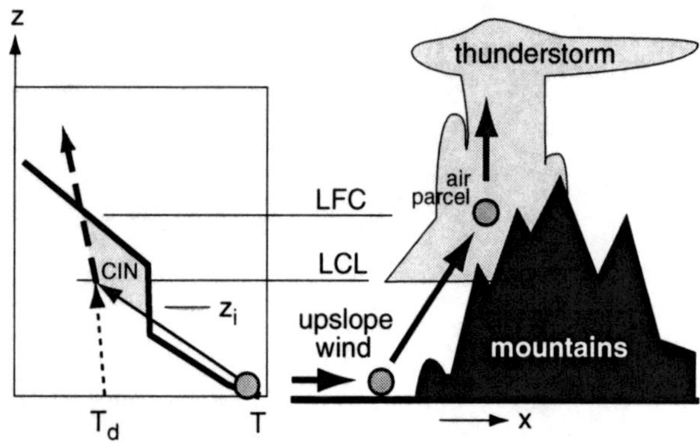

Figure 14.73
Wind hitting a mountain can be forced upslope, triggering thunderstorms called orographic thunderstorms.

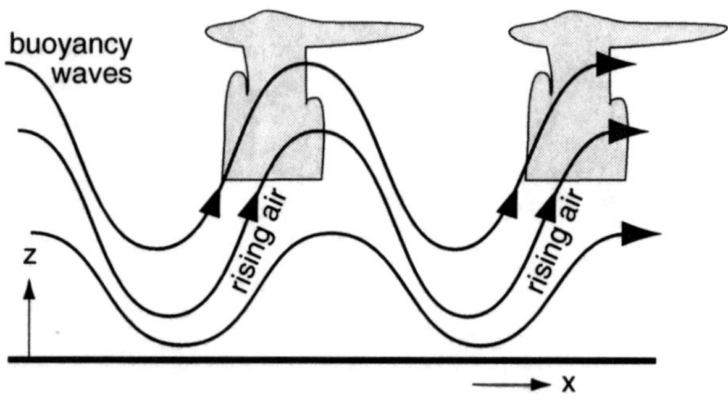

Figure 14.74
Thunderstorms generated in the updraft portions of buoyancy (gravity) waves.

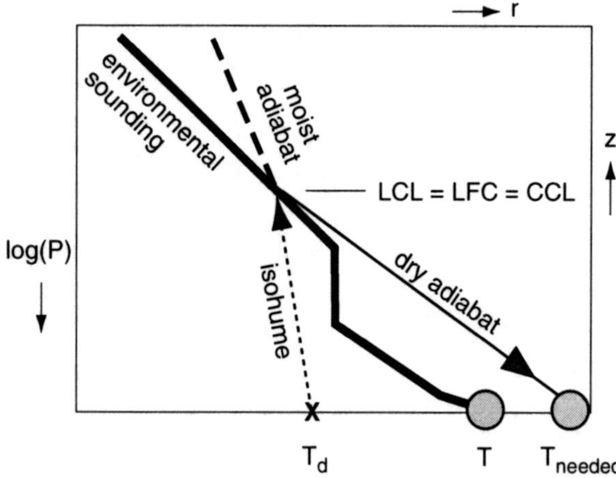

Figure 14.75
The convective temperature needed so that an air parcel has no CIN, and can reach its LFC under its own buoyancy. CCL = convective condensation level.

lines in the Doppler wind field, and as lines of weakly enhanced reflectivity due to high concentrations of insects (converging there with the wind) or a line of small cumulus clouds. When two or more airmass boundaries meet or cross, the chance of triggering thunderstorms is greater such as over Illinois and Texas in the case-study example of Fig. 14.72.

Horizontal winds hitting a mountain slope can be forced upward (Fig. 14.73), and can trigger **orographic thunderstorms**. Once triggered by the mountains, the storms can persist or propagate (trigger daughter storms via their gust fronts) as they are blown away from the mountains by the steering-level winds.

Sometimes when the air flows over mountains or is pushed out of the way by an advancing cold front, vertical oscillations called **buoyancy waves** or **gravity waves** (Fig. 14.74) can be generated in the air (see the Regional Winds chapter). If the lift in the updraft portion of a wave is sufficient to bring air parcels to their LFC, then thunderstorms can be triggered by these waves.

Another trigger is possible if the air near the ground is warmed sufficiently by its contact with the sun-heated Earth. The air may become warm enough to rise as a thermal through the cap and reach the LFC under its own buoyancy. The temperature needed for this to happen is called the **convective temperature**, and is found as follows (Fig. 14.75):

(1) Plot the environmental sounding and the surface dew-point temperature.
(2) From T_d, follow the isohume up until it hits the sounding. This altitude is called the **convective condensation level** (**CCL**).
(3) Follow the dry adiabat back to the surface to give the convective temperature needed.

You should forecast airmass thunderstorms to start when the surface air temperature is forecast to reach the convective temperature, assuming the environmental sounding above the boundary layer does not change much during the day. Namely, determine if the forecast high temperature for the day will exceed the convective temperature.

Often several triggering mechanisms work in tandem to create favored locations for thunderstorm development. For example, a jet streak may move across North America, producing upward motion under its left exit region and thereby reducing the strength of the cap as it passes (see the "Layer Method for Static Stability" section in the Atmospheric Stability chapter). If the jet streak crosses a region with large values of CAPE at the same time as the surface temperature reaches its maximum value, then the combination of these processes may lead to thunderstorm development.

Sample Application

Use the sounding in Fig. 14.25 to find the convective temperature and the convective condensation level. Are thunderstorms likely to be triggered?

Find the Answer

Given: Fig. 14.25.
Find: P_{CCL} = ? kPa, T_{needed} = ? °C , yes/no Tstorms

Step 1: Plot the sounding (see next column).

Step 2: From the surface dew-point temperature, follow the isohume up until it hits the sounding to find the CCL. Thus: $P_{CCL} \approx$ **63 kPa** (or $z_{CCL} \approx$ **3.7 km**).

Step 2: Follow the dry adiabat down to the surface, and read the temperature from the thermo diagram. $T_{needed} \approx$ **47 °C** is the convective temperature

Check: Values agree with figure.
Exposition: Typical high temperatures in the prairies of North America are unlikely to reach 47°C.

(continues in next column)

Sample Application *(continuation)*

Thus, the **triggering of a thunderstorm by a rising thermal is unlikely** for this case. However, thunderstorms could be triggered by an airmass boundary such as a cold front, if the other conditions are met for thunderstorm formation.

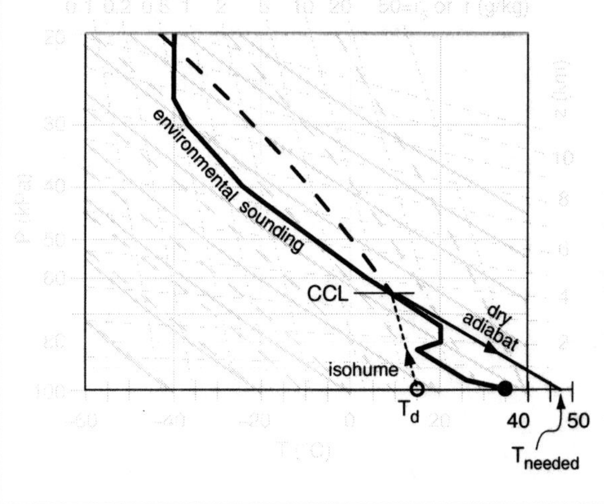

Finally, a word of caution. As with each of the other requirements for a thunderstorm, triggering (lift) by itself is insufficient to create thunderstorms. Also needed are the instability, abundant moisture, and wind shear. To forecast thunderstorms you must forecast all four of the key ingredients. That is why thunderstorm forecasting is so difficult.

THUNDERSTORM FORECASTING

Forecasting thunderstorms is not easy. Thunderstorms are very nonlinear (i.e., they grow explosively), and are extremely dependent on initial conditions such as triggering, shear, and static stability. Individual storms are relatively short lived (15 to 30 min), and are constantly changing in intensity and movement during their lifetimes. Intense thunderstorms also modify their environment, making the relationships between pre-storm environments and storm evolution even more challenging to apply.

Much appears to depend on random chance, since we are unable to observe the atmosphere to the precision needed to describe it fully. Will a boundary-layer thermal or an airmass boundary happen to be strong enough to be the first to break through the capping inversion? Will a previous thunderstorm create a strong-enough downburst to trigger a new

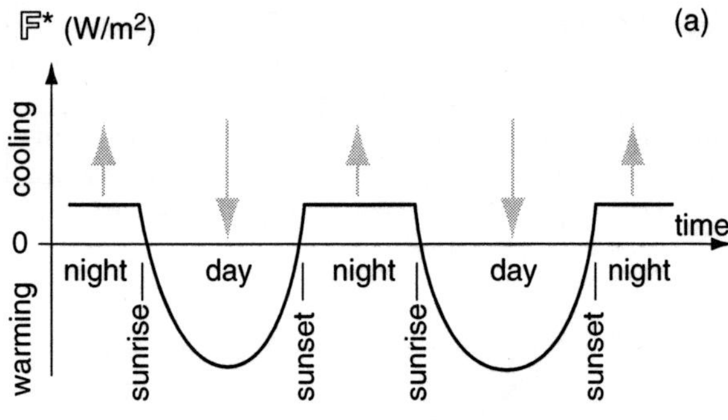

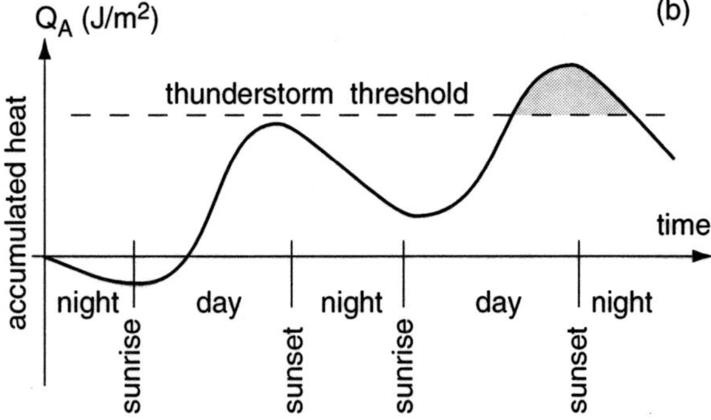

Figure 14.76
(a) Simplification of daily (diurnal) cycle of upward net radiation $\mathbb{F}^*$, which combines solar heating during the day and infrared cooling all the time, for summer over land. Grey arrows represent direction of net flux at the surface. (b) Accumulated (sensible plus latent) heats Q_A in the boundary layer. During summer, days are longer than nights, allowing accumulated heat to increase from day to day. Dash line thunderstorm threshold schematically represents the net effect of all the complex thunderstorm-genesis processes reviewed in this chapter. Grey shading shows that thunderstorms are most likely to form around sunset, plus and minus several hours

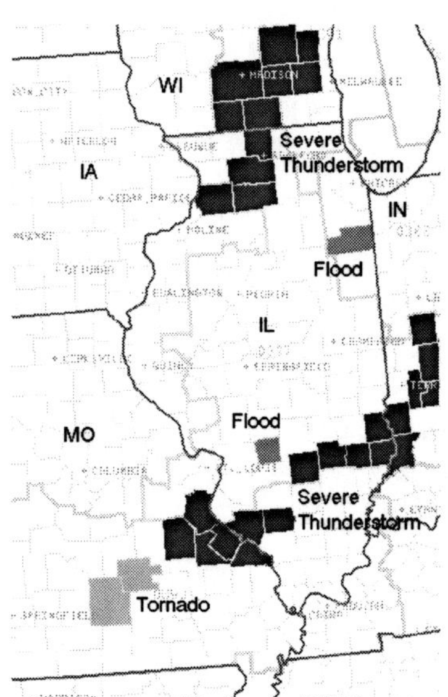

Figure 14.77
Map of tornado (light grey), flood (medium grey), and severe thunderstorm (black) **warnings** *associated with the 24 May 2006 case study in the USA. The shaded boxes indicate which counties were warned. State abbreviations are: IL = Illinois, MO = Missouri, IA = Iowa, IN = Indiana, WI = Wisconsin.*

thunderstorm with its gust front? Will high clouds happen to move over a region, shading the ground and thereby reducing the instability?

Tornadoes, hail, heavy rain, lightning, and strong straight-line wind events may or may not occur, and also vary in intensity, track, and duration within the constraints of the parent thunderstorm. Thunderstorms interact in complex ways with other existing thunderstorms and with the environment and terrain, and can trigger new thunderstorms.

Luckily, most thunderstorms over land have a very marked diurnal cycle, because atmospheric instability is strongly modulated by heating of the ground by the sun. Thunderstorms usually form in mid to late afternoon, are most frequent around sunset, and often dissipate during the night. As sketched in Fig. 14.76, the reason is that the greatest accumulation of heat (and moisture) in the boundary layer occurs not at noon, but about a half hour before sunset.

There are notable exceptions to this daily cycle. For example, thunderstorms can be triggered by the Rocky Mountains in late afternoon, and then propagate eastward all night to hit the Midwest USA with the greatest frequency around midnight to early morning.

Outlooks, Watches & Warnings

A thunderstorm is defined as "**severe**" in the USA if it has one or more of the following:

- tornadoes
- damaging winds (& any winds ≥ 25 m s^{-1})
- hail with diameter ≥ 1.9 cm

Severe weather (thunderstorm caused or not) also includes heavy rain that could cause flash flooding. All thunderstorms have lightning, by definition, so lightning is not included in the list of severe weather elements even though it kills more people than tornadoes, hail, or winds.

In the USA, severe convective weather forecasting is broken into three time spans: outlook, watch, and warning. These are defined as follows:

- **outlook**: 6 to 72 h forecast guidance for broad regions. Types:
  - **convective outlooks**: very technical
  - **public severe weather outlooks**: plain language

- **watch**: 0.5 to 6 h forecast that severe weather is favorable within specific regions called **watch boxes**. Watch types include:
  - **severe thunderstorm watch** (includes tornadoes)
  - **tornado watch**

• **warning:** 0 to 1 h forecast that severe weather is already occurring and heading your way. Warned are specific towns, counties, transportation corridors, and small regions called warning boxes.
Warning types include:
- **severe thunderstorm warning**
- **tornado warning**

Forecast methods for **warnings** are mostly **nowcasts**. Namely, wait until the severe weather is already occurring as observed by human spotters or radar signatures, or find other evidence indicating that severe weather is imminent. Then anticipate its movement in the next few minutes, and warn the towns that are in the path.

These warnings are delivered by activating civil-defense warning sirens and weather-radio alert messages, by notifying the news media, and by talking with local emergency planners, police and fire agencies. Although the warnings are the most useful because of the details about what, where, and when (Fig 14.77), they are also the shortest term forecasts, giving people only a few minutes to seek shelter. The safest tornado and outflow wind shelters are underground in a basement or a ditch, or in a specially designed reinforced-concrete above-ground **safe room**.

Methods for forecasting **watches** are essentially the methods described earlier in this chapter. Namely, short term (0 to 24 h) numerical weather prediction (NWP) model output, soundings from rawinsondes and satellites, and mesoscale analyses of surface weather-station data are analyzed. These analyses focus on the four elements needed for thunderstorm formation later in the day: high humidity in the boundary layer; nonlocal conditional instability; strong wind shear; and a trigger mechanism to cause the initial lifting. The indices and parameters described earlier in this chapter speed the interpretation of the raw data. To handle some of the uncertainty in thunderstorm behavior, probabilistic severe weather forecasts are produced using ensemble methods (see the NWP Chapter).

Nonetheless, watches are somewhat vague, with no ability to indicate which specific towns will be hit and exactly when they will be threatened (the INFO box has an example of a tornado watch and the associated **watch box** graphic). People in the watch area can continue their normal activities, but should listen to weather reports and/or watch the sky in case storms form nearby. Watches also help local officials prepare for events through changes in staffing of emergency-management and rescue organizations (fire departments, ambulance services, hospitals, police), and deploying storm spotters.

INFO • A Tornado Watch (WW)

```
Urgent - immediate broadcast requested
   Tornado watch number 387
   NWS Storm Prediction Center, Norman OK
   245 PM CDT Wed May 24 2006

   The NWS Storm Prediction Center has issued
a tornado watch for portions of
          Eastern Iowa
          Central into northeast Illinois
          Far east central Missouri
Effective this Wednesday afternoon and eve-
ning from 245 PM until  1000 PM CDT.

   Tornadoes...Hail to 3 inches in diameter...
Thunderstorm wind gusts to 70 mph...And dan-
gerous lightning are possible in these ar-
eas.

    The tornado watch area is approximately
along and 75 statute miles east and west of a
line from 30 miles east northeast of Dubuque
Iowa to 25 miles southwest of Salem Illinois
[see Fig. below].  For a complete depiction
of the watch see the associated watch outline
update (WOUS64 KWNS WOU7).

  Remember...A tornado watch means conditions
are favorable for tornadoes & severe thunder-
storms in & close to the watch area. Persons
in these areas should be on the lookout for
threatening weather conditions & listen for
later statements and possible warnings.

     Other watch info...Continue...WW 385...WW
386...

  Discussion...50+ knot westerly mid level jet
will spread across far southern Iowa/north-
ern Missouri through the evening and increase
shear atop moderate instability already in
place across WW.  Supercells and organized mul-
ticell clusters/lines are expected to evolve
and shift from west to east through the mid
evening.  Large hail could become quite large
with stronger cores...With additional isolated
tornado threat accompanying more persistent
supercells through the evening.  Wind damage
is also likely.

  Aviation...Tornadoes and a few severe thun-
derstorms with hail surface and aloft to 3
inches. Extreme turbulence and surface wind
gusts to 60 knots. A few cumulonimbi with
maximum tops to 50,000 ft. Mean storm motion
vector from 260° at 25 knots.
```

[Courtesy of the Storm Prediction Center, US National Weather Service, NOAA.]

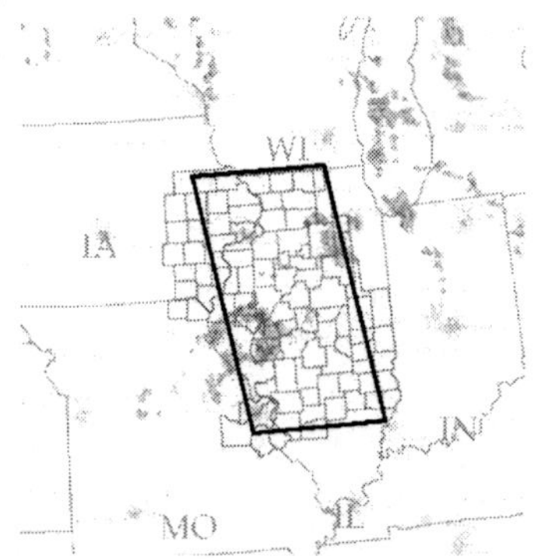

Figure 14.c
Example of a tornado watch box.

Table 14-5. Alphabetical listing of some thunderstorm (CB) stability indices, compiled from both this chapter and the next chapter.

Abbr.	Full Name	Use to anticipate
BRN	Bulk Richardson Number	CB type
BRN Shear	Bulk Richardson Number Shear	tornado likelihood
CAPE	Convective Available Potential Energy (SB=surface based; ML = mean layer; MU = most unstable)	CB intensity
CIN	Convective Inhibition	strength of cap
DCAPE	Downdraft CAPE	downburst and gust front intensity
EHI	Energy Helicity Index	supercell intensity tornado intensity
ML LCL	Mean-layer Lifting Condensation Level	moisture availability & CB intensity
S	Swirl Ratio	multi-vortex tornadoes
SCP	Supercell Composite Parameter	supercell & tornado intensity
SHIP	Significant Hail Parameter	large-hail likelihood
SRH	Storm-Relative Helicity	mesocyclone rotation
STP	Significant Tornado Parameter	tornado intensity

The **convective outlooks** include a general statement about the level of risk that severe weather will occur in broad regions spanning roughly 350 x 350 km, many hours or days into the future. These risk levels are:

• **Slight** (SLGT): well-organized severe thunderstorms are expected in small numbers and/or low coverage. Specifically: 5 to 29 reports of hail ≥ 2.5 cm, and/or 3 to 5 tornadoes, and/or 5 to 29 wind events.

• **Moderate** (MDT): a greater concentration of severe thunderstorms, and in most situations, greater magnitude of severe weather. Details: ≥ 30 reports of hail ≥ 2.5 cm, or 6 to 19 tornadoes, or numerous wind events (30 that might be associated with a squall line, bow echo or derecho).

• **High** (HIGH): a major severe-weather outbreak is expected, with great coverage of severe weather and enhanced likelihood of extreme weather (i.e., violent tornadoes or extreme convective wind events over a large area). Details: ≥ 20 tornadoes with at least two rated ≥ EF3, or an extreme derecho causing ≥ 50 widespread high wind events with numerous higher winds (≥ 35 m s^{-1}) and structural damage reports.

Stability Indices for Thunderstorms

As you have already seen, meteorologists have devised a wide variety of indices and parameters to help forecast thunderstorm existence, strength, and type. Table 14-5 summarizes indices that were discussed earlier in this chapter as well as indices from the next chapter. Many of these indices integrate over large portions of the environmental sounding or hodograph, and are automatically calculated by computer programs for display next to the computer-plotted sounding.

Because no single index has proved to be the best, new indices are devised, modified, and tested every year. Some day a single best index might be found.

The use of indices to aid severe weather forecasting has a long tradition. Many older indices were devised for calculation by hand, so they use data at a few key altitudes, rather than integrating over the whole sounding. Table 14-6 lists some of the older indices, and Table 14-7 give the associated forecast guidelines.

Table 14-8 compares the values of the different indices with respect to the forecasted weather elements. Beware that this table is highly simplified, and that the boundaries between different severity of storm are not as sharp as the table suggests. Nonetheless, you can use it as a rough guide for interpreting the index values that are often printed with plotted soundings or weather maps.

Table 14-6. Definition of some older thunderstorm stability indices. Notation:
T = environmental temperature (°C);
Td = environmental dew-point temperature (°C);
$Tp_{s\text{->}e}$ = final temperature of an air parcel that started with average conditions at height s, and which rose to ending height e following dry adiabat up to LCL, and moist adiabat above. M = wind speed (m s^{-1}). α = wind direction (degrees). Subscripts give pressure altitude.

Abbr.	Full Name	Values & Interpretation
K or KI	K Index	$K = T_{85kPa} + Td_{85kPa} + Td_{70kPa} - T_{70kPa} - T_{50kPa}$
LI	Lifted Index	$LI = T_{50kPa} - Tp_{95\text{->}50kPa}$
SSI	Sho-walter Stability Index	$SSI = T_{50kPa} - Tp_{85\text{->}50kPa}$
SWEAT	Severe Weather Threat Index	$SWEAT = 12{\cdot}Td_{85kPa} + 20{\cdot}(TT - 49) + 4{\cdot}M_{85kPa} + 2{\cdot}M_{50kPa} + 125{\cdot}[0.2 + \sin(\alpha_{50kPa} - \alpha_{85kPa})]$ where TT = total totals index. Note: more rules set some terms=0
TT	Total Totals Index	$TT = T_{85kPa} + Td_{85kPa} - 2{\cdot}T_{50kPa}$

Table 14-7. Interpretation of older thunderstorm stability indices. Notation: CB = thunderstorms.

Abbr.	Full Name	Values & Interpretation
K or KI	K Index	< 20 CB unlikely 20 to 30 Chance of scattered CB 30 to 40 Many CB; heavy rain > 40 CB; very heavy rain
LI	Lifted Index	> 2 CB unlikely 0 to 2 CB only if strong trigger -3 to 0 Weak CB possible -6 to -3 Moderate CB probable < -6 Severe CB likely
SSI	Sho-walter Stability Index	> 3 CB unlikely 1 to 3 Weak showers possible -3 to 0 Severe CB possible -6 to -4 Severe CB probable < -6 Severe CB likely
SWEAT	Severe Weather Threat Index	< 300 CB unlikely 300-400 Chance isolated severe CB 400-500 Severe CB likely; & tornado possible 500-800 Severe CB & tornado likely
TT	Total Totals Index	< 45 CB unlikely 45 to 50 Scattered CB possible 50 to 55 CB likely; some severe 55 to 60 Severe CB likely; tornado

Table 14-8. Approximate relationship between storm indices and storm intensity.

	Thunderstorm (CB) & Tornado (EF0 - EF5) Severity						
Index	**No CB**	**Ordinary CB**	**Marginal Supercell**	**Supercell, no tornado**	**Supercell & EF0-EF1**	**Supercell & EF2-EF5**	**Units**
BRN		150	70	30	30	30	
BRN Shear		7	30	45	55	70	m^2 s^{-2}
CAPE (ML)		950	1205	1460	1835	2152	J·kg^{-1}
CAPE (MU)		1750	1850	1950	2150	2850	J·kg^{-1}
CIN		18		35	12		J·kg^{-1}
EHI (ML 0-1km)		0.1	0.5	0.8	1.4	2.1	
K	15	25	35	45			°C
LCL (ML)		1.75	1.47	1.34	1.18	1.00	km
LI	+1.5	–1.5	–4.5	–7.5			°C
SCP		0	1.1	3.5	5.9	11.1	
Shear (0-6km)		8	15	22	23	24	m s^{-1}
SRH (0-1km)		20	70	115	155	231	m^2 s^{-2}
SRH-effective		16	60	117	166	239	m^2 s^{-2}
SSI	+4.5	+1.5	–1.5	–4.5	–7.5		°C
STP		0	0	0.4	0.9	2.7	
SWEAT	300	350	400	450	500		
TT	42	47	52	57	62		°C

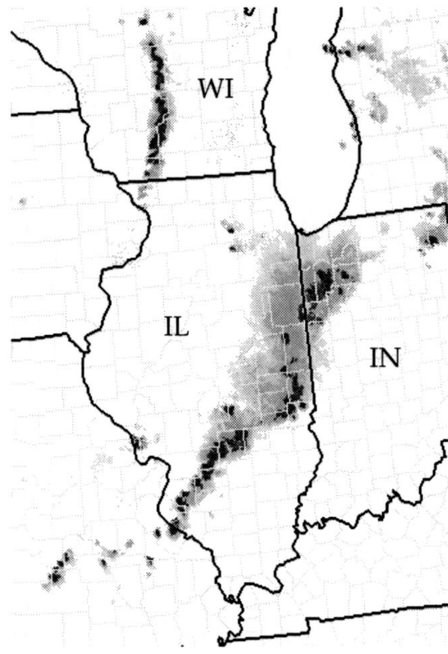

Figure 14.78
Weather radar composite reflectivity image of squall lines in Indiana (IN), Illinois (IL) and Wisconsin (WI), USA, at 2313 UTC on 24 May 2006. Darkest colors correspond to 60 to 65 dBZ reflectivities.

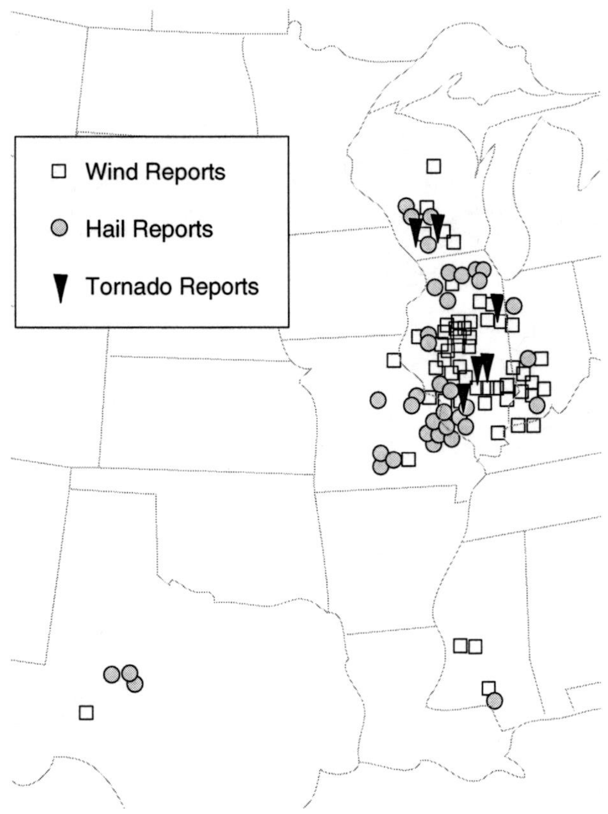

Figure 14.79
All storm reports during 24 May 2006 of damaging wind, hail, and tornadoes in the central USA.

Storm Case Study

Throughout this chapter and the next chapter are examples of weather maps for the severe-storm event of 24 May 2006. These maps presented indices and trigger locations that could be used to help forecast severe weather. The weather maps were created by the US National Weather Service and served over the internet in real time on that day, giving up-to-date weather information to meteorologists, public officials, storm chasers, radio and TV stations, and the general public.

On that day, warm humid air from the Gulf of Mexico was streaming northward into the Mississippi Valley, reaching as far north as the Midwest USA. This provided fuel in the boundary layer, ahead of a cold front that was located from Illinois to Texas. An additional trigger was a west-Texas dry line that merged with the cold front. Cold dry air aloft from the west contributed to the instability of the soundings, and wind shear near Illinois enabled supercells to form.

The US Storm Prediction Center anticipated that severe weather would form that afternoon, and issued a series of watches and warnings that successfully covered the severe weather area. Most of the severe weather was expected to occur that day in the Midwest, although there was also a chance of storms in Texas near the dry line.

Fig. 14.78 shows a radar snapshot of the squall line that moved across the Midwest USA at 22 UTC (5 PM local daylight time). Fig. 14.79 shows a map of the resulting **storm reports**, which included:

- 6 tornado reports
- 97 damaging wind reports & 1 wind >32 m s^{-1}
- 84 hail reports + 3 reports of hail > 5 cm diam.

A total of 187 reports of severe weather were compiled for this storm case. This was a typical severe weather day.

Now that we are finished with the sections on the formation conditions and characteristics of thunderstorms we will proceed into the next chapter on Thunderstorm Hazards. These include heavy rain, hail, downbursts of wind, gust fronts, lightning, thunder, and tornadoes.

REVIEW

Thunderstorms (cumulonimbi) are violent convective clouds that fill the depth of the troposphere. Thunderstorms look like mushrooms or anvils. The most violent thunderstorms are called supercells, in which the whole thunderstorm rotates as a mesocyclone. Other storm organizations include airmass thunderstorms, multicell storms, orographic storms, mesoscale convective systems, squall lines, and bow echoes.

Four conditions are needed to form thunderstorms: high humidity; instability; wind shear, and a trigger to cause lifting. Thermodynamic diagrams and hodographs are used to determine the moisture availability, static stability and shear. A variety of stability and shear indices have been devised to aid thunderstorm forecasting. Trigger mechanisms are often mountains or airmass boundaries such as synoptic fronts, dry lines, sea-breeze fronts, and gust fronts, such as determined from weather radar, satellite, and surface weather analyses.

Thunderstorms are like gigantic engines that convert fuel (moist air) into motion and precipitation via the process of condensation and latent heat release. Once triggered, thunderstorms can often sustain themselves within a favorable environment for 15 minutes to several hours. They are most frequent in the late afternoon and evening over land. The explosive growth of thunderstorms, relatively small diameters, and sensitivity to initial conditions make it difficult to forecast thunderstorms.

HOMEWORK EXERCISES

Broaden Knowledge & Comprehension

B1. Search the internet for (and print the best examples of) photographs of:
- a. airmass thunderstorms
- b. multicell thunderstorms
- c. orographic thunderstorms
- d. squall-line thunderstorms
- e. supercell thunderstorms
  - (i) classic
  - (ii) low precipitation
  - (iii) high precipitation
- f. mammatus clouds
- g. wall clouds
- h. beaver tail clouds
- i. flanking lines
- j. tail cloud

(Hint: search on "storm stock images photographs".) Discuss the features of your resulting image(s) with respect to information you learned in this chapter.

B2. Search the internet for (and print the best examples of) radar reflectivity images of
- a. airmass thunderstorms
- b. multicell thunderstorms
- c. squall line thunderstorms
- d. mesoscale convective systems
- e. bow echoes

Discuss the features of your resulting image(s) with respect to information you learned in this chapter.

B3. Same as the previous question, but for radar Doppler velocity images.

B4. Search the internet for (and print the best examples of) satellite images of:
- a. airmass thunderstorms
- b. multicell thunderstorms
- c. squall line thunderstorms
- d. mesoscale convective systems
- e. dry lines

Do this for visible, IR, and water vapor channels. Discuss the features of your resulting image(s) with respect to information you learned in this chapter.

B5. Search the internet for plotted soundings. In particular, find web sites that also list the values of key stability indices and key altitudes along with the sounding. From this site, get a sounding for a location in the pre-storm environment toward which thunderstorms are moving. Print the sounding and the stability indices, and use the sounding and indices to discuss the likelihood and severity of thunderstorms. (Hint: To find regions where thunderstorms are likely, first search the web for a weather radar image showing where the thunderstorms are. Alternately, find via web search the regions where the weather service has issued thunderstorm watches.)

B6. Search the internet for plotted hodographs. (Hint: Some sites plot hodographs along with plotted soundings, while others have separate hodograph plots.) Find a web site that also shows key wind-shear parameters associated with the hodograph, and perhaps also gives expected storm motion and/or storm-relative winds. Do this for a location into which thunderstorms are moving (see hint from previous exercise). Print and discuss the results.

B7. Search the internet for (and print the best examples of) real-time weather-map analyses or forecasts (from numerical models) showing plotted fields of the four conditions needed for convective storm formation. (Hint, you might find these 4 conditions on 4 separate maps, so you should print and discuss all 4 maps.) Discuss.

B8. Search the internet for maps showing real-time thunderstorm and tornado watches. Print and discuss the results. If you also did the previous exercise, then discuss the amount of agreement of the watch areas with the regions that satisfy the conditions for thunderstorms.

B9. Use the internet to find thunderstorm tutorials. (Hints: some elementary tutorials are at the "university of Illinois online meteorology guide" site. More advanced tutorials are at the "ucar meted" site. Much miscellaneous info is at the "wikipedia" site.) After reading the tutorial, print and discuss one new aspect of thunderstorms that wasn't in this textbook, but which you find interesting and/or important.

B10. Search the web for newer thunderstorm indices than were discussed in this chapter. Summarize and print key information (e.g., definitions for, equations governing, advantages compared to older indices) for one of these new indices, and if possible show a real-time map of the values of this index.

B11. Search the internet for a recent sounding close to your present location. Get and print this sounding data in text form. Then manually plot the temperatures and dew points on a thermo diagram, and plot the winds on a hodograph. Manually calculate a few key thunderstorm indices, and make your own forecast about whether thunderstorms are likely.

B12. Same as the previous exercise, but for a sounding just ahead of a severe thunderstorm location.

B13. Search the internet for a "storm spotter glossary", and list 10 new thunderstorm-related terms and their definitions that were not discussed in this chapter, but which you feel are important. Justify your choice of terms.

B14. Search the internet for (& print good examples of) new research that is being done on thunderstorms or any of the topics discussed in this chapter.

B15. Search the internet for four key storm components that spotters analyze to estimate thunderstorm strength, type, and stage of evolution.

B16. Search the internet to explain the difference between "storm chasers" and "storm spotters".

B17. Search the internet to find which agency or agencies within your own country are responsible for making the national forecasts of severe thunderstorms. Print their web address, along with basic information about their mission and location. Also, find and print a list of qualifications they want in new meteorologists that they hire.

B18. For the following country, where on the internet can you find up-to-date tornado/thunderstorm watches/warnings?

a. Canada b. Japan c. China
d. Australia e. Europe f. USA
g. or a teacher-assigned country?

Apply

A1(§). Plot the pressure difference ($\Delta P = P_{MCS} - P_{std.\,atm.}$) vs. height within a mesoscale convective system (MCS), given the temperature differences ($\Delta T = T_{MCS} - T_{std.atm.}$) below. Also plot the given temperature difference vs. height. The surface pressure $P_{MCS\,sfc}$ under the MCS is also given below. For the standard atmosphere (std. atm.), assume the surface pressure is 101.325 kPa. Use a vertical resolution of Δz = 250 m.

Exercise Part	a.	b.	c.	d.
	ΔT (K)			
0 ≤ z < 2 (km)	–10	–8	–13	–15
2 ≤ z < 4 (km)	0	0	0	0
4 ≤ z < 9 (km)	+6.5	+5	+10	+9
9 ≤ z < 11 (km)	–10	–7.7	–15	–13
11 ≤ z ≤ 12 (km)	0	0	0	0
$P_{MCS\,sfc}$ (kPa)	101.8	101.7	101.9	102.05

A2. Using your answer for ΔP at 2 km altitude from the previous exercise, calculate the rear-inflow jet (RIJ) speed into the mid-tropospheric meso-low within an MCS. Assume the RIJ is geostrophic, for a latitude of 40°N. Assume the MCS has a horizontal radius (km) of:

a. 80 b. 100 c. 120 d. 140 e. 160
f. 180 g. 200 h. 220 i. 240 j. 260

A3. Given the following prestorm sounding. Plot it on a thermo diagram (use a skew-T, unless your instructor specifies a different one). Find the mixed-layer height, tropopause height, lifting condensation level, level of free convection, and equilibrium level, for an air parcel rising from the surface. Use a surface (P = 100 kPa) dew-point temperature (°C) of:

a. 22 b. 21 c. 20 d. 19 e. 18 f. 17 g. 16
h. 15 i. 14 j. 13 k. 12 l. 11 m. 10 n. 9
o. 8 p. 7 q. 6 r. 5 s. 4

Sounding:

P (kPa)	T (°C)
20	–45
25	–45
30	–40
40	–30
50	–19
70	+5
80	15
88	21
92	21
98	26
100	30

A4. Using the result from the previous exercise, forecast the intensity (i.e., category) of thunderstorm and possibly tornadoes that are likely, given the lifting condensation level you found.

A5. Find the median and interquartile range for the following data sets:

a. 6 3 9 7 2 1 6 0 8
b. 8 5 9 8 1 1 3 2 6 8 4 9 7 0
c. 3 5 2 7 9 4 7 7
d. 9 6 8 0 9 1 3 2 9 7 8
e. 5 8 2 2 1 6 4 3 7 9
f. 4 5 6 7 8 9 0 1 2 3
g. 1 2 3 4 5 6 7 8 9
h. 9 8 7 6 5 4 3 2 1
i. 0 0 0 0 3 6 8 3 4 8 9 0 6

A6. Find the mean and standard deviation for the data set from the previous exercise.

A7. For the sounding from exercise A3, find the value of surface-based CAPE, using:
(1) the height (z) tiling method.
(2) the pressure (P) tiling method.
Assume $T_v \approx T$. Also, on the plotted sounding, shade or color the CAPE area.

A8. Start with the sounding from exercise A3. Assume that the forecast high temperature for the day is 2°C warmer than the surface temperature from the sounding. Find the value of surface-based CAPE. Assume $T_v \approx T$.

A9. For the sounding from exercise A3, find the value of mean-layer CAPE. Assume $T_v \approx T$. For the air parcel that you conceptually lift, assume its initial temperature equals the average temperature in the bottom 1 km of the sounding, and its initial dew-point is the average dew-point temperature in the bottom 1 km of the sounding (assuming that the mixing ratio is constant with height in this 1 km mixed layer).

A10. Using the result from the previous exercise, forecast the intensity of thunderstorms and possibly tornadoes that are likely, and indicate the uncertainty (i.e., the range of possible storm intensities) in this forecast, given the MLCAPE you found.

A11. Forecast the thunderstorm and/or tornado intensity, and indicate the uncertainty (i.e., the range of possible storm intensities) in this forecast, given a ML CAPE ($J \cdot kg^{-1}$) value of:

a. 1000 b. 1250 c. 1500 d. 1750 e. 2000
f. 2250 g. 2500 h. 2750 i. 3000 j. 3250
k. 3500 l. 250 m. 500 n. 750

A12. Given the sounding from exercise A3, except with a surface temperature of T = 24°C at P = 100 kPa. Assume the surface dew point from exercise A3 defines a mixing ratio that is uniform in the bottom 2 layers of the sounding.

Find the most unstable CAPE considering air parcels that start their rise from just the bottom two levels of the sounding (i.e., for air parcels that start at P = 100 and 98 kPa). Assume $T_v \approx T$.

A13. Using the result from the previous exercise, forecast the intensity (i.e., category) of thunderstorm and possibly tornadoes that are likely, given the MUCAPE you found, and indicate the uncertainty (i.e., the range of possible storm intensities) in this forecast.

A14. Forecast the thunderstorm and/or tornado intensity, and indicate the uncertainty (i.e., the range of possible storm intensities) in this forecast, given a MU CAPE ($J \cdot kg^{-1}$) value of:

a. 1000 b. 1250 c. 1500 d. 1750 e. 2000

f. 2250 g. 2500 h. 2750 i. 3000 j. 3250
k. 3500 l. 3750 m. 4000 n. 4250 n. 4500

A15. Using the SBCAPE you found in exercise A7, and the sounding and key altitudes from exercise A3, find the value of normalized CAPE, indicate of the CAPE is tall thin or short wide, and suggest whether thunderstorms would favor heavy precipitation or tornadoes.

A16. Estimate the max likely updraft speed in a thunderstorm, given a SB CAPE (J·kg^{-1}) value of:
a. 1000 b. 1250 c. 1500 d. 1750 e. 2000
f. 2250 g. 2500 h. 2750 i. 3000 j. 3250
k. 3500 l. 3750 m. 4000 n. 500 n. 750

A17. Given a wind of $(U, V) = (10, 5)$ m s^{-1} at z = 1 km. Find the wind difference magnitude, shear direction, and shear magnitude between z = 1 and 2 km, given winds at 2 km of (U, V) (m s^{-1}) of:
a. (20, –3) b. (25, –15) c. (30, 0) d. (25, 5)
e. (10, 15) f. (5, 20) g. (–15, 15) h. (–5, 0)
i. (–20, –30) j. (–5, 8) k. (–10, –10) l. (0, 0)

A18. Plot the following wind sounding data on a hodograph. (Make copies of the blank hodograph from Fig. 14.51 to use for all the hodograph exercises.)

z (km)	In each cell: wind direction (°), speed (m s^{-1})			
	Exercise			
	a	b	c	d
0	calm	100, 5	120, 8	150, 10
1	120, 5	120, 10	150, 5	180, 5
2	150, 8	160, 15	210, 5	240, 5
3	180, 12	220, 25	240, 10	260, 10
4	210, 15	240, 30	250, 15	260, 20
5	240, 25	250, 33	258, 25	250, 30
6	260, 40	250, 33	260, 35	240, 40

A19. On your plotted hodograph from the previous exercise, draw wind vectors from the origin to each wind data point. Discuss the relationship between these wind vectors and the original hodograph that connected just the tips of all the vectors.

A20. Using your hodograph plot from exercise A18, use the hodograph (not equations) to find the (U, V) components (m s^{-1}) at each altitude in the sounding.

A21. Graphically, using your hodograph plot from exercise A18, plot the local shear vector (m s^{-1}) across the z = 2 to 3 km layer.

A22. Same as the previous exercise, except solve for the local shear vector (m s^{-1}) mathematically.

A23. Graphically, using your hodograph plot from exercise A18, plot the 0 to 6 km mean shear vector (m s^{-1}).

A24. Same as exercise A23, except solve for the 0 to 6 km mean shear vector (m s^{-1}) mathematically.

A25. Using your hodograph plot from exercise A18, graphically find the 0 to 6 km total shear magnitude (m s^{-1}). Also, predict the likely thunderstorm and possibly tornado intensity based on this parameter, and indicate the uncertainty (i.e., the range of possible storm intensities) in this forecast.

A26(§). Same as the previous exercise, except solve for the 0 to 6 km total shear magnitude (m s^{-1}) mathematically. Also, predict the likely thunderstorm and possibly tornado intensity based on this parameter, and indicate the uncertainty (i.e., the range of possible storm intensities) in this forecast.

A27. Predict the thunderstorm and possibly tornado intensity based on the following total shear magnitude (m s^{-1}) across the 0 to 6 km layer, and indicate the uncertainty (i.e., the range of possible storm intensities) in this forecast.
a. 4 b. 6 c. 8 d. 10 e. 12 f. 14 g. 16
h. 18 i. 20 j. 22 k. 24 l. 26 m. 28 n. 30

A28. Graphically, using your hodograph plot from exercise A18, find the mean environmental wind direction (°) and speed (m s^{-1}), for normal storm motion.

A29(§). Same as the previous exercise, except solve mathematically for the mean environmental wind direction (°) and speed (m s^{-1}), for normal storm motion.

A30. Given the hodograph shape from exercise A18, indicate whether right or left-moving supercells would dominate. Also, starting with the "normal storm motion" from exercise A28 (based on hodograph from exercise A18), use the Internal Dynamics method on your hodograph to graphically estimate the movement (i.e., direction and speed) of
(1) Right-moving supercell thunderstorms
(2) Left-moving supercell thunderstorms

A31. Using your result from exercise A28 for normal storm motion (based on exercise A18), and assuming CAPE = 2750 J·kg^{-1}:

(1) calculate the shear ΔM (m s^{-1}) between the mean (0-6 km) environment winds and the average winds at z = 0.5 km.

(2) calculate the bulk Richardson number shear (BRN shear) in m^2·s^{-2}.

(3) calculate the bulk Richardson number (BRN).

(4) forecast the thunderstorm likelihood, thunderstorm type, and tornado intensity.

A32. Given the following value of bulk Richardson number, determine the likely thunderstorm type:

a. 7 b. 12 c. 17 d. 22 e. 27 f. 32 g. 37
h. 42 i. 47 j. 52 k. 57 l. 62 m. 67 n. 72

A33. Given the following value of bulk Richardson number (BRN) shear (m^2·s^{-2}), determine the likely thunderstorm and tornado intensity:

a. 5 b. 10 c. 20 d. 30 e. 40 f. 45 g. 50
h. 55 i. 60 j. 65 k. 70 l. 75 m. 80 n. 90

A34. Given the sounding from exercise A3, calculate the value of the convective inhibition (CIN) in J·kg^{-1} for an air parcel lifted from the surface. Assume $T_v \approx T$. Also, shade or color the CIN area on the plotted sounding.

A35. Given your answer to the previous exercise, interpret the strength of the cap and how it affects thunderstorm triggering.

A36. Calculate the mean-layer CIN using the sounding from exercise A3. Temperatures in the bottom 10 kPa of atmosphere are as plotted from the sounding, and assume that the mixing ratio is constant over this layer and equal to the mixing ratio that corresponds to the surface dew point.

A37. For the sounding from exercise A3:

(1) To what altitude must a surface air parcel be lifted by an airmass boundary or a mountain, to trigger thunderstorms?

(2) To what temperature must the surface air be heated in order to convectively trigger thunderstorms (i.e., find the convective temperature), and what is the value of the convective condensation level?

A38. What values of K index, Lifted Index, Showalter Stability Index, SWEAT Index, and Total-totals Index would you anticipate for the following intensity of thunderstorm (CB)?

a. no CB
b. ordinary CB
c. marginal supercell
d. supercell with no tornado
e. supercell with EF0 - EF1 tornado
f. supercell with EF2 - EF5 tornado

A39. What value of the Bulk Richardson Number, BRN-Shear, ML CAPE, MU CAPE, CIN, Energy Helicity Index, Lifting Condensation Level, Supercell Composite Parameter, 0-6 km Shear, Storm-Relative Helicity, effective Storm-Relative Helicity, and Significant Tornado Parameter would you anticipate for the following intensity of thunderstorm (CB)?

a. no CB b. ordinary CB
c. marginal supercell
d. supercell with no tornado
e. supercell with EF0 - EF1 tornado
f. supercell with EF2 - EF5 tornado

Evaluate & Analyze

E1. Identify as many thunderstorm features as you can, from the photo below. (Image "wea00106" courtesy of NOAA photo library).

E2. Why does a thunderstorm have a flat (anvil or mushroom) top, instead of a rounded top such as cumulus congestus clouds? (Ignore the overshooting dome for this question.)

E3. Almost all clouds associated with thunderstorms are caused by the lifting of air. List each of these clouds, and give their lifting mechanisms.

E4. Consider Figs. 14.4a & b. If you were a storm chaser, and were off to the side of the storm as indicated below, sketch which components of the storm and associated clouds would be visible (i.e., could be seen if you had taken a photo). Label the key cloud features in your sketch. Assume you are in the following direction from the storm:

a. northeast b. southwest c. northwest

For example, Fig. 14.4a shows the sketch for the view from southeast of the storm.

E5. Consider Fig. 14.4b. If a Doppler radar were located near the words in that figure given below, then sketch the resulting color patterns on the Doppler PPI wind display for near-surface winds.

a. "FFD" b. "RFD" c. "Anvil Edge" d. "Cu con"
e. "Gust Front" f. "Beaver Tail"

g. "Outflow Boundary Layer Winds"
h. "Inflow Boundary Layer Winds"
i. "Main Updraft" j. "Storm Movement"
k. "Forward Flank Downdraft"

E6. Circle the stations in the weather map below that are reporting thunderstorms. (Image courtesy of the Storm Prediction Center, NWS, NOAA.)

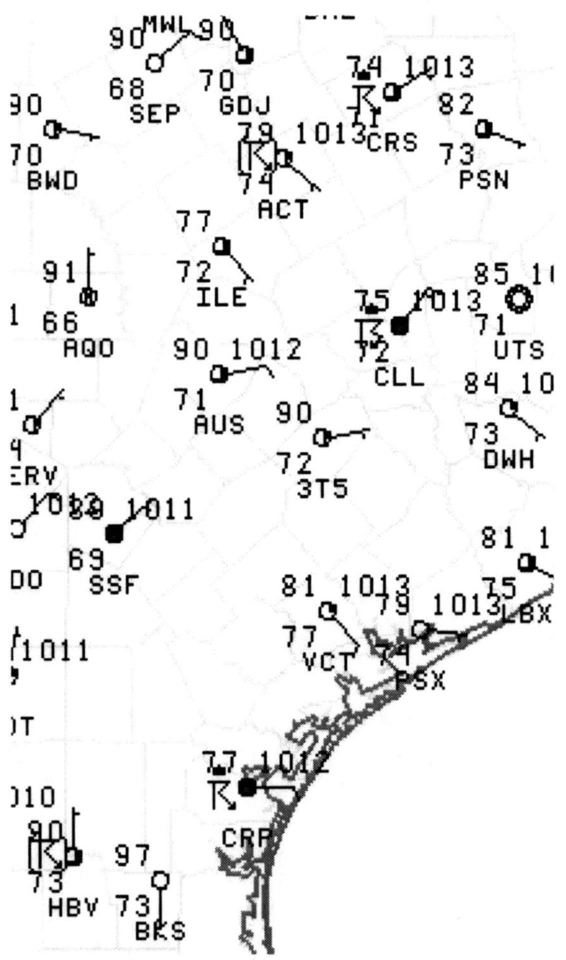

E7. Can a thunderstorm exist without one or more cells? Explain.

E8. For each thunderstorm type or organization, explain which triggered mechanism(s) would have most likely initiated it.

E9. For each thunderstorm type or organization, explain how the phenomenon would differ if there had been no wind shear.

E10. What is a derecho, and what causes it? Also, if strong straight-line winds can cause damage similar to that from a weak tornado, what are all the ways that you could use to determine (after the fact) if a damaged building was caused by a tornado or derecho?

E11. Summarize the different types of supercells, and list their characteristics and differences. Explain what factors cause a supercell to be of a specific type, and where you would most likely find it.

E12. In a thunderstorm, there is often one or more updrafts interspersed with one or more downdrafts. Namely, up- and down-drafts are often adjacent. Do these adjacent up- and down-drafts interfere or support each other? Explain.

E13. List the 4 conditions needed for thunderstorm formation. Then, consider a case where one of the conditions is missing, and explain why thunderstorms would be unlikely, and what would form instead (if anything). Do this for each of the 4 conditions separately.

E14. If thunderstorms normally occur at your town, explain how the 4 conditions needed for thunderstorm formation are satisfied for your region. Namely, where does the humid air usually come from, what conditions contribute to shear, what trigger mechanisms dominate in your region, etc.

E15. If thunderstorms are rare in your region, identify which one or more of the 4 conditions for thunderstorms is NOT satisfied, and also discuss how the other conditions are satisfied for your region.

E16. If the cap inhibits thermals from rising, why is it considered a good thing for thunderstorms?

E17. If your national weather service were to make only one upper-air (rawinsonde) sounding per day, when would you want it to happen, in order to be most useful for your thunderstorm forecasts?

E18. In the section on convective conditions and key altitudes, one thermo diagram was presented with the LCL above z_i (the mixed layer top), while the Sample Application showed a different situation with the LCL below z_i. Both are frequently observed, and both can be associated with thunderstorms.

Discuss and justify whether it is possible to have an environment favorable for thunderstorms if:

a. z_i is above the LFC
b. z_i is above the tropopause
c. LCL is at the Earth's surface
d. LFC is at the Earth's surface
e. LCL is above the LFC

E19. The tops of thunderstorms are often near the tropopause.

a. Why is that?

b. Why are thunderstorm tops usually NOT exactly at the tropopause.

E20. Based on tropopause info from earlier chapters, how would you expect thunderstorm depth to vary with:
a. latitude
b. season

E23. Same as the previous exercise, but for the sounding given in the Sample Application after Fig. 14.26.

E24. Fig. 14.22 shows schematically the information that is in Fig. 14.25. Explain how these two figures relate to each other.

E25. Consider Fig. 14.22. Why is the following environmental condition conducive to severe thunderstorms?
a. cold air near the top of the troposphere
b. dry air near the middle of the troposphere
c. a stable cap at the top of the boundary layer
d. a warm humid boundary layer
e. wind shear
f. strong winds aloft.
(Hint, consider info from the whole chapter, not just the section on convective conditions.)

E26. Fig. 14.26 shows a nonlocally conditionally un-stable environment up to the EL, yet the sounding in the top of that region (i.e., between the tropopause and the EL) is locally statically stable. Explain how that region can be both stable and unstable at the same time.

E27. How much energy does an airmass thunderstorm release (express your answer in units of megatons of TNT equivalent)? Assume all the water vapor condenses. Approximate the cloud by a cylinder of radius 5 km, with base at P = 90 kPa and top at P = 30 kPa. The ABL air has depth 1 km, and dew-point temperature (°C) of:
a. 17 b. 16 c. 15 d. 14 e. 13 f. 12 g. 11
h. 18 i. 19 j. 20 k. 22 l. 23 m. 24 n. 25

E28. Why do thunderstorms have (nearly) flat bases? (Hint, what determines the height of cloud base in a convective cloud?)

E29. Compare the different ways to present information about "high humidity in the boundary layer", such as the different weather maps shown in that section. What are the advantages and disadvantages of each?

Normally, meteorologists need use only one of the humidity metrics to get the info they need on moisture availability. Which one moisture variable would you recommend using, and why?

E30. Look up the definitions of the residual layer and nocturnal stable layer from the Atmospheric Boundary Layer chapter. Consider atmospheric conditions (temperature, humidity) in the residual layer, relative to the conditions in the mixed layer from the afternoon before. Use this information to explain why strong thunderstorms are possible at night, even after the near-surface air temperature has cooled significantly.

E31. Often statistical data is presented on a **box-and-whisker diagram**, such as sketched below.

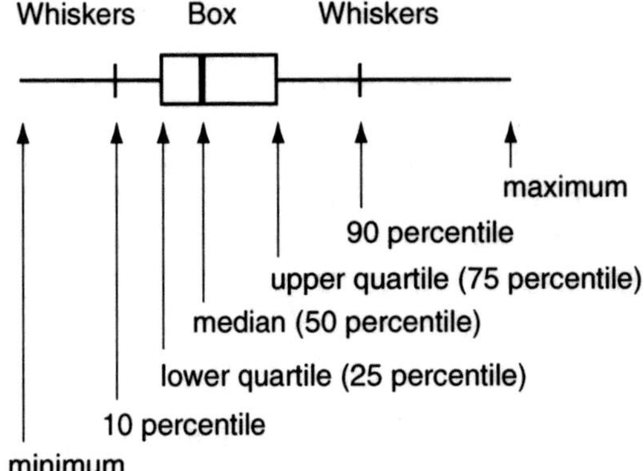

For example, the data such as in Fig. 14.33 is often presented in the journal literature as shown below (where the max and min values are not shown for this illustration, but are often given in the literature).

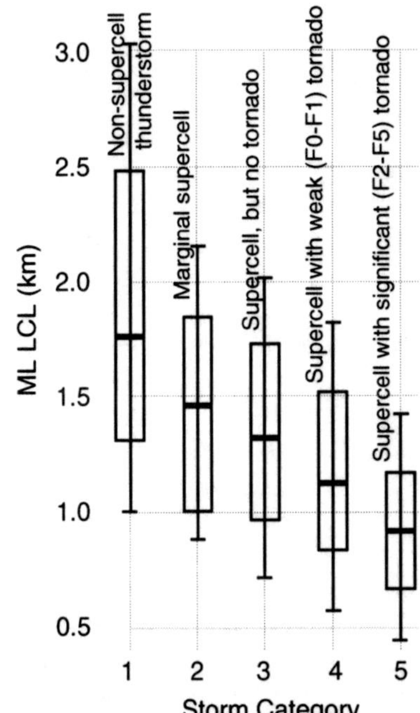

Create box-and-whisker diagrams, using data in this textbook for:

a. ML CAPE
b. MU CAPE
c. TSM
d. BRN Shear
e. 0-1 km SRH
f. 0-1 km EHI

E32. Look at a blank thermo diagram. At colder temperatures, the moist adiabats don't deviate far from the dry adiabats. Use this characteristic to help explain why CAPE is smaller for colder environments, and thus why strong thunderstorms are less likely.

E33. Thunderstorms are much more prevalent in the tropics than near the poles. Use all the info from this chapter to help explain this. Also relate it to info in other chapters, such as: radiation, global circulation, boundary layers etc.

E34. If you were given a file of ASCII text data of temperature and dew point vs. height in a sounding, show schematically (i.e., with a flow chart) a computer program that you could write to compute the CAPE. Don't actually write the program, just show the main steps, procedures, and/or data structures that you would use. (Hint: the method in Fig. 14.35 might be easier than the method in Fig. 14.36, especially for very small Δz.)

a. Do this based on eq. (14.2) [gives an approximate answer]

b. Do this based on eq. (14.1) [this is more accurate]

E35. Describe all the factors in a plotted sounding that could contribute to larger CAPE values.

E36. Compare and contrast the different parcel-origin methods for computing CAPE (SB, forecast SB, ML, and MU). Recommend which methods would be best for different situations.

E37. Derive eq. (14.5) from eq. (14.4) by using the hypsometric equation. Show your steps.

E38. Critically analyze eq. (14.6). Discuss the behavior of the equation as the height difference between the LFC and EL approaches zero.

E39. Consider CAPE in a prestorm sounding.

a. Can all the CAPE be consumed by the storm?

b. Do the CAPE eqs. have limitations? Discuss.

c. Are there reasons why thunderstorm vertical velocity might not get as large as given in eq. (14.8)?

E40. a. Is wind shear a vector or scalar quantity? Why?

b. Is it possible to have non-zero shear between two altitudes where the winds have the same speed? Explain.

E41. Suppose that Fig. 14.47 gives the winds relative to the ground. Redraw that figure showing the storm-relative winds instead. Also, explain why storm-relative winds are important for thunderstorm dynamics.

E42. On a hodograph, why is it important to list the altitude of the winds next to each wind data point?

E43. Draw a hodograph showing an environment where winds ____ with height. a. veer b. back

E44. On blank hodographs such as Fig. 14.51, why are the printed wind directions 180° out of phase with the normal compass directions?

E45. In Fig. 14.52, notice that the dark vertical and horizontal lines don't look straight. This is an optical illusion caused by the background circles of the hodograph. In fact, they are straight, as you can see by tilting the page so you are looking almost edge-on, sighting down each line. Based in this info, discuss why you should never draw the vertical or horizontal lines by eye, when trying to determine U and V components of wind.

E46. Three different types (i.e., ways of measuring) of shear vectors are shown in Fig. 14.53. Explain and contrast them.

E47. Create a table listing all the different types of shear that are discussed in this chapter, and add a column to the table that indicates what thunderstorm characteristics can be estimated from each shear type.

E48. A thunderstorm index or parameter is useful only if it can discriminate between different storm types or intensities. Using Fig. 14.56, critically evaluate the utility of TSM in estimating storm and tornado intensity.

E49. Explain the difference between the mean <u>shear</u> vector (as in Fig. 14.53) and the mean <u>wind</u> vector (as in Fig. 14.58), and discuss their significance.

E50. In Fig. 14.59, the mean wind is estimated as the center of mass of the shaded area. Is the actual mean wind exactly at the center of mass of the shaded area, or only close? [Hint, to help think about this question, create some very simple hodographs with simple shapes for the shaded area [such as a rectangle] for which you can calculate the exact center of mass. Then see if you can extend your argument to more complex, or arbitrary shapes. This scientific approach is called inductive reasoning (see p107).]

E51. Using information from Fig. 14.61, identify where in Fig. 14.60 are the right and left moving mesocyclones.

E52. Although Fig. 14.62c shows a hodograph that is concave up, the whole curve is above the origin. Suppose that, on a different day, the hodograph has the same shape, but is shifted to be all below the origin. Assume N. Hemisphere, mid-latitudes.

Would such a hodograph still favor left-moving supercell storms? Justify your answer.

E53. Both the bulk Richardson number (BRN) and the Supercell Composite Parameter (SCP) include CAPE and shear. However, BRN is CAPE/shear, while SCP is CAPE·shear (times a third factor that we can ignore for this Exposition). Which (BRN or SCP) would you anticipate to give the best storm intensity forecasts, and which is most physically justified? Explain.

E54. Figs. 14.67 - 14.69 show 3 different ways to calculate CIN. Discuss the advantages and disadvantages of each, and for which conditions each method is most appropriate.

E55. Table 14-4 relates CIN to the difficulty in triggering thunderstorms. For the "strong" and "intense" cap categories, can you think of a trigger mechanism that would be powerful enough to trigger thunderstorms? Explain.

E56. Figs. 14.71 and 14.73 show how thunderstorms can be triggered by warm, humid air forced to rise over an obstacle (cold air, or a mountain). In both situations, sometimes after the thunderstorms are triggered and develop to maturity, environmental wind shear causes them to move away from the triggering location. After they have moved away, what is needed in order for the supercell thunderstorm to continue to exist (i.e., not to die immediately after moving away from the triggering area)? Do you think this is possible in real life, and if so, can you give an example?

E57. Consider Fig. 14.74. The Regional Winds chapter states that buoyancy (gravity) waves need statically <u>stable</u> air to exist without damping. But thunderstorms need nonlocally conditionally <u>unstable</u> air to exist. These seems contradictory. Explain how it might be possible.

E58. Fig. 14.75 shows a way to estimate T_{needed} for convection, given T and T_d from a sounding earlier in the day. Although this is an easy method, it has a major flaw. Critically discuss this method. (Hint: what was assumed in order to use this method?)

E59. Step back from the details of thunderstorms, and look at the big picture. After the thunderstorm has occurred, some precipitation has fallen out, and there has been net heating of the air. Explain how these two processes affect the overall stability and entropy (in the sense of randomness) in the troposphere.

E60. Re-draw Fig. 14.76, but for a winter case over mid-latitude land where nocturnal cooling is longer duration than daytime heating, and for which the total accumulated heating during the day (because the sun is lower in the sky) is less than the accumulated cooling at night. Use you resulting figure to discuss the likelihood of thunderstorms in winter.

E61. Recall from other chapters that the ocean has a much large heat capacity than land, and thus has less temperature change during day and night. Redraw Fig. 14.76 for a location at a tropical ocean, and discuss how the resulting figure relates to thunderstorm occurrence there.

E62. Some of the older stability indices in Table 14-6 are very similar in basis to some of the newer indices in Table 14-5, as described throughout this chapter. The main differences are often that the old indices consider meteorological conditions at just a few levels, whereas the newer indices integrate (sum) over many levels. Create a table matching as many old indices to as many new indices as possible.

E63. Based on what you learned so far, which subsets of indices from Table 14-8 would you prefer to utilize in your own storm forecasting efforts, to forecast:

a. hail
b. heavy rain
c. lots of lightning
d. tornadoes
e. strong straight-line winds
f. downbursts

Justify your choices.

Synthesize

S1. Suppose that thunderstorms were typically 30 km deep. How would the weather and general circulation differ, if at all?

S2. How would general circulation differ if no thunderstorms exist in the atmosphere?

S3. Draw a supercell diagram similar to Fig. 14.4b, but for mid-latitudes in the southern hemisphere.

S4. Why do thunderstorms have flat, anvil-shaped tops, while cumulus congestus do not? Both reach their equilibrium level.

S5. Suppose that N. Hemisphere mid-latitude thunderstorms exist in an environment that usually has the same lower-tropospheric geographic arrangement of heat and humidity relative to the storm as our real atmosphere, but for which the winds aloft are from the east. How would thunderstorms differ, if at all?

S6. Suppose that thunderstorms never move relative to the ground. Could long-lasting supercells form and exist? Explain.

S7. In multicell storms, new cells usually form on the south side of the storm complex, closest to the supply of warm humid boundary-layer air. What changes in the environment or the thunderstorm might allow new cells to form on the north side of the storm? Assume N. Hemisphere.

S8. If orographic thunderstorms need the mountain-triggered lifting to be initiated, why can the storms persist after being blown away from the mountain?

S9. If the Earth's surface were smooth (i.e., no mountain ranges), then describe changes in the nature of thunderstorms. Could this alter the weather in your region? Explain.

S10. Suppose a 100 km diameter circular region of warm humid boundary layer air existed, surrounding by much colder air. If the cold air all around the circle started advancing toward the center of the circle and thunderstorms were triggered along this circular cold front, then describe the evolution of the thunderstorms.

S11. In Fig. 14.14 showing a vertical slice through a mesoscale convective system, the arrows show winds being drawn in toward low pressure in the mid troposphere, and other arrows blowing outward from high-pressure in the upper troposphere. Explain how this could happen, because normally we would expect winds to circulate around highs and lows due to Coriolis force, and not to converge or diverge (assuming no frictional drag because we are not in the boundary layer).

S12. Could a classic supercell change into a low-precipitation or a high-precipitation supercell? Explain what factors might cause this, and how the storm would evolve.

S13. If there was never a cap on the atmospheric boundary layer, explain how thunderstorms would differ from those in our real atmosphere, if at all.

S14. Start with the sounding in Fig. 14.24. Modify the dew point in the mid troposphere to create a new sounding that would support a layer of altocumulus castellanus (accus) clouds.

S15. Suppose that thunderstorm downdrafts could never penetrate downward through the cap at the top of the boundary layer. Explain how thunderstorms would differ, if at all?

S16. Suppose that shading of the ground by clouds would cause the Earth's surface to get warmer, not cooler during daytime. How would thunderstorms and climate differ, if at all?

S17. Suppose the atmospheric sounding over your town showed conditions nearly, but not quite, favorable for the existence of thunderstorms. If you could cause the surface energy balance over land to be partitioned differently between sensible and latent heat, how would you partition it in order to generate a thunderstorm? Note that given a fixed input of energy from the sun, increasing either the temperature or the humidity would decrease the other variable.

S18. When water vapor condenses in thunderstorms, suppose that the air cools instead of warms. Would thunderstorms occur? If so, describe their behavior.

S19. In the Exposition of rising air parcels in a nonlocally conditionally unstable environment, we assumed that the surrounding environment was not changing. But if there are many air parcels rising in a thunderstorm updraft, then there must be compensating subsidence in the environment that advects downward the temperature and moisture layers. How would this alter our description of the evolution of thunderstorms?

S20. Under what conditions would the median of a distribution exactly equal the mean? Under what conditions would half the interquartile range exactly equal the standard deviation?

S21. Consider Fig. 14.34. Suppose that when the air parcel is above its LFC, that it entrains environmental air at such a high rate that the air parcel follows a thermodynamic path that is exactly half way between the environment and the moist adiabats that passes through it at each height during its rise. Sketch the resulting path on a copy of Fig. 14.34, and discuss how thunderstorms would differ, if at all.

S22. Consider Fig. 14.34. Suppose that there is no frictional drag affecting the rising air parcel. All of the CAPE would lead to kinetic energy of the updraft. Once the air parcel reaches its EL, inertia would cause it to continue to rise (i.e., overshoot) until its negative potential energy (by being colder than the environment) balanced the initial kinetic energy. Assuming an isothermal sounding above the tropopause in that figure, determine exactly how high the overshooting air parcel would rise. Also, discuss whether such behavior is likely in real thunderstorms.

S23. Suppose that no wind shear existed in the environment. How would thunderstorms differ, if at all?

S24. Suppose that the wind profile of Fig. 14.53 (based on the wind data tabulated in a Sample Application a couple pages earlier) corresponded to the same environment as the thermodynamic sounding data in Fig. 14.24. Draw "phase-space" plots as follows:

a. T vs. U b. T vs. V c. T vs. M
d. T_d vs. U e. T_d vs. V f. T_d vs. M
g. T vs. T_d

(Hint: "phase-space" plots are explained in the Numerical Weather Prediction chapter as a way to help analyze the chaos of nonlinear-dynamics systems.)

Also, speculate on whether phase-diagram plots of real thunderstorm environments could be used to help forecast different types of thunderstorms.

S25. Suppose that moist adiabats curve concave upward instead of concave downward. How would thunderstorms be different, if at all?

S26. Suppose that the data plotted in the hodograph of Fig. 14.62b were everywhere below the origin of the hodograph. How would thunderstorms differ, if at all?

S27. Suppose that right-moving supercells altered the environment so that it favored left-moving supercells, and left-moving supercells altered the environment to favor right movers. How would thunderstorms differ, if at all?

S28. Suggest 3 or more thunderstorm trigger mechanisms beyond what was already discussed in this chapter.

S29. From the Precipitation chapter, recall the Wegener-Bergeron-Findeisen (WBF) cold-cloud process for forming precipitation. Describe the nature of thunderstorms if the WBF precipitation did not occur.

S30. Suppose that accurate thunderstorm (including hail, lightning, and tornado) warnings could be issued 2 days in advance. How would society and commerce change, if at all.

15 THUNDERSTORM HAZARDS

Contents

The basics of thunderstorms were covered in the previous chapter. Here we cover thunderstorm hazards:
- hail and heavy rain,
- downbursts and gust fronts,
- lightning and thunder,
- tornadoes and mesocyclones.

Two other hazards were covered in the previous chapter: turbulence and vigorous updrafts.

In spite of their danger, thunderstorms can also produce the large-diameter rain drops that enable beautiful rainbows (Fig. 15.1).

PRECIPITATION AND HAIL

Heavy Rain

Thunderstorms are deep clouds that can create:
- **large raindrops** (2 - 8 mm diameter), in
- **scattered showers** (order of 5 to 10 km diameter rain shafts moving across the ground, resulting in brief-duration rain [1 - 20 min] over any point), of
- **heavy rainfall rate** (10 to over 1000 mm h^{-1} rainfall rates).

The Precipitation Processes chapter lists world-record rainfall rates, some of which were caused by thunderstorms. Compare this to

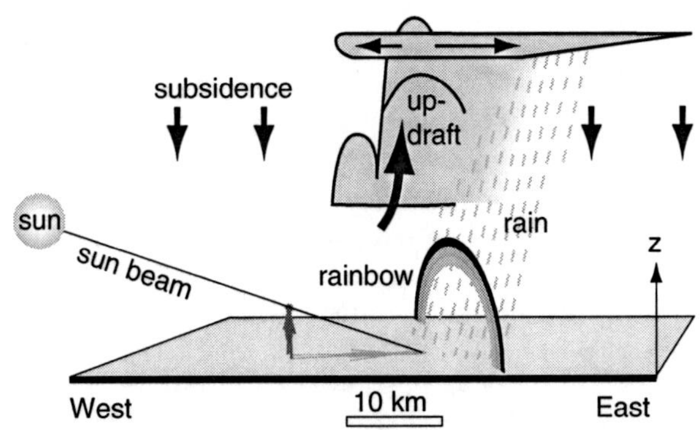

Figure 15.1
Rainbow under an evening thunderstorm. Updraft in the thunderstorm is compensated by weak subsidence around it to conserve air mass, causing somewhat clear skies that allow rays of sunlight to strike the falling large raindrops.

nimbostratus clouds, that create smaller-size **drizzle drops** (0.2 - 0.5 mm) and **small rain drops** (0.5 - 2 mm diameter) in widespread regions (namely, regions hundreds by thousands of kilometers in size, ahead of warm and occluded fronts) of light to moderate rainfall rate that can last for many hours over any point on the ground.

Why do thunderstorms have large-size drops? Thunderstorms are so tall that their tops are in very cold air in the upper troposphere, allowing cold-cloud microphysics even in mid summer. Once a spectrum of different hydrometeor sizes exists, the heavier ice particles fall faster than the smaller ones and collide with them. If the heavier ice particles are falling through regions of supercooled liquid cloud droplets, they can grow by **riming** (as the liquid water instantly freezes on contact to the outside of ice crystals) to form dense, conical-shaped **snow pellets** called **graupel** (< 5 mm diameter). Alternately, if smaller ice crystals fall below the 0°C level, their outer surface partially melts, causing them to stick to other partially-melted ice crystals and grow into miniature fluffy snowballs by a process called **aggregation** to sizes as large as 1 cm in diameter.

The snow aggregates and graupel can reach the ground still frozen or partially frozen, even in summer. This occurs if they are protected within the cool, saturated downdraft of air descending from thunderstorms (downbursts will be discussed later). At other times, these large ice particles falling through the warmer boundary layer will melt completely into large raindrops just before reaching the ground. These rain drops can make a big splat on your car windshield or in puddles on the ground.

Why scattered showers in thunderstorm? Often large-size, cloud-free, rain-free subsidence regions form around and adjacent to thunderstorms due to air-mass continuity. Namely, more air mass is pumped into the upper troposphere by thunderstorm updrafts than can be removed by in-storm precipitation-laden downdrafts. Much of the remaining excess air descends more gently outside the storm. This subsidence (Fig. 15.1) tends to suppress other incipient thunderstorms, resulting in the original cumulonimbus clouds that are either isolated (surrounded by relatively cloud-free air), or are in a thunderstorm line with subsidence ahead and behind the line.

Why do thunderstorms often have heavy rainfall?

- First, the upper portions of the cumulonimbus cloud is so high that the rising air parcels become so cold (due to the moist-adiabatic cooling rate) that virtually all of the water vapor carried by the air is forced to condense, deposit, or freeze out.
- Second, the vertical stacking of the deep cloud allows precipitation forming in the top of the storm to grow by collision and coalescence or accretion as it falls through the middle and lower parts of the cloud, as already mentioned, thus sweeping out a lot of water in a short time.
- Third, long lasting storms such as supercells or orographic storms can have continual inflow of humid boundary-layer air to add moisture as fast as it rains out, thereby allowing the heavy rainfall to persist. As was discussed in the previous chapter, the heaviest precipitation often falls closest to the main updraft in supercells (see Fig. 15.5).

Rainbows are a by-product of having large numbers of large-diameter drops in a localized region surrounded by clear air (Fig. 15.1). Because thunderstorms are more likely to form in late afternoon and early evening when the sun angle is relatively low in the western sky, the sunlight can shine under cloud base and reach the falling raindrops. In North America, where thunderstorms generally move from the southwest toward the northeast, this means that rainbows are generally visible just after the thundershowers have past, so you can find the rainbow looking toward the east (i.e., look toward your shadow). Rainbow optics are explained in more detail in the last chapter.

Any rain that reached the ground is from water vapor that condensed and did not re-evaporate. Thus, rainfall rate (RR) can be a surrogate measure of the rate of latent-heat release:

$$H_{RR} = \rho_L \cdot L_v \cdot RR \tag{15.1}$$

where H_{RR} = rate of energy release in the storm over unit area of the Earth's surface (J·s^{-1}·m^{-2}), ρ_L is the density of pure liquid water, L_v is the latent heat of vaporization (assuming for simplicity all the precipitation falls out in liquid form), and RR = rainfall rate. Ignoring variations in the values of water density and latent heat of vaporization, this equation reduces to:

$$H_{RR} = a \cdot RR \qquad \bullet(15.2)$$

where a = 694 (J·s^{-1}·m^{-2}) / (mm·h^{-1}), for rainfall rates in mm h^{-1}.

The corresponding warming rate averaged over the tropospheric depth (assuming the thunderstorm fills the troposphere) was shown in the Heat chapter to be:

$$\Delta T / \Delta t = b \cdot RR \tag{15.3}$$

where b = 0.33 K (mm of rain)$^{-1}$.

From the Water Vapor chapter recall that **precipitable water**, d_w, is the depth of water in a rain

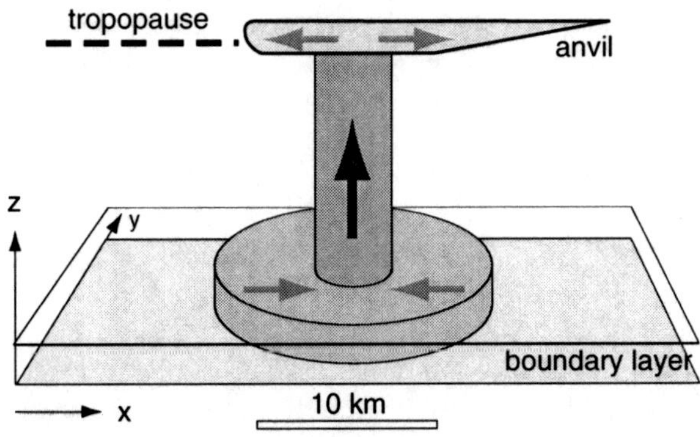

Figure 15.2
The thunderstorm updraft draws in a larger area of warm, humid boundary-layer air, which is fuel for the storm.

gauge if all of the moisture in a column of air were to precipitate out. As an extension of this concept, suppose that pre-storm boundary-layer air of mixing ratio 20 g kg^{-1} was drawn up into a column filling the troposphere by the action of convective updrafts (Fig. 15.2). If cloud base was at a pressure altitude of 90 kPa and cloud top was at 30 kPa, and if **half** of the water in the cloudy domain were to condense and precipitate out, then eq. (4.33) says that the depth of water in a rain gauge is expected to be d_w = 61 mm.

The ratio of amount of rain falling out of a thunderstorm to the inflow of water vapor is called **precipitation efficiency**, and ranges from 5 to 25% for storms in an environment with strong wind shear to 80 to 100% in weakly-sheared environments. Most thunderstorms average 50% efficiency. Processes that account for the non-precipitating water include anvil outflow of ice crystals that evaporate, evaporation of hydrometeors with entrained air from outside the storm, and evaporation of some of the precipitation before reaching the ground (i.e., **virga**).

Extreme precipitation that produce rainfall rates over 100 mm h^{-1} are unofficially called **cloudbursts**. A few cloudbursts or **rain gushes** have been observed with rainfall rates of 1000 mm h^{-1}, but they usually last for only a few minutes. As for other natural disasters, the more intense rainfall events occur less frequently, and have **return periods** (average time between occurrence) of order hundreds of years (see the Rainfall Rates subsection in the Precipitation chapter).

For example, a stationary **orographic thunderstorm** over the eastern Rocky Mountains in Colorado produced an average rainfall rate of 76 mm h^{-1} for 4 hours during 31 July 1976 over an area of about 11 x 11 km. A total of about 305 mm of rain fell into the catchment of the Big Thompson River, producing a flash flood that killed 139 people in the **Big Thompson Canyon**. This amount of rain is equiv-

Sample Application

A thunderstorm near Holt, Missouri, dropped 305 mm of rain during 0.7 hour. How much net latent heat energy was released into the atmosphere over each square meter of Earth's surface, and how much did it warm the air in the troposphere?

Find the Answer

Given: RR = 305 mm / 0.7 h = 436 mm h^{-1}.
Duration Δt = 0.7 h.
Find: $H_{RR} \cdot \Delta t$ = ? (J·m^{-2}) ; ΔT = ? (°C)

First, use eq. (15.2):
H_{RR} = [694 (J·s^{-1}·m^{-2})/(mm·h^{-1})]·[436 mm h^{-1}]·[0.7 h]·[3600s/h] = **762.5** MJ·m^{-2}

Next, use eq. (15.3):
$\Delta T = b \cdot RR \cdot \Delta t$ = (0.33 K mm^{-1})·(305 mm)
= **101 °C**

Check: Units OK, but values seem too large???

Exposition: After the thunderstorm has finished raining itself out and dissipating, why don't we observe air that is 101°C warmer where the storm used to be? One reason is that in order to get 305 mm of rain out of the storm, there had to be a continual inflow of humid air bringing in moisture. This same air then carries away the heat as the air is exhausted out of the anvil of the storm.

Thus, the warming is spread over a much larger volume of air than just the air column containing the thunderstorm. Using the factor of 5 number as estimated by the needed moisture supply, we get a much more reasonable estimate of (101°C)/5 ≈ 20°C of warming. This is still a bit too large, because we have neglected the mixing of the updraft air with additional environmental air as part of the cloud dynamics, and have neglected heat losses by radiation to space. Also, the Holt storm, like the Big Thompson Canyon storm, were extreme events — many thunderstorms are smaller or shorter lived.

The net result of the latent heating is that the upper troposphere (anvil level) has warmed because of the storm, while the lower troposphere has cooled as a result of the rain-induced cold-air downburst. Namely, the thunderstorm did its job of removing static instability from the atmosphere, and leaving the atmosphere in a more stable state. This is a third reason why the first thunderstorms reduce the likelihood of subsequent storms.

In summary, the **three reasons why a thunderstorm suppresses neighboring storms** are: (1) the surrounding environment becomes stabilized (smaller CAPE, larger CIN), (2) sources of nearby boundary-layer fuel are exhausted, and (3) subsidence around the storm suppresses other incipient storm updrafts. But don't forget about other thunderstorm processes such as the gust front that tend to trigger new storms. Thus, competing processes work in thunderstorms, making them difficult to forecast.

alent to a tropospheric warming rate of 25°C h^{-1}, causing a total latent heat release of about $9.1x10^{16}$ J. This **thunderstorm energy** (based only on latent heat release) was equivalent to the energy from 23 one-megaton nuclear bomb explosions (given about $4x10^{15}$ J of heat per **one-megaton nuclear bomb**).

This amount of rain was possible for two reasons: (1) the continual inflow of humid air from the boundary layer into a well-organized (long lasting) orographic thunderstorm (Fig 14.11), and (2) the weakly sheared environment allowed a precipitation efficiency of about 85%. Comparing 305 mm observed with 61 mm expected from a single troposphere-tall column of humid air, we conclude that the equivalent of about 5 troposphere-thick columns of thunderstorm air were consumed by the storm.

Since the thunderstorm is about 6 times as tall as the boundary layer is thick (in pressure coordinates, Fig. 15.2), conservation of air mass suggests that the Big Thompson Canyon storm drew boundary-layer air from an area about 5·6 = 30 times the cross-sectional area of the storm updraft (or 12 times the updraft radius). Namely, a thunderstorm updraft core of 5 km radius would ingest the fuel supply of boundary-layer air from within a radius of 60 km. This is a second reason why subsequent storms are less likely in the neighborhood of the first thunderstorm. Namely, the "fuel tank" is empty after the first thunderstorm, until the fuel supply can be regenerated locally via solar heating and evaporation of surface water, or until fresh fuel of warm humid air is blown in by the wind.

Figure 15.3
Large hailstones and damage to car windshield.

Hail

Hailstones are irregularly shaped balls of ice larger than 0.5 cm diameter that fall from severe thunderstorms. The event or process of hailstones falling out of the sky is called **hail**. The damage path on the ground due to a moving hail storm is called a **hail swath**.

Most hailstones are in the 0.5 to 1.5 cm diameter range, with about 25% of the stones greater than 1.5 cm. While rare, hailstones are called **giant hail** (or **large** or **severe hail**) if their diameters are between 1.9 and 5 cm. Hailstones with diameters ≥ 5 cm are called **significant hail** or **enormous hail** (Fig. 15.3) One stone of diameter 17.8 cm was found in Nebraska, USA, in June 2003. The largest recorded hailstone had diameter 20.3 cm and weighed 878.8 g — it fell in Vivian, S. Dakota, USA on 23 July 2010.

Hailstone diameters are sometimes compared to standard size balls (ping-pong ball = 4 cm; tennis ball ≈ 6.5 cm). They are also compared to nonstandard sizes of fruit, nuts, and vegetables. One such classification is the TORRO Hailstone Diameter relationship (Table 15-1).

Table 15-1. TORRO Hailstone Size Classification.

Size Code	Max. Diameter (cm)	Description
0	0.5 - 0.9	Pea
1	1.0 - 1.5	Mothball
2	1.6 - 2.0	Marble, grape
3	2.1 - 3.0	Walnut
4	3.1 - 4.0	Pigeon egg to golf ball
5	4.1 - 5.0	Pullet egg
6	5.1 - 6.0	Hen egg
7	6.1 - 7.5	Tennis ball to cricket ball
8	7.6 - 9.0	Large orange to soft ball
9	9.1 - 10.0	Grapefruit
10	> 10.0	Melon

Hail Damage

Large diameter hailstones can cause severe damage to crops, tree foliage, cars, aircraft, and sometimes buildings (roofs and windows). Damage is often greater if strong winds cause the hailstones to move horizontally as they fall. Most humans are smart enough not to be outside during a hail storm, so deaths due to hail in North America are rare, but animals can be killed. Indoors is the safest place for people to be in a hail storm, although inside a metal-roofed vehicle is also relatively safe (but stay away from the front and rear windows, which can break).

The terminal fall velocity of hail increases with hailstone size, and can reach magnitudes greater than 50 m s^{-1} for large hailstones. An equation for hailstone terminal velocity was given in the Precipitation chapter, and a graph of it is shown here in Fig. 15.4. Hailstones have different shapes (smooth and round vs. irregular shaped with protuberances) and densities (average is ρ_{ice} = 900 kg m^{-3}, but varies depending on the amount of air bubbles). This causes a range of air drags (0.4 to 0.8, with average 0.55) and a corresponding range of terminal fall speeds. Hailstones that form in the updraft vault region of a supercell thunderstorm are so heavy that most fall immediately adjacent to the vault (Fig. 15.5).

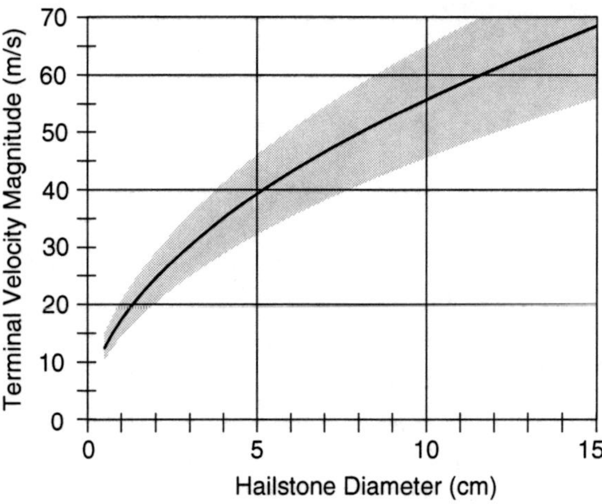

Figure 15.4
Hailstone fall-velocity magnitude relative to the air at pressure height of 50 kPa, assuming an air density of 0.69 kg m^{-3}.

Hail Formation

Two stages of hail development are embryo formation, and then hailstone growth. A hail **embryo** is a large frozen raindrop or graupel particle (< 5 mm diameter) that is heavy enough to fall at a different speed than the surrounding smaller cloud droplets. It serves as the nucleus of hailstones. Like all normal (non-hail) precipitation, the embryo first rises in the updraft as a growing cloud droplet or ice crystal that eventually becomes large enough (via collision and accretion, as discussed in the Precipitation chapter) to begin falling back toward Earth.

While an embryo is being formed, it is still so small that it is easily carried up into the anvil and out of the thunderstorm, given typical severe thunderstorm updrafts of 10 to 50 m s^{-1}. Most potential embryos are removed from the thunderstorm this way, and thus cannot then grow into hailstones.

The few embryos that do initiate hail growth are formed in regions where they are not ejected from the storm, such as: (1) outside of the main updraft in the flanking line of cumulus congestus clouds or in other smaller updrafts, called **feeder cells**; (2) in a side eddy of the main updraft; (3) in a portion of the main updraft that tilts upshear, or (4) earlier in the evolution of the thunderstorm while the main updraft is still weak. Regardless of how it is formed, it is believed that the embryos then move or fall into the main updraft of the severe thunderstorm a second time.

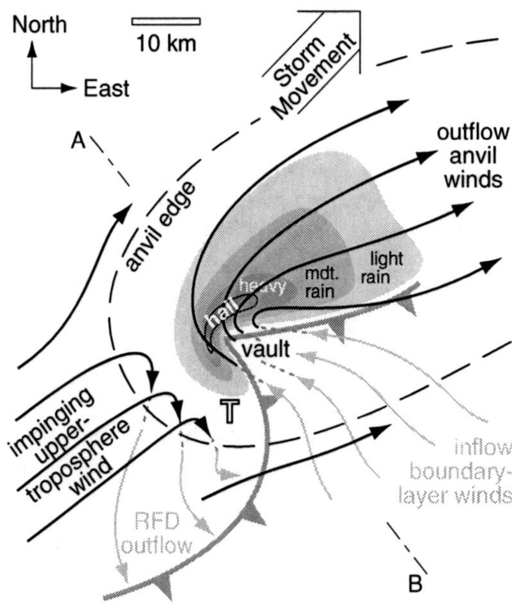

Figure 15.5
Plan view of classic (CL) supercell in the N. Hemisphere (copied from the Thunderstorm chapter). Low altitude winds are shown with light-grey arrows, high altitude with black, and ascending/descending with dashed lines. T indicates tornado location. Precipitation is at the ground. Cross section A-B is used in Fig. 15.10.

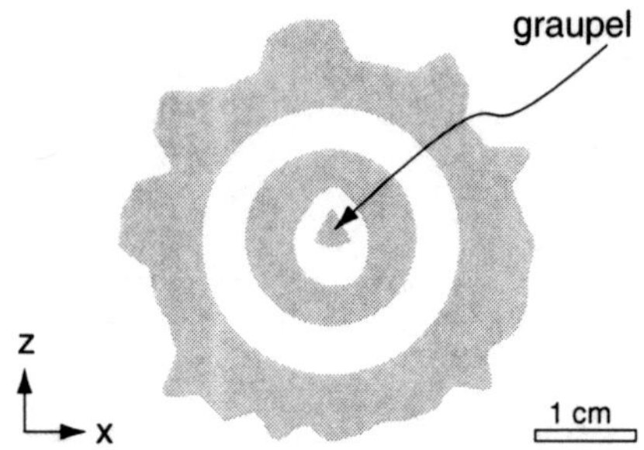

Figure 15.6
Illustration of slice through a hailstone, showing a graupel embryo surrounded by 4 layers of alternating clear ice (indicated with grey shading) and porous (white) ice.

Sample Application

If a supercooled cloud droplet of radius 50 µm and temperature –20°C hits a hailstone, will it freeze instantly? If not, how much heat must be conducted out of the droplet (to the hailstone and the air) for the droplet to freeze?

Find the Answer

Given: r = 50 µm = $5\text{x}10^{-5}$ m, T = –20°C
Find: ΔQ_E = ? J, ΔQ_H = ? J,
Is $\Delta Q_E < \Delta Q_H$? If no, then find $\Delta Q_E - \Delta Q_H$.
Use latent heat and specific heat for liquid water, from Appendix B.

Assume a spherical droplet of mass
$m_{liq} = \rho_{liq} \cdot Vol = \rho_{liq} \cdot (4/3) \cdot \pi \cdot r^3$
$= (1000 \text{ kg m}^{-3}) \cdot (4/3) \cdot \pi \cdot (5\text{x}10^{-5} \text{ m})^3 = 5.2\text{x}10^{-10}$ kg
Use eq. (3.3) to determine how much heat must be removed to freeze the whole droplet ($\Delta m = m_{liq}$):
$\Delta Q_E = L_f \cdot \Delta m = (3.34\text{x}10^5 \text{ J kg}^{-1}) \cdot (5.2\text{x}10^{-10} \text{ kg})$
$= 1.75\text{x}10^{-4}$ J .
Use eq. (3.1) to find how much can be taken up by allowing the droplet to warm from –20°C to 0°C:
$\Delta Q_H = m_{liq} \cdot C_{liq} \cdot \Delta T$
$= (5.2\text{x}10^{-10} \text{ kg}) \cdot [4217.6 \text{ J (kg·K)}^{-1}] \cdot [0°\text{C}-(-20°\text{C})]$
$= 0.44\text{x}10^{-4}$ J .
Thus $\Delta Q_E > \Delta Q_H$, so the sensible-heat deficit associated with –20°C is not enough to compensate for the latent heat of fusion needed to freeze the drop. The droplet will **NOT** freeze instantly.

The amount of heat remaining to be conducted away to the air or the hailstone to allow freezing is:
$\Delta Q = \Delta Q_E - \Delta Q_H = (1.75\text{x}10^{-4} \text{ J}) - (0.44\text{x}10^{-4} \text{ J})$
$= \mathbf{1.31\text{x}10^{-4}}$ J .

Check: Units OK. Physics OK.
Exposition: During the several minutes needed to conduct away this heat, the liquid can flow over the hailstone before freezing, and some air can escape. This creates a layer of clear ice.

The hailstone grows during this second trip through the updraft. Even though the embryo is initially rising in the updraft, the smaller surrounding supercooled cloud droplets are rising faster (because their terminal fall velocity is slower), and collide with the embryo. Because of this requirement for abundant supercooled cloud droplets, hail forms at altitudes where the air temperature is between –10 and –30°C. Most growth occurs while the hailstones are floating in the updraft while drifting relatively horizontally across the updraft in a narrow altitude range having temperatures of –15 to –20°C.

In pockets of the updraft happening to have relatively low liquid water content, the supercooled cloud droplets can freeze almost instantly when they hit the hailstone, trapping air in the interstices between the frozen droplets. This results in a porous, brittle, white layer around the hailstone. In other portions of the updraft having greater liquid water content, the water flows around the hail and freezes more slowly, resulting in a hard clear layer of ice. The result is a hailstone with 2 to 4 visible layers around the embryo (when the hailstone is sliced in half, as sketched in Fig. 15.6), although most hailstones are small and have only one layer. Giant hail can have more than 4 layers.

As the hailstone grows and becomes heavier, its terminal velocity increases and eventually surpasses the updraft velocity in the thunderstorm. At this point, it begins falling relative to the ground, still growing on the way down through the supercooled cloud droplets. After it falls into the warmer air at low altitude, it begins to melt. Almost all strong thunderstorms have some small hailstones, but most melt into large rain drops before reaching the ground. Only the larger hailstones (with more frozen mass and quicker descent through the warm air) reach the ground still frozen as hail (with diameters > 5 mm).

Hail Forecasting

Forecasting large-hail potential later in the day is directly tied to forecasting the maximum updraft velocity in thunderstorms, because only in the stronger updrafts can the heavier hailstones be kept aloft against their terminal fall velocities (Fig. 15.4). CAPE is an important parameter in forecasting updraft strength, as was given in eqs. (14.7) and (14.8) of the Thunderstorm chapter. Furthermore, since it takes about 40 to 60 minutes to create hail (including both embryo and hail formation), large hail would be possible only from long-lived thunderstorms, such as supercells that have relatively steady organized updrafts (which can exist only in an environment with appropriate wind shear).

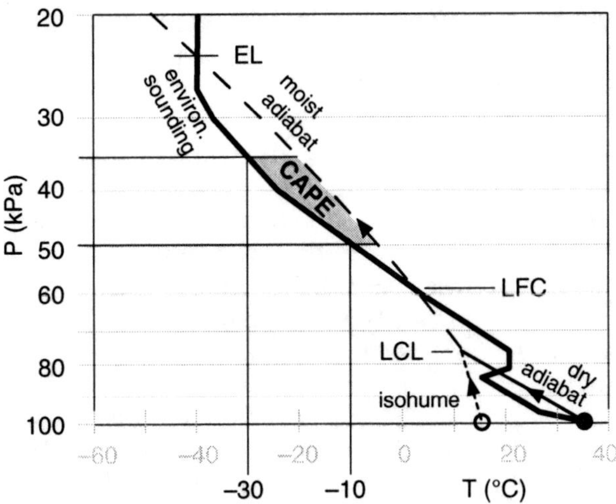

Figure 15.7
Shaded grey is the portion of CAPE area between altitudes where the environment is between –10 and –30°C. Greater areas indicate greater hail likelihood.

However, even if all these conditions are satisfied, hail is not guaranteed. So national forecast centers in North America do not issue specific hail watches, but include hail as a possibility in severe thunderstorm watches and warnings.

To aid in hail forecasting, meteorologists sometimes look at forecast maps of the portion of CAPE between altitudes where the environmental air temperature is $-30 \le T \le -10$°C, such as sketched in Fig. 15.7. Larger values (on the order of 400 J kg^{-1} or greater) of this portion of CAPE are associated with more rapid hail growth. Computers can easily calculate this portion of CAPE from soundings produced by numerical forecast models, such as for the case shown in Fig. 15.8. Within the shaded region of large CAPE on this figure, hail would be forecast at only those subsets of locations where thunderstorms actually form.

Weather maps of freezing-level altitude and wind shear between 0 to 6 km are also used by hail forecasters. More of the hail will reach the ground without melting if the freezing level is at a lower altitude. Environmental wind shear enables longer-duration supercell updrafts, which favor hail growth.

Research is being done to try to create a single forecast parameter that combines many of the factors favorable for hail. One example is the **Significant Hail Parameter (SHIP)**:

$$SHIP = \{ MUCAPE(\text{J kg}^{-1}) \cdot r_{MUP}(\text{g kg}^{-1}) \cdot \gamma_{70\text{-}50kPa}(^\circ\text{C km}^{-1}) \cdot [-T_{50kPa}(^\circ\text{C})] \cdot TSM_{0\text{-}6km}(\text{m s}^{-1}) \} \; / \; a \qquad (15.4)$$

Sample Application

What is the largest size hailstone that could be supported in a thunderstorm having CAPE = 1976 J kg^{-1}? Also give its TORRO classification.

Find the Answer

Given: CAPE = 1976 J kg^{-1}.
Find: d_{max} = ? cm (max hailstone diameter)

From Appendix A, note that units J kg^{-1} = m^2 s^{-2}. First, use eqs. (14.7) and (14.8) from the Thunderstorm chapter to get the likely max updraft speed. This was already computed in a Sample Application near those eqs:

$$w_{max\ likely} = 31 \text{ m s}^{-1}$$

Assume that the terminal fall velocity of the largest hailstone is just balanced by this updraft.

$$w_T = -w_{max\ likely} = -31 \text{ m s}^{-1}$$

where the negative sign implies downward motion. Then use Fig. 15.4 to find the diameter.

$$d_{max} \approx \underline{3.1 \text{ cm}}$$

From Table 15-1, the TORRO hail **size code is 4**, which corresponds to **pigeon egg to golf ball size**.

Check: Units OK. Physics OK.
Exposition: Hail this size would be classified as large or giant hail, and could severely damage crops.

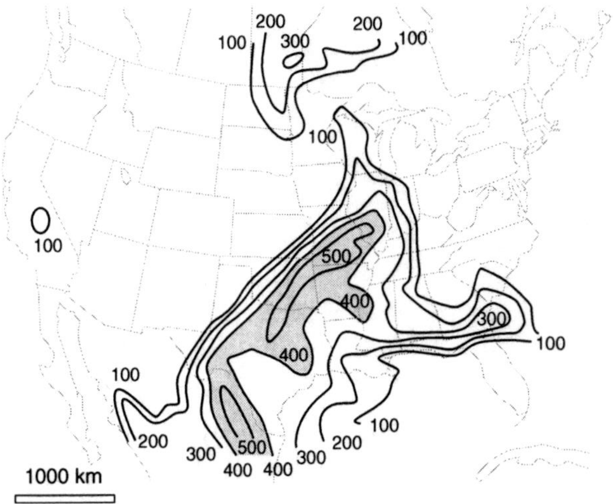

Figure 15.8
Portion of CAPE (J kg^{-1}) between altitudes where the environment is between –10 and –30°C. Larger values indicate greater hail growth rates. Case: 22 UTC on 24 May 2006 over the USA and Canada.

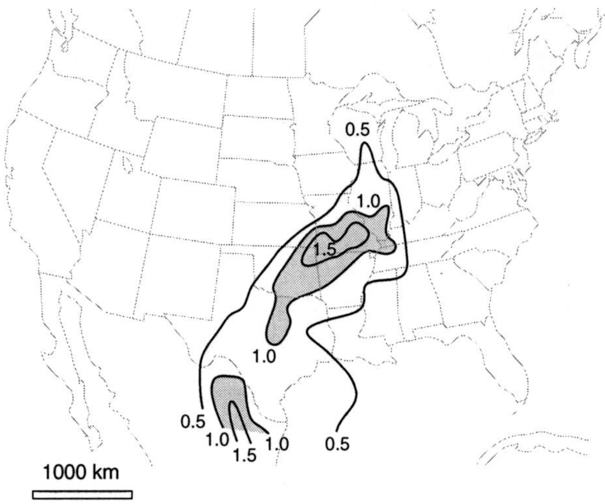

Figure 15.9
Values of significant hail parameter (SHIP) over the USA for the same case as the previous figure. This parameter is dimensionless.

Sample Application
Suppose a pre-storm environmental sounding has the following characteristics over a corn field:

$MUCAPE = 3000$ J kg^{-1},
$r_{MUP} = 14$ g kg^{-1},
$\gamma_{70\text{-}50kPa} = 5$ °C km^{-1},
$T_{50kPa} = -10$°C
$TSM_{0\text{-}6km} = 45$ m s^{-1}

If a thunderstorm forms in this environment, would significant hail (with diameters ≥ 5 cm) be likely?

Find the Answer
Given: values listed above
Find: $SHIP = ?$.

Use eq. (15.4):

$$SHIP = [(3000 \text{ J kg}^{-1}) \cdot (14 \text{ g kg}^{-1}) \cdot (5 \text{ °C km}^{-1}) \cdot (10\text{°C}) \cdot (45 \text{ m s}^{-1})] / (44\text{x}10^6)$$
$$= 99.5\text{x}10^6 / (44\text{x}10^6) = \underline{2.15}$$

Check: Units are dimensionless. Value reasonable.
Exposition: Because SHIP is much greater than 1.0, significant (tennis ball size or larger) hail is indeed likely. This would likely totally destroy the corn crop. Because hail forecasting has so many uncertainties and often short lead times, the farmers don't have time to take action to protect or harvest their crops. Thus, their only recourse is to purchase crop insurance.

where r_{MUP} is the water vapor mixing ratio for the most-unstable air parcel, $\gamma_{70\text{-}50kPa}$ is the average environmental lapse rate between pressure heights 70 and 50 kPa, T_{50kPa} is the temperature at a pressure height of 50 kPa, $TSM_{0\text{-}6km}$ is the total shear magnitude between the surface and 6 km altitude, and empirical parameter $a = 44\text{x}10^6$ (with dimensions equal to those shown in the numerator of the equation above, so as to leave *SHIP* dimensionless).

This parameter typically ranges from 0 to 4 or so. If $SHIP > 1$, then the prestorm environment is favorable for significant hail (i.e., hail diameters ≥ 5 cm). Significant hail is frequently observed when $1.5 \le SHIP$. Fig. 15.9 shows a weather map of *SHIP* for the 22 UTC 24 May 2006 case study.

Nowcasting (forecasting 1 to 30 minutes ahead) large hail is aided with weather radar:

- Large hailstones cause very large radar reflectivity (order of 60 to 70 dBZ) compared to the maximum possible from very heavy rain (up to 50 dBZ). Some radar algorithms diagnose hail when it finds reflectivities ≥ 40 dBZ at altitudes where temperatures are below freezing, with greater chance of hail for ≥ 50 dBZ at altitudes above the –20°C level.
- Doppler velocities can show if a storm is organized as a supercell, which is statistically more likely to support hail.
- Polarimetric methods (see the Satellites & Radar chapter) allow radar echoes from hail to be distinguished from echoes from rain or smaller ice particles.
- The updrafts in some supercell thunderstorms are so strong that only small cloud droplets exist, causing weak (<25 dBZ) radar reflectivity, and resulting in a **weak-echo region (WER)** on the radar display. Sometimes the WER is surrounded on the top and sides by strong precipitation echoes, causing a **bounded weak-echo region (BWER)**, also known as an **echo-free vault**. This enables very large hail, because embryos falling from the upshear side of the bounding precipitation can re-enter the updraft, thereby efficiently creating hail (Fig. 15.10).

Hail Locations

The hail that does fall often falls closest to the main updraft (Figs. 15.5 & 15.10), and the resulting **hail shaft** (the column of falling hailstones below cloud base) often looks white or invisible to observers on the ground. Most hail falls are relatively short lived, causing small (10 to 20 km long, 0.5 to 3 km wide) damage tracks called **hailstreaks**. Sometimes long-lived supercell thunderstorms can create longer **hailswaths** of damage 8 to 24 km wide and

160 to 320 km long. Even though large hail can be extremely damaging, the mass of water in hail at the ground is typically only 2 to 3% of the mass of rain from the same thunderstorm.

In the USA, most giant hail reaching the ground is found in the central and southern plains, centered in Oklahoma (averaging 6 to 9 giant-hail days yr^{-1}), and extending from Texas north through Kansas and Nebraska (3 or more giant-hail days yr^{-1}). Hail is also observed less frequently (1 to 3 giant-hail days yr^{-1}) eastward across the Mississippi valley and into the southern and mid-Atlantic states.

Although hail is less frequent in Canada than in the USA, significant hail falls are found in Alberta between the cities of Calgary and Edmonton, particularly near the town of Red Deer. Hail is also found in central British Columbia, and in the southern prairies of Saskatchewan and Manitoba.

In the S. Hemisphere, hail falls often occur over eastern Australia. The 14 April 1999 hailstorm over Sydney caused an estimated AUS$ 2.2 billion in damage, the second largest weather-related damage total on record for Australia. Hailstorms have been observed over North and South America, Europe, Australia, Asia, and Africa.

Hail Mitigation

Attempts at **hail suppression** (mitigation) have generally been unsuccessful, but active hail-suppression efforts still continue in most continents to try to reduce crop damage. Five approaches have been suggested for suppressing hail, all of which involve **cloud seeding** (adding particles into clouds to serve as additional or specialized hydrometeor nuclei), which is difficult to do precisely:

- **beneficial competition** - to create larger numbers of embryos that compete for supercooled cloud water, thereby producing larger numbers of smaller hailstones (that melt before reaching the ground). The methods are cloud seeding with **hygroscopic** (attracts water; e.g., salt particles) cloud nuclei (to make larger rain drops that then freeze into embryos), or seeding with **glaciogenic** (makes ice; e.g., silver iodide particles) ice nuclei to make more graupel.
- **early rainout** - to cause precipitation in the cumulus congestus clouds of the flanking line, thereby reducing the amount of cloud water available before the updraft becomes strong enough to support large hail. The method is seeding with ice nuclei.
- **trajectory altering** - to cause the embryos to grow to greater size earlier, thereby following a lower trajectory through the updraft where the temperature or supercooled liquid water con-

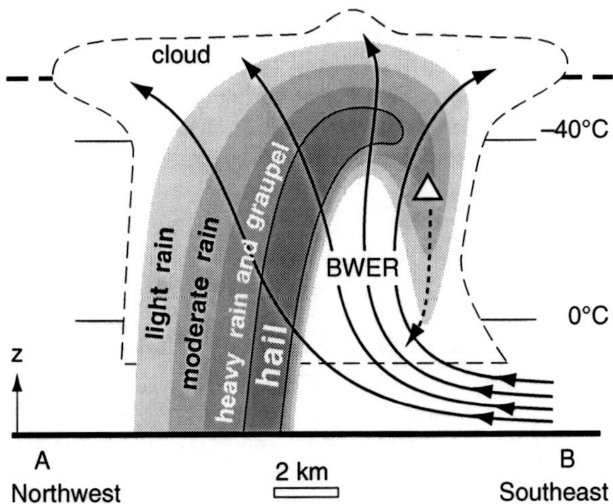

Figure 15.10
Vertical cross section through a classic supercell thunderstorm along slice A-B from Fig. 15.5. Thin dashed line shows visible cloud boundary, and shading indicates intensity of precipitation as viewed by radar. BWER = bounded weak echo region of supercooled cloud droplets. White triangle represents graupel on the upshear side of the storm, which can fall (dotted line) and re-enter the updraft to serve as a hail embryo. Thick dashed line is the tropopause. Isotherms are thin solid lines. Curved thick black lines with arrows show air flow.

INFO • Hail Suppression

For many years there has been a very active cloud seeding effort near the town of Red Deer, Alberta, Canada, in the hopes of suppressing hail. These activities were funded by some of the crop-insurance companies, because their clients, the farmers, demanded that something be done.

Although the insurance companies knew that there is little solid evidence that hail suppression actually works, they funded the cloud seeding anyway as a public-relations effort. The farmers appreciated the efforts aimed at reducing their losses, and the insurance companies didn't mind because the cloud-seeding costs were ultimately borne by the farmers via increased insurance premiums.

Sample Application

During cloud seeding, how many silver iodide particles need to be introduced into a thunderstorm to double the number of ice nuclei? Assume the number density of natural ice nuclei is 10,000 per cubic meter.

Find the Answer

Given: $n_{ice\ nuclei}/Volume$ = 10,000 m^{-3}
Find: N_{total} = ? total count of introduced nuclei

To double ice nuclei, the count of introduced nuclei must equal the count of natural nuclei:

$$N_{total} = (n_{ice\ nuclei}/Volume) \cdot Volume$$

Estimate the volume of a thunderstorm above the freezing level. Assume freezing level is at 3 km altitude, and the anvil top is at 12 km. Approximate the thunderstorm by a box of bottom surface area 12 x 12 km, and height 9 km (= 12 – 3).

$$Volume \approx 1300 \text{ km}^3 = 1.3\text{x}10^{12} \text{ m}^3$$

Thus:

$$N_{total} = (n_{ice\ nuclei}/Volume) \cdot Volume = (10{,}000 \text{ m}^{-3}) \cdot (1.3\text{x}10^{12} \text{ m}^3) = \underline{1.3\text{x}10^{16}}$$

Check: Units OK. Physics OK.

Exposition: Cloud seeding is often done by an aircraft. For safety reasons, the aircraft doesn't usually fly into the violent heart of the thunderstorm. Instead, it flies under the rain-free portion of cloud base, releasing the silver iodide particles into the updraft in the hopes that the nuclei get to the right part of the storm at the right time. It is not easy to do this correctly, and even more difficult to confirm if the seeding caused the desired change. Seeding thunderstorms is an uncontrolled experiment, and one never knows how the thunderstorm would have changed without the seeding.

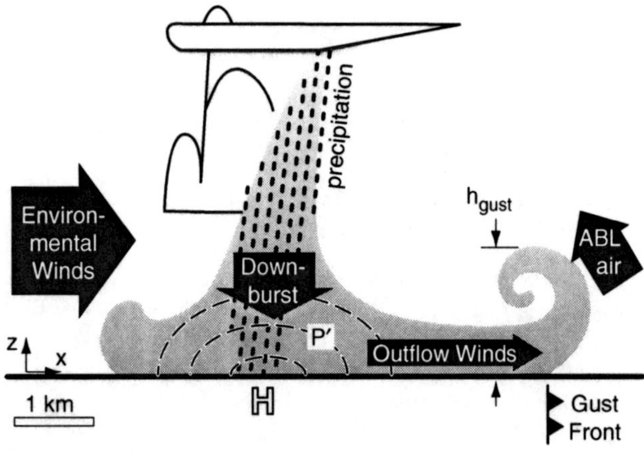

Figure 15.11
Vertical slice through a thunderstorm downburst and its associated gust front. Grey shading indicates the rain-cooled air. Behind the gust front are non-tornadic (i.e., straight-line) outflow winds. H indicates location of meso-high pressure at the surface, and isobars of positive pressure perturbation P' are dashed lines. Dotted lines show precipitation. ABL is the warm, humid environmental air in the atmospheric boundary layer.

tent is not optimum for large hail growth. This method attempts to increase rainfall (in drought regions) while reducing hail falls.

- **dynamic effects** - to consume more CAPE earlier in the life cycle of the updraft (i.e., in the cumulus congestus stage), thereby leaving less energy for the main updraft, causing it to be weaker (and supporting only smaller hail).
- **glaciation of supercooled cloud water** - to more quickly convert the small supercooled cloud droplets into small ice crystals that are less likely to stick to hail embryos and are more likely to be blown to the top of the storm and out via the anvil. This was the goal of most of the early attempts at hail suppression, but has lost favor as most hail suppression attempts have failed.

GUST FRONTS AND DOWNBURSTS

Attributes

Downbursts are rapidly descending (w = –5 to –25 m s^{-1}) downdrafts of air (Fig. 15.11), found below clouds with precipitation or virga. Downbursts of 0.5 to 10 km diameter are usually associated with thunderstorms and heavy rain. Downburst speeds of order 10 m s^{-1} have been measured 100 m AGL. The descending air can hit the ground and spread out as strong **straight-line winds** causing damage equivalent to a weak tornado (up to EF3 intensity). Smaller mid-level clouds (e.g., altocumulus with virga) can also produce downbursts that usually do not reach the ground.

Small diameter (1 to 4 km) downbursts that last only 2 to 5 min are called **microbursts**. Sometimes a downburst area will include one or more imbedded microbursts.

Acceleration of downburst velocity w is found by applying Newton's 2nd law of motion to an air parcel:

$$\frac{\Delta w}{\Delta t} \approx -\frac{1}{\rho}\frac{\Delta P'}{\Delta z} + |g| \cdot \left[\frac{\theta_v'}{\theta_{ve}} - \frac{C_v}{C_p}\frac{P'}{P_e} \right] \qquad \bullet(15.5)$$

Term: (A) (B) (C)

where w is negative for downdrafts, t is time, ρ is air density, $P' = P_{parcel} - P_e$ is pressure perturbation of the air parcel relative to the environmental pressure P_e, z is height, $|g|$ = 9.8 m s^{-2} is the magnitude of gravitational acceleration, $\theta_v' = \theta_{v\ parcel} - \theta_{ve}$ is the deviation of the parcel's virtual potential temperature from that of the environment θ_{ve} (in Kelvin), and C_v/C_p = 0.714 is the ratio of specific heat of air at constant volume to that at constant pressure.

Remember that the virtual potential temperature (from the Heat chapter) includes liquid-water and ice loading, which makes the air act as if it were colder, denser, and heavier. Namely, for the air parcel it is:

$$\theta_{v\ parcel} = \theta_{parcel} \cdot (1 + 0.61 \cdot r - r_L - r_I)_{parcel} \qquad (15.6)$$

where θ is air potential temperature (in Kelvin), r is water-vapor mixing ratio (in g g^{-1}, not g kg^{-1}), r_L is liquid water mixing ratio (in g g^{-1}), and r_I is ice mixing ratio (in g g^{-1}). For the special case of an environment with no liquid water or ice, the environmental virtual potential temperature is:

$$\theta_{ve} = \theta_e \cdot (1 + 0.61 \cdot r)_e \qquad (15.7)$$

Equation (15.5) says that three forces (per unit mass) can create or enhance downdrafts. (A) Pressure-gradient force is caused when there is a difference between the pressure profile in the environment (which is usually hydrostatic) and that of the parcel. (B) Buoyant force combines the effects of temperature in the evaporatively cooled air, precipitation drag associated with falling rain drops or ice crystals, and the relatively lower density of water vapor. (C) Perturbation-pressure buoyancy force is where an air parcel of lower pressure than its surroundings experiences an upward force. Although this last effect is believed to be small, not much is really known about it, so we will neglect it here.

Evaporative cooling and precipitation drag are important for initially accelerating the air downward out of the cloud. We will discuss those factors first, because they can create downbursts. The vertical pressure gradient becomes important only near the ground. It is responsible for decelerating the downburst just before it hits the ground, which we will discuss in the "gust front" subsection.

Precipitation Drag on the Air

When **hydrometeors** (rain drops and ice crystals) fall at their terminal velocity through air, the drag between the hydrometeor and the air tends to pull some of the air with the falling precipitation. This **precipitation drag** produces a downward force on the air equal to the weight of the precipitation particles. For details on precipitation drag, see the Precipitation Processes Chapter.

This effect is also called **liquid-water loading** or **ice loading**, depending on the phase of the hydrometeor. The downward force due to precipitation loading makes the air parcel act heavier, having the same effect as denser, colder air. As was discussed in the Atmos. Basics and Heat Budgets chapters, use virtual temperature T_v or virtual potential temperature θ_v to quantify this effective cooling.

Sample Application

10 g kg^{-1} of liquid water exists as rain drops in saturated air of temperature 10°C and pressure 80 kPa. The environmental air has a temperature of 10°C and mixing ratio of 4 g kg^{-1}. Find the: (a) buoyancy force per mass associated with air temperature and water vapor, (b) buoyancy force per mass associated with just the precipitation drag, and (c) the downdraft velocity after 1 minute of fall, due to only (a) and (b).

Find the Answer

Given: Parcel: r_L = 10 g kg^{-1}, T = 10°C, P = 80 kPa,
Environ.: r = 4 g kg^{-1}, T = 10°C, P = 80 kPa,
Δt = 60 s. Neglect terms (A) and (C).

Find: (a) Term($B_{\text{due to } T \,\&\, r}$) = ? N kg^{-1},
(b) Term($B_{\text{due to } r_L \,\&\, r_I}$) = ? N kg^{-1} (c) w = ? m s^{-1}

(a) Because the parcel air is saturated, $r_{parcel} = r_s$. Using a thermo diagram (because its faster than solving a bunch of equations), $r_s \approx$ 9.5 g kg^{-1} at P = 80 kPa and T = 10°C. Also, from the thermo diagram, $\theta_{parcel} \approx$ 28°C = 301 K. Thus, using the first part of eq. (15.6):

$\theta_{v\ parcel} \approx$ (301 K)·[1 + 0.61·(0.0095 g g^{-1})] ≈ 302.7 K

For the environment, also θ ≈ 28°C = 301 K, but r = 4 g kg^{-1}. Thus, using eq. (15.7):

$\theta_{ve} \approx$ (301 K)·[1 + 0.61·(0.004 g g^{-1})] ≈ 301.7 K

Use eq. (15.5):

Term($B_{\text{due to } T \,\&\, r}$) = $|g|$·[($\theta_{v\ parcel} - \theta_{ve}$) / θ_{ve}]
= (9.8 m s^{-2})·[(302.7 – 301.7 K) / 301.7 K]
= 0.032 m s^{-2} = <u>**0.032 N kg^{-1}**</u>.

(b) Because r_I was not given, assume r_I = 0 everywhere, and r_L = 0 in the environment.

Term($B_{\text{due to } r_L \,\&\, r_I}$) = $-|g|$·[$(r_L + r_I)_{parcel} - (r_L + r_I)_e$].
= – (9.8 m s^{-2})·[0.01 g g^{-1}] = <u>**–0.098 N kg^{-1}**</u>.

(c) Assume initial vertical velocity is zero.
Use eq. (15.5) with only Term B:
($w_{final} - w_{initial}$)/ Δt = [Term($B_{T \,\&\, r}$) + Term($B_{r_L \,\&\, r_I}$)]
w_{final} = (60 s)·[0.032 – 0.098 m s^{-2}] ≈ <u>**–4 m s^{-1}**</u>.

Check: Units OK. Physics OK.

Exposition: Although the water vapor in the air adds buoyancy equivalent to a temperature <u>in</u>crease of 1°C, the liquid-water loading decreases buoyancy, equivalent to a temperature <u>de</u>crease of 3°C. The net effect is that this saturated, liquid-water laden air acts 2°C colder and heavier than dry air at the same T.

CAUTION. The final vertical velocity assumes that the air parcel experiences constant buoyancy forces during its 1 minute of fall. This is NOT a realistic assumption, but it did make the exercise a bit easier to solve. In fact, if the rain-laden air descends below cloud base, then it is likely that the rain drops are in an <u>un</u>saturated air parcel, not saturated air as was stated for this exercise. We also neglected turbulent drag of the downburst air against the environmental air. This effect can greatly reduce the actual downburst speed compared to the idealized calculations above.

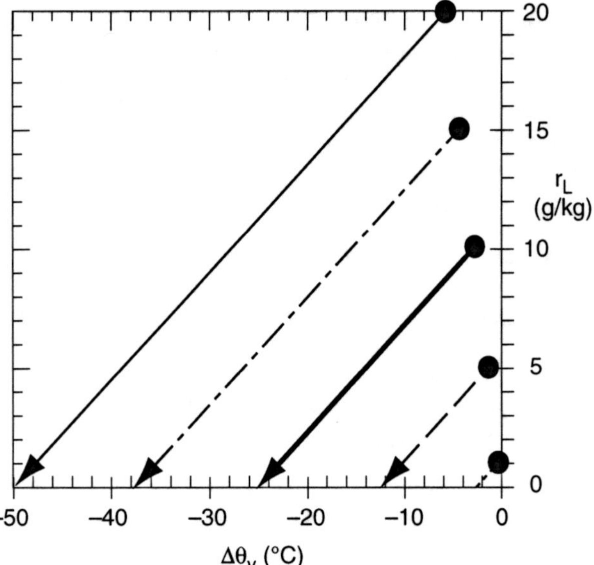

Figure 15.12
Rain drops reduce the virtual potential temperature of the air by both their weight (precipitation drag) and cooling as they evaporate. $\Delta\theta_v$ is the change of virtual potential temperature compared to air containing no rain drops initially. r_L is liquid-water mixing ratio for the drops in air. For any initial r_L along the vertical axis, the black dot indicates the $\Delta\theta_v$ due to only liquid water loading. As that same raindrop evaporates, follow the diagonal line down to see changes in both r_L and $\Delta\theta_v$.

Sample Application

For data from the previous Sample Application, find the virtual potential temperature of the air if:
a) all liquid water evaporates, and
b) no liquid water evaporates, leaving only the precipitation-loading effect.
c) Discuss the difference between (a) and (b)

Find the Answer

Given: $r_L = 10$ g kg^{-1} initially. $r_L = 0$ finally.
Find: (a) $\Delta\theta_{parcel} = ?$ K (b) $\Delta\theta_{v\ parcel} = ?$ K

(a) Use eq. (15.8):
$\Delta\theta_{parcel} = [2.5 \text{ K·kg}_{air}\text{·(g}_{water})^{-1}] \cdot (-10 \text{ g kg}^{-1}) = \underline{\mathbf{-25\ K}}$

(b) From the Exposition section of the previous Sample Application:
$\Delta\theta_{v\ parcel\ precip.\ drag} = -3$ K initially.
Thus, the <u>change</u> of virtual potential temperature (between before and after the drop evaporates) is
$\Delta\theta_{v\ parcel} = \Delta\theta_{v\ parcel\ final} - \Delta\theta_{v\ parcel\ initial}$
$= -25\text{K} - (-3\text{ K}) = \underline{\mathbf{-22\ K}}$.

Check: Units OK. Physics OK.
Exposition: (c) The rain is more valuable to the downburst if it all evaporates.

Cooling due to Droplet Evaporation

Three factors can cause the rain-filled air to be unsaturated. (1) The rain can fall out of the thunderstorm into drier air (namely, the rain moves <u>through</u> the air parcels, not <u>with</u> the air parcels). (2) As air parcels descend in the downdraft (being dragged downward by the rain), the air parcels warm adiabatically and can hold more vapor. (3) Mixing of the rainy air with the surrounding drier environment can result in a mixture with lowered humidity.

Raindrops can partially or totally evaporate in this unsaturated air, converting sensible heat into latent heat. Namely, air cools as the water evaporates, and cool air has negative buoyancy. Negatively buoyant air sinks, creating downbursts of air. One way to quantify the cooling is via the change of potential temperature associated with evaporation of Δr_L ($\text{g}_{liq.water}\ \text{kg}_{air}^{-1}$) of liquid water:

$$\Delta\theta_{parcel} = (L_v/C_p) \cdot \Delta r_L \qquad \bullet(15.8)$$

where $(L_v/C_p) = 2.5$ K·kg$_{air}$·(g$_{water}$)$^{-1}$, and where $\Delta r_L = r_{L\ final} - r_{L\ initial}$ is negative for evaporation. This parcel cooling enters the downdraft-velocity equation via θ_{parcel} in eq. (15.6).

Precipitation drag is usually a smaller effect than evaporative cooling. Fig. 15.12 shows both the precipitation-drag effect for different initial liquid-water mixing ratios (the black dots), and the corresponding cooling and liquid-water decrease as the drops evaporate. For example, consider the black dot corresponding to an initial liquid water loading of 10 g kg^{-1}. Even before that drop evaporates, the weight of the rain decreases the virtual potential temperature by about 2.9°C. However, as that drop evaporates, it causes a much larger amount of cooling to due latent heat absorption, causing the virtual-potential-temperature to decrease by 25°C after it has completely evaporated.

Evaporative cooling can be large in places where the environmental air is dry, such as in the high-altitude plains and prairies of the USA and Canada. There, raining convective clouds can create strong downbursts, even if all the precipitation evaporates before reaching the ground (i.e., for **virga**).

Downbursts are hazardous to aircraft in two ways. (1) The downburst speed can be faster than the climb rate of the aircraft, pushing the aircraft towards the ground. (2) When the downburst hits the ground and spreads out, it can create hazardous changes between headwinds and tailwinds for landing and departing aircraft (see the "Aircraft vs. Downbursts" INFO Box in this section.) Modern airports are equipped with Doppler radar and/or wind-sensor arrays on the airport grounds, so that warnings can be given to pilots.

Downdraft CAPE (DCAPE)

Eq. (15.5) applies at any one altitude. As precipitation-laden air parcels descend, many things change. The descending air parcel cools and looses some of its liquid-water loading due to evaporation, thereby changing its virtual potential temperature. It descends into surroundings having different virtual potential temperature than the environment where it started. As a result of these changes to both the air parcel and its environment, the θ_v' term in eq. (15.5) changes.

To account for all these changes, find term B from eq. (15.5) at each depth, and then sum over all depths to get the accumulated effect of evaporative cooling and precipitation drag. This is a difficult calculation, with many uncertainties.

An alternative estimate of downburst strength is via the **Downdraft Convective Available Potential Energy** (**DCAPE**, see shaded area in Fig. 15.13):

$$DCAPE = \sum_{z=0}^{z_{LFS}} |g| \cdot \frac{\theta_{vp} - \theta_{ve}}{\theta_{ve}} \cdot \Delta z \qquad \bullet(15.9)$$

where $|g|$ = 9.8 m s^{-2} is the magnitude of gravitational acceleration, θ_{vp} is the parcel virtual potential temperature (including temperature, water vapor, and precipitation-drag effects), θ_{ve} is the environment virtual potential temperature (Kelvin in the denominator), Δz is a height increment to be used when conceptually covering the DCAPE area with tiles of equal size.

The altitude z_{LFS} where the precipitation laden air first becomes negatively buoyant compared to the environment is called the **level of free sink** (**LFS**), and is the downdraft equivalent of the level of free convection. If the downburst stays negatively buoyant to the ground, then the bottom limit of the sum is at z = 0, otherwise the downburst would stop at a **downdraft equilibrium level** (**DEL**) and not be felt at the ground. DCAPE is negative, and has units of J kg^{-1} or m^2 s^{-2}.

By relating potential energy to kinetic energy, the downdraft velocity is approximately:

$$w_{max\ down} = -(2 \cdot |DCAPE|)^{1/2} \qquad (15.10)$$

Air drag of the descending air parcel against its surrounding environmental air could reduce the likely downburst velocity w_d to about half this max value (Fig. 15.14):

$$w_d = w_{max\ down} / 2 \qquad \bullet(15.11)$$

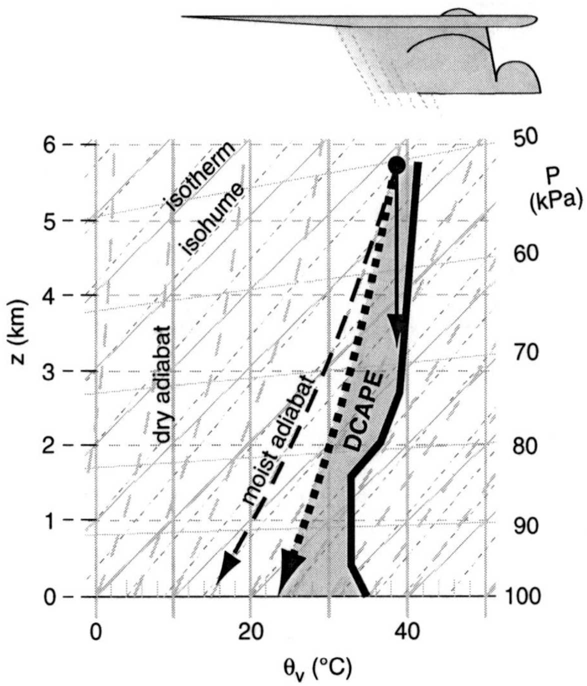

Figure 15.13
Thermo diagram (Theta-z diagram from the Stability chapter) example of Downdraft Convective Available Potential energy (DCAPE, shaded area). Thick solid line is environmental sounding. Black dot shows virtual potential temperature after top of environmental sounding has been modified by liquid-water loading caused by precipitation falling into it from above. Three scenarios of rain-filled air-parcel descent are shown: (a) no evaporative cooling, but only constant liquid water loading (thin solid line following a dry adiabat); (b) an initially saturated air parcel with evaporative cooling of the rain (dashed line following a moist adiabat); and (c) partial evaporation (thick dotted line) with a slope between the moist and dry adiabats.

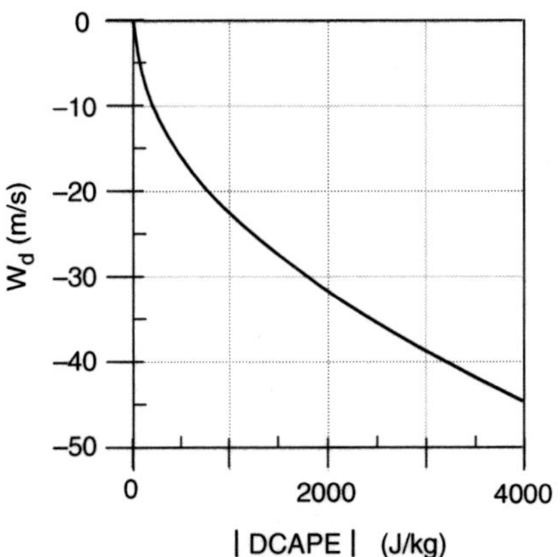

Figure 15.14
Downburst velocity w_d driven by DCAPE

Sample Application

For the shaded area in Fig. 15.13, use the tiling method to estimate the value of DCAPE. Also find the maximum downburst speed and the likely speed.

Find the Answer

Given: Fig. 15.13, reproduced here.
Find: $DCAPE$ = ? $m^2 s^{-2}$,
$w_{max\ down}$ = ? $m\ s^{-1}$ w_d = ? $m\ s^{-1}$

The DCAPE equation (15.9) can be re-written as

$$\text{DCAPE} = -[|g|/\theta_{ve}]\cdot(\text{shaded area})$$

Method: Overlay the shaded region with tiles:

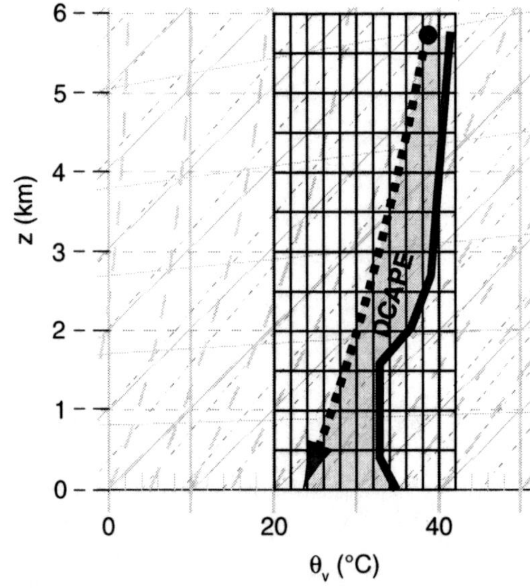

Each tile is 2°C = 2 K wide by 0.5 km tall (but you could pick other tile sizes instead). Hence, each tile is worth 1000 K·m. I count approximately 32 tiles needed to cover the shaded region. Thus:

(shaded area) = 32 x 1000 K·m = 32,000 K·m.

Looking at the plotted environmental sounding by eye, I estimate the average θ_{ve} = 37°C = 310 K.

$$DCAPE = -[(9.8\ m\ s^{-2})/310K]\cdot(32,000\ K\cdot m) = \underline{-1012}\ m^2\ s^{-2}$$

Next, use eq. (15.10):

$$w_{max\ down} = -[\ 2\cdot|-1012\ m^2\ s^{-2}\ |]^{1/2} = \underline{-45\ m\ s^{-1}}$$

Finally, use eq. (15.11):

$$w_d = w_{max\ down}/2 \ = \ \underline{-22.5\ m\ s^{-1}}.$$

Check: Units OK. Physics OK. Drawing OK.

Exposition: While this downburst speed might be observed 1 km above ground, the speed would diminish closer to the ground due to an opposing pressure-perturbation gradient. Since the DCAPE method doesn't account for the vertical pressure gradient, it shouldn't be used below about 1 km altitude.

Stronger downdrafts and associated straight-line winds near the ground are associated with larger magnitudes of DCAPE. For example, Fig. 15.15 shows a case study of DCAPE magnitudes valid at 22 UTC on 24 May 2006.

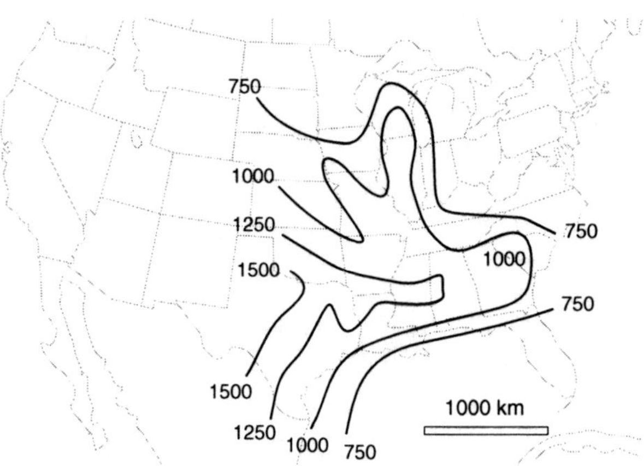

Figure 15.15
Example of downdraft DCAPE magnitude (J kg^{-1}) for 22 UTC 24 May 06 over the USA.

INFO • CAPE vs. DCAPE

Although DCAPE shares the same conceptual framework as CAPE, there is virtually no chance of practically utilizing DCAPE, while CAPE is very useful. Compare the two concepts.

For CAPE: The initial state of the rising air parcel is known or fairly easy to estimate from surface observations and forecasts. The changing thermodynamic state of the parcel is easy to anticipate; namely, the parcel rises dry adiabatically to its LCL, and rises moist adiabatically above the LCL with vapor always close to its saturation value. Any excess water vapor instantly condenses to keep the air parcel near saturation. The resulting liquid cloud droplets are initially carried with the parcel.

For DCAPE: Both the initial air temperature of the descending air parcel near thunderstorm base and the liquid-water mixing ratio of raindrops are unknown. The raindrops don't move with the air parcel, but pass through the air parcel from above. The air parcel below cloud base is often NOT saturated even though there are raindrops within it. The temperature of the falling raindrops is often different than the temperature of the air parcel it falls through. There is no requirement that the adiabatic warming of the air due to descent into higher pressure be partially matched by evaporation from the rain drops (namely, the thermodynamic state of the descending air parcel can be neither dry adiabatic nor moist adiabatic).

Unfortunately, the exact thermodynamic path traveled by the descending rain-filled air parcel is unknown, as discussed in the INFO box. Fig. 15.13 illustrates some of the uncertainty in DCAPE. If the rain-filled air parcel starting at pressure altitude of 50 kPa experiences no evaporation, but maintains constant precipitation drag along with dry adiabatic warming, then the parcel state follows the thin solid arrow until it reaches its DEL at about 70 kPa. This would not reach the ground as a down burst.

If a descending saturated air parcel experiences just enough evaporation to balance adiabatic warming, then the temperature follows a moist adiabat, as shown with the thin dashed line. But it could be just as likely that the air parcel follows a thermodynamic path in between dry and moist adiabat, such as the arbitrary dotted line in that figure, which hits the ground as a cool but unsaturated downburst.

Pressure Perturbation

As the downburst approaches the ground, its vertical velocity must decelerate to zero because it cannot blow through the ground. This causes the dynamic pressure to increase (P' becomes positive) as the air stagnates.

Rewriting Bernoulli's equation (see the Regional Winds chapter) using the notation from eq. (15.5), the maximum stagnation pressure perturbation P'_{max} at the ground directly below the center of the downburst is:

$$P'_{\max} = \rho \cdot \left[\frac{w_d^{\,2}}{2} - \frac{|g| \cdot \theta_v' \cdot z}{\theta_{ve}} \right] \qquad \bullet(15.12)$$

Term: (A) (B)

where ρ is air density, w_d is likely peak downburst speed at height z well above the ground (before it feels the influence of the ground), $|g|$ = 9.8 m s^{-2} is gravitational acceleration magnitude, and virtual potential temperature depression of the air parcel relative to the environment is $\theta_v' = \theta_{v\,parcel} - \theta_{ve}$.

Term (A) is an inertial effect. Term (B) includes the added weight of cold air (with possible precipitation loading) in increasing the pressure [because θ_v' is usually (but not always) negative in downbursts]. Both effects create a mesoscale high (**mesohigh**, **H**) pressure region centered on the downburst. Fig. 15.16 shows the solution to eq. (15.12) for a variety of different downburst velocities and virtual potential

Sample Application

A downburst has velocity –22 m s^{-1} at 1 km altitude, before feeling the influence of the ground. Find the corresponding perturbation pressure at ground level for a downburst virtual potential temperature perturbation of (a) 0, and (b) –5°C.

Find the Answer:
Given: w_d = –22 m s^{-1}, z = 1 km, θ_v' = (a) 0, or (b) –5°C
Find: P'_{max} = ? kPa

Assume standard atmosphere air density at sea level ρ = 1.225 kg m^{-3}
Assume θ_{ve} = 294 K. Thus $|g|/\theta_{ve}$ = 0.0333 m·s^{-2}·K^{-1}
Use eq. (15.12):
(a) P'_{max} = (1.225 kg m^{-3})·[(–22 m s^{-1})^2/2 – 0]
= 296 kg·m^{-1}·s^{-2} = 296 Pa = **0.296 kPa**

(b) θ_v' = –5°C = –5 K because it represents a temperature difference.
P'_{max} = (1.225 kg m^{-3})·
[(–22 m s^{-1})^2/2 – (0.0333 m·s^{-2}·K^{-1})·(–5 K)·(1000 m)]
P'_{max} = [296 + 204] kg·m^{-1}·s^{-2} = 500 Pa = **0.5 kPa**

Check: Units OK. Physics OK. Agrees with Fig. 15.16.
Exposition: For this case, the cold temperature of the downburst causes P' to nearly double compared to pure inertial effects. Although P'_{max} is small, P' decreases to 0 over a short distance, causing large $\Delta P'/\Delta x$ & $\Delta P'/\Delta z$. These large pressure gradients cause large accelerations of the air, including the rapid deceleration of the descending downburst air, and the rapid horizontal acceleration of the same air to create the outflow winds.

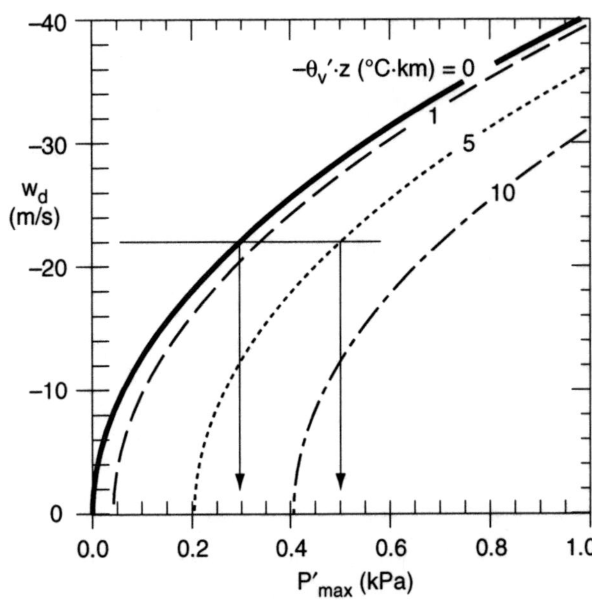

Figure 15.16
Descending air of velocity w_d decelerates to zero when it hits the ground, causing a pressure increase P'_{max} at the ground under the downburst compared to the surrounding ambient atmosphere. For descending air of the same temperature as its surroundings, the result from Bernoulli's equation is plotted as the thick solid line. If the descending air is also cold (i.e., has some a virtual potential temperature deficit $-\theta_v'$ at starting altitude z), then the other curves show how the pressure increases further.

temperature deficits. Typical magnitudes are on the order of 0.1 to 0.6 kPa (or 1 to 6 mb) higher than the surrounding pressure.

As you move away vertically from the ground and horizontally from the downburst center, the pressure perturbation decreases, as suggested by the dashed line isobars in Figs. 15.11 and 15.17. The vertical gradient of this pressure perturbation decelerates the downburst near the ground. The horizontal gradient of the pressure perturbation accelerates the air horizontally away from the downburst, thus preserving air-mass continuity by balancing vertical inflow with horizontal **outflow** of air.

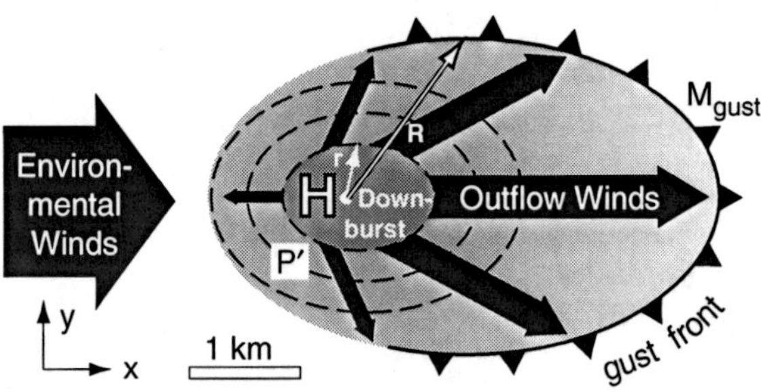

Figure 15.17
Horizontal slice through the air just above the ground, corresponding to Fig. 15.11. Shown are the downburst of average radius r (dark grey), gust front of average radius R (black arc, with cold-frontal symbols [triangles]), cool air (gradient shaded), and outflow winds (thick black arrows) flowing in straight lines. H shows the location of a meso-high-pressure center near the ground, and the dashed lines show isobars of positive perturbation pressure P′.

Outflow Winds & Gust Fronts

Driven by the pressure gradient from the mesohigh, the near-surface **outflow** air tends to spread out in all directions radially from the downburst. It can be enhanced or reduced in some directions by background winds (Fig. 15.17). **Straight-line** outflow winds (i.e., non-tornadic; non-rotating) behind the gust front can be as fast as 35 m s^{-1}, and can blow down trees and destroy mobile homes. Such winds can make a howling sound called **aeolian tones**, as wake eddies form behind wires and twigs.

The outflow winds are accelerated by the perturbation-pressure gradient associated with the downburst mesohigh. Considering only the horizontal pressure-gradient force in Newton's 2nd Law (see the Forces & Winds Chapter), you can estimate the acceleration from

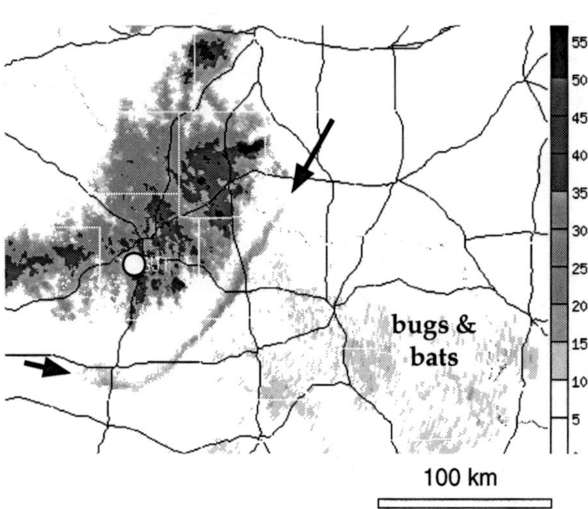

Figure 15.18
Radar reflectivity from the San Angelo (SJT), Texas, USA, weather radar. White circle shows radar location. Thunderstorm cells with heavy rain (darker greys) are over and northeast of the radar. Arrows show ends of gust front. Scale at right is radar reflectivity in dBZ. Radar elevation angle is 0.5°. Both figs. courtesy of the National Center for Atmospheric Research, based on National Weather Service radar data, NOAA.

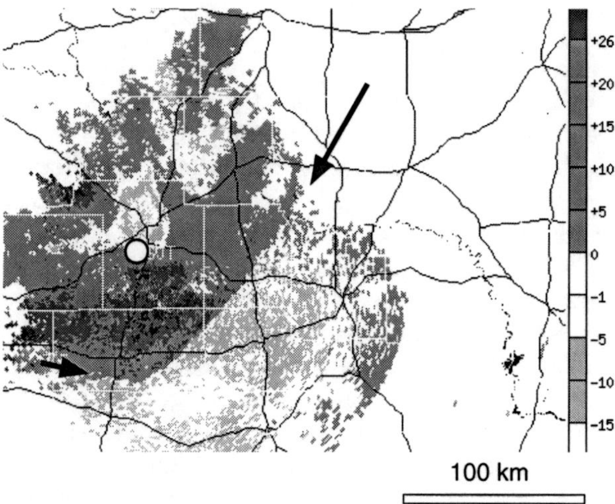

Figure 15.19
Same as 15.18, but for Doppler velocity. Medium and dark greys are winds away from the radar (white circle), and light greys are winds towards. Scale at right is speed in knots.

$$\frac{\Delta M}{\Delta t} = \frac{1}{\rho} \frac{P'_{max}}{r} \qquad \bullet(15.13)$$

where ΔM is change of outflow wind magnitude over time interval Δt , ρ is air density, P'_{max} is the pressure perturbation strength of the mesohigh, and r is the radius of the downburst (assuming it roughly equals the radius of the mesohigh).

The horizontal **divergence signature** of air from a downburst can be detected by Doppler weather radar by the couplet of "toward" and "away" winds, as was shown in the Satellites & Radar chapter. Figs. 15.18 and 15.19 show a downburst divergence signature and gust front.

The intensity of the downburst (Table 15-2) can be estimated by finding the maximum radial wind speed M_{max} observed by Doppler radar in the divergence couplet, and finding the max change of wind speed ΔM_{max} along a radial line extending out from the radar at any height below 1 km above mean sea level (MSL).

Gust front is the name given to the leading edge of the cold **outflow** air (Figs. 15.11 & 15.17). These fronts act like shallow (100 to 1000 m thick) cold fronts, but with lifetimes of only several minutes to a few hours. Gust fronts can advance at speeds ranging from 5 to 15 m s^{-1}, and their length can be 5 to 100 km. The longer-lasting gust fronts are often associated with squall lines or supercells, where downbursts of cool air from a sequence of individual cells can continually reinforce the gust front.

At fixed weather stations, temperature drops of 1 to 3 °C can be recorded as the gust front passes over. As this cold, dense air plows under warmer, less-dense humid air in the pre-storm environmental atmospheric boundary layer (ABL), the ABL air can be pushed up out of the way. If pushed above its LCL, clouds can form in this ABL air, perhaps even triggering new thunderstorms (a process called **propagation**). The new thunderstorms can develop their own gust fronts that can trigger additional thunderstorms, resulting in a storm sequence that can span large distances.

Greater gust front speeds are expected if faster downbursts pump colder air toward the ground. The continuity equation tells us that the vertical inflow rate of cold air flowing down toward the ground in the downburst must balance the horizontal outflow rate behind the gust front. If we also approximate the outflow thickness, then we can estimate the speed M_{gust} of advance of the gust front relative to the ambient environmental air as

$$M_{gust} = \left[\frac{0.2 \cdot r^2 \cdot |w_d \cdot g \cdot \Delta T_v|}{R \cdot T_{ve}} \right]^{1/3} \qquad (15.14)$$

Sample Application

For a mesohigh of max pressure 0.5 kPa and radius 5 km, how fast will the outflow winds become during 1 minute of acceleration?

Find the Answer

Given: P'_{max} = 0.5 kPa, r = 5 km, Δt = 60 s
Find: ΔM = ? m s^{-1}

Solve eq. (15.13) for M. Assume ρ = 1.225 kg m^{-3}.

$$\Delta M = \frac{(60\,\text{s})}{(1.225\,\text{kg/m}^3)} \cdot \frac{(0.5\,\text{kPa})}{(5\,\text{km})} = \underline{\mathbf{4.9\ m\ s^{-1}}}$$

Check: Physics OK. Units OK. Magnitude OK.
Exposition: In real downbursts, the pressure gradient varies rapidly with time and space. So this answer should be treated as only an order-of-magnitude estimate.

Table 15-2. Doppler-radar estimates of thunderstorm-cell downburst intensity based on outflow winds.

Intensity	Max radial-wind speed (m s^{-1})	Max wind difference along a radial (m s^{-1})
Moderate	18	25
Severe	25	40

Sample Application

A thunderstorm creates a downburst of speed of 4 m s^{-1} within an area of average radius 0.6 km. The gust front is 3°C colder than the environment air of 300 K, and is an average of 2 km away from the center of the downburst. Find the gust-front advancement speed and depth.

Find the Answer

Given: $\Delta T_v = -3$ K, $T_{ve} = 300$ K, $w_d = -4$ m s^{-1},
$r = 0.6$ km, $R = 2$ km,

Find: $M_{gust} = ?$ m s^{-1}, $h_{gust} = ?$ m

Employ eq. (15.14): $M_{gust} =$

$$\left[\frac{0.2\cdot(600\text{ m})^2\cdot|(-4\text{ m/s})\cdot(-9.8\text{ m}\cdot\text{s}^{-2})\cdot(-3\text{ K})|}{(2000\text{ m})\cdot(300\text{ K})}\right]^{1/3}$$

$= \underline{\mathbf{2.42}}$ m s^{-1}

Employ eq. (15.15): $h_{gust} =$

$$0.85\left[\left(\frac{(-4\text{m/s})\cdot(600\text{m})^2}{(2000\text{ m})}\right)^2\frac{(300\text{ K})}{|(-9.8\text{m}\cdot\text{s}^{-2})\cdot(-3\text{K})|}\right]^{1/3}$$

$= \underline{\mathbf{174\text{ m}}}$

Check: Units OK. Physics OK.

Exposition: Real gust fronts are more complicated than described by these simple equations, because the outflow air can turbulently mix with the environmental ABL air. The resulting mixture is usually warmer, thus slowing the speed of advance of the gust front.

where r is average downburst radius, R is average distance of the gust front from the downburst center, w_d is downburst speed, ΔT_v is virtual temperature difference between the cold-outflow and environmental air, and T_{ve} (K) is the environmental virtual temperature.

In the expression above, the following approximation was used for gust-front thickness h_{gust}:

$$h_{gust} \approx 0.85\cdot\left[\frac{T_{ve}}{|g\cdot\Delta T_v|}\left(\frac{w_d\cdot r^2}{R}\right)^2\right]^{1/3} \qquad (15.15)$$

As the gust front spreads further from the downburst, it moves more slowly and becomes shallower. Also, colder outflow air moves faster in a shallower outflow than does less-cold outflow air.

In dry environments, the fast winds behind the gust front can lift soil particles from the ground (a process called **saltation**). While the heavier sand particles fall quickly back to the ground, the finer dust particles can become **suspended** within the air. The resulting dust-filled outflow air is called a **haboob** (Fig. 15.20), **sand storm** or **dust storm**. The airborne dust is an excellent tracer to make the gust front visible, which appears as a turbulent advancing wall of brown or ochre color.

In moister environments over vegetated or rain-wetted surfaces, instead of a haboob you might see **arc cloud (arcus)** or **shelf cloud** over (not within) an advancing gust front. The arc cloud forms when the warm humid environmental ABL air is pushed upward by the undercutting cold outflow air. Sometimes the top and back of the arc cloud are connected or almost connected to the thunderstorm, in which case the cloud looks like a shelf attached to the thunderstorm, and is called a **shelf cloud**.

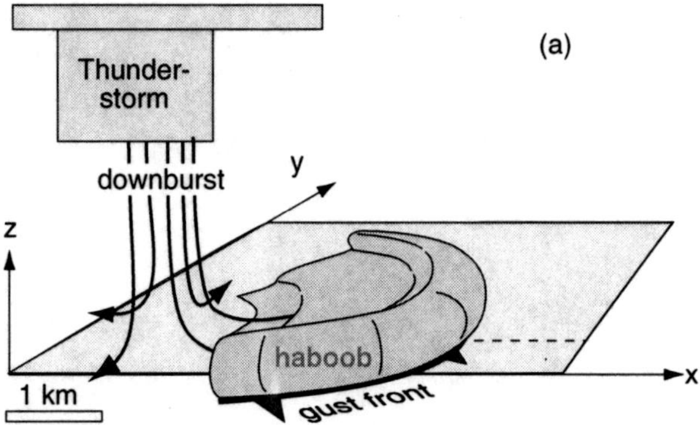

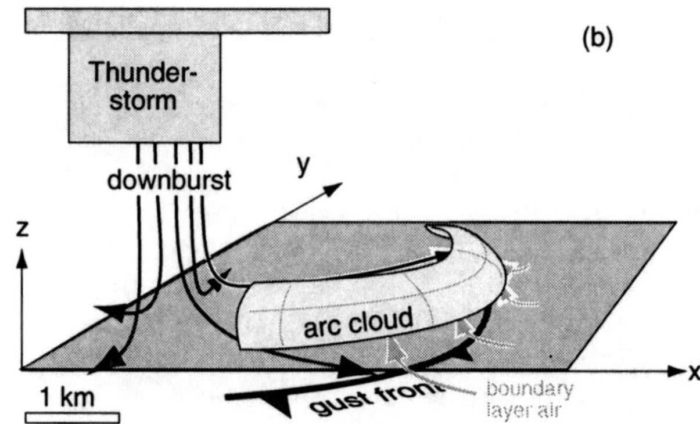

Figure 15.20
Sketches of (a) a haboob (dust storm) kicked up within the outflow winds, and (b) an arc cloud in the boundary-layer air pushed above the leading edge of the outflow winds (i.e., above the gust front). Haboobs occur in drier locations, while arc clouds occur in more humid locations.

When the arc cloud moves overhead, you will notice sudden changes. Initially, you might observe the warm, humid, fragrant boundary-layer air that was blowing toward the storm. After gust-front passage, you might notice colder, gusty, sharp-smelling (from ozone produced by lightning discharges) downburst air.

LIGHTNING AND THUNDER

Lightning (Fig. 15.21) is an electrical discharge (spark) between one part of a cloud and either:

- another part of the same cloud [**intracloud (IC) lightning**],
- a different cloud [**cloud-to-cloud (CC) lightning**, or **intercloud lightning**],
- the ground or objects touching the ground [**cloud-to-ground (CG) lightning**], or
- the air [**air discharge (CA)**].

Other weak high-altitude electrical discharges (blue jets, sprites and elves) are discussed later.

The lightning discharge heats the air almost instantly to temperatures of 15,000 to 30,000 K in a **lightning channel** of small diameter (2 to 3 cm) but long path (0.1 to 10 km). This heating causes:

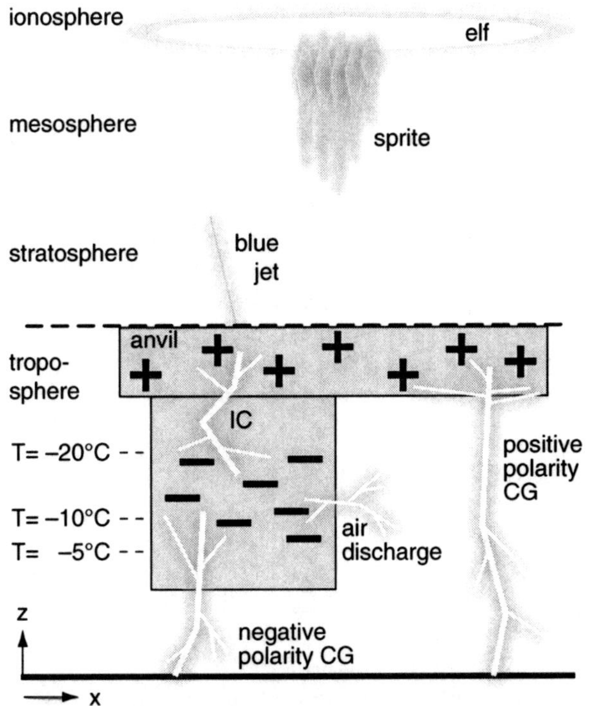

Figure 15.21
Types of lightning. Grey rectangles represent the thunderstorm cloud. (Vertical axis not to scale.)

INFO • Aircraft vs. Downbursts

113 people died in 1975 when a commercial jet flew through a downburst while trying to land at J. F. Kennedy airport near New York City.

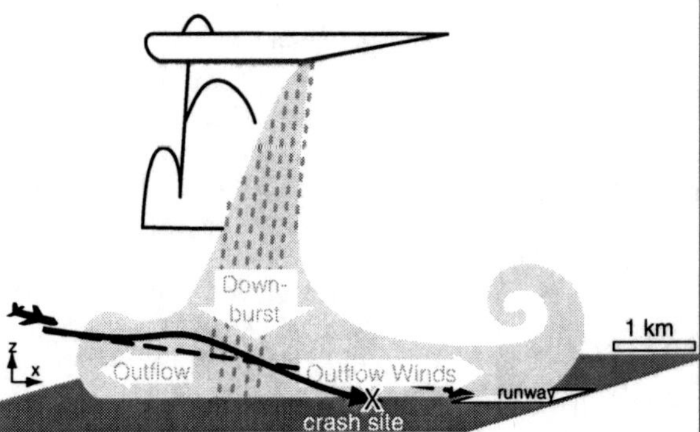

Figure 15.a
Sketch of the actual (solid line) and desired (dashed line) approach of the aircraft to a runway at JFK airport during a downburst event. Dotted lines represent rain; light blue shading represents the downburst and outflow air.

The airliner first encountered outflow headwinds while trying to gradually descend along the normal glideslope (dashed line) toward the runway. Due to the aircraft's forward inertia, the headwinds caused more air to blow over wings, generating more lift. This extra lift caused the aircraft to unintentionally climb (or not descend as fast), as shown by the solid line representing the actual aircraft path.

To compensate, the pilot slowed the engines and pointed the aircraft more downward, to try to get back down to the desired glideslope. But then the plane flew through the downburst, pushing the aircraft downward below the glideslope. So the pilot had to compensate in the opposite direction, by throttling up the engines to full power and raising the nose of the aircraft to try to climb back up to the glideslope.

By this time, the aircraft had reached the other side of the downburst, where the outflow winds were moving in the same direction as the aircraft. Relative to the aircraft, there was less air flowing over the wings and less lift, allowing gravity to pull the aircraft down even faster. The pilot increased pitch and engine power, but the turbine engines took a few seconds to reach full power. The inertia of the heavy aircraft caused its speed to increase too slowly. The slow airspeed and downward force of the downburst caused the aircraft to crash short of the runway.

That crash motivated many meteorological field experiments to enable better detection and forecasts of downbursts and gust fronts. Airports have special equipment (anemometers and Doppler radar) to detect dangerous wind shears; air-traffic controllers have a protocol to alert pilots and direct them to safety; and pilots are trained to carry extra speed and make earlier **missed-approach** decisions.

INFO • Electricity in a Channel

Electricity is associated with the movement of electrons and ions (charged particles). In metal channels such as electrical wires, it is usually only the electrons (negative charges) that move. In the atmospheric channels such as lightning, both negative and positive ions can move, although electrons can move faster because they are smaller, and carry most of the current. Lightning forms when **static electricity** in clouds discharges as **direct current** (DC).

Each electron carries one elementary negative charge. One **coulomb** (C) is an **amount of charge** (Q) equal to 6.24x10^{18} elementary charges. [Don't confuse C (coulombs) with °C (degrees Celsius).] The **main charging zone** of a thunderstorm is between altitudes where $-20 \le T \le -5$°C (Fig. 15.21), where typical thunderstorms hold 10 to 100 coulombs of static charge.

The movement of 1 C of charge per 1 second (s) is a **current** (I) of 1 A (**ampere**).

$$I = \Delta Q \, / \, \Delta t$$

The median current in lightning is 25 kA.

Most substances offer some **resistance** (R) to the movement of electrical charges. Resistance between two points along a wire or other conductive channel is measured in **ohms** (Ω).

An electromotive force (V, better known as the **electrical potential difference**) of 1 V (**volt**) is needed to push 1 A of electricity through 1 ohm of resistance.

$$V = I \cdot R$$

[We use italicized V to represent the variable (electrical potential), and non-italicized V for its units (volts).]

The **power** P_e (in watts W) spent to push a current I with voltage V is

$$P_e = I \cdot V$$

where 1 W = 1 V · 1 A. For example, lightning of voltage 1x10^9 V and current 25 kA dissipates 2.5x10^{13} W.

A lightning stroke might exist for Δt = 30 µs, so the energy moved is $P_e \cdot \Delta t$ = (2.5x10^{13} W)·(0.00003 s) = 7.5x10^8 J; namely, about 0.75 billion Joules.

- incandescence of the air, which you see as a bright flash, and
- a pressure increase to values in the range of 1000 to 10,000 kPa in the channel, which you hear as thunder.

On average, there are about 2000 thunderstorms active at any time in the world, resulting in about 45 flashes per second. Worldwide, there are roughly 1.4x10^9 lightning flashes (IC + CG) per year, as detected by optical transient detectors on satellites. Africa has the greatest amount of lightning, with 50 to 80 flashes km^{-2} yr^{-1} over the Congo Basin. In North America, the region having greatest lightning frequency is the Southeast, having 20 to 30 flashes km^{-2} yr^{-1}, compared to only 2 to 5 flashes km^{-2} yr^{-1} across most of southern Canada.

On average, only 20% of all lightning strokes are CG, as measured using ground-based lightning detection networks, but the percentage varies with cloud depth and location. The fraction of lightning that is CG is less than 10% over Kansas, Nebraska, the Dakotas, Oregon, and NW California, and is about 50% over the Midwest states, the central and southern Rocky Mountains, and eastern California.

CG is the type of lightning that causes the most deaths, and causes power surges or disruptions to electrical transmission lines. In North America, the southeastern states have the greatest density of CG lightning [4 to 10 flashes km^{-2} yr^{-1}], with Tampa, Florida, having the greatest CG flash density of 14.5 flashes km^{-2} yr^{-1}.

Most CG lightning causes the transfer of electrons (i.e., negative charge) from cloud to ground, and is called **negative-polarity** lightning. About 9% of CG lightning is **positive-polarity**, and usually is attached to the thunderstorm anvil (Fig. 15.21) or from the extensive stratiform region of a mesoscale convective system. Because positive-polarity lightning has a longer path (to reach between anvil and ground), it requires a greater voltage gradient. Thus, positive CG lightning often transfers more charge with greater current to the ground, with a greater chance of causing deaths and **forest fires**.

Origin of Electric Charge

Large-scale (macroscale) **cloud electrification** occurs due to small scale (microphysical) interactions between individual cloud particles. Three types of particles are needed in the same volume of cloud:

- small ice crystals formed directly by deposition of vapor on to ice nuclei;
- small supercooled liquid water (cloud) droplets;
- slightly larger graupel ice particles.

An updraft of air is also needed to blow the small particles upward relative to the larger ones falling down.

These three conditions can occur in cumulonimbus clouds between altitudes where the temperature is 0°C and –40°C. However, most of the electrical charge generation is observed to occur at heights where the temperature ranges between –5 and –20°C (Fig. 15.21).

The details of how charges form are not known with certainty, but one theory is that the falling graupel particles intercept lots of supercooled cloud droplets that freeze relatively quickly as a glass (i.e., too fast for crystals to grow). Meanwhile, separate ice nuclei allow the growth of ice crystals by direct deposition of vapor. The alignment of water molecules on these two types of surfaces (glass vs. crystal) are different, causing different arrangement of electrons near the surface.

If one of the small ice crystals (being blown upward in the updraft because of its small terminal velocity) hits a larger graupel (falling downward relative to the updraft air), then about 100,000 electrons (i.e., a charge of about 1.5×10^{-14} C) will transfer from the small ice crystal (leaving the ice crystal positively charged) to the larger glass-surfaced graupel particle (leaving it negatively charged) during this one collision (Fig. 15.22). This is the microphysical electrification process.

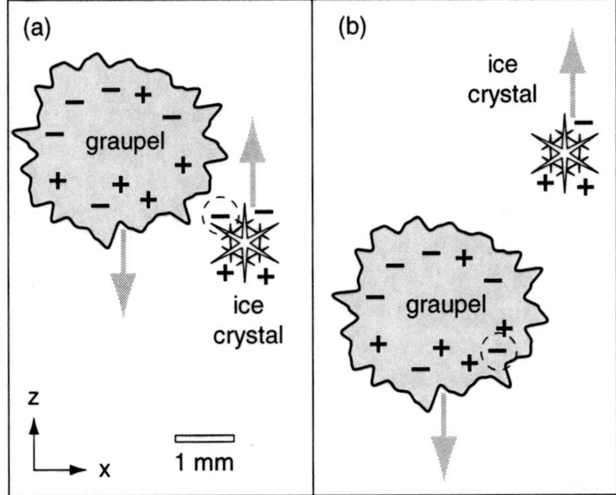

Figure 15.22
Illustration of charge transfer from a small, neutrally-charged, rising ice crystal to a larger, neutrally-charged, falling graupel or hail stone. The electron being transferred during the collision is circled with a dashed line. After the transfer, the graupel has net negative charge that it carries down toward the bottom of the thunderstorm, and the ice crystal has net positive charge that it carries up into the anvil.

INFO • Electricity in a Volume

The **electric field strength** (E, which is the magnitude of the **electric field** or the gradient of the **electric potential**) measures the electromotive force (voltage V) across a distance (d), and has units of V m^{-1} or V km^{-1}.

$$E = V / d .$$

Averaged over the whole atmosphere, $|E| \approx 1.3 \times 10^5$ V km^{-1} in the vertical. A device that measures electric field strength is called a **field mill**.

Near thunderstorms, the electric field can increase because of the charge build up in clouds and on the ground surface, yielding electric-potential gradients ($E = 1 \times 10^9$ to 3×10^9 V km^{-1}) large enough to ionize air along a narrow channel, causing lightning. When air is ionized, electrons are pulled off of the originally neutral molecules, creating a **plasma** of charged positive and negative particles that is a good conductor (i.e., low resistivity).

Electrical **resistivity** (ρ_e) is the resistance (R) times the cross-section area ($Area$) of the substance (or other conductive path) through which electricity flows, divided by the distance (d) across which it flows, and has units of Ω·m.

$$\rho_e = R \cdot Area \,/\, d$$

Air near sea level is not a good electrical conductor, and has a resistivity of about $\rho_e = 5 \times 10^{13}$ Ω·m. One reason why its resistivity is not infinite is that very energetic protons (**cosmic rays**) enter the atmosphere from space and can cause a sparse array of paths of ionized particles that are better conductors. Above an altitude of about 30 km, the resistivity is very low due to ionization of the air by sunlight; this conductive layer (called the **electrosphere**) extends into the ionosphere.

Pure water has $\rho_e = 2.5 \times 10^5$ Ω·m, while seawater is an even better conductor with $\rho_e = 0.2$ Ω·m due to dissolved salts.

Vertical **current density** (J) is the amount of electric current that flows vertically through a unit horizontal area, and has units A m^{-2}.

$$J = I \,/\, Area$$

In clear air, typical background current densities are 2×10^{-12} to 4×10^{-12} A m^{-2}.

Within a volume, the electric field strength, current density, and resistivity are related by

$$E = J \cdot \rho_e$$

Sample Application

What voltage difference is necessary to create a lightning bolt across the dry air between cloud base (at 2 km altitude) and the ground?

Find the Answer

Given: $\Delta z = 2$ km, $B = 3\text{x}10^9$ V km^{-1}

Find: $\Delta V_{lightning}$ = ? V

Use eq. (15.16):

$$\Delta V_{lightning} = (3\text{x}10^9 \text{ V km}^{-1})\cdot(2 \text{ km}) = 6\text{x}10^9 \text{ V}$$

Check: Units OK. Physics OK.

Exposition: Six billion volts is more than enough to cause cardiac arrest, so it is wise to avoid being struck by lightning. High-voltage electrical transmission lines are often about $3.5\text{x}10^5$ V. Lightning is nearly tied with floods as being the most fatal weather hazard in North America.

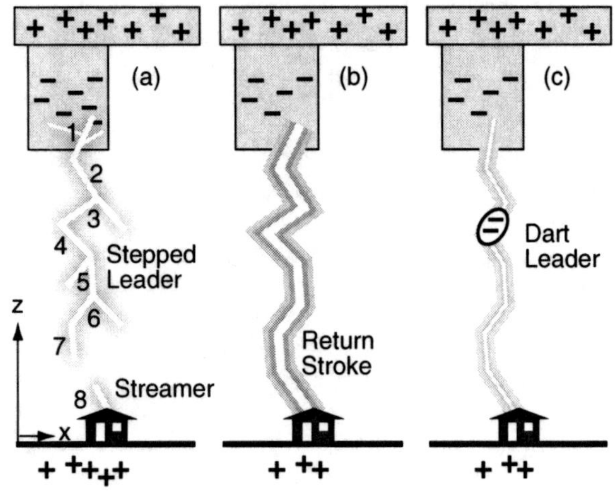

Figure 15.23

Sequence of events during lightning. (a) Stepped leader (1 to 7) moving rapidly downward from thunderstorm, triggers upward streamer (8) from objects on ground. (b) Intense return stroke transferring negative charge from the bottom of the thunderstorm to the ground. (c) Dart leader of negative charge following old ionized path toward ground, followed by another return stroke (not shown).

In 1 km^3 of thunderstorm air, there can be on the order of $5\text{x}10^{13}$ collisions of graupel and ice crystals per minute. The lighter-weight ice crystals carry their positive charge with them as they are blown in the updraft to the top of the thunderstorm, leading to the net positive charge in the anvil. Similarly, the heavier graupel carry their negative charges to the middle and bottom of the storm. The result is a macroscale (Fig. 15.21) charging rate of the thunderstorm cloud of order 1 C km^{-3} min^{-1}. As these charges continue to accumulate, the electric field (i.e., voltage difference) increases between the cloud and the ground, and between the cloud and its anvil.

Air is normally a good insulator in the bottom half of the troposphere. For dry air, a voltage gradient of $B_{dry} = 3\text{x}10^9$ V km^{-1} (where V is volts) is needed to ionize the air to make it conductive. For cloudy air, this **breakdown potential** is $B_{cloud} = 1\text{x}10^9$ V km^{-1}. Ionization adds or removes electrons to/from the air molecules. If lightning (or any spark) of known length Δz occurs, then you can use the breakdown potential to calculate the voltage difference $\Delta V_{lightning}$ across the lightning path:

$$\Delta V_{lightning} = B\cdot\Delta z \qquad \bullet(15.16)$$

where B is the dry-air or cloudy-air breakdown potential, depending on the path of the lightning.

Lightning Behavior & Appearance

When sufficient charge builds up in the cloud to reach the breakdown potential, an ionized channel called a **stepped leader** starts to form. It steps downward from the cloud in roughly 50 m increments, each of which takes about 1 μs to form, with a pause of about 50 μs between subsequent steps (Fig. 15.23). While propagating down, it may branch into several paths. To reach from the cloud to the ground might take hundreds of steps, and take 50 ms duration. For the most common (negative polarity) lightning from the middle of the thunderstorm, this stepped leader carries about 5 C of negative charge downward.

When it is within about 30 to 100 m of the ground or from the top of any object on the ground, its strong negative charge repels free electrons on the ground under it, leaving the ground strongly positively charged. This positive charge causes ground-to-air discharges called **streamers**, that propagate upward as very brief, faintly glowing, ionized paths from the tops of trees, poles, and tall buildings. When the top of a streamer meets the bottom of a stepped leader, the conducting path between the cloud and ground is complete, and there is a very strong (bright) **return stroke** from the ground to the cloud. During this return stroke, electrons drain downward first from the bottom of the stepped leader, and then drain downward from successively higher portions of the channel, producing the flash of light that you see. Taken together, the stepped leader and return stroke are called a lightning **stroke**.

Although thunderstorm winds and turbulence tend to rip apart the ionized channel, if the remaining negative charges in the cloud can recollect within about 20 to 50 ms, then another stroke can happen down the same conducting path. This second stroke (and subsequent strokes) forms as a **dart leader** carrying about 1 C of charge that moves smoothly (not in steps) down the existing channel (with no branches) to the ground, triggering another return stroke. Ten strokes can easily happen along that one ionized channel, and taken together are called a **lightning flash**. The multiple return strokes are what makes a lightning flash appear to flicker.

IC and CG lightning can have different appearances that are sometimes given colloquial names. **Anvil crawlers** or **spider lightning** is IC lightning that propagates with many paths (like spider legs) along the underside of anvils or along the bottom of the stratiform portion of mesoscale convective systems. Some spider lightning is exceptionally long, exceeding 100 km path lengths.

If an IC lightning channel is completely hidden inside a cloud, then often observers on the ground see the whole cloud illuminated as the interior light scatters off the cloud drops (similar to the light emitted from a frosted incandescent light bulb where you cannot see the filament). This is called **sheet lightning**. Lightning from distant thunderstorms may illuminate hazy cloud-free sky overhead, causing dim flashing sky glow called **heat lightning** in the warm, prestorm environment.

A **bolt from the blue** is a form of CG **anvil lightning** that comes out laterally from the side of a storm, and can travel up to 16 km horizontally into clear air before descending to the ground. To people near where this lightning strikes, it looks like it comes from clear blue sky (if they cannot see the thunderstorm that caused it).

Ball lightning has been difficult to study, but has been observed frequently enough to be recognized as a real phenomenon. It is rare, but seems to be caused by a normal CG strike that creates a longer lasting (many seconds to minutes) glowing, hissing plasma sphere that floats in the air and moves.

After the last return stroke of CG lightning, the rapidly dimming lightning channel sometimes exhibits a string of bright spots that don't dim as fast as the rest of the channel, and thus appear as a string of glowing beads, called **bead lightning**.

A SCIENTIFIC PERSPECTIVE • Be Safe *(part 3)*

More chase guidelines from Charles Doswell III (continued from the previous chapter):

The #2 Threat: Lightning

1. Stay inside your car if cloud-to-ground (CG) lightning is less than 1 mile away.
2. CG threat is high when rain first reaches you.
3. CG can strike well away from the storm. Move to safety. If you can't, then:
   a. Avoid being the tallest object around.
   b. Don't stand close to fences or power poles that are connected to wires going closer to the storm.
   c. Make yourself low (i.e., squat), but don't lay, sit, or kneel on the ground, and don't have your hands touching anything connected to the ground.
   d. You might not have any warning signs (hear thunder, hear hissing or crackling discharges of electricity nearby, or feel your hair stand on end) before you are struck.
   e. You and your chase-team members should be trained on how to do **CPR** (cardiopulmonary resuscitation), and don't be afraid to use it immediately if appropriate.
   f. Non-metalic camera tripods and insulated shoes don't reduce your threat of being struck by lightning.

(continues later in this chapter)

INFO • Lightning Burns the Air

The initial high temperature and pressure inside the lightning channel cause the oxygen in the air to react with the other gases.

Nitrogen, which makes up 78% of the atmosphere (see Chapter 1), oxidizes inside the ionized lightning path to become various oxides of nitrogen (NO_x). During rainout, the NO_x can fall as acid rain (nitric acid), which hurts the plants and acidifies streams and lakes on the short term. But over the long term, the NO_x rained out can help fertilize the soil to encourage plant growth.

Even the oxygen molecules (O_2) can be oxidized within the lightning channel to become ozone (O_3), which we smell in the air as a sharp or fresh odor. Sometimes this odor is carried down and out from the thunderstorm by the downburst and outflow winds, which we can smell when the gust front passes just before the thunderstorm arrives.

Because of all this oxidation, we can say that lightning causes the air to burn.

INFO • Lightning in Canada

In Canada, lightning currents greater than 100 kA are rare (approximately 1% of all cloud-to-ground CG strokes). Summer is when these large-current flashes occur. Large-current flashes with negative polarity can have 10 or more strokes, and are found mostly in western Canada. First-stroke peak currents are strongest in winter, and strongest in northern Canada.

In eastern Canada, about 11% of the CG flashes have positive polarity, and in western Canada about 17% have positive polarity. In British Columbia and Yukon about 25% of the CG flashes have positive polarity.

In western Canada, 89% of the positive CG lightning flashes had only one stroke, while 48% of the negative CG lightning had only one stroke. The average number of strokes per CG flash is 2 to 2.4 for negative polarity CG lightning flashes, and about 1 stroke for positive CG.

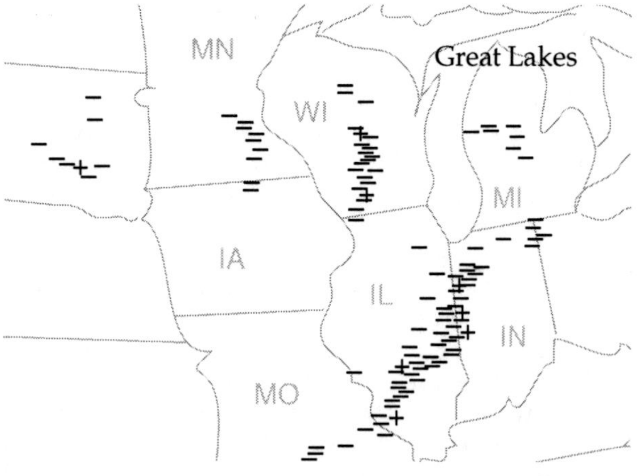

When CG multiple return strokes happen along a lightning channel that is blowing horizontally in the wind, the human eye might perceive the flash as a single broad ribbon of light called **ribbon lightning**. Lightning with a very brief, single return stroke is called **staccato lightning**.

Above the top of strong thunderstorms can be very brief, faint, electrical discharges that are difficult to see from the ground, but easier to see from space or a high-flying aircraft (Fig. 15.21). A **blue jet** is a vertical anvil-to-air discharge into the stratosphere. **Red sprites** can spread between cloud top and about 90 km altitude (in the mesosphere), and have diameters of order 40 km. **Elves** are extremely faint glowing horizontal rings of light at 90 km altitude with radii that increases at the speed of light, centered above strong lightning strokes in thunderstorms. Most people never see these.

Lightning Detection

During a lightning stroke, the changing flow of electricity along the lightning channel creates and broadcasts electromagnetic waves called **sferics.** A broad range of frequencies is transmitted, including radio waves that you hear as static or snaps on AM radio. Detectors on the ground receive these radio signals and accurately measure their strength and time of arrival. Other types of lightning sensors are based on magnetic direction finders.

To pinpoint the location of each lightning strike, a continent-wide array of multiple ground stations observe signals from the same lightning discharge. These ground stations either have direction-finding capability or relative time-of-arrival capability (to infer the range of the strike from the station). Regardless of the station capabilities, the strike is located by triangulating directions or ranges from all the stations that received the signal. All the strike locations during a time interval (5 minutes to a hour) are then plotted on a map (Fig. 15.24). Such an array of detectors on the Earth's surface and associated communication and computer equipment is called a **lightning detection network** (LDN).

Some of the sferics are generated at **very low frequencies (VLF)**. Some LDN systems measure these VLF at a frequency of about 10 kHz (wavelength ≈ 30 km). The advantage of these VLF waves is that they can travel large distances — trapped in a waveguide between the ionosphere and the ground.

When a VLF wave from lightning passes over an LDN ground station, it modulates the electric field near the station (see the INFO on "Electricity in a Volume"). When multiple stations measure the same wave, the distance D (m) between the lightning and the stations can be estimated, and the peak electric field E (V m^{-1}) measured at any one station

Figure 15.24
Map of negative (–) and positive (+) lightning strikes over the US Midwest from a (simulated) lightning detection network, for the 24 May 2006 case.

can be used to find the approximate peak current I_{max} (A) flowing in the lightning return stroke:

$$I_{\max} = -2\pi \cdot \varepsilon_o \cdot c^2 \cdot E \cdot D / v_L \qquad (15.17)$$

where ε_o = 8.854 x 10^{-12} A·s·(V·m)$^{-1}$ is the permittivity of free space, c = 3.00986 x 10^8 m s^{-1} is the speed of light, and v_L = 1.0 to 2.2 x 10^8 m s^{-1} is the return-stroke current velocity.

The USA has a **National Lightning Detection Network** (NLDN) that typically detects more than 20 million CG flashes per year. Peak electrical currents as high as 400 kA have been rarely observed, but the median peak current is about 25 kA. The average duration of a lightning stroke is about 30 μs, and the average peak power per stroke is about 10^{12} W. One to ten return strokes (with 50 - 300 ms pauses between strokes) can occur in the same ionized path before winds break the path apart.

Satellite systems also detect lightning. Low-light-level video and digital cameras have been on board some satellites and manned space vehicles, and have observed lightning at night from the flashes of light produced. An **optical transient detector (OTD)** has also been deployed that measures the changes in light leaving the portion of atmosphere viewed.

Lightning Hazards and Safety

When lightning strikes electric power lines it can cause power **surges** (transient spikes in voltage and current in the line). Based on observations of many such surges, the probability *Prob* that surge current will be greater than *I* (kA) in the power line is:

$$Prob = \exp\left\{-0.5 \cdot \left[\frac{\ln((I + I_o)/I_1)}{s_1}\right]^2\right\} \qquad (15.18)$$

for $I \geq (I_1 - I_o)$, where the empirical probability-distribution parameters are I_o = 2 kA, I_1 = 3.5 kA, and s_1 = 1.5. This curve is plotted in Fig. 15.25 — showing that 50% of the these surges exceed about 20 kA.

When a lightning-created surge travels down an electric power line, the voltage (and current) e at any point rapidly increases to its peak value e_{max} and then slowly decays. Electrical power engineers approximate this by:

$$e = e_{\max} \cdot a \cdot \left[\exp\left(-\frac{t}{\tau_1}\right) - \exp\left(-\frac{t}{\tau_2}\right)\right] \qquad (15.19)$$

The nominal constants are: τ_1 = 70 μs, τ_2 = 0.15 μs, and a = 1.014. Fig. 15.26 shows a surge that reaches its peak in 1 μs, and by 50 μs has decayed to half.

When lightning strikes sandy ground, it can melt and fuse the sand along its path into a long narrow

Sample Application

A lightning stroke of max intensity 10 kA occurs 100 km from a detection station. The station would likely measure what electric field value?

Find the Answer

Given: D = 100 km, I_{max} = 10 kA,
Find: E = ? V m^{-1}
Assume: v_L = 1.5 x 10^8 m s^{-1}

Rearrange eq. (15.17) to solve for E:

$E = [-v_L / (2\pi \cdot \varepsilon_o c^2)] \cdot (I_{\max} / D)$

E =[(– 1.5 x 10^8 m s^{-1}) / (2π· (8.854 x 10^{-12} A·s·(V·m)$^{-1}$)· (3.00986 x 10^8 m s^{-1})^2)] · [(10 kA)/100 km)] ·
= **–2.98 V m^{-1}**

Check: Physics and units are OK.
Lightning current flows from high to low voltage (i.e., it flows opposite to the voltage gradient), which is why the answer has a negative sign.
Exposition: Lightning can have positive or negative polarity (i.e., the charge that goes down to the ground). LDN detectors can measure this as well as the waveform of the lightning signal, and are thus able to discriminate between CG and cloud-to-cloud lightning. The net result is that LDNs can provide much valuable information to utility companies and forest fire fighters, including lightning intensity, location, polarity, and type (CG or other).

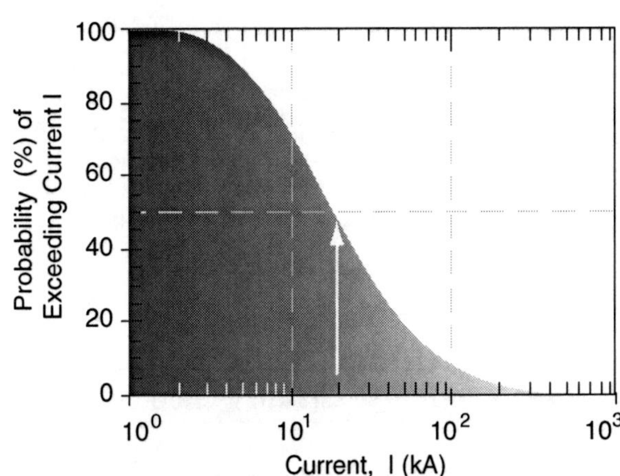

Figure 15.25
This curve (found from eq. 15.18) shows the probability that a surge of current I will be exceeded in an electrical power line.

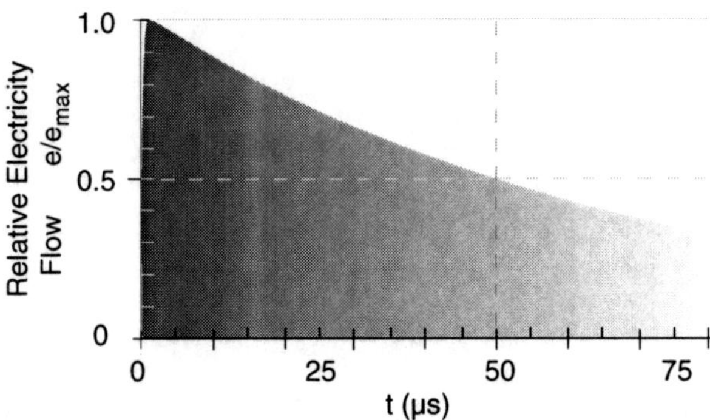

Figure 15.26
The surge of electricity in a power line struck by lightning, where e can be applied to current or voltage.

Sample Application

If lightning strikes a power line, what is the probability that the surge current will be 60 kA or greater? When will this surge decay to 5 kA?

Find the Answer

Given: I = 60 kA and I/I_{max} = 5/60 = 0.0833
Find: $Prob$ = ? %, t = ? μs after the surge starts

Use eq. (15.18):

$$Prob = \exp\left\{-0.5\cdot\left[\frac{\ln(\,(60+2\text{kA})\,/\,(3.5\text{kA})\,)}{1.5}\right]^2\right\}$$

$$= \exp(-1.8) = \underline{\mathbf{16\%}}$$

Use I/I_{max} in place of e/e_{max} in eq. (15.19). As time increases, the last exponential becomes small relative to the first exponential, and can be neglected. Use eq. (15.19) without the last exponential and solve it for time:

$$t = -\tau_1\cdot\ln[(1/a)\cdot e/e_{\max}] = -\tau_1\cdot\ln[(1/a)\cdot I/I_{\max}]$$

$$= -(70\ \mu\text{s})\cdot\ln[\ (1/1.014)\cdot 0.0833\ ] = \underline{\mathbf{175.\ \mu s}}$$

Check: Agrees with Figs 15.25 & 15.26. Physics & units OK.

Exposition: The brief intense power surge can open circuit breakers, blow fuses, and melt electric power transformers. The resulting disruption of power distribution to businesses and homes can cause computers to malfunction, files to be lost, and peripherals to be destroyed.

INFO • The 30-30 Lightning Safety Rule

For the non-storm-chaser, use the 30-30 Rule of lightning safety: If you are outdoors and the time between the lightning flash and thunder is 30 s or less, then seek shelter immediately. Wait 30 minutes after hearing the last thunder clap before going outdoors.

tube called a **fulgurite**. When it strikes trees and flows down the trunk to the ground, it can cause the moisture and sap in the tree to instantly boil, causing the bark to splinter and explode outward as lethal wooden shrapnel. Sometimes the tree trunk will split, or the tree will ignite.

The electrons flowing in lightning all have a negative charge and try to repel each other. While moving along the narrow lightning channel, the electrons are constrained to be close together. However, if the lightning hits a metal-skinned airplane, the electrons push away from each other so as to flow along the outside surfaces of the airplane, thus protecting the people inside. Such a **Faraday cage** effect also applies to metal-skinned cars. Other than the surprisingly loud noise and bright lightning flash, you are well protected if you don't touch any metal. Where the lightning attaches to the car or aircraft, a pinhole can be burned through the metal.

Dangerous activities/locations associated with lightning strikes are:

1. Open fields including sports fields.
2. Boating, fishing, and swimming.
3. Using heavy farm or road equipment.
4. Playing golf.
5. Holding a hard-wired telephone.
6. Repairing or using electrical appliances.

You should take precautions to avoid being struck by lightning. Avoid water and metal objects. Get off of high ground. Avoid open fields. Stay away from solitary trees or poles or towers (which might attract lightning). Go indoors or inside a metal-skinned car, van or truck if lightning is within 10 km (i.e., 30 seconds or less between when you see a lightning flash and when you hear its thunder). Even if you don't see lightning or hear thunder, if the hair on your head, neck or arms stands on end, immediately go inside a building or car. If indoors, avoid using hard-wired phones, hair driers, or other appliances, and avoid touching metal plumbing or water in your sink, shower, or tub.

If you are outside and no shelter is available, crouch down immediately in the lowest possible spot with your feet together and your hands over your ears. Do not lie down, because once lightning strikes the ground or a tree, it often spreads out along the surface of the ground and can still hit you. Do not put both your hands and feet on the ground, because then the lightning circuit could go through your heart.

If people near you are struck by lightning and fall down, do not assume they are dead. In many cases, lighting will cause a person to stop breathing or will stop their heart, but they can often be revived. If a person isn't breathing then try performing mouth-to-mouth resuscitation. If a person has no pulse, and if you have the training, apply cardiopulmonary resuscitation (CPR).

Cardiac arrest (stopped heart) is the main cause of death from a lightning strike. Surprisingly, there is usually very little burning of the skin. Other immediate effects include tinnitus (ringing in the ears due to the loud thunder), blindness, amnesia, confusion, cardiac arrhythmias, and vascular instability. Later problems include sleep disturbances, anxiety attacks, peripheral nerve damage, pain syndromes, fear of storms, personality changes, irritability, short-term memory difficulties, and depression. Lightning injures 800 to 1000 people per year in the USA, and kills 75 to 150 people per year. Support groups exist for lightning survivors.

Thunder

When lightning heats the air to T = 15,000 to 30,000 K, it instantly increases the air pressure to P = 1,000 to 10,000 kPa along the ionized lightning path, creating a **shock front** that moves at speeds up to ten times the speed of sound (i.e., Ma = 10, where Ma is **Mach number**). By 7 μs later, the supersonic shock-front speed has decreased to Mach 5, and has spread only about 1.5 cm from the edge of the lightning channel (Fig. 15.27). By 0.01 s after the lightning, the shock front has spread about 4 m in all directions around the lightning, and has a speed (Ma = 1.008) that is almost equal to the speed of sound (Ma = 1). Namely, it becomes a **sound wave** that continues to spread at the speed of sound, which you hear as thunder. So to understand thunder, we will study shock fronts first, and then sound.

Shock Front

A shock front in air is created by a pressure discontinuity or pressure step. The thickness of this pressure step is only a few micrometers. This shock front advances supersonically at speed C through the air toward the lower pressure. It is NOT like a piston that pushes against the ambient air in front of it. Instead it moves THROUGH the background low-pressure air, modifies the thermodynamic and dynamic state of the air molecules it overtakes, and leaves them behind as the front continues on. This modification of the air is irreversible, and causes entropy to increase.

To analyze the shock, picture an idealized vertical lightning channel (Fig. 15.28) of radius r_0. Assume the background air is calm (relative to the speed of the shock) and of uniform thermodynamic state. For simplicity, assume that air is an ideal gas, which is a bad assumption for the temperatures and pressures inside the lightning channel.

Because the shock front will expand as a cylinder of radius r around the axis of the lightning chan-

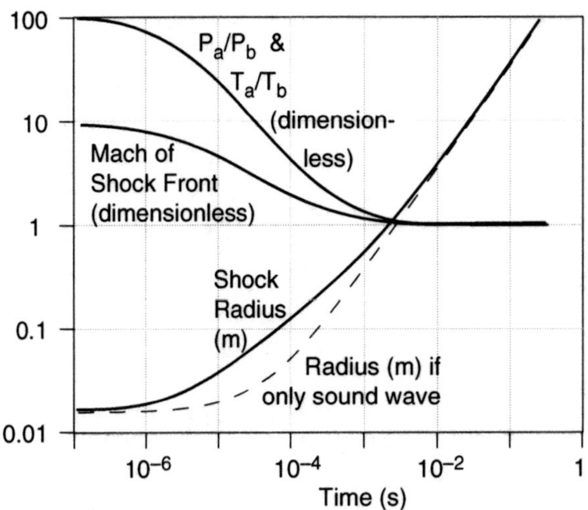

Figure 15.27
Evolution of initial stages of thunder as it propagates as a supersonic shock front. P_a/P_b is the ratio of average pressure behind the shock front to background pressure ahead of the front. T_a/T_b is similar ratio for absolute temperature. Propagation speed of the shock front is given by its Mach number. Radius of the shock front from the lightning axis is compared to radius if only sound waves had been created.

> **INFO • Force of Thunder**
>
> One time when I was driving, lightning struck immediately next to my car. The shock wave instantly pushed the car into the next lane, without breaking any windows or causing damage. I was amazed at the power of the shock wave, and happy to be alive.
> - *R. Stull*

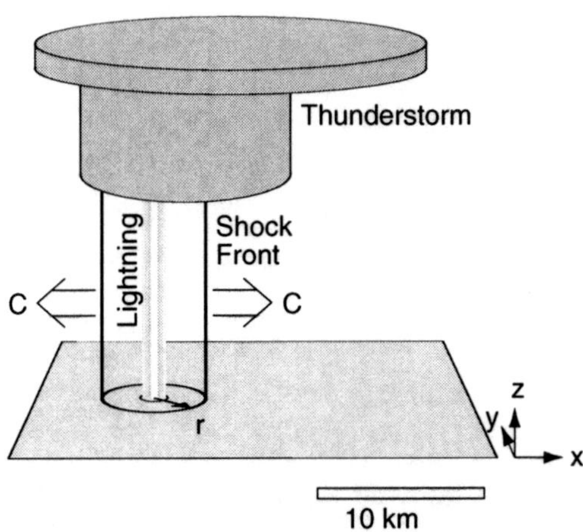

Figure 15.28
Sketch of idealized vertical lightning discharge that generates a cylindrical shock front of radius r that expands at speed C.

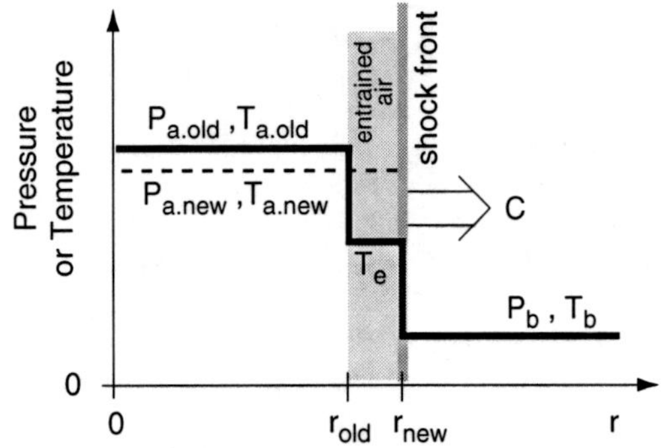

Figure 15.29
Characteristics of the thunder shock front as a function of radial distance r from the lightning-channel axis (at r = 0). P_a and T_a are the average pressure and temperature behind the shock front, relative to background conditions ahead of the front P_b and T_b. As the shock front overtakes background air molecules, their temperature is modified to be T_e, which is assumed to mix with the old conditions behind the front to create new average conditions. $r_{new} - r_{old} \approx$ a few µm.

nel, ignore the vertical (because no net change in the vertical), and use the circular symmetry to treat this as a 1-D **normal-shock** problem in the horizontal. Namely, the movement of the shock front across the air molecules is perpendicular to the face of the shock front.

Use subscript $_b$ to indicate background air (not yet reached by the shock), and subscript $_e$ to indicate entrained air (namely, background air that has been modified by the shock-front passage). For simplicity, assume that the entrained air instantly mixes with the rest of the air inside the shock-front circle, and use subscript $_a$ to indicate the resulting average conditions inside that circle (Fig. 15.29).

For a normal-shock, the Mach number (dimensionless) of the shock wave expanding into the background air is

$$Ma = \left[\frac{(P_a / P_b)\cdot(k-1)+k-1}{2k}\right]^{1/2} \qquad (15.20)$$

where P is pressure, and $k = C_p/C_v$. For air, $k = 1.40$, allowing the previous equation to be simplified to:

$$Ma = \left[\frac{(P_a / P_b)\cdot 6+1}{7}\right]^{1/2} \qquad (15.21)$$

One equation for the speed of sound (s) in air is

$$s = (k\cdot\Re\cdot T)^{1/2} \qquad •(15.22)$$

where $\Re = C_p - C_v = 287 \text{ m}^2 \text{ s}^{-2} \text{ K}^{-1}$ is the gas constant, and T is absolute air temperature. For background air, this simplifies to

$$s_b = a_s \cdot (T_b)^{1/2} \qquad (15.23)$$

where $a_s = 20 \text{ (m s}^{-1})\cdot\text{K}^{-1/2}$ is a constant for air, and T_b is background air temperature (in Kelvin). For example, if the background air has temperature 27°C (= 300 K), then sound speed is $s_b = 346.41 \text{ m s}^{-1}$.

By definition, the Mach number is the speed of the object (or the shock) divided by sound speed. Thus the speed C of the shock front through the calm background air is

$$C = Ma \cdot s_b \qquad •(15.24)$$

During a small time interval Δt, the radius r of the shock circle expands as:

$$r_{new} = r_{old} + C \cdot \Delta t \qquad (15.25)$$

The thin layer of air immediately behind the shock front is warmed (due to compression) to:

$$T_e = T_b \cdot \frac{[2+(k-1)\cdot Ma^2][2k\cdot Ma^2 - k + 1]}{(k+1)^2 Ma^2} \qquad (15.26)$$

or

$$T_e = T_b \cdot \frac{[5+Ma^2][7Ma^2 - 1]}{36 Ma^2} \qquad (15.27)$$

where T_b is background air temperature and Ma is the Mach number of the shock front. T_e is the temperature of the air entrained inside the shock circle.

By keeping track of the average temperature T_a of all air enclosed by the shock circle, use geometry to find how that average changes as the entrained air is added

$$T_{a.new} = T_e + \frac{(T_{a.old} - T_e)}{(r_{new} / r_{old})^2} \qquad (15.28)$$

Lightning-formed shocks are quite different from shocks caused by chemical explosives (i.e., bombs). Conventional bombs explosively release large amounts of gas via chemical reactions, which increases the pressure, temperature, and density of the atmosphere by quickly adding gas molecules that were not there before. As the resulting shock front expands, there is a net out rush of gas behind the shock as the gas density decreases toward background values.

Lightning, however, is a constant density (**isopycnal**) process, because no extra air molecules are added to the lightning channel. Namely, lightning starts with existing background air molecules and energizes whichever ones happen to be within the ionized channel. Furthermore, just outside the shock front nothing has changed in the background air (i.e., no density changes, and no movement of molecules toward or away from the shock front).

Therefore, by mass conservation, the average density ρ_a of air enclosed by the shock front is also constant and equal to the original (pre-lightning) background value ρ_b:

$$\rho_a = \rho_b = \text{constant} = P_b / \Re \cdot T_b \qquad \bullet(15.29)$$

using the ideal gas law with P_b as background air pressure, and T_b as background air temperature (in Kelvin). Immediately behind the shock front, the entrained air has higher density and pressure, but this is compensated by lower-density lower-pressure air closer to the lightning axis, resulting in constant average density as shown above.

Finally, the average pressure P_a of all air enclosed by the shock circle is found using the ideal gas law with constant density:

$$P_a / P_b = T_a / T_b \qquad \bullet(15.30)$$

where $T_a = T_{a.new}$. Equations (15.20) through (15.30) can be solved iteratively to find how conditions change as the shock evolves. Namely, P_a can be used back in eq. (15.20), and the calculations repeated. This assumes that the background thermodynamic state T_b and P_b of the undisturbed air is known.

To iterate, you need the initial conditions. Start with r_{old} = radius of the ionized lightning channel, although there is evidence that the incandescent region of air is about 10 to 20 times larger radius than this (so you could try starting with this larger value). Because of the isopycnal nature of lightning, if you know the initial lightning temperature T_a in Kelvin, use eq. (15.30) to find the initial pressure ratio P_a/P_b, as sketched in Fig. 15.30.

An iterative approach is demonstrated in a Sample Application, the results from which were used to create Fig. 15.27. The background air state was T_b = 300K and P_b = 100 kPa, giving ρ_b = 1.1614 kg m^{-3}. For the lightning, I used initial conditions of r_{old} = 1.5 cm = 0.015 m, $T_{a.old}$ = 30,000 K, P_a = 10,000 kPa.

Anyone who has been very close (within 1 m or less) of a lightning strike (but not actually hit by the lightning itself) feels a tremendous force that can instantly throw your body (or your car if you are driving) many meters horizontally (as well as rupturing your ear drums). This is the combined effect of the pressure difference across the shock front as it passes your body or your car, and the dynamic effect of a supersonic wind in the thin layer of entrained air immediately behind the shock front.

Assuming a normal shock, the extremely brief, outward-directed wind M_e in the entrained air immediately behind the shock is:

$$M_e = C - [C^2 - 2 \cdot C_p \cdot (T_e - T_b)]^{1/2} \qquad (15.31)$$

Sample Application

Background atmospheric conditions are P_b = 100 kPa, T_b = 300 K, and calm winds. If lightning heats the air in the lightning channel to 15,000 K, what is the initial speed of the shock front, and the initial speed of the air behind the shock front?

Find the Answer

Given: P_b = 100 kPa, T_b = 300 K, M_b = 0 m s^{-1}.
T_a = 15,000 K
Find: C = ? m s^{-1} , M_e = ? m s^{-1}

The shock speed equation requires the Mach speed and the speed of sound, which must be calculated first. Mach speed, in turn, depends on the initial pressure ratio. Use eq. (15.30) to find the pressure ratio:

$P_a/P_b = T_a/T_b$ = (15000 K)/(300 K) = 50

Next, use eq (15.21) to find the Mach speed of the shock front through the background air:

$Ma = \{[(50)\cdot 6 + 1]/ 7\}^{1/2} = 6.56$

The background speed of sound from eq. (15.23) is:

s_b = [20 (m s^{-1})·K$^{-1/2}$] · (300 K)$^{1/2}$ = 346.4 m s^{-1}

Thus, using eq. (15.24):

C = 6.56 · (346.4 m s^{-1}) = __2,271.6__ m s^{-1}

The equation for wind speed behind the shock requires the temperature behind the shock. Use eq. (15.27):

T_e = (300 K)· { [5 + 6.56^2]·[7· 6.56^2 – 1] / (36 · 6.56^2) }
= (300 K)·{ 48 · 300 / 1548 } = 2791 K

Finally, using eq. (15.31):

M_e = (2,271.6 m s^{-1}) – [(2,271.6 m s^{-1})^2 – 2·(1005 m^2 s^{-2} K^{-1})·(2791K – 300 K)]$^{1/2}$
= 2,271.6 – 392.3 m s^{-1} = __1,879.3__ m s^{-1}

Check: Units OK. Physics OK.

Exposition: Both the initial shock front and the initial wind behind the shock are supersonic, but the wind is slower than the shock front.

Nonetheless, the initial shock-front speed (about 2.3x10^3 m s^{-1}) is still much slower than the speed of light (3x10^8 m s^{-1}). Hence, we will see the lightning before we hear the thunder (except if lightning strikes you).

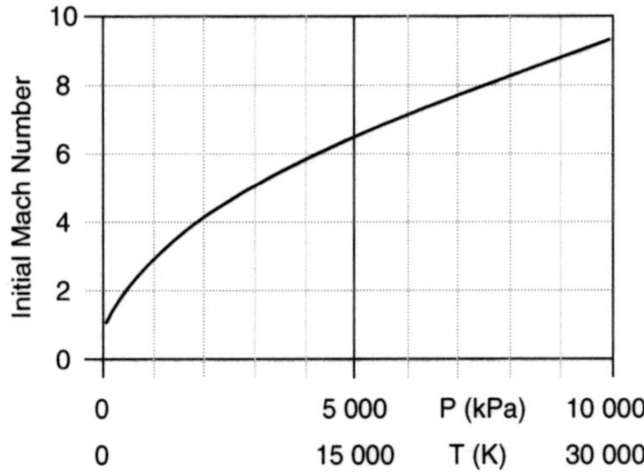

Figure 15.30
Higher temperature T in the lightning channel creates higher pressure P, which generates a shock front that initially moves with greater Mach number. (Mach 1 is the speed of sound).

Sample Application (§)

Background air is calm, with temperature 300 K and pressure 100 kPa. If lightning heats the air to 30,000 K within a vertical lightning channel of radius 1.5 cm, then find and plot the evolution of average temperature inside the shock circle (relative to background temperature), average relative pressure, Mach of the shock front, and shock radius. (Namely, produce the data that was plotted in Fig. 15.27.)

Find the Answer

Given: P_b = 100 kPa, T_b = 300 K, M_b = 0 m s^{-1}.
T_a = 30,000 K initially
Find: T_a/T_b = ? , P_a/P_b = ? , Ma = ? , r = ? m
and how they vary with time.

This is easily done with a spreadsheet. Because conditions vary extremely rapidly initially, and vary slower later, I will not use a constant time step for the iterations. Instead, I will use a constant ratio of

$$r_{new}/r_{old} = 1.05 \qquad \text{(a)}$$

Namely, I will redo the calculation for every 5% increase in shock radius. Thus,

$$t_{new} = t_{old} + (r_{new} - r_{old})/C_{old} \qquad \text{(b)}$$

Procedure: First, enter the given background air values in cells on the spreadsheet, and compute the speed of sound in the background air.

Next, create a table in the spreadsheet that holds the following columns: r, t, T_a, P_a/P_b, Ma, C, and T_e.

In the first row, start with r_{old} = 0.015 m at t_{old} = 0, and initialize with T_a = 30,000 K. Then compute the ratio T_a/T_b, and use that ratio in eq. (15.30) to find P_a/P_b. Use this pressure ratio to find Ma (using eq. 15.21) and C (using eq. 15.24 and knowing background sound speed). Finally, the last column in the first row is T_e found using eq. (15.27).

The second row is similar to the first, except estimate the new r using eq. (a), and the new t using eq. (b). The new T_a can be found using eq. (15.28). The other columns can then be filled down into this second row. Finally, the whole second row can be filled down to as many rows as you want (be careful: do not complete the table by filling down from the first row). Some results from that spreadsheet are shown in the table in the next column.

Fig. 15.27 shows a plot of these results.

Check: Units OK. Physics OK. Some decimal places have been dropped to fit in the table on this page.

Exposition: To check for accuracy, I repeated these calculations using smaller steps (1% increase in shock radius), and found essentially the same answer.

The equations in this section for shock-front propagation are not exact. My assumption of constant density, while correct when averaged over large scales, is probably not correct at the very small scale at the shock front. Thus, my equations are an oversimplification.

(table is in next column)

where C_p = 1005 m^2 s^{-2} K^{-1} is the specific heat of air at constant pressure. Although initially very fast, these winds are slower than the speed of the shock front. Initial supersonic post-shock winds are about 2500 m s^{-1} while the shock radius is still small (1.5 cm), but they quickly diminish to subsonic values of about 10 m s^{-1} as the shock front expands past 2 m radius.

The resulting sequence of winds at any point not on the lightning axis is: (1) no lightning-created winds prior to shock front passage; (2) near instantaneous increase in outward-directed winds M_e immediately after the shock front passes; which is quickly followed by (3) weaker inward-directed winds (never supersonic) drawn back toward the lower pressure along the lightning axis in order to conserve mass. A similar sequence of events has been observed with shock fronts from atmospheric nuclear-bomb explosions just above ground.

Sample Application *(continuation)*

r (m)	t (s)	T_a (K)	P_a / P_b	Ma	C (m s^{-1})	T_e (K)
0.0150	0	30000	100.0	9.27	3210	5291
0.0158	2.34E-07	27703	92.3	8.90	3085	4908
0.0165	4.89E-07	25584	85.3	8.56	2965	4555
0.0174	7.68E-07	23629	78.8	8.23	2849	4229
0.0182	1.07E-06	21825	72.7	7.91	2739	3928
0.0191	1.41E-06	20161	67.2	7.60	2632	3651
0.0201	1.77E-06	18626	62.1	7.30	2530	3395
0.0211	2.17E-06	17210	57.4	7.02	2433	3159
0.0222	2.60E-06	15904	53.0	6.75	2339	2941
0.0233	3.07E-06	14699	49.0	6.49	2249	2740
0.0244	3.59E-06	13587	45.3	6.24	2162	2555
0.0257	4.16E-06	12561	41.9	6.00	2079	2384
0.0269	4.77E-06	11615	38.7	5.77	2000	2226
0.0283	5.45E-06	10742	35.8	5.55	1924	2081
0.0297	6.18E-06	9937	33.1	5.34	1850	1946
0.0312	6.98E-06	9194	30.6	5.14	1780	1822
• • •						
0.0958	6.73E-05	1729	5.8	2.25	781	572
0.1006	7.35E-05	1621	5.4	2.19	757	553
0.1056	8.01E-05	1522	5.1	2.12	734	536
0.1109	8.73E-05	1430	4.8	2.06	712	520
0.1164	9.51E-05	1346	4.5	2.00	692	506
0.1222	1.03E-04	1268	4.2	1.94	672	492
• • •						
3.5424	0.0094	307	1.02	1.01	350	302
3.7195	0.0099	306	1.02	1.01	350	302
3.9055	0.0105	306	1.02	1.01	349	302
4.1007	0.0110	306	1.02	1.01	349	302
4.3058	0.0116	305	1.02	1.01	349	301
4.5210	0.0122	305	1.02	1.01	349	301
• • •						
35.091	0.100	300	1.00	1.00	347	300
36.845	0.105	300	1.00	1.00	347	300
38.687	0.111	300	1.00	1.00	347	300
• • •						

Sound

By about 0.1 s after the lightning stroke, the shock wave has radius 35 m, and has almost completely slowed into a sound wave. Because this happens so quickly, and so close to the lightning channel, ignore the initial shock aspects of thunder in this subsection, and for simplicity assume that the sound waves are coming directly from the lightning channel.

The speed of sound relative to the air depends on air temperature T — sound travels faster in warmer air. But if the air also moves at wind speed M, then the total speed of sound s relative to the ground is

$$s = s_o \cdot \left(\frac{T}{T_o}\right)^{1/2} + M \cdot \cos(\phi) \qquad \bullet(15.32)$$

where s_o = 343.15 m s^{-1} is a reference sound speed at T_o = 293 K (i.e., at 20°C), and ϕ is the angle between the direction of the sound and the direction of the wind. Namely, a head-wind causes slower propagation of sound waves.

Because light travels much faster than sound, you can estimate the range to a lightning stroke by timing the interval Δt between when you see lightning and hear thunder. Because sound travels roughly 1/3 km/s, divide the time interval in seconds by 3 to estimate the range in km. For range in statute miles, divide the time interval by 5 instead. These approximations are crude but useful.

Because sound wave speed depends on temperature, the portion of a wave front in warmer air will move faster than the portion in cooler air. This causes the wave front to change its direction of advance. Thus, its propagation path (called a **ray** path) will bend (**refract**).

Consider horizontal layers of the atmosphere having different temperatures T_1 and T_2. If a sound wave is moving through layer one at elevation angle α_1, then after passing into layer two the new ray elevation angle will be α_2.

To quantify this effect, define an **index of refraction** for sound in calm air as:

$$n = \sqrt{T_o / T} \qquad (15.33)$$

where the reference temperature is T_o = 293 K. Snell discovered that

$$n \cdot \cos(\alpha) = \text{constant} \qquad (15.34)$$

When applied to a sound ray moving from one layer to another, **Snell's law** can be rewritten as:

$$\cos(\alpha_2) = \sqrt{T_2 / T_1} \cdot \cos(\alpha_1) \qquad \bullet(15.35)$$

See the Atmospheric Optics chapter for more info.

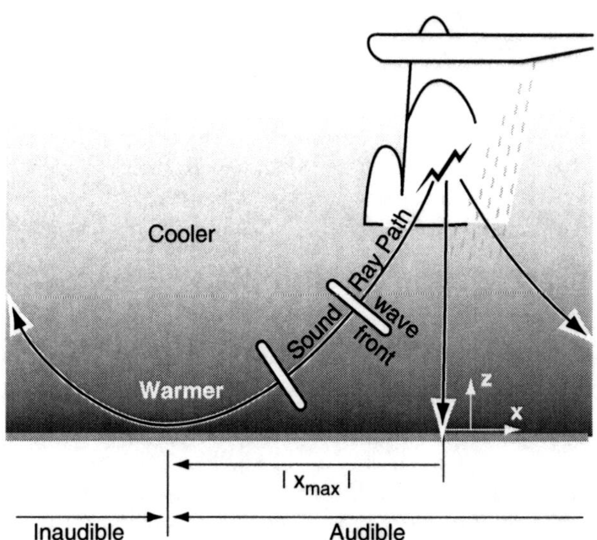

Figure 15.31
Wave fronts, ray paths, and audible range of thunder.

Sample Application

You see a thunderstorm approaching from the southwest. It is warm out (T = 35°C). Facing toward the storm, you see a lightning flash and hear the thunder 12 s later, and you feel a 10 m s^{-1} wind on your back. What is the distance to the lightning stroke?

Find the Answer

Given: With the wind at your back, this means that the wind blowing opposite to the direction that sound must travel to reach you; hence, ϕ = 180°. Also, Δt = 12 s, T = 35°C = 308 K, M = 10 m s^{-1}.
Find: Δx = ? km

Light speed is so fast that it is effectively instantaneous. So the time interval between "flash" and "bang" depends on sound speed:
Use eq. (15.32):
s = (343.15 m s^{-1})·[308K/293K]$^{1/2}$ + (10 m s^{-1})·cos(180°)
= 351.8 – 10.0 = 341.8 m s^{-1}
But speed is distance per time ($s = \Delta x/\Delta t$). Rearrange:
$\Delta x = s \cdot \Delta t$ = (341.8 m s^{-1})·(12 s) = **4.102 km**

Check: Physics and units OK.
Exposition: Typical wind speeds are much smaller than the speed of sound; hence, the distance calculation is only slightly affected by wind speed.

The approximate method of dividing the time interval by 3 is sometimes called the **3 second rule**. This rule is simple enough to do in your head while watching the storm, and would have allowed you to estimate $\Delta x \approx$ (12 s) / (3 s km^{-1}) = 4 km. Close enough.

If the time interval is 30 s or less, this means the storm is 10 km or closer to you, so you should immediately seek shelter. See the lighting safety INFO boxes.

Sample Application (§)

Suppose lightning occurs at 4 km altitude in a thunderstorm. Assume $\Delta T/\Delta z = -8$ K/km = constant. (a) How far horizontally from the lightning can you hear the thunder? (b) For the ray path that is tangent to the ground at that x_{max} point, plot the ray path backwards up to the lightning. $T = 308$ K near the ground.

Find the Answer

Given: $z = 4$ km, $T = 308$ K, $\gamma = 8$ K km^{-1}
Find: x_{max} = ? km, and plot (x, z) from $z = 0$ to 4 km

Use eq. (15.38): x_{max} =
$2 \cdot [(308\text{ K})\cdot(4\text{ km}) / (8\text{ K km}^{-1})]^{1/2}$ = **24.8 km**

Pick a small $\Delta x = 0.5$ km, and use eq. (15.36):
$\Delta\alpha = (8\text{K km}^{-1})\cdot(0.5\text{ km}) / [2\cdot(308\text{K})] = 0.00649$ radians

Use a spreadsheet to solve eqs. (15.37) with the constant value of $\Delta\alpha$ calculated above. We know the sound ray is tangent to the ground ($\alpha = 0$ radians) at the inaudibility point. Define $x = 0$ km and $z = 0$ km at this starting point. Then iterate eqs. (15.37) up to the altitude of the lightning.

x (km)	α (rad)	Δz (km)	z (km)
0	0	0	0
0.5	0.00649	0.00325	0.00325
1.0	0.01299	0.00649	0.00974
1..5	0.01948	0.00974	0.00195
2.0	0.02597	0.01299	0.03247
...	...	...	...
24.5	0.31818	0.16469	4.0477

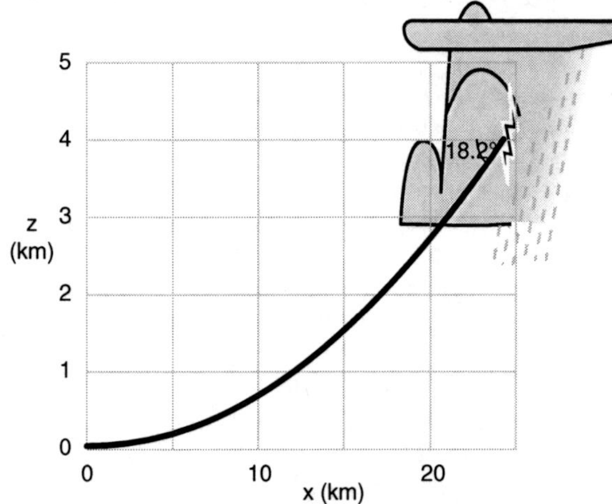

Check: Physics and units OK.
Exposition: $x = 0$ in the graph above is distance x_{max} from the lightning, where x_{max} was well approximated by eq. (15.38). For these weather conditions, of all the sound rays that radiated outward from the lightning origin, the one that became tangent to the ground was the one that left the storm with elevation angle 18.2° (= 0.31818 radians) downward from horizontal.

The previous expressions for Snell's law assumed a finite step change in temperature between layers that caused a sharp kink in the ray path. But if there is a gradual temperature change with distance along the ray path, then Snell's law for calm winds says there is a gradual bending of the ray path:

$$\Delta\alpha = \frac{\gamma}{2\cdot T}\cdot\Delta x \qquad (15.36)$$

where $\Delta\alpha$ is a small incremental change in ray elevation angle (radians) for each small increment of horizontal distance Δx traveled by the light. The vertical temperature variation is expressed as a lapse rate $\gamma = -\Delta T/\Delta z$, where T is the absolute temperature of background air. As a case study, we can assume γ is constant with height, for which case the ratio in eq. (15.36) is also nearly constant because T typically varies by only a small fraction of its magnitude.

You can solve eq (15.36) iteratively. Start with a known ray angle α at a known (x, z) location, and set a small fixed Δx value for your horizontal increment. Then, solve the following equations sequentially:

$$\begin{aligned} x_{new} &= x_{old} + \Delta x \\ \alpha_{new} &= \alpha_{old} + \Delta\alpha \\ \Delta z &= \Delta x\cdot\tan(\alpha) \\ z_{new} &= z_{old} + \Delta z \end{aligned} \qquad (15.37)$$

Continue solving eqs. (15.36 & 15.37) for more steps of Δx, using the "new" values output from the previous step as the "old" values to input for the next step. Save all the x_{new} and z_{new} values, because you can plot these to see the curved ray path (Fig. 15.31).

Thunderstorms usually happen on days when the sun has heated the ground, which in turn heated the bottom of the atmosphere. Thus, temperature often decreases with increasing height on thunderstorm days. Since sound waves bend toward air that is cooler, it means the thunder ray paths tend to be concave up (Fig 15.31).

This curvature can be significant enough that there can be a max distance x_{max} beyond which you cannot hear thunder (i.e., it is inaudible):

$$x_{\max} \cong 2\cdot\sqrt{T\cdot z/\gamma} \qquad (15.38)$$

where the sound has originated at height z, and where calm winds were assumed.

With wind, Snell's equation for the ray path is:

$$\frac{n\cdot(\cos\alpha)\left[1-(Ma\cdot n\cdot\sin\alpha)^2\right] - Ma\cdot(n\cdot\sin\alpha)^2}{1+Ma\cdot n\cdot(\cos\alpha)\left[1-(Ma\cdot n\cdot\sin\alpha)^2\right]^{1/2} - (Ma\cdot n\cdot\sin\alpha)^2}$$

$$= \text{constant} \qquad (15.39)$$

where $Ma = M/s_o$ is the Mach number of the wind.

TORNADOES

Tornadoes are violently rotating, small-diameter columns of air in contact with the ground. Diameters range from 10 to 1000 m, with an average of about 100 m. In the center of the tornado is very low pressure (order of 10 kPa lower than ambient).

Tornadoes are usually formed by thunderstorms, but most thunderstorms do not spawn tornadoes. The strongest tornadoes come from supercell thunderstorms. Tornadoes have been observed with a wide variety of shapes (Fig. 15.32).

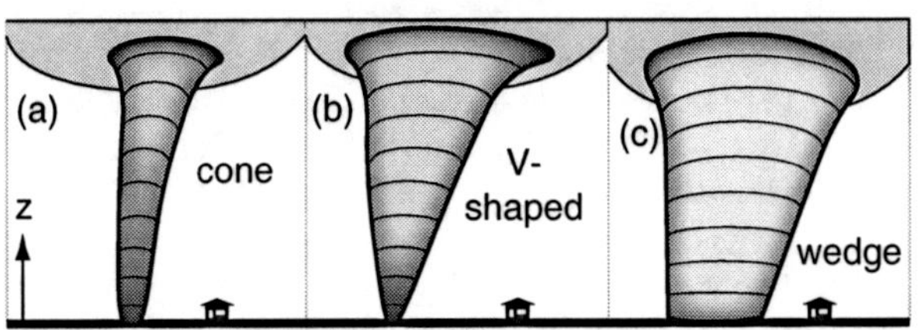

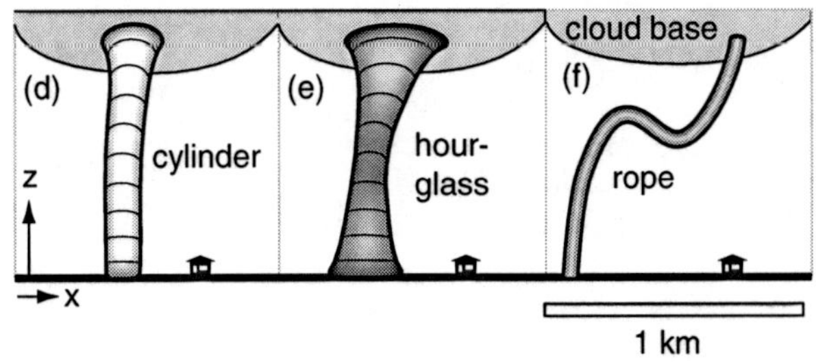

Figure 15.32
Illustration of some of the different tornado shapes observed.

Tangential Velocity & Tornado Intensity

Tangential velocities around tornadoes range from about 18 m s^{-1} for weak tornadoes to greater than 140 m s^{-1} for exceptionally strong ones. Tornado rotation is often strongest near the ground (15 to 150 m AGL), where upward vertical velocities of 25 to 60 m s^{-1} have been observed in the outer wall of the tornado. This combination of updraft and rotation can rip trees, animals, vehicles and buildings from the ground and destroy them. It can also loft trucks, cars, and other large and small objects, which can fall outside the tornado path causing more damage.

Often a two-region model is used to approximate tangential velocity M_{tan} in a tornado, with an internal **core** region of radius R_o surrounded by an external region. R_o corresponds to the location of fastest tangential velocity $M_{tan\ max}$ (Fig. 15.33), which is sometimes assumed to coincide with the outside edge of the visible funnel. Air in the core of the tornado rotates as a solid-body, while air outside the core is irrotational (has no relative vorticity as it moves around the tornado axis), and conserves angular momentum as it is drawn into the tornado. This model is called a **Rankine combined vortex (RCV)**.

The pressure deficit is $\Delta P = P_\infty - P$, where P is the pressure at any radius R from the tornado axis, and P_∞ is ambient pressure far away from the tornado (for $P_\infty \geq P$). At R_o, match the inner and outer tangential wind speeds, and match the inner and outer pressure deficits:

Core Region ($R < R_o$):

$$\frac{M_{\tan}}{M_{\tan\max}} = \frac{R}{R_o} \qquad \bullet(15.40)$$

$$\frac{\Delta P}{\Delta P_{\max}} = 1 - \frac{1}{2}\left(\frac{R}{R_o}\right)^2 \qquad \bullet(15.41)$$

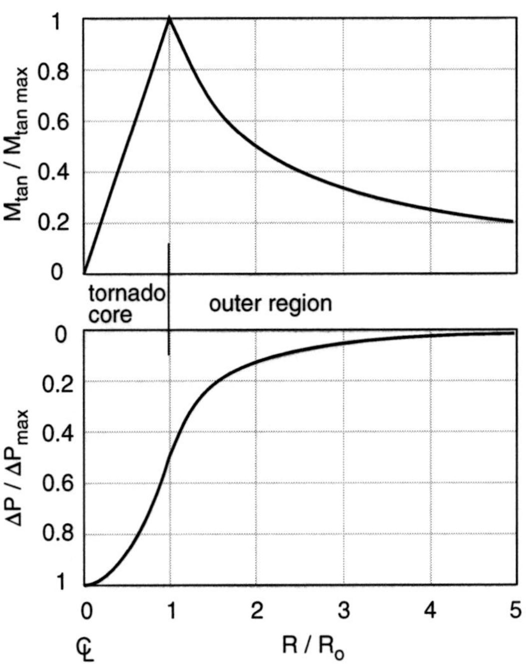

Figure 15.33
A Rankine-combined-vortex (RCV) model for tornado tangential velocity and pressure. The pressure deficit is plotted with reversed ordinate, indicating lower pressure in the tornado core.

Sample Application

If the max pressure deficit in the center of a 20 m radius tornado is 10 kPa, find the max tangential wind speed, and the wind and pressure deficit at R = 50 m.

Find the Answer

Given: R_o = 20 m, ΔP_{max} = 10 kPa, R = 50 m
Find: $M_{tan\ max}$ = ? m s^{-1}.
Also M_{tan} = ? m s^{-1} and ΔP = ? kPa at R = 50 m.
Assume: ρ = 1 kg m^{-3}.

Use eq. (15.44):

$M_{tan\ max} = [\Delta P_{max} / \rho]^{1/2} =$
$[(10{,}000 \text{ Pa}) / (1 \text{ kg m}^{-3})]^{1/2}$ = **100 m s^{-1}**

Because $R > R_o$, use outer-region eqs. (15.42 & 15.43):

M_{tan} = (100 m s^{-1}) · [(20 m)/(50 m)] = **40 m s^{-1}**

ΔP = 0.5·(10 kPa) · [(20 m)/(50 m)]^2 = **0.8 kPa**.

Check: Physics OK. Units OK. Agrees with Fig. 15.33.
Exposition: 10 kPa is quite a large pressure deficit in the core — roughly 10% of sea-level pressure. However, most tornadoes are not this violent. Typical tangential winds of 60 m s^{-1} or less would correspond to core pressure deficits of 3.6 kPa or less.

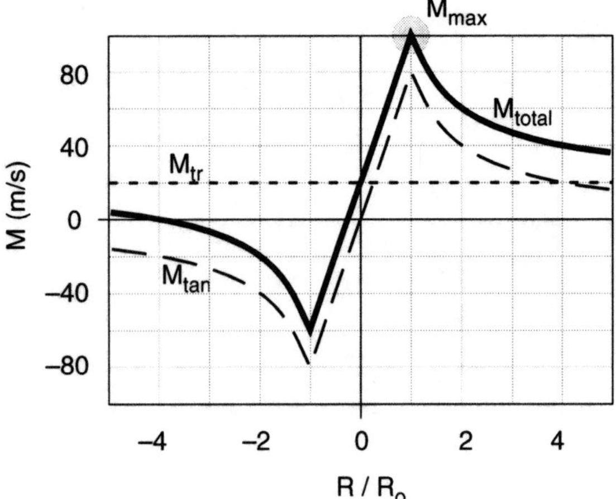

Figure 15.34
Illustration of sum of relative rotational (M_{tan}) wind and tornado translation (M_{tr}) to yield total winds (M_{total}) measured at the ground. Tornado intensity is classified based on the maximum wind speed (M_{max}) anywhere in the tornado.

Outer Region ($R > R_o$):

$$\frac{M_{\tan}}{M_{\tan \max}} = \frac{R_o}{R} \qquad \bullet(15.42)$$

$$\frac{\Delta P}{\Delta P_{\max}} = \frac{1}{2}\left(\frac{R_o}{R}\right)^2 \qquad \bullet(15.43)$$

These equations are plotted in Fig. 15.33, and represent the wind relative to the center of the tornado.

Max tangential velocity (at $R = R_o$) and max **core pressure deficit** (ΔP_{max}, at R = 0) are related by

$$\Delta P_{\max} = \rho \cdot (M_{\tan \max})^2 \qquad \bullet(15.44)$$

where ρ is air density. This equation, derived from the **Bernoulli equation**, $M_{\tan \max}$ can also be described as a **cyclostrophic wind** as explained in the Forces & Winds chapter (namely, it is a balance between centrifugal and pressure-gradient forces). Forecasting these maximum values is difficult.

Near the Earth's surface, frictional drag near the ground slows the air below the cyclostrophic speed. Thus there is insufficient centrifugal force to balance pressure-gradient force, which allows air to be sucked into the bottom of the tornado. Further away from the ground, the balance of forces causes zero net radial flow across the tornado walls; hence, the tornado behaves similar to a vacuum-cleaner hose.

The previous 5 equations gave tangential wind speed relative to the center of the tornado. But the tornado also moves horizontally (**translates**) with its parent thunderstorm. The total wind M_{total} at any point near the tornado is the vector sum of the rotational wind M_{tan} plus the translational wind M_{tr} (Fig. 15.34). The max wind speed M_{max} associated with the tornado is

$$|M_{max}| = |M_{tan\ max}| + |M_{tr}| \qquad \bullet(15.45)$$

and is found on the right side of the storm (relative to its translation direction) for cyclonically (counterclockwise, in the Northern Hemisphere) rotating tornadoes.

Most tornadoes rotate cyclonically. Less than 2% of tornadoes rotate the opposite direction (anticyclonically). This low percentage is due to two factors: (1) Coriolis force favors mesocyclones that rotate cyclonically, and (2) friction at the ground causes a turning of the wind with increasing height (Fig. 15.42, presented in a later section), which favors right-moving supercells in the Northern Hemisphere with cyclonically rotating tornadoes.

One **tornado damage scale** was devised by the Tornado and Storm Research Organization in Europe, and is called the **TORRO** scale (**T**). Another scale was developed for North America by Ted Fujita, and is called the **Fujita** scale (**F**).

In 2007 the Fujita scale was revised into an **Enhanced Fujita** (**EF**) scale (Table 15-3), based on better measurements of the relationship between winds and damage for 28 different types of structures. It is important to note that the EF intensity determination for any tornado is based on a damage survey AFTER the tornado has happened.

For example, consider modern, well-built single-family homes and duplexes, typically built with wood or steel studs, with plywood roof and outside walls, all covered with usual types of roofing, sidings, or brick. For this structure, use the following damage descriptions to estimate the EF value:

- If threshold of visible damage, then EF0 or less.
- If loss of gutters, or awnings, or vinyl or metal siding, or less than 20% of roof covering material, then EF0 - EF1.
- If broken glass in doors and windows, or roof lifted up, or more than 20% of roof covering missing, or chimney collapse, or garage doors collapse inward, or failure of porch or carport, then EF0 - EF2.
- If entire house shifts off foundation, or large sections of roof structure removed (but most walls remain standing), then EF1 - EF3.
- If exterior walls collapse, then EF2 - EF3.
- If most walls collapse, except small interior rooms, then EF2 - EF4.
- If all walls collapse, then EF3 - EF4.
- If total destruction & floor slabs swept clean, then EF4 - EF5.

Similar damage descriptions for the other 27 types of structures (including trees) are available from the USA Storm Prediction Center.

For any EF range (such as EF = 4), the lower threshold of maximum tangential 3-second-gust wind speed M_{max} is approximately:

$$M_{max} = M_o + a \cdot (\text{EF})^{1.2} \qquad (15.46)$$

where M_o = 29.1 m s^{-1} and a = 8.75 m s^{-1}.

The "derived" gust thresholds listed in Table 15-3 are often converted to speed units familiar to the public and then rounded to pleasing integers of nearly the correct value. Such a result is known as an **Operational Scale** (see Table 15-3).

Table 15-3. Enhanced Fujita scale for tornado-damage intensity. *(Derived-scale speeds from NOAA Storm Prediction Center.)*

Scale	Derived Max Tangential 3 s Gust Speed (m/s)	Operational Scales		Damage Classification Description (from the old Fujita F scale)	Relative Frequency	
		EF Scale (stat. miles/h)	Old F Scale (km/h)		USA	Canada
EF0	29.1 – 38.3	65 – 85	64 – 116	Light damage; some damage to chimneys, TV antennas; breaks twigs off trees; pushes over shallow-rooted trees.	29%	45%
EF1	38.4 – 49.1	86 – 110	117 – 180	Moderate damage; peels surface off roofs; windows broken; light trailer homes pushed or turned over; some trees uprooted or snapped; moving cars pushed off road.	40%	29%
EF2	49.2 – 61.6	111 – 135	181 – 252	Considerable damage; roofs torn off frame houses leaving strong upright walls; weak buildings in rural areas demolished; trailer houses destroyed; large trees snapped or uprooted; railroad boxcars pushed over; light object missiles generated; cars blown off roads.	24%	21%
EF3	61.7 – 75.0	136 – 165	253 – 330	Severe damage; roofs and some walls torn off frame houses; some rural buildings completely destroyed; trains overturned; steel-framed hangars or warehouse-type structures torn; cars lifted off of the ground; most trees in a forest uprooted or snapped and leveled.	6%	4%
EF4	75.1 – 89.3	166 – 200	331 – 417	Devastating damage; whole frame houses leveled leaving piles of debris; steel structures badly damaged; trees debarked by small flying debris; cars and trains thrown some distance or rolled considerable distances; large wind-blown missiles generated.	2%	1%
EF5	≥ 89.4	> 200	418 – 509	Incredible damage; whole frame houses tossed off foundation and blown downwind; steel-reinforced concrete structures badly damaged; automobile-sized missiles generated; incredible phenomena can occur.	< 1%	0.1%

Sample Application
Find Enhanced Fujita & TORRO intensities for M_{max} = 100 m s^{-1}.

Find the Answer
Given: M_{max}= 100 m s^{-1}.
Find: EF and T intensities

Use Tables 15-3 and 15-4: ≈ **EF5** , **T8** .

Exposition: This is a violent, very destructive, significant tornado.

Tornado intensity varies during the life-cycle of the tornado, so different levels of destruction are usually found along the damage path for any one tornado. Tornadoes of strength EF2 or greater are labeled **significant tornadoes**.

The TORRO scale (Table 15-4) is defined by maximum wind speed M_{max}, but in practice is estimated by damage surveys. The lower threshold of wind-speed for any **T** range (e.g., T7) is defined approximately by:

$$M_{max} \approx a \cdot (\mathbf{T} + 4)^{1.5} \tag{15.47}$$

where a = 2.365 m s^{-1} and **T** is the TORRO tornado intensity value. A weak tornado would be classified as T0, while an extremely strong one would be T10 or higher.

Any tornado-damage scale is difficult to use and interpret, because there are no actual wind measurements for most events. However, the accumulation of tornado-damage-scale estimates provides valuable statistics over the long term, as individual errors are averaged out.

Table 15-4. TORRO tornado scale. *(from www.torro.org.uk/site/tscale.php)*

Scale	Max. Speed (m s^{-1})	Tornado Intensity & Damage Description *(abridged from the Torro web site)* *[UK "articulated lorry" ≈ USA "semi-trailer truck" or "semi"]*
T0	17 – 24	**Light**. Loose light litter raised from ground in spirals. Tents, marquees disturbed. Exposed tiles & slates on roofs dislodged. Twigs snapped. Visible damage path through crops.
T1	25 – 32	**Mild**. Deck chairs, small plants, heavy litter airborne. Dislodging of roof tiles, slates, and chimney pots. Wooden fences flattened. Slight damage to hedges and trees.
T2	33 – 41	**Moderate**. Heavy mobile homes displaced. Semi's blown over. Garden sheds destroyed. Garage roofs torn away. Damage to tile and slate roofs and chimney pots. Large tree branches snapped. Windows broken.
T3	42 – 51	**Strong**. Mobile homes overturned & badly damaged. Light semis destroyed. Garages and weak outbuilding destroyed. Much of the roofing material removed. Some larger trees uprooted or snapped.
T4	52 – 61	**Severe**. Cars lifted. Mobile homes airborne & destroyed. Sheds airborne for considerable distances. Entire roofs removed. Gable ends torn away. Numerous trees uprooted or snapped.
T5	62 – 72	**Intense**. Heavy motor vehicles lifted (e.g., 4 tonne trucks). More house damage than T4. Weak buildings completely collapsed. Utility poles snapped.
T6	73 – 83	**Moderately Devastating**. Strongly built houses lose entire roofs and perhaps a wall. Weaker built structures collapse completely. Electric-power transmission pylons destroyed. Objects imbedded in walls.
T7	84 – 95	**Strongly Devastating**. Wooden-frame houses wholly demolished. Some walls of stone or brick houses collapsed. Steel-framed warehouse constructions buckled slightly. Locomotives tipped over. Noticeable debarking of trees by flying debris.
T8	96 – 107	**Severely Devastating**. Cars hurled great distances. Wooden-framed houses destroyed and contents dispersed over large distances. Stone and brick houses irreparably damaged. Steel-framed buildings buckled.
T9	108 – 120	**Intensely Devastating**. Many steel-framed buildings badly demolished. Locomotives or trains hurled some distances. Complete debarking of standing tree trunks.
T10	121 – 134	**Super**. Entire frame houses lifted from foundations, carried some distances & destroyed. Severe damage to steel-reinforced concrete buildings. Damage track left with nothing standing above ground.

Appearance

Two processes can make tornadoes visible: water droplets and debris (Fig. 15.35). Sometimes these processes make only the bottom or top part of the tornado visible, and rarely the whole tornado is invisible. Regardless of whether you can see the tornado, if the structure consists of a violently rotating column of air, then it is classified as a tornado.

Debris can be formed as the tornado destroys things on the Earth's surface. The resulting smaller fragments (dirt, leaves, grass, pieces of wood, bugs, building materials and papers from houses and barns) are drawn into the tornado wall and upward, creating a visible **debris cloud**. (Larger items such as whole cars can be lifted by the more intense tornadoes and tossed outward, some as much as 30 m.) If tornadoes move over dry ground, the debris cloud can include dust and sand. Debris clouds form at the ground, and can extend to various heights for different tornadoes, including some that extend up to wall-cloud base.

The water-condensation funnel is caused by low pressure inside the tornado, which allows air to expand as it is sucked horizontally toward the core. As the air expands it cools, and can reach saturation if the pressure is low enough and the initial humidity of the is air great enough. The resulting cloud of water droplets is called a **funnel cloud**, and usually extends downward from the thunderstorm cloud base. Sometimes this condensation funnel cloud can reach all the way to the ground. Most strong tornadoes have both a condensation funnel and a debris cloud (Fig. 15.35).

Because the tornado condensation funnel is formed by a process similar to the lifting condensation level (LCL) for normal convective cloud base, you can use the same LCL equation (see the Water Vapor chapter) to estimate the pressure P_{cf} at the outside of the tornado condensation funnel, knowing the ambient air temperature T and dew point T_d at ambient near-surface pressure P. Namely, $P_{cf} = P_{LCL}$.

$$P_{cf} = P \cdot \left[1 - b \cdot \left(\frac{T - T_d}{T} \right) \right]^{C_p/\Re} \qquad \bullet(15.48)$$

where $C_p/\Re = 3.5$ and $b = 1.225$, both dimensionless, and where T in the denominator must be in Kelvin.

Because both the condensation funnel and cloud base indicate the same pressure, the isobars must curve downward near the tornado (Fig. 15.36). Thus, the greatest horizontal pressure gradient associated with the tornado is near the ground (near "A" in Fig. 15.36). Drag at the ground slows the wind a bit there, which is why the fastest tangential winds in a tornado are found 15 to 150 m above ground.

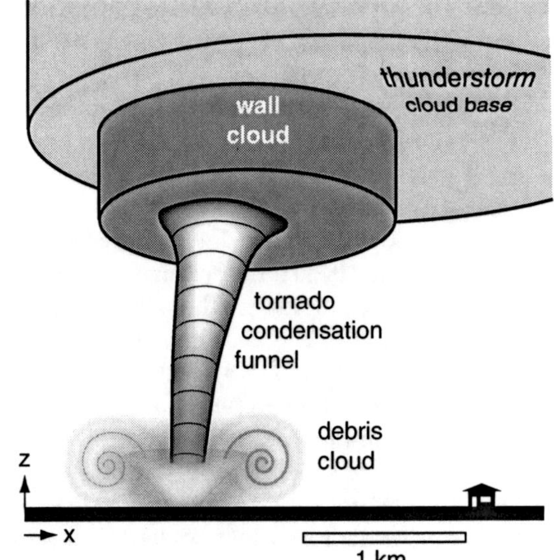

Figure 15.35
Condensation funnel and debris cloud.

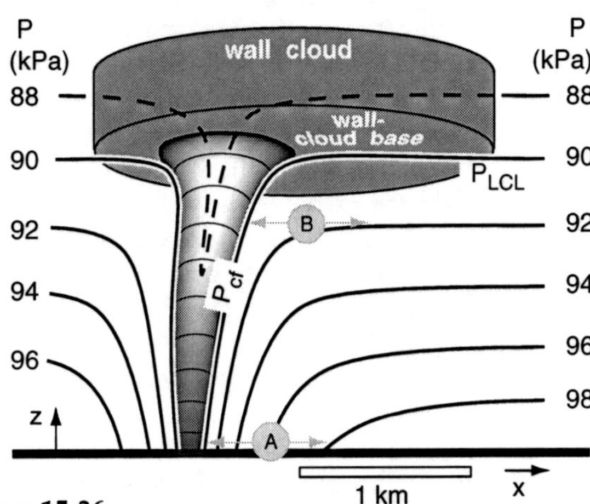

Figure 15.36
Relationship between lifting-condensation-level pressure (P_{LCL}), cloud base, and pressure at the condensation funnel (P_{cf}). Horizontal pressure gradient at point "A" is 4 times that at "B".

Sample Application

Under a tornadic thunderstorm, the temperature is 30°C and dew point is 23°C near the ground where pressure is 100 kPa. Find the near-surface pressure at the outside edge of the visible condensation funnel.

Find the Answer

Given: $T = 30°C = 303K$, $T_d = 23°C$, $P = 100$ kPa
Find: $P_{cf} = ?$ kPa

Use eq. (15.48): $P_{cf} = (100\text{kPa})\cdot[1 - 1.225\cdot(30-23)/303]^{3.5}$
$= \underline{\mathbf{90.4\ kPa}}$ at the tornado funnel-cloud edge.

Check: Units OK. Physics OK.
Exposition: The tornado core pressure can be even lower than at the edge of the condensation funnel.

Types of Tornadoes & Other Vortices

We will compare six types of vortices (Fig 15.37):

- supercell tornadoes
- landspout tornadoes
- waterspouts
- cold-air funnels
- gustnadoes
- dust devils, steam devils, firewhirls

Recall from the Thunderstorm chapter that a **mesocyclone** is where the whole thunderstorm updraft (order of 10 to 15 km diameter) is slowly rotating (often too slowly to see by eye). This rotation can last for 1 h or more, and is one of the characteristics of a **supercell** thunderstorm. Only a small percentage of thunderstorms are supercells with mesocyclones, but it is from these supercells that the strongest tornadoes can form. Tornadoes rotate faster and have smaller diameter (~100 m) than mesocyclones.

Supercell tornadoes form under (and are attached to) the main updraft of supercell thunderstorms (Figs. 15.32a-c, 15.35, & 15.37) or under a cumulus congestus that is merging into the main supercell updraft from the flanking line. It can be the most violent tornado type — up through EF5 intensity. They move horizontally (i.e., **translate**) at nearly the same speed as the parent thunderstorm (on the order of 5 to 40 m s^{-1}). These tornadoes will be discussed in more detail in the next subsections.

Landspouts are weaker tornadoes (EF0 - EF2, approximately) not usually associated with supercell thunderstorms. They are often cylindrical, and look like hollow soda straws (Fig. 15.32d). These short-lived tornadoes form along strong cold fronts. Horizontal wind shear across the frontal zone provides the rotation, and vertical stretching of the air by updrafts in the **squall-line thunderstorms** along the front can intensify the rotation (Fig. 15.37).

Waterspouts (Fig. 15.37) are tornadoes that usually look like landspouts (hollow, narrow, 3 to 100 m diameter cylinders), but form over water surfaces (oceans, lakes, wide rivers, bays, etc.). They are often observed in subtropical regions (e.g., in the waters around Florida), and can form under (and are attached to) cumulus congestus clouds and small thunderstorms. They are often short lived (usually 5 to 10 min) and weak (EF0 - EF1). The waterspout life cycle is visible by eye via changes in color and waves on the water surface: (1) dark spot, (2) spiral pattern, (3) spray-ring, (4) mature spray vortex, and (5) decay. Waterspouts have also been observed to the lee of mountainous islands such as Vancouver Island, Canada, where the initial rotation is caused by wake vortices as the wind swirls around the sides of mountains.

Unfortunately, whenever any type of tornado moves over the water, it is also called a waterspout.

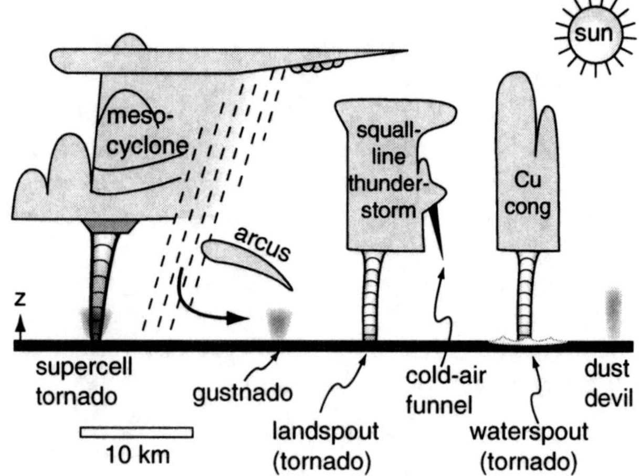

Figure 15.37
Illustration of tornado and vortex types.

Thus, supercell tornadoes (EF3 - EF5) would be called waterspouts if they moved over water. So use caution when you hear waterspouts reported in a weather warning, because without other information, you won't know if it is a weak classical waterspout or a strong tornado.

Cold-air funnels are short vortices attached to shallow thunderstorms with high cloud bases, or sometimes coming from the sides of updraft towers (Fig. 15.37). They are very short lived (1 - 3 minutes), weak, and usually don't reach the ground. Hence, they usually cause no damage on the ground (although light aircraft should avoid them). Cold-air funnels form in synoptic cold-core low-pressure systems with a deep layer of unstable air.

Gustnadoes are shallow (order of 100 m tall) debris vortices touching the ground (Fig. 15.37). They form along the gust-front from thunderstorms, where there is shear between the outflow air and the ambient air. Gustnadoes are very weak (EF0 or weaker) and very short lived (a few minutes). The arc clouds along the gust front are not usually convective, so there is little or no updraft to stretch the air vertically, and hence no mechanism for accelerating the vorticity. There might also be rotation or a very small condensation funnel visible in the overlying arc clouds. Gustnadoes translate with the speed of advance of the gust front.

Dust devils are not tornadoes, and are not associated with thunderstorms. They are fair-weather phenomena, and can form in the clear-air thermals of warm air rising from a heated surface (Fig. 15.37). They are weak (less than EF0) debris vortices, where the debris can be dust, sand, leaves, volcanic ash, grass, litter, etc. Normally they form during daytime in high-pressure regions, where the sun heats the ground, and are observed over the desert or other arid locations. They translate very slowly or not at all, depending on the ambient wind speed.

When formed by arctic-air advection over an unfrozen lake in winter, the resulting **steam-devils** can happen day or night. Smoky air heated by forest fires can create **firewhirls**. Dust devils, steam devils, firewhirls and gustnadoes look very similar.

Evolution as Observed by Eye

From the ground, the first evidence of an incipient supercell tornado is a **dust swirl** near the ground, and sometimes a rotating wall cloud protruding under the thunderstorm cloud base (Fig. 15.38). Stage 2 is the **developing stage**, when a condensation funnel cloud begins to extend downward from the bottom of the wall cloud or thunderstorm base, and the debris cloud becomes larger with well-defined rotation.

Stage 3 is the **mature stage**, when there is a visible column of rotating droplets and/or debris extending all the way from the cloud to the ground. This is the stage when tornadoes are most destructive. During stage 4 the visible tornado weakens, and often has a slender rope-like shape (also in Fig. 15.32f). As it dissipates in stage 5, the condensation funnel disappears up into the cloud base and the debris cloud at the surface weakens and disperses in the ambient wind. Meanwhile, a cautious storm chaser will also look to the east or southeast under the same thunderstorm cloud base, because sometimes new tornadoes form there.

Tornado Outbreaks

A **tornado outbreak** is when a single synoptic-scale system (e.g., cold front) spawns ten or more tornadoes during one to seven days (meteorologists are still debating a more precise definition). Tornado outbreaks have been observed every decade in North America for the past couple hundred years of recorded meteorological history. Sometimes outbreaks occur every year, or multiple times a year.

The following list highlights a small portion of the outbreaks in North America:

- 25 May - 1 June 1917: 63 tornadoes in Illinois killed 383 people.
- 18 March 1925 (tri-state) Tornado: Deadliest tornado(es) in USA, killing 695 people on a 350 km track through Missouri, Illinois & Indiana.
- 1 - 9 May 1930: 67 tornadoes in Texas killed 110.
- 5 - 6 April 1936: 17 tornadoes in Tupelo, Mississippi, and Gainesville, Georgia, killed 454.
- 15 - 24 May 1949: 74 torn. in Missouri killed 66.
- 7 - 11 April 1965 (Palm Sunday): 51 F2 - F5 tornadoes, killed 256.

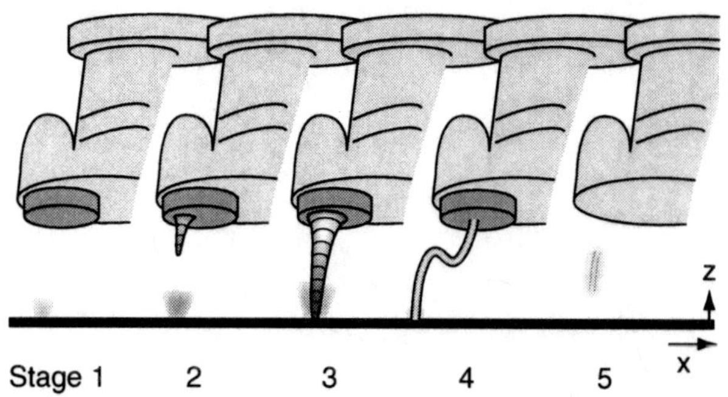

Figure 15.38
Stages in a supercell-tornado life cycle.

A SCIENTIFIC PERSPECTIVE • Be Safe *(part 4)*

More chase guidelines from Charles Doswell III.

The #3 Threat: The Storm

1. Avoid driving through the heaviest precipitation part of the storm (known as "**core punching**").
2. Avoid driving under, or close to, rotating wall clouds.
3. Don't put yourself in the path of a tornado or a rotating wall cloud.
4. You must also be aware of what is happening around you, as thunderstorms and tornadoes develop quickly. You can easily find yourself in the path of a new thunderstorm while you are focused on watching an older storm. Don't let this happen — be vigilant.
5. For new storm chasers, find an experienced chaser to be your mentor. (Work out such an arrangement ahead of time; don't just follow an experienced chaser uninvited.)
6. Keep your engine running when you park to view the storm.
7. Even with no tornado, straight-line winds can move hail or debris (sheet metal, fence posts, etc.) fast enough to kill or injure you, and break car windows. Move away from such regions.
8. Avoid areas of rotating curtains of rain, as these might indicate that you are in the dangerous center of a mesocyclone (called the "**bear's cage**").
9. Don't be foolhardy. Don't be afraid to back off if your safety factor decreases.
10. Never drive into rising waters. Some thunderstorms such as HP supercell storms can cause flash floods.
11. Always have a clear idea of the structure, evolution, and movement of the storm you are viewing, so as to anticipate safe courses of action.

(continues on next page)

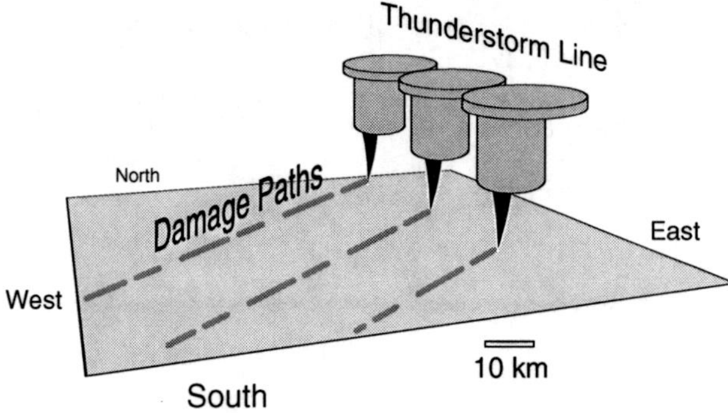

Figure 15.39
Sketch of parallel damage paths from a line of supercell thunderstorms during a tornado outbreak.

> **A SCIENTIFIC PERSPECTIVE • Be Safe** *(part 5)*
>
> More chase guidelines from Charles Doswell III.
>
> **The #3 Threat: The Storm** *(continuation)*
>
> 12. Plan escape routes in advance.
> 13. Although vehicles offer safety from lightning, they are death traps in tornadoes. If you can't drive away from the tornado, then abandon your vehicle and get into a ditch or culvert, or some other place below the line-of-fire of all the debris.
> 14. In open rural areas with good roads, you can often drive away from the tornado's path.
> 15. Don't park under bridge overpasses. They are <u>not</u> safe places if a tornado approaches.
> 14. Avoid chasing at night. Some difficulties include:
>   a. Don't trust storm movement as broadcast on radio or TV. Often, the media reports the heavy precipitation areas, not the action (dangerous) areas of the mesocyclone and tornado.
>   b. Storm info provided via various wireless data and internet services can be several minutes old or older
>   c. It is difficult to see tornadoes at night. Flashes of light from lightning & exploding electrical transformers (known as "**power flashes**") are often inadequate to see the tornado. Also, not all power flashes are caused by tornadoes.
>   d. If you find yourself in a region of strong inflow winds that are backing (changing direction counterclockwise), then you might be in the path of a tornado.
>   e. Flooded roads are hard to see at night, and can cut-off your escape routes. Your vehicle could hydroplane due to water on the road, causing you to lose control of your vehicle.
>   f. Even after you stop chasing storms for the day, dangerous weather can harm you on your drive home or in a motel.
>
> On his web site, Doswell offers many more tips and recommendations for responsible storm chasing.

- 3 - 4 April 1974: 148 tornadoes, killed 306 people in Midwest USA, and 9 in Canada.
- 31 May 1985: 41 tornadoes in USA & Canada, killed 76 in USA and 12 in Canada.
- 13 March 1990: 59 tornadoes in central USA, killed 2.
- 3 May 1999: 58 tornadoes in Oklahoma & Kansas, killed 44.
- 3 - 11 May 2003: 401 tornadoes in tornado alley killed 48.
- September 2004 in Hurricanes Francis & Ivan: 220 tornadoes.
- 26 - 31 August 2005 in Hurricane Katrina: 44 tornadoes in southeast USA.
- 5 - 6 February 2008 (Super Tuesday): 87 tornadoes in central USA killed 57.
- 22 - 25 May 2008: 234 tornadoes in central USA killed 10.
- 25 - 28 April 2011: 358 tornadoes in E. USA, killed 324, causing about $10 billion in damage.
- 21 - 26 May 2011: 242 tornadoes in midwest USA killed 180.

Outbreaks are often caused by a line or cluster of supercell thunderstorms. Picture a north-south line of storms, with each thunderstorm in the line marching toward the northeast together like troops on parade (Fig. 15.39). Each tornadic supercell in this line might create a sequence of multiple tornadoes (called a **tornado family**), with very brief gaps between when old tornadoes decay and new ones form. The aftermath are parallel tornado damage paths like a wide (hundreds of km) multi-lane highway oriented usually from southwest toward northeast.

Storm-relative Winds

Because tornadoes translate with their parent thunderstorms, the winds that influence supercell and tornado rotation are the environmental wind vectors <u>relative to a coordinate system that moves with the thunderstorm</u>. Such winds are called **storm-relative winds**.

First, find the storm motion vector. If the thunderstorm already exists, then its motion can be tracked on radar or satellite (which gives a vector based on its actual speed and direction of movement). For forecasts of future thunderstorms, recall from the previous chapter that many thunderstorms move in the direction of the mean wind averaged over the 0 to 6 km layer of air, as indicated by the "X" in Fig. 15.40. Some supercell storms split into two parts: a right moving storm and a left moving storm, as was shown in the Thunderstorm chapter. Namely, if tornado formation from a right-moving supercell is of concern, then use a mean storm vector associated with the "R" in Fig. 14.61 of the previous chapter (i.e., do not use the "X").

Next, to find storm-relative winds, take the vector difference between the actual wind vectors and the storm-motion vector. On a hodograph, draw the storm-relative wind vectors from the storm-motion point to each of the original wind-profile data points. This is illustrated in Fig. 15.40a, based on the hodograph and normal storm motion "X". After (optionally) repositioning the hodograph origin to coincide with the mean storm motion (Fig. 15.40b), the result shows the directions and speed of the storm-relative environmental wind vectors.

The algebraic components (U_j', V_j') of these storm-relative horizontal vectors are:

$$U_j' = U_j - U_s \qquad \bullet(15.49)$$

$$V_j' = V_j - V_s \qquad \bullet(15.50)$$

where (U_j, V_j) are the wind components at height index j, and the storm motion vector is (U_s, V_s). For a supercell that moves with the 0 to 6 km mean wind: (U_s, V_s) = ($\overline{U}, \overline{V}$) from the previous chapter. The vertical component of storm-relative winds $W_j' = W_j$, because the thunderstorm does not translate vertically.

Sample Application

For the right-moving supercell of Fig. 14.61 from the previous chapter plot the storm-relative wind vectors on a hodograph.

Find the Answer

Given: Fig. 14.61. Storm motion indicated by "R".
Find: Hodograph of storm-relative winds.
Method: Copy Fig. 14.61, draw relative vectors on it, and then re-center origin of hodograph to be at R:

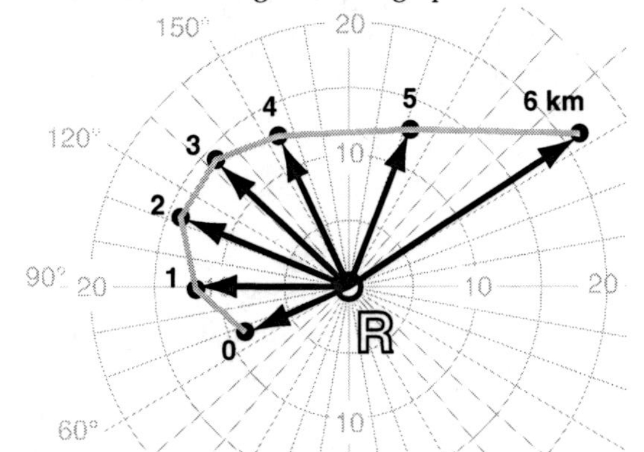

Check: Similar to Fig. 15.40b.
Exposition: Compared to storm X, storm R has less directional shear, but more inflow at middle altitudes.

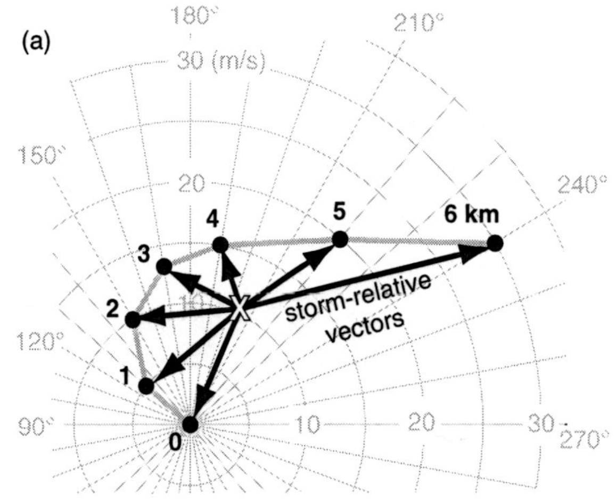

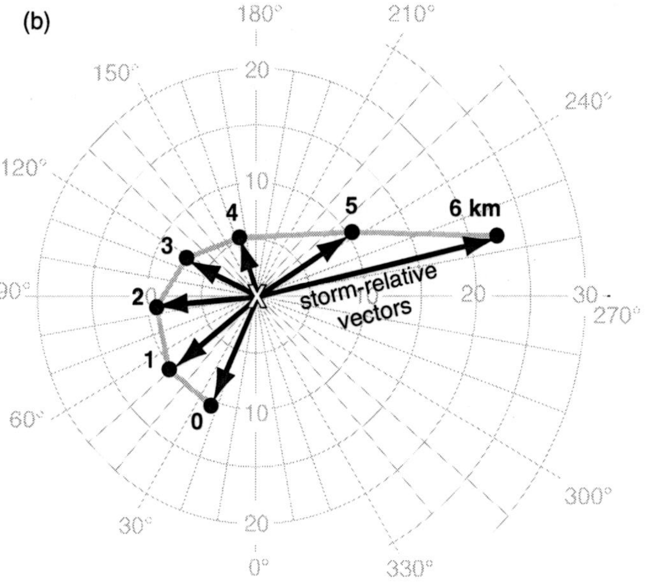

Figure 15.40
(a) Hodograph showing wind vector differences between the winds relative to a fixed coordinate system (grey line hodograph) and the mean storm motion (X), based on the data from Figs. 15.31 & 14.59. (b) Same data, but with the hodograph origin shifted to "X" to give **storm-relative** *wind vectors (black arrows). Some people find version (b) difficult to interpret, so they prefer to use version (a).*

Origin of Tornadic Rotation

Because tornado rotation is around a vertical axis, express this rotation as a **relative vertical vorticity** ζ_r. Relative vorticity was defined in the General Circulation chapter as $\zeta_r = (\Delta V/\Delta x) - (\Delta U/\Delta y)$, and a forecast (tendency) equation for it was given in the Extratropical Cyclone chapter in the section on cyclone spin-up.

Relative to the thunderstorm, there is little horizontal or vertical advection of vertical vorticity, and the beta effect is small because any one storm moves across only a small range of latitudes during its lifetime. Thus, mesocyclone and tornadic vorticity are affected mainly by tilting, stretching, and turbulent drag:

$$\underbrace{\frac{\Delta\zeta_r}{\Delta t}}_{spin-up} \approx \underbrace{\frac{\Delta U'}{\Delta z}\cdot\frac{\Delta W}{\Delta y} - \frac{\Delta V'}{\Delta z}\cdot\frac{\Delta W}{\Delta x}}_{tilting} \quad \bullet(15.51)$$

$$+ \underbrace{(\zeta_r + f_c)\cdot\frac{\Delta W}{\Delta z}}_{stretching} - \underbrace{C_d\cdot\frac{M}{z_{TornBL}}\cdot\zeta_r}_{turb.drag}$$

where the storm-relative wind components are (U', V', W), the Coriolis parameter is f_c, the tangential wind speed is approximately M, drag coefficient is C_d, and z_{TornBL} is the depth of the tornado's boundary layer (roughly z_{TornBL} = 100 m).

This simplified vorticity-tendency equation says that rotation about a vertical axis can increase (i.e., **spin up**) if horizontal vorticity is **tilted** into the vertical, or if the volume of air containing this vertical vorticity is **stretched** in the vertical (Fig. 15.41). Also, cyclonically rotating tornadoes (namely, rotating in the same direction as the Earth's rotation, and having positive ζ_r) are favored slightly, due to the Coriolis parameter in the stretching term. Rotation decreases due to **turbulent drag**, which is greatest at the ground in the tornado's boundary layer.

Most theories for tornadic rotation invoke tilting and/or stretching of vorticity, but these theories disagree about the origin of rotation. There is some evidence that different mechanisms might trigger different tornadoes.

Two theories focus on rotation about a horizontal axis in the atmospheric boundary layer. One theory suggests that **streamwise vorticity** (rotation about a horizontal axis aligned with the mean wind direction) exists in ambient (outside-of-the-storm) due to vertical shear of the horizontal wind $\Delta U/\Delta z$ in Fig. 15.42a. Once this inflow air reaches the thunderstorm, the horizontal streamwise vorticity is tilted by convective updrafts to create rotation around a

Sample Application

Given storm-relative horizontal wind shear in the environment of $\Delta U'/\Delta z$ of 15 m s^{-1} across 3 km of height, find the vorticity spin-up if vertical velocity increases from 0 to 10 m s^{-1} across Δy=10 km.

Find the Answer

Given: $\Delta U'/\Delta z$ = (15 m s^{-1})/(3 km) = 5x10^{-3} s^{-1}

$\Delta W/\Delta y$ = (10 m s^{-1})/(10 km) = 1x10^{-3} s^{-1}

Find: $\Delta\zeta_r/\Delta t$ = ? s^{-2}

Assume: All other terms in eq. (15.51) are negligible.

$\Delta\zeta_r/\Delta t$ = (5x10^{-3} s^{-1})·(1x10^{-3} s^{-1}) = **5x10^{-6}** s^{-2}

Check: Units OK. Physics OK.

Exposition: Tilting by itself might create a mesocyclone, but is too weak to create a tornado.

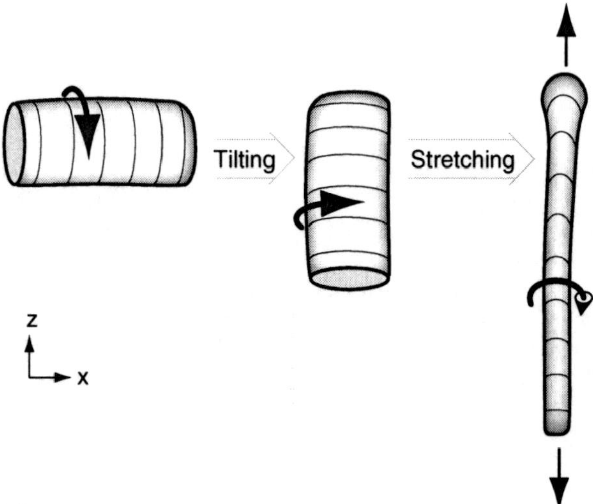

Figure 15.41

Tilting can change rotation about a horizontal axis to rotation about a vertical axis, to create a mesocyclone (a rotating thunderstorm). Stretching can then intensify the rotation into a tornado.

vertical axis. Another considers shears in vertical velocity that develop near the ground where cool precipitation-induced downdrafts are adjacent to warm updrafts. The result is vorticity about a nearly horizontal axis (Fig. 15.42b), which can be tilted towards vertical by the downdraft air near the ground.

Two other theories utilize thunderstorm updrafts to stretch existing cyclonic vertical vorticity. One suggests that the precipitation-cooled downdraft will advect the mesocyclone-base downward, thereby stretching the vortex and causing it to spin faster. Such a mechanism could apply to mesoscale convective vortices (MCVs) in the mid troposphere. The other theory considers thunderstorm updrafts that advect the top of the mesocyclone upward — also causing stretching and spin-up.

Yet another theory suggests the large (synoptic cyclone) scale rotation about a vertical axis can cascade down to medium (mesocyclone) scales and finally down to small (tornadic) scales. All of these previous theories are for supercell tornadoes.

Weaker tornadoes are suggested to form at boundaries between cold and warm airmasses near the ground. The cold airmass could be the result of precipitation-cooled air that creates a downburst and associated outflow winds. At the airmass boundary, such as a cold front or gust front, cold winds on one side of the boundary have an along-boundary component in one direction while the warmer winds on the other side have an along-boundary component in the opposite direction. The vertical vorticity associated with these shears ($\Delta U/\Delta y$ and $\Delta V/\Delta x$) can be stretched to create **landspouts** and **gustnadoes.**

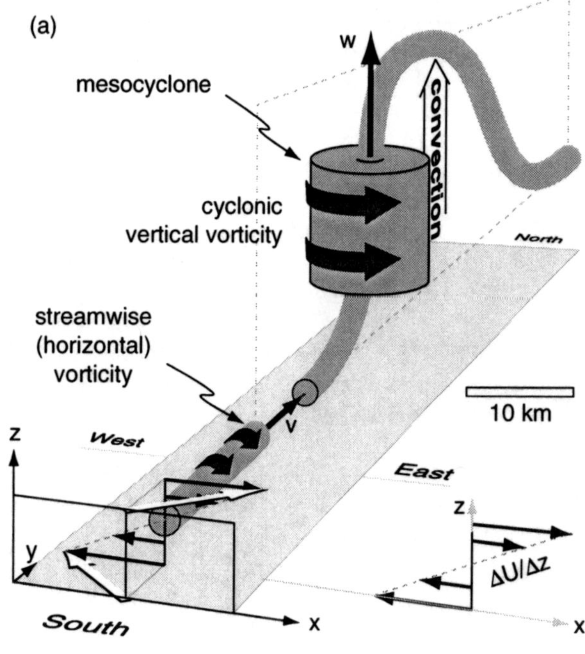

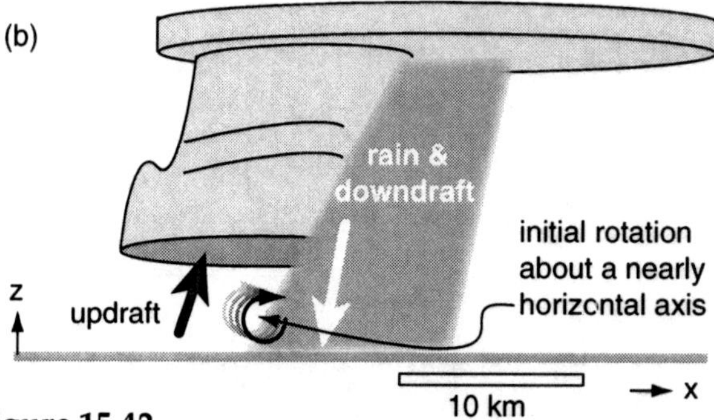

Figure 15.42
Theories for creation of initial rotation about a horizontal azis. (a) Vertical shear of the horizontal wind (ΔU/Δz) causes streamwise horizontal vorticity (dark slender horizontal cylinder), which can be tilted by convective updrafts to create mesocyclones (fat vertical cylinder). This wind profile is typical of the prairies in central North America. (b) Shear between thunderstorm downdraft and updraft creates rotation close to the ground.

Helicity

Many of the previous theories require a mesocyclone that has both rotation and updraft (Fig. 15.43a). The combination of these motions describes a helix (Fig. 15.43b), similar to the shape of a corkscrew.

Define a scalar variable called **helicity,** *H*, at any one point in the air that combines rotation around some axis with mean motion along the same axis:

$$H = U_{avg} \cdot \left[\frac{\Delta W}{\Delta y} - \frac{\Delta V}{\Delta z} \right] + V_{avg} \cdot \left[-\frac{\Delta W}{\Delta x} + \frac{\Delta U}{\Delta z} \right] + W_{avg} \cdot \left[\frac{\Delta V}{\Delta x} - \frac{\Delta U}{\Delta y} \right] \tag{15.52}$$

where $U_{avg} = 0.5 \cdot (U_{j+1} + U_j)$ is the average *U*-component of wind speed between height indices *j* and *j*+1. V_{avg} and W_{avg} are similar. $\Delta U/\Delta z = (U_{j+1} - U_j)/(z_{j+1} - z_j)$, and $\Delta V/\Delta z$ and $\Delta W/\Delta z$ are similar. Helicity units are $m \cdot s^{-2}$, and the differences Δ should be across very small distances.

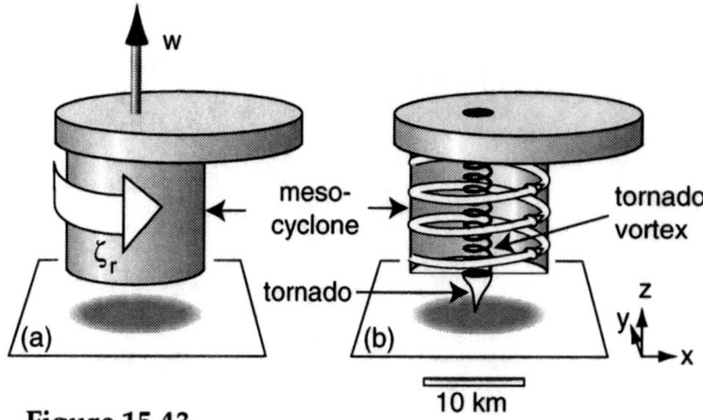

Figure 15.43
(a) Sketch of a supercell thunderstorm showing both mesocyclone rotation (white arrow) and updraft (black arrow). (b) Cross-section showing helical motions (white arrows) in the mesocyclone and smaller-diameter but faster rotation of the tornado vortex (black helix). ζ_r *is relative vorticity, and w is vertical velocity.*

Sample Application

(a) Given the velocity sounding below, what is the associated streamwise helicity?

z (km)	U (m s^{-1})	V (m s^{-1})
1	1	7
2	5	9

(b) If a convective updraft of 12 m s^{-1} tilts the streamwise vorticity from (a) into the vertical while preserving its helicity, what is the vertical vorticity value?

Find the Answer

Given: (a) in the table above (b) w = 12 m s^{-1}

Find: (a) H = ? m·s^{-2} (b) ζ_r = ? s^{-1}

(a) First find the average wind between the two given heights [U_{avg}=(1+5)/2 = 3 m s^{-1}. Similarly V_{avg} = 8 m s^{-1}]. Then use eq. (15.53):

$$H \approx (8\text{m/s})\cdot\frac{(5\text{–}1\text{m/s})}{(2000\text{–}1000\text{m})} - (3\text{m/s})\cdot\frac{(9\text{–}7\text{m/s})}{(2000\text{–}1000\text{m})}$$

$$= 0.032 + 0.006 \text{ m·s}^{-2} = \underline{0.038 \text{ m·s}^{-2}}$$

(b) Use this helicity in eq. (15.54):

$(0.038 \text{ m·s}^{-2}) = (12\text{m·s}^{-1}) \cdot \zeta_r$. Thus, $\zeta_r = \underline{0.0032 \text{ s}^{-1}}$

Check: Physics & units OK.

Exposition: Time lapse photos of mesocyclones show rotation about 100 times faster than for synoptic lows.

(a)

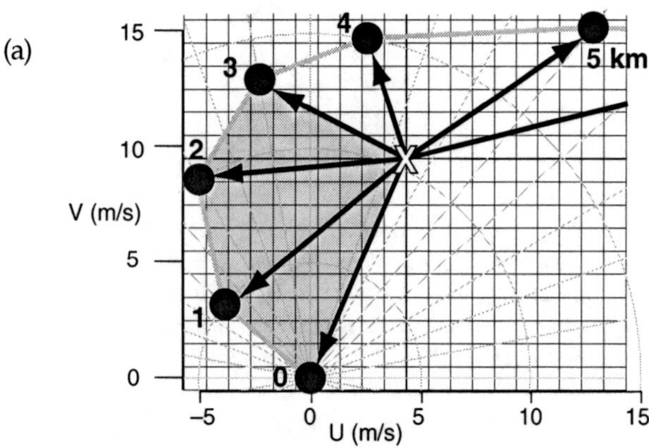

(b)

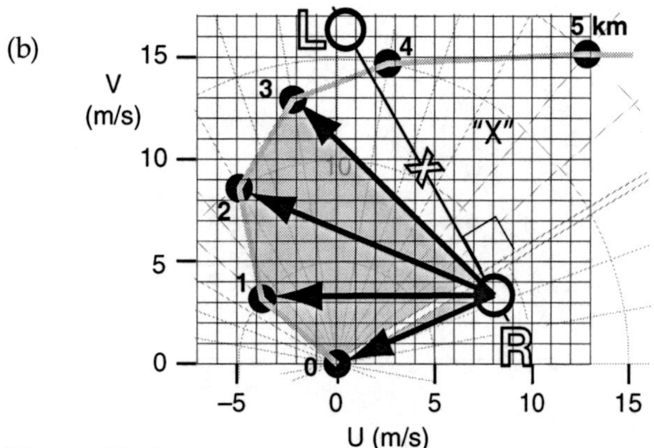

Figure 15.44

Storm-relative helicity (SRH) between the surface and z = 3 km is twice the shaded area. (a) For "normal" storm motion "X", based on hodograph of Figs. 14.61 & 15.40a. (b) For right-moving storm motion "R", based on the hodograph of Fig. 14.61.

If the ambient environment outside the thunderstorm has only vertical shear of horizontal winds, then eq. (15.52) can be simplified to be:

$$H \approx V_{avg}\cdot\frac{\Delta U}{\Delta z} - U_{avg}\cdot\frac{\Delta V}{\Delta z} \qquad \bullet(15.53)$$

which gives only **streamwise-vorticity** contribution to the total helicity. Alternately if there is only rotation about a vertical axis, then eq. (15.52) can be simplified to give the vertical-vorticity contribution to total helicity:

$$H = W_{avg}\cdot\left[\frac{\Delta V}{\Delta x} - \frac{\Delta U}{\Delta y}\right] = W_{avg}\cdot\zeta_r \qquad \bullet(15.54)$$

If this helicity H is preserved while thunderstorm up- and downdrafts tilt the streamwise vorticity into vertical vorticity, then you can equate the H values in eqs. (15.53 and 15.54). This allows you to forecast mesocyclone rotation for any given shear in the pre-storm environment. Greater values of streamwise helicity in the environment could increase the relative vorticity of a mesocyclone, making it more **tornadogenic** (spawning new tornadoes).

More useful for mesocyclone and tornado forecasting is a **storm relative helicity (SRH)**, which uses storm-relative environmental winds (U', V') to get a relative horizontal helicity contribution H'. Substituting storm-relative winds into eq. (15.53) gives:

$$H' \approx V'_{avg}\cdot\frac{\Delta U'}{\Delta z} - U'_{avg}\cdot\frac{\Delta V'}{\Delta z} \qquad (15.55)$$

where $U'_{avg} = 0.5\cdot(U'_{j+1} + U'_j)$ is the average U'-component of wind within the layer of air between height indices j and j+1, and V'_{avg} is similar.

To get the overall effect on the thunderstorm, SRH then sums H' over all atmospheric layers within the inflow region to the thunderstorm, times the thickness of each of those layers.

$$SRH = \sum H'\cdot\Delta z \qquad \bullet(15.56)$$

$$= \sum_{j=0}^{N-1}\left[(V'_j\cdot U'_{j+1}) - (U'_j\cdot V'_{j+1})\right] \qquad \bullet(15.57)$$

where N is the number of layers. j = 0 is the bottom wind observation (usually at the ground, z = 0), and $j = N$ is the wind observation at the top of the inflow region of air. Normally, the inflow region spans all the atmospheric layers from the ground to 1 or 3 km altitude. Units of SRH are m^2·s^{-2}.

On a hodograph, the SRH is twice the area swept by the storm-relative wind vectors in the inflow region (Fig. 15.44).

SRH is an imperfect indicator of whether thunderstorms are likely to be supercells, and might form tornadoes, hail, and strong straight-line winds. Fig. 15.45 shows the relationship between SRH values and tornado strength. Recent evidence suggests that the 0 to 1 km SRH (Fig. 15.46) works slightly better than SRH over the 0 to 3 km layer.

Sample Application

For the hodograph of Fig. 15.44a, graphically find SRH for the (a) z = 0 to 3 km layer, and (b) 0 to 1 km layer. (c, d) Find the SRHs for those two depths using an equation method. (e) Discuss the potential for tornadoes?

Find the Answer:

Given Fig. 15.44a.

Find: 0-3 km & 0-1 km SRH = ? $m^2 \cdot s^{-2}$, both graphically & by eq.

a) Graphical method: Fig. 15.44a is copied below, and zoomed into the shaded region. Count squares in the shaded region of the fig., knowing that each square is 1 m s^{-1} by 1 m s^{-1}, and thus spans 1 $m^2 \cdot s^{-2}$ of area.

When counting squares, if a shaded area (such as for square #2 in the fig. below) does not cover the whole square, then try to compensate with other portions of shaded areas (such as the small shaded triangle just to the right of square #2).

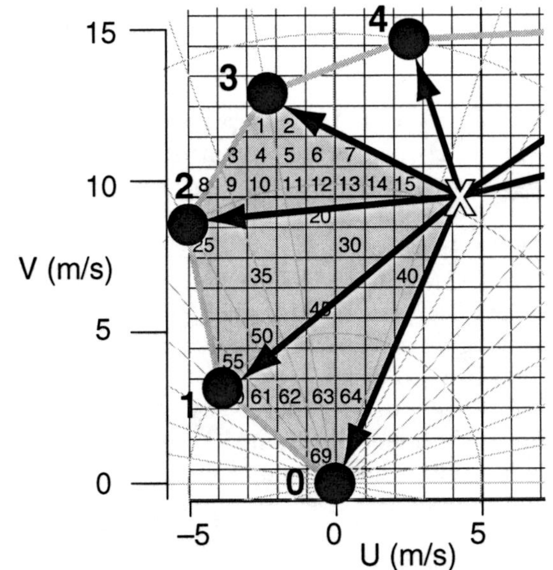

Thus, for 0 to 3 km SRH:

SRH = 2·(# of squares)·(area of each square)

SRH =2 · (69 squares) · (1 $m^2 \cdot s^{-2}$/square) = **138 $m^2 \cdot s^{-2}$**.

(continues in next column)

Sample Application *(SRH continuation)*

b) For 0 to 1 km: SRH:

SRH =2 · (25 squares) · (1 $m^2 \cdot s^{-2}$/square) = **50 $m^2 \cdot s^{-2}$**.

c) Equation method: Use eq. (15.57). For our hodograph, each layer happens to be 1 km thick. Thus, the index *j* happens to correspond to the altitude in km of each wind observation, for this fortuitous situation.

For the 0-3 km depth, eq. (15.57) expands to be:

$$\begin{aligned} SRH &= V'_0 \cdot U'_1 - U'_0 \cdot V'_1 && \text{(for } 0 \le z \le 1 \text{ km)} \\ &+ V'_1 \cdot U'_2 - U'_1 \cdot V'_2 && \text{(for } 1 \le z \le 2 \text{ km)} \\ &+ V'_2 \cdot U'_3 - U'_2 \cdot V'_3 && \text{(for } 2 \le z \le 3 \text{ km)} \end{aligned}$$

Because the figure at left shows storm-relative winds, we can pick off the (*U'*, *V'*) values by eye for each level:

z (km)	U' (m s^{-1})	V' (m s^{-1})
0	–4.3	–9.5
1	–8.1	–6.2
2	–9.3	–1.0
3	–7.7	+3.3

Plugging these values into the eq. above gives:

$$\begin{aligned} SRH &= (-9.5)\cdot(-8.1) - (-4.3)\cdot(-6.2) \\ &+ (-6.2)\cdot(-9.3) - (-8.1)\cdot(-1.0) \\ &+ (-1.0)\cdot(-7.7) - (-9.3)\cdot(+3.3) \end{aligned}$$

$$\begin{aligned} SRH &= 76.95 - 26.66 + 57.66 - 8.1 + 7.7 + 30.69 \\ &= \mathbf{138.2\ m^2 \cdot s^{-2}} \text{ for 0 - 3 km} \end{aligned}$$

d) 0-1 km SRH = 76.95 – 26.66 = **50.3 $m^2 \cdot s^{-2}$**.

e) Using Fig. 15.45 there is a **good chance of supercells and EF0 to EF1 tornadoes. Slight chance of EF2 tornado.** However, Fig. 15.46 suggests a **supercell with no tornado.**

Check: Units OK. Physics OK. Magnitudes OK.

Exposition: The graphical and equation methods agree amazingly well with each other. This gives us confidence to use either method, whichever is easiest.

The disagreement in tornado potential between the 0-1 and 0-3 km SRH methods reflects the tremendous difficulty in thunderstorm and tornado forecasting. Operational meteorologists often must make difficult decisions quickly using conflicting indices.

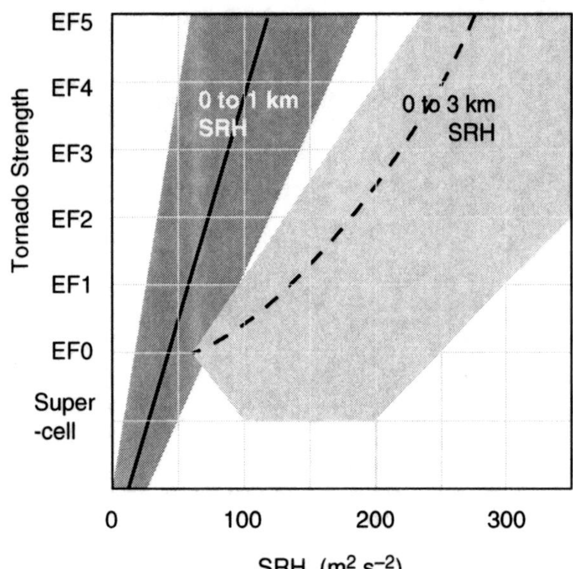

Figure 15.45
Approximate relationship between storm-relative helicity (SRH) and tornado strength on the Enhanced Fujita scale (EF0 to EF5), for North America. "Supercell" indicates a non-tornadic mesocyclonic thunderstorm. Caution: the shaded domain boundaries are not as sharp as drawn. Solid and dashed lines are medians.

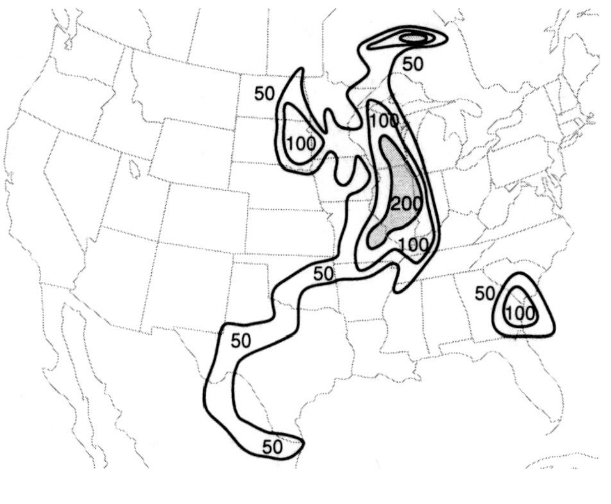

Figure 15.47
A case-study example of effective storm-relative helicity (eSRH) in units of $m^2 s^{-2}$ at 23 UTC on 24 May 2006, over the USA. Values of eSRH greater than 200 $m^2 s^{-2}$ are shaded, and indicate locations of greatest likelihood for tornadic supercells.

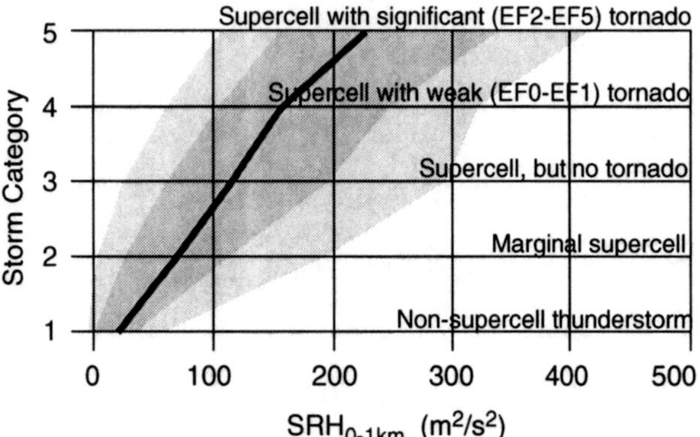

Figure 15.46
Statistics of thunderstorm intensity vs. storm-relative helicity (SRH) across the 0 to 1 km layer of air. For several hundred storms in central N. America, the black line is the median (50 percentile); dark grey shading spans 25 to 75 percentiles (the interquartile range); and light grey spans 10 to 90 percentiles.

Because actual thunderstorms do not necessarily draw in air from the 0 to 1 km layer, an **effective Storm Relative Helicity (eSRH)** has been proposed that is calculated across a range of altitudes that depends on CAPE and CIN of the environmental sounding. The bottom altitude is found as the lowest starting altitude for a rising air parcel that satisfies two constraints: CAPE ≥ 100 J kg^{-1} when lifted to its EL, <u>and</u> CIN ≥ –250 J kg^{-1} (i.e., is less negative than –250 J kg^{-1}). The top altitude is the lowest starting height (above the bottom altitude) for which rising air-parcel CAPE ≤ 100 J kg^{-1} <u>or</u> CIN ≤ –250 J kg^{-1}.

The eSRH calculation is made only if the top and bottom layer altitudes are within the bottom 3 km of the atmosphere. eSRH is easily found using computer programs. Fig. 15.47 shows a map of eSRH for the 24 May 2006 case.

eSRH better discriminates between non-tornadic and tornadic supercells than SRH. eSRH values are usually slightly smaller than SRH values. It works even if the residual-layer air ingested into the thunderstorm lies on top of a shallow stable layer of colder air, such as occurs in the evening after sunset.

One difficulty with SRH is its sensitivity to storm motion. For example, hodographs of veering wind as plotted in Fig. 15.44 would indicate much greater SRH for a right moving ("R") storm compared to a supercell that moves with the average 0 to 6 km winds ("X").

As sketched in Fig. 15.43a, both updraft and rotation (vorticity) are important for mesocyclone formation. You might anticipate that the most violent supercells have both large CAPE (suggesting strong updrafts) and large SRH (suggesting strong rotation). A composite index called the **Energy Helicity Index (EHI)** combines these two variables:

$$EHI = \frac{CAPE \cdot SRH}{a} \qquad \bullet(15.58)$$

where $a = 1.6\text{x}10^5\ \text{m}^4{\cdot}\text{s}^{-4}$ is an arbitrary constant designed to make EHI dimensionless and to scale its values to lie between 0 and 5 or so. Large values of EHI suggest stronger supercells and tornadoes.

CAPE values used in EHI are always the positive area on the sounding between the LFC and the EL. MLCAPE SRH values can be either for the 0 to 1 km layer in the hodograph, or for the 0 to 3 km layer. For this reason, EHI is often classified as 0-1 km EHI or 0-3 km EHI. Also, EHI can be found either from actual rawinsonde observations, or from forecast soundings extracted from numerical weather prediction models.

Table 15-5 indicates likely tornado strength in North America as a function of 0-3 km EHI. Caution: the EHI ranges listed in this table are only approximate. If 0-1 km EHI is used instead, then significant tornadoes can occur for EHI as low as 1 or 2, such as shown in Fig. 15.48. To illustrate EHI, Fig. 15.49 shows a forecast weather map of EHI for 24 May 2006.

Sample Application

Suppose a prestorm environmental sounding is analyzed, and shows that the convective available potential energy is 2000 J kg^{-1}, and the storm relative helicity is 150 m^2 s^{-2} in the bottom 1 km of atmosphere. Find the value of energy helicity index, and forecast the likelihood of severe weather.

Find the Answer:

Given: $CAPE$ = 2000 J kg^{-1} = 2000 m^2 s^{-2},
$SHR_{0\text{-}1km}$ = 150 m^2 s^{-2}

Find: EHI = ? (dimensionless)

Use eq. (15.58):

EHI = (2000 m^2 s^{-2}) · (150 m^2 s^{-2}) / (1.6x10^5 m^4 s^{-4})
= **1.88**

Check: Units dimensionless. Value reasonable.

Exposition: Using this value in Fig. 15.48, there is a **good chance for a tornadic supercell thunderstorm**, and the tornado could be significant (EF2 - EF5). However, there is a slight chance that the thunderstorm will be non-tornadic, or could be a marginal supercell.

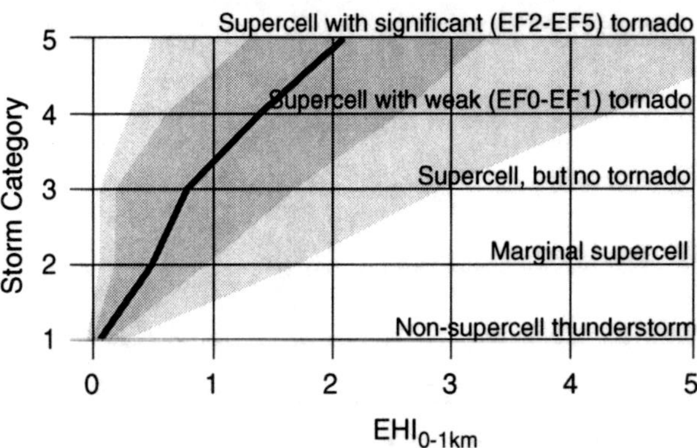

Figure 15.48
Statistics of thunderstorm intensity vs. energy helicity index (EHI) across the 0 to 1 km layer of air. For several hundred storms in central N. America, the black line is the median (50 percentile); dark grey shading spans 25 to 75 percentiles (the interquartile range); and light grey spans 10 to 90 percentiles.

Table 15-5. Energy Helicity Index (0-3 km EHI) as an indicator of possible tornado existence and strength.

EHI	Tornado Likelihood
< 1.0	Tornadoes & supercells unlikely.
1.0 to 2.0	Supercells with weak, short-lived tornadoes possible. Non-supercell tornadoes possible (such as near bow echoes).
2.0 to 2.5	Supercells likely. Mesocyclone-induced tornadoes possible.
2.5 to 3.0	Mesocyclone-induced supercell tornadoes more likely.
3.0 to 4.0	Strong mesocyclone-induced tornadoes (EF2 and EF3) possible.
> 4.0	Violent mesocyclone-induced tornadoes (EF4 and EF5) possible.

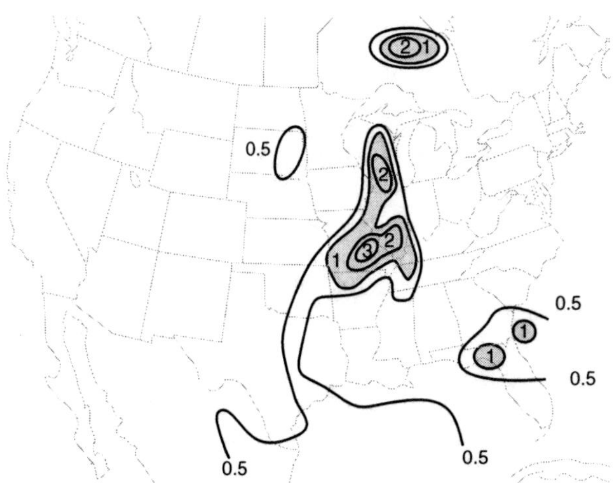

Figure 15.49
Map of energy helicity index (EHI) over the 0 to 1 km layer, for the 23 UTC case on 24 May 2006, over the USA. Dimensionless values greater than 1 are shaded, and suggest stronger supercell storms with greater likelihood of tornadoes.

Because none of the forecast parameters and indices give perfect forecasts of supercells and tornadoes, researchers continue to develop and test new indices. One example is the **Supercell Composite Parameter** (**SCP**), which is a normalized product of MUCAPE, eSRH, and effective bulk shear. Non-supercell thunderstorms have SCP values near zero, marginal and elevated supercells have values between 0 and 6, while strong surface-based supercells have values between 2 and 11.

Another experimental parameter is the **Significant Tornado Parameter** (**STP**), which is a normalized product of MLCAPE, surface-based effective bulk shear, eSRH, MLLCL, and MLCIN. Values between 0 and 1 are associated with non-tornadic supercells, while values between 1 and 5 indicate supercells likely to have significant (EF2 - EF5) tornadoes.

Multiple-vortex Tornadoes

If conditions are right, a single parent tornado can develop multiple mini daughter-tornadoes around the parent-tornado perimeter near the ground (Fig. 15.50c). Each of these daughter vortices can have strong tangential winds and very low core pressure. These daughter tornadoes are also known as **suction vortices** or **suction spots**. **Multiple-vortex tornadoes** can have 2 to 6 suction vortices. The process of changing from a single vortex to multiple vortices is called **tornado breakdown**.

A ratio called the **swirl ratio** can be used to anticipate tornado breakdown and the multi-vortex nature of a parent tornado:

$$S = M_{tan} / W \qquad \bullet(15.59)$$

where W is the average updraft speed in a mesocyclone, and M_{tan} is the tangential wind component around the mesocyclone. If we idealize the mesocyclone as being cylinder, then:

$$S = \frac{R_{MC} \cdot M_{tan}}{2 z_i \cdot M_{rad}} \qquad \bullet(15.60)$$

where M_{rad} is the inflow speed (i.e., radial velocity component,) R_{MC} is the radius of the mesocyclone, and the z_i is the depth of the atmospheric boundary-layer. For the special case of $R_{MC} \approx 2 \cdot z_i$, then:

$$S \approx M_{tan} / M_{rad} \,. \qquad \bullet(15.61)$$

When the swirl ratio is small (0.1 to 0.3), tornadoes have a single, well-defined, smooth-walled funnel (Fig. 15.50a), based on laboratory simulations. There is low pressure at the center of the tornado (**tornado core**), and the core contains updrafts at all heights.

Sample Application

At radius 2.5 km, a mesocyclone has a radial velocity of 1 m s^{-1} and a tangential velocity of 3 m s^{-1} . If the atmospheric boundary layer is 1.5 km thick, find the swirl ratio.

Find the Answer

Given: M_{rad} = 1 m s^{-1}, M_{tan} = 3 m s^{-1},
R_{MC} = 2500 m, and z_i = 1.5 km

Find: S = ? (dimensionless)

Although eq. (15.61) is easy to use, it contains some assumptions that might not hold for our situation. Instead, use eq. (15.60):

$$S = \frac{(2500 \text{ m}) \cdot (3 \text{ m/s})}{2 \cdot (1500 \text{ m}) \cdot (1 \text{ m/s})} = \underline{2.5} \text{ (dimensionless)}$$

Check: Physics and units OK.
Exposition: Because this swirl ratio exceeds the critical value, multiple vortices are likely.

Although the fundamental definition of the swirl ratio uses the average vertical velocity W in the mesocyclone, often W is not known and hard to measure. But using mass continuity, the vertical velocity in eq. (15.59) can be estimated from the radial velocity for eq. (15.60), which can be easier to estimate.

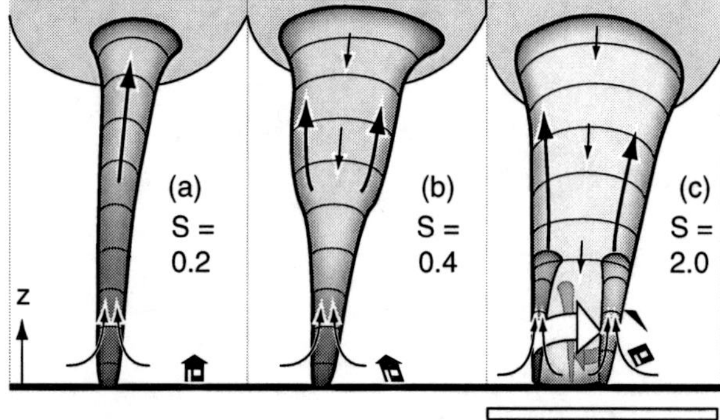

Figure 15.50
Sketch of the evolution of a tornado and how the bottom can separate into multiple vortices as the swirl ratio (S) increases. The multiple suction vortices move around the perimeter of the parent tornado

Core radius R_o of the tornado is typically 5% to 25% of the updraft radius, R_{MC}, in the mesocyclone.

If conditions change in the mesocyclone to have faster rotation and slower updraft, then the swirl ratio increases. This is accompanied by a turbulent downdraft in the top of the tornado core (Fig. 15.50b). The location where the core updraft and downdraft meet is called the **breakdown bubble**, and this stagnation point moves downward as the swirl ratio increases. The tornado is often wider above the stagnation point.

As the swirl ratio increases to a value of $S^* \approx 0.45$ (**critical swirl ratio),** the breakdown bubble gradually moves downward to the ground. The core now has a turbulent downdraft at all altitudes down to the ground, while around the core are strong turbulent updrafts around the larger-diameter tornado. At swirl ratios greater than the critical value, the parent tornado becomes a large-diameter helix of rotating turbulent updraft air, with a downdraft throughout the whole core.

At swirl ratios of about $S \geq 0.8$, the parent tornado creates multiple daughter vortices around its perimeter (Fig 15.50c). As S continues to increase toward 3 and beyond, the number of multiple vortices increases from 2 to 6.

Wind speeds and damage are greatest in these small suction vortices. The individual vortices not only circulate around the perimeter of the parent tornado, but the whole tornado system translates over the ground as the thunderstorm moves across the land. Thus, each vortex traces a cycloidal pattern on the ground, which is evident in post-tornado aerial photographs as cycloidal damage paths to crops (Fig. 15.51).

When single-vortex tornadoes move over a forest, they blow down the trees in a unique pattern that differs from tree blowdown patterns caused by straight-line downburst winds (Fig. 15.52). Sometimes these patterns are apparent only from an aerial vantage point.

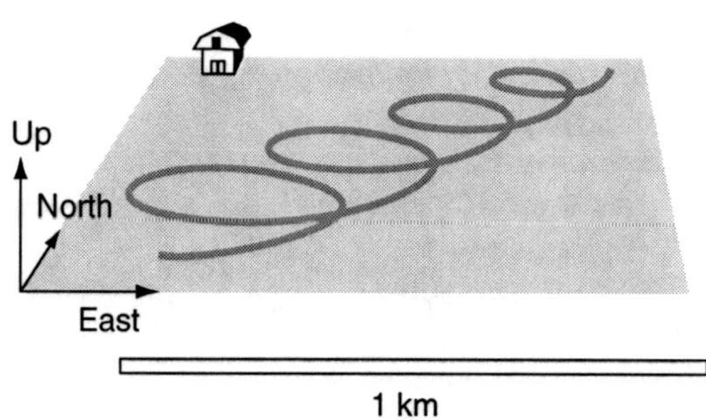

Figure 15.51
Cycloidal damage path in a farm field caused by a tornadic suction vortex imbedded in a larger diameter parent tornado.

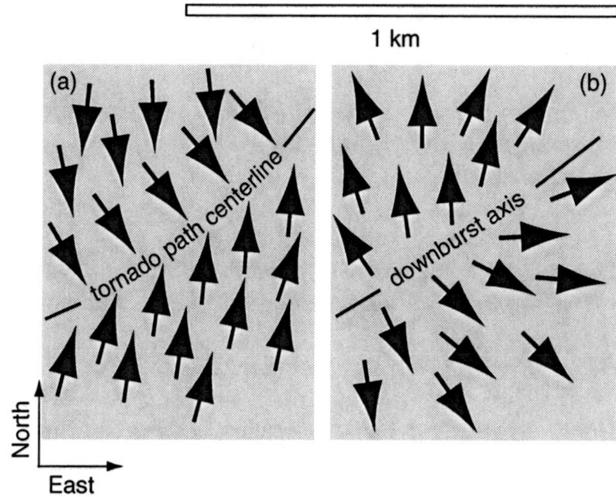

Figure 15.52
Damage signatures of (a) tornadoes and (b) downbursts in a forest, as seen in aerial photographs of tree blowdown.

REVIEW

Supercell thunderstorms can create tornadic winds, straight-line winds, downbursts, lightning, heavy rain, hail, and vigorous turbulence.

Larger hail is possible in storms with stronger updraft velocities, which is related to CAPE. When precipitation falls into drier air, both precipitation drag and evaporative cooling can cause acceleration of downdraft winds. When the downburst wind hits the ground, it spreads out into straight-line winds, the leading edge of which is called a gust front. Dust storms (haboobs) and arc clouds are sometimes created by gust fronts.

As ice crystals and graupel particles collide they each transfer a small amount of charge. Summed over billions of such collisions in a thunderstorm, sufficient voltage gradient builds up to create lightning. The heat from lightning expands air to create the shock and sound waves we call thunder. Horizontal wind shear can be tilted into the vertical to create mesocyclones and tornadoes.

HOMEWORK EXERCISES

Broaden Knowledge & Comprehension

B1. Search the web for (and print the best examples of) photographs of:
- a. tornadoes
  - (i) supercell tornadoes
  - (ii) landspouts
  - (iii) waterspouts
  - (iv) gustnadoes
- b. gust fronts and arc clouds
- c. hailstones, and hail storms
- d. lightning
  - (i) CG
  - (ii) IC (including spider lightning, also known as lightning crawlers)
  - (iii) a bolt from the blue
- e. damage caused by intense tornadoes

(Hint: search on "storm stock images photographs".) Discuss the features of your resulting photos with respect to information you learned in this chapter.

B2. Search the web for (and print the best examples of) radar reflectivity images of
- a. hook echoes
- b. gust fronts
- c. tornadoes

Discuss the features of your resulting image(s) with respect to information you learned in this chapter.

B3. Search the web for (and print the best example of) real-time maps of lightning locations, as found from a lightning detection network or from satellite. Print a sequence of 3 images at about 30 minute intervals, and discuss how you can diagnose thunderstorm movement and evolution from the change in the location of lightning-strike clusters.

B4. Search the web for discussion of the health effects of being struck by lightning, and write or print a concise summary. Include information about how to resuscitate people who were struck.

B5. Search the web for, and summarize or print, recommendations for safety with respect to:
- a. lightning b. tornadoes
- c. straight-line winds and derechoes
- d. hail
- e. flash floods
- f. thunderstorms (in general)

B6. Search the web (and print the best examples of) maps that show the frequency of occurrence of:
- a. tornadoes & tornado deaths
- b. lightning strike frequency & lightning deaths
- c. hail d. derechos

B7. Search the web for, and print the best example of, a photographic guide on how to determine Fujita or Enhanced Fujita tornado intensity from damage surveys.

B8. Search the web for (and print the best examples of) information about how different building or construction methods respond to tornadoes of different intensities.

B9. Search the web for, and print and discuss five tips for successful and safe tornado chasing.

B10. Use the internet to find sites for tornado chasers. The National Severe Storms Lab (NSSL) web site might have related info.

B11. During any 1 year, what is the probability that a tornado will hit any particular house? Try to find the answer on the internet.

B12. Search the web for private companies that provide storm-chasing tours/adventures/safaris for paying clients. List 5 or more, including their web addresses, physical location, and what they specialize in.

B13. Blue jets, red sprites, and elves are electrical discharges that can be seen as very brief glows in the mesosphere. They are often found at 30 to 90 km altitude over thunderstorms. Summarize and print info from the internet about these discharges, and included some images.

B14. Search the web for, and print one best example of, tornadoes associated with hurricanes.

B15. Search the web for a complete list of tornado outbreaks and/or tornado outbreak sequences. Print 5 additional outbreaks that were large and/or important, but which weren't already included in this chapter in the list of outbreaks. Focus on more-recent outbreaks.

B16. With regard to tornadoes, search the web for info to help you discuss the relative safety or dangers of:
- a. being in a storm shelter
- b. being in an above-ground tornado safe room
- c. being in a mobile home or trailer
- d. hiding under bridges and overpasses
- e. standing above ground near tornadoes

B17. Search the web for examples of downburst, gust front, or wind-shear sensors and warning systems at airports, and summarize and print your findings.

Apply

A1. If a thunderstorm cell rains for 0.5 h at the precipitation rate (mm h^{-1}) below, calculate both the net latent heat released into the atmosphere, and the average warming rate within the troposphere.

a. 50 b. 75 c. 100 d. 125 e. 150
f. 175 g. 200 h. 225 i. 250 j. 275
k. 300 l. 325 m. 350 n. 375 o. 400

A2. Indicate the TORRO hail size code, and descriptive name, for hail of diameter (cm):

a. 0.6 b. 0.9 c. 1.2 d. 1.5 e. 1.7
f. 2.1 g. 2.7 h. 3.2 i. 3.7 j. 4.5
k. 5.5 l. 6.5 m. 7.5 n. 8.0 o. 9.5

A3. Graphically estimate the terminal fall velocity of hail of diameter (cm):

a. 0.6 b. 0.9 c. 1.2 d. 1.5 e. 1.7
f. 2.1 g. 2.7 h. 3.2 i. 3.7 j. 4.5
k. 5.5 l. 6.5 m. 7.5 n. 8.0 o. 9.5

A4. A supercooled cloud droplet of radius 40 μm hits a large hailstone. Using the temperature (°C) of the droplet given below, is the drop cold enough to freeze instantly (i.e., is its temperature deficit sufficient to compensate the latent heat of fusion released)? Based on your calculations, state whether the freezing of this droplet would contribute to a layer of clear or white (porous) ice on the hailstone.

a. –40 b. –37 c. –35 d. –32 e. –30
f. –27 g. –25 h. –22 i. –20 j. –17
k. –15 l. –13 m. –10 n. –7 o. –5

A5. Given the sounding in exercise A3 of the previous chapter, calculate the portion of SB CAPE between altitudes where the environmental temperature is –10 and –30°C. Also, indicate if rapid hail growth is likely.

A6. Given the table below of environmental conditions, calculate the value of Significant Hail Parameter (SHIP), and state whether this environment favors the formation of hailstone > 5 cm diameter (assuming a thunderstorm indeed forms).

Exercise	a	b	c	d
MUCAPE (J kg^{-1})	2000	2500	3000	3500
r_{MUP} (g kg^{-1})	10	12	14	16
$\gamma_{70\text{-}50kPa}$ (°C km^{-1})	2	4	6	8
T_{50kPa} (°C)	–20	–15	–10	–5
$TSM_{0\text{-}6km}$ (m s^{-1})	20	30	40	50

A7. For the downburst acceleration equation, assume that the environmental air has temperature 2°C and mixing ratio 3 g kg^{-1} at pressure 85 kPa. A cloudy air parcel at that same height has the same temperature, and is saturated with water vapor and carries liquid water at the mixing ratio (g kg^{-1}) listed below. Assume no ice crystals.

(1) Find portion of vertical acceleration due to the combination of temperature and water vapor effects.

(2) Find the portion of vertical acceleration due to the liquid water loading only.

(3) By what amount would the virtual potential temperature of an air parcel change if all the liquid water evaporates and cools the air?

(4) If all of the liquid water were to evaporate and cool the air parcel, find the new vertical acceleration.

The liquid water mixing ratios (g kg^{-1}) are:

a. 20 b. 18 c. 16 d. 14 e. 12 f. 10 g. 9
h. 8 i. 7 j. 6 k. 5 l. 4 m. 3 n. 2

A8. Given the sounding from exercise A3 of the previous chapter, assume a descending air parcel in a downburst follows a moist adiabat all the way down to the ground. If the descending parcel starts at the pressure (kPa) indicated below, and assuming its initial temperature is the same as the environment there, plot both the sounding and the descending parcel on a thermo diagram, and calculate the value of downdraft CAPE.

a. 80 b. 79 c. 78 d. 77 e. 76 f. 75 g. 74
h. 73 i. 72 j. 71 k. 70 l. 69 m. 68 n. 67

A9. Find the downdraft speed if the DCAPE (J kg^{-1}) for a downburst air parcel is:

a. –200 b. –400 c. –600 d. –800 e. –1000
f. –1200 g. –1400 h. –1600 i. –1800 j. –2000
k. –2200 l. –2400 m. –2600 n. –2800 o. –3000

A10. If a downburst has the same potential temperature as the environment, and starts with vertical velocity (m s^{-1}, negative for descending air) given below, use Bernoulli's equation to estimate the maximum pressure perturbation at the ground under the downburst.

a. –2 b. –4 c. –6 d. –8 e. –10
f. –12 g. –14 h. –16 i. –18 j. –20
k. –22 l. –24 m. –26 n. –28 o. –30

A11. Same as the previous exercise, but in addition to the initial downdraft velocity, the descending air parcel is colder than the environment by the following product of virtual potential temperature depression and initial altitude (°C·km):

(1) –0.5 (2) –1 (3) –1.5 (4) –2
(5) –2.5 (6) –3 (7) –3.5 (8) –4
(9) –4.5 (10) –5 (11) –5.5 (12) –6
(13) –6.5 (14) –7 (15) –7.5 (16) –8

A12. Find the acceleration (m s^{-2}) of outflow winds from under a downburst, assuming a maximum mesohigh pressure (kPa) perturbation at the surface as given below, and a radius of the mesohigh of 3 km.
a. 0.1 b. 0.2 c. 0.3 d. 0.4 e. 0.5
f. 0.6 g. 0.7 h. 0.8 i. 0.9 j. 1.0
k. 1.1 l. 1.2 m. 1.3 n. 1.4 o. 1.5

A13. How fast will a gust front advance, and what will be its depth, at distance 6 km from the center of a downburst. Assume the downburst has radius 0.5 km and speed 9 m s^{-1} , and that the environmental around the downburst is 28°C. The magnitude of the temperature deficit (°C) is:
a. 1 b. 1.5 c. 2 d. 2.5 e. 3 f. 3.5
g. 4 h. 4.5 i. 5 j. 5.5 k. 6 l. 6.5
m. 7 n. 7.5 o. 8 p. 8.5 q. 9 r. 9.5

A14(§). Draw a graph of gust front depth and advancement speed vs. distance from the downburst center, using data from the previous exercise.

A15. Given a lightning discharge current (kA) below and a voltage difference between the beginning to end of the lightning channel of 10^{10} V, find (1) the resistance of the ionized lightning channel and (2) the amount of charge (C) transferred between the cloud and the ground during the 20 μs lifetime of the lightning stroke.
a. 2 b. 4 c. 6 d. 8 e. 10 f. 15 g. 20
h. 40 i. 60 j. 80 k. 100 l. 150 m. 200 n. 400

A16. To create lightning in (1) dry air, and (2) cloudy air, what voltage difference is required, given a lightning stroke length (km) of:
a. 0.2 b. 0.4 c. 0.6 d. 0.8 e. 1 f. 1.2 g. 1.4
h. 1.6 i. 1.8 j. 2.0 k. 2.5 l. 3 m. 4 n. 5

A17. For an electrical potential across the atmosphere of $1.3\text{x}10^5$ V km^{-1}, find the current density if the resistivity (Ω·m) is:
a. $5\text{x}10^{13}$ b. $1\text{x}10^{13}$ c. $5\text{x}10^{12}$ d. $1\text{x}10^{12}$ e. $5\text{x}10^{11}$
f. $1\text{x}10^{11}$ g. $5\text{x}10^{10}$ h. $1\text{x}10^{10}$ i. $5\text{x}10^{9}$ j. $1\text{x}10^{9}$
k. $5\text{x}10^{8}$ l. $1\text{x}10^{8}$ m. $5\text{x}10^{7}$ n. $1\text{x}10^{7}$ o. $5\text{x}10^{6}$

A18. What is the value of peak current in a lightning stroke, as estimated using a lightning detection network, given the following measurements of electrical field E and distance D from the ground station.

	$-E$ (V m^{-1})	D (km)
a.	1	10
b.	1	50
c.	2	10
d.	2	100
e.	3	20
f.	3	80
g.	4	50
h.	4	100
i.	5	50
j.	5	200
k.	6	75
l.	6	300

A19. For a power line struck by lightning, what is the probability that the lightning-generated current (kA) is greater than:
a. 2 b. 4 c. 6 d. 8 e. 10 f. 15 g. 20
h. 40 i. 60 j. 80 k. 100 l. 150 m. 200 n. 400

A20. When lightning strikes an electrical power line it causes a surge that rapidly reaches its peak but then slowly decreases. How many seconds after the lightning strike will the surge have diminished to the fraction of the peak surge given here:
a. 0.1 b. 0.15 c. 0.2 d. 0.25 e. 0.3
f. 0.35 g. 0.4 h. 0.45 i. 0.5 j. 0.55
k. 0.6 l. 0.65 m. 0.7 n. 0.75 o. 0.8

A21(§). If lightning heats the air to the temperature (K) given below, then plot (on a log-log graph) the speed (Mach number), pressure (as ratio relative to background pressure), and radius of the shock front vs. time given ambient background pressure of 100 kPa and temperature 20°C.
a. 16,000 b. 17,000 c. 18,000 d. 19,000
e. 20,000 f. 21,000 g. 22,000 h. 23,000
i. 24,000 j. 25,000 k. 26,000 l. 27,000
m. 28,000 n. 29,000 o. 30,000

A22. What is the speed of sound in calm air of temperature (°C):
a. –20 b. –18 c. –16 d. –14 e. –12
f. –10 g. –8 h. –6 i. –4 j. –2
k. 0 l. 2 m. 4 n. 6 o. 8
p. 10 q. 12 r. 14 s. 16 t. 18

A23(§). Create a graph with three curves for the time interval between the "flash" of lightning and the "bang" of thunder vs. distance from the lightning. One curve should be zero wind, and the other two are for tail and head winds of magnitude (m s^{-1}) given below. Given $T_{environment}$ = 295 K.
a. 2 b. 4 c. 6 d. 8 e. 10 f. 12 g. 14
h. 16 i. 18 j. 20 k. 22 l. 24 m. 26 n. 28

A24(§). For a lightning stroke 2 km above ground in a calm adiabatic environment of average temperature 300 K, plot the thunder ray paths leaving downward from the lightning stroke, given that they arrive at the ground at the following elevation angle (°).

a. 5 b. 6 c. 7 d. 8 e. 9 f. 10 g. 11
h. 12 i. 13 j. 14 k. 15 l. 16 m. 17 n. 18
o. 19 p. 20 q. 21 r. 22 s. 23 t. 24 u. 25

A25. What is the minimum inaudibility distance for hearing thunder from a sound source 7 km high in an environment of T = 20°C with no wind. Given a lapse rate (°C km^{-1}) of:

a. 9.8 b. 9 c. 8.5 d. 8 e. 7.5 f. 7 g. 6.5
h. 6 i. 5.5 j. 5 k. 4.5 l. 4 m. 3.5 n. 3
o. 2.5 p. 2 q. 1.5 r. 1 s. 0.5 t. 0 u. –1

A26. How low below ambient 100 kPa pressure must the core pressure of a tornado be, in order to support max tangential winds (m s^{-1}) of:

a. 20 b. 30 c. 40 d. 50 e. 60 f. 70 g. 80
h. 90 i. 100 j. 110 k. 120 l. 130 m. 140 n. 150

A27(§). For a Rankine Combined Vortex model of a tornado, plot the pressure (kPa) and tangential wind speed (m s^{-1}) vs. radial distance (m) out to 125 m, for a tornado of core radius 25 m and core pressure deficit (kPa) of:

a. 0.1 b. 0.2 c. 0.3 d. 0.4 e. 0.5
f. 0.6 g. 0.7 h. 0.8 i. 0.9 j. 1.0
k. 1.1 l. 1.2 m. 1.3 n. 1.4 o. 1.5

A28. If the max tangential wind speed in a tornado is 100 m s^{-1}, and the tornado translates at the speed (m s^{-1}) given below, then what is the max wind speed (m s^{-1}), and where is it relative to the center of the tornado and its track?

a. 2 b. 4 c. 6 d. 8 e. 10 f. 12 g. 14
h. 16 i. 18 j. 20 k. 22 l. 24 m. 26 n. 28

A29. What are the Enhanced Fujita and TORRO intensity indices for a tornado of max wind speed (m s^{-1}) of

a. 20 b. 30 c. 40 d. 50 e. 60 f. 70 g. 80
h. 90 i. 100 j. 110 k. 120 l. 130 m. 140 n. 150

A30 Find the pressure (kPa) at the edge of the tornado condensation funnel, given an ambient near-surface pressure and temperature of 100 kPa and 35°C, and a dew point (°C) of:

a. 30 b. 29 c. 28 d. 27 e. 26 f. 25 g. 24
h. 23 i. 22 j. 21 k. 20 l. 19 m. 18 n. 17

A31. For the winds of exercise A18 (a, b, c, or d) in the previous chapter, first find the storm movement for a

(1) normal supercell
(2) right-moving supercell
(3) left-moving supercell

Then graphically find and plot on a hodograph the storm-relative wind vectors.

A32. Same as previous exercise, except determine the (U_s, V_s) components of storm motion, and then list the (U_j', V_j') components of storm-relative winds.

A33. A mesocyclone at 38°N is in an environment where the vertical stretching ($\Delta W/\Delta z$) is (20 m s^{-1}) / (2 km). Find the rate of vorticity spin-up due to stretching only, given an initial relative vorticity (s^{-1}) of

a. 0.0002 b. 0.0004 c. 0.0006 d. 0.0008 e. 0.0010
f. 0.0012 g. 0.0014 h. 0.0016 i. 0.0018 j. 0.0020
k. 0.0022 l. 0.0024 m. 0.0026 n. 0.0028 o. 0.0030

A34. Given the hodograph of storm-relative winds in Fig. 15.40b. Assume that vertical velocity increases with height according to $W = a{\cdot}z$, where a = (5 m s^{-1})/km. Considering only the tilting terms, find the vorticity spin-up based on the wind-vectors for the following pairs of heights (km):

a. 0,1 b. 1,2 c. 2,3 d. 3,4 e. 4,5 f. 5,6 g. 0,2
h. 1,3 i. 2,4 j. 3,5 k. 4,6 l. 1,4 m. 2,5 n. 3,6
o. 1,5 p. 2,6 q. 1,6

A35. Same as the previous exercise but for the storm-relative winds in the hodograph of the Sample Application in the "Storm-relative Winds" subsection of the tornado section.

A36. Given the hodograph of winds in Fig. 15.40a. Assume W = 0 everywhere. Calculate the helicity H based on the wind-vectors for the following pairs of heights (km):

a. 0,1 b. 1,2 c. 2,3 d. 3,4 e. 4,5 f. 5,6 g. 0,2
h. 1,3 i. 2,4 j. 3,5 k. 4,6 l. 1,4 m. 2,5 n. 3,6
o. 1,5 p. 2,6 q. 1,6

A37. Same as the previous exercise, but use the storm-relative winds from Fig. 15.40b to get the storm-relative helicity H'. (Hint, don't sum over all heights for this exercise.)

A38. Given the hodograph of winds in Fig. 15.40a. Assume that vertical velocity increases with height according to $W = a{\cdot}z$, where a = (5 m s^{-1})/km. Calculate the vertical contribution to helicity (eq. 15.54) based on the wind-vectors for the following pairs of heights (km):

a. 0,1 b. 1,2 c. 2,3 d. 3,4 e. 4,5 f. 5,6 g. 0,2
h. 1,3 i. 2,4 j. 3,5 k. 4,6 l. 1,4 m. 2,5 n. 3,6
o. 1,5 p. 2,6 q. 1,6

A39. Use the storm-relative winds in the hodograph of the Sample Application in the "Storm-relative Winds" subsection of the tornado section. Calculate the total storm-relative helicity (SRH) graphically for the following height ranges (km):
a. 0,1 b. 0,2 c. 0,3 d. 0,4 e. 0,5 f. 0,6 g. 1,2
h. 1,3 i. 1,4 j. 1,5 k. 1,6 l. 2,3 m. 2,4 n. 2,5
o. 2,6 p. 3,5 q. 3,6

A40. Same as the previous exercise, but find the answer using the equations (i.e., NOT graphically).

A41. Estimate the intensity of the supercell and tornado (if any), given a 0-1 km storm-relative helicity ($m^2 s^{-2}$) of:
a. 20 b. 40 c. 60 d. 80 e. 100 f. 120 g. 140
h. 160 i. 180 j. 200 k. 220 l. 240 m. 260 n. 280
o. 300 p. 320 q. 340 r. 360 s. 380 t. 400

A42. Given a storm-relative helicity of 220 , find the energy-helicity index if the CAPE ($J kg^{-1}$) is:
a. 200 b. 400 c. 600 d. 800 e. 1000 f. 1200
g. 1400 h. 1600 i. 1800 j. 2000 k. 2200
l. 2400 m. 2600 n. 2800 o. 3000

A43. Estimate the likely supercell intensity and tornado intensity (if any), given an energy-helicity index value of:
a. 0.2 b. 0.4 c. 0.6 d. 0.8 e. 1.0 f. 1.2 g. 1.4
h. 1.6 i. 1.8 j. 2.0 k. 2.2 l. 2.4 m. 2.6 n. 2.8
o. 3.0 p. 3.2 q. 3.4 r. 3.6 s. 3.8 t. 4.0

A44. If the tangential winds around a mesocyclone updraft are 20 $m s^{-1}$, find the swirl ratio of the average updraft velocity ($m s^{-1}$) is:
a. 2 b. 4 c. 6 d. 8 e. 10 f. 12 g. 14
h. 16 i. 18 j. 20 k. 22 l. 24 m. 26 n. 28
o. 30 p. 32 q. 34 r. 36 s. 38 t. 40

A45. Given a mesocyclone with a tangential velocity of 20 $m s^{-1}$ around the updraft region of radius 1000 m in a boundary layer 1 km thick. Find the swirl ratio and discuss tornado characteristics, given a radial velocity ($m s^{-1}$) of:
a. 1 b. 2 c. 3 d. 4 e. 5 f. 6 g. 7
h. 8 i. 9 j. 10 k. 11 l. 12 m. 13 n. 14
o. 15 p. 16 q. 17 r. 18 s. 19 t. 20

Evaluate & Analyze

E1. Why cannot hook echoes be used reliably to indicate the presence of a tornado?

E2. Cases of exceptionally heavy rain were discussed in the "Precipitation and Hail" section of this chapter and in the Precipitation chapter section on Rainfall Rates. However, most of those large rainfall rates occurred over exceptionally short durations (usually much less than an hour). Explain why longer-duration extreme-rainfall rates are unlikely.

E3. Use the info in Fig. 15.4 and the relationship between max likely updraft speed and CAPE, to plot a new graph of max possible hailstone diameter vs. total CAPE.

E4. In Fig. 15.7, what is the advantage to ignoring a portion of CAPE when estimating the likelihood of large hail? Explain.

E5. Explain why the various factors in the SHIP equation (15.4) are useful for predicting hail?

E6. Figures 15.5 and 15.10 show top and end views of the same thunderstorm, as might be seen with weather radar. Draw a side view (as viewed from the southeast by a weather radar) of the same thunderstorm. These 3 views give a blueprint (mechanical drawing) of a supercell.

E7. Cloud seeding (to change hail or rainfall) is a difficult social and legal issue. The reason is that even if you did reduce hail over your location by cloud seeding, an associated outcome might be increased hail or reduced rainfall further downwind. So solving one problem might create other problems. Discuss this issue in light of what you know about sensitive dependence of the atmosphere to initial conditions (the "butterfly effect"), and about the factors that link together the weather in different locations.

E8. a. Confirm that each term in eq. (15.5) has the same units.
b. Discuss how terms A and C differ, and what they each mean physically.
c. In term A, why is the numerator a function of $\Delta P'$ rather than ΔP?

E9. a. If there were no drag of rain drops against air, could there still be downbursts of air?
b. What is the maximum vertical velocity of large falling rain drops relative to the ground, knowing that air can be dragged along with the drops as a downburst? (Hint: air drag depends on the velocity of the drops relative to the air, not relative to the ground.)
c. Will that maximum fall velocity relative to the ground be reached at the ground, or at some height well above ground? Why?

E10. A raindrop falling through unsaturated air will cool to a certain temperature because some of the drop evaporates. State the name of this temperature.

E11. Suppose that an altocumulus (mid-tropospheric) cloud exists within an environment having a linear, conditionally unstable, temperature profile with height. Rain-laden air descends from this little cloud, warming at the moist adiabatic rate as it descends. Because this warming rate is less than the conditionally unstable lapse rate of the environment, the temperature perturbation of the air relative to the environment becomes colder as it descends.

But at some point, all the rain has evaporated. Descent below this altitude continues because the air parcel is still colder than the environment. However, during this portion of descent, the air parcel warms dry adiabatically, and eventually reaches an altitude where its temperature equals that of the environment. At this point, its descent stops. Thus, there is a region of strong downburst that does NOT reach the ground. Namely, it can be a hazard to aircraft even if it is not detected by surface-based wind-shear sensors.

Draw this process on a thermo diagram, and show how the depth of the downburst depends on the amount of liquid water available to evaporate.

E12. Demonstrate that eq. (15.10) equates kinetic energy with potential energy. Also, what assumptions are implicit in this relationship?

E13. Eqs. (15.12) and (15.13) show how vertical velocities (w_d) are tied to horizontal velocities (M) via pressure perturbations P'. Such coupling is generically called a circulation, and is the dynamic process that helps to maintain the continuity of air (namely, the uniform distribution of air molecules in space). Discuss how horizontal outflow winds are related to DCAPE.

E14. Draw a graph of gust-front advancement speed and thickness vs. range R from the downburst center. Do what-if experiments regarding how those curves change with

a. outflow air virtual temperature?

b. downburst speed?

E15. Fig. 15.21 shows large accumulations of electrical charge in thunderstorm clouds. Why don't the positive and negative charge areas continually discharge against each other to prevent significant charge accumulation, instead of building up such large accumulations as can cause lightning?

E16. At the end of the INFO box about "Electricity in a Channel" is given an estimate of the energy dissipated by a lightning stroke. Compare this energy to:

a. The total latent heat available to the thunderstorm, given a typical inflow of moisture.

b. The total latent heat actually liberated based on the amount of rain falling out of a storm.

c. The kinetic energy associated with updrafts and downbursts and straight-line winds.

d. The CAPE.

E17. Look at both INFO boxes on electricity. Relate:

a. voltage to electrical field strength

b. resistance to resistivity

c. current to current density

d. power to current density & electrical potential.

E18. If the electrical charging process in thunderstorms depends on the presence of ice, then why is lightning most frequently observed in the tropics?

E19. a. Lightning of exactly 12 kA occurs with what probability?

b. Lightning current in the range of 8 to 12 kA occurs with what probability?

E20. How does the shape of the lightning surge curve change with changes of parameters τ_1 and τ_2?

E21. Show why eqs. (15.22) and (15.32) are equivalent ways to express the speed of sound, assuming no wind.

E22. Do you suspect that nuclear explosions behave more like chemical explosions or like lightning, regarding the resulting shock waves, pressure, and density? Why?

E23. The equations for shock wave propagation from lightning assumed an isopycnal processes. Critique this assumption.

E24. In Earth's atmosphere, describe the conditions needed for the speed of sound to be zero relative to a coordinate system fixed to the ground. How likely are these conditions?

E25. What might control the max distance from lightning that you could hear thunder, if refraction was not an issue?

E26. Show how the expression of Snell's law in an environment with gradually changing temperature (eq. 15.36) is equivalent to, or reduces to, Snell's law across an interface (eq. 15.35).

E27. Show that eq. (15.39) for Snell's Law reduces to eq. (15.34) in the limit of zero wind.

E28. Use Bernoulli's equation from the Regional Winds chapter to derive the relationship between tornadic core pressure deficit and tangential wind speed. State all of your assumptions. What are the limitations of the result?

E29(§). Suppose that the actual tangential velocity in a tornado is described by a **Rankine combined vortex (RCV)**. Doppler radars, however, cannot measure radial velocities at any point, but instead observe velocities averaged across the radar beam width. So the Doppler radar sees a smoothed version of the Rankine combined vortex. It is this smoothed tangential velocity shape that is called a **tornado vortex signature (TVS)**, and for which the Doppler-radar computers are programmed to recognize. This exercise is to create the tangential velocity curve similar to Fig. 15.34, but for a TVS.

Let ΔD be the diameter of radar beam at some range from the radar, and R_o be the core radius of tornado. The actual values of ΔD and R_o are not important: instead consider the dimensionless ratio $\Delta D/R_o$. Compute the TVS velocity at any distance R from the center of the tornado as the average of all the RCV velocities between radii of $[(R/R_o)-(\Delta D/2R_o)]$ and $[(R/R_o)+(\Delta D/2R_o)]$, and repeat this calculation for many values of R/R_o to get a curve. This process is called a **running average**. Create this curve for $\Delta D/R_o$ of:

a. 1.0 b. 1.5 c. 2.0 d. 2.5 e. 3.0 f. 3.5 g. 4.0
h. 4.5 i. 5.0 j. 0.9 k. 0.8 l. 0.7 m. 0.6 n. 0.5

Hint: Either do this by analytically integrating the RCV across the radar beam width, or by brute-force averaging of RCV values computed using a spreadsheet.

E30(§). For tornadoes, an alternative approximation for tangential velocities M_{tan} as a function of radius R is given by the **Burgers-Rott Vortex** (BRV) equation:

$$\frac{M_{\tan}}{M_{\tan\max}} = 1.398 \cdot \left(\frac{R_o}{R}\right) \cdot \left[1 - e^{-(1.12 \cdot R/R_o)^2}\right] \qquad (15.62)$$

where R_o is the core radius.

Plot this curve, and on the same graph replot the Rankine combined vortex (RCV) curve (similar to Fig. 15.33). Discuss what physical processes in the tornado might be included in the BRV that are not in the RCV, to explain the differences between the two curves.

E31. For the Rankine combined vortex (RCV), both the tangential wind speed and the pressure deficit are forced to match at the boundary between the tornado core and the outer region. Does the pressure <u>gradient</u> also match at that point? If not, discuss any limitations that you might suggest on the RCV.

E32. Suppose a suction vortex with max tangential speed $M_{s\ tan}$ is moving around a parent tornado of tangential speed $M_{p\ tan}$, and the parent tornado is translating at speed M_{tr}. Determine how the max speed varies with position along the resulting cycloidal damage path.

E33. The TORRO scale is related to the **Beaufort** wind scale (**B**) by:

$$\mathbf{B} = 2 \cdot (\mathbf{T} + 4) \qquad (15.63)$$

The Beaufort scale is discussed in detail in the Hurricane chapter, and is used to classify ocean storms and sea state. Create a graph of Beaufort scale vs. TORRO scale. Why cannot the Beaufort scale be used to classify tornadoes?

E34. Volcanic eruptions can create blasts of gas that knock down trees (as at Mt. St. Helens, WA, USA). The **air burst** from astronomical meteors speeding through the atmosphere can also knock down trees (as in Tunguska, Siberia). Explain how you could use the TORRO scale to classify these winds.

E35. Suppose that tangential winds around a tornado involve a balance between pressure-gradient force, centrifugal force, and Coriolis force. Show that anticyclonically rotating tornadoes would have faster tangential velocity than cyclonic tornadoes, given the same pressure gradient. Also, for anticyclonic tornadoes, are their any tangential velocity ranges that are excluded from the solution of the equations (i.e., are not physically possible)? Assume N. Hem.

E36. Does the outside edge of a tornado condensation funnel have to coincide with the location of fastest winds? If not, then is it possible for the debris cloud (formed in the region of strongest winds) radius to differ from the condensation funnel radius? Discuss.

E37. Given a fixed temperature and dew point, eq. (15.48) gives us the pressure at the outside edge of the condensation funnel.

a. Is it physically possible (knowing the governing equations) for the pressure deficit at the tornado axis to be higher than the pressure deficit at the visible condensation funnel? Why?

b. If the environmental temperature and dew point don't change, can we infer that the central pressure deficit of a large-radius tornado is lower than that for a small-radius tornado? Why or why not?

E38. Gustnadoes and dust devils often look very similar, but are formed by completely different mechanisms. Compare and contrast the processes that create and enhance the vorticity in these vortices.

E39. From Fig. 15.40a, you see that the winds at 1 km above ground are coming from the southeast. Yet, if you were riding with the storm, the storm relative winds that you would feel at 1 km altitude would be from the northeast, as shown for the same data in Fig. 15.40b. Explain how this is possible; namely, explain how the storm has boundary layer inflow entering it from the northeast even though the actual wind direction is from the southeast.

E40. Consider Fig. 15.42.

a. What are the conceptual (theoretical) differences between streamwise vorticity, and the vorticity around a local-vertical axis as is usually studied in meteorology?

b. If it is the streamwise (horizontal axis) vorticity that is tilted to give vorticity about a vertical axis, why don't we see horizontal-axis tornadoes forming along with the usual vertical tornadoes in thunderstorms? Explain.

E41. Compare eq. (15.51) with the full vorticity-tendency equation from the Extratropical Cyclone chapter, and discuss the differences. Are there any terms in the full vorticity equation that you feel should not have been left out of eq. (15.51)? Justify your arguments.

E42. Show that eq. (15.57) is equivalent to eq. (15.56). (Hint: use the average-wind definition given immediately after eq. (15.55).)

E43. Show mathematically that the area swept out by the storm-relative winds on a hodograph (such as the shaded area in Fig. 15.44) is indeed exactly half the storm-relative helicity. (Hint: Create a simple hodograph with a small number of wind vectors in easy-to-use directions, for which you can easily calculate the shaded areas between wind vectors. Then use inductive reasoning and generalize your approach to arbitrary wind vectors.)

E44. What are the advantages and disadvantages of eSRH and EHI relative to SRH?

E45. Suppose the swirl ratio is 1 for a tornado of radius 300 m in a boundary layer 1 km deep. Find the radial velocity and core pressure deficit for each tornado intensity of the

a. Enhanced Fujita scale. b. TORRO scale.

E46. One physical interpretation of the denominator in the swirl ratio (eq. 15.60) is that it indicates the volume of inflow air (per crosswind distance) that reaches the tornado from outside. Provide a similar interpretation for the swirl-ratio numerator.

E47. The cycloidal damage sketched in Fig. 15.51 shows a pattern in between that of a true cycloid and a circle. Look up in another reference what the true cycloid shape is, and discuss what type of tornado behavior would cause damage paths of this shape.

Synthesize

S1. Since straight-line outflow winds exist surrounding downbursts that hit the surface, would you expect similar hazardous outflow winds where the updraft hits the tropopause? Justify your arguments.

S2. Suppose that precipitation loading in an air parcel caused the virtual temperature to increase, not decrease. How would thunderstorms differ, if at all?

S3. If hailstones were lighter than air, discuss how thunderstorms would differ, if at all.

S4. If all hailstones immediately split into two when they reach a diameter of 2 cm, describe how hail storms would differ, if at all.

S5. Are downbursts equally hazardous to both light-weight, small private aircraft and heavy fast commercial jets? Justify your arguments?

S6. Suppose that precipitation did not cause a downward drag on the air, and that evaporation of precipitation did not cool the air. Nonetheless, assume that thunderstorms have a heavy precipitation region. How would thunderstorms differ, if at all?

S7. Suppose that downbursts did not cause a pressure perturbation increase when they hit the ground. How would thunderstorm hazards differ, if at all?

S8. Suppose that downbursts sucked air out of thunderstorms. How would thunderstorms differ, if at all?

S9. Do you think that lightning could be productively utilized? If so, describe how.

S10. Suppose air was a much better conductor of electricity. How would thunderstorms differ, if at all?

S11. Suppose that once a lightning strike happened, the resulting plasma path that was created through air persists as a conducting path for 30 minutes. How would thunderstorms differ, if at all.

S12. Suppose that the intensity of shock waves from thunder did not diminish with increasing distance. How would thunderstorm hazards differ, if at all?

S13. Refraction of sound can make noisy objects sound quieter, and can amplify faint sounds by focusing them. For the latter case, consider what happens to sound waves traveling different paths as they all reach the same focus point. Describe what would happen there.

S14. The air outside of the core of a Rankine-combined-vortex (RCV) model of a tornado is moving around the tornado axis. Yet the flow is said to be **irrotational** in this region. Namely, at any point outside the core, the flow has no vorticity. Why is that? Hint: consider aspects of a flow that can contribute to relative vorticity (see the General Circulation chapter), and compare to characteristics of the RCV.

S15. What is the max tangential speed that a tornado could possibly have? What natural forces in the atmosphere could create such winds?

S16. Do anticyclonically rotating tornadoes have higher or lower core pressure than the surrounding environment? Explain the dynamics.

S17. Consider Fig. 15.36. If the horizontal pressure gradient near the bottom of tornadoes was weaker than that near the top, how would tornadoes be different, if at all?

S18. If a rapidly collapsing thunderstorm (nicknamed a **bursticane,** which creates violent downbursts and near-hurricane-force straight-line winds at the ground) has a rapidly sinking top, could it create a tornado above it due to stretching of the air above the collapsing thunderstorm? Justify your arguments.

S19. Positive helicity forms not only with updrafts and positive vorticity, but also with downdrafts and negative vorticity. Could the latter condition of positive helicity create mesocyclones and tornadoes? Explain.

16 TROPICAL CYCLONES

Contents

Intense synoptic-scale cyclones in the tropics are called **tropical cyclones**. As for all cyclones, tropical cyclones have low pressure in the cyclone center near sea level. Also, the low-altitude winds rotate cyclonically (counterclockwise in the N. Hemisphere) around the storm and spiral in towards the center.

Tropical cyclones are called **hurricanes** over the Atlantic, Caribbean, Gulf of Mexico (Fig. 16.1), and eastern Pacific Oceans. They are called **typhoons** over the western Pacific. Over the Indian Ocean and near Australia they are called **cyclones**. In this chapter we use "tropical cyclone" to refer to such storms anywhere in the world.

Tropical cyclones differ from mid-latitude cyclones in that tropical cyclones do not have fronts. Tropical cyclones have warm cores while mid-latitude cyclones have cold cores. Tropical cyclones can persist two to three times longer than typical mid-latitude cyclones. To help explain this behavior, we start by describing tropical cyclone structure.

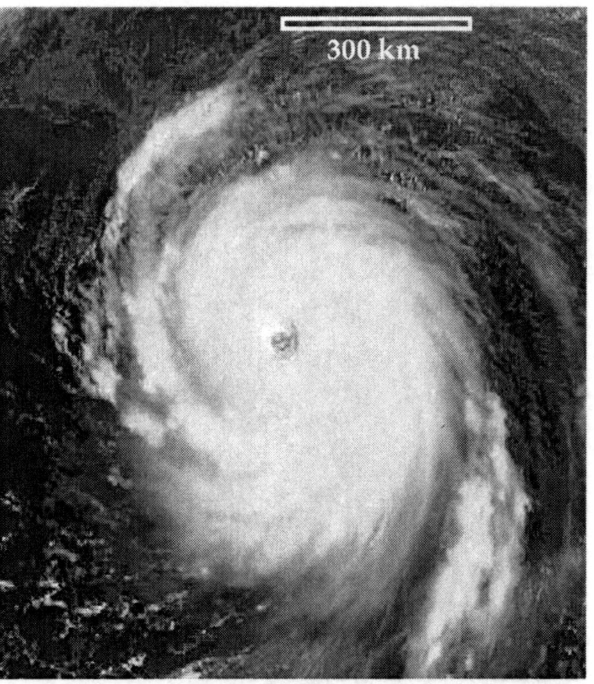

Figure 16.1
Visible-spectrum satellite picture of Hurricane Katrina over the Gulf of Mexico, taken 28 Aug 2005 at 1545 UTC. (GOES image courtesy of US DOC/NOAA.)

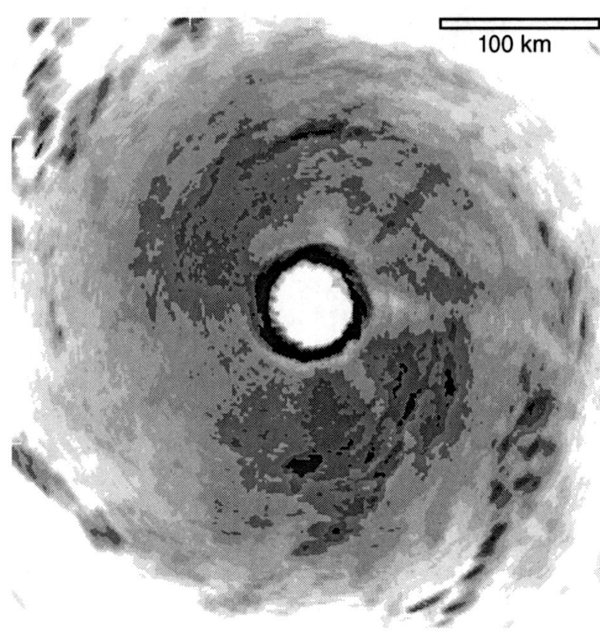

Figure 16.2
Radar reflectivity PPI image of Hurricane Katrina, taken from a research aircraft at altitude 2325 m on 28 Aug 2005 at 1747 UTC while it was over the Gulf of Mexico. Darkest regions in the eyewall correspond to 35 to 40 dBZ radar reflectivity. (Image courtesy of US DOC/NOAA/AOML/ Hurricane Research Division.)

TROPICAL CYCLONE STRUCTURE

Tropical cyclones are made of thunderstorms. Near the center (**core**) of the tropical cyclone is a ring or circle of thunderstorms called the **eyewall**. This is the most violent part of the storm with the heaviest rain and the greatest radar reflectivity (Fig. 16.2). The tops of these thunderstorms can be in the lower stratosphere: 15 to 18 km high. Thunderstorm bases are very low: in the boundary layer. Thus, Tropical cyclones span the tropical-troposphere depth.

The anvils from each of the thunderstorms in the eyewall merge into one large roughly-circular **cloud shield** that is visible by satellite (Fig. 16.1). These anvils spread outward 75 to 150 km away from the eye wall. Hence, tropical cyclone diameters are roughly 10 to 20 times their depth (Fig. 16.3), although the high-altitude outflow from the top of some asymmetric tropical cyclones can reach 1000s of km (Fig. 16.4).

In the middle of this eyewall is a calmer region called the **eye** with warm temperatures, subsiding (sinking) air, and fewer or no clouds. Eye diameter at sea level is 20 to 50 km. The eye is conical, with the larger diameter at the storm top (Fig. 16.5).

Spiraling out from the eye wall can be zero or more bands of thunderstorms called **spiral bands**.

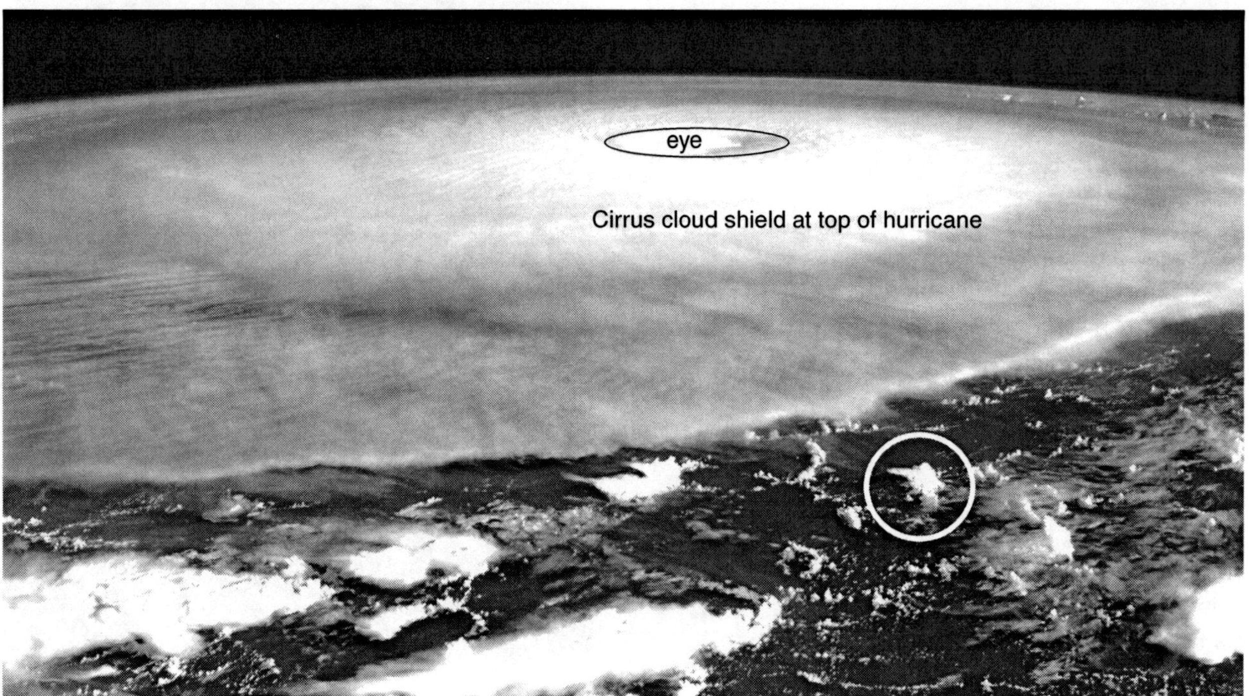

Figure 16.3
Hurricane Isabel as viewed from the International Space Station on 13 Sep 2003. The hurricane eye is outlined with the thin black oval, while the outflow from the eyewall thunderstorms fills most of this image. For scale comparison, the white circle outlines a nearby thunderstorm of depth and diameter of roughly 15 km. [Image courtesy of NASA - Johnson Spaceflight Center.]

Figure 16.4
NOAA GOES satellite -derived 3-D rendering of hurricane Floyd at 2015 UTC on 15 Sep 1999, showing asymmetric outflow of this Category 4 storm extending 1000s of km north, past the Great Lakes into Canada. The eye is in the NW Atlantic just to the east of the Georgia-Florida border, USA. The sea-surface temperature for this storm is in Fig. 16.51 at the end of this chapter. [Image by Hal Pierce is courtesy of the Laboratory for Atmospheres, NASA Goddard Space Flight Center.]

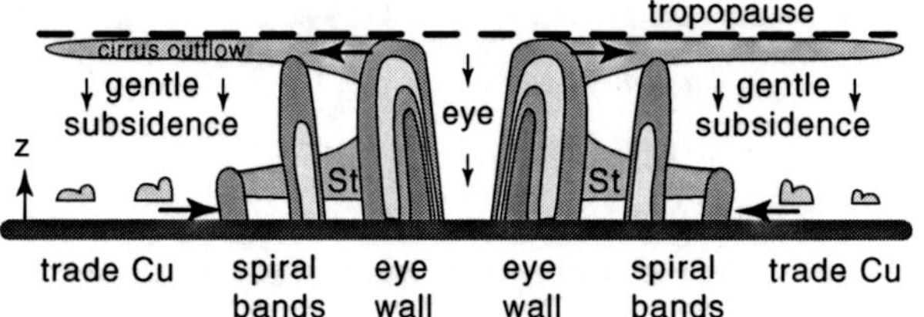

Figure 16.5
Vertical slice through a tropical cyclone. Green, yellow, orange, and red colors suggest the moderate to heavy rainfall as seen by radar. Light blue represents stratiform (St) clouds. The lighter rain that falls from the lower stratiform cloud deck of real tropical cyclones is not shown here. Tropopause altitude ≈ 15 km. Tropical-cyclone width ≈ 1500 km. Fair-weather trade-wind cumulus (Cu) clouds are sketched with grey shading. Arrows show radial and vertical wind directions.

Sometimes these spiral **rain bands** will merge to form a temporary second eyewall of thunderstorms around the original eye wall (Fig. 16.6). The region of lighter rain between the two eyewalls is called the **moat**.

During an **eyewall replacement cycle**, in very strong tropical cyclones, the inner eyewall dissipates and is replaced by the outer eyewall. When this happens, tropical cyclone intensity sometimes diminishes temporarily, and then strengthens again when the new outer eyewall diameter shrinks to that of the original eyewall.

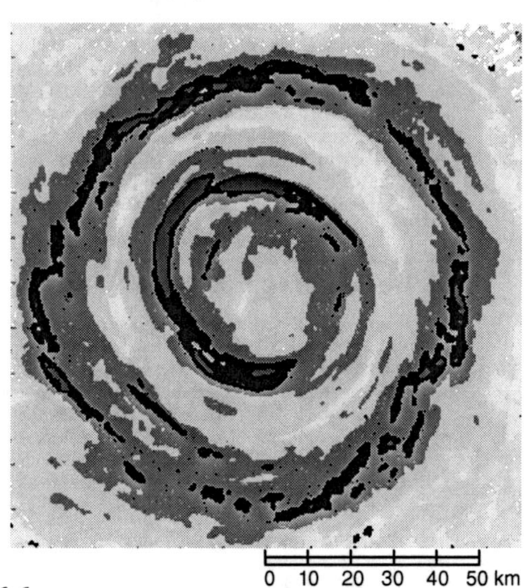

Figure 16.6
Zoomed view of double eyewall in Hurricane Rita, as viewed by airborne radar at 1 km altitude on 22 Sep 2005 at 1801 UTC. Darker greys show two concentric rings of heavier rain. Darkest grey corresponds to roughly 40 dBZ. [Modified from original image by Michael Bell and Wen-Chau Lee.]

INTENSITY AND GEOGRAPHIC DISTRIBUTION

Lower sea-level pressures in the eye and faster winds in the lower troposphere indicate stronger tropical cyclones. Several different tropical cyclone scales have been devised to classify tropical cyclone strength, as summarized below.

Table 16-1. Saffir-Simpson Hurricane Wind Scale — Definition. Based on max 1-minute sustained wind speed measured at standard anemometer height of 10 m. Applies in the Atlantic Ocean, and in the eastern Pacific from the coast of the Americas to 180°W.

Category	Wind Speed			
	m s^{-1}	km h^{-1}	knots	miles h^{-1}
1	33 - 42	119 - 153	64 - 82	74 - 95
2	43 - 49	154 - 177	83 - 95	96 - 110
3	50 - 58	178 - 209	96 - 113	111 - 130
4	59 - 69	210 - 249	114 - 135	131 - 155
5	> 69	> 249	> 135	> 155

Table 16-2. Saffir-Simpson Hurricane Wind Scale — Description of Expected Damage — *Examples.*

Category	Concise Statement Damage Expected *Examples*
1	**Very dangerous winds will produce some damage.** People, livestock, and pets struck by flying or falling debris could be injured or killed. Older (mainly pre-1994 construction) mobile homes could be destroyed, especially if they are not anchored properly as they tend to shift or roll off their foundations. Newer mobile homes that are anchored properly can sustain damage involving the removal of shingle or metal roof coverings, and loss of vinyl siding, as well as damage to carports, sunrooms, or lanais. Some poorly constructed frame homes can experience major damage, involving loss of the roof covering and damage to gable ends as well as the removal of porch coverings and awnings. Unprotected windows may break if struck by flying debris. Masonry chimneys can be toppled. Well-constructed frame homes could have damage to roof shingles, vinyl siding, soffit panels, and gutters. Failure of aluminum, screened-in, swimming pool enclosures can occur. Some apartment building and shopping center roof coverings could be partially removed. Industrial buildings can lose roofing and siding especially from windward corners, rakes, and eaves. Failures to overhead doors and unprotected windows will be common. Windows in high-rise buildings can be broken by flying debris. Falling and broken glass will pose a significant danger even after the storm. There will be occasional damage to commercial signage, fences, and canopies. Large branches of trees will snap and shallow rooted trees can be toppled. Extensive damage to power lines and poles will likely result in power outages that could last a few to several days. *Hurricane Dolly (2008) brought Category 1 winds and impacts to South Padre Island, Texas.*
2	**Extremely dangerous winds will cause extensive damage.** There is a substantial risk of injury or death to people, livestock, and pets due to flying and falling debris. Older (mainly pre-1994 construction) mobile homes have a very high chance of being destroyed and the flying debris generated can shred nearby mobile homes. Newer mobile homes can also be destroyed. Poorly constructed frame homes have a high chance of having their roof structures removed especially if they are not anchored properly. Unprotected windows will have a high probability of being broken by flying debris. Well-constructed frame homes could sustain major roof and siding damage. Failure of aluminum, screened-in, swimming pool enclosures will be common. There will be a substantial percentage of roof and siding damage to apartment buildings and industrial buildings. Unreinforced masonry walls can collapse. Windows in high-rise buildings can be broken by flying debris. Falling and broken glass will pose a significant danger even after the storm. Commercial signage, fences, and canopies will be damaged and often destroyed. Many shallowly rooted trees will be snapped or uprooted and block numerous roads. Near-total power loss is expected with outages that could last from several days to weeks. Potable water could become scarce as filtration systems begin to fail. *Hurricane Frances (2004) brought Category 2 winds and impacts to coastal portions of Port St. Lucie, Florida.*
3	**Devastating damage will occur.** There is a high risk of injury or death to people, livestock, and pets due to flying and falling debris. Nearly all older (pre-1994) mobile homes will be destroyed. Most newer mobile homes will sustain severe damage with potential for complete roof failure and wall collapse. Poorly constructed frame homes can be destroyed by the removal of the roof and exterior walls. Unprotected windows will be broken by flying debris. Well-built frame homes can experience major damage involving the removal of roof decking and gable ends. There will be a high percentage of roof covering and siding damage to apartment buildings and industrial buildings. Isolated structural damage to wood or steel framing can occur. Complete failure of older metal buildings is possible, and older unreinforced masonry buildings can collapse. Numerous windows will be blown out of high-rise buildings resulting in falling glass, which will pose a threat for days to weeks after the storm. Most commercial signage, fences, and canopies will be destroyed. Many trees will be snapped or uprooted, blocking numerous roads. Electricity and water will be unavailable for several days to a few weeks after the storm passes. *Hurricane Ivan (2004) brought Category 3 winds and impacts to coastal portions of Gulf Shores, Alabama.*
4	**Catastrophic damage will occur.** There is a very high risk of injury or death to people, livestock, and pets due to flying and falling debris. Nearly all older (pre-1994) mobile homes will be destroyed. A high percentage of newer mobile homes also will be destroyed. Poorly constructed homes can sustain complete collapse of all walls as well as the loss of the roof structure. Well-built homes also can sustain severe damage with loss of most of the roof structure and/or some exterior walls. Extensive damage to roof coverings, windows, and doors will occur. Large amounts of windborne debris will be lofted into the air. Windborne debris damage will break most unprotected windows and penetrate some protected windows. There will be a high percentage of structural damage to the top floors of apartment buildings. Steel frames in older industrial buildings can collapse. There will be a high percentage of collapse to older unreinforced masonry buildings. Most windows will be blown out of high-rise buildings resulting in falling glass, which will pose a threat for days to weeks after the storm. Nearly all commercial signage, fences, and canopies will be destroyed. Most trees will be snapped or uprooted and power poles downed. Fallen trees and power poles will isolate residential areas. Power outages will last for weeks to possibly months. Long-term water shortages will increase human suffering. Most of the area will be uninhabitable for weeks or months. *Hurricane Charley (2004) brought Category 4 winds and impacts to coastal portions of Punta Gorda, Florida.*
5	**Catastrophic damage will occur.** People, livestock, and pets are at very high risk of injury or death from flying or falling debris, even if indoors in mobile homes or framed homes. Almost complete destruction of all mobile homes will occur, regardless of age or construction. A high percentage of frame homes will be destroyed, with total roof failure and wall collapse. Extensive damage to roof covers, windows, and doors will occur. Large amounts of windborne debris will be lofted into the air. Windborne debris damage will occur to nearly all unprotected windows and many protected windows. Significant damage to wood roof commercial buildings will occur due to loss of roof sheathing. Complete collapse of many older metal buildings can occur. Most unreinforced masonry walls will fail which can lead to the collapse of the buildings. A high percentage of industrial buildings and low-rise apartment buildings will be destroyed. Nearly all windows will be blown out of high-rise buildings resulting in falling glass, which will pose a threat for days to weeks after the storm. Nearly all commercial signage, fences, and canopies will be destroyed. Nearly all trees will be snapped or uprooted and power poles downed. Fallen trees and power poles will isolate residential areas. Power outages will last for weeks to possibly months. Long-term water shortages will increase human suffering. Most of the area will be uninhabitable for weeks or months. *Hurricane Andrew (1992) brought Category 5 winds and impacts to coastal portions of Cutler Ridge, Florida.*

From an online report "The Saffir-Simpson Hurricane Wind Scale", National Hurricane Center, National Weather Service, NOAA. 2010.

Saffir-Simpson Hurricane Wind Scale

Herbert Saffir and Robert Simpson created a scale for Atlantic hurricane strength that came to be known as the **Saffir-Simpson Hurricane Scale**. This scale has been used since the 1970s to inform disaster-response officials about hurricane strength.

In 2010, the US National Hurricane Center updated this scale (now called the **Saffir-Simpson Hurricane Wind Scale**), and defined hurricane intensity on wind speed only. The scale ranges from category 1 for a weak hurricane to category 5 for the strongest hurricane (Table 16-1). A description of the expected damage for each category is in Table 16-2.

Typhoon Intensity Scales

For typhoons in the western North Pacific Ocean, storm intensity has been classified three different ways by three different organizations: the **Japan Meteorological Agency (JMA)**, the **Hong Kong Observatory (HKO)**, and the **US Joint Typhoon Warning Center (JTWC)**. See Table 16-3.

Table 16-3. Typhoon (tropical cyclone) intensity scale, based on max sustained winds during 10 minutes. Defined by JMA, unless otherwise noted. Applies between 100°E to 180°E.

Category	Wind Speed		
	m s^{-1}	km h^{-1}	knots
Tropical Depression	< 17	< 61	< 33
Tropical Storm	17 - 24	62 - 88	35 - 48
Severe Tropical Storm	25 - 32	89 - 117	48 - 63
Typhoon	33 - 41	118 - 149	64 - 79
Severe Typhoon	42 - 51 (HKO)	150 - 184 (HKO)	80 - 99 (HKO)
Super Typhoon	> 51 (HKO) > 67 (JTWC*)	>185 (HKO) > 241 (JTWC*)	>100 (HKO) > 130 (JTWC*)

* *Based on 1-minute average max sustained winds.*

Other Tropical-Cyclone Scales

Additional tropical cyclone intensity scales (with different wind-speed definitions and category names) have been defined by tropical cyclone agencies in:

- Australia (**Australian Bureau of Meteorology**, for S. Hemisphere east of 90°E),
- India (**Regional Specialized Meteorological Center in New Delhi**, for N. Hemisphere Indian Ocean between 45°E and 100°E), and by
- **Météo-France** (for S. Hemisphere west of 90°E).

Even more confusing, some agencies use 1-minute average (sustained) winds, some use 10-minute average (sustained) winds, and some use gusts.

Geographic Distribution and Movement

Fig. 16.7 outlines the regions of greatest frequency of tropical cyclones and shows typical storm tracks.

Tropical cyclones are steered mostly by the large-scale global circulation. Most tropical cyclones form between 10° and 30° latitude, which is the trade-wind region. Hence, most tropical cyclones are steered from east to west initially.

Later, these storms tend to turn poleward under the influence of monsoon circulations. For example, the Bermuda High (also known as the Azores High) over the North Atlantic Ocean has a clockwise circulation around it (see Fig. 11.31 in the General Circulation chapter), which turns tropical cyclones northward in the Western North Atlantic Ocean. The tracks of tropical cyclones and former tropical

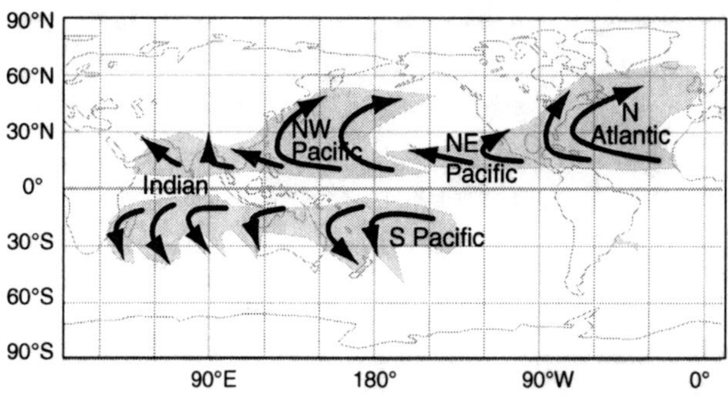

Figure 16.7
Map of tropical-cyclone locations (grey shading), with a general sense of the cyclone tracks (black arrows).

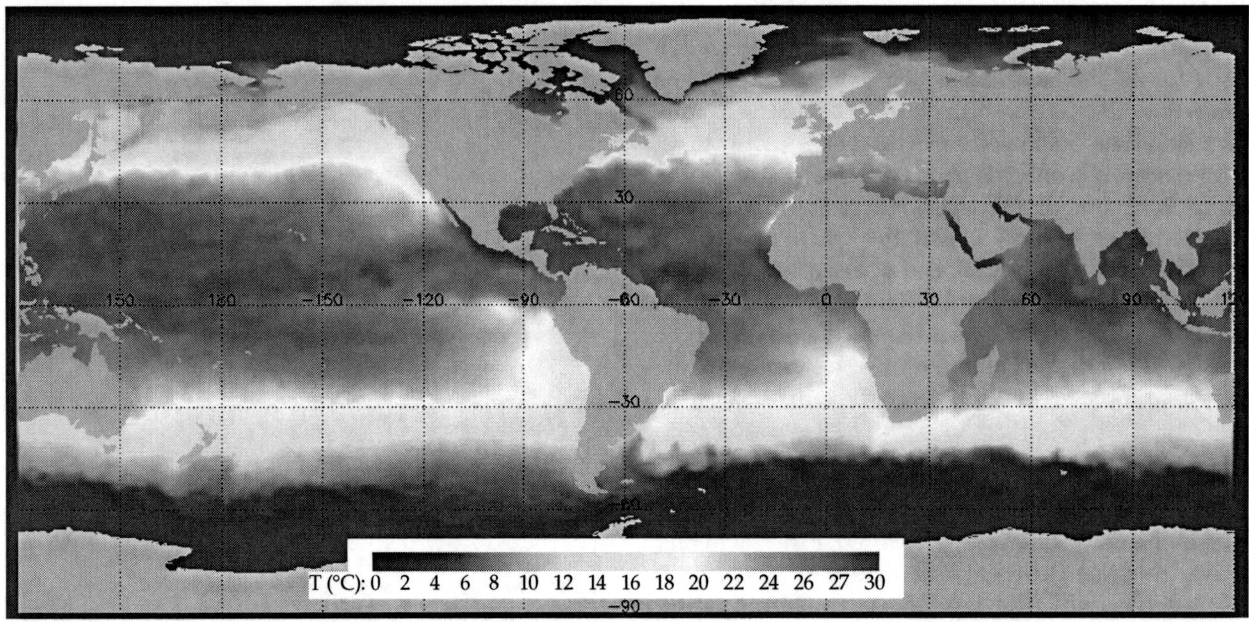

Figure 16.8
Sea surface temperatures (°C) at 00 UTC on 27 Sep 2003. [Courtesy of the USA Fleet Numerical Meteorology and Oceanography Center (FNMOC).]

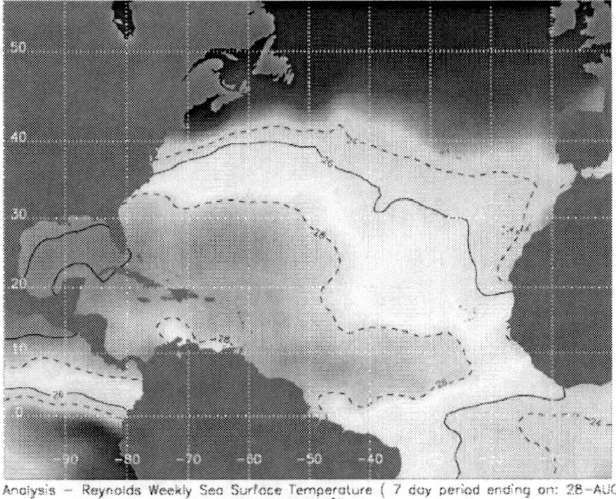

Figure 16.9
Analysis of Atlantic-ocean sea-surface temperatures (°C) averaged over 7 days ending on 28 Aug 2004. Hurricane Francis occurred during this time period . [Adapted from a USA National Hurricane Center image using NCEP/NOAA data.]

cyclones continue to turn toward the northeast as they proceed further into the mid-latitudes and encounter west winds in the global circulation.

A striking observation in Fig. 16.7 is that no tropical cyclones form at the equator. Also, none cross the equator. The next section explains why.

EVOLUTION

Requirements for Cyclogenesis

Seven conditions are necessary for tropical cyclones to form: a warm sea surface, non-zero Coriolis force, nonlocal conditional instability, high humidity in the mid troposphere, weak ambient wind shear, enhanced synoptic-scale vorticity, and a trigger.

Warm Sea Surface

The **sea surface temperature** (**SST**) must be approximately 26.5°C or warmer (Figs. 16.8 and 16.9 on this page, and 16.51 at the end of this chapter), and the warm surface waters must be at least 50 m deep. This warm temperature is needed to enable strong evaporation and heat transfer from the sea surface into the boundary layer. The warmer, more-humid boundary-layer air serves as the fuel for thunderstorms in the tropical cyclone.

The fast winds in tropical cyclones create large waves that stir the top part of the ocean. If the warm waters are too shallow, then this turbulent mixing will stir colder deeper water up to the surface (see

INFO Box two pages later). When this happens, the sea-surface temperature decreases below the 26.5°C threshold, and the tropical cyclone kills itself.

Coriolis Force

Tropical cyclones cannot exist within about 500 km of the equator (i.e., ≤ 5° latitude), because Coriolis force is near zero there (and is exactly zero right at the equator). Rarely, very small-diameter tropical cyclones have been observed closer to the equator, but none are observed right at the equator. Tropical cyclones cannot form at the equator, and existing cyclones cannot cross the equator (Fig. 16.7).

Without Coriolis force, boundary-layer air would be sucked directly into the eye by the low pressure there. Thus, air molecules would accumulate in the eye, causing pressure to increase towards ambient values. The result is that the low pressure would disappear, winds would die, and the tropical cyclone would cease to exist in less than a day. This is what happens to tropical cyclones that approach the equator (Fig. 16.10).

But with Coriolis force, winds in the bottom half of the troposphere are forced around the eye at gradient- or cyclostrophic-wind speeds. Namely, most of the air is flowing tangentially around the eye rather than flowing radially into it. Only close to the ground does drag change the winds into boundary-layer gradient winds, resulting in a small amount of convergence toward the eye. Thus, the tropical cyclone can persist for many days.

Nonlocal Conditional Instability

Because tropical cyclones are made of thunderstorms, the tropical environment must have sufficient nonlocal conditional instability to support deep thunderstorm convection. Namely, there must be a stable layer (i.e., a cap) above a warm humid boundary layer, and the mid-troposphere must be relatively cool compared to the boundary layer.

The warm humid boundary layer is achieved via strong heat and moisture fluxes from the warm sea surface into the air. The cap is the trade-wind inversion that was discussed in the General Circulation chapter. These conditions, combined with a cool mid troposphere, lead to large values of convective available potential energy (CAPE), as was thoroughly explained in the Thunderstorm chapter. Hence, large values of CAPE imply sufficient nonlocal conditional instability for tropical cyclones.

High Humidity in the Mid-troposphere

In a deep layer of air centered at roughly 5 km above sea level, humidity must be high. Otherwise, the incipient thunderstorms cannot continue to grow and become organized into tropical cyclones.

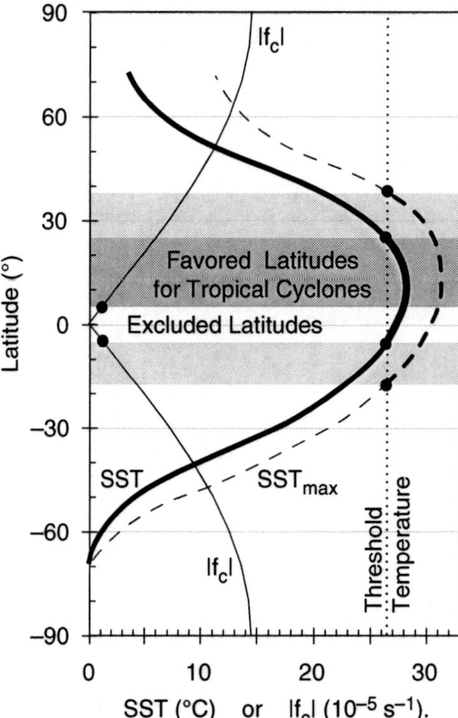

Figure 16.10
Approximate zonal average sea-surface temperature SST (°C, thick solid line) averaged around the globe during 3 Aug 2009. Also shown is the max SST (°C, thin dashed line) on that day at each latitude. The magnitude of the Coriolis parameter f_c (10^{-5} s^{-1}) is the thin solid line. Dotted line is the SST threshold of 26.5°C for tropical cyclones. Those latitudes that exceed the temperature threshold and have Coriolis parameter that is not near zero are favored locations for tropical cyclones (shaded medium grey). Tropical cyclones can also occur in limited regions (shaded light grey) at other latitudes where the max SST exceeds the threshold. "Excluded latitudes" are those for which Coriolis force is too small.

As the seasons progress, the location of peak SST shifts, causing the favored latitudes to differ from this particular August example.

Note that this differs from mid-latitude thunderstorms, where a drier mid-troposphere allows more violent thunderstorms. When dry environmental air is entrained into the sides of mid-latitude thunderstorms, some of the storm's hydrometeors evaporate, causing the strong downdrafts that define supercell storms and which create downbursts, gust fronts, and can help trigger tornadoes.

In the tropics, such downbursts from any one thunderstorm can disrupt neighboring thunderstorms, reducing the chance that neighboring thunderstorms can work together in an incipient eyewall. Also, the cold downburst air accumulates at the bottom of the troposphere, thereby increasing static stability and reducing deep convection.

INFO • Trop. Cyclone-induced Currents

Near-surface winds cause net ocean transport 90° to the right of the wind vector in the N. Hemisphere, as explained later in the Storm Surge section. Given tangential winds that circle a tropical cyclone, we expect the net Ekman transport (average ocean currents induced by the tropical cyclone) to be outward (Fig. 16.a)

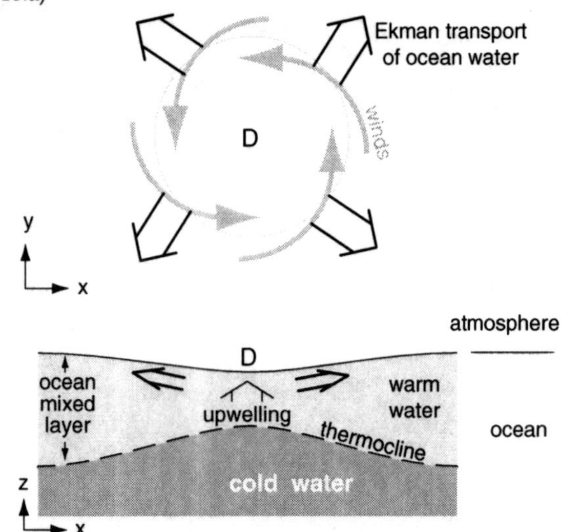

Fig. 16.a. *Ocean currents (double-shaft arrows) induced by tropical-cyclone winds (grey arrows).*

Such horizontal divergence (*D*) of the ocean-surface water does three things: (1) lowers the sea surface by removing sea water; (2) causes upwelling of water toward the surface; and (3) causes the colder deep water to be brought closer to the surface (Fig. 16.a-bottom). The **thermocline** is the interface between cold deep water and warmer mixed-layer above.

The tropical cyclone brings cold water closer to the surface, where ocean mixed-layer turbulence caused by breaking waves can further mix the warm and cold waters. Thus, tropical cyclones will kill themselves by cooling the sea-surface temperature unless the pre-storm warm ocean mixed-layer is deep enough (≥ 50 m).

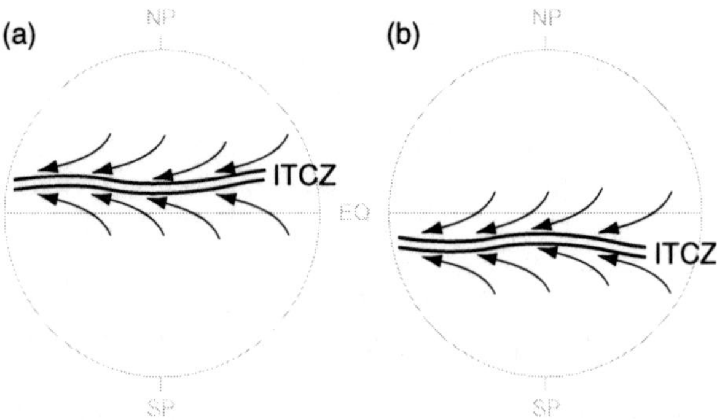

Figure 16.11. *Sketch of the Earth showing the Intertropical Convergence Zone (ITCZ, double solid line) for the early Autumn tropical-cyclone seasons during (a) August and (b) February. NP = North Pole. SP = South Pole. EQ = equator.*

Weak Ambient Wind Shear

Wind shear within four degrees of latitude of the incipient storm must be weak (ΔM < 10 m s^{-1} between pressure levels 80 and 25 kPa) to enable thunderstorm clusters to form. These clusters are the precursors to tropical cyclones.

If the shear is too strong, the updrafts in the thunderstorms become tilted, and latent heating due to water-vapor condensation is spread over a broader area. This results in less-concentrated warming, and a reduced ability to create a low-pressure center at sea level around which the thunderstorms can become organized into a tropical cyclone.

This requirement differs from that for mid-latitude thunderstorms. At mid-latitudes, strong shear in the ambient environment is good for storms because it encourages the creation of mesocyclones and supercell thunderstorms. Apparently, in the tropics such rotation of individual thunderstorms is bad because it interferes with the collaboration of many thunderstorms to create an eyewall.

Synoptic-scale Vorticity

A relative maximum of relative vorticity in the bottom half of the troposphere can help organize the thunderstorms into an incipient tropical cyclone. Otherwise, any thunderstorms that form would act somewhat independently of each other.

Tropical Cyclone Triggers

Even if all six previous conditions are satisfied, a method to trigger the tropical cyclone is also needed. A trigger is anything that creates synoptic-scale horizontal convergence in the atmospheric boundary layer. Such horizontal convergence forces upward motion out of the boundary-layer top as required by mass conservation (see the Atmospheric Forces & Winds chapter). Synoptic-scale upward motion can initiate and support an organized cluster of thunderstorms — incipient tropical cyclones.

Some triggers are: the ITCZ, easterly waves, Monsoon trough, mid-latitude fronts that reach the tropics, and Tropical Upper Tropospheric Troughs.

ITCZ

The **Intertropical Convergence Zone (ITCZ**, see General Circulation chapter) is the region where the northeasterly trade winds from the Northern Hemisphere meet the southeasterly trade winds from the Southern Hemisphere. This convergence zone shifts between about 10°N during August and September (end of N. Hemisphere summer, Fig. 16.11), and 10°S during February and March (end of S. Hemisphere summer). During these months of maximum asymmetry in global heating, the ITCZ is far enough from the equator for sufficient Coriolis force to enable tropical cyclones.

African Easterly Wave

Although equatorial Africa is hot, the Sahara Desert (roughly spanning 15° to 30°N) and Arabian Peninsula are even hotter. This temperature excess is within an atmospheric boundary layer that can reach 5 km depth by late afternoon, although average depth over 24 h is about 3 km. Thus, along the southern edge of the Sahara and the Arabian Peninsula is a strong temperature gradient (Fig. 16.12).

As explained in the thermal-wind section of the General Circulation chapter, a north-south temperature gradient creates a north-south pressure gradient that increases with altitude (up to the top of the boundary layer for this scenario). For Africa, higher pressure aloft is over the Sahara Desert and lower pressure aloft is further south over the Sudan and Congo.

This pressure gradient drives a geostrophic wind from the east over the African Sahel (at about 15°N). These winds reach maximum speeds of 10 to 25 m s^{-1} at an altitude of about 3 km in late summer. This is called the **African Easterly Jet (AEJ)**.

Like the subtropical and mid-latitude jets, the AEJ is barotropically and baroclinically unstable. This means that the jet tends to meander north and south while blowing from east to west. The mountains in eastern Africa (Ethiopia Highlands and Darfur Mountains) can trigger such oscillations.

The oscillations in this low-altitude jet are called waves, and are known as **Easterly Waves** because the wave troughs and crests move from the east at speeds of 7 to 8 m s^{-1}. These waves have wavelength of about 2,000 to 2,500 km, and have a wave period of 3 to 4 days. Although these waves are created over Africa, they continue to propagate west across the tropical North Atlantic, the Caribbean, central America, and finally into the tropical Northeast Pacific (Fig. 16.13).

Easterly waves can trigger tropical cyclones because of the low-altitude convergence that occurs east of the wave troughs. These waves are found predominantly in a latitude band between 10°N and 30°N, where there is lower pressure to the south and the subtropical high-pressure belt to the north. Thus, a trough of low pressure in an easterly wave corresponds to a region where the isobars meander to the north (Fig. 16.14) in the N. Hemisphere.

Because the pressure gradient is weak and sometimes difficult to analyze from sparsely spaced weather stations in this region, tropical meteorologists will often look at the streamlines instead of isobars. Streamlines are weather-map lines that are everywhere parallel to the flow at any instant in time

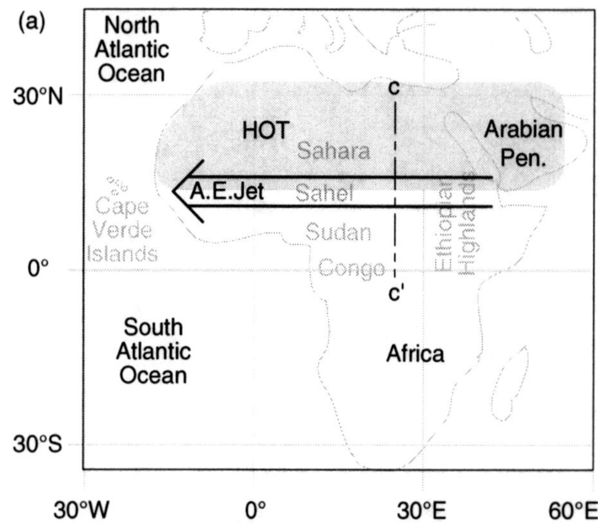

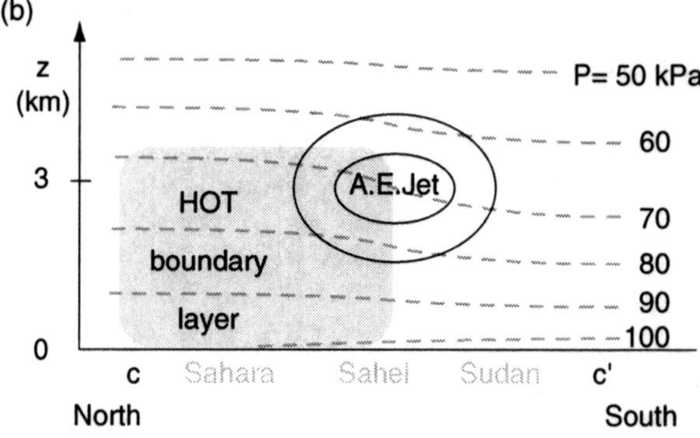

Figure 16.12
African Easterly (A.E.) Jet. (a) Map of Africa. (b) Vertical cross section along c - c' in (a), where thin black ovals represent isotachs around the core of the AEJ. Grey area indicates hot air. Note the low altitude of this A.E. jet.

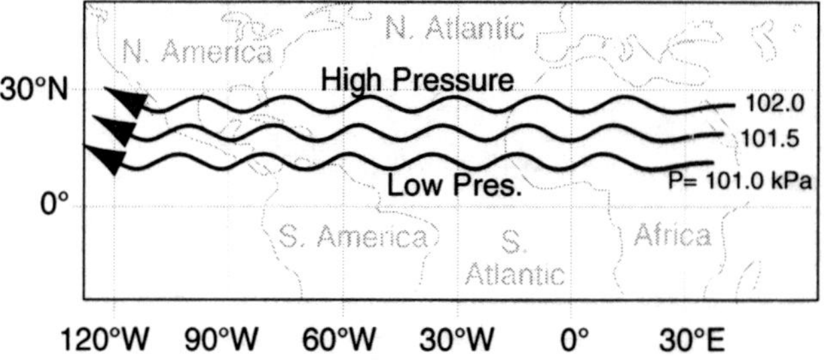

Figure 16.13
Easterly waves near the bottom of the troposphere. Wavelength is roughly 20° to 25° of longitude (≈ 2000 to 2500 km). Isobars of sea-level pressure are solid lines with arrows. Wave troughs propagate toward the west.

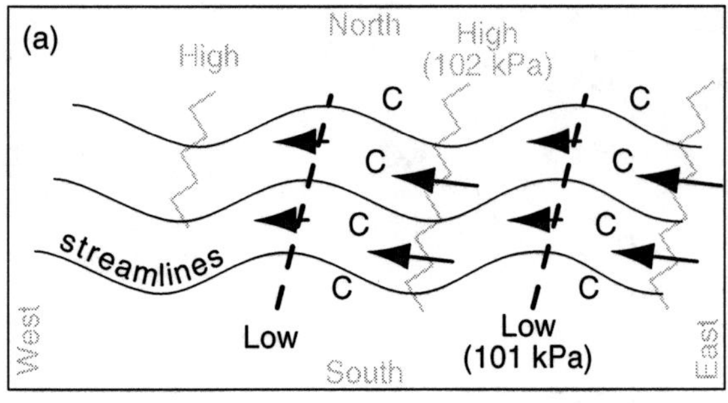

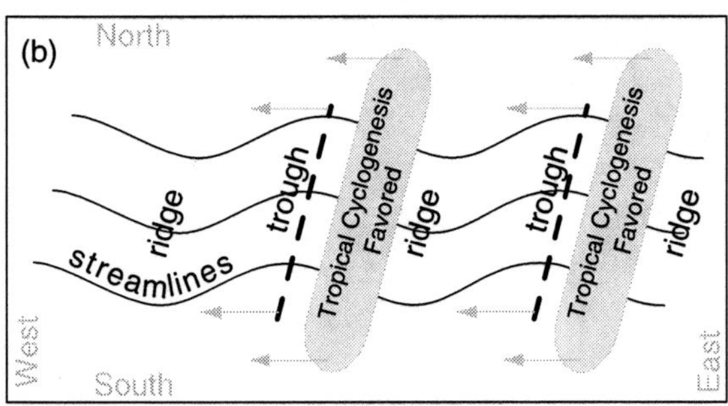

Figure 16.14
Characteristics of easterly waves in lower tropospheric flow over the tropical North Atlantic. (a) Black arrows indicate winds; longer arrows denote faster winds. C = convergence (b) Grey arrows show how the easterly waves and the tropical cyclogenesis regions propagate toward the west.

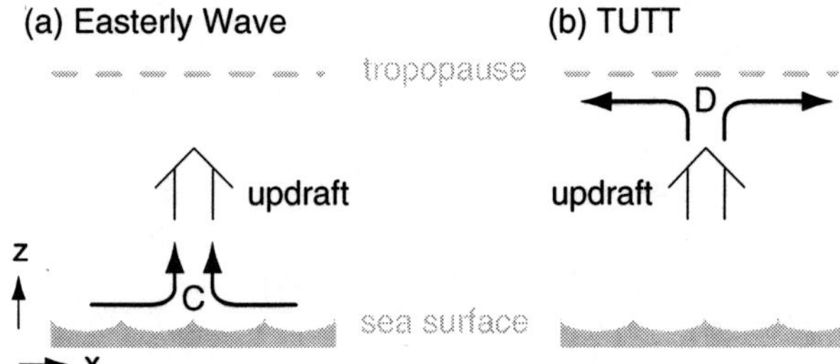

Figure 16.15
Mechanisms that create updrafts to trigger thunderstorm clusters: (a) C = lower-tropospheric horizontal convergence in an easterly wave; (b) D = horizontal divergence aloft in a TUTT.

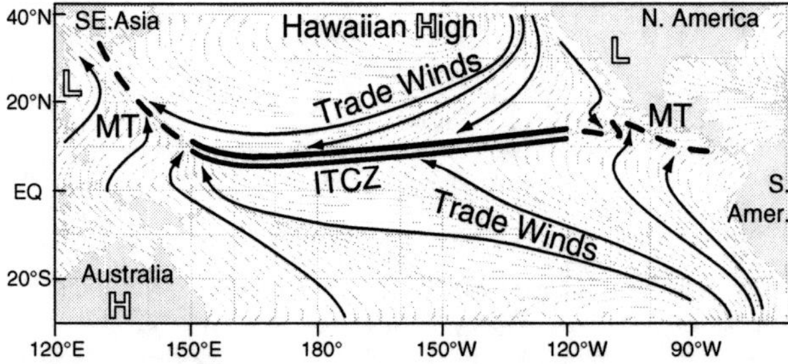

Figure 16.16
Streamline analysis (thin grey & thin black lines with arrows) of near-surface winds, averaged during August. MT denotes Monsoon Trough, shown by the thick dashed lines. 𝕃 *and* ℍ *are monsoon lows and highs. [modified from Naval Research Lab. image]*

(see the Regional Winds chapter). They more-or-less follow the isobars. Thus, regions where streamlines for low-altitude winds meander to the north can also be used to locate troughs in the easterly waves.

Recall that mid-latitude cyclones are favored in the updrafts under divergence regions in the polar jet (centered at the top of the troposphere). However, for tropical cyclogenesis we also look for updrafts above the convergence regions of the jet near the bottom of the troposphere (Fig. 16.15a).

Where are these convergence regions in easterly waves? According to the Atmospheric Forces & Winds chapter, winds are slower-than-geostrophic around troughs, and faster around ridges. This rule continues to hold in the tropics and subtropics. Thus, the convergence zone with fast inflow and slow outflow is east of the trough axis (Fig. 16.14a).

Also, since the trough axis often tilts eastward with increasing altitude, it means that a mid-tropospheric trough is often over a lower-troposphere convergence regions. This mid-troposphere trough has the cyclonic vorticity that encourages organization of thunderstorms into incipient tropical cyclones.

It is these regions east of troughs in easterly waves where the thunderstorms of incipient tropical cyclones can be triggered (Fig. 16.14b). As the easterly waves propagate from east to west, so move the convergence regions favoring tropical cyclone formation. This convergence mechanism is important for tropical cyclone triggering in both the North Atlantic and Northeast Pacific. Roughly 85% of the intense Atlantic hurricanes are triggered by easterly waves.

Monsoon Trough

A **monsoon trough** (MT) is where the ITCZ merges into a monsoon circulation. For example, Fig. 16.16 shows a monsoon low pressure (𝕃) region over southeast Asia (left side of Fig.), with winds rotating counterclockwise around it. Near the dashed line labeled MT in the western Pacific, monsoon winds from the southwest converge with easterly trade winds from the Hawaiian (Pacific) High.

Thus, the monsoon trough in the western Pacific is a convergence region that can trigger thunderstorms for incipient typhoons. Similar monsoon troughs have been observed in the Indian Ocean and the Eastern Pacific. In Fig. 16.16, the region where the monsoon trough in the western Pacific meets the ITCZ has enhanced convergence, and is a more effective trigger for tropical cyclones.

Mid-latitude Frontal Boundary

Sometimes cold fronts from midlatitudes can move sufficiently far equatorward to reach the subtropics. Examples include cold fronts reaching southern Florida (Fig. 16.17). Although these fronts are often weak and slow moving by the time the reach the subtropics, they nonetheless have convergence across them that can trigger tropical cyclone thunderstorms in the Caribbean and the subtropical western North Atlantic.

TUTT

TUTT stands for **Tropical Upper Tropospheric Trough**. This is a high-altitude cold-core low-pressure system that can form in the subtropics. East of the TUTT axis is a region of horizontal divergence aloft, which creates upward motion below it in order to satisfy air mass continuity (Fig. 16.15b). This upward motion can help trigger thunderstorms as precursors to tropical cyclones.

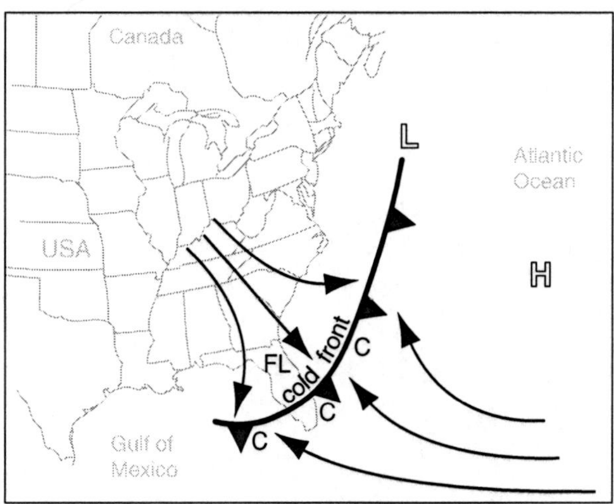

Figure 16.17
Lower tropospheric convergence (C) along a cold front over Florida (FL), USA.

Life Cycle

At locations where all of the necessary conditions are met (including any one trigger), the incipient tropical cyclones usually progress through the following intensification stages: tropical disturbance, tropical depression, tropical storm, tropical cyclone.

Tropical Disturbance

The US National Hurricane Center defines a **tropical disturbance** as "a discrete tropical weather system of apparently organized convection - generally 200 to 600 km in diameter — originating in the tropics or subtropics, having a nonfrontal migratory character, and maintaining its identity for 24 hours or more." Namely, it is a cluster of thunderstorms that stay together as they move across the ocean.

This thunderstorm cluster is visible by satellite as consisting distinct thunderstorms with their own anvils and separate precipitation regions. There is no eye, and little or no rotation visible. Some tropical disturbances form out of **Mesoscale Convective Systems** (**MCSs**), particularly those that develop a **Mesoscale Convective Vortex** (**MCV**, see the Thunderstorm chapter). Some are identified as the thunderstorm clusters that move with easterly waves.

Most tropical disturbances do not evolve into tropical cyclones. Hence, tropical disturbances are not usually named or numbered. However, tropical meteorologists watch them carefully as they evolve, as potential future tropical cyclones.

As condensation and precipitation continue in this thunderstorm cluster, more and more latent heat is released in those storms. Namely, the mid-tropospheric air near this cluster becomes warmer than the surrounding ambient air. According to the hypsometric relationship, pressure decreases more slowly with increasing altitude in this warm region. Thus, a synoptic-scale high-pressure starts to form near the top of the troposphere in the region of this cluster.

Tropical Depression (TD)

The high-pressure aloft begins to create a thermal circulation (see General Circulation chapter), where air aloft is driven horizontally outward — down the pressure gradient toward the lower pressure outside of the cluster. This diverging air aloft also starts to rotate anticyclonically.

The outward moving air aloft removes air molecules from the thunderstorm cluster region of the troposphere, thereby lowering the surface pressure under the cluster. The "depressed" pressure at sea level is why this stage gets its name: **tropical depression**. Tropical depressions are given an identification number, starting with number 1 each year.

As this weak surface low forms, it creates a pressure gradient that starts to suck winds horizontally toward the center of the low. This inflowing air begins rotating cyclonically due to Coriolis force. The surface low pressure is usually too weak to measure directly. However, this tropical depression stage is defined by measurable near-surface winds of 17 m s^{-1} or less (Table 16-4) turning cyclonically as a closed circulation around the cluster. There is still no eye at this stage, however the thunderstorms are becoming aligned in spiral rain bands (Fig. 16.18). Most tropical depressions do not strengthen further.

However, for the few storms that do continue to strengthen, the near-surface winds are boundary-layer gradient winds that spiral in towards the center of the cluster. This radial inflow draws in more warm humid boundary-layer air, thereby refueling

Figure 16.18
Tropical Depression 10 (in center of photo), over the tropical Atlantic Ocean south-southeast of the Cape Verde Islands. Africa is visible in the right quarter of the photo. (Visible satellite image taken 11:45 UTC on 2 Sep 2008, courtesy of NOAA.)

Table 16-4. Stages leading to tropical cyclones. M_{max} = near-surface wind-speed maximum. n/a = not applicable.

Stage	M_{max}		
	(m s^{-1})	(km h^{-1})	(knots)
Tropical Disturbance	n/a	n/a	n/a
Tropical Depression	< 17	< 61	≤ 33
Tropical Storm	17 - 32	61 - 118	33 - 63
Tropical Cycl.(Hurricane)	≥ 33	≥ 119	≥ 64

the thunderstorms and allowing them to persist and strengthen. Condensation and precipitation increase, causing a corresponding increase in latent-heat release.

This warm core of the storm further strengthens the high pressure aloft via the hypsometric relationship, which drives more outflow aloft and removes more air molecules from the troposphere near the cluster. Thus, the surface low pressure can continue to deepen in spite of the boundary-layer inflow. Namely, the high-pressure aloft creates an exhaust system, while the boundary layer is the intake system to the incipient cyclone.

Tropical Storm (TS)

When the surface low is deep enough to drive winds faster than 17 m s^{-1} (but less than 33 m s^{-1}) in a closed cyclone circulation, then the system is classified as a **tropical storm**. On weather maps, it is indicated with symbol 6.

At this point, the thunderstorm rain bands have nearly completely wrapped into a circle — the future eyewall (Fig. 16.19d). The anvils from the thunderstorms have usually merged in to a somewhat circular **central dense overcast (CDO)**, which is clearly visible as a large-diameter cold high cloud. There is no eye at this stage. Tropical storms are organized sufficiently to be able to modify their local environment to allow them to persist, without relying so much on a favorable ambient environment.

At this stage the storm is given a name for identification (and its former tropical-depression number is dropped). The same name is used if the storm further strengthens into a tropical cyclone, and for the rest of its life cycle.

Hurricanes (Atlantic and Eastern Pacific) and Cyclones (Indian Ocean and Southwestern Pacific near Australia) are usually given names of men or wom-

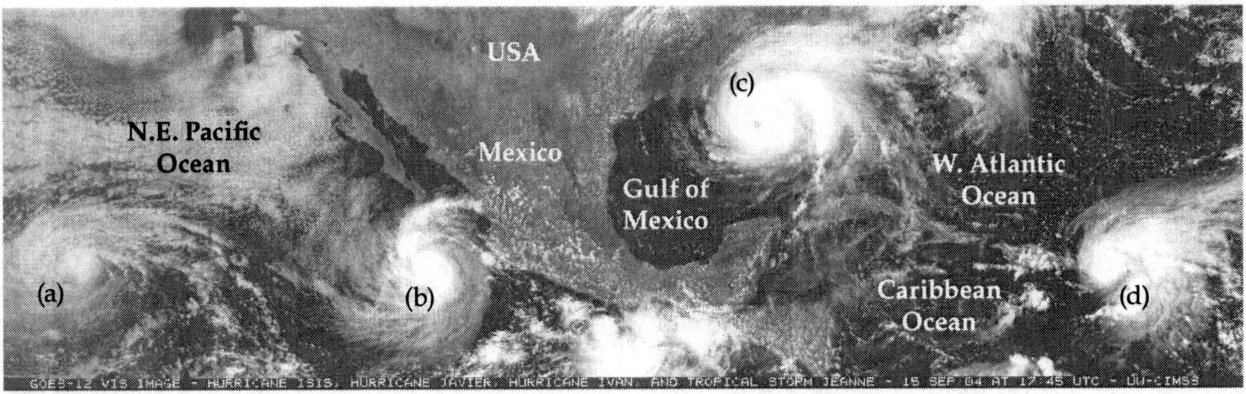

Figure 16.19
Visible satellite image at 17:45 UTC on 15 Sep 2004, showing four tropical systems. From left to right: (a) Hurricane Isis, (b) Hurricane Javier, (c) Hurricane Ivan, (d) Tropical Storm Jeanne. This image covers the eastern N. Pacific Ocean, Central America, Gulf of Mexico, Caribbean Ocean, and western N. Atlantic. (Image courtesy of University of Wisconsin–Madison CIMSS/ SSEC.)

en. These are assigned in alphabetical order according to lists that have been set in advance by the **World Meteorological Organization (WMO)**. Typhoons are given names of flowers, animals, birds, trees, foods (not in alphabetical order but in an order set by the WMO's Typhoon Committee).

Tropical-storm and tropical-cyclone names are re-used in a six-year cycle. However the names of the strongest, most destructive tropical cyclones are "retired" and never used again (they are replaced with other names starting with the same letter).

Tropical Cyclone (Hurricane, Typhoon)

Roughly half of the tropical storms in the Atlantic continue to strengthen into tropical cyclones. By this stage there is a well defined eye surrounded by an eyewall of thunderstorms. Max wind speeds are 33 m s^{-1} or greater in a closed circulation around the eye. The central dense overcast usually has a cloud-free eye. The weather map symbol for a tropical cyclone is 𝟗.

There is heavy precipitation from the eyewall thunderstorms, and the core of the storm is significantly warmer than the environment outside the storm. Sea-level pressure decrease in the eye is measurable and significant, and a rise of sea level into a storm surge is possible. The storm organization allows it to persist for days to weeks as it moves across ocean basins (Fig. 16.19a-c).

Movement/Track

Steering by the General Circulation

Tropical cyclones are moved by the larger-scale winds near them. These winds are variable, causing a wide variety of hurricane tracks and translation speeds. However, by focusing on the climatological average trade winds and the monsoon circulations (Figs. 16.7 & 16.20), we can then anticipate average movement of tropical cyclones.

During late Summer and early Fall, monsoon high-pressure regions form over the oceans, as was discussed in the General Circulation chapter. Winds rotate (clockwise; counterclockwise) around these highs in the (Northern; Southern) Hemisphere. Equatorward of the high center the monsoon winds are in the same direction as the trade winds — from east to west. Thus, average tropical cyclone tracks in the eastern and central longitudes of an ocean basin are usually zonal — from east to west.

However, near the west sides of the ocean basins, the monsoon winds turn poleward (Fig. 16.20), and thus steer the tropical cyclones away from the equator. Many of these tropical cyclones travel along the east coast of continents in a poleward direction. The tracks vary from year to year, and vary within any one year (Fig. 16.20). Some of the tracks in the Atlantic might never hit N. America. Other tracks can be along the East Coast of North America, causing significant damage to coastal cities. Yet other tracks continue further westward across the Caribbean Sea and into the Gulf of Mexico before reaching the mainland.

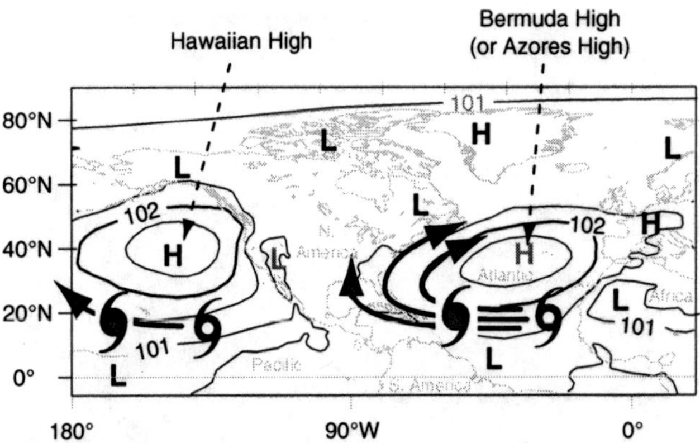

Figure 16.20
Monsoon sea-level pressure (kPa) pattern for July (copied from the General Circulation chapter), with isobars as thin black lines. Winds and average tropical cyclone tracks (dark arrows) tend to follow the isobars around the monsoon highs, although actual tracks are quite variable.

Extratropical Transition

As the monsoon circulations move the tropical cyclones poleward, the cyclones leave the tropics and subtropics and enter the mid-latitudes. Here, prevailing winds are from the west in the global circulation. Also, winds on the poleward side of the monsoon highs are from the west. Thus, many tropical cyclone tracks eventually turn towards the east at midlatitudes.

Extratropical transition is the name given to the movement of tropical cyclones out of the tropics and into the midlatitudes. At these latitudes the storms are generally moving over colder water, causing them to rapidly die. Nonetheless, these former tropical cyclones can change into extratropical cyclones and still cause damaging floods, wind, and waves if over coastal areas. Even dying tropical cyclones well away from a coastline can influence mid-latitude weather downwind by pumping copious amounts of moisture into the air. This humid air can later serve as the fuel for mid-latitude cyclones thousands of kilometers downwind, creating heavy rain and possible flooding.

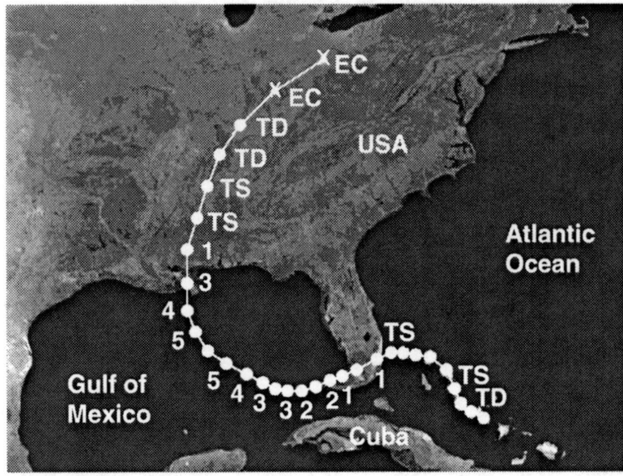

Figure 16.21
Track of hurricane Katrina, showing its intensity during 23 - 31 Aug 2005. 1 to 5 is Saffir-Simpson category. TS = tropical storm. TD = tropical depression. EC = extratropical cyclone.

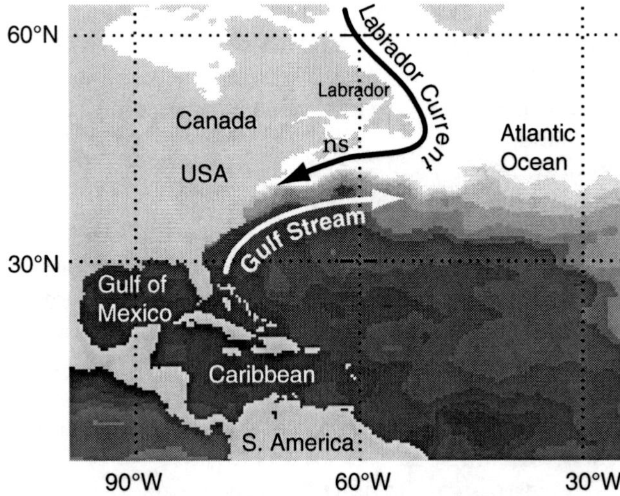

Figure 16.22
Enlargement of part of Fig. 16.8, showing sea-surface temperatures (warm = darker grey fill; cold = white fill). The Labrador Current and Gulf Stream are shown. ns = Nova Scotia.

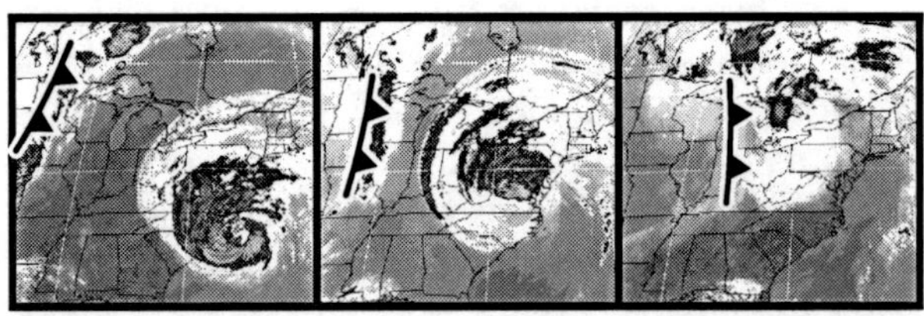

Figure 16.23
Collision of an extratropical cold front with former Hurricane Isabel in Sep 2003. (a) A cold front is over Minnesota, USA, while the hurricane eye makes landfall on the N. Carolina coast. (b) Later. (c) The front and hurricane collide over the Great Lakes. (IR GOES satellite images courtesy of NOAA.)

Tropical Cyclolysis

Tropical cyclones can exist for weeks because they have the ability to create their own fuel supply of warm humid boundary-layer air. They create this fuel by tapping the heat and water stored in the upper ocean. Tropical cyclones die (**tropical cyclolysis**) when they move to a location where they cannot create their own fuel, or if they are destroyed by larger-scale weather systems, as described next.

Movement over Land

If the global circulation steers a tropical cyclone over land, it begins to die rapidly because it is unable to tap the warm ocean for more fuel (Fig. 16.21). If the land is a small island or narrow isthmus, then the weakened tropical cyclone can possibly re-intensify when it again moves over warm ocean water.

Movement over Cold Water

If the global circulation steers a tropical cyclone over water colder than 26.5°C, then it is unable to create new warm humid air as fast the eyewall thunderstorms consume this fuel. As a result, tropical cyclones quickly weaken over cold water.

This often happens when tropical cyclones are steered poleward toward mid-latitudes, where sea surface temperatures (**SST**) are colder (Fig. 16.8). In the Atlantic, the Gulf Stream of warm ocean water moving north along the east coast of the USA allows tropical cyclones there to survive. However, further north, the very cold Labrador Current (Fig. 16.22) moving south from the Arctic along the east coast of Canada causes most tropical cyclones to diminish below tropical cyclone force before making landfall (see INFO Box on next page).

Tropical cyclolysis can also occur if the layer of warm ocean water is too shallow, such that tropical cyclone-induced ocean waves stir colder deep water to the surface. Also, if a tropical cyclone crosses the path of a previous tropical cyclone, the ocean surface along this previous path is often colder due to the wave stirring.

Interaction with Mid-latitude Lows

When tropical cyclones move into mid-latitudes, they can encounter the larger, more powerful mid-latitude cyclones (Fig. 16.23) that were discussed in a previous chapter. Strong mid-latitude cold fronts can inject cold air into the tropical cyclone, causing the tropical cyclone to die. The fronts and associated jet stream aloft often have strong wind shear, and can rip apart the tropical cyclones. Sometimes tropical cyclones will move into the path of a mid-latitude cyclone, allowing the mid-latitude cyclone to capture some of the tropical cyclone's energy and consume the tropical cyclone as the two systems merge.

DYNAMICS

Origin of Initial Rotation

Define **absolute angular momentum** (AAM) as the sum of a relative component of angular momentum associated with tropical-cyclone rotation plus a background component due to the rotation of the Earth:

$$AAM = M_{\tan} \cdot R + 0.5 \cdot f_c \cdot R^2 \qquad \bullet(16.1)$$

where the Coriolis parameter is f_c (≈ 0.00005 s^{-1} at 20° latitude), and the tangential component of velocity at distance R from the eye center is M_{tan}.

As near-surface air converges toward the weak low-pressure of an incipient tropical cyclone, absolute angular momentum is conserved for frictionless flow (i.e., for no drag against the sea surface). Eq. (16.1) indicates that there is nonzero AAM even if the incipient tropical cyclone has no rotation yet (i.e., even if $M_{tan.initial}$ = 0). Equating the resulting initial AAM for air starting at distance R_{init} from the eye center, with the final AAM after the air has moved closer to the eye (i.e., smaller R_{final}) yields:

$$M_{\tan} = \frac{f_c}{2} \cdot \left(\frac{R_{init}^2 - R_{final}^2}{R_{final}} \right) \qquad (16.2)$$

For real tropical cyclones you should not neglect frictional drag. Thus, you should anticipate that actual winds will be slower than given by eq. (16.2).

Subsequent Spin-up

As the winds accelerate around a strengthening tropical cyclone, centrifugal force increases. Recall from the Atmospheric Forces & Winds chapter that the equation for **gradient-wind** in cylindrical coordinates (eq. 10.32) is:

$$\frac{1}{\rho} \cdot \frac{\Delta P}{\Delta R} = f_c \cdot M_{\tan} + \frac{M_{\tan}^2}{R} \qquad (16.3)$$

where air density is ρ and the pressure gradient in the radial direction is $\Delta P / \Delta R$. In eq. (16.3) the last term gives the centrifugal force. The gradient wind applies at all radii from the center of the storm, and at all altitudes except near the bottom (in the boundary layer) and near the top (in the anvil or cirrus shield region).

Some researchers find it convenient to neglect Coriolis force closer to the center of the stronger tropical cyclones where the winds are faster. For

---

INFO • Hurricane Juan Hits Canada

The Canadian East Coast only rarely gets hurricanes — one every 3 to 4 years on average. But it gets many former hurricanes that have weakened to Tropical Storm or lower categories during passage over the cold Labrador current (for Atlantic Canada), or due to passage over land (for Central Canada).

On 29 Sep 2003, Hurricane Juan hit Nova Scotia, Canada, as a category 2 hurricane. Two factors made this possible: (1) warm (but colder than 26.5°C) sea-surface water was relatively close to Nova Scotia; and (2) the hurricane was translating north so quickly that it did not have time to spin down over the narrow cold-water region before making landfall.

This storm began as Tropical Depression #15 about 470 km southeast of Bermuda at noon on 25 Sep. Six hours later it became a Tropical Storm, and by noon 26 Sep it became a Hurricane while 255 km east of Bermuda. During the next 3 days it traveled northward over the Gulf Stream, with intensity of category 2 for most of its journey.

When it reached Canada, it dropped 25 to 40 mm of rain in Nova Scotia. The storm surge was 1.5 m, and the maximum wave height was roughly 20 m. Hundreds of thousands of people lost electrical power due to trees falling on power lines, and 8 people were killed in storm-related accidents.

Before Juan, the most remembered "former hurricane" to strike Canada was Hurricane Hazel. It transitioned into a strong extratropical cyclone before reaching Canada, but caused 81 deaths and severe flooding in southern Ontario in mid October 1954.

Since 1887, there were two hurricanes of category 3, five of category 2, and 26 of category 1. Info about Canadian hurricanes is available from the **Canadian Hurricane Centre**, Halifax, Nova Scotia.

Sample Application

Suppose that air at a latitude of 12° initially has no rotation as it is drawn toward a low-pressure center 500 km away. When the air reaches 200 km from the low center, what will be its relative tangential velocity?

Find the Answer

Given: R_{init} = 500 km, ϕ = 12°, R_{final} = 200 km
Find: M_{tan} = ? m s^{-1}

Compute the Coriolis parameter:
$f_c = 2 \cdot \omega \cdot \sin(\phi) = (1.458\text{x}10^{-4}\ \text{s}^{-1}) \cdot \sin(12°)$
$= 0.0000303\ \text{s}^{-1}$

Use eq. (16.2): M_{tan} =

$$\frac{(0.0000303\text{s}^{-1})}{2} \cdot \left(\frac{(500\text{km})^2 - (200\text{km})^2}{200\text{km}} \right) = \underline{15.9\ \text{m s}^{-1}}$$

Check: Units OK. Physics OK.
Exposition: This is halfway toward the 32 m s^{-1} that defines a Category 1 tropical cyclone. Tropical cyclones cannot exist at the equator because f_c = 0 there.

Sample Application

Given a tropical cyclone with: sea-level pressure difference between the eye and far outside of ΔP_{max} = 11.3 kPa, eyewall outside radius of R_o = 25 km, latitude = 14°N, and ρ = 1 kg m^{-3}. From the tropical cyclone model presented later in this chapter, assume the pressure gradient is roughly: $\Delta P/\Delta R = (4/5)\cdot\Delta P_{max}/R_o$.
a) Find the magnitude of the gradient wind just outside of the eye wall (i.e., at $R = R_o$).
b) Compare the relative importance of the Coriolis and centrifugal terms in the gradient wind eq.
c) Compare this gradient wind with the max expected surface wind of 67 m s^{-1}, and explain any difference.

Find the Answer

Given: ΔP_{max} = 11.3 kPa, R_o = 25 km, latitude = 14°N, ρ = 1 kg m^{-3}.
Find: M_{tan} = ? m s^{-1}. Compare terms.

a) Use: $\Delta P/\Delta R = (4/5)\cdot\Delta P_{max}/R_o$
Find f_c = (1.458x10^{-4} s^{-1})·sin(14°) = 3.5x10^{-5} s^{-1}.
Use eq. (16.3) and neglect translation speed:

$$\frac{(4/5)}{\rho}\cdot\frac{\Delta P_{max}}{R_o} = f_c \cdot M_{tan} + \frac{M_{tan}^2}{R}$$

Pressure Gradient = Coriolis + Centrifugal

Using trial and error (i.e., trying different values of M_{tan} in a spreadsheet until the left and right sides of the eq balance): M_{tan} = **<u>94.65 m s^{-1}</u>**.

b) From my spreadsheet, the terms in the eq. are:

0.361 = 0.003 + 0.358

Pressure Gradient = Coriolis + Centrifugal

Thus, the **<u>Coriolis force is about 2 orders of magnitude smaller than centrifugal force</u>**, at this location of strongest tangential wind just outside the eye wall. At this location, the cyclostrophic assumption (of neglecting Coriolis force) is OK.

c) The max expected <u>surface</u> wind of 67 m s^{-1} includes the effect of **<u>drag against the sea surface</u>**, which is why it is smaller than our answer from (a) M_{tan} = 95 m s^{-1}.

Check: Units OK. Physics OK.
Exposition: This scenario is similar to the composite tropical cyclone model presented later in this chapter.

Sample Application

Suppose a tropical cyclone has a pressure gradient of 1 kPa/15 km at radius 60 km. What is the value of the cyclostrophic wind? Given ρ = 1 kg m^{-3}.

Find the Answer

Given: R = 60 km, ΔP = 1 kPa, ΔR = 15 km,
Find: M_{cs} = ? m s^{-1}

Use eq. (16.4): $M_{cs} = \sqrt{\frac{(60\text{km})}{(1\text{kg}\cdot\text{m}^{-3})}\cdot\frac{(1000\text{Pa})}{(15\text{km})}}$ = **<u>63 m s^{-1}</u>**

Check: Physics & Units OK.
Exposition: This is a Category 4 tropical cyclone.

this situation, the tangential tropical cyclone winds can be crudely approximated by the **cyclostrophic wind**, M_{cs} (see the Atmospheric Forces & Winds Chapter):

$$M_{cs} = M_{tan} = \sqrt{\frac{R}{\rho}\cdot\frac{\Delta P}{\Delta R}} \qquad (16.4)$$

Nonetheless, the gradient-wind equation (16.3) is the most appropriate equation to use for tropical cyclones, at <u>middle</u> altitudes of the storm.

Inflow and Outflow

At the <u>bottom</u> of the tropical cyclone (Fig. 16.24), drag against the sea surface causes the winds to spiral in towards the eyewall. The **boundary-layer gradient wind** equation (see the Atmospheric Forces & Winds chapter) describes this flow well. Without this drag-related inflow, the eyewall thunderstorms would not get sufficient warm humid air to survive, causing the tropical cyclone to die.

Explaining the outflow at the <u>top</u> of a tropical cyclone is trickier, because drag forces are so small that we cannot invoke the boundary-layer gradient-wind equation. One important process is that the thunderstorm updrafts in the eyewall rapidly move air upward and deposit enough air molecules at the

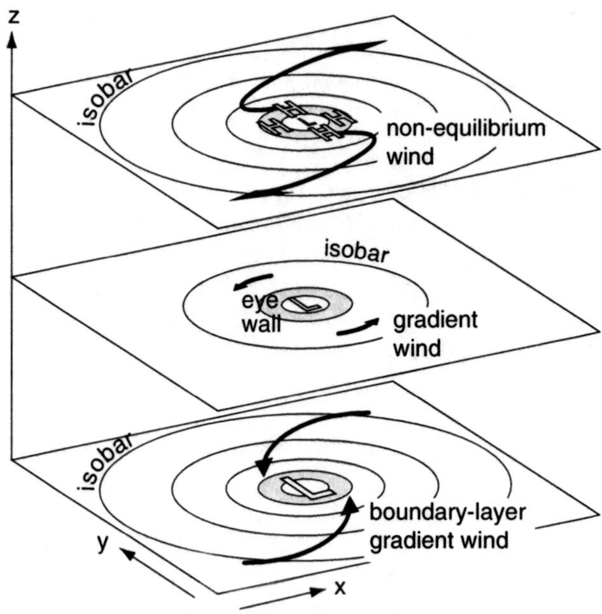

Figure 16.24
*Wind dynamics at the bottom, middle and top of a tropical cyclone. Thick arrows represent wind vectors. Low pressure **L** at storm bottom is associated with spiraling inflow winds. High pressure **H** at storm top is associated with outflow winds. Thus, storm top is the exhaust system of the tropical cyclone engine, and storm bottom is the intake system.*

top of the storm to contribute to high pressure there. Thus, the outflow is related to two factors.

(1) The cyclonically moving air from the boundary layer is brought to the top of the tropical cyclone by eyewall-thunderstorm updrafts so quickly that its inertia prevents it from instantly changing to anticyclonic outflow. Namely, the outflow is initially moving the wrong way (cyclonically) around the high (Fig. 16.24). The outflow must change direction and accelerate, and thus is not in steady state.

(2) It is physically impossible to create a balanced gradient-wind flow around a high-pressure area that is surrounded by an excessively strong pressure gradient, as was explained in the Atmospheric Forces & Winds chapter (see HIGHER MATH in the next column). But the thunderstorm updrafts help create such an excessive high pressure at storm top. In this situation, pressure-gradient force exceeds the compensating Coriolis force (Fig. 16.25), resulting in a net outward force that causes the air to accelerate outward from the eyewall as a non-equilibrium wind.

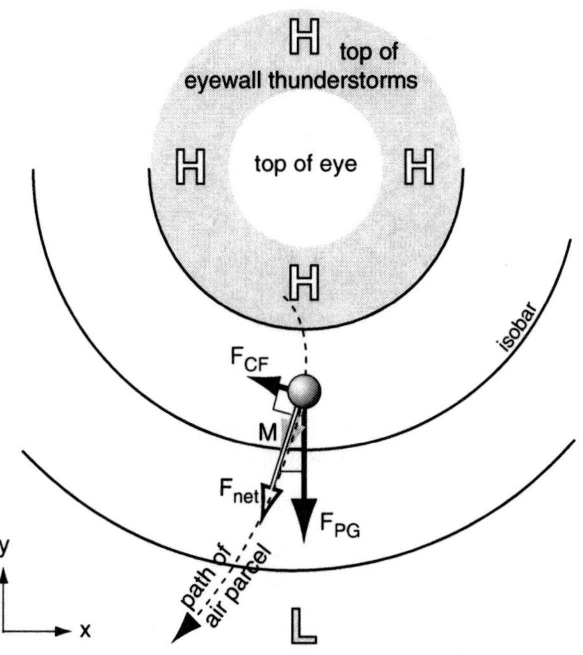

Figure 16.25
Isobars on a constant height surface near the top of a tropical cyclone. Grey ring represents the eyewall thunderstorms. H and L are high and low pressure. M is non-equilibrium wind speed. F_{CF} is Coriolis force, F_{PG} is pressure-gradient force, and F_{net} is the vector sum of forces.

HIGHER MATH • Non-equilibrium winds at the top of tropical cyclones

Recall from the Atmospheric Forces & Winds chapter that solutions for the gradient wind around an anticyclone are physically realistic only for curvature Rossby numbers $Ro_c \le 1/4$. But $Ro_c = (\partial P/\partial R)/(\rho \cdot f_c^2 \cdot R)$. Setting $Ro_c = 1/4$, separating variables, and integrating from $P = P_o$ at $R = 0$, to P at R gives:

$$P = P_o - a \cdot R^2 \qquad \text{(a)}$$

where $a = \rho \cdot f_c^2/8$. This equation gives the max change of pressure with radial distance R that is allowed for a gradient wind that is in equilibrium. But real tropical cyclones can have greater-than-equilibrium pressure gradients at the storm top.

For example, at tropical cyclone top ($z \approx 17$ km) suppose P_o = 8.8kPa in the eye and ρ = 0.14 kg m^{-3}. Thus, $a \approx 2.5\text{x}10^{-8}$ kPa km^{-1} at 15° latitude. The strongest horizontal pressure gradient possible for a gradient wind around the high-pressure center at the top of the eye is plotted below as the slope of the solid line (from equation a). This is a miniscule pressure gradient, due in part to the very small Coriolis force at 15° latitude.

But actual horizontal pressure gradients can be much stronger (dashed line). Thus, steady-state gradient winds (winds that follow the curved isobars) are not possible at tropical-cyclone top. Instead, the non-equilibrium winds are accelerating outward (crossing this isobars at a large angle) from the high due to the strong pressure gradient, and only gradually develop some anticyclone rotation further from the eye.

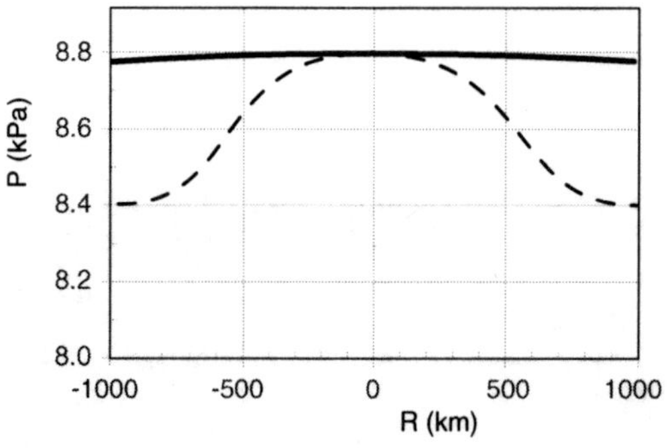

Fig. 16.b. *Horizontal pressure gradients at the top of a tropical cyclone. R = distance from cyclone center, P = atmospheric pressure. The slope of the solid line shows the max pressure gradient that admits a gradient-wind solution. The dashed line is hypothetical and focuses on the pressure gradient from the eyewall outward.*

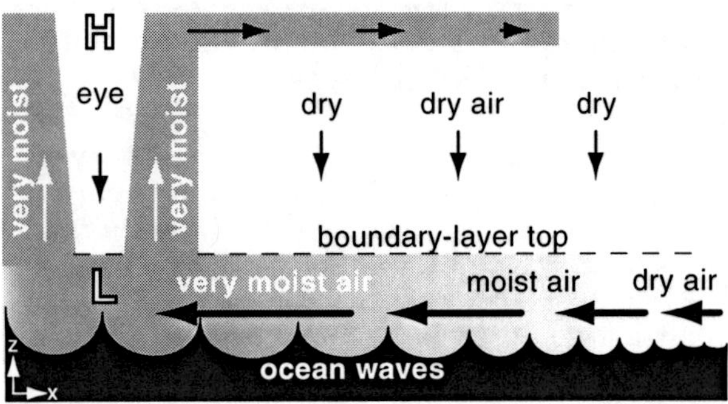

Figure 16.26
The intake system in the bottom of a tropical cyclone. As boundary-layer winds accelerate, they cause larger ocean waves that add more moisture (darker grey indicates higher humidity) to the inflowing air (thick arrows). **L** *is the low-pressure center at sea level. Air is exhausted out of the top of the storm in the central dense overcast composed of thunderstorm anvils.*

> **INFO• Tropical Cyclone Condensation Energy**
>
> One way to estimate the energy of tropical cyclones is by the amount of latent heat released during condensation. Rainfall rate is a measure of the net condensation.
>
> An average Atlantic Hurricane produces about 1.5 cm of rain day^{-1}, which when accumulated over the area (of radius 665 km) covered by hurricane rain is equivalent to **2.1×10^{10}** m^3 day^{-1}.
>
> This converts to **6×10^{14}** W (= 5.2×10^{19} J day^{-1}). Namely, it is about 200 times the daily electricity generation capacity of the whole world.

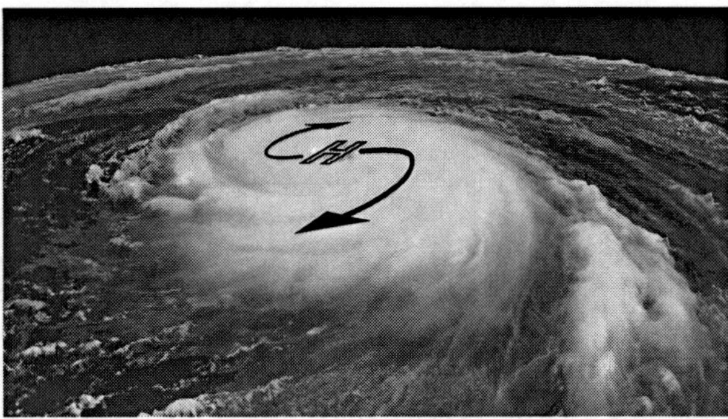

Figure 16.27
Exhaust system at the top of a tropical cyclone, where **H** *is the high-pressure center near the tropopause and arrows are outflow winds. Computer-enhanced 3-D satellite image of Hurricane Katrina over the Gulf of Mexico, taken 28 Aug 2005 at 1545 UTC. Visible satellite imagery is mapped into a 3-D surface of cloud-top altitudes estimated from satellite IR data. (Image courtesy of US DOC/NOAA.)*

THERMODYNAMICS

Tropical cyclones work somewhat like engines. There is an intake system (the atmospheric boundary layer) that draws in the fuel (warm, humid air). The engine (thunderstorms) converts heat into mechanical energy (winds and waves). And there is an exhaust system (precipitation fallout for water and anvil blowout for air) for the spent fuel.

Fuel Creation and Intake

In midlatitudes, thunderstorms can last for hours. However, tropical cyclones (which are made of thunderstorms) can last for weeks. Why the difference it longevity?

The main reason is that mid-latitude thunderstorms use warm humid air from only the nearby boundary layer, and after this nearby fuel is consumed the thunderstorms die. Even supercells and squall-line storms rely on an existing fuel supply of warm humid air, which they can utilize as they move across the countryside.

However, tropical cyclones create their own fuel of warm humid boundary-layer air. They do this via the fast near-surface winds that create large ocean waves (Fig. 16.26). These violent waves break and foam, causing rapid evaporation of sea-spray water into the air, and efficient transfer of heat from the ocean surface to the air. By the time this air reaches the base of the eyewall, it has an air temperature nearly equal to the sea-surface temperature and a relative humidity of nearly 100%.

Namely, tropical cyclones extract heat from the ocean. The ocean is a giant heat reservoir that has been absorbing sunlight all Summer and early Fall. This is the reason why warm, deep sea-surface temperatures are needed for tropical cyclones.

Exhaust

The large volume of inflowing boundary-layer air is good and bad for the tropical cyclone. It is good because this air carries the sensible- and latent-heat fuel for the storm. It is bad because it also brings massive amounts of other air molecules (nitrogen, oxygen) into the core of the storm.

The problem is that if these air molecules were to accumulate in the center of the storm, their weight would cause the sea-level air pressure to rise. This would weaken the surface low pressure, which would reduce the radial pressure gradient near sea level, causing the inflow winds and waves to diminish. With reducing inflow, the eyewall thunderstorms would run out of fuel and die, causing the tropical cyclone to disintegrate in half a day or so.

However, we know that tropical cyclones can exist for weeks. Thus, there must be some other mechanism that removes air from the tropical cyclone core as fast as it enters. That other mechanism is the strong outflow winds in the thunderstorm anvils at the top of the eyewall. This outflow exists because of high pressure at the top of the troposphere in the eye and eyewall regions (Figs. 16.26 & 16.27), which drives winds outward down the pressure gradient.

But what causes this high pressure aloft? We already discussed a non-hydrostatic process — the air deposited aloft from eyewall thunderstorm updrafts. But another process is hydrostatic — related to the excessive warmth of the center or **core** portion of the tropical cyclone.

Warm Core

Air rises moist adiabatically in the eyewall thunderstorms, releasing a lot of latent heat along the way. But then after losing water due to precipitation from the eyewall, some of the air warms as it descends dry adiabatically in the eye. Thus, the **core** (both the eye and the eye wall) of the tropical cyclone has much warmer temperature than the ambient air outside the storm.

Typical temperature excesses of the **warm core** are 0 to 4°C near sea level, and 10 to 16°C warmer at 12 to 16 km altitude (Fig. 16.28). An INFO box illustrates such a warm-core system.

This radial temperature gradient (warm core vs. cool exterior) causes a radial pressure-gradient reversal with increasing altitude (Fig. 16.29), as is explained by the thermal-wind expression. To determine this pressure gradient, define the sea-level pressure in the eye as $P_{B\ eye}$, and that of the distant surroundings at sea level as $P_{B\ \infty}$. Near the top of the tropical cyclone, define the core pressure as $P_{T\ eye}$ and that of its distant surroundings as $P_{T\ \infty}$ at the same altitude. Subscripts B and T denote bottom and top of the troposphere.

Suppose a tropical cyclone is approximately $z_{max} \approx 15$ km deep, has a temperature in the eye averaged over the whole tropical cyclone depth of $\overline{T_{eye}} = 273$ K, and has ambient surface pressure distant from the storm of $P_{B\ \infty} = 101.3$ kPa. The pressure difference at the top of the tropical cyclone ($\Delta P_T = P_{T\ \infty} - P_{T\ eye}$) is approximately related to the pressure difference at the bottom ($\Delta P_B = P_{B\ \infty} - P_{B\ eye}$) by:

$$\Delta P_T \approx a \cdot \Delta P_B - b \cdot \Delta T \qquad \bullet(16.5)$$

where $a \approx 0.15$ (dimensionless), $b \approx 0.7$ kPa K^{-1}, and $\Delta T = T_{eye} - T_{\infty}$ is the temperature difference averaged over the troposphere. Note that ΔP_T is negative.

Next, consider the tangential winds. These winds spiral cyclonically around the eye at low al-

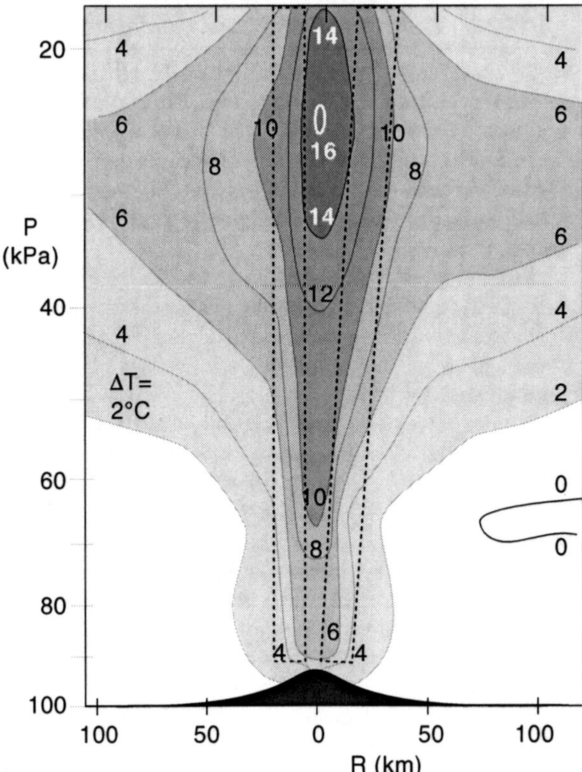

Figure 16.28
The exceptionally warm core of Hurricane Inez. Vertical cross section through the eye, showing the temperature excess ΔT compared to the environmental temperature at the same altitude. The approximate eyewall location is outlined with dotted lines. R is radial distance from the center of the eye, and P is pressure. Inez existed during 22 Sep to 11 Oct 1996 in the eastern Caribbean, and reached category 4 intensity. The black area at the bottom of the figure masks atmospheric pressures that did not exist at sea level, because of the very low surface pressure in and near the eye. (Based on data from Hawkins and Imbembo, 1976.)

Sample Application

The surface pressure in the eye of a tropical cyclone is 90 kPa, while the surrounding pressure is 100 kPa. If the core is 10 K warmer than the surroundings, find the pressure difference at the top of the tropical cyclone between the eye and surroundings.

Find the Answer

Given: $\Delta P_B = 100 - 90 = 10$ kPa, $\Delta T = 10$ K
Find: $\Delta P_T = ?$ kPa

Use eq. (16.5):
$\Delta P_T = (0.15)\cdot(10\text{kPa}) - (0.7\text{kPa K}^{-1})\cdot(10\text{K}) = \underline{\mathbf{-5.5\ kPa}}$

Check: Units OK. Physics OK.
Exposition: This answer is negative, meaning that the eye has higher pressure than the surroundings, aloft. This pressure reversal drives the outflow aloft.

INFO • Warm vs. Cold Core Cyclones

For a cyclone to survive and intensify, air must be constantly withdrawn from the top. This removal of air mass from the cyclone center (**core**) counteracts the inflow of boundary-layer air at the bottom, which always happens due to surface drag. The net result is low surface pressure in the core, which drives the near-surface winds.

Warm-core cyclones (e.g., tropical cyclones) and **cold-core cyclones** (e.g., extratropical lows) differ in the way they cause horizontal divergence to remove air from the top of the cyclones [see Fig. 16.c, parts (a) and (b)].

Tropical cyclones are vertically stacked, with the eye of the tropical cyclone near the top of the tropopause almost directly above the eye near the surface [Fig. 16.c(a)]. Intense latent heating in the tropical cyclone warms the whole depth of the troposphere near the core, causing high pressure aloft because warm layers of air have greater thickness than cold. This high aloft causes air to diverge horizontally at the top of the tropical cyclone, which is why visible and IR satellite loops show cirrus and other high clouds flowing away from the tropical cyclone center.

Extratropical lows are not vertically stacked, but have low pressure that tilts westward with increasing height [Fig 16.c(b)]. As the surface circulation around the cyclone advects in cold, polar, boundary-layer air on the west side of the cyclone, the small thicknesses in that sector cause pressure to decrease more rapidly with height. The net result is an upper-level trough west of, and a ridge east of, the surface low. A jet stream meandering through this trough-ridge system would cause horizontal divergence (D), as was shown in the Extratropical Cyclone chapter.

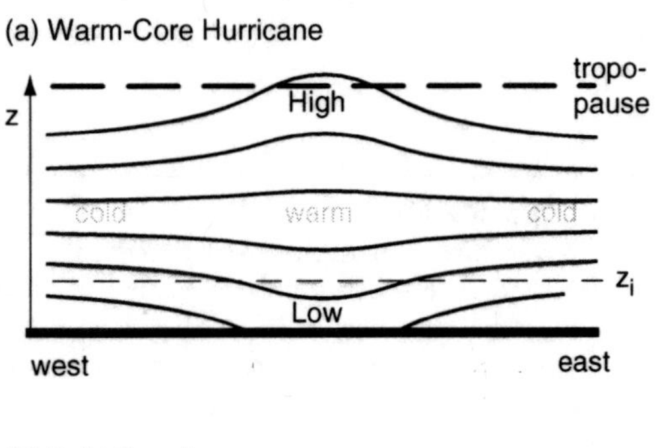

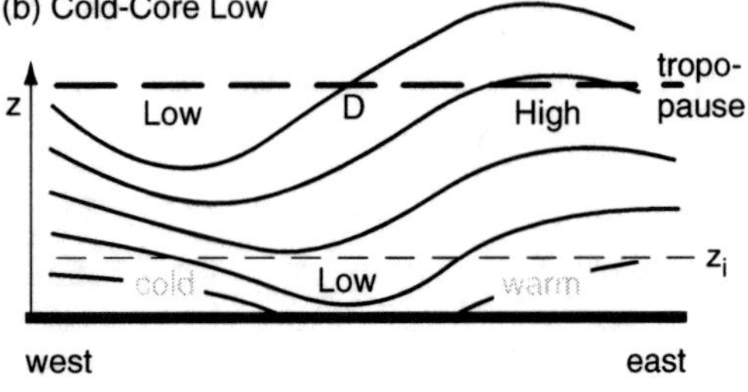

Figure 16.c.
Vertical cross section of (a) warm, and (b) cold core cyclones. Thin lines are isobars, and z_i is ABL top.

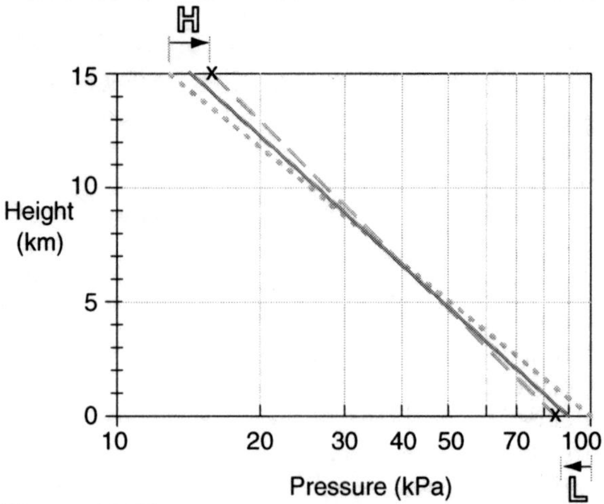

Figure 16.29
Blue dotted line: hydrostatic pressure decrease for cool air outside the tropical cyclone. Red solid line: hydrostatic vertical pressure gradient in the warm core. The core has higher pressure relative to its environment near the top of the storm, even though the core has lower pressure near sea level (z = 0), as explained by the hypsometric eq. Additional nonhydrostatic P variation (green dashed line between two x's) in the horizontal is due to strong updrafts in eyewall thunderstorms, removing low-altitude air and depositing it at storm top. This can cause non-hydrostatic descent of warm air in the eye. (This is a semi-log graph.)

titudes, but spiral anticyclonically near the top of the troposphere well away from the eye. Hence, the tangential velocity must decrease with altitude and eventually change sign. The ideal gas law can be used with the gradient wind eq. (16.3) to show how tangential wind component M_{tan} varies with altitude z :

$$\left(\frac{2\cdot M_{\tan}}{R}+f_c\right)\cdot\frac{\Delta M_{\tan}}{\Delta z}=\frac{|g|}{\overline{T}}\cdot\frac{\Delta T}{\Delta R} \qquad \bullet(16.6)$$

Sample Application

Suppose at a radius of 40 km the tangential velocity decreases from 50 m s^{-1} at the surface to 10 m s^{-1} at 10 km altitude. Find the radial temperature gradient at a latitude where the Coriolis parameter is 0.00005 s^{-1}. Assume $|g|/T = 0.0333$ m·s^{-2}·K^{-1} .

Find the Answer

Given: R = 40 km, M_{tan} = 50 m s^{-1} at z = 0,
M_{tan} = 10 m s^{-1} at z = 10 km, f_c = 0.00005 s^{-1} ,
$|g|/T$ = 0.0333 m·s^{-2}·K^{-1}
Find: $\Delta T/\Delta R$ = ? K km^{-1}

Use eq. (16.6):

$$\frac{|g|}{\overline{T}}\cdot\frac{\Delta T}{\Delta R}=\left(\frac{2\cdot(30\text{m/s})}{40{,}000\text{m}}+0.00005\text{s}^{-1}\right)\cdot\frac{(10-50\text{m/s})}{(10{,}000\text{m})}$$

$$= -6.2\text{x}10^{-6}\ \text{s}^{-2} .$$

Thus, $$\frac{\Delta T}{\Delta R}=\frac{-6.2\times10^{-6}\text{s}^{-2}}{0.0333\text{m}\cdot\text{s}^{-2}\cdot\text{K}^{-1}} = \underline{-0.19\ \text{K km}^{-1}}.$$

Check: Units OK. Physics OK.

Exposition: From the center of the eye, the temperature decreases about 7.4 K at a radius of 40 km, for this example. Indeed, the core is warm.

HIGHER MATH • Pressure Reversal

Derivation of eq. (16.5):

Use the hypsometric equation to relate the pressure at the top of the tropical cyclone eye to the pressure at the bottom of the eye. Do the same for the surroundings. Use those equations to find the pressure difference at the top:

$$\Delta P_T = P_{B\infty}\cdot\exp[-|g|\cdot z_{max} / (\Re\cdot\bar{T}_\infty)] \ - P_{B.eye}\cdot\exp[-|g|\cdot z_{max} / (\Re\cdot\bar{T}_{eye})]$$

Then, using $P_{B.eye} = P_{B\,\infty} - \Delta P_B$, collecting the exponential terms that are multiplied by $P_{B\,\infty}$, and finally using a first-order series approximation for those exponentials, one gets eq. (16.5), where

$a = \exp[-|g|\cdot z_{max} / (\Re\cdot\bar{T}_{eye})] \approx 0.15$,
$b = -(|g|\cdot z_{max}\cdot P_{B\infty}) / (\Re\cdot\bar{T}_{eye}\cdot\bar{T}_\infty) \approx 0.7$ kPa K^{-1},
$|g| = 9.8$ m·s^{-2} is gravitational acceleration magnitude, and
$\Re = 287.04$ m^2·s^{-2}·K^{-1} is the gas constant for dry air.

HIGHER MATH • Warm Core Winds

Derivation of eq. (16.6):

There are 3 steps: (1) scale analysis; (2) differentiation of eq. (16.3) with respect to height z, and (3) simplification of the pressure term.

(1) Scale Analysis

Differentiate the ideal gas law $P = \rho\cdot\Re\cdot T$ with respect to z, and use the chain rule of calculus:

$$\frac{\partial P}{\partial z} = \rho\Re\frac{\partial T}{\partial z} + \Re T\frac{\partial \rho}{\partial z}$$

Then divide this by the ideal gas law:

$$\frac{1}{P}\frac{\partial P}{\partial z} = \frac{1}{T}\frac{\partial T}{\partial z} + \frac{1}{\rho}\frac{\partial \rho}{\partial z}$$

Between $z = 0$ to 20 km, typical variations of the variables are: $P = 101$ to 5.5 kPa, $T = 288$ to 216 K, and $\rho = 1.23$ to 0.088 kg m^{-3}. Thus, the temperature term in the eq. above varies only 1/4 as much as the other two terms. Based on this scale analysis, we can neglect the temperature term, which leaves:

$$\frac{1}{P}\frac{\partial P}{\partial z} \approx \frac{1}{\rho}\frac{\partial \rho}{\partial z} \qquad \text{(a)}$$

(2) Differentiate eq. (16.3) with respect to z:

$$\frac{\partial}{\partial z}\left[\frac{1}{\rho}\frac{\partial P}{\partial R}\right] = f_c\cdot\frac{\partial M_{tan}}{\partial z} + \frac{2M_{tan}}{R}\cdot\frac{\partial M_{tan}}{\partial z}$$

Upon switching the left and right sides:

$$\left[\frac{2M_{tan}}{R} + f_c\right]\cdot\frac{\partial M_{tan}}{\partial z} = \frac{\partial}{\partial z}\left[\frac{1}{\rho}\frac{\partial P}{\partial R}\right] \qquad \text{(b)}$$

(3) Simplify the pressure term:

First, use the chain rule:

$$\frac{\partial}{\partial z}\left[\frac{1}{\rho}\frac{\partial P}{\partial R}\right] = \frac{1}{\rho}\cdot\frac{\partial}{\partial z}\left[\frac{\partial P}{\partial R}\right] + \frac{\partial P}{\partial R}\cdot\frac{\partial}{\partial z}\left[\rho^{-1}\right]$$

R and z are independent, allowing the order of differentiation to be reversed in the first term on the RHS:

(continues in next column)

HIGHER MATH • Warm Core Winds 2

Derivation of eq. (16.6) *(continuation)*

$$\frac{\partial}{\partial z}\left[\frac{1}{\rho}\frac{\partial P}{\partial R}\right] = \frac{1}{\rho}\cdot\frac{\partial}{\partial R}\left[\frac{\partial P}{\partial z}\right] - \frac{1}{\rho^2}\frac{\partial P}{\partial R}\cdot\frac{\partial \rho}{\partial z}$$

Use the hydrostatic eq. $\partial P/\partial z = -\rho\cdot|g|$ in the first term on the RHS:

$$\frac{\partial}{\partial z}\left[\frac{1}{\rho}\frac{\partial P}{\partial R}\right] = \frac{1}{\rho}\cdot\frac{\partial}{\partial R}\left[-\rho\cdot|g|\right] - \frac{1}{\rho}\frac{\partial P}{\partial R}\cdot\frac{1}{\rho}\frac{\partial \rho}{\partial z}$$

But $|g|$ is constant. Also, substitute (a) in the last term:

$$\frac{\partial}{\partial z}\left[\frac{1}{\rho}\frac{\partial P}{\partial R}\right] = -\frac{|g|}{\rho}\cdot\frac{\partial \rho}{\partial R} - \frac{1}{\rho}\frac{\partial P}{\partial R}\cdot\frac{1}{P}\frac{\partial P}{\partial z}$$

Substitute the ideal gas law in the first term on the RHS:

$$\frac{\partial}{\partial z}\left[\frac{1}{\rho}\frac{\partial P}{\partial R}\right] = -\frac{|g|}{\rho\cdot\Re}\cdot\frac{\partial(P\cdot T^{-1})}{\partial R} - \frac{1}{P}\frac{\partial P}{\partial R}\cdot\frac{1}{\rho}\frac{\partial P}{\partial z}$$

Use the chain rule on the first term on the right, and substitute the hydrostatic eq. in the last term:

$$\frac{\partial}{\partial z}\left[\frac{1}{\rho}\frac{\partial P}{\partial R}\right] = -\frac{P\cdot|g|}{\rho\cdot\Re}\cdot\frac{\partial(T^{-1})}{\partial R} - \frac{|g|}{\rho\cdot\Re\cdot T}\cdot\frac{\partial P}{\partial R} + \frac{|g|}{P}\frac{\partial P}{\partial R}$$

Substitute the ideal gas law in the 2nd term on the right:

$$\frac{\partial}{\partial z}\left[\frac{1}{\rho}\frac{\partial P}{\partial R}\right] = \frac{P\cdot|g|}{\rho\cdot\Re\cdot T^2}\cdot\frac{\partial T}{\partial R} - \frac{|g|}{P}\cdot\frac{\partial P}{\partial R} + \frac{|g|}{P}\frac{\partial P}{\partial R}$$

But the last two terms cancel. Using the ideal gas law in the remaining term leaves:

$$\frac{\partial}{\partial z}\left[\frac{1}{\rho}\frac{\partial P}{\partial R}\right] = \frac{|g|}{T}\cdot\frac{\partial T}{\partial R} \qquad \text{(c)}$$

(4) Completion:

Finally, equate (b) and (c):

$$\boxed{\left[\frac{2M_{tan}}{R} + f_c\right]\cdot\frac{\partial M_{tan}}{\partial z} = \frac{|g|}{T}\cdot\frac{\partial T}{\partial R}} \qquad (16.6)$$

which is the desired answer, when converted from derivatives to finite differences.

Table 16-5. Example of thermodynamic states within a tropical cyclone, corresponding to the points (Pt) circled in Fig. 16.30. Reference: $T_o = 273$ K, $P_o = 100$ kPa.

Pt	P (kPa)	T (°C)	T_d (°C)	r (g/kg)	s [J/(K·kg)]
1	100	28	–70	≈ 0	98
2	90	28	28	28	361
3	25	–18	–18	3.7	366
4	20	–83	–83	≈ 0	98

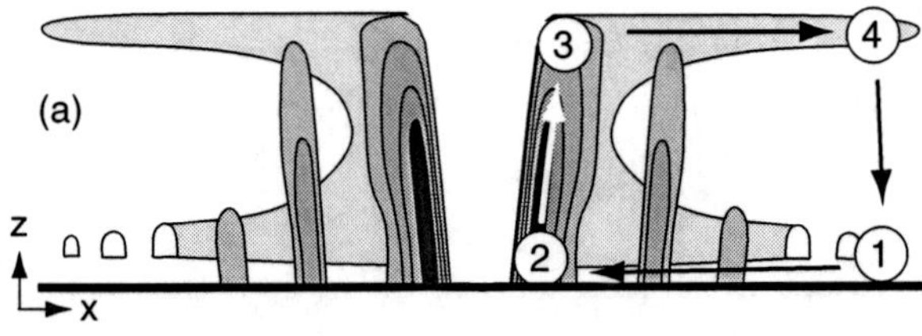

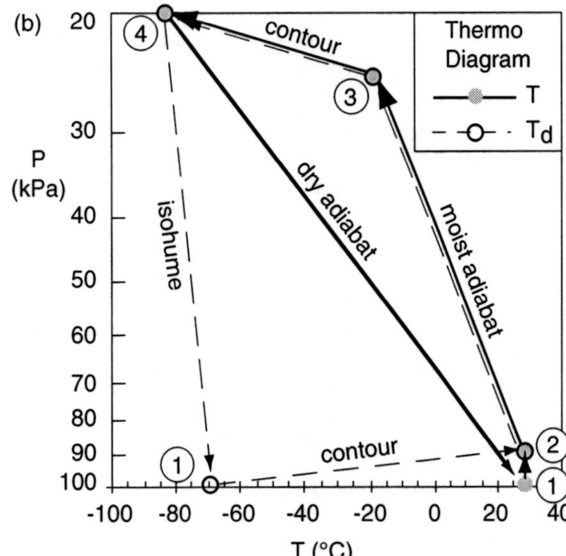

Figure 16.30
(a) Circulation of air through the tropical cyclone. (b) Thermodynamic diagram showing tropical cyclone processes for the same numbered locations as in (a). Solid lines show the temperature changes, dashed lines show dew point temperature changes.

where T is absolute temperature, and the overbar denotes an average over depth. Eqs. (16.5 & 16.6 are overly simplistic because they ignore nonhydrostatic and non-equilibrium effects.

Carnot-cycle Heat Engine

Tropical cyclones are analogous to **Carnot heat engines** in that they convert thermal energy into mechanical energy. One measure of the energy involved is the **total entropy** s per unit mass of air:

$$s = C_p \cdot \ln\left(\frac{T}{T_o}\right) + \frac{L_v \cdot r}{T} - \Re \cdot \ln\left(\frac{P}{P_o}\right) \qquad \bullet(16.7)$$

where C_p = 1004 J·kg^{-1}·K^{-1} is the specific heat of air at constant pressure, T is absolute temperature, L_v = 2500 J $g_{water\ vapor}^{-1}$ is the latent heat of vaporization, r is mixing ratio, $\Re$ = 287 J·kg^{-1}·K^{-1} is the gas-law constant, P is pressure. T_o = 273 K and P_o = 100 kPa are arbitrary reference values.

A thermo diagram is used to illustrate tropical cyclone thermodynamics (Fig. 16.30). Because of limitations in range and accuracy of thermo diagrams, you might find slightly different answers than those given in Table 16-5. For simplicity, assume constant sea surface temperature of T_{SST} = 28°C.

As an initial condition at Point 1 in Fig. 16.30, consider relatively warm (T = 28°C) dry air ($r \approx 0$, T_d = –70°C) in the boundary layer (z = 0; P = 100 kPa), but outside of the tropical cyclone. Using eq. (16.7), the initial entropy is $s_1 \approx$ 98 J·kg^{-1}·K^{-1}.

In the first part of the tropical cyclone's Carnot cycle, air in the boundary layer spirals in from Point 1 toward the eye wall of the tropical cyclone (Point 2) isothermally (T = 28°C, because of heat transfer with the sea surface) at constant height (z = 0 = sea level). Pressure decreases to P = 90 kPa as the air approaches the low-pressure eye. Evaporation from the sea surface increases the mixing ratio to saturation ($r \approx$ 28 g kg^{-1}, $T_d \approx$ 28°C), thereby causing entropy to increase to $s_2 \approx$ 361 J·kg^{-1}·K^{-1}. This evaporation is the major source of energy for the storm.

From Point 2, air rises moist adiabatically to Point 3 in the thunderstorms of the eye wall. The moist-adiabatic process conserves entropy (s_3 = 366 J·kg^{-1}·K^{-1} , within the accuracy of the thermo diagram, thus $s_2 \approx s_3$). During this process, temperature drops to $T \approx$ –18°C and mixing ratio decreases to about $r \approx$ 3.7 g kg^{-1}. However, the decrease of pressure (P = 90 to 25 kPa) in this rising air compensates to maintain nearly constant entropy.

Once the cloudy air reaches the top of the troposphere at Point 3, it flows outward to Point 4 at roughly constant altitude. (Sorry, the thermo diagrams at the end of the Atmospheric Stability chapter do not go high enough to simulate a real tropical cyclone of 15 km depth, so we will use $z \approx 10$ km here.) The divergence of air is driven by a pressure gradient of P = 25 kPa in the eye to 20 kPa outside the tropical cyclone.

During this high-altitude outflow from Points 3 to 4, air rapidly loses heat due to infrared radiation, causing its temperature to decrease from $T = T_d = -18°C$ to $-83°C$. The air remains saturated, and mixing ratio decreases from $r \approx 3.7$ to near 0. The cooling also converts more water vapor into precipitation. Entropy drops to 98 J·kg^{-1}·K^{-1}.

Finally, the air subsides dry adiabatically from Points 4 to 1, with no change of mixing ratio ($r \approx 0$; $T_d \approx -70°C$). Temperature increases adiabatically to T = 28°C, due to compression as the air descends into higher pressure (P = 100 kPa). This dry adiabatic process also preserves entropy, and thus is called an **isentropic** process (and dry adiabats are also known as **isentropes**). The final state of the air is identical to the initial state, at Point 1 in Fig. 16.30.

This Carnot process is a closed cycle – the air can recirculate through the tropical cyclone. However, during this cycle, entropy is gained near the sea surface where the temperature is warm, while it is lost near cloud top where temperatures are colder.

The gain of entropy at one temperature and loss at a different temperature allows the Carnot engine to produce mechanical energy ME according to

$$ME = (T_B - T_{T.avg}) \cdot (s_{eyewall} - s_\infty)_B \qquad \bullet(16.8)$$

where subscripts B and T denote bottom and top of the troposphere, eyewall denotes boundary-layer air under the eye wall, and ∞ denotes the ambient conditions at large distances from the tropical cyclone (e.g.: $P_\infty \approx 101.3$ kPa). This mechanical energy drives the tropical cyclone-force winds, ocean waves, atmospheric waves, and mixing of both the atmospheric and ocean against buoyant forces.

If all of the mechanical energy were consumed trying to maintain the tropical cyclone force winds against the frictional drag in the boundary layer (an unrealistic assumption), then the tropical cyclone could support the following maximum pressure ratio at the surface:

$$\ln\left(\frac{P_\infty}{P_{eye}}\right)_B = \frac{ME}{T_B \cdot \Re} \qquad (16.9)$$

Sample Application

Suppose air in the eye of a tropical cyclone has the following thermodynamic state: P = 70 kPa, r = 1 g kg^{-1}, T = 15°C. Find the entropy.

Find the Answer

Given: P = 70 kPa, r = 1 g kg^{-1}, T = 288 K.
Find: s = ? J·kg^{-1}·K^{-1} .

Use eq. (16.7):

$$s = \left(1004 \frac{J}{kg_{air} \cdot K}\right) \cdot \ln\left(\frac{288K}{273K}\right) + \left(2500 \frac{J}{g_{water}}\right) \cdot \left(1 \frac{g_{water}}{kg_{air}}\right) \frac{1}{288K} - \left(287 \frac{J}{kg_{air} \cdot K}\right) \cdot \ln\left(\frac{70kPa}{100kPa}\right)$$

s = **165 J·kg^{-1}·K^{-1}**

Check: Units OK. Physics OK.
Exposition: The actual value of entropy is meaningless, because of the arbitrary constants T_o and P_o. However, the difference between two entropies is meaningful, because the arbitrary constants cancel out.

Sample Application

Find the mechanical energy and minimum possible eye pressure that can be supported by the tropical cyclone of Table 16-5.

Find the Answer

Given: $s_{eyewall}$ = 361 J·kg^{-1}·K^{-1}, s_∞ = 98 J·kg^{-1}·K^{-1}
T_B = 28°C, $T_{T\,avg} \approx$ 0.5·(–18–83) = –50.5°C.
Find: ME = ? kJ kg^{-1}, P_{eye} = ? kPa

Use eq. (16.8): ME = (28 + 50.5)(K)·(361–98)(J·kg^{-1}·K^{-1}) = **20.6 kJ kg^{-1}**

Use eq. (16.9):

$$\ln\left(\frac{P_\infty}{P_{eye}}\right)_B = \frac{(20{,}600 J \cdot kg^{-1})}{(301K) \cdot (287 J \cdot kg^{-1} \cdot K^{-1})} = 0.238$$

Solving for P_{eye} gives:
P_{eye} =(101.3kPa)/exp(0.238) = **79.8 kPa**.

Check: Units OK. Physics OK.
Exposition. This eye pressure is lower than the actual eye pressure of 90 kPa. The difference is related to the ME of winds and waves.

INFO • The Power of Tropical Cyclones

Another measure of tropical cyclone power is the rate of dissipation of kinetic energy by wind drag against the sea surface. This is proportional to the ABL wind speed cubed, times the sea-surface area over which that speed is valid, summed over all areas under the tropical cyclone. K. A. Emanuel (1999, *Weather*, **54**, 107-108) estimates dissipation rates of 3 x 10^{12} W for a typical Atlantic hurricane (with max winds of 50 m s^{-1} at radius 30 km), and 3 x 10^{13} W for a Pacific typhoon (with max winds of 80 m s^{-1} at radius 50 km).

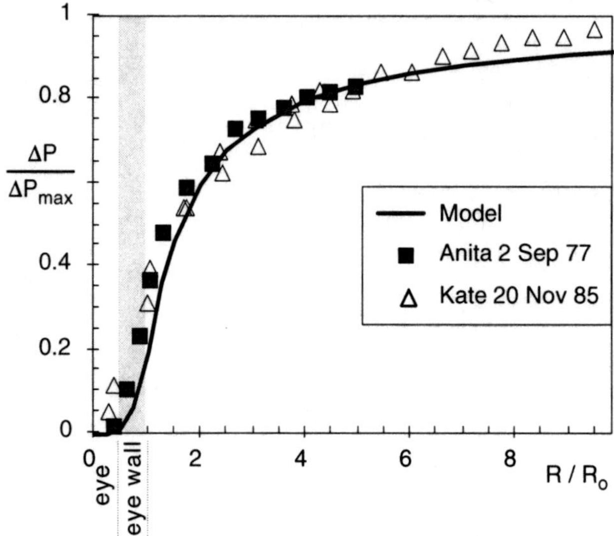

Figure 16.31
Surface pressure distribution across a tropical cyclone. Model is eq. (16.11).

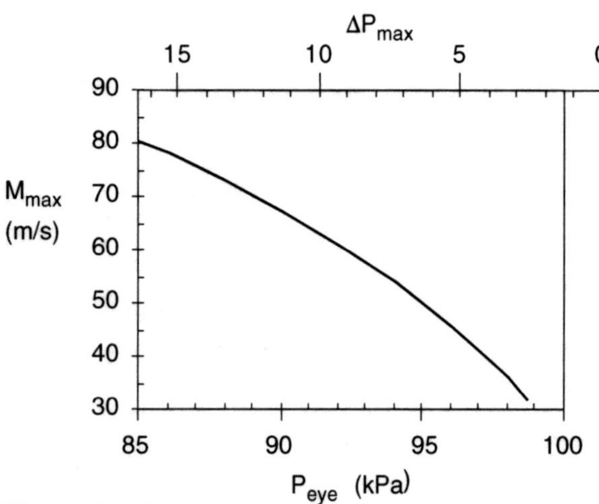

Figure 16.32
Maximum sustained winds around a tropical cyclone increase as sea-level pressure in the eye decreases.

Sample Application
What max winds are expected if the tropical cyclone eye has surface pressure of 95 kPa?

Find the Answer
Given: $P_{B\,eye}$ = 95 kPa, Assume $P_{B\,\infty}$ = 101.3 kPa
Find: M_{max} = ? m s^{-1}.
Assume translation speed can be neglected.

Use eq. (16.12):
M_{max} = [20 (m s^{-1})·kPa$^{-1/2}$]·(101.3–95kPa)$^{1/2}$ = **50 m s^{-1}**

Check: Units OK. Physics OK.
Exposition: This is a category 3 tropical cyclone on the Saffir-Simpson tropical cyclone Wind Scale.

A TROPICAL CYCLONE MODEL

Although tropical cyclones are quite complex and not fully understood, we can build an idealized model that mimics some of the real features.

Pressure Distribution

Eye pressures of 95 to 99 kPa at sea level are common in tropical cyclones, although a pressure as low as P_{eye} = 87 kPa has been measured. One measure of tropical cyclone strength is the pressure difference ΔP_{max} between the eye and the surrounding ambient environment P_∞ = 101.3 kPa:

$$\Delta P_{\max} = P_\infty - P_{eye} \qquad (16.10)$$

The surface pressure distribution across a tropical cyclone can be approximated by:

$$\frac{\Delta P}{\Delta P_{\max}} = \begin{cases} \frac{1}{5}\cdot\left(\frac{R}{R_o}\right)^4 & \text{for } R \le R_o \\ \left[1 - \frac{4}{5}\cdot\frac{R_o}{R}\right] & \text{for } R > R_o \end{cases} \qquad (16.11)$$

where $\Delta P = P(R) - P_{eye}$, and R is the radial distance from the center of the eye. This pressure distribution is plotted in Fig 16.31, with data points from two tropical cyclones.

R_o is the critical radius where the maximum tangential winds are found. In the tropical cyclone model presented here, R_o is twice the radius of the eye. Eyes range from 4 to 100 km in radius, with average values of 15 to 30 km radius. Thus, one anticipates average values of critical radius of: $30 < R_o < 60$ km, with an observed range of $8 < R_o < 200$ km.

Tangential Velocity

To be classified as a tropical cyclone, the sustained winds (averaged over 1-minute) must be 33 m s^{-1} or greater near the surface. While most anemometers are unreliable at extreme wind speeds, maximum tropical cyclone winds have been reported in the 75 to 95 m s range^{-1}.

As sea-level pressure in the eye decreases, maximum tangential surface winds M_{max} around the eye wall increase (Fig. 16.32). An empirical approximation for this relationship, based on **Bernoulli's equation** (see the Regional Winds chapter), is :

$$M_{\max} = a\cdot(\Delta P_{\max})^{1/2} \qquad •(16.12)$$

where a = 20 (m s^{-1})·kPa$^{-1/2}$.

If winds are assumed to be cyclostrophic (not the best assumption, because drag against the sea surface and Coriolis force are neglected), then the previous approximation for pressure distribution (eq. 16.11) can be used to give a distribution of tangential velocity M_{tan} (relative to the eye) in the boundary layer:

$$\frac{M_{\tan}}{M_{\max}} = \begin{cases} (R/R_o)^2 & \text{for } R \le R_o \\ (R_o/R)^{1/2} & \text{for } R > R_o \end{cases} \qquad (16.13)$$

where the maximum velocity occurs at critical radius R_o. This is plotted in Fig. 16.33, with data points from a few tropical cyclones.

For the tropical cyclones plotted in Fig. 16.33, the critical radius of maximum velocity was in the range of R_o = 20 to 30 km. This is a rough definition of the outside edge of the **eye wall** for these tropical cyclones, within which the heaviest precipitation falls. The maximum velocity for these storms was M_{max} = 45 to 65 m s^{-1}.

Winds in Fig. 16.33 are relative to the eye. However, the whole tropical cyclone including the eye is often moving. tropical cyclone **translation** speeds (movement of the center of the storm) can be as slow as M_t = 0 to 5 m s^{-1} as they drift westward in the tropics, and as fast as 25 m s^{-1} as they later move poleward. Average translation speeds of the tropical cyclone over the ocean are M_t = 10 to 15 m s^{-1}.

The total wind speed relative to the surface is the vector sum of the translation speed and the rotation speed. On the right quadrant of the storm relative to its direction of movement in the Northern Hemisphere, the translation speed adds to the rotation speed. Thus, tropical cyclone winds are fastest in the tropical cyclone's right quadrant (Fig. 16.34). On the left the translation speed subtracts from the tangential speed, so the fastest total speed in the left quadrant is not as strong as in the right quadrant (Fig. 16.35).

Total speed relative to the surface determines ocean wave and surge generation. Thus, the right quadrant of the storm near the eye wall is most dangerous. Also, tornadoes are likely there.

Radial Velocity

For an idealized tropical cyclone, boundary-layer air is trapped below the top of the boundary layer as the air converges horizontally toward the eye wall. Horizontal continuity in cylindrical coordinates requires:

$$M_{rad} \cdot R = \text{constant} \qquad (16.14)$$

where M_{rad} is the radial velocity component, negative for inflow. Thus, starting from far outside the tropi-

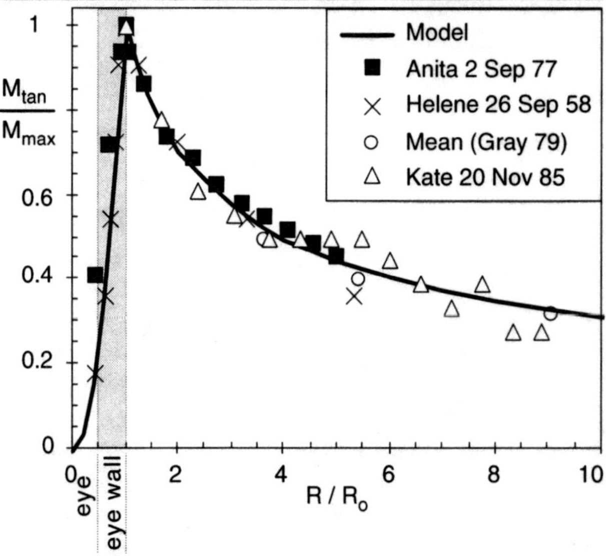

Figure 16.33
Tangential winds near the surface at various radii around tropical cyclones. Model is eq. (16.13).

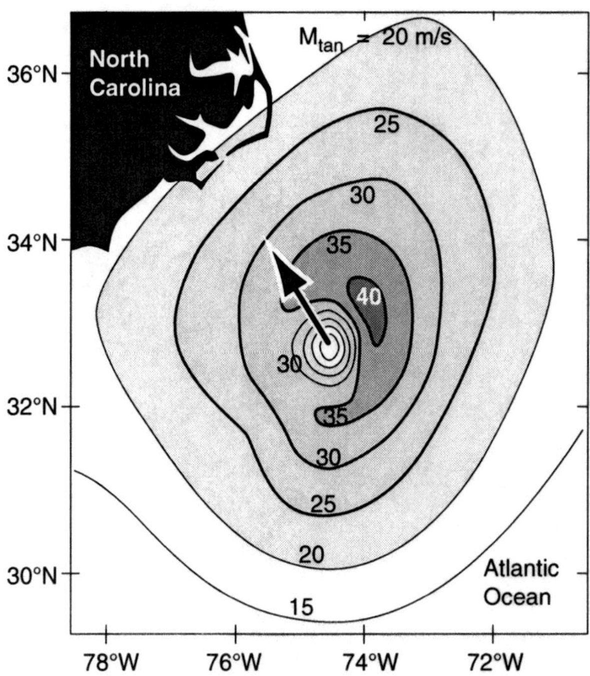

Figure 16.34
Max 1-minute total sustained surface winds (m s^{-1}) around Hurricane Isabel, 0730 UTC 18 Sep 2003. $M_{max\ total}$ = 42 m s^{-1}, and P_{eye} ≈ 95.7 kPa. Arrow shows translation toward 325° at 5.5 m s^{-1}. The Saffir-Simpson Hurricane Wind Scale is based on the fastest total sustained wind (rotational + translational) found anywhere in the hurricane.

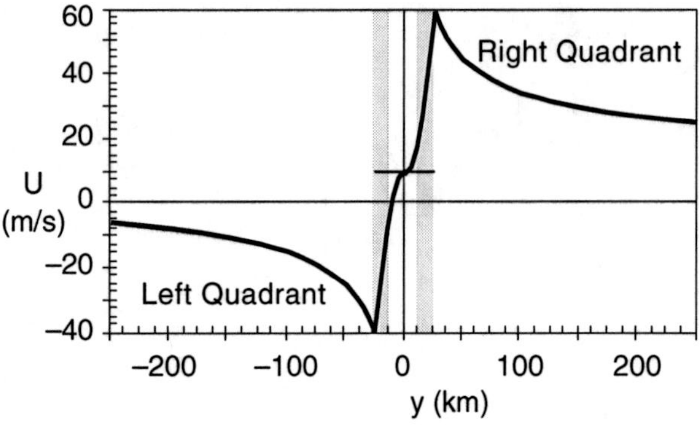

Figure 16.35
Sum of the modeled tangential winds relative to the eye and the translation speed (10 m s^{-1}) of the eye, for a hypothetical tropical cyclone. Assumes M_{max} = 50 m s^{-1} relative to the eye, R_o = 25 km, and uses a coordinate system with the x-axis aligned in the same direction as the translation vector.

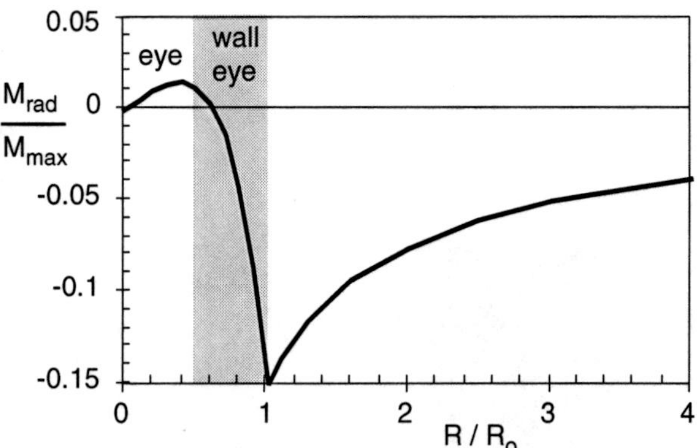

Figure 16.36
Radial winds in a tropical cyclone boundary layer. (Negative values indicate motion inward, converging toward R/R_o = 0.)

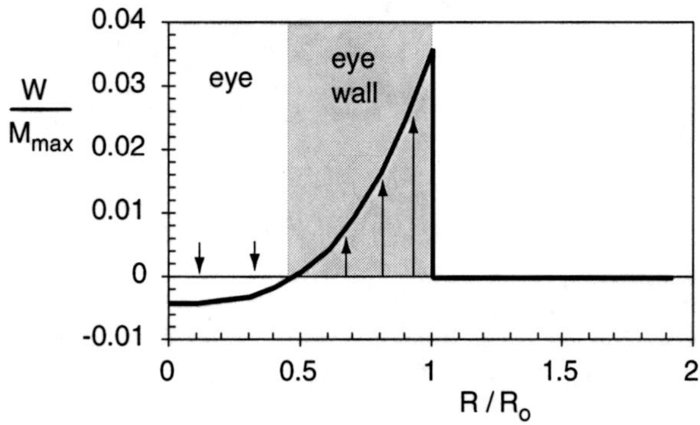

Figure 16.37
Vertical velocity at various radii around tropical cyclones, out of the top of the boundary layer. W_s = −0.2 m s^{-1} in the eye.

cal cyclone, as R decreases toward R_o, the magnitude of inflow must increase. Inside of R_o, thunderstorm convection removes air mass vertically, implying that horizontal continuity is no longer satisfied.

As wind velocities increase inward toward the eye wall from outside (see previous subsections), wave height and surface roughness also increase. The resulting turbulent drag against the ocean surface tends to couple the radial and tangential velocities, which we can approximate by $M_{rad} \propto M_{tan}{}^2$. Drag-induced inflow such as this eventually converges and forces ascent via the **boundary-layer pumping** process (see the Atmospheric Forces & Winds chapter).

The following equations utilize the concepts above, and are consistent with the tangential velocity in the previous subsection:

(16.15)

$$\frac{M_{rad}}{M_{\max}} = \begin{cases} -\frac{R}{R_o}\cdot\left[\frac{1}{5}\left(\frac{R}{R_o}\right)^3 + \frac{1}{2}\frac{W_s}{M_{\max}}\frac{R_o}{z_i}\right] & \text{for } R \le R_o \\ -\frac{R_o}{R}\cdot\left[\frac{1}{5} + \frac{1}{2}\cdot\frac{W_s}{M_{\max}}\cdot\frac{R_o}{z_i}\right] & \text{for } R > R_o \end{cases}$$

where W_s is negative, and represents the average subsidence velocity in the eye. Namely, the horizontal area of the eye, times W_s, gives the total kinematic mass flow downward in the eye. The boundary-layer depth is z_i, and M_{max} is still the maximum tangential velocity.

For example, Fig. 16.36 shows a plot of the equations above, using z_i = 1 km, R_o = 25 km, M_{max} = 50 m s^{-1}, and W_s = −0.2 m s^{-1}. In the eye, subsidence causes air to weakly diverge (positive M_{rad}) toward the eye wall. Inside the eye wall, the radial velocity rapidly changes to inflow (negative M_{rad}), reaching an extreme value of −7.5 m s^{-1} for this case. Outside of the eye wall, the radial velocity smoothly decreases as required by horizontal mass continuity.

Vertical Velocity

At radii less than R_o, the converging air rapidly piles up, and rises out of the boundary layer as thunderstorm convection within the eye wall. The vertical velocity out of the top of the boundary layer, as found from mass continuity, is

(16.16)

$$\frac{W}{M_{\max}} = \begin{cases} \left[\frac{z_i}{R_o}\left(\frac{R}{R_o}\right)^3 + \frac{W_s}{M_{\max}}\right] & \text{for } R < R_o \\ 0 & \text{for } R > R_o \end{cases}$$

For simplicity, we are neglecting the upward motion that occurs in the spiral rain bands at $R > R_o$.

As before, W_s is negative for subsidence. Although subsidence acts only inside the eye for real tropical cyclones, the relationship above applies it every where inside of R_o for simplicity. Within the eye wall, the upward motion overpowers the subsidence, so our simplification is of little consequence.

Using the same values as for the previous figure, the vertical velocity is plotted in Fig. 16.37. The maximum upward velocity is 1.8 m s^{-1} in this case, which represents an average around the eye wall. Updrafts in individual thunderstorms can be much faster.

Subsidence in the eye is driven by the non-hydrostatic part of the pressure perturbation (Fig. 16.29). Namely, the pressure gradient (shown by the dashed line between X's in that Fig.) that pushes air upward is weaker than gravity pulling down. This net imbalance forces air downward in the eye.

Temperature

Suppose that pressure difference between the eye and surroundings at the top of the tropical cyclone is equal and opposite to that at the bottom. From eq. (16.5) the temperature T averaged over the tropical cyclone depth at any radius R is found from:

$$\Delta T(R) = c \cdot \left[\Delta P_{\max} - \Delta P(R)\right] \tag{16.17}$$

where c = 1.64 K kPa^{-1}, the pressure difference at the bottom is $\Delta P = P(R) - P_{eye}$, and the temperature difference averaged over the whole tropical cyclone depth is $\Delta T(R) = T_{eye} - T(R)$. When used with eq. (16.11), the result is:

$$\frac{\Delta T}{\Delta T_{\max}} = \begin{cases} 1 - \frac{1}{5} \cdot \left(\frac{R}{R_o}\right)^4 & \text{for } R \le R_o \\ \frac{4}{5} \cdot \frac{R_o}{R} & \text{for } R > R_o \end{cases} \tag{16.18}$$

where $\Delta T_{max} = T_{eye} - T_{\infty} = c \cdot \Delta P_{max}$, and c = 1.64 K kPa^{-1}. This is plotted in Fig. 16.38.

Composite Picture

A coherent picture of tropical cyclone structure can be presented by combining all of the idealized models described above. The result is sketched in Fig. 16.39. For real tropical cyclones, sharp cusps in the velocity distribution would not occur because of vigorous turbulent mixing in the regions of strong shear.

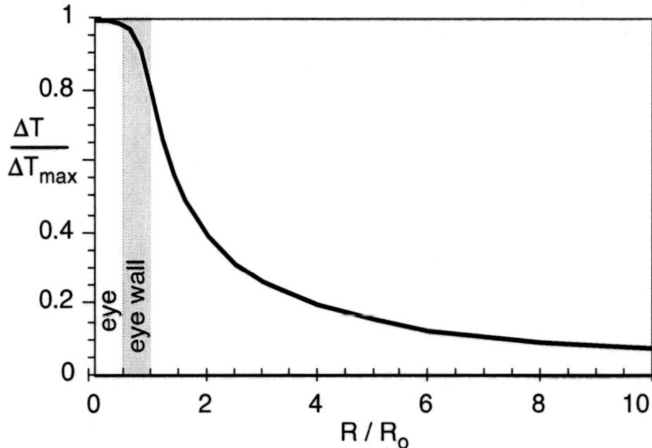

Figure 16.38
Temperature distribution, averaged over a 15 km thick tropical cyclone, showing the warm core.

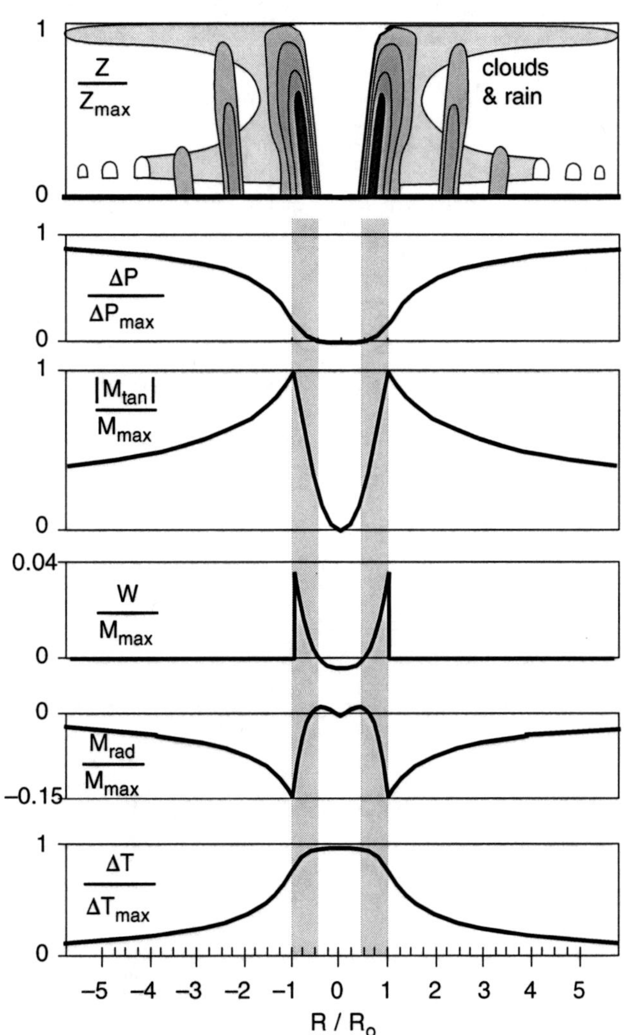

Figure 16.39
Composite tropical cyclone structure based on the idealized model. Pressure differences are at sea level. All horizontal velocities are in the boundary layer, while vertical velocity is across the top of the boundary layer. Temperature differences are averaged over the whole tropical cyclone depth. Grey vertical bands indicate eyewall locations.

Sample Application

A tropical cyclone of critical radius $R_o = 25$ km has a central pressure of 90 kPa. Find the wind components, vertically-averaged temperature excess, and pressure at radius 40 km from the center. Assume $W_s = -0.2$ m s^{-1} in the eye, and $z_i = 1$ km.

Find the Answer

Given: $P_{B\,eye} = 90$ kPa, $R_o = 25$ km, $R = 40$ km, $W_s = -0.2$ m s^{-1}, and $z_i = 1$ km.

Find: $P = ?$ kPa, $T = ?$ °C, $M_{tan} = ?$ m s^{-1}, $M_{rad} = ?$ m s^{-1}, $W = ?$ m s^{-1}

Assume $P_{B\,\infty} = 101.3$ kPa

Figure: Similar to Fig. 16.39. Note that $R > R_o$.

First, find maximum values:

$\Delta P_{max} = (101.3 - 90\text{kPa}) = 11.3$ kPa

Use eq. (16.12):

$M_{max} = [(20\text{m s}^{-1})\cdot\text{kPa}^{-1/2}]\cdot(11.3\text{kPa})^{1/2} = 67$ m s^{-1}

$\Delta T_{max} = c\cdot\Delta P_{max} = (1.64\text{ K kPa}^{-1})\cdot(11.3\text{kPa}) = 18.5$°C

Use eq. (16.11):

$$\Delta P = (11.3\text{kPa})\cdot\left[1 - \frac{4}{5}\cdot\frac{25\text{km}}{40\text{km}}\right] = 5.65\text{ kPa}$$

$P = P_{eye} + \Delta P = 90$ kPa $+ 5.65$ kPa $=$ **95.65 kPa**.

Use eq. (16.18):

$$\Delta T = (18.5°\text{C})\cdot\frac{4}{5}\cdot\frac{(25\text{km})}{(40\text{km})} = \mathbf{9.25°C}$$

averaged over the whole tropical cyclone depth.

(continues in next column)

Sample Application *(continuation)*

Use eq. (16.13):

$$M_{\tan} = (67\text{m/s})\cdot\sqrt{\frac{25\text{km}}{40\text{km}}} = \mathbf{53\ m\ s^{-1}}$$

Use eq. (16.15):

$$M_{rad} = -(67\text{m/s})\cdot\frac{(25\text{km})}{(40\text{km})}\cdot\left[\frac{1}{5} + \frac{1}{2}\cdot\frac{(-0.2\text{m/s})}{(67\text{m/s})}\cdot\frac{(25\text{km})}{(1\text{km})}\right]$$

$$M_{rad} = \mathbf{-6.8\ m\ s^{-1}}$$

Use eq. (16.16): $W =$ **0 m s^{-1}**

Check: Units OK. Physics OK.

Exposition: Using P_{eye} in Table 16-7, the approximate Saffir-Simpson category of this tropical cyclone is borderline between levels 4 and 5, and thus is very intense.

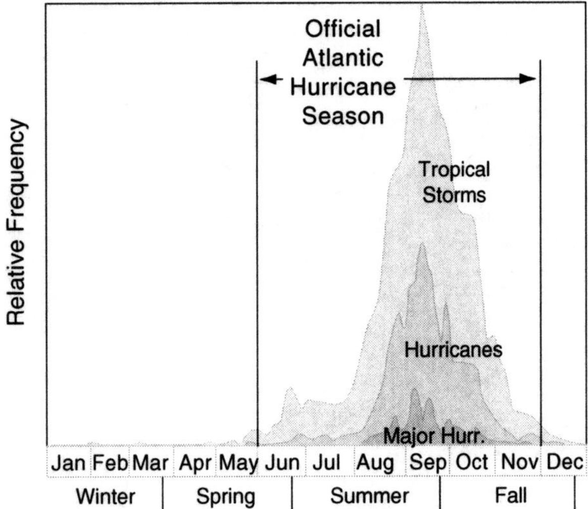

Figure 16.40
Relative frequency of Atlantic Tropical Storms, Hurricanes, and **Major Hurricanes** *(Saffir-Simpson categories 3 - 5).*

CLIMATOLOGY

Seasonality

Tropical cyclones are most frequent in late Summer and Fall of their respective hemisphere. This is because the sun has been highest in the sky, causing the top layers of the tropical ocean to accumulate the most heat. Fig. 16.40 shows the frequency of Atlantic Hurricanes. Table 16-6 shows periods of frequent tropical cyclones in all the ocean basins.

Table 16-6. Tropical Cyclone Seasons. Start and end dates are for the major portion of the storm season, but some storms occur outside the major season.

Location	Start	Peak	End
Atlantic	1 June	mid Sep	30 Nov
NE Pacific	15 May	late Aug /early Sep	30 Nov
NW Pacific*	July	late Aug /early Sep	Nov
N. Indian	Apr	2 peaks: Apr-Jun, & late Sep-early Dec	Dec
S Indian	Oct	2 peaks: mid-Jan, & mid Feb - early Mar	May
Australia & SW Pacific	Oct	late Feb /early Mar	May

* *The NW Pacific has typhoons all year.*

Locations of Strongest Cyclones

The largest number of strongest tropical cyclones is in the northwestern Pacific. The reason is that the Pacific is a larger ocean with warmer sea-surface temperatures, allowing typhoons more opportunity to organize and strengthen. Activity ranges from 17 typhoons in a slow year (1998) to 35 in an active year (1971). Also, the larger **fetch** (distance that wind blows over the ocean) allows larger ocean waves, which can cause more destruction (and better surfing further from a Pacific typhoon).

The opposite extreme is the South Atlantic, which has had only 2 recorded tropical cyclones in the past century. One was Cyclone Catarina, which struck Brazil in March 2004. The other was a Tropical Storm that formed west of Congo in April 1991. There might have been other tropical cyclones in the South Atlantic that were not recorded historically.

There are two reasons for the dearth of tropical cyclones in the South Atlantic. One is the weaker and sometimes nonexistent ITCZ, which reduces tropical cyclone triggering because of less convergence and less initial vorticity. The other is strong wind shear in the upper troposphere, which rips apart thunderstorm clusters before they can become tropical cyclones.

Natural Cycles & Changes in Activity

Atlantic hurricanes have a very large natural variability from year to year. For example, in the Atlantic there were only 4 hurricanes recorded in 1983, and 19 in 1994. An active hurricane year was 2005, with 14 hurricanes and 13 other tropical storms.

Hidden behind this large annual variability are longer-time-period variations of weaker amplitude, making them more difficult to detect and confirm. One is a **natural 40-year cycle** in Atlantic hurricane power (based on wind-speed cubed accumulated over the lifetime of all Atlantic hurricanes). This power was relatively high during the 1950s and 1960s, and was weaker during the 1970s, 1980s, and early 1990s. Since the late 1990s and 2000s the hurricane power has increased again.

Another is the **El Niño/La Niña**, which is an irregular 3 to 5 year cycle. During El Niño, there is a tendency for reduced tropical cyclone activity, and some displacement of these storms closer to the equator. The reason is that stronger west winds aloft during El Niño cause stronger wind shear across the troposphere, thereby inhibiting tropical cyclogenesis. Conversely, tropical cyclone activity is enhanced during La Niña.

Of concern these days is human-caused **global warming**. While most scientists suspect that there will be some effect on tropical cyclones if global warming continues, they have not yet found a clear signal. Debate continues.

HAZARDS

Human Population and Zoning

The most important factor causing increased deaths and destruction from tropical cyclones is the increase in **global population**. As population expands, more people live more densely in coastal areas threatened by tropical cyclones. Even if tropical-cyclone activity remains relatively constant, the human impact increases as population increases.

As more structures are built in vulnerable areas, so increase the property losses caused by tropical cyclone destruction. It also becomes more difficult to evacuate people along inadequate transportation networks. As more people move from farms to cities, there are increased fatalities due to urban flooding caused by heavy tropical-cyclone rainfall.

In highly developed countries, an easy solution would be proper land-use **zoning**. Namely, governments should not let people live and work in threatened areas. Instead, these areas can be used for parks, floodable farmland, wildlife refuges, etc. However, zoning is a political activity that sometimes results in poor decisions in the face of pressure from real-estate developers who want to build more waterfront properties. An example of such a poor decision is the reconstruction of New Orleans, Louisiana, USA, at its original location after being destroyed by hurricane Katrina in 2005.

Also, people are unfortunately encouraged to live in threatened areas because of the existence of hurricane **insurance** and hurricane **disaster relief**. Namely, some individuals choose to make these poor decisions on where to live because they do not have to bear the full costs of reconstruction — instead the cost is borne by all taxpayers.

In less developed, highly populated, low-lying countries such as Bangladesh, an additional problem is an inadequate **warning** system. Even when tropical cyclone tracks are successfully predicted, sometimes the warning does not reach rural poor people, and often there is inadequate transportation to enable their evacuation. Bangladesh has suffered terribly from tropical cyclones: 400,000 deaths in Nov 1970, 140,000 deaths in Apr 1991, and 10,000 deaths in May 1985.

So these aspects of tropical-storm hazards are social (cultural, political, religious, etc.). Do not assume that all problems can be ameliorated by technical solutions (more dams; higher levies). Instead, some tough decisions need to be made on zoning, transportation, and population growth.

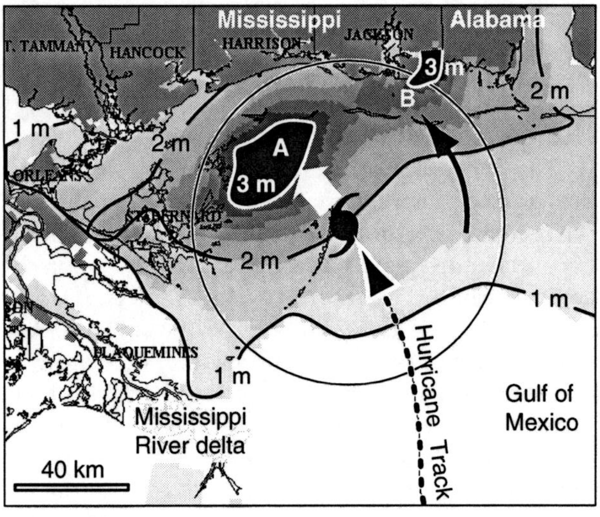

Figure 16.41
Simulated storm surge [rise (m) in sea level] associated with a hypothetical landfalling hurricane. (After NOAA/HRD).

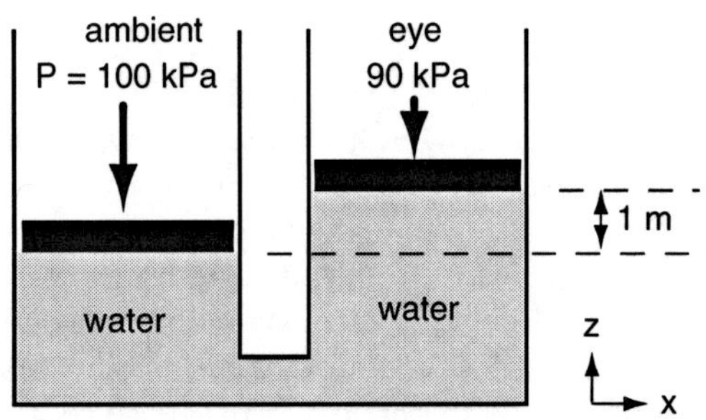

Figure 16.42
Reduced atmospheric pressure in the eye allows sea-level to rise.

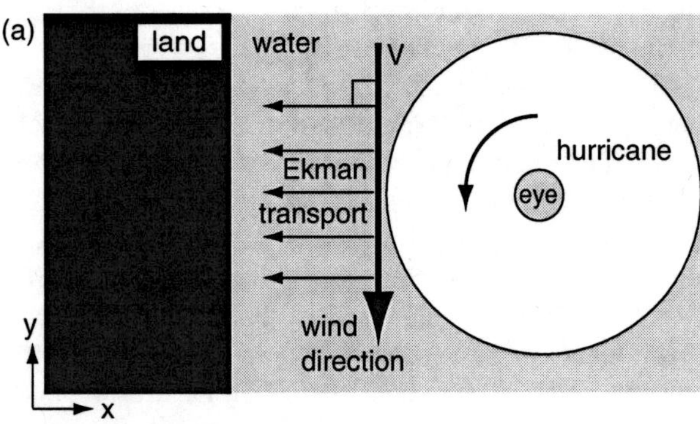

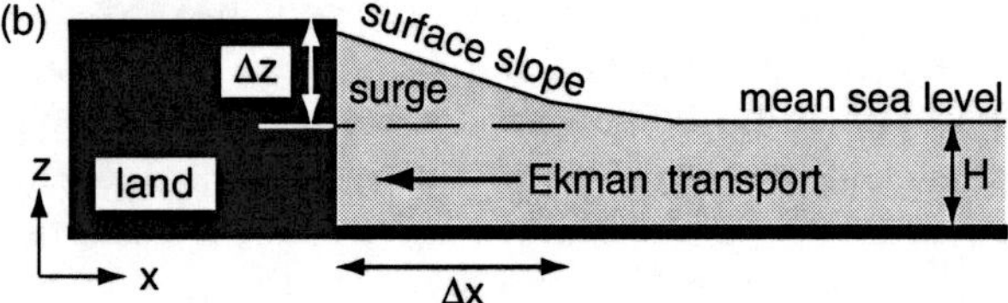

Figure 16.43
Onshore Ekman transport prior to tropical cyclone landfall, in the Northern Hemisphere. (a) Top view. (b) View from south.

Storm Surge

Much of the tropical cyclone-caused damage results from inundation of coastal areas by high seas (Fig. 16.41). The rise in sea level (i.e., the **storm surge**) is caused by the reduced atmospheric pressure in the eye, and by wind blowing the water against the coast to form a large propagating surge called a Kelvin wave. Table 16-7 gives typical storm-surge heights. High tides and high surface-waves can exacerbate the damage.

Atmospheric Pressure Head

In the eye of the tropical cyclone, atmospheric surface pressure is lower than ambient. Hence the force per area pushing on the top of the water is less. This allows the water to rise in the eye until the additional head (weight of fluid above) of water compensates for the reduced head of air (Fig. 16.42).

The amount of rise Δz of water in the eye is

$$\Delta z = \frac{\Delta P_{\max}}{\rho_{liq} \cdot |g|} \qquad \bullet(16.19)$$

where $|g|$ is gravitational acceleration magnitude (9.8 m s^{-2}), ρ_{liq} = 1025 kg m^{-3} is the density of sea water, and ΔP_{max} is the atmospheric surface pressure difference between the eye and the undisturbed environment.

To good approximation, this is

$$\Delta z \approx a \cdot \Delta P_{\max} \qquad (16.20)$$

where a = 0.1 m kPa^{-1}. Thus, in a strong tropical cyclone with eye pressure of 90 kPa (causing $\Delta P_{max} \approx$ 10 kPa), the sea level would rise 1 m.

Ekman Transport

Recall from the General Circulation chapter that ocean currents are generated by wind drag on the sea surface. The Ekman spiral describes how the current direction and speed varies with depth. The net **Ekman transport**, accumulated over all depths, is exactly perpendicular to the surface wind direction.

In the Northern Hemisphere, this net transport of water is to the right of the wind, and has magnitude:

$$\frac{Vol}{\Delta t \cdot \Delta y} = \frac{\rho_{air}}{\rho_{water}} \cdot \frac{C_D \cdot V^2}{f_c} \qquad (16.21)$$

where *Vol* is the volume of water transported during time interval Δt, Δy is a unit of length of coastline parallel to the mean wind, V is the wind speed near the surface (actually at 10 m above the surface), C_D is the drag coefficient, f_c is the Coriolis parameter, ρ_{air} = 1.225 kg·m^{-3} is the air density at sea level, and ρ_{water} = 1025 kg·m^{-3} is sea-water density.

As a tropical cyclone approaches the eastern coast of continents, the winds along the front edge of the tropical cyclone are parallel to the coast, from north to south in the Northern Hemisphere. Hence, there is net Ekman transport of water directly toward the shore, where it begins to pile up and make a storm surge (Fig. 16.43).

If the tropical cyclone were to hover just offshore for sufficient time to allow a steady-state condition to develop, then the Ekman transport toward the shore would be balanced by downslope sloshing of the surge. The surge slope $\Delta z/\Delta x$ for that hypothetical equilibrium is:

$$\frac{\Delta z}{\Delta x} \approx \frac{\rho_{air}}{\rho_{water}} \cdot \frac{C_D \cdot V^2}{|g| \cdot H} \qquad (16.22)$$

where H is the unperturbed ocean depth (e.g., 50 m) near the coast, and the other variables are the same as for eq. (16.21).

Kelvin Wave

Because the tropical cyclone has finite size, Ekman transport is localized to the region immediately in front of the tropical cyclone. Thus, water is piled higher between the tropical cyclone and the coast than it is further north or south along the coast. When viewed from the East, the surge appears as a long-wavelength wave, called a **Kelvin wave** (Fig. 16.44).

The propagation speed of the wave, called the **phase speed** c, is

$$c = \sqrt{|g| \cdot H} \qquad •(16.23)$$

which is also known as the **shallow-water wave speed**, where $|g|$ is gravitational acceleration magnitude and H is average water depth. Typical phase speeds are 15 to 30 m s^{-1}. These waves always travel with the coast to their right in the Northern Hemisphere, so they propagate southward along the east coast of continents, and northward along west coasts.

As the wave propagates south along the East Coast, it will hug the coast and inundate the shore immediately south of the tropical cyclone. Meanwhile, Ekman transport continues to build the surge in the original location. The net result is a continuous surge of high water along the shore that is closest to, and south of, the tropical cyclone. Typical surge depths can be 2 to 10 m at the coast, with extreme values of 13 to 20 m.

Table 16-7. Storm surge height S and sea-level pressure P_{eye}, from the old Saffir-Simpson classification. CAUTION: Actual storm-surge heights can vary significantly from these typical values. This is one of the reasons why the new Saffir-Simpson scale doesn't use S.

Category	P_{eye} (kPa)	S (m)
1	≥ 98.0	1.2 - 1.6
2	97.9 - 96.5	1.7 - 2.5
3	96.4 - 94.5	2.6 - 3.9
4	94.4 - 92.0	4.0 - 5.5
5	< 92.0	> 5.5

Sample Application

For tropical cyclone-force winds of 40 m s^{-1}, over a continental shelf portion of ocean of depth 50 m, find the volume transport rate and equilibrium surge slope. Assume C_D = 0.01 .

Find the Answer

Given: ρ_{air} = 1.225 kg·m^{-3}, ρ_{water} = 1025 kg·m^{-3}
M = 40 m s^{-1}, H = 50 m, C_D = 0.01
Find: $Vol/(\Delta t \cdot \Delta y)$ = ? m^2 s^{-1}, and $\Delta z/\Delta x$ = ?
Assume: f_c = 0.00005 s^{-1}

Use eq. (16.21):

$$\frac{Vol}{\Delta t \cdot \Delta y} = \frac{(1.225\text{kg/m}^3)}{(1025\text{kg/m}^3)} \cdot \frac{(0.01)\cdot(40\text{m/s})^2}{(0.00005\text{s}^{-1})} = \underline{382 \text{ m}^2 \text{ s}^{-1}}$$

Use eq. (16.22):

$$\frac{\Delta z}{\Delta x} \approx \frac{(1.225\text{kg/m}^3)}{(1025\text{kg/m}^3)} \cdot \frac{(0.01)\cdot(40\text{m/s})^2}{(9.8\text{m/s}^2)\cdot(50\text{m})} = \underline{3.9\text{x}10^{-5}}$$

Check: Units OK. Physics OK.
Exposition: The flow of water is tremendous. As it starts to pile up, the gradient of sea-level begins to drive water away from the surge, so it does not continue growing. The slope corresponds to 4 cm rise per km distance toward the shore. Over 10s to 100s km, the rise along the coast can be significant.

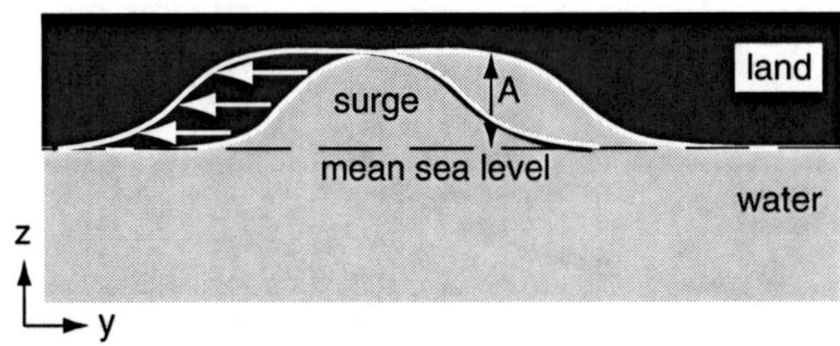

Figure 16.44
The surge, viewed from the east, is a Kelvin wave that propagates south.

Sample Application

Using info from the previous Sample Application, find the Kelvin wave phase speed, and the growth rate if the tropical cyclone follows the wave southward.

Find the Answer

Given: (same as previous example)
Find: $c = ?\ \text{m s}^{-1}$, $\Delta A/\Delta t = ?\ \text{m s}^{-1}$

Use eq. (16.23):

$$c = \sqrt{(9.8\text{m/s}^2)\cdot(50\text{m})} = \underline{\mathbf{22.1\ m\ s^{-1}}}$$

Use eq. (16.24):

$$\frac{\Delta A}{\Delta t} \approx \frac{(1.225\text{kg/m}^3)}{(1025\text{kg/m}^3)} \cdot \frac{(0.01)\cdot(40\text{m/s})^2}{\sqrt{(9.8\text{m/s}^2)\cdot(50\text{m})}}$$
$$= \underline{\mathbf{0.00086\ m\ s^{-1}}}$$

Check: Units OK. Physics OK.
Exposition: Tropical cyclones usually turn northward. If tropical cyclones were to translate southward with speed 22.1 m s^{-1}, matching the Kelvin wave speed, then an exceptionally dangerous situation would develop with amplification of the surge by over 3 m h^{-1}.

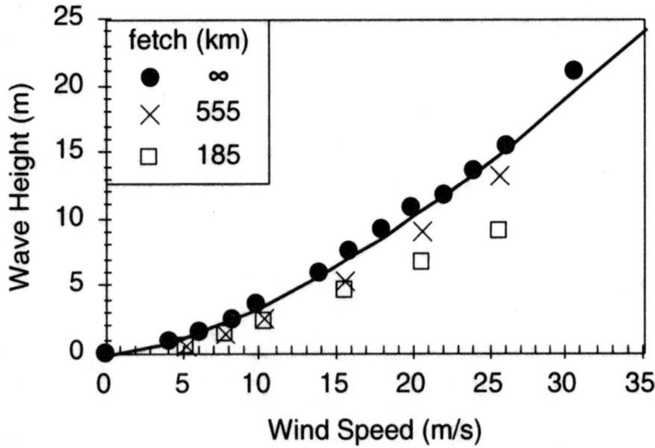

Figure 16.45
Wave height of surface wind-generated waves. Solid line is eq. (16.25) for unlimited fetch and duration. Data points are ocean observations with different fetch.

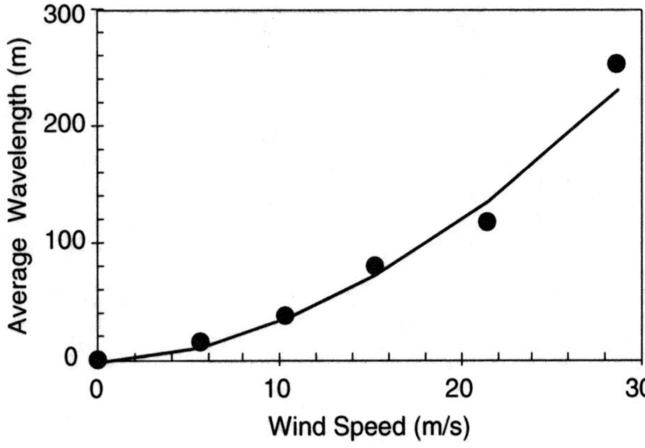

Figure 16.46
Wavelength of wind waves. Solid line is eq. (16.26).

Should the tropical cyclone move southward at a speed nearly equal to the Kelvin wave speed, then Ekman pumping would continue to reinforce and build the surge, causing the amplitude of the wave A to grow according to:

$$\frac{\Delta A}{\Delta t} \approx \frac{\rho_{air}}{\rho_{water}} \cdot \frac{C_D \cdot V^2}{\sqrt{|g|\cdot H}} \qquad (16.24)$$

where the amplitude A is measured as maximum height of the surge above mean sea level, and the other variables are the same as for eq. (16.21).

Surface Wind-waves

Waves are generated on the sea surface by action of the winds. Greater winds acting over longer distances (called **fetch**) for greater time durations can excite higher waves. High waves caused by tropical cyclone-force winds are not only a hazard to shipping, but can batter structures and homes along the coast.

Four coastal hazards of a tropical cyclone are:

- wave scour of the beach under structures,
- wave battering of structures,
- surge flooding, and
- wind damage.

The first two hazards exist only right on the coast, in the beach area. Also, the surge rapidly diminishes by 10 to 15 km inland from the coast.

For wind speeds M up to tropical cyclone force, the maximum-possible wave height (for unlimited fetch and duration) can be estimated from:

$$h = h_2 \cdot \left(\frac{M}{M_2}\right)^{3/2} \qquad (16.25)$$

where h_2 = 4 m and M_2 = 10 m s^{-1}. Wave heights are plotted in Fig. 16.45.

As winds increase beyond tropical cyclone force, the wave tops become partially chopped off by the winds. Thus, wave height does not continue to increase according to eq. (16.25). For extreme winds of 70 m s^{-1}, the sea surface is somewhat flat, but poorly defined because of the mixture of spray, foam, and chaotic seas that appear greenish white during daytime.

Wavelengths of the wind-waves also increase with wind speed. Average wavelengths λ can be approximated by:

$$\lambda = \lambda_2 \left(\frac{M}{M_2}\right)^{1.8} \qquad (16.26)$$

where λ_2 = 35 m, and M_2 = 10 m s^{-1}. Wavelengths are plotted in Fig. 16.46.

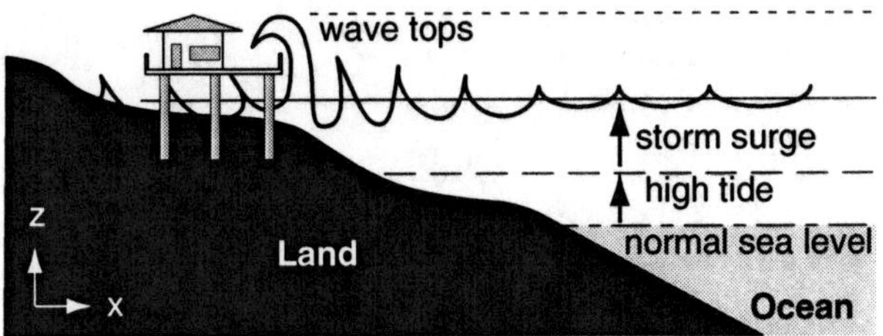

Figure 16.47
Tides, storm surge, and waves are additive.

The longest wavelength waves are called swell, and can propagate large distances, such as across whole oceans. Hence, a tropical cyclone in the middle tropical Atlantic can cause large surf in Florida well before the storm reaches the coast.

Wind and waves as affect mariners are classified according to the Beaufort scale. The INFO Box with Table 16-8 gives an historical description of the Beaufort scale, and the modern description is in Table 16-9.

Tides, storm surges, and wind waves are additive (Fig. 16.47). Namely, if the storm surge and high waves happen to occur during high tide according to routine astronomic tide tables, then the coastal destruction is likely to be greatest.

For safety, houses at coasts threatened with storm surges and tsunami are usually built on top of concrete or steel piles. These piles are driven very deep into the land to survive wave scour and erosion of the land. While the floor deck is above expected storm-surge plus high tide levels in this figure, the waves can still batter and damage the structure.

Sample Application

Find the maximum possible wave height (assuming unlimited fetch) and wavelength for tropical cyclone force winds of 35 m s^{-1}.

Find the Answer

Given: $M = 35$ m s^{-1}

Find: $h = ?$ m, $\lambda = ?$ m

Use eq. (16.25):

$$h = (4\text{m})\cdot\left(\frac{35\text{m/s}}{10\text{m/s}}\right)^{3/2} = \underline{\mathbf{26.2\ m}}$$

Use eq. (16.26):

$$\lambda = (35\text{m})\cdot\left(\frac{35\text{m/s}}{10\text{m/s}}\right)^{1.8} = \underline{\mathbf{334\ m}}$$

Check: Units OK. Physics OK. Agrees with Figs.

Exposition: Wavelengths are much longer than wave heights. Thus, wave slopes are small — less that 1/10. Only when these waves reach shore does wave slope grow until the waves break as surf.

INFO • Beaufort Wind-Scale History

In 1805, Admiral Beaufort of the British Navy devised a system to estimate and report wind speeds based on the amount of canvas sail that a full-rigged frigate could carry. It was updated in 1874, as listed in Table 16-8 for historical interest. Modern descriptors for the Beaufort wind scale are in Table 16-9.

Table 16-8. Legend: B = Beaufort number; D = modern classification; M = wind speed in knots (2 knots ≈ 1 m s^{-1}), S1 = speed through smooth water of a well-conditioned man-of-war with all sail set, and clean full; S2 = sails that a well-conditioned man-of-war could just carry in chase, full and by; S3 = sails that a well-conditioned man-of-war could scarcely bear.

B	D	M (kt)	Deep Sea Criteria
0	Calm	0 - 1	S1 = Becalmed
1	Light Air	1 - 3	S1 = Just sufficient to give steerageway
2	Slight Breeze	4 - 6	S1 = 1 - 2 knots
3	Gentle Breeze	7 - 10	S1 = 3 - 4 knots
4	Moderate Breeze	11 - 16	S1 = 5 - 6 knots
5	Fresh Breeze	17 - 21	S2 = Royals, etc.
6	Strong Breeze	22 - 27	S2 = Topgallant sails
7	High Wind	28 - 33	S2 = Topsails, jib, etc.
8	Gale	34 - 40	S2 = Reefed upper topsails and courses
9	Strong Gale	41 - 48	S2 = Lower topsails and courses
10	Whole Gale	49 - 55	S3 = lower main topsail and reefed foresail
11	Storm	56 - 65	S3 = storm staysails
12	Hurricane	> 65	S3 = no canvas

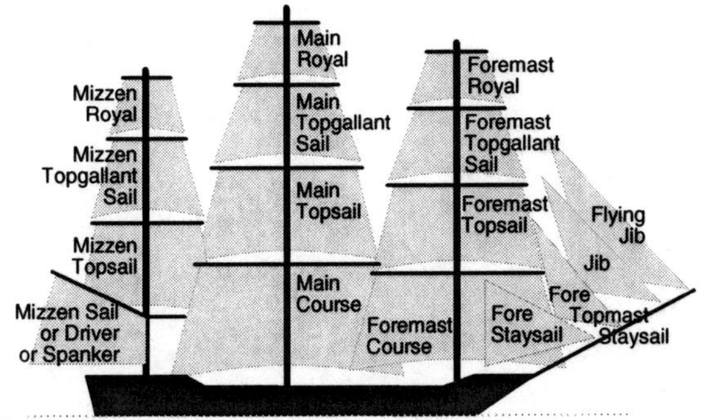

Figure 16.d. *Frigate sails.*

Table 16-9. Beaufort wind scale. B = Beaufort Number. (See INFO Box for historical info.)

B	Descrip-tion	Wind Speed (km/h)	(m/s)	Wave Height (m)	Sea Conditions (in deep ocean)	Land Conditions
0	Calm	< 1	< 0.3	0	Flat.	Calm. Smoke rises vertically.
1	Light Air	1 - 5	0.3 - 1.5	0 -00.2	Ripples without crests.	Wind motion visible in smoke.
2	Light Breeze	6 - 11	1.5 - 3.3	0.2 - 0.5	Small wavelets. Crests of glassy appearance, not breaking.	Wind felt on exposed skin. Leaves rustle.
3	Gentle Breeze	12 - 19	3.3 - 5.5	0.5 - 1	Large wavelets. Crests begin to break; scattered whitecaps.	Leaves and smaller twigs in constant motion.
4	Moderate Breeze	20 - 28	5.5 - 8.0	1 - 2	Small waves with breaking crests. Fairly frequent white horses.	Dust and loose paper raised. Small branches begin to move.
5	Fresh Breeze	29 - 38	8.0 - 11	2 - 3	Moderate waves of some length. Many white horses. Small amounts of spray.	Branches of a moderate size move. Small trees begin to sway.
6	Strong Breeze	39 - 49	11 - 14	3 - 4	Long waves begin to form. White foam crests are very frequent. Some airborne spray is present.	Large branches in motion. Whistling heard in overhead wires. Umbrella use becomes difficult. Empty plastic garbage cans tip over.
7	High Wind, Moderate Gale, Near Gale	50 - 61	14 - 17	4 - 5.5	Sea heaps up. Some foam from breaking waves is blown into streaks along wind direction. Moderate amounts of airborne spray.	Whole trees in motion. Effort needed to walk against the wind. Swaying of skyscrapers may be felt, especially by people on upper floors.
8	Gale, Fresh Gale	62 - 74	17 - 20	5.5 - 7.5	Moderately high waves with breaking crests forming spindrift. Well-marked streaks of foam are blown along wind direction. Considerable airborne spray.	Some twigs broken from trees. Cars veer on road. Progress on foot is seriously impeded.
9	Strong Gale	75 - 88	21 - 24	7 - 10	High waves whose crests sometimes roll over. Dense foam is blown along wind direction. Large amounts of airborne spray may begin to reduce visibility.	Some branches break off trees, and some small trees blow over. Construction/temporary signs and barricades blow over. Damage to circus tents and canopies.
10	Storm, Whole Gale	89 - 102	25 - 28	9 - 12.5	Very high waves with overhanging crests. Large patches of foam from wave crests give the sea a white appearance. Considerable tumbling of waves with heavy impact. Large amounts of airborne spray reduce visibility.	Trees are broken off or uprooted, saplings bent and deformed. Poorly attached asphalt shingles and shingles in poor condition peel off roofs.
11	Violent Storm	103 - 117	29 - 32	11.5 - 16	Exceptionally high waves. Very large patches of foam, driven before the wind, cover much of the sea surface. Very large amounts of airborne spray severely reduce visibility.	Widespread vegetation damage. Many roofing surfaces are damaged; asphalt tiles that have curled up and/or fractured due to age may break away completely.
12	Hurricane	≥ 118	≥ 33	≥14	Huge waves. Sea is completely white with foam and spray. Air is filled with driving spray, greatly reducing visibility.	Very widespread damage to vegetation. Some windows may break; mobile homes and poorly constructed sheds and barns are damaged. Debris may be hurled about.

Inland Flooding

In developed countries such as the USA, storm-surge warning and evacuation systems are increasingly successful in saving lives of coastal dwellers. However, inland flooding due to heavy rain from decaying tropical storms is increasingly fatal — causing roughly 60% of the tropical cyclone-related deaths in the USA during the past 30 years.

Streams and storm drains overflow, trapping people in cars and on roof tops. For this reason, people should not be complacent about former tropical cyclones reaching their inland homes, because these dying tropical cyclones contain so much tropical moisture that they can cause record-setting rainfalls.

The inland flooding hazard can affect people hundreds of kilometers from the coast. Of the 56 people who died in Hurricane Floyd (1999) in the eastern USA, 50 drowned in inland floods caused by heavy rains. Tropical storm Alberto dropped 53 cm of rain over Americus, Georgia, and 33 people drowned in 1994. Over 200 people drowned in Pennsylvania, New York, and New England from Hurricane Diane in 1955.

More recently, many unnecessary drowning deaths are caused by people driving into water of unknown depth covering the road. For people comfortable in driving their usual roads day after day, many find it hard to believe that their roads can become impassable. They unknowingly drive into deep water, causing the engine to stall and the car to stop in the middle of the water. If the water continues to rise, the car and passengers can be carried away. An easy solution is to approach each flooded road with caution, and be prepared to interrupt your journey and wait until the flood waters subside.

Thunderstorms, Lightning & Tornado Outbreaks

Because tropical cyclones are made of thunderstorms, they contain all the same hazards as thunderstorms. These include lightning, downbursts, gust fronts, downpours of rain, and tornadoes (see the Thunderstorm chapter for hazard details). Lightning is somewhat infrequent in the eyewall (about 12 cloud-to-ground strikes per hour), compared to about 1000 strikes per hour for midlatitude MCSs.

Tropical cyclones can cause **tornado outbreaks**. For example, Hurricanes Cindy and Katrina each spawned 44 tornadoes in the USA in 2005. Hurricane Rita spawned 101 tornadoes in Sep 2005. Hurricanes Frances and Ivan spawned 103 and 127 tornadoes, respectively, in Sep 2004.

Most tornadoes are weak (≤ EF2), and occur in the right front quadrant of tropical cyclones in the N. Hemisphere. Although some form near the eyewall, most tornadoes form in the thunderstorms of outer rain bands 80 to 480 km from the cyclone center.

TROPICAL CYCLONE FORECASTING

Prediction

The most important advance in tropical cyclone prediction is the weather satellite (see the Satellites & Radar chapter). Satellite images can be studied to find and track tropical disturbances, depressions, storms, and cyclones. By examining loops of sequential images of tropical-cyclone position, their past track and present translation speed and direction can be determined. Satellites can be used to estimate rainfall intensity and storm-top altitudes.

Research aircraft (**hurricane hunters**) are usually sent into dangerous storms to measure pressure, wind speed, temperature, and other variables that are not easily detected by satellite. Also, they can fly transects through the middle of the tropical storm to precisely locate the eye (Fig. 16.48). The two organizations that do this for Atlantic Hurricanes are the US Air Force Reserves **53rd Reconnaissance Squadron**, and the Aircraft Operations Center of the US **National Oceanic and Atmospheric Administration** (**NOAA**).

Forecasting future tracks and intensity of tropical cyclones is more difficult, and has lots of error. Computer codes (called models) describing atmospheric physics and dynamics (see the Numerical Weather Prediction (NWP) chapter) are run to fore-

Figure 16.48
Photo inside the eye of Hurricane Rita on 21 Sep 2005, taken from a NOAA P3 aircraft. Rita was category 5 at max intensity. (Image courtesy of NOAA/AOC.)

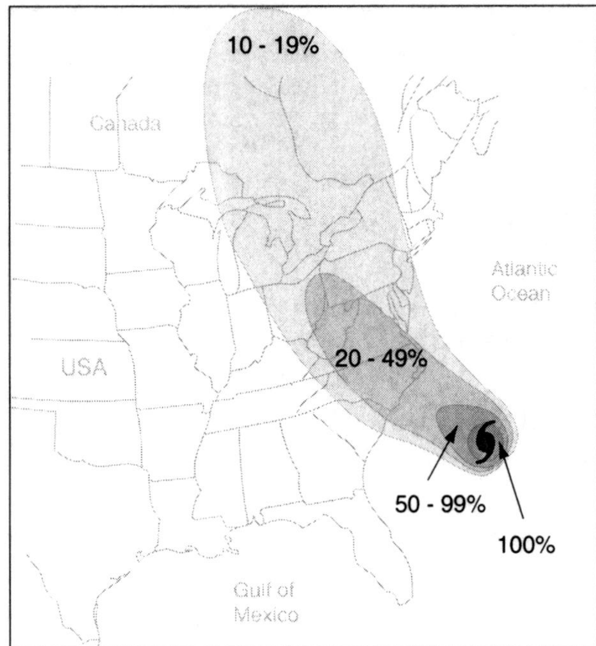

Figure 16.49
Probability forecast that the eye of Hurricane Isabel would pass within 120 km of any point on the map during the 72 h starting 1800 UTC on 17 Sep 2003. The actual location of the hurricane at this time is indicated with the hurricane map symbol.

cast the weather. But different models yield different forecast tracks. So human forecasters consider all the NWP model forecasts, and issue probability forecasts (Fig. 16.49) on the likelihood that any tropical cyclone will strike different sections of coastline. Local government officials and emergency managers then make the difficult (and costly) decision on whether to evacuate any sections of coastline.

Predicting tropical cyclone intensity is even more difficult. Advances have been made based on sea-surface temperatures, atmospheric static stability, ambient wind shear, and other factors. Also, processes such as eyewall replacement and interaction with other tropical and extratropical systems are considered. But much more work needs to be done.

Different countries have their own organizations to issue forecasts. In the USA, the responsible organization is the **National Hurricane Center**, also known as the **Tropical Prediction Center**, a branch of NOAA. This center issues the following forecasts:

- **Tropical Storm Watch** - tropical storm conditions are possible within 36 h for specific coastal areas. (Includes tropical storms as well as the outer areas of tropical cyclones.)
- **Tropical Storm Warning** - tropical storm conditions are expected within 12 h or less for specific coastal areas. (Includes tropical storms as well as the outer areas of tropical cyclones.)
- **Hurricane Watch** - hurricane conditions are possible within 36 h for specific coastal areas.
- **Hurricane Warning** - hurricane conditions are expected within 24 h or less for specific coastal areas. Based on the worst of expected winds, high water, or waves.

In Canada, the responsible organization is the **Canadian Hurricane Center,** a branch of the **Meteorological Service of Canada**.

Safety

The US National Hurricane Center offers these recommendations for a Family Disaster Plan:

- Discuss the type of hazards that could affect your family. Know your home's vulnerability to storm surge, flooding and wind.
- Locate a safe room or the safest areas in your home for each hurricane hazard. In certain circumstances the safest areas may not be your home but could be elsewhere in your community.
- Determine escape routes from your home and places to meet. These should be measured in tens of miles rather than hundreds of miles.
- Have an out-of-state friend as a family contact, so all your family members have a single point of contact.
- Make a plan now for what to do with your pets if you need to evacuate.
- Post emergency telephone numbers by your phones and make sure your children know how and when to call 911.
- Check your insurance coverage — flood damage is not usually covered by homeowners insurance.
- Stock non-perishable emergency supplies and a Disaster Supply Kit.
- Use a NOAA weather radio. Remember to replace its battery every 6 months, as you do with your smoke detectors.
- Take First Aid, cardiopulmonary resuscitation (**CPR**) and disaster-preparedness classes.

If you live in a location that receives an evacuation order, it is important that you follow the instructions issued by the local authorities. Some highways are designated as evacuation routes, and traffic might be rerouted to utilize these favored routes. Get an early start, because the roads are increasingly congested due to population growth. Before you evacuate, be sure to board-up your home to protect it from weather and looters.

Many people who live near the coast have storm shutters permanently installed on their homes, which they can close prior to tropical cyclone arrival. For windows and storefronts too large for shutters, have large pieces of plywood on hand to screw over the windows.

If a tropical cyclone overtakes you and you cannot escape, stay indoors away from windows. Street signs, corrugated metal roofs, and other fast-moving objects torn loose by tropical cyclone-force winds can slice through your body like a guillotine.

REVIEW

Hurricanes are tropical cyclones. They have low-pressure centers, called eyes, and rotation is cyclonic (counterclockwise in the Northern Hemisphere) near the surface. The tropical cyclone core is warm, which causes high pressure to form in the eye near the top of the storm. This high pressure drives diverging, anticyclonic winds out of the tropical cyclone.

Tropical cyclones are born over tropical oceans with temperature ≥26.5°C over 50 m or more depth. Evaporation from the warm ocean into the windy boundary layer increases the energy in the storm, which ultimately drives its circulation similar to a Carnot-cycle heat engine. Tropical cyclones die over cold water and over land, not due to the extra drag caused by buildings and trees, but due mostly to the lack of strong evaporation from the ocean.

Because tropical cyclones are born in the trade-wind regions of the global circulation, they are initially blown westward. Many eventually reach the eastern shores of continents where the global circulation turns them poleward. At the equator there is no Coriolis force; hence there is no rotation available to be concentrated into tropical cyclones. Thus, tropical cyclones are most likely to form between 10° to 30° latitude during autumn.

Updrafts are strongest in the eye wall of thunderstorms encircling the clear eye. Rotation is initially gathered from the absolute angular momentum associated with the Earth's rotation. As the storm develops and gains speed, centrifugal force dominates over Coriolis force within about 100 km of the eye, causing winds that are nearly cyclostrophic in the bottom third of the troposphere. Simple analytical models can be built to mimic the velocities, temperature, and pressure across a tropical cyclone.

While near the shore, tropical cyclones can cause damage due to storm-surge flooding, wind-wave battering, beach erosion, wind damage, heavy rain, tornadoes, and lightning. The surge is caused by Ekman transport of water toward shore, and by the reduced atmospheric pressure head.

Science Graffito

In 1989, category 5 hurricane Hugo moved directly over the US. Virgin Islands in the Caribbean Sea. When the hurricane reached the island of St. Croix, it temporarily stopped its westward translation, allowing the intense eye wall to blast the island with violent winds for hours. The following is an eyewitness account.

"It had been many years since St. Croix was in the path of a major storm. Hurricane Hugo reached into the Lesser Antilles with a deliberate vengeance. St. Croix was somewhat prepared. Many hundreds of people had moved into schools and churches to take refuge. But no one was ready for what happened next. By 1800 hours winds were a steady 50 kts [25 m s^{-1}] with gusts up to 70 kts [35 m s^{-1}] from the northwest. I was on the top floor of the wooden Rectory at the St. Patrick's Church in Frederiksted with my husband and 8 month old son."

"By 2000 hours it was apparent that our comfortable room with a view was not going to provide a safe haven. The electricity had been out for some time and a very big gust from the north blew the air conditioner out of the window, landing at the foot of our bed. We evacuated with only one diaper change and bottle of baby juice, leaving behind the playpen, high chair, and bundles of accessories brought from home. We followed Fr. Mike down the wooden staircase. Drafts were everywhere and glass doors exploded just as we passed on our way to Fr. O'Connor's main living quarters on the first floor, where the walls were made of thick coral blocks."

"We settled in again in spite of the persistent crashing and banging against the heavy wooden shutters. We had to shout to hear each other across the room and our ears were popping. In the bathroom, the plumbing sounded like a raging sea. The water in the toilet bowl sloshed around and vibrated. Mercifully the baby slept."

"Soon the thick concrete walls and floors were vibrating accompanied by a hum that turned into the 'freight train howl'. The banging intensified and persisted for the next 4 hours. By 0100 hours we were tense and sweaty and wondering if it would ever end and if there was anything left outside. Fr. O'Connor was praying and feared that many people must be dead. He got up to open a closet door and a wall of water flowed into the bedroom. At that point we moved to the dining room with a group of 8 other people trapped in the rectory and waited."

"There was concern that the rest of the roof would go and it was decided we would make a run for the schoolhouse made of 2-foot [0.61 m] thick concrete walls. I held the baby in my arms and with flimsy flip-flops [sandals], just about skated across the cement courtyard dodging flying branches and sheets of galvanized aluminum. The window was opened for us as a big, old mahogany tree blocked the door."

"Shortly after, the eye was over us. The thick wooden shutters were flung open and about 100 people outside climbed in the window. The housing project nearby had been stripped of its north and east walls. The eye remained over us for 2 hours then the wind started up with the same intensity coming from the southwest. Only now the room was packed. Strangers were sharing the same mattresses. People slept in desks and chairs made for elementary children, and it was hot. Toddlers and infants wailed."

(continues on next page)

Science Graffito

(continuation)

"There was no generator, only an occasional flashlight could be seen. Fears of [storm] surges were on everyone's' minds. We were only 200 meters from the west shoreline. Hugo had slowed down its eastward track to 4 mph [1.8 m s^{-1}] and the eye passed straight through the middle of the 23-mile [37 km] long island of St. Croix. It seems like the storm was in a fixed permanent position."

"When dawn broke, the winds still howled. By 0800 it was safe to open the windows and the landscape made me burst into tears. There was not a leaf left on a tree, there was not a tree left standing, just tangled branches lying sideways everywhere and not one blade of green grass. The wind had burned the ground and turned everything brown. The gray skies, light rain, and brown landscape persisted for several weeks."

"There were only 2 deaths reported but within weeks several dozen people died from heart attacks, strokes, electrocutions and other accidents associated with reconstruction. The majority of the island residents functioned without power for 3 to 6 months, using generators or candle power and gas stoves."

– Susan Krueger Allick Beach, 1999

HOMEWORK EXERCISES

Broaden Knowledge & Comprehension

B1. a. Search for satellite images and movies for a few recent tropical cyclones. Discuss similarities and differences of the tropical cyclone appearance.
b. Same, but for radar images during landfall.
c. Same, but for photos from hurricane hunters.

B2. Search for web sites that show tropical cyclone tracks for: a. the current tropical cyclone season.
b. past tropical cyclone seasons.

B3. What names will be used for tropical cyclones in the next tropical cyclone season, for the ocean basin assigned by your instructor?

B4. How many NWP models are available for forecasting tropical cyclone track, & how are they used?

B5. What is the long range forecast for the number of tropical cyclones for the upcoming season, for an ocean basin assigned by your instructor. (Or, if the season is already in progress, how are actual numbers and intensities comparing to the forecast.)

B6. Find maps of "tropical cyclone potential", "tropical cyclone energy", "intensity", or "sea-surface temperature". What are these products based on, and how are they used for tropical cyclone forecasting?

B7. Search the web for photos and info on **hot convective towers**. How do they affect tropical cyclones?

Apply

A1. At 10° latitude, find the absolute angular momentum ($m^2\ s^{-1}$) associated with the following radii and tangential velocities:

	R (km)	M_{tan} (m s^{-1})
a.	50	50
b.	100	30
c.	200	20
d.	500	5
e.	1000	0
f.	30	85
g.	75	40
h.	300	10

A2. If there is no rotation in the air at initial radius 500 km and latitude 10°, find the tangential velocity (m s^{-1} and km h^{-1}) at radii (km):
a. 450 b. 400 c. 350 d. 300
e. 250 f. 200 g. 150 h. 100

A3. Assume ρ = 1 kg m^{-3}, and latitude 20° Find the value of gradient wind (m s^{-1} and km h^{-1}) for:

	R (km)	$\Delta P/\Delta R$ (kPa/100 km)
a.	100	5
b.	75	8
c.	50	10
d.	25	15
e.	100	10
f.	75	10
g.	50	20
h.	25	25

However, if a gradient wind is not possible for those conditions, explain why.

A4. For the previous problem, find the value of cyclostrophic wind (m s^{-1} and km h^{-1}).

A5. Plot pressure vs. radial distance for the max pressure gradient that is admitted by gradient-wind theory at the top of a tropical cyclone for the latitudes (°) listed below. Use z = 17 km, P_o = 8.8 kPa.
a. 5 b. 7 c. 9 d. 11 e. 13 f. 17 g. 19
h. 21 i. 23 j. 25 k. 27 m. 29 n. 31 o. 33

A6. At sea level, the pressure in the eye is 93 kPa and that outside is 100 kPa. Find the corresponding pressure difference (kPa) at the top of the tropical cyclone, assuming that the core (averaged over the tropical cyclone depth) is warmer than surroundings by (°C):

a. 5 b. 2 c. 3 d. 4
e. 1 f. 7 g. 10 h. 15

A7. At radius 50 km the tangential velocity decreases from 35 m s^{-1} at the surface to 10 m s^{-1} at the altitude (km) given below:

a. 2 b. 4 c. 6 d. 8
e. 10 f. 12 g. 14 h. 16

Find the radial temperature gradient (°C/100km) in the tropical cyclone. The latitude = 10°, and average temperature = 0°C.

A8. Find the total entropy (J·kg^{-1}·K^{-1}) for:

	P (kPa)	T (°C)	r (g kg^{-1})
a.	100	26	22
b.	100	26	0.9
c.	90	26	24
d.	80	26	0.5
e.	100	30	25
f.	100	30	2.0
g.	90	30	28
h.	20	–36	0.2

A9. On a thermo diagram of the Atmospheric Stability chapter, plot the data points from Table 16-5. Discuss.

A10. Starting with saturated air at sea-level pressure of 90 kPa in the eye wall with temperature of 26°C, calculate (by equation or by thermo diagram) the thermodynamic state of that air parcel as it moves to:

a. 20 kPa moist adiabatically, and thence to
b. a point where the potential temperature is the same as that at 100 kPa at 26°C, but at the same height as in part (a). Thence to
c. 100 kPa dry adiabatically and conserving humidity. Thence to
d. Back to the initial state.
e to h: Same as a to d, but with initial T = 30°C.

A11. Given the data from Table 16-5, what would be the mechanical energy (J) available if the average temperature at the top of the tropical cyclone were

a. –18 b. –25 c. –35 d. –45
e. –55 f. –65 g. –75 h. –83

A12. For the previous problem, find the minimum possible eye pressure (kPa) that could be supported.

A13. Use P_∞ = 100 kPa at the surface. What maximum tangential velocity (m s^{-1} and km h^{-1}) is expected for an eye pressure (kPa) of:

a. 86 b. 88 c. 90 d. 92
e. 94 f. 96 g. 98 h. 100

A14. For the previous problem, what are the peak velocity values (m s^{-1} and km h^{-1}) to the right and left of the storm track, if the tropical cyclone translates with speed (m s^{-1}):

(i) 2 (ii) 4 (iii) 6 (iv) 8 (v) 10
(vi) 12 (vii) 14 (viii) 16 (ix) 18 (x) 20

A15. For radius (km) of:

a. 5 b. 10 c. 15 d. 20
e. 25 f. 30 g. 50 h. 100

find the tropical cyclone-model values of pressure (kPa), temperature (°C), and wind components (m s^{-1}), given a pressure in the eye of 95 kPa, critical radius of R_o = 20 km, and W_s = –0.2 m s^{-1}. Assume the vertically-averaged temperature in the eye is 0°C.

A16. (§) For the previous problem, plot the radial profiles of those variables between radii of 0 to 200 km.

A17. Use P_∞ = 100 kPa at the surface. Find the pressure-head contribution to rise of sea level (m) in the eye of a tropical cyclone with central pressure (kPa) of:

a. 86 b. 88 c. 90 d. 92
e. 94 f. 96 g. 98 h. 100

A18. Find the Ekman transport rate [km^3/(h·km)] and surge slope (m km^{-1}) if winds (m s^{-1}) of:

a. 10 b. 20 c. 30 d. 40
e. 50 f. 60 g. 70 h. 80

in advance of a tropical cyclone are blowing parallel to the shore, over an ocean of depth 50 m. Use C_D = 0.005 and assume a latitude of 30°.

A19. What is the Kelvin wave speed (m s^{-1} and km h^{-1}) in an ocean of depth (m):

a. 200 b. 150 c. 100 d. 80
e. 60 f. 40 g. 20 h. 10

A20. For the previous problem, find the growth rate of the Kelvin wave amplitude (m h^{-1}) if the tropical cyclone tracks south parallel to shore at the same speed as the wave.

A21. Find the wind-wave height (m) and wavelength (m) expected for wind speeds (m s^{-1}) of:

a. 10 b. 15 c. 20 d. 25
e. 30 f. 40 g. 50 h. 60

A22. For the previous problem, give the:

(i) Beaufort wind category, and give a modern description of the conditions on land and sea

(ii) Saffir-Simpson tropical cyclone Wind category, its
corresponding concise statement, and
describe the damage expected.

Evaluate & Analyze

E1. In Fig. 16.2, if the thin ring of darker shading represents heavy precipitation from the eyewall thunderstorms, what can you infer is happening in most of the remainder of the image, where the shading is lighter grey? Justify your inference.

E2. Thunderstorm depths nearly equal their diameters. Explain why tropical cyclone depths are much less than their diameters. (Hint, consider Fig. 16.3)

E3. If you could see movie loops of satellite images for the same storms shown in Figs. 16.1 and 16.4 in the Northern Hemisphere, would you expect these satellite loops to show the tropical cyclone clouds to be rotating clockwise or counterclockwise? Why?

E4. Speculate on why tropical cyclones can have eyes, but mid-latitude supercell thunderstorms do not?

E5. If a tropical cyclone with max sustained wind speed of 40 m s^{-1} contains tornadoes that do EF4 damage, what Saffir-Simpson Wind Scale category would you assign to it? Why?

E6. Why don't tropical cyclones or typhoons hit the Pacific Northwest coast of the USA and Canada?

E7. The INFO box on tropical cyclone-induced Currents in the ocean shows how Ekman transport can lower sea level under a tropical cyclone. However, we usually associate rising sea level with tropical cyclones. Why?

E8. Consider Fig. 16.12. Should there also be a jet along the north edge of the hot Saharan air? If so, explain its characteristics. If not, explain why.

E9. Fig. 16.14a shows wind moving from east to west through an easterly wave. Fig. 16.14b shows the whole wave moving from east to west. Can both these figures be correct? Justify your answer.

E10. In the Extratropical Cyclone chapter, troughs were shown as southward meanders of the polar jet stream. However, in Fig. 16.14 the troughs are shown as northward meanders of the trade winds. Explain this difference. Hint, consider the General Circulation.

E11. Compare and contrast a TUTT with mid-latitude cyclone dynamics as was discussed in the Extratropical Cyclone chapter.

E12. Consider Fig. 16.16. If a typhoon in the Southern Hemisphere was blown by the trade winds toward the Northern Hemisphere, explain what would happen to the tropical cyclone as it approaches the equator, when it is over the equator, and when it reaches the Northern Hemisphere. Justify your reasoning.

E13. Fig. 16.17 suggests that a cold front can help create a tropical cyclone, while Fig. 16.23 suggests that a cold front will destroy a tropical cyclone. Which is correct? Justify your answer. If both are correct, then how would you decide on tropical cyclogenesis or cyclolysis?

E14. In the life cycle of tropical cyclones, Mesoscale Convective Systems (MCSs) are known to be one of the possible initial stages. But MCSs also occur over the USA, as was discussed in the Thunderstorm chapter. Explain why the MCSs over the USA don't become tropical cyclones.

E15. Based on the visible satellite image of Fig. 16.19, compare and contrast the appearance of the tropical cyclone at (c) and the tropical storm at (d). Namely, what clues can you use from satellite images to help you decide if the storm has reached full tropical cyclone strength, or is likely to be only a tropical storm?

E16. What are the other large cloud areas in Fig. 16.19 that were not identified in the caption? Hint, review the Satellites & Radar chapter.

E17. The Bermuda High (or Azores High) as shown in Fig. 16.20 forms because the ocean is cooler than the surrounding continents. However, tropical cyclones form only when the sea-surface temperature is exceptionally warm. Explain this contradiction.

E18. If global warming allowed sea-surface temperatures to be warmer than 26.5°C as far north as 60°N, could Atlantic tropical cyclones reach Europe? Why?

E19. Could a tropical cyclone exist for more than a month? Explain.

E20. a. What is the relationship between angular momentum and vorticity?

b. Re-express eq. (16.1) as a function of vorticity.

E21. From the Atmospheric Forces & Winds chapter, recall the equation for the boundary-layer gradient wind. Does this equation apply to tropical cyclone boundary layers? If so what are it's limitations and characteristics.

E22. The top of Fig. 16.24 shows weak low pressure in the eye surrounding by high pressure in the eye-wall at the top of the storm. Explain why the storm cannot have high pressure also in the eye.

E23. Fig. 16.25 shows the wind vector M parallel to the acceleration vector F_{net}. Is that realistic? If not, sketch the likely vectors for M and F_{net}. Explain.

E24. The words "**supergradient winds**" means winds faster that the gradient wind speed. Are the outflow winds at the top of a tropical cyclone super-gradient? Justify your answer.

E25.(§) Using relationships from Chapter 1, plot an environmental pressure profile vs. height across the troposphere assuming an average temperature of 273 K. Assume the environmental sea-level pressure is 100 kPa. On the same graph, plot the pressure profile for a warm tropical cyclone core of average temperature (K): a. 280 b. 290 c. 300 assuming a sea-level pressure of 95 kPa. Discuss.

E26. Suppose that the pressure-difference magnitude between the eye and surroundings at the tropical cyclone top is only half that at the surface. How would that change, if at all, the temperature model for the tropical cyclone? Assume the sea-level pressure distribution is unaltered.

E27. Create a table that has a column listing attributes of mid-latitude cyclones, and another column listing attributes of tropical cyclones. Identify similarities and differences.

E28. The circles in Fig. 16.29 illustrate the hypsometric situation applied to the warm core of a tropical cyclone. Why cannot the hypsometric equation explain what drives subsidence in the eye?

E29. In Fig. 16.30b, why does the T_d line from Point 1 to Point 2 follow a contour that slopes upward to the right, even though the air parcel in Fig 16.30a is moving horizontally, staying near sea level?

E30. a. Re-express eq. (16.7) in terms of potential temperature.

b. Discuss the relationship between entropy and potential temperature.

E31. What factors might prevent a tropical cyclone from being perfectly efficient at extracting the maximum possible mechanical energy for any given thermodynamic state?

E32. For the tropical cyclone model given in this chapter, describe how the pressure, tangential velocity, radial velocity, vertical velocity, and temperature distribution are consistent with each other, based on dynamic and thermodynamic relationships. If they are not consistent, quantify the source and magnitude of the discrepancy, and discuss the implications and limitations. Consider the idealizations of the figure below. (ABL = atmos. boundary layer.)

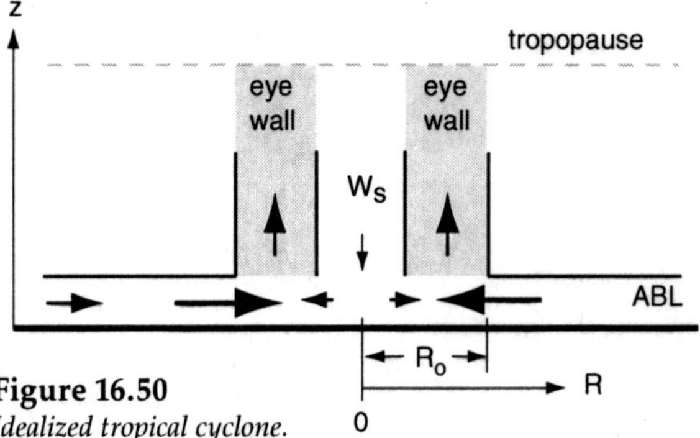

Figure 16.50
Idealized tropical cyclone.

Hints: a. Use the cyclostrophic relationship to show that tangential velocity is consistent with the pressure distribution.

b. Use mass conservation for inflowing air trapped within the boundary layer to show how radial velocity should change with R, for $R > R_o$.

c. Assuming wave drag causes radial velocity to be proportional to tangential velocity squared, show that the radial velocity and tangential velocity equations are consistent. For $R < R_o$, use the following alternate relationship based on observations in tropical cyclones: $M_{tan}/M_{max} = (R/R_o)^2$.

d. Rising air entering the bottom of the eye wall comes from two sources, the radial inflow in the boundary layer from $R > R_o$, and from the subsiding air in the eye, which hits the ground and is forced to diverge horizontally in order to conserve mass. Combine these two sources of air to compute the average updraft velocity within the eye wall.

e. Use mass continuity in cylindrical coordinates to derive vertical velocity from radial velocity, for $R < R_o$. (Note, for $R > R_o$, it was already assumed in part (a) that air is trapped in the ABL, so there is zero vertical velocity there.)

f. Use the hypsometric relationship, along with the simplifications described in the temperature subsection of the tropical cyclone model, to relate the radial temperature distribution (averaged over the whole tropical cyclone depth) to the pressure distribution.

E33. An article by Willoughby and Black (1996: tropical cyclone Andrew in Florida: dynamics of a disaster, *Bull. Amer. Meteor. Soc.*, **77**, 543-549) shows tangential wind speed vs. radial distance.

a. For their Figs. 3b and 3d, compare their observations with the tropical cyclone model in this chapter.

b. For their Figs. 3e - 3g, determine translation speed of the tropical cyclone and how it varied with time as the storm struck Florida.

E34. To reduce fatalities from tropical cyclones, argue the pros and cons of employing better mitigation technology and the pros and cons of population control. Hint: Consider the whole world, including issues such as carrying capacity (i.e., the finiteness of natural resources and energy), sustainability, politics and culture.

E35. In Fig. 16.41, one relative maximum (A) of storm-surge height is directly in front of the storm's path, while the other (B) is at the right front quadrant. Explain what effects could cause each of these surges, and explain which one might dominate further from shore while the other might dominate when the tropical cyclone is closer to shore.

E36. Fig. 16.43 for a storm surge caused by Ekman transport is for a tropical cyclone just off the east coast of a continent. Instead, suppose there was a tropical cyclone-force cyclone just off the west coast of a continent at mid-latitudes.

a. Would there still be a storm surge caused by Ekman transport?

b. Which way would the resulting Kelvin wave move (north or south) along the west coast?

E37. Devise a mathematical relationship between the Saffir-Simpson tropical cyclone Wind scale, and the Beaufort wind scale.

Synthesize

S1. What if the Earth rotated twice as fast. Describe changes to tropical cyclone characteristics, if any.

S2. What if the average number of tropical cyclones tripled. How would the momentum, heat, and moisture transport by tropical cyclones change the global circulation, if at all?

S3. Some science fiction novels describe "supercanes" with supersonic wind speeds. Are these physically possible? Describe the dynamics and thermodynamics necessary to support such a storm in steady state, or use the same physics to show why they are not possible.

S4. Suppose the tropical tropopause was at 8 km altitude, instead of roughly 16 km altitude. How would tropical cyclone characteristics change, if at all?

S5. What if the Earth's climate was such that the tropics were cold and the poles were hot, with sea surface temperature greater than 26°C reaching from the poles to 60° latitude. Describe changes to tropical cyclone characteristics, if any.

S6. Suppose that the sea surface was perfectly smooth, regardless of the wind speed. How would tropical cyclone characteristics change, if at all?

S7. Is it possible to have a tropical cyclone without a warm core? Be aware that in the real atmosphere, there are tropical cyclone-force cyclones a couple times a year over the northern Pacific Ocean, during winter.

S8. Suppose that the thermodynamics of tropical cyclones were such that air parcels, upon reaching the top of the eye wall clouds, do not loose any heat by IR cooling as they horizontally diverge away from the top of the tropical cyclone. How would the Carnot cycle change, if at all, and how would that affect tropical cyclone intensity?

S9. Is it possible for two tropical cyclones to merge into one? If so, explain the dynamics and thermodynamics involved. If so, is it likely that this could happen? Why?

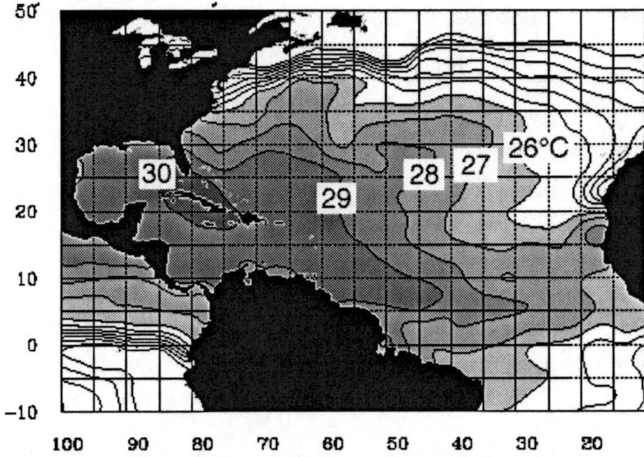

Figure 16.51
Sea surface temperature (°C) in the Atlantic Ocean averaged over the week of 5 to 11 Sep 1999, just prior to category-4 hurricane Floyd (see Fig. 16.4). Shaded regions are warmer than 26°C. The isotherm increment is 1°C, and the isotherm range is 20 to 30°C. [Adapted from a National Hurricane Ctr. image.]

17 REGIONAL WINDS

Contents

Each locale has a unique landscape that creates or modifies the wind. These local winds affect where we choose to live, how we build our buildings, what we can grow, and how we are able to travel.

During synoptic high pressure (i.e., fair weather), some winds are generated locally by temperature differences. These gentle circulations include thermals, anabatic/katabatic winds, and sea breezes.

During synoptically windy conditions, mountains can modify the winds. Examples are gap winds, boras, hydraulic jumps, foehns/chinooks, and mountain waves.

WIND FREQUENCY

Wind-speed Frequency

Wind speeds are rarely constant. At any one location, wind speeds might be strong only rarely during a year, moderate many hours, light even more hours, and calm less frequently (Fig. 17.1). The number of times that a range ΔM of wind speeds occurred in the past is the **frequency** of occurrence. Dividing the frequency by the total number of wind measurements gives a **relative frequency**. The expectation that this same relative frequency will occur in the future is the **probability** (*Pr*).

The probability distribution of mean wind speeds *M* is described by the **Weibull distribution**:

$$Pr = \frac{\alpha \cdot \Delta M \cdot M^{\alpha-1}}{M_o{}^{\alpha}} \cdot \exp\left[-\left(\frac{M}{M_o}\right)^{\alpha}\right] \tag{17.1}$$

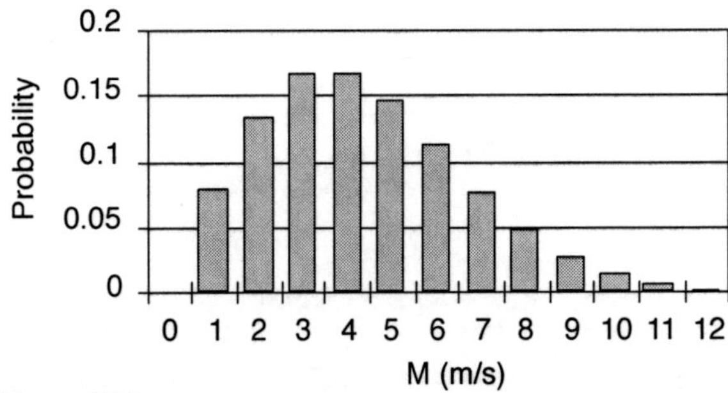

Figure 17.1
Wind-speed M probability (relative frequency) for a Weibull distribution with parameters $\alpha = 2$ and $M_o = 5$ m s^{-1}.

Sample Application

Given M_o = 5 m s^{-1} and α = 2, find the probability that the wind speed will be between 5.5 & 6.5 m s^{-1}?

Find the Answer

Given: M_o = 5 m s^{-1}, α = 2, M= 6 m s^{-1}
ΔM = 6.5 – 5.5 m s^{-1} = 1 m s^{-1},
Find: Pr = ? (dimensionless)

Use eq. (17.1):

$$Pr = \frac{2\cdot(1\text{m/s})\cdot(6\text{m/s})^{2-1}}{(5\text{m/s})^2}\cdot\exp\left[-\left(\frac{6\text{m/s}}{5\text{m/s}}\right)^2\right]$$

$= 0.114 = \underline{\mathbf{11.4\%}}$

Check: Units OK. Physics OK.
Exposition: This agrees with Fig. 17.1 at M = 6 m s^{-1}, which had the same parameters as this example. To get a sum of probabilities that is very close to 100%, use a smaller bin size ΔM and be sure not to cut off the tail of the distribution at high wind speeds.

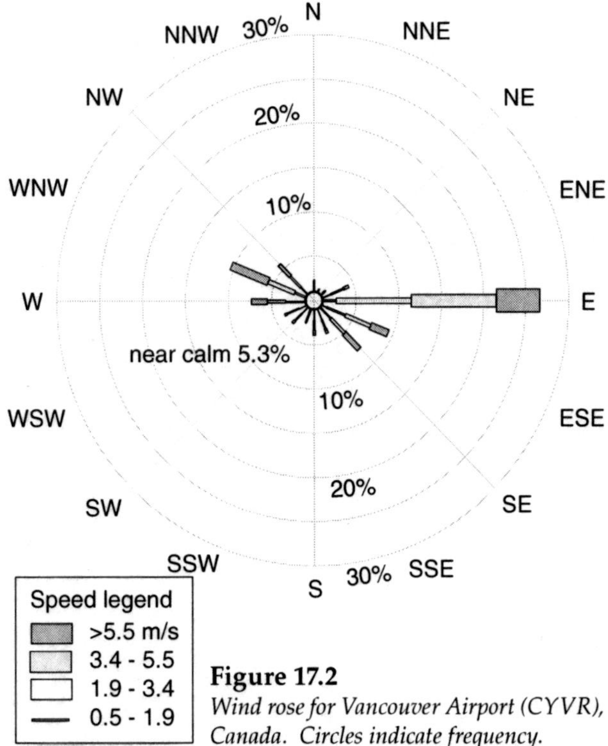

Figure 17.2
Wind rose for Vancouver Airport (CYVR), Canada. Circles indicate frequency.

Sample Application

How frequent are east winds at Vancouver airport?

Find the Answer

Use Fig. 17.2. Frequency ≈ **26%** .

Exposition: This is the sum of 2% for $0.5 < M \le 1.9$, plus 9% for $1.9 < M \le 3.4$, plus 10% for $3.4 < M \le 5.5$, plus 5% for $M > 5.5$ m s^{-1}.

where Pr is the probability (or relative frequency) of wind speed $M \pm 0.5\cdot\Delta M$. Such wind-speed variations are caused by synoptic, mesoscale, local and boundary-layer processes.

Location parameter M_o is proportional to the mean wind speed. For **spread parameter** α, smaller α causes wider spread of winds about the mean. Values of the parameters and the corresponding distribution shape vary from place to place.

The **bin size** or **resolution** is ΔM. For example, the column plotted in Fig. 17.1 for M = 3 m s^{-1} is the probability that the wind is between 2.5 and 3.5 m s^{-1}. The width of each column in the histogram is ΔM = 1 m s^{-1}. The sum of probabilities for all wind speeds should equal 1, meaning there is a 100% chance that the wind speed is between zero and infinity. Use this to check for errors. Eq. (17.1) is only approximate, so the sum of probabilities almost equals 1.

Use wind-speed distributions when estimating electrical-power generation by wind turbines, and when designing buildings and bridges to withstand extreme winds.

You can express extreme-wind likelihood as a **return period** (RP), which is equal to the total period of measurement divided by the number of times the wind exceeded a threshold. For example, if winds exceeding 30 m s^{-1} occurred twice during the last century, then the return period for 30 m s^{-1} winds is RP = (100 yr)/2 = 50 years. Faster winds occur less frequently, and have greater return periods.

Wind-direction Frequency

By counting the frequency of occurrence that winds came from each compass direction (N, NNE, NE, etc.) over a period such as 10 years, and then plotting that frequency on a polar graph, the result is called a **wind rose**. For example, Fig. 17.2 shows the wind rose for Vancouver Airport (CYVR).

The total length of each wind line gives the total frequency of any wind speed from that direction, while the width (or color) of the line subdivides that frequency into the portions associated with various wind speeds. (Not all wind roses are subdivided by wind speed.) The frequency of calm winds is usually written in the center of the circle if it fits, or is indicated off to the side. The sum of all the frequencies (including calm) should total 100%. At a glance, the longest lines indicate the predominant wind directions for any location.

For example, at Vancouver Airport, East (E) winds (winds from the east) are most frequent, followed by winds from the WNW and then from the ESE. Aircraft take-offs and landings are safer — and require shorter distances — if they are done into the wind. Hence, airports are built with their runways aligned parallel to the predominant wind directions (within reason, as dictated by property boundaries and obstacles).

Fig. 17.3 shows that the runways at Vancouver Airport are appropriate for the wind climatology of the previous figure. The end of each runway is labeled with the magnetic compass direction (in tens of degrees; e.g., 12 means 120° magnetic) towards which the aircraft is flying when approaching that end of the runway from outside the airport. Thus, aircraft will use runway 30 for winds from 300°. Parallel runways are labeled as left (L) or right (R).

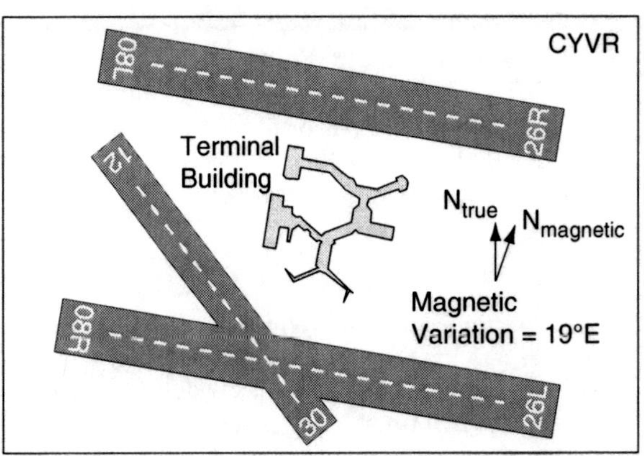

Figure 17.3
Plan view of runways at Vancouver International Airport (CYVR), Canada.

WIND-TURBINE POWER GENERATION

Kinetic energy of the wind is proportional to air mass times wind-speed squared. The rate at which this energy is blown through a wind turbine equals the wind speed. Thus, the theoretical power available from the wind is proportional to speed cubed:

$$Power = (\pi / 2) \cdot \rho \cdot E \cdot R^2 \cdot M_{in}^{\ 3} \qquad (17.2)$$

where R is the turbine-blade radius, M_{in} is incoming wind speed, and ρ is air density. Turbine efficiencies are E = 30% to 45%.

Faster winds and larger radii turbines allow greater power generation. Modern large wind turbines have a hub height (center of the turbine) of 80 m or more, to reach the faster winds higher above the surface. Turbines with radius of 30 m can generate up to 1.5 MW (mega Watts) of electricity, while blades of 40 m radius can generate up to 2.5 MW.

To see how a wind turbine works, consider Fig. 17.4 with an incoming wind speed M_{in}. Even before the wind reaches the disk swept out by the turbine blades, it feels the increased drag (higher pressure) from the turbine and begins to slow. It slows further while passing through the turbine (because the turbine is extracting energy from the wind), and slows more just behind the turbine due to the suction drag. Because the exit speed M_{out} is slower than the entrance speed, and because air-volume flow rate (= M · cross-section area) is conserved, the diameter of the air that feels the influence of the turbine must increase as wind speed decreases.

Zero exit speed is impossible, because the exited air would block subsequent in-flow, preventing power production. Also, if the exit speed equals the entrance speed, then power production is zero because no energy is extracted from the wind. Thus, wind turbines are designed to have an optimum wind-speed decrease of M_{out}/M_{in} = 1/3 (see HIGHER MATH box on **Betz' Law**). Albert Betz showed

Sample Application

A wind turbine at sea level uses a 30 m radius blade to convert a 10 m s^{-1} wind into electrical power at 40% efficiency. What is the theoretical power output?

Find the Answer

Given: ρ =1.225 kg·m^{-3}, R=30 m, M_{in} = 10 m s^{-1}, E = 0.4
Find: *Power* = ? kW (Appendix A defines Watt, W)

Use eq. (17.2):
$Power = (\pi/2)\cdot(1.225 \text{ kg·m}^{-3})\cdot(0.4)\cdot(30\text{m})^2\cdot(10\text{m s}^{-1})^3$
$= 6.93\text{x}10^5 \text{ kg·m}^2\text{·s}^{-3}$ = **693 kW**

Check: Units OK. Physics OK.
Exposition: To estimate annual wind turbine power, use the Weibull distribution to find the power for each wind speed separately, and add all these power increments. Do not use the annual average wind speed.

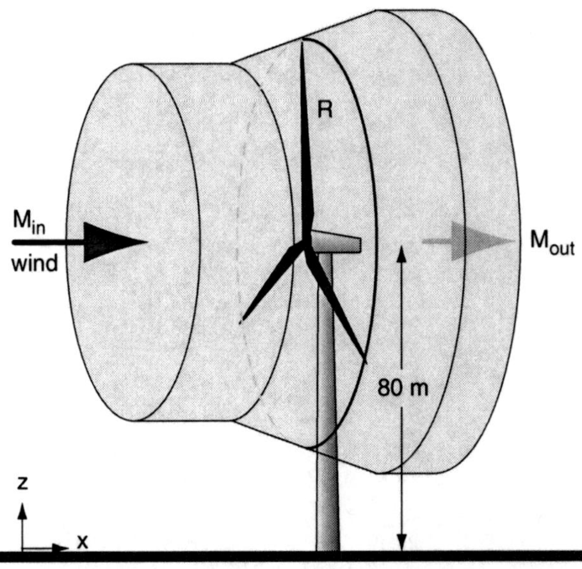

Figure 17.4
Wind turbine for electrical-power generation. Grey region shows the air that transfers some of its energy to the wind turbine.

HIGHER MATH • Betz' Law

In 1919 Albert Betz reasoned that the energy extracted by the turbine is the difference between incoming and outgoing kinetic energies:

$$Energy = 0.5 \cdot m \cdot M_{in}^2 - 0.5 \cdot m \cdot M_{out}^2 \qquad (1)$$

where m is air mass and M is wind speed. The amount of air mass moving through the disk swept by the turbine during time interval Δt is the air density ρ times disk area ($A = \pi R^2$) times average speed:

$$m = \rho \cdot A \cdot 0.5(M_{in} + M_{out}) \cdot \Delta t \qquad (2)$$

Plug this into the previous eq. and divide by Δt to get the power that the turbine can produce:

$$Power = 0.25 \cdot \rho \cdot A \cdot (M_{in} + M_{out}) \cdot (M_{in}^2 - M_{out}^2) \qquad (3)$$

Divide this power by the power of the incoming wind $0.5 \cdot \rho \cdot A \cdot M_{in}^3$ to get a theoretical efficiency E_o:

$$E_o = (1/2) \cdot [1 - (M_{out}/M_{in})^2] \cdot [1 + (M_{out}/M_{in})] \qquad (4)$$

Solve eq. (4) on a spreadsheet and plot E_o vs. M_{out}/M_{in} (see Fig. below) The peak in the curve gives

$$E_{max} = \underline{0.593} \quad \text{at} \quad M_{out}/M_{in} = \underline{1/3}.$$

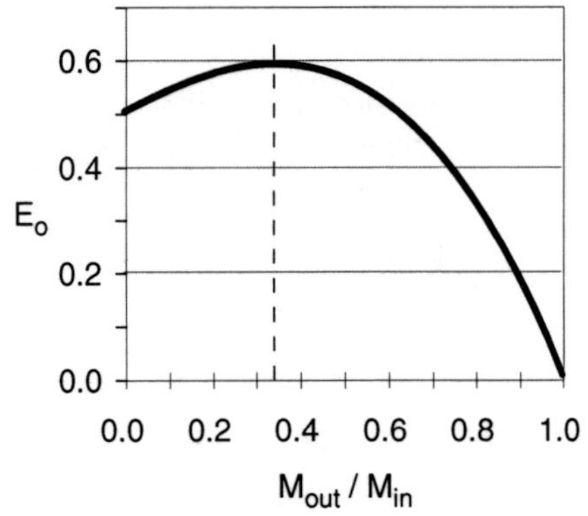

Fig. 17.a *Theoretical turbine efficiencies using Betz' Law.*

We can get the precise answer using calculus. Let the ratio of wind speeds be $r = M_{out}/M_{in}$, to simplify the notation. Use r in eq. (4):

$$E_o = (1/2) \cdot (1 - r^2) \cdot (1 + r) = (1/2) \cdot [1 + r - r^2 - r^3] \qquad (5)$$

Differentiate E_o and set $dE_o/dr = 0$ to find the value of r at max E_o:

$$3 r^2 + 2r - 1 = 0 \qquad (6)$$

Solving this quadratic eq for r gives $r = 1/3$ and $r = -1$, for which the only physically reasonable answer is

$$r = M_{out}/M_{in} = \underline{1/3} \ .$$

Finally, plug this r into eq. (5) to get the max E_o:

$$E_{max} = (1/2) \cdot [1 + (1/3) - (1/9) - (1/27)] \qquad (7)$$

Using a common denominator of 27, we find

$$E_{max} = (1/2) \cdot [32/27] = 16/27 = 0.593 = \underline{59.3\%}.$$

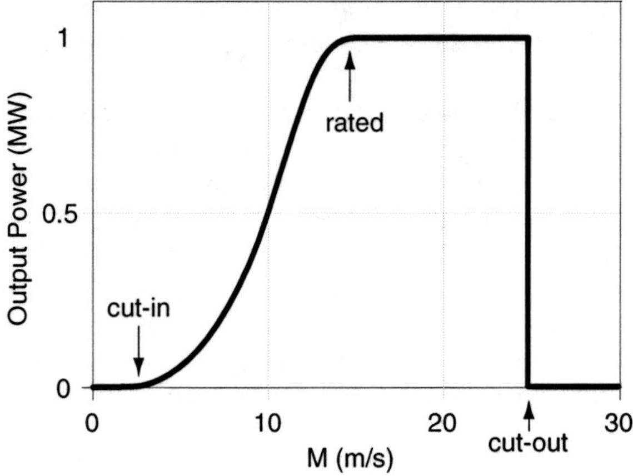

Figure 17.5
Typical power-output curve for a 1 MW wind turbine as a function of wind speed M.

that the theoretical maximum turbine efficiency at this optimum speed is $E_{max} = 16/27 = 59.3\%$, which is known as **Betz' Limit**.

Wind turbines need a wind speed of at least 3 to 5 m s^{-1} to start turning. This is called the **cut-in** speed. As wind speed increases, so increases the amount of power generated. At its **rated** wind speed (8 to 15 m s^{-1}), the turbine is producing the maximum amount of electricity that the generators can handle. As wind speeds increase further, the aerodynamics of the blades are designed to change (via feathering the blades to reduce their pitch, or causing aerodynamic stalling) to keep the shaft rotation rate and electrical power generation nearly constant. Namely, the efficiency is intentionally reduced to protect the equipment. Finally, for wind speeds at or above a **cut-out** wind speed (25 - 30 m s^{-1}), turbine rotation is stopped to prevent damage. Fig. 17.5 shows the resulting idealized power output curve for a wind turbine.

THERMALLY DRIVEN CIRCULATIONS

Thermals

Thermals are warm updrafts of air, rising due to their buoyancy. Thermal diameters are nearly equal to their depth, z_i (Fig. 17.6).

A rising thermal feels drag against the surrounding environmental air (not against the ground). This drag is proportional to the square of the thermal updraft velocity relative to its environment. Neglecting advection and pressure deviations, the equation of vertical velocity W from the Atmospheric Forces & Winds chapter reduces to:

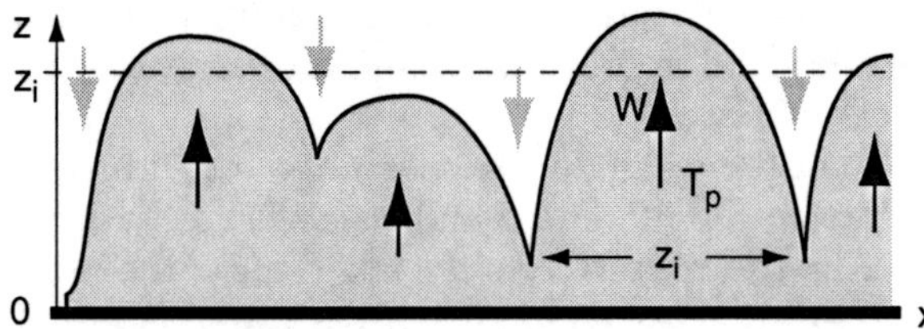

Figure 17.6
Thermals in a convective mixed layer of depth z_i. Black arrows show W updrafts in each thermal, and T_p is temperatue of the thermal. Grey arrows show free-atmosphere air being entrained down into the convective mixed layer between the thermals.

$$\underset{tendency}{\frac{\Delta W}{\Delta t}} = \underset{buoyancy}{\frac{\theta_{vp} - \theta_{ve}}{\overline{T}_{ve}} \cdot |g|} - \underset{turb.\ drag}{C_w \frac{W^2}{z_i}} \qquad (17.3)$$

where z_i is the mixed-layer (boundary-layer) depth, $C_w \approx 5$ is vertical drag coefficient, θ_v is virtual potential temperature, subscripts p and e indicate the air parcel (the thermal) and the environment, $\overline{T_{ve}}$ is average absolute virtual temperature of the environment, and $|g| = 9.8$ m·s^{-2} is gravitational acceleration.

At steady state, the acceleration is near zero ($\Delta W/\Delta t \approx 0$). Eq. (17.3) can be solved for the updraft speed of buoyant thermals (i.e., of warm air parcels):

$$W = \sqrt{\frac{|g| \cdot z_i}{C_w} \frac{(\theta_{vp} - \theta_{ve})}{\overline{T}_{ve}}} \qquad (17.4)$$

Thus, warmer thermals in deeper boundary layers have greater updraft speeds.

This equation also applies to deeper convection at the synoptic and mesoscales, such as weak updrafts in thunderstorms that rise to the top of the troposphere. For that case, z_i is the depth of the troposphere, and the temperature difference is that between the mid-cloud and the surrounding environment at the same height. For stronger updrafts and downdrafts in thunderstorms, the pressure deviation term of the equation of vertical motion must also be included (see the Thunderstorm chapters).

Cross-valley Circulations

(Circulations perpendicular to the valley axis.)

Anabatic Wind

During daytime in synoptically calm conditions (high-pressure center) with mostly clear skies, the sunlight heats mountain slopes. The warm mountain surface heats the neighboring air, which then rises. However, instead of rising vertically like thermals, the rising air hugs the slope as it rises. This warm turbulent air rising upslope is called an **anabatic wind** (Fig. 17.7). Typical speeds are 3 to

Sample Application

Find the steady-state updraft speed in the middle of (a) a thermal in a boundary layer that is 1 km thick; and (b) a thunderstorm in a 11 km thick troposphere. The virtual temperature excess is 2°C for the thermal and 5°C for the thunderstorm, and $|g|/\overline{T_{ve}} = 0.0333$ m·s^{-2}·K^{-1}.

Find the Answer

Given: $|g|/\overline{T_{ve}} = 0.0333$ m·s^{-2}·K^{-1},
(a) $z_i = 1000$ m, $T_{vp} - T_{ve} = 2$°C
(b) $z_i = 11{,}000$ m, $T_{vp} - T_{ve} = 5$°C,
Find: $W = ?$ m s^{-1}

Use eq. (17.4): (a) For the thermal:

$$W = \sqrt{\frac{(0.0333\text{m}\cdot\text{s}^{-2}\text{K}^{-1})(1000\text{m})(2\text{K})}{5}} = \underline{\mathbf{3.65\ m\ s^{-1}}}$$

(b) For the thunderstorm:

$$W = \sqrt{\frac{(0.0333\text{m}\cdot\text{s}^{-2}\text{K}^{-1})(11000\text{m})(5\text{K})}{5}} = \underline{\mathbf{19.1\ m\ s^{-1}}}$$

Check: Units OK. Physics OK.
Exposition: Actually, neither thermal nor thunderstorm updrafts maintain a constant speed. However, this gives us an order-of-magnitude estimate. In convection, these updrafts must have downdrafts between them, but usually of larger diameter and slower speeds. Air-mass continuity requires that mass flow up must balance mass flow of air down, across any arbitrary horizontal plane. This would cause quite a bumpy ride in an airplane, which is why most aircraft pilots try to avoid areas of **convection** (regions filled with thermal or thunderstorm up- and downdrafts).

When I pilot a small plane, I try to stay above the atmospheric boundary layer for the whole flight (except take-off and landing) to have a smooth ride, and I avoid thunderstorms by flying around them.

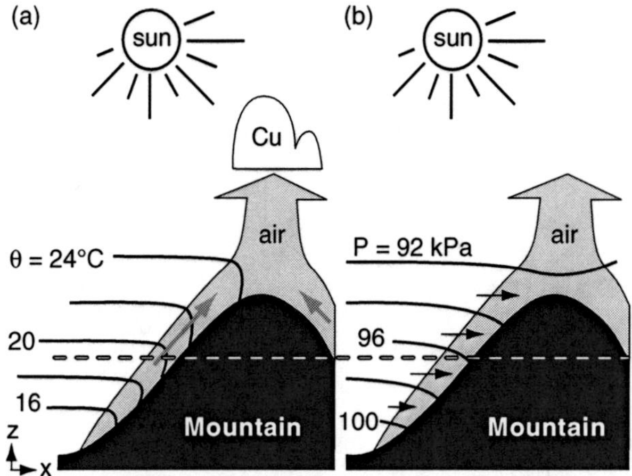

Figure 17.7
Anabatic winds (shaded grey). (a) Isentropes. (b) Isobars. Cu = cumulus cloud. θ = potential temperature. P = pressure.

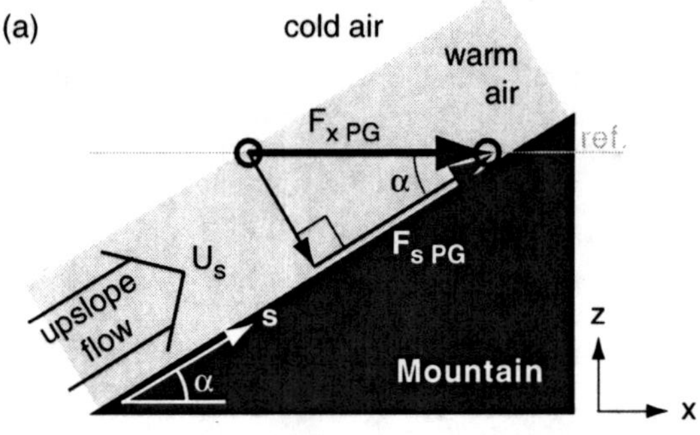

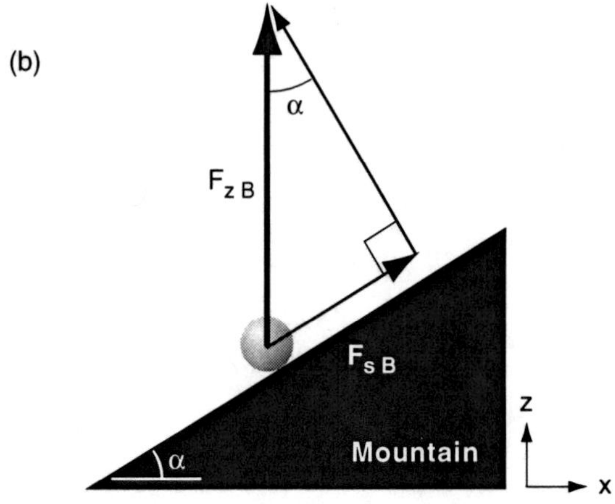

Figure 17.8
Geometry of anabatic (upslope) flows. F is force, subscript s is in the up-slope direction, subscript PG is pressure gradient, and subscript B is buoyancy. $F_{sB} = F_{sPG}$.

Sample Application
Anabatic flow is 5°C warmer than the ambient environment of 15°C. Find the horizontal and along-slope pressure-gradient forces/mass, for a 30° slope.

Find the Answer
Given: T_e = 15°C + 273 = 288 K, ΔT = 5 K, α = 30°
Find: (a) F_{xPG}/m = ? m·s^{-2}, (b) F_{sPG}/m = ? m·s^{-2}

$|g|\cdot\Delta T/T_e$ = (9.8m·s^{-2})·(5K)/(288K) = 0.17 m·s^{-2}

Use eq. (17.5):
$F_{x\ PG}/m$ = (0.17 m·s^{-2})·tan(30°) = **0.098 m·s^{-2}**

Use eq. (17.6):
$F_{s\ PG}/m$ = (0.17 m·s^{-2})·sin(30°) = **0.085 m·s^{-2}**

Check: Units OK, because m·s^{-2} = N kg^{-1}.
Exposition: Glider and hang-glider pilots use the anabatic updrafts to soar along mountain slopes and mountain tops.

5 m s^{-1}, and depths are hundreds of meters. The anabatic wind is the rising portion of a **cross-valley circulation**.

When the warm air reaches ridge top, it breaks away from the mountain and rises vertically, often joined by the updraft from the other side of the same mountain. Cumulus clouds called **anabatic clouds** can form just above ridge top in this updraft.

The dashed line in Fig. 17.7 is at a constant height above sea level. Following the line from left to right in Fig. 17.7a, potential temperatures of about 19°C are constant until reaching anabatic air near the mountain, where the potential temperature rises to about 21°C in this idealized illustration.

The temperature difference between the warmed air near the mountain and the cooler ambient air creates a small horizontal pressure gradient force (exaggerated in Fig. 17.7b) that holds the warm rising air against the mountain. To find this horizontal pressure-gradient force per unit mass m, use the hypsometric equation (see INFO box on next page):

$$\frac{F_{x\ PG}}{m} = -\frac{1}{\rho}\frac{\Delta P}{\Delta x} = |g|\frac{\Delta T}{T_e}\cdot\tan(\alpha) \qquad (17.5)$$

where Δx is horizontal distance (positive in the uphill direction), $|g|$= 9.8 m·s^{-2} is gravitational acceleration magnitude, $\Delta T = T_p - T_e$ is temperature difference between the air near and far from the slope, T_p is temperature of the warm near-mountain air, T_e is temperature of the cooler environmental air, and α is the mountain slope angle. Use absolute temperature in the denominator of the equation above.

The horizontal pressure difference across the anabatic flow is very small compared to the vertical pressure difference of air in hydrostatic balance. However, the horizontal pressure difference occurs across a short horizontal distance (tens of meters), yielding a modest pressure gradient that drives a measurable anabatic wind.

The portion of this pressure-gradient force in the along-slope direction (s) is $F_{s\ PG} = F_{x\ PG}\cdot\cos(\alpha)$, as can be seen from Fig. 17.8a. Combining this with the previous equation, and using the trigonometric identity $\tan(\alpha)\cdot\cos(\alpha) = \sin(\alpha)$, gives:

$$\frac{F_{s\ PG}}{m} = |g|\frac{\Delta T}{T_e}\cdot\sin(\alpha) \qquad (17.6)$$

But recall from the vertical equation of motion that the vertical buoyancy force is $F_{zB}/m = |g|\cdot(\Delta T/T_e)$. Thus, eq. (17.6) can also be interpreted as the component of vertical buoyancy force in the up-slope direction (s), as can be seen from Fig. 17.8b.

INFO • Anabatic Slope Flow

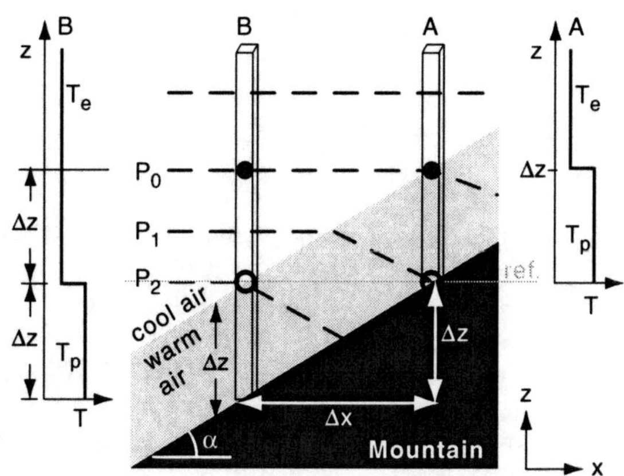

Fig. 17.b. *Dashed lines are isobars. Soundings at left and right correspond to columns of air B and A.*

Derivation of Horizontal Pressure Gradient

Consider an idealized situation of isothermal environmental air of temperature T_e and warmer near-mountain air (shaded grey) of temperature T_p with uniform vertical depth Δz, as sketched in Fig. 17.b. Consider two air columns: A and B. If a reference height *ref* is set at the base of column A (thin grey line), then place column B such that the same *ref* height is at the top of the warm-air layer.

For both columns A and B, start at the same pressure (P_o, at the solid black dots in Fig. 17.b). As you descend distance Δz, the pressure P increase depends on air temperature T, as given by the hypsometric equation (from Chapter 1):

$$\ln(P) = \ln(P_o) + \frac{\Delta z}{a \cdot T} \tag{1}$$

where $a = \Re_d/|g| = 29.3$ m K^{-1}. Thus, the ln(P) increases linearly with decreasing altitude (Fig. 17.c).

In column B, the temperature is uniformly cool between the solid black dot and the reference height, so pressure increases rapidly as you descend (Fig. 17.c). However, in column A, the temperature is uniformly

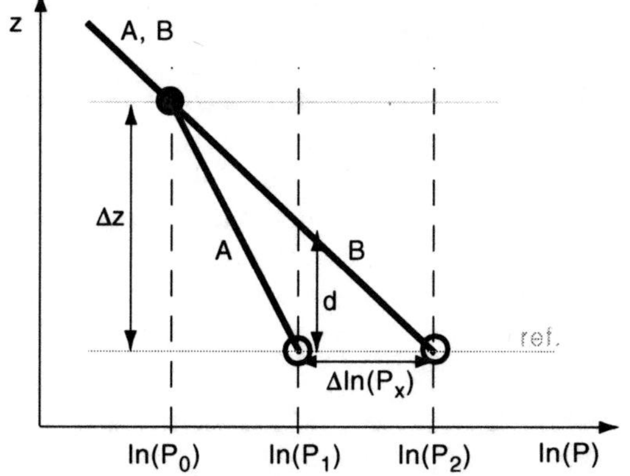

Fig. 17.c. *Change of pressure with height, as given by the hypsometric equation for cool air in column B and warm air in column A. (not to scale)*

(continues in next column)

INFO • Anabatic (continuation)

warmer, so the pressure doesn't increase as fast. By the time you have descended distance Δz from the black dot to the reference height, the pressure in the cold air has increased to P_2, but in the warm air has increased a smaller amount to P_1. Thus, at the reference height, there is a horizontal pressure difference $\Delta P = P_2 - P_1$ pointing toward the mountain slope.

To quantify this effect, start with the hypsometric equation, separately for columns A and B:

Col. B at ref.: $\ln(P_2) = \ln(P_o) + \Delta z/(a \cdot T_e)$

Col. A at ref: $\ln(P_1) = \ln(P_o) + \Delta z/(a \cdot T_p)$

where $a = \Re_d/|g| = 29.3$ m K^{-1} from Chapter 1. Subtract equation A from B

$$\ln(P_2) - \ln(P_1) = \frac{\Delta z}{a} \cdot \left(\frac{1}{T_e} - \frac{1}{T_p} \right)$$

Create a common denominator in the parentheses, and combine the ln() terms:

$$\ln\left(\frac{P_2}{P_1}\right) = \frac{\Delta z}{a} \cdot \left(\frac{T_p - T_e}{T_e \cdot T_p}\right)$$

Take the exponential of both sides, solve for P_1:

$$P_1 = P_2 \cdot \exp\left(-\frac{\Delta z}{a} \cdot \frac{\Delta T}{T_e^2} \right)$$

where $\Delta T = T_p - T_e$, and where $T_e^2 \approx T_e \cdot T_p$ because both are absolute temperatures. Next, subtract P_2 from both sides, and let $\Delta P = P_1 - P_2$ be the pressure change in the positive x-direction at the reference height:

$$\Delta P = -P_2 \cdot \left[1 - \exp\left(-\frac{\Delta z}{a} \cdot \frac{\Delta T}{T_e^2} \right) \right]$$

But $\exp(-y) \approx 1 - y + y^2/2 - \ldots$ Thus, the pressure decrease along the *ref.* height from B to A is

$$\Delta P \approx -P_B \cdot \left[\frac{\Delta z}{a} \cdot \frac{\Delta T}{T_e^2} \right] \tag{2}$$

Expanding *a* and using the ideal gas law gives:

$$-\frac{\Delta P}{\rho} \approx |g| \cdot \Delta z \cdot \frac{\Delta T}{T_e}$$

Divide both sides by Δx to give the horizontal pressure-gradient force per unit mass $F_{x\,PG}/m$:

$$\boxed{\frac{F_{x\,PG}}{m} = -\frac{1}{\rho}\frac{\Delta P}{\Delta x} = |g| \cdot \frac{\Delta T}{T_e} \cdot \frac{\Delta z}{\Delta x}}$$

Substituting $\Delta z/\Delta x = \tan(\alpha)$ gives the desired eq. (17.5), where α is the mountain slope angle.

$$\frac{F_{x\,PG}}{m} = -\frac{1}{\rho}\frac{\Delta P}{\Delta x} = |g| \frac{\Delta T}{T_e} \cdot \tan(\alpha) \tag{17.5}$$

Also, in Fig. 17.c, $d = z(P_{1\ at\ B}) - z(P_{1\ at\ A})$ is the deflection distance of the near-mountain end of isobar P_1 in Fig. 17.b.

Sample Application

For the scenario in the previous Sample Application, suppose a steady-state is reached where the only two forces are buoyancy and drag. Find the anabatic wind speed, assuming an anabatic flow depth of 50 m and drag coefficient of 0.05.

Find the Answer

Given: buoyancy term = 0.085 m·s^{-2} from previous Sample Application. C_D = 0.05, Δz = 50m

Find: U_s = ? m s^{-1}

Solve eq. (17.7) for U_s, considering only buoyancy and drag terms:

$$U_s = [(\Delta z/C_D)\cdot \text{buoyancy term})]^{1/2}$$

$$= [\ ((50\text{m})/(0.05)) \cdot (0.085\ \text{m·s}^{-2})\ ]^{1/2} = \underline{\mathbf{9.2\ m\ s^{-1}}}$$

Check: Units OK. Magnitude too large.

Exposition: In real anabatic flows, the temperature excess (5°C in this example) exists only close to the ground, and decreases to near zero by 50 to 100 m away from the mountain slope. If we had applied eq. (17.7) over the 5 m depth of the temperature excess, then a more-realistic answer of 2.9 m s^{-1} is found.

A better approach is to use the average temperature excess over the depth of the anabatic flow, not the maximum temperature excess measured close to the mountain slope.

Turbulence is strong during convective conditions such as during anabatic winds, which increases the turbulent drag against the ground.

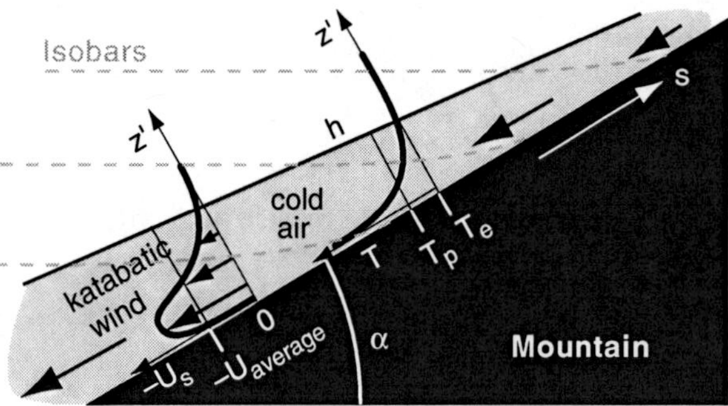

Figure 17.9

Katabatic winds (shaded light gray). Superimposed (thick black lines) are the along-slope wind-speed profile (U_s vs. z') and temperature profile (T vs z'), where z' is normal to the surface s. T_p is the average temperature in the katabatic-flow air, and T_e is the ambient environmental temperature just outside of the katabatic flow. The katabatic wind is indicated with the large arrows. Isobars (dashed grey lines) tilt upward in the cold air.

To forecast the speed of the upslope flow, apply the equation of horizontal motion to a sloping surface and use the approximation above:

$$\underbrace{\frac{\Delta U_s}{\Delta t}}_{\text{tendency}} = \underbrace{-U_s\frac{\Delta U_s}{\Delta s} - V\frac{\Delta U_s}{\Delta y}}_{\text{advection}} + \underbrace{|g|\frac{\Delta\theta_v}{T_{ve}}\sin(\alpha)}_{\text{buoyancy}} + \underbrace{f_c\cdot V}_{\text{Coriolis}} - \underbrace{C_D\frac{U_s^{\,2}}{\Delta z}}_{\text{drag}} \tag{17.7}$$

where U_s is along-slope (up- or down-slope) flow, V is across-slope flow, $\Delta\theta_v$ is virtual potential temperature difference between the slope-flow air and the environment (where $\Delta\theta_v = \Delta\theta = \Delta T$ for dry air), T_v is absolute virtual temperature, $f_c = 2\Omega\cdot\sin(latitude)$ is the Coriolis parameter, $2\Omega = 1.458423\text{x}10^{-4}\ \text{s}^{-1}$, C_D is a drag coefficient against the ground, and Δz is the vertical depth of the slope flow.

Katabatic Wind

During anticyclonic conditions of calm or light synoptic-scale winds at night, air adjacent to a cold mountain slope can become colder than the surrounding air. This cold, dense air flows downhill under the influence of gravity (buoyancy), and is called a **katabatic wind** (Fig. 17.9). It can also form in the daytime over snow- or ice-covered slopes.

The katabatic wind is shallowest at the top of the slope, and increases in thickness and speed further downhill. Typical depths are 10 to 100 m, where the depth is roughly 5% of the vertical drop distance from the hill top. Typical speeds are 3 to 8 m s^{-1}. Katabatic flows are shallower and less turbulent than anabatic flows.

Equation (17.7) also applies for katabatic flows, but with negative $\Delta\theta_v$. A horizontal pressure gradient drives the katabatic wind, but with hydrostatic air pressure increasing near the slope due to the cold dense air. Namely, isobars bend upward near the mountain during katabatic flows (Fig. 17.9).

The virtual potential temperature in a katabatic flow is generally coldest at the ground, and smoothly increases with height (Fig. 17.9). For the difference $\Delta\theta_v = (\theta_{vp} - \theta_{ve})$, use the <u>average</u> temperature in the katabatic flow θ_{vp} minus the environmental temperature at the same altitude θ_{ve}.

If there were no friction against the surface, then the fastest downslope winds would be where the air is the coldest; namely, closest to the ground. However, winds closest to the ground are slowed due to turbulent drag, leaving a nose of fast winds just above ground level (Fig. 17.9).

In Antarctica where downslope distances are hundreds of kilometers, the katabatic wind speeds are 3 to 20 m s^{-1} with extreme cases up to 50 m s^{-1}, and durations are many days. These attributes are sufficiently large that Coriolis force turns the equilibrium wind direction 30 to 50° left of the fall line.

However, for most smaller valleys and slopes you can neglect Coriolis force and the *V*-wind, allowing eq. (17.7) to be solved for some steady-state situations, as shown next.

Initially the wind (averaged over the depth of the katabatic flow) is influenced mostly by buoyancy and advection. It accelerates with distance *s* downslope:

$$|U_{average}| = \left[\left| g \cdot \frac{\Delta\theta_v}{T_{ve}} \cdot s \cdot \sin(\alpha) \right| \right]^{1/2} \qquad (17.8)$$

where $|g|$ = 9.8 m·s^{-2}, T_{ve} is absolute temperature in the environment at the height of interest, α is the mountain slope angle, and *s* is distance downslope.

The average katabatic wind eventually approaches an equilibrium where drag balances buoyancy:

$$|U_{eq}| = \left[\left| g \cdot \frac{\Delta\theta_v}{T_{ve}} \cdot \frac{h}{C_D} \cdot \sin(\alpha) \right| \right]^{1/2} \qquad (17.9)$$

where C_D is the total drag against both the ground and against the slower air aloft, and *h* is depth of the katabatic flow.

Sample Application (§)

Air adjacent to a 10° slope averages 10°C cooler over its 20 m depth than the surrounding air of virtual temperature 10°C. Find and plot the wind speed vs. downslope distance, and the equilibrium speed. C_D = 0.005 .

Find the Answer

Given: $\Delta\theta_v$= –10°C, T_{ve}=283 K, α= 10°, C_D= 0.005
Find: U (m s^{-1}) vs. x (km), where $s \approx x$.

Use eq. (17.9) to find final equilibrium value:

$$|U_{eq}| = \left[\left| \frac{(9.8\text{m}\cdot\text{s}^{-2})\cdot(-10°\text{C})}{283\text{K}} \cdot \frac{(20\text{m})}{0.005} \cdot \sin(10°) \right| \right]^{1/2}$$

$|U_{eq}|$ = **15.5 m s^{-1}**

Use eq. (17.8) for the initial variation.

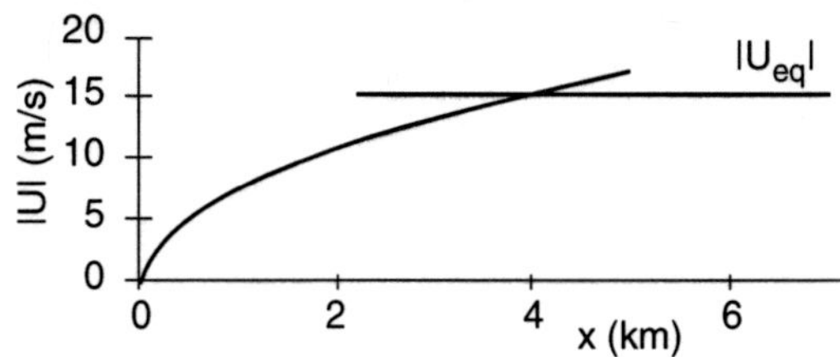

Check: Units OK. Physics OK.
Exposition: Although these two curves cross, the complete solution to eq. (17.7) smoothly transitions from the initial curve to the final equilibrium value.

Along-valley Winds

Katabatic and anabatic winds are part of larger circulations in the valley.

Night

At night, the katabatic winds from the bottom of the slopes drain into the valley, where they start to accumulate. This pool of cold air is often stratified like a layer cake, with the coldest air at the bottom and less-cold air on top. Katabatic winds that start higher on the slope often do not travel all the way to the valley bottom (Fig. 17.10). Instead, they either spread out at an altitude where they have the same buoyancy as the stratified pool in the valley, or they end in a turbulent eddy higher above the valley floor. This leads to a relatively mild **thermal belt** of air at the mid to upper portions of the valley walls — a good place for vineyards and orchards because of fewer frost days.

Meanwhile, the cold pool of air in the valley bottom flows along the valley axis in the same direction that a stream of water would flow. As this cold air drains out of the valley onto the lowlands, it is known as a **mountain wind**. This is part of an **along-valley circulation**. A weak return flow aloft (not drawn), called the **anti-mountain wind**, flows up-valley, and is the other part of this along-valley circulation.

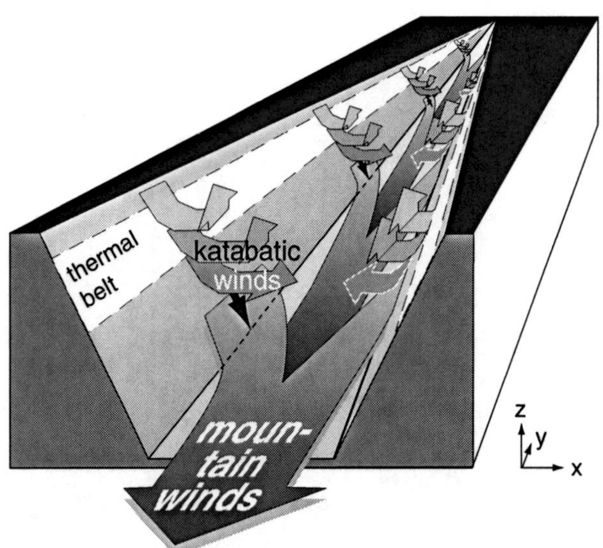

Figure 17.10
Katabatic winds are cross-valley flows that merge into the along-valley mountain winds draining down valley. Relatively warm air can exist in the "thermal belt" regions.

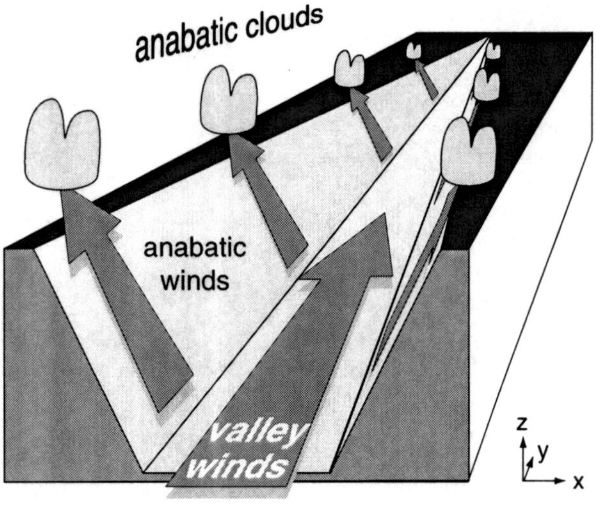

Figure 17.11
Anabatic winds are warm upslope flows that are resupplied with air by valley winds flowing upstream.

> **INFO • On Naming Local Winds**
>
> Local winds are often named by where they come from. Winds from the mountains are called mountain winds. Winds from the valley are called valley winds. A breeze from the sea is called a sea-breeze. The opposite is a land breeze.

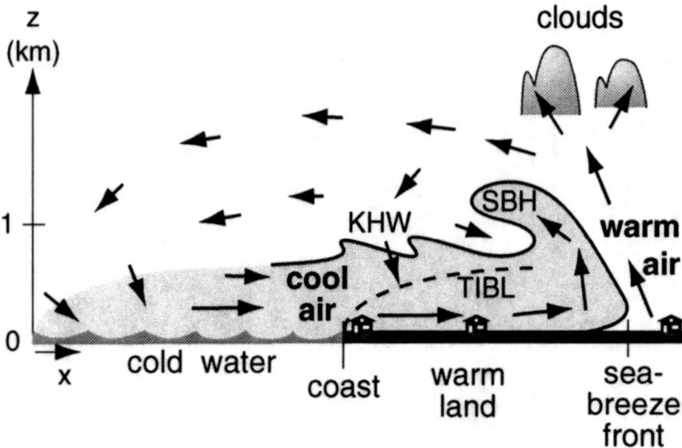

Figure 17.12
Vertical cross section through a sea-breeze circulation. KHW = Kelvin-Helmholtz waves. SBH = sea-breeze head. TIBL = thermal internal boundary layer.

Day

During anabatic conditions, there is often a weak **cross-valley horizontal circulation** (not drawn) from the tops of the anabatic updrafts back toward the center of the valley. Some of this air sinks (**subsides**) over the valley center, and helps trap air pollutants in the valley.

During daytime, the upslope anabatic winds along the valley walls remove air from the valley floor. This lowers the pressure very slightly near the valley floor, which creates a pressure gradient that draws in replacement air from lower in the valley. These upstream flowing winds are called **valley winds** (Fig. 17.11), and are part of an **along-valley circulation**. A weak return flow aloft (heading down valley; not drawn in Fig. 17.11) is called the **anti-valley wind**.

Transitions

Because the sun angle relative to the mountain slope determines the solar heating rate of that slope, the anabatic winds on one side of the valley are often stronger than on the other. At low sun angles near sunrise or sunset, the sunny side of the valley might have anabatic winds while the shady side might have katabatic winds.

Sea breeze

A **sea breeze** is a shallow cool wind that blows onshore (from sea to land) during daytime (Fig. 17.12). It occurs in large-scale high-pressure regions of weak or calm geostrophic wind under mostly clear skies. Similar flows called **lake breezes** form along lake shorelines, and **inland sea breezes** form along boundaries between adjacent land regions with different land-use characteristics (e.g., irrigated fields of crops adjacent to drier land with less vegetation).

The sea breeze is caused by a 5 °C or greater temperature difference between the sun-heated warm land and the cooler water. It is a surface manifestation of a thermally driven mesoscale circulation called the **sea-breeze circulation**, which often includes a weak return flow aloft from land to sea.

For warm air over land, the hypsometric equation states that hydrostatic pressure does not decrease as rapidly with increasing height as it does in the cooler air over the sea (Fig. 17.13). This creates a pressure gradient aloft between higher pressure over land and lower pressure over the sea, which initiates a wind aloft. This wind moves air molecules from over land to over sea, causing surface pressure over the warm land to decrease because fewer total molecules in the warm-air column cause less weight of air at the base of the column. Similarly, the surface pressure over the water increases due to molecules added to the cool-air column. This creates a pressure gradient near the surface that drives the bottom portion of the sea-breeze circulation. Such **hydrostatic thermal circulations** were explained in the General Circulation chapter.

A **sea-breeze front** marks the leading edge of the advancing cool marine air and behaves similarly to a weak advancing cold front or a thunderstorm gust front. If the updraft ahead of the front is humid enough, a line of cumulus clouds can form along the front, which can grow into a line of thunderstorms if the atmosphere is convectively unstable.

The raised portion of cool air immediately behind the front, called the **sea-breeze head**, is analogous to the head at the leading edge of a gust front. The sea-breeze head is roughly twice as thick as the subsequent portion of the **feeder** cool onshore flow (which is 0.5 to 1 km thick). The top of the sea-breeze head often curls back in a large horizontal roll eddy over warmer air from aloft.

Vertical wind shear at the density interface between the low-level sea breeze and the return flow aloft can create **Kelvin-Helmholtz (KH) waves**. These breaking waves in the air have wavelength of 0.5 to 1 km. The KH waves increase turbulent drag on the sea breeze by entraining low-momentum air from above the interface. A slowly subsiding return flow occurs over water and completes the circulation as the air is again cooled as it blows landward over the cold water.

As the cool marine air flows over the land, a **thermal internal boundary layer (TIBL)** forms just above the ground (Fig. 17.12). The TIBL grows in depth z_i with the square root of distance x from the shore as the marine air is modified by the heat flux F_H (in kinematic units K·m s^{-1}) from the underlying warm ground:

$$z_i = \left[\frac{2\cdot F_H}{\gamma\cdot M}\cdot x\right]^{1/2} \tag{17.10}$$

where $\gamma = \Delta\theta/\Delta z$ is the gradient of potential temperature in the air just before reaching the coast, and M is the wind speed.

In early morning, the sea-breeze circulation does not extend very far from the coast, but progresses further over land and water as the day progresses. Advancing cold air behind the sea-breeze front behaves somewhat like a **density current** or **gravity current** in which a dense fluid spreads out horizontally beneath a less dense fluid. When this is simulated in water tanks, the speed M_{SBF} of advance of the sea-breeze front, is

$$M_{SBF} = k\cdot\sqrt{|g|\cdot\frac{\Delta\theta_v}{T_v}\cdot d} \tag{17.11}$$

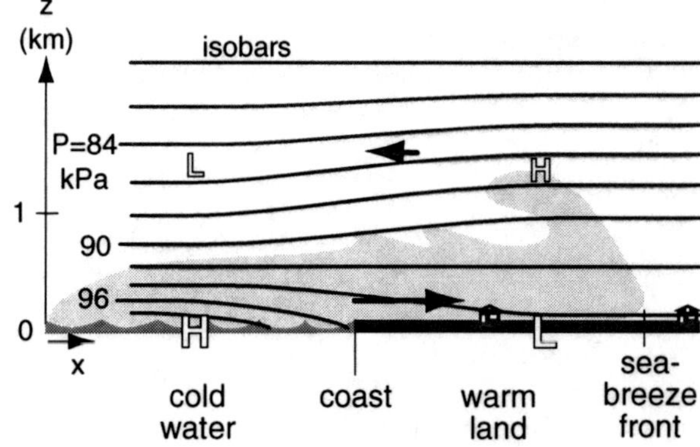

Figure 17.13
Vertical cross section through a sea-breeze circulation, showing isobars. H and L at any one altitude indicate relatively higher or lower pressure. The horizontal pressure gradient is greatly exaggerated in this illustration. Thick arrows show the coast-normal component of horizontal winds.

Sample Application
What horizontal pressure difference is needed in the bottom part of the sea-breeze circulation to drive a onshore wind that accelerates from 0 to 6 m s^{-1} in 6 h?

Find the Answer
Given: $\Delta M/\Delta t$ = (6 m s^{-1})/(6 h) = 0.000278 m·s^{-2}
Find: $\Delta P/\Delta x$ = ? kPa km^{-1}

Neglect all other terms in the horiz. eq. of motion:
$\Delta M/\Delta t = -(1/\rho)\cdot\Delta P/\Delta x$ (10.23a)
Assume air density is ρ = 1.225 kg m^{-3} at sea level.
Solve for $\Delta P/\Delta x$:
$\Delta P/\Delta x = -\rho\cdot(\Delta M/\Delta t)$ = −(1.225kg m^{-3})·(0.000278m·s^{-2})
= −0.00034 kg·m^{-1}·s^{-2}/m = −0.00034 Pa m^{-1}
= **−0.00034 kPa km^{-1}**

Check: Units OK. Magnitude OK.
Exposition: Only a small pressure gradient is needed to drive a sea breeze.

Sample Application
For a surface kinematic heat flux of 0.2 K m s^{-1}, wind speed of 5 m s^{-1}, and γ = 3K km^{-1}, find the TIBL depth 5 km from shore.

Find the Answer
Given: F_H=0.2 K·m s^{-1}, M=5 m s^{-1}, γ =3K km^{-1}, x=5 km
Find: z_i = ? m

Use eq. (17.10):
z_i = [2·(0.2K·m s^{-1})·(5km) / {(3K km^{-1})·(5m s^{-1})}]$^{1/2}$
= **0.365 km**

Check: Units OK. Magnitude reasonable.
Exposition: Above this height, the air still feels the marine influence, and not the warmer land.

Sample Application

Marine-air of thickness 500 m and virtual temperature 16°C is advancing over land. The displaced continental-air virtual temperature is 20°C. Find the sea-breeze front speed, and the sea-breeze wind speed.

Find the Answer

Given: $\Delta\theta_v = \Delta T_v = 20 - 16°C = 4°C$, $d = 500$ m
$T_{v\ average} = (16+20°C)/2 = 18°C = 291K$
Find: $M_{SBF} = ?$ m s^{-1}, $M = ?$ m s^{-1}

For speed-of-advance of the front, use eq. (17.11):
$M_{SBF} = (0.62) \cdot [(9.8\text{m·s}^{-2})\cdot(4\text{K})\cdot(500\text{m})/(291\text{K})]^{1/2}$
$= \underline{5.1 \text{ m s}^{-1}}$

For the sea-breeze wind speed, use eq. (17.12):
$M = 1.15 \cdot (5.1\text{m s}^{-1}) = \underline{5.9 \text{ m s}^{-1}}$

Check: Units OK. Magnitude OK.
Exposition: Because the sea-breeze circulation extends 20 to 250 km over the sea, mariners used these reliable breezes to sail north-south along the Atlantic coasts of Europe and Africa centuries ago. Upon reaching the latitude of the easterly trade winds in the tropical global circulation, they then sailed west towards the Americas. Upon reaching the Americas, they again used the sea-breezes to sail north and south along the East Coasts of North and South America. For the return trip to Europe, they sailed in the mid-latitude westerlies. Thus, by using both local sea-breeze winds and the global circulation, they achieved an effective commercial trade route.

Sample Application

Assuming calm synoptic conditions (i.e., no large-scale winds that oppose or enhance the sea-breeze), what maximum distance inland would a sea-breeze propagate. Use data from the previous Sample Application, for a latitude of 45°N.

Find the Answer:

Given: $M_{SBF} = 5.1$ m s^{-1} from previous example.
latitude = 45°N.
Find: $L = ?$ km

First, find the Coriolis parameter:
$f_c = (1.458\text{x}10^{-4} \text{ s}^{-1})\cdot\sin(45°) = 1.03\text{x}10^{-4} \text{ s}^{-1}$
Use eq. (17.13):
$L = (5.1 \text{ m s}^{-1})$ /
$[(7.27\text{x}10^{-5} \text{ s}^{-1})^2 - (1.03\text{x}10^{-4} \text{ s}^{-1})^2]^{1/2} = \underline{70 \text{ km}}$

Check: Units OK. Magnitude reasonable.
Exposition: At 30° latitude the denominator of eq. (17.13) is zero, causing $L = \infty$. But this is physically unreasonable. Thus, we expect eq. (17.13) to not be reliable at latitudes near 30°.

where $\Delta\theta_v$ is the virtual potential temperature difference between the cool marine sea-breeze air and the warmer air over land that is being displaced, T_v is an absolute average virtual temperature, $|g| = 9.8$ m·s^{-2} is gravitational acceleration magnitude, d is depth of the density current, and constant $k \approx 0.62$.

When fully developed, surface (10 m height) wind speeds in the marine, inflow portion of the sea breeze at the coast are 1 to 10 m s^{-1} with typical values of 6 m s^{-1}. The relationship between sea-breeze wind speed M at the coast and speed of the sea-breeze front is:

$$M \approx 1.15 \cdot M_{SBF} \qquad (17.12)$$

The sea-breeze front can advance L = 10 to 200 km inland by the end of the day, although typical advances are L = 20 to 60 km unless inhibited by mountains or by opposing synoptic-scale winds. Even without mountain barriers, the sea breeze will eventually turn away from its advance due to Coriolis force. For latitudes ≠ 30°, L is roughly

$$L \approx \frac{M_{SBF}}{\left|\omega^2 - f_c^{\,2}\right|^{1/2}} \qquad (17.13)$$

where M_{SBF} is given by the previous equation, $\omega = 2\pi$ day^{-1} = $7.27\text{x}10^{-5}$ s^{-1} is the frequency of the daily heating/cooling cycle, and f_c = $(1.458\text{x}10^{-4} \text{ s}^{-1})\cdot\sin(\textit{latitude})$ is the Coriolis parameter. As the front advances, prefrontal waves may cause wind shifts ahead of the front.

At the end of the day, the sea-breeze circulation dissipates and a weaker, reverse circulation called the **land-breeze** forms in response to the nighttime cooling of the land surface relative to the sea. Sometimes, the now-disconnected sea-breeze front from late afternoon continues to advance farther inland during the night as a **bore** (the front of dense fluid advancing under less-dense fluid; also described as a propagating solitary wave with characteristics similar to the **hydraulic jump**). In Australia, such a bore and its associated cloud along the wave crest are known as the **Morning Glory**.

In the vertical cross section normal to the coastline (as in Fig. 17.12), the surface wind oscillates back and forth between onshore and offshore, reversing directions during the morning and evening hours. The Coriolis force induces an oscillating along-shore wind component that lags the onshore-offshore component by 6 h (or 1/4 of a daily cycle). Hence, the horizontal wind vector rotates throughout the course of the day. Rotation is clockwise in the northern hemisphere and counterclockwise in the southern hemisphere.

The idealized sea-breeze hodograph has an elliptical shape (Fig. 17.14). For example, along a meridional coastline with the ocean to the west in the Northern Hemisphere, the diurnal component of the surface wind tends to be westerly (onshore) during the mid-day, northerly (alongshore) during the evening, easterly near midnight, and southerly (alongshore) near sunrise.

The sea-breeze wind and the mean (24 h average) synoptic-scale surface wind are additive. If the synoptic-scale wind in the above example is blowing, say, from the north, the surface wind speed will tend to be higher around sunset when the mean wind and the diurnal component are in the same direction, than around sunrise when they oppose each other.

Many coasts have complex shaped coastlines with bays or mountains, resulting in a myriad of interactions between local flows that distort the sea breeze and create regions of enhanced convergence and divergence. The sea breeze can also interact with boundary-layer thermals, and urban circulations, causing complex dispersion of pollutants emitted near the shore. If the onshore synoptic-scale geostrophic wind is too strong, only a TIBL develops with no sea-breeze circulation.

In regions such as the west coast of the Americas, where major mountain ranges lie within a few hundred kilometers of the coast, sea breezes and terrain-induced winds appear in combination.

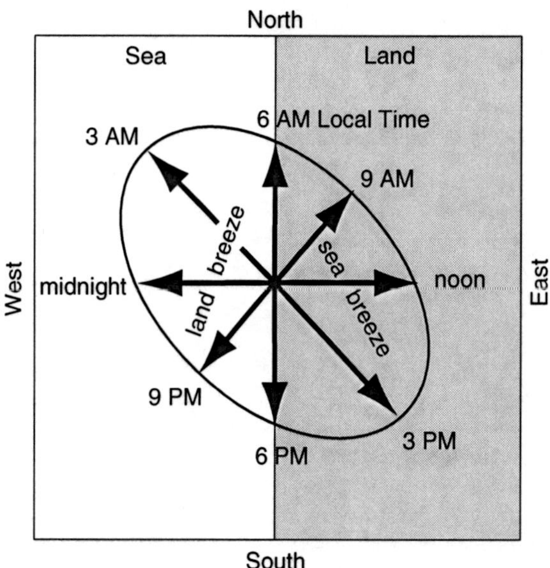

Figure 17.14
Idealized hodograph of surface (z = 10 m) wind vectors during a diurnal cycle for a sea breeze. Assumes Northern Hemisphere latitude of Europe, and fair-weather anticyclonic conditions of light to calm large-scale winds. (In this hodograph, compass directions show the direction toward which the wind blows. Also, the vectors are for different times, not different altitudes.)

OPEN-CHANNEL HYDRAULICS

Sometimes a dense cold-air layer lies under a less-dense warmer layer, with a relatively sharp temperature discontinuity ($\Delta T = \Delta\theta$) between the two layers (Fig. 17.15a and b). This temperature jump marks the **density interface** between the two layers. Examples of such a **two-layer system** include arctic air advancing behind a cold front and sliding under warmer air, cold gust fronts from thunderstorms, and cool marine air moving inland under warmer continental air.

These two-layer systems behave similarly to water in an **open channel** — a two layer system of dense water under less-dense air. Hence, you can apply **hydraulics** (applications of liquid flow based on its mechanical properties) to the atmosphere, for cases where air compressibility is not significant.

Sometimes the cold air can be stably stratified (Figs. 17.15c & d) as idealized here with constant lapse rate, where $\Delta\theta/\Delta z = \Delta T/\Delta z + \Gamma_d$, using the dry adiabatic lapse rate Γ_d = 9.8 °C km^{-1}. You can use modified hydraulic theory for these cases.

Hydraulic theory depends on the speed of waves on the interface between cold and warm air.

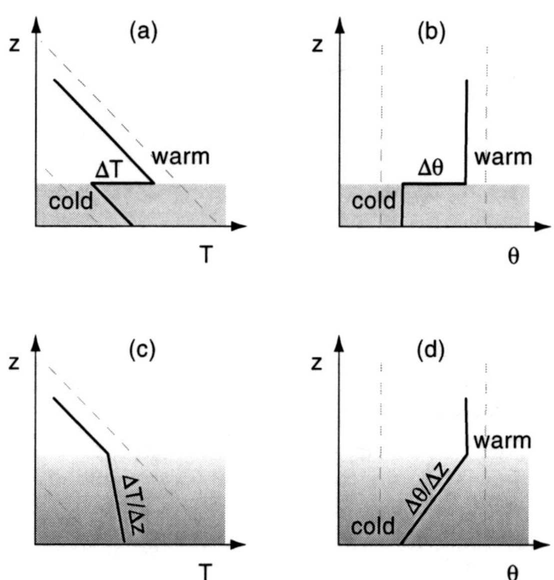

Figure 17.15
Idealized stratification situations in the atmosphere. The dashed lines indicate dry adiabats (lines of constant potential temperature θ) (a) and (b) are two ways of plotting the same situation: two-layers each having uniform potential temperature θ, and with a temperature jump between them. (c) and (d) both show a different situation: a linear change of T with height (i.e., constant stratification) instead of a temperature jump.

Because hydraulics depends on density, it is more accurate to use virtual temperature instead of temperature, to include the effect of humidity on air density. Namely, use T_v instead of T, use ΔT_v instead of ΔT, and $\Delta\theta_v$ instead of $\Delta\theta$.

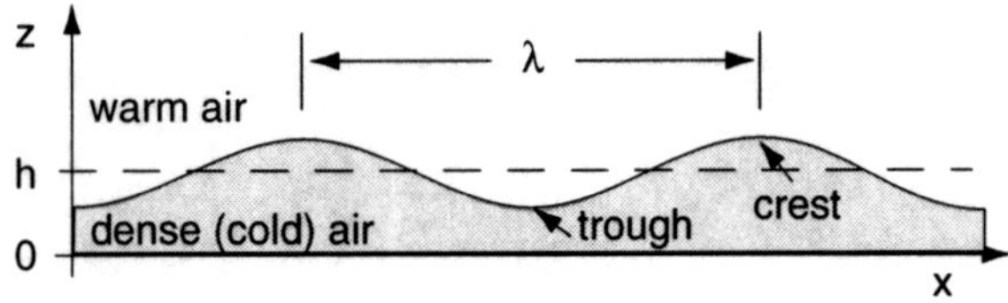

Figure 17.16
Sketch of waves on the interface between cold and warm air layers. λ is wavelength, and h is average depth of cold air.

Sample Application

For a two-layer air system with 5°C virtual potential temperature difference across the interface and a bottom-layer depth of 20 m, find the intrinsic group speed, and compare it to the speed of a water wave.

Find the Answer

Given: $\Delta\theta_v$ = 5°C = 5K, h = 20 m. Assume T_v = 283 K
Find: c_g = ? m s^{-1}, for air and for water

For a 2-layer air system, use eqs. (17.15 & 17.16):
$c_g = c_o = [(9.8 \text{ m·s}^{-2})·(5\text{K})·(20\text{m})/(283\text{K})]^{1/2}$ = **1.86 m s^{-1}**

For water under air, use eq. (17.14 & 17.16)):
$c_g = c_o = [(9.8 \text{ m·s}^{-2})·(20\text{m})]^{1/2}$ = **14 m s^{-1}**

Check: Units OK. Magnitude OK.
Exposition: Atmospheric waves travel much slower than channel or ocean waves, and have much longer wavelengths. This is because of the reduced gravity $|g'|$ for air, compared to the full gravity $|g|$ for water.

Sample Application

Find the internal-wave horizontal group speed in air for a constant virtual potential-temperature gradient of 5°C across a stable-layer depth of 20 m.

Find the Answer

Given: $\Delta\theta_v$ = 5°C = 5K, $\Delta z = h$ = 20 m. Let T_v = 283 K
Find: u_g = ? m s^{-1}

Use eq. (17.17): u_g =
$[(9.8 \text{ m·s}^{-2})·(5\text{K})/(283\text{K} · 20\text{m})]^{1/2} ·(20\text{m})$ = **1.86 m s^{-1}**

Check: Units OK. Magnitude OK.
Exposition: For the example here with $\Delta z = h$, the equation for u_g in a stably-stratified fluid is identical to the equation for c_g in a two layer system. This is one of the reasons why we can often use hydraulics methods for a stably-stratified atmosphere.

Wave Speed

Waves (vertical oscillations that propagate horizontally on the density interface) can exist in air (Fig. 17.16), and behave similarly to water waves. For hydraulics, if water in a channel is shallow, then long-wavelength waves on the water surface travel at the intrinsic "shallow-water" **phase speed** c_o of:

$$c_o = \sqrt{|g|\cdot h} \tag{17.14}$$

where $|g|$ = 9.8 m s^{-2} is gravitational acceleration magnitude, and h is average water depth. **Phase speed** is the speed of propagation of any wave crest. **Intrinsic** means relative to the mean fluid motion. **Absolute** means relative to the ground.

For a two-layer air system, the **reduced gravity** $|g'| = |g|\cdot\Delta\theta_v/T_v$ accounts for the cold-air buoyancy relative to the warmer air. For a shallow bottom layer, the resulting intrinsic phase speed of **surface waves** on the interface between the cold- and warm-air layers is:

$$c_o = \sqrt{|g'|\cdot h} \;=\; \left(|g|\cdot\frac{\Delta\theta_v}{T_v}\cdot h\right)^{1/2} \tag{17.15}$$

where $\Delta\theta_v$ is the virtual potential temperature jump between the two air layers, T_v is an average absolute virtual temperature (in Kelvin), h is the depth of the cold layer of air, and $|g|$ = 9.8 m·s^{-2} is gravitational acceleration magnitude.

The speed that wave energy travels through a fluid is the **group speed** c_g. Group speed is the speed that hydraulic information can travel relative to the mean flow velocity, and it determines how the upstream flow reacts to downstream flow changes. For a two-layer system with bottom-layer depth less than 1/20 the wavelength, the group speed equals the phase speed

$$c_g = c_o \tag{17.16}$$

For a statically-stable atmospheric system with constant lapse rate, there is no surface (no interface) on which the waves can ride. Instead, **internal waves** can exist that propagate both horizontally and vertically inside the statically-stable region. Internal waves reflect from solid surfaces such as the ground, and from statically neutral layers.

For internal waves, the horizontal component of group velocity u_g depends on both vertical and horizontal wavelength λ. To simplify this complicated situation, focus on infinitely-long waves in the horizontal (which propagate the fastest in the horizontal), and focus on a wave for which the vertical wavelength is proportional to the depth h of the statically stable layer of air. Thus:

$$u_g = N_{BV}\cdot h \;=\; \left(\frac{|g|}{T_v}\cdot\frac{\Delta\theta_v}{\Delta z}\right)^{1/2}\cdot h \tag{17.17}$$

where N_{BV} is the Brunt-Väisälä frequency, and $\Delta\theta_v/\Delta z$ is the vertical gradient of virtual potential temperature (a measure of static-stability strength).

Froude Number - Part 1

The ratio of the fluid speed (M) to the wave group velocity (c_g, the speed that energy and information travels) is called the **Froude number** Fr.

$$Fr = M / c_g \tag{17.18}$$

At least three different Froude numbers can be defined, depending on the static stability and the flow situation. We will call these Fr_1, Fr_2, and Fr_3, the last of which will be introduced in a later section.

For surface (interfacial) waves in an idealized atmospheric two-layer system where $c_g = c_o = (|g'|\cdot h)^{1/2}$, the Froude number is $Fr_1 = M/c_o$:

$$Fr_1 = \frac{M}{\left(|g|\cdot\frac{\Delta\theta_v}{T_v}\cdot h\right)^{1/2}} \tag{17.19}$$

where h is the depth of the bottom (cold) air layer, $\Delta\theta_v$ is the virtual potential temperature jump between the two air layers, T_v is an average absolute virtual temperature (in Kelvins), and $|g| = 9.8$ m·s^{-2} is gravitational acceleration magnitude.

For the other situation of a statically stable region supporting internal waves, the Froude number is $Fr_2 = M/u_g$:

$$Fr_2 = \frac{M}{\left(\frac{|g|}{T_v}\cdot\frac{\Delta\theta_v}{\Delta z}\right)^{1/2}\cdot h} \tag{17.20}$$

The nature of the flow is classified by the value of the Froude number:

- **subcritical (tranquil)** for $Fr < 1$
- **critical** for $Fr = 1$
- **supercritical (rapid)** for $Fr > 1$

For subcritical flow, waves and information travel upstream faster than the fluid is flowing downstream, thus allowing the upstream flow to "feel" the effect of both upstream conditions and downstream conditions such as flow constrictions. For supercritical flow, the fluid is moving so fast that no information can travel upstream (relative to a fixed location); hence, the upstream fluid does not "feel" the effects of downstream flow constrictions until it arrives at the constriction. For airflow, the words (upwind, downwind) can be used instead of (upstream, downstream).

Sample Application

For a two-layer air system with 5°C virtual potential temperature difference across the interface and a bottom-layer depth of 20 m, find the Froude number if the average flow speed is 5 m s^{-1}. Discuss whether the flow is critical.

Find the Answer

Given: $\Delta\theta_v$ = 5°C = 5K, h = 20 m. M = 5 m s^{-1}.
Assume T_v = 283 K

Find: Fr_1 = ? (dimensionless)

For a 2-layer air system, use eq. (17.19):

Fr_1 = (5 m s^{-1}) / [(9.8 m·s^{-2})·(5K)·(20m)/(283K)]$^{1/2}$
= (5 m s^{-1}) / (1.86 m s^{-1}) = <u>**2.69**</u>

Check: Units OK. Magnitude OK.

Exposition: This flow is <u>supercritical</u>. Hence, the air blows the wave downwind faster than it can propagate upwind against the mean flow.

Sample Application

For a smooth virtual potential-temperature gradient of 5°C across a stable-layer depth of 20 m, find the Froude number if the average flow speed is 5 m s^{-1}. Discuss whether the flow is critical.

Find the Answer

Given: $\Delta\theta_v$ = 5°C = 5K, $\Delta z = h$ = 20 m.
Assume T_v = 283 K

Find: Fr_2 = ? (dimensionless)

Use eq. (17.20). But since $\Delta z = h$, this causes eq. (17.20) to reduce to eq. (17.19). Hence, we get the same answer as in the previous Sample Application:

Fr_2 = <u>**2.69**</u>

Check: Units OK. Magnitude OK.

Exposition: This flow is also <u>supercritical</u>. Although this example was contrived to give the same virtual potential temperature gradient across the whole fluid depth as before, often this is not the case. So always use eq. (17.20) for a constant stable stratification flow, and don't assume that it always reduces to eq. (17.19).

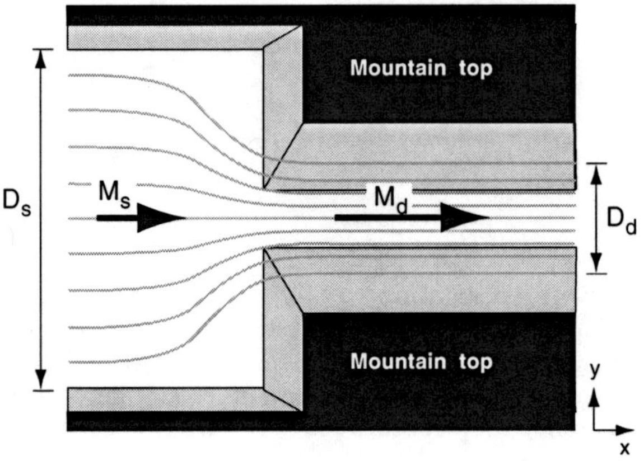

Figure 17.17
Acceleration of air through a constriction. View looking down from above.

Sample Application
If a 20 km wide band of winds of 5 m s^{-1} must contract to pass through a 2 km wide gap, what is the wind speed in the gap?

Find the Answer
Given: D_s = 20 km, D_d = 2 km, M_s = 5 m s^{-1}.
Find: M_d = ? m s^{-1}

Use eq. (17.23): M_d = [(20km)/(2km)]·(5m s^{-1}) = **50 m s^{-1}**

Check: Units OK. Physics OK.
Exposition: The actual winds would be slower, because turbulence would cause significant drag.

Conservation of Air Mass

Consider a layer of well-mixed cold air flowing at speed M_s along a wide valley of width D_s. If it encounters a constriction where the valley width shrinks to D_d (Fig. 17.17), the winds will accelerate to M_d to conserve the amount of air mass flowing. If the depth h of the flow is constant (not a realistic assumption), then **air-mass conservation** gives:

$$\textit{mass flowing } \underline{\textit{out}} = \textit{mass flowing } \underline{\textit{in}} \qquad (17.21)$$

$$\rho \cdot \textit{volume flowing out} = \rho \cdot \textit{volume flowing in}$$

$$\rho \cdot M_d \cdot h \cdot D_d = \rho \cdot M_s \cdot h \cdot D_s \qquad (17.22)$$

Thus,

$$M_d = \frac{D_s}{D_d} \cdot M_s \qquad \bullet(17.23)$$

under the assumptions of negligible changes in air density ρ. Thus, the flow must become faster in the narrower valley.

Similarly, suppose air is flowing downhill through a valley of constant-width D. Cold air of initial speed M_s might accelerate due to gravity to speed M_d further down the slope. Air-mass conservation requires that the depth of the flow h_d in the high speed region be less than the initial depth h_s in the lower-speed region.

$$h_d = \frac{M_s}{M_d} \cdot h_s \qquad (17.24)$$

Hydraulic Jump

Consider a layer of cold air flowing supercritically in a channel or valley. If the valley geometry or slope changes at some downstream location and allows wind speed M to decrease to its critical value (Fr = 1), there often occurs a sudden increase in flow depth h and a dramatic increase in turbulence. This transition is called an **hydraulic jump**. Downstream of the hydraulic jump, the wind speed is slower and the flow is subcritical.

In Fig. 17.18, you can consider the hydraulic jump as a wave that is trying to propagate upstream. However, the cold air flowing downslope is trying to wash this wave downstream. At the hydraulic jump, the wave speed exactly matches the opposing wind speed, causing the wave to remain stationary relative to the ground.

For example (Fig. 17.18), consider cold air flowing down a mountain slope. It starts slowly, and has Fr < 1. As gravity accelerates the air downslope and causes its depth to decrease, the Froude can eventually reach the critical value Fr = 1. As the air continues to accelerate downhill the flow can become supercritical (Fr > 1). But once this supercritical flow

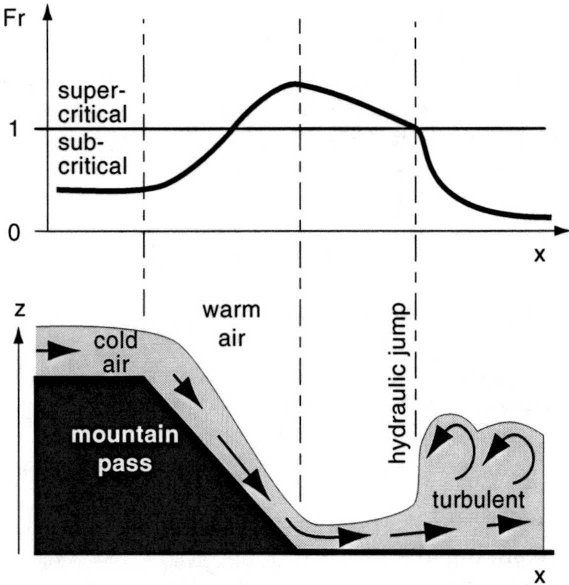

Figure 17.18
Variation of Froude number (Fr) with downwind distance (x), showing a hydraulic jump where the flow changes from supercritical to subcritical.

reaches the bottom of the slope and begins to decelerate due to turbulent drag across the lowland, its velocity can decrease and the flow depth gradually increases. At some point downstream the Froude number again reaches its critical value $Fr = 1$. An hydraulic jump can occur at this point, and turbulent drag increases. If this descending cold air is foggy or polluted, the hydraulic jump can be visible.

GAP WINDS

Basics

During winter in mountainous regions, sometimes the synoptic-scale weather pattern can move very cold air toward a mountain range. The cold air is denser than the warm overlying air, so buoyancy opposes rising motions in the cold air. Thus, the mountain range is a barrier that dams cold air behind it (Fig. 17.20).

However, river valleys, fjords, straits, and passes (Fig. 17.19) are mountain gaps through which the cold air can move as **gap winds** (Fig. 17.20). Gap wind speeds of 5 to 25 m s^{-1} have been observed, with gusts to 40 m s^{-1}. Temperature jumps at the top of the cold-air layer in the 5 to 10°C range are typical, while extremes of 15°C have been observed. Gap flow depths of 500 m to over 2 km have been observed. In any locale, the citizens often name the gap wind after their town or valley.

The cold airmass dammed on one side of the mountain range often has high surface pressure ($\mathbb{H}$), as explained in the Hydrostatic Thermal Circulation section of the General Circulation chapter. When synoptic low-pressure centers ($\mathbb{L}$) approach the opposite side of the mountain range, a pressure gradient of order (0.2 kPa)/(100 km) forms across the range that can drive the gap wind. Gap winds can also be driven by gravity, as cold air is pulled downslope through a mountain pass.

Divide gap flow into two categories based on the gap geometry: (1) short gaps, and (2) long gaps. Long gaps are ones with a gap width (order of 2 - 20 km) that is much less than the gap length (order of 100 km). Coriolis force is important for flow through long gaps, but is small enough to be negligible in short gaps.

Short-gap Winds

Use open-channel hydraulics for short gaps, and neglect Coriolis force. Although the gap-wind speed could range from subcritical to supercritical, observations suggest that one or more hydraulic jumps (Fig. 17.21) are usually triggered in the supercritical

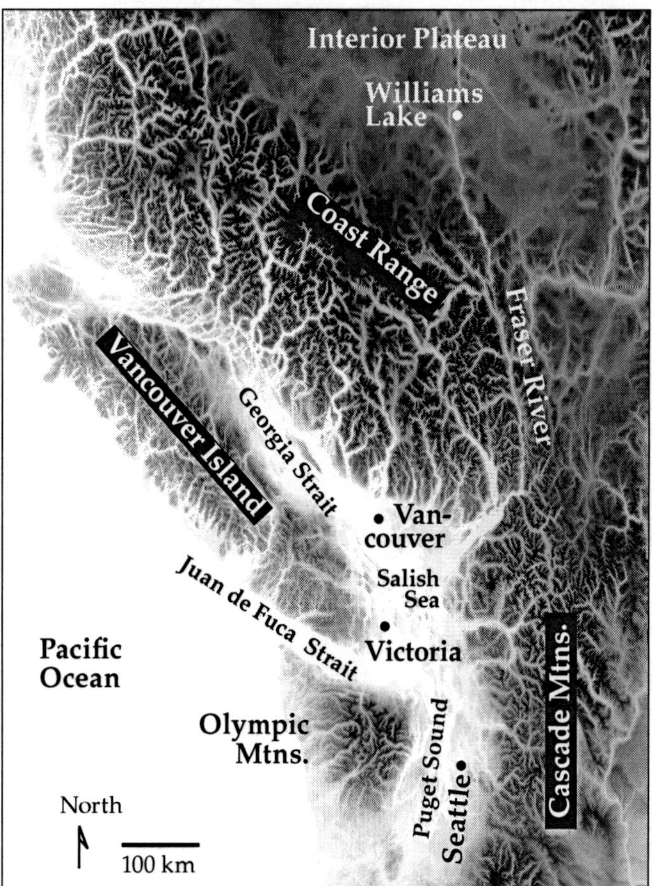

Figure 17.19
Geography of southwestern British Columbia, Canada, and northwestern Washington, USA, illustrating mountain gaps. Higher elevations are shown as darker greys, with the highest peaks 3,000 to 4,000 m above sea level. Ocean and very-low-elevation-land areas are white. Lower-elevation fjords, straits, and river valleys (i.e., gaps) appear as filaments of white or light-grey across the dark-shaded mountain range. In winter, sometimes very cold arctic air can pool in the Interior Plateau northeast of the Coast Mountains.

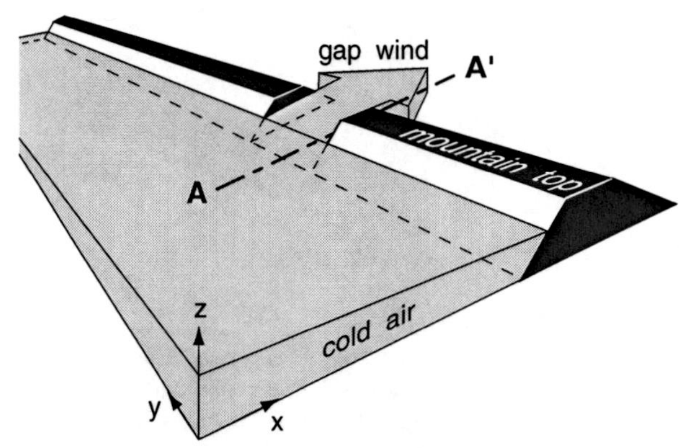

Figure 17.20
Cold air flow through a short gap.

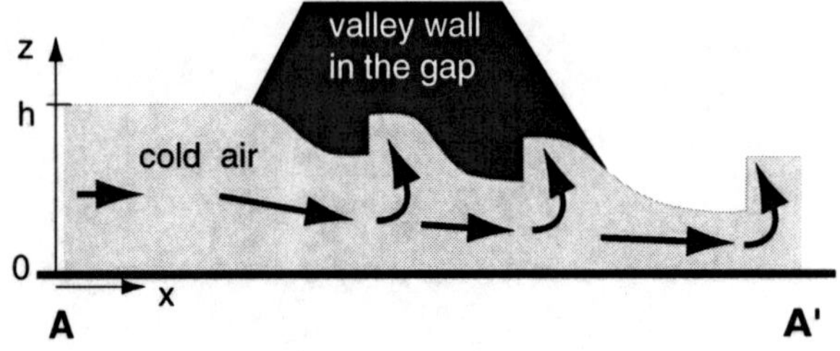

Figure 17.21
Cold air flow through a short gap. Vertical slice inside the gap, along section **A** - **A'** *from the previous figure. A series of 3 hydraulic jumps are shown in this idealization.*

> **Sample Application**
> Cold winter air of virtual potential temperature –5°C and depth 200 m flows through an irregular short mountain pass. The air above has virtual potential temperature 10°C. Find the max likely wind speed through the short gap.
>
> **Find the Answer**
> Given: $\Delta\theta_v$ = 15°C = 15K, h = 200 m.
> Find: $M_{gap\ max}$ = ? m s^{-1}
>
> For a 2-layer air system, use eq. (17.25):
> $M_{gap\ max}$ = [(9.8 m·s^{-2})·(15K)·(200m)/(267K)]$^{1/2}$
> = **10.5** m s^{-1}
>
> **Check**: Units OK. Magnitude OK.
> **Exposition**: Because of the very strong temperature jump across the top of the cold layer of air and the correspondingly faster wave group speed, faster gap flow speeds are possible. Speeds any faster than this would cause an hydraulic jump, which would increase turbulent drag and slow the wind back to this wind speed.

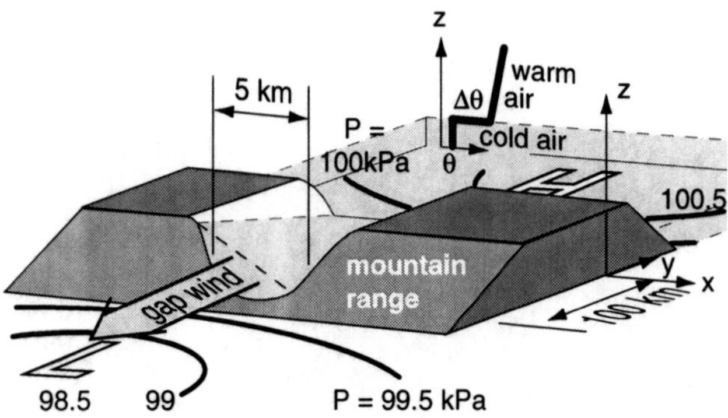

Figure 17.22
Scenario for gap winds in the N. Hemisphere in long valleys, with cold air (light grey) dammed behind a mountain range (dark grey). High pressure (H) is in the cold air, and low pressure (L) is at the opposite side of the mountains. Thick curved lines are sea-level isobars around the synoptic-scale pressure centers. The vertical profile of potential temperature θ *is plotted.*

regions due to irregularities in the valley shape, or by obstacles. The resulting turbulence in the hydraulic jumps causes extra drag, slowing the wind to its critical value.

The net result is that many gap winds are likely to have maximum speeds nearly equal to their critical value: the speed that gives Fr = 1. Using this in the definition of the Froude number allows us to solve for the likely maximum gap wind speed through gaps short enough that Coriolis force is not a factor:

$$M_{gap\ max} = \left[|g| \cdot \frac{\Delta\theta_v}{T_v} \cdot h \right]^{1/2} \tag{17.25}$$

Long-gap Winds

For long gaps, examine the horizontal forces (including Coriolis force) that act on the air. Consider a situation where the synoptic-scale isobars are nearly parallel to the axis of the mountain range (Fig. 17.22), causing a pressure gradient across the mountains.

For long narrow valleys this synoptic-scale cross-mountain pressure gradient is unable to push the cold air through the gap directly from high to low pressure, because Coriolis force tends to turn the wind to the right of the pressure gradient. Instead, the cold air inside the gap shifts its position to enable the gap wind, as described next.

Fig. 17.23 idealizes how this wind forms in the N. Hemisphere. Cold air (light grey in Fig. 17.23a) initially at rest in the gap feels the synoptically imposed pressure-gradient force F_{PGs} along the valley axis, and starts moving at speed M (shown with the short dark-grey arrow in Fig. 17.23a') toward the imposed synoptic-scale low pressure (**L**) on the opposite side of the gap. At this slow speed, both Coriolis force (F_{CF}) and turbulent drag force (F_{TD}) are correspondingly small (Fig. 17.23a'). The sum (black-and-white dotted arrow) of all the force vectors (black) causes the wind to turn slightly toward its right in the N. Hemisphere and to accelerate into the gap.

This turning causes the cold air to "ride up" on the right side of the valley (relative to the flow direction, see Fig. 17.23b). It piles up higher and higher as the gap wind speed M increases. But cold air is denser than warm. Thus slightly higher pressure (small dark-grey H) is under the deeper cold air, and slightly lower pressure (small dark-grey L) is under the shallower cold air. The result is a cross-valley mesoscale pressure-gradient force F_{PGm} per unit mass m (Fig. 17.23b') at the valley floor of:

$$\frac{F_{PGm}}{m} = -\frac{1}{\rho} \cdot \frac{\Delta P_m}{\Delta x} = -|g| \cdot \frac{\Delta\theta_v}{T_v} \cdot \frac{\Delta z}{\Delta x} \tag{17.26}$$

where $\Delta z/\Delta x$ is the cross-valley slope of the top of the cold-air layer, $\Delta P_m/\Delta x$ is the mesoscale pressure gradient across the valley, $\Delta\theta_v$ is the virtual potential temperature difference between the cold and warm air, T_v is an average virtual temperature (Kelvin), $|g| = 9.8 \text{ m·s}^{-2}$ is the magnitude of gravitational acceleration, and ρ is the average air density.

When this new pressure gradient force is vector-added to the larger drag and Coriolis forces associated with the moderate wind speed M, the resulting vector sum of forces (dotted white-and-black vector; Fig. 17.23b') begins to turn the wind to become almost parallel to the valley axis.

Gap-wind speed M increases further down the valley (Fig. 17.23c and c'), with the gap-wind cold air hugging the right side of the valley. In the along-valley direction (the $-y$ direction in Figs. 17.23), the synoptic pressure gradient force F_{PGs} is often larger than the opposing turbulent drag force F_{TD}, allowing the air to continue to accelerate along the valley, reaching its maximum speed near the valley exit. The **antitriptic wind** results from a balance of drag and synoptic-scale pressure-gradient forces. Similar gap winds are sometimes observed in the Juan de Fuca Strait (Fig. 17.19), with the fastest gap winds near the west exit region of the strait.

In the cross-valley direction, Coriolis force F_{CF} nearly balances the mesoscale pressure gradient force F_{PGm}. These last two forces define a mesoscale **"gap-geostrophic wind"** speed G_m parallel to the valley axis:

$$G_m = \left| \frac{g}{f_c} \cdot \frac{\Delta\theta_v}{T_v} \cdot \frac{\Delta z}{\Delta x} \right| \tag{17.27}$$

The actual gap wind speeds are of the same order of magnitude as this gap-geostrophic wind.

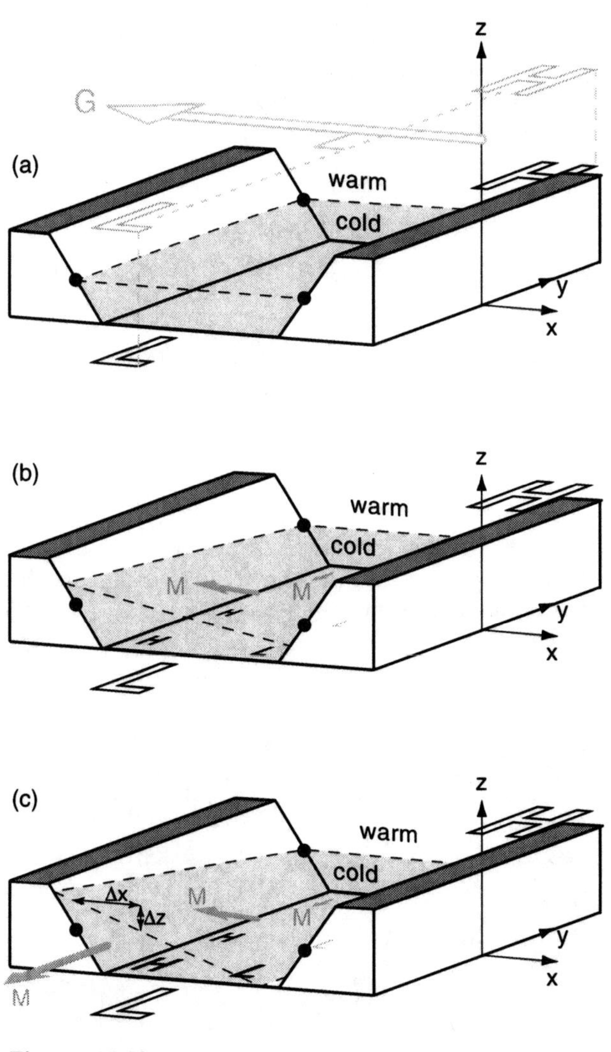

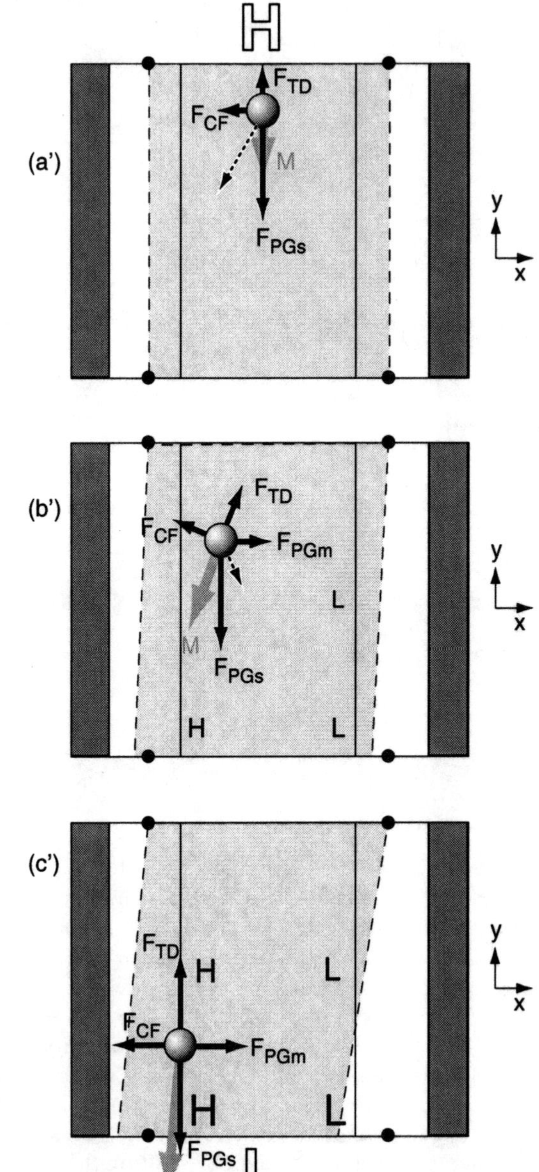

Figure 17.23
An enlargement of just the mountain gap portion of the previous figure. (a)-(c) Oblique view showing evolution of the cold air (light grey) in the gap. (a')-(c') Plan view (looking down from above) showing the corresponding forces acting on a cold air parcel as a gap wind forms (see text for details).

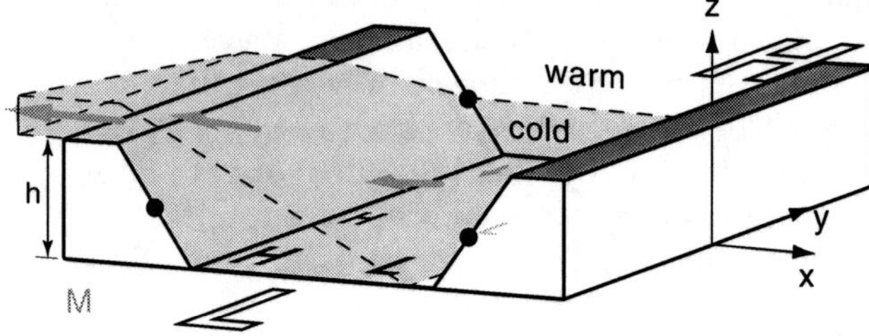

Figure 17.24
Gap winds can overtop the valley side walls if these walls are too low, or if the temperature difference between cold and warm layers of air is too small.

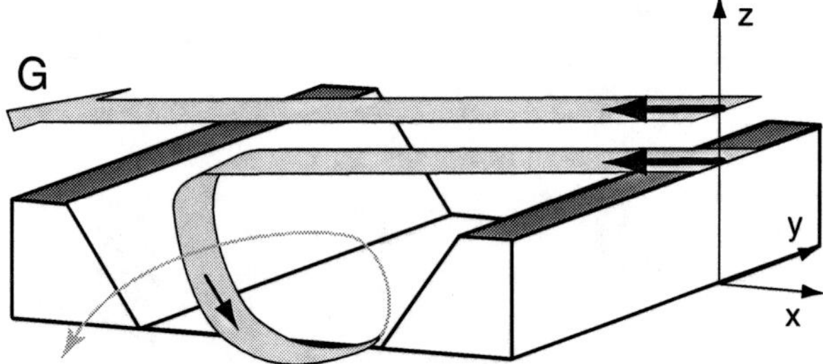

Figure 17.25
Without a layer of cold air near the surface, the cross-valley component of synoptic-scale geostrophic wind can create turbulent corkscrew motions and channeling of wind in the valley.

Sample Application
What long-gap wind speed can be supported in a strait 10 km wide through mountains 0.5 km high? The cold air is 4°C colder than the overlying 292 K air.

Find the Answer
Given: $h = \Delta z = 0.5$ km, $\Delta x = 10$ km, $\Delta\theta_v = 4°C = 4$ K,
$T_v = 0.5\cdot(288\text{K} + 292\text{K}) = 290\text{K}$.
Find: $G_m = ?$ m s^{-1}

Assume: $f_c = 10^{-4}$ s^{-1}.
Use eq. (17.27):
$G_m = |(9.8 \text{ m}\cdot\text{s}^{-2}/ 10^{-4} \text{ s}^{-1})\cdot(4\text{K}/290\text{K})\cdot(0.5\text{km}/10\text{km})|$
$= \underline{\mathbf{67 \text{ m s}^{-1}}}$.

Check: Units OK. Magnitude seems too large.
Exposition: The unrealistically large magnitude might be reached if the strait is infinitely long. But in a finite-length strait, the accelerating air would exit the strait before reaching this theoretical wind speed.

We implicitly assumed that the air was relatively dry, allowing $T_v \approx T$, and $\Delta\theta_v \approx \Delta\theta$. If humidity is larger, then you should be more accurate when calculating virtual temperatures. Also, as the air accelerates within the valley, mass conservation requires that the air depth Δz decreases, thereby limiting the speed according to eq. (17.27).

The synoptic-scale geostrophic wind on either side of the mountain range is nearly equal to the synoptic-scale geostrophic wind well above the mountain (G, in Fig. 17.23a). These synoptic winds are at right angles to the mesoscale gap-geostrophic wind.

The maximum possible gap wind speed is given by the equation above, but with Δz replaced with the height h of the valley walls above the valley floor. If either h or $\Delta\theta_v$ are too small, then some of the cool air can ride far enough up the valley wall to escape over top of the valley walls (Fig. 17.24), and the resulting gap winds are weaker. Gap winds occur more often in winter, when cold valley air causes large $\Delta\theta_v$.

For a case where $\Delta\theta_v$ is near zero (i.e., near-neutral static stability), the synoptic-scale geostrophic wind dominates. The vector component of G along the valley axis can appear within the valley as a **channeled wind** parallel to the valley axis. However, the cross-valley component of G can create strong turbulence in the valley (due to cavity, wake, and mountain-wave effects described later in this chapter). These components combine to create a turbulent corkscrew motion within the valley (Fig. 17.25).

COASTALLY TRAPPED LOW-LEVEL (BARRIER) JETS

Coriolis force is also important in locations such as the eastern Pacific Ocean, where low-altitude wind jets form parallel to the west coast of N. America. The dynamics describing these jets are very similar to the dynamics of long-gap winds.

Consider situations where synoptic-scale low-pressure systems reach the coast of N. America and encounter mountain ranges. Behind the approaching cyclone is cold air, the leading edge of which is the cold front (Fig. 17.26a). The cold air stops advanc-

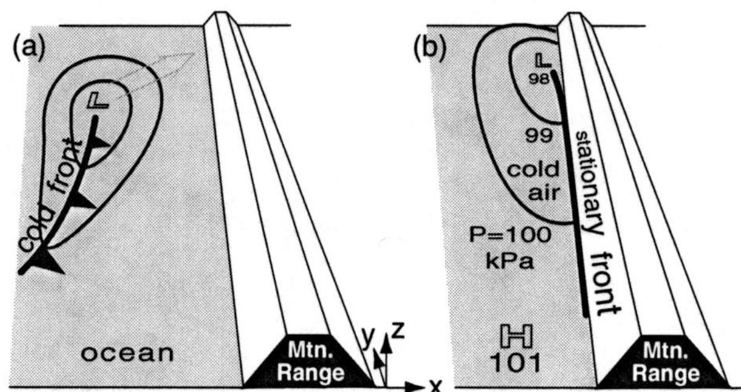

Figure 17.26
(a) Precursor synoptic conditions, as a low **L** *center approaches a coastal mountain (Mtn.) range in the N. Hemisphere. Curved lines are isobars at sea level. (b) Conditions later when the low reaches the coast, favoring coastally trapped low-level jets.*

ing eastward when it hits the mountains, causing a stationary front along the ocean side of the mountain range (Fig. 17.26b).

A pressure gradient forms parallel to the mountain range (Fig. 17.26b), between the low (𝕃) center to the north and higher pressure (ℍ) to the south. Fig. 17.27 shows a zoomed view of the resulting situation close to the mountains. The isobars are approximately perpendicular to the mountain-range axis, not parallel as was the case for gap winds.

As the synoptic-scale pressure gradient (*PGs*) accelerates the cold air from high towards low, Coriolis force (*CF*) turns this air toward the right (in the N. Hem.) causing the cold air to ride up along the mountain range. This creates a mesoscale pressure gradient (*PGm*) pointing down the cold-air slope (Fig. 17.28). Eventually an equilibrium is reached where turbulent drag (TD) nearly balances the synoptic pressure-gradient force, and Coriolis force is balanced by the mesoscale pressure gradient.

The end result is a low-altitude cold wind parallel to the coast, just west of the mountain range. The jet-core height is centered about 1/3 of the distance from the ocean (or lowland floor) to the ridge top. Jet core altitudes of 50 to 300 m above sea level have been observed along the west coast of N. America, while altitudes of about 1 km have been observed in California west of the Sierra-Nevada mountain range. Maximum speeds of 10 to 25 m s^{-1} have been observed in the jet core.

Width of the coastal jet is on the order of 100 to 150 km. This width is roughly equal to the **Rossby radius of deformation**, λ_R, which is a measure of the upstream region of influence of the mountain range on a flow that is in geostrophic balance. For a cold marine layer of air capped by a strong inversion as sketched in Fig. 17.28, the **external Rossby radius of deformation** is

$$\lambda_R = \frac{\sqrt{|g| \cdot h \cdot \Delta\theta_v / T_v}}{f_c} \qquad (17.28)$$

where $|g|$ = 9.8 m·s^{-2} is gravitational acceleration magnitude, h is mountain range height, f_c is the Coriolis parameter, $\Delta\theta_v$ is the jump of virtual potential temperature at the top of the marine air layer, and T_v is an average absolute virtual temperature of the air. For a statically-stable layer of air having a linear increase of potential temperature with height instead of a step discontinuity, an **internal Rossby radius of deformation** is

$$\lambda_R = \frac{N_{BV} \cdot h}{f_c} \qquad (17.29)$$

where $N_{BV} = [(|g|/T_v) \cdot \Delta\theta_v/\Delta z]^{1/2}$ is the Brunt-Väisälä frequency.

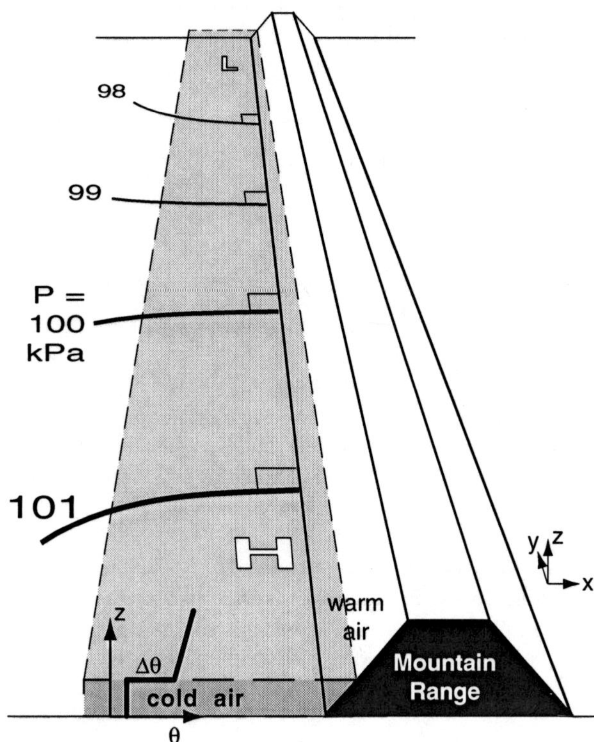

Figure 17.27
Synoptic conditions that favor creation of a mesoscale low-level jet parallel to the coast. Curved lines are isobars, ℍ and 𝕃 are high and low-pressure centers, and θ is potential temperature.

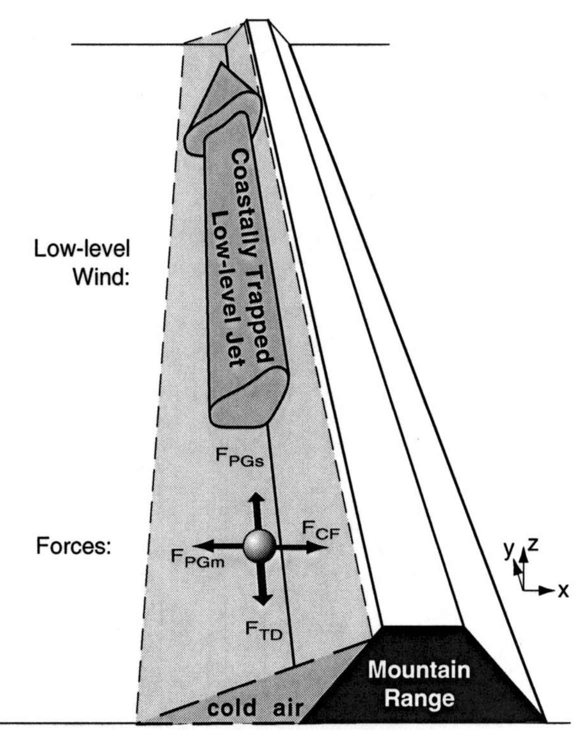

Figure 17.28
Final force (F) balance and cold-air location during a coastally trapped low-level jet (fast winds parallel to the coast). Vertical scale is stretched roughly 100:1 relative to the horizontal scale

Sample Application

The influence of the coast mountains extends how far to the west of the coastline in Fig. 17.28? The cold air has virtual potential temperature 8°C colder than the neighboring warm air. The latitude is such that the Coriolis parameter is 10^{-4} s^{-1}. Mountain height is 2000 m.

Find the Answer

Given: h = 2000 m, $\Delta\theta_v$ = 8°C = 8 K, $f_c = 10^{-4}$ s^{-1}.
Find: λ_R = ? km, external Rossby deformation radius

Assume: $|g|/T_v$ = 0.03333 m·s^{-2}·K^{-1}
The region of influence extends a distance equal to the Rossby deformation radius. Thus, use eq. (17.28):

λ_R = [(0.03333 m·s^{-2}·K^{-1})·(2000 m)·(8 K)]$^{1/2}$ /(10^{-4} s^{-1})
= 231,000 m = **231 km**

Check: Units OK. Magnitude OK.
Exposition: Even before fronts and low-pressure centers hit the coastal mountains, the mountains are already influencing these weather systems hundreds of kilometers offshore.

Sample Application (§)

Find and plot the path of air over a mountain, given: z_1 = 500 m, M = 30 m s^{-1}, b = 3, $\Delta T/\Delta z$ = –0.005 K m^{-1}, T = 10°C, and T_d = 8°C for the streamline sketched in Fig. 17.29.

Find the Answer

Given: (see above). Thus T = 283 K
Find: N_{BV} = ? s^{-1}, λ = ? m, and plot z vs. x

From the Stability chapter:

$$N_{BV} = \left[\frac{9.8\text{m}\cdot\text{s}^{-2}}{283\text{K}}\cdot(-0.005+0.0098)\right]^{1/2}$$

= 0.0129 s^{-1}

Use eq. (17.30)

$$\lambda = \frac{2\pi\cdot(30\text{m/s})}{0.0129\text{s}^{-1}} = 14.62\text{ km}$$

Solve eq. (17.31) on a spreadsheet to get the answer:

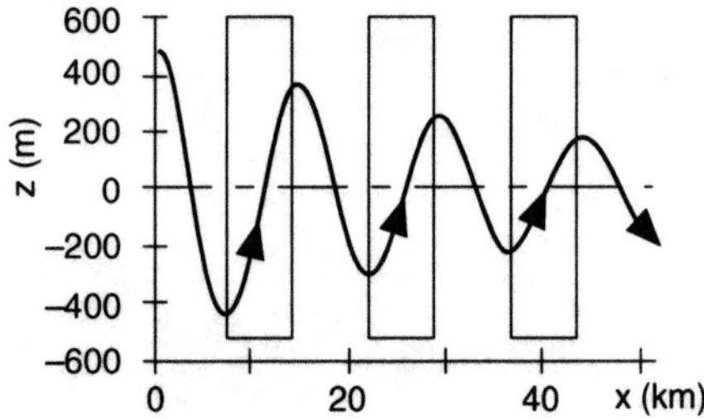

Check: Units OK. Physics OK. Sketch OK.
Exposition: Glider pilots can soar in the updraft portions of the wave, highlighted with white boxes.

MOUNTAIN WAVES

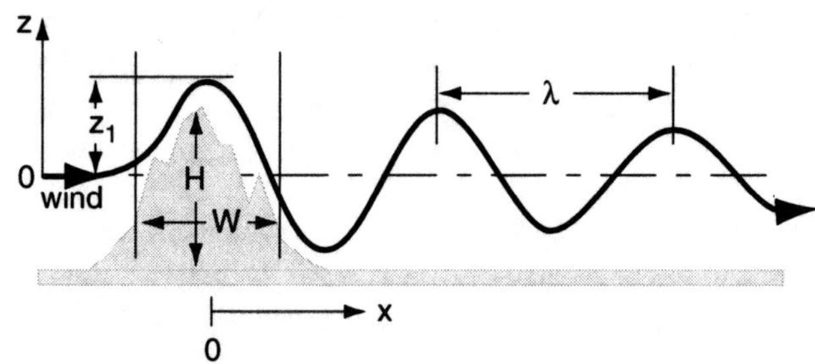

Figure 17.29
Mountain-wave characteristics.

Natural Wavelength

When statically stable air flows with speed M over a hill or ridge, it is set into oscillation at the Brunt-Väisälä frequency, N_{BV}. The natural wavelength λ is

$$\lambda = \frac{2\pi\cdot M}{N_{BV}} \qquad \bullet(17.30)$$

Longer wavelengths occur in stronger winds, or weaker static stabilities.

These waves are known as **mountain waves, gravity waves, buoyancy waves**, or **lee waves.** They can cause damaging winds, and interesting clouds (see the Clouds chapter).

Friction and turbulence damp the oscillations with time (Fig. 17.29). The resulting path of air is a damped wave:

$$z = z_1\cdot\exp\left(\frac{-x}{b\cdot\lambda}\right)\cdot\cos\left(\frac{2\pi\cdot x}{\lambda}\right) \qquad (17.31)$$

where z is the height of the air above its starting equilibrium height, z_1 is the initial amplitude of the wave (based on height of the mountain), x is distance downwind of the mountain crest, and b is a damping factor. Wave amplitude reduces to 1/e at a downwind distance of b wavelengths (that is, $b\cdot\lambda$ is the e-folding distance).

Lenticular Clouds

In the updraft portions of mountain waves, the rising air cools adiabatically. If sufficient moisture is present, clouds can form, called **lenticular clouds.** The first cloud, which forms over the mountain crest, is usually called a **cap cloud** (see Clouds chapter).

The droplet sizes in these clouds are often quite uniform, because of the common residence times of air in the clouds. This creates interesting optical phenomena such as **corona** and **iridescence** when the sun or moon shines through them (see the Atmospheric Optics chapter).

Knowing the temperature and dew point of air at the starting altitude before blowing over the mountain, a lifting condensation level (LCL) can be calculated using equations from the Water Vapor chapter. Clouds will form in the crests of those waves for which $z > z_{LCL}$.

Froude Number - Part 2

For individual hills not part of a continuous ridge, some air can flow around the hill rather than over the top. When less air flows over the top, shallower waves form.

The third variety of **Froude Number** Fr_3 is a measure of the ability of waves to form over hills. It is given by

$$Fr_3 = \frac{\lambda}{2 \cdot W} \qquad \bullet(17.32)$$

where W is the hill width, and λ is the natural wavelength. Fr_3 is dimensionless.

For strong static stabilities or weak winds, $Fr_3 << 1$. The natural wavelength of air is much shorter than the width of the mountain, resulting in only a little air flowing over the top of the hill, with small waves (Fig. 17.30a). If H is the height of the hill (Fig. 17.29), then wave amplitude $z_1 < H/2$ for this case. Most of the air is **blocked** in front of the ridge, or flows around the sides for an isolated hill.

For moderate stabilities where the natural wavelength is nearly equal to twice the hill width, $Fr_3 \approx 1$. The air **resonates** with the terrain, causing very intense waves (Fig. 17.30b). These waves have the greatest chance of forming lenticular clouds, and pose the threat of violent turbulence to aircraft. Extremely fast near-surface winds on the downwind (lee) side of the mountains cause **downslope wind storms** that can blow the roofs off of buildings. Wave amplitude roughly equals half the hill height: $z_1 \approx H/2$. Sometimes rotor circulations and **rotor clouds** will form near the ground under the wave crests (Fig. 17.30b; also see the Clouds chapter).

For weak static stability and strong winds, the natural wavelength is much greater than the hill width, $Fr_3 >> 1$. Wave amplitude is weak, $z_1 < H/2$. A turbulent **wake** will form downwind of the mountain, sometimes with a **cavity** of reverse flow near the ground (Fig. 17.30c). The cavity and rotor circulations are driven by the wind like a bike chain turning a gear.

For statically neutral conditions, $Fr = \infty$. A large turbulent wake occurs (Fig. 17.30d). These wakes are hazardous to aircraft.

Sample Application (§)
Replot the results from the previous Sample Application, indicating which waves have lenticular clouds.

Find the Answer
Given: (see previous Sample Application).
Find: z_{LCL} = ? m.

From the Moisture chapter: $z_{LCL} = a \cdot (T - T_d)$.
z_{LCL} = (125m °C^{-1})·(10°C – 8°C) = **250 m**
above the reference streamline altitude. From the sketch below, we find 1 cap cloud and 2 lenticular clouds.

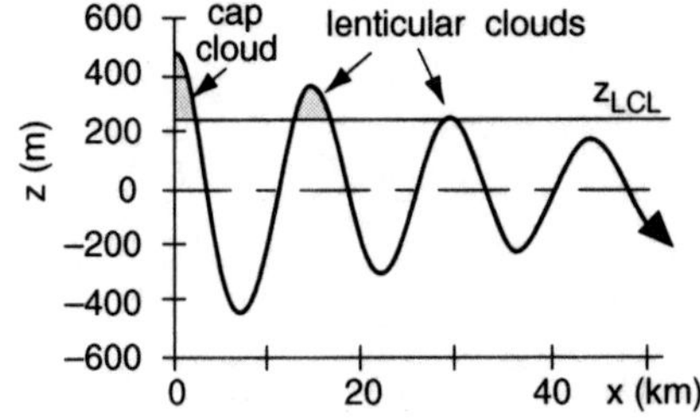

Check: Units OK. Physics OK. Sketch OK.
Exposition: Most clouds are blown with the wind, but **standing-lenticular clouds** are stationary while the wind blows through them!

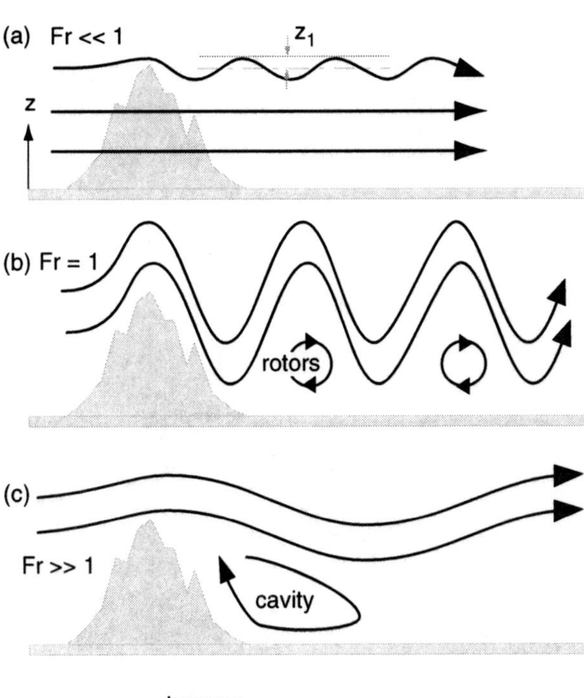

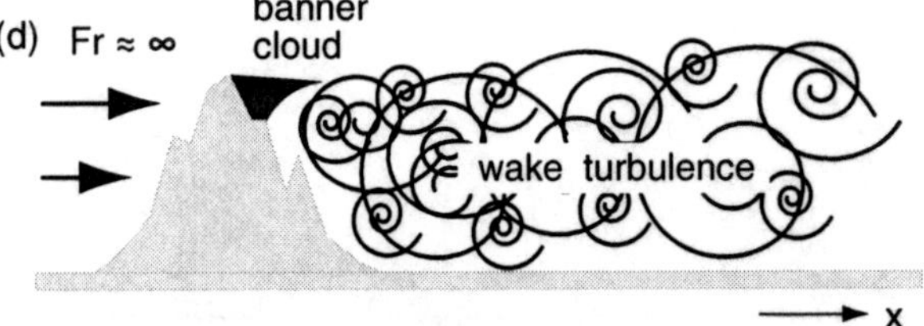

Figure 17.30
Mountain wave behavior vs. Froude number, Fr_3.

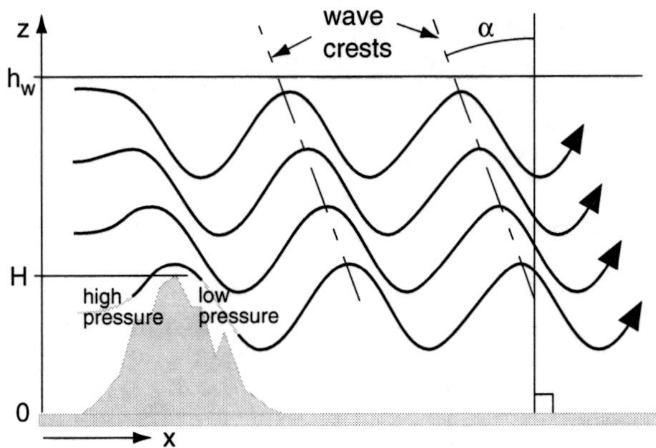

Figure 17.31
Vertical wave propagation, tilting crests, and wave drag.

Sample Application

For Fr_3 = 0.8, find the angle of the wave crests and the wave drag force over a hill of height 800 m. The Brunt-Väisälä frequency is 0.01 s^{-1}, and the waves fill the 11 km thick troposphere.

Find the Answer

Given: $Fr_3 = 0.8$, $H = 800$ m, $N_{BV} = 0.01$ s^{-1}.
Find: α = ?°, $F_{x\,WD}/m$ = ? m·s^{-2}

Use eq. (17.33): $\alpha = \cos^{-1}(0.8)$ = **36.9°**
Use eq. (17.34):

$$\frac{F_{xWD}}{m} = -\frac{[(800\text{m})\cdot(0.01\text{s}^{-1})]^2}{8\cdot(11{,}000\text{m})}\cdot 0.8\cdot[1-(0.8)^2]^{1/2}$$

$$= \mathbf{3.5x10^{-4}}\ \text{m·s}^{-1}.$$

Check: Units OK. Physics OK.
Exposition: This is of the same order of magnitude as the other forces in the equations of motion.

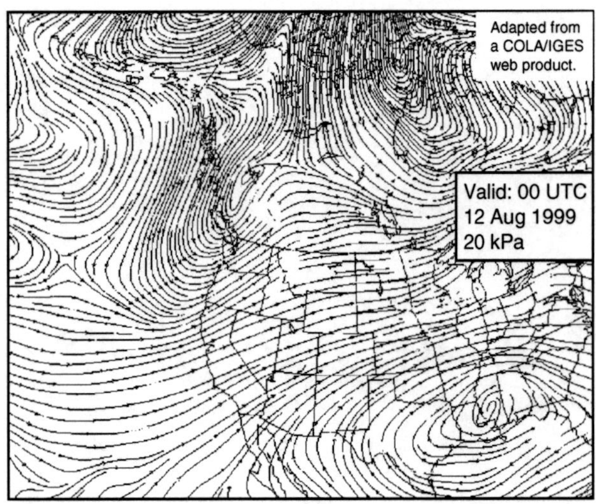

Figure 17.32
Streamlines near the tropopause, over N. America.

Sample Application

For a natural wavelength of 10 km and a hill width of 15 km, describe the type of waves.

Find the Answer

Given: λ = 10 km, W = 15 km.
Find: Fr = ? (dimensionless)

Use eq. (17.32): Fr_3 = (10km)/[2· (15km)] = **0.333**

Check: Units OK. Magnitude OK.
Exposition: Waves as in Fig. 17.30a form off the top of the hill, because $Fr_3 < 1$. Some air also flows around sides of hill.

Mountain-wave Drag

For $Fr_3 < 1$, wave crests tilt upwind with increasing altitude (Fig. 17.31). The angle α of tilt relative to vertical is

$$\cos(\alpha) = Fr_3 \qquad (17.33)$$

For this situation, slightly lower pressure develops on the lee side of the hill, and higher pressure on the windward side. This pressure gradient opposes the mean wind, and is called wave drag. Such wave drag adds to the skin drag. The **wave drag** (WD) force per unit mass near the ground is:

$$\frac{F_{xWD}}{m} = -\frac{H^2\cdot N_{BV}{}^2}{8\cdot h_w}\cdot Fr_3\cdot\left[1-Fr_3{}^2\right]^{1/2} \qquad (17.34)$$

where h_w is the depth of air containing waves.

Not surprisingly, higher hills cause greater wave drag. The whole layer of air containing these waves feels the wave drag, not just the bottom of this layer that touches the mountain.

STREAMLINES, STREAKLINES, AND TRAJECTORIES

Streamlines are conceptual lines that are everywhere parallel to the flow at some instant (i.e., a snapshot). This is an Eulerian point of view. Fig. 17.32 shows an example of streamlines on a weather map. Streamlines never cross each other except where the speed is zero, and the wind never crosses streamlines. Streamlines can start and end anywhere, and can change with time. They are often not straight lines.

Streaklines are lines deposited in the flow during a time interval by continuous emission of a tracer from a fixed point. Examples can be seen in aerial photographs of smoke plumes emitted from smokestacks, or volcanic ash clouds.

Trajectories, also called **path lines**, trace the route traveled by an air parcel during a time interval. This is the Lagrangian point of view. For **stationary** (not changing with time) flow, streamlines, streaklines, and trajectories are identical.

For nonstationary flow (flow that changes with time), there can be significant differences between them. For example, suppose that initially the flow is steady and from the north. Later, the wind suddenly shifts to come from the west.

Fig. 17.33 shows the situation shortly after this wind shift. Streamlines (thin solid lines in this figure) are everywhere from the west in this example. The streakline caused by emission from a smokestack is the thick grey line. The black dashed line shows the path followed by one air parcel in the smoke plume.

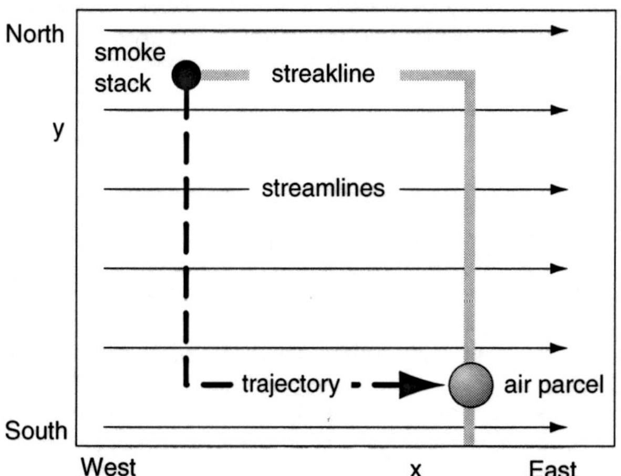

Figure 17.33
Streamlines, streaklines (smoke plumes), and trajectories (path lines) in nonstationary flow.

BERNOULLI'S EQUATION

Principles

Consider a **steady-state** flow (flow that does not change with time), but which can have different velocities at different locations. If we follow an air parcel as it flows along a streamline, its velocity can change as it moves from one location to another. For wind speeds $M \leq 20$ m s^{-1} at constant altitude, the air behaves as if it is nearly incompressible (namely, constant density ρ).

Incompressible Flow

For the special case of incompressible, steady-state, **laminar** (nonturbulent) motion with no drag, the equations of motion for an air parcel following a streamline can be simplified into a form known as **Bernoulli's equation**:

$$\frac{1}{2}M^2 + \frac{P}{\rho} + |g| \cdot z = C_B \qquad \bullet(17.35)$$

energy: kinetic + flow + potential = constant

where M is the total velocity along the streamline, P is static air pressure, ρ is air density, $|g|$ = 9.8 m·s^{-2} is gravitation acceleration magnitude, and z is height above some reference.

C_B is an arbitrary constant called **Bernoulli's constant** or **Bernoulli's function**. C_B is constant

Sample Application
Environmental air outside a hurricane has sea-level pressure 100 kPa. Find the rise in sea level at the eye, where sea-level pressure is 90 kPa. Neglect currents and wind waves.

Find the Answer
Given: P_{env} = 100 kPa, z_{env} = 0 m, P_{eye} = 90 kPa, $M \approx 0$
Find: z_{eye} = ? m , where z is height of sea level.

Consider a streamline in the water at the sea surface. ρ = 1025 kg m^{-3} for sea water.

Use Bernoulli's eq. (17.35) to find C_B for the environment, then use it for the eye:

Env: 0.5·(0m s^{-1})^2 + (100,000Pa)/(1025kg m^{-3}) + (9.8 m s^{-2})·(0m) = C_B = 97.6 m^2 s^{-2}

Eye: 0.5·(0m s^{-1})^2 + (90,000Pa)/(1025kg m^{-3}) + (9.8 m s^{-2})·(z_{eye}) = C_B = 97.6 m^2 s^{-2}

z_{eye} = {(97.6 m^2 s^{-1})–[(90,000Pa)/(1025kg m^{-3})]}/(9.8 m s^{-2})
= 1.0 m.

Check: Units OK. Magnitude OK.
Exposition: Such a rise in sea level is a hazard called the **storm-surge**.

HIGHER MATH • Bernoulli Derivation

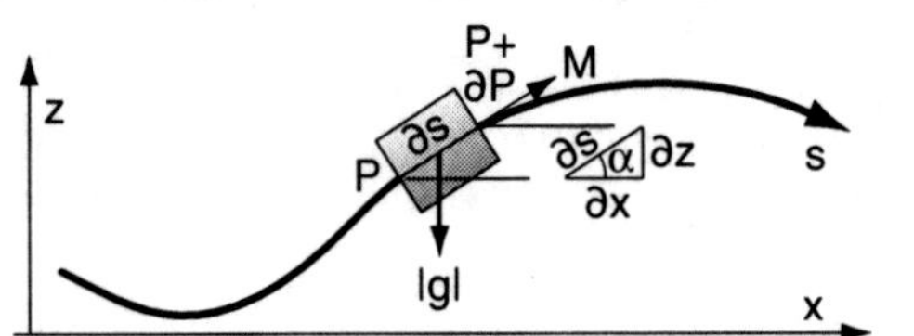

Fig. 17.d. *Forces on air parcel following streamline.*

To derive Bernoulli's equation, apply Newton's second law ($a = F/m$) along a streamline s: Acceleration is the total derivative of wind speed: $a = dM/dt = \partial M/\partial t + M\cdot\partial M/\partial s$. Consider the special case of flow that is steady at any location ($\partial M/\partial t = 0$) even though the flow can be different at different locations ($\partial M/\partial s \neq 0$). Thus

$$M\cdot\partial M/\partial s = F/m$$

The forces per unit mass F/m acting on a fluid parcel along the direction of the streamline are pressure-gradient force and the component of gravity along the streamline [$|g|\cdot\sin(\alpha) = |g|\cdot\partial z/\partial s$]:

$$M\cdot\partial M/\partial s = -(1/\rho)\cdot\partial P/\partial s - |g|\cdot\partial z/\partial s$$

or

$$(1/2)\, dM^2 + (1/\rho)\cdot dP + |g|\cdot dz = 0$$

For incompressible flow, ρ = constant. Integrate the equation above to get Bernoulli's equation:

$$(1/2)\, M^2 + P/\rho + |g|\cdot z = C_B$$

where C_B is the constant of integration.

This result applies only to steady incompressible flow along a streamline. Do not use it for situations where additional forces are important, such as turbulent drag, or across wind turbines or fans.

Sample Application

Cold-air flow speed 10 m s^{-1} changes to 2 m s^{-1} after passing a hydraulic jump. This air is 10°C colder than the surroundings. How high can the hydraulic-jump jump?

Find the Answer

Given: $M_1 = 10$ m s^{-1}, $M_2 = 2$ m s^{-1}, $\Delta T = 10$K, $z_1 = 0$.
Find: z_2 = ? m above the initial z.

Assume: $\Delta\theta_v = \Delta T$ and $|g|/T_v = 0.0333$ m·s^{-2}·K^{-1}. $P \approx$ constant on a streamline along the top of the cold air.

Use eq. (17.36), noting that $C_B - P/\rho$ is constant :

- At point 1: $0.5\cdot(10\text{m s}^{-1})^2 + 0 = C_B - P/\rho = 50\ \text{m}^2\,\text{s}^{-2}$
- At 2: $0.5\cdot(2\text{m s}^{-1})^2 + (0.0333\text{m}\cdot\text{s}^{-2}\cdot\text{K}^{-1})\cdot(10\text{K})\cdot z_2 = 50\ \text{m}^2\,\text{s}^{-2}$

Solve for z_2: $z_2 = [M_1^2 - M_2^2]\cdot T_v/(2\cdot|g|\cdot\Delta\theta_v)$ = **<u>144 m</u>**

Check: Units OK. Magnitude OK.

Exposition: Hydraulic jumps are very turbulent and would dissipate some of the mechanical energy into heat. So the actual jump height would be less.

along any one streamline, but different streamlines can have different C_B values.

Bernoulli's equation focuses on **mechanical-energy** conservation along a streamline. The first term on the left is the **kinetic energy** per unit mass. The middle term is the work done on the air (sometimes called **flow energy** per unit mass) that has been stored as pressure. The last term on the left is the **potential energy** per mass. Along any one streamline, energy can be converted from one form to another, provided the sum of these energies is constant.

In hydraulics, the gravity term is given by the change in depth of the water, especially when considering a streamline along the water surface. In meteorology, a similar situation occurs when cold air rises into a warmer environment; namely, it is a dense fluid rising against gravity. However, the gravity force felt by the rising cold air is reduced because of its buoyancy within the surrounding air. To compensate for this, the gravity factor in Bernoulli's equation can be replaced with a **reduced gravity** $g' = |g|\cdot\Delta\theta_v/T_v$, yielding:

$$\frac{1}{2}M^2 + \frac{P}{\rho} + |g|\frac{\Delta\theta_v}{T_v}\cdot z = C_B \qquad \bullet(17.36)$$

where $\Delta\theta_v$ is the virtual potential temperature difference between the warm air aloft and the cold air below, and T_v is absolute virtual air temperature (K). Thus, the gravity term is nonzero when the streamline of interest is surrounded by air of different virtual potential temperature, for air flowing up or down.

To use eq. (17.36), first measure all the terms in the left side of the equation at some initial (or upstream) location in the flow. Call this point 1. Use this to calculate the initial value of C_B. Then, at some downstream location (point 2) along the same streamline use eq. (17.36) again, but with the known value of C_B from point 1. The following equation is an expression of this procedure of equating final to initial flow states:

$$\frac{1}{2}M_2^{\,2} + \frac{P_2}{\rho} + |g|\frac{\Delta\theta_v}{T_v}\cdot z_2 = \frac{1}{2}M_1^{\,2} + \frac{P_1}{\rho} + |g|\frac{\Delta\theta_v}{T_v}\cdot z_1 \qquad \bullet(17.37)$$

Another way to consider eq. (17.36) is if any one or two terms increase in the equation, the other term(s) must decrease so that the sum remains constant. In other words, the sum of changes of all the terms must equal zero:

$$\Delta\left[\frac{1}{2}M^2\right] + \Delta\left[\frac{P}{\rho}\right] + \Delta\left[|g|\cdot\frac{\Delta\theta_v}{T_v}\cdot z\right] = 0 \qquad \bullet(17.38)$$

Caution: $\Delta[(0.5)\cdot M^2] \neq (0.5)\cdot[\Delta M]^2$.

The Bernoulli equations above do NOT work:

- anywhere the flow is turbulent
- behind obstacles that create turbulent wakes or which cause sudden changes in the flow
- at locations of heat input or loss
- at locations of mechanical-energy input (such as a fan) or loss (such as a wind turbine)
- near the ground where drag slows the wind
- where flow speed > 20 m s^{-1}
- where density is not approximately constant

Hence, there are many atmospheric situations for which the above equations are too simplistic.

Compressible Flow

For many real atmospheric conditions where winds can be any speed, you should use a more general form of the Bernoulli equation that includes thermal processes.

For an **isothermal** process, the equation becomes:

$$\frac{1}{2}M^2 + \Re_d\cdot T_v\cdot\ln(P) + |g|\frac{\Delta\theta_v}{T_v}\cdot z = C_B \qquad (17.39)$$

where $\Re_d$ = 287 m^2·s^{-2}·K^{-1} is the ideal gas constant for dry air, and where C_B is constant during the process (i.e., initial C_B = final C_B).

For **adiabatic** (isentropic; no heat transfer) flow, the Bernoulli equation is

$$\frac{1}{2}M^2 + \left(\frac{C_p}{\Re}\right)\frac{P}{\rho} + |g|\frac{\Delta\theta_v}{T_v}\cdot z = C_B \qquad (17.40)$$

where C_p is the specific heat at constant pressure, and $\Re$ is the ideal gas constant. For dry air, $C_{pd}/\Re_d$ = 3.5 (dimensionless).

Using the ideal gas law, this last equation for adiabatic flow along a streamline becomes

$$\frac{1}{2}M^2 + C_p\cdot T + |g|\frac{\Delta\theta_v}{T_v}\cdot z = C_B \qquad \bullet(17.41)$$

kinetic energy + sensible heat + potential energy = constant

where $C_p\cdot T$ is the **enthalpy** (also known as the **sensible heat** in meteorology), and C_p = 1004 m^2·s^{-2}·K^{-1} for dry air. In other words:

Sample Application

Wind at constant altitude decelerates from 15 to 10 m s^{-1} while passing through a wind turbine. What opposing net pressure difference would have caused the same deceleration in laminar flow?

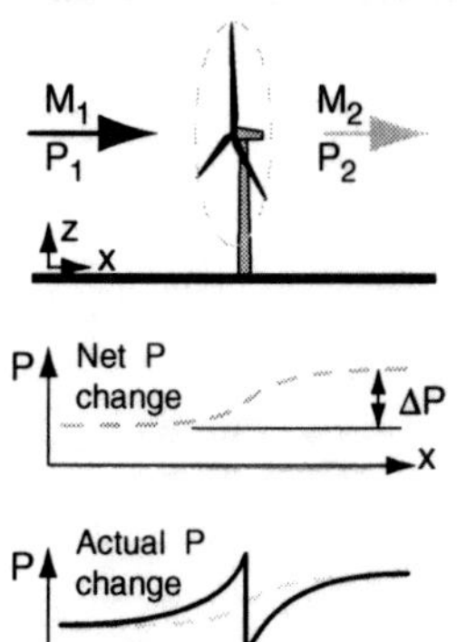

Find the Answer

Given: $\Delta z = 0$,
$M_1 = 15$ m s^{-1} , $M_2 = 10$ m s^{-1}.
Find: ΔP = ? Pa

Assume ρ = 1 kg m^{-3} = constant.
Solve eq. (17.38) for ΔP: $\Delta P = -\rho\cdot\Delta[(0.5)\cdot M^2]$
$\Delta P = -(1 \text{ kg m}^{-3})\cdot(0.5)\cdot[(10\text{m s}^{-1})^2 - (15\text{m s}^{-1})^2]$
= **<u>62.4 Pa</u>**

Check: Units OK. Magnitude OK.
Exposition: The process of extracting mechanical energy from the wind has the same affect as an opposing pressure difference. This pressure difference is small compared to ambient atmospheric pressure P = 100,000 Pa.

The actual pressure change across a wind turbine is shown in bottom figure.

Sample Application

Air with pressure 100 kPa is initially at rest. It is accelerated over a flat 0°C snow surface as it is sucked toward a household ventilation system. If the final speed is 10 m s^{-1}, what is the air pressure at the fan entrance?

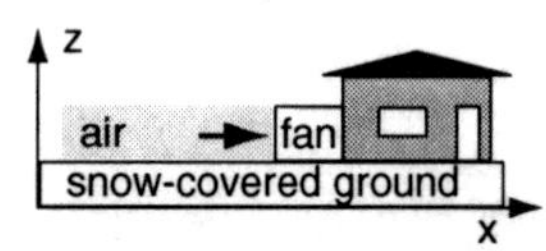

Find the Answer

Given: P_1 = 100 kPa, M_1 = 0, M_2 = 10 m s^{-1}, $\Delta z = 0$.
Find: P_2 = ? kPa

Assume the snow keeps their air at constant T_v = 0°C.
Use eq. (17.39) for an isothermal process:
$[0.5M_2^2 - 0] + \Re_d\cdot T_v\cdot[\ln(P_2) - \ln(P_1)] = 0$
Use $\ln(a) - \ln(b) = \ln(a/b)$. Then take exp of both sides, and rearrange: $P_2 = P_1\cdot\exp[-0.5M_2^2 / (\Re_d\cdot T_v)]$
P_2 =
= (100kPa)·exp[–0.5(10m s^{-1})^2/{(287m^2·s^{-2}·K^{-1})·(273K)}]
= (100kPa)·exp[–0.000638] = (100kPa)· (0.9994)
= **<u>99.94 kPa</u>**.

Check: Units OK. Magnitude OK.
Exposition: This decrease of about 0.06 kPa is small compared to ambient P = 100 kPa. The air-pressure decrease is expected because of the suction caused by the fan.

Sample Application

A short distance behind the engine of a jet aircraft flying in level flight, the exhaust temperature is 400 °C and the jet-blast speed is 200 m s^{-1}. After the jet exhaust decelerates to zero, what is the final exhaust air temperature, neglecting conduction & turbulent mixing.

Find the Answer

Given: M_1 = 200 m s^{-1}, T_1 =400°C = 673K, M_2 = 0, Δz=0
Find: T_2 = ? °C. Assume adiabatic process.

Rearrange eq. (17.42): $T_2 = T_1 + M_1^2/(2C_p)$
T_2 = (673K) + (200m s^{-1})^2/(2· 1004 m^2·s^{-2}·K^{-1})
= 673K + 19.9K = 693 K = **420°C**

Check: Units OK. Magnitude OK.
Exposition: Jet exhaust is turbulent and mixes quickly with the cooler ambient air, so it is not appropriate to use Bernoulli's equation. See the "dynamic warming" section later in this chapter for more info.

Sample Application

A 75 kW electric wind machine with a 2.5 m radius fan is used in an orchard to mix air to reduce frost damage on fruit. The fan horizontally accelerates air from 0 to 5 m s^{-1}. Find the temperature change across the fan, neglecting mixing with environmental air.

Find the Answer

Given: *Power* = 75 kW = 75000 kg·m^2·s^{-3}, R = 2.5 m, M_1 = 0, M_2 = 5 m s^{-1}, Δz = 0
Find: ΔT = ? °C

Assume that all the electrical energy used by the fan motor goes into a combination of heat and shaft work.

The mass flow rate through this fan is:
$\rho \cdot M_2 \cdot \pi \cdot R^2$ = (1.225kg m^{-3})·(5m s^{-1})·π·(2.5m)^2 =120 kg s^{-1}
Thus: $\Delta q + \Delta SW$ = *Power/(Mass Flow Rate)* = 624 m^2·s^{-2}
Use eq. (17.45): $\Delta T = (1/C_p)\cdot[\Delta q + \Delta SW - 0.5\cdot(M_2^2 - M_1^2)]$
= (1/1004 m^2·s^{-2}·K^{-1})·[(624 m^2·s^{-2}) – 0.5·(5m s^{-1})^2]
= 0.61 K = **0.61°C**

Check: Units OK. Magnitude OK.
Exposition: In spite of the large energy consumption of the electric motor, the heating is spread into a very large volume of air that passes through the fan. Hence, the amount of temperature change is small.

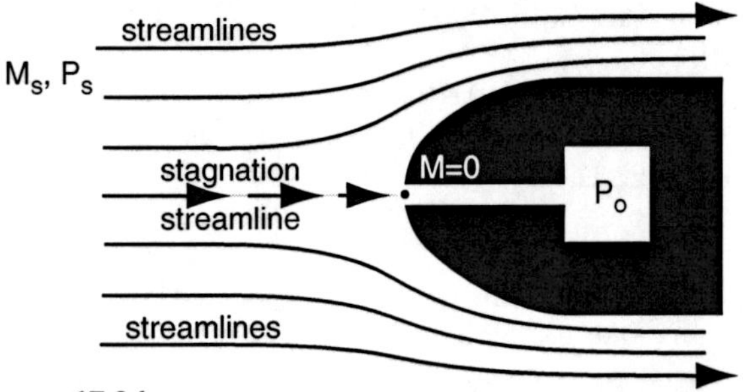

Figure 17.34
Streamlines, showing stagnation as air approaches the obstacle.

$$\frac{1}{2}M_2^2 + C_p \cdot T_2 + |g|\frac{\Delta\theta_v}{T_v}\cdot z_2 = \frac{1}{2}M_1^2 + C_p \cdot T_1 + |g|\frac{\Delta\theta_v}{T_v}\cdot z_1 \quad (\bullet 17.42)$$

where subscript 2 denotes final state, and subscript 1 denotes initial state. Equation (17.41) is also sometimes written as

$$\frac{1}{2}M^2 + C_v \cdot T + \frac{P}{\rho} + |g|\frac{\Delta\theta_v}{T_v}\cdot z = C_B \quad (17.43)$$

energy: kinetic + internal + flow + potential = constant

Again, C_B is constant during the adiabatic process.

Energy Conservation

Because these several previous equations also consider temperature, we cannot call them Bernoulli equations. They are **energy conservation equations** that consider mechanical and thermal energies following a streamline.

If we extend this further into an **energy budget equation**, then we can add the effects of net addition of **thermal energy** (heat per unit mass) Δq via radiation, condensation or evaporation, conduction, combustion, etc. We can also include **shaft work per unit mass** ΔSW done on the air by a fan, or work extracted from the air by a wind turbine.

$$\Delta\left[\frac{1}{2}M^2\right] + \Delta\left[\frac{P}{\rho}\right] + \Delta\left[C_v \cdot T\right] + \Delta\left[|g|\cdot\frac{\Delta\theta_v}{T_v}\cdot z\right] = \Delta q + \Delta SW \quad (17.44)$$

or

$$\Delta\left[\frac{1}{2}M^2\right] + \Delta\left[C_p \cdot T\right] + \Delta\left[|g|\cdot\frac{\Delta\theta_v}{T_v}\cdot z\right] = \Delta q + \Delta SW \quad \bullet(17.45)$$

Some Applications

Dynamic & Static Pressure & Temperature

Free-stream atmospheric pressure away from any obstacles is called the **static pressure** P_s. Similarly, let T be the free stream (initial) temperature.

When the wind approaches an obstacle, much of the air flows around it, as shown in Fig. 17.34. However, for one streamline that hits the obstacle, air decelerates from an upstream speed of $M_{initial} = M_s$ to an ending speed of M_{final} = 0. This ending point is called the **stagnation point**.

As air nears the stagnation point, wind speed decreases, and both the air pressure and temperature increase. The increased pressure is called the **dynamic pressure** P_{dyn}, and the increased temperature is called the **dynamic temperature** T_{dyn}. At the stagnation point where velocity is zero, the final dynamic pressure is called the **stagnation pressure** P_o, and the associated dynamic temperature is given the symbol T_o. Think of subscript "o" as indicating zero wind speed relative to the obstacle.

To find the dynamic effects at stagnation, use the energy conservation equation (17.42) for wind blowing horizontally (i.e., no change in z), and assume a nearly adiabatic process:

$$C_p \cdot T_o = \frac{1}{2} M_s^{\,2} + C_p \cdot T \qquad (17.46)$$

Solving for the dynamic temperature T_o gives:

$$T_o = T + \frac{M_s^{\,2}}{2 \cdot C_p} \qquad \bullet(17.47)$$

where C_p = 1004 m^2·s^{-2}·K^{-1} for dry air. Eq. (17.47) is valid for subsonic speeds (see INFO box).

This effect is called **dynamic warming** or **dynamic heating**, and must be avoided when deploying thermometers in the wind, because the wind stagnates when it hits the thermometer. As shown in Fig. 17.35, dynamic warming ($\Delta T = T_o - T$) is negligible ($\Delta T \le 0.2$°C) for flow speeds of $M_s \le 20$ m s^{-1}.

However, for thermometers on an aircraft moving 100 m s^{-1} relative to the air, or for stationary thermometers exposed to tornadic winds of M_s = 100 m s^{-1}, the dynamic warming is roughly $\Delta T \approx$ 5°C. For these extreme winds you can correct for dynamic warming by using the dynamic temperature T_o measured by the thermometer, and using the measured wind speed M_s, and then solving eq. (17.47) for free-stream temperature T.

In the Heat Budgets chapter is a relationship between temperature and pressure for an adiabatic, compressible process. Using this with the equation above allows us to solve for the **stagnation pressure**:

$$P_o = P_s \cdot \left(\frac{T_o}{T} \right)^{C_p/\Re} \qquad (17.48)$$

or

$$P_o = P_s \cdot \left(1 + \frac{0.5 \cdot M_s^{\,2}}{C_p \cdot T} \right)^{C_p/\Re} \qquad (17.49)$$

where $C_p/\Re \approx C_{pd}/\Re_d = 3.5$ for air, P_s is static (free-stream) pressure, M_s is free-stream wind speed along the streamline, $C_p \approx C_{pd}$ = 1004 J·kg^{-1}·K^{-1}, T is free-stream temperature, and subscript d denotes dry air. Fig. 17.35 shows stagnation-pressure increase.

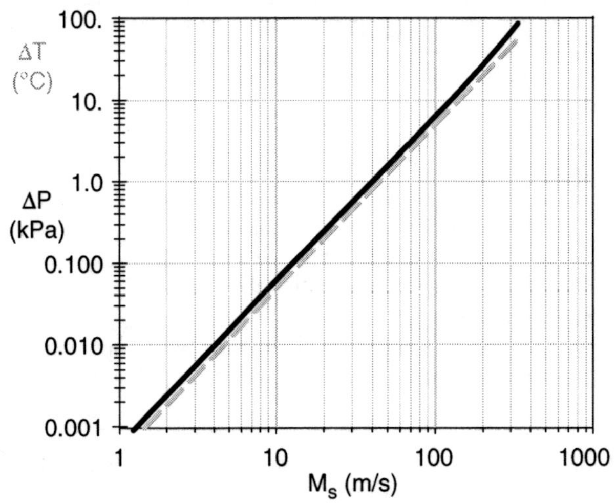

Figure 17.35
Increase in pressure (ΔP, black solid line) and temperature (ΔT, grey dashed line) due to dynamic warming when air of speed M_s hits an obstacle and stagnates. Ambient free-stream conditions for this calculation are T = 290K and P_s = 100 kPa, giving a speed of sound of c = 341 m s^{-1}.

Sample Application

Tornadic winds of 100 m s^{-1} and 30°C blow into a garage and stagnate. Find stagnation T & P. What net force pushes against a 3 x 5 m garage wall?

Find the Answer

Given: M_s =100 m s^{-1}, T=30°C, Wall area A=3x5 =15 m^2
Find: F_{net} = ? N. Assume: ρ = 1.2 kg m^{-3}, P = 100 kPa.

Use eq. (17.47):
T_o = (30°C)+[(100m s^{-1})^2/(2 · 1004m^2·s^{-2}·K^{-1}) = <u>35°C</u>.

Use eq. (17.48):
P_o = (100 kPa)·[(35+273)/(30+273)]$^{3.5}$ = **<u>105.9 kPa</u>**.

Compare with eq. (17.50): P_o = (100kPa) + [(0.5 · 1.2kg m^{-3})·(100m s^{-1})^2]·(1 kPa/1000Pa)= **<u>106 kPa</u>**

F_{net} = ΔP·A = (6 kN m^{-2})·(15 m^2) = **<u>90 kN</u>**

Check: Units OK. Physics OK.
Exposition: This force is equivalent to the weight of more than 1000 people, and acts on all walls and the roof. It is strong enough to pop the whole roof up off of the house. Then the walls blow out, and the roof falls back down onto the floor.

Hide in the basement. Quickly.

INFO • Speed of Sound

The speed c of sound in air is

$$c = [k \cdot \Re \cdot T]^{1/2} \quad (17.42)$$

where $k = C_p/C_v$ is the ratio of specific heats for air, $\Re$ is the ideal gas law constant, and T is absolute temperature. For dry air, the constants are: $k = C_{pd}/C_{vd} = 1.4$, and $\Re_d = 287.053$ (m² s⁻²)·K⁻¹. Thus, the speed of sound increases with the square root of absolute temperature.

The speed M of any object such as an aircraft or an air parcel can be compared to the speed of sound:

$$Ma = M / c \quad (17.43)$$

where Ma is the dimensionless **Mach number**. Thus, an object moving at Mach 1 is traveling at the speed of sound.

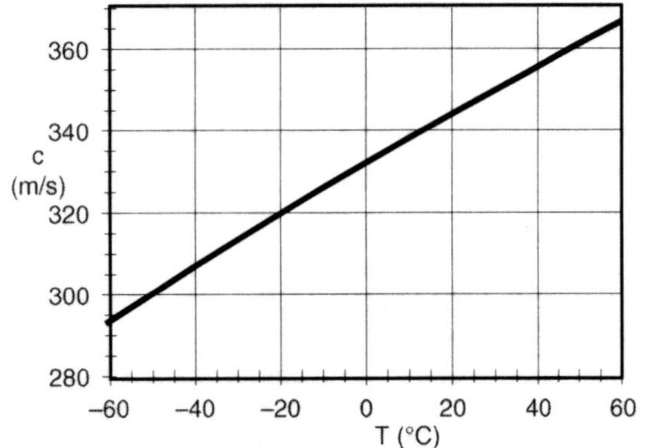

Fig. 17.e. *Speed of sound in dry air.*

For wind speeds of $M_s < 100$ m s⁻¹, the previous equation is very well approximated by the simple Bernoulli equation for incompressible flow:

$$P_o = P_s + \frac{\rho}{2} M_s^2 \quad \bullet(17.50)$$

where ρ is air density. Do not use eq. (17.50) to find dynamic heating when combined with the ideal gas law, because it neglects the large density changes that occur in high-speed flows that stagnate.

The previous 3 equations show that the pressure increase due to stagnation ($\Delta P = P_o - P_s$) is small ($\Delta P < 0.25$ kPa) compared to ambient atmospheric pressure ($P_s = 100$ kPa) for wind speeds of $M_s < 20$ m s⁻¹.

Dynamic effects make it difficult to measure static pressure in the wind. When the wind hits the pressure sensor, it decelerates and causes the pressure to increase. For this reason, static pressure instruments are designed to minimize flow deceleration and dynamic errors by having pressure ports (holes) along the sides of the sensor where there is no flow toward or away from the sensor.

Dynamic pressure can be used to measure wind speed. An instrument that does this is the **pitot tube**. Aircraft instruments measure stagnation pressure with the pitot tube facing forward into the flow, and static pressure with another port facing sideways to the flow to minimize dynamic effects. The instrument then computes an "indicated airspeed" from eq. (17.49 or 17.50) using the **pitot – static** pressure difference.

During tornadoes and hurricanes, if strong winds encounter an open garage door or house window, the wind trying to flow into the building causes pressure inside the building to increase dynamically. As is discussed in a Sample Application, the resulting pressure difference across the roof and walls of the building can cause them to blow out so rapidly that the building appears to explode.

Venturi Effect

Bernoulli's equation says that if velocity increases in the region of flow constriction, then pressure decreases. This is called the **Venturi effect.**

For gap winds of constant depth, eq. (17.37) can be written as

$$\frac{1}{2} M_s^2 + \frac{P_s}{\rho} = \frac{1}{2} M_d^2 + \frac{P_d}{\rho} \quad (17.51)$$

which can be combined with eq. (17.23) to give the Venturi pressure decrease:

$$P_d - P_s = \frac{\rho}{2} \cdot M_s^2 \cdot \left[1 - \left(\frac{D_s}{D_d} \right)^2 \right] \quad (17.52)$$

Sample Application

If a 20 km wide band of winds of 5 m s⁻¹ must contract to pass through a 2 km wide gap, what is the pressure drop in the gap compared to the non-gap flow?

Find the Answer

Given: $D_s = 20$ km, $D_d = 2$ km, $M_s = 5$ m s⁻¹.
Find: $M_d = ?$ m s⁻¹.

Assume: ρ = 1.2 kg m⁻³.

Use eq. (17.52):
$\Delta P = (1.2 \text{ kg·m}^{-3}/2)\cdot(5\text{m s}^{-1})^2\cdot[1 - ((20 \text{ km})/(2 \text{ km}))^2]$
$= -1485 \text{ kg·m}^{-1}\text{·s}^{-2} = \underline{-1.5 \text{ kPa}}$

Check: Units OK. Physics OK.
Exposition: This is a measurable drop in atmospheric pressure. The pressure measured at weather stations in such gaps should be compensated for this venturi effect to calculate the effective static pressure, such as could be used in analyzing weather maps.

Unfortunately, gap flows are often not constant depth, because the temperature inversion that caps these flows are not rigid lids.

DOWNSLOPE WINDS

Consider a wintertime situation of a layer of cold air under warm, with a synoptic weather pattern forcing strong winds toward a mountain range. Flow over ridge-top depends on cold-air depth, and on the strength of the temperature jump between the two layers. If conditions are right, fast winds, generically called **fall winds**, can descend along the lee slope. Sometimes these downslope winds are fast enough to cause significant destruction to buildings, and to affect air and land transportation. **Downslope wind storms** can be caused by mountain waves (previously discussed), bora winds, and foehns.

Bora

If the fast-moving cold air upstream is deeper than the ridge height H (Fig. 17.36), then very fast (hurricane force) cold winds can descend down the lee side. This phenomena is called a **Bora**. The winds accelerate in the constriction between mountain and the overlying inversion, and pressure drops according to the Venturi effect. The lower pressure upsets hydrostatic balance and draws the cold air layer downward, causing fast winds to hug the slope.

The overlying warmer air is also drawn down by this same pressure drop. Because work must be done to lower this warm air against buoyancy, Bernoulli's equation tells us that the Bora winds decelerate slightly on the way down. Once the winds reach the lowland, they are still destructive and much faster than the winds upstream of the mountain, but are slower than the winds at ridge top. See the Sample Application for Bernoulli's equation and Bora.

Boras were originally named for the cold fall wind along the Dalmatian coast of Croatia and Bosnia in winter, when cold air from Russia crosses the mountains and descends southwest toward the Adriatic Sea. The name Bora is used generally now for any cold fall wind having similar dynamics.

For situations where the average mountain ridge height is greater than z_i but the mountain pass is lower than z_i, boras can start in the pass (as a gap wind) and continue down the lee slope.

The difference between katabatic and Bora winds is significant. Katabatic winds are driven by the local thermal structure, and form during periods of weak synoptic forcing such as in high-pressure ar-

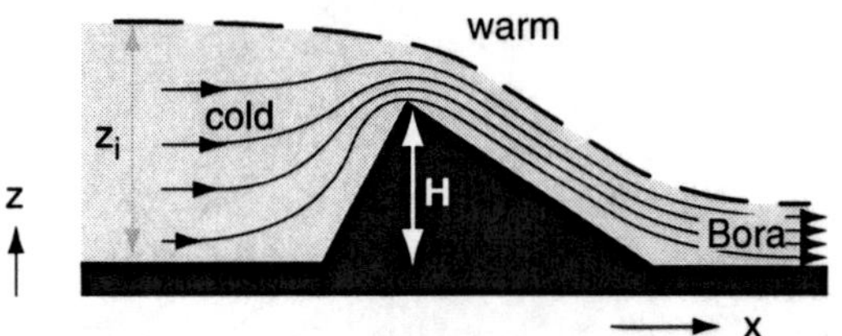

Figure 17.36
Cold Bora winds, during synoptic weather patterns where strong winds are forced toward the ridge from upstream. Thin lines are streamlines. Thick dashed line is a temperature inversion.

Sample Application

For the Bora situation of Fig. 17.36, the inversion of strength 6°C is 1200 m above the upstream lowland. Ridge top is 1000 m above the valley floor. If upstream winds are 10 m s^{-1}, find the Bora wind speed in the lee lowlands.

Find the Answer

Given: $H = 1$ km, $z_i = 1.2$ km, $\Delta\theta_v = 6$°C,
$M_s = 10$ m s^{-1}. Assume $|g|/T_v = 0.0333$ m·s^{-2}·K^{-1}
Find: M_{Bora} = ? m s^{-1}

Volume conservation similar to eq. (17.23) gives ridge top winds M_d:

$$M_d = \frac{z_i}{z_i - H}\cdot M_s = 60 \text{ m/s}$$

Assume Bora thickness = constant = $z_i - H$.
Follow the streamline indicated by the thick dashed line in Fig. 17.36. Assume ending pressure equals starting pressure on this streamline.
Use Bernoulli's eq. (17.37):

$$\left[\frac{1}{2}M^2 + |g|\frac{\Delta\theta_v}{T_v}\cdot z\right]_{ridgetop} = \left[\frac{1}{2}M^2 + |g|\frac{\Delta\theta_v}{T_v}\cdot z\right]_{Bora}$$

Combine the above eqs. Along the streamline, $z_{ridgetop} = z_i$, and $z_{Bora} = z_i - H$. Thus $\Delta z = H$. Solve for M_{Bora}:

$$M_{Bora} = \left[\left(\frac{z_i}{z_i - H}\right)^2\cdot M_s^{\,2} - 2\cdot\frac{|g|}{T_v}\cdot\Delta\theta_v\cdot H\right]^{1/2}$$

Finally, we can plug in the numbers: M_{Bora} =

$$\sqrt{\left(\frac{1.2\text{km}}{0.2\text{km}}\right)^2\left(10\frac{\text{m}}{\text{s}}\right)^2 - 2\cdot\left(0.033\frac{\text{m}}{\text{s}^2\text{K}}\right)\cdot(6\text{K})\cdot(1000\text{m})}$$

$$= \sqrt{(3600-400)\text{m}^2\text{s}^{-1}} = \underline{\mathbf{56.6\ m\ s^{-1}}}$$

Check: Units OK. Physics OK.
Exposition: Winds at ridge top were 60 m s^{-1}. Although the Bora winds at the lee lowlands are weaker than ridge top, they are still strong and destructive. Most of the wind speed up was due to volume conservation, with only a minor decrease given by Bernoulli's equation. This decrease is because kinetic energy associated with wind speed must be expended to do work against gravity by moving warm inversion air downward.

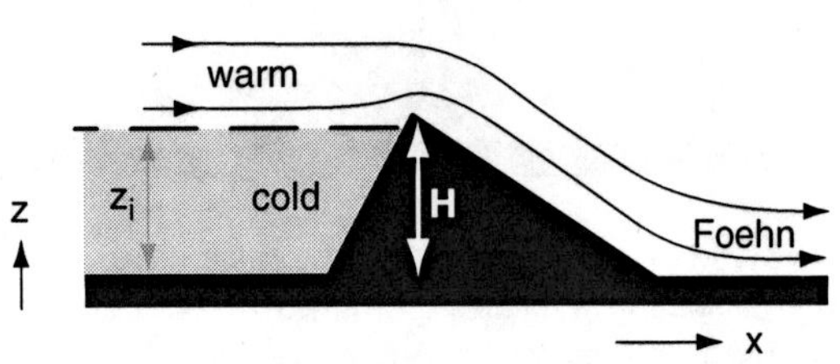

Figure 17.37
One mechanism for creating warm Foehn winds, given synoptic weather patterns where strong winds are forced toward the mountain ridge from upstream. Thin lines are streamlines. Thick dashed line is a temperature inversion.

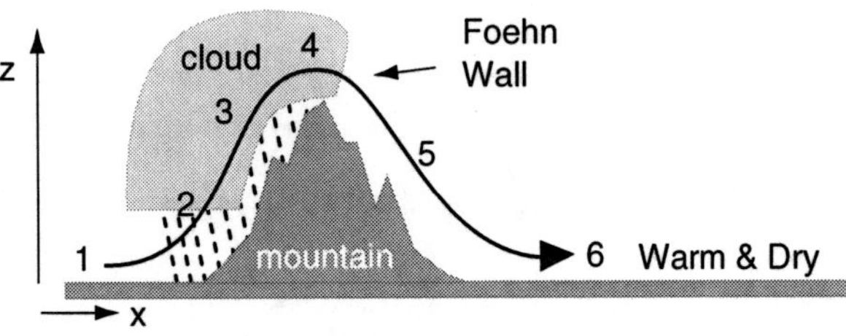

Figure 17.38
Another mechanism for creating warm Foehn winds, where net heating occurs due to formation and fallout of precipitation (dashed lines). Black curved line shows one streamline.

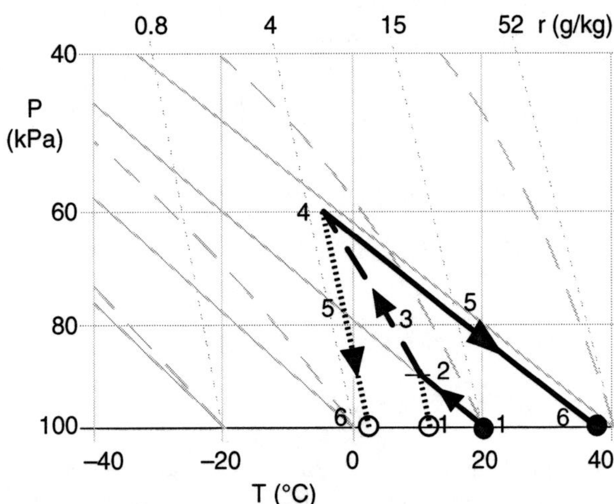

Figure 17.39
Foehn thermodynamics, plotted on a Stüve thermo diagram. P is pressure, T is temperature, r is water-vapor mixing ratio. Black dots and thick black lines indicate air temperature, while dashed lines and open circles indicate humidity. The numbers 1 - 6 correspond to the numbered locations in the previous figure.

eas of fair weather and light winds. Over mid-latitude land, katabatic winds exist only at night and are usually weak, while on the slopes of Antarctica they can exist for days and can become strong. Boras are driven by the inertia of strong upstream winds that form in regions of low pressure and strong horizontal pressure gradient. They can last for several days. Although both phenomena are cold downslope winds, they are driven by different dynamics.

Foehns and Chinooks

One mechanism for creating Foehn winds <u>does not</u> require clouds and precipitation. If the mountain height H is greater than the thickness z_i of cold air upstream, then the cold air is dammed behind the mountain and does not flow over (Fig. 17.37). The strong warm winds aloft can flow over the ridge top, and can warm further upon descending adiabatically on the lee side. The result is a <u>warm</u>, dry, downslope wind called the **Foehn**.

Foehn winds were originally named for the southerly winds from Italy that blow over the Alps and descend in Austria, Germany, and Switzerland. A similar downslope wind is called the **chinook** just east of the Rocky Mountains in the USA and Canada. Other names in different parts of the world are **zonda** (Argentina), **austru** (Romania), **aspre** (France). The onset of foehn winds in winter can be accompanied with a very rapid temperature increase at the surface. If the warm and very dry air flows over snowy ground, it rapidly melts and sublimes the snow, and is nicknamed "snow eater".

Another Foehn mechanism <u>is</u> based on net latent heating associated with condensation and precipitation on the upwind side of the mountain range. Consider an air parcel before it flows over a mountain, such as indicated at point (1) in Fig. 17.38. Suppose that the temperature is 20°C and dew point is 10°C initially, as indicated by the filled and open circles at point (1) in Fig. 17.39. This corresponds to about 50% relative humidity.

As the air rises along the windward slopes, it cools dry-adiabatically while conserving mixing ratio until the lifting condensation level (LCL) is reached (2). Further lifting is moist adiabatic (3) within the **orographic cloud** (a cloud caused by the terrain). Suppose that most of the condensed water falls out as precipitation on the windward slopes.

Over the summit (4), suppose that the air has risen to a height where the ambient pressure is 60 kPa. The air parcel now has a temperature of about –8°C. As it begins to descend down the lee side, any residual cloud droplets will quickly evaporate in the adiabatically warming air. The trailing edge of the orographic cloud is called a **Foehn wall**, because it looks like a wall of clouds when viewed from the lowlands downwind.

Continued descent will be dry adiabatic (5) because there are no liquid water drops to cause evaporative cooling. By the time the air reaches its starting altitude on the lee side (6), its temperature has warmed to about 35°C, with a dew point of about –2°C. This is roughly 10% relative humidity.

The net result of this process is: clouds and precipitation form on the windward slopes of the mountain, a Foehn wall forms just downwind of the mountain crest, and there is warming and drying in the lee lowlands.

CANOPY FLOWS

Forests and Crops

The leaf or needle layer of a crop or forest is called a **canopy**. Individual plants or trees in these crops or forests each cause drag on the wind. The average winds in the air space between these **plant-canopy** or **forest-canopy** obstacles is the canopy flow.

Just above the top of the canopy, the flow is approximately logarithmic with height (Fig. 17.40a), as is described in more detail in the Atmospheric Boundary-Layer chapter for statically neutral conditions in the surface layer:

$$M = \frac{u_*}{k} \ln\left(\frac{z-d}{z_o}\right) \qquad \text{for } z \geq h_c \quad (17.53)$$

where M is wind speed, z is height above ground, u_* is the **friction velocity** (a measure of the drag force per unit surface area of the ground), $k \approx 0.4$ is the **von Kármán constant**, d is the **displacement distance** ($0 \leq d \leq h_c$), and z_o is the **roughness length**, for an average canopy-top height of h_c.

If you can measure the actual wind speed M at 3 or more heights z within 20 m above the top of the canopy, then you can use the following procedure to find d, z_o, and u_*: (1) use a spreadsheet to plot your M values on a linear horizontal axis vs. their $[z–d]$ values on a logarithmic vertical axis; (2) experiment with different values of d until you find the one that aligns your wind points into a straight line; (3) extrapolate that straight line to $M = 0$, and note the resulting intercept on the vertical axis, which gives the roughness length z_o. (4) Finally, pick any point exactly on the plotted line, and then plug in its M and z values, along with the d and z_o values just found, to calculate u_*. using eq. (17.53).

If you do not have measurements of wind speed above the canopy top, you can use the following

Sample Application

Air from the Pacific Ocean (T =5°C, T_d = 3°C, z = 0) flows over the Coast Mountains ($z \approx$ 3000 m), and descends toward the interior plateau of British Columbia, Canada ($z \approx$1000 m). Fig. 17.19 shows the topography of this region. Find the final T and T_d.

Find the Answer

Given: T = 5 °C, T_d = 3 °C, z = 0 m initially.
Find: T = ? °C, T_d = ? ° at z = 1000 m finally.

Use the thermo diagram from chapter 5:

Use the thermo diagram to find that the initial mixing ratio is $r \approx 5$ g kg^{-1}. Clouds form as the air rises over the mountains, with base at z_{LCL} = (0.125 km °C^{-1})·(5–3°C) = 0.25 km. From there to z = 3000 m, the air follows a moist adiabat, reaching $T = T_d \approx$ –17°C.

Assuming all condensate falls out as precipitation on the windward side, the air then descends dry-adiabatically to the town of Williams Lake. The final state is: $T \approx$ **4°C** at z = 1000 m, and $T_d \approx$ **–13°C**.

Check: Units OK. Physics OK.
Exposition: The air is much drier ($RH \approx$ 25%) on the lee side, but nearly the same temperature as initially. This is typical of a foehn wind.

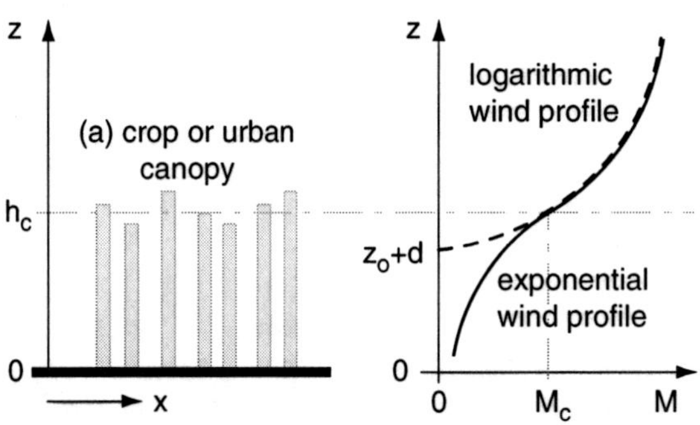

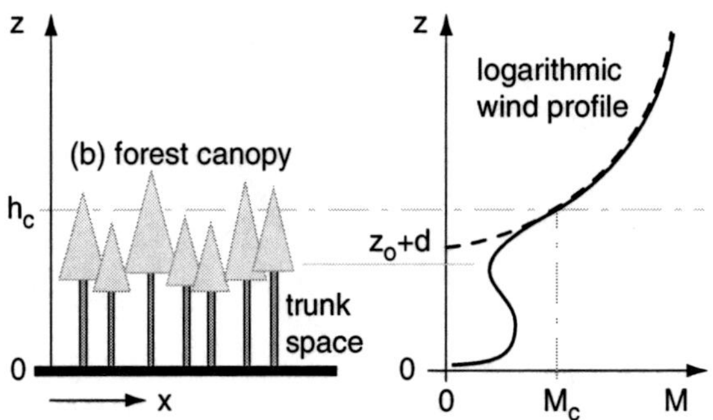

Figure 17.40
Canopy flows for (a) crops or urban canopies, and (b) forest canopies having a relatively open trunk space. Left: sketch of the canopy objects. Right: wind profile (solid line). The dashed black line shows a logarithmic profile extrapolated to zero wind speed M. Average top of the canopy is at height h_c.

Sample Application (§)

Given these wind measurements over a 2 m high corn crop: [z (m), M (m s^{-1})] = [5, 3.87] , [10, 5.0] , [20, 6.01]. Find the displacement distance, roughness length, and friction velocity. If the attenuation coefficient is 2.5, plot wind speed M vs. height over $0.5 \text{ m} \le z \le 5$ m.

Find the Answer

Given: h_c =2m, [z (m), M (m s^{-1})] listed above, a = 2.5.
Find: d = ? m, z_o = ? m, u_* = ? m s^{-1}, and plot M vs. z.

Guess d = 0, and plot M vs. log(z–d) on a spreadsheet. This d is too small (see graph below), because the curve is concave up. Guess d = 4, which is too large, because curve is concave down. After other guesses (some not shown), I find that <u>**d = 1.3 m**</u> gives the straightest line.

Next, extrapolate on the semi-log graph to M = 0, which gives an intercept of <u>**z_o = 0.2 m**</u>.

Solve eq. (17.53) for $u_* = k{\cdot}M/\ln[(z-d)/z_o]$
<u>**u_* =**</u> 0.4(5m s^{-1})/ln[(10–1.3)/0.2] = <u>**0.53 m s^{-1}**</u>

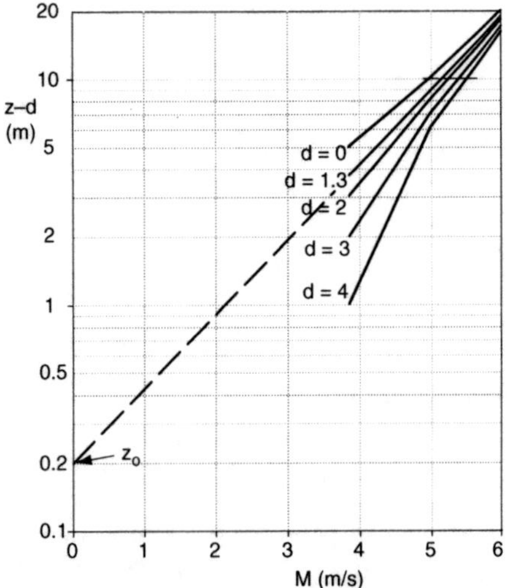

Solve eq. (17.53) for M_c at $z = h_c$ = 2 m:
M_c = [(0.53m s^{-1})/0.4] · ln[(2–1.3m)/0.2] = 1.66 m s^{-1}

Use eq. (17.54) to find M for a range of heights below h_c, and use eq. (17.53) for heights above h_c:

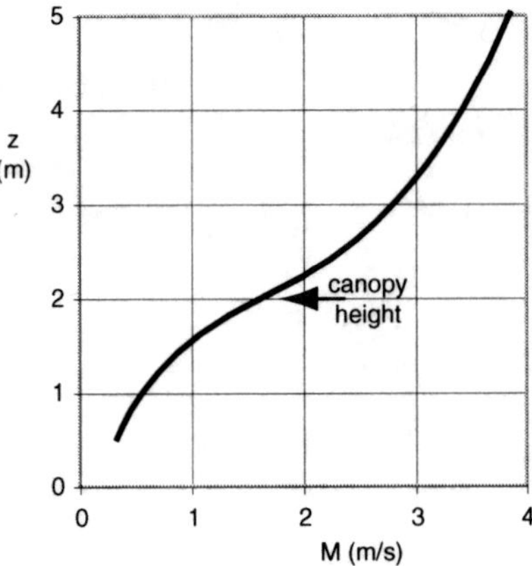

Check: Shape of curve looks reasonable.
Exposition: For this exercise, $z_o = 0.1\ h_c$, and $d = 0.65\ h_c$. Namely, the crude approximations are OK.

crude approximations to estimate the needed parameters: $d \approx 0.65{\cdot}h_c$ and $z_o \approx 0.1{\cdot}h_c$. Methods to estimate u_* are given in the Boundary-Layer chapter.

The average wind speed at the average canopy-top height is M_c. For the crude approximations above, we find that $M_c \approx 3.13\ u_*$.

Within the top 3/4 of canopy, an exponential formula describes the average wind-speed M profile:

$$M = M_c{\cdot}\exp\left[a{\cdot}\left(\frac{z}{h_c}-1\right)\right] \qquad \text{for } 0.5h_c \le z \le h_c \tag{17.54}$$

where a is an **attenuation coefficient** that increases with increasing leaf area and decreases as the mean distance between individual plants increases. Typical values are $a \approx$ 2.5 - 2.8 for oats and wheat; 2.0 - 2.7 for mature corn; 1.3 for sunflowers, 1.0 - 1.1 for larch and small evergreen trees, and 0.4 for a citrus orchard. The exponential and log-wind speeds match at the average canopy top h_c.

For a forest with relatively open trunk space (i.e., only the tree trunks without many leaves, branches, or smaller underbrush), the previous equation fails. Instead, a weak relative maximum wind speed can occur (Fig. 17.40b). In such forests, if the canopy is very dense, then the sub-canopy (trunk space) flow can be relatively disconnected from the flow above the tree tops. Weak katabatic flows can exist in the trunk space day and night.

Cities

The collection of buildings and trees that make up a city is sometimes called an **urban canopy**. These obstacles cause an <u>average</u> canopy-flow wind similar to that for forests and crops (Fig. 17.40a).

However, winds at any one location in the city can be quite different. For example, the street corridors between tall buildings can channel flow similar to the flow in narrow valleys. Hence these corridors are sometimes called **urban canyons**. Also, taller buildings can deflect down to the surface some of the faster winds aloft. This causes much greater wind speeds and gusts near the base of tall buildings than near the base of shorter buildings.

Cities can be 2 - 12°C degrees warmer than the surrounding rural countryside — an effect called the **urban heat island** (UHI, Fig. 17.41). Reasons include the abundance of concrete, glass and asphalt, which capture and store the solar heat during daytime and reduce the IR cooling at night. Also, vegetated areas are reduced in cities, and rainwater is channeled away through storm drains. Hence, there is less evaporative cooling. Also, fuel and electrical consumption by city residents adds heat via heating, air conditioning, industry, and transportation.

The city–rural temperature difference ΔT_{UHI} is greatest during clear calm nights, because the city stays warm while rural areas cool considerably due to IR radiation to space. The largest values ΔT_{UHI_max} occur near the city center (Fig. 17.42), at the location of greatest density of high buildings and narrow streets. For clear, calm nights, this relationship is described by

$$\Delta T_{UHI\_max} \approx a + b \cdot \ln(H/W) \tag{17.55}$$

where a = 7.54°C and b = 3.97°C. H is the average height of the buildings in the downtown city core, W is the average width of the streets at the same location, and H/W is dimensionless.

Temperature difference is much smaller during daytime. When averaged over a year (including windy and cloudy periods of minimal UHI), the average ΔT_{UHI} at the city center is only 1 to 2°C.

During periods of fair weather and light synoptic-scale winds, the warm city can generate circulations similar to sea breezes, with inflow of low-altitude rural air toward the city, and rising air over the hottest parts of town. These circulations can enhance clouds, and trigger or strengthen thunderstorms over and downwind of the city. With light to moderate winds, the UHI area is asymmetric, extending much further from the city in the downwind direction (Fig. 17.41), and the effluent (heat, air pollution, odors) from the city can be observed downwind as an **urban plume** (Fig. 17.43).

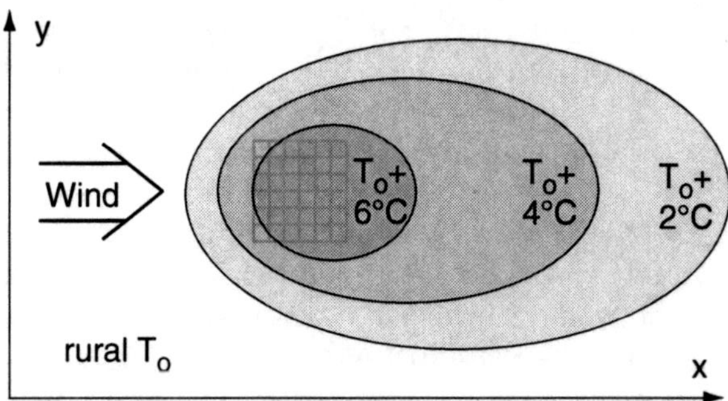

Figure 17.41
An urban heat island at night, where T_o is the rural air temperature. The grid represents city streets.

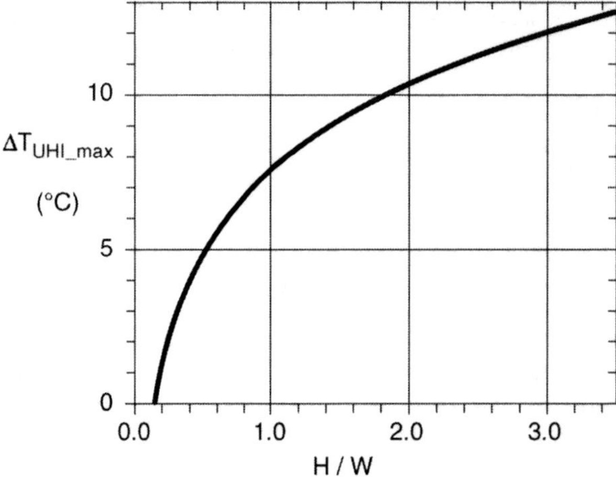

Figure 17.42
Maximum urban-heat-island temperature difference ΔT_{UHI_max} increases with average aspect ratio H/W of the urban canyons near the built-up city center. Based on data from North America, Europe, and Australasia. [Adapted from Oke, Mills, Christen, & Voogt, 2016: Urban Climates, submitted to Cambridge Univ. Press. (used with lead author's permission).]

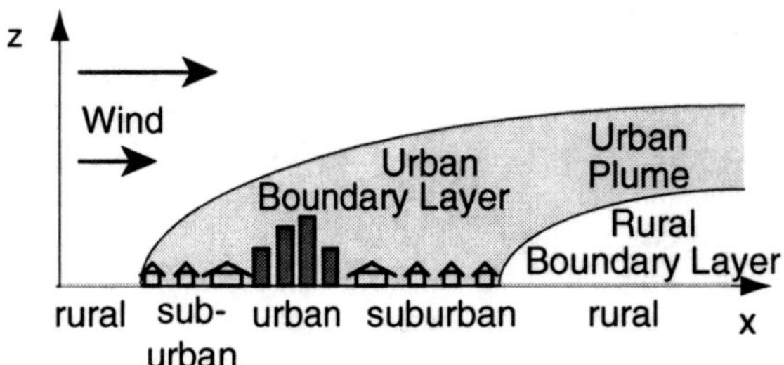

Figure 17.43
Sketch of an urban plume blowing downwind.

REVIEW

The probability of any wind speed can be described by a Weibull distribution, and the distribution of wind directions can be plotted on a wind rose. Regions with greater probability of strong winds are ideal for siting turbines for wind power.

During weak synoptic forcing (weak geostrophic winds), local circulations can be driven by thermal forcings. Examples include anabatic (warm upslope) and katabatic (cold downslope) winds, mountain and valley winds, and sea breezes near coastlines.

During strong synoptic forcing, winds can be channeled through gaps, can form downslope windstorms, and can create mountain waves and wave drag. The winds in short gaps can be well described by open-channel hydraulics and Bernoulli's equation. Winds in longer gaps and fjords are influenced by Coriolis force.

The Bora is a cold downslope wind — driven dynamically by the synoptic scale flow. Foehn winds are also driven dynamically, but are warm, and can be enhanced by orographic precipitation and latent heating. Hydraulic jumps occur downstream of bora winds and in some gap flows as the flow re-adjusts to hydrostatic equilibrium.

Wind and temperature instruments are constructed to minimize dynamic pressure and heating errors. Wind speed is reduced inside plant and urban canopies. The urban-heat-island effect of cities can induce local circulations.

HOMEWORK EXERCISES

Broaden Knowledge & Comprehension

B1. Search the web for wind-rose graphs for a location (weather station or airport) near you, or other location specified by your instructor.

B2. Search the web for wind-speed distributions for a weather station near you. Relate this distribution to extreme or record-breaking winds.

> **A SCIENTIFIC PERSPECTIVE • Simple is Best**
>
> Fourteenth century philosopher William of Occam suggested that "the simplest scientific explanation is the best". This tenant is known as **Occam's Razor**, because with it you can cut away the bad theories and complex equations from the good.
>
> But why should the simplest or most elegant be the best? There is no law of nature that says it must be so. It is just one of the philosophies of science, as is the scientific method of Descartes. Ultimately, like any philosophy or religion, it is a matter of faith.
>
> I suggest an alternative tenant: **"a scientific relationship should not be more complex than needed**." This is motivated by the human body — an amazingly complex system of hydraulic, pneumatic, electrical, mechanical, chemical, and other physical processes that works exceptionally well. In spite of its complexity, the human body is not more complex than needed (as determined by evolution).
>
> Although this alternative tenant is only subtly different from Occam's Razor, it admits that sometimes complex mathematical solutions to physical problems are valid. This tenant is used by a data-analysis method called **computational evolution** (or **gene-expression programming**). This approach creates a population of different algorithms that compete to best fit the data, where the best algorithms are allowed to persist with mutation into the next generation while the less-fit algorithms are culled via computational natural selection.

B3. Find on the web climatological maps giving locations of persistent, moderate winds. These are favored locations for wind turbine farms. Also search the web for locations of existing turbine farms.

B4. Search the web for a weather map showing vertical velocities over your country or region. Sometimes, these vertical velocities are given as omega (ω) rather than w, where ω is the change of pressure with time experienced by a vertically moving parcel, and is defined in the Extratropical Cyclone chapter. What is the range of vertical velocities on this particular day, in m s^{-1}?

B5. Search the web for the highest resolution (hopefully 0.5 km or better resolution) visible satellite imagery for your area. Which parts of the country have rising thermals, based on the presence of cumulus clouds at the top of the thermals?

B6. Search the web for lidar (laser radar) images of thermals in the boundary layer.

B7. Search the web for the highest resolution (hopefully 0.5 km or better resolution) visible satellite imagery for your area. Also search for an upper-air sounding (i.e., thermo diagram) for your area. Does the depth of the mixed layer from the thermo diagram agree with the diameter of thermals (clouds) visible in the satellite image? Comment.

B8. Access IR high resolution satellite images over cloud-free regions of the Rocky Mountains (or Cascades, Sierra-Nevada, Appalachians, or other significant mountains) for late night or early morning during synoptic conditions of high pressure and light winds. Identify those regions of cold air in valleys, as might have resulted from katabatic winds. Sometimes such regions can be identified by the fog that forms in them.

B9. Search the web for weather station observations at the mouth of a valley. Plotted meteograms of wind speed and direction are best to find. See if you can find evidence of mountain/valley circulations in these station observations, under weak synoptic forcing.

B10. Same as the previous problem, but to detect a sea breeze for a coastal weather station.

B11. Search the web for satellite observations of the sea breeze, evident as changes in cloudiness parallel to the coastline.

B12. Search the web for information on how tsunami on the ocean surface travel at the shallow-water wave speed as defined in this chapter, even when the waves are over the deepest parts of the ocean. Explain.

B13. Search the web for images of hydraulic jump in the atmosphere. If you can't find any, then find images of hydraulic jump in water instead.

B14. Access images from digital elevation data, and find examples of short and long gaps through mountain ranges for locations other than Western Canada.

B15. Search the web for news stories about dangerous winds along the coast, but limit this search to only coastally-trapped jets. If sufficient information is given in the news story, relate the coastal jet to the synoptic weather conditions.

B16. Access visible high-resolution satellite photos of mountain wave clouds downwind of a major mountain range. Measure the wavelength from these images, and compare with the wind speed accessed from upper air soundings in the wave region. Use those data to estimate the Brunt-Väisälä frequency.

B17. Access from the web photographs taken from ground level of lenticular clouds. Also, search for **iridescent** clouds on the web, to find if any of these are lenticular clouds.

B18. Access from the web pilot reports of turbulence, chop, or mountain waves in regions downwind of mountains. Do this over several days, and show how these reports vary with wind speed and static stability.

B19. Access high-resolution visible satellite images from the web during clear skies, that show the smoke plume from a major source (such as Gary, Indiana, or Sudbury, Ontario, or a volcano, or a forest fire). Assume that this image shows a streakline. Also access the current winds from a weather map corresponding roughly to the altitude of the smoke plume, from which you can infer the streamlines. Compare the streamlines and streaklines, and speculate on how the flow has changed over time, if at all. Also, draw on your printed satellite photo the path lines for various air parcels within the smoke plume.

B20. From the web, access weather maps that show streamlines. These are frequently given for weather maps of the jet stream near the tropopause (at 20 to 30 kPa). Also access from the web weather maps that plot the actual upper air winds from rawinsonde observations, valid at the same time and altitude as the streamline map. Compare the instantaneous winds with the streamlines.

B21. From the web, access a sequence of weather maps of streamlines for the same area. Locate a point on the map where the streamline direction has changed significantly during the sequence of maps. Assume that smoke is emitted continuously from that point. On the last map of the sequence, plot the streakline that you would expect to see. (Hint, from the first streamline map, draw a path line for an air parcel that travels until the time of the next streamline map. Then, using the new map, continue finding the path of that first parcel, as well as emit a new second parcel that you track. Continue the process until the tracks of all the parcels end at the time of the last streamline map. The locus of those parcels is a rough indication of the streakline.)

B22. Access from the web information for aircraft pilots on how the pitot tube works, and/or its calibration characteristics for a particular model of aircraft.

B23. Access from the web figures that show the amount of destruction for different intensities of tornado winds. Prepare a table giving the dynamic pressures and forces on the side of a typical house for each of those different wind categories.

B24. Access from the web news stories of damage to buildings or other structures caused by Boras, mountain waves, or downslope windstorms.

B25. Access from the web data or images that indicate typical height of various mature crops (other than the ones already given at the end of the Numerical exercises).

B26. Access from the web the near-surface air temperature at sunrise in or just downwind of a large city, and compare with the rural temperature.

B27. Use info from the web to estimate the urban canopy H/W aspect ratio for the city center nearest to you. Then use Fig. 17.42 to estimate ΔT_{UHI_max}.

Apply

A1.(§) Plot the probability of wind speeds using a Weibull distribution with a resolution of 0.5 m s^{-1}, and M_o = 8 m s^{-1}, for α =

a. 1.0 b. 1.5 c. 2.0 d. 2.5 e. 3.0 f. 3.5 g. 4.0
h. 4.5 i. 5.0 j. 5.5 k. 6.0 m. 7 n. 8 o. 9

A2. A wind turbine of blade radius 25 m runs at 35% efficiency. At sea level, find the theoretical power (kW) for winds (m s^{-1}) of:
a. 1 b. 2 c. 3 d. 4 e. 5 f. 6 g. 7 h. 8
i. 9 j. 10 k. 11 m. 12 n. 13 o. 14 p. 15

A3. Find the equilibrium updraft speed (m s^{-1}) of a thermal in a 2 km boundary layer with environmental temperature 15°C. The thermal temperature (°C) is: a. 16 b. 16.5 c. 17 d. 17.5 e. 18 f. 18.5
g. 19 h. 19.5 i. 20 j. 20.5 k. 21 m. 21.5

A4. Anabatic flow has a temperature excess of 4°C. Find the buoyant along-slope pressure gradient force per unit mass for a slope of angle (°):
a. 10 b. 15 c. 20 d. 25 e. 30 f. 35 g. 40
h. 45 i. 50 j. 55 k. 60 m. 65 n. 70 o. 75

A5(§). Plot katabatic wind speed (m s^{-1}) vs. downslope distance (m) if the environment is 20°C and the cold katabatic air is 15°C. The slope angle (°) is:
a. 10 b. 15 c. 20 d. 25 e. 30 f. 35 g. 40
h. 45 i. 50 j. 55 k. 60 m. 65 n. 70 o. 75

A6. Find the equilibrium downslope speed (m s^{-1}) for the previous problem, if the katabatic air is 5 m thick and the drag coefficient is 0.002.

A7. Find the depth (m) of the thermal internal boundary layer 2 km downwind of the coastline, for an environment with wind speed 8 m s^{-1} and γ = 4 K km^{-1}. The surface kinematic heat flux (K·m s^{-1}) is
a. 0.04 b. 0.06 c. 0.08 d. 0.1 e. 0.12
f. 0.14 g. 0.16 h. 0.18 i. 0.2 j. 0.22

A8. Assume T_v = 20°C. Find the speed (m s^{-1}) of advance of the sea-breeze front, for a flow depth of 700 m and a temperature excess $\Delta\theta$ (K) of:
a. 1.0 b. 1.5 c. 2.0 d. 2.5 e. 3.0 f. 3.5 g. 4.0
h. 4.5 i. 5.0 j. 5.5 k. 6.0 m. 7 n. 8 o. 9

A9. For the previous problem, find the sea-breeze wind speed (m s^{-1}) at the coast.

A10. For a sea-breeze frontal speed of 5 m s^{-1}, find the expected maximum distance (km) of advance of the sea-breeze front for a latitude (°) of
a. 10 b. 15 c. 20 d. 25 e. 80 f. 35 g. 40
h. 45 i. 50 j. 55 k. 60 m. 65 n. 70 o. 75

A11. What is the shallow-water wave phase speed (m s^{-1}) for a water depth (m) of:
a. 2 b. 4 c. 6 d. 8 e. 10 f. 15 g. 20
h. 25 i. 30 j. 40 k. 50 m. 75 n. 100 o. 200

A12. Assume $|g|/T_v$ = 0.0333 m·s^{-2}·K^{-1}. For a cold layer of air of depth 50 m under warmer air, find the surface (interfacial) wave phase speed (m s^{-1}) for a virtual potential temperature difference (K) of:
a. 1.0 b. 1.5 c. 2.0 d. 2.5 e. 3.0 f. 3.5 g. 4.0
h. 4.5 i. 5.0 j. 5.5 k. 6.0 m. 7 n. 8 o. 9

A13. For the previous problem, find the value of the Froude number Fr_1. Also, classify this flow as subcritical, critical, or supercritical. Given M = 15 m s^{-1}.

A14. Assume $|g|/T_v$ = 0.0333 m·s^{-2}·K^{-1}. Find the internal wave horizontal group speed (m s^{-1}) for a stably stratified air layer of depth 400 m, given $\Delta\theta_v/\Delta z$ (K km^{-1}) of:
a. 1.0 b. 1.5 c. 2.0 d. 2.5 e. 3.0 f. 3.5 g. 4.0
h. 4.5 i. 5.0 j. 5.5 k. 6.0 m. 7 n. 8 o. 9

A15. For the previous problem, find the value of the Froude number Fr_2. Also, classify this flow as subcritical, critical, or supercritical.

A16. Winds of 10 m s^{-1} are flowing in a valley of 10 km width. Further downstream, the valley narrows to the width (km) given below. Find the wind speed (m s^{-1}) in the constriction, assuming constant depth flow.
a. 1.0 b. 1.5 c. 2.0 d. 2.5 e. 3.0 f. 3.5 g. 4.0
h. 4.5 i. 5.0 j. 5.5 k. 6.0 m. 7 n. 8 o. 9

A17. Assume $|g|/T_v$ = 0.0333 m·s^{-2}·K^{-1}. For a two-layer atmospheric system flowing through a short gap, find the maximum expected gap wind speed (m s^{-1}). Flow depth is 300 m, and the virtual potential temperature difference (K) is:
a. 1.0 b. 1.5 c. 2.0 d. 2.5 e. 3.0 f. 3.5 g. 4.0
h. 4.5 i. 5.0 j. 5.5 k. 6.0 m. 7 n. 8 o. 9

A18. Find the long-gap geostrophic wind (m s^{-1}) at latitude 50°, given $|g|/T_v$ = 0.0333 m·s^{-2}·K^{-1} and $\Delta\theta_v$ = 3°C, and assuming that the slope of the top cold-air surface is given by the height change (km) below across a valley 10 km wide.
a. 0.3 b. 0.4 c. 0.5 d. 0.6 e. 0.7
f. 0.8 g. 0.9 h. 1.0 i. 1.1 j. 1.2
k. 2.4 m. 2.6 n. 2.8 o. 3

A19. Find the external Rossby radius of deformation (km) for a coastally trapped jet that rides against a mountain range of 2500 m altitude at latitude (°) given below, for air that is colder than its surroundings by 10°C. Assume $|g|/T_v$ = 0.0333 m·s^{-2}·K^{-1}.
a. 80 b. 85 c. 20 d. 25 e. 30 f. 35 g. 40
h. 45 i. 50 j. 55 k. 60 m. 65 n. 70 o. 75

A20. Assume $|g|/T_v$ = 0.0333 m·s^{-2}·K^{-1}. Find the natural wavelength of air, given

a. M = 2 m s^{-1}, $\Delta T/\Delta z$ = 5 °C km^{-1}
b. M = 20 m s^{-1}, $\Delta T/\Delta z$ = –8 °C km^{-1}
c. M = 5 m s^{-1}, $\Delta T/\Delta z$ = –2 °C km^{-1}
d. M = 20 m s^{-1}, $\Delta T/\Delta z$ = 5 °C km^{-1}
e. M = 5 m s^{-1}, $\Delta T/\Delta z$ = –8 °C km^{-1}
f. M = 2 m s^{-1}, $\Delta T/\Delta z$ = –2 °C km^{-1}
g. M = 5 m s^{-1}, $\Delta T/\Delta z$ = 5 °C km^{-1}
h. M = 2 m s^{-1}, $\Delta T/\Delta z$ = –8 °C km^{-1}

A21. For a mountain of width 25 km, find the Froude number Fr_3 for the previous problem. Draw a sketch of the type of mountain waves that are likely for this Froude number.

A22. For the previous problem, find the angle of the wave crests, and the wave-drag force per unit mass. Assume H = 1000 m and h_w = 11 km.

A23(§). Plot the wavy path of air as it flows past a mountain, given an initial vertical displacement of 300 m, a wavelength of 12.5 km, and a damping factor of

a. 1.0 b. 1.5 c. 2.0 d. 2.5 e. 3.0 f. 3.5 g. 4.0
h. 4.5 i. 5.0 j. 5.5 k. 6.0 m. 7 n. 8 o. 9

A24(§) Given a temperature dew-point spread of 1.5°C at the initial (before-lifting) height of air in the previous problem, identify which wave crests contain lenticular clouds.

A25. Cold air flow speed 12 m s^{-1} changes to 3 m s^{-1} after a hydraulic jump. Assume $|g|/T_v$ = 0.0333 m·s^{-2}·K^{-1}. How high can the hydraulic jump rise if the exit velocity (m s^{-1}) is

a. 1.0 b. 1.5 c. 2.0 d. 2.5 e. 3.0 f. 3.5 g. 4.0
h. 4.5 i. 5.0 j. 5.5 k. 6.0 m. 7 n. 8 o. 9

A26. Assuming standard sea-level density and streamlines that are horizontal, find the pressure change given the following velocity (m s^{-1}) change:

a. 1.0 b. 1.5 c. 2.0 d. 2.5 e. 3.0 f. 3.5 g. 4.0
h. 4.5 i. 5.0 j. 5.5 k. 6.0 m. 7 n. 8 o. 9

A27. Wind at constant altitude decelerates from 12 m s^{-1} to the speed (m s^{-1}) given below, while passing through a wind turbine. What opposing net pressure difference (Pa) would have caused the same deceleration in laminar flow?

a. 1.0 b. 1.5 c. 2.0 d. 2.5 e. 3.0 f. 3.5 g. 4.0
h. 4.5 i. 5.0 j. 5.5 k. 6.0 m. 7 n. 8 o. 9

A28. Air with pressure 100 kPa is initially at rest. It is accelerated isothermally over a flat 0°C snow surface as it is sucked toward a household ventilation system. What is the final air pressure (kPa) just before entering the fan if the final speed (m s^{-1}) through the fan is:

a. 1 b. 2 c. 3 d. 4 e. 5 f. 6 g. 7 h. 8
i. 9 j. 10 k. 11 m. 12 n. 13 o. 14 p. 15

A29. A short distance behind the jet engine of an aircraft flying in level flight, the exhaust temperature is 500°C. After the jet exhaust decelerates to zero, what is the final exhaust air temperature (°C), neglecting conduction and mixing, assuming the initial jet-blast speed (m s^{-1}) is:

a. 100 b. 125 c. 150 d. 175 e. 200 f. 210 g. 220
h. 230 i. 240 j. 250 k. 260 m. 270 n. 280 o. 290

A30. An 85 kW electric wind machine with a 3 m radius fan blade is used in an orchard to mix air so as to reduce frost damage on fruit. The fan horizontally accelerates the air from calm to the speed (m s^{-1}) given below. Find the temperature change (°C) across the fan, neglecting mixing with the environmental air.

a. 6 b. 6.5 c. 7 d. 7.5 e. 8 f. 8.5 g. 9
h. 9.5 i. 10 j. 10.5 k. 11 m. 12 n. 13 o. 14

A31. Tornadic air of temperature 25°C blows with speed (m s^{-1}) given below, except that it stagnates upon hitting a barn. Find the final stagnation temperature (°C) and pressure change (kPa).

a. 100 b. 125 c. 150 d. 175 e. 200 f. 210 g. 220
h. 230 i. 240 j. 250 k. 260 m. 270 n. 280 o. 290

A32. Find the speed of sound (m s^{-1}) and Mach number for M_{air} = 100 m s^{-1}, given air of temperature (°C):

a. –50 b. –45 c. –40 d. –35 e. –30
f. –25 g. –20 h. –15 i. –10 j. –5
k. 0 m. 5 n. 10 o. 15 p. 20

A33. Water flowing through a pipe with speed 2 m s^{-1} and pressure 100kPa accelerates to the speed (m s^{-1}) given below when it flows through a constriction. What is the fluid pressure (kPa) in the constriction? Neglect drag against the pipe walls.

a. 6 b. 6.5 c. 7 d. 7.5 e. 8 f. 8.5 g. 9
h. 9.5 i. 10 j. 10.5 k. 11 m. 12 n. 13 o. 14

A34. For the bora Sample Application, redo the calculation assuming that the initial inversion height (km) is: a. 1.1 b. 1.15 c. 1.25 d. 1.3 e. 1.35

f. 1.4 g. 1.45 h. 1.5 i. 1.55 j. 1.6
k. 1.65 m. 1.7 n. 1.75 o. 1.8 p. 1.85

A35. Use a thermodynamic diagram. Air of initial temperature 10°C and dew point 0°C starts at a height where the pressure (kPa) is given below. This air rises to height 70 kPa as it flows over a mountain, during which all liquid and solid water precipitate out. Air descends on the lee side of the mountain to an altitude of 95 kPa. What is the temperature, dew point, and relative humidity of the air at its final altitude? How much precipitation occurred on the mountain? [Hint: use a thermo diagram.]

a. 104 b. 102 c. 100 d. 98 e. 96 f. 94 g. 92
h. 90 i. 88 j. 86 k. 84 m. 82 n. 80

A36. Plot wind speed vs. height, for heights between 0.25 h_c and 5 h_c, where h_c is average plant canopy height. Given:

	Plant	h_c(m)	u_*(m s^{-1})	attenuation coef.
a.	Wheat	1.0	0.5	2.6
b.	Wheat	1.0	0.75	2.6
c.	Soybean	1.0	0.5	3.5
d.	Soybean	1.0	0.75	3.5
e.	Oats	1.5	0.5	2.8
f.	Oats	1.5	0.75	2.8
g.	Corn	2.0	0.5	2.7
h.	Corn	2.0	0.75	2.7
i.	Corn	2.5	0.5	2.2
j.	Corn	2.5	0.75	2.2
k.	Sunflower	2.75	0.5	1.3
m.	Sunflower	2.75	0.75	1.3
n.	Pine	3.0	0.5	1.1
o.	Pine	3.0	0.25	1.1
p.	Orchard	4.0	0.5	0.4
q.	Orchard	4.0	0.25	0.4
r.	Forest	20.	0.5	1.7
s.	Forest	20.	0.25	1.7

A37. Estimate the max urban heat island temperature excess compared to the surrounding rural countryside, for a city with urban-canyon aspect ratio (H/W) of:

a. 0.5 b. 0.75 c. 1.0 d. 1.25 e. 1.5 f. 1.75
g. 2.0 h. 2.25 i. 2.5 j. 2.75 k. 3.0 m. 3.25

Evaluate & Analyze

E1. For a Weibull distribution, what is the value of the probability in any one bin as the bin size becomes infinitesimally small? Why?

E2(§). Create a computer spreadsheet with location and spread parameters in separate cells. Create and plot a Weibull frequency distribution for winds by referencing those parameters. Then try changing the parameters to see if you can get the Weibull distribution to look like other well-known distributions, such as Gaussian (symmetric, bell shaped), exponential, or others.

E3. Why was an asymmetric distribution such as the Weibull distribution chosen to represent winds?

E4. What assumptions were used in the derivation of Betz' Law, and which of those assumptions could be improved?

E5. To double the amount of electrical power produced by a wind turbine, wind speed must increase by what percentage, or turbine radius increase by what percentage?

E6. For the Weibull distribution as plotted in Fig. 17.1, find the total wind power associated with it.

E7. In Fig. 17.5, what determines the shape of the wind-power output curve between the cut-in and rated points?

E8. List and explain commonalities among the equations that describe the various thermally-driven local flows.

E9. If thermals with average updraft velocity of W = 5 m s^{-1} occupy 40% of the horizontal area in the boundary layer, find the average downdraft velocity.

E10. What factors might affect rise rate of the thermal, in addition to the ones already given in this chapter?

E11. Anabatic and lenticular clouds were described in this chapter. Compare these clouds and their formation mechanisms. Is it possible for both clouds to occur simultaneously over the same mountain?

E12. Is the equation describing the anabatic pressure gradient force valid or reasonable in the limits of 0° slope, or 90° slope. Explain.

E13. Explain in terms of Bernoulli's equation the horizontal pressure gradient force acting on anabatic winds.

E14. What factors control the shape of the katabatic wind profile, as plotted in Fig. 17.9?

E15. The Sample Application for katabatic wind shows the curves from eqs. (17.8) and (17.9) as crossing. Given the factors that appear in those equations, is a situation possible where the curves never cross? Describe.

E16. Suppose a mountain valley exits right at a coastline. For synoptically weak conditions (near zero geostrophic wind), describe how would the mountain/valley circulation and sea-breeze circulation interact. Illustrate with drawings.

E17. The thermal internal boundary layer can form both during weak- and strong-wind synoptic conditions. Why?

E18. For stronger land-sea temperature contrasts, which aspects of the sea-breeze would change, and which would be relatively unchanged? Why?

E19. At 30° latitude, can the sea-breeze front advance an infinite distance from the shore? Why?

E20. In the Southern Hemisphere, draw a sketch of the sea-breeze-vs.-time hodograph, and explain it.

E21. For what situations would open-channel hydraulics NOT be a good approximation to atmospheric local flows? Explain.

E22. Interfacial (surface) wave speed was shown to depend on average depth of the cold layer of air. Is this equation valid for any depth? Why?

E23. In deriving eq. (17.17) for internal waves, we focused on only the fastest wavelengths. Justify.

E24. In what ways is the Froude number for incompressible flows similar to the Mach number for compressible flows?

E25. If supercritical flows tend to "break down" toward subcritical, then why do supercritical flows exist at all in the atmosphere?

E26. Is it possible to have supercritical flow in the atmosphere that does NOT create an hydraulic jump when it changes to subcritical? Explain?

E27. Contrast the nature of gap winds through short and long gaps. Also, what would you do if the gap length were in between short and long?

E28. For gap winds through a long gap, why are they less likely to form in summer than winter?

E29. Can coastally trapped jets form on the east coast of continents in the N. Hemisphere? If so, explain how the process would work.

E30. It is known from measurements of the ionosphere that the vertical amplitude of mountain waves increases with altitude. Explain this using Bernoulli's equation.

E31. What happens to the natural wavelength of air for statically unstable conditions?

E32. Why are lenticular clouds called **standing lenticular**?

E33. Compare and contrast the 3 versions of the Froude number. Do they actually describe the same physical processes? Why?

E34. Is there any max limit to the angle a of mountain wave crests (see Fig. 17.31)? Comment.

E35. If during the course of a day, the wind speed is constant but the wind direction gradually changes direction by a full 360°, draw a graph of the resulting streamline, streakline, and path line at the end of the period. Assume continuous emissions from a point source during the whole period.

E36. Identify the terms of Bernoulli's equation that form the hydrostatic approximation. According to Bernoulli's equation, what must happen or not happen in order for hydrostatic balance to be valid?

E37. Describe how the terms in Bernoulli's equation vary along a mountain-wave streamline as sketched in Fig. 17.29.

E38. If a cold air parcel is given an upward push in a warmer environment of uniform potential temperature, describe how the terms in Bernoulli's equation vary with parcel height.

E39. For compressible flow, show if (and how) the Bernoulli equations for isothermal and adiabatic processes reduce to the basic incompressible Bernoulli equation under conditions of constant density.

E40. In the Sample Application for the pressure variation across a wind turbine, hypothesize why the actual pressure change has the variation that was plotted.

E41. In Fig. 17.34, would it be reasonable to move the static pressure port to the top center of the darkly shaded block, given no change to the streamlines drawn? Comment on potential problems with a static port at that location.

E42. Design a thermometer mount on a fast aircraft that would not be susceptible to dynamic warming. Explain why your design would work.

E43. In Fig. 17.34, speculate on how the streamlines would look if the approaching flow was supersonic. Draw your streamlines, and justify them.

E44. Comment on the differences and similarities of the two mechanisms shown in this Chapter for creating Foehn winds.

E45. For Bora winds, if the upwind cold air was over an elevated plateau, and the downwind lowland was significantly lower than the plateau, how would Bora winds be different, if at all? Why?

E46. If the air in Fig. 17.38 went over a mountain but there was no precipitation, would there be a Foehn wind?

E47. Relate the amount of warming of a Foehn wind to the average upstream wind speed and the precipitation rate in mm h^{-1} .

E48. How sensitive is the solution for wind speed above a plant canopy? [Hint: see the Sample Application in the canopy flow section.] Namely, if you have a small error in estimating displacement distance d, are the resulting errors in friction velocity u_* and roughness length z_o relatively small or large?

Synthesize

S1. Suppose that in year 2100 everyone is required by law to have their own wind turbine. Since wind turbines take power from the wind, the wind becomes slower. What effect would this have on the weather and climate, if any?

S2. If fair-weather thermals routinely rose as high as the tropopause without forming clouds, comment on changes to the weather and climate, if any.

S3. Suppose that katabatic winds were frictionless. Namely, no turbulence, no friction against the ground, and no friction against other layers of air. Speculate on the shape of the vertical wind profile of the katabatic winds, and justify your arguments.

S4. If a valley has two exists, how would the mountain and valley winds behave?

S5. Suppose that katabatic winds flow into a bowl-shaped depression instead of a valley. Describe how the airflow would evolve during the night.

S6. If warm air was not less dense than cold, could sea breezes form? Explain.

S7. Why does the cycling in a sea-breeze hodograph not necessarily agree with the timing of the pendulum day?

S8. What local circulations would disappear if air density did not vary with temperature? Justify.

S9. Can a Froude number be defined based on deep-water waves rather than shallow-water waves? If so, write an equation for the resulting Froude number, and suggest applications for it in the atmosphere.

S10. What if waves could carry no information and no energy. How would the critical nature of the flow change, if at all?

S11. If the Earth did not rotate, compare the flow through short and long gaps through mountains.

S12. If no mountains existing along coasts, could there ever be strong winds parallel to the coast?

S13. If mountain-wave drag causes the winds to be slower, does that same drag force cause the Earth to spin faster? Comment.

S14. Suppose that mountain-wave drag worked oppositely, and caused winds to accelerate aloft. How would the weather & climate be different, if at all?

S15. Is it possible for a moving air parcel to not be traveling along a streamline? Comment.

S16. Suppose that Bernoulli's equation says that pressure decreases as velocity decreases along a streamline of constant height. How would the weather and climate be different, if at all? Start by commenting how Boras would be different, if at all.

S17. Suppose you are a 2 m tall person in a town with average building height of 8 m. How would the winds that you feel be different (if at all) than the winds felt by a 0.2 m tall cat in a young corn field of average height 0.8 m?

S18. If human population continued to grow until all land areas were urban, would there be an urban heat island? Justify, and relate to weather changes.

18 ATMOSPHERIC BOUNDARY LAYER

Contents

Sunrise, sunset, sunrise. The daily cycle of radiative heating causes a daily cycle of sensible and latent heat fluxes between the Earth and the air, during clear skies over land. These fluxes influence only the bottom portion of the troposphere — the portion touching the ground (Fig. 18.1).

This layer is called the **atmospheric boundary layer** (ABL). It experiences a **diurnal** (daily) cycle of temperature, humidity, wind, and pollution variations. Turbulence is ubiquitous in the ABL, and is one of the causes of the unique nature of the ABL.

Because the boundary layer is where we live, where our crops are grown, and where we conduct our commerce, we have become familiar with its daily cycle. We perhaps forget that this cycle is not experienced by the rest of the atmosphere above the ABL. This chapter examines the formation and unique characteristics of the ABL.

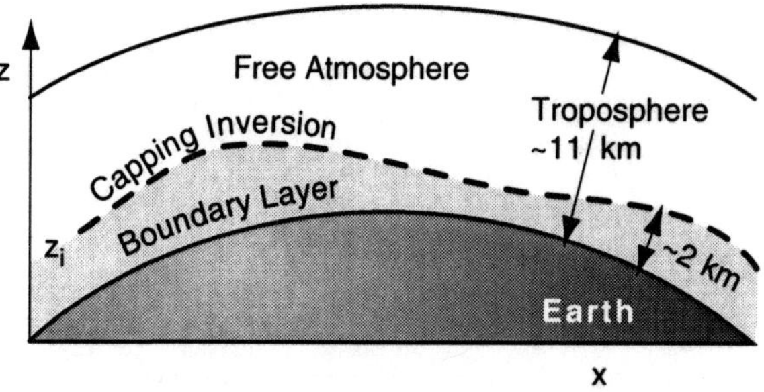

Figure 18.1
Location of the boundary layer, with top at z_i.

STATIC STABILITY — A REVIEW

Explanation

Static stability controls formation of the ABL, and affects ABL wind and temperature profiles. Here is a quick review of information from the Atmospheric Stability chapter.

If a small blob of air (i.e., an **air parcel**) is warmer than its surroundings at the same height or pressure, the parcel is positively buoyant and rises. If cooler, it is negatively buoyant and sinks. A parcel

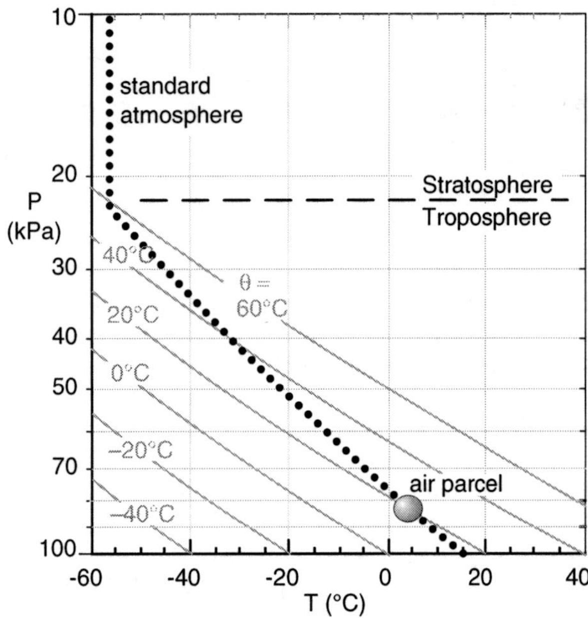

Figure 18.2
Standard atmosphere (dotted black line) plotted on a thermodynamic diagram. The circle represents a hypothetical air parcel. Diagonal grey lines are dry adiabats.

Sample Application

Find the vertical gradient of potential temperature in the troposphere for a standard atmosphere.

Find the Answer

Given: $\Delta T/\Delta z = -6.5°C\ km^{-1}$ from Chapter 1 eq. (1.16)
Find: $\Delta\theta/\Delta z = ?\ °C\ km^{-1}$

Use eq. (3.11) from the Heat chapter:

$$\theta(z) = T(z) + \Gamma_d \cdot z \qquad (3.11)$$

where the dry adiabatic lapse rate is $\Gamma_d = 9.8°C\ km^{-1}$
Apply this at heights z_1 and z_2, and then subtract the z_1 equation from the z_2 equation:

$$\theta_2 - \theta_1 = T_2 - T_1 + \Gamma_d \cdot (z_2 - z_1)$$

Divide both sides of the equation by $(z_2 - z_1)$. Then define $(z_2 - z_1) = \Delta z$, $(T_2 - T_1) = \Delta T$, and $(\theta_2 - \theta_1) = \Delta\theta$ to give the algebraic form of the answer:

$$\Delta\theta/\Delta z = \Delta T/\Delta z + \Gamma_d$$

This eq. applies to any vertical temperature profile.

If we plug in the temperature profile for the standard atmosphere:

$$\Delta\theta/\Delta z = (-6.5°C\ km^{-1}) + (9.8°C\ km^{-1}) = \mathbf{3.3°C\ km^{-1}}$$

Check: Units OK. Agrees with Fig. 18.2, where θ increases from 15°C at the surface to 51.3°C at 11 km altitude, which gives (51.3–15°C)/(11km) = 3.3°C km^{-1}.
Exposition: θ gradually increases with height in the troposphere, which as we will see tends to gently oppose vertical motions. Although the standard-atmosphere (an engineering specification similar to an average condition) troposphere is statically stable, the real troposphere at any time and place can have layers that are statically stable, neutral, or unstable.

with the same temperature as its surrounding environment experiences zero buoyant force.

Figure 18.2 shows the standard atmosphere from Chapter 1, plotted on a thermodynamic diagram from the Stability chapter. Let the standard atmosphere represent the environment or the background air. Consider an air parcel captured from one part of that environment (plotted as the circle). At its initial height, the parcel has the same temperature as the surrounding environment, and experiences no buoyant forces.

To determine static stability, you must ask what would happen to the air parcel if it were forcibly displaced a small distance up or down. When moved from its initial capture altitude, the parcel and environment temperatures could differ, thereby causing buoyant forces.

If the buoyant forces on a displaced air parcel push it back to its starting altitude, then the environment is said to be **statically stable**. In the absence of any other forces, statically stable air is **laminar**. Namely, it is smooth and non-turbulent.

However, if the displaced parcel is pulled further away from its starting point by buoyancy, the portion of the atmosphere through which the air parcel continues accelerating is classified as **statically unstable**. Unstable regions are **turbulent** (gusty).

If the displaced air parcel has a temperature equal to that of its new surroundings, then the environment is **statically neutral**.

When an air parcel moves vertically, its temperature changes adiabatically, as described in previous chapters. Always consider such adiabatic temperature change before comparing parcel temperature to that of the surrounding environment. The environment is usually assumed to be **stationary**, which means it is relatively unchanging during the short time it takes for the parcel to rise or sink.

If an air parcel is captured at P = 83 kPa and T = 5°C (as sketched in Fig. 18.2), and is then is forcibly lifted dry adiabatically, it cools following the θ = 20°C adiabat (one of the thin diagonal lines in that figure). If lifted to a height where the pressure is P = 60 kPa, its new temperature is about T = –20°C.

This air parcel, being colder than the environment (thick dotted line in Fig. 18.2) at that same height, feels a downward buoyant force toward its starting point. Similarly if displaced downward from its initial height, the parcel is warmer than its surroundings at its new height, and would feel an upward force toward its starting point.

Air parcels captured from any initial height in the environment of Fig. 18.2 always tend to return to their starting point. Therefore, the standard atmosphere is statically stable. This stability is critical for ABL formation.

Rules of Thumb for Stability in the ABL

Because of the daily cycle of radiative heating and cooling, there is a daily cycle of static stability in the ABL. ABL static stability can be anticipated as follows, without worrying about air parcels for now.

Unstable air adjacent to the ground is associated with light winds and a surface that is warmer than the air. This is common on sunny days in fair-weather. It can also occur when cold air blows over a warmer surface, day or night. In unstable conditions, thermals of warm air rise from the surface to heights of 200 m to 4 km, and turbulence within this layer is vigorous.

At the other extreme are **stable** layers of air, associated with light winds and a surface that is cooler than the air. This typically occurs at night in fair-weather with clear skies, or when warm air blows over a colder surface day or night. Turbulence is weak or sometimes nonexistent in stable layers adjacent to the ground. The stable layers of air are usually shallow (20 - 500 m) compared to the unstable daytime cases.

In between these two extremes are **neutral** conditions, where winds are moderate to strong and there is little heating or cooling from the surface. These occur during overcast conditions, often associated with bad weather.

BOUNDARY-LAYER FORMATION

Tropospheric Constraints

Because of buoyant effects, the vertical temperature structure of the troposphere limits the types of vertical motion that are possible. The standard atmosphere in the troposphere is not parallel to the dry adiabats (Fig. 18.2), but crosses the adiabats toward warmer potential temperatures as altitude increases.

That same standard atmosphere is replotted as the thick dotted grey line in Fig. 18.3, but now in terms of its potential temperature (θ) versus height (z). The standard atmosphere slopes toward warmer potential temperatures at greater altitudes. Such a slope indicates statically stable air; namely, air that opposes vertical motion.

The ABL is often turbulent. Because turbulence causes mixing, the bottom part of the standard atmosphere becomes homogenized. Namely, within the turbulent region, warmer potential-temperature air from the standard atmosphere in the top of the ABL is mixed with cooler potential-temperature air from near the bottom. The resulting mixture has a

INFO • Engineering Boundary Layers

In wind tunnel experiments, the layer of air that turbulently "feels" frictional drag against the bottom wall grows in depth indefinitely (Fig. 18.a). This engineering boundary-layer thickness h grows proportional to the square root of downstream distance x, until hitting the top of the wind tunnel.

On an idealized rotating planet, the Earth's rotation imposes a dynamical constraint on ABL depth (Fig. 18.b). This maximum depth is proportional to the ratio of wind drag (related to the friction velocity u_*, which is a concept discussed later in this chapter) to Earth's rotation (related to the Coriolis parameter f_c, as discussed in the Atmos. Forces & Winds chapter). This dynamic constraint supersedes the turbulence constraint.

For the real ABL on Earth, the strong capping inversion at height z_i makes the ABL unique (Fig. 18.c) compared to other fluid flows. It constrains the ABL thickness and the eddies within it to a maximum size of order 200 m to 4 km. This stratification (thermodynamic) constraint supersedes the others. It means that the temperature structure is always very important for the ABL.

(a) Engineering Boundary Layers:

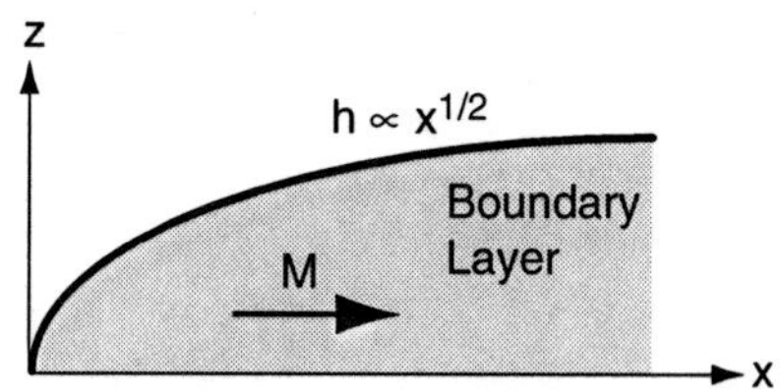

(b) Boundary Layers on a Rotating Planet:

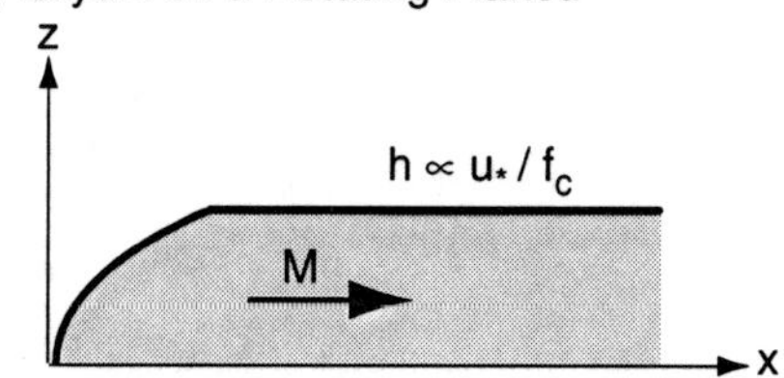

(c) Boundary Layers in Earth's Stratified Atmosphere:

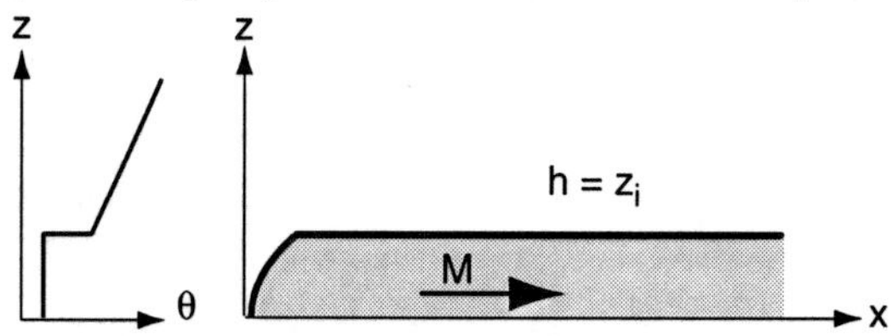

Figure 18.a-c
Comparison of constraints on boundary layer thickness, h. M is mean wind speed away from the bottom boundary, θ is potential temperature, z is height, and x is downwind distance.

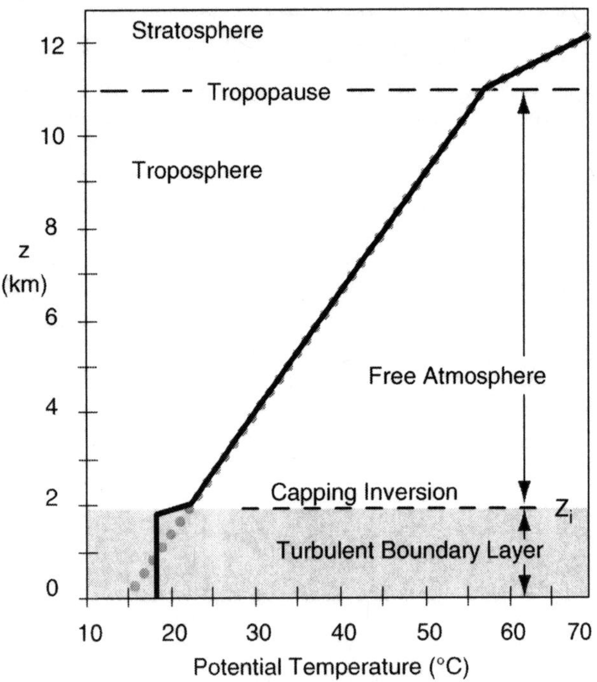

Figure 18.3
Restriction of ABL depth by tropospheric temperature structure during fair weather. The standard atmosphere is the grey dotted line. The thick black line shows an idealized temperature profile after the turbulent boundary layer modifies the bottom part of the standard atmosphere.

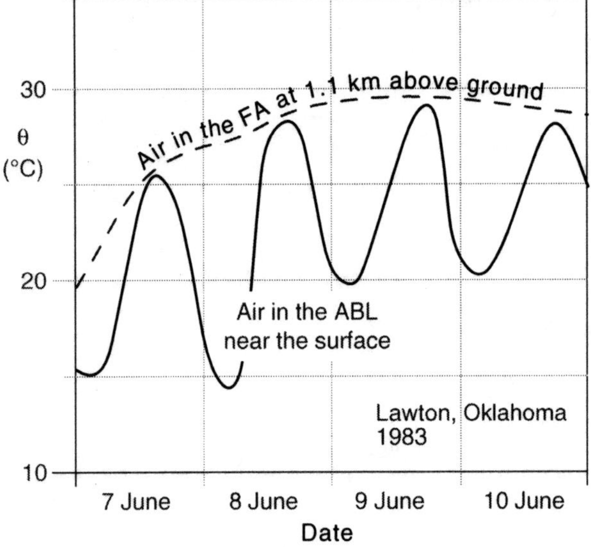

Figure 18.4
Observed variations of potential temperature in the ABL (solid line) and the free atmosphere (FA) (dashed line). The daily heating and cooling cycle that we are so familiar with near the ground does not exist above the ABL.

medium potential temperature that is uniform with height, as plotted by the thick black line in Fig. 18.3. In situations of vigorous turbulence, the ABL is also called the **mixed layer** (**ML**).

Above the mixed layer, the air is usually un-modified by turbulence, and retains the same temperature profile as the standard atmosphere in this idealized scenario. This tropospheric air above the ABL is known as the **free atmosphere** (**FA**).

As a result of a turbulent mixed layer being adjacent to the unmixed free atmosphere, there is a sharp temperature increase at the mixed layer top. This transition zone is very stable, and is often a **temperature inversion**. Namely, it is a region where temperature increases with height. The altitude of the middle of this inversion is given the symbol z_i, and is a measure of the depth of the turbulent ABL.

The temperature inversion acts like a lid or cap to motions in the ABL. Picture an air parcel from the mixed layer in Fig. 18.3. If turbulence were to try to push it out of the top of the mixed layer into the free atmosphere, it would be so much colder than the surrounding environment that a strong buoyant force would push it back down into the mixed layer. Hence, air parcels, turbulence, and any air pollution in the parcels, are trapped within the mixed layer.

There is always a strong stable layer or temperature inversion capping the ABL. As we have seen, turbulent mixing in the bottom of the statically-stable troposphere creates this cap, and in turn this cap traps turbulence below it.

The capping inversion breaks the troposphere into two parts. Vigorous turbulence within the ABL causes the ABL to respond quickly to surface influences such as heating and frictional drag. However, the remainder of the troposphere does not experience this strong turbulent coupling with the surface, and hence does not experience frictional drag nor a daily heating cycle. Fig. 18.4 illustrates this.

In summary, the bottom 200 m to 4 km of the troposphere is called the atmospheric boundary layer. ABL depth is variable with location and time. Turbulent transport causes the ABL to feel the direct effects of the Earth's surface. The ABL exhibits strong diurnal (daily) variations of temperature, moisture, winds, pollutants, turbulence, and depth in response to daytime solar heating and nighttime IR cooling of the ground. The name "boundary layer" comes from the fact that the Earth's surface is a boundary on the atmosphere, and the ABL is the part of the atmosphere that "feels" this boundary during fair weather.

Synoptic Forcings

Weather patterns such as high (**H**) and low (**L**) pressure systems that are drawn on weather maps are known as **synoptic** weather. These large diameter (≥ 2000 km) systems modulate the ABL. In the N. Hemisphere, ABL winds circulate clockwise and spiral out from high-pressure centers, but circulate counterclockwise and spiral in toward lows (Fig. 18.6). See the Dynamics chapter for details on winds.

The outward spiral of winds around highs is called **divergence**, and removes ABL air horizontally from the center of highs. Conservation of air mass requires **subsidence** (downward moving air) over highs to replace the horizontally diverging air (Fig. 18.5). Although this subsidence pushes free atmosphere air downward, it cannot penetrate into the ABL because of the strong capping inversion. Instead, the capping inversion is pushed downward closer to the ground as the ABL becomes thinner. This situation traps air pollutants in a shallow ABL, causing air stagnation and air-pollution episodes.

Similarly, horizontally converging ABL air around lows is associated with upward motion (Fig. 18.5). Often the synoptic forcings and storms associated with lows are so powerful that they easily lift the capping inversion or eliminate it altogether. This allows ABL air to be deeply mixed over the whole depth of the troposphere by thunderstorms and other clouds. Air pollution is usually reduced during this situation as it is diluted with cleaner air from aloft, and as it is washed out by rain.

Because winds in high-pressure regions are relatively light, ABL air lingers over the surface for sufficient time to take on characteristics of that surface. These characteristics include temperature, humidity, pollution, odor, and others. Such ABL air is called an **airmass**, and was discussed in the chapter on Airmasses and Fronts. When the ABLs from two different high-pressure centers are drawn toward each other by a low center, the zone separating those two airmasses is called a **front**.

At a frontal zone, the colder, heavier airmass acts like a wedge under the warm airmass. As winds blow the cold and warm air masses toward each other, the cold wedge causes the warm ABL to peel away from the ground, causing it to ride up over the colder air (Figs. 18.7a & b). Also, thunderstorms can vent ABL air away from the ground (Figs. 18.7a & b). It is mainly in these stormy conditions (statically stable conditions at fronts, and statically unstable conditions at thunderstorms) that ABL air is forced away from the surface.

Although an ABL forms in the advancing airmass behind the front, the warm humid air that was pushed aloft is not called an ABL because it has lost contact with the surface. Instead, this rising warm air cools, allowing water vapor to condense and make the clouds that we often associate with fronts.

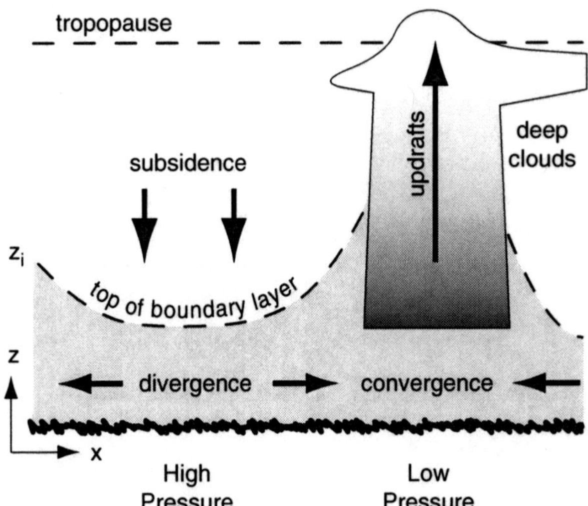

Figure 18.5
Influence of synoptic-scale vertical circulations on the ABL.

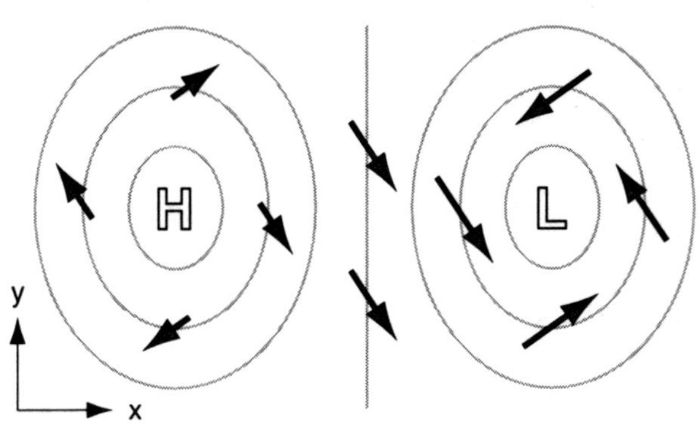

Figure 18.6
Synoptic-scale horizontal winds (arrows) in the ABL near the surface. Thin lines are isobars around surface high (H) and low (L) pressure centers.

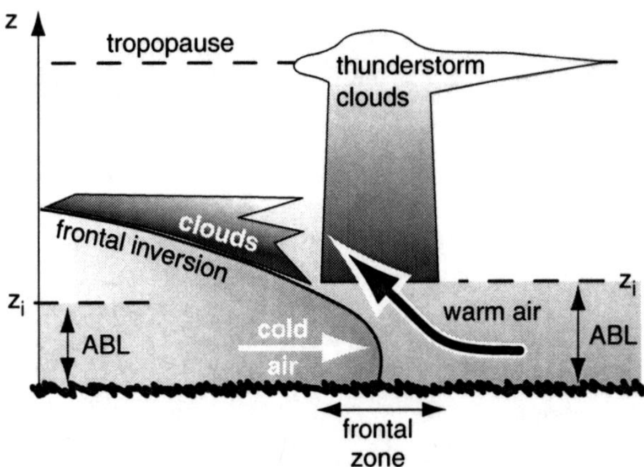

Figure 18.7a
Idealized ABL modification near a frontal zone.

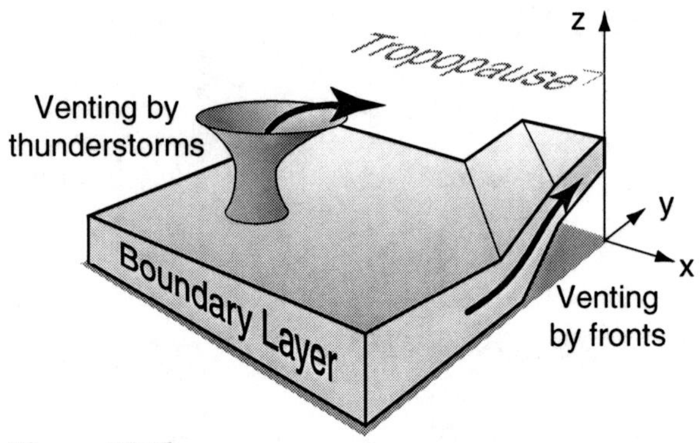

Figure 18.7b
Idealized venting of ABL air away from the surface during stormy weather.

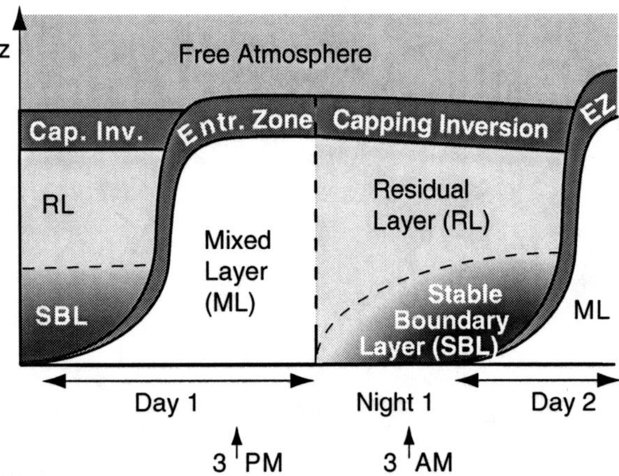

Figure 18.8
Components of the boundary layer during fair weather in summer over land. White indicates nonlocally statically unstable air, light grey (as in the RL) is neutral stability, and darker greys indicate stronger static stability.

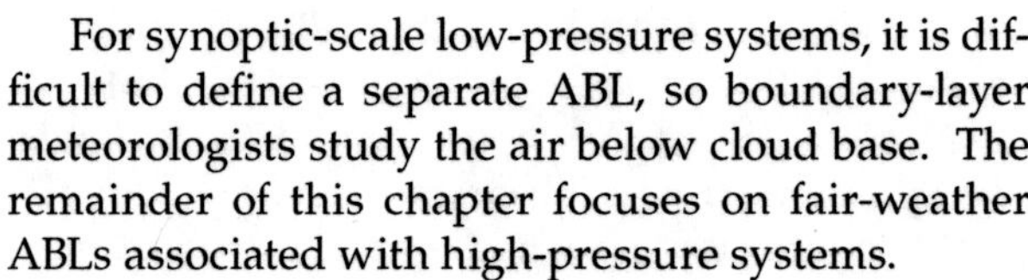

For synoptic-scale low-pressure systems, it is difficult to define a separate ABL, so boundary-layer meteorologists study the air below cloud base. The remainder of this chapter focuses on fair-weather ABLs associated with high-pressure systems.

ABL STRUCTURE AND EVOLUTION

The fair-weather ABL consists of the components sketched in Fig. 18.8. During daytime there is a statically-unstable **mixed layer (ML)**. At night, a statically **stable boundary layer (SBL)** forms under a statically neutral **residual layer (RL)**. The residual layer contains the pollutants and moisture from the previous mixed layer, but is not very turbulent.

The bottom 20 to 200 m of the ABL is called the **surface layer** (SL, Fig. 18.9). Here frictional drag, heat conduction, and evaporation from the surface cause substantial variations of wind speed, temperature, and humidity with height. However, turbulent fluxes are relatively uniform with height; hence, the surface layer is known as the **constant flux layer**.

Separating the **free atmosphere** (FA) from the mixed layer is a strongly stable **entrainment zone (EZ)** of intermittent turbulence. Mixed-layer depth z_i is the distance between the ground and the middle of the EZ. At night, turbulence in the EZ ceases, leaving a nonturbulent layer called the **capping inversion (CI)** that is still strongly statically stable.

Typical vertical profiles of temperature, potential temperature, humidity mixing ratio, and wind speed are sketched in Fig. 18.9. The "day" portion of Fig. 18.9 corresponds to the 3 PM time indicated in Fig. 18.8, while "night" is for 3 AM.

Next, look at ABL temperature, winds, and turbulence in more detail.

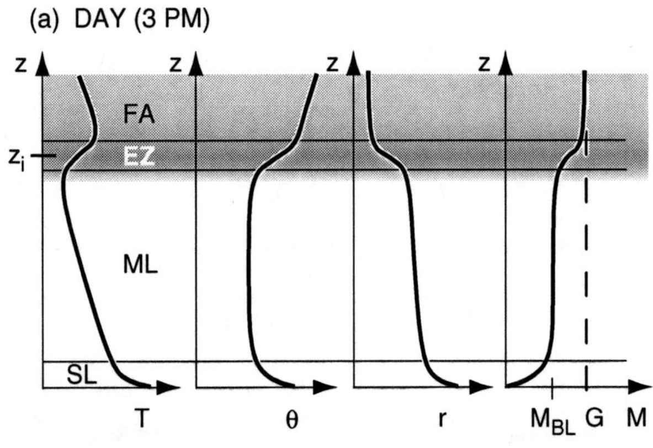

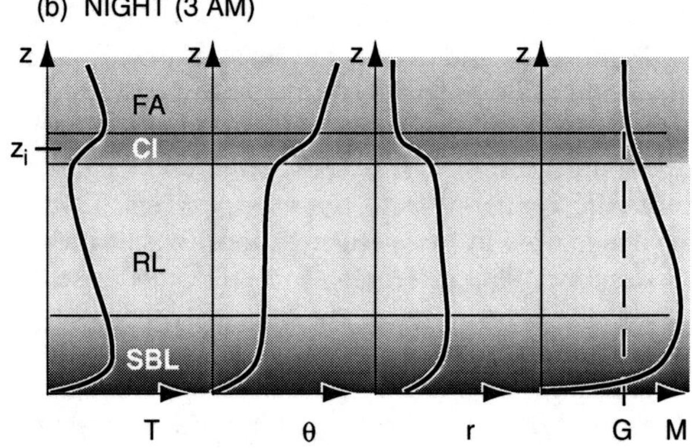

Figure 18.9
Typical vertical profiles of temperature (T), potential temperature (θ), mixing ratio (r, see the Water Vapor chapter), and wind speed (M) in the ABL. The dashed line labeled G is the geostrophic wind speed (a theoretical wind in the absence of surface drag, see the Atmos. Forces & Winds chapter). M_{BL} is average wind speed in the ABL. Shading corresponds to shading in Fig. 18.8, with white being statically unstable, and black very stable.

TEMPERATURE

The capping inversion traps in the ABL any heating and water evaporation from the surface. As a result, heat accumulates within the ABL during day, or whenever the surface is warmer than the air. Cooling (actually, heat loss) accumulates during night, or whenever the surface is colder than the air. Thus, the temperature structure of the ABL depends on the accumulated heating or cooling.

Cumulative Heating or Cooling

The cumulative effect of surface heating and cooling is more important on ABL evolution than the instantaneous heat flux. This cumulative heating or cooling Q_A equals the area under the curve of heat flux vs. time (Fig. 18.10). We will examine cumulative nighttime cooling separately from cumulative daytime heating.

Nighttime

During clear nights over land, heat flux from the air to the cold surface is relatively constant with time (dark shaded portion of Fig. 18.10). If we define t as the time since cooling began, then the accumulated cooling per unit surface area is:

$$Q_A = \mathbb{F}_{H\ night} \cdot t \qquad (18.1a)$$

For night, Q_A is a negative number because $\mathbb{F}_H$ is negative for cooling. Q_A has units of J m^{-2}.

Dividing eq. (18.1a) by air density and specific heat ($\rho_{air} \cdot C_p$) gives the kinematic form

$$Q_{Ak} = F_{H\ night} \cdot t \qquad (18.1b)$$

where Q_{Ak} has units of K·m.

For a night with variable cloudiness that causes a variable surface heat flux (see the Solar & IR Radiation chapter), use the average value of $\mathbb{F}_H$ or F_H.

Daytime

On clear days, the nearly sinusoidal variation of solar elevation and downwelling solar radiation (Fig. 2.13 in the Solar & IR Radiation chapter) causes nearly sinusoidal variation of surface net heat flux (Fig. 3.9 in the Heat chapter). Let t be the time since $\mathbb{F}_H$ becomes positive in the morning, D be the total duration of positive heat flux, and $\mathbb{F}_{H\,max}$ be the peak value of heat flux (Fig. 18.10). These values can be found from data such as in Fig. 3.9. The accumulated daytime heating per unit surface area (units of J m^{-2}) is:

$$Q_A = \frac{\mathbb{F}_{H\max} \cdot D}{\pi} \cdot \left[1 - \cos\left(\frac{\pi \cdot t}{D}\right)\right] \qquad (18.2a)$$

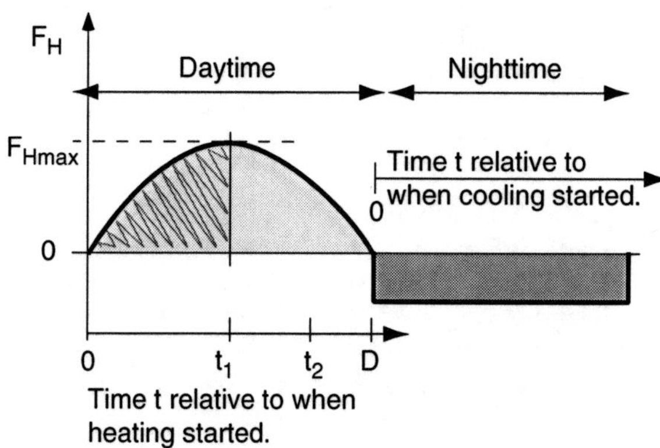

Figure 18.10
Idealization of the heat flux curve over land during fair weather, from Fig. 3.9 of the Heat chapter. Light shading shows total accumulated heating during the day, while the hatched region shows the portion of heating accumulated up to time t_1 after heating started. Dark shading shows total accumulated cooling during the night.

HIGHER MATH • Cumulative Heating

Derivation of Daytime Cumulative Heating

Assume that the kinematic heat flux is approximately sinusoidal with time:

$$F_H = F_{H\max} \cdot \sin(\pi \cdot t / D)$$

with t and D defined as in Fig. 18.10 for daytime. Integrating from time $t = 0$ to arbitrary time t :

$$Q_{Ak} = \int_{t'=0}^{t} F_H dt' = F_{H\max} \cdot \int_{t'=0}^{t} \sin(\pi \cdot t' / D) dt'$$

where t' is a dummy variable of integration.

From a table of integrals, we find that:

$$\int \sin(a \cdot x) = -(1/a) \cdot \cos(a \cdot x)$$

Thus, the previous equation integrates to:

$$Q_{Ak} = \frac{-F_{H\max} \cdot D}{\pi} \cdot \cos\left(\frac{\pi \cdot t}{D}\right)\Bigg|_0^t$$

Plugging in the two limits gives:

$$Q_{Ak} = \frac{-F_{H\max} \cdot D}{\pi} \cdot \left[\cos\left(\frac{\pi \cdot t}{D}\right) - \cos(0)\right]$$

But the cos(0) = 1, giving the final answer (eq. 18.2b):

$$Q_{Ak} = \frac{F_{H\max} \cdot D}{\pi} \cdot \left[1 - \cos\left(\frac{\pi \cdot t}{D}\right)\right]$$

Sample Application

Use Fig. 3.9 from the Heat chapter to estimate the accumulated heating and cooling in kinematic form over the whole day and whole night.

Find the Answer

By eye from Fig. 3.9:

Day: $\mathbb{F}_{H\,max} \approx$ 150 W·m^{-2} , D = 8 h

Night: $\mathbb{F}_{H\,night} \approx$ –50 W·m^{-2} (averaged)

Given: Day: $t = D$ = 8 h = 28800. s

Night: t = 24h – D = 16 h = 57600. s

Find: Q_{Ak} = ? K·m for day and for night

First convert fluxes to kinematic form by dividing by $\rho_{air}\cdot C_p$:

$F_{H\,max} = \mathbb{F}_{H\,max} / \rho_{air}\cdot C_p$

= (150 W·m^{-2}) / [1231(W·m^{-2})/(K·m s^{-1})]

= 0.122 K·m s^{-1}

$F_{H\,night}$ = (–50W·m^{-2})/[1231(W·m^{-2})/(K·m s^{-1})]

= –0.041 K·m s^{-1}

Day: use eq. (18.2b):

$$Q_{Ak} = \frac{(0.122 \text{ K}\cdot\text{m/s})\cdot(28800 \text{ s})}{3.14159}\cdot[1-\cos(\pi)]$$

= **2237 K·m**

Night: use eq. (18.1b):

Q_{Ak} = (–0.041 K·m s^{-1})·(57600. s) = **–2362 K·m**

Check: Units OK. Physics OK.

Exposition: The heating and cooling are nearly equal for this example, implying that the daily average temperature is fairly steady.

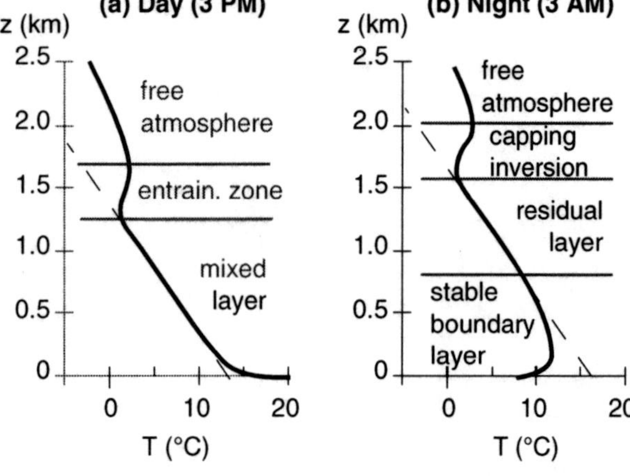

Figure 18.11

Examples of boundary-layer temperature profiles during day (left) and night (right) during fair weather over land. Adiabatic lapse rate is dashed. The heights shown here are illustrative only. In the real ABL the heights can be greater or smaller, depending on location, time, and season.

In kinematic form (units of K·m), this equation is

$$Q_{Ak} = \frac{F_{H\max}\cdot D}{\pi}\cdot\left[1-\cos\left(\frac{\pi\cdot t}{D}\right)\right] \tag{18.2b}$$

Temperature-Profile Evolution

Idealized Evolution

A typical afternoon temperature profile is plotted in Fig. 18.11a. During the daytime, the environmental lapse rate in the mixed layer is nearly adiabatic. The unstable surface layer (plotted but not labeled in Fig. 18.11) is in the bottom part of the mixed layer. Warm blobs of air called **thermals** rise from this surface layer up through the mixed layer, until they hit the temperature inversion in the entrainment zone. Fig. 18.12a shows a closer view of the surface layer (bottom 5 - 10% of ABL).

These thermal circulations create strong turbulence, and cause pollutants, potential temperature, and moisture to be well mixed in the vertical (hence the name **mixed layer**). The whole mixed layer, surface layer, and bottom portion of the entrainment zone are statically unstable.

In the entrainment zone, free-atmosphere air is incorporated or **entrained** into the mixed layer, causing the mixed-layer depth to increase during the day. Pollutants trapped in the mixed layer cannot escape through the EZ, although cleaner, drier free atmosphere air is entrained into the mixed layer. Thus, the EZ is a one-way valve.

At night, the bottom portion of the mixed layer becomes chilled by contact with the radiatively-cooled ground. The result is a stable ABL. The bottom portion of this stable ABL is the surface layer (again not labeled in Fig. 18.11b, but sketched in Fig. 18.12b).

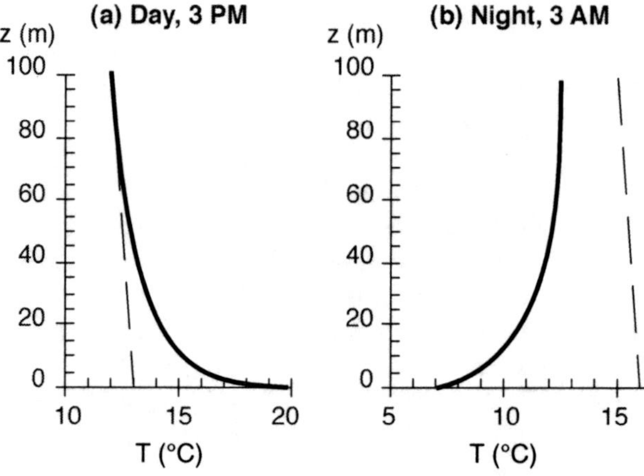

Figure 18.12

Examples of surface-layer temperature profiles during day (left) and night (right). Adiabatic lapse rate is dashed. Again, heights are illustrative only. Actual heights might differ.

Above the stable ABL is the residual layer. It has not felt the cooling from the ground, and hence retains the adiabatic lapse rate from the mixed layer of the previous day. Above that is the capping temperature inversion, which is the nonturbulent remnant of the entrainment zone.

Seasonal Differences

During summer at mid- and high-latitudes, days are longer than nights during fair weather over land. More heating occurs during day than cooling at night. After a full 24 hours, the ending sounding is warmer than the starting sounding as illustrated in Figs. 18.13 a & b. The convective mixed layer starts shallow in the morning, but rapidly grows through the residual layer. In the afternoon, it continues to rise slowly into the free atmosphere. If the air contains sufficient moisture, cumuliform clouds can exist. At night, cooling creates a shallow stable ABL near the ground, but leaves a thick residual layer in the middle of the ABL.

During winter at mid- and high-latitudes, more cooling occurs during the long nights than heating during the short days, in fair weather over land. Stable ABLs dominate, and there is net temperature decrease over 24 hours (Figs. 18.13 c & d). Any nonfrontal clouds present are typically stratiform or fog. Any residual layer that forms early in the night is quickly overwhelmed by the growing stable ABL.

Fig. 18.14 shows the corresponding structure of the ABL. Although both the mixed layer and residual layer have nearly adiabatic temperature profiles, the mixed layer is nonlocally unstable, while the residual layer is neutral. This difference causes pollutants to disperse at different rates in those two regions.

If the wind moves ABL air over surfaces of different temperatures, then ABL structures can evolve in space, rather than in time. For example, suppose the numbers along the abscissa in Fig. 18.14b represent distance x (km), with wind blowing from left to right over a lake spanning $8 \le x \le 16$ km. The ABL

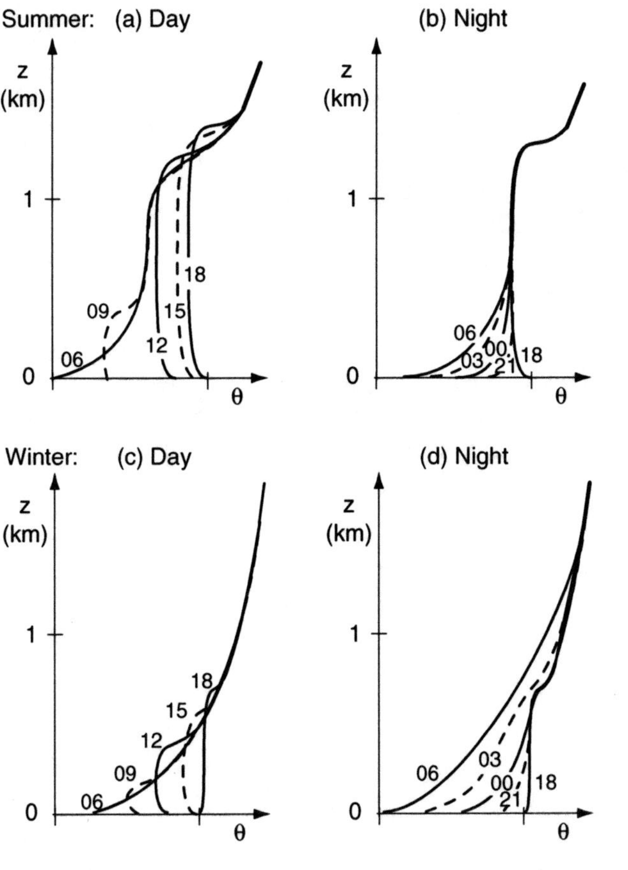

Figure 18.13
Evolution of potential temperature θ profiles for fair-weather over land. Curves are labeled with local time in hours. The nighttime curves begin at 18 local time, which corresponds to the ending sounding from the previous day.

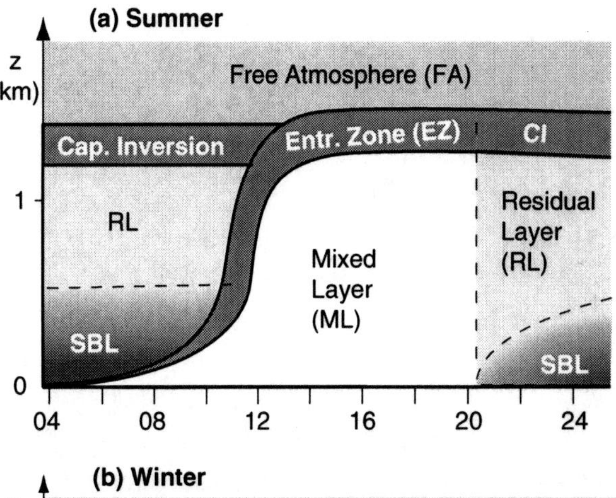

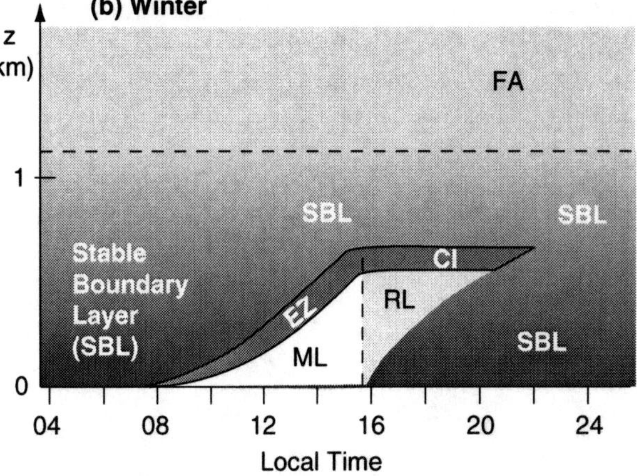

Figure 18.14
Daily evolution of boundary-layer structure by season, for fair weather over land. CI = Capping Inversion. Shading indicates static stability: white = unstable, light grey = neutral (as in the RL), darker greys indicate stronger static stability.

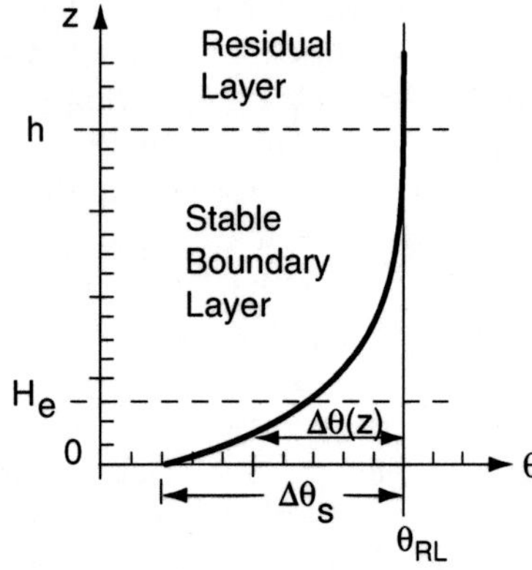

Figure 18.15
Idealized exponential-shaped potential temperature profile in the stable (nighttime) boundary layer.

Sample Application (§)
Estimate the potential temperature profile at the end of a 12-hour night for two cases: windy (10 m s^{-1}) and less windy (5 m s^{-1}). Assume Q_{Ak} = –1000 K·m.

Find the Answer
Given: Q_{Ak} = –1000 K·m, t = 12 h
(a) M_{RL} = 10 m s^{-1}, (b) M_{RL} = 5 m s^{-1}
Find: θ vs. z

Assume: flat prairie. To plot profile, need H_e & $\Delta\theta_s$.
Use eq. (18.4):

$$H_e \approx (0.15\text{m}^{1/4}\cdot\text{s}^{1/4})\cdot(10\text{m}\cdot\text{s}^{-1})^{3/4}\cdot(43200\text{s})^{1/2}$$
(a) H_e = 175 m
$$H_e \approx (0.15\text{m}^{1/4}\cdot\text{s}^{1/4})\cdot(5\text{m}\cdot\text{s}^{-1})^{3/4}\cdot(43200\text{s})^{1/2}$$
(b) H_e = 104 m

Use eq. (18.5):
(a) $\Delta\theta_s$ = (–1000K·m)/(175) = –5.71 °C
(b) $\Delta\theta_s$ = (–1000K·m)/(104) = –9.62 °C
Use eq. (18.3) in a spreadsheet to compute the potential temperature profiles, using H_e and $\Delta\theta_s$ from above:

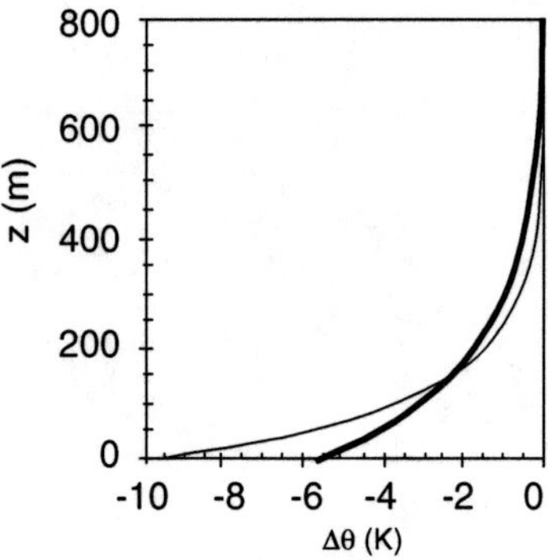

Check: Units OK, Physics OK. Graph OK.
Exposition: The windy case (thick line) is not as cold near the ground, but the cooling extends over a greater depth than for the less-windy case (thin line).

structure in Fig. 18.14b could occur at midnight in mid-latitude winter for air blowing over snow-covered ground, except for the unfrozen lake in the center. If the lake is warmer than the air, it will create a mixed layer that grows as the air advects across the lake.

Don't be lulled into thinking the ABL evolves the same way at every location or at similar times. The most important factor is the temperature difference between the surface and the air. If the surface is warmer, a mixed layer will develop regardless of the time of day. Similarly, colder surfaces will create stable ABLs.

Stable-ABL Temperature

Stable ABLs are quite complex. Turbulence can be intermittent, and coupling of air to the ground can be quite weak. In addition, any slope of the ground causes the cold air to drain downhill. Cold, downslope winds are called **katabatic winds**, as discussed in the Regional Winds chapter.

For a simplified case of a contiguously-turbulent stable ABL over a flat surface during light winds, the potential temperature profile is approximately exponential with height (Fig. 18.15):

$$\Delta\theta(z) = \Delta\theta_s \cdot e^{-z/H_e} \qquad (18.3)$$

where $\Delta\theta(z) = \theta(z) - \theta_{RL}$ is the potential temperature difference between the air at height z and the air in the residual layer. Δθ(z) is negative.

The value of this difference near the ground is defined to be $\Delta\theta_s = \Delta\theta(z{=}0)$, and is sometimes called the **strength** of the stable ABL. H_e is an **e-folding height** for the exponential curve. The actual **depth** h of the stable ABL is roughly $h = 5{\cdot}H_e$.

Depth and strength of the stable ABL grow as the cumulative cooling Q_{Ak} increases with time:

$$H_e \approx a \cdot M_{RL}{}^{3/4} \cdot t^{1/2} \qquad (18.4)$$

$$\Delta\theta_s = \frac{Q_{Ak}}{H_e} \qquad (18.5)$$

where a = 0.15 m$^{1/4}$·s$^{1/4}$ for flow over a flat prairie, and where M_{RL} is the wind speed in the residual layer. Because the cumulative cooling is proportional to time, both the depth $5{\cdot}H_e$ and strength $\Delta\theta_s$ of the stable ABL increase as the square root of time. Thus, fast growth of the SBL early in the evening decreases to a much slower growth by the end of the night.

Mixed-Layer (ML) Temperature

The shape of the potential temperature profile in the mixed layer is simple. To good approximation it is uniform with height. Of more interest is the evolution of mixed-layer average θ and z_i with time.

Use the potential temperature profile at the end of the night (early morning) as the starting sounding for forecasting daytime temperature profiles. In real atmospheres, the sounding might not be a smooth exponential as idealized in the previous subsection. The method below works for arbitrary shapes of the initial potential temperature profile.

A graphical solution is easiest. First, plot the early-morning sounding of θ vs. z. Next, determine the cumulative daytime heating Q_{Ak} that occurs between sunrise and some time of interest t_1, using eq. (18.2b). This heat warms the air in the ABL; thus the area under the sounding equals the accumulated heating (and also has units of K·m). Plot a vertical line of constant θ between the ground and the sounding, so that the area (hatched in Fig. 18.16a) under the curve equals the cumulative heating.

This vertical line gives the potential temperature of the mixed layer $\theta_{ML}(t_1)$. The height where this vertical line intersects the early-morning sounding defines the mixed-layer depth $z_i(t_1)$. As cumulative heating increases with time during the day ($Area_2$ = total grey-shaded region at time t_2), the mixed layer becomes warmer and deeper (Fig. 18.16a). The resulting potential temperature profiles during the day are sketched in Fig. 18.16b. This method of finding mixed layer growth is called the **encroachment method**, or **thermodynamic method**, and explains roughly 90% of typical mixed-layer growth on sunny days with winds less than 10 m s^{-1}.

Entrainment

As was mentioned earlier, the turbulent mixed layer grows by entraining non-turbulent air from the free atmosphere. One can idealize the mixed layer as a **slab model** (Fig. 18.17a), with constant potential temperature in the mixed layer, and a jump of potential temperature (Δθ) at the EZ.

Entrained air from the free atmosphere has warmer potential temperature than air in the mixed layer. Because this warm air is entrained downward, it corresponds to a negative heat flux $F_{H\,zi}$ at the top of the mixed layer. The heat-flux profile (Fig. 18.17b) is often linear with height, with the most negative value marking the top of the mixed layer.

The entrainment rate of free atmosphere air into the mixed layer is called the **entrainment velocity**, w_e, and can never be negative. The entrainment velocity is the volume of entrained air per unit horizontal area per unit time. In other words it is a volume flux, which has the same units as velocity.

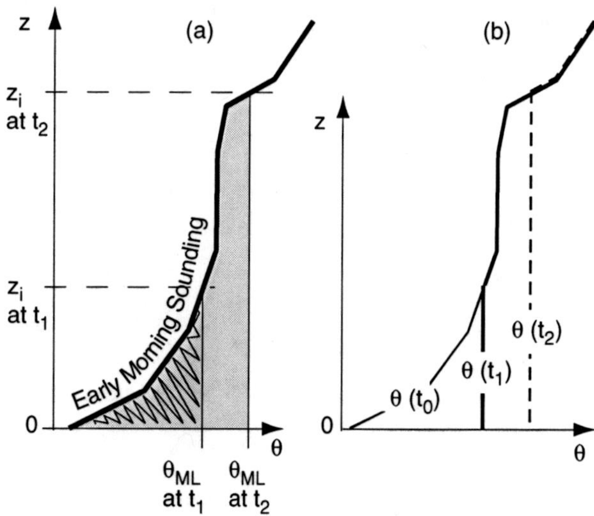

Figure 18.16
Evolution of the mixed layer with time, as cumulative heating increases the areas under the curves.

Sample Application

Given an early morning sounding with surface temperature 5°C and lapse rate Δθ/Δz = 3 K km^{-1}. Find the mixed-layer potential temperature and depth at 10 AM, when the cumulative heating is 500 K·m.

Find the Answer

Given: θ_{sfc} = 5°C, Δθ/Δz = 3 K km^{-1},
Q_{Ak} =0.50 K·km
Find: θ_{ML} = ? °C, z_i = ? km
Sketch:

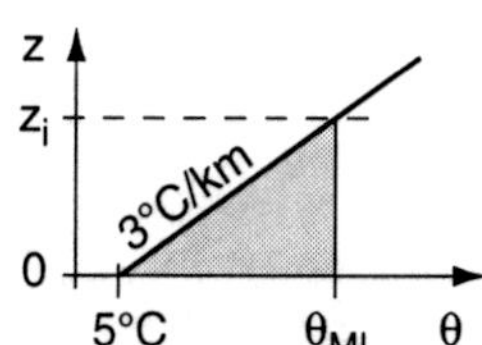

The area under this simple sounding is the area of a triangle: *Area* = 0.5·[*base*]·(*height*) ,
Area = 0.5·[(Δθ/Δz)·z_i]·(z_i)
0.50 K·km = 0.5·(3 K km^{-1})·z_i^2
Rearrange and solve for z_i:
z_i = **0.577 km**

Next, use the sounding to find the θ_{ML}:
$\theta_{ML} = \theta_{sfc} + (\Delta\theta/\Delta z)\cdot z_i$
θ_{ML} = (5°C) + (3°C km^{-1})·(0.577 km) = **6.73 °C**

Check: Units OK. Physics OK. Sketch OK.
Exposition: Arbitrary soundings can be approximated by straight line segments. The area under such a sounding consists of the sum of areas within trapezoids under each line segment. Alternately, draw the early-morning sounding on graph paper, count the number of little grid boxes under the sounding, and multiply by the area of each box.

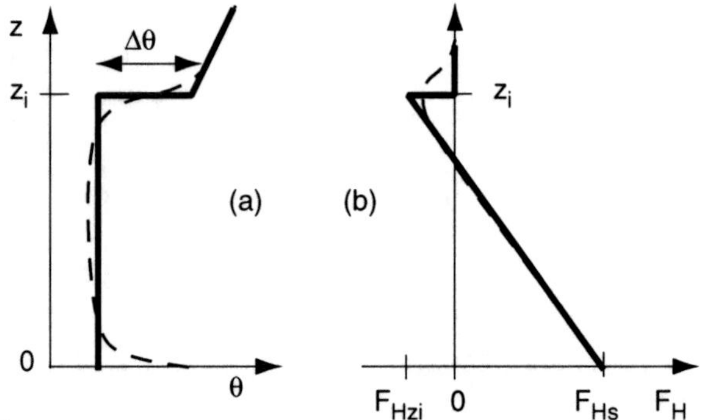

Figure 18.17
(a) Slab idealization of the mixed layer (solid line) approximates a more realistic potential temperature (θ) profile (dashed line). (b) Corresponding real and idealized heat flux (F_H) profiles.

Sample Application
The mixed layer depth increases at the rate of 300 m h^{-1} in a region with weak subsidence of 2 mm s^{-1}. If the inversion strength is 0.5°C, estimate the surface kinematic heat flux.

Find the Answer
Given: w_s= –0.002 m s^{-1}, $\Delta z_i/\Delta t$= 0.083 m s^{-1}, Δθ = 0.5°C
Find: $F_{H\,sfc}$ = ? °C·m s^{-1}

First, solve eq. (18.6) for entrainment velocity:
$w_e = \Delta z_i/\Delta t - w_s$ = [0.083 – (–0.002)] m s^{-1} = 0.085 m s^{-1}

Next, use eq. (18.7):
F_{Hzi} = (–0.085 m s^{-1})·(0.5°C) = –0.0425 °C·m s^{-1}

Finally, rearrange eq. (18.8):
$F_{H\,sfc}$= $-F_{Hzi}/A$= –(–0.0425 °C·m s^{-1})/0.2 = **0.21°C·m** s^{-1}

Check: Units OK. Magnitude and sign OK.
Exposition: Multiplying by ρ·C_p for the bottom of the atmosphere gives a dynamic heat flux of $\mathbb{F}_H$ = 259 W m^{-2}. This is roughly half of the value of max incoming solar radiation from Fig. 2.13 in the Solar & IR Radiation chapter.

Sample Application
Find the entrainment velocity for a potential temperature jump of 2°C at the EZ, and a surface kinematic heat flux of 0.2 K·m s^{-1}.

Find the Answer
Given: $F_{H\,sfc}$ = 0.2 K·m s^{-1}, Δθ = 2 °C = 2 K
Find: w_e = ? m s^{-1}

Use eq. (18.9):

$$w_e \cong \frac{A \cdot F_{H\,sfc}}{\Delta\theta} = \frac{0.2 \cdot (0.2\,\mathrm{K \cdot m/s})}{(2\,\mathrm{K})} = \mathbf{0.02\ m}\ \mathrm{s^{-1}}$$

Check: Units OK. Physics OK.
Exposition: While 2 cm s^{-1} seems small, when applied over 12 h of daylight works out to z_i = 864 m, which is a reasonable mixed-layer depth.

The rate of growth of the mixed layer during fair weather is

$$\frac{\Delta z_i}{\Delta t} = w_e + w_s \qquad \bullet(18.6)$$

where w_s is the synoptic scale vertical velocity, and is negative for the subsidence that is typical during fair weather (recall Fig. 18.5).

The kinematic heat flux at the top of the mixed layer is

$$F_{Hzi} = -w_e \cdot \Delta\theta \qquad \bullet(18.7)$$

where the sign and magnitude of the temperature jump is defined by Δθ = θ(just above z_i) – θ(just below z_i). Greater entrainment across stronger temperature inversions causes greater heat-flux magnitude. Similar relationships describe entrainment fluxes of moisture, pollutants, and momentum as a function of jump of humidity, pollution concentration, or wind speed, respectively. As for temperature, the jump is defined as the value above z_i minus the value below z_i. Entrainment velocity has the same value for all variables.

During free convection (when winds are weak and thermal convection is strong), the entrained kinematic heat flux is approximately 20% of the surface heat flux:

$$F_{H\,zi} \cong -A \cdot F_{H\,sfc} \qquad (18.8)$$

where A = 0.2 is called the **Ball ratio**, and $F_{H\,sfc}$ is the surface kinematic heat flux. This special ratio works only for heat, and does not apply to other variables. During windier conditions, A can be greater than 0.2. Eq. (18.8) was used in the Heat chapter to get the vertical heat-flux divergence (eq. 3.40 & 3.41), a term in the Eulerian heat budget.

Combining the two equations above gives an approximation for the entrainment velocity during free convection:

$$w_e \cong \frac{A \cdot F_{H\,sfc}}{\Delta\theta} \qquad \bullet(18.9)$$

Combining this with eq. (18.6) gives a mixed-layer growth equation called the **flux-ratio method**. From these equations, we see that stronger capping inversions cause slower growth rate of the mixed layer, while greater surface heat flux (e.g., sunny day over land) causes faster growth. The flux-ratio method and the thermodynamic methods usually give equivalent results for mixed layer growth during free convection.

For stormy conditions near thunderstorms or fronts, the ABL top is roughly at the tropopause. Alternately, in the Atmos. Forces & Winds chapter (Mass Conservation section) are estimates of z_i for bad weather.

WIND

For any given weather condition, there is a theoretical equilibrium wind speed, called the **geostrophic wind** *G*, that can be calculated for frictionless conditions (see the Atmos. Forces & Winds chapter). However, steady-state winds in the ABL are usually slower than geostrophic (i.e., **subgeostrophic**) because of frictional and turbulent drag of the air against the surface, as was illustrated in Fig. 18.9a.

Turbulence continuously mixes slower air from close to the ground with faster air from the rest of the ABL, causing the whole ABL to experience drag against the surface and to be subgeostrophic. This vertically averaged steady-state ABL wind M_{BL} is derived in the Atmos. Forces & Winds chapter. The actual ABL winds are nearly equal to this theoretical M_{BL} speed over a large middle region of the ABL.

Winds closer to the surface (in the **surface layer, SL**) are even slower (Fig. 18.9a). Wind-profile shapes in the SL are empirically found to be similar to each other when scaled with appropriate length and velocity scales. This approach, called **similarity theory**, is described later in this section.

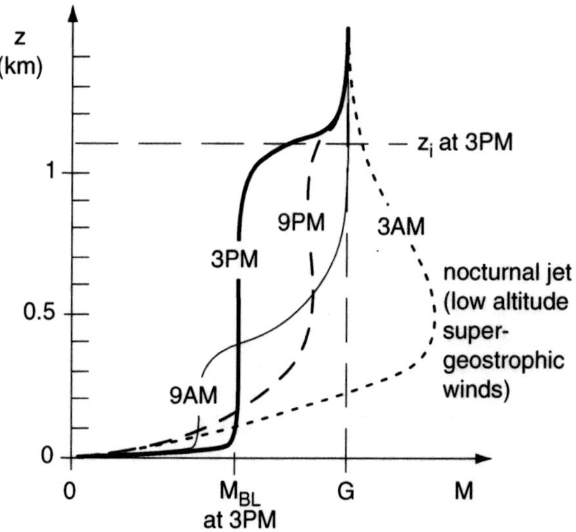

Figure 18.18
Typical ABL wind-profile evolution during fair weather over land. G is geostrophic wind speed. M_{BL} is average ABL wind speed and z_i is the average mixed-layer depth at 3 PM local time. The region of **supergeostrophic** *(faster-than-geostrophic: M > G) winds is called a nocturnal jet.*

Wind Profile Evolution

Over land during fair weather, the winds often experience a diurnal cycle as illustrated in Fig. 18.18. For example, a few hours after sunrise, say at 9 AM local time, there is often a shallow mixed layer, which is 300 m thick in this example. Within this shallow mixed layer the ABL winds are uniform with height, except near the surface where winds approach zero.

As the day progresses, the mixed layer deepens, so by 3 PM a deep layer of subgeostrophic winds fills the ABL. Winds remain moderate near the ground as turbulence mixes down faster winds from higher in the ABL. After sunset, turbulence intensity usually diminishes, allowing surface drag to reduce the winds at ground level. However, without turbulence, the air in the mid-ABL no longer feels drag against the surface, and begins to accelerate.

By 3 AM, the winds a few hundred meters above ground can be supergeostrophic, even though the winds at the surface might be calm. This low-altitude region of supergeostrophic winds is called a **nocturnal jet**. This jet can cause rapid horizontal transport of pollutants, and can feed moisture into thunderstorms. Then, after sunrise, turbulence begins vertical mixing again, and mixes out the jet with the slower air closer to the ground.

For measurements made at fixed heights on a very tall tower, the same wind-speed evolution is shown in Fig. 18.19. Below 20 m altitude, winds are

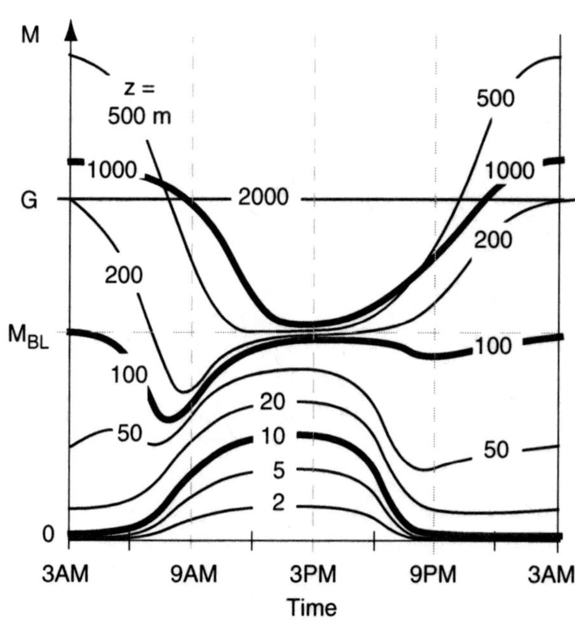

Figure 18.19
Typical ABL wind speed evolution at different heights. G is geostrophic wind speed, M_{BL} is average ABL wind, and the vertical time lines correspond to the profiles of Fig. 18.18.

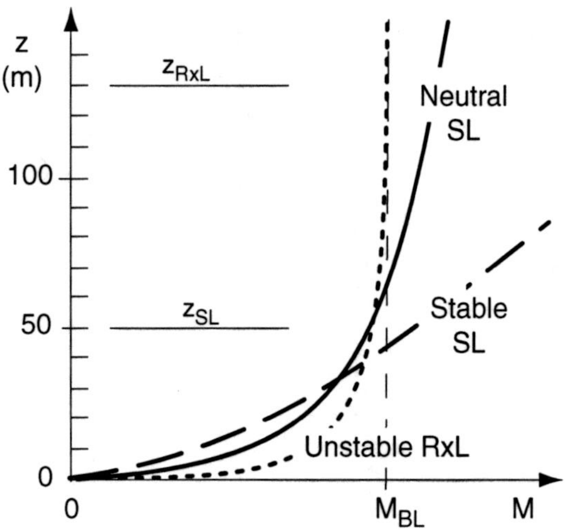

Figure 18.20
Typical wind speed profiles in the surface layer (SL, bottom 5% of the ABL) and radix layer (RxL, bottom 20% of the ABL), for different static stabilities. z_{RxL} and z_{SL} give order-of-magnitude depths for the radix layer and surface layer.

Table 18-1. The Davenport-Wieringa roughness-length z_o (m) classification, with approximate drag coefficients C_D (dimensionless).

z_o (m)	Classification	C_D	Landscape
0.0002	sea	0.0014	sea, paved areas, snow-covered flat plain, tide flat, smooth desert
0.005	smooth	0.0028	beaches, pack ice, morass, snow-covered fields
0.03	open	0.0047	grass prairie or farm fields, tundra, airports, heather
0.1	roughly open	0.0075	cultivated area with low crops & occasional obstacles (single bushes)
0.25	rough	0.012	high crops, crops of varied height, scattered obstacles such as trees or hedgerows, vineyards
0.5	very rough	0.018	mixed farm fields and forest clumps, orchards, scattered buildings
1.0	closed	0.030	regular coverage with large size obstacles with open spaces roughly equal to obstacle heights, suburban houses, villages, mature forests
≥ 2	chaotic	≥0.062	centers of large towns and cities, irregular forests with scattered clearings

often calmer at night, and increase in speed during daytime. The converse is true above 1000 m altitude, where winds are reduced during the day because of turbulent mixing with slower near-surface air, but become faster at night when turbulence decays.

At the ABL bottom, near-surface wind speed profiles have been found **empirically** (i.e., experimentally). In the bottom 5% of the statically neutral ABL is the **surface layer**, where wind speeds increase roughly logarithmically with height (Fig. 18.20).

For the statically stable surface layer, this logarithmic profile changes to a more linear form (Fig. 18.20). Winds close to the ground become slower than logarithmic, or near calm. Winds just above the surface layer are often not in steady state, and can temporarily increase to faster than geostrophic (**supergeostrophic**) in a process called an **inertial oscillation** (Figs. 18.9b and 18.20).

The bottom 20% of the convective (unstable) ABL is called the **radix layer** (RxL). Winds in the RxL have an exponential power-law relationship with height. The RxL has faster winds near the surface, but slower winds aloft than the neutral logarithmic profile. After a discussion of drag at the ground, these three wind cases at the bottom of the ABL will be described in more detail.

Drag, Stress, Friction Velocity, and Roughness Length

The frictional force between two objects such as the air and the ground is called **drag**. One way to quantify drag is by measuring the force required to push the object along another surface. For example, if you place your textbook on a flat desk, after you first start it moving you must continue to push it with a certain force (i.e., equal and opposite to the drag force) to keep it moving. If you stop pushing, the book stops moving.

Your book contacts the desk with a certain surface area. Generally, larger contact area requires greater force to overcome friction. The amount of friction force per unit surface contact area is called **stress**, τ, where for stress the force is parallel to the area. Contrast this with pressure, which is defined as a force per unit area that is perpendicular to the area. Units of stress are N m^{-2}, and could also be expressed as Pascals (Pa) or kiloPascals (kPa).

Stress is felt by both objects that are sliding against each other. For example, if you stack two books on top of each other, then there is friction between both books, as well as between the bottom book and the table. In order to push the bottom book in one direction without moving the top book, you must apply a force to the top book in the opposite direction as the bottom book.

Think of air within the ABL as a stack of layers of air, much like a stack of books. Each layer feels stress from the layers above and below it. The bottom layer feels stress against the ground, as well as from the layer of air above. In turn, the surface tends to be pushed along by air drag. Over the ocean, this **wind stress** drives the **ocean currents**.

In the atmosphere, stress caused by turbulent motions is many orders of magnitude greater than stress caused by molecular viscosity. For that reason, we often speak of **turbulent stress** instead of frictional stress, and **turbulent drag** rather than frictional drag. This turbulent stress is also called a **Reynolds stress**, after Osborne Reynolds who related this stress to turbulent gust velocities in the late 1800s.

Because air is a fluid, it is often easier to study the stress per unit density ρ of air. This is called the **kinematic stress**. The kinematic stress against the Earth's surface is given the symbol u_*^2, where u_* is called the **friction velocity**:

$$u_*^2 = |\tau / \rho| \qquad \bullet(18.10)$$

Typical values range from $u_* = 0$ during calm winds to $u_* = 1$ m s^{-1} during strong winds. Moderate-wind values are often near $u_* = 0.5$ m s^{-1}.

For fluid flow, turbulent stress is proportional to wind speed squared. Also stress is greater over rougher surfaces. A dimensionless **drag coefficient** C_D relates the kinematic stress to the wind speed M_{10} at z = 10 m.

$$u_*^2 = C_D \cdot M_{10}^2 \qquad \bullet(18.11)$$

The drag coefficient ranges from C_D = 2x10^{-3} over smooth surfaces to 2x10^{-2} over rough or forested surfaces (Table 18-1). It is similar to the bulk heat-transfer coefficient of the Heat chapter.

The surface roughness is usually quantified as an **aerodynamic roughness length** z_o. Table 18-1 shows typical values of the roughness length for various surfaces. Rougher surfaces such as sparse forests have greater values of roughness length than smoother surfaces such as a frozen lake. Roughness lengths in this table are not equal to the heights of the houses, trees, or other roughness elements.

For statically neutral air, there is a relationship between drag coefficient and aerodynamic roughness length:

$$C_D = \frac{k^2}{\ln^2(z_R / z_o)} \qquad (18.12)$$

where k = 0.4 is the **von Kármán constant**, and z_R = 10 m is a reference height defined as the standard

Sample Application

Find the drag coefficient in statically neutral conditions to be used with standard surface winds of 5 m s^{-1}, over (a) villages, and (b) grass prairie. Also, find the friction velocity and surface stress.

Find the Answer

Given: z_R = 10 m for "standard" winds
Find: C_D = ? (dimensionless), u_* = ? m s^{-1}, τ = ? N m^{-2}

Use Table 18-1:
(a) z_o = 1 m for villages. (b) z_o = 0.03 m for prairie

Use eq. (18.12) for drag coefficient:

(a) $C_D = \frac{0.4^2}{\ln^2(10\text{m} / 1\text{m})}$ = **0.030** (dimensionless)

(b) $C_D = \frac{0.4^2}{\ln^2(10\text{m} / 0.03\text{m})}$ = **0.0047** (dimensionless)

Use eq. (18.11) for friction velocity:
(a) $u_*^2 = C_D \cdot M_{10}^2 = 0.03 \cdot (5\text{m s}^{-1})^2 = 0.75 \text{ m}^2 \cdot \text{s}^{-2}$
Thus u_* = **0.87 m** s^{-1}.

(b) Similarly, u_* = **0.34 m** s^{-1}.

Use eq. (18.10) for surface stress, and assume ρ = 1.2 kg m^{-3}:

(a) $\tau = \rho \cdot u_*^2 = (1.2 \text{ kg m}^{-3}) \cdot (0.75\text{m}^2 \text{ s}^{-2}) =$
$\tau = 0.9 \text{ kg} \cdot \text{m}^{-1} \cdot \text{s}^{-2}$ = **0.9 Pa** (using Appendix A)

(b) $\tau = 0.14 \text{ kg} \cdot \text{m}^{-1} \cdot \text{s}^{-2}$ = **0.14 Pa**

Check: Units OK. Physics OK.
Exposition: The drag coefficient, friction velocity, and stress are smaller over smoother surfaces.

In this development we examined the stress for fixed wind speed and roughness. However, in nature, greater roughness and greater surface drag causes slower winds (see the Atmos. Forces & Winds chapter).

Sample Application

If the wind speed is 20 m s^{-1} at 10 m height over an orchard, find the friction velocity.

Find the Answer

Given: M_{10} = 20 m s^{-1} at z_R = 10 m, z_o = 0.5 m
Find: u_* = ? m s^{-1}

Use eq. (18.13):
u_* = (0.4)·(20 m s^{-1})/ln(10m/0.5m) = **2.67 m** s^{-1}

Check: Units OK. Magnitude OK.
Exposition: This corresponds to a large stress on the trees, which could make the branches violently move, causing some fruit to fall.

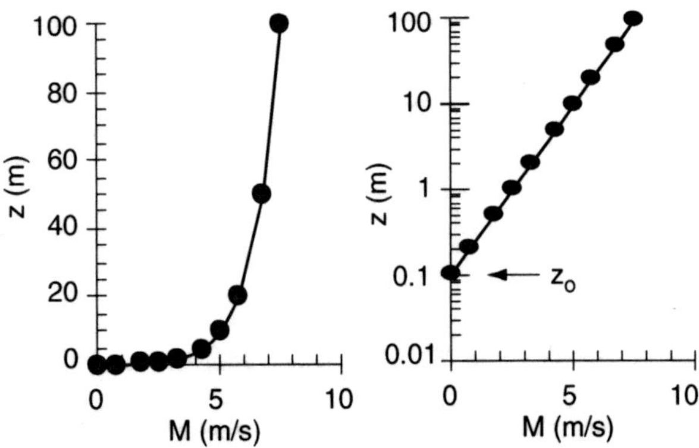

Figure 18.21
Wind-speed (M) profile in the statically-neutral surface layer, for a roughness length of 0.1 m. (a) linear plot, (b) semi-log plot.

Sample Application

On an overcast day, a wind speed of 5 m s^{-1} is measured with an anemometer located 10 m above ground within an orchard. What is the wind speed at the top of a 25 m smoke stack?

Find the Answer

Given: M_1 = 5 m s^{-1} at z_1 = 10 m
Neutral stability (because overcast)
z_o = 0.5 m from Table 18-1 for an orchard
Find: M_2 = ? m s^{-1} at z_2= 25 m

Sketch:

z
M_2
z_2
M_1
anemometer
z_1
0

Use:
eq. (18.14b):

$$M_2 = 5(\text{m/s})\cdot\frac{\ln(25\text{m}/0.5\text{m})}{\ln(10\text{m}/0.5\text{m})} = \mathbf{6.53\ m}\ \text{s}^{-1}$$

Check: Units OK. Physics OK. Sketch OK.
Exposition: Hopefully the anemometer is situated far enough from the smoke stack to measure the true undisturbed wind.

anemometer height for measuring **"surface winds"**. The drag coefficient decreases as the air becomes more statically stable. For unstable air, roughness is less important, and alternative approaches are given in the Heat chapter and the Atmos. Forces & Winds chapter.

Combining the previous two equations gives an expression for friction velocity in terms of surface wind speed and roughness length:

$$u_* = \frac{k \cdot M_{10}}{\ln[z_R / z_o]} \tag{18.13}$$

The physical interpretation is that faster winds over rougher surfaces causes greater kinematic stress.

Log Profile in the Neutral Surface Layer

Wind speed M is zero at the ground (more precisely, at a height equal to the aerodynamic roughness length). Speed increases roughly logarithmically with height in the statically-neutral surface layer (bottom 50 to 100 m of the ABL), but the shape of this profile depends on the surface roughness:

$$M(z) = \frac{u_*}{k}\ln\left(\frac{z}{z_o}\right) \qquad \bullet\text{(18.14a)}$$

Alternately, if you know wind speed M_1 at height z_1, then you can calculate wind speed M_2 at any other height z_2 :

$$M_2 = M_1 \cdot \frac{\ln(z_2 / z_o)}{\ln(z_1 / z_o)} \tag{18.14b}$$

Many weather stations measure the wind speed at the standard height z_1 = 10 m.

An example of the **log wind profile** is plotted in Fig. 18.21. A perfectly logarithmic wind profile (i.e., eq. 18.14) would be expected only for **neutral** static stability (e.g., overcast and windy) over a uniform surface. For other static stabilities, the wind profile varies slightly from logarithmic.

On a semi-log graph, the log wind profile would appear as a straight line. You can determine the roughness length by measuring the wind speeds at two or more heights, and then extrapolating the straight line in a semi-log graph to zero wind speed. The z-axis intercept gives the roughness length.

Log-Linear Profile in Stable Surf. Layer

During statically stable conditions, such as at nighttime over land, wind speed is slower near the ground, but faster aloft than that given by a logarithmic profile. This profile in the surface layer is empirically described by a **log-linear profile** formula with both a logarithmic and a linear term in z:

$$M(z) = \frac{u_*}{k}\left[\ln\left(\frac{z}{z_o}\right) + 6\frac{z}{L}\right] \qquad \bullet(18.15)$$

where M is wind speed at height z, $k = 0.4$ is the von Kármán constant, z_o is the aerodynamic roughness length, and u_* is friction velocity. As height increases, the linear term dominates over the logarithmic term, as sketched in Fig. 18.20.

An **Obukhov length** L is defined as:

$$L = \frac{-u_*^3}{k \cdot (|g| / T_v) \cdot F_{Hsfc}} \qquad \bullet(18.16)$$

where $|g| = 9.8$ m s^{-2} is gravitational acceleration magnitude, T_v is the absolute virtual temperature, and F_{Hsfc} is the kinematic surface heat flux. L has units of m, and is positive during statically stable conditions (because F_{Hsfc} is negative then). The Obukhov length can be interpreted as the height in the stable surface layer below which shear production of turbulence exceeds buoyant consumption.

Profile in the Convective Radix Layer

For statically unstable ABLs with vigorous convective thermals, such as occur on sunny days over land, wind speed becomes uniform with height a short distance above the ground. Between that uniform wind-speed layer and the ground is the **radix layer** (RxL). The wind speed profile in the radix layer is:

$$M(z) = M_{BL} \cdot \left(\zeta_*^D\right)^A \cdot \exp\left[A \cdot \left(1 - \zeta_*^D\right)\right] \quad \text{for } 0 \le \zeta_* \le 1.0 \qquad \bullet(18.17a)$$

and

$$M(z) = M_{BL} \quad \text{for } 1.0 \le \zeta_* \qquad (18.17b)$$

where $\zeta_* = 1$ defines the top of the radix layer. In the bottom of the RxL, wind speed increases faster with height than given by the log wind profile for the neutral surface layer, but becomes tangent to the uniform winds M_{BL} in the mid-mixed layer (Fig. 18.20).

The dimensionless height in the eqs. above is

$$\zeta_* = \frac{1}{C} \cdot \frac{z}{z_i} \cdot \left(\frac{w_*}{u_*}\right)^B \qquad (18.18)$$

where w_* is the **Deardorff velocity**, and the empirical coefficients are $A = 1/4$, $B = 3/4$, and $C = 1/2$. $D = 1/2$ over flat terrain, but increases to near $D = 1.0$ over hilly terrain.

The Deardorff velocity (eq. 3.39) is copied here:

$$w_* = \left[\frac{|g|}{T_v} \cdot z_i \cdot F_{Hsfc}\right]^{1/3} \qquad \bullet(18.19a)$$

Sample Application (§)

For a friction velocity of 0.3 m s^{-1}, aerodynamic roughness length of 0.02 m, average virtual temperature of 300 K, and kinematic surface heat flux of –0.05 K·m s^{-1} at night, plot the wind-speed profile in the surface layer. (Compare profiles for statically stable and neutral conditions.)

Find the Answer

Given: $u_* = 0.3$ m s^{-1}, $z_o = 0.02$ m, $T_v = 300$ K, $F_{Hsfc} = -0.05$ K·m s^{-1}

Find: $M(z) = ?$ m s^{-1}

Use eq. (18.16):

$L = -(0.3\text{m s}^{-1})^3/[0.4\cdot(9.8\text{m}\cdot\text{s}^{-2})\cdot(-0.05\text{K}\cdot\text{m s}^{-1})/(300\text{K})]$

= **41.3 m**

Use eq. (18.14a) for M in a neutral surface layer.
For example, at $z = 50$ m:
$M = [(0.3\text{m s}^{-1})/0.4] \cdot \ln(50\text{m}/0.02\text{m})$ = **5.9 m** s^{-1}

Use eq. (18.15) for M in a stable surface layer.
For example, at $z = 50$ m: $M = [(0.3\text{m s}^{-1})/0.4] \cdot [\ln(50\text{m}/0.02\text{m}) + 6\cdot(50\text{m}/41.3\text{m})]$ = **11.3 m** s^{-1}

Use a spreadsheet to find M at the other heights:

z (m)	M(m s^{-1})$_{neutral}$	M (m s^{-1})$_{stable}$
0.02	0.0	0.0
0.05	0.7	0.7
0.1	1.2	1.2
0.2	1.7	1.7
0.5	2.4	2.5
1	2.9	3.0
2	3.5	3.7
5	4.1	4.7
10	4.7	5.7
20	5.2	7.4
50	5.9	11.3
100	6.4	17.3

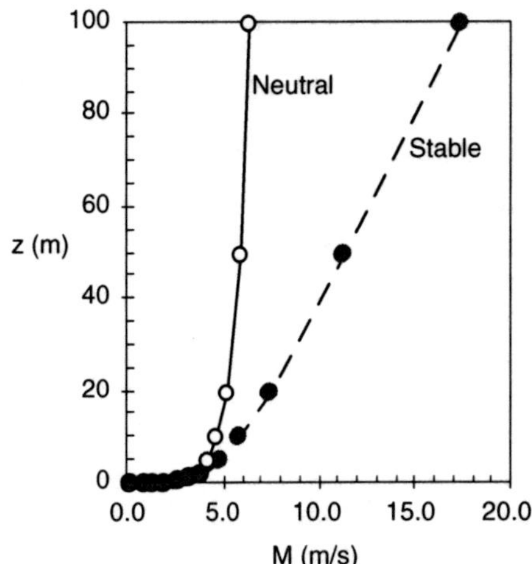

Check: Units OK. Physics OK. Plot OK.

Exposition: Open circles are for neutral, solid are for statically stable. The linear trend is obvious in the wind profile for the stable boundary layer.

Sample Application (§)

For a 1 km deep mixed layer with surface heat flux of 0.3 K·m s^{-1} and friction velocity of 0.2 m s^{-1}, plot the wind speed profile using a spreadsheet. Terrain is flat, and mid-ABL wind is 5 m s^{-1}.

Find the Answer

Given: F_{Hsfc} =0.3 K·m s^{-1}, u_* = 0.2 m s^{-1}, z_i = 1000 m, M_{BL} = 5 m s^{-1}, D = 0.5 .

Find: $M(z)$ = ? m s^{-1} .

The ABL is statically unstable, because F_{Hsfc} is positive. First, find w_* = ? m s^{-1} using eq. (18.19a).

Assume: $|g|/T_v$ = 0.0333 m·s^{-2}·K^{-1} (typical).

$$w_* = \left[\left(0.0333\frac{\text{m}}{\text{s}^2\text{K}}\right)\cdot(1000\text{m})\cdot\left(0.3\frac{\text{K}\cdot\text{m}}{\text{s}}\right)\right]^{1/3}$$

$$= (10 \text{ m}^3 \text{ s}^{-3})^{1/3} = 2.15 \text{ m s}^{-1}$$

Use eq. (18.18) in a spreadsheet to get ζ_* at each z, then use eq. (18.17) to get each M. For example, at z = 10 m:

ζ_* = 2·(10m/1000m)·[(2.15m s^{-1})/(0.2m s^{-1})]$^{3/4}$
= **0.1187** , &

M =(5m s^{-1})·(0.119$^{1/2}$)$^{1/4}$·exp[0.25·(1– 0.119$^{1/2}$)]
= **4.51m** s^{-1}

z (m)	ζ_*	M (m s^{-1})
0	0.000	0.00
0.1	0.001	2.74
0.2	0.002	2.98
0.5	0.006	3.32
1.0	0.012	3.59
2	0.024	3.87
5	0.059	4.24
10	0.119	4.51
15	0.178	4.66
20	0.237	4.75

etc.

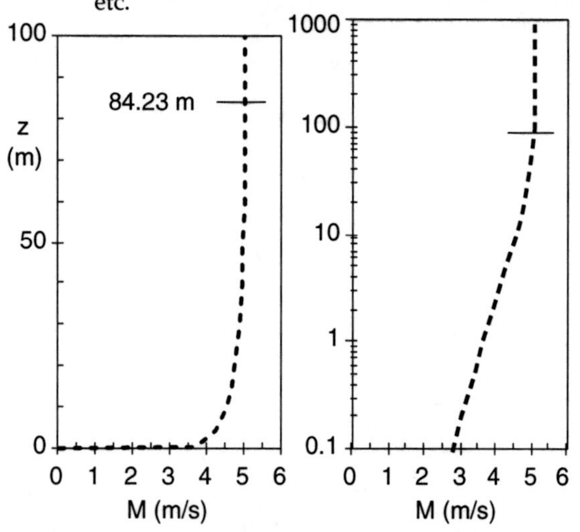

Check: Units OK. Physics OK. Sketch OK.

Exposition: This profile smoothly merges into the uniform wind speed in the mid-mixed layer, at height ζ_* = 1.0, which is at $z = C{\cdot}z_i{\cdot}(u_*/w_*)_B$ = 84.23 m from eq. (18.18).

where $|g|$ = 9.8 m s^{-2} is gravitational acceleration magnitude, T_v is absolute virtual temperature, z_i is depth of the ABL (= depth of the mixed layer), and F_{Hsfc} is the kinematic sensible heat flux (units of K·m s^{-1}) at the surface. Typical values of w_* are on the order of 1 m s^{-1}. The Deardorff velocity and buoyancy velocity w_B (defined in the Heat chapter) are both convective velocity scales for the statically unstable ABL, and are related by:

$$w_* \approx 0.08\, w_B \tag{18.19b}$$

To use eq. (18.17) you need to know the average wind speed in the middle of the mixed layer M_{BL}, as was sketched in Fig. 18.9. The Atmos. Forces & Winds chapter shows how to estimate this if it is not known from measurements.

For both the free-convection radix layer and the forced-convection surface layer, turbulence transports momentum, which controls wind-profile shape, which then determines the shear (Fig. 18.22). However, differences between the radix layer and surface layer are caused by differences in feedback.

In the neutral surface layer (Fig. 18.22a) there is strong feedback because wind shear generates the

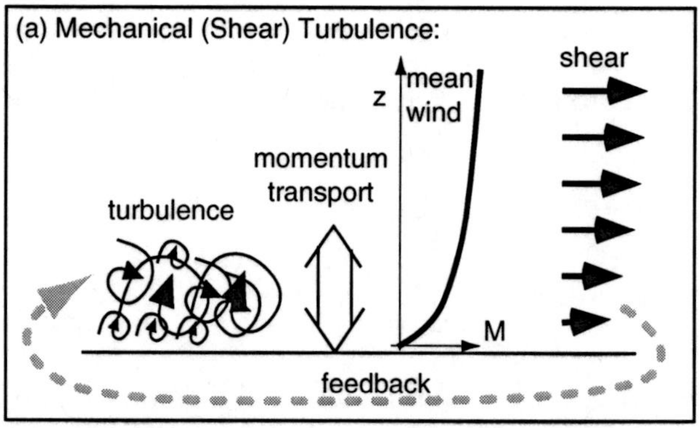

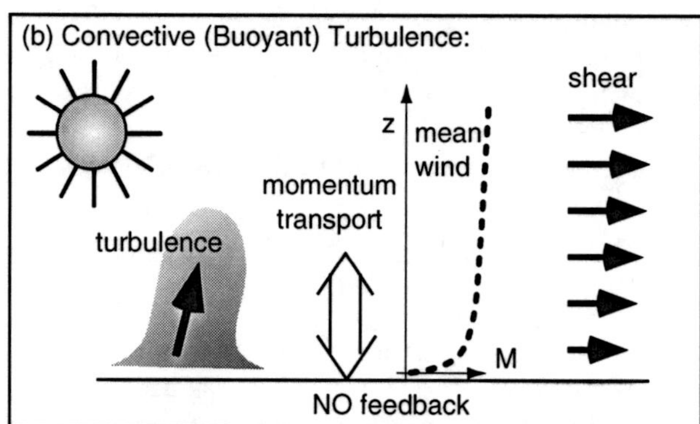

Figure 18.22

(a) Processes important for the log-wind profile in the surface layer dominated by mechanical turbulence (forced convection; neutral stability). (b) Processes important for the radix-layer wind profile during convective turbulence (free convection; statically unstable).

turbulence, which in turn controls the wind shear. However, such feedback is broken for convective turbulence (Fig. 18.22b), because turbulence is generated primarily by buoyant thermals, not by shear.

TURBULENCE

Mean and Turbulent Parts

Wind can be quite variable. The total wind speed is the superposition of three types of flow:

mean wind – relatively constant, but varying slowly over the course of hours

waves – regular (linear) oscillations of wind, often with periods of ten minutes or longer

turbulence – irregular, quasi-random, non-linear variations or gusts, with durations of seconds to minutes

These flows can occur individually, or in any combination. Waves are discussed in the Regional Winds chapter. Here, we focus on mean wind and turbulence.

Let $U(t)$ be the x-direction component of wind at some instant in time, t. Different values of $U(t)$ can occur at different times, if the wind is variable. By averaging the **instantaneous wind** measurements over a time period, P, we can define a **mean wind** $\overline{U}$, where the overbar denotes an average. This mean wind can be subtracted from the instantaneous wind to give the **turbulence** or gust part u' (Fig. 18.23).

Similar definitions exist for the other wind components (U, V, W), temperature (T) and humidity (r):

$$u'(t) = U(t) - \overline{U} \tag{18.20a}$$

$$v'(t) = V(t) - \overline{V} \tag{18.20b}$$

$$w'(t) = W(t) - \overline{W} \tag{18.20c}$$

$$T'(t) = T(t) - \overline{T} \tag{18.20d}$$

$$r'(t) = r(t) - \overline{r} \tag{18.20e}$$

Thus, the wind can be considered as a sum of mean and turbulent parts (neglecting waves for now).

The averages in eq. (18.20) are defined over time or over horizontal distance. For example, the mean temperature is the sum of all individual temperature measurements, divided by the total number N of data points:

$$\overline{T} = \frac{1}{N}\sum_{k=1}^{N} T_k \qquad \bullet(18.21)$$

Science Graffito

"I am an old man now, and when I die and go to heaven there are two matters on which I hope for enlightenment. One is quantum electrodynamics, and the other is the turbulent motion of fluids. And about the former I am rather optimistic."
– *Sir Horace Lamb* (1932)

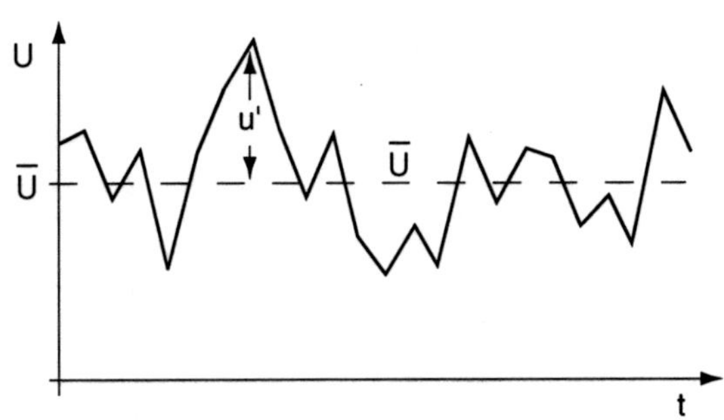

Figure 18.23
The instantaneous wind speed U shown by the zigzag line. The average wind speed $\overline{U}$ is shown by the thin horizontal dashed line. A gust velocity u' is the instantaneous deviation of the instantaneous wind from the average.

Sample Application

Given the following measurements of total instantaneous temperature, T, find the average $\overline{T}$. Also, find the T' values.

t (min)	T (°C)	t (min)	T (°C)
1	12	6	13
2	14	7	10
3	10	8	11
4	15	9	9
5	16	10	10

Find the Answer

As specified by eq. (18.21), adding the ten temperature values and dividing by ten gives the average $\overline{T}$ = 12.0°C. Subtracting this average from each instantaneous temperature gives:

t (min)	T '(°C)	t (min)	T '(°C)
1	0	6	1
2	2	7	–2
3	–2	8	–1
4	3	9	–3
5	4	10	–2

Check: The average of these T' values should be zero, by definition, useful for checking for mistakes.
Exposition: If a positive T' corresponds to a positive w', then warm air is moving up. This contributes positively to the heat flux.

where k is the data-point index (corresponding to different times or locations). The averaging time in eq. (18.21) is typically about 0.5 h. If you average over space, typical averaging distance is 50 to 100 km.

Short term fluctuations (described by the primed quantities) are associated with small-scale swirls of motion called **eddies**. The superposition of many such eddies of many sizes makes up the **turbulence** that is imbedded in the mean flow.

Molecular viscosity in the air causes friction between the eddies, tending to reduce the turbulence intensity. Thus, turbulence is NOT a conserved quantity, but is **dissipative**. Turbulence decays and disappears unless there are active processes to generate it. Two such production processes are **convection**, associated with warm air rising and cool air sinking, and **wind shear**, the change of wind speed or direction with height.

Normally, weather forecasts are made for mean conditions, not turbulence. Nevertheless, the net effects of turbulence on mean flow must be included. Idealized average turbulence effects are given in the chapters on Thermodynamics, Water Vapor, and Atmos. Forces and Winds.

Meteorologists use statistics to quantify the net effect of turbulence. Some statistics are described next. In this chapter we will continue to use the overbar to denote the mean conditions. However, we drop the overbar in most other chapters in this book to simplify the notation.

Sample Application (§)

(a) Given the following V-wind measurements. Find the mean wind speed, and standard deviation.
(b) If the standard deviation of vertical velocity is 1 m s^{-1}, is the flow isotropic?

t (h)	V (m s^{-1})
0.1	2
0.2	–1
0.3	1
0.4	1
0.5	–3
0.6	–2
0.7	0
0.8	2
0.9	–1
1.0	1

Find the Answer

Given: Velocities listed at right
$\sigma_w = 1$ m s^{-1}.
Find: $\overline{V}$ = ? m s^{-1}, σ_v = ? m s^{-1}, isotropy = ?

(a) Use eq. (18.21), except for V instead of T:

$$\overline{V}(z) = \frac{1}{n}\sum_{i=1}^{n} V_i(z) = \frac{1}{10}(0) \quad = \underline{\mathbf{0\ m}}\ \mathrm{s}^{-1}$$

Use eq. (18.22), but for V:

$$\sigma_v^{\ 2} = \frac{1}{n}\sum_{i=1}^{n}(V_i - \overline{V})^2$$

σ_v^2 =(1/10)·(4+1+1+1+9+4+0+4+1+1) = 2.6 m^2 s^{-2}

Finally, use eq. (18.23), but for v:

$$\sigma_v = \sqrt{2.6 m^2 \cdot s^{-2}} \quad = \underline{\mathbf{1.61\ m}}\ \mathrm{s}^{-1}$$

Check: Units OK. Physics OK.
Exposition: (b) Anisotropic, because $\sigma_v > \sigma_w$ (see next subsection). This means that an initially spherical smoke puff would become elliptical in cross section as it disperses more in the horizontal than the vertical.

Variance and Standard Deviation

The **variance** σ^2 of vertical velocity is an overall statistic of gustiness:

$$\sigma_w^{\ 2} = \frac{1}{N}\sum_{k=1}^{N}(W_k - \overline{W})^2 = \frac{1}{N}\sum_{k=1}^{N}(w_k{}')^2 = \overline{w'^{\,2}} \qquad \bullet(18.22)$$

Similar definitions can be made for $\sigma_u^{\ 2}$, $\sigma_v^{\ 2}$, $\sigma_\theta^{\ 2}$, etc. Statistically, these are called "biased" variances. Velocity variances can exist in all three directions, even if there is a mean wind in only one direction.

The **standard deviation** σ is defined as the square-root of the variance, and can be interpreted as an average gust (for velocity), or an average turbulent perturbation (for temperatures and humidities, etc.). For example, standard deviations for vertical velocity, σ_w, and potential temperature, σ_θ, are:

$$\sigma_w = \sqrt{\sigma_w^{\ 2}} = \overline{(w')^2}^{\,1/2} \qquad (18.23a)$$

$$\sigma_\theta = \sqrt{\sigma_\theta^{\,2}} = \overline{(\theta')^2}^{\,1/2} \tag{18.23b}$$

Larger variance or standard deviation of velocity means more intense turbulence.

For statically **stable** air, standard deviations in an ABL of depth h have been empirically found to vary with height z as:

$$\sigma_u = 2 \cdot u_* \cdot \left[1 - (z/h)\right]^{3/4} \tag{18.24a}$$

$$\sigma_v = 2.2 \cdot u_* \cdot \left[1 - (z/h)\right]^{3/4} \tag{18.24b}$$

$$\sigma_w = 1.73 \cdot u_* \cdot \left[1 - (z/h)\right]^{3/4} \tag{18.24c}$$

where u_* is friction velocity. These equations work when the stability is weak enough that turbulence is not suppressed altogether.

For statically **neutral** air:

$$\sigma_u = 2.5 \cdot u_* \cdot \exp(-1.5 \cdot z/h) \tag{18.25a}$$

$$\sigma_v = 1.6 \cdot u_* \cdot \left[1 - 0.5 \cdot (z/h)\right] \tag{18.25b}$$

$$\sigma_w = 1.25 \cdot u_* \cdot \left[1 - 0.5 \cdot (z/h)\right] \tag{18.25c}$$

For statically **unstable** air:

$$\sigma_u = 0.032 \cdot w_B \cdot \left(1 + \left[1 - (z/z_i)\right]^6\right) \tag{18.26a}$$

$$\sigma_v = 0.032 \cdot w_B \tag{18.26b}$$

$$\sigma_w = 0.11 \cdot w_B \cdot (z/z_i)^{1/3} \cdot \left[1 - 0.8 \cdot (z/z_i)\right] \tag{18.26c}$$

where z_i is the mixed-layer depth, w_B is buoyancy velocity (eq. 3.38 or 18.19b). These relationships are important for air-pollution dispersion, and are used in the Air Pollution chapter. These equations are valid only within the boundary layer (i.e., from z = 0 up to h or z_i).

Isotropy

If turbulence has nearly the same variance in all three directions, then turbulence is said to be **isotropic**. Namely:

$$\sigma_u^{\,2} = \sigma_v^{\,2} = \sigma_w^{\,2} \tag{•18.27}$$

Sample Application (§)

Plot the vertical profiles of standard deviation of vertical velocity for statically stable and neutral situations for which the friction velocity is 0.5 m s^{-1}. The boundary-layer depth is 300 m.

Find the Answer

Given: u_* = 0.5 m s^{-1}, h = 300 m

Find: σ_w = ? vs. z for neutral and stable conditions

Use eqs. (18.24c and 18.25c). For example, for z = 100m:

Stable:

σ_w = 1.73·(0.5m s^{-1})·[1–(100m/300m)]$^{0.75}$ = **0.638 m** s^{-1}

Neutral:

σ_w = 1.25·(0.5m s^{-1})·[1–0.5·(100m/300m) = **0.521 m** s^{-1}

Using a spreadsheet with a range of heights yields:

	sigma w (m/s)	
z (m)	Stable	Neut.
0.0	0.865	0.625
1.0	0.863	0.624
10	0.843	0.615
100	0.638	0.521
150	0.514	0.469
200	0.379	0.417
250	0.266	0.365
300	0.0	0.313

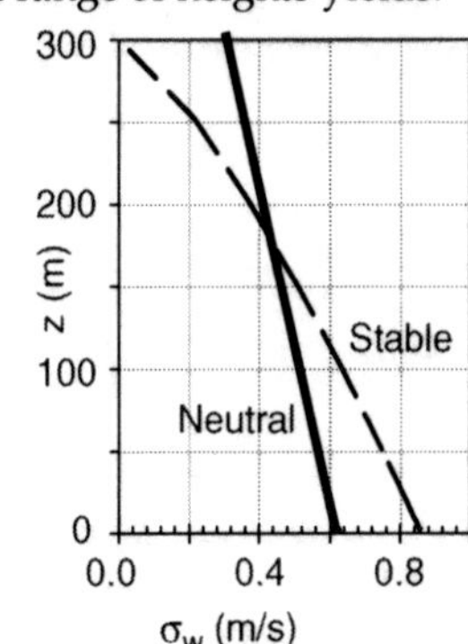

Check: Units OK. Magnitudes OK.

Exposition: Turbulence diminishes rapidly with increasing height for statically stable conditions.

Sample Application (§)

For a 1 km thick unstable ABL, compare the vertical profiles of σ_u & σ_w. Assume w_B = 30 m s^{-1}.

Find the Answer

Given: z_i = 1000 m , w_B = 30 m s^{-1}

Find: σ_u & σ_w vs. z .

Use eqs. (18.26a and 18.26c) in a spreadsheet, and plot.

	sigma (m/s)	
z (m)	σ_u	σ_w
0	0.865	0.625
50	0.863	0.624
100	0.843	0.615
200	0.638	0.521
300	0.514	0.469
500	0.379	0.417
800	0.266	0.365
1000	0.0	0.313

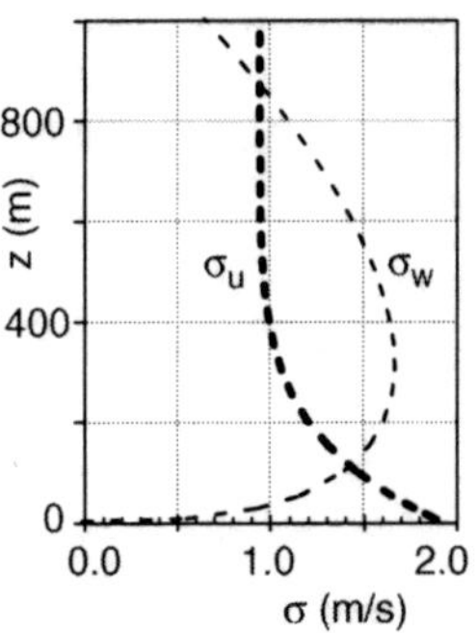

Check: Units OK. Magnitude reasonable.

Exposition: The flow is anisotropic at most heights. Near both the surface and z_i, horizontal turbulence is greater than vertical. But in the middle of the mixed layer, vertical turbulence is greater than horizontal.

Sample Application

Find u_*, the velocity standard deviations, and TKE in statically stable air at height 50 m in an ABL that is 200 m thick. Assume $C_D = 0.002$, and the winds at height 10 m are 5 m s^{-1}.

Find the Answer

Given: $z = 50$ m, $h = 200$ m, $C_D = 0.002$,
$M =$ 5 m s^{-1}. Statically stable.
Find: $u_*, \sigma_u, \sigma_v, \sigma_w$ = ? m s^{-1}. TKE = ? m^2 s^{-2}.

Use eq. (18.11):
$u_*^2 = 0.002 \cdot (5\text{m s}^{-1})^2 = 0.05$ m^2 s^{-2}. $u_* =$ **0.22 m** s^{-1}

Use eqs. (18.24a-c):
$\sigma_u = 2 \cdot (0.22\text{m s}^{-1}) \cdot [1-(50\text{m}/200\text{m})]^{3/4}$ = **0.35 m** s^{-1}

$\sigma_v = 2.2 \cdot (0.22\text{m s}^{-1}) \cdot [1-(50\text{m}/200\text{m})]^{3/4}$ = **0.39 m** s^{-1}

$\sigma_w = 1.73 \cdot (0.22\text{m s}^{-1}) \cdot [1-(50\text{m}/200\text{m})]^{3/4}$ = **0.31 m** s^{-1}

Use eq. (18.28b):

$TKE = 0.5 \cdot [0.35^2 + 0.39^2 + 0.31^2]$ = **0.185** m^2 s^{-2}.

Check: Units OK. Physics OK.
Exposition: In statically stable air, vertical turbulence is generally less than horizontal turbulence. Also, turbulence intensity increases with wind speed. If the atmosphere is too stable, then there will be no turbulence (see "dynamic stability" in the Atmospheric Stability chapter).

Science Graffito

"Big whirls have little whirls that feed on their velocity, and little whirls have lesser whirls and so on to viscosity – in the molecular sense."
– *L.F. Richardson*, 1922: "Weather Prediction by Numerical Process". p66.

[CAUTION: Do not confuse this word with "isentropic", which means adiabatic or constant entropy.]

Turbulence is **anisotropic** (not isotropic) in many situations. During the daytime over bare land, rising thermals create stronger vertical motions than horizontal. Hence, a smoke puff becomes **dispersed** (i.e., spread out) in the vertical faster than in the horizontal. At night, vertical motions are very weak, while horizontal motions can be larger. This causes smoke puffs to **fan** out horizontally with only little vertical dispersion in statically stable air.

Turbulence Kinetic Energy

An overall measure of the intensity of turbulence is the **turbulence kinetic energy** per unit mass (TKE):

$$TKE = 0.5 \cdot \left[\overline{(u')^2} + \overline{(v')^2} + \overline{(w')^2} \right] \qquad \bullet(18.28a)$$

$$TKE = 0.5 \cdot \left[\sigma_u^2 + \sigma_v^2 + \sigma_w^2 \right] \qquad \bullet(18.28b)$$

TKE is usually produced at the scale of the boundary-layer depth. The production is made mechanically by wind shear and buoyantly by thermals.

Turbulent energy cascades through the **inertial subrange**, where the large-size eddies drive medium ones, which in turn drive smaller eddies. Molecular viscosity continuously damps the tiniest (**microscale**) eddies, dissipating TKE into heat. TKE is not conserved.

The tendency of TKE to increase or decrease is given by the following TKE budget equation:

$$\frac{\Delta TKE}{\Delta t} = A + S + B + Tr - \varepsilon \qquad \bullet(18.29)$$

where A is advection of TKE by the mean wind, S is shear generation, B is buoyant production or consumption, Tr is transport by turbulent motions and pressure, and ε is viscous dissipation rate. For **stationary** (steady-state) turbulence, the tendency term on the left side of eq. (18.29) is zero.

Mean wind blows TKE from one location to another. The **advection** term is given by:

$$A = -U \cdot \frac{\Delta TKE}{\Delta x} - V \cdot \frac{\Delta TKE}{\Delta y} - W \cdot \frac{\Delta TKE}{\Delta z} \qquad (18.30)$$

Thus, turbulence can increase (or decrease) at any location if the wind is blowing in greater (or lesser) values of TKE from somewhere else.

Wind shear generates turbulence near the ground according to:

$$S = u_*^2 \cdot \frac{\Delta M}{\Delta z} \quad (18.31a)$$

in the surface layer, where u_* is the friction velocity, and $\Delta M/\Delta z$ is the wind shear. To good approximation for near-neutral static stability:

$$S \approx a \cdot M^3 \quad (18.31b)$$

where $a = 2\text{x}10^{-4}\ \text{m}^{-1}$ for wind speed M measured at a standard height of z = 10 m. Greater wind speeds near the ground cause greater wind shear, and generate more turbulence.

Buoyancy can either increase or decrease turbulence. When thermals are rising from a warm surface, they generate *TKE*. Conversely, when the ground is cold and the ABL is statically stable, buoyancy opposes vertical motion and consumes *TKE*.

The rate of **buoyant production or consumption** of *TKE* is:

$$B = \frac{|g|}{T_v} \cdot F_{H\,sfc} \quad (18.32)$$

where $|g|$ = 9.8 m·s^{-2} is gravitational acceleration magnitude, T_v is the absolute virtual air temperature near the ground, and $F_{H\,sfc}$ is the kinematic effective surface heat flux (positive when the ground is warmer than the air). Over land, $F_{H\,sfc}$ and B are usually positive during the daytime, and negative at night.

Turbulence can advect or **transport** itself. For example, if turbulence is produced by shear near the ground (in the **surface layer**), then turbulence motions will tend to move the excess *TKE* from the surface layer to locations higher in the ABL. Pressure fluctuations can have a similar effect, because turbulent pressure forces can generate turbulence motions. This pressure term is difficult to simplify, and will be grouped with the turbulent transport term, *Tr*, here.

Molecular viscosity dissipates turbulent motions into heat. The amount of heating is small, but the amount of damping of *TKE* is large. The **dissipation** is always a loss:

$$\varepsilon \approx \frac{(TKE)^{3/2}}{L_\varepsilon} \quad (18.33)$$

where $L_\varepsilon \approx 50$ m is a **dissipation length scale**.

The ratio of buoyancy to shear terms of the *TKE* equation is called the **flux Richardson number**, R_f:

$$R_f = \frac{-B}{S} \approx \frac{-(|g|/T_v) \cdot F_{H\,sfc}}{u_*^2 \cdot \frac{\Delta M}{\Delta z}} \quad (18.34a)$$

HIGHER MATH • Shear Generation

To get *TKE* shear-generation eq. (18.31b), start with eq. (18.31a):

$$S = u_*^2 \cdot (\partial M/\partial z) \quad (18.31a)$$

But

$$(\partial M/\partial z) = u_*/(k{\cdot}z) \quad \text{from eq. (18.14a).}$$

Thus,

$$S = u_*^3 / (k{\cdot}z)$$

But

$$u_*^2 = C_D \cdot M^2 \quad \text{from eq. (18.11)}$$

which gives

$$u_*^3 = C_D^{3/2} \cdot M^3$$

Thus:

$$S = [C_D^{3/2}/(k{\cdot}z)] \cdot M^3$$

or

$$S = a \cdot M^3 \quad (18.31b)$$

where

$$a = [C_D^{3/2}/(k{\cdot}z)]$$

For $C_D \approx 0.01$, k = 0.4 is von Kármáns constant, and z = 10 m, the result is $a = 2.5\text{x}10^{-4}\ \text{m}^{-1}$, but which can vary by an order of magnitude depending on the drag coefficient and height.

Sample Application

Assume steady state, and neglect advection and transport. What equilibrium TKE is expected in the surface layer with a mean wind of 5 m s^{-1} and surface heat flux of –0.02 K·m s^{-1}? The ambient temperature is 25°C, and the air is dry.

Find the Answer

Given: M = 5 m s^{-1}, $F_{H\,sfc}$ = –0.02 K·m s^{-1}, A = 0, Tr = 0, $\Delta TKE/\Delta t = 0$ for steady state, T_v = 298 K.

Find: TKE = ? m^2 s^{-2}.

Rearrange eq. (18.29).

$$\varepsilon = S + B$$

But ε depends on *TKE*, thus, we can rearrange eq. (18.33) to solve for $TKE = (L_\varepsilon \cdot [\varepsilon])^{2/3}$ and then plug in the eq. above:

$$TKE = \{ L_\varepsilon \cdot [S + B] \}^{2/3}$$

Use eqs. (18.31b and 18.32) to find S and B, and plug into the equation above:

$$TKE = \left\{ L_\varepsilon \cdot \left[a \cdot M^3 + (|g|/T_v) \cdot F_{H\,sfc} \right] \right\}^{2/3} = \left\{ (50\text{m}) \cdot \right.$$

$$\left. \left[\left(2 \times 10^{-4}\text{m}^{-1}\right)\left(5\frac{\text{m}}{\text{s}}\right)^3 + \frac{9.8\text{ms}^{-2}}{298\text{K}}(-0.02\text{Km/s}) \right] \right\}^{2/3}$$

$$= \{1.25 - 0.033\ \text{m}^3\,\text{s}^{-3}\}^{2/3} = \underline{\mathbf{1.14}}\ \text{m}^2\,\text{s}^{-2}$$

Check: Units OK. Physics OK.

Exposition: This turbulence intensity is weak, as is typical at night when heat fluxes are negative. Also, eq. (18.31b) for S is not accurate for statically stable conditions.

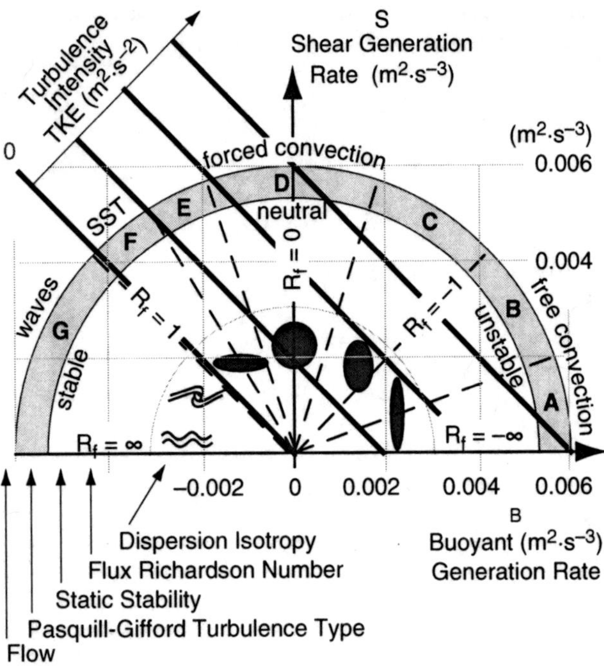

Figure 18.24
Rate of generation of TKE by buoyancy (abscissa) and shear (ordinate). Shape and rates of plume dispersion (dark spots or waves). Dashed lines separate sectors of different Pasquill-Gifford turbulence type (A - G). Isopleths of TKE intensity (dark diagonal lines). R_f is flux Richardson number. SST is stably-stratified turbulence.

Sample Application
For the previous Sample Application, determine the nature of convection (free, forced, etc.), the Pasquill-Gifford (PG) turbulence type, and the flux Richardson number. Assume no clouds.

Find the Answer
Given: (see previous Sample Application)
Find: $S = ?\ \mathrm{m^2\,s^{-3}}$, $B = ?\ \mathrm{m^2\,s^{-3}}$, $R_f = ?$, PG = ?

Use eq. (18.31b):

$$S \approx \left(2\times10^{-4}\mathrm{m^{-1}}\right)\left(5\frac{\mathrm{m}}{\mathrm{s}}\right)^3 = 0.025\ \mathrm{m^2\,s^{-3}}$$

Use eq. (18.32):

$$B = \frac{9.8\mathrm{ms^{-2}}}{298\mathrm{K}}(-0.02\mathrm{Km/s}) = -0.00066\ \mathrm{m^2\,s^{-3}}$$

Because the magnitude of B is less than a third of that of S, we conclude convection is <u>**forced**</u>.

Use eq. (18.34): $R_f = -(-0.00066) / 0.025 =$ <u>**0.0264**</u> and is dimensionless.

Use Fig. 18.24. Pasquill-Gifford Type = <u>**D**</u> (but on the borderline near E).

Check: Units OK. Physics OK.
Exposition: The type of turbulence is independent of the intensity. Intensity is proportional to S+B.

$$R_f \approx \frac{-(|g|/T_v)\cdot F_{H\,sfc}}{a\cdot M^3} \tag{18.34b}$$

with $a \approx 2\mathrm{x}10^{-4}\ \mathrm{m^{-1}}$. R_f is approximately equal to the **gradient or bulk Richardson number**, discussed in the Stability chapter. Generally, turbulence dies if $R_f > 1$.

Free and Forced Convection

The nature of turbulence, and therefore the nature of pollutant dispersion, changes with the relative magnitudes of terms in the *TKE* budget. Two terms of interest are the shear S and buoyancy B terms.

When $|B| < |S/3|$, the atmosphere is said to be in a state of **forced convection** (Fig. 18.24). These conditions are typical of windy overcast days, and are associated with near **neutral static stability**. Turbulence is nearly **isotropic**. Smoke plumes disperse at nearly equal rates in the vertical and lateral, which is called **coning**. The sign of B is not important here — only the magnitude.

When B is positive and $|B| > |3{\cdot}S|$, the atmosphere is said to be in a state of **free convection**. **Thermals** of warm rising air are typical in this situation, and the ABL is **statically unstable** (in the nonlocal sense; see the Stability chapter). These conditions often happen in the daytime over land, and during periods of cold-air advection over warmer surfaces. Turbulence is **anisotropic**, with more energy in the vertical, and smoke plumes loop up and down in a pattern called **looping**.

When B is negative and $|B| > |S|$, static stability is so strong that turbulence cannot exist. During these conditions, there is virtually no dispersion while the smoke blows downwind. **Buoyancy waves (gravity waves)** are possible, and appear as waves in the smoke plumes. For values of $|B| \approx |S|$, breaking **Kelvin-Helmholtz waves** can occur (see the Stability chapter), which cause some dispersion.

For B negative but $|B| < |S|$, weak turbulence is possible. These conditions can occur at night. This is sometimes called **stably-stratified turbulence (SST)**. Vertical dispersion is much weaker than lateral, causing an **anisotropic** condition where smoke spreads horizontally more than vertically, in a process called **fanning**.

Fig. 18.24 shows the relationship between different types of convection and the terms of the *TKE* equation. While the ratio of B/S determines the nature of convection, the sum $S + B$ determines the intensity of turbulence. A Pasquill-Gifford turbulence type (Fig. 18.24) can also be defined from the relative magnitudes of S and B, and is used in the Air Pollution to help estimate pollution dispersion rates.

Turbulent Fluxes and Covariances

Rewrite eq. (18.22) for variance of w as

$$\text{var}(w) = \frac{1}{N}\sum_{k=1}^{N}(W_k - \overline{W})\cdot(W_k - \overline{W}) \qquad (18.35)$$

By analogy, a **covariance** between vertical velocity w and potential temperature θ can be defined as:

$$\text{covar}(w,\theta) = \frac{1}{N}\sum_{k=1}^{N}(W_k - \overline{W})\cdot(\theta_k - \overline{\theta})$$
$$= \frac{1}{N}\sum_{k=1}^{N}(w_k{}')\cdot(\theta_k{}') \qquad \bullet(18.36)$$
$$= \overline{w'\theta'}$$

where the overbar still denotes an average. Namely, one over N times the sum of N terms (see middle line of eq. 18.36) is the average of those items. Comparing eqs. (18.35) with (18.36), we see that variance is just the covariance between a variable and itself.

Covariance indicates the amount of common variation between two variables. It is positive where both variables increase or decrease together. Covariance is negative for opposite variation, such as when one variable increases while the other decreases. Covariance is zero if one variable is unrelated to the variation of the other.

The **correlation coefficient** $r_{a,b}$ is defined as the covariance between a and b normalized by the standard deviations of the two variables a and b. Using vertical velocity and potential temperature for illustration:

$$r_{w,\theta} = \frac{\overline{w'\theta'}}{\sigma_w \cdot \sigma_\theta} \qquad \bullet(18.37)$$

By normalized, we mean that $-1 \le r_{a,b} \le 1$. A correlation coefficient of +1 indicates a perfect correlation (both variables increase or decrease together proportionally), –1 indicates perfect opposite correlation, and zero indicates no correlation. Because it is normalized, $r_{a,b}$ gives no information on the absolute magnitudes of the variations.

In the ABL, many turbulent variables are correlated. For example, in the statically unstable ABL (Fig. 18.25a), parcels of warm air rise and while other cool parcels sink in convective circulations. Warm air ($\theta' = +$) going up ($w' = +$) gives a positive product [$(w'\theta')_{up} = +$]. Cool air ($\theta' = -$) going down ($w' = -$) also gives a positive product [$(w'\theta')_{down} = +$].

The average of those two products is also positive [$\overline{w'\theta'} = 0.5\cdot((w'\theta')_{up} + (w'\theta')_{down}) = +$]. The result gives positive correlation coefficients $r_{w,\theta}$ during free convection, which is typical during daytime.

Science Graffito

Seen on a bumper sticker:
"**Lottery**: A tax on people who are bad at math."

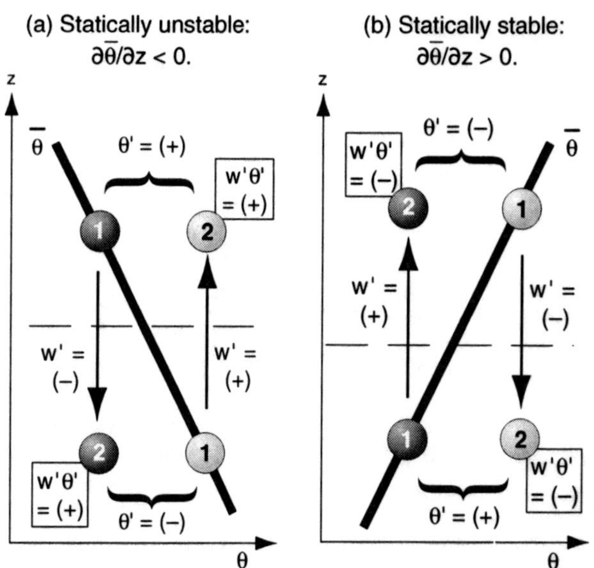

Figure 18.25
(a) Relationship between turbulent potential temperature θ and vertical velocity w for a statically unstable environment (e.g., daytime over land with clear skies) . (b) Same, but for a statically stable environment (e.g., nighttime over land with clear skies). In both figures, the thick line represents the ambient environment, circles represent air parcels, with light grey being the warm air parcel and dark gray being cool. Numbers 1 and 2 indicate starting and ending positions of each air parcel during time interval Δt.

Sample Application (§)

Fast-response measurements of potential temperature θ, water-vapor mixing ratio r, and u and w components of wind are given below as a function of time t. For θ and w, find their means, variances, and standard deviations. Also find the covariance, correlation coefficient, kinematic heat flux, and the heat flux (W m^{-2}).

Columns C and D are not used in this example, but will be used in some of the homeworks.

	A	B	C	D	E
1	Given:				
2	t (s)	θ (°C)	r (g/kg)	U (m/s)	W (m/s)
3	0	21	6.0	10	-5
4	0.1	28	9.5	6	4
5	0.2	29	10.0	7	3
6	0.3	25	8.0	3	4
7	0.4	22	6.5	5	0
8	0.5	28	9.5	15	-5
9	0.6	23	7.0	12	-1
10	0.7	26	8.5	16	-3
11	0.8	27	9.0	10	2
12	0.9	24	7.5	8	-4
13	1.0	21	6.0	14	-4
14	1.1	24	7.5	10	1
15	1.2	25	8.0	13	-2
16	1.3	27	9.0	5	3
17	1.4	29	10.0	7	5
18	1.5	22	6.5	11	2
19	1.6	30	10.5	2	6
20	1.7	23	7.0	15	-1
21	1.8	28	9.5	13	3
22	1.9	21	6.0	12	-3
23	2.0	22	6.5	16	-5
24	avg =	25			0

Find the Answer:

Given: Data above in rows 2 through 23.

Find: $\overline{W}$ =? m s^{-1}, $\overline{\theta}$ = ? °C, σ_w^2 = ? m^2 s^{-2}, σ_θ^2 = ? °C^2, σ_w =? m s^{-1}, σ_θ = ? °C, $\overline{w'\theta'}$ = ? K·m s^{-1}, $r_{w,\theta}$ = ? (dimensionless) F_H = ? K·m s^{-1}, $\mathbb{F}_H$ = ? W m^{-2} .

First, use eq. (18.21) to find the mean values. These answers are already shown in row 24 above.

$\overline{\theta}$ = **25 °C**, $\overline{W}$ = **0 m** s^{-1}

Next, use eqs. (18.20) to find the deviation from the mean, for each of the observations. The results are tabulated in columns G and H on the next page. Then square each of those perturbation (primed) values, as tabulated in columns I and J on the next page.

Use eq. (18.22), averaging the squared perturbations to give the variances (row 24, columns I and J):

σ_θ^2 = **8.48 °C^2**, σ_w^2 = **12.38** m^2 s^{-2}.

The square root of those answers (eq. 18.23) gives the standard deviations in row 25, columns I and J:

σ_θ = **2.91 °C**, σ_w = **3.52 m** s^{-1} .

(continues on next page)

Similarly, for statically stable conditions (Fig. 18.25b) where wind-shear-induced turbulence drives vertical motions against the restoring buoyant forces (see the Stability chapter), one finds cold air moving up, and warm air moving down. (Parcel warmness or coldness is measured relative to the ambient mean potential temperature $\overline{\theta}$ at the same ending height as the parcel.) This gives $\overline{w'\theta'}$ = – , which is often the case during night.

More important than the statistics are the physical processes they represent. Covariances represent fluxes. Look at the air parcels crossing the horizontal dashed line in Fig. 18.25a. Pretend that the dashed line is an edge view of a horizontal area that is 1 m^2.

During time interval Δt, the warm (light-grey shaded) air parcel moves warm air upward through that area in Fig. 18.25a. Heat flux is defined as heat moved per area per time. Thus this rising warm air parcel contributes to a positive heat flux. Similarly, the cold sinking air parcel (shaded dark grey) contributes to a positive heat flux through that area (because negative w' times negative θ′ is positive). Both parcels contribute to a positive heat flux.

This implies that covariance between vertical velocity and potential temperature is a turbulent kinematic heat flux, F_H [= $F_{z\ turb}(\theta)$ in the notation of the Heat chapter]:

$$\overline{w'\theta'} = F_H \qquad \bullet(18.38a)$$

Similarly, the covariance between vertical velocity w and water vapor mixing ratio, r (see the Water Vapor chapter), is a kinematic moisture flux, $F_{z\ turb}(r)$:

$$\overline{w'r'} = F_{z\ turb}(r) \qquad \bullet(18.38b)$$

Momentum flux is even more interesting. Recall from physics that **momentum** is mass times velocity. Units would be kg·m·s^{-1}. Therefore, momentum flux (momentum per area per time) would have units of (kg·m·s^{-1})·m^{-2}·s^{-1} = (kg·m·s^{-2})·m^{-2} = N·m^{-2}. Appendix A was used to find the equivalent units for a force of 1 Newton (N). But N·m^{-2} is a force per unit area, which is the definition of stress, τ. Thus, stress and momentum flux are physically the same in a fluid such as air.

A kinematic momentum flux is the momentum flux divided by air density ρ, which from the paragraph above is equal to τ/ρ. But this is just the definition of friction velocity squared u_*^2 (eq. 18.10).

As in Fig. 18.25, if vertically moving air parcels (w') transport air with different horizontal velocities (u') across a horizontal area per unit time, then the covariance between vertical velocity and horizontal velocity is a kinematic momentum flux. From the paragraph above, the magnitude is also equal to the friction velocity squared. Thus:

$$\left|\overline{w'u'}\right| = \left|F_{z\,turb}(\text{momentum})\right| = \left|\tau / \rho\right| = u_*^{\,2} \qquad \bullet(18.38c)$$

where $\overline{w'u'}$ is called a **Reynolds stress**.

In Fig. 18.25, air mass is conserved. Namely, each rising air parcel is compensated by a descending air parcel with the same air mass. Thus, as seen from the discussion above, turbulence can cause a net vertical transport of heat, moisture, and momentum, even though there is no net transport of air mass.

Turbulent fluxes given by eqs. (18.38) are called **eddy-correlation fluxes**. They can be measured with fast response velocity, humidity, and temperature sensors, sampling at about 10 Hz for 30 minutes. Turbulent fluxes can also be parameterized, as discussed next.

Turbulence Closure

To forecast the weather (see the NWP chapter), we need to solve the Eulerian conservation equations for temperature, humidity, and wind:

- temperature forecasts ← heat conservation eq. ← First Law of Thermodynamics (see the Thermodynamics chapter)
- humidity forecasts ← water conservation eq. ← Eulerian water-budget equation (see the Water Vapor chapter)
- wind forecasts ← momentum conservation eq. ← Newton's Second Law (see the Atmos. Forces & Winds chapter)

For example, the Eulerian net heat-budget equation from the Thermodynamics chapter (eq. 3.51) is:

$$\frac{\Delta T}{\Delta t} = Advection + Radiation + LatentHeat - \frac{\Delta F_{z\,turb}(\theta)}{\Delta z} \qquad (18.39a)$$

where the last term is the turbulence term. But from eq. (18.38a), we recognize the turbulent heat flux as a covariance. Thus, eq. (18.39a) can be rewritten as:

$$\frac{\Delta \overline{T}}{\Delta t} = (otherPhysics) - \frac{\Delta \overline{w'\theta'}}{\Delta z} \qquad (18.39b)$$

The derivation of this equation is shown in the HIGHER MATH box on the next page.

Because the heat flux $\overline{w'\theta'}$ is needed in eq. (18.39b), we need to get a forecast equation for it:

$$\frac{\Delta \overline{w'\theta'}}{\Delta t} = (otherPhysics) - \frac{\Delta \overline{w'w'\theta'}}{\Delta z} \qquad (18.40)$$

Sample Application (§) *(continuation)*

Use eq. (18.36) and multiply each w' with θ' to give in column K the values of $w'\theta'$. Average those to get the covariance: $\overline{w'\theta'} = F_H =$ **6.62 K·m** s^{-1}, which is the kinematic heat flux by definition.

Use eq. (18.37) and divide the covariance by the standard deviations to give the correlation coef: $r_{w,\theta}$ = **0.65** (dimensionless).

Use eq. (2.11) to give heat flux, with $\rho{\cdot}C_p$ =1231 (W m^{-2})/(°C·m s^{-1}) from Appendix B, yielding $\mathbb{F}_H = \rho{\cdot}C_p{\cdot}F_H$ = **8150 W m^{-2}**.

	G	H	I	J	K	L
1	θ'	w'	θ'^{2}	w'^{2}	w'θ'	
2	(°C)	(m/s)	(°C^2)	(m/s)^2	°C·(m/s)	
3	–4	–5	16	25	20	
4	3	4	9	16	12	
5	4	3	16	9	12	
6	0	4	0	16	0	
7	–3	0	9	0	0	
8	3	–5	9	25	–15	
9	–2	–1	4	1	2	
10	1	–3	1	9	–3	
11	2	2	4	4	4	
12	–1	–4	1	16	4	
13	–4	–4	16	16	16	
14	–1	1	1	1	–1	
15	0	–2	0	4	0	
16	2	3	4	9	6	
17	4	5	16	25	20	
18	–3	2	9	4	–6	
19	5	6	25	36	30	
20	–2	–1	4	1	2	
21	3	3	9	9	9	
22	–4	–3	16	9	12	
23	–3	–5	9	25	15	
24	**0**	**var.=**	8.48	12.38	**6.62=**	**covar**
25		**st.dev.=**	2.61	3.52	**0.65=**	$r_{w,\theta}$

Check: The average is zero of the singled primed values, as they always should be. Units OK.

Exposition: The magnitude of $\mathbb{F}_H$ is unrealistically big for this contrived data set.

Sample Application

If $\overline{w'\theta'}$ = 0.2 K·m s^{-1} at the surface, and is 0 at the top of a 1 km thick layer, find the warming rate.

Find the Answer

Given: $\overline{w'\theta'}$ = 0.2 K·m s^{-1} at z = 0, $\overline{w'\theta'}$ =0 at z = 1km

Find: $\Delta\overline{T}/\Delta t$ = ? K h^{-1}

Use eq. (18.39b):

$\Delta\overline{T}/\Delta t$ = – (0 – 0.2 K·m s^{-1}) /(1000m – 0)

= 2x10^{-4} K s^{-1} = **0.72 K** h^{-1}.

Check: Units OK. Magnitude small.

Exposition: Over 12 hours of daylight, this surface flux would warm the thick layer of air by 8.6°C.

HIGHER MATH • Turbulence Terms

Why does a turbulence covariance term appear in the forecast equation for average temperature, $\overline{T}$? To answer, consider the vertical advection term (3.31) in the heat-budget equation (3.17) as an example:

$$\frac{\partial T}{\partial t} = \cdots - W\frac{\partial \theta}{\partial z}$$

For each dependent variable (T, W, θ), describe them by their mean plus turbulent parts:

$$\frac{\partial(\overline{T}+T')}{\partial t} = \cdots - (\overline{W}+w')\frac{\partial(\overline{\theta}+\theta')}{\partial z}$$

or

$$\frac{\partial \overline{T}}{\partial t} + \frac{\partial T'}{\partial t} = \cdots - \overline{W}\frac{\partial \overline{\theta}}{\partial z} - \overline{W}\frac{\partial \theta'}{\partial z} - w'\frac{\partial \overline{\theta}}{\partial z} - w'\frac{\partial \theta'}{\partial z}$$

Next, average the whole equation. But the average of a sum is the same as the sum of the averages:

$$\overline{\frac{\partial \overline{T}}{\partial t}} + \overline{\frac{\partial T'}{\partial t}} = \cdots - \overline{\overline{W}\frac{\partial \overline{\theta}}{\partial z}} - \overline{\overline{W}\frac{\partial \theta'}{\partial z}} - \overline{w'\frac{\partial \overline{\theta}}{\partial z}} - \overline{w'\frac{\partial \theta'}{\partial z}}$$

[**Aside**: Let A be any variable. Expand into mean and turbulent parts: $A = \overline{A} + a'$, then average the whole eq: $\overline{A} = \overline{\overline{A}} + \overline{a'}$. But the average of an average is just the original average: $\overline{A} = \overline{A} + \overline{a'}$. This equation can be valid only if $\overline{a'} = 0$. Thus, the average of any term containing a single primed variable (along with any number of unprimed variables) is zero.]

Thus the heat budget becomes:

$$\overline{\frac{\partial \overline{T}}{\partial t}} = \cdots - \overline{\overline{W}\frac{\partial \overline{\theta}}{\partial z}} - \overline{w'\frac{\partial \theta'}{\partial z}}$$

The term on the left and the first term on the right are average of averages, and can be rewritten by just the original averages. The last term can be transformed into **flux form** using the turbulent continuity equation (which works if you apply it to the turbulent advection in all 3 directions, but which is not shown here). The end result is:

$$\frac{\partial \overline{T}}{\partial t} = \cdots - \overline{W}\frac{\partial \overline{\theta}}{\partial z} - \frac{\partial \overline{w'\theta'}}{\partial z}$$

This says that to forecast the average temperature, you need to consider not only the average advection by the mean wind (first term on the right), but you also need to consider the turbulence flux divergence (last term on the right).

Similar terms appear for advection in the x and y directions. Also, similar terms appear in the forecast equations for moisture and wind. Thus, the effects of turbulence cannot be neglected.

To simplify the notation in almost all of this book, the overbar is left off of the terms for mean temperature, mean wind, etc. Also, earlier in this chapter, and in other chapters, the turbulence flux divergence term has already been parameterized directly as a function of non-turbulent (average) wind, temperature, humidity, etc. Such parameterizations are turbulence closure approximations.

But this contains yet another unknown $\overline{w'w'\theta'}$. A forecast equation for $\overline{w'w'\theta'}$ would yield yet another unknown. Hence, we need an infinite number of equations just to forecast air temperature. Or, if we use only a finite number of equations, then we have more unknowns than equations.

Hence, this set of equations is mathematically not closed, which means they cannot be solved. To be a **closed system of equations**, the number of unknowns must equal the number of equations.

One reason for this **closure problem** is that it is impossible to accurately forecast each swirl and eddy in the wind. To work around this problem, meteorologists **parameterize** the net effect of all the eddies; namely, they use a finite number of equations, and approximate the unknowns as a function of known variables. Such an approximation is called **turbulence closure**, because it mathematically closes the governing equations, allowing useful weather forecasts and engineering designs.

Turbulence Closure Types

For common weather situations with mean temperature, wind and humidity that are nearly horizontally uniform, turbulent transport in any horizontal direction nearly cancels transport in the opposite direction, and thus can be neglected. But vertical transport is significant. Medium and large size turbulent eddies can transport air parcels from many different source heights to any destination height within the turbulent domain, where the smaller eddies mix the parcels together.

Different approximations of turbulent transport consider the role of small and large eddies differently. **Local closures**, which neglect the large eddies, are most common. This gives turbulent heat fluxes that flow down the local gradient of potential temperature, analogous to molecular diffusion or conduction (see the Heat and Air Pollution chapters). One such turbulence closure is called **K-theory**.

A **nonlocal closure** alternative that accounts for the superposition of both large and small eddies is called **transilient turbulence theory** (T3). While this is more accurate, it is also more complicated. There are many other closures that have been proposed. K-theory is reviewed here.

K-Theory

One approximation to turbulent transport considers only small eddies. This approach, called **K-theory**, **gradient transport theory**, or **eddy-diffusion theory**, models turbulent mixing analogous to molecular diffusion Using heat flux F_H for example:

$$F_H = \overline{w'\theta'} = -K \cdot \frac{\Delta\overline{\theta}}{\Delta z} \qquad \bullet(18.41a)$$

This parameterization says that heat flows down the gradient of potential temperature, from warm to cold. The rate of this turbulent transfer is proportional to the parameter K, called the **eddy viscosity** or **eddy diffusivity**, with units $m^2 \cdot s^{-1}$.

Similar expressions can be made for moisture flux as a function of the mean mixing-ratio (r) gradient, or momentum flux as a function of the shear in horizontal wind components (U, V):

$$\overline{w'r'} = -K \frac{\Delta \bar{r}}{\Delta z} \quad (18.41b)$$

$$\overline{w'u'} = -K \frac{\Delta \overline{U}}{\Delta z} \quad (18.41c)$$

$$\overline{w'v'} = -K \frac{\Delta \overline{V}}{\Delta z} \quad (18.41d)$$

K is expected to be larger for more intense turbulence. In the surface layer, turbulence is generated by wind shear. **Prandtl** made a **mixing-length** suggestion that:

$$K = k^2 \cdot z^2 \cdot \left| \frac{\Delta \overline{M}}{\Delta z} \right| \quad (18.42)$$

where k = 0.4 is von Kármán's constant (dimensionless), z is height above ground, and $\Delta M/\Delta z$ is mean shear of the horizontal wind M.

When K-theory used in the Eulerian heat budget equation, neglecting all other terms except turbulence, the result gives the heating rate of air at height z due to **turbulent flux divergence** (i.e., change of flux with height):

$$\frac{\Delta \bar{\theta}(z)}{\Delta t} = K \cdot \left[\frac{\bar{\theta}(z+\Delta z) - 2\bar{\theta}(z) + \bar{\theta}(z-\Delta z)}{(\Delta z)^2} \right] \quad (18.43)$$

{For those of you who like calculus, the ratio in square brackets is an approximation to the second derivative $[\partial^2 \bar{\theta} / \partial z^2]$. Namely, it is equal to the curvature of the potential temperature vertical profile.} Although the example above was for heat flux, you can also use it for moisture or momentum flux by substituting $\bar{r}$ or $\overline{U}$, $\overline{V}$ in place of $\bar{\theta}$.

K-theory works best for windy surface layers, where turbulent eddy sizes are relatively small. Fig. 18.26 shows that heat flux flows "down" the temperature gradient from warm to colder potential temperature, which gives a negative (downward) heat flux in the statically stable surface layer. Fig. 18.25 is also a small-eddy (K-theory-like) illustration.

K-theory does not apply at the solid ground, but only within the air where turbulence exists. For heat fluxes at the surface, use approximations given in earlier in this chapter, and in the Heat chapter.

A SCIENTIFIC PERSPECTIVE • Parameterization Rules

For some situations in science and engineering, the governing equations are not known or are so complicated as to be unwieldy. Nonetheless, an approximation might be good enough to give useful answers.

A **parameterization** is an approximation to an unknown term, made using one or more known terms or factors. Because these approximations do not come from first principles, one or more fudge factors are often included in the substitute term to make it have the correct behavior or order-of-magnitude. These fudge factors are called **parameters**, and are not known from first principles. They must be found **empirically** (from field or laboratory experiments, or from numerical simulations).

Different scientists and engineers can invent different parameterizations for the same unknown term. While none are perfect, each would have different advantages and disadvantages. However, every parameterization must follow certain rules.

The parameterization should:

(1) be physically reasonable,
(2) have the same dimensions as the unknown term,
(3) have the same scalar or vector properties,
(4) have the same symmetries,
(5) be invariant under an arbitrary transformation of coordinate system,
(6) be invariant under an **inertial** or **Newtonian transformation** (e.g., a coordinate system moving at constant speed and direction),
(7) satisfy the same constraints & budget equations.

Even if the parameterization satisfies the above rules, it often will be successful for only a limited range of conditions. An example from this book is the transport velocity, w_T. One parameterization was developed for statically neutral boundary layers, while a different one worked better for statically unstable mixed layers. Thus, every parameterization should state the limitations on its use, including accuracy and range of validity.

In summary, parameterizations only approximate nature. As a result, they will never work perfectly. But they can be designed to work satisfactorily.

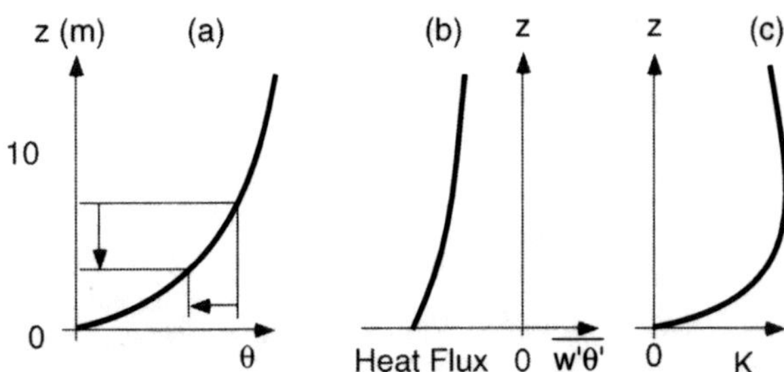

Figure 18.26
Typical profiles of (a) potential temperature; (b) heat flux; and (c) eddy diffusivity in the statically stable surface layer.

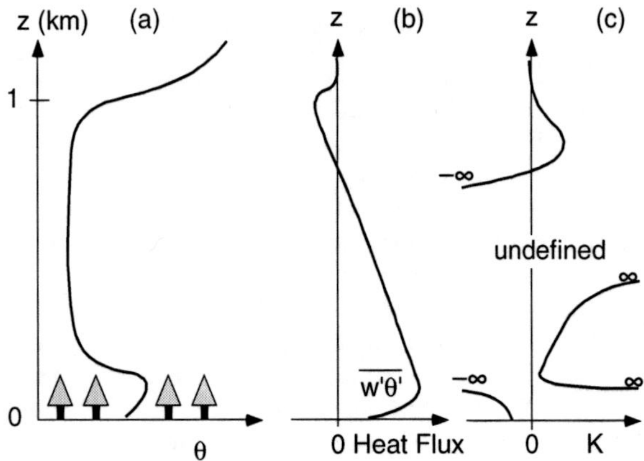

Figure 18.27
The value (c) of K needed to get (b) the observed heat flux F_H from (a) the observed potential temperature θ over a forest for an unstable (convective) ABL.

Sample Application
Instruments on a tower measure θ = 15°C and M = 5 m s^{-1} at z = 4 m, and θ = 16°C and M = 8 m s^{-1} at z = 10 m. What is the vertical heat flux?

Find the Answer:
Given:

z (m)	θ (°C)	M (m s^{-1})
10	16	8
4	15	5

Find: F_H = ? K·m s^{-1}

First, use eq. (18.42), at average z = (10+4)/2 = 7 m

$$K = k^2 \cdot z^2 \cdot \left|\frac{\Delta M}{\Delta z}\right| = [0.4 \cdot (7\text{m})]^2 \left|\frac{(8-5)\text{m/s}}{(10-4)\text{m}}\right| = 3.92 \text{ m}^2 \text{ s}^{-1}$$

Next, use eq. (18.41):
$F_H = -K(\Delta\theta/\Delta z) = -(3.92 \text{ m}^2 \text{ s}^{-1})\cdot(16\text{–}15°\text{C})/(10\text{–}4\text{m})$
= **−0.65 K·m** s^{-1}

Check: Units OK. Physics OK.
Exposition: The negative sign means a downward heat flux, from hot to cold. This is typical for statically stable ABL.

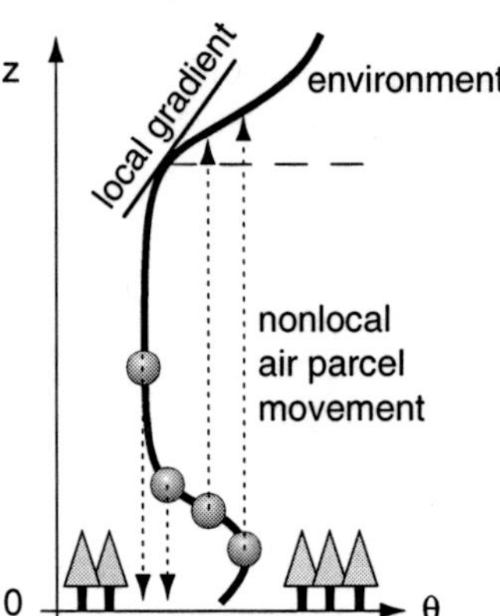

Figure 18.28
Typical potential temperature θ profile in the ABL over a forest, showing nonlocal convective movement of some of the air parcels that are moving across relatively large vertical distances during free convection.

K-theory has difficulty for convective ABLs and should not be used there. Figs. 18.27a and b illustrate these difficulties, giving typical values in the atmosphere, and the resulting K values backed out using eq. (18.42). Negative and infinite K values are unphysical.

Nonlocal Closure

Instead of looking at local down-gradient transport as was done in K-theory, you can look at all ranges of distances across which air parcels move during turbulence (Fig. 18.28). This is an approach called **nonlocal closure**, and is useful for a statically unstable ABL having free convection.

For heat flux at the altitude of the dashed line in Fig. 18.28, K-theory (small-eddy theory) would utilize the local gradient of θ at that altitude, and conclude that the heat flux should be downward and of a certain magnitude. However, if the larger-size eddies are also included (as in nonlocal closure), such as the parcel rising from tree-top level, we see that it is bringing warm air upward (a positive contribution to heat flux). This could partially counteract, or even overwhelm, the negative contribution to flux caused by the local small eddies.

As you can probably anticipate, a better approach would be to consider all size eddies and nonlocal air-parcel movement. One such approach is called **transilient turbulence theory** (T3). This approach uses arrays to account for eddies of each different size, and combines them to find the average effect of all eddy sizes. It is more complex, and will not be described here.

REVIEW

We live in a part of the atmosphere known as the boundary layer (ABL), the bottom 200 m to 4 km of the troposphere. Tropospheric static stability and turbulence near the Earth's surface combine to create this ABL, and cap it with a temperature inversion.

Above the boundary layer is the free atmosphere, which is not turbulently coupled with the ground (except during stormy weather such as near low pressure centers, fronts, and thunderstorms). Thus, the free atmosphere does not normally experience a strong diurnal cycle.

Within the ABL are significant daily variations of temperature, winds, static stability, and turbulence over land under mostly clear skies. These variations are driven by the diurnal cycle of radiative heating of the ground during daytime and cooling at night.

During daytime under fair weather (i.e., in anticyclonic or high-pressure regions), vigorous turbulence mixes potential temperature, humidity, wind speed, and pollutants to be nearly uniform with height. This turbulence creates a well-mixed layer that grows due to entrainment of free atmosphere air from above.

At night in fair weather, there is a shallow stable boundary layer near the ground, with a nearly neutral residual layer above. Turbulence is weak and sporadic. Winds often become calm near the surface, but can be very fast a few hundred meters above ground.

The bottom 5 to 10% of the ABL is called the surface layer. Surface drag causes the wind to be zero near the ground, and to increase with height. The shape of this wind profile is somewhat logarithmic, but depends on the roughness of the surface, and on convection.

Turbulence is a quasi-random flow phenomenon that can be described by statistics. Covariance of vertical velocity with another variable represents the vertical kinematic flux of that variable. Heat fluxes, moisture fluxes, and stress can be expressed as such an eddy-correlation statistic.

Velocity variances represent components of turbulent kinetic energy (*TKE*) per unit mass, a measure of the intensity of turbulence. *TKE* is produced by wind shear and buoyancy, is advected from place to place by the mean and turbulent winds, and is dissipated into heat by molecular viscosity.

The relative magnitudes of the shear and buoyant production terms determine whether convection is free or forced. The sum of those terms is proportional to the intensity of turbulence. The ratio gives the flux Richardson number for determining whether turbulence can persist.

Turbulence is so complex that it cannot be solved exactly for each swirl and eddy. Instead, parameterizations are devised to allow approximate solutions for the net statistical effect of all turbulent eddies. Parameterizations, while not perfect, are acceptable if they satisfy certain rules.

One type of local parameterization, called K-theory, neglects the large eddies, but gives good answers for special regions such as the surface layer in the bottom 10% of the atmospheric boundary layer. It is popular because of its simplicity. Another type of parameterization is called transilient turbulence theory (T3), which is a nonlocal closure that includes all eddy sizes. It is more accurate, more complicated, and works well for free convection.

HOMEWORK EXERCISES

Broaden Knowledge & Comprehension

B1. Access the upper-air soundings every 6 or 12 h for a rawinsonde station near you (or other site specified by your instructor). For heights every 200 m (or every 2 kPa) from the surface, plot how the temperature varies with time over several days. The result should look like Fig. 18.4, but with more lines. Which heights or pressure levels appear to be above the ABL?

B2. Same as the previous question, but first convert the temperatures to potential temperatures at those selected heights. This should look even more like Fig. 18.4.

B3. Access temperature profiles from the web for the rawinsonde station closest to you (or for some other sounding station specified by your instructor). Convert the resulting temperatures to potential temperatures, and plot the resulting θ vs z. Can you identify the top of the ABL? Consider the time of day when the sounding was made, to help you anticipate what type of ABL exists (e.g., mixed layer, stable boundary layer, neutral layer.)

B4. Access a weather map with surface wind observations. Find a situation or a location where there is a low pressure center. Draw a hypothetical circle around the center of the low, and find the average inflow velocity component across this circle. Using volume conservation, and assuming a 1 km thick ABL, what vertical velocity to you anticipate over the low?

B5. Same as previous question, but for a high-pressure center.

B6. Access a number of rawinsonde soundings for stations more-or-less along a straight line that crosses through a cold front. Identify the ABL top both ahead of and behind the front.

B7. Access the current sunrise and sunset times at your location. Estimate a curve such as in Fig. 18.10.

B8. Find a rawinsonde station in the center of a clear high pressure region, and access the soundings every 6 h if possible. If they are not available over N. America, try Europe. Use all of the temperature, humidity, and wind information to determine the evolution of the ABL structure, and create a sketch similar to Fig. 18.8, but for your real conditions. Consider seasonal affects, such as in Figs. 18.13 & 18.14.

B9. Access a rawinsonde sounding for nighttime, and find the best-fit parameters (height, strength) for an exponential potential temperature profile that best fits the sounding. If the exponential shape is a poor fit, explain why.

B10. Get an early morning sounding for a rawinsonde site near you. Calculate the anticipated accumulated heating, based on the eqs. in the Solar & IR Radiation & Heat chapters, and predict how the mixed layer will warm & grow in depth as the day progresses.

B11. Compare two sequential rawinsonde soundings for a site, and estimate the entrainment velocity by comparing the heights of the ABL. What assumptions did you make in order to do this calculation?

B12. Access the wind information from a rawinsonde site near you. Compare the wind speed profiles between day and night, and comment on the change of wind speed in the ABL.

B13. Access the "surface" wind observations from a weather station near you. Record and plot the wind speed vs. time, using wind data as frequent as are available, getting a total of roughly 24 to 100 observations. Use these data to calculate the mean wind, variance, and standard deviation.

B14. Same as the previous exercise, but for *T*.

B15. Same as the previous exercise, except collect both wind and temperature data, and find the covariance and correlation coefficient.

B16. Search the web for **eddy-correlation sensors**, **scintillometers**, **sonic anemometers**, or other instruments used for measuring vertical turbulent fluxes, and describe how they work.

B17. Search the web for eddy viscosity, eddy diffusivity, or K-theory values for *K* in the atmosphere.

Apply

A1(§). Calculate and plot the increase of cumulative kinematic heat (cooling) during the night, for a case with kinematic heat flux (K·m s^{-1}) of:
a. –0.02 b. –0.05 c. – 0.01 d. –0.04 e. –0.03
f. – 0.06 g. – 0.07 h. –0.10 i. –0.09 j. –0.08

A2 (§). Calculate and plot the increase of cumulative kinematic heat during the day, for a case with daytime duration of 12 hours, and maximum kinematic heat flux (K·m s^{-1}) of:
a. 0.2 b. 0.5 c. 0.1 d. 0.4 e. 0.3
f. 0.6 g. 0.7 h. 1.0 i. 0.9 j. 0.8

A3(§). For a constant kinematic heat flux of –0.02 K·m s^{-1} during a 12-hour night, plot the depth and strength of the stable ABL vs. time. Assume a flat prairie, with a residual layer wind speed (m s^{-1}) of:
a. 2 b. 5 c. 8 d. 10 e. 12
f. 15 g. 18 h. 20 i. 22 j. 25

A4(§). For the previous problem, plot the vertical temperature profile at 1-h intervals.

A5. Find the entrained kinematic heat flux at the top of the mixed layer, given a surface kinematic heat flux (K·m s^{-1}) of
a. 0.2 b. 0.5 c. 0.1 d. 0.4 e. 0.3
f. 0.6 g. 0.7 h. 1.0 i. 0.9 j. 0.8

A6. Find the entrainment velocity for a surface heat flux of 0.2 K·m s^{-1}, and a capping inversion strength of (°C): a. 0.1 b. 0.2 c. 0.3 d. 0.5 e. 0.7
f. 1.0 g. 1.2 h. 1.5 i. 2.0 j. 2.5

A7. For the previous problem, calculate the increase in mixed-layer depth during a 6 h interval, assuming subsidence of – 0.02 m s^{-1}.

A8. Calculate the surface stress at sea level for a friction velocity (m s^{-1}) of:
a. 0.1 b. 0.2 c. 0.3 d. 0.4 e. 0.5
f. 0.6 g. 0.7 h. 0.8 i. 0.9 j. 1.0

A9. Find the friction velocity over a corn crop during wind speeds (m s^{-1}) of:
a. 2 b. 3 c. 4 d. 5 e. 6 f. 7 g. 8
h. 9 i. 10 j. 12 k. 15 m. 18 n. 20 o. 25

A10. Find the roughness length and standard drag coefficient over: a. sea b. beach
c. tundra d. low crops e. hedgerows
f. orchards g. village h. city center

A11. Same as previous problem, but find the friction velocity given a 10 m wind speed of 5 m s^{-1}.

A12(§). Given M_1 = 5 m s^{-1} at z_1 = 10 m, plot wind speed vs. height, for z_o (m) =
a. 0.001 b. 0.002 c. 0.005 d. 0.007 e. 0.01
f. 0.02 g. 0.05 h. 0.07 i. 0.1 j. 0.2
k. 0.5 m. 0.7 n. 1.0 o. 2.0 p. 2.5

A13. Find the drag coefficients for problem A12.

A14(§). For a neutral surface layer, plot wind speed against height on linear and on semi-log graphs for friction velocity of 0.5 m s^{-1} and aerodynamic roughness length (m) of:

a. 0.001	b. 0.002	c. 0.005	d. 0.007	e. 0.01
f. 0.02	g. 0.05	h. 0.07	i. 0.1	j. 0.2
k. 0.5	m. 0.7	n. 1.0	o. 2.0	p. 2.5

A15. An anemometer on a 10 m mast measures a wind speed of 8 m s^{-1} in a region of scattered hedgerows. Find the wind speed at height (m):

a. 0.5	b. 2	c. 5	d. 30
e. 1.0	f. 4	g. 15	h. 20

A16. The wind speed is 2 m s^{-1} at 1 m above ground, and is 5 m s^{-1} at 10 m above ground. Find the roughness length. State all assumptions.

A17. Over a low crop with wind speed of 5 m s^{-1} at height 10 m, find the wind speed (m s^{-1}) at the height (m) given below, assuming overcast conditions:

a. 0.5	b. 1.0	c. 2	d. 5	e. 15	f. 20
g. 25	h. 30	i. 35	j. 40	k. 50	m. 75

A18. Same as previous problem, but during a clear night when friction velocity is 0.1 m s^{-1} and surface kinematic heat flux is –0.01 K·m s^{-1}. Assume $|g|/T_v$ = 0.0333 m·s^{-2}·K^{-1}.

A19(§). Plot the vertical profile of wind speed in a stable boundary layer for roughness length of 0.2 m, friction velocity of 0.3 m s^{-1}, and surface kinematic heat flux (K·m s^{-1}) of:

a. –0.02	b. –0.05	c. – 0.01	d. –0.04	e. –0.03
f. – 0.06	g. – 0.07	h. –0.10	i. –0.09	j. –0.08

Assume $|g|/T_v$ = 0.0333 m·s^{-2}·K^{-1}.

A20. For a 1 km thick mixed layer with $|g|/T_v$ = 0.0333 m·s^{-2}·K^{-1}, find the Deardorff velocity (m s^{-1}) for surface kinematic heat fluxes (K·m s^{-1}) of:

a. 0.2	b. 0.5	c. 0.1	d. 0.4	e. 0.3
f. 0.6	g. 0.7	h. 1.0	i. 0.9	j. 0.8

A21§). For the previous problem, plot the wind speed profile, given u_* = 0.4 m s^{-1}, and M_{BL} = 5 m s^{-1}.

A22(§). For the following time series of temperatures (°C): 22, 25, 21, 30, 29, 14, 16, 24, 24, 20

a. Find the mean and turbulent parts
b. Find the variance
c. Find the standard deviation

A23(§). Using the data given in the long Sample Application in the "Turbulent Fluxes and Covariances" section of this Chapter:

a. Find the mean & variance for mixing ratio *r*.
b. Find the mean and variance for *U* .
c. Find the covariance between *r* and *W*.
d. Find the covariance between *U* and *W*.
e. Find the covariance between *r* and *U*.
f. Find the correlation coefficient between *r* & *W*.
g. Find the correlation coef. between *u* & *W*.
h. Find the correlation coefficient between *r* & *U*.

A24(§). Plot the standard deviations of *U*, *V*, and *W* with height, and determine if and where the flow is nearly isotropic, given u_* = 0.5 m s^{-1}, w_B = 40 m s^{-1}, *h* = 600 m, and z_i = 2 km, for air that is statically:

a. stable b. neutral c. unstable

Hint: Use only those "given" values that apply to your stability.

A25(§). Same as exercise A24, but plot *TKE* vs. *z*.

A26(§). Plot wind standard deviation vs. height, and determine if and where the flow is nearly isotropic, for all three wind components, for

a. stable air with *h* = 300 m, u_* = 0.1 m s^{-1}
b. neutral air with *h* = 1 km, u_* = 0.2 m s^{-1}
c. unstable air with z_i = 2 km, w_B = 0.3 m s^{-1}

A27(§). Plot the turbulence kinetic energy per unit mass vs. height for the previous problem.

A28. Given a wind speed of 20 m s^{-1}, surface kinematic heat flux of –0.1 K·m s^{-1}, *TKE* of 0.4 m^2 s^{-2}, and $|g|/T_v$ = 0.0333 m·s^{-2}·K^{-1}, find

a. shear production rate of *TKE*
b. buoyant production/consumption of *TKE*
c. dissipation rate of *TKE*
d. total tendency of *TKE* (neglecting advection and transport)
e. flux Richardson number
f. the static stability classification?
g. the Pasquill-Gifford turbulence type?
h. flow classification

A29. Given the following initial sounding

z (m)	θ (°C)	M (m s^{-1})
700	21	10
500	20	10
300	18	8
100	13	5

Compute the

a. value of eddy diffusivity at each height.
b. turbulent heat flux in each layer, using K-theory
c. the new values of θ after 30 minutes

of mixing.

d. turbulent kinematic momentum flux $\overline{w'u'}$ in each layer, using K-theory. (Hint, generalize the concepts of heat flux. Also, assume $M = U$, and $V = 0$ everywhere.)

e. the new values of M after 30 minutes of mixing.

Evaluate & Analyze

E1. Can the ABL fill the whole troposphere? Could it extend far into the stratosphere? Explain.

E2. Estimate the static stability near the ground now at your location?

E3. If the standard atmosphere was statically neutral in the troposphere, would there be a boundary layer? If so, what would be its characteristics.

E4. It is nighttime in mid winter with snow on the ground. This air blows over an unfrozen lake. Over the lake, what is the static stability of the air, and what type of ABL exists there.

E5. It is daytime in summer during a solar eclipse. How will the ABL evolve during the eclipse?

E6. Given Fig. 18.3, if turbulence were to become more vigorous and cause the ABL depth to increase, what would happen to the strength of the capping inversion? How would that result affect further growth of the ABL?

E7. The ocean often has a turbulent boundary layer within the top tens of meters of water. During a moderately windy, clear, 24 h period, when do you think stable and unstable (convective) ocean boundary layers would occur?

E8. Fig. 18.4 shows the free-atmosphere line touching the peaks of the ABL curve. Is it possible during other weather conditions for the FA curve to cross through the middle of the ABL curve, or to touch the minimum points of the ABL curve? Discuss.

E9. It was stated that subsidence cannot penetrate the capping inversion, but pushes the inversion down in regions of high pressure. Why can subsidence not penetrate into the ABL?

E10. Fig. 18.7a shows a cold front advancing. What if it were retreating (i.e., if it were a warm front). How would the ABL be affected differently, if at all?

E11. Draw a sketch similar to Fig. 18.8, except indicating the static stabilities in each domain.

E12. Copy Fig. 18.9, and then trace the nighttime curves onto the corresponding daytime curves. Comment on regions where the curves are the same, and on other regions where they are different. Explain this behavior.

E13. Use the solar and IR radiation equations from the Solar & IR Radiation chapter to plot a curve of daytime radiative heat flux vs. time. Although the resulting curve looks similar to half of a sine wave (as in Fig. 18.10), they are theoretically different. How would the actual curve change further from a sine curve at higher or lower latitudes? How good is the assumption that the heat flux curve is a half sine wave?

E14. Similar to the previous question, but how good is the assumption that nighttime heat flux is constant with time?

E15. In the explanation surrounding Fig. 18.10, the accumulated heating and cooling were referenced to start times when the heat flux became positive or negative, respectively. We did not use sunrise and sunset as the start times because those times did not correspond to when the surface heat flux changes sign. Use all of the relevant previous equations in the chapter (and in the Solar & IR Radiation chapter) to determine the time difference today at your town, between:

a. sunrise and the time when surface heat flux first becomes positive.

b. sunset and the time when surface heat flux first becomes negative.

E16. At nighttime it is possible to have well-mixed layers if wind speed is sufficiently vigorous. Assuming a well-mixed nocturnal boundary layer over a surface that is getting colder with time, describe the evolution (depth and strength) of this stable boundary layer. Also, for the same amount of cooling, how would its depth compare with the depth of the exponentially-shaped profile?

E17. Derive an equation for the strength of an exponentially-shaped nocturnal inversion vs. time, assuming constant heat flux during the night.

E18. Use eq. (18.3) and the definition of potential temperature to replot Fig. 18.15 in terms of actual temperature T vs. z. Discuss the difference in heights between the relative maximum of T and relative maximum of θ.

E19. For a linear early morning sounding (θ increases linearly with z), analytically compare the growth rates of the mixed layer calculated using the thermodynamic and the flux-ratio methods. For some idealized situations, is it possible to express one in terms of the other? Were any assumptions needed?

E20. Given an early morning sounding as plotted in Fig. 18.16a. If the daytime heat flux were constant with time, sketch a curve of mixed-layer depth vs. time, assuming the thermodynamic method. Comment on the different stages of mixed layer growth rate.

E21. Use the early-morning (6 AM) sounding given below with surface temperature 5°C. Find the mixed-layer potential temperature and depth at 11 AM, when the cumulative heating is 800 K·m. Assume the early-morning potential temperature profile is: a. $\Delta\theta/\Delta z = 2 \text{ K km}^{-1}$ = constant

b. $\Delta\theta(z) = (8°\text{C})\cdot\exp(-z/150\text{m})$

Hint: Use the encroachment method.

E22. If the heating rate is proportional to the vertical heat-flux divergence, use Fig. 18.17b to determine the heating rate at each z in the mixed layer. How would the mixed-layer T profile change with time?

E23. Assume that equations similar to eq. (18.7) applies to moisture fluxes as well as to heat fluxes, but that eq. (18.9) defines the entrainment velocity based only on heat. Combine those two equations to give the kinematic flux of moisture at the top of the mixed layer as a function of potential temperature and mixing ratio jumps across the mixed-layer top, and in terms of surface heat flux.

E24. If the ABL is a region that feels drag near the ground, why can the winds at night accelerate, as shown in Fig. 18.18?

E25. Use eqs. (18.11) and (18.14a) to show how eq. (18.12) is derived from the log wind profile.

E26. Given the moderate value for u_* that was written after eq. (18.10), what value of stress (Pa) does that correspond to? How does this stress compare to sea-level pressure?

E27. Derive eq. (18.14b) from (18.14a).

E28. Given the wind-speed profile of the Sample Application in the radix-layer subsection, compare the bottom portion of this profile to a log wind profile. Namely, find a best fit log-wind profile for the same data. Comment on the differences.

E29. Using eq. 18.21, find the average value (or simplify the notation) of $c{\cdot}T$, where c is a constant.

E30. Abbreviation "rms" means "root mean square". Explain why "rms" can be used to describe eq. (18.23).

E31. For fixed w_B and z_i, plot eqs. (18.26) with z, and identify those regions with isotropic turbulence.

E32. Using eqs. (18.24 – 18.26), derive expressions for the TKE vs. z for statically stable, neutral, and unstable boundary layers.

E33. Derive an expression for the shear production term vs. z in the neutral surface layer, assuming a log wind profile.

E34. Knowing how turbulent heat flux varies with z (Fig. 18.17b) in a convective mixed layer, comment on the TKE buoyancy term vs. z for that situation.

E35. Given some initial amount of TKE, and assuming no production, advection, or transport, how would TKE change with time, if at all? Plot a curve of TKE vs. time.

E36. Use K-theory to relate the flux Richardson number to the gradient Richardson number ($R_i = [(|g|/T_v)\cdot\Delta\theta\cdot\Delta z] \,/\, \Delta M^2$, see the Stability chapter).

E37. If shear production of TKE increases with time, how must the buoyant term change with time in order to maintain a constant Pasquill-Gifford stability category of E?

E38. Create a figure similar to Fig. 18.25, but for the log wind profile. Comment on the variation of momentum flux with z in the neutral surface layer.

E39. A negative value of eddy-correlation momentum flux near the ground implies that the momentum of the wind is being lost to the ground. Is it possible to have a positive eddy-correlation momentum flux near the surface? If so, under what conditions, and what would it mean physically?

Synthesize

S1. At the end of one night, assume that the stable ABL profile of potential temperature has an exponential shape, with strength 10°C and e-folding depth of 300 m. Using this as the early-morning sounding, compute and plot how the potential temperature and depth of the mixed layer evolve during the morning. Assume that D = 12 hr, and F_{Hmax} = 0.1 K·m s^{-1}. (Hint: In the spirit of the "Seek Solutions" box in a previous chapter, feel free to use graphical methods to solve this without calculus. However, if you want to try it with calculus, be very careful to determine which are your dependent and independent variables.)

S2. Suppose that there was no turbulence at night. Assume, the radiatively-cooled surface cools only the bottom 1 m of atmosphere by conduction. Given typical heat fluxes at night, what would be the resulting air temperature by morning? Also describe how the daytime mixed layer would evolve from this starting point?

S3. On a planet that does not have a solid core, but has a gaseous atmosphere that increases in density as the planet center is approached, would there be a boundary layer? If so, how would it behave?

S4. What if you were hired by an orchard owner to tell her how deep a layer of air needs to be mixed by electric fans to prevent frost formation. Create a suite of answers based on different scenarios of initial conditions, so that she can consult these results on any given day to determine the speed to set the fans (which we will assume can be used to control the depth of mixing). State and justify all assumptions.

S5. What if the Earth's surface was a perfect conductor of heat, and had essentially infinite heat capacity (as if the Earth were a solid sphere of aluminum, but had the same albedo as Earth). Given the same solar and IR radiative forcings on the Earth's surface, describe how ABL structure and evolution would differ from Fig. 18.8, if at all?

S6. What if all the air in the troposphere were saturated & cloudy. How would the ABL be different?

S7. Suppose that for some reason, the actual 2 km thick ABL in our troposphere was warm enough to have zero capping inversion on one particular day, but otherwise looked like Fig. 18.3. Comment on the evolution of the ABL from this initial state. Also, how might this initial state have occurred?

S8. If the ABL were always as thick as the whole troposphere, how would that affect the magnitude of our diurnal cycle of temperature?

S9. It was stated in this chapter that entrainment happens one way across the capping inversion, from the nonturbulent air into the turbulent air. Suppose that the capping inversion was still there, but that both the ABL below the inversion and the layer of air above the inversion were turbulent. Comment on entrainment and the growth of the mixed layer.

S10. Suppose that *TKE* was not dissipated by viscosity. How would that change our weather and climate, if at all?

S11. Suppose that wind shear destroyed *TKE* rather than generated it. How would that change our weather and climate, if at all?

S12. Suppose there were never a temperature inversion capping the ABL. How would that change our weather and climate, if at all?

S13. Suppose that the winds felt no frictional drag at the surface of the Earth. How would that change our weather and climate, if at all?

S14. Verify that K-theory satisfies the rules of parameterization.

S15. Positive values of K imply down-gradient transport of heat (i.e., heat flows from hot to cold, a process sometimes called the Zeroth Law of Thermodynamics). What is the physical interpretation of negative values of K?

19 POLLUTANT DISPERSION

Contents

Every living thing pollutes. Life is a chemical reaction, where input chemicals such as food and oxygen are converted into growth or motion. The reaction products are waste or pollution.

The only way to totally eliminate pollution is to eliminate life — not a particularly appealing option. However, a system of world-wide population control could stem the increase of pollution, allowing residents of our planet to enjoy a high quality of life.

Is pollution bad? From an anthropocentric point of view, we might say "yes". To do so, however, would deny our dependence on pollution. In the Earth's original atmosphere, there was very little oxygen. Oxygen is believed to have formed as pollution from plant life. Without this pollutant, animals such as humans would likely not exist now.

However, it is reasonable to worry about other **chemicals that threaten our quality of life**. We call such chemicals **pollutants**, regardless of whether they form naturally or **anthropogenically** (man-made). Many of the natural sources are weak emissions from large area sources, such as forests or swamps. Anthropogenic sources are often concentrated at points, such as at the top of smoke stacks (Fig. 19.1). Such high concentrations are particularly hazardous, and been heavily studied.

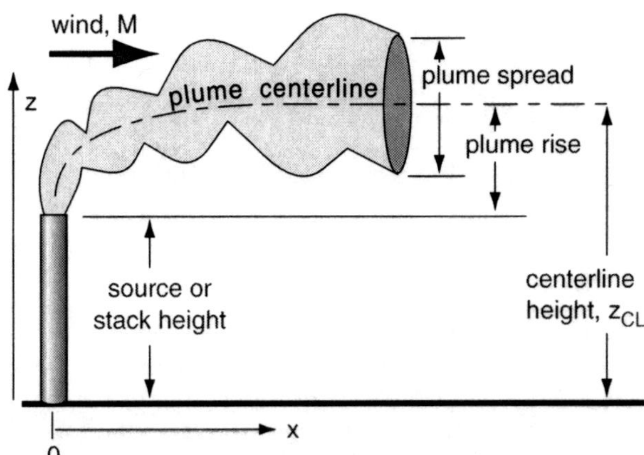

Figure 19.1
Pollutant plume characteristics.

INFO • Pollutant Concentration Units

The amount of a pollutant in the air can be given as a **fraction** or **ratio**, q. This is the amount (moles) of pollution divided by the total amount (moles) of all constituents in the air. For air quality, the ratios are typically reported in **parts per million (ppm)**. For example, 10 ppm means 10 parts of pollutant are contained within 10^6 parts of air. For smaller amounts, **parts per billion (ppb)** are used.

Although sometimes the ratio of masses is used (e.g., **ppmm** = parts per million by mass), usually the ratio of volumes is used instead (**ppmv** = parts per million by volume).

Alternately, the amount can be given as a **concentration**, c, which is the mass of pollutant in a cubic meter of air. For air pollution, units are often micrograms per cubic meter (μg m^{-3}). Higher concentrations are reported in milligrams per cubic meter, (mg m^{-3}), while lower concentrations can be in nanograms per cubic meter (ng m^{-3}).

The conversion between fractions and concentrations is

$$q(\text{ppmv}) = \frac{a \cdot T}{P \cdot M_s} \cdot c(\mu\text{g/m}^3)$$

where T is absolute temperature (Kelvin), P is total atmospheric pressure (kPa), M_s is the molecular weight of the pollutant, and a = 0.008314 kPa·K^{-1}·(ppmv)·(μg m^{-3})$^{-1}$.

For a standard atmosphere at sea level, where temperature is 15°C and pressure is 101.325 kPa, the equation above reduces to

$$q(\text{ppmv}) = \frac{b}{M_s} \cdot c(\mu\text{g/m}^3)$$

where b = 0.02363 (ppmv) / (μg m^{-3}).

For example, nitrogen dioxide (NO_2) has a molecular weight of M_s = 46.01 (see Table 1-2 in Chapter 1). If concentration c = 100 μg m^{-3} for this pollutant, then the equation above gives a volume fraction of q = (0.02363/46.01) · (100) = 0.051 ppmv.

Science Graffito

"The solution to pollution is dilution."

– Anonymous.

This aphorism was accepted as common sense during the 1800s and 1900s. By building taller smoke stacks, more pollutants could be emitted, because the pollutants would mix with the surrounding clean air and become dilute before reaching the surface.

However, by 2000, society started recognizing the global implications of emitting more pollutants. Issues included greenhouse gases, climate change, and stratospheric ozone destruction. Thus, government regulations changed to include total emission limits.

DISPERSION FACTORS

The stream of polluted air downwind of a smoke stack is called a **smoke plume**. If the plume is buoyant, or if there is a large effluent velocity out of the top of the smoke stack, the center of the plume can rise above the initial emission height. This is called **plume rise**.

The word "plume" in air pollution work means a long, slender, nearly-horizontal region of polluted air. However, the word "plume" in atmospheric boundary-layer (ABL) studies refers to the relatively wide, nearly vertical updraft portion of buoyant air that is convectively overturning. Because smoke plumes emitted into the boundary layer can be dispersed by convective plumes, one must take great care to not confuse the two usages of the word "plume".

Dispersion is the name given to the spread and movement of pollutants. Pollution dispersion depends on

- wind speed and direction
- plume rise, and
- atmospheric turbulence.

Pollutants disperse with time by mixing with the surrounding cleaner air, resulting in an increasingly dilute mixture within a spreading smoke plume.

Wind and **turbulence** are characteristics of the ambient atmosphere, as were described in earlier chapters. While emissions out of the top of the stack often have strong internal turbulence, this quickly decays, leaving the ambient atmosphere to do the majority of the dispersing.

The direction that the effluent travels is controlled by the local, synoptic, and global-scale winds. Pollutant destinations from known emission sources can be found using a **forward trajectory** along the mean wind, while source locations of polluted air that reach receptors can be found from a **backward trajectory**.

The goal of calculating dispersion is to predict or diagnose the pollutant concentration at some point distant from the source. **Concentration** c is often measured as a mass per unit volume, such as μg m^{-3}. It can also be measured as volume ratio of pollutant gas to clean air, such as parts per million (**ppm**). See the INFO box for details about units.

A **source - receptor** framework is used to relate emission factors to predicted downwind concentration values. We can examine pollutants emitted at a known rate from a **point source** such as a **smoke stack**. We then follow the pollutants as they are blown downwind and mix with the surrounding air. Eventually, the mixture reaches a receptor such as a sensor, person, plant, animal or structure, where we can determine the expected concentration.

In this chapter, we will assume that the mean wind is known, based on either weather observations, or on forecasts. We will focus on the plume rise and dispersion of the pollutants, which allows us to determine the concentration of pollutants downwind of a known source.

AIR QUALITY STANDARDS

To prevent or reduce health problems associated with air pollutants, many countries set air quality standards. These standards prescribe the maximum average concentration levels in the ambient air, as allowed by law. Failure to satisfy these standards can result in fines, penalties, and increased government regulation.

In the USA, the standards are called **National Ambient Air Quality Standards (NAAQS)**. In Canada, they are called Canadian Ambient Air Quality Standards (CAAQS). The European Union (EU) also sets Air Quality Standards. Other countries have similar names for such goals. Table 19–1 lists standards for a few countries. Governments can change these standards.

In theory, these average concentrations are not to be exceeded anywhere at ground level. In practice, meteorological events sometimes occur, such as light winds and shallow ABLs during anticyclonic conditions, that trap pollutants near the ground and cause concentration values to become undesirably large.

Also, temporary failures of air-pollution control measures at the source can cause excessive amounts of pollutants to be emitted. Regulations in some of the countries allow for a small number of concentration **exceedances** without penalty.

To avoid expensive errors during the design of new factories, smelters, or power plants, **air pollution modeling** is performed to determine the likely pollution concentration based on expected emission rates. Usually, the greatest concentrations happen near the source of pollutants. The procedures presented in this chapter illustrate how concentrations at receptors can be calculated from known emission and weather conditions.

By comparing the predicted concentrations against the air quality standards of Table 18–1, engineers can modify the factory design as needed to ensure compliance with the law. Such modifications can include building taller smoke stacks, removing the pollutant from the stack effluent, changing fuels or raw materials, or utilizing different manufacturing or chemical processes.

Table 19-1. Air quality concentration standards for the USA (US), Canada (CAN), and The European Union (EU) for some of the commonly-regulated chemicals, as of Oct 2015. Concentrations represent averages over the time periods listed. For Canada, both the 2015 and 2020 standards are shown.

Avg. Time	US	CAN	EU
Sulfur Dioxide (SO_2)			
1 day			125 µg m^{-3}
3 h	1300 µg m^{-3} or 0.5 ppm		
1 h	75 ppb		350 µg m^{-3}
Nitrogen Dioxide (NO_2)			
1 yr	100 µg m^{-3} or 53 ppb		40 µg m^{-3}
1 h	100 ppb		200 µg m^{-3}
Carbon Monoxide (CO)			
8 h	10,000 µg m^{-3} or 9 ppm		10,000 µg m^{-3}
1 h	40,000 µg m^{-3} or 35 ppm		
Ozone (O_3)			
8 h	0.070 ppm	63 -> 62 ppb	120 µg m^{-3}
Fine Particulates, diameter < 10 µm (PM_{10})			
1 yr			40 µg m^{-3}
1 day	150 µg m^{-3}		50 µg m^{-3}
Very Fine Particulates, diam. < 2.5 µm ($PM_{2.5}$)			
1 yr	12 µg m^{-3}	10->8.8 µg m^{-3}	25 µg m^{-3}
1 day	35 µg m^{-3}	28->27 µg m^{-3}	
Lead (Pb)			
1 yr			0.5 µg m^{-3}
3 mo	0.15 µg m^{-3}		
Benzene (C_6H_6)			
1 yr			5 µg m^{-3}
Arsenic (As)			
1 yr			6 ng m^{-3}
Cadmium (Cd)			
1 yr			5 ng m^{-3}
Nickel (Ni)			
1 yr			20 ng m^{-3}
PAH (Polycyclic Aromatic Hydrocarbons)			
1 yr			1 ng m^{-3}

TURBULENCE STATISTICS

For air pollutants emitted from a point source such as the top of a smoke stack, mean wind speed and turbulence both affect the pollutant concentration measured downwind at ground level. The mean wind causes pollutant transport. Namely it blows or advects the pollutants from the source to locations downwind. However, while the plume is advecting, turbulent gusts acts to spread, or **disperse**, the pollutants as they mix with the surrounding air. Hence, we need to study both mean and turbulent characteristics of wind in order to predict downwind pollution concentrations.

Science Graffito

"If I had only one day left to live, I would live it in my statistics class — it would seem so much longer."

– Anonymous. [from C.C Gaither (Ed.), 1996: *Statistically Speaking: A Dictionary of Quotations*, Inst. of Physics Pub., 420 pp].

Review of Basic Definitions

Recall from the Atmospheric Boundary-Layer (ABL) chapter that variables such as velocity components, temperature, and humidity can be split into mean and turbulent parts. For example:

$$M = \overline{M} + M' \tag{19.1}$$

where M is instantaneous speed in this example, $\overline{M}$ is the mean wind speed [usually averaged over time (≈ 30 min) or horizontal distance (≈ 15 km)], and M' is the instantaneous deviation from the mean value.

The mean wind speed at any height z is

$$\overline{M}(z) = \frac{1}{N}\sum_{i=1}^{N} M_i(z) \tag{19.2}$$

where M_i is the wind speed measured at some time or horizontal location index i, and N is the total number of observation times or locations. Use similar definitions for mean wind components $\overline{U}$, $\overline{V}$, and $\overline{W}$.

Smoke plumes can spread in the vertical direction. Recall from the ABL chapter that the ABL wind speed often varies with height. Hence, the wind speed that affects the pollutant plume must be defined as an average speed over the vertical thickness of the plume.

If the wind speeds at different, equally spaced layers, between the bottom and the top of a smoke plume are known, and if k is the index of any layer, then the average over height is:

$$\overline{\overline{M}} = \frac{1}{K}\sum_{k=1}^{K} \overline{M}(z_k) \tag{19.3}$$

where the sum is over only those layers spanned by the plume. K is the total number of layers in the plume.

Sample Application (§)

Given an x-axis aligned with the mean wind U = 10 m s^{-1}, and the y-axis aligned in the crosswind direction, V. Listed at right are measurements of the V-component of wind.

a. Find the V mean wind speed and standard deviation.

b. If the vertical standard deviation is 1 m s^{-1}, is the flow isotropic?

t (h)	V (m s^{-1})
0.1	2
0.2	–1
0.3	1
0.4	1
0.5	–3
0.6	–2
0.7	0
0.8	2
0.9	–1
1.0	1

Find the Answer

Given: V speeds above, σ_w = 1 m s^{-1}, $\overline{U}$ = 10 m s^{-1}

Find: $\overline{V}$ = ? m s^{-1}, σ_v = ? m s^{-1}, isotropic (yes/no) ?

Assume V wind is constant with height.

Use eq. (19.2), except for V instead of M:

$\overline{V}(z) = \frac{1}{N}\sum_{i=1}^{N} V_i(z)$ = (1/10)·(0) = **0** m s^{-1}

Use eq. (19.5), but for V: $\sigma_V^2 = \frac{1}{N}\sum_{k=1}^{N}(V_k - \overline{V})^2$

= (1/10) · [4 + 1 + 1 + 1 + 9 + 4 + 0 + 4 + 1 + 1]
= 2.6 m^2 s^{-2}

Use eq. (19.6)

σ_v = [2.6 m^2 s^{-2}]$^{1/2}$ = **1.61** m s^{-1}

Use eq. (19.7): (σ_v = 1.61 m s^{-1}) > (σ_w = 1.0 m s^{-1}), therefore **Anisotropic** in the y-z plane (but no info on σ_u here).

Check: Units OK. Physics OK.

Exposition: The sigma values indicate the rate of plume spread. In this example is greater spread in the crosswind direction than in the vertical direction, hence dispersion looks like the "statically stable" case plotted in Fig. 19.2. By looking at the spread of a smoke plume, you can estimate the static stability.

This works for nearly horizontal plumes that have known vertical thickness. For the remainder of this chapter, we will use just one overbar (or sometimes no overbar) to represent an average over both time (index i), and vertical plume depth (index k).

The coordinate system is often chosen so that the x-axis is aligned with the mean wind direction, averaged over the whole smoke plume. Thus,

$$\bar{M} = \bar{U} \tag{19.4}$$

There is no lateral (crosswind) mean wind ($\bar{V} \approx 0$) in this coordinate system. The mean vertical velocity is quite small, and can usually be neglected ($\bar{W} \approx 0$, except near hills) compared to plume dispersion rates. However, $u' = U - \bar{U}$, $v' = V - \bar{V}$, and $w' = W - \bar{W}$ can be non-zero, and are all important.

Recall from the ABL chapter that **variance** $\sigma_A{}^2$ of any quantity A is defined as

$$\sigma_A{}^2 = \frac{1}{N}\sum_{k=1}^{N}(A_k - \bar{A})^2 = \frac{1}{N}\sum_{k=1}^{N}(a'^2) = \overline{a'^2} \tag{19.5}$$

The square root of the variance is the **standard deviation**:

$$\sigma_A = \left(\sigma_A{}^2\right)^{1/2} \tag{19.6}$$

The ABL chapter gives estimates of velocity standard deviations.

Isotropy (again)

Recall from the ABL chapter that turbulence is said to be **isotropic** when:

$$\sigma_u{}^2 = \sigma_v{}^2 = \sigma_w{}^2 \tag{19.7}$$

As will be shown later, the rate of smoke dispersion depends on the velocity variance. Thus, if turbulence is isotropic, then a smoke puff would tend to expand isotropically, as a sphere; namely, it would expand equally in all directions.

There are many situations where turbulence is **anisotropic** (not isotropic). During the daytime over bare land, rising thermals create stronger vertical motions than horizontal. Hence, a smoke puff would **loop** up and down and disperse more in the vertical. At night, vertical motions are very weak, while horizontal motions can be larger. This causes smoke puffs to **fan** out horizontally at night, and for other stable cases.

Similar effects operate on smoke plumes formed from continuous emissions. For this situation, only the vertical and lateral velocity variances are relevant. Fig. 19.2 illustrates how isotropy and anisotropy affect average smoke plume cross sections.

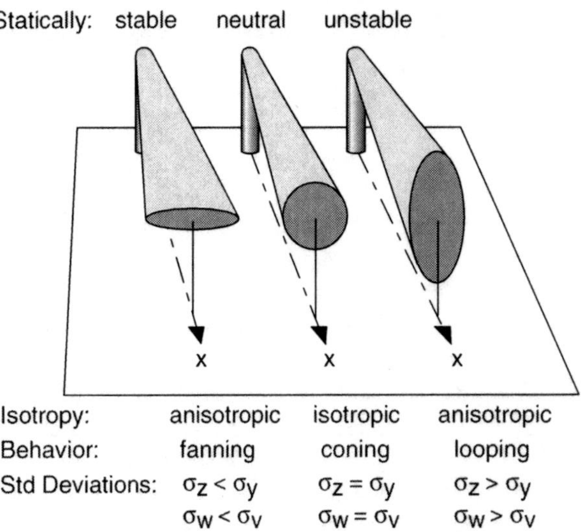

Figure 19.2
Isotropic and anisotropic dispersion of smoke plumes. The shapes of the ends of these smoke plumes are also sketched in Fig. 19.3, along the arc labeled "dispersion isotropy".

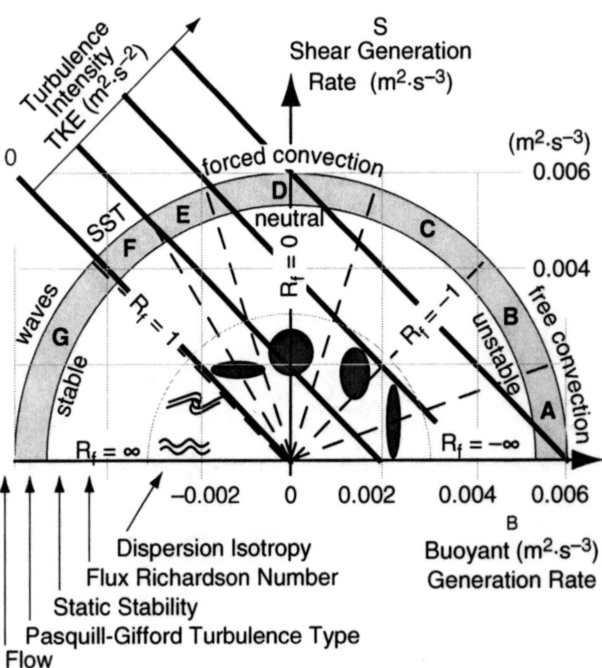

Figure 19.3
Rate of generation of TKE by buoyancy (abscissa) and shear (ordinate). Shape and rates of plume dispersion (dark spots or waves). Dashed lines separate sectors of different Pasquill-Gifford turbulence type. Isopleths of TKE intensity (dark diagonal lines). R_f is flux Richardson number. SST is stably-stratified turbulence. (See the Atmospheric Boundary Layer chapter for turbulence details.)

Table 19-2a. Pasquill-Gifford turbulence types for **Daytime**. *M* is wind speed at z = 10 m.

M (m s^{-1})	Insolation (incoming solar radiation)		
	Strong	Moderate	Weak
< 2	A	A to B	B
2 to 3	A to B	B	C
3 to 4	B	B to C	C
4 to 6	C	C to D	D
> 6	C	D	D

Table 19-2b. Pasquill-Gifford turbulence types for **Nighttime**. *M* is wind speed at z = 10 m.

M (m s^{-1})	Cloud Coverage	
	≥ 4/8 low cloud or thin overcast	≤ 3/8
< 2	G	G
2 to 3	E	F
3 to 4	D	E
4 to 6	D	D
> 6	D	D

Sample Application

Determine the PG turbulence type during night with 25% cloud cover, and winds of 5 m s^{-1}.

Find the Answer

Given: M = 5 m s^{-1}, clouds = 2/8 .
Find: PG = ?
Use Table 18–2b. <u>PG = "D"</u>

Check: Units OK. Physics OK.
Exposition: As wind speeds increase, the PG category approaches "D" (statically neutral), for both day and night conditions. "D" implies "forced convection".

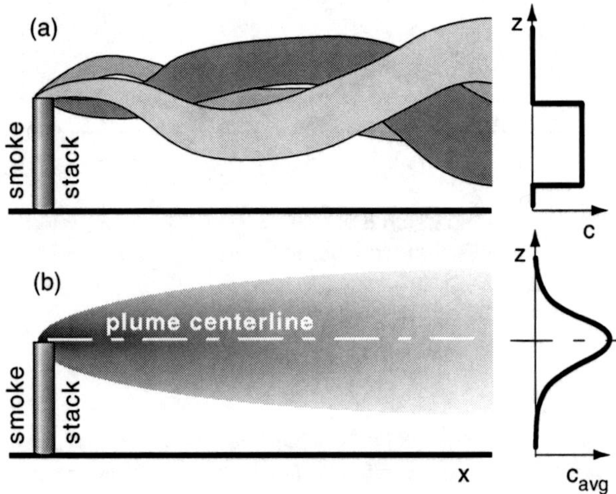

Figure 19.4
(a) Snapshots showing an instantaneous smoke plume at different times, also showing a concentration c profile for the dark-shaded plume. (b) Average over many plumes, with the average concentration c_{avg} profile shown at right.

Pasquill-Gifford (PG) Turbulence Types

During weak advection, the nature of convection and turbulence are controlled by the wind speed, incoming solar radiation (**insolation**), cloud shading, and time of day or night. Pasquill and Gifford (PG) suggested a practical way to estimate the nature of turbulence, based on these forcings.

They used the letters "A" through "F" to denote different turbulence types, as sketched in Fig. 19.3 (reproduced from the ABL chapter). "A" denotes free convection in statically unstable conditions. "D" is forced convection in statically neutral conditions. Type "F" is for statically stable turbulence. Type "G" was added later to indicate the strongly statically stable conditions associated with meandering, wavy plumes in otherwise <u>non</u>turbulent flow. PG turbulence types can be estimated using Tables 19-2.

Early methods for determining pollutant dispersion utilized a different plume spread equation for each Pasquill-Gifford type. One drawback is that there are only 7 discrete categories (A – G); hence, calculated plume spread would suddenly jump when the PG category changed in response to changing atmospheric conditions.

Newer air pollution models do not use the PG categories, but use the fundamental meteorological conditions (such as shear and buoyant TKE generation, or values of velocity variances that are continuous functions of wind shear and surface heating), which vary smoothly as atmospheric conditions change.

DISPERSION STATISTICS

Snapshot vs. Average

Snapshots of smoke plumes are similar to what you see with your eye. The plumes have fairly-well defined edges, but each plume wiggles up and down, left and right (Fig. 19.4a). The concentration c through such an instantaneous smoke plume can be quite variable, so a hypothetical vertical profile is sketched in Fig. 19.4a.

A time exposure of a smoke-stack plume might appear as sketched in Fig. 19.4b. When averaged over a time interval such as an hour, most of the pollutants are found near the centerline of the plume. Average concentration decreases smoothly with distance away from the centerline. The resulting profile of concentration is often bell shaped, or Gaussian. Air quality standards in most countries are based on averages, as was listed in Table 18–1.

Center of Mass

The **plume centerline** height z_{CL} can be defined as the location of the **center of mass** of pollutants. In other words, it is the average height of the pollutants. Suppose you measure the concentration c_k of pollutants at a range of equally-spaced heights z_k through the smoke plume. The center of mass is a weighted average defined by:

$$z_{CL} = \overline{z} = \frac{\sum_{k=1}^{K} c_k \cdot z_k}{\sum_{k=1}^{K} c_k} \qquad \bullet(19.8)$$

where K is the total number of heights, and k is the height index, and the overbar denotes a mean.

For passive tracers with slow exit velocity from the stack, the plume centerline is at the same height as the top of the stack. For buoyant plumes in hot exhaust gases, and for smoke blown at high speed out of the top of the stack, the centerline rises above the top of the stack.

A similar center of mass can be found for the crosswind (lateral) location, assuming measurements are made at equal intervals across the plume. Passive tracers blow downwind. Thus, the center of mass of a smoke plume, when viewed from above such as from a satellite, follows a mean wind trajectory from the stack location (see the discussion of streamlines, streaklines, and trajectories in the Regional Winds chapter).

Standard Deviation – Sigma

For time-average plumes such as in Figs. 19.4b and 19.5, the plume edges are not easy to locate. They are poorly defined because the bell curve gradually approaches zero concentration with increasing distance from the centerline. Thus, we cannot use edges to measure plume spread (depth or width).

Instead, the standard deviation σ_z of pollutant location is used as a measure of plume spread, where standard deviation is the square root of the variance σ_z^2. The vertical-position deviations must be weighted by the pollution concentration to find sigma, as shown here:

$$\sigma_z = \left[\frac{\sum_{k=1}^{K} c_k \cdot (z_k - \overline{z})^2}{\sum_{k=1}^{K} c_k} \right]^{1/2} \qquad \bullet(19.9)$$

where $\overline{z} = z_{CL}$ is the average height found from the previous equation.

A similar equation can be defined for lateral standard deviation σ_y. The vertical and lateral dis-

ON DOING SCI. • Data Misinterpretation

Incomplete data can be misinterpreted, leading to expensive erroneous conclusions. Suppose an air-pollution meteorologist/engineer erects a tall tower at the site of a proposed 75 m high smoke stack. On this tower are electronic thermometers at two heights: 50 and 100 m. On many days, these measurements give temperatures that are on the same adiabat (Fig. a).

Regarding static stability, one interpretation (Fig. b) is that the atmosphere is statically <u>neutral</u>. Another interpretation (Fig. c) is that it is a statically <u>unstable</u> convective mixed layer. If neutral stability is erroneously noted on a day of static instability, then the corresponding predictions of dispersion rate and concentrations will be embarrassingly wrong.

To resolve this dilemma, the meteorologist/engineer needs additional info. Either add a third thermometer near the ground, or add a net radiation sensor, or utilize manual observations of sun, clouds, and wind to better determine the static stability.

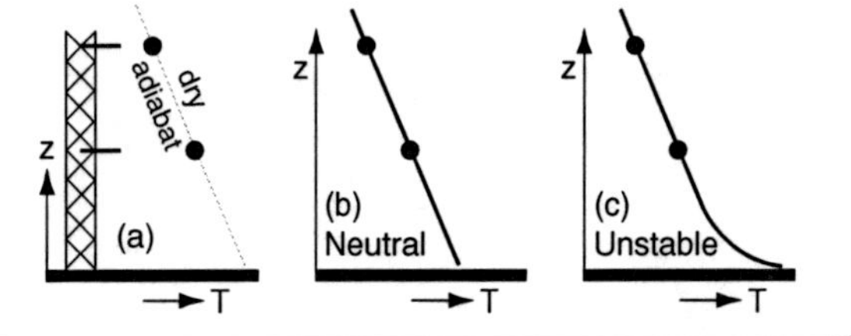

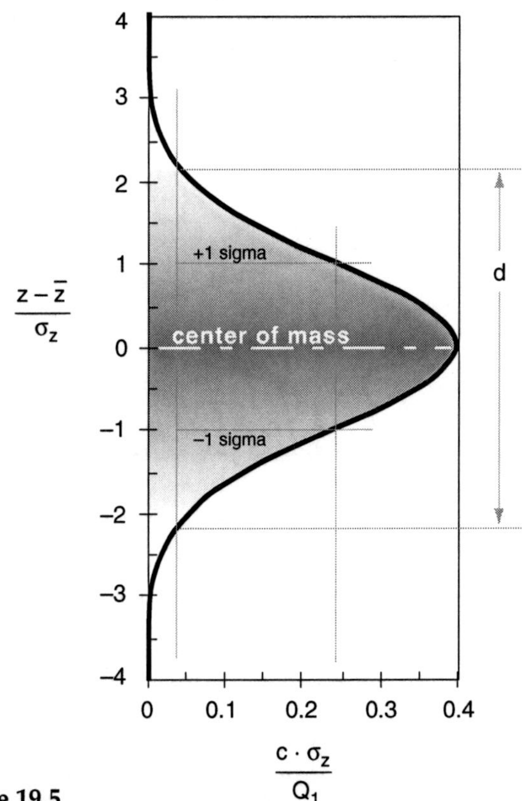

Figure 19.5
Gaussian curve.
Each unit along the ordinate corresponds to one standard deviation away from the center. Nominal 1-D plume width is d.

persion need not be equal, because the dispersive nature of turbulence is not the same in the vertical and horizontal when turbulence is anisotropic.

When the plume is compact, the standard deviation and variance are small. These statistics increase as the plume spreads. Hence we expect sigma to increase with distance downwind of the stack. Such spread will be quantified in the next main section.

Sample Application (§)

Given the following 1-D concentration measurements. Find the plume centerline height, standard deviation of height, and nominal plume depth. Plot the best-fit curve through these data points.

z (km)	c(μg m^{-1})
2.0	0
1.8	1
1.6	3
1.4	5
1.2	7
1.0	6
0.8	2
0.6	1
0.4	0
0.2	0
0.0	0

Find the Answer

Given: Δz = 0.2 km, with concentrations above

Find: $\bar{z}$ = ? km, σ_z = ? km, Q_1 = ? km·μg m^{-1}, d = ? km, and plot $c(z)$ = ? μg m^{-1}

Use eq. (19.8) to find the plume centerline height:

$\bar{z}$ = (30.2 km·μg·m^{-1}) / (25 μg·m^{-1}) = **1.208 km**

Use eq. (19.9):

σ_z^2 = (1.9575 km^2·μg·m^{-1})/(25 μg·m^{-1}) = 0.0783 km^2

$\sigma_z = [\sigma_z^2]^{1/2}$ = [0.0783 km^2]$^{1/2}$ = **0.28 km**

Use eq. (19.12):

d = 4.3 · (0.28 km) = **1.2 km**

Use eq. (19.11):

Q_1 = (0.2 km) · (25 μg·m^{-1}) = **5.0 km·μg m^{-1}**

Use eq. (19.10) to plot the best-fit curve:

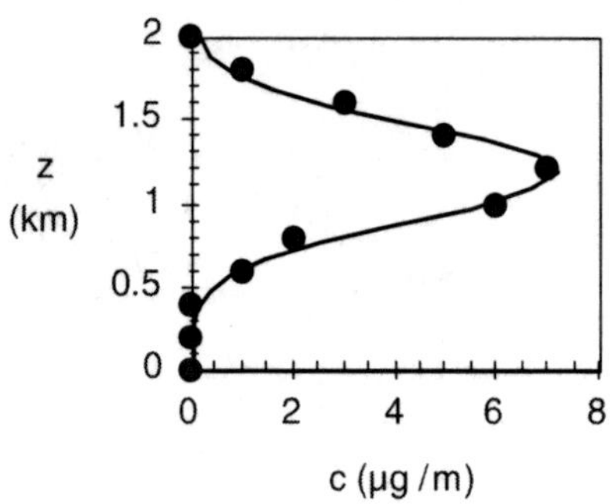

Check: Units OK. Physics OK. Sketch OK.

Exposition: The curve is a good fit to the data. Often the measured data will have some scatter due to the difficulty of making concentration measurements, but the equations in this section are able to find the best-fit curve. This statistical curve-fitting method is called the **method of moments**, because we matched the first two statistical moments (mean & variance) of the Gaussian distribution to the corresponding moments calculated from the data.

Gaussian Curve

The **Gaussian** or "**normal**" curve is bell shaped, and is given in one-dimension (1-D) by:

$$c(z) = \frac{Q_1}{\sigma_z\sqrt{2\pi}} \cdot \exp\left\{-0.5\left[\frac{z-\bar{z}}{\sigma_z}\right]^2\right\} \qquad •(19.10)$$

where $c(z)$ is the one-dimensional concentration (g m^{-1}) at any height z, and Q_1 (g) is the total amount of pollutant emitted.

This curve is symmetric about the mean location, and has tails that asymptotically approach zero as z approaches infinity (Fig. 19.5). The area under the curve is equal to Q_1, which physically means that pollutants are conserved. The **inflection points** in the curve (points where the curve changes from concave left to concave right) occur at exactly one σ_z from the mean. Between $\pm 2 \cdot \sigma_z$ are 95% of the pollutants; hence, the Gaussian curve exhibits strong central tendency.

Eq. (19.10) has three parameters: Q_1, $\bar{z}$, and σ_z. These parameters can be estimated from measurements of concentration at equally-spaced heights through the plume, in order to find the best-fit Gaussian curve. The last two parameters are found with eqs. (19.8) and (19.9). The first parameter is found from:

$$Q_1 = \Delta z \cdot \sum_{k=1}^{K} c_k \qquad •(19.11)$$

where Δz is the height interval between neighboring measurements.

Nominal Plume Edge

For practical purposes, the edge of a Gaussian plume is defined as the location where the concentration is 10% of the centerline concentration. This gives a plume spread (e.g., depth from top edge to bottom edge) of

$$d \approx 4.3 \cdot \sigma_z \qquad •(19.12)$$

TAYLOR'S STATISTICAL THEORY

Statistical theory explains how plume dispersion statistics depend on turbulence statistics and downwind distance.

Passive Conservative Tracers

Many pollutants consist of gases or very fine particles. They passively ride along with the wind, and trace out the air motion. Hence, the rate of dispersion depends solely on the air motion (wind and turbulence) and not on the nature of the pollutant. These are called **passive tracers**. If they also do not decay, react, fall out, or stick to the ground, then they are also called **conservative tracers,** because all pollutant mass emitted into the air is conserved.

Some pollutants are not passive or conservative. Dark soot particles can adsorb sunlight to heat the air, but otherwise they might not be lost from the air. Thus, they are active because they alter turbulence by adding buoyancy, but are conservative. Radioactive pollutants are both nonconservative and active, due to radioactive decay and heating.

For passive conservative tracers, the amount of dispersion (σ_y or σ_z) depends not only on the intensity of turbulence (σ_v or σ_w, see the ABL chapter), but on the distribution of turbulence energy among eddies of different sizes. For a plume of given spread, eddies as large as the plume diameter cause much greater dispersion than smaller-size eddies. Thus, dispersion rate increases with time or downwind distance, as shown below.

Dispersion Equation

G.I. Taylor theoretically examined an individual passive tracer particle as it moved about by the wind. Such an approach is **Lagrangian**, as discussed in the Thermodynamics chapter. By averaging over many such particles within a smoke cloud, he derived a statistical theory for turbulence.

One approximation to his result is

•(19.13a)

$$\sigma_y^2 = 2\cdot\sigma_v^2\cdot t_L^2\cdot\left[\frac{x}{M\cdot t_L} - 1 + \exp\left(-\frac{x}{M\cdot t_L}\right)\right]$$

•(19.13b)

$$\sigma_z^2 = 2\cdot\sigma_w^2\cdot t_L^2\cdot\left[\frac{x}{M\cdot t_L} - 1 + \exp\left(-\frac{x}{M\cdot t_L}\right)\right]$$

where x is distance downwind from the source, M is wind speed, and t_L is the Lagrangian time scale.

HIGHER MATH • Diffusion Equation

The Gaussian concentration distribution is a solution to the diffusion equation, as is shown here.

For a conservative, passive tracer, the budget equation says that concentration c in a volume will increase with time t if greater tracer flux F_c enters a volume than leaves. In one dimension (z), this is:

$$\frac{dc}{dt} = -\frac{\partial F_c}{\partial z} \qquad \text{(a)}$$

If turbulence consists of only small eddies, then turbulent flux of tracer flows down the mean tracer gradient:

$$F_c = -K\frac{\partial c}{\partial z} \qquad \text{(b)}$$

where K, the eddy diffusivity, is analogous to a molecular diffusivity (see K-Theory in the Atmospheric Boundary Layer chapter), and F_c is in kinematic units (concentration times velocity).

Plug eq. (b) into (a), and assume constant K, which gives the 1-D **diffusion equation**:

$$\boxed{\frac{dc}{dt} = K\frac{\partial^2 c}{\partial z^2}} \qquad \text{(c)}$$

This parabolic differential equation can be solved with initial conditions (IC) and boundary conditions (BC). Suppose a smoke puff of mass Q grams of tracer is released in the middle of a vertical pipe that is otherwise filled with clean air at time $t = 0$. Define the vertical coordinate system so that $z = 0$ at the initial puff height (and $\bar{z} = 0$). Dispersion up and down the pipe is one-dimensional.

IC: $c = 0$ at $t = 0$ everywhere except at $z = 0$.

BC1: $\int c\,dz = Q$, at all t, where integration is $-\infty$ to ∞

BC2: c approaches 0 as z approaches $\pm\infty$, at all t.

The solution is:

$$c = \frac{Q}{(4\pi K t)^{1/2}}\exp\left(\frac{-z^2}{4Kt}\right) \qquad \text{(d)}$$

You can confirm that this is a solution by plugging it into eq. (c), and checking that the LHS equals the RHS. It also satisfies all initial & boundary conditions.

Comparing eq. (d) with eq. (19.10), we can identify the standard deviation of height as

$$\sigma_z = \sqrt{2Kt} \qquad \text{(e)}$$

which says that tracer spread increases with the square root of time, and greater eddy-diffusivity causes faster spread rate. Thus, the solution is Gaussian:

$$\boxed{c = \frac{Q}{\sqrt{2\pi}\cdot\sigma_z}\exp\left[-\frac{1}{2}\left(\frac{z}{\sigma_z}\right)^2\right]} \qquad \text{(19.10)}$$

Finally, using **Taylor's hypothesis** that $t = x/M$, we can compare eq. (e) with the σ_z version of eq. (19.15), and conclude that:

$$K = \sigma_w^2\cdot t_L \qquad \text{(f)}$$

showing how K increases with turbulence intensity.

Sample Application (§)

Plot vertical and horizontal plume spread σ_z and σ_y vs. downwind distance x, using a Lagrangian time scale of 1 minute and wind speed of 10 m s^{-1} at height 100 m in a neutral boundary layer of depth 500 m. There is a rough surface of varied crops.

a) Plot on both linear and log-log graphs.

b) Also plot the short and long-distance limits of σ_y on the log-log graph.

Find the Answer

Given: z = 100 m, M = 10 m s^{-1}, t_L = 60 s, h = 500 m

Find: σ_z and σ_y (m) vs. x (km).

Refer to Atmospheric Boundary Layer chapter to calculate the info needed in the eqs. for Taylor's statistical theory.

Use Table 18-1 for rough surface of varied crops: aerodynamic roughness length is z_o = 0.25 m

Use this z_o in eq. (18.13) to get the friction velocity:
u_* = [0.4·(10m s^{-1})] / ln(100m/0.25m) = 0.668 m s^{-1}

Use this u_* in eq. (18.25b) to get the velocity variance
σ_v = 1.6·(0.668m s^{-1})·[1 – 0.5·(100m/500m)] = 0.96 m s^{-1}

Similarly, use eq. (18.25c): σ_w = 0.75 m s^{-1}

Use σ_v and σ_w in eqs. (19.13a & b) in a spreadsheet to calculate σ_y and σ_z, and plot the results on graphs:

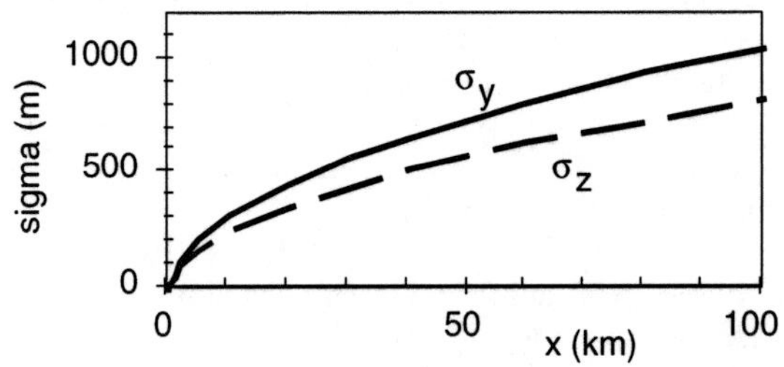

A linear graph is shown above, and log-log below.

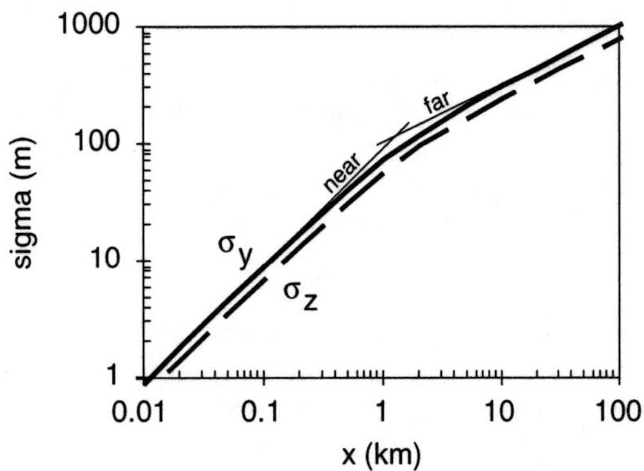

Eqs. (19.14) & (19.15) for the near and far approximations are plotted as the thin solid lines.

Check: Units OK. Physics OK. Sketch OK.

Exposition: Plume spread increases with distance downwind of the smoke stack. $\sigma_y \approx \sigma_z$ at any x, giving the nearly isotropic dispersion expected for statically neutral air. The cross-over between short and long time limits is at $x \approx 2{\cdot}M{\cdot}t_L$ = 1.2 km.

Thus, the spread (σ_y and σ_z) of passive tracers increases with turbulence intensity (σ_v and σ_w) and with downwind distance x.

The **Lagrangian time scale** is a measure of how quickly a variable becomes uncorrelated with itself. For very small-scale atmospheric eddies, this time scale is only about 15 seconds. For convective thermals, it is on the order of 15 minutes. For the synoptic-scale high and low pressure systems, the Lagrangian time scale is on the order of a couple days. We will often use a value of $t_L \approx$ 1 minute for dispersion in the boundary layer.

Dispersion Near & Far from the Source

Close to the source (at small times after the start of dispersion), eq. (19.13a) reduces to

$$\sigma_y \approx \sigma_v \cdot \left(\frac{x}{M}\right) \qquad \bullet(19.14)$$

while far from the source it can be approximated by:

$$\sigma_y \approx \sigma_v \cdot \left(2 \cdot t_L \cdot \frac{x}{M}\right)^{1/2} \qquad \bullet(19.15)$$

There are similar equations for σ_z as a function of σ_w.

Thus, we expect plumes to initially spread linearly with distance near to the source, but change to square-root with distance further downwind.

DISPERSION IN NEUTRAL & STABLE BOUNDARY LAYERS

To calculate pollutant concentration at the surface, one needs to know both the height of the plume centerline, and the spread of pollutants about that centerline. **Plume rise** is the name given to the first issue. **Dispersion** (from Taylor's statistical theory) is the second. When they are both used in an expression for the average spatial distribution of pollutants, pollution concentrations can be calculated.

Plume Rise

Ground-level concentration generally decreases as plume-centerline height increases. Hence, plume rise above the physical stack top is often desirable. The centerline of plumes can rise above the stack top due to the initial momentum associated with exit velocity out of the top of the stack, and due to buoyancy if the effluent is hot.

Neutral Boundary Layers

Statically neutral situations are found in the residual layer (not touching the ground) during light winds at night. They are also found throughout the bottom of the boundary layer (touching the ground) on windy, overcast days or nights.

The height z_{CL} of the plume centerline above the ground in neutral boundary layers is:

$$z_{CL} = z_s + \left[a \cdot l_m^2 \cdot x + b \cdot l_b \cdot x^2 \right]^{1/3} \qquad \bullet(19.16)$$

where a = 8.3 (dimensionless), b = 4.2 (dimensionless), x is distance downwind of the stack, and z_s is the physical stack height. This equation shows that the plume centerline keeps rising as distance from the stack increases. It ignores the capping inversion at the ABL top, which would eventually act like a lid on plume rise and upward spread.

A momentum length scale, l_m, is defined as:

$$l_m \approx \frac{W_o \cdot R_o}{M} \qquad \bullet(19.17)$$

where R_o is the stack-top radius, W_o is stack-top exit velocity of the effluent, and M is the ambient wind speed at stack top. l_m can be interpreted as a ratio of vertical emitted momentum to horizontal wind momentum.

A buoyancy length scale, l_b, is defined as:

$$l_b \approx \frac{W_o \cdot R_o^2 \cdot |g|}{M^3} \cdot \frac{\Delta\theta}{\theta_a} \qquad \bullet(19.18)$$

where $|g|$ = 9.8 m s^{-2} is gravitational acceleration magnitude, $\Delta\theta = \theta_p - \theta_a$ is the temperature excess of the effluent, θ_p is the initial stack gas potential temperature at stack top, and θ_a is the ambient potential temperature at stack top. l_b can be interpreted as a ratio of vertical buoyancy power to horizontal power of the ambient wind.

Stable Boundary Layers

In statically stable situations, the ambient potential temperature increases with height. This limits the plume-rise centerline to a final equilibrium height z_{CLeq} above the ground:

$$z_{CLeq} = z_s + 2.6 \cdot \left(\frac{l_b \cdot M^2}{N_{BV}^2} \right)^{1/3} \qquad \bullet(19.19)$$

where the Brunt-Väisälä frequency N_{BV} is used as a measure of static stability (see the Atmospheric Stability chapter).

Sample Application (§)

At stack top, effluent velocity is 20 m s^{-1}, temperature is 200°C, emission rate is 250 g s^{-1} of SO_2. The stack is 75 m high with 2 m radius at the top. At stack top, the ambient wind is 5 m s^{-1}, and ambient potential temperature is 20°C. For a neutral boundary layer, plot plume centerline height vs. downwind distance.

Find the Answer

Given: W_o = 20 m s^{-1}, Q = 250 g s^{-1}, z_s = 75 m, M = 5 m s^{-1}, θ_a = 293 K, θ_p = 473 K + (9.8 K km^{-1})·(0.075 km) = 474 K

Find: $z_{CL}(x)$ = ? m.

Use eq. (19.17): l_m = (20m s^{-1})·(2m) / (5m s^{-1}) = 8 m

Use eq. (19.18):

$$l_b \approx \frac{(20\text{m/s}) \cdot (2\text{m})^2 \cdot (9.8\text{m·s}^{-2})}{(5\text{m/s})^3} \cdot \frac{(474 - 293\text{K})}{293\text{K}} = 3.87 \text{ m}$$

Use eq. (19.16):

$$z_{CL} = (75\text{m}) + \left[8.3 \cdot (8\text{m})^2 \cdot x + 4.2 \cdot (3.87\text{m}) \cdot x^2 \right]^{1/3}$$

This is shown as the solid line on the plot below:

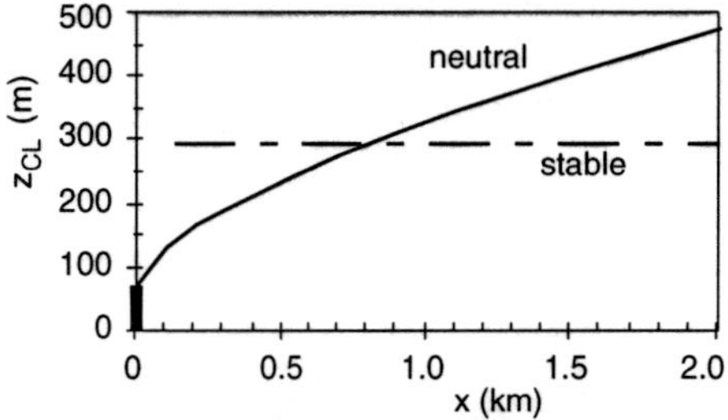

Check: Units OK. Physics OK. Sketch OK.

Exposition: In neutral conditions, the plume continues to rise with downwind distance. However, real plumes usually hit an elevated inversion & stop rising. Use a thermo diagram to locate the inversions aloft.

Sample Application

Same as previous example, but for a stable boundary layer with $\Delta\theta_a/\Delta z$ = 5°C km^{-1}. Find z_{CLeq}.

Find the Answer

Given: W_o = 20 m s^{-1}, Q = 250 g s^{-1}, z_s = 75 m, M = 5 m s^{-1}, θ_p = 474 K, θ_a = 293 K, $\Delta\theta_a/\Delta z$ = 5°C km^{-1}

Find: Δz_{CLeq} = ? m, then plot z_{CLeq} vs. x

Use eq. (5.4b) to find the Brunt-Väisälä frequency^2:

N_{BV}^2 = [(9.8m·s^{-2})/293K]·[5K/1000m] =1.67x10^{-4} s^{-2}

Use eq. (19.19):

$$z_{CLeq} = (75\text{m}) + 2.6 \cdot \left(\frac{(3.87\text{m}) \cdot (5\text{m/s})^2}{1.67 \times 10^{-4} \text{s}^{-2}} \right)^{1/3}$$

$z_{CLeq} = z_s + \Delta z_{CL\,eq}$ = 75 m + 216.7 m = **291.75 m**

See dashed line in the previous Sample Application.

Check: Units OK. Physics OK. Sketch OK.

Exposition: The actual plume centerline does not reach the equilibrium height instantly. Instead, it approaches it a bit slower than the neutral plume rise curve plotted in the previous solved ex.

Sample Application (§)

Given a "surface" wind speed of 10 m s^{-1} at 10 m above ground, neutral static stability, boundary layer depth 800 m, surface roughness length 0.1 m, emission rate of 300 g s^{-1} of passive, non-buoyant SO_2, wind speed of 20 m s^{-1} at plume centerline height, and Lagrangian time scale of 1 minute.

Plot isopleths of concentration at the ground for plume centerline heights of: (a) 100m, (b) 200m

Find the Answer

Given: M = 10 m s^{-1} at z = 10 m, z_o = 0.1 m,
M = 20 m s^{-1} at z = 100 m = z_{CL}, neutral,
Q = 300 g s^{-1} of SO_2, t_L = 60 s, h = 800 m

Find: c (μg m^{-3}) vs. x (km) and y (km), at z = 0.

Assume z_{CL} is constant.

Use eq. (18.13) from the Atmospheric Boundary Layer (ABL) chapter:

u_* = 0.4·(10 m s^{-1})/ln(10 m/0.1 m) = 0.869 m s^{-1}

(a) Use eqs. (18.25b) & (18.25c) from the ABL chapter:

σ_v = 1.6·(0.869m s^{-1})·[1–0.5(100/800)] = 1.3 m s^{-1}

σ_w = 1.25·(0.869m s^{-1})·[1–0.5(100/800)] = 1.02 m s^{-1}

Use eq. (19.13a & b) in a spreadsheet to get σ_y and σ_z vs. x. Then use eq. (19.21) to find c at each x and y:

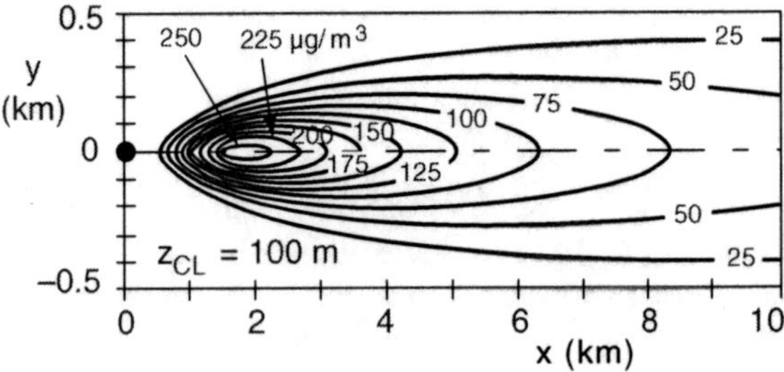

(b) Similarly, for a higher plume centerline:

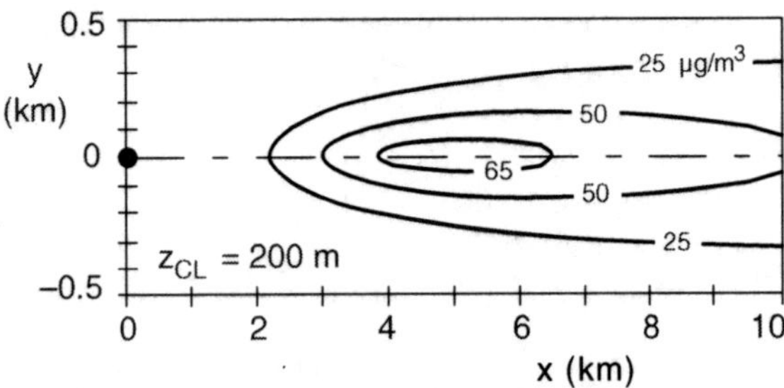

Check: Units OK. Physics OK.

Exposition: These plots show the pollutant **footprints**. Higher plume centerlines cause lower concentrations at the ground. That is why engineers design tall smoke stacks, and try to enhance buoyant plume rise.

Faster wind speeds also cause more dilution. Because faster winds are often found at higher altitudes, this also favors tall stacks for reducing surface concentrations.

Gaussian Concentration Distribution

For neutral and stable boundary layers (PG types C through F), the sizes of turbulent eddies are relatively small compared to the depth of the boundary layer. This simplifies the problem by allowing turbulent dispersion to be modeled analogous to molecular diffusion. For this situation, the average concentration distribution about the plume centerline is well approximated by a 3-D Gaussian bell curve:

•(19.20)

$$c = \frac{Q}{2\pi\sigma_y\sigma_z M}\cdot\exp\left[-0.5\cdot\left(\frac{y}{\sigma_y}\right)^2\right]\cdot \left\{\exp\left[-0.5\cdot\left(\frac{z-z_{CL}}{\sigma_z}\right)^2\right]+\exp\left[-0.5\cdot\left(\frac{z+z_{CL}}{\sigma_z}\right)^2\right]\right\}$$

where Q is the source emission rate of pollutant (g s^{-1}), σ_y and σ_z are the plume-spread standard deviations in the crosswind and vertical, y is lateral (crosswind) distance of the receptor from the plume centerline, z is vertical distance of the receptor above ground, z_{CL} is the height of the plume centerline above the ground, and M is average ambient wind speed at the plume centerline height.

For receptors at the ground (z = 0), eq. (19.20) reduces to:

•(19.21)

$$c = \frac{Q}{\pi\sigma_y\sigma_z M}\cdot\exp\left[-0.5\cdot\left(\frac{y}{\sigma_y}\right)^2\right]\cdot\exp\left[-0.5\cdot\left(\frac{z_{CL}}{\sigma_z}\right)^2\right]$$

The pattern of concentration at the ground is called a **footprint**.

The above two equations assume that the ground is flat, and that pollutants that hit the ground are "reflected" back into the air. Also, they do not work for dispersion in statically unstable mixed layers.

To use these equations, the turbulent velocity variances σ_v^2 and σ_w^2 are first found from the equations in the Atmospheric Boundary Layer chapter. Next, plume spread σ_y and σ_z is found from Taylor's statistical theory (eqs. 19.13). Plume centerline heights z_{CL} are then found from the previous subsection. Finally, they are all used in eqs. (19.20) or (19.21) to find the concentration at a receptor.

Recall that Taylor's statistical theory states that the plume spread increases with downwind distance. Thus, σ_y, σ_z, and z_{CL} are functions of x, which makes concentration c a strong function of x, in spite of the fact that x does not appear explicitly in the two equations above.

DISPERSION IN UNSTABLE BOUNDARY LAYERS (CONVECTIVE MIXED LAYERS)

During conditions of light winds over an underlying warmer surface (PG types A & B), the boundary layer is **statically unstable** and in a state of **free convection**. Turbulence consists of thermals of warm air that rise from the surface to the top of the mixed layer. These vigorous updrafts are surrounded by broader areas of weaker downdraft. The presence of such large turbulent structures and their asymmetry causes dispersion behavior that differs from the usual Gaussian plume dispersion.

As smoke is emitted from a point source such as a smoke stack, some of the emissions are by chance emitted into the updrafts of passing thermals, and some into downdrafts. Thus, the smoke appears to **loop** up and down, as viewed in a snapshot. However, when averaged over many thermals, the smoke disperses in a unique way that can be described deterministically. This description works only if variables are normalized by free-convection scales.

The first step is to get the meteorological conditions such as wind speed, ABL depth, and surface heat flux. These are then used to define the ABL convective scales such as the Deardorff velocity w_*. Source emission height, and downwind receptor distance are then normalized by the convective scales to make dimensionless distance variables.

Next, the dimensionless (normalized) variables are used to calculate the plume centerline height and vertical dispersion distance. These are then used as a first guess in a Gaussian equation for cross-wind-integrated concentration distribution, which is a function of height in the ABL. By dividing each distribution by the sum over all distributions, a corrected cross-wind-integrated concentration can be found that has the desirable characteristic of conserving pollutant mass.

Finally, the lateral dispersion distance is estimated. It is used with the cross-wind-integrated concentration to determine the dimensionless Gaussian concentration at any lateral distance from the plume centerline. Finally, the dimensionless concentration can be converted into a dimensional concentration using the meteorological scaling variables.

Although this procedure is complex, it is necessary, because non-local dispersion by large convective circulations in the unstable boundary layer works completely differently than the small-eddy dispersion in neutral and stable ABLs. The whole procedure can be solved on a spreadsheet, which was used to produce Figs. 19.7 and 19.8.

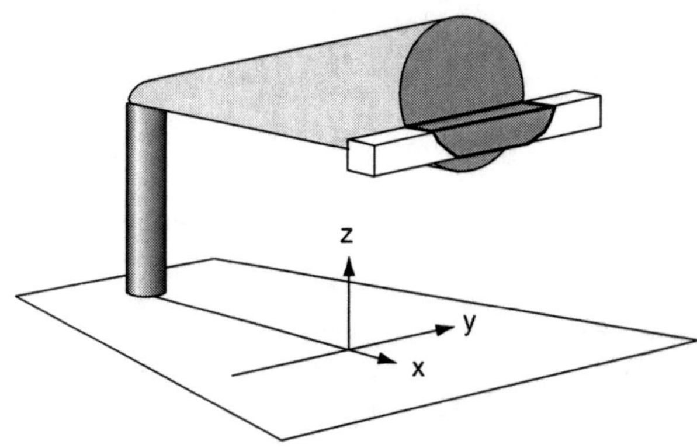

Figure 19.6
Crosswind integrated concentration c_y is the total amount of pollutants in a thin conceptual box (1 m^2 at the end) that extends crosswind (y-direction) across the smoke plume. This concentration is a function of only x and z.

Relevant Variables

Physical Variables:

c = concentration of pollutant (g m^{-3})

c_y=**crosswind-integrated concentration** (g m^{-2}), which is the total amount of pollutant within a long-thin box that is 1 m^2 on each end, and which extends laterally across the plume at any height z and downwind location x (see Fig. 19.6)

Q = emission rate of pollutant (g s^{-1})

x = distance of a receptor downwind of the stack (m)

z = height of a receptor above ground (m)

z_{CL} = height of the plume centerline (center of mass) above the ground (m)

z_s = source height (m) after plume-induced rise

σ_y = lateral standard deviation of pollutant (m)

σ_z = vertical standard deviation of pollutant (m)

σ_{zc} = vertical standard deviation of crosswind-integrated concentration of pollutant (m)

Mixed-Layer Scaling Variables:

F_H = effective surface kinematic heat flux (K·m s^{-1}), see Surface Fluxes section of Thermo. chapter.

M = mean wind speed (m s^{-1})

$$w_* = \left[\frac{|g| \cdot z_i \cdot F_H}{T_v}\right]^{1/3}$$ = **Deardorff velocity** (m s^{-1}) (19.22)

$\approx 0.08 \cdot w_B$, where w_B is the buoyancy velocity.

z_i = depth of the convective mixed layer (m)

Dimensionless Scales:

These are usually denoted by uppercase symbols (except for M and Q, which have dimensions).

$$C = \frac{c \cdot z_i^2 \cdot M}{Q}$$ = dimensionless concentration •(19.23)

$$C_y = \frac{c_y \cdot z_i \cdot M}{Q}$$ = dimensionless crosswind-integrated concentration (19.24)

$$X = \frac{x \cdot w_*}{z_i \cdot M}$$ = dimensionless downwind distance of receptor from source •(19.25)

$Y = y / z_i$ = dimensionless crosswind (lateral) distance of receptor from centerline (19.26)

$Z = z / z_i$ = dimensionless receptor height (19.27)

$Z_{CL} = z_{CL} / z_i$ = dimensionless plume centerline height (19.28)

$Z_s = z_s / z_i$ = dimensionless source height (19.29)

$\sigma_{yd} = \sigma_y / z_i$ = dimensionless lateral standard deviation (19.30)

$\sigma_{zdc} = \sigma_{zc} / z_i$ = dimensionless vertical standard deviation of crosswind-integrated concentration (19.31)

As stated in more detail earlier, to find the pollutant concentration downwind of a source during convective conditions, three steps are used: (1) Find the plume centerline height. (2) Find the crosswind integrated concentration at the desired x and z location. (3) Find the actual concentration at the desired y location.

Plume Centerline

For neutrally-buoyant emissions, the dimensionless height of the center of mass (= centerline Z_{CL}) varies with dimensionless distance downwind X: •(19.32)

$$Z_{CL} \approx 0.5 + \frac{0.5}{1 + 0.5 \cdot X^2} \cdot \cos\left[2\pi \frac{X}{\lambda} + \cos^{-1}(2 \cdot Z_s - 1)\right]$$

where Z_s is the dimensionless source height, and the dimensionless wavelength parameter is $\lambda = 4$.

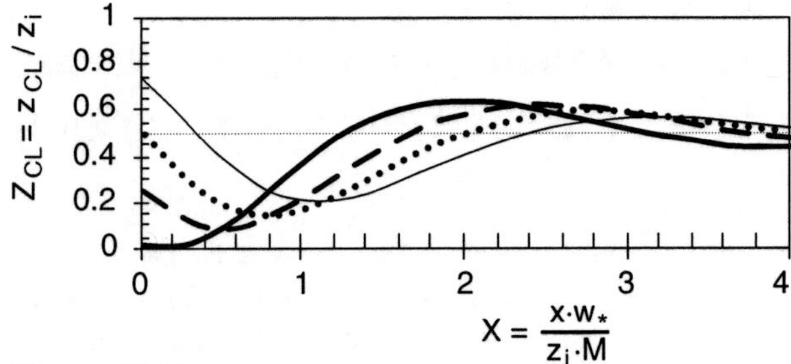

Figure 19.7
Height of the averaged pollutant centerline z_{CL} with downwind distance x, normalized by mixed-layer scales. Dimensionless source heights are $Z_s = z_s/z_i = 0.025$ (thick solid line); 0.25 (dashed); 0.5 (dotted); and 0.75 (thin solid). The plume is neutrally buoyant.

The centerline tends to move down from elevated sources, which can cause high concentrations at ground level (see Fig. 19.7). Then further downwind, they rise a bit higher than half the mixed-layer depth, before reaching a final height at $0.5 \cdot z_i$. For buoyant plumes, the initial downward movement of the centerline is much weaker, or does not occur.

Crosswind Integrated Concentration

The following algorithm provides a quick approximation for the crosswind integrated concentration. Find a first guess dimensionless C_y' as a function of dimensionless height Z using a Gaussian approach for vertical dispersion:

$$C_y' = \exp\left[-0.5 \cdot \left(\frac{Z - Z_{CL}}{\sigma_{zdc}'}\right)^2\right] \qquad \bullet(19.33)$$

where the prime denotes a first guess, and where the vertical dispersion distance is:

$$\sigma_{zdc}' = a \cdot X \qquad (19.34)$$

with $a = 0.25$. This calculation is done at K equally-spaced heights between the ground to the top of the mixed layer.

Next, find the average over all heights $0 \le Z \le 1$:

$$\overline{C_y'} = \frac{1}{K} \sum_{k=1}^{K} C_y' \qquad (19.35)$$

where index k corresponds to height z. Finally, calculate the revised estimate for dimensionless crosswind integrated concentration at any height:

$$C_y = C_y' / \overline{C_y'} \qquad \bullet(19.36)$$

Examples are plotted in Fig. 19.8 for various source heights.

Concentration

The final step is to assume that lateral dispersion is Gaussian, according to:

$$C = \frac{C_y}{(2\pi)^{1/2} \cdot \sigma_{yd}} \exp\left[-0.5 \cdot \left(\frac{Y}{\sigma_{yd}}\right)^2\right] \qquad \bullet(19.37)$$

The dimensionless standard deviation of lateral dispersion distance from an elevated source is

$$\sigma_{yd} \approx b \cdot X \qquad (19.38)$$

where $b = 0.5$.

At large downwind distances (i.e., at $X \geq 4$), the dimensionless crosswind integrated concentration always approaches $C_y \rightarrow 1.0$, at all heights. Also, directly beneath the plume centerline, $Y = 0$.

REVIEW

Pollutants emitted from a smoke stack will blow downwind and disperse by turbulent mixing with ambient air. By designing a stack of sufficient height, pollutants at ground level become sufficiently dilute as to not exceed local environmental air-quality standards. Additional buoyant plume rise above the physical stack top can further reduce ground-level concentrations.

Air-quality standards do not consider instantaneous samples of pollutant concentration. Instead, they are based on averages over time. For such averages, statistical descriptions of dispersion must be used, including the center of mass (plume centerline) and the standard deviation of location (proportional to plume spread).

For emissions in the boundary layer, the amount of dispersion depends on the type of turbulence. This relationship can be described by Taylor's statistical theory.

During daytime conditions of free convection, thermals cause a peculiar form of dispersion that often brings high concentrations of pollutants close to the ground. At night, turbulence is suppressed in the vertical, causing little dispersion. As pollutants remain aloft for this case, there is often little hazard at ground level. Turbulent dispersion is quite anisotropic for these convective and stable cases.

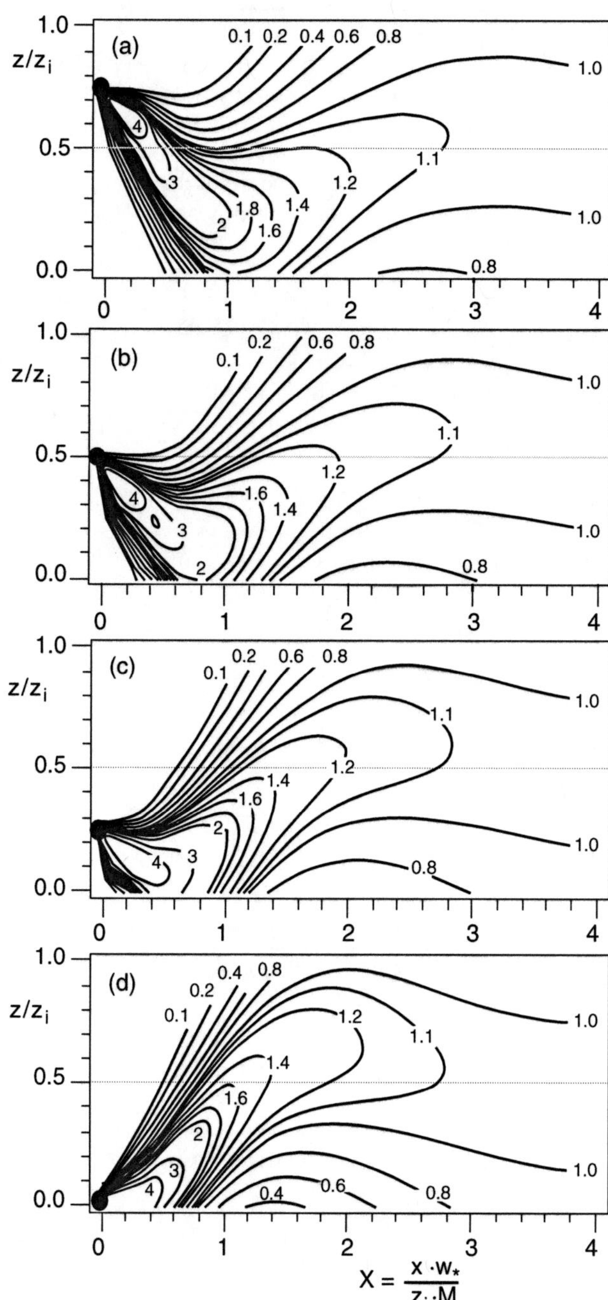

Figure 19.8 (at right)
Isopleths of dimensionless crosswind-integrated concentration $C_y = c_y \cdot z_i \cdot M / Q$ in a convective mixed layer, where c_y is crosswind integrated concentration, z_i is depth of the mixed layer, M is mean wind speed, and Q is emission rate of pollutants. Source heights are $Z_s = z_s/z_i =$ (a) 0.75, (b) 0.5, (c) 0.25, (d) 0.025, and are plotted as the large black dot at left.

Science Graffito

"The service we render to others is really the rent we pay for our room on the Earth."
– Sir Wilfred Grenfell.

Sample Application

Source emissions of 200 g s^{-1} of SO_2 occur at height 150 m. The environment is statically unstable, with a Deardorff convective velocity of 1 m s^{-1}, a mixed layer depth of 600 m, and a mean wind speed of 4 m s^{-1}.

Find the concentration at the ground 3 km downwind from the source, directly beneath the plume centerline.

Find the Answer

Given: $Q = 200$ g s^{-1}, $z_s = 150$ m, $z_i = 600$ m, $M = 4$ m s^{-1}, $w_* = 1$ m s^{-1}

Find: c (µg m^{-3}) at $x = 3$ km, $y = z = 0$.

Use eq. (19.25):

$$X = \frac{(3000\text{m})\cdot(1\text{m/s})}{(600\text{m})\cdot(4\text{m/s})} = 1.25$$

Use eq. (19.29): $Z_s = (150\text{m}) / (600\text{m}) = 0.25$

From Fig. 19.8c, read $C_y \approx 0.9$ at $X = 1.25$ and $Z = 0$.

Use eq. (19.38): $\sigma_{yd} \approx 0.5 \cdot (1.25) = 0.625$

Use eq. (19.37) with $Y = 0$:

$$C = \frac{0.9}{\sqrt{2\pi}\cdot 0.625} = 0.574$$

Finally, use eq. (19.23) rearranging it first to solve for concentration in physical units:

$$c = \frac{C\cdot Q}{z_i^2 \cdot M} = \frac{(0.574)\cdot(200\text{g/s})}{(600\text{m})^2\cdot(4\text{m/s})} = 79.8 \ \mu\text{g m}^{-3}$$

Check: Units OK. Physics OK.

Exposition: We were lucky that the dimensionless source height was 0.25, which allowed us to use Fig. 19.8c. For other source heights not included in that figure, we would have to create new figures using equations (19.33) through (19.36).

In statically neutral conditions of overcast and strong winds, turbulence is more isotropic. Smoke plumes disperse at roughly equal rates in the vertical and lateral directions, and are well described by Gaussian formulae.

Various classification schemes have been designed to help determine the appropriate turbulence and dispersion characteristics. These range from the detailed examination of the production of turbulence kinetic energy, through examination of soundings plotted on thermo diagrams, to look-up tables such as those suggested by Pasquill and Gifford.

Finally, although we used the words "smoke" and "smoke stack" in this chapter, most emissions in North America and Europe are sufficiently clean that particulate matter is not visible. This clean-up has been expensive, but commendable.

A SCIENTIFIC PERSPECTIVE • Citizen Scientist

Scientists and engineers have at least the same responsibilities to society as do other citizens. Like our fellow citizens, we ultimately decide the short-term balance between environmental quality and material wealth, by the goods that we buy and by the government leaders we elect. Be informed. Take a stand. Vote.

Perhaps we have more responsibility, because we can also calculate the long-term consequences of our actions. We have the ability to evaluate various options and build the needed technology. Take action. Discover the facts. Design solutions.

HOMEWORK EXERCISES

Broaden Knowledge & Comprehension

B1. Search the web for the government agency of your country that regulates air pollution. In the USA, it is the Environmental Protection Agency (EPA). Find the current air pollution standards for the chemicals listed in Table 19-1.

B2. Search the web for an air quality report for your local region (such as town, city, state, or province). Determine how air quality has changed during the past decade or two.

B3. Search the web for a site that gives current air pollution readings for your region. In some cities, this pollution reading is updated every several minutes, or every hour. If that is the case, see how the pollution reading varies hour by hour during a typical workday.

B4. Search the web for information on health effects of different exposures to different pollutants.

B5. Air-pollution models are computer codes that use equations similar to the ones in this chapter, to predict air-pollution concentration. Search the web for a list of names of a few of the popular air-pollution models endorsed by your country or region.

B6. Search the web for inventories of emission rates for pollutants in your regions. What are the biggest polluters?

B7. Search the web for an explanation of **emissions trading**. Discuss why such a policy is or is not good for industry, government, and people.

B8. Search the web for information on **acid rain**. What is it? How does it form? What does it do?

B9. Search the web for information on **forest death** (**waldsterben**) caused by pollution or acid rain.

B10. Search the web for instruments that can measure concentration of the chemicals listed in Table 19-1.

B11. Search the web for "web-cam" cameras that show a view of a major city, and discuss how the visibility during fair weather changes during the daily cycle on a workday.

B12. Search the web for information of plume rise and/or concentration predictions for complex (mountainous) terrain.

B13. Search the web for information to help you discuss the relationship between "good" ozone in the stratosphere and mesosphere, vs. "bad" ozone in the boundary layer.

B14. For some of the major industry in your area, search the web for information on control technologies that can, or have, helped to reduce pollution emissions.

B15. Search the web for satellite photos of emissions from major sources, such as a large industrial complex, smelter, volcano, or a power plant. Use the highest-resolution photographs to look at lateral plume dispersion, and compare with the dispersion equations in this chapter.

B16. Search the web for information on forward or backward trajectories, as used in air pollution. One example is the Chernobyl nuclear accident, where radioactivity measurements in Scandinavia were used with a back trajectory to suggest that the source of the radioactivity was in the former Soviet Union.

B17. Search the web for information on chemical reactions of air pollutants in the atmosphere.

B18. Search the web for satellite photos and other information on an **urban plume** (the pollutant plume downwind of a whole city).

B19. To simplify the presentation of air-quality data to the general public, many governments have created an **air-quality index** that summarizes with a simple number how clean or dirty the air is. For your national government (or for the USA if your own government doesn't have one), search the web for info about the air-quality index. How is it defined in terms of concentrations of different pollutants? How do you interpret the index value in terms of visibility and/or health hazards?

Apply

A1. Given the following pollutant concentrations in $\mu m\ m^{-3}$, convert to volume fraction units ppmv assuming standard sea-level conditions:

a. SO_2 1300	b. SO_2 900	c. SO_2 365
d. SO_2 300	e. SO_2 80	f. SO_2 60
g. NO_2 400	h. NO_2 280	i. NO_2 200
j. NO_2 150	k. NO_2 40	m. CO 40,000
n. CO 35,000	o. CO 20,000	p. CO 15,000

q. O_3 235   r. O_3 160   s. O_3 157
t. O_3 100   u. O_3 50   v. O_3 30

A2. Same as previous exercise, but for a summer day in Denver, Colorado, USA, where T = 25°C and P = 82 kPa.

A3. Create a table similar to Table 19-1, but where all the ppm values of volume fraction have been converted into concentration units of µg m^{-3}.

A4. Given wind measurements in the table below.

a. Find the mean wind speed component in each direction
b. Create a table showing the deviation from the mean at each time for each wind component.
c. Find the velocity variance in each direction.
d. Find the standard deviation of velocity for each wind direction.
e. Determine if the turbulence is isotropic.
f. Speculate on the cross-section shape of smoke plumes as they disperse in this atmosphere.

t (min)	U (m s^{-1})	V (m s^{-1})	W (m s^{-1})
1	8	1	0
2	11	2	–1
3	12	0	1
4	7	–3	1
5	12	0	–1

A5. Determine the Pasquill-Gifford turbulence type

a. Strong sunshine, clear skies, winds 1 m s^{-1}
b. Thick overcast, winds 10 m s^{-1}, night
c. Clear skies, winds 2.5 m s^{-1}, night
d. Noon, thin overcast, winds 3 m s^{-1}.
e. Cold air advection 2 m s^{-1} over a warm lake.
f. Sunset, heavy overcast, calm.
g. Sunrise, calm, clear.
h. Strong sunshine, clear skies, winds 10 m s^{-1}.
i. Thin overcast, nighttime, wind 2 m s^{-1}.
j. Thin overcast, nighttime, wind 5 m s^{-1}.
k. Thin overcast, 9 am, wind 3.5 m s^{-1}.

A6. Given turbulence kinetic energy (TKE) buoyant generation (B) and shear generation (S) rates in this table (both in units of m^2·s^{-3}), answer questions (i) - (vi) below.

	B	S		B	S
a.	0.004	0.0	k.	0.0	0.004
b.	0.004	0.002	m.	0.0	0.006
c.	0.004	0.004	n.	–0.002	0.0
d.	0.004	0.006	o.	–0.002	0.002
e.	0.002	0.0	p.	–0.002	0.004
f.	0.002	0.002	q.	–0.002	0.006
g.	0.002	0.004	r.	–0.004	0.0
h.	0.002	0.006	s.	–0.004	0.002
i.	0.0	0.0	t.	–0.004	0.004
j.	0.0	0.002	u.	–0.004	0.006

(i) Specify the nature of flow/convection
(ii) Estimate the Pasquill-Gifford turbulence type.
(iii) Classify the static stability (from strongly stable to strongly unstable)
(iv) Estimate the **flux Richardson number** $R_f = -B/S$
(v) Determine the dispersion isotropy
(vi) Is turbulence intensity (TKE) strong or weak?

A7. Given the table below with pollutant concentrations c (µg m^{-3}) measured at various heights z (km), answer these 5 questions.

(i) Find the height of center of mass.
(ii) Find the vertical height variance.
(iii) Find the vertical height standard deviation.
(iv) Find the total amount of pollutant emitted.
(v) Find the nominal plume spread (depth)

z (km)	c (µg m^{-3}) Question: a	b	c	d	e
1.5	0	0	0	0	0
1.4	0	10	0	86	0
1.3	5	25	0	220	0
1.2	25	50	0	430	0
1.1	20	75	0	350	0.04
1.0	45	85	0	195	0.06
0.9	55	90	2	50	0.14
0.8	40	93	8	5	0.18
0.7	30	89	23	0	0.13
0.6	10	73	23	0	0.07
0.5	0	56	7	0	0.01
0.4	0	30	3	0	0
0.3	0	15	0	0	0
0.2	0	5	0	0	0
0.1	0	0	0	0	0

A8.(§) For the previous problem, find the best-fit Gaussian curve through the data, and plot the data and curve on the same graph.

A9. Using the data from question A7, find the nominal plume width from edge to edge.

A10. Given lateral and vertical velocity variances of 1.0 and 0.5 m^2 s^{-2}, respectively. Find the variance of plume spread in the lateral and vertical, at distance 3 km downwind of a source in a wind of speed 5 m s^{-1}. Use a Lagrangian time scale of:

a. 15 s	b. 30 s	c. 1 min	d. 2 min
e. 5 min	f. 10 min	g. 15 min	h. 20 min
i. 5 s	j. 45 s	m. 12 min	n. 30 min

A11.(§) For a Lagrangian time scale of 2 minutes and wind speed of 10 m s^{-1}, plot the standard deviation of vertical plume spread vs. downwind distance for a vertical velocity variance (m^2 s^{-2}) of:

a. 0.1 b. 0.2 c. 0.3 d. 0.4 e. 0.5
f. 0.6 g. 0.8 h. 1.0 i. 1.5 j. 2
k. 2.5 m. 3 n. 4 o. 5 p. 8

A12.(§) For the previous problem, plot σ_z if
(i) only the near-source equation
(ii) only the far source equation
is used over the whole range of distances.

A13. Given the following emission parameters:

	W_o (m s^{-1})	R_o (m)	$\Delta\theta$ (K)
a.	5	3	200
b.	30	1	50
c.	20	2	100
d.	2	2	50
e.	5	1	50
f.	30	2	100
g.	20	3	50
h.	2	4	20

Find the momentum and buoyant length scales for the plume-rise equations. Assume $|g|/\theta_a \approx 0.0333$ m·s^{-2}·K^{-1} , and M = 5 m s^{-1} for all cases.

A14.(§) For the previous problem, plot the plume centerline height vs. distance if the physical stack height is 100 m and the atmosphere is statically neutral.

A15. For buoyant length scale of 5 m, physical stack height 10 m, environmental temperature 10°C, and wind speed 2 m s^{-1}, find the equilibrium plume centerline height in a statically stable boundary layer, given ambient potential temperature gradients of $\Delta\theta/\Delta z$ (K km^{-1}):

a. 1 b. 2 c. 3 d. 4 e. 5 f. 6 g. 7
h. 8 i. 9 j. 10 k. 12 m. 15 n. 18 o. 20

A16. Given $\sigma_y = \sigma_z = 300$ m, $z_{CL} = 500$ m, $z = 200$ m, $Q = 100$ g s^{-1}, $M = 10$ m s^{-1}. For a neutral boundary layer, find the concentration at y (km) =

a. 0 b. 0.1 c. 0.2 d. 0.3 e. 0.4 f. 0.5 g. 0.7
h. 1 i. 1.5 j. 2 k. 3 m. 4 n. 5 o. 6

A17(§). Plot the concentration footprint at the surface downwind of a stack, given: $\sigma_v = 1$ m s^{-1}, $\sigma_w = 0.5$ m s^{-1}, $M = 2$ m s^{-1}, Lagrangian time scale = 1 minute, $Q = 400$ g s^{-1} of SO_2, in a stable boundary layer. Use a plume equilibrium centerline height (m) of:

a. 10 b. 20 c. 30 d. 40 e. 50 f. 60 g. 70
h. 15 i. 25 j. 35 k. 45 m. 55 n. 65 o. 75

A18. Calculate the dimensionless downwind distance, given a convective mixed layer depth of 2 km, wind speed 3 m s^{-1}, and surface kinematic heat flux of 0.15 K·m s^{-1}. Assume $|g|/T_v \approx 0.0333$ m·s^{-2}·K^{-1} . The actual distance x (km) is:

a. 0.2 b. 0.5 c. 1 d. 2 e. 3 f. 4
g. 5 h. 7 i. 10 j. 20 k. 30 m. 50

A19. If $w_* = 1$ m s^{-1}, mixed layer depth is 1 km, wind speed is 5 m s^{-1}, $Q = 100$ g s^{-1}, find the
a. dimensionless downwind distance at $x = 2$ km
b. dimensionless concentration if $c = 100$ µg m^{-3}
c. dimensionless crosswind integrated concentration if $c_y = 1$ mg m^{-2}

A20.(§) For a convective mixed layer, plot dimensionless plume centerline height with dimensionless downwind distance, for dimensionless source heights of:

a. 0 b. 0.01 c. 0.02 d. 0.03 e. 0.04
f. 0.05 g. 0.06 h. 0.07 i. 0.08 j. 0.09
k. 0.1 m. 0.12 n. 0.15 o. 0.2 p. 0.22

A21.(§) For the previous problem, plot isopleths of dimensionless crosswind integrated concentration, similar to Fig. 19.8, for convective mixed layers.

A22. Source emissions of 300 g s^{-1} of SO_2 occur at height 200 m. The environment is statically unstable, with a Deardorff convective velocity of 1 m s^{-1}, and a mean wind speed of 5 m s^{-1}.

Find the concentration at the ground at distances 1, 2, 3, and 4 km downwind from the source, directly beneath the plume centerline. Assume the mixed layer depth (km) is:

a. 0.5 b. 0.75 c. 1.0 d. 1.25 e. 1.5 f. 1.75
g. 2.0 h. 2.5 i. 3.0 j. 3.5 k. 4.0 m. 5.0

(Hint: Interpolate between figures if needed, or derive your own figures.)

Evaluate & Analyze

E1. Compare the two equations for variance: (19.5) and (19.9). Why is the one weighted by pollution concentration, and the other not?

E2. To help understand complicated figures such as Fig. 19.3, it helps to separate out the various parts. Using the info from that figure, produce a separate sketch of the following on a background grid of B and S values:
a. TKE (arbitrary relative intensity)
b. R_f c. Flow type d. Static stability
e. Pasquill-Gifford turbulence type
f. Dispersion isotropy (plume cross section)

g. Suggest why these different variables are related to each other.

E3. Fig. 19.3 shows how dispersion isotropy can change as the relative magnitudes of the shear and buoyancy TKE production terms change. Also, the total amount of spread increases as the TKE intensity increases. Discuss how the shape and spread of smoke plumes vary in different parts of that figure, and sketch what the result would look like to a viewer on the ground.

E4. Eq. (19.8) gives the center of mass (i.e., plume centerline height) in the vertical direction. Create a similar equation for plume center of mass in the horizontal, using a cylindrical coordinate system centered on the emission point.

E5. In eq. (19.10) use $Q_1 = 100$ g m^{-1} and $\bar{z} = 0$. Plot on graph paper the Gaussian curve using σ_z (m) =

a. 100 b. 200 c. 300 d. 400

Compare the areas under each curve, and discuss the significance of the result.

E6. Why does a "nominal" plume edge need to be defined? Why cannot the Gaussian distribution be used, with the definition that plume edge happens where the concentration becomes zero. Discuss, and support your arguments with results from the Gaussian distribution equation.

E7. The Lagrangian time scale is different for different size eddies. In nature, there is a superposition of turbulent eddies acting simultaneously. Describe the dispersion of a smoke plume under the influence of such a spectrum of turbulent eddies.

E8. While Taylor's statistical theory equations give plume spread as a function of downwind distance, x, these equations are also complex functions of the Lagrangian time scale t_L. For a fixed value of downwind distance, plot curves of the variation of plume spread (eq. 19.13) as a function of t_L. Discuss the meaning of the result.

E9. a. Derive eqs. (19.14) and (19.15) for near-source and far-source dispersion from Taylor's statistical theory equations (19.13).

b. Why do the near and far source dispersion equations appear as straight lines in a log-log graph (see the Sample Application near eq. (19.15)?

E10. Plot the following sounding on the boundary-layer $\theta - z$ thermo diagram from the Atmospheric Stability chapter. Determine the static stability vs. height. Determine boundary-layer structure, including location and thickness of components of the boundary layer (surface layer, stable BL or convective mixed layer, capping inversion or entrainment zone, free atmosphere). Speculate whether it is daytime or nighttime, and whether it is winter or summer. For daytime situations, calculate the mixed-layer depth. This depth controls pollution concentration (shallow depths are associated with periods of high pollutant concentration called **air-pollution episodes**, and during calm winds to **air stagnation events**). [Hint: Review how to nonlocally determine the static stability, as given in the ABL and Stability chapters.]

z (m)	a. T (°C)	b. T (°C)	c. T (°C)	d. T (°C)
2500	–11	8	–5	5
2000	–10	10	–5	0
1700	–8	8	–5	3
1500	–10	10	0	5
1000	–5	15	0	10
500	0	18	5	15
100	4	18	9	20
0	7	15	10	25

E11. For the ambient sounding of the previous exercise, assume that a smoke stack of height 100 m emits effluent of temperature 6°C with water-vapor mixing ratio 3 g kg^{-1}. (Hint, assume the smoke is an air parcel, and use a thermo diagram.)

(i). How high would the plume rise, assuming no dilution with the environment?

(ii). Would steam condense in the plume?

E12. For plume rise in statically neutral conditions, write a simplified version of the plume-rise equation (19.16) for the special case of:

a. momentum only b. buoyancy only

Also, what are the limitations and range of applicability of the full equation and the simplified equations?

E13. For plume rise in statically stable conditions, the amount of rise depends on the Brunt-Väisälä frequency. As the static stability becomes weaker, the Brunt-Väisälä frequency changes, and so changes the plume centerline height. In the limit of extremely weak static stability, compare this plume rise equation with the plume rise equation for statically neutral conditions. Also, discuss the limitations of each of the equations.

E14. In eq. (19.20), the "reflected" part of the Gaussian concentration equation was created by pretending that there is an imaginary source of emissions an equal distance underground as the true source is above ground. Otherwise, the real and imaginary sources are at the same horizontal location and have the same emission rate.

In eq. (19.20), identify which term is the "reflection" term, and show why it works as if there were emissions from below ground.

E15. In the Sample Application in the Gaussian Concentration Distribution subsection, the concentration footprints at ground level have a maximum value neither right at the stack, nor do concentrations monotonically increase with increasing distances from the stack. Why? Also, why are the two figures in that Sample Application so different?

E16. Show that eq. (19.20) reduces to eq. (19.21) for receptors at the ground.

E17. For Gaussian concentration eq. (19.21), how does concentration vary with:

a. σ_y b. σ_z c. M

E18. Give a physical interpretation of crosswind integrated concentration, using a different approach than was used in Fig. 19.6.

E19. For plume rise and pollution concentration in a statically unstable boundary layer, what is the reason for, or advantage of, using dimensionless variables?

E20. If the Deardorff velocity increases, how does the dispersion of pollutants in an unstable boundary layer change?

E21. In Fig. 19.8, at large distances downwind from the source, all of the figures show the dimensionless concentration approaching a value of 1.0. Why does it approach that value, and what is the significance or justification for such behavior?

Synthesize

S1. Suppose that there was not a diurnal cycle, but that the atmospheric temperature profile was steady, and equal to the standard atmosphere. How would local and global dispersion of pollutants from tall smoke stacks be different, if at all?

S2. In the present atmosphere, larger-size turbulent eddies often have more energy than smaller size one. What if the energy distribution were reversed, with the vigor of mixing increasing as eddy sizes decrease. How would that change local dispersion, if at all?

S3. What if tracers were not passive, but had a special magnetic attraction only to each other. Describe how dispersion would change, if at all.

S4. What if a plume that is rising in a statically neutral environment has buoyancy from both the initial temperature of the effluent out of the top of the stack, and also from additional heat gained while it was dispersing.

A real example was the black smoke plumes from the oil well fires during the Gulf War. Sunlight was strongly absorbed by the black soot and unburned petroleum in the smoke, causing solar warming of the black smoke plume.

Describe any resulting changes to plume rise.

S5. Suppose that smoke stacks produced smoke rings, instead of smoke plumes. How would dispersion be different, if at all?

S6. When pollutants are removed from exhaust gas before the gas is emitted from the top of a smoke stack, those pollutants don't magically disappear. Instead, they are converted into water pollution (to be dumped into a stream or ocean), or solid waste (to be buried in a dump or landfill). Which is better? Why?

S7. Propose methods whereby life on Earth could produce zero pollution. Defend your proposals.

S8. What if the same emission rate of pollutions occurs on a fair-weather day with light winds, and an overcast rainy day with stronger winds. Compare the dispersion and pollution concentrations at the surface for those situations. Which leads to the least concentration at the surface, locally? Which is better globally?

S9. Suppose that all atmospheric turbulence was extremely anisotropic, such that there was zero dispersion in the vertical , but normal dispersion in the horizontal.

a. How would that affect pollution concentrations at the surface, for emissions from tall smoke stacks?

b. How would it affect climate, if at all?

S10. What if ambient wind speed was exactly zero. Discuss the behavior of emission plumes, and how the resulting plume rise and concentration equations would need to be modified.

S11. What if pollutants that were emitted into the atmosphere were never lost or removed from the atmosphere. Discuss how the weather and climate would be different, if at all?

S12. If there were no pollutants in the atmosphere (and hence no cloud and ice nuclei), discuss how the weather and climate would be different, if at all.

S13. Divide the current global pollutant emissions by the global population, to get the net emissions per person. Given the present rate of population increase, discuss how pollution emissions will change over the next century, and how it will affect the quality of life on Earth, if at all.

20 NUMERICAL WEATHER PREDICTION (NWP)

Contents

Most weather forecasts are made by computer, and some of these forecasts are further enhanced by humans. Computers can keep track of myriad complex nonlinear interactions among winds, temperature, and moisture at thousands of locations and altitudes around the world — an impossible task for humans. Also, data observation, collection, analysis, display and dissemination are mostly automated.

Fig. 20.1 shows an automated forecast. Produced by computer, this **meteogram** (graph of weather vs. time for one location) is easier for non-meteorologists to interpret than weather maps. But to produce such forecasts, the equations describing the atmosphere must first be solved.

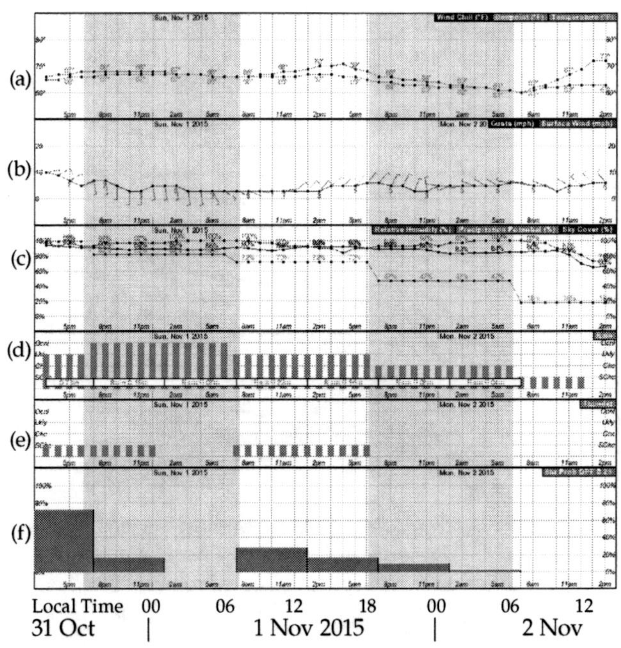

Figure 20.1
Two-day weather forecast for Jackson, Mississippi USA, plotted as a meteogram (time series), based on initial conditions observed at 12 UTC on 31 Oct 2015. (a) Temperature & dewpoint (°F), (b) winds, (c) humidity, precipitation, cloud-cover, (d) rainfall amounts, (e) thunderstorm likelihood, (f) probability of precipitation > 0.25 inch. Produced by US NWS.

INFO • Alternative Vertical Coordinate

Eqs. (20.1-20.7) use z as a vertical coordinate, where z is height above mean sea level. But local terrain elevations can be higher than sea level. The atmosphere does not exist underground; thus, it makes no sense to solve the meteorological equations of motion at heights below ground level.

To avoid this problem, define a **terrain-following coordinate** σ **(sigma)**. One definition for σ is based on the hydrostatic pressure $P_{ref}(z)$ at any height z relative to the hydrostatic pressure difference between the earth's surface ($P_{ref\,bottom}$) and a fixed pressure ($P_{ref\,top}$) representing the top of the atmosphere:

$$\sigma = \frac{P_{ref}(z) - P_{ref\,top}}{P_{ref\,bottom} - P_{ref\,top}}$$

$P_{ref\,bottom}$ varies in the horizontal due to terrain elevation (see Fig. 20.A) and varies in space and time due to changing surface weather patterns (high- and low-pressure centers). The new vertical coordinate σ varies from 1 at the earth's surface to 0 at the top of the domain.

The figure below shows how this **sigma coordinate** varies over a mountain. **Hybrid coordinates** (Fig. 20.5) are ones that are terrain following near the ground, but constant pressure aloft.

If σ is used as a vertical coordinate, then (U, V) are defined as winds along a σ surface. The vertical advection term in eq. (20.1) changes from $W \cdot \Delta U/\Delta z$ to $\dot{\sigma} \cdot \Delta U / \Delta \sigma$, where sigma dot is analogous to a vertical velocity, but in sigma coordinates. Similar changes must be made to most of the terms in the equations of motion, which can be numerically solved within the domain of $0 \leq \sigma \leq 1$.

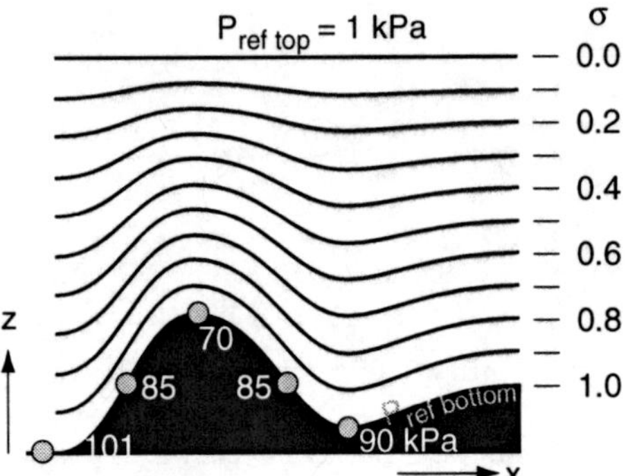

Figure 20.A.
Vertical cross section through the atmosphere (white) and earth (black). White numbers represent surface air pressure at the weather stations shown by the grey dots. For the equation above, $P_{ref\,bottom}$ = 70 kPa at the mountain top, which differs from $P_{ref\,bottom}$ = 90 kPa in the valley.

Although sigma coordinates avoid the problem of coordinates that go underground, they create problems for advection calculations due to small differences between large terms. To reduce this problem, stair-step terrain-following coordinates have been devised — known as **eta coordinates** (η).

SCIENTIFIC BASIS OF FORECASTING

The Equations of Motion

Numerical weather forecasts are made by solving Eulerian equations for U, V, W, T, r_T, ρ and P.

From the Forces & Winds chapter are forecast equations for the three wind components (U, V, W) (modified from eqs. 10.23a & b, and eq. 10.59):

$$\frac{\Delta U}{\Delta t} = -U\frac{\Delta U}{\Delta x} - V\frac{\Delta U}{\Delta y} - W\frac{\Delta U}{\Delta z} - \frac{1}{\rho}\cdot\frac{\Delta P}{\Delta x} + f_c \cdot V - \frac{\Delta F_{z\,turb}(U)}{\Delta z} \tag{20.1}$$

$$\frac{\Delta V}{\Delta t} = -U\frac{\Delta V}{\Delta x} - V\frac{\Delta V}{\Delta y} - W\frac{\Delta V}{\Delta z} - \frac{1}{\rho}\cdot\frac{\Delta P}{\Delta y} - f_c \cdot U - \frac{\Delta F_{z\,turb}(V)}{\Delta z} \tag{20.2}$$

$$\frac{\Delta W}{\Delta t} = -U\frac{\Delta W}{\Delta x} - V\frac{\Delta W}{\Delta y} - W\frac{\Delta W}{\Delta z} - \frac{1}{\rho}\frac{\Delta P'}{\Delta z} + \frac{\theta_{vp} - \theta_{ve}}{T_{ve}}\cdot|g| - \frac{\Delta F_{z\,turb}(W)}{\Delta z} \tag{20.3}$$

From the Heat Budgets chapter is a forecast equation for temperature T (modified from eq. 3.51):

$$\frac{\Delta T}{\Delta t} = -U\frac{\Delta T}{\Delta x} - V\frac{\Delta T}{\Delta y} - W\left[\frac{\Delta T}{\Delta z} + \Gamma_d\right] - \frac{1}{\rho \cdot C_p}\frac{\Delta \mathbb{F}^*_{z\,rad}}{\Delta z} + \frac{L_v}{C_p}\frac{\Delta r_{condensing}}{\Delta t} - \frac{\Delta F_{z\,turb}(\theta)}{\Delta z} \tag{20.4}$$

From the Water Vapor chapter is a forecast equation (4.44) for total-water mixing ratio r_T in the air:

$$\frac{\Delta r_T}{\Delta t} = -U\frac{\Delta r_T}{\Delta x} - V\frac{\Delta r_T}{\Delta y} - W\frac{\Delta r_T}{\Delta z} + \frac{\rho_L}{\rho_d}\frac{\Delta Pr}{\Delta z} - \frac{\Delta F_{z\,turb}(r_T)}{\Delta z} \tag{20.5}$$

From the Forces & Winds chapter is the continuity equation (10.60) to forecast air density ρ:

$$\frac{\Delta \rho}{\Delta t} = -U\frac{\Delta \rho}{\Delta x} - V\frac{\Delta \rho}{\Delta y} - W\frac{\Delta \rho}{\Delta z} - \rho\left[\frac{\Delta U}{\Delta x} + \frac{\Delta V}{\Delta y} + \frac{\Delta W}{\Delta z}\right] \tag{20.6}$$

For pressure P, use the equation of state (ideal gas law) from Chapter 1 (eq. 1.23):

$$P = \rho \cdot \Re_d \cdot T_v \tag{20.7}$$

In these seven equations: f_c is Coriolis parameter, P' is the deviation of pressure from its hydrostatic value, θ_{vp} and θ_{ve} are virtual potential temperatures of the air parcel and the surrounding environment, T_{ve} is virtual temperature of the environment, $|g|$ = 9.8 m s^{-2} is the magnitude of gravitational acceleration, Γ_d = 9.8 K km^{-1} is the dry adiabatic lapse rate, $\mathbb{F}^*_{z\ rad}$ is net radiative flux, $L_v \approx 2.5\text{x}10^6$ J kg^{-1} is the latent heat of vaporization, $C_p \approx 1004$ J·kg^{-1}·K^{-1} is the specific heat of air at constant pressure, $\Delta r_{condensing}$ is the increase in liquid-water mixing ratio associated with water vapor that is condensing, $\rho_L \approx 1000$ kg·m^{-3} and ρ_d are the densities of liquid water and dry air, Pr is precipitation rate (m s^{-1}) of water accumulation in a rain gauge at any height z, $\Re_d = 287$ J·kg^{-1}·K^{-1} is the gas constant for dry air, and T_v is the virtual temperature. For more details, see the chapters cited with those equations.

Notice the similarities in eqs. (20.1 - 20.6). All have a **tendency term** (rate of change with time) on the left. All have **advection** as the first 3 terms on the right. Eqs. (20.1 - 20.5) include a **turbulence flux divergence** term on the right. The other terms describe the special forcings that apply to individual variables. Sometimes the hydrostatic equation (Chapter 1, eq. 1.25b) is also included in the set of forecast equations:

$$\frac{\Delta P_{ref}}{\Delta z} = -\rho \cdot |g| \tag{20.8}$$

to serve as a reference state for the definition of $P' = P - P_{ref}$, as used in eq. (20.3).

Equations (20.1) - (20.7) are the **equations of motion**. They are also known as the **primitive equations**, because they forecast fundamental (primitive) variables rather than derived variables such as vorticity. The first six equations are **budget equations**, because they forecast how variables change in response to inputs and outputs. Namely, the first three equations describe **momentum conservation** per unit mass of air. Eqs. (20.4 - 20.5) describe **heat conservation** and **moisture conservation** per unit mass of air. Eq. (20.6) describes **mass conservation**.

The first six equations are **prognostic** (i.e., forecast the change with time), and the seventh (the ideal gas law) is **diagnostic** (not a function of time). The third equation includes **non-hydrostatic** processes, the fourth equation includes **diabatic** processes (non-adiabatic heating), and the sixth equation includes **compressible** processes.

These equations of motion are **nonlinear**, because many of the terms in these equations consist of products of two or more dependent variables. Also, they are **coupled** equations, because each equation contains variables that are forecast or diagnosed

INFO • Alternative Horizontal Coord.

Spherical Coordinates

For the Cartesian coordinates used in eqs. (20.1-20.8), the coordinate axes are straight lines. However, on Earth we prefer to define x to follow the Earth's curvature toward the East, and define y to follow the Earth's curvature toward the North. If U and V are defined as velocities along these spherical coordinates, then <u>add</u> the following terms to the right side of the horizontal momentum equations (20.1-20.3):

$$+\frac{U \cdot V \cdot \tan(\phi)}{R_o} - \frac{U \cdot W}{R_o} - [2 \cdot \Omega \cdot W \cdot \cos(\phi)] \tag{20.1b}$$

$$-\frac{U^2 \cdot \tan(\phi)}{R_o} - \frac{V \cdot W}{R_o} \tag{20.2b}$$

$$+\frac{U^2 + V^2}{R_o} + [2 \cdot \Omega \cdot U \cdot \cos(\phi)] \tag{20.3b}$$

where $R_o \approx 6371$ km is the average Earth radius, ϕ is latitude, and $\Omega = 0.7292\text{x}10^{-4}$ s^{-1} is Earth's rotation rate. The terms containing R_o are called the **curvature terms**. The terms in square brackets are small components of Coriolis force (see the INFO Box "Coriolis Force in 3-D" from the Forces & Winds chapter).

Map Factors

Suppose we pick (x, y) to represent horizontal coordinates on a map projection, such as shown in the INFO Box on the next page. Let (U, V) be the horizontal components of winds in these (x, y) directions. [Vertical velocity W applies unchanged in the z (up) direction.] One reason why meteorologists use such map projections is to avoid singularities, such as near the Earth's poles where meridians converge.

The equations of motion can be rewritten for any map projection. For example, eq. (20.1) can be written for a **polar stereographic projection** as:

$$\begin{aligned}\frac{\Delta U}{\Delta t} = & -m_o \cdot U \frac{\Delta U}{\Delta x} - m_o \cdot V \frac{\Delta U}{\Delta y} - W \frac{\Delta U}{\Delta z} \\ & - m_x \cdot V^2 + m_y \cdot U \cdot V - \frac{U \cdot W}{R_o} + [Cor] \\ & - \frac{m_o}{\rho} \cdot \frac{\Delta P}{\Delta x} + f_c \cdot V - \frac{\Delta F_{z\ turb}(U)}{\Delta z}\end{aligned} \tag{20.1c}$$

where $[Cor]$ is a 3-D Coriolis term, and the **map factors** (m) are:

$$m_o = \frac{1 + \sin(\phi_o)}{1 + \sin(\phi)} = \frac{L^2 + x^2 + y^2}{2 \cdot R_o \cdot L}$$

and

$$m_x = x / (R_o \cdot L) \quad , \quad m_y = y / (R_o \cdot L) \; ,$$

where $L = R_o \cdot [1 + \sin(\phi_o)]$, $R_o \approx 6371$ km is the average Earth radius, ϕ is latitude, and ϕ_o is the reference latitude for the map projection (see INFO Box).

Thus, the equation has extra terms, and many of the terms are scaled by a map factor. Eqs. (20.2 - 20.6) have similar changes when cast on a map projection.

Sample Application (§)

Plot the given coordinates: (a) on a lat-lon grid, and (b) on a polar stereographic grid with $\phi_o = 60°$.

Find the Answer

Given: Latitudes (ϕ) & longitudes (λ) of N. America				
Each column holds [ϕ(°) λ(°)]. λ is positive eastward				
50 -125	9 -76	38 -77	55 -82	0 0
40 -125	11 -84	46 -65	58 -95	0 180
23 -110	15 -84	43 -66	68 -82	
24 -110	15 -88	46 -60	70 -140	0 -45
30 -115	22 -87	45 -65	73 -157	0 135
32 -114	22 -90	50 -65	65 -168	
22 -106	18 -91	50 -60	58 -158	0 45
20 -106	18 -96	53 -56	53 -170	0 -135
7 -80	22 -98	48 -59	60 -146	
9 -78	27 -97	47 -52	60 -140	0 0
4 -76	30 -85	53 -56	50 -125	0 10
0 -80	28 -83	60 -65		0 20
	25 -81	58 -68		0 30
0 -48	26 -80	64 -78	0 -90	0 etc.
10 -63	30 -82	52 -79	0 90	0 350
12 -73	35 -76	53 -83		0 360

Hint: In Excel, copy these numbers into 2 long columns: the first for latitudes and the second for longitudes. Leave blank rows in Excel corresponding to the blank lines in the table, to create discontinuous plotted lines.

(a) Lat-Lon Grid: To save space, only the portion of the grid near North America is plotted.

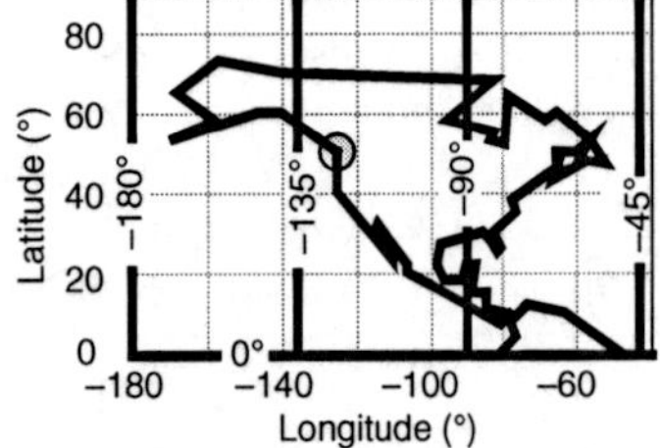

Fig. 20.B1.

(b) Polar Stereographic Grid:

Hint: In Excel, don't forget to convert from (°) to (radians).

To demonstrate the Excel calculation for the first coordinate (near Vancouver): $\phi = 50°$, $\lambda = -125°$:

$L = (6371 \text{ km})\cdot[1 + \sin(60°\cdot\pi/180°)] = 11,888 \text{ km}.$

$r = (11888 \text{ km})\cdot\tan[0.5\cdot(90° - 50°)\cdot\pi/180°] = 4327 \text{ km}$

$x = (4327 \text{ km})\cdot\cos(-125°\cdot\pi/180°) =$ **-2482 km**

$y = (4327 \text{ km})\cdot\sin(-125°\cdot\pi/180°) =$ **-3545 km**

That point is circled on the maps above and below:

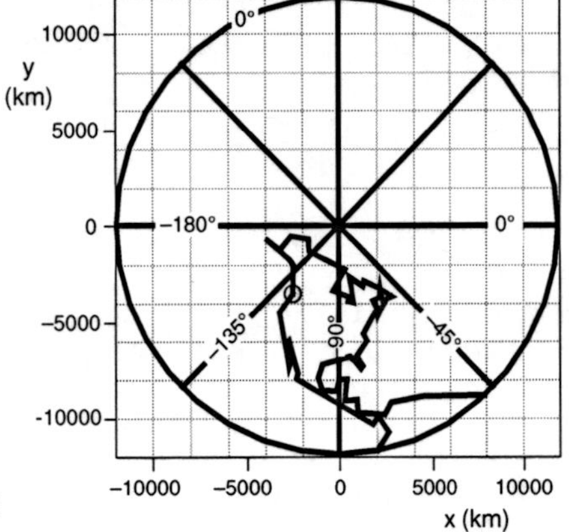

Fig. 20.B2

INFO • Map Projections

A map displays the 3-D Earth's surface on a 2-D plane. On maps you can also: (1) create perpendicular (x, y) coordinates; and (2) rewrite the equations of motion within these map coordinates. You can then solve these eqs. to make numerical weather forecasts.

Create a map by projecting the spherical Earth on to a plane (**stereographic** projection), a cylinder (**Mercator** projection), or a cone (**Lambert** projection), where the cylinder and cone can be "unrolled" after the projection to give a flat map. Although other map projections are possible, the 3 listed above are **conformal**, meaning that the angle between two intersecting curves on the Earth is equal to the angle between the same curves on the map.

For stereographic projections, if the projector is at the North or South Pole, then the result is a **polar stereographic** projection (Fig. 20.C). For any latitude (ϕ) longitude (λ, positive eastward) coordinates on Earth, the corresponding (x, y) map coordinates are:

$$x = r\cdot\cos(\lambda) \quad , \quad y = r\cdot\sin(\lambda) \qquad \text{(F20.1)}$$

$$r = L\cdot\tan[0.5\cdot(90° - \phi)] \quad , \quad L = R_o\cdot[1+\sin(\phi_o)] \qquad \text{(F20.2)}$$

$R_o = 6371$ km = Earth's radius, and ϕ_o is the latitude intersected by the projection plane. The Fig. below has $\phi_o = 60°$, but often $\phi_o = 90°$ is used instead.

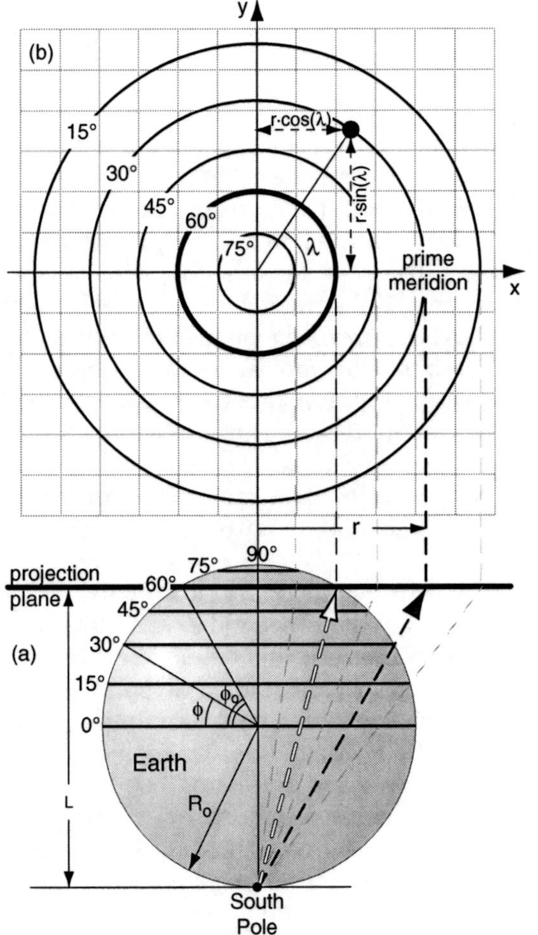

Fig. 20.C. *Polar stereographic map projection.*

from one or more of the other equations. Hence, all 7 equations must be solved together.

Unfortunately, no one has yet succeeded in solving the full governing equations analytically. An **analytical solution** is itself an algebraic equation or number that can be applied at every location in the atmosphere. For example, the equation $y^2 + 2xy = 8x^2$ has an analytical solution $y = 2x$, which allows you to find y at any location x.

Approximate Solutions

To get around this difficulty of no analytical solution, three alternatives are used. One is to find an exact analytical solution to a simplified (approximate) version of the governing equations. A second is to conceive a simplified physical model, for which exact equations can be solved. The third is to find an approximate **numerical solution** to the full governing equations (the focus of this chapter).

(1) An atmospheric example of the first method is the geostrophic wind, which is an exact solution to a highly simplified equation of motion. This is the case of steady-state (equilibrium) winds above the boundary layer where friction can be neglected, and for regions where the isobars are nearly straight.

(2) Early **numerical weather prediction** (**NWP**) efforts used the 2nd method, because of the limited power of early computers. Rossby derived simplified equations by modeling the atmosphere as if it were one layer of water surrounding the Earth. Charney, von Neumann, and others extended this work and wrote a program for a one-layer **barotropic** atmosphere (Fig. 20.2a) for the ENIAC computer in 1950. These earliest programs forecasted only vorticity and geopotential height at 50 kPa.

(3) Modern NWP uses the third method. Here, the full primitive equations are solved using finite-difference approximations for full **baroclinic** scenarios (Fig. 20.2b), but only at discrete locations called **grid points**. Usually these grid points are at regularly-spaced intervals on a map, rather than at each city or town.

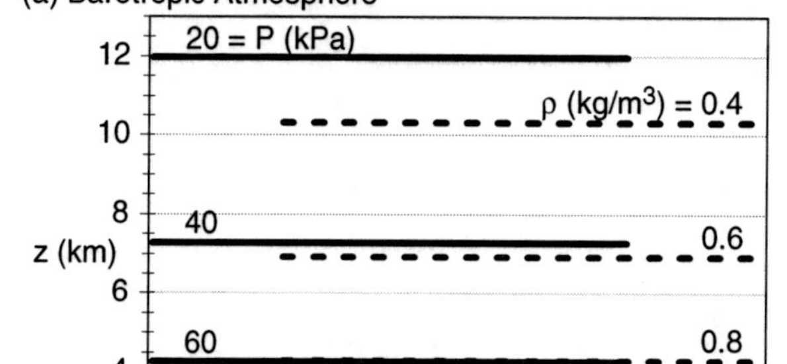
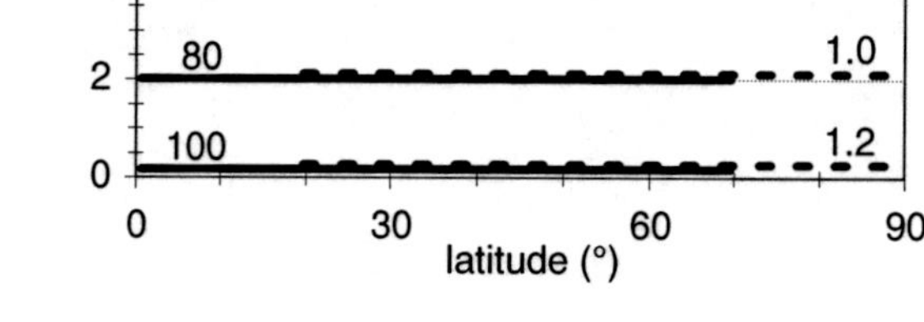

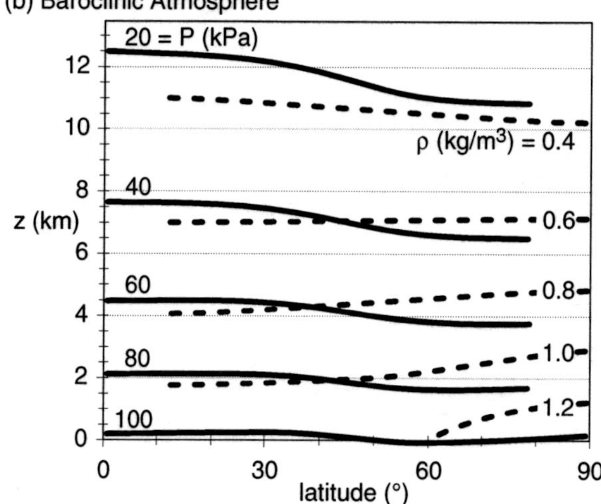

Figure 20.2
(a) Barotropic idealization, based on the standard atmosphere from Chapter 1. (b) Baroclinic example, based on data from the General Circulation chapter.

INFO • Barotropic vs. Baroclinic

In a **barotropic** atmosphere (Fig. 20.2a), the **isobars** (lines of equal pressure) do not cross the **isopycnics** (lines of equal density). This would occur for a situation where there are no variations of temperature in the horizontal. Hence, there could be no thermal-wind effect.

In a **baroclinic** atmosphere (Fig. 20.2b), isobars can cross isopycnics. Horizontal temperature gradients contribute to the tilt of the isopycnics. These temperature gradients also cause changing horizontal pressure gradients with increasing altitude, according to the thermal-wind effect.

The real atmosphere is baroclinic, due to differential heating by the sun (see the General Circulation chapter). In a baroclinic atmosphere, potential energy associated with temperature gradients can be converted into the kinetic energy of winds.

Dynamics, Physics and Numerics

If computers had infinite power, then we could: forecast the movement of every air molecule, forecast the growth of each snowflake and cloud droplet, precisely describe each turbulent eddy, consider atmospheric interaction with each tree leaf and blade of grass, diagnose the absorption of radiation for an infinite number of infinitesimally fine spectral bands, account for every change in terrain elevation, and could even include the movement and activities of each human as they affect the atmosphere. But it might be a few years before we can do that. At pres-

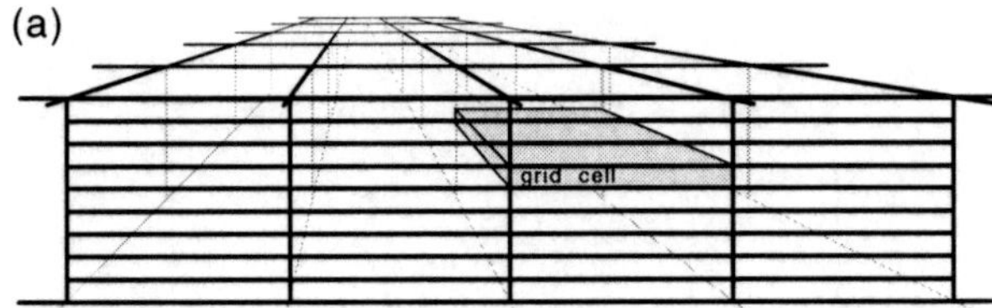

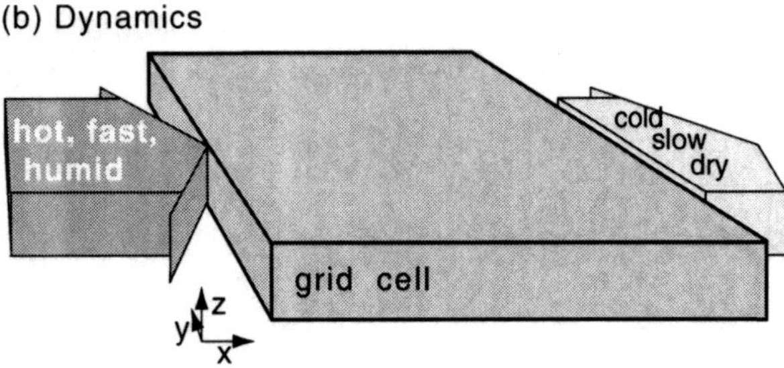

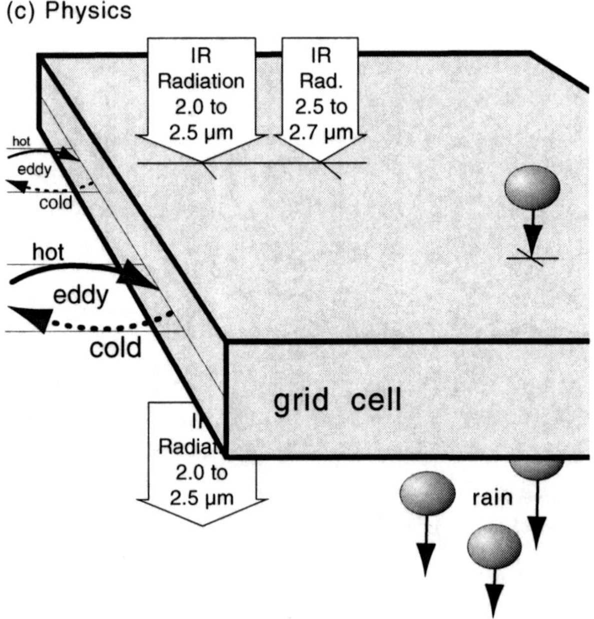

Figure 20.3
(a) The forecast **domain** *(the portion of atmosphere we wish to forecast) is split into discrete grid cells, such as the shaded one. The 3-D grid cells are relatively thin, with sizes on the order of 10s m in the vertical, and 10s km in the horizontal.*
(b) Enlargement of the shaded grid cell, illustrating one **dynamics** *process (advection in the x-direction). Namely, the resolved U wind is blowing in hot, fast, humid air from the upwind neighboring grid cell, and is blowing out colder, slower, drier air into the downwind neighboring cell. Simultaneously, advection could be occurring by the V and W components of wind (not shown). Not shown are other resolved forcings, such as Coriolis and pressure-gradient forces.*
(c) Further enlargement, illustrating **physics** *such as turbulence, radiation, and precipitation. Turbulence is causing a net heat flux into the left side of the grid cell in this example, even though the turbulence has no net wind (i.e., the wind-gust arrows moving air into the grid cell are balanced by gusts moving the same amount of air out of the grid cell). Two of the many radiation bands are shown, where infrared (IR) wavelengths in the 2.0 to 2.5 μm "window" band shine through the grid cell, while wavelengths in the 2.5 to 2.7 band are absorbed by water vapor and carbon dioxide (see the Satellites & Radar chapter), causing warming in the grid cell. Some liquid water is falling into the top of the grid cell from the cell above, but even more is falling out the bottom into the grid cell below, suggesting a removal of water and net latent heating due to condensation.*

ent, we must make compromises to our description of the atmosphere.

Numerics: The main compromise is the process of **discretization**, where:

(1) we split the continuum of space into a finite number of small volumes called **grid cells** (Fig. 20.3a), within which we forecast average conditions;
(2) we approximate the smooth progression of time with finite **time steps**; and
(3) we replace the elegant equations of motion with **numerical approximations**.

These topics are generically known as **numerics**, as will be discussed in detail later. Numerics also include the **domain** being forecast, the mapping and coordinate systems, and the representation of data.

The word **"dynamics"** refers to the **governing equations.** It applies to only the resolved portions of motions, thermodynamic states, and moisture states (Fig. 20.3b) for the particular discretization used. A variable or process is said to be **resolved** if it can be represented by the average state within a grid cell. The dynamics described by eqs. (20.1 - 20.7) depend on sums, differences, and products of these resolved grid-cell average values.

The word **"physics"** refers to other processes (Fig. 20.3c and Table 20-1) that:

(1) are not forecast by the equations of motion, or
(2) are not well understood even though their effects can be measured, or
(3) involve motions or particles that are too small to resolve (called **subgrid scale**), or
(4) have components that are too numerous (e.g., individual cloud droplets or radiation bands), or
(5) are too complicated to compute in finite time.

However, unresolved processes can combine to produce resolved forecast effects. Because we cannot neglect them, we parameterize them instead.

A **parameterization** is a physical or statistical approximation to a physical process by one or more known terms or factors. Parameterization rules are given in an "A SCIENTIFIC PERSPECTIVE" box in the Atmos. Boundary Layer chapter. In NWP, the "knowns" are the resolved state variables in the grid cells, and any imposed boundary conditions such as solar radiation, surface topography, land use, ice coverage, etc. Empirically estimated factors called **parameters** tie the knowns to the approximated physics.

Sample Application

Suppose subgrid-scale cloud coverage C is parameterized by

$C = 0$ for $RH \le RH_o$
$C = [(RH - RH_o) / (1 - RH_o)]^2$ for $RH_o \le RH < 1$
$C = 1$ for $RH \ge 1$

RH is the grid-cell average relative humidity. Parameter $RH_o \approx 0.8$ for low and high clouds, and $RH_o \approx 0.65$ for mid-level cloud. Plot parameterized cloud coverage vs. resolved relative humidity.

Find the Answer

Given: info above
Find: C vs. RH

Spreadsheet solution is graphed at right Grey curve: mid-level clouds. Black curve: low and high clouds.

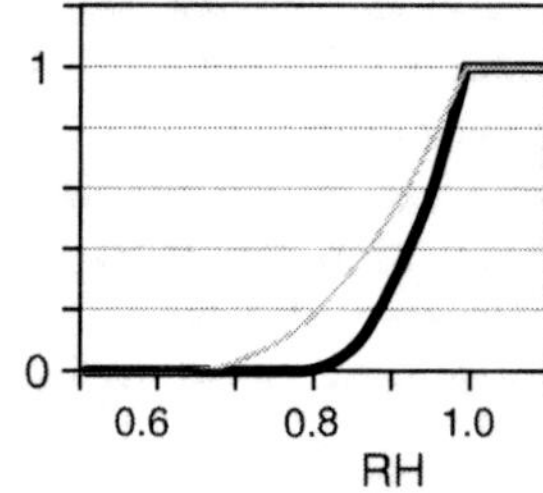

Check: Coverage bounded between clear & overcast.
Exposition: Partial cloud coverage is important for computing how much radiation reaches the ground.

Because parameterizations are only approximations, no single parameterization is perfectly correct. Different scientists might propose different parameterizations for the same physical phenomenon. Different parameterizations might perform better for different weather situations.

Models

The computer code that incorporates one particular set of dynamical equations, numerical approximations, and physical parameterizations is called a **numerical model** or **NWP model**. People developing these extremely large sets of computer code are called **modelers**. It typically takes teams of modelers (meteorologists, physicists, chemists, and computer scientists) several years to develop a new numerical weather model.

Different forecast centers develop different numerical models containing different dynamics, physics and numerics. These models are given names and acronyms, such as the Weather Research and Forecasting (**WRF**) model, the Global Environmental Multiscale (**GEM**) model, or the Global Forecast System (**GFS**). Different models usually give slightly different forecasts.

Table 20-1. Some physics parameterizations in NWP.

Process	Approximation Methods
Cloud Coverage	• Subgrid-scale cloud coverage as a function of resolved relative humidity. Affects the radiation budget.
Precipitation & Cloud Microphysics	Considers conversions between water vapor, cloud ice, snow, cloud water, rain water, and graupel + hail. Affects large-scale condensation, latent heating, and precipitation based on resolved supersaturation. Methods: • bulk (assumes a size distribution of hydrometeors); or • bin (separate forecasts for each subrange of hydrometeor sizes).
Deep Convection	• Approximations for cumuliform clouds (including thunderstorms) that are narrower than grid-cell width but which span many grid layers in the vertical (i.e., are unresolved in the horizontal but resolved in the vertical), as function of moisture, stability and winds. Affects vertical mixing, precipitation, latent heating, & cloud coverage.
Radiation	• Impose solar radiation based on Earth's orbit and solar emissions. Include absorption, scattering, and reflection from clouds, aerosols and the surface. • Divide IR radiation spectrum into small number of wide wavelength bands, and track up- and down-welling radiation in each band as absorbed and emitted from/to each grid layer. Affects heating of air & Earth's surface.
Turbulence	Subgrid turbulence intensity as function of resolved winds and buoyancy. Fluxes of heat, moisture, momentum as function of turbulence and resolved temperature, water, & winds. Methods: • local down-gradient eddy diffusivity; • higher-order local closure; or • nonlocal (transilient turb.) mixing.
Atmospheric Boundary Layer (ABL)	Vertical profiles of temperature, humidity, and wind as a function of resolved state and turbulence, based on forecasts of ABL depth. Methods: • bulk; • similarity theory.
Surface	• Use albedo, roughness, etc. from statistical average of varied land use. • Snow cover, vegetation greenness, etc. based on resolved heat & water budget.
Sub-surface heat & water	• Use climatological average. Or forecast heat conduction & water flow in rivers, lakes, glaciers, subsurface, etc.
Mountain-wave Drag	• Vertical momentum flux as function of resolved topography, winds and static stability.

INFO • Moore's Law & Forecast Skill

Gordon E. Moore co-founded the **integrated-circuit** (computer-chip) manufacturer Intel. In 1965 he reported that the maximum number of transistors that were able to be inexpensively manufactured on integrated circuits had doubled every year. He predicted that this trend would continue for another decade.

Since 1970, the rate slowed to about a doubling every two years. This trend, known as **Moore's Law**, has continued for over 4 decades.

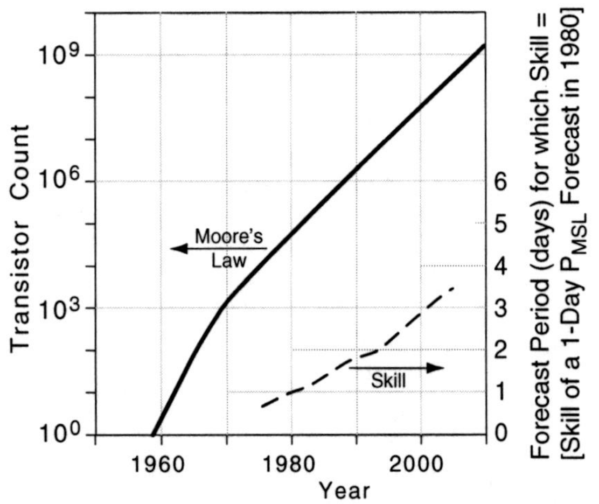

Fig. 20.D. *Moore's Law and forecast skill vs. time.*

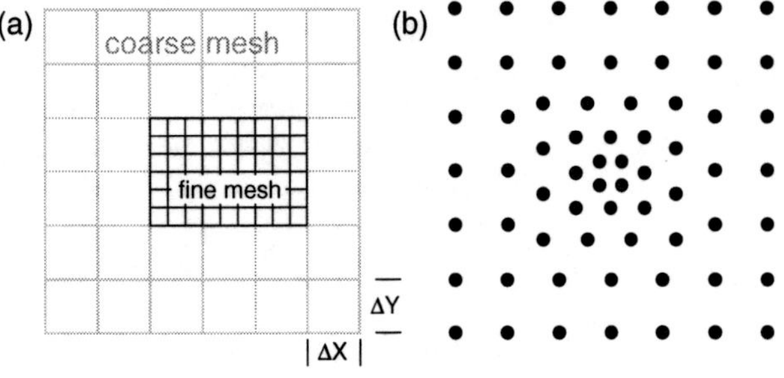

Figure 20.4
Horizontally nested grids. (a) Discrete meshes (shown as grid cells). (b) Variable mesh (shown as grid points).

GRID POINTS

Define the size of a grid cell in the three Cartesian directions as ΔX, ΔY, and ΔZ (Fig. 20.6A). Typical values are ΔX = ΔY = one to hundreds of kilometers, while ΔZ = one to hundreds of meters. Small-size grid cells give **fine-resolution** (or **high-resolution**) forecasts, and large-size cells give **coarse-resolution** (or **low-resolution**) forecasts.

Because we forecast only the average condition of weather variables at each grid cell, we can represent these average values as being physically located at a **grid point** (Fig. 20.6A) in each cell. The distance between grid points is the same as the grid-cell size: ΔX, ΔY, ΔZ. More closely spaced grid points have finer resolution (see a later INFO Box on Resolution).

Finer resolution requires more grid cells to span your forecast domain. Each cell requires a certain number of numerical calculations to make the forecast. Thus, more cells require more total calculations. Hence, finer resolution forecasts take longer to compute, but often give more accurate forecasts.

Thus, your choice of domain and grid size is a compromise between forecast timeliness and accuracy, based on the computer power available. As computer power has improved over the past 6 decades, so have weather-forecast resolution and skill (see INFO Box on Moore's Law and Forecast Skill). **Skill** is the forecast improvement relative to some reference such as climatology.

Nested and Variable Grids

Alternatives exist to the domain-size vs. resolution trade off. In the horizontal, use a fast-running coarse-grid over a large domain to span large-scale weather systems, and nest inside that a smaller-horizontal-domain finer-mesh grid (Fig. 20.4a). Such **nested grids** reduce overall run time while capturing finer-scale features where they are needed most. Typically, the fine mesh has a horizontal grid size (ΔX) of 1/3 of the coarse-mesh grid size, although ratios of 1/5 have sometimes been used. Nesting can continue with successively finer nests. The author's research team has run nested grids with grid sizes ΔX = 108, 36, 12, 4, 1.33, and 0.44 km.

Nested grids can employ **one-way nesting**, where the coarse grid is solved first, and its output is applied as time-varying boundary conditions to the finer grid. For **two-way nesting**, both grids are solved together, and features from each grid are fed into the other at each time step. Two-way nesting often gives better forecasts, but are more complicated to implement.

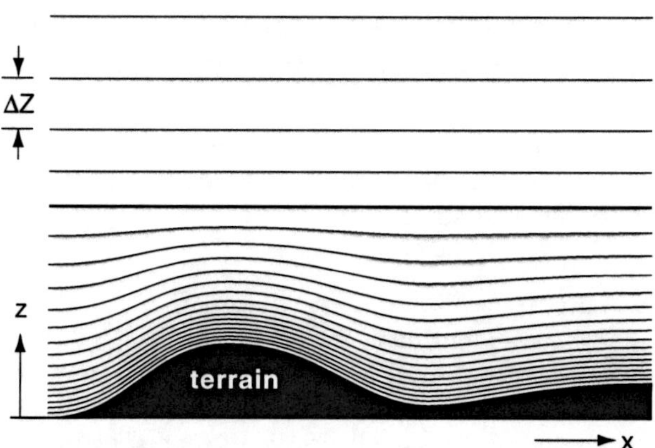

Figure 20.5
Illustration of variable grid increments (ΔZ) in the vertical.

An alternative to discrete nested grids in the horizontal is a **variable-mesh grid** (Fig. 20.4b), which uses smoothly varying grid spacings. Again, the finer mesh is positioned over the region of interest.

In the vertical, fine resolution (i.e., small ΔZ) is needed near the Earth's surface and in the boundary layer, because of important small-scale motions and strong vertical gradients. To reduce the computation time, coarse resolution (i.e., larger ΔZ) is acceptable higher above the surface — in the stratosphere and upper troposphere. Variable mesh vertical grids (i.e., smoothly changing ΔZ values) are often used for this reason (Fig. 20.5). For models using pressure or sigma as a vertical coordinate, ΔP or $\Delta\sigma$ varies smoothly with height. As an alternative, some NWP models use discrete vertical nests.

Staggered Grids

You could represent all the cell-average variables at the same grid point, as in Fig. 20.6A (called an A-Grid). But this has some undesirable characteristics: wavy motions do not disperse properly, some wave energy gets stuck in the grid, and some weather variables oscillate about their true value.

Instead, grid points are often arranged in a **staggered grid** arrangement within the cell, with different variables being represented by points at different locations in the grid (Fig. 20.6 Grids B - E). Staggered Grid D has many of the same problems as unstaggered Grid A. Grids B and C have fewer problems.

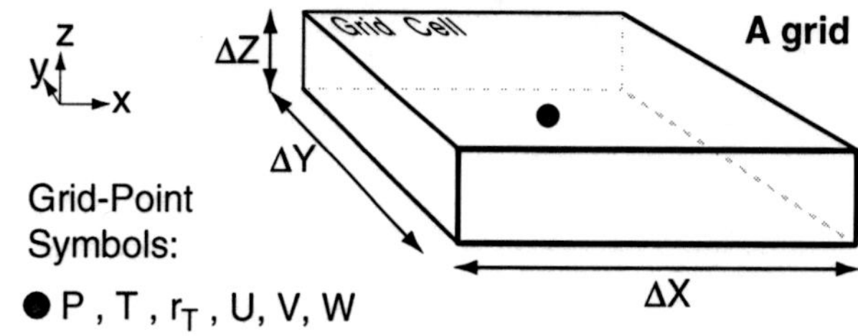

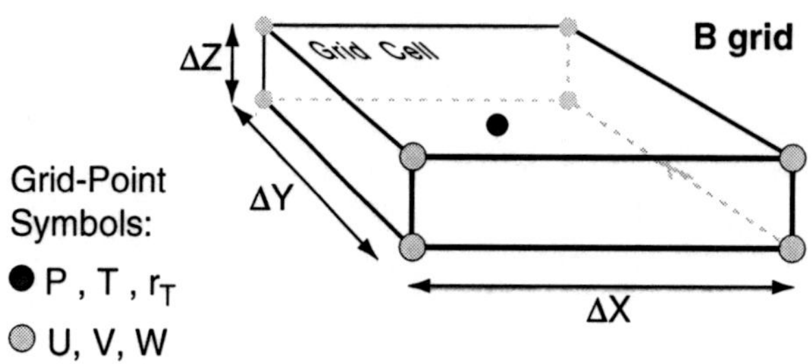

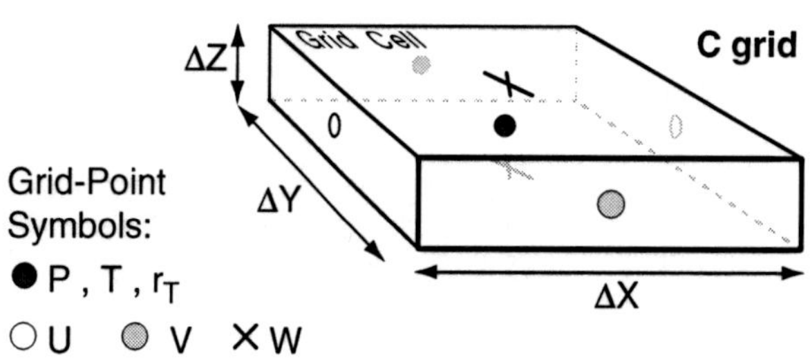

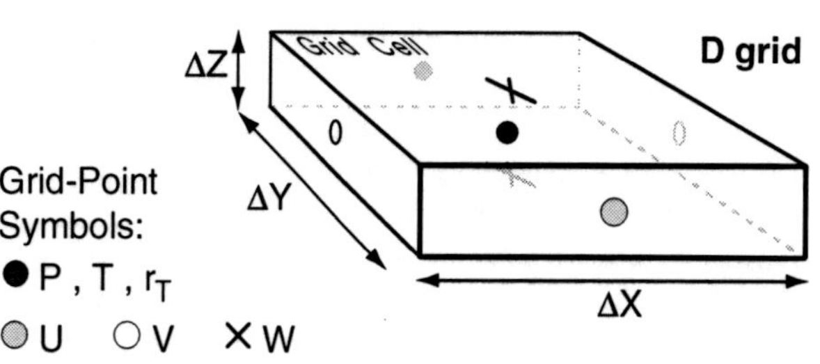

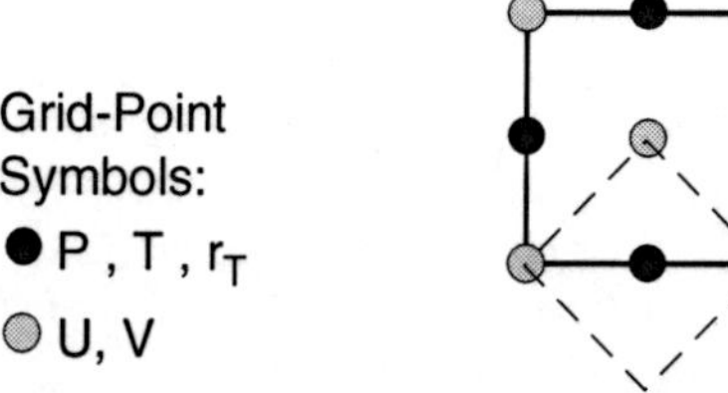

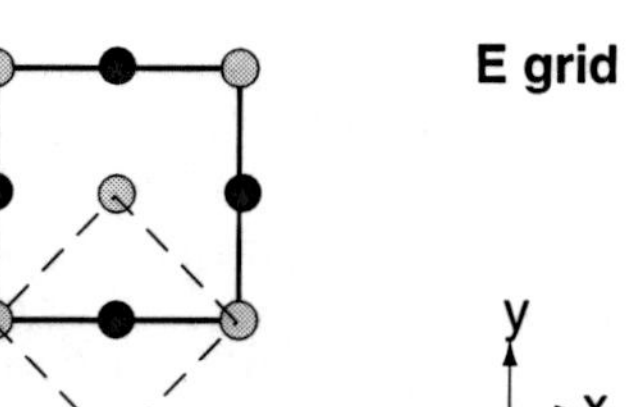

Figure 20.6 (in right column)
Akio Arakawa (1972) identified 5 grid arrangements. Grid A is an unstaggered grid (where all variables are at the same grid point). All the others are staggered grids. Only 2 dimensions are shown for grid E, which is a rotated version of Grid B. Grids C and D differ in their locations of the U and V winds.

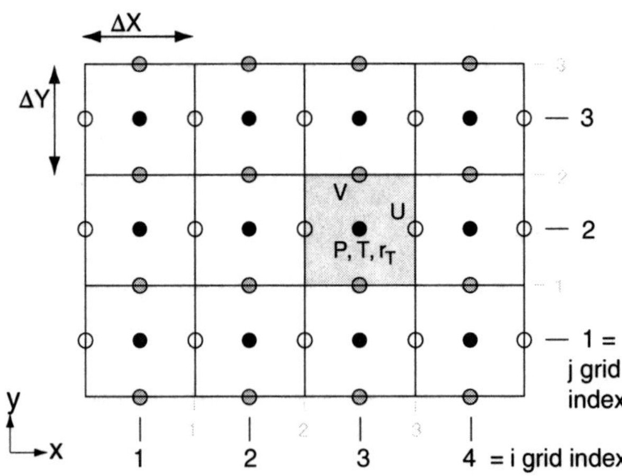

Figure 20.7
Arrangement and indexing of grid cells for a 2-dimensional C-Grid. Each variable in the shaded cell has indices i = 3, j = 2. For variables located at grid-cell edges, some models use whole-index numbering, as shown by the grey numbers. Other models use half indices, as in Fig. 20.9.

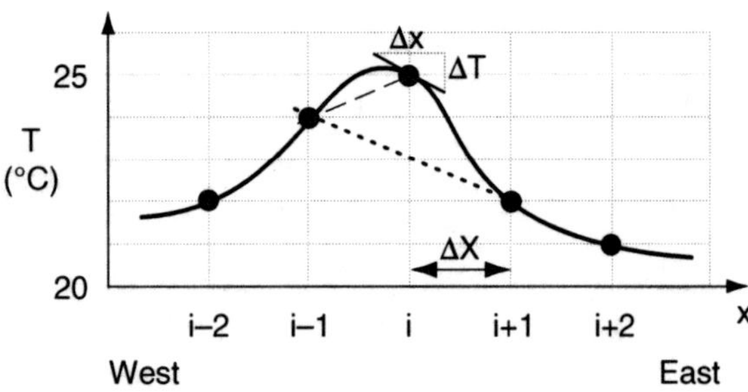

Figure 20.8
The slope (thin unbroken black line) of the temperature (T) curve (thick black line) at grid point i is represented by ΔT/Δx. The lower-case "x" is used in Δx to represent an infinitesimally small increment of distance, while upper-case "X" is used in ΔX to indicate the spacing between grid points (black dots), where i is a grid-point index. Thin dashed and dotted lines are various finite-difference approximations to the slope at i.

FINITE-DIFFERENCE EQUATIONS

Here we see how to find discrete numerical approximations to the equations of motion (20.1 - 20.7) as applied to grid cells.

Notation

Cells are identified by a set of indices (*i, j, k*) that indicate their (*x, y, z*) positions within the domain. Fig. 20.7 shows a two-dimensional example. By using these indices as subscripts, we can specify any variable at any grid-point location. For example, $T_{3,2}$ is the temperature in the center of the shaded grid cell, at *x*-location $i = 3$, and *y*-location $j = 2$. For a 3-D grid, you can use 3 indices or subscripts.

[CAUTION: Throughout this book, we have used ratios of differences (such as ΔT/Δx) instead of derivatives (∂T/∂x) to represent the local slope or local gradient of a variable. While this allowed us to avoid calculus, it causes problems in this chapter because Δx now has two conflicting meanings: (1) Δx is an infinitesimal increment of distance, such as used to find the local slope of a curve at point i in Fig. 20.8. (2) ΔX is a finite distance between grid points, such as between points i and i+1 in Fig. 20.8.

To artificially discriminate between these two meanings, we will use lower-case "x" in Δx to represent an infinitesimal distance increment, and upper-case "X" in ΔX to represent a finite distance between grid points.]

Approximations to Spatial Gradients

The equations of motion (20.1 - 20.6) contain many terms involving local gradients, such as the horizontal temperature gradient $\Delta T/\Delta x$. So to solve these equations, we need a way to approximate the local gradients as a function of things that we know — e.g., values of *T* at the discrete grid points.

But when the local gradient of an analytical variable is represented at one grid point as a function of its values at other grid points, the result is an infinite sum of terms — each term of greater power of ΔX or Δt. This is a **Taylor series** (see the HIGHER MATH box). The most important terms in the series are the first ones — the ones of lowest power of ΔX (said to be of **lowest order**).

However, the **higher-order** terms do slightly improve the accuracy. For practical reasons, the numerical forecast can consider only the first few terms from the Taylor series. Such a series is said to be **truncated**; namely, the highest-order terms are cut from the calculation. For example, a second-order approximation to T' (= $\Delta T/\Delta x$) has an error of about $\pm T'/6$, while a third-order approximation has an error of about $\pm T'/24$.

Different approximations to the local gradients have different **truncation errors**. Such approximations can be applied to the local gradient of any weather variable — the illustrations below focus on temperature (T) gradients. Assuming a mean wind from the west, an **upwind first-order difference** approximation is:

$$\left.\frac{\Delta T}{\Delta x}\right|_i \approx \frac{(T_i - T_{i-1})}{\Delta X} \tag{20.9}$$

which applies at grid point i. But first-order approximations to the gradient (shown by slope of the dashed line in Fig. 20.8) can have large errors relative to the actual gradient (shown by the slope of the thin black line).

A **centered second-order difference** gives a better approximation to the gradient at i:

$$\left.\frac{\Delta T}{\Delta x}\right|_i \approx \frac{(T_{i+1} - T_{i-1})}{2\Delta X} \tag{20.10}$$

as sketched by the dotted line in Fig. 20.8.

An even-better **centered fourth-order difference** for the gradient at i is:

$$\left.\frac{\Delta T}{\Delta x}\right|_i \approx \frac{1}{12\Delta X}\left[8(T_{i+1} - T_{i-1}) - (T_{i+2} - T_{i-2})\right] \tag{20.11}$$

as shown by the thin solid line in Fig. 20.8.

Use similar equations for gradients of other variables (U, V, W, r_T, ρ). Orders higher than fourth-order are also used in some numerical models.

Sample Application

Find $\Delta T/\Delta x$ at grid point i in Fig. 20.8 using 1, 2, & 4th order gradients, for a horizontal grid spacing of 5 km.

Find the Answer

Given: T_{i-2} = 22, T_{i-1} = 24, T_i = 25, T_{i+1} = 22, T_{i+2} = 21°C from the data points in Fig. 20.8. ΔX = 5 km.

Find: $\Delta T/\Delta x$ = ? °C km^{-1}

For Upwind 1st-order Difference, use eq. (20.9):

$\Delta T/\Delta x \approx$ (25 – 24°C)/(5 km) = **0.2 °C km^{-1}**

For Centered 2nd-order Difference, use eq. (20.10):

$\Delta T/\Delta x \approx$ (22 – 24°C) /[2·(5 km)] = **–0.2 °C km^{-1}**

For Centered 4th-order Difference, use eq. (20.11):

$\Delta T/\Delta x \approx$ [8·(22–24°C) – (21–22°C)]/[12·(5 km)]

$\approx$ [(–16 + 1)°C] /(60 km) = **–0.25 °C km^{-1}**

Check: Units OK. Agrees with lines in Fig. 20.8.

Exposition: Higher-order differences are better approximations, but none give the true slope exactly.

HIGHER MATH • Taylor Series

The equations of motion have terms such as $U\cdot\partial T/\partial x$. We can use a Taylor series to approximate derivative $\partial T/\partial x$ as a function of discrete grid-point values. [Notation: use T' for $\partial T/\partial x$, use T'' for $\partial^2 T/\partial x^2$.]

Any analytic function such as temperature vs. distance $T(x)$ can be expanded into an infinite series called a **Taylor series** if the derivatives (T', T'', etc.) are well behaved near x. To find the value of T at $(x + \Delta X)$, where ΔX is a small finite distance from x, use a Taylor series of the form: (20.BA1)

$$T(x+\Delta X) \approx T(x) + \frac{(\Delta X)^1}{1!}\cdot T'(x) + \frac{(\Delta X)^2}{2!}\cdot T''(x) + \frac{(\Delta X)^3}{3!}\cdot T'''(x) + \frac{(\Delta X)^4}{4!}\cdot T''''(x) + \dots$$

Apply the Taylor series to grid points (Fig. 20.8), where the spatial position is indicated by an index i: (20.BA2)

$$T_{i+1} \approx T_i + \Delta X\cdot T_i' + \frac{(\Delta X)^2}{2}\cdot T_i'' + \frac{(\Delta X)^3}{6}\cdot T_i''' + \frac{(\Delta X)^4}{24}\cdot T_i'''' + \dots$$

Similarly, by using $-\Delta X$ in the Taylor expansion, you can estimate T upwind, at grid index i–1: (20.BA3)

$$T_{i-1} \approx T_i - \Delta X\cdot T_i' + \frac{(\Delta X)^2}{2}\cdot T_i'' - \frac{(\Delta X)^3}{6}\cdot T_i''' + \frac{(\Delta X)^4}{24}\cdot T_i'''' - \dots$$

For practical reasons, **truncate** the series to a finite number of terms. The more terms you keep, the smaller the **truncation error**. The lowest power of the ΔX term not used defines the **order** of the truncation. Higher-order truncations have less error.

- For a simple upwind difference (with poor, first-order error in ΔX), solve eq (20.BA3) for T':

$$T_i' = \frac{(T_i - T_{i-1})}{\Delta X} + O(\Delta X) \tag{20.BA4}$$

where the last term indicates the truncation error. This T' value gives the dashed-line slope in Fig. 20.8.

- For a centered difference (with moderate, second-order error in ΔX), subtract eq. (20.BA3) from (20.BA2) and solve the result for T':

$$T_i' = \frac{(T_{i+1} - T_{i-1})}{2\Delta X} + O(\Delta X)^2 \tag{20.BA5}$$

This T' value gives the dotted-line slope in Fig. 20.8.

- For an even-better, 4th-order, centered difference, use: (20.BA6)

$$T_i' = \frac{1}{12\Delta X}\left[8(T_{i+1} - T_{i-1}) - (T_{i+2} - T_{i-2})\right] + O(\Delta X)^4$$

which is a slightly better fit to the true slope at i.

In this chapter, we use $\Delta T/\Delta x$ in place of T'. Hence, the bullets above give approximations to $\Delta T/\Delta x$.

Grid Computation Rules

(1) When multiplying or dividing any two variables, both of those variables must be at the same point in space. The result applies at that same point.

(2) When adding, averaging, or subtracting any two variables, if both those variables are at the same point in space, then the result applies at the same point.

(3) However, when adding, averaging, or subtracting two variables at different locations in space, the sum, average, or difference applies at a physical location halfway between the locations of the original variables.

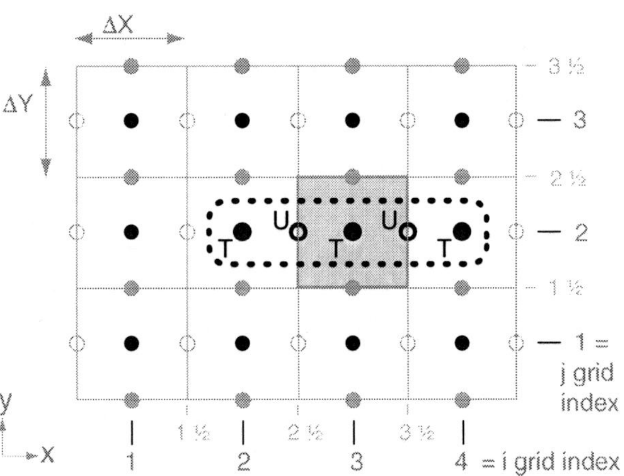

Figure 20.9
Sketch of a two-dimensional C grid. Consider the computation of temperature advection by the U wind, as contributes to the temperature tendency at the one grid point centered in the shaded cell. The grid points needed to make that calculation are outlined with the dotted line, and their arrangement is called a stencil.

Sample Application
What is the warming rate at grid point (i=3, j=2) in Fig. 20.9 due to temperature advection in the x-direction, given $T_{2,2}$ = 22°C, $T_{3,2}$ = 23°C, $T_{4,2}$ = 24°C, $U_{2½,2}$ = –5 m s^{-1}, $U_{3½,2}$ = –7 m s^{-1}, ΔX = 10 km?

Find the Answer
Given: T and U values above. ΔX = 10 km
Find: $\Delta T/\Delta t = -U \cdot \Delta T/\Delta x$ = ? °C h^{-1}.

Use eq. (20.12):
$\Delta T/\Delta t \approx -0.5\cdot(-7-5 \text{ m s}^{-1}) \cdot [\, 0.5\cdot(24 - 22°\text{C})/(10^4\text{m})\,]$
$\approx (6 \text{ m s}^{-1}) \cdot [\, 1°\text{C}/(10^4\text{m})\,] \cdot (3600 \text{ s h}^{-1})$ = **2.16 °C h^{-1}**

Check: Units OK. Sign OK. Magnitude OK.
Exposition: Winds are advecting in warmer air from the East, causing advective warming.

NWP Corollary 1: The forecast at any one point is affected by ALL other points in the forecast domain.

Grid Computation Rules

For mathematical and physical consistency, the grid computation rules at left must be obeyed when making calculations with grid-point values. Rule 3 is handy because you can use it to "move" values to locations where you can then multiply by other variables while obeying Rule 1.

For example, consider the temperature forecast equation (20.4) for grid point (i = 3, j = 2), for the C-grid in Fig. 20.9. The first term on the right side of eq. (20.4) is temperature advection in the x-direction. If we choose to use second-order difference eq. (20.10) at location (i,j) = (3,2), we have a mismatch because we do not have wind at that same location. Rule 1 says we can not multiply the wind times the T gradient.

However, we can use Rule 3 to average the U-winds from the right and left of the temperature point, knowing that this average applies halfway between the two U points. The average thus spatially coincides with the temperature gradient, so we can multiply the two factors together.

For that one grid point (i,j) = (3,2), the result is:

$$\left\{-U\cdot\frac{\Delta T}{\Delta x}\right\}_{3,2} \approx -\left(\frac{U_{3½,2}+U_{2½,2}}{2}\right)\cdot\left[\frac{T_{4,2}-T_{2,2}}{2\cdot\Delta X}\right] \tag{20.12}$$

where the ½ grid index numbering method was used for values at the edges of the grid cell (Fig. 20.9). The parentheses hold the average U, and the square brackets hold the centered second-order difference approximation for the local T gradient.

Similarly, for any grid point (i,j), the result is:

$$\left\{-U\cdot\frac{\Delta T}{\Delta x}\right\}_{i,j} \approx -\left(\frac{U_{i+½,j}+U_{i-½,j}}{2}\right)\cdot\left[\frac{T_{i+1,j}-T_{i-1,j}}{2\cdot\Delta X}\right] \tag{20.13}$$

The spatial arrangement of all grid points used in any calculation is called a **stencil**. Fig. 20.9 shows the stencil for eq. (20.13). Different grid arrangements (Grids A - E) and different approximation orders will result in different stencils.

Significantly, the forecast for any one grid point (such as i,j) depends on the values at other nearby grid points [such as (i–1,j) , (i–½,j) , (i+½,j) , (i+1,j)]. In turn, forecasts at each of these points depends on values at their neighbors. This interconnectivity is summarized as NWP Corollary 1, at left.

For grid points near the edges of the domain, special stencils using one-sided difference approximations must be used, to avoid referencing grid points that don't exist because they are outside of the domain. Alternately, a **halo** of **ghost-cell** grid points outside the forecast domain can be specified using values found from a larger coarser domain or from imposed **boundary conditions** (**BCs**; i.e., the state of the air along the edges of the forecast domain).

Time Differencing

The smooth flow of time implied by the left side of the equations of motion can be approximated by a sequence of discrete time steps, each of duration Δt. For example, the temperature tendency term on the left side of eq. (20.4) can be written as a centered time difference:

$$\frac{\Delta T}{\Delta t} \approx \frac{T(t+\Delta t) - T(t-\Delta t)}{2 \cdot \Delta t} \tag{20.14}$$

When combined with the right side of eq. (20.4), the result gives the temperature at some future time as a function of the temperatures and winds at earlier times:

$$T_{3,2}(t+\Delta t) = T_{3,2}(t-\Delta t) + 2\Delta t \cdot [\text{RHS of eq. } 20.4] \tag{20.15}$$

Typical time-step durations Δt are on the order of a few seconds to tens of minutes, depending on the grid size (see the section on numerical stability).

The equation above is a form of the **leapfrog** scheme. It gets its name because the forecast starts from the previous time step (t–Δt) and leaps over the present step (t) to make a forecast for the future (t+Δt). Although it leaps over the present step, it utilizes the present conditions to determine the future conditions. Fig. 20.10 shows a sketch of this scheme.

The two leapfrog solutions (one starting at t–Δt and the other starting at t, illustrated above and below the time line in Fig. 20.10) sometimes diverge from each other, and need to be occasionally averaged together to yield a consistent forecast. Without such averaging the solution would become unstable, and would numerically blow up (see next section).

There are other numerical solutions that work better than the leapfrog method. One example is the **Runge-Kutta** method, described in the INFO Box.

By combining eqs. (20.12 & 20.15), we get a temperature forecast equation that includes only the U-advection forcing:

$$T_{3,2}(t+\Delta t) = T_{3,2}(t-\Delta t) + \frac{\Delta t}{\Delta X} \cdot \left\{ (U_{3\frac{1}{2},2} + U_{2\frac{1}{2},2}) \cdot \tfrac{1}{2} \left[T_{4,2} - T_{2,2} \right] \right\}_t \tag{20.16}$$

where the subscript t at the very right indicates that all of the terms inside the curly brackets are evaluated at time t. So the future temperature (at t+Δt), depends on the current temperature and winds (at t) and on the past temperature (at t–Δt). The concept of a **stencil** can be extended to include the 4-D arrangement of locations and times needed to forecast one aspect of physics for any grid point.

Generalizing the previous equation, and recalling NWP Corollary 1, we infer that: the forecast Δt

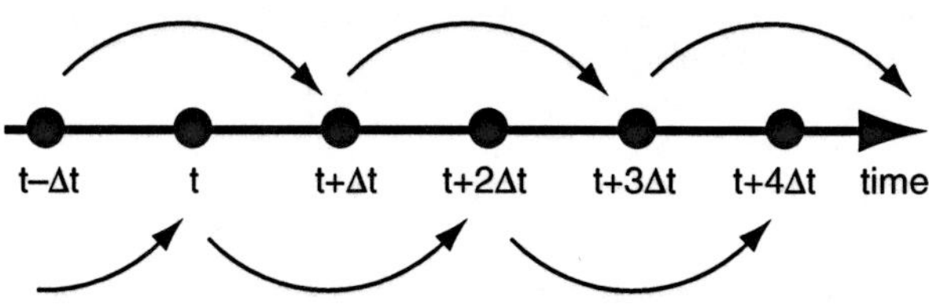

Figure 20.10
Time line illustrating the "leapfrog" time-differencing scheme.

INFO • Time Differencing Methods

The prognostic equations of motion (20.1 - 20.6) can be written in a generic form:

$$\Delta A / \Delta t = f[A, x, t] \, ,$$

where A = any dependent variable (e.g., U, V, W, T, etc.), f is a function that describes all the dynamics and physics of the equations of motion, and x represents all other independent variables (x, y, z) as indicated by grid-point indices (i, j, k). Knowing present (at time t) and all past values at the grid points, how do we make a small time step Δt into the future?

One of the simplest methods is called the **Euler method** (also known as the Euler-forward method):

$$A(t+\Delta t) = A(t) + \Delta t \cdot f[A(t), x, t]$$

But this is only first-order accurate, and is never used because errors accumulate so quickly that the numerical forecast often **blows up** (forecasts values of ± infinity) causing the computer to **crash** (premature termination due to computation errors).

The **leapfrog method** was already given in the text, and is second-order accurate.

$$A(t+\Delta t) = A(t-\Delta t) + 2\Delta t \cdot f[A(t), x, t]$$

Higher-order accuracy has less error.

Also popular is the **fourth-order Runge-Kutta method**, which has even less error, but requires intermediate steps done in the order listed:

(1) $k_1 = f[\, A(t) \, , x \, , t \,]$

(2) $k_2 = f[\, A(t) + \tfrac{1}{2}\Delta t \cdot k_1 \, , \, x \, , \, t + \tfrac{1}{2}\Delta t \,]$

(3) $k_3 = f[\, A(t) + \tfrac{1}{2}\Delta t \cdot k_2 \, , \, x \, , \, t + \tfrac{1}{2}\Delta t \,]$

(4) $k_4 = f[\, A(t) + \Delta t \cdot k_3 \, , \, x \, , \, t + \Delta t \,]$

(5) $A(t+\Delta t) = A(t) + (\Delta t / 6) \cdot [\, k_1 + 2k_2 + 2k_3 + k_4 \,]$

Sample Application (§)

Given a 1-D array consisting of 10 grid points in the x-direction with the following initial temperatures (°C):
T_i (t = 0) = 21.76 22.85 22.85 21.76 20.00 18.24 17.15 17.15 18.24 20.00 for i = 1 to 10.

Assume that the lateral boundaries are cyclic, so that this number sequence repeats outside this primary domain. Grid spacing is 3 km and wind speed from the west is 10 m s^{-1}. For a 250 s time step, forecast the temperature at each point for the first 6 time steps, using leapfrog temporal and 4th-order centered spatial differences.

Find the Answer

Given: ΔX = 3 km , Δt = 250 s , U = 10 m s^{-1}. Initially:

i =	1	2	3	4	5	6	7	8	9	10
T =	21.76	22.85	22.85	21.76	20.00	18.24	17.15	17.15	18.24	20.00

Find: T_i at t = 250 s, 500 s, etc. out to 1500 s

We can use leapfrog for every time step except the first step, because for the first step we have no temperatures before time zero. So I will use an Eulerian time difference for the first step. The resulting eqs. are:

For 1st time step:

$$T_i(t+\Delta t) = T_i(t=0) - \frac{\Delta t \cdot U}{\Delta X \cdot 12}[8\cdot(T_{i+1}-T_{i-1})-(T_{i+2}-T_{i-2})]\Big|_{t=0}$$

For all other time steps:

$$T_i(t+\Delta t) = T_i(t-\Delta t) - \frac{\Delta t \cdot U \cdot 2}{\Delta X \cdot 12}[8\cdot(T_{i+1}-T_{i-1})-(T_{i+2}-T_{i-2})]\Big|_{t}$$

Solving these in a spreadsheet gives, the following, where each row is a new time step and each column is a different grid point:

t(s)	[i = 1	2	3	4	5	6	7	8	9	10]
0	[21.76	22.85	22.85	21.76	20.00	18.24	17.15	17.15	18.24	20.00]
250	[20.50	22.37	23.34	23.03	21.56	19.50	17.63	16.66	16.97	18.44]
500	[18.28	20.34	22.27	23.34	23.13	21.72	19.66	17.73	16.66	16.87]
750	[17.43	18.83	20.68	22.27	22.99	22.57	21.17	19.32	17.73	17.01]
1000	[16.66	17.46	19.22	21.29	22.86	23.34	22.54	20.78	18.71	17.14]
1250	[17.15	16.56	17.29	19.05	21.17	22.85	23.44	22.71	20.95	18.83]
1500	[18.67	17.34	17.02	17.84	19.49	21.33	22.66	22.98	22.16	20.51]

These are plotted as Fig. 20.11b, showing a temperature pattern that is advected by the wind toward the East.

Check: Units OK. Fig. 20.11b looks reasonable

Exposition: The **Courant Number** [$\Delta t \cdot U/\Delta x$] is (250 s) · (10 m s^{-1}) / (3000 m) = 0.833 (dimensionless). Since this number is less than 1, it says that the solution could be numerically stable (see the Numerical Error section).

The boxes in the table above show which numbers are used in the calculation. For example, the temperature forecast at i = 4 and t = 1500 s used data from the grey boxes near it, based on leapfrog in time and 4th-order centered in space. For i = 4 and t = 250 s, the Euler forward time difference was used. For i = 9 and t = 750 s, one of the grey boxes wrapped around, due to the cyclic boundary conditions.

into the future of any one variable at one location can depend on the current state of ALL other variables at ALL other locations. Thus, ALL other variable at ALL other locations must be stepped forward the same one Δt into the future, based on current values. Only after they all have made this step can we proceed to the next step, to get to time $t + 2\Delta t$ into the future. This characteristic is summarized as NWP Corollary 2:

NWP Corollary 2: ALL variables at ALL grid points must march in step into the future*.

**Some terms (e.g., for acoustic waves) and some parameterizations require very short time steps for numerical stability. They can be programmed to take many "baby" steps for each "adult" time step Δt in the model, to enable them to remain synchronized (holding hands) as they advance toward the future.*

To start the whole NWP, we need **initial conditions (ICs)**. These ICs are estimated by merging weather observations with past forecasts (see the Data Assimilation section). ICs are often named by the **synoptic time** when most of the observations were made, such as the "00 UTC analysis", the "00 UTC initialization", or the "00 UTC model run". Modern assimilation schemes can also incorporate **asynoptic** (off-hour) observations.

Discretized Equations of Motion

In summary, the physical equations of motion, which are essentially smooth analytical functions, must be discretized to work at grid points. For example, if we use leapfrog time differencing with second-order spatial differencing on a C-grid, the temperature forecast equation (20.4) becomes:

(20.17)

$$T_{i,j,k}(t+\Delta t) = T_{i,j,k}(t-\Delta t) + 2\Delta t \cdot \Bigg\{ -\left(\frac{U_{i+\frac{1}{2},j,k}+U_{i-\frac{1}{2},j,k}}{2}\right)\cdot\left[\frac{T_{i+1,j,k}-T_{i-1,j,k}}{2\cdot\Delta X}\right] -\left(\frac{V_{i,j+\frac{1}{2},k}+V_{i,j-\frac{1}{2},k}}{2}\right)\cdot\left[\frac{T_{i,j+1,k}-T_{i,j-1,k}}{2\cdot\Delta Y}\right] -\left(\frac{W_{i,j,k+\frac{1}{2}}+W_{i,j,k-\frac{1}{2}}}{2}\right)\cdot\left[\frac{T_{i,j,k+1}-T_{i,j,k-1}}{2\cdot\Delta Z}\right] -\frac{\Re\cdot T_{i,j,k}}{P_{i,j,k}\cdot C_p}\cdot\left[\frac{\mathbb{F}_{z\,rad\;i,j,k+\frac{1}{2}}-\mathbb{F}_{z\,rad\;i,j,k-\frac{1}{2}}}{\Delta Z}\right] +\frac{L_v}{C_p}\cdot\left[\frac{r_{cond\;i,j,k}(t+\Delta t)-r_{cond\;i,j,k}(t-\Delta t)}{2\Delta t}\right] -\left[\frac{F_{z\,turb\;i,j,k+\frac{1}{2}}-F_{z\,turb\;i,j,k-\frac{1}{2}}}{\Delta Z}\right] \Bigg\}$$

Finite-difference equations that are used to forecast winds and humidity are similar. If we had used higher-order differencing, and included the curvature terms and mapping factors, the result would have contained even more terms.

Although the equation above looks complicated, it is trivial for a digital computer to solve because it is just algebra. Computing this equation takes a finite time — perhaps a few microseconds. Similar computation time must be spent for all the other grid points in the domain. These computations must be repeated for a succession of short time steps to reach forecast durations of several days. Thus, for many grid points and many time steps, the total computer run-time accumulates and can take many minutes to several hours on powerful computers.

NUMERICAL ERRORS & INSTABILITY

Causes of NWP errors include **round-off error**, **truncation error**, **numerical instability**, and **dynamical instability**. Dynamical instability related to initial-condition errors will be discussed later in the section on chaos. Additional errors not considered in this section are coding bugs, computer viruses, user errors, numerical or physical approximations, simplifications and parameterizations.

Round-off Error

Round-off error exists because computers represent numbers by a limited number of binary bits (e.g., 32, 64, 128 bits). As a result, some real decimal numbers can be only approximately represented in the computer. For example, a 32-bit computer can resolve real numbers that are different from each other by about 3×10^{-8} or greater. Any finer differences are missed.

To demonstrate, I wrote a computer program to start with x = 0.0, and then repeatedly add 0.1 to x (printing x at each step) until it reaches x = 3.0, at which point I programmed it to stop. When I used single precision (32-bits), my program never stopped. After 30 additions it had found x = 2.9999993, but since this was not exactly equal to 3.0, the program kept adding 0.1 in an **infinite loop** (i.e., ran forever). When I tried it again using double precision (64 bits) it also never stopped, getting only as close to 3.0 as x = 3.0000000000000013 .

Namely, the slight error between decimal and binary representations of a number can accumulate, or can cause unexpected outcomes of conditional tests ("if" statements). Most modern computers use many bits to represent numbers. Nonetheless, always consider round-off errors when you write programs.

INFO • Early History of NWP

The first equations of fluid mechanics were formulated by Leonhard Euler in 1755, using the differential calculus invented by Isaac Newton in 1665, Gottfried Wilhelm Leibniz in 1675, and using partial derivatives devised by Jean le Rond d'Alembert in 1746.

Terms for molecular viscosity were added by Claude-Louis Navier in 1827 and George Stokes in 1845. The equations describing fluid motion are often called the **Navier-Stokes equations**. These primitive equations for fluid mechanics were refined by Herman von Helmholtz in 1888.

About a decade later Vilhelm Bjerknes in Norway suggested that these same equations could be used for the atmosphere. He was a very strong proponent of using physics, rather than empirical rules, for making weather forecasts.

In 1922, Lewis Fry Richardson in England published a book describing the first experimental numerical weather forecast — which he made by solving the primitive equations with mechanical desk calculators. His book was very highly regarded and well received as one of the first works that combined physics and dynamics in a thorough, interactive way.

It took him 6 weeks to make a 6 h forecast. Unfortunately, his forecast of surface pressure was off by an order of magnitude compared to the real weather. Because of the great care that Richardson took in producing these forecasts, most of his peers concluded that NWP was not feasible. This discouraged further work on NWP until two decades later.

John von Neumann, a physicist at Princeton University's Institute for Advanced Studies, and Vladimir Zworykin, an electronics scientist at RCA's Princeton Laboratories and key inventor of television, proposed in 1945 to initiate NWP as a way to demonstrate the potential of the recently-invented electronic computers. Their goal was to simulate the global circulation. During the first few years they could not agree on how to approach the problem.

Von Neumann formed a team of theoretical meteorologists including Carl-Gustav Rossby, Arnt Eliassen, Jule Charney, and George Platzman. They realized the need to simplify the full primitive equations in order to focus their limited computer power on the long waves of the global circulation. So Charney and von Neumann developed a simple one-layer barotropic model (conservation of absolute vorticity).

Their first electronic computer, the ENIAC, filled a large room at Princeton, and used vacuum tubes that generated tremendous heat and frequently burned out. The research team had to translate the differential equations into discrete form, write the code in machine language (FORTRAN and C had not yet been invented), decide how large a forecast domain was necessary, and do many preliminary calculations using slide rules and mechanical calculators.

Their first ENIAC forecasts were made in March-April 1950, for three weather case studies over North America. This was the start of modern NWP.

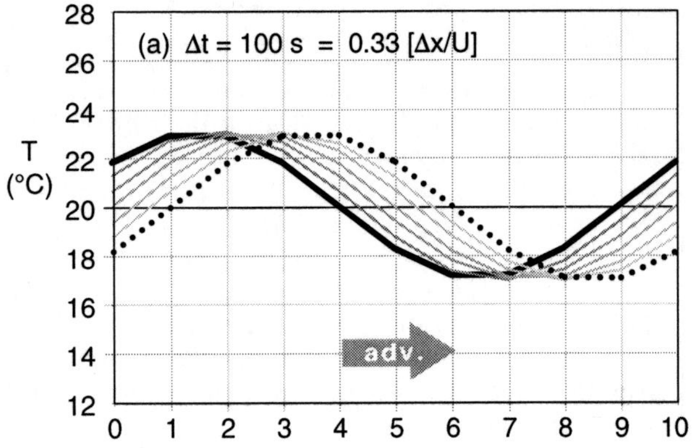

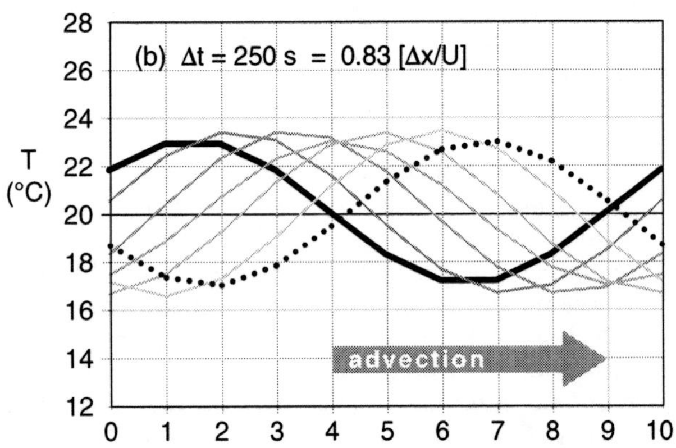

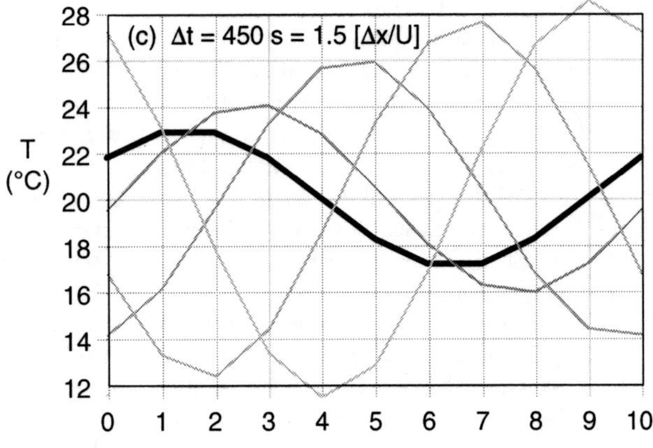

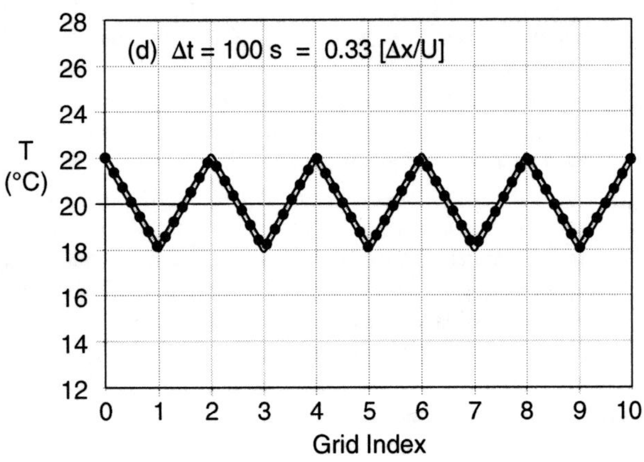

Truncation Error

Truncation error was already discussed, and refers to the neglect (i.e., truncation) of higher-order terms in a Taylor series approximation to local gradients. If we retain more terms in the Taylor series, then the result is a higher-order solution that is more accurate, but which takes longer to run because there are more terms to compute. If we truncate the series at lower order, the numerical solution is faster but less accurate. In NWP, time and space difference schemes are chosen as a compromise between accuracy and speed.

Numerical Instability

Numerical instability causes forecasts to **blow up**. Namely, the numerical solution rapidly diverges from the true solution, can have incorrect sign, and can approach unphysical values (±∞). Truncation error is one cause of numerical instability.

Numerical instability can also occur if the wind speeds are large, the grid size is small, and the time step is too large. For example, eq. (20.16) models advection by using temperature in neighboring grid cells. But what happens if the wind speed is so strong that temperature from a more distant location in the real atmosphere (beyond the neighboring cell) can arrive during the time step Δt? Such a physical situation is not accounted for in the numerical approximation of eq. (20.16). This can create numerical errors that amplify, causing the model to blow up (see Fig. 20.11).

Such errors can be minimized by taking a small enough time step. The specific requirement for stability of advection processes in one dimension is

$$\Delta t \le \frac{\Delta X}{|U|} \qquad \bullet(20.18)$$

with similar requirements in the y and z directions. This is known as the **Courant-Friedrichs-Lewy (CFL) stability criterion**, or the **Courant condition.** When modelers use finer mesh grids with smaller ΔX values, they must also reduce Δt to preserve numerical stability. The combined effect greatly increases model run time on the computer. For example, if ΔX and ΔY are reduced by half, then

Figure 20.11 (at left)
Examples of numerical stability for advection, with $\Delta X = 3$ km and $U = 10$ m s^{-1}. Thick black line is initial condition, and the forecast after each time step is shown as lighter grey, with the last (6th) step dotted. A temperature signal of wavelength $10 \cdot \Delta X$ is numerically stable for time steps Δt of (a) 100 s and (b) 250 s, but (c) = 450 s exceeds the CFL criterion, and the solution blows up (i.e., the wave amplitude increases without bound). (d) A $2 \cdot \Delta X$ wave does not advect at all (i.e., is unphysical).

INFO • Resolution vs. Grid Spacing

Theoretically, the smallest horizontal wavelength you can resolve with data at discrete grid points is $2{\cdot}\Delta X$. However, the finite-difference equations that are used to describe advection and other dynamics in NWP models are unable to handle $2{\cdot}\Delta X$ waves. Namely, these waves either do not advect at all (Fig. 20.11d), or they are numerically unstable.

To avoid such unphysical behavior, small wavelength waves are numerically filtered out of the model. As a result, the smallest waves that are usually retained in NWP models are about 5 to $7{\cdot}\Delta X$.

Hence, the actual **resolution** (i.e., the smallest weather features that can be modeled) are about 7 times the **grid spacing**. Stated another way, if you know the size of the smallest weather system or terrain-related flow that you want to be able to forecast, then you need to design your NWP model with horizontal grid spacing ΔX smaller than 1/7 of that size.

so must Δt, thereby requiring 8 times as many computations to complete the forecast.

For other physical processes such as diffusion and wave propagation, there are other requirements for numerical stability. To preserve overall stability in the model, one must satisfy the most stringent condition; that is, the one requiring the smallest time step. Some high-resolution NWP models use time steps of Δt = 5 s or less.

For advection, one way to avoid the time-step limitation above is to use a **semi-Lagrangian** method. This scheme uses the wind at each grid point to calculate a **backward trajectory**. The backward trajectory indicates the source location for air blowing into the grid cell of interest. This source location need not be adjacent to the grid-cell of interest. By carrying the values of meteorological variables from the source to the destination during the time step, advection can be successfully modeled (i.e., be numerically stable) even for long time steps.

INFO • Lipschitz Continuity

A semi-Lagrangian numerical approach can be numerically stable if the velocity and advected variables A are limited in how fast they vary along the back trajectory path s. Namely, a graph of A vs. s must not cross into the grey shaded cone of Fig. 20.E, for a double cone centered anywhere along s.

This smoothness requirement is called the **Lipschitz condition.**

In the example shown in Fig. 20.E, the curve at (1) is OK, but at (2) is bad because the curve crosses the grey cone.

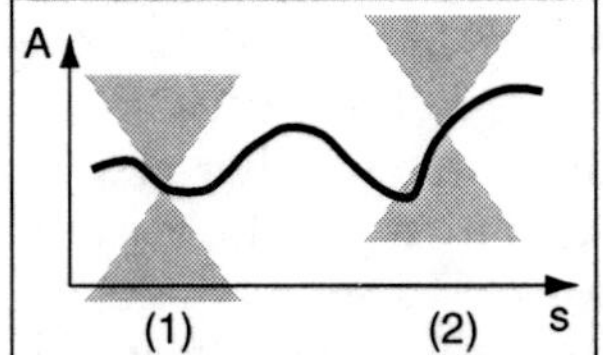

Fig. 20.E

Sample Application

What grid size, domain size, number of grid points, and time steps would you use for a numerical model of a hurricane, and how many computations would be needed to make a 3-day forecast? How fast should your computer be? [Hint: Use info from the Hurricane chapter.]

Find the Answer

This is an example of how you **design an NWP system**, including both the software and hardware.

Assume the smallest feature you want to resolve is a thunderstorm in the eyewall. If tropical thunderstorms are about 14 km in diameter, then you would want ΔX = (14 km)/7 = 2 km to horizontally resolve it.

Hurricanes can be 300 km in diameter. To model the whole hurricane and a bit of its surrounding environment, you might want a horizontal domain of 500 km by 500 km. This works out to (500 km / 2 km) = 250 grid points in each of the x and y directions, giving $(250)^2$ = 62,500 grid points in the horizontal. If you want a model with 50 vertical levels, then you need (50) · (62,500) = 3,125,000 grid points total.

If you want to be able to forecast hurricanes up through category 5 (wind speed > 69 m s^{-1}), then design for a max wind of 80 m s^{-1}. The CFL criterion (eq. 20.18) gives Δt = (2000 m)/(80 m s^{-1}) = 25 s. Thus, a 3-day forecast would require (72 h) · (3600 s h^{-1}) / (25 s) = 10,368 time steps.

The temperature forecast eq (20.17) has about 43 arithmetic operations (adds, subtracts, multiplies, divides). We have 7 equations of motion, so this gives about (7 ·43 ≈) 300 operations. You must do these operations at each grid point for each time step, giving a total = 3,125,000 x 10,368 x 300 ≈ 10^{13} operations.

But we haven't included the calculations for all the other physics (clouds, turbulence, precipitation, radiation, etc.) that must be done at each grid point. As a quick estimate, round up to 10^{15} **floating-point** (real-number) operations.

But you need to complete all these calculations quickly, in order to be useful as a forecast to warn people to evacuate. Suppose you design the model to finish within 3 h (=10,800 s) of computer run time. Thus, your computer must be powerful enough to compute at the rate of (10^{15} operations)/(10,800 s) ≈ 10^{11} = 100 Giga flops (where 1 **flops** = 1 **floating-point operation per second**).

Check: Units OK. Physics OK.

Exposition: The number of calculations needed to make a hurricane forecast is tremendous, and requires a powerful computer. As computer power increases, NWP modelers strive for finer horizontal and vertical resolution spanning larger domains, and including more accurate (and complex) physics parameterizations.

Namely, NWP modelers always tend to fully use all the computer power available, and dream of even more powerful computers.

A SCIENTIFIC PERSPECTIVE • Mathematics

"To have to translate one's verbal statement into mathematical formulae compels one to scrutinize the ideas therein expressed. Next the possession of formulae makes it much easier to deduce the consequences. In this way absurd implications, which might have passed unnoticed in a verbal statement, are brought clearly into view & stimulate one to amend the formula.

Mathematical expressions have, however, their special tendencies to pervert thought: the definiteness may be spurious, existing in the equation but not in the phenomena to be described; and the brevity may be due to the omission of the more important things, simply because they cannot be mathematized. Against these faults we must constantly be on our guard."

– L.F. Richardson, 1919

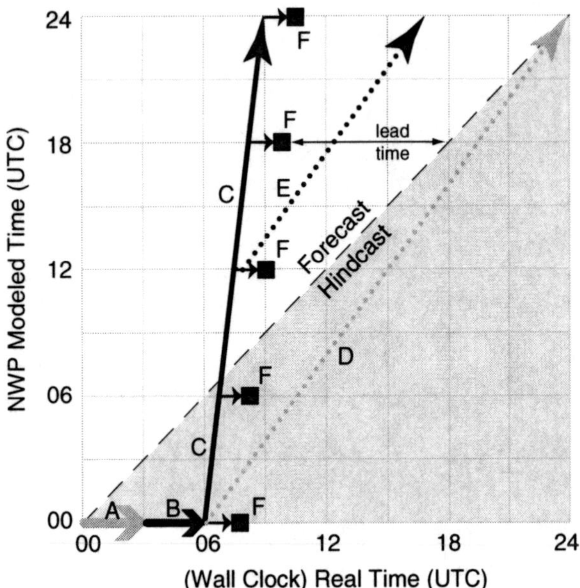

Figure 20.12
Hypothetical forecast schedule, for a 00 UTC initialization.
A: wait for weather observations to arrive.
B: data assimilation to produce the analysis (ICs).
C: coarse-mesh forecast.
D: fine-mesh forecast, initialized from 00 UTC.
E: fine-mesh forecast initialized from coarse forecast at 12 h.
F: postprocessing and creation of products (e.g., weather maps).

THE NUMERICAL FORECAST PROCESS

Weather forecasting is an **initial value problem**. As shown in eq. (20.17), you must know the **initial conditions** on the right hand side in order to forecast the temperature at later times ($t + \Delta t$). Thus, to make forecasts of real weather, you must start with observations of real weather.

Weather-observation platforms and instruments were already discussed in the Weather Reports chapter. Data from these instruments are communicated to central locations. Government forecast centers use these weather data to make the forecasts.

There are three phases of this forecast process. First is pre-processing, where weather observations from various locations and times around the world are assimilated into a regular grid of initial conditions known as an **analysis**. Second is the actual computerized NWP forecast, where the finite-difference approximations of the equations of motion are iteratively stepped forward in time. Finally, **post-processing** is performed to refine and correct the forecasts, and to produce additional secondary products tailored for specific users.

Fig. 20.12 shows a hypothetical forecast schedule, for a weather forecast initialized from 00 UTC synoptic observations. First, it takes a few hours (timeline A in Fig. 20.12) for all the data to be communicated from around the world to the weather forecast center (WFC). This step includes quality control, and rejection of suspected bad data.

Next, the data assimilation programs run for a few hours (B) to create a gridded analysis field. This is the optimum initial condition for the NWP model. At this point, we are ready to start making the forecast, but the initial conditions are already 6 h old compared to the present weather.

So the first part of forecast (C) is spent trying to catch up to "present". This wasted initial forecast period is not lamented, because startup problems associated with the still-slightly-imbalanced initial conditions yield preliminary results that should be discarded anyway. Forecasts that occur AFTER the weather has already happened are known as **hindcasts**, as shown by the shaded area in Fig. 20.12.

The computer continues advancing the forecast (C) by taking small time steps. As the NWP forecast reaches key times, such as 6, 12, 18, and 24 (=00) UTC, the forecast fields are saved for post-processing and display (F). **Lead time** is how much the forecast is ahead of real time. For example, for coarse-mesh model (C), weather-map products (F) that are produced for a **valid time** of 18 UTC appear with a lead time of about 8 h before 18 UTC actually happens, in this hypothetical illustration.

Fig. 20.12 shows a coarse-mesh model (C) that takes 3 h of computation for each 24 h of forecast, as indicated by the slope of line (C). A finer-mesh model might take longer to run (with gentler slope). Model (D) takes 18 h to make a 24 h forecast, and if initialized from the 00 UTC initial conditions, might never catch up to the real weather during Day 1. Hence, it would be useless as a forecast — it would never escape from the hindcast domain.

But for one-way nesting, a fine-mesh forecast (E) could be initialized from the 12 UTC coarse-mesh forecast. This is analogous to a multi-stage rocket, where the coarse mesh (C) blasts the forecast from the past to the future, and then the finer-mesh (E) can remain in the future even though E has the same slope as D.

NWP meteorologists always have the need for speed. Faster computers allow most phases of the forecast process to run faster, allowing finer-resolution forecasts over larger domains with more accuracy and greater lead time. Speed-up can also be achieved computationally by making the dynamics and physics subroutines run faster, by utilizing more processor cores, and by utilizing special computer chips such as **Graphics Processing Units** (**GPU**s). However, tremendous speed-up of a few subroutines might cause only a small speed-up in the overall run time of the NWP model, as explained by **Amdahl's Law** (see INFO Box).

The actual duration of phases (A) through (F) vary with the numerical forecast center, depending on their data-assimilation method, model numerics, domain size, grid resolution, and computer power. Details of the forecast phases are explained next.

Balanced Mass and Flow Fields

Over the past few decades it was learned by hard experience that numerical models give bad forecasts if they are initialized with the raw observed data. One reason is that the in-situ observation network has large gaps, such as over the oceans and in much of the Southern Hemisphere. Also, while there are many observations at the surface, there are fewer in-situ observations aloft. Remote sensors on satellites do not observe many of the needed dynamic variables (U, V, W, T, r_T, ρ) directly, and have very poor vertical resolution. Observations can also contain errors, and local flow around mountains or trees can cause observations that are not representative of the larger-scale flow.

The net effect of such gaps, errors, and inconsistencies is that the numerical representation of this initial condition is **imbalanced**. By imbalanced, we mean that the observed winds disagree with the theoretical winds, where theoretical winds such as

INFO • Amdahl's Law

Computer architect Gene Amdahl described the overall speedup factor S_{ALL} of a computer program as a function of the speedup S_i of individual subroutines, where P_i is the portion of the total computation done by subroutine i:

$$S_{ALL} = \left[\sum (P_i / S_i) \right]^{-1}$$

and where $\Sigma P_i = 1$.

Special programs called **profilers** can find how much time it takes to run each component of an NWP model, such as for the implementation of the WRF model shown below.

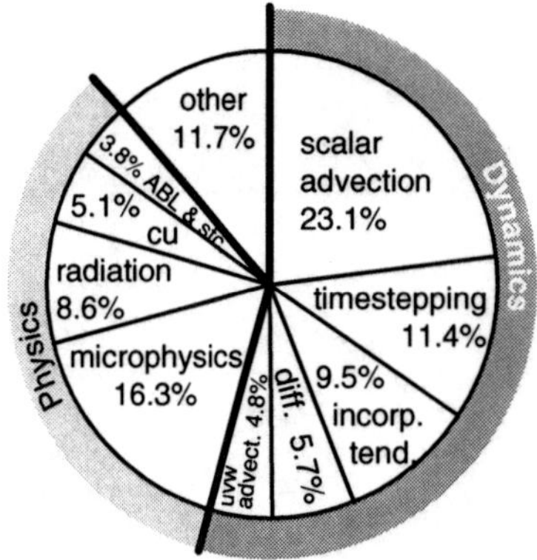

Fig. 20.F. *Portion of total run time of the WRF model for some of the major components.*
"incorp. tend." = incorporation of tendencies.
"diff." = diffusion.
"uvw advect." = advection of U, V, and W wind components.
"microphysics" = hydrometeor parameterizations.
"cu" = parameterizations for convective clouds.
"ABL & sfc." = boundary layer and surface parameterizations.

For example, if **graphics-processing units** (**GPUs**) speed-up the microphysics 20 times and speedup scalar advection by 1.8 times (i.e., an 80% speedup), and the remaining 60.6% of WRF has no speedup, then overall:

$$S_{ALL} = [0.163/20 + 0.231/1.8 + 0.606/1]^{-1} = \underline{\mathbf{1.35}}$$

Namely, even though the microphysics portion of the model is sped up 2000%, the overall speedup of WRF is 35% in this hypothetical example.

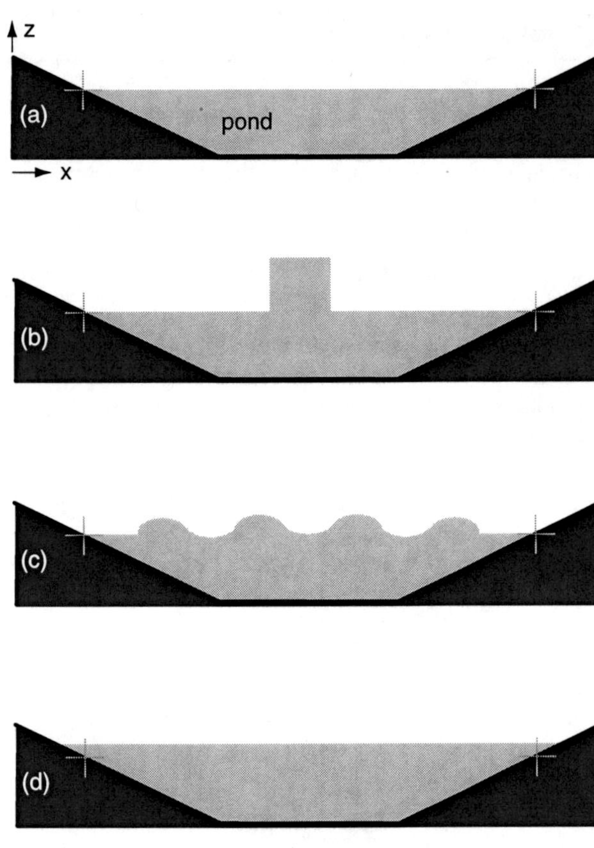

Figure 20.13
Demonstration of a dynamic system becoming balanced. (a) Balanced initial state of a pond of water (shaded grey), with no waves and no currents. (b) Extra water added in center of pond, causing the water-mass distribution to not be in equilibrium with the waves and currents. (c) Wave generation as the pond adjusts itself toward a new balanced state. (d) Final balanced state with slightly higher water everywhere, but no waves and no currents.

the geostrophic wind are based on temperature and pressure fields via atmospheric dynamics.

Balanced and imbalanced flows can be illustrated with a pond of water. Suppose initially the water-level is everywhere level, and the water currents and waves are zero (Fig. 20.13a). This flow system is **balanced**, because with a level pond surface we indeed expect no currents or waves. Next, add extra water to the center of the pond (Fig. 20.13b). This **mass field** (i.e., the distribution of water mass in the pond) is not balanced with the **flow field** (i.e., the motions or circulations within the pond, which for Fig. 20.13b are assumed to be zero). This imbalance causes waves and currents to form (Fig. 20.13c), which help to redistribute mass. These transient waves and currents decay, leaving the pond in a new balanced state (i.e., level water surface, no waves, no currents), but with slightly greater water depth.

Consider what happens to a numerical model of the pond if observation errors are incorporated into the initial conditions. Suppose that the water level in the center of the numerical pond is erroneously "observed" to be 1 m higher than the level everywhere else (Fig. 20.13b). Namely, the "true" initial conditions might be like Fig. 20.13a, but observation errors might cause the "modeled" initial conditions to be like Fig. 20.13b.

A well-designed numerical model of a pond would simulate the dynamical behavior of an actual pond. Hence, the modeled pond would respond as in Figs. 20.13c & d, even though the actual pond would remain motionless as in Fig. 20.13a.

The transient waves and currents are an artifact of the poor initial conditions in the model, and are not representative of the true flow in the real pond. Hence, the forecast results are not to be trusted during the first few minutes of the forecast period while the model is adjusting itself to a balanced state.

Numerical forecasts of the atmosphere have the same problem, but on a longer time scale than a pond. Namely, the first 0.5 to 3 hours of a weather forecast are relatively useless while the model adjusts to imbalances in the initial conditions (see the Data Assimilation section). During this startup period, simulated atmospheric waves are bouncing around in the model, both vertically and horizontally.

After the first 3 to 12 h of forecast, the dynamics are fairly well balanced, and give essentially the same forecast as if the fields were balanced from the start. However, spurious waves in the model might also cause unjustified rejection of good data during data assimilation (see next subsection).

Also, the erroneous waves can generate erroneous clouds that cause erroneous precipitation, etc. The net result could be an unrealistic loss of water from the model that could reduce the chance of future cloud formation and precipitation. Change of water content is just one of many **irreversible processes** that can permanently harm the forecast.

In summary, initialization problems cause a transient period of poor forecast quality, and can permanently degrade longer-term forecast skill or cause rejection of good data. Hence, data-assimilation methods to reduce startup imbalances, such as described next, are highly desirable.

Data Assimilation and Analysis

The technique of incorporating observations into the model's initial conditions is called **data assimilation**. Most assimilation techniques capitalize on the tendency of NWP models to create a balanced state during their forecasts.

One can utilize the balanced state from a previous forecast as a **first guess** of the initial conditions for a new forecast. When new weather observations are incorporated with the first guess, the result is called a weather **analysis**.

To illustrate the initialization process, suppose a forecast was started using initial conditions at 00 UTC, and that a 6-hour forecast was produced, valid at 06 UTC. This 06 UTC forecast could serve as the first guess for new initial conditions, into which the new 06 UTC weather observations could be incorporated. The resulting 06 UTC analysis could then be used as the initial conditions to start the next forecast run. The process could then be repeated for successive forecasts started every 6 h.

Although the analysis represents current or recent-past weather (not a forecast), the analyzed field is usually not exactly equal to the raw observations because the analysis has been smoothed and partially balanced. Observations are used as follows.

First, an automated initial screening of the raw data is performed. During this **quality control** phase, some observations are rejected because they are unphysical (e.g., negative humidities), or they disagree with most of the surrounding observations. In locations of the world where the observation network is especially dense, neighboring observations are averaged together to make a smaller number of statistically-robust observations.

When incorporating the remaining weather observations into the analysis, the raw data from various sources are not treated equally. Some sources have greater likelihood of errors, and are weighted less than those observations of higher quality. Also, observations made slightly too early or too late, or made at a different altitude, are weighted less. In some locations such as the tropics where Coriolis force and pressure-gradients are weak, more weight can be given to the winds than to the pressures.

We focus on two data-assimilation methods here: optimum interpolation and variational. Both are

INFO • The Pacific Data Void

One hazard of data assimilation is that the resulting analysis does not represent truth, because the analysis includes a previous forecast as a first guess. If the previous forecast was wrong, then the subsequent analysis is poor.

Even worse are situations where there are little or no observation data. For data-sparse regions, the first-guess from the previous forecast dominates the "analysis". This means that future forecasts start from old forecasts, not from observations. Forecast errors tend to accumulate and amplify, causing very poor forecast skill further downstream.

One such region is over the N.E. Pacific Ocean. From Fig. 9.23 in the Weather Reports & Map Analysis chapter, there are no rawinsonde observations (RAOBs) in that region to provide data at the dynamically important mid-tropospheric altitudes. Ships and buoys provide some surface data, and aircraft and satellites provide data near the tropopause, but there is a sparsity of data in the middle. This is known as the **Pacific data void**.

Poor forecast skill is indeed observed downstream of this data void, in British Columbia, Canada, and Washington and Oregon, USA. The weather-forecast difficulty there is exacerbated by the complex terrain of mountains and shoreline.

INFO • Giving a Weather Briefing

NWP forecast maps make up an important part of most weather briefings, but they should not be the only part. Bosart (2003), Snelling (1982), West (2011), and others recommend the following:

Your discussion should answer 6 questions:

- What has happened?
- Why has it happened?

- What is happening?
- Why is it happening?

- What will happen?
- Why will it happen?

Identify forecast issues throughout your briefing:

- Difficult/tricky forecast details.
- Significant/interesting weather.

Go from the large-scale to the smaller scales.
Verify your previous forecast.
Encourage questions, discussion, and debate.

For the past and current weather portions of the briefing, show satellite images/animations, radar images/animations, soundings, and weather analyses.

Speak clearly, concisely, loudly, and with confidence. No forecast is perfect, but do the best you can. Your audience will appreciate your sincerity.

Table 20-2.
Standard deviation σ_o of observation errors.

Sensor Type	σ_o
Wind errors in the lower troposphere:	(m s^{-1})
Surface stations and ship obs	3 to 4
Drifting buoy	5 to 6
Rawinsonde, wind profiler	0.5 to 2.7
Aircraft and satellite	3
Pressure errors:	(kPa)
Surface weather stations & Rawinsonde	0.1
Ship and drifting buoy	0.2
S. Hemisphere manual analysis	0.4
Geopotential height errors:	(m)
Surface weather stations	7
Ship and drifting buoy	14
S. Hemisphere manual analysis	32
Rawinsonde	13 to 26
Temperature errors:	(°C)
ASOS surface automatic weather stn.	0.5 to 1.0
Rawinsonde upper-air obs at z < 15 km	0.5
at altitudes near 30 km	< 1.5
Humidity errors:	
ASOS surface weather stations: T_d (°C)	0.6 to 4.4
Rawinsonde in lower troposph. RH (%)	5
near tropopause: RH (%)	15

Sample Application

A drifting buoy observes a wind of M = 10 m s^{-1}, while the first guess for the same location gives an 8 m s^{-1} wind with 2 m s^{-1} likely error. Find the analysis wind speed.

Find the Answer

Given: M_O = 10 m s^{-1}, M_F = 8 m s^{-1}, σ_f = 2 m s^{-1}
Find: M_A = ? m s^{-1}

Use Table 20-2 for Wind Errors: Drifting buoy:
σ_o = 6 m s^{-1}

Use eq. (20.20) for wind speed M:

$$M_A = M_F \cdot \frac{\sigma_o{}^2}{\sigma_f{}^2 + \sigma_o{}^2} + M_O \cdot \frac{\sigma_f{}^2}{\sigma_f{}^2 + \sigma_o{}^2}$$

$$M_A = (8\text{m/s}) \cdot \frac{(6\text{m/s})^2}{(2\text{m/s})^2 + (6\text{m/s})^2} + (10\text{m/s}) \cdot \frac{(2\text{m/s})^2}{(2\text{m/s})^2 + (6\text{m/s})^2}$$

= (8 m s^{-1})·(36/40) + (10 m s^{-1})·(4/40) = <u>8.2 m s^{-1}</u>

Check: Units OK. Physics OK.

Exposition: Because the drifting buoy has such a large error, it is given very little weight in producing the analysis. If it had been given equal weight as the first guess, then the average of the two would have been 9 m s^{-1}. It might seem disconcerting to devalue a real observation compared to the artificial value of the first guess, but it is needed to avoid startup problems.

objective analysis methods in the sense that they are calculated by computer based on equations, in contrast to a "subjective analysis" by a human (such as was demonstrated in the Map Analysis chapter).

Objective optimum interpolation

Let σ_o be the standard deviation of raw-observation errors from a sensor such as a rawinsonde (Table 20-2). Larger σ indicates larger errors.

Let σ_f be the standard deviation of errors associated with the **first guess** from a previous forecast. These are also known as **background errors**. Generally, error increases with increasing forecast range. For example, some global NWP models have the following errors for geopotential height Z of the 50-kPa isobaric surface:

$$\sigma_{Zf} = a \cdot t \tag{20.19}$$

where $a \approx$ 11 m day^{-1} and t is forecast time range. Namely, a first guess from a 2-day forecast is less accurate (has greater error) than a first guess from a 1-day forecast.

An **optimum interpolation** analysis weights the first guess F and the observation O according to their errors to produce an analysis field A:

$$A = F \cdot \frac{\sigma_o{}^2}{\sigma_f{}^2 + \sigma_o{}^2} + O \cdot \frac{\sigma_f{}^2}{\sigma_f{}^2 + \sigma_o{}^2} \tag{20.20}$$

where A, F, O , and σ all apply to the same one weather element, such as pressure, temperature, or wind. If the observation has larger errors than the first guess, then the analysis weights the observation less and the first-guess more.

The equation above can be used to define a **cost function** J:

$$J(A) = \frac{1}{2}\left[\frac{(F-A)^2}{\sigma_f^2} + \frac{(O-A)^2}{\sigma_o^2}\right] \tag{20.21}$$

where the optimum analysis from eq. (20.20) gives the minimum cost for eq. (20.21).

Optimum interpolation is "local" in the sense that it considers only the observations near a grid point when producing an analysis for that point. Optimum interpolation is not perfect, leaving some imbalances that cause atmospheric gravity waves to form in the subsequent forecast. A **normal-mode initialization** modifies the analysis further by removing the characteristics that might excite gravity waves.

Variational data assimilation

Another scheme, called **variational analysis,** attempts to match secondary characteristics calculated from the analysis field to observations so as to minimize the cost function. For example, the radiation emitted by air of the analyzed temperature is compared to radiance measured by satellite, allowing corrections to be made to the temperature analysis as appropriate.

Eq. (20.21) can be modified to utilize such secondary observations:

$$J(A) = \frac{1}{2}\left[\frac{(F-A)^2}{\sigma_f^2} + \frac{(Y-H(A))^2}{\sigma_{yo}^2}\right] \qquad (20.22)$$

where H is an operator that converts from the analysis variable A to the secondary observed variable Y, and σ_{yo} is the standard deviation of observation errors for variable Y. The "best" analysis A is the one that minimizes the value of the cost function J. This minimum can be found by an iterative approach, or by trial and error.

For example, suppose a satellite radiometer looks toward Paris, and measures the upwelling radiance L for an infrared wavelength λ corresponding to the water-vapor channel of Fig. 8.8 in the Satellites & Radar chapter. This channel has the strongest returns at about 8 km altitude. For this example, Y is the measured radiance L. $H(A)$ is the Planck blackbody radiance function (eq. 8.1), and A is the analysis temperature at 8 km over Paris. To find the best analysis: guess different values of A, calculate the associated cost function values, and iterate towards the value of A that yields the lowest value of cost function.

The variational approach allows you to consider all worldwide observations at the same time — a method called **3DVar**. To do this, the first-guess (F), analysis (A), and observation (O) factors in eq. (20.22) must be replaced by vectors (arrays of numbers) containing all grid points in the whole 3-D model domain, and all observations worldwide made at the analysis time. Also, the first-guess and observation error variances must be replaced by covariance matrices. Although the resulting matrix-equation version of (20.22) contains millions of elements, large computers can iterate towards a "best" analysis.

An extension is **4DVar**, where the additional dimension is time. This allows off-time observations to be incorporated into the variational analysis. 4DVar is even more computationally expensive than 3DVar. Although the role of any analysis method is to create the initial conditions for an NWP forecast, often it takes more computer time and power to create the optimum initial conditions than to run the subsequent numerical forecast.

Sample Application

The water vapor channel of a satellite observes a radiance of 1.53 W·m^{-2}·µm^{-1}·sr^{-1} over Paris. On average, this channel has an error standard deviation of 0.05 W·m^{-2}·µm^{-1}·sr^{-1}. Output from a previous NWP run gives a temperature of –35°C at altitude 8 km over Paris, which we can use as a balanced first guess for the new analysis. This forecast has an error standard deviation of 1°C. The upper troposphere over Paris is very humid. For your analysis, find the "best" estimate of the temperature at 8 km over Paris.

Find the Answer

Given: L_o = 1.53 W·m^{-2}·µm^{-1}·sr^{-1}, T_F = –35°C at 8 km
σ_{Lo} = 0.05 W·m^{-2}·µm^{-1}·sr^{-1}, σ_{Tf} = 1°C, water-vapor channel: λ = 6.7 µm from the Satellites & Radar chapter with peak weighting function at $z \approx 8$ km. High humidity implies blackbody emissions.
Find: T_A = ? °C at 8 km altitude over Paris.

Use eq. (20.22), & use Planck's Law eq. 8.1 for $H(T_A)$

$$J(T_A) = \frac{1}{2}\left\{\frac{(T_F - T_A)^2}{\sigma^2_{Tf}} + \frac{1}{\sigma^2_{Lo}}\left[L_o - \frac{c_{1B}\cdot\lambda^{-5}}{\exp\left(\frac{c_2}{\lambda\cdot T_A}\right)-1}\right]^2\right\}$$

with c_{1B} = 1.191043x10^8 W·m^{-2}·µm^4·sr^{-1} and c_2 = 14387.75 µm K from near eq. (8.1) in The Satellites & Radar chapter. [Hint: don't forget to convert T_A in Kelvin to use in Planck's Law.]

Calculate $J(T_A)$ for various values of T_A. I will start with a std. atmosphere guess of T_A = –37°C at 8 km.

J = 0.5 · { [(–35°C)–(–37°C)]^2/(1°C)^2 +
(1/(0.05u)^2 ·[1.53u – (8822u/(exp(9.09926) – 1))]^2 }

J = 0.5 ·{ 4 + 118.4 } = 61.2 (dimensionless)

where u are units of W·m^{-2}·µm^{-1}·sr^{-1}.

Repeating the calculation in a spreadsheet for other values of T_A in the range of –37 to –25°C gives:

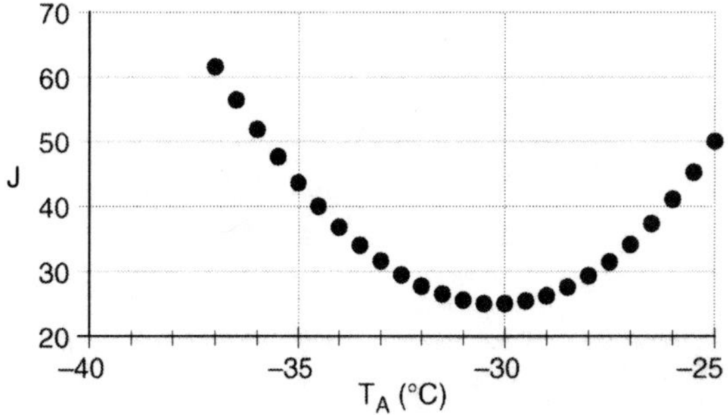

The best analysis is at minimum cost J, near T_A= <u>–30°C</u>

Check: Units OK. T_A value reasonable.

Exposition: The satellite radiance corresponds to a brightness temperature of –25°C. An error propagation calculation (see Appendix A) for eq. (8.1) shows that a radiance error of σ_{Lo} = 0.05 W·m^{-2}·µm^{-1}·sr^{-1} corresponds to a brightness-temperature error of 1°C. Thus, both the first guess and the observed radiance have the same effective error, giving T_A equally weighted between the satellite value –25°C and the first-guess value –35°C. Try different error values in the spreadsheet to see how T_A varies between –35 & –25°C.

Table 20-3. Hierarchy of operational numerical weather prediction (NWP) models.

Forecast Type	Fcst. Duration & (Fcst. Cycle)	Domain & (ΔX)
nowcasts	0 to 3 h (re-run every few minutes)	local: town, county (100s of m)
short-range	3 h to 3 days (re-run every few hours)	regional: state, national, continental (1 to 5 km)
medium-range	3 to 7 days (re-run daily)	continental to global (5 to 25 km)
long-range	7 days to 1 month (re-run daily or weekly)	global (25 to 100 km)
seasonal	1 to 12 months (re-run monthly)	global (100 to 500 km)
GCM*	1 to 1000 years [non-operational (not run routinely); focus instead on case studies & hypothetical scenarios.]	global (100 to 500 km)

** GCM = Global Climate Model -or- General Circulation Model.*

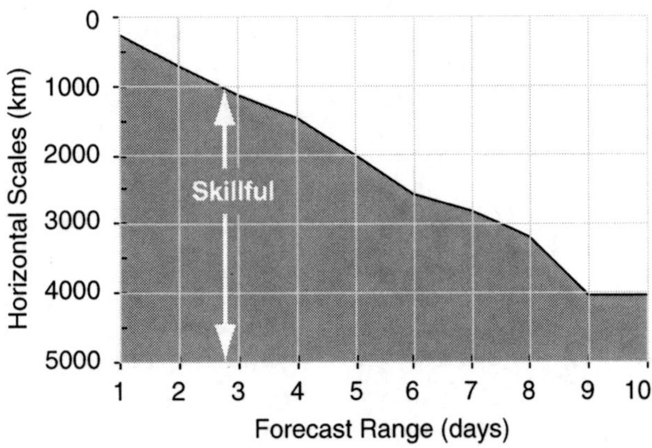

Figure 20.14
Range of horizontal scales having reasonable forecast skill (shaded) for various forecast durations. [from the European Centre for Medium Range Weather Forecasts (ECMWF), 1999]

Forecast

The next phase of the forecast process is the running of the NWP models. Recall that weather consists of the superposition of many different scales of motion (Table 10-6), from small turbulent eddies to large Rossby waves. Different NWP models focus on different time and spatial scales (Table 20-3).

Unfortunately, the forecast quality of the smaller scales deteriorates much more rapidly than that for the larger scales. For example, cloud forecasts might be good out to 2 to 12 hours, frontal forecasts might be good out to 12 to 36 hours, while the Rossby-wave forecasts might be useful out to several days. Fig. 20.14 indicates the ranges of horizontal scales over which the forecast is reasonably skillful.

Don't be deceived when you look at a weather forecast, because all scales are superimposed on the weather map regardless of the forecast duration. Thus, when studying a 5 day forecast, you should try to ignore the small features on the weather map such as thunderstorms or frontal positions. Even though they exist on the map, they are probably wrong. Only the positions of the major ridges and troughs in the jet stream might possess any forecast skill at this forecast duration. Maps in the next section illustrate such deterioration of small scales.

Case Study: 22-25 Feb 1994

Figures 20.15 show the weather valid at 00 UTC on 24 February 1994. Fig. 20.15a gives the **verifying analysis**; namely, a smoothed fit to the actual weather measured at 00 UTC on 24 Feb 1994.

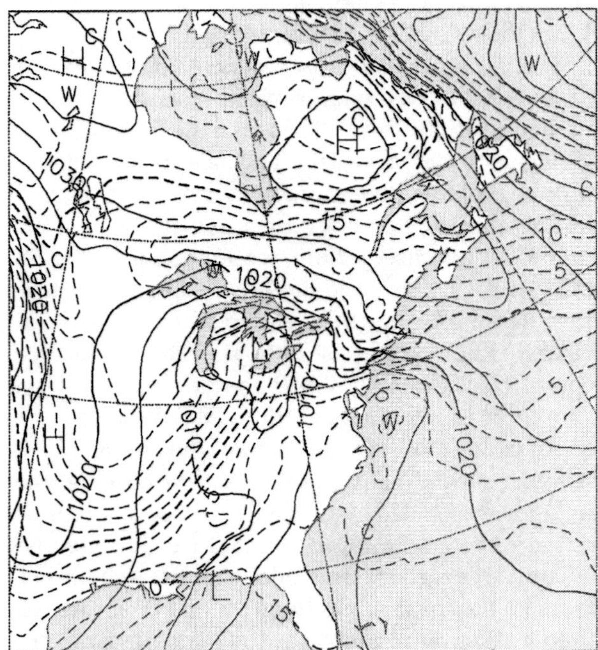

Figure 20.15a
Analysis of 85 kPa temperature and mean-sea-level pressure, valid 00 UTC 24 Feb 94. (Courtesy of ECMWF.)

Figs. 20.15b-e give the weather forecasts valid for the same time, but initialized 1.5, 3.5, 5.5, and 7.5 days earlier. For example, Fig. 20.15b was initialized with weather observations from 12 UTC on 22 Feb 94, and the resulting 1.5 day forecast valid at 00 UTC on 24 Feb 94 is shown in the figure. Fig. 20.15c was initialized from 12 UTC on 20 Feb 94, and the resulting 3.5 day forecast is shown in the figure. Thus, each succeeding figure is the result of a longer-range forecast, which started with earlier observations, but ended at the same time.

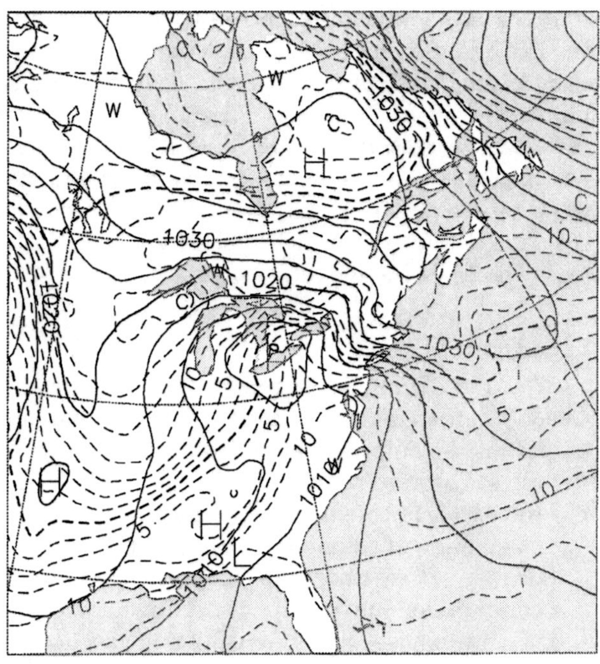

Figure 20.15b
1.5-day forecast, valid 00 UTC 24 Feb 94, started from initialization data at 12 UTC on 22 Feb 94. (From ECMWF.)

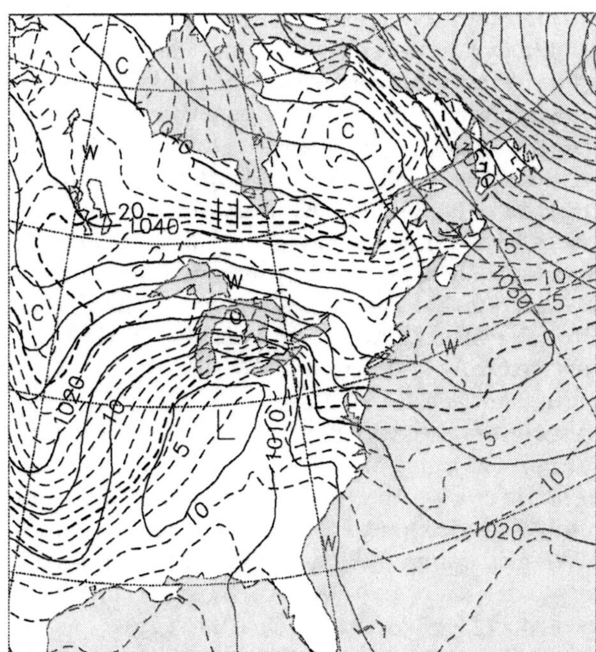

Figure 20.15c
3.5-day forecast, valid 00 UTC 24 Feb 94, started from initialization data at 12 UTC on 20 Feb 94. (From ECMWF.)

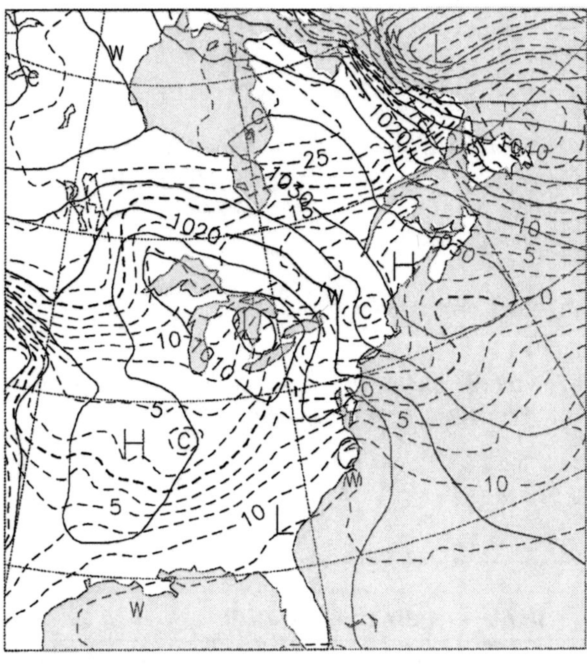

Figure 20.15d
5.5-day forecast, valid 00 UTC 24 Feb 94, started from initialization data at 12 UTC on 18 Feb 94. (From ECMWF.)

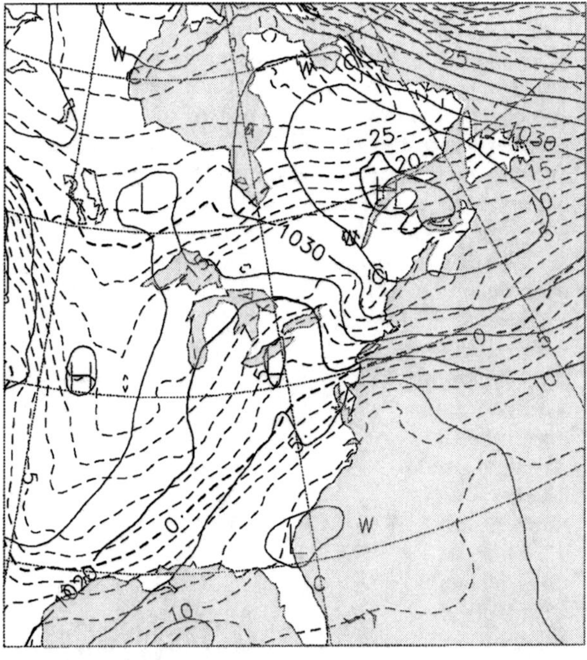

Figure 20.15e
7.5-day forecast, valid 00 UTC 24 Feb 94, started from initialization data at 12 UTC on 16 Feb 94. (From ECMWF.)

INFO • Linear Regression

Suppose that y represents a weather element observed at a weather station. Let x be the corresponding forecast by a NWP model. Over many days, you might accumulate many (N) data points (x_i, y_i) of forecasts and corresponding observations, where i is the data-point index.

If you anticipate that the relationship between x and y is linear, then that relationship can be described by:

$$y = a_o + a_1 \cdot x$$

where a_o is an unknown bias (called the **intercept**), and a_1 is an unknown trend (called the **slope**).

The best-fit (in the **least-squared** error sense) coefficients are:

$$a_o = \frac{\left(\overline{x}\right)\cdot\overline{xy} - \overline{x^2}\cdot\left(\overline{y}\right)}{\left(\overline{x}\right)^2 - \overline{x^2}} \qquad a_1 = \frac{\left(\overline{x}\right)\cdot\overline{y} - \overline{xy}}{\left(\overline{x}\right)^2 - \overline{x^2}}$$

The overbar indicates an average over all data points of the quantity appearing under the overbar; e.g.:

$$\overline{x} = \frac{1}{N}\sum_{i=1}^{N} x_i \qquad \overline{xy} = \frac{1}{N}\sum_{i=1}^{N}\left(x_i \cdot y_i\right) \qquad \overline{x^2} = \frac{1}{N}\sum_{i=1}^{N}\left(x_i^{\,2}\right)$$

INFO • Kalman Filter (KF)

Rudolf Kalman suggested a method that we can modify to estimate the bias x in tomorrow's forecast. It uses the observed bias y in today's forecast, and also uses yesterday's estimate for today's bias x_{old}:

$$x = x_{old} + \beta \cdot \left(y - x_{old}\right)$$

The **Kalman gain** β depends on ratio $r = \sigma^2_{PL}/\sigma^2_{NWP}$, where σ^2_{PL} is the "predictability-limit" error variance associated with the chaotic nature of a "perfect" weather-forecast model, and σ^2_{NWP} is the error variance of the operational NWP model. If those error variances are steady, then $\beta = 0.5\cdot[(r^2+4r)^{1/2} - r]$. The e-folding response time (days) is $\tau = -1/[\ln(1-\beta)]$.

Midlatitude weather is more variable and less predictable in winter. As a result, useful values are:

- Winter: $r \approx 0.06$, $\beta = 0.217$, $\tau = 4$ days.
- Summer: $r \approx 0.02$, $\beta = 0.132$, $\tau = 7$ days.

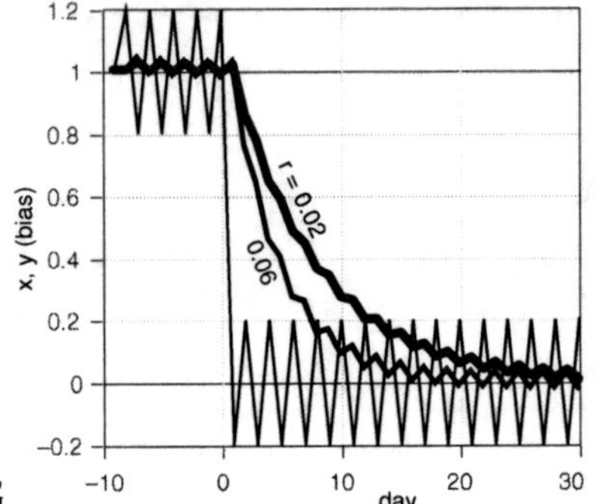

This Fig. shows a noisy input y (thin line) and KF responses x (thicker lines) for different values of the ratio r. The KF adapts to changes, and is recursive.

Fig. 20.G

Solid isobars are MSL pressure in mb (1000 mb = 100 kPa), plotted every 5 mb. Dashed isotherms are 85-kPa temperatures, plotted every 2.5°C, with the 0°C line darker. The map domain covers eastern North America, and is centered on Lake Ontario.

These figures demonstrate the inconsistency of forecasts started at different times with different initial conditions. Such inconsistency is inherent in all forecasts, and illustrates the limits of predictability. The analysis (Fig. 20.15a) shows a low centered near Detroit, Michigan, with a cold front extending southwest toward Arkansas. The 1.5-day forecast (Fig. 20.15b) is reasonably close, but the 3.5-day forecast (Fig. 20.15c) shows the low too far south and the cold front too far west. The 5.5 and 7.5-day forecasts (Fig. 20.15d & e) show improper locations for the fronts and lows, but the larger scales are good.

Post-processing

After the dynamical computer model has completed its forecast, additional **post-processing** computations can be made with the saved output. Postprocessing can include:

- forecast refinement to correct biases,
- calculation of secondary weather variables,
- drawing of weather maps and other graphics,
- compression into databases & climatologies, &
- verification (see the Forecast Quality section).

Forecast Refinement

Forecasts often contain **biases** (systematic errors; see Appendix A), due to: the NWP model formulation; the initial conditions used; and characteristics of different locales. For example, towns might be located in valleys or near coastlines. These are landscape features that can modify the local weather, but which might not be captured by a coarse-mesh numerical model. A number of automated statistical techniques (e.g., **linear regression**, **Kalman filtering**) can be applied as post-processing to reduce the biases and to tune the model output toward the climatologically-expected or observed local weather.

Two classical statistical methods are the **Perfect Prog Method (PPM)** and **Model Output Statistics (MOS)**. Both methods use a best-fit statistical regression (see INFO boxes) to relate input fields (**predictors**) to different output fields (**predictands**). An example of a predictand is surface temperature at a weather station, while predictors for it might include values interpolated from the nearest NWP grid points. PPM uses observations as predictors to determine regression coefficients, while MOS uses model forecast fields. Once the regression coefficients are found, both methods then use the model forecast fields as the predictors to find the surface-temperature forecast for that weather station.

Best fit regressions are found using multi-year sets of predictors and predictands. The parameters of the resulting best fit regression equations are held constant during their subsequent usage.

The PPM method has the advantage that it does not depend on the particular forecast model, and can be used immediately after changing the forecast model. The PPM produces best predictand values only when the model produces perfect predictor forecasts, which is rare.

The MOS advantage is that any systematic model errors can be compensated by the statistical regression. A disadvantage of MOS is that a multi-year set of model output must first be collected and statistically fit, before the resulting regression can be used for future forecasts. Both MOS and PPM have a disadvantage that the statistical parameters are fixed.

Alternative methods include the **Kalman Filter** (KF; see INFO box) and **Updateable MOS**, which continually refine their statistical parameters each day. They share the advantage of MOS in that they use model output for the predictors. They learn from their mistakes (i.e., are adaptive), and can automatically and quickly retune themselves after any changes in the numerical model or in the synoptic conditions. They are recursive (tomorrow's bias correction depends on today's bias correction, not on many years of past data), which significantly reduces the data-storage requirements. A disadvantage is that the KF cannot capture rare, extreme events.

Calculation of Secondary Variables

Fundamental output from the NWP forecast include winds, temperature, pressure or height, mixing ratio, and precipitation. Additional weather variables can be created for human forecasters, for the general public, and for specific industries such as agriculture, transportation, and utilities. Some of these secondary variables (such as relative humidity) can be calculated directly from the primary fields using their defining equations. Other secondary variables (such as visibility) can be estimated statistically via regression.

Secondary thermodynamic variables include: potential temperature, virtual potential temperature, liquid-water or equivalent potential temperature, wet-bulb temperature, near-surface (z = 2 m) temperature, surface skin temperature, surface heat fluxes, surface albedo, wind-chill temperature, static stability, short- and long-wave radiation, and various storm-potential indices such as CAPE.

Secondary moisture variables include: relative humidity, cloudiness (altitudes and coverage), precipitation type and amount, visibility, near-surface dew-point (z = 2 m), soil wetness, and snowfall.

Sample Application (§)

Given the following set of past data (T_{fcst} vs. T_{obs}):
(a) Find the best-fit straight line. Namely, train MOS using linear regression (see INFO box).
(b) If the NWP forecast for tomorrow is T_{fcst} = 15°C, then post-process it to estimate the bias-corrected T ?

Find the Answer

Given: The first 3 columns of data below.
Find: a_o = ? °C, a_1 = ? , T_{fcst} = ? °C

(a) Train MOS using past data to find the coefficients.

i	$x=T_{fcst}$	$y=T_{obs}$	$x \cdot y$	x^2
1	-10	-11.4	114.0	100
2	-8	-7.9	62.8	64
3	-6	0.7	-4.0	36
4	-4	-2.7	10.9	16
5	-2	-1.4	2.8	4
6	0	3.8	0.0	0
7	2	5.0	10.1	4
8	4	3.8	15.2	16
9	6	6.7	40.0	36
10	8	14.7	117.5	64
11	10	13.9	139.1	100
12	12	17.9	214.3	144
13	14	17.7	247.8	196
14	16	24.8	397.0	256
15	18	22.0	396.4	324
16	20	28.4	568.0	400
17	22	27.2	599.5	484
18	24	34.3	822.0	576
19	26	34.7	902.7	676
20	28	29.7	831.1	784
21	30	36.7	1100.3	900
avg=	10.0	14.2	313.7	246.7

a_o = (10 · 313.7 – 246.7 · 14.2) / (10^2 – 246.7) = **2.52 °C**
a_1 = (10 · 14.2 – 313.7) / (10^2 – 246.7) = **1.17** (dim'less)

The original data points and the best-fit line are:

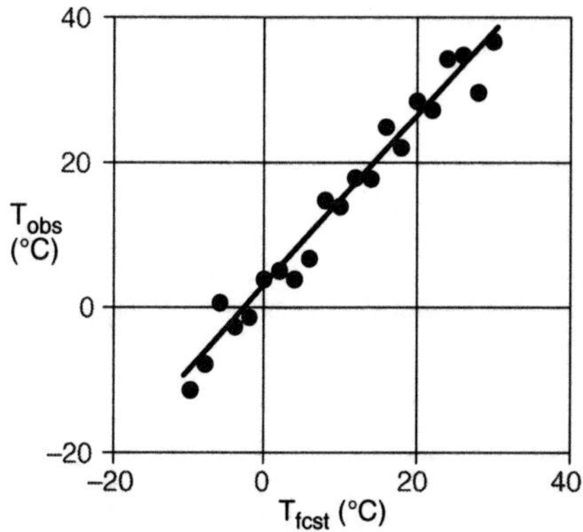

(b) Use MOS to correct the forecast: $T = a_o + a_1 \cdot T_{fcst}$.
T = (2.52°C) + 1.17·(15°C) = **20.1 °C**.

Check: Units OK. Line fits data nicely.
Exposition: The predictor (x) and predictand (y) need not have the same units. For example, MOS could be trained to use model forecasts of relative humidity (%) to estimate observed values of visibility (km).

If you draw a vertical line from each data point to the regressed line, then the length of each line is a measure of the error between the data and the regression. Square each error value and sum them to give the total error. Linear regression is a "best fit" in the sense of finding the "least squared error".

Sample Application

Given the following simplified MOS regression:

$$T_{min} = -295 + 0.4 \cdot T_{15} + 0.25 \cdot \Delta TH + 0.6 \cdot T_d$$

for daily minimum temperature (K) in winter at Madison, Wisconsin, where T_{15} = observed surface temperature (K) at 15 UTC, ΔTH = model forecast of 100-85 kPa thickness (m), and T_d = model fcst. dew point (K). Predict T_{min} given NWP model forecasts of: T_{15} = 273 K, ΔTH = 1,200 m, and T_d = 260 K.

Find the Answer

Given: T_{15} = 273 K, ΔTH = 1,200 m, and T_d = 260 K.
Find: T_{min} = ? K

$$T_{min} = -295 + 0.4\cdot(273) + 0.25\cdot(1{,}200) + 0.6\cdot(260)$$
$$= 270.2 \approx \underline{-3°C}.$$

Check: Units OK. Physics OK.
Exposition: Chilly, but typical for winter in Madison. Note that MOS regressions can be made for any variables in any units. Thus, (1) units might not be consistent from term to term in the regression, but (2) you MUST use the same units for each variable as was used when the MOS regression was created.

INFO • Human Contributions

NWP forecasts are rarely perfect. Thus, humans have the opportunity to improve the forecasts.

By comparing the current model forecast to recent observations, you can discover biases early in the forecast that you can use to correct future forecasts. For example, if the NWP forecast for your location was 2 °C too cold during the past few hours, perhaps it will continue to be 2 °C too cold in the next few hours. Errors in timing or position of fronts and cyclones early in the forecast can be extrapolated to anticipate what corrections are needed for wind and precipitation forecasts later.

You can incorporate knowledge of local effects that might not be resolved by the model — such as the effects of local terrain, coastlines, land use, and urban centers. You may have noticed that some NWP models do not perform well in certain weather patterns, so you can anticipate similar errors when those same weather patterns happen in the future.

Humans can also tailor the weather forecasts to individual users and industries. Examples are forecasts for road or rail transportation, aviation, shipping/fishing/recreational boating, servicing of offshore oil platforms, electrical utility companies (including hydro and wind-energy), air-quality regulatory agencies, avalanche forecast centers, flood and landslide prediction, forest-fire fighting, ski resorts, emergency services, etc.

Humans are also rarely perfect. Nonetheless, studies have shown that the best forecasts are achieved by a combination of NWP and human experience.

Secondary dynamic variables include streamlines, trajectories, absolute vorticity, potential vorticity, isentropic potential vorticity, dynamic tropopause height, vorticity advection, Richardson number, dynamic stability, near-surface winds (z = 10 m), surface stress, surface roughness, mean-sea-level pressure, and turbulence.

While many of the above variables are computed at central numerical-computing facilities, additional computations can be made by separate organizations. Local forecast offices of National Weather Services can tailor the numerical guidance to produce local forecasts of maximum and minimum temperature, precipitation, cloudiness, and storm and flood warnings for the neighboring counties.

Consulting firms, broadcast companies, utility companies, and airlines, for example, acquire the fundamental and secondary fields via data networks such as the internet. From these fields they compute products such as computerized flight plans for aircraft, crop indices and threats such as frost, hours of sunshine, and heating- or cooling-degree days for energy consumption.

Universities also acquire the primary and secondary output fields, to use for teaching and research. Some of the applications result in weather maps that are put back on the internet.

Weather Maps and Other Graphics

The fundamental and secondary variables that are output from the NWP and from forecast refinement are arrays of numbers. To make these data easier to use and interpret by humans, the numbers can be converted into weather-map graphics and animations, **meteograms** (plots of a weather variable vs. time), sounding profiles, cross-sections, text forecasts, and other output forms. Computation of these outputs can take hours, depending on the graphical complexity and the number of products, and thus cannot be neglected in the forecast schedule.

Some visualization programs for NWP output include: GrADS, Vis5D, MatLab, NCAR RIP, GEMPAK, unidata IDV, AWIPS, and NinJo.

Compression into Databases and Climatologies

It is costly to save the terabytes of output produced by operational NWP models every day for every grid point, every level, and every time step. Instead, only key weather fields at RAOB **mandatory levels** in the atmosphere are saved. These WMO **standard isobaric surfaces** are: surface, 100, 92.5, 85, 70, 50, 40, 30, 25, 20, 15, 10, 7, 5, 3, 2, & 1 kPa. Output files can be converted from model-specific formats to standard formats (NetCDF, Vis5D, SQL, GRIB). Forecasts at key locations such as weather stations can be accumulated into growing climatologies.

NONLINEAR DYNAMICS AND CHAOS

Predictability

Recall that NWP is an initial-value problem, where these initial values are partially based on observed weather conditions. Unfortunately, the observations include instrumentation, sampling, and representativeness errors. We have already examined how such errors cause startup problems due to imbalanced flow conditions. How do these errors affect the long-range predictability?

Lorenz suggested that the equations of motion (which are **nonlinear** because they contain products of dependent variables, such as U and T in $U \cdot \Delta T/\Delta x$) are **sensitive to initial conditions**. Such sensitivity means that small differences in initial conditions can grow into large differences in the forecasts.

This is unfortunate. Initial conditions will always have errors. Thus our forecasts will always become less accurate with increasing forecast time. Thus, there is a **limit to the predictability of weather** that is related to **instability of the dynamics**.

A simple physical illustration of **sensitive dependence to initial conditions** is a toy balloon. Inflate one with air and then let it go to fly around the room. Repeat the experiment, being careful to inflate the balloon the same amount and to point it in the same direction. You probably know from experience that the path and final destination of the balloon will differ greatly from flight to flight. In spite of how simple a toy balloon seems, the dynamical equations describing its flight are extremely sensitive to initial conditions, making predictions of flight path virtually impossible.

Lorenz Strange Attractor

Another illustration of sensitive dependence to initial conditions was suggested by Lorenz. Suppose we examine 2-D convection within a tank of water, where the bottom of the tank is heated (Fig. 20.16). The vertical temperature gradient from bottom to top drives a circulation of the water, with warm fluid trying to rise. The circulation can modify the temperature distribution within the tank.

A very specialized, highly-simplified set of equations that approximates this flow is:

$$\frac{\Delta C}{\Delta t} = \sigma \cdot (L - C)$$

$$\frac{\Delta L}{\Delta t} = r \cdot C - L - C \cdot M \qquad (20.23)$$

$$\frac{\Delta M}{\Delta t} = C \cdot L - b \cdot M$$

A SCIENTIFIC PERSPECTIVE • Scientific Revolutions

In the late 1950s and early 1960s, Ed Lorenz was making numerical forecasts of convection to determine if statistical forecasts were better than NWP forecasts using the nonlinear dynamical equations. One day he re-ran a numerical forecast, but entered slightly different initial data. He got a strikingly different answer.

He was curious about this effect, and allowed himself to become sidetracked from his original investigations. This lead to his description of chaos, discovery of a strange attractor, and realization of the sensitive dependence of some equations to initial conditions.

He published his results in 1963 ("Deterministic non-periodic flow". *J. Atmos. Sci.*, **20**, p 130-141). From 1963 to about 1975, this innovative paper was rarely cited by other scientists — a clue that it was not yet accepted by his colleagues. However, between 1975 to 1980, researchers were becoming increasingly aware of his work. From 1980 to present, this paper has been cited on the order of 100 times per year.

About a decade and a half elapsed before this new theory gained wide acceptance, which is typical of many paradigm shifts. Namely, it takes about one human generation for a scientific revolution to mature, because typically the older scientists (who hold the power) are not willing to make the shift. The scientific revolution occurs when this group retires and younger scientists (with the newer ideas) take their place.

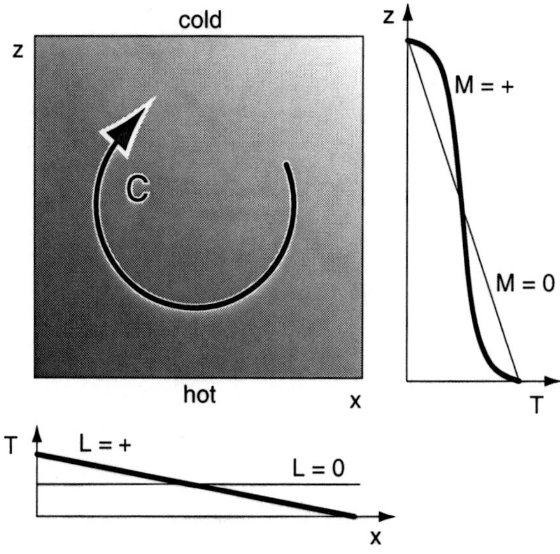

Figure 20.16
Tank of fluid (shaded), showing circulation C. The vertical M and horizontal L distributions of temperature are also shown.

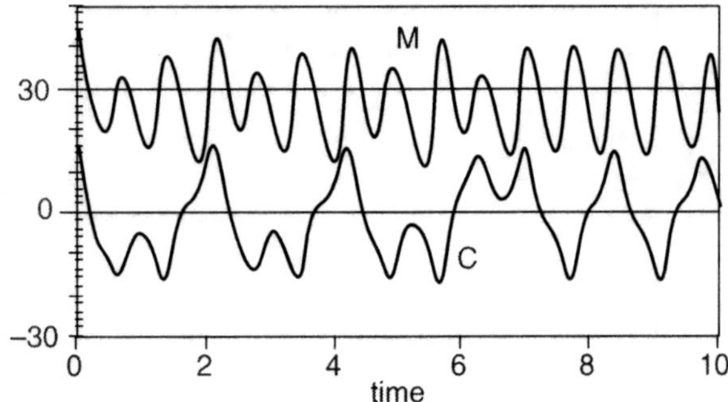

Figure 20.17
Time evolution of circulation C and mixing M.

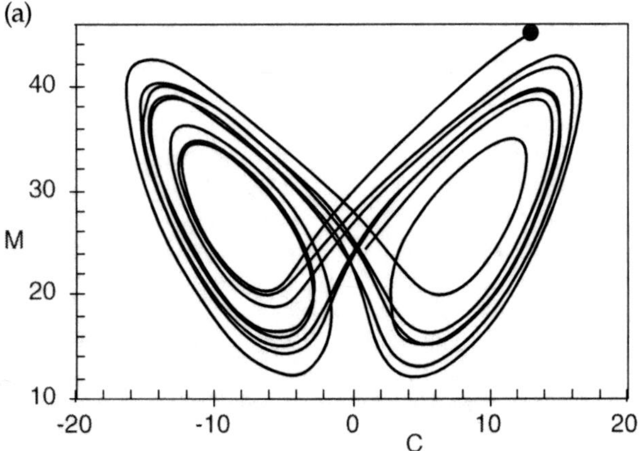

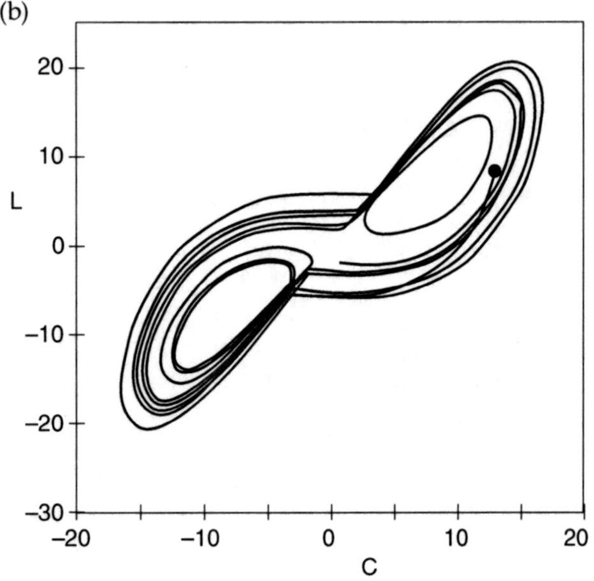

Figure 20.18
"Butterfly" showing evolution of the solution to the Lorenz equations in phase space. Solid dot indicates initial condition. (a) M vs. C. (b) L vs. C.

where C gives the circulation (positive for clockwise, and greater magnitude for a more vigorous circulation), L gives the left-right distribution of temperature (positive for warm water on the left), and M indicates the amount of vertical mixing (0 for a linear temperature gradient, and positive when temperature is more uniformly mixed within the middle of the tank). Each of these dependent variables is dimensionless. Terms $C{\cdot}M$ and $C{\cdot}L$ are nonlinear.

Fig. 20.17 shows forecasts of C and M vs. time, made with parameter values:

$$\sigma = 10.0, \;\; b = 8/3, \;\; \text{and } r = 28$$

and initial conditions:

$$C(0) = 13.0, \;\; L(0) = 8.1, \;\; \text{and } M(0) = 45.$$

Note that all three variables were forecast together, even though L was not plotted to reduce clutter. From Fig. 20.17 it is apparent that the circulation changes direction chaotically, as indicated by the change of sign of C. Also, the amount of mixing in the interior of the tank increases and decreases, as indicated by chaotic fluctuations of M.

When one dependent variable is plotted against another, the result is a **phase-space** plot of the solution. Because the Lorenz equations have three dependent variables, the phase space is three-dimensional. Figs. 20.18 shows two-dimensional views of the solution, which looks like a butterfly.

This solution exhibits several important characteristics that are similar to the real atmosphere. First, it is irregular or chaotic, meaning that it is impossible to guess the solution in the future. Second, the solution is bounded within a finite domain:

$$-20 < C < 20, \quad -30 < L < 30, \quad 0 \le M < 50.$$

which implies that the solution will always remain physically reasonable. Third, the solution (M vs. C) appears to flip back and forth between two favored regions, (i.e., the separate wings of the butterfly). These wings tend to attract the solution toward them, but in a rather strange way.

Hence, they are called **strange attractors.** Fourth, the exact solution is very dependent on the initial conditions, as illustrated in the Sample Applications next. Yet, the eventual solution remains attracted to the same butterfly.

The atmosphere has many more degrees of freedom (i.e., is more complex) than the simple Lorenz model. So we anticipate that the atmosphere is **intrinsically unpredictable** due to its nonlinear chaotic nature. In other words, **there is a limit to how well we can predict the weather**.

Sample Application (§)

Solve the Lorenz equations for the parameters and initial conditions listed previously in this chapter. Use a dimensionless time step of Δt = 0.01, and forecast from t = 0 to t = 10.

Find the Answer

Given: C(0) = 13.0, L(0) = 8.1, and M(0) = 45, and σ = 10.0, b = 8/3, and r = 28.

Find: C(t) = ?, L(t) = ?, M(t) = ?

First, rewrite eqs. (20.23) in the form of a forecast:

$$C(t+\Delta t) = C(t) + \Delta t \cdot [\sigma \cdot (L(t) - C(t))]$$

$$L(t+\Delta T) = L(t) + \Delta t \cdot [r \cdot C(t) - L(t) - C(t) \cdot M(t)]$$

$$M(t+\Delta T) = M(t) + \Delta t \cdot [C(t) \cdot L(t) - b \cdot M(t)]$$

As an example, for the 1st step:

C(0.01) = 13.0 + 0.01·[10.0·(8.1-13.0)] = 12.51

Next, set this up on a spreadsheet, a portion of which is reproduced below.

t	C	L	M
0.00	13.00	8.1	45.00
0.01	12.51	5.809	44.85
0.02	11.84	3.643	44.38
0.03	11.02	1.666	43.63
0.04	10.08	-0.07	42.65
0.05	9.069	-1.55	41.51
0.06	8.007	-2.76	40.26
0.07	6.931	-3.71	38.96
0.08	5.866	-4.44	37.67
0.09	4.836	-4.96	36.4
0.10	3.856	-5.32	35.19

Note that your answers might be different than these, due to different round-off errors and mathematical libraries on the spreadsheets.

Plots. These answers are already plotted in Figs. 20.17 - 20.18.

Check: Units dimensionless. Physics OK.

Exposition: The forecast equations above use an Euler time-differencing scheme, which is the least accurate. Nevertheless, it illustrates the Lorenz attractor.

Science Graffiti

Sensitive Dependence on Initial Conditions

"Does the flap of a butterfly's wings in Brazil set off a tornado in Texas?" – E. Lorenz & P. Merilees, 1972.

"Can a man sneezing in China cause a snow storm in New York? – George R. Stewart, 1941: *Storm*.

"Did the death of a prehistoric butterfly change the outcome of a US presidential election?" – Ray Bradbury, 1952, 1980, *A Sound of Thunder*.

Sample Application (§)

Repeat the previous Sample Application, but for a slightly different initial condition: M(0) = 44.

Find the Answer

Given: C(0) = 13.0, L(0) = 8.1, and M(0) = 44, and σ = 10.0, b = 8/3, r = 28.

Find: C(t) = ?, L(t) = ?, M(t) = ?

As in the previous Sample Application.

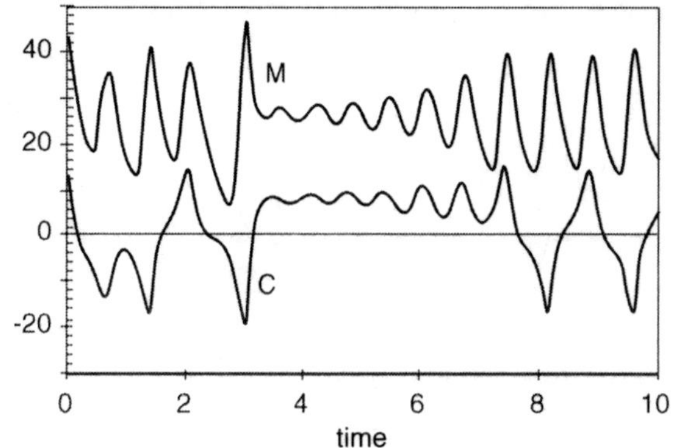

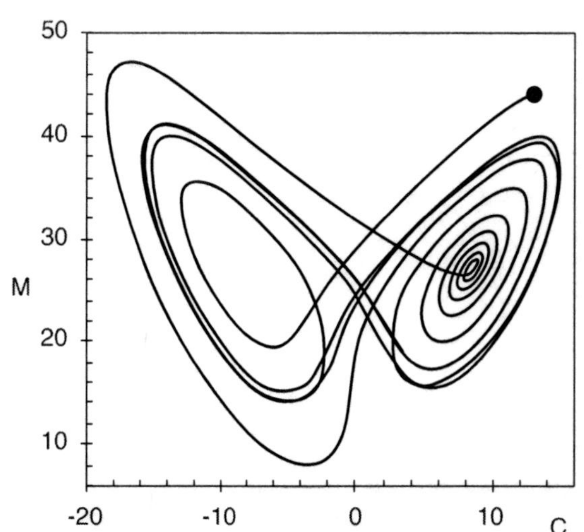

Check: Units OK. Physics OK.

Exposition: These are quite different from Figs. 20.17 & 20.18, demonstrating sensitive dependence to initial conditions.

Science Graffito

"Computations indicate that a perfect model should produce:

- three-day forecasts ... which are generally good;
- one-week forecasts ... which are occasionally good;
- and two-week forecasts ... which, although not very good, may contain some useful information"

– E. Lorenz, 1993: *The Essence of Chaos*. Univ. of Washington Press, Seattle. 227 pp.

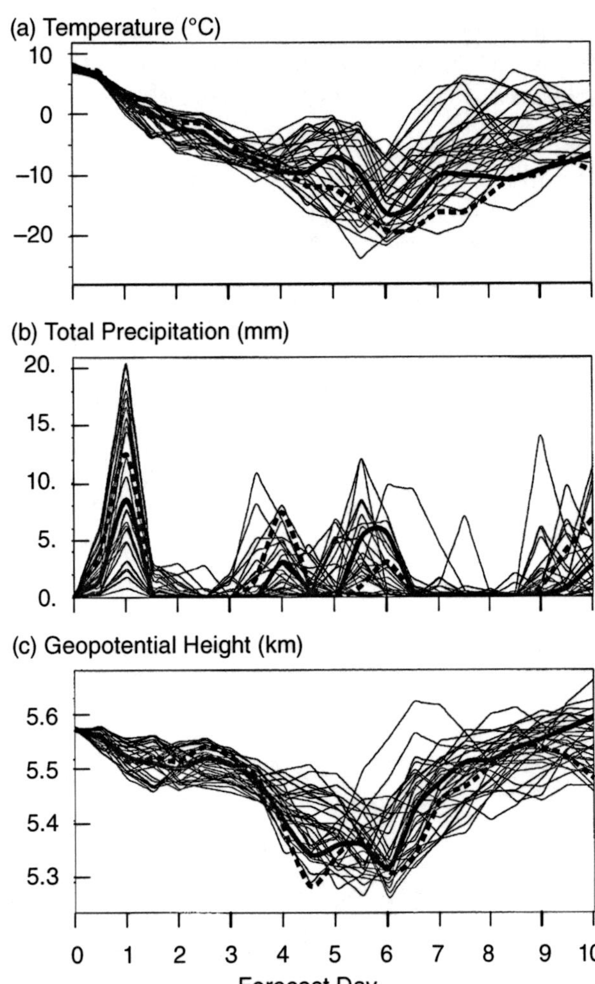

Figure 20.19
Ten-day ensemble forecasts for Des Moines, Iowa, starting from 19 Feb 1994. (a) Temperature (°C) at 85 kPa. (b) Total precipitation (mm). (c) Geopotential height (km) of the 50 kPa surface. Thick solid line is a coarse-resolution forecast that was started from the analyzed initial conditions. Thin lines are course-resolution forecasts, each started from a slightly different initial condition created by adding a different small perturbation to the analysis. The thick dotted line shows a single high-resolution forecast started from the analyzed initial conditions. (Courtesy of ECMWF, 1999).

Ensemble Forecasts

Some forecast centers repeatedly forecast the same time period, but for different conditions. These differences can be created by using different initial conditions, physical parameterizations, numerics, and/or NWP models. Such a procedure yields an **ensemble** of forecasts, which reveal the sensitive dependence of weather forecasts on those different conditions. Fig. 20.19 shows 10-day ensemble forecasts of 85-kPa temperature, precipitation, and 50-kPa geopotential heights at Des Moines, Iowa, for the same case-study period as discussed previously.

The heavy dotted line in these figures shows the single forecast run made with the high-resolution model, starting from the "best" initial conditions. This is the "official" forecast produced by ECMWF. The thick solid line starts from the same initial conditions, but is made with a coarser grid resolution to save forecast time. Note that the forecast changes substantially when the grid resolution changes.

The thin lines are forecasts from slightly different initial conditions. To save computer time, these multiple forecasts are also made with the lower grid-resolution. The forecasts of temperature and geopotential height start out quite close, and diverge slowly during the first 3.5 days. At that time (roughly when the cyclone reaches Des Moines late 22 Feb 94), the solution rapidly diverges, signaling a sudden loss in forecast skill that is never regained.

The spread of the ensemble members informs you about the **uncertainty** of the forecast. Unfortunately, you have no way of knowing which of the ensemble members will be closest to reality.

By averaging all the ensemble members together, you can find an **ensemble average** forecast that is usually more skillful than any individual member. This is one of the strengths of ensemble forecasting.

After several days into the forecast (Fig. 20.19), the ensemble forecasts seem chaotic. Yet this chaotic solution is bounded within a finite region — perhaps a "strange attractor" such as studied by Lorenz.

Studies of chaos often focus on the eventual state of the solution, at times far from the initial condition. At these long times, the dynamics have forgotten the initial state. Although this eventual state is somewhat useless as a weather forecast, it can provide some insight into the range of possible climatic conditions that are allowed by the dynamical equations in the model (i.e., the **model's climate**).

Ensemble forecasts can also suggest which conditions are unlikely — valuable information for some users. For example, if none of the ensemble forecasts give temperatures below freezing on a particular day, then a categorical "no freeze" forecast could be made. Better **confidence** in such forecasts is possible by calibrating the ensemble spread into a probabilistic forecast, discussed next.

Probabilistic Forecasts

Due to the inherent unpredictability of the atmosphere, we cannot confidently make **deterministic forecasts** such as "the temperature tomorrow at noon will be exactly 19.0°C". However, it is possible to routinely create **probabilistic forecasts** similar to "there is a 60% chance the temperature tomorrow at noon will be between 17°C and 21°C, and an 80% chance the it will be between 14 and 24°C."

Ensemble forecasts are increasingly used to create probabilistic forecasts. Various methods can be used to convert the distribution of ensemble members into calibrated probabilistic forecasts, such as illustrated in Fig. 20.20.

The **spread** of the probability distribution (i.e., the **uncertainty** in the forecast) depends on the season (greater spread in winter), the location (greater spread downwind of data-void regions), the climate (greater spread in parts of the globe where weather is more variable), and on the accuracy of the NWP models. A perfect forecast would have no spread.

Often probability forecasts are given as a cumulative probability *CP* that some **threshold** will be met. In Fig. 20.20, the lowest solid line represents a cumulative probability of *CP* = 10% (there is a 10% chance that the observed temperature will be colder than this forecast temperature) and the highest solid black line is for *CP* = 90% (there is a 90% chance that the observation will be colder than this forecast).

For example, using Fig. 20.20 we could tell a farmer that there is 35% chance the low temperature will be below freezing ($T_{threshold}$ = 0°C) on 6 - 7 Jan 2011.

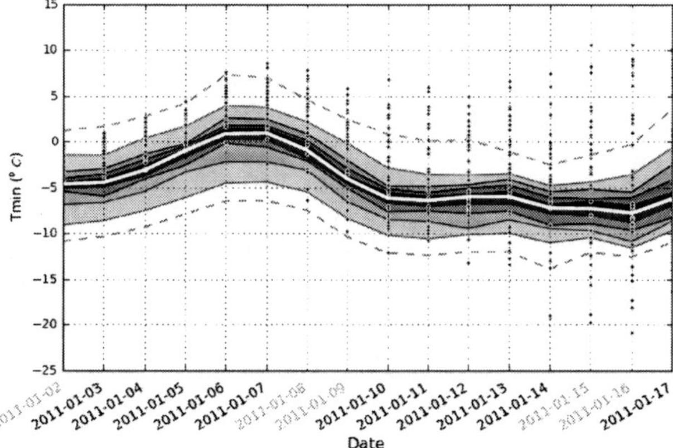

Figure 20.20
Calibrated probabilistic forecast for the minimum temperature (T_{min} °C) at Vancouver Airport (CYVR) for 2 to 17 Jan 2011. The thin white line surrounded by black is the median (50% of the observations should fall below this line, and 50% above). The black, dark grey, medium grey, and light grey regions span ±10%, 20%, 30%, and 40% of cumulative probability around the median. The dashed lines are ±49% around the median — namely, 98% of the time the observations should fall between the dashed lines. The dots are the raw forecasts from each of the 42 members of the North American Ensemble Forecast System.

FORECAST QUALITY & VERIFICATION

NWP forecasts can have both **systematic error** and **random error** (see Appendix A). By making ensemble forecasts you can reduce random errors caused by the chaotic nature of the atmosphere. By postprocessing each ensemble member using Model Output Statistics, you can reduce systematic errors (biases) before computing the ensemble average. After making these corrections, the forecast is still not perfect. How good is the forecast?

Verification is the process of determining the quality of a forecast. Quality can be measured in different ways, using various statistical definitions.

One of the least useful measures of quality is forecast **accuracy**. For example, in Vancouver, Canada, clouds are observed 327 days per year, on average. If I forecast clouds every day of the year, then my accuracy (= number of correct forecasts / total number of forecasts) will be 327/365 = 90% on the average. Although this accuracy is quite high, it shows no skill. To be skillful, I must beat climatology to successfully forecast which days will be cloudless.

Skill measures forecast improvement above some reference such as the climatic average, persistence, or a random guess. On some days the forecast is better than others, so these measures of skill are usually averaged over a long time (months to years) and over a large area (such as all the grid points in the USA, Canada, Europe, or the world).

Methods to calculate verification scores and skills for various types of forecasts are presented next.

Continuous Variables

An example of a continuous variable is temperature, which varies over a wide range of values. Other variables are bounded continuous. For example, precipitation is bounded on one side — it cannot be less than zero. Relative humidity is bounded on two sides — it cannot be below 0% nor greater than 100%, but otherwise can vary smoothly in between.

First, define the terms. Let:

A = initial analysis (based on observations)
V = verifying analysis (based on later obs.)
F = deterministic forecast
C = climatological conditions
n = number of grid points being averaged

An **anomaly** is defined as the difference from climatology at any instant in time. For example

$F - C$ = predicted anomaly
$A - C$ = persistence anomaly
$V - C$ = verifying anomaly

A **tendency** is the change with time:
$F - A$ = predicted tendency
$V - A$ = verifying tendency

An **error** is the difference from the observations (i.e., from the verifying analysis):
$F - V$ = forecast error
$A - V$ = persistence error
The first error is used to measure forecast accuracy. Note that the persistence error is the negative of the verifying tendency.

As defined earlier in this book, the overbar represents an average. In this case, the average can be over n times, over n grid points, or both:

$$\overline{X} = \frac{1}{n}\sum_{k=1}^{n} X_k \qquad \bullet(20.24)$$

where k is an arbitrary grid-point or time index, and X represents any variable.

Mean Error

The simplest quality statistic is the **mean error (ME)**:

$$ME = \overline{(F - V)} = \overline{F} - \overline{V} \qquad \bullet(20.25)$$

This statistic gives the mean **bias** (i.e., a mean difference) between the forecast and verification data. [CAUTION: For precipitation, sometimes the bias is given as a ratio $\overline{F}/\overline{V}$.]

A **persistence forecast** is a forecast where we say the future weather will be the same as the current or initial weather (A) — namely, the initial weather will persist. Persistence forecasts are excellent for very short range forecasts (minutes), because the actual weather is unlikely to have changed very much during a short time interval. The quality of persistence forecasts decreases exponentially with increasing lead time, and eventually becomes worse than climatology. Analogous to mean forecast error, we can define a **mean persistence error**:

$$\overline{(A - V)} = \overline{A} - \overline{V} = \text{mean persistence error} \qquad (20.26)$$

Other persistence statistics can be defined analogous to the forecast statistics defined below, by replacing F with A.

For forecast ME, positive errors at some grid points or times can cancel out negative errors at other grid points or times, causing ME to give a false impression of overall error.

Mean Absolute Error

A better alternative is the **mean absolute error (MAE)**:

$$MAE = \overline{|F - V|} \qquad \bullet(20.27)$$

Thus, both positive and negative errors contribute positively to this error measure.

Mean Squared Error and RMSE

Another way to quantify error that is independent of the sign of the error is by the **mean squared error (MSE)**

$$MSE = \overline{(F - V)^2} \qquad (20.28)$$

A similar mean squared error for climatology (MSEC) can be defined by replacing F with C in the equation above. This allows a **mean squared error skill score (MSESS)** to be defined as

$$MSESS = 1 - \frac{MSE}{MSEC} \qquad (20.29)$$

which equals 1 for a perfect forecast, and equals 0 for a forecast that is no better than climatology. Negative skill score means the forecast is worse than climatology.

The MSE is easily converted into a **root mean square error (RMSE)**:

$$RMSE = \sqrt{\overline{(F - V)^2}} \qquad \bullet(20.30)$$

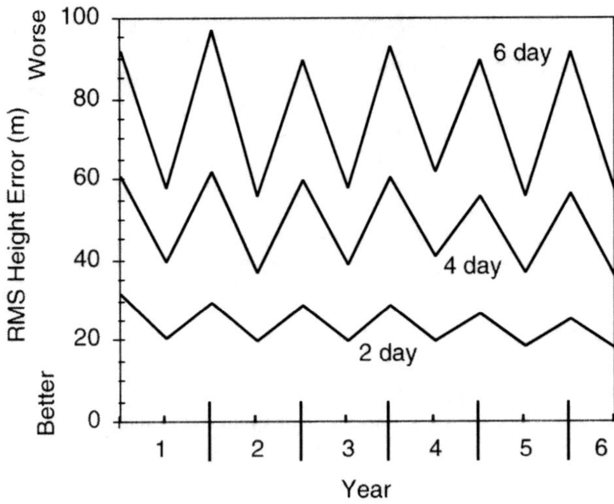

Figure 20.21
Maximum (winter) and minimum (summer) RMS error of 50-kPa heights over the Northern Hemisphere, for forecast durations of 2, 4, and 6 days. (a hypothetical case)

This statistic not only includes contributions from each individual grid point or time, but it also includes any mean bias error.

RMSE increases with forecast duration (i.e., lead time, or forecast range). It is also greater in winter than summer, because summer usually has more quiescent weather. Fig. 20.21 shows the maximum (winter) and minimum (summer) RMS errors of the 50 kPa heights in the Northern Hemisphere, for hypothetical forecasts.

Correlation Coefficient

To discover how well the forecast and verification vary together (e.g., hotter than average on the same days, colder than average on the same days), you can use the Pearson product-moment **correlation coefficient**, *r*:

$$r = \frac{\overline{F'V'}}{\sqrt{\overline{(F')^2}} \cdot \sqrt{\overline{(V')^2}}} \qquad \bullet(20.31)$$

where

$$F' = F - \overline{F} \quad \text{and} \quad V' = V - \overline{V} \qquad (20.32)$$

The correlation coefficient is in the range $-1 \le r \le 1$. It varies from +1 when the forecast varies perfectly with the verification (e.g., forecast hot when verification hot), to –1 for perfect opposite variation (e.g., forecast hot when verification cold), and is zero for no common variation.

Anomaly Correlation

Anomaly correlations indicate whether the forecast (or persistence) is varying from climatology in the same direction as the observations. For example, if the forecast is for warmer-than-normal temperatures and the verification confirms that warmer-than-normal temperatures were observed, then there is a positive correlation between the forecast and the weather. At other grid points where the forecast is poor, there might be a negative correlation. When averaged over all grid points, one hopes that there are more positive than negative correlations, giving a net positive correlation.

By dividing the correlations by the standard deviations of forecast and verification anomalies, the result is normalized into an **anomaly correlation coefficient**. This coefficient varies between 1 for a perfect forecast, to 0 for an awful forecast. (Actually, for a really awful forecast the correlation can reach a minimum of –1, which indicates that the forecast is opposite to the weather. Namely, the model forecasts warmer-than-average weather when colder weather actually occurs, and vice-versa.)

Sample Application (§)

Given the following synthetic analysis (*A*), NWP forecast (*F*), verification (*V*), and climate (C) fields of 50-kPa height (km). Each field represents a weather map (North at top, East at right).

Analysis:

5.3	5.3	5.3	5.4
5.4	5.3	5.4	5.5
5.5	5.4	5.5	5.6
5.6	5.5	5.6	5.7
5.7	5.6	5.7	5.7

Forecast:

5.5	5.2	5.2	5.3
5.6	5.4	5.3	5.4
5.6	5.5	5.4	5.5
5.7	5.6	5.5	5.6
5.7	5.7	5.6	5.6

Verification:

5.4	5.3	5.3	5.3
5.5	5.4	5.3	5.4
5.5	5.5	5.4	5.5
5.6	5.6	5.5	5.6
5.6	5.7	5.6	5.7

Climate:

5.4	5.4	5.4	5.4
5.4	5.4	5.4	5.4
5.5	5.5	5.5	5.5
5.6	5.6	5.6	5.6
5.7	5.7	5.7	5.7

Find the mean error of the forecast and of persistence. Find the forecast MAE and MSE. Find MSEC and MSESS. Find the correlation coefficient between the forecast and verification. Find the RMS errors and the anomaly correlations for the forecast and persistence.

Find the Answer

Use eq. (20.25): $ME_{forecast}$ = 0.01 km = **10 m**
Use eq. (20.26): $ME_{persistence}$ = **15 m**
Use eq. (20.27): MAE = **40 m**
Use eq. (20.28): $MSE_{forecast}$ = 0.004 km^2 = **4000 m^2**
Use eq. (20.28): $MSEC$ = 0.0044 km^2 = **4500 m^2**
Use eq. (20.29): $MSESS$ = 1 – (4000/4500) = **0.11**
Use eq. (20.30): $RMSE_{forecast}$ = **63 m**
Use eq. (20.30): $RMSE_{persistence}$ = **87 m**
Use eq. (20.31): r = **0.92** (dimensionless)
Use eq. (20.33): forecast anomaly correlation = **81.3%**
Use eq. (20.34): persist. anomaly correlation = **7.7%**

Check: Units OK. Physics OK.
Exposition: Analyze (i.e., draw height contour maps for) the analysis, forecast, verification, and climate fields. The analysis shows a Rossby wave with ridge and trough, and the verification shows this wave moving east. The forecast amplifies the wave too much. The climate field just shows the average of higher heights to the south and lower heights to the north, with all transient Rossby waves averaged out.

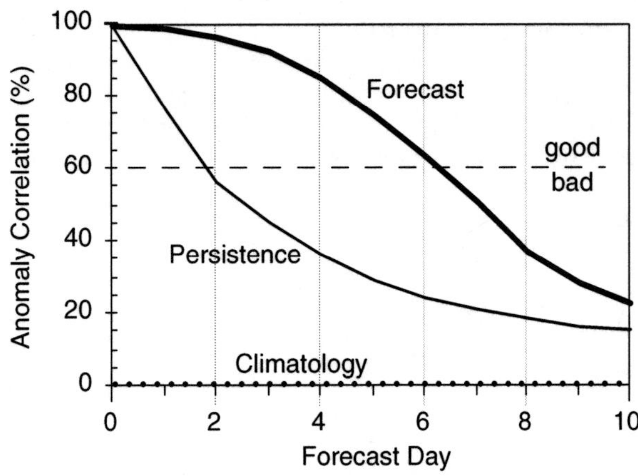

Figure 20.22
NWP forecast skill (thick line) of 50-kPa heights for a hypothetical forecast model. A persistence forecast (thin line) is one where the weather is assumed not to change with time from the initial conditions. A value of 100% is a perfect forecast, while a value of 0 is no better than climatology (dotted line).

(a)

		Observation Yes	Observation No
Forecast	Yes	Hit	False Alarm
Forecast	No	Miss	Correct Rejection or Correct Negative

(b)

		Observation Yes	Observation No
Forecast	Yes	a	b
Forecast	No	c	d

(c)

		Observation Yes	Observation No
Forecast	Yes	Mitigated Loss (C)	Cost (C)
Forecast	No	Loss (L)	No Cost (0)

Figure 20.23
Contingency table for a binary (Yes/No) situation. "Yes" means the event occurred or was forecast to occur. (a) Meaning of cells. (b) Counts of occurrences, where a + b + c + d = n. (c) Expense matrix, where C = cost for taking protective action (i.e., for mitigating the loss), and L = loss due to an unmitigated event.

The definitions of these coefficients are:

anomaly correlation coef. for the forecast =

$$\frac{\overline{\left[(F-C)-\overline{(F-C)}\right]\cdot\left[(V-C)-\overline{(V-C)}\right]}}{\sqrt{\overline{\left[(F-C)-\overline{(F-C)}\right]^2}\cdot\overline{\left[(V-C)-\overline{(V-C)}\right]^2}}} \tag{20.33}$$

anomaly correlation coef. for persistence =

$$\frac{\overline{\left[(A-C)-\overline{(A-C)}\right]\cdot\left[(V-C)-\overline{(V-C)}\right]}}{\sqrt{\overline{\left[(A-C)-\overline{(A-C)}\right]^2}\cdot\overline{\left[(V-C)-\overline{(V-C)}\right]^2}}} \tag{20.34}$$

Fig. 20.22 compares hypothetical persistence and NWP-forecast anomaly correlations for 50-kPa geopotential height. One measure of forecast skill is the vertical separation between the forecast and persistence curves in Fig. 20.22.

This figure shows that the forecast beats persistence over the full 10 days of forecast. Also, using 60% correlation as an arbitrary measure of quality, we see that a "good" persistence forecast extends out to only about 2 days, while a "good" NWP forecast is obtained out to about 6 days, for this hypothetical case for 50 kPa geopotential heights. For other weather elements such as precipitation or surface temperature, both solid curves decrease more rapidly toward climatology.

Binary / Categorical Events

A binary event is a yes/no event, such as snow or no snow. Continuous variables can be converted into binary variables by using a **threshold**. For example, is the temperature below freezing (0°C) or not? Does precipitation exceed 25 mm or not?

A **contingency table** (Fig. 20.23a) has cells for each possible outcome of forecast and observation. "Hit" means the event was successfully forecast. "Miss" means it occurred but was not forecast. "False Alarm" means it was forecast but did not happen. "Correct Rejection" means that the event was correctly forecast not to occur.

After making a series of categorical forecasts, the forecast outcomes can be counted into each cell of a contingency table. Let a, b, c, d represent the counts of all occurrences as shown in Fig. 20.23b. The total number of forecasts is n:

$$n = a + b + c + d \tag{20.35}$$

The **bias score** B indicates over- or under-prediction of the frequency of event occurrence:

$$B = \frac{a+b}{a+c} \tag{20.36}$$

The **portion correct** PC (also known as **portion of forecasts correct** PFC) is

$$PC = \frac{a+d}{n} \tag{20.37}$$

But perhaps a portion of PC could have been due to random-chance (dumb but lucky) forecasts. Let E be this "random luck" portion, assuming that you made the same ratio of "YES" to "NO" forecasts:

$$E = \left(\frac{a+b}{n}\right)\cdot\left(\frac{a+c}{n}\right) + \left(\frac{d+b}{n}\right)\cdot\left(\frac{d+c}{n}\right) \tag{20.38}$$

We can now define the portion of correct forecasts that was actually skillful (i.e., not random chance), which is known as the **Heidke skill score** (HSS):

$$HSS = \frac{PC - E}{1 - E} \tag{20.39}$$

The **hit rate** H is the portion of actual occurrences (obs. = "YES") that were successfully forecast:

$$H = \frac{a}{a+c} \qquad \bullet(20.40)$$

It is also known as the **probability of detection** POD.

The **false-alarm rate** F is the portion of non-occurrences (observation = "NO") that were incorrectly forecast:

$$F = \frac{b}{b+d} \qquad \bullet(20.41)$$

Don't confuse this with the **false-alarm ratio** FAR, which is the portion of "YES" forecasts that were wrong:

$$FAR = \frac{b}{a+b} \tag{20.42}$$

A **true skill score** TSS (also known as **Peirce's skill score** PSS, and as **Hansen and Kuipers' score**) can be defined as

$$TSS = H - F \tag{20.43}$$

which is a measure of how well you can discriminate between an event and a non-event, or a measure of how well you can detect an event.

A **critical success index** CSI (also known as a **threat score** TS) is:

$$CSI = \frac{a}{a+b+c} \tag{20.44}$$

Sample Application

Given the following contingency table, calculate all the binary verification statistics.

		Observation	
		Yes	No
Forecast	Yes:	90	50
	No	75	150

Find the Answer:

Given: $a = 90$, $b = 50$, $c = 75$, $d = 150$

Find: $B, PC, HSS, H, F, FAR, TSS, CSI, GSS$

First, use eq. (20.35): $n = 90 + 50 + 75 + 150 = 365$

So apparently we have daily observations for a year.

Use eq. (20.36): B = (90 + 50) / (90 + 75) = **0.85**

Use eq. (20.37): PC = (90 + 150) / 365 = **0.66**

Use eq. (20.38):

E = [(90+50)·(90+75) + (150+50)·(150+75)]/(365^2)

E = 68100 / 133225 = **0.51**

Use eq. (20.39): HSS = (0.66 – 0.51) / (1 – 0.51) = **0.31**

Use eq. (20.40): H = 90 / (90 + 75) = **0.55**

Use eq. (20.41): F = 50 / (50 + 150) = **0.25**

Use eq. (20.42): FAR = 50 / (90 + 50) = **0.36**

Use eq. (20.43): TSS = 0.55 – 0.25 = **0.30**

Use eq. (20.44): CSI = 90 / (90 + 50 + 75) = **0.42**

Use eq. (20.45): a_r = [(90+50)·(90+75)]/365 = 63.3

Use eq. (20.46):

GSS = (90–63.3) / (90–63.3+50+75) = **0.18**

Check: Values reasonable. Most are dimensionless.

Exposition: No verification statistic can tell you whether the forecast was "good enough". That is a subjective decision to be made by the end user. One way to evaluate "good enough" is via a cost/loss model (see next section).

Sample Application (§)

Given the table below of k = 1 to 31 forecasts of the probability p_k that the temperature will be below threshold 20°C, and the verification $o_k = 1$ if indeed the observed temperature was below the threshold.

(a) Find the Brier skill score. (b) For probability bins of width $\Delta p = 0.2$, plot a reliability diagram, and find the reliability Brier skill score.

k	p_k	o_k	BN	j	k	p_k	o_k	BN	j
1	0.43	0	0.18	2	16	0.89	1	0.01	4
2	0.98	1	0.00	5	17	0.13	0	0.02	1
3	0.53	1	0.22	3	18	0.92	1	0.01	5
4	0.33	1	0.45	2	19	0.86	1	0.02	4
5	0.50	0	0.25	3	20	0.90	1	0.01	5
6	0.03	0	0.00	0	21	0.83	0	0.69	4
7	0.79	1	0.04	4	22	0.00	0	0.00	0
8	0.23	0	0.05	1	23	1.00	1	0.00	5
9	0.20	1	0.64	1	24	0.69	0	0.48	3
10	0.59	1	0.17	3	25	0.36	0	0.13	2
11	0.26	0	0.07	1	26	0.56	1	0.19	3
12	0.76	1	0.06	4	27	0.46	0	0.21	2
13	0.17	0	0.03	1	28	0.63	0	0.40	3
14	0.30	0	0.09	2	29	0.10	0	0.01	1
15	0.96	1	0.00	5	30	0.40	1	0.36	2
					31	0.73	1	0.07	4

Find the Answer

Given: The white portion of the table above.
Find: BSS = ? , $BSS_{reliability}$ = ?, and plot reliability.

(a) Use eq. (20.47). The grey-shaded column labeled BN shows each contribution to the numerator $(p_k - o_k)^2$ in that eq. The sum of BN = 4.86. The sum of o_k = 16. Thus, the eq is: $BSS = 1 - [4.86 / \{16 \cdot (31-16)\}] =$ **0.98**

(b) There are J = 6 bins, with bin centers at p_j = 0, 0.2, 0.4, 0.6, 0.8, and 1.0. (Note, the first and last bins are one-sided, half-width relative to the nominal "center" value.) I sorted the forecasts into bins using j = round($p_k/\Delta p$, 0), giving the grey j columns above.

For each j bin, I counted the number of forecasts n_j falling in that bin, and I counted the portion of those forecasts that verified n_{oj}. See table below:

j	p_j	n_j	n_{oj}	n_{oj}/n_j	num
0	0	2	0	0	0
1	0.2	6	1	0.17	0.04
2	0.4	6	2	0.33	0.16
3	0.6	6	3	0.50	0.36
4	0.8	6	5	0.83	0.04
5	1.0	5	5	1	0

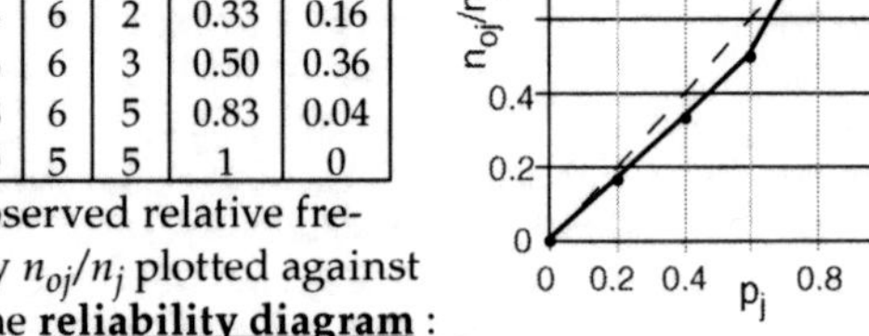

The observed relative frequency n_{oj}/n_j plotted against p_j is the **reliability diagram** :

Use eq. (20.48). The contribution to the numerator from each bin is in the num column above, which sums to 0.6 . Thus: $BSS_{reliability} = 0.6 / \{16 \cdot (31-16)\}$ = **0.0025**

Check: Small $BSS_{reliability}$ agrees with the reliability diagram where the curve nearly follows the diagonal.

Exposition: N = 31 is small, so these statistics are not very robust. Large BSS suggests good probability forecasts. Also, the forecasts are very reliable.

which is the portion of hits given that the event was forecast, or observed, or both. This is often used as a forecast-quality measure for rare events.

Suppose we consider the portion of hits that might have occurred by random chance a_r:

$$a_r = \frac{(a+b)\cdot(a+c)}{n} \tag{20.45}$$

Then we can subtract this from the actual hit count to modify CSS into an **equitable threat score** ETS, also known as **Gilbert's skill score** GSS:

$$GSS = \frac{a - a_r}{a - a_r + b + c} \tag{20.46}$$

which is also useful for rare events.

For a perfect forecasts (where $b = c = 0$), the values of these scores are: B = 1, PC = 1, HSS = 1, H = 1, F = 0, FAR = 0, TSS = 1, CSI = 1, GSS = 1.

For totally wrong forecasts (where $a = d = 0$): B = 0 to ∞, PC = 0, HSS = negative, H = 0, F = 1, FAR = 1, TSS = –1, CSI = 0, GSS = negative.

Probabilistic Forecasts

Brier Skill Score

For calibrated probability forecasts, a **Brier skill score** (BSS) can be defined relative to climatology as

$$BSS = 1 - \frac{\sum_{k=1}^{N} (p_k - o_k)^2}{\left(\sum_{k=1}^{N} o_k\right)\cdot\left(N - \sum_{k=1}^{N} o_k\right)} \tag{•(20.47)}$$

where p_k is the forecast probability ($0 \le p_k \le 1$) that the threshold will be exceeded (e.g., the probability that the precipitation will exceed a precipitation threshold) for any one forecast k, and N is the number of forecasts. The verifying observation $o_k = 1$ if the observation exceeded the threshold, and is set to zero otherwise.

BSS = 0 for a forecast no better than climatology. BSS = 1 for a perfect deterministic forecast (i.e., the forecast is $p_k = 1$ every time the event happens, and $p_k = 0$ every time it does not). For probabilistic forecasts, $0 \le BSS \le 1$. Larger BSS values are better.

Reliability

How **reliable** are the probability forecasts? Namely, when we forecast an event with a certain probability, is it observed with the same relative frequency? To determine this, after you make each forecast, sort it into a forecast probability bin (j) of probability width Δp, and keep a tally of the number of forecasts (n_j) that fell in this bin, and count how many of the forecasts verified (n_{oj}, for which the corresponding observation satisfied the threshold).

For example, if you use bins of size $\Delta p = 0.1$, then create a table such as:

bin index	bin center	fcst. prob. range	n_j	n_{oj}
$j = 0$	$p_j = 0$	$0 \le p_k < 0.05$	n_0	n_{o0}
$j = 1$	$p_j = 0.1$	$0.05 \le p_k < 0.15$	n_1	n_{o1}
$j = 2$	$p_j = 0.2$	$0.15 \le p_k < 0.25$	n_2	n_{o2}
etc.	...	...	...	...
$j = 9$	$p_j = 0.9$	$0.85 \le p_k < 0.95$	n_9	n_{o9}
$j = 10 = J$	$p_j = 1.0$	$0.95 \le p_k \le 1.0$	n_{10}	n_{o10}

A plot of the observed relative frequency (n_{oj}/n_j) on the ordinate vs. the corresponding forecast probability bin center (p_j) on the abscissa is called a **reliability diagram**. For perfect reliability, all the points should be on the 45° diagonal line.

A Brier skill score for **relative reliability** ($BSS_{reliability}$) is:

$$BSS_{reliability} = \frac{\sum_{j=0}^{J}\left[(n_j \cdot p_j) - n_{oj}\right]^2}{\left(\sum_{k=1}^{N} o_k\right) \cdot \left(N - \sum_{k=1}^{N} o_k\right)} \qquad (20.48)$$

where $BSS_{reliability} = 0$ for a perfect forecast.

ROC Diagram

A **Relative Operating Characteristic (ROC)** diagram shows how well a probabilistic forecast can **discriminate** between an event and a non-event. For example, an event could be heavy rain that causes flooding, or cold temperatures that cause crops to freeze. The probabilistic forecast could come from an ensemble forecast, as illustrated next.

Suppose that the individual NWP models of an $N = 10$ member ensemble made the following forecasts of 24-h accumulated rainfall R for Day 1:

NWP model	R (mm)	NWP model	R (mm)
Model 1	8	Model 6	4
Model 2	10	Model 7	20
Model 3	6	Model 8	9
Model 4	12	Model 9	5
Model 5	11	Model 10	7

Consider a precipitation threshold of 10 mm. The ensemble above has 4 models that forecast 10 mm or more, hence the forecast probability is $p_1 = 4/N = 4/10 = 40\%$. Supposed that 10 mm or more of precipitation was indeed observed, so the observation flag is set to one: $o_1 = 1$.

On Day 2, three of the 10 models forecast 10 mm or more of precipitation, hence the forecast probability is $p_2 = 3/10 = 30\%$. On this day precipitation did NOT exceed 10 mm, so the observation flag is set to zero: $o_2 = 0$. Similarly, for Day 3 suppose the forecast probability is $p_3 = 10\%$, but heavy rain was observed, so $o_3 = 1$. After making ensemble forecasts every day for a month, suppose the results are as listed in the left three columns of Table 20-4.

An end user might need to make a decision to take action. Based on various economic or political reasons, the user decides to use a probability threshold of $p_{threshold} = 40\%$; namely, if the ensemble model forecasts a 40% or greater chance of daily rain exceeding 10 mm, then the user will take action. So we can set forecast flag $f = 1$ for each day that the ensemble predicted 40% or more probability, and $f = 0$ for the other days. These forecast flags are shown in Table 20-4 under the $p_{threshold} = 40\%$ column.

Other users might have other decision thresholds, so we can find the forecast flags for all the other probability thresholds, as given in Table 20-4. For an N member ensemble, there are only $(100/N) + 1$ discrete probabilities that are possible. For our example with $N = 10$ members, we can consider only 11 different probability thresholds: 0% (when no members exceed the rain threshold), 10% (when 1 out of the 10 members exceeds the threshold), 20% (etc.), . . . 90%, 100%.

For each probability threshold, create a 2x2 contingency table with the elements a, b, c, and d as shown in Fig. 20.23b. For example, for any pair of observation and forecast flags (o_j, f_j) for Day j, use

a = count of days with hits $(o_j, f_j) = (1, 1)$.
b = count of days with false alarms $(o_j, f_j) = (0, 1)$.
c = count of days with misses $(o_j, f_j) = (1, 0)$.
d = count of days: correct rejection $(o_j, f_j) = (0, 0)$.

For our illustrative case, these contingency-table elements are shown near the bottom of Table 20-4.

Next, for each probability threshold, calculate the hit rate $H = a/(a+c)$ and false alarm rate $F = b/(b+d)$, as defined earlier in this chapter. These are shown in the last two rows of Table 20-4 for our example. When each (F, H) pair is plotted as a point on a graph, the result is called a **ROC diagram** (Fig. 20.24).

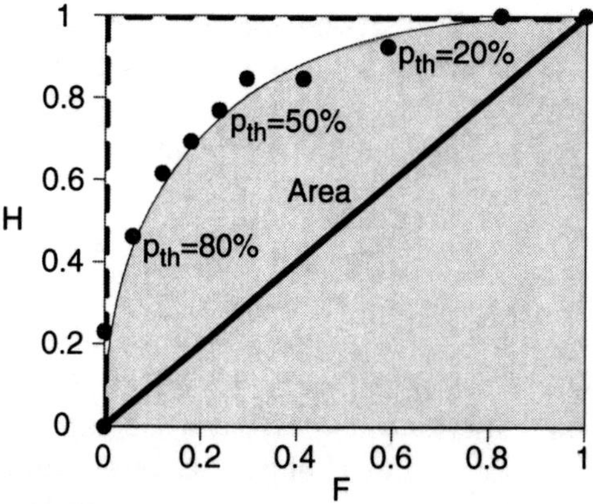

Figure 20.24
ROC diagram plots hit rate (H) vs. false-alarm rate (F) for a range of probability thresholds (p_{th}).

Table 20-4. Sample calculations for a ROC diagram. Forecast flags *f* are shown under the probability thresholds.

Day	*o*	*p* (%)	Probability Threshold $p_{threshold}$ (%)										
			0	10	20	30	40	50	60	70	80	90	100
1	1	40	1	1	1	1	1	0	0	0	0	0	0
2	0	30	1	1	1	1	0	0	0	0	0	0	0
3	1	10	1	1	0	0	0	0	0	0	0	0	0
4	1	50	1	1	1	1	1	1	0	0	0	0	0
5	0	60	1	1	1	1	1	1	1	0	0	0	0
6	0	30	1	1	1	1	0	0	0	0	0	0	0
7	0	40	1	1	1	1	1	0	0	0	0	0	0
8	1	80	1	1	1	1	1	1	1	1	1	0	0
9	0	50	1	1	1	1	1	1	0	0	0	0	0
10	1	20	1	1	1	0	0	0	0	0	0	0	0
11	1	90	1	1	1	1	1	1	1	1	1	1	0
12	0	20	1	1	1	0	0	0	0	0	0	0	0
13	0	10	1	1	0	0	0	0	0	0	0	0	0
14	0	10	1	1	0	0	0	0	0	0	0	0	0
15	1	70	1	1	1	1	1	1	1	1	0	0	0
16	0	70	1	1	1	1	1	1	1	1	0	0	0
17	1	60	1	1	1	1	1	1	1	0	0	0	0
18	1	90	1	1	1	1	1	1	1	1	1	1	0
19	1	80	1	1	1	1	1	1	1	1	1	0	0
20	0	80	1	1	1	1	1	1	1	1	1	0	0
21	0	20	1	1	1	0	0	0	0	0	0	0	0
22	0	10	1	1	0	0	0	0	0	0	0	0	0
23	0	0	1	0	0	0	0	0	0	0	0	0	0
24	0	0	1	0	0	0	0	0	0	0	0	0	0
25	1	70	1	1	1	1	1	1	1	1	0	0	0
26	0	10	1	1	0	0	0	0	0	0	0	0	0
27	0	0	1	0	0	0	0	0	0	0	0	0	0
28	1	90	1	1	1	1	1	1	1	1	1	1	0
29	0	20	1	1	1	0	0	0	0	0	0	0	0
30	1	80	1	1	1	1	1	1	1	1	1	0	0
Contingency Table Values		*a* =	13	13	12	11	11	10	9	8	6	3	0
		b =	17	14	10	7	5	4	3	2	1	0	0
		c =	0	0	1	2	2	3	4	5	7	10	13
		d =	0	3	7	10	12	13	14	15	16	17	17
Hit Rate: $H = a/(a+c)$ =			1.00	1.00	0.92	0.85	0.85	0.77	0.69	0.62	0.46	0.23	0.00
False Alarm Rate: $F=b/(b+d)$			1.00	0.82	0.59	0.41	0.29	0.26	0.18	0.12	0.06	0.00	0.00

The area *A* under a ROC curve (shaded in Fig. 20.24) is a measure of the overall ability of the probability forecast to discriminate between events and non-events. Larger area is better. The dashed curve in Fig. 20.24 illustrates perfect discrimination, for which $A = 1$. A random probability forecast with no ability to discriminate between events has a curve following the thick diagonal line, for which $A = 0.5$. Since this latter case represents no skill, a **ROC skill score** SS_{ROC} can be defined as

$$SS_{ROC} = (2 \cdot A) - 1 \qquad (20.49)$$

where $SS_{ROC} = 1$ for perfect skill, and $SS_{ROC} = 0$ for no discrimination skill.

Although a smooth curve is usually fit through the data points in a ROC diagram, we can nonetheless get a quick estimate of the area by summing the trapezoidal areas under each pair of data points. For our illustration, $A \approx 0.84$, giving $SS_{ROC} \approx 0.68$.

Cost / Loss Decision Models

Weather forecasts are often used to make decisions. For example, a frost event might cause *L* dollars of economic loss to a citrus crop in Florida. However, you could save the crop by deploying orchard fans and smudge pots, at a cost of *C* dollars.

Another example: A hurricane or typhoon might sink a ship, causing *L* dollars of economic loss. However, you could save the ship by going around the storm, but the longer route would cost *C* dollars in extra fuel and late-arrival fees.

No forecast is perfect. Suppose the event is forecast to happen. If you decide to spend *C* to mitigate the loss by taking protective action, but the forecast was bad and the event did not happen, then you wasted *C* dollars. Alternately, suppose the event is NOT forecast to happen, so you decide NOT to take protective action. But the forecast was bad and the event actually happened, causing you to lose *L* dollars. How do you decide which action to take, in consideration of this forecast uncertainty?

Suppose you did NOT have access to forecasts of future weather, but did have access to climatological records of past weather. Let o be the climatological (base-rate) frequency ($0 \le o \le 1$) that the event occurred in the past, where $o = 0$ means it never happened, and $o = 1$ means it always happened.

Assume this same base rate will continue to occur with that frequency in the future. If your $C \le o \cdot L$, then it would be most economical to always mitigate, at cost C. Otherwise, it would be cheaper to never mitigate, causing average losses of $o \cdot L$. The net result is that the expense based only on climate data is

$$E_{climate} = \min(C, o \cdot L) \qquad (20.50)$$

But you can possibly save money if you use weather forecasts instead of climatology. The expected (i.e., average) expense associated with a sometimes-incorrect deterministic forecast $E_{forecast}$ is the sum of the cost of each contingency in Fig. 20.23c times the relative frequency that it occurs in Fig. 20.23b:

$$E_{forecast} = \frac{a}{n}C + \frac{b}{n}C + \frac{c}{n}L \qquad (20.51)$$

This assumes that you take protective action (i.e., try to mitigate the loss) every time the event is forecast to happen. For some of the individual events the forecast might be bad, causing you to respond inappropriately (in hindsight). But in the long run (over many events) your taking action every time the event is forecast will result in a minimum overall expense to you.

If forecasts were perfect, then you would never have any losses because they would all be mitigated, and your costs would be minimal because you take protective action only when needed. On average you would need to mitigate at the climatological frequency o, so the expenses expected with perfect forecasts are

$$E_{perfect} = o \cdot C \qquad (20.52)$$

Combine these expenses to find the **economic value** V of deterministic forecasts relative to climatology:

$$V = \frac{E_{climate} - E_{forecast}}{E_{climate} - E_{perfect}} \qquad (20.53)$$

V is an economic skill score, where $V = 1$ for a perfect forecast, and $V = 0$ if the forecasts are no better than using climatology. V can be negative if the forecasts are worse than climatology, for which case your best course of action is to ignore the forecasts and choose a response based on the climatological data, as previously described.

Define a **cost/loss ratio** r_{CL} as

$$r_{CL} = C / L \qquad (20.54)$$

Different people and different industries have different protective costs and unmitigated losses. So you should estimate your own cost/loss ratio for your own situation.

The economic-value equation can be rewritten using hit rate H, false-alarm rate F, and the cost/loss ratio r_{CL}:

$$V = \frac{\min(r_{CL}, o) - [F \cdot r_{CL} \cdot (1-o)] + [H \cdot (1 - r_{CL}) \cdot o] - o}{\min(r_{CL}, o) - o \cdot r_{CL}} \qquad (20.55)$$

You can save even more money by using probabilistic forecasts. If the probabilistic forecasts are perfectly calibrated (i.e., are reliable), then you should take protective action whenever the forecast probability p of the event exceeds your cost/loss ratio:

$$p > r_{CL} \qquad (20.56)$$

The forecast probability of the event is the portion of cumulative probability that is beyond the event-threshold condition. For the frost example, it is the portion of the cumulative distribution of Fig. 20.20 below a $T_{threshold} = 0°C$.

Sample Application

Given forecasts having the contingency table from the Sample Application 4 pages ago. Protective cost is \$75k to avoid a loss of \$200k. Climatological frequency of the event is 40%. Find the value of the forecast.

Find the Answer:

Given: a=90, b=50, c=75, d=150, C=\$75k, L=\$200k, o=0.4

Find: E_{clim} = \$?, E_{fcst} = \$?, E_{perf} = \$?, V = ?

Use eq. (20.50): E_{clim} = min(\$75k, 0.4·\$200k) = \$75k.

Use eq. (20.35): n = 90+50+75+150 = 365.

Use eq. (20.51):

E_{fcst} =(90+50)·\$75k/365 + (75/365)·\$200k = \$70k

Use eq. (20.52): E_{perf} = 0.4 · \$75k = \$30k.

Use eq. (20.53): V = (\$75k – \$70k)/(\$75k – \$30k) = <u>**0.11**</u>

Check: Units are OK. Values are consistent.

Exposition: The forecast is slightly more valuable than climatology.

r_{CL} = \$75k/\$200k = <u>**0.375**</u> from eq. (20.54).

If you are lucky enough to receive probabilistic forecasts, then based on eq. (20.56) you should take protective action when $p > 0.375$.

REVIEW

The atmosphere is a fluid. It obeys the laws of fluid mechanics, thermodynamics, and conservation of mass and water. If we can solve these equations, then we can forecast the weather.

Unfortunately, no one has yet found an analytical solution to these equations. Instead, we can approximate these continuous differential equations with finite-difference equations (i.e., algebra) that we can solve on a computer. One approach is to divide the spatial domain into finite-size grid cells, and to forecast the average conditions for a grid point within each cell.

If we know (or can approximate) the initial condition of each grid point by assimilating new observations into past forecasts, then we can make an iterative forecast by taking finite-size time steps into the future. To do this, all grid points must be marched together one step at a time.

Some aspects of atmospheric physics cause resolvable changes to the forecast, even though the grid resolution is not fine enough to resolve all the physical details. Hence, physical processes such as turbulence, radiation, clouds, and precipitation must be parameterized as a simplified function of variables that can be resolved in the model (winds, temperature, etc.).

The finite-difference equations suffer from truncation, round-off, numerical instability, and dynamic instability errors. Round-off errors are smaller when more bits are used to represent numbers in a computer. Truncation errors are smaller when more of the higher-order terms are retained in Taylor-series approximations of derivatives. Numerical instability is reduced when the time step is sufficiently small relative to the grid size. Dynamic instability refers to the sensitive dependence of the forecast on initial conditions and model parameters. Dynamic instability can be reduced with better weather analyses, but it cannot be eliminated.

No numerical forecast is perfect. For any specific location, forecasts might have a consistent bias or systematic error. Most of these biases can be removed by using statistics such as model output statistics (MOS) to post-process the NWP output. Random errors associated with the chaotic nonlinear-dynamic nature of the atmosphere can be estimated and/or reduced by making multiple forecasts (called ensemble forecasts) with different initial conditions or parameterizations. The ensemble forecasts can be averaged to give a deterministic forecast, and they can be used to make probabilistic forecasts.

Unfortunately, we have not discovered a way to reduce all errors. Thus, forecast error usually increases with increasing forecast time (i.e., how far into the future you forecast). Often forecast skill is defined relative to some baseline or reference, such as climatology. Short-range (out to 3 days) forecasts show significant skill, while medium-range forecasts show modest skill out to about 10 days. In addition to deterministic forecasts, NWP is increasingly used to make probabilistic forecasts. Methods exist to verify deterministic, binary, and probabilistic forecasts.

The resulting forecasts can be analyzed and graphed to reveal cyclones, fronts, airmasses, and other weather systems. Of all the tools (satellites, radar, weather balloons, etc.) that meteorologists use to forecast the weather, only NWP gives the future weather with a skill that is better than persistence.

HOMEWORK EXERCISES

Broaden Knowledge & Comprehension

B1. Search the web for info about each of the following operational weather forecast centers. Describe the full title, location, computers that they use, and models that they run. Also answer any special questions indicated below for these forecast centers.

a. CMC (and list the branches of CMC).
b. NCEP (and list the centers that make up NCEP)
c. ECMWF
d. FNMOC
e. UK Met Office

B2. Search the web for the government forecast centers in Germany, Japan, China, Australia, or any other country specified by your instructor. Describe the NWP models they run.

B3. Based on web searches for each of the numerical models listed below:

a. Define the full title.
b. At which centers or universities are they run?
c. Find the max forecast duration for each run.
d. Find the domain (i.e., world, N. Hem. N. America, Canada, Oklahoma, etc.).
e. What is the finest horizontal grid resolution?
f. How many model layers are in the vertical?
g. Which type of grid arrangement (A, B, etc.) is used?
h. What order spatial and time differencing schemes are used?

GEM
NAM
GFS
AVN
NOGAPS
COAMPS
ECMWF
MC2
UW-NMS
MM5
RAMS
ARPS
WRF-NMM
WRF-ARW

B4. Search the web for models in addition to those listed in the previous exercise, which are being run operationally. Describe the basic characteristics of these models.

B5. Search the web for a discussion of MOS. What is it, and why is it useful to forecasting?

B6. Find on the web different forecast models that produce precipitation forecasts for Vancouver, Canada (or other city specified by your instructor). Do this for as many models as possible that are valid at the same time and place. Specify the date/time for your discussion. Try to pick an interesting day when precipitation is starting or ending, or a storm is passing. Compare the forecasts from the different models, and if possible search the web for observation data of precipitation against which to validate the forecasts.

B7. At which web sites can you find forecast sea states (e.g., wave height, etc.)?

B8. Based on results of a web search, discuss different ways that ensemble forecasts can be presented via images and graphs.

B9. What types of publicly available daily forecasts are being made by a university (not a government operational center) closest to your location?

B10. What are the broad categories of observation data that are used to create the analyses (the starting point for all forecasts). Hint, see the ECMWF data coverage web site, or similar sites from NCEP or the Japanese forecast agency.

B11. Search the web for verification scores for the national weather forecast center that forecasts for your location. How have the scores changed by season, by year? How do the anomaly correlation scores vary with forecast day, compared to the results from ECMWF shown in this chapter?

B12. Find a web site that shows plots of the Lorenz "butterfly", similar to Fig. 20.18. Even better, search the web for a 3-D animation, showing how the solution chaotically shifts from wing to wing.

B13. Search the web for other equations that have different strange attractors. Discuss how the equations and attractors differ from those of Lorenz.

B14. Examine from the web the forecast maps that are produced by various forecast centers. Instead of looking at the quality of the forecasts, look at the quality of the weather map images that are served on the web. Which forecast centers produce the maps that are most attractive? Which are easiest to understand? Which are most useful? What geographic map projections are used for your favorite maps?

B15. Search the web for a meteogram of the weather forecast for your town (or for a town near you, or a town specified by the instructor). What are the advantages and disadvantages of using meteograms to present weather forecasts, rather than weather maps?

B16. Search the web for a summary of different options that are used for physics parameterizations in WRF, or other model selected by your instructor.

B17. Search the web for images/photos of some of the earliest computers used in weather forecasting, such as the ENIAC computer. Discuss how computers have changed.

B18. Find examples of probabilistic forecasts on the web. Print a few examples and discuss.

B19. Computational fluid dynamics (CFD) is the name for numerical methods used to solve fluid dynamics equations in engineering. Search the web for images of CFD forecasts for the finest grid resolution that you can find. What flow situation is it solving? What are the grid spacings and time steps?

Apply

A1. An NWP model has a bottom hydrostatic pressure of 95 kPa over the mountains, and a top pressure of 5 kPa. Find the sigma coordinate value for a height where the pressure (kPa) is:
a. 90 b. 85 c. 80 d. 75 e. 70 f. 65 g. 60 h. 55
i. 50 j. 45 k. 40 m. 35 n. 30 o. 25 p 20

A2. For a polar stereographic map projection with a reference latitude of 60°, find the (x, y) coordinates on the map that corresponds to the lat & lon at:
a. Montreal b. Boston c. New York City
d. Philadelphia e. Baltimore f. Washington DC
g. Atlanta h. Miami i. Toronto j. Chicago
k. St. Louis m. New Orleans n. Minneapolis
o. Kansas City p. Oklahoma City q. Dallas
r. Denver s. Phoenix t. Vancouver
u. Seattle v. San Francisco w. Los Angeles
x. A location specified by your instructor

A3. For a polar stereographic map projection with a reference latitude of 90°, find the map factors m_o, m_x, and m_y for latitudes (°) of:
a. 90 b. 85 c. 80 d. 75 e. 70 f. 65 g. 60 h. 55
i. 50 j. 45 k. 40 m. 35 n. 30 o. 25 p 20

A4. Estimate subgrid-scale cloud coverage of low and high clouds for a grid-average RH (%) of:
a. 70 b. 75 c. 80 d. 85 e. 90 f. 95 g. 100

A5. Given the following temperature T values (°C) at the indicated grid-point indices (i) for ΔX = 10 km:

i:	1	2	3	4	5	6	7	8	9	10
T:	30	27	26	28	24	20	18	23	25	25

Find the gradient $\Delta T/\Delta x$ for upwind first-order difference, centered second-order difference, and centered fourth-order difference, at grid index i:
a. 3 b. 4 c. 5 d. 6 e. 7 f. 8

A6. Same as the previous exercise, except find the advection term $-U{\cdot}\Delta T/\Delta x$ for a C grid with U values (m s^{-1}) of:

i:	1.5	2.5	3.5	4.5	5.5	6.5	7.5	8.5	9.5
U:	3	3	4	5	7	10	14	19	25

A7(§). Suppose that an equation of motion in some strange universe has the form $\Delta U/\Delta t = U{\cdot}t^2/\tau_o^3$ where constant τ_o = 1 min, and variable wind U = 1 m s^{-1} initially (t = 0). With time steps of Δt = 1 min, use the (a) Euler method, (b) leapfrog method, and (c) fourth-order Runge-Kutta method to forecast the value of U at t = 5 min. Compare your results to the analytical solution of $U = U_o \cdot \exp[\,(1/3){\cdot}(t/\tau_o)^3]$ where U_o = 1 m s^{-1} is the initial condition. [Hint: for the leapfrog method, you will need to use the Euler forward method for the first step.] Show your work and the results at each time step.

A8(§). Given a 1-D array consisting of 12 grid points in the x-direction with the following initial temperatures (°C). $T_i(t{=}0)$ =
20 24 20 16 20 24 20 16 20 24 20 16 20
for i = 1 to 12. Assume that the lateral boundaries are cyclic, so that the number sequence repeats outside this primary domain. Grid spacing is 5 km, and wind speed is from the west at 10 m s^{-1}. Use the leapfrog time-step method (except for the first time step) and 4th-order spatial differencing. Make enough time steps to forecast out to t = 5000 s. Use time steps of size Δt (s) =
a. 100 b. 200 c. 300 d. 400 e. 500 f. 600
Plot the temperature graph at each time step, and comment on the numerical stability.

A9. Given the following pairs of [grid spacings (km), wind speeds (m s^{-1})], find the largest time step that satisfies the CFL criterion.
a. [0.1, 50] b. [0.2, 30] c. [0.5, 75] d. [1.0, 50]
e. [2, 40] f. [3, 80] g. [5, 50] h. [10, 100]
i. [15, 75] j. [20, 20] k. [33, 50] m. [50, 75]

A10. A program has 3 subprograms that each take 1/3 of the running time of the whole program. If the first subprogram is sped up 10 times, the second subprogram is sped up 40%, and the third one is sped up as

indicated, what is the total speedup of the program?
a. 10% b. 50% c. 75% d. 100% e. 3 times
f. 5 times g. 10 times h. 20 times i. 50 times

A11. A surface weather station reports a temperature of 20°C with an observation error of σ_T = 1°C. An NWP model forecasts temperature of 24°C at the same point. For optimum interpolation, find the analysis temperature and the cost function if the NWP forecast error (°C) is
a. 0.2 b. 0.4 c. 0.6 d. 0.8 e. 1.0 f. 1.2
g. 1.5 h. 2.0 i. 3 j. 4 k. 5

A12. Suppose the first guess pressure in an optimum interpolation is 100 kPa, with an error of 0.2 kPa. Find the analysis pressure if an observation of P = 102 kPa was observed by:
a. surface weather station b. ship
c. Southern Hemisphere manual analysis

A13(§). A surface weather station at P = 100 kPa reports a dew-point temperature of T_d = 10°C with an observation error of σ_{Td} = 3°C. An NWP model forecasts a mixing ratio of r = 12 g kg^{-1} at the same point. For variational data assimilation, find the analysis mixing ratio (in g kg^{-1}, and plot the variation of the cost function with mixing ratio) if the NWP mixing-ratio forecast error (g kg^{-1}) is:
a. 0.1 b. 0.2 c. 0.4 d. 0.5 e. 0.7 f. 1.0
g. 1.2 h. 1.5 i. 2 j. 2.5 k. 3 m. 4 n. 5
[Hint: Use (4.15b) from the Water Vapor chapter as your "H" function to convert from r to T_d.]

A14. Using Fig. 20.14 estimate at what forecast range (days) do we lose the ability to forecast:
a. tornadoes b. hurricanes
c. fronts d. cyclones
e. Rossby waves f. thunderstorms
g. Boras h. lenticular clouds

A15. Given the following pairs of x, y values. Use linear regression to find the slope and intercept of the best-fit straight line.

	x=	1	2	3	4	5	6	7	8	9
a.	y=	0.1	0.4	0.2	0.6	0.3	0.3	0.5	0.8	0.7
b.	y=	2	4	7	7	10	12	16	14	18
c.	y=	7	9	12	12	15	17	21	19	23
d.	y=	–3	–1	2	2	5	7	11	9	13
e.	y=	10	7	9	8	6	3	3	3	2
f.	y=	–20	–25	–30	–35	–40	–45	–50	–55	–60

A16. Given the forecast bias input (y, thin line) in Fig. 20.G, plot the Kalman filter estimate x of tomorrow's bias for error variance ratios (r) of:
a. 0.001 b. 0.002 c. 0.005 d. 0.01 e. 0.03
f. 0.04 g. 0.05 h. 0.07 i. 0.08 j. 0.09 k. 0.1

A17. Using the MOS regression from the Sample Application in this chapter, calculate the predictand if each of the predictors based on forecast-model output increased by
a. 1% b. 2%. c. 3% d. 4% e. 5%
f. 6%. g. 7% h. 8% i. 9% j. 10%

A18.(§) For the Lorenz equations, with the same parameters and initial conditions as used in this chapter, reproduce the results similar to the first Sample Application, except for all 1000 time steps. Also:
a. Plot L and C on the same graph vs. time.
b. Plot M vs. L c. Plot L vs C.

A19.(§) Given the following fields of 50-kPa height (km). Find the:
a. mean forecast error
b. mean persistence error
c. mean absolute forecast error
d. mean squared forecast error
e. mean squared climatology error
f. mean squared forecast error skill score
g. RMS forecast error
h. correlation coefficient between forecast and verification
i. forecast anomaly correlation
j. persistence anomaly correlation
k. Draw height contours by hand for each field, to show locations of ridges and troughs.

Each field (i.e., each weather map) below covers an area from North to South and West to East.

Analysis:

5.2	5.3	5.4	5.3
5.3	5.4	5.5	5.4
5.4	5.5	5.6	5.5
5.5	5.6	5.7	5.6
5.6	5.7	5.8	5.7

Forecast:

5.3	5.4	5.5	5.4
5.5	5.4	5.5	5.6
5.6	5.6	5.6	5.6
5.8	5.7	5.6	5.7
5.9	5.8	5.7	5.8

Verification:

5.3	5.3	5.3	5.4
5.4	5.3	5.4	5.5
5.5	5.4	5.5	5.5
5.7	5.5	5.6	5.6
5.8	5.7	5.6	5.6

Climate:

5.4	5.4	5.4	5.4
5.4	5.4	5.4	5.4
5.5	5.5	5.5	5.5
5.6	5.6	5.6	5.6
5.7	5.7	5.7	5.7

A20. Given the following contingency table, calculate all the binary verification statistics.

		Observation	
		Yes	No
Forecast	Yes:	150	65
	No:	50	100

A21. Given forecasts having the contingency table of exercise N20. Protective cost is \$5k to avoid a loss of \$50k. Climatological frequency of the event is 50%. (a) Find the value of the forecast. (b) If you can get probabilistic forecasts, then what probability would you want in order to decide to take protective action?

A22. Given the table below of k = 1 to 20 forecasts of probability p_k that 24-h accumulated precipitation will be above 25 mm, and the verification $o_k = 1$ if the observed precipitation was indeed above this threshold. (a) Find the Brier skill score. (b) For probability bins of width $\Delta p = 0.2$, plot a reliability diagram, and (c) find the reliability Brier skill score.

k	p_k	o_k	k	p_k	o_k
1	0.9	1	11	0.4	0
2	0.85	1	12	0.35	0
3	0.8	0	13	0.3	1
4	0.75	1	14	0.25	0
5	0.7	1	15	0.2	0
6	0.65	1	16	0.15	1
7	0.6	0	17	0.1	0
8	0.55	1	18	0.05	0
9	0.5	0	19	0.02	0
10	0.45	1	20	0	0

A23. For any <u>one</u> part of this exercise (Ex a to Ex d) of this problem, a 10-member ensemble forecast system forecasts probabilities that 24-h accumulated rainfall will exceed 5 mm. The observation flags (o) and forecast probabilities (p) are given in the table (in the next column) for a 30-day period. Calculate the hit rate and false-alarm rate for the full range of allowed probability thresholds, and plot the result as a ROC diagram. Also find the area under the ROC curve and find the ROC skill score.

Data for calculations of a ROC diagram:

		(Ex a)	(Ex b)	(Ex c)	(Ex d)			(Ex a)	(Ex b)	(Ex c)	(Ex d)
Day	o	p(%)	p(%)	p(%)	p(%)	Day	o	p(%)	p(%)	p(%)	p(%)
1	1	50	10	100	0	16	0	60	30	20	40
2	0	20	0	0	10	17	1	70	60	60	50
3	1	20	30	90	20	18	1	90	70	60	60
4	1	60	40	90	30	19	1	80	80	60	70
5	0	50	30	0	40	20	0	70	70	30	80
6	0	20	40	0	50	21	0	10	80	30	90
7	0	30	50	10	60	22	0	10	90	30	100
8	1	90	80	80	70	23	0	0	0	40	10
9	0	40	70	10	80	24	0	0	10	40	20
10	1	30	100	80	90	25	1	80	40	50	30
11	1	100	100	70	100	26	0	0	30	40	40
12	0	10	0	10	0	27	0	0	40	0	50
13	0	0	0	20	10	28	1	100	70	50	60
14	0	10	10	20	20	29	0	10	60	0	70
15	1	80	40	70	30	30	1	90	10	50	0

Evaluate & Analyze

E1. Use the meteogram of Fig. 20.1.

a. After 20 Feb, when does the low pass closest to Des Moines, Iowa?

b. During which days does it rain, and which does it snow?

c. During which days is there cold-air advection?

d. Based on the wind direction, does the low center pass north or south of Des Moines.

e. After 20 Feb, when does the cold front pass Des Moines?

f. What is the total amount of precipitation that fell during the midweek storm?

g. How does this forecast, which was initialized with data from 19 Feb, compare with the actual observations (refer to a previous chapter)?

E2(§). Reproduce the polar stereographic map from the Sample Application for map projections. Then add:

a. Greenland b. Europe c. Asia

d. your location if in the N. Hemisphere.

E3. If Moore's law continues to hold, and if forecast skill continues to improve as it has in the past, then estimate the transistor count on an integrated circuit, and the skillful forecast period (days) for the year: a. 2010 b. 2015 c. 2020 d. 2025 e. 2030
Also, comment on what factors might cause errors in your estimate.

E4. Speculate on the capability of weather forecasting if digital computers had not been invented.

E5. Critique the validity of a statement that "A variable mesh grid is analogous to a number of discrete nested grids."

E6. What procedure (i.e., what equations and how they are manipulated) would you use to create the 4th-order centered difference of eq. (20.BA6).

E7. If NWP Corollaries 1 and 2 did not exist, describe tricks that you could use to increase the speed of numerical weather forecasts.

E8. Write a finite-difference equation similar to eq. (20.13) for the shaded grid cell of Fig. 20.9, but for:
a. vertical advection
b. advection in the y-direction

E9. Draw the stencil of grid points used for computing horizontal advection, but for the ___ grid for 2nd-order spatial differencing.
a. A b. B c. D

E10. If the atmosphere is balanced, and if observations of the atmosphere are perfectly accurate, why would numerical models of the atmosphere start out imbalanced?

E11. For the case study forecast of Figs. 20.15, first photocopy the figures. Then, on each map
a. Draw the likely location for fronts.
b. Indicate the locations of low centers
c. Comment on the forecast accuracy for fronts, cyclones, and the large-scale flow for this case.

E12. Suppose you are making weather forecasts for Pittsburgh, Pennsylvania, which is close to the intersection of the 40°N parallel and 80°W meridian, shown by the intersection of latitude and longitude lines in Figs. 20.15 just south of Lake Erie. During the 7.5 days prior to 00 UTC 24 Feb 94, your temperature forecasts for 00 UTC 24 Feb would likely change as you received newer updated forecast maps.

What is your temperature forecast for 00 UTC 24 Feb, if you made it ___ days in advance from the ECMWF forecast charts of Fig. 20.15?
a. 7.5 b. 5.5 c. 3.5 d. 1.5
e. and which forecast was closest to the actual verifying analysis?

E13.(§) Suppose the Lorenz equations were modified by assuming that $C = L$. For the second two Lorenz equations, replace every C with L, and recalculate for the first 1000 time steps.
a. Plot L and M vs. time on the same graph.
b. Plot M vs. L.

Note that the solution converges to a steady-state solution. On the graph of M vs. L, this is called a **fixed point**. This fixed point is an **attractor**, but not a strange attractor.
c. Describe what type of physical circulation is associated with this solution.

E14. Experiment with the Lorenz equations on a spreadsheet. Over what range of values of the parameters σ, b, and r, do the solutions still exhibit chaotic solutions similar to that shown in Fig. 20.18?

E15. A pendulum swings with a regular oscillation.
a. Plot the position of the pendulum vs. time.
b. Plot the velocity of the pendulum vs. time.
c. Plot the position vs. velocity. This is a called a **phase diagram** according to chaos theory. How does it differ from the phase diagram (i.e., the butterfly) of the Lorenz strange attractor?

E16. Use the ensemble forecast for Des Moines in Fig. 20.19.
a. What 85-kPa temperature forecast, and with what reliability, would you make for forecast day: 1, 3, 5, 7, and 9 ?
b. Which temperature ranges would you be confident to forecast would NOT occur, for day: 1, 3, 5, 7, and 9 ?
c. In spite of the forecast uncertainty, are you confident about the general trends in temperature?
d. Could you confidently forecast when rain is most likely? If so, how much rain would you predict?

E17. Fit an exponential curve to the persistence data of Fig. 20.22. What is the e-folding time?

E18. In what order are weather maps presented in the weather briefing given by your favorite local TV meteorologist? What are the advantages and disadvantages of this approach compared to the order suggested in this chapter?

Synthesize

S1. Learn what an analog computer is, and how it differs from a digital computer. If automated weather forecasts were made with analog rather than digital computers, how would forecasts be different, if at all?

S2. a. Suppose that there were no weather observations in the western half of N. America. How would the forecast quality over Washington, DC, and Ottawa, Canada, be different, if at all? Given that national legislators live in those cities, speculate on the changes that they would require of the national weather services in the USA and Canada in order to improve the forecasts.

b. Extending the discussion from part (a), suppose that weather observations are back to normal in N. America, but the seats of government were moved to Seattle and Vancouver. Given what you know about the Pacific data void, speculate on the changes that they would require of the national weather services in the USA and Canada in order to improve the forecasts.

S3. How many grid points are needed to forecast over the whole world with roughly 1-m grid spacing? When do you anticipate computer power will have the capability to do such a forecast? What, if any, are the advantages to such a forecast?

S4. Design a grid arrangement different from that in Fig. 20.9, but which is more efficient (i.e., involves fewer calculations) or utilizes a smaller stencil.

S5. a. Suppose one person developed and ran a NWP model that gave daily forecasts with twice the skill as those produced by any other NWP model run operationally around the world. What power and wealth could that person accumulate, and how would they do it? What would be the consequences, and who would suffer?

b. Same question as part (a), but for one country rather than one person.

S6. Look up the Runge-Kutta finite-difference method in a book or internet site on numerical methods. Find equations for a Runge-Kutta method that is higher order than fourth-order. Can you implement this method on a computer spreadsheet? Try it.

S7. Suppose that there was not a CFL numerical stability criterion that restricted the time step that can be used for NWP. How would NWP be different, if at all? Even without a numerical stability criterion, would there be any other restrictions on the time step? If so, discuss.

S8. Speculate on the ability of national forecast centers to make timely weather forecasts if a computer hacker destroyed the internet and other world-wide data networks.

S9. If greater spread of ensemble members in an ensemble forecast means greater uncertainty, then is greater spread desirable or undesirable in an ensemble forecast?

S10. Which would likely give more-accurate forecasts: a categorical model with very fine grid spacing, or an average of ensemble runs where each ensemble member has coarse grid spacing? Why?

S11. Same as the previous question, but specifically for over steep mountainous terrain. Discuss.

S12. Finite-difference equations are approximations to the full, differential equations that describe the real atmosphere. However, such finite-difference equations can also be thought of as exact representations of a numerical atmosphere that behaves according to different physics. How is this numerical atmosphere different from the real atmosphere? How would physical laws differ for this numerical atmosphere, if at all? When the NWP model runs for a long time, if it approaches steady state, is this state equal to the real climate or to the **"model" climate**?

S13. Suppose that electricity did not exist. How would you make numerical weather forecasts? Also, how would you disseminate the results to customers?

S14. How good must a numerical forecast be, to be good enough? Should the quality and value of a weather forecast be determined by meteorologists, computer scientists, or end users? Discuss.

S15. Lorenz suggested that there is a limit to predictability. Is that a "hard" limit, or might it be possible to make skillful forecasts beyond that limit? Discuss.

S16. Comment on the interconnectivity of the atmosphere, as expressed by NWP corollary 1.

21 NATURAL CLIMATE PROCESSES

Contents

The dominant climate process is radiation to and from the Earth-ocean-atmosphere system. Increased absorption of solar (shortwave) radiation causes the climate to warm, which is compensated by increased infrared (IR, longwave) radiation out to space. This strong negative feedback has allowed the absolute temperature at Earth's surface to vary by only 4% over millions of years.

But small changes have indeed occurred. Recent changes are associated with human activity such as greenhouse-gas emissions and land-surface modification. Other changes are natural — influenced by astronomical and tectonic factors. These primary influences can be amplified or damped by changes in clouds, ice, vegetation, and other feedbacks.

To illustrate dominant processes, consider the following highly simplified "toy" models.

RADIATIVE BALANCE

Consider an Earth with no atmosphere (Fig. 21.1). Given the near-constant climate described above, assume a balance of radiation input and output.

Incoming solar radiation minus the portion reflected, multiplied by the area intercepted by Earth, gives the total radiative input:

$$Radiation\ In = (1-A)\cdot S_o \cdot \pi \cdot R_{Earth}^2 \qquad \bullet(21.1)$$

where the fraction reflected (A = 0.3) is called the **global albedo** (see INFO Box). The annual-average **total solar irradiance (TSI)** over all wavelengths

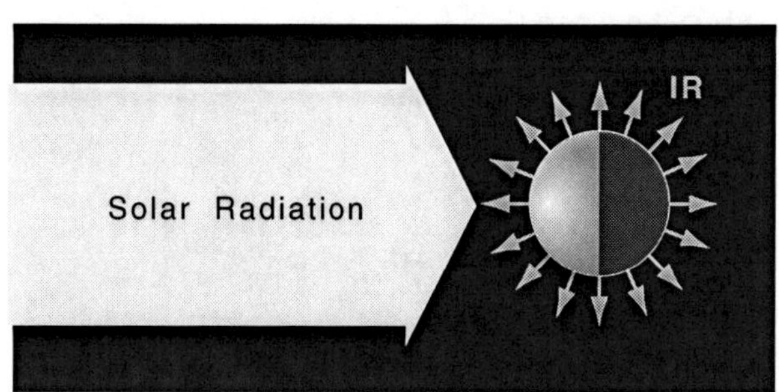

Figure 21.1
Earth's radiation balance of solar energy in and infrared (IR) radiation out.

INFO • Albedo

The average global reflectivity (portion of incoming sunlight that is reflected back to space) for Earth is called the **albedo**, A. But the exact value of A is difficult to measure, due to natural variability of cloud and snow cover, and due to calibration errors in the satellite radiation sensors.

Trenberth, Fasullo & Kiehl (2009, *Bul. Amer. Meteor. Soc.*) report a range of values of $27.9\% \le A \le 34.2\%$. They suggest $A = 29.8\%$. We use $\mathbf{A \approx 30\%}$ here.

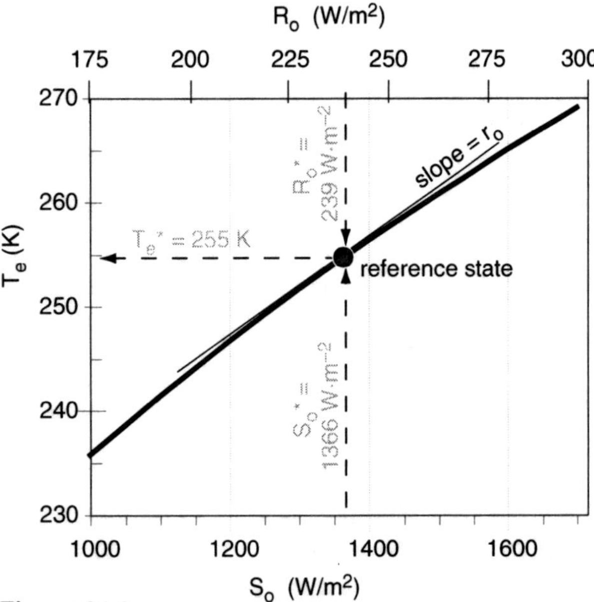

Figure 21.2
Thick curved line shows how the effective radiation temperature T_e for Earth varies with changes in solar irradiance S_o, given global albedo A = 0.3. Dashed lines show present conditions, which will be used as a reference point or reference state (R_o, T_e*). Thin black straight line shows the slope r_o of the curve at the reference point. $R_o = (1–A)·S_o/4$ is net solar input for each square meter of the Earth's surface (see INFO Box, next page).*

Sample Application

Show the calculations leading to $T_e = -18°C$.

Find the Answer:

Given: $T_e = -18°C$, $S_o = 1366\ W·m^{-2}$, $A = 0.30$

Find: Show the calculation

Use eq. (21.3):

$$T_e = \left[\frac{(1-0.3)\cdot(1366\ W\cdot m^{-2})}{4\cdot(5.67\times10^{-8}\ W\cdot m^{-2}\cdot K^{-4})}\right]^{1/4}$$

$= 254.8\ K \approx 255\ K =$ **−18°C**

Check: Physics & units are OK.

Exposition: Earth's radius appears in both the radiation-in and -out equations, but is not in eq. (21.3).

as measured by satellites is $S_o = 1366\ W·m^{-2}$. The interception area is the same as the area of Earth's shadow — a disk of area $\pi R_{Earth}{}^2$, where R_{Earth} = 6371 km is the average radius of Earth. The TSI has fluctuated about ±0.5 $W·m^{-2}$ (the thickness of the vertical dashed line in Fig. 21.2) during the past 33 years due to the average sunspot cycle. See the Solar & IR Radiation chapter for TSI details.

IR radiation is emitted from every point on Earth's surface, which we will assume to behave as a radiative blackbody (emissivity = 1). For this toy model, assume that the earth is spherical with surface area $4·\pi R_{Earth}{}^2$. If IR emissions/area are given by the Stefan-Boltzmann law (eq. 2.15), then the total outgoing radiation is:

$$Radiation\ Out = 4\pi\cdot R_{Earth}^2\cdot\sigma_{SB}\cdot T_e^4 \qquad \bullet(21.2)$$

where the Stefan-Boltzmann constant is σ_{SB} = $5.67\times10^{-8}\ W·m^{-2}·K^{-4}$.

The real Earth has a heterogeneous surface with cold polar ice caps, warm tropical continents, and other temperatures associated with different land and water geographic features. Define an **effective radiation emission temperature** (T_e) such that the emissions from a hypothetical earth of uniform surface temperature T_e is the same as the total heterogeneous-Earth emissions.

For a steady-state climate (i.e., no temperature change with time), outputs must balance the inputs. Thus, equate outgoing and incoming radiation from eqs. (21.1 and 21.2), and then solve the resulting equation for T_e:

$$T_e = \left[\frac{(1-A)\cdot S_o}{4\cdot\sigma_{SB}}\right]^{1/4} = \left[\frac{R_o}{\sigma_{SB}}\right]^{1/4} \qquad \bullet(21.3)$$

$$\approx 254.8\ K \approx -18°C$$

where $R_o = (1–A)·S_o/4$.

Fig. 21.2 shows this **reference state** at 254.8°C (≈255°C), given an albedo of 0.3. It also shows how changes of solar irradiance along the abscissa of the graph need only small changes in effective temperature along the ordinate in order to reach a new radiative equilibrium. Namely, the radiation balance causes a **negative feedback** (a stable climate) rather than runaway global warming.

The geological record illustrates the steadiness of Earth's past climate (Fig. 21.3). About 50 M years ago, the absolute temperature of Earth's surface averaged roughly 2% warmer (i.e., 6 K = 6°C warmer) than present. For the most recent 10 k years, the geologic record indicates temperature oscillations on the order of ±1°C relative to our current average surface temperature of 15°C. Nonetheless, these small temperature changes (with ΔT range of 10°C over the past 500 Myr) can cause significant changes in ice-cap coverage and sea level.

But our simple toy model is perhaps too simple, because the modeled –18°C temperature colder than the +15°C actual surface temperature (from Chapter 1). To improve the modeled temperature, we must include some additional physics.

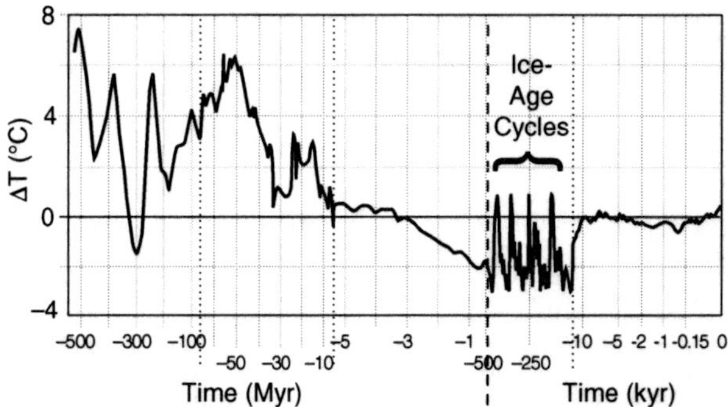

Figure 21.3
Paleotemperature estimate (smoothed) as a difference ΔT from the 1961 to 1990 temperature average. Time is relative to calendar year 2000. Scale breaks at the dotted and dashed lines. Glacial/interglacial oscillations (i.e., ice-age cycles) prior to 500,000 yr ago have been smoothed out.

The So-called "Greenhouse Effect"

To make our Earth-system model be slightly more realistic, add an atmosphere of uniform absolute temperature T_A everywhere (Fig 21.4). Suppose this atmosphere is opaque (not transparent) in the IR, but is perfectly transparent for visible light.

For this idealized case, sunlight shines through the atmosphere and reaches the Earth's surface, where a portion is reflected and a portion is absorbed, causing the surface to warm. IR emitted from the warm Earth is totally absorbed by the atmosphere, causing the atmosphere to warm. But the atmosphere also emits radiation: some upward to space, and some back down to the Earth's surface (a surface that absorbs 100% of the IR that hits it).

This downward IR from the atmosphere adds to the solar input, thus allowing the Earth's surface to have a greater surface temperature T_s than with no atmosphere. This is called the **greenhouse effect**, even though greenhouses do not work this way. Water vapor is one of many "greenhouse gases" that absorbs and emits IR radiation.

Assume the atmosphere-Earth system has existed for sufficient years to reach a **steady state**. Such an equilibrium state requires outgoing IR from the atmosphere-Earth system to balance absorbed incoming solar radiation. Namely:

$$T_A = T_e = 255 \text{ K} = -18°\text{C} \tag{21.4}$$

Thus, the temperature of the atmosphere must equal the **effective emission temperature** for the atmosphere-Earth system.

You can use the Stefan-Boltzmann law $\sigma_{SB} \cdot T_A{}^4$. to find the radiation emitted both up and down from the atmosphere, which is assumed to be thin relative to Earth's radius. For the atmosphere to be in steady-state, these two streams of outgoing IR radiation must balance the one stream of incoming IR from the Earth's surface $\sigma_{SB} \cdot T_s{}^4$, which requires:

$$\sigma_{SB} \cdot T_s^4 = 2 \cdot \sigma_{SB} \cdot T_e^4 \tag{21.5}$$

Thus

$$T_s = 2^{1/4} \cdot T_e \tag{•21.6}$$

$$= 303 \text{ K} = 30°\text{C}$$

While the no-atmosphere case was too cold, this

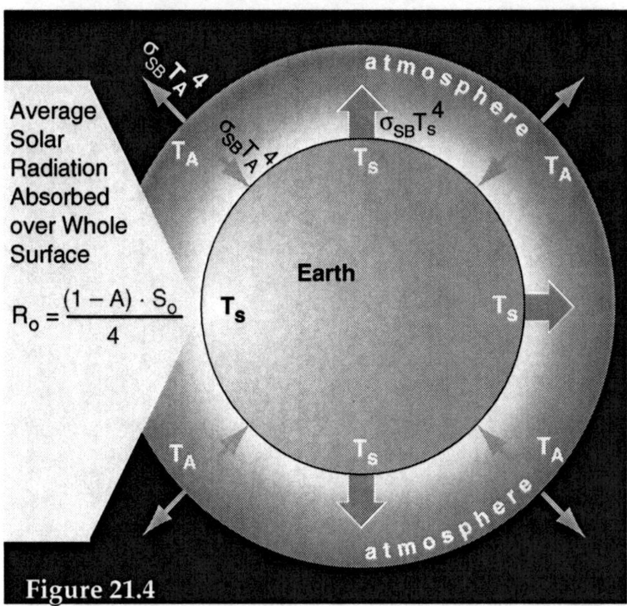

Figure 21.4
Simplified atmosphere-Earth system with greenhouse effect. Atmosphere thickness is exaggerated. R_o is the net solar input.

INFO • Why does $R_o = (1–A)\cdot S_o / 4$?

In Figs. 21.4 and 21.5, the **net solar input** R_o is given as $(1–A)\cdot S_o/4$ instead of S_o. The $(1–A)$ factor is because any sunlight reflected back to space is not available to heat the Earth system. But why also divide by 4? The reason is due to geometry.

Of the radiation streaming outward from the sun, the shadow cast by the Earth indicates how much radiation was intercepted. The shadow is a circle with radius equal to the Earth's radius R_{Earth}, hence the interception area is $\pi \cdot R_{Earth}{}^2$.

But we prefer to study the energy fluxes relative to each square meter of the Earth's surface. The surface area of a sphere is $4 \cdot \pi \cdot R_{Earth}{}^2$. Namely, the surface area is four times the area intercepted by the sun.

Hence, the average net solar input that we allocate to each square meter of the Earth's surface is:
$R_o = (1–A)\cdot S_o/4 = (1–0.3)\cdot(1366 \text{ W·m}^{-2})/4 = \mathbf{239\ W·m^{-2}}$.

Sample Application

What temperature is the Earth's surface for the one-layer atmosphere of the previous sample applic.

Find the Answer

Given: $T_e = 254.8$ K
Find: $T_s = ?$ K

Apply eq. (21.6): $T_s = 2^{1/4}\cdot(254.8\text{K}) = 302.96 \text{ K} = \underline{29.95°C}$

Check: Physics and units are OK.
Exposition: Twice the radiation at Earth's surface, but not twice the surface absolute temperature.

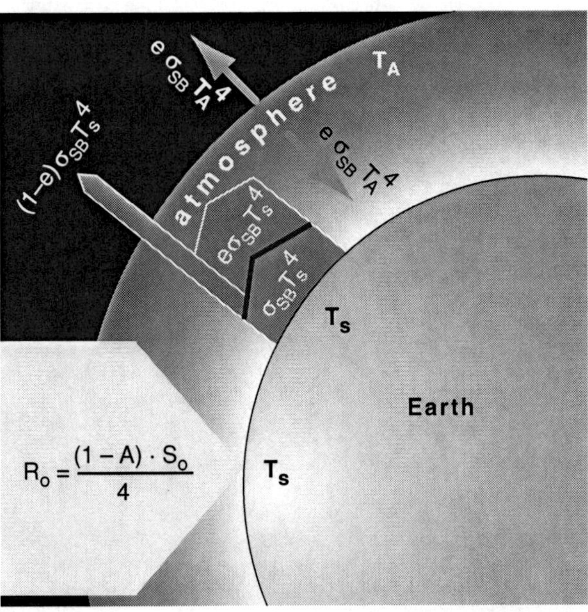

Figure 21.5
Idealized single-layer atmosphere having an IR dirty window.

Sample Application

Given an idealized case similar to the previous Sample Application, but with a dirty atmospheric window having $e_\lambda = 0.899$. Find T_A and T_s.

Find the Answer

Given: $e_\lambda = 0.899$, and $T_e = 254.8$ K,
Find: $T_A = ?$ K, $T_s = ?$ K

Apply eq. (21.10):

$$T_s = \left(\frac{2}{2-0.899}\right)^{1/4}\cdot(254.8\text{K}) = 295.8 \text{ K} = \underline{22.8°C}$$

From Eq. (21.7) we see that the effective emission temperature of the atmosphere is:

$$T_A = (0.5)^{1/4}\cdot T_s = 0.84\cdot(295.8 \text{ K}) \approx 249 \text{ K} = \underline{-24°C}$$

Check: Physics & units are OK.
Exposition: This simple "dirty window" model is better, but still does not give the desired T_s of 15°C.

opaque-atmosphere case is too warm (recall from Chapter 1 that the standard atmosphere T_s = 15°C). Perhaps the air is not fully opaque in the IR.

Atmospheric Window

As discussed in the Satellites & Radar chapter, Earth's atmosphere is partially transparent (i.e., is a dirty **atmospheric window**) for the 8 to 14 μm range of infrared radiation (IR) wavelengths and is mostly opaque at other wavelengths. Of all the IR emissions upward from Earth's surface, suppose that 89.9% is absorbed and 10.1% escapes to space (Fig. 21.5). Thus, the atmosphere will not warm as much, and in turn, will not re-emit as much radiation back to the Earth's surface.

At each layer (Earth's surface, atmosphere, space) outgoing and incoming radiative fluxes balance for an Earth-system in equilibrium. Also, fluxes out of one layer are input to an adjacent layer. For example, the radiative balance of the atmosphere becomes

$$e_\lambda \cdot \sigma_{SB} \cdot T_s^{\,4} = 2 \cdot e_\lambda \cdot \sigma_{SB} \cdot T_A^{\,4} \tag{21.7}$$

while at the Earth's surface it is

$$R_o + e_\lambda \cdot \sigma_{SB} \cdot T_A^{\,4} = \sigma_{SB} \cdot T_s^{\,4} \tag{21.8}$$

From Kirchhoff's law (see the Solar & IR Radiation chapter), recall that absorptivity equals emissivity, which for this idealized case is $e_\lambda = a_\lambda$ = 89.9%. Eqs. (21.3, 21.7 & 21.8) can be combined to solve for key absolute temperatures:

$$T_A^4 = \frac{T_e^4}{2-e_\lambda} \qquad \bullet(21.9)$$

$$T_s^4 = \frac{2\cdot T_e^4}{2-e_\lambda} \qquad \bullet(21.10)$$

where T_e is still given by eq. (21.4). The result is:

$$T_A \approx 249 \text{ K} = -24°\text{C}$$

and

$$T_s \approx 296 \text{ K} = 23°\text{C} \tag{21.11}$$

The temperature at Earth's surface is more realistic, but is still slightly too warm.

A by-product of human industry and agriculture is the emission of gases such as methane (CH_4), carbon dioxide (CO_2), freon CFC-12 ($C\,Cl_2\,F_2$), and nitrous oxide (N_2O) into the atmosphere. These gases, known as **anthropogenic greenhouse gases**, can absorb and re-emit IR radiation that would otherwise have been lost out the dirty window. The resulting increase of IR emissions from the atmosphere to Earth's surface could cause **global warming**.

Average Energy Budget

Radiation is not the only physical process that transfers energy. Between the Earth and bottom layer of the atmosphere can also be sensible-heat transfer via conduction (see the Heat Budgets chapter), and latent-heat transfer via evapotranspiration (see the Water Vapor chapter). Once this heat and moisture are in the air, they can be transported further into the atmosphere by the mean wind, by deep convection such as thunderstorms, by shallow convection such as thermals, and by smaller turbulent eddies in the atmospheric boundary layer.

One estimate of the annual mean of energy fluxes between all these components, when averaged over the surface area of the globe, is sketched in Fig. 21.6. According to the Stefan-Boltzmann law, the emitted IR radiation from the Earth's surface (396 W·m^{-2}) corresponds to an average surface temperature of 16°C, which is close to observed temperature of 15°C.

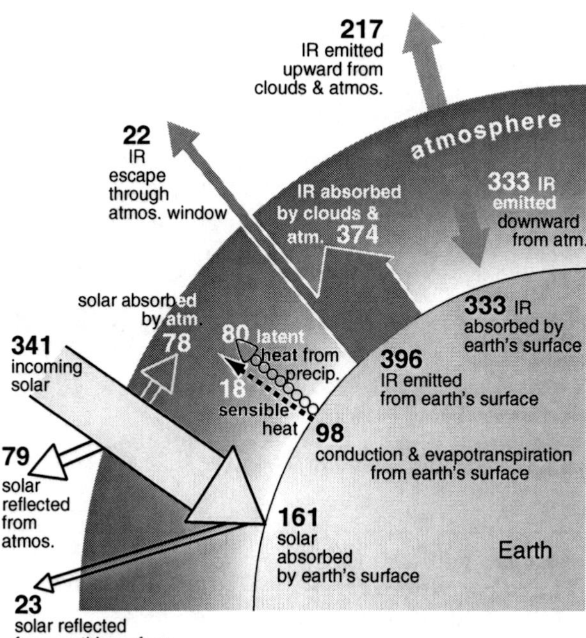

Figure 21.6
Approximate global annual mean energy budget for the Earth system, based on data from 2000-2005. The "atmosphere" component includes the air, clouds, and particles in the air. Numbers are approximate energy fluxes in units of W·m^{-2}. Errors are of order 3% for fluxes at the top of the atmosphere, and errors are 5 to 10% for surface fluxes. The incoming solar radiation, when averaged over all the Earth's surface, is $S_o/4$ = 341 W·m^{-2}. After subtracting the outgoing (reflected) solar radiation (341 – 79 – 23 = 239 W·m^{-2}), the result is the same R_o that we used before. (Based partially on info from Trenberth, Fasullo and Kiehl 2009 & Trenberth & Fasullo 2012.)

ASTRONOMICAL INFLUENCES

If the incoming solar radiation (**insolation**) changes, then the radiation budget of the Earth will adjust, possibly altering the climate. Two causes of average-daily-insolation $\overline{E}$ change at the top of the atmosphere are changes in solar output S_o and changes in Earth-sun distance R (recall eq. 2.21).

Milankovitch Theory

In the 1920s, astrophysicist Milutin Milankovitch proposed that slow fluctuations in the Earth's orbit around the sun can change the sun-Earth distance. Distance changes alter insolation due to the inverse square law (eq. 2.16), thereby contributing to climate change. Ice-age recurrence (inferred from ice and sediment cores) was later found to be correlated with orbital characteristics, supporting his hypothesis. Milankovitch focused on three orbital characteristics: eccentricity, obliquity, and precession.

Eccentricity

The shape of Earth's elliptical orbit around the sun (Fig. 21.7a) slowly fluctuates between nearly circular (**eccentricity** $e \approx 0.0034$) and slightly elliptical ($e \approx 0.058$). For comparison, a perfect circle has $e = 0$. Gravitational pulls by other planets (mostly Jupiter and Saturn) cause these eccentricity fluctuations.

Earth's eccentricity at any time (past or future) can be approximated by a sum of cosine terms:

Sample Application
Confirm that energy flows in Fig. 21.6 are balanced.

Find the Answer
Solar: incoming = [reflected] + (absorbed)
341 = [79 + 23] + (78 + 161)
341 = 341 W·m^{-2}

Earth's surface: incoming = outgoing
(161 + 333) = (98 + 396)
494 = 494 W·m^{-2}

Atmosphere: incoming = outgoing
(78 + 18 + 80 + 374) = (217 + 333)
550 = 550 W·m^{-2}

Earth System: in from space = out to space
341 = (23 + 79 + 22 + 217)
341 = 341 W·m^{-2}

Check: Yes, all budgets are balanced.
Exposition: The solar budget must balance because the Earth-atmosphere system does not create/emit solar radiation. However, IR radiation is created by the sun, Earth, and atmosphere; hence, the IR fluxes by themselves need not balance.

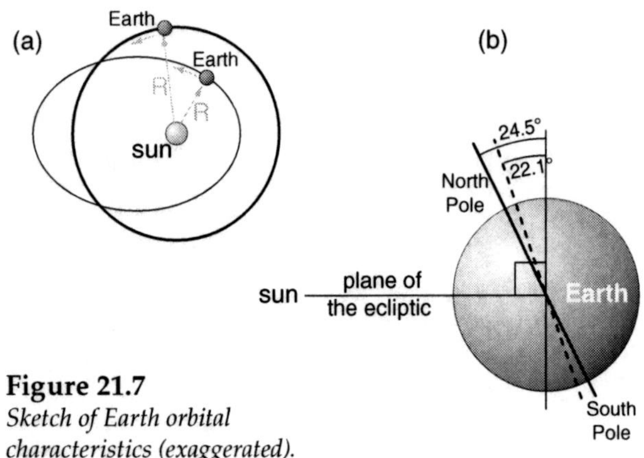

Figure 21.7
Sketch of Earth orbital characteristics (exaggerated).
(a) Eccentricity variations. (b) Obliquity variations.

Sample Application
Plot ellipses for eccentricities of 0, 0.005, 0.058 & 0.50. Hints: the (x, y) coordinates are given by $x = a\cdot\cos(t)$, and $y = b\cdot\sin(t)$, where t varies from 0 to 2π radians. Assume a semi-major axis of $a = 1$, and get the semi-minor axis from $b = a\cdot(1 - e^2)^{1/2}$.

e = 0, 0.005, 0.058
e = 0.5
a
b

Find the Answer
Plotted at right:

Exposition. The first 3 curves (solid lines) look identical. Thus, for the full range of Earth's eccentricities ($0.0034 \le e \le 0.58$), we see that Earth's orbit is nearly circular ($e = 0$).

$$e \approx e_o + \sum_{i=1}^{N} A_i \cdot \cos\left[C \cdot \left(\frac{t}{P_i} + \frac{\phi_i}{360°} \right) \right] \qquad \bullet(21.12)$$

where $e_o = 0.0275579$, A_i are amplitudes, t is time in years relative to calendar year 2000, P_i are oscillation periods in years, ϕ_i are phase shifts in degrees, and i is an index that counts the 1 to N terms. C should be either 360° or $2\cdot\pi$ radians, depending on the type of trigonometric argument required by your calculator, spreadsheet, or programming language.

Table 21-1 shows values for these orbital factors for $N = 5$ terms in this series. [CAUTION: *A more accurate description requires N =20 or more terms. The first 10 terms are listed in Table 21-1b near exercise E9 at the end of this chapter.*] I used a spreadsheet to solve these 5 terms for times ranging from 1 million years in the past to 1 million years into the future (see results in Fig. 21.8).

By eye, we see two superimposed oscillations of eccentricity with periods of about 100 and 400 kyr (where "kyr" = kilo years = 1000 years). The first oscillation correlates very well with the roughly 100 kyr period for cool events associated with **ice ages**. The present eccentricity is about 0.0167. Short-term forecast: eccentricity will change little during the next 100,000 years. Variations in e affect Earth-sun distance R via eq. (2.4) in the Solar & IR Radiation chapter.

Sample Application.
Find the Earth-orbit eccentricity 600,000 years ago (i.e., 600,000 years before year 2000).

Find the Answer.
Given: $t = -600{,}000$ years (negative sign for past).
Find: e (dimensionless).

Use eq. (21.12) with data from Table 21-1:

$e \approx 0.0275579$
$+ 0.010739\cdot\cos[2\pi\cdot(-600000\text{yr}/405091\text{yr} + 170.739°/360°)]$
$+ 0.008147\cdot\cos[2\pi\cdot(-600000\text{yr}/\ 94932\text{yr} + 109.891°/360°)]$
$+ 0.006222\cdot\cos[2\pi\cdot(-600000\text{yr}/123945\text{yr} - 60.044°/360°)]$
$+ 0.005287\cdot\cos[2\pi\cdot(-600000\text{yr}/\ 98857\text{yr} - 86.140°/360°)]$
$+ 0.004492\cdot\cos[2\pi\cdot(-600000\text{yr}/130781\text{yr} + 100.224°/360°)]$

$e \approx 0.0275579 + 0.0107290 + 0.0081106 + 0.0062148$
$- 0.0019045 - 0.0016384 \approx$ **<u>0.049</u>** (dimensionless)

Check: $0.0034 \le e \le 0.058$, within the expected eccentricity range for Earth. Agrees with Fig. 21.8.
Exposition: An **interglacial** (non-ice-age) period was ending, & a new **glacial** (ice-age) period was starting.

Table 21-1. Factors in orbital series approximations.

index	A	P (years)	ϕ (degrees)
Eccentricity:			
i= 1	0.010739	405,091.	170.739
2	0.008147	94,932.	109.891
3	0.006222	123,945.	-60.044
4	0.005287	98,857.	-86.140
5	0.004492	130,781.	100.224
Obliquity:			
j= 1	0.582412°	40,978.	86.645
2	0.242559°	39,616.	120.859
3	0.163685°	53,722.	-35.947
4	0.164787°	40,285.	104.689
Climatic Precession:			
k= 1	0.018986	23,682.	44.374
2	0.016354	22,374.	-144.166
3	0.013055	18,953.	154.212
4	0.008849	19,105.	-42.250
5	0.004248	23,123.	90.742

Simplified from Laskar, Robutel, Joutel, Gastineau, Correia & Levrard, 2004: A long-term numerical solution for the insolation quantities of the Earth. "Astronomy & Astrophysics", ***428****, 261-285.*

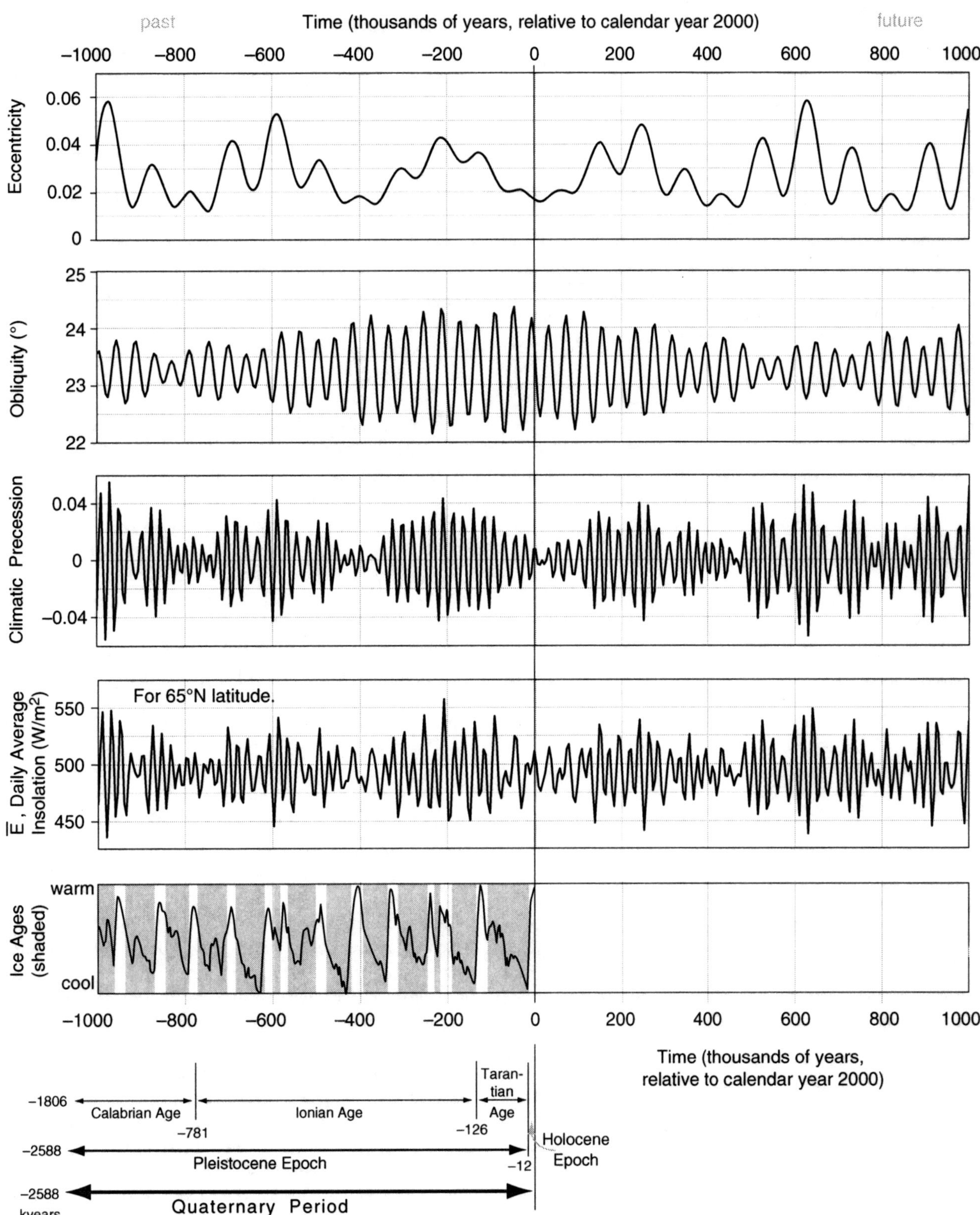

Figure 21.8
Past and future orbital characteristics in Milankovitch theory, as calculated using only the 4 or 5 highest-order terms (see Table 21-1) in the series approximations of eqs. (21.12 - 21.20). Bottom curve shows relative temperature changes estimated from ice and sediment cores. Ice ages (glacial periods) are shaded grey. Precession is for the summer solstice (i.e., only the e·sin(ϖ) term is used).

Sample Application
Estimate Earth's obliquity 90,000 yrs in the future (relative to calendar year 2000).

Find the Answer
Given: t = 90,000 yr (positive sign indicates future)
Find: $\varepsilon = ?°$

Use eq. (21.13) with coefficients from Table 21-1.
ε = 23.2545°
+ 0.582412°·[2π·(90000yr/40978yr + 86.645°/360°)]
+ 0.242559°·[2π·(90000yr/39616yr + 120.859°/360°)]
+ 0.163685°·[2π·(90000yr/53722yr – 35.947°/360°)]
+ 0.164787°·[2π·(90000yr/40285yr + 104.689°/360°)]

ε=23.2545° –0.537349° –0.189273° –0.145639° –0.162777°
= **22.219°**

Check: Units OK. Within Earth range 22.1°≤ ε ≤24.5°.
Exposition: Agrees with Fig. 21.8. At this small obliquity, seasonal contrasts are smaller than at present, allowing glaciers to grow due to cooler summers.

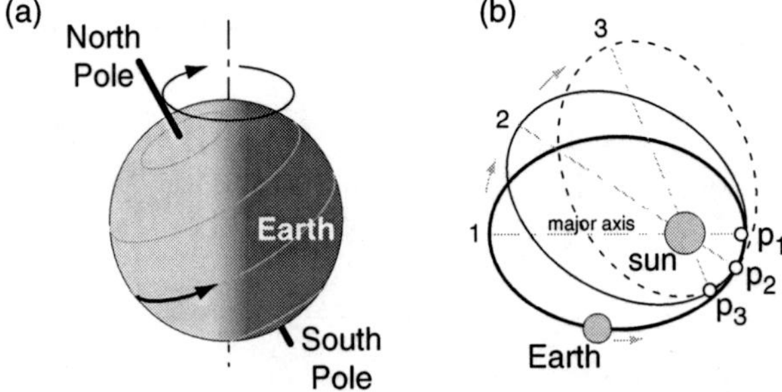

Figure 21.9
Sketch of additional Earth orbital characteristics. (a) Precession of Earth's axis. (b) Precession of the orbit-ellipse major axis (aphelion precession) at successive times 1 - 3. Small white dots p_1, p_2, p_3 show locations of the perihelion (point on Earth's orbit closest to sun). All the orbit ellipses are in the same plane.

Obliquity

The tilt (obliquity) of Earth's axis slowly fluctuates between 22.1° and 24.5°. This tilt angle is measured from a line perpendicular to Earth's orbital plane (the ecliptic) around the sun (Fig. 21.7b).

A series approximation for obliquity ε is:

$$\varepsilon = \varepsilon_o + \sum_{j=1}^{N} A_j \cdot \cos\left[C \cdot \left(\frac{t}{P_j} + \frac{\phi_j}{360°} \right) \right] \qquad \bullet(21.13)$$

where ε_o = 23.254500°, and all the other factors have the same meaning as for the eq. (21.12). Table 21-1 shows values for these orbital factors for N = 4 terms in this series.

Fig. 21.8 shows that obliquity oscillates with a period of about 41,000 years. The present obliquity is 23.439° and is gradually decreasing.

Obliquity affects insolation via the solar declination angle (eq. 2.5 in the Solar & IR Radiation chapter, where a different symbol Φ_r was used for obliquity). Recall from the Solar & IR Radiation chapter that Earth-axis tilt is responsible for the seasons, so greater obliquity would cause greater contrast between winter and summer. One hypothesis is that ice ages are more likely during times of lesser obliquity, because summers would be cooler, causing less melting of glaciers.

Precession

Various precession processes affect climate.

Axial Precession. Not only does the tilt magnitude (obliquity) of Earth's axis change with time, but so does the tilt direction. This tilt direction rotates in space, tracing one complete circle in 25,680 years (Fig. 21.9a) relative to the fixed stars. It rotates opposite to the direction the Earth spins. This precession is caused by the gravitational pull of the sun and moon on the Earth, because the Earth is an oblate spheroid (has larger diameter at the equator than at the poles). Hence, the Earth behaves somewhat like a spinning toy top.

Aphelion Precession. The direction of the major axis of Earth's elliptical orbit also precesses slowly (with a 174,000 to 304,00 year period) relative to the fixed stars (Fig. 21.9b). The rotation direction is the same as the axial-precession, but the precession speed fluctuates due to the gravitational pull of the planets (mostly Jupiter and Saturn).

Equinox Precession. Summing axial and aphelion precession rates (i.e., adding the inverses of their periods) gives a total precession with fluctuating period of about 22,000 years. This combined precession describes changes to the angle (measured at the sun) between Earth's orbital positions at the perihelion (point of closest approach) and at the moving vernal (Spring) equinox (Fig. 21.10). This fluctuating angle (ϖ, a math symbol called "variant pi") is the equinox precession.

For years when angle ϖ is such that the winter solstice is near the perihelion (as it is in this century), then the cooling effect of the N. Hemisphere being tilted away from the sun is slightly moderated by the fact that the Earth is closest to the sun; hence winters and summers are not as extreme as they could be. Conversely, for years with different angle ϖ such that the summer solstice is near the perihelion, then the N. Hemisphere is tilted toward the sun at the same time that the Earth is closest to the sun — hence expect hotter summers and colder winters. However, seasons near the perihelion are shorter duration than seasons near the aphelion, which moderates the extremes for this latter case.

Climatic Precession. ϖ by itself is less important for insolation than the combination with eccentricity e: $e{\cdot}\sin(\varpi)$ and $e{\cdot}\cos(\varpi)$. These terms, known as the climatic precession, can be expressed by series approximations, where both terms use the same coefficients from Table 21-1, for $N = 5$.

$$e \cdot \sin(\varpi) \approx \sum_{k=1}^{N} A_k \cdot \sin\left[C \cdot \left(\frac{t}{P_k} + \frac{\phi_k}{360^\circ}\right)\right] \qquad \bullet(21.14)$$

$$e \cdot \cos(\varpi) \approx \sum_{k=1}^{N} A_k \cdot \cos\left[C \cdot \left(\frac{t}{P_k} + \frac{\phi_k}{360^\circ}\right)\right] \qquad (21.15)$$

Other factors are similar to those in eq. (21.12).

A plot of eq. (21.14), labeled "climatic precession", is shown in Fig. 21.8. It shows high frequency (22,000 year period) waves with amplitude modulated by the eccentricity curve from the top of Fig. 21.8. Climatic precession affects sun-Earth distance, as described next.

Insolation Variations

Eq. (2.21) from the Solar & IR Radiation chapter gives average daily insolation $\overline{E}$. It is repeated here:

$$\overline{E} = \frac{S_o}{\pi} \cdot \left(\frac{a}{R}\right)^2 \cdot \left[h_o{'} \cdot \sin(\phi) \cdot \sin(\delta_s) + \cos(\phi) \cdot \cos(\delta_s) \cdot \sin(h_o)\right] \qquad \bullet(21.16)$$

where S_o =1366 W m^{-2} is the solar irradiance, a = 149.457 Gm is the semi-major axis length, R is the sun-Earth distance for any day of the year, $h_o{'}$ is the hour angle in radians., ϕ is latitude, and δ_s is solar declination angle. Factors in this eq. are given next.

Instead of solving this formula for many latitudes, climatologists often focus on one key latitude:

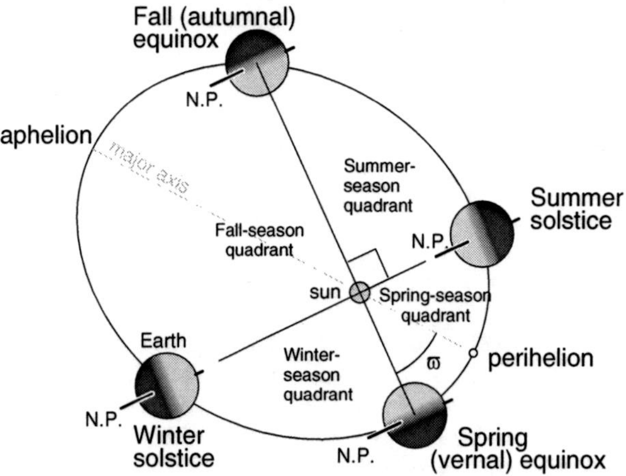

Figure 21.10
Illustration for sometime in the future, showing the angle (ϖ) of the perihelion from the moving vernal equinox. N.P. = North Pole. Seasons are for the N. Hemisphere.

Sample Application
Estimate the sine term of the climatic precession for year 2000.

Find the Answer
Given: $t = 0$
Find: $e{\cdot}\sin(\varpi)$ = ? (dimensionless)

Use eq. (21.14) with coefficients from Table 21-1.
$e{\cdot}\sin(\varpi)$ =
+ 0.018986·[2π·(0yr/23682yr + 44.374°/360°)]
+ 0.016354·[2π·(0yr/22374yr – 144.166°/360°)]
+ 0.013055·[2π·(0yr/18953yr + 154.212°/360°)]
+ 0.008849·[2π·(0yr/19105yr – 42.250°/360°)]
+ 0.004248·[2π·(0yr/23123yr + 90.742°/360°)]

$e{\cdot}\sin(\varpi)$ = 0.0132777 – 0.0095743 + 0.0056795 – 0.0059498 + 0.0042476 = **0.00768**

Check: Agrees with curve plotted in Fig. 21.8.
Exposition: The small eccentricity in year 2000 (i.e., almost at present) causes small climatic precession.

Table 21-2. Key values of λ, the angle between the Earth and the moving vernal equinox. Fig. 21.10 shows that solstices and equinoxes are exactly 90° apart.

Date	λ (radians, °)	sin(λ)	cos(λ)
Vernal (Spring) equinox	0 , 0	0	1
Summer solstice	π/2 , 90°	1	0
Autumnal (Fall) equinox	π , 180°	0	–1
Winter solstice	3π/2 , 270°	–1	0

Sample Application

For the summer solstice at 65°N latitude, find the average daily insolation for years 2000 and 2200.

Find the Answer

Given: $\lambda = \pi/2$, $\phi = 65°N$, (a) $t = 0$, (b) $t = 200$ yr
Find: $\overline{E} = ?$ W m^{-2} (Assume $S_o = 1366$ W m^{-2})

(a) From previous statements and Sample Applications:

$e = 0.0167$, $\delta_s = \varepsilon = 23.439°$, $e{\cdot}\sin(\varpi) = 0.00768$

Use eq. (21.17): $h_o = \arccos[-\tan(65°)\cdot\tan(23.439°)]$ $= 158.4° = 2.7645$ radians $= h_o'$.

Eq. (21.20): $(a/R) = (1+0.00768)/[1-(0.0167)^2] = 1.00796$

Plugging all these into eq. (21.16):

$\overline{E} = [(1366 \text{ W m}^{-2})/\pi]\cdot(1.00796)^2 \cdot [2.7645 \cdot\sin(65°) \cdot\sin(23.439°) + \cos(65°)\cdot\cos(23.439°)\cdot\sin(158.4°)]$

$\overline{E}$ = **503.32** W m^{-2} for $t = 0$.

(b) Eq. (21.12) for $t = 200$: $e \approx 0.0168$

Eq. (21.13): $\varepsilon = 23.2278°$. Eq. (21.14): $e{\cdot}\sin(\varpi) = 0.00730$

Eq. (21.20): $a/R = 1.00758$. Eq. (21.16): $\overline{E}$ = **499.3** W m^{-2}

Check: Units OK. Magnitude ≈ that in Fig. 2.11 (Ch2).

Exposition: Milankovitch theory indicates that insolation will decrease slightly during the next 200 yr.

INFO • Sun-Earth Distance

Recall from Chapter 2 (eq. 2.4) that the sun-Earth distance R is related to the semi-major axis a by:

$$\frac{a}{R} = \frac{1 + e\cdot\cos(\nu)}{1 - e^2} \qquad (21.19)$$

where ν is the true anomaly (position of the Earth from the perihelion), and e is the eccentricity from eq. (21.12). By definition, $\nu = \lambda - \varpi$ (21.a)
where λ is position of the Earth from the moving vernal equinox, and ϖ is position of the perihelion from the moving vernal equinox. Thus,

$$\frac{a}{R} = \frac{1 + e\cdot\cos(\lambda - \varpi)}{1 - e^2} \qquad (21.b)$$

Using trigonometric identities, the cosine term is:

$$\frac{a}{R} = \frac{1 + \cos(\lambda)\cdot[e\cdot\cos(\varpi)] + \sin(\lambda)\cdot[e\cdot\sin(\varpi)]}{1 - e^2} \qquad (21.c)$$

The terms in square brackets are the **climatic precessions** that you can find with eqs. (21.14) and (21.15).

To simplify the study of climate change, researchers often consider a special time of year; e.g., the summer solstice ($\lambda = \pi/2$ from Table 21-2). This time of year is important because it is when glaciers may or may not melt, depending on how warm the summer is. At the summer solstice, the previous equation simplifies to:

$$\frac{a}{R} = \frac{1 + [e\cdot\sin(\varpi)]}{1 - e^2} \qquad (21.20)$$

Also, during the summer solstice, the solar declination angle from eq. (21.18) simplifies to: $\delta_s = \varepsilon$

$\phi = 65°N$. This latitude crosses Alaska, Canada, Greenland, Iceland, Scandinavia, and Siberia, and is representative of genesis zones for ice caps.

The hour angle h_o at sunrise and sunset is given by the following set of equations:

$$\alpha = -\tan(\phi)\cdot\tan(\delta_s) \qquad (21.17a)$$

$$\beta = \min[1, (\max(-1, \alpha)] \qquad (21.17b)$$

$$h_o = \arccos(\beta) \qquad (21.17c)$$

Although the hour angle h_o can be in radians or degrees (as suits your calculator or spreadsheet), the hour angle marked with the prime (h_o') must be in radians.

The solar declination angle, used in eq. (21.16), is

$$\delta_s = \arcsin[\sin(\varepsilon)\cdot\sin(\lambda)] \approx \varepsilon\cdot\sin(\lambda) \qquad (21.18)$$

where ε is the obliquity from eq. (21.13), and λ is the **true longitude** angle (measured at the sun) between the position of the Earth and the position of the moving vernal equinox. Special cases for λ are listed in Table 21-2 on the previous page.

Recall from the Solar & IR Radiation chapter (eq. 2.4) that the sun-Earth distance R is related to the semi-major axis a by:

$$\frac{a}{R} = \frac{1 + e\cdot\cos(\nu)}{1 - e^2} \qquad (21.19)$$

where ν is the true anomaly (position of the Earth from the perihelion), and e is the eccentricity. For the special case of the summer solstice (see the Sun-Earth Distance INFO Box), this becomes:

$$\frac{a}{R} = \frac{1 + [e\cdot\sin(\varpi)]}{1 - e^2} \qquad (21.20)$$

where e is from eq. (21.12) and $e{\cdot}\sin(\varpi)$ is from eq. (21.14)

I used all the information above to solve eq. (21.16) on a spreadsheet for average daily insolation at latitude 65°N during the summer solstice for 1 Myr in the past to 1 Myr in the future. This results in the curve for $\overline{E}$ plotted in Fig. 21.8. This is the climatic signal resulting from Milankovitch theory. Some scientists have found a good correlation between it and the historical temperature curve plotted at the bottom of Fig. 21.8, although the issue is still being debated.

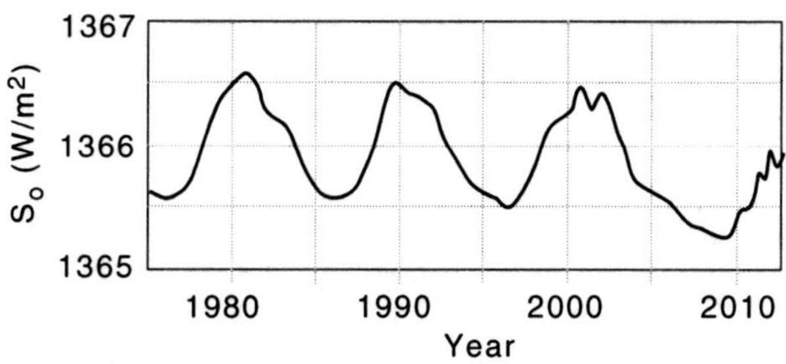

Figure 21.11
Variation of annual average total solar irradiance S_o at the top of Earth's atmosphere during 3 recent sunspot cycles. (based on 2013 data from Phys.-Meteor. Obs. Davos/World Rad. Ctr.)

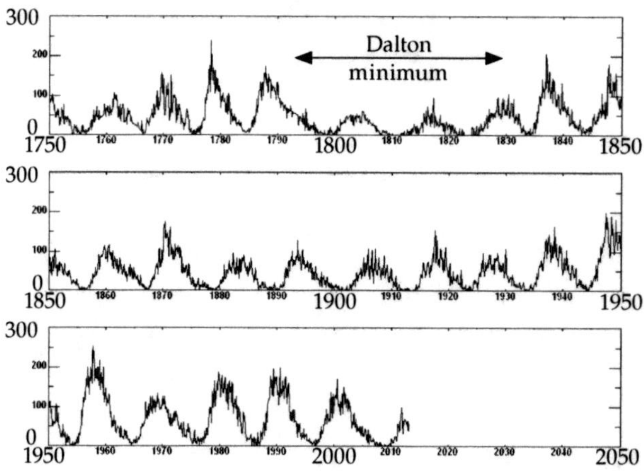

Figure 21.12
Monthly average sunspot number during 1750 - 2012. NASA.

Solar Output

The average total solar irradiance (TSI) currently reaching the top of Earth's atmosphere is S_o = 1366 W·m^{-2}, but the magnitude fluctuates daily due to solar activity, with extremes of 1362 and 1368 W·m^{-2} measured by satellite since 1975. Satellite observations have calibration errors on the order of ±7 W·m^{-2}.

Sunspot Cycle

When averaged over each year, the resulting smoothed curve of TSI varies with a noticeable 11-yr cycle (Fig. 21.11), corresponding to the 9.5 to 11-yr sunspot cycle (Fig. 21.12) as observed by telescope. Because the TSI variation is only about 0.1% of its total magnitude, the sunspot cycle is believed to have only a minor (possibly negligible) effect on recent climate change.

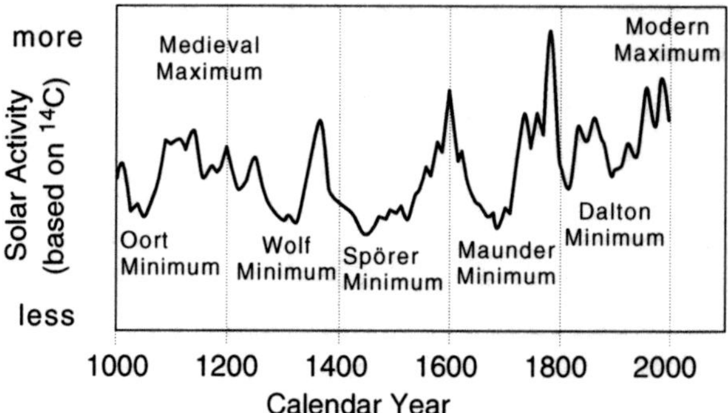

Figure 21.13
Carbon 14 concentration as a proxy for solar activity. Approximate dates for these minima: **Oort Minimum:** *1040 - 1080,* **Wolf Minimum:** *1280 - 1350,* **Spörer Minimum:** *1450 - 1550,* **Maunder Minimum:** *1645 - 1715, and* **Dalton Minimum:** *1790 - 1830. The 11-year sunspot cycle still exists during this time span, but was smoothed out for this plot. (based on data from Muscheler et al, 2007, Quat.Sci.Rev.,* ***26****.)*

Longer-term Variations in Solar Output

To get solar output information for the centuries before telescopes were invented, scientists use **proxy** measures. Proxies are measurable phenomena that vary with solar activity, such as beryllium-10 concentrations in Greenland ice cores, or carbon-14 concentration in tree rings (**dendrochronology**). Radioactive carbon-14 dating suggests solar activity for the past 1,000 years as plotted in Fig. 21.13. Noticeable is the 200-yr cycle in solar-activity minima.

Going back 11,000 yr in time, Fig. 21.14 shows a highly-smoothed carbon-14 estimate of solar/sunspot activity. There is some concern among scientists that Earth-based factors might have confounded the proxy analysis of tree rings, such as extensive volcanic eruptions that darken the sky and reduce tree growth over many decades.

Even further back (3 to 4 Myr ago), the sun was younger and weaker, and emitted only about 70% of what it currently emits. About 5 billion years into the future, the sun will grow into a red-giant star (killing all life on Earth), and later will shrink into a white dwarf star.

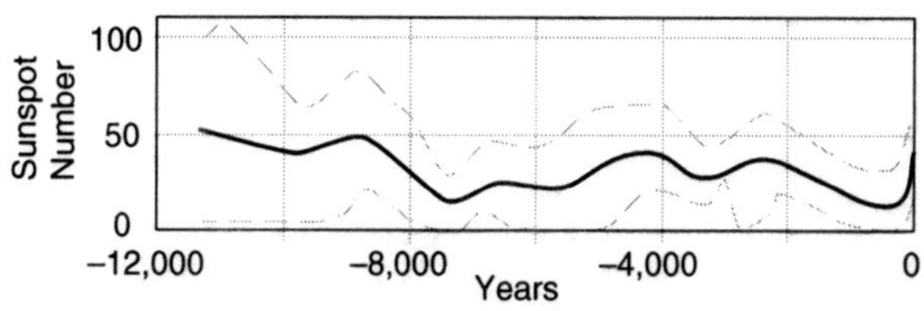

Figure 21.14
Smoothed proxy estimate of sunspots. Time is relative to calendar year 2000. Dashed lines show envelope of fluctuations.

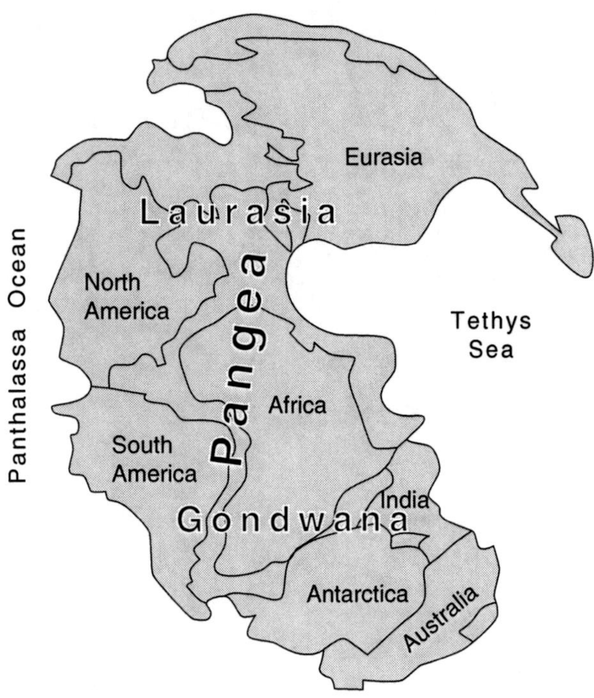

Figure 21.15
Location of continents about 225 million years ago to form the supercontinent Pangea (also spelled as Pangaea and Pangæa). This supercontinent spanned north-south from pole to pole.

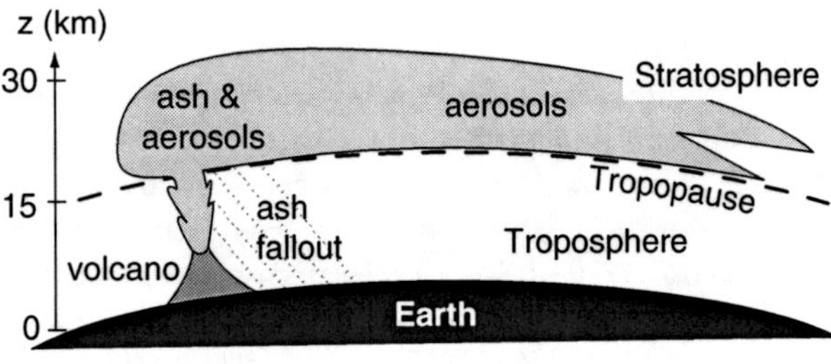

Figure 21.16
Eruption of volcanic ash and aerosols into the stratosphere. Average times for particles to fall 1 km are: 36 yr for 0.1 μm diameter particles, 3.2 yr for 0.5 μm, 328 days for 1 μm, 89 days for 2 μm, 14.5 days for 5 μm, and 3.6 days for 10 μm.

TECTONIC INFLUENCES

Continent Movement

As Earth-crust tectonic plates move, continents form, change, merge and break apart. About 225 million years ago the continents were merged into the **Pangea** supercontinent (Fig. 21.15). The interior of such large continents are sufficiently far from the moderating influence of oceans that extreme daily and seasonal temperature variations can occur. Large monsoon circulations can form due to the temperature contrasts between continent and ocean. The more recent (200 million years ago) supercontinents of Gondwanaland and Laurasia would likely have had similar monsoon circulations, and extreme temperature variations in their interiors.

Colliding tectonic plates can cause **mountain building** (**orogenesis**). Large mountain ranges can block portions of the global circulation, changing the Rossby waves in the jet stream, and altering cyclogenesis regions. Mountains can partially block the movement of airmasses, changing temperature and precipitation patterns.

Changes in continent location and ocean depths cause changes in ocean circulations. **Thermohaline** circulations are driven by water-density differences caused by temperature and salt-concentration variations. These large, slow circulations connect deep water and surface waters of all oceans, and are important for redistributing the excess heat from the equator toward the poles. Climatic changes in ocean-surface temperatures can alter atmospheric temperature, humidity, winds, cloudiness, and precipitation distribution.

Volcanism

When volcanoes erupt explosively, the eruptive heat and vertical momentum ejects tons of tiny rock particles called **volcanic ash** into the upper troposphere and lower stratosphere (Fig. 21.16). During the few weeks before this ash falls out of the air, it can block some sunlight from reaching the surface.

Gases such as **sulfur dioxide** (SO_2) are also emitted in large volumes (0.1 to 6000 Mtons from the larger eruptions), and are much more important for climate change. SO_2 oxidizes in the atmosphere to form sulfuric acid and sulfate **aerosols**, tiny droplets that fall so slowly they seem suspended in the air.

Those sulfate particles in the stratosphere can remain suspended in the atmosphere for a few years as they are spread around the world by the winds. The sulfates absorb and reflect some of the incoming solar radiation, allowing less to be transmitted

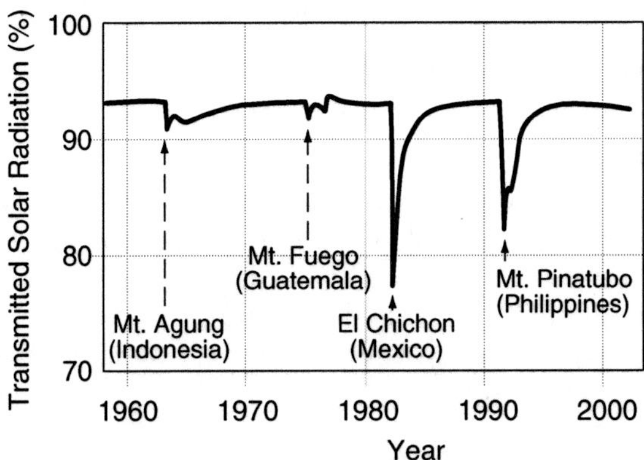

Figure 21.17
Reduction in transmitted shortwave radiation from the sun due to aerosols and particulates erupted into the atmosphere from volcanoes, as observed at Muana Loa.

toward the ground (Fig. 21.17). Also, these aerosols absorb upwelling IR radiation from the Earth and troposphere.

The net result for aerosol particle diameters less than 0.5 µm is that the stratosphere warms while the Earth's surface and lower troposphere cool. This cooling causes short-term (2 - 3 yr) climate change known as **volcanic winter** (analogous to the effects of nuclear winter). The surface cooling can be so great as to ruin vegetation and crops around the world, as was experienced throughout history (Table 21-3). For a brief explosive eruption, the stratospheric effects are greatest initially and gradually diminish over a 3-year period as the aerosols fall out. For violent continuous eruptions, the climatic effects can last decades to millennia.

Volcanoes also erupt hundreds of millions tons of **greenhouse gases** (gases that absorb IR radiation as can alter the global heat budget), including carbon dioxide (CO_2) and water vapor (H_2O). These gases can cause longer term global warming after the aerosol-induced cooling has ended.

After the large eruptions, you can see amazingly beautiful sunsets and sunrises with a bright **red glow** in the sky. Sometimes you can see a **blue moon** at night. These phenomena are due to scattering of sunlight by sulfate aerosols with diameters ≤ 0.5 µm (see the Atmospheric Optics chapter).

Table 21-3. Some notable volcanic eruptions.

Volcano (Date, Location) • Effect
Siberian Traps, creating floods of basaltic lava (~250 Myr ago, Russia) - Lasted a million years, emitting tremendous volumes of sulfates and greenhouse gases into the atmosphere that caused a mass extinction at the **Permian-Triassic** geological-period boundary. Known as the "**Great Dying**", where 96% of all marine species and 70 percent of terrestrial vertebrate species disappeared. Major volcanic eruptions likely contributed to other mass extinctions: End-Ordovician (~450 Myr ago) and Late Devonian (~375 Myr ago).
Deccan Traps, creating floods of basaltic lava (~65.5 Myr ago, India) • The Cretaceous–Paleogene extinction event (K–Pg event, formerly known as Cretaceous–Tertiary or **K-T event**) was when the **dinosaurs** disappeared (except for those that evolved into modern-day birds). The sulfate and greenhouse-gas emissions from this massive eruption are believed to be a contributing factor to the K-T event, which was exacerbated by a massive asteroid impact in Mexico.
Yellowstone Caldera (640 kyr ago, USA) • Emissions reduced N. Hem. temperatures by about 5°C
Lake Toba (~90 kyr ago, Indonesia) • Emissions killed most of the humans on Earth, possibly causing a **genetic bottleneck** of reduced genetic diversity. Stratospheric aerosol load ≈ 1000 Mtons. Optical depth $\tau \approx 10$ (see Atmospheric Optics chapter for definition of τ).
Mt. Tambora (1815, Indonesia) • Emissions caused so much cooling in the N. Hemisphere that 1816 is known as the "**Year Without Summer**". Widespread famine in Europe and N. America due to crop failure and livestock deaths. Followed by an exceptionally cold winter. Stratospheric aerosol load ≈ 200 Mtons. Optical depth $\tau \approx 1.3$.
Krakatau (1883, Indonesia) • Violent explosion heard 3,500 km away. Ash reached 80 km altitude. Global temperature decrease ≈ 0.3 - 1.2°C. 50 Mtons of aerosols into stratosphere. $\tau \approx 0.55$.

Darkness

"I had a dream, which was not all a dream.
The bright sun was extinguish'd, and the stars
Did wander darkling in the eternal space,
Rayless, and pathless, and the icy Earth
Swung blind and blackening in the moonless air;
Morn came and went — and came, and
brought no day,
And men forgot their passions in the dread
Of this their desolation; and all hearts
Were chill'd into a selfish prayer for light."

- Lord Byron (1816 –the Year Without Summer.)

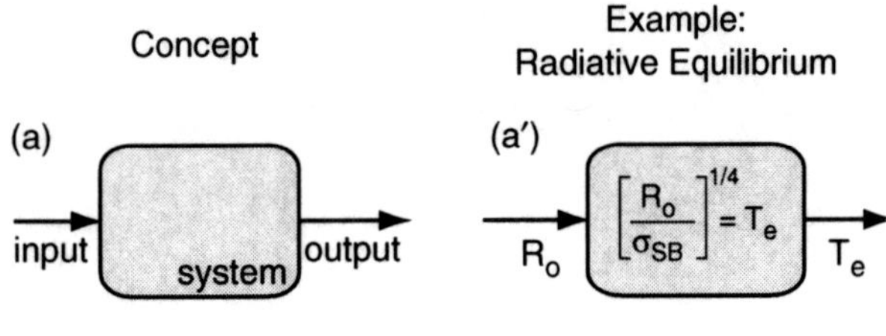

Linearize:

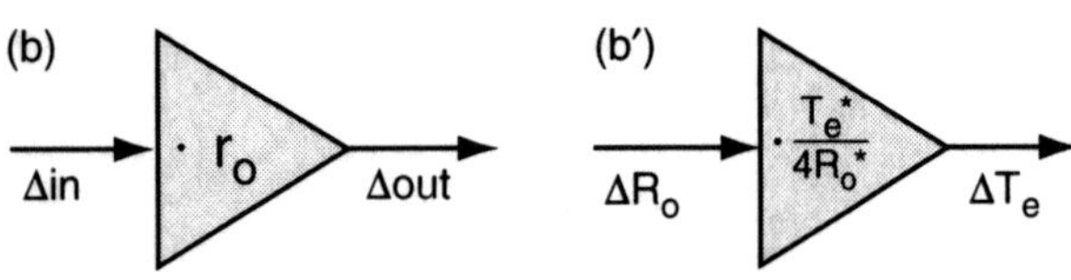

Add Feedback:

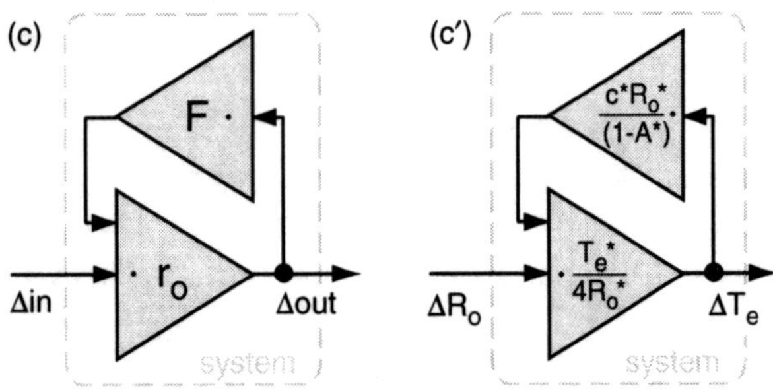

Figure 21.18
Left column: feedback concepts. Right column: example of ice-albedo feedback within the radiative-equilibrium process, as described by nonlinear eq. (21.3). The reference state is (R_o, T_e*, A*, c*).*

> **Sample Application**
> Suppose r_o = 0.1 and F = –3. Find the feedback factor, gain, and compare system responses with and without feedback.
>
> **Find the Answer**
> Given: F = –3, no-feedback system response r_o = 0.1
> Find: f = ? , G = ? , r = ? = response with feedback
>
> Use eq. (21.24):
> f = (0.1) · (–3) = **<u>–0.3</u>** is the feedback factor
> Use eq. (21.26):
> G = 1 / [1 – (–0.3)] = **<u>0.77</u>** is the gain
> Use eq. (21.28):
> r = (0.77)·(0.1) = **<u>0.077</u>** is the system response <u>with</u> feedback, compared to r_o = 0.1 without.
>
> **Check**: Units all dimensionless. Magnitudes OK.
> **Exposition**: This system is <u>damped</u>. It has <u>negative feedback</u>. This feedback diminishes the response. For many systems, the output units differ from the input units, in which case r_o , F, and r would have units.

FEEDBACKS

Concept

Consider a system or physical process with an input and output (Fig. 21.18a). This system might be **linear** (described by a straight <u>line</u> when output is plotted vs. input) or **nonlinear** (described by a curved <u>line</u>). Various values of input give various values of output.

Choose one particular set of input and output values as a **reference state** of the system. Find the slope r_o of the line at this reference point. This slope describes the **response** of the output (Δout) to a small change of input (Δin). Namely,

$$\Delta out = r_o \cdot \Delta in \qquad (21.21)$$

as sketched in Fig. 21.18b. The slope is a straight line. Thus, even if the original system is nonlinear, this slope describes a **linearized approximation** to the system response at the reference point.

Feedback (Fig. 21.18c) is where the output signal is added to the input via some feedback process F:

$$\Delta out = r_o \cdot [\ \Delta in + F \cdot \Delta out] \qquad (21.22)$$

A linear approximation is also used for this feedback. The <u>feedback is internal to the system</u> (Fig. 21.18c). Thus the amount of feedback responds to changes in the system. Contrast that to <u>inputs</u>, which are considered **external forcings** that do not respond to changes within the system.

The previous feedback equation can be rewritten as

$$\Delta out = r_o \cdot \Delta in + f \cdot \Delta out \qquad \bullet(21.23)$$

where the **feedback factor** f is given by

$$f = r_o \cdot F \qquad (21.24)$$

Solving eq. (21.33) for Δout gives:

$$\Delta out = \frac{1}{(1-f)} \cdot r_o \cdot \Delta in \qquad (21.25)$$

Eq. (21.25) looks similar to eq. (21.21) except for the new factor $1/(1–f)$. In electrical engineering, this factor is called the **gain**, G:

$$G = \frac{1}{(1-f)} \qquad (21.26)$$

[*CAUTION: Some climate researchers use a different definition of gain than electrical engineers. See the INFO Box on the next page.*]

If we rewrite eq. (21.25) as

$$\Delta\text{out} = r \cdot \Delta\text{in} \qquad (21.27)$$

then we see that the linearized process WITH feedback (eq. 21.27) looks identical to that without (eq. 21.21) but with a different **response**:

$$r = G \cdot r_o \qquad (21.28)$$

If $f \geq 1$, then the feedback is so strong that the response increases without limit (i.e., a **runaway** response), and no equilibrium exists. If $0 < f < 1$, then there is **positive feedback** with gain $G > 1$ that leads to an amplified but stable new equilibrium. If $f < 0$, then there is **negative feedback** with gain $G < 1$ that damps the response toward a different stable equilibrium.

For N feedback processes that are independent of each other and additive, the total gain is

$$G = \left[1 - \sum_{i=1}^{N} f_i \right]^{-1} \qquad (21.29)$$

where f_i is the feedback factor for any one process i.

Idealized Example

To illustrate feedback, use the simple no-atmosphere blackbody radiative-equilibrium system of Fig. 21.1. The net solar input per square meter of the Earth's surface is

$$R_o = (1 - A) \cdot S_o / 4 \qquad (21.30)$$

(see the INFO box on the 3rd page of this chapter). The IR-radiation output is described by the Stefan-Boltzmann law, resulting in the following energy balance of input = output:

$$R_o = \sigma_{SB} \cdot T_e^4 \qquad (21.31)$$

as indicated in Fig. 21.18a'.

The dashed lines in Fig. 21.2 show how an input of R_o = 239 W m^{-2} (S_o = 1366 W m^{-2}) results in an output $T_e \approx 255$ K. Use this blackbody Earth-equilibrium condition as the **reference state**, and denote the reference values with asterisks (R_o*, T_e*). The thick curved line in Fig. 21.2 shows how T_e varies nonlinearly for different inputs of R_o, assuming A = 0.3 is constant (i.e., no feedback other than the background outgoing IR).

No Feedback

Focus on this no-feedback example first (Fig. 21.18b'). For small solar-input variations ΔR_o about its reference state, the corresponding temperature

INFO • Gain — Different Definitions

Some climate researchers define the **gain** g as

$$g = \frac{\Delta out - \Delta out_o}{\Delta out}$$

where Δout_o is the background response with NO feedbacks. Thus:

$$g = \frac{G \cdot r_o \cdot \Delta in - r_o \cdot \Delta in}{G \cdot r_o \cdot \Delta in} = 1 - \frac{1}{G}$$

But $G = 1/(1-f)$ from eq. (21.26). Thus

$$g = f$$

Namely, the "gain" used in some climate studies is the feedback factor used in this chapter and in electrical engineering.

HIGHER MATH • No-feedback Response

Start with no feedback, as described by the energy balance of eq. (21.31).

$$R_o = \sigma_{SB} \cdot T_e^4$$

Take the derivative:

$$dR_o = 4\,\sigma_{SB} \cdot T_e^3 \; dT_e$$

Rearrange to find the change of output with input (dT_e/dR_o),

$$\frac{\partial T_e}{\partial R_o} = \frac{1}{4 \cdot \sigma_{SB} \cdot T_e^3}$$

Finally, convert to finite differences:

$$r_o = \frac{\Delta T_e}{\Delta R_o} = \frac{1}{4 \cdot \sigma_{SB} \cdot T_e^3} \qquad (21.33)$$

Another form of this answer can be found by multiplying the numerator and denominator by T_e, and then substituting eq. (21.31) in the denominator:

$$r_o = \frac{T_e}{4 \cdot R_o}$$

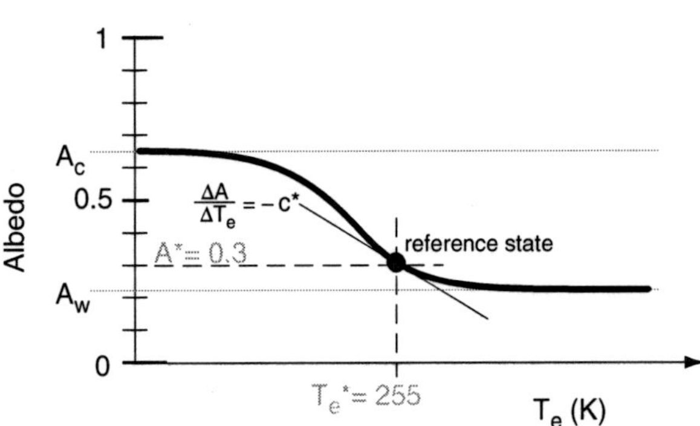

Figure 21.19
Thick line is a sketch of response of Earth's albedo to changes in radiative equilibrium temperature T_e, associated with changes in snow- and ice-cover. The thin line of slope $-c^*$ corresponds to the slope of the thick line at the reference point. Values for A_c and A_w are not accurate — only for illustration.

> **Sample Application**
> For ice-albedo radiative-equilibrium with a blackbody-Earth reference state, find the feedback factor and gain. Compare system responses with and without feedback.
>
> **Find the Answer**
> Given: T_e*= 255 K, R_o*= 239 W·m^{-2}, A*= 0.3, c*= 0.01 K^{-1}
> Find: r_o = ? (K/W·m^{-2}), f = ? , G = ? , r = ? (K/W·m^{-2})
>
> Use eq. (21.33) for the no-feedback system response:
> r_o = (255K)/(4 · 239W·m^{-2}) = **0.267** K/(W·m^{-2})
>
> Use eq. (21.36) for the feedback factor:
> f = (0.01 K^{-1})·(255K) / [4 · (1 – 0.3)] = **0.91** (dim'less)
>
> Use eq. (21.26): G = 1/(1 – 0.91) = **11.1** (dimensionless)
>
> Use eq. (21.28):
> r = (11.1) · [0.267 K/(W·m^{-2})] = **2.96** K/(W·m^{-2})
>
> **Check**: Response is reasonable. Units OK.
> **Exposition**: This illustrates positive feedback; namely, the response is **amplified** by the feedback. As an example, $\Delta S_o \approx 1$ W·m^{-2} due to sunspot cycle (Fig. 21.11) implies ΔR_o = 0.175 W·m^{-2}, giving $\Delta T_e = r \cdot \Delta R_o$ = 0.52°C variation in equilibrium radiation temperature due to sunspots.

change ΔT_e can be approximated by a straight line tangent to the curve at the reference point (Fig. 21.2), as defined by

$$r_o = \Delta T_e \,/\, \Delta R_o \tag{21.32}$$

Based on eq. (21.31) the slope r_o of this straight line is (see Higher Math box on the previous page):

$$r_o = \frac{1}{4 \cdot \sigma_{SB} \cdot T_e^{*3}} = \frac{T_e^*}{4 \cdot R_o^*} = 0.267 \text{ K/(W·m}^{-2}) \tag{21.33}$$

Ice Albedo Feedback

Next, add **ice-albedo feedback** by allowing the albedo to vary with temperature. Changes in temperature T_e alter the areal extent of highly reflective ice caps and glaciers, thereby changing the albedo (A) between two extremes (Fig. 21.19). A warm limit A_w assumes 0% snow cover, and is based on the reflectivities of the ground, crops, cities, and oceans. A cold-limit A_c represents 100% snow cover, causing a highly reflective Earth. The blackbody Earth state is useful as a reference point (T_e* = 255 K, A* = 0.3).

Use the slope (thin line in Fig. 21.19) of the A vs. T_e curve at the reference point as a **linear approximation** of the nonlinear feedback (thick line). Numerical simulations of the global climate suggest this slope is:

$$\frac{\Delta A}{\Delta T} = -c^* \tag{21.34}$$

where c^* = 0.01 K^{-1} is the magnitude of the slope at the reference point. The ice-albedo feedback described here is highly oversimplified, for illustrative purposes.

For this idealized ice-albedo radiative-equilibrium case, the feedback equation (21.23) becomes:

$$\Delta T_e = r_o \cdot \Delta R_o + f \cdot \Delta T_e \tag{21.35}$$

where r_o was given in eq. (21.33), and where

$$f = \frac{c^* \cdot T_e^*}{4 \cdot (1 - A^*)} \tag{21.36}$$

(see the Higher Math box on the next page).

The resulting gain is

$$G = \frac{4 \cdot (1 - A^*)}{4 \cdot (1 - A^*) - c^* \cdot T_e^*} \tag{21.37}$$

After multiplying r_o by the gain to give r, we find that the temperature response to solar-input variations (when feedbacks ARE included) is:

$$r = \frac{(1 - A^*) \cdot T_e^*}{R_o^* \cdot [4 \cdot (1 - A^*) - c^* \cdot T_e^*]} \tag{21.38}$$

Fig. 21.20 shows how the total response r with feed-

HIGHER MATH • Feedback Example

For ice-albedo feedback, infinitesimal variations in solar irradiance dS_o and albedo dA cause infinitesimal changes in equilibrium temperature dT_e :

$$dT_e = \frac{\partial T_e}{\partial S_o} dS_o + \frac{\partial T_e}{\partial A} dA \tag{21.a}$$

To find the factors that go into this equation, start with a simple radiative-equilibrium balance:

$$T_e = \left[\frac{(1-A)\cdot S_o}{4\cdot\sigma_{SB}} \right]^{1/4} \tag{21.3}$$

and take the derivative $\partial T_e/\partial S_o$:

$$\frac{\partial T_e}{\partial S_o} = \frac{1}{4}\left[\frac{(1-A)\cdot S_o}{4\cdot\sigma_{SB}} \right]^{-3/4} \cdot \left[\frac{1-A}{4\cdot\sigma_{SB}} \right]$$

Substituting eq. (21.3) into this gives:

$$\frac{\partial T_e}{\partial S_o} = \frac{1}{4}[T_e]^{-3} \cdot \left[\frac{1-A}{4\cdot\sigma_{SB}} \right]$$

which we can rewrite as:

$$\frac{\partial T_e}{\partial S_o} = \frac{1}{4}\cdot T_e \cdot \left[\frac{1-A}{4\cdot\sigma_{SB}\cdot T_e^4} \right]$$

Solve (21.3) for S_o and substituting into the eq. above :

$$\frac{\partial T_e}{\partial S_o} = \frac{1}{4}\cdot T_e \cdot \left[\frac{1}{S_o} \right]$$

You can similarly find $\partial T_e/\partial A$ from eq. (21.3) to be:

$$\frac{\partial T_e}{\partial A} = \frac{-T_e}{4\cdot(1-A)}$$

When these derivatives are plugged into eq. (21.a) and applied at the reference point (S_o*, T_e*, A*), the result is:

$$dT_e = \frac{T_e*}{4\cdot S_o*} dS_o - \frac{T_e*}{4\cdot(1-A*)} dA$$

Albedo decreases when Earth's temperature increases, due to melting snow and retreating glacier coverage:

$$dA = -c*\cdot dT_e$$

where $c*$ is the magnitude of this variation at the reference point. Plugging this into the previous eq. gives:

$$dT_e = \frac{T_e*}{4\cdot S_o*} dS_o + \frac{c*\cdot T_e*}{4\cdot(1-A*)} dT_e \tag{21.b}$$

(continues in next column)

HIGHER MATH • Feedback (continued)

Next, take the derivative of eq. (21.30):

$$dR_o = [(1-A)/4]\cdot dS_o$$

Dividing this by eq. (21.30) shows that:

$$\frac{dR_o}{R_o} = \frac{dS_o}{S_o}$$

Plug this into eq. (21.b), and convert from calculus to finite differences:

$$\boxed{\Delta T_e = \frac{T_e*}{4\cdot R_o*}\cdot\Delta R_o + \frac{c*\cdot T_e*}{4\cdot(1-A*)}\cdot\Delta T_e}$$

Compare this with eq. (21.23) to realize that:

$$r_o = \frac{T_e*}{4\cdot R_o*} \quad \text{(eq. 21.33)} \quad \text{and} \quad f = \frac{c*\cdot T_e*}{4\cdot(1-A*)}$$

Thus,

$$F = \frac{f}{r_o} = \frac{c*\cdot R_o*}{(1-A*)} \quad \text{and} \quad G = \frac{4\cdot(1-A*)}{4\cdot(1-A*) - c*\cdot T_e*}$$

For the idealized (toy model) example in this section, the reference values are:

T_e* = 255 K = radiative equilibrium temperature
S_o* = 1366 W·m^{-2} = solar irradiance
R_o* = 239 W·m^{-2} = solar input / m^2 Earth's surface
A* = 0.3 (dimensionless) = albedo
c* = 0.01 K^{-1} =albedo response to temperature change

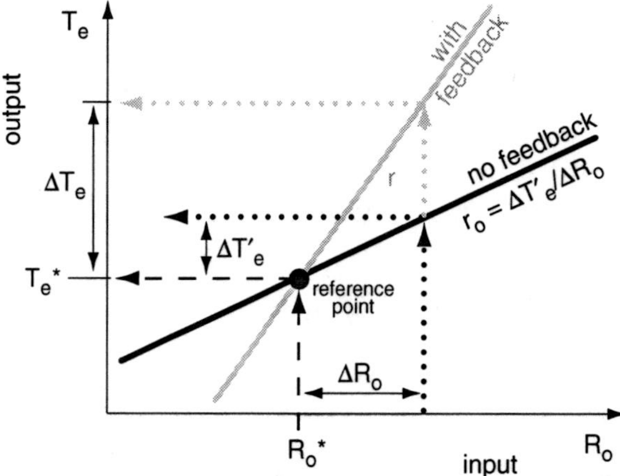

Figure 21.20
Response of system with feedback (grey) and without (black). The prime ' denotes the no-feedback response.

Table 21-4. Feedbacks affecting Earth's climate.
(+) = positive, (–) = negative feedbacks.
F = Feedback parameter $(W·m^{-2})·K^{-1}$.
f = feedback factor (dimensionless)
G = Gain (dimensionless)
r = system response $K/(W·m^{-2})$. This is also known as the **sensitivity**, λ.

Processes for gradual climate change:				
Feedback	**F**	**f** = $r_o·F$	**G** = 1/ (1–f)	**r** = $G·r_o$
IR Radiative (–) • blackbody Earth • realistic Earth				r_o = 0.27 0.31
Water-vapor (+)	1.8 ±0.18	0.48	1.93	0.51
Lapse-rate (–)	–0.84 ±0.26	–0.22	0.82	0.22
Cloud (+)	0.69 ± 0.38	0.18	1.2	0.33
Surface albedo (+)	0.26 ±0.08	0.07	1.1	0.29
Ocean CO_2 (+)				
Biological (–)				
Processes for Abrupt climate change:				
• Meridional Overturning Ocean Circulation • Fast Antarctic & Greenland Ice-sheet Collapse • Volcanoes • Methane Release (Hydrate Instability & Permafrost) • Biogeochemical				

Based on IPCC AR4, 2007.

Sample Application

Given the feedbacks listed in Table 21-4, what is the total response.

Find the Answer

Given: Table 21-4. Use realistic Earth for r_o.
Find: r = ? $K/(W·m^{-2})$.
Assume: total response is based on only those feedback processes for which numbers are given in the table.

Use eq. (21.29): $G = 1 / [1 - \Sigma f]$
$G = 1 / [1 - (0.48 - 0.22 + 0.18 + 0.07)]$
$G = 1 / [1 - 0.51] = 2.04$

Use eq. (21.28): $r = G \cdot r_o$
$r = 2.04 \cdot 0.31 =$ **0.63** $K/(W·m^{-2})$.

Check: Units OK. Physics OK.
Exposition: For each $\Delta R = 1\ W·m^{-2}$ change in solar input, the Earth-system response with all feedbacks is $\Delta T = 0.63$ K, which is twice the response (0.31 K) without feedback (other than the background IR feedback). [CAUTION: The total response is NOT equal to the sum of the individual responses r. Instead, use the sum of feedback factors f to get the net feedback.]

back differs from the r_o without. With feedback, eq. (21.32) becomes:

$$\Delta T_e = r \cdot \Delta R_o \qquad (21.39)$$

Table 21-4 lists some feedbacks that can affect Earth's climate. These are briefly discussed next.

Infrared Radiative (IR) Feedback

IR is the strongest feedback — it dominates all others. It was already discussed as the "radiation out" term in eq. (21.2). This important negative feedback allows the Earth to have an equilibrium state. Even a small change in Earth's temperature results in a strong compensating energy loss because IR radiation is to the 4th power of T_e in the Stefan-Boltzmann equation. This strong damping keeps Earth's climate relatively steady.

As we have seen, the reference response to changes in net solar input is $r_o = 0.267\ K/(W·m^{-2})$. Global climate models that include more realistic atmospheric absorption and emission of IR radiation find $r_o = 0.30$ to $0.32\ K/(W·m^{-2})$.

Water-vapor Feedback

Warmer air can hold more water vapor, which absorbs and re-radiates more IR radiation back to the surface and reduces the atmospheric window sketched in Fig. 21.5. The warmer surface warms the air, resulting in a very strong positive feedback (see Table 21-4). The upper troposphere is where this feedback is most effective.

Water vapor is an important natural greenhouse gas. If the climate warms and the upper tropospheric water-vapor content increases to the point where the atmospheric window is mostly closed (net atmospheric emissivity $e \approx 1$), then we return to the negative feedback sketched in Fig. 21.4 and steady Earth temperature of $T_s = 30°C$ from eq. (21.6).

Lapse-rate Feedback

The simple model of Fig. 21.4 uses the same air temperature T_A to determine IR radiation both up to space and down toward the Earth. However, in the real atmosphere the temperature varies with altitude — decreasing with lapse rate 6.5 °C·km^{-1} in the troposphere. Thus, colder temperatures at the top of the troposphere emit less upward IR radiation to space than is emitted downward from the warmer bottom of the troposphere. This difference in upward and downward IR radiation is part of our normal climate, causing our normal temperatures.

Global warming tends to reduce the lapse rate. Namely, the temperature difference diminishes between the top and bottom of the troposphere. Relative to each other, IR radiation out to space will increase and IR radiation down toward the Earth will decrease. The net effect is to cool the Earth system. Thus, lapse-rate feedback is negative (Table 21-4).

Water-vapor and lapse-rate feedbacks are closely related, and are often considered together. Their combined effect is a modest positive feedback.

Cloud Feedback

Cloud feedback is likely important, but the details are poorly understood. Relative to clear skies, clouds at all heights cause cooling by reflecting incoming solar radiation. But clouds also cause warming by trapping (absorbing and re-radiating back towards the surface) upwelling IR radiation. The small difference between these large opposing forcings can have very large error, depending on cloud type and altitude. For our present climate, the net effect of clouds is cooling.

Table 21-4 indicates a positive feedback for clouds, based on global climate modeling. Namely, an increase in radiative input to the Earth system would decrease the cloud cover, thus causing less cooling. But much more research must be done to gain more confidence in cloud-feedback processes.

Sometimes **aerosol feedback** is grouped with cloud feedback. Aerosols (microscopic solid or liquid particles in the air) can have similar effects as cloud droplets. However, some aerosols such as sulfates are darker than pure water droplets, and can absorb more solar radiation to cause atmospheric warming. As already discussed, volcanoes can inject sulfate aerosols into the stratosphere. **Phytoplankton** (microscopic plant life) in the ocean can release into the atmospheric boundary layer a chemical called dimethyl sulfide, which can later be oxidized into sulfate aerosols.

Ice–albedo (Surface) Feedback

This positive feedback was discussed earlier in the idealized feedback example, where we found a response of r = 2.96 K/(W·m^{-2}). More realistic estimates from Global Climate Model simulations are r = 0.27 to 2.97 K/(W·m^{-2}). At our current reference temperature, this feedback mostly affects the climate at high latitudes.

This positive-feedback sensitivity applies only over the limited range of Fig. 21.19 for which albedo can vary with temperature. At the cold extreme of a completely snow covered Earth (**snowball Earth**), the albedo is constant at its cold-limit value A_c. At the warm extreme the Earth has no ice caps or snow cover, and albedo is constant at A_w.

At these two extremes, the negative feedback of IR radiation again dominates, leading to steady Earth

Sample Application

Use Table 21-4 to estimate the combined water-vapor and lapse-rate response, $r_{wv.lr}$.

Find the Answer

Given Table 21-4: $f_{water.vapor}$ = 0.48, $f_{lapse.rate}$ = –0.22
Find: $r_{wv.lr}$ = ? K/(W·m^{-2})
Assume: r_o = 0.267 K/(W·m^{-2}) from eq. (21.33)

For multiple feedbacks, we must use eq. (21.29):
$G = 1 / [1 - (0.48 - 0.22)] = 1.35$
Use eq. (21.28):
$r_{wv.lr}$ = 1.35 · [0.267 K/(W·m^{-2})] = <u>0.361 K/(W·m^{-2})</u>

Check: Units OK. Magnitude OK.
Exposition: This response is midway between water vapor alone (r = 0.51) and lapse-rate alone (r = 0.22).

INFO • Greenhouse Gases

A **greenhouse gas** is an atmospheric gas that absorbs and emits IR radiation. Of particular concern are gases that absorb and emit within the 8 to 14 µm wavelength range of the **atmospheric window**. As the concentrations of such gases increase, the atmospheric window can close due to the increased atmospheric emissivity (Fig. 21.5), resulting in global warming.

Greenhouse gases, ranked by importance, are:
(1) water vapor (H_2O),
(2) carbon dioxide (CO_2),
(3) methane (CH_4),
(4) nitrous oxide (N_2O),
(5) ozone (O3),
(6) halocarbons [e.g., freon CFC-12 ($C\,Cl_2\,F_2$)]

Except for (6), most greenhouse gases have both natural and **anthropogenic** (man-made) sources. CO_2 is of particular concern because it is released by humans burning fossil fuels (coal, oil) for energy.

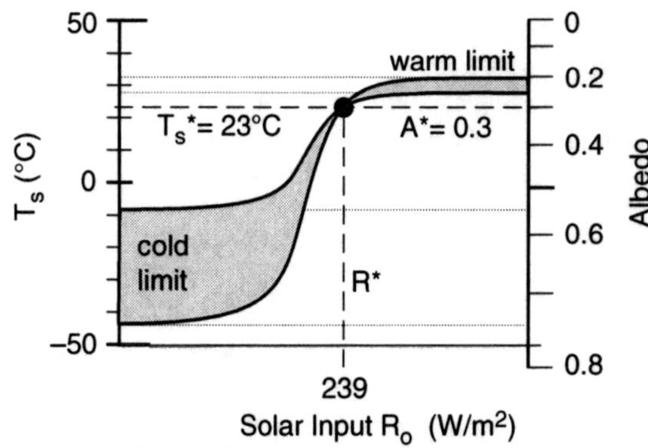

Figure 21.21
Illustration of how Earth surface temperature is bounded once the Earth becomes fully snow covered (cold limit) or fully snow-free (warm limit). Grey shading shows temperature uncertainty due to uncertainty in the limiting albedo values.

INFO • Effervescence and CO_2

Carbonated beverages retain their tart flavor when cold. The tartness or sourness comes from carbonic acid (H_2CO_3) due to CO_2 dissolved in water (H_2O). But as the beverage warms, it cannot hold as much dissolved CO_2, forcing some of the CO_2 to **effervesce** (bubble out). Namely, the beverage loses its fizz, causing it to taste "flat".

The same phenomenon happens to the oceans. If oceans warm, they lose some of their fizz.

INFO • Climate Sensitivity

The Intergovernmental Panel on Climate Change (**IPCC**) states that **climate sensitivity** is "a measure of the climate system response to sustained radiative forcing. It is defined as the equilibrium global average surface warming" (ΔT_s) "following a doubling of CO_2 concentration."

Based on research by international teams using global climate models, IPCC suggests in their 4th Assessment Report (AR4, 2007) that "this sensitivity is likely to be in the range of" ΔT_s = "2 to 4.5°C with a best estimate of 3°C" for a doubling of the greenhouse gas CO_2. This temperature rise would be equivalent to a **radiative forcing** (ΔR_o in Fig. 21.18b′ and c′; abbreviated as *RF* by IPCC) of 3.7 to 4.0 $W \cdot m^{-2}$.

But CO_2 has not doubled yet. What are the actual forcings so far. The IPCC is particularly concerned about anthropogenic forcings, because these can be mitigated by people. Based on estimates of anthropogenic changes between years 1750 (the start of the industrial age) and 2005, they find that people have inadvertently altered the net radiative forcing of the Earth system as follows:

Process	ΔR_o or RF ($W \cdot m^{-2}$)
CO_2 increase	1.66
CH_4 increase	0.48
N_2O increase	0.16
Halocarbons	0.34
O_3 in stratosphere	–0.05
O_3 in troposphere	0.35
H_2O from CH_4 in stratosphere	0.07
Surface albedo/land use	–0.2
Surface albedo/carbon on snow	0.1
Aerosols/ direct	-0.5
Aerosols/ via cloud albedo	-0.7
Jet contrails	0.01
Total anthropogenic	**1.6**
	range [0.6 to 2.4]

Compare this to a natural process:

Solar irradiance change	**0.12**

The large anthropogenic effect relative to the natural solar variations is what motivates policy makers to take action.

temperatures. For snowball Earth, cold-limit albedo estimates are in the range of $0.55 \le A_c \le 0.75$. When used in the radiation balance of eq. (21.3) the cold-limit blackbody temperature is $197 \text{ K} \le T_e \le 228 \text{ K}$, yielding a surface temperature ($T_s = 1.161 \cdot T_e$, from eq 21.10) of $-44°C \le T_s \le -8°C$.

For the warm limit (assuming that clouds having 37% albedo still cover roughly 70% of the Earth) $0.20 \le A_w \le 0.25$, and the equilibrium blackbody temperature is $259 \text{ K} \le T_e \le 263 \text{ K}$. This corresponds to a surface-temperature of $28°C \le T_s \le 33°C$ using eq. (21.10). Thus, runaway global warming or cooling is not possible beyond these bounds (see Fig. 21.21).

Ocean CO_2 Feedback

Warmer seawater can hold less dissolved carbon dioxide (CO_2) than colder water (see INFO box). Hence global warming will cause oceans to release CO_2 into the atmosphere. Greater atmospheric CO_2 concentrations reduce the atmospheric window due to an increase in net atmospheric emissivity *e*, and thus cause the atmosphere to re-radiate more IR radiation back toward the Earth surface. The net result is positive feedback.

Biological CO_2 Feedback

As plants grow, they consume CO_2 to make hydrocarbons and carbohydrates via photosynthesis. As a result, the carbon is **sequestered** (stored) as the body of the plant. Some of the carbohydrates (e.g., sugars) are consumed by the plant, and are converted back to CO_2 and **transpired** (exhaled by the plant) into the atmosphere.

When plants die and decay, or when they burn in a **wild fire**, they release their remaining carbon as CO_2 back to the atmosphere. If plants are buried under sediment before they decay or burn, their carbon can be fossilized into coal, oil, natural gas (**fossil fuels**). Carbonate rocks can also sequester carbon from the atmosphere, but the carbon can later be released through weathering of the rocks.

The change of carbon from CO_2 to other forms is called the **global carbon cycle**. It is extremely difficult to model the global carbon cycle, which leads to large uncertainty in estimates of climate feedback. Also, anthropogenic effects of fossil-fuel and biomass burning affect the cycle. At present, it is believed that biological CO_2 causes a positive feedback. Namely, global warming will cause more CO_2 to be released than sequestered.

Processes of **abrupt climate change** in Table 21-4 are left to the reader to explore (via an Evaluate & Analyze exercise). But next we introduce the intriguing gaia hypothesis.

GAIA HYPOTHESIS & DAISYWORLD

James Lovelock proposed a thought-provoking hypothesis: life regulates the climate to create an environment that favors life. Such self-maintained stability is called **homeostasis**. His hypothesis is called **gaia** — Greek for "mother Earth".

Lovelock and Andrew Watson illustrate the "biological homeostasis of the global environment" with **daisyworld**, a hypothetical Earth containing only light and dark colored daisies. If the Earth is too cold, the dark daisies proliferate, increasing the absorption of solar radiation. If too warm, light-colored daisies proliferate, reflecting more sunlight by increasing the global albedo (Fig. 21.22).

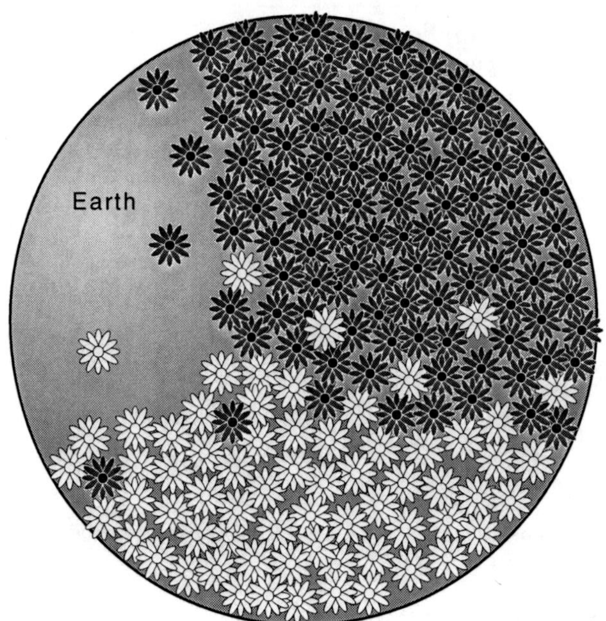

Figure 21.22
A hypothetical daisyworld with only light and dark colored flowers, or bare ground.

Physical Approximations

We modify Lovelock's toy model here to enable an easy solution of daisyworld on a spreadsheet. Given are two constants: the Stefan-Boltzmann constant $\sigma_{SB} = 5.67\text{x}10^{-8}$ W·m^{-2}·K^{-4} and solar irradiance $S_o = 1366$ W·m^{-2}. Small variations in solar output are parameterized by luminosity, L, where $L = 1$ corresponds to the actual value for Earth. These factors are combined into a solar-forcing ratio:

$$q = \frac{L \cdot S_o}{4 \cdot \sigma_{SB}} \tag{21.40}$$

Adjustable parameters include:
$A_L = 0.75$ is the albedo of light-colored daisies
$A_D = 0.25$ is the albedo of dark-colored daisies
$A_G = 0.5$ is the bare-ground albedo (no flowers).
$D = 0.3$ is death rate for both light and dark daisies.

$Tr = 0.6$ is an efficiency for horizontal heat transport. For example, $Tr = 1$ implies efficient spreading of heat around the globe, causing the local temperature to be controlled by the global-average albedo. Conversely, $Tr = 0$ forces the local temperature to be a function of only the local albedo in any one patch of daisies.

The model forecasts the fraction of the globe covered by light daisies (C_L) and the fraction covered by dark daisies (C_D). Some locations will have no daisies, so the bare-ground fraction (C_G) of Earth's surface is thus:

$$C_G = 1 - C_D - C_L \tag{21.41}$$

To start the calculation, assume an initial state of $C_L = C_D = 0$.

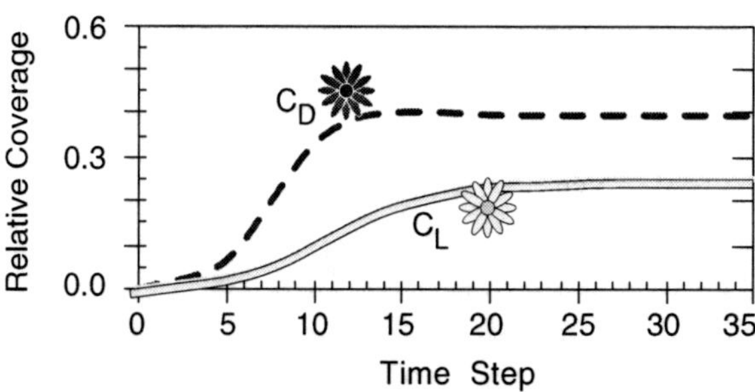

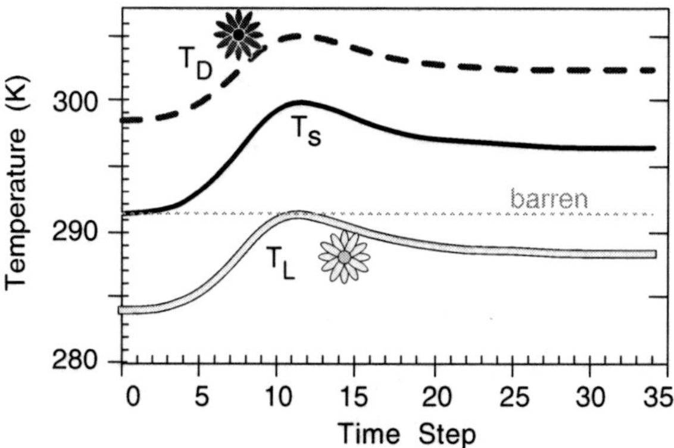

Figure 21.23
Evolution of daisy coverage and temperatures as daisyworld evolves toward an equilibrium, given a luminosity of L = 1.2 .

Sample Application (§)

Compute and plot the evolution of temperatures and daisy coverage for a daisyworld having: bare ground albedo = 0.45, light-colored daisy albedo = 0.85, dark daisy albedo = 0.2, daisy death rate = 0.25, transport parameter = 0.65, luminosity of the sun = 1.5, seed coverage = 0.01, and time step = 0.5 .

Find the Answer

Given: D=0.25, A_L=0.85, A_D=0.20, A_G=0.45, Tr=0.65, L=1.5, C_s = 0.01

Find: T_L = ? K, T_D = ? K, T_s = ? K, C_L = ? , C_D = ?

Find the solar forcing ratio from eq. (21.40):

$$q = \frac{1.5\cdot(1366\text{W}\cdot\text{m}^{-2})}{4\cdot(5.67\times10^{-8}\text{W}\cdot\text{m}^{-2}\cdot\text{K}^{-4})} = 9.036\text{x}10^9 \text{ K}^4$$

Because we are not told an initial coverage of daisies, assume $C_L = C_D = 0$.

Use eqs. (21.41 - 21.46) in a spreadsheet, and iterate forward in time from t = 0 using Δt = 0.5 .

t	C_D	C_L	C_G	A	T_s (K)	T_D (K)	T_L (K)	β_D	β_L
0	0.00	0.00	1.00	0.45	315.7	321.8	305.1	0.00	0.70
0.5	0.01	0.01	0.98	0.45	315.5	321.6	304.9	0.00	0.71
1	0.01	0.01	0.98	0.45	315.4	321.5	304.8	0.00	0.72
1.5	0.01	0.01	0.98	0.45	315.2	321.4	304.7	0.00	0.73
2	0.01	0.02	0.97	0.45	315.0	321.3	304.5	0.00	0.74
...									
7	0.01	0.19	0.80	0.52	304.6	313.2	295.0	0.00	1.00
7.5	0.01	0.24	0.75	0.54	301.2	310.7	291.9	0.25	0.96
...									
36.5	0.18	0.45	0.37	0.59	294.1	305.4	285.5	0.68	0.67
37	0.18	0.45	0.37	0.59	294.1	305.4	285.5	0.68	0.67

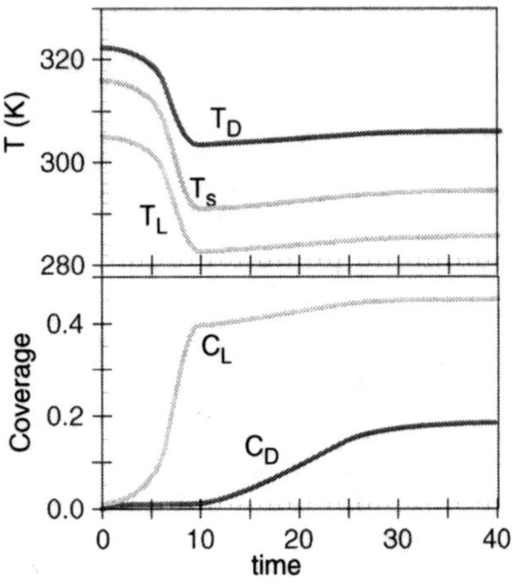

Check: Physics & units are OK.

Exposition: The solar luminosity is large, causing a planetary temperature that is initially too warm for any dark-colored daisies (except for the seed coverage). But as light-colored daisies proliferate, the world cools to the point where some dark-colored daisies can grow. The final steady surface temperature is not very different from that on the previous page, even though this world has more light than dark daisies.

Define a planetary albedo (A) as the coverage-weighted average of the individual daisy albedoes (A_i):

$$A = C_G\cdot A_G + C_D\cdot A_D + C_L\cdot A_L \qquad (21.42)$$

Use this albedo with the solar forcing ratio to find the world's effective-radiation absolute temperature:

$$T_e^4 = q\cdot(1-A) \qquad (21.43)$$

For a one-layer atmosphere with no window, the average surface temperature is

$$T_s^4 = 2\cdot T_e^4 \qquad (21.44)$$

But the patches of dark and light daisies have the following temperatures:

$$T_D^4 = (1-Tr)\cdot q\cdot(A-A_D)+T_s^4 \qquad (21.45a)$$

$$T_L^4 = (1-Tr)\cdot q\cdot(A-A_L)+T_s^4 \qquad (21.45b)$$

Suppose that daisies grow only if their patch temperatures are between 5°C and 40°C, and that daisies have the fastest growth rate (β) near the middle of this range (where T_o = 295.5 K = 22.5°C):

$$\beta_D = 1-b\cdot(T_o-T_D)^2 \qquad (21.46a)$$

$$\beta_L = 1-b\cdot(T_o-T_L)^2 \qquad (21.46b)$$

where negative values of β are truncated to zero. The growth factor is $b = 0.003265 \text{ K}^{-2}$.

Because dark and light daisies interact via their change on the global surface temperature, you must iterate the coverage equations together as you step forward in time

$$C_{D\,new} = C_D + \Delta t\cdot C_D\cdot(C_G\cdot\beta_D - D) \qquad (21.47a)$$

$$C_{L\,new} = C_L + \Delta t\cdot C_L\cdot(C_G\cdot\beta_L - D) \qquad (21.47b)$$

Namely, given the daisy coverages at any one time step, insert these on the right side of the above two equations, and step forward in time to find the new coverages one time step Δt later. The time units are arbitrary, so you could use Δt = 1 or Δt = 0.5. To get daisies to grow if none are on the planet initially, assume a seed coverage of C_s = 0.01 and force $C_{L\,new} \geq C_s$ and $C_{D\,new} \geq C_s$

With the new coverages, eqs. (21.41 - 21.47) can be computed again for the next time step. Repeat for many time steps. For certain values of the parameters, the solution (i.e., coverages and temperatures) approach a steady-state, which results in the desired homeostasis equilibrium.

Steady-state and Homeostasis

Suppose $L = 1.2$ with all other parameters set as described in the Physical Approximations subsection. After iterating, the final steady-state values are $C_L = 0.25$, $C_D = 0.39$ and $T_s = 296.75$ K (see Fig. 21.23). Compare this to the surface temperature of a planet with no daisies: $T_{s\,barren} = 291.65$ K, found by setting $A = A_G$ in eqs. (21.43 - 21.44).

Suppose you rerun daisyworld with all the same parameters, but with different luminosity. The resulting steady-state conditions are shown in Fig. 21.24 for a variety of luminosities. For luminosities between about 0.94 to 1.70, the daisy coverage is able to adjust so as to maintain a somewhat constant T_s — namely, it achieves homeostasis. Weak incoming solar radiation (insolation) allows dark daisies to proliferate, which convert most of the insolation into heat. Strong insolation is compensated by increases in light daisies, which reflect the excess energy.

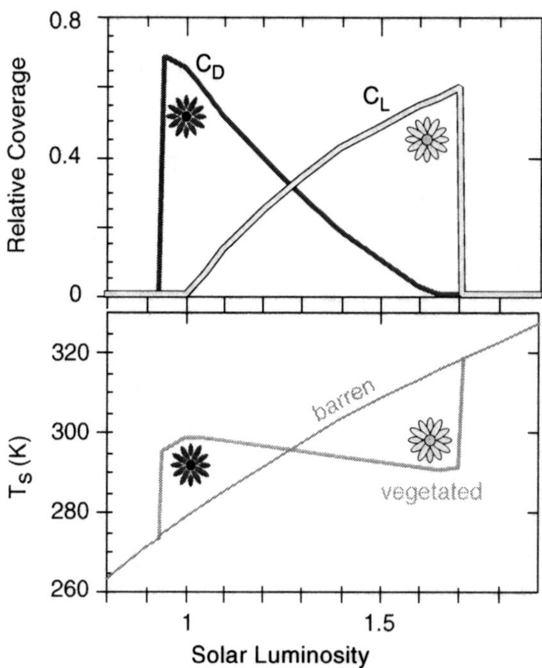

Figure 21.24
Final steady-state values of average surface temperature T_s and daisy coverages (C_D & C_L), as a function of luminosity. The bottom graph compares the lushly flowered (vegetated) world with an unflowered (barren) world.

GCMS

Some computer codes for numerical weather prediction (see the NWP chapter) are designed to forecast global climate. These codes, called **GCMs** (global climate models) have a somewhat-coarse 3-D arrangement of grid points filling the global atmosphere. At each grid point the governing equations for heat, momentum, mass, moisture, and various chemicals are repeatedly solved as the GCM iterates forward in time.

Other GCMs use spectral rather than grid-point numerics in the horizontal, where a sum of sine waves of various wavelengths are used to approximate the spatial structure of the climate. Regardless of the numerical scheme, the GCM is designed to forecast decades or even centuries into the future.

Such a long-duration forecast is computationally expensive. Improved climate forecasts are possible by coupling forecasts of the atmosphere, ocean, biosphere, cryosphere (ice) and soil — but this further increases computational expense. To speed-up the forecast, compromises are made in the numerics (e.g., coarse grid spacing in the horizontal; reduced number of vertical layers), dynamics, or physics (e.g., simplified parameterizations of various feedbacks).

One disadvantage of coarse grid spacings is that some dynamical and physical phenomena (e.g., thunderstorms, thermals, clouds, etc.) become **subgrid scale**, meaning they are not directly resolvable. As an example, a GCM with horizontal grid spacing of 1° of latitude would be able to resolve phenomena as small as 7° of latitude (≈ 777 km); hence, it could not "see" individual thunderstorms or clusters of thun-

Human- & Climate-Change Timescales

"We tend to predict impacts on people by assuming a static human society. But humans are adapting quickly, and time scales of human change are faster than the climate change. There's no unique perfect temperature at which human societies operate. Run them a little warmer, and they'll be fine; run them a little cooler, they'll be fine. People are like roaches — hard to kill. And our technology is changing rapidly, and we're getting richer, internationally and individually, at an absurd rate. We're rapidly getting less and less sensitive to climatic variations."

– David Keith

(from an interview published in *Discover*, Sep 2005.)

Table 21-5. Both OPMs (short-range prognostic models for daily operational forecasts) and GCMs (global-climate models) are NWP (numerical weather prediction) models.

Characteristic	OPM	GCM
desired outcome	deterministic forecasts (of the actual weather)	forecasts that are statistically skillful (e.g., forecast the right number of cyclones, even though their locations and times are wrong.
forecast horizon	days	decades
horizontal grid spacing	10 m to 10 km	10 km to 100 km
boundary conditions	less important	more important
initial conditions	very important	less important
approximations	partially parameterized	extensively parameterized

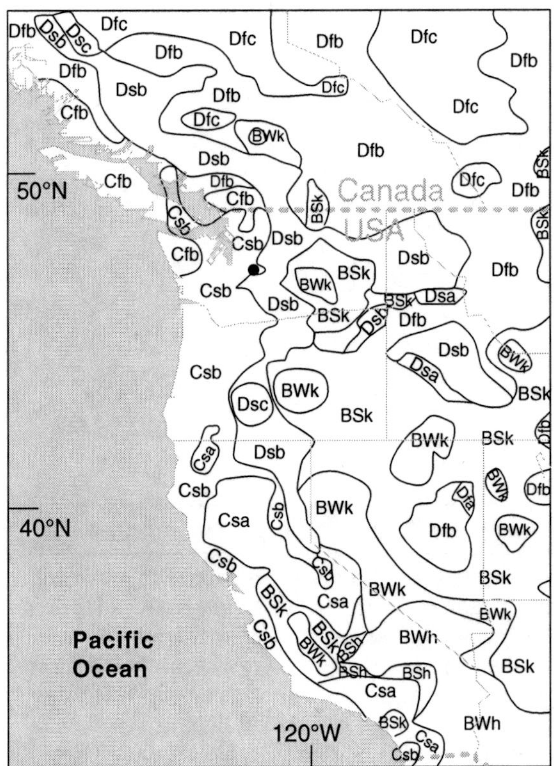

Figure 21.25
Köppen climate map for western North America. See Tables 21-6 & 21-8 for explanations of the climate code letters. [Color maps of Köppen climate classification for the whole world are available on the internet.]

derstorms. Although unresolved, subgrid-scale thunderstorms still affect the global climate forecast. Therefore their effects must be approximated via **parameterizations.**

Even for resolvable phenomena such as midlatitude cyclones, GCMs are designed to forecast the correct number of cyclones, although their locations and dates are usually incorrect (Table 21-5). The motivation is that the correct number of cyclones will transport the correct amount of heat, moisture and momentum when averaged over long time periods, thus increasing the likelihood that the climate simulation is reasonable.

Other phenomena, such as solar input or deep-soil temperatures, are imposed as boundary conditions (BCs). By very careful prescription of these BCs, the GCM can be designed so that its numerical climate does not drift too far from the actual climate. This is often validated by running the GCM for past years and comparing the result to the observed climate, giving confidence that the same accuracy is possible for future years.

Initial conditions (ICs) are less important for GCMs, because the ability to forecast the exact location of individual cyclones decreases to nil within a week or so (see the Forces and Winds Chapter).

GCMs are often used to address what-if questions, such as "what if the concentration of CO_2 were to double", or "what if deforestation were to reduce vegetation coverage." But comparing GCM runs with and without CO_2 doubling or deforestation, we can estimate the effects of such changes on future climate. But such predictions carry a lot of uncertainty because of the many parameterizations and approximations used in GCMs.

PRESENT CLIMATE

So far, we focused on processes that create and modify global climate. But before we leave this chapter, let's briefly summarize current climate.

Definition

Climate describes the slowly varying characteristics of the atmosphere, as part of the atmosphere/hydrosphere/cryosphere/biosphere/land-surface system. Climate is found by averaging or filtering the hourly or daily weather data over 30 years, with the last year ending in zero (e.g., 1971-2000, or 1981-2010). Mean values of weather, temporal and spatial variability, and extremes are the statistics that are recorded as components of the climate signal.

Table 21-6. Köppen climate codes and defining criteria. *(Peel et al, 2007). Also see Table 21-8 near end of chapter.*

1st	2nd	3rd	Description	Criteria*
A			Tropical	$T_{cold} \geq 18$
	f		Rainforest	$P_{dry} \geq 60$
	m		Monsoon	[not Af] and $[P_{dry} \geq (100 - MAP/25)]$
	w		Savannah	[not Af] and $[P_{dry} < (100 - MAP/25)]$
B			Arid	$MAP < [10 \cdot P_{threshold}]$
	W		Desert	$MAP < [5 \cdot P_{threshold}]$
	S		Steppe	$MAP \geq [5 \cdot P_{threshold}]$
		h	Hot	$MAT \geq 18$
		k	Cold	$MAT < 18$
C			Temperate	$[T_{hot} \geq 10]$ and $[0 < T_{cold} < 18]$
	s		Dry summer	$[P_{sdry} < 40]$ and $[P_{sdry} < (P_{wwet}/3)]$
	w		Dry winter	$P_{wdry} < [P_{swet}/10]$
	f		Without dry season	[not Cs] and [not Cw]
		a	Hot summer	$T_{hot} > 22$
		b	Warm summer	[not a] and $[N_{monT10} \geq 4]$
		c	Cold summer	[not (a or b)] and $[1 \leq N_{monT10} < 4]$
D			Cold	$[T_{cold} \leq 0]$ and $[T_{hot} > 10]$
	s		Dry summer	$[P_{sdry} < 40]$ and $[P_{sdry} < (P_{wwet}/3)]$
	w		Dry winter	$P_{wdry} < [P_{swet}/10]$
	f		Without dry season	[not Ds] and [not Dw]
		a	Hot summer	$T_{hot} \geq 22$
		b	Warm summer	[not a] and $[T_{mon10} \geq 4]$
		c	Cold summer	[not (a or b or d)]
		d	Very cold winter	[not (a or b)] and $T_{cold} < -38$
E			Polar	$T_{hot} < 10$
	T		Tundra	$T_{hot} > 0$
	F		Frost	$T_{hot} \leq 0$

Köppen Climate Classification

In 1884, Wladimir Köppen used temperature and precipitation to define different types of climate. He drew maps, classifying every point on land according to these climates. The climate at each location controls the dominant type of plants that naturally grow there. The Köppen climate map was modified by Rudolf Geiger in the early 1900s, and was significantly updated in 2007 by Peel, Finlayson and McMahon (*Hydrol. Earth Syst. Sci.*, **11**, 1633-1644) using modern weather data.

A two-or-three-letter code is used to identify each climate type (also see Table 21-8 in the Homework Exercises section). The criteria used to classify the climate types are in Table 21-6. Fig. 21.25 shows a small portion of the global climate map.

Table 21-6 *(continuation)*. *Footnotes

MAP = mean annual precipitation (mm·yr^{-1})
MAT = mean annual temperature (°C)
N_{monT10} = number of months where the temperature is above 10 (°C)
P_{dry} = precipitation of the driest month (mm·month^{-1})
P_{sdry} = precipitation of the driest month in summer (mm·month^{-1})
P_{swet} = precipitation of the wettest month in summer (mm·month^{-1})
P_{wdry} = precipitation of the driest month in winter (mm·month^{-1})
P_{wwet} = precipitation of the wettest month in winter (mm/month)
T_{hot} = average temperature of the hottest month (°C)
T_{cold} = average temperature of the coldest month (°C)

$P_{threshold}$ (mm) = is given by the following rules:
If (70% of MAP occurs in winter) then
$P_{threshold} = (2 \cdot MAT)$
Elseif (70% of MAP occurs in summer) then
$P_{threshold} = (2 \cdot MAT) + 28$
Else $P_{threshold} = (2 \cdot MAT) + 14$

Summer (winter) is defined as the warmer (cooler) six month period of [1 Oct. - 31 Mar.] and [1 Apr. - 30 Sep.].

Table 21-7. Climate oscillations. SST = sea-surface temperature, SLP = sea-level pressure.

Name	Period (yr)	Key Place	Key Variable
El Niño (EN)	2 to 7	Tropical E. Pacific	SST
Southern Oscillation (SO)	2 to 7	Tropical Pacific	SLP
Pacific Decadal Oscillation (PDO)	20 to 30	Extratrop. NE Pacific	SST
North Atlantic Oscillation (NAO)	variable	Extratrop. N. Atlantic	SLP
Arctic Oscillation (AO)	variable	N. half of N. hem.	Pressure
Madden-Julian Oscillation (MJO)	30 to 60 days	Tropical Indian & Pacific	Convective Precipitation

INFO • What is an Oscillation?

Oscillations are regular, repeated variations of a measurable quantity such as temperature (T) in space x and/or time t. They can be described by sine waves: $T = A \cdot \sin[C \cdot (t - t_o) / \tau]$
where A is a constant **amplitude** (signal strength), τ is a constant **period** (duration one oscillation), t_o is a constant **phase shift** (the time delay after $t = 0$ for the signal to first have a positive value), and C is 2π radians or 360° (depending on your calculator).

True oscillations are **deterministic** — the signal T can be predicted exactly for any time in the future. The climate "oscillations" discussed here are not deterministic, cannot be predicted far in the future, and do not have constant amplitude, period, or phase.

NATURAL OSCILLATIONS

The 30-year-average climate is called the **normal climate**. Any shorter-term (e.g., monthly) average weather that differs or varies from the climate norm is called an **anomaly**. Natural climate "oscillations" are recurring, multi-year anomalies with irregular amplitude and period, and hence are not true oscillations (see the INFO Box).

Table 21-7 lists some of these oscillations and their major characteristics. Although each oscillation is associated with, or defined by, a location or place on Earth (Fig. 21.26), most of these oscillations affect climate over the whole world. At the defining place(s), a combination of key weather variables is used to define an "**index**", which is the varying climate signal. Fig. 21.27 shows some of these oscillation indices.

El Niño - Southern Oscillation (ENSO)

(**El Niño**; **La Niña**) occurs when the tropical sea-surface temperature (**SST**) in the central and eastern Pacific ocean is (warmer; cooler) than the multi-decadal climate average.

The El Niño index plotted in Fig. 21.27a is the Pacific SST anomaly in the tropics (averaged within ±20° latitude of the equator) minus the Pacific SST anomaly outside the tropics (averaged over more poleward latitudes). Alternative El Niño indices are also useful.

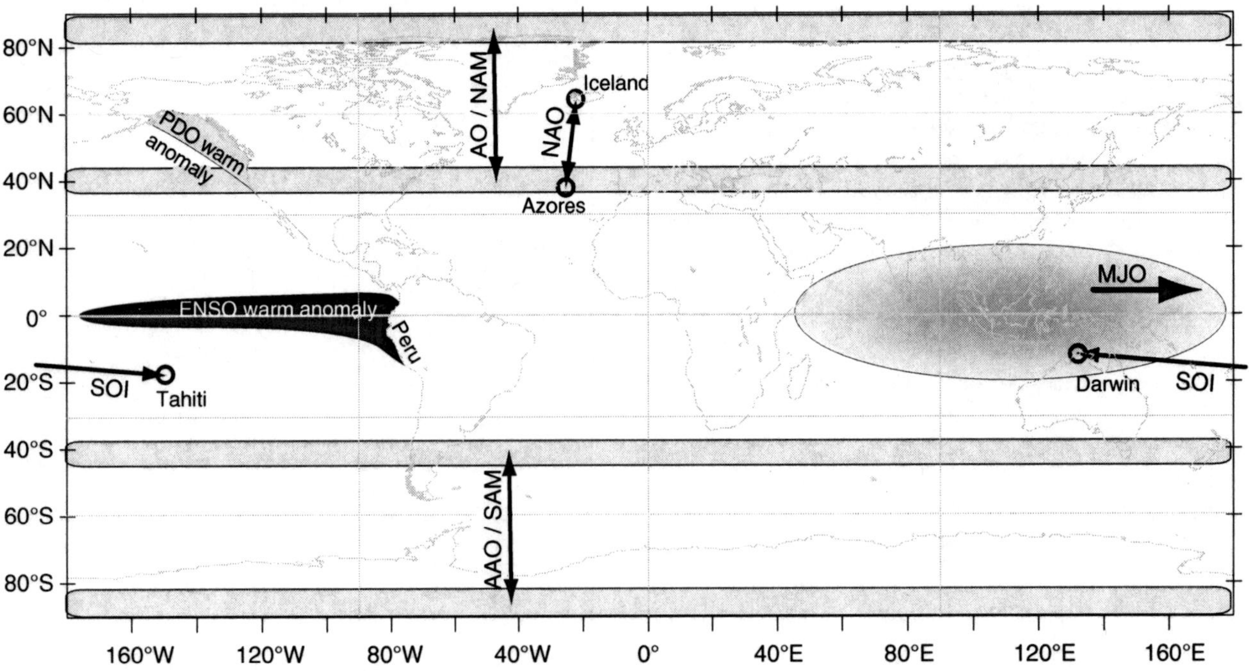

Figure 21.26 *Places and key variables that define some climate oscillations. Acronyms are defined in the text.*

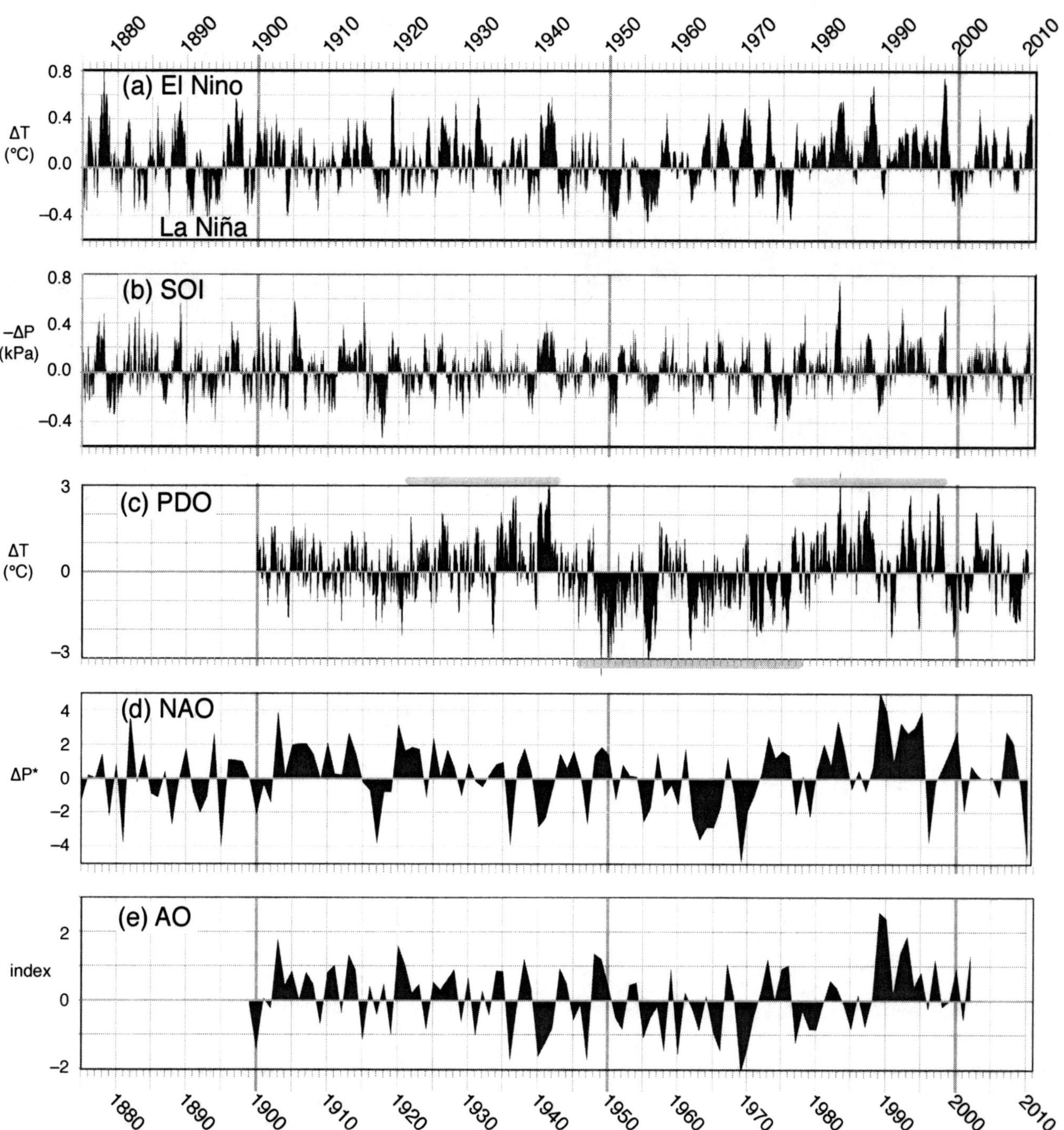

Figure 21.27
Natural oscillations of some climate indices.

(a) **El Niño** *index, based on average sea-surface temperature (SST) anomalies* ΔT *[data courtesy of Todd Mitchell of the Univ. of Washington Joint Inst. for Study of the Atmosphere and Ocean (US JISAO)].*

(b) *Negative of the* **Southern Oscillation Index** *(***SOI***), based on sea-level pressure (SLP) difference:* $-\Delta P = P_{Darwin} - P_{Tahiti}$ *[data courtesy of the Climate Analysis Section of the National Center for Atmospheric Research's (NCAR) Climate and Global Dynamics Division (NCAR/CGD/CAS)].*

(c) **Pacific Decadal Oscillation** *(***PDO***) index, based on the leading principal component for monthly Pacific SST variability north of 20° latitude [data courtesy of Nate Mantua and Steven Hare of UW JISAO]. Grey bands show warm and cool phases.*

(d) **North Atlantic Oscillation** *(***NAO***) index, based on the difference of normalized SLP (defined on p 824):* $\Delta P^* = P^*_{Azores} - P^*_{Iceland}$ *for winter (Dec - Mar) [data courtesy of Jim Hurrell, NCAR/CGD/CAS].*

(e) **Arctic Oscillation** *(***AO***), based on the leading principal component of winter SLP in the N. Hemisphere for winter (Jan-Mar) [courtesy of Todd Mitchell of UW JISAO].*

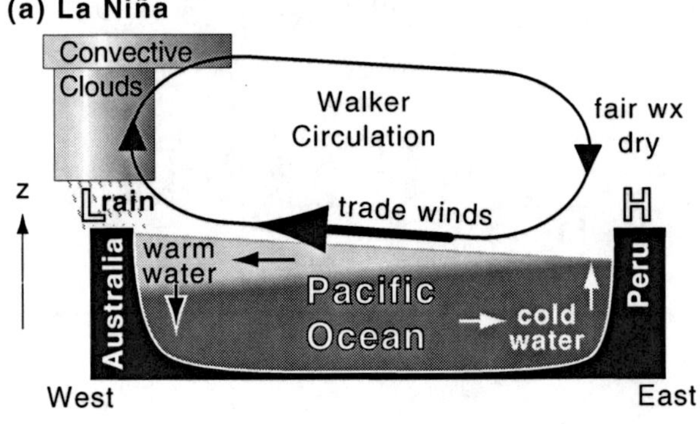

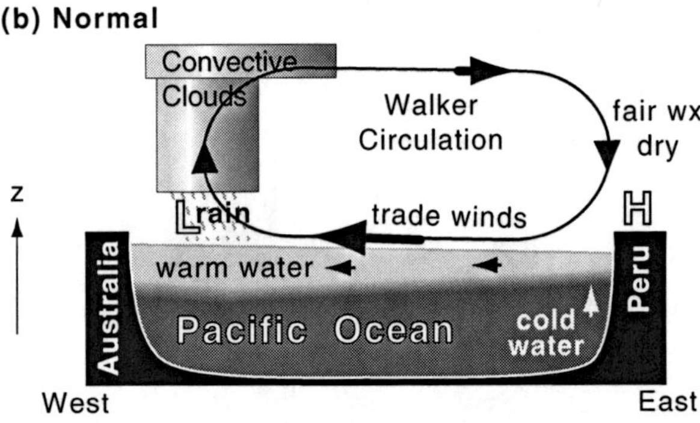

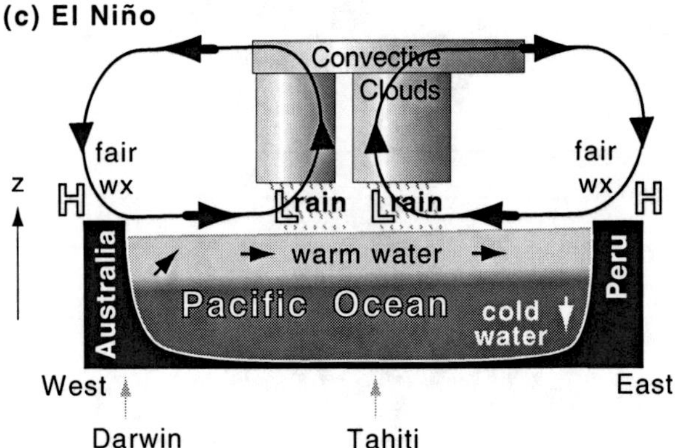

Figure 21.28
Processes that drive ENSO. (a) La Niña conditions, where cold upwelling nutrient-rich water off the coast of Peru enables increases in fish populations. Also, cool equatorial water upwells near the central Pacific. Strong trade winds. (b) Normal conditions, with modest trade winds. (c) El Niño conditions, associated with a warm-water anomaly in the eastern Pacific, and a decline in fish stocks there. H and L indicate centers of high and low atmospheric pressure, respectively.

The **Southern Oscillation** refers to a change in sign of the east-west atmospheric pressure gradient across the tropical Pacific and Indian Oceans. One indicator of this pressure gradient is the sea-level pressure difference between Tahiti (an island in the central equatorial Pacific, Fig. 21.26) and Darwin, Australia (western Pacific). This indicator ($\Delta P = P_{Tahiti} - P_{Darwin}$) is called the **Southern Oscillation Index (SOI)**.

The negative of the SOI, plotted in Fig. 21.27b, agrees with the El Niño index (Fig. 21.27a). The reason for this agreement is that El Niño (EN) and the Southern Oscillation (SO) are the ocean and atmospheric signals of the same coupled phenomenon, abbreviated as **ENSO**.

During normal and La Niña conditions, the east-to-west decrease in air pressure drives the easterly trade winds, which drag much of the warm ocean-surface waters to the west side of the Pacific basin (Fig. 21.28a). The trade winds are part of the **Walker Circulation**, which includes cloudy rainy updrafts over the warm SST anomaly in the western tropical Pacific, a return flow aloft, and fair-weather (fair wx) dry downdrafts. Heavy rain can fall over Australia. Other changes are felt around the world, including cool winters over northern North America, with drier warmer winters over the southern USA.

During El Niño, this pressure gradient breaks down, causing the trade winds to fail, and allowing anomalously warm western Pacific waters to return to the eastern Pacific (Fig. 21.28c). Thunderstorms and rain tend to be strongest over the region of warmest SST. As the SST anomaly shifts eastward, droughts (allowing wild fires) occur in Australia, and heavy rains (causing floods and landslides) occur in Peru. Other changes are felt around the world, including warm winters over northern N. America and cool, wet winters over southern N. America.

Pacific Decadal Oscillation (PDO)

The PDO is a 20 to 30 year variation in the extratropical SST near the American coast in the NE Pacific. It has a warm or positive phase (i.e., warm in the key place sketched in Fig. 21.26) and a cool or negative phase. Phase durations are highlighted by the thick grey lines in Fig. 21.27c. The PDO signal is found using a statistical method called **principal component analysis (PCA**, also called **empirical orthogonal function analysis**, **EOF**; see the INFO box on the next three pages.)

The PDO affects weather in northwestern North America as far east as the Great Lakes, and affects the salmon production between Alaska and the Pacific Northwest USA. There is some debate among scientists whether the PDO is a statistical artifact associated with the ENSO signal.

INFO • PCA and EOF

Principal Component Analysis (**PCA**) and **Empirical Orthogonal Function** (**EOF**) analysis are two names for the same statistical analysis procedure. It is a form of data reduction designed to tease out dominant patterns that naturally occur in the data. Contrast that with Fourier analysis, which projects the data onto sine waves regardless of whether the data naturally exhibits wavy patterns or not.

You can use a spreadsheet for most of the calculations, as is illustrated here with a contrived example.

1) The Meteorological Data to be Analyzed

Suppose you have air temperature T measurements from different locations over the North Pacific Ocean. Although PCA works for any number of locations, we will use just 4 locations for simplicity:

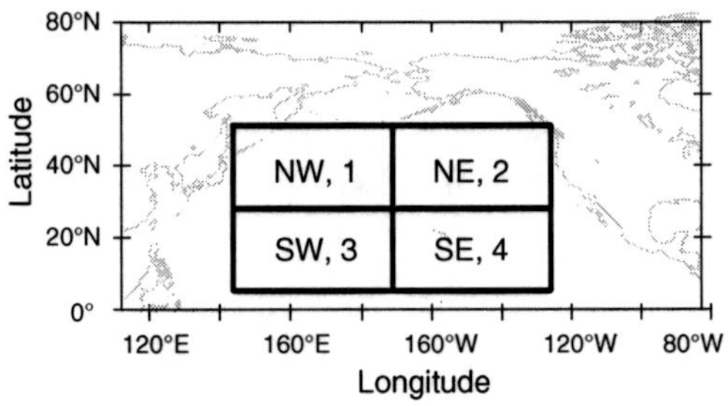

Fig. 21.a. *Locations for weather data.*

These 4 locations might have place names such as Northwest quadrant, Northeast quadrant, etc., but we will index them by number: j = 1 to 4. You can number your locations in any order.

For each location, suppose you have annual average temperature for N = 10 years, so your raw <u>input data</u> on a spreadsheet might look this:

Location Index:	j = 1	2	3	4
Location:	NW	NE	SW	SE
Year	Air Temperature (°C)			
i = 1	0	3	8	5
2	4	7	6	8
3	0	6	6	7
4	4	9	5	8
5	0	8	5	6
6	0	9	6	6
7	4	2	10	4
8	0	4	9	4
9	4	6	8	5
10	0	7	5	5
Average:	1.6	6.1	6.8	5.8

(continues)

INFO • PCA and EOF *(continuation)*

These meteorological data are plotted below:

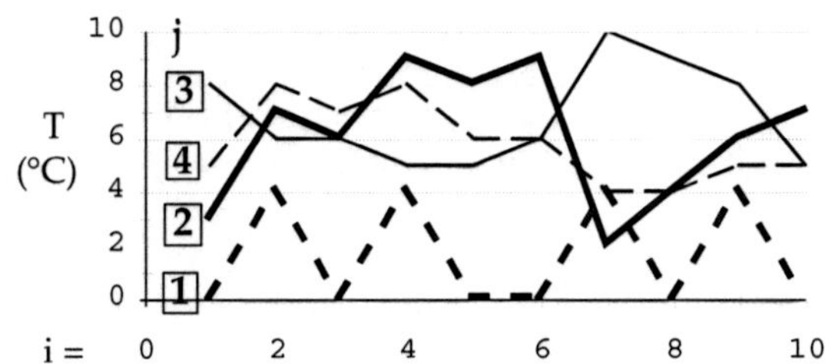

Fig. 21.b. *Graphs of raw data at each of 4 locations, where numbers in boxes indicate the j index for each line.*

Looking at this data by eye, we see that location 2 has a strong signal. Location 4 varies somewhat similarly to 2 (i.e., has positive correlation); location 3 varies somewhat oppositely (negative correlation). Location 1 looks uncorrelated with the others.

2) Preliminary Statistical Computations

Next, use your spreadsheet to find the average temperature at each location (shown at the bottom of the previous table). Then subtract the average from the individual temperatures at the same locations, to make a table of temperature **anomalies** T':

$$T'_i = T_i - T_{avg} \tag{21.48}$$

Location	j= 1	2	3	4
Year	Air Temperature Anomalies T' (°C)			
i= 1	-1.6	-3.1	1.2	-0.8
2	2.4	0.9	-0.8	2.2
3	-1.6	-0.1	-0.8	1.2
4	2.4	2.9	-1.8	2.2
5	-1.6	1.9	-1.8	0.2
6	-1.6	2.9	-0.8	0.2
7	2.4	-4.1	3.2	-1.8
8	-1.6	-2.1	2.2	-1.8
9	2.4	-0.1	1.2	-0.8
10	-1.6	0.9	-1.8	-0.8

[Check: The average of each T′ column should = 0.]

The covariance between variables A and B is:

$$\text{covar}(A,B) = \frac{1}{N}\sum_{i=1}^{N} A'_i \cdot B'_i \tag{21.49}$$

where covar(A,A) is the variance. Use a spreadsheet to compute a table of covariances, where T1 are the temperatures at location j=1, T2 are from location 2, etc.:

<u>covar(T1,T1)</u>	covar(T1,T2)	covar(T1,T3)	covar(T1,T4)
covar(T1,T2)	**<u>covar(T2,T2)</u>**	covar(T2,T3)	covar(T2,T4)
covar(T1,T3)	covar(T2,T3)	**<u>covar(T3,T3)</u>**	covar(T3,T4)
covar(T1,T4)	covar(T2,T4)	covar(T3,T4)	**<u>covar(T4,T4)</u>**

(continues)

INFO • PCA and EOF *(continuation)*

The numbers in that table are symmetric around the main diagonal, so you need to compute the covariances for only the black cells, and then copy the appropriate covariances into the grey cells.

This table is called a **covariance matrix**. When I did it in a spreadsheet, I got the values below:

3.84	-0.16	0.72	0.72
-0.16	**5.29**	-3.48	2.22
0.72	-3.48	**2.96**	-1.74
0.72	2.22	-1.74	**1.96**

The underlined values along the main diagonal are the variances. The sum of these diagonal elements (known as the **trace of the matrix**) is a measure of the total variance of our original data (= 14.05 °C^2).

3) More Statistical Computations

Next, find the **eigenvalues** and **eigenvectors** of the covariance matrix. You cannot do that in most spreadsheets, but other programs can do it for you.

I used a website: http://www.arndt-bruenner.de/mathe/scripts/engl_eigenwert2.htm where I copied my covariance matrix into the web page. It gave back the following info, which I pasted into my spreadsheet. Under each eigenvalue is the corresponding eigenvector.

k=	1	2	3	4
	Eigenvalues (largest first)			
	8.946	4.118	0.740	0.247
	Eigenvectors (columns under evalues)			
j=1	-0.049	0.959	-0.144	-0.238
2	0.749	0.031	-0.579	0.323
3	-0.548	0.126	-0.267	0.782
4	0.369	0.251	0.757	0.477

The sum of all eigenvalues = 14.05 °C^2 = the total variance of the original data. Those eigenvectors with the largest eigenvalues explain the most variance. Eigenvectors are dimensionless.

4) Find the Principal Components

Next, compute the **principal components** (PCs) — one for each eigenvector (*evect*). The number of eigenvectors (K = 4 in our case) equals the number of locations (J = 4 in our case).

Each PC will be a column of N numbers (i = 1 to 10 in our case) in our spreadsheet. Compute each of these N numbers in your spreadsheet using:

$$PC(k,i) = \sum_{j=1}^{J} [evect(k,j) \cdot T(j,i)] \qquad (21.50)$$

For each data table (T data; eigenvectors; PCs) used in eq. (21.50), the first index indicates the column and the second indicates the row.

(continues)

INFO • PCA and EOF *(continuation)*

Namely:
• k indicates which PC it is (k = 1 for the first PC, k = 2 for the second PC, up through k = 4 in our case) and also indicates the eigenvector (*evect*).
• i is the observation index (i = 1 to 10 yrs in our case), and indicates the corresponding element in any PC.
• j is the data variable index (1 for NW quadrant, etc.), and indicates an element within each eigenvector.

For example, focus on the first PC (i.e., k = 1); i.e., the first column in the table below. The first row (i = 1) is computed as:

$PC(1,1) = -0.049{\cdot}0 + 0.749{\cdot}3 - 0.548{\cdot}8 + 0.369{\cdot}5 = -0.29$

For the second row (i = 2), it is:

$PC(1,2) = -0.049{\cdot}4 + 0.749{\cdot}7 - 0.548{\cdot}6 + 0.369{\cdot}8 = 4.71$

Each PC element has the same units as the original data (°C in our case). The resulting PCs are:

k =	1	2	3	4
	PC1	PC2	PC3	PC4
i=1	-0.29	2.36	-0.08	9.61
2	4.71	6.82	-0.17	9.81
3	3.79	2.70	0.23	9.97
4	6.76	6.75	-1.06	9.68
5	5.46	2.38	-1.42	9.35
6	5.66	2.54	-2.26	10.46
7	-2.70	6.16	-1.37	9.42
8	-0.46	2.26	-1.68	10.24
9	1.76	6.29	-2.39	9.63
10	4.34	2.10	-1.60	8.55
evalue:	8.946	4.118	0.740	0.247

Each PC applies to the whole data set, not just to one location. These PCs, plotted below, are like alternative T data:

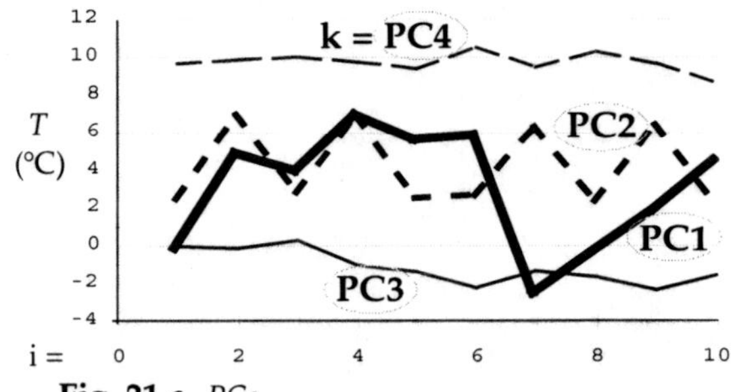

Fig. 21.c. *PCs.*

• PC1 captures the dominant pattern that is in the raw data for locations 2, 3, and 4. This 1st PC explains 8.946/14.05 = 63.7% of the original total variance.
• PC2 captures the oscillation that is in the raw data for location 1. Since location 1 was uncorrelated with the other locations, it couldn't be captured by PC1, and hence needed its own PC. The 2nd PC explains 4.118/14.05 = 29.3% of the variance.
• PC3 captures a nearly linear trend in the data.
• PC4 captures a nearly constant offset.

(continues)

INFO • PCA and EOF *(continuation)*

5) Synthesis

Now that we've finished computing the PCs, we can see how well each PC explains the original data. Let T_k be an approximation to the original data, reconstructed using only the kth single principal component. Such reconstruction is known as **synthesis**.

$$T_k(j,i) = evect(k,j) \cdot PC(k,i) \qquad (21.51)$$

using the same (column, row) notation as before.

For k = 1 (the dominant PC), the reconstructed T_1 data is:

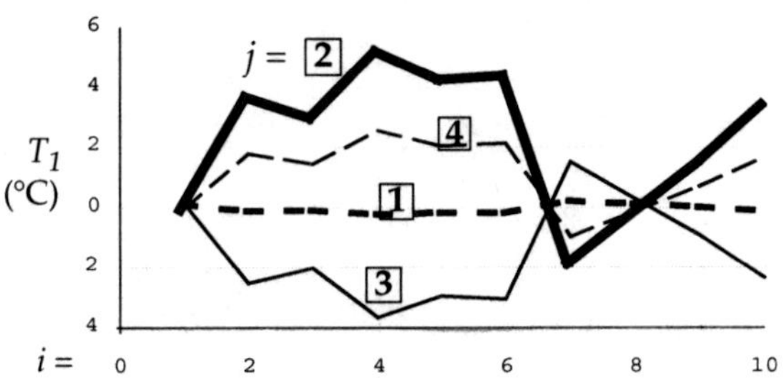

Fig. 21.d. *Reconstruction of temperature for each geographic location j = 1 to 4 (boxed numbers), but using only the 1st PC.*

It is amazing how well just one PC can capture the essence (including positive and negative correlations) of the signal patterns in the raw data.

The spatial pattern associated with PC1 can be represented on a map by the elements of the first eigenvector (arbitrarily scaled by 100 here & rounded):

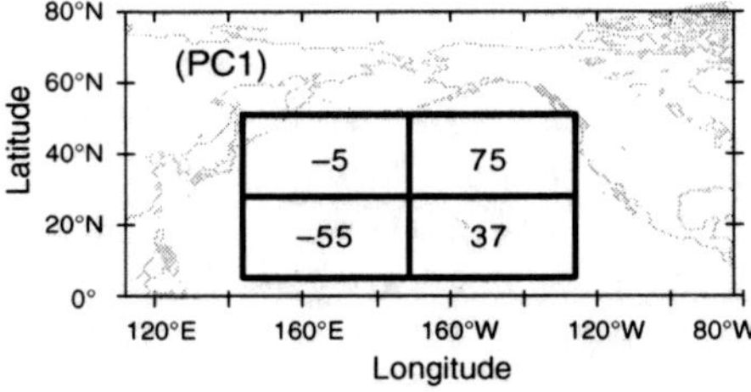

Fig. 21.e. *Geographic pattern of influence of PC1.*

6) Interpretation:

A dominant pattern in the NE quadrant extends in diminished form to the SE quadrant, with the SW quadrant varying oppositely. But this pattern is not experienced very much in the NW quadrant. Similar maps can be created for the other eigenvectors. When data from more than 4 stations are used, you can contour the eigenvector data when plotted on a map.

Eq. (21.51) can also be used to partially reconstruct the data based on only PC2 (i.e., k = 2). As shown in Fig. 21.f, you won't be surprised to find that it almost completely captures the signal in the NW quadrant (j = 1), but doesn't explain much in the other regions.

(continues)

INFO • PCA and EOF *(continuation)*

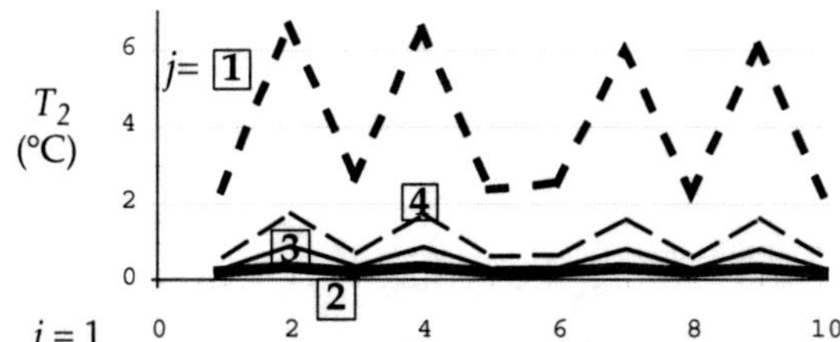

Fig. 21.f. *Similar to Fig. 21.d, but using only PC2.*

Plotting the elements of the second eigenvector scaled by 100 on a map at their geographic locations:

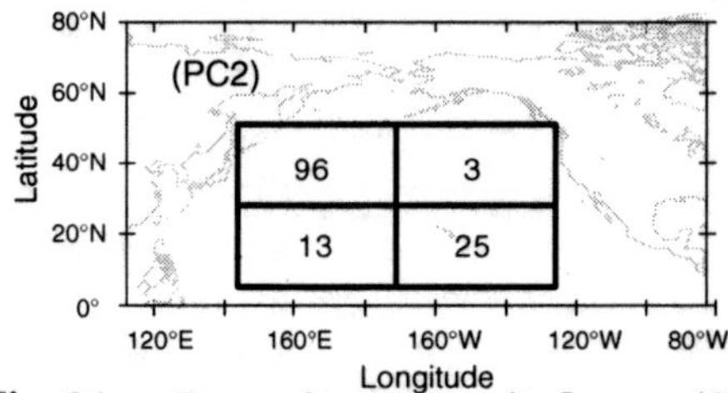

Fig. 21.g. *Geographic pattern of influence of PC2.*

7) Data Reduction

Suppose that you compute graphs similar to Figs. 21.d & 21.f for the other PCs, and then add all location 1 curves from each PC. Similarly, add all location 2 curves together, etc. The result would have perfectly synthesized the original data as was plotted in Fig. 21.a. Although the last 2 PCs don't explain much of the variance (fluctuations of the raw data), they are important to get the mean temperature values.

But what happens if you synthesize the temperature data with only the first 2 PCs (see plot below). These first 2 PCs explain 93% of the total variance. Indeed, the graph below captures most of the variation in the original signals that were in Fig. 21.a, including positive, negative, and near-zero correlations. This allows **data reduction**: We can use only the dominant 2 PCs to explain most of the signals, instead of using all 4 columns of raw data.

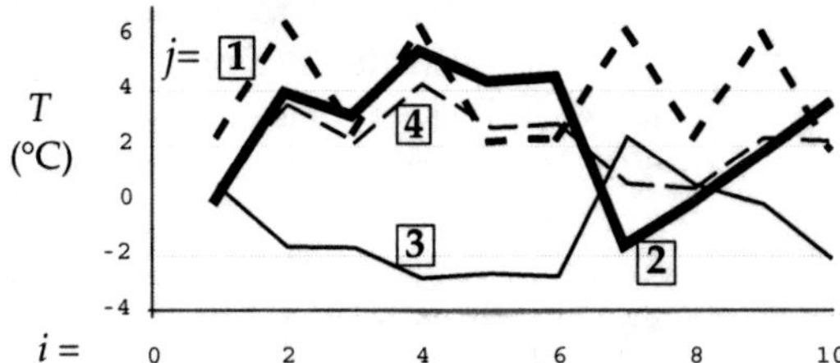

Fig. 21.h. *Reconstruction of temperature using only the first 2 PCs.*

8) Conclusions

For our small data set, the data reduction might not seem like much, but for larger data sets with hundreds of geographic points, this can be an important way to find dominant patterns in noisy data, and to describe these patterns efficiently. For this reason, PCA/EOF is valued by climate researchers.

North Atlantic Oscillation (NAO)

In the General Circulation chapter (Fig. 11.31) we identified an Iceland low and Azores (or Bermuda) high on the map of average sea-level pressure (SLP). The NAO is a variation in the strength of the pressure difference between these two pressure centers (Fig. 21.26).

Greater differences (called the positive phase of the NAO) correspond to stronger north-south pressure gradients that drive stronger west winds in the North Atlantic — causing mid-latitude cyclones and associated precipitation to be steered toward N. Europe. Weaker differences (negative phase) cause weaker westerlies that tend to drive the extratropical cyclones toward southern Europe.

An NAO index is defined as $\Delta P^* = P^*_{Azores} - P^*_{Iceland}$, where $P^* = (P - P_{clim})/\sigma_P$ is a normalized pressure that measures how many standard-deviations (σ_P) the pressure is from the mean (P_{clim}). The winter average (Dec - Mar) is often most relevant to winter storms in Europe, as plotted in Fig. 21.27d.

Arctic Oscillation (AO)

The AO is also known as the **Northern Annular Mode (NAM)**. It is based on the leading PCA mode for sea-level pressure in the N. Hemisphere, and varies over periods of weeks to decades. Fig. 21.27e shows the winter (Jan-Mar) averages.

The AO acts over the whole tropospheric depth, and is associated with the **polar vortex** (a belt of mid-troposphere through lower-stratosphere westerly winds circling the pole over the polar front). It compares the average pressure anomaly over the whole Arctic with pressure anomaly averaged around the 37 to 45°N latitude band. Thus, the AO pattern is zonal, like a bulls-eye centered on the N. Pole (Fig. 21.26). You can think of the NAO as a regional portion of the AO.

The positive, high-index, warm phase corresponds to greater pressure difference between mid-latitudes and the North Pole. This is correlated with faster westerly winds, warmer wetter conditions in N. Europe, warmer drier winters in S. Europe, and warmer winters in N. America. The negative, low-index, cool phase has weaker pressure gradient and winds, and opposite precipitation and temperature patterns.

Madden-Julian Oscillation (MJO)

The MJO is 45-day variation (±15 days) between enhanced and reduced convective (CB) precipitation in the tropics. It is the result of a region of enhanced convection (sketched as the grey shaded oval in Fig. 21.26) that propagates toward the east at roughly 5 to 10 m s^{-1}.

The region of enhanced tropical convection consists of thunderstorms, rain, high anvil cloud tops (cold, as seen by satellites), low SLP anomaly, convergent U winds in the boundary layer, and divergent U winds aloft. MJO can enhance tropical cyclone and monsoonal-rain activity. Outside of this region is reduced convection, clearer skies, subsiding air, less or no precipitation, divergent U winds in the boundary layer, convergent U winds aloft, and reduced tropical-cyclone activity.

The convective regions are strongest where they form over the Indian Ocean and as they move eastward over the western and central Pacific Ocean. By the time they reach the eastern Pacific and the Atlantic Oceans, they are usually diminished, although they can still have observable effects.

For example, when the thunderstorm region approaches Hawaii, the moist storm outflow can merge with a southern branch of the jet stream to feed abundant moisture toward the Pacific Northwest USA and SW Canada. This **atmospheric river** of moist air is known as the **Pineapple Express**, and can cause heavy rains, flooding, and landslides due to orographic uplift of the moist air over the coastal mountains (see the Extratropical Cyclone chapter).

A useful analysis method to detect the MJO and other propagating phenomena is to plot key variables on a **Hovmöller diagram** (see INFO box). Eastward propagating weather features would tilt down to the right in such a diagram. Westward moving ones would tilt down to the left. The slope of the tilt indicates the zonal propagation speed.

Other Oscillations

Other climate oscillations have been discovered:

- **Antarctic Oscillation (AAO**, also known as the **Southern Annular Mode (SAM)** in S. Hem. SLP (sketched in Fig. 21.26). 2 - 7 yr variation.
- **Arctic Dipole Anomaly** (ADA), 2 - 7 yr variation with high SLP over northern Canada and low SLP over Eurasia.
- **Atlantic Multi-decadal Oscillation (AMO)**, 30 - 40 yr variations in SST over the N. Atlantic
- **Indian Ocean Dipole (IOD)**. 7.5 yr oscillation between warm east and warm west SST anomaly in the Indian Ocean.
- **Interdecadal Pacific Oscillation (IDPO** or **IPO)** for SST in the N. and S. Hemisphere extratropics. 15-30 yr periods.
- **Pacific North American Pattern (PNA)** in SLP and in 50 kPa geopotential heights.
- **Quasi-biennial Oscillation (QBO)** in stratospheric equatorial zonal winds
- **Quasi-decadal Oscillation (QDO)** in tropical SST, with 8-12 yr periods.

Other climate variations may also exist.

REVIEW

Climate is weather averaged over 30 yr. Earth's climate is controlled by both external and internal processes. External processes include the balance between incoming solar radiation and outgoing infrared (IR) radiation. Internal processes include changes to the composition or structure of the atmosphere that alter how energy is distributed within the Earth-ocean-atmosphere-cryosphere-biosphere system.

Externally, changes in Earth's orbit around the sun as predicted by Milankovitch can alter the amount of sunlight reaching Earth, and changes in the tilt of Earth's axis can alter the severity of the seasons. Solar output can also vary. Changes in Earth's average albedo (via snow cover, cloud cover, or coverage of vegetation) can modulate the amount of radiation lost from the Earth system.

Internally, normal chemicals (water vapor, carbon dioxide, etc.) in the atmosphere absorb IR radiation, causing the air to become warmer and re-radiate a large amount of IR radiation back toward the Earth's surface. This greenhouse effect does not have perfect efficiency, because some wavelength bands of IR radiation can escape directly to space through the so-called "atmospheric window." Anthropogenic and natural changes to these greenhouse gases can alter the greenhouse efficiency and shift the climate equilibrium — an effect called "climate change."

Natural internal processes such as volcanic eruptions can also change the climate via emission of sulfates into the stratosphere. Movement of tectonic plates can shift the locations of continents and oceans relative to the equator, altering global-scale and continental-scale (monsoon) circulations and climate.

Some processes in the atmosphere cause negative feedbacks that tend to stabilize the climate, while other positive-feedback processes tend to amplify climate variations. Many of these feedbacks apply to external, internal, natural, and anthropogenic forcings. Overall, our climate is remarkable steady.

Short-term (few years to few decades) variations in Earth's climate are also observed. El Niño is one example out of many such "oscillations" in the climate signal.

Researchers have classified the current climate (Köppen). They also utilize tools including global climate models (GCMs), principal component analysis (PCA), Hovmöller diagrams, and conceptual models (daisyworld) to study climate variability.

INFO • Hovmöller Diagram

A Hovmöller diagram is a plot of a meteorological variable as a function of date and longitude. Usually date is along the ordinate — increasing downward. Longitude is along the abscissa. The meteorological variable is contoured with grey or color fill. The latitude or latitude band of these meteorological observations is fixed, and is specified in the figure legend. These diagrams are useful for detecting the zonal movement of weather features over time.

For example, suppose that on 1 January you flew an airplane around the world along the equator (0° latitude), and remotely measured the sea surface temperature (SST). After subtracting the average SST, you would be left with a temperature anomaly T' at each point around the equator. Suppose the first graph below shows what you measured.

Later, on 15 January, make the same flight, and plot your measurements (second graph below). Repeat the flights for many different dates.

Instead of plotting each circumnavigation on a separate graph, we can write all the T' numbers on the same diagram of time vs. longitude. Then analyze the result as you learned in the Map Analysis chapter, by drawing isotherms of T'. The result is a Hovmöller diagram (third diagram below). You can use any meteorological variable, not just T'.

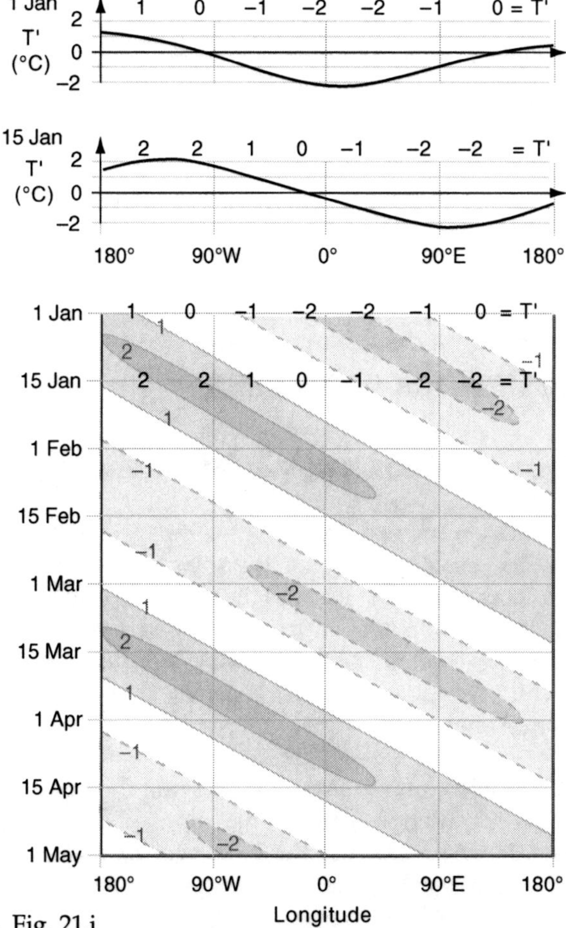

Fig. 21.i

PERSPECTIVES • Scientific Ethics

It is a worthy human trait to strive to be the best. But for scientists measuring their success by the papers they publish, the temptation to cheat can unfortunately drive people to be unethical.

Such unethical issues have been discussed by C. J. Sindermann (1982: *Winning the Games Scientists Play*. Plenum. 290 pp) and by the Sigma Xi scientific research society (1984: *Honor in Science*. Sigma Xi Pubs., 41 pp):

"• **Massaging** – performing extensive transformations... to make inconclusive data appear to be conclusive;

• **Extrapolating** – developing curves [or proposing theories] based on too few data points, or predicting future trends based on unsupported assumptions about the degree of variability measured;

• **Smoothing** [or **trimming**] – discarding data points too far removed from expected or mean values;

• **Slanting** [or **cooking**]– deliberately emphasizing and selecting certain trends in the data, ignoring or discarding others which do not fit the desired or preconceived pattern;

• **Fudging** [or **forging**]– creating data points to augment incomplete data sets or observations; and

• **Manufacturing** – creating entire data sets de novo, without benefit of experimentation or observation."

HOMEWORK EXERCISES

Broaden Knowledge & Comprehension

B1. Search the web for graphs comparing different satellite measurements of the solar irradiance (i.e., the solar constant). Discuss the relative magnitudes of the variation of solar irradiance with time vs. the errors associated with measuring solar irradiance. [Hint: Recall info about accuracy and precision from Appendix A.]

B2. Search the web for maps showing the ice cover on Earth, and how ice cover has changed during the past decades. Discuss these changes in relation to ice-albedo feedback.

B3. Use info from the web to discuss the difference between **Bond albedo** and **geometric albedo**. Which one of these albedos is most appropriate to use for studying the radiation budget of other planets?

B4. Search the web for animations or graphs showing/predicting the changes in eccentricity, obliquity, and precession of the Earth over time. Discuss the simplifications made in these diagrams/animations to communicate this info to the general public.

B5. Search the web for more accurate Earth orbital parameters, to improve and expand the data that were in Tables 21-1 and 21-1b. Save and print this table.

B6. Search the web for accurate graphs of ice ages. Discuss what ice-age details are missing or poorly resolved in the ice-age graph in Fig. 21.8.

B7. Search the web for graphs of solar activity or sunspot number, going as far back in time as you can find. Discuss the regular and irregular components of the sunspot cycle.

B8. Search the web for animations or a sequence of still-frames showing how the continents moved to create Pangea, and how they continued to move into their current configuration. Discuss how monsoonal circulations change and alter the continental climates as the continents move.

B9. Search the web for photos of one recent volcanic eruptions that show: (a) ash fallout close to the volcano, and (b) the plume of aerosols injected into the stratosphere, showing their track around the Earth. Discuss the relative amount of global impact due to the one volcanic eruption.

B10. Search the web for photos of beautiful sunsets that were enhanced by volcanic aerosols in the stratosphere. Read ahead in the Atmospheric Optics chapter to see why you can see red skies and blue moons after some volcanic eruptions.

B11. Search the web for info about other major volcanic eruptions that were not listed in Table 21-3. Discuss the impacts of these eruptions on species.

B12. Search the web for the most recent Assessment Report from the Intergovernmental Panel on Climate Change (IPCC), and identify new information and understanding that updates information in this Chapter.

B13. Search the web for additional climate feedback processes that were not listed in Table 21-4, and discuss their relative importance to the ones already listed.

B14. Search the web for data on actual amounts of global warming during the past two decades, and compare to the predictions presented in older IPCC reports. How accurate are climate predictions?

B15. Search the web for animations of Daisyworld evolution. Enjoy.

B16. Lovelock, the inventor of Daisyworld, wrote a more recent book called "The Revenge of Gaia." If you don't have time to read the book, search the web for a summary of its key points, and argue either for or against his thesis.

B17. Search the web (particularly IPCC Assessment Reports) that summarize the attributes of the various global climate models (GCMs) that were used for the most recent IPCC climate report. Summarize the commonalities and differences between these various GCMs.

B18. Search the web for color maps showing Köppen climate classification for the whole world. Focus on where you live, and discuss how the climate classification does or does not agree with the climate you observed in your area.

B19. Search the web for color diagrams illustrating the processes involved in these climate oscillations:
 a. El Niño or ENSO, including the relationship between trade winds, upwelling water, and SST
 b. PDO c. NAO d. AO e. MJO

B20. Often climate variations are indicated on **anomaly maps**. An example frequently used for ENSO is the sea-surface-temperature anomaly. (a) Define "anomaly map". (b) Show or give the URL for two other climate anomaly maps on the internet. (c) What is the relationship between sea-surface-temperature anomalies and rainfall shifts or cloud-coverage changes?

B21. Search the web for maps showing the geographic distribution of dominant principal components for any one of the climate oscillations.

B22. Search the web for a Hovmöller diagram for climate oscillations such as MJO in the most recent 6 or 12 months, and analyze it to explain the direction and movement of features.

B23. Based on an internet search, what does the word **teleconnection** mean with respect to phenomena in the global circulation?

B24. Identify the most-recent major El Niño event, and use the internet to identify or catalog 5 of the worldwide consequences (such as storms or fires).

B25. NCDC is a US-government center. What does the acronym mean, and what can this center do for you? List 8 of the main data types you can access.

B26. Search the web graphs showing observed changes in many of the greenhouse gases.

Apply

A1. Find the radiation input to the Earth system for an albedo of:
 a. 0.1 b. 0.15 c. 0.2 d. 0.25 e. 0.35 f. 0.4
 g. 0.45 h. 0.5 i. 0.55 j. 0.6 k. 0.65 m. 0.7

A2. Use the albedoes from exercise A1 to calculate the effective radiation temperature of Earth.

A3. For a global albedo of 0.3, find the radiation input to the Earth system for the following total solar irradiances (W m^{-2}):
 a. 1350 b. 1352 c. 1354 d. 1356 e. 1358
 f. 1360 g. 1362 h. 1364 i 1365 j. 1367
 k. 1368 m. 1370 n. 1372 o. 1374

A4. Use the solar irradiance values from exercise A3 to calculate the effective radiation temperatures of Earth. Assume the global albedo is 0.3 .

A5. Find the net radiation out from the Earth system, for an Earth effective radiation emission temperature (°C) of
 a. –19 b. –17 c. –15 d. –13 e. –11
 f. –9 g. –7 h. –5 i. –3 j. –1
 k. 1 m. 3 n. 5 o. 10 p. 15

A6. Given the planets and dwarf planets listed below, find their effective radiation temperatures.

Planet	Albedo	Planet	Albedo
a. Mercury	0.119	b. Venus	0.90
c. Mars	0.25	d. Jupiter	0.343
e. Saturn	0.342	f. Uranus	0.30
g. Neptune	0.29	h. Pluto	0.40
i. Eris	0.96		

A7. Given an idealized atmosphere having one layer that is totally opaque to infrared radiation, what is the temperature of the planetary surface for:
 (i) exercise A2 (ii) exercise A4 (iii) exercise A6

A8. Given an Earth with an idealized atmosphere having one layer, but with an infrared window, what is the temperature at Earth's surface. Use 255 K as the effective emission temperature, and assume the emissivity is:
a. 0.70 b. 0.73 c. 0.76 d. 0.80 e. 0.83
f. 0.86 g. 0.88 h. 0.91 i. 0.93 j. 0.95
k. 0.97 m. 0.99

A9. Use Fig. 21.6 to answer these questions:
a. What fraction of energy absorbed at the Earth's surface comes from solar radiation?
b. What fraction of energy leaving the Earth's surface consists of (latent + sensible) heat?
c. What is the net flux of IR radiation between the Earth's surface and the atmosphere, and how does this compare to the flux of (latent + sensible) heat?
d. What is the total amount of upwelling energy (solar, IR, sensible, latent) that reaches the bottom of the idealized atmosphere in that figure?
e. Why is the 79 W m^{-2} of "solar reflected from atmos." not included in the energy budget of the atmosphere as one of the outgoing fluxes (Hint: See the Sample Application after that figure)?
f. Why is the (sensible + latent) heat flux upward from the top of the atmosphere not shown?

A10(§). On a very large graph, carefully plot two ellipses. One has eccentricity of 0.0 (i.e., is a circle), and the other has an eccentricity of:
a. 0.05 b. 0.1 c. 0.15 d. 0.2 e. 0.25 f. 0.3
g. 0.35 h. 0.4 i. 0.6 j. 0.7 k. 0.8 m. 0.9

A11(§). Using the data from Table 21-1, find the Earth-orbit eccentricity for the following time:
a. 400 kyr ago b. 1.1 Myr ago c. 1.2 Myr ago
d. 1.4 Myr ago e. 1.6 Myr ago f. 2 Myr ago
g. 625 kyr in the future h. 1.1 Myr in future
i. 1.3 Myr in future j. 2 Myr in future

A12.(§). Same as for A11, but find the obliquity (°).

A13.(§). Same as for A11, but find the climatic precession at the summer solstice (i.e., the $e \cdot \sin(\varpi)$ term).

A14(§). Same as for A11 (and A13), but find the ratio of semi-major axis to sun-Earth distance: a/R.

A15(§). Same as A11 (and using your results from A12 - A14), but find the average daily insolation at 65°N during the summer solstice.

A16. Find the sun-Earth distance R (Gm) given an eccentricity of 0.05 and a climatic precession value of: a. −0.05 b. −0.04 c. −0.03 d. −0.02 e. −0.01
f. 0.01 g. 0.02 h. 0.03 i. 0.04 j. 0.05

A17. Use the definition of optical depth from the Beer's Law section of the Radiation chapter to estimate the optical depth for background "clean" air, which transmits about 94% of incident solar radiation. Also, use data from Fig. 21.17 on recent volcanic eruptions, and compare the "clean" optical depth to the optical depth for:
a. Mt. Agung (Indonesia)
b. Mt. Fuego (Guatemala)
c. El Chichon (Mexico)
d. Mt. Pinatubo (Philippines)

A18. Find the feedback factor, gain, and system response, given a no-feedback response of $r_o = 0.2$, and a feedback value (F) of:
a. −8 b. −6 c. −4 d. −2 e. 1 f. 2 g. 3
h. 4 i. 5 j. 6 k. 7 m. 8 n. 9 o. 10

A19. Based on the answer to A18, is the feedback negative, stable positive, or run-away positive?

A20. For a no-feedback reference state for Earth, find the no-feedback climate response r_o for a blackbody Earth equilibrium radiative temperature of T_e^* (K):
a. 200 b. 210 c. 220 d. 230 e. 240 f. 250
g. 260 h. 270 i. 280 j. 290 k. 300 m. 310

A21. Given a reference Earth blackbody radiative temperature of $T_e^* = 255$ K and a reference albedo of $A^* = 0.3$. Find the feedback factor, gain, and total system response if ice-albedo feedback behaves such that the change of Earth albedo (associated with ice coverage changes) with surface temperature ($\Delta A/\Delta T = -c$), where c (K^{-1}) has a value of:
a. 0.002 b. 0.005 c. 0.008 d. 0.012 e. 0.015
f. 0.02 g. 0.03 h. 0.05 i. 0.10 j. 0.2

A22. Suppose that doubling of CO_2 in the global atmosphere had the same effect as increasing the incoming solar radiation by $\Delta R_o = 4$ W m^{-2} (due to the increased absorption of solar radiation, as explained in the INFO box on Climate Sensitivity). Given Table 21-4, how much would Earth's equilibrium temperature change (i.e., find ΔT_e in °C) for a reference response with no feedbacks, and for the following feedback process:
a. water-vapor only b. lapse-rate only
c. cloud only d. surface-albedo only
e. all of a - d combined

A23(§). Program the Daisyworld equations into your spreadsheet. Calculate the surface temperature (K), coverage of white daisies, and coverage of black daisies, for luminosity of:
a. 0.9 b. 1.0 c. 1.1 d. 1.2 e. 1.3
f. 1.4 g. 1.5 h. 1.6 i. 1.7 j. 1.8

A24. Describe the climate at
a. Oregon coast
b. Olympic Mtns (NW corner of Washington)
c. central Vancouver Island (British Columbia)
d. at the USA-Canada border labels in Fig. 21.25
e. north-central California
f. north-central Idaho
g. western Nevada
h. extreme SE California
i. central part of Alberta (portion shown in map)
j. south central Idaho

Evaluate & Analyze

E1. Derive an equation for the temperature at Earth's surface vs. the effective radiation temperature, assuming two concentric atmospheric layers that are both opaque to infrared radiation. Hint, consider the methods in the Greenhouse-effect subsection.

E2. Modify the derivations from the Greenhouse-effect subsection to calculate the Earth-surface temperature, given a volcanic-ash-filled single-layer atmosphere that absorbs all IR radiation and absorbs 25% of the solar radiation that tries to move through it. Assume surface albedo is unchanged.

E3. Looking at estimates of past temperatures, the global climate 102 Myr ago was about 2 - 5 °C warmer than present, while 2 Myr ago the global climate was 1.5 - 3.0°C cooler. Is our current "global warming" approaching the normal climate, or is it deviating further away. Discuss.

E4. Explain how the "atmospheric window" in Fig. 21.5 relates to Figs. 8.3 and 8.4 in the Satellites & Radar chapter.

E5. Consider the heat/energy fluxes shown in Fig. 21.6. Assume an atmosphere with a single opaque layer. What if the sun suddenly gets hotter, causing incoming solar radiation to increase by 50%.

Which of the fluxes would change the quickest and which the slowest? How would the new climate equilibrium compare to Earth's present climate?

E6. In music there is a concept of "**beat frequency**". Namely, when two sounds are produced that have nearly the same tone (frequency), you can hear a third tone that depends on the difference between the two frequencies. Look-up the definition of "beat frequency", and use that info to explain why the Climatic Precession curve in Fig. 21.8 looks the way it does.

E7. a. Given the max and min values of climatic precession plotted in Fig. 21.8, what are the max and min of values of (a/R); namely, the ratio of semi-major axis to sun-Earth distance?

b. Simplify the equation (21.18) for the solar declination angle for the special case of the summer solstice.

c. Given the max and min of obliquities plotted in Fig. 21.8, what are the max and min of solar declination angles (°) for the summer solstice?

d. Combine the info in Figs. 21.11 - 21.13 to estimate the max and min values of solar irradiance (W m^{-2}) during the past 1000 years.

e. Given your answers to (a) - (d) above, which has the largest effect on the value of daily-average insolation at 65°N during the summer equinox: climatic precession, obliquity, eccentricity, or solar irradiance?

E8. In Fig. 21.8, the Daily Average Insolation graph is based on a combination of the previous three graphs in that figure. Looking at the nature of the signals in all four top graphs in Fig. 21.8, which of the top 3 graphs best explains the Insolation, and which of the top 3 least explain the Insolation? Why?

E9(§). Table 21-1 shows only the first $N = 4$ or 5 terms for Earth's orbital factors. Use the first $N = 10$ terms (in Table 21-1b, on the next page) to more accurately re-calculate the first 3 graphs in Fig. 21.8. [The original paper by Laskar et al has over 20 terms for each orbital series.]

E10. Use the orbital data for Earth to answer the following. How many years in the future is the first time when the perihelion will coincide with the summer solstice? This is important because it means that summers will be hotter and winters colder in the N. Hemisphere, because the N. Hemisphere will be tilted toward the sun at the same date that the Earth is closest to the sun.

E11 Show that the right-hand approximation in eq. (21.18) for the solar declination is the same as eq. (2.5). [Hints: Consider that in summer, Earth is 90° further along its orbit than it was in Spring. Use trig identities that relate sines to cosines. Consider the definitions of various orbital angles, as summarized in the INFO box on Sun-Earth Distance.]

E12. Explain why Pangea's climate would be different than our present climate. [Hint: Consider global circulations and monsoonal circulations, as discussed in the General Circulation chapter.]

Table 21-1b. Factors in orbital series approximations.

index	A	P (years)	ϕ (degrees)
Eccentricity:			
i= 1	0.010739	405,091.	170.739
2	0.008147	94,932.	109.891
3	0.006222	123,945.	-60.044
4	0.005287	98,857.	-86.140
5	0.004492	130,781.	100.224
6	0.002967	2,373,298.	-168.784
7	0.002818	977,600.	57.718
8	0.002050	105,150.	49.546
9	0.001971	486,248.	148.744
10	0.001797	688,038.	137.155
Obliquity:			
j= 1	0.582412°	40,978.	86.645
2	0.242559°	39,616.	120.859
3	0.163685°	53,722.	-35.947
4	0.164787°	40,285.	104.689
5	0.095382°	41,697.	-112.872
6	0.094379°	41,152.	60.778
7	0.087136°	9,572,151.	39.928
8	0.064348°	29,842.	-15.13
9	0.072451°	28,886.	-155.175
10	0.080146°	40,810.	-70.983
Climatic Precession:			
k= 1	0.018986	23,682.	44.374
2	0.016354	22,374.	-144.166
3	0.013055	18,953.	154.212
4	0.008849	19,105.	-42.250
5	0.004248	23,123.	90.742
6	0.002742	19,177.	61.600
7	0.002386	19,025.	74.660
8	0.001796	22,570.	-145.502
9	0.001908	19,260.	119.333
10	0.001496	16,465.	141.244

*Simplified from Laskar, Robutel, Joutel, Gastineau, Correia & Levrard, 2004: A long-term numerical solution for the insolation quantities of the Earth. "Astronomy & Astrophysics", **428**, 261-285.*

E13. a. Use the definitions of optical depth and visual range from the Beer's Law section of the Solar & IR Radiation chapter to find a relationship for visibility (km) as a function of optical depth τ.

b. Use your answer from (a) with the data from Table 21-3 to estimate the atmospheric visibilities associated with the following volcanic eruptions:

(i) Lake Toba (ii) Mt. Tambora (iii) Krakatau

E14. For every type of climate feedback, explain why runaway climate change is not possible. [Or if it is possible, show how runaway climate changes applies for only a limited change of Earth temperatures before a new stable climate equilibrium is reached. Can you recall any Hollywood movies that showed runaway climate change, with or without reaching a stable equilibrium?]

E15. For the feedback interconnections shown in Fig. 21.18, the feedback outcomes are only: negative (damped), positive (amplified), or runaway (increasing without limit). Can you think of a different type of interconnection or process that could cause the system response to oscillate? Explain.

E16. Ice-albedo processes were used for an idealized illustration of feedback. Try to develop your own feedback concepts that account for the following (and ignores all other feedbacks except the fundamental radiative feedback). Suppose there is a CO_2 feedback process whereby warmer Earth temperatures cause the oceans to release more CO_2 into the air. But more CO_2 in the air acts to partially close the IR window sketched in Fig. 21.5.

a. What info or what relationships do you need to be able to figure out the feedback? [Just list what info you need here — do not actually try to find these relationships.]

b. If you are good at calculus and if assigned by your instructor, then try to create a governing equation similar to that used for ice-albedo feedback, then take the derivatives as was done in the Higher Math Feedback Example box. There is no single correct answer — many different approximations and solutions might be reasonable. Justify your methods.

E17. For feedbacks in the Earth climate system, use error-propagation methods as discussed in Appendix A, to propagate the errors listed in the feedback (F) column of Table 21-4 into the other columns for feedback factor (f), gain (G), & system response (r).

E18. Do outside readings to explain how the following abrupt climate-change processes work (Table 21-4). Also, explain if feedback is positive or negative.

a. meridional overturning ocean circulation
b. fast Antarctic & Greenland ice-sheet collapse
c. volcanoes d. biogeochemical
e. methane release due to hydrate instability and permafrost)

E19.(§) For a daisyworld

a. Redo the calculation shown in this chapter.
b. What happens to the temperature over black and white daisies if transport Tr changes to 0 or 1?
c What happens if you start with initial conditions of 100% coverage by white or black daisies?
d. What happens if you take a time step increment of 2, 4, or 8?
e. What parameter values prevent homeostasis from occurring (i.e., eliminate the nearly constant temperature conditions on a fecund daisyworld?
f. Is it possible for homeostasis to occur on Earth, given the large fraction of area covered by oceans?

Table 21-8. Names of Köppen climate classes.

Exer-cise	Code	Name
a	Af	equatorial climate
b	Am	monsoon climate
c	Aw	tropical savanna climate
d	BWh	warm desert climate
e	BWk	cold desert climate
f	BSh	warm semi-arid climate (warm steppe clim.)
g	BSk	cold semi-arid climate (cold steppe climate)
h	Csa	warm mediterranean climate
i	Csb	temperate mediterranean climate
j	Cwa	warm humid subtropical climate
k	Cwb	temperate humid subtropical climate
l	Cwc	cool humid subtropical climate
m	Cfa	warm oceanic/marine; or humid subtropical
n	Cfb	temperate oceanic/marine climate
o	Cfc	cool oceanic/marine climate
p	Dsa	warm continental (mediterranean continental)
q	Dsb	temperate continental (mediter. continental)
r	Dsc	cool continental climate
s	Dsd	cold continental climate
t	Dwa	warm humid continental climate
u	Dwb	temperate humid continental climate
v	Dwc	cool continental clim.; cool subarctic climate
w	Dwd	cold continental clim.; cold subarctic climate
x	Dfa	warm humid continental climate
y	Dfb	temperate humid continental climate
z	Dfc	cool continental clim.; cool subarctic climate
aa	Dfd	cold continental clim.; cold subarctic climate
ab	ET	tundra climate
ac	EF	ice cap climate

E20(§). In Fig. 21.24 for Daisyworld, homeostasis is possible for luminosities spanning only a certain range. What changes to the daisy growth parameters or equations would permit homeostasis for luminosities spanning a greater range? Confirm your results by solving the resulting Daisyworld equations with your new parameters.

E21. Where in the world (but NOT in any region plotted in Fig. 21.25) would you expect to find the climate in Table 21-8 above. (Hint, consider info in the General Circulation chapter, and also use the Köppen climate definitions in Table 21-6.)

E22(§-difficult). Find the covariance matrix, the eigenvectors and eigenvalues (using some other resource such as a web-based eigenvector calculator), leading 2 principal components (PCs), interpret the results, and reconstruct (synthesize) the original data as approximated using the dominant 2 PCs, given time series in Table 21-9. Assume that the locations for these time series correspond to the geographic locations shown in Fig. 21.a of the **PCA** INFO box.

Table 21-9. Time series for PCA exercise.

Location Index:	j = 1	2	3	4
Location:	NW	NE	SW	SE
Year	Air Temperature (°C)			
i = 1	9	1	1	8
2	7	2	1	8
3	5	3	2	7
4	7	4	5	6
5	9	5	9	5
6	7	6	9	5
7	5	5	5	5
8	7	4	5	6
9	9	3	4	7
10	7	2	1	8

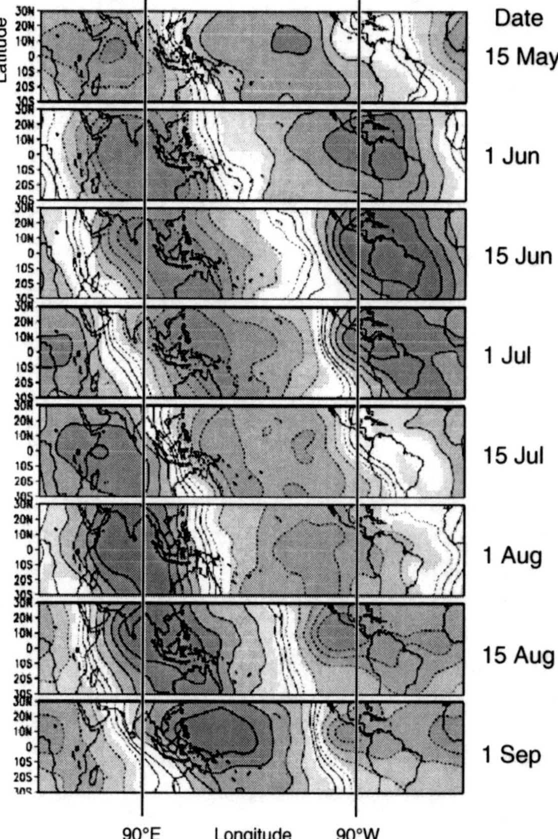

Figure 21.29
Hovmöller Diagram showing 20 kPa winds. (Shading is arbitrary for this exercise.) Modified from data courtesy of the Climate Prediction Center, National Weather Service, National Oceanic and Atmospheric Admin.

E23. Given the Hovmöller diagram in Fig. 21.29, determine the zonal speed and direction that features move.

Synthesize

S1. What if emitted IR radiation were constant — not a function of temperature. How would Earth's climate be different, if at all?

S2. Create an energy balance to determine T_s and T_e for an atmosphere with:
a. 2 opaque atmospheric layers (no window)
b. 3 opaque atmospheric layers (no window)
c. extrapolate your theory so it works with more layers

S3. Suppose the same side of the Earth always faced the sun, and the atmosphere was stationary over the Earth. Assume no heat transport between Earth's cold and warm sides. Write the equations (along with your justification) for energy balances for each side of the Earth, for the following scenarios:
a. a single-layer opaque atmosphere.
b. a single-layer atmosphere, but having an IR window.

S4. Suppose radioactive decay deep inside the Earth caused as much heat transport into the bottom of the atmosphere as is incoming solar radiation into the top of Earth's atmosphere. Work out the energy budgets, and calculate steady-state Earth-surface temperatures. Discuss feedbacks. (Hint, assume a single-layer opaque atmosphere.)

S5. Estimate the variation in insolation if Earth's orbit were to have eccentricity of 0.8.

S6. Given idealized ozone holes centered over each pole, where each hole extends 25° of latitude from the pole. In each hole, assume that the atmosphere is totally transparent to IR radiation. Assume that Earth's climate can be divided in to latitude bands of uniform surface temperature: hot near the equator, temperate at mid latitudes, and cold directly under the ozone holes. (a) Devise energy-balance equations for each latitude band, (b) find the average surface temperature within each band, and (c) find the overall Earth-surface average temperature. Justify any assumptions you make.

S7. Suppose a long-lasting solar storm ionizes the air and totally closes the infrared window on Earth's sunny side. However, Earth's shady side is not affected, allowing an atmospheric window to exist in the infrared. (a) What is the steady-state Earth-surface temperature (averaged over the whole globe)? (b) Show your budget equations justify your assumptions to support your answer.

S8. For Daisyworld, start with global coverages of 50% dark and 50% light daisies. Assume that no bare ground is allowed at any time, although relative coverage of light and dark daisies can change. (a) Devise equations to describe this scenario, and (b) compare your results with the Daisyworld equations and results earlier in this chapter. (c) Does the new Daisyworld allow homeostasis? (d) Is the new Daisyworld more or less sensitive to changes in luminosity?

S9. If runaway climate change caused the Earth to reach a new warmer equilibrium, would the Earth's climate be able to recover to the climate we currently have? What positive and/or negative feedback processes would encourage such recovery?

S10. Would global warming alter:
a. the global circulation?
b. thunderstorm severity?
c. hurricane severity?
Explain how.

S11. Discuss the scientific accuracy of recent Hollywood movies where the theme was weather or climate disasters.

S12. Repeat the principal component analysis from the INFO box, but when you calculate the table of principal components (PC), use the temperature anomaly T' in eq. (21.50) instead of T. Then finish all the remaining calculations using these new PCs.

What is the sum of elements in each PC (i.e., the sum of each column in the PC table)? What is the relationship between your plotted PCs and Fig. 21.c? When you synthesize the signal using only PC1, how does your result compare to Fig. 21.d?

Outcome: you are synthesizing T' and not T. Most researchers use PCA to focus on the anomaly.

22 ATMOSPHERIC OPTICS

Contents

Light can be considered as photon particles or electromagnetic waves, either of which travel along paths called **rays**. To first order, light rays are straight lines within a uniform transparent medium such as air or water, but can **reflect** (bounce back) or **refract** (bend) at an interface between two media. Gradual refraction (curved ray paths) can also occur within a single medium containing a smooth variation of optical properties.

The beauty of nature and the utility of physics come together in the explanation of rainbows, halos, and myriad other atmospheric optical phenomena.

RAY GEOMETRY

When a **monochromatic** (single color) light ray reaches an interface between two media such as air and water, a portion of the incident light from the air can be reflected back into the air, some can be refracted as it enters the water (Fig. 22.1), and some can be absorbed and changed into heat (not sketched). Similar processes occur across an air-ice interface.

Reflection

The angle θ_3 of the reflected ray always equals the angle θ_1 of the incident ray, measured with respect to a line normal (perpendicular) to the interface:

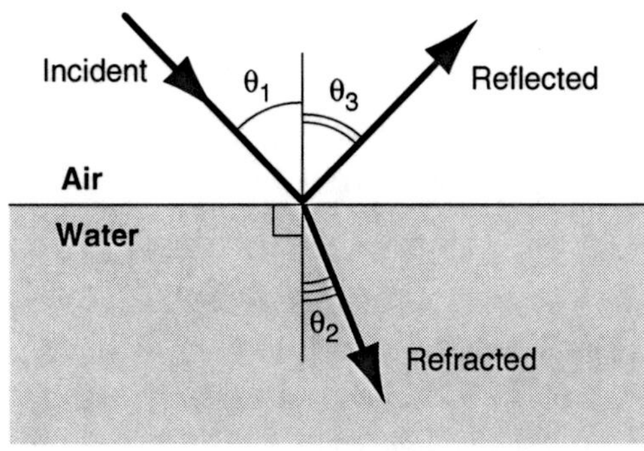

Figure 22.1
Geometric optics at an air-water interface. Black arrows are light-ray paths.

Sample Application

Rays of red and violet light in air strike a water surface, both with incident angle of 60°. Find the angle of refraction for each color, given T = 20°C, P = 101 kPa?

Find the Answer

Given: $\theta_1 = 60°$, $T = 20°C$, $P = 101$ kPa

Find: $\theta_2 = ?°$ for red (λ=0.7 µm) and violet (λ=0.4 µm).

Use eq. (22.5):

$\theta_2 = \arcsin[(n_1/n_2)\cdot\sin(\theta_1)] = \arcsin[\mu_{12}\cdot\sin(\theta_1)]$

with μ_{12} from eq. (22.3) and refractive indices from Table 22-1.

For red: $\mu_{12} = n_1/n_2 = 1.0002704/1.3305 = 0.7518$

$\theta_2 = \arcsin[0.7518\cdot\sin(60°)]$

θ_2 = **40.62°**

Similarly for violet: θ_2 = **40.15°**

Check: Units OK. Physics OK.

Exposition: Had there been no bending, then both answers would have been 60°. Angles closer to 60° for this example correspond to less bending. The answers above confirm the statement that red light is bent less than violet. The **amount of bending** (difference between incident and refracted angles) is large: 60° – 40° = 20°.

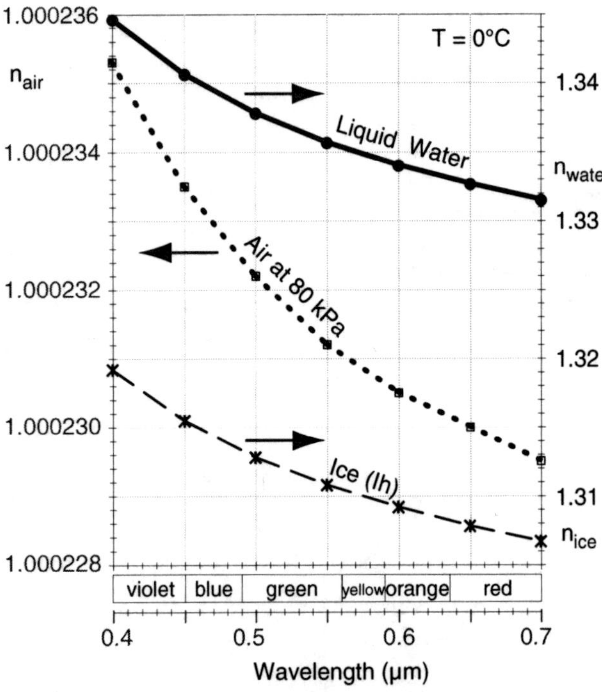

Figure 22.2

Refractive index n for various wavelengths of visible light at T = 0°C. See Table 22-1 for details. The wavelength bands that we perceive as different **colors** *is approximate, and is slightly different for each person. In this chapter we will use red ≈ 0.7 µm, orange ≈ 0.62 µm, yellow ≈ 0.58 µm, green ≈ 0.53 µm, blue ≈ 0.47 µm, and violet ≈ 0.4 µm wavelength.*

$$\theta_1 = \theta_3 \qquad \bullet(22.1)$$

The reflected angle does not depend on color (i.e., is not a function of the wavelength of light).

Refraction

Refractive index

The **refractive index** n_i for medium i relative to a vacuum is defined as:

$$n_i = \frac{c_o}{c_i} \qquad \bullet(22.2)$$

where c_i is the speed of light through medium i and $c_o = 3 \times 10^8$ m s^{-1} is the speed of light in a vacuum.

The ratio of refractive indices is sometimes defined as

$$\mu_{12} = \frac{n_1}{n_2} \qquad \bullet(22.3)$$

where subscripts 1 and 2 refer to the media containing the incident ray and refracted rays, respectively. Different colors and different media have different refractive indices, as indicated in Table 22-1 and Fig. 22.2.

Snell's Law

The relationship between the incident angle θ_1 and refracted angle θ_2 (Fig. 22.1) is called **Snell's Law**:

$$\frac{\sin\theta_1}{\sin\theta_2} = \frac{c_1}{c_2} = \frac{n_2}{n_1} = \frac{1}{\mu_{12}} \qquad \bullet(22.4)$$

Incident and refracted rays are always in the same plane. This plane includes the line that is normal to the surface. It is the plane that gives the smallest angle between the incident ray and the surface.

Solving Snell's Law for the refracted ray angle gives:

$$\theta_2 = \arcsin[\mu_{12}\cdot\sin(\theta_1)] \qquad (22.5)$$

Red light is bent less than violet light as it passes through an interface. Thus, refraction causes white light to be spread into a **spectrum** of colors. This phenomenon is called **dispersion**.

A viewer looking <u>toward</u> the incoming light ray (Fig. 22.3) would see a **light point** in the celestial sphere overhead at angle θ_2 from normal.

Table 22-1a. Refractive index (n_{air}) for **air** with pressure P = 101.325 kPa and relative humidity RH = 75%. *From NIST.*

λ (µm)	Temperature (°C)				
	−40	−20	0	20	40
0.4 violet	1.0003496	1.0003219	1.0002982	1.0002773	1.0002583
0.45	1.0003469	1.0003194	1.0002958	1.0002751	1.0002562
0.5	1.0003450	1.0003176	1.0002942	1.0002736	1.0002548
0.55	1.0003436	1.0003163	1.0002930	1.0002725	1.0002537
0.6	1.0003425	1.0003154	1.0002921	1.0002716	1.0002529
0.65	1.0003417	1.0003146	1.0002914	1.0002710	1.0002523
0.7 red	1.0003410	1.0003140	1.0002908	1.0002704	1.0002518

Table 22-1b. Refractive index (n_{air}) for **air** with P = 80 kPa and RH = 75%. *Based on data from NIST modified Edlen Eq. calculator.*

λ (µm)	Temperature (°C)				
	−40	−20	0	20	40
0.4 violet	1.0002760	1.0002541	1.0002353	1.0002188	1.0002036
0.45	1.0002738	1.0002521	1.0002335	1.0002171	1.0002019
0.5	1.0002723	1.0002507	1.0002322	1.0002158	1.0002008
0.55	1.0002712	1.0002497	1.0002312	1.0002150	1.0001999
0.6	1.0002704	1.0002489	1.0002305	1.0002143	1.0001993
0.65	1.0002697	1.0002483	1.0002300	1.0002138	1.0001988
0.7 red	1.0002692	1.0002479	1.0002295	1.0002134	1.0001984

Table 22-1c. Refractive index (n_{air}) for **air** with P = 40 kPa and RH = 75%. *Based on data from NIST modified Edlen Eq. calculator.*

λ (µm)	Temperature (°C)				
	−40	−20	0	20	40
0.4 violet	1.0001379	1.0001270	1.0001176	1.0001091	1.0001009
0.45	1.0001369	1.0001260	1.0001166	1.0001082	1.0001000
0.5	1.0001361	1.0001253	1.0001160	1.0001076	1.0000994
0.55	1.0001355	1.0001248	1.0001155	1.0001071	1.0000990
0.6	1.0001351	1.0001244	1.0001151	1.0001068	1.0000987
0.65	1.0001348	1.0001241	1.0001149	1.0001066	1.0000985
0.7 red	1.0001346	1.0001239	1.0001147	1.0001064	1.0000983

Table 22-1d. Refractive index (n_{water}) for **liquid water**. *Based on data from IAPWS 1997 release and CRC Handbook.*

λ (µm)	Temperature (°C)				
	−40	−20	0	20	40
0.4 violet	(values below	1.3429	1.3446	1.3436	1.3411
0.45	0°C are for	1.3390	1.3406	1.3396	1.3371
0.5	supercooled	1.3362	1.3378	1.3368	1.3344
0.55	water)	1.3341	1.3357	1.3347	1.3323
0.6		1.3324	1.3340	1.3330	1.3306
0.65		1.3310	1.3326	1.3317	1.3293
0.7 red		1.3299	1.3315	1.3305	1.3281
density (kg m^{-3}):		**993.547**	**999.840**	**998.207**	**992.200**

Table 22-1e. Refractive index (n_{ice}) for **ice (Ih)**. *Based on data from S.G. Warren (1984) and V.F. Petrenko & R.W. Whitworth (1999).*

λ (µm)	Temperature (°C)				
	−40	−20	0	20	40
0.4 violet	1.3206	1.3199	1.3192	(values at 0°C were extrapolated)	
0.45	1.3169	1.3162	1.3155		
0.5	1.3142	1.3135	1.3128	(See INFO Box in Precipitation chapter for info on ice phases such as Ih.)	
0.55	1.3122	1.3115	1.3108		
0.6	1.3106	1.3099	1.3092		
0.65	1.3092	1.3085	1.3078		
0.7 red	1.3081	1.3074	1.3067		

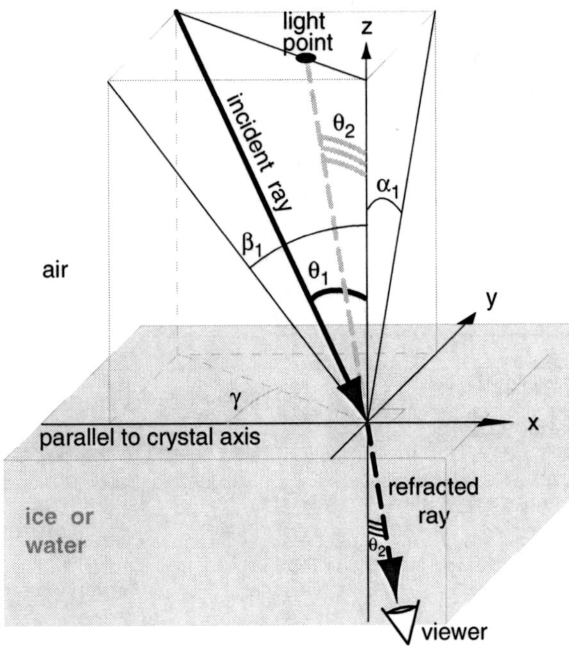

Figure 22.3
Components of refraction geometry. Incident ray is heavy solid arrow. Refracted ray is heavy dashed arrow. The tail (thick grey dashed line) of the refracted ray is extended to show that it lies in the same plane as the incident ray.

Sample Application
A ray of red light in air strikes a water surface, with incidence angle components of $\alpha_1 = 45°$ and $\beta_1 = 54.74°$. What are the corresponding component angles of the refracted ray? Assume $T = 20°C$ and $P = 101.325$ kPa.

Find the Answer
Given: $\alpha_1 = 45°$, $\beta_1 = 54.74°$.
Find: $\alpha_2 = ?°$, $\beta_2 = ?°$

From Table 22-1 for red light:
$\mu_{12}{}^2 = (1.0002704/1.3305)^2 = (0.7518)^2 = 0.5652$
Also: tan(45°) = 1, and tan(54.74°) = 1.414
Next, solve eq. (22.8):

$$b_{\alpha\beta} = \frac{0.5652}{1+(1-0.5652)\cdot(1+2)} = 0.2453$$

Then use eqs. (22.7):
$\tan^2(\alpha_2) = 0.2453\cdot[\tan^2(45°)] = 0.2453$
$\tan^2(\beta_2) = 0.2453\cdot[\tan^2(54.74°)] = 0.4906$

Thus: $\alpha_2 = \arctan[(0.2453)^{0.5}] =$ **26.35°**
$\beta_2 = \arctan[(0.4906)^{0.5}] =$ **35.01°**

Check: Units OK. Physics OK.
Exposition: Using eq. (22.6) with the incident angle components of 45° and 54.74°, we find that the incident ray angle is $\theta_1 = \arctan[(1+2)^{0.5}] = 60°$. This is the same as a previous Sample Application. Using eq. (22.6) on the answers above, we find $\theta_2 = \arctan[(0.2453+0.4906)^{0.5}]$ = **40.62°**. This is also the same as a previous Sample Application, which verifies eqs. (22.6 - 22.9).

Sample Application
Find the speed of red light through liquid water and through air at $T = 20°C$ and $P = 101.325$ kPa.

Find the Answer
Given: $\lambda = 0.7$ µm, $T = 20°C$ and $P = 101.325$ kPa.
Find: $c_{air} = ?$ m s^{-1} , $c_{water} = ?$ m s^{-1}

Use eq. (22.2) with refractive indices from Table 22-1:
$c_{air} = c_o/n_{air} = (3\times10^8 \text{m s}^{-1})/1.0002704 =$ **2.999×10^8 m s^{-1}**
$c_{water} = c_o/n_{water} = (3\times10^8 \text{m s}^{-1})/1.3305 =$ **2.255×10^8 m s^{-1}**

Check: Speeds reasonable. Units OK.
Exposition: The difference in speeds of light is useful for understanding Huygens' principle.

Snell's Law in 3 Dimensions

Sometimes it is easier to work with x and y components of the incident ray, where the x-axis might be aligned with the axis of a columnar ice crystal, for example, and the y-axis might be on the crystal prism surface (Fig. 22.3). The relationship between the component angles and the incident angle is:

$$\tan^2\theta_1 = \tan^2\alpha_1 + \tan^2\beta_1 \qquad (22.6)$$

This relationship also applies to refracted angles (θ_2, α_2, β_2), and will be used extensively later in this chapter to discuss ice-crystal optics.

Component angles α and β of the refracted and incident rays do NOT individually obey Snell's law (eq. 22.4). Nevertheless, Snell's law can be reformulated in terms of components as follows:

$$\tan^2\alpha_2 = b_{\alpha\beta}\cdot\tan^2\alpha_1 \qquad (22.7a)$$

$$\tan^2\beta_2 = b_{\alpha\beta}\cdot\tan^2\beta_1 \qquad (22.7b)$$

where α_2 and β_2 are the components of the refracted ray (analogous to α_1 and β_1), and

$$b_{\alpha\beta} = \frac{\mu_{12}{}^2}{1+(1-\mu_{12}{}^2)\cdot\left\{\tan^2\alpha_1+\tan^2\beta_1\right\}} \qquad (22.8)$$

These equations are abbreviated as

•(22.9b)

$$\alpha_2 = S_\alpha(\alpha_1,\beta_1,,\mu_{12}) = \arctan\left[\left(b_{\alpha\beta}\cdot\tan^2\alpha_1\right)^{1/2}\right]$$

•(22.9b)

$$\beta_2 = S_\beta(\alpha_1,\beta_1,,\mu_{12}) = \arctan\left[\left(b_{\alpha\beta}\cdot\tan^2\beta_1\right)^{1/2}\right]$$

where S represents Snell's law for components.

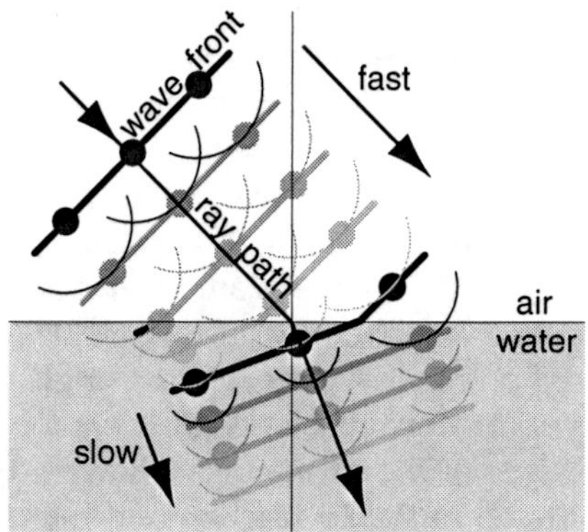

Figure 22.4
Propagation of wave fronts across an interface.

Huygens' Principle

Huygens suggested that every point along a wave front acts like a generator of new spherical secondary wavelets. Wave-front position after some time interval is located at the tangent to all of the new wavelets. Thus, when a portion of a wave front encounters a medium with a slower light velocity, then that portion of the wave slows, causing the whole wave front to turn into the medium (Fig. 22.4).

Critical Angle

When Snell's law is applied to light rays moving from a denser medium (having slower light velocity) to a less-dense medium (having faster light velocity), there is a **critical angle** at which light is bent so much that it follows the interface. At angles greater than this critical incidence angle, light cannot refract out of the dense medium at all. Instead, all of the light reflects (Fig. 22.5).

The critical angle θ_c is found from:

$$\sin(\theta_c) = \frac{n_2}{n_1} \qquad \bullet(22.10)$$

where n_1 is the refractive index for the incident ray (i.e., in the denser medium with slower light velocity). An example is a light ray moving out of water and into air (i.e., $n_2 < n_1$). For red light the critical angle is about 48.7°, while for violet light the it is about 48.1°. There is no critical angle for light moving into a denser medium (i.e., $n_2 > n_1$).

LIQUID-DROP OPTICS

Most rain-drop water optics are found by looking away from the sun. **Rainbows** are circles or portions of circles (Fig. 22.6) that are centered on the **antisolar** point, which is the point corresponding to the shadow of your head or camera. (As viewed

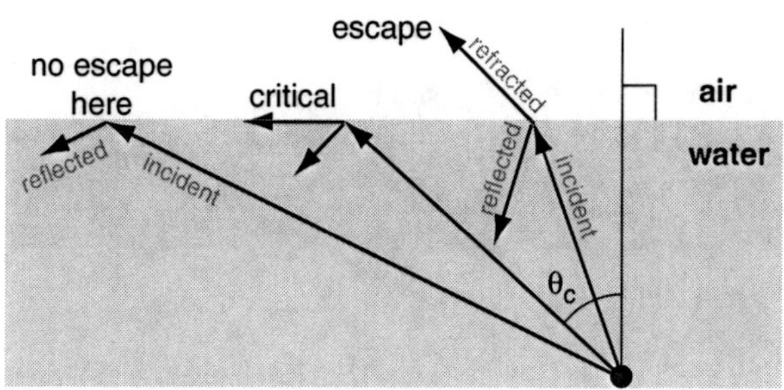

Figure 22.5
Angles greater than critical θ_c do not allow light to escape from a dense to less dense medium.

Sample Application

Find the critical angle for red light going from a cirrus-cloud ice crystal to air, where the air state is T = –20°C & P = 40 kPa.

Find the Answer

Given: red light λ = 0.7 µm, T = –20°C & P = 40 kPa.
Find: θ_c = ? °

From Table 22-1: n_{ice} = 1.3074 and n_{air} = 1.0001239.
Use eq. (22.10):
θ_c = arcsin(1.0001239/1.3074) = <u>**49.9°**</u>.

Check: Magnitude OK. Units OK.
Exposition: This critical angle is close to that for liquid water, because the refractive indices for liquid and solid water are nearly the same.

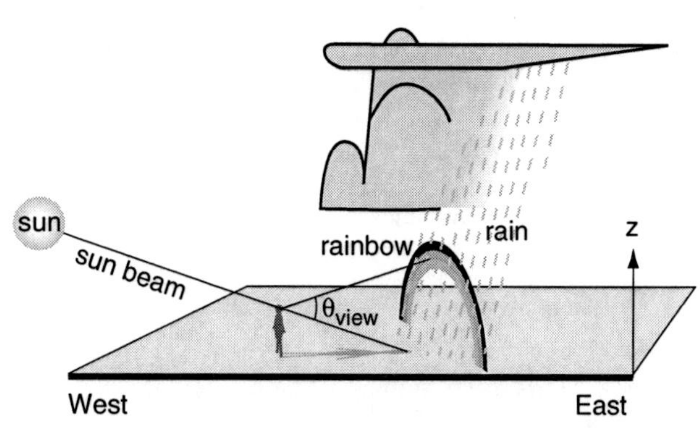

Figure 22.6
Orientation of rainbows relative to the observer and the sun.

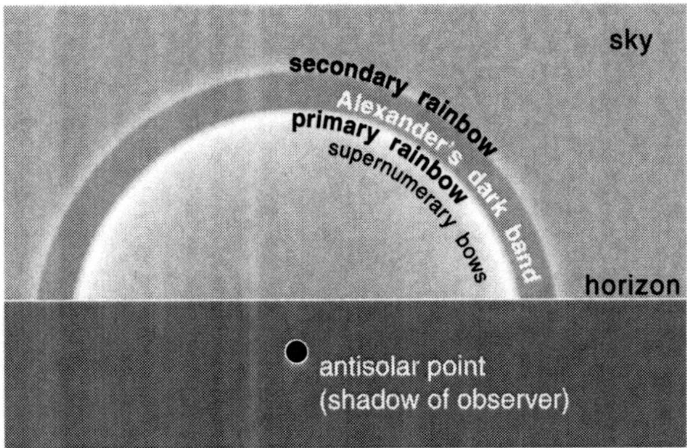

Figure 22.7
Rainbows.

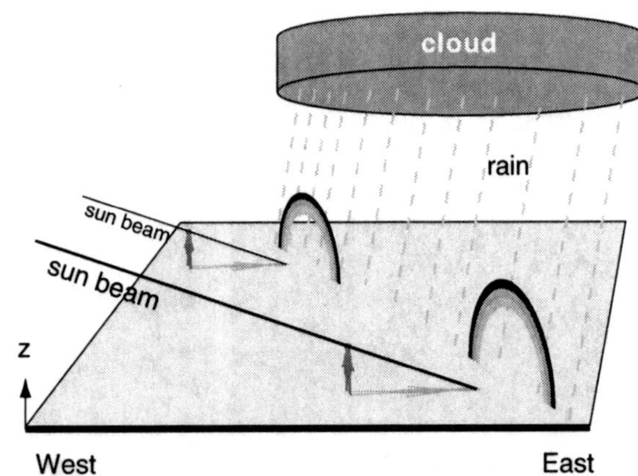

Figure 22.8
Different observers see different rainbows.

by the observer, the antisolar point is also called the sun's **antipodal point**.)

Primary rainbows have red on the outside of the circle, and are the brightest and most easily seen (Fig. 22.7). The **viewing angle** (θ_{view} = angle between two lines: the line from your eye to the rainbow and the line from your eye and the antisolar point) is about 42°. **Secondary rainbows** have red on the inside at viewing angle of about 50°. **Supernumerary bows** are very faint, and touch the inside of the primary rainbow — these will be discussed in the Diffraction and Interference section.

At any one time for any one rain storm, different observers see different rainbows caused by different light rays interacting with different rain drops (Fig. 22.8). Hence, one observer in a region of bright sun and large drops might see a vivid rainbow, while another observer a half kilometer away might see only a weak or partial rainbow in light rain with smaller drops.

We can explain these rainbow phenomena using geometric optics, assuming spherical rain drops. First, we will consider the portion of incident sunlight that is lost due to reflection off the outside of the drop. Then we will look at the remainder that is refracted and reflected in the drop to make the rainbow.

Reflection from Water

Rays of light that hit a water surface at a shallow angle are reflected more than those hitting straight on. Reflectivity r vs. elevation angle Ψ above the water surface is approximately:

$$r = r_o + r_1 \cdot e^{-\Psi/a} \qquad (22.11)$$

where $r_o = 0.02$, $r_1 = 0.98$, and $a = 9.3°$. This is plotted in Fig. 22.9.

Sample Application

Just after sunrise, sunlight hits a smooth lake surface at elevation angle 5°. Find the reflectivity.

Find the Answer
Given: $\Psi = 5°$
Find: $r = ?$ (dimensionless)

Use eq. (22.11):
$r = 0.02 + 0.98 \cdot \exp(-5°/9.3°) = \underline{0.59}$

Check: Units dimensionless. Magnitude agrees with Fig. 22.9.
Exposition: 59% of incident sunshine in this case is reflected away, and is not available to heat the lake.

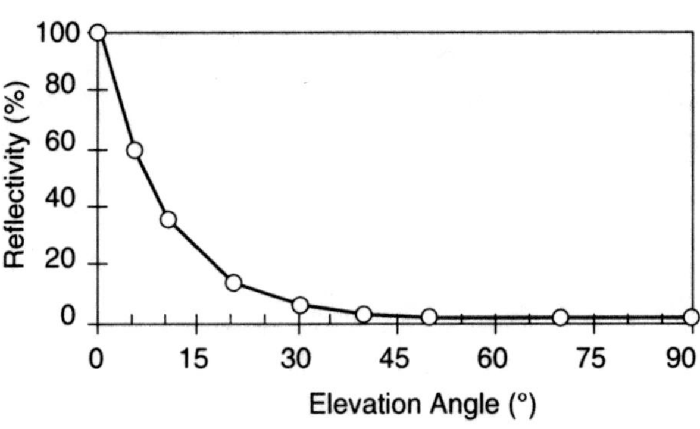

Figure 22.9
Reflectivity from a water surface as a function of ray elevation angle.

Sample Application

For a light ray hitting a spherical rain drop at a distance from the center of 90% of the drop radius, find the elevation angle and the fraction of light reflected.

Find the Answer:
Given: $x/R = 0.9$
Find: $\Psi = ?°$, and $r = ?$ dimensionless

Use eq. (22.12): $\Psi = \arccos(0.9) = 25.8°$
Use eq. (22.11): $r = 0.02 + 0.98 \cdot \exp(-25.8°/9.3°) = \underline{\mathbf{0.081}}$

Check: Units OK. Magnitude agrees with Fig. 22.9.
Exposition: Even at this large impact parameter, only 8% of the light is reflected off the outside.

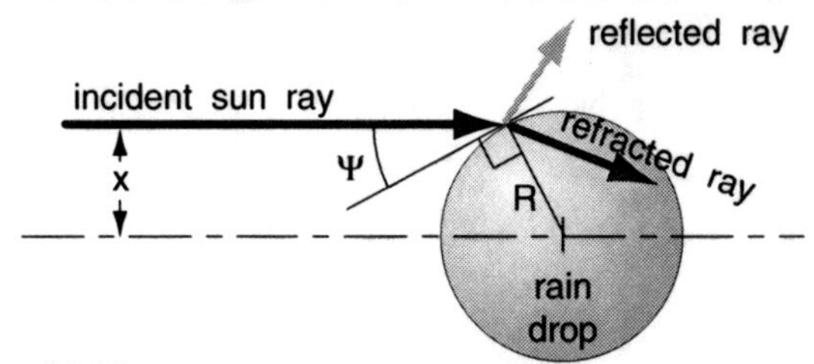

Figure 22.10
Fate of incident ray after first impact with drop.

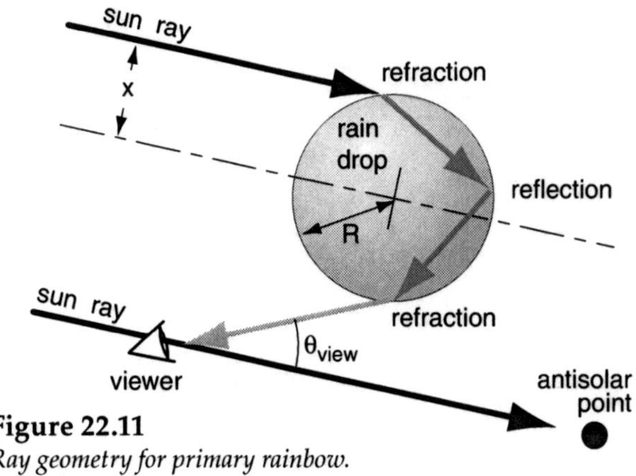

Figure 22.11
Ray geometry for primary rainbow.

Rays of sunlight can enter a spherical drop at any distance x from the centerline (Fig. 22.10). The ratio of this distance to the drop radius R is called the **impact parameter** (x/R). Because of curvature of the drop surface, rays arriving at larger impact parameters strike the drop at smaller elevation angles:

$$\psi = \arccos\left(\frac{x}{R}\right) \qquad \bullet(22.12)$$

Thus, the amount of reflected light from a drop increases with increasing impact parameter.

Primary Rainbow

Rays that are not reflected from the outside of the drop can make zero or more reflections inside the drop before leaving. Those entering rays that make one reflection (in addition to the two refractions during entry and exit) cause the **primary rainbow** (Fig. 22.11). White light is dispersed by the refractions such that reds are bent less and appear on the outside of the rainbow circle, while violets are bent more and appear on the inside (Fig. 22.12).

Variations of impact parameter cause a range of output viewing angles θ_{view} according to:

$$\theta_{view} = 4 \cdot \arcsin\left(\frac{n_{air}}{n_{water}} \cdot \frac{x}{R}\right) - 2 \cdot \arcsin\left(\frac{x}{R}\right) \qquad \bullet(22.13)$$

where x/R is the impact parameter, and n is refractive index. In other words, there is not a single magic angle of 42° for the primary rainbow. Instead, there is a superposition of many rays of different colors with a wide range of viewing angles.

To learn how these rays interact to form the rainbow, we can solve eq. (22.13) on a spreadsheet for a large number of evenly spaced values of the impact parameter, such as intervals of 0.02 for x/R. Output

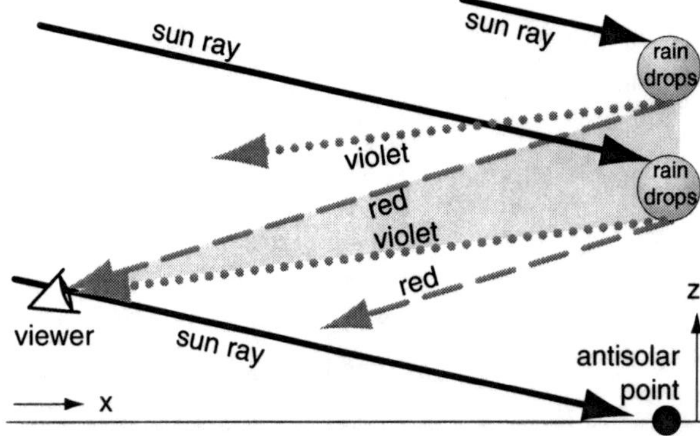

Figure 22.12
Viewer sees primary rainbow (shaded) with red light coming from drops further from the antisolar point than for violet. Thus, the different colors you see come from different drops.

Sample Application

For red light calculate the primary-rainbow viewing angle for x/R= 0.70 in air of P= 80 kPa and T= 0°C.

Find the Answer
Given: $x/R = 0.70$
$n_{air} = 1.0002295$, $n_{water} = 1.3315$ from Table 22-1.
Find: $\theta_{view} = ?°$
Use eq. (22.13) for the primary rainbow:
$\theta_{view} = 4 \cdot \arcsin(0.7512 \cdot 0.70) - 2 \cdot \arcsin(0.70) = \underline{\mathbf{38.05°}}$

Check: Units OK. Physics OK. Agrees with Fig. 22.13.
Exposition: This lights inside the primary rainbow.

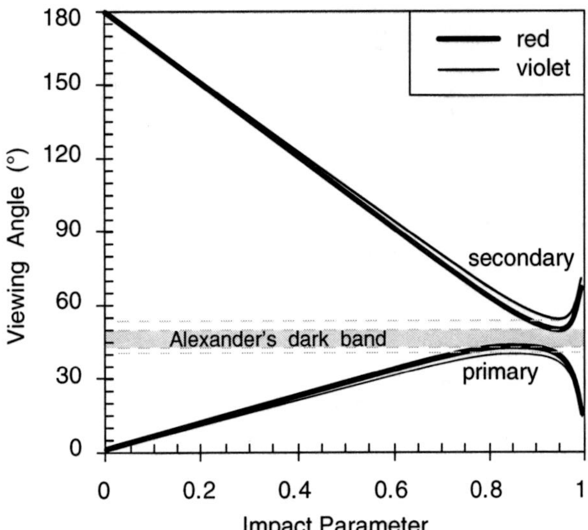

Figure 22.13a
Viewing angles for rainbow rays.

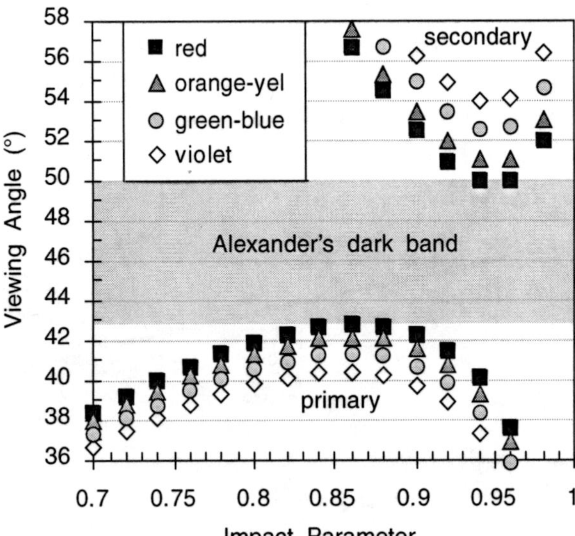

Figure 22.13b
Viewing angles for rainbow rays (expanded version).

> **Sample Application**
> For red light calculate the secondary-rainbow viewing angle for x/R= 0.90, with P= 80 kPa and T= 0°C.
>
> **Find the Answer**
> Given: $x/R = 0.90$
> $n_{air} = 1.0002295$, $n_{water} = 1.3315$ from Table 22-1.
> Find: $\theta_{view} = ?°$
>
> Use eq. (22.14) for the primary rainbow:
> $\theta_{view} = 180° + 2\cdot\arcsin(0.90) - 6\cdot\arcsin(0.7512 \cdot 0.90)$
> $= \underline{53.09°}$
>
> **Check**: Units OK. Physics OK.
> **Exposition**: This almost agrees with Fig. 22.13, which was created for a slightly different T and P.

viewing angles, measured from the antisolar point, range between 0° and 42.7°, as plotted in Fig. 22.13a. This is why the sky is brighter inside the primary rainbow (i.e., for viewing angles of 0 to 42.7°), than just outside of it.

Fig. 22.13b is a blow-up of Fig. 22.13a. Three of the data points have a viewing angle of about 42.5°, and these all correspond to rays of red light. Hence, the primary rainbow looks red at that viewing angle. At 42.0°, there are mostly orange-yellow data points, and just a couple red points. Hence, at this angle the rainbow looks orange. Similar arguments explain the other colors. Inside the primary rainbow the sky looks white, because roughly equal amounts of all colors are returned at each viewing angle.

Secondary Rainbow

Entering light rays can also make two reflections before leaving the raindrop, as sketched in Fig. 22.14. The extra reflection reverses the colors, putting red on the inside of the circle of a **secondary rainbow**. At each internal reflection some light is lost (refracted) out of the drop. Thus, the secondary rainbow is faint -- only 43% as bright as the primary rainbow.

The relationship between impact parameter and viewing angle for a secondary rainbow is:

•(22.14)

$$\theta_{view} = 180° + 2\cdot\arcsin\left(\frac{x}{R}\right) - 6\cdot\arcsin\left(\frac{n_{air}}{n_{water}}\cdot\frac{x}{R}\right)$$

where x/R is the impact parameter, and n is refractive index. Again, a range of output viewing angles occurs because light enters over the full range of impact parameters. In this case, the sky is dark inside of the secondary rainbow (about 50° viewing angle), and bright outside, as sketched in Figs. 22.13a & b, based on spreadsheet calculations.

Higher-order rainbows (having more internal reflections) are theoretically possible, but are rarely seen because they are so faint:

- **Tertiary Rainbow**:
  3 internal reflections
  $\theta_{view} = 137.52°$
  intensity ≈ 24% as bright as the primary
- **Quaternary Rainbow**:
  4 internal reflections
  $\theta_{view} = 137.24°$
  intensity ≈ 15% as bright as the primary
- **5th Order Rainbow**:
  5 internal reflections
  $\theta_{view} = 52.9°$
  very faint

Viewing angles for tertiary and quaternary rainbows are 43° from the <u>sun</u> (i.e., not from the antisolar point). Hence, their faint presence is washed out

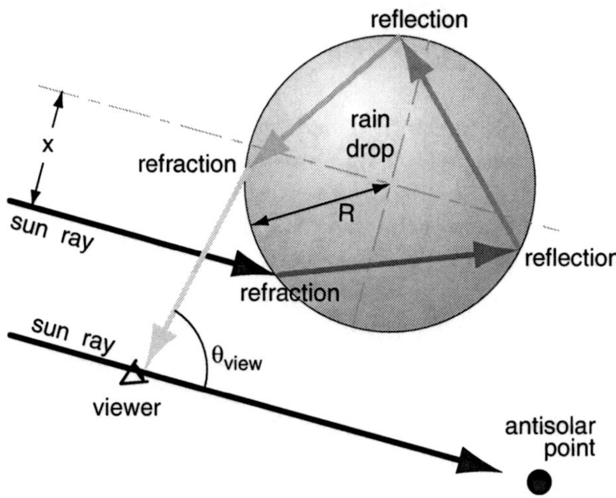

Figure 22.14
Ray geometry for secondary rainbow.

by the intense zero-order glow from the sun. **Zero-order glow** is light from the sun passing through a raindrop with no internal reflections while en route to the observer.

Alexander's Dark Band

Neither the primary nor secondary rainbows return light in the viewing-angle range between 42.7° and 50°. Hence, the sky is noticeably darker between these rainbows (Figs. 22.7 & 22.13). This dark region is called **Alexander's dark band**, after Greek philosopher Alexander of Aphrodisias who described it during the third century A.D.

Other Rainbow Phenomena

Larger diameter raindrops cause more vivid, colorful rainbows. But as discussed in the Satellite & Radar chapter, large raindrops are often not spherical. A mixture of large non-spherical drops (oblate spheroids the shape of hamburger buns) and moderate size spherical drops causes cause **twinned rainbows**, where the top of the rainbow circle splits into two rainbows (Fig. 22.15a).

Sea-spray rainbows are about 0.8° smaller radius than normal rainbows, because of greater refraction by salt water. Near sunset, the blue-diminished sunlight illuminates raindrops with reddish light, creating **red bows** with enhanced red and orange bands. Faint rainbows (**moon bows**) can also be created by moonlight at night.

Reflection rainbows form when the incident sunlight bounces off a lake before hitting the raindrops (Fig. 22.15b). This bow is centered on the **anthelic point**, which is as high above the horizon as the antisolar point is below, but at the same azimuth.

INFO • Rainbows and the Renaissance

German monk Theodoric von Freiberg created a physical model of rainbow physics in 1304 by filling a glass sphere with water. By carefully shining light through it at different impact parameter values, he could measure the various output angles. However, the physics behind these observations were not explained until the renaissance, after Willebrod Snell van Royen had discovered the law for refraction in the early 1600's.

Twenty years after Snell's discovery, Descartes used it to demonstrate the capabilities of his scientific method by mathematically solving the rainbow problem. His calculations were very similar to our spreadsheet calculations, except that he manually performed calculations for a large number of different impact parameters. Through these repetitive "brute-force" calculations he verified the magic number of 42° for the brightest returned light, and 50° for the secondary rainbow.

About 30 years later, Newton applied his knowledge of color dispersion to explain why the rainbow has colors. Also, he was able to use his invention of calculus to elegantly derive the magic angle of 42°.

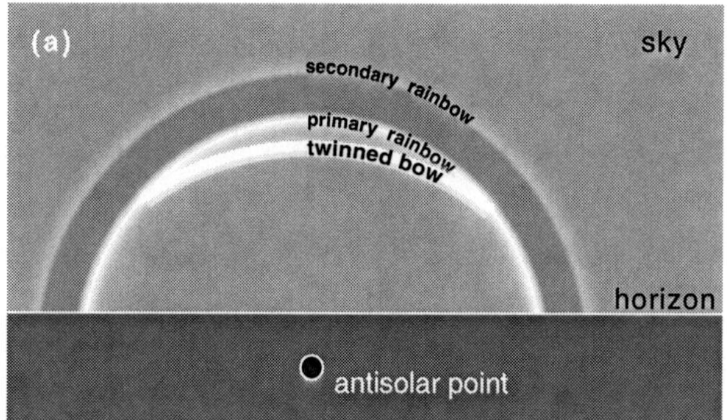

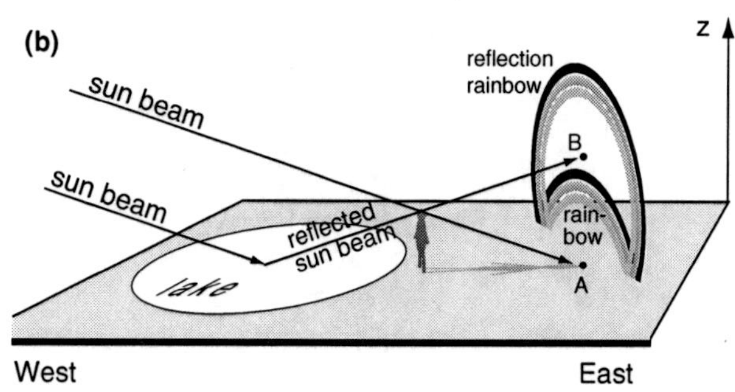

Figure 22.15
(a) Twinned primary rainbow (twinned bow) inside the normal primary rainbow, caused by large oblate rain drops. (b) Reflection rainbow caused by sunlight reflecting from a lake or ocean before reaching the falling raindrops. The reflection bow is a circle centered on the anthelic point (B), while the normal primary rainbow circle is centered on the antisolar point (A).

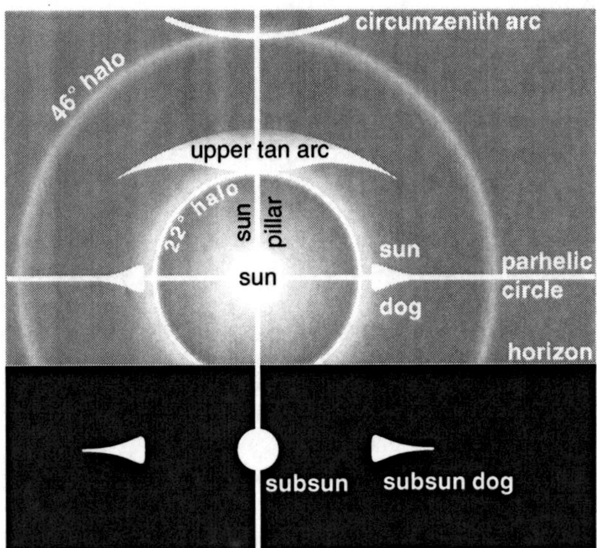

Figure 22.16
Some of the more common halos and other optical phenomena associated with ice crystals in air.

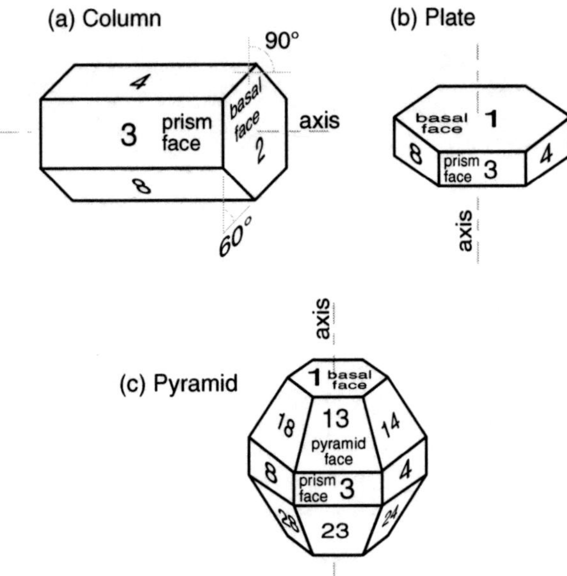

Figure 22.17
Ice-crystal geometry relevant to atmospheric optics. The angles highlighted in (a) are known as **wedge angles**, β_c, *which correspond to the angles of a prism that would yield the refraction as discussed in this section. There are many different wedge angles (not shown here) between different faces of the pyramid.*

ICE CRYSTAL OPTICS

Most of the common ice-crystal optical phenomena are seen by looking more-or-less toward the sun (Fig. 22.16). To observe them, shield your eyes from the direct rays of the sun to avoid blinding yourself. Also, wear sunglasses so your eyes are not dazzled by the bright sky close to the sun.

Subsuns, sun pillars, and parhelic circles are caused by simple reflection from the outside surface of ice crystals. Reflections from inside the ice crystal can also contribute to the intensity of these phenomena.

Sundogs, halos, circumzenith arcs, and tangent arcs are caused by refraction through the ice crystals, with the red color closest to the sun.

Subsun dogs are caused by both refraction and reflection in the ice. There are many other optical phenomena related to ice crystals, only a few of which we cover here.

Ice crystals (types of snow) can form as high-altitude cirrus clouds, can be naturally seeded within and fall from the base of mid-altitude water-droplet clouds, can grow within low- or mid-altitude cloudless cold air of high relative humidity, or can be stirred up by the wind from fresh surface snow. For all these situations, the best optical displays occur with a diffuse concentration of crystals in the air. Namely, the air looks mostly or partially clear, except for sparkles of light reflected from the sparsely spaced crystals (nicknamed **diamond dust**).

The crystals have a variety of shapes, most with hexagonal cross section (see Figs. 7.12 and 7.13 in the Precipitation chapter). The two shapes most important for atmospheric optical phenomena are the **hexagonal column** and **hexagonal plate** (Fig. 22.17). The plate is just a very short column. **Hexagonal pyramid**-capped columns cause rarer halos.

Plates or columns, if larger than about 30 µm, gently fall through the air with the **orientation** relative to horizontal as sketched in Fig. 22.17. Namely, the crystal axis of large columns is nearly parallel to the ground, and the axis of large plates is nearly vertical. Surprisingly, this natural orientation is associated with relatively high aerodynamic drag on the falling crystal. Smaller crystals tend to tumble as they fall, giving random orientations.

Crystal faces are numbered as shown in Fig. 22.17, where **basal faces** numbered 1 and 2 mark the ends of the hexagonal column, and **prism faces** 3 - 8 wrap sequentially around the sides. In naming the relative positions of prism faces: 4 and 3 are examples of **adjacent** faces, 4 and 8 are **alternate** faces, while 4 and 7 are **opposite** faces. **Pyramid faces** are numbered as double digits, where the first digit indicates the number of the adjacent basal face, and the second digit indicates the number of the adjacent prism face. The numbering is useful to track ray paths.

As shown in Fig. 22.17, the most important **wedge angles** (β_c) of plate and column crystals are 60° (such as between faces 3 and 7, or 4 and 8) and 90° (such as between faces 3 and 1, or between 4 and 2). Angles for pyramids are discussed later.

Sun Pillar

A sun pillar is a vertical column of light that appears to come out of the top (and sometimes the bottom) of the sun (Fig. 22.16). It forms when sunlight reflects off the outside of any of the following nearly-horizontal faces:

- prism faces of large oriented column ice crystals (such as off of face 8 in Fig. 22.18a),
- basal faces of fluttering large oriented plates (such as off face 2 in Fig. 22.18b), or
- basal faces of large oriented **dendrites** (6-armed snowflake plates) (such as off the bottom of the snowflake in Fig. 22.18d).

These crystals act like myriad tiny mirrors.

Internal reflections can contribute to the intensity of sun pillars. In the example of Fig. 22.18c, light enters the ice crystal (with refraction) through face 5, then reflects from the inside of the crystal at face 1, and leaves (with refraction) from face 8.

An end view of this phenomenon for reflection from oriented hexagonal columns is sketched in Fig. 22.19. Column crystals are free to rotate about their column axis as they gently fall to Earth, and flat plates and dendrites have random wobble of their axes causing them to flutter like leafs when they fall. As a result, some of the ice crystals have faces that by chance reflect the sunlight to your eye, while other ice crystals reflect the sunlight elsewhere.

Relatively few sunbeams reflect to your eye, causing the sun pillar to be very faint. The best time to observe sun pillars is at sunrise or sunset when the sun is hidden just below the horizon, but is still able to illuminate the ice precipitation. Sun pillars often appear to be red not because of refraction in the crystal, but because light from the rising or setting sun has lost much of its blue components due to scattering from air molecules before it reaches the ice crystals.

Similar pillars can form above and below:

- bright street lights
- the moon and bright planets (e.g., Venus) and
- other bright lights, such as aircraft lights,

and these pillars often have the same color as the light source.

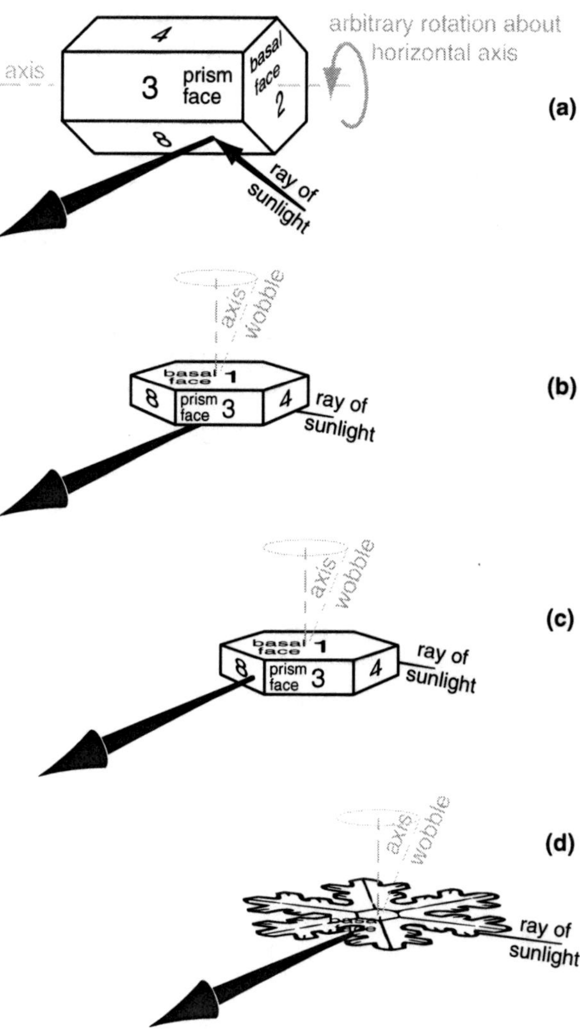

Figure 22.18
Sun-pillar optics caused by reflection of sunlight off the faces of large (a) oriented ice columns, (b & c) oriented plates, and (d) oriented dendrites.

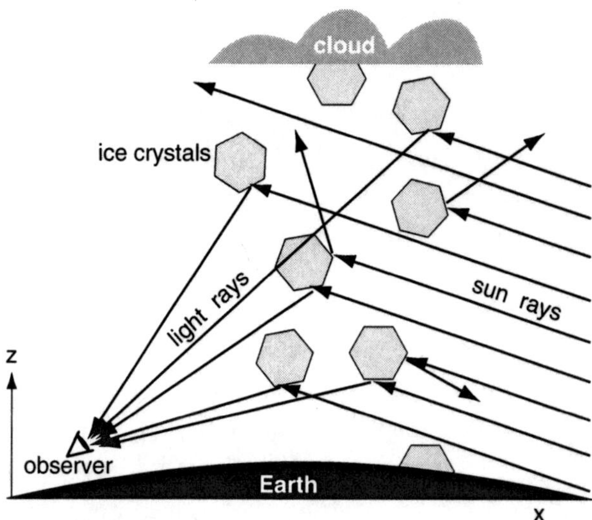

Figure 22.19
Geometry of sun pillar optics (end view) reflecting from oriented large hexagonal-column ice crystals (size is exaggerated).

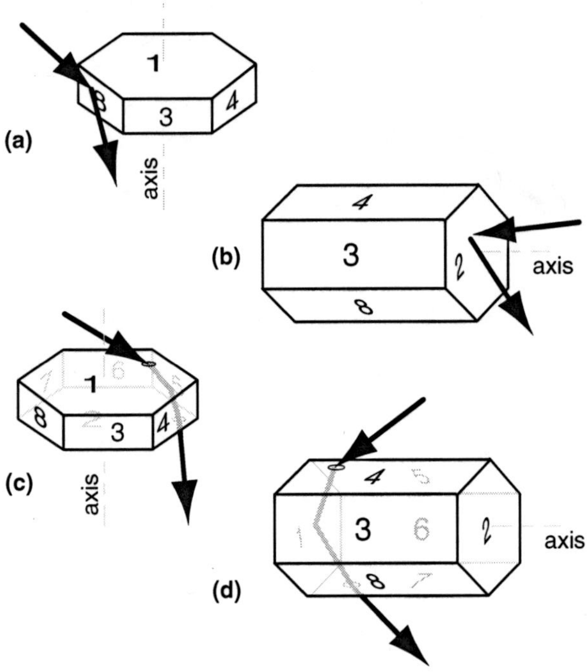

Figure 22.20
Ray geometry of parhelic-circle optics, showing some of the possible external and internal reflections. Small ovals indicate points where light rays enter or leave the ice crystal.

Parhelic Circle

The parhelic circle is a horizontal band of white light extending left and right through the sun (Fig. 22.16). It can be formed by external reflection off of the vertical faces of large hexagonal plates (e.g., face 8 in Fig. 22.20a) or large hexagonal columns (e.g., face 2 in Fig. 22.20b). It can also be formed by internal reflection off of similar vertical faces, such as the ray path touching faces 1 5 2 in Fig. 22.20c, or the path touching faces 4 1 7 in Fig. 22.20d.

These crystals can have any rotation orientation about their axis, causing reflections from many different ice crystals to reach your eye from many different angles left and right from the sun (Fig. 22.21). In other words, they act as a large number of small mirrors.

Subsun

The subsun is a spot of white light seen below the horizon (Fig. 22.16), as viewed looking down from a bridge, mountain, or aircraft. It forms by external reflection of sunlight off of the top surfaces of large hexagonal plates (face 1 in Fig. 22.22a), and by internal reflection (face 2 in Fig. 22.22b; namely, the ray touches faces 7 2 4). If all the hexagonal plates are oriented horizontally, they act as mirrors to produce a simple reflection of the sun (Fig. 22.23). As given by eq. (22.1), the subsun appears the same angle θ_3 below the horizon that the sun is above θ_1.

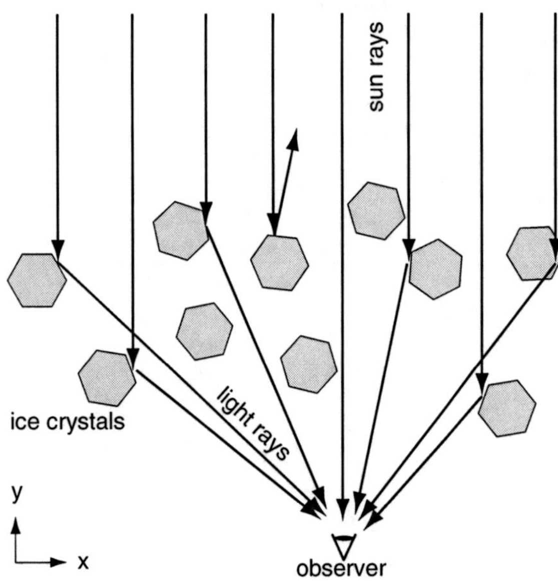

Figure 22.21
Reflection from many large hexagonal plates (size is exaggerated) to form a parhelic circle (top or plan view).

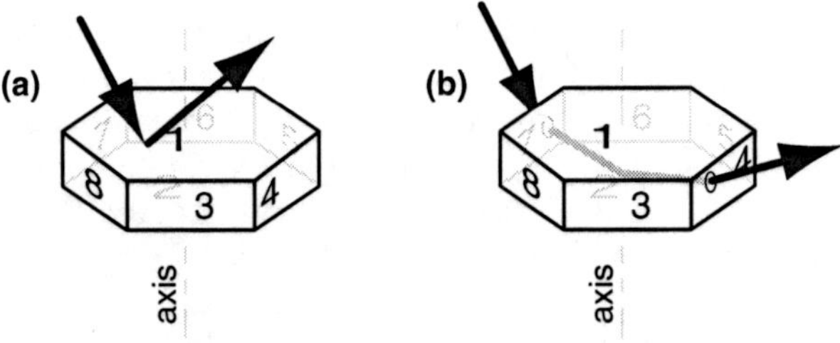

Figure 22.22
Ray geometry of subsun optics. Small ovals indicate points where light rays enter or leave the ice crystal.

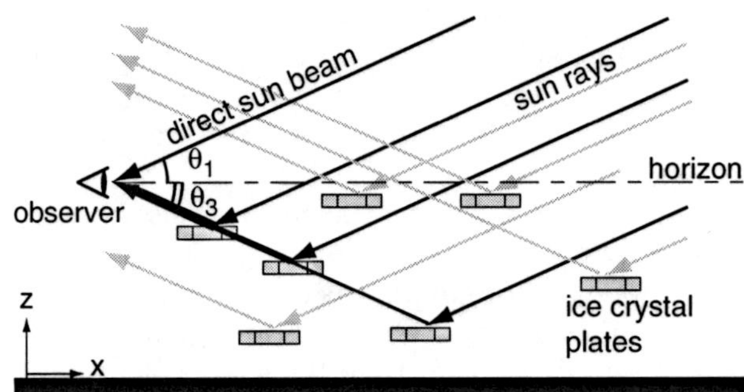

Figure 22.23
Reflection from many ice-crystal plates to form a subsun (side view). Grey rays reach someone else's eyes.

22° Halo

Optics for 22° halos (Fig. 22.16) are not as simple as might be expected. Small ice-crystal columns are free to tumble in all directions. As a result, light rays can enter the crystal at a wide range of angles, and can be refracted over a wide range of sky.

The brightest light that we identify as the halo is caused by refraction through those ice crystals that happen to be oriented with their column axis perpendicular to the sun rays (Fig. 22.24). We will study this special case. Different crystals can nevertheless have different rotation angles about their axis (the dark point in the center of the ice crystal of Fig. 22.24). As a result, the angle of incidence θ_1 can vary from crystal to crystal, thereby causing different viewing (output) angle θ_2 values.

For hexagonal crystals (for which the **wedge angle** is $\beta_c = 60°$) and for the special case of light rays approaching normal to the column axis, Snell's law yields:

$$\theta_2 = \theta_1 - \beta_c + \arcsin\left\{\frac{n_{ice}}{n_{air}} \cdot \sin\left[\beta_c - \arcsin\left(\frac{n_{air}}{n_{ice}} \cdot \sin\theta_1\right)\right]\right\} \qquad \bullet(22.15)$$

where n is refractive index.

The 22° halo observed by a person is the result of superposition of rays of light from many ice crystals possessing many different rotation angles. Hence, it is not obvious why 22° is a magic number for θ_2. Also, if refraction is involved, then why are the red to yellow colors seen most vividly, while the blues and violets are fading to white?

To answer these questions, suppose ice crystals are randomly rotated about their axes. There would be an equal chance of finding an ice crystal having any incidence angle θ_1. We can simulate this on a computer spreadsheet by doing calculations for a large number of evenly-spaced incidence angles (for example, every 2°), and then calculating the set of θ_2 angles as an outcome. The result is plotted in Fig. 22.25 for two of the colors (because n_{air}/n_{ice} varies with color).

Colors and angles of a 22° halo can be explained using Fig. 22.26 (which is a zoomed view of Fig. 22.25). For viewing (output) angles less than about 21.6°, there is no refracted light of any color; hence, the halo is dark inside (the edge closest to the sun).

In the range of viewing angles averaging 22° (i.e., from 21.6° to 22.6°), there are 29 data points plotted in Fig. 22.26, which cause the bright ring of light we call the halo. There are also data points at larger angles, but they are more sparse — causing the brightness of the halo to gradually fade at greater viewing angles from the sun.

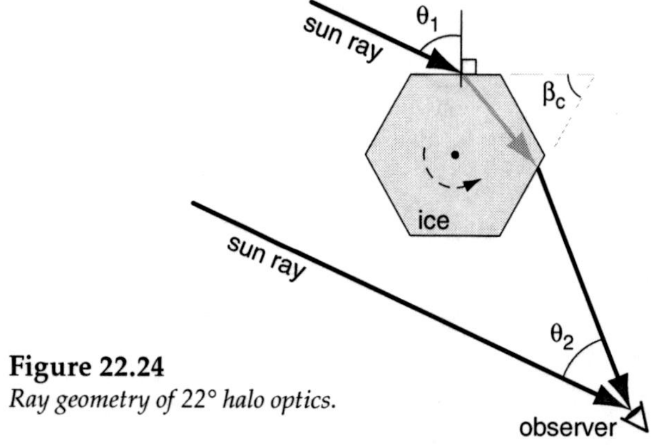

Figure 22.24
Ray geometry of 22° halo optics.

Sample Application

What is θ_2 if $\theta_1 = 80°$ and $\beta_c = 60°$ for red light in a 22° halo? $T = -20°C$, $P = 80$ kPa, $RH = 75\%$.

Find the Answer

Given: $\theta_1 = 80°$. $\beta_c = 60°$. $\lambda = 0.7$ µm for red light.
$n_{air} = 1.0002479$, $n_{ice} = 1.3074$ from Table 22-1.
Find: $\theta_2 = ?°$

Use eq. (22.15) with $\beta_c = 60°$ for hexagonal crystal:

$$\theta_2 = 80° - 60° + \text{asin}\left\{\frac{1.3074 \cdot \sin}{1.0002479}\left[60° - \text{asin}\left(\frac{1.0002479}{1.3074} \cdot \sin 80°\right)\right]\right\}$$

$$= 80° - 60° + 14.59° = \underline{\mathbf{34.59°}}$$

Check: Units OK. Physics OK.
Exposition: Quite different from 22°, but agrees with Fig. 22.25. Namely, this is one of the light rays that contributes to the bright glow outside the 22° halo, as sketched in Fig. 22.16.

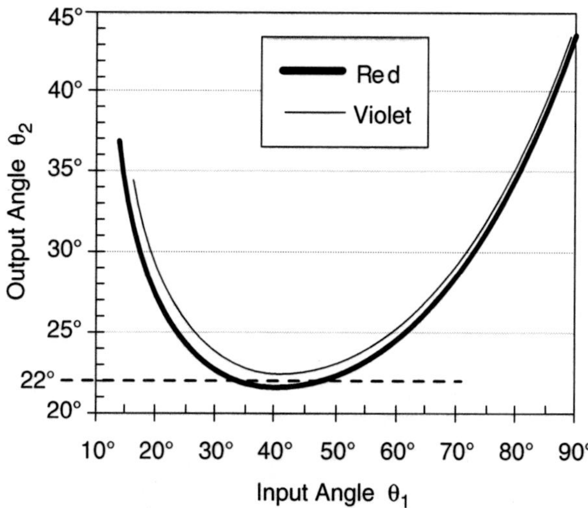

Figure 22.25
Viewing (output) angles θ_2 from hexagonal ice crystals in a 22° halo, for various incident angles θ_1 of sunlight.

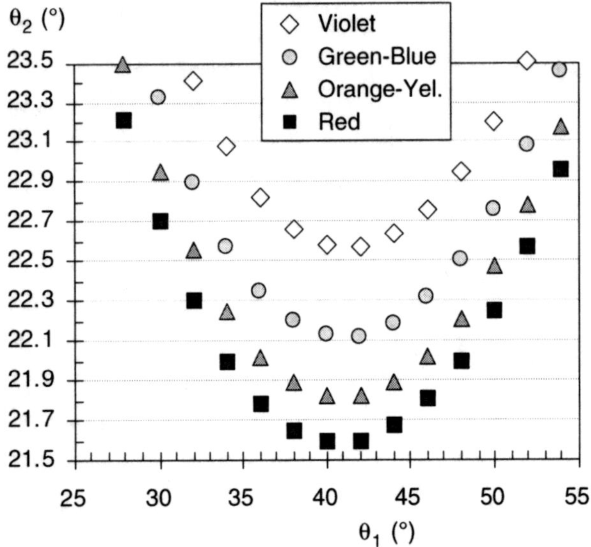

Figure 22.26
Blow-up of viewing (output) angles θ_2 from hexagonal ice crystals in a 22° halo, for various incident angles θ_1.

> **Sample Application**
> Find θ_{2min} for green-blue light for wedge angle 60°.
>
> **Find the Answer**
> Given: $\beta_c = 60°$, $n_{ice} = 1.3135$ for blue-green (λ=0.5 µm).
> Find: θ_{2min} = ? °. (Assume T = –20°C & $n_{air} \approx 1.0$.)
>
> Use eq. (22.16): θ_{2min} = 2·arcsin[1.3135 ·sin(30°)] – 60°
> = 82.10° – 60° = **22.10°**
>
> **Check**: Agrees with Table 22-2.
> **Exposition**: The different colors of the 22° halo are not all at exactly 22°.
> CAUTION: If you do these angle calculations on a spreadsheet, don't forget to convert to/from radians for the sine/arcsine terms.

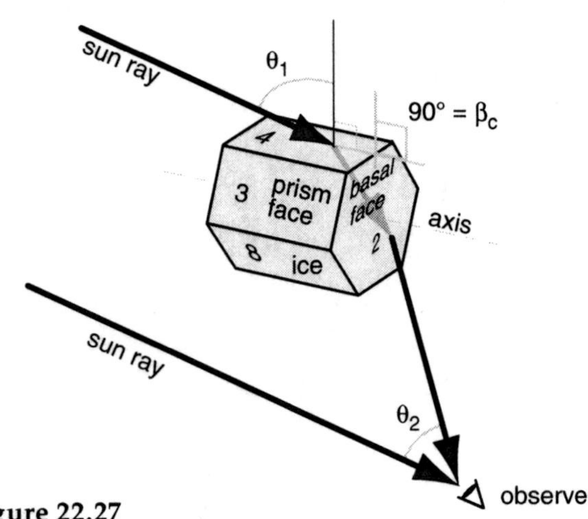

Figure 22.27
Ray geometry for 46° halo.

For a viewing angle range of θ_2 = 21.5° to 21.7°, we see four "red light" data points in Fig. 22.26, but no other colors. Thus, the portion of the halo closest to the sun looks bright red. In the next range of viewing angles (21.7° - 21.9°), we see four "yellow-orange" and two "red" data points. Thus, in this range of viewing angles we see bright orange light.

In the angle range 22.1 - 22.3°, there are four blue-green data points, and two orange-yellows and two reds. These colors combine to make a bluish white color. By an angle of 23° , there are roughly equal portions of all colors, creating white light. Thus, the colors of a 22° halo can have bright reds, oranges, and yellows on the inside, fading to light green and bluish-white further away from the sun.

The **minimum viewing angle** is the angle for which the halo gets its name, and can be found from:

$$\theta_{2\,min} = 2 \cdot \arcsin\left[\frac{n_{ice}}{n_{air}} \cdot \sin\left(\frac{\beta_c}{2}\right)\right] - \beta_c \qquad (22.16)$$

where β_c is wedge angle and n is refractive index. This is at the bottom of the curves in Fig. 22.26.

46° Halo

The 46° halo (Fig. 22.16) forms by sun rays shining through a prism face and a basal end of short hexagonal columns (Fig. 22.27). Equations (22.15 and 22.16) apply, but with an ice wedge angle β_c = 90°. The resulting viewing angles θ_2 are shown in Fig. 22.28 for a variety of input angles θ_1. Except for the different radius, this halo has visual characteristics similar to those of the 22° halo. But it is faint, and less frequently seen.

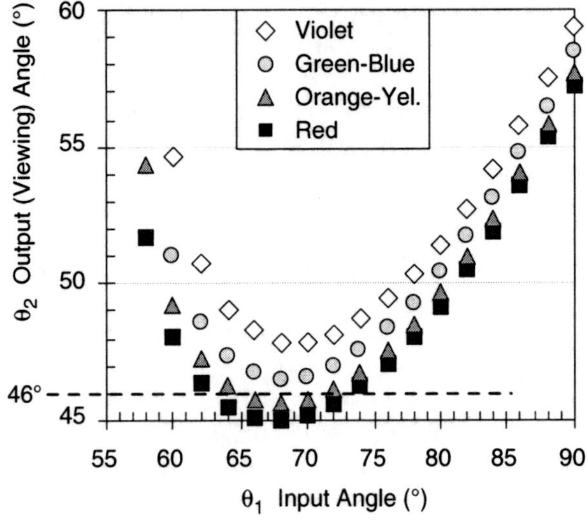

Figure 22.28
Viewing (output) angles for a 46° halo.

Sample Application

Find the viewing angle θ_2 for a 46° halo, given input angle θ_1 = 80° and β_c = 90° for red light.

Find the Answer

Given: $\theta_1 = 80°$, $\beta_c = 90°$. λ = 0.7 µm for red light.
Find: $\theta_2 = ?°$
Assume: n_{air} = 1.0002479, n_{ice} = 1.3074 from Table 22-1.

Use eq. (22.15): $\theta_2 = 80° - 90° +$

$$\text{asin}\left\{\frac{1.3074\cdot \sin}{1.0002479}\left[90° - \text{asin}\left(\frac{1.0002479}{1.3074}\cdot \sin 80°\right)\right]\right\}$$

= **49.25°**

Check: Units OK. Physics OK.
Exposition: This agrees with Fig. 22.28.

Halos Associated with Pyramid Crystals

Ice crystals with pyramid ends (Fig. 22.17c) have a wide variety of wedge angles depending on the path of the light ray through the crystal. This can cause additional halos of differing radii centered on the sun (Fig. 22.29).

Table 22-2 lists the ray path, wedge angle, and minimum viewing angle (from eq. 22.16) for some of these odd-radius halos. The 9° halo is often hard to see against the glare close to the sun, while the 18°, 20°, 23°, and 24° halos might be mis-reported as 22° halos unless accurate measurements are made. Nonetheless, most of the rare halos have been observed and photographed.

Table 22-2. Halo viewing angles (θ_{2min}) found from eq. (22.16). Ray paths refer to the face numbers in the Fig. at right. β_c = wedge angle of ice crystal (from *W. Tape & J. Moilanen, 2006: "Atm. Halos & the Search for Angle x"*. $n_{air} \approx 1.0$

		color =	red	yellow orange	green blue	violet
		wavelength (µm) =	0.7	0.6	0.5	0.4
(For T = –20°C)		n_{ice} =	1.3074	1.3099	1.3135	1.3199
Halo Name	β_c (°)	Ray Path	Minimum **Halo Viewing Angles** (°)			
9°	25	18 · 5	8.88	8.95	9.06	9.24
18°	52.4	18 · 24	18.11	18.27	18.49	18.89
20°	56	18 · 15	19.73	19.90	20.14	20.58
22°	60	8 · 4	21.64	21.83	22.10	22.59
23°	62	18 · 7	22.65	22.85	23.14	23.66
24°	63.8	18 · 4	23.60	23.81	24.11	24.65
35°	80.2	18 · 14	34.53	34.87	35.37	36.26
46°	90	1 · 3	45.18	45.71	46.49	47.91

Science Graffito

"The skipper he stood beside the helm,
His pipe was in his mouth,
And he watched how the veering flaw did blow
The smoke now West, now South.
Then up and spake an old Sailor
Had sailed to the Spanish Main,
'I pray thee, put into yonder port,
For I fear a Hurricane.
Last night **the moon had a golden ring**,
And tonight no moon we see!'
The skipper he blew a whiff from his pipe,
And a scornful laugh laughed he."
– by Henry Wadsworth Longfellow, from *The Wreck of the Hesperus.*

INFO • Estimating 22° Angles

For most people, if you extend your arm fully and open your hand as widely as possible, then the angle that you see subtended by your hand is about 22°. This works for people of different sizes, because hand size and arm length are often proportional.

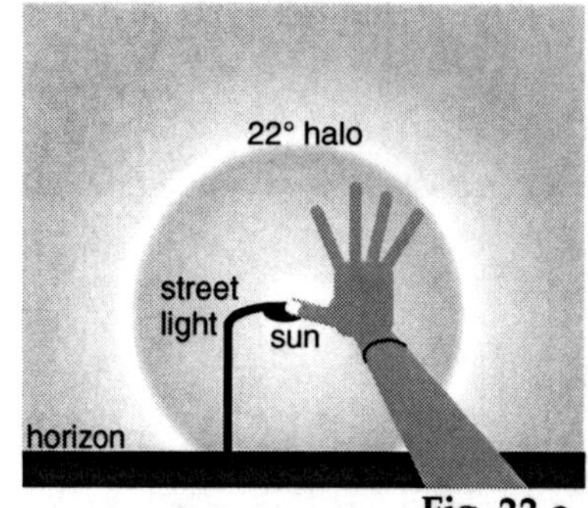

Fig. 22.a

To view halos, first move so that the sun is hidden behind an obstacle such as a street light, stop sign, or a corner of a building, so you don't blind yourself. Then extend your arm and stretch your hand, and position the tip of your thumb over the location where the center of the sun would be (hidden behind the obstacle). The tip of your little finger should just touch the halo if it is indeed a 22° halo.

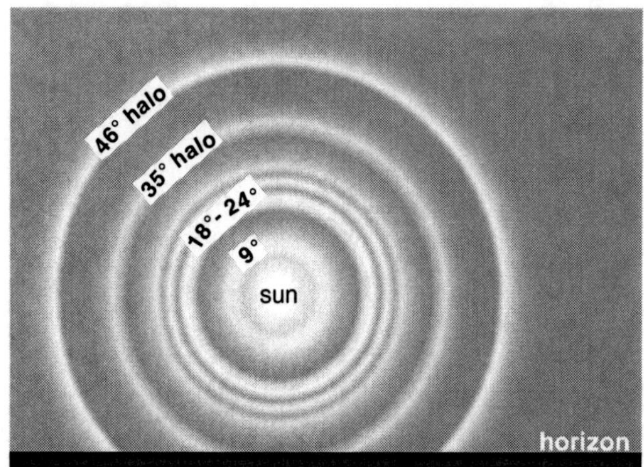

Figure 22.29
Halos. *(It is unusual to see all these halos at the same time.)*

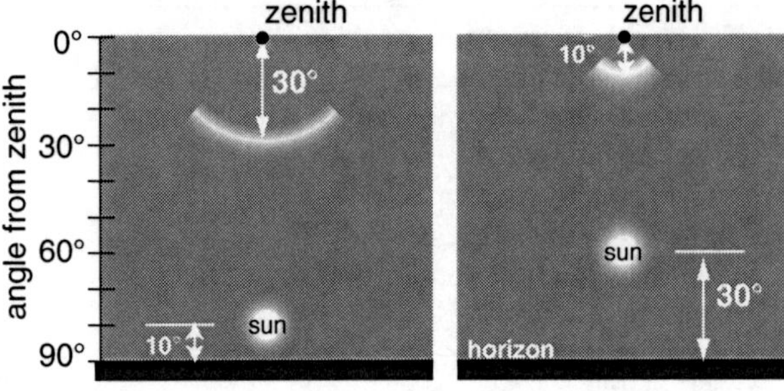

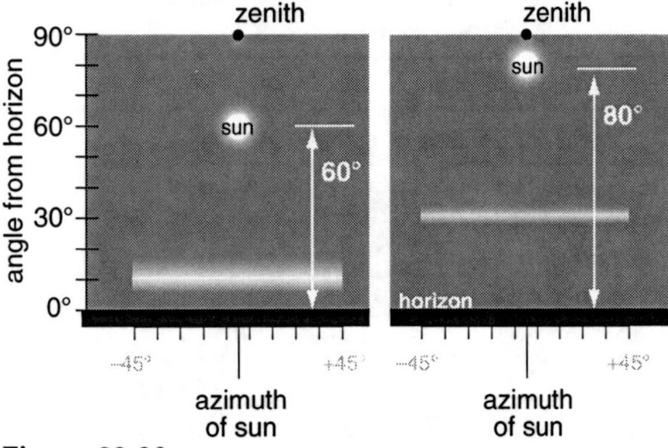

Figure 22.30
Although both circumzenith and circumhorizon arcs are portions of circles around the zenith and both are parallel to the horizon, we perceive them differently. Circumzenith arcs look like a part of a circle around the zenith point, as sketched in (a). Circumhorizon arcs look like horizontal lines just above the horizon, as sketched in (b). Angles above are approximate.

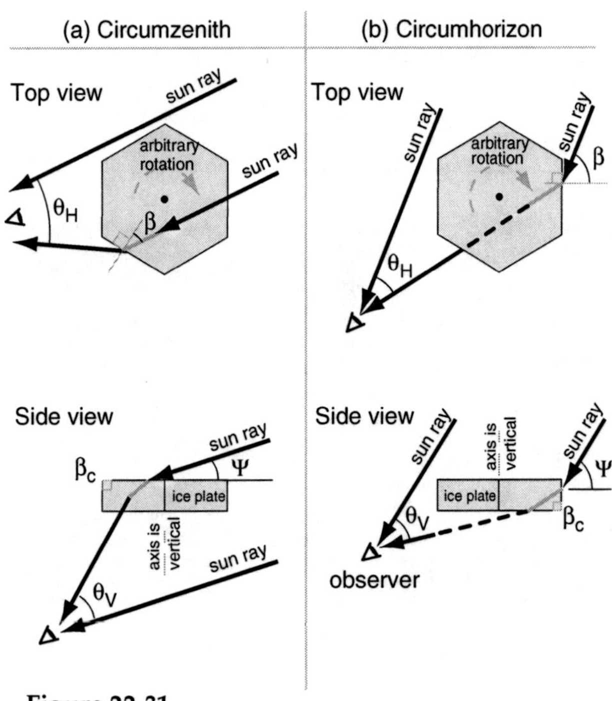

Figure 22.31
Ray geometry for (a) circumzenith & (b) circumhorizon arcs.

Circumzenith & Circumhorizon Arcs

Circumzenith arcs and circumhorizon arcs are similar (Fig. 22.30). They are both portions of circles at constant viewing angle above the horizon, and thus at constant viewing angle from the zenith (the point directly above the viewer). They are lines parallel to the horizon that partially encircle the viewer. Only the portion of the circle within about ±45° azimuth of the sun's azimuth from the viewer are visible; hence, they are seen as arcs, not circles.

Both arcs are caused by light refracted through the $\beta_c = 90°$ wedge angle of large oriented hexagonal plates (Fig. 22.31). For circumzenith arcs the light ray enters the top of the plate (through basal face 1) and exits through the side (such as through prism face 8 in Fig. 22.17b). For circumhorizon arcs the ray enters a side (such as face 5) and exits the bottom (face 2). Because refraction is involved, these arcs can be colorful, with red closest to the sun.

The arc's viewing-angle from the zenith (i.e., the apparent radius of the arc) varies with solar elevation (Fig. 22.32). The circumzenith arc is seen high in the sky (above the sun) only when the sun is low in the sky (for solar elevation angles of $0° < \Psi < 32°$). The circumzenith arc is always outside the 46° halo (Fig. 22.16).

The circumhorizon arc is seen low in the sky (below the sun) only when the sun is high in the sky ($58° < \Psi \leq 90°$). At high latitudes the circumhorizon arc is never visible because the sun never gets high enough in the sky.

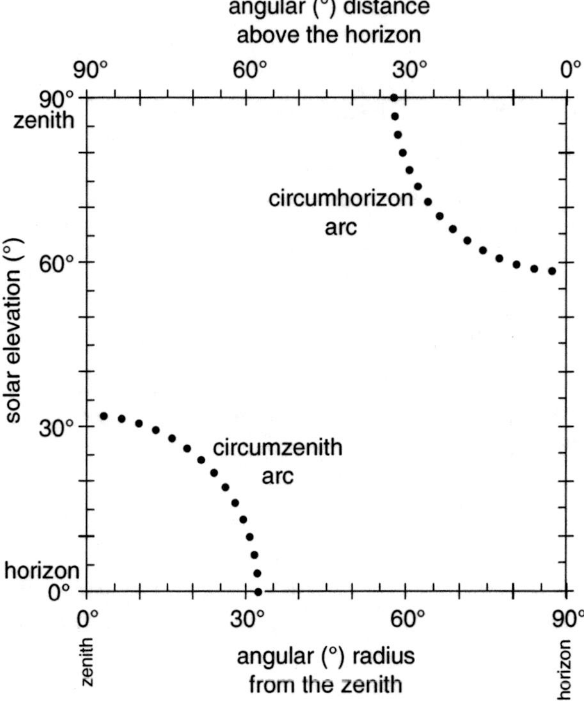

Figure 22.32
Radius of circumzenith and circumhorizon arcs about the zenith point, as viewed from the ground.

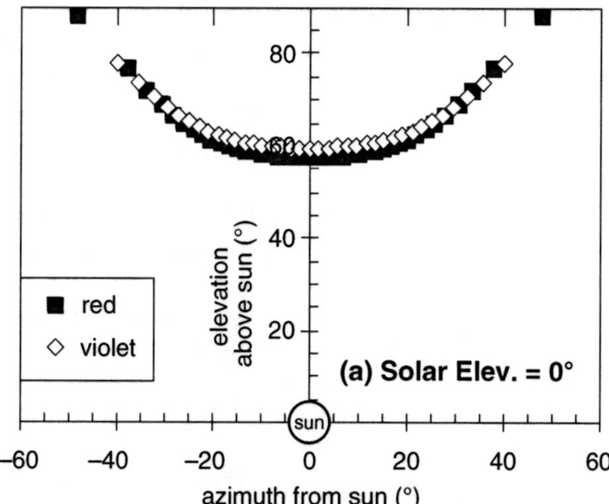

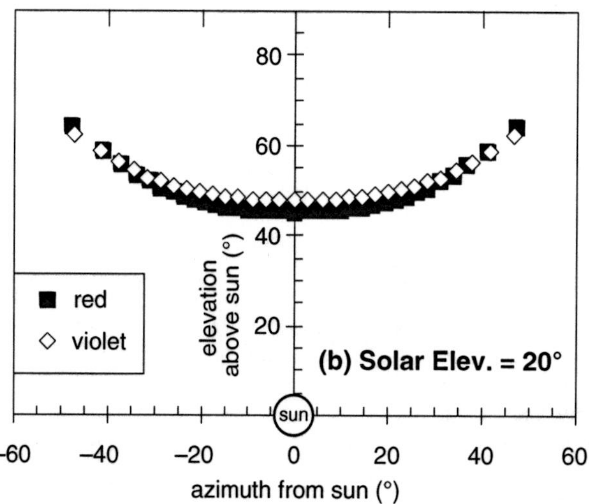

Figure 22.33
Circumzenith arcs found by solving eq. (22.17) for a variety of crystal rotation angles β, at solar elevation angles of (a) 0°, at left, and (b) 20°, above.

To illustrate how these arcs form, we will focus on just one of them — the circumzenith arc. The large hexagonal plates can assume any rotation angle β about their vertically-oriented column axis. At different times and locations, there are different solar elevation angles ψ. The ray equations for vertical viewing angle θ_V (elevation above the sun) and horizontal angle θ_H (azimuth from the sun) are:

$$\theta_V = S_\alpha[\arccos(\mu_{ai}\cdot\cos\psi), \beta, \mu_{ia}] - \psi \qquad (22.17)$$

$$\theta_H = S_\beta[\arccos(\mu_{ai}\cdot\cos\psi), \beta, \mu_{ia}] - \beta$$

Subscripts *"ai"* are for rays going from air to ice (e.g.; $\mu_{ai} = n_{air}/n_{ice}$), while subscripts *"ia"* are from ice to air (e.g.; $\mu_{ia} = n_{ice}/n_{air}$).

A spreadsheet can be used with a large number of evenly spaced values of β to simulate the superposition of rays from a large number of randomly-rotated ice plates. The resulting locus of output viewing angles traces the circumzenith arc in Fig. 22.33.

Eq. (22.17) does not include reflection of some or all of the rays from the outside of the crystal as a function of incidence angle, nor does it include the critical angle for rays already inside the crystal. Thus, some of the points plotted in Fig. 22.33 might not be possible, because the sun rays corresponding to those points either cannot enter the crystal, or cannot leave it.

Science Graffito

"It is clear and cold... And all about is snow and silence. And in the sky are three suns, and all the air is flashing with the dust of diamonds."
– Jack London, on a trip down the Yukon River in search of gold in Alaska.

Sample Application

For a circumzenith arc, what are the elevation and azimuth viewing angles (relative to the sun)? Solar elevation is 10°. Rotation angle is β = 20°. Assume nearly red light with: n_{air} = 1.0002753 and n_{ice} = 1.307 .

Find the Answer

Given: Ψ = 10°, β = 20°, n_{air} = 1.0002753, n_{ice} = 1.307
Find: θ_V = ?°, and θ_H = ?°

First, find the refraction parameters:

$\mu_{ai} = n_{air}/n_{ice} = 0.7653$, and $\mu_{ia} = 1/\mu_{ai} = 1.3067$

Next: $\arccos[\mu_{ai}\cos(\Psi)] = \arccos(0.7653\cdot\cos 10°) = 41.09°$

Then, use eqs. (22.17):

$\theta_V = S_\alpha[41.09°, 20°, 1.3067] - 10°$
$\theta_H = S_\beta[41.09°, 20°, 1.3067] - 20°$.

To find the S values, we first need to use eq. (22.8):

$$b_{\alpha\beta} = \frac{(1.3067)^2}{1+[1-(1.3067)^2]\cdot\left\{[\tan(41.09°)]^2+[\tan(20°)]^2\right\}}$$
$= 4.634$

Then use eq. (22.9):

$$S_\alpha(\alpha_1, \beta_1, \mu_{12}) = \arctan\left[\left\{4.634\cdot[\tan(41.09)]^2\right\}^{1/2}\right]$$
$S_\alpha = 61.95°$

Similarly: $S_\beta = 38.08°$
Finally: $\theta_V = 61.95° - 10° =$ **<u>51.95°</u>**
$\theta_H = 38.08° - 20° =$ **<u>18.08°</u>**

Check: Units OK. Physics OK.
Exposition: This answer agrees with Fig. 22.33b. Obviously this calculation is tedious for more than one ray, but it is easy on a spreadsheet.

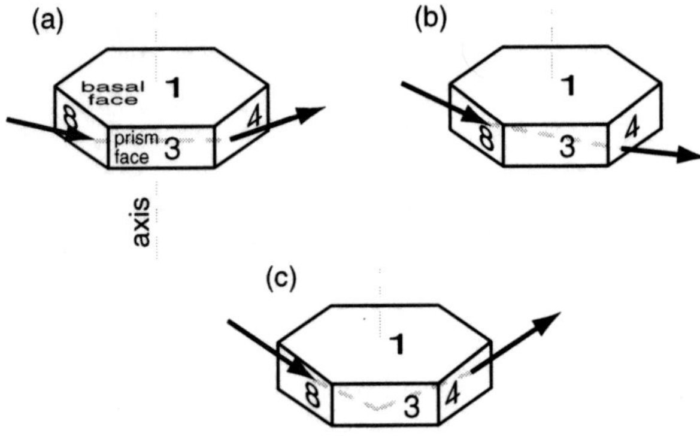

Figure 22.34
Geometry of sun dog optics (oblique side view), for (a) low sun elevation angle, (b) higher sun angle, and for (c) subsun dogs.

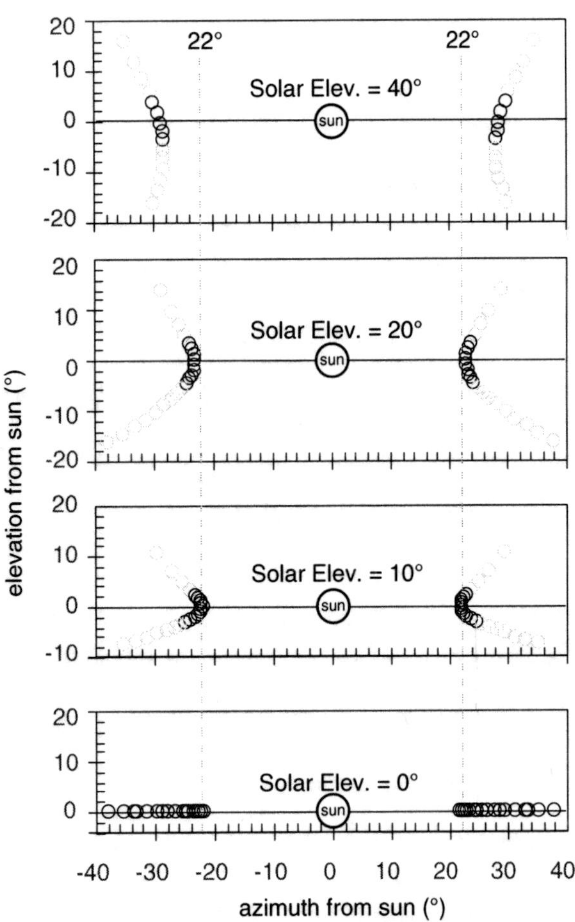

Figure 22.35
Variation of parhelia (sun-dog) appearance with solar elevation angle. Vertical dashed lines are 22° from sun. Greater density of data points indicates brighter parhelia. Grey data points, although solutions to eqs. (22.18), are unphysical because those equations don't consider the finite size of the ice crystals.

Sun Dogs (Parhelia)

Parhelia are bright spots of light to the right and left of the sun, roughly 22° from the sun (Fig. 22.16). As the sun moves through the sky, the parhelia move with the sun as faithful companions — earning their nickname **sun dogs**.

They are formed when light shines through the sides (prism faces) of large horizontal ice-crystal plates that gently fall through the air with their column axes vertical. The crystals are free to rotate about their column axes, so there is no favored rotation angle.

Because of the preferred orientation of the column axis, sun-dog optics depend on the solar elevation angle. When the sun is on the horizon, sun rays shine through the crystal as sketched in Fig. 22.34a, with a nearly horizontal path through prism faces such as 8 4. This causes the parhelia at a viewing angle of 22° from the sun. If a 22° halo is also present due to other smaller ice crystals in the air, then each parhelion appears as a bright spot on the halo to the right or left of the sun.

For higher sun angles (Fig. 22.34b), ice-crystal geometry causes the sun dogs to move horizontally further away from the sun. For example, at a solar elevation angle of $\Psi = 47°$, the sun dogs are at a horizontal viewing angle of 31° from the sun; namely, they are found outside the 22° halo. As the solar elevation angle becomes greater than about 60°, the sun dog is so faint and so far from the 22° halo that it effectively ceases to exist.

For a single ray input from the sun, let β be the rotation angle of the ice-crystal plate about its column axis. The ray equation for a sun dog is:

(22.18)

$$\theta_V = S_\alpha[S_\alpha(\psi,\beta,\mu_{ai}),\ 60° - S_\beta(\psi,\beta,\mu_{ai}),\ \mu_{ia}] - \psi$$

$$\theta_H = S_\beta[S_\alpha(\psi,\beta,\mu_{ai}),\ 60° - S_\beta(\psi,\beta,\mu_{ai}),\ \mu_{ia}] + \beta - 60°$$

where θ_V is the viewing angle of the output ray (elevation above the sun), and θ_H is the viewing angle (azimuth to the right of the sun).

The sun dog as viewed by an observer is the superposition of rays from many ice crystals having a variety of rotation angles. For randomly rotated crystals, there is an equal chance of any β between 0° and 90°. This is simulated on a spreadsheet by solving the sun-dog equation for a large number of equally-spaced β angles. The results are plotted in Fig. 22.35. For a finite size ice crystal with aspect ratio typical of hexagonal plates, ray paths with large vertical viewing angle are not physically possible because the light ray would exit or reflect from a basal face before reaching the required prism face.

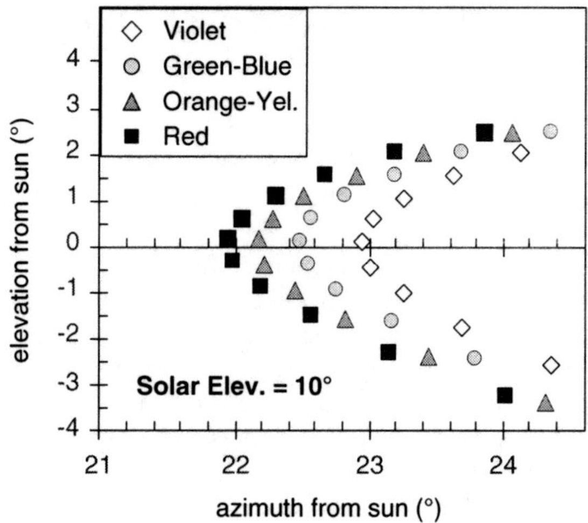

Figure 22.36
Dispersion of colors in a sun dog.

Sun dog colors are similar to the halo, with bright red on the inside (closest to the sun), then orange and yellow, but with green, blue, and violet fading to white (Fig. 22.36 is an expanded view of Fig. 22.35.)

Subsun Dogs (Subparhelia)

Subparhelia (spots of light to the left and right of the subsun) are sketched in Fig. 22.16. Ray geometry for a subsun dog is identical to that for a sun dog, except that the ray within the ice reflects from the inside of the bottom (basal) face (face 2 in Fig. 22.34c), in addition to refractions through prism faces when entering and exiting the crystal.

Tangent Arcs

Tangent arcs are caused when light rays are refracted through large hexagonal column ice crystals, which often fall with their column axis nearly horizontal (Fig. 22.17a). Three phenomena in this tangent-arc category include the **upper-tangent arc**, the **lower-tangent arc**, and the **circumscribed halo**. We will examine the optics of upper-tangent arcs in detail, and then briefly discuss the others.

Upper-Tangent Arc

In Fig. 22.16, the upper tangent arc is just above (and is tangent to) the 22° halo. The shape of this arc changes with solar elevation angle, Ψ.

In the absence of wind shear, the ice-crystal column axes can point in any compass direction within that horizontal plane. Thus, the crystal axis can have any directional orientation γ with respect to the compass direction of the incoming sun ray. The crystal can also have any rotation angle α_1 about the column axis (see Fig. 22.37).

Sample Application

Find the vertical and horizontal viewing angles of a sun-dog ray of red light for solar elevation 40° and crystal rotation β = 45°.

Find the Answer

Given: $\Psi = 40°$, $\beta = 45°$

Assume for red light: $n_{air} \approx 1.0002753$, $n_{ice} \approx 1.307$

Find: $\theta_V = ?°$, $\theta_H = ?°$

Use eq. (22.18), with $\mu_{ai} = 0.7653$ from before:

$$\theta_V = S_\alpha[S_\alpha(40°, 45°, 0.7653), \; 60° - S_\beta(40°, 45°, 0.7653), \; 1.3067] - 40°$$

$$\theta_H = S_\beta[S_\alpha(40°, 45°, 0.7653), \; \{60° - S_\beta(40°, 45°, 0.7653)\}, \; 1.3067] + 45° - 60°$$

For the S_α and S_β inside the square brackets:
$b_{\alpha\beta} = 0.3433$, $S_\alpha = 26.18°$, $S_\beta = 30.37°$

For outside the square brackets: $b_{\alpha\beta} = 2.8448$, because
$S_\alpha[26.18°, (60°–30.37°), 1.3067] =$
$S_\alpha[26.18°, 29.63°, 1.3067] = 39.67°$

$S_\beta[26.18°, 29.63°, 1.3067] = 43.81°$

Finally:
$\theta_V = 39.67° - 40° = \underline{\mathbf{-0.332°}}$
$\theta_H = 43.81° - 15° = \underline{\mathbf{28.81°}}$

Check: Units OK. Physics OK.
Exposition: This portion of the sun dog is nearly at the same elevation as the sun, but is 28.81° to the side of the sun. This is 6.81° outside of the 22° halo.

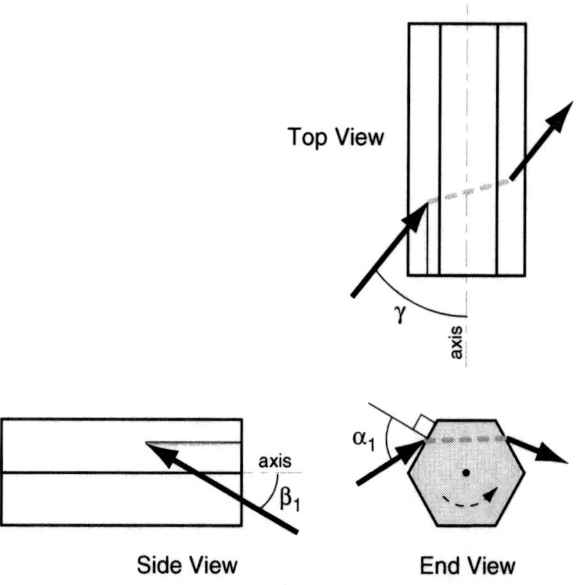

Figure 22.37
Ice crystal geometry associated with upper-tangent arcs.

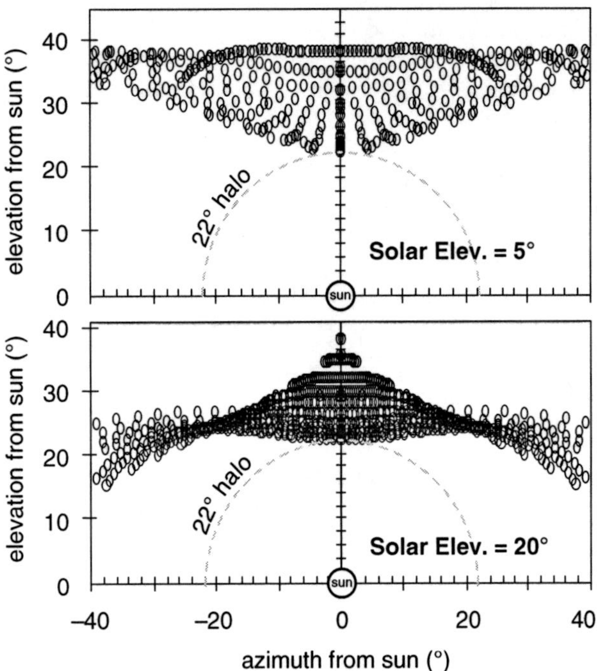

Figure 22.38
The locus of data points (from the solution of eqs. 22.19) shows the upper tangent arc for low solar elevation angles. Data points that were sparser (corresponding to fainter illumination) have been removed to clarify this illustration.

> **Sample Application**
> Find the vertical and horizontal viewing angles of a ray of red light in an upper tangent arc for solar elevation Ψ = 10°, crystal rotation α_1 = 76°, and crystal axis direction γ = 45°.
>
> **Find the Answer**
> Given: $\Psi = 10°$, $\alpha_1 = 76°$, $\gamma = 45°$.
> Assume: $n_{air} \approx 1.0002753$, $n_{ice} \approx 1.307$
> Find: $\theta_V = ?°$, $\theta_H = ?°$
>
> Use eqs. (22.19). Various intermediate results are:
> $\beta_1 = 76°$
> $b_{\alpha\beta} = 0.0402$ (1st calculation)
> $\alpha_3 = 21.2°$, $\beta_3 = 38.8°$
> $b_{\alpha\beta} = 3.9142$ (2nd calculation)
> $\alpha_5 = 22.50°$, $\beta_5 = 57.84°$
> $G = 2.701$, $\varepsilon = 75.4°$, $\phi = -30.4°$
>
> Finally:
> θ_V = **25.20°**
> θ_H = **39.25°**
>
> **Check**: Units OK. Physics OK.
> **Exposition**: This one data point within the upper tangent arc is above the sun and above the top of the 22° halo, but is off to the right. These calculations would need to be repeated for many other values of input angles to create an image similar to Fig. 22.38

Because there are three input angles for this case, the solution of ray paths is quite nasty. The equations below show how one can calculate the ray output vertical viewing angle θ_V (elevation above the sun), and the horizontal viewing angle θ_H (azimuth from the sun), for any input ray. All other variables below show intermediate steps.

As before, one must examine rays entering many crystals having many angles in order to generate the locus of points that is the upper tangent arc. Even on a spreadsheet, this solution is tedious. Again, no consideration was made for the finite size of the crystal, or rays that cannot enter the crystal due to reflection, or rays that are trapped inside because of the critical angle.

Given: Ψ, α_1, and γ:

$$\beta_1 = \arctan\left[\cos(\gamma)\cdot\tan(90° - \psi)\right] \tag{22.19}$$

$$\alpha_3 = 60° - S_\alpha(\alpha_1, \beta_1, \mu_{ai})$$

$$\beta_3 = S_\beta(\alpha_1, \beta_1, \mu_{ai})$$

$$\alpha_5 = 60° - S_\alpha(\alpha_3, \beta_3, \mu_{ia}) - \alpha_1 + \arctan\left[\sin(\gamma)\cdot\tan(90° - \psi)\right]$$

$$\beta_5 = S_\beta(\alpha_3, \beta_3, \mu_{ia})$$

$$G = [\tan(\alpha_5)]^2 + [\tan(\beta_5)]^2$$

$$\varepsilon = \arctan\left[\frac{\tan(\beta_5)}{\tan(\alpha_5)}\right]$$

$$\phi = 90° - \gamma - \varepsilon$$

$$\theta_V = 90° - \psi - \arctan\left[G^{1/2}\cdot\cos(\phi)\right]$$

$$\theta_H = \arctan\left[\sin(\phi)\cdot\left(\frac{G}{1 + G\cdot[\cos(\phi)]^2}\right)^{1/2}\right]$$

To keep the spreadsheet calculations to a finite size, we used γ in the range of 55° to 90°, and α_1 in the range of 0° to 90°. The results for two solar elevations are shown in Fig. 22.38.

The locus of points of the upper tangent arc looks like the wings of a bird. For low solar elevations, the wings are up in the air. As the sun rises in the sky, the wings gently lower. At solar elevations of about 45° or more the wings are wrapped closely around the 22° halo. Above about 60° solar elevation, there is no solution to the upper tangent arc that is physically realistic.

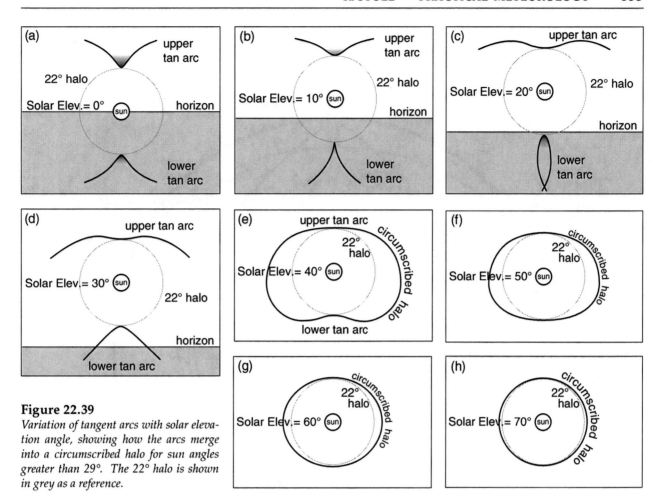

Figure 22.39
Variation of tangent arcs with solar elevation angle, showing how the arcs merge into a circumscribed halo for sun angles greater than 29°. The 22° halo is shown in grey as a reference.

Lower Tangent Arcs

Figs. 22.39 show lower tangent arcs, which occur during the same conditions as upper tangent arcs. Lower tangent arcs are tangent to the bottom of the 22° halo. Upper and lower tangent arcs have similar ray paths (Fig. 22.40) through large hexagonal column ice crystals. Eqs. (22.19) also apply to lower tangent arcs.

For solar elevations of 22° or less, you cannot see the lower tangent arc unless you look down into ice-crystal filled air. You can have such a view from aircraft, mountain peaks, bridges or tall buildings.

Circumscribed Halo

At solar elevations of 29° and greater, the upper and lower tangent arcs merge to form a circumscribed halo (Figs. 22.39). At solar elevations above 70°, the circumscribed halo is so tight around the 22° halo that they are often indistinguishable.

Other Halos

Fig. 22.41 and Table 22-3 show most of the halos and arcs.

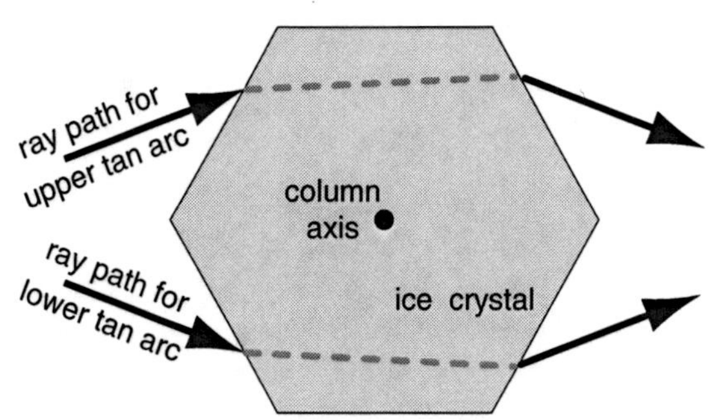

Figure 22.40
Ray paths for upper and lower tangent arcs. This is an end view of a hexagonal column ice crystal oriented with its column axis horizontal (out of the page).

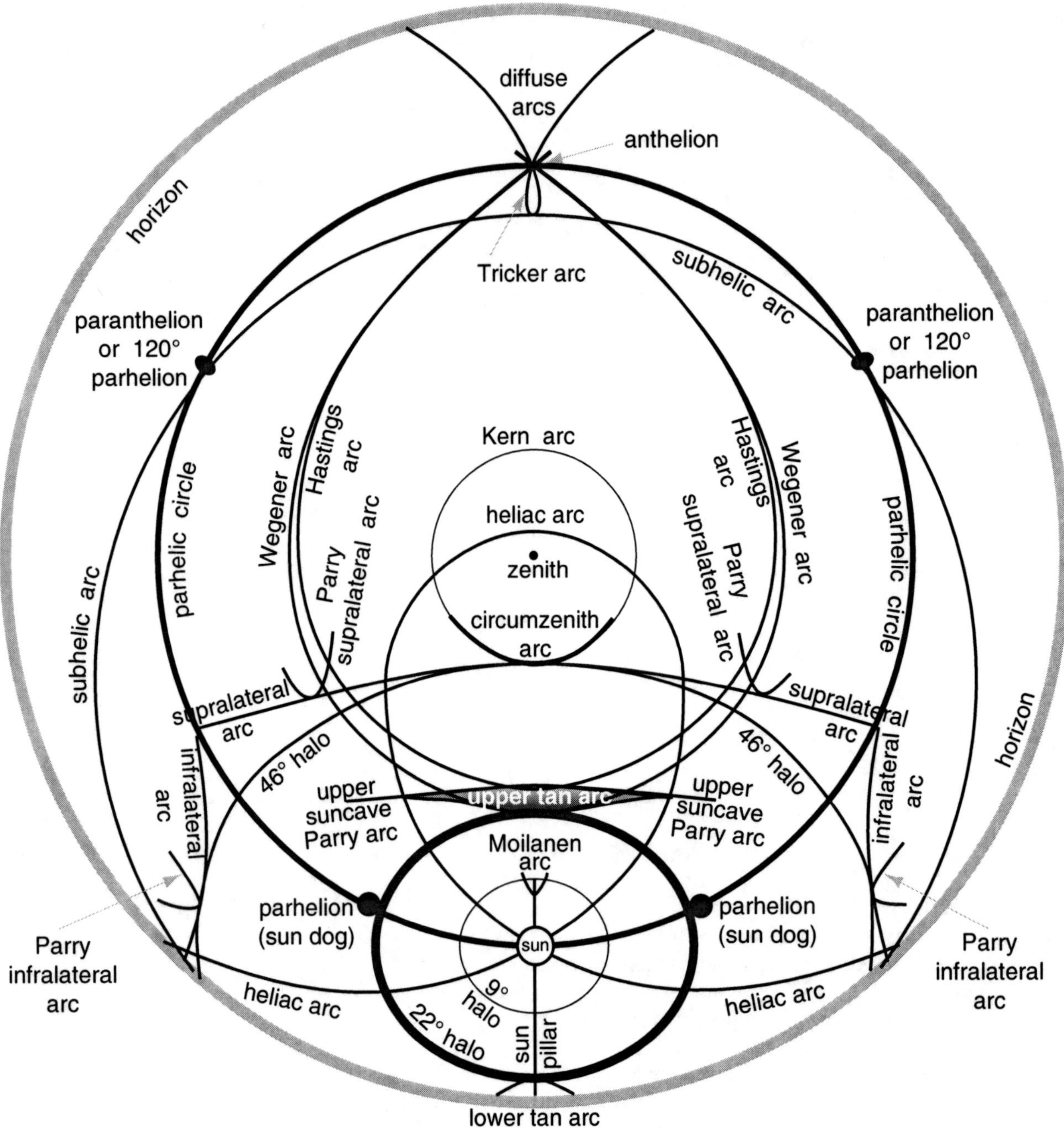

Figure 22.41

Halos in the sky, for a solar elevation of 26°. This view is looking straight up toward the zenith, as what could be photographed using an all-sky camera (with a fish-eye lens). The outermost thick grey circle represents the horizon, and the zenith point is marked as a dot in the center of this circle. Some of these halos are extremely rare. Most would not be seen at the same time, because they require ice crystals of different shapes and having perfect orientation. This fish-eye view distorts the 9°, 22°, and 46° halos, which would appear as circles to the observer. The halos of other diameters (including the 28° Scheiner's halo) are not plotted. [Based on data from Les Cowley's website (viewed Jan 2010) http://www.atoptics.co.uk/ , and from W. Tape, 1994: "Atmospheric Halos", and W. Tape and J. Moilanen, 2006: "Atmospheric Halos and the Search for Angle x", published by AGU.]

Table 22-3. Alphabetical catalog of halos. Fig. 22.41 shows most of these halos.

Legend:

Ice crystals:
- A = large oriented hexagonal pyramid
- a = small hex. pyramid (random orientations)
- C = large hexagonal column (col. axis horiz.)
- c = small hex. column (random orientations)
- L = Lowitz large hex. plate (see INFO Box)
- P = large hexagonal plate (col. axis vertical)

Angles:
- 22° ≈ angle of phenomenon from sun, associated with 60° ice wedge angle.
- 46° ≈ angle of phenomenon from sun, associated with 90° ice wedge angle.

Optics:
- m = (mirror) = one or more reflections
- r = one or more refractions

9° halo, a, r
9° Parry arc, A, r
18° halo, a, r
20° halo, a, r
22° halo, c, r
22° Lowitz arcs (upper & lower), L, r (rare)
22° Parry arcs (upper & lower; suncave & sunvex), C, r
23° halo, a, r
24° halo, a, r
35° halo, a, r
44° parhelia (sun dogs of sun dogs, formed when light passes through 2 hex. plates), 2P, 22°, r
46° halo, c, r
46° Lowitz arcs, L, r (extremely rare)
46° Parry arcs (infralateral, supralateral), C, r
120° parhelia (left & right), P, r, m
Anthelic arc, C, r, m (= Antisolar arc)
Anthelion (a crossing point of diffuse & anthelic arcs)
Antisolar arc, C, r, m (= Anthelic arc)
Circumhorizon arc, P, 46°, r (also Parry)
Circumscribed halo, C, 22°, r
Circumzenith arc, P, 46°, r (also Parry)
Diffuse arcs, C, r, m
Halos (see 9° through 46° halos above)
Hastings arcs (of upper suncave Perry arc; and of lower suncave Perry arc), r, m
Heliac arc, C, m or (r, m)
Infralateral arc, C, 46°, r (also Parry)
Kern arc, P, 46°, r, m (rare)
Lower suncave Parry arc, C, 22°, r
Lower sunvex Parry arc, C, 22°, r
Lower tangent arc, C, 22°, r
Lower Wegener arc, C, r, m
Lowitz arcs (upper & lower), L, 22°, r (rare)
Moilanen arc (rare; seen in ice crystals from snow-making machines)
Paranthelia (= 120° parhelia, left & right), P, r, m
Parhelia (sun dogs) (left & right), P, 22°, r
Parhelic circle, [P, m or (r, m)] or [C, m or (r, m)] (also Parry)
Parry arcs (Many arcs. See separate listings for: 9°, upper suncave, upper sunvex, lower suncave, lower sunvex, circumzenith, circumhorizon, infralateral, supralateral, Hastings, helic, subhelic, anthelic, subanthelic)
Scheiner's 28° halo (from cubic Ic ice crystals), r
Subanthelic arc, C, r, m
Subcircumzenith arc, P, 46°, r, m (not yet observed)
Subhelic arc, C, r, m
Subhorizon halo (halos and arcs below the horizon, including portions of diffuse, Tricker, subhelic, lower Wegener, and lower tangent arcs)
Subparhelia (subsun dogs) (left & right), P, 22°, r, m
Subparhelic circle, [P, r, m] or [C, r, m]
Subsun, P, r
Subsun dogs (subparhelia) (left & right), P, 22°, r, m
Sun dogs (parhelia) (left & right), P, 22°, r
Sun pillar, C or P, r
Supralateral arc, C, 46°, r (also Parry)
Tangent arc (upper & lower), C, 22°, r
Tricker arc, C, r, m
Upper suncave Parry arc, C, 22°, r
Upper sunvex Parry arc, C, 22°, r
Upper tangent arc, C, 22°, r
Upper Wegener arc, C, r, m
Wegener arcs (upper & lower), C, r, m

INFO • Special Crystal Orientations

A **Parry oriented column** crystal has both a horizontal column axis and horizontal top and bottom prism faces (see Fig. at right). Resulting **Parry arcs** (Fig. 22.41) are observed 1% as frequently as 22° halos.

top prism face is horizontal
column axis is horizontal
bottom prism face is horizontal

Fig. 22.b

A **Lowitz** oriented plate crystal has an unusual orientation. **Lowitz arcs** are extremely rare & faint (not shown in Fig. 22.41).

Lowitz axis is horizontal

Fig. 22.c

INFO • Scheiner's Halo

In year 1629, Christopher Scheiner observed a rare 28° halo, which has been observed only about 6 times since then. One explanation is that sun rays pass through an octahedral crystal made of cubic ice Ic. Cubic ice is discussed in a INFO Box in the Precipitation chapter. The Fig. below shows a ray path through a truncated octahedron.

The wedge angle of 70.528° yields a minimum viewing angle of 27.46° according to eq. (22.16), assuming a refractive index of 1.307. But this cubic-ice explanation is still being debated.

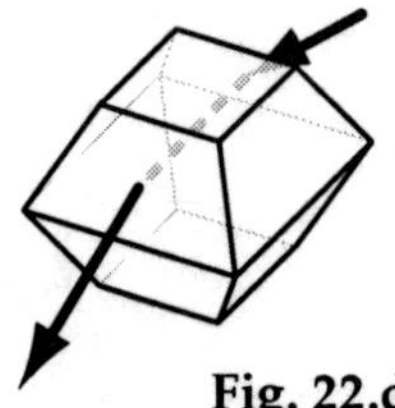

Fig. 22.d

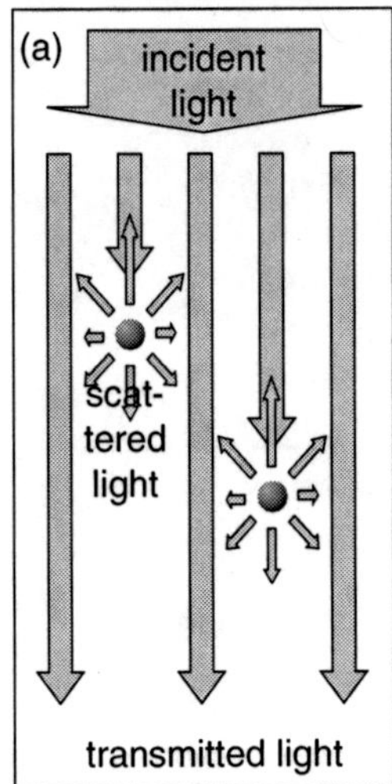

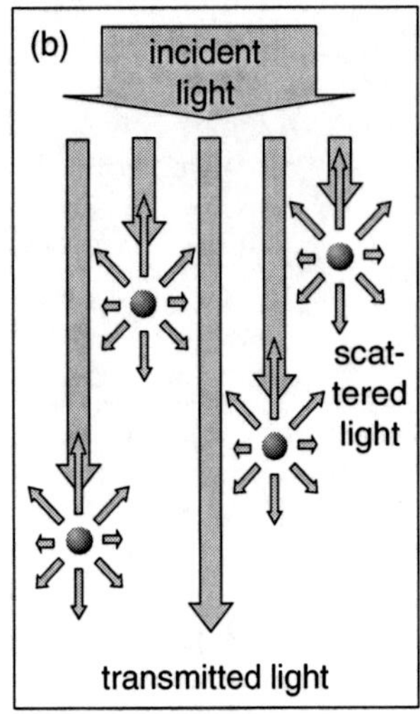

Figure 22.42
Incident light (long downward arrows) can be scattered (small arrows in all directions) by particles (black dots) that are in the path of the light. Paths with few particles (a) scatter less light and transmit more light than paths with many particles (b).

Sample Application
What fraction of light is scattered for $\tau = 2$?

Find the Answer
Given: $\tau = 2$
Find: $I_{scat}/I_o = ?$

Use eq. (22.21): $I_{scat}/I_o = 1 - e^{-2} = 0.865$ = <u>86.5%</u>

Check: Units OK. Magnitudes agree with text.
Exposition: A large portion of light is scattered. This optical depth can occur in polluted air.

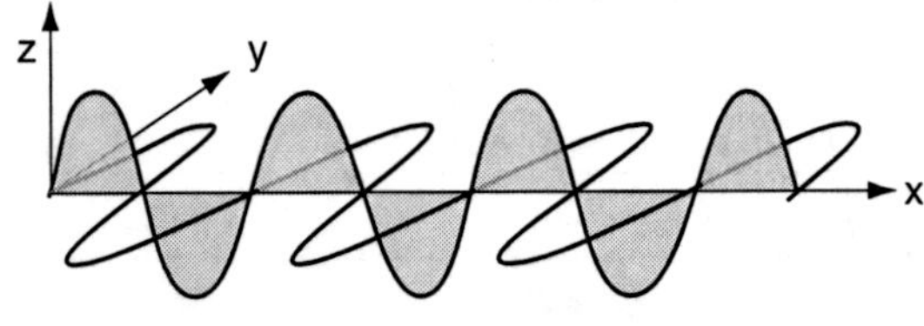

Figure 22.43
Unpolarized light consists of two cross-polarized parts.

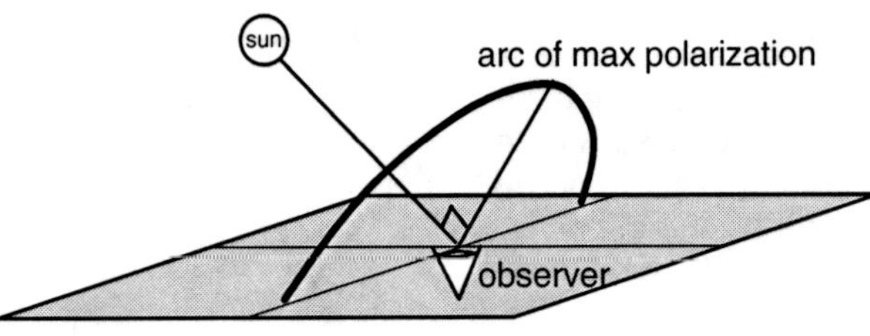

Figure 22.44
Arc of max polarization in the sky due to Rayleigh scattering.

SCATTERING

Background

Optical thickness

Light can scatter off of air molecules, pollutant particles, dust, and cloud droplets. More particles in the air cause more of the light to be scattered (Fig. 22.42). The ratio of transmitted (I_{tran}, non-scattered) to incident (I_o) light can be quantified using the **optical thickness** or **optical depth** τ (dimensionless) in Beer's Law:

$$\frac{I_{tran}}{I_o} = e^{-\tau} \qquad (22.20)$$

Thus, the relative amount of light scattered (I_{scat}) is:

$$\frac{I_{scat}}{I_o} = 1 - \frac{I_{tran}}{I_o} = 1 - e^{-\tau} \qquad (22.21)$$

Zero optical thickness means no light is scattered, while $\tau > 5$ implies that virtually all the light is scattered. Optical thickness increases for longer paths and for higher particle concentrations. Examples are τ = 0.03 to 0.3 for a clean dry atmosphere, and $\tau > 10$ for thick clouds.

Polarization

Light propagating in the *x*-direction can be thought of as having oscillations in the *y* and *z* directions (Fig. 22.43). This is unpolarized light. Polarizing filters eliminate the oscillations in one direction (for example, the shaded curve) while passing the other oscillations. What remains is **polarized** light, which has half the intensity of the unpolarized ray.

Sunlight becomes polarized when it is scattered from air molecules. The maximum amount of polarization occurs along an arc in the sky that is 90° from the sun, as viewed from the ground (Fig. 22.44).

If the sky scatters light with one polarity, and a polarizing camera filter or polarized sunglasses are rotated to eliminate it, then virtually no sky light reaches the camera or observer. This makes the sky look very deep blue, providing a very striking background in photographs of clouds or other objects. If you want to maximize this effect, pick a camera angle looking toward the 90° arc from the sun.

Scattering by small dust particles in ice also explains why glaciers are often magnificently blue.

Types of Scattering

The type and efficiency of scattering depends on the ratio of particle diameter D to the wavelength λ of the light. Table 22-4 summarizes the types of scattering. Actual size ranges of cloud droplets and aerosols are wider than indicated in this table.

Table 22-4. Scattering of visible light. D = particle diameter. λ = wavelength of light.

D / λ	Particles	Diameter, D (µm)	Type	Phenomena	Scattering Varies with: λ	Direction	Polarization
< 1	air molecules	0.0001 to 0.001	Rayleigh	blue sky, red sunsets	X		X
≈ 1	aerosols (smog)	0.01 to 1.0	Mie	brown smog	X	X	X
> 1	cloud droplets	10 to 100	geometric	white clouds		X	

Rayleigh Scattering

Air molecules have sizes of $D \approx 0.0001$ to 0.001 µm, which are much smaller than the wavelength of light (λ = 0.4 to 0.7 µm). These particles cause **Rayleigh scattering**. The ratio of scattered intensity of radiation I_{scat} to the incident radiation intensity I_o is:

$$\frac{I_{scat}}{I_o} \approx 1 - \exp\left[-\frac{a \cdot (n_{air} - 1)^2}{\rho \cdot \lambda^4} \cdot x \right] \qquad (22.22)$$

where a = 1.59x10^{-23} kg, ρ is air density, n_{air} is the refractive index, and x is the path length of light through the air.

Fig. 22.45a shows the relative amount of scattering vs. wavelength. Because of the λ^{-4} dependence, shorter wavelengths such as blue and violet are scattered much more (about a factor of 10) than red light, which makes the **sky blue**. For light shining vertically through a clean dry atmosphere, τ = 0.3 for violet light and 0.03 for red.

Sunlight intensity varies according to Planck's law (Fig. 22.45b and the Solar & Infrared Radiation chapter). The product of the top 2 curves shows the amount of sunlight that is scattered in the atmosphere (Fig. 22.45c). All curves have been normalized to have a maximum of 1.0.

The curve in Fig. 22.45c peaks in the ultraviolet portion of the spectrum. Although this is not visible to the naked eye, the scattered ultraviolet light can affect photographic films to produce a picture that looks hazier than the view by eye. Haze filters on cameras can filter out this unwanted scattered ultraviolet light.

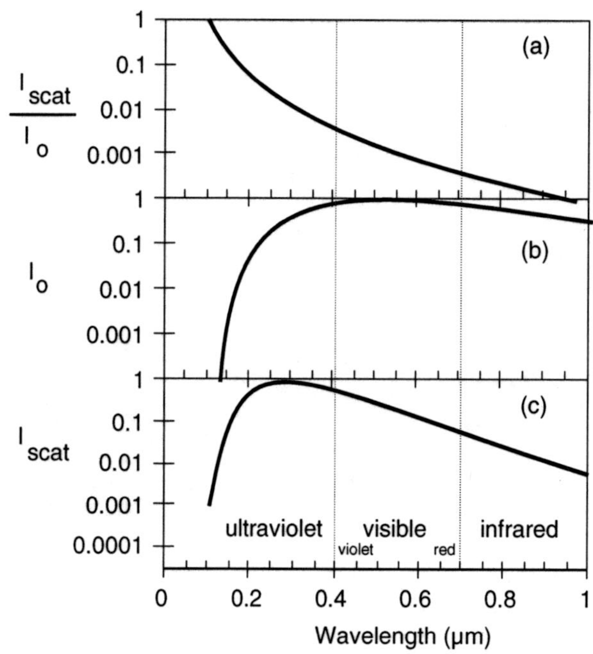

Figure 22.45
Top: Relative Rayleigh scattering. Middle: Planck spectrum for the sun. Bottom: product of the top two curves, indicating the amount of sunlight scattered in clean air.

Geometric Scattering

For large particles such as cloud droplets, light reflects from them according to geometric optics. Because all wavelengths are reflected equally, **clouds look white** when illuminated by white sunlight.

There is, however, a directional dependence. Light that is scattered in the same direction that the incident ray is pointing is called **forward scattering**. The opposite is **backward scattering**. For clouds, forward scattering is usually greater than backscattering. Thus, clouds you see in the direction of the sun look bright white, while those in the direction of your shadow look slightly more grey.

Sample Application

What fraction of incident violet light is scattered by air molecules along a 20 km horizontal ray path near the Earth?

Find the Answer

Given: x = 5x10^6 m, λ = 4x10^{-7} m for violet
Find: I_{scat}/I_o = ?
Assume: ρ = 1 kg·m^{-3} for simplicity.

Use eq. (22.22): I_{scat}/I_o =

$$1 - \exp\left[-\frac{(1.59\times10^{-23}\text{kg})\cdot(0.0002817)^2}{(1\ \text{kg}\cdot\text{m}^{-3})\cdot(4\times10^{-7}\text{m})^4} \cdot (2\times10^4\text{m}) \right]$$

= <u>0.627</u> = 62.7% scattered

Check: Units OK. Physics OK.
Exposition: Looking toward the horizon, distant objects are difficult to see because some of the light is lost. Vertical rays experience less scattering, because air density decreases with height.

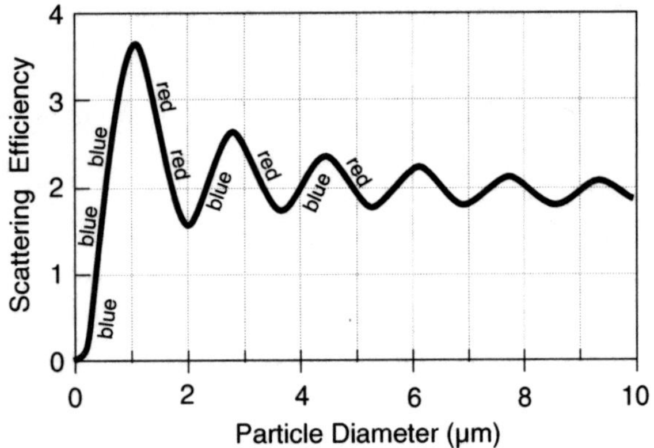

Figure 22.46
Mie scattering efficiency and dominant color scattered out of white light for various particle diameters.

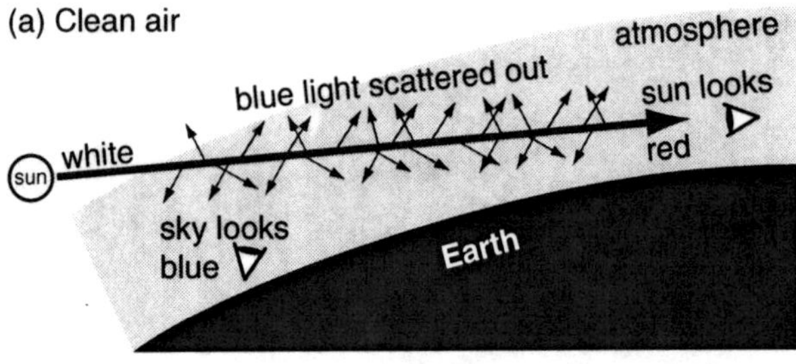

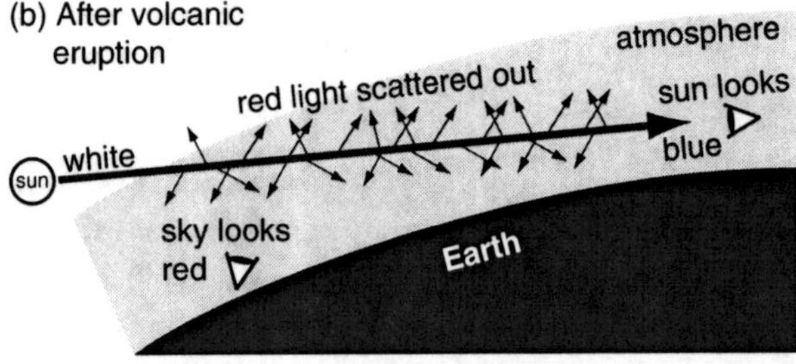

Figure 22.47
Scattering of sunlight as it travels through a long path through the atmosphere. (a) Clear skies. (b) Air hazy due to tiny sulfuric-acid aerosol droplets suspended in the stratosphere after a volcanic eruption.

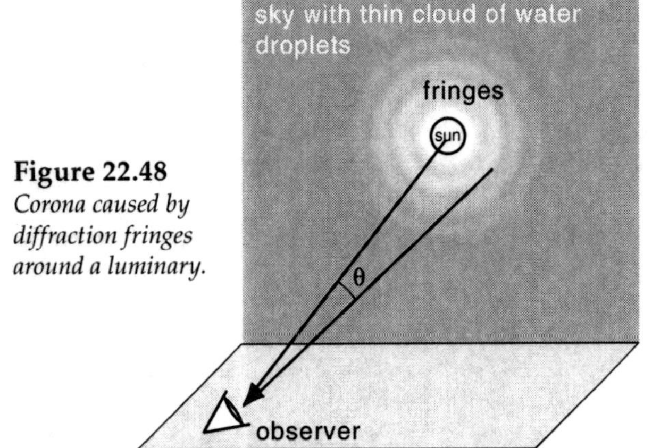

Figure 22.48
Corona caused by diffraction fringes around a luminary.

Mie Scattering

Gustov Mie proposed a comprehensive theory that describes reflection, scattering, polarization, absorption, and other processes for all size particles. The theory reduces to Rayleigh scattering for particles smaller than the wavelength, and to geometric scattering for larger particles. Aerosol particles are middle size, so no simplification of Mie theory is possible.

Fig. 22.46 shows that aerosols with diameters greater than about 0.1 µm are 10 to 1000 times more efficient light scatterers than air molecules, which is why polluted air looks hazy and has low visibility. Uniform aerosol smog particles can produce bluish or reddish-brown colors, depending on the dominant aerosol diameter.

Most aerosol particles have diameters smaller than 1 µm. Fig. 22.46 shows that these aerosols still scatter more blue light than red, causing blue haze. When the sun is low in the sky, the light rays travel through a very long path of air en route to the observer. For this situation, so much blue light is scattered out (making the **sky blue**) by these small aerosols and air molecules that it leaves more red in the remaining direct beam — giving a beautiful **red sun** at sunrise or sunset (Fig. 22.47a).

However, some acid droplet aerosols have diameters of about 1.5 µm, which scatters more red than blue. In some polluted regions where combustion processes emit SO_2 and NO_2 into the air, these pollutants can react with water in the air to form sulfuric acid and nitric acid. Scattering in this condition creates a **brown cloud** haze.

Another source of 1.5 µm diameter acid aerosols is volcanic eruptions, where sulfur emissions change into sulfuric acid droplets in the stratosphere. When the sun is low in the sky, so much red light is scattered out that beautiful **crimson sunset skies** are often observed after volcanic eruptions (Fig. 22.47b). Also, this leaves the remaining direct beam of sunlight in the setting sun slightly blue, which is hard to see because it is so bright. However, direct moonlight from a rising or setting moon can often be blue — causing a **blue moon** after a volcanic eruption.

For aerosols, forward scattering is usually greater than backscattering. Scattering can be polarized.

DIFFRACTION & INTERFERENCE

When sunlight or moonlight passes through a thin cloud of water droplets, diffraction and interference can produce a disk of bright sky centered on the luminary, surrounded by one or more colored rings or **fringes** (Fig. 22.48).

Recall from physics that when wave fronts hit the edge of an object, part of the wave bends around the edge — a process called **diffraction**. This is consistent with Huygens' principle, where every point on an incident wave front can be thought of as a source of new wavelets radiating away, and that the subsequent position of the wave front depends on the superposition of the wavelets.

Fig. 22.49a sketches cloud droplets and the wave fronts of incident light from the sun or moon. The edge of each droplet is marked with an "x". Consider the wavelets generated from each edge, sketched as the thin grey concentric circles. Superposition of these wavelets, indicated with the thick black wavy line, can **constructively interfere** to create a new wave train. Fig. 22.49a shows one of the resulting wave trains — the one for light that is transmitted straight through the cloud to the observer. Other angles have **destructive interference**.

However, there are many other angles where the wavelet fronts constructively interfere, one of which is sketched in Fig. 22.49b. This has the net effect of producing an additional train of wave fronts that leaves the cloud at a different angle. Because this angle of light leaving the cloud is the same as the angle observed by the observer, the observer sees the constructive interference (i.e., bright light) from all those different drops that happen to be the same viewing angle away from the sun. The different angles that can be produced from a set of drops produces many rings of different radii, called fringes.

The angle θ to each ring depends on the radius R of the droplet and the wavelength λ of light:

$$\theta = \arcsin\left(\frac{m \cdot \lambda}{R}\right) \tag{22.23}$$

where m is a dimensionless diffraction parameter given in Table 22-5.

For any one color, there are faint rings of light separated by darker gray background. The different colors of the spectrum have different fringe radii, causing the fringes of any one color (e.g. blue-green) to appear in the dark spaces between fringes of another color (e.g., red). Fringes further from the sun (with higher fringe index) are less bright, as given in Table 22-5 and plotted in Fig. 22.50.

Corona, Iridescence and Glory

Wave clouds (standing **lenticular** clouds) have extremely uniform drop sizes. This means that all the drops in the cloud produce the same fringe angles, for any one color. Hence, the diffraction from all the droplets reinforce each other to produce bright colorful fringes (rings) called **corona**. Most

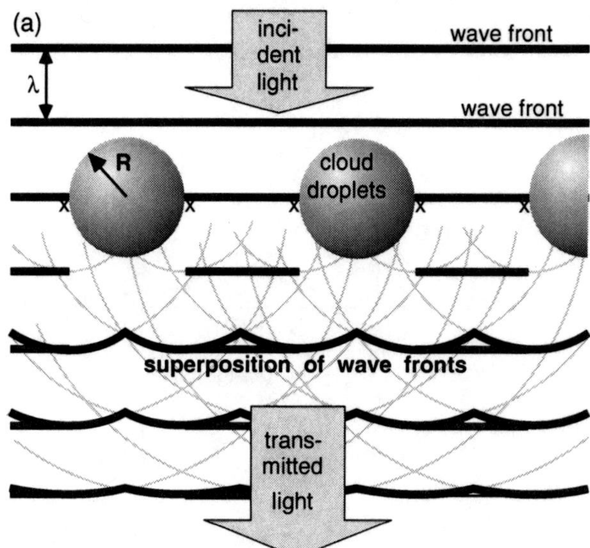

Figure 22.49a
Diffraction of light from edges of cloud droplets.

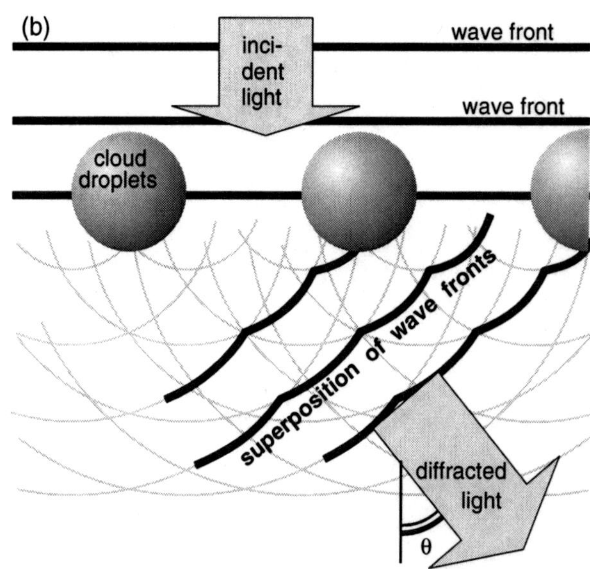

Figure 22.49b
Another direction where constructive interference happens.

Table 22-5. Diffraction fringes.

Fringe Index	m	Relative Intensity
1	0	1.0
2	0.819	0.01745
3	1.346	0.00415
4	1.858	0.00165
5	2.362	0.00078
6	2.862	0.00043
7	3.362	0.00027
8	3.862	0.00018

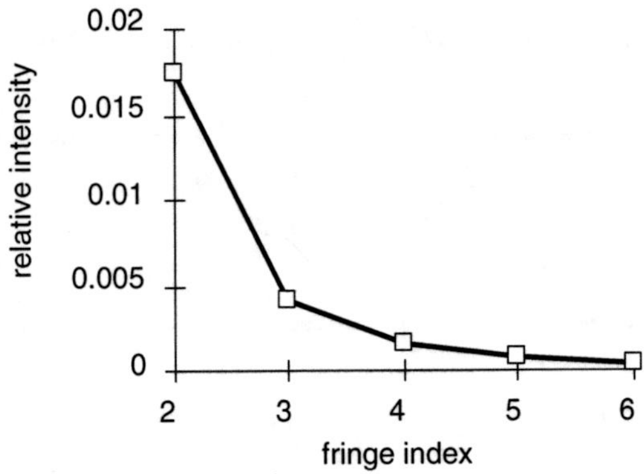

Figure 22.50
Diffraction fringe brightness. Larger indices indicate fringes that are further from the luminary.

Sample Application
The first visible fringe (index = 2) of red is 10° from the moon. What is the cloud drop radius?

Find the Answer
Given: $\theta = 10°$, $\lambda = 0.7$ µm.
$m = 0.819$ from Table 19-3 for fringe #2.
Find: $R = ?$ µm
Assume $n_{air} \approx 1.0002753$ for red light.

Rearrange eq. (22.23):

$$R = \frac{m \cdot \lambda}{\sin\theta} = \frac{0.819 \cdot (0.7 \mu m)}{\sin(10°)} = \underline{3.3\ \mu m}$$

Check: Units OK. Physics OK.
Agrees with Fig. 22.51.
Exposition: Smaller than typical cloud-droplet size, perhaps associated with a wave cloud.

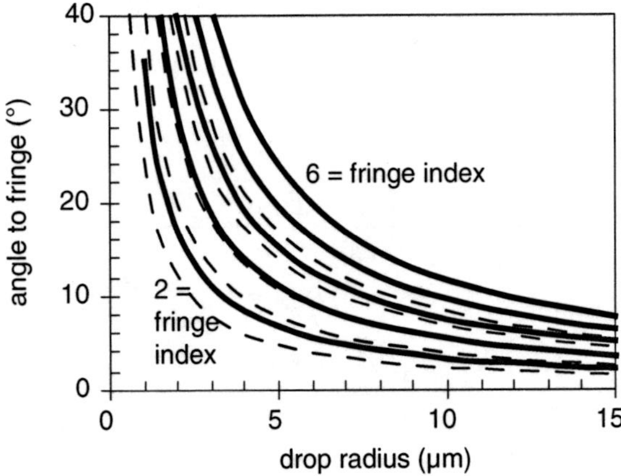

Figure 22.51
Viewing angle vs. cloud droplet size, for red (solid) and blue-green (dashed) diffraction fringes.

other clouds contain drops with a wide range of sizes, causing the colors to smear together to form a whitish disk (called an **aureole**) touching the sun or moon.

Eq. (22.23) can be solved on a spreadsheet for various cloud droplet sizes and colors. The results in Fig. 22.51 show that smaller droplets produce larger-diameter fringes. For reference, typical cloud droplets have 10 µm radii, and the viewing angle subtended by the sun is 0.534°.

To help discriminate between corona and halos, remember that corona are bright disks of light that touch the sun or moon, while halos have a dark region between the luminary and the halo ring. Also, corona are formed from liquid water droplets, while halos are formed from ice crystals.

Droplets near the edges of clouds (particularly wave clouds) can be very small as they form or evaporate, and can cause fringes of very large angular radius. As a result, the edges of clouds near the sun or moon are often colorful, a phenomenon called **iridescence**.

When looking down from above a cloud, diffraction patterns are sometimes seen around the shadow of the observer. This is called **glory**, and is associated with both diffraction and reflection from the cloud droplets. When viewed from an aircraft flying above the clouds, the glory is a circle of light centered on the shadow of the airplane.

Supernumerary Bows

As sketched in Fig. 22.7, supernumerary bows are closely spaced faint pinkish-purple and turquoise diffraction arcs touching the inside of the primary rainbow arc. The viewing-angle width of this bow (containing 2 to 4 sequences of fringes) is 1 to 2°. Supernumerary bows are most visible when the rain drop diameter is smaller than about 1 mm, during a rain shower when most of the falling drops have nearly the same size (i.e., a narrow drop-size distribution).

When the rain shower has a wider drop-size distribution, the greater flattening of the larger drops causes many drops to have nearly the same vertical thickness, even though their horizontal diameters vary widely. Thus, supernumerary bows are often most visible inside the top of the rainbow arc, and less visible inside the sides. Rainbows and supernumerary bows are also visible when the sun shines on dew drops on grass, and in water sprays from irrigation systems. Very faint supernumerary bows can sometimes be seen outside the secondary rainbow.

To explain the optics, look at Fig. 22.13b and consider red light at viewing angle 40°, for example. We see that light at this viewing angle comes from two different paths through the rain drop: one at impact parameter 0.74 and the other at impact parameter 0.94. These two different paths travel different distances through the drop, hence their wave fronts can get out of phase with each other. At certain angles for certain colors the wave fronts interfere with each other, canceling to yield no light of that color. At other angles for other colors, the wave fronts are in phase and reinforce each other, yielding a bright color. This **constructive and destructive interference** of wave fronts creates supernumeraries.

MIRAGES

Refractive index n_{air} varies with air density ρ:

$$n_{air} - 1 \approx (n_{ref} - 1) \cdot \rho / \rho_{ref} \tag{22.24}$$

where *ref* denotes a reference condition (such as given for air at any of the *T* and *P* values in Table 22-1). Knowing *T* and *P*, you can use the ideal gas law to find density [$\rho = P/(\Re \cdot T)$ for $\Re = 0.287$ (kPa K^{-1})·(m^3 kg^{-1}); see Chapter 1 for details]. Thus, you can rewrite eq. (22.24) as:

$$n_{air} - 1 = (n_{ref} - 1) \cdot \frac{T_{ref}}{T} \cdot \frac{P}{P_{ref}} \tag{22.25}$$

A sharp change in density between two media causes a sharp kink in the ray path (Fig. 22.1). A gradual change of density causes a smoothly curving ray path (Fig. 22.52). The radius of curvature R_c (positive for concave up) is:

$$R_c \approx \frac{\rho_{ref}}{(n_{ref} - 1) \cdot (\cos\alpha) \cdot (\Delta\rho / \Delta z)} \tag{22.26}$$

where α is the angle of the ray above horizontal, and the gradient of density $\Delta\rho/\Delta z$ is assumed to be perpendicular to the Earth's surface.

Substituting the ideal gas law and the hydrostatic relationship into eq. (22.26) yields

$$R_c \approx \frac{-(T / T_{ref}) \cdot (P_{ref} / P)}{(n_{ref} - 1) \cdot (\cos\alpha) \cdot \left\{ \frac{1}{T} \left[\frac{\Delta T}{\Delta z} + a \right] \right\}} \tag{22.27}$$

where $a = 0.0342$ K·m^{-1}.

Because average density decreases with height in the atmosphere (see Chapter 1), eq. (22.27) gives a negative radius of curvature in the presence of weak temperature gradients. In other words, the ray is bent downward in a standard atmosphere. This

Sample Application
During winter with cold (T = –30°C) high-pressure (P = 104 kPa), find the refractive index for violet light.

Find the Answer
Given: T = –30°C = 243 K, P = 104 kPa, λ = 0.4 µm.
Find: n_{air} = ? (dimensionless)

Use Table 22-1: For T_{ref}= –20°C = 253K & P_{ref}= 101.325 kPa, the table gives n_{ref} = 1.0003219, assuming *RH*=75%.

Use eq. (22.25):
n_{air} = 1 + (0.0003219)·(253/243)·(104/101.325)
= **1.0003440**

Check: Value reasonable compared to Table 22-1.
Exposition: By the ideal gas law, air density increases when temperature decreases and pressure increases. Higher density gives greater refractive index.

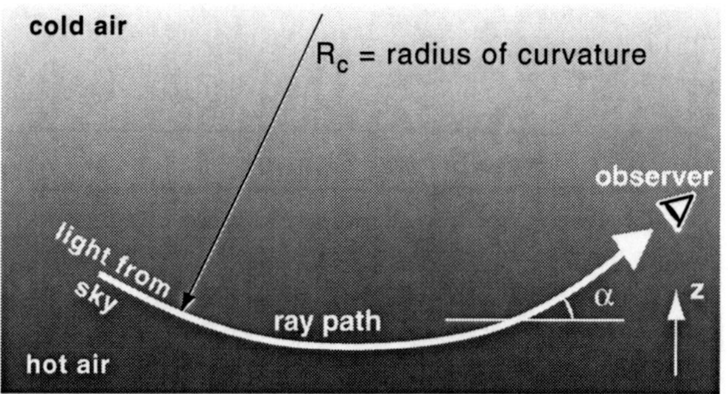

Figure 22.52
Ray curvature within a vertical gradient of density, creating an inferior mirage.

Sample Application
What is the radius of curvature of a horizontal indigo (λ ≈ 0.45 µm from Table 2-3 in the Radiation chapter) ray near the ground in a standard atmosphere?

Find the Answer
Given: z = 0, α = 0°,
n_{air} ≈ 1.00028 from Table 22-1 for indigo light.
Find: R_c = ? km

Assume T = 288K and P = 101.3 kPa in a standard atmosphere near the ground. Assume the vertical temperature gradient is negligible. Rearrange eq. (22.27):

$$R_c \approx \frac{-1}{(0.00028) \cdot (\cos 0°) \cdot (0.0342 \mathrm{K \cdot m^{-1}} / 288\mathrm{K})}$$

= **–30,000 km**

Check: Units OK. Physics OK.
Exposition: The large negative radius implies very weak downward curvature, which agrees with the small angle of the **apparent sunrise** (Table 2-2).

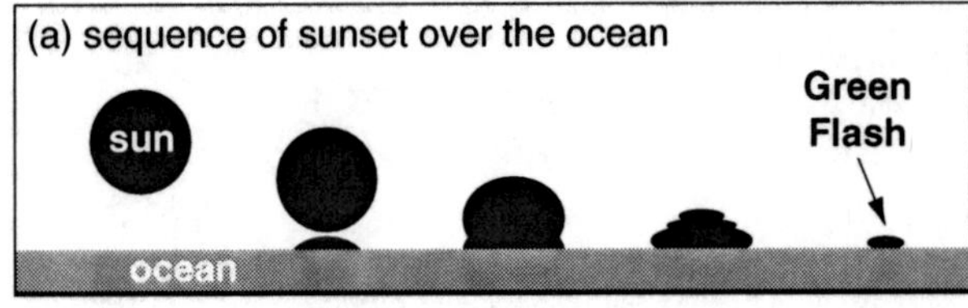

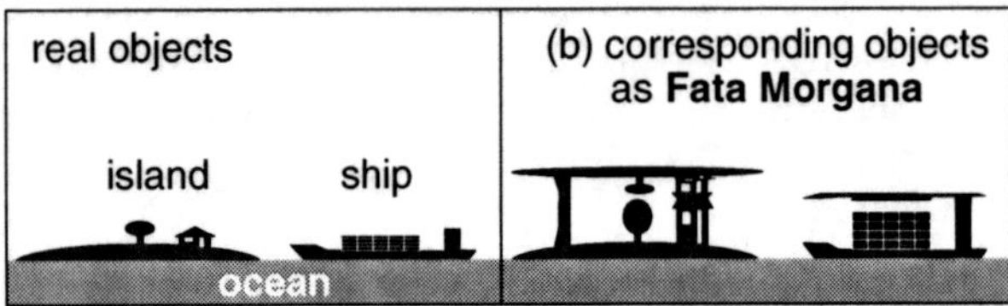

Figure 22.53
(a) Appearance of the sun setting over the ocean in inferior mirage conditions. (b) Distortion of distant objects during Fata Morgana mirage.

INFO • Newton and Colors

To confirm his laws of motion, Isaac Newton wanted to view the motions of the planets. He built his own telescopes for this purpose. However, the images he observed through his lenses were blurry. For example, images of stars were spread (**dispersed**) into a streak of colors.

After experimenting with different lenses, he concluded that neither the glass nor the construction was flawed. He realized that there must be some unknown physics causing this optical phenomenon. Like many great scientists, he allowed himself to get side-tracked to study this phenomenon in detail.

One of his experiments was to obtain a triangular prism, and to allow sunlight to pass through it. He observed that the white sunlight is composed of a spectrum of colors: red, orange, yellow, green, blue, indigo, and violet. Newton must have had a unique sense of color, because most people cannot discriminate between indigo and violet in the spectrum.

Newton concluded that the refraction of light through a lens inevitably causes color **dispersion**. Thus, a pinpoint of white starlight would be spread into a smear of colors. His solution to the telescope problem was to design a telescope without glass lenses. Instead he invented a reflecting telescope using curved mirrors, because reflection does not cause separation of light into colors.

agrees with Huygens' principle, that says light rays are bent toward media of higher density.

For ray paths near horizontal, the total bending through the whole atmosphere is about 0.567°, which is why **apparent sunrise** occurs before geometric sunrise (see the Radiation chapter). In everyday life, we rarely notice refraction associated with the standard atmosphere. However, in the presence of strong temperature gradients we can see mirages.

For rays to bend up instead of down, the term in square brackets in eq. (22.27) must be negative. This is possible when $\Delta T/\Delta z < -a$. A temperature decrease of at least 3.5°C per 10 cm height rise is necessary. Such a strong gradient is possible on hot sunny days in the air touching strongly-absorbing (black) surfaces. This condition of warm air under cold causes **inferior mirages**, where the objects appear lower than they really are. Warm under cold air is statically unstable (see the Atmospheric Stability chapter), causing turbulence that makes these mirages shimmer.

Inferior mirages are common above black roads on hot sunny days. Looking down toward the road, you see light rays refracted from the sky (Fig. 22.52). As these shimmer in the convective turbulence, the mirages look similar to reflections from the surface of water puddles.

Inferior mirages can also form over the ocean, if a thin layer of warm air is created by heat conduction from a warm sea surface, but colder air exists aloft. As the sun sets over the ocean, the base of the sun sometimes appears to spread out (Fig. 22.53a), causing the shape of the sun's outline to look like the Greek letter omega (Ω). Later, just as the top of the sun sets, the inferior-mirage effect causes the top of the sun to briefly appear emerald green just before it disappears below the horizon. This phenomenon is called **green flash**. *[CAUTION: to avoid damaging your eyes, do not look directly at the sun.]*

Cold air under warm air (as during early morning over land, or in arctic regions over ice, or where warm air flows over cold seas) causes **superior mirages**, where objects appear higher than they are. This accentuates the downward bending of rays, causing the image of the object to **loom** or stretch vertically upward.

The **Fata Morgana** mirage is caused by light rays passing through one or more elevated temperature inversions (a sharp interface between cold air below and warmer air aloft). Some ray paths bend as in superior mirages, and others bend as in inferior mirages. This causes portions of normal objects to appear stretched in the vertical, and other portions to appear compressed (Fig. 22.53b).

The results are fanciful images of mountains, ramparts, and castle turrets (some of which appear to float above ground) where none exist in reality.

The name of this mirage comes from Fairy Morgan (or Fata Morgana in Italian) of the King Arthur legend, an enchantress/magician who could create illusions of floating castles, etc.

REVIEW

Atmospheric optical phenomena can be caused by reflection, refraction, scattering, diffraction and interference. Some of these processes are caused by the difference in density between hydrometeors (liquid and solid water particles) and air.

Common ice-crystal optical phenomena include the sun pillar, parhelic circle, subsun, 22° halo, 46° halo, circumzenith arc, sun dogs, subsun dogs and tangent arcs. They are usually found by looking toward the sun. The hexagonal column shape of many ice crystals creates 60° and 90° prismatic effects. The somewhat random orientation of these crystals causes sunlight to be returned in different directions from different crystals. The phenomena we see are the superposition from many crystals. Many other rarer halos can occur, some of which are associated with hexagonal pyramid crystals.

Rain-drop phenomena are a bit easier to describe, because the approximate spherical shape eliminates orientational dependence. Rainbows are seen by looking away from the sun. Cloud droplets produce corona, iridescence, glory, and aureole. Air molecules cause blue sky and red sunsets. Refraction in air causes amazing mirages.

This brings you to the end of this book. Along the way, you examined the physics of atmospheric pressure, wind, temperature, and humidity. You saw how these concepts weave together to make cyclones, precipitation, thunderstorms, fronts, and hurricanes that we experience as weather. You have the quantitative wherewithal to critically evaluate weather-related issues, and to make sound decisions that affect society and the future of our planet.

If this is your last course on meteorology, be confident that you have the background to delve into the meteorology journals to learn the latest details that might help you in your engineering or scientific work. If you plan to continue your study of meteorology, the overview that you have gained here will help you put into context the advanced material that you will soon be learning.

As the Irish might bless: "May the wind be always at your back, may the sun shine warm upon your face, and may the rains fall soft upon your fields."

Sincerely, *Roland Stull*

A SCIENTIFIC PERSPECTIVE • Great Scientists Can Make Big Mistakes Too

Isaac Newton explained many atmospheric optical phenomena in his books on optics. Not all his theories were correct. One of his failures was his explanation of the blue sky.

He thought that the sky was blue for the same reason that soap bubbles or oil slicks have colors. For blue sky, he thought that there is interference between light reflecting from the backs of small water droplets and the light from the front of the drops.

Although Newton's theory was accepted for about 175 years, eventually observations were made of the polarization of sky light that were inconsistent with the theory. Lord Rayleigh proposed the presently accepted theory in 1871.

Like Newton, many great scientists are not afraid to propose radical theories. Although most radical theories prove to be wrong, the few correct theories are often so significant as to eventually create **paradigm shifts** (radical changes) in scientific thought. Unfortunately, "publish or perish" demands on modern scientists discourage such "high risk, high gain" science.

HOMEWORK EXERCISES

Broaden Knowledge & Comprehension

B1. Search the web for images of the following atmospheric optical phenomena:

a. 22° halo b. 46° halo
c. sun dogs d. sub sun dogs
e. sub sun f. sun pillar
g. supersun h. upper tangent arc
i. parhelic circle j. lower tangent arc
k. white clouds l. circumzenith arc
m. Parry arcs n. primary rainbow
o. red sunset p. secondary rainbow
q. blue sky r. Alexander's dark band
s. corona t. crepuscular rays
u. iridescence v. anti-crepuscular rays
w. glory x. fata morgana
y. green flash z. mirages
aa. halos other than 22° and 46°
ab. circumhorizon arc ac. circumscribed halo
ad. Lowitz arcs ae. Scheiner's halo
af. Moilanen arc ag. Hastings arc
ah. Wegener arc ai. supralateral arc

aj. infralateral arcs ak. subhelic arc
am. heliac arc an. diffuse arcs
ao. Tricker arc ap. 120° parhelia (120° sun dogs)
aq. various subhorizon arcs
ar. fog bow as. blue moon
at. reflection rainbow au. twinned rainbow
av. supernumerary bows

B2. Search the web for images and descriptions of additional atmospheric phenomena that are not listed in the previous question.

B3. Search the web for microphotographs of ice crystals for ice Ih. Find images of hexagonal plates, hexagonal columns, and hexagonal pyramids.

B4. Search the web for diagrams or microphotographs of ice crystals for cubic ice Ic.

B5. Search the web for lists of indices of refraction of light through different materials.

B6. Search the web for highway camera or racetrack imagery, showing inferior mirages on the roadway.

B7. Search the web for computer programs to simulate atmospheric optical phenomena. If you can download this software, try running it and experimenting with different conditions to produce different optical displays.

B8. Search the web for images that have exceptionally large numbers of atmospheric optical phenomena present in the same photograph.

B9. Search the web for information on linear vs. circular polarization of sky light. Polarizing filters for cameras can be either circular or linear polarized. The reason for using circular polarization filters is that many automatic cameras loose the ability to auto focus or auto meter light through a linear polarizing filter. Compare how clear sky would look in a photo using circular vs. linear polarization filters.

B10. Search the web for literature, music, art, or historical references to atmospheric optical phenomena, other than the ones already listed in this chapter.

B11. Search the web for info on the history of scientific understanding of atmospheric optical phenomena. (E.g., Snell's law, etc.)

Apply

A1. Calculate the angle of reflection, given the following angles of incidence:

a. 10° b. 20° c. 30° d. 40° e. 45
f. 50° g. 60° h. 70° i. 80° j. 90°

A2. Find the refractive index for the following (medium, color) for $T = 0°C$ and $P = 80$ kPa.

a. air, red b. air, orange c. air, yellow
d. air, green e. air, blue f. air, violet
g. water, red h. water, orange i. water, yellow
j. water, green k. water, blue
m. ice, red n. ice, orange o. ice, yellow
p. ice, green q. ice, blue r. ice, violet

A3. Find the speed of light for the medium and color of the previous exercise.

A4. Find the ratio μ of refractive indices for the following pairs of medium from exercise A2:

a. (a, g) b. (a, m) c. (g, m) d. (b, h) e. (b, n)
f. (h, n) g. (c, i) h. (c, o) i. (i, o) j. (d, j)
k. (d, p) m. (j, p) n. (e, k) o. (e, q) p. (k, q)

A5. For the conditions in Fig. 22.2, find the angle of refraction in water, given an incident angle of 45° in air for λ (μm) of

a. 0.4 b. 0.45 c. 0.5 d. 0.55 e. 0.6
f. 0.65 g. 0.7 h. 0.42 i. 0.47 j. 0.52
k. 0.57 m. 0.62 n. 0.67

A6. For green light, calculate the angle of refraction for: (i) water, and (ii) ice, given the following incident angles in air, for conditions in Fig. 22.2:

a. 5° b. 10° c. 15° d. 20° e. 25° f. 30° g. 35°
h. 40° i. 45° j. 50° k. 55° m. 60° n. 65° o. 70°
p. 75° q. 80° r. 85°

A7. For Snell's law in 3-D, given the following component angles in degrees (α_1, β_1), find the total incident angle (θ_1) in degrees.

a. 5, 20 b. 5, 35 c. 5, 50 d. 5, 65 e. 5, 80
f. 20, 35 g. 20, 50 h. 20, 65 i. 20, 80 j. 35, 50
k. 35, 65 m. 35, 80 n. 50, 65 o. 50, 80 p. 65, 80

A8. Using the incident components in air from the previous problem, find the refraction components of blue-green light in ice. Check your answers using Snell's law for the total refraction angle. Assume the refractive index in air is 1.0 and in ice is 1.31 .

A9. Use the data in Fig. 22.2 to calculate the critical angles for light moving from ice to air, for λ (μm) of

a. 0.4 b. 0.45 c. 0.5 d. 0.55 e. 0.6
f. 0.65 g. 0.7 h. 0.42 i. 0.47 j. 0.52
k. 0.57 m. 0.62 n. 0.67

A10. Same as the previous exercise, but for light moving from water to air.

A11. Find the reflectivity from spherical rain drops, given impact parameters of:
a. 0.0 b. 0.1 c. 0.2 d. 0.3 e. 0.4 f. 0.5
g. 0.6 h. 0.7 i. 0.75 j. 0.8 k. 0.85 m. 0.9
n. 0.95 o. 0.98

A12. What value(s) of the impact parameter gives a primary-rainbow viewing angle of θ_1 = 42.0° for light of the following wavelength (µm)?
a. 0.4 b. 0.45 c. 0.5 d. 0.55 e. 0.6
f. 0.65 g. 0.7 h. 0.42 i. 0.47 j. 0.52
k. 0.57 m. 0.62 n. 0.67

A13. Find the output viewing angle for a primary rainbow with the conditions in Fig. 22.2 for green light and the following impact parameters.
a. 0.5 b. 0.55 c. 0.6 d. 0.65 e. 0.7 f. 0.75
g. 0.8 h. 0.85 i. 0.9 j. 0.95

A14.(§) Use a spreadsheet to calculate and plot primary-rainbow viewing angles for a full range of impact parameters from 0 to 1, for the conditions in Fig. 22.2. Do this for light of wavelength (µm):
a. 0.4 b. 0.45 c. 0.5 d. 0.55 e. 0.6
f. 0.65 g. 0.7 h. 0.42 i. 0.47 j. 0.52
k. 0.57 m. 0.62 n. 0.67

A15.(§) Same as previous exercise but for a secondary rainbow.

A16. For yellow light in a 22° halo, what is the viewing angle for the following input angles? Use the conditions in Fig. 22.2.
a. 15° b. 20° c. 25° d. 30° e. 35°
f. 40° g. 45° h. 50° i. 55° j. 60°
k. 65° m. 70° n. 75° o. 80° p. 85°

A17. Same as the previous exercise, but for blue light.

A18.(§) Use a spreadsheet to generate curves for green light, similar to those drawn in Fig. 22.25 for the 22° halo. Use the conditions in Fig. 22.2.

A19.(§) Same as previous exercise, but for the 46° halo.

A20. For yellow light in a 46° halo, find the viewing angle for the following input angles. Use the conditions in Fig. 22.2.
a. 58° b. 60° c. 62° d. 64° e. 66° f. 68° g. 70°
h. 72° i. 74° j. 76° k. 78° m. 80° n. 82°
o. 84° p. 86° q. 88°

A21. Same as the previous exercise, but for blue light.

A22. Find the minimum viewing angle (i.e., the nominal halo angle) for green light using the conditions in Fig. 22.2 for the following ice-crystal wedge angle:
a. 25° b. 52.4° c. 56° d. 60° e. 62° f. 63.8°
g. 70.5° h. 80.2° i. 90° j. 120° k. 150°

A23. For a solar elevation of 10° and conditions as in Fig. 22.2, calculate the viewing angle components for a circumzenith arc for red light with a crystal rotation of β =
a. 10° b. 12° c. 14° d. 16° e. 18° f. 20°
g. 22° h. 24° i. 26° j. 28° k. 30°
m. 32° n. 34°

A24.(§) For conditions as in Fig. 22.2, calculate and plot circumzenith arcs similar to Fig. 22.33 for red and violet light, but for solar elevation angles of
a. 10° b. 12° c. 14° d. 16° e. 18° f. 20°
g. 22° h. 24° i. 26° j. 28° k. 30° m. 0°
n. 2° o. 4° p. 6° q. 8°

A25. For a solar elevation of 10°, calculate the viewing angle components for a sun dog for red light with the following crystal rotations β. Use the conditions in Fig. 22.2.
a. 15° b. 20° c. 25° d. 30° e. 35°
f. 40° g. 45° h. 50° i. 55° j. 60°
k. 65° m. 70° n. 75° o. 80° p. 85°

A26.(§) Calculate and plot sun dog arcs similar to Fig. 22.35, but for the following solar elevation angles. Use the conditions of Fig. 22.2 for green light.
a. 5° b. 15° c. 25° d. 30°
e. 35° f. 45° g. 50° h. 55° i. 60°

A27.(§) For γ = 80°, plot upper tangent arc elevation and azimuth angles for a ray of orange-yellow light, for a variety of evenly-space values of rotation angle. Use a solar elevation of:
a. 10° b. 12° c. 14° d. 16° e. 18° f. 8°
g. 22° h. 24° i. 26° j. 28° k. 30°
m. 32° n. 34° o. 36° p. 38° q. 40°

A28. Find the fraction of incident light that is scattered for an optical thickness of:
a. 0.01 b. 0.02 c. 0.05 d. 0.10 e. 0.2
f. 0.5 g. 1. h. 2 i. 3 j. 4
k. 5 m. 7 n. 10 o. 20 p. 50

A29. What fraction of incident red light is scattered from air molecules along a horizontal path near the Earth's surface, with path length (km) of

a. 0.5 b. 1 c. 2 d. 5 e. 10 f. 20 g. 50
h. 100 i. 200 j. 500 k. 1000 m. 2000

A30.(§) Calculate and plot the relative fraction of scattering as a function of wavelength, for Rayleigh scattering in air.

A31. For cloud droplets of 5 μm radius, find the corona fringe viewing angles for the 2nd through 8th fringes, for light of the following wavelength (μm):
a. 0.4 b. 0.45 c. 0.5 d. 0.55 e. 0.6
f. 0.65 g. 0.7 h. 0.42 i. 0.47 j. 0.52
k. 0.57 m. 0.62 n. 0.67

A32.(§) Calculate and plot the viewing angle vs. wavelength for the:
a. second corona fringe.
b. third corona fringe.
c. fourth corona fringe.

A33. (§) Calculate and plot the ratio of refractive index in air to a reference refractive index in air, for a variety of values of ratio of air density to reference air density.

A34. If the temperature decreases 10°C over the following altitude, find the mirage radius of curvature of a light ray. Assume a standard atmosphere at sea level.
a. 1 mm b. 2 mm c. 3 mm d. 4 mm
e. 5 mm f. 6 mm g. 7 mm h. 8 mm
i. 9 mm j. 1 cm k. 2 cm m. 3 cm
n. 4 cm o. 5 cm p. 10 cm q. 20 cm

Evaluate & Analyze

E1. If light is coming from the water towards the air in Fig. 22.1 (i.e., coming along the θ_2 path but in the opposite direction), then sketch any incident and reflected rays that might occur.

E2. Salt water is more dense than fresh water. Sketch how light wave fronts behave as they approach the salt-water interface from the fresh water side.

E3. Is it possible for a material to have a refractive index less than 1.0 ? Explain.

E4. In general, how does the refractive index vary with temperature, pressure, and density?

E5. Use geometry and trig to derive eq. (22.6).

E6. Consider Fig. 22.5. If you were in a boat near the words "no escape here" and were looking into the water towards an object in the water at the black dot near the bottom of that figure, what would you see?

E7. Speculate on why rainbows are common with rain from cumulonimbus clouds (thunderstorms) but not with rain from nimbostratus clouds.

E8. If you were standing on a mountain top in a rain shower, and the sun was so low in the sky that sun rays were shining at a small angle upward past your position, could there be a rainbow? If so, where would it be, and how would it look?

E9. Use the relationship for reflection of light from water to describe the variations of brightness of a wavy sea surface during a sunny day.

E10. How does the intensity and color of light entering the rain drop affect the brightness of colors you see in the rainbow?

E11. Draw a sketch of a sun ray shining through a rain drop to make a primary rainbow. If red light is bent less than violet as a sun ray refracts and reflects through a rain drop, why isn't red on the inside of the primary rainbow instead of the outside?

E12. Use geometry to derive eq. (22.13).

E13. Use geometry to derive eq. (22.14).

E14. Draw a sketch of a sun ray path through a raindrop for a:
a. Tertiary rainbow b. Quaternary rainbow.
Your sketch must be consistent with the viewing angle listed for these rainbows.

E15. During the one reflection or two reflections of light inside a raindrop for primary and secondary rainbows, what happens to the portion of light that is not reflected? Who would be able to see it, and where must they look for it?

E16. Why are higher-order rainbows fainter than lower-order ones?

E17. If neither the primary or secondary rainbows return light within Alexander's dark band, why is it not totally black?

E18. Compare and contrast a twinned rainbow and supernumerary bows.

E19. Often hexagonal column ice crystals in the real atmosphere have indentations in their basal faces. These are known as hollow columns. How might that affect ice-crystal optical phenomena?

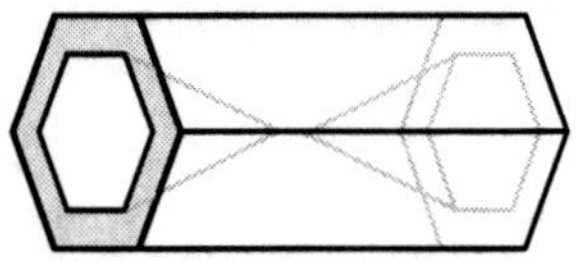

E20. For hexagonal columns, what other crystal angles exist besides 60° and 90°? For these other crystal angles, at what viewing angles would you expect to see light from the crystal?

E21. In Fig. 22.18, why are plate-crystal wobble and column-crystal rotation necessary to explain a sun pillar?

E22. Fig. 22.18d shows a dendrite ice crystal. In addition to the sun pillar, for what other halos & optical phenomena might such dendrites be important?

E23. Fig. 22.22a has only reflection while Fig. 22.22b also has two refractions. How might this affect the color of subsuns?

E24. Eq. (22.15) does not consider the finite size of a hexagonal column. What range of input angles would actually allow rays to exit from the face sketched in Fig. 22.24 for 22° halos?

E25. Use geometry to derive eq. (22.15) for the 22° halo.

E26. Derive equation (22.16) for the minimum viewing angle for halos. (You might need calculus for this).

E27.(§) Suppose a halo of 40° was discovered. What wedge angle β_c for ice would cause this?

E28. When the sun is on the horizon, what is the angle between the 22° halo and the bottom of the circumzenith arc?

E29. Is it possible to see both a circumzenith and a circumhorizon arc at the same time? Explain.

E30 Why might only one of the sun dogs appear?

E31. Using geometry, derive the minimum thickness to diameter ratio of hexagonal plates that can create sun dogs, as a function of solar elevation angle.

E32. For both sun dogs and rainbows, the red color comes via a path from the sun to your eyes that does not allow other colors to be superimposed. However, for both phenomena, as the wavelength gets shorter, wider and wider ranges of colors are superimposed at any viewing angle. Why, then, do rainbows have bright, distinct colors from red through violet, but sundogs show only the reds through yellows, while the large viewing angles yield white color rather than blue or violet?

E33. Why are the tangent arcs always tangent to the 22° halo?

E34. Sometimes it is possible to see multiple phenomena in the sky. List all of the optical phenomena associated only with

a. large hexagonal plates
b. large hexagonal columns
c. small hexagonal columns
d. small hexagonal plates
e. large hexagonal pyramids

E35. If optical depth is defined as the optical thickness measured vertically from the top of the atmosphere, then sketch a graph of how optical depth might vary with height above ground for a standard atmosphere.

E36. If you take photographs using a polarizing filter on your camera, what is the angle between a line from the subject to your camera, and a line from the subject to his/her shadow, which would be in the proper direction to see the sky at nearly maximum polarization? By determining this angle now, you can use it quickly when you align and frame subjects for your photographs.

E37 In the Solar & Infrared Radiation chapter, Beer's law was introduced, which related incident to transmitted light. Assume that transmitted light is 1 minus scattered light.

a. Relate the Rayleigh scattering equation to Beer's law, to find the absorption coefficient associated with air molecules.

b. If **visibility** is defined as the distance traveled by light where the intensity has decreased to 2% of the incident intensity, then find the visibility for clean air molecules.

c. Explain the molecular scattering limit in the atmospheric transmittance curve of Fig. 8.4a in the Satellite & Radar chapter.

E38. Discuss the problems and limitations of using visibility measurements to estimate the concentration of aerosol pollutants in air.

E39. For atmosphere containing lots of 1.5 μm diameter sulfuric acid droplets, describe how the sky and sun would look during a daily cycle from sunrise past noon to sunset.

E40. To see the brightest fringes furthest from the sun during a corona display, what size cloud droplets would be best?

E41. If you were flying above a cloud during daytime with no other clouds above you, describe why you might not be able to see glory for some clouds.

E42. Contrast and compare the refraction of light in mirages, and refraction of sound in thunder (see the Thunderstorm chapter). Discuss how the version of Snell's law in the Thunderstorm chapter can be applied to light in mirages.

E43. Microwaves are refracted by changes in atmospheric humidity. Use Snell's law or Huygens' principle to describe how ducting and trapping of microwaves might occur, and how it could affect weather radar and air-traffic control radar.

E44. What vertical temperature profile is needed to see the Fata Morgana mirage?

E45. Green, forested mountains in the distance sometimes seem purple or black to an observer. Also, the mountains sometimes seem to loom higher than they actually are. Discuss the different optical processes that explain these two phenomena.

E46. Paint on traffic signs and roadway lines is often sprinkled with tiny glass spheres before the paint dries, in order to make the signs more reflective to automobile headlights at night. Explain how this would work. Also, would it be possible to see other optical phenomena from these glass spheres, such as rainbows, halos, etc? Justify your answer.

Synthesize

S1. Suppose atmospheric density were (a) constant with height; or (b) increasing with height. How would optical phenomena be different, if at all?

S2. How would optical phenomena be different if ice crystals were octagonal instead of hexagonal?

S3. Suppose the speed of light through liquid and solid water was faster than through air. How would optical phenomena be different, if at all?

S4. After a nuclear war, if lots of fine Earth debris were thrown into the atmosphere, what optical phenomena would cockroaches (as the only remaining life form on Earth) be able to enjoy?

S5. Using the data in Fig. 22.2, fit separate curves to the refractive index variation with wavelength for air, water, and ice. Can you justify any or all of the resulting equations for these curves based on physical principles?

S6. Knowing the relationship between optical phenomena and the cloud (liquid or water) microphysics, and the relationship between clouds and atmospheric vertical structure, cyclones and fronts, create a table that tells what kind of weather would be expected after seeing various optical phenomena.

S7. If large raindrops were shaped like thick disks (short cylinders) falling with their cylinder axis vertical, how would rainbows be different, if at all?

S8. If large ice crystals were shaped like cylinders falling with their cylinder axis horizontal, how would ice-crystal optics be different, if at all?

S9. Consider Fig. 22.5. Could a thin layer of different fluid be floated on the liquid water to allow all light ray angles to escape from liquid to air? Justify your proposal.

S10. If there were two suns in the sky, each with light rays going to the same raindrop but arriving from different angles, is there a special angle between the arriving sun rays such that the two separate rainbows reinforce each other to make a single brighter rainbow?

S11. Design an ice-crystal shape that would cause:
a. different halos and arcs than exist naturally.
b. a larger number of natural halo and arcs to occur simultaneously.

S12. Suppose rain drops and ice crystals could cause refraction but not reflection or diffraction. What atmospheric optical phenomena could still exist? Justify your answers.

S13. Suppose that spacecraft landing on other planets can photograph optical phenomena such as halos, arcs, bows, etc. Describe how you could analyze these photographs to determine the chemicals and/or temperature of these planetary atmospheres.

S14. Design equipment to be used in the Earth's atmosphere that could determine the vertical temperature profile by measuring the characteristics of mirages.

APPENDIX A • SCIENTIFIC TOOLS

Contents

Many physical sciences including atmospheric science share the same fundamental definitions and analysis techniques. These fundamentals include problem-solving methods, standard units, ways of expressing relationships, formats for plotting the results, measurement uncertainty, and error propagation. The fundamentals reviewed here are used throughout this book.

A SCIENTIFIC PERSPECTIVE • Problem Solving

The following method aids understanding the problem, speeds solution, and helps to avoid errors. This method is used throughout the book in the various Sample Applications.

1) List the "Given" variables with their symbols, values and units.
2) List the unknown variables to "Find", with units.
3) Sketch the objects, velocities, etc. if appropriate.
4) Determine which equation(s) contains the unknown variable as a function of the knowns. This equation might need to be rearranged to solve for the unknown. If the solution equation contains more unknowns, find additional equations for them.
5) Make assumptions, if necessary, for any of the unknowns for which you have no equations. Clearly state your assumptions, and justify them.
6) Solve the equations using the known or assumed values, being sure to carry along the units. Show your intermediate steps.
7) Identify the final answer by putting a box around it, underlining it, or making it bold face.
8) Check your answer. If the solved units don't match the desired units of the unknown, then either a mistake was made, or unit conversion might be needed (e.g., convert from knots to m s^{-1}). Also, certain functions such as "ln" and "exp" require arguments that are dimensionless, while trig functions like "sine" need an argument in degrees or radians. These are clues to help catch mistakes. Also, compare your answer with your sketch, to check if it is physically reasonable. Check other physical constraints (e.g., humidities cannot be negative, speeds cannot be infinite).
9) Discuss the significance of the answer.

Table A-1. Basic dimensions and their SI units.

Dimension	Unit	Abbrev.
length	meter	m
mass	kilogram	kg
time	second	s
electrical current	ampere	A
temperature (thermodynamic)	kelvin	K
amount of substance	mole	mol
luminous intensity	candela	cd

Table A-2. Some derived dimensions & their SI units.

Dimension	Unit (Abbrev.)	Composition
force	newton (N)	$kg \cdot m \cdot s^{-2}$
energy	joule (J)	$kg \cdot m^2 \cdot s^{-2}$
power	watt (W)	$kg \cdot m^2 \cdot s^{-3}$
pressure, stress	pascal (Pa)	$kg \cdot m^{-1} \cdot s^{-2}$
temperature	degree Celsius (°C)	T(K) – 273.15
frequency	hertz (Hz)	s^{-1}
electric charge	coulomb (C)	$s \cdot A$
electric potential	volt (V)	$m^2 \cdot kg \cdot s^{-3} \cdot A^{-1}$
electric resistance	ohm (Ω)	$m^2 \cdot kg \cdot s^{-3} \cdot A^{-2}$
plane angle	radian (rad)	$m \cdot m^{-1}$
solid angle	steradian (sr)	$m^2 \cdot m^{-2}$

Table A-3. Prefixes. (*Size is in terms of USA "short scale" designations. Some international "long scale" designations are shown in *italics*, if different.)

Multiplier	Size*	Unit	Abbrev.
10^{24}	septillion (*quadrillion*)	yotta	Y
10^{21}	sextillion (*trilliard*)	zetta	Z
10^{18}	quintillion (*trillion*)	exa	E
10^{15}	quadrillion (*billiard*)	peta	P
10^{12}	trillion (*billion*)	tera	T
10^{9}	billion (*milliard*)	giga	G
10^{6}	million	mega	M
10^{3}	thousand	kilo	k
10^{2}	hundred	hecto	h
10^{1}	ten	deka	da
10^{-1}	tenth	deci	d
10^{-2}	hundredth	centi	c
10^{-3}	thousandth	milli	m
10^{-6}	millionth	micro	µ
10^{-9}	billionth	nano	n
10^{-12}	trillionth	pico	p
10^{-15}	quadrillionth	femto	f
10^{-18}	quintillionth	atto	a
10^{-21}	sextillionth	zepto	z
10^{-24}	septillionth	yocto	y

DIMENSIONS AND UNITS

Standards

There are seven basic dimensions in science, from which all other dimensions are derived (Table A-1). The first letters of the first three units are 'm, k, s'; hence, this system of units is sometimes called the **MKS system**. It has been adopted as the international system (**SI**) of units.

Derived units are formed from combinations of basic units. Examples of derived units that are used frequently in meteorology are listed in Table A-2.

A **prefix** can be added in front of these units to indicate larger or smaller values, such as kilometer (km), which is 1000 meters. The most commonly used prefixes are given in Table A-3. In this book, units and prefixes are NOT italicized, while *variables are italicized*. Other units are listed in Table A-4.

When we quantify things in nature, the value has four parts (but magnitude and prefix are optional):

•(A.1)

value = number · magnitude · prefix · units

A **scientific notation** example is:

distance value = 5x10² km

where number = 5, magnitude = 10^2 (= 100), prefix = k (=1000), and units = m. This is equivalent to 500,000 m. The value must include the units (except for a few things that are truly dimensionless). The number can include a negative sign (–) if needed.

Table A-4. Some other units. (* unofficial symbol)

Name	Symbol	Value in SI units
minute	min	1 min = 60 s
hour	h	1 h = 60 min = 3600 s
day	d	1 d = 24 h = 86,400 s
degree (angle)	°	1° = (π/180) rad
liter (or litre)	L	1 L = 10^{-3} m^3
metric ton (or tonne)	t	1 t = 10^3 kg
astronomical unit	ua	1 ua ≈ 1.49598 x 10^{11} m
ångström	Å	1 Å = 10^{-10} m
nautical mile	nm*	1 naut. mile = 1852 m
knot	kt*	1 knot = (1852 m / 3600 s)
hectare	ha	1 ha = 10^4 m^2
bar	bar	1 bar = 100 kPa = 10^5 Pa

INFO • Binary multiples

The IEEE Standards Board officially adopted the SI units of Table A-3. For example, 1 kilobit = 1000 bits.

CAUTION: Historically, manufacturers have called 2^{10} bits (=1,024 bits) as 1 kilobit (NOT SI standard), 2^{20} bits (=1,048,576) as 1 megabit (NOT SI standard), 2^{30} bits (=1,073,741,824) as 1 gigabit (NOT SI std.).

Unit Conversion

Create ratios of equivalent values to enable easy unit conversion. For example, a velocity of 1 knot equals 0.51 m s^{-1}. Because these two values equal, their ratio must be one:

$$1 = \frac{(1 \text{ knot})}{(0.51 \text{ m s}^{-1})} \qquad \text{(A.2)}$$

[*CAUTION: Although the* *numbers* *1 and 0.51 do NOT equal each other (and their ratio does NOT equal 1), the values include the units, and hence the ratio of values does equal 1.*]

The inverse of 1 is 1, thus, it makes no difference which value appears in the numerator. For example:

$$1 = \frac{(0.51 \text{ m/s})}{(1 \text{ knot})} \qquad \text{(A.3)}$$

When any quantity is multiplied by 1, its value does not change. Hence, by multiplying a velocity by the ratio in eqs. (A.2) or (A.3), we can change the units without changing the physics or the value.

Even if a conversion relationship is not known, you can sometimes figure it out if you know equivalent values that apply to a situation. For example, what is the conversion between pressure in "pounds per square inch" (PSI) and in "inches of mercury" (in Hg). Perhaps you might already know that the average pressure at sea level is 14.7 PSI. You might also know that standard sea-level pressure is 29.92 in Hg. Thus, these two values are equivalent at sea level, and their ratio gives the conversion between them:

$$1 = \frac{(14.7 \text{ PSI})}{(29.92 \text{ in Hg})} = 0.49 \frac{(\text{PSI})}{(\text{in Hg})}$$

or 1 in Hg = 0.49 PSI.

Ratios can also be formed to add or remove prefixes and magnitudes to units. For example, 1 milligram (mg) equals 0.001 grams, by definition. Thus their ratio is one. We can use this ratio to find 5 x 10^7 mg in units of kg. For example:

$$(5\text{x}10^7 \cancel{\text{mg}}) \cdot \frac{0.001 \cancel{\text{g}}}{1 \cancel{\text{mg}}} \cdot \frac{1 \text{ kg}}{1000 \cancel{\text{g}}} = 50 \text{ kg}$$

This trick of forming ratios to do conversions works only when both units have the same zero point. In the example above, 0 PSI = 0 inHg. Similarly, 0 m s^{-1} = 0 knots. This trick fails for temperature conversions, because °F, °C, and K all have different zero points. For temperature conversions, you must use special conversion formulae, as described in the "Relationships and Graphs" section.

Sample Application

Convert a wind speed of 10 knots into m s^{-1}.

Find the Answer:

Given: M = 10 kt wind speed

Find: M = ? m s^{-1}

Multiply wind speed by 1 in the form of eq. (A.3):

$$M = 10 \text{ kt} = (10 \text{ kt}) \cdot 1$$

$$= (10 \text{ kt}) \cdot \left(\frac{0.51 \text{ m/s}}{1 \text{ kt}}\right)$$

$$= \left(\frac{10 \cdot 0.51}{1}\right) \cdot \left(\frac{\cancel{\text{kt}} \cdot (\text{m/s})}{\cancel{\text{kt}}}\right) \qquad \text{grouping}$$

$$= \underline{\mathbf{5.1 \text{ m s}^{-1}}}$$

Check: Units OK. Physically reasonable.

Exposition: Note how we grouped the numbers separately from the units. You can calculate the number group using your calculator. You can reduce the units group by canceling identical units in the numerator and denominator (such as knots, in this example).

How do you know whether to use (A.2) or (A.3)? Given a value with knots in the numerator, we want to multiply it by a ratio that has knots in the denominator, so that the knots will cancel. Eq. (A.3) will work.

Sample Application

Sometimes you can use units to guess the form of an equation. For example, metabolic heat-production rate by humans sitting quietly is about 100 watts. Find the number of calories produced in half a day.

Find the Answer

Given: C = 100 W, Δt = 0.5 d

Find: B = ? cal heat production

Eqs: From tables of unit conversion in other books

1 W = 14.3353 cal min^{-1}, 1 h = 60 min, & 1 d = 24 h

By forming each of these equivalences as ratios, you can convert from watts into calories, and then from minutes into hours into days:

$$C = (100 \cancel{\text{W}}) \cdot \left(\frac{14.3353 \text{ cal}/\cancel{\text{min}}}{1 \cancel{\text{W}}}\right) \cdot \left(\frac{60 \cancel{\text{min}}}{1 \cancel{\text{h}}}\right) \cdot \left(\frac{24 \cancel{\text{h}}}{1 \text{ d}}\right)$$

$$= 2.06 \text{ x } 10^6 \text{ cal d}^{-1}$$

By looking at the units of the last line it is obvious that if we multiply it by the time Δt in days, then we will be left with our desired units of calories. Thus, the final equation is:

$$B(\text{cal}) = C(\text{cal day}^{-1}) \cdot \Delta t(\text{day})$$

$$= (2.06 \text{ x } 10^6 \text{ cal d}^{-1}) \cdot (0.5 \text{ d}) = \underline{\mathbf{1.03 \text{ x } 10^6 \text{ cal}}}$$

Check: Units OK. Physically reasonable.

Exposition: Over a million calories of heat is given off by a human sitting still for half a day. The number of calories of food we eat should be sufficient to replace those calories burned metabolically. Caution: "calories" listed on food packages are really kilocalories.

Sample Application

Suppose the air at the ground has a temperature of 20°C, while the air at height 500 m is 15 °C. Find the vertical temperature gradient.

Find the Answer

Given:

z_2 = 500 m top altitude

z_1 = 0 m ground altitude

$T_2 = T(z_2)$ = 15 °C temperature at 500 m altitude

$T_1 = T(z_1)$ = 20 °C temperature at ground

Find: $\Delta T/\Delta z$ = ? °C m^{-1} temperature gradient

Sketch:

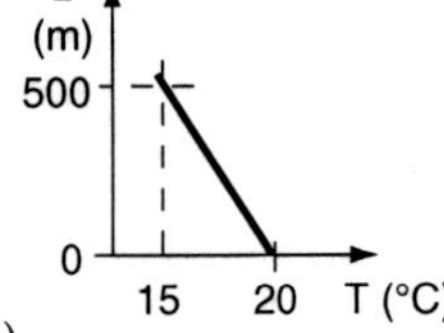

Use definition of a gradient:

$\Delta T/\Delta z = (T_2 - T_1) / (z_2 - z_1)$

$= (15°C - 20°C) / (500\text{ m} - 0\text{ m})$

$= (-5\text{ °C}) / (500\text{ m}) =$ **-0.01°C m^{-1}**

Check: Units OK. Sketch OK. Sign negative.

Exposition: If this gradient is constant with height, then the temperature at your head is 0.02°C colder than at your toes, assuming you are roughly 2m tall.

CAUTION: Suppose you erroneously had computed the gradient as (20 – 15°C) / (500 – 0 m). Then your answer would have had the wrong sign, because you had erroneously computed $(T_1 - T_2)/(z_2 - z_1)$ instead of the desired $(T_2 - T_1)/(z_2 - z_1)$. Always form your differences in the same order (point 2 – point 1) in both the numerator and denominator.

HIGHER MATH • Calculus

In theoretical meteorology, the physics of the atmosphere is described by differential equations. Outside of these "HIGHER MATH" boxes, we utilize the following approximations to avoid calculus.

A derivative can be approximated as:

$$\frac{\partial T}{\partial z} \approx \frac{\Delta T}{\Delta z} \qquad \text{for small } \Delta z$$

A total derivative:

$$\frac{dT}{dt} = \frac{\partial T}{\partial t} + U\frac{\partial T}{\partial x} + V\frac{\partial T}{\partial y} + W\frac{\partial T}{\partial z}$$

is approximated as:

$$\approx \frac{\Delta T}{\Delta t} + U\frac{\Delta T}{\Delta x} + V\frac{\Delta T}{\Delta y} + W\frac{\Delta T}{\Delta z}$$

Also, you can approximate an integral by a sum:

$$\int T\, dz \approx \sum T(z)\cdot \Delta z$$

where T is temperature, z is height, t is time, (U, V, W) are wind components in the (x, y, z) directions, which are positive toward the (east, north, up).

FUNCTIONS AND FINITE DIFFERENCE

As in other fields of science and engineering, functional relationships describe how one variable (the **dependent** variable) changes when one or more other variables (the **independent** variables) change. Suppose that P_2 is the pressure at time t_2, and P_1 is the pressure at time t_1. Pressure "varies" with time, or "is a function of time". Such functional dependence is written generically as $P(t)$. For a single value of pressure that occurs at specific time, such as at t = 50 s, the notation P(50 s) is used.

The symbol Δ means change or difference. Differences must always be taken in the same direction relative to the independent variable. For example, if temperature T and pressure P both vary with time t, then ΔT and ΔP are defined as their values at the later time minus their values at the earlier time. For example, $\Delta T = T(t_2) - T(t_1)$, and $\Delta P = P(t_2) - P(t_1)$, where t_2 is later than t_1. The notation is sometimes simplified to be $\Delta T = T_2 - T_1$ and $\Delta P = P_2 - P_1$ for $\Delta t = t_2 - t_1$.

In a different example, let temperature T depend on independent variable height z. Then

$$\Delta T = T(z_2) - T(z_1) = T_2 - T_1$$

and

$$\Delta z = z_2 - z_1$$

where z_2 is higher than z_1. Furthermore, a ratio such as $\Delta T/\Delta z$ is equivalent to $(T_2 - T_1)/(z_2 - z_1)$, or $[T(z_2) - T(z_1)]/(z_2 - z_1)$, where the differences in the numerator and denominator must always be taken in the same direction (e.g., point 2 – point 1). Don't be deceived into subtracting the smaller value from the larger one when you compute a finite difference, because if you do, the sign of your answer might be wrong (see the Sample Application on this page).

The change of something with distance is called a **gradient**. Thus, $\Delta T/\Delta z$ is a **vertical temperature gradient**.

Although calculus is a useful mathematical tool for studying the physics of the atmosphere, this book is designed for an audience who might not have had calculus. In the place of differential calculus we will use finite differences, Δ. In place of integral calculus, we will use sums or graphically examine the area under curves.

However, for those students with a calculus background, "HIGHER MATH" boxes are scattered here and there in the book to provide a taste of theoretical meteorology. These "HIGHER MATH" boxes are surrounded by a double line as shown at left, and may be safely skipped by students wishing to avoid calculus.

RELATIONSHIPS AND GRAPHS

Although $T(z)$ says that there is some functional relationship between temperature T and height z, it does not specify what that relationship is. Temperature could increase as height increases. It could decrease as height increases. It could increase with the square of height. It could vary logarithmically with height. It could be invariant with height.

The particular form of the function might be governed by some underlying physics, and have a fixed functional form that is sometimes called a physical "law" or a mathematical definition. Other relationships can be found by plotting observational data.

Linear, **semi-log** and **log-log** graphs are frequently used to discover and display relationships between dependent and independent variables.

Linear

A **linear** relationship yields a straight line on a **linear graph** (values along both axes increase at constant rates with distance from the origin).

For example, the relationship between temperature in degrees Celsius and Fahrenheit is linear; namely, temperature in Fahrenheit is proportional to the first power of temperature in Celsius

$$T_{°F} = a \cdot T_{°C} + b \qquad \bullet(A.4)$$

where the parameters in this equation are $a = 9/5$ (°F/°C) and $b = 32$ °F, see Fig. A.1.

The slope (change in values along the vertical axis per change of values along the horizontal axis) of the line equals the factor a, which is 9/5 in this case. As shown in Fig. A.1, the nonzero parameter b causes the plotted line to cross the vertical axis not at the origin ($T_{°C}$, $T_{°F}$) = (0, 0), but at an intercept of T = 32°F [i.e., at point (0°C, 32°F].

The relationship between temperature in degrees Celsius and absolute temperature in kelvins is also linear:

$$T(K) = T(°C) + 273 \qquad \bullet(A.5)$$

If plotted on a linear graph, the slope would be 1 and the intercept is 273 K. Most equations using temperature require the use of absolute temperature.

Logarithmic

An **exponential** or **logarithmic** relationship gives a straight line when plotted on a **semi-log graph** (one axis is linear, the other is logarithmic).

For example, the decrease of pressure with height is logarithmic in atmospheres where the temperature is constant with height:

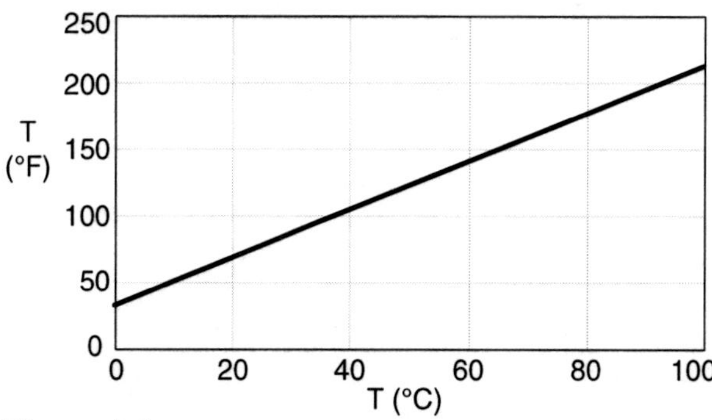

Figure A.1
Relationship between Celsius (°C) and Fahrenheit (°F) temperature (T), plotted on a linear graph.

Sample Application
Convert 10°C to Fahrenheit.

Find the Answer:
Given: $T_{°C}$ = 10°C. Find: $T_{°F}$ = ? °F

Sketch: (see Fig A.1). Use eq. (A.4):
$T_{°F} = a \cdot T_{°C} + b$
= (9/5 °F/°C)·(10°C) + 32°F
= 18°F + 32°F = **50°F**

Check: Units OK. Sketch OK. Physics OK.
Exposition: As the temperature in Celsius increases by equal amounts of 10°C, the corresponding Fahrenheit values increase in equal amounts of 18°F. Such a constant rate of increase of the dependent variable for a constant rate of increase of independent variable indicates a linear relationship.

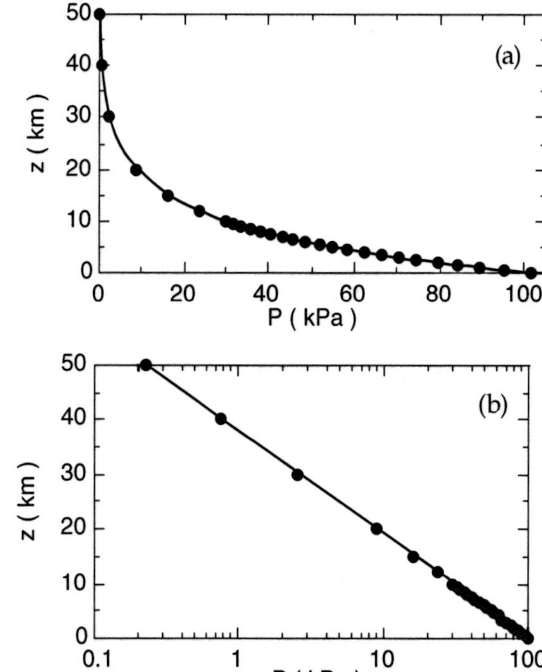

Figure A.2
Height z vs. pressure P in the atmosphere, plotted on (a) linear; and (b) semi-log graphs. (Copied from Chapter 1.)

Sample Application
At what height does P = 50 kPa, given T = 0°C?

Find the Answer
Given: P = 50 kPa, T = 0 + 273.15 = 273.15 K.
Find: z = ? km

Solve eq. (A.6) for z: $z = (T/a) \cdot \ln(P_o/P)$
z = [(273.15K) / (0.0342 K m^{-1})] · ln(101.3kPa/50kPa)
= (7987 m) · ln(2.026) = 5639 m = **5.6 km**.

Check: Units OK. Agrees with Fig. A.2.
Exposition: At this height the air is so thin that people would die of **hypoxia** (lack of oxygen) unless they breath pressurized oxygen.

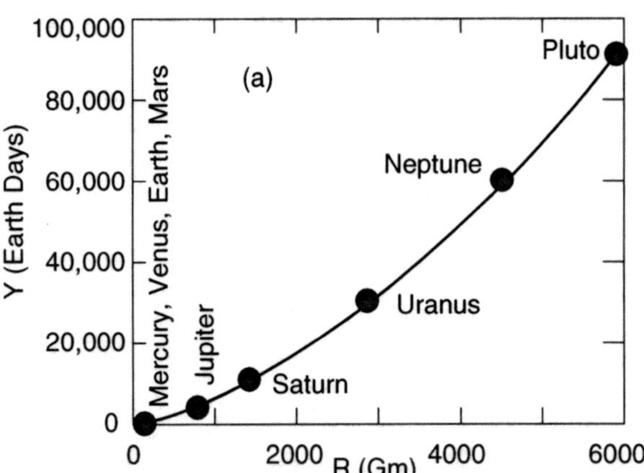

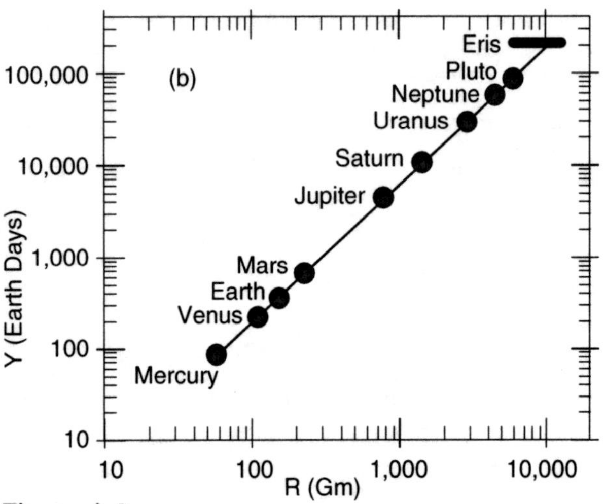

Figure A.3
Plot of planetary orbital periods in Earth days versus distance from sun in billions of meters, on (a) linear, (b) log-log graphs.

Sample Application
Find the orbital period of Mercury (R = 58 Gm).
Find the Answer
Given: R = 58 Gm. Find: Y = ? d
Use eq.(A.7): Y = [0.1996 d·(Gm)$^{-3/2}$] · (58 Gm)$^{3/2}$ = **88d**
Check: Units OK. Agrees with Fig. A.3
Exposition: 4 orbits of Mercury per each Earth orbit.

$$\ln\left(\frac{P}{P_o}\right) = \frac{-a}{T} \cdot z \qquad \text{(A.6)}$$

where a = 0.0342 K m^{-1}, P_o = 101.3 kPa for Earth, and average temperature T must be in Kelvin.

[CAUTION: symbol a in this equation has a different value than in the previous equation. I often re-use symbols in this book, because there are not enough symbols for all the variables.]

Fig. A.2 plots the relationship between P and z in both linear and semi-log graphs, for T = 280 K. Pressure decreases rapidly with height near the ground, but decreases more slowly at higher altitude.

From Fig. A.2a we see that the portion of the curve in the lowest 3 km of the atmosphere is almost a straight line, and conclude that the variation of pressure with height is nearly linear in that region. Near the ground, the pressure decreases roughly 10 kPa with each 1 km height gain.

Fig. A.2b plots P with a logarithmic scale along the bottom axis, and plots z with a linear scale on the vertical axis. All of the data points at all heights fall on a straight line on this graph. This is evidence that the logarithm of P is proportional to z.

Power

If the dependent variable is proportional to a **power** of the independent variable, then the data will appear as a straight line on a **log-log graph** (both axes are logarithmic).

For example, Johannes Kepler, the 17th century astronomer, discovered that planets in the solar system have elliptical orbits around the sun, and that the time period Y of each orbit is related to the average distance R of the planet from the sun by:

$$Y = a \cdot R^{3/2} \qquad \text{(A.7)}$$

Parameter $a \approx$ 0.1996 d·(Gm)$^{-3/2}$, where d is the abbreviation for Earth days and Gm is gigameters (= 10^6 km).

Using a table of the average distance of the planets from the sun (see Numerical Exercise N1 in the Radiation chapter), you can calculate orbital period using eq. (A.7). These are plotted in Fig. A.3a on a linear-linear graph, and in Fig. A.3b on a log-log graph (copied from the Radiation chapter).

The slope of the straight line on the log-log graph (Fig. A.3b) equals the power of the exponent. Namely, the range of orbital periods between Mercury and Pluto is about 3 decades (i.e., 100 to 1,000 to 10,000 to 100,000), while the range of distances from the sun is 2 decades. Thus, the slope is 3 to 2, as indicated in eq. (A.7).

ERRORS

Error is the difference between a measured (or estimated) value and the true (or reference) value.

Systematic Error & Accuracy

If part of the error is **systematic** or **repeatable** (namely, you get the same error each time you make a measurement), then the difference between the average measurement and the true value is called the **bias**. Smaller bias magnitude (i.e., lower systematic errors) corresponds to greater **accuracy**. Namely, accuracy indicates how close your **average** observations are to truth (Fig. A4).

Systematic errors can be due to **errors in instrument calibration**, **personal errors** (such as parallax error in reading a dial), **erroneous experimental conditions** (such as not shielding a thermometer from sunlight), and **imperfect technique** (such as breathing on a thermometer before you read it).

If you can calculate or otherwise know the bias, then you can remove this bias from your observations to correct for systematic error. Namely, you can easily make your corrected observations more accurate.

Random Error & Precision

After removing systematic errors, you might find that your observations still have some unexplained variability from measurement to measurement. These are called **random errors** (Fig. A.4). Experiments with smaller random errors are said to have higher **precision**; namely, are more precise. The **standard deviation** (or spread) of your observations is a measure of the random error — greater standard deviation indicates greater random error and lower precision.

Random errors can be due to **errors in judgement** (such as by manually reading a dial with poor resolution), **fluctuating conditions** (such as trying to determine sea level on a wavy ocean), **small disturbances** (such mechanical vibrations of an instrumented tower in high winds), and **errors in definition** (such as measuring the dimension of a fractal-shaped cloud, which depends on the size of the measuring stick).

Unfortunately, the probabilistic nature of random errors makes them difficult to remove after the fact. Often, the only recourse is to repeat the experiment under better controlled conditions and with higher quality instruments, and be sure to take a large number of observations to improve the statistical robustness of your results.

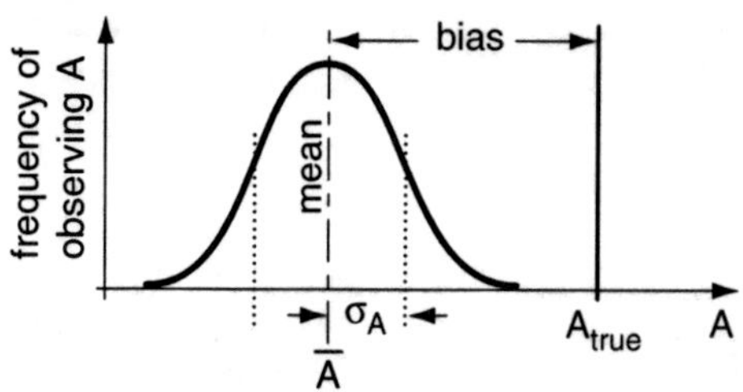

Figure A.4
The dark curve is the frequency that different A values are observed. Accuracy is related to bias. Precision is related to spread of curve (described by standard deviation σ_A). $\overline{A}$ is the mean.

Reporting Observations

For any variable A that you have measured N times to yield a **data set** $(A_1, A_2, A_3, ..., A_N)$, let $\overline{A}$ be the **mean value**, and σ_A be the **standard deviation**. These are defined as:

$$\overline{A} = \frac{1}{N}\sum_{i=1}^{N} A_i \qquad \bullet\text{(A.8)}$$

and

$$\sigma_A = \left[\frac{1}{(N-1)}\sum_{i=1}^{N}(A-\overline{A})^2\right]^{1/2} \qquad \bullet\text{(A.9)}$$

where i is a **dummy index** that points to a single **data element** A_i in your data set.

After removing any known biases, the resulting observation is usually reported or written as a mean (average) value plus or minus (±) the standard deviation:

$$A = \overline{A} \pm \sigma_A \qquad \bullet\text{(A.10)}$$

where the precision or **standard uncertainty** is given by the standard deviation σ_A.

Sample Application

Given these T observations: (15, 13, 20, 12, 10, 17, 18)°C. T_{true} = 8°C. Find the mean, bias & std dev.

Find the Answer

Given: T_{true} = 8°C, and the data set above.
Find: $\overline{T}$ = ? °C, σ_T = ? °C, bias = ? °C.

Use eq. (A.8): $\overline{T}$ = (1/7)(15+13+20+12+10+17+18) = **15°C**
Define the deviation from mean as: $T' = T - \overline{T}$
Thus, our observation have T' = (0, –2, 5, –3, –5, 2, 3)°C
Use eq. (A.9), rewritten as $\sigma_T = [(N–1)^{-1} \Sigma(A'^2)]^{1/2}$
Thus: $\sigma_T = [(1/6) \cdot (0+4+25+9+25+4+9)(°C)^2]^{1/2}$ = **3.56°C**
From the raw observations: **T = 15 ± 3.56 °C** .
Bias = $\overline{T} - T_{true}$ = 15 – 8 °C = **7°C** .

Check: Units OK. We seem to have a warm bias.
Exposition: Our observations are **not accurate** (large bias) and are **not precise** (large σ_T).

HIGHER MATH • Error Propagation

Suppose D is a function of A, B and C, where C is not a constant. Namely: $D(A, B, C)$.

If the error standard deviations σ_A, σ_B, and σ_C for A, B & C are known, then the **propagation of errors** into the standard deviation σ_D of variable D is:

$$\sigma_D = \left[\left(\frac{\partial D}{\partial A}\right)^2 \cdot \sigma_A^2 + \left(\frac{\partial D}{\partial B}\right)^2 \cdot \sigma_B^2 + \left(\frac{\partial D}{\partial C}\right)^2 \cdot \sigma_C^2 + 2r_{AB} \cdot \left(\frac{\partial D}{\partial A}\right) \cdot \left(\frac{\partial D}{\partial B}\right) \cdot \sigma_A \cdot \sigma_B + 2r_{AC} \cdot \left(\frac{\partial D}{\partial A}\right) \cdot \left(\frac{\partial D}{\partial C}\right) \cdot \sigma_A \cdot \sigma_C + 2r_{BC} \cdot \left(\frac{\partial D}{\partial B}\right) \cdot \left(\frac{\partial D}{\partial C}\right) \cdot \sigma_B \cdot \sigma_C \right]^{1/2} \quad \text{(A.a)}$$

The correlation coefficients r are defined as

$$r_{AB} = \frac{1}{(N-1)\cdot\sigma_A\cdot\sigma_B} \cdot \sum_{i=1}^{N} \left[(A_i - \overline{A})\cdot(B_i - \overline{B})\right]$$

and correlations r_{BC} and r_{AC} are defined similarly. If A and B are independent, the $r_{AB} = 0$, and similarly for the other correlation coefficients.

For example, suppose you measure air density $(\rho = \overline{\rho} \pm \sigma_\rho)$ and temperature $(T = \overline{T} \pm \sigma_T)$, and calculate pressure (P) using the ideal gas law $P = \rho\cdot\Re\cdot T$, where $\Re$ is a constant. Thus, from calculus: $\partial P/\partial\rho = \Re\cdot T$, and $\partial P/\partial T = \rho\cdot\Re$. Assume ρ and T are independent, thus the correlation coefficient $r_{\rho T} = 0$.

Our best estimate of pressure is

$$\overline{P} = \overline{\rho}\cdot\Re\cdot\overline{T} .$$

To estimate the pressure error σ_P, use eq. (A.a) to propagate the other errors into the pressure error:

$$\sigma_P = \left[\left(\frac{\partial P}{\partial \rho}\right)^2 \cdot \sigma_\rho^2 + \left(\frac{\partial P}{\partial T}\right)^2 \cdot \sigma_T^2\right]^{1/2}$$

$$\sigma_P = \left[(\Re\cdot T)^2\cdot\sigma_\rho^2 + (\Re\cdot\rho)^2\cdot\sigma_T^2\right]^{1/2}$$

Multiply the right side by 1 in the form of $\overline{P}/(\overline{\rho}\cdot\Re\cdot\overline{T})$

$$\sigma_P = \overline{P}\cdot\left[\left(\frac{1}{\overline{\rho}}\right)^2 \cdot \sigma_\rho^2 + \left(\frac{1}{\overline{T}}\right)^2 \cdot \sigma_T^2\right]^{1/2}$$

or

$$\sigma_P = \overline{P}\cdot\left[(\sigma_\rho/\overline{\rho})^2 + (\sigma_T/\overline{T})^2\right]^{1/2}$$

where we use the averages as our best estimates of P, ρ, and T. This last result looks like eq. (A.14). In fact, we could have used eq. (A.14) directly and avoided all the calculus.

Thus, we would report our calculated pressure as:

$$P = \overline{P} \pm \sigma_P$$

Similarly, the mean and standard deviation of some other variable B would be $\overline{B}$ and σ_B.

For example, Newton's constant of gravitation G is reported (CODATA 2006) as:

$$G = 6.67428\text{x}10^{-11} \pm 0.00067\text{x}10^{-11} \text{ m}^3\text{ kg}^{-1}\text{ s}^{-2} .$$

Error Propagation

Error propagation tells us how the errors in A and B affect the error of D, where D depends on A and B according to some equation. Namely, how can we estimate σ_D knowing σ_A and σ_B. Assume the errors in A and B are independent of each other.

For a simple sum or difference (e.g., $D = A + B$, or $D = A - B$), then

$$\sigma_D = [\sigma_A{}^2 + \sigma_B{}^2]^{1/2} \quad \text{(A.11)}$$

For $D = c\cdot A$ where c is a constant, then

$$\sigma_D = c \cdot \sigma_A \quad \text{(A.12)}$$

Similarly, if $D = c_A\cdot A \pm c_B\cdot B$ where c_A and c_B are different constants, then

$$\sigma_D = [c_A{}^2\cdot\sigma_A{}^2 + c_B{}^2\cdot\sigma_B{}^2]^{1/2} \quad \text{(A.13)}$$

For a simple product $D = c\cdot A\cdot B$ or quotient $D = c\cdot A/B$, then

$$\sigma_D = \overline{D} \cdot [\,(\sigma_A/\overline{A}\,)^2 + (\sigma_B/\overline{B}\,)^2\,]^{1/2} \quad \text{(A.14)}$$

where $\overline{D}$ is the average of D, $\overline{A}$ is the average of A, and $\overline{B}$ is the average of B.

For a simple power relationship $D = c\cdot A^m$ where m is a fixed constant, then

$$\sigma_D = \overline{D} \cdot m \cdot (\sigma_A/\overline{A}\,) \quad \text{(A.15)}$$

For the general case of a product of factors raised to various fixed (errorless) powers $D = c\cdot A^m\cdot B^q$, then

$$\sigma_D = \overline{D} \cdot [\,m^2\cdot(\sigma_A/\overline{A}\,)^2 + q^2\cdot(\sigma_B/\overline{B}\,)^2\,]^{1/2} \quad \text{(A.16)}$$

For a logarithm such as $D = \ln(c\cdot A)$, where c is a constant, then

$$\sigma_D = (\sigma_A/\overline{A}\,) \quad \text{(A.17)}$$

For an exponential such as $D = e^{c\cdot A}$ where c is a constant, then

$$\sigma_D = c \cdot \overline{D} \cdot \sigma_A \quad \text{(A.18)}$$

For more complicated relationships, the rules above can be combined or used sequentially (or see the HIGHER MATH box).

Sample Application
Observations give $P_1 = 100 \pm 0.1$ kPa, $P_2 = 50 \pm 0.5$ kPa, and $\overline{T_v} = 260 \pm 5$ K. Use hypsometric eq. to find Δz.

Find the Answer
Given: $\overline{P}_1 = 100$ kPa, $\sigma_{P1} = 0.1$ kPa, $\overline{P}_2 = 50$ kPa,
$\sigma_{P2} = 0.5$ kPa, $\overline{T_v} = 260$ K, $\sigma_T = 5$ K.
Hyp. eq.(1.26a): $\Delta z = a \cdot \overline{T_v} \cdot \ln(P_1/P_2)$, a=29.3m K^{-1}.
Find: $\Delta z = ? \pm ?$ m. Namely, find $\overline{\Delta z} = ?$m, $\sigma_{\Delta Z} = ?$ m

Method: Use error propagation rules sequentially.
For (P_1/P_2): *Average(P/P)* = (100kPa)/(50kPa) = 2
Use eq. (A.14): $\sigma_{P/P} = 2 \cdot [(0.1/100)^2 + (0.5/50)^2]^{1/2} = 0.02$

For $a \cdot \ln(P_1/P_2)$: *Average* =(29.3m K^{-1})·ln(2) = 20.31 m K^{-1}
Use eq. (A.17): $\sigma_{a \cdot ln}$ =(29.3m K^{-1})·(0.02/2) = 0.293 m K^{-1}

For $\overline{T_v} \cdot a \cdot \ln(P_1/P_2)$:*Average*=(20.31m K^{-1})·(260K)= **5281m**
Use eq. (A.14): $\sigma_{\Delta Z}$=(5281m)·[(5/260)^2+(0.293/20.31)^2]$^{1/2}$
= **127 m**
Thus: Δz = **5281 ± 127 m**

Exposition: Notice that error-propagation eqs. (A.11 - A.18) are dimensionally consistent. A good check.

A SCIENTIFIC PERSPECTIVE • Have Passion

The best scientists and engineers need more than the good habits of diligence and meticulousness. They need passion for their field, and they need creativity. In this regard, they are kindred spirits to artists, composers, musicians, authors, and poets.

While an observation is something that can usually be quantified, the explanation or theory for it comes from the minds of people. For example, does light consist of particles (photons) or waves? Probably it consists of neither, but those are two theories from the creative imagination of scientists that have proved useful in explaining the observations.

The joy that a scientist feels after successfully explaining an observation, and pride that an engineer feels for making something work within the constraints of physics and economics, are no less intense than the joy and pride felt by an artist who has just completed his or her masterpiece.

Approach your work with passion, evaluate your result objectively, and enjoy your travel through life as you help society.

A SCIENTIFIC PERSPECTIVE

Science is a philosophy. It is faith in a set of principles that guide the actions of scientists. It is a faith based on observation. Scientists try to explain what they observe. Theories not verified by observations are discarded. This philosophy applies to **atmospheric science**, also known as **meteorology**.

A good theory is one that works anywhere, anytime. Such a theory is said to be **universal**. Engineers utilize universal theories with the expectation they will continue working in the future. The structures, machines, circuits, and chemicals designed by engineers that we use in every-day life are evidence of the success of this philosophy.

But we scientists and engineers are people, and share the same virtues and foibles as others. Those of you planning to become scientists or engineers might appreciate learning some of the pitfalls so that you can avoid them, and learning some of the tools so that you can use them to good advantage.

For this reason, scattered throughout the book are boxes called "A SCIENTIFIC PERSPECTIVE", summarized in Table A–5. These go beyond the mathematical preciseness and objective coldness that is the stereotype of scientists. These boxes cover issues and ideas that form the fabric of the philosophy of science. As such, many are subjective. While they give you one scientist's (my) perspectives, I encourage you to discuss and debate these issues with other scientists, colleagues, and teachers.

Table A-5. Guidelines & issues for scientists. Chapters & topics of the "A SCIENTIFIC PERSPECTIVE" boxes.

Chapter	Issues	Page
1	Descartes and the Scientific Method	2
1	Check for Errors	16
1	Be Meticulous	20
1	Give Credit	24
2	Scientific Laws — The Myth	38
2	Seek Solutions	46
3	Expert vs. Novice	72
4	Look for Patterns	107
6	Cargo Cult Science	167
7	Consequences	218
9	Creativity in Engineering	288
10	Be Creative	293
11	Toy Models	330
11	Residuals	340
11	The Scientific Method Revisited	343
11	Model Sensitivity	350
12	Math Clarity	393
13	Truth vs. Uncertainty	470
14	Be Safe (many parts)	485, 517
15	Be Safe (continuation)	567, 583, 584
17	Simple is Best	680
18	Parameterization Rules	715
19	Data Misinterpretation	729
19	The Citizen Scientist	738
20	Mathematics	762
20	Scientific Revolutions	773
21	Ethics and Data Abuse	826
22	Great Scientists Make Big Mistakes Too	863
A	Problem Solving	869
A	Have Passion	877

REVIEW

SI units are used by atmospheric scientists. Functional relationships between variables can sometimes be discerned when the data is plotted on linear, semi-log, or log-log graphs. An organized approach to problem solving is recommended, including error propagation. The philosophy of science blends the passions of people with the objective analysis of observations. These principles are used throughout this book.

HOMEWORK EXERCISES

Broaden Knowledge & Comprehension

B1. Search the web for recommended units to use in meteorology, and list 3 that were not given in this chapter. [Hint, on the American Meteorological Society web site, search for a "Guide for Authors".]

B2. For 5 of the universal constants listed in Appendix B, search the web to find their precision (i.e., the $\pm\sigma$ value). [Hint: try NIST (U.S. National Institute of Standards and Technology.]

Apply

A1. Convert the values on the left to the units at right, using a table of conversions.

a. 50 miles =? km b. 15 knots = ? m s^{-1}
c. 30 lb in^{-2} =? kPa d. 5000 kW =? horsepower
e. 150 lb_{Mass} =? kg f. 150 lb_{Force} =? N
g. 12 ft = ? m h. 50 km h^{-1} = ? m s^{-1}

A2. Solve the expression on the left, and give the answer in the units at right, using a table of conversions, and the basic definitions of units:

a. (55. knots) x (36. inches) =? $m^2 \cdot s^{-1}$
b. (14. lb_F in^{-2}) x $(2.5\ m)^2$ = ? N
c. (120. lb_M) x (3. knots) / day =? mN
d. (15. inHg) x (2. ft^3) = ? J
e. (500. mb) x (3. knots) x $(5.\ in)^2$ =? kW
f. (9.8 $m \cdot s^{-2}$) x (6 kg) / (ft^2) = ? mb
g. (4200 $J \cdot kg^{-1} \cdot K^{-1}$) x (5°C) x (3.3 g) = ? ergs
h. $(2\ ha)^{1/2}$ / 3 weeks = ? m s^{-1}

A3. Find $\Delta T/\Delta z$ between the height at assigned letter (a - e) and the height immediately above it?

	z(m)	T(°C)
	1000	10
e.	500	15
d.	200	17
c.	100	17
b.	50	15
a.	0	10

A4. Convert the following temperatures:

a. 15°C = ? K b. 50°F = ? °C c. 70 °F =? K
d. 48°C = ? °F e. 400 K = ? °F f. 250 K=?°C

A5(§). Plot the following relationships on linear, semi-log, and log-log graphs. Use a spreadsheet (§) on a personal computer to make this easier. Any variable with subscript "o" represents a constant.

a. $I = I_o \cdot (R_o/R)^2$ b. $U = U_o \cdot \ln(z/z_o)$
c. $E = E_o \cdot \exp(-z/z_o)$ d. $f = c_o / \lambda$
e. $q = e / e_o$ f. $c = c_o \cdot \exp[-(z/z_o)^2]$
g. $w = [2 \cdot (F/m)_o \cdot z]^{1/2}$ h. $\Delta P/\Delta P_o = (1/5) \cdot (R/R_o)^4$

A6. Given: A = 4000±20 m, B = 300±5 K, C = 80±2 m Also, k = 5 is a constant. Find $D = \overline{D} \pm \sigma_D$ for:

a. $D = A + C$ b. $D = A - C$ c. $D = 3C + A$
d. $D = k \cdot B$ e. $D = k^2 \cdot A$ f. $D = A/B$
g. $D = A \cdot C$ h. $D = A \cdot B/C$ i. $D = (C+A)/B$
j. $D = A^k$ k. $D = B^{-k}$ m. $D = C^k \cdot B^{1/3}$
n. $D = k \cdot \ln(A/25\text{m})$ o. $D = k \cdot (C-A)/\ln(B/273\text{K})$

Evaluate & Analyze

E1. Is anything dimensionally wrong with the following equations, given: P = 100 kPa, z = 2 km, T = 30°C, W = 0.5 m s^{-1}? If so, why?

a. $\log(P) = z$ b. $\sin(T) = W$
c. $\arccos(P) = T$ d. $\exp(-P/z) = 1$
e. $\cos(W) = 2$ f. $\ln(0) = z$
g. $\ln(-10) = T$ h. $\exp(0) = 1$

E2. What is the difference between a good assumption and a bad one? What can you do to detect bad assumptions?

Synthesize

S1. Suppose that you discovered a new physical characteristic of nature, and that you devised a new dimension to explain it. However, also assume that the list of basic dimensions in Table A–1 is still valid, which means that your new dimension must be able to be described in terms of the basic dimensions. Describe the steps that you could take to determine the relationship between your new physical dimension and the basic units.

S2. Suppose science did not involve human creativity. What physical "laws" might have been described differently than they are now, or might have not been discovered at all?

B CONSTANTS & CONVERSION FACTORS

Contents

UNIVERSAL CONSTANTS

[from US National Institute of Standards and Technology (NIST), based on 2014 CODATA]

c_o = 299,792,458. m/s = speed of light in a vacuum

c_1 = 3.741 771 790 x10^8 W $\cdot$m^{-2} $\cdot\mu$m^4 = first radiation constant (in Planck's law)

c_{1B} = 1.191 042 953 x10^8 W $\cdot$m^{-2} $\cdot\mu$m^4 $\cdot$sr^{-1} = first radiation constant for spectral radiance

c_2 = 1.438 777 36 x10^4 μm $\cdot$K = second radiation constant (in Planck's law)

G = 6.674 08 x10^{-11} m^3 $\cdot$s^{-2} $\cdot$kg^{-1} = Newtonian gravitational constant

h = 6.626 070 040 x10^{-34} J $\cdot$s = Planck constant

k_B = 1.380 648 52 x10^{-23} J $\cdot$K^{-1} $\cdot$molecule^{-1} = Boltzmann constant

N_A = 6.022 140 857 x10^{23} mol^{-1} = Avogadro constant

σ_{SB} = 5.670 367 x10^{-8} W $\cdot$m^{-2} $\cdot$K^{-4} = Stefan-Boltzmann constant

T = –273.15°C = 0 K = absolute zero (not considered a true universal constant)

MATH CONSTANTS

[from *CRC Handbook of Chemistry and Physics*]

e = 2.718 281 828 459 = base of natural logarithms

$1/e$ = 0.367 879 441 = e-folding ratio

π = 3.141 592 653 589 793 238 462 643 = pi

sqrt(2) = 1.414 213 562 373 095

EARTH CHARACTERISTICS

1° latitude = 111 km = 60 nautical miles (nm) [Caution: This relationship does NOT hold for degrees longitude.]

a = 149.598 Gm = semi-major axis of Earth orbit

A = 0.306 = Bond albedo (NASA 2015)

A = 0.367 = visual geometric albedo (NASA 2015)

b = 149.090 Gm = semi-minor axis of Earth orbit

d = 149.597 870 7 Gm = average sun-Earth distance = 1 Astronomical Unit (AU) (NASA 2015)

$d_{aphelion}$ = 152.10 Gm = furthest sun-Earth distance, which occurs about 4 July (NASA 2015)

$d_{perihelion}$ = 147.09 Gm = closest sun-Earth distance, which occurs about 3 January (NASA 2015)

d_r = 173 = 22 June = approx. day of summer solstice

e = 0.0167 = eccentricity of Earth orbit around sun

g = –9.806 65 m $\cdot$s^{-2} = average gravitational acceleration on Earth at sea level (negative = downward). (from 2014 CODATA)

$|g| = g_o \cdot [1 + A \cdot \sin^2(\phi) - B \cdot \sin^2(2\phi)] - C \cdot H$
= variation of gravitational-acceleration magnitude with latitude ϕ and altitude H(m) above mean sea level. g_o = 9.780 318 4 m$\cdot$s^{-2}, A = 0.005 3024, B = 0.000 0059, C = 3.086x10^{-6} s^{-2}.

M = 5.9726 x10^{24} kg = mass of Earth (NASA 2015)

P_{earth} = 365.256 days = Earth orbital period (2015)

P_{moon} = 27.3217 days = lunar orbital period (2015)

$P_{sidereal}$ = 23.934 469 6 h = sidereal day = period for one revolution of the Earth about its axis, relative to fixed stars

R_{earth} = 6371.0 km = volumetric average Earth radius (from NASA 2015)
= 6378.1 km = Earth radius at equator
= 6356.8 km = Earth radius at poles

S = 1367.6 W $\cdot$m^{-2} = solar irradiance (solar constant) at top of atmosphere (NASA 2015)
≈ 1.125 K $\cdot$m $\cdot$s^{-1} = kinematic solar constant (based on mean sea-level density)

T_e = 254.3 K = effective radiation emission black-body temperature of Earth system (NASA 2015)

Φ_r = 23.44° = 0.4091 radians = tilt of Earth axis = obliquity relative to the orbital plane (2015)

Ω = 0.729 210 7 x10^{-4} s^{-1} = sidereal rotation frequency of Earth (NASA 2015)

$2\cdot\Omega$ = 1.458 421 x10^{-4} s^{-1} = Coriolis factor

$2\cdot\Omega$ / R_{earth} = 2.289 x10^{-11} m^{-1} $\cdot$s^{-1} = beta factor

AIR AND WATER CHARACTERISTICS

$a = 0.0337$ (mm/day) $\cdot$(W/m^2)$^{-1}$ = water depth evaporation per latent heat flux
$B = 3 \times 10^9$ V/km = breakdown potential for dry air
$C_{vd} = 717$ J$\cdot$kg$^{-1}\cdot$K^{-1} = specific heat for dry air at constant volume
$C_{pd} = 1003$ J$\cdot$kg$^{-1}\cdot$K^{-1} = specific heat for dry air at constant pressure at –23°C
$C_{pd} = 1004$ J$\cdot$kg$^{-1}\cdot$K^{-1} = specific heat for dry air at constant pressure at 0°C
$C_{pd} = 1005$ J$\cdot$kg$^{-1}\cdot$K^{-1} = specific heat for dry air at constant pressure at 27°C
$C_{pv} = 1850$ J$\cdot$kg$^{-1}\cdot$K^{-1} = specific heat for water vapor at constant pressure at 0°C
$C_{pv} = 1875$ J$\cdot$kg$^{-1}\cdot$K^{-1} = specific heat for water vapor at constant pressure at 15°C
$C_{liq} = 4217.6$ J$\cdot$kg$^{-1}\cdot$K^{-1} = specific heat of liquid water at 0°C
$C_{ice} = 2106$ J$\cdot$kg$^{-1}\cdot$K^{-1} = specific heat of ice at 0°C
$D = 2.11 \times 10^{-5}$ m$^2\cdot$s^{-1} = molecular diffusivity of water vapor in air in standard conditions
$e_o = 0.611$ kPa = reference vapor pressure at 0°C
$k = 0.0253$ W$\cdot$m$^{-1}\cdot$K^{-1} = molecular conductivity of air at sea level in standard conditions
$L_d = 2.834 \times 10^6$ J$\cdot$kg^{-1} = latent heat of deposition at 0°C
$L_f = 3.34 \times 10^5$ J$\cdot$kg^{-1} = latent heat of fusion at 0°C
$L_v = 2.501 \times 10^6$ J$\cdot$kg^{-1} = latent heat of vaporization at 0°C
$n = 3.3 \times 10^{28}$ molecules$\cdot$m^{-3} for liquid water at 0°C
$n_{air} \approx 1.000\,277$ = index of refraction for air
$n_{water} \approx 1.336$ = index of refraction for liquid water
$n_{ice} \approx 1.312$ = index of refraction for ice
$P_{STP} = 101.325$ kPa = standard sea-level pressure (STP = Standard Temperature & Pressure)
$\Re_d = 0.287\,053$ kPa$\cdot$K$^{-1}\cdot$m$^3\cdot$kg^{-1} = $C_{pd} - C_{vd}$
= 287.053 J$\cdot$K$^{-1}\cdot$kg^{-1} = gas constant for dry air
$\Re_v = 461.5$ J$\cdot$K$^{-1}\cdot$kg^{-1} = water-vapor gas constant
= 4.61×10^{-4} kPa$\cdot$K$^{-1}\cdot$m$^3\cdot$g^{-1}
$Ri_c = 0.25$ = critical Richardson number
$s_o = 343.15$ m/s = sound speed in standard, calm air
$T_{STP} = 15$°C = standard sea-level temperature
$\varepsilon = 0.622\ g_{water}\cdot g_{air}{}^{-1} = \Re_d / \Re_v$ = gas-constant ratio
$\gamma = 0.0004\ (g_{water}\cdot g_{air}{}^{-1})\cdot K^{-1} = C_p / L_v$
= $0.4\ (g_{water}\cdot kg_{air}{}^{-1})\cdot K^{-1}$ = psychrometric constant
$\Gamma_d = 9.75$ K$\cdot$km^{-1} = $|g|/C_p$ = dry adiabatic lapse rate
$\rho_{STP} = 1.225$ kg$\cdot$m^{-3} = standard sea-level air density
$\rho_{avg} = 0.689$ kg$\cdot$m^{-3} = air density averaged over the troposphere (over z = 0 to 11 km)

$\rho_{liq} = 999.84$ kg$\cdot$m^{-3} = density of liquid water at 0°C
= 1000.0 kg$\cdot$m^{-3} = density of liquid water at 4°C
= 998.21 kg$\cdot$m^{-3} = density of liquid water at 20°C
= 992.22 kg$\cdot$m^{-3} = density of liquid water at 40°C
= 983.20 kg$\cdot$m^{-3} = density of liquid water at 60°C
= 971.82 kg$\cdot$m^{-3} = density of liquid water at 80°C
= 958.40 kg$\cdot$m^{-3} = density of liquid water at 100°C
$\rho_{sea\text{-}water} = 1025$ kg$\cdot$m^{-3} = avg. density of sea water, (sea water contains 34.482 g of salt ions per kg of water, on average)
$\rho_{ice} = 916.8$ kg$\cdot$m^{-3} = density of ice at 0°C
$\sigma = 0.076$ N$\cdot$m^{-1} = surface tension of pure water 0°C

CONVERSION FACTORS & COMBINED PARAMETERS

$C_{pd} / C_{vd} = k = 1.400$ (dimensionless)
= specific heat ratio
$C_{pd} / |g| = 102.52$ m$\cdot$K^{-1}
$C_{pd} / L_v = 0.0004\ (g_{water}\cdot g_{air}{}^{-1})\cdot K^{-1} = \gamma$
= $0.4\ (g_{water}\cdot kg_{air}{}^{-1})\cdot K^{-1}$
= psychrometric constant
$C_{pd} / \Re_d = 3.50$ (dimensionless)
$C_{vd} / C_{pd} = 1/k = 0.714$ (dimensionless)
$|g|/C_{pd} = \Gamma_d = 9.8$ K$\cdot$km^{-1} = dry adiabatic lapse rate
$|g|/\Re_d = 0.0342$ K$\cdot$m^{-1} = 1/(hypsometric constant)
$L_v / C_{pd} = 2.5\ K / (g_{water}\cdot kg_{air}{}^{-1})$
$L_v / \Re_v = 5423$ K = Clausius-Clapeyron parameter for vaporization
$\Re_d / C_{pd} = 0.28571$ (dimensionless) = potential-temperature constant
$\Re_d / \Re_v = \varepsilon = 0.622\ g_{water}\cdot g_{air}{}^{-1}$ = gas-constant ratio
$\Re_d / |g| = 29.29$ m$\cdot$K^{-1} = hypsometric constant
$\rho_{air}\cdot C_{pd\,air} = 1231$ (W$\cdot$m^{-2}) / (K$\cdot$m$\cdot$s^{-1}) at sea level
= 12.31 mb$\cdot$K^{-1} at sea level
= 1.231 kPa$\cdot$K^{-1} at sea level
$\rho_{air}\cdot|g| = 12.0$ kg$\cdot$m$^{-2}\cdot$s^{-2} at sea level
= 0.12 mb$\cdot$m^{-1} at sea level
= 0.012 kPa$\cdot$m^{-1} at sea level
$\rho_{air}\cdot L_v = 3013.5$ (W$\cdot$m^{-2}) / [$(g_{water}\cdot kg_{air}{}^{-1})\cdot$(m$\cdot$s^{-1})] at sea level
$\rho_{liq}\cdot C_{liq} = 4.295 \times 10^6$ (W$\cdot$m^{-2}) / (K$\cdot$m$\cdot$s^{-1})
1 megaton nuclear explosion $\approx 4 \times 10^{15}$ J
2π radians = 360°
$(1-\varepsilon)/\varepsilon = 0.61$ = virtual temperature constant

INDEX

B

C

D

E

F

G

H

I

J

K

L

M

N

O

P

Q

R

T

U

V

W

X

Y

Z

CPSIA information can be obtained
at www.ICGtesting.com
Printed in the USA
LVOW09s0452181116
513539LV00001B/1/P